STRUCTURAL GEOLOGY

STRUCTURAL GEOLOGY

SECOND EDITION

Robert J. Twiss

Eldridge M. Moores

University of California at Davis

W. H. FREEMAN AND COMPANY

New York

Publisher:	Susan Finnemore Brennan
Acquisitions Editor:	Valerie Raymond
Marketing Manager:	Scott Guile
Editorial Assistant:	Whitney Clench
Media and Supplements Editor:	Amy Thorne
Photo Editor:	Bianca Moscatelli
Cover Designer:	Vicki Tomaselli
Text Designer:	Cambraia Fernandes
Project Editor:	J. Carey Publishing Service
Illustrations:	Network Graphics
Illustration Coordinator:	Bill Page
Production Coordinator:	Ellen Cash
Composition:	Matrix Publishing Services
Printing and Binding:	RR Donnelley

Library of Congress Control Number: 2006931628

ISBN-13: 978-0-7167-4951-6
ISBN-10: 0-7167-4951-3

Printed in the United States of America

Second printing

W. H. Freeman and Company
41 Madison Avenue
New York, NY 10010
Houndmills, Basingstoke RG21 6XS, England

www.whfreeman.com

Contents

Preface *xi*

CHAPTER 1 Introduction 1

1.1 What Are Structural Geology and Tectonics? 1
1.2 Structural Geology, Tectonics, and the Use of Models 2
1.3 The Interior of the Earth and of Other Terrestrial Bodies 7
1.4 The Earth's Crust and Plate Tectonics: Introduction 9
1.5 Ocean Basins 14
1.6 The Structure of Continental Crust 17
1.7 Precambrian Shields 19
1.8 Phanerozoic Regions 25
1.9 Summary and Preview 30
Box 1-1 *The Scientific Method* *4*

PART I BRITTLE DEFORMATION 35

CHAPTER 2 Fractures and Joints 37

2.1 Classification of Extension Fractures 38
2.2 Geometry of Fracture Systems in Three Dimensions 43
2.3 Features of Fracture Surfaces 52
2.4 Timing of Fracture Formation 54
2.5 Relationships of Fractures to Other Structures 57
Box 2-1 *Fractals and the Description of Joint Patterns* *46*

CHAPTER 3 Introduction to Faults 61

3.1 Types of Faults 61
3.2 Recognition of Faults 63
3.3 Determination of Fault Displacement 70
3.4 Fault Geometry 81
3.5 Balanced Cross Sections 88
3.6 Summary and Preview 88

CHAPTER 4 Normal Faults 91

4.1 Characteristics of Normal Faulting 91
4.2 Shape and Displacement of Normal Faults 93
4.3 Structural Associations of Normal Faults 95
4.4 Kinematic Models of Normal Fault Systems 107
4.5 Determination of Extension associated with Normal Faults 111

CHAPTER 5 Thrust or Reverse Faults 115

5.1 Recognition of Thrust Faults 115
5.2 Shape and Displacement of Thrust Faults 117
5.3 Structural Environments of Thrust Faults 119
5.4 Kinematic Models of Thrust Fault Systems 127
5.5 Geometry and Kinematics of Thrust Systems in the Hinterland 128
5.6 Analysis of Displacement on Thrust Faults 131

CHAPTER 6 Strike-Slip Faults 135

6.1 Characteristics of Strike-Slip Faults 136

6.2 Shape, Displacement, and Related Structures 139
6.3 Structural Associations of Strike-Slip Faults 141
6.4 Kinematic Models of Strike-Slip Fault Systems 146
6.5 Analysis of Displacement 148
6.6 Balancing Strike-Slip Faults 149

CHAPTER 7 Stress 151

7.1 Force, Traction, and Stress 151
7.2 The Mohr Diagram for Two-Dimensional Stress 168
7.3 Terminology for States of Stress 173
7.4 A Closer Look at the Mohr Circle for Two-Dimensional Stress 175
7.5 The Stress Tensor 180
Box 7-1 *What Is a Vector: A Brief Review* *154*
Box 7-2 *The Mohr Diagram for Three-Dimensional Stress* *178*
Box 7-3 *What Is a Tensor?* *184*
Box 7-4 *Sign Conventions Galore: A Cautionary Note* *187*
Box 7-5 *Derivation of Principal Stresses in Two Dimensions* *189*
Appendix 7-A *Illustrative Problem 1* *193*
Appendix 7-B *Illustrative Problem 2* *201*

CHAPTER 8 Mechanics of Fracturing and Faulting: Experiment and Theory 209

8.1 Experimental Fracturing of Rocks 209
8.2 A Fracture Criterion for Tension Fractures 210
8.3 The Coulomb Fracture Criterion for Confined Compression 212
8.4 Effects of Confining Pressure on Fracturing and Frictional Sliding 216
8.5 Effects of Pore Fluid Pressure on Fracturing and Frictional Sliding 220
8.6 Effects on Fracturing of Anisotropy, the Intermediate Principal Stress, Temperature, and Scale 221
8.7 Limitations of the Coulomb Fracture Criterion 225
8.8 The Griffith Theory of Fracture 226
Box 8-1 *The Coulomb Fracture Criterion in Terms of Principal Stresses* *214*

CHAPTER 9 Mechanics of Natural Fractures and Faults 231

9.1 Elastic Deformation 231
9.2 Techniques for Determining Stress in the Earth 233
9.3 Mechanisms of Stressing the Earth's Crust 236
9.4 Stress in the Earth 238
9.5 Stress Histories and the Origin of Joints 241
9.6 The Spacing of Extension Fractures 250
9.7 Distinguishing Extension Fractures from Shear Fractures 251
9.8 Fractures Associated with Faults 252
9.9 Fractures Associated with Folds 253
9.10 Stress Distributions and Faulting 254
9.11 The Mechanics of Large Overthrusts 258
9.12 Cause and Effect: A Word of Caution 267
Box 9-1 *The Effect of Burial and Uplift on Stress in the Crust* *244*
Box 9-2 *Simplified Model of a Thrust Sheet* *264*

PART II DUCTILE DEFORMATION 271

CHAPTER 10 The Description of Folds 273

10.1 Geometric Parts of Folds 274
10.2 Fold Scale and Attitude 280
10.3 The Elements of Fold Style 282
10.4 The Order of Folds 290
10.5 Common Styles and Structural Associations of Folding 292

CHAPTER 11 Foliations and Lineations in Deformed Rocks 297

11.1 Tectonites 297
11.2 Compositional Foliations 298
11.3 Disjunctive Foliations 299
11.4 Crenulation Foliations 302
11.5 Continuous Foliations 302
11.6 The Relationship of Foliations to Other Structures 303
11.7 Special Types of Foliation and Nomenclature 309
11.8 Structural Lineations 309
11.9 Mineral Lineations 313
11.10 Associations of Lineations with Other Structures 316

CHAPTER 12 Geometry of Homogenous Strain 319

12.1 Measures of Strain 321
12.2 The State of Strain 324
12.3 Special States of Strain 328
12.4 Progressive Deformation 333
12.5 Progressive Stretch of Material Lines 336
12.6 Homogeneous and Inhomogeneous Deformation 339
12.7 The Representation of Three-Dimensional Strain States and Progressive Strains 339
12.8 Tensor Representations of Strain 350
12.9 Finite Strain of an Arbitrary Line Segment and the Mohr Circle 355
12.10 Applications of Strain Analysis 361
Box 12-1 *Other Measures of Linear Strain* *322*
Box 12-2 *Terminology of Strain Compared with Stress: Beware!* *346*
Box 12-3 *A More Quantitative View of Strain* *352*

CHAPTER 13 Kinematic Analysis of Folds 363

13.1 Flexural Folding of a Layer 364
13.2 Passive Shear Folding of a Layer 369
13.3 Volume-Loss Folding of a Layer 371
13.4 Homogeneous Flattening of Folds in a Layer 374
13.5 Folding of Multilayers 376
13.6 Formation of Kink and Chevron Folds 379
13.7 Fault-Bend and Fault-Propagation Folding of a Multilayer 383
13.8 Drag Folds and Hansen's Method for Slip-Line Determination 390
13.9 Superposed Folding 390
13.10 Diapiric Flow 395

CHAPTER 14 Analysis of Foliations and Lineations 399

14.1 Material and Nonmaterial Foliations and Lineations 399
14.2 Mechanisms of Formation of Foliations and Lineations and their Relationships to Strain 400
14.3 Interpretation of the Morphological Types of Foliation 405
14.4 Steady-State Foliations 412
14.5 Foliations and Shear Planes 413
14.6 Interpretation of Morphological Types of Lineation 415
14.7 Lineations on Folds 421

CHAPTER 15 Observations of Strain in Deformed Rocks 423

15.1 Measuring Strain in Rocks 423
15.2 Relationship of Strain to Foliations and Lineations 426
15.3 Measurement of Strain in Folds 427
15.4 Strain in Shear Zones 432
15.5 Deformation History 446
Box 15-1 *Brittle Strain Inferred from Fault Systematics* *439*
Appendix 15-A *Common Techniques for Measuring Strain* *449*

PART III RHEOLOGY 457

CHAPTER 16 Macroscopic Aspects of Rock Deformation: Rheology and Experiment 459

16.1 Continuum Models of Material Behavior 460
16.2 Experiments on Friction and Cataclastic Flow: Implications for Faulting 466
16.3 Experimental Investigation of Ductile Flow 475
16.4 Steady-State Creep 477
16.5 The Effects of Pressure, Grain Size, Chemical Environment, and Partial Melt on Steady-State Creep 481
16.6 Application of Experimental Rheology to Natural Deformation 487
Box 16-1 *Measures of Strain Rate* *465*
Box 16-2 *The Rate- and State-Dependent Friction Law* *471*
Box 16-3 *Experimental Determination of the Material Constants in the High-Temperature Creep Equation* *485*
Box 16-4 *Constitutive Equations in Three Dimensions* *491*

CHAPTER 17 Microscopic Aspects of Ductile Deformation: Mechanisms and Fabrics 495

17.1 Mechanisms of Low-Temperature Deformation 497
17.2 Twin Gliding 499
17.3 Diffusion and Solution Creep 500
17.4 Linear Crystal Defects: The Geometry and Motion of Dislocations 503

17.5 Mechanisms of Dislocation Creep 513
17.6 Microstructural Fabrics associated with Dislocation Creep 518
17.7 Preferred Orientation Fabrics of Dislocation Creep 524
17.8 Symmetry Principles in the Interpretation of Deformed Rocks 535
Box 17-1 *Rheologies Inferred from Mechanisms of Ductile Deformation* *517*
Box 17-2 *Inferring the Orientation and Magnitude of Paleostresses in Deformed Rocks* *520*

CHAPTER 18 Scale Models and Quantitative Models of Rock Deformation 543

18.1 Constraints on Physical Models 544
18.2 The Theory of Scale Models 546
18.3 Scale Models of Folding 548
18.4 Scale Models of Gravity-Driven Deformation 551
18.5 Plastic Slip-Line Field Theory and Faulting 553
18.6 Analytic Solution for the Viscous Buckling of a Competent Layer in an Incompetent Matrix 561
18.7 Numerical Models of Buckling and the Effects of Different Rheologies 563
Box 18-1 *Formulation of a Mathematical Model with Application to the Problem of Viscous Deformation* *573*

PART IV REGIONAL ASSOCIATIONS OF STRUCTURES 579

CHAPTER 19 Development of Structures at Active Plate Margins 581

19.1 Divergent Margins on the Continents: Continental Rifting 581
19.2 Divergent Margins in Ocean Basins 590
19.3 Major Strike-Slip Faults: Transform Faults and Megashears 601
19.4 Convergent Margins 615
19.5 Active Collisions 626
Box 19-1 *Structures of Convergent and Divergent Strike-Slip along the Boundaries of the Sierran Microplate* *611*

CHAPTER 20 Anatomy and Tectonics of Orogenic Belts 639

20.1 Introduction 639
20.2 The Foredeep or Foreland Basin 640
20.3 The External Thrust Complex: Foreland Fold and Thrust Belt, Slate Belt, Ophiolites, and Sutures 644
20.4 The Crystalline Core Zone: Metamorphism 648
20.5 The Crystalline Core Zone: Structure and Lithology 653
20.6 Extensional Deformation and Low-Angle Detachments 660
20.7 High-Angle Fault Zones 665
20.8 Minor Structures and Strain in the Interpretation of Orogenic Zones 666
20.9 Tectonics, Topography, and Erosion 670
20.10 Tectonics and Metamorphism 671
20.11 Simple Models of Orogenic Deformation 673
20.12 A Two-Dimensional Plate Tectonic Model of Orogeny 677
20.13 The "Wilson Cycle" and Plate Tectonics 685
20.14 Terrane Analysis 688

APPENDIX 1 The Orientation and Representation of Structures 693

A1.1 The Attitude of Planes and Lines 693
A1.2 Graphical Presentation of Orientation Data 695
A1.3 Geologic Maps 697
A1.4 Cross Sections: Portrayal of Structures in Three Dimensions 698

APPENDIX 2 Geophysical Techniques 701

A2.1 Seismic Studies 701
A2.2 Analysis of Gravity Anomalies 708
A2.3 Geomagnetic Studies 709

APPENDIX 3 Units and Constants

Basic SI (Système Internationale; mks) Units 713
Table of SI Multiples 713
Other Systems of Units 713
Units and Constants Used in the Book and Some Common Equivalents 714

INDEX 717

TABLES

Table 3.1 Fault Rock Terminology 64
Table 7.1 Development of the Concept of Stress 152
Table 7.2 Components for the Two-Dimensional Stress σ in the Mohr Circle Sign Convention 165
Table 7.3 Components for the Three-Dimensional Stress σ in the Mohr Circle Sign Convention 168
Table 7.4 Notation for Stress 191
Table 7-4.1 Sign Conventions for Stress Components 187
Table 9.1 Stress Interpretation of Fractures in Folds 254
Table 9-1.1 Mechanical Properties of Sediment during Burial and Uplift 244
Table 9-2.1 Relationships among Fracture Angle, Coefficient of Internal Friction, and K 265
Table 10.1 Elements of Fold Style 283
Table 10.2 Aspect Ratio 285
Table 10.3 Tightness of Folding 286
Table 10.4 Bluntness of Folds 287
Table 10.5 Style of a Folded Layer 289
Table 12.1 Extensional Strain of a Material Line 322
Table 12-2.1 Strain versus Stress Terminology 347
Table 12-2.2 Antonym Pairs: Strain versus Stress Terminology 348
Table 16.1 Examples of Material Constants for Steady-State Power-Law Flow of Selected Rocks in the Moderate-Stress Regime 488
Table 17.1 Dominant Slip Systems of Some Common Minerals 510
Table 18.1 Scale Factors for Selected Variables in Mechanics 547

Preface

NEED FOR A NEW EDITION

It has been fourteen years, as we write this Preface, since the publication of the first edition of *Structural Geology*. We are grateful for both the reception that the book has received and the apparent demand for a new edition. We also appreciate the comments and feedback we have received about the book. It is largely with these comments in mind, and with the intent of updating some of the material and indicating some of the new and exciting directions of research, that we have undertaken this revision.

PHILOSOPHY OF THE BOOK

Our philosophy in writing the book, and our overall aims, are as follows:

For the student:

To engage the students' primary fascination with geology and the structures present in rocks;

To build on that fascination by showing how our understanding of the structures progresses from careful observation and description, to the use of theory (stress, strain) and experiment (brittle fracturing, ductile flow) in producing models, and then to the testing of the models against observation and the interpretation of the structures;

To provide students an easily comprehended discussion of the basics of structural geology, as well as a clear and progressive path from that basic level to an accessible introduction to more advanced material;

To reinforce, through the organization of the material, the understanding of the primary method of scientific investigation, which proceeds from observation and description, through the development of kinematic and mechanical models based on theory and experiment, to testing, interpretation, and explanation;

To introduce areas of exciting ongoing research in the field that highlight the dynamic nature of structural geology.

For the instructor:

To accommodate an instructor's choice for the appropriate level of presentation, especially of the more technical material;

To accommodate an instructor's preference for the organization of a course in structural geology.

Structural geology and tectonics represent a continuum in the scale at which we study deformation in the Earth, with structural geology concerned largely with the microscopic to regional scales and tectonics concerned largely with the regional to global scales. Studies at the different scales are interdependent, however, and in fact, studies at one scale invariably inform studies at other scales. To emphasize this connection, we have placed the major topics in structural geology (Chapters 2–18) between global tectonic bookends, so to speak — Chapter 1 at one end and Chapters 19 and 20 at the other. Moreover, throughout the book, we use examples of structures that develop on a local as well as a regional scale. In this way, we intend to indicate the interdependence and interconnection between the small and large scales of studying deformation in the Earth.

These goals have not changed from the first edition, but in this second edition, we have tried to enhance the clarity of the presentation by improving the organization of the material in the chapters. The level at which different instructors choose to teach this material varies substantially, and different instructors may choose to pursue different parts of the subject to different levels. We have tried to refine the organization of the second edition to accommodate better these different choices.

FLEXIBLE ORGANIZATION AND LEVEL OF INSTRUCTION

The broad organizational features of our text that make it amenable to a variety of choices in organization and levels of instruction can be summarized as follows:

The introductory chapter (Chapter 1) includes a discussion of the principal tectonic features of the Earth, which constituted Chapter 21 in the first edition. Our intent here is to provide, at the outset, a global context for the study of deformation in the Earth.

Chapters 2 to 15 constitute the core of most courses in structural geology. Although we prefer a linear progress through these chapters, we have adopted a division of topics into chapters, which allows the material easily to be taught in different orders, depending on the preferences of the individual instructor.

These chapters are organized into Part I on brittle deformation (Chapters 2–9) and Part II on ductile deformation (Chapters 10–15). The discussion in each of these parts proceeds in a manner comparable to actual scientific investigation. First we describe the observed structures, then we introduce the theory and experiment necessary to understand the structures and to develop models for the formation of the structures, and finally we compare the models against the observations to evaluate the models and to interpret the origins and significance of the structures.

Each chapter in Part III (Chapters 16–18) and Part IV (Chapters 19 and 20) stands alone and can be chosen to provide more advanced material in a variety of topics, as time permits and inclination dictates. The sole exception is that Section 16.1 is a useful prerequisite for Chapters 17 and 18. We discuss rheology and experimental rock deformation (Chapter 16), deformation mechanisms and fabrics (Chapter 17), mechanical modeling (Chapter 18), formation of structures at active plate margins (Chapter 19), and orogenic belts (Chapter 20).

Chapters 19 and 20 complement the introduction to large-scale tectonics of Chapter 1 and emphasize the interface between large-scale tectonic environments and the smaller-scale structures that develop in them. Our intent is to explore the integral relationship between structural geology and tectonics and to reinforce the recognition of that relationship.

We do not expect that the entire book can be covered in a one-quarter, or even a one-semester, structural geology course. In fact, we would expect instructors to select sections from Chapters 1 to 15 (Parts I and II) that conform to their choice of level of presentation and then possibly to select additional chapters or sections from Parts III and IV (Chapters 16–20) to support the topics and level consistent with the instructor's choices and the time available.

CHOICE OF LEVEL AND TOPIC COVERAGE

We have reorganized substantial parts of the book to make a more consistent progression within individual chapters from simple to complex. This new structure should make it easier for instructors to choose the level of complexity and detail at which they wish to teach the course, and for students to progress more consistently from the introductory to the more advanced subject matter.

Throughout the book, boxes generally provide introductions to more advanced topics. These ancillary topics can be incorporated into the course, used for self-study by more advanced students, or used as an introduction for independent research projects and papers.

We assume students have had only an introductory course in physical geology and at least high school courses in mathematics, including algebra, geometry, and trigonometry, and in physics (particularly mechanics). Some of the boxes also assume familiarity with introductory calculus. Many universities require first-year calculus and college-level physics and chemistry as part of a major in geology. We feel students should have some opportunity to see why these courses are important to the study of geology, and so we have included some appropriate material. Other colleges, however, do not require as much advanced preparation, and we have tried to structure the book so that it is easy to use in a structural geology course that emphasizes a more descriptive approach.

Because of the depth of discussion and the boxes on more advanced topics, however, we hope the book will retain its usefulness and value beyond a basic course in structural geology.

Specific features of the major reorganization in this new edition of the book are as follows:

The first chapter now includes a discussion of the principal tectonic features of the Earth, whereas that material was presented near the end of the first edition.

The section on techniques for specifying and representing the orientation of structures, and the section on geophysical techniques, have been moved to an appendix.

In the chapter on stress (Chapter 7), the first three sections use only the Mohr circle notation and sign convention for stress. These sections provide a sufficient basis for the remainder of the book, except for some of the advanced material in boxes.

Two sample problems on calculating stresses appear as appendices to Chapter 7. The first uses only two-dimensional analysis and Mohr circle notation (relying only on Sections 7.1–7.3). The second shows the connection between the Mohr circle and the components of the three-dimensional stress tensor, expressed in tensor component notation (relying on Sections 7.4 and 7.5).

The variety of sign conventions used for stress have been collected together in a new easily-referenced box (Box 7-4 and Table 7-4.1). Although all these conventions are used in the geologic literature, our book uses only the geologic Mohr circle convention, except for the most advanced material in some of the boxes, which requires the geologic tensor convention.

In the chapter on strain (Chapter 12), the first sections (12.1–12.6) focus on two-dimensional strain, largely from a geometric point of view. These sections are adequate to understand the remainder of the book, except for the material in some boxes. Three-dimensional strain appears in Section 12.7, and the strain tensor and tensor notation are introduced in Section 12.8, toward the end of the chapter. The finite strain Mohr circle is discussed in Section 12.9 (see "New Additions").

The chapters on kinematic models for the formation of folds (Chapter 13) and foliations and lineations (Chapter 14) follow the chapter on strain (Chapter 12), whereas in the first edition they preceded the strain chapter. These chapters now incorporate strain in a more unified discussion of models of formation of folds, foliations, and lineations.

The detailed description of techniques for measuring strain in rocks is now presented in an appendix to Chapter 15 (Appendix 15-A).

The chapter on scale and quantitative models of rock deformation (Chapter 18) has been reorganized so the sections on the more intuitive scale modeling appear first, followed by discussion of results of numerical modeling.

The discussion of plastic slip-line fields is now in the scale modeling section (18.5), and it includes the Tapponnier plasticene model for southeast Asia as an example.

The entire discussion of formulation of a mathematical model of viscous deformation has been incorporated into Box 18-1.

NEW ADDITIONS

We have updated the text, incorporating new and exciting areas of research and making a connection between the discussion of small-scale structures and the tectonic environments and processes that result in their development. Major new additions include the following:

A new box on the scientific method (Box 1-1).

A new box introducing fractal geometry and its use in describing systems of fractures (Box 2-1).

New material discussing the relationship between fault length and displacement (Section 3.3(vi)) and fault length distributions (Section 3.4(i)).

New material in the chapter on normal faults (Section 4.3(iii)) on the structure of metamorphic core complexes.

New material in the chapter on thrust faults (Section 5.3(iii)) on back-thrusts and tectonic wedging.

A new section discussing the limitations of the Coulomb fracture criterion (Section 8.7).

A new discussion on the origin of curved joints and joint intersections (Sections 9.5(vii) and 9.5(viii)).

A refined discussion of three important features of a folded surface—the aspect ratio, tightness, and bluntness (Section 10.3(i)), including clear diagrams of these different features.

New material (Section 12.7(iii)) on triaxial deformation in convergent and divergent strike-slip zones ("transpression" and "transtension").

A new section (12.9) using only geometry, trigonometry, and algebra to show how the finite deformation of an arbitrarily oriented line segment is described by a finite-strain Mohr circle.

A new box (Box 12-2), with tables, that discusses the distinction between the terminology for stress and for strain, the confusions that often arise, and the distinctions that should be maintained.

A new discussion of the trishear model for fault-propagation folding (Section 13.7(ii)).

A new box on determining the average extension in faulted regions from the systematics of fault geometry and displacement (Box 15-1).

An enhanced discussion of friction, faulting, and cataclastic flow (Section 16.2), including a new box on state- and rate-dependent friction (Box 16-2) and its relationship to earthquakes.

A new section on the use of symmetry principles in interpreting the fabrics of deformed rocks (Section 17.8).

New material on the results of power-law and visco-elastic models for folding (Section 18.7).

A new chapter (Chapter 19) on the development of structures at active plate boundaries. Our intent with this chapter and Chapter 20 on the anatomy and tectonics of orogenic belts is to connect the material on individual structures discussed throughout most of the book with the tectonic environments in which these structures form.

A new discussion of extensional deformation and low-angle detachments in orogenic zones (Section 20.6).

A new section on tectonics, topography, and erosion (Section 20.9).

A new section discussing a self-consistent two-dimensional plate tectonic model of collisional orogeny, including underplating, nappe formation, exhumation, ultra-high-pressure metamorphism, root formation, and gravitational collapse of topographic highs and orogenic roots.

ACKNOWLEDGMENTS

We owe a tremendous debt of gratitude to our colleagues and students for educating us with their research, sharing with us their insights and criticisms, and permitting us the use of their figures. We remain indebted to the reviewers of the first edition, including Roy Dokka, Roy Kligfield, William MacDonald, Stephen Marshak, Peter Mattson, Cris Mawer, Sharon Mosher, Raymond Siever, Carol Simpson, and Doug Walker. In particular, William MacDonald, Sharon Mosher, and Carol Simpson put in a large amount of work to provide us with exceptionally detailed comments for the first edition, for which we remain most grateful indeed.

Reviewers who shared their insights into how the first edition could be improved included Phillip A. Armstrong and Nancy Lindsley-Griffin, who provided us with particularly detailed and helpful comments on the whole book, and Hassan Babaie and Sarah E. Tindall, who gave us very helpful suggestions on the first and second halves of the book, respectively. We benefited also from careful reviews on selected chapters and sections from Magali I. Billen, Randall Marrett, and Don Turcotte. We also appreciate reviews from Gary Solar, Gary Lash, and Cathy J. Busby.

We are also deeply indebted to all those authors and their publishers who have permitted us to reproduce figures from their published works and who in many cases have provided us with original prints. They are too numerous to list here, but we list the credits in the individual figure captions. We could not have done this without their kind and generous cooperation. We must single out John Ramsay for special thanks, however, for permitting us to use an unusually large number of figures from his many seminal published works, whether in their original or modified form. Our reliance on his work is testimony to the huge impact he has had on the development of modern structural geology from an almost purely descriptive science to one based firmly in the process of detailed analysis and modeling.

Obviously, then, this book would not have been possible without all this feedback and help from our colleagues. We cannot thank them enough.

Last but not least, we wish to thank everyone at W. H. Freeman and Company who helped us get the first edition into print and prodded us into completing the revisions for the second edition. John Staples started us on this project, and Jeremiah Lyons, publisher, and Christine Hastings, project editor, brought the first edition to fruition. Editor Holly Hodder started the ball rolling on the second edition. Editor Valerie Raymond was adept at patiently but persistently pushing us through the process of actually producing the revised manuscript and, with project editor Jennifer Carey and the staff at Freeman, finally getting it into print. To all involved with this project, we offer our sincere thanks.

Chapter 1

INTRODUCTION

1.1 WHAT ARE STRUCTURAL GEOLOGY AND TECTONICS?

The Earth is a dynamic planet. The evidence is all around us. Earthquakes and volcanic eruptions regularly jar many parts of the world. Many rocks exposed at the Earth's surface reveal a continuous history of such activity, and some have been uplifted from much deeper levels in the crust where they were broken, bent, and contorted. These processes, however, proceed in extremely slow motion on the scale of a human lifetime or even on the scale of human history. The "continual" eruption of a volcano can mean that it erupts once in one or more human lifetimes. The "continual" shifting and grinding along a fault in the crust means that a major earthquake might occur in the same place once every 50 to 150 years. At the almost imperceptible rate of a few millimeters per year (about the rate at which your toenails grow!), high mountain ranges can be uplifted in the geologically short span of only a million years. A 5-mm/yr uplift rate, for example, would produce a 5-kilometer-high mountain in a million years, if erosion did not reduce the altitude at the same time. A million years, however, is already more than two orders of magnitude longer than the whole of recorded history, and these processes have been going on for hundreds of millions of years, an extent of time that so stretches our imaginations that it has been called "deep time." The Earth's crust, however, preserves a record of this constant dynamic activity, and if we can learn to read and decipher this record, we can learn much about the history of our planet and how it has evolved through deep time.

Structural geology and tectonics are two branches of geology that are closely related in both their subject matter and their approach to the study of the evolution of the Earth. They are concerned with reconstructing the inexorable motions that have shaped the evolution of the Earth's outer layers. The terms **structural geology** and **tectonics** are derived from similar roots. **Structure** comes from the Latin word *struere*, which means "to build," and **tectonics** from the Greek word *tektos*, which means "builder," the reference being to the motions and processes that build the crust of the Earth. The motion may be simply a **rigid-body motion** that transports a body of rock from one place to another causing no change in its size or shape and, therefore, leaving no permanent imprint. The motion also may be a **deformation** that breaks a rock to form fractures or faults or makes solid rock flow, thus changing its shape or size. Such motions leave a permanent record in the form of structures that can be observed in the rock. Our intent in this book is to present our current understanding of how, and under what conditions, the different types of structures form, and to show how we can use that understanding to reconstruct the history of the crust.

For example, the Earth's crust may break along faults, and the two pieces slide past one another. Sections of continental crust may pull apart as oceans open, and they subsequently collide with each other as oceans close. Such events result in bending and breaking of rocks in the shallow crust and in the puttylike flow of solid rocks at greater depth. Mountain ranges are uplifted and subsequently eroded, exposing the deeper levels of the crust. The breaking, bending, and flowing of rocks all produce permanent structures such as fractures, faults, and folds that we can use as clues to reconstruct the deformation that produced them. Even on a much smaller scale, the preferred alignment of platy and elongate mineral grains in the rocks and the submicroscopic imperfections in crystalline structure of the mineral grains all help us to reconstruct and trace the course of the deformation.

The fields of structural geology and tectonics are both concerned with the study of the history of active or past motions and deformation in the Earth's crust and upper mantle. They differ in that structural geology predominantly deals with the study of deformation in rocks at a scale ranging from the submicroscopic to the regional, whereas tectonics predominantly deals with a regional to global scale. The two realms of study are interdependent, and at the regional scale, structural geology and tectonics overlap. Our interpretation of the history of large-scale motions must be consistent with the observations of deformation that has occurred at a small to local scale in the rocks. Conversely, the origins of local deformation need to be understood in the context of the history of the large-scale motions that we deduce from plate tectonics.

Structural geology and tectonics have undergone a period of rapid development since the 1960s. Structural geology has changed from an almost purely descriptive discipline to an increasingly quantitative one. New insights into the processes of deformation and the formation of structures at a wide variety of scales have become possible through the application of theoretical principles of **continuum mechanics.**[1] Our ability to measure the actual tectonic movements of the Earth's crust over a period of just a few years, to deform rocks directly in the laboratory, and to study deformed minerals at the submicroscopic level has provided insights that have been used in many field-based investigations vastly to improve our understanding of naturally deformed rocks.

In that same period of time, the revolution in tectonics was based largely on the development of the theory of plate tectonics. This theory now provides the framework for study of almost all large-scale motions and deformation affecting the Earth's crust and upper mantle. Field-based studies have taken on new meaning because plate tectonic theory has given us a new basis for interpreting the tectonic significance of structures and for inferring the history of regional deformation.

[1]**Continuum mechanics** is the study of how bodies of different materials deform when subjected to forces. Materials are idealized mathematically as **continua**, which are characterized by having no structure or discontinuities of any sort at any scale. Real materials, however, all have discontinuities at some scale. Rocks, for example, are composed of grains that at the grain boundaries are discontinuous in crystallographic orientation, structure, and/or composition; at the atomic scale the discontinuities are even more profound. Because we wish to describe the material behavior only in volumes that are very large compared to the grain size or the interatomic distances, however, these discontinuities are relatively insignificant. Whatever effect they might have can be assumed to average out over a relatively large volume. Thus a continuum often approximates well the behavior of such materials. This approximation enables us to describe the deformation of rock in simple mathematical terms.

Geophysics has become increasingly important to both structural geology and tectonics, as indicated by the number of diagrams throughout this book that present geophysical data. Seismic and gravity studies provide information on the geometry of large-scale structures at depth, which adds the critical third spatial dimension to our observations, and seismic focal mechanism solutions provide information on the ongoing deformation within plate boundaries. In addition, studies of rock magnetism and paleomagnetism provide data on past and present motions of the plates, which are essential for reconstructing global tectonic patterns.

Satellite imagery has also developed tremendously since about the 1960s. Satellites, such as the Global Positioning System (GPS), provide powerful new tools for analyzing active motions of parts of the Earth's crust relative to one another. Satellites also have been crucial to advances in our knowledge of the surface of other planets.

Research in structural geology and tectonics also depends on other branches of geology. Petrology and geochemistry provide data on temperature, pressure, and ages of deformation and metamorphism, which have become essential for the accurate interpretation of deformation and its tectonic significance. Sedimentology and paleontology are also important in reconstructing the patterns and ages of structural and tectonic events. These topics, however, are generally beyond the scope of this book.

In the remainder of this chapter, we set the stage for our investigation of structures in rocks (Section 1.2) by reviewing briefly the method of geological investigation and considering how we learn about processes in the Earth and interpret the records that those processes leave behind in the rock. This method involves the interaction between observation and the use of models, and it provides an essential organizing principle for our later discussions. Next (Section 1.3), we review the major features of the Earth's interior and the effect of the dynamics of the interior on the outer crust, which is the part of the Earth that we study with structural geology and tectonics. Finally (Sections 1.4–1.8), we review the major tectonic features of the Earth's crust and how they fit together as a dynamic system characterized by the processes of plate tectonics. These processes provide the context within which structures in the rocks develop.

1.2 STRUCTURAL GEOLOGY, TECTONICS, AND THE USE OF MODELS

All field studies in structural geology and many in tectonics rely on observations of deformed rocks at the Earth's surface (see Box 1-1). These studies generally begin with observations of features at outcrop scale—that is, a scale of a few millimeters to several meters. They may then proceed "down-scale" to observations made at the microscopic or even electron-microscopic level of microns[2] or "up-scale" to more regional observations at a scale of hundreds to thousands of kilometers. At the largest scale, observations are generally based on a compilation of observations from smaller scales. *None of these observations alone provides a complete view of all structural and tectonic processes.* Our understanding increases as we integrate our observations of the Earth over all scales.

In order to derive understanding from direct observations of the Earth, we observe the behavior of rocks under the controlled conditions of laboratory experiments. We use all these observations, as well as mathematical calculations, to devise and constrain models of the processes. These models represent our hypotheses concerning how structures develop, and by comparing the characteristics of the models with our observations, we can test and improve our understanding of the processes occurring in the Earth.

Good models are important for understanding structural and tectonic features, whether they be as concrete as a geologic map or as abstract as a set of numerical calculations on a computer. It is important to remember, however, that models are approximations of real situations; they are not themselves the geologic reality. When reading about the structural development of a feature or an area, a structural geologist should keep in mind such questions as: How strongly do the data constrain the model? What assumptions are implicit in the model? Are the assumptions justified? Are other models equally plausible? Are the models testable, and has the researcher provided any reproducible tests? Do the available data satisfy the tests? Another way of putting this is that a good model includes the potential that it can be disproven by comparing its characteristics with observations of the real world. We present many models of processes in this book. They should all be evaluated critically and with skepticism. Only such a critical approach to scientific interpretation can lead to efficient advances in our understanding of the Earth and, ultimately, of other planets.

Our first task in trying to unravel the deformation of the Earth's crust and the history of that deformation is to observe and record, carefully and systematically, the structures in the rock, including such features as lithologic contacts, fractures, faults, folds, and preferred orientations of mineral grains. In general, this process consists of determining the **geometry** of the structures. Where are the structures located in the rocks? What are their characteristics? How are they oriented in space and with respect to one another? How many times in the past have the rocks been deformed? Which structures belong to which episodes of deformation? Answering these and similar questions constitutes the initial phase of any structural investigation.

[2]One micron is one micrometer, or one-millionth (10^{-6}) of a meter, or one-thousandth (10^{-3}) of a millimeter.

In some circumstances, determining the geometry of rock structures is an end in itself; it is important, for example, in the location of economic deposits. To understand the processes that occur in the Earth, however, we need an explanation for the geometry. Initially, we want to know the **kinematics** of formation of the structures—that is, the motions that have occurred in producing them. Beyond that, we want to understand the **mechanics** of formation—that is, the interaction between forces and motions and how these lead to the geometry of the observed structures.

In large part, we improve our understanding by making conceptual models of the structures and how they form and then testing the predictions derived from the models against observation. **Geometric models** are three-dimensional interpretations of the distribution and orientation of structures within the Earth. They are based on mapping, geophysical data, and any other observational information we have. Typically, we present such models as geologic maps and as vertical cross sections along a particular line through an area.

Kinematic models prescribe a specific history of motion that could have carried the system from the undeformed to the deformed state, or from one configuration to another. These models are not concerned with why or how the motion occurred or what the physical properties of the system were. The plate tectonics model is a good example of a kinematic model. We can assess the validity of such models by comparing the geometries of the motion and deformation observed in the Earth with those deduced from the model.

Mechanical models are based on our understanding of basic laws of continuum mechanics, such as the conservation of mass, momentum, angular momentum, and energy, and on our knowledge of the **constitutive equations**, which are equations that describe the relationships between the forces and the motions that occur during deformation and which depend on the material properties of the specific rocks in question. Information on constitutive equations and material properties comes largely from laboratory experiments in which rocks are deformed under conditions that reproduce, as nearly as possible, the conditions within the Earth. Using mechanical models, we can calculate the hypothetical deformation of a body of rock that is subjected to a prescribed set of physical conditions such as forces, displacements, temperatures, and pressures. A model of the driving forces of plate tectonics based on the mechanics of convection in the mantle is an example of a mechanical model. Such models represent a deeper level of analysis and understanding than kinematic models, because the motions of the model, and the resulting geometry of the deformation, are not assumed but must be a consequence of the mechanical properties and physical conditions that we use in the model to represent the behavior of the rocks.

We use geometric, kinematic, and mechanical models to help us understand deformation on all scales from

BOX 1-1 The Scientific Method

i. The Science of Geology

Geology is the study of the history of Earth's evolution. It is based, first, on observations of the Earth itself, and increasingly on observations of other planetary bodies as well, but such sciences as biology, chemistry, physics, and materials science also are required to understand the processes we observe. Geology differs from these other sciences, however, in at least three ways.

First, geology is fundamentally a historically oriented science dealing with processes that for the most part occur on a time scale that is immense compared with human lives. The geologically short time of a million years is already more than two orders of magnitude longer than all of recorded human history! Thus, it is impossible to observe an entire process directly; we can see only what is happening at a single geological instant in time. Because of this constraint, the inference of geologic processes relies heavily on the *fundamental assumption* that *spatial variation can be interpreted in terms of temporal evolution*. In other words, we assume that the same process can be found in various stages of advancement in different places and that therefore we can piece together observations made in different places to infer a temporal evolution of that process.

Second, geology deals with large-scale and complex systems for which it is difficult if not impossible to construct controlled experiments. Thus, the observation and description of natural features acquire proportionately more importance than they have in most other sciences.

Third, the fact that geologic evidence is fragmentary and incomplete makes many of the inferences drawn from the data nonunique and highly dependent on our intuition and experience.

Despite these differences, the methods employed by geologists to investigate the Earth are philosophically similar to those used in other realms of science. In the remainder of this Box, we explore some of the philosophical underpinnings of our science that are usually understood implicitly and imparted at best by osmosis.

ii. From Hypotheses to Laws

In discussions of structural and tectonic features of the Earth, we refer repeatedly to *models* of the processes that result in the formation of structures such as fractures, faults, folds, foliations, lineations, and the production of crystallographic preferred orientations. This continual use of models prompts the question: What do scientific models have to do with reality? To answer this question, and to understand a scientific discussion, it is necessary to understand the meanings of the terms *fact*, *hypothesis*, *model*, *theory*, and *law* as they are used in science.

A **fact** is an objective observation or measurement, empirically verifiable by any trained observer. The rock type that occurs at a given outcrop, or the orientation of a dipping bed are two examples of objective facts. Facts are as near to "truth" as one gets in science, but because error always exists, the precision and accuracy[1] of a measurement or observation are never perfect, and in many cases, particularly in laboratory experiments, a 'direct observation' is in fact mediated by complex machinery and electronics..

An **hypothesis** is a proposed explanation of one or more observations. The term *model* is often synonymous with hypothesis. The mere proposal of an hypothesis or model does not, of course, imply that it is useful. For example, the ancient Egyptian belief that frogs formed from the mud of the Nile did not contribute much to the understanding of biology.

To be useful, an hypothesis must satisfy three criteria:

1. The hypothesis must be *testable*, which means that it must provide predictions that, at least in principle, can be either verified or proven false by observation. Ideally, the predictions of the model should indicate some previously unrecognized or unexpected aspect or behavior of the process or system under investigation.
2. The hypothesis must be *powerful*, which means that it must be capable of explaining a large number of disparate observations.
3. The hypothesis must be *parsimonious*, which means that it must use a minimum of assumptions compared to the amount of data that it explains. A model of a process that requires new assumptions to explain the behavior observed under each new set of conditions is not parsimonious. Moreover, the hypothesis should satisfy the principle of *Occam's Razor*, which requires the elimination of all nonessential features of an hypothesis.[2]

The criterion of testability is absolutely crucial. A useful model need not be right or "True," but it does need to be testable, which means it must include the potential for being disproved by observations of the physical world. If an hypothesis or model is not testable—if it does not provide the possibility for being disproven, at least in principle—it is not a scientific proposition, it cannot be a part of any scientific inquiry, and it cannot become part of the body of scientific knowledge and understanding. Disproof of an hypothesis can

[1]*Precision* is the closeness of one measurement to the mean value of many measurements of the same quantity. The standard deviation, for example, is a statistical measure of precision. *Accuracy* is the closeness of a measurement to the true value of the quantity. Thus, one can measure the attitude of a smooth planar bedding surface more precisely than a rough irregular one. If in doing so one used a compass with the wrong magnetic declination, the measurement would be inaccurate even though it may have been precise.

[2]After the English philosopher William of Occam (or Ockham), 1280–1349, who expressed it thus: "Multiplicity ought not to be posited without necessity."

be absolute, because in principle, a single observation can show that an hypothesis does not work. Proof, on the other hand, is never absolute, because no matter how successful a model has been in accounting for the characteristics of the observed world, it is impossible to guarantee that there will never be an inconsistent observation. In practice, of course, scientists tend to rationalize away one or a few inconsistent observations by questioning the methods or ability of the observer or by simply ignoring them as currently unexplainable anomalies. So disproof of an hypothesis typically is accepted only after the buildup of a significant quantity of anomalous observations, and the hypothesis is abandoned only after a better one is devised. Moreover, the longer an hypothesis has been widely accepted, the more resistant scientists typically are to accepting its demise.

The other two criteria provide only a qualitative standard, because there is no objective way of quantifying power or parsimony. Nevertheless, different hypotheses can usually be compared against one another on the basis of these principles.

Those hypotheses or models that best satisfy these three criteria are considered "elegant." Although elegant models are extremely useful in understanding our physical world, they necessarily are simplifications and idealizations of reality and, for all their success, should not be confused with "Truth" with a capital T.

The restrictions on what constitutes an acceptable hypothesis are quite severe, and they impose a stringent boundary on the sorts of questions that scientists can legitimately pose and that science can be expected to answer. Thus science is not an approach that can be used to answer all questions about everything. For example, most moral, ethical, religious, artistic, and legal questions simply are not within the realm of science. Investigating such issues requires different systems of inquiry and thought, which should be understood to coexist with the scientific philosophy rather than to compete with it.

A model that is very broad in scope and that dominates the framework of assumptions from which all observations are interpreted is termed a *paradigm*. A major shift in paradigm constitutes a scientific revolution, and such a shift often is accomplished only when the careers of the older generation of scientists end and a younger generation that does not have its careers invested in the old paradigm comes to dominate a field. The plate tectonic revolution is an outstanding example of such a shift in paradigm. It replaced a view of the Earth as a basically static place dominated at most by vertical tectonic movements, with a view of the Earth as an extremely mobile system dominated by large and constant horizontal motions of crustal plates and continental blocks.

A *theory*, in scientific terminology, is a coherent group of general propositions that, through repeated testing and verification, has achieved widespread acceptance as an explanation of a class of observed phenomena. Unfortunately, this word has another definition that is not entirely consistent with the scientific usage. In the common vernacular, "theory" can be used to mean a proposed explanation of a phenomenon that is still conjectural. In scientific language, we use the word "hypothesis" to express this meaning. The discrepancy between the scientific use of "theory," meaning a well-established set of propositions, and the contradictory vernacular use, meaning a proposition that is still conjectural, is commonly a source of great confusion in the public mind. Darwin's Theory of Evolution by Natural Selection comes to mind as an egregious case in point, at least in the United States.

A *law*, in scientific terminology, is a statement of a basic relationship among phenomena in the physical world that is invariable; to be considered a law, such a relationship must have been shown so often to be consistent with observation that its validity is no longer in serious question. The main difference between a theory and a law is that the term *theory* is usually applied to a broader set of propositions, whereas *law* implies a very specific proposition. In both cases, however, the propositions are widely accepted as explanations of the physical world. Even scientific laws, however, can change. Although Newton's laws of motion, for example, are still useful for everyday applications, they do not work at velocities approaching the speed of light; for these conditions, they have been supplanted by the relativistic laws of motion based on Einstein's Theory of Relativity.

iii. Deductive and Inductive Science

Scientific inquiry proceeds by two distinctive methods, the *inductive* (or *Baconian*) method, named for the English philosopher Sir Francis Bacon (1561–1626), who first espoused it, and the *model-deductive* (or *Darwinian*) method, named after the English naturalist Charles Darwin (1809–1892), who championed and practiced it. The Baconian method involves the collection of observations without regard to theory, with the expectation that explanations and natural laws will become apparent from the organization and synthesis of large amounts of observational data. The Darwinian method presumes that the scientist can devise a model that accounts for a set of observations. The model is then used to make predictions about nature, which then must be tested by comparing the predictions against objective observation of nature. By a continual iterative process of devising models, testing them through prediction and observation, modifying them to eliminate inconsistencies between prediction and observation, and further testing them, a model emerges that accounts well for the observations.

The Baconian method is intended to provide a systematic and objective approach to scientific study. Although its application results in the amassing of large quantities of data, as a system of scientific investigation it is ultimately inefficient, as it lacks a means of deciding what data are important to gather. The data that could be gathered are in principle unlimited, and we could easily miss a critical observation simply

(continued)

BOX 1-1 The Scientific Method *(continued)*

because it never occurred to us that it might be useful. This problem is embodied in the common expression, "The eye seldom sees what the mind does not anticipate." The historian and philosopher of science, Thomas H. Kuhn, expressed the problem with the Baconian method as follows:

> *since any description must be partial, the typical natural history often omits from its immensely circumstantial accounts just those details that later scientists will find sources of important illumination. (Kuhn, 1970, p. 16)*

Thus it is impossible to make a set of observations of natural phenomena that are sufficiently complete to serve all the needs of future investigators. In addition, despite the best efforts to the contrary, every scientist's work is affected to some extent by the prejudices of the society, including its scientists, the limits of current knowledge, and the brain's limitations in dealing with the unfamiliar.

Nevertheless, the Baconian method is useful in the early stages of study of a subject when data gathering helps define the field and provides the basis on which preliminary models can be formulated. Examples of the application of this method in geology include the reconnaissance mapping of a poorly known area, and the accumulation of basic information that generally follows the application of a new technology to the study of some aspect of the Earth.

The strength of the Darwinian method is that it provides a system for focusing the search for data. The method works best if a scientist can invent more than one model to account for a given set of observations, because the scientist thereby avoids the common pitfall of becoming psychologically wedded to a particular model and ignoring inconsistent data that would tend to refute it. We call this technique the *method of multiple working hypotheses*. The predictions of a model, or contradictory predictions of two different models, direct the scientist's attention toward gathering specific data that are critical for the evaluation of the models.

Even if an individual scientist does not succeed in devising multiple hypotheses, however, others in the scientific community will, and fierce arguments in the scientific literature between proponents of different models are not unusual. The net result, however, is that this method is more efficient at focusing our investigations and improving our understanding.

It is a common but egregious misunderstanding of scientists and the scientific method, however, to believe that scientists are strictly rational and unintuitive individuals and that science progresses only by the application of strict principles of logic and rational thought. The Baconian method, for example, does not specify how to carry out the process of induction that leads from the amassed data to natural laws and an understanding of nature; the Darwinian method also provides no method for the development of hypotheses or models that are required for the process of prediction and testing against observation. In the end, we must admit that both processes rely on the nonrational and nonlogical creativity, imagination, and intuition of the scientist, and that therefore these "scientific methods" are not entirely an objective and rational approach to understanding our surroundings.

It is, in fact, precisely this nonrational and nonlogical aspect of scientific inquiry that is fundamental to the progress of science. This aspect of science can never be taught by a teacher to a student because we cannot teach how to get good ideas or how to find a flash of insight. The importance of this aspect of science also means that science never can be totally systematized in any epistemology.[3] The drive of curiosity, the flash of insight, the leap of intuition are the expressions of a process that we must accept as a part of our human nature but that we do not as yet understand and cannot predict or control. It is the factor that unites the work of scientists with all the creative endeavors of human beings, whether the result of the creativity is a scientific hypothesis, a piece of music, a dance, a painting, a piece of sculpture, a poem, a novel, a play, an invention, or any result of the myriad of creative human activities that lead us in new directions and to new understanding.

For scientists, however, the results of this creative process must be passed through the filter of rational objective analysis before they can become an accepted part of the body of scientific knowledge and understanding. The filtering process requires that predictions be derived from the models and that those predictions be tested by comparing them against the objective observation of nature. Those hypotheses that fail the tests are ultimately discarded. This is the part of the sci-

[3] *Epistemology* is the branch of philosophy that considers how we know what we know and what the limits of our knowing are.

the smallest to the global scale. It is important to realize, however, that even though we may be able to invent some model whose properties resemble observations of part of the Earth, such a model is not necessarily a good one. Predictions based on a model tell us *only* about the properties and characteristics of the *model*, not the actual conditions in the Earth. The relevance of a model for understanding the Earth, therefore, must always be tested against observation.

Observations guide the formulation of models. The models in turn provide predictions that can be compared with reality, thereby stimulating new observations of the real world. Comparisons of model predictions with observations of the Earth constitute tests of the model's relevance. New observations that confirm the predictions support the model, and to that extent, we accept the model as a reasonable representation of the processes occurring in the Earth. If at any time observations con-

entific process that is rational and logical and that usually receives the most attention. It is the part that teachers try to teach, that students try to learn, and that philosophers try to systematize. But the irony is that this part of the scientific process is the part that comes after the fundamental insight, and it is really the insight that is at the core of the process.

In the final analysis, the practice of science is the practice of phenomenology: It is a practice of finding what works as an explanation and a predictor of nature, not of finding what is "True," whatever that may be. If we find a model that works consistently, we use it to improve our ability to understand, predict, and control our surroundings; if the model does not work, we discard it and look for a better one. It is ultimately a totally practical and utilitarian endeavor, not a philosophical one. Ideas about how it should or should not be accomplished, or how it was or was not accomplished, or what the significance of the results might be in terms of "Truth" are philosophical in nature, but scientists will simply continue to do whatever they can think of to find models that work, to discover how well they work, and to apply them to further our understanding of our physical world.

Good models are important for understanding structural and tectonic features, whether they be as basic as a geologic map or as theoretical as a set of numerical calculations run on a computer. When reading about the structural development of a feature or an area, a structural geologist should keep in mind such questions as: What are the assumptions inherent in the model? How strongly do the data constrain the model? Are other models equally plausible? Are the models testable, and has the researcher provided any tests? Do the available data satisfy the tests? Only such a critical approach to scientific interpretation can lead to efficient advances in our understanding of the Earth and, ultimately, of other planets.

iv. Correlation versus Causation

One of the most difficult problems in gathering and interpreting data, such as is encountered with the Baconian approach to science, is the problem of whether a correlation between two sets of observations should be interpreted as indicating a causative relationship in the processes by which they formed. We state categorically: *Correlation does not imply causation*. The erroneous assumption that a correlation implies a causative relation constantly appears in the popular press reports of science and is a major source of misunderstanding of scientific results.

In geology, we frequently are faced with the problem because of the observational nature of our science and the difficulty of performing experiments that can test whether a causative relationship actually exists. Correlation of the orientations of two different structures, for example, such as two different sets of joints, or a set of fractures and a set of faults, does not necessarily imply that the two structures are genetically, much less causally, related. Such a correlation, for example, could arise from the development of different structures in the same rock at different times or from the development of different structures by different and unrelated mechanisms.

Even the consistency of a relationship is not adequate proof of causation. If one structure always occurs in the same relationship to another, that may confirm only a common cause of the two phenomena rather than a causative relation between the two. Ultimately, one of the best ways to justify the assumption of a causative relationship is to develop a model of a process that predicts such a relationship. To the extent that such a model can be tested and confirmed, the assumption of a causative relationship is supported. Without such a model, it can be extremely difficult to determine whether a causative relationship exists, and the inference of a causative relationship from a correlation is highly suspect.

An example of such a problem is the correlation of various fracture orientations with the orientations of the layers across a fold in which the fractures occur, as discussed in Section 2.5(ii). Such a correlation raises questions such as: Did the fractures precede the folding and become reoriented by the folding of the layers? Did the folding precede the fractures and did the geometry of the pre-existing structure govern the location and orientation of subsequent fracture formation? Or did different orientations of fracture develop at different stages of fold development and record the evolution of the deformational regime during the folding? A model of the mechanics of folding can help us evaluate which of these possibilities is most likely and what evidence we might look for to try to confirm the applicability of the model.

tradict the predictions, we must refine the model or reject it and devise a new one. Our understanding of structural and tectonic processes improves gradually by a continual repetition of the processes of making observations, formulating models based on those observations, deriving predictions from the models, testing the models with new observations, and fine-tuning the models accordingly. This process, in fact, is common to all science (see Box 1-1 for more discussion of the scientific method).

1.3 THE INTERIOR OF THE EARTH AND OF OTHER TERRESTRIAL BODIES

Although most structural and tectonic processes discussed in this book are present either at the surface or within the outermost layer of the Earth (the crust), the large scale over which these motions are consistent indicates that they reflect deeper, interior processes. Moreover, Earth is not alone in the solar system. In this age

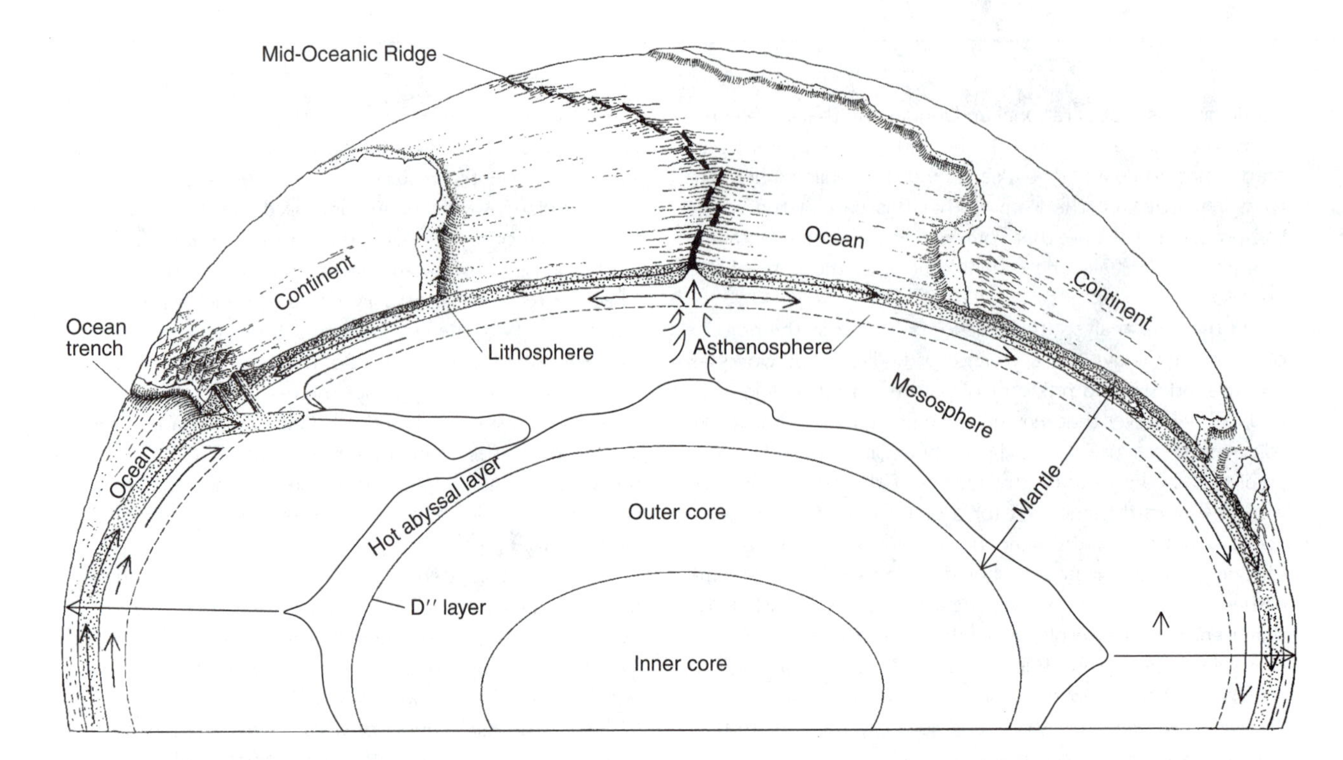

FIGURE 1.1 Diagrammatic cross section of the Earth showing the inner and outer core; the core–mantle transitional layers, labeled *D''* and *Hot abyssal layer*; the mantle with its three main layers–the lithospheric mantle, the asthenosphere, and the mesosphere; and the crust, which is the top part of the lithosphere. The spreading centers and subduction zones of plate tectonics are also indicated. New lithosphere is produced by upwelling along oceanic spreading centers. Subduction zones recycle lithospheric plates back into the interior. Some plates descend to the lower mantle. (After Wyllie, 1975 and Kellogg et al., 1999)

of space exploration, the observations and models we make of the dynamic processes in the Earth are relevant to our understanding of other planets, and our observations of other planets provide constraints on how all planets, including our own, must function and evolve.

According to current models, the Earth is divided into three approximately concentric shells; these are, from the center outward, the core, the mantle, and the crust (Figure 1.1). The **core** is composed of very dense material believed to be predominantly an iron–nickel alloy. It includes a solid inner core and a liquid outer core. Seismic data suggest that the inner core consists of crystals that are strongly aligned with respect to one another. Surrounding the core is the **mantle**, a thick shell of much lower density than the core and composed largely of solid magnesium–iron silicates. The **crust** is a thin layer that surrounds the mantle; the crust is composed predominantly of relatively low-density minerals. It is mostly made up of igneous rocks of granitic to basaltic composition, sediments and sedimentary rocks, and the metamorphic equivalents of these rocks. Chemically, the crust is composed predominantly of sodium, potassium, and calcium alumino-silicates, but it also includes other low-density minerals such as carbonates.

The mantle is itself subdivided into three main zones. The top of the mantle is a relatively cold, dense, strong layer called the **lithospheric mantle** The lithosphere as a whole actually includes the crust and the upper mantle to a depth of about 100 kilometers under ocean basins and to depths of up to 200 or 300 kilometers under continents. Beneath the lithosphere is the **asthenosphere**, which is a less-dense mantle layer that is closer to its melting temperature than the lithosphere and therefore considerably weaker. The **mesosphere** is the stronger, high-density lower mantle that consists principally of denser crystalline phases of magnesium–iron silicates or oxides.

The temperature of the Earth increases with depth with a gradient of approximately 25°C per kilometer in the crust and upper mantle, and with considerably smaller gradients deeper within the Earth. Several sources of heat account for this increase of temperature with depth: Residual heat was trapped during the original accretion of the Earth approximately 4500 million years ago. Heat is produced continually by the spontaneous decay of radioactive elements within the Earth. Heat may also be added by the latent heat of crystallization from slow solidification of the liquid outer core and by the dissipation of tidal energy resulting from the gravitational interaction among the Earth, Moon, and Sun.

The increase of temperature with increasing depth results in a flow of thermal energy toward the surface. For many reasons, we accept that this energy is transported by at least two systems of convection within the Earth.

One convection system transports heat out of the liquid core,[3] and this flow is part of the process that generates the Earth's magnetic field. This heat is transferred by conduction to the lower mantle. A separate system of convection in the solid mantle carries the heat transferred from the core plus heat generated by radioactive decay of isotopes within the mantle toward the surface. This convection operates by the flow of the solid crystalline mantle material. Heat escapes from the Earth by conduction through the cold lithospheric boundary layer, by advection of heat in intrusive and eruptive magmas, and by upwelling of asthenosphere at oceanic spreading centers. Partial melting of the asthenosphere takes place during this upwelling, and the melts rise to the ocean bottom where they crystallize to form the rocks of the oceanic crust. The return flow for the mantle convection occurs largely at subduction zones, where cold lithosphere descends as slabs back into the mantle. Detailed seismic and geochemical investigations suggest that some lithospheric slabs penetrate as deep as the bottom of the mantle.

Thus we could view the Earth as a vast sphere of seething convective flow. Our studies of structural geology and tectonics focus only on the motion and deformation of the outer 20 to 30 kilometers of this enormous system. It puts into perspective the task of understanding the dynamics of the Earth to realize that what we have available to study is the thin rind (some might call it the scum) of relatively cold but low-density crust that rides passively on top of the lithospheric boundary layer to the mantle convection currents. Nevertheless, that thin layer preserves a long history of this convective activity, which is available for study nowhere else. So we must learn to interpret it and to integrate its lessons with the results of other branches of geology to come to a better understanding of the Earth's evolution.

The Earth is one of a class of similar objects in the solar system called the **terrestrial bodies**. They are the innermost four planets (Mercury, Venus, Earth, and Mars), plus the Earth's Moon and several moons of Jupiter and Saturn, including Io, Titan, and Europa. All the terrestrial bodies apparently consist of a central core of very dense material that is probably an iron–nickel alloy, a mantle of less-dense silicates, and a thin crust of even lower-density silicates. The inner four planets differ in their total size and their relative volumes of core. The core is proportionately greatest for Mercury and progressively smaller for Venus, Earth, and Mars. The core of the Moon is proportionately smaller still.

Although our study of the other terrestrial bodies is based primarily on information from orbiting spacecraft and a few landings on the Moon, Venus, and Mars, the current state of our knowledge invites fascinating comparisons with the Earth and speculation about the origin of the various similarities and differences. As our understanding of the other terrestrial bodies increases, it will provide tests for models we have devised to explain dynamic processes occurring within the Earth. These models will have to account for the presence or absence of such processes on the basis of the size, structure, and internal physical conditions of the other terrestrial bodies.

1.4 THE EARTH'S CRUST AND PLATE TECTONICS: INTRODUCTION

The crust of the Earth is broadly divided into continental crust of approximately granodioritic composition and oceanic crust of roughly basaltic composition. Land—that part of the Earth's surface above sea level—is principally continental, the exceptions being islands in the oceans. At the present time, 29.22 percent of the Earth's surface is land and 70.78 percent is oceans and seas. Continental crust makes up 34.7 percent of the total area of the Earth, and it underlies most of the land area as well as the continental shelves and continental regions covered by shallow seas, such as Hudson's Bay and the North Sea. The remaining 65.3 percent of the Earth's surface is oceanic crust (Figure 1.2).

The surface elevation of the Earth shows a strongly bimodal distribution. Most of the continental surface lies within a few hundred meters of sea level, and most of the ocean floor lies approximately 5 kilometers below the sea surface. This distribution is evident from the two types of **hypsometric[4] diagrams** shown in Figure 1.3. Figure 1.3*A* is a cumulative plot that shows the total percentage of surface above a given elevation; Figure 1.3*B* is a histogram that shows the percentage of the surface within a given elevation interval. The difference in elevation between continent and ocean floor (Figure 1.3*B*) results from a number of factors, including the thickness and density differences between the continental and oceanic crusts, tectonic activity, erosion, sea level, and the ultimate strength of continental rocks, which determines their ability to maintain an unsupported slope above oceanic crust.

Available information from other terrestrial bodies, such as Venus, Mars, and Earth's Moon, indicate that their distributions of elevation are unimodal, not bimodal, as with Earth. The difference is thought to be because these other terrestrial bodies lack the contrast between continental and oceanic crust that exists on Earth.

The characteristics of the Earth's crust are largely the direct or indirect result of motions of the lithosphere. The theory of plate tectonics describes these motions and accounts for most observable tectonic activity in the Earth, as well as the tectonic history recorded in the ocean basins. The theory holds that the Earth's lithosphere is divided at present into seven major and several minor plates that are in motion with respect to one another

[3]The heat presumably arises from crystallization of the solid inner core. The inner core may have reached a significant size only within the last 1 Ga (10^9 years) or so of Earth history.

[4]After the Greek words *hypsos*, "elevation," and *metron*, "measure."

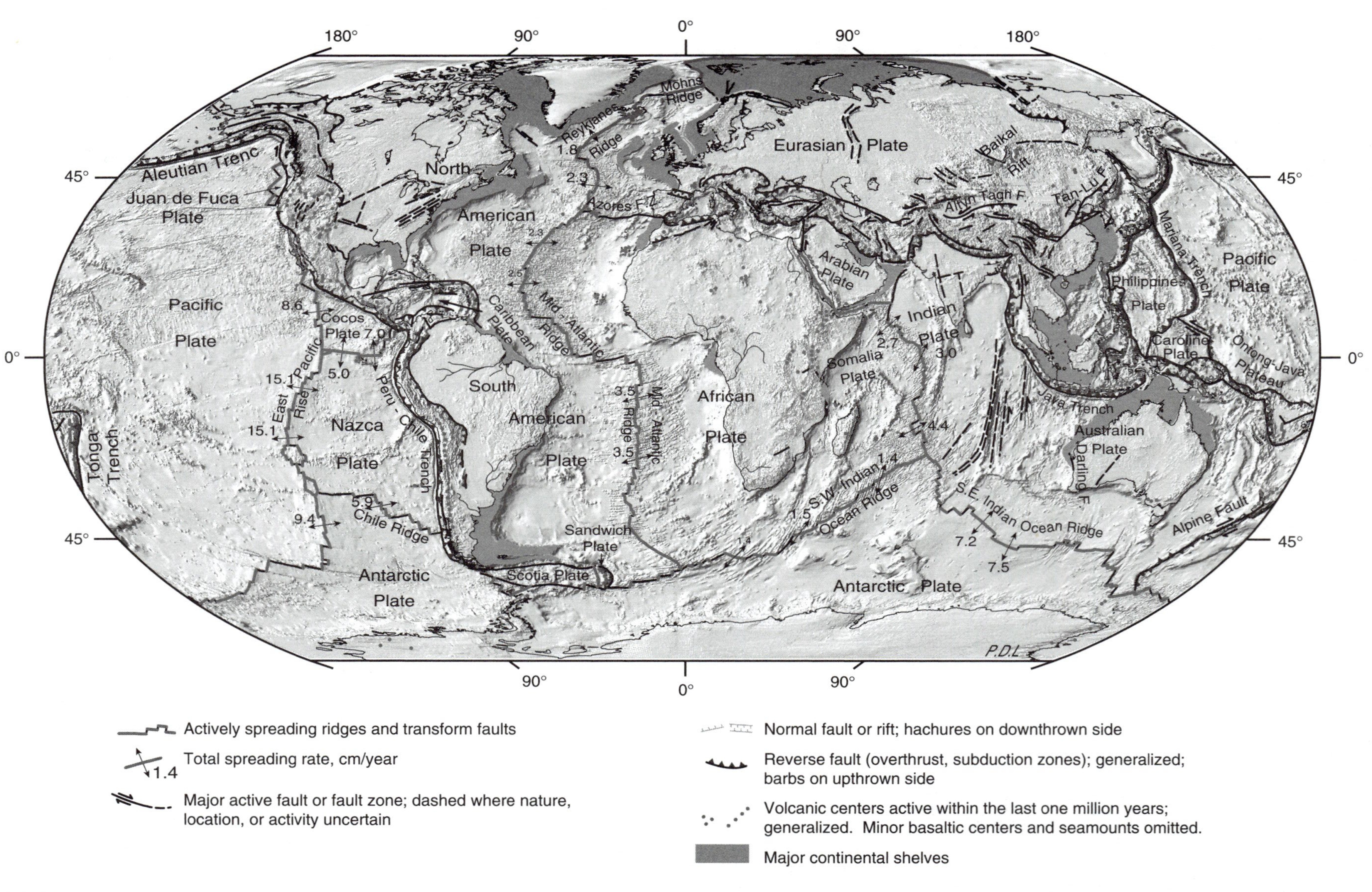

FIGURE 1.2 Distribution of land, continental shelves, ocean basins, major plate boundaries, and tectonic plates on the surface of the Earth. Selected average relative plate velocities are illustrated. (After NASA, 2002)

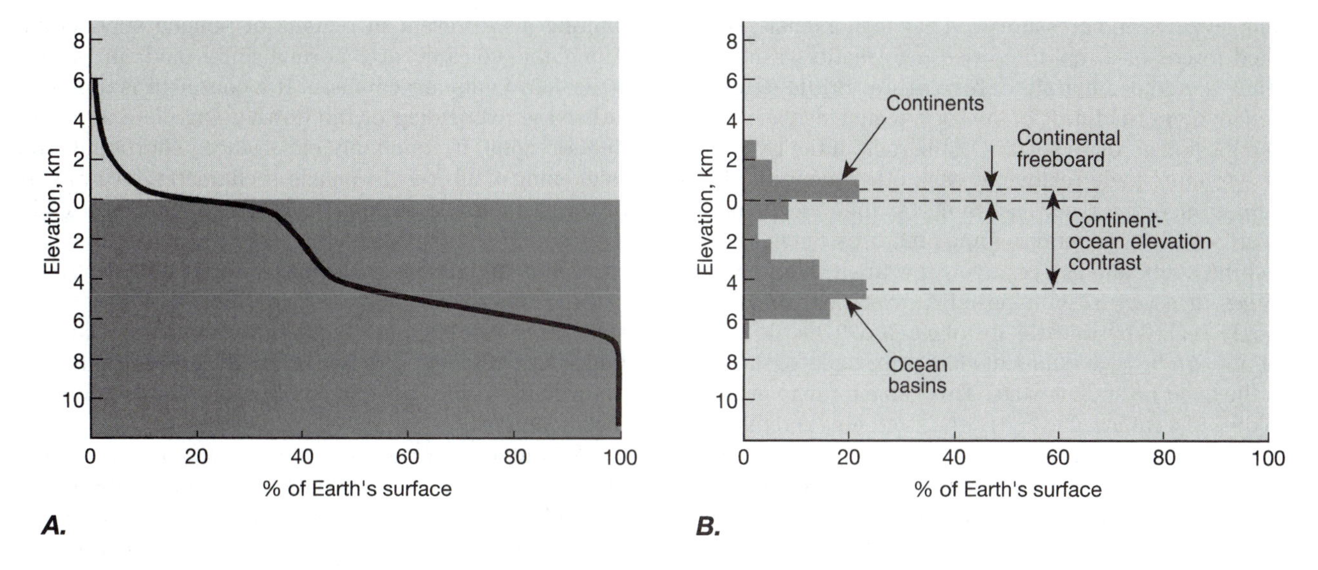

FIGURE 1.3 Distribution of topographic elevations on the Earth. *A.* Cumulative curve showing the percentage of the Earth's surface that is above a particular elevation. *B.* Histogram showing the percentage of the Earth's surface that falls within each 1-kilometer interval of elevation. The continental freeboard is the difference between the mean elevation of the continents and mean sea level. The continent-ocean elevation contrast is the difference between the mean elevation of the continents and that of the ocean basins.

(Figures 1.1 and 1.2) and that the motion of each plate is, to a first approximation, a rigid-body motion. This does not necessarily imply, however, that the plates are rigid; it simply means that the interiors of the plates are not deforming, that is, they are not changing shape. Deformation of the plates is concentrated primarily in belts tens to hundreds of kilometers wide along the plate boundaries. In a few regions, however, deformation extends deep into plate interiors.

The different types of boundaries between these plates include divergent boundaries, convergent or consuming boundaries, and conservative or transform fault boundaries. At divergent boundaries, which are mainly midoceanic ridges, two plates move away from each other and lithosphere is created. At convergent or consuming boundaries, also called subduction zones, two plates move toward each other and one descends beneath the other, recycling lithosphere back into the mantle. At conservative or transform fault boundaries, two plates move horizontally past each other without creation or destruction of lithosphere. The asthenosphere apparently is weak and deforms relatively easily; it is the layer along which the plates slide relative to the deeper mantle. Earthquakes indicate that subducted plates descend to at least 700 kilometers in some instances; geophysical data suggest that some plates even descend into the lower mantle (see Figure 1.1).

The most direct evidence for plate tectonic processes and seafloor spreading comes from the oceanic crust, where divergent motion at midoceanic ridges adds new material to lithospheric plates. As indicated in Figure 1.4, however, the maximum age of the oceanic crust limits this evidence to the last 180 million years—that is, to the last 4 percent of Earth history. Any evidence of plate tectonic processes for the preceding 96 percent of Earth history must come from the continental crust, which contains a much longer record of the Earth's activity. We must therefore learn to understand the large-scale tectonic significance of deformation in the continental crust so that we can see further back into the history of the Earth's dynamic activity.

In the geologic record, highly deformed continental rocks tend to be concentrated in long linear belts comparable to the belts of deformation associated with current plate boundaries. This observation suggests that belts of deformation in the continental crust record the existence and location of former plate boundaries. If this hypothesis is correct, and if we can learn what structural characteristics of deformation correspond to the different types of plate boundaries, we can use these structures in ancient continental rocks to infer the pattern and processes of former tectonic activity. In this sense, the plate tectonic model has united the disciplines of structural geology and tectonics and made them interdependent.

The types of structures that develop in rocks during deformation (characteristically along plate boundaries) depend on a number of factors, including the orientation and intensity of the forces applied to the rocks, or the motions to which the rocks are subjected; the physical conditions, such as the temperature and pressure, under which the rocks are deformed; and the mechanical properties of the rocks, which are strongly affected by the physical conditions and the type of rock. At relatively low temperatures and pressures and at a high intensity of applied forces or a rapid imposed deformation, rock generally undergoes **brittle deformation**. Brittle deformation involves fracturing of the rock, which involves a loss of cohesion across a surface in the rock. If the two sides of a fracture slide relative to each other along the fracture surface, the result is a fault. On the other hand, at relatively high temperatures and pressures, but below the melting point, and at a relatively low intensity of applied forces or a very slow imposed deformation, rock commonly reacts by **ductile deformation.**[5] Ductile deformation is a flow, or coherent change in shape, of the rock in the solid crystalline state. This behavior may produce folding of stratigraphic layers, the stretching and thinning of layers, and the parallel alignment of mineral grains in the rock to form pervasive planar and linear preferred orientations. Description of the structures formed from the brittle and ductile modes of deformation are the focus, respectively, of Parts I and II of this book.

In the belts of deformation along the plate boundaries, the relative motions of adjacent plates, which can be divergent, convergent, or conservative, largely determine the style of deformation. Differences between oceanic and continental crust also affect the nature of deformation along plate boundaries.

At divergent boundaries, material flowing upward in the mantle accommodates the separation of the plates at depth. Partial melting of the upwelling mantle forms basaltic magma, and intrusion and extrusion of these basalts produce the new oceanic crust. The separation of the plates may also result in horizontal stretching and vertical thinning of the crust by normal faulting near the surface and ductile thinning at deeper levels. When a divergent plate boundary develops within a continent, the associated horizontal stretching and vertical thinning lower the mean elevation of the boundary zone sufficiently that it becomes flooded by the sea. Such stretched and thinned continent commonly underlies the wide continental shelves (Figure 1.2).

Subduction zones at convergent boundaries are the places where a lithospheric plate plunges back into the interior of the Earth, recycling the oceanic crust at its top back into the mantle. Sediments on the down-going plate may be subducted with the down-going plate or scraped off, in varying proportions, and partial melting of the down-going plate and sediments, or of the mantle wedge above it, produces characteristic volcanic arcs on the over-riding plate. Structures at the plate boundary are predominantly systems of thrust faults with strike-slip faulting also present in regions of oblique subduction. Along the volcanic arc, normal faults and, in places, strike-slip faults are common. If a continent is a part of either the over-riding or the down-going plate at a subduction zone, it commonly experiences shortening and thickening of its crust by means of characteristic systems of thrust faults. In regions of continent–continent collision, complex deformation results in faulting of all types, as well as folding, metamorphism, and igneous activity.

At conservative or transform fault boundaries, the structures that form are typically systems of strike-slip faults or, at deeper levels, vertical zones of ductile deformation that have a subhorizontal direction of displacement.

A variety of secondary structures also develop in any of these tectonic environments, so the presence of any particular structure *per se* is not necessarily diagnostic of the type of boundary at which it developed. The genesis can be inferred only after careful study of the regional pattern of the structures and their associations.

Structural and tectonic processes profoundly influence other Earth processes as well. For example, the continents have varied in number, size, and geographic position as a result of plate tectonic processes. As plate tectonics changed the shape and distribution of continents and ocean basins, the patterns of oceanic and atmospheric circulation changed accordingly. The resulting changes in environmental conditions have affected both the patterns of sedimentary environments, as revealed by studies of sediments and stratigraphy, and the patterns of natural selection and evolution, as revealed by studies of the fossil record.

Because plate boundaries are sites of major thermal anomalies in the crust and upper mantle, these areas

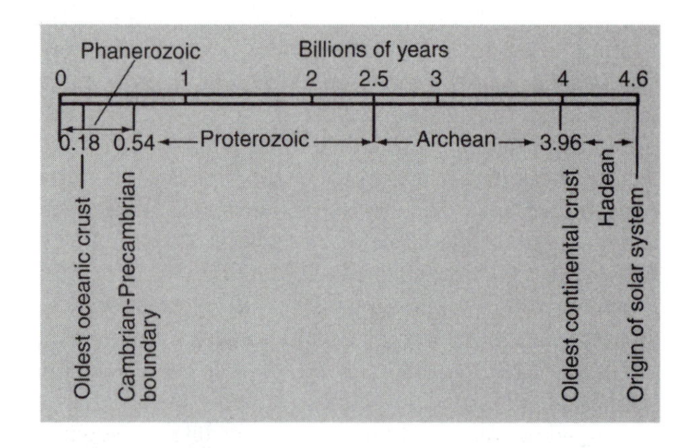

FIGURE 1.4 Time line showing different events in Earth's history and the ages of the oldest oceanic and continental crusts. The "eons" of Earth time—the Phanerozoic including the Paleozoic, Mesozoic, and Cenozoic eras; the Proterozoic including the Paleoproterozoic, Mesoproterozoic, and Neoproterozoic eras; the Archean; and the Hadean are shown. The last is the interval of time between the formation of the solar system and the oldest rocks on Earth.

[5]We use the term *ductile* to imply coherent nonrecoverable deformation that occurs in the solid state without loss of cohesion (brittle fracturing) at the scale of crystal grains or larger. The term has broader significance in other contexts, but no other word adequately describes this behavior. In particular, the term *plastic* has other specific connotations that do not accurately reflect the behavior we wish to describe. See the introduction to Part II for a more detailed discussion of these terms.

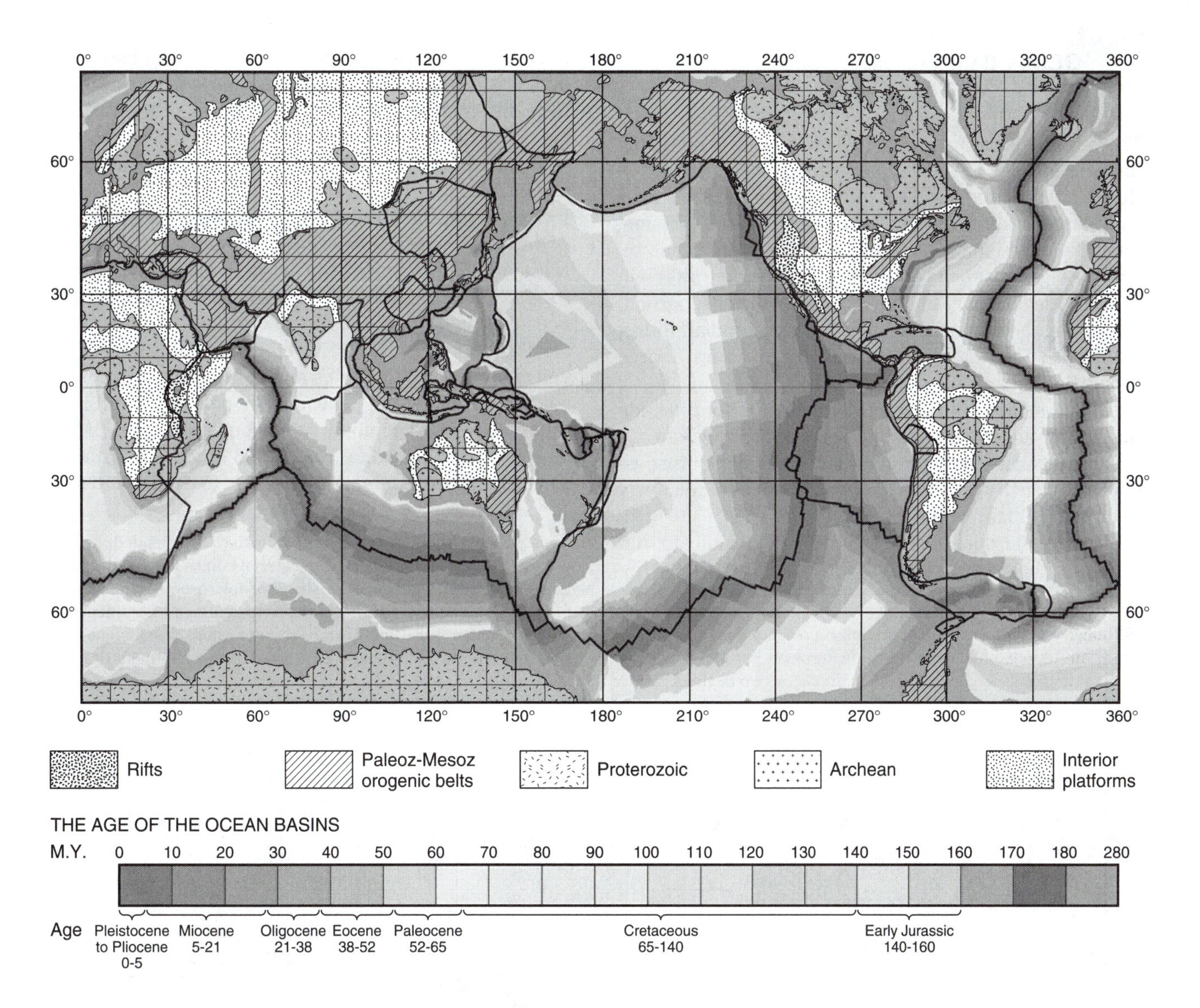

FIGURE 1.5 Generalized world map showing major features of continental crust and age of oceanic crust. Continental features include: Precambrian shields with Archean and Proterozoic areas, interior lowlands, orogenic belts, rifts, and margins. (Modified after Anonymous, 1950; Pitman, 1974; Stanley, 1999, figure 8-25, p. 227; Sloss, 1996)

control the occurrence and distribution of igneous and metamorphic rocks, which are studied in "hard rock" petrology. Similarly, the formation, concentration, and preservation of mineral deposits are profoundly affected by structures and their tectonic environments, as well as by the thermal anomalies at plate boundaries. As exploitation inexorably depletes the Earth's resources, increasingly sophisticated and subtle exploration strategies are required in order to find and develop new deposits. Structural geology and tectonics are assuming an increasingly crucial role in the search for metal and hydrocarbon deposits.

Although oceans occupy most of the Earth's surface area, oceanic crust is substantially younger than continental crust. Relatively young oceanic crust near active spreading centers occupies a larger area than the oldest crust, most of which has disappeared down subduction zones (Figure 1.5). Thus most oceanic crust has a relatively simple history that reflects only the most recent events, within the last 4 percent of the Earth's tectonic history.

Although subordinate in area to oceanic crust, the continents are better exposed, better studied, and better known. Continental crust ranges in age from 0 to at least 3.96 Ga, which encompasses the last 88 percent of Earth's tectonic history (Figure 1.4). Thus, for most of Earth's history, the only record of active geologic processes is preserved in continental rocks. The geology of continental rocks is far more complex than that of oceanic crust (Figure 1.5), and the record of the oldest history is obscured by subsequent events, which may overprint, destroy, or bury the older evidence.

1.5 OCEAN BASINS

Vast areas of the oceanic bottom are flat or nearly flat. The oceanic crust underlying these areas is remarkably uniform in thickness and composition. Oceanic crust is thin relative to continental crust, ranging in thickness from 3 to 10 kilometers and averaging about 7 kilometers. Compositionally, oceanic crust consists predominantly of igneous rocks of basaltic composition, although in some magma-starved slow-spreading regions, particularly along the mid-Atlantic and Southwest Indian Ridge, abundant serpentinized mantle rock makes up a large proportion of the oceanic crust.

These nearly flat areas of the ocean bottom include the ridges[6] and abyssal plains. Scattered throughout the ocean basins are plateaus of anomalously thick crust, island arc–trench systems, and aseismic ridges of relatively thick crust (Figure 1.6).

Gravity measurements over the oceans indicate that generally the free air anomaly is nearly zero (see Appendix A2.2). Thus, for the most part, ocean basins are in isostatic equilibrium, and differences in elevation reflect differences in density or thickness of the underlying crust and/or mantle.

An average layered model for the oceanic crust, shown in Figure 1.7*A*, is based on the seismic P-wave velocity (V_p) measurements. The lithologic interpretation of these layers results from direct sampling of the oceanic crust and from comparison with on-land exposures of rock sequences (**ophiolites**) thought to represent old oceanic crust.

The uppermost layer, Layer 1, has a seismic P-wave velocity (V_p) of 3 to 5 km/s and is interpreted as unconsolidated sediment of pelagic, hemi-pelagic, or turbiditic origin.[7] Layer 2, commonly subdivided into Layers 2A, 2B, and 2C with V_p ranging from 4 to 6 km/s, is interpreted as predominantly submarine basaltic extrusive and shal-

[6]Although the midocean ridge system represents one of the most important topographic and tectonic features on Earth, the average slope of its flanks is generally less than one or two degrees.

[7]Pelagic sediments form by settling of material suspended throughout the ocean water column. The material comes either from windborne dust from land or from shells of microscopic animals and plants. Hemi-pelagic sediments contain significant continental or volcanic material. Turbidite is a sediment formed by sediment-laden bottom currents and is generally derived from a continent or island source.

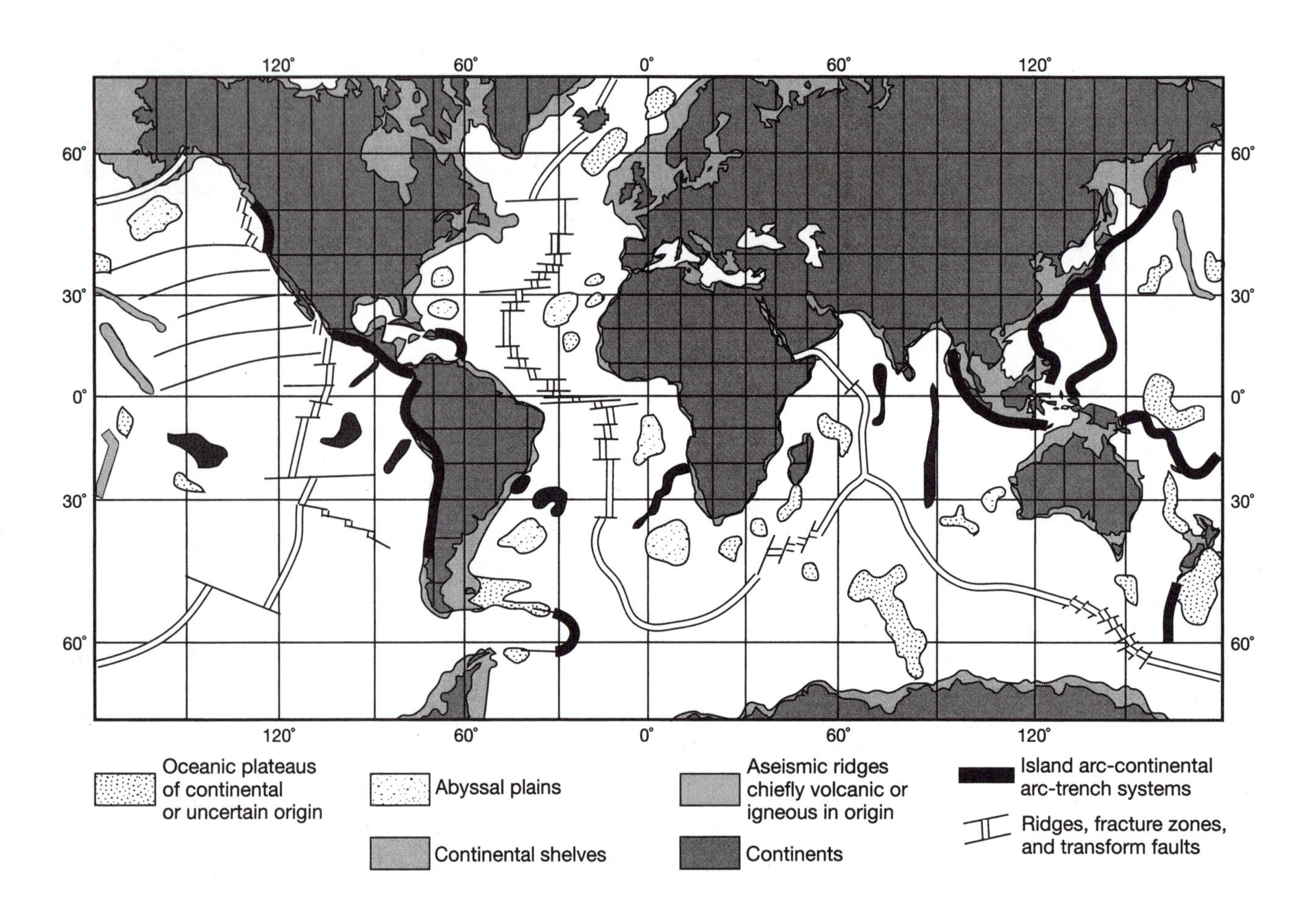

FIGURE 1.6 Generalized world map showing major oceanic features: ridges, transform faults and fracture zones, oceanic plateaus, aseismic ridges, and island or continental arc-trench systems. (Modified from Bally, 1980; Uyeda, 1978)

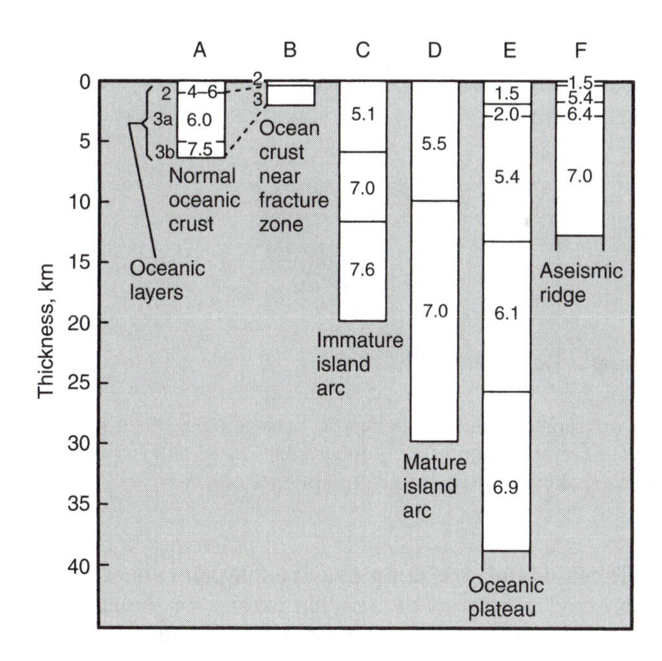

FIGURE 1.7 Layered models of seismic velocity for typical oceanic crust and other oceanic features. P-wave velocities in km/s are indicated for the different model layers.

low intrusive rocks. Layer 2A has a relatively low velocity that increases rapidly with depth; Layer 2B is a layer of relatively constant velocity; and Layer 2C displays a rapid increase of seismic velocity with depth. Layer 3, with common subdivisions 3A and 3B and V_p ranging from 6 to 7.5 km/s, is thought to represent mafic-ultramafic[8] plutonic rocks and/or serpentinized mantle peridotite. Layer 3 subdivisions may also reflect varying quantities of olivine in plutonic rocks.

For descriptive purposes, we divide the principal features of oceanic crust into features characteristic of plate margins and those characteristic of plate interiors.

i. Features of Oceanic Plate Margins

Divergent plate margins are topographically high regions characteristically in the middle of the ocean basins (except for the eastern Pacific Ocean and northwestern Indian Ocean). These **midoceanic ridges** form a continuous world-girdling topographic swell approximately 40,000 km long, 2.5 km high above the abyssal floors of the ocean basins on either side, and 1000 to 3000 km wide. Structures on ridges are predominantly active normal faults, as revealed by the morphology of the ocean floor and by the first motion studies of earthquakes (Figure 1.8; see also Appendix A2.1 and Box A2-4). The faulting is consistent with extension perpendicular to the trend of the ridge and parallel to the inferred relative plate motion.

Transform fault boundaries in the oceans are the seismically active portions of **fracture zones**—great

[8]"Mafic" and "ultramafic" are terms used to indicate the composition of the rocks in question. "Mafic" rocks have one or more Mg-Fe–bearing mineral, such as amphibole, pyroxene, or olivine. "Ultramafic" rocks consist chiefly of Mg-Fe–bearing minerals. Rocks of basaltic or gabbroic compositions are mafic, whereas peridotites and serpentinites are ultramafic.

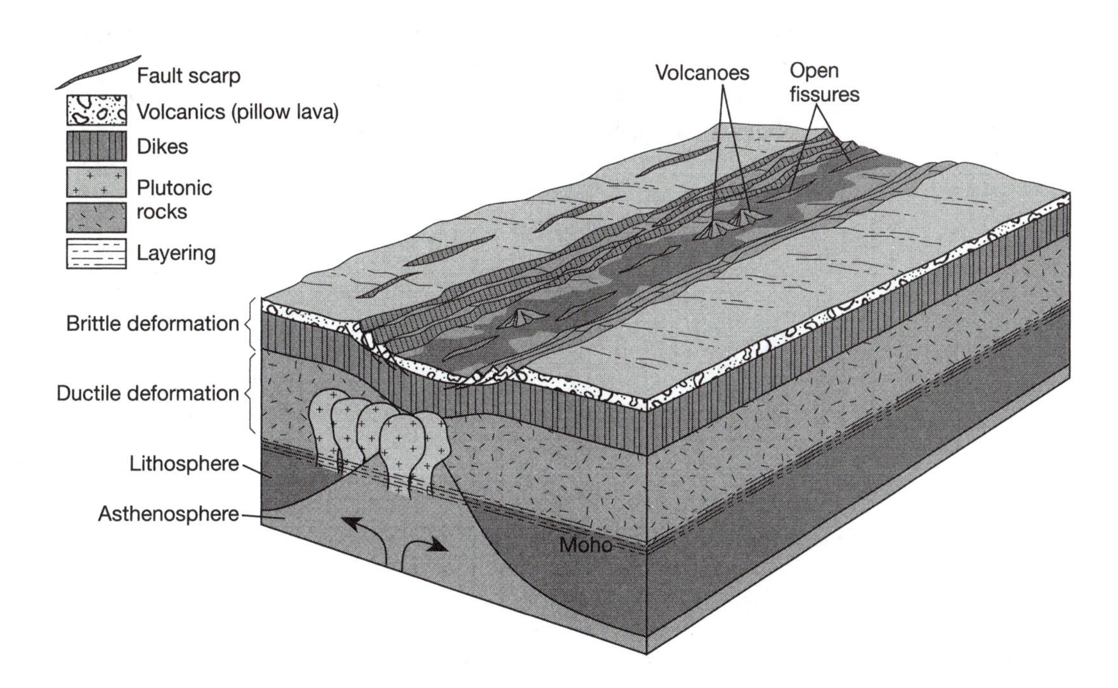

FIGURE 1.8 Block diagram schematically illustrating principal features of a midoceanic divergent plate margin. Extensional (normal fault) structures at the surface pass downward into a zone of magmatic intrusion and ductile stretching. Lithosphere thickens away from the plate margin. Not to scale.

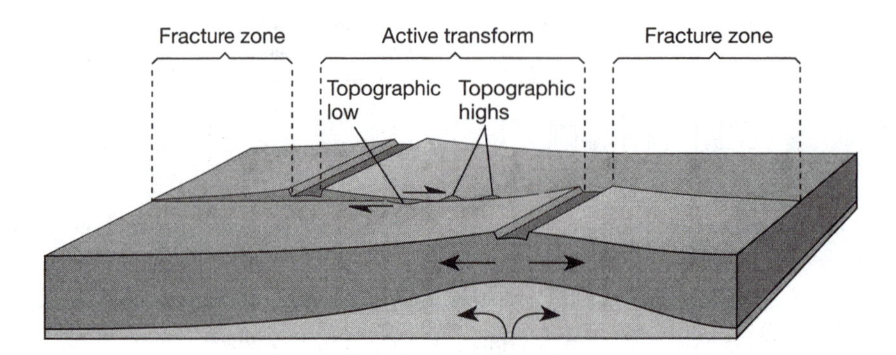

FIGURE 1.9 Block diagram schematically illustrating a conservative, or transform fault, boundary in oceanic crust offsetting a divergent margin (ridge). Structures of each offset portion of the ridge are as in Figure 1.8. "Topographic highs" and "topographic lows" schematically represent sharp ridges and troughs that are present along transform fault zones. Not to scale.

rectilinear fracture systems within the oceanic crust. They are characterized by pronounced differential topographic relief, commonly sharp ridge and trough topography, steeply dipping faults, and deformed oceanic rocks (Figure 1.9). They range in length up to 10,000 km. Although they commonly are fairly narrow features, some fast-slipping transform faults (principally in the east Pacific) are up to 100 km or more in width and contain short spreading segments. The seismically inactive portions of these fracture zones represent fossil transform faults. First motion studies of the earthquakes along the active transform fault portions indicate mostly strike-slip faults with characteristic horizontal relative motion *opposite* in sense to that of the apparent offset of the ridge crest. The thickness of oceanic crust near fracture zones and transform faults tends to be less than average for the interfault portions of the oceanic crust (Figure 1.7*B*).

Convergent plate margins in the oceans exhibit chains of volcanic islands accompanied by parallel trenches, which are the deepest parts of the ocean basins. These **island arc–deep sea trench** pairs generally are arrayed in a series of arcs that extend for thousands of kilometers and join at cusps. The volcanic islands are spaced approximately 70 to 80 km apart and rise above submerged ridges that tend to be a few hundred kilometers wide. Trenches are up to 12 km deep and approximately 100 km wide. The landward side of trenches is characterized by systems of active thrust faults and possibly by strike-slip faults in regions of oblique subduction. Island arcs and regions behind the arcs exhibit active normal faults and in some cases strike-slip faults (Figure 1.10). Trenches are associated with pronounced negative Bouguer gravity anomalies, indicating a marked mass deficiency below the seafloor.

The crust of island arc regions averages 25 km in thickness, considerably thicker than that of normal oceanic crust. It is rather variable, however, thinning abruptly to oceanic thicknesses on either side of the arc axis (Figure 1.7*C*, *D*). Younger immature arcs tend to have a thinner crust (Figure 1.7*C*) than older mature ones (Figure 1.7*D*).

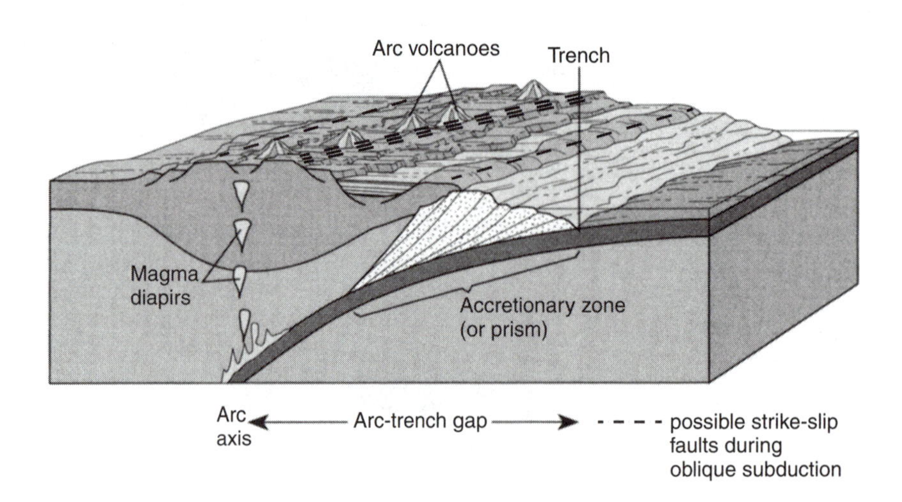

FIGURE 1.10 Block diagram schematically illustrating principal features (island arc and deep sea trench) in an intra-oceanic convergent plate margin or subduction zone. One plate descends beneath another along a marginal zone of thrust faults. Partial melting of downgoing crust produces blobs of magma that rise and become volcanoes. Not to scale.

ii. Features of Oceanic Plate Interiors

Away from plate margins, the deepest regions of the ocean (about 5 km below sea level) are vast areas of very flat ocean floor, the **abyssal plains**. These plains represent areas of normal oceanic crust covered by pelagic sediments and turbidites, sediments deposited from turbidity currents.

Broad elevated regions, or **oceanic plateaus**, have a variety of origins. Some apparently are continental rocks, others are inactive volcanic arcs. The origin of others is unclear. They range in area from a few hundred to many thousands of square kilometers, and they stand 1 to 4 km above the normal ocean floor. Crustal thicknesses are generally more similar to those of continental crust than to oceanic crust (Figure 1.7*E*).

Linear ridges characterized by high elevation, anomalously thick oceanic crust, and a general lack of associated seismic activity, are called **aseismic ridges**. Their lack of seismic activity and more limited dimensions, as illustrated in Figure 1.6, set them apart from the mid-oceanic ridges. In most cases, they represent linear constructional ridges formed by chains of basaltic volcanoes. The Hawaiian Islands–Emperor Seamount chain, extending northwest from Hawaii to Midway Island and north to the Kamchatka trench (Figure 1.6) forms the most famous example of this type of crustal feature. Crustal thickness of aseismic ridges is considerably larger than normal oceanic crust and comparable to that of island arcs (Figure 1.7*F*).

1.6 THE STRUCTURE OF CONTINENTAL CRUST

Continents are most conveniently subdivided into regions according to their large-scale tectonic features. It is also useful to distinguish Precambrian from Phanerozoic regions of continental crust. Thus we distinguish Precambrian shields from Phanerozoic regions consisting of interior lowlands, orogenic belts, and continental rifts and margins. These subdivisions are the basis of our discussion in Sections 1.7 and 1.8.

Compared with oceanic crust, continental crust is more complex in structure, because in general it is older and has experienced a much longer history of tectonic activity. Continental crust is also thicker and less dense than oceanic crust and has lower seismic velocity. Typical features of continental crust are illustrated in Figure 1.11, which is an idealized east–west cross section of North America.

The average thickness of continental crust is about 35 km, although there is considerable variation from that average, depending on location and tectonic setting (Figure 1.12). The crust tends to be thickest under mountainous regions (up to 70 km or more), about average under sedimentary platforms, and relatively thin under rifts such as the Basin and Range province, under continental margins, and under Precambrian shields. The crustal seismic velocity tends to increase with depth, but layers in which the velocity decreases with increasing depth, known as **velocity inversions**, are reported from some regions, such as the Basin and Range province and the Tibetan Plateau of central Asia.

The nature of the deep continental crust can be observed directly where rocks formed at depth have been uplifted and exposed by erosion. Exposures of shallow crustal rocks characteristically show metamorphosed sedimentary and igneous rocks that have been deformed by faulting and folding and intruded by igneous rocks. Deep crustal rocks characteristically consist of granulites, which are very high-grade metamorphic rocks, and granitic gneisses, and these rocks also contain very complex structure (for examples, see Figures 1.14 and 1.15). Such structures are relatively common in many

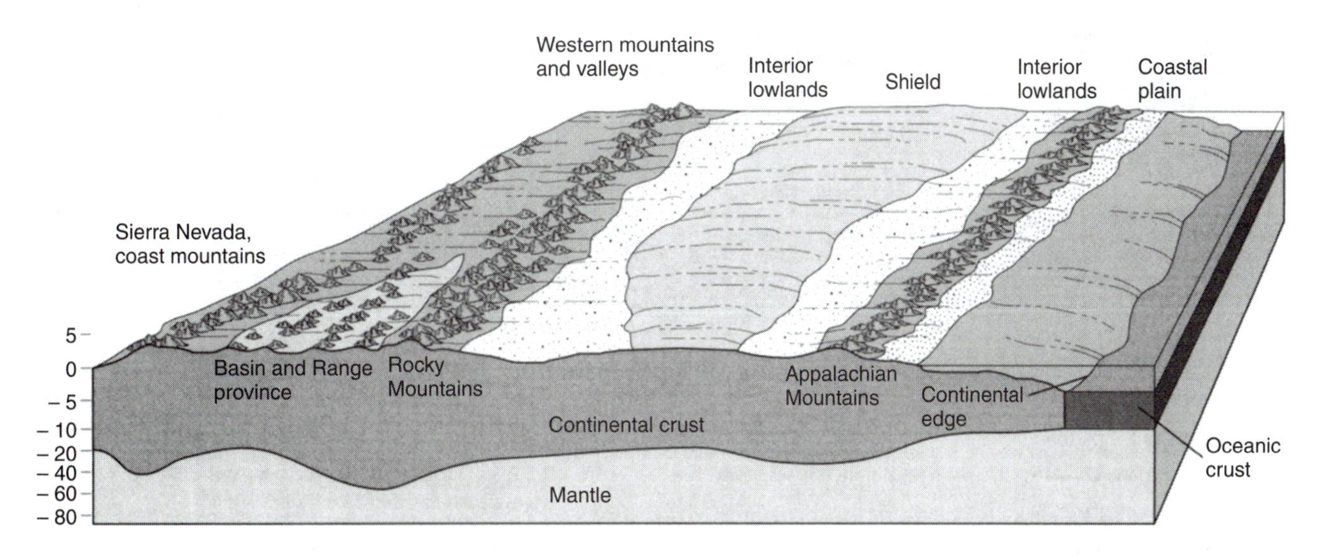

FIGURE 1.11 Generalized block diagram of the North American continent with an East-West section showing variation in crustal thickness in kilometers for the various tectonic provinces. Note the change in vertical exaggeration at about 10 km.

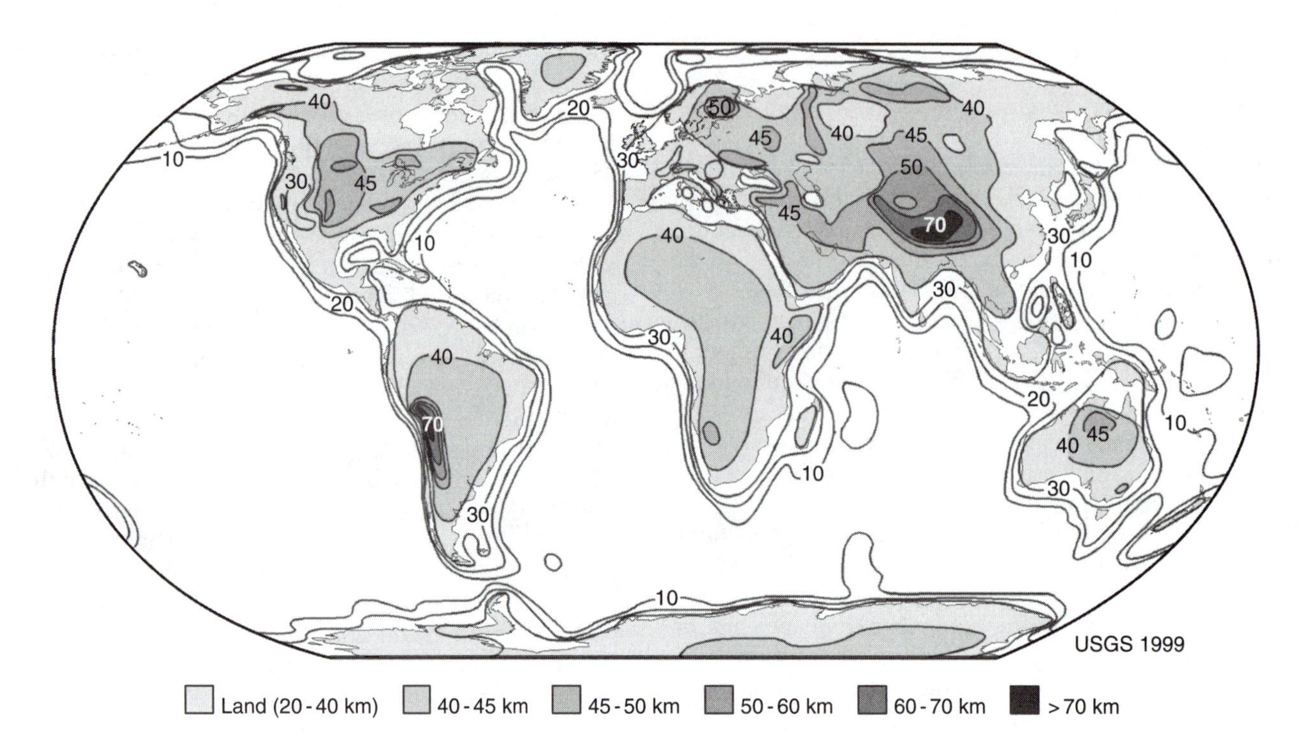

FIGURE 1.12 Generalized map of the world showing continental crustal thickness in kilometers. Note that the continental margins are approximately outlined by the 30 km contour. (US Geological Survey contour map based on the crustal model CRUST 5.1 [Mooney et al., 1998] http://quake.usgs.gov/research/structure/CrustalStructure/index.html)

Precambrian shields and in deeply eroded central zones of Phanerozoic orogenic belts. It seems reasonable to infer that they are common throughout the lower crust.

Correlation of field data with the seismic velocity structure of the continental crust suggests the generalized petrologic and seismic model shown in Figure 1.13. The upper levels of the crust, beneath the sedimentary cover, consist of metamorphosed sedimentary and volcanic rocks intruded in places by granitic rocks. The middle crust is composed mostly of migmatite, a felsic rock that was partially melted and strongly deformed during metamorphism, and that is characterized by a slightly

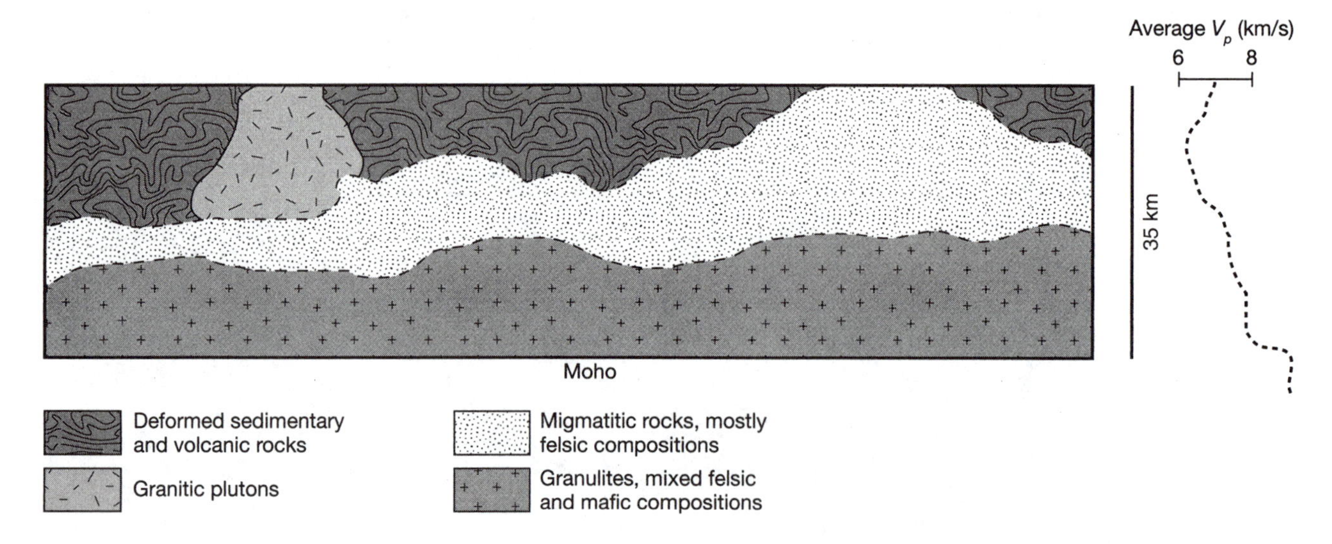

FIGURE 1.13 Generalized crustal model showing lateral and vertical inhomogeneities to account for observations in deeply eroded regions and for observed variations in seismic velocity. The diagram to the right shows idealized average P-wave velocities (V_P) through the crust. (Modified from Smithson et al., 1979, p. 263)

lower seismic velocity than the upper crust. The lower crust consists of highly folded rocks commonly exhibiting granulite facies[9] of metamorphism, intruded by mafic and silicic plutonic rocks. These rocks have a slightly higher seismic velocity than the upper crust. Scattered, highly folded remnants of gneissic ultramafic rocks are also probably present, based on inferred lower crustal exposures in regions such as the Italian Alps, southwestern New Zealand, and central Canada.

The Moho discontinuity under continents is generally less sharp than underneath oceans. It occurs over a depth range of several kilometers and marks a gradual transition from mostly mafic to mostly ultramafic rocks. Some subcontinental mantle may be mafic rocks in the form of **eclogite**, a high-pressure metamorphic rock chemically comparable to basalt, but composed of Na-rich pyroxene and Mg-rich garnet. It has a high density and a high seismic velocity that are comparable to mantle values.

This model of the crust includes great lateral and vertical heterogeneity, which is consistent with the available information. A great amount of detailed exploration of the continental crust is taking place around the world, and we can expect much to change in our knowledge and concepts in the years ahead.

1.7 PRECAMBRIAN SHIELDS

All continents exhibit large areas where Precambrian rocks greater than 600 million years old (Ma) are exposed at the surface (Figure 1.5). These regions commonly form topographically rolling uplands that stand higher than the lowlands surrounding them, thus giving rise to the name *Precambrian shield*.

We subdivide Precambrian shields into Archean and Proterozoic[10] terranes based on the age of the rocks. This subdivision has tectonic significance that is of worldwide utility. Archean rocks are mostly greater than 2500 Ma, and Proterozoic rocks range in age from 2500 Ma to approximately 540 Ma (see Figure 1.4). Archean regions display evidence of greater crustal instability or mobility than Proterozoic regions. The tectonic distinction is not universal or abrupt, however, and the transition from one tectonic style to the other varies by a few hundred million years or so from place to place.

i. Archean Terranes

Rocks in Archean terranes are divisible on the basis of their metamorphic grade into **high-grade gneissic regions**, exhibiting amphibolite or granulite facies of metamorphism, and **greenstone belts**, characterized by rocks at greenschist or lower grades of metamorphism.[11] Both types are characteristically intruded by younger granitic plutons. Figure 1.14 shows a map of a typical region, part of the Kalahari craton of southern Africa (Figure 1.14*A*) that displays the typical division into greenstone belt, gneiss, and granitic rocks (Figure 1.14*B*).

High-grade gneisses form the bulk of Archean regions. They consist predominantly of quartzo-feldspathic gneisses derived by metamorphism of felsic igneous rocks, but they also contain subordinate metasedimentary rocks, including metamorphosed quartzites, volcanogenic sediments, iron formations, and carbonate rocks (Figure 1.14*C*). Deformed mafic-ultramafic complexes form the rest of the gneissic regions.

The high-grade gneissic regions are complexly mixed on a scale of tens to hundreds of kilometers with lower-grade greenstone belts (Figure 1.14*B*) that contain mafic to silicic volcanic rocks and shallow intrusive bodies, volcanogenic sediments of similar composition, and subordinate flows and shallow sills of olivine-rich magmas (Figure 1.14*D*).

Three tectonic and structural features are common to all Archean terranes. First, most rocks are highly deformed and display more than one generation of folds (Figures 1.14*D* and 1.15) as well as deformed and folded low-angle faults. Figure 1.14*D* shows a complex pattern of folding in the Barberton Mountain Land in the Kalahari craton of Swaziland and South Africa. Figure 1.15 shows a map and cross section of a complexly folded mafic-ultramafic complex, the Fiskenaesset complex, and surrounding gneiss and amphibolite in southwestern Greenland.

Second, the contacts between greenstone belts and high-grade gneissic areas are complex. In some places, the contacts are shear zones that mask the original relationship. Elsewhere, rocks are deposited on older gneissic basement and are subsequently metamorphosed to the greenschist facies. In still other areas, gneissic granitic rocks intrude rocks of the greenstone belt.

Third, the sedimentary rock types fall into one of two broad categories: Either they are immature volcanogenic sediments that are characteristic of the greenstone belts and parts of the gneissic terranes, or they are a quartzite-carbonate-iron-formation assemblage associated in many

---

[9]The term *metamorphic facies* refers to a distinctive assemblage of metamorphic minerals that are characteristic of certain conditions of pressure and temperature. The granulite facies of metamorphism generally comprises an assemblage of garnet, pyroxene, and feldspar. It usually indicates high temperature (above 650°C) and high pressure (0.25–1 GPa; 1GPa = 10^3 MPa) (see Figure 20.8).

[10]Archean comes from the Greek word *archi*, which means "beginning"; Proterozoic comes from the Greek words *proteron*, and *zoe*, which mean, respectively, "before" and "life," an allusion to geologists' original but erroneous impression that these rocks bore no fossils.

[11]The greenschist facies is characterized by the presence of chlorite and actinolite. The amphibolite facies typically includes hornblende and may or may not include aluminosilicate minerals and garnet. Pressure and temperature conditions for greenschist facies are approximately 250–450°C and 0.2–0.9 GPa; for amphibolite facies they are about 450–700°C and 0.2–1 GPa (see Figure 20.8)

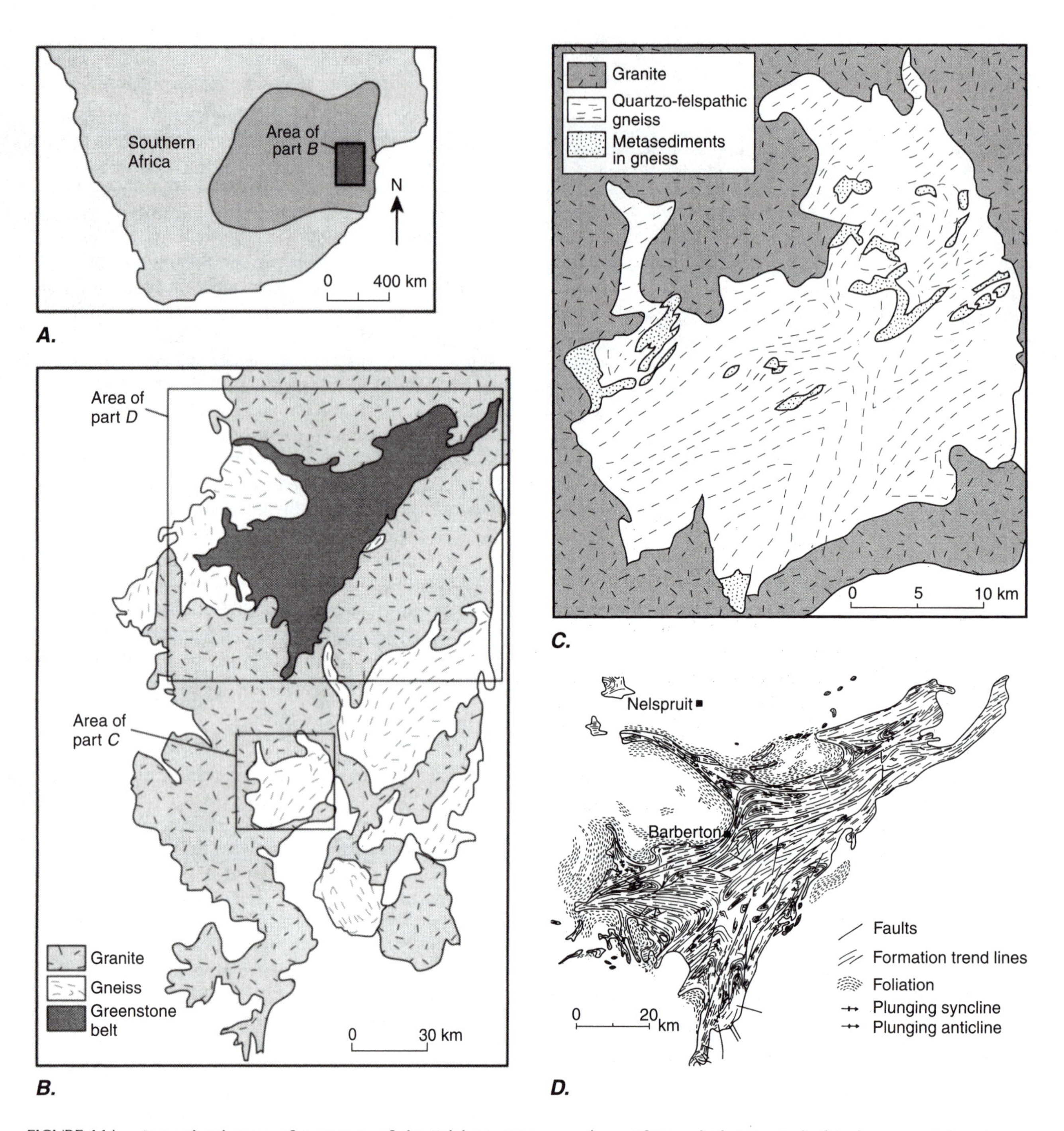

FIGURE 1.14 Generalized maps of a portion of the Kalahari craton, southern Africa, which is typical of Archean crustal regions, showing gneissic terranes with metasedimentary units, granitic rocks, and a greenstone belt. *A.* Regional map of southern Africa; the shaded region is the Kalahari craton. *B.* Regional map of a portion of the Kalahari craton showing areas of granite, gneiss, and the Barberton greenstone belt. *C.* The Mankayane inlier showing infolds in gneiss of metamorphosed igneous and sedimentary rocks consisting of a mafic-ultramafic unit structurally overlain by a sedimentary unit of metaquartzite, schists of quartzofeldspathic, pelitic, and calcareous composition, and iron formations. (Modified after Jackson, 1984) *D.* Detailed map of Barberton greenstone belt showing internal structure. (After K. Anhaueuser, 1984)

areas with multiply deformed mafic-ultramafic layered igneous complexes, found only in gneissic terranes (Figure 1.14*C*).

The study of Archean tectonics is fairly new, because numerous large-scale detailed maps and sufficiently precise radiometric dating techniques became available only in about 1975. The worldwide presence of these characteristic sedimentary and structural associations implies that similar sedimentary and tectonic conditions occurred globally during Archean time and that these con-

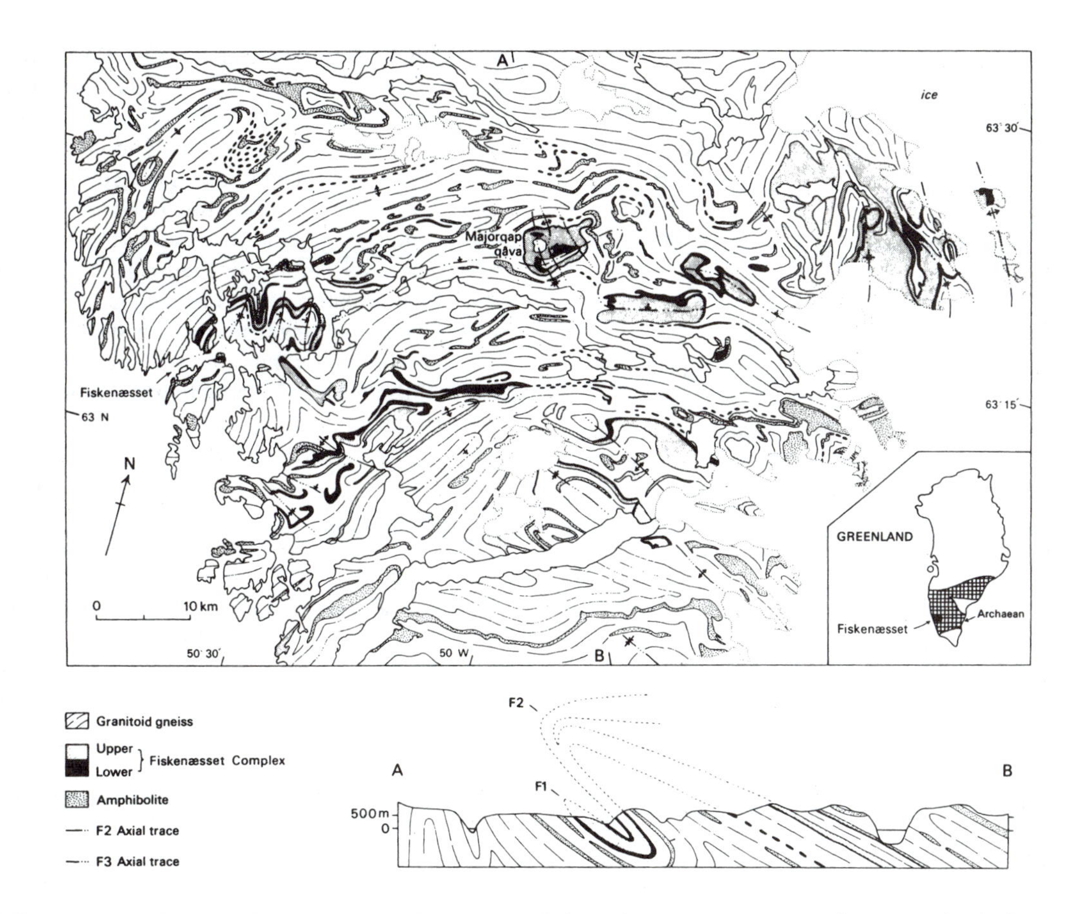

FIGURE 1.15 Generalized map and cross section of a portion of the Fiskenaesset region, southern Greenland, showing refolded folds. The Fiskenaesset complex is a mafic-ultramafic stratiform sequence; the lower unit is peridotite, the upper unit is gabbroic. (After J. S. Myers in Kröner and Greiling, 1984)

ditions differ markedly from those characteristic of Phanerozoic time. In particular, the widespread metamorphism and the presence of ultramafic magmatic rocks indicate higher temperatures in the earth during Archean times. The petrology of the ultramafic magmas imply that they formed by about 50% melting of a mantle source at temperatures of approximately 1500°C. Theoretical heat-budget calculations suggest that the rate of increase of temperature with depth in the Earth (the geothermal gradient) was approximately 2 or 3 times the present one, although some studies of metamorphic rocks have suggested geothermal gradients similar to present ones.

The question of whether plate tectonics operated in the Archean is controversial. Many authors argue that plate tectonics was prescnt, although plates may have been more numerous and smaller on average than at present. In particular, seismic reflection profiling in the Precambrian Canadian Shield by the "Lithoprobe" project clearly points to older Archean cratons separated by younger (but still Archean) **sutures**, which are regions of deformed oceanic material thought to represent the remnants of disappeared oceans. This evidence strongly indicates the presence of horizontal movement akin to modern plate tectonics (see Figure 1.16).

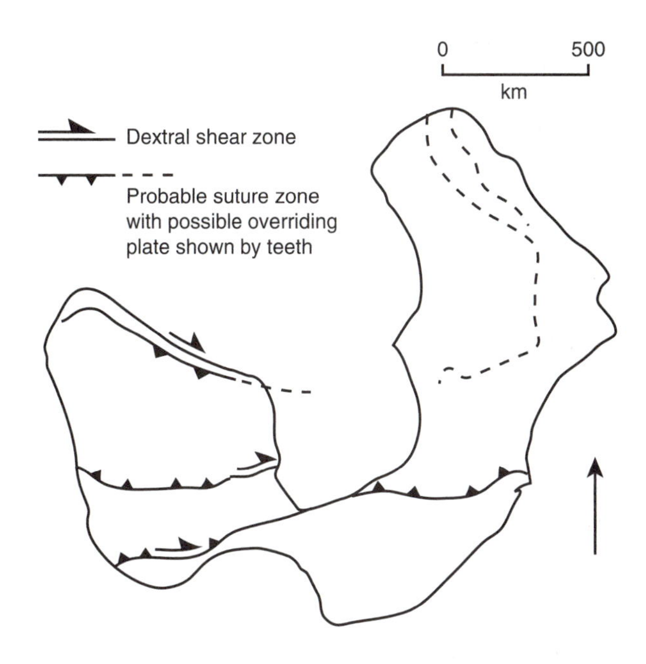

FIGURE 1.16 Generalized map of the Archean Superior Province, Canada, showing principal probable and possible suture zones between separate crustal blocks (see Figure 1.5 for location). (After Percival et al., 2004)

Oceanic crust probably was considerably thicker in Archean times, perhaps resembling modern oceanic plateau thicknesses. Continents, however, were apparently smaller and less numerous in Archean times. Archean terrestrial sediments are relatively rare, a notable exception being gold-bearing quartzose sandstones and conglomerates in regions such as South Africa, Western Australia, Canada, and eastern Siberia.

ii. Proterozoic Terranes

Proterozoic terranes include both slightly deformed stable regions and highly deformed mobile areas, which contrasts with the evidence for nearly ubiquitous mobility displayed by Archean terranes. Tectonically stable regions of the Earth's crust, called **cratons,**[12] first became abundant in Proterozoic time. In these areas, vast deposits of weakly deformed, relatively unmetamorphosed Proterozoic sediments typically overlie a basement of deeply eroded, deformed, and metamorphosed Archean rocks. These sediments characteristically consist of areally extensive stratigraphic units of mature sediments such as quartzites and quartz-pebble conglomerates, which indicate a tectonically stable environment. Quartzites are commonly intercalated with abundant iron formations, composed of interstratified iron-rich oxides, iron carbonates, and iron silicates. They host many vast Precambrian placer gold and uranium deposits, as well as most of the world's iron ore deposits.

Proterozoic deformed belts are of two general types. One type includes multiply deformed regions rich in volcanic rocks, reminiscent of Archean terranes as well as the core zones of many Phanerozoic volcanic-rich orogenic[13] belts. The other type is characterized by thick sedimentary sequences deposited in linear troughs, presumably along ancient continental margins, and subsequently deformed to form linear fold and thrust belts. These are similar to the fold and thrust belts that occur along the flanks of Phanerozoic orogenic belts.

Proterozoic igneous rocks also display distinctive differences when compared with older and younger terranes. Regionally extensive Proterozoic dike swarms of basaltic composition commonly cut Archean terranes (Figure 1.17). Several Proterozoic dike systems are associated with extensive mafic-ultramafic stratiform complexes, such as the Muskox and Bird River complexes in Canada, the Stillwater complex in the United States, and the Bushveld Complex in South Africa (Figure 1.17). These complexes are essentially undeformed masses of layered igneous rocks hundreds to tens of thousands of square kilometers in area. They resemble the layered igneous complexes of the Archean but differ in that they are only weakly deformed. Figure 1.18 shows an example of such a feature, the Bushveld Complex. The vertical columnar sections from three widely separated locations in the complex illustrate the amazingly continuous nature of the different distinctive layers in the complex. The dark layers in the columnar sections in Figure 1.18 represent individual layers of chromite within a sequence of gabbroic cumulate rocks that can be traced for tens of kilometers.

There are other distinctive igneous rock suites in Proterozoic regions. Large intrusive massifs of anorthosite[14] are present in regions of mid-Proterozoic age (1000–2000 Ma) (indicated by the symbol "A" in Figure 1.17). These rocks are more Na-rich than anorthositic layers in stratiform mafic-ultramafic complexes. In some cases, they are clearly igneous in origin, but in other cases deformation and recrystallization have so modified the primary texture and structures of the rock that their origin is difficult to decipher.

Little-deformed "anorogenic magmatic suites" are widespread in Scandinavia, southern Greenland, and south-central to eastern North America. These suites include thick, extensive silicic volcanic sequences, as well as a distinctive kind of granitic rock called "rapakivi[15] granite," which characteristically contains ovoid phenocrysts[16] with cores of orthoclase that are surrounded by rims of plagioclase. These rocks are tens to hundreds of millions of years younger than deformed sequences into which they intrude. They are thought to have formed well after local tectonic activity had ceased, by the melting of continental crust that had been thickened by earlier orogenic activity.

Associated with many Proterozoic fold and thrust belts are a series of smaller, linear, sediment-filled grabens, called **aulacogens**[17] that generally strike at high angles to the trend of the deformed belts. Those that have been mapped in the North American continent are shown in Figure 1.19*A*. Careful mapping of several aulacogens shows that their sediments are correlative with the thick sediments of the associated deformed belt, as well as with thinner undeformed platform sediments on either

[12]From the Greek word *kratos*, which means "power."

[13]The term *orogenic* is derived from the Greek words *oros*, which means "mountain" and *genesis*, which means "origin" or "birth." We use the term to refer to areas that are major belts of pervasive deformation. The term *mobile belt* also means approximately the same thing; it denotes regions that have been tectonically mobile. *Mountain belt* is a geomorphic term that refers to areas of high and rugged topography. Most mountain belts are also orogenic belts, so the two terms are often used interchangeably. Not all orogenic belts, however, are mountainous; although they probably were at one time, they have been eroded down to a gently undulating surface.

[14]An igneous rock composed almost completely of calcium-rich plagioclase feldspar.

[15]Derived from the Finnish word for "crumbly stone."

[16]Phenocrysts are large crystals that have crystallized from a melt.

[17]From the Greek word *avlax*, which means "furrow."

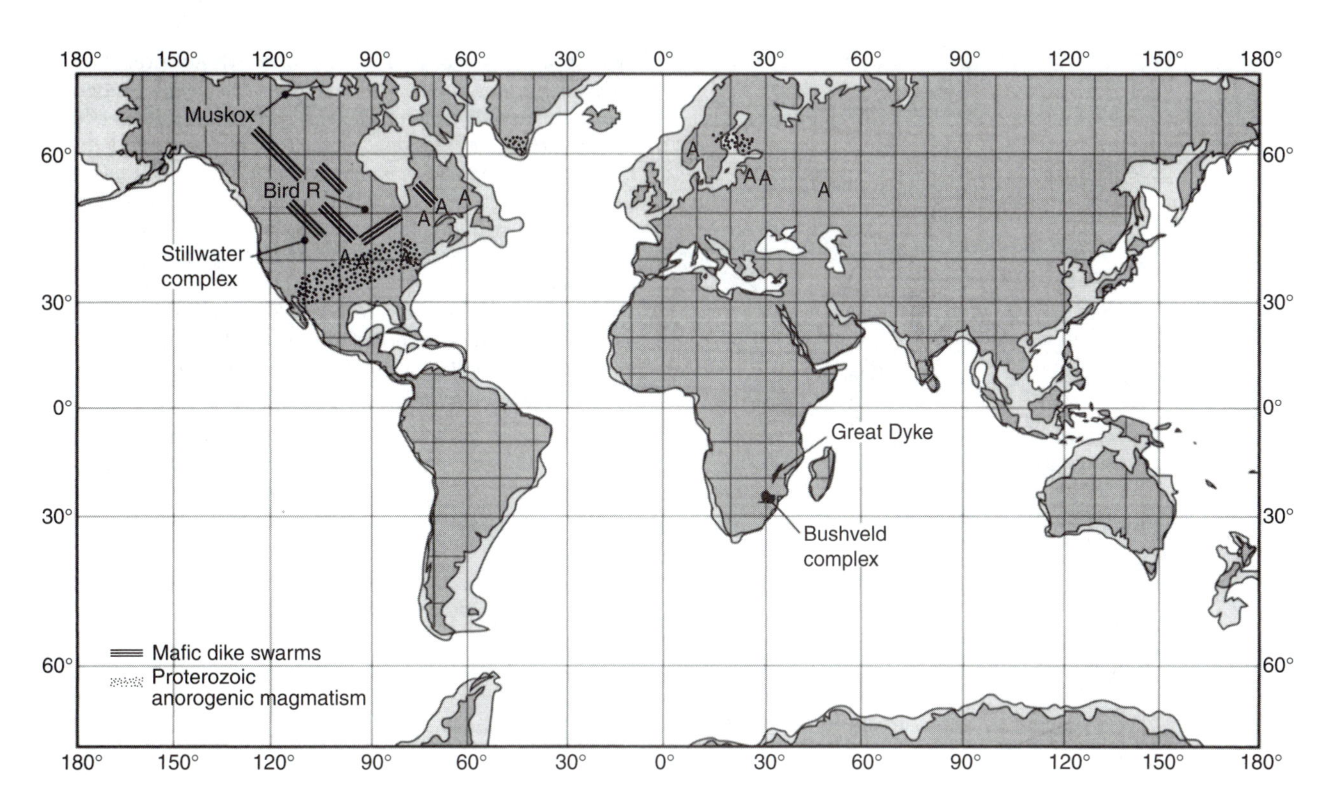

FIGURE 1.17 World map showing distribution of several mafic-ultramafic stratiform complexes (black dots and blobs), mafic dike swarms, anorogenic magmatic provinces, and anorthosite complexes (indicated by symbols A), all of Proterozoic age. (Modified after Stanton, 1972; Windley, 1993)

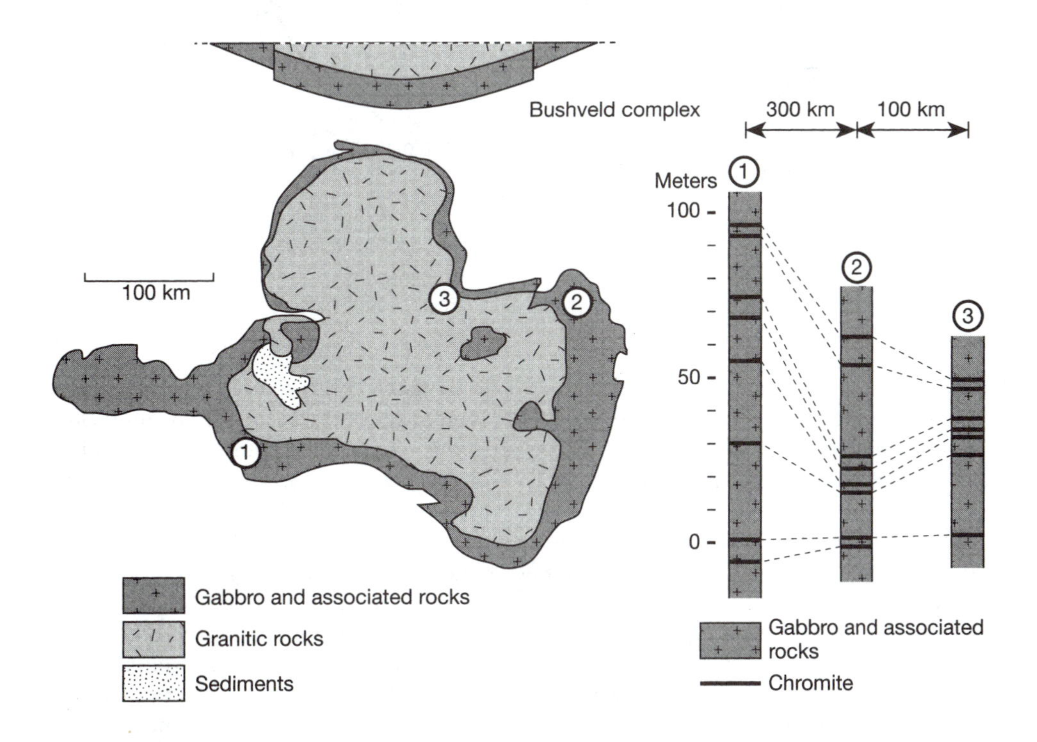

FIGURE 1.18 Simplified map and cross section of the Bushveld complex, South Africa, with selected stratigraphic sections. Dark lines are chromite layers. Note the similarity in the sections over great distances. (Modified after Stanton, 1972, pp. 313, 316)

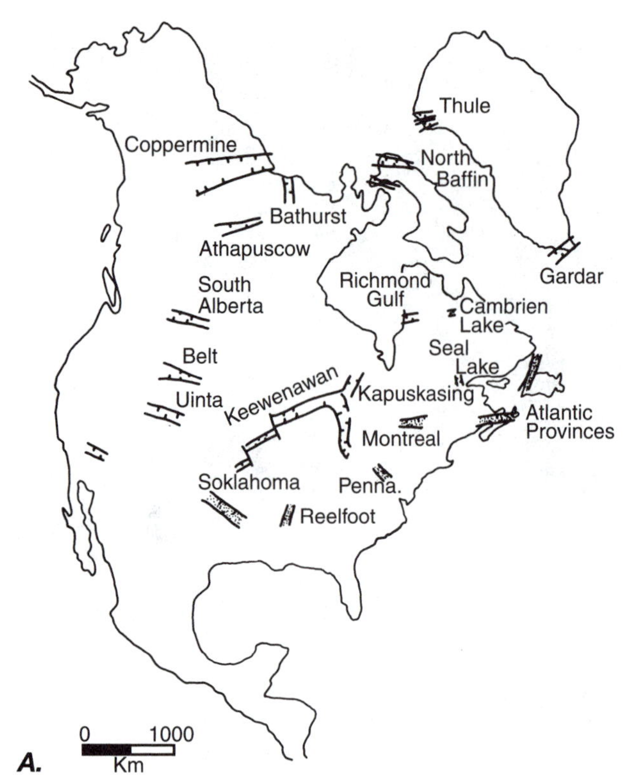

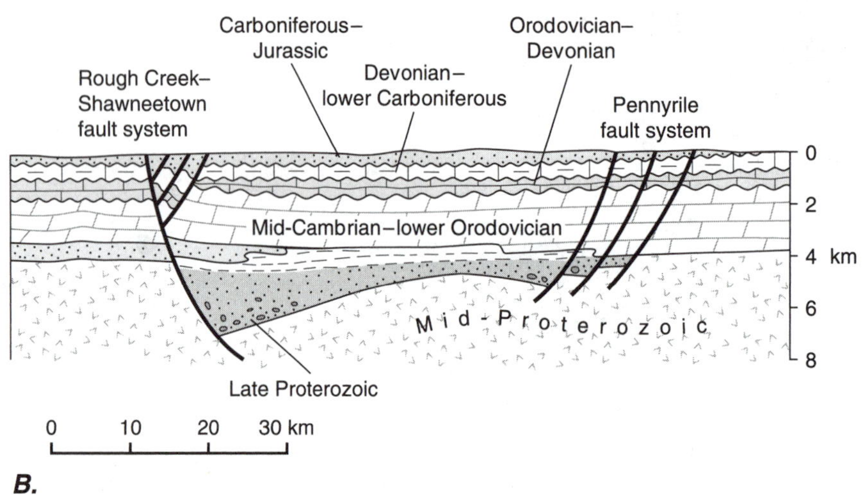

FIGURE 1.19 Aulacogens. *A*. Map of North America, showing inferred and documented Precambrian aulacogens. (Modified after Burke, 1981) *B*. Generalized cross section of the Reelfoot Rift, midcontinental United States. The rift formed in late Precambrian time and was mostly active during the middle Cambrian–lower Ordovician time. Minor thrust separation in parts of the stratigraphy on the Rough Creek–Shawneetown fault system may result from subsequent, post-depositional movement. Renewed movement on this rift is thought by some authors to be the cause of seismic activity in the U.S. midcontinental region, including the very large New Madrid earthquakes of 1811–1812, the largest so far recorded in the United States. (After Kolata and Nelson, 1997)

side (Figure 1.19*B*). The sediments commonly are undeformed or only slightly folded, with fold axes trending parallel to the axis of the trough.

Thus tectonic conditions during the Proterozoic apparently differed from those in the Archean. Extensive undeformed platform sequences indicate the widespread existence in Proterozoic times of large stable continental regions. Regional dike swarms and linear sediment troughs, such as aulacogens, indicate that these regions were capable of undergoing brittle extension. In this respect Proterozoic tectonics may have closely resembled processes that operated in the Phanerozoic rather than the Archean, although oceanic crust may have been considerably thicker in mid-lower Proterozoic and Archean times than in the late Proterozoic–Phanerozoic. Some workers have even suggested that the Archean–Proterozoic transition is the single most important tectonic transition in all of Earth history. What caused this transition, and what does it indicate? Why was the Archean so different from later times? How did the Proterozoic differ from later times? What was the nature of global tectonics during Proterozoic and Archean times, and how did it differ from Phanerozoic plate tectonics? Addressing these provocative questions is beyond the scope of this book, but we focus on understanding the formation of the structures associated with deformation by which some of these questions must be addressed. It is clear that although the thermal state of the Earth and the composition of igneous rocks in Archean times differed from those in Proterozoic and Phanerozoic times, the rock-forming and rock deforming processes that operated in the early Earth were comparable with those in later times.

1.8 PHANEROZOIC REGIONS

As we consider Phanerozoic (Cambrian and younger) features of the Earth, the available evidence increases vastly, so that we can obtain a much more detailed picture of the structural characteristics of the younger parts of continents than we can of the Precambrian areas. We briefly describe features of cratonic platforms, of orogenic belts, of continental rifts, and finally of continental margins.

i. Interior Lowlands and Cratonic Platforms

All continents contain regions of interior lowlands and cratonic platforms where relatively thin sequences of sedimentary rocks overlie Precambrian rocks that are subsurface continuations of the shields. With minor exceptions, these sedimentary rocks are flat-lying and composed of lithologic units that are continuous over vast areas larger than the Precambrian shields themselves. For the most part, these regions are plains that stand a few hundred meters above sea level. To a structural geologist they are relatively monotonous. Yet the economic wealth of these regions in coal, petroleum, mineral deposits, and agricultural resources is such that a large body of knowledge about them has accumulated, giving rise to what the late American structural geologist P. B. King called "the science of gently dipping strata."

Most interior platform sedimentary sequences begin with middle Cambrian or younger deposits; lower Cambrian or late Precambrian rocks are generally found only at the edges of the platforms. Throughout much of the world, and especially in North America, the contact with the underlying Precambrian shield rocks is a profound unconformity commonly called a great unconformity that marks a worldwide transgression of the sea over older continental interiors. In most places, a time gap of tens to hundreds of millions of years is represented by this unconformity. Exceptions to this generalization do exist, however, especially in parts of former Gondwanaland,[18] which was still in the process of consolidation and which experienced orogenic activity into the Ordovician.

Most platform sediments are marine and represent deposition in epeiric[19] seas. A major exception is the platform sequence of much of Gondwanaland, which is mostly nonmarine in origin. The marine sediments record

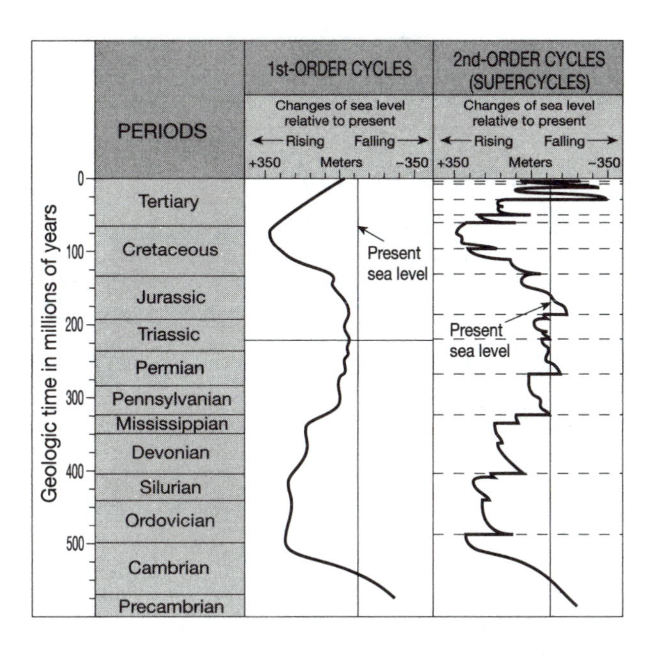

FIGURE 1.20 Transgression regression curves for the world's continents. Curves show major long-term changes in sea level (first-order cycles) and more detailed changes (second-order cycles). (Modified after Cloetingh, 1986; Vail et al., 1977)

major periods of transgression and regression throughout the Phanerozoic, which in turn reflect major fluctuations in the level of the oceans relative to that of the continent (Figure 1.20). The main structural characteristic of these platforms is a group of cratonic basins separated by intervening domes or arches. Figure 1.21 shows the distribution of cratonic basins throughout the world and identifies the basins and arches of North America. Many of these features exhibit evidence of vertical movement of the crust lasting intermittently for tens to hundreds of millions of years. Arches served as sources of sediment during some stratigraphic intervals and were covered in others, but with thinner stratigraphic sequences than the surrounding platforms. Basins may contain thicker sequences deposited in deeper water. In times of general regression, these basins show evidence of restricted circulation and even dessication.

North America provides numerous good examples of these features (Figure 1.21). The Transcontinental Arch is a region that stood high relative to the surrounding area throughout most of Paleozoic time. The sedimentary facies of some stratigraphic intervals shows that at some times it actually was emergent in an otherwise flooded continental region. Conversely, the Michigan and Illinois Basins are areas that were relatively depressed features throughout most of the Paleozoic. During times of high sea level, sediments in these basins were of deeper water origin and thicker than sediments on the surrounding platforms. During times of low sea level, basin sediments record evidence of restricted circulation. During some regressive periods, evaporite deposits developed.

[18]The supercontinent formed by India, Africa, Australia, and Antarctica.

[19]From the Greek word *epiros*, which means "continent."

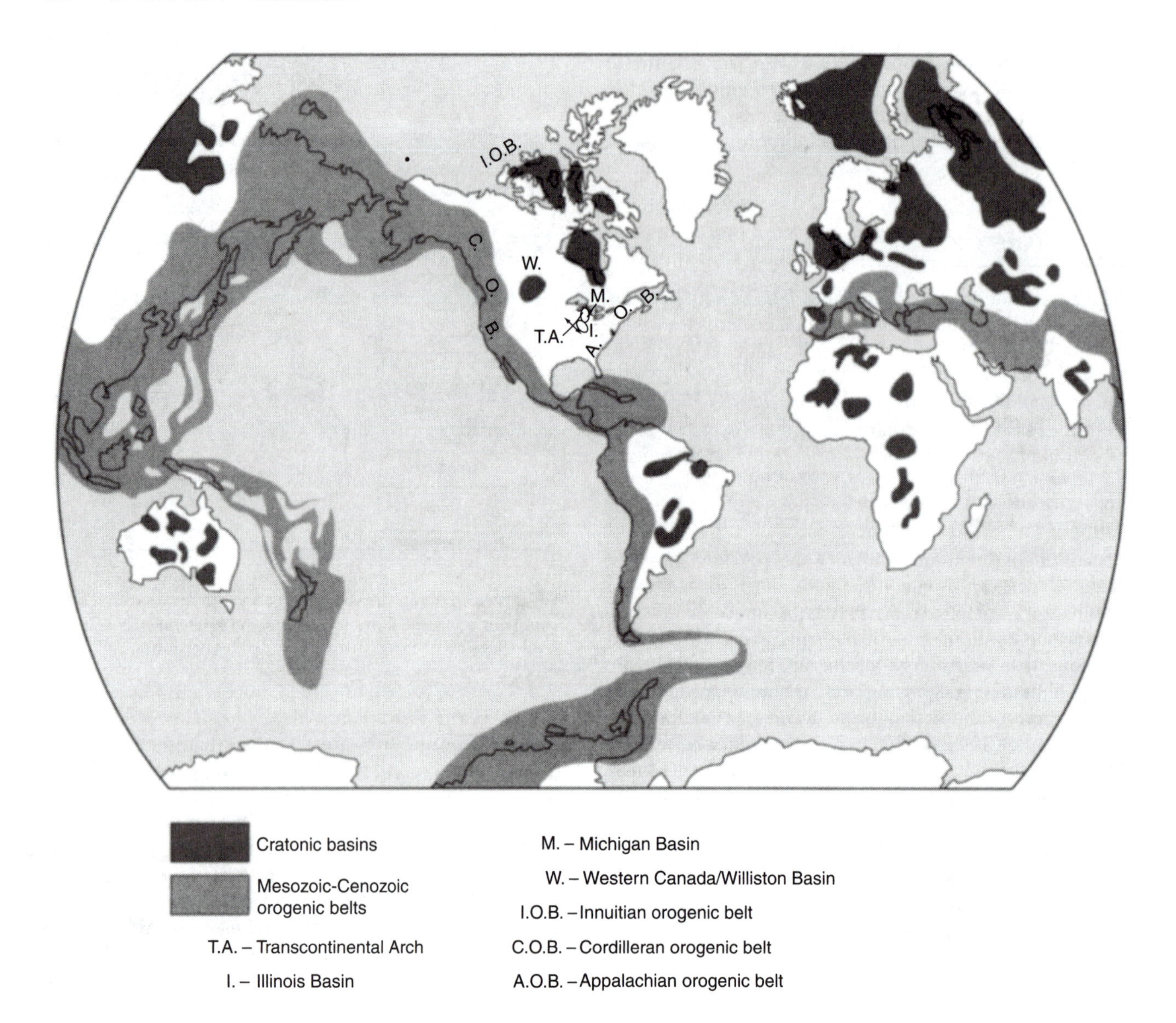

FIGURE 1.21 Map of the world showing the distribution of cratonic basins. On North America, basins and arches of the interior platform are identified. (Modified after Bally et al., 1979)

The existence of domes and basins on the continental platform and the reflection of fluctuations of sea level in platform sediments have been known for decades. Even in the light of our current understanding of plate tectonics and its operation during part or all of Phanerozoic time, it remains something of a puzzle to explain what tectonic processes might have caused these domes and basins to form, and how those processes relate to plate tectonics.

ii. Orogenic Belts

Orogenic belts are one of the most prominent tectonic features of continents, and they have been the primary focus of work in structural geology for the past century. We discuss these belts in more detail in Chapter 20; here we wish only to mention them briefly for the sake of completeness. These belts are formed characteristically of thick sequences of shallow-water sandstones, limestones, and shales deposited on continental crust, as well as oceanic deposits characterized by deep-water turbidites and pelagic sediments, commonly with volcaniclastic sediments and volcanic rocks. Orogenic belts characteristically have been deformed and metamorphosed to varying degrees and intruded by plutonic rocks, chiefly of granitic affinity.

Structurally, many—but not all—orogenic belts display a crude bilateral symmetry that is manifest by a linear central core area of thick deformed and metamorphosed sedimentary and/or volcanic accumulations bordered on either side by low-grade metamorphic to nonmetamorphosed rocks, typically consisting of continental shelf deposits such as sandstones, limestones, and shales. These regions are folded and faulted, but not as extensively deformed as the core areas, and they in turn are bordered by undeformed regions of either oceanic or continental rocks. In the past, much significance was attached to the symmetrical nature of orogenic belts. Recent work, however, has demonstrated that the symmetry is more apparent than real, as in many cases the structures on the two sides of the center are of different ages.

The application of the plate tectonic model to the study of orogenic belts has revolutionized ideas on the origin of these belts. We now believe that orogenic belts form at convergent margins as a result of the collision[20] of a continental block, an island arc, or other thick crust of oceanic origin on a subducting plate with another continent or an island arc on the over-riding plate. Different types of orogenic belts form depending on the nature of these colliding blocks. Noncollisional mountain belts can form, for example, as a continental volcanic arc over a subduction zone or in response to a component of convergence along a continental transform boundary, but strictly speaking, these are not considered orogenic belts.

Thus the tectonic history of an orogenic belt may record some aspects of the history of plate tectonic activity. By studying the structural and tectonic history of young orogenic belts, we can discover the relationship between orogenic structures and associated plate tectonic activity. Similar structures in inactive or older orogenic belts can then be used to infer the existence of similar plate tectonic activity in the geologic past.

iii. Continental Rifts

Continental rifts are areas marked by abundant normal faulting, shallow earthquake activity, and mountainous topography. The North American Basin and Range province and the East African Rift Valleys are examples. Both regions exhibit north-trending systems of normal faults bounding blocks of uplifted or down-dropped crust that extend over an area approximately 100 to 600 kilometers from east to west and 2000 kilometers from north to south (see Figure 4.9). In such regions, the continental crust is undergoing extension, and in the geologic record, this characteristically has preceded the breakup of continents and the formation of new ocean basins.

[20]Much confusion exists in the literature about the terms *convergent* and *collision*. Some writers conflate the two terms, using the term *collisional margin* for convergent margins. It is important to distinguish between the terms, however. A convergent margin can evolve smoothly for millions of years as oceanic crust is dragged down a subduction zone. A collision occurs when a crustal block of continent, oceanic plateau, or island arc is carried on a downgoing plate into a subduction zone. A collision causes an abrupt change in the relative plate motion or in the plate configuration.

iv. Modern Continental Margins

The margins of the present continents are apparently marked by a relatively sharp transition from continental to oceanic crust. That transition, however, is poorly exposed and difficult to resolve with common exploration geophysical techniques. Seismic refraction, which generally assumes models characterized by continuous subhorizontal layers, is difficult to apply where the layers are discontinuous, as at margins of continents. Seismic reflection techniques can penetrate the thick marginal sedimentary sequences and produce images of the continent–oceanic crust transition, but only since the 1970s has this technique been publicized in the geologic literature. Consequently, continental margins have been traditionally less well known than the interiors of continents or even the ocean basins. Nevertheless, a great deal of information has been gained from geophysical studies and scientific drilling by the Ocean Drilling Project (ODP) and the International Ocean Drilling Project (IODP).

Four types of continental margins are recognizable, based on their tectonic environment (Figure 1.22): passive, or Atlantic-style, margins; convergent, or Andean-style, margins; transform, or California-style, margins; and back-arc, or Japan Sea–style, margins. The geographic name that is sometimes used to refer to each style represents a region where it is characteristically developed.

Passive margins, also called **rifted margins** or **Atlantic-style margins**, are present on both sides of the Atlantic, as well as around the Indian and Arctic Oceans and Antarctica. They are belts of horizontally lengthened and vertically thinned continental crust that form as continents rift apart to form new ocean basins. They initiate at a divergent plate boundary, but as spreading proceeds and the ocean basin widens, they end up in a midplate position far from the divergent plate boundary (see Figure 1.23).

Passive margins include a coastal plain and a submarine topographic shelf of variable width, generally underlain by a thick (10–15 km) sequence of shallow-water mature clastic or biogenic sediments. Along some margins, an outer ridge is present in the thick sedimentary sequence, generally at the point where the shelf passes into a steeper topographic slope toward the ocean basin. A relatively thick sequence of sediments (roughly 10 km) is generally present along the continental rise and slope (Figure 1.23). Normal faults, including those active during sedimentation, are the most characteristic structural features found in the sediments along these margins. Many margins are volcanic-rich, exhibiting extensive amounts of intrusive and extrusive rocks that erupted during rifting and now occur near the base of the rifted margin sediments.

Convergent margins, or **Andean-style margins**, are present where consuming plate boundaries are located along a continental margin (Figure 1.24). They exhibit an abrupt topographic change from a deep sea

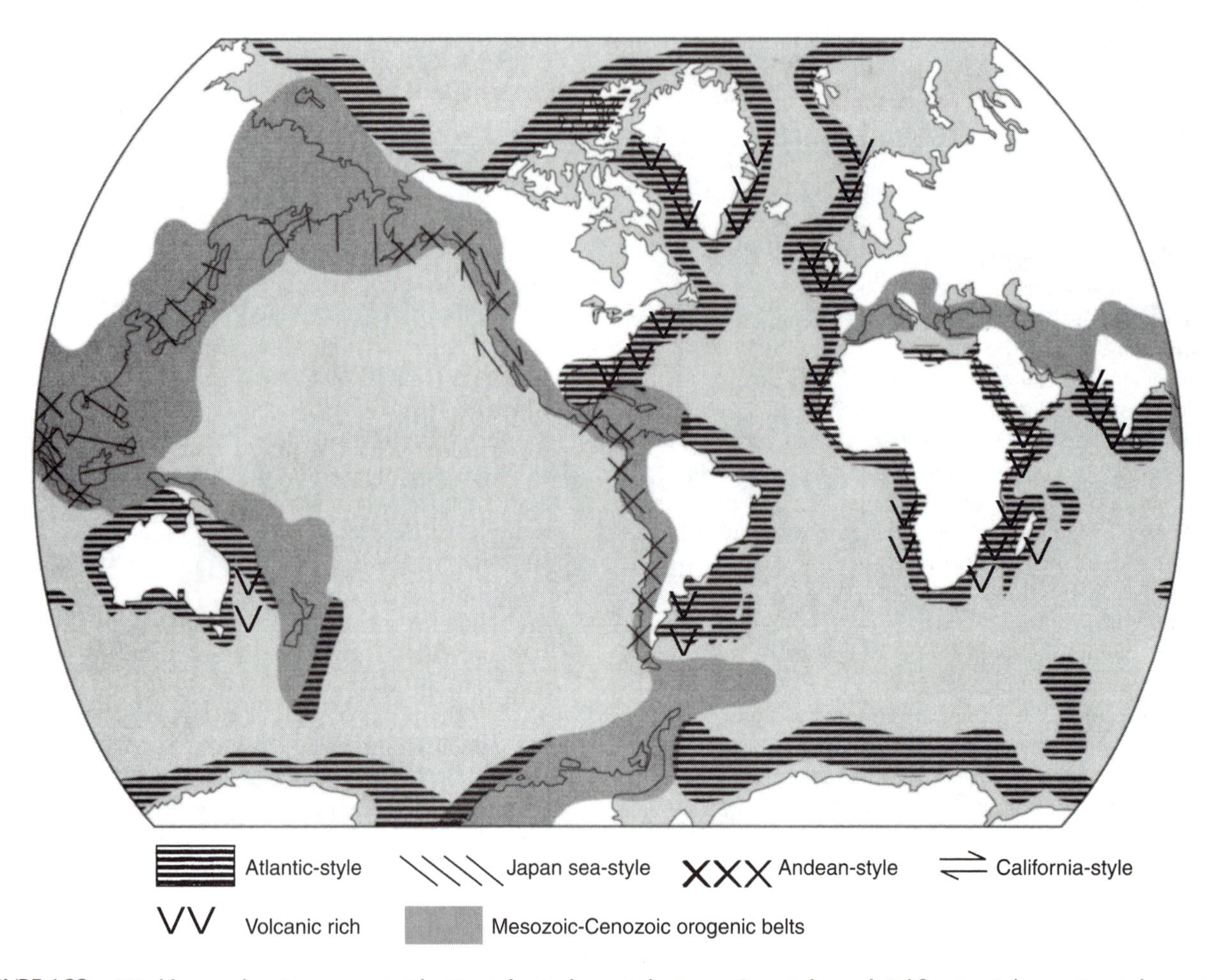

FIGURE 1.22 World map showing present Atlantic-style, Andean-style, Japan Sea-style, and California-style continental margins, with areas of major volcanic-rich margins indicated. (Modified after Bally et al., 1979; Menzies et al., 2002)

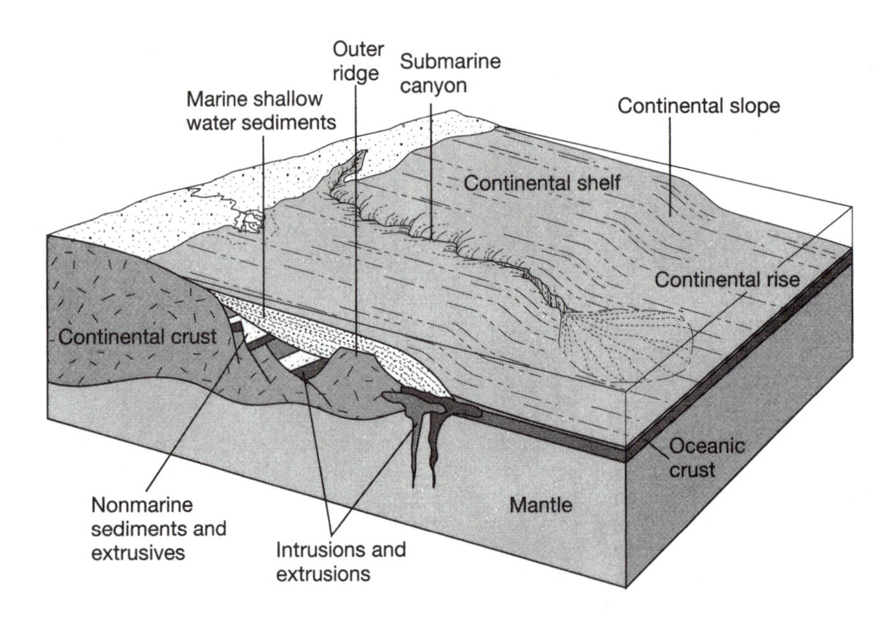

FIGURE 1.23 Generalized block diagram of passive, or Atlantic-style, continental margin. Not to scale.

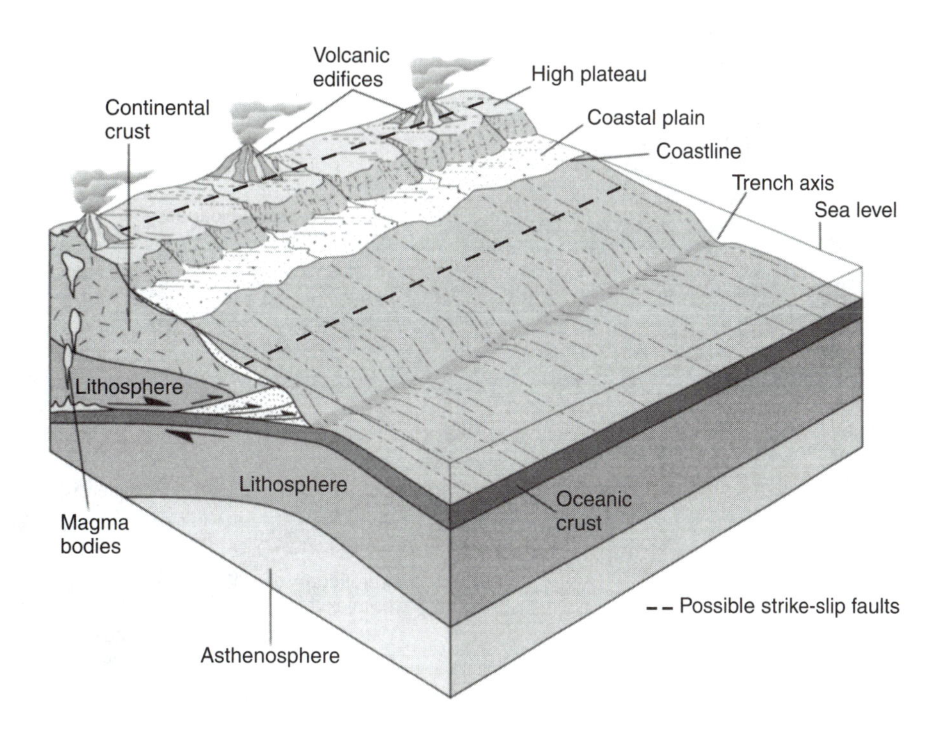

FIGURE 1.24 Generalized block diagram of convergent, or Andean-style, continental margin. Not to scale.

trench offshore to a high belt of mountains within 100 to 200 kilometers of the coast. Continental shelves tend to be narrow or absent. The mountains along these margins are characterized by a chain of active stratovolcanoes of principally andesitic composition (Figure 1.24). Subduction results in thrust complexes near the trench. In regions of oblique subduction, strike-slip faults and high-angle normal faults form near the volcanic axis, and either normal or thrust faults form between the volcanic axis and the continent. Strike-slip faults may be present near the trench, along the volcanic axis, or in the back-arc region, depending on the angle between the subduction direction and the continental margin.

Transform margins, or **California-style margins**, are also characterized by sharp topographic differences between ocean and continent. They are marked by active strike-slip faulting, sharp local topographic relief, a poorly developed shelf, irregular ridge and basin topography, and many deep sedimentary basins. Figure 1.25 shows schematically the development of such topography by strike-slip displacement on two faults along an irregular continental margin. As the faults move, they progressively displace portions of the continent from each other (Figure 1.25*B*), thereby producing an alternation in places of narrow ocean basins and continental fragments. The Pacific margin of the United States is the type example, and the many faults of the San Andreas fault system have produced a ridge-and-basin topography off southern California.

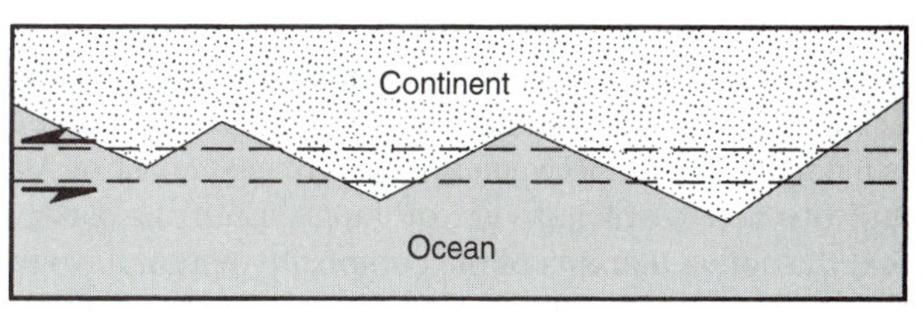

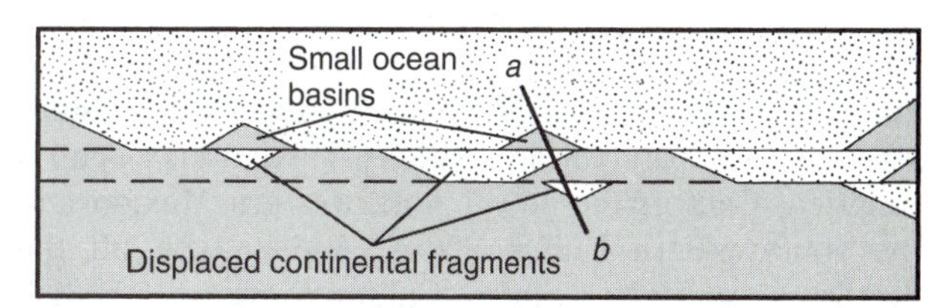

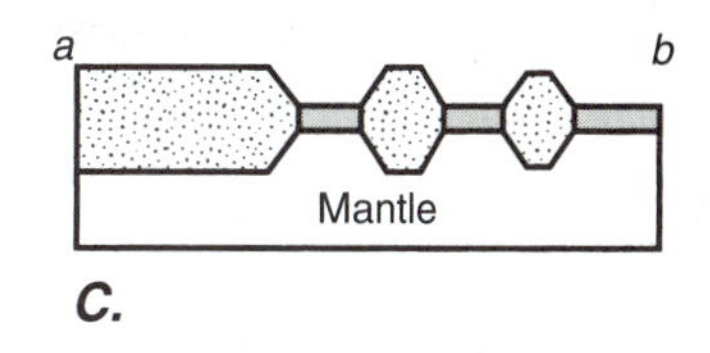

FIGURE 1.25 Generalized maps and cross section illustrating development of a California-style or transform continental margin. Not to scale. *A*. Irregular continental margin and a two-fault strike-slip system. *B*. After motion on both faults of the system, portions of the continent are displaced to new positions. *C*. Cross section *ab* in *B* showing ridge-and-basin structure.

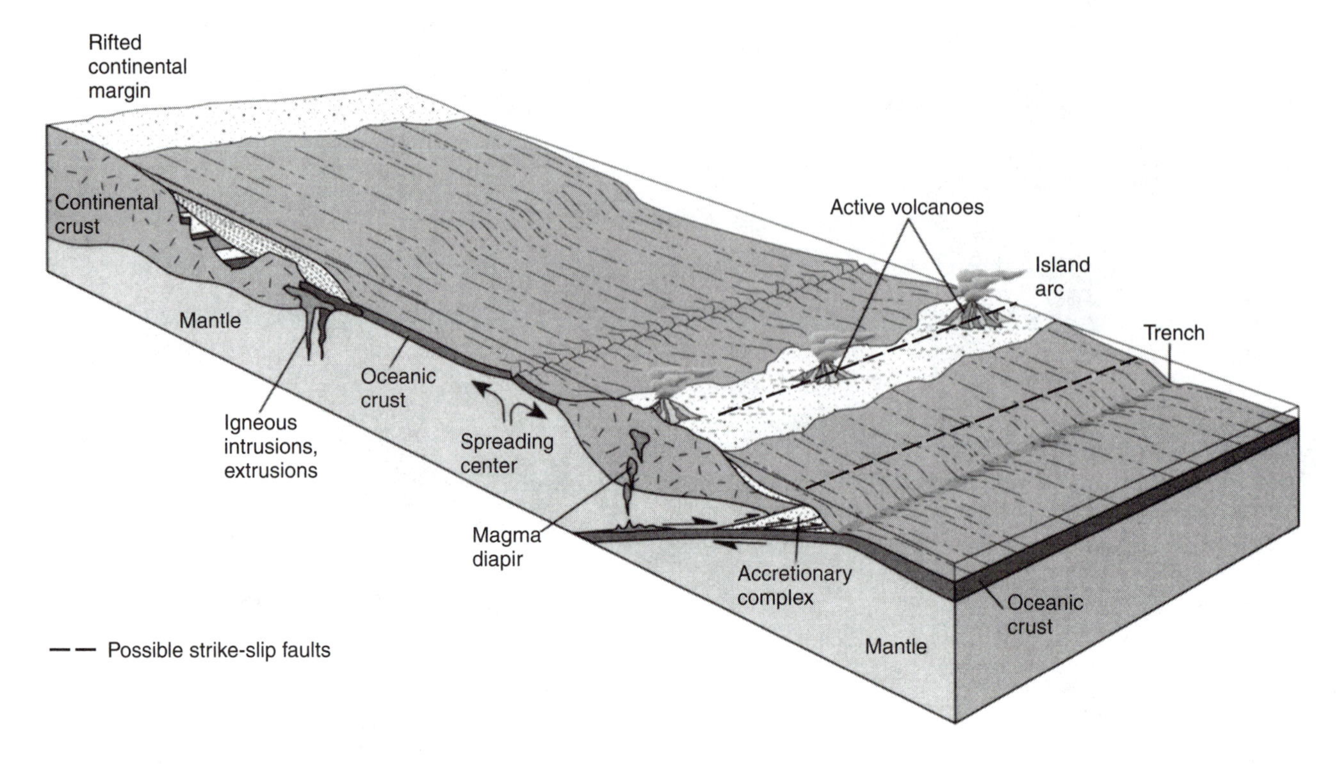

FIGURE 1.26 Generalized block diagram of Japan Sea–style margin. Not to scale.

Back-arc margins, or **Japan Sea–style margins**, are composite margins consisting of a combination of a passive Atlantic-style margin separated by a narrow oceanic region from an active island arc. The Japan Sea is a narrow ocean between the passive east coast of Asia and the active volcanic arc of Japan. Both the passive and the active margins of the composite margin have the features of the individual margins as described earlier (Figure 1.26).

These margin types are "end-members" of a continuous variation. The nature of a continental margin can change along a strike. Thus North America's western margin is Andean-type in southern Mexico and Central America, California-type off northwestern Mexico and the southwestern United States, Andean-type off the northwestern United States, California-type off British Columbia and southeastern Alaska, Andean- and then Japan Sea–type off southern and southwestern Alaska.

1.9 SUMMARY AND PREVIEW

In a sense, the study of Earth deformation processes is a detective exercise. As in all other branches of geology, our evidence is usually incomplete, and we must use all available paths of investigation to limit the uncertainties. Thus we study modern processes to help us understand the results of past deformations. We use indirect geophysical observation to detect structures that lie beneath the surface where we cannot see them. We make observations on all scales, from the submicroscopic to the regional, and we try to integrate them into a unified model. We perform laboratory experiments to study the behavior of rocks under conditions that at least partially reproduce those found in the Earth. And we use mechanical modeling, in which we apply the principles of continuum mechanics to calculate the expected behavior of rocks under different conditions.

At the level of this book, we cannot hope to cover all these aspects in detail. Our aim, rather, is to provide a thorough basis for field observation of geologic structures and to introduce the various paths of investigation that can add valuable data to our observations and lead to deeper understanding of structural and tectonic processes. We hope also to instill an appreciation for the interdependence and essential unity of the disciplines of structural geology and tectonics.

We have chosen to omit from this book material that is traditionally taught in a structural geology laboratory. There are good laboratory manuals available that cover these important topics of field technique and measurement of structures, and for reasons of space, we focus on the description and interpretation of the structures. In Appendix 1, however, we provide a very brief review of some basic structural techniques, and in Appendix 2 we briefly discuss basic geophysical techniques that arise repeatedly in our discussions.

We have arranged the topics to be covered into four major parts following this introductory chapter: Part I (Chapters 2–9) covers the structures typically associated with brittle deformation; Part II (Chapters 10–15) discusses structures formed during ductile deformation; Part III (Chapters 16–18) deals with rheology, or the characteristics and mechanisms of ductile flow in rocks; and Part IV (Chapters 19–20) discusses tectonics and the relationshipships among plate tectonics, crustal deformation, and the structures formed in different tectonic settings.

Our approach in Parts I and II is first to describe the characteristics and geometry of the different types of structures that we can observe in the field. For each class of structures, we then introduce relevant concepts from continuum mechanics and pertinent results from laboratory experiments, and then we apply these concepts to understand the structures initially described. We introduce the concept of stress in Part I to describe the intensity of forces, so that we can explain the origin of fractures and faults in rocks. We introduce the concept of strain in Part II to describe deformation so that we can better understand the structures formed during ductile deformation.

The manner in which deformation of rock depends on the intensity of forces applied to it is determined by the relationships between the stress and either the strain or the strain rate. These relationships are the subject of Part III, where we also discuss the mechanisms that give rise to the flow of rocks and the characteristic microstructures that result from the operation of those mechanisms.

By applying these ideas to the observable deformation in the Earth, we can understand the conditions necessary for the formation of different structures, and this in turn helps us to determine the deformational processes and tectonic environments in which structures form. Our presentation generally follows the process of research and interpretation, which must start with the geometric description and analysis of the structures that exist in the rocks and then ideally proceeds to kinematic and mechanical interpretation of those structures.

Finally, in Part IV, we discuss the connection between small- to regional-scale structures and the tectonic processes that gave rise to them, paying particular attention to orogenic belts. For two centuries, orogenic belts have fascinated those who study the Earth. They preserve much of the information that exists about the interaction of plates through geologic history, and the challenge is to reconstruct those interactions from the structures we can observe.

Although the current theory of plate tectonics provides a unifying model within which we can understand much of the deformation of the Earth's crust, it does not answer all the questions we have about the structural and tectonic evolution of the Earth; and, of course, tectonic processes have not necessarily remained the same throughout the Earth's entire history. One of the challenges of modern structural geology and tectonics is to study ancient deformation to see whether, in fact, models based on modern tectonics are appropriate or if the observed structural patterns and associations require different models for the various stages in the Earth's evolution and thus indicate an evolution in tectonic processes. Plate tectonic theory is a major advance in our understanding, but it is itself evolving. The problems that remain are generating provocative research questions. Answering them will lead to further advances and, undoubtedly, more questions.

Tectonic processes, of course, are on-going, and they produce hazards for society, both for the population directly and for the structures such as buildings, dams, mines, and waste repositories on which society depends. To understand the stability or instability of the Earth's crust requires that we understand the deformational processes to which it is subject, and thus that we understand structural geology.

We hope this book, and its companion volume, *Tectonics* (Moores and Twiss), will stimulate the curiosity and ambition of a new generation of geologists to explore in greater detail the various paths of investigation we introduce and, ultimately, to create new approaches to enhance our understanding.

REFERENCES AND ADDITIONAL READINGS

Anhausser, K. 1984. Structural elements of Archaean granite-greenstone terranes as exemplified by the Barberton Mountain Land, southern Africa. In A. Kröner and Greiling, R. , eds., *Precambrian Tectonics Illustrated*, Schweizerbartsche, Stuttgart, p. 57–78.

Anonymous. 1950. Der Bau der Erde, Gotha, Justus Perthes, 1:40,000,000.

Bally, A. W. 1980. Basins and subsidence: A summary. In A. W. Bally, P. L. Bender, T. R. McGetchin, and R. I. Walcott, eds., *Dynamics of Plate Interiors*, eds.,. Geodynamics Series, Vol. 1., American Geophysical Union, Washington, DC, and Geological Society of America, Boulder, CO: 5–20.

Blundell, D., R. Freeman, and S. Mueller, eds. 1992. *Continent Revealed: The European Geotraverse*. Cambridge University Press, New York, 275 pp.

Burchfiel, B. C. 1983. The continental crust. In R. Siever, ed., *The Dynamic Earth*. New York Scientific American.

Burke, K. 1980. Intracontinental rifts and aulacogens. In *Geophysics Study Committee, Continental Tectonics*, Washington, D.C., National Academy of Sciences: 42–50.

Chamberlin, T. C. 1897. The method of multiple working hypotheses. *Journal of Geology*, 5: 837–848.

Cloetingh, S. 1986. Intraplate stress: A new tectonic mechanism for fluctuations of relative sea level. *Geology* 14: 617–620.

Cloos, M. 1993. Lithospheric bouyancy and collisional orogenesis: Subduction of oceanic plateaus, continental margins, island arcs, spreading ridges, and seamounts. *Geol. Soc. Amer. Bull.* 105: 715–737.

Cox, A., and R. B. Hart. 1986. *Plate Tectonics: How It Works*. Blackwell, London, 392 pp.

Francheteau, J. 1983. The oceanic crust. In *The Dynamic Earth*. New York Scientific American.

Gardner, M., 1981. *Science: Good, Bad and Bogus*. Buffalo, N. Y.: Prometheus Books.

Geological Society of America. 1988–94. *Decade of North American Geology*. Boulder, CO Geological Society of America.

Goodwin, A. M. 1981. Archean plates and greenstone belts. In A. Kroner, ed., *Precambrian Plate Tectonics*, Amsterdam, Netherlands, Elsevier.

Goodwin, A. M. 1991. *Precambrian Geology: The Dynamic Evolution of the Continental Crust*. Academic Press, San Diego, 666 pp.

Hoffman, P. 1988. United Plates of America. *Ann. Rev. Earth and Planet. Sci.*: 16 pp. Palo Alto, CA., Annual Reviews, Inc. p. 543–603.

Hoffman, P., Dewey, J. F., and Burke, K. 1974. Aulacogens and their genetic relations to geosynclines, with a Proterozoic example from Great Slave Lake, Canada, in R. H. Dott, ed., Modern and Ancient Geosynclinal Sedimentation. *SEPM Special Publ.* 19, American Association of Petroleum Geologists, Tulsa. pp. 38–55.

Intermargins. 2005. Map of crustal thickness: http://www.intermargins.org/images/maps/crustal_thickness

Jackson, M. P. A. 1984. Archean structural styles in ancient gneiss complex of Swaziland, southern Africa. In A. Kröner and R. Greiling, eds, *Precambrian Tectonics Illustrated*, Schweizerbart'sche, Stuttgart: 1–18.

Kellogg, L. H. 1992. Mixing in the mantle. *Ann. Rev. Earth and Planet. Sci.* 20: 365–388.

Kellogg, L. H., B. H. Hager, and R. van der Hilst. 1999. Compositional stratification in the deep mantle. *Science* 283: 1881–1884.

Kolata, D. R., and W. J. Nelson. 1997. Role of the Reelfoot Rift/Rough Creek Graben in the evolution of the Illinois Basin. *Geol. Soc. Amer. Spec. Paper* 312, 287–298.

Kosso, P.. 1992. *Reading the Book of Nature : An Introduction to the Philosophy of Science* Cambridge [England]; New York, NY, USA: Cambridge University Press.

Kröner, A., and Greiling, R. , eds. *Precambrian Tectonics Illustrated*, Schweizerbartsche, Stuttgart, 419 p.

Kuhn, T. S. 1970. The Structure of Scientific Revolutions. Second Edition. University of Chicago Press.

Langmuir. I. and R.N. Hall. 1989. Pathological Science. *Physics Today* 42(10): 36–48.

Laudan, Larry; 1981; *Science and Hypothesis, Historical Essays on Scientific Methodology; The University of Western Ontario Series in Philosophy of Science, v.19*; D. Reidel Publishing Co., Dordrecht, Holland; Boston, U.S.A.; London, England; 258 pp.

Mahoney, J. J., and M. F. Coffin, eds. 1997. *Large Igneous Provinces: Continental, Oceanic, and Planetary Flood Volcanism*. Washington, D.C. American Geophysics Union, 100 pp.

Menzies, M., S. L. Klemperer, C. J. Ebinger, and J. Baker, eds. 2002. *Volcanic Rifted Margins*. Geological Society of America Boulder, CO Spec. Paper 362. 230 pp.

Mooney, W. D., G. Laske, and T. G. Masters. 1998. CRUST 5.1; a global crustal model at 5 degrees × 5 degrees. *Jour. Geophys. Res.* 103: 727–747.

Moores, E. M. 2002. Pre-1Ga (pre-Rodinian) ophiolites: Their tectonic and environmental implications. *Geol. Soc. Amer. Bull.* 114: 80–95.

Müller, R. D., W. R. Roest., J.-Y. Royer, L. M. Gahan, and J. G. Sclater. 1996. Digital isochrons of the world's oceans, June 17, 1996. *J. Geophys. Res.* 102(B2): 3211–3214; www.geosci.usyd.edu.au/research/marinegeophysics/Resprojects/Agegrid/digit_isochrons.html

NASA. 2002. Digital Tectonic Activity Map of the Earth, NASA Map DTAM-1: http://denali.gsfc.nasa.gov/dtam/

National Academy of Sciences–National Research Council. 1980. *Continental Tectonics*. Washington, DC: National Academy of Sciences.

National Research Council. 1979. *Continental Margins: Geological and Geophysical Research Needs and Problems*. Washington, D.C. National Academy of Sciences: 302 pp.

Nisbet, E. G. 1987. *The Young Earth: An Introduction to Archaean Geology*. Allen and Unwin, Boston, 402 pp.

Oreskes, N. 1999. *The Rejection of Continental Drift: Theory and Method in American Earth Science*. Oxford University Press, New York, 420 pp.

Oreskes, N. ed., with H. Le Grand. 2001, *Plate Tectonics*. Westview, Boulder, CO, 424 pp.

Percival, J. A., W. Bleeker, F. A. Cook, T. Rivers, G. Ross, and C. van Staal. 2004. PanLITHOPROBE Workshop IV: Intra orogen correlations and comparative orogenic anatomy. *Geosci. Canada* 31: 23–39.

Pitman, W. 1974. Age of ocean floor. Geol. Soc. Amer. Map and Chart MC-6.

Platt, J. R. 1964. Strong inference. *Science* 146: 347–353.

Press, F., and R. Siever. 1986. *Earth, Fourth Edition*. Freeman, New York.

Rousseau, Denis L. 1992. Case studies in pathological science. *American Scientist*, 80: 54–63. Comment and reply in *Am. Sci.* 80: 107–110.

Siever, R., ed. 1983. *The Dynamic Earth.* New York, Scientific American.

Sleep, N. J. 1992. Hotspot volcanism and mantle plumes. *Ann. Rev. Earth and Planet. Sci.* 20: 19–44.

Sloss, L. L. 1996. Sequence stratigraphy on the craton:Caveat emptor. Geological Society of America, Spec. Paper. 306: 425–434.

Smithson, S. B., P. N. Shive, and S. K. Brown. 1977. Seismic velocity, reflection, and structure of the crystalline crust. In J. G. Heacock, ed., *The Earth's Crust, Its Nature and Physical Properties*, American Geophysical Union Monograph 20: 254–270.

Stanley, S. 1986. Earth and Life Through Time. W. H. Freeman and Co., New York, 538 pp.

Stanley, S. 1999. *Earth System History.* Freeman, New York, 615 pp.

Stanton, R. L. 1972. *Ore Petrology.* McGraw-Hill, New York, 713 pp.

Trendall, A. P. 1968. Three great basins of Precambrian banded iron formation deposition: A systematic comparison. *Geol. Soc. Amer. Bull.* 79: 1527–1544.

Uyeda, S. 1978. *The New View of the Earth.* Freeman, New York, 217 pp.

Vail, P. R., R. M. Michum, Jr., and S. Thompson. 1977. Seismic Stratigraphy and global changes of sea level, part 4: global cycles of relative changes of sea level. In C. E. Paxton, ed. Seismic Stratigraphy—Application to Hydrocarbon Exploration, *AAPG Mem.* 26: 83–97.

Windley, B. F. 1993a. *The Evolving Continents, Third Edition*, Wiley, New York, 526 pp.

Windley, B. F. 1993b. Proterozoic anorogenic magmatism and its orogenic connections. J. *Geol. Soc. London* 150: 39–50.

Windley, B. F., F. C. Bishop, and J. V. Smith. 1981. Metamorphosed layered igneous complexes in Archean granulite-gneiss belts. *Ann. Rev. Earth and Planet. Sci* 9: 175–198.

Witzke, B .J., G. A. Ludvigson, and J. Day. 1996. Paleozoic sequence stratigraphy: Views from the North American Craton. Geological Society of America Spec. Pap. 306: 425–434.

Wyllie, P. J. 1975. *The Way the Earth Works: An Introduction to the New Global Geology and Its Revolutionary Development.* Wiley, New York, 296 pp.

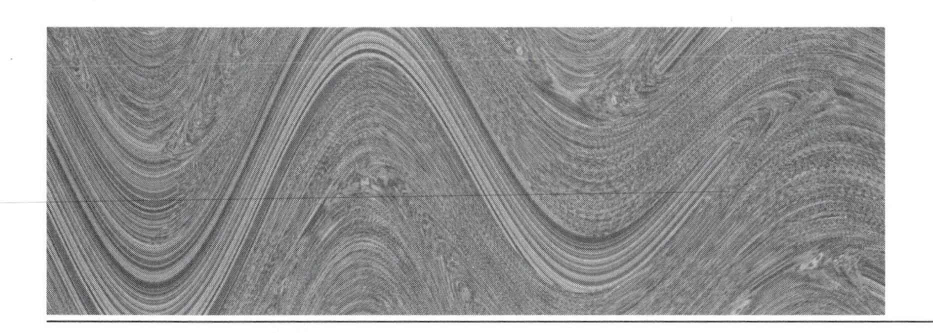

PART I

BRITTLE DEFORMATION

Part I of this book focuses on the structures formed in rocks predominantly by brittle deformation—that is, the breaking of rocks along well-defined fracture planes or zones. Depending on the relative motion that occurs across the fracture plane, the fractures are either extension fractures or shear fractures. We describe the general characteristics of extension fractures in Chapter 2. Faults are shear fractures in rocks generally at the scale of an outcrop or greater. Larger faults are commonly structures of major tectonic importance. We introduce the general characteristics of faults in Chapter 3, and in Chapters 4, 5, and 6, we discuss the characteristics and tectonic significance of each of the three major types of faults: normal, thrust, and strike-slip.

Having described the structures of predominantly brittle origin that we observe in rocks, we next turn our attention to understanding how and why these structures form. Rocks break when they are subjected to an excessive amount of force, and in Chapter 7, we introduce the concept of stress as a measure of the intensity of forces applied to a material. In Chapter 8, we review experimental evidence and theory about how the stress imposed on a rock is related to the types of fractures that form and to the mechanism of formation. With this background, we return, in Chapter 9, to the interpretation of brittle structures that we find in the Earth. By understanding the mechanisms by which fractures form, and by being able to interpret the evidence we observe in the rocks, we can deduce the physical conditions that prevailed in the rock during this fracturing process, thereby opening another window on the tectonic evolution of the Earth's crust and the dynamic processes that drive that evolution.

Beyond their use in investigating the tectonic evolution of the Earth's crust, fractures are of major importance to our environment and to the continued viability of our society. First, because fractures often serve as conduits for groundwater, they are the site of preferential weathering and thereby control the form of much of the Earth's topography. Indeed, some of the world's most inspiring landforms, such as Yosemite Valley in California, the Grand Canyon, the Alps, the islands of the Mediterranean, and Uluru (Ayer's Rock) in Australia owe much of their form to preferential erosion caused by the presence of fractures.

Furthermore, because fractures provide conduits for the migration of fluids through solid rock, they are of great significance in the migration of groundwater, of hydrocarbons, and of hydrothermal and metamorphic fluids. Thus they are significant in the fields of hydrogeology, oil and gas migration and recovery, and geothermal heat extraction. In addition, fractures affect the location of hydrothermal mineral deposits and are of major importance in the integrity of nuclear waste disposal sites, which must safely contain their lethal waste for 10,000 years or more. As a consequence of this association with the world economy and the safety of future generations, understanding the characteristics of fractures and the conditions of their formation is of very real social importance.

Finally, because the cohesion of the rocks is lost across fracture surfaces, they are planes of weakness in the rock. This inherent weakness must be accounted for in the building of dams, bridge abutments, tunnels, mines, and similar engineering projects.

Chapter 2

FRACTURES AND JOINTS

Fractures[1] are surfaces along which rocks or minerals have broken, creating two free surfaces where none existed before; they are therefore surfaces across which the material has lost cohesion.

Fractures are among the most common of all geologic features; hardly any outcrop of rock exists that does not have some fractures through it. They are significant both for the information they provide regarding the sequence of tectonic events during which the fractures formed and for the physical characteristics they impart to the rock in which they occur. Fractures are critical, for example, in determining the permeability of rock to fluid flow and the fluid storage properties of the rock, which are both significant in evaluating aquifer characteristics, contaminant transport, and the migration of oil and gas. They also strongly affect the mechanical properties of the rock, which are important in the design of structures such as dams and tunnels.

Because the outcrop scale is easy to observe and is the basis of all field geology, we emphasize the descriptive characteristics of fractures at the outcrop scale. Studies of fractures in rocks, however, show that the fracture geometry is **self-similar**, which means that the fractures have the same geometric pattern and spatial distribution regardless of whether the scale at which they are viewed is a microscopic scale, an outcrop scale, or a regional scale (see Box 2-1). This characteristic is important for understanding how fractures determine the physical characteristics of the rocks in which they occur.

We distinguish two basic types of fracture, **extension** and **shear** fractures, according to the relative motion that has occurred across the fracture surface during formation. For **extension fractures**, the relative motion, as the fracture propagates, is perpendicular to the fracture walls, which is referred to as **mode I** propagation (Figure 2.1*A*). For **shear fractures**, the relative motion during propagation is parallel to the surface. Two end-member modes of shear fracture propagation are possible: **Mode II** propagation occurs if the sliding motion is perpendicular to the propagating tip, or edge, of the fracture (Figure 2.1*B*); **mode III** propagation occurs if the sliding motion is parallel to the propagating tip (Figure 2.1*C*). Thus a shear fracture whose propagating edge is an expanding closed loop involves both mode II and mode III propagation, as well as some combination of both, depending on the local orientation of the fracture tip relative to the direction of shearing on the fracture. A fracture that has components of displacement both perpendicular and parallel to the fracture surface is an **oblique extension fracture**.

The presumed mode of propagation is sometimes used as a basis for classifying fractures, but for understanding the formation and significance of fractures, a descriptive classification based on the relative displacement across the fracture surface is more useful than a genetic classification based on interpretations of how the fractures formed.

The descriptive criteria used to classify fractures include the orientation, relative to the fracture surface, of the displacement on the fracture, and the geometry of the fractures, including their orientation, the extent of individual fractures, and the distinctive patterns formed by associated fractures. The terminology applied to fractures reflects a recognition of various associations of field characteristics, but it does not reflect a systematic

[1]From the Latin *fractus*, which means "broken."

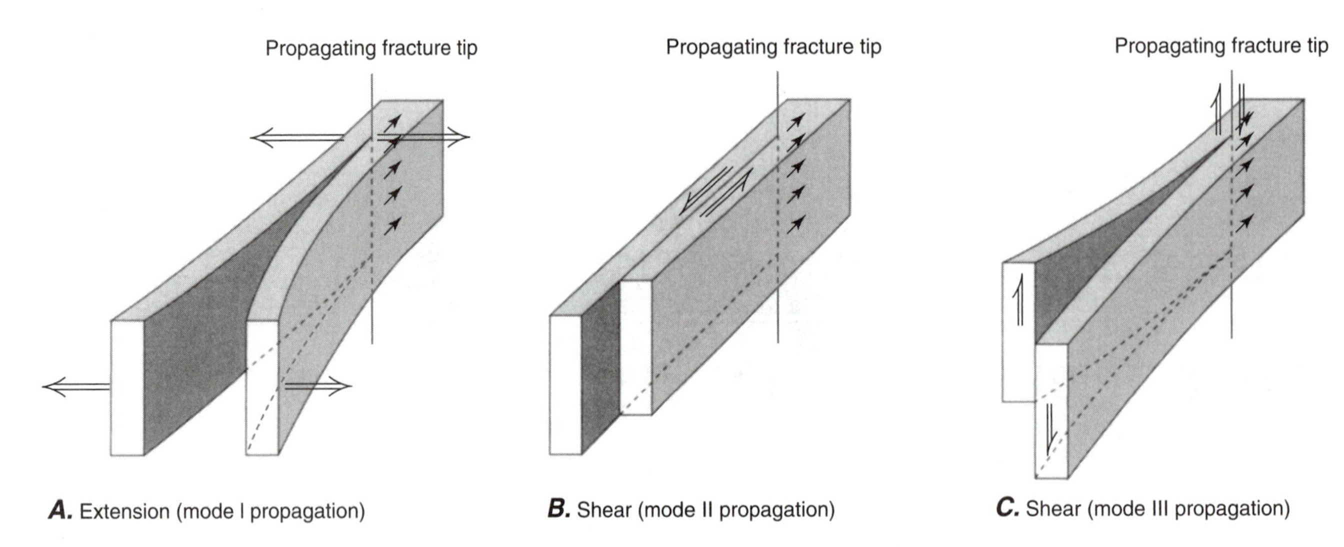

FIGURE 2.1 The distinctions among the major types of fractures are based on the relative displacement of the material on opposite sides of the fracture. Arrows on the fracture tip show the fracture propagation direction. Double-shafted arrows show relative motion across the fracture plane. *A*. Extension fracture; propagation is by mode I fracturing, for which the relative displacement is perpendicular to the fractures. *B*. Shear fracture; propagation is by mode II fracturing, for which relative displacement is a sliding parallel to the fracture and perpendicular to the edge of the fracture. *C*. Shear fracture; propagation is by mode III fracturing for which relative displacement is a sliding parallel to the fracture and parallel to the edge of the fracture. (After Kulander, Barton, and Dean, 1979)

application of these criteria to the definition of a formal classification system. Thus beyond very general criteria, a systematic classification of fractures is difficult. When in doubt about whether a particular classification term is appropriate, we recommend using the simple and direct term "fracture" with appropriate modifiers.

The study of fractures comprises four general categories of observations: (1) the distribution and geometry of the fracture system; (2) the surface features of the fractures; (3) the relative timing of the formation of different fractures; and (4) the geometric relationship of fractures to other structures.

The geologic history of fractures is notoriously difficult to interpret. Evidence bearing on the mode of fracture formation and the relative time of formation of different fractures is often ambiguous. As planes of weakness in the rock, fractures are subject to reactivation in later tectonic events, so some of the observable features of a fracture may be completely unrelated to the time and mode of its formation. Careful study of fractures, however, has led to major progress in understanding their origins and significance, a topic to which we return in Chapter 9.

2.1 CLASSIFICATION OF EXTENSION FRACTURES

If many fractures occur in the same area and have a similar orientation and arrangement, they are referred to as a **set** of fractures. We refer to individual extension fractures that show very small displacement normal to their surfaces and no, or very little, displacement parallel to their surfaces, as **joints**[2]; a group of them is called a **joint set**. A fracture with a small shear displacement, however, may be an extension fracture on which shear displacement accumulated after formation.

Systematic joints have the attributes of roughly planar fracture surfaces, regular parallel orientations, and regular spacing (Figures 2.2 and 2.3*A*). **Nonsystematic joints** are curved and irregular in geometry, although they may be present in distinct aerially persistent sets and are distinguished by nearly always terminating against older joints that belong to a systematic set (Figure 2.3*B*). The term "joint" or "joint set" alone, however, usually refers to systematic joints unless specifically indicated otherwise. A **joint zone** is a quasi-continuous joint that is composed of a series of closely associated parallel fractures and that extends much further than any of the individual fractures (Figure 2.3*A*, *C*). In practice, such a joint zone is also called simply a joint. Two or more joint sets affecting the same volume of rock constitute a **joint system**[3] (Figures 2.2 and 2.3*D*; see also

[2]Unfortunately, there is no universally accepted definition of the term joint. The definition given here is conservative in that fractures satisfying this definition would be called joints by every other definition of the term.

[3]Note that the terms "joint system" and "systematic joint" have different meanings and should not be confused.

FIGURE 2.2 Joints. *A*. Outcrop showing a joint system made up of three distinct sets of joints. *B*. Joints of different orientations terminating against lithologic contacts. At the contact between the layers, the parent joint in the upper layer changes abruptly to twist hackle in the lower layer (see top of Figure 2.14*A*). (From Engelder, 1985)

Figure 2.15*A* and Figure 2-1.3 in Box 2-1). Joint sets and systems are nearly ubiquitous in rock outcrops, and they may persist over hundreds to thousands of square kilometers, each set displaying a constant or only gradually varying orientation (Figure 2.4; see also Figure 2.12). Such systems can show up as linear features, or **lineaments**, on high-altitude photographic and radar images (Figure 2.4).

Sheet joints, **sheeting**, or **exfoliation joints** are extension fractures that are smoothly curved at a scale on the order of hundreds of meters. They are subparallel to the topography and result in a characteristic smooth, rounded topography (Figure 2.5). Sheet joints may be found in many kinds of rocks, but the characteristic topography is best displayed in plutonic rocks in mountainous regions where the joints appear to cut the rock into sheets like the layers of an onion. Many sheet joints apparently formed later than other joint sets, although in some cases they predate late phases of intrusive activity, as indicated by dikes present along the joints.

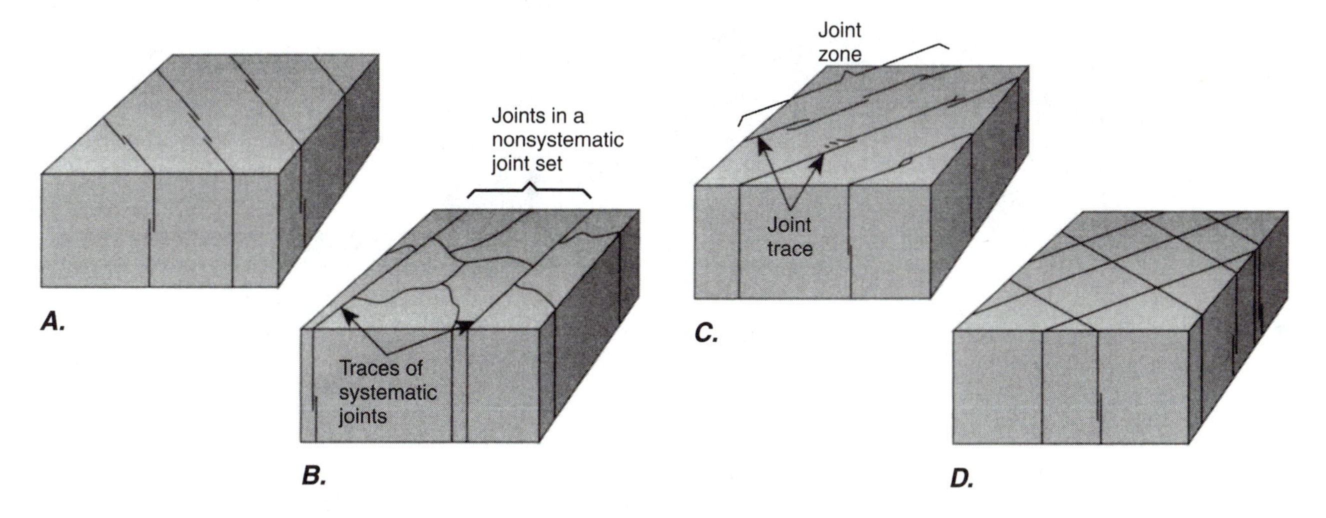

FIGURE 2.3 Diagrammatic views of joint sets and joint systems. *A*. Geometry of a systematic joint set. *B*. Typical pattern of nonsystematic joints and their characteristic termination against systematic joints. *C*. Joint zones forming quasi-continuous joints of much larger extent than the individual fractures. *D*. Two sets of mutually intersecting joints. Joints in each set cut joints of the other set. There is no consistent relationship whereby joints of one set terminate on joints of the other set. (After Hodgson, 1961b, figures 12, 16, 17)

FIGURE 2.4 Fracture controlled topography is evident in the southern tip of the Sinai Peninsula, Northern Red Sea, between the Gulf of Suez (left) and the Gulf of Aqaba (right). (From Short and Blair, 1986, plate 37, p. 111. Apollo 7 7-5-1623)

Columnar joints are extension fractures characteristic of shallow tabular igneous intrusions, dikes or sills, or thick extrusive flows. The fractures separate the rock into roughly hexagonal or pentagonal columns (Figure 2.6), which are often oriented perpendicular to the contact of the igneous body with the surrounding rock.

A variety of terms exist that refer to a characteristic orientation of joints in space or in relation to other structures. These characteristics are mainly useful in distinguishing among different fracture sets, but in general, they do not describe essential characteristics of a classification system that would relate to the mechanism of fracture formation. **Strike joints** and **dip joints** are vertical joints parallel to the strike or dip of the bedding, respectively, and **bedding joints** are parallel to the bedding. **Cross joints** are systematic joints of a set that

FIGURE 2.5 Sheet joints in granitic rock at Little Shuteye Pass, Sierra Nevada, California. (Photo by N. K. Hubert in Twidale, 1982)

either consistently terminate against the joints of another set or that cut a fold or some other linear feature at high angles; joints having other orientations relative to linear structures are **oblique joints** or **diagonal joints**.

Extension fractures commonly develop in association with shear zones in deformed rocks. **Pinnate fractures**, or **feather fractures**, are extension fractures that form ***en echelon*** arrays along brittle shear fractures (Figure 2.7), which means the extension fractures are parallel to one another but offset from one another along the trend of the shear fracture, which is oblique to the extension fracture plane. The sense of rotation through the acute angle from the fault plane to the extension fracture plane is the same as the shear sense on the fault plane (see also the discussion in Section 3.3(iii); Figure 3.17 *A*, *E*).

Gash fractures are extension fractures, usually mineral-filled, that form *en echelon* sets along zones of ductile shear in the same orientation as the pinnate fractures. They are generally S- or Z-shaped, depending on the sense of shear along the zone. The photograph in Figure 2.8 shows two *en echelon* sets of gash fractures (white veins) arrayed along crossing shear zones. The orientation of gash fractures relative to the shear zone can be used in the same way as feather fractures to determine the sense of shear on the associated shear zone. For distinctly S- or Z-shaped gash fractures, the sense of shear is also indicated by the sense of rotation of the central part of the fracture relative to the fracture tips. The

A.

B.

FIGURE 2.6 Columnar jointing. Devil's Postpile National Monument, California. *A*. Columnar jointing in an andesitic flow. Note that the orientation of the columns varies from vertical to a shallow plunge. *B*. Cross section of the columnar joints.

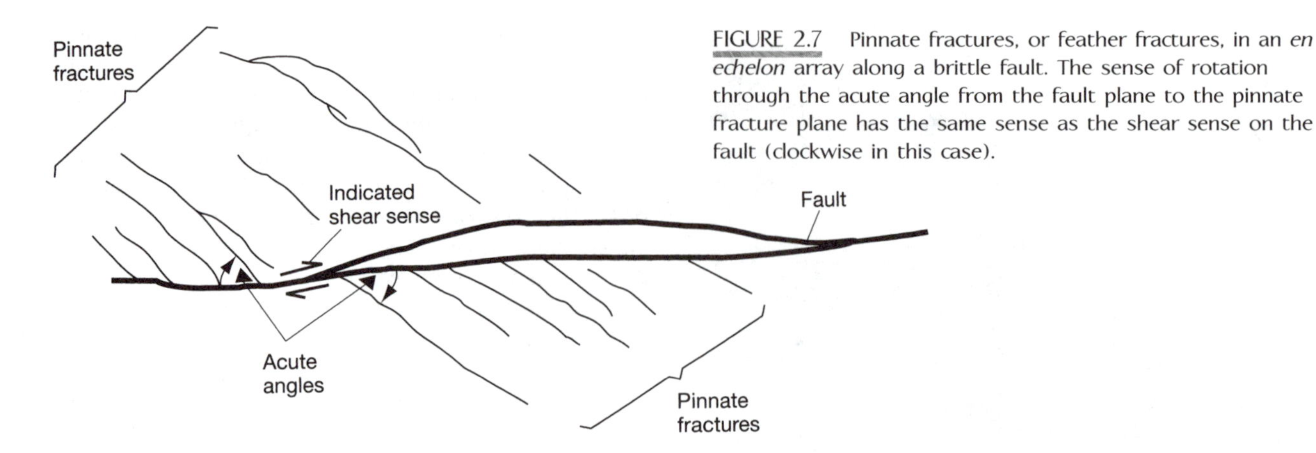

FIGURE 2.7 Pinnate fractures, or feather fractures, in an *en echelon* array along a brittle fault. The sense of rotation through the acute angle from the fault plane to the pinnate fracture plane has the same sense as the shear sense on the fault (clockwise in this case).

Z-shaped and S-shaped veins in the left and right shear zones in Figure 2.8, respectively, show the "top-down" sense of shear along each zone of fractures (see Figure 15.18). Extension fractures may also be associated with other structures, including folds and igneous intrusions, as described in Section 2.5.

Veins are extension fractures that are filled with mineral deposits. The deposit may be massive or composed of fibrous crystal grains of such minerals as quartz or calcite. The fibrous fillings can be very useful in interpreting the deformation associated with opening of the vein, as we discuss in detail in Sections 11.9 and 14.6(v).

FIGURE 2.8 Gash fractures (white veins) are extension fractures that commonly develop in a shear zone. Dark seams are solution features. Gash fractures are aligned along differently oriented planar shear zones that make an angle of approximately 50° with each other. The tips of the fractures curve toward the bisector of the angle between these shear zones. (Photo courtesy of Richard Sibson)

2.2 GEOMETRY OF FRACTURE SYSTEMS IN THREE DIMENSIONS

In studying the origin of fractures in rocks, we collect data on the spatial pattern and distribution of the fractures in each fracture set, which include the orientation of the fractures, the size of the fractures, the spacing of the fractures, and the relationship of the spacing to lithology and bed thickness. Systems of fractures, however, are commonly characterized by a self-similar geometry, which means the geometry of the fracture system is scale-invariant, i.e., it appears the same regardless of the scale of resolution of the observation. Thus the actual size and spacing one would measure depend on the scale, or the resolution, at which the fracture system is observed. In such a system, characteristic size and spacing are difficult to define precisely, except perhaps in terms of **fractal geometry**[4] (see Box 2-1). We discuss each of these characteristics below.

i. Orientation of Fractures

Many fractures tend to develop in sets characterized by a consistent fracture orientation. The determination of the preferred orientations of different fracture sets in a rock is one of the most common means of studying fracture systems (Figure 2.9). To be objective, we should collect the orientations of all the fractures visible in an area of outcrop that is large relative to the spacing between fractures of the most widely spaced set. To establish regional patterns we measure numerous exposures distributed over a large area. We correlate fracture sets, and especially joints, from one outcrop to another, assuming that fracture sets are related if their orientations are the same or are smoothly and consistently varying. Other criteria, however, such as evidence establishing the relative timing of fractures, direction of fracture propagation, and whether the fractures are limited to individual strata (cf. Figure 2.9*B*), can also be important in establishing correlations.

Such a program of orientation measurement may present difficulties. Many fracture planes are curved or twisted, making it difficult to decide what orientations are representative and what range of orientations are associated with a particular fracture set. Introducing subjective judgment into the process of choosing orientations to measure, however, may introduce bias into the sampling that could distort the description of the fracture geometry and could ignore fracture orientations that do not fit neatly into one of the sets with a clear preferred orientation. Some geologists produce detailed maps of all

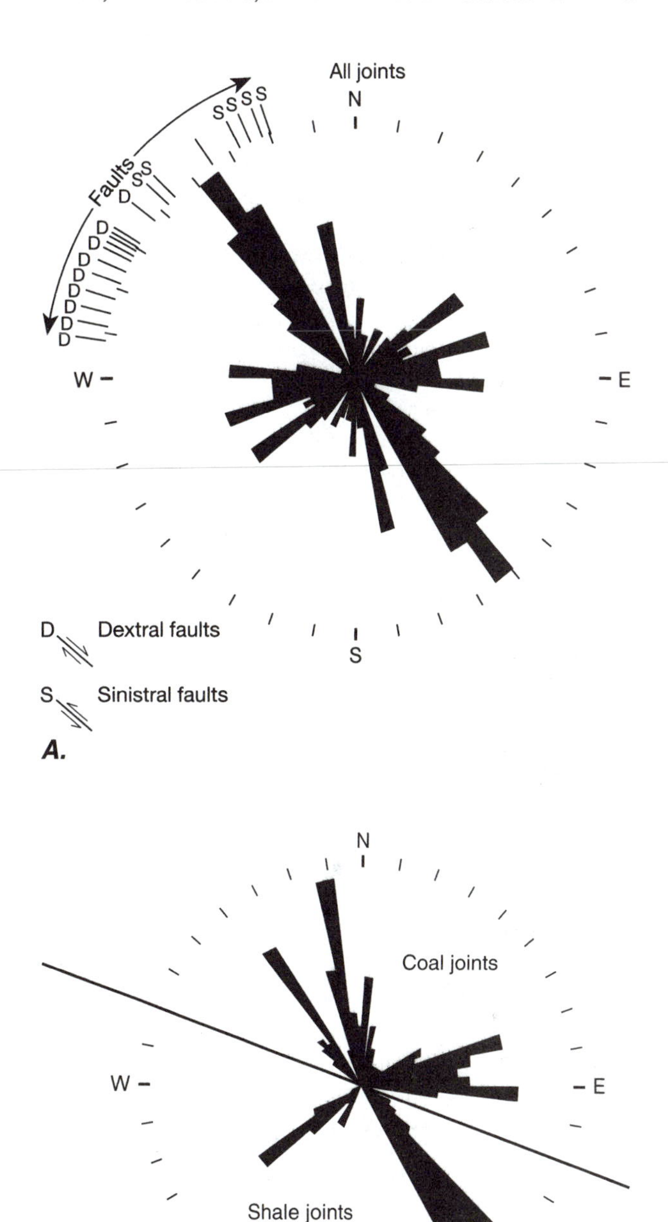

FIGURE 2.9 Rose diagrams showing the strikes of joints in a coal/shale sequence. Azimuth intervals are 5°; the length of the longest radius represents 40 percent of the measurements. *A*. All joints plotted together. Notice the diagram has a twofold axis of symmetry, i.e., a 180° rotation of the diagram is identical to the original. *B*. A split rose diagram showing the strikes of joints in coal in the upper half of the diagram and those in shale in the lower half. Note the difference in orientation of the joints in the two rock types. (After Nichelsen and Hough, 1967, figure 2)

[4]**Fractal geometry** is a branch of mathematics that identifies and quantifies how patterns repeat from one scale to another in a system for which the geometry is independent of scale. Fractures are features that have a fractal geometry.

fractures exposed on a given area of outcrop (see Figure 2.15*A*, Box 2-1, Figure 2-1.3) to analyze the fracture patterns. This method, however, produces just a two-dimensional pattern and does not record the three-dimensional orientations of the individual fractures and fracture sets. At this point in our understanding, the type of orientation analysis performed must be chosen to be consistent with the use intended for the data, and the geologist must keep in mind the limitations and potential biases of different data gathering methods.

Although orientation data on fractures are very useful, interpretations that rely too heavily on orientations can be misleading or incomplete. For example, because shear fractures commonly intersect one another at an angle of roughly 60°, it is often incorrectly assumed that all fractures that intersect at such an angle are shear fractures. Similarly, the consistent orientation of joints relative to other structures is often taken to indicate a genetic relationship. Although such interpretations cannot be ruled out *a priori*, they are unreliable unless corroborated by other evidence (see Box 1-1(iv)).

More than one orientation of fracture may be associated with a single fracturing event. Genetically related fractures may differ in orientation as a result of segmentation and twisting of the fracture plane, curving of the plane, reorientation of the fracture into parallelism with a local planar weakness in the rock, or branching of the fracture into two or more orientations. Some fractures may be of only local extent and may even result from human activity such as excavation or blasting. Careful study is required to identify the significant data.

The orientation of genetically related fractures may differ from one lithology to another; on the other hand, the fractures in layers of different rock types may result from different events. Figure 2.9*A*, *B*, for example, contains the same data, except that in the split rose diagram of Figure 2.9*B*, the joints in coal are plotted in the top half, and those in the shale are in the lower half. Note the different joint patterns from the two lithologies. Careful investigation has led to the conclusion that the joints in the coal formed first. Gentle folding of the region then broadened the distribution of the earlier joint orientations, and finally the joints in the shale formed. Separating the joints by lithology, therefore, can be important in deciphering the history of joint formation.

Orientation data on fractures are conveniently collected and compared by using orientation histograms, rose diagrams, or spherical projections, all of which we discuss in Appendix A1.2.

ii. Scale and Shape of Fractures

Individual fracture planes have definite tip lines where the fracture ends. Field observations indicate that a joint may terminate by simply dying out (Figure 2.10*A*), by curving and dying out (Figure 2.10*B*), by kinking and dying out (Figure 2.10*C*), by twisting and segmenting into an *en echelon* set of small extension fractures (Figure 2.10*D*; compare Figure 2.14*A*), by branching and dying out (Figure 2.10*E*), or by curving into a pre-existing joint (Figure 2.10*F*, *G*, *H*). These intersecting and branching

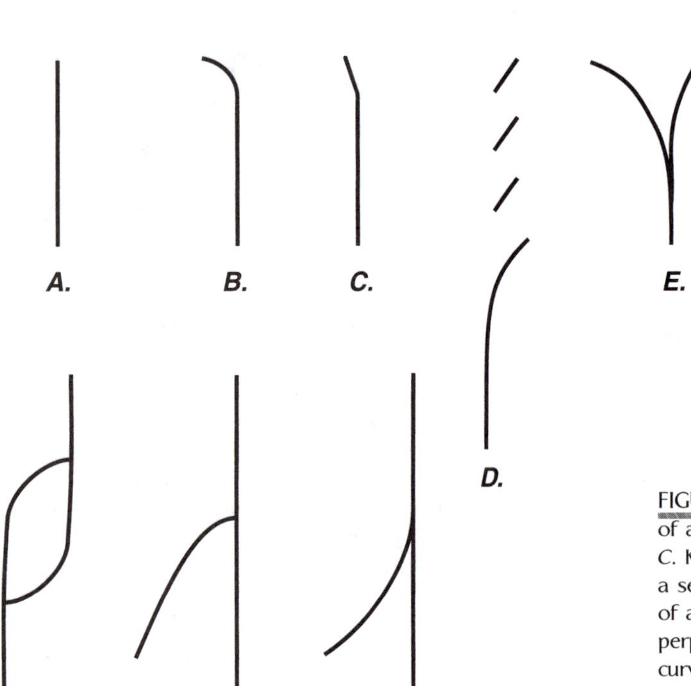

FIGURE 2.10 Terminations of individual joints. *A*. Dying out of a straight joint. *B*. Curving and dying out of a joint. *C*. Kinking and dying out of a joint. *D*. Dying out of a joint in a series of *en echelon* fractures. *E*. Branching and dying out of a joint. *F*. Two overlapping joints, each curving toward a perpendicular intersection with the other. *G*. One joint curving toward a perpendicular intersection with another. *H*. One joint curving toward a parallel intersection with another. (In part after Hodgson, 1961b)

relationships may result from joints developed at different times. The amount of displacement across the joint decreases toward the joint termination. In a given joint set, individual joint traces or joint zones (see Figure 2.3*C*) range in length from a few centimeters to many meters—and even up to kilometers in the case of **master joints**. Fractures also exist at scales as small as the microscopic level; such fractures are better referred to as **microfractures** than as joints.

The shape of individual joints depends largely on the rock type and its structure. In uniform rock, such as granite, argillite, or thin-bedded rocks of uniform composition, the boundary of an individual joint plane tends to be roughly circular to elliptical in shape, with the long axis horizontal. In sedimentary sequences involving rocks of highly different mechanical properties, such as interbedded sandstone and shale, one dimension of a joint is commonly constrained by the upper and lower contacts of the bed in which the joint forms, and the joint tends to be of much greater extent parallel to the bedding than across it. Joints in individual beds of one lithology often end against beds of another lithology (Figure 2.2*B*).

The shape of master joints is not well known because it is difficult to see the third dimension. In areas of great vertical relief, however, joints can be traced to a depth of more than 1 kilometer.

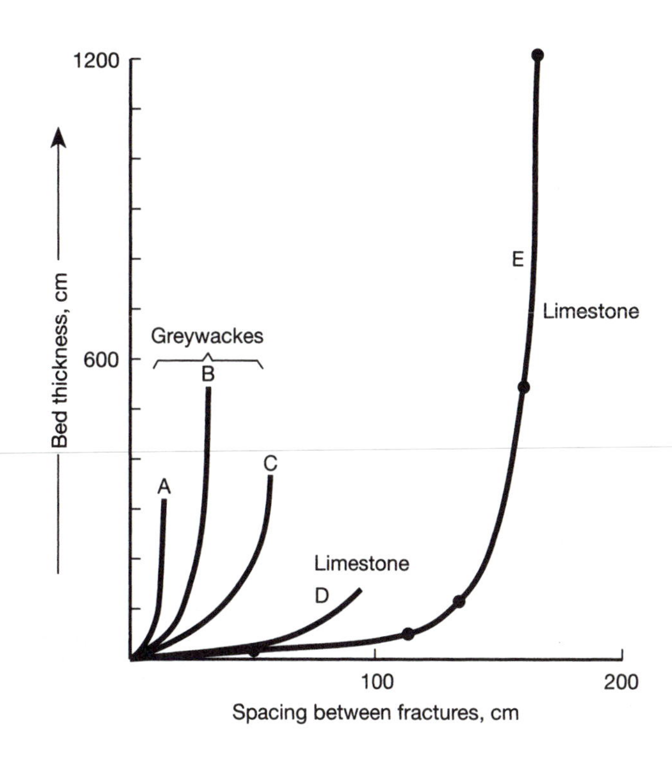

FIGURE 2.11 Relationship between bed thickness and fracture spacing measured in sandy layers in wackes (data sets A, B, and C) and in limestones (data sets D and E). For data set A, the shale interlayers are less than 5 cm thick, and for sets B and C they are greater than 5 cm thick. (Data from Ladeira and Price, 1981)

iii. Spacing of Fractures

The spacing of fractures in a systematic set can be measured either as the average perpendicular distance between fractures or the average number of fractures found in a convenient standard distance normal to the fractures. Such spacing measurements should be done only on joints of similar scale, because in at least some systems, the spacing is dependent on the size (length) of the joints (see Box 2-1). The average spacing of joints of a given scale tends to be remarkably consistent, and it depends in part on the rock type and the thickness of the bed in which the fractures are developed (Figure 2.11). Data sets A through C are measurements made in the sandy layers of a sequence of wackes from several locations with different thicknesses of shale interbeds. Data sets D and E are from different limestones. Two features of the plot are significant. First, the spacing between adjacent joints increases with increasing bed thickness, up to a maximum value beyond which spacing is independent of the thickness. Second, the maximum spacing is considerably greater for the limestones than for the wackes, which demonstrates the effect of lithology independent of the thickness of the beds.

iv. Spatial Pattern and Distribution of Fracture Systems

The most useful method of studying the pattern and distribution of fracture sets is to plot maps of the location and orientation of the fractures. In areas of very good exposure, it may be possible to map joints individually and trace out the relationship of joints to one another and to lithology, and to analyze the anisotropy of the fracture pattern, the connectivity of the fractures, or their geometry. On such maps, we can also plot the strikes and dips of the fractures, their relationship to other local structures, and the amount and direction of shear (if any) on the fractures. Examples of detailed maps of complex fracture systems are given in Figure 2.15 and Figure 2-1.3 (Box 2-1).

In most cases, there is neither enough exposed rock nor enough time available to permit such detailed mapping. Usually data from outcrops scattered over a large area are plotted on a map. From these data one constructs **form lines**, or **trajectories**, of the individual joint sets by correlating and matching the sets from one outcrop to the next and assuming the strikes of joints in the same set vary smoothly if at all from one outcrop to the next. Figure 2.12 shows such a map for an area on the Appalachian plateau. The consistency of orientation of joints over such large areas indicates that they record regional tectonic conditions.

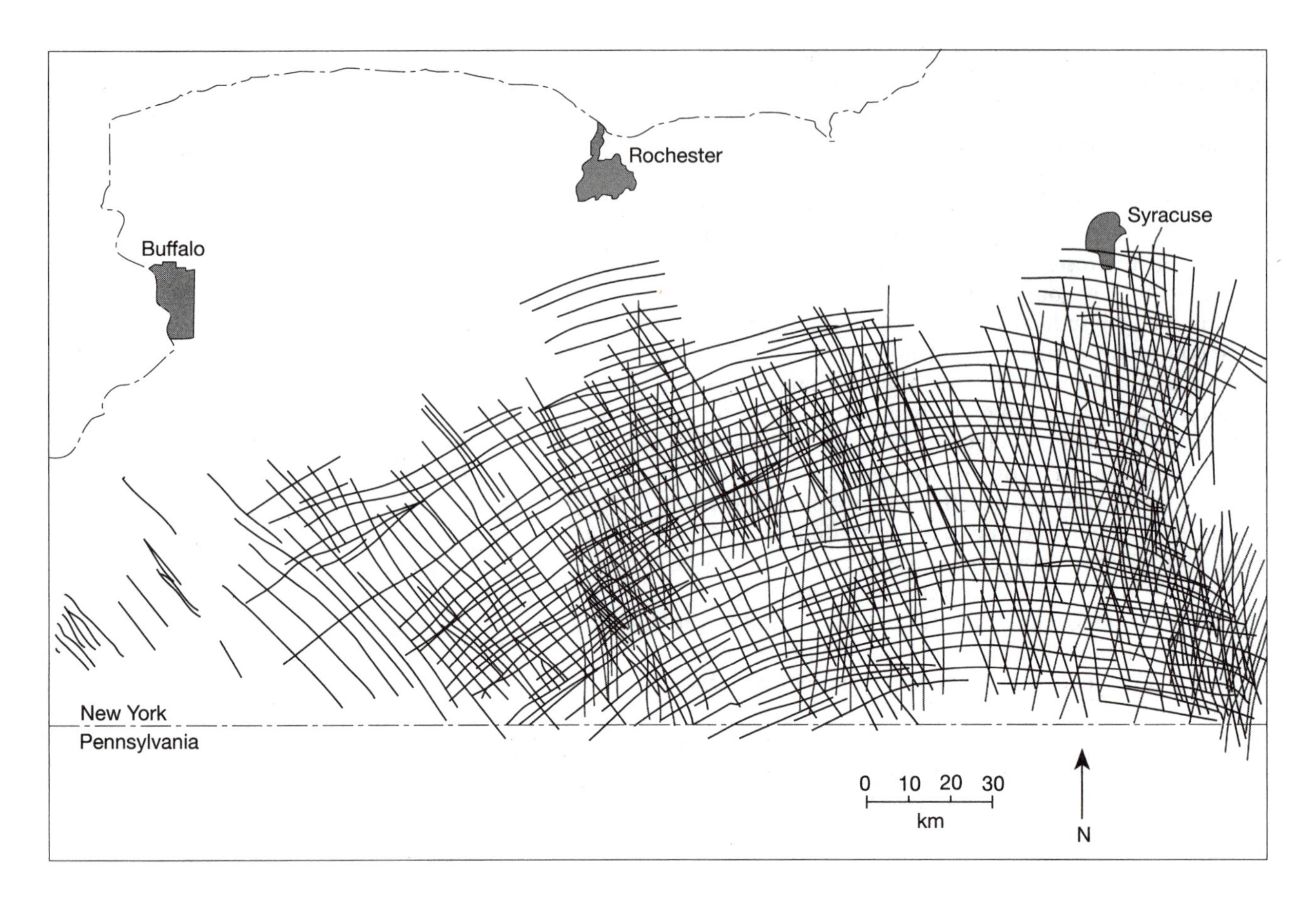

FIGURE 2.12 Map of joints in the Appalachian plateau in New York State. The trend lines are constructed parallel to the dominant strike of joints measured in local outcrops throughout the area. (From Engelder and Geiser 1980)

BOX 2-1 Fractals and the Description of Joint Patterns

Joints occur in rocks on a vast range of scales, from master joints on the scale of kilometers or tens of kilometers down to small fractures on the scale of centimeters or less. The spacing between joints occurs with a corresponding range of scales. This range of scales has made the characterization of joint systems very difficult. The ability to describe accurately the characteristics of joint patterns, however, is important in predicting such phenomena as the fluid flow and storage properties of rock, necessary in calculating groundwater flow and oil flow in reservoir rock, and the bulk mechanical properties of rock, important to numerous types of construction projects. This complexity has led to the proposal that the geometry of joint systems is **fractal**. We first introduce briefly what a fractal object is, and then we describe how the concept of fractals might be applicable to joint patterns. A variety of features of a physical system such as a joint system can have fractal characteristics. These might include, for example, the fracture length or the displacement on the fracture, which for extension fractures determines the fracture opening or aperture. We concentrate in this discussion on the fractal geometry, or the spatial distribution of fractures, which defines the extent to which the fractures fill the space in which they are embedded.

i. What Is a Fractal?

An object having a fractal geometry is one for which the geometry is **scale-invariant**. The object is also said to be **self-similar**, which means that any part of the object viewed at a particular scale looks similar to the object viewed at any other scale. Thus, for example, the pattern of joints viewed from an airplane at the scale of a quadrangle would look the same as the pattern of joints as seen close-up in an outcrop. This idea is actually quite familiar to geologists, who understand the necessity of placing a familiar object such as a ruler or a rock hammer in photographs of geologic structures so that the size of the structure is apparent, because the structures themselves, commonly being self-similar in nature, do not provide any information as to the scale.

One way to obtain an intuitive grasp of a fractal geometry is to consider a line embedded in a plane (Figure 2-1.1A). The line has a dimension of 1, and the plane has a dimension

(continued)

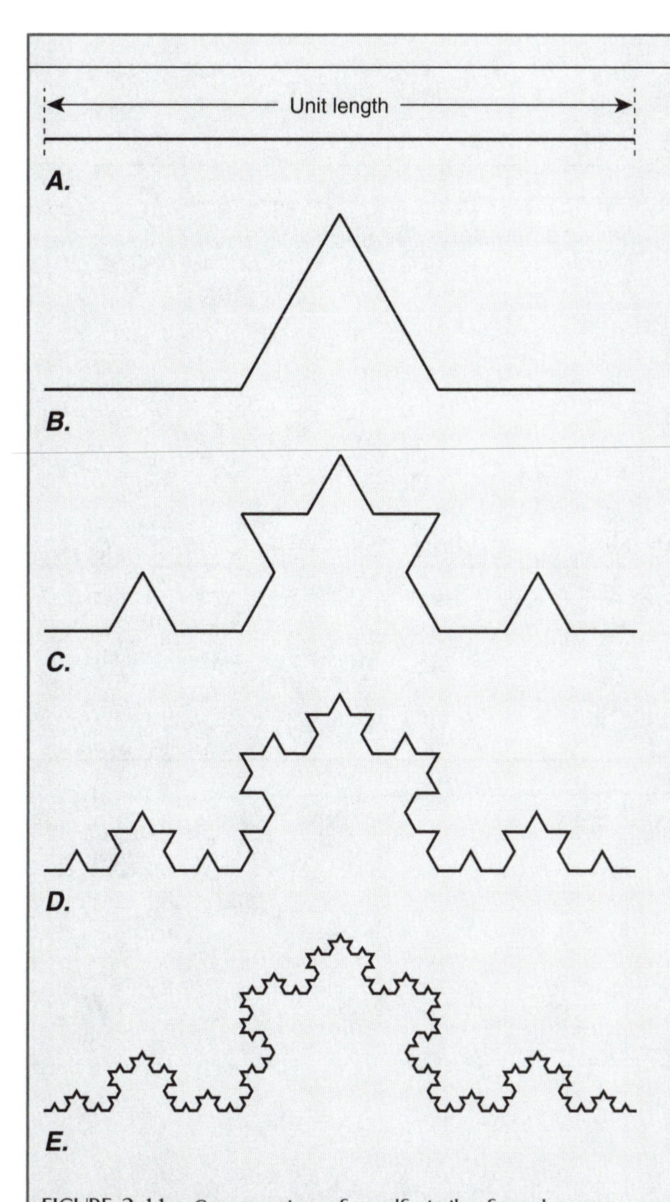

FIGURE 2-1.1 Construction of a self-similar fractal curve, the von Koch curve. The middle third of each line segment is replaced by two segments, each of which has the same length as the segment that is replaced. *A*. A normal line of length 1. *B–E*. Generations 1 through 4 in the construction of the von Koch curve. (After Malinverno, 1995, with permission from Springer Science and Business Media)

of 2. Assume the length of the line is 1. Now let us make that line jagged by replacing the middle third of the line length with two line segments, each with length equal to 1/3 (Figure 2-1.1*B*). Each of the resulting line segments can in turn be replaced by a jagged line with segments having a length of 1/3 of that initial line segment. In principle, this procedure can continue *ad infinitum* (Figure 2-1.1*C–E*). The resulting object, called the **von Koch curve**, has a fractal geometry. Regardless of what scale, i.e., what strength of magnification, we use to look at the line, we will always see the same pattern; thus the curve is said to be self-similar, or its geometry is scale-independent.

An attribute of a fractal geometry is that we cannot characterize it using normal geometric measures. For example, we cannot characterize the von Koch curve by its length, because, as we show next, the length of the curve depends on the length of the ruler we use to measure it. Imagine measuring the length of a line by opening a divider up to span a length r and then walking the divider along the line from one end to the other and counting how many steps n are required to reach the end. This is called the **divider** or **ruler method** of measuring an object. The length of the line is just nr, the number of steps n multiplied by the span of the divider r.

For a normal one-dimensional line, for any length of r, we always get the same total length of the line (Figure 2-1.2*A*). Thus the length of the line in Figure 2-1.2*A* is

$$nr = 1 \tag{2-1.1}$$

for any value of r. For the van Koch curve, however, if we use a span of length $r = 1$ for the divider, then the line in Figure 2-1.1*A* is spanned in one step $n = 1$ and thus has a length of $nr = (1)(1) = 1$. If the divider spans a length $r = 1/3$, however, then, since the von Koch curve has four segments of length 1/3 (Figure 2-1.1*B*; cf. Figure 2-1.2*B*), it requires $n = 4$ steps to measure the line length of the object, which must be $nr = 4\ (1/3) = 1.333$. If our divider spans a length $r = 1/9$, then because there are $n = 16$ segments of that length (Figure 2-1.1*C*), the length of the von Koch curve at that scale must be $nr = 16\ (1/9) = 1.778$. This process can go on indefinitely. The first few steps are summarized in the following table:

Lengths of the von Koch Curve

Span, r	Number of increments, n	Length	Figure 2-1.1
1	1	1	*A*
1/3	4	1.333	*B*
1/9	16	1.778	*C*
1/27	64	2.370	*D*
1/81	256	3.160	*E*
. . .	. . .	. . .	

Thus the shorter the span r we use to measure the length of the fractal line, the longer the line appears to be. Evidently, the line has no characteristic length that we can use to specify its geometry.

In two dimensions, the area of a square with a side of length 1 that is divided into n parts by a scale of length r is given by an equation comparable to Equation (2-1.1):

$$nr^2 = 1 \tag{2-1.2}$$

For example, a square whose side has length 1 has an area of $(1)(1) = 1$. If it is divided into four smaller squares (Figure 2-1.2*C*),

(continued)

BOX 2-1 Fractals and the Description of Joint Patterns *(continued)*

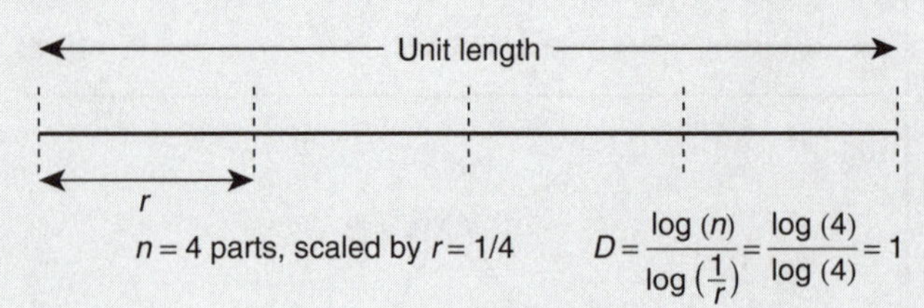

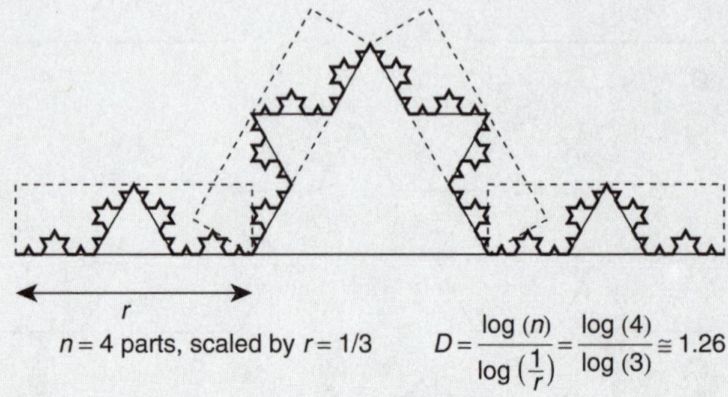

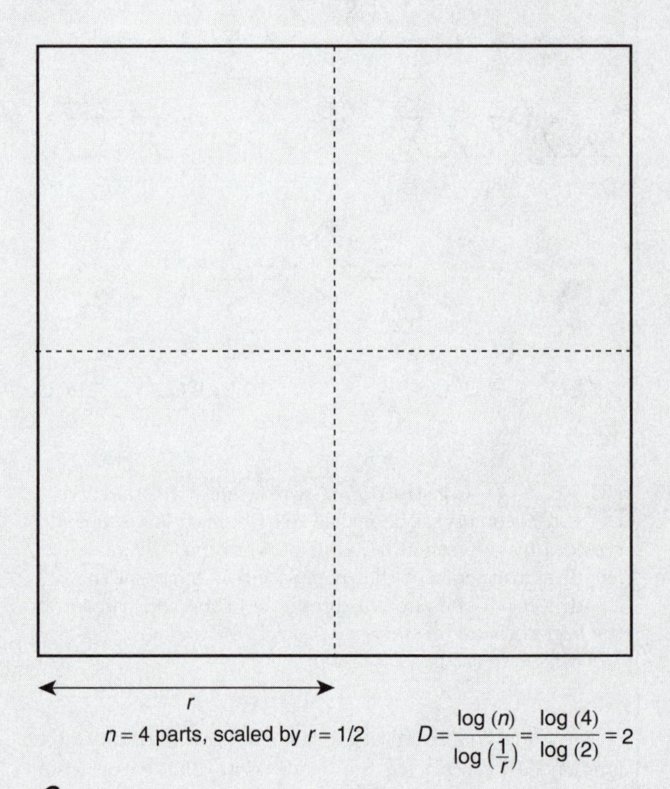

FIGURE 2-1.2 Determination of the dimension of an object. *A.* A normal line has dimension $D = 1$. Its length, given by the number of parts n multiplied by the length r of each part, is always the same, regardless of the size of r. *B.* The length of the von Koch curve increases as the scale of the measure r decreases. This property is characterized by the fractal dimension $D = 1.26$. *C.* A normal plane has dimension $D = 2$. Its area, given by the number of parts n multiplied by the area of each part r^2, is always the same, regardless of the size of r. (After Malinverno, 1995, with permission from Springer Science and Business Media)

each with a side $r = 1/2$, each has an area $r^2 = 1/4$, so the total area of the original square is (4) (1/4) = 1. Thus for any dimension D, our usual understanding of the Euclidean size C of a geometric object is given by the general equation

$$n\,r^D = C \quad \text{or} \quad n = \frac{C}{r^D} \qquad (2\text{-}1.3)$$

where $D = 1$ for a one-dimensional line, $D = 2$ for a two-dimensional plane, and $D = 3$ for a three-dimensional block. C is a scaling constant that defines how many parts of dimension $r = 1$ there are in the object; in Equations (2-1.1) and (2-1.2), C is equal to 1.

We see from the first Equation (2-1.3) that D is determined by

$$\log n + D \log r = \log C \qquad (2\text{-}1.4)$$

$$\log n = -D \log r + \log C = D \log\left(\frac{1}{r}\right) + \log C \qquad (2\text{-}1.5)$$

We use the results for two different scales to eliminate the constant C from Equation (2-1.4)

$$\log n_1 + D \log r_1 = \log n_2 + D \log r_2 \qquad (2\text{-}1.6)$$

$$D = \frac{-\Delta\log n}{\Delta\log r} = \frac{\Delta\log n}{\Delta\log\left(\frac{1}{r}\right)} \qquad (2\text{-}1.7)$$

where we define

$$\Delta\log n \equiv \log n_2 - \log n_1 = \log\frac{n_2}{n_1} \qquad (2\text{-}1.8)$$

$$\Delta\log r \equiv \log r_2 - \log r_1 = \log\frac{r_2}{r_1} \qquad (2\text{-}1.9)$$

$$\Delta\log\left(\frac{1}{r}\right) \equiv \log\left(\frac{1}{r_2}\right) - \log\left(\frac{1}{r_1}\right) = \log\left(\frac{r_1}{r_2}\right) \qquad (2\text{-}1.10)$$

Applying the second of the equivalent equations (2-1.7) to the von Koch curve with Equations (2-1.8) to (2-1.10), and using the second and third row in the above table (any two rows would give the same result), we discover that

$$D = \frac{\Delta\log n}{\Delta\log\left(\frac{1}{r}\right)} = \frac{\log\left(\frac{n_2}{n_1}\right)}{\log\left(\frac{r_1}{r_2}\right)} = \frac{\log\left(\frac{16}{4}\right)}{\log\left(\frac{9}{3}\right)} = \frac{\log 4}{\log 3} = 1.26 \qquad (2\text{-}1.11)$$

This result shows that the fractal von Koch curve has a dimension $D = 1.26$, which is part way between that of a normal line and a normal plane. This number is in general not an integer, but a fractional dimension and thus has been termed the **fractal dimension**. Regardless of the scale r at which we examine the von Koch curve, the fractal dimension is always the same, and thus it characterizes an important aspect of the geometry of this object. The power-law relationship of the type in Equations (2-1.3) is characteristic of a fractal, or self-similar, system.

A coastline is an example of a natural object that has the characteristics of a fractal geometry. An oft-quoted example is Mandlebrot's analysis of the west coast of Britain, which has a fractal dimension very close to that of the von Koch curve. The von Koch curve, however, is very regular and precisely self-similar, whereas the west coast of Britain is very irregular and only statistically self-similar. Nevertheless, the fractal dimension characterizes a quality of roughness that is similar in the von Koch curve and in the coastline. Natural-looking objects such as coastlines can be generated by including a degree of randomness in the fractal construction.

Of course, in the real world, there are always upper and lower bounds, called the **fractal limits**, beyond which the fractal dimension ceases to describe the object. In the case of a coastline, the upper fractal limit obviously must be no larger than the dimension of the island or continent bounded by the coast, and the lower fractal limit is perhaps no smaller than a grain of sand.

Different fractal objects can have different fractal dimensions. The fractal dimension defines the extent to which a fractal object fills the space in which it is embedded. An object with a dimension between 1 and 2 occupies more area than a line, but does not completely occupy the area of the plane in which it is embedded. Equation (2-1.5) shows that if we measure the length of a fractal line with a wide variety of scales r and then make a plot of log n versus log $(1/r)$, the plot should result in a straight line whose slope is the fractal dimension D. This gives us a way of testing the fractal nature of a physical system. A number of geologic systems have been described using fractals, including joint patterns, fracture surface roughness, and fault gouge grain size. Here, our interest is in fracture patterns, and in the following we look briefly at their fractal nature.

ii. The Fractal Nature of Joint Systems

If joints have a fractal geometry, then the fractures occur on a very wide range of scales with smaller fractures being much more numerous than the larger ones. What does this have to do with describing joints? It has been proposed that joint systems have the quality of being self-similar, so that regardless of the scale at which you look at a joint system, within the fractal limits, it looks the same. It is impractical to try to map a system of joints in three dimensions, but it can be shown that if a volumetric fractal structure is isotropic[1] and random, the fractal dimensions determined from a planar section through the volume will be 1 less than the fractal dimension determined for the volume, and the fractal dimension of a linear section through the volume will be 2 less. Thus if we determine the fractal dimension of a joint system on a plane to be D_2 and if that system is isotropic and random, then the fractal dimension of a linear section determined from a drill hole through the structure would be $D_l = D_2 - 1$ and the dimension of the three-dimensional structure would be $D_3 = D_2 + 1$. This provides some justification for examining the pattern that a joint system makes on an outcrop surface. Figure 2-1.3 shows four maps of joint systems. Note the wide variety of scales and the fact that nothing about the joint patterns gives any indication of the scale at which the pattern was mapped.

In order to determine whether such joint patterns are fractal, we use a measurement method called the **box method**, which is comparable to, but more convenient than, the divider method. The box method measures the length of the object by counting the number of boxes of dimension r that are required to completely cover the object, rather than using a divider spanning the length r to step out its length. To use the box method, we construct a square grid with a grid spacing of r, place the grid over the object (in this case a system of joints), and count the number of boxes n in the grid in which the object is found. Repeating this procedure for a large number of values of r and plotting the results on a graph of log n versus log $(1/r)$ (Equation (2-1.5)) should, for a fractal object, give a straight line with a slope equal to the fractal dimension D (Figure 2-1.4).

Figure 2-1.5 shows the results of applying the box method to the joint system maps in Figures 2.15A and 2-1.3, with the lower and upper fractal limits defined, respectively, by the shortest line length in the map and the length at which the number of occupied boxes equals the total number of boxes. The lines are the least-squares fit to all the points measured, but points are plotted for only three of the lines, and only a selection of the points that were measured is plotted. The data all give an excellent fit to a line, and the fractal dimension for the four maps varies from 1.5 to 1.7, with an accuracy of about ± 0.02, so the differences in the value of D are significant. The different intercepts of the lines simply reflect the different scaling constants C (Equation (2-1.5)) used in the measurement of the different maps. Thus the geometry of a joint system can be characterized, at least in part, by its fractal dimension. The fractal dimension for each map has been determined over less than an order of magnitude range of scale, which is smaller than would be ideal for such a measurement. Taken together, however, the scales in Figure 2-1.3 span a range of more than three orders of magnitude.

Analysis of fracture patterns formed by different fracturing mechanisms shows that the fractal dimension is not characteristic of particular mechanisms. The fractal dimension also does not characterize all the geometric properties of a fracture system. For example, although the fractal dimension increases with increasing fracture density, it seems to be rather

[1]**Isotropic** means that the properties are the same in all directions.

(continued)

BOX 2-1 Fractals and the Description of Joint Patterns *(continued)*

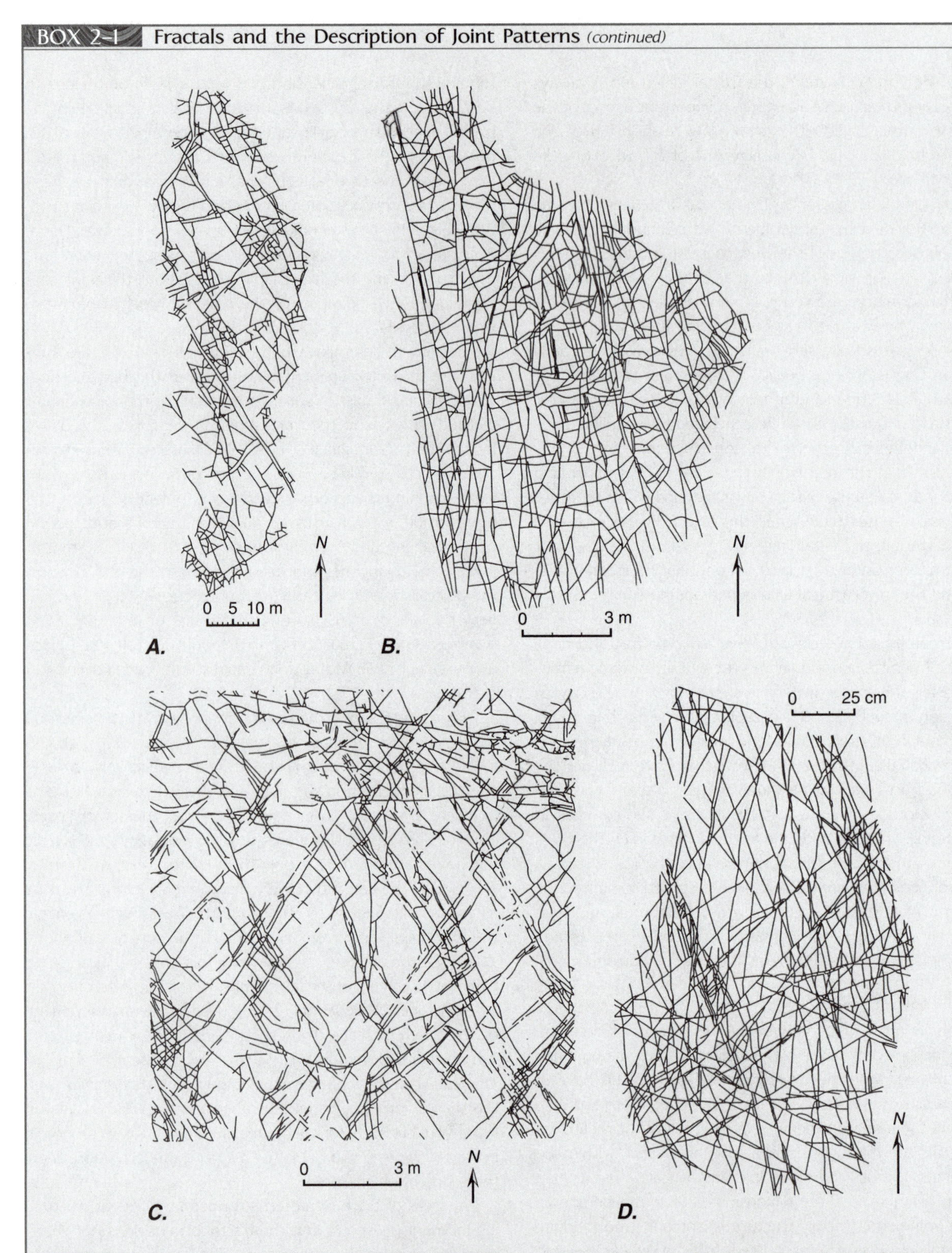

FIGURE 2-1.3 Maps of joint systems and their fractal dimensions. The accuracy in the value of the fractal dimension is about ±0.02. (Berkowitz and Hadad, 1997) *A.* Paintbrush tuff, Yucca Mt., Nevada; $D = 1.61$. *B.* Paintbrush tuff, Yucca Mt., Nevada; $D = 1.70$. *C.* Niagaran dolomite, Lannon, Wisconsin; $D = 1.60$. *D.* Lyons sandstone, Morrison, Colorado; $D = 1.50$. (All maps presented in Barton, 1995, figure 8.7d, g, i, and j, respectively. *A*, *B*, and *D* mapped by Barton, 1995; *C* mapped by LaPointe and Hudson, 1985. *A*, *B*, and *D* reproduced with permission from Springer Science and Business Media; *C* reproduced with permission from the Geological Society of America)

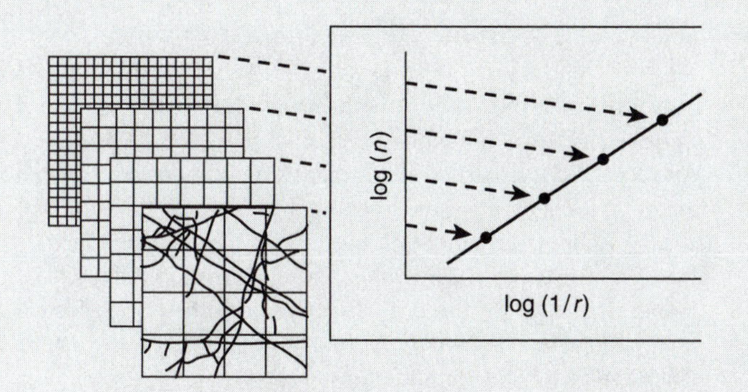

FIGURE 2-1.4 The box method of measuring fractal dimension. A grid of squares each with a side of length r is superimposed on the fracture map, and the number of squares n occupied by fractures is counted. Grids of successively smaller grid size r are used to generate a series of points on a plot of $\log(1/r)$ versus $\log n$. For a fractal object, the plot is a straight line whose slope is the fractal dimension D. (After Barton, 1995, with permission from Springer Science and Business Media)

insensitive to the distribution of lengths and orientations of the fractures in the system. It also is insensitive to tendencies for clustering or scattering (anticlustering) of fractures, which determines the extent to which fractures are clustered together or are evenly spaced. Moreover, fracture sets commonly develop sequentially (Figure 2.15), and thus the fractal dimension for the joint system would change with time. The fractal dimension for the geometry of the joint system would then describe just the final fracture pattern, regardless of when different sets of fractures formed.

iii. A Model for Fragmentation

With these caveats in mind, we present as an example a model for fragmentation, or **comminution**, that results in a fractal geometry of fractures. One hypothesis for the process of comminution proposes that when two grains of comparable size come into contact, one of them will fragment, but a large fragment is unlikely to break if it is in contact only with small fragments, and a small fragment is unlikely to break if it is in contact with a large fragment. The result is that, in general, no two fragments of roughly equal size should end up in contact with each other.

A model for producing a fractal structure with these characteristics is illustrated in Figure 2-1.6. An initial cube is divided into eight pieces by three mutually perpendicular bisecting planes. Of those eight, two cubes on a diagonal of the initial cube are left intact, and the other six are each subdivided into eight other cubes. This process continues indefinitely, always at each scale leaving two cubes on a diagonal intact and subdividing the remaining cubes according to the same scheme. If the length of an edge of the initial cube is $r_0 = 1$, then the number of cubes N_k of size r_k is given in the following table:

Step No., k	Cube Size, r	Number of Cubes, n
0	1	1
1	1/2	2
2	1/4	12
3	1/8	72
4	1/16	432
...	...	...

The fractal dimension of the fragmented cube is given from Equations (2-1.7)–(2-1.10):

$$D = \frac{\Delta \log n}{\Delta \log\left(\frac{1}{r}\right)} = \frac{\log\left(\frac{12}{2}\right)}{\log\left(\frac{4}{2}\right)} = \frac{\log 6}{\log 2} = 2.58$$

The dimension for the fractures on a surface of the cube is one less than for the corresponding three-dimensional structure, or 1.58. This dimension is very close to the dimensions observed for the joint patterns in Figure 2-1.3. The joint

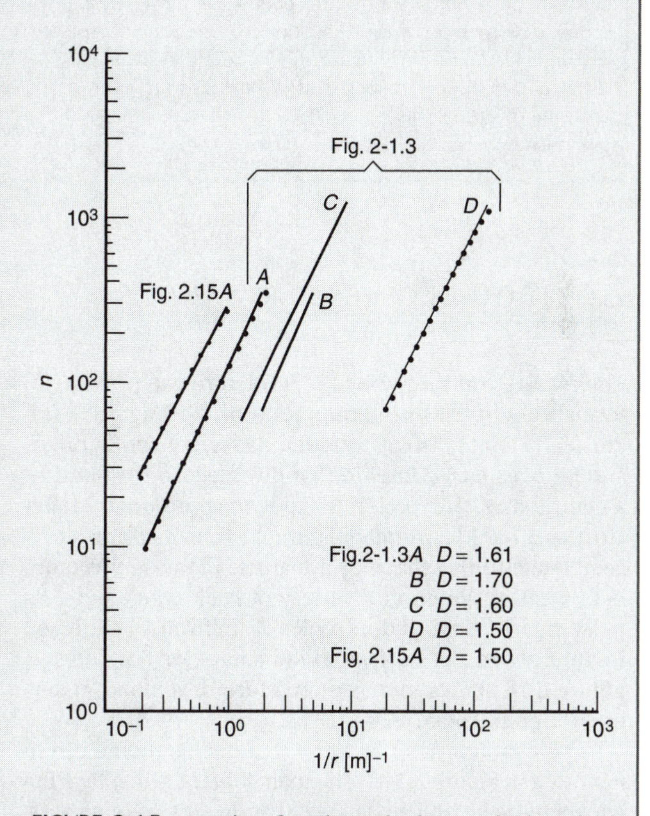

FIGURE 2-1.5 Results of applying the box method to the joint maps of Figure 2-1.3 and 2.15A. For the lines with plotted data points, the plot shows every third point generated from the analysis. The correlation coefficient for the fits to these lines is better than 0.99, where 1 indicates a perfect fit. (After Barton, 1995, with permission from Springer Science and Business Media)

BOX 2-1 Fractals and the Description of Joint Patterns *(continued)*

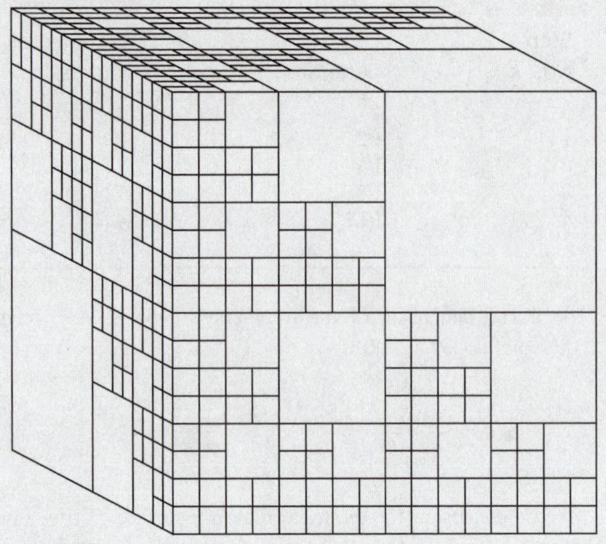

FIGURE 2-1.6 A fractal model for fragmentation. At each scale, a cube is divided into eight smaller cubes by three mutually perpendicular bisecting planes. Of these eight cubes, two that occur on a diagonal of the original cube are left undivided, and the remaining six are subdivided. The result is that no two cubes of the same size are ever adjacent to each other. (After Turcotte and Huang, 1995, with permission from Springer Science and Business Media)

systems with the closest fractal dimension are in Figure 2-1.3*A* ($D = 1.61$) and Figure 2-1.3*C* ($D = 1.60$). Of the four, these two look most as if they satisfy the comminution model requiring, in general, that two fragments of equal size should not be in contact. Of the other two joint maps, the fracture pattern in Figure 2-1.3*B* has a higher fractal dimension ($D = 1.70$), and that in Figure 2-1.3*D* (and in Figure 2.15*A*) has a lower dimension ($D = 1.50$). Although higher fractal dimension indicates a higher density of fractures, it is perhaps not obvious just from inspection that this is true, probably because of the fractal limits that apply for the different maps.

The foregoing discussion of the cube describes the comminution model in action. The fractal dimensions for the joint maps in Figure 2-1.3 are close to that of the model fractured cube and therefore define some characteristic similarity in their geometries. Nevertheless, it is not obvious that this comminution model is an appropriate explanation for the jointing process, especially in view of the caveats noted above.

In summary, we conclude that the fractal dimension is a way of describing certain fundamental characteristics of the geometry of self-similar systems. It does not encompass, however, all significant geometric characteristics, and it may be difficult to infer from this description anything specific about the mechanism by which the system developed. It is a subject of ongoing research.

2.3 FEATURES OF FRACTURE SURFACES

The features on the surface of a fracture can provide information critical to interpretation of the fracture's origin. Many joints display a regular pattern of subtle ridges and grooves called **hackle** that diverges from a point or a central axis. The pattern is known as **plumose structure** or a **hackle plume** (Figure 2.13), named for its resemblance to the shape of a feather. Plumose structure is present on joints in a variety of rock types, but it is most clearly displayed in rocks of uniform fine-grained texture, when the surface is illuminated at low angles. Figure 2.13 shows plumose structure developed in several different rock types.

The characteristic features of a hackle plume are illustrated in Figure 2.14*A*. The main joint face displays the hackle plume with hackle lines that diverge from an axis. Toward the edges of the main joint face, the joint plane may segment into a series of planes that are slightly twisted from the main joint face. The twisting may increase gradually away from the axis (Figure 2.13*C*; lower half of Figure 2.14*A*), or it may develop abruptly at a shoulder (Figure 2.2*B*; top of Figure 2.14*A*). In either case, the results is a **hackle fringe** composed of a set of extension fractures, or **fringe faces**, aligned *en echelon* along the trend of the main joint face and connected to one another by curving fringe steps. The fringe faces themselves may show second-order hackle plumes with associated fringes.

Fringe faces at the edge of a joint should not be confused with pinnate fractures and gash fractures, even though they are all extension fractures that form *en echelon* arrays. Fringe faces usually make a considerably smaller angle with the main joint face than pinnate or gash fractures make with the shear surface. Moreover, in three dimensions, fringe faces are restricted to the edge of a joint surface, which commonly displays plumose structure, whereas pinnate and gash fractures occur along the entire shear fracture.

In some cases, curvilinear features called **rib marks** and **ripple marks** cross the lines of hackle on the fracture face. The rib marks are either cuspate in cross section (Figure 2.14*B*) or composed of smoothly curved ramps connecting adjacent parallel surfaces of the joint face (Figure 2.14*C*; see also Figure 2.13*B*). They tend to be perpendicular to the hackle lines. The ripple marks are rounded in cross section and oblique to the hackle lines (Figure 2.14*D*). Hackle plumes form a variety of different patterns (compare photos in Figure 2.13), which can characterize particular sets of joints and which reflect important differences in the fracturing process.

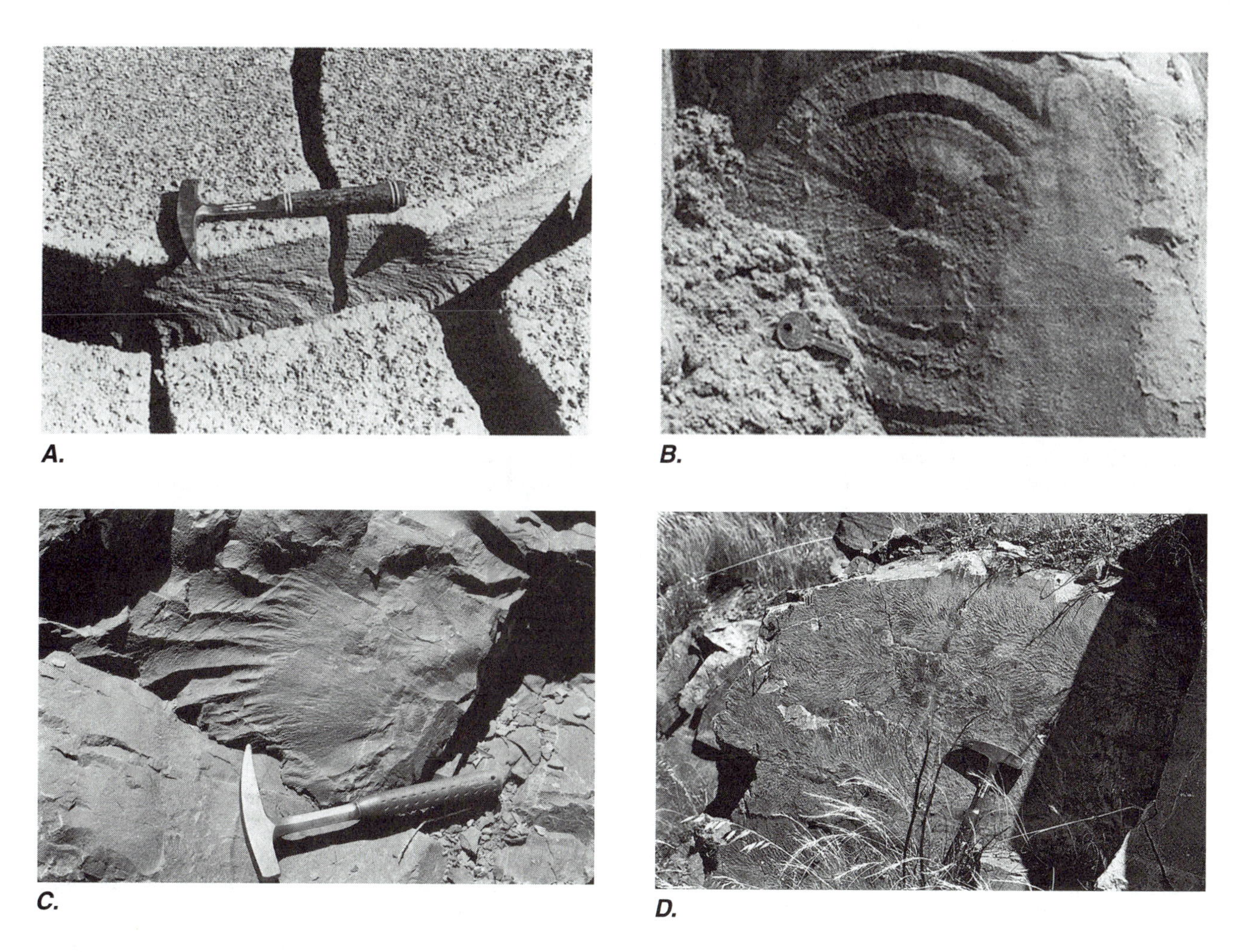

FIGURE 2.13 Plumose structure, or hackle plumes, in *A*. a mudstone. *B*. a chalk. *C*. a basalt D. a basalt. (A and B from Bahat, 1979, plates 1B, 4)

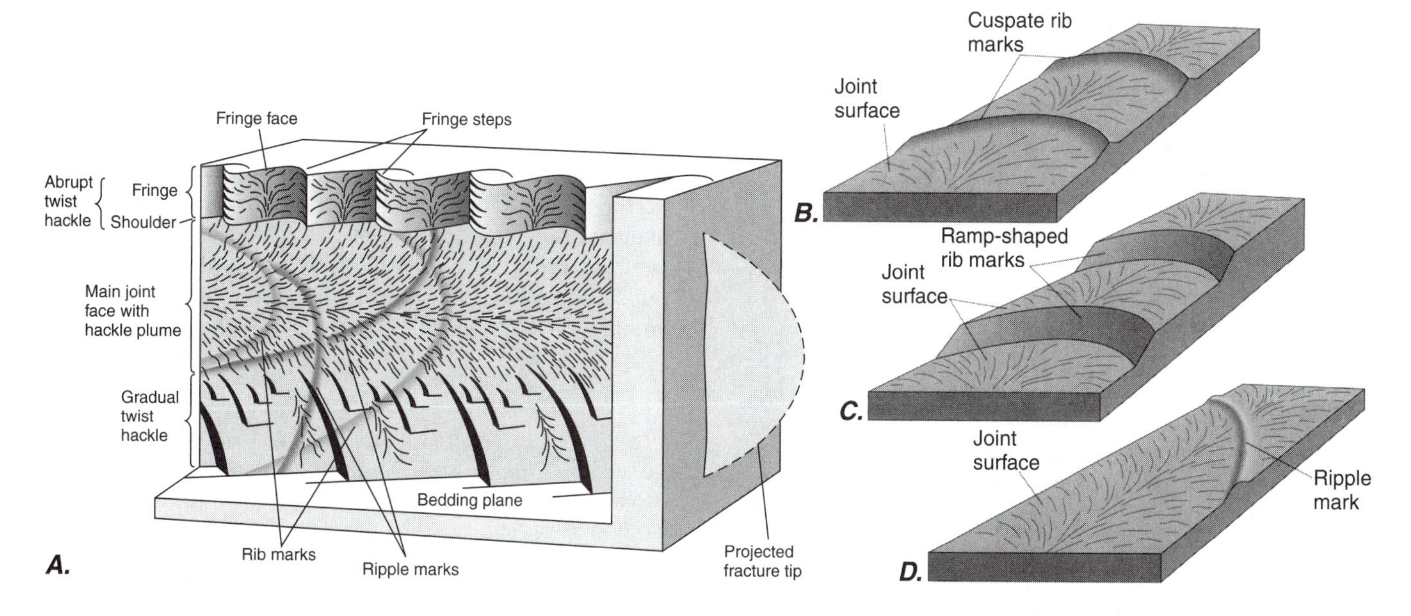

FIGURE 2.14 Schematic block diagrams illustrating markings on joint faces. *A*. Features that characteristically appear on the surface of a systematic joint, including a hackle plume, rib marks, ripple marks, and joint fringe. (After Hodgson, 1961a, figure 1) *B*. Cuspate rib marks. *C*. Ramp-shaped rib marks. *D*. Ripple marks. (After Kulander, Barton, and Dean, 1979, figure 77)

Plumose structure is a unique feature of brittle extension fractures that distinguishes them from shear fractures. The direction of divergence of the hackle lines is the direction in which the fracture propagated, and the lines of hackle form normal to the fracture front and parallel to the local propagation direction of the fracture front. When traced back along the plume axis, the hackle is usually found to radiate from a single point, which is the point of origin of the fracture. Rib marks are interpreted to be arrest lines where fracture propagation halted temporarily. Ripple marks are interpreted to form during very rapid fracture propagation, in which case they are called **Walner lines**. By careful study of the surfaces of joints, therefore, we can learn a great deal about where fractures initiated and how they propagated. The use of hackle plumes is treated in greater detail in the next section, and we discuss the interpretation of joints further in Section 9.5, after we have examined the mechanics and mechanisms of fracture formation.

In some cases, a fracture displays **slickenside lineations** on its surface, indicating that shear has occurred on the fracture (see Sections 11.8, 11.9, and 14.6). Slickenside lineations occur as parallel sets of ridges and grooves, light and dark streaks of fine-grained pulverized rock, or linear mineral fibers (see Figures 3.8, 11.25, and 11.26). Because extension fractures commonly accumulate small amounts of shear displacement during tectonic movements subsequent to their formation, such displacements may, but do not necessarily, indicate that the fracture formed by shearing.

Joints and other fractures may have a thin mineral deposit—such as quartz, feldspar, calcite, zeolite, chlorite, or epidote—along their surfaces. These mineral layers indicate either that the fracture was open or that fluids under pressure were able to force it open, flow along the fracture, and deposit minerals from solution. In some cases, a fracture is clearly associated with a zone of alteration in the surrounding rock, indicating diffusion of material into or out of the rock surrounding the fracture. Some joints have been affected by dissolution, resulting in open fissures.

2.4 TIMING OF FRACTURE FORMATION

The interpretation of the development of fracture sets relies on determination of the timing of their formation relative to other fracture sets and structures. Although these relationships are often ambiguous and difficult to sort out, especially for extension fractures, we can make a few generalizations.

Where more than one set of joints are developed, younger joints must terminate against older joints, because an extension fracture cannot propagate across a free surface such as another unsealed extension fracture. In Figure 2.3*B*, for example, the nonsystematic joints are clearly younger than the systematic ones. Many such terminations are at a high angle, forming T-shaped intersections, and the younger extension fracture may curve toward a high-angle intersection where it approaches an older fracture (Figure 2.10*G*). Low-angle intersections also occur in some joint systems (Figure 2.10*H*). Where two fractures each curve toward a high-angle intersection with the other (Figure 2.10*F*), the fractures must be coeval.

Analyzing the abutting relations of extension fractures in a complex fracture pattern can reveal the sequence of development of fractures in the system. Such an analysis is illustrated in Figure 2.15. In the complete fracture map (Figure 2.15*A*), the abutting relations are used to assign each fracture to the oldest possible generation. The inferred sequence of fracture development is illustrated in Figure 2.15*B*–*G*; in each figure the newest formed fractures are shown in black and the pre-existing fractures are shown in gray. Note that the earliest generation of fractures forms a highly ordered pattern of systematic fractures that are long, parallel, and poorly interconnected (Figure 2.15*B*). Subsequent generations of fractures are shorter, less systematic, and tend to abut older fractures to form polygonal blocks. Younger generations of fractures progressively increase the connectivity of the fracture system.

This type of analysis only works for systems of extension fractures in which there has been no healing of fractures by mineral deposition. The comparison of the progressive fragmentation of the rock by the addition of subsequent generations of fractures, as in Figure 2.15, with the progressive development of several sets of systematic joints, as in Figure 2-1.3*D*, suggests that in the formation of the latter, healing of earlier fractures may be important.

In many cases, joints cross-cut one another, a relationship that cannot be interpreted in terms of relative timing of fracture formation. This relationship can arise in several ways. If the first-formed joint is closed and has a high pressure keeping the joint faces together when a later joint forms, the later joint may be able to propagate across the earlier closed joint. If the first-formed joint is cemented by mineral deposits, it no longer acts as a free surface, and a later joint can cut across the older one. Subsequent dissolution of the mineral deposit leaves an ambiguous cross-cutting relationship. These situations

FIGURE 2.15 A history of fracture formation, determined by assigning fractures to the oldest possible generation, assuming that younger fractures abut older fractures (quartz monzonite, Cedar City Utah). *A*. Map of the complete fracture pattern (the fractal dimension is 1.50; see Box 2-1). *B*-*G*. Each diagram shows the successive addition of the next-youngest set of fractures (black lines) to an existing set of older fractures (gray lines). The fractal dimension increases from 1.29 (*B*) to 1.50 (*G*) (see Box 2-1). (From Barton, 1995, after Kolb et al., 1970, with permission from Springer Science and Business Media)

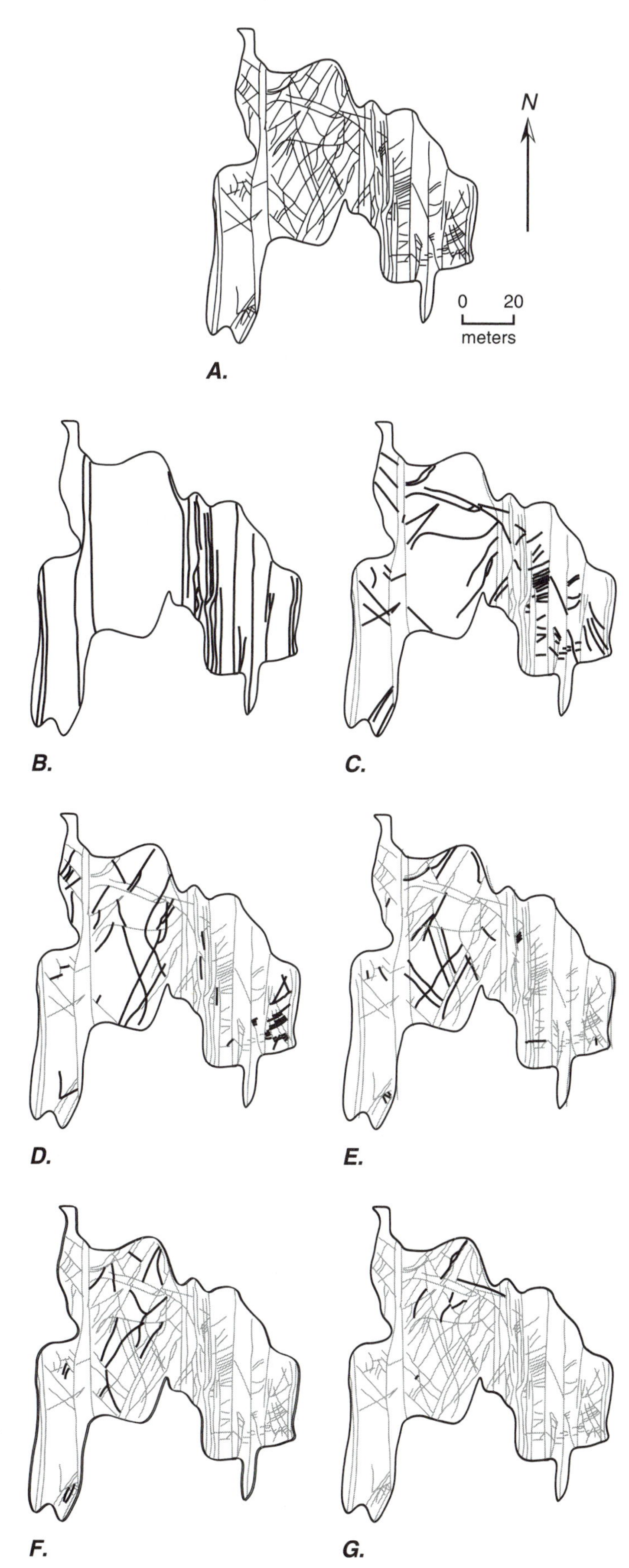

violate the assumptions used to infer the sequence of fracturing events in Figure 2.15. An early shallow joint in a layer may be cut by a later deeper joint that propagates up to one side of the older joint, continues around the edge of the older joint, and propagates back to the opposite side, leaving a joint intersection for which the interpretation of relative time of formation is ambiguous (Figure 2.16*A*). The relative age can be deciphered only if the hackle on the younger joint can be examined to determine the propagation direction of the fracture front. Two joints of the same orientation may also initiate from the same place on opposite sides of an older fracture (Figure 2.16*B*). Those two joints appear as a continuous joint, leaving ambiguous the interpretation of the intersection with the earlier fracture.

Two joint sets could, in principle, also form during the same fracturing event. For example, one set of joints could originate in one orientation at the top of a layer and propagate down, while another set of joints could originate in another orientation at the bottom of a layer and propagate up (Figure 2.17*C*). Their intersection would then show inconsistent relative-age relationships, because at different points along the intersection, the hackle lines would indicate that different joints had formed earliest. A similar fracture geometry would be created if two pairs of coplanar joints formed at the same time and place (Figure 2.16*D*). It is possible to distinguish the histories illustrated in Figure 2.16*B*, *D* only by examining the hackle plume geometry on the joint surfaces. Inconsistent relationships can also result when shear displacement occurs on a later fracture. The offset first fracture can then appear to be a younger fracture terminating against the second fracture (Figure 2.16*E*). For all the cases illustrated in Figure 2.16, the hackle plume geometry is of critical importance in interpreting the fracturing history of joints.

Several structures indicate that extension fractures can form in sediments before they have consolidated into rock. When such fractures form before the deposition of overlying sediments, the open fractures may be filled by the sediment subsequently deposited on top. Mudcracks are one obvious example. If a steeply dipping fracture forms in uncompacted sediments and becomes mineralized before compaction is complete, the mineralized fracture may form a series of folds to accommodate the shortening associated with the subsequent compaction of the sediment. Extension fracturing of unconsolidated sediment in the presence of high pore fluid pressure can result in the formation of **clastic dikes**. The opening of the fracture creates a low-pressure area into which pore fluid rushes, carrying unconsolidated material of contrasting lithology. The existence of such structures proves that some joints, at least, can form very early in the history of a rock.

Fractures that cross-cut a geologic boundary or a geologic structure clearly postdate the formation of that boundary or structure. For example, a joint set that cuts across an intrusive contact is younger than the intrusive

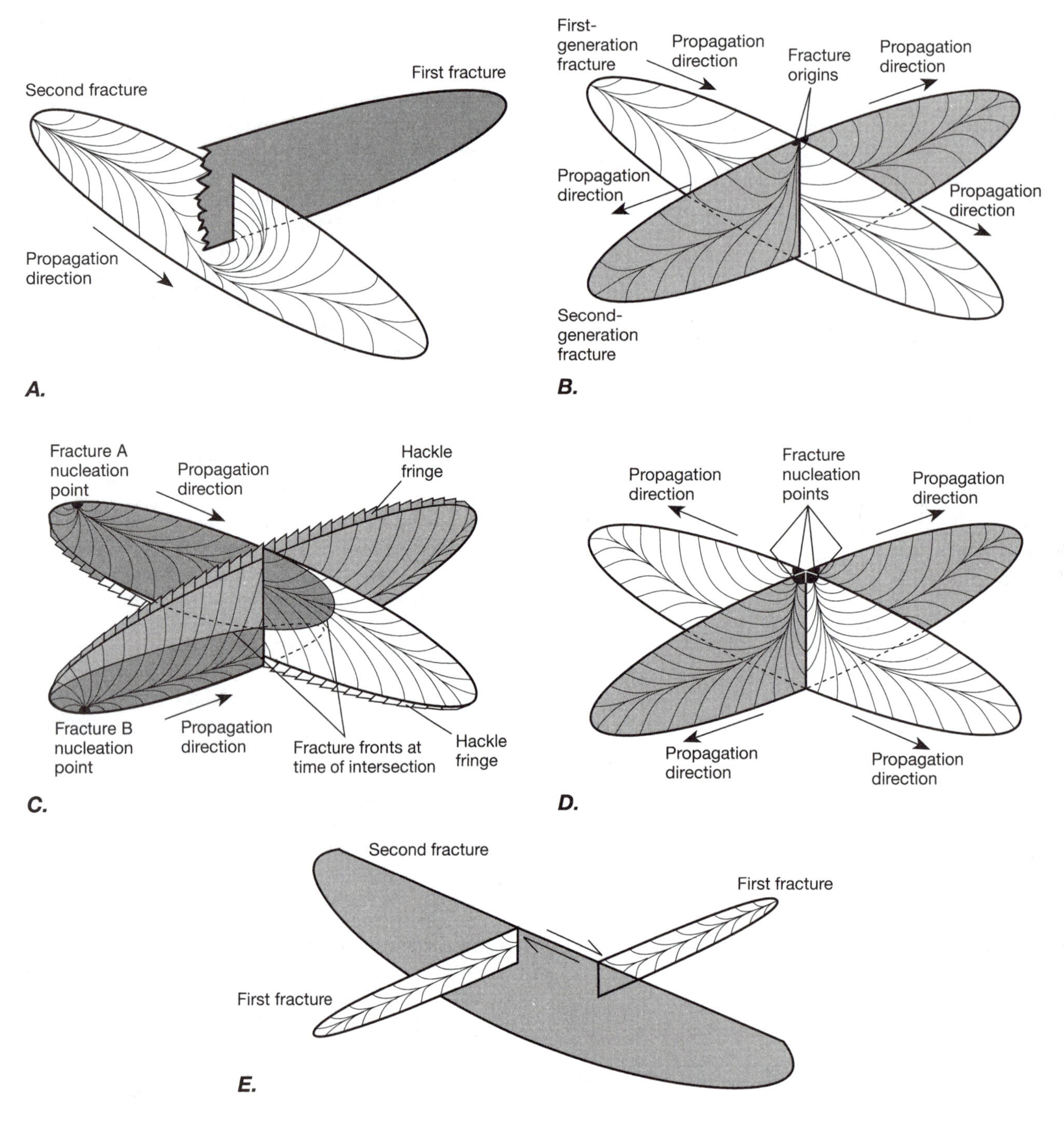

FIGURE 2.16 Origin of fracture intersections. Solid arrows indicate the general direction of fracture propagation. *A.* An early shallow fracture (shaded gray) cut by a later deeper fracture. The intersection relationship would be ambiguous were it not for the distinctive hackle pattern. *B.* Two coplanar fractures (shaded gray) originate at adjacent points on opposite sides of an earlier fracture to produce an ambiguous intersection. The hackle patterns resolve the ambiguity at the intersection and make the origins of the fractures clear. *C.* Two intersecting fractures that originated in different orientations at the top and bottom of the layer. The darkly shaded portions of the fracture surfaces and the fracture fronts indicate the geometry of the propagating fractures when the two fracture fronts intersected. Hackle patterns indicate that the top of fracture A and the bottom of fracture B were the earliest to form at the intersection. *D.* Two pairs of coplanar joints originate at the same point producing an intersection indistinguishable from the one shown in (*B*), except for the tell-tale hackle pattern. *E.* Shear offset of an early fracture on a later shear fracture produces an apparent termination of the older fracture against the younger. (After Kulander, Barton, and Dean, 1979)

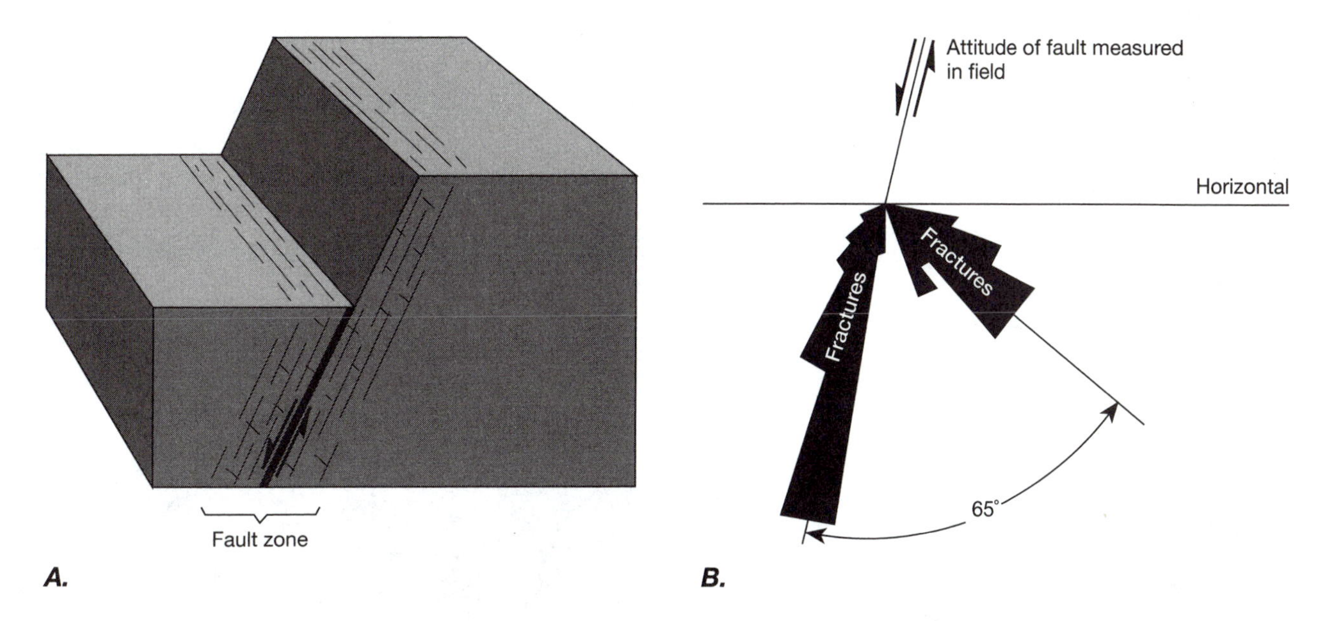

FIGURE 2.17 Shear fractures associated with faulting. *A*. Normal fault with dominant parallel shear fractures (long lines) and subordinate conjugate shear fractures (short lines). *B*. Rose diagram plotted in the vertical plane, showing the distribution of dips of two sets of fractures associated with a normal fault. (From Stearns, 1968)

event, and joints that maintain a constant orientation across folded layers must have formed after the folding. Fractures that are clearly affected by a geologic structure are older than that structure. A joint set that changes orientation over a fold but everywhere maintains the same angular relationship with the bedding could either predate or be synchronous with the folding, but it is not likely to be younger.

If one set of joints is consistently mineralized or has igneous rocks injected along the fractures, then it must be older than the mineralizing or intrusive event. If a second set of joints in the same rocks is free of the mineralization or intrusion, then it probably formed after the mineralizing or intrusive event.

Applying criteria such as these has shown clearly that joints can form at any time in the history of a rock—from the earliest time, when the sediment has not yet consolidated, to the latest time when the joints postdate all other structures in the rock. It is likely, therefore, that more than one mechanism produce joints. We discuss possible mechanisms in Chapter 9.

2.5 RELATIONSHIPS OF FRACTURES TO OTHER STRUCTURES

i. Fractures Associated with Faults

Fractures often form as subsidiary features spatially related to other structures. If such a relationship can be documented, the fractures can provide information about the origin of the associated structure.

In some cases, faults are accompanied by two sets of small-scale shear fractures at an angle of approximately 60° to each other with opposite senses of shear. These are called **conjugate shear fractures**. Figure 2.17 shows data for a system of conjugate fractures that developed in an area closely associated with a known fault. The rose diagram in Figure 2.17*B* is plotted in the vertical plane normal to the strike of the fault, and the distribution of fracture dips is plotted below the horizontal line. The orientation of the fault is also indicated on the figure. The major set of fractures is clearly parallel to the fault; the second and less-well-developed set is approximately 65° from the first set.

Extension fractures associated with faulting include pinnate fractures and gash fractures, which were described in Section 2.1 (Figures 2.7 and 2.8).

ii. Fractures Associated with Folds

Fractures often develop in rocks in association with folding. A variety of orientations, related symmetrically to the fold, have been reported. It is convenient to refer the orientations to a mutually orthogonal system of coordinates (a, b, c) related to the fold geometry and the bedding. The b axis is parallel to the fold axis, which in general is the line about which the bedding planes are folded (Figure 2.18). A line parallel to the fold axis is parallel to the bedding regardless of where on the fold it lies. Thus the b axis has a constant orientation for all the (a, b, c) axes. The c axis is everywhere perpendicular to the bedding, and the a axis lies in the bedding plane perpendicular to the fold axis (b) and the c axis.

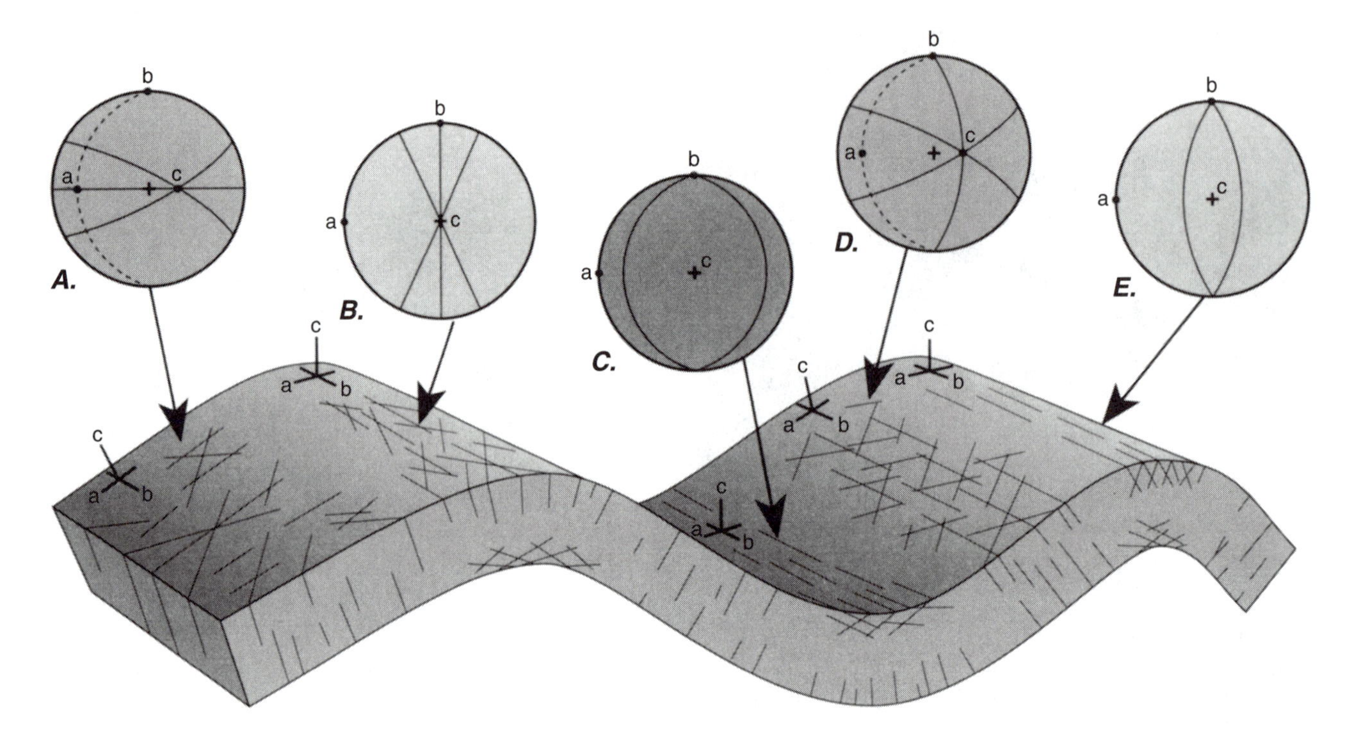

FIGURE 2.18 Fractures associated with folds. The stereographic projections show the orientations of the reference coordinate system (*a*, *b*, *c*), the bedding where it is not horizontal (dotted great circles), and the fractures (solid great circles). (After Price, 1966; Stearns, 1968)

Figure 2.18 is a diagrammatic illustration of the orientations of fractures that have been reported from folds. Fractures parallel to the plane of the *a* and *c* axes and to the plane of the *b* and *c* axes are called ***ac* fractures** and ***bc* fractures**, respectively. The fractures shown in sets (A), (B), and (D) in Figure 2.18 are all perpendicular to the bedding. In sets (A) and (B), the *ac* fractures or the *bc* fractures bisect the acute angle between the two other fracture sets, which are **oblique fractures**. In set (D), the *bc* fractures bisect the obtuse angle between the oblique fractures. The inclined fractures in sets (C) and (E) are parallel to the fold axis *b*. Those in set (C) make a low angle with bedding, and those in set (E) make a high angle with bedding.

Fractures in sets (A) and (D) are particularly common on fold limbs. Sets (B) and (E) tend to be associated with the convex sides of a fold where the curvature is strongest. Set (C) occurs on the strongly curved concave sides of folds.

It is possible that the fractures of all these orientations formed in association with folding, that the *ac* and *bc* fractures are extension fractures, and that the oblique and inclined fractures are shear fractures. That interpretation is not justified, however, simply on the basis of fracture pattern and orientation. Such fractures have been shown to predate the folding in some cases, and to postdate it in others, and as we have remarked before, the presence of shear displacement on a fracture does not necessarily mean that the fracture originated as a shear fracture. In order to make a well-documented interpretation of complex fracture systems, it is critical to describe all the characteristics that we have discussed for the various fracture sets. This includes citing specific evidence for extensional or shear displacement and fracturing, the spatial distribution of fractures, and evidence suggesting the relative sequence of formation of the fractures in different sets.

iii. Fractures Associated with Igneous Intrusions

Fractures can form in association with igneous intrusions, and some types occur *only* within igneous rock. We have already described columnar jointing in sills and thick flows (Figure 2.6) and sheet joints or exfoliation joints in plutonic rocks (Figure 2.5). In many cases, the internal structure of plutonic rocks is related in a simple manner to the orientations of other fractures that develop. Especially near the margins of plutonic bodies, platy minerals such as mica and tabular mineral grains may be aligned parallel to one another, creating a planar structure in the rock called a **foliation**. Elongate mineral grains also may be aligned parallel to one another within the foliation, creating a linear structure called a **lineation** (see Chapter 11).

We can describe the orientations of fractures with respect to these structures by again using coordinate axes (a, b, c), where a is parallel to the lineation, b lies in the foliation perpendicular to a, and c is perpendicular to the foliation. Joints commonly form parallel to c and thus perpendicular to the foliation. If they are also parallel to the lineation, they are referred to as ***ac* joints**; if they are perpendicular to the lineation, they are called **cross joints** or ***bc* joints**. Diagonal joints also occur, usually at an angle of about 45° to the lineation and normal to the foliation. Cross joints (*bc*) typically contain pegmatite dikes or hydrothermal deposits.

REFERENCES AND ADDITIONAL READINGS

Bahat, D. 1979. Theoretical considerations on mechanical parameters of joint surfaces based on studies of ceramics. *Geol. Mag.* 116: 81–166.

Bahat, D. 1990. Genetic classification of joints in chalk and their corresponding fracture characteristics. In J. B. Burland (preface), *Chalk: Proceedings of the 1989 International Chalk Symposium*, Thomas Telford, London: 79–86.

Bahat, D. 1991. *Tectonofractography.* Springer-Verlag, New York, 354 pp.

Bahat, D., and T. Engelder. 1984. Surface morphology on crossfold joints of the Appalachian plateau, New York and Pennsylvania. *Tectonophysics* 104: 299–313.

Barton, C.C. 1995. Fractal analysis of scaling and spatial clustering of fractures. In C. C. Barton and P. R. LaPointe, *Fractals in Earth Sciences*, Plenum Press, New York: chapter 8, 141–178.

Beach, A. 1975. The geometry of en-echelon vein arrays. *Tectonophysics* 28: 245–263.

Berkowitz, B., and A. Hadad. 1997. Fractal and multifractal measures of natural and synthetic fracture networks. *J. Geophys. Research*, B, Solid Earth and Planets, 102: 12,205–12,218.

Cosgrove, J. W., and T. Engelder. 2004. *The Initiation, Propagation, and Arrest of Joints and Other Fractures, Sp. Pub. 231.* Geological Society of London.

Engelder, T. 1985. Loading paths to joint propagation during a tectonic cycle: An example from the Appalachian plateau, U.S.A.. *J. Struct. Geol.* 7: 459–476.

Engelder, T. 2004. Tectonic implications drawn from differences in the surface morphology on two joint sets in the Appalachian Valley and Ridge, Virginia. *Geology* 32: 413–416.

Engelder, T., and P. Geiser. 1980. On the use of regional joint sets as trajectories of paleostress fields during the development of the Appalachian plateau, New York. *J. Geophys. Research* 85: 6319–6341.

Hodgson, R. A. 1961a. Classification of structures on joint surfaces. *Amer. J. Sci.* 259: 493–502.

Hodgson, R. A. 1961b. Regional study of jointing in Comb Ridge–Mavarre Mountain area, Arizona and Utah. *Amer. Ass. Petrol. Geol. Bull.* 45: 1–38.

Johnston, J. D., and K. J. W. McCaffrey. 1996. Fractal geometries of vein systems and the variation of scaling relationships with mechanism. *J. Struct. Geol.* 18: 349–358.

Kolb, C. R., W. J. Farrell, R. W. Hunt, and J. R. Curro, Jr. 1970. Geological investigation of the mine shaft sites, Cedar City, Utah. U.S. Army Engineer Waterways Experiment Station, report MS2170.

Kulander, B. R., C. C. Barton, and S. L. Dean. 1979. The application of fractography to core and outcrop fracture investigations. U.S. Dept. of Energy, METC/SP-79/3; National Technical Information Service, U.S. Dept. of Commerce, Springfield, VA 22161.

Kulander, B. R., and S. L. Dean. 1985. Hackle plume geometry and joint propagation dynamics. In Ove Stephansson, ed. *Proceedings of the International Symposium on Fundamentals of Rock Joints, Bjorkliden, 15–20 Sept. 1985.* Swedish Natural Science Research Council, Centek Publishers, Lulea, Sweden.

Ladeira, F. L., and N. J. Price. 1981. Relationship between fracture spacing and bed thickness. *J. Struct. Geol.* 3: 179–183.

LaPointe, P. R., and J. A. Hudson. 1985. Characterization and interpretation of rock mass joint patterns. *Geol. Soc. Amer. Sp. Paper 199.*

Malinverno, A. 1995. Fractals and ocean floor topography: A review and a model. In C. C. Barton and P. R. LaPointe, eds., *Fractals in the Earth Sciences.* Plenum Press, New York: 107–130.

Marrett, R. 1996. Aggregate properties of fracture populations. *J. Struct. Geol.* 18: 169–178.

Nichelsen, R. P., and V. D. Hough. 1967. Jointing in the Appalachian plateau of Pennsylvania. *Geol. Soc. Amer. Bull.* 78: 609–630.

Price, N. J. 1966. Fault and joint development in brittle and semibrittle rock. Pergamon Press, New York.

Rabinovitch, A., and D. Bahat. 1999. Model of joint spacing distribution based on the shadow compliance. *J. Geophys. Research* 104: 4877–4886.

Rives, T., M. Razack, J.-P. Petit, and K. D. Rawnsley. 1992. Joint spacing: Analogue and numerical simulations. *J. Struct. Geol.* 14: 925–937.

Short, N. M, and R. W. Blair. 1986. *Geomorphology from Space*, NASA SP 486.

Stearns, D. W. 1968. Certain aspects of fractures in naturally deformed rocks. In R. E. Riecker, ed., *NSF Advanced Science Seminar in Rock Mechanics for College Teachers of Structural Geology*. Terrestrial Sciences Laboratory, Air Force Cambridge Research Laboratories. Bedford, MA: 97–118.

Turcotte, D. L. 1997. *Fractals and Chaos in Geology and Geophysics, Second Edition*. Cambridge Univ. Press, New York, 398 pp.

Turcotte, D. L., and J. Huang. 1995. Fractal distributions in geology, scale invariance, and deterministic chaos. In C. C. Barton and P. R. La Pointe, eds., *Fractals in the Earth Sciences*. Plenum Press, New York: 1–40.

Twidale, C. R. 1982. *Granite Landforms*. Elsevier Sci. Publ. Co., Amsterdam, Netherlands, 395 pp.

Wu, H., and D. D. Pollard. 1991. Fracture spacing, density, and distribution in layered rock masses: Results from a new experimental technique. In: J.-C. Roegiers, ed., *Rock Mechanics as a Multidisciplinary Science*. Balkema, Rotterdam: 1175–1184.

Younes, A. I., and T. Engelder. 1999. Fringe cracks: Key structures for the interpretation of the progressive Alleghanian deformation of the Appalachian plateau. *Geol. Soc. Amer. Bull.* 111: 219–239.

Chapter 3

INTRODUCTION TO FAULTS

A **fault** is a surface or narrow zone in the Earth's crust along which one side has moved relative to the other in a direction parallel to the surface or zone. Most faults that we see at the Earth's surface are brittle shear fractures (Figure 3.1*A*) or zones of closely spaced shear fractures (Figure 3.1*B*), but some are narrow shear zones of ductile deformation where movement took place without loss of cohesion at the outcrop scale (Figure 3.1*C*). We generally use the term *fault* for shear fractures or zones that extend over distances of meters or larger. Features at the scale of centimeters or less are called **shear fractures**, and shear fractures at the scale of a millimeter or less, which may be visible only under a microscope, are sometimes called **microfaults**. Faults are often structural features of first-order importance on the Earth's surface and in its interior. They affect blocks of the Earth's crust thousands or millions of square kilometers in area, and they include major plate boundaries hundreds or even thousands of kilometers long.

This large range of scales over which faults and shear fractures develop reflects the self-similar geometry of these structures (see Box 2-1(i)). Characteristically, large faults consist of a network of smaller faults, each of which consists of a network of still smaller shear fractures, and so on. At each scale, the geometric characteristics of the fractures are similar and the characteristic fault structures we describe in this chapter and in Chapters 4 through 6 generally occur at a wide variety of scales.

The word *fault* is derived from an eighteenth- and nineteenth-century mining term for a surface across which coal layers are offset. Many such mining terms were transferred to geology in its early days, despite the fact that the mining lexicon was often complex and ambiguous. The past century has seen a number of attempts to rationalize and systematize this terminology, although there is still no agreement on precise definitions for some words. We try to employ only those terms that are the most prevalent and useful in describing faults.

3.1 TYPES OF FAULTS

A fault divides the rocks it cuts into two **fault blocks**. For an inclined fault, geologists have adopted the miners' terms **hanging wall** for the bottom surface of the upper fault block and **footwall** for the top surface of the lower fault block (Figures 3.1*A*; 3.3*A*, *B*). In a tunnel, these surfaces literally hang overhead or lie under foot. The fault block above the fault is the **hanging wall block**, and the block below the fault is the **footwall block**. For a vertical fault, of course, these distinctions do not apply, and the sides of the fault are named in accordance with geographic directions: the northwest side and the southeast side, for instance.

Faults are classified in terms of the attitude of the fault surface. If a fault dip is more than 45°, it is a **high-angle fault**; if it is less than 45°, it is a **low-angle fault**.

We also divide faults into three categories depending on the orientation of the **relative displacement**, or **slip**, which is the net distance and direction that the hanging wall block has moved with respect to the footwall block (Figure 3.2). On **dip-slip faults**, the slip is approximately parallel to the dip of the fault surface; on **strike-slip faults**, the slip is approximately horizontal, parallel to the strike of the fault surface; and on **oblique-slip faults**, the slip is inclined obliquely on the fault surface. An oblique-slip vector can always be described as the sum of a strike-slip component and a dip-slip component,

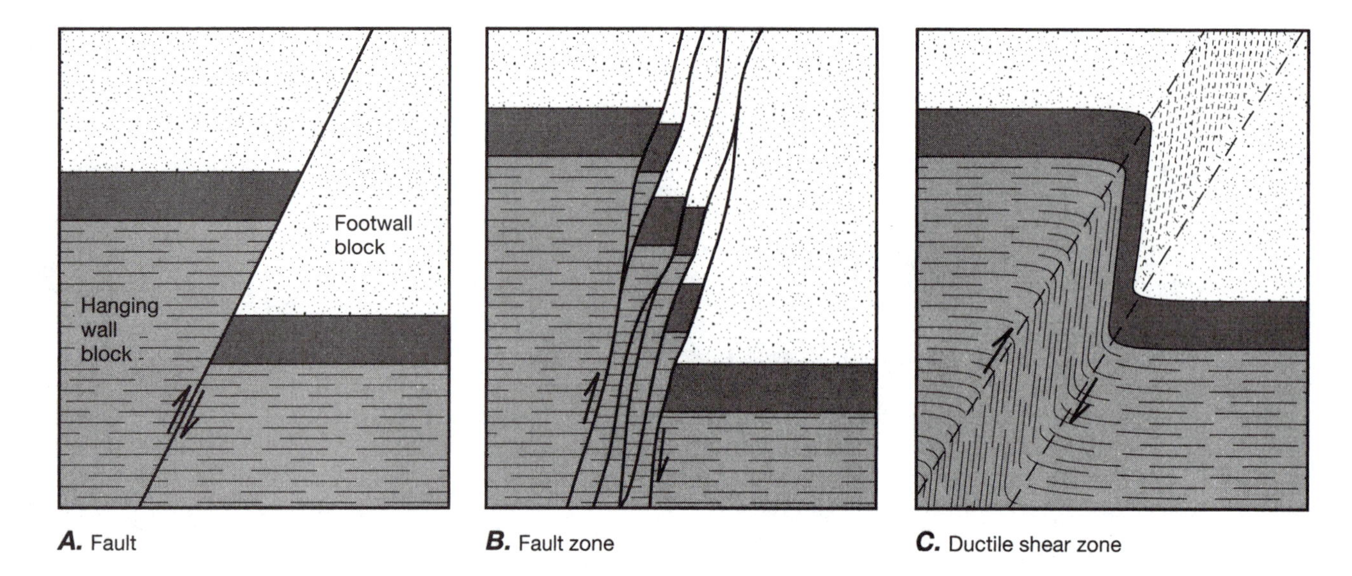

FIGURE 3.1 Three styles of faulting. *A*. A single fault consists of a single shear fracture. *B*. A fault zone may comprise a set of associated shear fractures. *C*. A fault zone may be a zone of ductile shear.

or as the sum of a horizontal component and a vertical component (Figure 3.2). The dip-slip component may in turn be described as the sum of a vertical component and a horizontal component, which are sometimes called the **throw** and the **heave**, respectively.

We subdivide faults further in terms of the relative movement, or shear sense, along them. Inclined dip-slip faults on which the hanging wall block moves down relative to the footwall block are **normal faults** (Figure 3.3*A*). Those on which the hanging wall block moves up relative to the footwall block are **thrust faults** (Figure 3.3*B*). High-angle thrust faults are often called **reverse faults**. Vertical faults characterized by dip-slip motion, of course, cannot be classified as either normal or reverse faults, so we simply specify which side of the fault has moved up or down. Strike-slip faults are **right-lateral**, or **dextral**, if the fault block across the fault from the observer moved to the right (Figure 3.3*C*); they are **left-lateral**, or **sinistral**, if that block moved to the left (Figure 3.3*D*). Oblique-slip faults may be described according to the nature of the strike-slip and dip-slip components. Figure 3.3*E*, for example, shows sinistral normal slip, and Figure 3.3*F* shows sinistral reverse slip. For rotational faults the slip changes rapidly with horizontal distance along the fault (Figure 3.3*G*). Such faults are sometimes referred to as **scissor faults**.

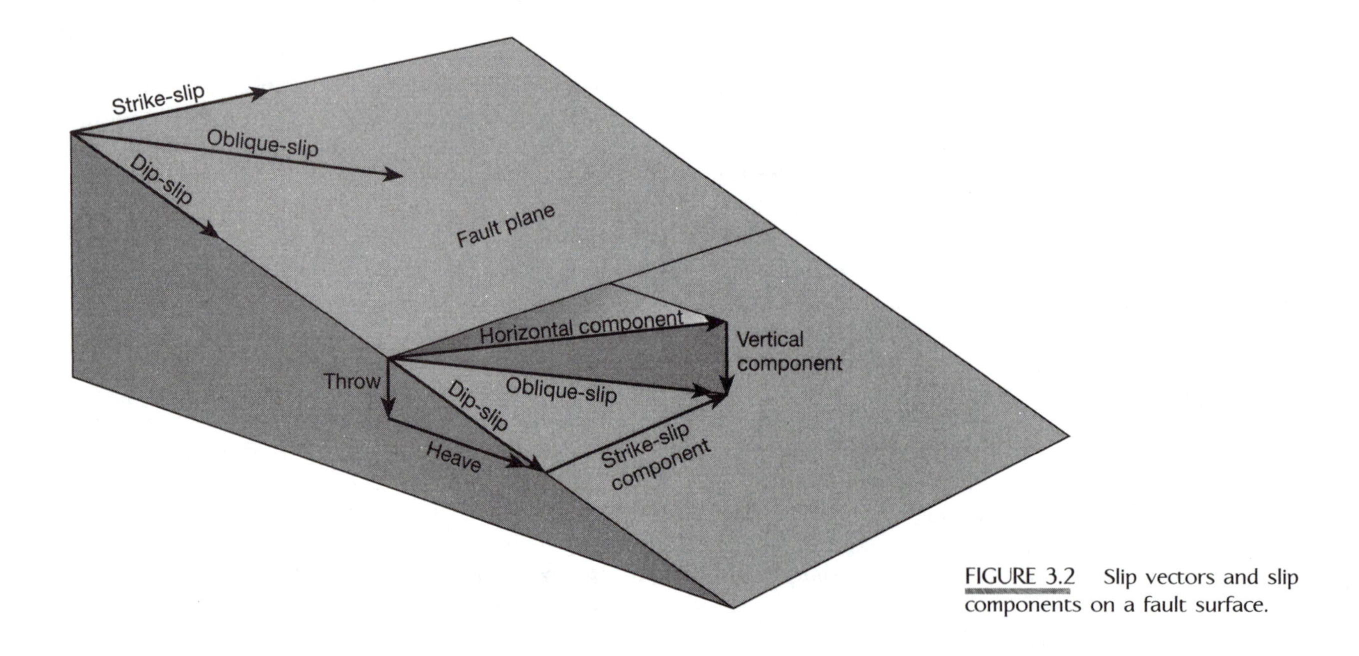

FIGURE 3.2 Slip vectors and slip components on a fault surface.

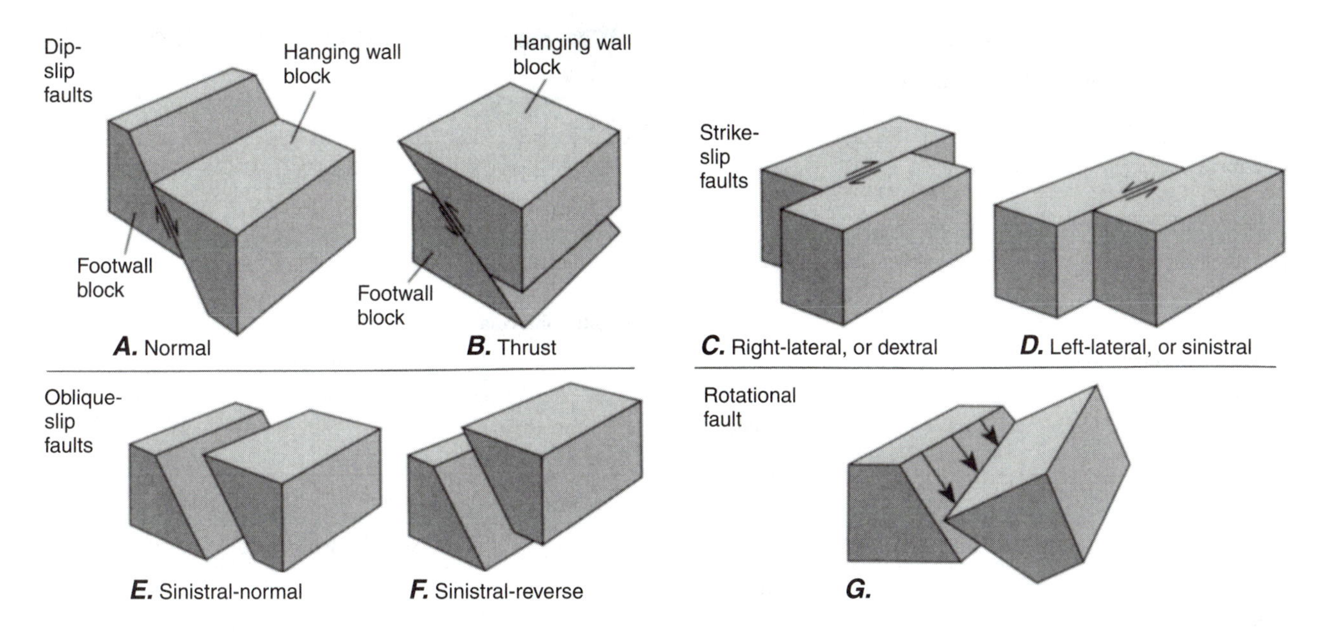

FIGURE 3.3 Faulted blocks showing the characteristic displacement for the different classes of faults.

3.2 RECOGNITION OF FAULTS

The criteria for recognizing faults comprise three broad categories: first, features intrinsic to faults themselves; second, effects on geologic or stratigraphic units; and third, effects on physiographic features. We briefly consider each of these categories.

i. Features Intrinsic to Faults

Faults can often be recognized by the characteristic textures and structures that develop in rocks as a result of shearing (Table 3.1). These textures and structures vary with the amount and rate of shear and with the physical conditions under which the faulting took place, including temperature and pressure, which typically are functions of the depth of faulting (Figure 3.4).

Faults formed at depths less than about 10 to 15 kilometers typically have **cataclastic rocks** present in the fault zone. These rocks have been fractured into clasts or ground into powder during brittle deformation. Individual fragments are generally sharp, angular, and internally fractured. Cataclastic rocks usually lack any internal planar or linear structure, although foliated gouge and cataclasite also are observed. Friable cataclastic rocks are typical of faulting above depths of 1 to 4 kilometers. Cohesive cataclasites may form at depths up to 10 to 15 kilometers.

Fault rock terminology and classification are not universally agreed on (see Table 3.1). We divide cataclastic rocks into four main categories: the **breccia series**, **gouge**, **cataclasites**, and **pseudotachylite**. The abundance of fine-grained matrix distinguishes rocks in the breccia series (less than about 30%; Figure 3.5*A*) from cataclasites (more than about 30%; Figure 3.5*B*). We further subdivide the breccia series into **megabreccia** (Figure 3.5*A*), **breccia**, and **microbreccia**. In megabreccia and breccia, the clasts are predominantly rock fragments. In microbreccia the clasts are principally mineral grain fragments. Gouge is essentially a continuation of the breccia series to finer clast size. In outcrop, it appears as a finely ground, whitish rock powder. Cataclasites include a range of clast sizes and vary from 30% fine-grained matrix up to 100 percent matrix (Figure 3.5*B*). They are generally cohesive rocks.

Remarkably, cataclastic rocks are self-similar—the size, shape, and arrangement of the grains look very much the same over a wide range of scales (compare Figure 3.5*A*, *B*). The distribution of fragment sizes has a fractal geometry, that is, it is self-similar, and it accords with a model of fragmentation (see Box 2-1(iii)) by which clasts of the same size tend not to be in direct contact with one another (Figure 3.5*B*). The fractal nature of fragment size distribution means that the definition of "clasts" and "matrix" is scale-dependent (fragments that are part of the matrix in megabreccia are clasts in the finer-scale view; Table 3.1). Although breccia, microbreccia, and gouge are generally noncohesive, deposition of silica during or subsequent to formation can turn them into hard, cohesive, silicified fault rock.

Pseudotachylite (Figure 3.6) is a massive rock that frequently appears in microbreccias or surrounding rocks as dark veins of glassy or cryptocrystalline material. It characteristically contains a matrix of crystals less than 1 μm in diameter and/or small amounts of glass or devitrified glass cementing a mass of fractured material together. Under a petrographic microscope, the matrix appears isotropic; that is, between crossed polarizers no

TABLE 3.1 Fault Rock Terminology*

Cataclastic Rocks					
Fabric	**Texture**	**Name**		**Clast Size**	**Matrix**
Generally no preferred orientations	Cataclastic: Sharp angular fragments	Breccia Series	Megabreccia	> 0.5 m	$< 30\%$
			Breccia	1–500mm	$< 30\%$
			Microbreccia	< 1 mm	$< 30\%$
May be foliated		Gouge		< 0.1 mm	$< 30\%$
May be foliated		Cataclasite		generally $\leq \sim 10$mm	$> 30\%$
May be foliated		Pseudotachylite			glass, or grain size $\leq 1\ \mu$m

Mylonitic Rocks					
Fabric	**Texture**	**Name**		**Matrix Grain Size**	**Matrix**
Foliated and lineated	Metamorphic: Interlocking grain boundaries, sutured to polygonal	Mylonitic gneiss		$> 50\ \mu$m	
		Mylonite Series	Protomylonite	$< 50\ \mu$m	$< 50\%$
			Mylonite	$< 50\ \mu$m	50%–90%
			Ultramylonite	$< 10\ \mu$m	$> 90\%$

*The terminology applied to fault rocks is by no means generally agreed upon. The definitions of the different categories, and the quantitative boundaries we have placed on them, should therefore be understood as guidelines to present usage, which, however, can vary from one geologist to another. We believe, for example, that what we have defined as mylonite would fit anyone's definition, but other geologists use mylonite in a broader sense, even to include what we call mylonitic gneiss.

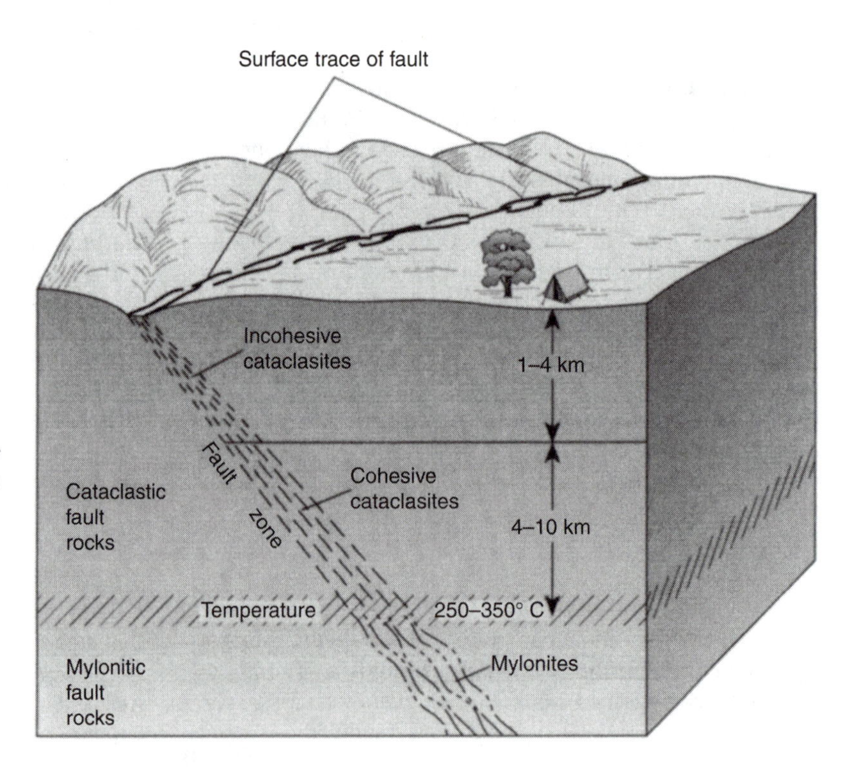

FIGURE 3.4 Schematic block diagram of a portion of the Earth's crust, showing the surface trace of a fault zone (i.e., its exposure on the Earth's surface) and the variation with depth of the type of fault rock within the fault zone. Incoherent cataclasites (plus pseudotachylite if dry) characterize depths above 1 to 4 kilometers. Below that, coherent cataclasites (plus pseudotachylite if dry) are present at depths of up to 15 kilometers. Mylonites are present at depths greater than 10 to 15 kilometers and temperatures greater than 250°C to 350°C. (After Sibson 1977)

A.

B.

FIGURE 3.5 Cataclastic rocks. *A*. Megabreccia composed of very large fragments of limestone (Titus Canyon, Death Valley National Monument). *B*. Cataclasite from the Whipple Mountain detachment, southeast California.

light is transmitted. This behavior is characteristic of glass and of extremely fine-grained material. During an earthquake under dry conditions at depths generally less than 10 to 15 kilometers, frictional heating can be sufficient to melt small portions of the rock. The resulting material may intrude through fractures in the adjacent rock before quenching to form veins of pseudotachylite. Thus the presence of pseudotachylite is one of the few definite indicators of paleo-seismic activity.

Cataclastic rocks occur in zones ranging from a few millimeters in thickness up to extensive zones one or more kilometers thick. In general, the greater the thickness and the smaller the grain size, the greater the amount of displacement that has accumulated on the fault.

Fault zones formed at depths exceeding about 10 to 15 kilometers are characterized by another type of very fine-grained rock called **mylonitic rocks** (Figure 3.7). These rocks form only as a result of ductile deformation, which occurs in crustal rocks at temperatures generally in excess of 250°C to 350°C.[1] Mylonitic rocks have a matrix of very fine grains that is derived by reduction of grain size from the original rock. Variable amounts of relict coarse mineral grains, called **porphyroclasts**, may be present, surrounded by the fine-grained matrix. The fine grains show an interlocking grain boundary texture characteristic of metamorphic rocks; the grain boundaries themselves may be polygonal, forming 120° triple junctions, or they may be highly sutured. Mylonitic rocks exhibit a strong planar and linear internal structure, called *foliation* and *lineation*, respectively (see Chapter 11). These structures tend to be oriented at small angles to the fault zone, that is, parallel to subparallel to it.

FIGURE 3.6 Veins of dark pseudotachylite cutting a light-colored gneiss. Actual width of view shown is about 6 centimeters.

Mylonitic rocks form as a result of the recrystallization of mineral grains during rapid ductile deformation. Their polygonal to sutured grain boundaries are different from the sharp angular shapes characteristic of brittle fracturing that occurs in cataclasites.

If the grain size is reduced from the original grain size but is coarser than about 50 μm, the rock is a **mylonitic gneiss**. If the matrix grain size is less than 50 μm, the rock belongs to the **mylonite series**, which we subdivide on the basis of increasing percentage of fine-grained matrix (Table 3.1) into **protomylonite**, **mylonite**, and **ultramylonite**. In ultramylonites, the characteristic grain size of less than 10 μm causes the matrix to appear glassy in a hand sample.

Mylonites are generally present in ductile shear zones ranging in thickness from a few millimeters to several meters. Some mylonites, however, are kilometer-thick

[1]We discuss in detail the structures and processes of ductile deformation in Parts II and III of this book. We include here a brief description of some of these features because they characterize many fault zones.

FIGURE 3.7 Mylonitic quartzite showing large feldspar porphyroclast in a much finer-grained matrix of strongly recrystallized quartz grains.

bodies that apparently define wide shear zones. All transitional stages, from original country rock through mylonitic gneisses to ultramylonites, may be present in such a zone.

Where exposed, fault planes commonly are smooth, polished surfaces called **slickensides**,[2] which form in response to shearing on the fault planes or in the fault gouge. Fault surfaces, including slickensides, typically contain strongly oriented linear features, known as **slickenlines**, **slickenside lineations**, or **striations**, that are parallel to the direction of slip. These lineations are of three types: **ridges and grooves**, **mineral streaks**, and **mineral fibers**, or **slickenfibers**. Ridges and grooves may result from scratching and gouging of the fault surface (Figure 3.8*A*), from the accumulation of gouge behind hard protrusions or **asperities**, from the development of irregularities in the fracture surface itself forming **ridge-in-groove lineations** or **fault mullions** (Figure 3.8*B*), or from growth of slickenfibers (Figure 3.8*C*). Slickenfibers are long, single-crystal mineral fibers that grow parallel to the direction of fault displacement. They fill gaps that develop along the fault during gradual shearing (see Figures 3.8*C*, 3.16, and 11.26*A* and Sections 3.3, 11.9(ii), and 14.6(v)). Mineral streaks are streaks on slickensides that result from the pulverization and shearing out of mineral grains within the gouge (Figure 11.25).

Faults that develop at relatively shallow depths are **dilatant**, which means their volume increases during faulting due to the formation and accumulation of open fractures in the rock. Dilatant fault zones provide pathways for the flow of groundwater and hydrothermal fluids. As a result, many fault zones contain secondary deposits of minerals, including calcite (see Figure 3.5*A*) and silica (quartz, opal, or chalcedony) as vein deposits or as cement for the pre-existing fault gouge or breccia. Many economically valuable ore deposits form by precipitation of ore minerals from hydrothermal fluids flowing along fault zones.

ii. Effects of Faulting on Geologic or Stratigraphic Units

Displacement along faults generally juxtaposes rocks that do not belong together in undisturbed geologic sequences. The resulting discontinuities provide some of the best evidence for the presence of a fault.

A break in an otherwise continuous geologic feature, such as sedimentary bedding, may indicate the presence of a fault. A stratigraphic discontinuity, however, may also result from an unconformity or an intrusive contact, and it is important to distinguish such features from faults. Characteristic features of unconformities include

[2]Confusion exists about the exact meaning of this term. Many authors use "slickenside" to refer to the *lineations* that occur on the fault surfaces. This usage is not consistent with the original definition, however, which refers to a polished *surface* that may, or may not, be lineated.

A.

B.

C.

FIGURE 3.8 Lineations on fault surfaces formed during fault slip. *A*. Lineations formed by scratching and gouging of the fault surface. *B*. Ridge-in-groove lineations, or fault mullions. *C*. Calcite slickenfiber lineations. (Sample: D. Benner; photo: M. Graziose)

fossil soil horizons, erosional channels, basal conglomerates, depositional contacts, and the parallelism or near parallelism of the strata that overlie the unconformity. Distinctive features of intrusive contacts (Figure 3.9) include metamorphism in the adjacent country rocks, fragments of country rock suspended in the intrusion (xenoliths), and dikes or veins of igneous rock cutting the country rock adjacent to the intrusion.

The presence of **horses**, or **fault slices**, along a discontinuity is clear evidence of a fault. *Horses* are volumes of rock surrounded on all sides by faults (see Figure 3.27). They are sliced from either the footwall or the hanging wall block by a branch of the fault and are displaced a significant distance from their original position. Thus they may appear markedly out of place stratigraphically. If the local stratigraphy is known, identification of the original stratigraphic position of the rocks in a horse provides a constraint on the direction and amount of movement. Along faults that separate similar rock types, a horse of a different lithology may be the only observable evidence for the fault.

Repetition of strata or omission of strata in a known stratigraphic sequence is another possible indication of a fault. This criterion is especially important in the interpretation of subsurface geology, where often the only data available are from drill holes. Figure 3.10 shows a diagrammatic cross section of a region of horizontal bedding with drill-hole sections that show either **repeated**

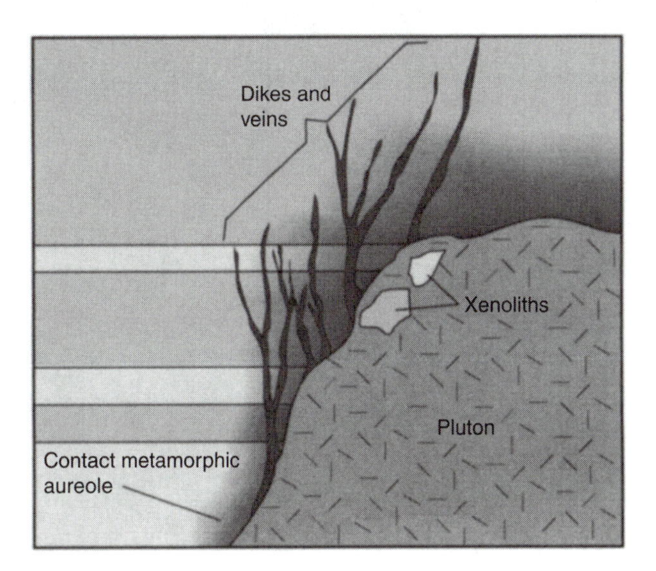

FIGURE 3.9 Characteristics of a plutonic intrusive contact in sedimentary rocks.

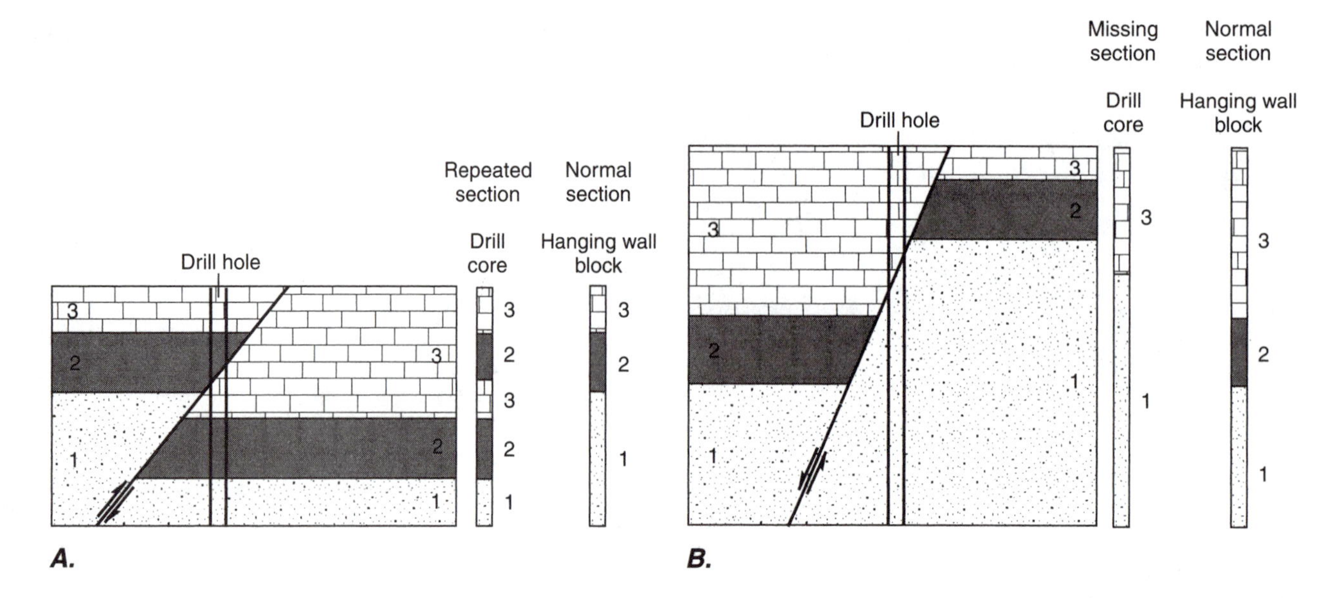

FIGURE 3.10 *A*. Thrust fault resulting in repeated section in a vertical drill hole. *B*. Normal fault resulting in missing section in a vertical drill hole.

section (Figure 3.10*A*) or **missing section** (Figure 3.10*B*). If enough information is present, it is possible to map a fault in the subsurface solely on the basis of information obtained by drilling.

As in the truncation of structures, it is important to make sure that the omission of strata does not result from an unconformity and that the repetition of strata does not result from a facies change associated with alternating transgressions and regressions. The distinction between faults and facies changes can be subtle, and failure to distinguish them correctly has resulted in some spectacular geologic errors.

Bedding surfaces near faults may have been bent in the direction of motion of the opposite fault block. These

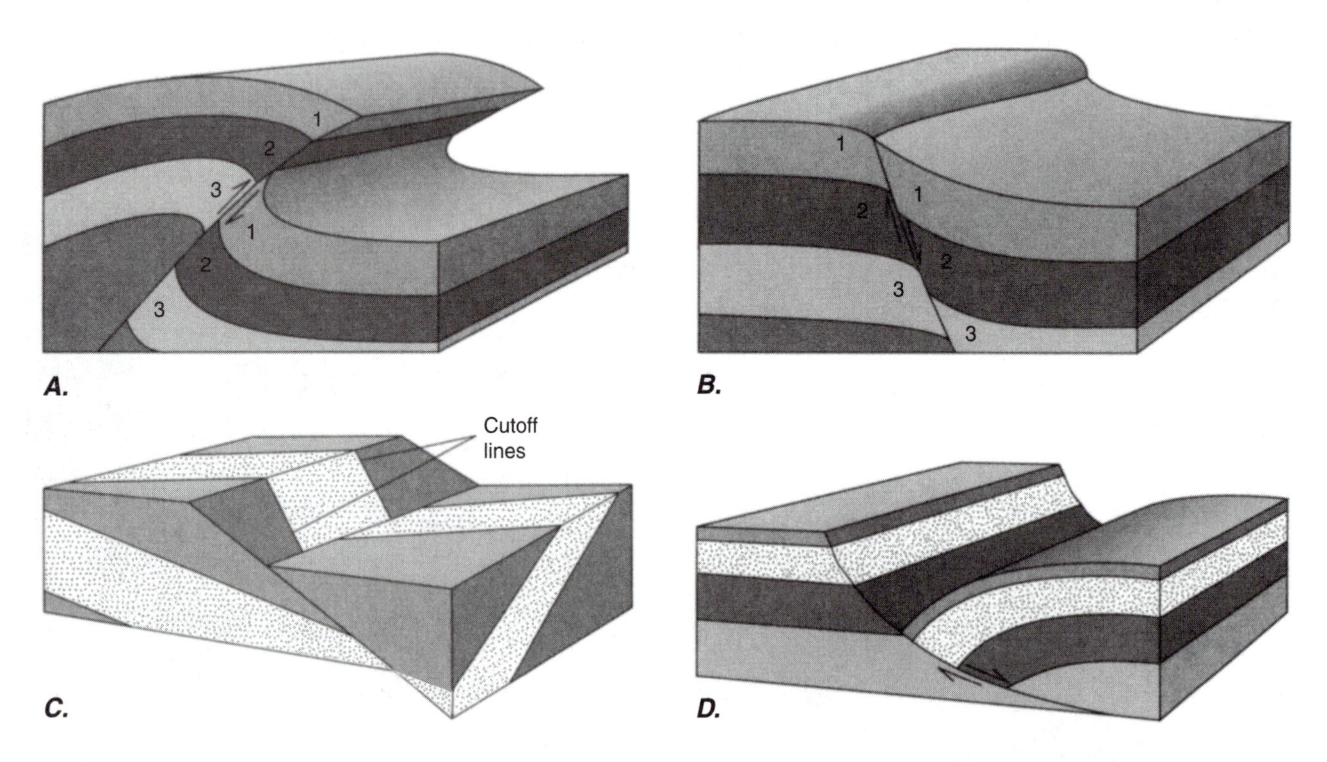

FIGURE 3.11 Drag folds in sedimentary layers along faults. *A*. A thrust fault. *B*. A normal fault. *C*. If the cutoff line of the bedding makes a small angle with the displacement direction on the fault surface, the formation of drag folds is less likely. *D*. Rollover anticline in the hanging wall of a normal fault.

bends are called **drag folds**. They are most likely to develop where the traces of the sedimentary layers on the fault plane—that is, the **cutoff lines**—are at a high angle to the slip direction on the fault (Figure 3.11*A*, *B*). Drag folds are less likely to form if the cutoff lines are nearly parallel to the slip direction (Figure 3.11*C*). In some cases, if not most, formation of a drag fold precedes the development of the fault (cf. Figure 5.10*A*; see also Figures 13.31, 13.33, 13.34).

Drag folds are especially well developed along many thrust faults (Figure 3.11*A*). Along normal faults, **rollover anticlines**, which develop in the hanging wall block, are more common (Figure 3.11*D*). The direction of bending in these folds is opposite to that found in drag folds, and they reflect the deformation necessary to accommodate the hanging wall block to a curved fault surface (see Section 4.4).

iii. Physiographic Criteria for Faulting

Many active and inactive faults have pronounced effects on topography, stream channels, and groundwater flow. Because these effects frequently suggest the existence of a fault, they are useful in geologic mapping.

Scarps are linear features characterized by sharp increases in the topographic slope; they suggest the presence of faults. There are two types of fault-related scarps: **Fault scarps** are continuous linear breaks in slope that result directly from displacement of topography by a fault (Figure 3.12*A*), and **fault line scarps** are erosional features that are characteristic of both active and inactive faults. Figure 3.12 illustrates three steps in the progressive erosion at a fault. Initially, the upthrown footwall block (Figure 3.12*A*) forms a fault scarp. Erosion carves valleys in the fault scarp, leading to formation of **faceted spurs** along the mountain front (Figure 3.12*B*). Eventually the upthrown block is eroded down to the same level as the downthrown hanging wall block (Figure 3.12*C*). Subsequent erosion exposes the thin layer in the hanging wall block, which is more resistant than the surrounding layers. Further erosion occurs most rapidly in the least resistant rocks (Figure 3.12*D*), leaving a fault line scarp that, as in this case, does not necessarily indicate the sense of displacement on the fault.

Fault benches are linear topographic features characterized by an anomalous decrease in slope. They form where a fault displaces an originally smooth slope so that a strip of shallower slope results, or where erosion of the less resistant rocks in a fault zone produces a shallower slope than is supported by the surrounding, more resistant rocks. Fault benches may be associated with any of the different fault types.

Ridges, valleys, or streams may be offset along a fault. Figure 3.13 shows two **offset streams** that have been displaced by strike-slip motion on a fault during continued stream activity. The deflection of the stream channels may indicate the sense of slip on the fault, but if the fault displacement is sufficiently large, original

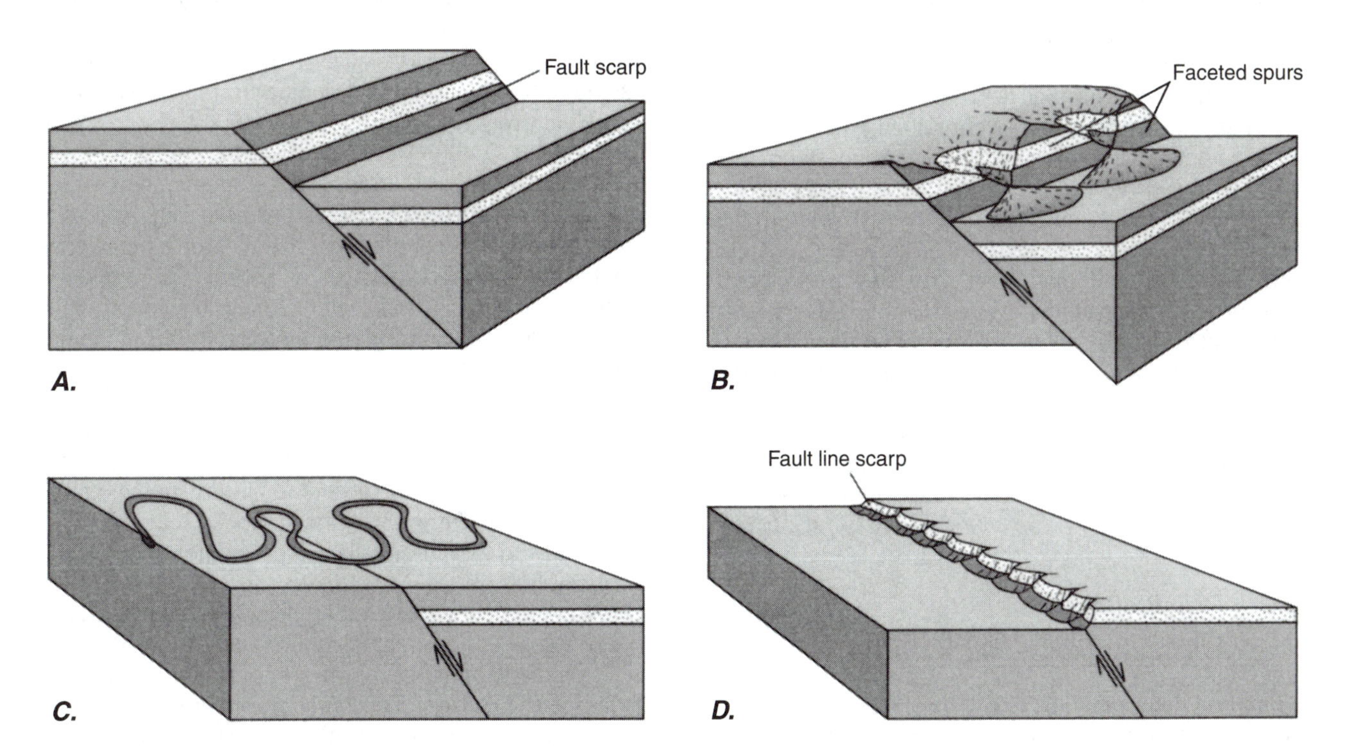

FIGURE 3.12 Erosion of fault scarps. *A*. Faulting produces a fault scarp. *B*. Erosion of valleys in the fault scarp produces faceted spurs. *C*. Erosion wears away the thin resistant layer in the topographically high footwall block and levels the topography. *D*. Erosion reaches the level of the resistant layer in the hanging wall block. More rapid erosion in the less-resistant layers in the footwall block leaves a topographic step, a fault line scarp.

FIGURE 3.13 Photograph of San Andreas fault, central California, showing dextral stream offset along the trace of the fault. (Courtesy of J. Shelton)

stream channels may be abandoned and new ones formed downstream from the fault. In this case, the "dog leg" in the channel may not correspond to the sense of displacement.

A fault surface or fault zone may act as either a conduit or a barrier for groundwater, depending on the permeability of the material both in the fault and on either side of the fault. A breccia zone forms an excellent conduit for water, but a thick gouge zone containing abundant clay minerals may act as a barrier to flow. If faulting offsets an aquifer or juxtaposes an impermeable rock, such as a plutonic or metamorphic rock, against a good aquifer, it may also significantly alter the flow of groundwater. Thus fault traces are often characterized by springs and by water-filled depressions called **sag ponds**.

A stream tends to form a consistent equilibrium profile characterized by a slope that gradually steepens toward the headwaters. Such a profile can change because of fault movements or because of the variable erodability of the bedrock. Any sharp changes in the profile, referred to as a "nickpoint," or a change in the shape of a stream valley that cannot be ascribed to the change in erosional resistance of the bedrock may betray a fault, but further evidence is necessary to confirm its existence.

Faulting often juxtaposes rocks of different composition. Different rocks produce soils of different composition, which in turn often support differences in vegetation. Thus changes in vegetation are often a clue to the location of a fault.

3.3 DETERMINATION OF FAULT DISPLACEMENT

Complete determination of the displacement on a fault requires knowledge of the **magnitude** and **direction** of its displacement. Some features indicate the total displacement; others permit a partial or approximate determination; still others only place a constraint on the possible displacements. Determination of the original orientation of a fault and of the associated displacement is an important part of understanding faults, as these provide two of the major features used for classifying faults and for interpreting their tectonic significance. Faults may be rotated into different attitudes by later tectonic activity, which, if not recognized, can give rise to inaccurate interpretations. For example, it is possible for a thrust fault with the hanging wall up to be rotated to such an extent about a horizontal axis roughly parallel to the fault strike that it resembles a normal fault with the hanging wall down.

In systems of related faults, a subsidiary fault is said to be a **synthetic fault** if the fault has a similar dip direction and the same shear sense as that of the main fault; it is an **antithetic fault** if the fault has an opposite dip direction and the opposite shear sense to that of the main fault.

i. Relative versus Absolute Displacement

Discussion of fault movement generally concerns only the relative movement of the rock on opposite sides of a fault. An absolute sense of movement can rarely be obtained from fieldwork, because we generally do not have a reference point independent of both of the fault blocks. Thus, for example, it is generally impossible to distinguish whether the hanging wall of a normal fault has decreased in elevation because of faulting or if the footwall has increased in elevation.

In special cases, however, it may be possible to determine absolute motions with respect to an independent reference. Investigations of active deformation using GPS (satellite-based Global Positioning System) surveys can resolve current motions of fault blocks relative to a globally defined satellite reference frame. In the 1994 Northridge earthquake in California, for example, increases in elevation of the hanging wall of the thrust fault amounted to several tens of centimeters. Radar interferometry, using two satellite radar images of the same area taken at different times, can also detect small changes in elevation, on the order of a few millimeters to a few tens of millimeters.

Sea level may also provide an independent reference frame from which to determine absolute motions of faults. Faulting associated with the Alaska earthquake of 1964 near Anchorage provided one interesting example. Measurement of motion of fault blocks relative to sea level showed that in one place where a normal fault had formed, both sides had moved upward. Figure 3.14*A* shows part of the area affected by the fault and gives contours of the amount of uplift that occurred during faulting. Note that the uplift contours show that both sides of the Patton Bay Fault moved up but that the northwest side moved up farther than the southeast side. This effect is more clearly shown in Figure 3.14*B*, which is a plot of the amount of uplift along the line AA′ that crosses the fault. This interesting result on a historical earthquake makes one cautious about assuming that the

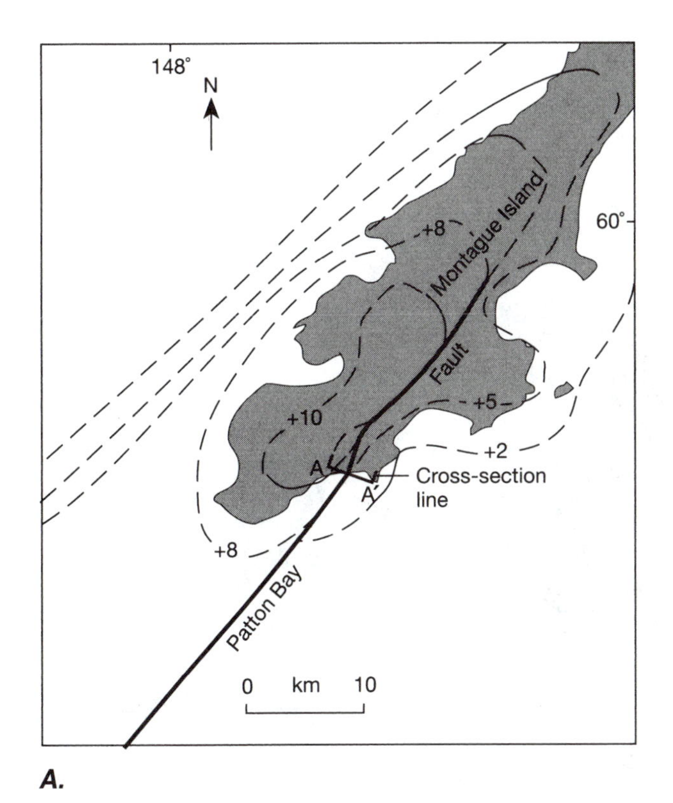

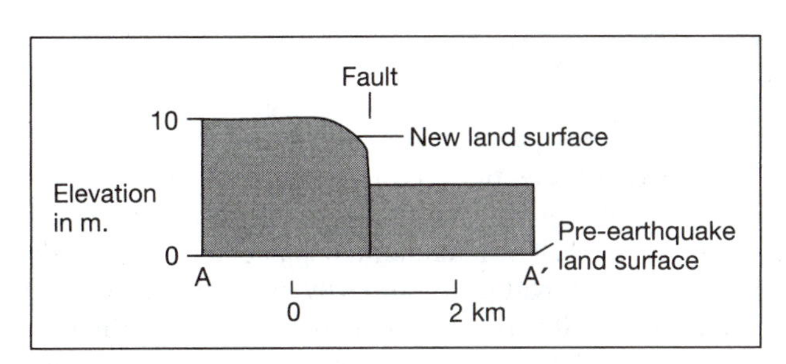

FIGURE 3.14 Faulting of Montague Island, southern Alaska, associated with the 1964 Alaska earthquake. The map shows that the uplift contours are positive on both sides of the Patton Bay fault. Contours are marked in meters of uplift. The cross section shows a plot of the uplift magnitudes along the line A-A′ that crosses the Patton Bay fault. The fault is a normal fault resulting from uplift by different amounts on opposite sides of the fault. (After Plafker 1965)

relative motion of the hanging wall is also the absolute motion.

ii. Complete Determination of Displacement

The complete determination of displacement on a fault requires identification of a particular pre-existing linear feature that intersects the fault surface and is displaced by it. A faulted linear feature produces two **piercing points** where the feature intersects the fault surface in the hanging wall block and in the footwall block, respectively (Figure 3.15). Before faulting, the two piercing points were adjacent to each other. After faulting, therefore, the vector connecting the two points uniquely defines the direction and magnitude of fault displacement, and the relative positions of the linear feature on opposite sides of the fault give the sense of shear of the fault.

Several linear geologic features provide piercing points on fault surfaces. The intersection of two distinct planes always defines a unique line that, when cut by a fault, can be used to determine the fault displacement (Figure 3.15*A*). Examples include two older intersecting faults; two differently oriented veins or dikes; a bedding plane intersecting with a fault, vein, or dike; and the intersection of an unconformity with a geologic contact such as a bedding plane or intrusive contact. The line of maximum curvature, or hinge line, on a folded surface also provides a unique line by which to determine the displacement (Figure 3.15*B*). Buried river channels and linear sandstone bodies (shoestring sands) are linear stratigraphic features that can be used, and cylindrical bodies such as volcanic necks and some ore deposits can serve in the same manner. Although piercing points generally are defined by the intersection of linear structures in the rock with the fault plane, geodetic lines that are surveyed across a fault both before and after a faulting event, satellite interferometry, or GPS data may also provide evidence for the complete determination of displacement.

iii. Partial Determination of Displacement from Small-Scale Structures

In many cases where a fault or ductile shear zone is identified, it is possible to determine the orientation of the displacement vector and the sense of shear but not the magnitude of the displacement. This type of information can be obtained by examining features at the microscopic to outcrop scale.

As we have noted, slickenside lineations form parallel to the direction of displacement on a fault (see Figure 3.8), but the magnitude of the displacement is more difficult to obtain. Ridge-in-groove lineations that form during propagation of the fracture may be longer than the displacement vector on the fault. Mineral streaks may result from comminution and smearing out of mineral grains and may give a minimum estimate of the displacement magnitude, although this correlation has never been proved.

During brittle faulting, numerous asymmetric features develop on fault surfaces that can be used as kinematic indicators to define the sense of shear on the fault. These features can only be observed on the fault surface if the rock on one side of the fault has been removed to expose the surface. The direction of relative motion of the removed block we refer to as the "downstream direction" on the fault; the "upstream direction" is the direction opposite to that motion. Several features form asymmetric steps on the fault surface with long steps inclined at a small angle to the fault and short risers at a high angle to the fault. If the risers face downstream, they

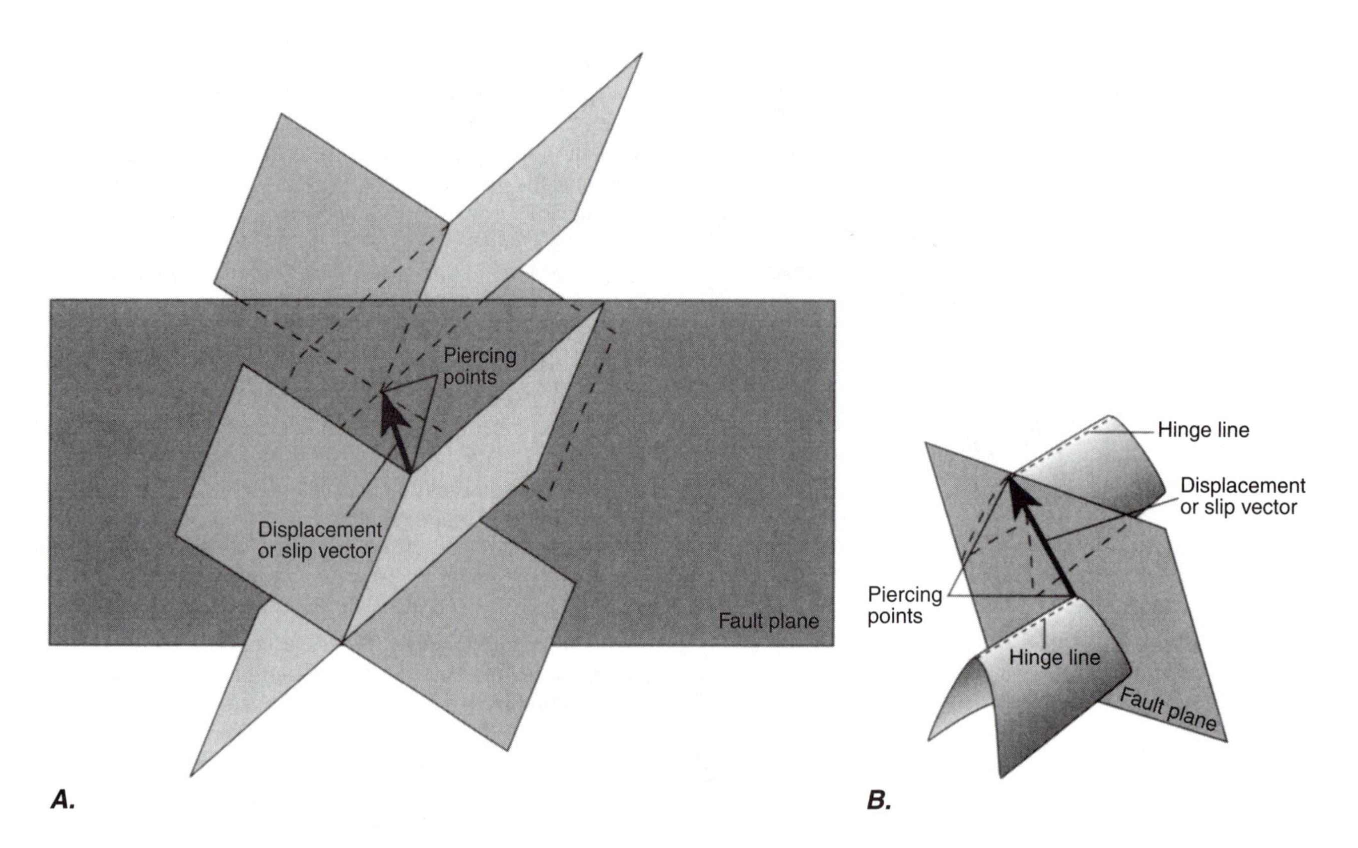

FIGURE 3.15 Complete determination of the displacement or slip vector from the offset of a unique linear feature cut by a fault. *A*. The intersection of two planar features. *B*. The hinge line of a fold.

are **congruous steps**; if they face upstream they are **incongruous steps**. The most common kinematic indicators for the relative motion on a fault result from five processes: 1) precipitation of secondary mineral grains or dissolution of material on the fault surface; 2) secondary fracturing; 3) gouging or plucking of the fault surface; 4) frictional wear, damage, or surface polishing; and 5) the accumulation and dragging of gouge material along the fault surface. We discuss each of these processes in order below.

1) If a facet of a fault surface is oriented so that it tends to open during shearing, that is, it is dilational, minerals can fill the opening by precipitating from solution to form sheaves of elongate mineral fibers known as a **slickenfiber lineation** (Figures 3.8*C*; 3.16*A*, *B*). The fibers grow at a small angle to the shear fracture boundary such that an arrow pointing along a fiber from its point of attachment on one fault block indicates the direction of relative motion of the opposite block (Figure 3.16*C*). Because fiber growth is a relatively slow phenomenon, the slickenfibers must form during slow aseismic movement (creep) on a fault, so that fiber growth can keep pace with the displacement. Opposite ends of the same mineral fiber should join points that were adjacent when the fiber started growing. Thus if the attachment points of fibers on both fault blocks are preserved, as in a cross section of the fault (Figure 3.16*C*), the length of the fiber is a measure of the displacement magnitude on a particular fracture for the period of fiber growth. The maximum displacement magnitudes recorded on individual fractures by these fibers are rather small—generally a few millimeters to 20 centimeters. Fibers much longer than this size are either not formed or not preserved. In principle, the minimum total displacement across a fault zone should be the sum of the fiber lengths on all shear fractures in a cross section of the zone, but where displacements on the order of meters or more are involved, this measurement would be time-consuming, if not impossible, to make.

Slickenfibers on fault surfaces are generally exposed by the erosional removal of part of one of the fault blocks. The crystals grow at a low angle to the fault surface and tend to break off either along the fibers or at a high angle to them. They therefore form a set of congruous steps on the fault surface (Figure 3.16*B*) such that the step risers face downstream, and the surface feels smoothest to the hand when rubbed in the downstream direction, that is, the direction of relative motion of the missing block.

If contraction occurs across a facet of a fault surface, the shortening may be accommodated by solution of the rock, especially in limestones. Solution occurs along highly irregular surfaces called **stylolites**. Where formed along a shear fracture, the peaks on the stylolites are oriented at a low angle to the fault surface and are some-

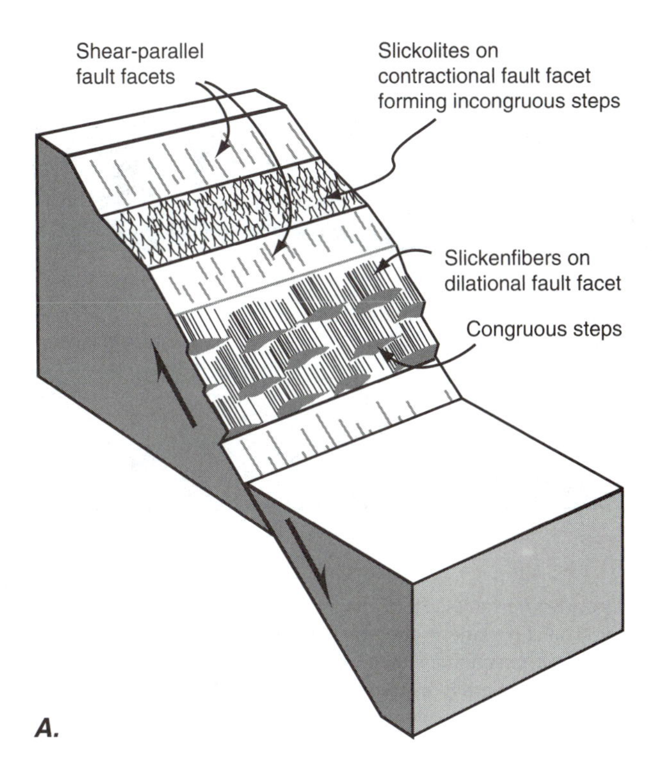

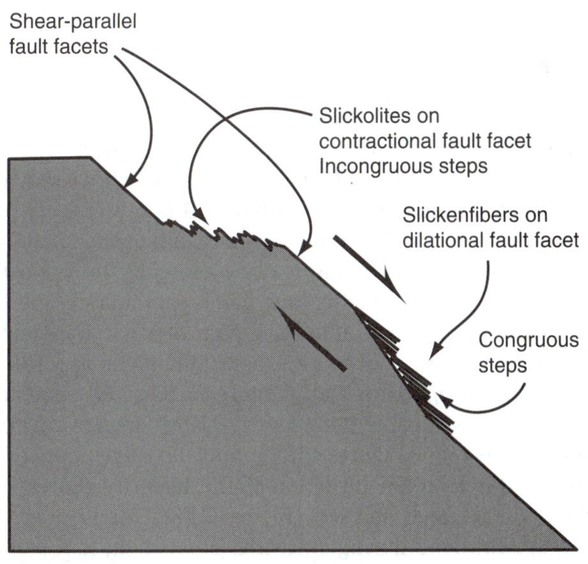

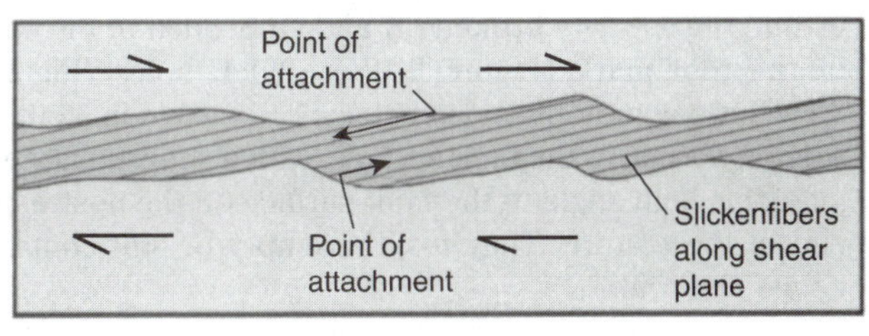

FIGURE 3.16 *A.* Block diagram. *B.* Cross section showing an irregular fault surface with shear sense indicators that form on differently oriented facets of the surface: Slickolites are solution structures that form on contractional facets (especially in limestones and marbles); slickenfibers are depositional sheaves of mineral fibers that form on dilational facets. *C.* Slickenfibers as indicators of shear sense and minimum displacement. Fibers extend from where they are attached to one wall in the direction of motion of the opposite block of the fault. An arrow along a slickenfiber attached to one wall of the fault points in the direction of relative slip of the opposite wall. The length of the fiber from one wall to the other is a measure of the minimum displacement on that fault. With one of the fault blocks eroded away, slickenfibers form congruous steps on the fault surface with step risers facing in the downstream direction. The fault surface is smoothest when rubbed in the downstream direction, that is, the direction of relative slip of the missing block. (N.B. Using the smoothest-feeling direction to infer the relative sense of displacement does not work for slickolites.)

times called **slickolites**. These peaks point in the upstream direction and may form incongruous steps. Irregularities on faults may form alternating facets across which components of dilation or contraction occur, where alternating zones of slickenfibers and slickolites, respectively, may form (Figure 3.16*B*).

2) Secondary fractures that develop during faulting may be either extension or shear fractures. Both fracture types can form at many angles to the main fault surface, but most form such that in a given fault block, the fracture plane cuts into the fault surface in the downstream direction (Figure 3.17). In general, secondary extension fractures are not striated and may be filled with secondary minerals; secondary shear fractures are generally striated. Because secondary fractures can form both congruous and incongruous steps on the fault surface, the direction that feels smoothest when rubbing one's hand along the fault surface is not a good criterion for the sense of shear. Secondary fractures provide five criteria that are useful for determining the sense of shear on the fault surface.

a) As viewed on an exposed fault surface, extension fractures cut into the fault surface at an angle of 30° to 50° in the direction of movement of the missing fault block (Figure 3.17*A*). These extension fractures are essentially the same as the pinnate fractures shown in a section normal to the fault in Figure 2.7 (Section 2.1). Sets of these fractures may form *en echelon* arrays of gash fractures, which are apparent when viewed in a section looking parallel to the fault and perpendicular to the slip vector.

b) If the extension fractures are crescent-shaped as exposed on the fault surface, they are concave downstream, in the direction of motion of the missing fault block (Figure 3.17*B*). Such fractures

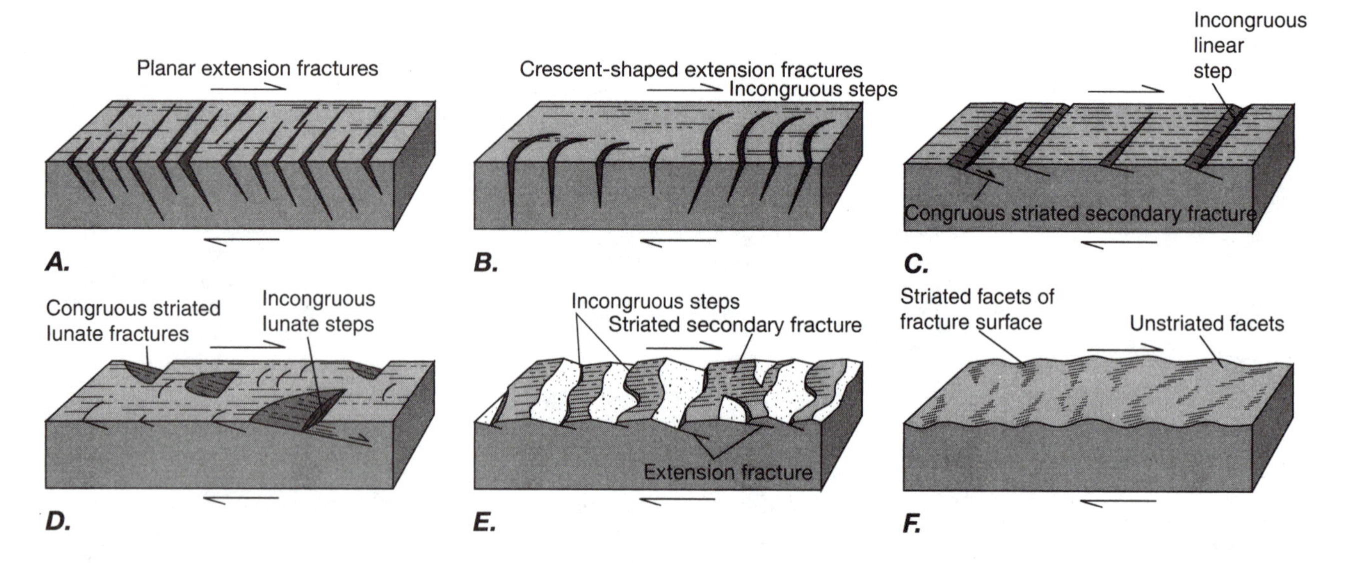

FIGURE 3.17 Secondary fracturing that provides shear sense criteria on brittle faults. Block diagrams show the relationship between secondary fractures and the sense of shear on a brittle fault. The top plane is the shear plane; relative motion is indicated by arrows. Extension fractures are unstriated and may be filled with secondary minerals. Striated fracture surfaces are shear fractures. (N.B. The relative movement of the fault in these features is generally in the rough, rather than the smooth direction.) (After Petit 1987)

may actually accumulate a very minor amount of antithetic slip (opposite to the shear sense on the main fault) to form incongruous steps, as if the small blocks between the fractures had rotated slightly in the sense of shear on the main fault (see the right-hand end of Figure 3.17*B*).

c) Extension fractures at high angles to the fault surface may form congruous steps if fragments are detached from the downstream side of the fracture wall and shifted downstream.

d) If striated secondary fractures cut into the wall of the main fault plane, they are oriented so they cut into the surface in the direction of motion of the missing fault block (Figure 3.17*C*). Fracturing of the acute wedge of rock between the secondary shear and the fault surface produces incongruous steps in the surface that face upstream opposite to the motion of the missing fault block. The steps may be predominantly linear (Figure 3.17*C*), or they may have a lunate morphology (Figure 3.17*D*).

e) Some striated secondary shears do not cut below the fault surface. They may alternate with unstriated secondary extension fractures that do cut below the fault surface (Figure 3.17*E*), or they may be simply the facets of irregularities in the fault surface itself (Figure 3.17*F*). In these cases, the striated facets face upstream. If the acute angle wedge between the secondary shear and extension fractures breaks off, then incongruous steps form, facing upstream (Figure 3.17*E*).

The secondary shear fractures shown in Figure 3.17*C*, *D*, and *E* are examples of **Riedel shears**, which we discuss in Chapters 6 and 8 (see Figures 6.4*A* and 8.7).

3) Gouging of the fault surface results from a hard asperity on one side of the surface plowing the opposite side during shear, leaving long linear tool marks. The asperities can be hard mineral grains or rock fragments. The edges of the tool mark are parallel if the asperity is fixed in the opposite wall (Figure 3.18*A*); the edges diverge downstream if the asperity plows deeper into the gouged surface (Figure 3.18*B*); and the edges converge in the downstream direction if the asperity plows out from the gouged surface (Figure 3.18*C*) or is progressively destroyed during shearing. Thus using tool marks as shear sense indicators requires caution unless one can either identify the congruous step at the start of the tool mark from which the asperity was plucked (Figure 3.18*A*) or find the asperity broken off and embedded in the end of the tool mark (Figure 3.18*B*). Pluck holes (Figure 3.18*D*) are commonly asymmetrically concave in a cross section normal to the fault surface, with a congruous step making a high angle to the fault surface on the upstream end of the feature. They may be empty or still contain gouge material.

4) Irregularities in the fault surface may show damage such as crushing on the facets that face upstream. These irregularities may also have polished or striated facets facing upstream (Figure 3.17*E*).

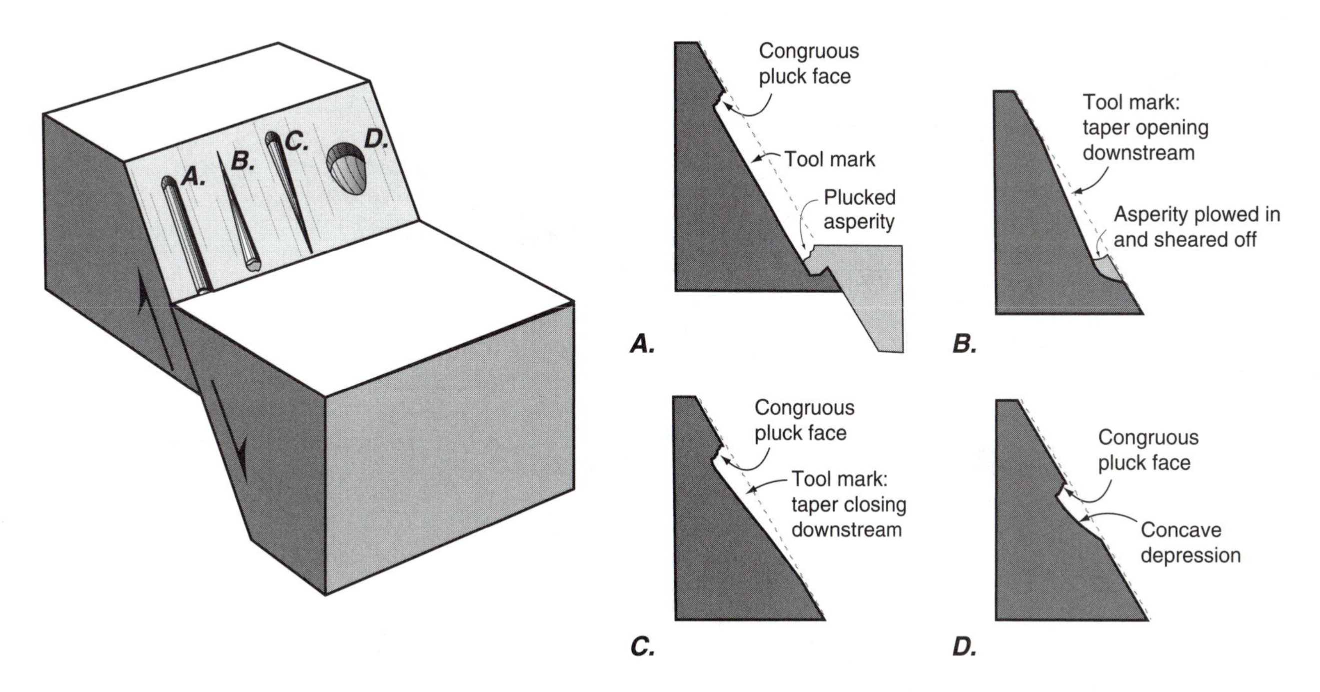

FIGURE 3.18 Gouging or plucking of the fault surface provides shear sense criteria. *A.* Tool mark gouged into the fault surface by a hard asperity. *B.* Tool mark from asperity plowing into the fault surface. The taper of the tool mark opens downstream with asperity sheared off at the end. *C.* Tool mark from a plucked asperity plowing out from the fault surface leaving a tool mark with a taper closing downstream. *D.* A concave pluck scar with a congruous pluck face on the upstream side.

5) Gouge may accumulate downstream from a protuberance in the fault plane to form a conical ridge that narrows in the downstream direction. Gouge may be dragged downstream and plastered over the upstream ends of concavities in the fault surface to form delicate overhanging lips of gouge. In some cases, gouge may be observed trailing out of a pluck hole and plastered onto the fault surface downstream from the hole.

This list of fault surface features is not exhaustive. Other less common asymmetric features have also been used to identify the slip directions on brittle faults. There are no simple rule-of-thumb means for determining relative fault displacement, so careful examination of individual situations and rock types is necessary, and this may reveal other useful criteria.

Ductile shear zones may contain a number of small-scale structures that indicate the shear sense. Platy minerals may become aligned to form a foliation (shown by dashed lines in all parts of Figure 3.19 except part *B*, and labeled *S* in Figure 3.19*A*). At the boundaries of a shear zone, the foliation makes an angle of about 45° with the boundary, and near the center of the shear zone it becomes roughly parallel to the shear zone (Figure 3.19*A*). The sigmoidal curvature of the foliation defines the sense of shear as indicated in the figure.

Ductile faults also characteristically exhibit extended tube-shaped folds in layering that are called **sheath folds** (Figure 3.19*B*). The long dimension of these folds is approximately parallel to the direction of slip on the ductile fault, and the complex folding results in sheaths that can close in either direction.

Many rocks in ductile shear zones contain large crystals. Some are relict crystals, or **porphyroclasts**[3] that survived the shearing and reduction in grain size from the original rock. Others are **porphyroblasts**,[4] which are mineral grains that grow to a relatively large size in a rock during metamorphism and deformation.[5]

Porphyroclasts found in mylonites may have asymmetric "tails" composed of very fine grains that are recrystallized from the edges of the porphyroclast itself. The sense of asymmetry of the tails defines the sense of shear in the deformed rock. We distinguish two different tail morphologies, the σ-type and the δ-type. On σ-porphyroclasts, the tails extend from each side of the grain in the downstream direction of the relative shear in the matrix, and the tails do not cross the line parallel to the foliation through the center of the grain (Figure 3.19*C*). The δ-type is derived from the σ-type by rotation of the porphyroclast in a sense consistent with the shear,

[3]After the Latin word *porphyry,* which means "purple," and the Greek word *klastos,* which means "broken."

[4]After *porphyry* and the Greek word *blastos,* which means "growth."

[5]The use of *porphyry* to refer to a rock with large crystals comes from the fact that statues of Roman emperors were carved from purple volcanic rock containing large feldspar phenocrysts.

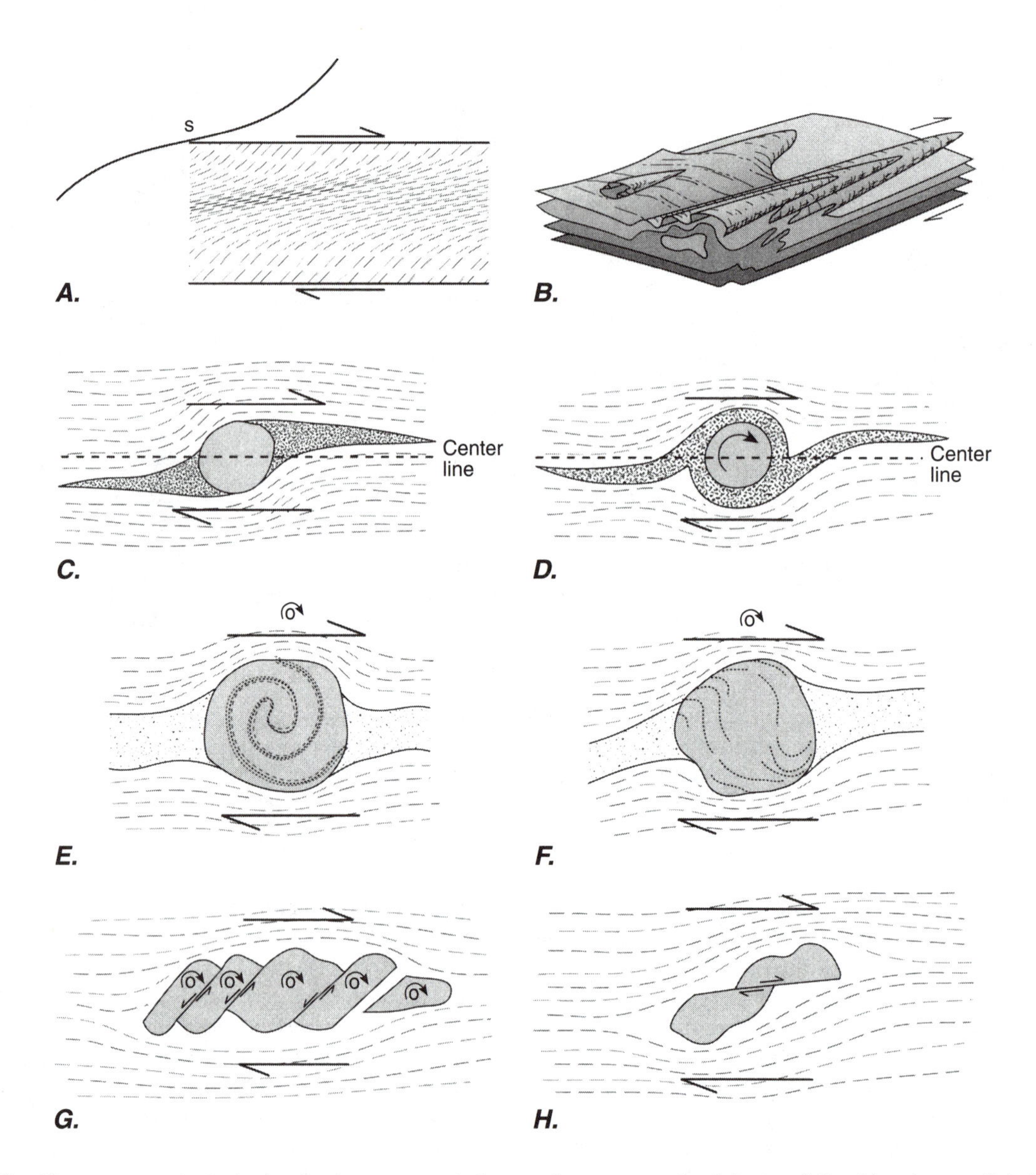

FIGURE 3.19 Shear sense criteria in ductile shear zones. *A.* Sense of curvature of a foliation defined by the parallel alignment of platy mineral grains. *B.* The orientation and asymmetry of sheath folds shown with the surrounding layers of rock removed. Although sheath folds are, for convenience, shown here closing to the right, in fact they can close in either direction. *C.* The sense of asymmetry of σ-type "tails" of recrystallized porphyroclastic material relative to the porphyroclast. *D.* δ-type "tails" develop from rotation of the σ-type "tails." *E.* The sense of rotation of a helical train of mineral grains included within a growing porphyroblastic crystal grain such as a garnet or staurolite. *F.* An inclusion train that indicates only a small amount of rotation is not, by itself, adequate evidence of shear sense. *G, H.* The sense of shear on fractures in mineral grains such as feldspar or mica depends on the shear sense of the fault and the angle between the fracture and the shear plane.
(*C, E, F, G* after Simpson and Schmidt 1983; *D* after Simpson 1986)

and the tails do cross the center line (Figure 3.19*D*). **Pressure shadows**, or **strain shadows**, are volumes of mineral grains that have crystallized as overgrowths in zones on opposite sides of a porphyroclast or porphyroblast. They form an elongate structure that is generally parallel to the foliation and may show an asymmetric shape similar to the asymmetric tails of recrystallized porphyroclast material. It is difficult to infer unambiguously a sense of shear from these overgrowths, however, and they should not be confused with the tails composed of recrystallized porphyroclast mineral.

Porphyroblasts do not deform with the rest of the rock but rotate as rigid grains during ductile deformation of the matrix. Common minerals that form porphyroblasts include garnet and staurolite. As they grow during deformation, they enclose adjacent minerals, such as micas or quartz, from the matrix. Continued rotation and growth of a porphyroblast result in a helical train of inclusions that defines the sense of rotation of the grain and thereby the sense of shear in the rock (Figure 3.19*E*). Interpretation of shear sense from inclusion trains that indicate only small amounts of rotation, however, is unreliable because such inclusion trains may actually be crenulations preserved from an earlier foliation rather than a record of porphyroblast rotation (Figure 3.19*F*).

Some porphyroclastic minerals, such as mica and feldspar, tend to shear on discrete fractures or crystallographic planes to accommodate ductile deformation in the surrounding matrix. If the fractures initially make a high angle with the shear plane, then the individual mineral fragments rotate with the imposed shear, and the shear sense on the fracture planes is opposite to that in the surrounding matrix (Figures 3.19*G* and 3.7). If, on the other hand, the fractures make a relatively low angle with the shear plane, then the shear sense on the fractures is the same as it is in the matrix (Figures 3.19H).

iv. Partial Determination of Displacement from Large-Scale Structures

In regions where fault displacement measures tens to hundreds of kilometers, large-scale geologic features that have been offset can be used to determine the displacement direction and shear sense and to estimate the displacement magnitude of the fault. Such features include shorelines, sides of sedimentary basins, the source and depositional site of distinctive sediments, and paleomagnetic measurements. Figure 3.20 shows a paleogeographic map of an Oligocene sedimentary basin in western California that has been cut by the San Andreas Fault. The different map patterns distinguish marine from nonmarine deposits. Offset of the shoreline along the fault from A to B suggests a right-lateral component of displacement of approximately 300 kilometers since Oligocene time.

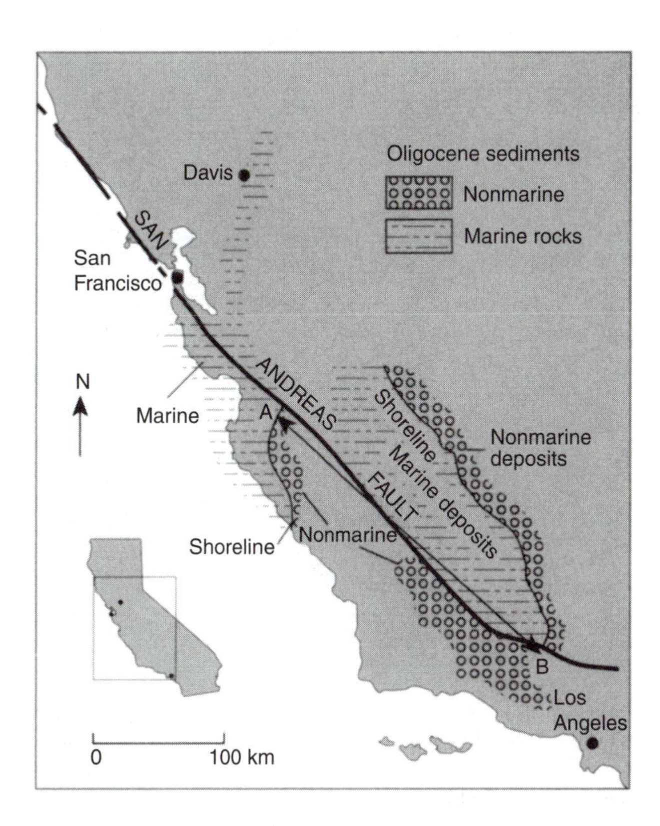

FIGURE 3.20 Shorelines of an Oligocene sedimentary basin offset about 300 kilometers along the San Andreas fault, California. Points A and B were originally adjacent to each other. (After Addicott 1968)

Isopach maps[6] are maps showing the contours of equal thickness of a geologic unit. If there is a regular variation in the thickness of a layer, fault offset of the layer (Figure 3.21*A*) may show up as a discontinuity on an isopach map (Figure 3.21*B*). Each isopach is a unique line of constant layer thickness, and matching isopachs across the discontinuity should in principle make it possible to determine the horizontal component (*H*) or the strike-slip component (*S*) of the displacement. If the data are from well logs, however, they are not very closely spaced. Thus the locations of isopachs and their intersection with the fault may be only approximate. Moreover, unless the fault strike is known, the strike-slip component of displacement cannot be determined accurately. Because isopach maps do not include elevation information, we cannot use them to determine the dip-slip and the vertical components of displacement.

Structure contour maps are maps of elevation contours on a particular geologic surface at depth, generally a stratigraphic horizon. Faults in the subsurface can be identified from discontinuities in structure contours (Figure 3.22). If the structure contours display a linear structure, such as a fold hinge that can be matched uniquely across the fault (Figure 3.22*A*), the fault displacement can be determined. The horizontal component of fault displacement (*H*) (Figure 3.22*B*) is determined by matching the map pattern of the structure contours across the fault; the vertical component is determined from the different elevations of two initially adjacent structural features. If the strike of the fault can be determined, we can use trigonometry to find 1) the strike-slip component of displacement (*S*) from the horizontal component (*H*) and 2) the actual displacement (*D*) from the strike-slip (*S*) and dip-slip (*d*) components (Figure 3.22*A*). As with isopachs, however, the determination is subject to the accuracy of the contours themselves.

In some cases, it is possible to use the geomagnetic field as a means to determine the relative latitudinal motions of two sides of a fault. Sedimentary or igneous rocks commonly preserve a paleomagnetic record of the orientation of the magnetic field at the time the rocks form. If those rocks are displaced substantially in latitude, for example, along a large-scale strike-slip fault, the difference between the paleomagnetic field orientation preserved in the rocks on either side of the fault can be used to determine the relative displacement along the fault. Studies of paleomagnetism cannot detect longitudinal

[6] After the Greek words *isos*, which means "equal," and *pachos*, which means "thickness."

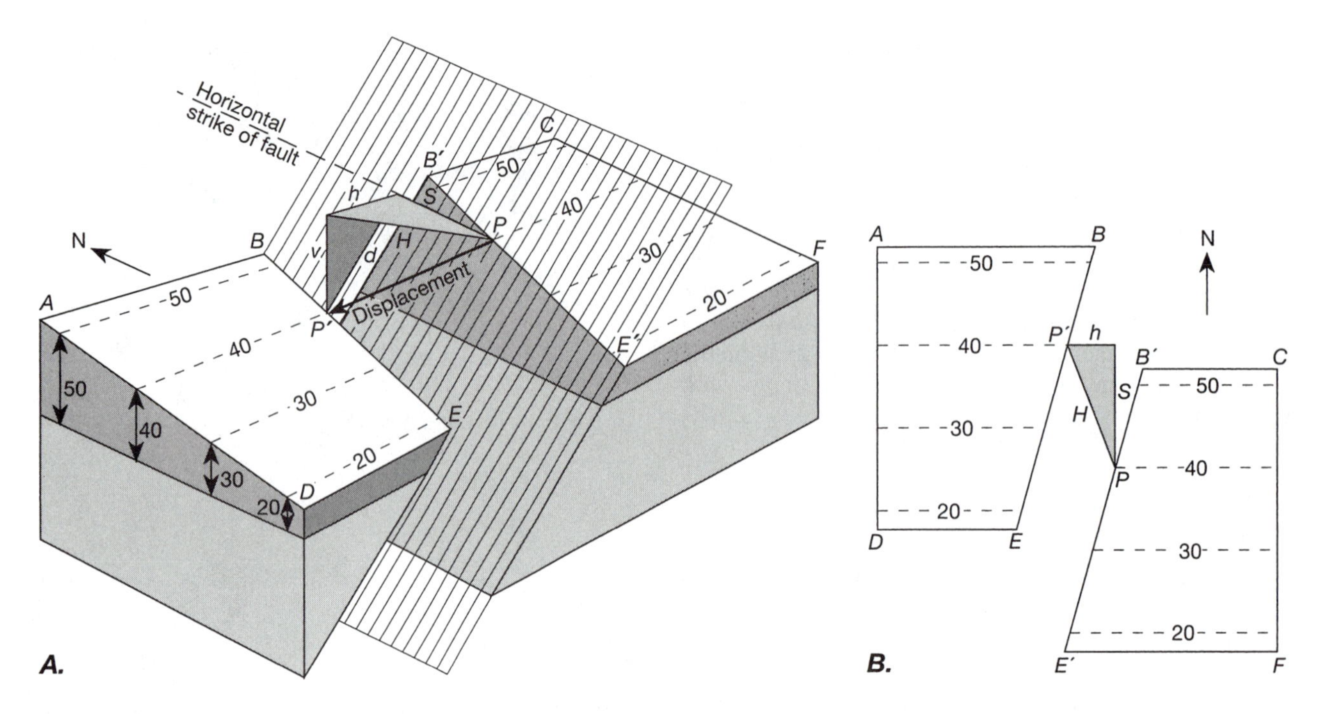

FIGURE 3.21 Interpretation of a fault from a discontinuity on an isopach map. *A*. Block diagram showing a bed of varying thickness cut by a fault. Contours of equal thickness (isopach lines) are drawn on the top surface of the bed. The true displacement and the strike-slip (*S*), dip-slip (*d*), horizontal (*H*), and vertical (*v*) components are shown. For ease of interpreting the three-dimensional drawing, we show a special case for which the isopachs are parallel to the strike of the layer surface and perpendicular to the strike of the fault (*h* is the horizontal component perpendicular to the dip of the fault, or the horizontal dip-slip component). *B*. Isopach map of the structure shown in *A*. The horizontal component of displacement (*H*) is determined by connecting the map projections of two points *P* and *P′* that mark where the same isopach on opposite sides of the fault intersects the fault surface. The strike-slip component of displacement (*S*) is determined by connecting the extensions of equal isopach lines with a line parallel to the strike of the fault.

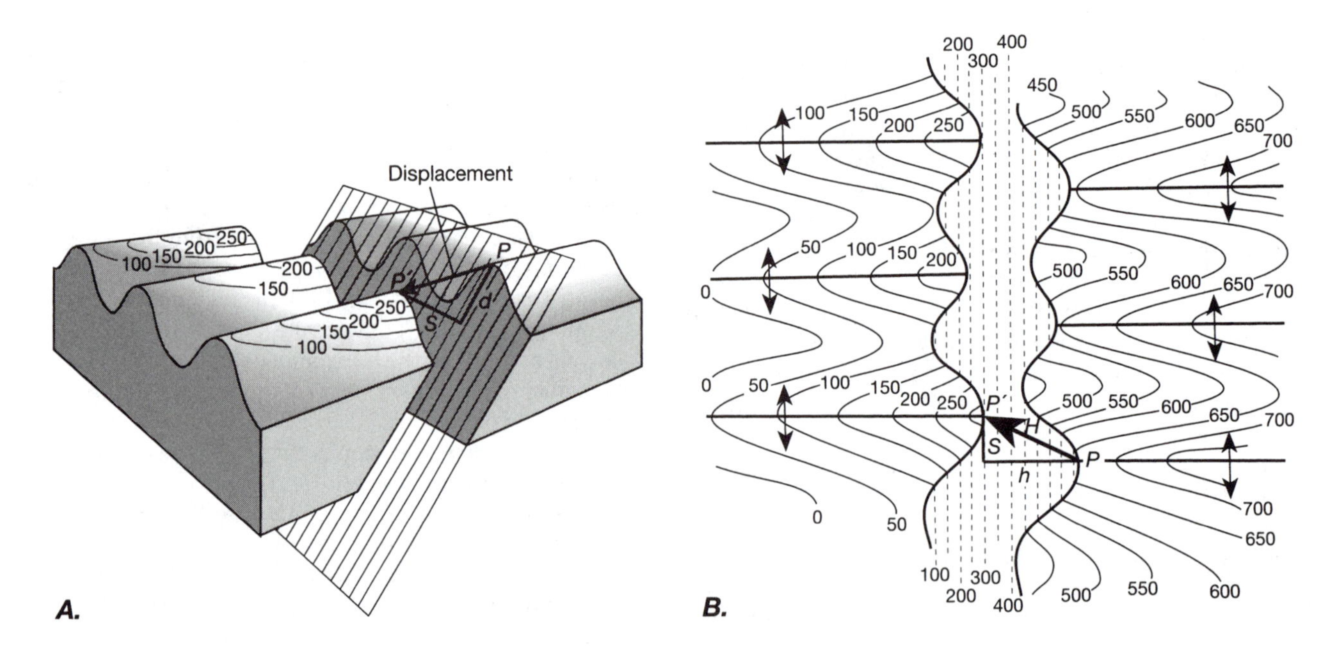

FIGURE 3.22 Interpretation of a fault from a structure contour map. *A*. Three-dimensional diagram showing the folded surface of a stratigraphic contact that has been cut and displaced by a fault. The contact has contours of equal elevation (structure contours) drawn on it. *B*. A structure contour map of the same structure shown in *A*. The horizontal component of displacement (*H*) is determined by joining the points on the map that are the vertically projected piercing points *P* and *P′* of the fold hinge on opposite sides of the fault. The strike-slip component (*S*) is parallel to the fault strike, and the horizontal dip-slip component (*h*) is normal to the fault strike.

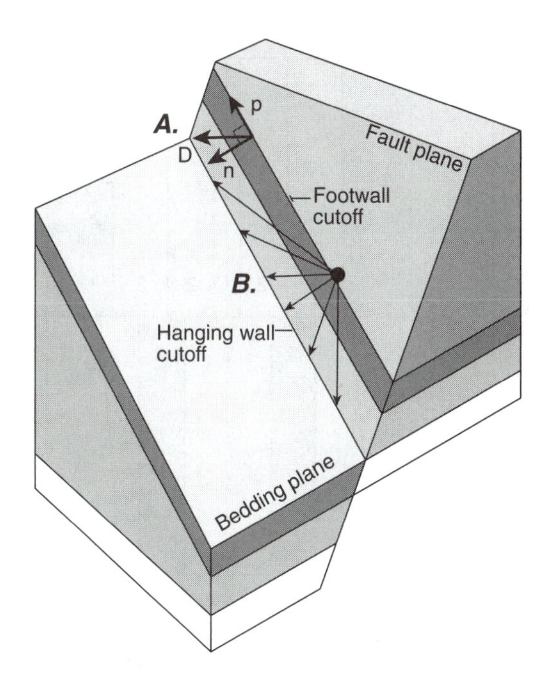

FIGURE 3.23 Faulted planar features are nonunique indicators of fault displacement. *A*. The true displacement of the hanging wall block on the fault (*D*) can be specified as the sum of the component normal to the cutoff line of the planar feature (*n*) and the component parallel to the cutoff line (*p*). *B*. Because the parallel component of displacement (*p*) does not produce any offset of the bedding plane, any value of *p* results in the same geometry for the faulted plane, and the displacement *D* is therefore not uniquely defined.

motions, and the latitudinal motion generally must be more than 10° to be beyond the limits of error.

v. Nonunique Constraints on Displacement

Frequently, the principal evidence for a fault consists of the offset of a planar structure, typically sedimentary bedding. It is very important to realize that this offset alone can never define the displacement on the fault, regardless of the appearance of the outcrop pattern. The reason for this is not difficult to understand (see Figure 3.23). We express the true displacement vector (**D**) as the sum of its component vectors in the fault plane that are normal (**n**) and parallel (**p**) to the **cutoff line**, which is the intersection of the planar feature with the fault (Figure 3.23*A*). The normal component (**n**) is the perpendicular distance between the matching cutoff lines on the hanging wall and the footwall, respectively, and this component can be measured easily. The component (**p**) parallel to the cutoff line, however, produces no change in the orientations or locations of the cutoff lines, and thus it cannot be observed. The complete description of the displacement (**D**) is therefore indeterminate. Figure 3.23*B* shows six of the infinite number of displacement vectors that could produce the same geometry for the faulted bedding plane. Among these six vectors are those with components of normal, reverse, dextral, and sinistral displacement. Each vector has the same component of displacement (**n**) normal to the cutoff line but a different component (**p**) parallel to the cutoff.

Thus, if the only information available about a fault is the offset of parallel planar features, we cannot talk about the slip or displacement on the fault because we cannot determine it. We speak instead of the **separation**, which is the distance measured in a specified direction between the same planar feature on opposite sides of the fault (Figure 3.24). The separation enables us to determine only the component of displacement normal to the cutoff line; any displacement component parallel to that line is indeterminate. Illustrated in the figure are two of the common separations and the directions in which each is measured—the **strike separation**, measured parallel to the strike of the fault, and the **dip separation**, measured parallel to the dip line of the fault. Other separations used in some cases include the **stratigraphic separation**, measured normal to a bedding plane; the **vertical separation**, measured in a vertical direction (it is the vertical component of both the dip and the stratigraphic separations); and the **horizontal separation**, measured in a plane normal to the strike of the fault (it is the horizontal component of the dip separation).

On the same fault, a separation measured for one plane is different from the separation measured in the same direction for a plane of a different orientation. In fact, if the two planar features are appropriately oriented, one strike separation can be right-lateral and the other left-lateral. For example, opposite limbs of a faulted fold can have very different separations (Figure 3.25). Similarly, opposite senses of dip separation can develop for appropriately oriented planes and displacements.

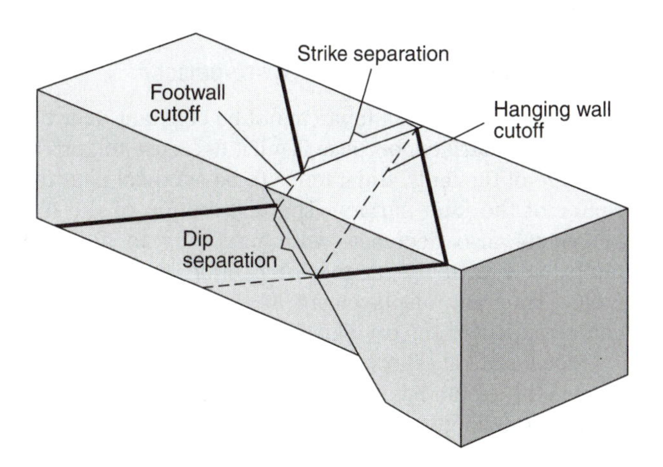

FIGURE 3.24 Block diagram indicating the strike separation and the dip separation of a faulted layer on a dip-slip normal fault where the footwall block has been eroded down to the same elevation as the hanging wall block.

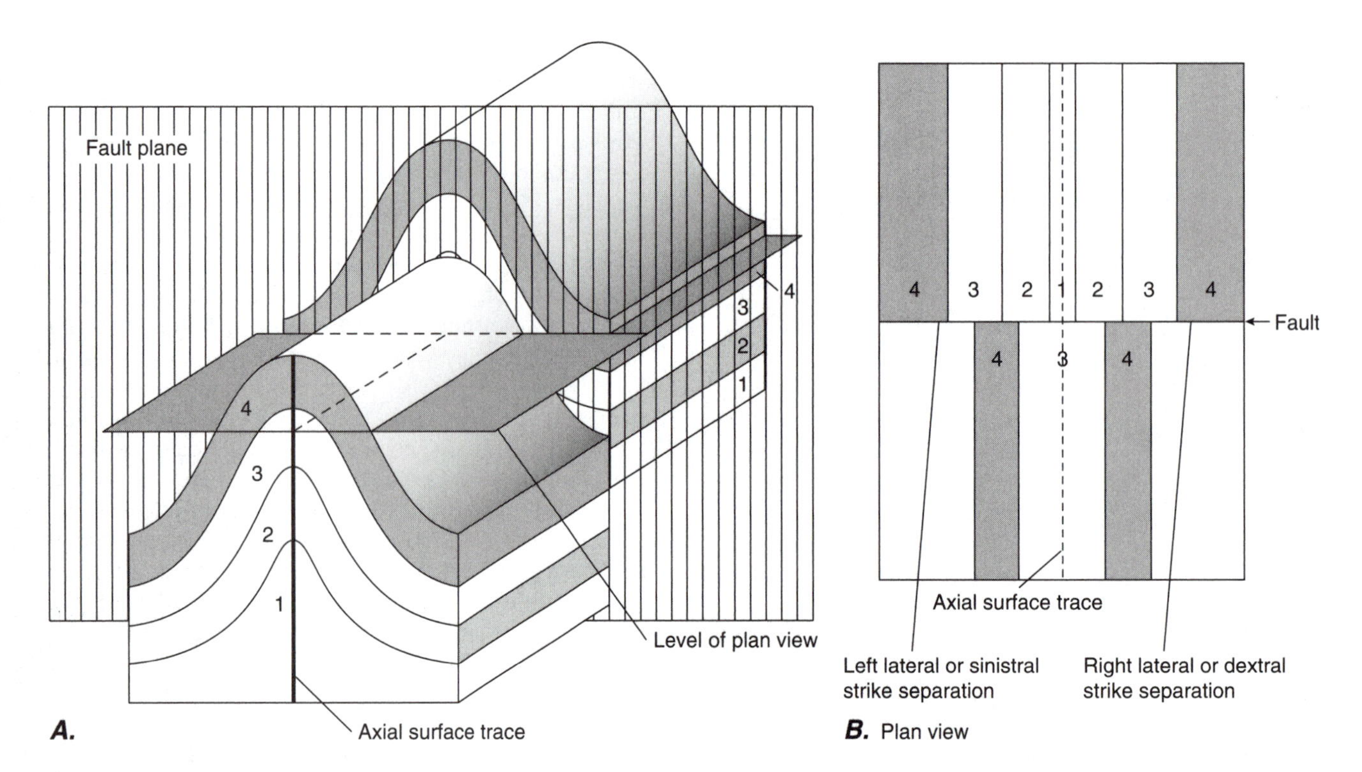

FIGURE 3.25 Examples of different separations produced by the faulting of folded layers. *A*. Fold offset by dip-slip movement along a vertical fault. *B*. Map view at the level of the horizontal plane in *A*, showing the opposite separations obtained on opposite sides of the fold.

These examples emphasize the nonuniqueness of the separation and the difficulty, or impossibility, of using it to characterize a fault. If, however, the separation in a particular direction can be determined for two planes of different orientation offset by a fault, then it is possible to determine a unique magnitude and direction for the displacement vector. In effect, the intersection of the two planes defines a line that intersects the fault at unique piercing points on the hanging wall and footwall (see Figure 3.15*A*).

vi. Displacement—Length Systematics

The displacement on a fault cannot be constant over the whole fault surface, because it must decrease to zero at the edge of the fault, and it tends to be maximal near the center of the fault surface. The distribution of the displacement also decreases with proximity to adjacent faults, and with the degree of overlap with adjacent faults. Thus any displacement measurement on a fault can be expected to vary along the fault.

Measurements show, however, that the maximum displacement, or the mean displacement, δ on a fault is related to the length L_f of the fault by an equation of the form

$$\delta = \frac{L_f^p}{B} \quad \text{or} \quad \log \delta = -\log B + p \log L_f \qquad (3.1)$$

where B and p are empirical constants. Figure 3.26 shows a compilation of eleven different data sets that collectively span about six orders of magnitude in fault length. Although the intercepts $\log B$ of the different sets are not the same, and they tend to be larger for the longer faults, the slopes of the individual data sets are statistically all consistent with a value of $p = 1$. Thus, on average, these data indicate that the maximum displacement on a fault increases linearly with the length of the fault. Other researchers have proposed that the best-fit slope is approximately 1.5. For the data in Figure 3.26, such a value could result from fitting a single line to the different data sets that have $\log L_f > -1$, but such a fit does not account for the bias introduced by the larger intercepts ($\log B$) associated with the longer faults (Figure 3.26).

In addition to inherent variability in the systematics, scatter occurs in these data because of the variability of the displacement across the fault surface and the difficulty of determining where the maximum displacement should be measured. Scatter also results from the uncertainty that the length of a fault measured on a topographic surface, for example, is an accurate measure of the fault length. If a fault is envisioned as an elliptical area, the length of the fault observed on a topographic surface depends on exactly where the topographic surface cuts through the ellipse, on how the ellipse is oriented relative to the topographic surface, and on how the faults are distributed in the volume.

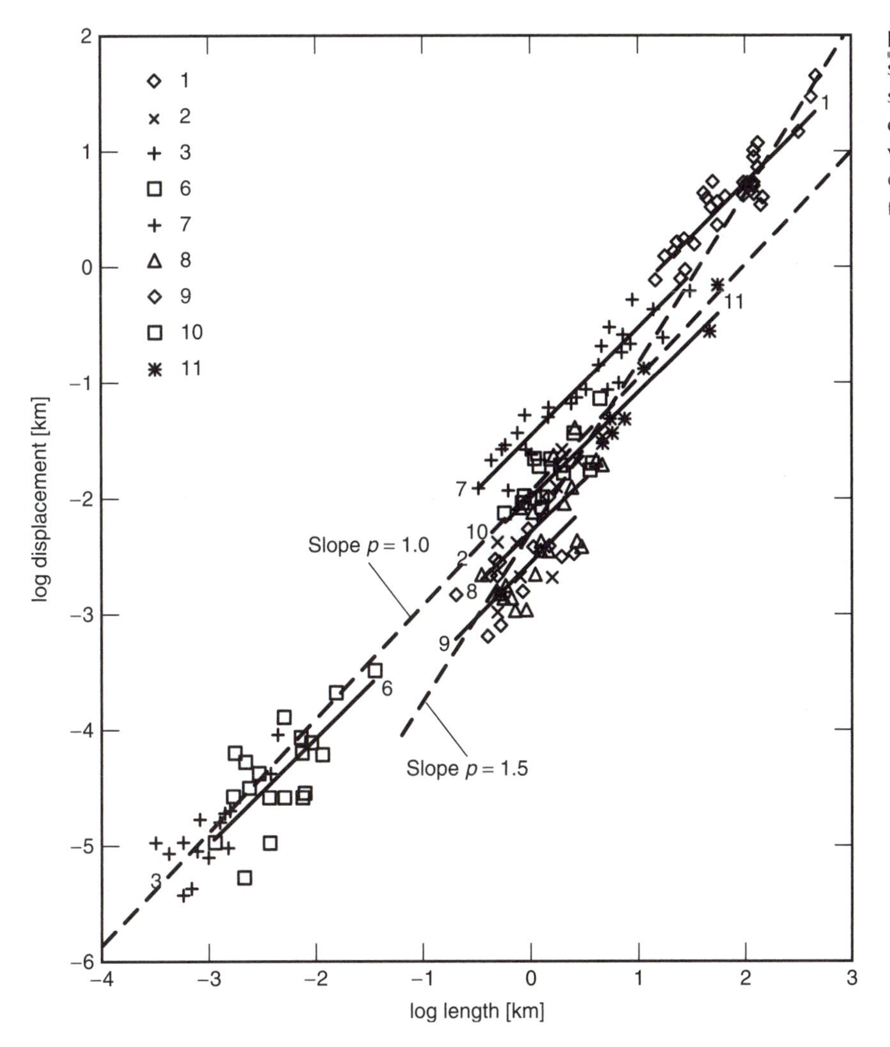

FIGURE 3.26 Displacement-length systematics for faults. The different symbols and numbers identify data from different faults. Dashed lines show lines with slopes of $p = 1.0$ and 1.5. (From Clark and Cox 1996, figure 1, with permission from Elsevier)

3.4 FAULT GEOMETRY

All faults are somewhat irregular surfaces of finite extent in three-dimensional space. Faults are usually viewed, however, on two-dimensional outcrop or topographic surfaces, or on maps or cross sections, as lines or zones of discontinuity having widely varying, but finite lengths (Figure 3.27*A*). The depiction and analysis of faults on two-dimensional surfaces (Figure 3.27*A*, *B*) encourage us to ignore the three-dimensional aspects of fault geometry (Figure 3.27*C*), which are important to understanding the kinematics of faults and the associated deformation. We consider these various characteristics of fault geometry in this section.

i. Fault-Length Scaling

The representation of faults on a map should give an indication of the lengths of the various fault segments. Where faults and associated shear fractures have been mapped over a wide range of length scales, we find that shorter faults are far more numerous than longer ones (Figure 3.28*A*). Plots of the cumulative number N of faults having a length greater than or equal to L_f suggest a power-law scaling relationship (Figure 3.28*B*), indicating that the fault lengths of a fault system show a fractal, or self-similar, distribution (compare Box 2-1).[7]

$$N = \frac{K}{L_f^m} \quad \text{or} \quad \log N = \log K - m \log L_f \qquad (3.2)$$

[7]Note that in Box 2-1 we considered the fractal geometry of fracture networks, which we evaluated in terms of the distribution of fracture traces on a two-dimensional plane. In that case, the fractal dimension describes the extent to which fracture traces fill the plane in which they are embedded. In the present case, however, we are discussing the distribution only of fracture lengths, which is a one-dimensional characteristic that does not address how the fracture traces are distributed in two dimensions. Thus, although the distribution of fracture traces in a plane and the distribution of fracture lengths both obey a power-law distribution (compare the first Equation (3.2) with the second Equation (2-1.3)), they describe characteristics of the fracture set in different dimensions.

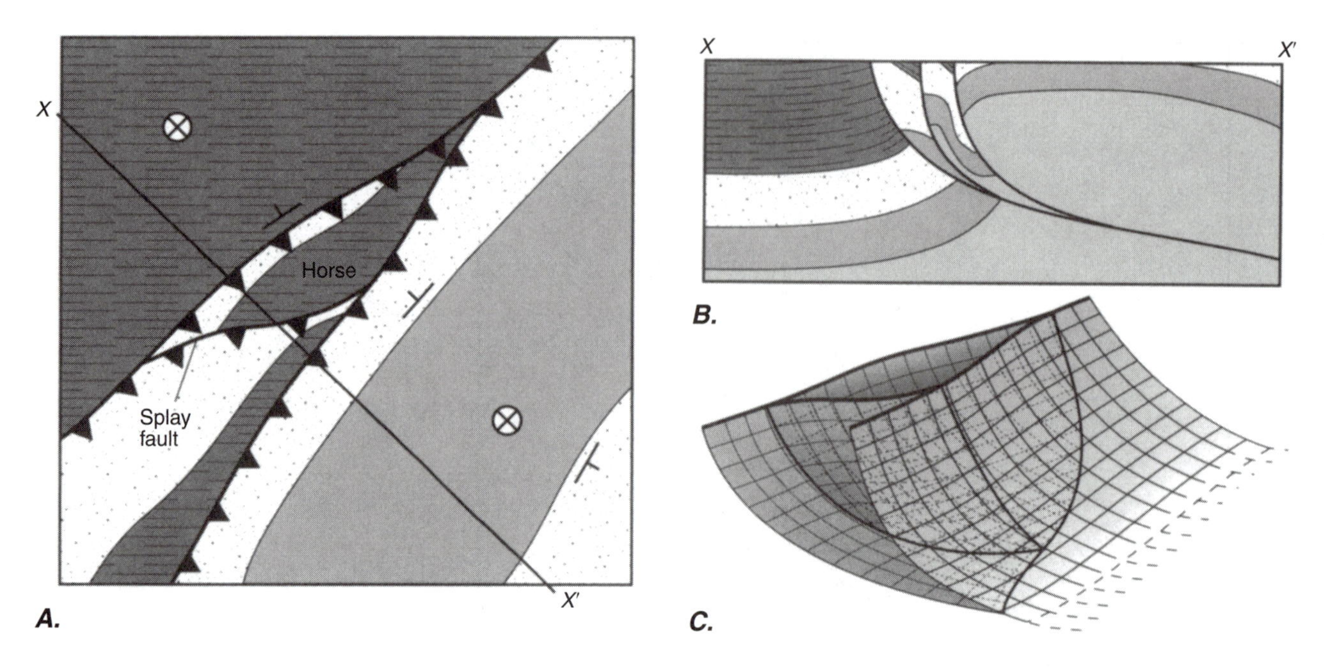

FIGURE 3.27 Three-dimensional representation of faults. *A*. Geologic map of branching imbricate thrust faults connected by a subsidiary splay fault that isolates a horse. *B*. Cross section along XX′ through the fault system in *A*. *C*. Portrayal of the three-dimensional geometry of the faults in *A* and *B*.

where K and m are empirical constants.[8] This relationship indicates that within a population of faults and associated shear fractures, one can expect to find a large number of very small shear fractures, but only a few very large faults.[9]

[8]Ideally, the constants in this equation would be determined by counting the total number of faults of any given length within a volume. In practice, it is impossible to actually see and count fractures in a three-dimensional volume of the crust, so the measurements are usually done by determining the lengths of faults exposed on a two-dimensional surface, as in Figure 3.28. Because a short fault has a smaller probability of intersecting a given surface of observation than a large fault, the exponent m determined from a two-dimensional survey should be 1 less than the exponent that would be determined from a full survey of the volume (compare the discussion at the beginning of Box 2-1(ii)).

Note that the cumulative number N of faults of length $\geq L_f$ is related to the number n of faults of length L_f by $n = dN / dL_f$, if the number distribution can be approximated by a continuous function. This simply means that the number of faults n having a length between L_f and $L_f + dL_f$ is just the increase in the cumulative number of faults dN from the N for the length L_f to $N + dN$ for the length $L_f + dL_f$.

[9]This relationship could be summarized by the ditty:

Big faults have lots of little faults
That slip and slide beside 'em,
And little faults have littler faults
And so *ad infinitum*.

Except, of course, that real scaling relations have upper and lower limits beyond which they cease to describe the physical system, so they do not apply *ad infinitum*.

ii. Termination Lines

Every fault surface, no matter what its type, must end in every direction, and the end is marked by a **termination line**. A termination line must be continuous and must form a closed line about the fault surface; it cannot simply end. It has different features, depending on the geometry of the termination.

The termination of a fault at the surface of the Earth is the **fault trace** on the topographic surface (Figures 3.29–3.32). It may be the original boundary of the fault or the intersection of an originally deeper part of the fault with a surface of erosion. It is in essence the cutoff line of the Earth's surface on the fault.

At a brittle-fluid or brittle-ductile interface, the displacement discontinuity on a fault is easily accommodated by the flow of the fluid or ductile material. Thus the discontinuity cannot extend beyond the brittle material, and the cutoff line of the interface defines the termination of the fault.

If one fault terminates against another fault of the same age, the intersection line of the faults must be parallel to the displacement direction on both faults (Figure 3.29*A*). If a younger fault terminates against an older fault, the displacement vector on the younger fault must parallel the termination line (Figure 3.29*B*). If an older fault is cut and offset by a younger fault, the termination line of the older fault against the younger has no relationship to the slip direction on either fault (Figure 3.29*C*). For a strike-slip fault with a very large displacement (Figure 3.29*C*), one might only observe fault intersections that would resemble those in Figure 3.29*B*,

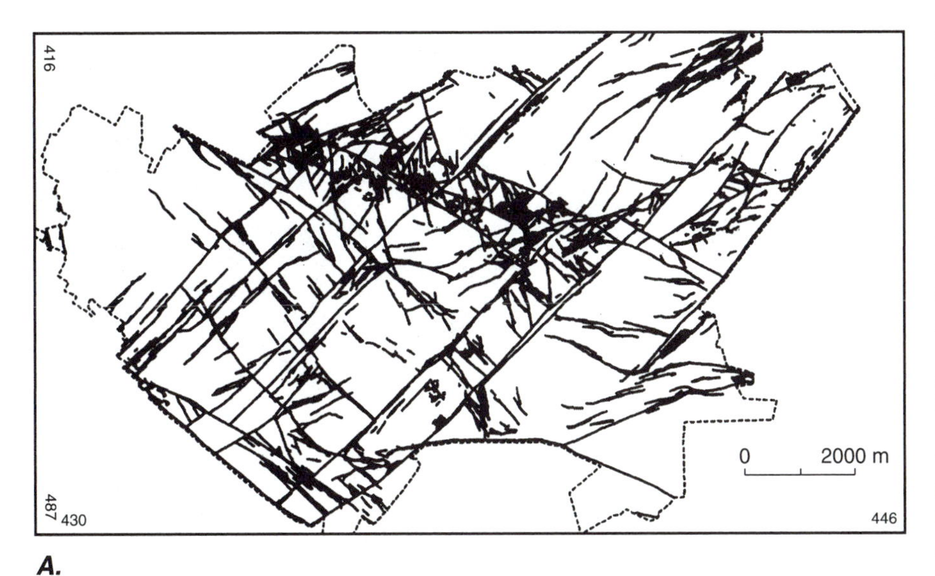

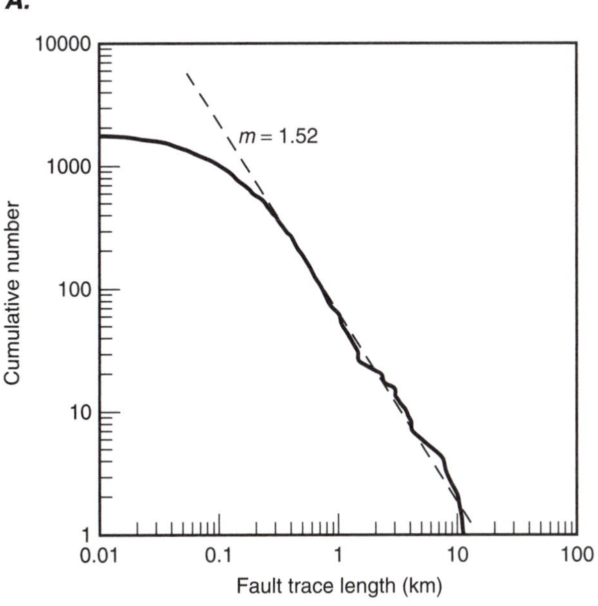

FIGURE 3.28 Fault-length systematics for a fault set mapped in the South Yorkshire coal fields in the United Kingdom. (From Waterson et al. 1996, figures 2a, 10c, with permission from Elsevier) *A.* Map of faults in the coal field. The area of the map is 87 square kilometers and includes 2257 fault traces with lengths between 10 meters and 12 kilometers. The faults striking northeast and northwest are normal faults, the northeast-striking set being the earliest-formed; the faults striking west-northwest are later dextral strike-slip fault zones. North-south and east-west boundaries of the figure are labeled by U.K. National Grid Reference numbers. *B.* Cumulative frequency versus fault trace length for the two sets of normal faults (striking northeast and northwest) but excluding the strike-slip faults (striking west-northwest) mapped in *A*. Note the self-similar geometry indicated by the straight-line portion of the plot and the limits at the ends of the plot where the plot ceases to be linear. Above lengths of about 5 to 10 kilometers, the plot is biased by faults that intersect the boundaries of the map and thus are not represented by their true lengths (censoring). Below lengths of about 0.25 kilometer, the plot is biased by the limits on the lengths of the faults that were actually mapped (truncation). The slope of the linear portion of the plot is negative and has a value of approximately $-m = -1.52$ (Equation (3.2)). Analyzing the two fault sets separately gives slopes of -1.36 and -1.87 for the northeast- and northwest-striking faults, respectively.

which could result in an erroneous interpretation of the slip direction and relative age of the faults.

A look at any geologic map of a faulted terrane demonstrates that the traces of individual faults are of limited extent. The termination line of a fault is a **tip line** where the fault displacement has decreased to the extent that it can be accommodated by coherent deformation distributed through the solid rock (Figures 3.30 and 3.31). If a fault trace ends without running into another fault, it must end at a tip line. In Figure 3.30 the tip line is parallel to the displacement vector on the fault where it intersects the horizontal surface of the block, and it is perpendicular to the displacement where it intersects the vertical side. Because tip lines of faults are generally roughly elliptical in shape at depth, all relative orientations of displacement and tip line between these two end-member types occur on a single fault. Thus, the tip line for a single fault has propagated with the characteristics of both mode II and mode III fractures (compare Figure 2.1), depending on the relative orientation of the displacement and the tip line.

Below the surface, then, a fault can be bounded on all sides by a continuous tip line that connects at both ends with the surface trace of the fault. A **blind fault**, however, does not break the surface anywhere and thus is completely surrounded by a termination line that is either a tip line or a branch line (see below). Erosion may subsequently eliminate part of the tip line to create a trace of the fault on the Earth's surface.

A **branch line** is a line of intersection where either a fault surface splits into two fault surfaces of the same type or two fault surfaces of the same type merge into one. All segments of the fault shown in Figure 3.31*A* are completely surrounded by a termination line that is a fault trace at the surface, a tip line, or a branch line, except, of course, for the trace of the fault on the verti-

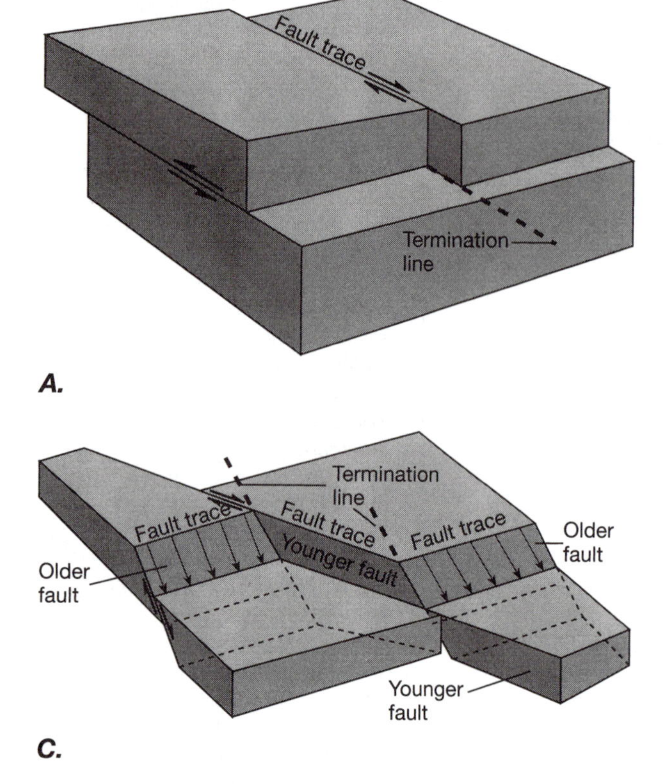

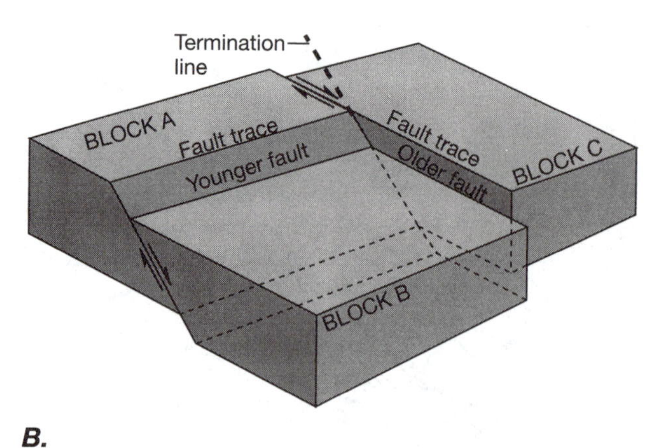

FIGURE 3.29 Block diagrams showing the geometry of termination lines at fault intersections. *A.* A right lateral strike-slip fault terminates against a horizontal fault of the same age. The termination line is parallel to the slip direction. *B.* A younger normal fault terminates against an older vertical fault. The termination line is parallel to the displacement vector on the terminated normal fault. *C.* An older normal fault is offset by, and terminates against, a younger strike-slip fault. The termination lines do not parallel the displacement vectors of either fault.

cal left side of the block, which is an artificial line on the cross section. In Figure 3.27*C*, the horse is bounded on all edges below the surface by branch lines.

Faults of all types commonly die out in a set of **splay faults**, which are smaller, subsidiary faults that branch off from the main fault (Figure 3.31*A*). Where splay faults branch off from the main fault at fairly regular intervals and have comparable geometries, they form an **imbricate fan**, which can be either extensional (Figure 3.31*B*) or contractional (Figure 3.31*C*). In Figure 3.27*C*, the horse is bounded on all edges below the surface by branch lines.

Two circumstances occur in which a fault can end, in a sense, without being bounded by a termination line. First, if the fault surface curves, the nature of the fault may change completely, but no termination line exists. Figure 3.32 illustrates a normal fault that changes orientation to become a vertical oblique-slip fault. Second, a fault may extend deep into the crust or even into the upper mantle. With increasing depth in the Earth, the temperature and pressure rise, and if they are sufficiently high, rocks become ductile. This ability of rock to flow ultimately limits the depth to which a fault can maintain its identity as a shear zone. As we trace a fault deeper into the crust, we expect it first to change from a zone of brittle deformation into a ductile shear zone. At some depth, which we do not know well, the zone of deformation must spread out, until the nature of the fault as a discrete shear zone is lost, and the displacement is accommodated by the slow, widespread flow of the

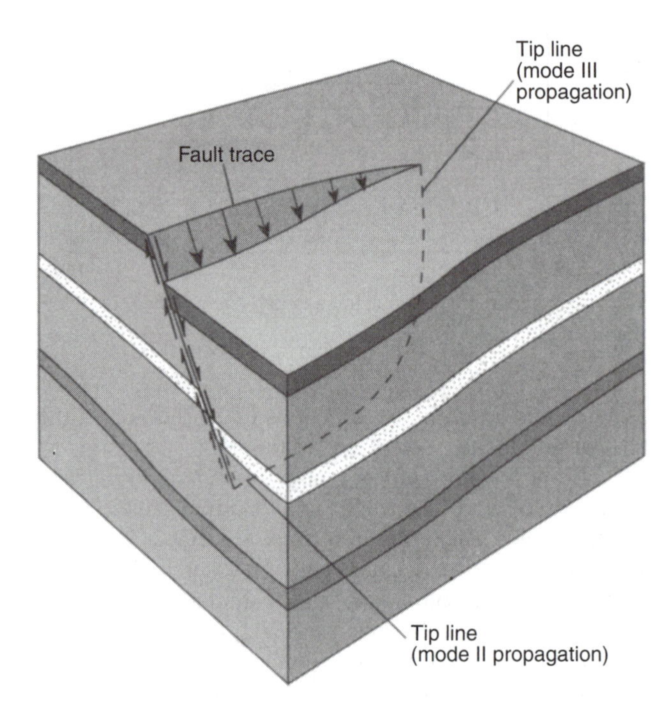

FIGURE 3.30 Geometry of tip lines. Displacement on a normal fault dies out along the strike and down the dip. The tip line is a continuous quasi-elliptical line. The angle it makes with the displacement vector changes around the perimeter of the fault and defines the mode of propagation of that part of the fault. Mode III propagation (tip line parallel to displacement); mode II propagation (tip line perpendicular to displacement) (see Figure 2.1).

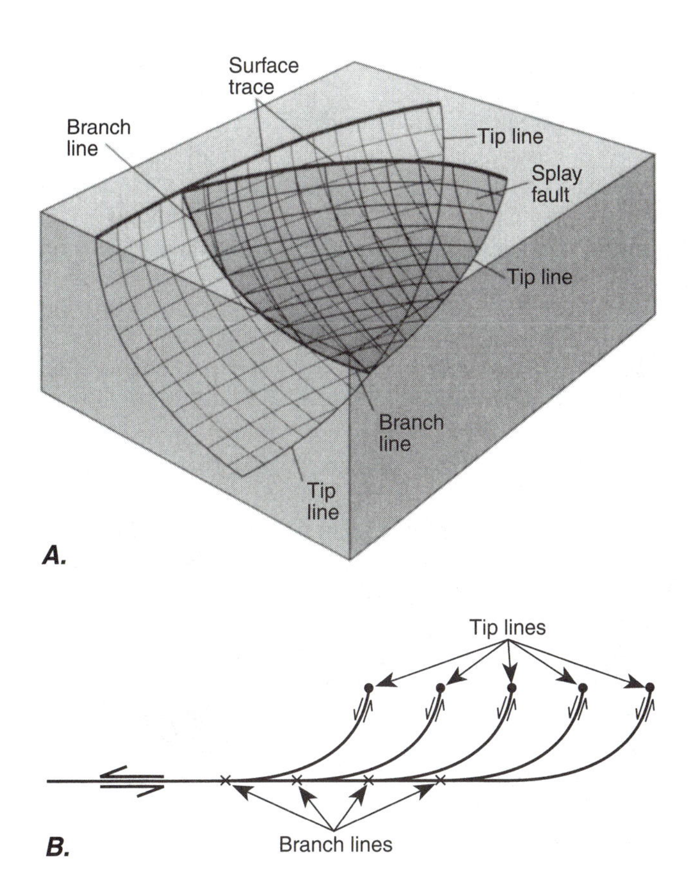

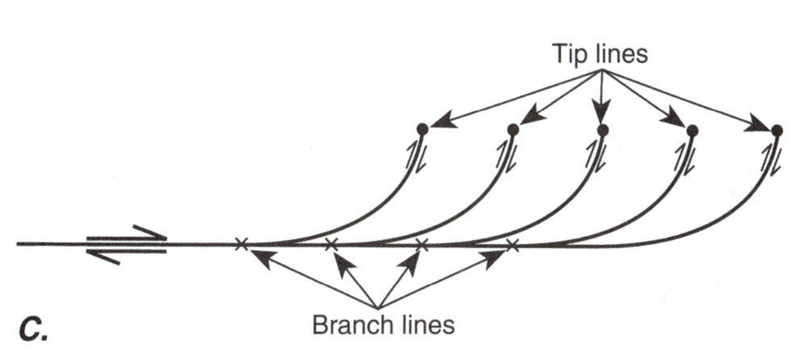

FIGURE 3.31 Splay faults and the geometry of branch lines. *A.* The three-dimensional geometry of a splay fault shows how the fault surface is completely bounded by a surface trace, a branch line, or a tip line. *B.* Extensional imbricate fans. *C.* Contractional imbricate fans.

rocks. In this circumstance, the boundary of the fault is indistinct.

The depth at which a fault loses its identity is not well known, but it probably depends in part on the magnitude and rate of the displacement on the fault as well as on the composition of the rock being faulted. Brittle fracturing is replaced by ductile flow at depths of about 10 to 20 kilometers. Fault zones have been traced by seismic reflection techniques into the lower crust to depths of about 25 kilometers, and in some places, slices of the upper mantle are exposed at the surface along faults, suggesting that faults have extended at least to the Moho. Subduction zones, in fact, are major thrust faults, some of which can be traced hundreds of kilometers into the mantle, although we do not expect most crustal faults to extend to such depths. We have limited opportunity to observe rocks that have deformed near and below the base of the crust, however, so our knowledge is very poor.

iii. Ramps, Jogs, Duplexes, and Transfer Zones

An individual fault surface generally is not a flat or smoothly curved surface but instead may have **fault ramps** connecting segments of the fault (Figure 3.33*A*). A **frontal ramp** is oriented such that its intersection with the main fault surface is approximately perpendicular to the displacement direction on the fault. On strike-slip faults, frontal ramps are called **jogs** or **bends**. A **lateral ramp** is oriented so that its intersection with the main fault surface is oblique or parallel to the displacement direction on the fault. If it is parallel, it is also a **step** or **sidewall ramp**; otherwise, it is an **oblique ramp**. If a fault is segmented but the segments are not connected by a distinct ramp fracture, the structure is a **step-over**.

In general, displacement of a fault block over a ramp induces in the block a deformation whose characteristics depend on the orientation of the fault and the displacement. A frontal ramp or jog can be extensional (Figure 3.33*B*) or contractional (Figure 3.33*C*), depending on whether the material is pulled apart or pushed together across the ramp by the dominant shear on the fault zone.

During faulting at a ramp, the location of the ramp can migrate as the fault surface cuts in discrete jumps into one fault block or the other. The result is a **fault duplex**, characterized by a stack of horses bounded by traces of the main fault. Duplexes can be thrust (Figure 3.34*A*), normal (Figure 3.34*B*), or strike-slip (Figure 3.34*C*). We discuss the characteristics of duplex formation that are specific to the different fault types in Chapters 4 through 6.

Large fault systems are made up of a system of major generally parallel faults, none of which extends for the total length of the system. Systems of normal, thrust,

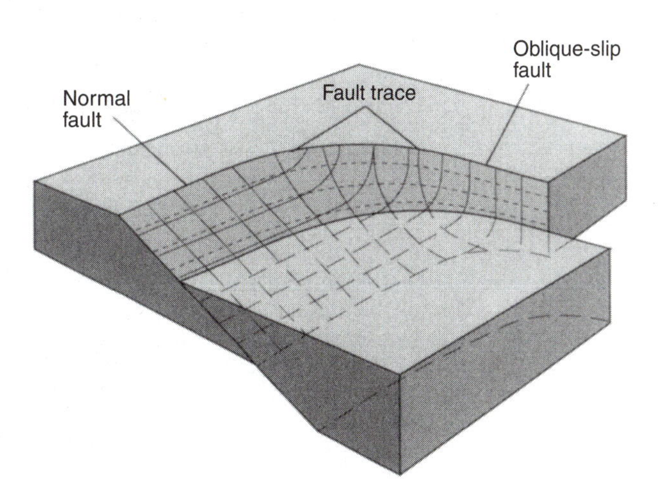

FIGURE 3.32 A change in orientation of a fault surface modifies a normal fault into a vertical oblique-slip fault. No termination line can be identified.

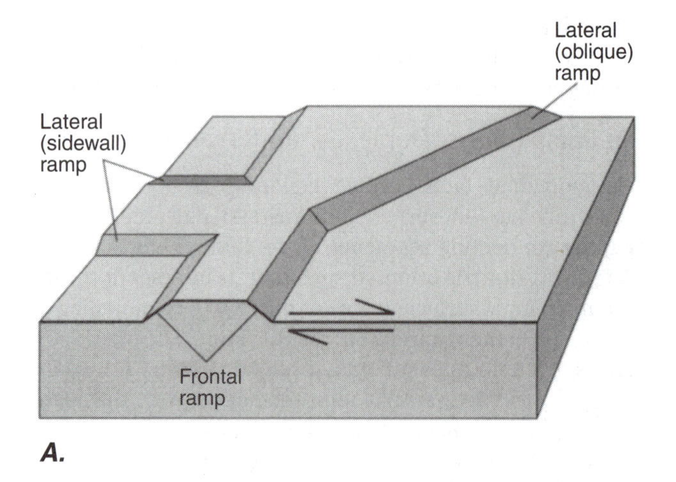

FIGURE 3.33 Geometry of fault ramps. Any of the diagrams can be oriented arbitrarily and therefore can apply to any fault type. *A*. Schematic shape of a fault surface. *B*. An extensional frontal ramp or jog. The dashed arrows indicate that material tends to be pulled apart at the ramp. *C*. A contractional frontal ramp or jog. The dashed arrows indicate that material tends to be pushed together at the ramp.

and strike-slip faults are all common. Because individual faults die out along the fault system, the total displacement across the system is maintained by the transfer, or relay, of slip from the end of one fault to an adjacent parallel fault.

A **transfer zone** is a local structure that accommodates the transfer of slip between two parallel adjacent larger faults. The structure can be a single fault, referred to as a **transfer fault**, also known as a **relay fault**, or it can be a zone of distributed deformation called a **transfer zone** or a **relay zone**. Within a transfer zone, the deformation may be distributed either on a set of parallel faults oblique to the major structures or in a zone of less organized distributed brittle deformation (see the discussion of damage zones in Section 3.4(iv)). The transfer zone may also be a zone of ductile folding (see Figure 5.16). A fault ramp (Figure 3.33) is in essence a transfer fault, or a **relay ramp**, and a fault duplex (Figure 3.34) is in essence a transfer zone. The term *transfer fault*, however, is most commonly applied to strike-slip faults that transfer slip between larger adjacent and parallel normal faults (see Figures 4.11 and 4.12) or thrust faults. Geometrically, however, thrust faults or normal faults can act to transfer slip from one strike-slip fault to an adjacent subparallel strike-slip fault.

iv. Damage Zones

Many faults are depicted in an overly simplified manner as single planes of discontinuity that cut through a body of rock (see Figure 3.1*A*). Any detailed look at a real fault zone, however, reveals a much more complex structure. A brittle fault is generally a zone of anastomosing interlaced fractures, which may be dominated by a zone of cataclastic rocks in which most of the slip has accumulated. On either side of the fault and at the fault tips, however, is a **damage zone** consisting of secondary fractures in a variety of orientations and displaying various senses of slip. This damage zone can be of variable extent, but in some cases it can be considerably wider than the fault zone itself.

Displacement on a fault typically is a maximum near the central part of the fault surface and decreases to a minimum at the fault tips. The changes in displacement along a fault, and the residual displacement at its tip line, must be accommodated in the adjacent rock by distributed deformation. The intensity of deformation in the accommodation zone is greatest near the fault, and it decreases with increasing distance from the fault. Where the accommodation deformation is most intense, however, secondary fracturing may occur, creating the damage zone. The damage may accumulate along the walls of an active fault or at the tips of the fault, but in either case it serves to distribute the accommodation deformation through a larger volume of rock surrounding the

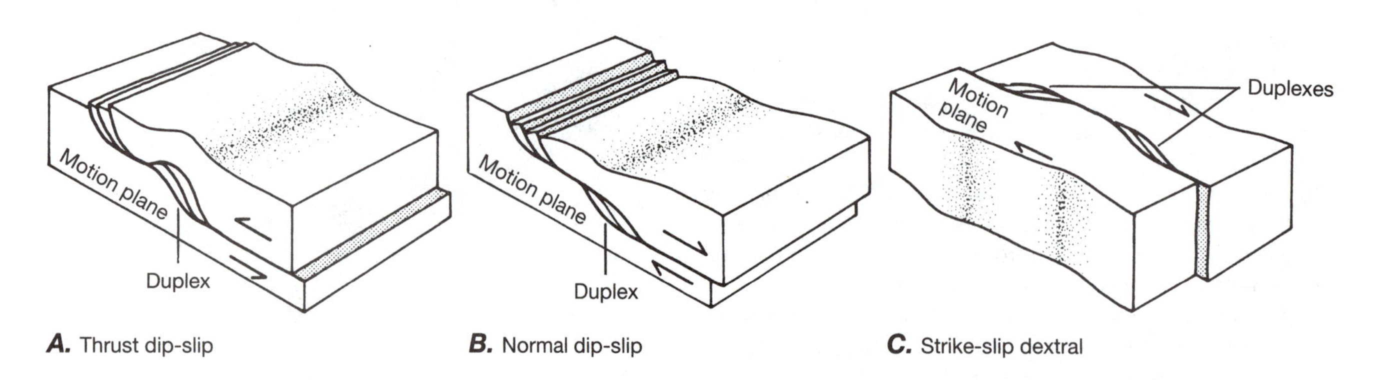

FIGURE 3.34 Block diagram illustrating duplexes. The motion plane contains the slip direction and the line perpendicular to the fault. (Woodcock and Fischer 1986)

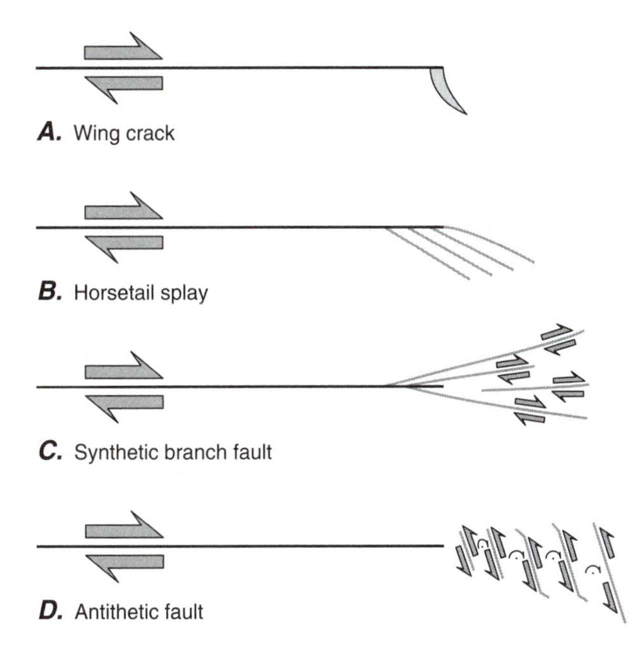

FIGURE 3.35 Damage zones at the tips of faults show a variety of structures. (Kim et al. 2004, with permission from Elsevier) *A*. Wing cracks develop on small shear fractures. *B*. Horsetail splay of secondary "synthetic" faults, each of which has the same sense of shear as the parent fault. *C*. Synthetic branch faults, each with the same shear sense as the parent fault. *D*. Antithetic faults bound blocks that rotate in the same sense as the parent fault, thus giving the bounding faults the opposite sense of shear.

main fault zone, thereby decreasing its intensity on individual faults.

Characteristics of the secondary fracturing at a fault tip depend in part on the orientation of the displacement vector relative to the tip line. Damage zones that develop at mode II tip lines (see Figures 2.1 and 3.30) may occur as a set of extension cracks, splay faults, or branch faults (Figure 3.35). **Wing cracks** (Figure 3.35*A*) are extension fractures typically associated with small amounts of displacement on small fractures. On larger faults, comparable deformation results in pinnate fractures (Figure 2.7) or a **horsetail splay** with synthetic or oblique-synthetic slip on the secondary shears (Figure 3.35*B*). Deformation at a fault tip may be accommodated by a set of **synthetic branch faults** (Figure 3.35*C*), which have the same shear sense as the main fault. The damage zone at a fault tip may also consist of a set of **antithetic faults** (Figure 3.35*D*), which have a shear sense opposite to that of the main fault and which accommodate rotation of blocks, where the rotation is in the same sense as the shear on the main fault. More complex damage zones comprise mixtures of these types of secondary fractures.

If the main fault subsequently propagates through the damage zone, the already-existing damage becomes a feature of the rock on either side of the fault. Damage along the walls of faults also may reflect a distributed shear that extends beyond the main fault itself, or it could reflect deformation required to accommodate irregularities in the main fault surface. Thus damage zones are characteristic features adjacent to any part of a fault, and the secondary fractures that indicate sense of motion on faults, such as are illustrated in Figure 3.17, are themselves part of the damage zone.

Damage also accumulates in a fault step-over, where one fault ends and an adjacent parallel fault begins. The step-over can become a transfer zone within which the deformation accommodates the transfer of slip from the

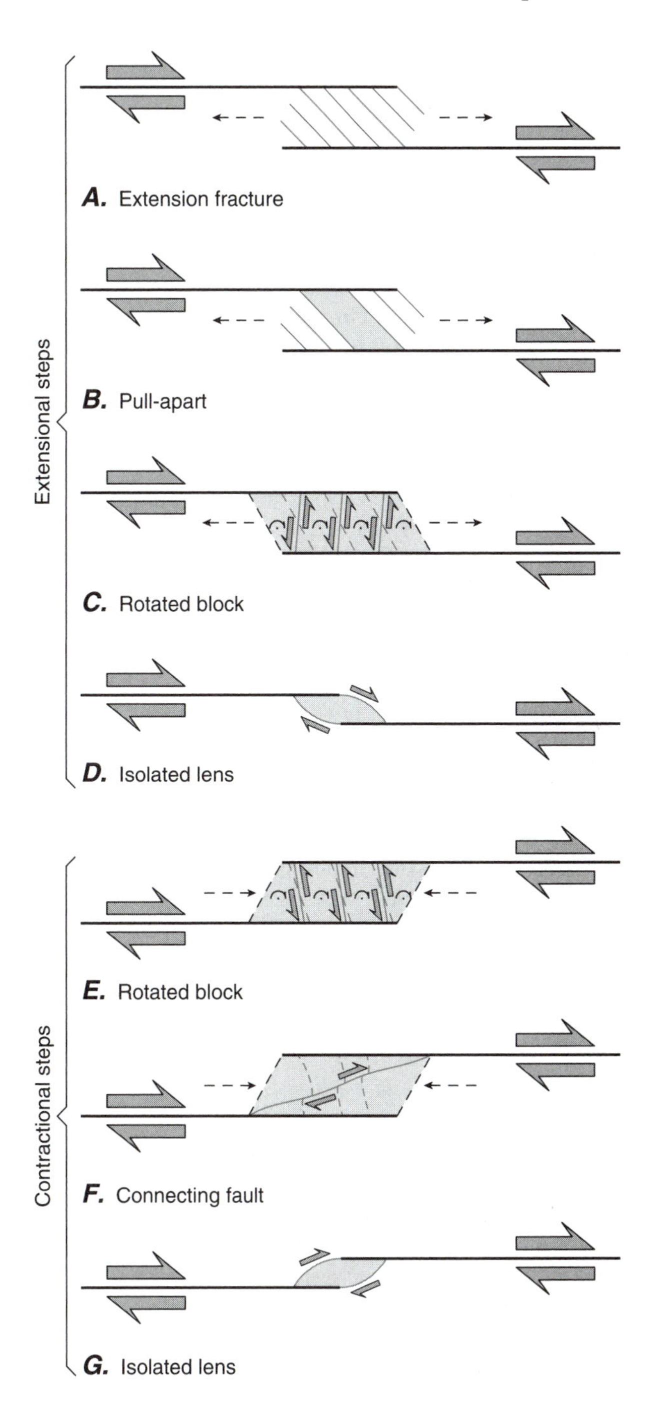

FIGURE 3.36 (*A–D*) Structures of damage zones at extensional step-overs and (*E–G*) contractional step-overs. (Kim et al. 2004, with permission from Elsevier)

end of one fault to the beginning of an adjacent parallel fault (Figure 3.36). A step-over may be either extensional (Figure 3.36*A–D*), if the shear sense on the faults requires the material in the step-over to lengthen, or contractional, if the material in the step-over must shorten (Figure 3.36*E–G*). These zones may subsequently evolve into a duplex structure (Figure 3.34) or in some cases become the locus of high permeability zones that lead to formation of important ore deposits.

The formation of smaller shear fractures in a damage zone around a larger fault are part of the process by which the power-law scaling of fault lengths (Equation (3.2), Section 3.4 (i)) develops.

3.5 BALANCED CROSS SECTIONS

Cross sections in faulted sedimentary sequences that have not undergone significant deformation perpendicular to the plane of the section must preserve the original cross-sectional area of the undeformed section. Such cross sections are called **balanced cross sections**. The requirement that cross-sectional area be preserved places significant constraints on possible interpretations of structure at depth, because any proposed interpretation must make it possible to restore the fault to an acceptable undeformed state characterized by continuous layers with no gaps or overlaps in the stratigraphic section. The cross section must be balanced between two **pinning points**, which are vertical reference lines chosen to pass through undeformed sections of a stratigraphic sequence. Thus the shape of these reference lines is assumed to have been unchanged by the deformation.

If the deformation has not changed the thicknesses of units in the stratigraphic section, then balancing the section can be achieved by line balancing, which requires that the length of each contact between the pinning points be the same before and after deformation. If deformation has shortened and thickened units within the plane of the section, however, the lengths of contact lines can change, and balancing must be done by area balancing, which requires that the area of each unit between pinning points be conserved during deformation.

Although the balancing requirement imposes an important constraint on the construction of cross sections of unmetamorphosed sedimentary rocks, in many geologic situations there may be significant deformation normal to any possible cross section, for example, along a strike-slip fault perpendicular to the section. Under such conditions, material moves in or out of the cross section plane, and the fundamental assumption of constant cross-sectional area, which is the basis for constructing balanced cross sections, is no longer valid. Some three-dimensional techniques have been developed to account roughly for this deformation, but the process requires large amounts of detailed information, and balancing cross sections under these conditions becomes an increasingly complicated and approximate procedure.

3.6 SUMMARY AND PREVIEW

In this chapter, we have described many features common to faults in general, but we have limited our discussion to individual fault surfaces or to simple sets of faults. In the next three chapters, we complete our description of faults, dealing separately with each of the major types of faults. We concentrate on those features that are unique to each fault type, including the structures associated with the different faults, the geometry of complicated fault systems, and the tectonic settings in which the different types of faults are found. Despite the differences in tectonic setting of the three major types of faults, the geometric characteristics discussed in this chapter are remarkably consistent for all fault systems. The differences that do exist are largely attributable to the difference in orientation of the faults with respect to the Earth's surface.

REFERENCES AND ADDITIONAL READINGS

Addicott, W. O. 1968. In W. R. Dickinson and A. Grantz, eds. *Proceedings of the Conference on Geologic Problems of San Andreas Fault System (1967: Stanford University)*. Stanford University Publications.Geological Sciences, v. 11, Stanford Calif., School of Earth Sciences, Stanford University.

Angelier, J. 1994. Fault slip analysis and paleostress reconstruction. In P. L. Hancock, ed. *Continental Deformation*. Pergamon Press, Oxford: 53–100.

Clark, R. M., and S. J. D. Cox. 1996. A modern regression approach to determining fault displacement–length scaling relationships. *J. Struct. Geol.* 18(2/3): 147–152.

Clifton, A. E., R. W. Schlische, M.O. Withjack, and R. V. Ackermann. 2000. Influence of rift obliquity on fault-population systematics: Results of experimental clay models. *J. Struct. Geol.* 22: 1491–1509.

Doblas, M., V. Mahecha, M. Hoyos, and J. Lopez-Ruiz. 1997. Slickenside and fault surface kinematic indicators on active normal faults on the Alpine Betic Cordilleras, Granada, southern Spain. *J. Struct. Geol.* 19(2): 159–170.

Fleuty, M. J. 1975. Slickensides and slickenlines. *Geol. Mag.* 112: 319–322.

Groshong, R. H. Jr. 1994. Area balance, depth to detachment, and strain in extension. *Tectonics* 13(6): 1488–1497.

Kim, Y.-S., D. C. P. Peacock, and D. J. Sanderson. 2004. Fault damage zones. *J. Struct. Geol.* 26(3): 503–517.

Mount, V. S; J. Suppe, and S. C. Hook. 1990. A forward modeling strategy for balancing cross sections. *Bull. Amer. Assoc. of Petrol. Geol.* 74(5): 521–531.

Petit, J. P. 1987. Criteria for the sense of movement on fault surfaces in brittle rocks. *J. Struct. Geol.* 9: 597–608.

Plafker, G. 1965. Tectonic deformation associated with the 1964 Alaska earthquake. *Science* 148(3678): 1675–1687.

Sammis, C. G., and R. L. Biegel. 1989. Fractals, fault-gouge, and friction. *Pure and Appl. Geophys.* 131: 255-271.

Sammis, C. G; and S. J. Steacy. 1995. Fractal fragmentation in crustal shear zones. In C. C. Barton and P. R. La Pointe, eds., *Fractals in the Earth Sciences.* Plenum Press, New York: 179–204.

Sibson, R. H. 1977. Fault rocks and fault mechanisms. *J. Geol. Soc. Lond.* 133: 190–213.

Sibson, R. H. 1983. Continental fault structure and the shallow earthquake source. *J. Geol. Soc. Lond.* 140: 741–767.

Simpson, C. 1986. Determination of movement sense in mylonites. *J. Geol. Educ.* 34: 246–261.

Simpson, C., and S. M. Schmidt. 1983. An evaluation of criteria to deduce the sense of movement in sheared rocks. *Geol. Soc. Amer. Bull.* 94: 1281–1288.

Waterson, J., J. J. Walsh, P. A. Gillespie, and S. Easton. 1996. Scaling systematics of fault sizes on a large-scale range fault map *J. Struct. Geol.* 18(2/3): 199–214.

Wheeler, J., and R. W. H. Butler. 1994. Criteria for identifying structures related to true crustal extension in orogens. *J. Struct. Geol.* 16: 1023–1027.

Wise, D. U., D. E. Dunn, J. T. Engelder, P. A. Geiser, R. D. Hatcher, S. A. Kish, A. L. Odom, and S. Schamel. 1984. Fault-related rocks: Suggestions for terminology. *Geology* 12: 391–394.

Woodcock, N. J., and M. Fischer. 1986. Strike-slip duplexes. *J. Struct. Geol.* 8: 725–735.

Woodward, N. B., S. E. Boyer, and J. Suppe. 1989. Balanced geological cross-sections: An essential technique in geological research and exploration. *Short Course in Geology*, v.6. American Geophysical Union, Washington, DC.

Chapter 4

NORMAL FAULTS

Normal faults[1] (Figure 4.1) are inclined dip-slip faults along which the hanging wall block has moved down with respect to the footwall block. Generally, they emplace younger rocks on top of older rocks, and in a vertical section through the fault, stratigraphic section is missing. Most normal faults have steep dips of about 60°, but many have lower dips, some approaching horizontal. As a result of the hanging-wall-down motion, normal faults accommodate a lengthening, or extension, of the Earth's crust.

4.1 CHARACTERISTICS OF NORMAL FAULTING

i. Separation and Normal Faulting

The separations produced by normal faulting parallel to the strike and the dip depend on the relative orientations of the fault and the stratigraphic layering. As noted in Section 3.3(v), separation can be quite misleading as an indication of the nature of a fault. For example, Figure 4.2 shows a series of block diagrams of a normal fault cutting various orientations of bedding, none of which is overturned. In all the diagrams on the left, pure normal dip-slip motion on the fault displaces the stratigraphy and produces a scarp on the footwall block. Each diagram on the right shows the same geometry as the diagram to its left except that the fault scarp has been eroded away, leaving a horizontal planar surface.

In Figure 4.2*A*, the fault cuts horizontal beds, leaving a simple stratigraphic discontinuity. In Figure 4.2*B–F*, the bedding is inclined at various angles to the fault, resulting in some potentially confusing separations. In Figure 4.2*B*, for example, on a horizontal plane the stratigraphy is repeated across the fault, although on the vertical section, stratigraphy is missing. This example emphasizes the fact that for a normal fault, the characteristic of missing stratigraphic section only applies to a vertical section through the fault. If, however, on a vertical section normal to the strike of the fault the apparent dip of the beds is in the same direction as, but steeper than, the fault (Figure 4.2*C*), a vertical hole through the fault would reveal *repeated* stratigraphy, and if the beds are not overturned, the normal fault places older beds on top of younger ones. Thus this is an exception to the usual relationship for normal faults of missing stratigraphy in a vertical section and of younger-over-older relationships across the fault.

In Figure 4.2*D* and *E*, the stratigraphic pattern is characterized by strike separation on the horizontal plane and normal dip separation on a vertical section. In *D*, the strike separation is right lateral, and in *E* it is left lateral, even though both faults have identical, pure normal dip-slip displacement.

In Figure 4.2*F*, the fault and the displacement are both parallel to the bedding, leaving no separation visible on any bedding plane. In fact, for any situation, including all those shown in Figure 4.2, any component of displacement parallel to the cutoff line of the bedding on the fault surface produces no effect on the separation in any

[1]The terms *normal* and *reverse* as applied to faults stem from eighteenth- and nineteenth-century mining. The "normal" situation in the coal mines of Britain was that the coal seams in the hanging wall block moved down on the fault relative to the same seam in the footwall block. If the hanging wall block had moved up, the "reverse" of the normal situation existed.

A.

B.

FIGURE 4.1 Normal faults. *A*. Small-scale normal faults, including examples with opposite directions of dip in beds of volcanic ash exposed in a road cut approximately 5 meters high near Klamath Falls, Oregon. (Photograph by N. Lindsley-Griffin) *B*. A normal fault bounds the east side of the Stillwater Range in Nevada, separating the rugged topography of the range from the flat valley floor. Notice the sinuous trace of the fault and the faceted spurs, along the length of the fault. (Photograph by J. Stewart)

section through the fault (compare Figure 3.23). Thus, because of the indeterminate magnitude of this component of displacement, none of the patterns of separation in Figure 4.2 is unique to normal dip-slip motion on the fault.

A surface across which the metamorphic grade of the rocks changes abruptly from high-grade rocks below to lower-grade or unmetamorphosed rocks above may be a normal fault. The cutting out of metamorphic grades is comparable geometrically to the cutting out of stratigraphy.

ii. Folds Associated with Normal Faults

In areas where flat-lying beds are deformed by normal faults, rollover folds in the hanging wall block are common, illustrated by the deeper strata in Figure 4.3 (compare Figure 3.11*D*). In these folds, the beds in the hanging wall block tilt down into the fault, which is opposite to the direction of tilt on drag folds (compare Figure 3.11*B*). They form on **listric**[2] **normal faults**, which are concave upward faults, or fault surfaces whose dip decreases with increasing depth (see Section 4.2). As the hanging wall block slips on the fault, it deforms to maintain contact with the footwall block across the fault, thereby producing a bend in the layering (see Figure 4.19).

On normal faults, drag folds tend to be smaller-scale features than rollover folds and may be less common. In these folds, the beds in the hanging wall block tilt up against the fault, and those in the footwall block tilt down into the fault, as shown by the shallow strata in Figure 4.3 and the small-scale folds against the eastern fault of Railroad Valley in Figure 4.6. Where they can be definitely recognized, they indicate the sense of relative displacement across the fault.

Folds may also develop during normal faulting if the fault has ramps or flats at depth, as we discuss below (see Figure 4.5 and the associated discussion).

iii. Features of Fault Surfaces

Like all faults, normal faults exist at all levels in the crust. Fault surface features are variable and depend on the shape of the fault, the depth at which movement on the fault occurred, and whether faulting was accommodated by brittle fracture and frictional sliding, or ductile deformation.

Like all faults at shallow levels, normal faults develop cataclastic rocks (see Figure 3.5), slickensides, and slickenside lineations (Figure 3.8) along their surfaces. Some gently dipping normal faults in regions such as the Basin and Range province of the United States are characterized by large thicknesses of breccia and megabreccia (see Figure 3.5*A*). In many cases, such breccias develop in hanging wall blocks from the pervasive fracturing and in-

[2]After the Greek word *listron*, meaning "shovel."

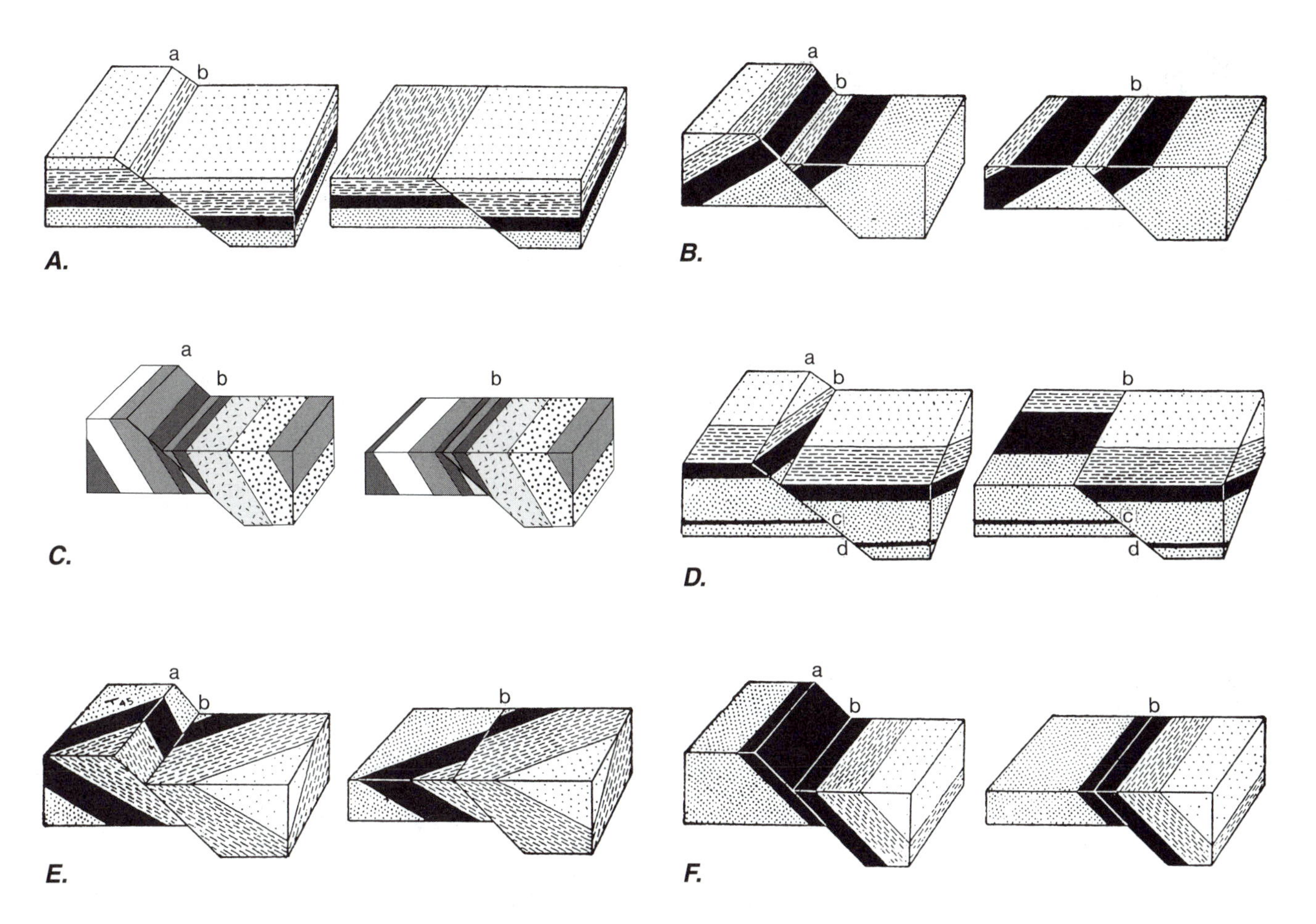

FIGURE 4.2 Separations of stratigraphy created by dip-slip normal faults cutting different attitudes of bedding. In the diagrams on the left, the fault blocks have been displaced, leaving scarps on the footwall block. In the diagrams on the right, the scarp has been eroded down to a level even with the hanging wall block. (*A*, *B*, and *D–F* after Billings 1954)

ternal faulting that accommodate brittle, layer-parallel extension above shallow normal faults. In other cases, they may be associated with large low-angle landslides.

At deeper structural levels, normal faults develop features associated with ductile deformation, including mylonitic textures (see Figure 3.7), which may be present in shear zones tens to hundreds of meters in thickness.

4.2 SHAPE AND DISPLACEMENT OF NORMAL FAULTS

i. The Shape of the Surface Trace

The sharp planar discontinuities that we commonly use to represent normal faults, such as those in Figure 4.2, are idealized representations of the structures we actually observe in nature. The surface trace of a normal fault is generally not a straight line, but instead may be a sinuous curve or a series of connected, roughly-straight, line segments (Figure 4.1*B*; see also Figure 4.12 and Section 3.4(iii)). Although surface trace irregularities may result in part from intersection of inclined faults with irregular topography, in many places the faults themselves must be nonplanar surfaces.

ii. The Shape at Depth

Normal faults need not maintain a constant dip with increasing depth, as is shown in Figure 4.3. Some listric normal faults turn into or join a detachment fault at depth (Figure 4.4). A **detachment fault** is a low-angle fault that marks a major boundary between unfaulted rocks below and a hanging wall block above that is commonly deformed and faulted. Normal faults in the hanging wall block may form a set of **imbricate faults**, which are closely spaced parallel faults of the same type that either terminate against the detachment fault (left end of Figure 4.4) or merge with it (right end of Figure 4.4).

Lateral ramps on the fault show up at the surface as an irregular fault trace (Figure 4.1*B*), and frontal ramps may also occur (Figures 3.33*A*, 4.5). Thus a map or a cross section alone does not provide a complete picture of the geometry of a normal fault at depth, and the potential complications of the third dimension must be kept in mind when interpreting a simple two-dimensional view.

FIGURE 4.3 Seismic reflection profile of a listric normal fault. The hanging wall block is cut by a set of subsidiary "synthetic" normal faults, which are faults having the same dip direction and sense of shear as the major fault. Note the small-scale drag of the layers in both blocks along the steep part of the fault and the larger-scale rollover fold in the layers of the hanging wall block along the shallowly dipping parts. (After A.G. Wintershalle in Wernicke and Burchfiel 1982) *A.* Seismic reflection profile. *B.* Interpretation of bedding and faulting on the seismic section.

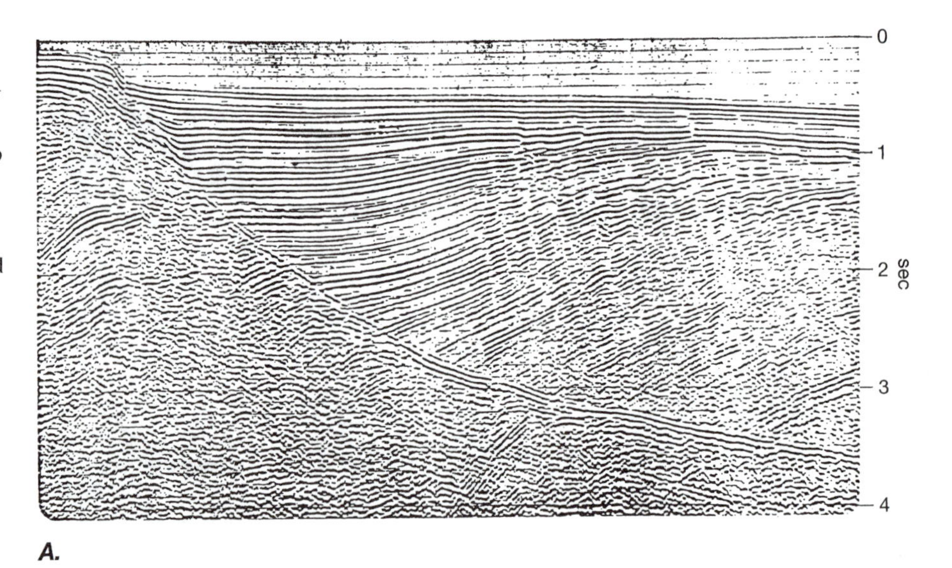

B.

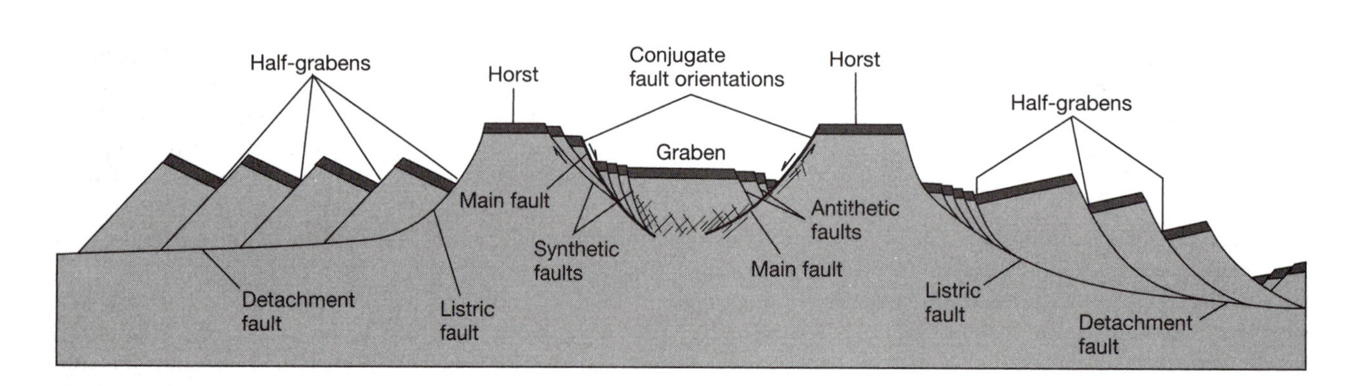

FIGURE 4.4 Systems of normal faults commonly are characterized by a main fault with associated subsidiary faults and by low-angle detachment faults with imbricate fault blocks in the hanging wall block.

iii. Displacement on Normal Faults

By definition, displacement on ideal normal faults is parallel to the dip of the fault surface. If the strike of the fault varies, however, rigid movement of the hanging wall block relative to the footwall block cannot everywhere be down the dip of the fault. This fact is illustrated in Figure 3.32, in which, as the fault trace curves, the displacement on the fault varies from pure normal dip slip to oblique slip. Thus the complex shape of real faults requires that they depart from our idealized models.

Movement on normal faults can be either nonrotational or rotational, depending on whether the orientations of the fault blocks remain constant or change as a result of the faulting. Both conditions exist in nature. If the apparent dip of the fault measured in the direction of the displacement does not change with depth (for example, in Figures 4.2, 3.32), and if the fault itself is not rotated during faulting, then in general the orientations of the fault blocks do not change during slip. Horizontal beds in the blocks remain horizontal; inclined beds maintain the same strike and dip, although rotational (scissor) faults (Figure 3.3*G*) are an exception to this rule. If, however, the dip changes with depth (e.g., the fault is listric or has a ramp-flat geometry), then slip must result in rotation or deformation of the hanging wall block. Ideally, the rotation takes place about an axis parallel to the strike of the fault plane, and originally horizontal bedding ends up dipping toward the fault on which the fault block rotates. The angle between the bedding and the fault remains constant (see right and left ends of Figure 4.4).

A hanging wall block moving over a fault with ramp-flat geometry must in general deform internally. If a ramp connects two more shallowly dipping segments of the fault, slip on the fault produces a fault-ramp syncline (Figure 4.5*A*). If a flat connects two more steeply dipping segments of the fault, slip produces a fault-bend anticline (Figure 4.5*B*), which is comparable in part to a rollover anticline. These folds must parallel the associated ramp or flat, whether it intersects the main fault surface in a line perpendicular or oblique to the displacement direction. Deformation in the hanging wall block may take place by ductile bending (Figure 4.5), distributed faulting that may be either synthetic to the main fault (Figures 4.3*B*, 4.4, left side of the "graben") or antithetic to it (Figure 4.4, right side of the "graben"; see also Figure 4.19*D*), or a combination of both.

4.3 STRUCTURAL ASSOCIATIONS OF NORMAL FAULTS

Normal faults generally are present as systems of many associated faults. In many cases, the orientations of the faults fall into two groups, which are referred to as conjugate orientations; they have comparable dip angles but

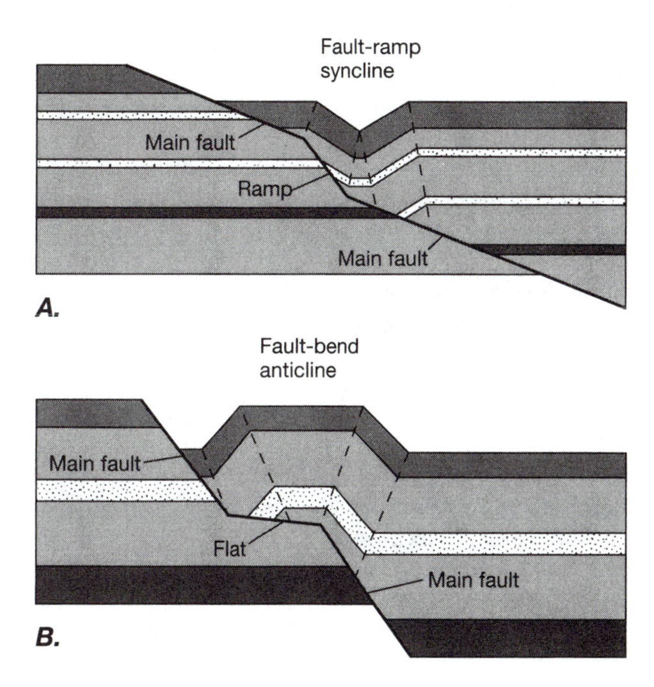

FIGURE 4.5 Displacement on normal faults with a ramp-flat geometry. *A*. A fault-ramp syncline. *B*. A fault-bend anticline.

opposite dip directions and opposite senses of shear (conjugate faults bound the graben in Figure 4.4).

Commonly in such systems, some of the faults have a major amount of displacement along them and accommodate the major deformation, whereas other faults have a relatively small amount of displacement and provide the minor adjustments required for the large-scale displacements. If the smaller-scale faults are parallel to the major fault and have the same sense of shear, they are **synthetic faults**; if they are in the conjugate orientation, that is, if they have comparable dips but in the opposite dip direction from the main faults, they are **antithetic faults** (Figure 4.4, opposite sides of the "graben").

A **graben**[3] is a down-dropped block bounded on both sides by conjugate normal faults that dip toward the down-dropped block on both sides (Figure 4.4). A **half-graben** is a down-dropped tilted block bounded on only one side by a major normal fault. A **horst**[4] is a relatively uplifted block bounded by two conjugate normal faults that dip away from the uplifted block on both sides. These terms may refer either to the topographic feature formed by the faulting or to the structural feature of relatively down-dropped or uplifted fault blocks (Figure 4.1*A*). Alternating uplifted and down-dropped fault blocks are called a **horst-and-graben structure**. In some cases, a series of half-grabens result in tilted fault blocks that

[3]After the German word *grabe*, meaning "ditch."

[4]After the German word *horst* meaning "a retreat" or "aerie," the nest of a bird of prey, typically built on a high cliff.

also form alternating topographic highs and lows (Figure 4.4, left and right sides).

In regions of active normal faulting, horsts and the higher ends of tilted fault blocks provide the sediments that accumulate in the basins formed by the grabens and the lower ends of tilted blocks. As faulting continues during deposition of the basin sediments, the sediments themselves often become involved in the faulting. Study of these faults and the age, composition, thickness, and distribution of the sediments can reveal when the major periods of uplift occurred, as well as the sequence and the time of exposure of the different rock types in the uplifted fault blocks, known as the **unroofing sequence**.

Railroad Valley (Figure 4.6) in east-central Nevada illustrates the nature of an individual graben. This structure clearly is down-dropped on both sides, though more so on the east. It has a valley fill of late Tertiary and Quaternary sediments approximately 6 km thick that includes coarse to very coarse alluvial deposits shed from the surrounding ranges, playa lake sediments, and landslide deposits. These sediments record the interplay between faulting and concurrent sedimentation.

Systems of normal faults exist either on a local scale, subsidiary to other structures, or on a regional scale, where they dominate the structure. We consider each scale of structure briefly here.

i. Local Normal Faults Associated with Other Structures

Normal fault systems of local extent are generally associated with other structures whose geometry requires extension of crustal layers. Examples include domes, folds, cavities, and pull-apart structures on strike-slip faults.

Structural domes cut by a system of normal faults commonly result from the intrusion of bodies of salt or magma. The faults radiate from the center of the dome and may include a single major fault, one or two grabens (Figure 4.7*A*), or a Y-shaped set of grabens. The displacement on the faults is greatest at the center and dies

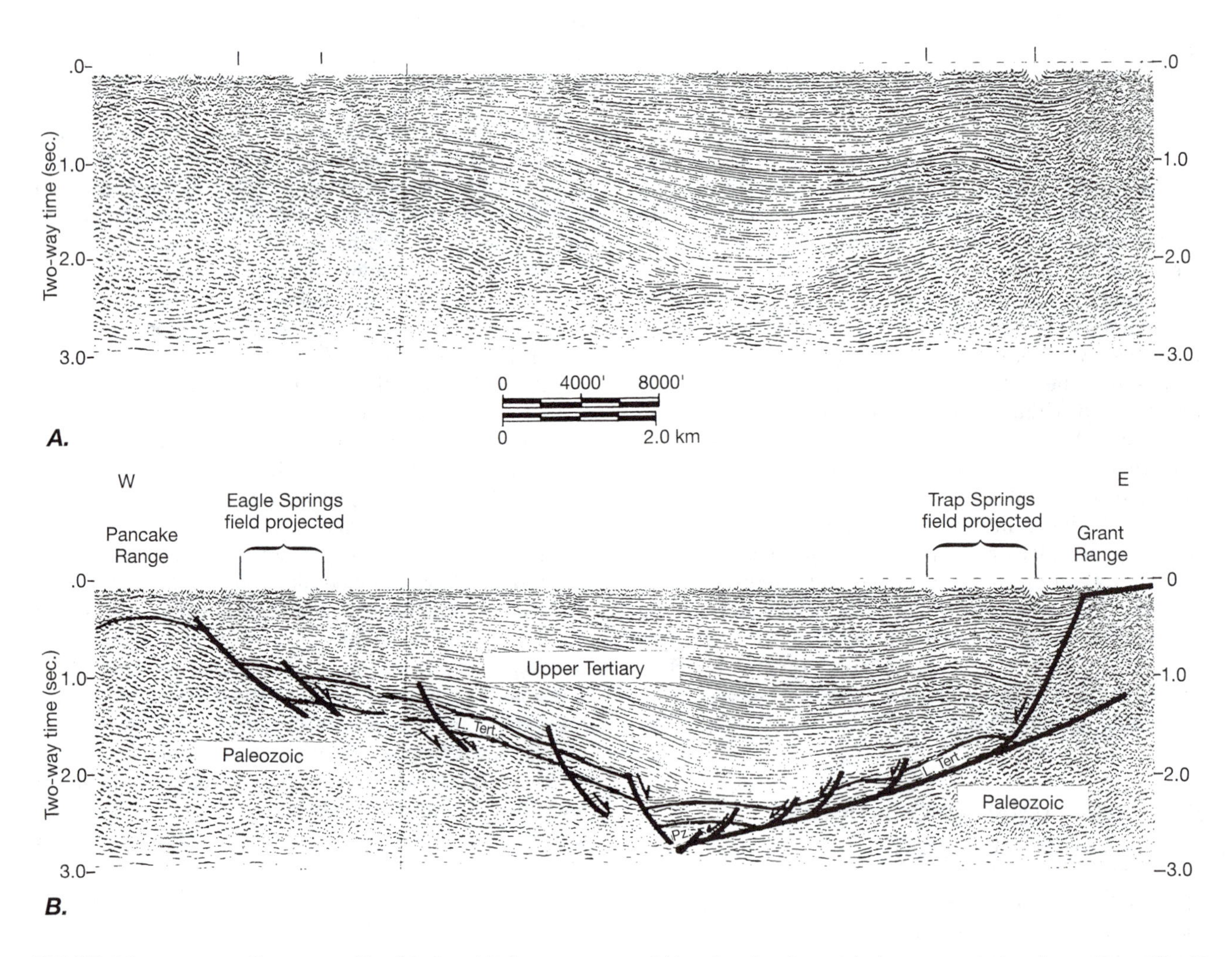

FIGURE 4.6 Seismic reflection profile of Railroad Valley, east-central Nevada, showing typical asymmetrical graben. (From Effimoff and Pinezich 1986) *A*. Seismic reflection profile. *B*. Interpretation of the faults on the seismic section.

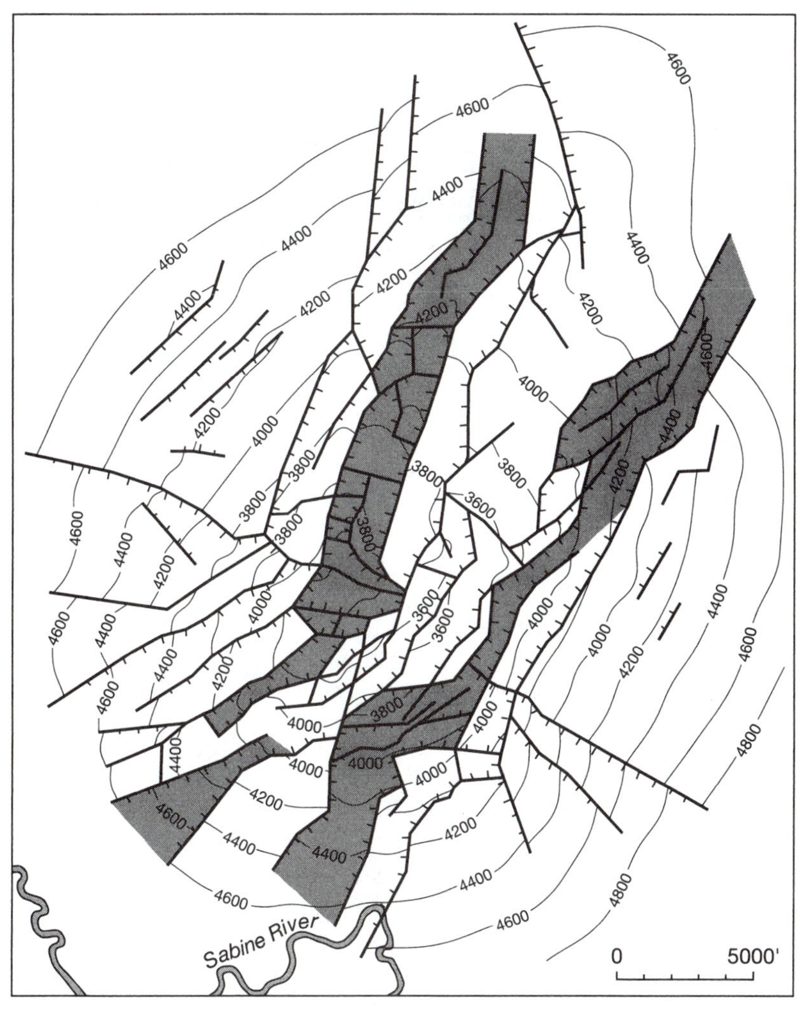

A.

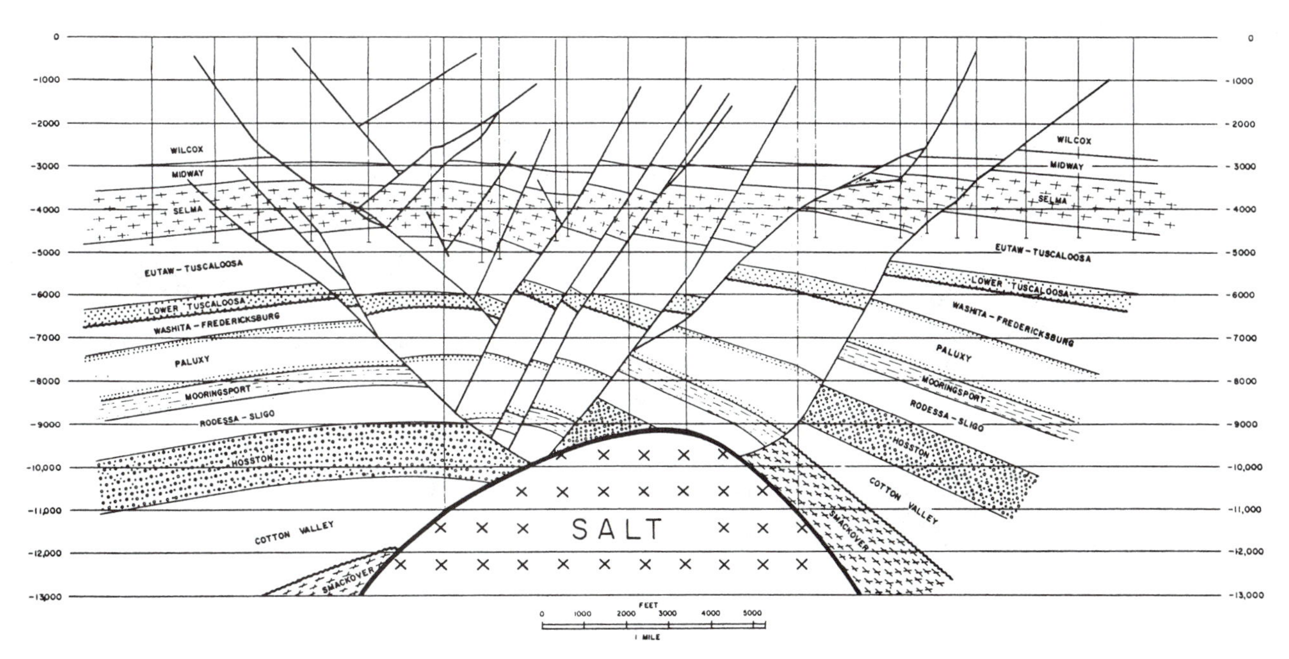

B.

FIGURE 4.7 Normal faults over structural domes. *A.* The uplift over a salt dome transected by a pair of grabens with radial faults. The central grabens are shaded for emphasis. (After Wentlandt 1951) *B.* Schematic east-west cross section of a graben and associated faults over a salt dome. (After Cloos 1968)

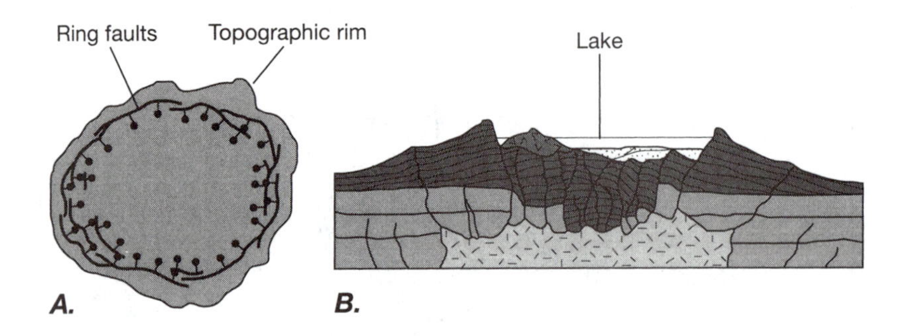

FIGURE 4.8 Normal faults associated with a caldera, a volcanic collapse structure. *A*. Schematic map of ring faults around a caldera at Crater Lake, Oregon. *B*. Schematic cross section of the caldera structure at Crater Lake, Oregon. (After Press and Siever 1988)

out at tip lines near the margin of the dome. At depth, the main faults terminate at or near the margin of the intrusive (Figure 4.7*B*); subsidiary faults may join or terminate against other faults. Elongate domes, described as doubly plunging anticlinal folds, commonly exhibit a comparable pattern of normal faults.

If a cavity forms at depth, surficial rocks commonly collapse into it along a set of concentric normal faults, forming a system of **ring faults**. Examples of such structures include calderas, which form by the collapse of surficial rocks into a magma chamber emptied during an explosive eruption (Figure 4.8), **diatremes**, which are volcanic pipes explosively blasted through crustal rocks, or collapse structures that result from the dissolution of limestone, salt, or gypsum at depth. Individual faults are not continuous around the circumference of such structures. Where one fault dies out, however, the displacement associated with the collapse is taken up by adjacent faults, thereby forming concentric rings of discontinuous normal faults (Figure 4.8*A*). At depth, the major faults must terminate at the cavity boundary (Figure 4.8*B*).

ii. Regional Systems of Normal Faults

Regional systems of normal faults define large distinct structural provinces in many parts of the world (Figure 4.9). Continental examples of such provinces include the Basin and Range province in western North America (see Figures 4.10, 4.12), the East African Rift region, the western Turkey–Aegean Sea region, and the Shaanxi Graben in China, all of which are currently active. Inactive continental provinces of normal faulting include the Triassic-Jurassic–age graben system of the eastern United

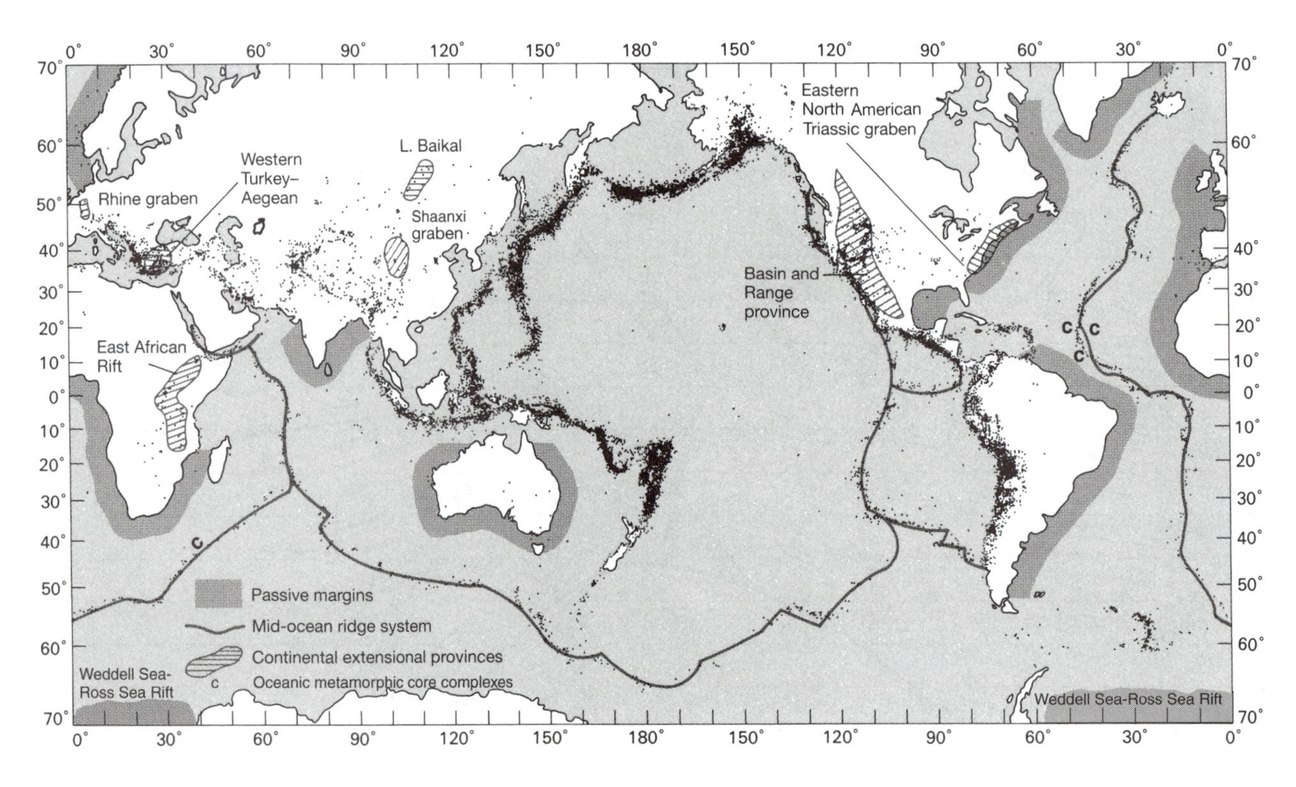

FIGURE 4.9 Map of the world showing regions dominated by extensional normal faulting. (Modified after Uyeda 1977)

States, the Tertiary Rhine Graben system in western Europe, and the Tertiary-Quaternary(?) Weddell Sea–Ross Sea rifted region of Antarctica. Systems of normal faults in basement rocks also characterize rifted, and now passive, continental margins such as along eastern North America, western Europe, western Africa, eastern South America, southern India, and western and southern Australia (Figure 4.9). The sedimentary cover rocks of many such margins also contain systems of active normal faults not necessarily associated with the underlying basement systems, such as in the Gulf Coast province of the southern United States (see Figure 4.16). In the oceanic crust, the midoceanic ridge system constitutes a world-encircling extensional province characterized by active normal faulting, which therefore characterizes the inactive structure of most of the ocean basins.

In all of these regions, normal faults form in conjugate sets that have faults with approximately the same strike, but with dips of varying magnitude in opposite directions (Figures 4.6, 4.16). The faults are always of limited length (Figure 4.10). Where one fault dies out at a tip line, regional extension is taken up by displacement

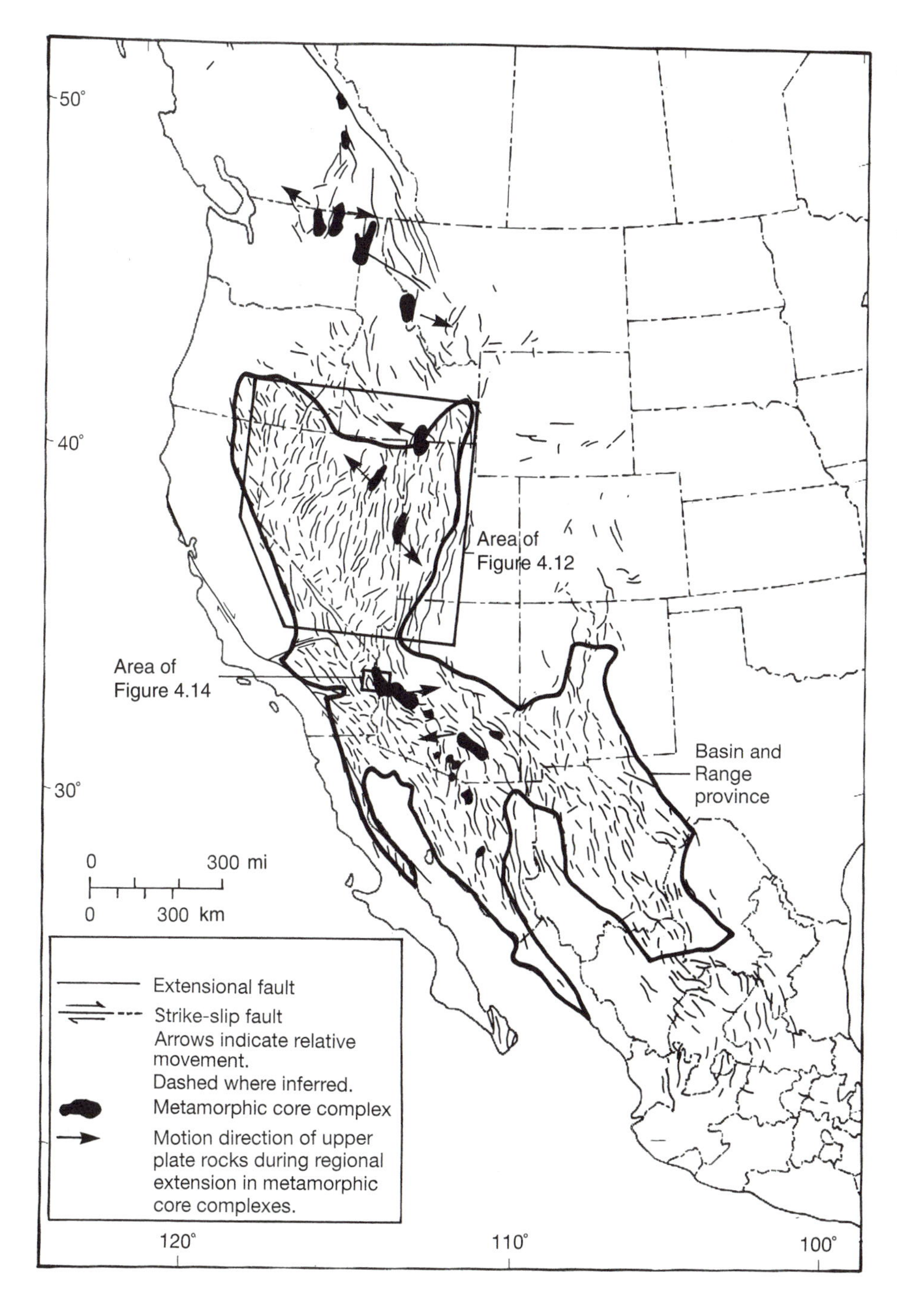

FIGURE 4.10 The extensional province of the North American Cordilera, showing the boundaries of the Basin and Range province (bold outline) and the distribution of metamorphic core complexes (areas in black). (After Coney 1980; Wust 1986; Stewart 1978)

on adjacent faults. Between these faults is commonly a **transfer zone** or **relay zone** within which deformation is accommodated by folding, faulting, and fracturing. In some cases, these transfer zones may be distinct strike-slip **transfer faults**. Transfer zones or faults may divide the extensional province into domains distinguished by different amounts of extension, different predominant orientation of faults, or different predominant directions of tilting. A schematic model of the geometry is shown in Figure 4.11.

Many rifted passive continental margins in the world originated as extensional terranes during the plate tectonic breakup of continental masses. Beneath layers of younger sediments, these margins are characterized by systems of normal faults with geometries similar to that shown schematically in Figure 4.11.

In the Great Basin area of the Basin and Range province, several strike-slip faults have been recognized that cut across the dominant normal fault system at high angles. Both dextral and sinistral faults occur. Other major discontinuities at a high angle to the dominant structure have been recognized as the boundaries of domains of similar dip of strata or similar tilt of the fault blocks, as structural discontinuities within domains of similar dip or tilt, or as boundaries of paleontologic associations (Figure 4.12). The direction of tilting of fault blocks tends to be consistent over large areas (Figure 4.12), which suggests a structural association at depth and requires some discontinuity between major domains. These boundaries may be transfer faults of the type shown in Figure 4.11.

iii. Metamorphic Core Complexes

In cases of extreme extension, normal faulting effectively strips off the shallower layers of rock to expose rocks that originally were deeper in the crust. This process allows us to examine rocks that were deep enough to undergo ductile faulting. In the Basin and Range province of western North America there are numerous regions called **metamorphic core complexes**, where the crust has been extended in a roughly east-west direction (Figure 4.10) by amounts on the order of 100% to 400% on major detachment faults. As a result, the metamorphic and plutonic rocks that lie beneath the detachment faults have been brought up to the surface from depths as great as 20 kilometers. The faults are characterized by extensive development of mylonite, commonly with an overprint of cataclastic deformation (see Section 3.2(i)).

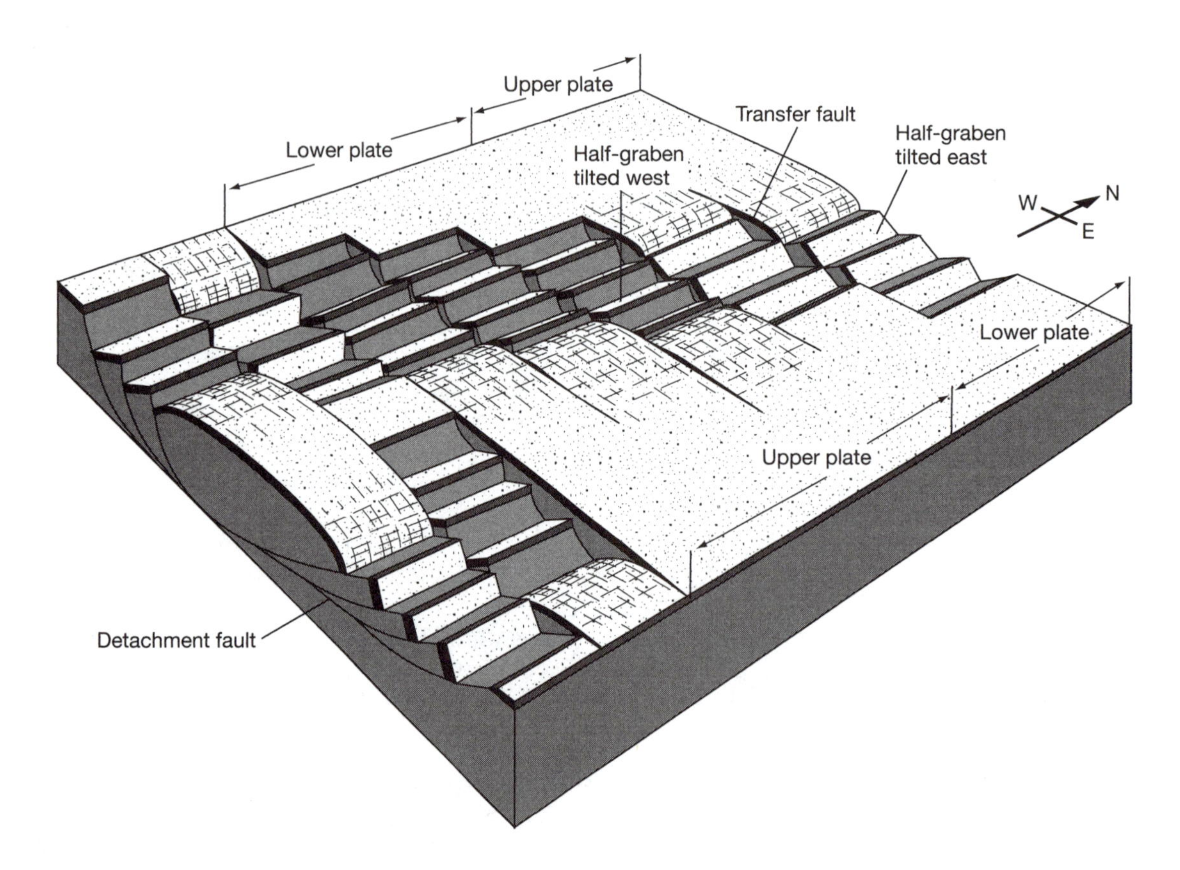

FIGURE 4.11 Model of the fault geometry in basement rocks of a continental extensional province. Different domains of normal faulting are separated by transfer faults. Some domains, such as the two on the left, may contain sets of oppositely dipping normal faults separated by an unfaulted block. (Modified after Lister et al. 1986)

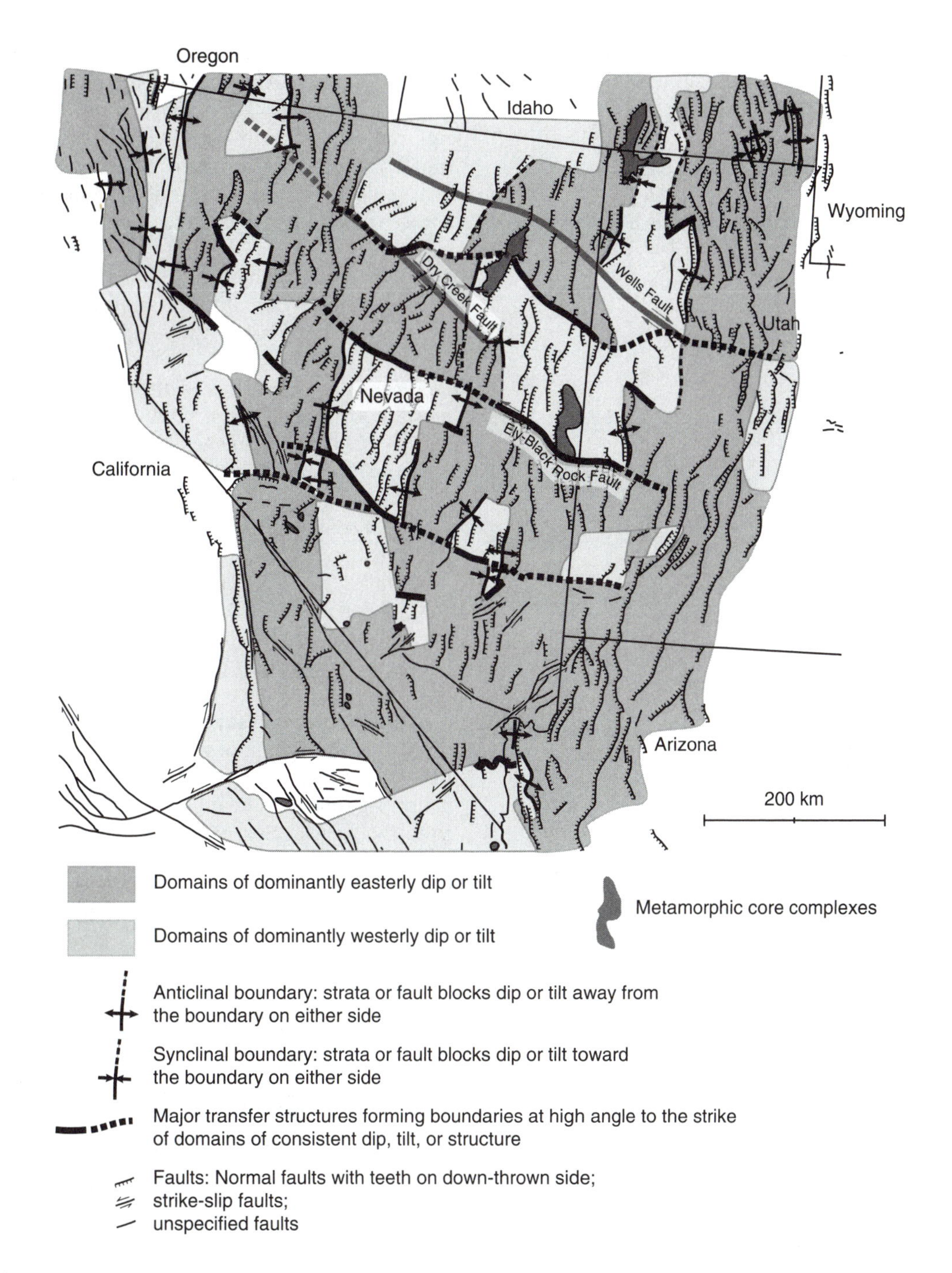

FIGURE 4.12 Structure of the Great Basin in the Basin and Range province of Nevada and neighboring states, showing the major faults, the domains of predominantly westerly or easterly dip of strata or tilt of major fault blocks, and the transfer zones and anticlinal and synclinal axes that bound these domains. (Modified after Faulds and Stewart 1998; Stewart 1978) Other strike-slip faults in northeastern Nevada (Wells, Dry Creek, and Ely–Black Rock faults) have been identified by the offset of stratigraphic and structural trends. (After Thorman and Ketner 1979) The Ely–Black Rock fault is similar in part to a transfer fault identified by Faulds and Stewart. (Figure location is shown in Figure 4.10.)

Figure 4.13 shows schematically one model for the development of such a complex. In some cases, rotations of brittlely faulted blocks can reach 90° or more. In the Whipple Mountains of southeastern California, for example (Figure 4.14), the rocks beneath the detachment fault are extensively mylonitized and have a gently dipping foliation. The detachment fault itself contains mylonitic rocks, which in turn have been deformed by cataclasis, reflecting the change from ductile to brittle deformation as normal faulting brought the deeper

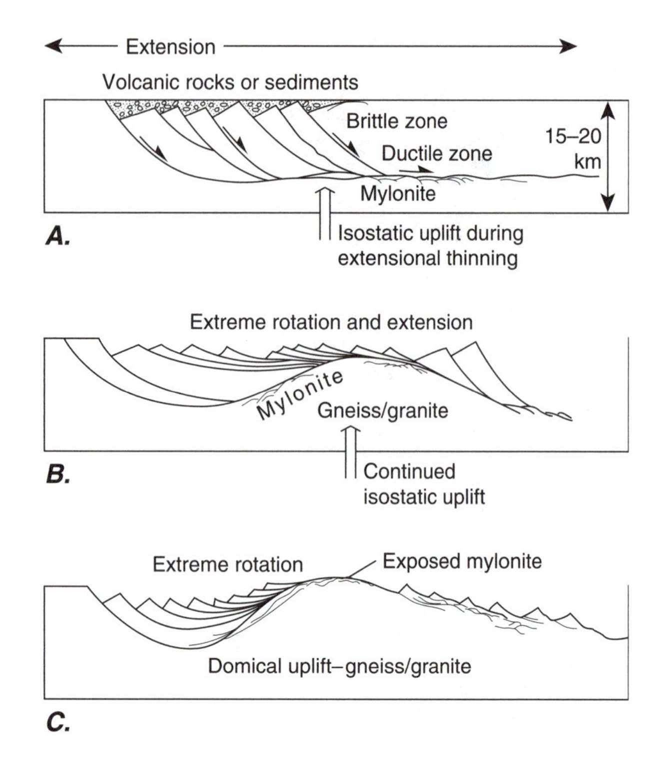

FIGURE 4.13 Diagrammatic cross section illustrating possible origin of a metamorphic core complex. *A.* Formation of listric normal faults with a ductile zone in midcrust. *B.* Structure after extension, with rotation and isostatic uplift in the most extended region where crust has been extremely thinned. *C.* Exposure of mylonite after erosion.

rocks up toward the surface and the temperature and pressure decreased. The "upper-plate" rocks above the detachment fault are unmetamorphosed and have been strongly rotated on an imbricate set of listric normal faults that merge at depth with the detachment fault (Figure 4.14*B*, *C*).

Thus an ideal metamorphic core complex contains, from top to bottom: (a) an unmetamorphosed sequence of volcanic and coarse clastic sedimentary rocks that has been faulted and variably tilted along curved half-graben systems of normal faults; (b) a zone of highly sheared rocks including cataclasites and mylonites (the cataclasites, which are characteristic of shallow, low-temperature brittle shearing, commonly cross-cut the mylonitized rocks, which contain many complex folds and structures indicating higher-temperature ductile shearing; see Section 3.3(iii) and Figure 3.19); and (c) a gradational change from mylonite to unsheared older metamorphosed basement rocks, which in places are intruded by granitic rocks similar in age to the mylonite and the normal faults.

The detachment fault commonly has a corrugated shape, with the axis of the corrugations parallel to the slip direction on the fault. These corrugations are not the result of folding but are an original structure of the fault that has been described as a **mega-mullion structure** (compare with the small-scale fault mullions in Figure 3.8*B*). The footwall basement rocks and the overlying detachment fault may form a structural dome, sometimes referred to as a **turtle-back**, with the mega-mullions continuous across it. Shear sense indicators in the mylonites give a consistent sense of displacement on the fault across such domes, indicating the fault has been folded, presumably by isostatic uplift of the thinned crust, as the shallower crust in the hanging wall block is stripped off the top of the footwall block, and the footwall block is extracted from beneath the hanging wall block (Figure 4.13).

These characteristics of the detachment on a metamorphic core complex structure are strikingly apparent on the shaded relief images of an inferred core complex at Dante's Domes near the end of a slow-spreading segment on the mid-Atlantic ridge (Figure 4.15). Only a thin veneer of sediment has covered the structure, and no erosion has damaged it, leaving a remarkable image of the detachment surface. Both the mega-mullion structure and the doming are apparent. Serpentinites dredged from the detachment indicate mantle rocks have been exhumed in the footwall block, as shown in the schematic cross section in Figure 4.15*B*.

The widespread development of core complexes in the Basin and Range province and the large amount of extension associated with them indicate that they are of major tectonic significance. Reports of similar features from the Aegean Sea, Papua–New Guinea, areas of the midoceanic ridges (particularly along the mid-Atlantic ridge (Figure 4.15), and the southwest Indian ridge suggest that core complexes have worldwide importance.

iv. Faulting along Continental Margins: The U.S. Gulf Coast

Buried normal fault systems along rifted continental margins resemble those of exposed rifted regions. The northern Gulf Coast region of the United States from Texas to Alabama is an example of such normal faulting along a modern continental margin. The area is well characterized by abundant seismic imaging because of its economic importance as a major hydrocarbon-bearing region.

The U.S. Gulf Coast region is characterized by thick accumulations of sediment, rapid subsidence, and an arcuate system of normal faults whose extent is closely associated with the extent of major Jurassic salt deposits (Figure 4.16*A*). Most faults dip southward (Figure 4.16*A*, *B*, *C*), although north-dipping faults also exist. South-dipping faults commonly show rollover anticlines, indicating the faults are listric, with gentler dips at depth (Figure 4.16*B*).

Many normal faults along the Gulf Coast are **growth faults**, also referred to as **regional contemporaneous faults**, that are active during sedimentation. These faults typically have stratigraphic sequences characterized by units that are thicker on the hanging wall block than they are on the footwall block (Figure 4.17; cf. Figure 4.16*B*).

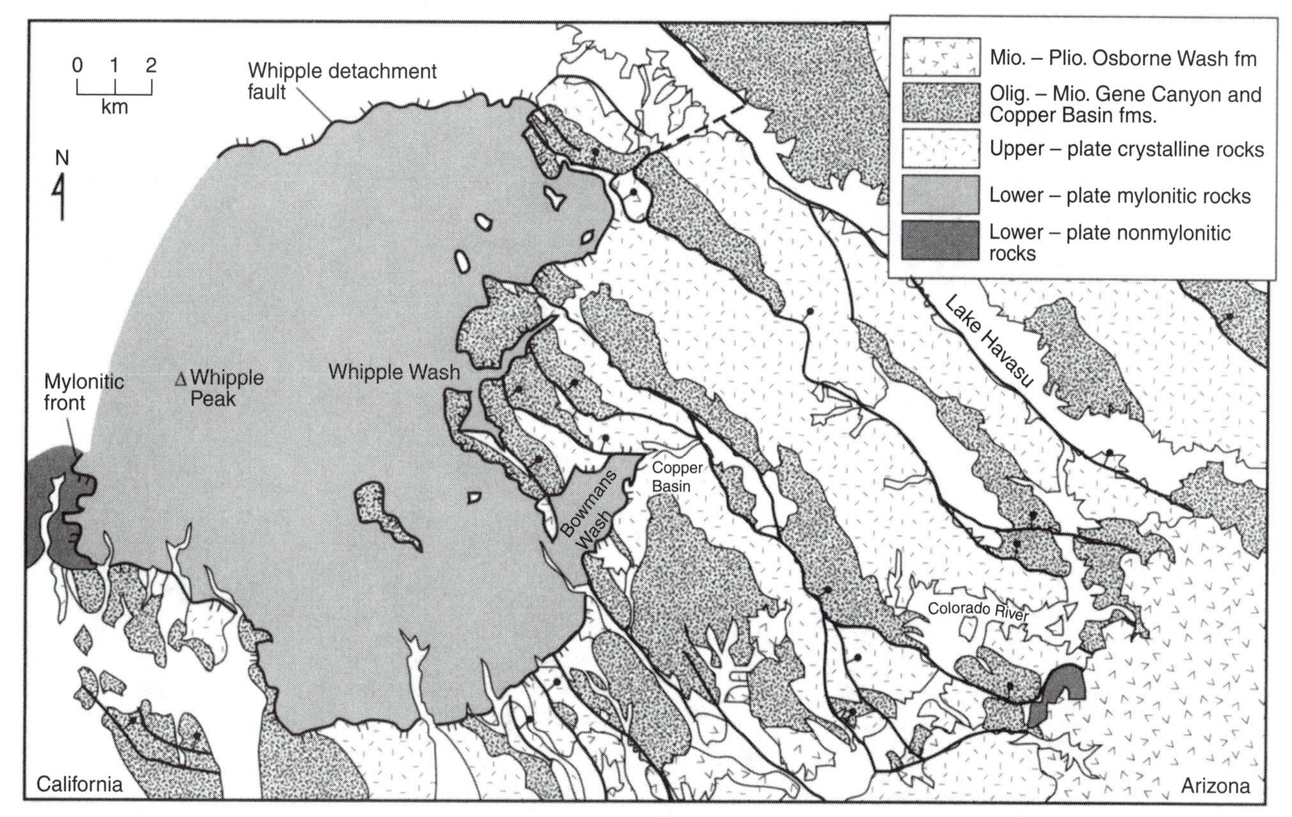

A.

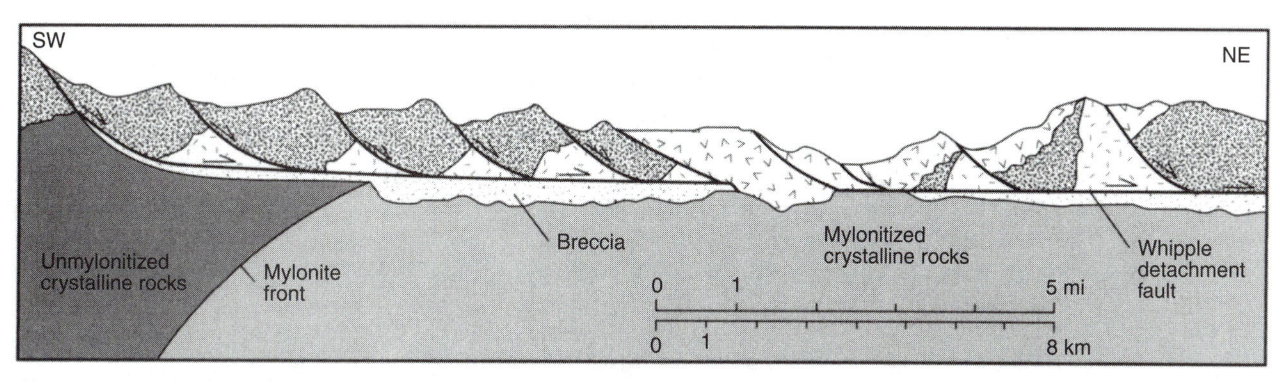

B.

C.

FIGURE 4.14 Faulting in the Whipple Mountain metamorphic core complex of the Basin and Range province, southeastern California. (*A* and *B* from Davis et al. 1980) *A*. Map of the Whipple Mountain metamorphic core complex (see Figure 4.10 for location). *B*. Diagrammatic cross section through the Whipple Mountains before uplift domed the detachment fault. *C*. The Whipple Mountain detachment fault is marked here by a topographic ledge of cataclastic rocks indicated by the white arrow (see Figure 3.5*B*) along which the dark-colored tilted Tertiary strata are faulted against the underlying lighter-colored mylonitic gneisses.

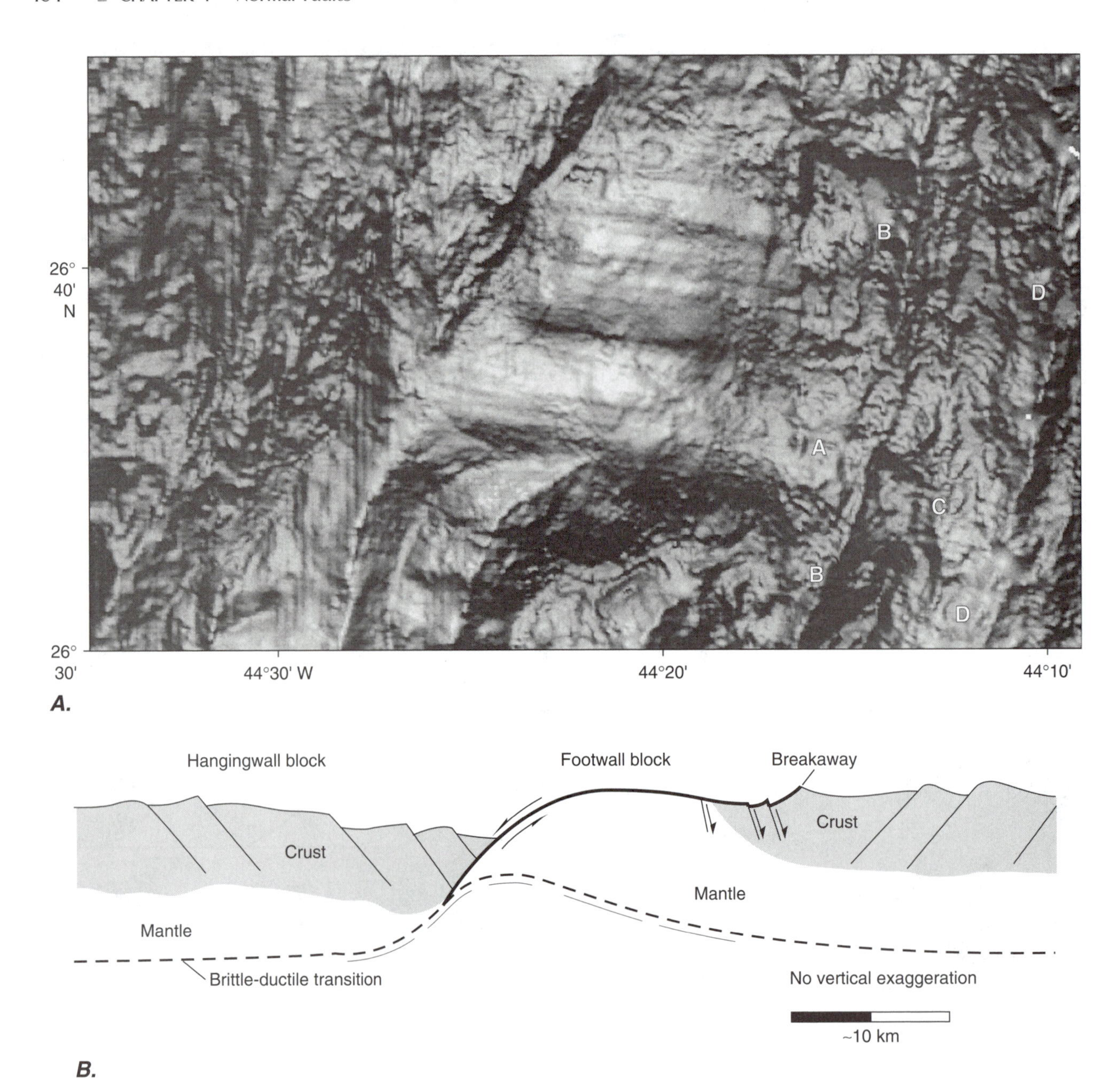

FIGURE 4.15 An inferred metamorphic core complex developed at Dante's Domes along the mid-Atlantic ridge. (From Tucholke et al. 2001) *A.* Shaded relief image of the exposed detachment fault. The detachment is domed from east to west and is characterized by corrugations aligned east-west that define a mega-mullion structure parallel to the slip direction. Hanging wall block to the west (left). Illumination is from the north. *B.* Schematic interpretive cross section of the detachment parallel to the mega-mullion structure. The normal fault cut down into the upper mantle to the brittle-ductile transition. The crust-mantle boundary has been exposed by slip along the detachment extracting the footwall block from under the hanging wall block and by isostatic doming of the thinned crust and upper mantle.

This disparity in thickness develops because as faulting continues, sediments accumulate most rapidly in the deepest parts of the basin, thereby keeping the surface of deposition approximately flat and horizontal. The down-faulted block thus accumulates a greater thickness of any particular unit. Older units show greater amounts of displacement and larger tilts than younger units, because they have experienced a longer history of faulting.

Growth faults apparently form in two ways: by differential compaction of shale layers in a sandstone-shale sequence, and by gravity sliding toward the basin. Coarser sands are generally deposited closer to shore than muds but they undergo less compaction during consolidation and lithification. The greater compaction of the shale can initiate a growth fault near the boundary with the less compactable sands. Growth faults can also develop by

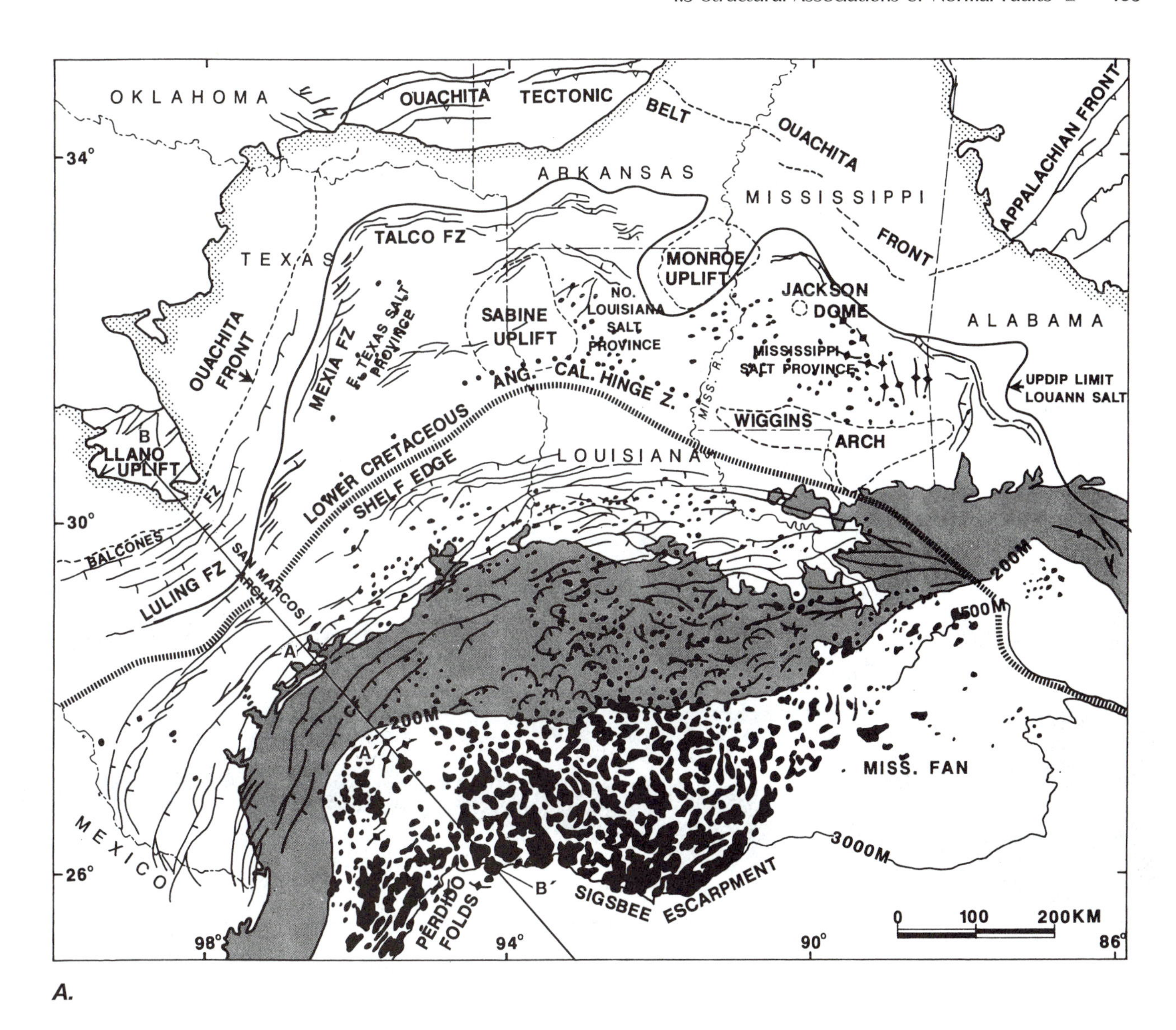

FIGURE 4.16 Normal fault province in the Gulf Coast area of the United States bounding the Gulf of Mexico. (After Worrall and Snelson 1989) *A.* Map of the Gulf Coast region showing major normal faults and salt structures (black). Major salt deposits occur south of the line marking the up-dip limit of the Louann salt, and this area is closely associated with the province of normal faulting. The area shaded gray marks the continental shelf to a depth of 200 m. *B.* Cross section A-A′ in A across the continental shelf of southwest Texas. Much of the area is believed to be underlain by salt deposits, which are not shown because of a lack of seismic resolution. Note the growth faults and salt structure on the right. There is no vertical exaggeration. *C.* Interpretive cross section of the Gulf Coast from the Llano uplift in the northwest to the Gulf abyssal plain in the southeast (B–B′ in A). Jurassic salt is believed to underlie much of the shallow structure and to form the major detachment zone, although the structure is not known. Note the salt nappe behind the Sigsbee escarpment and the underlying Perdido fold belt. Vertical exaggeration 5×.

(continued)

formation of a decollement at the base of a sequence, commonly in easily deformed shale or salt deposits. In the Gulf Coast region, accumulation of large thicknesses of sediment on the continental shelf has caused the thick underlying salt deposits (originally up to 1500 to 2100 m thick) to flow toward the basin. This flow causes the development of listric growth faults along the continental margin with a detachment in the salt layers and associated development of compressional structures in the basin, such as the huge salt nappe behind the Sigsbee escarpment and the Perdido fold belt (Figure 4.16*A*, *C*).

Similar faults are present along many rifted continental margins such as the Atlantic Ocean margins of North America, Europe, and Africa, as well as the Indian Ocean margins of Africa, India, and Australia. In all these regions, the fault systems and salt domes are important traps for hydrocarbons and thus are of great interest to the petroleum industry.

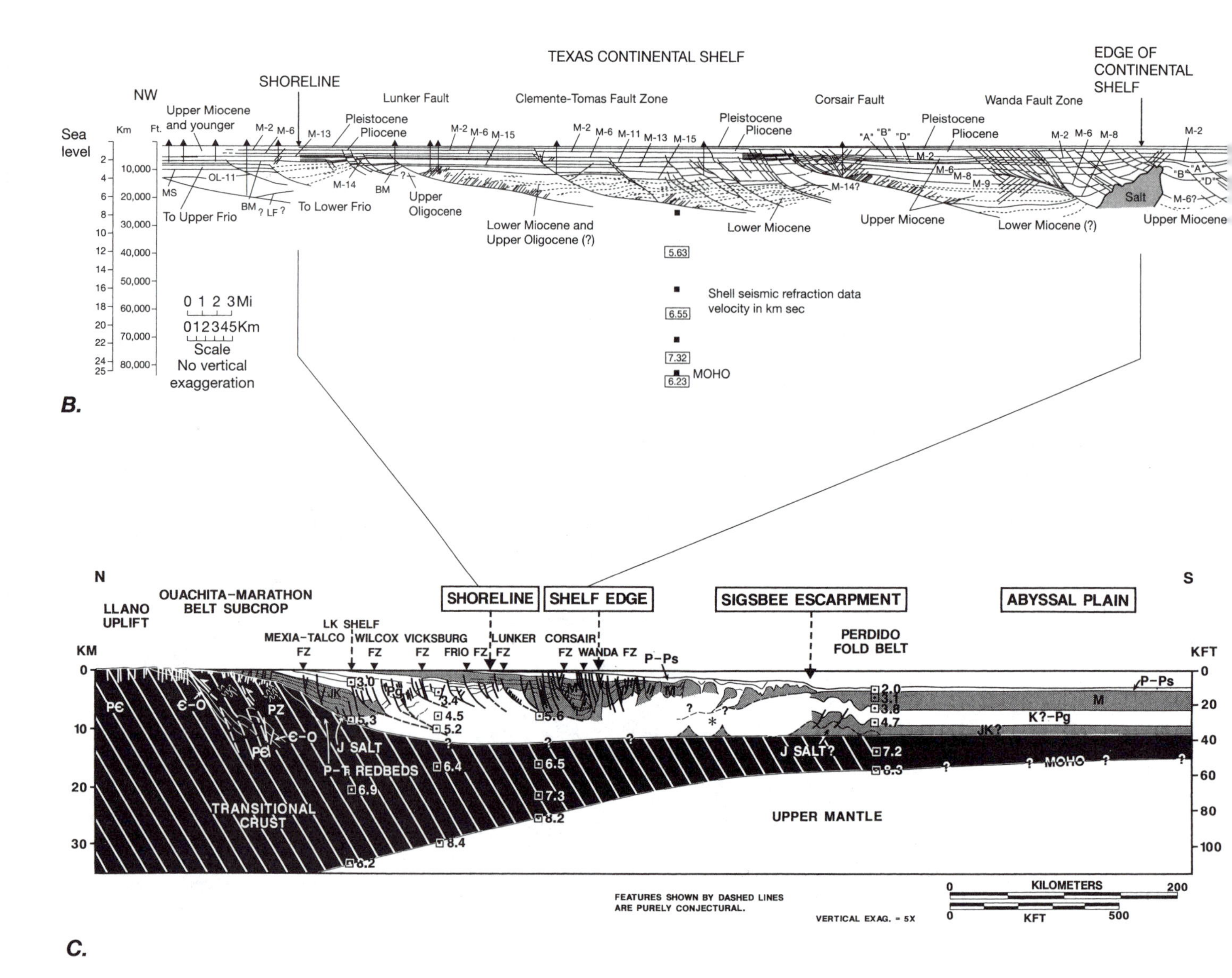

FIGURE 4.16 *(continued)*

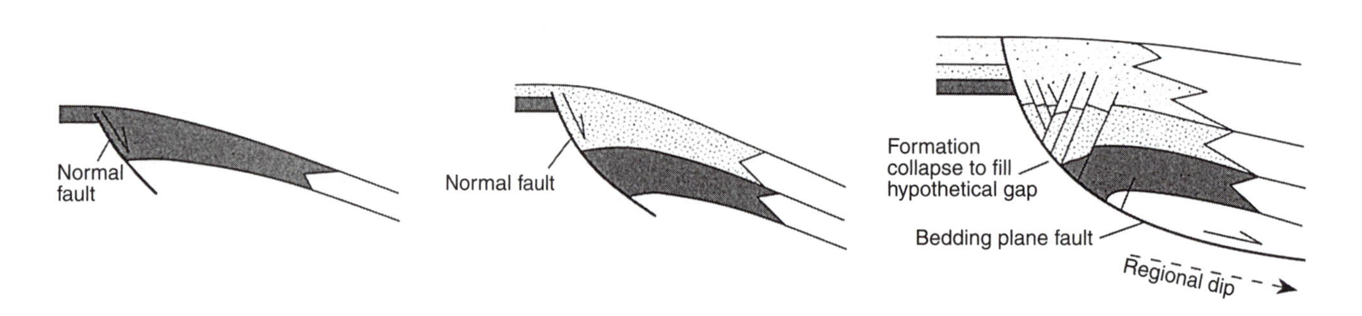

FIGURE 4.17 Development of growth faults. Displacement on a listric normal fault occurs during sedimentation, resulting in equivalent beds being thicker in the hanging wall block than in the footwall block. The fault passes into a bedding-plane fault at depth. (After Bruce 1973)

A single set of parallel normal faults can accommodate extension in only one direction, approximately perpendicular to the strike of the faults. In regions where extension occurs in two perpendicular horizontal directions, more than one orientation of normal fault is required, and commonly a rhombic pattern of fault traces develops. The angle between the faults depends on the relative magnitude of the extension in the two directions.

4.4 KINEMATIC MODELS OF NORMAL FAULT SYSTEMS

A kinematic model of any fault system is a description of the motions that have occurred on the faults in the system. A fundamental constraint on any model of faulting is that the volume of the blocks of rock involved in the faulting must be conserved. If the deformation was two-dimensional, the cross-sectional area of each unit must remain constant, and appropriate models or cross sections must be balanced.[5] Thus running the inferred fault motions backward from the present configuration must not produce overlaps of different fault blocks or large gaps between the different blocks. The model must also account for horizontal extension in the footwall block of major detachment faults.

We generally use cross sections of normal faults to display the geometry of faulting at depth, and any cross section inevitably implies some kinematic model of faulting, whether intended or not. Cross sections, however, are commonly incomplete in that they do not include all fault termination lines. This incompleteness may reflect a lack of data, which makes it impossible to determine how apparent geometrical problems are accommodated at depth, or it may be required by the scale of the section needed to portray important details of structure or stratigraphy. In any case, such cross sections make it easy to ignore the necessity to conserve volume. The result may be unbalanced cross sections that are geometrically impossible, that leave unresolved fundamental problems about the tectonics of an area, or that contain unintended implications about the kinematics of faulting.

Figure 4.18*A*, for example, shows an unbalanced cross section of a graben that is geometrically impossible. There is no way that the motion on the two faults can be reversed to produce an originally continuous layer without large gaps (Figure 4.18*B*) or overlaps (Figure 4.18*C*). Nor does this cross section specify what happens to the fault at depth. This type of inconsistency is commonly difficult to recognize in a cross section showing multiple intersecting faults. We note other examples of incomplete cross sections in the next section.

[5]Any component of motion out of the plane of the cross section, or volume loss due to solution, however, makes the balancing exercise unreliable (see Section 3.5).

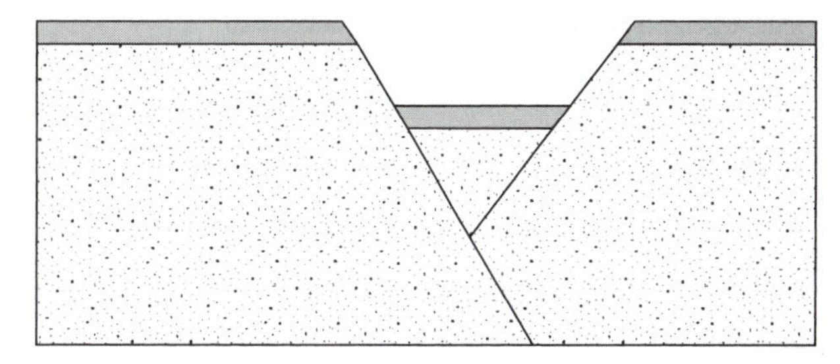

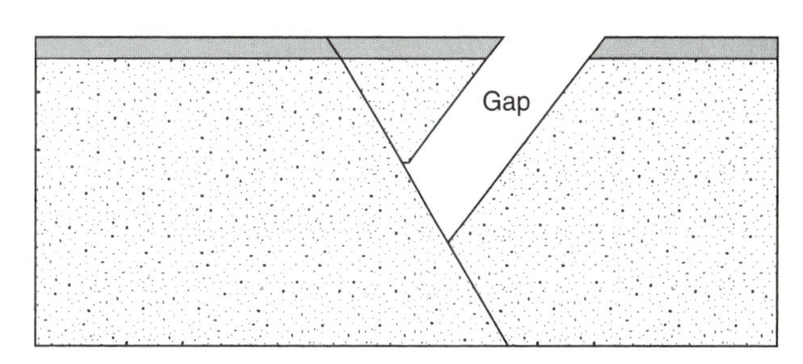

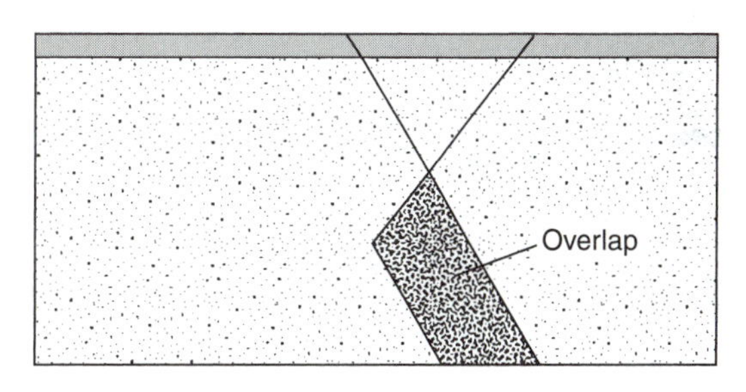

FIGURE 4.18 Geometrical constraints on cross sections. *A*. A geometrically impossible cross section of a graben. Attempted reversal of the fault motion leads to *B*. major gaps or to *C*. major overlap of fault blocks.

Normal fault provinces typically show tilted fault blocks (Figure 4.14*B*), and in some cases the rotations may approach 90° or more. Horizontal bedding typically is rotated about an axis roughly parallel to the strike of the fault so that the beds dip toward the fault. On geometrical grounds, this type of block rotation must imply either that the fault surfaces curve toward shallower dips with increasing depth or that planar faults rotated with the fault blocks during faulting. In the remainder of this section, we discuss three kinematic models for normal faulting that result in tilted fault blocks.

Figure 4.19 illustrates some geometrical problems inherent in accommodating extension on a listric normal fault. Horizontal extension of the block on a listric normal fault by an amount d opens a large gap between the hanging wall and footwall blocks (Figure 4.19*A*, *B*). If the bottom edge of the hanging wall block must conform to the shape of the listric fault while keeping constant the length L of the surface layer and the total area of the

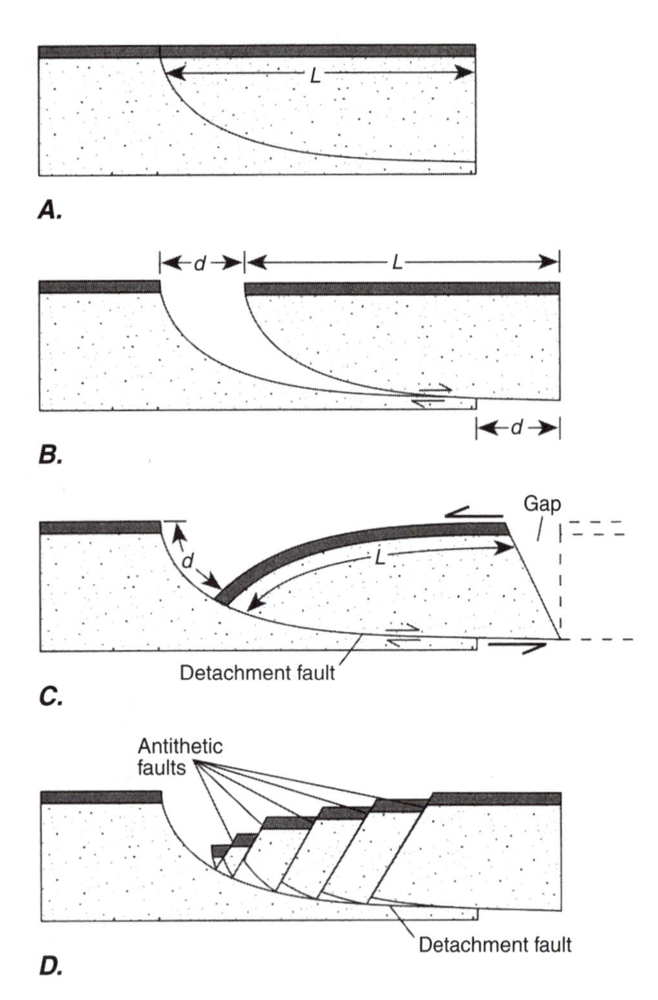

FIGURE 4.19 Model for the geometry of displacement on a listric normal fault accompanied by rollover folding or antithetic normal faulting. (In part after Hamblin 1965) *A.* Crustal block with the incipient fault. The length *L* of the hanging wall block is kept constant. *B.* Rigid displacement of the hanging wall block a distance *d* parallel to the horizontal part of the listric normal fault results in the opening of a geologically ridiculous gap. *C.* Deformation distributed through the hanging wall block allows contact to be maintained along the fault and results in rollover folding of the layers. The length *L* remains constant, resulting in the development of another gap problem in the hanging wall block. *D.* Distributed faulting on antithetic faults in the hanging wall block reduces the gap problem along the normal fault to small misfits along the listric fault.

block (Figure 4.19*C*), the process requires internal deformation of the hanging wall block. The resulting geometry is a rollover anticline commonly associated with listric normal faults.

One problem introduced by this model is that if there is no extension of the layers in the hanging wall block, a shearing must be distributed throughout the block, as indicated by the large arrows in Figure 4.19*C*. If the entire hanging wall block does not shear, a triangular gap must open up between the sheared and unsheared portions of the block (Figure 4.19*C*). Neither the shearing of the entire hanging wall block nor the triangular gap is geologically reasonable.

The problem illustrated in Figure 4.19*C* can be alleviated by allowing extension parallel to the layer in the hanging wall block. A set of antithetic faults cutting the hanging wall block (Figure 4.19*D*) permits the block to conform fairly well to the listric fault and effectively extends the block above the curved part of the detachment. As a result, the right edge of the block remains perpendicular to the base. Greater continuity and smaller gaps under the antithetic fault blocks can be obtained by closer spacing of the faults. The residual gaps are easily accommodated by local fracturing of the blocks.

A second model for slip on a listric fault requires the hanging wall block to maintain contact along the curved part of the listric fault by rotating as it slides. This mechanism can only work if the hanging wall block breaks up into a set of domino-like blocks along synthetic faults dipping in the same direction as the main fault (Figure 4.20*A*, *B*). Rotation of the fault blocks requires the synthetic fault planes to rotate as well, and the result is comparable to the collapse of a row of standing dominos. The triangular gap that opens at the right where the set of synthetic faults ends could be closed by a set of antithetic faults as shown in Figure 4.20*B*. Again the small gaps that occur below the synthetic fault blocks can be accommodated by closer spacing of the faults and by localized fracturing near the base of the fault blocks.

A third model of slip on listric normal faults requires slip of tapered fault blocks on a set of imbricate listric normal faults. As the fault blocks slip down the faults (Figure 4.21*A*, *B*), they must deform to conform to the shape of the fault. At large amounts of extension, the imbricate blocks are almost completely flattened out on the listric detachment fault, and bedding in the fault blocks is rotated into very steep dips. Rotation of the surface layer approaches a value equal to the initial dip of the fault where it cut the layer (Figure 4.21*A*). The unfaulted part of the hanging wall block, however, encounters the same geometrical problems as in Figure 4.19.

In principle, the latter two models might be distinguishable in the field. For the "domino block" model of planar rotational normal faults (Figure 4.20*B*), the dip of the bedding is constant across the entire hanging wall block above the detachment fault. For the imbricate listric fault model, however, if extension has not been too great (Figure 4.21*B*), the dip of the bedding should increase with distance in the direction of displacement on the detachment fault.

The models we have discussed are, of course, simplified idealizations of the natural world. The presence of triangular gaps in the models, for example, results from our implicit assumption that the fault blocks behave rigidly. Although this assumption is in accord with our intuitive experience with rock on a relatively small scale over short

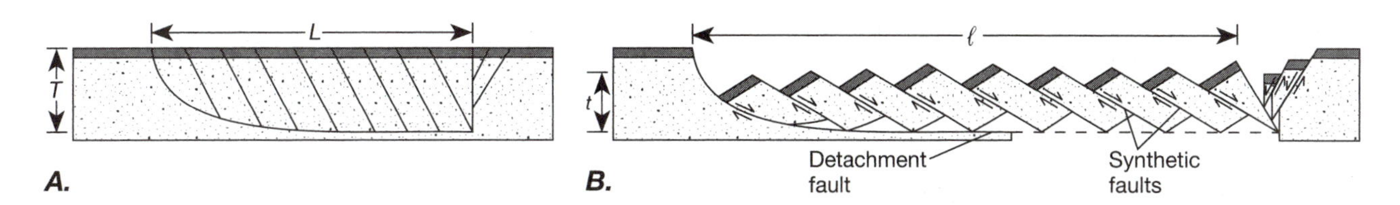

FIGURE 4.20 Model for the geometry of displacement on a listric normal fault accompanied by synthetic normal faulting. (After Wernicke and Burchfiel 1982) *A.* Rotation on the listric fault can be accommodated by forming a set of synthetic faults. L and T are the original length and thickness of the hanging wall block. *B.* After extension of the crust, the fault blocks have slipped and rotated on the system of normal faults synthetic to the listric normal fault. Here ℓ and t are the final length and average thickness of the hanging wall block.

periods of time, on the scale of tens of kilometers and millions of years, the mechanical behavior of rocks is quite different. When the natural behavior is scaled down to models we might make in the laboratory, the mechanical properties of rock are closer to those of sand or clay (see Sections 18.3 through 18.5). Thus the deformation in the hanging wall block required by the model in Figure 4.19*D* and the flattening of the imbricate fault blocks in Figure 4.21*B* are not outrageous propositions, and the small gaps that open up along the detachment fault in models such as Figures 4.19*D* and 4.20*B* could be easily accommodated by local deformation (see Figure 18.4).

The geometry of normal fault systems also is generally more complex than our model cross sections imply. It is common, for example, for listric normal faults to have a ramp-flat geometry and to cut progressively into either the hanging wall block or the footwall block as faulting proceeds. In some cases, later normal faults also commonly cross-cut earlier systems of normal faults in the same episode of extension. Figure 4.22, for example,

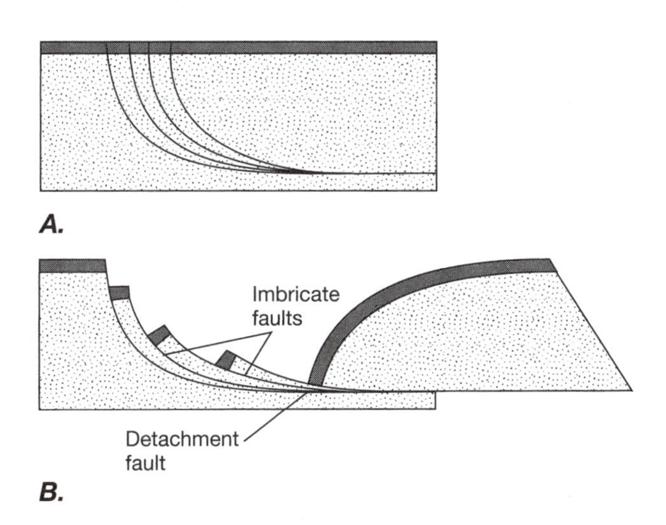

FIGURE 4.21 Model for the geometry of displacement on a set of imbricate listric normal faults. (After Wernicke and Burchfiel 1982) *A.* Geometry of incipient imbricate listric normal faults. *B.* As the imbricate fault blocks slip down the faults, they rotate and straighten out.

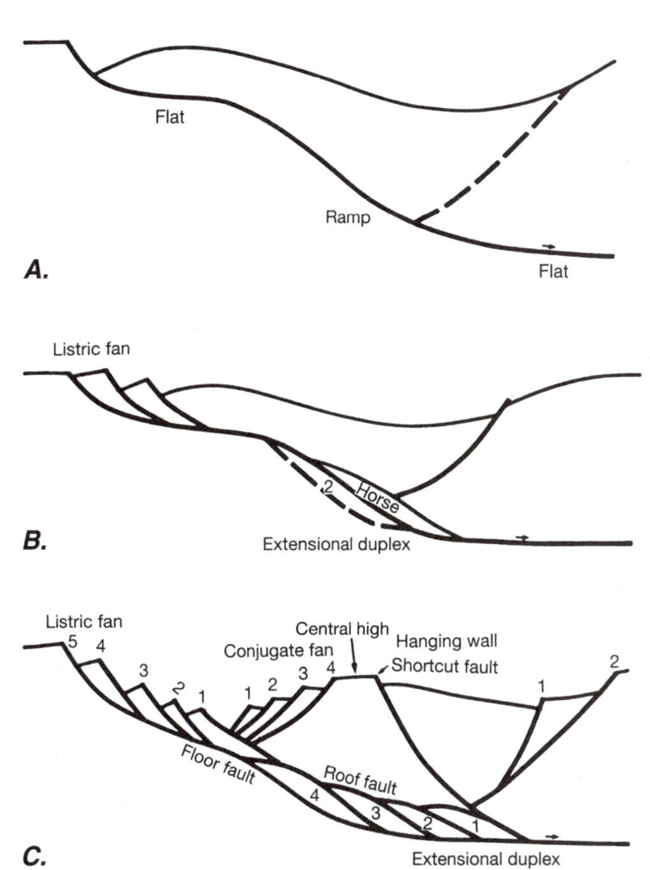

FIGURE 4.22 Model for the progressive development of a listric fan and an extensional duplex associated with a ramp and flat in a normal fault. (After Gibbs 1984) *A.* Listric normal fault with a ramp and flat geometry. A rollover anticline and a ramp syncline are developed in the hanging wall block. *B, C.* Progressive propagation of the fault into the footwall block produces a listric imbricate fan near the surface and an extensional duplex at depth. Eventually other faults, including the conjugate imbricate fan, develop to accommodate the deformation of the hanging wall block.

illustrates the structure resulting from the progressive cutting of the active fault back into the footwall block, as indicated by the number sequence. An imbricate listric fan of faults is formed at the surface, and with a set of conjugate faults, it defines a graben. At depth, an **extensional duplex** develops, characterized by a stack of horses that are progressively cut from the footwall block and transferred to the hanging wall block. The **floor fault**, which defines the bottom of the duplex, is the active fault, whereas the roof fault, which bounds the top of the duplex, is never active as a single fault.

Although all three models in Figures 4.19 through 4.21 can account for the rotation of fault blocks, all of them ignore a major tectonic problem implied by a horizontal or low-angle detachment fault. Normal faulting inherently increases the distance between two points on opposite sides of the fault and, on the average, must thin the faulted block if cross-sectional area is conserved (compare Figure 4.20*A* and *B*). Thus if extension is accommodated by normal faulting on a detachment surface and the rocks below that surface are not extended by the same amount, then there must exist a region in the hanging wall block where the extension is compensated by an equivalent shortening. This compensation is presumably the relationship in the Gulf Coast region between the system of normal growth faults and the Perdido fold belt and salt nappe structures (Figure 4.16*C*). Alternatively, the basement below the detachment must extend by the same amount, although perhaps by a mechanism other than brittle thinning. Stretching and thinning of the crust must in turn be accommodated in the mantle by a flow of mantle rock. Because the Earth is not expanding, horizontal stretching of the crust must be compensated for somewhere by crustal shortening or destruction of crust, for example, in an orogenic belt or at a subduction zone.

Figure 4.23 shows two models that account for the complete geometry of crustal normal fault systems. In Figure 4.23*A*, major normal faults in the upper crust join one of two symmetrically located detachment faults that become horizontal at the depth where deformation changes from brittle to ductile. The tip line for each detachment is at the same location in the middle of the faulted terrane. Below the detachments, the crust extends and thins by ductile deformation and the extension may also be accommodated to some extent by magmatic intrusion. Below the crust, the mantle accommodates the crustal extension by ductile inflow of rock.

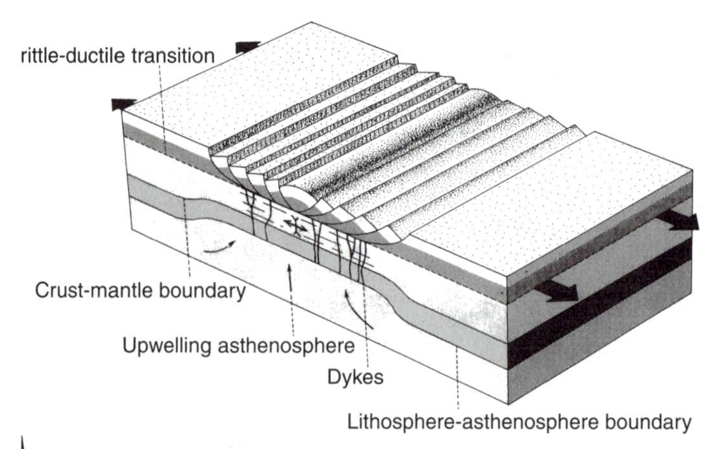

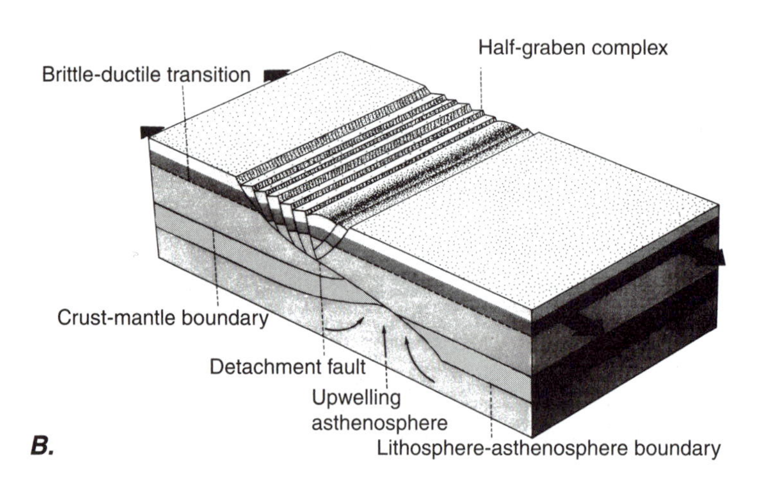

FIGURE 4.23 Complete cross sections accounting for the extension across the entire lithosphere in provinces of normal faulting. (After Lister and Davis 1989) *A*. The shallower crust extends by brittle normal faulting. The deeper crust extends and thins by ductile deformation. The extension is accommodated in the mantle by ductile inflow of material. Dip directions of normal faults are symmetrical about the center of the province. *B*. Extension occurs by displacement along a normal detachment fault that extends completely through the lithosphere. The brittle shallow crust extends by brittle imbricate listric normal faulting. Faulting on the detachment at depth is by ductile shear. The extension is accommodated in the mantle by ductile inflow of material. The dips of normal faults are predominantly in a direction synthetic to the detachment.

In Figure 4.23*B*, a major detachment fault extends completely through the lithosphere, changing from a brittle to a ductile fault at a depth of roughly 15 to 20 kilometers. Predominantly synthetic imbricate normal faulting in the hanging wall block produces an asymmetric normal fault province. Thinning of the crust by faulting on the detachment fault is accommodated in the mantle by ductile inflow of material. The termination of the fault is at the base of the lithosphere where the hanging wall block and the asthenosphere are both moving to the right at approximately the same rate. A large amount of extension would juxtapose deep crustal mylonites in the footwall block against the brittlely faulted blocks of the hanging wall block. Pervasive ductile extension of the crust in the footwall block is not required.

Both models include all termination lines of all the faults and account for all the required tectonic motions. The actual driving force for the extension could be the same in both cases. Structural aspects of each model that in principle could be tested include the predicted symmetry or asymmetry of normal faulting across the structural province, the extent of ductile extension of the crust below the detachment, and the geometry of the Moho. The tests are not easy to make, however, and which model is better remains to be determined through field and geophysical investigations.

4.5 DETERMINATION OF EXTENSION ASSOCIATED WITH NORMAL FAULTS

In studying normal fault systems, we wish to estimate quantitatively the amount of extension in a region. We define the **extension** e as the change in length in a given direction caused by the deformation, divided by the original length. Thus in Figure 4.20, for example, $e = (\ell - L/)L$, where ℓ is the deformed length, and L the initial undeformed length. Estimates of the extension can constrain our reconstructions of an area and help us better understand its tectonic history. We can estimate the amount of extension from fault geometry and by using map relationships to restore the stratigraphy to its original state.

i. Estimates of Extension Based on Fault Geometry

To evaluate the extension across a region using fault geometry, we must make a few simplifying assumptions. We assume that the fault strike is uniform and that the change in length of the region is the sum of the horizontal extensions on all the faults (Figure 4.24*A*). The extension is then this change in length divided by the original distance across the region, measured normal to the strike of the faults.

For example, let us take a simple cross-sectional model of planar nonrotating normal faults producing a

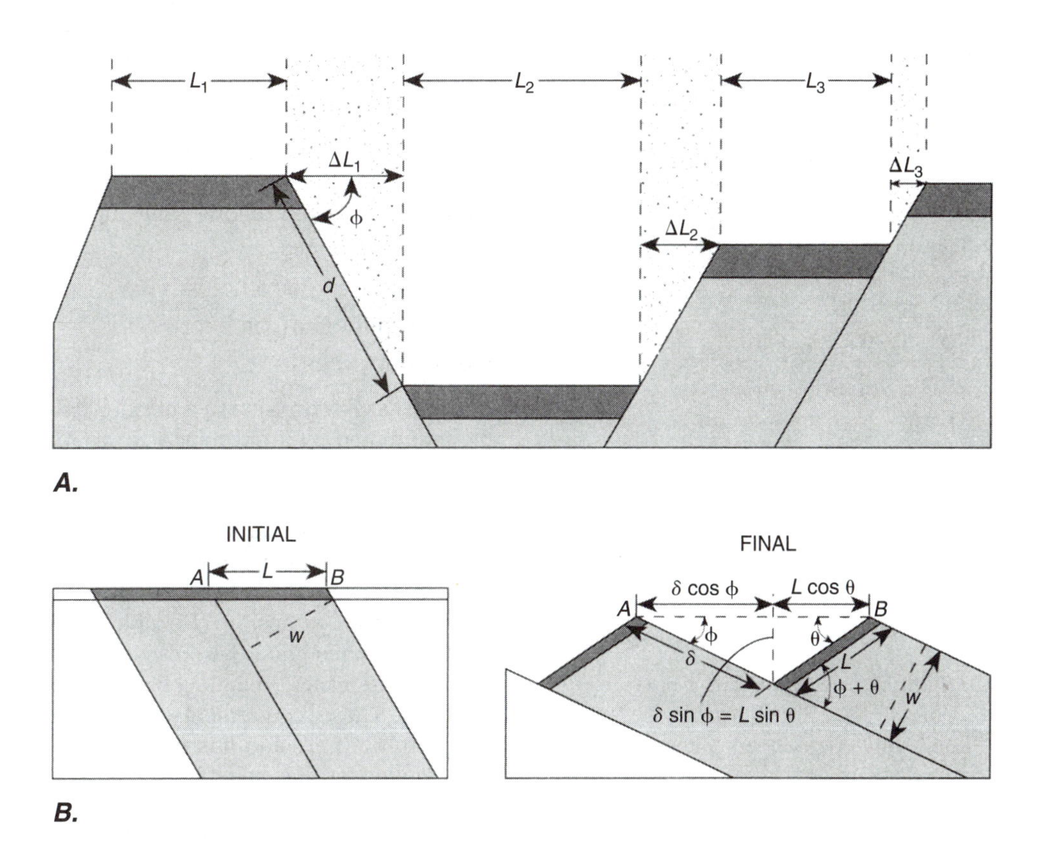

FIGURE 4.24 Determination of extension in a terrane faulted by planar normal faults. *A*. On nonrotating faults, the overall horizontal extension e is the ratio of the sum of the changes in horizontal length on each fault ΔL_i divided by the sum of the original lengths L_i of the strata in each fault block. *B*. Geometric relationship among the dip of the faults ϕ, the dip of the beds θ, and extension e, assuming equally spaced, planar, rotating normal faults above a detachment fault (see Figure 4.20*B*).

horst-and-graben structure (Figure 4.24*A*).[6] The segments of a particular stratigraphic layer labeled L_i (L_1, L_2, L_3, and so on), when summed together, equal the original length of the cross section. The segments labeled ΔL_i, when summed together, give the total change in horizontal length. The extension e is calculated for a total of N faults by the formula:

$$e = \frac{\sum_{i=1}^{N} \Delta L_i}{\sum_{i=1}^{N} L_i} \tag{4.1}$$

For an individual fault, the change in length ΔL is related to the dip-slip displacement δ and the dip angle of the fault ϕ by:

$$\Delta L = \delta \cos \phi \tag{4.2}$$

For a model of rotating planar normal faults (Figure 4.20*A*, *B*), if we assume that bedding is initially horizontal and that, on the average, the faults have the same orientation, spacing, and slip, we can easily derive a relationship between the extension e, the dip of the rotated bedding θ, and the dip of the rotated fault planes ϕ (Figure 4.24*B*):

$$e \equiv \frac{\overline{AB}_{\text{final}} - \overline{AB}_{\text{initial}}}{\overline{AB}_{\text{initial}}} \tag{4.3}$$

From the right diagram in Figure 4.24*B*, Equation (4.3) can be expressed as

$$e = \frac{(\delta \cos \phi + L \cos \theta) - L}{L}$$

$$e = \frac{\delta}{L} \cos \phi + \cos \theta - 1 \tag{4.4}$$

and

$$\delta \sin \phi = L \sin \theta \quad \text{or} \quad \frac{\delta}{L} = \frac{\sin \theta}{\sin \phi} \tag{4.5}$$

From a standard trigonometric identity, we have,

$$\sin(\theta + \phi) = \sin \theta \cos \phi + \sin \phi \cos \theta$$

or

$$\sin \theta \cos \phi = \sin(\theta + \phi) - \sin \phi \cos \theta \tag{4.6}$$

[6]Note that this model ignores the obvious problem of the fault geometry at depth (compare Figure 4.18).

Thus using the second Equation (4.5) in Equation (4.4), and then introducing the Equation (4.6), we get,

$$\begin{aligned} e &= \frac{\sin \theta}{\sin \phi} \cos \phi + \cos \theta - 1 \\ &= \frac{\sin(\theta + \phi) - \sin \phi \cos \theta}{\sin \phi} + \cos \theta - 1 \\ e &= \frac{\sin(\theta + \phi)}{\sin \phi} - 1 \end{aligned} \tag{4.7}$$

The assumptions limit the applicability of the model, but it can be used in some cases to give a rough approximation of the extension due to slip on the major faults.

The power-law distribution of fault lengths in a faulted terrane (Figure 3.28B; Equation (3.2), Section 3.4(i)) shows that by just accounting for the extension associated with the largest faults, as in Figure 4.24, we ignore the contribution to extension from all of the smaller faults. It is not immediately obvious whether the smaller displacements on these smaller faults (Figure 3.26; Equation (3.1), Section 3.3(vi)) contribute significantly to the total extension of an area, or whether it is negligible compared to the contribution of the largest faults. We examine the question in more detail in Box 15-1.

For the models of listric normal fault slip in Figures 4.19*C* and 4.21*B*, which involve deformation of the fault blocks above the detachment, there is no simple geometric relationship among fault dip, bedding dip, and displacement.

ii. Determination of Extension from Map Relations

In some cases extension in a normal-faulted terrane may be determined by a **palinspastic**[7] **restoration**, that is, by constructing a balanced cross section parallel to the slip direction and restoring the geology to its original configuration prior to deformation. Comparison of the restored cross-sectional length with the length after faulting, as measured in the field, makes it possible to calculate the extension. Reliable application of the method requires good data from the subsurface as well as from the surface, including firm constraints on the slip direction. These requirements severely limit the utility of the method. If erosion has removed much of the originally faulted terrane, and if large-displacement listric normal faults are prevalent, then the difficulty in knowing the original curvature of the faults also limits the reliability of the restoration.

[7]After the Greek words *palin*, which means "again," and *spasmos*, which means "contraction" or "breaking."

REFERENCES AND ADDITIONAL READINGS

Billings, M. P. 1954. *Structural Geology*. New York, Prentice-Hall.

Bruce, C. H. 1973. Pressured shale and related sediment deformation: Mechanism for development of regional contemporaneous faults. *Amer. Assoc. Petrol. Geol. Bull.* 57: 878–886.

Brun, J.-P., and P. Choukroune. 1983. Normal faulting, block tilting, and décollement in a stretched crust. *Tectonics* 2: 345–356.

Brun, J.-P., D. Sokoutis, and J. Van Den Driesche. 1994. Analogue modeling of detachment fault systems. *Geology* 22: 319–322.

Buck, W. R. 1991. Modes of continental lithospheric extension. *J. Geophys. Res.* 96: 20161–20178.

Buck, W. R., P. T. Delaney, J. A. Karson, and Y. Lagabrielle, eds. 1998. Faulting and magmatism at mid-ocean ridges. Amer. Geophys. Union. Geophysical Monograph 106: 348 pp.

Cloos, E. 1955. Experimental analysis of fracture patterns. *Geol. Soc. Amer. Bull.* 66(3): 241–256.

Cloos, E. 1968. Experimental analysis of Gulf Coast fracture patterns. *Amer. Assoc. Petrol. Geol. Bull.* 52: 420–444.

Coney, P. J. 1980. Cordilleran metamorphic core complexes: An overview. In P. J. Coney and G. H. Davis, eds., *Geol. Soc. Amer. Mem.* 153: 7–31.

Coward, M. P., J. F. Dewey, and P. L. Hancock, eds. 1987. *Continental Extensional Tectonics*. Geological Society of London Spec. Pub. 28: 637 pp.

Davis, G. A., J. L. Anderson, E. G. Frost, and T. J. Shackelford. 1980. Mylonitization and detachment faulting in the Whipple-Buckskin-Rawhide Mountains terrane, southeastern California and western Arizona. *Geol. Soc. Amer. Mem.* 153: 79–129.

Davis, G. H., and P. Coney. Geologic development of the Cordilleran metamorphic core complex. *Geology* 7: 120–124.

Donath, F. 1962. Analysis of basin-range structure, south central Oregon. *Geol. Soc. Amer. Bull.* 73: 1–16.

Effimoff, I., and A. R. Pinezich. 1986. Tertiary structural development of selected basins: Basin and range province, northeastern Nevada. In L. Mayer, ed. Geological Society of America Spec. Paper 208: 31–42.

Faulds, J. E., and J. H. Stewart., eds. 1998. *Accommodation Zones and Transfer Zones: The Regional Segmentation of the Basin and Range Province*, Geological Society of America Spec. Paper 323: 257 pp.

Gibbs, A. D. 1984. Structural evolution of extensional basin margins. *J. Geol. Soc. Lond.* 141: 609–620.

Hamblin, W. K. 1965. Origin of "reverse drag" on the downthrown side of normal faults. *Geol. Soc. Amer. Bull.* 76(10): 1145–1164.

Lister, G. S., and G. A. Davis. 1989. The origin of metamorphic core complexes and detachment faults formed during Tertiary continental extension in the northern Colorado River region, USA. *J. Struct. Geol.* 11(1/2): 65–94.

Lister, G. S., M. A. Etheridge, and P. A. Symonds. 1986. Detachment faulting and the evolution of passive continental margins. *Geology* 14: 246–250.

Murray, G. E. 1961. *Geology of the Atlantic and Gulf Coast Province of North America*. New York, Harper & Bros: 692 pp.

Press, F., and R. Siever. 1986. *Earth*. New York, W.H. Freeman & Co.

Reches, Z. 1979. Analysis of faulting in three-dimensional strain field. *Tectonophys.* 47(1/2): 109–129.

Rosenberg, C. L., J.-P. Brun, and D. Gapais. 2004. Indentation model of the Eastern Alps and the origin of the Tauern window. *Geology* 32: 997–1000.

Stewart, J. H. 1978. Basin-range structure in western North America: A review. In R. B. Smith and G. P. Eaton, eds., *Cenozoic Tectonics and Regional Geophysics of the Western Cordillera*, Geological Society of America Memoir 152: 1–32.

Stewart. J. H. 1998. Regional characteristics, tilt domains, and extensional history of the later Cenozoic Basin and Range Province, western North America. In J. E. Faulds and J. H. Stewart, eds., *Accommodation Zones and Transfer Zones: The Regional Segmentation of the Basin and Range Province*, Geological Society of America Spec. Paper 323: 47–74.

Stewart. J. H., et al. 1998. Map showing Cenozoic tilt domains and associated structural features, western North America. In J. E. Faulds and J. H. Stewart, eds., *Accommodation Zones and Transfer Zones: The Regional Segmentation of the Basin and Range Province*, Geological Society of America Spec. Paper 323: 257 pp.

Thorman, C. H. and K. B. Ketner. 1979. West-northwest strike-slip faults and other structures in allochthonous rocks in central and eastern Nevada and western Utah. In G. W. Newman and H. D. Goode, eds., *Basin and Range Symposium and Great Basin Field Conference*. Denver, CO, Rocky Mountain Association of Geologists: 121–133.

Tucholke, B. E., K. Fujioka, T. Ishihara, G. Hirth, and M. Kinoshita. 2001. Submersible study of an oceanic megamullion in the central North Atlantic. *J. Geophys. Res.* 106(B8): 16145–16161.

Uyeda, S. 1977. *New View of the Earth.* New York, W.H. Freeman & Co.

Wentlandt, E. A. 1951. Hawkins field, Wood County, Texas. Austin, University of Texas Pub. No. 153–158.

Wernicke, B., and B. C. Burchfiel. 1982. Modes of extensional tectonics. *J. Struct. Geol.* 4: 104–115.

Worrall, D. M., and S. Snelson. 1989. Evolution of the northern Gulf of Mexico, with emphasis on Cenozoic growth faulting and the role of salt. In A. W. Bally and A. R. Palmer, eds., *The Geology of North America: An Overview*, DNAG, The Geology of North America, Volume A. Boulder, CO, Geological Society of America: 97–138.

Wust, S. L. 1986. Regional correlation of extension directions in Cordilleran metamorphic core complexes. *Geology* 14(10): 828–830.

Zoback, M. L., R. E. Anderson, and G. A. Thompson. 1981. Cenozoic evolution of the state of stress and style of tectonism of the Basin-Range province of the western United States. *Phil. Trans. Roy. Soc. Lond.* 300: 407–434.

Chapter 5

THRUST OR REVERSE FAULTS

Thrust and reverse faults are dip-slip faults on which the hanging wall block has moved up relative to the footwall block. Generally, older rocks are emplaced over younger rocks, and in a vertical section through the fault stratigraphic section is duplicated (Figure 3.10*A*). These faults accommodate horizontal shortening of the Earth's crust. Reverse faults have dips greater than 45°, whereas thrust faults have dips less than 45°. We concentrate our discussion on thrust faults because they are much more abundant and tectonically more significant than reverse faults.

Thrust faults exist at all scales. They range from small ones with extents and displacements on the order of millimeters to meters (Figure 5.1*A*), through major low-angle thrusts on the scale of mountain ranges that show displacements on the order of tens to hundreds of kilometers (Figure 5.1*B*), up to the global-scale features of convergent plate margins, which are enormous complex zones of thrust faults having total displacements as large as thousands of kilometers.

A hanging wall block above a very low-angle thrust commonly extends over an area that is very large compared with its thickness and therefore is called a **thrust sheet** or a **nappe**.[1] A thrust sheet that has moved a large distance and is thus geologically out of place is an **allochthon**, and the rocks within it are **allochthonous**. A large region of rock that has not been moved and is close to its original location, such as the basement rocks in the footwall block of a thrust, is an **autochthon** and the rocks within it are **autochthonous**.[2] Figure 5.1*B* is a photo of the Keystone thrust in southern Nevada. The irregular dark/light contact is the trace of the low-angle thrust fault, which dips gently to the west (left) and is cut by irregular topography. The light rocks forming the cliff are autochthonous Jurassic sandstones, whereas the overlying dark rocks are an allochthonous Paleozoic sequence that extends from lower Paleozoic at the fault to upper Paleozoic in the snow-covered peaks in the background. These rocks have moved up to 20 kilometers on the thrust fault.

5.1 RECOGNITION OF THRUST FAULTS

Most thrust fault surfaces exhibit the deformation features intrinsic to faults, as well as the other features associated with faults, that are discussed in Section 3.2. In this section we confine our discussion to stratigraphic characteristics that are unique to thrust faults.

Thrust faults characteristically emplace older rocks on top of younger rocks. On a vertical section through a thrust fault, stratigraphic section is generally duplicated (Figure 5.2*A–D*; compare Figure 3.10*A*). These characteristics are illustrated in Figure 5.2, where each pair of block diagrams shows the results of thrust faulting of upright stratified rocks having a specific orientation relative to the fault. The left diagram of each pair shows the hanging wall block suspended over the footwall block, and the right diagram shows the same structure with the hanging wall block eroded down to the same level as the footwall block.

The horizontal separation across a thrust fault depends on the initial attitude of the layers relative to the fault, as illustrated in Figure 5.2; compare, for instance, Figure 5.2*C* and *D*, which show opposite senses of strike

[1] After the French word *nappe*, which means "sheet."

[2] After the Greek words *allo*, which means "other," *chthonous*, which means "ground" or "earth," and *auto*, which means "this."

FIGURE 5.1 The geometry and expression of thrust faults at different scales. (Photos courtesy of R. J. Varga) *A*. Thrust fault cutting carbonate strata, Valley and Ridge province of the Appalachian Mountains, Tennessee. *B*. The Keystone thrust, southern Nevada. For scale, the road is the four-lane interstate highway 115 west of Las Vegas.

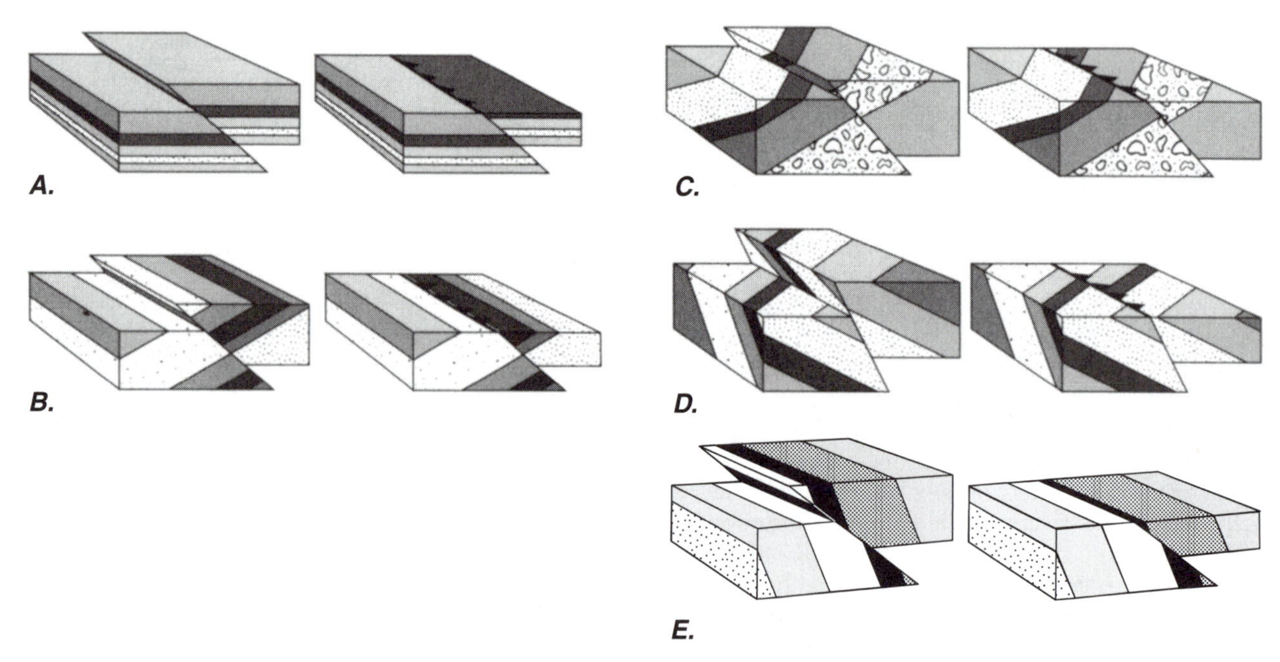

FIGURE 5.2 The effect of the dip of strata on the separation developed as a result of thrust faulting. The right diagram in each pair shows the hanging wall block eroded down to the same level as the footwall block. In a vertical section perpendicular to the strike of the fault, strata are normally repeated across the fault, and older rocks are emplaced over younger, except in *E*. The top surfaces show: *A*. a simple discontinuity. *B*. the cutting out of strata. *C*. left lateral separation. *D*. right lateral separation. *E*. If strata dip more steeply than, but in the same direction as, the fault, thrusting cuts out strata in a vertical section and emplaces younger rocks over older.

separation resulting from identical thrust displacement on the fault. Thus the separations on either the horizontal surface or the vertical section do not define uniquely the displacement on the fault (compare Figure 3.23). Generally, a vertical section through the faults shows the duplication of stratigraphy, with older rocks resting on top of younger rocks across the fault (Figure 5.2*A–D*). The exception to this generalization is if the strata are upright and dip more steeply than, but in the same direction as, the fault (Figure 5.2*E*). In this case, strata are missing in a vertical line through the fault, and younger rocks are emplaced over older. Of course, in areas of complex deformation, for example, where the stratigraphy is overturned or folded, the conventional relationships also need not hold.

Several types of stratigraphic contrasts may indicate the presence of a thrust fault on which the hanging wall block has been displaced from substantially deeper levels or from large horizontal distances.

Plutonic or high-grade metamorphic rocks are generally associated with deeper structural levels than unmetamorphosed or low-grade rocks. Thus if plutonic or high-grade metamorphic rocks overlie low-grade or unmetamorphosed sedimentary rocks in a footwall block, the normal structural sequence is inverted, suggesting that thrusting has occurred.

Some thrust faults separate stratigraphic sequences of essentially the same age but of markedly different sedimentary facies. Thrust faults of this nature commonly emplace allochthonous rocks of oceanic or deep-water environments, usually shales, cherts, or oceanic crustal rocks, on top of autochthonous shallow-water deposits such as limestone and sandstone, or even rocks of continental origin. The striking discontinuity in the sedimentary environments suggests that the contact between the two sequences is a thrust fault of large displacement.

In some cases, highly deformed allochthonous rocks overlie slightly or undeformed autochthonous rocks. If, for example, the rocks above and below the fault are the same stratigraphic layers, but those above are deformed by folds and those below are not, a thrust fault probably separates the two sequences.

All of these criteria are only general indicators: Each pattern conceivably could form in some other fashion, for example, from two or more episodes of deformation. The possibility of such structural complications means that we must pay careful attention to stratigraphic sequence and to evidence for sedimentary environment, conditions of igneous crystallization, and conditions of metamorphism. The geologic literature contains many examples of egregious errors committed by conscientious geologists who, when mapping a region, failed to take adequate account of these factors.

5.2 SHAPE AND DISPLACEMENT OF THRUST FAULTS

i. The Shape of Thrust Faults

Many map traces of thrust faults are highly irregular, a feature resulting from either the intersection of a shallow dipping fault with the irregularities of topography (Figure 5.1*B*) or the folding of the fault surface. In some

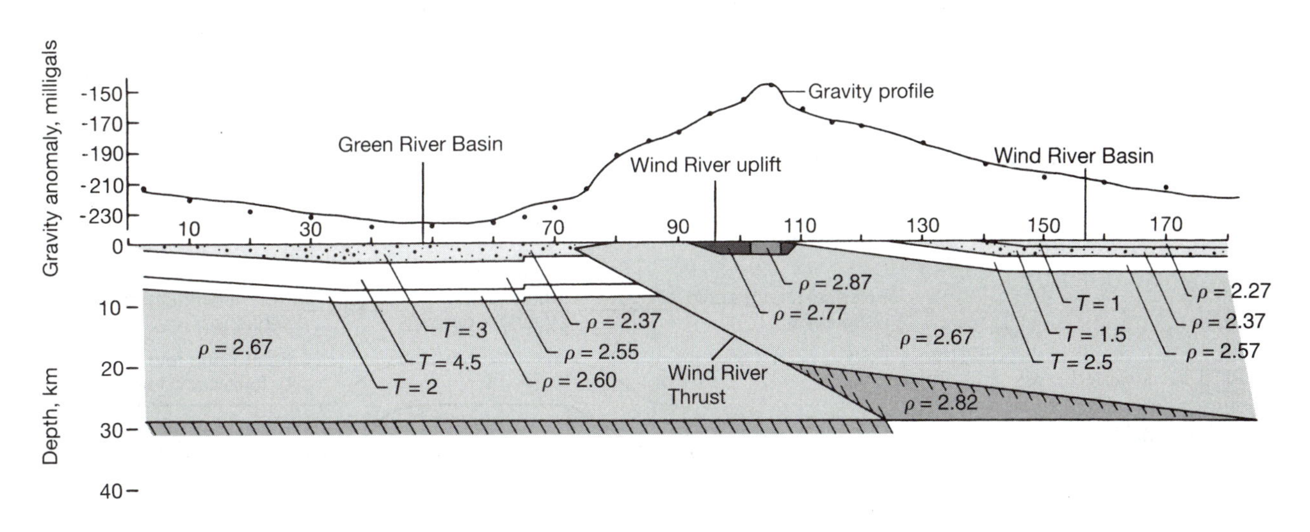

FIGURE 5.3 The shape of the Wind River thrust under the Wind River Mountains in Wyoming, which maintains almost a constant angle through the entire crust. This gravity model has been constrained by seismic reflection data. Values of T and ρ are, respectively, the thicknesses (in kilometers) and densities (in grams per cubic centimeter) of the different layers. On the gravity profile, dots are measured values of the gravity anomaly; the solid line is the anomaly computed from the model. (After Smithson et al. 1978, Fig. 5)

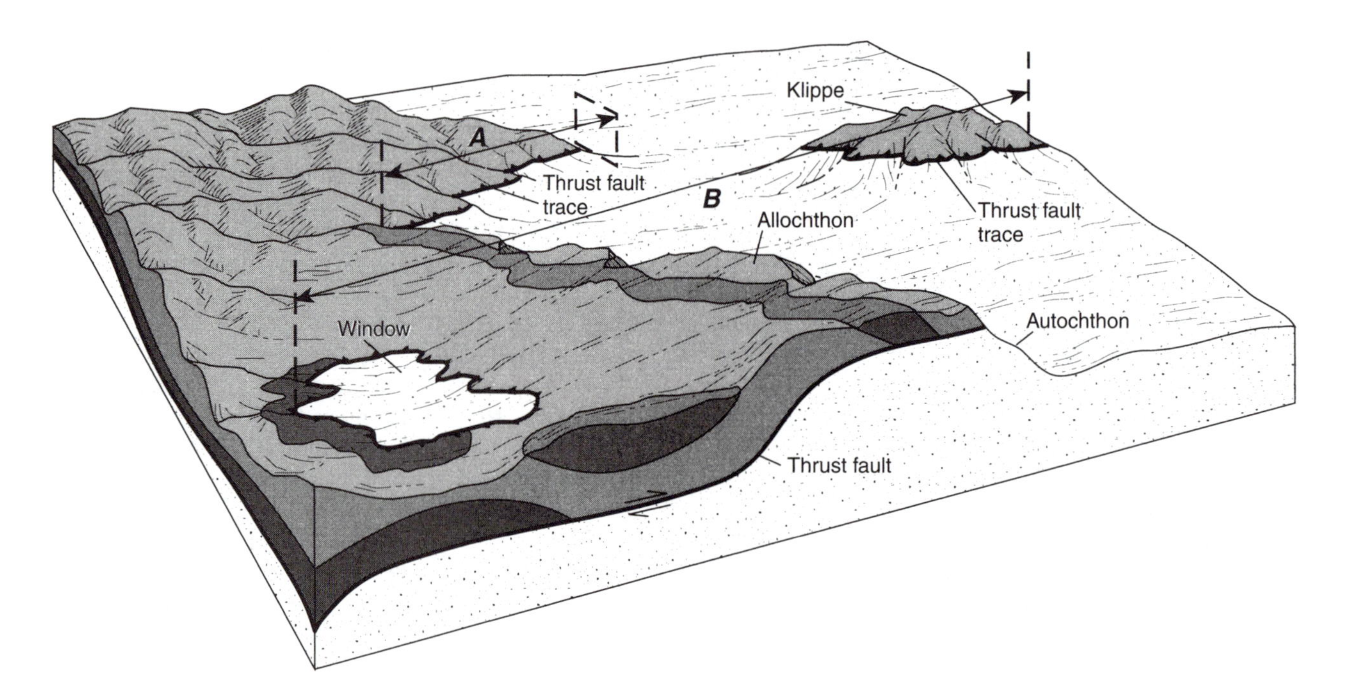

FIGURE 5.4 Block diagram illustrating a thrust surface, thrust sheet, thrust fault trace, window or fenster, klippe, allochthon, autochthon, and conventional thrust symbol with teeth on the hanging wall side. Minimum constraints on the displacement are given by (*A*) the sinuosity of the thrust fault trace and (*B*) the distance from the back of the window to the front of the klippe.

cases, however, an irregular fault trace on the surface may reflect the original irregular path cut by the fault through the stratigraphy.

At depth, thrust faults generally are listric faults that curve toward shallow or horizontal dips (see Figure 5.12). Some faults, however, continue at a dip of roughly 30° through most of the crust and into the mantle (Figure 5.3). Others become steeper with depth, such as where they accommodate compression at a jog in a strike-slip fault (see Sections 6.2(ii) and 6.3(ii) and Figures 6.7*C* and 6.8*B*) or where they end against an upwardly moving intrusion (see Figure 5.9*B*).

Erosion of a thrust sheet that lies above a shallowly dipping fault commonly leaves an isolated remnant of the allochthon, a **klippe**,[3] separated from the rest of the sheet (Figure 5.4). A klippe indicates a minimum extent of the original thrust sheet. In other cases, erosion can create a hole through the thrust sheet, a **window** or **fenster**,[4] and expose an isolated area of the rocks beneath the thrust (Figure 5.4). These rocks may be part of the autochthon or part of another underlying thrust sheet. A window provides a minimum constraint on how far the thrust sheet extends out over the underlying rocks.

[3]After the German word *klippe*, which means "cliff," reflecting the fact that most klippen in the Alps are bounded by cliffs.

[4]After the German word *fenster*, which means "window."

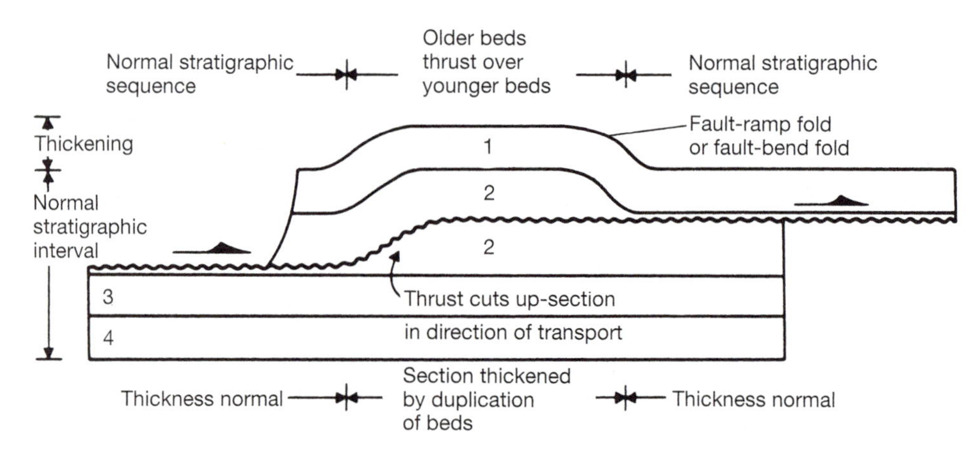

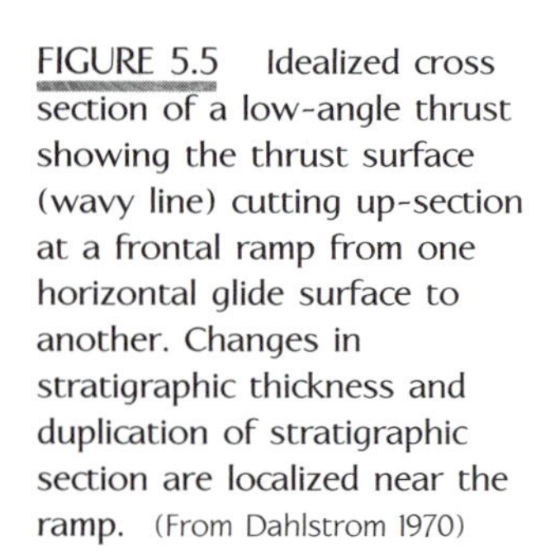
FIGURE 5.5 Idealized cross section of a low-angle thrust showing the thrust surface (wavy line) cutting up-section at a frontal ramp from one horizontal glide surface to another. Changes in stratigraphic thickness and duplication of stratigraphic section are localized near the ramp. (From Dahlstrom 1970)

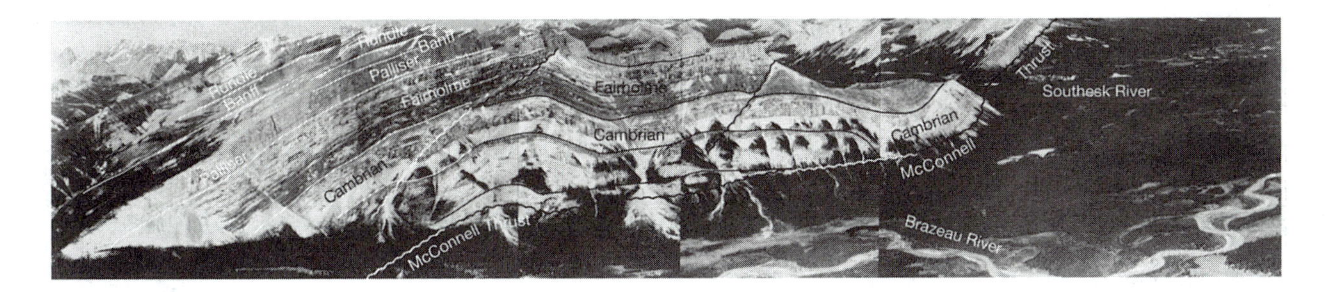

FIGURE 5.6 The McConnell thrust at the Brazeau River in Alberta. Composite photograph looking north at a cliff containing the McConnell thrust. The background and foreground to the cliff do not match up well across the joins between the frames because the vantage points for the photographs are different. The scale is given by the thickness of the Fairholme formation, which is about 1400 ft (425 m). The McConnell thrust cuts up-section from the lower Cambrian units (center) to the upper Cambrian (right) along the cliff (compare Figure 5.5). (Photos from Dahlstrom 1970)

A low-angle thrust fault generally does not form a smooth simple surface. Its fault plane characteristically cuts through the stratigraphy in steps, alternately following flat bedding planes or easily deformed layers such as shale or evaporite beds, and then cutting up section in the direction of displacement to form a **frontal ramp** (Figure 5.5; compare Section 3.4(ii) and Figures 3.33*A* and 3.34*A*). The fault surface thereby develops a characteristic **ramp-flat** geometry. Figure 5.5 also shows that in some places where the fault parallels the bedding, a normal stratigraphic sequence is preserved despite the presence of the fault.

Figure 5.6 is a composite photograph of the McConnell thrust in Alberta, Canada, that shows such a ramp. To the left, the thrust is parallel to bedding and located within a lower Cambrian unit. To the right, the lower and middle Cambrian units of the thrust sheet are truncated against the fault. Still further right, the fault is again parallel to bedding, but here it is located within the upper Cambrian unit. The geometry is comparable to the diagram in Figure 5.5, with the lower Cambrian corresponding roughly to layer 2.

Lateral ramps are also common features of low-angle thrust faults, as illustrated in Figure 5.7 for the Lewis thrust, which is part of the Canadian Rockies foreland fold and thrust belt in Montana (locations shown in Figure 5.11*B*). Both oblique and sidewall ramps may have dips as low as 15° to 20° (Figure 5.7*A*, *B*). If sidewall faults are steeply dipping, they are strike-slip faults called either **tear faults** or **transfer faults**, such as those that occur at the ends of the Pine Mountain thrust in the Appalachian Valley and Ridge province (Figure 5.7*C*, *D*; location shown in Figure 5.11*A*; see also Figure 5.13).

ii. Displacement on Thrust Faults

The direction of displacement of the hanging wall block on a thrust fault defines its **vergence**. For a thrust fault that has an eastward vergence, or that verges toward the east, for example, the hanging wall block has moved upward toward the east on the west-dipping thrust.

Although displacement on thrust faults is typically up the dip of the fault surface, on irregularities in the fault plane such as lateral ramps, the displacement must in general be oblique slip. Ramps in the fault surface also require that the thrust sheet deform as it moves. Movement of a thrust sheet over a ramp causes a fold to appear in the thrust sheet, called a **fault-ramp fold** or a **fault-bend fold** (Figures 5.5, 5.6). The trend of the fold reflects the trend of the ramp below the thrust sheet. Frontal ramps are steeper than the main fault surface and therefore produce **ramp anticlines** (Figure 5.5). On lateral ramps, if the displacement has a component up the ramp, then an anticline forms (Figure 5.8*A*, *B*); if displacement has a component down the ramp, then a syncline develops (Figure 5.8*A*, *C*).

5.3 STRUCTURAL ENVIRONMENTS OF THRUST FAULTS

Thrust faults exist as local faults, as sets of faults subsidiary to larger structures, or as large systems involving multiple thrusts and extending over whole mountain ranges. We consider each structural environment in turn.

i. Local Thrust Faults

Subsidiary thrust faults form wherever the geometry of other structures requires local convergence or shortening and the rocks react brittlely. We describe a number of such occurrences.

Diapiric structures involve less-dense rocks that move upward through more dense surroundings, such as salt intruding overlying sediments. In some cases, the diapir shoves the surrounding rocks upward and outward, and marginal thrust faults develop. One common example of such features is thrust faults marginal to some salt domes (Figure 5.9). In plan view, the thrust faults mimic the outline of the diapir. Because the cover rocks must also be stretched by this motion, normal faults develop over the top of the dome, as described in Section 4.3(i).

Thrust faults commonly are present where bends in strike-slip faults result in convergence of the rocks across the fault. We discuss this structural environment in greater detail in Chapter 6 (see Figures 6.7 and 6.12).

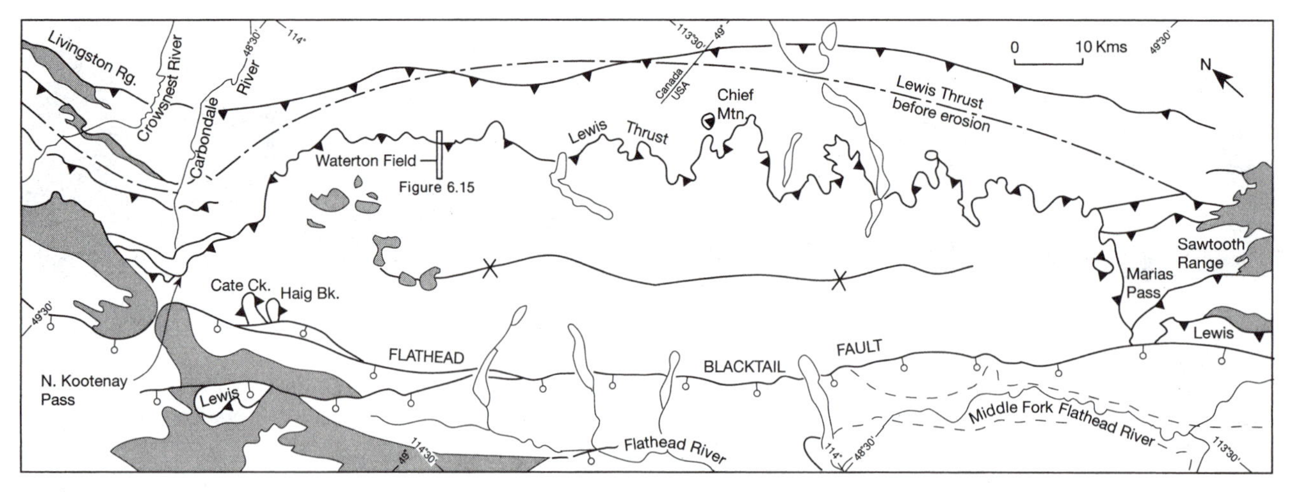

A.

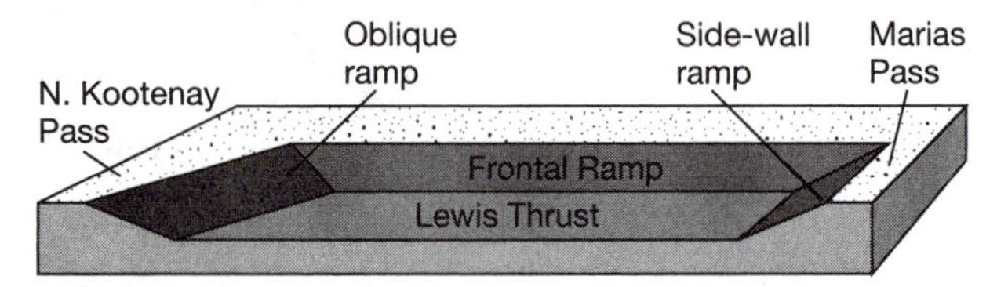

B.

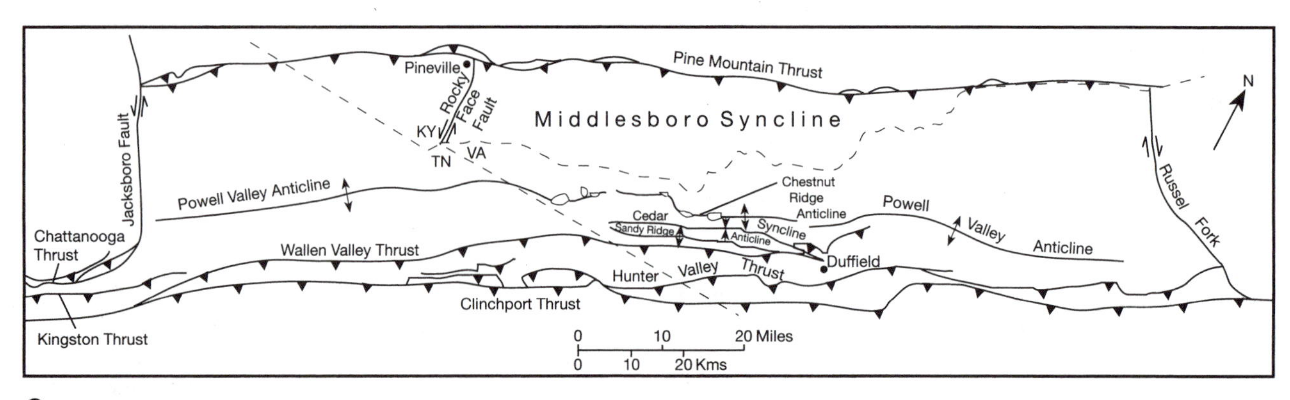

C.

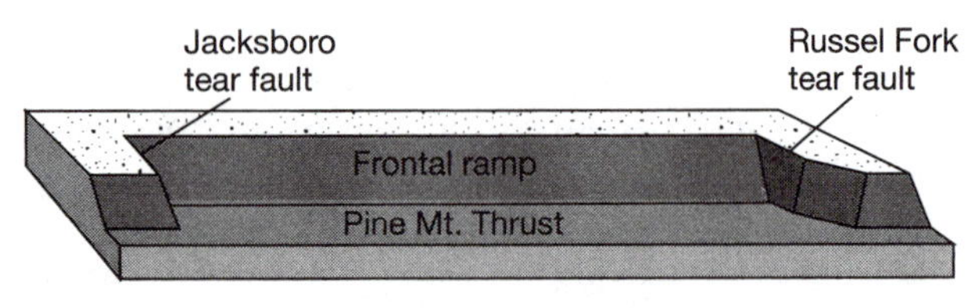

D.

FIGURE 5.7 Structure of the low-angle thrust faults. *A.* Map of the Lewis thrust near the Canada–United States border between Alberta and Montana (see the location marked on Figure 5.11*B*). The irregular nature of the thrust is a reflection of topography on the shallowly dipping fault surface. Note the Chief Mountain klippe near the border and the Cate Creek and Haig Brook windows near North Kootenay Pass. (From Boyer and Elliot 1982) *B.* Schematic block diagram showing the geometry of the Lewis thrust surface. Note in particular the frontal ramp that brings the fault up to the surface, the side-wall ramp near Marias Pass, and the oblique ramp near North Kootenay Pass. *C.* Map of the Pine Mountain thrust in the southern Appalachian Valley and Ridge province (see the location marked on Figure 5.11*A*). Tear faults mark the northeast and southwest ends of the Pine Mountain thrust sheet. (From Mitra 1988) *D.* Schematic block diagram showing the geometry of the Pine Mountain thrust surface. The tear faults bound the frontal ramp at either end.

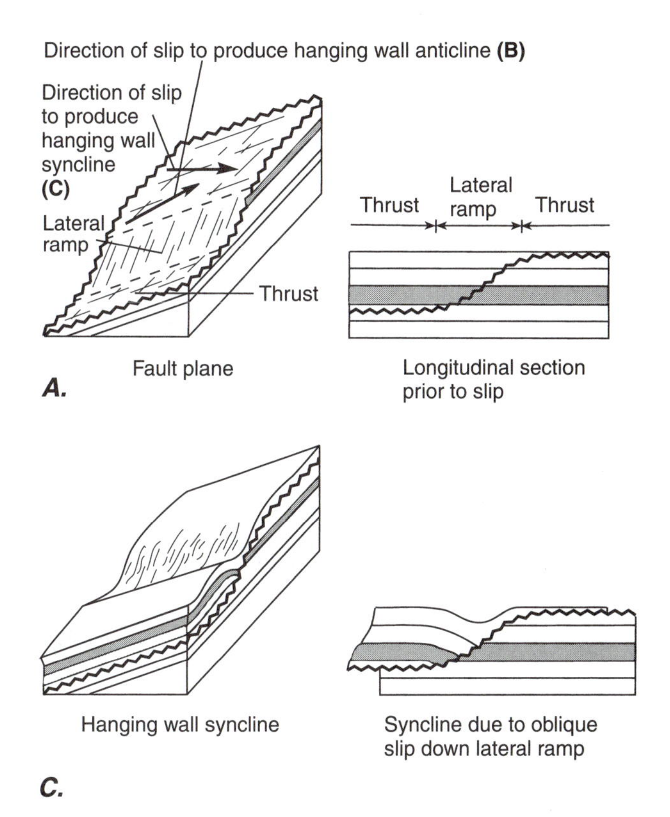

FIGURE 5.8 Effects of oblique displacement on sidewall ramps. The diagram on the left of each pair is a block diagram; the right diagram is the section shown as the right-hand face of the block. (After Dahlstrom 1970) *A*. Geometry of the thrust surface and the sidewall ramp with the hanging wall block removed. The arrows on the block diagram show directions of displacement leading to ramp folds shown in *B* and *C*. *B*. Hanging wall anticline produced by oblique slip up the sidewall ramp. *C*. Hanging wall syncline produced by oblique slip down the sidewall ramp.

ii. Thrust Faults Related with Folds

Thrust faults also are commonly associated with folds in four ways. First, as some folds develop, they reach a stage at which the sides, or limbs, of the fold cannot be rotated any closer together. Continued shortening of the layered sequence results in development of a thrust fault that generally cuts the steep or overturned limb of the fold (Figure 5.10*A*). Second, folds may develop as a result of thrusting to accommodate the deformation above the tip line of the thrust and are therefore called **fault-propagation folds**. As the displacement on the thrust increases, the tip line propagates through the layers and

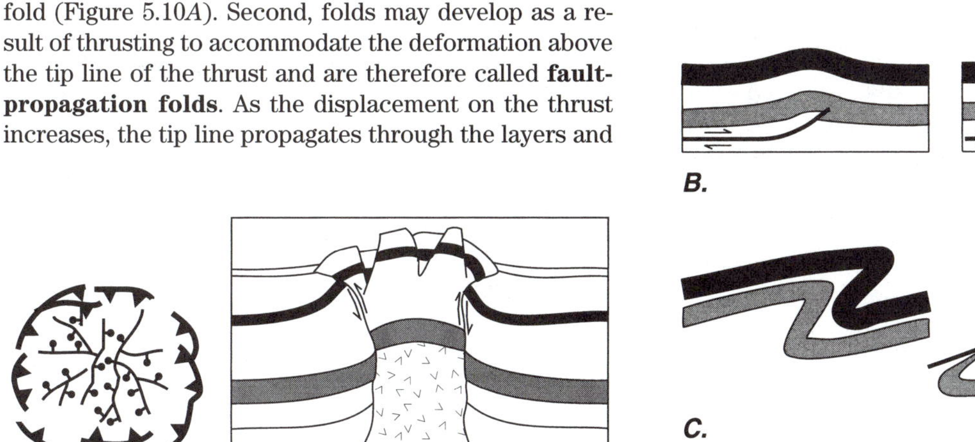

FIGURE 5.9 Peripheral thrust faults produced by diapiric intrusion. Normal faults in the central area accommodate extension associated with uplift. *A*. Schematic map. The balls on the ends of the tick marks on the faults indicate the down-dropped side of the normal faults. *B*. Schematic cross section.

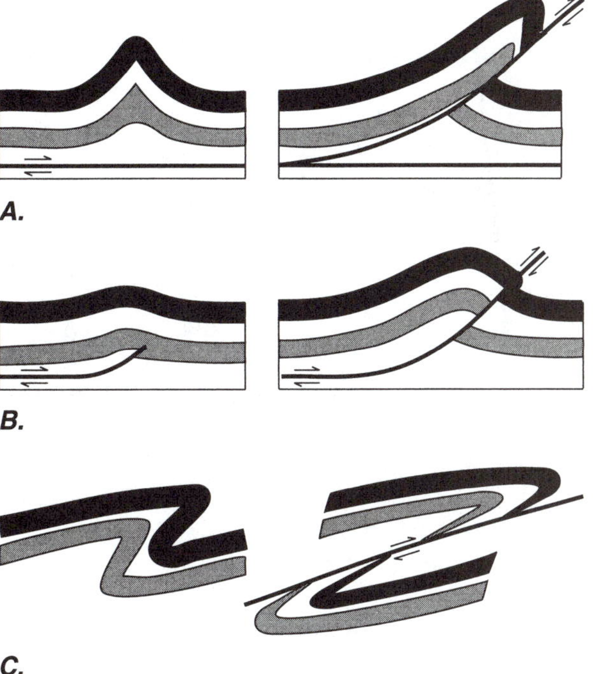

FIGURE 5.10 Diagrammatic cross sections illustrating relationships between folds and thrust faults (see also Figure 5.8). *A*. Thrust fault cuts up from the decollement through the foreland limb of a fold when the fold becomes too tight to accommodate further shortening. *B*. Fold forms in association with the propagation of a thrust fault. *C*. Formation of a fold by ductile flow can result in the shearing out of one limb to form a ductile thrust fault.

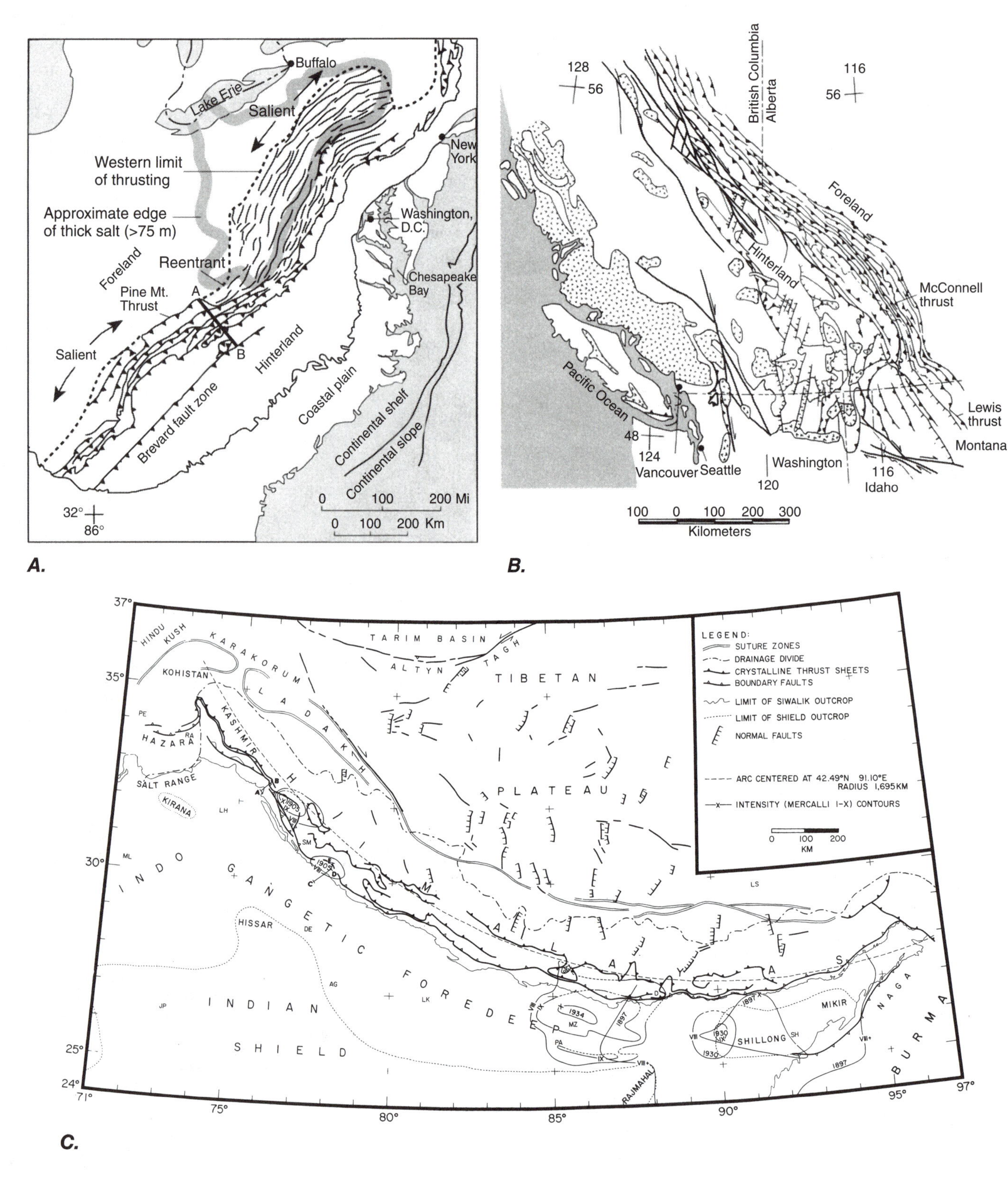

FIGURE 5.11 Major thrust systems showing the foreland, hinterland, salient, or virgation and re-entrant or syntaxis relative to the direction of movement for each fold and thrust belt. Teeth on the thrust faults are on the side of the hanging wall. *A.* Generalized map of the Appalachians. Plain lines are fold hinges; barbed lines are thrust faults with barbs on the hanging wall side. (After Harris and Bayer 1979) *B.* Generalized map of the Canadian Cordillera. (After Price and Hatcher 1983) *C.* Generalized map of the Himalaya, showing main thrusts, normal faults, the Indus-Tsangpo suture (northern boundary of the Himalaya proper), as well as regions of historic earthquakes. (After Seeber et al. 1981)

cuts the steep limb of the fold (Figure 5.10*B*; see Figures 13.31, 13.33, and 13.34). Third, folds may develop a steep or inverted limb that becomes progressively sheared and thinned until it is in effect a ductile thrust fault (Figure 5.10*C*). Fourth, where thrust faults have an alternating ramp-flat geometry, movements along the faults cause fault-bend folds to form in the hanging wall block (Figures 5.5, 5.6, and 5.8).

iii. Thrust Systems

The most common examples by far of large thrust systems on the continents are the thrust faults in **foreland fold and thrust belts**, which mark the margins of major orogenic belts. Orogenic belts commonly become the sites of subsequent continental rifting, and therefore such thrust systems also tend to occur along the edges of ancient rifted continental margins. Because of the economic importance of the major reserves of hydrocarbons found in these belts and because of their intrinsic interest as a major type of tectonic feature of the world, these systems have been the object of an enormous amount of research.

The geometry of such thrust systems is distinctive. In plan view, a foreland fold and thrust belt consists of a set of many thrust faults and folds, more or less parallel to one another, that extend for hundreds or even thousands of kilometers (Figure 5.11). The area in front of the thrusts toward which the thrust sheet moved is the **foreland**, and the region behind the thrusts is the **hinterland**. Although in some places these systems are nearly straight, generally they are curved, as illustrated in Figure 5.11. We describe the curvature by its relationship to the direction of relative motion of the thrust sheet. In a **salient** or **virgation**, faults and folds form an arcuate belt convex toward the foreland. In a **re-entrant** or **syntaxis**, the arcuate belt is concave toward the foreland. Figure 5.11 shows three examples of such thrust systems, from the Appalachians (Figure 5.11*A*), the Canadian Cordillera (Figure 5.11*B*), and the Himalaya (Figure 5.11*C*). Thrust systems also display differences in elevation along strike. Relatively high areas, or **culminations**, commonly are present along salients, and relatively low regions, or **depressions**, commonly accompany reentrants.

In cross section, fold and thrust belts consist of a set of low-angle listric thrust faults having the same general orientation and thus the same vergence (Figure 5.12), which asymptotically join a major low-angle fault at depth called a **decollement**.[5] Decollements characteristically are parallel to bedding and occur along weak layers in the stratigraphy, such as salt, gypsiferous rocks, or shales; they separate deformed rocks in the overlying thrust sheets from differently deformed or undeformed rocks below. The thrust system consists of a wedge-shaped package of deformed rocks that is thinnest toward the foreland and thickens toward the hinterland. This package overlies a basement of undeformed rocks at a gently dipping basal decollement, often referred to as a **sole fault** or a **detachment**,[6] which cuts up through the stratigraphic section toward the foreland (Figure 5.12).

Most thrust faults include frontal and lateral fault ramps as described in Section 5.2 (Figures 5.5 through 5.8; compare Figure 3.34*A*). As a result, fault ramp folds are a common feature. Not all folds in thrust sheets are

[5]After the French word *décollement*, which means "unsticking, loosening, or disengagement" and thus is comparable to "detachment."

[6]The term *detachment* is sometimes reserved for low-angle normal faults.

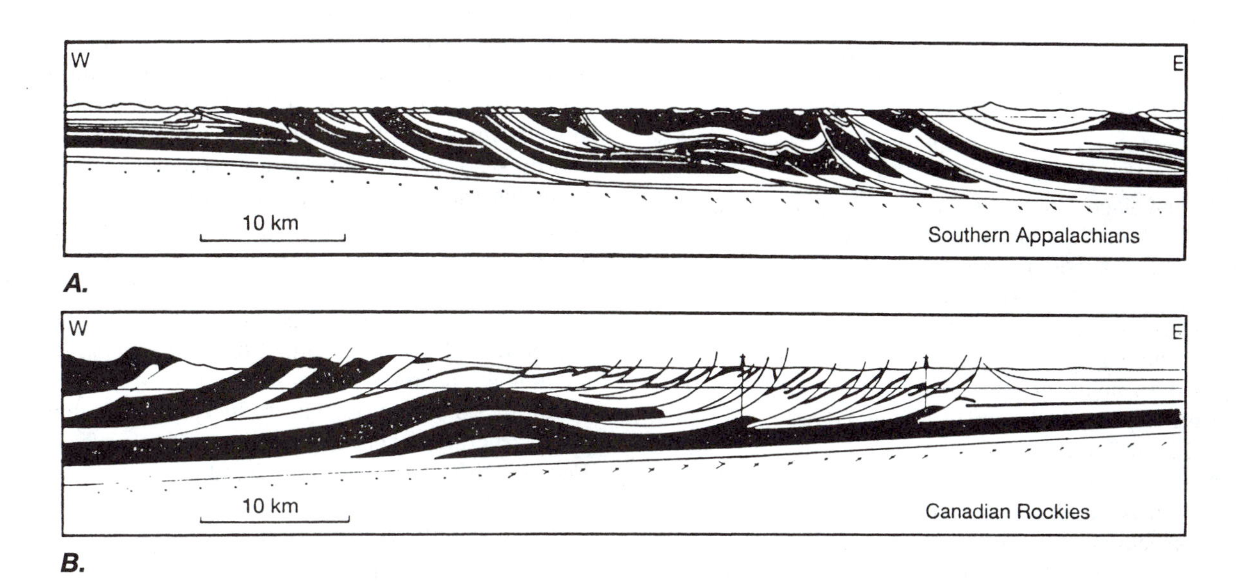

FIGURE 5.12 Cross sections of the major fold and thrust belts shown in Figure 5.11*A* and *B*. *A*. Southern Appalachians. (Davis et al. 1983, after Roeder et al. 1978) *B*. Canadian Cordillera. (Davis et al. 1983, after Bally et al. 1966)

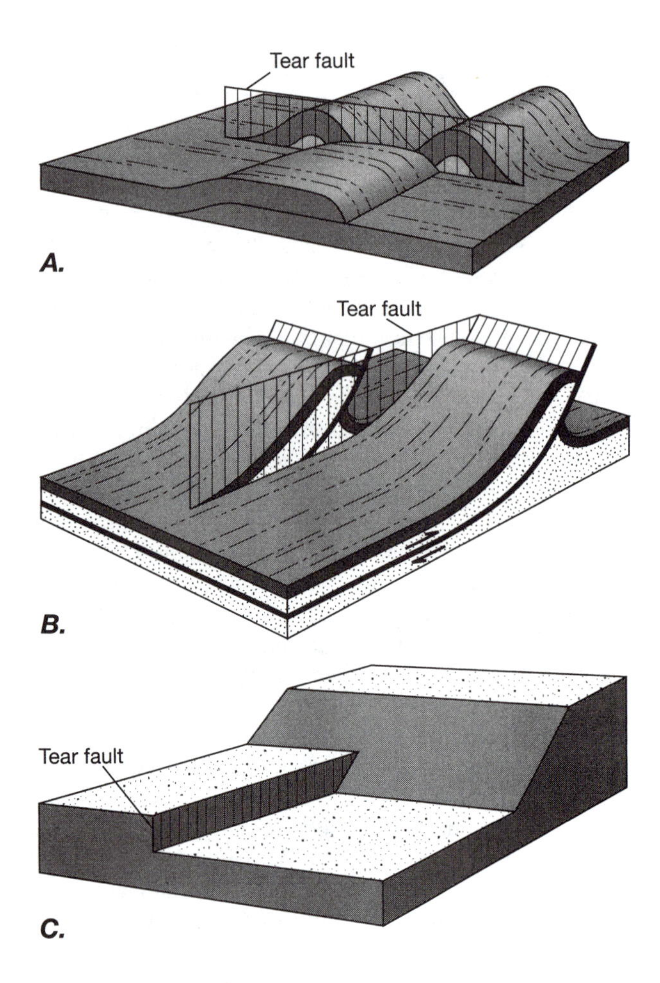

FIGURE 5.13 Thrust sheets segmented by tear faults. *A.* Shortening is accommodated by thrusting on one side of a tear fault and by folding on the other. *B.* Two noncoplanar imbricate thrusts are connected by a tear fault. *C.* Two segments of a thrust fault surface, each at a different structural level, are connected by a vertical sidewall ramp, or tear fault.

necessarily fault ramp folds, however; some may form to accommodate shortening of the thrust sheet above a flat portion of the decollement.

Thrust sheets are not structurally continuous features. Rather, they are segmented by tear or transfer faults, which accommodate differential displacement of different parts of the sheet, or connect parts of the active thrust that are not coplanar (Figure 5.13). For example, one part of the thrust sheet may shorten by faulting, and an adjacent part may shorten by folding. The discontinuity in displacement is then taken up by a tear fault (Figure 5.13*A*).

Many thrust systems include an **imbricate fan** or **schuppen zone**,[7] in which a number of individual listric thrust sheets, all with the same vergence, overlap like a series of roofing tiles (Figure 5.12). Faults in such imbricate systems generally are concave upward, with their dips becoming shallower with increasing depth and distance behind the thrust front. These listric thrust faults may cut the paleotopographic surface, or they may terminate upward at tip lines within the stratigraphic section. At depth, they terminate at branch lines along the decollement.

A **thrust duplex** is a system of imbricate thrust faults bounded below and above, respectively, by a floor thrust and a roof thrust. The imbricate thrusts form a stack of horses (Figure 5.14; cf. Figure 3.34*A*), with each thrust branching off from the floor thrust at a branch line and curving upward to define a roof thrust at the top of the stack. Unlike an imbricate fan that can break through to the surface, a duplex is by definition contained within the stratigraphic section. Like an imbricate fan, however, a duplex can develop along a frontal ramp on which the main thrust rises toward the foreland.

Duplexes exhibit a variety of forms that are distinguished by the arrangement of the horses and the geometry of the faults bounding them. Strata within the stack of horses of a duplex generally display characteristic asymmetrical anticline–syncline pairs as shown in Figure 5.14*B*, *D*. Beds above and below the duplex commonly parallel the roof and floor faults. The horses commonly dip toward the hinterland (hinterland-dipping) and form a zone of roughly constant thickness between the roof and the floor thrusts (Figure 5.14*B*). The horses may also form an antiformal stack over which the roof thrust curves through an antiform (Figure 5.14*C*). In still other cases, the horses dip toward the foreland (foreland-dipping), again defining a zone of roughly constant thickness between the roof and the floor thrusts (Figure 5.14*D*).

The Lewis thrust in the Waterton oil field in southern Alberta, Canada (Figure 5.7*A*), displays a more complex duplex structure (Figure 5.15*A*). There the Lewis thrust appears as the floor thrust and the Mount Crandell thrust is the roof thrust. The duplex geometry combines elements of a hinterland-dipping duplex and an antiformal stack. Higher thrust faults in the duplex, and the horses between them, are folded over fault ramps and associated horses lower in the section, indicating that slip on the higher thrusts must have occurred before the lower ones became active. Thus formation of the thrusts progressed in time downward in the duplex stack and toward the foreland (see Section 5.4). The Waterton field duplex illustrates the fact that in duplexes, earlier thrusts may be folded by displacement on later faults, resulting in culminations in the earlier thrust. Erosion of such culminations subsequently can produce windows exposing lower structural levels. Restoration of this cross section (Figure 5.15*B*) is discussed in detail in Section 5.5.

Individual thrust faults, like any other structure, are limited in extent; generally they are considerably shorter than the fold and thrust belt as a whole. If the horizontal shortening normal to the trend of a fold and thrust belt is relatively constant along the belt, then where a thrust fault dies out along strike at a tip line, its displacement must be transferred to another overlapping,

[7]After the German word *schuppe*, which means "scale," as in the scales on a fish.

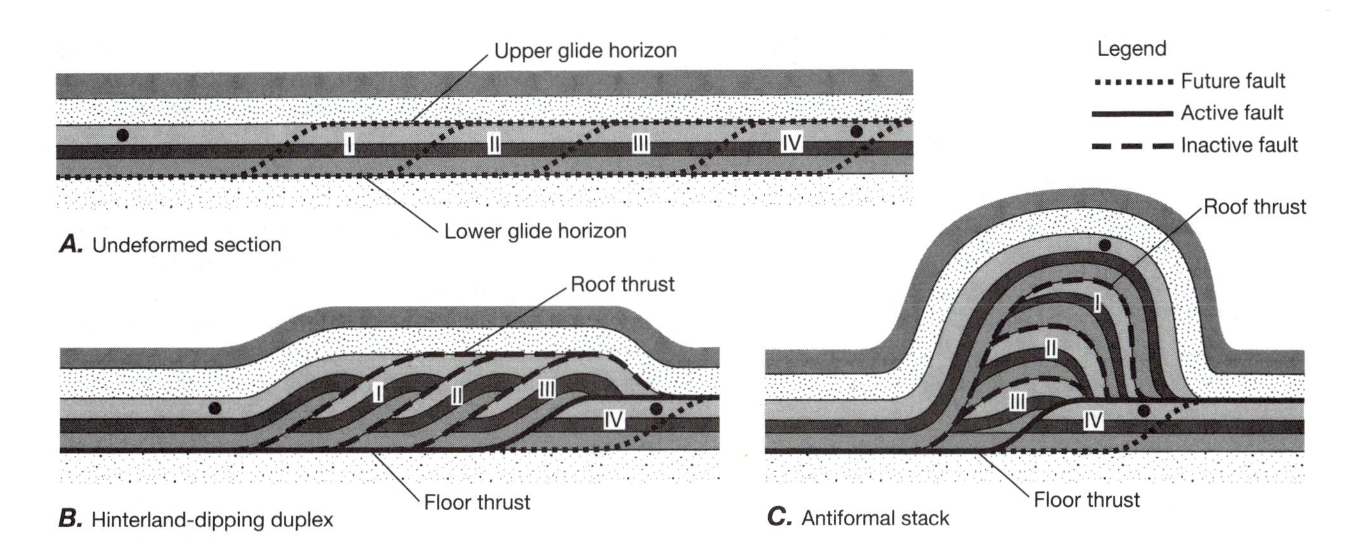

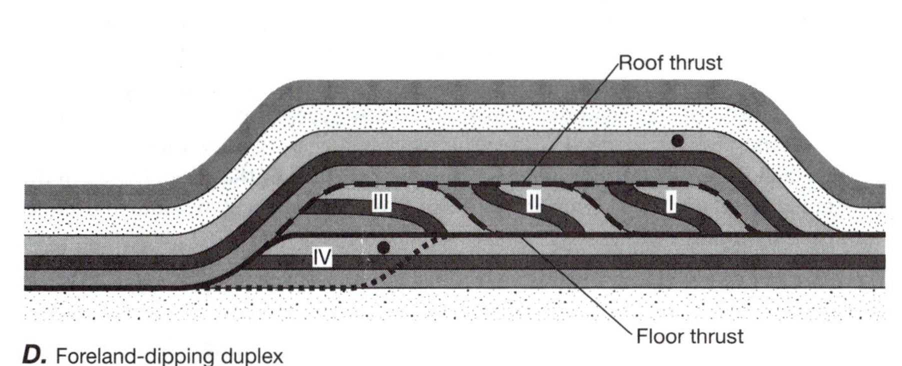

FIGURE 5.14 Schematic geometry of duplex structures resulting from the progressive cutting of the thrust fault into the footwall block. Thrust faults are marked by heavy lines; short-dashed lines are used for future faults; solid for active parts of the fault; long-dashed for inactive parts of the fault on which displacement has occurred. The large black dots in the upper layer mark the same two points in each diagram. The roman numerals mark the same horses in each diagram.

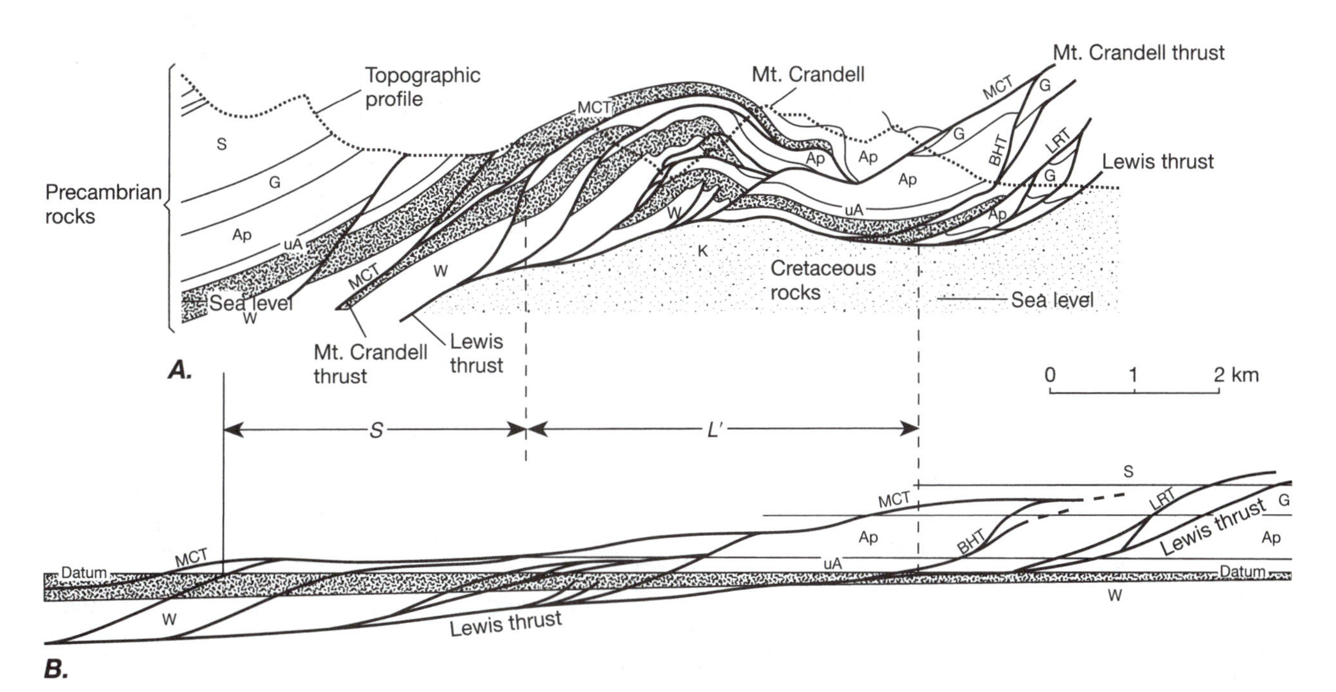

FIGURE 5.15 Cross section of duplex structure near the Waterton field in the Lewis thrust sheet near the Canada–United States border (compare Figure 5.7*A*). (From Boyer and Elliot 1982) *A*. Generalized cross section showing that the Lewis thrust is the floor of the duplex where Precambrian rocks are thrust over Cretaceous siliciclastics. The Mount Crandell thrust is the roof thrust. *B*. Palinspastic balanced cross section restoring the duplex in *A* to its original configuration.

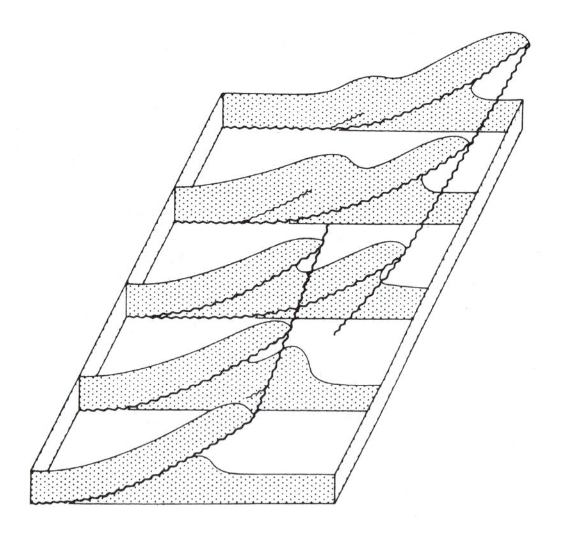

FIGURE 5.16 A simple transfer zone where one thrust fault dies out and the displacement is transferred through the transfer zone by folds to an *en echelon* thrust fault. (From Dahlstrom 1970)

or *en echelon*, thrust. Figure 5.16 shows an example of the **transfer zone** between two *en echelon* thrusts that merge into the same basal decollement. As the displacement on the upper thrust decreases, shortening is taken up first by a fold in its footwall block and then by a new thrust that cuts the fold (cf. Figure 5.10*A* or *B*). Finally the upper thrust decays into a fold in the hanging wall of the lower thrust, and the displacement is progressively transferred to the lower thrust.

Because many fold and thrust belts are developed in continental marginal sequences that overlie an older rifted continental margin, thrust faults may be localized along normal faults that initially were active during the rifting stage. In such cases, the stratigraphy of each thrust slice differs slightly from its neighbors. The original normal faults have been transformed into thrust faults by horizontal shortening of a formerly extended terrane. This change from one style of deformation to another, in this case from extensional to contractional deformation, is sometimes called **inversion tectonics**.

Thrust faults in a thrust system need not all have the same vergence as the sole fault that underlies the thrust complex. In a variety of situations, faults called **back thrusts** may develop that have opposite vergence. One process that results in the characteristic development of back thrusts is called **tectonic wedging**. Tectonic wedges, in cross section, are triangular packages of rock bounded above and below, respectively, by a roof thrust and a sole thrust that have opposite vergences (Figure 5.17). The sole and roof thrusts of the triangle join in the subsurface to form a tip that propagates into the undeformed rocks, wedging them apart.

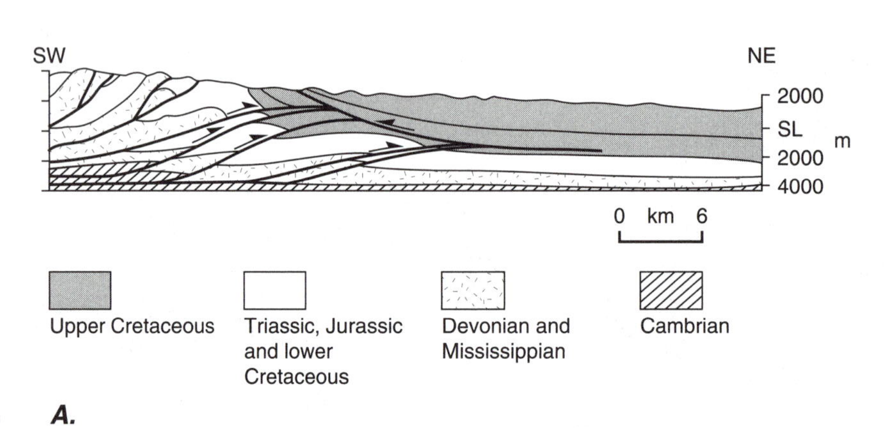

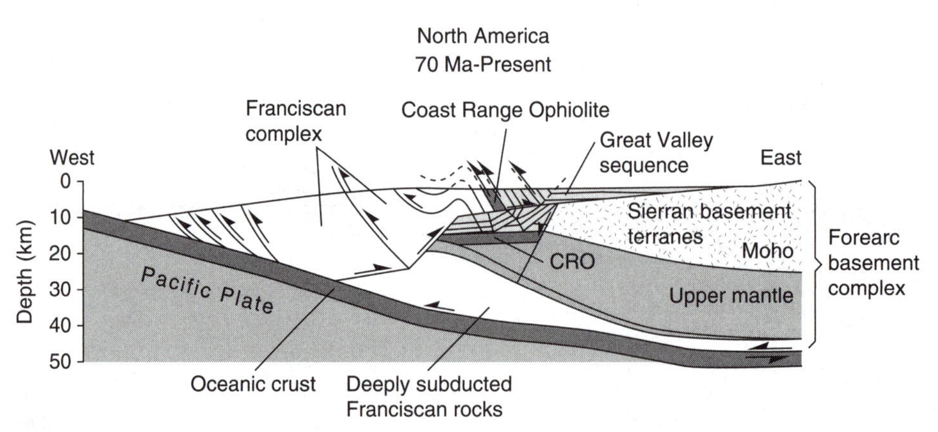

FIGURE 5.17 Tectonic wedging in subduction-thrust systems. *A*. Cross section of the eastern edge of the Canadian Rockies fold and thrust belt. (After Price 1986; Mountjoy 1980) *B*. Generalized cross section of the Pacific margin of North America in California, prior to development of the San Andreas fault system, showing antithetic thrusting of the Franciscan, the Coast Range Ophiolite (CRO), and the Great Valley Sequence against the western edge of the Sierra Nevada basement. (After Wakabayashi and Unruh 1994)

Figure 5.17*A* shows an example from the Canadian Cordillera in Alberta. The east vergent fold and thrust belt forms a tectonic wedge bounded above by the west-vergent roof thrust. The foreland stratigraphy is wedged apart at the tip of this structure, with the shallower strata riding west, up and over the triangle zone along the roof thrust. Multiple back thrusts such as this may develop in sequence as the tip of the tectonic wedge propagates into the foreland stratigraphic section. This type of structure is characteristic of the eastern margin of the Cordilleran fold and thrust belt for hundreds of kilometers from northwestern Alberta into the northern United States in Montana.

Tectonic wedging is not restricted to the edges of fold and thrust belts. Figure 5.17*B* shows an example from the Coast Range of northern California, where tectonic wedging occurs within the forearc of a subduction zone. In this case, the wedging is quite complex. The sole fault is the west-vergent, east-dipping subduction zone that existed off the west coast of California from about 165 Ma (million years ago) until some time between about 20 Ma and the present, depending on the latitude. The forearc basement complex, including Sierran basement terranes, an upper mantle wedge, and underlying deeply subducted Franciscan rocks, form a tectonic wedge between the down-going plate and the remaining Franciscan complex in the accretionary prism. The roof thrust to this wedge is an east-vergent thrust that rises eastward through the forearc and carries the overriding Franciscan complex eastward into the forearc basin. This roof thrust itself forms the sole thrust to a tectonic wedge farther east in the Great Valley sequence, which constitutes the sediments of the forearc basin. This wedge tilts and lifts the Great Valley rocks on a series of west-vergent roof thrusts that form major structures of the Coast Range. Thus the east-vergent roof thrust to the basement tectonic wedge is itself the sole fault to a tectonic wedge whose west-vergent roof thrust carries the forearc sediments westward over the accretionary complex.

5.4 KINEMATIC MODELS OF THRUST FAULT SYSTEMS

To understand how thrust systems form, we need to know the sequence of development of thrusts in the imbricate fans and duplexes, that is, whether new faults develop in front of (toward the foreland) or behind (toward the hinterland) the existing faults. The structure of duplexes is diagnostic of this sequence, and we discuss four models by which a duplex might form and describe the features that are characteristic of each.

The first model (Figure 5.14) assumes that the decollement cuts up section at a fault ramp and that the active ramp steps progressively into the footwall block (toward the foreland), cutting out a new horse from the footwall block and transferring it to the hanging wall block with each step. Figure 5.14*A* shows the undeformed section with short-dashed heavy lines indicating the location of future faults, and with the block numbers increasing in the order in which they will be transferred from footwall to hanging wall block. Initially, blocks I through IV are all part of the footwall, and block I is overridden by the strata to the left. After a certain displacement, the fault ramp steps a set distance into the footwall block, transferring horse I to the hanging wall block, which is then thrust up the ramp and over the remaining footwall block. This process of displacement on the fault followed by a stepping of the fault ramp into the footwall block successively transfers horses II and III from the footwall block into the hanging wall block, one after the other, and thrusts them successively over the remaining footwall block (Figure 5.14*B*–*D*). In Figure 15.14*B*, *C*, and *D*, block IV would be the next horse to be transferred to the hanging wall block and thrust up the footwall ramp.

Figure 5.14*B*–*D* shows the duplex structures that result for a different magnitude of the displacement that occurs after each step of the ramp into the footwall block. If the front end of the youngest horse is displaced only to the top of the ramp, a hinterland-dipping duplex results (Figure 5.14*B*). If the displacement emplaces the front end of the youngest horse near the point where the next frontal ramp will emerge, an antiformal stack of horses results (Figure 5.14*C*). If the youngest horse is emplaced so that only its rear end overlaps the point at which the next ramp will emerge, a foreland-dipping duplex results (Figure 5.14*D*).

The active thrust of a duplex is the floor thrust in each diagram of Figure 5.14, as indicated by the solid heavy line. The roof thrust is never active as a single fault. It is a composite fault consisting of fault segments on successive horses that were active at different stages in the development of the structure. In Figure 5.14*B*, the roof thrust appears horizontal because each horse is displaced only to the top of its ramp. If the displacement of each horse were less or more than this amount, the roof thrust would develop a more irregular shape.

Faults that become folded during the formation of a duplex must be older than unfolded faults (e.g., Figure 5.14*C*), because a folded shape makes faults unsuitable for slip. Moreover, faults become folded because of slip on fault ramps deeper in the structure, implying that the youngest faults are the deeper ones and thus that faulting has propagated toward the foreland. This model seems able to account for the geometry of duplexes found in nature. Duplexes with hinterland-dipping, antiformal, and foreland-dipping structures lie along a continuum that reflects progressively larger ratios of displacement on each ramp to the length of each horse in the structure.

The second and third of our four models of duplex development invoke a process by which ramps in the thrust fault cut progressively into the hanging wall block (toward the hinterland), transferring horses from the hanging wall block to the footwall block (Figure 5.18). In

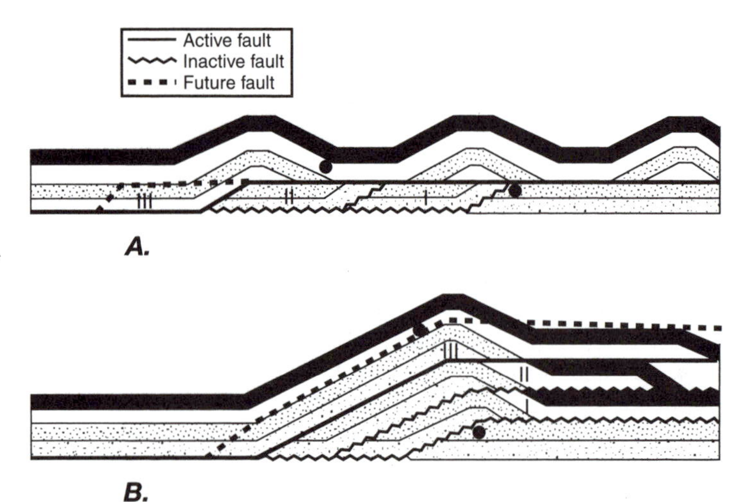

FIGURE 5.18 Two kinematic models for the formation of duplex structure in which the duplex develops by the stepwise retreat as the thrust fault frontal ramp steps back into the hanging wall block leaving horses in the footwall block. Thrust faults are marked by heavy lines: solid for active segments of the thrust system, and wavy for inactive segments of the thrust system. (After Boyer and Elliot 1982) *A*. The upper glide horizon remains the same with each stepwise retreat of the frontal ramp. *B*. The upper glide horizon steps up in the structure with each stepwise retreat of the frontal ramp.

the second model (Figure 5.18*A*), the upper glide horizon remains at the same structural level with each stepwise retreat of the frontal ramp. Block III will be the next horse to be transferred from the hanging wall block to the footwall block. In the third model (Figure 5.18*B*), the roof thrust steps up in the structure with each step of the ramp back into the hanging wall block. Again, block III will become the next horse to be transferred from the hanging wall block to the footwall block. Evidence for such structures is unusual, indicating that these second and third models do not represent common geologic processes.

In a fourth model, a duplex could form if a younger thrust fault forms the roof fault of a duplex by truncating a pre-existing imbricate thrust fan (Figure 5.19). The major characteristics of this model are that the roof thrust is the youngest fault in the system and is a single fault, active at the same time over its entire length, and that the anticlinal parts of the horses are offset, resulting in a truncated imbricate fan in the hanging wall block.

For imbricate thrust fans, it is difficult in many cases to determine the sequence of formation of the splay faults. The kinematic models are similar to those for duplexes shown in Figures 5.14 and 5.18, but because the splay faults eventually break the surface, no roof fault forms and a stack of imbricate thrust slices would form regardless of whether the new faults cut into the footwall or the hanging wall block. Thus the geometry of an imbricate thrust fan alone does not indicate whether the faulting progressed toward the foreland or the hinterland. Stratigraphic information, particularly in regions where sedimentation is active, can help resolve the ambiguity. In many thrust systems, such as the Idaho–Wyoming fold and thrust belt, the Canadian Rockies, and to some extent the Himalaya, the sediments caught up in the thrust wedge become progressively younger as one approaches the foreland, and distinctive sediments eroded off the thrust sheet are themselves cut by thrust faults that postdate the sedimentation. Inactive faults further toward the hinterland in the thrust wedge may be covered unconformably by sediments that are themselves cut by thrust faults toward the foreland. These observations indicate that the thrust faults have propagated by cutting progressively into the footwall block toward the foreland.

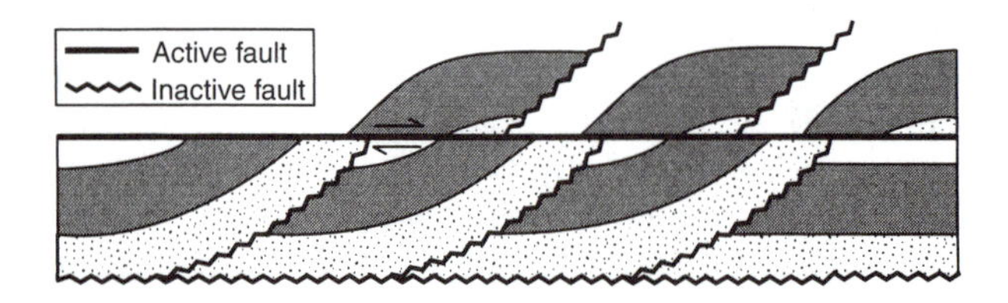

FIGURE 5.19 A duplex structure develops if an imbricate fan is truncated by a younger thrust, which then forms the roof thrust of the duplex. Heavy lines indicate faults: solid where the fault is active, and wavy where the fault is inactive. (After Boyer and Elliot 1982)

Most large fold and thrust systems are dominated by the progressive cutting of fault ramps and fault splays into the footwall block and thus toward the foreland (Figure 5.14). Examples of faults cutting progressively into the hanging wall block, and thus toward the hinterland, are present at least locally and are called **out-of-sequence thrusts**, or **back thrusts**, but they represent a minor proportion of faults in these thrust systems.

5.5 GEOMETRY AND KINEMATICS OF THRUST SYSTEMS IN THE HINTERLAND

None of the thrust system models discussed in the previous section deals with a complete cross section containing all fault termination lines (see Section 3.4(ii)). Because thrust systems accommodate substantial horizontal shortening of the crust and because these systems are composed of shallowly dipping faults, we must consider what happens to the continental crust below the sole fault. Moreover, to complete the model, we must consider what becomes of the sole fault beneath the hinterland. We discuss three models that provide a geometrically complete system.

First, the sole faults may return to the surface somewhere so that the horizontal shortening accommodated

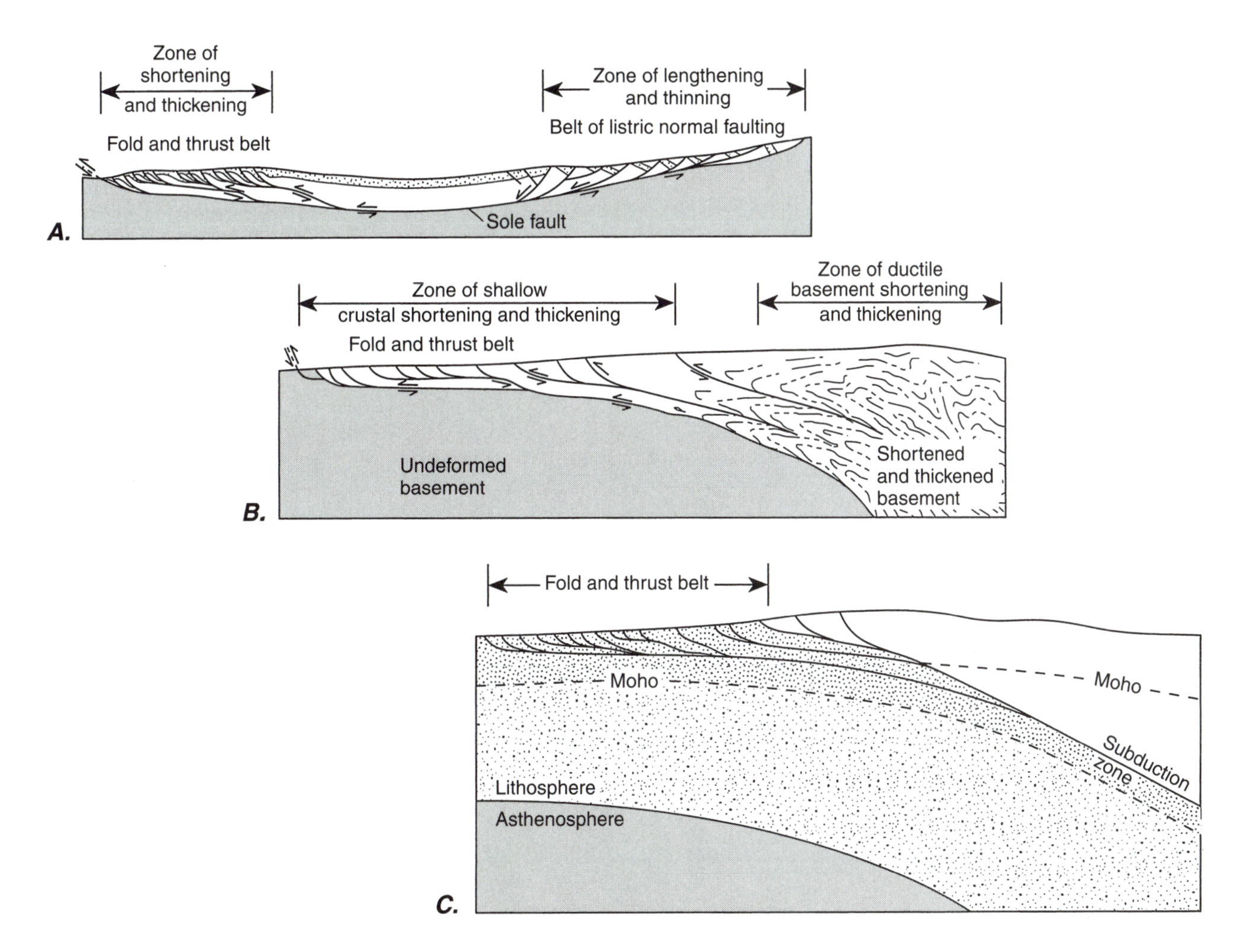

FIGURE 5.20 Hypothetical and schematic, but geometrically complete, models for fold and thrust belts. *A*. A fold and thrust belt paired with an extensional terrane. *B*. The root zone model. C. The subduction model. The patterned areas define the down-going plate.

by listric thrust faults in one area of the crust is balanced by horizontal lengthening in a system of listric normal faults in another region (Figure 4.16*A*). The implied pairing of a belt of shortening with a belt of lengthening may occur with shallow fault systems such as in the sediments of the Gulf Coast (Figure 4.16*C*). The scale of displacement there, however, is probably much less than observed in typical foreland fold and thrust belts, which have never been paired with an area of comparable extension.

Second, the basement rocks may be shortened by processes other than thrusting. The hinterland of an orogenic belt is characterized by high-grade metamorphic rocks that show abundant ductile deformation. Possibly sole faults of the foreland fold and thrust belts terminate in a so-called root zone of ductile deformation within the metamorphic core. The gravitational collapse of the topographic high created by the shortening and thickening of the metamorphic core could be responsible for compression in the shallow wedge-shaped fold and thrust belt on the margin of the orogenic belt (Figure 5.20*B*; compare the model experiment in Figure 18.2*B*).

Still a third possibility, and an intriguing one, is that the large displacements and shortening in fold and thrust belts reflect the involvement of continental crust in the largest type of thrust system that we know on Earth, a subduction zone (Figure 5.20*C*). According to seismic evidence, some subducted slabs are continuous down to depths of 1200 km or more, implying at least that much slip on the subduction thrust. Where continental crust is on a down-going plate, it can be carried into a subduction zone and subducted to a depth of as much as about 100 km. This implies roughly 140 km of slip on a subduction zone that dips at 45°, or 200 km for a 30° dip. Thus the ultimate sole fault to a foreland fold and thrust belt may be simply a subduction zone, in which case the thrust belt is a series of splay faults off a convergent plate boundary fault, and the driving force for thrusting would be that of subduction itself. Several current continental collision zones provide examples of this situation. The west-vergent West Taiwan fold and thrust belt (Figure 5.21*A*) is a direct result of a continent–island arc collision between the East Asian continent, which is being

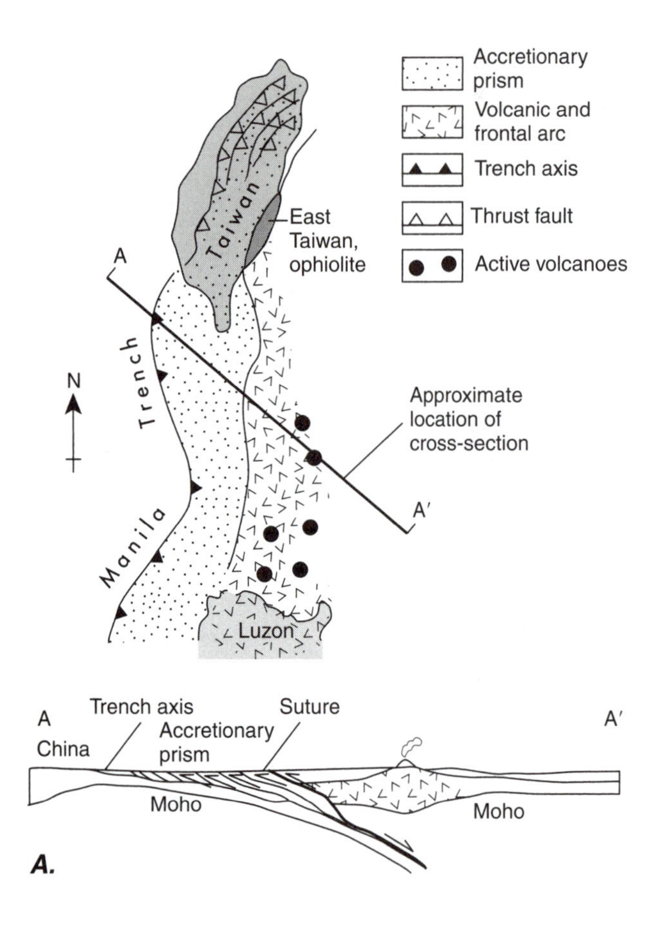

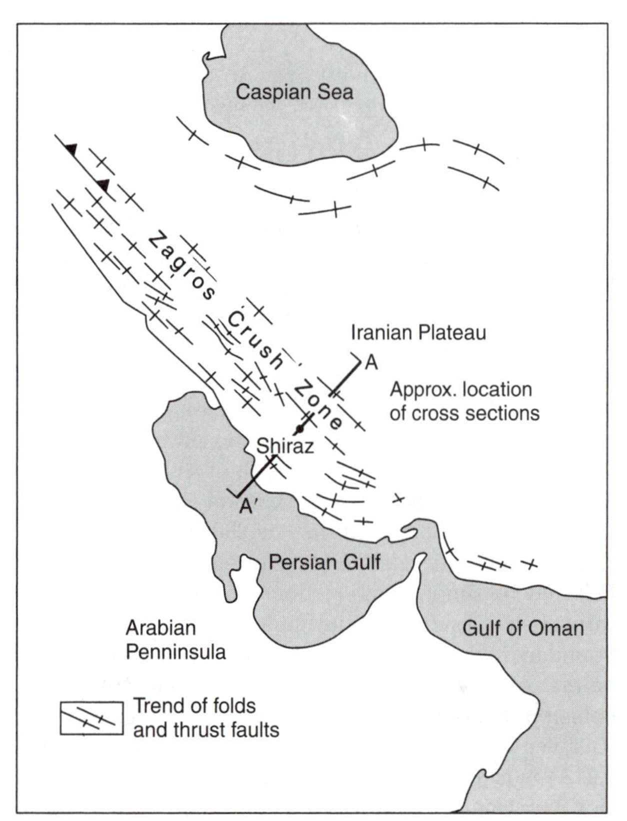

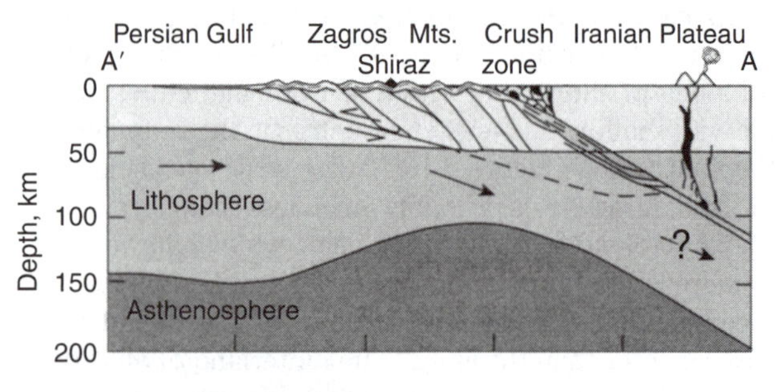

FIGURE 5.21 Thrust systems related to collision at a subduction zone. *A*. Map and cross section of the Taiwan region. Map shows the western fold and thrust belt in the accretionary prism, the eastern volcanic arc, and the presently active major faults. (After Liou et al. 1977) *B*. The Zagros fold and thrust belt, Iran. Map shows the Zagros Crush Zone and folded region of the Zagros Mountains. Cross section is an interpretation based on surface geology and the location of seismic events. (Map after Stöcklin 1974; cross section after Bird 1978)

subducted eastward under the West Luzon arc. The Zagros fold and thrust belt in southwest Iran (Figure 5.21*B*) is a continent–continent collision between the Arabian plate, which is being subducted northeastward beneath Iran on the Eurasian plate. Thus at least some foreland fold and thrust belts record places where the continental crust has been dragged into a subduction zone.

5.6 ANALYSIS OF DISPLACEMENT ON THRUST FAULTS

On most thrust faults, the existence of piercing points (Section 3.3(ii)) on both sides of the fault is rare, and it is necessary to resort to other methods to obtain an indication of the displacement. As with other faults, the three things we need to determine are the displacement direction, the shear sense, and the amount of the displacement.

i. Direction and Sense of Displacement

In addition to general features of faults that indicate shear sense and relative displacement direction (Sections 3.3, 5.1, and 5.2(ii)), some structures within the thrust sheet as well as the geometry of the thrust system itself can be used to constrain the thrust motion.

If, on a regional scale, the total amount of displacement on a thrust is generally the same along its length, and if erosion has not worn the sheet back unevenly, then the displacement is commonly taken to be approximately normal to the regional strike of the thrust fault or thrust system. Using this criterion, the thrust sheets in the southern Appalachian Mountains (Figure 5.11*A*) moved to the northwest, those in the Canadian Rockies (Figure 5.11*B*) moved to the northeast.

On many thrust faults it may not be obvious which direction the thrust sheet moved on the fault, especially on a local scale, and we must then rely on stratigraphic evidence. Although we tend to think of the hanging wall block as moving up the dip of the thrust fault, this may be an unreliable assumption because after thrusting, fault dips are commonly altered by folding. The slip direction of a thrust sheet is best indicated by the tendency of thrust faults to cut up-section in the direction of displacement (Figures 5.12 and 5.15).

The problem with a folded thrust fault is illustrated by the Mount Crandell thrust in Figure 5.15*A*. In the vicinity of Mount Crandell, the folded fault has a northeast dip, which is opposite to its dip further to the northeast and to the southwest. Based only on the exposures in Mount Crandell, an inference of up-dip displacement of the thrust sheet would indicate top to the southwest. A restoration of the initial structure of the Mount Crandell area (Figure 5.15*B*), however, shows that in fact the Mount Crandell thrust cuts gradually but consistently up-section from southwest to northeast throughout its length, indicating that the relative displacement of the thrust sheet is toward the northeast, consistent with the regional picture.

In imbricate thrust systems, the thrusts as a rule branch up from the basal decollement in the direction of relative movement of the thrust sheet (Figures 5.12 and 5.15), although in Figure 5.17*B*, the back thrusts that carry the Great Valley rocks over the Franciscan tectonic wedge have a vergence opposite to that of the sole fault. In general, however, in a complexly deformed region, identification of thrusts branching off the sole fault provides another indication of the sense of their relative displacement. The use of any of these criteria requires accurate data regarding the overall attitude of the faults, the nature of their intersections, and their relationships with the stratigraphy.

In many thrust sheets, asymmetric folds develop during thrusting. Such folds have one longer side, or limb, that dips at a relatively shallow angle, and one shorter limb that either dips steeply in the opposite direction or is overturned (e.g., Figure 5.10*C*). The tops of the folds can be thought of as leaning in the direction of the steep or overturned limb. This direction of leaning, the **vergence** of the fold, indicates the direction of relative motion of the thrust sheet. We discuss the geometry of folds in more detail in Chapter 10.

ii. The Amount of Displacement Determined from Maps

In order to gain a better understanding of large-scale tectonic processes and to predict possible sites for accumulation of economic deposits such as oil, we need to determine the amount of displacement along a thrust fault or system. Unfortunately, a map cannot provide unequivocal determinations of displacement, and it is therefore advisable to employ more than one method to obtain a more complete understanding of the thrusting.

In some cases, the irregularities in the thrust trace, including windows and klippes, provide a minimum estimate of displacement. The desired measurement is the distance between the exposures closest to and farthest from, the hinterland in the movement direction. Figure 5.4 illustrates the technique. Figure 5.4*A* shows the displacement determinable from irregularities in the trace of the thrust fault. Using the exposure of a window and a klippe (Figure 5.4*B*) provides a larger lower bound for the displacement.

For the Lewis thrust (Figure 5.7*A*), we assume a displacement direction of N55E perpendicular to the average trend of the fault trace. The minimum possible displacement determined from the fault trace irregularities (Figure 5.4*A*) is approximately 12 km. Measuring parallel to the same displacement direction but using the thrust trace and the Cate Creek window, however, gives a minimum possible displacement of about 40 km.

Figure 5.22 shows the application of this analysis to two very famous windows, the Engadine and Tauern windows of the eastern Alps. In each case, the upper thrust sheets or nappes, called the Austro-Alpine nappes, overlie a lower series called the Penninic nappes. The distance

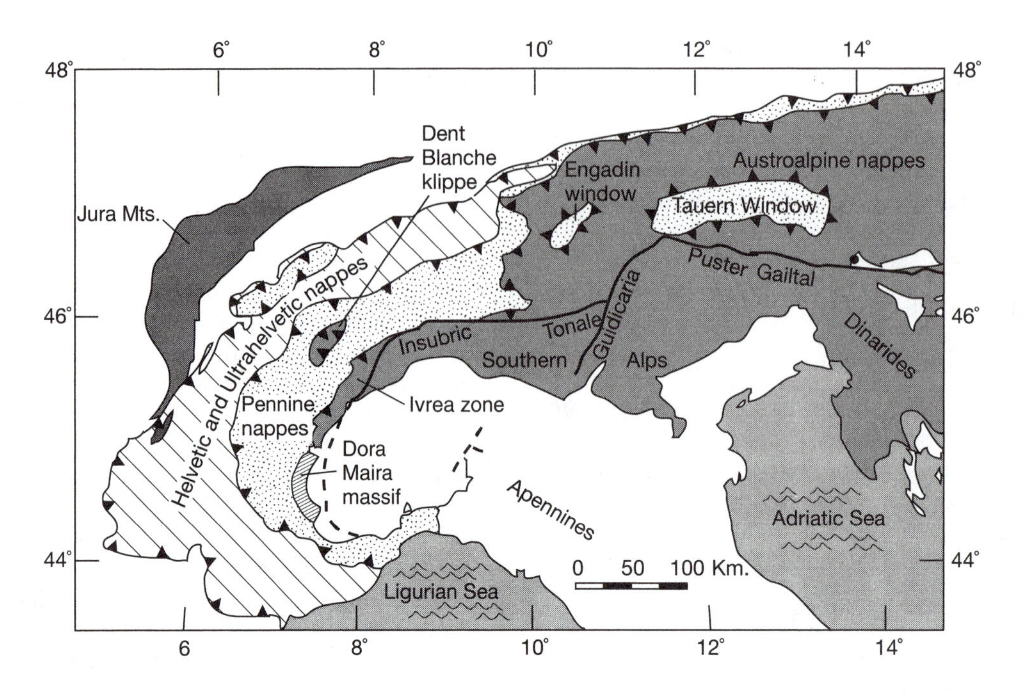

FIGURE 5.22 Generalized map of the Alpine region, Europe, showing three principal thrust complexes: Helvetic (diagonal ruling), Penninic (stipple pattern), and Austro-Alpine systems (medium grey). Two windows, the Tauern and the Engadin, show the Penninic nappes underneath the Austro-Alpine nappes. The Dent Blanche klippe is an erosional outlier of the Austro-Alpine nappes on top of the Penninic nappes; to the northwest is another klippe of the Penninic nappes on top of the Helvetic nappes. (Modified from Ernst 1973)

from the rear of these windows roughly northward to the front of the main thrust trace indicates a minimum possible displacement of as much as 100 km for the Austro-Alpine nappes.

iii. The Amount of Displacement Determined from Cross Sections

Cross sections of thrust faults can be used to determine the magnitude of the displacement if the section is parallel to the displacement direction on the fault. Figure 5.23 shows the simplest example of this situation in which the displacement and shortening are related by the dip angle (ϕ) of the fault, and the displacement (δ) is determined with a simple linear measurement.

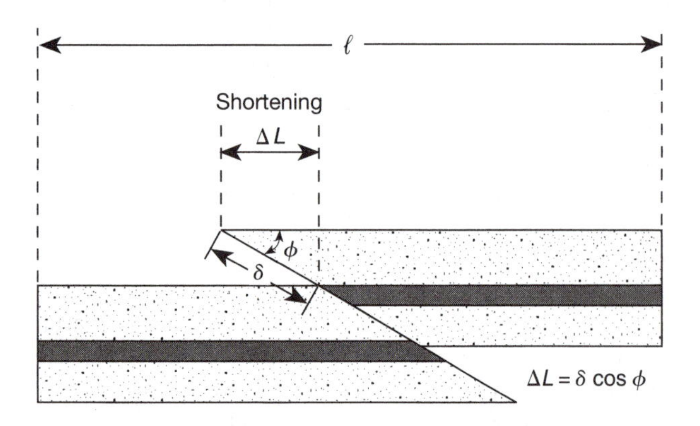

FIGURE 5.23 Shortening associated with thrust faulting, showing how the change in length ΔL is related to the displacement d and the dip angle of the fault ϕ for simple faults.

For more complicated structures, the determination of the total amount of displacement and shortening is more difficult. The cross section through the Lewis thrust system in Figure 5.15, for example, consists of a combination of imbricate and duplex faults, some of which have been folded above the younger thrusts. The original continuous stratigraphic sequence appears intact at the left side of the cross section.

In such cases, we construct a balanced cross section, using techniques described in Section 3.5. We concentrate for this example on the area between the Lewis thrust and the Mount Crandell thrust. The lower Altyn formation, shown as the shaded layer, is used as the reference layer because it is contained in most of the thrust wedges and horses of the thrust system. The pinning points must be to the northeast (right) of where the Lewis thrust cuts up through the stratigraphic section that is being balanced and to the southwest (left) of the duplex between the Mount Crandell and Lewis thrusts.

Figure 5.15*B* is the balanced **palinspastic**[8] cross section showing the undeformed stratigraphic sequence

[8]A "palinspastic map" is one that has restored the geology, geographic features, and tectonic features as nearly as possible to their configuration that preceded deformation by folding or faulting.

with the paths of the various thrust faults through the sequence. Two reference points at the top of the shaded lower Altyn unit in both the deformed and the palinspastic cross sections show that the amount of shortening (*S*) caused by the thrusting amounts to almost 3.5 km for a section originally only 8 km long. Thus this part of the section has been shortened by about 43%.

Across the Appalachian Valley and Ridge province from the Pine Mountain thrust to the Brevard fault (between points A and B in Figure 5.11*A*; see Figure 5.12*A*), the fold and thrust belt has accommodated roughly 280 km of shortening of the Earth's crust. "Retrodeforming" the thrust faults and taking eroded section into account reveal that the original width of the belt must have been about 435 km. A shortening of more than 60% has occurred!

REFERENCES AND ADDITIONAL READINGS

Allmendinger, R. W., L. D. Brown, J. E. Oliver, and S. Kaufman. 1983. COCORP deep seismic profiles across the Wind River Mountains, Wyoming. In A. W. Bally, ed., *Seismic Expression of Structural Styles*, volume 3, Tectonics of Compressional Provinces/Strike Slip Tectonics, AAPG Studies in Geology, v. 15, American Association of Petroleum Geologists, Tulsa, OK, USA: p. 3.2.1-29–3.2.1-33.

Bally, A. W., P. L. Gordy, and G. A. Stewart. 1966. Structure, seismic data and orogenic evolution of southern Canadian Rockies. *Canadian Petrol. Geol. Bull.* 14: 337–381.

Bird, P. 1978. Finite element modelling of lithosphere deformation: The Zagros collision orogeny. *Tectonophysics* 50: 307–336.

Boyer, S. E. 1995. Sedimentary basin taper as a factor controlling the geometry and advance of thrust belts. *Amer. J. Sci.* 295(10): 1220–1254.

Boyer, S. E., and D. Elliot. 1982. Thrust systems. *Amer. Assoc. Petrol. Geol. Bull.* 66: 1196–1230.

Brewer, J. A., S. B. Smithson, J. E. Oliver, S. Kaufman, and L. D. Brown. 1980. The Laramide orogeny: Evidence from COCORP deep crustal seismic profiles in the Wind River Mountains, Wyoming. *Tectonophysics* 62: 165–189.

Colpron, M., M. J. Warren, and R. A. Price. 1998. Selkirk fan structure, southeastern Canadian Cordillera: Tectonic wedging against an inherited basement ramp. *Geol. Soc. Amer. Bull.* 110: 1060–1074.

Coogan, J. C., and P. G. DeCelles. 1996. Extensional collapse along the Sevier Desert reflection, northern Sevier Desert basin, Western United States. *Geology* 24: 933–936.

Dahlstrom, C. D. A. 1970. Structural geology in the eastern margin of the Canadian Rocky Mountains. *Canad. Petrol. Geol. Bull.* 18: 332–406.

Davis, D., J. Suppe, and F. A. Dahlen. 1983. Mechanics of fold-and-thrust belts and accretionary wedges. *J. Geophys. Res.* 88: 1153–1172.

DeCelles, P. G. 2004. Late Jurassic to Eocene evolution of the Cordilleran thrust belt and foreland basin system, western US. *Amer. J. of Sci.* 304: 105–168.

DeCelles, P. G., and G. Mitra. 1995. History of the Sevier orogenic wedge in terms of critical taper models, northeast Utah and southwest Wyoming. *Geol. Soc. Amer. Bull.* 107: 454–462.

Ernst, G. 1973. Interpretative synthesis of metamorphism in the Alps. *Geol. Soc. Amer. Bull.* 84: 2053–2078.

Gansser, A. 1981. Himalaya: and overview. In H. K. Gupta and F. M. Delany, eds, *Zagros, Hindu Kush, Himalaya, Geodynamic Evolution*, Geodynamics series 3, 215–242.

Harris, L. D., and K. C. Bayer. 1979. Sequential development of the Appalachian orogen above a master decollement—a hypothesis. *Geology* 7: 568–572.

Koyi, H. A., K. Hessami, and A. Teixell. 2000. Epicenter distribution and magnitude of earthquakes in fold-thrust belts: Insights from sandbox models. *Geophys. Res. Lett.* 27: 273–276.

Lawton, D. C., D. A. Spratt, and J. C. Hopkins. 1994. Tectonic wedging beneath the Rocky Mountain foreland basin, Alberta, Canada. *Geology* 22: 519–522.

Liou, J. G., C.-Y. Lan, J. Suppe, and W. G. Ernst. 1977. The East Taiwan ophiolite, its occurrence, petrology, metamorphism, and tectonic setting. Taipei (Taiwan): Mining Research and Service Organization.

Liu, S., and J. M. Dixon. 1995. Localization of duplex thrust-ramps by buckling: analog and numerical modelling. *J. Struct. Geol.* 17(6): 875–886.

Macedo, J., and S. Marshak, 1999. Controls on the geometry of fold-thrust belt salients. *Geol. Soc. Amer. Bull.* 111: 1808–1822.

McQuarrie, N. 2004. Crustal scale geometry of the Zagros fold-thrust belt, Iran. *J. Struct. Geol.* 26: 519–535.

Merle, O. 1998. *Emplacement Mechanisms of Nappes and Thrust Sheets*. Boston, Dordrecht, 159 pp.

Mitra, S. 1988. Three dimensional geometry and kinematic evolution of the Pine Mountain thrust system, southern Appalachians. *Geol. Soc. Amer. Bull.* 100: 72–95.

Mitra, S., and G. W. Fisher, eds. 1992. *Structural Geology of Fold and Thrust Belts*. Baltimore, Johns Hopkins University Press.

Moores, E. M., and H. W. Day. 1984. Overthrust model for the Sierra Nevada. *Geology* 12: 416–419.

Mountjoy, E. W. 1980. In: *Geology, Mt-Robson, Alberta-British Columbia Geol. Surv. Can. Map 1499A* (Compiler).

Price, R. A. 1981. The Cordilleran foreland thrust and fold belt in the southern Canadian Rocky Mountains. In N. J. Price, ed, *Thrust and Nappe Tectonics*, Geological Society of London Special Publication 9.

Price, R. A. 1986. The southeastern Canadian Cordillera: Thrust faulting, tectonic wedging, and delamination of the lithosphere. *J. Struct. Geol.* 8(3/4): 239–254.

Price, R. A., and Hatcher, R. D. Jr. 1983. Tectonic significance of similarities in the evolution of the Alabama-Pennsylvania Appalachians and the Alberta–British Columbia Canadian cordellera. *Geol. Soc. Amer. Memoir* 158: 149–160.

Rodgers, J. 1970. *Tectonics of the Appalachians*. Wiley-Interscience, New York, 271 pp.

Roeder, D. H., and H. Bögel. 1978. Geodynamic interpretation of the Alps. In. H. Closs, D. Roeder, and K. Schmidt, eds., *Alps, Apennines, Hellenides: geodynamic investigation along geotraverses by an international group of geoscientists*, Scientific report—Inter-Union Commission on Geodynamics, Scientific Report, no. 38: Stuttgart: Schweizerbart, 191–212.

Seeber, L., J. G. Armbruster, and R. C. Quittmeyer. 1981. Seismicity and continental subduction in the Himalayan arc. In H. K. Kupta and F. M. Delany, eds., *Zagros-Hindu Kush-Himalaya Geodynamic Evolution*, AGU-GSA Geodynamics Series volume 3: 215–242.

Smithson, S. B., J. Brewer, S. Kaufman, J. Oliver, C. Hurich. 1978. Nature of the Wind River Thrust, Wyoming, from COCORP deep reflection data and from gravity data. *Geology* 6(11): 648–652.

Stöcklin, J. 1974. Possible ancient continental margins of Iran. In C. Burk and C. L. Drake, eds., *Geology of Continental Margins*. New York, Springer.

Unruh, J. R., V. R. Ramirez, S. P. Phipps, and E. M. Moores. 1991. Tectonic wedging beneath fore-arc basins: Ancient and modern examples from California and the Lesser Antilles. *GSA Today*, 1(9): 185–186.

Vergés, J., M. Marzo, and J. A. Muñoz. 2002. Growth strata in foreland settings. *Sed. Geol.* 146: 1–9.

Wakabayashi, J., and J. R. Unruh. 1994. Tectonic wedging, blueschist metamorphism and exposure of blueschists: Are they compatible? *Geology* 23: 85–88.

Williams, C. A., C. Conners, F. A. Dahlen, E. J. Price, and J. Suppe. 1994. Effect of the brittle-ductile transition on the topography of compressive mountain belts on Earth and Venus. *J. Geophys. Res.* 99(B10): 19947–19974.

Chapter 6

STRIKE-SLIP FAULTS

Most strike-slip faults are approximately vertical, at least near the surface of the Earth. As a result, their fault traces tend to be straight to gently curved lines on a map, even across rugged topography (Figure 6.1). Displacement on strike-slip faults is essentially horizontal, either a right or a left lateral shear, and thus it results in no net addition or subtraction of area to the crust. In some cases, oblique strike-slip motion that includes a component of reverse or normal slip results from a component of horizontal shortening or lengthening perpendicular to the fault trace in addition to the predominant strike-slip motion. Strike-slip faults exist on all scales in both oceanic and continental crust.

Tear faults are relatively small-scale local strike-slip faults, commonly subsidiary to other structures such as folds, thrust faults, or normal faults (e.g., Figures 5.7*C*, *D*, 5.13, and 6.10). They are steeply dipping and oriented subparallel to the regional direction of displacement. They occur in the hanging wall blocks of low-angle faults and accommodate differences in the amounts of displacement in the allochthon on either side of the fault. They also occur as vertical fault steps separating an allochthon from the adjacent autochthonous rocks (Figure 5.13*C*).

The term *transfer fault* (see Section 3.4(iii)) is applied to two different geometries of fault, one of which occurs in extensional or contractional terranes, and the other of which occurs in strike-slip terranes. In extensional or contractional terranes, transfer faults are strike-slip faults that tend to parallel the regional direction of displacement and to mark the boundaries of domains in which normal or thrust faults show different geometry and displacement (cf. Figures 4.11, 4.12, 5.13*A*, 5.16 and 6.10). Imbricate systems of normal faults or thrust faults—and possibly their detachments—terminate against such transfer faults, and in adjacent domains the normal or thrust faults may have different orientations and different amounts of displacement. There is no clear distinction between these strike-slip faults and tear faults described above.

In strike-slip terranes, transfer faults lie at a high angle to the regional displacement direction and connect adjacent or *en echelon* parallel strike-slip faults. They accommodate the transfer of displacement from one fault to the next, and slip on these faults is generally oblique. These transfer faults often develop into strike-slip duplexes, which we discuss in Section 6.2(ii) (Figures 6.6 and 6.7).

Transform faults and **transcurrent faults** are major regional strike-slip fault systems that generally comprise zones of many associated faults (Figure 6.2). Transform faults are strike-slip faults that form segments of lithospheric plate boundaries (Figure 6.2*A*). Transcurrent faults, on the other hand, are regional-scale strike-slip faults in continental crust that are not parts of the plate margin (Figure 6.2B).[1] Both types of faults may be many hundreds of kilometers long and may have accumulated relative displacements of up to several hundred kilometers.

---

[1]The specific usage of these two terms is not universally agreed on. The confusion arises in part from the fact that before the development of plate tectonics, *transcurrent fault* was used to refer to all major strike-slip faults, some of which are now recognized to be plate boundaries. Moreover, *transform fault* originally referred to faults connecting offset segments of oceanic spreading ridges. Its use has been generalized to include all plate boundary strike-slip faults. *Wrench fault* is another term used to refer to strike-slip faults in a variety of specific senses; we do not use this term.

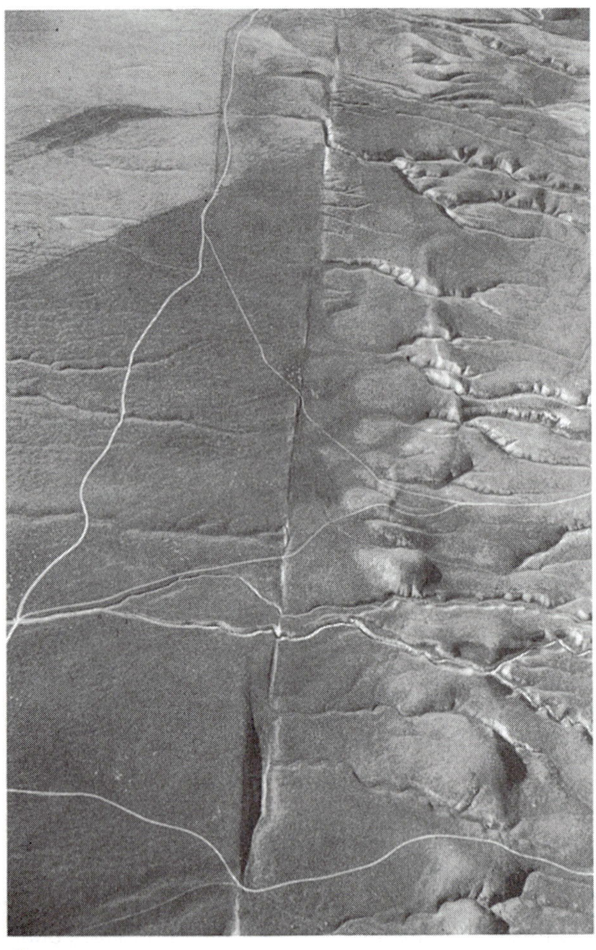

A.

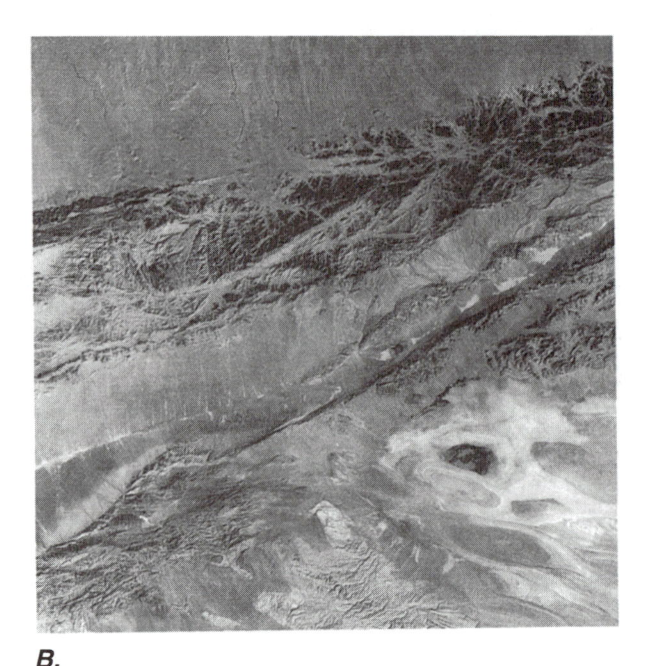

B.

FIGURE 6.1 Photographs of strike-slip faults illustrating rectilinear fault traces. *A*. San Andreas fault, California. Air view looking southwestward along trace of the San Andreas fault, 12 miles west of Taft, California. Length of trace is approximately 5 kilometers. (From Shelton 1966, figure 237) *B*. Landsat image of Altyn Tagh fault, China, showing through-going nature of the structure (see map in Figure 6.2*B*). (After Molnar and Tapponnier 1975)

At outcrop or local scale, transform and transcurrent faults are indistinguishable. One must identify them based on the regional plate tectonic environment and the tectonic role that each plays. For most plates, the recognition of a transform boundary is straightforward. In a few situations, however, such as in Asia (Figure 6.2*B*), the distinction between transform and transcurrent faults depends in part on how small a block one chooses to accept as a "tectonic plate."

The San Andreas fault system of California (Figures 6.1*A*, 6.2*A*), is a right-lateral transform fault system 1300 km long that connects two triple junctions, one south of the Gulf of California and the other at Cape Mendocino on the north coast of California. It consists of many roughly parallel faults in a zone as much as 100 to 150 km wide. It displays along its length many of the characteristic features of strike-slip faults, and because it has been exceptionally well studied, it furnishes numerous examples of structures that we describe in the following sections.

Central and eastern Asia contains a complex system of transcurrent faults (Figures 6.1*B*, 6.2*B*) dominated by left-lateral faults in eastern Tibet and right-lateral faults in an area extending from Lake Baikal in the northeast to the Herat fault in the southwest. Many workers attribute this complex system of faults to the effects of the northward-moving Indian plate indenting the Asian crustal block, and this model accounts for many of the observed features. Several examples of characteristic strike-slip fault structures that we discuss in the following sections come from this complex.

6.1 CHARACTERISTICS OF STRIKE-SLIP FAULTS

Many large strike-slip faults are marked by prominent continuous topographic features on the Earth's surface that are visible even from space (Figure 6.1*B*). The topographically high side of a strike-slip fault commonly changes from one side to the other along the fault trace. The topographic expression of the fault may result from minor components of vertical slip along segments of the fault that could be caused by: a component of shortening or lengthening across the fault; differences in temperature of the rocks across the fault; a juxtaposition of originally separate topographic features; or a juxtaposition of rocks having different resistance to erosion.

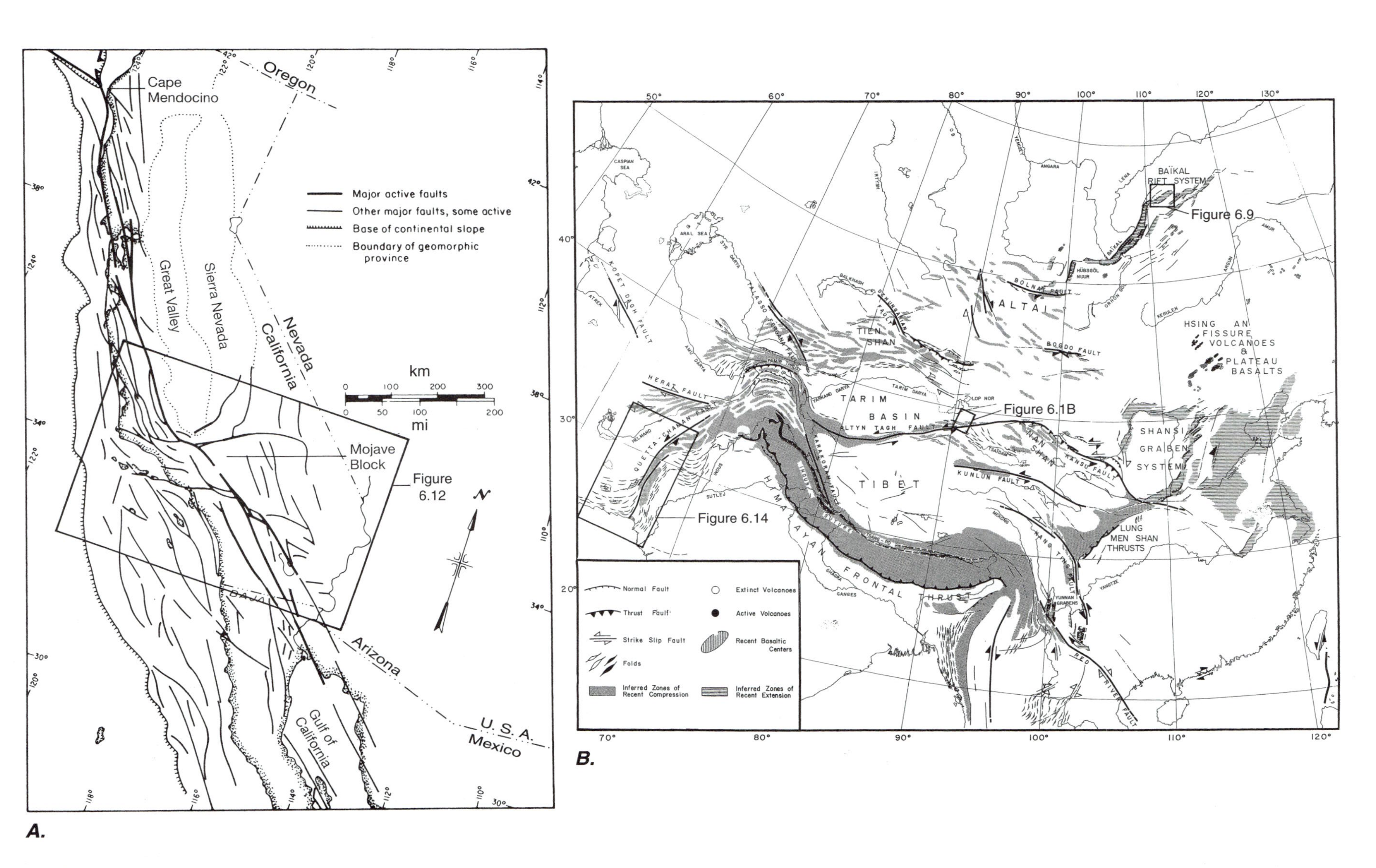

FIGURE 6.2 Major strike-slip fault systems. *A*. San Andreas fault system, California, showing multiple faults. (After Crowell 1979, figure 2) *B*. Simplified map of active tectonics in Asia. Heavy lines are faults; arrow pairs indicate sense of displacement. (After Tapponnier and Molnar 1977)

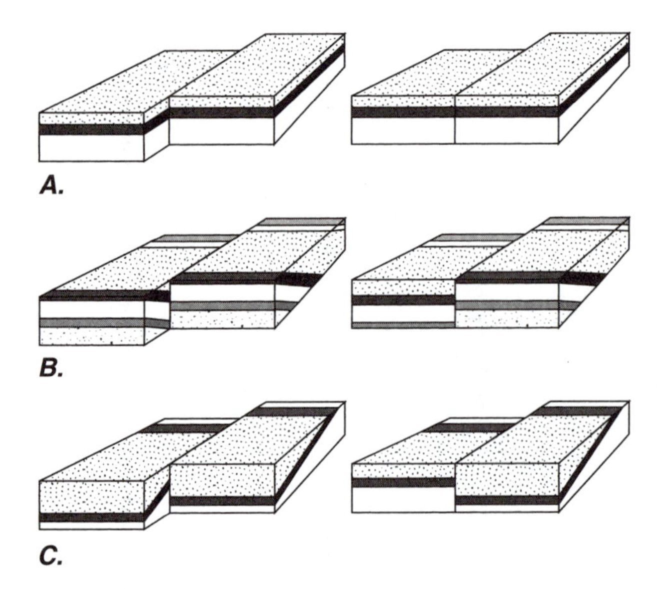

FIGURE 6.3 Separations of stratigraphic units as a result of left-lateral (sinistral) strike-slip faulting. Left diagrams show the strike-slip displacement; right diagrams show a planar cross section perpendicular to the fault after the displacement. *A.* Strike-slip faulting of horizontal beds produces no separation on a map or a vertical section. The displacement is parallel to the cutoff line of the strata on the fault surface. *B.* Beds are tilted so that the separation on a vertical section is right fault block up. *C.* Beds are tilted so that the separation on a vertical section is left fault block up.

The predominantly horizontal slip on strike-slip faults produces a horizontal separation that commonly is used as an indication of strike-slip faulting (Figure 6.3). If the cutoff line of the bedding on the fault is parallel to the displacement, however, no separation is evident (Figure 6.3*A*). If beds are inclined so that their cutoff lines are oblique to the displacement, the separation on a vertical cross section of the fault can be either right-side-up (Figure 6.3*B*) or left-side-up (Figure 6.3*C*), depending on the relative orientation of the beds and the fault and on the sense of displacement on the fault. Large strike separations of a planar boundary, such as a lithologic contact, amounting to many tens or hundreds of kilometers, constitute reasonable evidence for strike-slip faulting (e.g., Figure 3.20), although small strike separations can result from other types of fault slip (Figures 4.2*D*, *E* and 5.2*C*, *D*).

It is important to realize that the sense of shear on a transform fault is determined by the nature and arrangement of the plate boundaries connected by the transform, not by the apparent offset of those boundaries. For example, for a transform fault connecting two segments of a spreading ridge, the shear sense on the fault is opposite to the apparent offset of the ridge segments (see Figures 19.20*A* and 19.22). For a transform fault connecting two segments of a subduction zone, however, the apparent offset of the subduction zone is the same as the shear sense on the fault (see Figure 19.21). Nevertheless, the apparent offset is not an indication of the relative motion of the plate boundary segments connected by the transform fault. For example, it is only for the geometry shown in Figure 19.21*A* that the shear sense on the transform fault indicates the sense of relative motion of the two subduction zone segments. For the geometry of Figure 19.21*B*, the subduction zone segments approach each other, which is opposite to the shear sense on the fault; and for the geometry in Figure 19.21*C*, the subduction zone segments do not move relative to each other.

Strike-slip faults display the typical features that we discuss in Chapter 3. Slickenside lineations are subhorizontal. Drag folds may form along some strike-slip faults if the bedding is favorably oriented (Section 3.2(ii)), although folds reflecting shortening across the fault are more common (see Figure 6.4*B* and the associated discussion). Geomorphic features characteristic of strike-slip faults include linear erosional depressions (Figure 6.1*A*), sag ponds, springs, offset streams (Figure 3.13), and topography, including **shutter ridges**, which occur where horizontal slip on a strike-slip fault has placed a topographic ridge in front of a topographic valley, shutting off or displacing the normal drainage from the valley.

A variety of shear fractures, folds, and normal and thrust faults are found associated with strike-slip faults. The orientations of these structures relative to the main strike-slip fault are characteristic of the sense of shear on the fault (Figure 6.4). Subsidiary shear fractures, known as **Riedel shears** or **R shears**, are synthetic to the main fault, which means they develop at a small angle from the fault (roughly 10° to 20°) and have the same shear sense as the main fault (Figure 6.4*A*). They commonly develop in an *en echelon* array along the main fault, and they are oriented such that a rotation through the acute angle from the main fault trace to the trace of the R shear has the same sense as the shear sense on the main fault e.g. a clockwise angle for a right-lateral fault. Other subsidiary shears may also develop. **P shears** are synthetic to the main fault and are oriented symmetri-

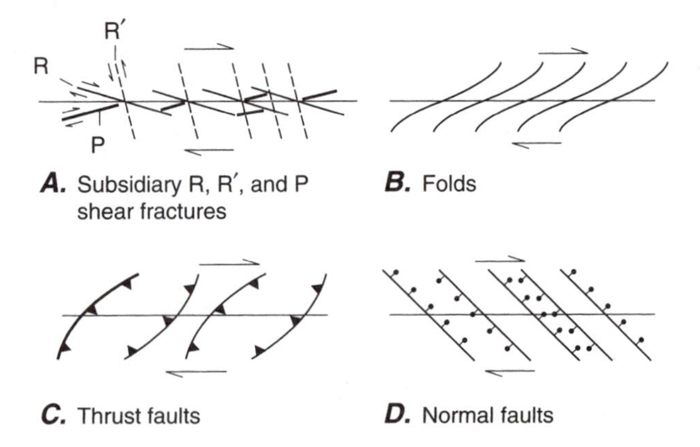

FIGURE 6.4 Structures associated with strike-slip faults and their orientations relative to the shear sense on the fault. (Modified after Sylvester 1988)

cally with respect to the main fault from the orientation of the R shears. Thus a rotation through the acute angle from the trace of the main fault to the trace of the P shear has the opposite sense from the shear sense on the main fault. Conjugate Riedel shears, or **R′ shears**, are antithetic shear fractures, which means they are oriented at high angles from the fault (roughly 70° to 80°) and have a shear sense opposite to that of the main fault. Synthetic shear fractures that are parallel to the main fault are sometimes referred to as **Y shears**. On a small scale, these various secondary shear fractures, plus extension fractures that bisect the acute angle between the R and R′ shears, are responsible for some of the brittle shear sense criteria that we discuss in Section 3.3 (see Figure 3.17). They are therefore commonly used to infer the kinematics of the main fault. On a large scale, these fractures can form a complex anastomosing network of faults that can become very difficult to interpret.

Folds and thrust faults commonly form in an *en echelon* arrangement above the tip lines or beside major strike-slip faults (Figure 6.4*B*, *C*). The trend of the fold hinges and the strike of the thrust faults are oriented at 135° or more from the strike-slip fault, where the angle is measured in the same sense as the shear sense on the fault, for example, clockwise for a right-lateral fault. These structures record a component of horizontal shortening that is oblique to the strike-slip fault, roughly perpendicular to the trends of the folds and the thrust faults.

Normal faults may also form *en echelon* arrays along strike-slip faults, and they are oriented at roughly 45° to the main fault, measured in the same sense as the shear sense on the fault. Thus the orientation of normal faults is close to perpendicular to the orientations characteristic of fold hinges and thrust faults (Figure 6.4*D*). These faults record a component of horizontal lengthening that is oblique to the strike-slip fault and normal to the shortening orientation recorded by folds and thrust faults.

Many of these associated structures develop as a result of the inherent geometry of strike-slip faults and the displacement along them, as we discuss in Section 6.2. Other structures reflect a distributed component of displacement along or across the fault, as we describe in Section 6.4.

6.2 SHAPE, DISPLACEMENT, AND RELATED STRUCTURES

i. Single Faults

At depth, strike-slip faults may terminate on another fault, commonly a low-angle detachment, or they may continue through the crust and lose their identity at depth in a zone of ductile deformation. Earthquakes along modern strike-slip faults typically are present only down to depths of about 15 to 20 km. Below this seismic zone, aseismic shear is probably accommodated by cataclastic flow in a transition zone, and below that by ductile flow. A strike-slip fault terminating against a horizontal decollement is geometrically equivalent to a dip-slip fault terminating against a vertical tear fault, in that the displacement on the fault is parallel to the termination line in both cases.

Although they are characteristically vertical with straight map traces, strike-slip faults also include **bends** (or **jogs**), and **step-overs** (or **offsets**) (Figure 6.5). Bends are curved parts of a continuous fault trace that connect two noncoplanar but approximately parallel segments of fault. Step-overs or offsets are regions where one fault segment ends and another *en echelon* fault segment of the same orientation begins. Bends and step-overs are described geometrically as being either right or left, depending on whether the bend or step is to the right or left as one progresses along the fault. This description is the same regardless of the sense of shear on the fault zone. Bends are geometrically equivalent to frontal ramps on dip slip faults in that the displacement on the bend or ramp is approximately perpendicular to the line of intersection of the bend or ramp with the main fault.

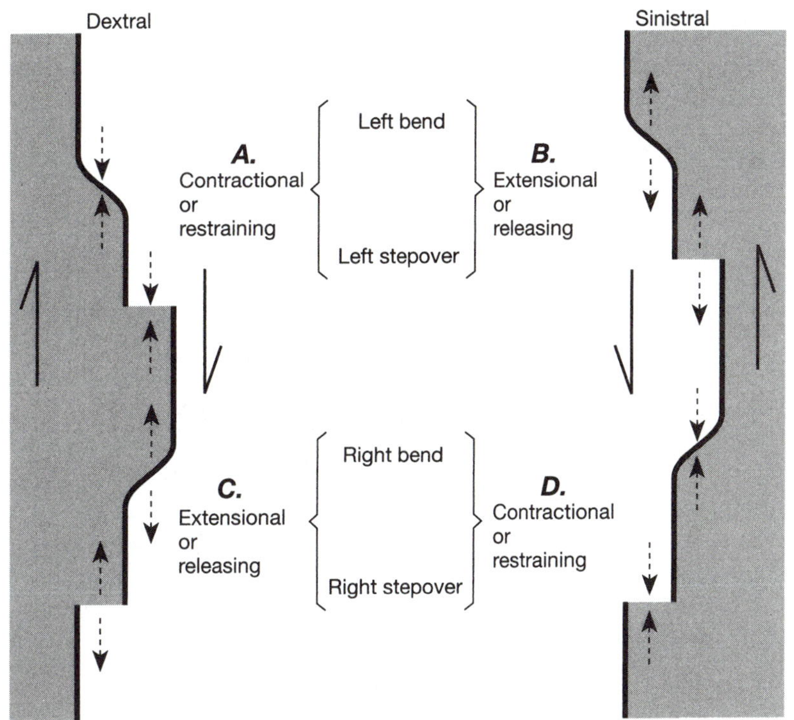

FIGURE 6.5 The geometry and terminology for right and left bends and stepovers on dextral and sinistral strike-slip faults. Large arrows show relative shear on the fault; pairs of dashed arrows indicate the extension or contraction across the bends and step-overs.

Displacement on strike-slip faults ideally is horizontal and therefore parallel to the strike of the fault. For a vertical fault, the trace of the fault on a map is straight, regardless of the topography, and thus is parallel to strike—and therefore also parallel to the ideal displacement direction. We describe a bend or step-over kinematically as **contractional** or **restraining**, if material is pushed together by the dominant fault shear (dashed arrow pairs, Figure 6.5*A*, *D*); the bend or step-over is **extensional**, **releasing**, or **dilatant** if material is pulled apart by the dominant shear (dashed arrow pairs, Figure 6.5 *B*, *C*). On dextral faults (Figure 6.5*A*, *C*), right bends and step-overs are extensional and left bends and step-overs are contractional, whereas on left-lateral faults (Figure 6.5*B*, *D*), left bends and step-overs are extensional and right bends and step-overs are contractional. Thus bends, step-overs, and broad curves in the trace of a strike-slip fault do not permit pure strike-slip motion but require some accommodating deformation.

ii. Strike-Slip Duplexes

Displacement along strike-slip faults that have bends or step-overs produces a complex zone of deformation. Commonly the result is a **strike-slip duplex**, which is a set of horizontally stacked horses bounded on both sides by segments of the main fault (see Figure 3.34*C*). Such a duplex may be extensional (Figure 6.6) or contractional (Figure 6.7), depending on whether it forms at an extensional or contractional bend or step-over.

Strike-slip duplexes must differ from duplexes that form along dip-slip faults because the different orientation of the shear plane places different constraints on the deformation. For dip-slip faults, the faulting accommodates either a thickening or a thinning of the crust, which results in a vertical displacement of the surface of the Earth, which is a free surface (Figure 3.34*A*, *B*). For strike-slip faults, however, the corresponding thickening or thinning would have to occur in a horizontal direction (Figure 3.34*C*), which is impossible because of the constraint imposed by the rest of the crust. There being no free vertical surface, the required thickening or thinning can only be accommodated by vertical motion of the free horizontal surface, and therefore slip on strike-slip duplex faults cannot be purely strike-slip, but must be oblique. To accommodate this component of motion, faults in a strike-slip duplex must have a different geometry from those in dip-slip duplexes.

The oblique slip on the faults bounding the horses in an extensional duplex must be a combination of strike-slip and normal slip (Figure 6.6*C*); on the faults in a contractional duplex it must be a combination of strike-slip and reverse slip (Figure 6.7*C*). The shortening associated with contractional duplexes also may be accommodated by folding subparallel to the reverse faults (see Figure 6.4). The deformation required at contractional or extensional bends provides one mechanism for producing the *en echelon* folds and the normal and thrust faults associated with strike-slip faults that we describe in Section 6.1 (Figure 6.4; see Figure 6.12).

In a strike-slip duplex, the shape of the faults on a vertical section normal to the main fault trace is referred to as a **flower structure**. If the dip-slip component is normal, the faults tend to be concave up and form a **nor-**

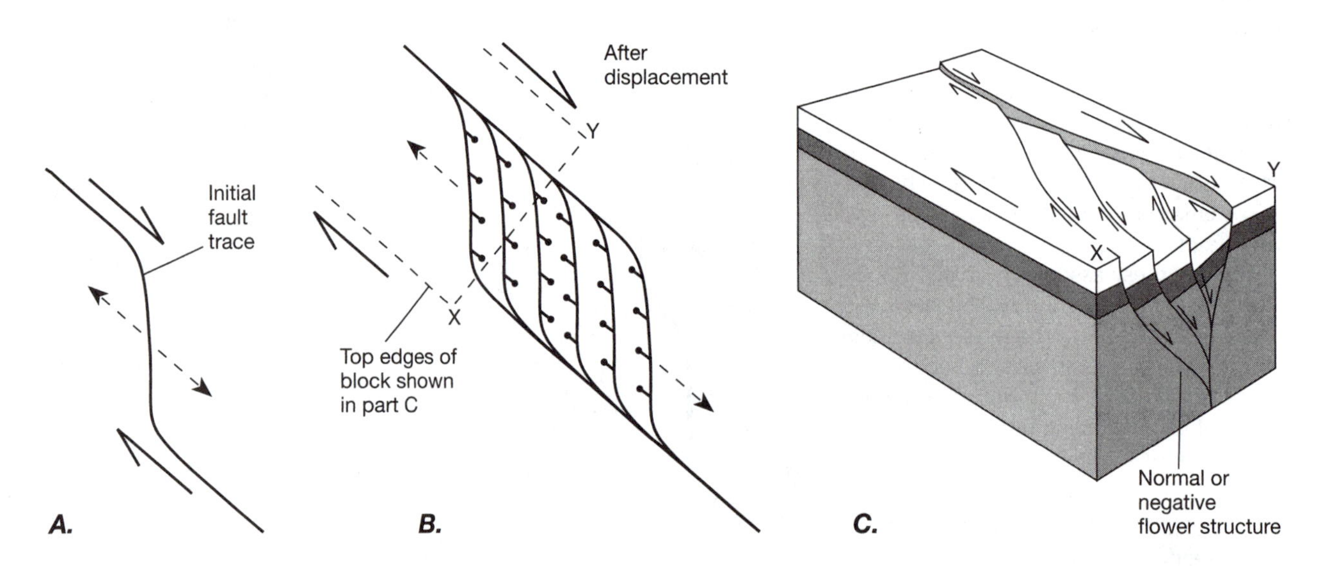

FIGURE 6.6 Formation of an extensional duplex at an extensional (or releasing) bend. Large arrows indicate the dominant shear sense of the fault zone; small arrows indicate the sense of strike-slip and normal components of motion on the fault splays. *A*. Extensional bend on a dextral strike-slip fault. *B*. An extensional duplex developed from the bend in *A*. *C*. A block diagram showing a normal negative flower structure in three dimensions. (After Woodcock and Fischer 1986, figure 12a)

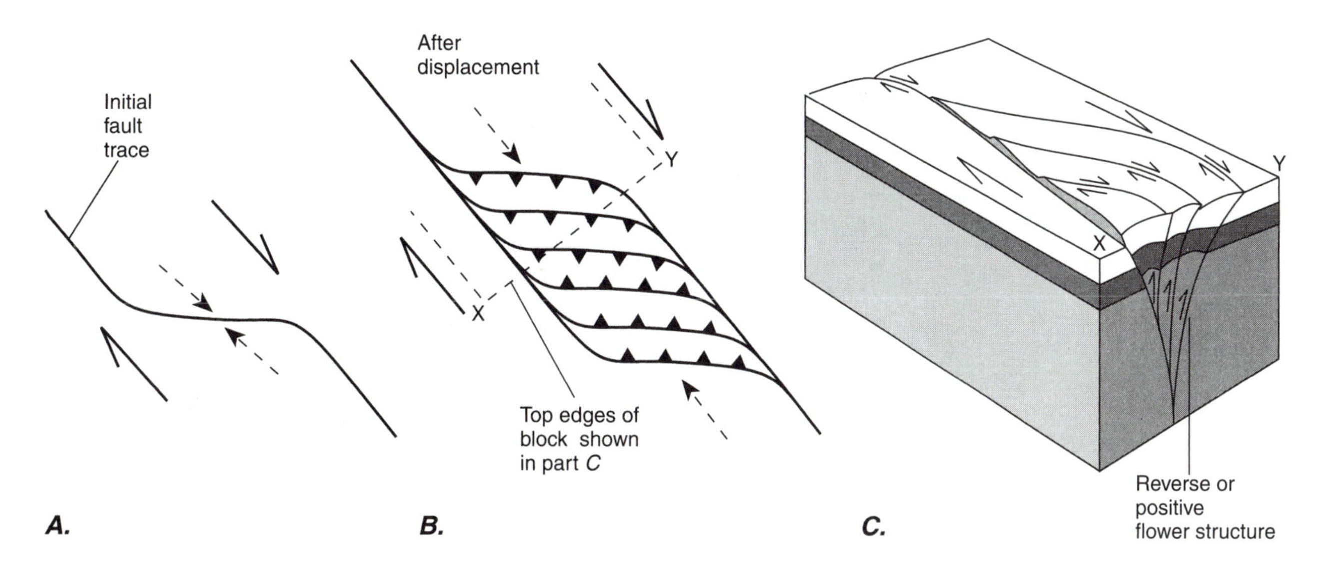

FIGURE 6.7 Formation of a contractional duplex at a contractional (or restraining) jog. Large arrows indicate the dominant shear sense of the fault zone; small arrows indicate the sense of strike-slip and reverse components of motion on the fault splays. *A*. Contractional bend on a dextral fault. *B*. A contractional duplex developed from the bend in *A*. *C*. A block diagram showing reverse, or positive, flower structure in three dimensions. The block faces are vertical planes along the dashed lines in *B*. (After Woodcock and Fischer 1986, figure 12b)

mal or **negative flower structure**, also known as a **tulip structure** (Figures 6.6*C* and 6.8*A*). If the dip-slip component is reverse, the faults tend to be convex up and form a **reverse** or **positive** flower structure, also referred to as a **palm tree structure** (Figures 6.7*C* and 6.8*B*). All these botanical names suggest the similarity in cross-sectional form between the plants and the faults, but given the difference in the third dimension, they are not particularly apt.

In actual cases, the slip on faults in strike-slip duplexes may vary along strike from having a normal component at one end to a thrust component at the other. Such faults, which sometimes are called **scissor faults**, accommodate the rotation of horse blocks in the duplex.

Examples of these two types of flower structure can be seen in seismic reflection profiles from the southern Andaman Sea (see Figure 6.8*A*) and from the Ardmore Basin in southern Oklahoma (Figure 6.8*B*). In Figure 6.8*A*, the graben-like offset of reflectors across the fault zone indicates a component of normal slip on the faults characteristic of a normal (negative) flower structure. In Figure 6.8*B*, the major faults show a component of thrusting characteristic of a reverse or positive flower structure.

Displacement at extensional bends and step-overs produces topographic depressions known as **pull-apart basins**, which commonly fill with water to produce sag ponds or lakes. On a large scale, pull-apart basins are commonly rhomb-shaped fault-bounded basins several kilometers or tens of kilometers in dimension (Figure 6.9). Faulting may be accompanied by volcanic eruptions that cover the floor of the basin. Because they form topographic depressions, pull-apart basins generally accumulate large thicknesses of either alluvial or lake deposits, or both. With continued displacement, the basin may eventually be split by a younger segment of the fault that separates opposite sides of the basin from each other.

6.3 STRUCTURAL ASSOCIATIONS OF STRIKE-SLIP FAULTS

i. Tear Faults

Tear faults are secondary structures generally associated with major faults and folds. These predominantly strike-slip faults characteristically develop in regions of normal faulting (see Figures 4.11 and 4.12) and in fold and thrust sheets (see Section 5.2(i); Figures 5.7*C*, *D* and 5.13). They accommodate different amounts of crustal lengthening or shortening in adjacent regions. The Jura Mountains of Switzerland (Figure 6.10; see location in Figure 5.22) are a classic example of a fold and thrust belt that is segmented by generally north- to northwest-trending tear faults. Fold hinges terminate laterally against the tear faults (compare Figure 5.13*A*, *B*), separating sections of the thrust sheet that have different magnitudes of displacement on the decollement.

ii. Bends, Step-Overs, and Duplexes

Transcurrent and transform faults never occur as simple planar faults through the crust. At all scales, they are characterized by complex zones of anastomosing, parallel, or

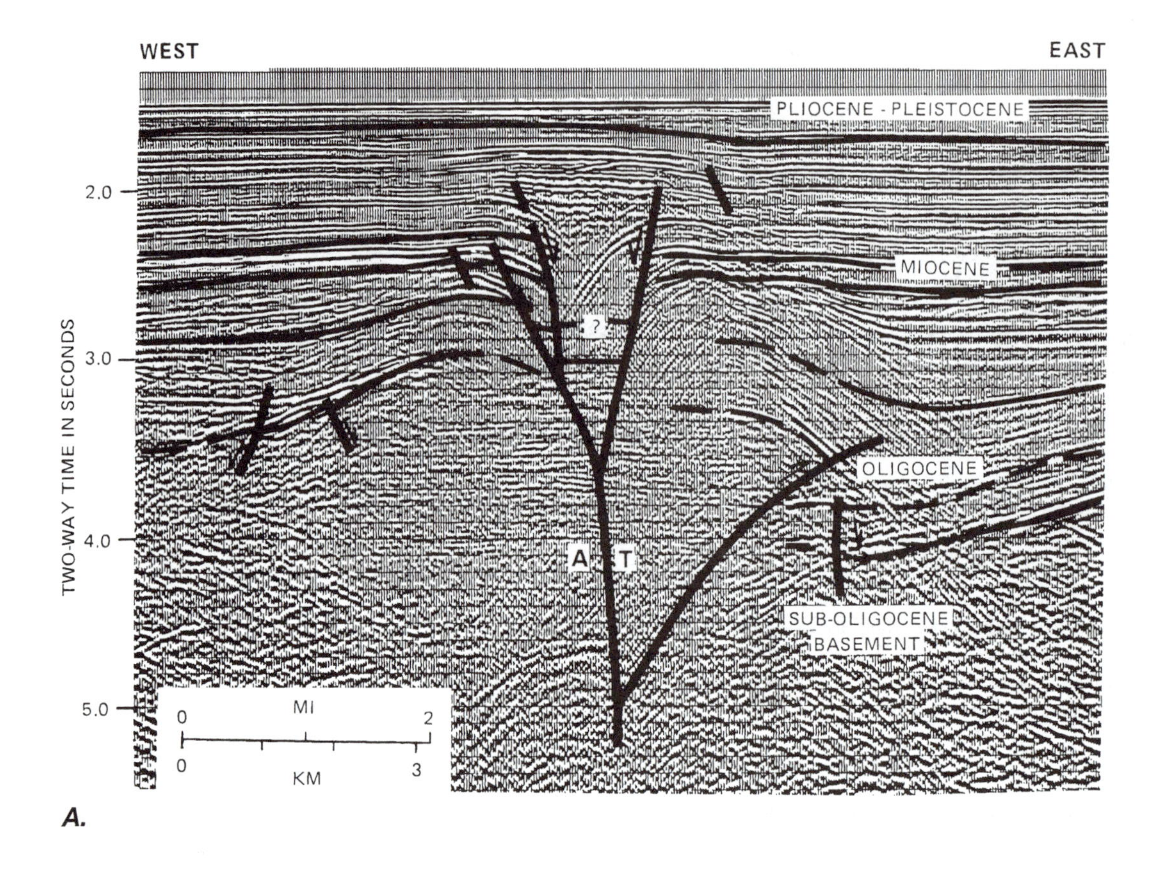

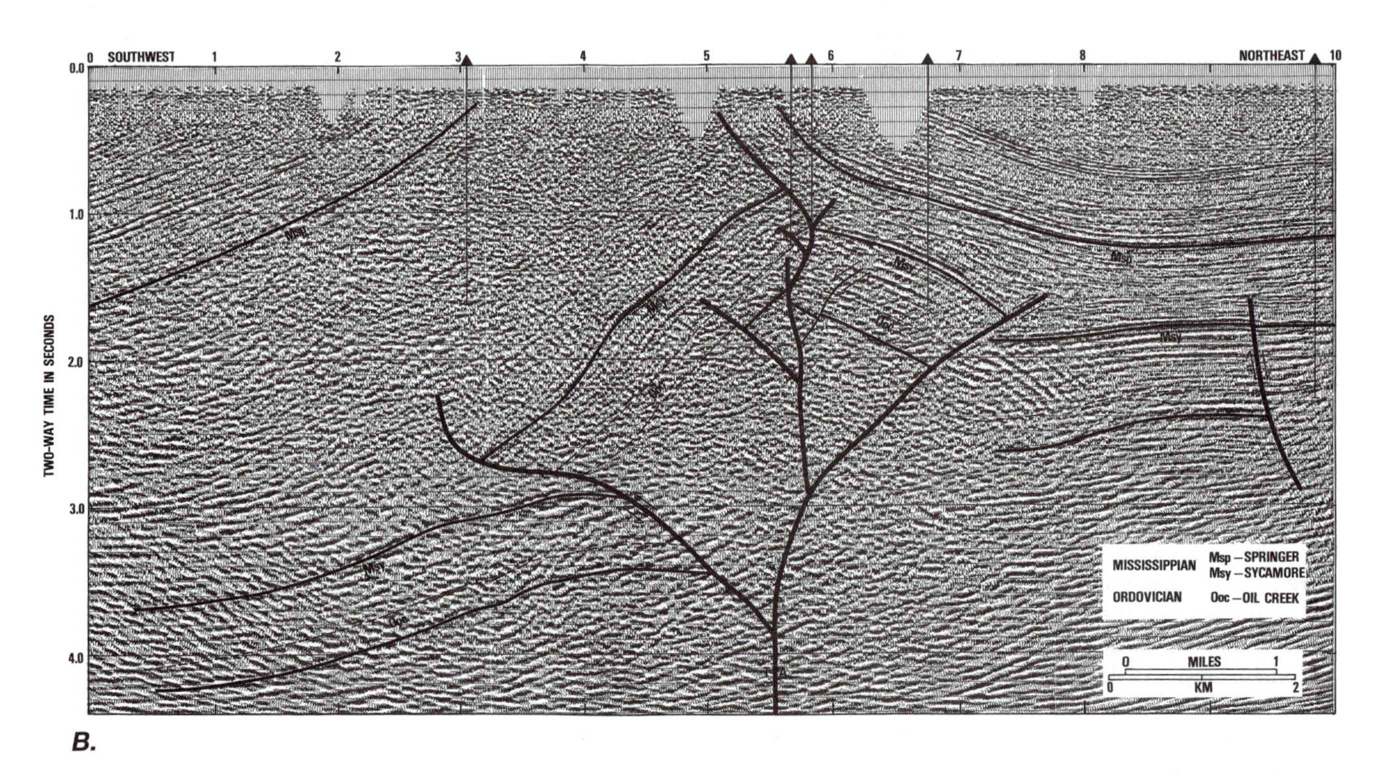

FIGURE 6.8 Seismic profiles of flower structures. *A*. Example of a negative flower structure from an extensional duplex on a dextral strike-slip fault from the Andaman Sea between India and the Malay Peninsula. Unmigrated seismic reflection profile. *B*. Example of positive flower structure from a contractional duplex on a sinistral strike-slip fault in the Ardmore Basin, Oklahoma. Migrated seismic profile. (After Harding 1985)

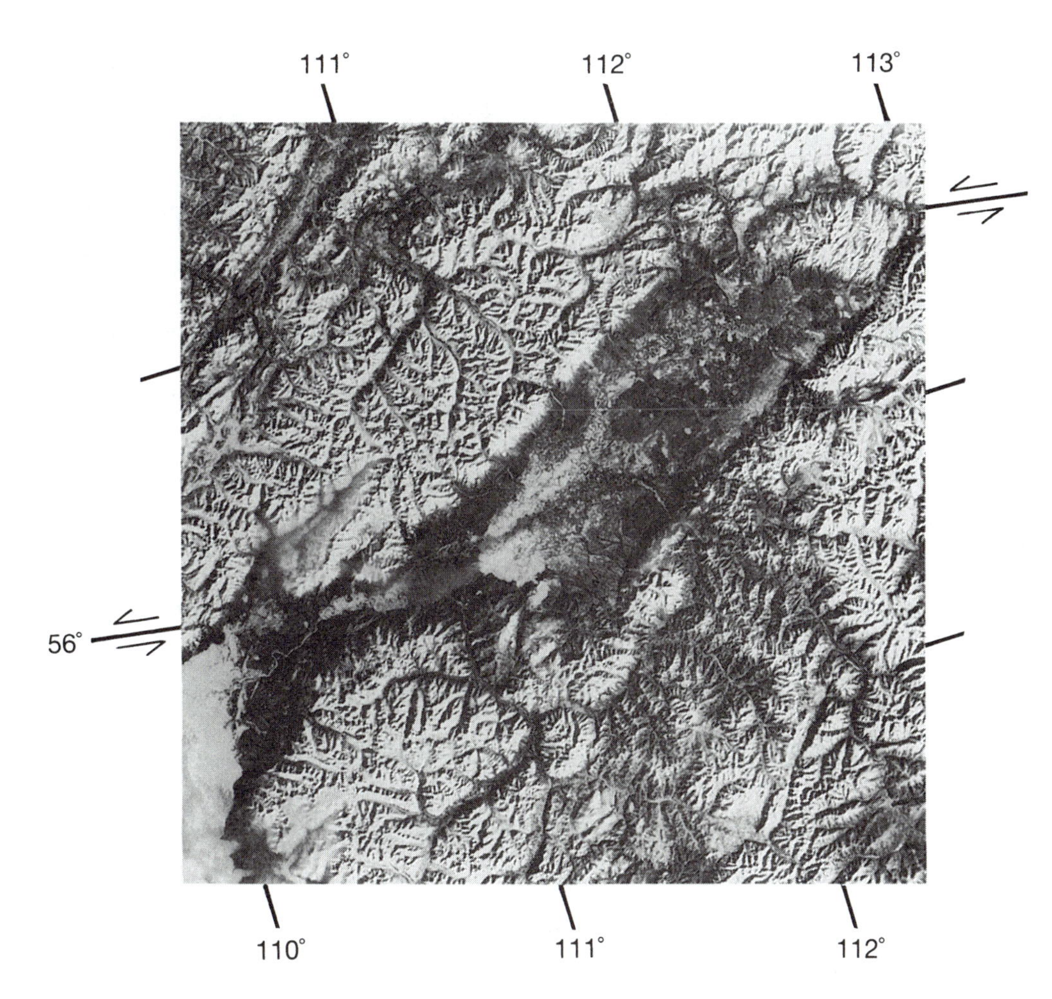

FIGURE 6.9 The Angara graben, a major pull-apart basin northeast of Lake Baikal in Siberia (see the map in Figure 6.2*B*). The basin formed at a left step-over in a left-lateral strike-slip fault, indicated at the upper right and lower left of the photo. The basin is bounded by northeast-trending normal fault scarps. (Landsat image 1584-03070-5; from Tapponnier and Molnar 1979)

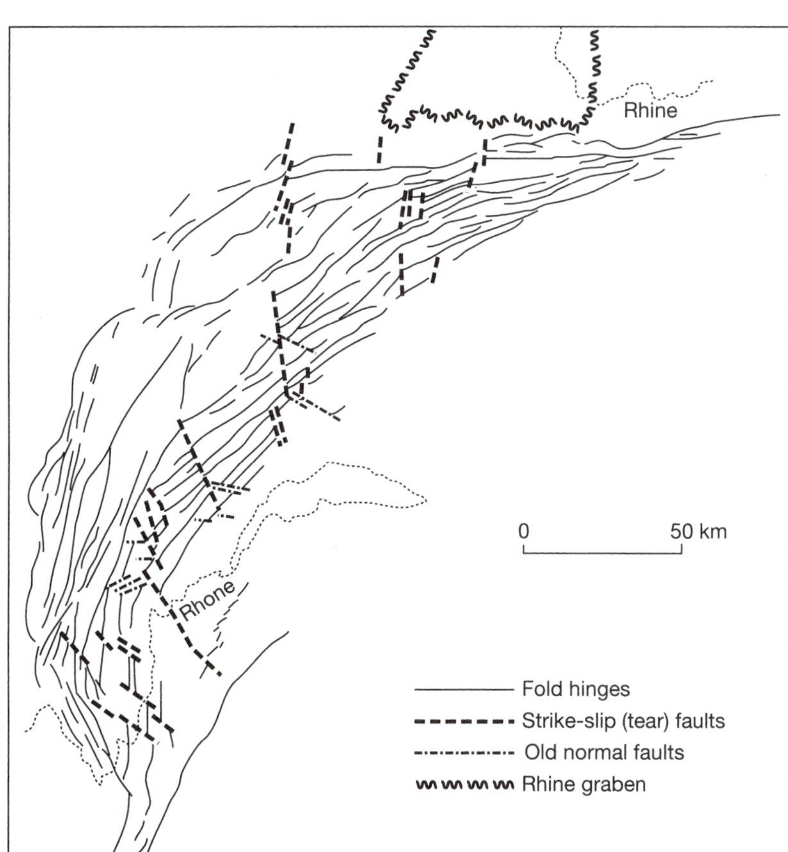

FIGURE 6.10 Tear faults in the Jura fold and thrust belt (see Figure 5.20 for location). The generalized map of the Jura Mountains shows the major fold axes, wrench faults, and the boundary of the Rhine graben. Note how the fold axes tend to terminate against the tear faults. (After Laubscher 1972, figure 3)

en echelon faults that are not perfectly straight (Figure 6.2*A*) and along which the strike-slip motion results in a variety of accommodation structures (Figure 6.4).

An excellent example of an extensional duplex occurs on the active Dasht-E Bayaz fault in northeastern Iran (Figure 6.11). The duplex is in the process of developing at a left bend on the left-lateral fault. The main trace of the fault trends obliquely through the middle of the duplex, defining what is essentially a transfer fault at a left step-over in the strike-slip fault. Subsidiary faults to the east and a dense concentration of fractures to the west outline two horses within which the fracture density is much lower. The inset in Figure 6.11 shows an idealization of the duplex geometry.

An even larger left bend in the San Andreas fault system, known as the Big Bend, occurs in southern California in the region where the Garlock fault intersects the San Andreas (Figures 6.2*A* and 6.12). Here the contraction expected at a left bend in a dextral fault is reflected by the Transverse Ranges, a block of crust that is rapidly rising on east-west–trending thrust faults (Figure 6.12). Along this contractional bend, however, extensional basins are also present, illustrating the complex interplay of extensional and contractional structures in major strike-slip fault systems. These basins, which are filled with Neogene sediments, probably represent remnants of pull-apart basins that originally formed in extensional duplexes, some of which have been displaced from their original location.

The bend at the Transverse Ranges coincides with the intersection of the right-lateral San Andreas fault and the left-lateral Garlock and Big Pine faults (Figure 6.12). The Mojave block between the Garlock and San Andreas faults contains northwest-trending dextral strike-slip faults as well as west-trending sinistral strike-slip faults, and parts of the block have experienced large amounts of roughly east-west extension. The active, complex, predominantly dextral Eastern California Shear Zone connects with the San Andreas fault south of the Big Bend, traverses the Mojave block (the zone extends approximately between the Helendale fault and the Pisgah-Bullion fault; Figure 6.12), crosses the Garlock fault just east of the Sierra Nevada, and continues north along the eastern side of the Sierra Nevada (between the Owens Valley fault and the Death Valley fault; Figure 6.12). All these faults are no older than Tertiary and many are currently active, producing large earthquakes. For example, historic earthquakes of magnitude greater than 6.5 along the San Andreas system and the Eastern California Shear Zone are plotted on Figure 6.12. Understanding such a complex mosaic of faults requires an understanding of the history of each individual fault in relation to all the others, and is indeed a challenge! In Section 6.4 we discuss one kinematic model for the faulting in this region that accounts for some aspects of the geology but is only a partial solution to the complex puzzle of deformation.

iii. Terminations

Strike-slip faults can terminate in the crust at a zone of either extensional or contractional deformation, depending on the location of the deformation zone relative to the slip vectors on the fault. Extension may be accommodated where strike-slip faults splay and turn into an imbricate fan of normal faults (Figure 6.13*A*, *B*). Sim-

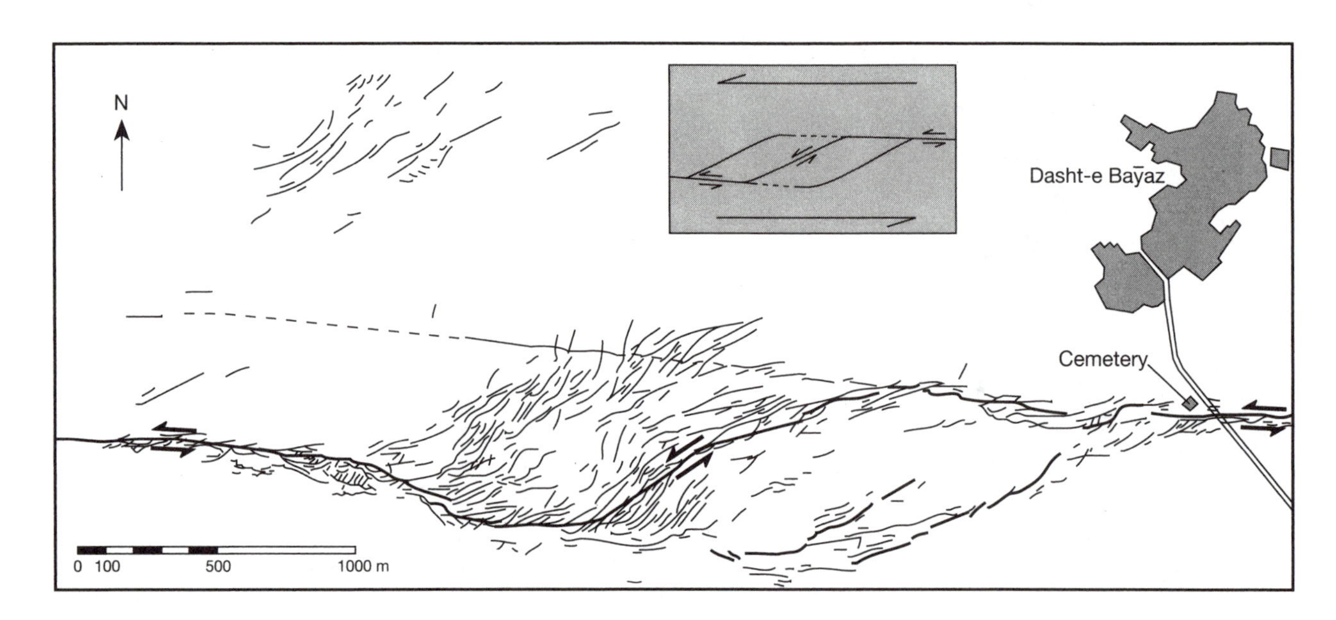

FIGURE 6.11 An extensional duplex developing on the active Dasht-E Bayaz fault in northeastern Iran. (After Tchalenko and Ambraseys 1970)

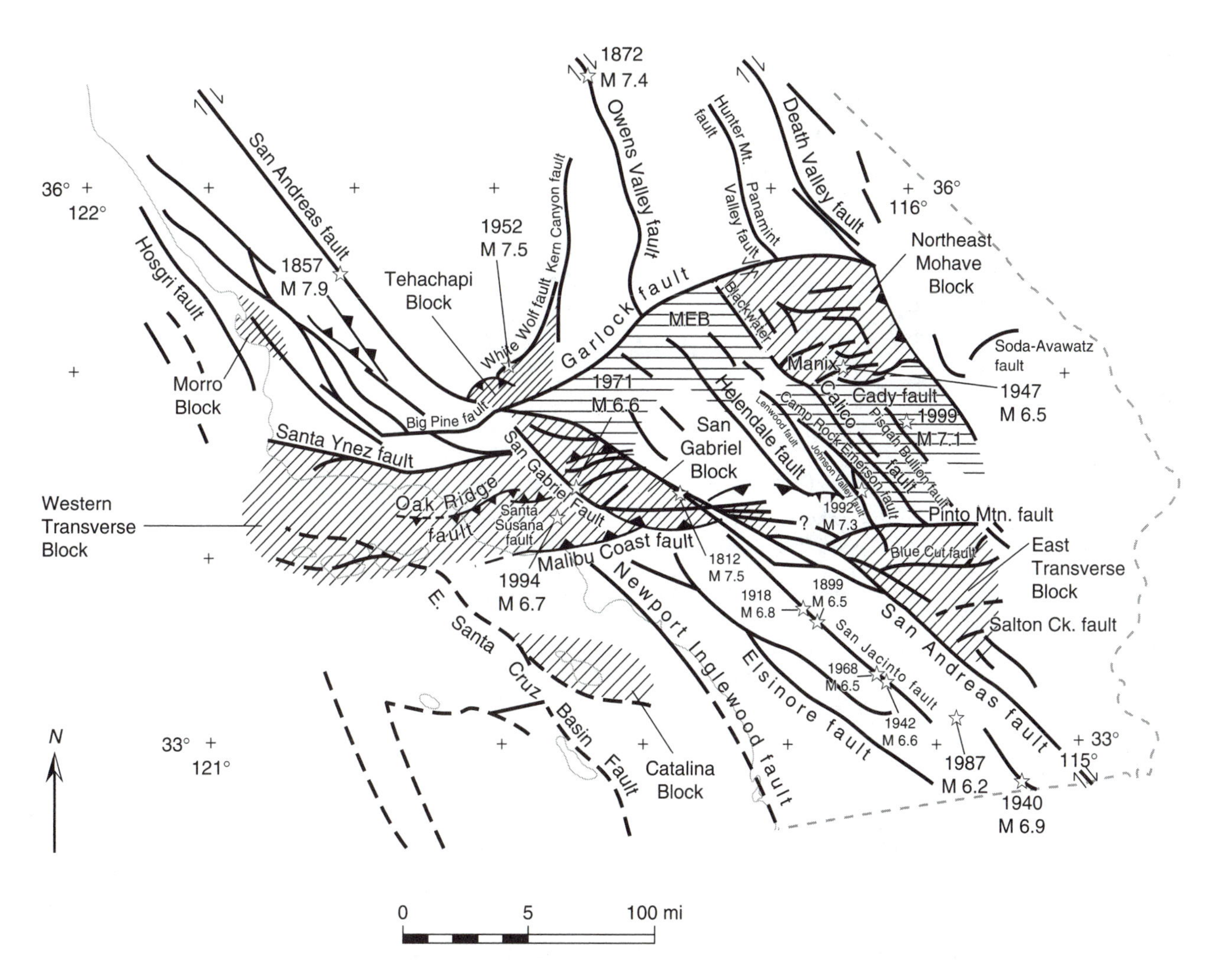

FIGURE 6.12 The San Andreas and Garlock fault systems in southern California. Diagonal ruled areas are regions of clockwise rotation; the horizontal-ruled area is a region of east-west extension. Stars indicate epicenters of historical earthquakes with magnitude (*M*) greater than 6.5: 1812, Wrightwood, *M* 7.5; 1857, Fort Tejon, *M* 7.9; 1872, Owens Valley, *M* 7.4; 1899, San Jacinto, *M* 6.5; 1918, San Jacinto, *M* 6.8; 1940, Imperial Valley, *M* 6.9; 1942, Fish Creek Mountains, *M* 6.6; 1947, Manix, *M* 6.5; 1952, Kern County, *M* 7.5; 1968, Borrego Mountains, *M* 6.5; 1971, San Fernando, *M* 6.6; 1987, Elmore Ranch, *M* 6.2; 1992, Landers, *M* 7.3; 1994, Northridge, *M* 6.7; and 1999, Hector Mine, *M* 7.1. Locations and magnitudes of earthquakes before the mid-twentieth century are approximate. (Tectonic map after Luyendyk 1991; earthquakes from Southern California Earthquake Data Center, http://www.data.scec.org/clickmap.html)

ilarly, contraction may be accommodated by an imbricate fan of thrust faults or folds (Figure 6.13*C*, *D*). Within such zones, the fault curves around, shallows in dip, and ends up trending approximately perpendicular to the direction of movement (Figure 6.13*A*–*D*). The strike-slip displacement diminishes progressively to zero along the fault.

The active Quetta-Chaman fault system of Pakistan is a good example of this structure (Figure 6.14; see the west edge of Figure 6.2*B*). There the left-lateral Quetta-Chaman fault system terminates southward into a series of thrust faults and folds that in fact are part of a modern convergent plate margin.

At its east end, the Garlock fault apparently turns south and becomes a thrust fault that dips westward underneath the Soda and Avawatz Mountains (Figure 6.12). The geology here, however, is complicated by the intersection of the Garlock fault with the Death Valley fault zone, which is a mostly right-lateral strike-slip fault that is part of the Eastern California Shear Zone.

In some cases, the fault may branch into a fan of strike-slip splay faults, called a **horsetail splay** (Figure 6.13*E*), which commonly curve toward the receding fault block. The displacement on any individual splay is relatively small, but the sum of the displacements on each fault in the splay equals the displacement on the main

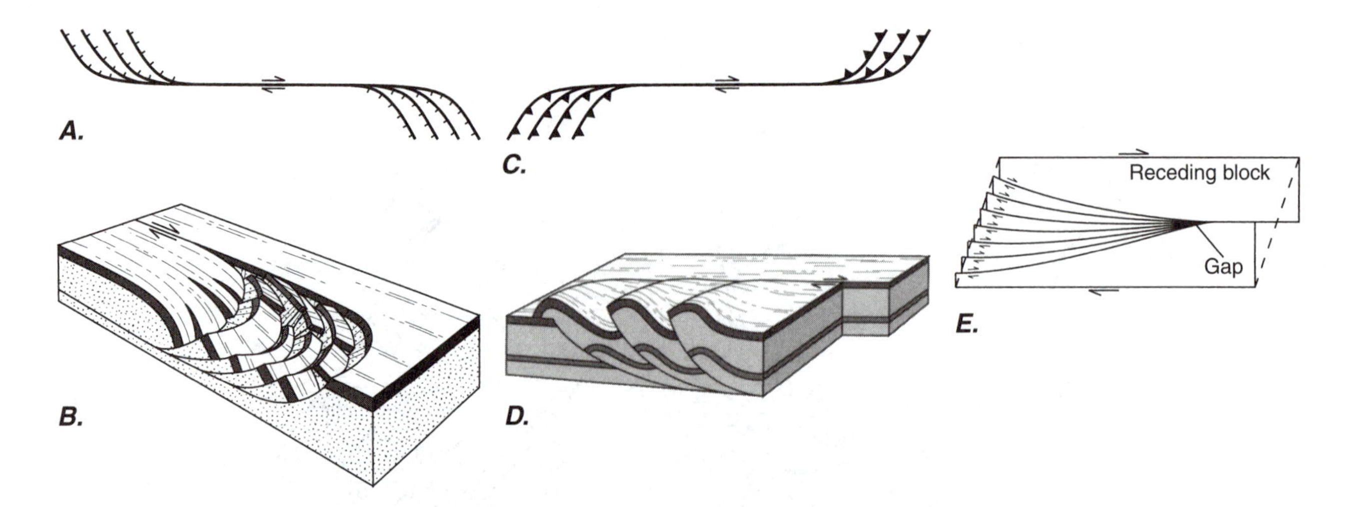

FIGURE 6.13 Termination of dextral strike-slip faults by formation of imbricate fans. (Similar diagrams can be drawn for a sinistral fault.) *A*. Geometry of extensional imbricate fans at the ends of a dextral fault. *B*. Extensional normal faulting at the termination of a dextral strike-slip fault. *C*. Geometry of contractional imbricate fans at the ends of a dextral fault. *D*. Contractional folding and thrust faulting at the termination of a dextral strike-slip fault. *E*. Geometry of a horsetail splay of strike-slip faults at the ends of a dextral strike-slip fault. The total displacement for the single fault on the right side of the block is equaled by the sum of small displacements on the individual splay faults on the left of the block. Splay faults tend to have a curvature that is concave toward the receding block. (After Freund 1974)

strike-slip fault. The fan thereby distributes the deformation through a large volume of crust. The geometry of a horsetail splay and the relative slip on a strike-slip fault is comparable to that of an imbricate fan of listric faults on a low-angle thrust or normal fault.

The Hope fault, which is one strand of the Alpine fault system in New Zealand, provides a good example of the termination of a fault at a horsetail splay (Figure 6.15). The fault splays out against the Alpine fault with relatively small amounts of displacement distributed among each of the splays, as indicated by the horizontal separation of the Pounamou formation (compare Figure 6.13*E*).

Transform faults terminate at major plate boundaries, where the relative slip on the fault is accommodated either by production of crust at a spreading center or destruction of crust at a subduction zone (see Section 19.3).

6.4 KINEMATIC MODELS OF STRIKE-SLIP FAULT SYSTEMS

As with other faults, it is useful to consider simplified kinematic models of strike-slip fault systems in order to gain insight into the possible complexities that can develop. In this section, we discuss models of distributed shear and of oblique strike slip that can account for some of the folds, thrust faults, and normal faults that develop near strike-slip faults. We also discuss a model that accounts for some aspects of the regional deformation associated with the fault systems in southern California.

Many of the structures that develop near strike-slip faults can be accounted for by assuming that part of the shearing is distributed through the rock on either side of the fault. This model is illustrated in Figure 6.16, which shows two squares inscribed across a strike-slip fault (Figure 6.16*A*) that becomes separated by motion on the fault and deformed into parallelograms by shearing distributed on either side of the fault (Figure 6.16*B*). Because of the distributed shear, one diagonal of the square becomes shorter, and the other diagonal becomes longer. This model suggests an explanation for the formation and orientation of folds and thrust faults, which trend perpendicular to the direction of shortening, and normal faults, which trend perpendicular to the direction of lengthening.

The orientation of major strike-slip faults is not necessarily exactly parallel to the direction of relative motion of the adjacent fault blocks. For transform faults, for example, minor changes in plate motion can result in a component of shortening or lengthening across the fault, which can only be accommodated by the development of other structures such as folds and thrust faults or normal faults, respectively. The contractional strike-slip model for the San Andreas fault may be the explanation for the uplift of the Coast Ranges on the west side of the Central Valley of California, and for a series of thrust faults and folds that are currently active along the west side of the valley.

One kinematic model of the San Andreas system in southern California is shown in Figure 6.17. It attempts to integrate the numerous strike-slip faults in the region into a rational pattern. The model idealizes the domains

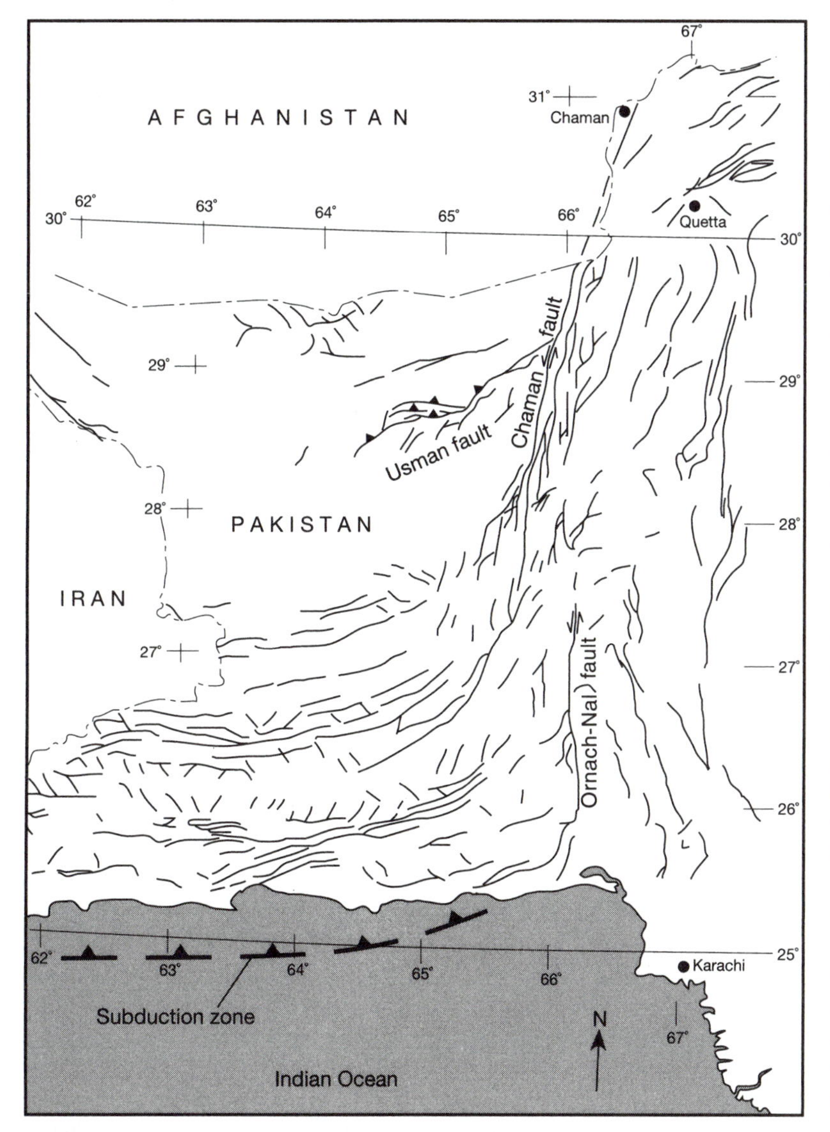

FIGURE 6.14 Map of faults in southern Pakistan. Major north-south-trending faults such as the Chaman and the Ornach-Nal faults are sinsitral strike-slip faults that pass southward into an east-west-trending fold and thrust belt. Most of the east-west-trending faults are interpreted to be thrust faults synthetic to the subduction zone that lies in the Indian Ocean to the south. The short northeast- and northwest-trending faults in the south may be conjugate orientations of tear faults. (From the geologic map of Pakistan, with interpretation after Lawrence and Yeats 1978; see the map in Figure 6.2B)

of roughly parallel faults by assuming that they are perfectly straight faults defining the boundaries of rigid fault blocks (compare Figures 6.2*A* and 6.12). The domains comprise either a set of right-lateral northwest-trending faults or a set of left-lateral faults originally trending north but now trending east-northeast. The model predicts that domains dominated by northwest-trending right-lateral faults should not have rotated significantly during the deformation, but that domains dominated by left-lateral faults, such as the Transverse Ranges block, should have rotated clockwise by as much as 80° (Figure 6.17*B*). Subsequent to rotation, the Transverse Ranges block has been split into eastern and western blocks by slip on the San Andreas fault (Figure 6.17*B*; compare Figure 6.12). Slip on the fault system produces a net north-south shortening and east-west lengthening of southern California.

The assumption of rigid fault blocks requires that numerous gaps open during shearing, especially along domain boundaries (shaded areas in Figure 6.17*B*). Although deformation around these gaps would certainly be complex, many of the model gaps can be correlated with deep basins filled with large thicknesses of young, locally derived sediments. In addition, paleomagnetic determinations of the orientation of the paleopole in parts of the region are consistent with the large rotations and with progressive rotation through time, as indicated by the model. These data permit only small counterclockwise rotation of the right-lateral fault domains, but large clockwise rotations up to 70° to 80° are indicated for the western Transverse Range block.

Despite its complexity, this controversial model is not complete. For example, it does not account for slip along the Garlock fault, for Basin and Range extension north of the Garlock fault, for east-west extension in the Mojave block, for development of the Eastern California Shear Zone, for the rise of the Transverse Ranges on thrust faults associated with the contractional duplex along the San Andreas fault, or for any nonrigid behavior of the various fault blocks. Moreover, it does not address the problem of the fault geometry and the displacement at depth. Nevertheless, such tentative models are useful because they provide testable predictions, and they focus attention on critical problems. More recently, several partial models have been proposed to account for some of the features unexplained in this model.

As with other faults, complete models of strike-slip faults must account for the termination of the faults at depth as well as along strike (see Section 6.2). For large fault systems, such as the San Andreas fault, the Alpine fault of New Zealand, and the Red River and Altyn Tagh faults in Asia, displacements of hundreds of kilometers have accumulated. The only crustal structures that seem capable of accommodating such enormous displacements are plate boundaries.

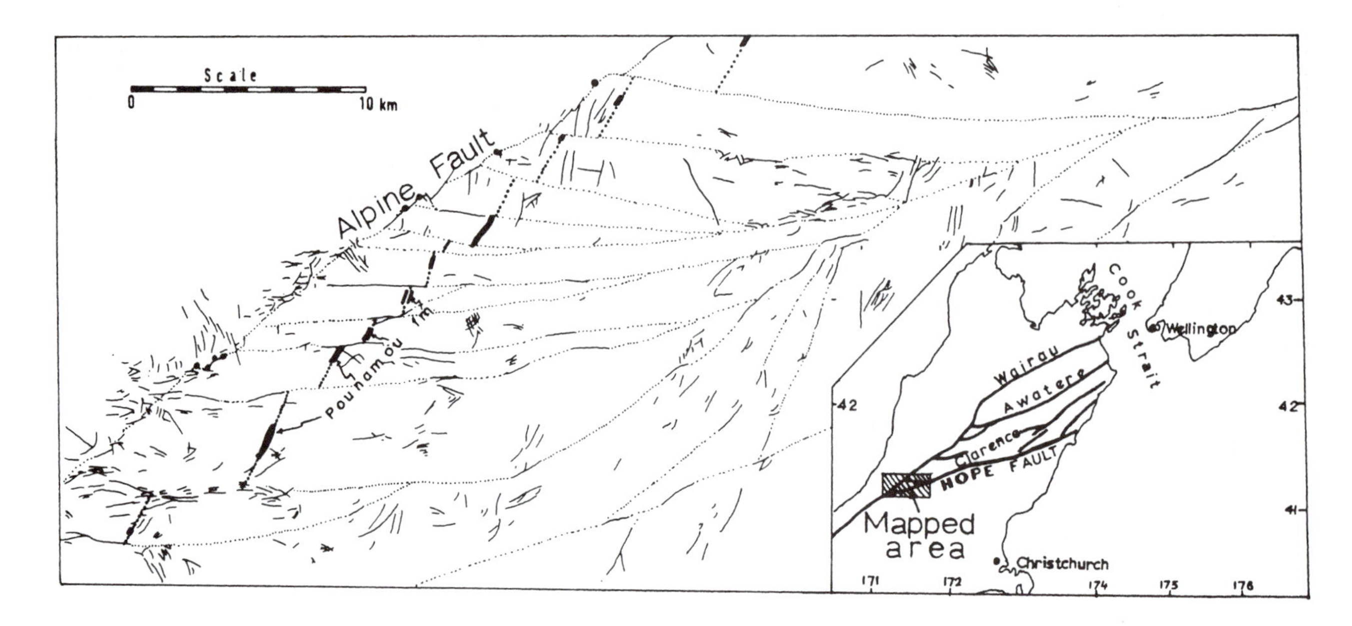

FIGURE 6.15 Termination of the Hope fault against the Alpine fault in New Zealand. Both splaying of the Hope fault and curving of the splays toward the receding fault block are evident. The displacement on the splays is defined by the offset of the Pounamou formation. (After Freund 1974)

The function of the San Andreas fault as a transform fault between triple junctions in the Gulf of California and off the northwest coast of California and Oregon is well recognized (see Section 19.3(v)). The association of the megashears in Asia with the collision of India seems clear, but except for the Quetta-Chaman faults, the structures themselves do not serve as transform faults, and their association with plate boundaries other than the collision zone is not obvious. Accommodation of such large displacements, however, would seem to require involvement of not just crustal, but also mantle rocks.

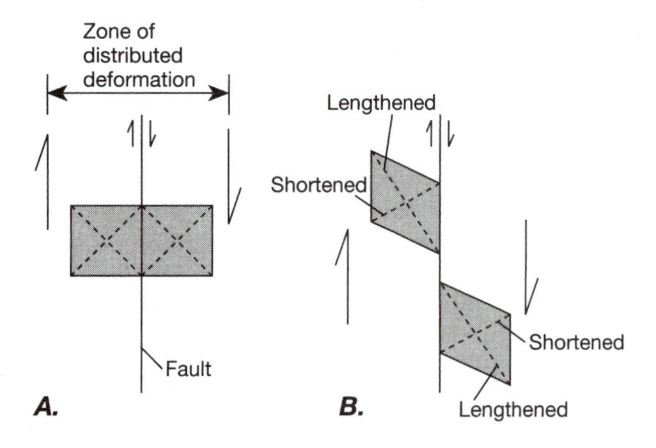

FIGURE 6.16 Kinematic model of shearing on a strike-slip fault where part of the shearing is distributed on either side of the fault. *A* and *B*. Before and after shearing on and near the fault, respectively. Folds and thrust faults form at 45° angles to the main fault perpendicular to the direction of shortening, and normal faults form at 45° angles to the main fault perpendicular to the direction of lengthening (see Figure 6.4).

6.5 ANALYSIS OF DISPLACEMENT

In Section 3.3, we discussed the most important methods of determining displacement on strike-slip faults. The matching of displaced geological features on opposite sides of the fault provides the most reliable determinations. For relatively small displacements, the problem of distinguishing the separation from the displacement is an important one. The possibility is rather remote, however, that horizontal separations on the order of a hundred kilometers are produced by displacement on a fault other than a strike-slip fault. Figure 3.20 shows an example of the determination of large displacement on the San Andreas fault.

The geometric model used for strike-slip faulting in Figure 6.17 is identical to the model for planar rotating normal faults (Figure 4.24*B*). One need only imagine Figure 4.24*B* to be a map view of strike-slip faults instead of a cross section of normal faults: In that case, θ is the angle of rotation of the blocks from the fixed boundary, ϕ is the angle between the faults and the fixed boundary, and the width of the fault block is $w = L\sin(\phi + \theta)$. The displacement δ and the other three parameters describing the deformation w, ϕ, and θ, are related to one another by the equation

$$\frac{\delta}{w} = \left(\frac{\delta}{L}\right)\left(\frac{L}{w}\right) = \left[\frac{\sin\theta}{\phi}\right]\left[\frac{1}{\sin(\theta + \phi)}\right] \tag{6.1}$$

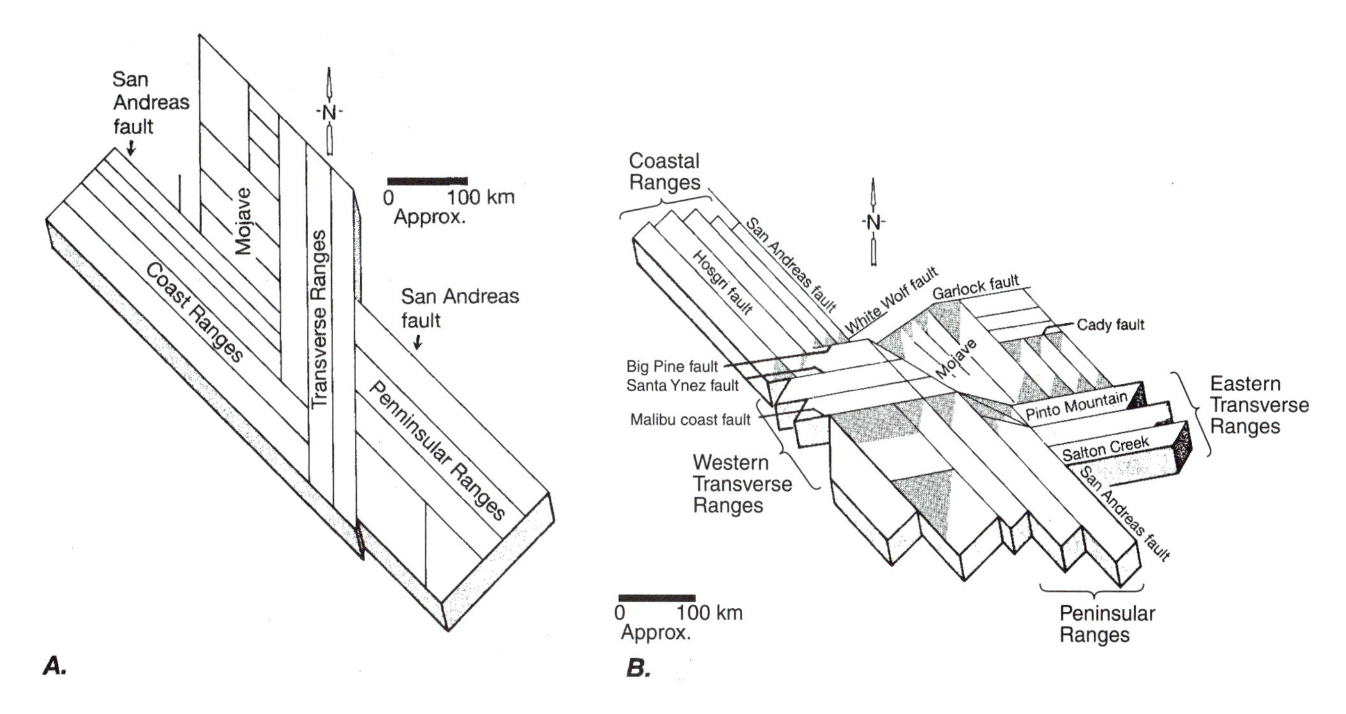

FIGURE 6.17 Rigid strike-slip fault block model for the development of the structures associated with the San Andreas fault system in southern California. (After Luyendyk et al. 1980; see also Luyendyk 1991). *A*. Configuration of faults before displacement in Oligocene time. *B*. Present configuration showing the right lateral offset of the Transverse Ranges. Shaded areas indicate predicted location of basins that would open up as a result of the sliding of the rigid blocks

which must be satisfied if the model is correct. In principle, all the parameters are measurable. We can determine the rotation of a crustal block θ by measuring the rotation of the paleomagnetic pole; we can measure the present angle of the faults ϕ from a fixed boundary and the width w of a particular fault block, and we can use the displacement of geological features on individual faults to determine the displacement δ. As discussed earlier, application of this test of the model to the southern California region has had some success.

6.6 BALANCING STRIKE-SLIP FAULTS

In Section 3.5 we discussed the technique of balancing cross sections of dip-slip (normal and thrust) faults. The assumptions used in balancing are valid only when the deformation has been essentially two-dimensional so that no net movement of material has taken place into or out of the plane of the cross section. Vertical cross sections perpendicular to strike-slip faults (e.g., Figures 6.6*C*, 6.7*C*, and 6.8) do not meet this requirement, and it is generally inappropriate to attempt to balance such cross sections.

The appropriate plane for possible balancing of strike-slip faults is the plane of the map that contains the fault-slip vector. This plane, however, is generally parallel to bedding, and therefore the boundaries that are used to measure lengths and areas in balancing dip-slip faults are commonly not available for strike-slip faults. Moreover, the vertical displacement that accompanies deformation at bends and step-overs violates the strict condition of two-dimensional deformation. Thus any valid balancing of strike-slip faults would have to account for the motion normal to the plane of balancing.

Because of these difficulties, the use of balanced sections as a method of analyzing strike-slip fault zones is not common.

REFERENCES AND ADDITIONAL READINGS

Aydin, A., and A. Nur. 1982. Evolution of pull-apart basins and their scale independence. *Tectonics* 1: 91–105.

Aydin, A., and B. M. Page. 1984. Diverse Pliocene-Quaternary tectonics in a transform environment, San Francisco Bay region, California. *Geol. Soc. Amer. Bull.* 95: 1303–1317.

Christie-Blick, N., and K. T. Biddle. 1985. Deformation and basin formation along strike-slip faults. In

K. T. Biddle and N. Christie-Blick, eds., *Strike-Slip Deformation, Basin Formation, and Sedimentation.* Society of Economic Paleontologists and Mineralogists Special Publication 37: 1–34.

Cowgill, E., A. Yin, T. M. Harrison, and W. Xiao-Feng. 2003. Reconstruction of the Altyn Tagh fault based on U-Pb geochronology: Role of back thrusts, mantle sutures, and heterogeneous crustal strength in forming the Tibetan Plateau. *J. Geophys. Res.* 108: 2346.

Cowgill, E., A. R. Yin, J. R. Arrowsmith, X. F. Wang, and S. Zhang. 2004. The Akato Tagh bend along the Altyn Tagh fault northwest Tibet 1: Smoothing by vertical-axis rotation and the effect of topographic stresses on bend-flanking faults. *Geol. Soc. Bull.* 116(11–12): 1423–1442.

Crowell, J. C. 1974. Sedimentation along the San Andreas fault, California, in *Modern and Ancient Geosynclinal Sedimentation; Successor Basin Assemblages, Special Publication—Society of Economic Paleontologists and Mineralogists,* 19: 292–303.

Crowell, J. C. 1979. The San Andreas fault through time. *Bull. Geol. Soc. London* 133: 292–302.

Cummings, D. 1976. Theory of plasticity applied to faulting, Mojave Desert, southern California. *Geol. Soc. Amer. Bull.* 87: 720–724.

Davis, G. A., and B. C. Burchfiel. 1973. Garlock fault—An intracontinental transform structure, southern California. *Geol. Soc. Amer. Bull.* 84: 1407–1422.

Freund, R. 1974. Kinematics of transform and transcurrent faults. *Tectonophysics* 21: 93–134.

Garfunkel, Z. 1974. Model for the late Cenozoic tectonic history of the Mojave Desert, California, and for its relation to adjacent regions. *Geol. Soc. Amer. Bull.* 85: 1931–1944.

Garfunkel, Z., and H. Ron. 1985. Block rotation and deformation by strike-slip faults, 2: The properties of a type of macroscopic discontinuous deformation. *J. Geophys. Res.* 90: 8589–8602.

Harding, T. P. 1985. Seismic characteristics and identification of negative flower structures, positive structures, and positive structural inversion. *Amer. Assoc. Petrol. Geol. Bull.* 69: 582–700.

Laubscher, H. P. 1972. Some overall aspects of Jura dynamics. *Amer. J. Sci.* 272: 293–304.

Luyendyk, B. P., M. J. Kamerling, and R. Terres. 1980. Geometric model for Neogene crustal rotations in southern California. *Geol. Soc. Amer. Bull.* Part I, 91: 211–217.

McClay, K., and M. Bonora. 2001. Analog models of restraining stepovers in strike-slip fault systems. *Amer. Assoc. Petrol. Geol. Bull.* 85: 233–260.

Molnar, P., and P. Tapponnier. 1975. Cenozoic tectonics of Asia: Effects of a continental collision. *Science* 189: 419–426.

Ron, H., and Y. Eyal. 1985. Intraplate deformation by block rotation and mesostructures along the Dead Sea transform, northern Israel. *Tectonics* 4: 85–105.

Ron, H., R. Freund, and Z. Garfunkel. 1984. Block rotation by strike-slip faulting: Structural and paleomagnetic evidence. *J. Geophys. Res.* 89: 6256–6270.

Ron, H., A. Nur, and Y. Eyal. 1990. Multiple strike-slip fault sets: A case study from the Dead Sea transform. *Tectonics* 9: 1421–1431.

Shelton, J. S. 1966. *Geology Illustrated.* San Francisco, W.H. Freeman and Company, 434 pp.

Sylvester, A. G. 1988. Strike-slip faults. *Geol. Soc. Amer. Bull.* 100(11): 1666–1703.

Tapponnier, P., and P. Molnar. 1976. Slip-line field theory and large scale continental tectonics. *Nature* 264: 319–323.

Tapponnier, P., and P. Molnar. 1977. Active faulting and tectonics in China. *J. Geophys. Res.* 82: 2905–2930.

Tapponnier, P., and P. Molnar. 1979. Active faulting and Cenozoic tectonics of the Tien Shan, Mongolia, and Baykal regions. *J. Geophys. Res.* 84: 3425–3459.

Teyssier, C., and B. Tikoff. 1998. Strike-slip partitioned transpression of the San Andreas fault system: A lithospheric-scale approach. In R. E. Holdsworth, R. A. Strachan, and J. F. Dewey, eds., *Continental Transpressional and Transtensional Tectonics.* Geol. Soc. Lond. Special Paper 135: 143–158.

Woodcock, N. J., and M. Fischer. 1986. Strike-slip duplexes. *J. Struct. Geol.* 8(7): 725–735.

Yin, A., and T. M. Harrison. 2000. Geologic evolution of the Himalayan-Tibetan orogen. *Ann. Rev. Earth and Planet. Sci.* 28: 211–280.

Chapter 7

STRESS

We describe in Chapters 2–6 many kinds of fractures and faults that form in rocks as a result of brittle deformation. We now want to explore why these structures exist. What caused them to form? What do they tell us about the processes operating in the Earth at the time they formed? To answer these questions, we must understand how and why rocks break, and the first step is to understand how forces affect the materials to which they are applied.

7.1 FORCE, TRACTION, AND STRESS

i. Preview

Our experience tells us that if we apply too much force to an object, it breaks. But what do we mean by "too much"? How can we use knowledge of the applied forces to predict when fracture will occur? We need to analyze what happens when forces are applied to a body of rock or to any other solid material. To this end, we first discuss force, then the concept of force intensity, or traction; this leads to the concept of surface stress, which finally allows us to describe the full state of stress. Before developing these ideas in detail, however, we give a brief introduction to help make the ideas more familiar and to serve as a guide to the subsequent discussion.

The concept of stress can be confusing at first, largely because we require a set of numbers to describe it (four numbers for a two-dimensional stress; nine numbers for a three-dimensional stress), and keeping track of all these numbers during the algebra needed to calculate stresses requires some getting used to. The fundamental concept of stress, however, is really quite straightforward.

We start from the idea of a force, which we all have experienced as a push or a pull applied to a body of some material (Table 7.1*A*, left). The **intensity** of the force on a body is the significant factor in determining how that body will react, and we express that intensity as the force divided by the area of the surface over which the force is distributed; we call this quantity the **traction**, and obviously it must have units of force per unit area (Table 7.1*B*, left). At equilibrium, any given surface must have equal and opposite forces acting on opposite sides of the surface, and hence the tractions acting on opposite sides of the surface also must be equal and opposite. This pair of equal and opposite tractions taken together is called the **surface stress** (Table 7.1*C*, left). In this way, the concept of a surface stress arises directly from our experience of a force applied across a surface of a given area.

It turns out that in two dimensions if we know the surface stresses on only two perpendicular surfaces that pass through a point, we can calculate the surface stress that acts on any surface of any other orientation through that point. To make this calculation in three dimensions, we need to know the surface stresses on three mutually perpendicular surfaces. These two (or three) sets of surface stresses then define the **state of stress at a point**, or more simply, the **stress** (Table 7.1*D*, left).

That is really all there is to it. So where do all the numbers come from that we have to deal with? Let us look at this question first for the two-dimensional case. Force is a vector quantity, which means it has both magnitude and direction (see Box 7-1). In two dimensions, a vector is represented in a particular coordinate system by two components, where each component is a number defining the length of the vector in a direction parallel to one of the coordinate axes (Table 7.1*A*, right). When we

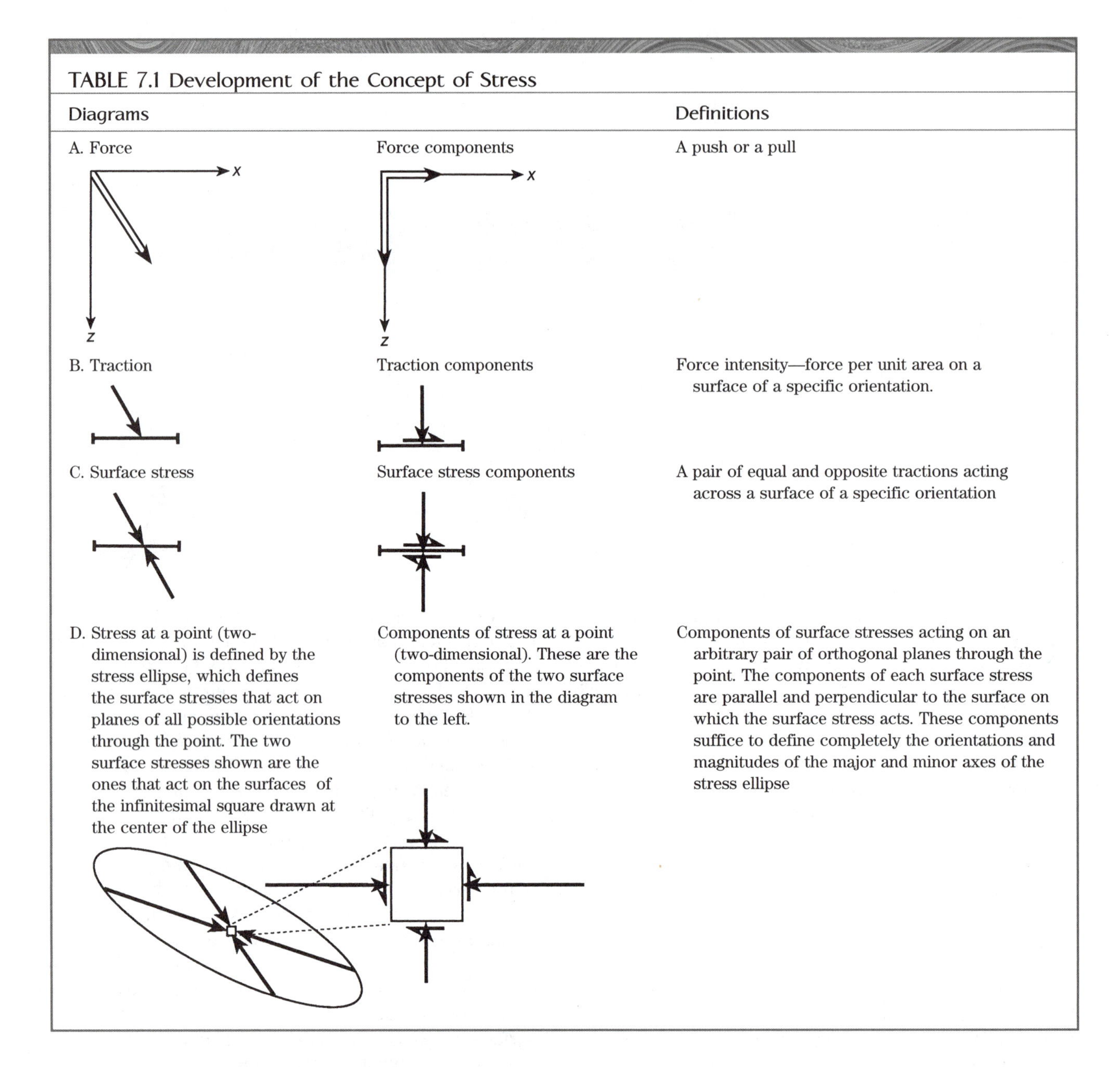

TABLE 7.1 Development of the Concept of Stress

Diagrams		Definitions
A. Force	Force components	A push or a pull
B. Traction	Traction components	Force intensity—force per unit area on a surface of a specific orientation.
C. Surface stress	Surface stress components	A pair of equal and opposite tractions acting across a surface of a specific orientation
D. Stress at a point (two-dimensional) is defined by the stress ellipse, which defines the surface stresses that act on planes of all possible orientations through the point. The two surface stresses shown are the ones that act on the surfaces of the infinitesimal square drawn at the center of the ellipse	Components of stress at a point (two-dimensional). These are the components of the two surface stresses shown in the diagram to the left.	Components of surface stresses acting on an arbitrary pair of orthogonal planes through the point. The components of each surface stress are parallel and perpendicular to the surface on which the surface stress acts. These components suffice to define completely the orientations and magnitudes of the major and minor axes of the stress ellipse

divide force by an area to get the traction, we still need two numbers to specify the resulting traction components, so the traction is also represented by two numbers (Table 7.1*B*, right). Because the tractions that act on opposite sides of a given surface are equal and opposite (Table 7.1*C*, right), we only need to specify one of the tractions in order to know them both. Therefore the surface stress is represented by one of the tractions and thus by a set of two numbers. Finally, to represent the state of stress, we require two surface stresses, and thus two pairs of tractions that act on two mutually perpendicular surfaces, with the tractions in each pair being equal and opposite (Table 7.1*D*, right). To describe the state of stress, therefore, we end up needing four numbers ([2 surface stresses] × [2 components per surface stress]). In three dimensions, a force, a traction, and a surface stress are each defined by three components. The state of stress is defined by the surface stresses on three mutually perpendicular surfaces and thus by nine components ([3 surface stresses] × [3 components per surface stress]).

The complexity, of course, is in the details. Let us now look more carefully at those details. To minimize the number of components we must keep track of, we consider the problem in two dimensions, and we progressively develop these concepts, which we introduce through a series of simple steps, to end up with a full description of the stress in a convenient notation. We need to use algebra in this development, but mathematics is only a shorthand for expressing physical relationships and for logically manipulating those relationships so that they can be easily interpreted in a way that makes physical sense. It is important to keep in mind the physical meaning behind the symbols and to understand the physical significance of the final form of the equations that we derive.

ii. Force

We start with the idea of **force**, because it is the basic concept, and we all have an intuitive grasp of what a force is from everyday experience. For example, we all are held on the Earth's surface by the Earth's gravitational force, and we all exert forces on objects every day by pushing or pulling on them. A force is a **vector quantity**, which means it has a magnitude (how strong the push or pull is) and a direction (which way the push or pull is). Graphically, a force is usually represented by an arrow (see Table 7.1*A*, left), where the arrow's length and orientation represent, respectively, the magnitude and direction of the force. In any coordinate system (e.g., in Table 7.1*A*, x might be north and z might be down), we can define a vector quantity in terms of its components parallel to each coordinate. In Table 7.1*A*, right, these components are F_x and F_z, and these symbols represent numbers that define the magnitudes of the force components in appropriate units such as Newtons in the x and z directions, respectively. In Box 7-1 we give a brief review of vectors and vector components.

Forces come in two types, internal forces and external forces, depending on whether they originate inside or outside of a body. **Internal forces** include the interatomic forces within a crystal lattice. These forces are balanced internally and thus do not cause motion or deformation of the body. They do, however, determine the **material properties**, such as the strength, hardness, or stiffness of a body. We account for the internal forces implicitly through the values of the material constants that appear in equations describing the material behavior; we discuss these equations further in Sections 9.1 and 9.4 and Chapters 16, 17, and 18.

External forces are the forces of explicit concern to us here as they produce the motion and deformation that are preserved as structures that we can study in rocks. There are two types of external forces:

1. **Body forces** act on each particle of mass, independent of its surroundings. The Earth's force of gravity is such a force. It affects every particle of rock in a volume, and its magnitude is proportional to the mass within that volume. Gravity is the most important body force in structural geology. Other body forces include electrostatic and magnetic forces, but they do not concern us here.
2. **Surface forces** result from the action of one body, or one part of a body, on another across a shared surface. For example, if we push on a block of rock with our hand, we apply a surface force to the block across the area of contact. We can also imagine a surface within a body of rock across which a surface force is applied by one part of the body on the other.

We focus our attention on surface forces.

iii. Traction: A Measure of Force Intensity

In Figure 7.1*A*, the tyrannosaur stands on a table with a wide support pillar having a cross-sectional area A_A. In Figure 7.1*B*, the table pedestal, made of the same material, has a much smaller cross-sectional area A_B, and the pedestal breaks under the applied force. Although the force is the same (the weight of the tyrannosaur), the effect is different, and the difference is the size of the pillar that supports the force. Evidently, the intensity of the

FIGURE 7.1 The intensity of an applied force increases as the area across which it is distributed decreases. *A*. A tyrannosaur is happy being securely supported on a large pillar of cross-sectional area A_A. *B*. The tyrannosaur, to its dismay, is too heavy for the table pedestal of smaller cross-sectional area A_B, which breaks under its weight.

BOX 7-1 What Is a Vector? A Brief Review

A **scalar** is a physical quantity that is characterized only by its magnitude. Examples of such quantities are temperature, mass, and density, each of which is represented by a single number (e.g., 35° C, 1000 kg, and 2500 kg/m^3).

A **vector** is a quantity that has both a magnitude and a direction. Familiar examples of vector quantities include velocity and force. We can define a vector quantity completely only by giving its magnitude and the direction in which it acts; for example, a plane travels 400 km/hr in a horizontal northeast direction. We commonly represent a vector diagrammatically by an arrow; the length of the arrow is proportional to the vector's magnitude, and the direction in which the arrow points indicates the direction of the vector, which is generally defined relative to a specified set of coordinates.

We can add two vectors by using the parallelogram rule. If, for example, we wish to add two forces **V** and **W** that act on a point *p*, we draw the arrows representing the forces tail to tail and construct a parallelogram with the arrows defining two adjacent sides (Figure 7-1.1*A*). The sum of the forces, called the **resultant force R**, is then the vector from the common origin to the diagonally opposite corner of the parallelogram. Thus the effect of applying the forces **V** and **W** to *p* is the same as if the resultant force **R** were applied to *p*. Note that an equivalent construction to find the resultant force **R** is to place the two arrows representing **V** and **W** tail to head and then to draw an arrow from the tail of the first to the head of the second (Figure 7-1.1*B*).

In order to specify a direction, we need to adopt some frame of reference, such as the geographic coordinates east, north, and up (which we used to describe the velocity of the airplane). In three dimensions we commonly use a mutually perpendicular system of coordinates known as Cartesian coordinates,[1] and we assume that we know their orientations.

[1]Named after the seventeenth-century French philosopher René Descartes.

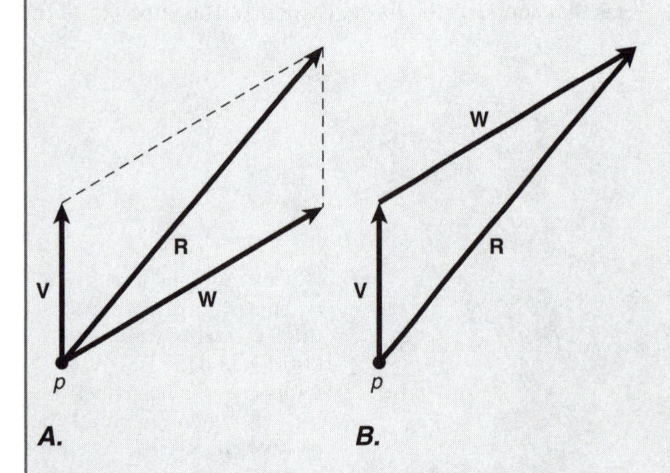

FIGURE 7-1.1 Vector addition. The resultant vector **R** is the sum of vectors **V** and **W**. *A*. The parallelogram rule for vector addition. *B*. The "head-to-tail" rule for vector addition.

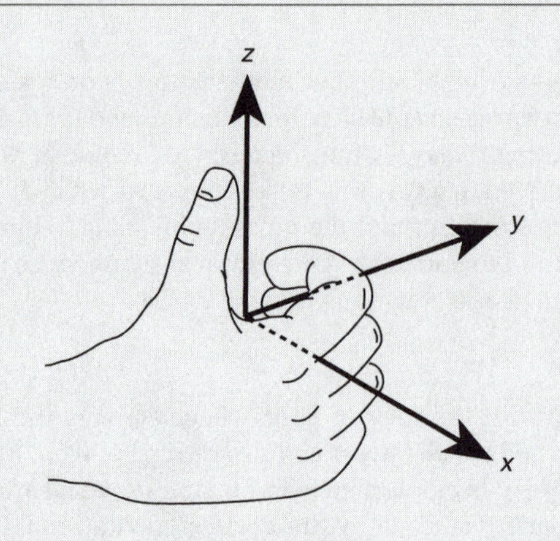

FIGURE 7-1.2 The right-hand rule defining the relative orientations of axes (*x*, *y*, *z*) in a right-handed orthogonal Cartesian coordinate system.

Here we label the axes *x*, *y*, and *z* according to the right-hand rule. By this rule, if the fingers of the right hand are oriented to curve along the direction of rotation from positive *x* to positive *y*, then the thumb points in the direction of positive *z* (Figure 7-1.2).

If a vector **V** represents, for example, a force in three-dimensional space (Figure 7-1.3), using the parallelogram rule shows that it can be considered the result of adding two forces: one, $\mathbf{V}_z$, parallel to the *z* axis and the other, **W**, lying in the (*x*, *y*)-plane:

$$\mathbf{V} = \mathbf{W} + \mathbf{V}_z \tag{7-1.1}$$

Using the parallelogram rule again for **W** shows that it can be

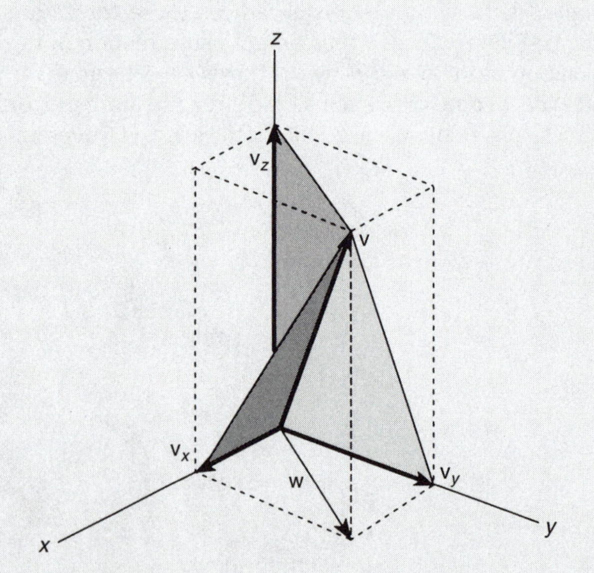

FIGURE 7-1.3 A vector **V** and its vector components ($\mathbf{V}_x$, $\mathbf{V}_y$, $\mathbf{V}_z$) in three dimensions. **W** is the projection of **V** on the (*x*, *y*)-plane. $\mathbf{V}_x$, $\mathbf{V}_y$, and $\mathbf{V}_z$ are the projections of **V** onto the three coordinate axes. $\mathbf{V}_x$ and $\mathbf{V}_y$ are also the projections of **W** on the *x* and *y* coordinate axes, respectively.

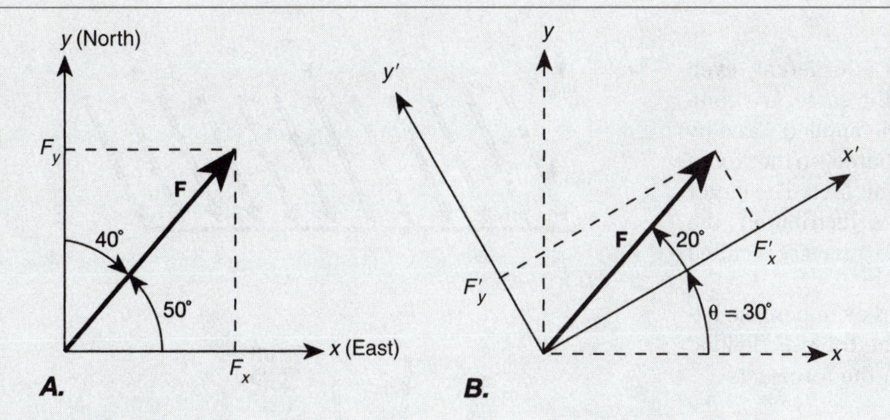

FIGURE 7-1.4 The dependence of the scalar components of a vector F on the orientation of the coordinate system. *A*. Components (F_x, F_y) of F in the (x, y) coordinate system. *B*. Components (F'_x, F'_y) of the same vector F in the (x', y') coordinate system.

considered the result of two forces $\mathbf{V}_x$ and $\mathbf{V}_y$, which parallel the x and y axes, respectively:

$$\mathbf{W} = \mathbf{V}_x + \mathbf{V}_y \tag{7-1.2}$$

Combining Equations (7-1.1) and (7-1.2) shows that the force **V** is the result of adding three forces, each acting parallel to one of the coordinate axes:

$$\mathbf{V} = \mathbf{V}_x + \mathbf{V}_y + \mathbf{V}_z \tag{7-1.3}$$

$\mathbf{V}_x$, $\mathbf{V}_y$, and $\mathbf{V}_z$ are the component vectors of **V**. If we designate their lengths by V_x, V_y, and V_z, respectively, then these are called the **scalar components**, or just the **components**, of the vector **V** in the given coordinate system. By convention, the components are always written in order. Thus the vector **V** can be represented by an ordered array of three scalar components

$$\mathbf{V} : [V_x, V_y, V_z] \tag{7-1.4}$$

For a fixed vectorial quantity such as a given force, the values of the components representing that vector quantity depend not only on the magnitude and direction of the quantity but also on the orientation of the coordinate system in which the components are defined. The problem is simpler to explain in two dimensions, for which the reference coordinates are, say, x positive due east and y positive due north. For example, if **F** is a force of 100 N (Newtons) acting 50° north of east (or, equivalently, 40° east of north) (Figure 7-1.4*A*), then the force vector is completely defined by its components (F_x, F_y) in the (x, y) coordinate system:

$$(F_x, F_y) = (64.3, 76.6)\ \text{N}$$

where

$$F_x = |F| \cos 50^\circ = (100)(0.643) = 64.3\ \text{N}$$
$$F_y = |F| \sin 50^\circ = (100)(0.766) = 76.6\ \text{N} \tag{7-1.5}$$

If, however, we use a different coordinate system, say, (x', y'), where x' is 30° counterclockwise from x (Figure 7-1.4*B*), then exactly the same force vector **F** has components given by

$$(F'_x, F'_y) = (94.0, 34.2)\ \text{N}$$

where

$$F'_x = |F| \cos 20^\circ = (100)(0.940) = 94.0\ \text{N}$$
$$F'_y = |F| \sin 20^\circ = (100)(0.342) = 34.2\ \text{N} \tag{7-1.6}$$

For a given vector **F**, the components in different coordinate systems are systematically related. If, in Figure 7-1.5, we use θ to designate the angle between x and x', as well as that between y and y' then using the sides of the shaded triangles, it is not difficult to show that

$$F'_x = F_x \cos\theta + F_y \sin\theta$$
$$F'_y = -F_x \sin\theta + F_y \cos\theta \tag{7-1.7}$$

The same situation exists in the more general three-dimensional case. Although the equations are slightly more complicated, the principle is the same: The vector **F** is the physical quantity, such as force, and the same force is represented by a different ordered set of components in each different coordinate system, where those sets of components are related by equations like Equations (7-1.7).

Because Equations (7-1.7) enable us to transform the component values from one known coordinate system to another, they are called the **transformation equations**. For a quantity to be a vector, its components must transform according to the rule given by these equations for two dimensions or by comparable equations for three dimensions.

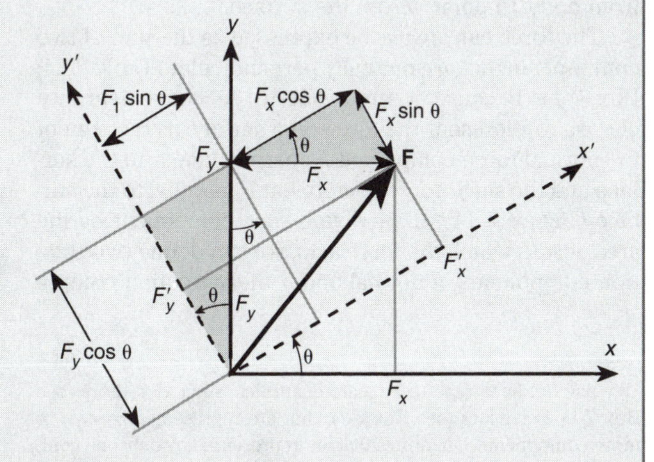

FIGURE 7-1.5 Geometric relationships between the scalar components of the same vector in two differently oriented coordinate systems. The sides of the shaded triangles can be used to deduce the values of the components (F'_x, F'_y) from the components (F_x, F_y) and the angle θ.

force experienced by the two pillars is different, even though the magnitude of the force is the same. We compute this force intensity by dividing the applied force by the area over which that force is distributed, so the "force intensity" is expressed as a force per unit area. The larger the area over which a given force is distributed, the smaller is the force intensity. This force intensity is called the **traction** (Table 7.1*B*).

We designate the traction by the Greek uppercase letter sigma $\boldsymbol{\Sigma}$, and we define it to be the force $\mathbf{F}$ divided by the area A of the surface on which the force acts[1]

$$\boldsymbol{\Sigma} \equiv \frac{\mathbf{F}}{A} \tag{7.1}$$

Thus the material in the pillar can withstand a traction equal to the tyrannosaur's weight divided by the cross-sectional area A_A, but not the greater traction equal to the tyrannosaur's weight divided by the smaller area A_B. The traction thus has physical units of force per unit area.[2]

The relationship in Equation (7.1) assumes that the force $\mathbf{F}$ is uniformly distributed over the surface A (Figure 7.2*A*). In many cases, the force is not uniformly distributed but varies in magnitude and direction across a surface, as indicated in Figure 7.2*B*. In such a case, we can only define the traction at a point, represented as an infinitesimal area dA of the surface on which the infinitesimal part of the total force $d\mathbf{F}$ acts (Figure 7.2*B*). In this case, we can express the traction at a point as

$$\boldsymbol{\Sigma} \equiv \frac{d\mathbf{F}}{dA} \tag{7.2}$$

Thus the magnitude and direction of the traction can vary from point to point across the surface.

The force can always be expressed as the sum of two components that are mutually perpendicular (Table 7.1*A*; Box 7-1). Because it turns out to be convenient, we choose to represent the force on a surface as the sum of the normal force component F_n perpendicular to the surface and the shear force component F_s parallel to the surface (Figure 7.3*A*). If we divide each component by the area across which the force acts, we can define two traction components, a normal and a shear traction component (Figure 7.3*B*; see Table 7.1*B*). We designate these traction components with a lowercase Greek sigma, σ, to which we add the subscripts n and s to indicate, respectively, the normal and shear components. Thus the traction $\boldsymbol{\Sigma}$ is defined by its normal and shear components (σ_n, σ_s), where

$$\sigma_n \equiv \frac{F_n}{A} \qquad \sigma_s \equiv \frac{F_s}{A} \tag{7.3}$$

These components are independent of each other, which means that changing one component does not require a change in the other component, although of course such a change reflects a change in the original traction.

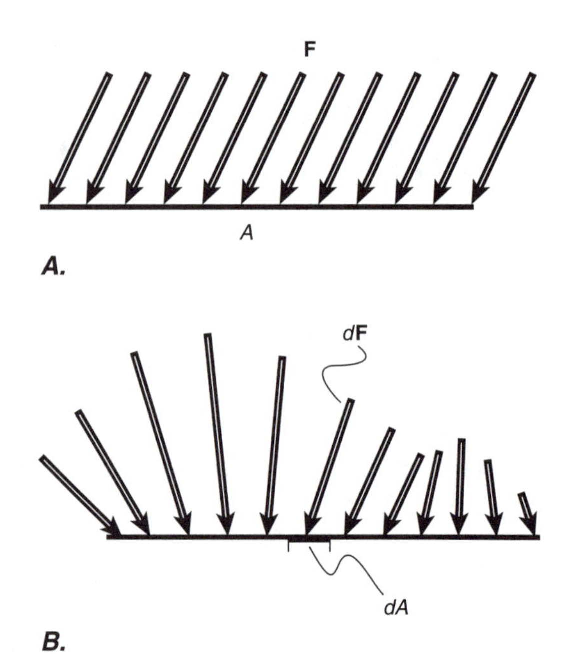

FIGURE 7.2 A force need not be constant in magnitude and orientation across a surface. *A*. A force F that is constant across a surface. *B*. A force that varies in magnitude and orientation across a surface. Each infinitesimal area *dA* has a force *dF* acting on it.

iv. Surface Stress

A **material surface** in a body is a surface defined by a particular set of particles of the material, or **material points**. We require that the material points making up a surface cannot accelerate away from the material points on either side of it. Thus, according to Newton's second law, a force on one side of a surface must be opposed by an equal and opposite force on the other side, and the sum of all the forces across the surface must be equal to zero. Figure 7.3*C* shows two forces

[1]We use boldface type to indicate quantities, such as vectors (see Box 7-1) or tensors (see Box 7-3), that are represented by two or more components. Quantities that are represented by only one component, such as temperature or density are called **scalars**, and they are written in italic type.

[2]Tractions thus have the same units as pressure. A pressure differs from a traction, however, in that a pressure is always perpendicular to the surface across which it acts, whereas a traction need not be.

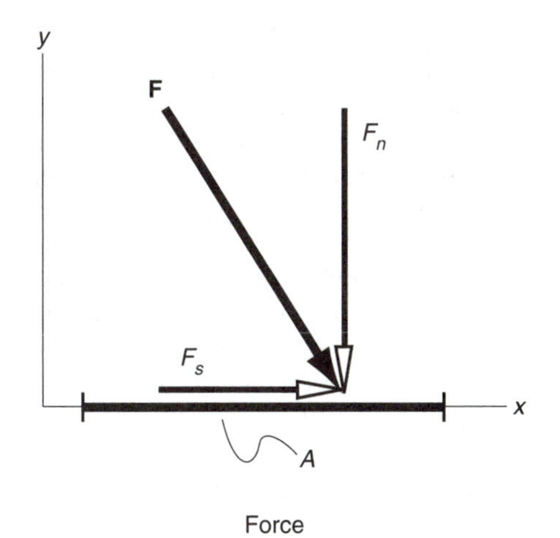

A.

y

Σ

σ_n

σ_s

x

A

Traction

B.

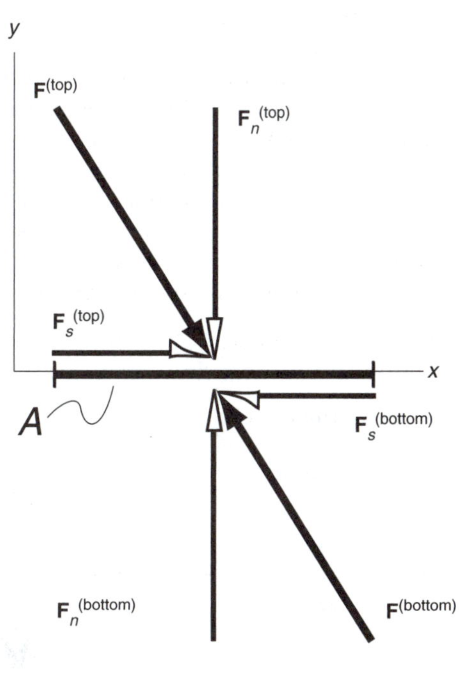

C.

y

$\Sigma^{(top)}$

$\sigma_n^{(top)}$

$\sigma_s^{(top)}$

X

A

$\sigma_s^{(bottom)}$

$\sigma_s^{(bottom)}$

$\Sigma^{(bottom)}$

Surface stress
(pairs of opposing
tractions)

D.

FIGURE 7.3 Force, traction, and surface stress. *A*. A force *F* acting on a surface *A* is represented by its components normal and parallel to the surface (F_n, F_s). *B*. The force *F* divided by the area *A* on which it acts gives the traction $\mathbf{\Sigma}$; similarly each force component, divided by the area, gives the components of the traction (σ_n, σ_s) normal and parallel to the surface. *C*. A force $F^{(top)}$ or its components ($F_n^{(top)}$, $F_s^{(top)}$) applied to the top of the surface of area *A* is balanced by an equal and opposite force $F^{(bottom)}$ or its components ($F_n^{(bottom)}$, $F_s^{(bottom)}$) on the bottom of the surface. *D*. The force intensity is given by the associated tractions $\mathbf{\Sigma}^{(top)}$ and $\mathbf{\Sigma}^{(bottom)}$, which are equal and opposite and together define the surface stress $\mathbf{\Sigma}$. Each traction can be expressed in terms of its normal and shear components. The balanced pairs of components ($\sigma_n^{(top)}$, $\sigma_n^{(bottom)}$) and ($\sigma_s^{(top)}$, $\sigma_s^{(bottom)}$) are, respectively, the normal surface stress component σ_n and the shear surface stress component σ_s.

$\mathbf{F}^{(top)}$ and $\mathbf{F}^{(bottom)}$ that are equal and opposite to each other

$$\mathbf{F}^{(top)} = -\mathbf{F}^{(bottom)} \quad \text{or} \quad \mathbf{F}^{(top)} + \mathbf{F}^{(bottom)} = 0 \tag{7.4}$$

Similarly, there must be two equal and opposite tractions acting on the surface as well (Figure 7.3*D*). We can express this force balance in terms of tractions if we divide the forces by the area across which they act

$$\frac{\mathbf{F}^{(top)}}{A} + \frac{\mathbf{F}^{(bottom)}}{A} = 0$$

$$\mathbf{\Sigma}^{(top)} + \mathbf{\Sigma}^{(bottom)} = 0 \quad \text{or} \quad \mathbf{\Sigma}^{(top)} = -\mathbf{\Sigma}^{(bottom)} \tag{7.5}$$

Because the components of the traction are independent of each other, this same relationship must also apply to each of the traction components separately:

$$\sigma_n^{(top)} = -\sigma_n^{(bottom)} \qquad \sigma_s^{(top)} = -\sigma_s^{(bottom)} \tag{7.6}$$

The **surface stress** $\mathbf{\Sigma}$ is defined to be the pair of equal and opposite tractions acting on any surface of a specific orientation, such as $\mathbf{\Sigma}^{(top)}$ and $\mathbf{\Sigma}^{(bottom)}$ in Figure 7.3*D* (see Table 7.1*C*). The surface stress can also be represented by its normal and shear components σ_n and σ_s, respectively. σ_n is defined by the pair of equal and opposite normal traction components $\sigma_n^{(top)}$ and $\sigma_n^{(bottom)}$, and σ_s is defined by the pair of equal and opposite shear traction components $\sigma_s^{(top)}$ and $\sigma_s^{(bottom)}$, sometimes referred to as a **shear couple** (Figure 7.3*D*).

Because a surface stress consists of a pair of equal and opposite tractions, if we know the traction on one side of the surface, we automatically know the traction on the other side, and thus we know what the surface stress is. Accordingly, we can use the same symbol for the surface stress as for the traction, $\mathbf{\Sigma}$, and we can use the same symbols for the normal and shear components of the surface stress as for the corresponding traction components, σ_n and σ_s, respectively. The numerical values for these components are the same for the traction and for the associated surface stress, except possibly for the sign (which we discuss in the following paragraphs). If we want specifically to indicate one or the other of the tractions that make up the surface stress, we use a superscript that identifies the particular side of the surface on which the traction acts, such as the (top) or (bottom) in Equations (7.5) and (7.6).

To define the sign of a traction component, we must use a coordinate system with axes parallel and perpendicular to the surface on which the traction acts, such as the axes (x, y) shown in Figure 7.3*B*. If the traction component points in a positive coordinate direction, such as σ_s in Figure 7.3*B* (which points in the positive x direction), the value of the traction is positive, and if it points in a negative coordinate direction, such as σ_n in Figure 7.3*B* (which points in the negative y direction), its value is negative.

We must use a different sign convention for the stress, however, because the normal and shear components of the surface stress always include a pair of tractions that point in opposite directions. Normal stresses are distinguished in sign according to whether they are compressive or tensile: If the normal traction components push the material together across the surface (the two normal tractions point toward each other; Figure 7.4*A*), the normal surface stress component is a **compressive stress**. If the normal traction pair pulls the material apart across the surface (the two normal tractions point away from each other; Figure 7.4*B*), the normal surface stress component is a **tensile stress**. The shear stress components can be distinguished by the sense of rotation of an imaginary ball placed between the arrows representing the shear traction components. The surface shear stress component is either counterclockwise (the sense of rotation of a ball in a sinistral strike-slip fault; Figure 7.4*C*) or clockwise (the sense of rotation of a ball in a dextral strike-slip fault; Figure 7.4*D*). Based on these distinctions we use the following conventions, which define the **geologic Mohr circle sign convention** for stress:

compressive normal stress	is	positive (Figure 7.4*A*)
tensile normal stress	is	negative (Figure 7.4*B*)

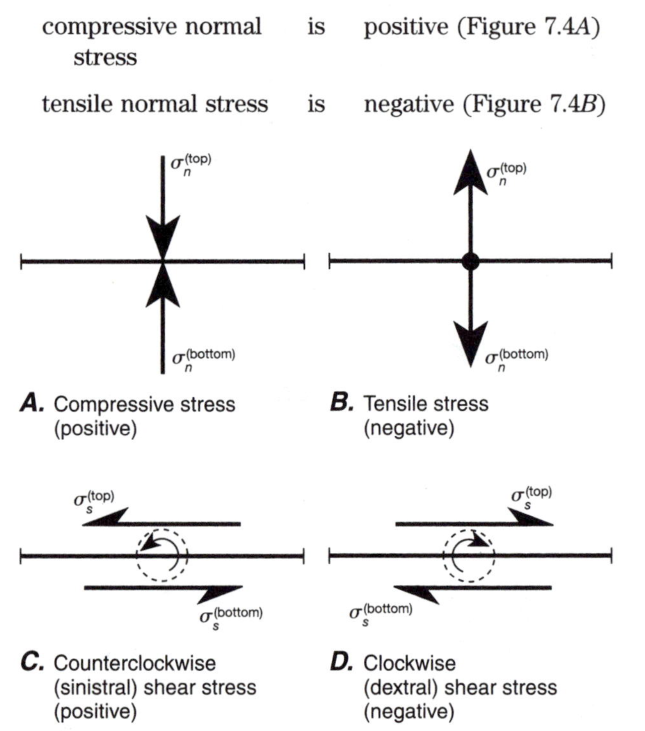

FIGURE 7.4 The Mohr circle sign conventions for the components of the stress at a point.

counterclockwise (sinistral) shear stress	is	positive (Figure 7.4*C*)
clockwise (dextral) shear stress	is	negative (Figure 7.4*D*).

Thus, the magnitudes, or absolute values, of the surface stress components are the same as the magnitudes of either traction in the pair that make up the surface stress, but the signs of the tractions and surface stresses are defined differently. To calculate a surface stress component, therefore, we can simply calculate one of the associated traction components and then adjust the sign to be consistent with our stress sign convention.

These sign conventions are appropriate for plotting stress components on the **Mohr diagram**, which is a common method of graphically displaying and solving stress problems that we discuss in Sections 7.2 and 7.4. The geologic sign convention defines compressive stresses to be positive because those are the most commonly encountered normal stresses in geological problems, and it is convenient to deal with positive numbers as much as possible. The sign convention we give for shear stress is required for plotting those stress components on the Mohr diagram, but it has the distinct disadvantage of being nonunique. In particular, a counterclockwise (positive) shear stress viewed from one side of a plane is a clockwise (negative) shear stress viewed from the opposite side, and vice versa. Its use therefore requires care and consistency in setting up and solving problems.

In this text, we use the geologic Mohr circle sign convention exclusively through Section 7.4 and in the rest of the book, with the exception of some boxes in Chapters 16 and 18, where we use the tensor sign convention and the specific notation for components in that convention, which we introduce in Section 7.5. It is important to be aware, however, that there are still other conventions in common use in the geologic literature. We discuss the geologic tensor sign convention in Section 7.5 and other sign conventions in Box 7-4. For reference, we summarize all these vexing conventions in Table 7-4.1 in Box 7-4.

v. Stress Across Planes of Different Orientations

The fracture plane in the broken table pedestal in Figure 7.1*B* is not parallel to the cross-sectional area of the pedestal A_B, but it is a common orientation for a shear fracture under these circumstances, as we show in Section 8.3. We need to know the surface stress on any plane of any orientation through the pedestal, because then we can determine which of those planes is most critically stressed and thus which will become the fracture plane. We start by using a simple example of determining the stress on a plane of any orientation. This example involves only a uniaxial stress (see Section 7.3), and the equations we derive therefore do not apply to a more general biaxial stress. We consider the more general case in Sections 7.2 and 7.5.

It is tempting to think of stress components as being similar to vector components such as the components of force. In fact, however, stress components behave very differently because stress combines the effects of both force and area (Equation (7.1)), both of which change as the orientation of the surface changes. In calculating the stress, therefore, *we must account for the changes in both the force components F_n and F_s, and the area A, as the orientation of the plane changes.* In the rest of this subsection, we look more carefully at how to account for both of these changes simultaneously.

First let us calculate the components of the surface stress acting on a cross-sectional plane perpendicular to the long dimension of the pedestal (Figure 7.5*A*, *B*). If the tyrannosaur has a weight of W and the cross-sectional area of the pedestal is A ($A = L \times L$ in Figure 7.5*A*), then from Equation (7.1), the magnitude of the surface stress $\boldsymbol{\Sigma}$ on the (horizontal) cross-sectional plane would be the dinosaur's weight divided by the area, or

$$|\boldsymbol{\Sigma}| \equiv \Sigma = \frac{W}{A} = \frac{W}{LL} \tag{7.7}$$

(From here on, we consider only the *magnitudes* of the forces and surface stresses, so we use symbols in italic type, not boldface).

In this special case, the normal stress component σ_n equals the magnitude of the surface stress itself Σ, and the shear stress component σ_s is zero, because the force acts exactly perpendicular to the surface A and there is no component of force parallel to the surface (Figure 7.5*B*). Thus,

$$\sigma_n = \Sigma = \frac{W}{A} \qquad \sigma_s = 0 \tag{7.8}$$

Suppose, now, that we wanted to calculate the magnitude of surface stress Σ' acting on another plane through the column that is inclined to the left at an angle θ. In this case, the area A' (Figure 7.5*C*) is different from that of the cross-sectional area A because one of the sides of the plane has a length ($L/\cos\theta$) instead of just L. Since the area is equal to the product of the lengths of the sides of the plane in Figure 7.5*C*, we have

$$A' = L\left(\frac{L}{\cos\theta}\right) = \frac{LL}{\cos\theta} \quad \text{or} \quad A' = \frac{A}{\cos\theta} \tag{7.9}$$

Notice that Equation (7.9) tells us that the area A' is greater than the area A because the cosine of any positive angle θ different from zero always has a value less

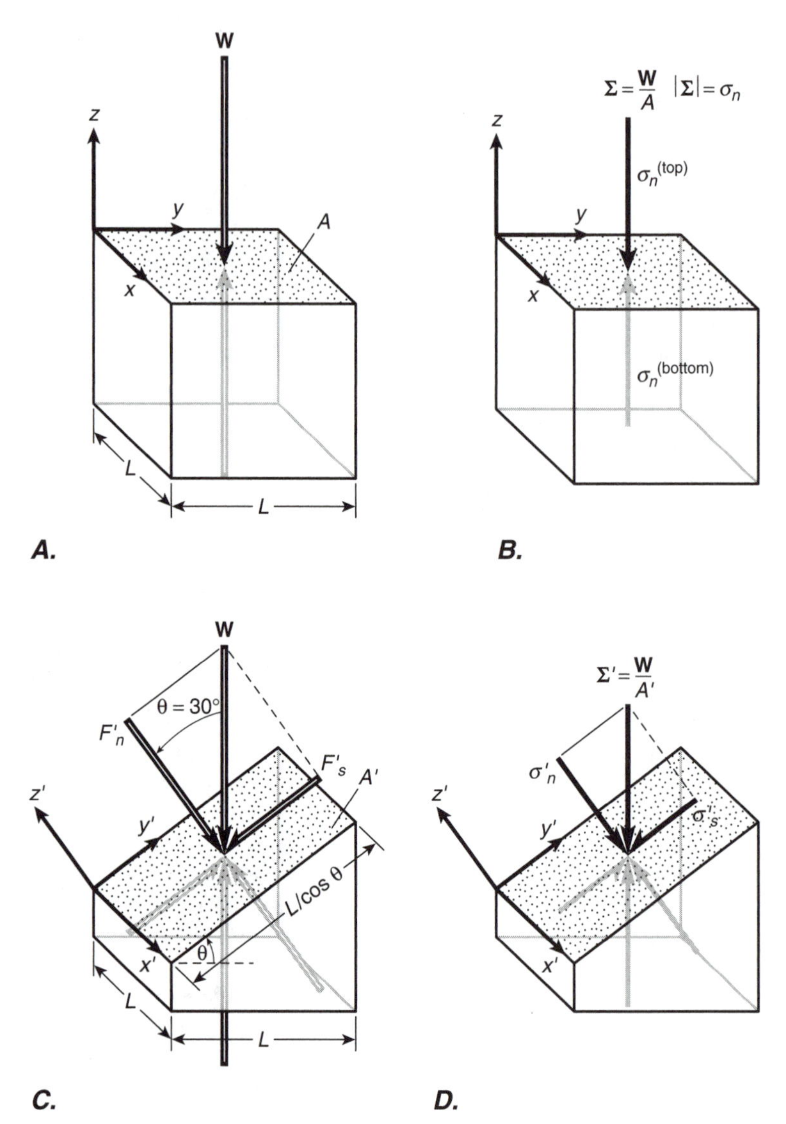

FIGURE 7.5 Determination of the stress components on surfaces of different orientation. *A*. In supporting the weight W of the tyrannosaur in Figure 7.1*A*, the upper part of the pillar exerts a force W on the lower part across a cross-sectional plane with an area *A* normal to the pillar axis. The lower part of the pillar exerts an equal and opposite force on the upper part. *B*. The magnitude of the surface stress $\boldsymbol{\Sigma}$ on the plane of area *A* is the force divided by the area. The force and the surface stress are normal to the surface, so there is no shear stress component. *C*. On a plane inclined at an angle $\theta = 30°$ and having an area A', the same force W acts in the same direction. Components of the force normal and parallel to the plane are F'_n and F'_s. *D*. The magnitude of the surface stress $\boldsymbol{\Sigma}'$ is the same force W divided by the larger area A'. The magnitude of the normal stress and shear stress components σ'_n and σ'_s are the normal and parallel force components divided by A'.

than 1. Using Equation (7.1) for the inclined plane, therefore, we have (Figure 7.5*D*)

$$\Sigma' = \frac{W}{A'} = \frac{W}{A/\cos\theta} = \frac{W}{A}\cos\theta = \Sigma \cos\theta \quad (7.10)$$

where we used Equation (7.8) to get the last equality. Thus *even though the magnitude W of the force is the same on both planes* (Figure 7.5*A*, *C*) *and even though the orientations of the force* **W** (Figure 7.5*A*, *C*) *and of both of the surface stresses* Σ *and* Σ' (Figure 7.5*B*, *D*) *are all vertical, the magnitude of the surface stress* Σ' *is less than that of* Σ*, because* A' *is bigger than A*. This shows that the surface stress behaves differently from the force, and thus it cannot be calculated in the same way.

It should come as no surprise that the *components* of the force and the *components* of the stress also behave differently from each other. Because **W** is perpendicular to the cross-sectional plane *A* (Figure 7.5*A*), the normal component of force on the plane is just *W*, and the shear component of force, which would act parallel to the plane, is zero. But **W** is not perpendicular to the plane A' (Figure 7.5*C*), so it has a normal component F'_n and a shear component F'_s on that plane, both of which are nonzero. As illustrated in Figure 7.5*C*, these components of the force vector equal, respectively, *W* times the cosine or sine of θ

$$F'_n = W\cos\theta \quad F'_s = W\sin\theta \quad (7.11)$$

These equations for the force components define how the normal and shear components of the force vector vary with the orientation θ of the plane (remember, however, that we are using a special simple example; see Box 7-1, Equations (7-1.7), for the more general form of these equations). These equations tell us that when $\theta = 0$ (Figure 7.5*A*), the normal force component is equal to *W* (because $\cos\theta = 1$), whereas the shear force component is zero (because $\sin\theta = 0$). $\theta = 0$ is just the special case when the area A is perpendicular to $\mathbf{W}$.

If Equation (7.11) tells us how the force components change when the orientation of the surface changes, how do the stress components change? From the preceding discussion, we know that the normal and shear stress components are simply the corresponding force components divided by the area across which they act (Figure 7.5*C*, *D*) or

$$\sigma'_n = \frac{F'_n}{A'} = \frac{W\cos\theta}{A'} = \Sigma'\cos\theta \qquad (7.12)$$

$$\sigma'_s = \frac{F'_s}{A'} = \frac{W\sin\theta}{A'} = \Sigma'\sin\theta \qquad (7.13)$$

where we used Equations (7.11) and where $\Sigma' = W/A'$ is the magnitude of the surface stress $\mathbf{\Sigma}'$.

We can now relate the surface stress components (σ'_n, σ'_s) on the plane A' to normal stress component σ_n on the plane A. To do so we simply substitute the value of A' in terms of A given in Equation (7.9) into Equations (7.12) and (7.13):

$$\sigma'_n = \frac{W\cos\theta}{A/\cos\theta} = \frac{W}{A}\cos^2\theta = \sigma_n\cos^2\theta \qquad (7.14)$$

$$\sigma'_s = \frac{W\sin\theta}{A/\cos\theta} = \frac{W}{A}\sin\theta\cos\theta = \sigma_n\sin\theta\cos\theta \quad (7.15)$$

Note that if $\theta = 0$, $\sigma'_n = \sigma_n$ and $\sigma'_s = 0$, which is again the special case when the area A is perpendicular to $\mathbf{W}$.

In the special case we have considered, Equations (7.11) are the **transformation equations** for the force vector components, and Equations (7.14) and (7.15) are the **transformation equations** for the stress components. They define how these different components change as the orientation of the plane on which they act changes. The important point to note here is that *the components of force and of stress behave differently: In the transformation equations for force components (Equation (7.11)), the trigonometric terms appear as single sine or cosine functions; in the transformation equations for stress components (Equations (7.14) and (7.15)), however, they appear as products or squares of the sine and cosine functions.* Thus there is a fundamental difference between force and stress. In Equations (7.11), the single trigonometric term comes from resolving the force vector into directions normal and parallel to the plane. In the first Equations (7.14) and (7.15), the trigonometric term in the numerator serves the same purpose, but the trigonometric term that is divided into A comes from resolving the area (Equation (7.9)), which does not have to be done for the force components.

Thus neither the traction nor the surface stress is a vector quantity, because they are both inseparable from the area on which they act, and that area changes with the orientation of the plane. A traction has the characteristics of a vector only if we consider a surface of fixed orientation and thus of fixed area.[3] The force vector, however, is independent of the area of the surface on which it acts. This difference is the most important distinction between stress (and the associated tractions) and force.

vi. Calculating Tractions and Stresses: A Numerical Example for a Special Case

It can be helpful in understanding stress to see a numerical example of how the magnitude of the traction components and their associated stress components are actually calculated. We use the simple case and the associated equations developed in the preceding subsection. We emphasize, however, that some of these equations are for the special case called a uniaxial stress (see Section 7.3), and these are not the equations one would use to solve general stress problems in two dimensions. The general equations include additional terms that account for a biaxial state of stress and are given in Section 7.2 (Equations (7.34)–(7.36)) and Section 7.4 (Equations (7.44) and (7.46)).

Let us consider the tyrannosaur standing on the table in Figure 7.1*A*, and find the traction and stress components, first on the horizontal cross-sectional plane through the pillar shown in Figure 7.5*A* and *B*, and then on the inclined plane shown in Figure 7.5*C* and *D*.

Consider the x and y coordinate axes to be horizontal and the z coordinate to be vertical and positive up (Figure 7.5*A*, *B*). Suppose a cross section of the pillar measures $L = 2$ meters on a side (almost 6.6 feet). A tyrannosaur was a rather large animal. Small ones might have weighed about 80,000 N (Newtons),[4] using the SI (Systeme Internationale) units that are the standard in science (see Appendix 3, Tables A3.1, and A3.2). In English units, this tyrannosaur would have weighed about 18,000 pounds, or 9 English tons.

What is the magnitude of the tractions on opposite sides of the horizontal plane, and of the surface stress on that plane defined by those tractions? The magnitude of each of the tractions, and of the associated surface stress, is just the force per unit area (Equation (7.7)). The force applied is the weight of the tyrannosaur (80,000 N) and the area is the cross-sectional area of the pillar,

[3]The traction is sometimes called the *stress vector*, but we eschew that terminology because of the confusing implication that the traction is a vector; it only behaves like a vector if the orientation of the surface on which it acts is held constant, which is not necessary.

[4]A Newton is the amount of force required to accelerate 1 kilogram of mass at 1 meter per second per second (1 N = 1 kg m/s^2). Appropriately enough, a force of 1 Newton is approximately equal to the weight of an apple (1 N = 0.225 lb).

4 m² (2 m × 2 m). The weight of the tyrannosaur acts in a negative coordinate direction on the pillar (negative z is down), and the pillar provides an equal and opposite force in a positive coordinate direction (up) on the tyrannosaur (Figure 7.5*A*). Thus the *magnitude* of the two tractions on opposite sides of the surface, and of the associated surface stress is[5] (Figure 7.5*B*; compare Equation (7.1)):

$$|\boldsymbol{\Sigma}^{(\text{bottom})}| = |\boldsymbol{\Sigma}^{(\text{top})}| = |\boldsymbol{\Sigma}| = \frac{W}{A} = \frac{80{,}000\ N}{4\ m^2}$$
$$= 20{,}000 \text{ Pa} = 0.02 \text{ MPa} \quad (7.16)$$

What are the normal and shear components of the tractions on the horizontal plane and the normal and shear components of the surface stress defined by them? In this special case, the applied force acts exactly perpendicular to the surface, and therefore it has no component parallel to the surface. Thus the normal traction components represent all the tractions on the surface, because the shear traction components are zero (Equation (7.3)). The sign is positive for the traction that points in the positive coordinate direction ($\sigma_n^{(\text{bottom})}$, the traction applied by the pillar to the tyrannosaur) and negative for the equal but opposite traction that points in the negative coordinate direction ($\sigma_n^{(\text{top})}$, the traction applied by the tyrannosaur to the pillar) (Figure 7.5*B*; Equation (7.6)). Thus, the traction components are:

$$\sigma_n^{(\text{bottom})} = -\sigma_n^{(\text{top})} = \frac{W}{A} = \frac{80{,}000 \text{ N}}{4 \text{ m}^2}$$
$$= 20{,}000 \text{ Pa} = 0.02 \text{ MPa} \quad (7.17a)$$
$$\sigma_s^{(\text{bottom})} = \sigma_s^{(\text{top})} = 0$$

The surface stress components have the same magnitude as their associated tractions (Equation (7.8)). The normal surface stress component σ_n is compressive and therefore positive. It therefore is identical in magnitude and sign to the traction acting on the bottom (negative side) of the xy coordinate surface. The shear surface stress component is zero.

$$\begin{aligned} \sigma_n &= \sigma_n^{(\text{bottom})} = 0.02 \text{ MPa} \\ \sigma_s &= \sigma_s^{(\text{bottom})} = 0 \text{ MPa} \end{aligned} \quad (7.17b)$$

[5]Newtons per square meter (N/m²), pascals (Pa), megapascals (MPa), bars (b), and kilobars (kb) are all units of stress—that is, force per unit area—but bars and kilobars belong in the cgs (centimeter-gram-second) system of units, whereas the others are part of the SI system. These units are related by the following:

$$10^6 \text{ N/m}^2 = 10^6 \text{ Pa} = 1 \text{ MPa} = 10 \text{ b} = 0.01 \text{ kb}$$

A pressure of 1 Pa is rather small. Because one apple weighs approximately 1 Newton (0.225 lb), if one made applesauce out of that apple and spread it in an even layer over an area of 1 m², the pressure created on that surface would be about 1 Pa. Atmospheric pressure (14.7 lb/in²) is approximately 10^5 Pa, 0.1 MPa, or 1 b.

What is the magnitude of the tractions on opposite sides of the inclined plane and of the surface stress on that plane defined by those tractions? We choose coordinates x' and y' to lie in the inclined plane and z' to be normal to the plane (Figure 7.5*C, D*), so that $x' = x$, and y' and z' are rotated by an angle $\theta = 30°$ from the horizontal y and vertical z axes, respectively. Again, the magnitude of the tractions is just the applied force per unit area. The force—which is the weight of the tyrannosaur—is unchanged. The area, however, is larger because the plane is inclined (the normal to the plane, z', is oriented at $\theta = 30°$ from the vertical z axis). Thus the area is (Equation (7.9)):

$$A' = L\left(\frac{L}{\cos\theta}\right) = 2\left(\frac{2}{\cos 30}\right) = \frac{4}{0.866} = 4.62 \text{ m}^2 \quad (7.18)$$

and the *magnitude* of the tractions on opposite sides of the surface and of the associated surface stress is (Equations (7.1) and (7.7)):

$$|\boldsymbol{\Sigma}'^{(\text{top})}| = |\boldsymbol{\Sigma}'^{(\text{bottom})}| = |\boldsymbol{\Sigma}'| = \frac{W}{A'} = \frac{80{,}000 \text{ N}}{4.62 \text{ m}^2}$$
$$\cong 17{,}300 \text{ Pa} \cong 0.017 \text{ MPa} \quad (7.19)$$

What are the normal and shear components of the two equal and opposite tractions on the inclined plane, and of the surface stress components defined by those tractions? The normal and shear components of the tractions are calculated from the normal and shear components of the force divided by the area of the inclined plane. The components of the force normal and parallel to the inclined plane are (Figure 7.5*C*; Equation (7.11)):

$$\begin{aligned} F'_n &= W\cos\theta = (80{,}000 \text{ N})\cos 30° = (80{,}000 \text{ N})(0.866) \\ F'_n &\cong 69{,}300 \text{ N} \\ F'_s &= W\sin\theta = (80{,}000 \text{ N})\sin 30° = (80{,}000 \text{ N})(0.5) \\ F'_s &= 40{,}000 \text{ N} \end{aligned} \quad (7.20)$$

The sign of each traction component is determined by the relation of the traction component to the local coordinate axes x', y', and z' (Figure 7.5*D*). If the traction component points in a positive coordinate direction, it is positive, and if it points in a negative coordinate direction, it is negative. Thus the normal and shear traction components are (Equations (7.1) and (7.6) with Equations (7.18) and (7.20)):

$$\begin{aligned} \sigma_n'^{(\text{bottom})} &= -\sigma_n'^{(\text{top})} = \frac{F'_n}{A'} = \frac{69{,}300 \text{ N}}{4.62 \text{ m}^2} \\ &\cong 15{,}000 \text{ Pa} \cong 0.015 \text{ MPa} \\ \sigma_s'^{(\text{bottom})} &= -\sigma_s'^{(\text{top})} = \frac{F'_s}{A'} = \frac{40{,}000 \text{ N}}{4.62 \text{ m}^2} \\ &\cong 8{,}660 \text{ Pa} \cong 0.0087 \text{ MPa} \end{aligned} \quad (7.21)$$

The normal stress is compressive and thus positive; looking in the negative x' direction, the shear stress is counterclockwise and thus also positive. Thus the normal and shear components of the surface stress have exactly the same magnitude and sign as the traction components acting on the bottom (negative side) of the $x'y'$ coordinate surface (compare Equations (7.12) and (7.13) with Equation (7.21).

$$\begin{aligned} \sigma'_n &= \sigma'^{(\text{bottom})}_n = +0.015 \text{ MPa} \\ \sigma'_s &= \sigma'^{(\text{bottom})}_s = +0.0087 \text{ MPa} \end{aligned} \tag{7.22}$$

We could also calculate the normal and shear stress components on the inclined plane from the stress components on the horizontal plane using Equations (7.14) and (7.15) with (7.17):

$$\begin{aligned} \sigma'_n &= \sigma_n \cos^2\theta = (20{,}000 \text{ Pa}) \cos^2 30° \\ \sigma'_n &= (20{,}000 \text{ Pa})(0.866)^2 \cong 15{,}000 \text{ Pa} \cong 0.015 \text{ MPa} \\ \sigma'_s &= \sigma_n \sin\theta \cos\theta = (20{,}000 \text{ Pa}) \sin 30° \cos 30° \\ \sigma'_s &= (20{,}000 \text{ Pa})(0.5)(0.866) \cong 8{,}660 \text{ Pa} \cong 0.0087 \text{ MPa} \end{aligned} \tag{7.23}$$

Of course, these same equations with $\theta = 0°$ would give the results for the surface stress components on the horizontal plane that are the same as we found in Equations (7.17b).

vii. The Two-Dimensional State of Stress at a Point

We know the state of stress $\boldsymbol{\sigma}$ at a point in a body if we can determine the normal stress and shear stress components, σ_n and σ_s, respectively, that act on a plane of *any* orientation passing through that point. There are, of course, an infinite number of such planes, so we need to know, what is the minimum amount of information that enables us to determine the stress components on any plane?

For the two-dimensional case, if the normal stress components on the planes are either all compressive or all tensile, the stress is particularly easy to visualize. The surface stress at a point that acts on any plane through the point can be plotted as a pair of opposing arrows. These arrows point either toward (for compressive stress) or away from (for tensile stress) the point, and the orientation and length of each arrow represent one of the tractions. If we plot the surface stresses on all planes of all possible orientations through a point, it turns out that the ends of all the arrows all fall on an ellipse, that is, the arrows are radii of the ellipse. We call this ellipse the **stress ellipse** (Figure 7.6*A*).[6] This ellipse must represent the state of stress at that point because it is a direct representation of the surface stresses on all possible orientations of plane through the point. This result is very convenient, because it shows us that if we have enough information to define the shape and orientation of the stress ellipse, we know the complete state of stress at the point. The stress ellipse is thus a graphic representation of a general two-dimensional state of stress.

The simplest way to describe the shape of any ellipse is to define the magnitudes of the major and minor axes of the ellipse and their orientations relative to a known reference frame. These axes on the stress ellipse are the two surface stresses called the **principal stresses**, because they are the maximum and minimum of the surface stresses acting on any plane through the point (Figure 7.6*A*). We label the maximum and minimum principal stresses[7] $\hat{\sigma}_1$ and $\hat{\sigma}_3$ (Figure 7.6*A*) such that, by standard convention,

$$\hat{\sigma}_1 \geq \hat{\sigma}_3 \tag{7.24}$$

We represent a point in a two-dimensional body as an infinitesimally small square whose sides are perpendicular to the coordinate axes and thus parallel to coordinate planes (Figure 7.6*B*). Opposite sides of the square represent the opposite sides of a single plane through the point, and the perpendicular pairs of sides of the square thus represent two perpendicular planes through the point. Thus different orientations of the square represent the coordinate surfaces at the same point in space for different orientations of coordinates. The planes on which the principal stresses act are the **principal planes**, and coordinate axes parallel to the principal stresses are the **principal coordinates** or **principal axes**[8] $\hat{x}_1$ and $\hat{x}_3$ (Figure 7.6*B*, *C*). The principal stresses are also unique in that they are normal to the planes on which they act, which means that the shear stress components on the principal planes are always zero (Figure 7.6*C*).

Note that a single principal stress by itself does not define the complete state of stress at a point, as is illustrated by the fact that the length and orientation of one of the principal stresses do not define the complete shape of the stress ellipse; to define the stress ellipse and the

[6]States of stress are possible in which the normal stress components on some orientations of plane are compressive while the normal stress components on other orientations of plane are tensile. This state of stress still plots as a stress ellipse, because an ellipse is a quadratic function in which the variables are raised to the second power, which eliminates differences in sign. In such a situation, however, the stress ellipse is difficult to interpret correctly. The possibility that such a state of stress exists in the Earth is small, however, and thus of little practical significance.

[7]Here and throughout the book, we always use "hats" (circumflexes) above symbols specifically to indicate principal values or principal coordinates. This is a convention of convenience we have adopted for this book: Whenever you see a symbol with a circumflex above it, you know it is referring to principal coordinates. It is not a general convention, however, and commonly the principal stresses, for example, are just written σ_1 and σ_3.

[8]We label the principal coordinates $\hat{x}_1$ and $\hat{x}_3$ instead of x and z so that they are associated directly, by the subscript, with the principal stresses $\hat{\sigma}_1$ and $\hat{\sigma}_3$. It is common in many discussions of stress to distinguish coordinate axes by different numerical subscripts. We describe the notation in more detail in Section 7.5.

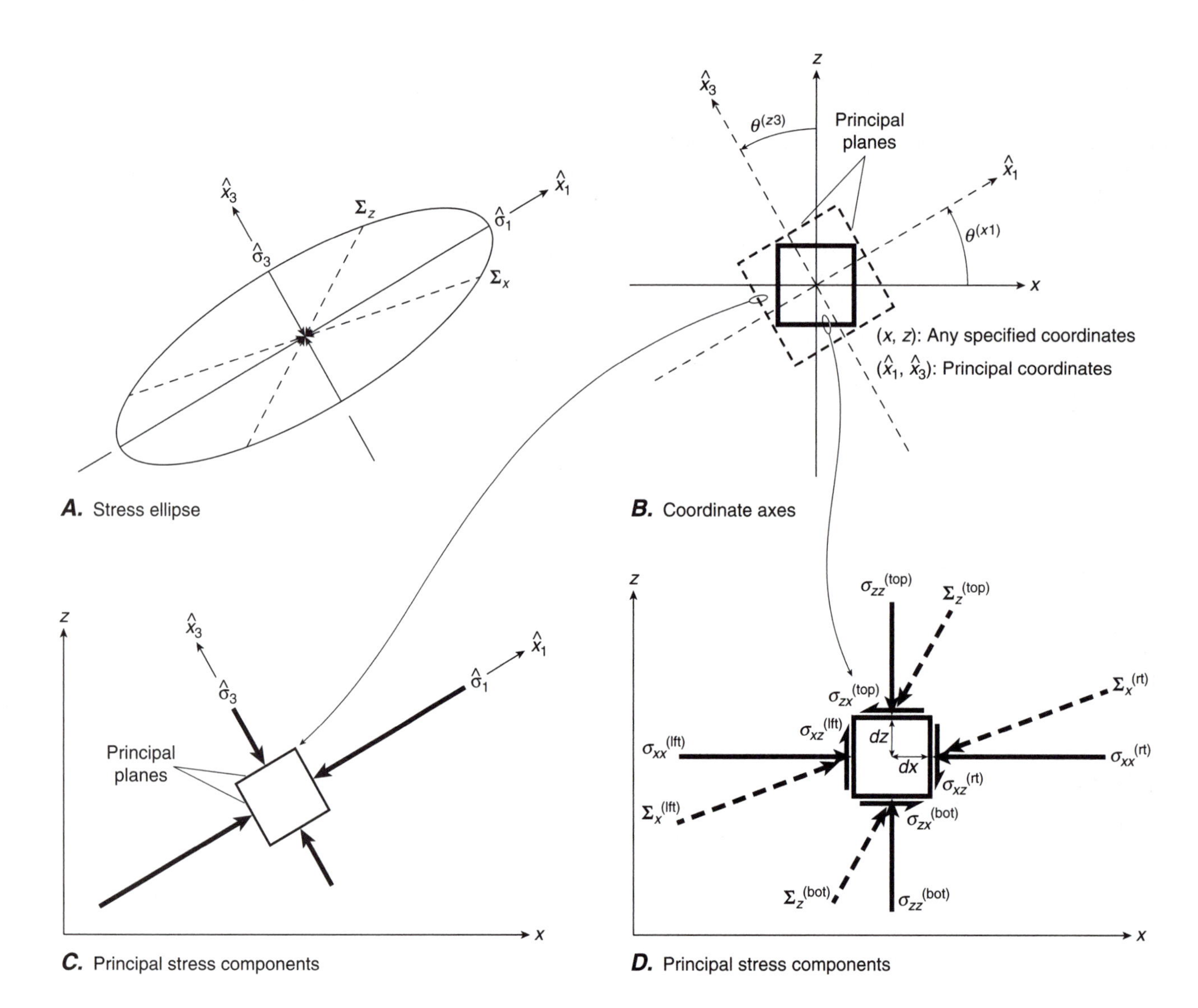

FIGURE 7.6 Representation of the state of two-dimensional stress at a point. *A*. The stress ellipse is the locus of all the surface stresses (plotted as pairs of opposing arrows) that act on planes of all orientations through the point. In two dimensions, those planes are all normal to the plane of the diagram. All normal stress components must have the same sign in order to interpret the stress ellipse easily. The principal stresses are the major and minor axes of the ellipse and are the maximum and minimum of all normal stresses on the planes. The principal coordinate axes are parallel to the principal stresses, and the planes normal to the principal stresses are the principal planes. *B*. Two different coordinate systems in which we can define the state of stress. The coordinate square in each system represents the opposite sides of two perpendicular coordinate planes through an infinitesimal point. Coordinates x and z could be geographic coordinates east and up, for example. Coordinates $\hat{x}_1$ and $\hat{x}_3$ are the principal coordinates. $\theta^{(x1)}$ and $\theta^{(z3)}$ are the angles between the axis pairs (x, $\hat{x}_1$) and (z, $\hat{x}_3$), respectively, as indicated by the superscripts. *C*. The stress at a point can be completely defined in the principal coordinate system ($\hat{x}_1$, $\hat{x}_3$) by the two principal stresses ($\hat{\sigma}_1$, $\hat{\sigma}_3$) that act on the two perpendicular principal coordinate planes (shear stress components are zero on these planes). *D*. The stress at a point can also be completely defined in any other coordinate system (x, z) by specifying the surface stresses ($\mathbf{\Sigma}_x$, $\mathbf{\Sigma}_z$) or their components (σ_{xx}, σ_{xz}), (σ_{zx}, σ_{zz}) on the two perpendicular coordinate planes. Shear stress components in general are not zero. Superscripts identify specific tractions and traction components.

state of stress we need two principal stresses. Here again we see how stress differs from force, because multiple forces acting on a point can always be represented as a single net force on the point. Note that for the special case of unaxial stress, which we considered in the preceding two subsections, the minimum principal stress is zero and thus the stress state cannot be represented by an ellipse.

The principal stresses are the surface stresses that act on a specific pair of perpendicular planes through the point (Figure 7.6*B*, *C*). But the mathematical properties of an ellipse show us that we can define its geometry, and thus define the state of stress at a point, if we know the surface stresses that act on *any pair* of perpendicular planes through the point (Figure 7.6*B*, *D*).

In order to discuss the general case of two-dimensional stress at a point, we define a pair of perpendicular reference coordinates x and z (Figure 7.6*B*, *D*). As an example, x could be east, and z could be up. We refer to planes perpendicular to x as x-planes and planes perpendicular to z as z-planes. We can then label each surface stress according to the coordinate plane on which it acts. Thus the surface stresses that define the state of stress in this coordinate system are $\mathbf{\Sigma}_x$ and $\mathbf{\Sigma}_z$, which act respectively on the x- and z-planes. (Note the surface stresses plotted as dashed radii of the stress ellipse in Figure 7.6*A* are the same surface stresses plotted as dashed arrows in Figure 7.6*D*. In Figure 7.6*D*, each traction and each traction component is labeled individually.) Each surface stress has a normal and a shear component, so we label each component according to the surface on which it acts and the coordinate to which it is parallel (Figure 7.6*D*). The components of the surface stress $\mathbf{\Sigma}_x$ therefore are labeled σ_{xx} and σ_{xz}, where the first subscript x shows that both components act on the x-plane, and the second subscript shows that the components are parallel to the x and z coordinates, respectively. Thus σ_{xx} is the normal stress component, and σ_{xz} is the shear stress component. Similarly, for the surface stress $\mathbf{\Sigma}_z$ acting on the z-plane, σ_{zz} is the normal stress component and σ_{zx} is the shear stress component. In general, a surface stress $\mathbf{\Sigma}$ is not perpendicular to the plane on which it acts (see the description in footnote 12), so both the normal stress and the shear stress components (σ_n and σ_s) on an arbitrary surface are nonzero.

Using this same reference coordinate system, the principal axis $\hat{x}_1$ makes an angle $\theta^{(x1)}$ with the x = east axis, where the superscripts identify the two axes between which the angle is measured. The principal axis $\hat{x}_3$, which is perpendicular to $\hat{x}_1$, is also oriented at an angle $\theta^{(z3)} = \theta^{(x1)}$ relative to the z axis (Figure 7.6*B*). These axes define the unique coordinate system in which the normal stresses on the coordinate planes are maximum or minimum, that is, they are the principal stresses $\hat{\sigma}_1$ and $\hat{\sigma}_3$, and the shear stresses on those planes are all zero. Thus the magnitudes of the principal surface stresses are completely defined by their normal stress components $\hat{\sigma}_1$ and $\hat{\sigma}_3$. By a longstanding convention used only for principal stresses, the second subscript on the normal stress component is omitted, because it must be the same as the first, and because the shear stresses, which have two different subscripts, must be zero. It is very useful to remember the following general property: *Any plane on which the shear stress is zero* (such as plane *A* in Figure 7.5*A*) *must be a principal plane, and the normal stress on that plane must be a principal stress.* Because the shear stresses are zero on the principal planes, using the principal stresses is a particularly simple way to define the stress at a point, but the orientation of the principal axes with respect to the reference frame axes $\theta^{(x1)} = \theta^{(z3)}$ must always be specified as well.

To summarize, then, the stress $\boldsymbol{\sigma}$ at a point is defined in a general reference coordinate system (x, z) by two surface stresses $\mathbf{\Sigma}_x$ and $\mathbf{\Sigma}_z$, each acting on one of the coordinate planes and each consisting of two components, a normal and a shear component (σ_{xx} and σ_{xz}) and (σ_{zz} and σ_{zx}), respectively.[9] In general, none of these components is zero. The same state of stress $\boldsymbol{\sigma}$ is also defined by the principal stresses $\hat{\sigma}_1$ and $\hat{\sigma}_3$ and their orientation $\theta^{(x1)} = \theta^{(z3)}$ relative to the (x, z) reference coordinates. These are the normal stresses on the principal planes, which are the maximum or minimum of all possible normal stresses on any possible plane; the shear stress components on the principal planes are zero. We summarize these relations in tabular form in Table 7.2. Our complete notation for stress components, and the different types of stress that we introduce in this chapter, is summarized for easy reference in Table 7.4 at the end of the chapter.

Because we are dealing with a continuum, each material point, represented by a coordinate square, must be in mechanical equilibrium. This requirement assures that the point cannot undergo an acceleration or an angular acceleration relative to its neighboring points. Thus the forces acting on the point parallel to each of the coordinate axes must sum to zero, and the moments of the

[9]The only cases for which one surface stress is sufficient to define the stress at a point are for a hydrostatic pressure, in which case the stress ellipse is a circle and all shear stresses on every orientation of plane are zero, and for a uniaxial stress, in which case one of the principal stresses is zero and a stress ellipse is not defined.

TABLE 7.2 Components for the Two-Dimensional Stress $\boldsymbol{\sigma}$ in the Mohr Circle Sign Convention

Coordinate System	Normal Components*	Shear Components*
Principal coordinates ($\hat{x}_1$, $\hat{x}_3$)	$\hat{\sigma}_1$, $\hat{\sigma}_3$	$(\hat{\sigma}_{13}, \hat{\sigma}_{31}) = (0, 0)$
Any specified coordinates (x, z)	σ_{xx}, σ_{zz}	$\sigma_{xz} = -\sigma_{zx}$

* The first subscript in each symbol defines the coordinate plane on which the stress component acts. The second subscript defines the coordinate axis to which the component is parallel. If there is only one subscript, it identifies both the plane on which the component acts, and the coordinate axis to which the component is parallel.

forces acting on the point about an axis normal to the plane of the diagram also must sum to zero.[10] In what follows, we show that this balance of forces and moments places constraints on the possible values of the tractions and surface stresses.

We know from Equation (7.6) that each normal traction and each shear traction occurs as one of a pair of equal and opposite components on opposite sides of a plane (Figure 7.3*D*). Accordingly, from Figure 7.6*D*,

$$\begin{aligned} \sigma_{xx}^{(\mathrm{rt})} &= -\sigma_{xx}^{(\mathrm{lft})} & \sigma_{zz}^{(\mathrm{top})} &= -\sigma_{zz}^{(\mathrm{bot})} \\ \sigma_{xz}^{(\mathrm{rt})} &= -\sigma_{xz}^{(\mathrm{lft})} & \sigma_{zx}^{(\mathrm{top})} &= -\sigma_{zx}^{(\mathrm{bot})} \end{aligned} \tag{7.25}$$

The product of a traction component and the area on which it acts is a force component acting on the coordinate square. Using Figure 7.6*D*, we sum all the force components that are parallel to the x axis, and separately we sum all force components that are parallel to the z axis. For equilibrium to exist, each of these sums must be equal to zero.

Forces parallel to the x axis:

$$\sigma_{xx}^{(\mathrm{rt})} A_x + \sigma_{xx}^{(\mathrm{lft})} A_x + \sigma_{zx}^{(\mathrm{top})} A_z + \sigma_{zx}^{(\mathrm{bot})} A_z = 0 \tag{7.26a}$$

Forces parallel to the z axis:

$$\sigma_{zz}^{(\mathrm{top})} A_z + \sigma_{zz}^{(\mathrm{bot})} A_z + \sigma_{xz}^{(\mathrm{rt})} A_x + \sigma_{xz}^{(\mathrm{lft})} A_x = 0 \tag{7.26b}$$

If we use Equations (7.25) to eliminate one of each of the traction component pairs from Equations (7.26), we obtain the identity $0 = 0$, which shows that Equations (7.25) are indeed the constraints that must be satisfied for the point to be in equilibrium.

We also require the sum of the moments of the forces about the point to be equal to zero. This sum involves only the shear traction components, because the moment arms for the normal traction components are all zero. From Figure 7.6*D*, the infinitesimal dimensions of the square representing the infinitesimal point are $2(dx)$ and $2(dz)$. Thus the moment arm of the shear traction components in each case equals dx or dz, as appropriate. Adding all the moments of the forces and requiring their sum to be zero gives

$$\begin{aligned} \sigma_{xz}^{(\mathrm{rt})} A_x dx + \sigma_{xz}^{(\mathrm{lft})} A_x(-dx) + \sigma_{zx}^{(\mathrm{top})} A_z dz \\ + \sigma_{zx}^{(\mathrm{bot})} A_z(-dz) = 0 \end{aligned} \tag{7.27}$$

Because $A_x = A_z$ and $dx = dz$, we can eliminate these quantities from the equation by division and, using Equations (7.25), show that the shear tractions are related by

$$\sigma_{xz}^{(\mathrm{lft})} = -\sigma_{zx}^{(\mathrm{bot})} \qquad \sigma_{xz}^{(\mathrm{rt})} = -\sigma_{zx}^{(\mathrm{top})} \tag{7.28}$$

and therefore the shear stress components are related by

$$\sigma_{xz} = -\sigma_{zx} \tag{7.29}$$

Thus *of the four stress components in the* (x, z) *coordinate system, only three are independent:* σ_{xx}, σ_{xz} $(= -\sigma_{zx})$, *and* σ_{zz}.

It might appear that we can define the state of stress with fewer numbers if we use the principal stresses than if we use the stress components in a general coordinate system, because the shear stress components must always be zero. However, the orientations of the principal axes are not known a priori. Thus in a given reference coordinate system (for example, geographic coordinates with x = east, z = up), the stress is completely defined if we know the three independent components of the stress σ_{xx}, σ_{xz} $(= -\sigma_{zx})$, and σ_{zz}. Alternatively, the stress is also defined if we know the magnitudes of the two principal stresses $\hat{\sigma}_1$ and $\hat{\sigma}_3$, plus an angle that specifies the orientation of the principal axes relative to the reference axes, such as $\theta^{(x1)}$, the angle between the x = east coordinate and $\hat{x}_1$ (Figure 7.6*B*). Three numbers therefore are required regardless of whether we describe the stress with components in an arbitrary reference coordinate system or with components in the principal coordinates.

viii. The Three-Dimensional State of Stress at a Point

The description of the stress in three dimensions is a direct extrapolation of its description in two dimensions. If all normal stress components have the same sign, we can graphically represent the stress at a point by a stress ellipsoid (Figure 7.7*A*). The major, intermediate, and minor principal axes of the ellipsoid are parallel to the principal coordinate axes $(\hat{x}_1, \hat{x}_2, \hat{x}_3)$. The orientations of these axes relative to the reference coordinates (x, y, z) are defined by three angles, such as $(\theta^{(x1)}, \theta^{(y2)}, \theta^{(z3)})$ (Figure 7.7*B*), where the superscripts identify the two axes between which the angle is measured. The principal axes of the ellipsoid represent the maximum, intermediate,[11] and minimum principal stresses, respectively, which we label in accordance with the convention

$$\hat{\sigma}_1 \geq \hat{\sigma}_2 \geq \hat{\sigma}_3 \tag{7.30}$$

The principal stresses are the surface stresses acting on the three mutually perpendicular principal planes through a point, which we represent as an infinitesimal cube with faces parallel to the principal planes and perpendicular to the principal axes; the two sides of each principal plane are represented by one pair of parallel faces of the cube (Figure 7.7*C*). On the three principal planes, normal stresses have extreme values and shear stresses are zero.

The stress ellipsoid can also be defined in the arbitrarily chosen reference coordinates (x, y, z) (Figure 7.7*D, E*). The coordinate planes, which are normal to the

[10]Remember that the moment of a force is the force multiplied by the length of the lever arm on which the force operates.

[11]The intermediate principal stress is known mathematically as a minimax, because it is a minimum in one of the two principal planes in which it lies, and a maximum in the other.

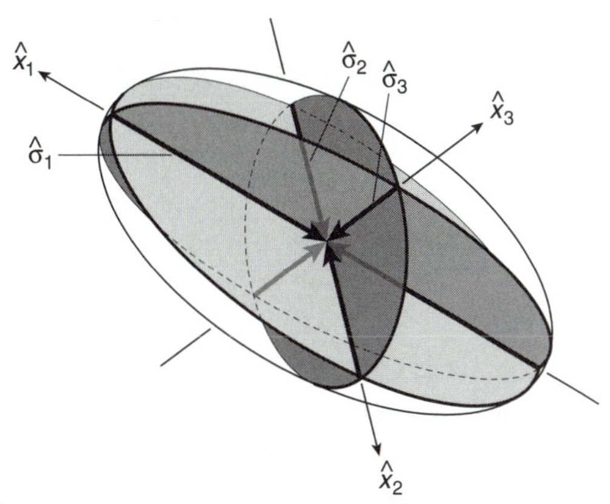

A. Stress ellipsoid

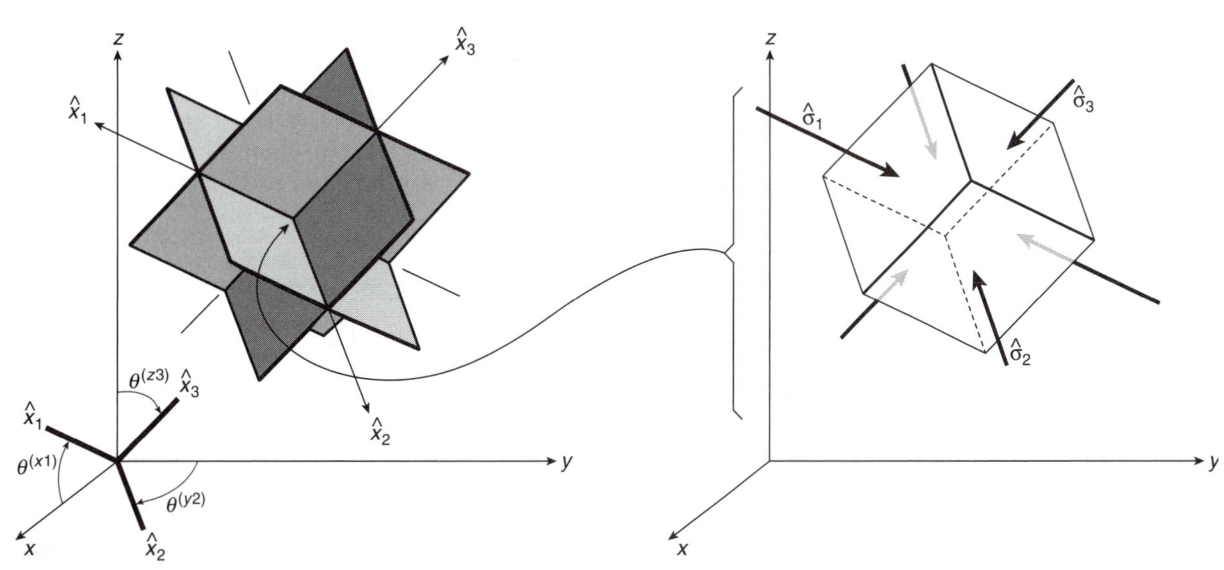

B. Principal coordinate axes and planes

C. Principal stress components

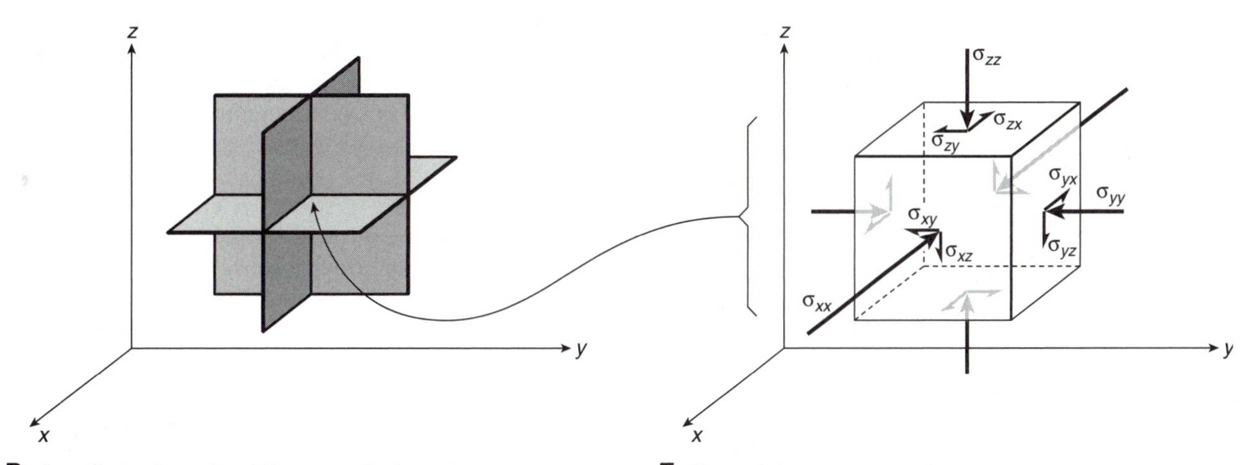

D. Coordinate planes in arbitrary coordinate system

E. General stress components

FIGURE 7.7 The state of three-dimensional stress at a point. *A*. The stress ellipsoid is defined by the surface stresses that act on planes of all possible orientations through a point. Shaded planes are the principal planes. Stress components illustrated are the principal stresses. *B*. The principal coordinates ($\hat{x}_1$, $\hat{x}_2$, $\hat{x}_3$) and the principal planes through an infinitesimal point relative to a general coordinate system (x, y, z). *C*. Principal stresses on the principal planes at the point in *B*. *D*. Coordinate planes of the general coordinate system (x, y, z) through the same infinitesimal point shown in *B*. *E*. Stress components on the coordinate planes of the general coordinate system (x, y, z) shown in *D*.

TABLE 7.3 Components for the Three-Dimensional Stress $\boldsymbol{\sigma}$ in the Mohr Circle Sign Convention

Coordinate System	Normal Components*	Shear Components*
Principal coordinates $(\hat{x}_1, \hat{x}_2, \hat{x}_3)$	$\hat{\sigma}_1, \hat{\sigma}_2, \hat{\sigma}_3$	$(\hat{\sigma}_{12}, \hat{\sigma}_{21}) = (0, 0)$ $(\hat{\sigma}_{13}, \hat{\sigma}_{31}) = (0, 0)$ $(\hat{\sigma}_{23}, \hat{\sigma}_{32}) = (0, 0)$
Any specified coordinates (x, y, z)	$\sigma_{xx}, \sigma_{yy}, \sigma_{zz}$	$\sigma_{xy} = -\sigma_{yx}$, $\sigma_{xz} = -\sigma_{zx}$, $\sigma_{yz} = -\sigma_{zy}$

* The first subscript in each symbol defines the coordinate plane on which the stress component acts. The second subscript defines the coordinate axis to which the component is parallel. If there is only one subscript, it identifies both the plane on which the component acts, and the coordinate axis to which the component is parallel.

x, y, and z axes, are again represented by an infinitesimal cube whose faces are parallel to those three planes. The surface stresses on these coordinate planes are, respectively, $\boldsymbol{\Sigma}_x$, $\boldsymbol{\Sigma}_y$, and $\boldsymbol{\Sigma}_z$. Each of these surface stresses has three components, one parallel to each of the coordinate axes, $(\sigma_{xx}, \sigma_{xy}, \sigma_{xz})$, $(\sigma_{yx}, \sigma_{yy}, \sigma_{yz})$, and $(\sigma_{zx}, \sigma_{zy}, \sigma_{zz})$. (Figure 7.7*E* shows only the components of the surface stresses.) Each stress component consists of a pair of equal and opposite traction components that act on opposite faces of the cube. We label the components of these three surface stresses by using the same convention we used for two-dimensional stress: The first subscript identifies the coordinate axis normal to the plane on which the stress component acts, and the second subscript identifies the coordinate axis to which the stress component is parallel. Thus each component with two identical subscripts is a normal stress; each component with two different subscripts is a shear stress. This notation is included in the summary of stress notation in Table 7.4 at the end of the chapter.

Thus the state of stress $\boldsymbol{\sigma}$ at a point, represented by the stress ellipsoid, is uniquely described by the three principal stresses $\hat{\sigma}_1 \geq \hat{\sigma}_2 \geq \hat{\sigma}_3$ and their orientations $(\theta^{(x1)}, \theta^{(y2)}, \theta^{(z3)})$ relative to the reference axes (Figure 7.7*B*, *C*). It is also described by three surface stresses $\boldsymbol{\Sigma}_x$, $\boldsymbol{\Sigma}_y$, $\boldsymbol{\Sigma}_z$, or their nine components, acting on any three mutually perpendicular surfaces of specified orientation through the point (Figure 7.7*D*, *E*). These relations are summarized in Table 7.3.

Figure 7.7*E* shows that, in general, nine components are required to define the state of stress at a point in three dimensions: three components for each of three surface stresses. As in the case for two dimensions, however, only half of the shear stress components are independent. The requirement that the sum of the moments of the forces about the point must equal zero, yields the following relationships among the six shear stress components (compare Equation (7.29)):

$$\sigma_{xy} = -\sigma_{yx} \qquad \sigma_{yz} = -\sigma_{zy} \qquad \sigma_{zx} = -\sigma_{xz} \tag{7.31}$$

We summarize these relations in Table 7.3. Because of Equations (7.31), we need only six numbers to specify the state of stress at a point in three dimensions. In a general coordinate system, those numbers are the three normal stress components and the three independent shear stress components. In principal coordinates, those six numbers are the magnitudes of the three principal stresses plus the three angles that uniquely define the orientations of the three principal axes relative to the reference axes.

7.2 THE MOHR DIAGRAM FOR TWO-DIMENSIONAL STRESS

From the preceding discussion and the characteristics of the stress ellipse (see Figure 7.6), we have seen that the normal stress and shear stress components on a plane change systematically with the orientation of the plane. From the stress ellipse, however, it is difficult to extract the relationship between the orientation of a plane and the values of normal stress and shear stress components on it.[12] That relationship is remarkably simple, however, when the stress is plotted on a **Mohr diagram**[13] for which the horizontal axis is the value of the normal stress σ_n and the vertical axis is the value of the shear stress σ_s. As before, we consider compressive normal stresses and counterclockwise shear stresses to be positive in this diagram.

For a given state of stress, we can show (see Section 7.4) that as a plane through a point progressively rotates through all possible orientations, the normal and shear components of stress on the plane, when plotted on a Mohr diagram, trace out a circle, called the **Mohr circle**,

[12]The components of the outward unit normal vector **n** to the plane on which a surface stress $\boldsymbol{\Sigma}^{(n)}$ acts can be found from the following relation, which for reference we simply state without derivation.

$$(n_1, n_2, n_3) = (\hat{\Sigma}_1^{(n)}/\hat{\sigma}_1,\ \hat{\Sigma}_2^{(n)}/\hat{\sigma}_2,\ \hat{\Sigma}_3^{(n)}/\hat{\sigma}_3)$$

where $(\hat{\Sigma}_1^{(n)}, \hat{\Sigma}_2^{(n)}, \hat{\Sigma}_3^{(n)})$ are the components, in the principal coordinates, of the surface stress.

[13]Named after Christian Otto Mohr (1835–1918), a German professor of mechanics and civil engineering.

whose center lies on the normal stress axis. Several characteristics of the Mohr circle show clearly how the stress at a point is related to the surface stresses on planes through the point. We number these characteristics in order to refer to them easily in subsequent sections. We recommend that students read this section to get an idea of how the Mohr circle works and then study the illustrative problem in Appendix 7A to gain a working understanding of how the Mohr circle can be applied to solve problems.

1) The Mohr Diagram

(i) The diagram has axes that are values of the stress components. It is therefore very important to distinguish the Mohr diagram from a diagram of "physical space" whose axes are spatial coordinates. It is always necessary to draw a separate diagram of physical space, along with the Mohr diagram, and to transfer data carefully from one diagram to the other (Figure 7.8) according to the rules relating the two types of diagram, which we present in the following discussion.

(ii) The Mohr circle is a complete representation of the state of stress at a point, because the points on the circle include the normal stress and shear stress components of the surface stress on planes of all possible orientations through the point. Each point on the circle represents the surface stress on a differently oriented plane.

2) Principal Stresses

(i) The maximum and minimum normal stresses $\hat{\sigma}_1$ and $\hat{\sigma}_3$ have values defined by the intersection of the Mohr circle with the σ_n axis (Figure 7.8*B*). Note that these two points are the only surface stresses on the Mohr circle for which the shear stress is zero. Note also that the principal stresses plot at the opposite ends of a diameter of the circle.

3) Surface Stress and the Orientation of Planes

(i) If we know the normal and shear stresses on two mutually perpendicular planes, we know the state of stress, and thus we can plot the Mohr circle, as described in item (v) below. Using the Mohr circle, we can then determine the surface stress on any other plane P, if we know how P is oriented relative to the planes on which we already know the stresses. *It is very important to remember that the orientation of a plane in physical space is defined by the angle between its normal* **n** *and a coordinate axis, not by the angle between the plane itself and that axis* (Figure 7.8*A*).

For example, let us assume that initially we know the orientations and surface stresses on the principal planes. If $\mathbf{n}^{(P)}$ is the normal to plane P and $\hat{x}_1$ is the normal to a principal plane, we can then specify the orientation of the plane P in physical space by the angle θ measured from $\hat{x}_1$ to

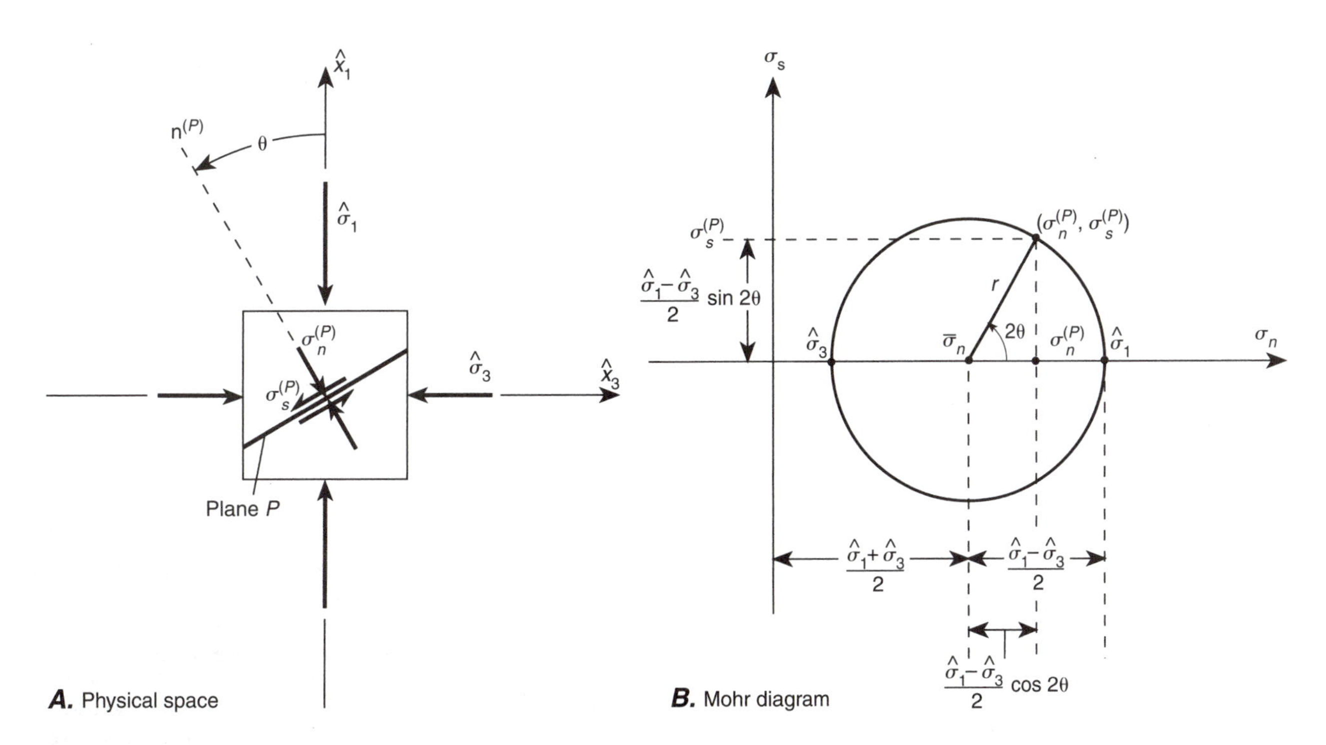

FIGURE 7.8 Using the Mohr diagram to plot the stress at a point. *A*. Diagram in physical space of the relationships between the stress components, the principal coordinate axes, $(\hat{x}_1, \hat{x}_3)$ and the plane P with its normal $n^{(P)}$. Superscripts (P) identify stress components acting on plane P. *B*. Stress at the point in *A* represented on a Mohr diagram by a Mohr circle. Superscripts (P) identify the stress components acting on the plane P.

$\mathbf{n}^{(P)}$ (Figure 7.8*A*), where we measure the angle from the known (we know the surface stress on the plane normal to $\hat{x}_1$) to the unknown (we do not know the surface stress on the plane normal to $\mathbf{n}^{(P)}$).

To find the surface stress on plane P from the Mohr circle, we then measure an angle 2θ between two radii of the circle, where the first radius touches the Mohr circle at the known surface stress $(\hat{\sigma}_1, 0)$ on the principal plane normal to $\hat{x}_1$, and the second radius touches the Mohr circle at the unknown surface stress $(\sigma_n^{(P)}, \sigma_s^{(P)})$ on the plane P normal to $\mathbf{n}^{(P)}$ (Figure 7.8*B*). We must use the same sense of rotation for 2θ that we used to measure θ in physical space from the known to the unknown quantity.

The angle θ is also necessarily the angle in physical space from $\hat{\sigma}_1$ to $\sigma_n^{(P)}$, because $\hat{\sigma}_1$ is parallel to $\hat{x}_1$ and $\sigma_n^{(P)}$ is parallel to $\mathbf{n}^{(P)}$. Thus it is the angle between the normal stress components on the $\hat{x}_1$ principal plane and the plane P. Figures 7.8 through 7.10 illustrate some important features of plotting stresses on the Mohr diagram, specifically:

(ii) Angles measured in physical space are doubled when plotted on the Mohr diagram, and they are measured in the same sense on the Mohr diagram as in physical space. Thus as θ takes on values from 0° to 180° in physical space, the angle 2θ plotted on the Mohr diagram takes on values from 0° to 360°, and the entire Mohr circle is swept out (Figure 7.8*B*). This relationship reflects the fact that all planes have two normals that are 180° apart. Therefore in physical space, the angles $180° \leq \theta < 360°$ are redundant because they merely duplicate the orientations of the plane defined by the angles $0° \leq \theta < 180°$.

(iii) The orientation in physical space of the normal stress component $\sigma_n^{(P)}$ on plane P is easily determined from the point on the Mohr circle where the normal and shear stress components $(\sigma_n^{(P)}, \sigma_s^{(P)})$ plot. Suppose that in physical space the normal stress $\sigma_n^{(P)}$ is oriented at an angle θ from the maximum principal stress $\hat{\sigma}_1$ (Figure 7.8*A*). On the Mohr circle, the normal and shear surface stress components on plane P $(\sigma_n^{(P)}, \sigma_s^{(P)})$ plot at the end of the radius that lies at an angle 2θ from the radius to the maximum principal stress $(\hat{\sigma}_1, 0)$ (Figure 7.8*B*).

(iv) Suppose there are two arbitrary planes in physical space P and P' whose normals are $\mathbf{n}$ and $\mathbf{n}'$ (Figure 7.9*A*). Suppose furthermore that the angle from $\hat{x}_1$ to $\mathbf{n}$ is a counterclockwise angle θ and the angle from $\mathbf{n}$ to $\mathbf{n}'$ is a counterclockwise angle α. Then on the Mohr diagram there are two points on the Mohr circle, $(\sigma_n^{(P)}, \sigma_s^{(P)})$ and $(\sigma_n^{(P')}, \sigma_s^{(P')})$, that define the normal stress and shear stress

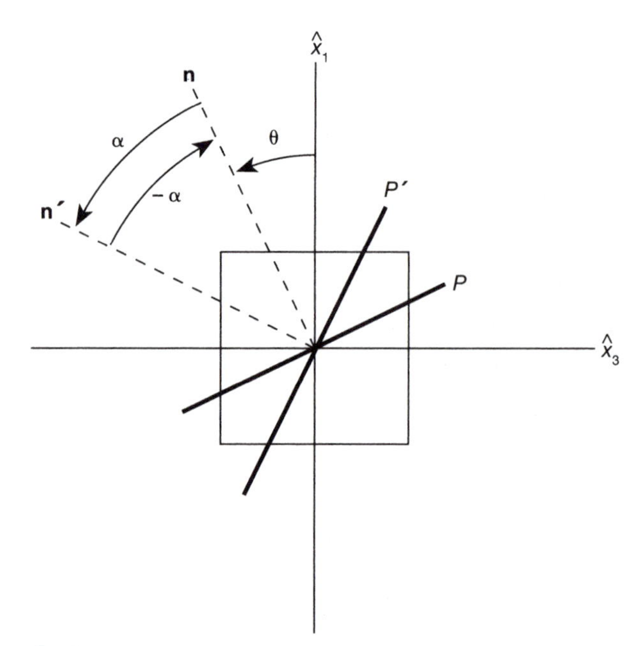

A. Physical space

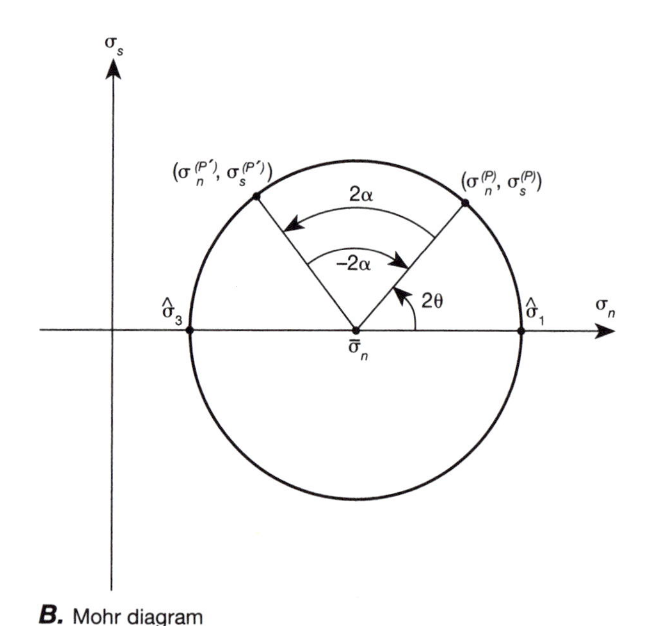

B. Mohr diagram

FIGURE 7.9 Using the Mohr diagram to represent the stress components on two planes at different orientations in physical space. *A*. Two-dimensional diagram in physical space of two planes P and P' and their respective normals n and n′. n is oriented relative to the principal axis $\hat{x}_1$ by an angle θ. The angle between n and n′, measured counterclockwise from n to n′, is α; measured clockwise from n′ to n, it is $-\alpha$. *B*. Mohr diagram representation of the surface stresses on the two planes shown in *A*. Note that the angles plotted are double the angles in physical space, but the sense of rotation for measuring the angles is the same.

components on P and P', respectively (Figure 7.9B). As shown in that figure, the angle between radii to those points is 2α, measured counterclockwise from $(\sigma_n^{(P)}, \sigma_s^{(P)})$ to $(\sigma_n^{(P')}, \sigma_s^{(P')})$.

Suppose, on the other hand, that we measure the angle in physical space from $\mathbf{n}'$ to $\mathbf{n}$. In this case it is a clockwise *negative* angle $-\alpha$. Thus we would plot a clockwise (negative) angle -2α on the Mohr circle from the radius at $(\sigma_n^{(P')}, \sigma_s^{(P')})$ to the radius at $(\sigma_n^{(P)}, \sigma_s^{(P)})$ (Figure 7.9B). Thus, to determine the unknown stress components on a plane from the Mohr circle, we need only know how that plane is oriented in physical space relative to some other plane on which we already know the stress components.

(v) From these properties of the Mohr circle, we now can see how to construct the circle. The surface stress components that act on two perpendicular planes in physical space ($\alpha = 90°$) must plot at opposite ends of a diameter of the Mohr circle ($2\alpha = 180°$). Thus the principal stresses $\hat{\sigma}_1$ and $\hat{\sigma}_3$, which act on perpendicular planes (inner square, Figure 7.10A), plot at opposite ends of a diameter of the circle (Figure 7.10B). Similarly, the two pairs of components $(\sigma_{xx}, \sigma_{xz})$ and $(\sigma_{zz}, \sigma_{zx})$ that specify the surface stresses acting on the perpendicular coordinate planes of any specified coordinate system (outer square, Figure 7.10A) also plot at opposite ends of a diameter of the Mohr circle (Figure 7.10B).

Fundamentally, this statement is a corollary of the fact that angles measured in physical space are doubled when plotted on the Mohr diagram (item (ii)). Thus, if we plot on a Mohr diagram the two surface stresses acting on any pair of perpendicular surfaces, we know they lie at opposite ends of a diameter of the Mohr circle, and therefore we can construct the whole circle. If we can do that, then we know the state of stress at the point because we know the surface stress components on planes of all orientations through the point.

4) Conjugate Planes of Maximum Shear Stress

(i) The stresses on the planes whose normals lie at $\theta = \pm 45°$ to the maximum principal stress $\hat{\sigma}_1$ in physical space (Figure 7.11A) occur on the Mohr circle at $2\theta = \pm 90°$, measured from $(\hat{\sigma}_1, 0)$ (Figure 7.11B). These planes plot at the top and bottom of the circle, and therefore on these planes, the absolute value of the shear stress $|\sigma_s|$ is a maximum. These planes are the **conjugate planes of**

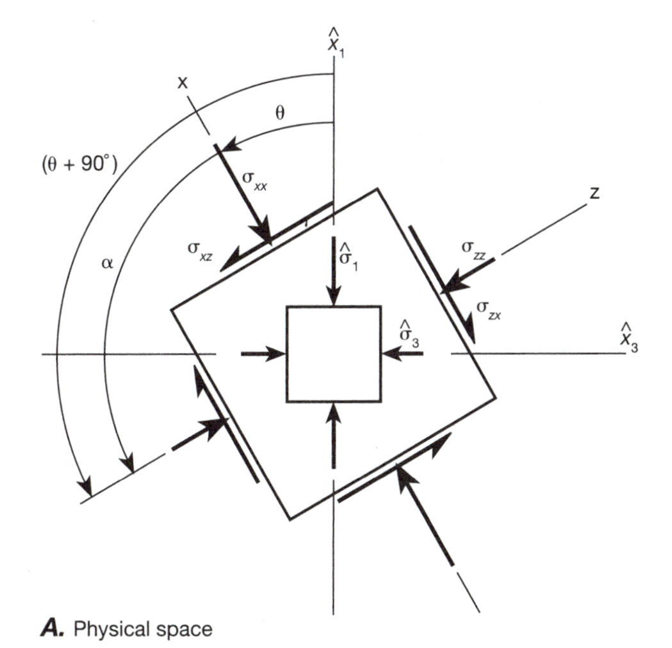

A. Physical space

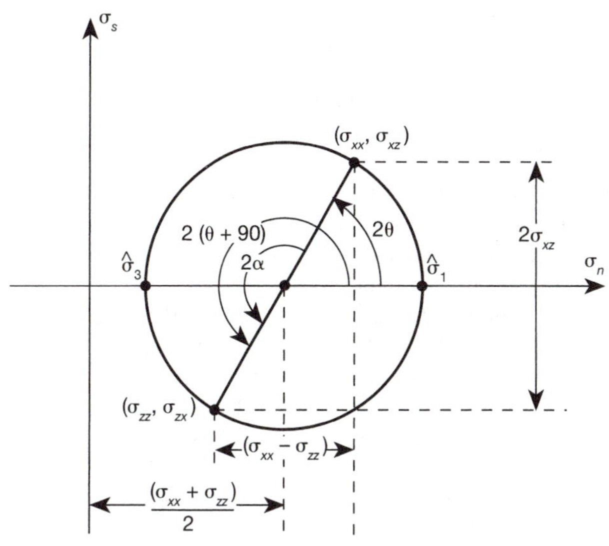

B. Mohr diagram

FIGURE 7.10 Transferring stress components from a diagram of physical space to a Mohr diagram. *A*. Diagram in physical space of components of the same state of stress in two separate coordinate systems: $(\hat{x}_1, \hat{x}_3)$, the principal coordinate system, and (x, z), a general coordinate system. The coordinate squares are drawn in different sizes for convenience, but both squares represent the two sides of the two coordinate surfaces through an infinitesimal point. *B*. Mohr diagram representation of the stress components in the two coordinate systems represented in *A*. Stress components acting on two perpendicular planes plot at opposite ends of a diameter of the Mohr circle. The principal stresses plot on the σ_n axis. The two scalar invariants of the stress are the center of the Mohr circle, defined as the mean of any two normal stresses that plot at opposite ends of a diameter, and the radius of the Mohr circle. The right triangle with sides of length $2\sigma_{xz}$ and $(\sigma_{xx} - \sigma_{zz})$ has a hypotenuse whose length is twice the radius.

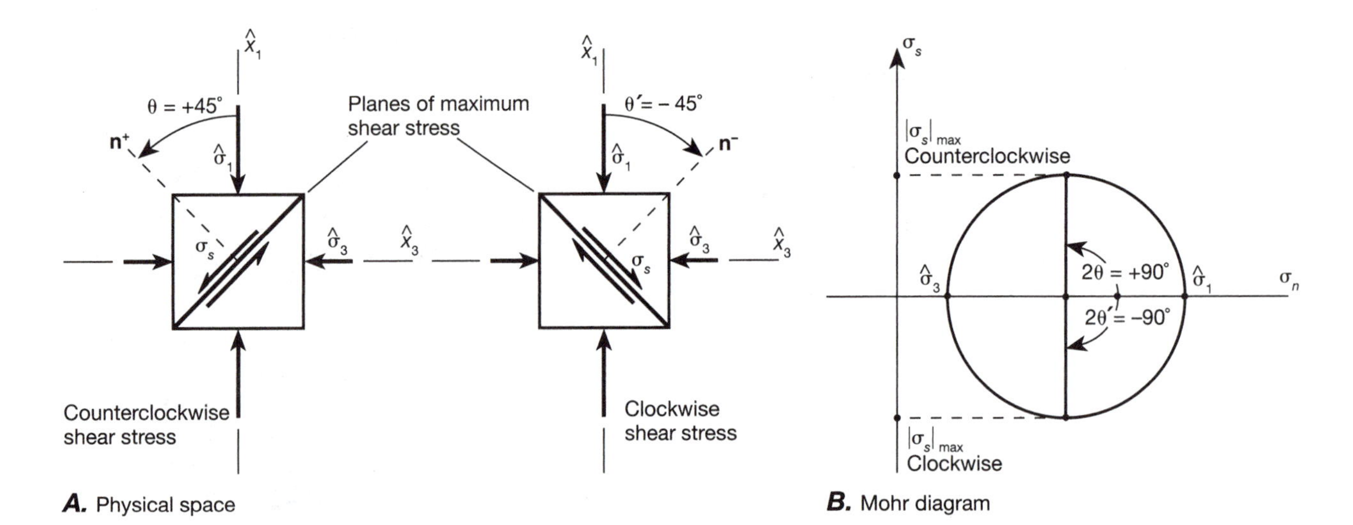

FIGURE 7.11 Planes of maximum shear stress. *A.* Two diagrams in physical space showing the angular separation of $\theta = \pm 45^\circ$ between the principal stresses and the normals to planes of maximum shear stress. The two planes are conjugate shear planes. *B.* Mohr diagram showing the plot of the surface stresses on the two planes of maximum shear stress. They plot, respectively, at the top and bottom of the Mohr circle at $2\theta = \pm 90^\circ$ from $\hat{\sigma}_1$ (and from $\hat{\sigma}_3$).

maximum shear stress, and in physical space the normals to these planes (and in this special case, the planes themselves) lie at $\pm 45^\circ$ to the maximum compressive stress $\hat{\sigma}_1$. The stresses on these planes plot at opposite ends of a vertical diameter of the Mohr circle (Figure 7.11*B*). Thus in physical space the normals to the planes are perpendicular, as are the planes themselves (Figure 7.11*A*).

5) Scalar Invariants of the Stress

(i) The magnitude of the state of stress at a point is uniquely characterized by two quantities called the **scalar invariants** of the stress. On the Mohr diagram these quantities are $\overline{\sigma}_n$, the value of σ_n at the center of the Mohr circle, and r, the length of the radius of the Mohr circle. By examining the Mohr circle, we see that these quantities are, respectively, half the sum and half the difference of the principal stresses (Figure 7.8*B*):

$$\overline{\sigma}_n = \frac{\hat{\sigma}_1 + \hat{\sigma}_3}{2} \qquad r = |\sigma_s|_{(\max)} = \frac{\hat{\sigma}_1 - \hat{\sigma}_3}{2} \tag{7.32}$$

The quantity $\overline{\sigma}_n$ is thus the **mean normal stress,** and $r = |\sigma_s|_{(\max)}$ is the **maximum shear stress,** which actually is the maximum absolute value of the shear stress (Figure 7.11*B*). These equations result from the fact that the principal stresses plot at opposite ends of a diameter of the Mohr circle. The same invariants can be calculated from any set of components $(\sigma_{xx}, \sigma_{xz})$, $(\sigma_{zz}, \sigma_{zx})$ at opposite ends of any diameter of the circle (Figure 7.10*B*):

$$\begin{aligned} \overline{\sigma}_n &= \frac{\sigma_{xx} + \sigma_{zz}}{2} \\ r &= |\sigma_s|_{(\max)} \\ &= \frac{1}{2}\sqrt{(\sigma_{xx} - \sigma_{zz})^2 + (2\sigma_{xz})^2} \end{aligned} \tag{7.33}$$

where we get the second equation by using the Pythagorean theorem[14] for the right triangle having a hypotenuse of $2r$ and sides $(\sigma_{xx} - \sigma_{zz})$ and $2\sigma_{xz}$ in Figure 7.10*B*. Equations (7.33) reduce to (7.32) when the state of stress is expressed in the principal coordinates, because then the normal stresses become the principal stresses and the shear stresses are zero. The mean normal stress and the maximum shear stress are quantities having magnitude but no direction, and therefore they are scalar quantities. These quantities are unique characteristics of a particular state of stress. Because we can calculate them from stress components defined in any coordinate system, as shown by Equation (7.33), they are invariant to changes in the coordinate axes and thus are called scalar invariants. If we know the scalar invariants, we can construct the whole Mohr circle, because we know the location of the center $(\overline{\sigma}_n)$ and the length of the radius $r = |\sigma_s|_{(\max)}$.

The scalar invariants of the stress describe fundamental geometric characteristics of the stress

[14]The Pythagorean theorem states that the square of the hypotenuse of a right triangle equals the sum of the squares of the other two sides.

ellipse (Figure 7.6*A*). The mean normal stress is proportional to the mean radius of the ellipse, and the difference between the squares of the two invariants is proportional to the area of the ellipse ($A_{(\text{ellipse})} = \pi\hat{\sigma}_1\hat{\sigma}_3$). Using Equations (7.32),

$$\overline{\sigma}_n^2 - r^2 = \left[\frac{\hat{\sigma}_1 + \hat{\sigma}_3}{2}\right]^2 - \left[\frac{\hat{\sigma}_1 - \hat{\sigma}_3}{2}\right]^2 = \hat{\sigma}_1\hat{\sigma}_3 = \frac{1}{\pi} A_{(\text{ellipse})}$$

The existence of two scalar invariants for the stress illustrates another way in which the stress differs from a vector quantity. The size of a vector is defined by one scalar invariant, its magnitude, represented by the length of an arrow; but the stress requires two scalar invariants to define its magnitude, the location of the center of the Mohr circle and the length of the radius.

6) Equations of the Mohr Circle

(i) Using the Mohr circle and the stress invariants, we can easily calculate the normal and shear stress components on any orientation of plane in physical space (Figure 7.8). For any such plane whose normal **n** is at an angle θ from the maximum principal stress $\hat{\sigma}_1$, the normal and shear stresses are as follows:

$$\sigma_n = \overline{\sigma}_n + r \cos 2\theta \qquad \sigma_s = r \sin 2\theta \tag{7.34}$$

$$\sigma_n = \left(\frac{\hat{\sigma}_1 + \hat{\sigma}_3}{2}\right) + \left(\frac{\hat{\sigma}_1 - \hat{\sigma}_3}{2}\right) \cos 2\theta \tag{7.35}$$

$$\sigma_s = \left(\frac{\hat{\sigma}_1 - \hat{\sigma}_3}{2}\right) \sin 2\theta \tag{7.36}$$

To get equations (7.35) and (7.36) we substituted Equations (7.32) for $\overline{\sigma}_n$ and r into Equation (7.34).

Thus the Mohr circle provides a very quick and convenient method for obtaining solutions to stress problems and for remembering the correct equations. We shall use the Mohr circle repeatedly in the applications of stress to the understanding of brittle deformation in rocks. In Appendix 7A we give examples of problems that can be solved by using the Mohr circle. We recommend study of these sample problems to obtain a better working knowledge of how to use the Mohr circle. Appendix 7B is another sample problem that presumes knowledge of the material in Box 7-2 and Section 7.5.

7.3 TERMINOLOGY FOR STATES OF STRESS

A number of terms that refer to certain specific states of stress are common in the literature. They all have special characteristics that are easy to describe in terms of the principal stresses and Mohr circle diagrams (Figure 7.12). Figure 7.12 also shows square arrays (matrices) of numbers, which are the components of the associated stress tensors that we define in Section 7.5. See Table 7.4 at the end of the chapter for a summary of the main symbols we use for stress notation.

Hydrostatic pressure, $\hat{\sigma}_1 = \hat{\sigma}_2 = \hat{\sigma}_3 = p$ (Figure 7.12*A*). All principal stresses are compressive and equal. No shear stresses exist on any plane, so all orthogonal coordinate systems are principal coordinates. The Mohr circle reduces to a point on the normal stress axis.

Uniaxial stress. The Mohr diagram for the three-dimensional stress is a single circle tangent to the ordinate at the origin. There are two possible cases:

1. *Uniaxial compression*, $\hat{\sigma}_1 > \hat{\sigma}_2 = \hat{\sigma}_3 = 0$ (Figure 7.12*B*). The only stress applied is a compressive stress in one direction. This geometry is commonly used in testing the strength of rock samples in the laboratory. (This is the stress state that we used for the simple examples in Subsections 7.1(v) and 7.1(vi))
2. *Uniaxial tension*, $0 = \hat{\sigma}_1 = \hat{\sigma}_2 > \hat{\sigma}_3$ (Figure 7.12*C*). The only stress applied is a tension in one direction. Engineers often use this geometry to test the mechanical properties of metals.

Axial compressive stress, axial compression, or confined compression, $\hat{\sigma}_1 > \hat{\sigma}_2 = \hat{\sigma}_3 > 0$ (Figure 7.12*D*). A uniaxial compression of magnitude $(\hat{\sigma}_1 - \hat{\sigma}_3)$ is added to a state of hydrostatic stress $(\hat{\sigma}_2 = \hat{\sigma}_3)$. This state is frequently used in laboratory experiments to determine the high-temperature high-pressure properties of rock.

Extensional stress, axial extension, or extension, $\hat{\sigma}_1 = \hat{\sigma}_2 > \hat{\sigma}_3 > 0$ (Figure 7.12*E*). A uniaxial tension of magnitude $(\hat{\sigma}_1 - \hat{\sigma}_3)$ is added to a hydrostatic stress $(\hat{\sigma}_1 = \hat{\sigma}_2)$. This state is also sometimes used in high-temperature high-pressure laboratory deformation experiments. *The terms* axial extension *and* extension *are unfortunate in this context and should be avoided, because they are used with different meanings when they refer to stress or to strain (see Box 12-2 and Table 12-2.2), and the distinction between stress and strain should always be kept clear.*

Triaxial stress, $\hat{\sigma}_1 > \hat{\sigma}_2 > \hat{\sigma}_3$ (Figure 7.12*F*). The principal stresses are all unequal and can be of either sign. The stress plots on the Mohr diagram as three distinct circles (see Box 7-2).

Pure shear stress or pure shear, $\hat{\sigma}_1 = -\hat{\sigma}_3$ *and* $\hat{\sigma}_2 = 0$ (Figure 7.12*G*). The maximum and minimum principal stresses are equal in magnitude and opposite in sign; the intermediate principal stress is zero. The normal stress on planes of maximum

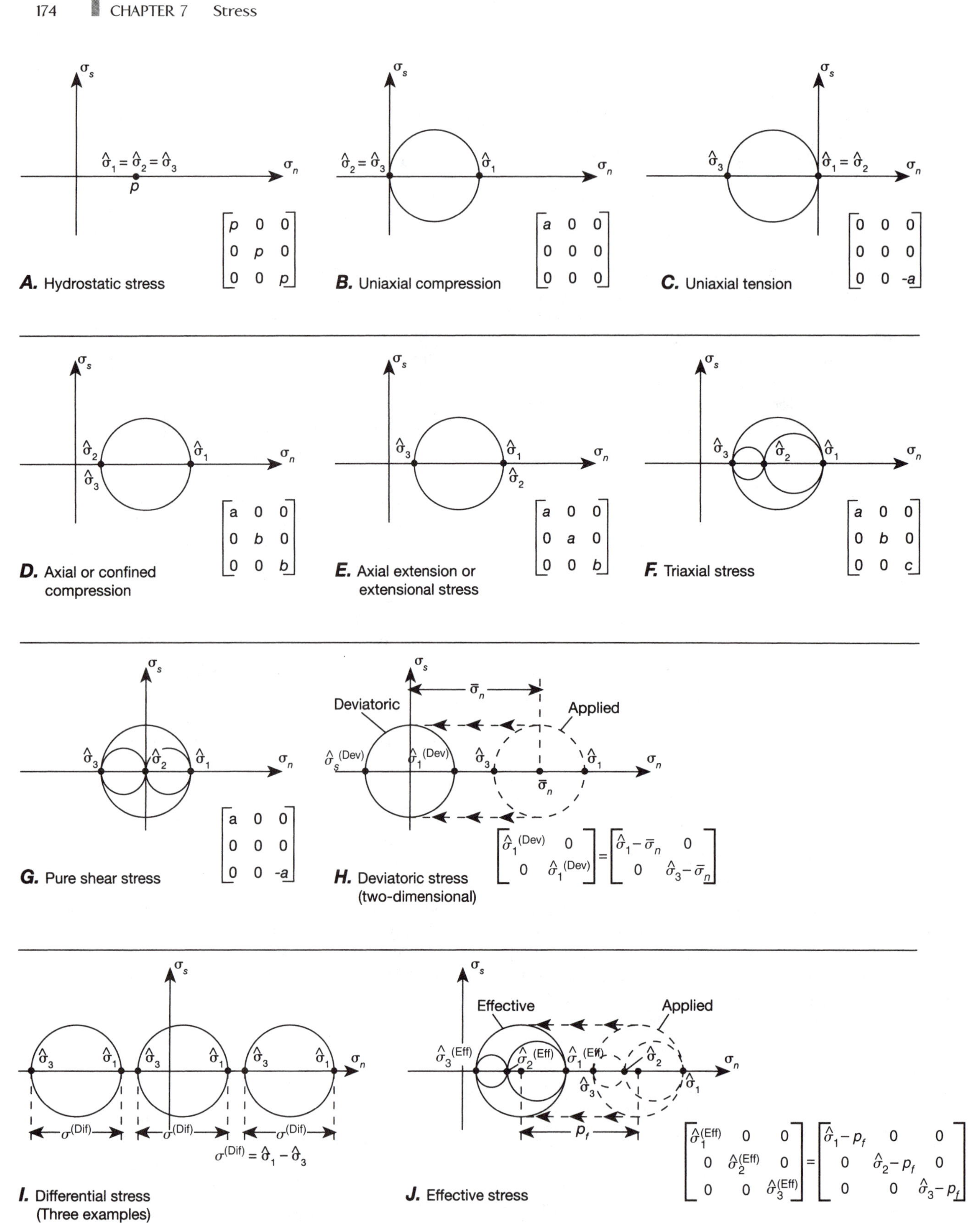

FIGURE 7.12 Mohr diagrams illustrating special states of stress. Each diagram also includes the arrays (matrices) of stress components in the principal coordinates, which are defined in Section 7.5. The three principal stresses occur along a diagonal of the array, written in order from the maximum at the top left to the minimum at the bottom right. Here p, a, b, and c are all positive values, and we assume $a > b > c$.

shear stress is zero—hence the name. The Mohr diagram is centered on the origin. *The term* pure shear *is an unfortunate contraction and should not be used to refer to stress, because it has different meanings when applied to stress or to strain (see Box 12-2 and Table 12-2.2); the ambiguity can cause confusion, so the distinction between stress and strain should always be kept clear.*

Deviatoric stress (Figure 7.12*H*). The deviatoric normal stress on a plane $\sigma_n^{(Dev)}$ is defined by subtracting the mean normal stress $\overline{\sigma}_n$ from any normal stress component σ_n, including, in particular, the principal stresses $\hat{\sigma}_1$ and $\hat{\sigma}_3$:

$$\begin{aligned} \sigma_n^{(Dev)} &\equiv \sigma_n - \overline{\sigma}_n \\ \hat{\sigma}_1^{(Dev)} &\equiv \hat{\sigma}_1 - \overline{\sigma}_n \\ \hat{\sigma}_3^{(Dev)} &\equiv \hat{\sigma}_3 - \overline{\sigma}_n \end{aligned} \tag{7.37}$$

where the mean normal stress is given for two dimensions by the first Equations (7.32) or (7.33). The deviatoric shear stresses are unchanged from their original values, so the Mohr circle remains the same diameter. For the deviatoric stress in two dimensions, the center of the Mohr circle is shifted to the origin of the graph so that it appears to be a two-dimensional pure shear stress (compare Figure 7.12*G* and *H*). The deviatoric stress is useful in describing material behavior that depends only on the size of the Mohr circle, which is a measure of the maximum shear stress, and not on the location of the Mohr circle along the normal stress axis, which is a measure of the average pressure.

Differential stress (Figure 7.12*I*). The differential stress $\sigma^{(Dif)}$ is the difference between the maximum and minimum principal stresses:

$$\sigma^{(Dif)} \equiv \hat{\sigma}_1 - \hat{\sigma}_3 \tag{7.38}$$

It is always a positive scalar quantity that is the diameter of the ($\hat{\sigma}_1$, $\hat{\sigma}_3$) Mohr circle (= $2r$; see the second Equation (7.32)) and therefore twice the maximum shear stress. As outlined in Section 7.2, property 5, the differential stress is therefore a scalar invariant of the stress. For an axial compressive stress or an axial extensional stress, the differential stress is the uniaxial part of the stress that is added to (see (Figure 7.12*D*) or subtracted from (see Figure 7.12*E*), the hydrostatic stress.

Effective stress (Figure 7.12*J*). The effective normal stress on a plane $\sigma_n^{(Eff)}$ is defined by subtracting the pore fluid pressure in the rock p_f from any normal stress component σ_n, including, in particular, the principal stresses $\hat{\sigma}_1$ and $\hat{\sigma}_3$:

$$\begin{aligned} &\sigma_n^{(Eff)} \equiv \sigma_n - p_f \\ &\hat{\sigma}_1^{(Eff)} \equiv \hat{\sigma}_1 - p_f \qquad \hat{\sigma}_3^{(Eff)} \equiv \hat{\sigma}_3 - p_f \end{aligned} \tag{7.39}$$

The shear stresses for the effective stress are unchanged from their original values, so the Mohr circle does not change diameter. As shown in Figure 7.12*J*, the effective stress is the result of shifting the Mohr circle toward lower normal stresses by an amount equal to the pore fluid pressure p_f. We discuss the effective stress in greater detail in Section 8.5, where we show that the mechanical behavior of a brittle material depends on the effective stress, not on the applied stress.

7.4 A CLOSER LOOK AT THE MOHR CIRCLE FOR TWO-DIMENSIONAL STRESS

In this section, we derive the equations for the Mohr circle. We restrict our discussion to two-dimensional problems. Manipulation of the equations for two dimensions is significantly less complex than for three dimensions, and retaining the third dimension adds little to intuitive understanding. An analysis in three dimensions is analogous, as summarized in Box 7-2 (readers should finish this section before reading the box).

We pose the following question: If we know the orientation of the principal axes and the values of the principal stresses at a point, how can we determine the normal and shear components of the surface stress that act on a plane of arbitrarily prescribed orientation through that point?

In order for us to analyze a stress problem in two dimensions, the plane of analysis must be perpendicular to one of the principal axes, usually the intermediate principal axis $\hat{x}_2$. That axis therefore must lie in all the planes on which we can calculate the surface stresses. Thus we consider the infinitesimal cube centered on the point in question with faces parallel to the principal planes and thus perpendicular to the principal axes (Figure 7.13*A*). The plane *P* is parallel to $\hat{x}_2$ but is otherwise of arbitrary orientation. With this geometry we can use a two-dimensional analysis to determine the surface stress on *P*, with the plane of analysis being the ($\hat{x}_1$, $\hat{x}_3$) plane and with the $\hat{x}_2$ axis being perpendicular to that plane (Figure 7.13*B*). All possible surface stresses, which include $\hat{\sigma}_1$ and $\hat{\sigma}_3$, will then lie on the stress ellipse that is the intersection of the three-dimensional stress ellipsoid with the ($\hat{x}_1$, $\hat{x}_3$) plane that passes through the center of the ellipsoid (Figure 7.7*A*).

In order to derive the correct equations for the normal and shear stress, we must be careful to set the diagram up so that all quantities are drawn in their positive sense. Thus we note the following conventions:

Convention 1. The orientation of the plane *P* is defined by the angle θ measured from the positive $\hat{x}_1$ axis to **n**, where **n** is the vector of unit length that is normal to plane *P*. Positive angles are measured counterclockwise, and we construct the diagram for this example so that the angle θ is positive (Figure 7.13*B*).

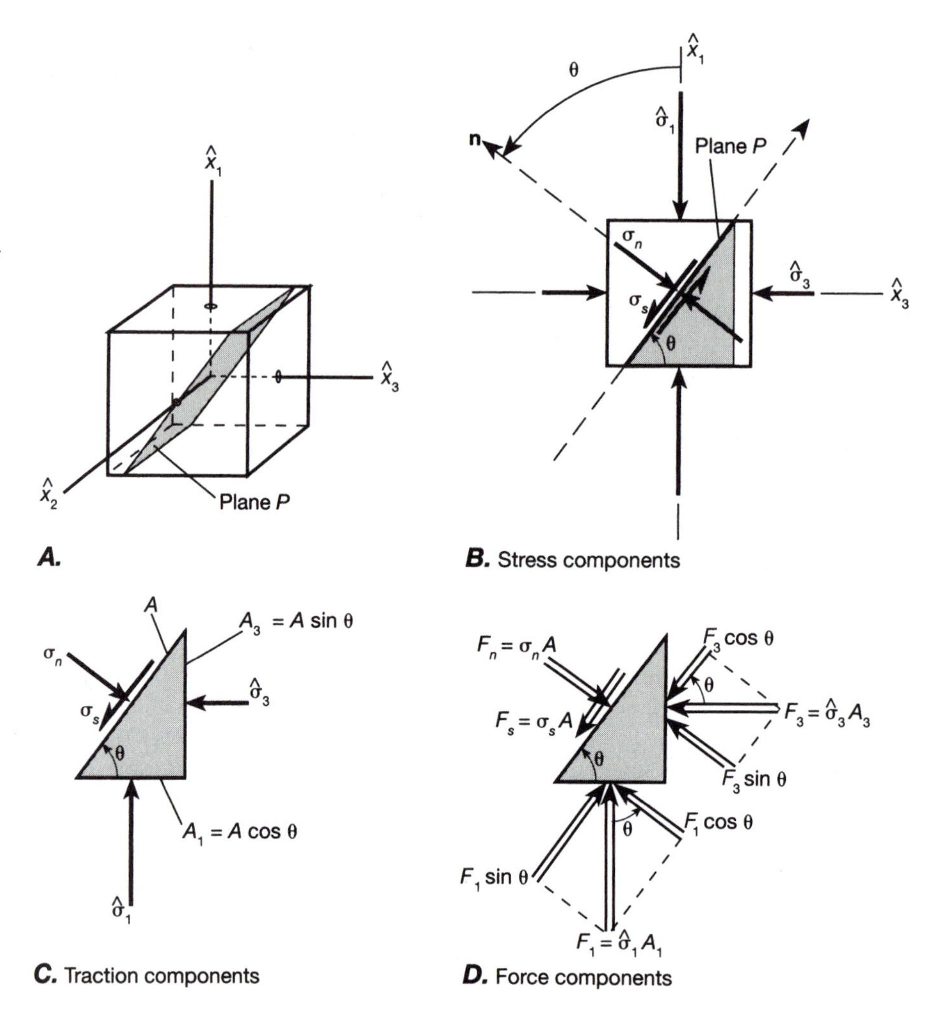

FIGURE 7.13 Geometry for determining the normal and shear stress components on a plane *P* through a point, oriented relative to the principal coordinates. *A*. The plane *P* through the infinitesimal cube is parallel to $\hat{x}_2$ but is otherwise arbitrary in its orientation. *B*. Two-dimensional view of *A* showing distribution of stress components. All stress components and angles are drawn as positive. *C*. A free-body diagram of the shaded triangular element in *B*, showing only those traction components that act on the exterior surfaces of the element. *D*. Forces derived from the traction components in *C* and the associated force components parallel and perpendicular to the inclined plane *P* of area *A*.

Convention 2. We draw the diagram with positive stress components, according to the geologic Mohr circle sign convention. Thus principal stress components on the coordinate planes are drawn as compressive stresses. The normal stress component on plane *P* is also drawn as a compressive normal stress, and the shear stress on plane *P* is drawn counterclockwise.

Now we want to determine the normal and shear components (σ_n, σ_s) of the surface stress acting on plane *P*. To this end, in Figure 7.13*C* we draw a free-body diagram of the shaded triangular element shown in Figure 7.13*B*, and we draw only those traction components that represent the action of the surrounding material on the triangle. Because the infinitesimal square in Figure 7.13*B* is in equilibrium, the triangular element in Figure 7.13*C* must also be in equilibrium, and we can determine the surface stress components on plane *P* by applying Newton's second law, which requires that the sum of all the forces exerted on the triangular element equal zero.

We convert the traction components into force components by multiplying each traction by the area of the surface on which it acts (Figure 7.13*D*). Although in this derivation, we are dealing with tractions on the outer surface of the triangular element, we will persist in using the components and sign convention for the surface stresses, taking care to account for the orientations of the tractions when we add or subtract the forces. In this way, our analysis will give the appropriate value for the surface stress on *P*. The areas of plane *P* and of the sides of the triangular element normal to $\hat{x}_1$ and $\hat{x}_3$ are, respectively, A, A_1, and A_3, so the forces acting on the triangular element are

$$\begin{aligned} F_n &= \sigma_n A \quad & F_s &= \sigma_s A \\ F_1 &= \hat{\sigma}_1 A_1 \quad & F_3 &= \hat{\sigma}_3 A_3 \end{aligned} \tag{7.40}$$

The force on A_1 can be resolved into a pair of components parallel to F_n and to F_s, which are, respectively, the normal and tangential forces on *P* (Figure 7.13*D*). The same is true of the force on A_3. Equilibrium of the triangular element is maintained if all forces perpendicular to *P* sum to zero and if all forces parallel to *P* sum to zero. From Figure 7.13*D*, these conditions imply that

$$\begin{aligned} F_n - F_1 \cos\theta - F_3 \sin\theta = 0 \\ F_s - F_1 \sin\theta + F_3 \cos\theta = 0 \end{aligned} \tag{7.41}$$

where forces acting in the same direction as F_n or F_s in Figure 7.13*D* are added, and those acting in the opposite direction are subtracted. Rearranging Equation (7.41) to isolate F_n and F_s on the left side and substituting for the force components from Equation (7.40), we get

$$\begin{aligned} \sigma_n A &= \hat{\sigma}_1 A_1 \cos\theta + \hat{\sigma}_3 A_3 \sin\theta \\ \sigma_s A &= \hat{\sigma}_1 A_1 \sin\theta - \hat{\sigma}_3 A_3 \cos\theta \end{aligned} \tag{7.42}$$

We can eliminate the area terms from these equations by substituting the following relationships (Figure 7.13*C*):

$$A_1 = A\cos\theta \quad A_3 = A\sin\theta \tag{7.43}$$

into Equation (7.42) and dividing through by A. By these manipulations, we express the force balance (Equations (7.41)) strictly in terms of the stress components so that they give the results we seek:

$$\begin{aligned} \sigma_n &= \hat{\sigma}_1 \cos^2\theta + \hat{\sigma}_3 \sin^2\theta \\ \sigma_s &= (\hat{\sigma}_1 - \hat{\sigma}_3)\sin\theta\cos\theta \end{aligned} \tag{7.44}$$

Note that all the terms with θ involve products of sine and cosine functions. One of the trigonometric terms comes from resolving the force vectors (Equation (7.41)) and the other from resolving the areas (Equation (7.43)). The need to resolve both of these quantities to determine stress components gives those components the characteristics of two directions, which distinguish them from the vector components such as force, which have a unidirectional character (compare Boxes 7-1 and 7-3).

Equations (7.44) are the equations for the general case, and they reduce to Equations (7.14) and (7.15), which we derived in Section 7.1 for the special case of uniaxial stress. To show the equivalence of the two results, we replace σ_n and σ_s on the left of Equations (7.44) with the symbols σ'_n and σ'_s, and on the right side of Equations (7.44) we set $\hat{\sigma}_1 = \sigma_n$ and $\hat{\sigma}_3 = 0$, which is the condition for uniaxial stress.

Thus given the orientation of any plane defined by θ, we can calculate the normal stress and shear stress components on that plane if we know only the principal stresses and their orientations. These equations, therefore, justify our earlier assumption that the components of the state of stress at a point are necessary and sufficient for determining the normal stress and shear stress components on a plane of any orientation through that point.

We can put the equations in a more easily interpreted form by using the standard trigonometric identities:

$$\cos^2\theta = 0.5(1 + \cos 2\theta) \qquad \sin^2\theta = 0.5(1 - \cos 2\theta)$$
$$\sin\theta\cos\theta = 0.5\sin 2\theta \tag{7.45}$$

Substituting Equations (7.45) into Equations (7.44) and rearranging gives

$$\begin{aligned} \sigma_n &= \left(\frac{\hat{\sigma}_1 + \hat{\sigma}_3}{2}\right) + \left(\frac{\hat{\sigma}_1 - \hat{\sigma}_3}{2}\right)\cos 2\theta \\ \sigma_s &= \left(\frac{\hat{\sigma}_1 - \hat{\sigma}_3}{2}\right)\sin 2\theta \end{aligned} \tag{7.46}$$

Here $(\hat{\sigma}_1 + \hat{\sigma}_3)/2$ is the mean normal stress, and $(\hat{\sigma}_1 - \hat{\sigma}_3)/2$ is the maximum possible shear stress, as evidenced from the fact that sin 2θ in the second equation can be no greater than 1.

Equations (7.46) are identical to Equations (7.35) and (7.36), which we deduced from the geometry of the Mohr circle. Thus Equations (7.46) are parametric equations for the Mohr circle, with σ_n and σ_s as the variables and θ as the parameter.

We can put the equations for the Mohr circle into a form that is perhaps more familiar for the equation of a circle by eliminating θ from Equations (7.46). We rewrite the first Equation (7.46) as

$$\sigma_n - \left(\frac{\hat{\sigma}_1 + \hat{\sigma}_3}{2}\right) = \left(\frac{\hat{\sigma}_1 - \hat{\sigma}_3}{2}\right)\cos 2\theta \tag{7.47}$$

We then square both sides of the second Equation (7.46) and of Equation (7.47) and add the resulting two equations together. Applying the trigonometric identity

$$\sin^2 2\theta + \cos^2 2\theta = 1 \tag{7.48}$$

yields the result

$$\left[\sigma_n - \left(\frac{\hat{\sigma}_1 + \hat{\sigma}_3}{2}\right)\right]^2 + \sigma_s^2 = \left(\frac{\hat{\sigma}_1 - \hat{\sigma}_3}{2}\right)^2 \tag{7.49}$$

This equation has the form

$$(x - a)^2 + y^2 = r^2 \tag{7.50}$$

which is the equation of a circle that has a radius r and its center at a point a along the x axis. Thus we have shown that the normal and shear stress on planes of all possible orientations plot as a circle on a graph whose axes are the normal stress and the shear stress. That circle has its center on the normal stress axis at a point whose value is half the sum of the principal stresses, or the mean normal stress $(\hat{\sigma}_1 + \hat{\sigma}_3)/2$. The radius of the circle is half the difference between the principal stresses, or the maximum shear stress $(\hat{\sigma}_1 - \hat{\sigma}_3)/2$. This circle is the Mohr circle. In the form of Equation (7.49), however, the relation between the normal and shear stress on a plane, and the orientation of that plane, is lost.

In general, stress is a three-dimensional quantity, although we have treated it in two dimensions because it is easier to understand, and because many problems can be reduced to two-dimensional problems. A Mohr diagram can, however, be defined for the full three-dimensional stress, although the simplicity of the two-dimensional Mohr diagram is lost. We review briefly some of the characteristics of the three-dimensional Mohr diagram in Box 7-2.

BOX 7-2 The Mohr Diagram for Three-Dimensional Stress

i. Description of the Three-Dimensional Mohr Circles

In Section 7.4 we discuss the two-dimensional problem of determining the surface stress acting on planes that are parallel to the $\hat{x}_2$ principal axis. In that circumstance, the two-dimensional stress components lie within the $(\hat{x}_1, \hat{x}_3)$ coordinate plane. Exactly the same properties of the Mohr circle that are discussed in Section 7.4 apply for the stresses in the other coordinate planes. Thus, in general, to find the stresses on planes parallel to any of the principal axes $\hat{x}_k$, a two-dimensional diagram of the $(\hat{x}_i, \hat{x}_j)$-plane is used, where (i, j, k) all have different values between 1 and 3 and where k designates the coordinate axis normal to the plane of analysis. To standardize the direction in which we view the principal plane defined by any of these pairs of axes, the axes must be plotted so that there is a clockwise rotation from the axis parallel to the larger of the two principal stresses, toward the axis parallel to the smaller one (Figure 7-2.1). This ensures that the signs of the shear stresses will be consistent. Thus, assuming $i < j$, (i, j, k) can take on the values (1, 3, 2) (Figure 7-2.1*A*), (1, 2, 3) (Figure 7-2.1*B*), or (2, 3, 1) (Figure 7-2.1*C*). Within any of these diagrams of the $(\hat{x}_i, \hat{x}_j)$-plane, we then give the orientation of any plane that is parallel to the third principal axis $\hat{x}_k$ by the angle θ_k, which is measured between $\hat{x}_i$ (the axis parallel to the larger principal stress in the plane) and the normal to the plane; this angle is taken to be positive counterclockwise in the plane of the diagram (Figure 7-2.1). With these conventions, the equations analogous to Equations (7.46) and (7.49) for each of the three principal planes can be written in general form as

for $(i,j,k) = (1,3,2), (1,2,3)$, or $(2,3,1)$ (7-2.1)

$$\sigma_n = \left(\frac{\hat{\sigma}_i + \hat{\sigma}_j}{2}\right) + \left(\frac{\hat{\sigma}_i - \hat{\sigma}_j}{2}\right)\cos 2\theta_k$$

$$\sigma_s = \left(\frac{\hat{\sigma}_i + \hat{\sigma}_j}{2}\right)\sin 2\theta_k \qquad (7\text{-}2.2)$$

$$\left[\sigma_n - \left(\frac{\hat{\sigma}_i + \hat{\sigma}_j}{2}\right)\right]^2 + \sigma_s^2 = \left(\frac{\hat{\sigma}_i + \hat{\sigma}_j}{2}\right)^2 \qquad (7\text{-}2.3)$$

When $(i, j, k) = (1, 3, 2)$, for example, we obtain Equations (7.46) and (7.49).

ii. Additional Properties of the Mohr Circle

To the main properties of a single Mohr circle discussed in Section 7.2, we append the following properties that apply to a Mohr diagram of three-dimensional stress. The numbered headings are the same as those in Section 7.2, and the items under each heading are numbered in Roman numerals consecutively, continuing on from the items under the same heading in Section 7.2. The application of these properties to solving some three-dimensional stress problems is illustrated in Appendix 7B.

1) The Mohr Diagram

(iii) The three-dimensional stress plots on a Mohr diagram as a set of three Mohr circles, each of which is a graph of the surface stress components on sets of planes that are parallel to one of the principal axes (Figure 7-2.2). The three circles are defined by Equations (7-2.2), with Equation (7-2.1), and each involves one pair of the principal stresses. In Figure 7-2.2*A*, the three angles θ_k for $k = 1, 2, 3$ are shown as negative (clockwise) angles. All the properties discussed in Section 7.2 apply to each of these circles.

2) Principal Stresses

(iii) All three principal stresses plot on the σ_n axis. Each principal stress plots at a point that is common to two of the Mohr circles. If all the principal stresses are unequal, there are no other common points among the circles. Each of the principal stresses forms one end of the diameters of two Mohr circles, with the other two principal stresses lying at the opposite ends of those diameters. This geometry is consistent with the fact that the three principal stresses in physical space act on

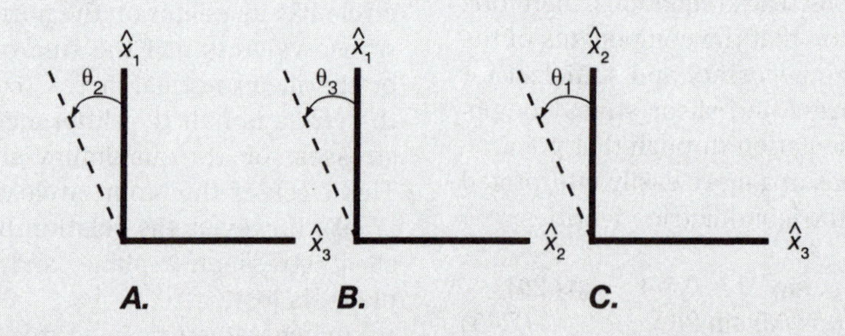

FIGURE 7-2.1 The pairs of principal coordinate axes for the three Mohr circles in three-dimensional stress. The axes are oriented so that in each case there is a clockwise rotation from the positive axis parallel to the larger normal stress toward the positive axis parallel to the smaller one. These orientations standardize the views of the coordinate planes and thus eliminate the ambiguity of the signs assigned to the shear stress components by the Mohr circle sign convention. The orientation of a plane is defined in each diagram by the angle its normal makes with the coordinate axis parallel to the larger normal stress. The diagrams show positive (counterclockwise) angles θ_2, θ_3, and θ_1 measured about the principal axis that is normal to each diagram, respectively, $\hat{x}_2$, $\hat{x}_3$, and $\hat{x}_1$.

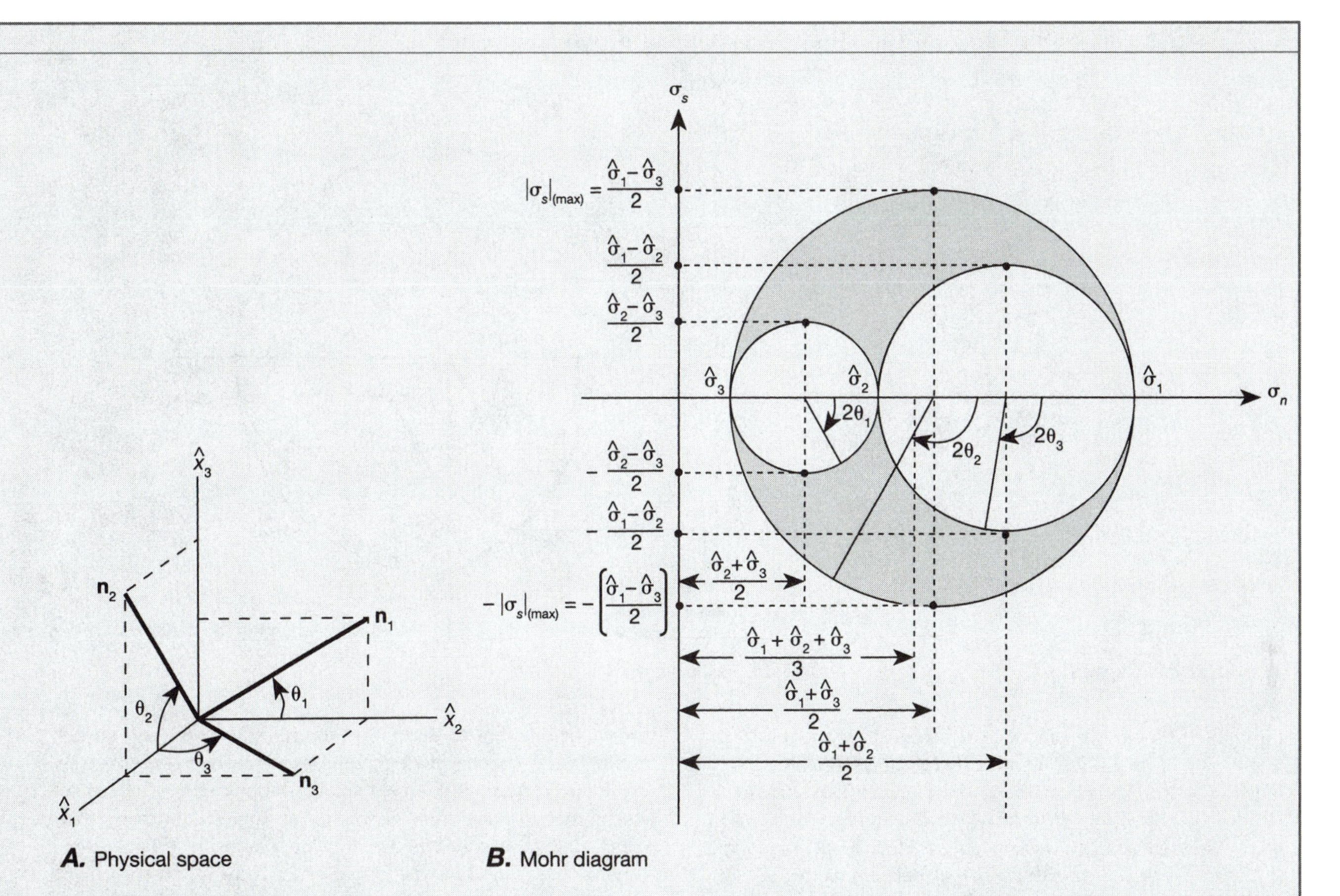

FIGURE 7-2.2 Geometry of three-dimensional stress on a Mohr diagram. *A.* Diagram of physical space showing the three principal coordinate axes $\hat{x}_1$, $\hat{x}_2$, and $\hat{x}_3$ and the normals n_1, n_2, and n_3 to three planes (not shown) that are parallel, respectively, to $\hat{x}_1$, $\hat{x}_2$, and $\hat{x}_3$. θ_1, θ_2, and θ_3 are the angles between the appropriate principal axis and the normals n_k, measured in one of the coordinate planes as shown. When viewed from the proper direction so that the axes are in the standardized orientation (Figure 7-2.1), all three angles are clockwise (negative) as shown. *B.* Mohr diagram. Mohr circles are shown for a three-dimensional stress. Each Mohr circle represents the two-dimensional stress in one of the principal coordinate planes. Angles plotted are shown in *A*. The surface stress components on the planes corresponding to the normals in *A* plot on the circumference of the appropriate Mohr circle.

three mutually perpendicular surfaces (cf. property 3(v) in Section 7.3).

3) Surface Stress and the Orientation of Planes

(vi) Planes that are not parallel to one of the principal axes have normals that do not lie in any of the principal coordinate planes (Figure 7-2.3*A*). The components of the surface stress on all such planes must plot on the Mohr diagram within the largest Mohr circle and outside the two smaller circles in the area shaded in Figure 7-2.2*B*. The construction on the Mohr diagram for determining the stress components on such a plane is indicated in Figure 7-2.3*B* for which the geometry in physical space is shown in Figure 7-2.3*A*. The complexity of such three-dimensional problems is beyond the scope of this book, and our interest is confined to problems involving planes that parallel one of the principal axes.

4) Conjugate Planes of Maximum Shear Stress

(ii) The maximum absolute values of the shear stress on any plane in three-dimensional space plot on the ($\hat{\sigma}_1$, $\hat{\sigma}_3$) Mohr circle at $2\theta_2 = \pm 90^\circ$ (Figure 7-2.4*A*). These stresses occur on a conjugate set of planes $P^{(+)}$ and $P^{(-)}$ in physical space that are parallel to $\hat{x}_2$ and that have normals $\mathbf{n}^{(+)}$ and $\mathbf{n}^{(-)}$ lying in the ($\hat{x}_1$, $\hat{x}_3$)-plane at $\theta = \pm 45^\circ$, respectively, from $\hat{x}_1$ (Figure 7-2.4*B*). Thus, although each Mohr circle individually has maximum absolute values of the shear stress (Figure 7-2.2*B*), the maxima for the ($\hat{\sigma}_1$, $\hat{\sigma}_2$) and the ($\hat{\sigma}_2$, $\hat{\sigma}_3$) Mohr circles are maxima only for the particular set of planes that are parallel to $\hat{x}_3$ and to $\hat{x}_1$, respectively. The true maxima for planes of all possible orientations occur only at the maxima for the ($\hat{x}_1$, $\hat{x}_3$) Mohr circle (Figure 7-2.4*A*).

(continued)

BOX 7-2 The Mohr Diagram for Three-Dimensional Stress *(continued)*

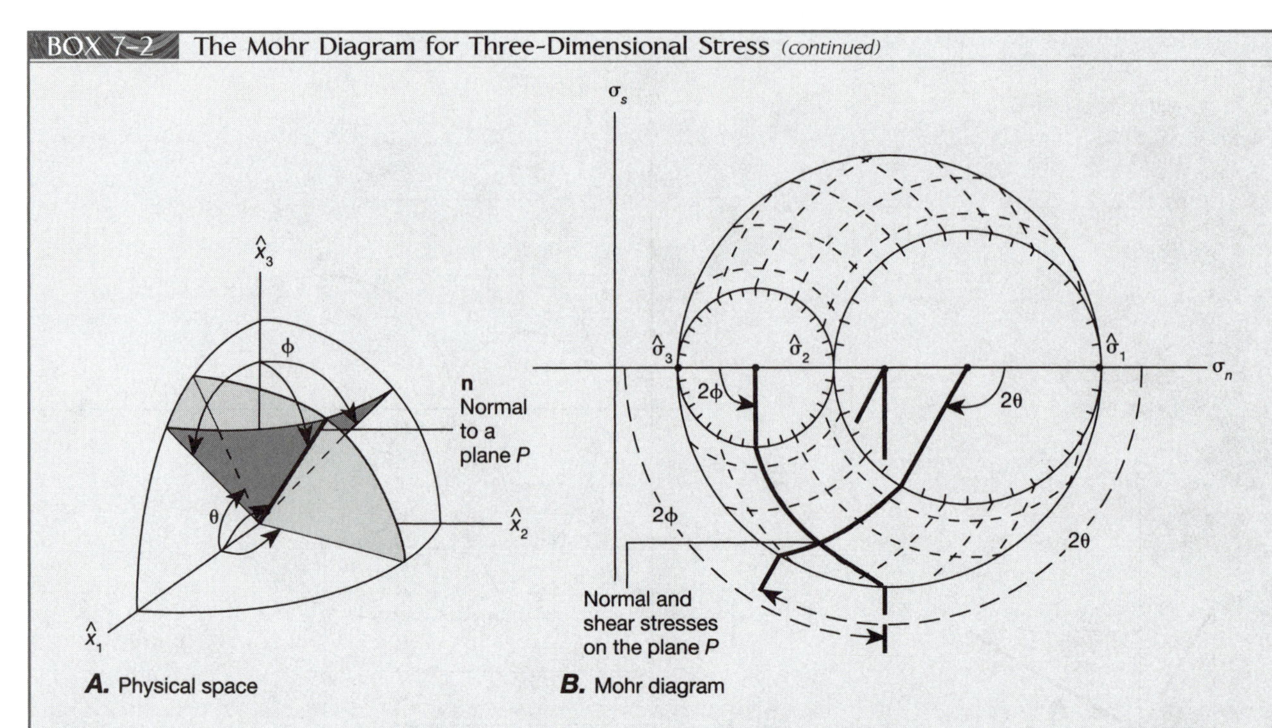

FIGURE 7-2.3 Mohr diagram for stress components on a plane of arbitrary orientation in three dimensions. *A.* Physical space: The normal n to the plane *P* (not shown) is defined by the angles θ from $\hat{x}_1$ and ϕ from $\hat{x}_3$. Thus n lies at the intersection of two cones, one of half opening-angle θ about the $\hat{x}_1$ axis and the other of half opening-angle ϕ about the $\hat{x}_3$ axis. The orientation of n is therefore described by the angle θ measured from $\hat{x}_1$ in both the ($\hat{x}_1$, $\hat{x}_2$)-plane and the ($\hat{x}_1$, $\hat{x}_3$)-plane, and by the angle ϕ measured from $\hat{x}_3$ in both the ($\hat{x}_1$, $\hat{x}_3$)-plane and the ($\hat{x}_2$, $\hat{x}_3$)-plane. Counterclockwise angles measured in the principal coordinate planes are positive when the coordinate axes are viewed according to Figure 7-2.1. The θ is negative (clockwise) in both the ($\hat{x}_1$, $\hat{x}_3$)- and ($\hat{x}_1$, $\hat{x}_2$)-planes; ϕ is positive (counterclockwise) in both the ($\hat{x}_1$, $\hat{x}_3$)- and ($\hat{x}_2$, $\hat{x}_3$)-planes. *B.* Mohr diagram: The angles in *A* are transferred to the Mohr diagram to determine the normal stress and shear stress acting on the plane *P*. Families of dashed curves are arcs concentric with the two smaller Mohr circles. The radii to the Mohr circles are constructed at angles defined by the cone angles measured in the principal planes in *A*: Radii are constructed at a clockwise angle of 2θ from $\hat{\sigma}_1$ on the ($\hat{\sigma}_1$,$\hat{\sigma}_2$) and ($\hat{\sigma}_1$,$\hat{\sigma}_3$) Mohr circles. The endpoints of these radii are connected by a dashed circular arc concentric with the ($\hat{\sigma}_2$,$\hat{\sigma}_3$) Mohr circle. Similarly, radii are constructed at a counterclockwise angle of 2ϕ from $\hat{\sigma}_3$ on the ($\hat{\sigma}_1$,$\hat{\sigma}_3$) and the ($\hat{\sigma}_2$,$\hat{\sigma}_3$) Mohr circles. The endpoints of these radii are connected by a dashed circular arc concentric with the ($\hat{\sigma}_1$,$\hat{\sigma}_2$) Mohr circle. The intersection of these two dashed arcs marks the normal and shear stresses on the plane *P*. (After Jaeger 1962)

5) Scalar Invariants of the Stress

(ii) In three dimensions there are three scalar invariants of the stress tensor that characterize the stress at a point. In terms of the stress ellipsoid (Figure 7. 7A), the three invariants are proportional, respectively, to the mean of the three principal radii of the ellipsoid, the sum of the areas of the three principal planes of the ellipsoid, and the volume of the ellipsoid. Each of these invariants provides an independent measure of the size of the ellipsoid, and thus of the magnitude of the stress.

7.5 THE STRESS TENSOR

For students interested in understanding the connection between the preceding material and the common treatments of continuum mechanics, we present in this section some details about the notation, sign conventions, characteristics, and behavior of the stress tensor and its components. This material is not necessary for the rest of the book, except that the component notation for the stress tensor is useful for following the discussion in Boxes 16-4 and 18-1. Thus this section can be skipped without compromising understanding of the rest of the text. A summary of the notation we use for the various stress components and sign conventions is included in Table 7.4 at the end of the chapter.

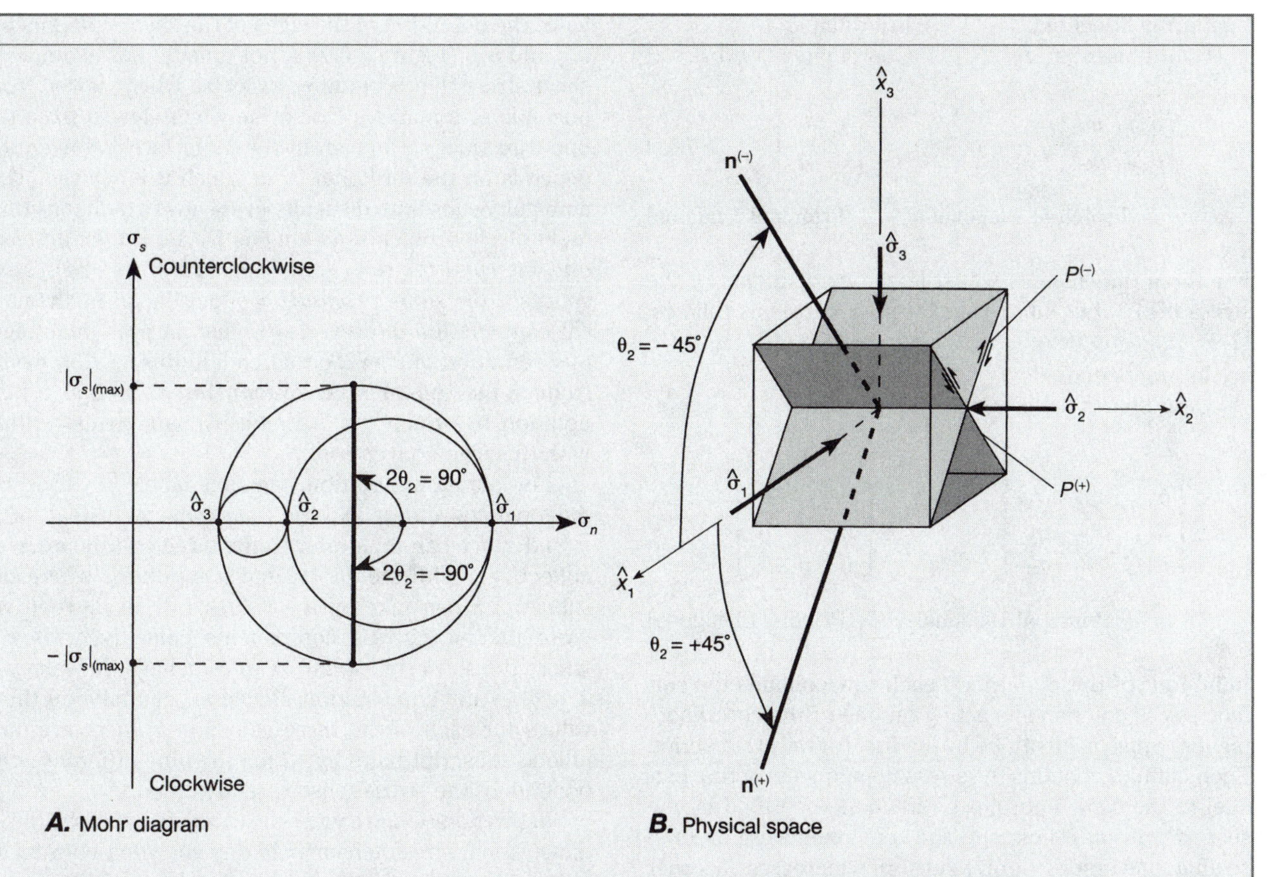

FIGURE 7-2.4 Planes of maximum shear stress in three dimensions. *A*. Mohr diagram showing maximum absolute values of the shear stress. *B*. Diagram of physical space showing the conjugate planes of maximum shear stress and their relationship to the principal stresses.

With the added complexity of the third dimension, none of the invariants has the simple geometric interpretation, in terms of the Mohr circle, that we found for the two invariants of the two-dimensional stress tensor. Thus, we consider briefly only the first invariant, which is still the mean normal stress defined by

$$\overline{\sigma}_n = \frac{\sigma_{xx} + \sigma_{yy} + \sigma_{zz}}{3} = \frac{\hat{\sigma}_1 + \hat{\sigma}_2 + \hat{\sigma}_3}{3} \qquad (7\text{-}2.4)$$

where the first expression is for components in a general coordinate system, and the second is for the components in the principal coordinates (Figure 7-2.2*B*). Note that for the general three-dimensional case, the mean normal stress is not the center of any of the three Mohr circles.

i. The Stress-Component Matrix

The stress at a point $\boldsymbol{\sigma}$ belongs to a group of mathematical quantities called **second-rank tensors**, and it is therefore called the **stress tensor** (see Box 7-3). In two dimensions, a second-rank tensor is represented in a given coordinate system by four scalar components (see Table 7.2). In three dimensions, a second-rank tensor is specified by nine scalar components (see Table 7.3). It is most convenient to list these components in systematic order in tabular form. Such a listing of components is called a **matrix** of stress components. For example, from Table 7.2, the components of the stress $\boldsymbol{\sigma}$ can be written as follows:

In Any Specified Coordinates (x, z): In Principal Coordinates $(\hat{x}_1, \hat{x}_3)$:

$$\begin{bmatrix} \sigma_{xx} & \sigma_{xz} \\ \sigma_{zx} & \sigma_{zz} \end{bmatrix} \qquad \begin{bmatrix} \hat{\sigma}_1 & 0 \\ 0 & \hat{\sigma}_3 \end{bmatrix} \tag{7.51}$$

Principal Diagonal Principal Diagonal

For three dimensions, we write the components of the stress $\boldsymbol{\sigma}$ in Table 7.3 in a similar matrix form as follows:

In Any Specified Coordinates (x, y, z): In Principal Coordinates $(\hat{x}_1, \hat{x}_2, \hat{x}_3)$:

$$\begin{bmatrix} \sigma_{xx} & \sigma_{xy} & \sigma_{xz} \\ \sigma_{yx} & \sigma_{yy} & \sigma_{yz} \\ \sigma_{zx} & \sigma_{zy} & \sigma_{zz} \end{bmatrix} \qquad \begin{bmatrix} \hat{\sigma}_1 & 0 & 0 \\ 0 & \hat{\sigma}_2 & 0 \\ 0 & 0 & \hat{\sigma}_3 \end{bmatrix} \tag{7.52}$$

Principal Diagonal Principal Diagonal

In all four of these matrices, each row contains the surface stress components acting on one of the coordinate planes, which is identified by the first (or only) subscript. Each column contains the stress components acting parallel to the same coordinate direction, identified by the second (or only) subscript. The components are ordered so that the first (or only) subscript increases in order from top to bottom in each column, and the second (or only) subscript increases in order from left to right in each row. Notice that with this ordering scheme, the normal stress components, for which both subscripts are the same, occur on a diagonal of the matrix, which is called the **principal diagonal**. The shear stress components are in the off-diagonal positions.

The second matrix in Equations (7.51) and (7.52) lists the stress components in the principal coordinate system (see Figure 7.6*A*, *B*, *C*; Figure 7.7*A*, *B*, *C*). Because the shear stress components are necessarily zero on the principal planes, however, we use a common contracted subscript convention in which we write only one of the two identical subscripts for the nonzero normal stresses.

In the first Equations (7.51) and (7.52) we use the notation that we have adopted for the Mohr circle sign convention. This is not a mathematically correct representation of the stress tensor, however, because the signs of the shear stress components in the Mohr circle sign convention are not unique, so the off-diagonal matrix elements are not adequately defined. We resolve this problem in the next section with a change to the tensor sign convention and a corresponding change in notation.

ii. Stress Tensor Notation and the Tensor Sign Convention

Although the stress components $(\sigma_{xx}, \sigma_{xz})$, $(\sigma_{zz}, \sigma_{zx})$, as defined in Section 7.2 are useful for Mohr circle problems, the definition of the signs of the shear components σ_{xz} and σ_{zx} (Figure 7.6*D*) is not unique. For example, a shear stress that is counterclockwise when viewed from one side of a plane is clockwise when viewed from the opposite side, so its sign in the Mohr circle convention depends on the direction from which it is viewed. This ambiguity causes no difficulty in solving two-dimensional Mohr circle problems, *as long as we set up the problem and interpret the solution using a diagram of physical space in the same orientation.* For general mathematical computation, however, we must have an unambiguous definition of sign. To that end, in this section we introduce the **tensor sign convention** as well as a new notation by which we distinguish components written with this sign convention.[15]

For our new notation, we first label the three orthogonal coordinate axes as x_1, x_2, and x_3 instead of x, y, and z. We can then refer to the three coordinate axes collectively with the short-hand notation x_k, where the subscript k can take on the values 1, 2, or 3. Then we write the nine stress components collectively as σ_{kl}, where the subscripts k and l can each have the values 1, 2, or 3 in any combination. Because l can take on three values for each of the three values of k, there are nine unique subscript pairs by which the nine different components of the stress tensor are identified.

In this notation, we write the components of the three-dimensional stress tensor $\boldsymbol{\sigma}$ in any specified coordinate system and in principal coordinates respectively as:

In Any Specified Coordinates (x_1, x_2, x_3): In Principal Coordinates $(\hat{x}_1, \hat{x}_2, \hat{x}_3)$:

$$\sigma_{kl} = \begin{bmatrix} \sigma_{11} & \sigma_{12} & \sigma_{13} \\ \sigma_{21} & \sigma_{22} & \sigma_{23} \\ \sigma_{31} & \sigma_{32} & \sigma_{33} \end{bmatrix} \qquad \sigma_{kl} = \hat{\sigma}_{kl} = \begin{bmatrix} \hat{\sigma}_1 & 0 & 0 \\ 0 & \hat{\sigma}_2 & 0 \\ 0 & 0 & \hat{\sigma}_3 \end{bmatrix} \tag{7.53}$$

Principal Diagonal Principal Diagonal

As before, the first (or only) subscript defines the coordinate axis that is perpendicular to the plane on which the stress component acts, and the second (or only) subscript defines the coordinate axis that is parallel to the stress component. The normal stress components, for which both subscripts are the same (or for which there is only one subscript), occur along the principal diagonal, and the shear stress components occur in the off-diagonal positions.

[15]Beware! There is no standardized stress notation in the literature. The different notations that we use here for the stress components defined with the Mohr circle sign convention and those defined with the tensor sign convention are adopted for clarity in this book, but they do not represent a generally accepted notation. Sometimes, for example, the notation we use for components in the Mohr circle sign convention is used for components in the tensor sign convention.

Each row of the component matrix thus contains the three components of one of the surface stresses, which acts on the coordinate plane identified by the first subscript. For example, the components in the first row of the first matrix, $[\sigma_{11}, \sigma_{12}, \sigma_{13}]$, are the normal stress and two shear stress components, respectively, of the surface stress $\mathbf{\Sigma}_1$ that acts on the coordinate plane perpendicular to the x_1 axis. Thus we see that the stress tensor also can be represented by an array of three surface stresses, each acting on a different coordinate plane.

In Any Specified Coordinates (x_1, x_2, x_3): In Principal Coordinates $(\hat{x}_1, \hat{x}_2, \hat{x}_3)$:

$$\begin{bmatrix} \mathbf{\Sigma}_1 \\ \mathbf{\Sigma}_2 \\ \mathbf{\Sigma}_3 \end{bmatrix} \qquad \begin{bmatrix} \hat{\mathbf{\Sigma}}_1 \\ \hat{\mathbf{\Sigma}}_2 \\ \hat{\mathbf{\Sigma}}_3 \end{bmatrix} \tag{7.54}$$

The new sign convention that we associate with this notation is illustrated in Figure 7.14, where, as before (Figure 7.7), the two sides of each of the three coordinate surfaces through a point are separated into the opposite sides of an infinitesimal cube. The positive and negative sides of the cube are labeled in Figure 7.14*A* and defined as follows: A vector of unit length that points out of the cube perpendicular to one of the cube faces is referred to as an **outward unit normal**. If the outward unit normal to a coordinate plane on the cube points in a positive coordinate direction, that face of the cube is labeled the positive side of the coordinate surface (Figure 7.14*A*; "(+)" signs on the visible cube faces); and if the outward unit normal points in a negative coordinate direction, the face is labeled the negative side of the coordinate surface (Figure 7.14*A*; "(−)" signs on the hidden cube faces).

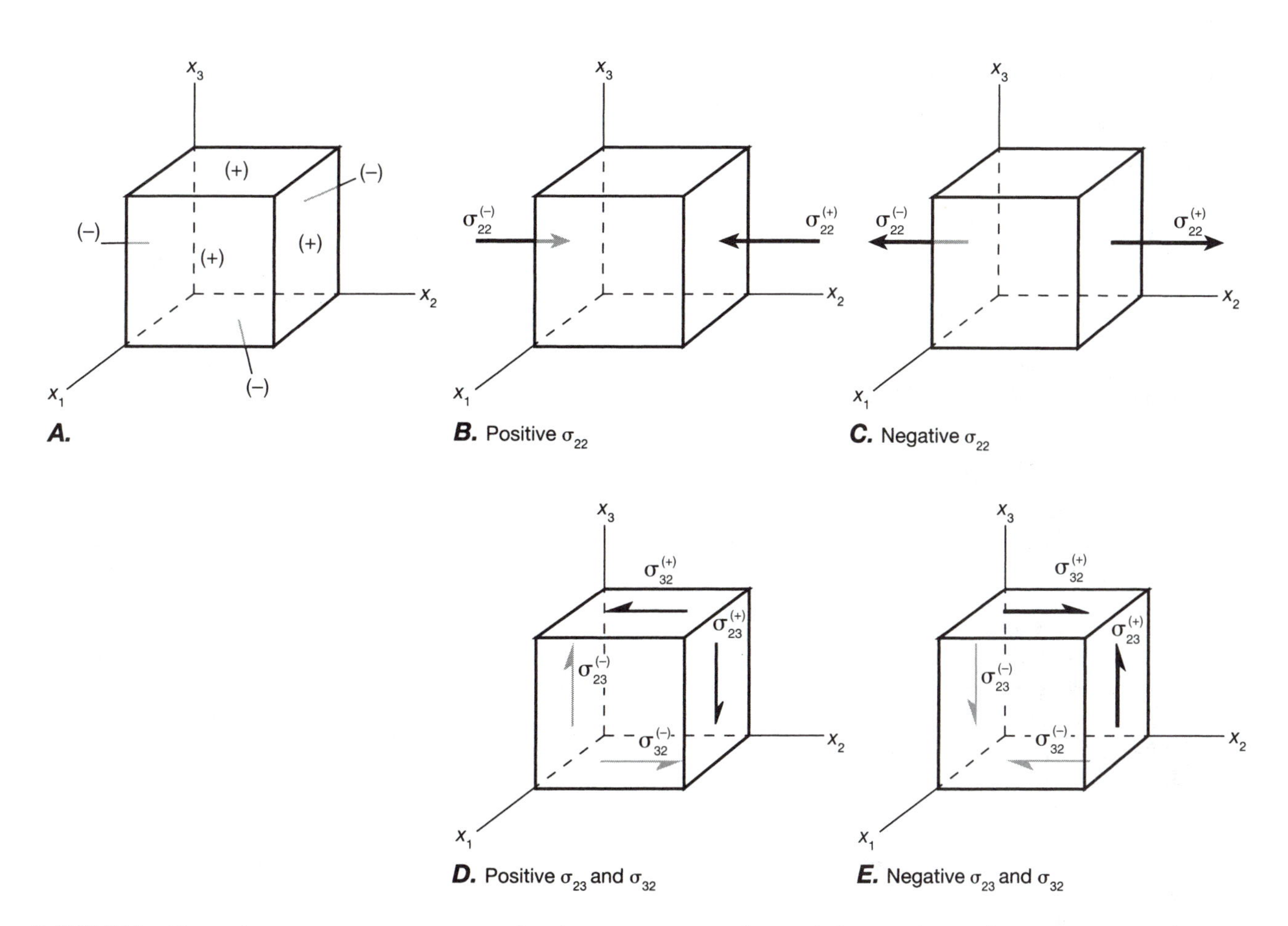

FIGURE 7.14 The geologic tensor sign convention for the stress components. The three coordinate surfaces intersecting at a point in space are represented by an infinitesimal cube with opposite faces of the cube representing the opposite sides of one coordinate plane. Traction components are labeled separately, with superscripts "(+)" or "(−)" on opposing traction components indicating whether they act on a positive or negative side of the coordinate surface, respectively. The stress components are defined to have the same values and signs as the traction components acting on the negative side of the coordinate surfaces. *A.* The positive and negative sides of an infinitesimal coordinate cube. *B.* Positive σ_{22}: a compressive stress. *C.* Negative σ_{22}: a tensile stress. *D.* Positive σ_{23} and σ_{32}: Traction components on negative faces of the cube act in the positive coordinate direction. *E.* Negative σ_{23} and σ_{32}: Traction components on negative faces of the cube act in the negative coordinate direction.

BOX 7-3 What Is a Tensor?

A **tensor** is a mathematical quantity that can be used to describe the physical state or the physical properties of a material. Some types of tensor quantities are given specific names. These include **scalars**, such as temperature or mass, and **vectors**, such as force or velocity. Other types of tensors are simply referred to as tensor quantities such as stress (and strain—see Chapter 12). The different types of tensor quantities can all be distinguished by their rank, as explained in the following discussion. Although tensor quantities are described in terms of their components in a particular coordinate system, they describe a physical quantity that is actually independent of the coordinate system. Think, for example, of velocity, which has a specific magnitude and orientation regardless of the coordinate system in which it is described.

The **rank** r of a tensor is a number that distinguishes different types of tensor quantities. The rank determines how many scalar components are required to describe the tensor completely; it equals the number of subscripts that are part of the symbol for the tensor components; and it indicates how many coordinate directions are associated with each component. We illustrate these properties for tensors of various rank.

The number of components c needed to describe a tensor quantity equals the dimension d of the physical space raised to the power given by the rank r:

$$c = d^r$$

In three-dimensional space ($d = 3$), for example, a **scalar** is a tensor of zero rank ($r = 0$) and so has $3^0 = 1$ component. A scalar is defined simply by its magnitude and thus is associated with no coordinate directions. We represent scalars mathematically by a single symbol, such as T for temperature and m for mass; we do not need to write any subscripts on these symbols, because there is only one value and no orientation associated with the physical quantity.

A vector is a first-rank tensor ($r = 1$), which in three-dimensional space ($d = 3$) is described with $3^1 = 3$ components (see Box 7-1). Force, velocity, and acceleration are all vector quantities. Vectors describe physical quantities that are characterized by a magnitude and a direction. We represent a vector with an arrow. The length of the arrow is proportional to the magnitude of the vector, and the orientation of the arrow, relative to a specified coordinate system, defines the direction of the vector. The vector components are represented mathematically by a symbol with a single subscript, such as for a force F_k. The subscript k is understood to take on the values 1, 2, and 3 in three-dimensional space, and the subscripted symbol represents the three components (F_1, F_2, F_3), each measured parallel to the corresponding coordinate x_k = (x_1, x_2, x_3). In two-dimensional space, there are two components, and k takes on just two values, such as 1 and 2. Each component is the magnitude of the vector parallel to one of the coordinate axes, and thus each is associated with one coordinate direction. Thus in a given three-dimensional coordinate system, a vector (a first-rank tensor) can be represented by an array of three zeroth-rank tensors (i.e., three scalars).

A second-rank tensor ($r = 2$) in three-dimensional space ($d = 3$) has $3^2 = 9$ components. In two dimensions ($d = 2$) second-rank tensors ($r = 2$) have $2^2 = 4$ components. Second-rank tensors describe physical quantities that are characterized by three magnitudes in three dimensions or two magnitudes in two dimensions. These are the scalar invariants, which for the stress are represented in two dimensions by the center of the Mohr circle and its radius, or equivalently by the mean radius and the area of the stress ellipse. In three dimensions, the scalar invariants of the stress tensor are related to the mean radius, the surface area, and the volume of the stress ellipsoid. We discuss these invariants further in Sections 7.2, property 5.i, and Box 7-2, property 5.ii. The components of a second-rank tensor, such as stress, are written with two subscripts σ_{kl} where each of the subscripts k and l takes on the values 1, 2, and 3 in three dimensions to define the nine stress components. In two-dimensional space, each of the subscripts k and l takes on just two values, such as 1 and 2 to define the four components. Each component of a second-rank tensor is associated with two coordinate directions. For the stress tensor, for example, these orientations are the normal to the plane on which the stress component acts and the orientation of the stress component acting on that plane.

Second-rank tensors can be represented by an array of three first-rank tensors (vectors), and thus by an array of nine scalars (three scalars to represent each of the three vectors). In structural geology the most important examples of second-rank tensors are stress, introduced in this chapter, and strain, introduced in Chapter 12. The array of three vectors

The simplest statement of the **geologic stress tensor sign convention** is that each stress tensor component has the same sign as the associated traction component that acts on the *negative* side of the coordinate surface. Figure 7.14*B* and *D* show stress tensor components (σ_{22}, σ_{23}, σ_{32}) that are positive because in each case, the traction on the negative side of the coordinate surface, indicated by the superscript minus sign ($\sigma_{22}^{(-)}$, $\sigma_{23}^{(-)}$, $\sigma_{32}^{(-)}$), points in a positive coordinate direction. Figure 7.14*C* and *E* show stress tensor components (σ_{22}, σ_{23}, σ_{32}) that are negative because in each case there, the traction on the negative side of the coordinate surface ($\sigma_{22}^{(-)}$, $\sigma_{23}^{(-)}$, $\sigma_{32}^{(-)}$) points in a negative coordinate direction.[16]

[16]A more general statement of the geologic tensor sign convention that can be applied to any traction component is that if the outward unit normal to a coordinate surface, and a traction component acting on that surface, point in opposite coordinate senses (i.e., one points in a positive coordinate direction and the other points in a negative coordinate direction), then the stress component associated with that traction component is positive; if the outward unit normal and the traction component point in the same coordinate sense, the associated stress component is negative.

that represent the stress tensor are the coordinate-plane tractions; because the coordinate planes have a constant orientation and area, these tractions have the characteristics of vectors. Thus any rank of tensor can be represented as an array of d tensors of one lower rank, where d is the dimension of the space.

A second-rank tensor **t** is called a **symmetric tensor** if, for any given pair of indices (i, j) it is true that $t_{ij} = t_{ji}$. These equations are nontrivial only if $i \neq j$. In three dimensions, there are three such equations, which means that only six of the nine components are independent; in two dimensions there is one such equation, and thus three of the four components are independent. Second-rank symmetric tensors can be represented diagrammatically by an ellipsoid in three dimensions or an ellipse in two dimensions. Stress and strain are both second-rank symmetric tensors.

The components of a tensor must preserve the characteristics of magnitude and orientation of the tensor in any coordinate system, and for that to be true, the components in different coordinate systems must be related to one another in a specific way. That relation is called a **transformation equation**. Different rank tensors are also characterized by different transformation equations. For scalars, the transformation equation is just the identity relation; scalars have the same value in any coordinate system. For vectors, the terms in the transformation equations (Equations (7-1.7)) involve the first power of the sine and cosine functions. For second-rank tensors, however, the terms in the transformation equations involve the products of sine and cosine functions (Equations (7.44) and (7.46); (7-2.2) and (7-2.3) in Box 7-2; or (7-5.3) and (7-5.4) in Box 7-5). The difference arises from the fact that for vector components, only the single direction associated with each component must be transformed, whereas for second-rank tensor components, the two directions associated with each component must be transformed. Because one set of components in one coordinate system can always be calculated from the set in any other coordinate system as long as the relative orientation of the two systems is known, the sets of components are not independent, and each set describes the same physical quantity.

For any vector, it is always possible to define a coordinate system in which all components of the vector are zero except one. This is the case when one coordinate axis is parallel to the vector and the others are perpendicular to it. The analogous situation for any second-rank symmetric tensor **t**, for which $t_{ij} = t_{ji}$, is that there is always a coordinate system of a particular orientation, called the **principal coordinates**, in which the only nonzero components of t_{ij} are those with subscripts $i = j$ (which lie on the principal diagonal of the component matrix). The values of these components are called the **principal values**. For the stress tensor, this means that on the principal planes, which are defined by any pair of the principal coordinates, the normal stresses are extremes—that is, maximum, minimum, or minimax[2]—and the shear stresses are all zero. The principal coordinates are parallel to the principal axes of the stress ellipsoid (Figure 7.7A); the principal stresses are the surface stresses on the principal planes and are parallel to the principal axes.[3]

Physical quantities that are described by tensors of higher rank also exist. For example, the piezoelectric material constants are represented by a third-rank tensor whose components can be symbolized by A_{ijk} and the elastic constants of a material are defined, in general form, by a fourth-rank tensor symbolized by A_{ijkm}. These tensor components are associated with three and four directions, respectively. In particular, the piezoelectric material constants describe the relationship between the stress on a material (two directions) and the associated electric field vector (one direction). The elastic constants describe the relationship between the stress on a material (two directions) and the associated strain (two directions). These higher-rank tensors do not concern us in this book.

[2]A minimax is a quantity that is simultaneously a minimum in the plane that contains the minimax and the maximum and a maximum in the plane that contains the minimax and the minimum. The intermediate principal axis of an ellipsoid is an example of a minimax.

[3]For students familiar with linear algebra, the principal stresses and principal directions are the eigenvalues and eigenvectors, respectively, for the matrix of stress components. Finding the principal values can be done by diagonalizing the matrix.

Thus for normal stress components, this convention defines compressive stresses to be positive and tensile stresses to be negative (Figure 7.14*B*, *C*), just as in the Mohr circle sign convention. If the coordinates (x_1, x_2, x_3) were parallel to the coordinates (x, y, z), respectively, then using our notation to differentiate components in the two sign conventions, we could write

$$\sigma_{11} = \sigma_{xx}, \quad \sigma_{22} = \sigma_{yy}, \quad \sigma_{33} = \sigma_{zz} \tag{7.55}$$

For shear stress components, however, the signs of the tensor components can be different from the sign of the Mohr circle components. Notice, for example, that in Figure 7.14*D* the shear stress components σ_{23} and σ_{32} are both positive even though one is a counterclockwise shear couple and the other is clockwise. We know from equations (7.29) and (7.31) that these two shear stresses must have the same absolute value, and thus, using the tensor sign convention, we can write:

$$\sigma_{12} = \sigma_{21} \quad \sigma_{13} = \sigma_{31} \quad \sigma_{23} = \sigma_{32} \tag{7.56}$$

Note the differences between this sign convention and that for the Mohr circle: (1) In the tensor sign convention, the definition for the signs of the shear stress components are unique and unambiguous, whereas in the Mohr circle sign convention, the signs of the shear stress components depend on the direction from which those components are viewed; (2) pairs of shear stress components that have the same numerical subscripts but arranged in opposite order always have the same magnitude and opposite shear senses. In the tensor sign convention they have the same sign (Equation (7.56)), whereas in the Mohr circle sign convention they have *opposite* signs (Equations (7.29) and (7.31)). In the stress tensor matrix, these pairs of equal shear stress components occur in symmetrical positions relative to the principal diagonal, and because of this, the matrix is called a **symmetric matrix** and the stress tensor is called a **symmetric tensor**.

We can now understand the inherent incompatibility of the tensor sign convention with the Mohr circle sign convention. The stress components in Figure 7.13*B* are all drawn to be positive in accordance with the Mohr circle sign convention. Thus if we choose two planes whose normals are at angles of θ and $\theta + 90°$, a counterclockwise shear stress will be considered positive on both of them. We know from considerations of equilibrium, however, that the shear stresses on two perpendicular planes must be equal and of opposite shear sense, for example, counterclockwise and clockwise, respectively, (e.g., Figure 7.10). Thus according to the Mohr circle sign convention, the signs of the shear stresses must be opposite (Equation (7.29)) (the counterclockwise one is positive, the clockwise one is negative). We also know, however, that in the tensor sign convention, the shear stresses on two perpendicular planes, which necessarily are of opposite shear sense, must have the same sign (Equation (7.56); Figure 7.14*D*, *E*). Thus, unavoidably, the shear stress sign convention for the Mohr circle is inconsistent with the convention for the stress tensor because of the different way in which the signs of these components are determined. The need to shift from one convention to the other, when plotting or determining stress tensor components on a Mohr circle, is a common source of error, which we discuss in Subsection 7.5(iv).

We use the sign conventions and notation for the Mohr circle in most discussions of stress in this book, although in a few instances, especially in the Boxes in Chapters 16 and 18 concerning the relationships between stress and strain, the stress tensor notation is more convenient. Unfortunately, other sign conventions also are in common use, and strictly for reference, we discuss and summarize all these vexing conventions in Box 7-4. Tables 7.2 and 7.3 present a summary of the Mohr circle notation we use for discussing stress, and Table 7.4 at the end of the chapter summarizes all the notation we use for stress.

iii. Stress Tensor Components in Two Dimensions

It is commonly convenient to analyze a three-dimensional problem in two dimensions. Not only is the amount of algebra reduced, but two-dimensional problems are particularly easy to graph and solve on a Mohr diagram. A two-dimensional analysis, however, is possible only if the plane of analysis is a principal plane and therefore if the planes on which we want to determine the stress are parallel to a principal axis. Most commonly, the plane of analysis is the $(\hat{x}_1, \hat{x}_3)$ plane, with the $\hat{x}_2$ axis perpendicular to the plane of analysis and parallel to the planes on which we want to determine the stress (Figure 7.15*A*). It is not necessary to know the orientation of the principal axes *a priori*, but from the geometry of the problem, we must have some reason to know the plane in which those axes should lie. Under those circumstances, we adopt a coordinate system x_k $(k = 1{:}3)$, such that $x_2 = \hat{x}_2$ (the intermediate principal stress axis), whereupon $\sigma_{22} = \hat{\sigma}_2$. The x_2-plane is then the $\hat{x}_2$-plane and thus is a principal plane, which is an appropriate plane in which to analyze the problem in two dimensions. The shear stress components on the $(x_2 = \hat{x}_2)$-plane must be zero because it is a principal plane, so $\sigma_{21} = \hat{\sigma}_{21} = 0$ and $\sigma_{23} = \hat{\sigma}_{23} = 0$. Because of the symmetry of the stress tensor (Equations (7.56)), two of the other shear stress components must also be zero ($\sigma_{12} = 0$; $\sigma_{32} = 0$). The matrix of stress components must therefore include at least one row and one column in which the two shear stress components are zero. Thus, in tensor notation, the stress tensor $\boldsymbol{\sigma}$ is defined by its tensor components σ_{kl} according to

$$\boldsymbol{\sigma} : \sigma_{kl} = \begin{bmatrix} \sigma_{11} & 0 & \sigma_{13} \\ 0 & \hat{\sigma}_{22} & 0 \\ \sigma_{31} & 0 & \sigma_{33} \end{bmatrix} \tag{7.57}$$

All the nonzero stress components are shown in Figure 7.15*A*, and all except $\sigma_{22} = \hat{\sigma}_{22} \equiv \hat{\sigma}_2$ lie in the (x_1, x_3)-plane. Thus all of the stress components with a "2" as one of its subscripts are either zero or do not affect the stress on the planes parallel to $x_2 = \hat{x}_2$. Under these conditions, the components of the stress tensor that lie in the (x_1, x_3)-plane completely define the surface stress on any plane parallel to $x_2 = \hat{x}_2$ (Figure 7.15*B*). This is the reason we are justified in reducing such a problem to a two-dimensional analysis.

We can obtain the matrix of components for the two-dimensional stress tensor simply by eliminating from the matrix in Equation (7.57) all components having a "2" as one of the subscripts, leaving

$$\boldsymbol{\sigma}\text{: } \sigma_{kl} = \begin{bmatrix} \sigma_{11} & \sigma_{13} \\ \sigma_{31} & \sigma_{33} \end{bmatrix} \tag{7.58}$$

The two-dimensional stress tensor expressed in principal coordinates is obtained from the second matrix in Equation (7.53) by deleting all stress components that have a "2" as a subscript.

$$\boldsymbol{\sigma}\text{: } \sigma_{kl} = \hat{\sigma}_{kl} = \begin{bmatrix} \hat{\sigma}_1 & 0 \\ 0 & \hat{\sigma}_3 \end{bmatrix} \tag{7.59}$$

BOX 7-4 Sign Conventions Galore: A Cautionary Note

In nearly all of this book, we use the geologic Mohr circle sign convention, which assigns compressive normal stress components and counterclockwise shear stress components a positive sign, and tensile normal stress components and clockwise shear stress components a negative sign. Recall, however, that this sign convention for the shear stress components is nonunique because any shear stress can be clockwise or counterclockwise, depending on the direction from which the shear couple is viewed (see Section 7.1(iv)). In Section 7.5, we introduce the geologic tensor sign convention for stress. This convention provides a unique definition of sign for all stress components, which is why all mathematically analytic treatments of stress use the tensor sign convention. The tensor sign convention, however, cannot be used to plot stresses on the Mohr diagram, because the tensor signs of the shear stresses are incompatible with the Mohr diagram. Thus we are stuck with at least these two different sign conventions, both of which are necessary in different circumstances.

It is important to be aware, however, that there are still other sign conventions. In particular, the engineering sign convention is the opposite of the geologic convention for both the Mohr circle and the tensor components. The **engineering Mohr circle sign convention** assigns positive values to tensile stresses and to clockwise shear stresses, and the **engineering tensor sign convention** assigns to a stress component the sign of the associated traction that acts on the positive side of a coordinate surface.[4] This results in tensile stresses being positive and shear stresses having the opposite sign from that of the geologic tensor sign convention.

This convention presumably arose because engineers often test materials in tension, and like everyone else, they find it easier to deal with positive numbers. The engineering convention, however, has become the standard not only in engineering, but also in most analytic applications of continuum mechanics, which include analytical structural geology, geophysics, and physics. As a result, both the geologic and the engineering sign conventions are common in the geologic literature.

Thus, there are four different sign conventions for stress: the geologic and engineering sign conventions, respectively, for the Mohr circle and for the stress tensor. The existence of four different sign conventions that are all in common use in different contexts provides an unending source of confusion and vexation, not the least because the literature is not always explicit as to which sign convention is being used. As if this is not confusing enough, at least one source uses the geologic convention for normal stress and the engineering convention for shear stress! Although it is not necessary to master all these conventions for the level at which this book is written, it is at least important to be aware of their existence because of the problems and confusion that can arise. We mention the engineering conventions here for the sake of completeness and for future reference, but we do not use them at all in this book.

For easy reference, Table 7-4.1 summarizes these four different sign conventions.

[4]More generally, for the engineering tensor sign convention, if the outward unit normal to a coordinate surface and a traction component acting on that surface point in the same coordinate senses (i.e., both point in a positive coordinate direction or both point in a negative coordinate direction), then the stress component associated with that traction component is positive; if the outward unit normal and the traction component point in opposite coordinate senses, then the associated stress component is negative. For the geologic tensor sign convention, the signs are exactly the opposite.

TABLE 7-4.1 Sign Conventions for Stress Components

Convention	Normal Stress	Sign	Shear Stress	Sign
Geologic Mohr Circle Convention	Compressive	+	Counterclockwise	+
	Tensile	−	Clockwise	−
Geologic Stress Tensor Convention	Compressive: Normal traction component on the *negative* side of a coordinate surface points in a positive coordinate direction	+	Shear traction component on the *negative* side of a coordinate surface points in a positive coordinate direction	+
	Tensile: Normal traction component on the *negative* side of a coordinate surface points in a negative coordinate direction	−	Shear traction component on the *negative* side of a coordinate surface points in a negative coordinate direction	−
Engineering Mohr Circle Convention	Tensile	+	Clockwise	+
	Compressive	−	Counterclockwise	−
Engineering Stress Tensor Convention	Tensile: Normal traction component on the *positive* side of a coordinate surface points in a positive coordinate direction	+	Shear traction component on the *positive* side of a coordinate surface points in a positive coordinate direction	+
	Compressive: Normal traction component on the *positive* side of a coordinate surface points in a negative coordinate direction	−	Shear traction component on the *positive* side of a coordinate surface points in a negative coordinate direction	−

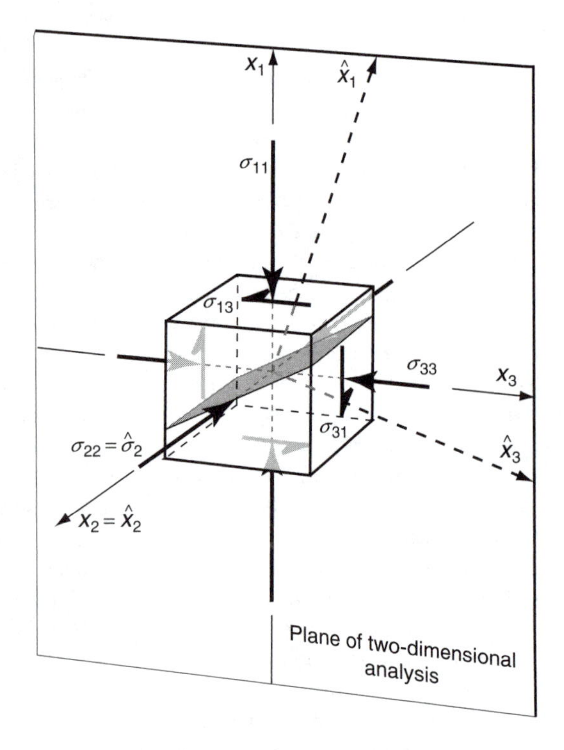

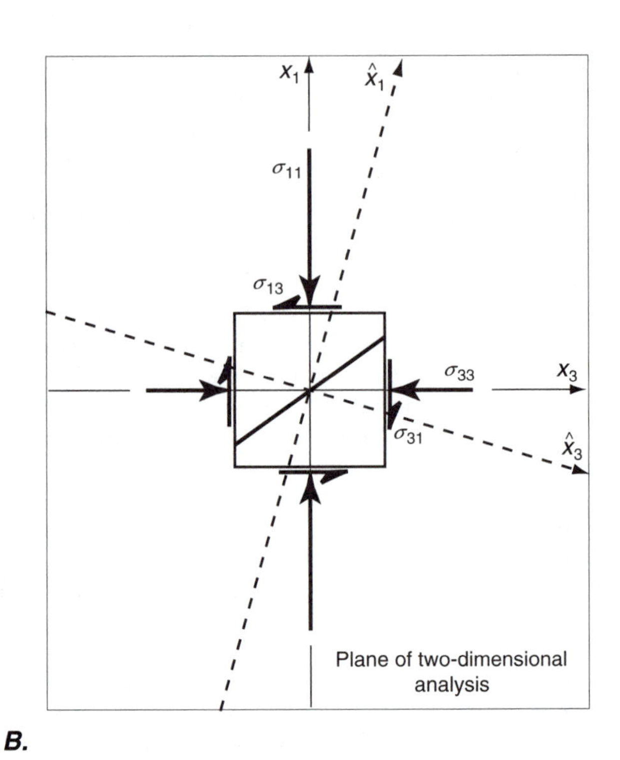

FIGURE 7.15 Geometry of stress in three dimensions that permits an analysis in two dimensions. *A.* A three-dimensional diagram showing one coordinate axis, for example, x_2, parallel to one of the principal coordinates, for example, $\hat{x}_2$. In this case, the (x_1, x_3) coordinate plane contains the principal axes $\hat{x}_1$ and $\hat{x}_3$, and one can make a two-dimensional analysis of the stress in the (x_1, x_3)-plane. *B.* Two-dimensional diagram showing the relevant geometry of *A*.

Equations (7.44) and (7.46) allow us to calculate the normal and shear stresses on any orientation of plane if we know the values of the principal stresses. But what if we do not know the principal stresses? It is most convenient to use the stress tensor sign convention to derive a generalization of these equations, which we give in Box 7-5. The resulting equations allow us to find the normal and shear stresses on any orientation of plane if we know the tensor stress components for the state of stress in any particular coordinate system. Moreover, from those equations, we can derive the equations that give the orientations and values of the principal stresses (see Box 7-5).

iv. The Stress Tensor and the Mohr Circle

It often proves useful to plot stress tensor data on the Mohr diagram. Intrinsically, this is not difficult to do, but it requires considerable care, first to deal with the incompatibility between the unique tensor sign convention for shear stresses and the nonunique Mohr circle sign convention, and second to change from one sign convention to the other before plotting the stress components. We discuss the procedures for dealing with these problems in this subsection. See Appendix 7B for an example of the application of the points covered in this subsection.

Most problems involving the Mohr circle are analyzed in two dimensions, so we focus on that aspect. To eliminate the ambiguity in the nonunique Mohr circle shear stress sign convention, we must establish a convention for viewing the geometry of the stress components and for constructing a two-dimensional diagram of the problem. To that end, we state the following:

Convention. The two-dimensional diagram of the stress components must be constructed so that there is a clockwise sense of rotation from the positive coordinate axis parallel to the largest normal stress component, toward the positive coordinate axis parallel to the smallest normal stress component.

If the plane of two-dimensional analysis is the $(\hat{x}_1, \hat{x}_3)$-plane, which contains the principal stresses $\hat{\sigma}_1$ and $\hat{\sigma}_3$, the effect of this convention is that the positive direction of the intermediate principal axis $\hat{x}_2$ is always pointing toward the viewer. If the plane of analysis is the $(\hat{x}_1, \hat{x}_2)$-plane or the $(\hat{x}_2, \hat{x}_3)$-plane, the convention requires, respectively, that the positive direction of the $\hat{x}_3$ or the $\hat{x}_1$ axis point away from the viewer. This convention therefore fixes the direction from which we view the diagram of stress components. Because the sign of the shear stress in the Mohr circle sign convention depends on the direction from which the shear couple is viewed, and because the tensor sign convention is independent of the viewing direction, this convention for viewing the coordinate axes ensures that the transfer of shear stress components from the tensor sign convention to the Mohr circle sign convention is consistent and will give the correct answer.

The same convention applies to the normal stress components in any given coordinate system, not just the principal coordinates, although the plane of analysis still must be perpendicular to one of the principal axes (see the discussion in the preceding subsection). Let us assume that we know the values of the tensor components for the two-dimensional stress as in Equation (7.58), and furthermore, let us assume that $\sigma_{11} > \sigma_{33}$. We construct Figure 7.15*B* so that x_1 is positive up and x_3 is positive to the right, giving a clockwise rotation from the positive coordinate axis x_1 and the largest normal stress σ_{11} toward the positive coordinate axis x_3 and the smallest normal stress σ_{33}. Had the relative values of the components been $\sigma_{33} > \sigma_{11}$, then we would have had to view the diagram in Figure 7.15*B* from the opposite side, so that x_3 would be positive to the left and x_1 would be positive up. This would give a clockwise rotation from the positive coordinate axis x_3 and the largest normal stress σ_{33} toward the positive coordinate axis x_1 and the smallest normal stress σ_{11}.

In solving problems using the Mohr circle when the stress components are given in the tensor sign convention, we recommend the following procedure to ensure that the signs of the stress components are correct (see the example given in Appendix 7B):

1. Draw a diagram of the coordinate square in physical space, with the coordinate axes oriented relative to each other according to the preceding Convention and with the stress components appropriately oriented according to the signs of the given components and the tensor sign convention (Figure 7.15*B*).
2. Make a table listing the appropriate tensor symbol for each stress component in the first column and the value of that component in the tensor sign convention in the second column.
3. In a third column opposite each component, list its value according to the Mohr circle sign convention.
    - Normal stress components have the same sign as the tensor components.
    - Determine the correct sign for the shear stress components by using the diagram of the coordinate square. A shear stress component in the Mohr circle sign convention is positive if it is a counterclockwise couple on the coordinate square, negative if it is a clockwise couple.
4. Finally, plot the values of the components thus determined on the Mohr diagram.

Obviously, the reverse of this procedure should be used to transfer solutions from the Mohr circle to tensor components.

BOX 7-5 Derivation of Principal Stresses in Two Dimensions

In order to show that principal stresses must exist for any stress tensor, we need to derive equations analogous to Equations (7.44) but expressed in terms of the stress components in a general coordinate system. The principles of the derivation are the same as those used to obtain Equations (7.44), so we merely outline the procedure here.

We limit ourselves to considering planes parallel to $\hat{x}_2$ so that our diagrams of physical space will necessarily include the $\hat{x}_1$ and $\hat{x}_3$ principal axes. We consider an infinitesimal coordinate square with its faces perpendicular to arbitrary coordinate axes x_1 and x_3 and with both normal and shear stresses acting on its faces (Figure 7-5.1*A*). The normal **n** to the plane *P* on which we want to determine the normal stress and shear stress components (σ_n, σ_s) makes an angle α with x_1. We draw all stress components as positive components using the tensor sign convention, and we construct the angle α to be a positive (counterclockwise) angle.

We make a free-body diagram of the shaded triangular element (Figure 7-5.1*B*) and construct a diagram of the forces acting on the triangular element (Figure 7-5.1*C*), where

$$F_{1n} = \sigma_{11}A_1 \quad F_{3n} = \sigma_{33}A_3 \quad F_n = \sigma_n A \qquad (7\text{-}5.1)$$
$$F_{1s} = \sigma_{13}A_1 \quad F_{3s} = \sigma_{31}A_3 \quad F_s = \sigma_s A$$

Here, the first subscript on the symbol for force indicates the normal to the plane on which the particular force acts, and the second subscript "*n*" or "*s*" designates a normal or shear force. The areas of the triangle are related by (Figure 7-5.1*B*):

$$A_1 = A\cos\alpha \quad A_3 = A\sin\alpha \qquad (7\text{-}5.2)$$

We resolve each of the four force vectors acting on the coordinate planes into two components parallel to F_n and F_s, respectively (dashed arrows, Figure 7-5.1*C*), and then we require equilibrium by setting the sum of the forces in each of these two directions equal to zero. Then, expressing forces in terms of stresses with Equations (7-5.1), rearranging the equations to isolate σ_n and σ_s on the left, substituting Equations (7-5.2) for A_1 and A_3, dividing through by *A* to eliminate it from the equations, and using the symmetry condition of the stress tensor Equations (7.56), we find that

$$\sigma_n = \sigma_{11}\cos^2\alpha - 2\sigma_{13}\sin\alpha\cos\alpha + \sigma_{33}\sin^2\alpha \qquad (7\text{-}5.3)$$

$$\sigma_s = (\sigma_{11} - \sigma_{33})\sin\alpha\cos\alpha + \sigma_{13}(\cos^2\alpha - \sin^2\alpha) \qquad (7\text{-}5.4)$$

(continued)

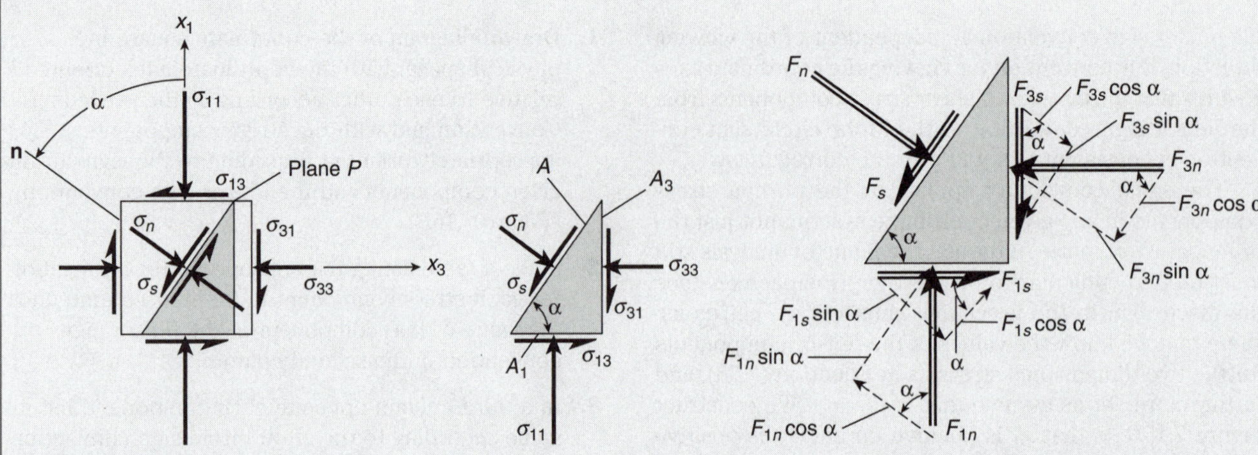

FIGURE 7-5.1 Geometric relationships used to deduce the transformation equations for components of two-dimensional stress. *A*. An infinitesimal coordinate square in an arbitrary coordinate system showing the components of stress on the coordinate surfaces and on an arbitrary plane *P*. All quantities are shown as positive quantities. *B*. The traction components acting on the exterior surfaces of the shaded triangle in *A*. Tractions are labeled with the associated stress components, because we want the transformation equations in terms of stress. Sign differences are accounted for in formulating the equations. Areas A, A_1, and A_3 can be thought of as the areas of the sides of a triangular prism of unit dimension normal to the diagram. *C*. Forces acting on the isolated triangular element, showing their components parallel and perpendicular to plane *P*.

We now wish to determine the orientation α_0 of planes on which σ_n is a maximum or a minimum. To this end, we differentiate Equation (7-5.3) with respect to α and set the result equal to zero:

$$\frac{d\sigma_n}{d\alpha} = 0$$
$$= (\sigma_{11} - \sigma_{33}) \sin \alpha_0 \cos \alpha_0 + \sigma_{13} (\cos^2 \alpha_0 - \sin^2 \alpha_0) \quad (7\text{-}5.5)$$

where we have used α_0 instead of α to indicate that the angle is no longer arbitrary. The right sides of Equations (7-5.5) and (7-5.4) are identical. This means that the condition for σ_n to be extreme is also the condition for σ_s to be zero, confirming that the shear stresses on any principal plane must be zero.

We solve Equation (7-5.5) for α_0 by using the trigonometric identities:

$$\begin{aligned} \cos 2\alpha_0 &= \cos^2 \alpha_0 - \sin^2 \alpha_0 \\ \sin 2\alpha_0 &= 2 \sin \alpha_0 \cos \alpha_0 \\ \tan 2\alpha_0 &= \frac{\sin 2\alpha_0}{\cos 2\alpha_0} \\ \tan 2\alpha_0 &= \tan 2(\alpha_0 + 90^\circ) \end{aligned} \quad (7\text{-}5.6)$$

The result is

$$\tan 2\alpha_0 = \tan 2(\alpha_0 + 90^\circ) = \frac{-2\sigma_{13}}{(\sigma_{11} - \sigma_{33})} \quad (7\text{-}5.7)$$

Thus Equations (7-5.4) and (7-5.5) show that for planes on which σ_n is a maximum or a minimum, σ_s is zero. Equations (7-5.7) show that there are two such planes. The normals to these planes make angles of α_0 and $(\alpha_0 + 90^\circ)$ with x_1. The planes are therefore perpendicular to each other (Figure 7-5.2), and their normals are the orientations of the principal axes of stress.

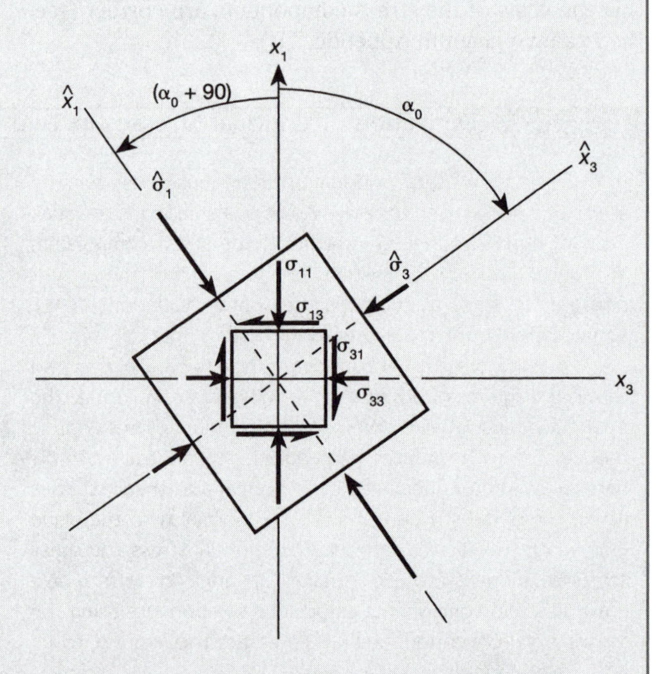

FIGURE 7-5.2 Stress components for a specific state of stress shown on infinitesimal coordinate squares for a general coordinate system (x_1, x_3) and for the principal coordinate system ($\hat{x}_1$, $\hat{x}_3$). Both squares are infinitesimal, but they are drawn with different sizes for clarity. The angle α_0 is obtained from Equation (7-5.7).

TABLE 7.4 Notation for Stress[1]

Symbol	Definition
$\boldsymbol{\Sigma}$	Traction or surface stress (for tractions, a superscript is added in places to define the side of the plane on which the particular traction acts)
σ_n, σ_s	Normal and shear components, respectively, of surface stress or traction (for tractions, a superscript is added in places to define the side of the plane on which the particular traction component acts)
$\boldsymbol{\Sigma}_x$, $\boldsymbol{\Sigma}_y$, $\boldsymbol{\Sigma}_z$	Surface stresses acting on planes normal to the x, y, and z coordinate axes
$\boldsymbol{\sigma}$	Stress at a point represented by the stress ellipse in two dimensions and the stress ellipsoid in three dimensions; a second-rank tensor
Two dimensions: $\begin{bmatrix} \sigma_{xx} & \sigma_{xz} \\ \sigma_{zx} & \sigma_{zz} \end{bmatrix}$ Three dimensions: $\begin{bmatrix} \sigma_{xx} & \sigma_{xy} & \sigma_{xz} \\ \sigma_{yx} & \sigma_{yy} & \sigma_{yz} \\ \sigma_{zx} & \sigma_{zy} & \sigma_{zz} \end{bmatrix}$	Components of stress at a point, in the Mohr circle sign convention, defining the stress at a point in the $[x, (y), z]$ coordinate system. Each row contains the components of one of the surface stresses $\boldsymbol{\Sigma}_x$, $(\boldsymbol{\Sigma}_y)$, or $\boldsymbol{\Sigma}_z$, respectively. The first subscript in each symbol defines the coordinate axis that is normal to the coordinate plane on which the stress component or surface stress acts. The second subscript defines the coordinate axis to which the stress component is parallel (only two coordinates and two surface stresses are used for two-dimensional stress)
$\hat{\sigma}_k$ Two dimensions: $[\hat{\sigma}_1, \hat{\sigma}_3]$ Three dimensions: $[\hat{\sigma}_1, \hat{\sigma}_2, \hat{\sigma}_3]$	Principal stresses, where $k = 1, 3$ in two dimensions, and $k = 1, 2, 3$ in three dimensions, representing the maximum, (intermediate), and minimum principal stresses parallel to coordinate axes $\hat{x}_k : [\hat{x}_1, (\hat{x}_2), \hat{x}_3]$, respectively, and therefore perpendicular to the $\hat{x}_1$, $(\hat{x}_2)$, $\hat{x}_3$ coordinate planes (only two coordinates are used for two-dimensional stress)
$\boldsymbol{\Sigma}_k$	Surface stress or traction on surface normal to x_k axis, where $k = 1, 3$ for two dimensions and $k = 1, 2, 3$ for three dimensions (for tractions, a superscript is added in places to define the side of the plane on which the particular traction acts)
σ_{kl} Two dimensions: $= \begin{bmatrix} \sigma_{11} & \sigma_{13} \\ \sigma_{31} & \sigma_{33} \end{bmatrix}$ Three dimensions: $= \begin{bmatrix} \sigma_{11} & \sigma_{12} & \sigma_{13} \\ \sigma_{21} & \sigma_{22} & \sigma_{23} \\ \sigma_{31} & \sigma_{32} & \sigma_{33} \end{bmatrix}$	Tensor components of the stress at a point, in the tensor sign convention, defining the stress at a point in the $[x_1, (x_2), x_3]$ coordinate system. Each row contains the components of one of the surface stresses $\boldsymbol{\Sigma}_1$, $(\boldsymbol{\Sigma}_2)$, or $\boldsymbol{\Sigma}_3$, respectively. The first subscript in each symbol defines the axis normal to the coordinate plane on which the stress component acts. The second subscript defines the coordinate axis to which the stress component is parallel (k and $l = 1, (2), 3$) (only two values of the subscript are used for two-dimensional stress)
$\bar{\sigma}_n$	Mean normal stress—the average of the normal stress components. A scalar invariant of the stress tensor
$\sigma^{(Dif)}$	Differential stress—a positive scalar quantity equal to the difference between the maximum and minimum principal stresses $(\hat{\sigma}_1 - \hat{\sigma}_3)$
$\sigma_n^{(Dev)}$	Deviatoric normal stress, equal to the normal stress minus the mean normal stress $(\sigma_n - \bar{\sigma}_n)$
$\sigma_{kl}^{(Dev)}$	Deviatoric stress—equal to the stress tensor components with each of the normal stress components reduced by the mean normal stress, where k and $l = 1, (2)$, or 3 (only two values of the subscripts are used for two-dimensional stress)
$\sigma_n^{(Eff)}$	Effective normal stress, equal to the normal stress minus the pore fluid pressure $(\sigma_n - p_f)$
$\sigma_{kl}^{(Eff)}$	Effective stress components, equal to the components of the stress tensor with each of the normal stress components reduced by the pore fluid pressure, where k and $l = 1, (2)$, or 3 (only two values of the subscript are used for two-dimensional stress)

[1]Boldface type, with or without subscripts, indicates vectors and tensors. Italic type with subscripts indicates scalars or scalar components of vectors and tensors.

REFERENCES AND ADDITIONAL READINGS

Eringen, A. C. 1967. *Mechanics of Continua.* New York, Wiley.

Fung, Y. C. 1965. *Foundations of Solid Mechanics.* Englewood Cliffs, NJ, Prentice-Hall.

Hubbert, M. K. 1972. *Structural Geology.* New York, Hafner Publishing Co.

Jaeger, J. C. 1962. *Elasticity, Fracture, and Flow, Second Edition*, London, Methuen & Co. Ltd.

Means, W. D. 1976. *Basic Concepts of Stress and Strain for Geologists.* New York, Springer-Verlag.

Appendix 7-A

Illustrative Problem 1

For this illustrative problem, we assume the reader has read Section 7.2 on the characteristics of the Mohr circle; we illustrate how to use the Mohr circle to solve some common types of two-dimensional stress problems that are relevant to geology. We use only the geologic Mohr circle sign convention for stress in this example.

Consider a fault block that is 5 km thick and rests on a horizontal detachment associated with a listric normal fault. Figure 7-A.1*A* shows a perspective view of the fault block with the coordinate system (x, y, z) that we use for reference. We analyze the problem in two dimensions in the (x, z)-plane. Figure 7-A.1*B* shows a two-dimensional block representing a section through part of the detachment sheet. Such a block is sometimes called a "free-body diagram" because the surrounding material is removed and its action on the free block is represented by the tractions on the outside surfaces of the block. These tractions arise from the force of gravity (the overburden), the applied tectonic stress (which we assume to be an east-west horizontal tensile stress), and the frictional resistance to sliding on the detachment.

In order to determine the two-dimensional stress at an arbitrary point along the detachment surface, we draw the free-body diagram such that the point in question is at the bottom left corner of the block (Figure 7-A.1*B*). We then determine the stress components σ_{zz}, σ_{xx}, and $\sigma_{zx} = -\sigma_{xz}$ at this point. We consider each of these components in turn. The point at this corner of the block is shown enlarged in Figure 7-A.1*C*.

The vertical normal stress σ_{zz} is the overburden stress due to gravity, and it equals the weight per unit area of the overlying rock. This is

$$\sigma_{zz} = \rho g h \qquad (7\text{-A}.1)$$

where ρ is the mass density (mass per unit volume) of the rock, g is the magnitude of the gravitational acceleration, and h is the depth from the top surface. The stress is compressive and therefore positive.

The horizontal normal stress σ_{xx} is the sum of the horizontal compressive stress associated with the overburden and the tectonically applied stress, represented by a positive number T. Because we assume the rock has some finite strength, the part of σ_{xx} due to the overburden is some fraction $\kappa < 1$ of the vertical normal stress (in a fluid, however, $\kappa = 1$).[1] We assume, furthermore, that T is a tensile tectonic stress, constant with depth, that tends to stretch the fault block in an east-west direction. Thus

$$\sigma_{xx} = \kappa(\rho g h) - T \qquad (7\text{-A}.2)$$

where we subtract T because it is a positive number that represents a tensile (negative) stress. σ_{xx} could, in principal, be positive or negative, depending on the relative values of the overburden and of T. As a general rule for the Earth, a tectonically imposed tensile stress would not be constant with depth, and the component of stress due to the compressive overburden would be larger than any tectonically imposed tensile stress. Thus with the possible exception of the shallowest levels in the crust, the normal components of stress in the Earth are all compressive.

The frictional shear stress along the detachment σ_{zx} is given by the product of the coefficient of friction μ and the normal stress across the sliding surface, a relationship known as Amontons' second law of friction. Because the block is being pulled to the left, the shear stress is counterclockwise, and therefore it must be positive in the geologic Mohr circle sign convention:

$$\sigma_{zx} = \mu \sigma_{zz} = \mu(\rho g h) \qquad (7\text{-A}.3)$$

Because of the relationships between the shear stress components acting on perpendicular planes (Equation (7.29)), we now have a means of determining all three independent components of the two-dimensional stress for this problem.

In order to introduce definite numbers into the analysis, we adopt the following geologically reasonable values for the symbols in the equations:

$$\begin{aligned} &\rho = 2700 \text{ kg/m}^3 && h = 5000 \text{ m} \\ &g = 9.8 \text{ m/s}^2 && T = 10 \text{ MPa} \\ &\kappa = 0.45 && \\ &\mu = 0.5 && \end{aligned} \qquad (7\text{-A}.4)$$

[1] A means of evaluating limits for this constant, for the condition when the principal stresses are horizontal and vertical, is discussed in Section 9.4(ii); see Equation (9.7).

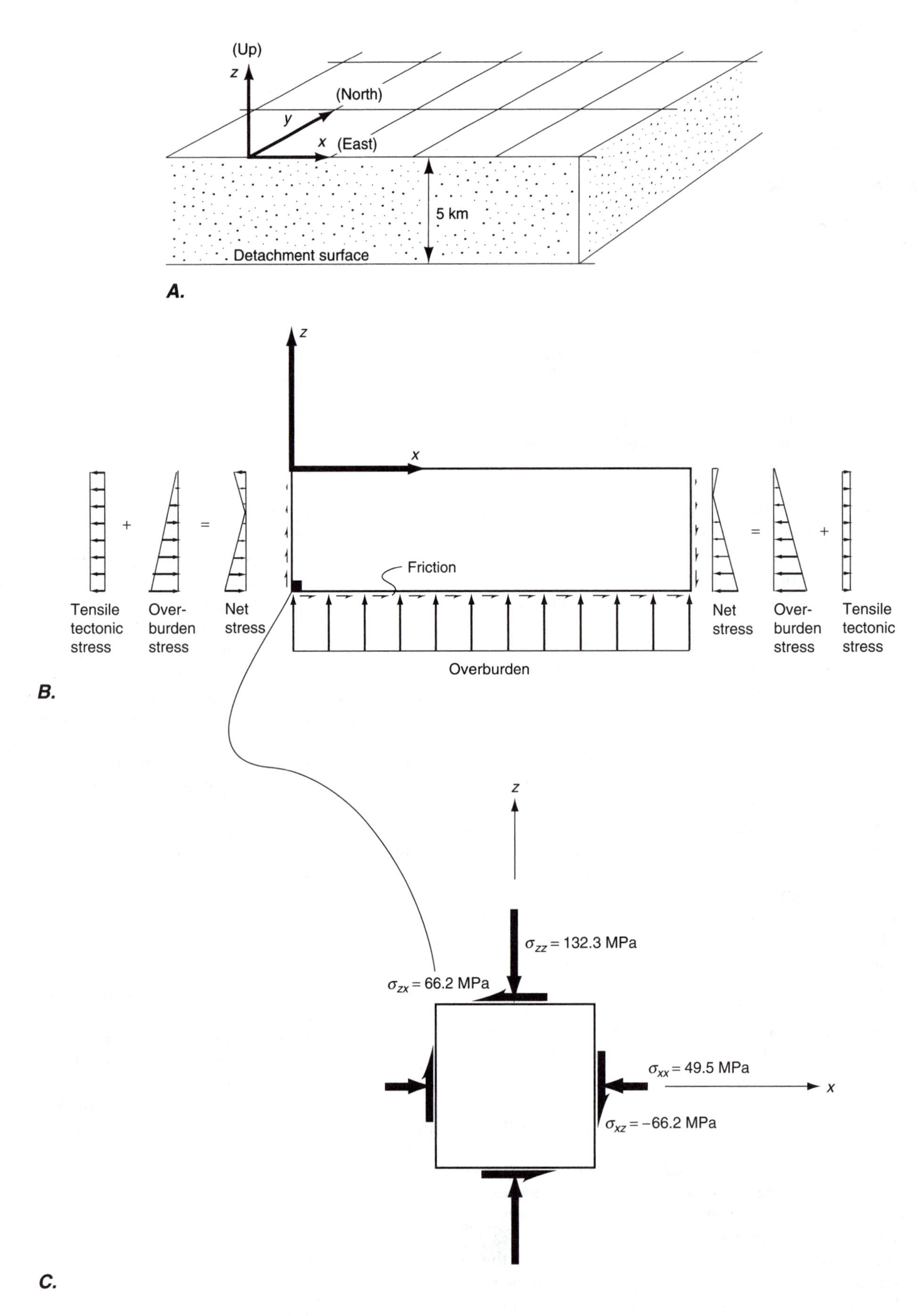

FIGURE 7-A.1 Simplified model of a fault block resting on a horizontal detachment and subjected to a tensile tectonic stress. *A*. A fault block bounded below by a horizontal detachment fault. The coordinate axes are used in the analysis. *B*. A free-body diagram of an isolated part of the fault block showing the tractions on the base due to the overburden stress and the frictional shear stress, and the tractions on the vertical sides due to the overburden, the added tensile tectonic stress, the sum of these two, and the shear stresses required by the symmetry of the stress tensor. *C*. The stresses on the coordinate square from the lower-left corner of the isolated body of the fault block in *B*.

Using these values with Equations (7-A.1) to (7-A.3) and using the relationships between shear stress components (Equations (7.29)), we obtain the following values for the components of the stress at the corner of the block:

$$\begin{aligned}\sigma_{xx} &= 49.5 \text{ MPa}\\ \sigma_{zz} &= 132.3 \text{ MPa}\\ \sigma_{zx} &= -\sigma_{xz} = 66.2 \text{ MPa}\end{aligned} \quad (7\text{-A}.5)$$

The assumptions we make here are admittedly oversimplified compared to what we expect in the real Earth, but they suffice for illustrative purposes. A similar but more detailed analysis is applied in Section 9.11 (Figure 9.22) to understand thrust faults (see also Box 9-2).

In the following discussion, we refer by number to the properties of the Mohr circle that we discussed in Section 7.2, and we do not duplicate those discussions here.

Question A1

Construct the Mohr circle for the two-dimensional stress that acts at this point on planes normal to the (x, z)-plane. Note that these planes are all parallel to the y axis and perpendicular to our two-dimensional diagram. These planes and the pertinent stress components are shown in Figure 7-A.1*C*.

Procedure

1. We have calculated the stress components using the Mohr circle sign convention, so it is simply a matter of plotting the information on the Mohr diagram. The two pairs of stress components $(\sigma_{xx}, \sigma_{xz}) = (49.5, -66.2)$ MPa and $(\sigma_{zz}, \sigma_{zx}) = (132.3, 66.2)$ MPa are the normal and shear components of the two surface stresses that act, respectively, on the x coordinate plane and the z coordinate plane. Each pair plots as a point on the Mohr diagram (Figure 7-A.2*A*).

2. A line connecting these two points on the Mohr diagram must be a diameter of the Mohr circle (property 3v), and the point where the diameter intersects the σ_n axis is the center of the circle (property 5i). The circle can then be constructed on the Mohr diagram with a drafting compass.

Discussion

As an alternative to plotting the diameter of the Mohr circle, we can use Equations (7.33) to calculate the center of the Mohr circle, which is the mean normal stress, and its radius. From these two quantities, the whole circle can be constructed

$$\overline{\sigma}_n = \frac{\sigma_{xx} + \sigma_{zz}}{2} = \frac{49.5 + 132.3}{2} = 90.9 \text{ MPa} \quad (7\text{-A}.6)$$

$$\begin{aligned} r &= 0.5[(\sigma_{xx} - \sigma_{zz})^2 + (2\sigma_{xz})^2]^{0.5}\\ &= 0.5[(49.5 - 132.3)^2 + 4(66.2)^2]^{0.5}\end{aligned}$$

$$r = 78.1 \text{ MPa} \quad (7\text{-A}.7)$$

Question A2

What are the values and orientations of the principal stresses in the (x, z)-plane? Draw a diagram of physical space showing the relationship between the (x, z) coordinates and the principal coordinates in that plane.

Procedure

The values of the principal stresses are read from Figure 7-A.2*A* at the points where the Mohr circle intersects the σ_n axis (property 2i). These values can also be obtained by adding and subtracting the magnitude of the radius of the Mohr circle (Equation (7-A.7)) to and from the value of $\overline{\sigma}_n$, the center of the circle (Equation (7-A.6)). We label the values in decreasing order such that the stress in principal coordinates is given by

$$\begin{aligned}\hat{\sigma}_1 &= \overline{\sigma}_n + r = 169.0 \text{ MPa}\\ \hat{\sigma}_3 &= \overline{\sigma}_n - r = 12.8 \text{ MPa}\end{aligned} \quad (7\text{-A}.8)$$

The orientations of the principal axes are also determined from Figure 7-A.2*A*, using properties 3i and 3ii of Section 7.2 and the same orientation of axes used in Figure 7-A.1. We recommend tabulating the measurements as shown in the following tables to avoid confusion and to ensure proper observance of the conventions. Measurements on the Mohr diagram are shown in Figure 7-A.2*A*, and the corresponding measurements in physical space are shown in Figure 7-A.2*B*.

Angular Relations on the Mohr Diagram

Angle	Sense of angle	Measured from	Measured to
$2\alpha_1 = 122°$	Counterclockwise	$(\sigma_{xx}, \sigma_{xz}) = (49.5, -66.2)$	$(\hat{\sigma}_1, 0) = (169.0, 0)$
$2\alpha_2 = -58°$	Clockwise	$(\sigma_{xx}, \sigma_{xz}) = (49.5, -66.2)$	$(\hat{\sigma}_3, 0) = (12.8, 0)$

Angular Relations in Physical Space

Angle	Sense of angle	Measured from	Measured to
$\alpha_1 = 61°$	Counterclockwise	x (or σ_{xx})	$\hat{x}_1$ (or $\hat{\sigma}_1$)
$\alpha_2 = -29°$	Clockwise	x (or σ_{xx})	$\hat{x}_3$ (or $\hat{\sigma}_3$)

Discussion

Measurements on the Mohr diagram and in physical space must always be made from the known to the unknown. We know initially the surface stress components on the x-plane, $(\sigma_{xx}, \sigma_{xz}) = (49.5, -66.2)$ MPa, and we know initially the orientation of the x axis. On the Mohr diagram, the angle $2\alpha_1 = 122°$ is measured from the radius at the initially known surface stress $(\sigma_{xx}, \sigma_{xz})$ to the radius at the unknown surface stress $(\hat{\sigma}_1, 0)$ (property 3i). It is twice the angle $\alpha_1 = 61°$ in physical space measured from the initially known coordinate axis x to the

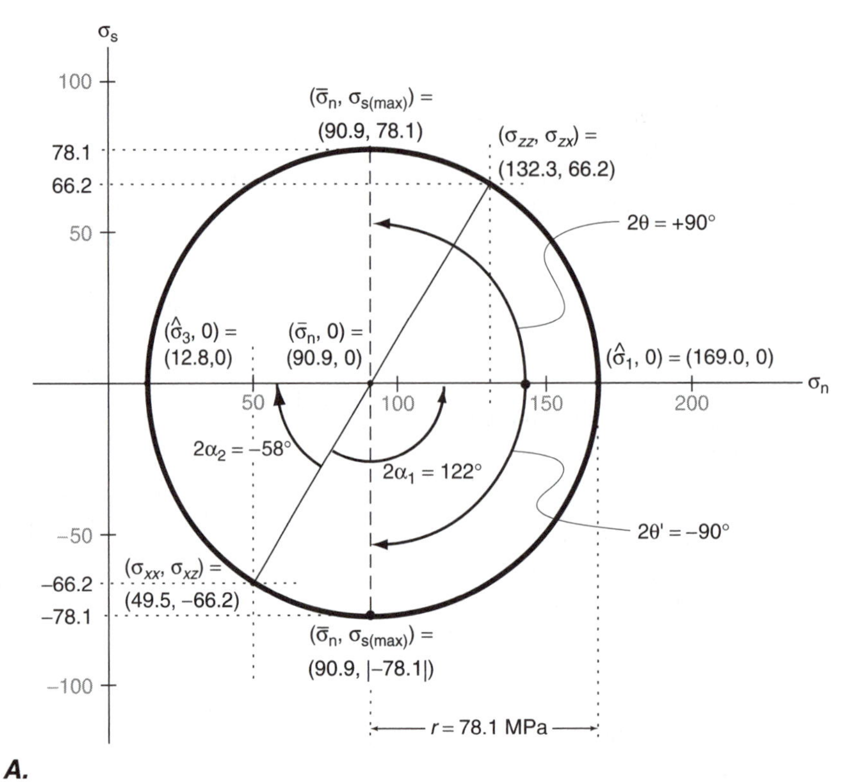

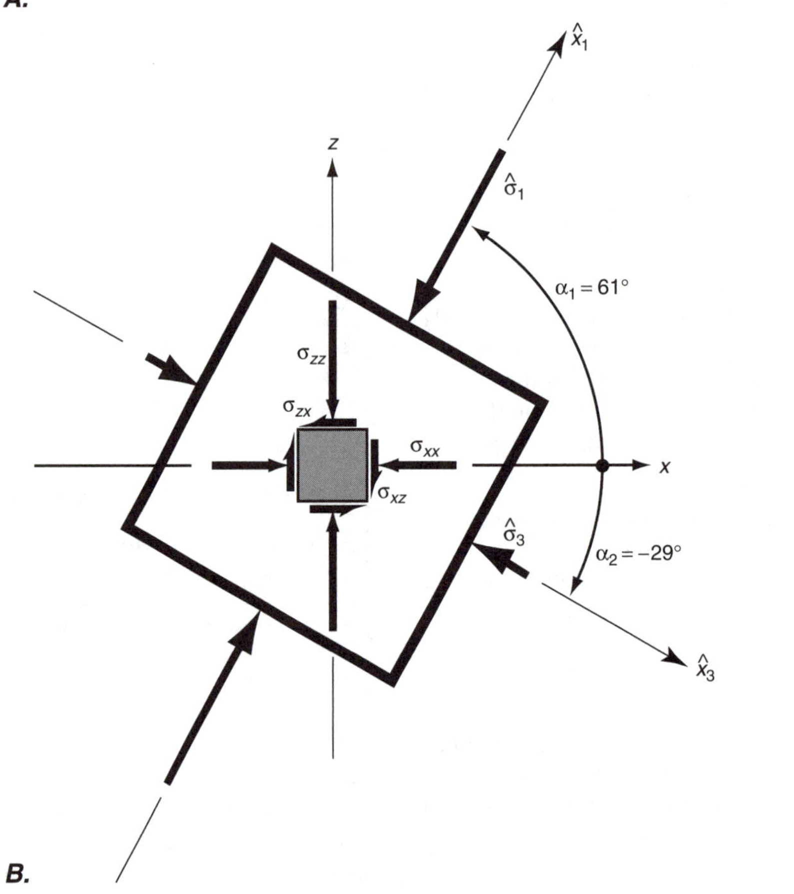

FIGURE 7-A.2 Construction of the Mohr circle for the problem illustrated in Figure 7-A.1. *A.* Mohr circle construction showing the plot of the calculated stresses at opposite ends of the oblique diameter, with the angles that define the location of the principal stresses and the maximum shear stresses. *B.* Determining the orientation of principal stresses in physical space from the angles on the Mohr circle in *A.* *C, D.* Determining the orientation in physical space of the two planes of maximum shear stress from the angles on the Mohr circle in *A.* *E.* Relation in physical space between the principal stresses and the planes of maximum shear stress.

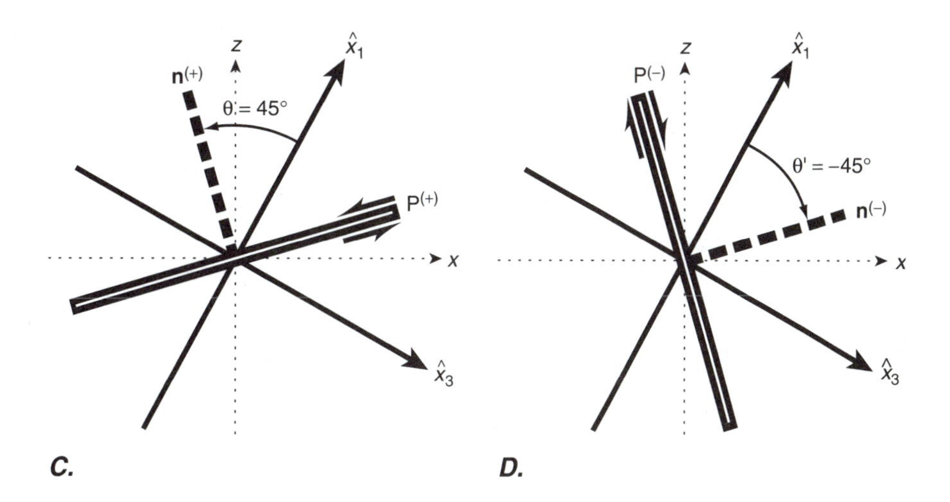

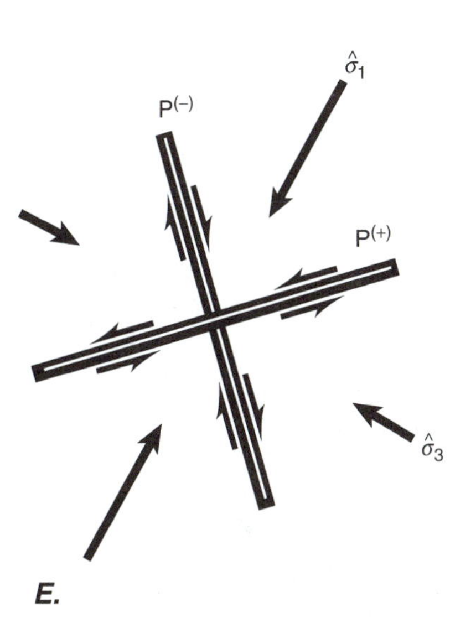

FIGURE 7-A.2 (continued)

unknown principal coordinate axis $\hat{x}_1$ (property 3ii). These two axes are the normals to the respective coordinate planes on which the normal stress components σ_{xx} and $\hat{\sigma}_1$ act.

Once we know the orientation of the $\hat{x}_1$ axis, we also know the orientation of the $\hat{x}_3$ axis, because it must be perpendicular to $\hat{x}_1$. We can also determine the orientation of the $\hat{x}_3$ axis from the Mohr circle by measuring the angle $2\alpha_2 = -58°$ from $(\sigma_{xx}, \sigma_{xz}) = (49.5, -66.2)$ MPa clockwise to $(\hat{\sigma}_3, 0)$, and then transferring half that angle $\alpha_2 = -29°$ to the diagram of physical space (Figure 7-A.2 *A*, *B*).

We could use the z axis as a reference in the same way, in which case the corresponding angles on the Mohr circle would be measured from the radius at $(\sigma_{zz}, \sigma_{zx}) = (132.3, 66.2)$ MPa, and the corresponding angles in physical space would be measured from the z axis.

We label the principal axes such that $\hat{x}_1$ and $\hat{x}_3$ are parallel to, respectively, the maximum and minimum compressive stresses (Figure 7-A.2*B*). These axes are necessarily perpendicular, because they are the normals to planes whose stress components plot at opposite ends of a diameter of the Mohr circle (Figure 7-A.2*A*) (properties 3v).

Note that at this corner of the fault block, the principal stresses are not horizontal and vertical; they are both inclined. This is necessary because the horizontal and vertical planes have a shear stress on them and thus cannot be principal planes.

Question A3

What is the mean normal stress for the two-dimensional stress in the (x, z)-plane?

Procedure

The mean normal stress for the two-dimensional state of stress that acts on the set of planes perpendicular to the (x, z)-plane (and parallel to the y axis) can be determined from the Mohr diagram (property 5i): Simply read off the value of the normal stress at the center of the Mohr circle (Figure 7-A.2). Alternatively, it can be calculated using the first Equation (7.32):

$$\overline{\sigma}_n = \frac{\hat{\sigma}_1 + \hat{\sigma}_3}{2} = \frac{169.0 + 12.8}{2} = 90.9 \text{ MPa} \quad (7\text{-A.9})$$

The first Equation (7.33), which we used in Equation (7-A.6), gives the same result.

Discussion

In the two-dimensional case, $\overline{\sigma}_n$ is the center of the Mohr circle, and it is the value of the normal stress on the planes of maximum shear stress. It should be apparent from the geometry of the circle that the mean normal stress is the mean of any pair of normal stresses that occur at opposite ends of any diameter of Mohr's circle, and these are the normal stresses that act on perpendicular planes in physical space. This relationship illustrates why the mean normal stress is a scalar invariant of the stress.

Question A4

What are the maximum absolute values of the shear stress acting on planes normal to the (x, z)-plane, and what are the orientations of the planes on which these values occur?

Procedure

The maximum absolute value of the shear stress is the length of the radius of the Mohr circle, and the planes on which this shear stress occurs are the planes on which the mean normal stress (the center of the Mohr circle) is the normal stress (properties 4 and 5i). The magnitude of the shear stress is read directly from the $\hat{\sigma}_1 = \hat{\sigma}_3$ Mohr circle (Figure 7-A.2*A*):

$$|\sigma_s|_{(\max)} = 78.1 \text{ MPa} \quad (7\text{-A.10})$$

This value is also half the difference in the principal stresses (the second Equation (7.32)):

$$|\sigma_s|_{(\max)} = \frac{\hat{\sigma}_1 - \hat{\sigma}_3}{2} = \frac{169.0 - 12.8}{2} = 78.1 \text{ MPa} \quad (7\text{-A.11})$$

which is the same result as from Equation (7-A.7).

The orientations of the normals to the planes of maximum shear stress are determined from the angles on the Mohr circle (Figure 7-A.2*A*) as tabulated in the following tables; they are plotted in a diagram of physical space in Figure 7-A.2*C*, *D*, and *E*:

Angular Relationships on the Mohr Diagram

Angle	Sense of angle	Measured from	Measured to
$2\theta = +90°$	Counterclockwise	$(\hat{\sigma}_1,0) = (169.0, 0)$	(90.9, 78.1)
$2\theta' = -90°$	Clockwise	$(\hat{\sigma}_1,0) = (169.0, 0)$	(90.9, −78.1)

Angular Relations in Physical Space

Angle	Sense of angle	Measured from	Measured to
$\theta = +45°$	Counterclockwise	$\hat{x}_1$	$\mathbf{n}^{(+)}$
$\theta' = -45°$	Clockwise	$\hat{x}_1$	$\mathbf{n}^{(-)}$

where $\mathbf{n}^{(+)}$ and $\mathbf{n}^{(-)}$ are, respectively, the normals to the planes $P^{(+)}$ and $P^{(-)}$ on which the maximum shear stress is, respectively, positive and negative in the Mohr circle sign convention (Figure 7-A.2*C*, *D*, *E*).

Discussion

The value of the maximum shear stress is simply given by the length of the radius of the appropriate Mohr circle. In the (x, z)-plane, $|\sigma_s|$ is a maximum on the Mohr circle at those points where the radius is normal to the σ_n axis. From Figure 7-A.2*A*, $2\theta = -2\theta' = 90°$. Thus in physical space, the normals to the planes of maximum shear stress are at $\theta = +45°$ (Figure 7-A.2*C*) and $\theta' = -45°$ (Figure 7-A.2*D*) from $\hat{\sigma}_1$ and $\hat{x}_1$. The planes of maximum shear stress are therefore perpendicular to each other. They are called the **conjugate planes of maximum shear stress**, and the relationship between these planes and the principal stresses is shown in Figure 7-A.2*E*.

The magnitude of the radius of the Mohr circle can be calculated directly from the components of stress on any two perpendicular surfaces, as illustrated by Equations (7-A.7) and (7-A.11) (see the second Equations (7.32) and (7.33)). This result illustrates why the radius is an invariant of the stress.

Question A5

What are the values of the normal and shear components of the surface stress that act on each of the following planes?

Plane A is perpendicular to the (x, z)-plane, and its normal is at an angle $\alpha_A = 35°$ from x (Figure 7-A.3*A*).

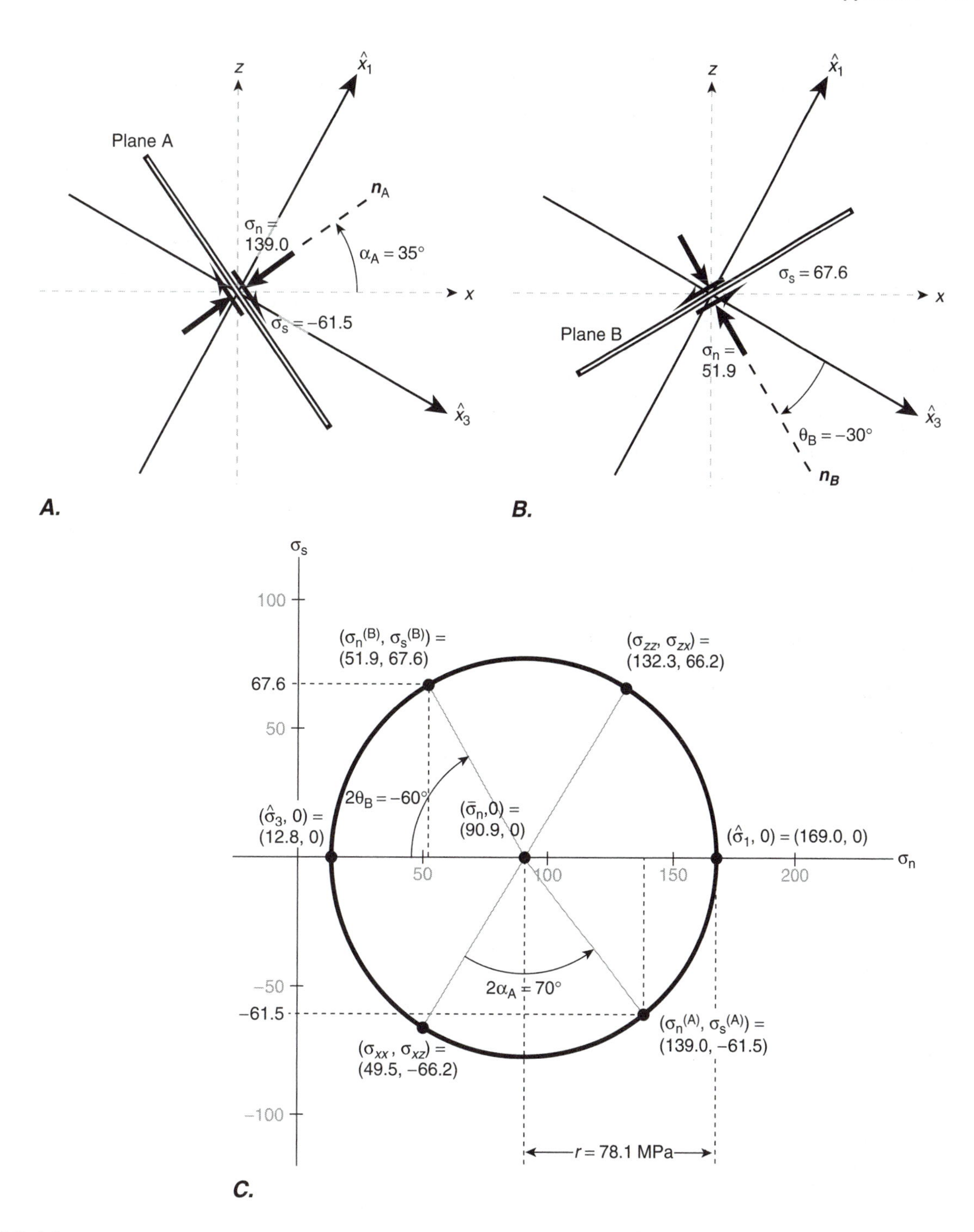

FIGURE 7-A.3 Determining the normal and shear stress components on planes of different orientations that are parallel to the y axis and the intermediate principal stress. *A*. Orientation of plane *A* relative to coordinate axes (x, z). *B*. Orientation of plane *B* relative to coordinate axes $(\hat{x}_1, \hat{x}_3)$. *C*. Determining the stress components on planes *A* and *B* using the Mohr circle.

Plane *B* is perpendicular to the (x, z)-plane, and its normal is at an angle $\theta_B = -30°$ from the minimum principal axis $\hat{x}_3$ (Figure 7-A.3*B*).

Procedure

In physical space, the planes *A* and *B* are perpendicular to the (x, z)-plane (which is the same as the $(\hat{x}_1, \hat{x}_3)$-plane and the $(\hat{\sigma}_1, \hat{\sigma}_3)$-plane) (Figure 7-A.3*A*, *B*). Thus the normals $\mathbf{n}_A$ and $\mathbf{n}_B$ to the planes *A* and *B* lie in the (x, z)-plane. The surface stresses acting on planes *A* and *B* must therefore plot on the $(\hat{\sigma}_1, \hat{\sigma}_3)$ Mohr circle. On the Mohr diagram, the construction that defines the stress components on the relevant planes is derived from properties 3i and 3ii. The relationships in physical space, summarized in the following table "Angular Relationships in Physical Space," are transferred to the Mohr circle as indicated in the following table "Angular Relationships on the Mohr Circle" and in Figure 7-A.3*C*.

In physical space (Figure 7-A.3*A*), the normal $\mathbf{n}_A$ to plane *A* is defined by the angle $\alpha_A = 35°$ measured counterclockwise from x. Here x is the normal to the plane on which the surface stress components are $(\sigma_{xx}, \sigma_{xz}) = (49.5, -66.2)$ MPa. Thus we can find the stress components on plane *A* by constructing a new radius on the Mohr circle at an angle $2\alpha_A = 70°$ counterclockwise from the radius at the point $(\sigma_{xx}, \sigma_{xz}) = (49.5, -66.2)$ MPa. The point on the Mohr circle at the end of the new radius is the surface stress on the plane *A*, and the surface stress components are read off the Mohr diagram, giving (Figure 7-A.3*C*):

$$(\sigma_n^{(A)}, \sigma_s^{(A)}) = (139.0, -61.5) \text{ MPa}$$

A similar procedure is used to find the stress components on plane *B* (Figure 7-A.3*C*), except that in this case the orientation of $\mathbf{n}_B$ in physical space is found by measuring the angle $\theta_B = -30°$ clockwise from $\hat{x}_3$ and the new radius of the Mohr circle is constructed at a clockwise angle $2\theta_B = -60°$ measured from $(\hat{\sigma}_3, 0) = (12.8, 0)$ MPa. The end of the new radius is the surface stress on plane *B*:

$$(\sigma_n^{(B)}, \sigma_s^{(B)}) = (51.9, 67.6) \text{ MPa}$$

Angular Relationships in Physical Space

Plane	Angle	Sense of angle	Measured from	Measured to
A	$\alpha_A = 35°$	Counter-clockwise	x	$\mathbf{n}_A$
B	$\theta_B = -30°$	Clockwise	$\hat{x}_3$	$\mathbf{n}_B$

Angular Relationships on the Mohr Diagram

Plane	Angle	Sense of angle	Measured from	Measured to
A	$2\alpha_A = 70°$	Counter-clockwise	$(\sigma_{xx}, \sigma_{xz}) = (49.5, -66.2)$	$(\sigma_n^{(A)}, \sigma_s^{(A)}) = (139.0, -61.5)$
B	$2\theta_B = -60°$	Clockwise	$(\hat{\sigma}_3, 0) = (12.8, 0)$	$(\sigma_n^{(B)}, \sigma_s^{(B)}) = (51.9, 67.6)$

Appendix 7-B

Illustrative Problem 2

For this illustrative problem, we assume that the reader is familiar with the matrix form of writing the stress components (Equations (7.51), (7.52)) and with the stress tensor notation and sign convention, as described in Section 7.5 (Equations (7.53)). Here we analyze a similar problem to that in Appendix 7A (Figure 7-A.1*A*), but we consider the problem in three dimensions, and instead of a tensile tectonic stress T, we consider that a compressive tectonic stress C, constant with depth, acts parallel to the x axis and is added to the stresses derived from the overburden (Figure 7-B.1*A*). The assumption that a tectonic stress would be constant with depth is again a major oversimplification that we adopt only for the purpose of illustration.

As before, we have the vertical stress defined by the overburden:

$$\sigma_{zz} = \rho g h \qquad (7\text{-B.}1)$$

In this case, we have two horizontal normal stress components acting on the x- and y-planes, respectively:

$$\sigma_{xx} = \kappa(\rho g h) + C \quad \sigma_{yy} = \kappa(\rho g h) + \nu C \quad (7\text{-B.}2)$$

Here, the normal stress parallel to x is a fraction κ of the overburden plus the compressive tectonic stress C, and the normal stress parallel to y is the same fraction κ of the overburden plus a fraction ν of the tectonic stress. The fractions κ and ν arise from the theory of elasticity (see Section 9.1, Equation (9.3), and Section 9.4(ii), Equation (9.7)[2]), where ν is the Poisson ratio and

$$\kappa \equiv \frac{\nu}{(1-\nu)} \qquad (7\text{-B.}3)$$

[2]We make a rough initial approximation that the principal stresses are horizontal and vertical, which in fact is only true at the surface, so this is an oversimplification. The dependence of $\hat{\sigma}_{xx}$ and $\hat{\sigma}_{yy}$ on the overburden is then given by Equation (9.7) (Section 9.4). To find the contribution of the tectonic stress C to the normal stress in the y direction, recast Equation (9.5) (Section 9.1) by cycling the subscripts, substituting y for z, x for y, and z for x so $\hat{e}_{yy}$ appears to the left of the equal sign. Set $\hat{e}_{yy} = 0$, because we require that no strain occur in the y direction (see Section 9.4). Set $\hat{\sigma}_{zz} = 0$ and $\hat{\sigma}_{xx} = C$, because we are considering only the horizontal tectonic stress. Finally, solve for $\hat{\sigma}_{yy}$ to find the contribution of the tectonic stress to this stress component.

The shear stresses on the z-plane are determined by the friction of the block on the horizontal fault plane and the normal stress across the plane:

$$\sigma_{zx} = -\mu\sigma_{zz} = -\mu(\rho g h) = -\sigma_{xz} \qquad (7\text{-B.}4)$$

σ_{zx} is negative because as the block is pushed toward the positive x direction, the friction resisting the motion generates a clockwise (negative) shear stress (Figure 7-B.1*B*). The last relation in Equation (7-B.4) comes from the necessary balance of shear stresses (Equation (7.31)). In the three-dimensional case, we must also determine the shear stress on the z-plane parallel to y. Since there is no component of the tectonic stress parallel to the y axis, the block will not move in that direction and there will therefore be no frictional resistance developed in that direction. Thus we can assume

$$\sigma_{zy} = -\sigma_{yz} = 0 \qquad (7\text{-B.}5)$$

where again we used the balance of the shear stresses (Equation (7.31)). For simplicity, we will also assume the shear stress on the y-plane in the x direction σ_{yx} is also zero. In effect, this means we assume the conditions on the fault block extend indefinitely in the y direction. We discuss this further below. Using the balance of the shear stresses (Equation (7.31)) also then gives the shear stress on the x-plane in the y direction:

$$\sigma_{yx} = -\sigma_{xy} = 0 \qquad (7\text{-B.}6)$$

We now have values for all nine of the stress components in the Mohr circle sign convention, and we can organize them in an array as:

$$\begin{bmatrix} \sigma_{xx} & \sigma_{xy} & \sigma_{xz} \\ \sigma_{yx} & \sigma_{yy} & \sigma_{yz} \\ \sigma_{zx} & \sigma_{zy} & \sigma_{zz} \end{bmatrix} = \begin{bmatrix} \kappa(\rho g h) + C & 0 & \mu(\rho g h) \\ 0 & \kappa(\rho g h) + \nu C & 0 \\ -\mu(\rho g h) & 0 & \rho g h \end{bmatrix} \qquad (7\text{-B.}7)$$

$$= \begin{bmatrix} 238.2 & 0 & 66.2 \\ 0 & 148.2 & 0 \\ -66.2 & 0 & 132.3 \end{bmatrix} \text{MPa}$$

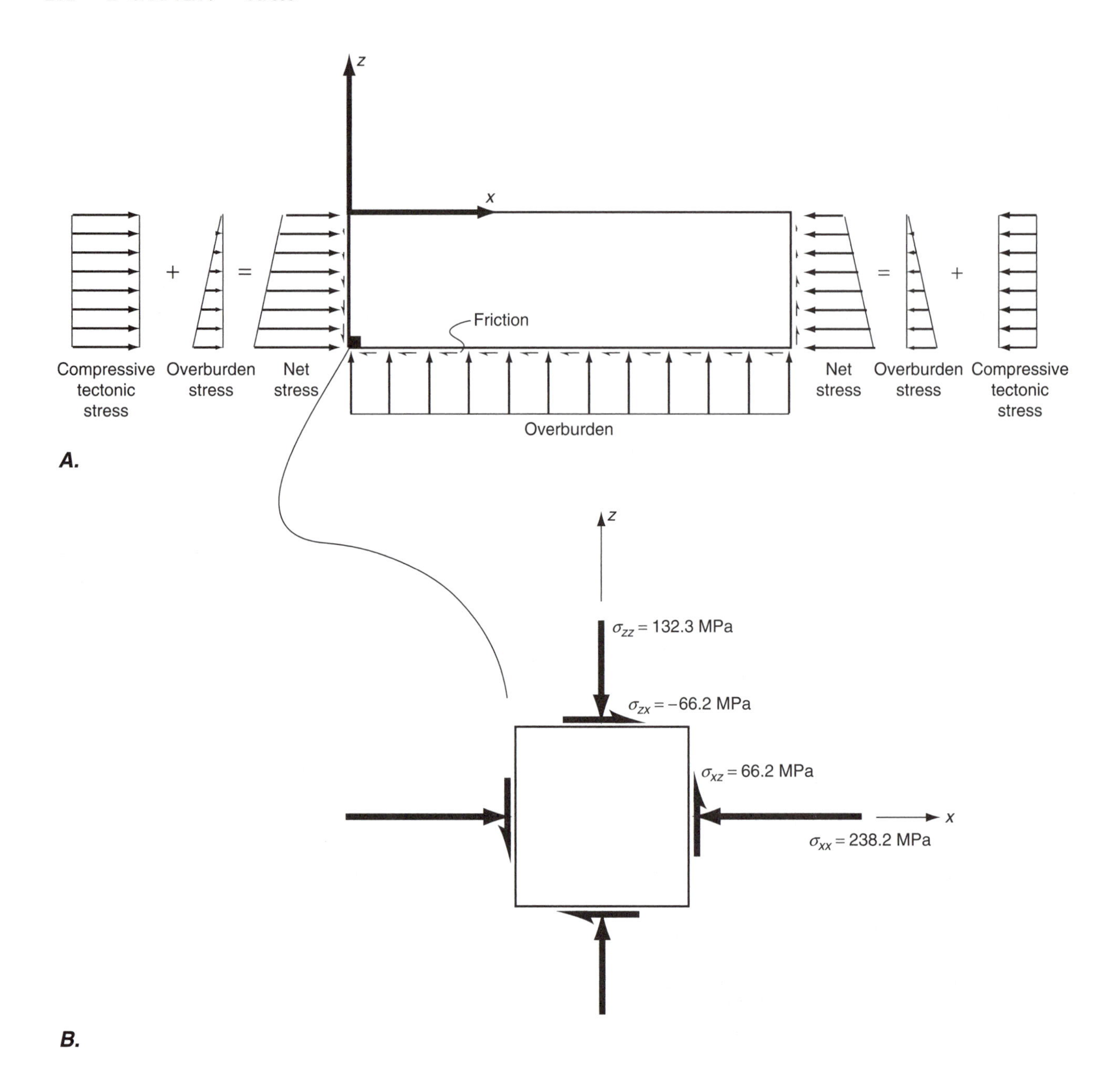

FIGURE 7-B.1 Simplified model of a fault block resting on a detachment, as in Figure 7-A.14, but here subjected to a compressive tectonic stress. *A*. A free-body diagram of an isolated part of the fault block showing the tractions on the base due to the overburden stress and the frictional shear stress, and the tractions on the vertical sides due to the overburden, the added compressive tectonic stress, the sum of these two, and the shear stresses required by the symmetry of the stress tensor. *B*. Stress components on the coordinate square from the lower-left corner of the isolated body of the fault block.

where, to find the values in the last matrix, we used the values in Equation (7-A.4), Equation (7-B.3), and

$$\nu = 0.4$$
$$C = 150 \text{ MPa}$$

The net result is that we have assumed the y-planes to be principal planes, because from Equations (7-B.5) and (7-B.6), both shear stresses on y-planes are zero, $\sigma_{yx} = \sigma_{yz} = 0$ (Section 7.2, property 2i).

This result for the stress allows us to use a two-dimensional analysis to determine stresses on any plane parallel to the y axis and thus perpendicular to the (x, z)-plane. Two-dimensional problems are simpler because we use only four stress components rather than all nine, which we must use for a three-dimensional analysis (see Equation (7.52)). Let us look briefly at why we are justified in using this simplification. In a two-dimensional analysis, the four stress components we consider must all lie in one plane, in this case the y-plane, which contains the x and z axes. In order to justify ignoring the other five stress components, they must contribute nothing to the sum of the forces (stress times area) parallel to the axes in this plane. In the current example, we see that the shear stresses on the y-plane parallel to the x and z axes are both zero (from Equations (7-B.6) and (7-B.5), respectively):

$$\sigma_{yx} = \sigma_{yz} = 0 \qquad \text{(7-B.8)}$$

This condition, along with Equations (7.31), give Equations (7-B.5) and (7-B.6), which means that we can ignore all four of these stress components because they are zero. Moreover, because the normal stress on the (x, z)-plane, σ_{yy}, acts parallel to the y axis, it does not affect the stress on planes that are themselves parallel to the y axis, and therefore we do not need to consider this stress in our two-dimensional analysis. Thus we are justified in ignoring all the five stress components that, in this case, have a "y" as one of the subscripts.

By induction from these results, we can state the general rule: *In order to analyze a problem in two dimensions, the axis parallel to the third dimension (y in this example) must be parallel to one of the principal stresses; and the plane perpendicular to that axis, that is, the plane in which we do the analysis (the (x, z)-plane in this example), must be a principal plane.*

This example implicitly assumes the stresses on the block are unchanged for an indefinite extent in the y direction. If that were not true, for example, if the block were bounded by a tear fault perpendicular to the y axis (Figure 5.13), then we would have to consider the shear stresses σ_{yx}, and possibly σ_{yz}, to be nonzero. In that case, Equations (7-B.5) and (7-B.6) would not be true, and the (x, z)-plane would not be a principal plane because the shear stresses on it would not all be zero. The principal stresses then would not lie in the (x, z)-plane but rather would be inclined at some angle, and we would have to use a full three-dimensional analysis. In practice, we can justify a two-dimensional analysis if the (x, z)-plane of our analysis is far enough away (e.g., more than twice the thickness of the block) from any boundary like a tear fault, so that the effect of the boundary is negligible.

The four stress components that are important to our two-dimensional analysis (σ_{xx}, σ_{xz}, σ_{zz}, σ_{zx}) can be written in a matrix, using the Mohr circle sign convention, as

$$\begin{bmatrix} \sigma_{xx} & \sigma_{xz} \\ \sigma_{zx} & \sigma_{zz} \end{bmatrix} = \begin{bmatrix} 238.2 & 66.2 \\ -66.2 & 132.3 \end{bmatrix} \text{MPa} \qquad \text{(7-B.9)}$$

Figure 7-B.1*B* is the diagram in the (x, z)-plane showing these stress components. We examine some questions that pertain particularly to the three-dimensional stress, which we did not cover in Appendix 7-A, and we refer by number to the properties of the Mohr circles discussed in Section 7.2 and Box 7-2.

Question B1

(a) Plot the Mohr circles for the three-dimensional stress and determine the magnitudes and orientations of the three principal stresses.

(b) Draw a two-dimensional diagram of physical space showing the relationships among the (x, z) coordinates and:

- **(i)** the surface stress components on the x and z coordinate planes,
- **(ii)** the principal coordinates $\hat{x}_1$ and $\hat{x}_3$ and the principal stresses on the principal planes.

Procedure

(a) Using values from Equation (7-B.9), we plot on the Mohr diagram the normal and shear components of the surface stress that act on the x- and z-planes, (238.2, 66.2) MPa and (132.3, −66.2) MPa, respectively (Figure 7-B.2A). The line between these points is a diameter of one of the Mohr circles (Section 7.2, property 3v) and the point at which the diameter crosses the σ_n axis is the center of the circle. Because we now know the location of the center of the circle and the length of its radius to one of the points on the circumference of the circle (i.e., we know the invariants of the two-dimensional stress (Section 7.2, property 5)), we can draw the Mohr circle with a drafting compass. Where the circle crosses the σ_n axis, the shear stresses are zero and the normal stresses are the two principal stresses, the maximum (270.1 MPa) and the minimum (100.5 MPa) (Section 7.2, property 2). We already know from our preceding discussion that σ_{yy} = 148.2 MPa is the third principal stress (Equation (7-B.7)), so we can draw the other two Mohr circles that represent the two-dimensional stresses in the other two principal planes, which gives us a plot of the three-dimensional state of stress (Figure 7-B.2A; Box 7-2, property 2). We then order the three principal stresses from maximum to minimum such that

$$\hat{\sigma}_1 = 270.1 \text{ MPa} \\ \hat{\sigma}_2 = 148.2 \text{ MPa} \qquad \text{(7-B.10)} \\ \hat{\sigma}_3 = 100.5 \text{ MPa}$$

(b) **(i)** On a diagram of physical space showing the x and z axes (Figure 7-B.2*B*), we draw the infinitesimal coordinate square (the small square shaded gray in Figure 7-B.2*B*), and using the Mohr circle sign convention and the convention for interpreting the meaning of the subscripts (Equation (7.52) and following

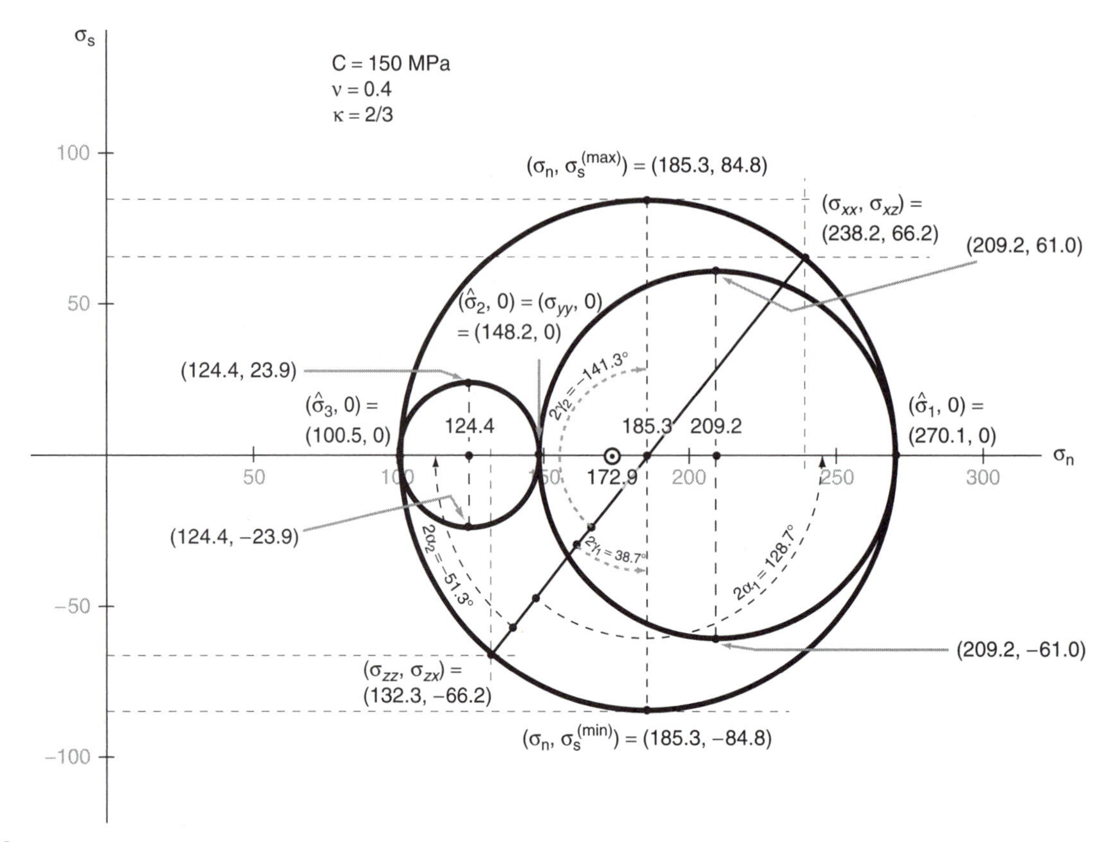

σ_s
C = 150 MPa
ν = 0.4
κ = 2/3
100
50
−50
−100
$(\sigma_n, \sigma_s^{(max)})$ = (185.3, 84.8)
$(\sigma_{xx}, \sigma_{xz})$ = (238.2, 66.2)
(209.2, 61.0)
$(\hat{\sigma}_2, 0) = (\sigma_{yy}, 0)$ = (148.2, 0)
(124.4, 23.9)
$(\hat{\sigma}_3, 0)$ = (100.5, 0)
124.4
$2\gamma_2 = -141.3°$
185.3
209.2
$(\hat{\sigma}_1, 0)$ = (270.1, 0)
σ_n
50
100
150
172.9
200
250
300
(124.4, −23.9)
$2\alpha_2 = -51.3°$
$2\gamma_1 = 38.7°$
$2\alpha_1 = 128.7°$
(209.2, −61.0)
$(\sigma_{zz}, \sigma_{zx})$ = (132.3, −66.2)
$(\sigma_n, \sigma_s^{(min)})$ = (185.3, −84.8)

A.

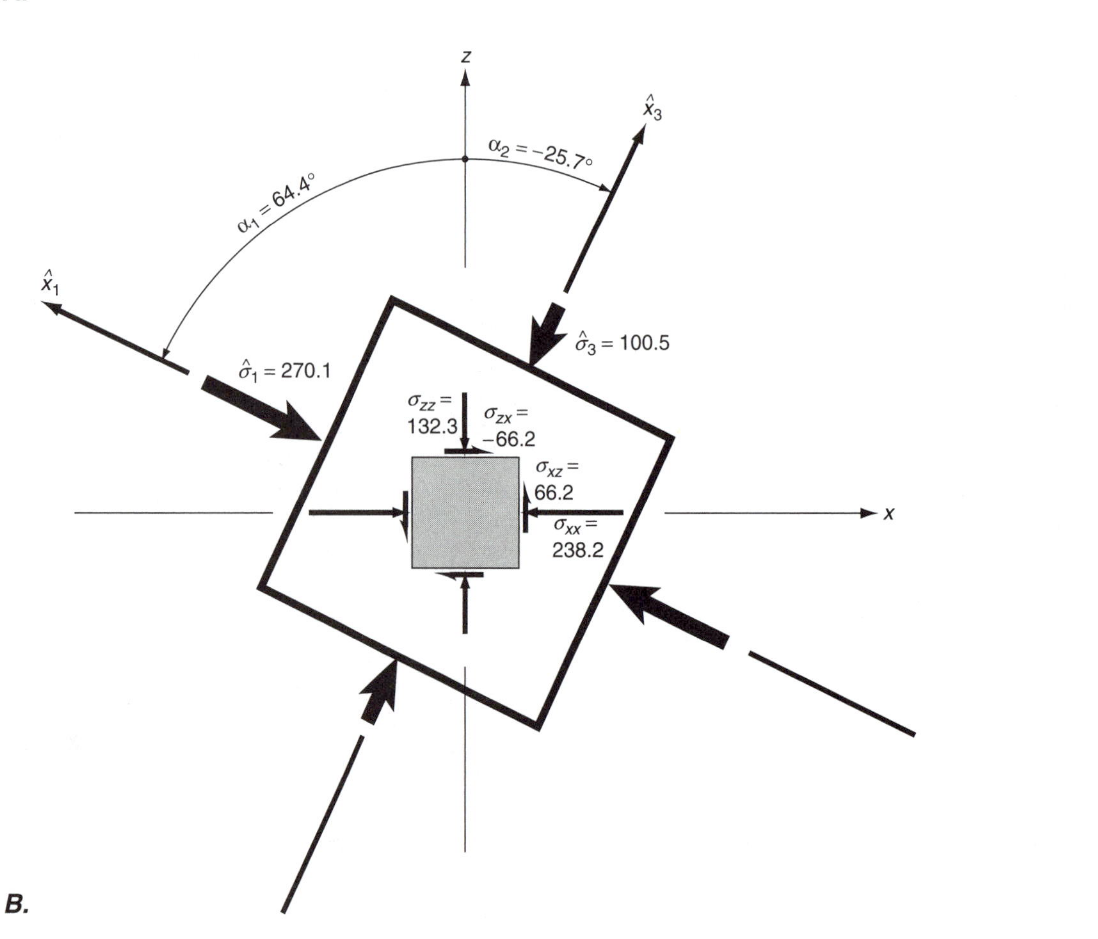

z
$\hat{x}_3$
$\alpha_2 = -25.7°$
$\alpha_1 = 64.4°$
$\hat{x}_1$
$\hat{\sigma}_3$ = 100.5
$\hat{\sigma}_1$ = 270.1
σ_{zz} = 132.3
σ_{zx} = −66.2
σ_{xz} = 66.2
σ_{xx} = 238.2
x

B.

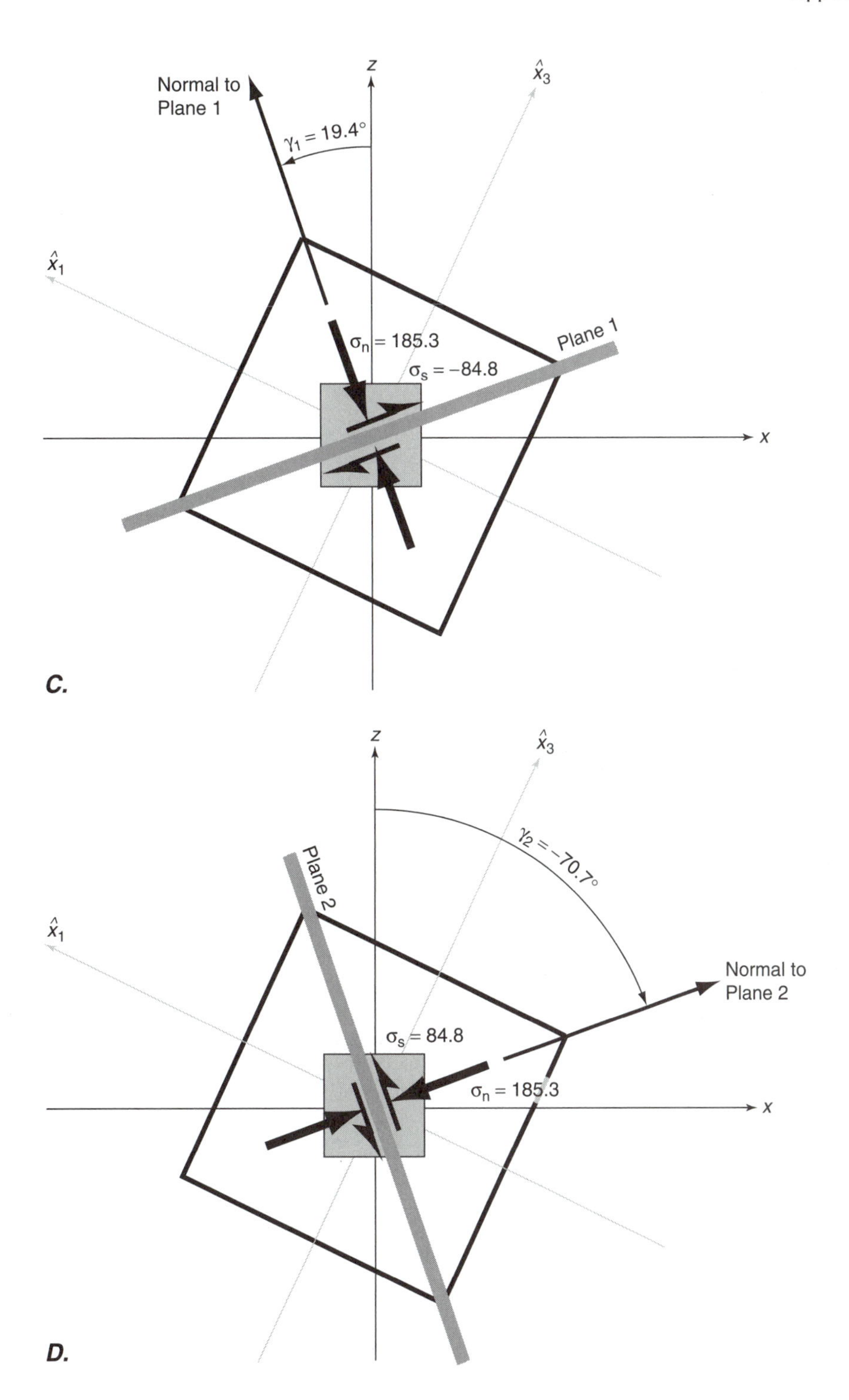

FIGURE 7-B.2 *A*. Three-dimensional Mohr circle diagram for the state of stress in the block in Figure 7-B.1. The oblique diameter of the circle is defined by the calculated stresses lying in the (x, z)-plane. The σ_n axis shows the centers for the three Mohr circles for the three-dimensional stress, as well as the three-dimensional mean normal stress. Angles define the locations of the principal stresses and the maximum shear stresses. *B*. Determining the orientation of principal stresses in physical space from the Mohr circle. *C, D*. Determining from the Mohr circles in *A*, the orientation in physical space of the two planes of maximum shear stress. Note that these two planes are perpendicular to each other and oriented 45° from the principal axes.

text), we draw the stress components given in Equation (7-B.9) on the appropriate coordinate surfaces.

(ii) We now determine the orientations of the principal axes and principal stresses relative to the x and z axes. Because we know both the orientation of the z axis and the surface stress on the z-plane, we measure on the Mohr circle the counterclockwise (positive) angle $2\alpha_1 = 128.7°$ from the radius at $(\sigma_{zz}, \sigma_{zx}) = (132.3, -66.2)$ MPa to the radius at the maximum principal stress $(\hat{\sigma}_1, 0) = (270.1, 0)$ MPa (Figure 7-B.2*A*). In physical space (Figure 7-B.2*B*) we measure half that angle $\alpha_1 = 64.4°$ counterclockwise from the z axis to find the orientation of $\hat{x}_1$. This is the principal axis normal to the plane on which the maximum compressive stress $\hat{\sigma}_1$ acts, and thus it is also the orientation of $\hat{\sigma}_1$.

Similarly, we can determine the orientation of $\hat{\sigma}_3$ by measuring on the Mohr circle (Figure 7-B.2*A*) the clockwise angle $2\alpha_2 = -51.3°$ from the radius at $(\sigma_{zz}, \sigma_{zx}) = (132.3, -66.2)$ MPa to the radius at the minimum principal stress at $(\hat{\sigma}_3, 0) = (100.5, 0)$ MPa. In physical space (Figure 7-B.2*B*) we measure half that angle $\alpha_2 = -25.7°$ from the z axis clockwise to $\hat{x}_3$, the axis normal to the plane on which the minimum compressive stress $\hat{\sigma}_3$ acts, and thus also the orientation of $\hat{\sigma}_3$.

Discussion

Of course, this last measurement was not strictly necessary, because we know that the two principal stresses plot on the Mohr circle at opposite ends of a diameter, and thus in physical space they must be perpendicular (Section 7.2, property 3v).

We could also just as well have measured the angles on the Mohr circle from the radius at $(\sigma_{xx}, \sigma_{xz}) = (238.2, 66.2)$ MPa to the radii to the two principal stresses, in which case we would find the orientations of the principal stresses in physical space by measuring half those angles in the same senses from the x axis. The answers must necessarily be the same.

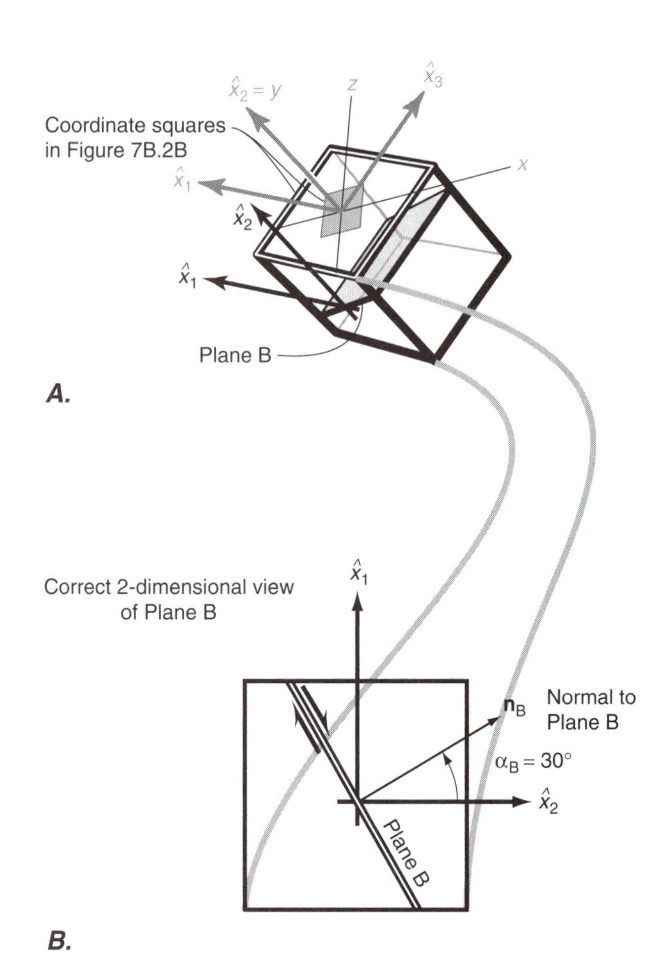

FIGURE 7-B.3 Using the three-dimensional Mohr circle to find stress components on a plane B that is parallel to the $\hat{x}_3$ axis. *A*. Choosing the correct two-dimensional view of plane B in physical space. The $(\hat{x}_1, \hat{x}_2)$ principal plane must be viewed looking in the positive $\hat{x}_3$ direction so that the rotation from positive $\hat{x}_1$ to positive $\hat{x}_2$ is clockwise. *B*. The correct view of plane B in physical space. *C*. Solving for the surface stress on plane B using the $(\hat{\sigma}_1, \hat{\sigma}_2)$ Mohr circle.

Question B2

In the three-dimensional solid, what are the absolute values of the maximum shear stress and the orientations of the planes on which these shear stresses occur?

Procedure

In order to answer this question, we must consider the Mohr diagram for the three-dimensional stress (Box 7-2). When all three Mohr circles are plotted (Figure 7-B.2*A*; Box 7-2, properties 1iii and 4ii), it is clear that the maximum absolute value of the shear stress (84.8 MPa) occurs at the top and bottom points on the largest Mohr circle, which is the $(\hat{\sigma}_1, \hat{\sigma}_3)$ circle representing surface stresses on planes parallel to the $\hat{x}_2$ axis (Box 7-2, property 4ii). The maximum absolute shear stress on the $(\hat{\sigma}_1, \hat{\sigma}_2)$ Mohr circle (61.0 MPa), and on the $(\hat{\sigma}_2, \hat{\sigma}_3)$ Mohr circle (23.9 MPa) are only maxima for the sets of planes that are parallel, respectively, to $\hat{x}_3$ and $\hat{x}_1$.

The orientations of the planes of maximum absolute shear stress relative to the reference axes x and z are determined on the Mohr circle (Figure 7-B.2*A*) by measuring the angles $2\gamma_1 = 38.7°$ and $2\gamma_2 = -141.3°$ from the radius at $(\sigma_{zz}, \sigma_{zx}) = (132.3, -66.2)$ MPa to the respective radii at $(\overline{\sigma}_n, \sigma_s^{(\min)}) = (185.3, -84.8)$ and at $(\overline{\sigma}_n, \sigma_s^{(\max)}) = (185.3, 84.8)$. In physical space (Figure 7-B.2*C*), measuring half these angles $\gamma_1 = 19.4°$ and $\gamma_2 = -70.7°$ from the z axis, in the sense indicated by the signs of the angles, defines the orientation of the *normals* to the planes 1 and 2 of maximum absolute shear stress. In Appendix 7-A we found that the planes of maximum absolute shear stress are oriented at 45° to the maximum and minimum principal stresses. We see that the differences in the angles

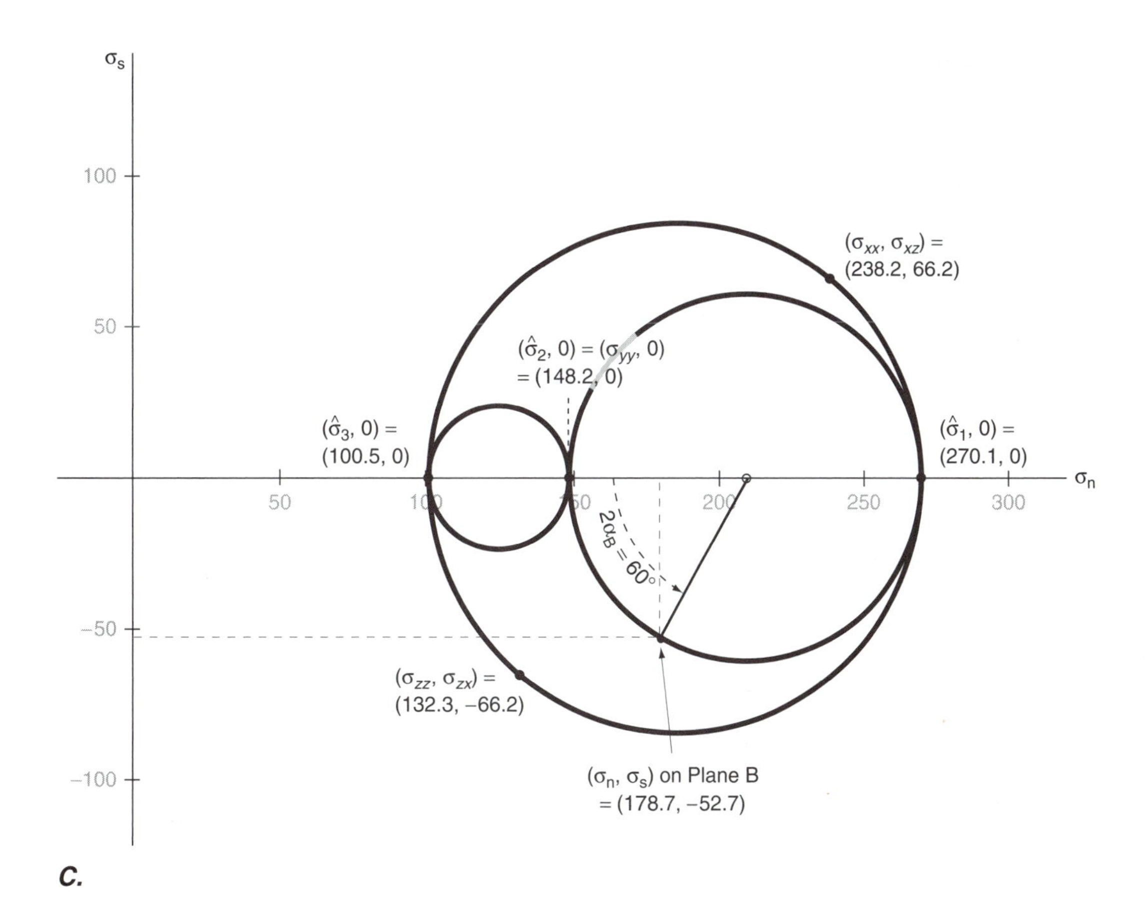

FIGURE 7-B.3 (continued)

given by $(\alpha_1 - \gamma_1)$ and $(\alpha_2 - \gamma_2)$ are both 45° (compare Figure 7-B.2*B* with Figure 7-B.2*C*, *D*), which shows that the planes of maximum shear stress deduced for this example give the same result.

The orientations of the two planes of maximum absolute shear stress in three-dimensional physical space are defined such that they are parallel to the $y = \hat{x}_2$ axis and their normals lie at 45° to the maximum and minimum principal stresses $\hat{\sigma}_1$ and $\hat{\sigma}_3$ (Figure 7-B.2*C*, *D*). For this special case only, in which the normals bisect the angle between the maximum and minimum principal stresses, the planes themselves also are oriented at angles of 45° to those principal stresses. Thus the normal to one shear plane lies in the other shear plane because the two planes of maximum shear stress are perpendicular to each other (Figure 7-B.2*C*, *D*).

Question B3

What is the mean normal stress for the two-dimensional stress and the three-dimensional stress? How is each related to the Mohr circles?

Procedure

In a two-dimensional stress, the mean normal stress is simply the location of the center of the Mohr circle defined by the two principal stresses that lie in the two-dimensional plane (Section 7.2, property 5i) (Figure 7-B.2*A*). For the three Mohr circles of the three-dimensional stress plotted in Figure 7-B.2*A*, from the largest to the smallest circle, the two-dimensional mean normal stresses are

For the $(\hat{\sigma}_1, \hat{\sigma}_3)$ Mohr circle:

$$\overline{\sigma}_n = \frac{\hat{\sigma}_1 + \hat{\sigma}_3}{2} = \frac{270.1 + 100.5}{2} = 185.3 \text{ MPa}$$

For the $(\hat{\sigma}_1, \hat{\sigma}_2)$ Mohr circle:

$$\overline{\sigma}_n = \frac{\hat{\sigma}_1 + \hat{\sigma}_2}{2} = \frac{270.1 + 148.2}{2} = 209.2 \text{ MPa}$$

For the $(\hat{\sigma}_2, \hat{\sigma}_3)$ Mohr circle:

$$\overline{\sigma}_n = \frac{\hat{\sigma}_2 + \hat{\sigma}_3}{2} = \frac{148.2 + 100.5}{2} = 124.4 \text{ MPa}$$

Thus the mean normal stresses for each of the three Mohr circles lie at the mean of the maximum and minimum principal stresses for each particular circle. For three dimensions, however, the mean normal stress is the mean of all three principal stresses and is not the center of any of the three Mohr circles (Box 7-2, property 5ii) (Figure 7-B.2*A*).

For the three-dimensional stress:

$$\overline{\sigma}_n = \frac{\hat{\sigma}_1 + \hat{\sigma}_2 + \hat{\sigma}_3}{3} = \frac{270.1 + 148.2 + 100.5}{3}$$
$$= 172.9 \text{ MPa}$$

Question B4

What are the normal and shear stresses acting on a plane B that is parallel to the principal axis $\hat{x}_3$ with its normal $\mathbf{n}_B$ at an angle $\alpha_B = 30°$ from $\hat{x}_2$ and thus from $\hat{\sigma}_2$, where the angle is measured when looking in the positive $\hat{x}_3$ direction?

Procedure

For the two-dimensional analysis of planes parallel to $\hat{x}_3$, we must use the $(\hat{x}_1, \hat{x}_2)$-plane and the $(\hat{\sigma}_1, \hat{\sigma}_2)$ Mohr circle. When drawing these relationships in a diagram of physical space, we must be careful to view the $(\hat{x}_1, \hat{x}_2)$-plane from the same direction as specified for the measurement of the angle. If we plotted the positive angle α_B when looking in the negative $\hat{x}_3$ direction, the plane B would be oriented in the wrong quadrant of the principal coordinate axes in physical space. The sign of the shear stress obtained by plotting a counterclockwise angle $2\alpha_B$ from $(\hat{\sigma}_2, 0)$ on the Mohr circle would be correct for the correctly oriented plane; for the incorrectly oriented plane, however, that shear stress would have the wrong sign. To view the $(\hat{x}_1, \hat{x}_2)$-plane from the correct direction, that is to look in the positive $\hat{x}_3$ direction, we must look up at the downward-facing side of the plane, as shown in Figure 7-B.3. From this point of view, if we plot the angle α_B as a counterclockwise (positive) angle measured from the $\hat{x}_2$ axis to $\mathbf{n}_B$ (Figure 7-B.3*B*), we find the correct orientation of plane B.

On the $(\hat{\sigma}_1, \hat{\sigma}_2)$ Mohr circle (Figure 7-B.3*C*), we plot twice the angle α_B in a counterclockwise (positive) direction from the radius at $(\hat{\sigma}_2, 0) = (148.2, 0)$ MPa to the radius at the normal and shear stresses on plane B, $(\sigma_n, \sigma_s) = (178.7, -52.7)$ MPa.

Question B5

Write the array of Mohr circle stress components (Equation (7-B.7)) as the mathematically correct matrix of stress tensor components.

Procedure

Assume the three coordinate axes (x, y, z) are relabeled, respectively, (x_1, x_2, x_3). We apply the geologic stress tensor sign convention to all the nonzero components, which are diagrammed in Figure 7-B.1*B*. That convention requires that the stress tensor component take the sign of its associated traction component that acts on the negative side of the coordinate surface. The two normal tractions on the negative coordinate surfaces both point in a positive coordinate direction, so the normal stress components are both positive (compressive). The two shear traction components on the negative sides of the coordinate surfaces both point in a negative coordinate direction, so both shear stress components are negative. Thus the stress tensor components are written in matrix form as:

$$\begin{bmatrix} \sigma_{11} & \sigma_{12} & \sigma_{13} \\ \sigma_{21} & \sigma_{22} & \sigma_{23} \\ \sigma_{31} & \sigma_{32} & \sigma_{33} \end{bmatrix} = \begin{bmatrix} 238.2 & 0 & -66.2 \\ 0 & 148.2 & 0 \\ -66.2 & 0 & 132.3 \end{bmatrix} \text{MPa}$$

Notice that this is a symmetric tensor, because the shear stress components in symmetric positions in the matrix relative to the principal diagonal are equal:

$$\sigma_{kl} = \sigma_{lk} \quad (k \neq l)$$

Chapter 8

MECHANICS OF FRACTURING AND FAULTING: EXPERIMENT AND THEORY

In this chapter, we investigate the experimental and theoretical relationships between stress and the formation of rock fractures. We wish to understand the conditions under which fractures develop in Earth materials as a guide to understanding how and why natural fractures and faults form.

In the laboratory, we can subject rock samples to many different stress states, which is how we discover how rocks behave. If the stress is large enough, the sample can fracture. Fracturing rocks in the laboratory under different states of stress and different conditions makes it possible to formulate **fracture criteria**, which define both the stress state at fracture and the orientations of the fractures relative to the principal stresses. Fracture criteria enable us to infer the conditions in the Earth when fractures formed in the rocks and to determine whether, for any given set of conditions, we can expect the rock to be stable or to fracture. Thus we devote much of the chapter to different fracture criteria and how they are affected by physical conditions such as confining pressure, pore pressure, rock anisotropy, the intermediate principal stress, and temperature. Finally, we consider a unifying model for brittle fracture, the **Griffith theory for brittle fracture**, which accounts for many observed effects at the microscopic and submicroscopic levels.

8.1 EXPERIMENTAL FRACTURING OF ROCKS

In a typical rock deformation experiment, a piece of rock[1] is cut into a cylinder with a diameter ranging from less than 1 centimeter to tens of centimeters in some cases, and a length typically two to four times the diameter. The sample is placed between two pistons of hardened steel or similar material, which are forced together by a device such as a hydraulic ram. The axial stress on the sample is then increased until the sample fails, and the experiment provides data on the stress state required to cause failure.

Failure of a sample occurs when the sample is unable to support a stress increase without permanent

[1]Experimentalists often select samples of rocks that are fine-grained, homogeneous (i.e., each sample is the same throughout), and isotropic (i.e., each sample has the same mechanical properties in all directions). Examples of commonly used rocks include the Solenhofen limestone (a Jurassic limestone from the Alps, which is the parent material of all *Archeopteryx* fossils so far discovered), the Yule marble (from Colorado, widely used as a building stone, including for the Lincoln Memorial in Washington, DC), Hasmark dolomite (from the southern Appalachians), Berea sandstone (from Kentucky), and Westerly granite (from New England). The locations of origin are interesting, but the rocks were chosen for their homogeneous isotropic properties.

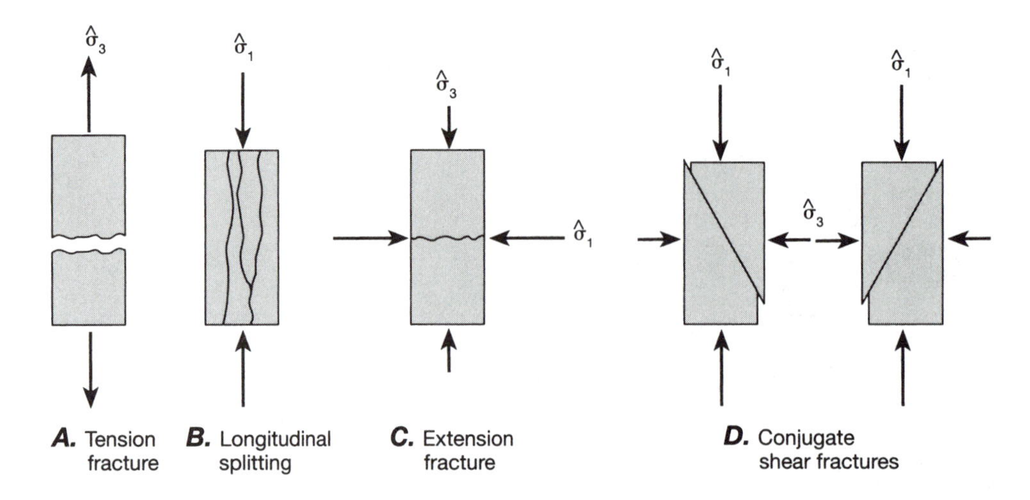

FIGURE 8.1 Types of fractures developed during experiments on brittle rocks.

deformation. The stress at which failure occurs is a measure of the strength of the material. Because failure can occur in a number of different ways, there are a variety of different measures of material strength. **Brittle failure** occurs with the formation of a **brittle fracture**, which is a surface or zone across which the material loses cohesion, that is, it breaks apart. **Ductile failure** occurs when the material deforms permanently without losing cohesion.

Experiments investigating the effects of pressure on failure employ a sample sealed in an impermeable jacket and surrounded in a pressure chamber by a fluid under pressure. The pressure of the surrounding fluid (the confining pressure) and that of the fluid in the pore spaces in the rock (the **pore fluid pressure**) can be controlled independently to determine the influence of each on the behavior of the rock samples. In experiments on temperature effects, the temperature is controlled with a small furnace, often an electrical coiled-wire heating element that surrounds the sample in the pressure chamber.

In such experiments, the applied forces generally are normal to the surfaces of the sample so that the principal axes of stress are either parallel to the cylinder axis (the axial stress) or perpendicular to the cylinder axis (the radial stress or confining pressure). Experiments usually are done in either uniaxial compression (Figure 7.12*B*) or confined compression (Figure 7.12*D*), the axial stress being the maximum compressive stress $\hat{\sigma}_1$. Axial extension experiments in which the axial stress is the minimum compressive stress $\hat{\sigma}_3$ (Figure 7.12*E*) are also common. Special modifications to the equipment permit experiments in uniaxial tension (Figure 7.12*C*), although these are uncommon in rock deformation experiments.

Experiments on brittle failure reveal two fundamentally different types of fracture: extension fractures and shear fractures. These two fracture types mimic natural fractures in rocks, which we described, respectively, in Chapters 2 and 3. Each type is characterized by a different direction of displacement relative to the fracture surface and a different orientation of the fracture plane relative to the principal stresses.

For **extension fractures** (Figure 8.1*A*, *B*, *C*), the fracture plane is perpendicular to the minimum principal stress $\hat{\sigma}_3$ and parallel to the maximum principal stress $\hat{\sigma}_1$. Displacement is approximately normal to the fracture surface. Extension fractures are tension fractures (Figure 8.1*A*) if the minimum principal stress $\hat{\sigma}_3$ is tensile, as in uniaxial tension (Figure 7.12*C*). They form by longitudinal splitting (Figure 8.1*B*) if the minimum principal stress is equal or close to zero and the maximum compressive stress $\hat{\sigma}_1$ is the axial stress, as in uniaxial compression (Figure 7.12*B*). Fractures that form by longitudinal splitting tend to be more irregular in orientation and shape than other extension fractures. Extension fractures may also form under conditions of extensional stress (Figure 8.1*C*) in which the unique axial stress is the minimum compressive stress $\hat{\sigma}_3$ (Figure 7.12*E*).

Shear fractures form in confined compression (Figure 7.12*D*) at angles of less than 45° to the maximum compressive stress $\hat{\sigma}_1$ (Figure 8.1*D*). Displacement is parallel to the fracture surface. If the state of stress is triaxial (Figure 7.12*F*), the shear fractures are parallel to the intermediate principal stress $\hat{\sigma}_2$, and the fracture planes form a conjugate pair of orientations at angles less than 45° on either side of the maximum compressive stress $\hat{\sigma}_1$. If $\hat{\sigma}_2 = \hat{\sigma}_3$ (Figure 7.12*D*), the possible orientations of shear fractures are tangent to a cone of less than 45° about the $\hat{\sigma}_1$ axis (see Section 8.3).

8.2 A FRACTURE CRITERION FOR TENSION FRACTURES

Experiments on rocks under uniaxial tension show that, for each rock type, there is a characteristic value of tensile stress (T_0) at which tension fracturing occurs. The

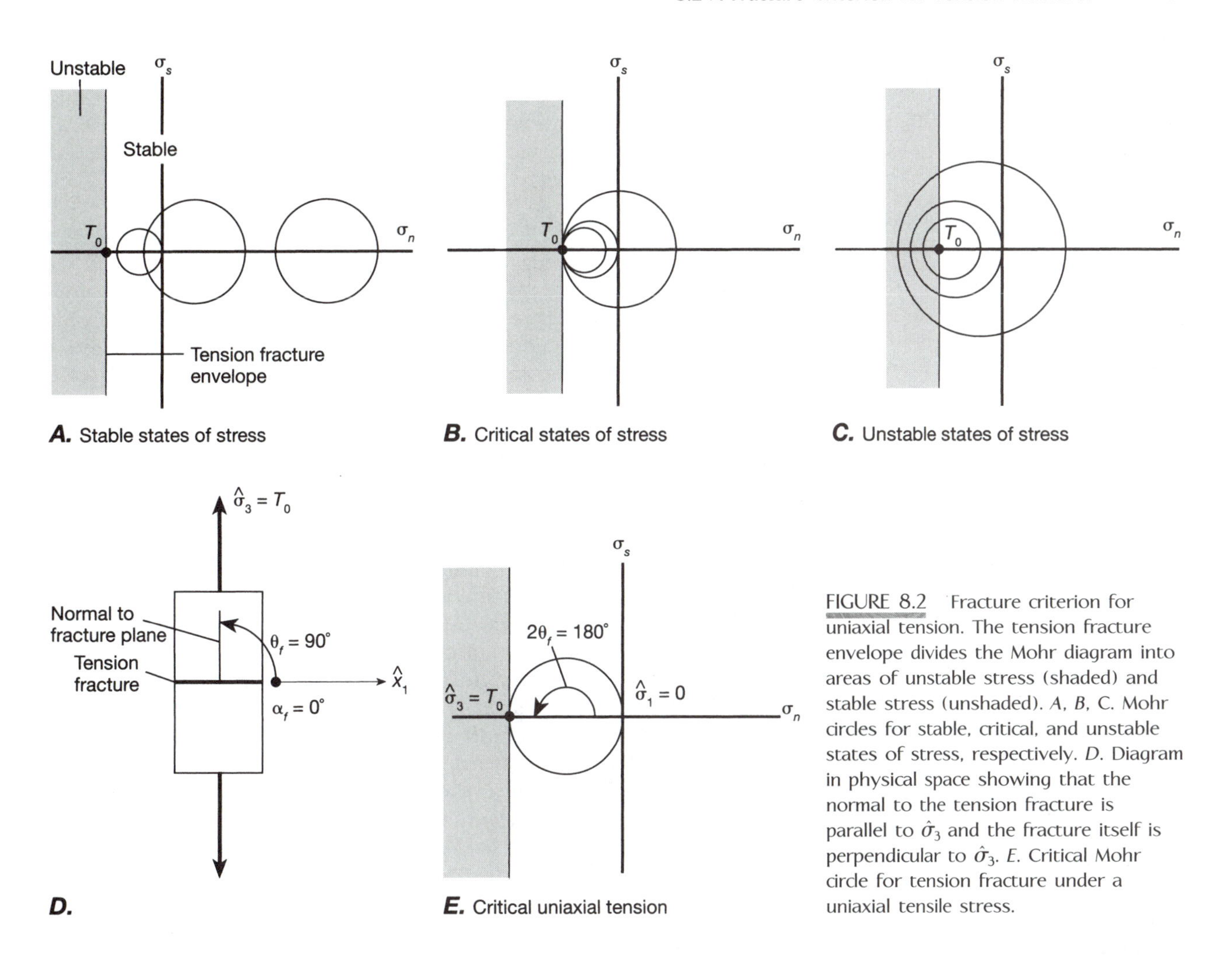

FIGURE 8.2 Fracture criterion for uniaxial tension. The tension fracture envelope divides the Mohr diagram into areas of unstable stress (shaded) and stable stress (unshaded). *A, B, C.* Mohr circles for stable, critical, and unstable states of stress, respectively. *D.* Diagram in physical space showing that the normal to the tension fracture is parallel to $\hat{\sigma}_3$ and the fracture itself is perpendicular to $\hat{\sigma}_3$. *E.* Critical Mohr circle for tension fracture under a uniaxial tensile stress.

rock is stable at tensile stresses smaller than T_0, but it cannot support larger tensile stresses. T_0 is the **tensile strength** of the material. On a Mohr diagram, the boundary between stable and unstable states of tensile stress is called the **tension fracture envelope** (Figure 8.2*A*). It is a line perpendicular to the σ_n axis at T_0 and is represented by the equation

$$\sigma_n^* = T_0 \tag{8.1}$$

where σ_n^* is the critical normal stress required to produce fracture. A Mohr circle that lies to the right of the line represents a stable stress state (Figure 8.2*A*). A Mohr circle tangent to the line (a critical Mohr circle) represents a state of stress that causes tension fracturing (Figure 8.2*B*). Mohr circles that cross the line represent unstable states of stress that the material cannot support (Figure 8.2*C*).

We can describe the orientation of a fracture plane relative to the principal stresses by the **fracture plane angle** α_f, which is the angle between the maximum principal stress $\hat{\sigma}_1$ and the fracture, or by the **fracture angle** θ_f, which is the angle between the maximum principal stress $\hat{\sigma}_1$ and the normal to the fracture plane. For a given plane, $|\theta_f - \alpha_f| = 90°$, and if both angles are acute, they are opposite in sign (compare Figures 8.2*D* and 8.3*D*, *E*). On a Mohr diagram, the angle we use to plot the surface stresses that act on a plane is the angle defined by the normal to the plane, relative to some known direction (Section 7.2, property 3i). Thus to plot the stresses on a fracture plane, we must use the fracture angle θ_f.

In experiments, the tension fracture plane is normal to the maximum tensile stress $\hat{\sigma}_3$. Thus the fracture plane angle α_f is 0° and the fracture angle θ_f is 90° (Figure 8.2*D*). On the Mohr diagram (Figure 8.2*E*), the stress on the fracture plane plots at an angle of $2\theta_f = 180°$ from $(\hat{\sigma}_1, 0)$. The normal stress and shear stress components on the fracture plane thus plot exactly at the point of tangency between the critical Mohr circle and the tension fracture envelope.

Equation (8.1) thus provides a fracture criterion, because it defines both the stress required for fracturing and the orientation of the fracture: A tension fracture forms on any plane in the material on which the normal stress reaches the critical value T_0, and the fracture plane is perpendicular to the maximum tensile stress $\hat{\sigma}_3$. This fracture criterion, however, applies only to tension fractures formed under conditions of tensile stress. It does not account for the occurrence of extension fractures that develop under conditions in which none of the principal stresses are tensile (Figure 8.1*B*, *C*), such as longitudinal splitting.

8.3 THE COULOMB FRACTURE CRITERION FOR CONFINED COMPRESSION

In confined compression experiments (Figure 7.12*D*), the relationship between the state of stress and the occurrence of shear fracturing is more complicated than for uniaxial tension. Fracture experiments on different samples of the same rock show that the initiation of fracturing depends on the differential stress ($\sigma^{(Dif)} \equiv \hat{\sigma}_1 - \hat{\sigma}_3$) and that the magnitude of the differential stress necessary to cause shear fracture increases with increasing confining pressure.[2] The fracture angle θ_f between $\hat{\sigma}_1$ and the normal to the fracture plane is generally around $\pm 60°$, so the fracture plane angle α_f between the fracture plane itself and the maximum compressive stress $\hat{\sigma}_1$ must be about $\pm 30°$ (Figure 8.1*D*).

Experimental data show that it is possible to construct on the Mohr diagram a shear fracture envelope that separates stable from unstable states of stress. This envelope commonly approximates a pair of straight lines that are symmetric across the σ_n axis (Figure 8.3*A*–*C*), although in fact the lines may be slightly concave toward that axis (see Figure 8.4*B*). Any Mohr circle contained between the two lines of the fracture envelope represents a stable stress state (Figure 8.3*A*). A Mohr circle tangent to the lines represents a critical state of stress that causes fracturing (Figure 8.3*B*). A Mohr circle that crosses the fracture envelope represents an unstable state of stress that cannot be supported by the material (Figure 8.3*C*); before the Mohr circle reaches this state, fracturing would already have occurred. On any critical Mohr circle, the points of tangency with the shear fracture envelope (Figure 8.3*B*) indicate the surface stress components on the actual fracture plane at the time of fracture.

The straight line approximation to the shear fracture envelope is the **Coulomb fracture criterion**, and it states that the critical shear stress $|\sigma_s^*|$ is equal to a constant c plus the tangent of the slope angle ϕ of the line times the normal stress σ_n, or

$$|\sigma_s^*| = c + \mu \sigma_n \tag{8.2}$$

where

$$\mu = \tan \phi \tag{8.3}$$

[2]Remember that in confined compression, the confining pressure p defines the magnitude of the radial stress that is exerted by the pressure medium, $p = \hat{\sigma}_2 = \hat{\sigma}_3$.

and where σ_s^* is the critical shear stress; μ and c are the slope and intercept of the lines, respectively; and ϕ is the slope angle of the line, taken to be positive (Figure 8.3*B*). Because the equation is written in terms of the absolute value of the critical shear stress, it describes both lines of the fracture criterion.

The two constants in Equation (8.2), c and μ, characterize the failure properties of the material, and they vary from one type of rock to another. The **cohesion** c is the resistance to shear fracture on a plane across which the normal stress is zero. We call μ the **coefficient of internal friction** and ϕ the **angle of internal friction** because of the similarity, when the cohesion is zero, between Equation (8.2) and Amontons' second law of frictional resistance (see Section 8.4(ii), Equation (8.7), and footnote 3). We can also express the Coulomb fracture criterion in terms of the principal stresses (see Box 8-1).

The Coulomb fracture criterion states that whenever the state of stress in a rock is such that on a plane of some orientation, the surface stress components (σ_n, σ_s) satisfy Equation (8.2), that plane can become a shear fracture. For any critical stress state, there are two points on the Mohr circle at angles of $\pm 2\theta_f$ from $\hat{\sigma}_1$ where the circle is tangent to the two lines given by Equation (8.2) (Figure 8.3*B*; note that $2\theta_f' = -2\theta_f$). These points define the stresses on two differently oriented planes called the **conjugate shear planes** (Figure 8.3*D*, *E*). On the Mohr circle (Figure 8.3*B*), the angles between the σ_n axis and the radii to the points of tangency with the fracture criterion are $\pm 2\theta_f$. Thus in physical space, the normals to the two conjugate shear fractures must be at angles of $\pm \theta_f$ to the $\hat{x}_1$ axis (and to $\hat{\sigma}_1$), and the fracture planes themselves make an angle of $\pm \alpha_f$ with $\hat{\sigma}_1$ (Figure 8.3*D*, *E*). The fracture criterion does not predict which orientation of fracture should actually form.

On the Mohr circle, the radius to the tangent point must be perpendicular to the fracture envelope, so the angles θ_f and α_f are related to the slope angle of the fracture envelope ϕ by the following equations:

$$\begin{aligned} |2\theta_f| &= (90 + \phi) \text{ degrees} = \left(\frac{\pi}{2} + \phi\right) \text{ radians} \\ |2\alpha_f| &= (90 - \phi) \text{ degrees} = \left(\frac{\pi}{2} - \phi\right) \text{ radians} \end{aligned} \tag{8.4}$$

Thus the fracture envelope defines both the critical stress required for fracture and the orientation of the shear fracture that develops.

If the three principal stresses are unequal, the line of intersection of the conjugate shear planes parallels the intermediate principal stress $\hat{\sigma}_2$. In uniaxial or confined compression where $\hat{\sigma}_2 = \hat{\sigma}_3$, there is no unique intermediate principal stress axis, and there are an infinite number of possible orientations for the shear fracture planes, which are distributed as tangents to a cone whose axis is parallel to $\hat{\sigma}_1$.

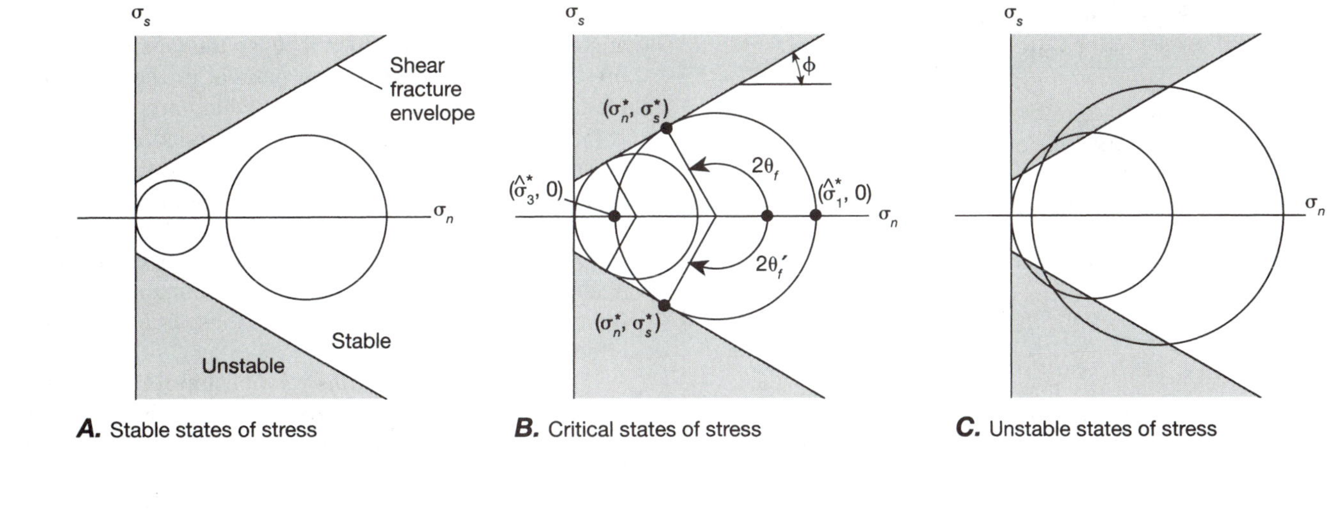

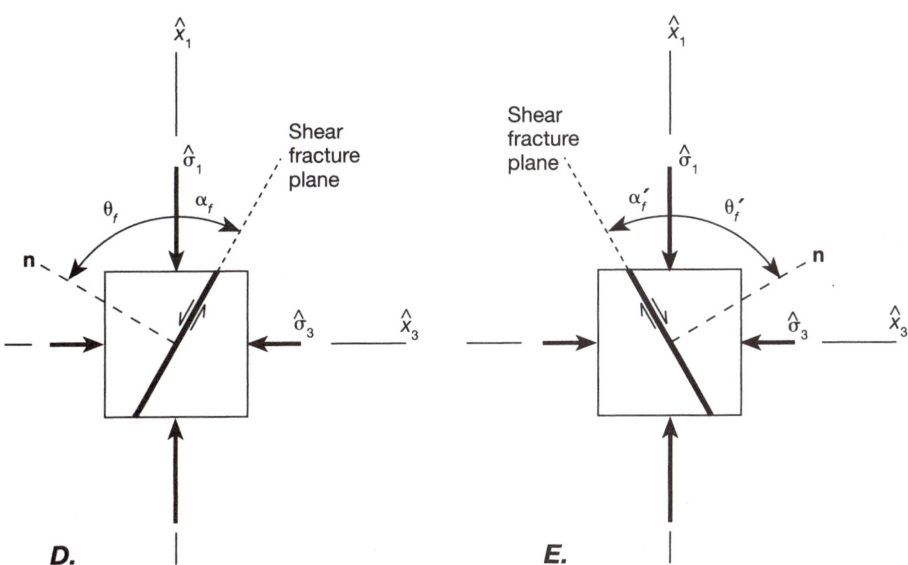

FIGURE 8.3 Coulomb fracture criterion for axial compression. The fracture criterion divides the Mohr diagram into areas of unstable stress (shaded) and stable stress (unshaded). *A*. Mohr circles representing stable states of stress. *B*. Mohr circles for critical states of stress; circles are tangent to the line bounding the shaded area (fracture envelope), and radii are drawn to the points of tangency. The angle of internal friction ϕ and twice the fracture angles $2\theta_f$ and $2\theta_f'$ are shown. *C*. Mohr circle illustrating unstable states of stress. *D*, *E*. Diagrams in physical space showing the orientations of the shear fractures indicated by the two fracture angles labeled in *B*. Note that the maximum compressive stress $\hat{\sigma}_1$ bisects the acute angle between the conjugate shear planes.

Most rocks develop shear fractures with planes oriented approximately $\alpha_f = 30°$ to the maximum principal stress direction. For example, the Coulomb fracture criterion that best fits the experimental data for shear fracturing of Berea sandstone is, in units of megapascals (MPa),

$$|\sigma_s^*| = 24.1 + 0.49\ \sigma_n \tag{8.5}$$

By comparing Equations (8.2) and (8.5) and using Equations (8.3) and (8.4), we find that $\phi = 26°$, $|\theta_f| = 58°$, and $|\alpha_f| = 32°$.

The relationships expressed in Equations (8.3) and (8.4) indicate that μ, ϕ, θ_f, and α_f all measure the same physical property. Figure 8.4*A* shows a histogram of experimentally determined values of α_f on a variety of rocks under widely differing experimental conditions. The mean of the fracture plane angles is about 29°, very close to the oft-quoted "constant" of 30°. These data suggest that the angle α_f between the shear fracture plane and the maximum compressive stress $\hat{\sigma}_1$ is about 30°. Thus the fracture angle θ_f between the normal to the shear fracture plane and the maximum compressive stress $\hat{\sigma}_1$ is about 60°.

BOX 8-1 The Coulomb Fracture Criterion in Terms of Principal Stresses

The Coulomb fracture criterion is sometimes expressed in terms of the critical principal stresses at fracture. We can derive this relationship from Equation (8.2), using the positive value of σ_s^*. We substitute for σ_n and σ_s from Equations (7.35) and (7.36) or Equation (7.46) with $\theta = \theta_f$ and use the relationship from Equation (8.3) and the first Equation (8.4):

$$\mu = \tan\phi = -\frac{1}{\tan 2\theta_f} = -\cot 2\theta_f \qquad (8\text{-}1.1)$$

After some algebraic manipulation, using the standard trigonometric relationships $\tan 2\theta_f = (\sin 2\theta_f / \cos 2\theta_f)$ and $\sin^2 2\theta_f + \cos^2 2\theta_f = 1$, we find that the maximum compressive stress required for fracture varies linearly with the minimum principal stress according to

$$\hat{\sigma}_1^* = S + K\hat{\sigma}_3 \qquad (8\text{-}1.2)$$

where

$$K = \frac{1-\cos 2\theta_f}{1+\cos 2\theta_f} = \left[\mu + \sqrt{1+\mu^2}\right]^2$$
$$S = \frac{2c\sin 2\theta_f}{1+\cos 2\theta_f} = 2c\left[\mu + \sqrt{1+\mu^2}\right] = 2cK^{1/2} \qquad (8\text{-}1.3)$$

μ is the coefficient of internal friction, and c is the cohesion. The asterisk on $\hat{\sigma}_1$ indicates the critical maximum compressive stress for a given value of the minimum compressive stress. Equation (8-1.2) shows that S is the fracture strength under uniaxial compression when $\hat{\sigma}_3 = 0$. Note, however, that setting $\hat{\sigma}_1 = 0$ does not give the tensile strength under uniaxial tension, because this equation is an expression of the Coulomb fracture criterion, which does not account for tensile fracture.

In this form, the fracture criterion is commonly plotted as a line on a graph of maximum versus minimum principal stress. Although the Mohr circle does not plot on such a graph, the difference between $\hat{\sigma}_1^*$ and $\hat{\sigma}_3$ is the diameter of the critical Mohr circle.

The large scatter in values of α_f in Figure 8.4*A* results partly from combining experiments performed under different conditions of stress, confining pressure, pore fluid pressure, temperature, and rock type. The data in the histogram indicated by dark shading are for jacketed dry samples of Berea sandstone. Here the mean fracture plane angle is 31°, and the scatter is reduced, although the angles still range between 26° and 38°. Temperature apparently has no effect on these data. The fracture plane angle tends to increase with confining pressure, however, as shown by Figure 8.4*B*, indicating a fracture envelope concave toward the σ_n axis, although the scatter is still considerable. Thus the experimental variation is considerable in the average value of 30° for α_f. We should expect a similar variation in nature.

It may seem strange that shear fractures form at an angle of approximately $\alpha_f = \pm 30°$ to the maximum compressive stress $\hat{\sigma}_1$ rather than parallel to the conjugate planes of maximum shear stress, which are oriented at ±45°. The lower angle results from competing effects of normal and shear stresses on a given fracture orientation. Both a minimum normal stress and a maximum shear stress will favor the development of a shear fracture, but the normal stress is not a minimum on the same plane for which the shear stress is a maximum. Thus the fracture angle is an optimization of these two competing effects.

Figure 8.5 illustrates this relationship for one of the fracture experiments on Berea sandstone. For this particular state of stress (inset, Figure 8.5*A*; the principal stresses are on the diagonal of the matrix of numbers), the two solid curves in Figure 8.5*A* show how the normal stress σ_n (Equation (7.35)) and the shear stress σ_s (Equation (7.36)) vary with changing orientation of a plane (see also Equations (7.44) and (7.46)). That orientation is defined by the angle θ between the normal to the plane and the $\hat{\sigma}_1$ direction (Figure 8.5*B*). Note that it is impossible to minimize the normal stress and maximize the shear stress on the same plane. We can calculate the curve for the critical shear stress $|\sigma_s^*|$ required to cause fracture on any particular orientation of plane from the Coulomb fracture criterion for Berea sandstone (Equation (8.5)) using values for σ_n from the curve in Figure 8.5*A*. Where the available shear stress equals the critical shear stress needed to cause fracture, the curves for σ_s and $|\sigma_s^*|$ touch, and that point defines the orientation of the fracture plane.

The predicted fracture angle where, within experimental error, the curves touch is $\theta_f = 58°$ ($\alpha_f = 32°$); the experimentally observed angle is $\theta_f = 55°$ ($\alpha_f = 35°$). The two curves, however, are almost parallel for angles of roughly ±5° on either side of the ideal θ_f. Thus within this range of angles, all the planes are very nearly at the critical shear stress. Under such conditions, minor heterogeneities must affect which plane ultimately becomes the fracture plane. This factor probably accounts for some of the scatter in the observed data (dark shaded histogram bars in Figure 8.4*A*).

By far the most common stress state in natural environments is triaxial stress (Figure 7.12*F*). Because faults are shear fractures, the Coulomb fracture criterion leads us to expect faults to form parallel to the intermediate principal stress and at angles of $\alpha_f \approx \pm 30°$ to the maximum compressive stress $\hat{\sigma}_1$. In nature, one orientation of the conjugate pair tends to be locally dominant, although both are present over large areas.

We can get some insight into how natural deformation might proceed by examining scale models in the laboratory. The mechanical properties of dry sand and wet clay provide appropriate scaled analogues for the properties of the Earth's crust (see Sections 18.1 through 18.5). Figure 8.6 shows one experiment in which a layer

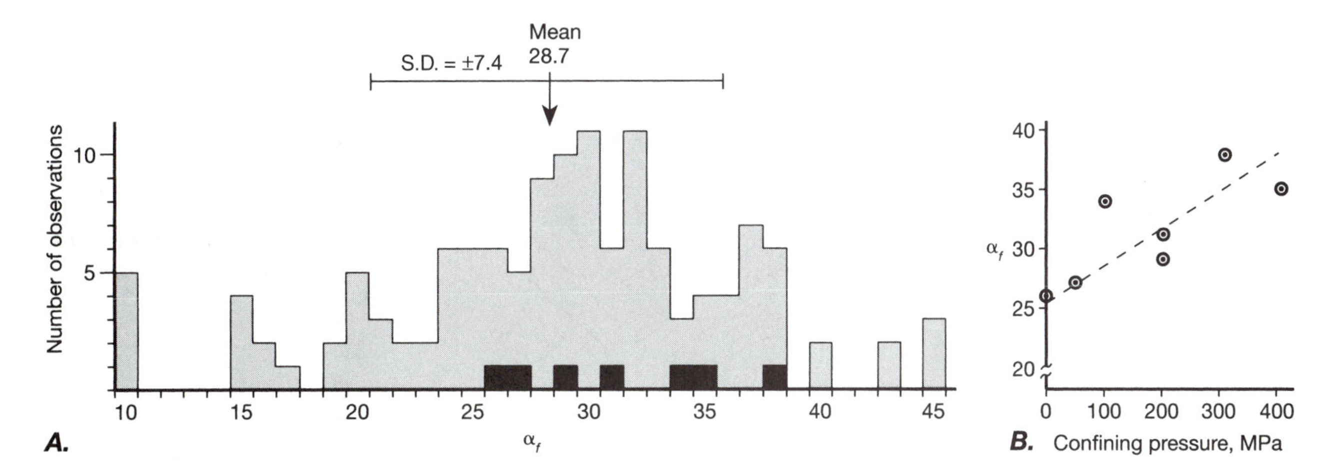

FIGURE 8.4 Experimental values of the fracture angle. *A.* Histogram of experimentally determined fracture plane angles α_f between the fracture plane and the maximum compressive stress $\hat{\sigma}_1$. The diagram includes data from jacketed and unjacketed plutonic, volcanic, sedimentary, and metamorphic rocks under confining pressures from about 0.5 to 500 MPa, temperatures from 20° C to 800° C, and with or without pore fluid. Dark shading accents data angles for jacketed dry samples of Berea sandstone. *B.* Variation of the fracture plane angle with confining pressure for jacketed dry samples of Berea sandstone (dark shaded data in *A*). Data from Handin (1966, table II-3)

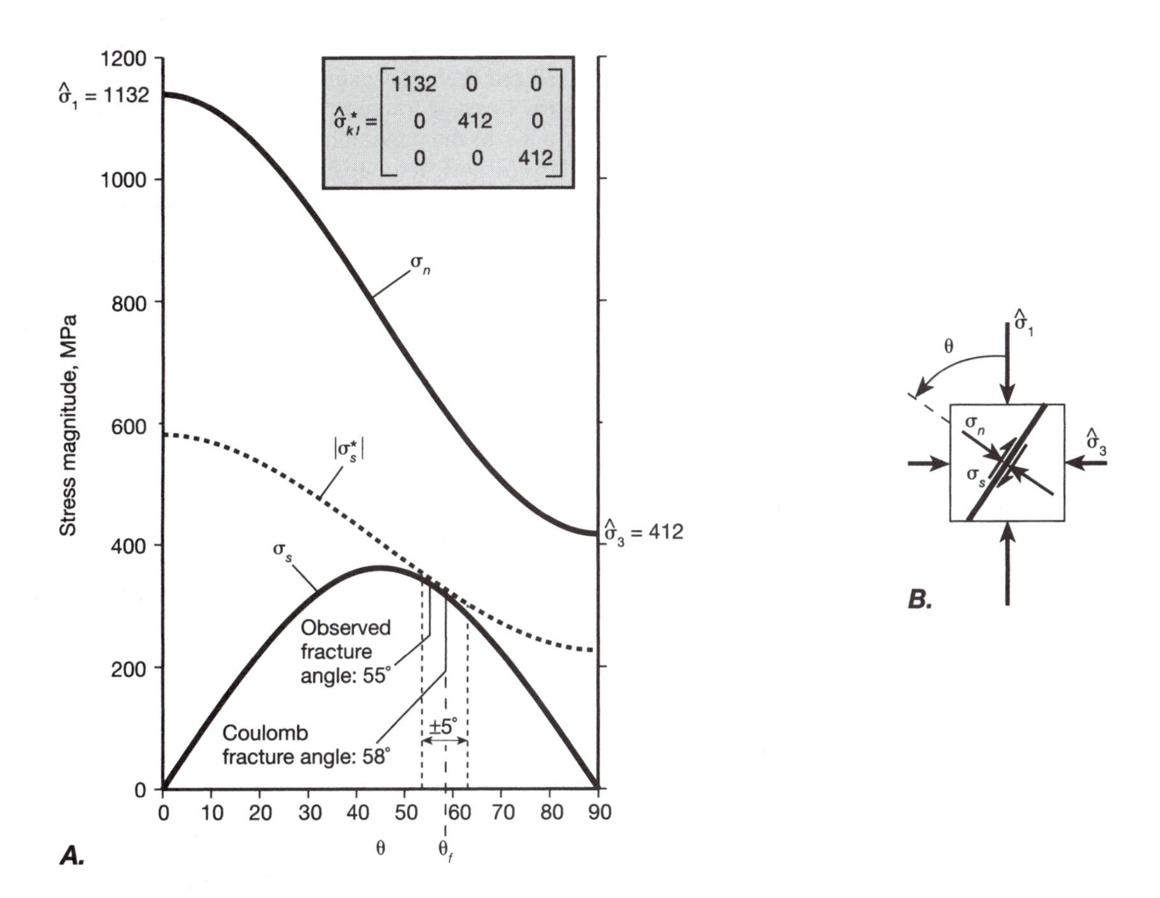

FIGURE 8.5 *A.* The surface stress components σ_n and σ_s and the critical shear stress σ_s^* needed for fracture are plotted versus the orientation θ of the surface for a particular critical state of stress on a sample of Berea sandstone (inset shows the principal stresses on the diagonal of the matrix, in MPa). The fracture angle is the one that minimizes the difference between $|\sigma_s^*|$ and σ_s, as shown on the graph. *B.* Relationship in physical space between the principal stresses, the orientation of the plane, and the surface stress components on the plane. (Fracture data for Berea sandstone from Handin 1966)

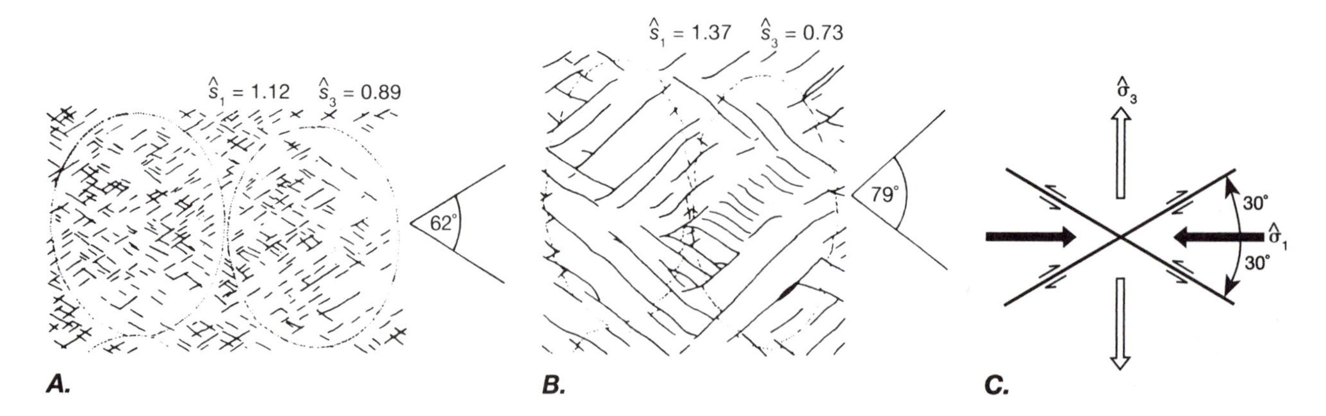

FIGURE 8.6 Pure shear in a clay block deformed by stretching of an underlying rubber sheet. Ellipses record the distortion of original circles. *A.* Shear fractures form in two conjugate orientations separated by an angle of about 60°. *B.* Domains develop in which displacement is concentrated on one orientation of faults or the other. The angle between the conjugate fault sets increases. *C.* Intrepretation of the faults in terms of the Coulomb fracture criterion. (After Hoeppner et al. 1969, referenced in Freund 1974)

of clay was shortened in one direction and stretched in a perpendicular direction by the deformation of an underlying sheet of rubber. The amount of deformation is indicated by the ellipses, which were circles in the undeformed state. Faults initially develop in conjugate orientations that are approximately 60° apart (Figure 8.6*A*). Slip on one set of faults offsets faults of the other orientation, however, and interferes with the deformation. Thus with continued deformation, domains develop in which faults of one orientation dominate (Figure 8.6*B*). The angle between the conjugate sets of faults increases with increasing deformation as a result of rotation of the original fracture planes. The initial orientations of the conjugate shear fractures are consistent with the Coulomb fracture criterion, as indicated in Figure 8.6*C*.

A different geometry of faults develops if the layer of clay is subjected to a shear parallel to its boundaries (Figure 8.7). Again two conjugate sets of faults labeled *R* and *R′* appear (Figure 8.7*A*). Neither is parallel to the imposed direction of shearing, however, and they are referred to as **secondary shears** or **Riedel shears**. *R* shears are "synthetic," having the same sense (sinistral in Figure 8.7*A*) as the imposed shear, and they are oriented about 15° from the plane of the imposed shear. *R′* shears are "antithetic," having a sense of shear (dextral in Figure 8.7*A*) opposite to that imposed, and they are oriented about 75° to 80° from the plane of the imposed shear. With increasing deformation, *R* shears rotate to smaller angles, and another set of secondary synthetic shears (labeled *P*) develops, oriented at about −10° from the imposed shear plane (Figure 8.7*B*; compare Figure 6.4*A*, which shows an imposed dextral shear).

We can account qualitatively for the *R* and the *R′* shears in terms of Coulomb fracture angles if we assume that the imposed shear results in a state of pure shear stress (Figure 7.12*G*) and that the plane of imposed shear is the plane of maximum shear stress at 45° to $\hat{\sigma}_1$ (Figure 8.7*C*). The state of pure shear stress results in zero normal stress on the plane of maximum shear stress, which implies that only a shear stress is imposed on the clay block. The orientations of the Riedel shears *R* and *R′* are then consistent with their being conjugate shear fractures at fracture plane angles of approximately $\alpha = \pm 30°$ to the maximum principal stress $\hat{\sigma}_1$, as is expected from the Coulomb fracture. The secondary shear fractures labeled *P* are not explained at all by this simple analysis.

The Coulomb fracture criterion is a reasonable and useful approximation to many data on shear fracturing under compressive stresses. Nevertheless, it applies only to states of stress with positive principal stresses and to critical differential stresses that are not too large (see Section 8.2), and it does not account for all observed shear fracturing. Thus we need to understand how other factors affect the development of shear fractures so that we might better predict the onset of fracturing and better interpret the observations we make of the Earth. In the following sections, therefore, we examine the effects of confining pressure (Section 8.4), pore fluid pressure (Section 8.5), and anisotropy, the intermediate principal stress, temperature, and scale (Section 8.6). We discuss limitations to the Coulomb fracture criterion in Section 8.7 (see also Section 9.12), and we describe the Griffith theory of fracture and its relation to observed fracture behavior in Section 8.8.

8.4 EFFECTS OF CONFINING PRESSURE ON FRACTURING AND FRICTIONAL SLIDING

i. Confining Pressure and Shear Failure

Experimental data show that the Coulomb fracture criterion does not apply in the tensile part of the Mohr di-

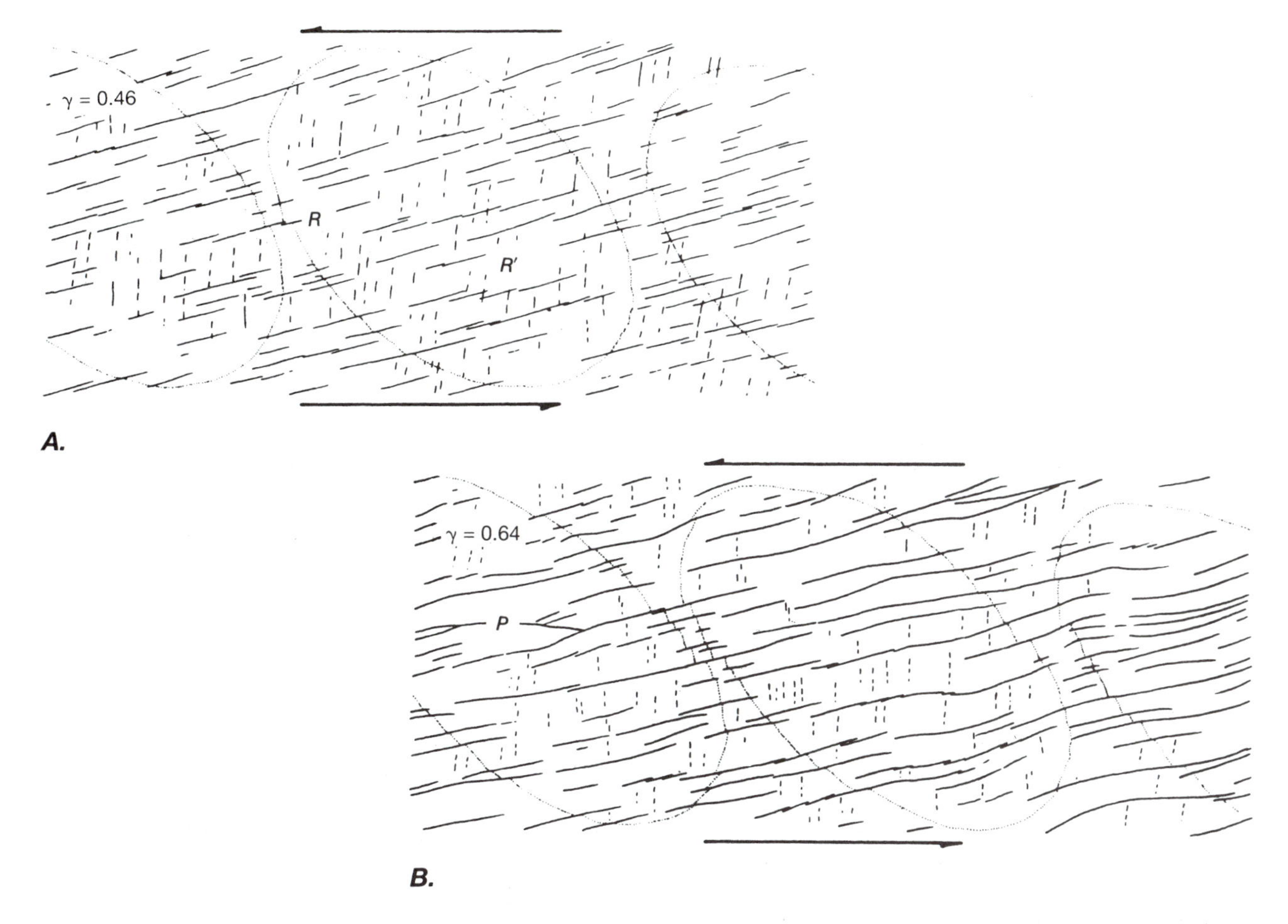

FIGURE 8.7 Sinistral simple shear in a clay block induced by shearing the substrate of the clay. *A.* Conjugate Riedel shears R and R′ form at angles of about 10° and 75°, respectively, to the imposed direction of shear. *B.* R' shears rotate counterclockwise and *R* shears rotate clockwise. *R* shears become dominant and wavy. *P* shears begin to form at an angle of roughly −10° to −15° to the imposed shear direction. *C.* Interpretation of *R* and R' in terms of the Coulomb fracture criterion. The plane of imposed shear is the plane of maximum shear stress oriented at a 45° angle from the maximum compressive stress. It is a plane of pure shear stress. This analysis does not account for *P* shears. (After Freund 1974)

agram. In fact, experiments indicate that rather than the straight-line fracture envelope shown in Figure 8.3*B*, a more comprehensive fracture criterion should actually approximate a parabolic curve on the tensile (negative) side of the normal stress axis that connects with the Coulomb fracture criterion on the compressive (positive) side (Figure 8.8). If the Mohr circle is tangent to the parabolic fracture envelope where the envelope crosses the σ_n axis, tension fractures form normal to the least principal stress ($2\theta_f = 180°$; Figure 8.8*A*). With increasing confining pressure, the critical Mohr circle shifts to the right and increases in diameter, but the point of tangency of the circle with the parabolic failure envelope does not predict accurately the observed failure angle. At higher confining pressures, shear fractures are oriented according to the Coulomb fracture criterion ($2\theta_f \cong 120°$; Figure 8.8*B*).

At still higher confining pressures, the fracture envelope becomes concave toward the normal stress axis and decreases in slope. As a consequence, $2\theta_f$ decreases with increasing confining pressure, and the fracture plane angle α_f increases (Figure 8.8*C*; compare Figure 8.4*B*). This change in fracture orientation coincides with a transition from brittle to ductile behavior. In the ductile region the Coulomb criterion no longer applies, and another failure criterion, the **von Mises criterion**, becomes applicable.

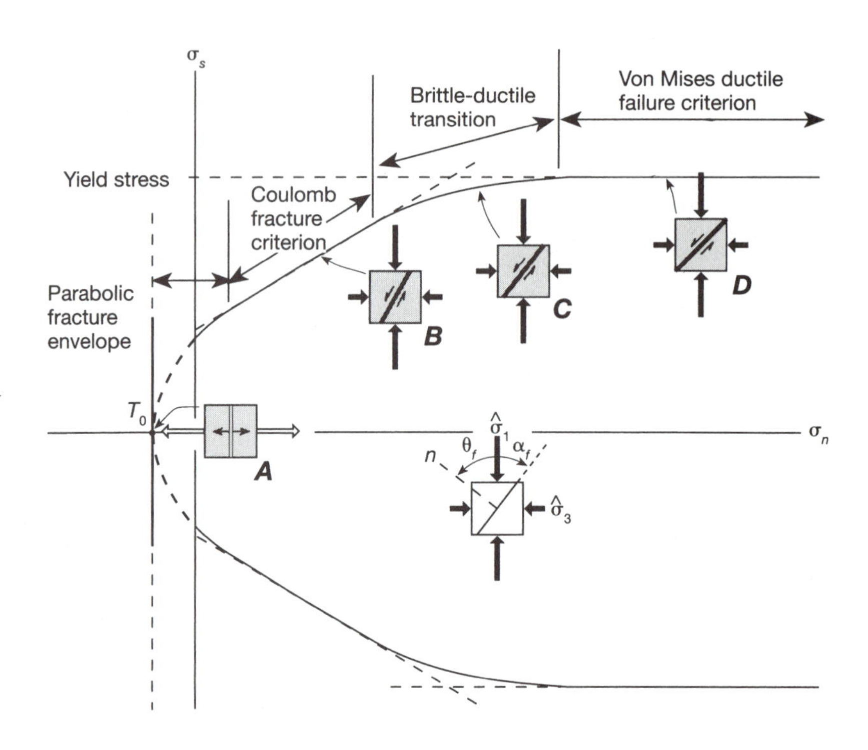

FIGURE 8.8 Mohr diagram showing schematic portrayal of failure envelopes and related fractures. For the dashed portion of the parabolic curve in the field of tensile normal stresses, the orientation of the fractures is not well predicted by the angle made by the radius of the critical Mohr circle to the point of tangency with the failure envelope. Shaded boxes are physical space diagrams showing the orientation of the failure plane at different points along the failure envelope. Double lines represent extension fracturing; solid lines indicate shear fractures. *A*. Tension fracture. *B*. Brittle shear fracture according to the Coulomb criterion. *C*. Shear fracture in the brittle-ductile transition. *D*. Ductile shear failure according to the von Mises criterion.

On a Mohr diagram, the von Mises criterion consists of a pair of parallel lines of constant shear stress symmetric about the normal stress axis (Figure 8.8). This criterion implies that ductile deformation begins at a critical shear stress, called the **yield stress**, which is independent of the confining pressure, and that planes of ductile failure are the planes of maximum shear stress ($2\theta_f = \pm 90°$; Figure 8.8*E*). The algebraic expression for the von Mises criterion is

$$|\sigma_s^*| = \text{constant} \tag{8.6}$$

Figure 8.9 illustrates the transition from brittle fracture to ductile deformation with increasing confining pressure. The figure shows the results of a series of room-temperature experiments on samples of Wombeyan marble from Australia. At a confining pressure of about 0.1 MPa (atmospheric pressure), longitudinal splitting occurs (Figure 8.9*A*); at 3.5 MPa, standard shear fractures form (Figure 8.9*B*); at 35 MPa, the deformation is transitional and is characterized by more pervasive fracturing and by the development of conjugate shears with a large fracture angle (Figure 8.9*C*). Finally, at 100 MPa confin-

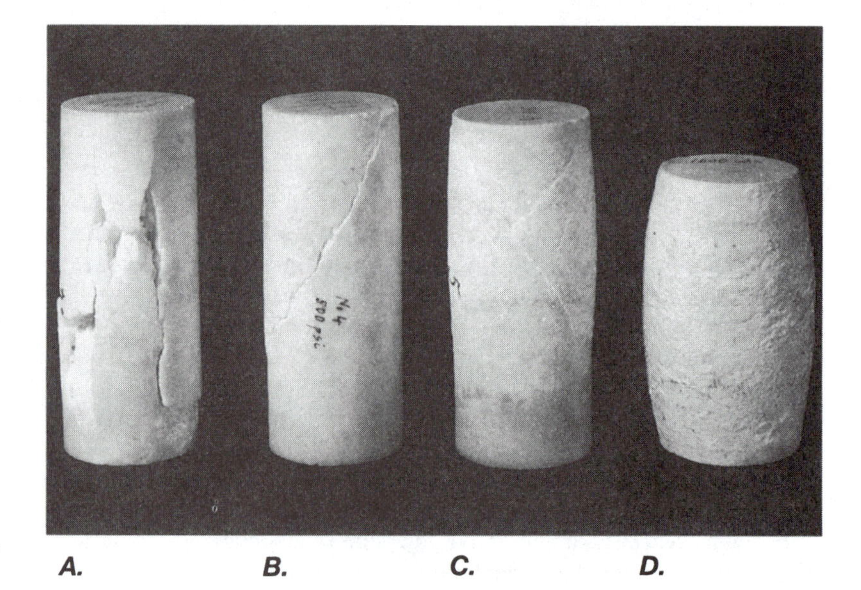

FIGURE 8.9 Effect of confining pressure p on mode of deformation. Experimental results on Wombeyan marble at room temperature. All samples had the same initial length. *A*. Longitudinal splitting at $p = 0.1$ MPa (atmospheric pressure). *B*. Single shear fracture at $p = 3.5$ MPa (≈ 100 m depth). *C*. Brittle-ductile transition at $p = 35$ MPa (≈ 1 km depth). *D*. Ductile flow at $p = 100$ MPa (1 kb ≈ 3 km depth). (From Patterson 1978, fig. 48)

ing pressure, the marble cylinder deforms ductilely into a smooth barrel shape (Figure 8.9D).

Thus we can account for the failure of rocks over a broad range of pressure only by means of a composite failure criterion, as illustrated schematically on the Mohr diagram in Figure 8.8. In the Earth, of course, the situation is even more complex, because an increase in pressure with depth goes hand in hand with an increase in temperature, which lowers the yield stress for ductile deformation.

ii. Confining Pressure, Frictional Sliding, and Cataclasis

Fault movement at the Earth's surface often occurs along pre-existing fractures, which implies that frictional sliding on a pre-existing fracture is easier than the formation of a new fracture. Because the Earth's crust is pervasively fractured, it is important to understand frictional sliding and how it competes with the process of brittle fracture. After a shear fracture develops in a rock at relatively low confining pressure, the fracture plane is a plane of weakness because the rock possesses no cohesion across it. Subsequent deformation at low confining pressures occurs by frictional sliding on the fracture. The onset of frictional sliding is described by a pair of lines on the Mohr diagram, similar to the Coulomb fracture criterion except that they pass through the origin (Figure 8.10). These lines are a plot of **Amontons' second law** of friction,[3] which states that the critical shear stress for frictional sliding is proportional to the normal stress, or

$$|\sigma_s^*| = \bar{\mu}\sigma_n \tag{8.7}$$

where $|\sigma_s^*|$ is the magnitude of the critical shear stress and $\bar{\mu}$ is the coefficient of sliding friction. The coefficient of sliding friction is generally greater than the coefficient of internal friction ($\bar{\mu} > \mu$) (cf. Equation (8.2)), so at low confining pressure, the differential stress $\sigma^{(Dif)} \equiv \hat{\sigma}_1 - \hat{\sigma}_3$ required to produce sliding is less than that needed to form another fracture. Immediately after fracture (Mohr circle I, Figure 8.10), the differential stress must drop to a level below the frictional sliding criterion (Mohr circle II, Figure 8.10). This behavior explains why faulting near the Earth's surface often results from sliding on pre-existing faults.

[3]Guillaume Amontons (1663–1705) was a French experimental physicist. He published his two laws of friction in 1699, although according to Scholz (1990), the same laws had been discovered 200 years earlier by Leonardo da Vinci but apparently had never been published. Amontons' first law (also known as the law of Leonardo da Vinci) states that *The frictional force is independent of the area of the surfaces in contact.* Amontons' second law (also known as the law of Euler and Amontons) states that *The friction is proportional to the normal load across the sliding surfaces.* The third law of friction, known as Coulomb's law, states that *The force of friction is independent of the velocity of sliding.* Friction is actually more complex than these laws imply, as we explain in Section 16.2 and Box 16-2.

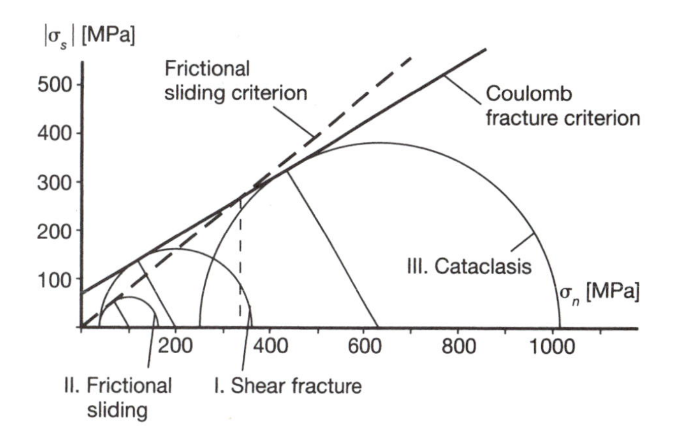

FIGURE 8.10 Example of Coulomb criterion for fracture (solid line): ($|\sigma_s^*| = 70 + 0.6\ {}_E\sigma_n$) and for frictional sliding on an existing fault surface (dashed line $|\sigma_s^*| = 0.81\ {}_E\sigma_n$). Data for Weber sandstone. Circle I: a critical stress for shear fracture. Circle II: a critical stress for frictional sliding on the fracture plane at constant confining pressure. Circle III: a critical stress for cataclastic flow during which fracturing requires a lower differential stress than frictional sliding at the same confining pressure $\hat{\sigma}_3$. (After Byerlee 1975)

At low confining pressure, frictional sliding occurs as a smooth, continuous motion called **stable sliding** (Figure 8.11*A*). As the compressive stress across the sliding surface increases with increasing confining pressure, the motion changes to **stick-slip** behavior (Figure 8.11*B*), which is characterized by "stick" intervals of no motion, during which the shear stress increases, alternating with "slip" intervals of rapid sliding that relieve the stress (Figure 8.11*B*). On a much larger scale than laboratory experiments, this same phenomenon is responsible for the episodic nature of many earthquakes, as suggested by a detailed fault slip history deduced from young deformed sediments along the San Andreas fault in southern California (Figure 8.11*C*). The "stick" parts of the cycle represent the periods of quiescence between earthquakes, and the "slip" parts represent the earthquakes themselves.

As confining pressure increases still further, the frictional sliding criterion and the fracture criterion cross, and less shear stress is required to form a new fracture in a rock than to slide along an existing one (Mohr circle III, Figure 8.10). The rock deforms by pervasive brittle fracturing and comminution of the grains, rather than by sliding on pre-existing cracks. This process of **cataclasis** during **cataclastic flow** results in a cataclasite, a rock typical of many fault zones that form in the shallow crust where brittle deformation dominates (see Figure 3.5*B* and Section 16.2).

The phenomenon of frictional sliding is actually much more complex than the first approximation that we have presented here, and that complexity has important implications for the behavior of active faults. See Section 16.2 and Box 16-2 for a more detailed look at friction.

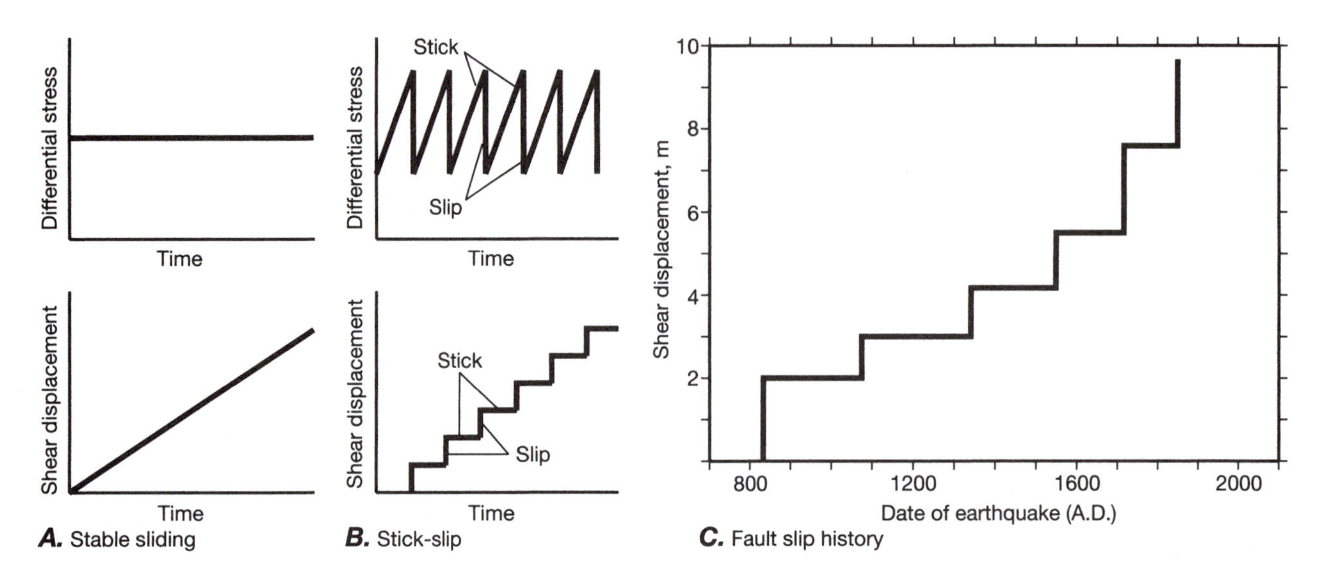

FIGURE 8.11 Variation of differential stress ($\sigma^{(Dif)} = \hat{\sigma}_1 - \hat{\sigma}_3$) and shear displacement with time. *A.* For stable sliding. *B.* For stick-slip. *C.* A record of displacement versus time for the San Andreas fault in southern California. (After Sieh 1982)

8.5 EFFECTS OF PORE FLUID PRESSURE ON FRACTURING AND FRICTIONAL SLIDING

The presence of pore fluid causes a rock to behave as though the confining pressure were lower by an amount equal to the pore fluid pressure. The mechanical behavior is described in terms of the effective stress (Section 7.3, Equation (7.39); see Figure 7.12*J*). The effective stress is the result of reducing all applied normal stress components by an amount equal to the pore fluid pressure p_f, while leaving the applied shear stress components unchanged. Thus the Mohr circle for the effective stress is the same size as for the applied stress, but it is shifted to the left along the normal stress axis toward smaller normal stresses by an amount equal to the pore fluid pressure (Figure 8.12).

The fracture criterion remains the same, except that the normal stress is replaced by the effective normal stress:

$$|\sigma_s^*| = c + \mu(\sigma_n^{(Eff)}) = c + \mu(\sigma_n - p_f) \qquad (8.8)$$

where

$$\sigma_n^{(Eff)} \equiv \sigma_n - p_f \qquad (8.9)$$

The effect of shifting the Mohr circle to the left is that states of stress that are stable at zero pore fluid pressure may become unstable if the pore pressure is sufficiently high. If the differential stress $\sigma^{(Dif)}$ (the diameter of the Mohr circle) is small, as is commonly the case in the Earth, and if the pore fluid pressure p_f exceeds the minimum compressive stress $\hat{\sigma}_3$ by an amount equal to the tensile strength of the rock (that is, $|\hat{\sigma}_3 - p_f| = T_0$), then extension fracturing can occur, even at great depths (Figure 8.12*A*). If $\sigma^{(Dif)}$ is relatively large, on the other hand, shear fracturing can result if the pore fluid pressure is sufficiently high (Figure 8.12*B*).

Pore pressure has exactly the same effect on frictional sliding. An increase in the pore fluid pressure causes a decrease in the effective normal stress across the sliding surface. Because frictional stress is proportional to the effective normal stress across the sliding surface, the critical shear stress necessary for sliding also decreases.

Thus pore fluid pressure is geologically important for several reasons. First, it lowers the differential stress necessary to cause failure and permits fracture at depths where the rock otherwise would be either stable or in the realm of ductile behavior. Second, it can shift the conditions of frictional sliding from those favoring stick-slip behavior to those favoring stable sliding. And third, it can shift deformation from cataclastic flow to frictional sliding. High pore fluid pressure is commonly an important factor in the development of joints and large-scale faults, as we discuss in Chapter 9.

The most obvious origins of pore fluid are water incorporated into a sediment during subaqueous deposition and fluids released from minerals by dehydration reactions during metamorphism. If a rock at a given depth is permeable all the way to the surface, the pore spaces are interconnected, and the fluid pressure cannot exceed the weight of a column of water extending from the sur-

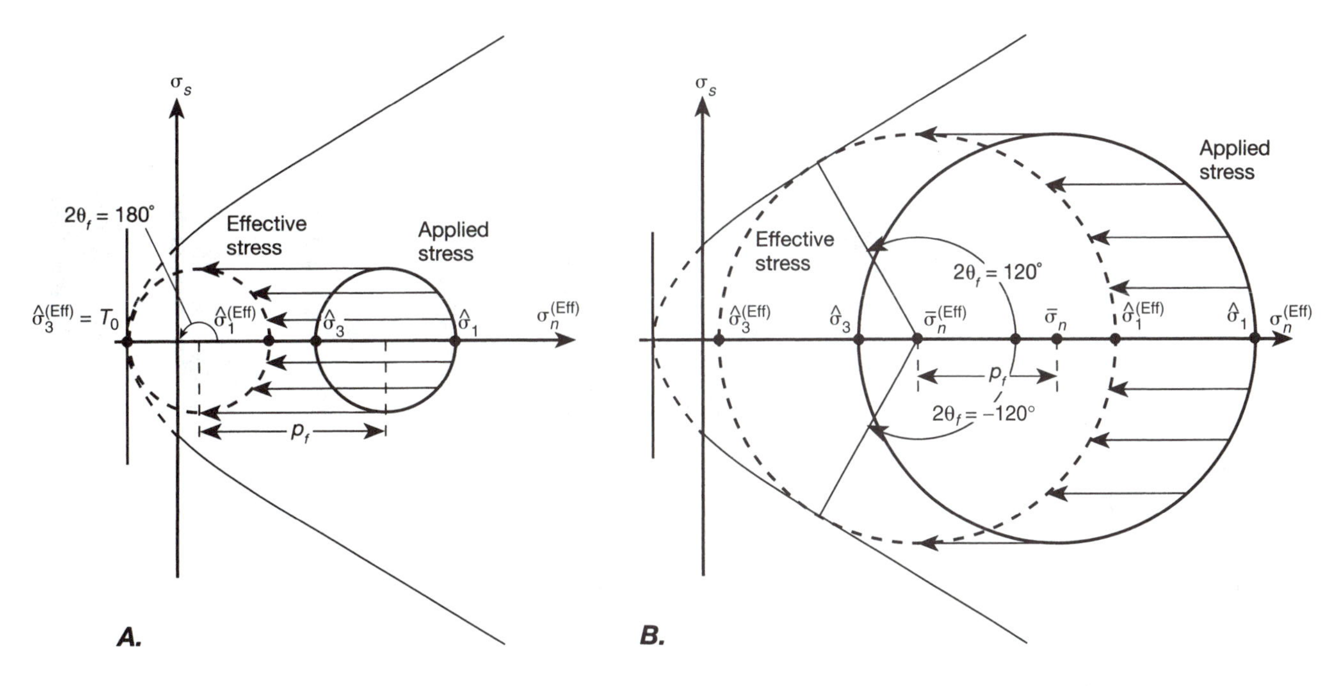

FIGURE 8.12 Mohr diagram with the effective normal stress plotted on the abscissa, showing the effect of pore fluid pressure on the fracture stability of rock. *A*. At small differential stress, an increase in pore pressure leads to tension fracture. *B*. At large differential stress an increase in pore pressure leads to shear fracture.

face to the given depth. The lithostatic pressure results from the weight of the overlying rock at a particular depth. Thus assuming complete hydraulic communication between the surface and depth, the hydrostatic pore fluid pressure (p_f) and the vertical lithostatic normal stress (σ_V) at a given depth are, respectively,

$$p_f = \rho_w g h \quad \text{and} \quad \sigma_V = \rho_r g h \tag{8.10}$$

where ρ_w and ρ_r are the densities of water and rock, respectively; g is the magnitude of the acceleration due to gravity; and h is the depth in question. If the density of water is 10^3 kg/m^3 and that of sediment is $(2.3)(10^3)$ kg/m^3, then the ratio λ of the hydrostatic pore pressure to the lithostatic pressure is

$$\lambda \equiv \frac{p_f}{\sigma_V} = \frac{\rho_w g h}{\rho_r g h} = 0.4 \tag{8.11}$$

It is perhaps surprising that values of λ approaching 1 are not uncommon in deep wells drilled in many sedimentary sequences, as well as in accretionary prisms along consuming plate margins. Thus in these areas, impermeable barriers must exist that prevent free communication of the fluid with the surface. These barriers and the resultant high fluid pressures that can accumulate beneath them may play a role in the localization and occurrence of earthquakes in seismically active regions.

8.6 EFFECTS ON FRACTURING OF ANISOTROPY, THE INTERMEDIATE PRINCIPAL STRESS, TEMPERATURE, AND SCALE

i. Effect of Anisotropy

So far in this discussion, we have assumed that rocks have the same mechanical properties in all directions—that is, that they are **mechanically isotropic**. In this case the fracture criterion is the same regardless of the orientation of the principal stresses in the rock. Many rocks, however, are **mechanically anisotropic**; that is, their strength is different in different directions. A mechanical anisotropy may result, for example, from a preferred planar alignment of platy minerals in a rock, called a foliation or cleavage, which is characteristic of slates and schists. Such rocks break easily, or cleave, along these planes of weakness. A pervasive joint set has the same effect at a larger scale. These planes of weakness dominate the strength of the rock and the orientation of the fractures that develop for a wide range of orientations of the principal stresses relative to the anisotropy.

A series of fracture experiments performed on the Martinsburg slate illustrate these effects. Figure 8.13*A* shows examples of the fractured samples in copper jackets for different orientations (δ) of the slaty cleavage plane relative to the maximum compressive stress $\hat{\sigma}_1$. Figure 8.13*B* shows the relationship between the fracture

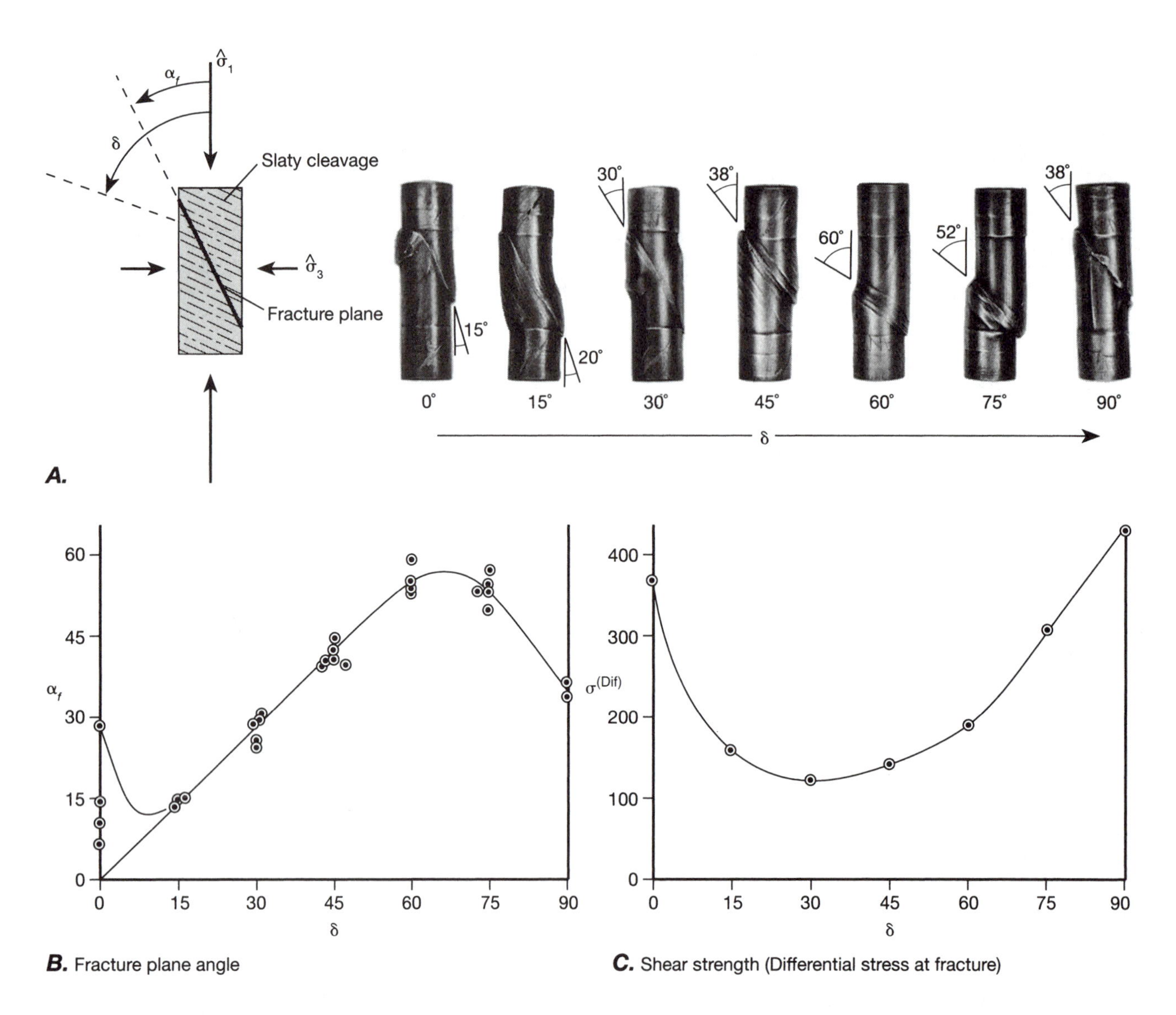

FIGURE 8.13 The effect of anisotropy on fracture. Experimental results of fracture of Martinsburg slate at confining pressures of 50, 100, and 200 MPa. α_f is the fracture angle, δ defines the angle between the cleavage and the maximum compressive stress $\hat{\sigma}_1$. *A*. Copper-jacketed samples of Martinsburg slate showing the angle of shear fracture developed for the different values of δ. Samples with $\delta = 45°$ and 75° deformed at 100 MPa confining pressure ($\approx$ 2.5- to 3-km depth); the other samples deformed at 200 MPa confining pressure ($\approx$ 5- to 6-km depth). *B*. Variation of the fracture plane angle α_f with the orientation of the cleavage. *C*. Variation of the rock strength, measured in terms of the differential stress at fracture, with orientation of the cleavage. (After Donath 1961)

plane angle α_f and the angle δ. If the cleavage plane and the maximum compressive stress $\hat{\sigma}_1$ are either parallel ($\delta = 0°$) or perpendicular ($\delta = 90°$), there is no resolved shear stress on the cleavage plane because the cleavage is parallel to a principal plane of stress. In these cases, fracture strength is a maximum (Figure 8.13*C*), and shear fractures form at the usual angle of about 30° (Figure 8.13*B*). If the cleavage plane and $\hat{\sigma}_1$ are parallel ($\delta = 0°$), however, there is also a tendency to develop longitudinal splitting at low confining pressures. At values of δ between 15° and 60°, shear fractures tend to develop parallel to the cleavage, and even if δ is as high as 75°, the cleavage still has a substantial influence on the fracture plane orientation. Shear strength is lower for those orientations of slaty cleavage that affect the formation of shear fractures (Figure 8.13*C*), and it is a minimum when the cleavage is parallel to the usual shear fracture plane at an angle of approximately 30° to the maximum compressive stress $\hat{\sigma}_1$.

In simple terms, two different fracture criteria are necessary to account for the behavior of the rock. One criterion, plotted as the outer pair of solid lines in Figure 8.14, applies to fractures that develop across the plane of weakness. The Mohr circle cannot cross this cri-

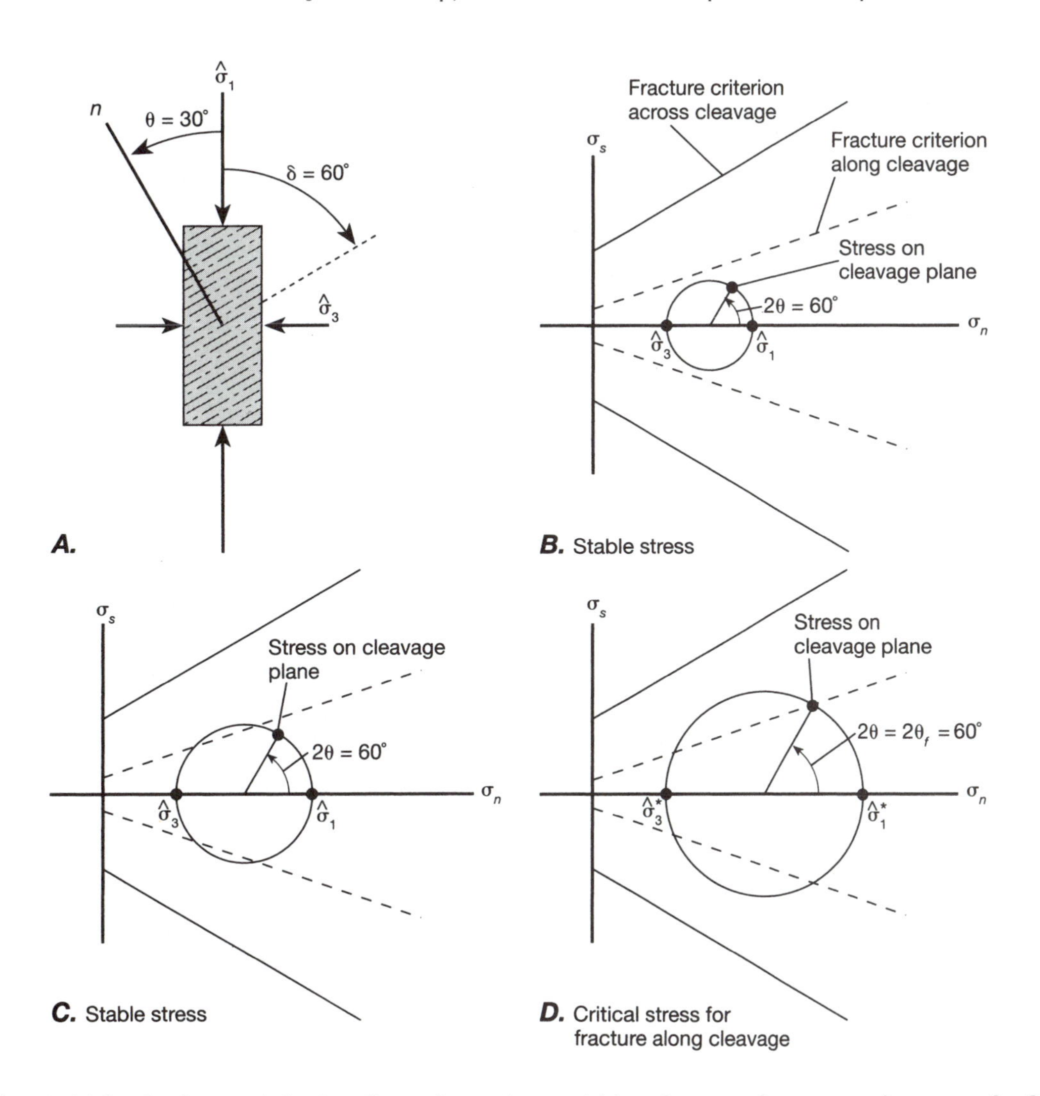

FIGURE 8.14 Model for the fracture behavior of an anisotropic material based on two fracture envelopes: one for fracture *across* the plane of weakness (outer pair of solid lines) and one for fracture *parallel* to the plane of weakness (inner pair of dashed lines). For these diagrams the normal to slaty cleavage is at an angle $\theta = 30^\circ$ from the maximum compressive stress (the cleavage plane itself is at an angle $\delta = 60^\circ$ from $\hat{\sigma}_1$). *A.* Diagram of physical space showing the relative orientations of the cleavage and the principal stresses. *B.* Stable state of stress. *C.* Stable state of stress. The inner envelope defines the critical stresses only for the surface stress *on the cleavage plane*, which in this case is in the stable zone. *D.* Critical state of stress. The surface stress on the cleavage plane reaches the critical condition defined by the inner envelope, and fracture occurs parallel to the cleavage.

terion, because a fracture forms at the usual shear angle as soon as the surface stress components on any plane reach the critical values. The second criterion, plotted as the inner pair of dashed lines, applies only to the surface stress components acting on the cleavage plane.

The stress is stable as long as the Mohr circle is within the outer fracture envelope, and the surface stress components on the cleavage plane plot within the inner fracture envelope (Figure 8.14*B*, *C*). Note that in Figure 8.14*C*, the stable Mohr circle can cross the inner fracture envelope as long as the surface stress acting on the cleavage plane remains in the stable field. Unstable stresses occur either when the surface stress on the cleavage plane reaches the inner fracture envelope (Figures 8.14*D* and 8.15*B*) or when the surface stress on the cleavage is stable but the Mohr circle intersects the outer fracture envelope (Figure 8.16*B*). In the former case, the fracture develops parallel to the cleavage; in the latter case the fracture develops across the cleavage.

The shear strength of a rock equals the differential stress at shear fracture (plotted in Figure 8.13*C*) and thus is measured as the diameter of the critical Mohr circle. If the cleavage is oriented such that the surface stress on the cleavage plane plots at the point where the Mohr circle is tangent to the inner fracture envelope (Figure 8.15*B*), the shear strength is a minimum. If the cleavage

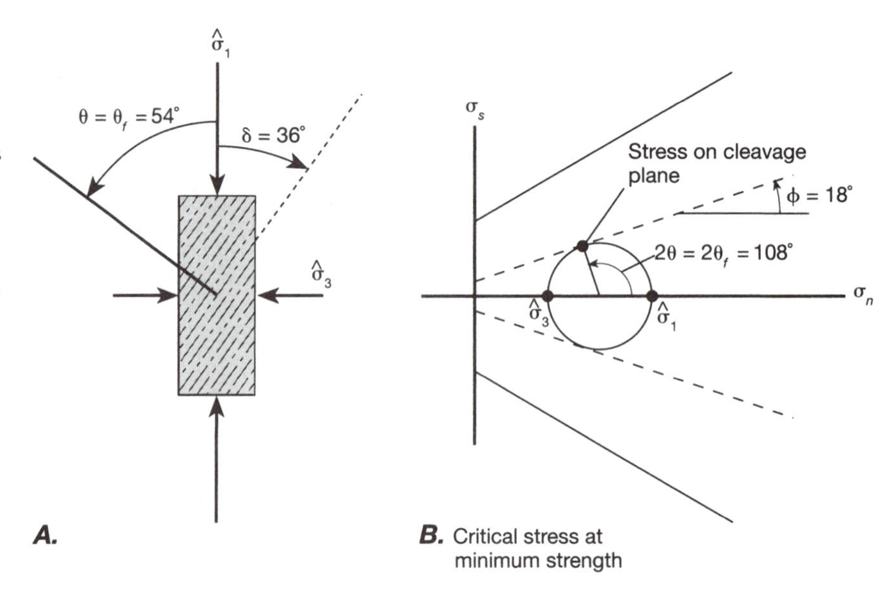

FIGURE 8.15 The minimum strength for a rock with a planar mechanical anisotropy. The fracture criteria are the same as in Figure 8.14. *A.* Diagram of physical space: The plane of weakness is oriented parallel to the direction of preferred shear fracture. *B.* Mohr diagram: The plane of weakness is oriented so that the surface stress on the plane plots at the point of tangency between the Mohr circle and the inner (dashed) fracture criterion.

is oriented such that the shear stress on the cleavage plane is very small, the rock fractures across the cleavage and the shear strength is a maximum (Figure 8.16*B*).

ii. Effect of the Intermediate Principal Stress

So far in our discussion of fracture criteria, we have assumed that for isotropic rocks, shear fractures develop parallel to the intermediate principal stress $\hat{\sigma}_2$. In that orientation, $\hat{\sigma}_2$ contributes nothing to the normal or shear stress on the fracture plane, so it should have no effect on the fracture strength. This assumption is only approximately valid, however, because experiments indicate that $\hat{\sigma}_2$ does have a small effect on a rock's fracture strength. The strength is highest and the fracture plane angle α_f is lowest when the intermediate principal stress equals the maximum compressive stress $\hat{\sigma}_2 = \hat{\sigma}_1$ (extensional stress; see Figure 7.12*E*). Conversely, the strength is lowest and the fracture plane angle is highest when the intermediate principal stress equals the minimum com-

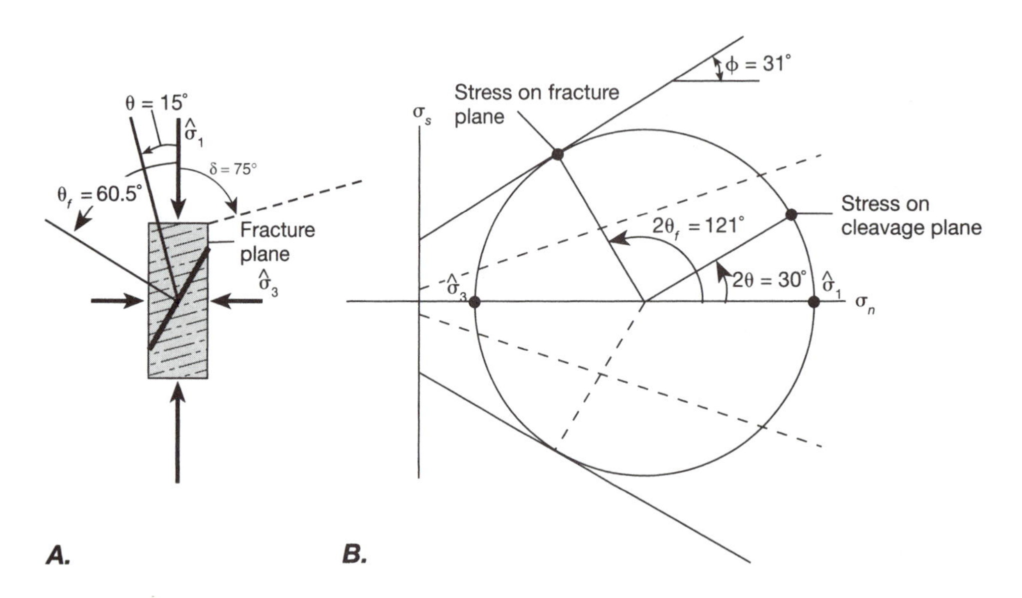

FIGURE 8.16 Shear fracture develops across the plane of weakness when stresses on that plane remain below the critical state. *A.* Stable orientation of the cleavage with respect to the principal stresses. Only one of the conjugate fracture planes is shown. *B.* Mohr diagram showing the same fracture criteria as in Figure 8.15. The orientation of the anisotropy is such that the surface stress on the plane of weakness is within the inner failure envelope—and is thus stable—when the Mohr circle intersects the envelope for fracture across the anisotropy. Thus the rock fractures across the anisotropy. The dashed radius indicates the stress on the conjugate fracture plane.

pressive stress $\hat{\sigma}_2 = \hat{\sigma}_3$ (confined compression; see Figure 7.12*D*). These relationships imply that the angle of internal friction ϕ for the fracture envelope is highest for extensional stress and lowest for confined compression.

For anisotropic rocks, the plane of weakness need not be parallel to the intermediate principal stress. If it is not parallel to $\hat{\sigma}_2$, then that stress component contributes to the normal and shear stresses on the plane of weakness and therefore affects the shear stress required for fracturing parallel to the plane.

iii. Effect of Temperature

The effect of temperature on brittle fracturing is difficult to investigate experimentally, because above temperatures ranging from 200°C to 500°C (depending on the composition of the rock) ductile deformation mechanisms become important. The increase in temperature lowers the von Mises yield stress for ductile behavior (Figure 8.8), thereby lowering the pressure of the brittle-ductile transition and reducing the field of brittle behavior. Experimental data suggest, however, that there is also a small decrease in the brittle shear strength with increasing temperature.

iv. Effect of Scale

Rock samples that are tested in the laboratory generally are homogeneous samples without flaws. In nature, however, such flaws as joints, faults, and compositional heterogeneities are a characteristic feature of large bodies of rock. Thus we should expect that the strengths determined from flawless samples in the laboratory might not describe the behavior of large bodies of rock. In pervasively jointed rock, for example, the strength may be determined more by the properties of the joints than by the rock material between them. In fact, experiments have demonstrated that as the scale of the sample tested increases, the measured strength decreases. Thus we must expect that the fracture strength of the Earth's brittle crust is less than that suggested by most measurements made on crustal rocks in the laboratory.

8.7 LIMITATIONS OF THE COULOMB FRACTURE CRITERION

The Coulomb fracture criterion is a two-dimensional criterion based on the assumption that stress is the cause of the deformation. This criterion does not apply in all situations, however, and it is important to understand its limitations. We have discussed the limits of applicability at low mean stresses where tensile stresses can occur and at high stresses where the von Mises ductile failure criterion becomes important, but under some circumstances, the linear Coulomb fracture criterion does not strictly apply.

We have used the Coulomb fracture criterion to understand the development of Riedel shears (Figure 8.7). In fact, however, the normal range of applicability of the linear Coulomb fracture criterion does not extend to the state of pure shear stress (Figure 7.12*G*) that seems applicable to this experiment because the point of tangency between the fracture criterion line and the pure shear stress Mohr circle would lie in the tensile region of the Mohr diagram. This would imply that the Riedel shears should open up during development, which is not observed. We thus have a fracture geometry that seems consistent with the expectations from the Coulomb fracture criterion but that in fact is inconsistent with the strict application of the criterion. The orientation of the P Riedel shears in the experiment illustrated in Figure 8.7 does not fit any prediction of the Coulomb fracture criterion either for fracture under the assumed state of stress or for the angular relation between P and the R and R' shears (Figure 8.7*C*). Thus the shear fractures in this experiment are actually examples for which the Coulomb fracture criterion does not apply, notwithstanding our analysis in Section 8.3.

We can gain some understanding of these shears from a strictly kinematic analysis. The R shears accommodate a component of the imposed shearing of the block, but they also act like normal faults that result in a thinning of the block normal to the imposed shear plane and lengthening of the block parallel to it (Figure 8.7). The R' and P shears, on the other hand, act like thrust faults that result in a shortening of the block parallel to the imposed shear plane and thickening perpendicular to it. Because the simple shearing constraints imposed on the block require no lengthening parallel to the shear zone and no thinning normal to it, the shearing on the R shears makes the deformation provided by shearing on R' and P shears geometrically necessary. This analysis, however, gives no answer to the question of why the shear fractures should develop in these particular orientations. We discuss this example in terms of the applied boundary conditions in Section 9.12, although the complete understanding of the origin of Riedel shears remains an unresolved problem.

In other cases, even within the nominal range of stresses where the Coulomb criterion should apply, it is not always successful. A particularly important example of this limitation is the case of three-dimensional deformation accommodated by brittle fracture. The conjugate shear fractures predicted by the Coulomb fracture criterion can only accommodate deformation in the $(\hat{x}_1, \hat{x}_3)$-plane, that is, the $(\hat{\sigma}_1, \hat{\sigma}_3)$ principal plane, because both the fractures are perpendicular to that principal plane and the slip direction on each fracture lies in that plane (Figure 8.17*A*): The average shortening of the sample is parallel to $\hat{x}_1$ and the average lengthening is parallel to $\hat{x}_3$. These fractures, however, cannot accommodate any change of dimension parallel to $\hat{x}_2$. If a sample is forced to deform in three dimensions so that there is also some change of length parallel to $\hat{x}_2$, then experiments show that a system of four fracture orientations develops, two pairs of conjugate fractures in which the fractures are oriented symmetrically about the principal axes but are not perpendicular to any of the principal planes (Figure 8.17*B*). This set of fractures can accommodate a three-dimensional deformation, but their orientations are not predicted by the Coulomb fracture criterion.

It would seem that the Coulomb fracture criterion must be a two-dimensional version of a more general criterion. One proposal for such a criterion is that fractures develop in orientations such that the work required to accommodate the imposed deformation is a minimum, but the resulting criterion is not simple to use. It may be that under the circumstances of three-dimensional deformation, the cause of the process is the displacements imposed on the body, and the stresses are one of the effects of the imposed displacements, so the conditions are not the same as is assumed for the Coulomb fracture criterion. We discuss cause and effect in more detail in Sections 9.12 and 18.1(iv).

8.8 THE GRIFFITH THEORY OF FRACTURE

i. Griffith Cracks

So far, we have discussed empirical fracture criteria that relate the initiation of fracturing to stress and other physical conditions. These criteria have been reasonably successful in accounting for the macroscopic brittle behavior of most geologic materials, but they contribute little to our understanding of the physical mechanism of fracturing on a microscopic or molecular level.

One can calculate the theoretical tensile strength of a solid material based on the strengths of the atomic bonds in the constituents of the solid. The strength derived in this manner, however, is generally about two orders of magnitude higher than the experimentally determined tensile strength of the material. In an attempt to account for this discrepancy, A. A. Griffith, in the early twentieth century, proposed that all solids contain a myriad of microscopic to submicroscopic randomly oriented cracks, now called **Griffith cracks** (see stage I in Figure 8.21*B*), that reduce substantially the strength of the material.

A Griffith crack is a small penny-shaped or slit-like crack that in cross section is much longer than it is thick and that has a very small radius of curvature at its tips. Griffith cracks may be imperfections within the crystal lattice of crystal grains in a rock, or they may be cracks within or between crystal grains in the rock. The cracks themselves are commonly modeled as extremely flattened ellipsoids, so the cross-sectional shape is elliptical with a large ratio of major axis to minor axis. (The diagrams of Griffith cracks given in Figures 8.18 through 8.20 show the minor axis $2c$ drawn very much larger in relation to the major axis $2a$ than is actually used in a model for a real Griffith crack.)

The ability of a Griffith crack to reduce substantially the strength of a material derives from the fact that an

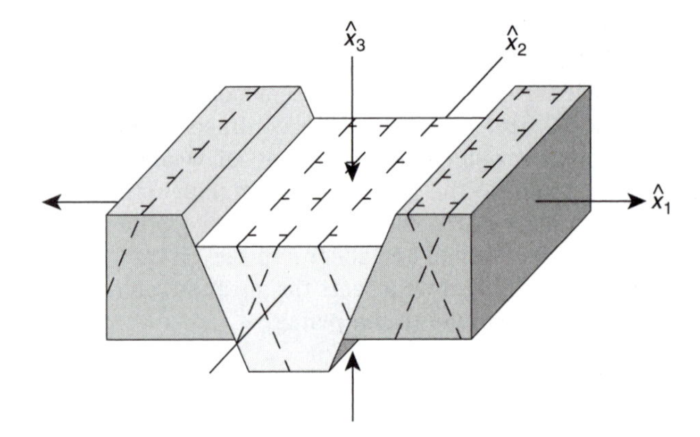

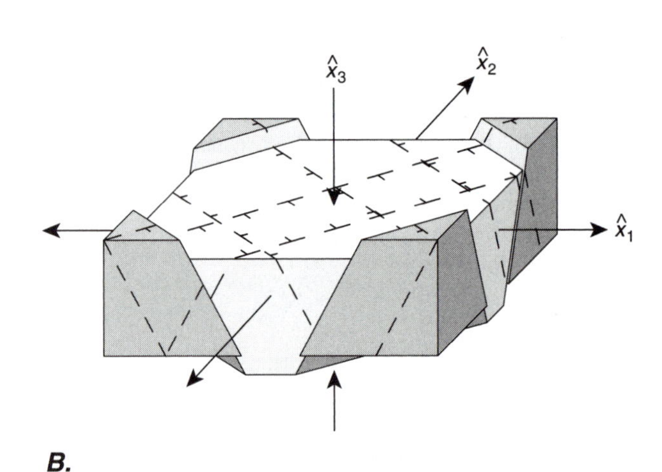

FIGURE 8.17 Sets of shear fractures needed to accommodate different deformation geometries. Tick marks on the horizontal traces of the fractures indicate the direction of dip. *A.* Conjugate normal faults predicted by the Coulomb fracture criterion can only accommodate plane strain, with lengthening parallel to $\hat{x}_1$, shortening parallel to $\hat{x}_3$, and no change of dimension parallel to $\hat{x}_2$. *B.* Four sets of faults, two sets of conjugate pairs, can accommodate a triaxial deformation, with lengthening parallel to $\hat{x}_1$ and $\hat{x}_2$, and shortening parallel to $\hat{x}_3$. The orientations of these fractures are not predicted by the Coulomb fracture criterion. (From Reches 1978)

applied stress in general produces a local high concentration of tensile stress near a crack tip. Thus we distinguish the *applied stresses*, which are determined by the forces per unit area applied to the surfaces of the body, from the *local stresses*, which describe the state of stress immediately adjacent to a Griffith crack. The smaller the radius of curvature at the crack tip and therefore the larger the ratio a/c of the ellipsoidal model of the crack, the higher the local concentration of tensile stress near the crack tip.

To understand the mechanism, there are two factors that we must consider. The first factor is the way the local stresses are distributed around the surface of a Grif-

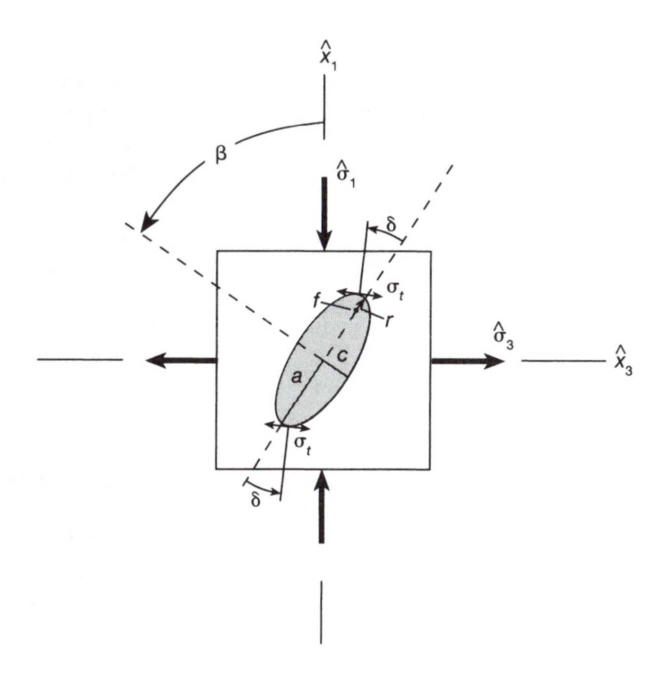

FIGURE 8.18 Schematic diagram of Griffith crack in two dimensions idealized as an ellipse with major semi-axis *a*, minor semi-axis *c*, and radius of curvature *r* at the crack tip measured from the focus *f*. σ_t is the *local* tensile normal stress that is parallel to the surface of the ellipse. It acts on a plane perpendicular to the surface at a point defined by the angle δ between the major axis of the ellipse and the normal to the ellipse. For the sake of clarity, the ratio of the minor axis to the major axis is much larger in the diagram than is expected for real Griffith cracks. β is the angle between the normal to the crack plane and the applied principal stress $\hat{\sigma}_1$.

fith crack. In general, the local tensile stress is a maximum near the crack tip at a point that we define by the angle δ between the normal to the ellipsoidal surface of the crack and the major axis of the crack (Figure 8.18). The second factor is the crack orientation β relative to the applied principal stresses, which determines the magnitude and location of the local stress maximum. The orientation β^* of the most severely stressed Griffith crack and the location δ^* and orientation of the local maximum tensile stress on that crack surface govern how a fracture develops from a Griffith crack.

ii. Formation of Tension Fractures

For a body under an applied tensile stress, the Griffith cracks are open. In this case, any crack surface is a free surface, which cannot support a shear stress. A crack surface therefore must be a principal plane of the local stress. A free surface also cannot support a normal tensile stress. Thus the local maximum tensile stress σ_t^{max} must be parallel to the elliptical crack surface.

The orientation of the most critically stressed Griffith crack is perpendicular to the maximum applied tensile stress ($\beta^* = 90°$, Figure 8.19). In this orientation, the location of the maximum stress concentration is exactly at the crack tip ($\delta^* = 0°$), and the orientation of the local tensile stress σ_t^{max} is parallel to the applied tensile stress $\hat{\sigma}_3$. If the ellipticity of such a crack is, for example, $a/c \approx 100$, which is reasonable for a Griffith crack, then the magnitude of the local tensile stress at the crack tip is approximately 200 times that of an applied uniaxial tensile stress ($\sigma_t^{max} \approx 200\ \hat{\sigma}_3$). Thus the stress at the crack tip can be at the theoretical strength for the material when the applied stress is still roughly two orders of magnitude lower than the theoretical strength.

When the true strength of the material is exceeded at the crack tip, the crack propagates in a plane normal to the local tensile stress σ_t^{max}, and in this case the plane of propagation is parallel to the plane of the crack itself. As the crack propagates, the ellipticity of the crack *a/c* increases, which in turn increases the stress concentration. At constant applied stress, the growth of the crack therefore leads to an instability and the crack propagates rapidly, causing a tensile fracture to form (Figure 8.1*A*). The crack stops propagating when it reaches a surface it cannot cross, such as the boundary of the sample or another open crack, or when the applied stress decreases to the point at which local stress concentrations are subcritical.

iii. Longitudinal Splitting

Under conditions of uniaxial compression, those Griffith cracks that are not essentially parallel to the compressive stress are closed by the component of normal stress across their surfaces. Cracks parallel to the applied stress

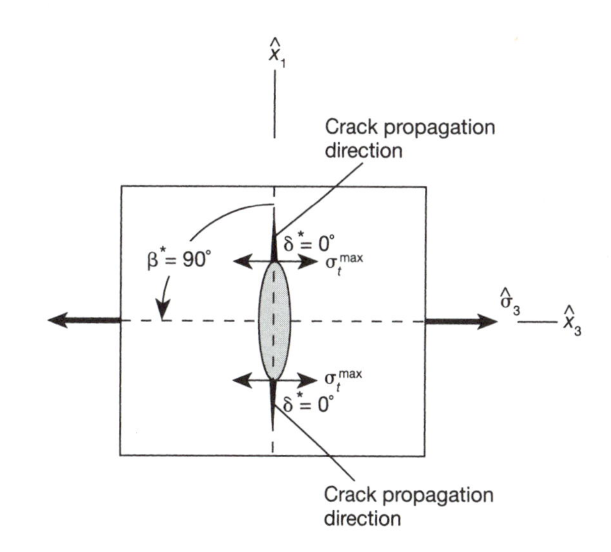

FIGURE 8.19 Orientation of the most critically stressed Griffith crack under applied uniaxial tensile stress ($\beta^* = 90°$). The maximum local tensile stress concentration is at the crack tips ($\delta^* = 0°$), and the orientation of the local maximum tensile stress σ_t^{max} is parallel to the applied tensile stress $\hat{\sigma}_3$. The crack grows perpendicular to the local maximum tensile stress σ_t^{max} and therefore perpendicular to $\hat{\sigma}_3$.

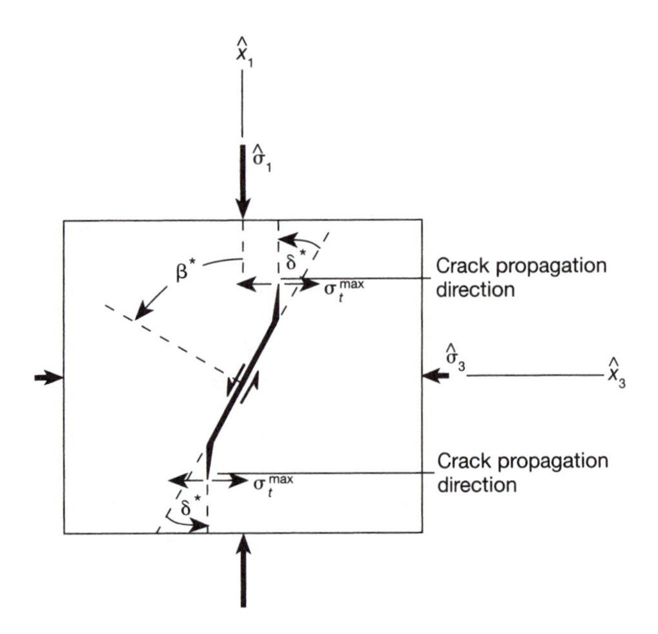

FIGURE 8.20 Orientation of the most critically stressed Griffith crack under applied confined compression. The crack is closed, and the orientation of the most critically stressed crack falls in the range $45^\circ \leq \beta^* \leq 90^\circ$. A local tensile stress concentration develops near, but not at, the crack tips and is a maximum at an angle $\delta^* > 0^\circ$. The local tensile stress maximum σ_t^{max} is oriented such that the crack grows progressively toward parallelism with σ_1. Crack growth must be accommodated by frictional sliding on the closed part of the crack surface.

need not close, however, and even though a compressive stress is applied, the local stress concentration at the crack tip is tensile and oriented normal to the applied compressive stress. The situation is identical to that shown in Figure 8.19, except that the applied stress is a compressive stress parallel to $\hat{x}_1$ rather than a tensile stress parallel to $\hat{x}_3$ (compare Figure 8.1*A*, *C*). For the most critically stressed Griffith crack in uniaxial compression, we again have $\beta^* = 90^\circ$ and $\delta^* = 0^\circ$ (Figure 8.19). If the ellipticity $a/c \approx 100$, then the local maximum tensile stress on the crack is roughly 25 times the magnitude of the applied *compressive* stress $\sigma_t^{max} \approx -25\,\hat{\sigma}_1$). When such cracks propagate, they grow roughly parallel to the applied compressive stress, leading to the development of longitudinal splitting (Figure 8.1*B*). This mechanism therefore accounts for the formation of extension fractures in compression at low to zero confining pressure. An increase in the confining pressure tends to close cracks of this orientation and to reduce the local stress concentrations that lead to this mode of fracturing.

iv. Formation of Shear Fractures

The behavior of Griffith cracks in confined compression is more complicated. The cracks in general are closed, because the applied stress produces a compressive normal stress across the crack surfaces. Friction on the closed crack surfaces and the applied compressive stress cause the distribution of local stress around the crack to differ from that around Griffith cracks under applied tensile stress. Shear along the closed cracks results in tensile stress near the crack tips that reaches a maximum at a point not, in general, exactly at the tip. The crack grows by the opening of tensile cracks (wing cracks) that accommodate the shear on the crack surface (Figure 8.20). The orientation of the most critically stressed Griffith crack in compression is at an angle β^* between 45° and 90°, depending on the relative values of $\hat{\sigma}_1$ and $\hat{\sigma}_3$. Thus the most critically stressed crack plane is between 45° and 0° to the maximum compressive stress, which is a range of orientations that includes those within which shear fractures generally form.

The location of the point of maximum tensile stress concentration, however, is at a point where the surface of the crack is more nearly perpendicular to the maximum compressive stress (δ^* in Figure 8.20). Thus when the crack grows, an instability does not develop because the new part of the crack grows toward parallelism with the maximum compressive stress, which is a more stable orientation (Figure 8.20; see stage III of Figure 8.21*B*). The onset of Griffith crack growth in compression, therefore, does not lead immediately to shear failure. The differential stress must be increased considerably beyond that required for initial crack growth.

The mechanism of shear fracture formation is illustrated in Figure 8.21. The data for a typical fracture experiment are shown diagrammatically in Figure 8.21*A*, where the axial extension (the change in length divided by the initial length, $\Delta L/L$) and the volumetric extension (the change in volume divided by the initial volume, $\Delta V/V$) are plotted against the differential stress. The physical states of the material for the five different stages of the stress-strain curves (labeled I to V) are illustrated schematically in Figure 8.21*B*. In stage I, the initial undeformed material is filled with open Griffith cracks, and the low slope of the stress-strain curves results from the relatively large deformation associated with the closing up of the cracks. In stage II, the cracks are closed and stable; no crack growth occurs with increasing stress. Crack growth begins in stage III, where the volume of the opening cracks begins to offset the normal volumetric decrease associated with the increase in stress. As the applied stress increases, less critically oriented cracks begin to grow. Eventually, in stage IV, the volume decrease caused by increasing compression is completely offset by the volume increase caused by growing cracks, and the volume of the material actually begins to increase (a phenomenon called **dilation**). At this stage, the local stress fields around the cracks start to interact, the cracks begin to join together, and finally, in stage V, a throughgoing shear fracture develops. Thus shear fracturing under compressive stresses actually depends on the growth of tension cracks in the material.

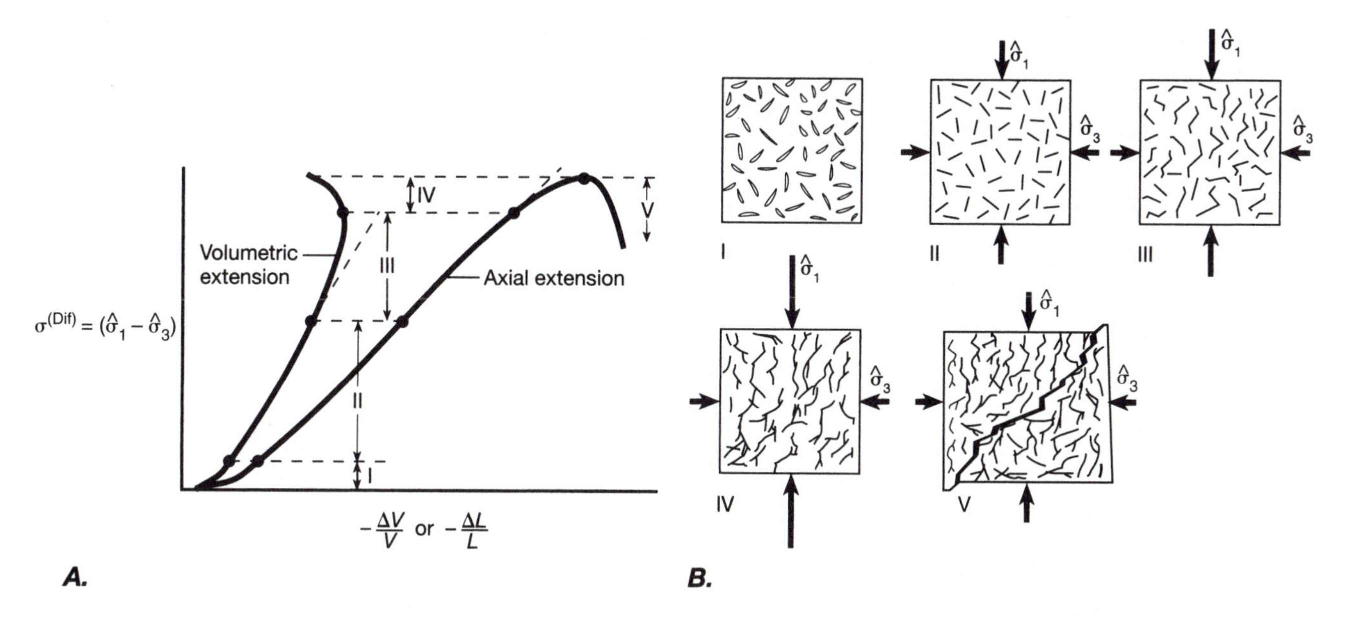

FIGURE 8.21 The process of shear fracturing in compression. *A.* Schematic plot of differential stress ($\sigma^{(Dif)}$) versus axial extension ($e_n = \Delta L/L$) and volumetric extension ($e_V = \Delta V/V$) for a characteristic shear fracture experiment in uniaxial compression. Stages I through V reflect changes in physical processes within the rock. (After Beniawski 1967) *B.* Diagrammatic stages in the formation of a brittle shear fracture corresponding to the stages labeled on the stress-strain curves in *A.* I: Griffith substance containing many open cracks. As stress increases, the cracks gradually close. II: The cracks are all closed and stable. Increase in stress does not cause crack growth. III: Under increasing load, the most critically oriented and the longest cracks begin to propagate toward the direction of maximum compressive stress. As the load increases, shorter and less favorably oriented cracks begin to grow. IV: The specimen is at this stage almost a granular solid. The local stress fields around cracks begin to interact. V: A through-going shear fracture forms by the coalescence of many small fractures. (Modified after Johnson 1970, fig. 10.20)

v. Griffith Fracture Criteria for Tensile and Compressive Stress

The Griffith crack models can be used to derive different theoretical failure criteria for conditions of tensile and compressive stress. In the tensile stress regime, the predicted fracture criterion is a parabolic envelope, which corresponds reasonably well with experimental observation. The form of the fracture criterion for compressive stresses is similar to the Coulomb fracture criterion, but the predicted constants are different from the experimentally determined ones. The discrepancy exists because the Griffith theory provides a criterion for the first initiation of Griffith crack growth (Figure 8.21, beginning of stage III), whereas the development of a shear fracture occurs at considerably higher differential stresses (at the beginning of stage V). The actual process of shear fracture development is very complex, and the details of the mechanism are poorly understood.

vi. The Griffith Theory and the Effects of Confining Pressure, Pore Fluid Pressure, and the Intermediate Principal Stress

The growth of tension cracks from Griffith cracks in a compressive stress regime is accompanied by frictional sliding on the Griffith crack surface (Figure 8.20). When confining pressure increases, the normal stress across the cracks increases, and the frictional resistance to sliding must also increase. To initiate crack growth, therefore, the shear stress on the crack must increase, and this requires an increase in the applied differential stress ($\sigma^{(Dif)} = \hat{\sigma}_1 - \hat{\sigma}_3$). Under higher confining pressures, then, higher differential stresses are required for shear fracturing, as specified by the Coulomb fracture criterion. Thus the angle of internal friction ϕ that defines the slope of the Coulomb fracture envelope is related to the friction on the surface of the Griffith cracks.

A confining pressure applied to a rock provides the same normal stress across Griffith cracks of all orientations. The presence of a pore fluid under pressure in the cracks directly counteracts the externally applied normal stress, so the net normal stress across a Griffith crack is exactly the effective normal stress defined in Section 7.3 (the first Equation (7.39)) and discussed in Section 8.5. In this manner, the Griffith crack theory accounts for the observed effect of pore pressure on brittle fracturing.

As the value of the intermediate principal stress $\hat{\sigma}_2$ varies from $\hat{\sigma}_1$ to $\hat{\sigma}_3$, the fracture strength of the rock decreases slightly (Section 8.6(ii)). The Griffith theory provides a model to account for this behavior. In a three-dimensional material, Griffith cracks are distributed in all orientations, and those that eventually coalesce into a shear fracture are not necessarily exactly parallel to $\hat{\sigma}_2$. On those cracks, $\hat{\sigma}_2$ contributes a small component of compressive stress. Thus the frictional resistance to sliding on the crack surface depends slightly on $\hat{\sigma}_2$, and it is higher for higher values of $\hat{\sigma}_2$. A higher frictional resistance on Griffith cracks results in the need for a larger applied differential stress to cause fracturing.

REFERENCES AND ADDITIONAL READINGS

Beniawski, Z. T. 1967. Mechanisms of brittle fracture of rock: Parts I, II, and III. *Int. J. Rock Mech. Mining Sci.* 4: 395–430.

Blanpied, M. L., D. A. Lockner, and J. D. Byerlee. 1991. Fault stability inferred from granite sliding experiments at hydrothermal conditions. *Geophys. Res. Lett.* 18: 609–612.

Byerlee, J. 1975. The fracture strength and frictional strength of Weber sandstone. *Int. J. Rock Mech. Mining Sci.* 12: 1–4.

Cox, S. J. D., and C. H. Scholz. 1988. Rupture initiation in shear fracture of rocks: An experimental study. *J. Geophys. Res.* 93(B4): 3307–3320.

Donath, F. 1961. Experimental study of shear failure in anisotropic rocks. *Geol. Soc. Amer. Bull.* 72: 985–990.

Engelder, T. 1999. Transitional-tensile fracture propagation: A status report. *J. Struct. Geol.* 21(8/9): 1049–1055.

Freund, R. 1974. Kinematics of transform and transcurrent faults. *Tectonophysics* 21: 93–134.

Handin, J. 1966. Strength and ductility. In S. P. Clark, Jr., ed., *Handbook of Physical Constants*, Geol. Soc. Amer. Mem. 97: 223–289.

Johnson, A. M. 1970. *Physical Processes in Geology.* San Francisco, Freeman Cooper.

Paterson, M. S. 1978. *Experimental Rock Deformation—The Brittle Field.* New York, Springer Verlag.

Reches, Z. 1978. Analysis of faulting in three-dimensional strain field. *Tectonophysics* 47(1/2): 109–129.

Reches, Z. 1983. Faulting of rocks in three-dimensional strain fields; II, Theoretical analysis. *Tectonophysics* 95(1–2): 133–156.

Reches, Z., and J. H. Dieterich. 1983. Faulting of rocks in three-dimensional strain fields; I, Failure of rocks in polyaxial, servo-control experiments. *Tectonophysics* 95(1–2): 111–132.

Scholz, C. H. 1990. *The Mechanics of Earthquakes and Faulting.* New York, Cambridge University Press.

Sieh, K. 1982. Late Holocene displacement history along the south-central reach of the San Andreas fault. Ph.D. diss., Stanford University.

Wei, K., and J.-C. De Bremaecker. 1994. Fracture growth under compression. *Jour. Geophys. Res.* 99(B8): 13781–13790.

Wu, H., and D. D. Pollard. 1995. An experimental study of the relationship between joint spacing and layer thickness. *J. Struct. Geol.* 17(6): 887–905.

Chapter 9

MECHANICS OF NATURAL FRACTURES AND FAULTS

In this chapter, we use our knowledge of stress and mechanics of brittle fracture from the preceding two chapters, with the introduction to elastic deformation presented in Section 9.1, to gain a deeper understanding of brittle deformation. With that background, we can reexamine fractures and faults, described in Chapters 2 through 6, and draw some conclusions about the conditions under which they form.

Bear in mind that like all geology, structural geology is a historical science for which much of the record of past activity is obscure or lost. In cases of complex deformation history, structures may be difficult to interpret, either because the relative timing of the formation of different structures is obscure or because structures formed under one set of conditions are reactivated under another. Thus, for example, it may be difficult to determine whether a set of fractures was formed before, during, or after folding. Fractures developed as extension fractures may undergo a subsequent shear fracture history. Preexisting thrust faults may be reactivated as normal faults. Such complex and multiple histories make the interpretation of structures challenging and often controversial.

With that caution in mind, then, we first present an introduction to elastic deformation, and we discuss some techniques for measuring stress within the Earth. Next we discuss the results of measuring the magnitude and orientation of stress within the Earth, discuss the possible origins of these stresses, and apply that knowledge to understanding the formation of extension fractures and faults.

9.1 ELASTIC DEFORMATION

During rock deformation experiments, if the stresses are small enough that fracture does not occur, the applied stress causes changes in the length, diameter, and volume of the sample that are proportional to the stress applied. Removal of the stress causes the strain completely to disappear, and the strain is therefore said to be **recoverable**. This type of deformation is called **elastic deformation**. The primary information that such experiments yield is the relationship between the axial force applied through the pistons and the associated change in dimensions of the sample (see Figure 8.21*A*).

One of the most common measures of deformation is the strain, which we discuss in detail in Chapter 12. One of the measures of strain is the extensional strain e_n or simply the extension. It is the change in length of the sample per unit of initial length—that is,

$$e_n = \frac{l - L}{L} = \frac{\Delta L}{L} \tag{9.1}$$

where L is the initial length, l is the deformed length, and ΔL is the change in length. We also express the extension as a percentage change in length by multiplying e_n by 100. We show in Chapter 12 that if a spherical object is subjected to a homogeneous deformation, it deforms into an ellipsoid called the strain ellipsoid. The principal axes of this ellipsoid are the directions of the principal extensions, which are the maximum, intermediate (minimax), and minimum extensions, and which we label $\hat{e}_1 \geq \hat{e}_2 \geq \hat{e}_3$. In two dimensions, an initial circle is deformed into an ellipse, called the strain ellipse, for which the principal axes are parallel to the principal extensions $\hat{e}_1 \geq \hat{e}_3$.

In a uniaxial state of stress, the magnitude of elastic extension parallel to the applied stress is directly proportional to the magnitude of the stress (see Figure 9.1):

$$\sigma_n = E\, e_n \quad e_n = \frac{\sigma_n}{E} \tag{9.2}$$

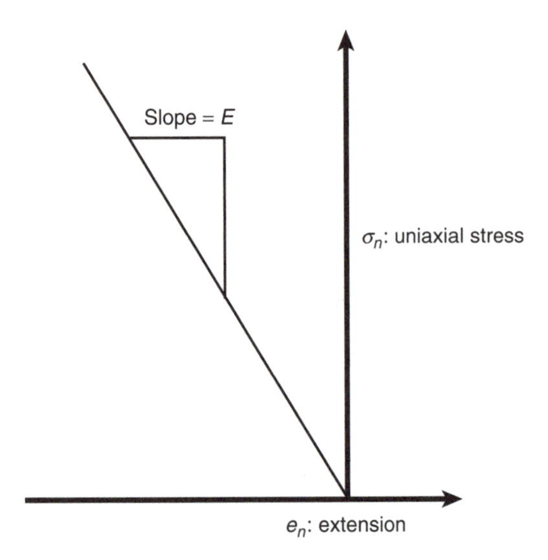

FIGURE 9.1 Diagram showing linear relationship between normal stress σ_n and extension e_n for perfectly elastic materials. The slope is the value of the Young modulus E, which is negative because our sign conventions result in a positive stress (compression), producing a negative extension (shortening).

where the constant of proportionality E is **Young's modulus**. Young's modulus is one of two elastic constants we need to characterize the elastic behavior of an isotropic material.[1] For rocks, E characteristically has values in the range of -0.5×10^5 to -1.5×10^5 MPa. The maximum extension that most rocks can reach before they fracture is generally quite small—a few percent at most and generally much less.

In uniaxial compression, the sample shortens parallel to the applied stress. If the material is incompressible, its volume is conserved, and the sample must therefore expand in a direction normal to the shortening (Figure 9.2*A*). Materials never are perfectly incompressible, however, so there is a net decrease in volume in any deformation caused by a compressive stress. The **Poisson ratio** ν is the absolute value of the ratio given by the extension $\hat{e}_\perp$ normal to an applied compressive stress, divided by the extension $\hat{e}_\parallel$ parallel to the applied compression:

$$\nu \equiv \left| \frac{\hat{e}_\perp}{\hat{e}_\parallel} \right| \tag{9.3}$$

The Poisson ratio is the second elastic constant that characterizes the behavior of an isotropic elastic material. If a material were perfectly incompressible, ν would equal 0.5. For most rocks, however, the Poisson ratio ranges from 0.25 to 0.33. The expansion normal to an applied compression is the **Poisson expansion**.

In a sample under confined compression, the magnitude of the axial extension depends not only on the axial stress but also, because of the Poisson expansion, on the radial stresses. An axial compressive stress parallel to $\hat{z}$, for example, produces an axial shortening parallel to $\hat{z}$ as well as radial Poisson expansions parallel to $\hat{x}$ and $\hat{y}$. Similarly, if a radial compressive stress is added for which the two principal stresses parallel to $\hat{x}$ and $\hat{y}$ are equal, each radial stress produces shortening parallel to the stress plus an expansion in the two directions perpendicular to it. Thus the axial shortening parallel to $\hat{z}$ is reduced by the Poisson expansion in that direction associated with each of the radial compressive stresses (Figure 9.4*B*). This relationship is expressed mathematically as

$$\hat{e}_{zz} = \frac{1}{E}\hat{\sigma}_{zz} - \nu\left(\frac{1}{E}\right)\hat{\sigma}_{xx} - \nu\left(\frac{1}{E}\right)\hat{\sigma}_{yy} \tag{9.4}$$

$$\hat{e}_{zz} = \frac{1}{E}\hat{\sigma}_{zz} - \frac{\nu}{E}(\hat{\sigma}_{xx} + \hat{\sigma}_{yy}) \tag{9.5}$$

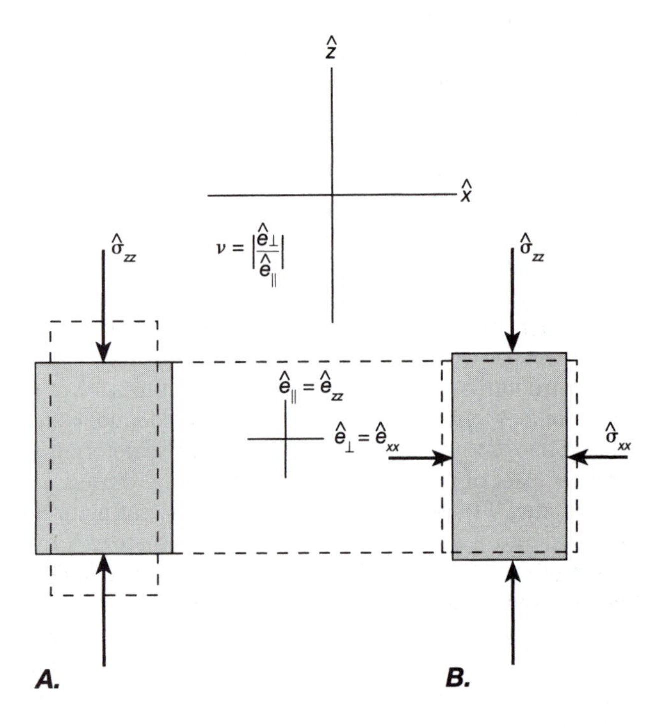

FIGURE 9.2 The Poisson effect. Dimensional changes are exaggerated for clarity. *A*. Comparison of the unstressed state (dashed rectangle) with the shape caused by a uniaxial compressive stress (shaded rectangle). The sample shortens parallel to the $\hat{z}$ axis (a negative extension, $\hat{e}_\parallel$) and lengthens perpendicular to the $\hat{z}$ axis (a positive extension $\hat{e}_\perp$) illustrating the Poisson expansion. *B*. If a radial pressure $p = \hat{\sigma}_{xx} = \hat{\sigma}_{yy}$ is added to the uniaxial stress $\hat{\sigma}_{zz}$ in *A*, the Poisson expansion due to the radial pressure (shaded rectangle) decreases the initial amount of shortening parallel to $\hat{\sigma}_{zz}$ (dashed rectangle).

[1]Young's modulus for the geologic sign convention is a negative number, because a positive stress (uniaxial compression) decreases the length of the specimen, thus producing a negative extension.

where for consistency in labeling stress and extension, we have relabeled the principal axes $\hat{x}$, $\hat{y}$, and $\hat{z}$.[2] Compare these relationships with the second Equation (9.2) for uniaxial stress. In Equation (9.4), the two terms containing ν give the Poisson expansion that would develop in the $\hat{z}$ direction if each radial stress component parallel to $\hat{x}$ and $\hat{y}$, respectively, were the only applied stress.

The volumetric extension e_V is the change in volume divided by the initial volume. If the principal extensions are very small, as is the case for elastic deformation in rocks (see Sections 12.1 and 12.2), they are related to the volumetric extension by the second Equation (12.11) and Equation (12.15):

$$e_V \equiv \frac{\Delta V}{V} \cong \hat{e}_1 + \hat{e}_2 + \hat{e}_3 \tag{9.6}$$

9.2 TECHNIQUES FOR DETERMINING STRESS IN THE EARTH

Techniques for determining the state of stress in the Earth have been developed in the geological engineering, mining, and energy industries. In mining, knowledge of the state of stress is important for the design of safe tunnels and stable open pits. In dam construction, the stresses in the abutments and foundations are measured before, during, and after construction to ensure safe design and operation, particularly to determine that they do not exceed the strength of the rocks in question. In petroleum and geothermal energy production, artificially fracturing rocks at depth can increase their permeability, thereby enhancing the yield from wells. Control of artificial fracturing requires knowledge of the state of stress at depth.

We also need to know the state of stress within the Earth in order to understand how and why plates move; why, where, and when earthquakes occur; and why and how deformational structures form. Because of these practical and fundamental needs, the determination of stress in the Earth has become a field of geologic investigation in its own right.

Techniques are available for determining the current state of stress in the Earth as well as for determining the "paleostress" that existed at some time in the geologic past. We can never actually measure stress directly, however, even an existing stress in rock. We always have to measure an observable effect produced by the stress, which is often an elastic deformation (such as the extension defined by Equation (9.1)), and then infer the stress by knowing the relation between the effect and the stress (such as the constitutive relation between stress and extension in Equation (9.2)). When such measurements are made using well-calibrated instruments like strain gauges or flat jacks, or when the stress depends on the local strain of rock whose elastic properties can be measured in the laboratory, such as in the overcoring technique (both of which we describe in Subsection 9.2(i)), the results for the stress are reliable. In some cases, however, we may be able to measure the strains associated with a deformation, but the inference of stress from those measurements would depend on large-scale mechanical properties of rock that are not well known. Such is the case with strains determined from seismic focal mechanism solutions (see Appendix 2), slickenline data on faults (see Section 15.4), or the orientation of large-scale deformational structures such as folds. In this situation, it is common practice to report the strain measurements as "stress" measurements, but the equivalence is not always justified, and the measurements are more reliably reported simply as measurements of strain.

With some techniques, we can completely determine the orientation and magnitude of the principal stresses; others provide only partial information on the state of stress. We discuss here techniques that rely on the effects of elastic or brittle deformation; in Chapter 17, we discuss others that rely on the effects of ductile deformation (see Box 17-2).

i. Stress Relief Measurements

Stress relief techniques for determining stress depend on the fact that the stress on an elastic material produces a linearly proportional strain (Section 9.1, Figure 9.1). Removal of the stress causes the strain to disappear. Thus measurement of the change in strain that accompanies unloading (stress relief) allows us to infer the original stresses if we know the elastic constants of the rock.

Overcoring is a common technique that involves drilling a hole in the rock, attaching strain gauges to the surface of the hole, and then drilling an annulus around the hole to form a hollow cylinder of rock no longer subjected to stress from the surrounding rock (Figure 9.3*A*). The release of stress causes elastic deformation of the cylinder, so its dimensions and its initial circular cross section change. Using the theory of elastic deformation (see Equation (9.5), for example), we can calculate the magnitude and direction of the original stresses from measurements of this deformation. The calculation is not simple, however, because the presence of the first hole

[2]Because of the conventions for the principal values of stress $\hat{\sigma}_1 \geq \hat{\sigma}_2 \geq \hat{\sigma}_3$ and extension $\hat{e}_1 \geq \hat{e}_2 \geq \hat{e}_3$, and the sign conventions that require both a compressive stress and a lengthening extension (Equation (9.1)) to be positive, we encounter the inconvenient fact that the stress component $\hat{\sigma}_1$ and the extension component $\hat{e}_3$ that parallel the same coordinate axis have different subscripts. To avoid this confusion, we relabel the principal axes $\hat{x}$, $\hat{y}$, and $\hat{z}$ and relabel the principal stresses and extensions so that

$$\begin{bmatrix} \hat{\sigma}_{xx} \\ \hat{\sigma}_{yy} \\ \hat{\sigma}_{zz} \end{bmatrix} = \begin{bmatrix} \hat{\sigma}_3 \\ \hat{\sigma}_2 \\ \hat{\sigma}_1 \end{bmatrix} \quad \text{and} \quad \begin{bmatrix} \hat{e}_{xx} \\ \hat{e}_{yy} \\ \hat{e}_{zz} \end{bmatrix} = \begin{bmatrix} \hat{e}_1 \\ \hat{e}_2 \\ \hat{e}_3 \end{bmatrix}$$

With this notation, the principal stress and the principal extension that parallel the same coordinate axis have the same subscripts, so for example, Equation (9.2) would be $\hat{\sigma}_{xx} = E\hat{e}_{xx}$.

changes the stress from its value in the solid rock (Figure 9.3*B*), and this effect must be taken into account. For technical reasons, the maximum practical depth of boreholes for overcoring is 30 to 50 meters.

The International Ocean Drilling Project uses the shape of drill holes to infer the orientation of horizontal stresses in a given region of the oceanic crust. When a hole is drilled into a material under compressive stress, the applied maximum compressive stress becomes concentrated at the surface of the hole where it is tangent to the hole (Figure 9.3*B*). This causes fractures to develop at this point, which are parallel to the sides of the hole and are comparable to the longitudinal splitting observed in experiments (Figures 8.1*B*, 8.9*A*). Flakes of rock spall off these areas, enlarging the hole diameter in a direction that is perpendicular to the maximum applied compressive stress. The orientation of the axis of enlargement of the hole can be determined by either of two instruments, a borehole televiewer (a TV camera lowered down the hole) or a formation microscanner, an instrument that measures the shape of the hole. The instrument is lowered into the hole to observe where the spalling occurs on the sides of the hole or to measure the shape of the hole and the orientation of its largest diameter. These measurements then define the orientation of the maximum compressive stress.

The flat jack is an instrument used in underground excavations to measure the normal component of stress acting on a plane of a particular orientation. Reference pins are inserted into the rock to form a rectangular grid, and the distances d_1 to d_6 between pins are measured (Figure 9.3*C*). Cutting a slot into the rock between the pins relieves the stress locally and changes the distances between the reference pins. A thin hollow steel plate, the flat jack, is inserted into the slot, and the remaining space in the slot is filled with grout. When the grout has hardened, oil is pumped into the hollow flat jack until the increased oil pressure causes the reference pins to return as closely as possible to their original relative positions. The measured oil pressure is then taken to be the normal component of stress acting in the rock across the plane of the flat jack, but the flat jack cannot reproduce any shear stresses that may have existed on the plane of the slot. Making several such measurements in different orientations, however, allows one to determine the complete state of stress in the rock.

ii. Hydraulic Fracturing ("Hydrofrac") Measurements

The hydrofrac technique uses fluid pressure to induce fracturing of rock in place. By monitoring the fluid pressure as it is increased to the failure point of the rock, and by making some simplifying assumptions, it is possible to determine both the magnitudes and orientations of the principal stresses. The technique was initially developed by the petroleum industry to increase the permeability and thereby the yield of underground petroleum deposits developed by drilling.

A section of a borehole is sealed off with two inflatable rubber packers (Figure 9.4*A*), and the fluid pressure between the pressure seals is pumped up until, at a critical pressure P_c, a tension fracture forms in the borehole. Immediately after fracturing, pumping ceases and the fluid is sealed in. The pressure drops slightly as the fluid flows into the crack, and as long as the fluid does not leak out of the crack into the surrounding rock, the pressure stabilizes at a value called the instantaneous shut-in pressure P_s, which is the pressure that is just sufficient to keep the fracture open.

Using these measurements, the magnitudes and orientations of the maximum and minimum horizontal principal stresses $\sigma_{H(\max)}$ and $\sigma_{H(\min)}$ and the vertical principal stress σ_V are determined as follows: Because the surface of the Earth is a free surface on which the shear stress must be zero, the principal stresses there must be perpendicular and parallel to the surface, or approximately vertical and horizontal. We assume the stresses at depth have the same orientation. If the borehole is vertical and the tension fracture is parallel to it,[3] the minimum horizontal compressive stress in the rock equals the instantaneous shut-in pressure; that is, $\sigma_{H(\min)} = P_s$. The orientation of $\sigma_{H(\min)}$ is taken to be perpendicular to the fracture, and the fracture orientation is determined by using a downhole televiewer or by making an oriented impression of the surface of the borehole (Figure 9.4*B*). In order to fracture the rock, the critical pressure P_c must equal the sum of the tensile strength T_0 of the rock and the minimum compressive stress tangent to the surface of the borehole. T_0 must be determined experimentally in the laboratory, and the minimum stress tangent to the borehole can be related to $\sigma_{H(\max)}$ and $\sigma_{H(\min)}$ through the elastic theory of a hole in a stressed solid (see for example Figure 9.3*B*). Thus, we can calculate $\sigma_{H(\max)}$ from these relationships if we know P_c, T_o, and $\sigma_{H(\min)}$. The orientation of $\sigma_{H(\max)}$ is horizontal and perpendicular to $\sigma_{H(\min)}$. The vertical normal stress σ_V is assumed to be equal to the overburden stress, which is caused by the weight of the overlying rock. The overburden stress must be greater than, or at least equal to, the measured P_C if the fracture in the borehole is not horizontal, whereby $\sigma_V = \rho_r g h \geq P_c$. Thus all three principal stresses and their orientations are determined. Measurements at depths of up to 5 kilometers beneath the ground surface have been achieved.

iii. "Stress" Orientations from Earthquake First-Motion Studies

Earthquakes are associated with regional stresses in the Earth, and they occur at depths ranging from near-surface down to approximately 700 km at some subduction zones. The radiation pattern of first motions of

[3]These conditions can be observed by downhole scanning instruments such as the borehole televiewer.

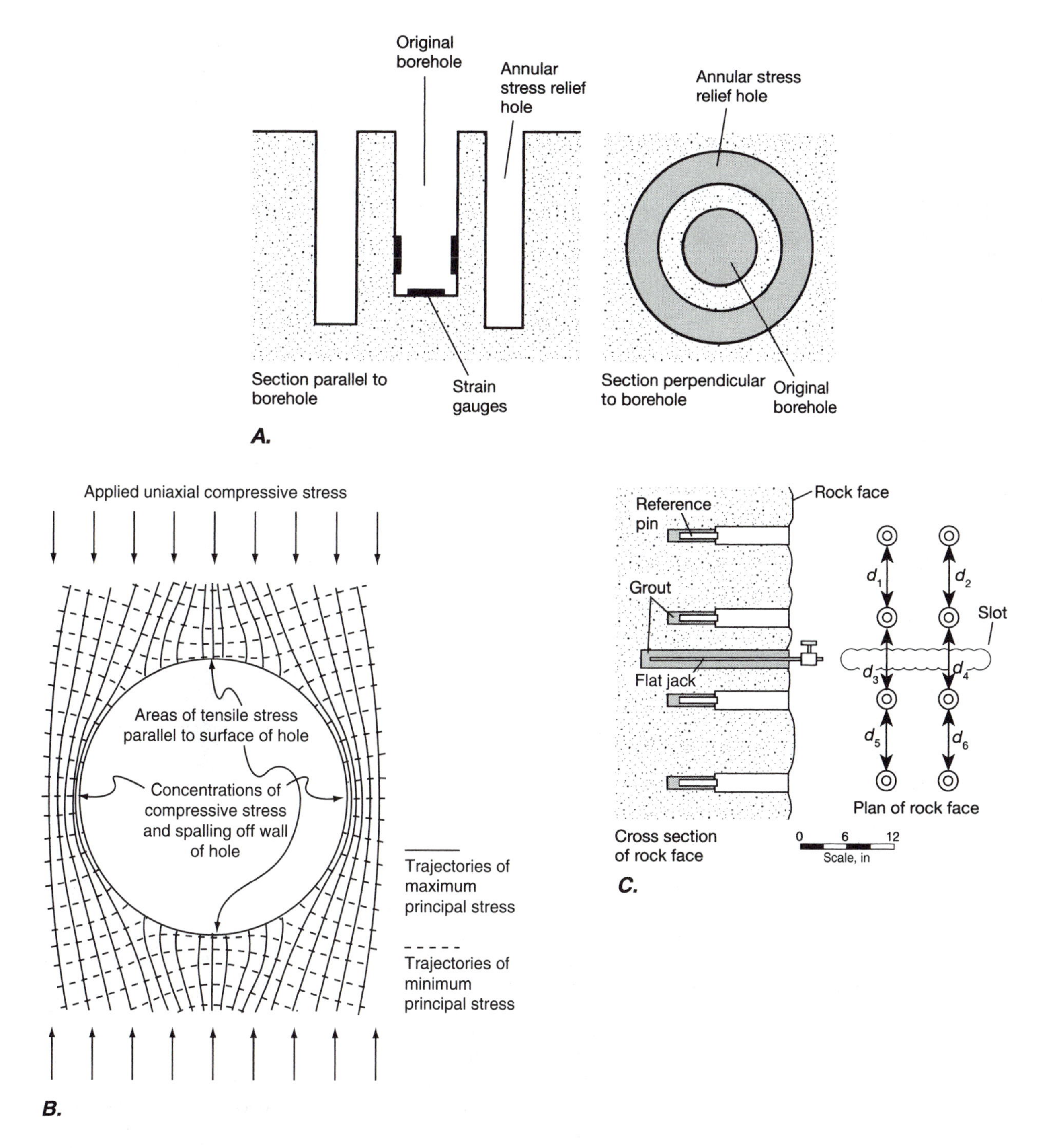

FIGURE 9.3 Stress relief techniques for measuring stress. *A.* Stress relief by the overcoring technique. A shallow borehole less than 50 meters deep is drilled, and strain gauges are glued to the surface of the hole. A larger annular hole concentric with the first hole releases the stresses on the intervening hollow cylinder of rock, which results in a change of shape of the cylinder. The resulting strain measured by strain gauges is then converted to stress by means of calibrations of the elastic behavior of the strain gauges. *B.* Stress trajectories around a circular hole in an elastic plate under uniaxial compression. Solid lines: trajectories of $\hat{\sigma}_1$; dashed lines: trajectories of $\hat{\sigma}_3$. Stresses are highest where trajectories are closest together. Spalling off the sides of the hole (by longitudinal splitting) where $\hat{\sigma}_1$ is concentrated elongates the diameter of the hole in a direction normal to the applied maximum compressive stress. (After Jaeger 1962) *C.* The flat jack technique is used to measure the component of stress normal to a plane of a particular orientation. An array of reference pins is inserted into the rock face, and the distances $d_1, d_2, \ldots, d_6$ between them are measured. A slot is cut, releasing the stress across the face of the slot and changing the distances between the pins. The flat jack is inserted and the remaining space in the slot is filled with a cement. Pumping hydraulic fluid into the flat jack restores the original distances when the fluid pressure equals the original normal stress across the plane of the flat jack. (After Merrill 1968)

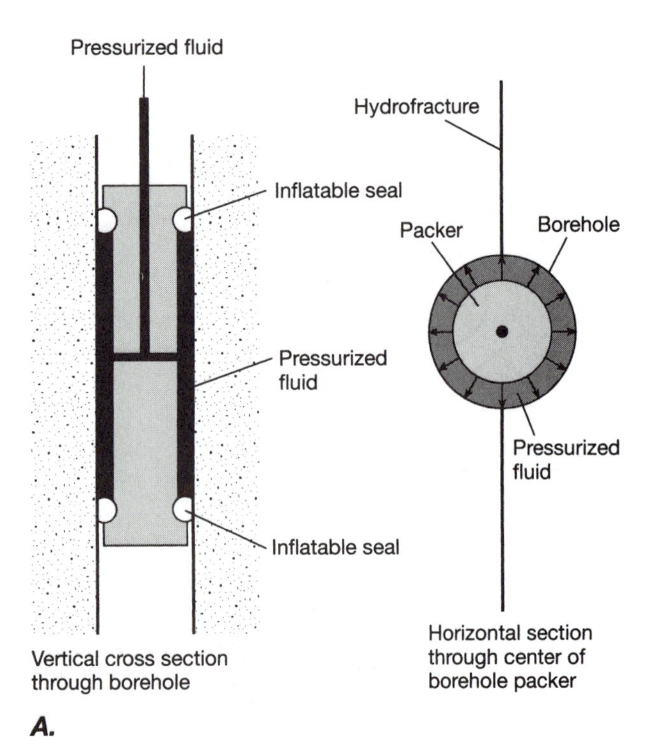

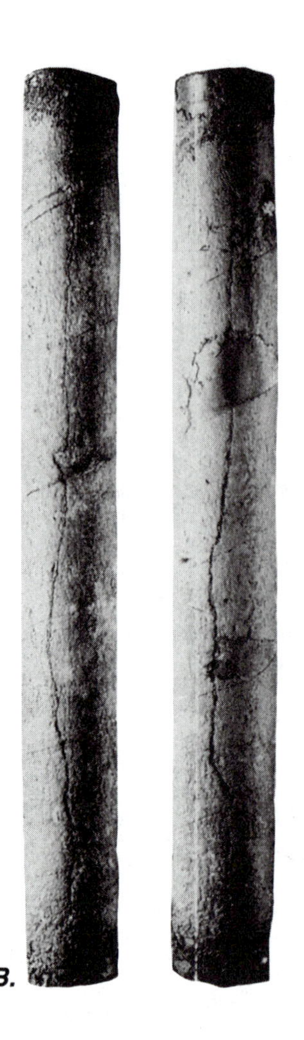

FIGURE 9.4 The hydraulic fracturing (hydrofracting) method of measuring stress. *A*. Vertical and horizontal sections through a borehole packer, which is used to isolate a section of the borehole. Fluid pressure is increased between the seals to induce hydrofracture, which occurs at the critical pressure P_c. *B*. Image of the walls of the borehole following hydraulic fracturing. The impression of the induced extension fracture is clearly visible approximately parallel to the borehole. (From Zoback et al. 1980)

seismic *P* waves for the earthquake defines the orientation of, and the sense of shear on, two perpendicular nodal planes at depth, one of which is the fault plane (see Appendix 2, Box A2-4),[4] although identifying which nodal plane is the actual fault plane is often not possible. It is commonly assumed that the maximum compressive stress $\hat{\sigma}_1$ bisects the angle between the two nodal planes in the *rarefaction* first-motion quadrant (see Box A2-4) and is perpendicular to the line of intersection of the nodal planes. This assumption allows the inference of stresses at depths in the Earth that are otherwise inaccessible for this purpose. The stress interpretation thus assumes that $\hat{\sigma}_1$ is at 45° to the shear plane. This angle is consistent with the von Mises criterion for ductile failure, but not with the Coulomb fracture criterion, which implies the maximum compressive stress should lie approximately 30° from the shear fracture plane. *This bisector of the nodal planes actually defines the axis of minimum principal infinitesimal extension* $\hat{e}_3$ (*maximum infinitesimal shortening*; see Chapter 12). Assuming the principal stress axes are parallel to the principal extension axes involves assumptions about the mechanical properties of the rock that are not necessarily correct, although it probably gives a reasonable first approximation to the orientation of the principal stresses. Because the measurement actually gives the orientations of the principal extension axes, however, and because the assumptions involved in the stress interpretation are not necessarily appropriate, we treat this topic in greater detail in our discussion of strain observed in rocks in Section 15.4(i).

[4]Field observation of faults and the slip directions on them, like seismic first-motion studies, can be used under favorable circumstances to infer the orientation of the principal strains accommodated by the faulting. We discuss this very useful technique further in Section 15.4(i).

9.3 MECHANISMS OF STRESSING THE EARTH'S CRUST

The many causes for stress or changes of stress in the Earth include the overburden (that is, the overlying rocks and fluid at any given point), the driving mechanisms of plate tectonics, horizontal and vertical tectonic motions, the inhomogeneous mechanical properties of the crust, changes (over space and time) in temperature and pressure, chemical changes in rocks, and pore fluid pressure. Once we understand the mechanisms by which stresses arise, we can begin to understand the actual state of stress in the Earth. In addition, we can gain insight into the possible origin of fractures in the Earth by modeling loading histories and their consequences. In this section, we examine some of the principal mechanisms by which stress originates or changes in the Earth's crust.

i. The Overburden

The overburden stress at any depth in the Earth results from the weight of the overlying column of material, which could be rock, fluid, or both. Major topographic variations of the surface tend to be isostatically compensated by crustal roots extending into the mantle. Above the depth of isostatic compensation, however, stress gradients associated with the topography can lead to ductile flow and the collapse of the surface topography and the associated mantle root. The greater the topographic relief, the greater the magnitude of the effect. Large-scale deviations from the geoid,[5] such as depressions that can result from the rapid melting of continental ice sheets that had reached isostatic equilibrium, can cause stress gradients that drive flow in the deep mantle. The influence on stress of variations in topography or the geoid dies out with increasing depth and is generally negligible at depths greater than the horizontal dimension of the variation.

Thickening the crust by deposition of sediments or by tectonic processes such as thrust faulting can increase the overburden pressure. Conversely, thinning the crust by erosion or tectonic denudation by widespread low-angle normal faulting can decrease the overburden pressure.

ii. Driving Processes of Tectonics

Stresses associated with plate motion are one of the major sources of regional stress in the lithosphere. Such stresses may arise from the pull of the cold down-going slab as it descends in a subduction zone, from the push of a midoceanic ridge associated with its relative topographic elevation above the adjacent seafloor, from the drag between the lithosphere and the underlying asthenosphere as the plates move relative to the underlying mantle, or from the interaction between adjacent plates. Stresses associated with convergent boundaries are particularly important to the formation of regional systems of thrust faults and folds during subduction of oceanic lithosphere and during collisions of continents or island arcs with subduction zones. Likewise, stresses associated with divergent and transform boundaries are important in the formation of regions characterized by normal faults or strike-slip faults.

[5]The geoid is an imaginary surface at some reference height on the Earth that is everywhere perpendicular to the gravitational force vector.

iii. Horizontal and Vertical Motions

Bending of the crust and of the lithosphere generates stresses whose spatial extent is comparable to the wavelength of the bending. Plates bend at subduction zones where the plate enters the trench. Plates bend during isostatic response to surface loads, such as the huge volume of volcanic rocks of the Hawaiian islands, the thick accumulations of ice in continental ice sheets, or the thick accumulations of sediment in sedimentary basins or along continental margins. Bending also occurs as a result of isostatic response to the unloading caused by tectonic denudation associated with major low-angle normal faults, erosion, or the melting of ice sheets. A deviatoric tensile stress should develop on the convex side of the bend and a deviatoric compressive stress on the concave side (Figure 9.5). Lithospheric plates also bend as they drift from one latitude to another, because the Earth is roughly ellipsoidal in shape and the surface has a greater curvature (a smaller radius of curvature) at the equator than at the poles. The plates must bend to accommodate this change in curvature.

Vertical motion on a roughly spherical body, such as results from isostatic adjustment, can also induce stresses in rocks simply by changing the radius of the body of rock from the Earth's center. As a segment of the crust is uplifted, for example, it should subtend a constant central angle. As the radial distance from the Earth's center increases, however, the arc length increases, thereby stretching the rock in both horizontal directions and inducing associated stresses.

iv. Thermal and Pressure Effects

Thermal expansion or contraction of rocks in response to changes in temperature induces stresses in the rocks if they are not free to expand or contract. The stress that is induced must be just sufficient to counteract the thermally-induced changes in dimension. Because different rocks have different coefficients of thermal expansion, a temperature change induces different stresses in two immediately adjacent but different rock types, such as a limestone and a sandstone. Stresses also arise where different amounts of temperature change occur in adjacent

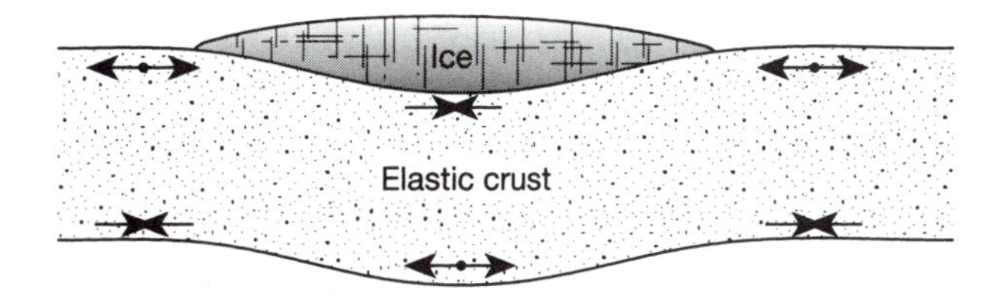

FIGURE 9.5 Bending of the elastic part of the lithosphere in response to loading by a continental ice sheet (or a volcanic pile or a sedimentary basin) causes deviatoric compression and tension on the concave and convex sides, respectively, of the bends.

rocks, such as where a hot magmatic intrusion cools while the cooler adjacent country rock warms.

Because different types of rocks have different elastic coefficients, changes in pressure associated with the addition or removal of overburden induce different amounts of strain under unconstrained conditions. If constraints impose the same deformation on rocks with different properties, stresses will vary from one rock type to another so that the deformation remains the same in the different rocks.

v. Pore Fluid Pressure

Finally, the existence of pore fluid pressure in rocks strongly affects their mechanical response and can cause extension fracturing even under conditions of purely compressive applied stresses (see Section 8.5 and Figure 8.12*A*). High pore fluid pressures can develop simply from the compaction of water-saturated impermeable sediments. As compaction decreases the pore volume, the pore fluid pressure must increase if the water cannot escape from the sediment.

Water has a higher coefficient of thermal expansion than sediment. If the pores are filled with water trapped by impermeable layers, the pore fluid pressure must increase with temperature, a phenomenon referred to as **aquathermal pressuring**.

Prograde metamorphic reactions, which occur under conditions of increasing temperature and pressure, are commonly dehydration or decarbonation reactions that release water or carbon dioxide, respectively, into the rock. Even at relatively low temperatures, the maturation of hydrocarbons can generate a high pore fluid pressure leading to extension fracturing. Most unfaulted crystalline rocks have a very low permeability. If fluids are produced by metamorphic reactions faster than they can migrate away through the rock, the pore fluid pressure must increase. Hydrofractures caused by such fluid overpressure deep in the crust are a common feature of metamorphic terranes, as evidenced by quartz veins found in many metamorphic rocks. Some workers have suggested that dehydration of deeply subducted serpentine in oceanic lithosphere may be a cause of deep-focus earthquakes in down-going lithospheric slabs.

Partial melting during very high-grade metamorphism in deep crustal regions may also create high pore fluid pressure. In such a situation, the first melts to form are fluid-rich and generally of granitic composition. If the fluid cannot escape, the pressure of the melt can become very high. Some dikes and veins in the deepest core regions of mountain belts may originate as fractures induced by the fluid pressure of such melts.

9.4 STRESS IN THE EARTH

We now wish to inquire what stresses are in fact observed in the Earth's crust. With our knowledge of the possible mechanisms for development of such stresses, we will be in a position to interpret the significance of the observations with respect to the processes that operate in the Earth.

i. Vertical Normal Stress

We often assume that the principal stresses are vertical and horizontal, because they must have that orientation at the horizontal surface of the Earth. If this assumption is correct, plotting principal stress orientations on a stereonet should produce a tight cluster of axes about the center of the net and a distribution of axes around the periphery. Figure 9.6*A* shows, for example, principal stress orientations determined in southern Africa. Although axes do cluster around the center and the periphery of the stereogram, there is a large amount of scatter, suggesting that the assumption is only a rough generalization.

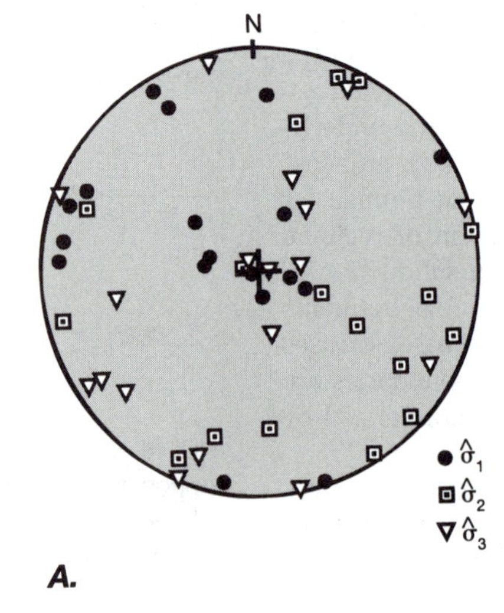

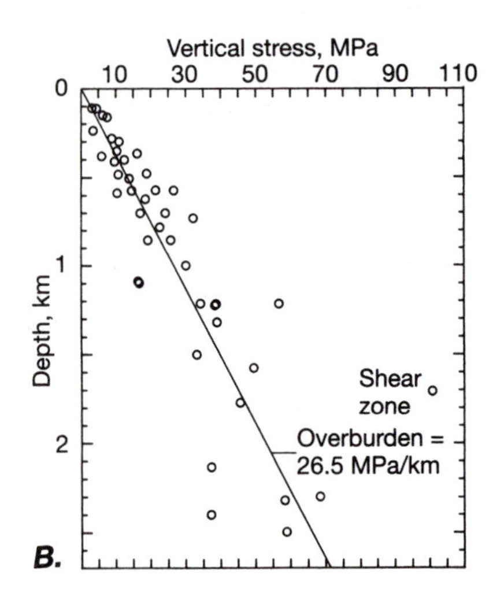

FIGURE 9.6 The orientation and magnitude of the vertical component of stress in the Earth. *A*. Plot of the orientations of the three principal stresses in southern Africa. Equal-area, lower hemisphere projection. *B*. Plot of the magnitude of the vertical component of stress. The line is the lithostatic load for a rock density of 2700 kg/m^3. (After McGarr and Gay 1978)

Another common assumption is that the vertical normal stress should be equal to the overburden, which is determined by the density of the rocks. Figure 9.6*B* shows a set of measurements of the vertical normal stress compared with the overburden stress for a mean rock density of 2700 kg/m^3. In fact, although the overburden stress is a good average of the vertical stresses, there is a great deal of variability, which again warns us that our common assumption is an oversimplification.

ii. Nontectonic Horizontal Normal Stress

In an undeformed sedimentary basin, we generally expect the overburden to dominate the state of stress. If our expectation is correct, the principal stresses should be vertical and horizontal. The vertical normal stress should be the maximum compressive stress and should equal the overburden. The horizontal stress, however, is more difficult to estimate. We can suppose that the sedimentary rocks in a basin behave as an elastic solid and that the geometry of the Earth requires that there can be no change in horizontal dimension of these rocks ($\hat{e}_{xx} = 0$), since the horizontal arc length of the sedimentary basin does not change. If unconstrained, however, the rocks would show a horizontal Poisson expansion in response to a vertical load. We can then calculate the magnitude of the horizontal stress σ_H that would exactly counteract that Poisson expansion. We proceed as follows:

Write the elasticity equation for $\hat{e}_{xx}$ from Equation (9.5) by changing subscripts z to x and x to z. Then set $\hat{\sigma}_{xx} = \hat{\sigma}_{yy} = \sigma_H$, $\hat{\sigma}_{zz} = \sigma_V$, and $\hat{e}_{xx} = 0$. Solving for the horizontal stress gives

$$\sigma_H = \frac{\nu}{(1 - \nu)} \sigma_V \tag{9.7}$$

For a Poisson ratio (ν) between 0.25 and 0.33, which are common values for rock, this equation implies that the horizontal stress should range from about one-third to one-half the vertical stress. The constant $\nu/(1 - \nu)$ is one possible value for the constant κ in Equations (7A.2) and (7B.2) (Appendices 7-A and 7-B) and in Equation (9.9) (Section 9.10(ii)).

Figure 9.7 shows the values of the minimum horizontal compressive stress in sedimentary basins in the United States, determined by the hydrofrac technique. For comparison, the different lines indicate the overburden stress, the hydrostatic pressure, and the minimum compressive stress predicted from the Poisson effect for three values of ν. Except for three measurements in granite, the stress calculated from the Poisson effect stress is too low, indicating that the assumptions we made for the calculation are not realistic.

If we had assumed that rocks were sufficiently ductile so that the flow would eliminate any differential stress, it would be equivalent to assuming that $\nu = 0.5$. In that case, the state of stress would be lithostatic and equal to the overburden ($\sigma_H = \sigma_V = \rho_r gh$). Figure 9.7 shows that this is not a realistic assumption either.

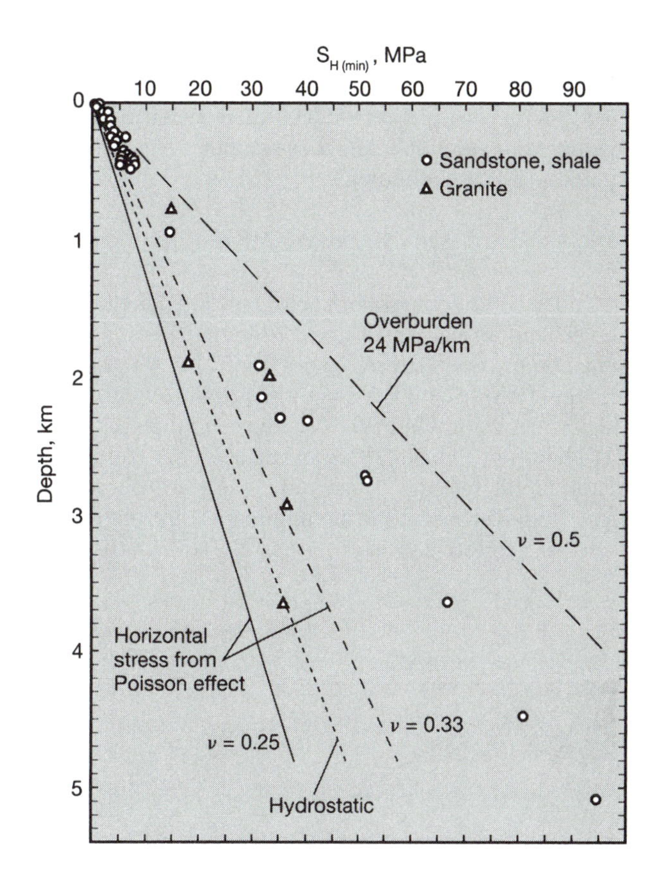

FIGURE 9.7 The minimum horizontal compressive stress measured by the hydrofract technique in sedimentary basins in the United States (data points) compared with the overburden pressure, the hydrostatic pressure, and the minimum horizontal stress predicted by the Poisson effect for $\nu = 0.25$ and 0.33. (After McGarr and Gay 1978)

The best that we can say is that the horizontal normal stress calculated from extreme values of the Poisson ratio gives maximum and minimum bounds for a nontectonic horizontal stress. Some other effect seems to be at work to give a value for that stress that lies between the calculated extremes.

iii. Tectonic Horizontal Normal Stress

The only constraint we can put on horizontal stresses of tectonic origin is that the differential stress (the diameter of the Mohr circle) must not exceed the strength of the rock. We assume the strength is determined by the Coulomb fracture criterion, which we express as a relationship between the maximum and minimum principal stresses at fracture (see Box 8-1, Equation (8-1.2)). We also assume that the principal stresses are horizontal and vertical and that the vertical stress is the overburden, although these assumptions are not necessarily accurate. We consider the cases for horizontal tectonic lengthening and horizontal tectonic shortening with a fracture angle $\theta_f = 60°$ and a cohesion $c = 10$ MPa (a common value for rocks), which gives a fracture strength under uniaxial compression of $S = 34.6$ MPa and the proportionality factor $K = 3$. Using these values we can rewrite Equation (8-1.2) as follows:

$$\hat{\sigma}_1^* = 34.6 + 3\hat{\sigma}_3 \text{ [MPa]} \tag{9.8}$$

Under conditions of horizontal tectonic lengthening, the vertical normal stress (the overburden) is the maximum compressive stress, or $\sigma_V = \hat{\sigma}_1$, and we can use Equation (9.8) to find the minimum possible value of $\hat{\sigma}_3$, which is shown in Figure 9.8*A* as the solid line to the left of the heavy line labeled "Overburden." Under conditions of horizontal tectonic shortening, the vertical normal stress (the overburden) is the minimum principal stress, or $\sigma_V = \hat{\sigma}_3$, and we can use Equation (9.8) to find the maximum possible value of $\hat{\sigma}_1$, which is shown in Figure 9.8*A* as the solid line to the right of the heavy line labeled "Overburden." The strength of the rock under these two different tectonic conditions is the differential stress $\sigma^{(Dif)}$ measured by the horizontal distance between the heavy "Overburden" line and one of the other solid lines.

To include the effects of pore fluid pressure on the strength of the rock, we replace the principal stresses $\hat{\sigma}_1$ and $\hat{\sigma}_3$ in Equation (9.8) with the effective principal stresses $\hat{\sigma}_1^{(Eff)} = \hat{\sigma}_1 - p_f$ and $\hat{\sigma}_3^{(Eff)} = \hat{\sigma}_3 - p_f$, respectively, where p_f is the pore fluid pressure. From Equation (8.11) we can write the pore fluid pressure as a function of the overburden, $p_f = \lambda\sigma_V$, and we can also substitute σ_V for the appropriate principal stress, as discussed in the preceding paragraph. The resulting equations for different values of λ are plotted in Figure 9.8*A* as the dashed lines for each of the two tectonic environments. Because

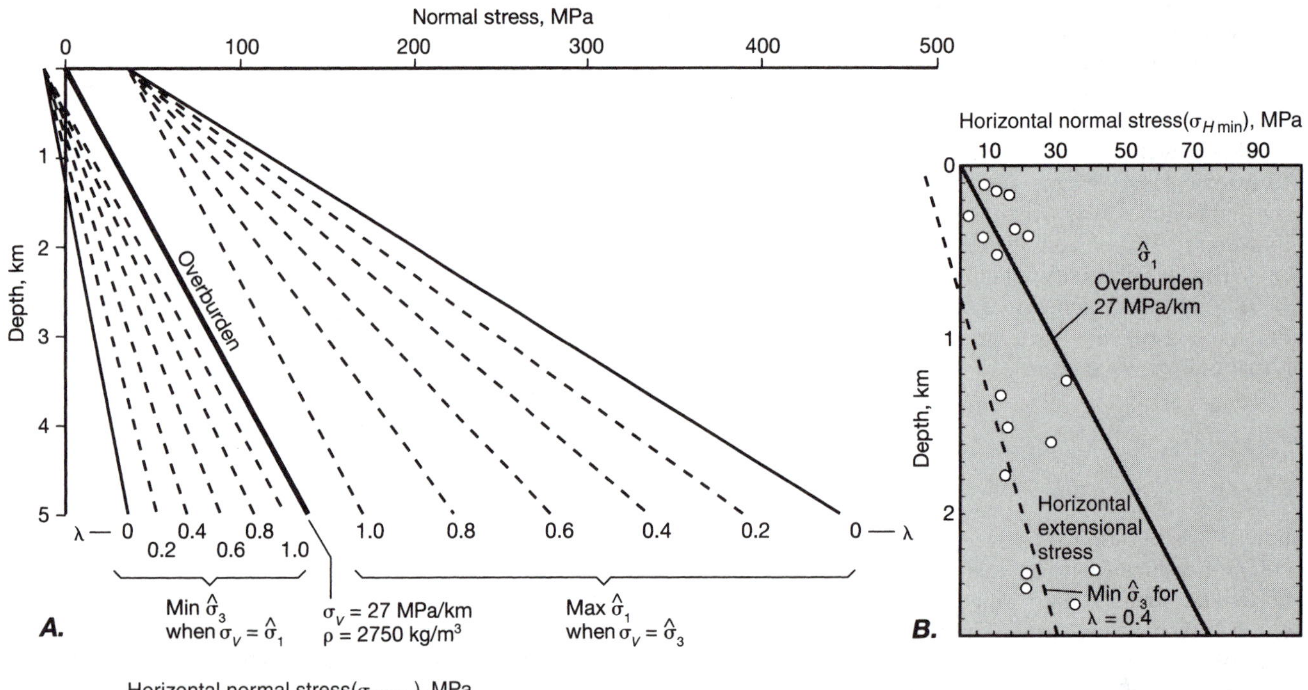

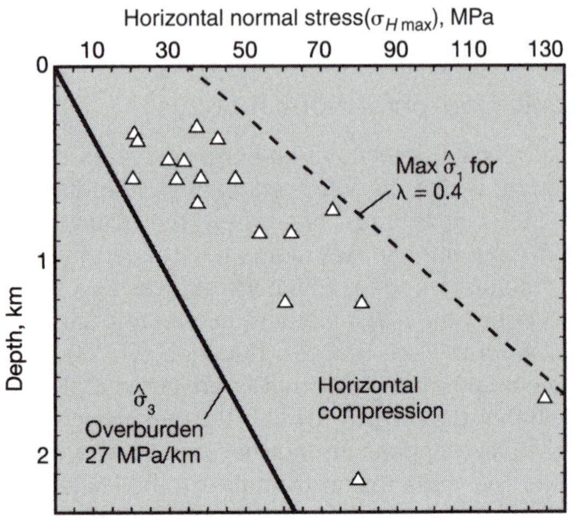

FIGURE 9.8 Constraints on maximum and minimum principal stresses based on the Coulomb fracture criterion and the effect of pore fluid pressure. *A*. Theoretical curves for the minimum possible value of $\hat{\sigma}_3$ when the vertical stress is $\hat{\sigma}_1$ (to the left of the "Overburden" line) and for the maximum possible value of $\hat{\sigma}_1$ when the vertical stress is $\hat{\sigma}_3$ (to the right of the "Overburden" line) for varying values of the pore fluid pressure ratio λ. Note that the minimum principal stress cannot be tensile below about 1 kilometer depth. *B*. Minimum horizontal stress measurements from an area of tectonic horizontal lengthening in southern Africa, showing the constraints provided by the overburden and the minimum principal stress for hydrostatic pore fluid pressure ratio $\lambda = 0.4$ (After McGarr and Gay 1978) *C*. Maximum horizontal stress measurements from an area of tectonic horizontal shortening in Canada showing constraints on stress for a pore fluid ratio $\lambda = 0.4$. (After McGarr and Gay 1978)

the distance between the dashed lines and the "Overburden" line decreases with increasing λ for both tectonic environments, the figure shows that the strength of the rock in each case decreases with increasing pore fluid pressure, as discussed in Section 8.5.

Figure 9.8*B* shows measurements of the minimum horizontal stress from an area of subsidence and normal faulting in southern Africa. The solid line is again the overburden, and the dashed line is the minimum possible stress for $\lambda = 0.4$, the value for permeable saturated rock (Equation (8.11)). At shallow depths, several values of the horizontal stress exceed the overburden. At greater depths, the predicted maximum and minimum stresses are better constraints to the data. Observed values that fall below the $\lambda = 0.4$ line could result from a lower value of λ. Alternatively, these values could result from a larger value of the cohesion c, and thus of the material constant S (Equation (8-1.2), second Equation (8-1.3)), which would have the effect of giving smaller calculated values for $\hat{\sigma}_3^{(Eff)}$.

Note that the plot of minimum values of $\hat{\sigma}_3$ in Figure 9.8*A* indicates that actual tensile stresses (negative values of the normal stress) cannot exist below a depth of about 1 kilometer. *In fact, tensile stresses have never been measured within the Earth.*

Measurements of the maximum horizontal stress from a region of folding and thrust faulting in Canada are illustrated in Figure 9.8*C*. The solid line is again the overburden stress, and the dashed line is the maximum possible compressive stress for $\lambda = 0.4$. All the measured stresses fall between the two lines.

This analysis shows that considerable variation exists in the Earth's crust, which indicates that application of theory gives only an approximation of the real state of affairs. This method of constraining the differential stress applies at best to the upper 15 to 20 kilometers of the crust, which is the depth range of brittle behavior. Below that, the increases in temperature and pressure induce a weakening of the rock because of the onset of ductile deformation processes (see Sections 16.3 to 16.6), and the Coulomb fracture criterion on which Equation (9.8) is based does not predict the strength of the rocks.

iv. Regional Distributions of Stress

Figure 9.9 summarizes the distribution of stresses inferred from the measurements of strains using the techniques discussed in Section 9.2. Although commonly referred to as a "stress map," the principal axes plotted are better interpreted in terms of the principal strains that are the basis of the measurements. Figure 9.9*A* shows the worldwide distribution of these strains, including the location of measurements, the orientation of principal extension axes, and the technique used for the measurement. Over large regions, there is reasonable agreement among the different measurement techniques, but the orientations of the principal axes change substantially even within a single plate, and differences among nearby measurements are probably larger than errors introduced by equating principal stress axes to principal extension axes. These strains reflect major tectonic processes in the Earth and provide important constraints on models of the driving forces for plate tectonics, which must account, at least approximately, for the observed strain distribution within the plates.

Figure 9.9*B* presents a more detailed summary of strain, or nominal stress orientation measurements, in the United States. The boundaries separate regions of roughly similar states of strain. These regions approximately correspond to geologic provinces in which the structures reflect distinct tectonic regimes.

9.5 STRESS HISTORIES AND THE ORIGIN OF JOINTS

Given the wide variety of mechanisms for inducing and changing stress conditions in the Earth's crust, it is not surprising that fractures in the crust have numerous possible origins. In this section, we look at possible loading histories that can lead to the formation of joints. Because joints are extension fractures, the tension fracture criterion applies to the explanation of their origin.

For sedimentary basins, we distinguish two principal sets of conditions: those that cause jointing during burial and those that cause jointing during uplift and erosion. The stress path associated with burial followed by uplift is not a reversible path, because the mechanical properties of the material change with time. During burial, the unconsolidated sediments gradually become compacted and lithified and may be affected by tectonic deformation or even recrystallization. Thus when rocks are uplifted, they are very different materials from when they were buried, and they have different mechanical properties. This difference affects the way stresses accumulate and fractures form (see Box 9-1).

All stresses that have been measured in the Earth are compressive; true tensile stresses have never been observed. For extension fractures to form, therefore, two conditions must be met: 1) pore fluid pressure must be large enough for the effective minimum principal stress to become tensile; and 2) the differential stress must be small enough so that at the critical pore fluid pressure, extension fractures form (Figure 8.12*A*) rather than shear fractures (Figure 8.12*B*).[6] Values of the tensile strength $|T_0|$ for small rock samples measured in the laboratory range from a few MPa for weak sedimentary rocks up to around 40 MPa for crystalline rocks. Widespread planes of weakness in crustal rocks, such as fractures and bedding planes, however, result in very low bulk-tensile

[6]According to the Griffith theory of fracture, the differential stress (the diameter of the Mohr circle) that can cause extension fracturing is limited by $\hat{\sigma}_1 - \hat{\sigma}_3 < 4|T_0|$, where T_0 is the tensile strength. This is the largest Mohr circle that can be tangent to the parabolic fracture criterion at the vertex of the parabola (see Figure 8.12).

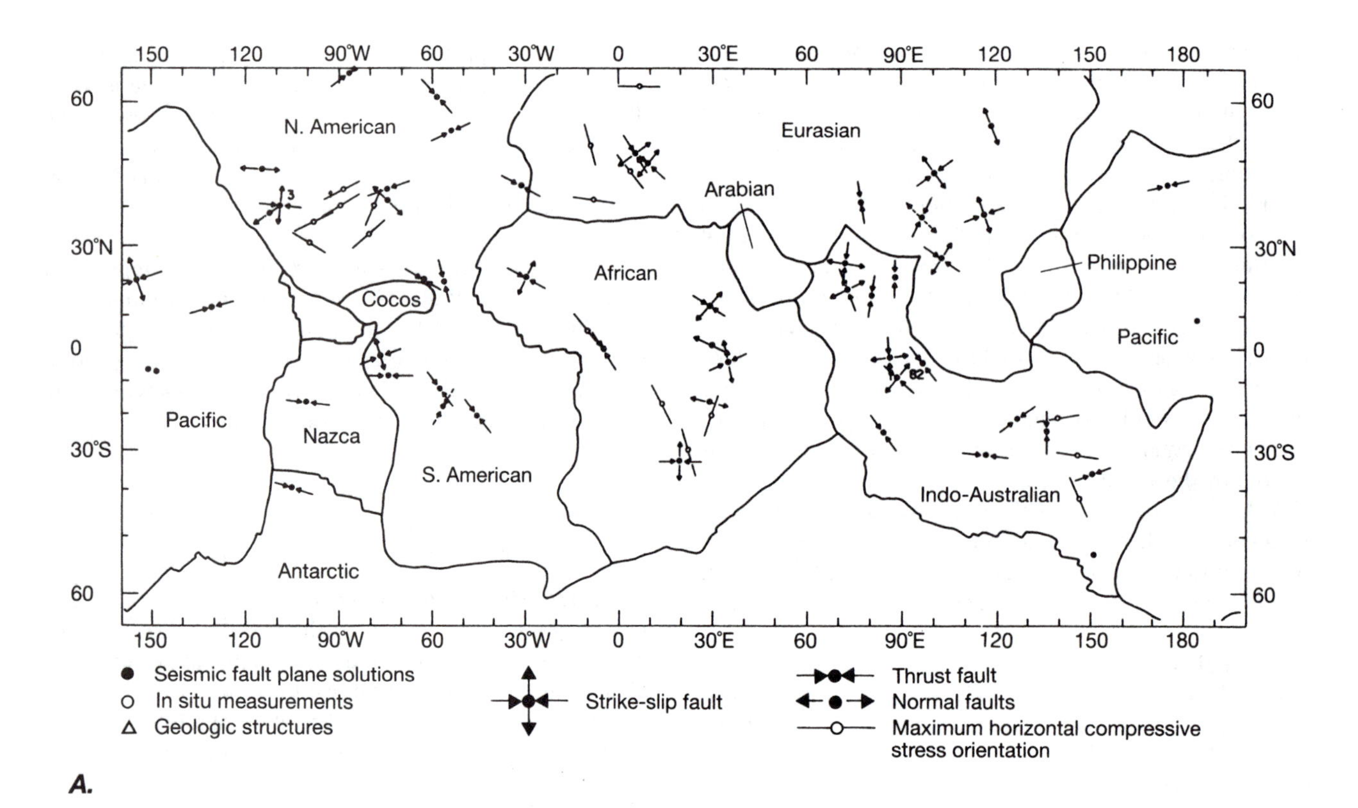

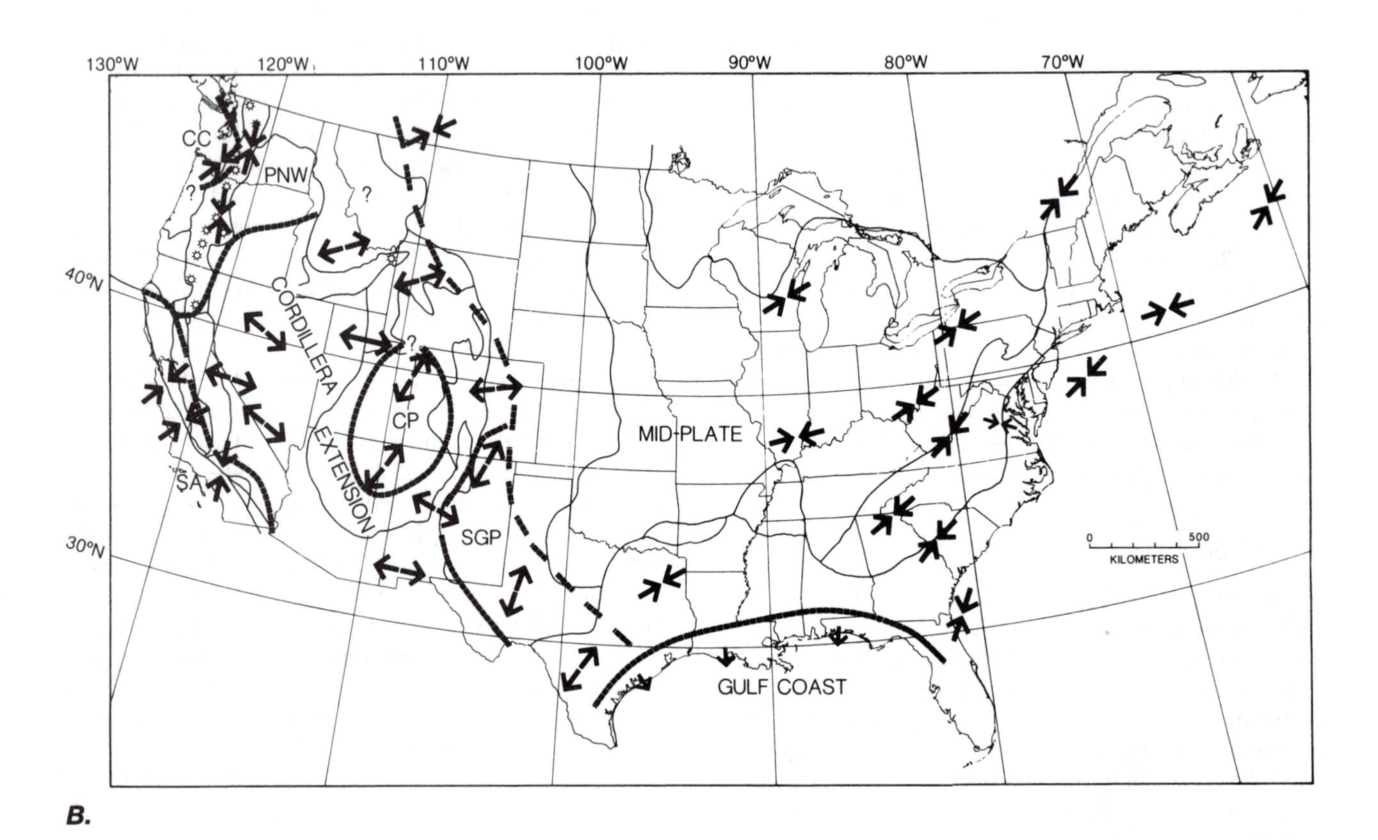

FIGURE 9.9 Regional stress fields inferred in general by equating principal stress axes to measured principal strain axes. *A*. Worldwide distribution of principal stress orientations. (After Richardson et al. 1979) *B*. Stress orientations within the coterminus United States. (After Zoback and Zoback 1980)

strengths. Measured differential stresses are generally small and tend to increase slightly with depth, being generally less than 20 MPa near the surface, and at a depth of 5 km, reaching values of no more than 50 MPa in sedimentary rocks and 70 MPa in crystalline rocks (Figure 9.8). Thus hydrofracture should be common in rocks, and it probably results in extension fractures.

i. Joint Formation During Burial

In tectonically quiescent sedimentary basins, measured fluid pressures at depths less than about 3 km are generally not greater than hydrostatic pressure. This fact suggests that the flow of fluids through the rock is unrestricted above depths of 3 km. With increasing depth of burial, flow becomes restricted, and compaction and aquathermal pressuring, or the maturation of hydrocarbons, can increase the fluid pressure more rapidly than the minimum compressive stress increases. Eventually, hydrofracture results.

Evidence of joint formation by hydrofracture before the sediments have solidified into rock is preserved in some cases as **clastic dikes**, which are intrusive dikes of sedimentary material that cross-cut bedding. They form in permeable rocks, when the sudden opening of a fracture results in a sudden local decrease of pore fluid pressure. The high-pressure gradient this creates causes a rapid flow of pore water into the fracture, and if the sediment is unconsolidated, some of it may be carried along into the fracture, producing a clastic dike.

The different mechanical properties of different rock types mean that in general, rocks do not fracture at the same time. As a simple example, consider an interlayered sandstone-shale sequence in which the overburden stress is the maximum compressive stress, $\hat{\sigma}_1 = \rho_r g h$. The sandstone is stronger than the shale, which means it can support a larger differential stress. Thus the horizontal normal stress, which is the minimum compressive stress $\hat{\sigma}_3$, can be smaller in the sandstone than in the shale (Figure 9.10), and a smaller pore fluid pressure would then be required to cause hydrofracturing in the sandstone than in the shale (see Box 9-1). As the pore pressure gradually increases during burial, therefore, hydrofractures would first develop only in the sandstone and would not extend into the shale. When pore pressure in the shale rises enough to cause it to hydrofracture, the sandstone could also fracture, and fractures therefore could propagate across the sandstone-shale contacts. Thus the extent of a set of joints with respect to lithology is an important factor in interpreting the history of joint development.

ii. Joint Formation During Uplift and Erosion

We assume that when rocks undergo uplift and erosion, the principal stresses are horizontal and vertical, and the vertical stress equals the overburden. The changes in horizontal stress components during uplift determine whether jointing occurs in this phase of the rock's history. The im-

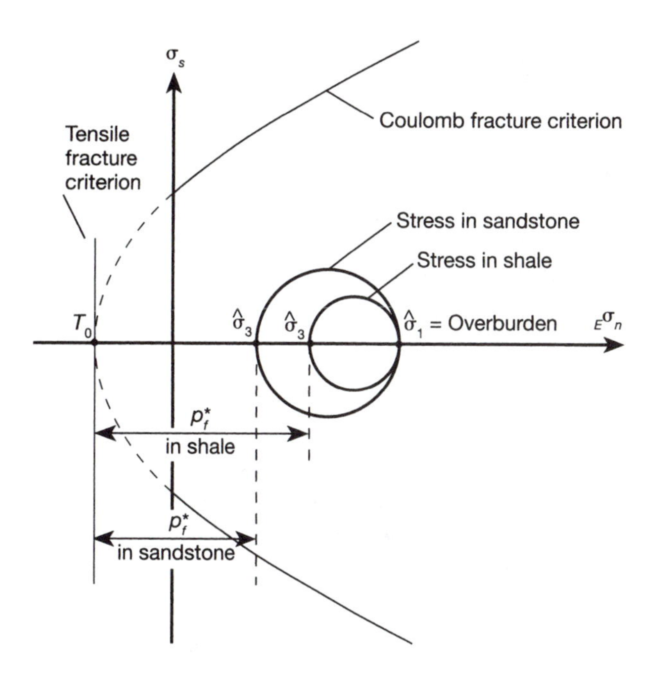

FIGURE 9.10 Stresses in interbedded sandstone and shale. The sandstone layers can support a larger differential stress than the shales. p_f^* is the critical pore pressure required for hydrofracture, and it is smaller for the sandstones than for the shale.

portant factors governing the evolution of the horizontal stresses are the Poisson and the thermal effects, which both tend to decrease the horizontal stress, and bending stresses, which are caused by bending of the rock layers during uplift. Without detailed knowledge of the geometry of bending, we cannot predict these stresses, so we will not include them in the following discussion, but that does not imply that they are unimportant.

If the rocks behave as an elastic material, the Poisson effect predicts that a decrease in the vertical load will cause expansion in the vertical direction and corresponding contraction in the horizontal direction (Figure 9.2A). The rocks are not free to change horizontal dimensions, however, so the horizontal components of stress must decrease sufficiently to offset exactly the Poisson contraction (Equation (9.7)). Starting, for example, from a lithostatic stress at the deepest point of burial, for which $\hat{\sigma}_1 = \hat{\sigma}_2 = \hat{\sigma}_3 = \rho_r g h$, uplift and erosion would result in a decrease in all components of compressive stress, but the horizontal stress would decrease less than the vertical component (Equation (9.7)). Thus the horizontal principal stress would end up as the maximum compressive stress (see Box 9-1).

The decrease in temperature associated with uplift causes a thermal contraction of the rock in both the vertical and horizontal directions. Because the rocks cannot change horizontal dimension, the horizontal compressive stress decreases by an amount that exactly offsets the thermal contraction. This decrease, added to the decrease

BOX 9-1 The Effect of Burial and Uplift on Stress in the Crust

We consider a simple model of the evolution of stress in rocks during burial, lithification, and uplift. The model includes only the overburden, the Poisson effect, and the thermal effect, and we calculate the stress required to maintain a horizontal extension of zero. The change in the maximum and minimum horizontal normal stresses $\Delta\sigma_{H(max)}$ and $\Delta\sigma_{H(min)}$ as a function of the changes in vertical stress $\Delta\sigma_V$ and temperature ΔT is given by

$$\Delta\sigma_{H(\max)} = \Delta\sigma_{H(\min)} = \left(\frac{\nu}{1-\nu}\right)\Delta\sigma_V - \left(\frac{E}{1-\nu}\right)\alpha\ \Delta T \quad (9\text{-}1.1)$$

where α is the coefficient of thermal expansion, which gives the extension (Equation (9.1)) per degree of temperature change. The first term on the right side of this equation gives the stress required to counteract the Poisson effect and comes from Equation (9.7). The second term gives the stress required to counteract the thermal effect and comes from the equations of elasticity similar to Equation (9.5). By switching subscripts z to x and x to z in Equation (9.5), we obtain

$$\hat{e}_{xx} = \frac{1}{E}\ \hat{\sigma}_{xx} - \frac{\nu}{E}\ (\hat{\sigma}_{yy} + \hat{\sigma}_{zz})$$

A change in temperature will only affect the horizontal stresses, so to calculate this effect, we set the vertical overburden stress to zero and assume the horizontal stresses are equal:

$$\hat{\sigma}_{zz} = 0 \quad \hat{\sigma}_{xx} = \hat{\sigma}_{yy}$$

Substituting these assumptions into the equation for $\hat{e}_{xx}$ and solving for $\hat{\sigma}_{xx}$ gives

$$\hat{\sigma}_{xx} = \left(\frac{E}{1-\nu}\right)\hat{e}_{xx}$$

We want to find the stress that will just counteract the horizontal thermal contraction on uplift. The thermal contraction is $\alpha\ \Delta T$, so the horizontal extension necessary to counteract that is

$$\hat{e}_{xx} = -\alpha\ \Delta T$$

Substitution into the preceding equation gives the stress $\hat{\sigma}_{xx}$ required to counteract the horizontal thermal contraction, which is then the second term on the right side of Equation (9-1.1).

The changes in stress and temperature indicated by the Δ in Equation (9-1.1) are the final values minus the initial values, and we can express $\Delta\sigma_V$ and ΔT as a function of the change in depth:

$$\Delta\sigma_{H(\max)} = \Delta\sigma_{H(\min)} \equiv \sigma_H^{(f)} - \sigma_H^{(i)} \quad (9\text{-}1.2)$$

$$\Delta\sigma_V \equiv \sigma_V^{(f)} - \sigma_V^{(i)} = \rho_r g\,(h^{(f)} - h^{(i)})$$

$$\Delta\sigma_V = (25\ \text{MPa/km})(h^{(f)} - h^{(i)}) \quad (9\text{-}1.3)$$

$$\Delta T = (25^\circ\text{C/km})(h^{(f)} - h^{(i)}) \quad (9\text{-}1.4)$$

where the superscripts (f) and (i) indicate "final" and "initial" values, respectively, and where h is the depth in kilometers.

TABLE 9-1.1 Mechanical Properties of Sediment during Burial and Uplift*

	Burial		Uplift	
	Sand	Clay	Sandstone	Shale
E [MPa]	-1.0×10^3	Small	-16.5×10^3	-4.9×10^3
ν	0.21	0.5	0.33	0.36
α [°C^{-1}]	10.0×10^{-6}	—	10.8×10^{-6}	10.0×10^{-6}

* Data assembled from various sources by Engelder (1985).

associated with the Poisson effect, can be sufficient to make the horizontal principal stress become the minimum principal stress.

In most examples of uplift and erosion, the net effect of the interplay of Poisson and thermal effects is that the horizontal stress is the minimum compressive stress, and the vertical stress is the maximum compressive stress (see Box 9-1). The formation of joints, however, requires a pore fluid pressure sufficient to make the minimum *effective* principal stress a tensile stress, and if that stress is horizontal, then vertical joints can form with an orientation normal to that stress component. Thus, as we discussed in Section 8.5, the pore fluid pressure plays a critical role in producing an *effective tensile* stress in a *compressive* stress regime.

The development of a set of vertical joints relieves the effective tensile stress normal to the set. If the other horizontal principal effective stress is also tensile, then it becomes the maximum effective tensile stress, and a second set of vertical joints may form orthogonal to the first set. Such systems of orthogonal vertical joints are a common feature, for example, of the flat-lying sediments

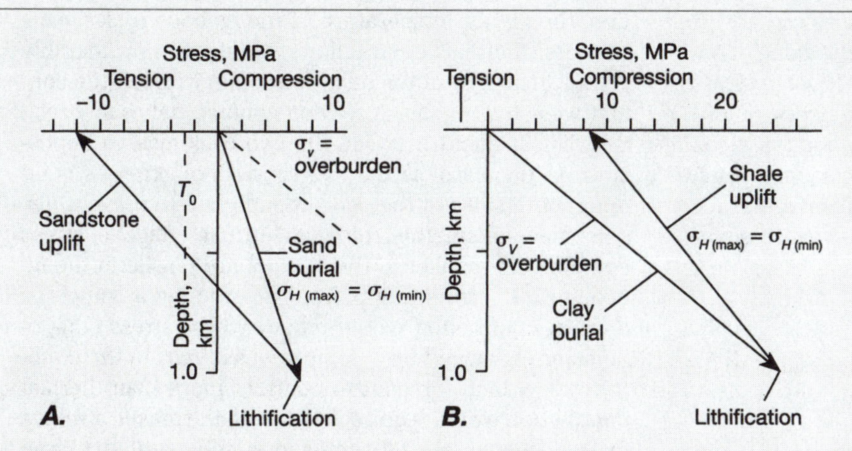

FIGURE 9-1.1 Graphs of stress histories during burial of a sediment, followed by lithification and uplift of a rock using the physical properties from Table 9.1.1. (After Engelder 1985) *A.* Sand/sandstone and *B.* clay/shale.

We then substitute Equations (9-1.2) through (9-1.4) into Equation (9-1.1).

With this model, we determine the history of stress for both a sandstone and a shale that are first buried as unconsolidated sediments to a depth of 1 km, then lithified at the maximum depth of burial, and finally uplifted back to the surface. The elastic constants and the coefficients of thermal expansion listed in Table 9-1.1 show that the clay and shale have different mechanical properties, as do the sand and sandstone. This fact ensures that the stress history during burial is different from that during uplift, as shown in Figure 9-1.1.

For the burial, we take $\sigma_H^{(i)} = 0$ MPa, $h^{(f)} = 1$ km, and we solve for $\sigma_H^{(f)}$ by using the constants for sand and clay listed in Table 9-1.1. At a depth of 1 km, the final horizontal stresses on the sand and clay, respectively, are 7 MPa and 25 MPa, and the vertical stress for both is 25 MPa (Figure 9-1.1). For uplift, we use the final horizontal stress from burial as the initial horizontal stress for uplift, $\sigma_H^{(i)} = 7$ or 25 MPa, we use $h^{(i)} =$ 1 km and $h^{(f)} = 0$ km, and we use the constants for sandstone and shale from Table 9-1.1. The final horizontal stresses at the surface are −12 MPa and 9 MPa for the sandstone and shale, respectively (Figure 9-1.1).

For the sand in this simple model, the horizontal stress is compressive during burial and is the minimum principal stress; the vertical stress is the overburden and is the maximum principal stress (Figure 9-1.1*A*). Because lithification from sand to sandstone changes the elastic properties of the material (Table 9-1.1), uplift of the sandstone carries it along a stress-depth path of shallower slope than for burial. Thus the horizontal stress on the sandstone decreases more rapidly during uplift than it increased in the sand during burial, and it actually becomes tensile during uplift. The tensile strength T_0 is exceeded after only about half the overburden has been removed (Figure 9-1.1*A*), at which point joints could form.

The assumption for clay that $\nu = 0.5$ results in all stress components being lithostatic along the stress-depth path for burial (Equation (9.7) and Figure 9-1.1*B*). After lithification, uplift causes a decrease in the horizontal stress, but this decrease is less than that of the overburden. Thus the horizontal stress remains compressive throughout the history and is the maximum principal stress during uplift (Figure 9-1.1*B*).

Different lithologies thus can have very different stress histories in response to the same externally applied conditions, and the same fractures do not necessarily develop in all rock types, even in the same location.

Tensile stresses have not been measured in rocks; lithification is not likely to occur only at the greatest depth of burial; and we have neglected the effects of pore fluid pressure; so this model is oversimplified. It does, however, illustrate some of the variability that is inherent in the evolution of stress at depth in different rocks.

in the North American midcontinent and the Colorado plateau regions of the United States. Because of the difficulty of determining the relative timing of different joint sets, this interpretation of origin of such orthogonal sets remains hypothetical at present.

iii. Tectonic Joints

If tectonic stresses are imposed on a rock during burial, then compaction and the restriction of pore fluid circulation may occur at shallower depths than would be the case under lithostatic loading alone. The resulting high pore fluid pressures can cause hydrofracturing at depths much shallower than 3 kilometers. In such cases, the orientation of the joints should reflect the orientation of the principal tectonic stresses. Tectonic stresses may be applied to the rocks either before or after the formation of burial joints. Because tectonic deformation is commonly accompanied by the development of a foliation in the rocks (see Chapter 11), the cross-cutting relationships between joints and foliations can be an important element in reconstructing the sequence of deformational events.

Tectonic stresses can affect rocks, of course, during uplift as well as during burial. Such stresses can govern the orientation of new joints by changing the value of one of the horizontal components of stress. If a horizontal stress became the minimum compressive stress, as proposed in the preceding subsection, and if a horizontal tensile tectonic stress were added, then vertical joints would form normal to the tectonic stress. If the horizontal tectonic stress were the maximum compressive stress, vertical joints could form parallel to it.

iv. The Origin of Sheet Joints

As noted in Chapter 2, sheet joints are subparallel to the topographic surface, and we mentioned two possible mechanisms for their formation:

1. The topography controls the orientation of the sheet joints. Suppose, for example, that a tectonic compression causes the maximum compressive stress to remain horizontal and the minimum compressive stress to be vertical during uplift. As rock approaches the surface, the vertical stress approaches zero. The topography, however, would affect the local orientation of the stress field, because the topographic surface is a free surface that must be a principal surface of stress, since it can support no shear stress. Thus the principal stresses locally must be perpendicular or parallel to the topography. The maximum compressive stress, then, would parallel the topographic surface, and joints parallel to topography could propagate in a manner similar to longitudinal splitting (Figures 8.1*B*, 8.9*A*).
2. As an alternative hypothesis, the orientation of sheet joints is controlled by pre-existing stresses in the rock, and the joints affect the evolution of the topography. For example, in a plutonic igneous body, cooling at depth concentric with the boundary of the pluton could produce residual thermal stresses within the body with the maximum compressive stress ($\hat{\sigma}_1$) subparallel to the boundary. As the minimum compressive stress decreases toward zero during uplift, sheet jointing could develop by longitudinal splitting, and the orientations of the joints would reflect the shape of the boundary or cooling surfaces in the pluton. Subsequent erosion is controlled by the orientations of the joints.

The interpretation of sheet joints is not clear-cut, and each hypothesis could be correct in different cases.

v. The Origin of Columnar Joints

Polygonal fracture patterns, or columnar joints, are common features of many igneous extrusions and shallow intrusions. These fractures result from thermal stresses set up by unequal cooling and thermal contraction between the igneous body and the country rock. After solidification, the higher temperature of the igneous rock means that its thermal contraction would be considerably greater than that of the adjacent country rock if the contact were free to slip. A welded contact makes any relative displacement between the two rock masses impossible. In this case, as the two rocks cool, stresses build up on both sides of the contact sufficient to prevent displacement along the contact. Normal effective stress components parallel to the contact are tensile in the igneous rock, preventing it from contracting as much as thermal contraction would require; these stress components are balanced by a compressive stress in the country rock, which forces it to contract more than thermal contraction would require. In general, the tensile stresses in the igneous rock become oriented parallel to the isothermal surfaces during cooling. Because rocks are weaker in tension than in compression, the igneous rocks tend to form tensile fractures perpendicular to the surfaces of equal temperature.

The origin of the hexagonal shape of the columns is not well understood. More than one set of fractures is required to relieve the tensile stress in two orthogonal directions. Such a system of fractures can fill a volume with close-packed fracture-bounded prisms if the prism cross section is triangular, rectangular, or hexagonal. Of these three configurations, the hexagonal prisms have the smallest fracture surface area per unit volume of prism. Thus fracture-bounded prisms with a hexagonal cross section require less energy to produce than other prism shapes, and this form of columnar joint is predominant. In principle, however, two sets of fractures should suffice to relieve tensile stresses in two orthogonal directions, and we do not yet understand the mechanism of development of the three sets of tensile fractures that define the hexagonal prisms.

A similar process must also account for the development of hexagonal mud cracks, which form during the desiccation and associated contraction of the surface layers of mud.

vi. Interpretation of Plumose Structure

Comparison of experiments on brittle extension fracture in ceramics with the morphology of joint surfaces provides a means of interpreting the significance of some of the features of plumose structure on joints.

Fast rupture propagation in a bed under high stress concentration at the crack tip results in the formation of a symmetrical plumose structure about a straight central axis. The rupture front propagates perpendicular to the hackle lines, and this morphology indicates a single rupture front. Rough hackle characteristically defines the plume under these circumstances (Figure 2.13*A*). Slow fracture propagation under low stress concentration at the fracture tip typically results in multiple wandering plume axes and associated hackle plumes, which indicates fracture propagation that splits into multiple fracture fronts (Figure 2.13*D*). These hackle plumes characteristically show much lower relief, forming smoother joint faces.

Rib marks are interpreted as arrest lines indicating where the fracture propagation has stopped and restarted. They therefore mark the location of the rupture front when the propagation stopped. Multiple sets of rib marks (Figures 2.13*B*, 2.14*B*, *C*) indicate an episodic propagation of the fracture. The episodic nature of the propagation could be explained in terms of a gradual buildup of pore fluid pressure necessary to initiate a fracturing event, followed by a drop in the pore pressure in response to the increase in volume provided by the opening of the fracture and a consequent stalling of the fracture propagation.

vii. Interpretation of Joint Curvature

Curvature or kinks of joints at the ends (Figure 2.10*B*, *C*) and the development of *en echelon* fringe fractures (Figures 2.10*D*, 2.13*C*, 2.14*A*) indicate a change in the orientation of the stress field in the rock volume into which the fracture propagates. The stress may rotate because of local structures, such as the presence of another fracture, which we discuss in Subsection 9.5(viii), or because of a difference in mechanical properties between adjacent beds. The principal stresses may also rotate in time as tectonic conditions change. Depending on whether the spatial change in principal stress orientation is gradual or abrupt, fringe fractures are, respectively, gradual (Figures 2.13*C*, 2.14*A* bottom half of the main joint face) or abrupt (Figures 2.2*B*, 2.14*A* top half of main joint face).

If an extension fracture propagates into a volume where the stress has changed orientation, then the original plane of the fracture must in general experience a shear stress. If propagation were to continue parallel to the original plane of the joint, it would have to occur under mixed-mode conditions. These conditions are represented by two end members, either mixed modes I and II if the shear stress is perpendicular to the fracture front (Figure 2.1*A*, *B*; right side of Figure 9.11) or mixed modes I and III if the shear stress is parallel to the fracture front (Figure 2.1*A*, *C*, top of Figure 9.11). In fact, however, the propagating fracture tends to reorient so that the fracture plane is a true extension fracture, oriented perpendicular to the minimum principal stress with zero shear stress on the fracture plane. Thus if fracture propagation by mixed modes I and II is required, the fracture tends to turn (Figure 2.10*B*) or kink (Figure 2.10*C*) into a new orientation normal to the new minimum principal stress (right side of Figure 9.11). If the fracture propagation by mixed modes I and III is required, the fracture plane tends to break down into an *en echelon* set of fringe fractures,

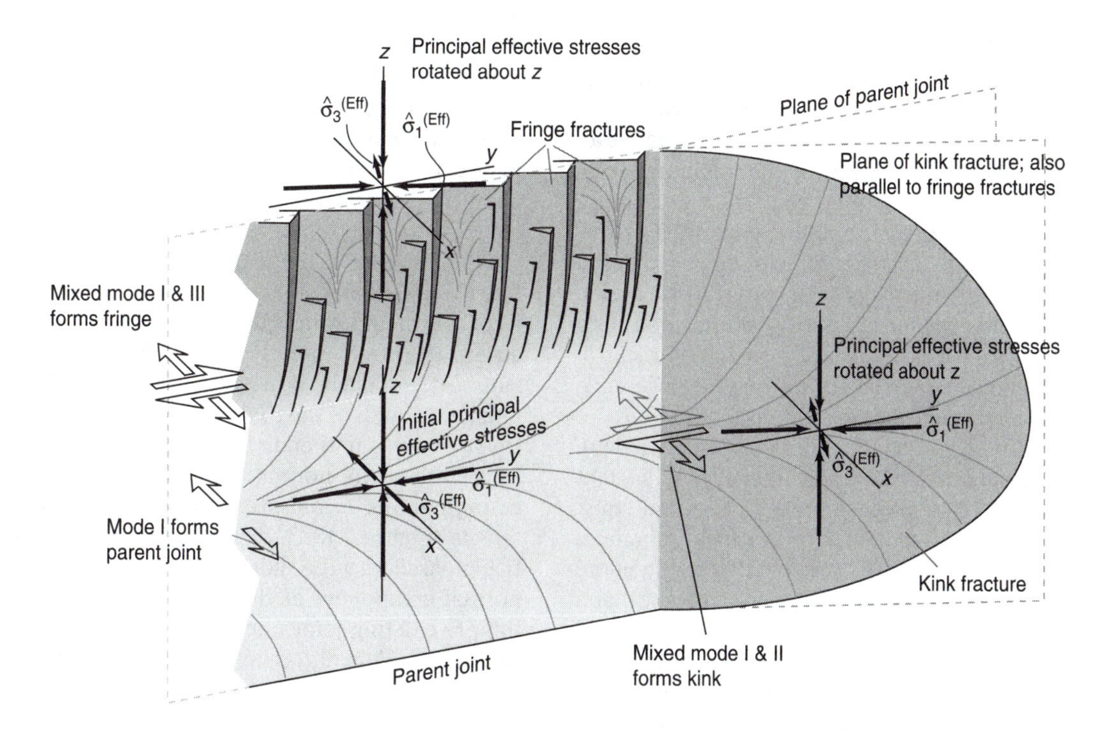

FIGURE 9.11 Hackle fringes and kink fractures on joints indicate that the joint has propagated into volumes of rock in which the principal stresses are rotated relative to those that cause the parent joint. A rotated stress field results in shear stresses on the plane parallel to the parent joint, and the fracture changes orientation to remain perpendicular to the minimum principal stress, forming either a kink fracture or an *en echelon* set of fringe fractures, depending, respectively, on whether the shear stress is perpendicular or parallel to the propagating fracture front. (After diagrams in Younes and Engelder 1999)

each of which is normal to the new minimum principal stress (Figure 2.10*D*; top of Figure 9.11).

viii. Interpretation of Joint Intersections

The origin of at least some of the geometries of joint intersections can be understood in terms of the propagation history of the joints and the manner in which the stress field at a distance from the joints becomes perturbed immediately adjacent to the joints. We look in particular at the joint intersections illustrated by Figure 2.10*F*, *G*, *H*. In the following discussion, when we refer to the stress, we actually mean the effective stress ($\hat{\sigma}_k^{(Eff)} \equiv \hat{\sigma}_k - p_f$), where p_f is the pore fluid pressure (see Figure 7.12*J* and Section 8.5), which we assume is the same everywhere. The stress field far from the influence of a joint is termed the *remote stress*, and the perturbed stress in the neighborhood of a joint is the *local stress*. We analyze the joints in the horizontal plane, and we assume that a joint always propagates so that it is oriented perpendicular to the local minimum principal stress. Thus if a joint is curved, it reflects a curvature in the principal stress trajectories through the rock (see the preceding subsection).

If each of a pair of parallel joints turns toward the other and intersects at a high angle (Figure 2.10*F*), we infer that the joints propagated toward each other at the same time and that they were close enough for the presence of each joint to affect the local stress field around the other. Figure 9.12*A* shows a crack subjected to a pore fluid pressure p_f and a remote stress field for which the minimum compressive stress $\hat{\sigma}_{zz}$ is normal to the crack face and the maximum compressive stress $\hat{\sigma}_{xx}$ is parallel. The pore fluid pressure exceeds the minimum compressive stress by an amount equal to the fracture strength of the rock. We can obtain an intuitive understanding of how the local stress affects the propagation of two parallel but offset cracks propagating toward each other by examining the locally perturbed stress around the tip of a single crack. The local principal stress trajectories about a crack depend strongly on the difference between the two remote principal stresses $\Delta\sigma = \hat{\sigma}_{xx} - \hat{\sigma}_{zz}$. Figure 9.12*B* shows one example of these trajectories about the right half of the crack. If a new crack (right side of Figure 9.12*B*) propagates toward the initial crack so that it remains perpendicular to the local minimum compressive stress, it will first turn away from the initial crack and subsequently turn toward it to intersect it at a high angle, as shown by the heavy dashed line extending from the end of the new crack. The details of the path depend strongly on the value of $\Delta\sigma$ and on the ratio of the crack length to the perpendicular distance between the two cracks. Prediction of the propagation path requires the calculation of the effect of both fracture tips on the local stress, and this stress changes as the cracks lengthen, thereby affecting the further propagation of each joint. Thus the actual propagation history and joint geometry can only be determined with a computer using a stepwise numerical calculation of each increment of propagation. Figure 9.12*C* shows several crack propagation paths calculated in this manner for four different values of $\Delta\sigma$. For $\Delta\sigma > 0$, the initial turn of the fractures away from each other does not occur, and as the magnitude of $\Delta\sigma$ increases, the curving of the fracture propagation paths is suppressed. The curvature of the paths is a maximum if $\Delta\sigma < 0$. Thus the pattern of interaction between joints can be interpreted as an indicator of the difference in the remote principal stresses when the joints formed.

The perpendicular and parallel intersections of joints in Figure 2.10*G*, *H* can be understood as an intersection of a younger joint with a pre-existing joint. In Figure 9.13*A*, *B*, the straight joint in each case formed first, oriented normal to the initial maximum tensile stress $\hat{\sigma}_3^{(Eff)}$. Subsequently, the stress field changed orientation, and in this new stress field, the younger joint developed, again with the joint oriented normal to the new $\hat{\sigma}_3^{(Eff)}$. The presence of a pre-existing joint perturbs the new remote stress orientations and magnitudes within a neighborhood whose dimension is of the order of the length of the joint. As the later joint propagates into this neighborhood, its orientation changes to reflect the orientation of the perturbed stresses (heavy dashed lines in Figure 9.13*A*, *B*).

If the younger joint intersects the older one at either 90° or 0°, the local principal stresses $\hat{\sigma}_k^{(Eff)}$ must be perpendicular and parallel to the older joint. This condition is satisfied if the older joint was an open fracture at the time of formation of the younger joint, because an open joint can support no shear stress, which means the joint plane must have been a principal plane for the local stress.

Figure 9.13*A*, *B* show the situation after the first joint formed and during the formation of the second joint as it propagates toward the neighborhood of perturbed stress. The remote minimum principal stress $\hat{\sigma}_3^{(Eff)}$ is tensile, perpendicular to the younger joint, and equal to the tensile strength of the rock T_0. If the younger joint curves toward being perpendicular to the older joint (Figure 2.10*G*), then the normal component of the local stress acting parallel to the pre-existing joint must have been tensile (Figure 9.13*A*). If the second joint curves toward being parallel to the older joint (Figure 2.10*H*), then the normal component of the local stress acting parallel to the pre-existing joint must have been compressive (Figure 9.13*B*). This difference is determined by the relative magnitude and orientation of the remote principal stresses.

If the older joint is closed during the formation of the younger joint, then it can support a shear stress, which, however, will be smaller than the shear stress that the solid rock can support. In this case, the local principal

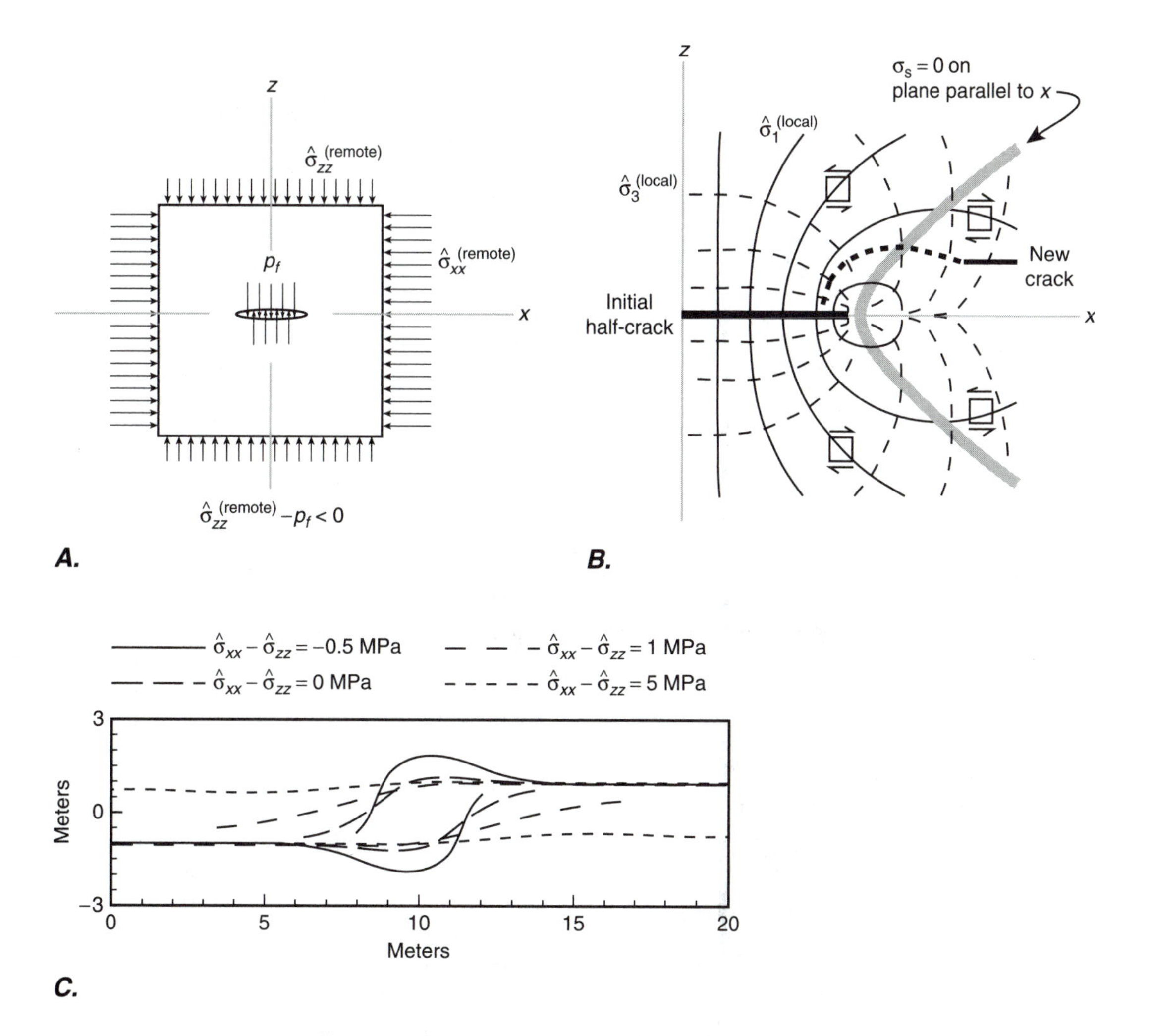

FIGURE 9.12 Interaction of propagating joints that are parallel but offset from each other. (After Olson and Pollard 1989) *A.* A crack in a block of rock subjected to an internal pore fluid pressure p_f and to a remote stress field for which $\hat{\sigma}_{xx}$ is the maximum compressive stress and $\hat{\sigma}_{zz}$ is the minimum compressive stress. For crack propagation to occur, p_f must exceed $\hat{\sigma}_{zz}$ by an amount equal to the fracture strength of the rock. *B.* An example of the trajectories of the local principal stresses about the right half of the crack in *A*, shown as the thick black line at the left of the diagram. Solid and dashed lines are the maximum and minimum principal stress trajectories, respectively. The thick gray line shows where the shear stress on a plane parallel to the initial crack is equal to zero and thus where the local principal stresses are perpendicular and parallel to the initial crack plane. The shear stress component on the plane parallel to the initial crack changes shear sense across the gray line, as indicated by the shear couples on the boxes. A new crack, shown as the thick black line on the right of the diagram, would propagate perpendicular to the local minimum principal stress, as shown by the thick dashed black line extending from the end of the new crack. To the right of the gray line, the new crack turns away from the initial crack; to the left of the gray line it turns toward the initial crack to intersect it at a high angle. *C.* Four different propagation paths calculated for four different sets of values of the remote principal stresses, where $\Delta\sigma = \hat{\sigma}_{xx} - \hat{\sigma}_{zz}$. The curvature of the propagation paths is a maximum for $\Delta\sigma < 0$ and is increasingly suppressed as the positive value of $\Delta\sigma$ increases. The propagation paths initially turn away from each other only for values of $\Delta\sigma \leq 0$. For this example, the ratio of the initial crack length to the initial perpendicular distance between the two cracks is 3 (two 12-meter cracks separated by 4 meters; only half of each crack is shown).

stresses do not have to be perpendicular and parallel to the old joint, and thus the intersection angle of the young joint can be different from 90° or 0°, the difference increasing with increasing magnitude of the shear stress that is supported by friction on the closed joint face.

These cases can be calculated rigorously using elasticity theory, but because the propagation of a joint into the neighborhood of perturbed stress changes the perturbation, the calculation must either use simplifying assumptions or be done in stepwise fashion to take account

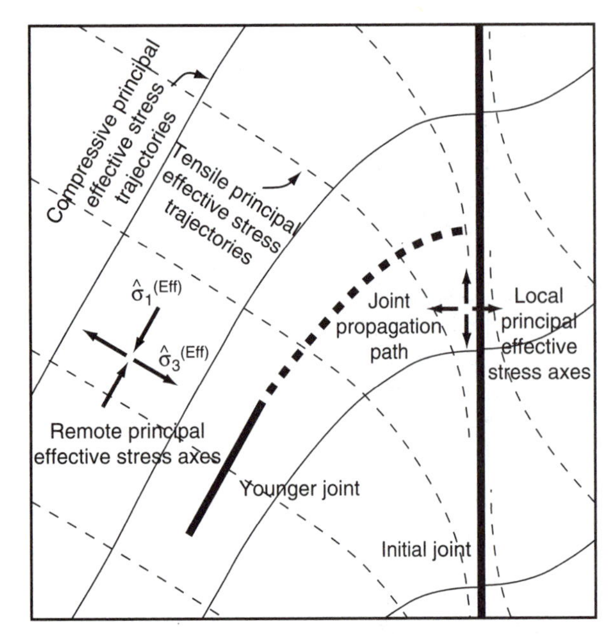

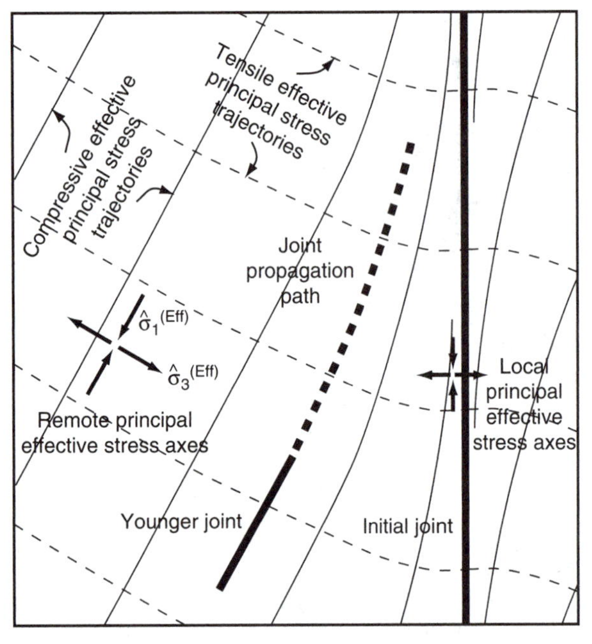

FIGURE 9.13 Schematic plot of the stress trajectories in the neighborhood of an open pre-existing fracture that is intersected by a younger fracture. *A*. The remote stress has a component of normal stress parallel to the pre-existing fracture that is tensile. The younger fracture curves to intersect the pre-existing fracture at an angle of 90°. *B*. The remote stress has a component of normal stress parallel to the pre-existing fracture that is compressive. The younger fracture curves to intersect the pre-existing fracture at an angle of 0°.

of the stress changes associated with each increment of propagation. Qualitatively, however, the effects we have just described, and the origin of curvature in joint orientations, are correct.

The bifurcation of the joint in Figure 2.10*E* may represent a propagation of the joint into a region where it must accommodate extension in two directions, which cannot be accomplished by a single extension fracture.

9.6 THE SPACING OF EXTENSION FRACTURES

The regular spacing of joints and the dependence of that spacing on layer thickness (Figure 2.11) are characteristics for which any proposed mechanism of formation must account. Several explanations have been proposed, but it is unclear which, if any, is correct.

One hypothesis involves the pore fluid pressure. When a fracture forms, the pore fluid pressure in the neighborhood of the fracture decreases as pore fluid flows into the open fracture. As the pore fluid pressure declines, the effective Mohr circle moves away from the failure criterion, so further fracture in the vicinity of the initial fracture is impossible. A second fracture can form in the rock only beyond the zone of reduced pore pressure, thereby defining the minimum spacing for the formation of hydrofractures. This distance must depend on the permeability of the rock at the time of fracture, so highly permeable rocks should have a larger fracture spacing than less permeable rocks.

The contact forces between adjacent layers can also produce fractures. To illustrate this process, consider three layers with welded contacts, with the mechanical properties of the central layer different from those of the outside layers (Figure 9.14). Suppose that with uplift the two outside layers extend more than the central layer. The normal component of stress parallel to the layers is compressive in the outside layers and tensile in the central layer. The force F_t resulting from the tensile stress across the thickness of the central layer must balance the forces F_s exerted by the shear stresses along the surfaces of the layer: $F_t = 2F_s$. F_s increases with the length l of the layer. Thus the spacing of the extension fractures that can form within the central layer is determined by the length of layer necessary to build up a tensile stress equal to the fracture strength, $F_t = T_0$. For a thicker layer, the fracture spacing should be larger because the force required to fracture the layer is larger. In principle, this also must be the type of process involved in the formation of columnar joints.

If a tensile fracture develops, the nearby tensile stress normal to the fracture surface is relieved. The stress relief diminishes away from the crack to a negligible amount at a distance about five to ten times the crack depth. Beyond that distance, another crack may

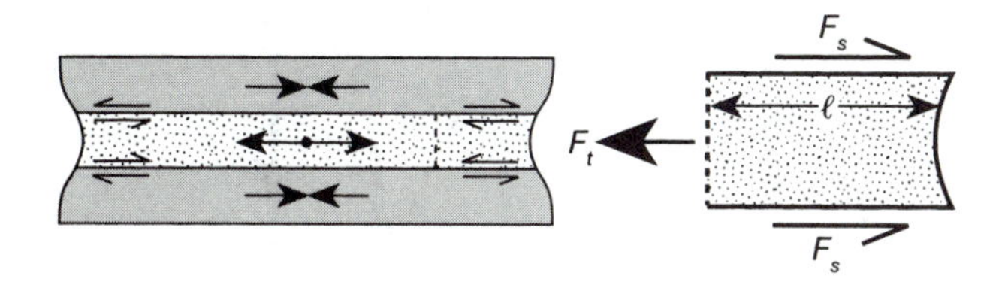

FIGURE 9.14 Changes in pressure, stress, or temperature can induce stresses in interlayered rocks if the elastic constants or coefficients of thermal expansion differ from one layer to the next. In a three-layer sequence, if the central layer expanded less than the enclosing layers, it would be in deviatoric tension. The force created by the shear stress on the boundaries of the central layer increases with length ℓ of the boundary. To cause an extension fracture, ℓ must be long enough so that the tensile force in the layer divided by its cross-sectional area equals its tensile strength. Under these conditions, ℓ is the minimum spacing of fractures.

develop. If fracture depth is limited by the thickness of a layer, this relationship suggests that fracture spacing should vary with layer thickness, as is indeed observed. Fracture spacing, however, is usually much less than that predicted by this relationship (Figure 2.11), indicating that this mechanism is not the only one affecting the rocks.

9.7 DISTINGUISHING EXTENSION FRACTURES FROM SHEAR FRACTURES

It is often difficult to tell the difference between fractures that have formed as extension fractures and those that have formed as shear fractures, unless some distinguishing characteristic of the mode of formation is present.

The presence of plumose structure on the fracture surface is clear evidence of formation by extension fracturing. Lack of any offset, even down to the microscopic scale, is also clear evidence of extension fracturing.

The presence of pinnate fractures along a fracture is good evidence that the fracture originated as a shear fracture. Pinnate fractures may be extensional cracks that form approximately parallel to the maximum compressive stress. They may also be secondary shear fractures (Riedel shears; Figure 8.7) that may form at Coulomb fracture angles under locally rotated orientations of the principal stresses relative to the main shear fracture. The orientation of such fractures is not necessarily a reliable indication of their origin. Fractures that display ridge-and-groove lineations (see Figures 3.8*A*, *B* and 14.16*A*, *B* and Section 14.6) also must have formed as shear fractures. Such features, however, commonly are not present or are not easily observed on all shear fracture surfaces.

The ambiguity in the interpretation of fracture origin is particularly troublesome for those fractures along which there is shear displacement. Such fractures may originate as shear fractures, or they may be extension fractures that are subsequently reactivated as shear planes. Reactivated fractures could even have mineral fiber or other slickenside lineations (see Chapter 14), although if shear displacement is very small, slickenside lineations might not develop.

The angular relationship between sets of fractures in rocks is not diagnostic of the origin of the fractures, although many interpretations in the literature assume otherwise. Two sets of fractures intersecting in an acute angle often are interpreted to be conjugate shear fractures, and sets of three fractures in which one set bisects the acute angle between the other two have been interpreted as sets of conjugate shear fractures bisected by an extension fracture. On the basis of the Coulomb fracture criterion and the tensile fracture criterion, the principal stresses commonly inferred from these fracture orientations are as follows (Figure 9.15): The maximum compressive stress $\hat{\sigma}_1$ bisects the acute angle between the conjugate shear fractures and parallels the extension fracture. The intermediate principal stress $\hat{\sigma}_2$ parallels both the line of intersection of the conjugate shear frac-

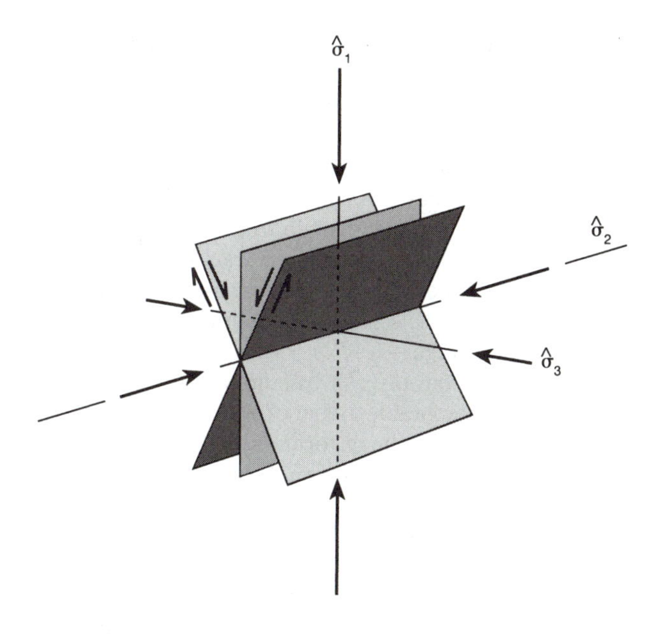

FIGURE 9.15 Principal stresses commonly inferred from fracture orientations. The maximum compressive stress $\hat{\sigma}_1$ bisects the acute angle between conjugate shear planes; the intermediate compressive stress $\hat{\sigma}_2$ is parallel to the intersection line of the conjugate shear planes; and the minimum compressive stress $\hat{\sigma}_3$ bisects the obtuse angle between the conjugate planes. The acute angle between conjugate shear planes is bisected by an extension fracture.

tures and the extension fracture itself; and the minimum compressive stress $\hat{\sigma}_3$ is perpendicular to the extension fracture and bisects the obtuse angle between the conjugate fractures.

Such an interpretation of the origin of fractures based on their relative orientations, however, is unjustified without independent evidence of the nature of the fractures and their relative times of formation. In several well-documented examples, careful investigation of the relative timing of joints has revealed that all the fractures forming a pattern similar to that expected for conjugate shear fractures are in fact extension fractures that developed at different times and under the influence of different orientations of stress (Figure 2.12). Thus the reader should be suspicious of all interpretations in which the angle between fractures is the only evidence cited for a shear fracture origin.

9.8 FRACTURES ASSOCIATED WITH FAULTS

The fractures that are parallel and conjugate to faults (Figure 2.17) may represent conjugate shear fractures corresponding to the two fracture orientations predicted by the Coulomb fracture criterion (Figure 8.3*B*, *D*, *E*). In such a case, the approximate stress orientations are as shown for the conjugate shear planes in Figure 9.15.

Many pinnate fractures that are arrayed *en echelon* along a shear fracture (Figures 2.7 and 3.17*A*) form as extension fractures during shearing. When formed, they are oriented approximately perpendicular to the minimum compressive stress. Some pinnate fractures may also originate as secondary shear fractures such as the R Riedel shears (Figures 8.7 and 3.17*C*, *D*) or the P secondary shears (Figures 8.7 and 3.17*E*, *F*). The sense of rotation through the acute angle from the fault plane to the planes of both the extension fractures and the R Riedel shears is the same as the shear sense on the fault. This accounts for the sense-of-shear criteria discussed in Section 3.3(iii) (Figure 3.17).

The relationship of gash fractures to the associated shear zone is comparable to that of pinnate (feather) fractures. Gash fractures may form as extension fractures perpendicular to the minimum compressive stress $\hat{\sigma}_3$ (Figure 9.16*A*). The gash fractures, however, may be rotated by ductile deformation during or after formation (Figures 9.16*B* and 2.8). Gash fractures that initiate at different times during the ductile shear should show different amounts of rotation (Figures 9.16*B* and 2.8; see also Figure 15.18). Because the minimum compressive stress $\hat{\sigma}_3$ is normal to any unrotated part of the gash fracture, either the tips of the sigmoidal fractures or the latest formed fractures provide the best estimate of the stress orientations.

In some cases, *en echelon* gash fractures occur parallel to a conjugate shear zone, an orientation that is not accounted for by this analysis (Figure 2.8). Such orientations may be consistent with the fracture criteria we have discussed and may record locally rotated stress axes. Their geometry may be better accounted for, however, by assuming that the fractures form perpendicular to the direction of greatest incremental extension, a possibility we discuss further in Section 15.4 (Figure 15.17). In other cases, each gash fracture may have been completely rotated by ductile deformation, or the set of fractures may have formed as hybrid shears, which have components of both extension and shear across their surfaces.

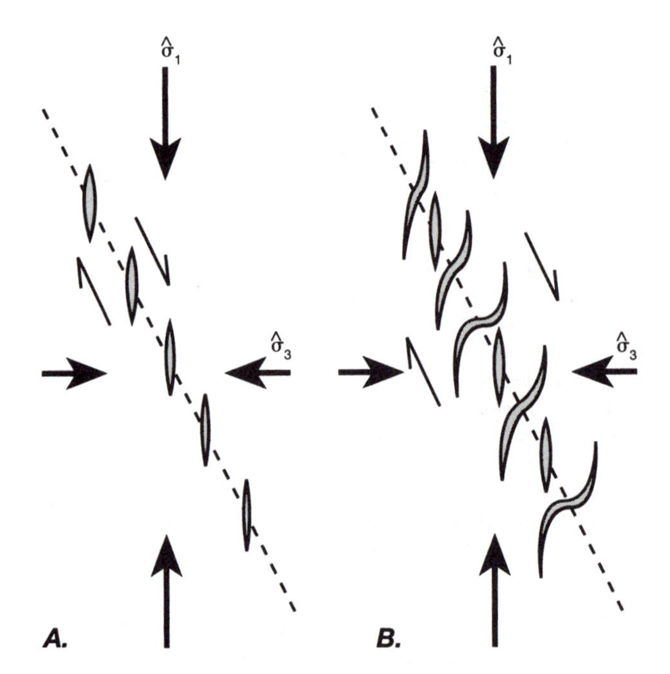

FIGURE 9.16 Extension fracture model for the formation of gash fractures. *A*. Gash fractures form an *en echelon* array along a shear zone with each fracture perpendicular to the minimum compressive stress. *B*. Ductile shearing along the shear zone rotates the central portions of the fractures, leaving a sigmoidal fracture with the tips of the fractures perpendicular to the minimum compressive stress $\hat{\sigma}_3$. Fractures formed at different times during the ductile shearing show different amounts of rotation, and the smallest and youngest fractures may not be rotated at all.

A word of caution is in order concerning the use of fracture orientations to infer the orientations of the principal stresses. Not all fractures near faults form at the same time as the faults. Fractures that predate a fault may actually influence its orientation, because they are pre-existing planes of weakness that give the rock a mechanical anisotropy. Such anisotropies are common and can lead to fractures at orientations different from those predicted by the Coulomb fracture criterion (Figure 8.13), which assumes an isotropic con-

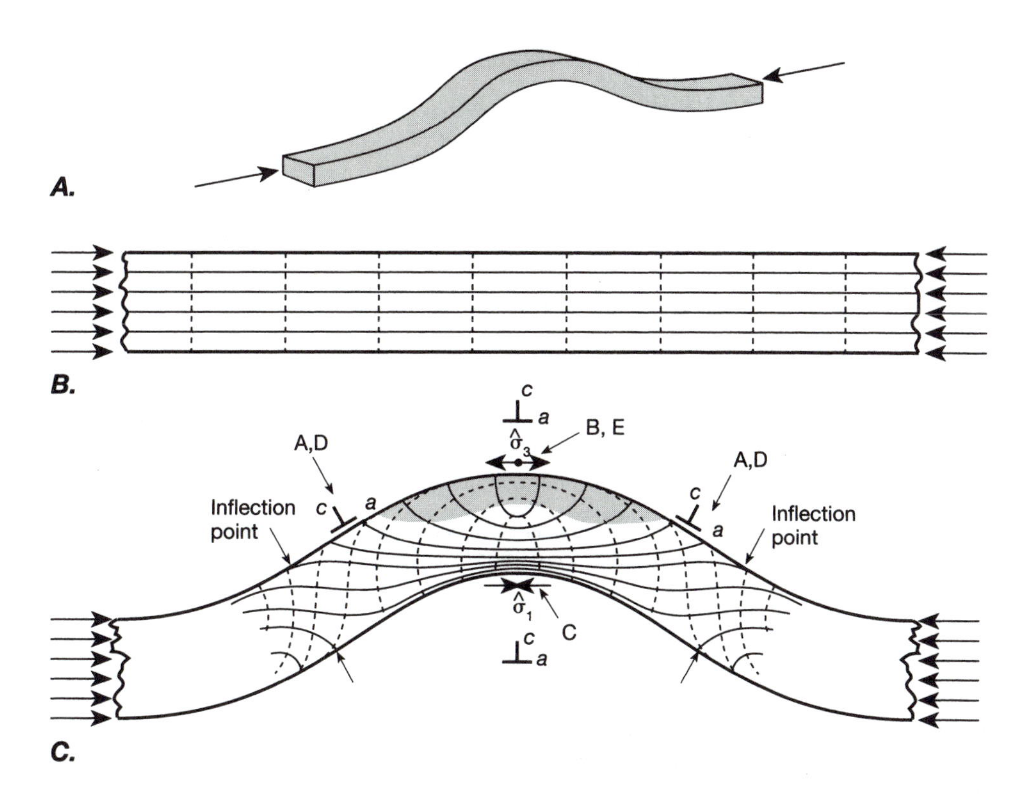

FIGURE 9.17 Stress distribution in a bar of gelatin undergoing buckling by layer-parallel compression. (After Currie et al. 1962) *A*. Perspective view of the folding experiment. *B*. Stress trajectories in the bar before folding. Solid lines parallel the maximum compressive stress $\hat{\sigma}_1$; dashed lines parallel the minimum compressive stress $\hat{\sigma}_3$. *C*. Stress trajectories in the bar after folding. The shaded area shows where the layer-parallel principal stress component is tensile. The letters A through E show locations where different fracture orientations can develop; they correspond to the fracture patterns shown in Figure 2.18 and listed in Table 9.1. Fracture patterns shown in Figure 9.15, when oriented with respect to the reference axes *a*, *b*, and *c* as indicated in Table 9.1, can account for the observed fracture patterns.

tinuum. On the other hand, other fractures might postdate an adjacent fault and can have an orientation totally unrelated to it.

9.9 FRACTURES ASSOCIATED WITH FOLDS

The fracture orientations associated with folds (Figure 2.18) have been interpreted as sets of conjugate shear fractures with or without a bisecting set of extension fractures (Figure 9.15). This interpretation seems to be based largely on the relative orientations of the different fracture planes, which in the last section we argued is not a reliable criterion. Some studies of fractures associated with folds, in fact, have found that at least some of the fractures existed in the rocks before the folding.

On the other hand, because the orientation and magnitude of stresses in layers undergoing folding vary radically, both from one place to another in the fold and through time as the fold develops, it is possible to account, at least qualitatively, for most of the observed fracture orientations.

Figure 9.17 shows the evolution of stress in an elastic layer folded by layer-parallel compression (Figure 9.17*A*).[7] The lines in Figure 9.17*B*, *C* are stress trajectories, which are everywhere parallel to the principal stresses. Solid lines are trajectories for the maximum principal stress $\hat{\sigma}_1$, and the dashed lines are the trajectories for the minimum principal stress $\hat{\sigma}_3$. The closer the trajectories are to one another, the greater the magnitude of the stress. Before the bar buckles (Figure 9.17*B*), the maximum principal stress is everywhere parallel to the length of the bar, and the minimum principal stress is everywhere perpendicular to the top and bottom of the bar. After buckling (Figure 9.17*C*), the stress orientations are more complex. The maximum principal

[7]The experiment was performed on a bar of gelatin illuminated from behind by plane-polarized light. Because gelatin is a "photoelastic" material, it rotates the plane of polarization by an amount proportional to the elastic strain. By observing the bar through a polarizer set perpendicular to the original plane of polarization, one can determine the amount of rotation of the polarized light and interpret it in terms of the strain magnitude, which by the equations of elasticity, is proportional to the magnitude of the stress.

TABLE 9.1 Stress Interpretation of Fractures in Folds

Fracture Set[b]	Principal Stress Parallel to Reference Axes[a]			Time of Formation[c]	Place of Formation
	a	*b*	c		
A	$\hat{\sigma}_1$	$\hat{\sigma}_3$	$\hat{\sigma}_2$	Before folding	Throughout fold
B	$\hat{\sigma}_3$	$\hat{\sigma}_1$	$\hat{\sigma}_2$	During folding	Convex areas of maximum curvature
C	$\hat{\sigma}_1$	$\hat{\sigma}_2$	$\hat{\sigma}_3$	During folding	Concave areas of maximum curvature
D (conjugate pair)	$\hat{\sigma}_1$	$\hat{\sigma}_3$	$\hat{\sigma}_2$	Before folding	Throughout fold
D (*bc* fractures)	$\hat{\sigma}_3$	$\hat{\sigma}_1$	$\hat{\sigma}_2$	During folding	Convex side
E	$\hat{\sigma}_3$	$\hat{\sigma}_2$	$\hat{\sigma}_1$	During folding	Convex areas of maximum curvature

[a]Here *c* is normal to bedding, *a* and *b* are in the plane of the bedding, *b* is parallel to the fold axis, and *a* is normal to *b* and *c* (see Figures 2.18 and 9.17*C*).
[b]The letters correspond to the fracture sets shown in Figure 2.18 and to the locations around the fold shown in Figure 9.17*C*.
[c]In general, "Before folding" corresponds to the stress state in Figure 9.17*B*, and "During folding" corresponds to the stress state in Figure 9.17*C*.

stress on the concave side of the fold is roughly parallel to the bar and considerably larger than the applied stress, but on the convex side, it is at a high angle to the bar. The minimum principal stress on the convex side is parallel to the bar and is actually a tensile stress in the shaded area. This model is really two-dimensional, because it does not account for changes in fold geometry along the fold axis perpendicular to the diagram in Figure 9.17*C*.

The important points to emphasize from this example are that the orientations of the principal stresses change through time during the buckling process and that the magnitudes—and even the signs—of the principal stresses also change. Because different stresses occur in folds in the same place at different times, and different stresses occur at the same time in different places, we can account for the variety of fracture orientations that are commonly observed in folds (Figure 2.18) in terms of extension fractures or Coulomb shear fractures. Table 9.1 summarizes the interpretation of the different observed fracture sets (Figure 2.18) in terms of the states of stress that could develop during folding (Figure 9.17). The labels of the different fracture sets A through E correspond to the same labels in Figure 2.18 and Table 9.1. They are used in Figure 9.17*C* to indicate the locations on the fold where the different fracture sets are commonly found. The reference axes *a*, *b*, and *c* are defined in Table 9.1 and Figure 2.18 and are shown in Figure 9.17*C*, where *b* is everywhere perpendicular to the plane of the diagram.

On the strongly convex side of the fold, the orientation of $\hat{\sigma}_3$ parallel to the *a* axis in the layer and of $\hat{\sigma}_1$ parallel to the *c* axis and perpendicular to the layer can account for the extension fractures (*bc*) in set B and possibly in set D and for the conjugate shear fractures of set E. The conjugate shear fractures in set B are also consistent with $\hat{\sigma}_3$ being parallel to the *a* axis in the layer as predicted for the convex side of the folds. For these fractures, however, $\hat{\sigma}_1$ and $\hat{\sigma}_2$ must exchange orientations from those in Figure 9.17*C* so that $\hat{\sigma}_1$ is parallel to *b* and $\hat{\sigma}_2$ is parallel to the *c* axis and perpendicular to the layer. This could happen, for example, if gentle folding occurred in the third dimension parallel to *b*.

On the strongly concave side of the fold, the orientations of the principal stresses reverse. $\hat{\sigma}_1$ becomes parallel to both the *a* axis and the layer, and $\hat{\sigma}_3$ becomes parallel to the *c* axis and perpendicular to the layer. In this orientation, the stresses can account for the conjugate shear fractures of set C. With the same orientation of $\hat{\sigma}_1$, parallel to the *a* axis, but with $\hat{\sigma}_3$ parallel to the *b* axis and $\hat{\sigma}_2$ parallel to the *c* axis and perpendicular to the layer, we can account for the conjugate shear fractures in sets A and D and the extension fractures in set A. This stress orientation is an exchange of the orientations of $\hat{\sigma}_2$ and $\hat{\sigma}_3$ from that implied in Figure 9.17*B*, *C*. This exchange could occur early in the folding process (Figure 9.17*B*) if some gentle folding of the layer occurred about the *a* axis, although as we discussed in Section 9.6, these two types of fracture are unlikely to have formed at the same time.

Thus the difference between sets B and E and between sets A and C is that the stresses parallel to *b* and *c* exchange positions, presumably depending on local deformation about the *a* axis in the direction parallel to *b*.

As this somewhat ad hoc interpretation of the fractures in folds indicates, this subject is currently in need of considerable study. We discuss the stress field associated with folding in more detail in Sections 18.6 and 18.7.

9.10 STRESS DISTRIBUTIONS AND FAULTING

i. Anderson's Theory of Faulting

The Coulomb fracture criterion provides a useful theoretical explanation for the threefold classification of faults into normal, thrust, and strike-slip faults. We call this explanation **Anderson's theory of faulting** after the British geologist E. M. Anderson, who proposed it. The theory depends on the fact that the surface of the Earth is a free surface, which can support no shear stress. The Earth's surface thus must be a principal plane of stress, and the principal stresses must be normal and parallel to it. The Coulomb fracture criterion requires that shear fracture planes contain the intermediate principal stress $\hat{\sigma}_2$ and that the fracture plane angle α_f between the fracture plane and the maximum compressive stress $\hat{\sigma}_1$ must be less than 45° (Figures 8.3*B*, *D*, *E*; 8.4; and 8.5). The type of fault that develops in a given situation depends on which of the three principal stresses is vertical.

We diagrammatically illustrate the various possibilities in Figure 9.18, where we assume a fracture plane angle of $\alpha_f = 30°$. If the maximum compressive stress $\hat{\sigma}_1$ is vertical, the faults that form should be normal faults with dips of 60° and hanging-wall-down sense of shear (Figure 9.18*A*). If the minimum compressive stress $\hat{\sigma}_3$ is vertical, the faults should be thrust faults that dip at 30° with a hanging-wall-up shear sense (Figure 9.18*B*). If the intermediate principal stress $\hat{\sigma}_2$ is vertical, faults should be vertical strike-slip faults with horizontal shear directions (Figure 9.18*C*).

The stress orientations measured in the Earth in regions of active tectonics are generally consistent with this interpretation. For example, the Basin and Range province in Nevada is characterized by roughly north-trending normal faults (Figure 4.12). The minimum horizontal stress is oriented approximately east-west (Figure 9.9*B*), which is consistent with the fault orientation and with the maximum compressive stress being vertical (Figure 9.18*A*). These faults, therefore, are consistent with the requirements of Anderson's theory of normal faults. The tectonics of the Himalaya is dominated by north-south-directed thrusting (Figure 5.11*C*). Near the northern boundary between the Indian and Asian plates, the maximum compressive stress is oriented approximately north-south (Figure 9.9*A*), which is consistent with the minimum compressive stress being vertical and thus with Anderson's model of thrust faults (Figure 9.18*B*). Right-lateral strike-slip motion occurs along the northwest-southeast-oriented San Andreas fault in California (Figure 6.2*A*). The maximum compressive stress measured in this area is oriented roughly northeast-southwest (Figure 9.9*B*), consistent with the intermediate compressive stress being vertical as required by Anderson's theory of strike-slip faults (Figure 9.18*C*). In this case, however, the maximum compressive stress appears to be at a much higher angle to the fault than Anderson's theory would predict. There is reason to believe, however, that the maximum compressive stress may actually be closer to north-south than indicated in this figure. The interpretation of the stress along the San Andreas fault is controversial and the subject of ongoing research.

ii. Faulting and the Distribution of Stress with Depth

The Coulomb fracture criterion provides a concise explanation for the existence of the three major types of faults observed at the Earth's surface. Strictly speaking, however, it applies only near the surface of the Earth, and it assumes strictly planar faults in isotropic material. Most faults are curved and are not just confined to the shallow parts of the crust. Most rocks are not isotropic, and the principal stresses may not be parallel to the horizontal and vertical directions (Figure 9.6*A*).

Thus it is of interest to examine theoretically how the orientation of the stress field might change with depth. To do so, we look at a two-dimensional case and adopt

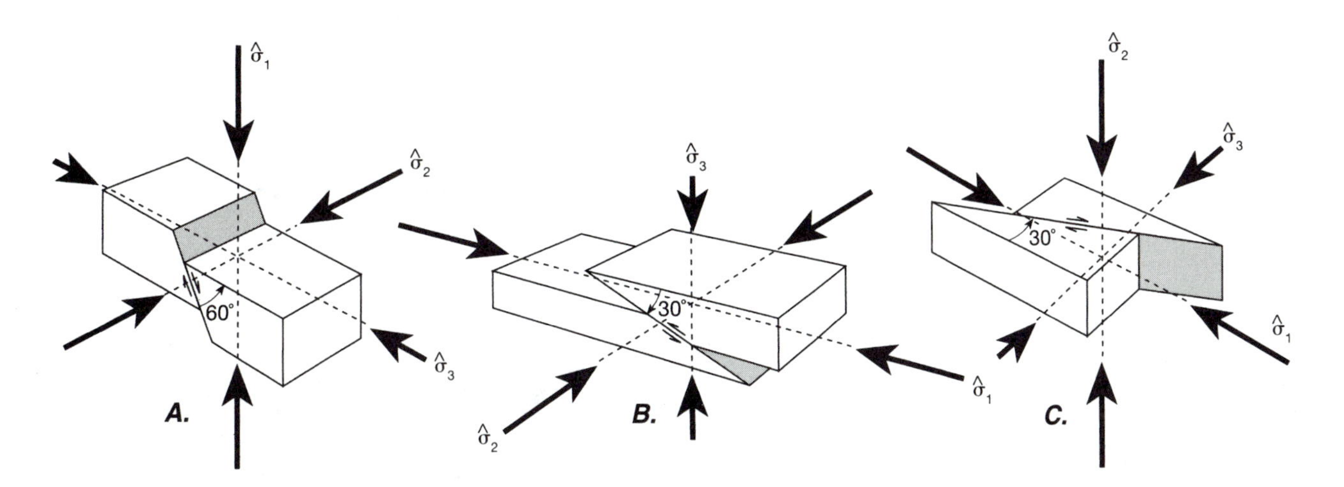

FIGURE 9.18 Anderson's theory of faulting, showing the relationship between the orientation of the principal stresses and the different ideal fault types. *A*. Normal fault with maximum compressive stress $\hat{\sigma}_1$ vertical. *B*. Thrust fault with minimum compressive stress $\hat{\sigma}_3$ vertical. *C*. Strike-slip fault with intermediate compressive stress $\hat{\sigma}_2$ vertical.

a coordinate system with the z axis positive down and the x axis horizontal and positive pointing to the right. We isolate a block of the Earth's crust and consider the distribution of stresses along the boundaries of the block and within it.

We begin by considering the stresses on rock that arise only from the overburden. The vertical normal stress σ_{zz} at any depth z in the block is simply the overburden

$$\sigma_{zz} = \rho_r\, g\, z \tag{9.8}$$

where ρ_r is the average density of the rock, g is the magnitude of the acceleration due to gravity, and z is the depth. The corresponding horizontal stress is equal to a fraction κ of the vertical stress

$$\sigma_{xx} = \kappa\, \rho_r\, g\, z \tag{9.9}$$

where κ is a factor less than 1 that depends on the effective Poisson ratio of the rock (see Section 9.3, Equation (9.7)).

Because no shear stresses exist at the surface and none are applied to any other surface of the block, the principal planes of stress must lie parallel and perpendicular to the sides of the block. In other words, the maximum compressive stress $\hat{\sigma}_1 = \sigma_{zz}$ is everywhere vertical, and the minimum compressive stress $\hat{\sigma}_3 = \sigma_{xx}$ is everywhere horizontal. The stress trajectories, which are lines everywhere parallel to the orientations of these principal stresses, are horizontal and vertical throughout the block. The state of stress arising only from the overburden is often called the **standard state**. We discuss various faulting situations by superimposing additional stresses on the standard state.

Consider, first of all, superposition of a tectonic horizontal compressive stress on the standard state that is adequate to cause faulting (Figure 9.19). In the figure, a supplementary horizontal compressive stress C is added to the standard state. If the added stress is sufficiently large, the horizontal stress becomes the maximum compressive stress $\hat{\sigma}_1 = \sigma_{xx}$, as shown in Figure 9.19*A*. The potential faults, shown in Figure 9.19*B*, are a conjugate set of thrust faults, and the geometry corresponds to that assumed in Anderson's theory (Figure 9.18*B*).

This model of the stress distribution is unrealistically simple. In particular, we have assumed that no shear stresses exist on the boundaries of the crustal block, and as a result, all the stress trajectories are straight lines. If a crustal block were extending or shortening, for example, we would expect shear stresses opposing the motion to be present along the base of the block.

We therefore consider next the effect of adding a horizontal shear stress that increases with depth and has a constant magnitude along the base of the block (Figure 9.20*A*):

$$\sigma_{zx} = k\, z \tag{9.10}$$

This assumption also is overly simple, but the results are interesting. The symmetry of the stress tensor requires that vertical shear stresses balance the horizontal ones, so that shear stresses σ_{xz} must exist on the vertical boundaries of the block. Moreover, the requirement that all horizontal forces must sum to zero means that the horizontal normal force on the right side of the block must be less than that on the left, the difference being made up by the force contributed by the shear stress on the base.

The fact that shear stresses exist on the sides and bottom of the block means that these boundaries are no longer principal planes of stress and that the principal axes in general are no longer horizontal and vertical. The

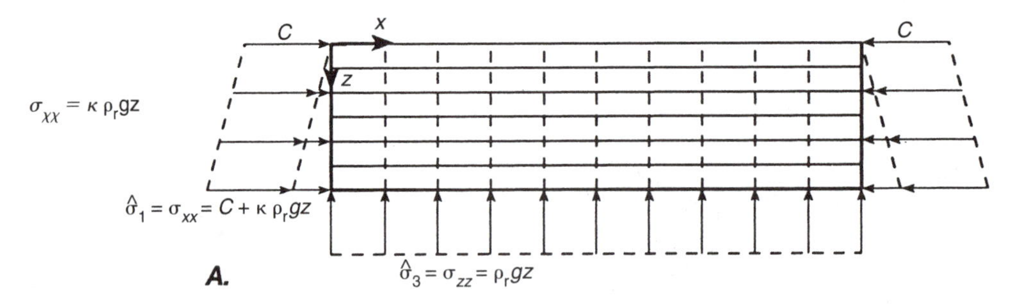

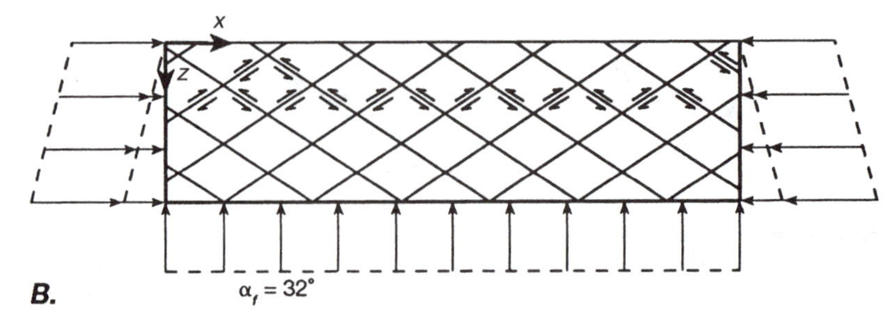

FIGURE 9.19 Free-body diagram for a horizontal compressive tectonic stress (C) constant with depth added to the standard state stress, which consists of a horizontal stress and the vertical overburden that increase with depth. *A*. Tractions and stress trajectories. Solid lines are trajectories of $\hat{\sigma}_1$; dashed lines are trajectories of $\hat{\sigma}_3$. *B*. Attitudes of potential shear fractures, assuming a fracture plane angle $\alpha_f = 32°$.

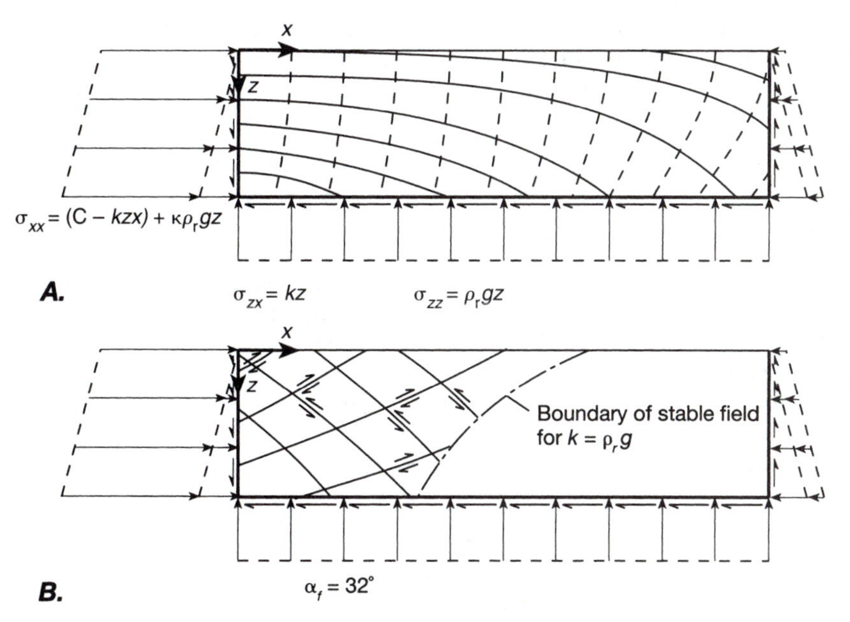

FIGURE 9.20 Free-body diagram for a compressive tectonic stress added to the standard state stress and including shear stresses on the boundaries of the block. *A*. Tractions and stress trajectories. Solid lines are trajectories of $\hat{\sigma}_1$; dashed lines are trajectories of $\hat{\sigma}_3$. *B*. Potential fault surfaces. The blank area indicates the region of stability where stresses are subcritical, as determined by the fracture criterion given in terms of the principal stresses by $\hat{\sigma}_1 = 4\,\hat{\sigma}_3 + 100$ MPa, where $\rho_r g = 25$ MPa/km and the value of $k = \rho_r g$ is chosen as an example. (After Hafner 1951)

top of the block, however, still supports no shear stress, so it must be a principal plane, and the vertical boundaries just at the free surface also must be principal planes. Thus the shear stress on both the horizontal plane, σ_{zx}, and the vertical plane, σ_{xz}, must diminish to zero at the surface, as indicated by Equation (9.10). The principal stress trajectories, therefore, are horizontal and vertical at the surface and curve with depth to provide the increasing shear stress with depth, on the vertical and horizontal surfaces (Figure 9.18*A*). With this stress distribution, the potential fault surfaces (Figure 9.20*B*) show a curvature comparable to that found on natural faults. Two possible directions of faulting are shown: one concave upward, reminiscent of many listric thrust faults (compare Figure 5.12); and one concave downward, reminiscent of faults along some basement uplifts (compare Figure 5.9).

Other possible boundary conditions, consisting of different stress distributions applied to the boundaries of the block would lead to different models of tectonic environments. The diagram in Figure 9.21, for example, shows a stress distribution along the base of the block consisting of a vertical normal stress that varies as a sine function across the length of the base, as well as a horizontal shear stress that varies as a cosine function across the same length. These illustrated stresses are added to the standard state of stress. The imposed stresses cause the block to bend, and the stress trajectories are comparable to those in the folded layer in Figure 9.17*C*.

This stress distribution is a possible model for an oceanic spreading center, where upwelling and laterally spreading material provide a vertical tectonic stress that decreases laterally from the spreading axis and a horizontal shear that increases away from the spreading axis. The potential fault surfaces form a conjugate set of normal faults symmetrically oriented about the center of the block. Listric normal faults dip toward the center on both sides, and the conjugate faults dip away from the center and get steeper with depth. The listric faults are comparable to faults observed on either side of the spreading

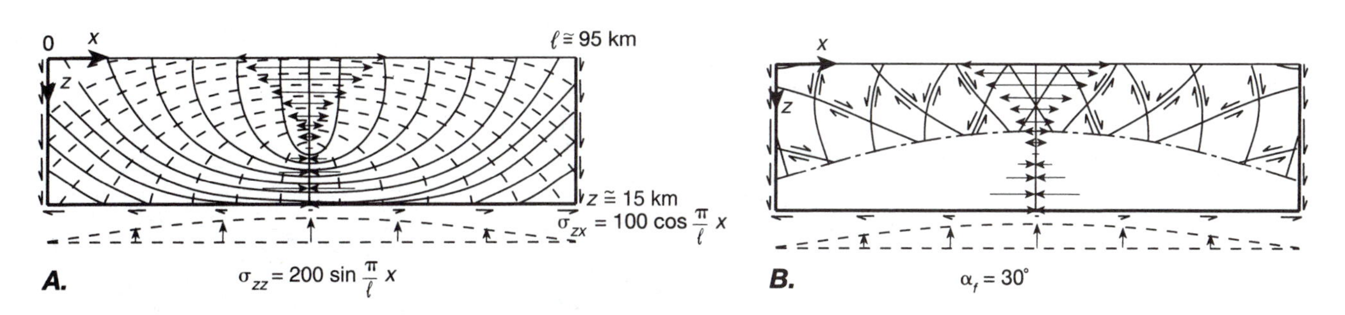

FIGURE 9.21 Free-body diagram for a block with a basal normal stress varying as a sine function and the shear stress varying as a cosine function from the left edge of the block. The standard stress state is not shown but is also assumed. *A*. Tractions and stress trajectories. Solid lines are trajectories of $\hat{\sigma}_1$; dashed lines are trajectories of $\hat{\sigma}_3$. *B*. Potential fault surfaces. The blank area shows the field of stability if the maximum value of σ_{zz} is 200 MPa. (After Hafner 1951)

axis of midoceanic ridges (compare with the distribution of fractures on folds: Figures 2.18*E* and 9.17*C* and Table 9.1).

9.11 THE MECHANICS OF LARGE OVERTHRUSTS

In Chapter 5 we discussed the existence of large overthrust sheets that extend for distances of up to hundreds of kilometers along strike and more than 100 kilometers across strike. Such large thrusts have been known since the end of the nineteenth century. Soon after they were recognized, however, it became clear that to push such a large mass would seem to require forces that the rocks would be unable to withstand.

i. The Thrust Sheet Problem

M. S. Smoluchowski first formulated the problem in elementary form in 1909. Consider a rectangular block (Figure 9.22) of height H (parallel to the z coordinate axis), width W parallel to the thrusting direction (and to the x coordinate axis), and length L perpendicular to the direction of motion (and parallel to the y coordinate axis). Its weight per unit volume is $\rho_r g$, and the coefficient of sliding friction on the base of the block is $\overline{\mu}$. The frictional force that resists the motion of the block F_f equals the normal force across the base F_n times the coefficient of friction $\overline{\mu}$ (Amontons' second law of friction; Equation (8.7); see Section 8.4(ii), footnote 3). That is,

$$F_f = \overline{\mu}\, F_n \tag{9.11}$$

$$F_f = \overline{\mu} \times (\text{normal force per unit area}) \times (\text{area}) \tag{9.12}$$

$$F_f = \overline{\mu}\, (\rho_r g H)(WL) \tag{9.13}$$

The driving force required to move the block must be greater than or equal to the frictional resistance F_f. If the driving force is applied across the back vertical face of the block, the stress on that face is the driving force per unit area,

$$\sigma_{xx} = \frac{F_f}{LH} = \overline{\mu}\, \rho_r g W \tag{9.14}$$

where to get the second equation, we introduced Equation (9.13) for F_f. This stress cannot exceed the fracture strength of the block. Choosing average values for the coefficients of friction, density, and strength

$$\overline{\mu} = 0.6 \quad \rho_r = 2500 \text{ kg/m}^3 \quad \sigma^*_{xx} = 250 \text{ MPa}$$

we can solve Equation (9.14) for W:

$$W = \frac{\sigma^*_{xx}}{\overline{\mu}\rho_r g} = 17{,}007 \text{ m} \cong 17 \text{ km} \tag{9.15}$$

Thus this model predicts that the maximum possible dimension of an overthrust sheet in the direction of thrusting is $W = 17$ km. For larger dimensions, the fracture strength of the rock is exceeded at the rear face of the sheet before the frictional resistance can be overcome. Large overthrusts, however, are known to have widths W of more than 100 km, so something must be wrong with this model. A more sophisticated analysis yielding more general but comparable results is given in Box 9-2.

There are several assumptions in this simple model that may be inappropriate for explaining the mechanics of emplacement of large thrust sheets: (1) The force of friction on the base of the thrust could be lower than we assumed. (2) The very assumption that resistance to motion is frictional in origin may be incorrect; for example, the shear along the decollement in some cases may be accommodated by ductile flow of weak rocks. (3) The thrust sheet may be driven not by a push from the rear but by gravitational forces. (4) Thrust sheets in general are not rectangular blocks, as assumed in our model, but

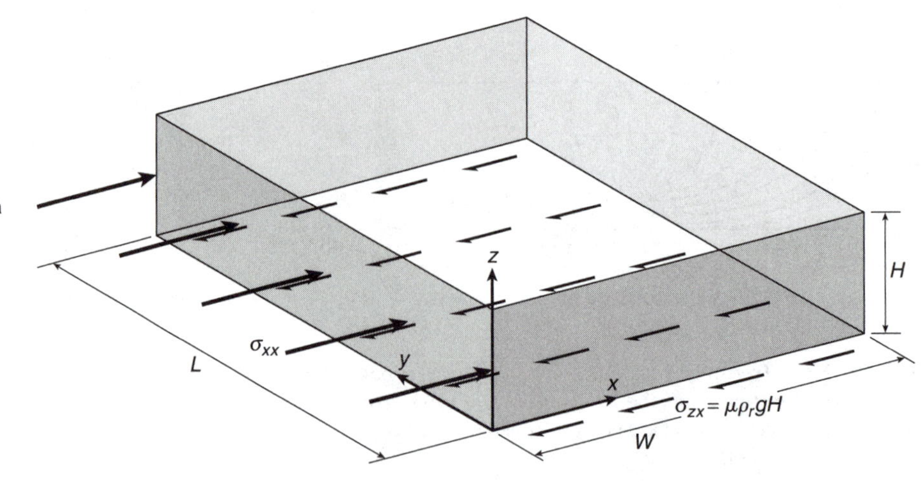

FIGURE 9.22 Simplified model of a thrust sheet of thickness H, width parallel to the thrust direction W, and length normal to the slip direction L. The force from the compressive stress σ_{xx} on the rear vertical face must balance the force on the base from the frictional resistance σ_{zx} without exceeding the strength of the rock.

instead taper to smaller thicknesses toward the foreland. (5) The thrust sheet itself may not behave as a Coulomb-type material but could behave ductilely. (6) Thrust sheets do not move en masse as a single sheet, but rather caterpillar-style, by the propagation of localized domains of slip along the fault. All of these factors may be important in explaining aspects of the mechanics of thrust sheets, and we discuss each in turn.

ii. Basal Friction

The force of frictional resistance can be reduced in two possible ways: The coefficient of friction $\bar{\mu}$ on the base can be significantly smaller than we assumed, either intrinsically or because of lubrication, or the effective normal stress across the decollement can be less than we assumed.

Laboratory measurements of the coefficient of friction of rock on rock consistently give values near $\bar{\mu} = 0.85$ and do not leave much possibility for significant reduction. The presence of water on a rock interface actually seems to increase the coefficient of friction; it does not act as a lubricant.

A high pore fluid pressure along the decollement (λ approaching 1) would reduce the effective normal stress across the surface and thereby lower the frictional resistance (Equation (9.11); see Section 8.5). If the frictional resistance decreases, then a horizontal normal stress σ_{xx} equal to the critical fracture stress can move a greater width of thrust sheet. For zero resistance, the possible width of the thrust sheet is unlimited. Sedimentary basins in active tectonic regions are prime locations for the formation of high pore fluid pressure (Section 9.4), and large overthrust sheets are common in such environments. This explanation has been accepted as a fundamental mechanism associated with the emplacement of large thrust sheets, but other mechanisms also operate to affect the mechanics of thrust emplacement and the internal structure of thrust sheets.

iii. Basal Ductile Flow

The assumption that friction on the base of the thrust sheet provides the resistance to motion may be wrong. Deformation at the base could be accommodated, for example, by ductile flow of rock with a yield stress that is significantly smaller than the frictional resistance.

Thrust faults commonly follow layers of weak rock in the stratigraphic section. Large accumulations of evaporites (such as halite, gypsum, and anhydrite) underlie many sedimentary basins, including the Gulf Coast of the United States, southwestern Iran, and the Appalachian plateau in western Pennsylvania and adjacent states, and these basins often are the sites of major thrusting. Evaporites, especially halite (common rock salt), are among the weakest rocks known. For conditions characteristic of geologic deformation, halite has a yield stress in the range of 0.1 to 1 MPa. Even at shallow depths and low temperatures, the differential stress at which halite flows is one to two orders of magnitude less than frictional stresses and the yield stresses of other rocks. Thus where thrust faults can occupy salt beds, the resistance to motion is significantly less than where the salt is absent, and our model would suggest that in those areas, the width of thrust sheets could extend much farther out toward the foreland. This explanation may account for the major salient in the northwestern Appalachians (Figure 5.11*A*). Here the large belt of very gentle folding in the Appalachian plateau northwest of the Valley and Ridge province is almost coincident with the extent of Silurian salt beds at depth.

Anhydrite and gypsum also have relatively low yield stresses, and strata rich in these minerals also commonly act as decollement zones. Where evaporites are not present, shales are generally the weakest rocks, and at greater depths and higher temperatures, limestone (marble) and even quartzite may be sufficiently weak to localize major zones of ductile shear in a decollement.

iv. Gravitational Driving Forces

One problem with our simple model arises from the need to drive the thrust sheet forward by means of a stress transmitted through the thrust sheet from the rear. If the force of gravity were the driving force, however, this restriction would not arise, because gravitational forces act independently on every point in a body.

Gravitational sliding occurs if the shear force provided by the force of gravity (F_s) is at least equal to the frictional resistance on the decollement ($F_s = F_f$; Figure 9.23*A*). If we know the resistance, we can determine the slope necessary to cause such a thrust sheet to slide. From Equation (9.11), therefore, we have

$$\bar{\mu} = \frac{F_s}{F_n} \tag{9.16}$$

If gravity were the only force driving the sheet, then the normal force across the decollement (F_n) and the shear force parallel to it (F_s) would be related to the dip δ of the thrust surface by

$$\tan \delta = \frac{F_s}{F_n} \tag{9.17}$$

Using Equation (9.16) and assuming that the coefficient of friction $\bar{\mu} = 0.6$, which is actually a low value compared with most experimental data, we find that

$$\begin{aligned} \tan \delta &= \bar{\mu} \approx 0.6 \\ \delta &= 31° \end{aligned} \tag{9.18}$$

Thus a slope of at least 31° is required to move the thrust block gravitationally against a conservatively selected value of the frictional resistance. A 100-kilometer thrust sheet would need to slide off a topographic high of at

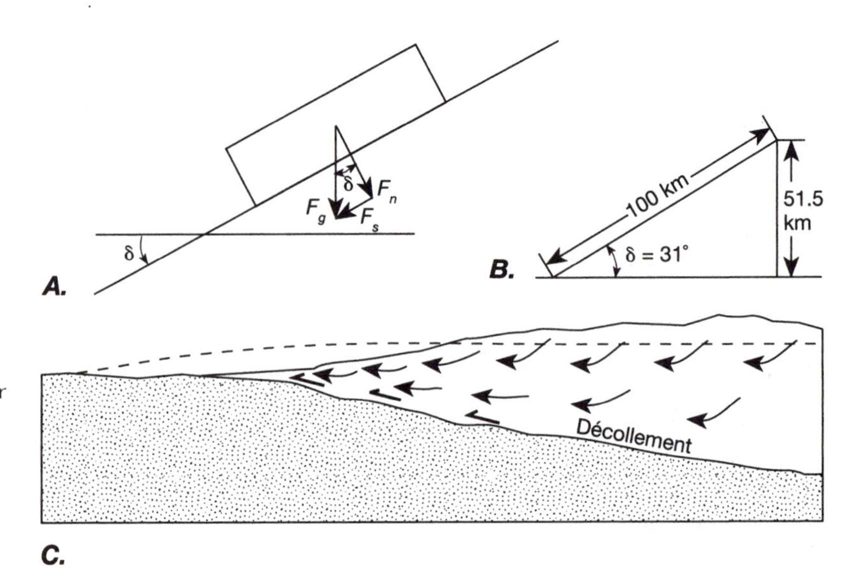

FIGURE 9.23 Models of gravitationally driven thrust sheets. A. Resolution of the gravitational force on a thrust sheet to determine the driving force available (F_s) and the normal force across the decollement (F_n). B. Normal rock friction would require too steep a slope to account for the observed size of thrust sheets and dips of decollement. C. Gravitational collapse of a tectonically produced topographic high by ductile flow within the thrust sheet. The solid lines indicate the tectonically uplifted topography; the dashed line indicates the topography after gravitational collapse of the uplift. Arrows indicate the general pattern of flow within the collapsing sheet.

least 51.5 kilometer altitude for this mechanism to explain some of the larger thrust sheets (Figure 9.23*B*). Given that Chomolungma (Mt. Everest) is less than 9 kilometers above sea level, this solution is not satisfactory. Moreover, evidence for steep dips over significant lengths of large thrust sheets is utterly lacking. This mechanism could account for the observations only if it were effective on slopes on the order of a few degrees at most. Such slopes imply a very small resistance along the decollement, and we must include in the model either high pore fluid pressure or ductile flow to make it acceptable.

If tectonic processes thicken the crust and create a topographic high, gravitational collapse of the thickened part of the crust could result in the formation of thrust sheets. This mechanism requires ductile flow throughout much of the thickened part of the crust, which spreads outward under its own weight rather like a mound of silicon putty spreads out into a puddle or, to draw an even more apt analogy, like a continental ice sheet spreads out from its center (Figure 9.23*C*). The driving force is provided by the topographic slope of the thickened region of crust, and the slope of the decollement is not restricted; it could even slope upward in the direction of thrusting, as is a common feature of thrust sheets.

Intuitively, gravitational forces may not seem strong enough to cause rocks to deform significantly. We must not forget, however, that ultimately, gravitational forces drive the whole plate tectonic machine through mantle convection. Given the great lengths of time available and the ability of rocks to creep slowly in response to relatively small differential stresses, emplacement of thrust sheets by gravitational collapse cannot be discounted.

v. Tapered Thrust Sheets: The Critical Coulomb Wedge Model

Active thrust sheets, such as occur in the Himalaya and in western Taiwan (Figure 9.24*A*), and active submarine accretionary prisms over subduction zones, which are also a type of thrust sheet (Figure 9.24*B*), are wedge-shaped rather than rectangular in cross section, with thickness increasing with increasing distance from the toe of the thrust sheet. This observation suggests that it is incorrect to model a thrust sheet as a rectangular block as we did at the beginning of this subsection (Figure 9.22). An analysis that accounts for the wedge shape largely removes the mechanical problems of emplacing large thrust sheets.

To account for tapered thrust sheets, we adopt the **critical Coulomb wedge** model, which describes the mechanics of a brittle, deformable thrust sheet undergoing frictional sliding on a basal decollement. In order to understand the various factors that control the shape of a thrust sheet, we examine part of the sheet as a free-body diagram (Figure 9.25). In such a diagram, we represent the mechanical effects of all the material surrounding the free-body portion of the thrust sheet by tractions (i.e., force per unit area) distributed across the boundaries of the body.

To set up the mechanical model for a thrust sheet (Figure 9.25), we make the following initial assumptions: We assume the basal decollement is a flat plane that slopes upward toward the foreland at an angle β. As we observe in nature (Figures 9.24 and 9.26), the surface of the thrust sheet must slope downward toward the foreland, and we label that slope angle α. We assume that the rocks in the thrust sheet are everywhere just at the critical stress for failure, as defined by the Coulomb fracture criterion, and thus that the stress everywhere in the thrust sheet is as large as possible. Finally, we assume that the acceleration is small enough to be negligible, in which case the equation of equilibrium (Newton's first law of motion) requires that the driving forces that tend to push the thrust wedge toward the foreland must be balanced by the resisting forces that tend to prevent motion. The problem then is to account for all the forces

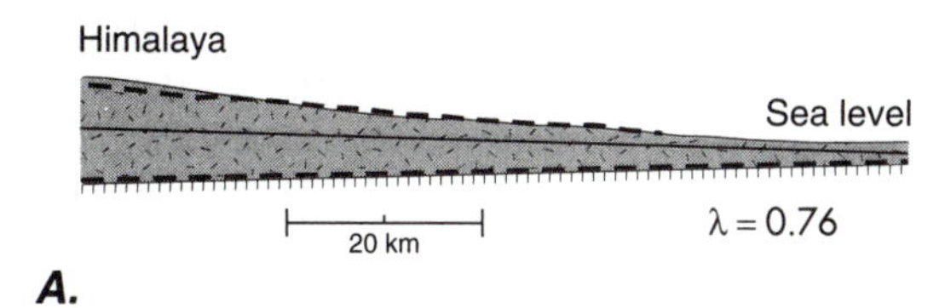

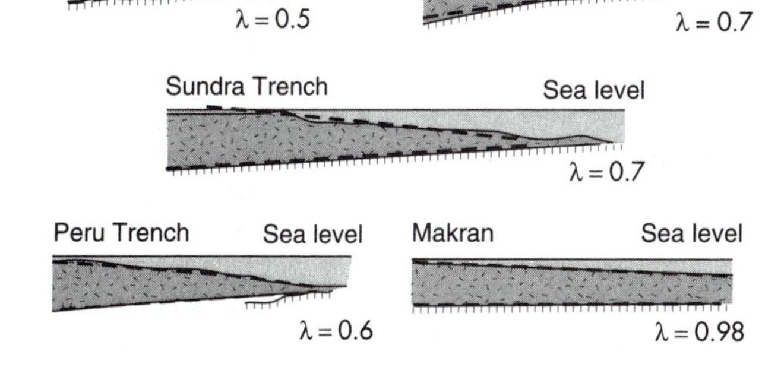

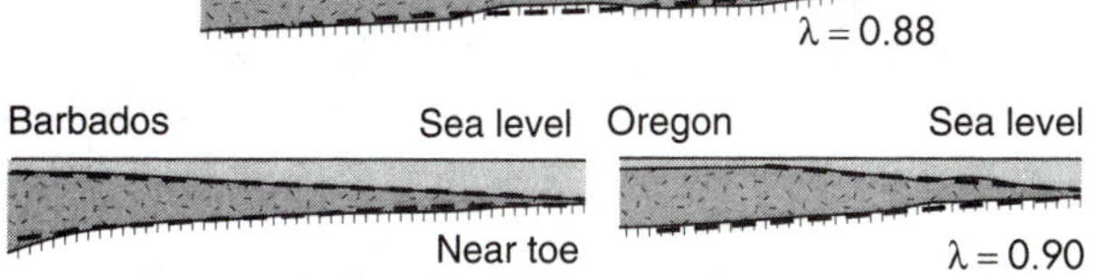

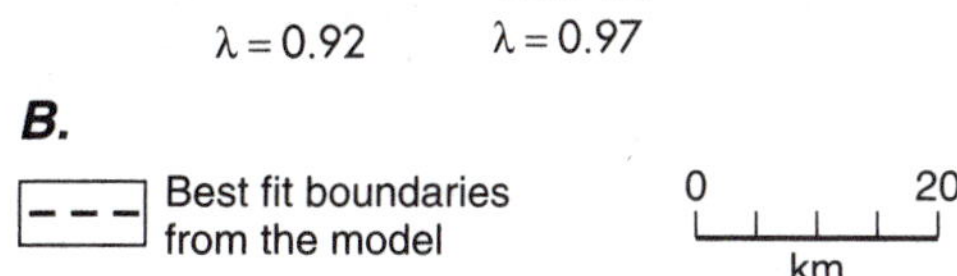

FIGURE 9.24 Cross sections of thrust wedges. Dashed lines show the fit of Equation (9.19) for the listed value of $\lambda = \lambda_i = \lambda_b$. *A.* Active subaerial thrust wedges: the Himalaya and western south-central Taiwan. *B.* Active submarine thrust wedges that occur as accretionary prisms overlying active subduction zones. (After Davies et al. 1983)

that can act on the thrust wedge and try to explain how the wedge-shaped geometry is a natural consequence of the mechanics.

The details of the calculations are presented in Box 9-2 for a thrust sheet with a horizontal decollement ($\beta = 0$, Figure 9.25). The broad outline of the argument, however, is not difficult to grasp. The main driving force is contributed by the horizontal normal traction σ_{xx} that acts on the rear vertical face of the free-body wedge (Figure 9.25). σ_{xx} is as large as the strength of the rock will allow, and it increases with depth because the strength of the rock increases with increasing pressure. The resisting force to thrusting comes predominantly from the frictional resistance to sliding on the decollement, which is given by the coefficient of friction $\overline{\mu}_b$ times the effective normal stress on the decollement $(\sigma_n - p_b)$ (Figure 9.25). Calculating these forces, and then equating the driving force to the force of resistance leads to an equation that accounts for the basic wedge shape of a thrust sheet (see Equation (9.19)).

It is not necessary to go through the calculations, however, in order to understand the results qualitatively. If we were to increase the length of the free-body part of the thrust sheet by moving the rear vertical face an increment in the positive x direction (Figure 9.25), the frictional resistance would increase because the area of the base of the thrust sheet would increase. To counteract this and maintain the possibility for the thrust sheet to

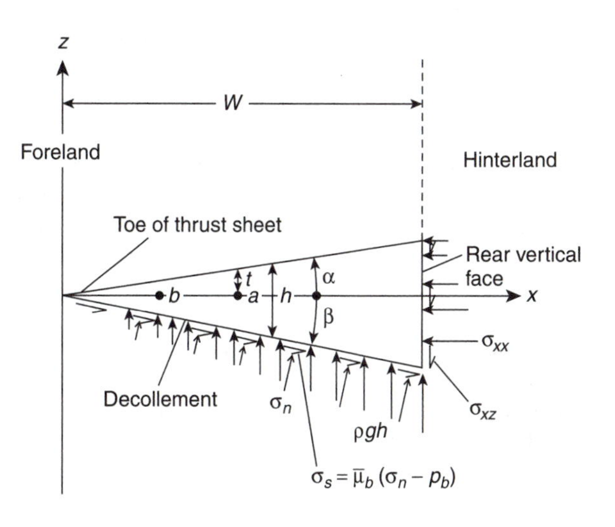

FIGURE 9.25 Free-body diagram of a portion of a thrust wedge. See text for definitions of the symbols.

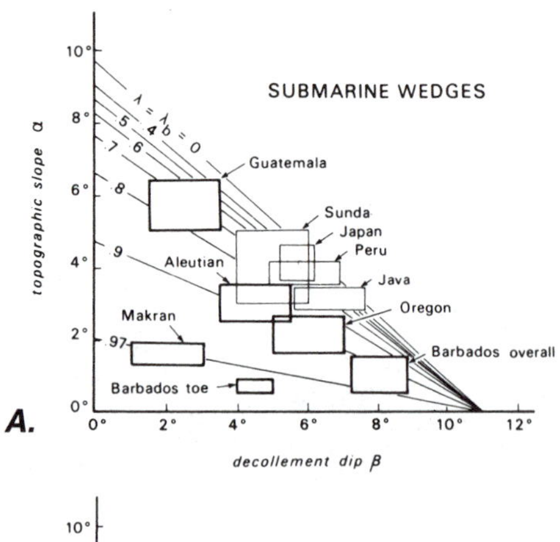

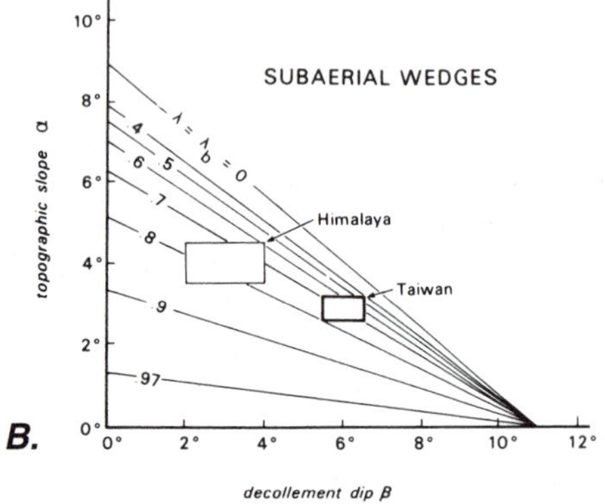

FIGURE 9.26 The measured geometry of thrust wedges compared with the theoretically predicted relationships among the topographic slope angle α, the slope of the decollement β, and the pore fluid pressure ratio λ. (N.B. The scale is the same in all diagrams, but the section for Taiwan crosses the entire orogen and thus is more complete than the other ones). A. Subaerial thrust wedges. B. Submarine accretionary prisms, which in fact are thrust wedges. (From Davies et al. 1983)

move, the driving force created by σ_{xx} must increase. But σ_{xx} itself cannot increase because it is already at the critical fracture strength, the maximum that the rock can withstand. Thus the only way to increase the magnitude of the driving force is to increase the area over which σ_{xx} acts, which means making the thrust sheet thicker. This simple argument shows that the thickness h of the thrust sheet must increase with increasing distance x from the toe, and this fundamental relationship thus defines the tapered shape of the thrust sheet. Accounting for all the details adds considerable complexity to the analysis (see Box 9-2), but the basic result still holds true.

The careful mechanical analysis of this problem leads to the following approximate equation for the surface slope angle α for subaerial wedges:

$$\alpha = \frac{\overline{\mu}_b(1-\lambda_b) - (K-1)(1-\lambda_i)\beta}{(K-1)(1-\lambda_i)+1} \tag{9.19}$$

where the angles α and β are measured in radians, β is the slope of the decollement, λ_b and λ_i are the ratios of pore fluid pressure to overburden pressure, respectively, along the decollement and internal to the thrust sheet, $\overline{\mu}_b$ is the coefficient of friction along the decollement, and K is predominantly a measure of the Coulomb fracture strength of the rock in the thrust sheet (see Box 8-1). A slight modification of the equation is necessary for subaqueous wedges to account for the pressure of the water column. We examine the factors that contribute to the form of Equation (9.19).

Friction and Rock Strength. Let us first assume that the decollement is horizontal ($\beta = 0$) and that pore fluid pressure is everywhere zero ($\lambda_b = \lambda_i = 0$). The surface slope is then defined from Equation (9.19) by

$$\alpha = \frac{\overline{\mu}_b}{K} \tag{9.20}$$

which is simply the ratio between coefficients that describe the frictional resistance and the Coulomb strength of the rock. Higher coefficients of friction on the decollement $\overline{\mu}_b$ result in higher slope angles α reflecting the need for a larger force on the rear vertical face to overcome the friction and drive the thrust sheet. Because the rock is always at the critical fracture point, the only way to increase the driving force is to increase the area of the rear vertical face, and that implies a higher slope angle. A higher strength of the rock in the thrust sheet, on the other hand, indicated by K, results in lower slope angles, because the rock can support a higher stress on the rear vertical face and therefore the area need not be as large to obtain the required driving force.

Pore Fluid Pressure. As the pore fluid pressure along the decollement increases, that is, λ_b increases from 0 to 1, the numerator of Equation (9.19) becomes smaller and the surface slope decreases. Physically, the pore fluid pressure on the decollement decreases the frictional resistance to thrusting by decreasing the effective normal stress on the decollement. As a result, the driving force necessary to move any given width of the thrust sheet decreases. Thus the change in height of the rear vertical face with increasing length of the thrust sheet that is required to maintain the driving force is reduced, and the surface slope is reduced.

Increasing pore fluid pressure within the thrust sheet, on the other hand, tends to increase the surface slope of the thrust wedge, because as λ_i increases from 0 to 1, the denominator of Equation (9.19) becomes smaller, decreasing toward a value of 1, whereas the numerator becomes bigger. Physically, pore fluid pressure within the thrust sheet makes the rock weaker and thus decreases the maximum possible horizontal stress σ_{xx} that can be applied to the rear vertical face of the wedge. To maintain the necessary driving force, the surface slope of the thrust sheet must be higher so that the rear vertical face at any point in the thrust sheet has a larger area (force = stress × area). Thus for larger values of λ_i the area of the rear vertical face must be larger, and this implies the surface slope angle of the thrust sheet must be larger.

Dip of the Decollement. The slope β of the decollement affects the resistance to motion of the thrust sheet in three ways. First, the steeper the decollement, the more work the driving force must do to raise the thrust sheet up the decollement. In effect, the driving force must not only overcome the frictional resistance, it must also overcome the component of the vertical overburden stress that acts down the dip of the decollement. This component increases with the slope β of the decollement as ($\rho gh \sin \beta$). Second, a higher slope of the decollement increases the component of normal stress on the decollement contributed by the driving force on the rear vertical face, thereby causing an increase in the frictional resistance. Third, an increase in the slope β of the decollement increases the area of the rear vertical face at any given distance from the toe, if the surface slope α remains constant, so the available driving force would increase. The net result of these effects is shown by the coefficient of β in Equation (9.19), which has the form $B/(B + 1)$, where $B = (K - 1)(1 - \lambda_i)$. This ratio must always be less than 1. Thus the surface slope α decreases with increasing slope of the decollement β, but the decrease in α is always less than the increase in β.

The Driving Force from the Surface Slope. Besides the main driving force on the rear vertical face (Figure 9.25), the surface slope itself creates a small driving force. The overburden stress (ρgt) at a given point at depth t below the surface (for example, at point a in Figure 9.25) decreases toward the toe of the thrust sheet (e.g., at point b) because along any horizontal line (e.g., $z = 0$), the height t of the sheet above that line decreases toward the toe. Because part of the horizontal normal component of the stress σ_{xx} is proportional to the overburden (Equation (9.7)), there is a horizontal gradient in σ_{xx} which is equivalent to a horizontal force per unit volume of material ([stress × distance^{-1}] = [force × area^{-1} × distance^{-1}] = [force × volume^{-1}]). This same force makes balls of silicone putty collapse into puddles if they are left standing unsupported on a table, and it drives the flow of continental ice sheets and the gravitational collapse of topographic highs (cf. Figure 9.23). Because the topographic slope is small, however, this component of driving force is not very large.

Discussion. The wedge described by Equation (9.19) is the thinnest body of Coulomb material that can be moved over the decollement. If material is added to the front, or toe, of the wedge, the whole wedge must deform to maintain the critical taper. The deformation takes the form of thrust faults, folds, and fault ramp folds internal to the wedge, all of which provide a net shortening and thickening of the wedge (see Figure 5.12). If the surface slope angle of the wedge becomes too large, then it will be decreased by the propagation of the thrust fault out in front of the wedge to lengthen the thrust sheet and by the thinning and lengthening of the thrust sheet, through slip on internal normal faults.

The mechanics of tapered thrust sheets has been compared to the mechanics of dirt or snow wedges that form in front of a bulldozer or snowplow blade. The analogy can be misleading, however, because for it to be applicable, the blade should be a flat vertical blade. In actual practice, the blades are vertically curved, a design that forces the snow or dirt to slide up the blade and fall forward to form a pile whose surface slope is the angle of repose of the material rather than the slope of the critical taper that we are discussing. The angle of repose is the steepest slope angle that loose material can support and is generally about 30°, whereas the steepest slopes predicted by the critical Coulomb wedge model are about 10° (Figure 9.26; cf. Figure 9-2.2 in Box 9-2).

A comparison of the critical Coulomb wedge model with observations of natural thrust wedges is shown in Figure 9.26, which plots the dip of the decollement β against the dip of the surface slope α. The lines are the theoretically predicted relationship for a variety of values of the pore fluid pressure ratio λ, assuming the λ ratio on the decollement equals that within the thrust wedge $\lambda_b = \lambda_i = \lambda$. The theoretical results for subaerial wedges are derived from Equation (9.19). The boxes indicate the approximate geometries of active wedges as labeled. It is clear from this figure that most thrust wedges require a value of λ considerably above the hydrostatic value of about 0.4, implying significant overpressure of the pore fluid. Such values of λ are consistent with measurements made in wells that penetrate into some of these wedges, which lends credence to the theory. Other models for thrust wedges are possible, however, which give different results, as discussed briefly in the next subsection.

vi. Non-Coulomb Theories for Thrust Wedges

An important assumption that could be questioned in the Coulomb wedge model of a thrust sheet is that the material in the thrust sheet behaves as a Coulomb material. The thrust sheet could, for example, behave as a plastic material with a yield strength that is independent of confining pressure, which would lead to different surface slopes for a thrust wedge. For a plastic thrust wedge, the strength of the material does not increase with depth, so the horizontal stress that could be supported on a vertical face would not increase with depth. In fact, because temperature increases with depth, and the yield stress for ductile flow generally decreases with increasing temperature, the horizontal stress that could be applied to a

BOX 9-2 Simplified Model of a Thrust Sheet

We adopt a simplified model of a thrust sheet composed of cohesionless material underlain by a horizontal decollement on which motion occurs by frictional sliding. In general, the height of the sheet as a function of the horizontal distance from the front of the thrust sheet, measured parallel to the direction of displacement x, is $h(x)$. At the maximum width of the sheet, $x_{(max)} = W$, we take $h(W) = H$ (Figure 9-2.1). The tractions acting on the external surfaces of the thrust sheet are as shown in Figure 9-2.1. In the following derivations, we use the notation for the associated stress components by dropping the superscripts "+" and "−," and we account for the orientations of the tractions when we sum the associated forces. The stress components include:

σ_T a horizontal tectonic compressive stress applied to the rear vertical face of the sheet,

σ_F a frictional shear stress applied along the bottom of the thrust sheet, and

σ_V a vertical compressive stress applied along the bottom of the thrust sheet (due to the overburden).

The total tectonic driving force and the frictional forces acting on the thrust sheet are just the appropriate stresses multiplied by the area of the surface on which each acts. Because the stresses vary across the surfaces, however, we must integrate the stress times an infinitesimal area, over the finite area. If we assume these areas have unit dimension in the y direction, we can integrate over just one dimension to obtain:

$$F_T = \int_0^H \sigma_T\, dz, \quad \text{at} \quad x = W \tag{9-2.1}$$

$$F_F = \int_0^W \sigma_F\, dx, \quad \text{at} \quad z = 0 \tag{9-2.2}$$

The sum of the horizontal tectonic force F_T driving the thrust and the total force of frictional resistance on the base F_F must be zero if the thrust wedge is moving as a block but not accelerating. These forces act in opposite coordinate directions, so

$$F_F + F_T = 0 \tag{9-2.3}$$

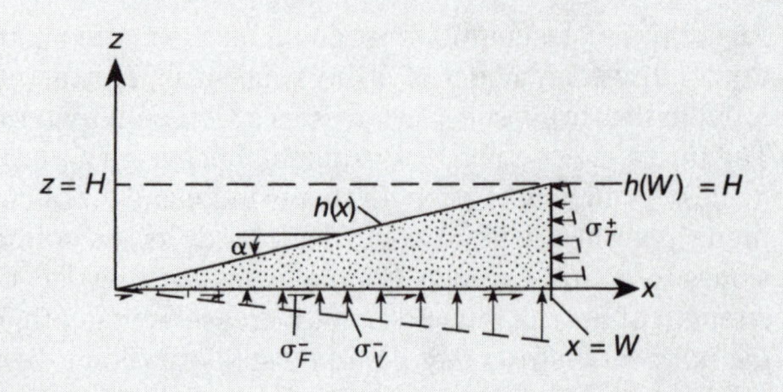

FIGURE 9-2.1 Free-body diagram for a model of a thrust sheet, showing the geometry of the wedge and the tractions acting on the surfaces (see text).

We now make the following assumptions:

1. The vertical normal stress, which equals the overburden pressure, is approximately the minimum compressive stress.
2. The horizontal tectonic normal stress is approximately the maximum compressive stress.
3. The horizontal stress is as large as possible and thus is given by the Coulomb fracture criterion for the material in the thrust sheet.
4. The pore fluid pressure on the decollement at the base of the thrust sheet $p_F^{(b)}$ is distinct from that in the internal part of the thrust sheet $p_F^{(i)}$.

Assumptions 1 and 2 imply that the principal stresses are everywhere horizontal and vertical, which cannot actually be true because there is a shear stress on the horizontal base of the sheet. Thus the stress trajectories should be inclined (see Figure 9.20A), but our simplifying assumption should be a reasonable first approximation if the shear stress is relatively small and therefore the inclination of the maximum principal stress is small.

Let us consider first how we can calculate the frictional resistance (Equation 9-2.2). The frictional traction on the base σ_F is related to the effective vertical traction $(\sigma_V - p_F^{(b)})$ by the coefficient of friction on the base $\overline{\mu}_b$ (Amontons' second law; see Section 8.4(ii), footnote 3):

$$\sigma_F = \overline{\mu}_b\,[\sigma_V - p_F^{(b)}], \quad \text{at} \quad z = 0 \tag{9-2.4}$$

The thickness of material overlying any given point at height z above the decollement is just $(h - z)$, so we express assumption 1 as

$$\sigma_V = \rho_r g(h - z) \cong \hat{\sigma}_3 \tag{9-2.5}$$

The pore fluid pressure along the base at $z = 0$ can be expressed as a fraction λ_b of the vertical stress given by Equation (9-2.5):

$$p_F^{(b)} = \lambda_b \sigma_V = \lambda_b \rho_r g h \tag{9-2.6}$$

Thus the force of frictional resistance is found by substituting Equations (9-2.5) and (9-2.6) into (9-2.4), and then substituting the result into Equation (9-2.2) to obtain

$$F_F = \overline{\mu}_b(1 - \lambda_b)\rho_r g \int_0^W h(x)\, dx \tag{9-2.7}$$

Now we calculate the horizontal tectonic force (Equation (9-2.1)). We express assumption 3, the Coulomb fracture criterion, in terms of the principal stresses by using Equation (8-1.2) with the effective principal stresses $\hat{\sigma}_k^{(Eff)} \equiv \hat{\sigma}_k - p_F^{(i)}$ substituted for the principal stresses $\hat{\sigma}_k$, where $k = 1$ or 3 and where $p_F^{(i)}$ is the pore fluid pressure internal to the thrust sheet. Thus

$$\hat{\sigma}_1^* - p_F^{(i)} = S + K(\hat{\sigma}_3 - p_F^{(i)})$$

The asterisk on $\hat{\sigma}_1$ indicates that it is the critical maximum principal stress for fracture at the given value of $\hat{\sigma}_3$. Setting S =

0 to represent the fracture criterion for a cohesionless material, rearranging the equation to solve for $\hat{\sigma}_1^*$, and using assumption 2, we find that

$$\sigma_T \cong \hat{\sigma}_1^* = K\hat{\sigma}_3 - (K-1)p_f^{(i)}, \quad \text{at} \quad x = W \qquad (9\text{-}2.8)$$

The pore fluid pressures within the thrust sheet can be expressed as a fraction λ_i of the vertical stress given by Equation (9-2.5). At $x = W$, $h(x) = H$, so we find at the rear vertical face of the thrust sheet:

$$p_f^{(i)} = \lambda_i \sigma_V = \lambda_i \rho_r g(H - z) \qquad (9\text{-}2.9)$$

The tectonic force driving the thrust sheet can now be found by substituting $\hat{\sigma}_3$ from Equation (9-2.5) and $p_f^{(i)}$ from Equation (9-2.9) into Equation (9-2.8), and substituting that result for σ_T into Equation (9-2.1) to obtain,

$$F_T = [K - (K-1)\lambda_i]\rho_r g \int_0^H (H - z)dz$$
$$= 0.5[K - (K-1)\lambda_i]\rho_r g H^2 \qquad (9\text{-}2.10)$$

Substituting Equations (9-2.7) and (9-2.10) into (9-2.3) and rearranging gives

$$0.5CH^2 = \int_0^W h\,dx \qquad (9\text{-}2.11)$$

where

$$C \equiv \frac{K - (K-1)\lambda_i}{\overline{\mu}_b(1-\lambda_b)} = \frac{(K-1)(1-\lambda_i)+1}{\overline{\mu}_b(1-\lambda_b)} \qquad (9\text{-}2.12)$$

Note that for Equation (9-2.12) if there is no pore fluid pressure, then λ_i and λ_b are both zero and C is the ratio of the fracture strength constant of the thrust sheet, K, to the frictional resistance on the base, $\overline{\mu}_b$. In general, then, C is just this ratio modified by the effects of pore fluid pressure. From the second Equation (8-1.3), we see that K depends on the coefficient of internal friction, which we here label μ_i. In order for sliding to occur on the base, rather than faulting to occur within the wedge, the coefficient of sliding friction on the base must be less than the coefficient of internal friction for the Coulomb fracture criterion; that is, $\overline{\mu}_b < \mu_i$.

Equation (9-2.11) can be interpreted in two different ways:

1. We can assume the maximum thickness H and the surface shape $h(x)$ of the thrust sheet are given, in which case Equation (9-2.11) determines the maximum possible width W for which the stress on the thrust sheet remains below the Coulomb fracture criterion or

2. We can take both H and W to be variables and use Equation (9-2.11) to determine the relation between these two variables and thus the shape of a thrust sheet that is everywhere at the critical stress for Coulomb failure. We discuss these interpretations in turn.

For interpretation 1, suppose the thrust sheet is a rectangular block,

$$h(x) = H, \qquad 0 \le x \le W \qquad (9\text{-}2.13)$$

where H is a constant. We considered this problem in a simplified way in Section 9.11(i). Using Equation (9-2.13) in (9-2.11), integrating, and rearranging, we find that

$$W = 0.5\,CH \qquad (9\text{-}2.14)$$

Thus for a given thickness H of a block-shaped thrust sheet, and for given values of K, $\overline{\mu}_b$, λ_i, and λ_b, which determine C, this relationship gives the maximum width W for a thrust sheet that can be moved over the decollement. We assume for simplicity that $\overline{\mu}_b = \mu_i$ and $\lambda_b = \lambda_i$, and we use Equation (9-2.14) and the values from Table 9-2.1 to graph the dependence of $C/2$ ($= W/H$) on λ (Figure 9-2.2A). For zero pore fluid pressure, $C/2$ is between 2.75 (for $\theta_f = 65°$) and 3.85 (for $\theta_f = 50°$). Thus for a thrust sheet of thickness $H = 5$ km, the width W must be between about 13.7 and 19.7 kilometers for the different values of the fracture angle θ_f. This result is of the same order as the more-approximate solution in Equation (9.15). Note that the possible length of the thrust sheet increases without limit as λ approaches 1.

For an alternative interpretation of Equation (9-2.11), we assume that the entire thrust sheet must be just at the critical Coulomb fracture stress, and we use the equation to determine the shape of the thrust sheet. On the left side of Equation (9-2.11), C is dimensionless, and H^2 has units of [length]^2. Thus Equation (9-2.11) can be satisfied only if $h(x)$ is a linear function of x, because the integral must have dimensions of [length]^2:

$$h(x) = A\,x + B \qquad (9\text{-}2.15)$$

We require that

$$h(W) = H \quad A = \tan\alpha \qquad (9\text{-}2.16)$$

where the first Equation (9-2.16) is implicit in the way the quantities are defined for the problem and α is defined as the surface slope of the wedge. Substituting this equation into Equation (9-2.15) shows that

$$H = AW + B \qquad (9\text{-}2.17)$$

TABLE 9-2.1 Relationships among Fracture Angle, Coefficient of Internal Friction, and K[a]

θ_f	μ_i	K
65°	0.84	4.60
60°	0.58	3.00
55°	0.36	2.04
50°	0.18	1.42
45°	0.00	1.00

[a] Given θ_f, the tabulated values of μ_i and K are calculated respectively from Equation (8-1.1) and the first Equation (8-1.3)

(continued)

BOX 9-2 Simplified Model of a Thrust Sheet *(continued)*

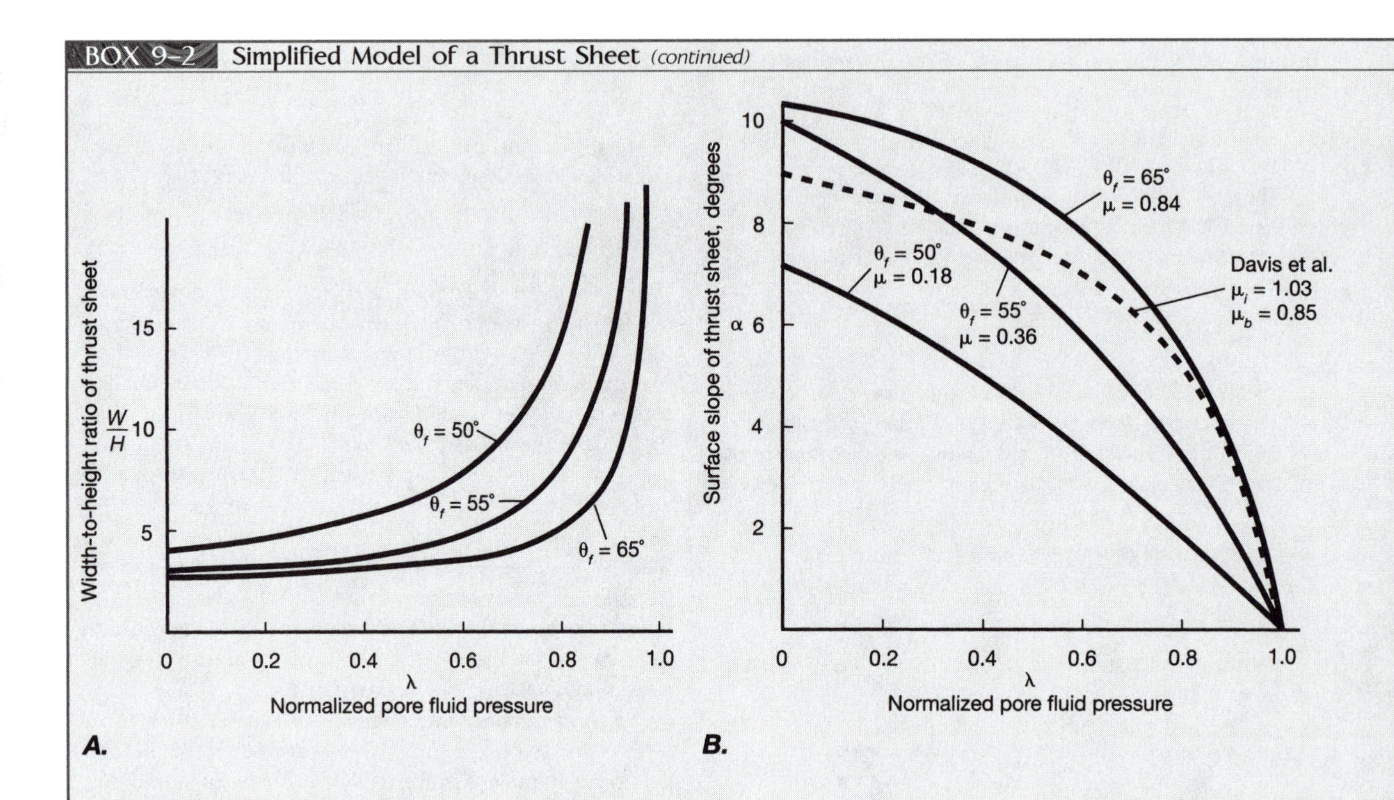

FIGURE 9-2.2 The permissible geometry of thrust sheets as a function of the pore fluid pressure ratio λ. *A.* Maximum possible ratio of width to height for a thrust sheet shaped like a rectangular block, plotted as a function of λ. *B.* Equilibrium surface slope α for a wedge-shaped thrust sheet with a horizontal decollement, plotted as a function of λ. Solid lines result from the current analysis; the dashed line results from the more detailed model of Davis et al. (1983).

vertical face could actually decrease with increasing depth. Thus if a ductile wedge were weaker on average than a Coulomb wedge and the resistance on the base were the same, the surface slope would have to be larger for a ductile wedge than for a Coulomb wedge.

Ductile deformation instead of frictional sliding on the base could offset the effects of a ductile, instead of a Coulomb, thrust wedge. Two effects are important. First, a weak material such as salt along the decollement would reduce the resistance to thrust motion substantially below that resulting from friction. Second, the ductile yield strength is relatively insensitive to the normal stress across the surface, unlike the frictional resistance. As a result, the resistance to thrust motion is the same under thicker parts of the thrust sheet as it is under thinner parts, whereas the frictional resistance increases as the thickness of the thrust sheet increases. This enhances the weakening effect of ductile flow at the decollement relative to frictional resistance and thus would tend to require a smaller surface slope to maintain the driving force for thrusting. This mechanism could account for some of the very low slopes observed (Figure 9.26).

Both temperature and pressure increase with increasing depth in the Earth. The plastic yield strength is relatively insensitive to changes in pressure but decreases dramatically with increasing temperature. The Coulomb strength and frictional resistance, on the other hand, increase significantly with increasing pressure but are relatively insensitive to changes in temperature. Thus at some depth, probably near 10 to 15 kilometers, we can expect plastic behavior to replace both Coulomb behavior and frictional resistance. If that depth is reached within a single thrust sheet, it should be apparent from a change in the surface slope of the thrust wedge.

vii. The Propagation of Slip Domains

The critical Coulomb wedge model for the mechanics of a thrust sheet assumes that the entire mass is at the critical stress for fracture as predicted by the Coulomb frac-

Using Equation (9-2.15), the right side of Equation (9-2.11) is

$$\int_0^W h\,dx = \int_0^W (Ax + B)dx = 0.5\,AW^2 + BW \quad (9\text{-}2.18)$$

Substituting Equations (9-2.17) and (9-2.18) into Equation (9-2.11) and simplifying, we get

$$0.5[A^2C - A]W^2 + [ABC - B]W + 0.5[B^2C] = 0 \quad (9\text{-}2.19)$$

This relationship must hold for a thrust wedge of any width W, and for this to be true, each of the coefficients in brackets must independently be zero:

$$A^2C - A = 0 \quad ABC - B = 0 \quad B^2C = 0 \quad (9\text{-}2.20)$$

To satisfy the third Equation (9-2.20), either $B = 0$ or $C = 0$. Taking $C = 0$ implies from Equation (9-2.12) that K is a function of λ_i or, through the definition of K in Equation (8-1.3), that the fracture angle θ_f is a function of the pore pressure ratio. Experimental work shows that this is a physically unacceptable solution, so we must choose $B = 0$. This result implies, from Equation (9-2.15), that the thrust wedge tapers to a point, which is physically reasonable for a material with no cohesion (Figure 9-2.1).

Both the first and second Equations (9-2.20) give exactly the same condition:

$$A = \frac{1}{C} \quad (9\text{-}2.21)$$

Introducing Equation (9-2.12) and the second Equation (9-2.16) into Equation (9-2.21), we find that

$$\tan\alpha \cong \alpha = \frac{\bar{\mu}_b(1 - \lambda_b)}{(K - 1)(1 - \lambda_i) + 1} = \frac{1}{C} \quad (9\text{-}2.22)$$

where α is the topographic slope angle of the thrust wedge. The first relation is the small-angle approximation to tan α, where α is given in radians. The second equation is the special case of Equation (9.19) when $\beta = 0$.

If $\lambda_b = \lambda_i = 0$, that is, if there is no pore fluid pressure, the surface slope of the thrust wedge is determined by $\bar{\mu}_b/K$, the ratio of frictional resistance on the base of the sheet to the fracture strength constant of the thrust wedge. As the pore fluid pressure internal to the wedge increases, the effect is to decrease the strength of the wedge. Thus for higher values of λ_i, the denominator of Equation (9-2.22) decreases, the ratio increases, and the surface slope of the thrust wedge increases. As the pore fluid pressure along the base increases, there is less resistance to frictional sliding. Thus for higher values of λ_b, the numerator in Equation (9-2.22) decreases, the ratio decreases, and the surface slope of the thrust wedge decreases.

For purposes of simplification, we assume $\mu_i = \bar{\mu}_b = \mu$ and $\lambda_b = \lambda_i = \lambda$. Figure 9-2.2*B* shows the relationships then predicted by the second Equation (9-2.22) between the topographic slope α of the thrust sheet and the magnitude of the pore fluid pressure ratio λ for values of the constants in Table 9-2.1. The predicted slopes are all less than about 10°. The slopes approach 0°, and the frictional resistance to sliding on the decollement decreases toward zero as λ approaches 1. These results are comparable to the angles calculated from more sophisticated analyses for the same angle of the decollement (Figure 9-2.2). More thorough analyses include the dip of the decollement as a variable.

ture criterion and that motion occurs continuously on the entire fault plane. These assumptions provide a simplified and only approximate model, however, because the entire thrust sheet neither moves as a rigid block nor undergoes pervasive deformation at one time. Rather, the deformation is accommodated by the discontinuous slip events over finite areas of faults within and at the base of the sheet. Such slip events, which are localized in time and space, commonly cause earthquakes, which we can observe. Only by averaging these events over a long period of time, on the order of perhaps tens of thousands to hundreds of thousands of years, would we see the pattern of pervasive deformation and the slip of the entire thrust sheet on the decollement that we assume for the model.

Physical models of sliding by localized events have been constructed by assuming that the frictional resistance on finite areas of a fault varies in magnitude in a random pattern. As the applied stress on the thrust sheet builds up, the weakest areas are the first to slip, and the slipping is limited to the area of weakness. The release of stress on the weak areas requires a concentration of stress in other areas, and these areas can then deform or slip at externally applied stresses that are lower than would be predicted from the strength of these areas. The net result is that the deformation can occur at applied stresses that are lower than would be predicted from an average value of the mechanical properties. The effect is similar to the effect of dislocations in a ductile crystalline material, which allow ductile deformation at stresses well below the theoretical strength predicted on the basis of atomic bonds (see Chapter 17).

9.12 CAUSE AND EFFECT: A WORD OF CAUTION

Throughout our discussion of the origin of brittle fractures, we have implicitly assumed that stress is the cause of the process and that deformation, whether strain or brittle fracture, is the result or effect. For experiments on fracturing, we have assumed, for example, that an axial stress is applied parallel to the cylinder axis of a rock sample, and a confining pressure is applied in the radial direction. We then infer that the rock fractures when the stress reaches a critical value, that is, that stress is the cause of the fracturing. This, however, may not always be an appropriate way to view the process. In some situations, it may be more appropriate to consider that a deformation is the cause of the process and that the stress develops in the sample an effect of the deformation. Under such circumstances, we should discuss the origin of different structures in terms of deformation, rather than stress.

In the experiment pictured in Figure 8.7, for example, although we interpreted the results in terms of stress, the experimental set-up actually imposed the simple shear deformation on the clay block, so the stress must be an effect of the deformation. Had we imposed the stress in Figure 8.7*C* on the boundaries of the clay block, we would not have obtained a simple shear deformation, but rather a deformation symmetric relative to the stress axes, as, for example, in Figure 8.6*A*.

Thus the cause of a process is determined by the constraints that are imposed on the boundaries of the system, which are called the **boundary conditions.** Stress and displacement boundary condition are not equivalent. For example, uniform stress boundary conditions imposed on an isotropic material can never lead to a simple shear deformation, because trying to impose one shear couple on a material will always result in the development of an equal conjugate shear couple (Equations (7.29), (7.31), 7.56)). Thus applying a uniform stress boundary condition to an isotropic material must always result in a deformation that is symmetric relative to the principal stress axes. Simple shear of an isotropic material can occur, however, if the appropriate displacements are imposed on the boundaries.

In Section 8.7, we argue that the Coulomb fracture criterion is not an appropriate model to explain the Riedel shears that developed in the clay block experiment shown in Figure 8.7. Instead, the fractures could be understood, at least in part, with reference to the geometric constraints of the imposed deformation. Thus in this case, the structures and the stress can be understood as being caused by the deformation.

It is relatively easy to identify the boundary conditions in a deformation experiment, but often much more ambiguous in the case of deformation in the Earth, for which no simple boundary conditions may in fact exist. The identification of cause and effect in that case may also be ambiguous. In some if not most cases, however, it may be more useful to consider structures as reflecting an imposed deformation rather than an imposed stress.

We discuss the use of strain to interpret structures in Chapters 13 through 15. The relationships between stress and deformation are the topic of Chapter 16. Finally, we revisit the role of boundary conditions and the problems of cause and effect and their relation to symmetry in Sections 18.1(iii) and 18.1(iv) and in Box 18-1(iv).

REFERENCES AND ADDITIONAL READINGS

i. Stress in the Plates

Brace, W. F., and D. L. Kohlstedt. 1980. Limits of lithospheric stress imposed by laboratory experiments. *J. Geophys. Res.* 85: 6248–6252.

Jaeger, J. C. 1962. *Elasticity , Fracture and Flow with Engineering and Geological Applications.* London, Methuen & Co. Ltd.; New York, John Wiley & Sons, 208 pp.

McGarr, A. 1980. Some constraints on the levels of shear stress in the crust from observations and theory. *J. Geophys. Res.* 85: 6231–6238.

McGarr, A., and N. C. Gay. 1978. State of stress in the Earth's crust. *Ann. Rev. Earth and Planet. Sci.* 6: 405–436.

Merrill, R. H. 1968. Measurement of in situ stress and strain in rock. Rock Mechanics Seminar, NSF Advanced Science Seminar in Rock Mechanics for College Teachers of Structural Geology, Boston College, 1967. Bedford, MA, Terrestrial Science Laboratory, Air Force Cambridge Research Laboratory, 151–199.

Richardson, R. M., S. C. Solomon, and N. H. Sleep. 1979. Tectonic stress in the plates. *Rev. Geophys. and Sp. Phys.* 17: 981–1019.

Zoback, M. D., H. Tsukahara, and S. Hickman. 1980. Stress measurement at depth in the vicinity of the San Andreas fault: Implications for the magnitude of shear stress at depth. *J. Geophys. Res.* 85: 6157–6173.

Zoback, M. L., and M. D. Zohack. 1980. State of stress in the coterminus United States. *J. Geophys. Res.* 85: 6113–6156.

Zoback, M.L., M. D. Zoback, J. Adams, M. Assumpcao, S. Bell, E. A. Bergman, P. Blümling, N. R. Brereton, D. Denham, J. Ding, K. Fuchs, N. Gay, S. Gregersen, H. K. Gupta, A. Gvishiani, K. Jacob, R. Klein, P. Knoll, M. Magee, J. L. Mercier, B. Müller, C. Paquin, K. Rajendran, O. Stephansson, G. Suarez, M. Suter, A. Udias, Z. H. Xu, and M. Zhizhin 1989. Global patterns of tectonic stress. *Nature* 341: 291–298.

ii. Formation and Interpretation of Joints

Bahat, D. 1979. Theoretical considerations on mechanical parameters of joint surfaces based on studies on ceramics. *Geol. Mag.* 116: 81–92.

Bahat, D. 1991. Plane stress and plane strain fracture in Eocene chalks around Beer Sheva. *Tectonophysics* 196: 6147.

Bahat, D., and T. Engelder. 1984. Surface morphology on joints of the Appalachian plateau, New York and Pennsylvania. *Tectonophysics* 104: 299–313.

Brown, E. T., and E. Hoek. 1978. Trends in relationships between measured in-situ stresses and depth. *Int. J. Rock Mech. Mining Sci. and Geomech. Abstr.* 15: 211–215.

Degraff, J. M., and A. Aydin. 1987. Surface morphology of columnar joints and its significance to mechanics and direction of joint growth. *Geol. Soc. Amer. Bull.* 99: 605–617.

Engelder, T. 1985. Loading paths to joint propagation during a tectonic cycle: An example from the Appalachian plateau, U.S.A. *J. Struct. Geol.* 7: 459–476.

Engelder, T. 2004. Tectonic implications drawn from differences in the surface morphology on two joint sets in the Appalachian Valley and Ridge, Virginia. *Geology* 32(5): 413–416.

Engelder, T., and P. Geiser. 1980. On the use of regional joint sets as trajectories of paleostress fields during the development of the Appalachian plateau, New York. *J. Geophys. Res.* 85: 6319–6341.

Haxby W. F., and D. L. Turcotte. 1976. Stresses induced by the addition or removal of overburden and associated thermal effects. *Geology* 4(3): 181–184.

Kulander, B. R., and S. L. Dean. 1985. Hackle plume geometry and joint dynamics. In *Fundamentals of Rock Joints. Proceedings of the International Symposium of Rock Joints*, Bjorkliden, Sweden, 85–94.

Ladeira, F. L., and N. J. Price. 1981. Relationship between fracture spacing and bed thickness. *J. Struct. Geol.* 3(2): 179–184.

Olson, J., and D. D. Pollard. 1989. Inferring paleostresses from natural fracture patterns: A new method. *Geology* 17: 345–348.

Pollard, D. D., and R. C. Fletcher. 2005. *Fundamentals of Structural Geology*. Cambridge, UK, and New York, Cambridge University Press, 500 pp.

Pollard, D. D., and P. Segall. 1987. Theoretical displacements and stresses near fractures in rock: With applications to faults, joints, veins, dikes, and solution surfaces. In B. K. Atkinson, ed., *Fracture Mechanics of Rock*, London, Academic Press, 277–349.

Pollard, D. D., P. Seagall, and P. T. Delaney. 1982. Formation and interpretation of dilatant *en echelon* cracks. *Geol. Soc. Amer. Bull.* 93: 1291–1303.

Younes, A. I., and T. Engelder. 1999. Fringe cracks: Key structures for the interpretation of the progressive Alleghanian deformation of the Appalachian plateau. *Geol. Soc. Amer. Bull.* 111: 219–239.

iii. Fractures Associated with Faults

Beach, A. 1975. The geometry of *en echelon* vein arrays. *Tectonophysics* 28: 245–263.

Conrad, R. E. II, and M. Friedman. 1976. Microscopic feather fractures in the faulting process. *Tectonophysics* 33: 187–198.

Engelder, T. 1974. Cataclasis and the generation of fault gouge. *Geol. Soc. Amer. Bull.* 85: 1515–1522.

Friedman, M., and J. M. Logan. 1970. Microscopic feather fractures. *Geol. Soc. Amer. Bull.* 81: 3417–3420.

iv. Fractures Associated with Folds

Currie, J. B., H. W. Parnode, and R. P. Trump. 1962. Development of folds in sedimentary strata. *Geol. Soc. Amer. Bull.* 73: 655–674.

Friedman, M. R., H. H. Hugman III, and J. Handin. 1980. Experimental folding of rocks under confining pressure, Part VIII: Forced folding of unconsolidated sand and of lubricated layers of limestone and sandstone. *Geol. Soc. Amer. Bull.* Part 1, 91(5): 307–312.

Handin, J., M. Friedman, K. D. Min, and L. J. Pattison. 1976. Experimental folding of rocks under confining pressure, Part II: Buckling of multilayered rock beams. *Geol. Soc. Amer. Bull.* 87: 1035–1048.

Norris, D. K. 1967. Structural analysis of the Queensway folds, Ottawa, Canada. *Canad. J. Earth Sci.* 4: 299–321.

Spang, J. H., and R. H. Groshong, Jr. 1981. Deformation mechanism and strain history of a minor fold from the Appalachian Valley and Ridge. *Tectonophysics* 72: 323–342.

Stearns, D. W. 1968. Certain aspects of fractures in naturally deformed rocks. In R. E. Riecker, ed., *NSF Advanced Science Seminar in Rock Mechanics for College Teachers of Structural Geology*, Bedford, MA, Terrestrial Sciences Laboratory, Air Force Cambridge Research Laboratories, 97–118.

v. Mechanics of Faulting and Thrust Sheets

Angelier, J. 1984. Tectonic analysis of fault slip data sets. *J. Geophys. Res.* 89(B7): 5835–5848.

Chapple, W. M. 1978. Mechanics of thin-skinned fold-and-thrust belts. *Geol. Soc. Amer. Bull.* 89(8): 1189–1198.

Dahlen, F. A. 1984. Noncohesive critical Coulomb wedges: An exact solution. *J. Geophys. Res.* 89: 10,125–10,133.

Dahlen, F. A., J. Suppe, and D. Davis. 1984. Mechanics of fold-and-thrust belts and accretionary wedges: Cohesive Coulomb theory. *J. Geophys. Res.* 89: 10,087–10,101.

Davis, D., and T. Engelder. 1985. The role of rock salt in fold-and-thrust belts. *Tectonophysics* 119: 67–88.

Davis, D., J. Suppe, and F. A. Dahlen. 1983. Mechanics of fold-and-thrust belts and accretionary wedges. *J. Geophys. Res.* 88: 1153–1172.

Elliot, D. 1976. The motion of thrust sheets. *J. Geophys. Res.* 81: 949–963.

Emerman, S., and D. Turcotte. 1983. A fluid model for the shape of accretionary wedges. *Earth and Planet. Sci. Lett.* 63: 379–384.

Hafner, W. 1951. Stress distributions and faulting. *Geol. Soc. Amer. Bull.* 62: 373–398.

Hubbert, M. K., and W. W. Rubey. 1959. Role of fluid pressure in mechanics of overthrust faulting. *Geol. Soc. Amer. Bull.* 70: 115–206.

Scholz, C. H. 2002. *The Mechanics of Earthquakes and Faulting, 2nd Edition.* New York, Cambridge University Press, 471 p.

Stockmal, G. S. 1983. Modeling of large-scale accretionary wedge formation. *J. Geophys. Res.* 88: 8271–8287.

Tickoff, B., and S. F. Wojtal. 1999. Displacement control of geologic structures. *J. Struct. Geol.* 21(8/9): 959–967.

Zheng, Y., T. Wang, M. Ma, and G. A. Davis. 2004. Maximum effective moment criterion and the origin of low-angle normal faults. *J. Struct. Geol.* 26: 271–285.

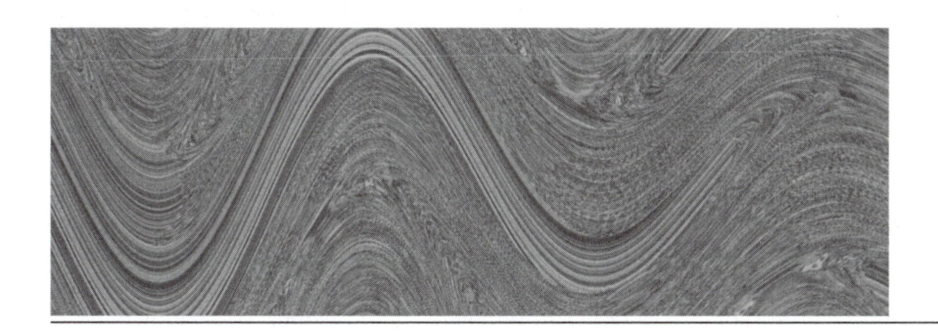

PART II

DUCTILE DEFORMATION

In Part II we consider structures in rocks that form as a result of ductile deformation, or flow of rocks in the solid state. We define ductile deformation as a permanent, coherent, solid-state deformation, for which there is no loss of cohesion on the scale of crystal grains or larger. Thus mineral grains deform coherently without the formation of open cracks inside the grains or along grain boundaries, and deformation is distributed on the scale of crystal grains and larger as opposed to being concentrated on discrete shear planes. Angular grain fragments that result from brittle fracturing are not present.

Many structures exist in rocks that could only form by flow of the rocks in the solid state. Solid-state flow at first may seem an oxymoron: Liquids flow, but do solids? In fact they do. Much of our modern use of metals, for example, depends on solid-state flow. Bars, rods, and sheets of steel, as well as copper and aluminum wire, are all produced by rolling out blocks of metal or drawing the metal through dies, essentially forcing it to flow in the solid state into the desired configuration. Glaciers also flow slowly yet inexorably downhill by processes that include solid-state flow. Metals and glacier ice, like rocks, are polycrystalline materials, that is, they are made up of an aggregate of crystals. By analogy, then, it may not seem so surprising that rocks also can undergo large amounts of solid-state flow when subjected to the appropriate conditions.

Our aims in Part II are to document the evidence for the solid-state flow of rocks; provide the means for objectively describing the characteristics of the resulting structures; introduce the concept of strain, by which we can measure ductile deformation; and begin to understand the processes by which the different types of structures form.

First we describe folds in rocks (Chapter 10), then foliations and lineations in rocks (Chapter 11). We then introduce the concept of strain, which is essential to an understanding of these features (Chapter 12). With this concept, we then analyze various models for the formation of folds (Chapter 13), and of foliation and lineations (Chapter 14). We end Part II by describing methods for actually measuring strain in rocks and discussing examples of its application to the study of natural structures (Chapter 15).

Terminology

The term *ductile deformation* is used in the literature in several different ways, which creates considerable confusion and misunderstanding. Much of the problem arises from the fact that there are at least three different criteria by which ductile deformation can be recognized. These criteria are: (1) the characteristic structures that are preserved in the rocks; (2) the rheology of the deformation—that is, the form of the relationship among stress, strain rate, pressure, and temperature; and (3) the microscopic mechanisms responsible for the deformation.

Many writers prefer to use the terms *plastic* or *crystal plastic* deformation. These terms imply a mechanism of deformation that may not be appropriate. For example, it is not clear that they would appropriately describe deformation accomplished by solution-diffusion phenomena. The term *plastic* carries implications of a specific type of rheological behavior that does not include, for example, dependence of the strain rate on the first power of the stress (see Part III).

The definition we adopt is specifically a descriptive and nongenetic one. The value of such a definition is that it depends on observable structures and characteristics, and differences in interpretation do not affect the use of a given word. We expect, however, that the identification of appropriate descriptive features should be useful for inferring some constraints on the possible rheology and mechanism of the deformation.

Thus in Part III, we discuss those characteristics of rheology—that is, the physical conditions and mechanisms of flow of rocks—that are associated with the ductile structures. In particular, we find that ductile deformation, in our descriptive sense of the term, is associated with specific rheological behavior that is characterized by: (1) a thermally activated process that occurs at elevated temperatures, usually above about half the absolute melting point of the material; (2) a dependence of strain rate on stress raised to a power generally between 1 and 5; and (3) a weak dependence on the confining pressure. In terms of mechanism, ductile deformation is accomplished by diffusion of material through the rock, including the diffusion through crystal lattices of lattice defects called vacancies, and by the motion through crystal lattices of lattice defects called dislocations, which can also be aided by diffusion. Thus, in practice, observational evidence for any of these conditions or phenomena may provide additional justification for applying the term *ductile deformation.*

Our definition of ductile deformation specifically excludes cataclastic flow, which many would include within the meaning of the term, but which is not a coherent deformation at the grain scale and which we consider to be characteristic of the brittle-ductile transition. It also excludes soft-sediment deformation, or granular flow, which is also not coherent at the grain scale. Both these types of deformation might be termed "quasi-ductile deformation," as on a large scale they can produce structures comparable to ductile deformation.

In the final analysis, there is no completely satisfactory and unambiguous term to use. The reader should bear in mind that the processes by which rocks deform fall within a complete gradation from brittle to ductile, and the imposition of boundaries on such a gradation, although useful for descriptive purposes, is somewhat arbitrary.

Chapter 10

THE DESCRIPTION OF FOLDS

Folds are wave-like undulations in initially planar structures in rocks, such as sedimentary strata, that develop during deformation. They are the most obvious and common structures that demonstrate the existence of ductile deformation in the Earth. In fact, as long ago as 1669, the Danish naturalist Nicholas Steno described folds and attributed them to Earth movements. Folds occur on all scales, ranging from huge bends in orogenic belts (Figure 10.1*A*), large features that dominate the structure of orogenic core zones (Figure 10.1*B*) and form entire mountain sides (Figure 10.1*C*), through mesoscopic folds on the scale of an outcrop (Figure 10.1*D*), to folds visible only under a microscope.

Orogenic belts are all characterized by a number of fold systems. Some regions, such as Alaska and the central Andes, display large bends of the entire orogen, known as **oroclines** (Figure 10.1*A*). Elsewhere the flanks of orogenic belts are generally marked by large fold and thrust belts in unmetamorphosed to lightly metamorphosed sedimentary rocks that overly major decollements (see Chapter 5). These belts, exemplified by the Appalachian Valley and Ridge province (Figures 10.2*A*, 5.11*A*, and 5.12*A*), the Canadian Rockies (Figures 5.11*B* and 5.12*B*), and the Jura mountains north of the Alps (Figures 10.2*B*, 5.22, 6.10), commonly contain folds that are continuous for tens or even hundreds of kilometers and that in cross section exhibit layers of relatively constant thickness.

In the central regions, or core zones, of orogenic belts, the exposed rocks were generally deformed at greater depth where the temperature is higher than in the outer fold and thrust belts. The deformation there is associated with pervasive metamorphism and recrystallization of the rocks, and the folding is more intense, resulting in folds with a different appearance from those in the fold and thrust belts (Figure 10.1*C*, *D*). Cross sections of the core zones of the Alps (Figure 10.1*B*) and of the New England Appalachians (see Figure 13.42) indicate the large-scale character of such fold systems. Shapes of folds similar to those in metamorphic rocks also typify deformed salt deposits (see Figure 13.41*B*) and glaciers, which undergo solid-state flow at much lower temperatures than silicate rocks. Such structures imply a high degree of mobility of the rocks and their component minerals during deformation.

Although most folds we observe are in bedding or former bedding surfaces, they also affect other types of layers, including dikes, veins, metamorphic or igneous compositional layering, or foliations.

Apart from their esthetic appeal, folds are usually studied to reveal their geometry. The shape, orientation, and extent of folds can be of critical importance in finding economically valuable deposits and in predicting continuations of known deposits. Oil and gas are commonly trapped in the up-bowed parts of folds. Ore deposits may be concentrated in the most sharply curved areas of folds (the hinge zone) or located in particular layers that have been folded.

Beyond their economic importance, however, folds provide a record of tectonic processes in the Earth. The great variety of fold shapes in rocks must reflect the physical conditions (such as stress, temperature, and pressure) and the mechanical properties of the rock that existed when the folds developed. If we could understand the significance of fold geometry, then we would have a valuable key to understanding conditions of deformation in the Earth.

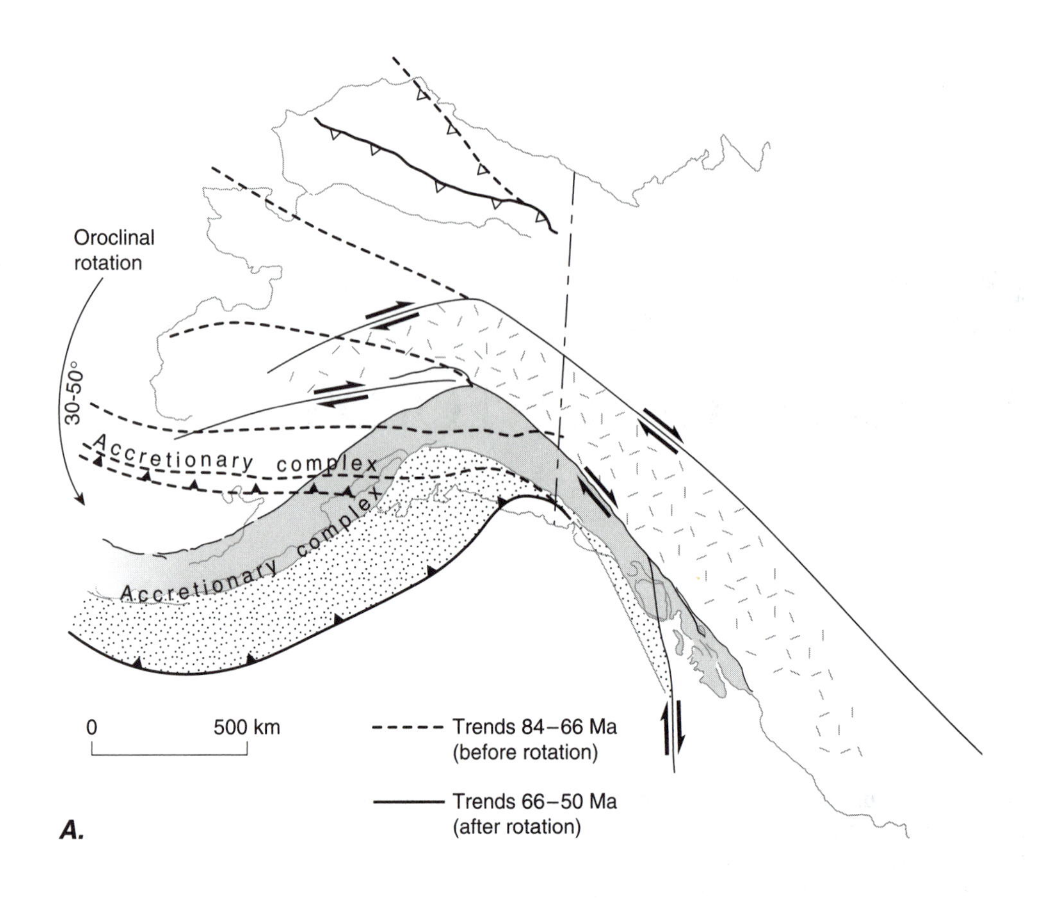

FIGURE 10.1 Scales of folding: from oroclines to folds in ductile metamorphic rocks. *A*. Generalized map of Alaska, showing trends of major tectonic features during latest Cretaceous (84 Ma to 66 Ma) and early Paleogene time (66 Ma to 50 Ma) after 30 to 50° counterclockwise rotation of western Alaska. (After Plafker and Berg 1994, figures 5F and 5G, pp. 1011–1012)

(continued)

The description of folds should be free of genetic implications, because genetic terms require interpretation of the fold origin, which may not be well understood. Ultimately, however, we wish to associate fold geometry with the mechanism of formation so that accurate description can lead to useful interpretation. Our aim is to understand how folds attain the wide variety of geometries that they display in nature and thereby to be able to infer the conditions within the Earth under which the folds formed. We must first describe what characteristics need to be explained, and thus the geometric description in this chapter serves as the basis for the discussion in subsequent chapters of the kinematics (Chapter 13) and the mechanics (Sections 18.3, 18.6, and 18.7) of fold formation.

The terminology for describing folds has evolved and accumulated over the past century or so of geologic investigation. It is extensive and not always consistent. We introduce the most useful terms in this chapter and present an objective system of describing fold geometry in terms of elements of fold style.

10.1 GEOMETRIC PARTS OF FOLDS

The simplest element of a fold that displays its characteristic geometry is a single folded surface such as a bedding surface, which is the interface between two layers of rock. A folded layer can be viewed as the volume contained between two such surfaces. Most folds consist of a stack of layers folded together, and they can be described as a nested set of folded surfaces. We discuss the geometry of folds by looking first at folded surfaces, then at folded layers and multilayers.

i. Parts of a Single Folded Surface

Figure 10.3 shows several features of folds in a single surface. A single fold is bounded on each side by an **in-**

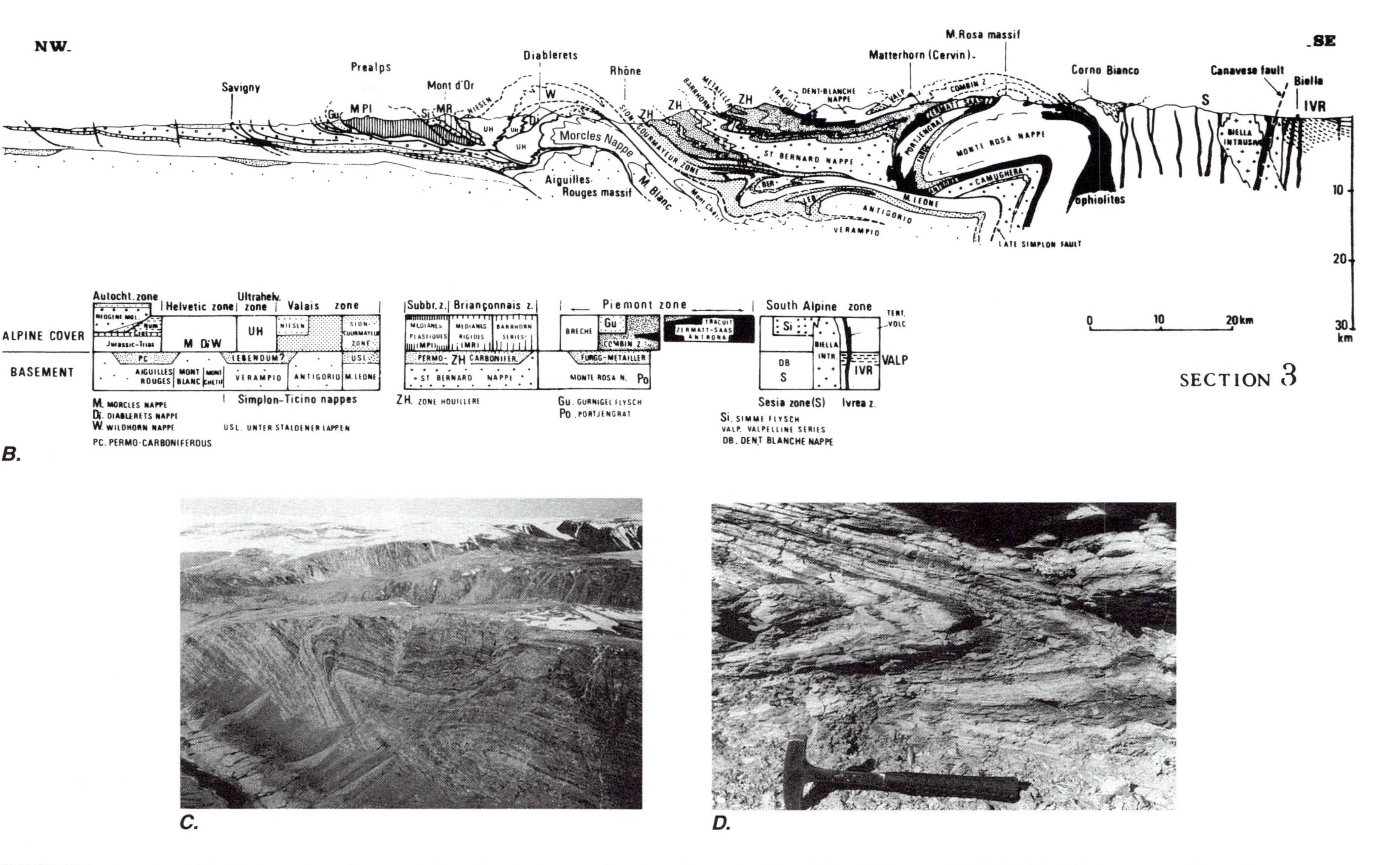

FIGURE 10.1 (continued) *B.* Cross section of metamorphosed rocks in central region of western Alps. (After Debelmas et al. 1983) *C.* This fold in metamorphic gneiss (Grandjeans-Fjord, east Greenland) is typical of folding in the metamorphic cores of orogenic belts. Height of the cliff is 800 meters. (Courtesy Haller collection of photographs of East Greenland, Geologisk Museum, Copenhagen, Denmark) *D.* Fold in banded marbles in the Snake Range detachment, western Nevada.

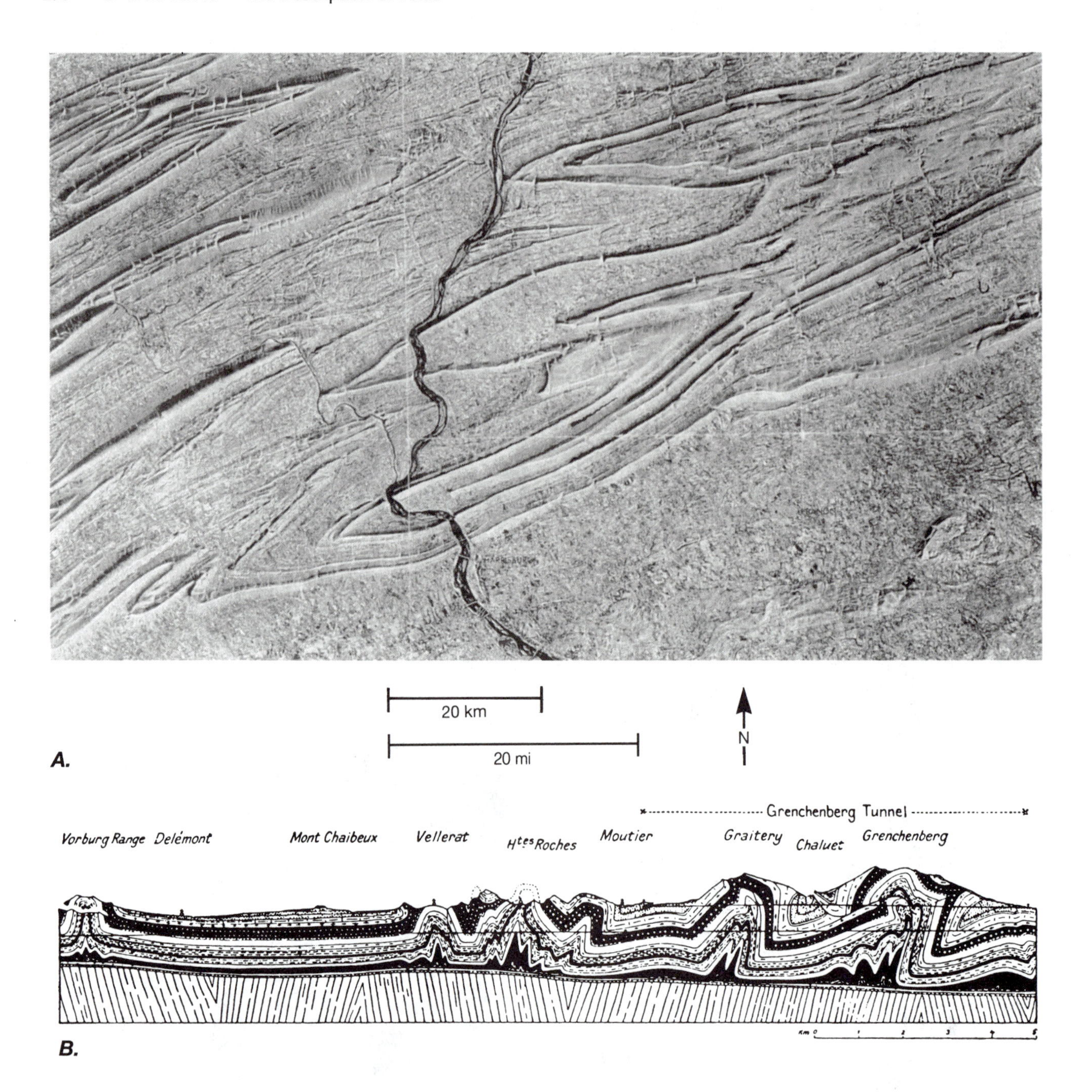

FIGURE 10.2 Folds in sedimentary rocks of fold and thrust belts. *A*. Folds in sedimentary rocks of Appalachians. Ridges are formed by erosion-resistant sandstones and conglomerates. Note the fold train of doubly plunging anticlines and synclines in the northwest. (Courtesy U.S. Geological Survey) *B*. Cross section of the Jura Mountains north of the Alps, showing the folds in the sedimentary layers (largely limestones) and the decollement, or sole fault, below the fold. Folds are class 1B and 1C. (After Buxdorf 1916)

flection line where the surface changes its sense of curvature—for example, from convex up to concave up (fold I, for example, is bounded by inflection lines i_1 and i_2). If the fold surface is planar in the region of the inflection, then by definition we take the inflection line to be the midline of the planar segment. A **fold train** is a series of folds having alternating senses of curvature. Folds that are convex upward (folds I and III) are **antiforms**, and folds that are concave upward (fold II) are **synforms**. A **fold system** is a set of folds of regional extent having a comparable geometry and presumably a common origin.

The **curvature** of any surface is a measure of the change of orientation per unit distance along the surface. A circular arc has constant curvature, and a flat plane has no curvature. In general the curvature measured along the folded surface from one inflection line to the

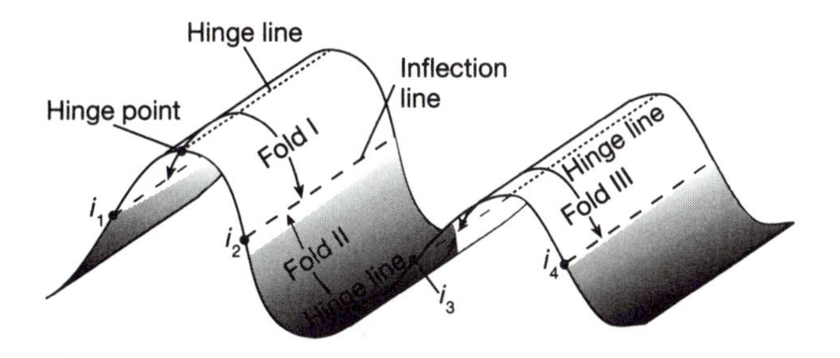

FIGURE 10.3 Features of a fold train in a single surface. Folds I and III are convex up; fold II is concave up. Folds I and III are unshaded; fold II and the incomplete parts of folds at either end are shaded. Inflection lines (dashed) delimit individual folds; points i_1, i_2, i_3, and i_4 are the inflection points on the cross section of the fold. Dotted lines are the hinge lines of each fold.

next is not constant, and the **hinge line**, or more simply the **hinge**, is the line in the folded surface along which the curvature is greatest (Figures 10.3 and 10.4*A*, *B*). A single fold may have more than one hinge (Figure 10.4*B*). If the maximum curvature is constant along an arc of finite length, then we take the midpoint of the arc to be the location of the hinge point (Figure 10.4*C*). The curvature may also vary in magnitude *along* any given hinge, and the hinge need not be a straight line (see, for instance, Figure 10.5*A*).

A fold with a single hinge closes where the limbs converge at the hinge zone (Figure 10.4*A*). On an outcrop pattern of such a fold, the **closure** is also sometimes called the **nose** of the fold. For a double-hinge fold, the closure is in the region of minimum curvature between the two hinges (Figure 10.4*B*).

The **hinge zone** is the most highly curved portion of a fold near the hinge line (Figure 10.4*A*); the **limbs** (sometimes called the **flanks**) are regions with the lowest curvature and include the inflection lines. Technically, the hinge zone is that portion of the folded surface having a greater curvature than the reference circle that is tangent to both limbs at the inflection points of the fold (Figure 10.4*A*). In the unusual case of a fold with constant curvature, the areas near the hinge and the inflection lines are still referred to loosely as the hinge zone and the limbs, respectively.

The **crest** and **trough lines** on a fold are the lines of highest and lowest elevation, respectively, on the folded surface (Figure 10.5). These lines do not necessarily coincide with the hinge (Figure 10.5*B*), and they need not be straight lines (Figure 10.5*A*). **Culminations** and **depressions** are areas where crest or trough lines go through maximum or minimum elevations, respectively.

We generally portray the form of a fold by its **profile**, which is the trace of the folded surface on a plane normal to the hinge line (Figure 10.6*A*). The profile is the form of the fold seen when it is viewed parallel to its hinge. The curvature of most folds is greatest along their profiles. The hinge line, inflection lines, and crest and trough lines appear on the profile, of course, as points.

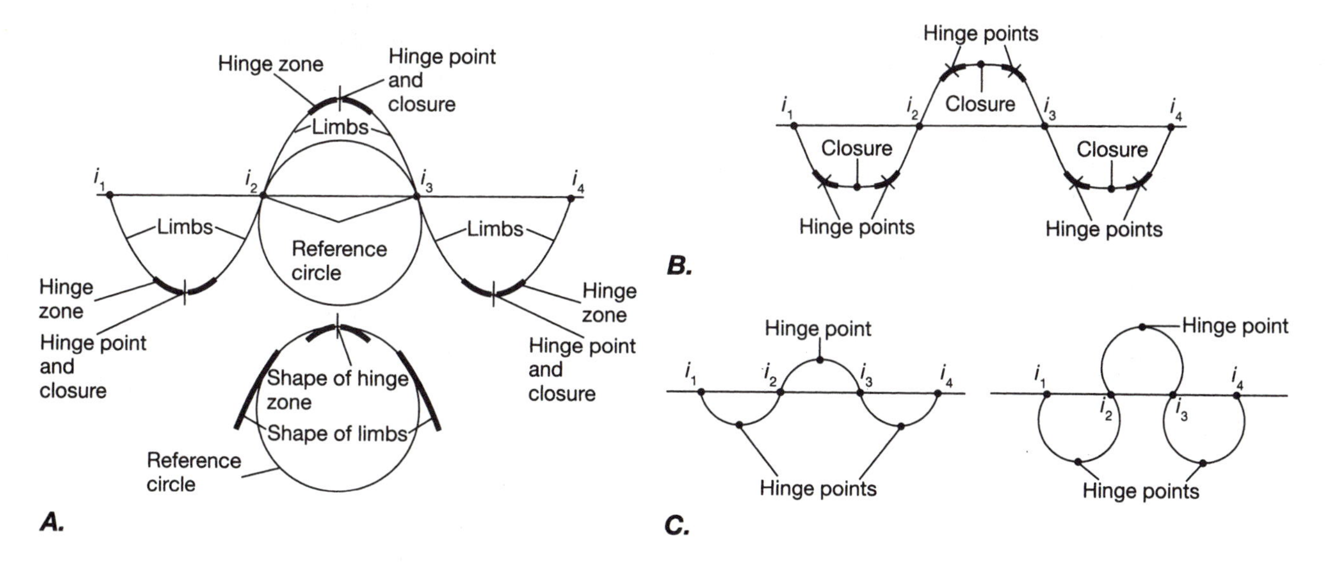

FIGURE 10.4 Definition of a hinge point, closure, hinge zone, and limb of a fold. *A*. The hinge points are points of maximum curvature. The closure point is the hinge point on a single-hinge fold. The *hinge zone* and *limb* are defined with reference to a circle that is tangent to both sides of the fold at two adjacent inflection points. The part of the fold that has a curvature greater than that of the reference circle is the hinge zone; the parts between the hinge zone and the inflection points that have a curvature less than that of the reference circle are the limbs. *B*. Individual folds may have two hinges. The closure point is the point of minimum curvature between the two hinges. *C*. Fold trains in which each fold has constant curvature and thus is the arc of a circle (perfect circular folds). Hinge points are the midpoints of each of the arcs.

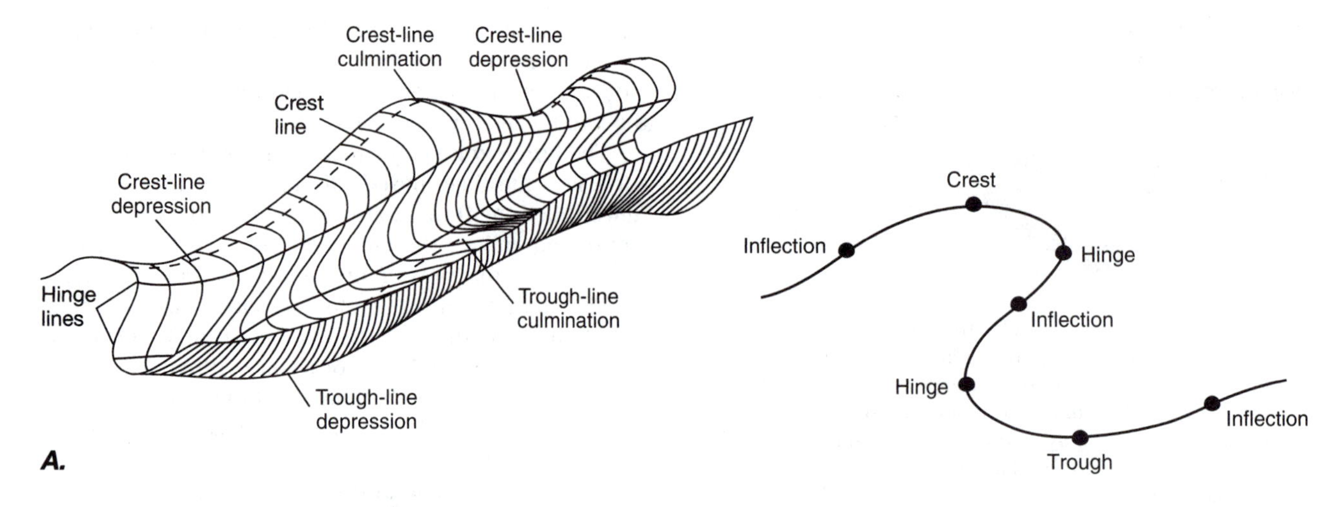

FIGURE 10.5 Crest and trough of a fold. *A*. Three-dimensional view of fold. *B*. Cross section normal to the hinge.

A **cylindrical fold** is one for which a line of constant orientation, called the **fold axis**, can be moved along the folded surface without losing contact with it at any point (Figure 10.6*A*). Thus it is a line of fixed orientation that is parallel to every orientation of the folded surface. Folds that do not possess this property are called **noncylindrical** folds. A **conical fold** is one whose surface is everywhere at a constant nonzero angle to a line of fixed orientation, which is also called the fold axis (Figure 10.7*A*). A fold axis thus is a theoretical geometric property of certain kinds of folds. Folds that are neither cylindrical nor conical in geometry do not, strictly speaking, possess a fold axis, but in general they do have a hinge.

Although natural folds are never geometrically perfect, many have approximately cylindrical geometry, at least locally, and thus they can be described by the orientation of an approximate fold axis. Even irregular folds generally can be divided into local segments, each of which is approximately cylindrical so the fold axis can be defined locally. The irregularity of the fold can be described by the variation in the fold axis orientation from place to place.

At the hinge of a cylindrical fold, the fold axis coincides with the hinge line, and for this reason the two terms are commonly used interchangeably. It is useful to maintain the distinction, however, because the *hinge line* is a physical linear feature defined by the line of maximum curvature on the folded surface. It thus has a specific orientation at a *specific location* on the folded surface. The fold axis, however, is a geometrical construct,

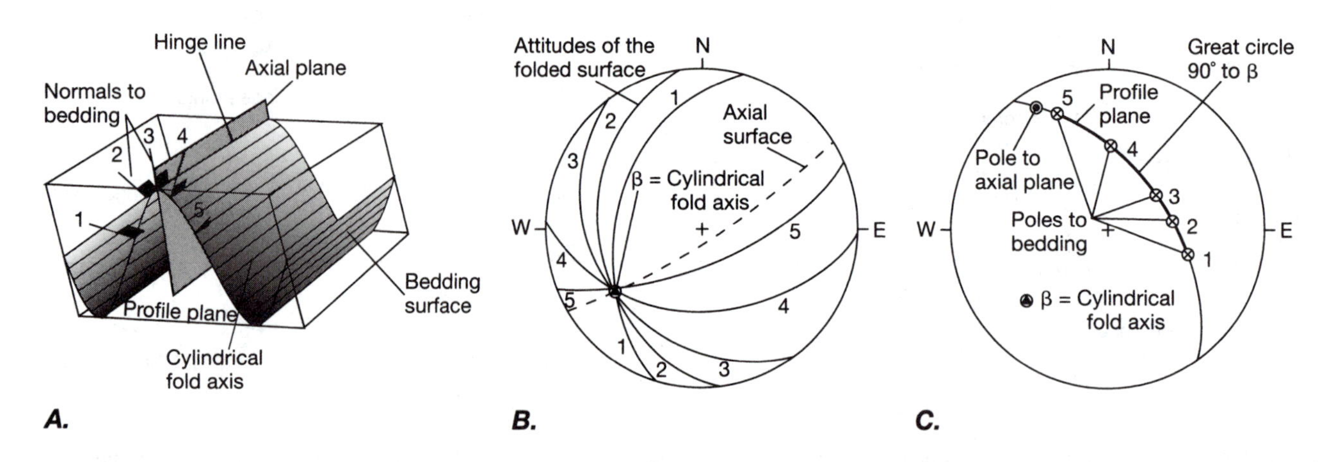

FIGURE 10.6 Geometry of a cylindrical fold in three dimensions and on a spherical projection. *A*. Diagram of a cylindrically folded surface showing the fold axis (parallel lines on the folded surface), the profile plane, and the perpendiculars to the folded surface, which are parallel to the profile plane. *B*. Schmidt net (equal area) plot of several orientations of the cylindrically folded surface in *A*, all of which intersect at the fold axis β. *C*. Schmidt net (equal area) plot of the poles to the surface orientations plotted in *B*. All the poles must lie along a great circle perpendicular to fold axis β. The great circle also defines the orientation of the profile plane.

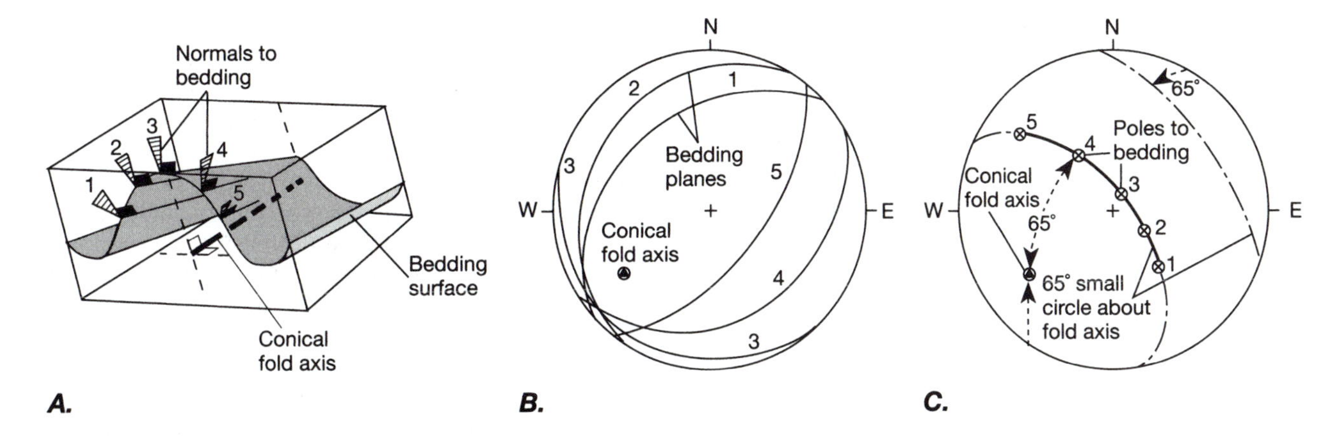

FIGURE 10.7 Geometry of a conical fold in three dimensions and on a spherical projection. *A*. Diagram of a conically folded surface. The fold axis is a line of constant orientation that is at a constant nonzero angle to the folded surface, in this case 25°. The front face of the block is perpendicular to the fold axis, not the hinge line, and therefore is not the profile plane. The perpendiculars to the bedding surface (1 through 5) are in this case all 25° from the plane normal to the fold axis and 65° from the fold axis itself. *B*. Schmidt net (equal area) plot of various attitudes of a conically folded surface. Note the great circles do not all meet at a single point. *C*. Schmidt net plot of the poles to the folded surface and the fold axis shown in *A*. The poles lie on a small circle around the fold axis.

most commonly defined with a stereographic projection, which is the orientation of line that is common to the orientations of all parts of a folded surface. It is therefore a line having no specific location on the fold but only a *specific orientation* that characterizes the geometry of the entire fold, at least locally. Moreover, noncylindrical folds have a hinge line but (with the exception of conical folds) no fold axis.

The geometry of a cylindrical fold and its fold axis has a particularly simple representation on a stereographic projection (see Appendix A1.2), which can be extremely useful in the analysis of folding in a region. A line of constant orientation plots as a point on a stereographic projection. A plane plots as a great circle. If the line lies in the plane, then on the projection, the point must lie somewhere on the great circle. The fold axis is by definition a common orientation to all attitudes of a cylindrically folded surface. Thus on a stereographic projection (Figure 10.6*B*) the point β representing the attitude of the fold axis must lie on each of the great circles representing attitudes of the folded surface. These great circles must all intersect at the fold axis orientation β (Figures 10.6*A*, *B*). Moreover, any line perpendicular to the folded surface must also be perpendicular to the fold axis (Figure 10.6*A*). On a stereographic projection, all lines perpendicular to a reference line must lie along the great circle normal to it. Thus the locus of lines plotted normal to the folded surface (called the poles to the surface) must lie along a great circle, and the normal to this great circle is therefore the fold axis (Figure 10.6*C*). This great circle also defines the orientation of the profile plane (Figure 10.6*A*).

These geometrical relations, and the fact that many folds are at least locally almost cylindrical, enable a field geologist to deduce the orientation of a fold axis from measurements of two or (preferably) more different attitudes of a folded surface. Plotting these attitudes as either great circles (Figure 10.6*B*) or their poles (Figure 10.6*C*) makes it possible to determine the orientation of the fold axis. This technique is especially useful in areas where, owing to scale or limited exposure, the hinge line is not directly observable.

For a conical fold (Figure 10.7*A*), the great circles representing orientations of the folded surface do not intersect in a point, and they do not possess an easily recognized relationship to the fold axis (Figure 10.7*B*). The poles to the folded surface, however, must lie along a small circle at a constant angle from the fold axis (Figure 10.11*C*).

ii. Parts of Folded Layers and Multilayers

The geometry of a folded layer or a stack of folded layers is equivalent to that of a nested set of two or more folded surfaces. A single multilayer fold is delimited by two **inflection surfaces** that join the inflection lines on adjacent folded surfaces in the nested stack (Figure 10.8).

The surface joining all hinge lines in a particular nested set of folds is variously called the **hinge surface**, the **axial surface**, or (if the surface is planar) the **axial plane**. In field studies, we usually recognize folds by the pattern of the folded layers on the surface of exposure, an outcrop or topographic surface. The intersection of the axial surface with a surface of exposure is a linear feature called the **axial surface trace**, which generally is very different from both the hinge and the fold axis and must not be confused with either (Figure 10.9). The orientation of the axial surface can be constructed on a

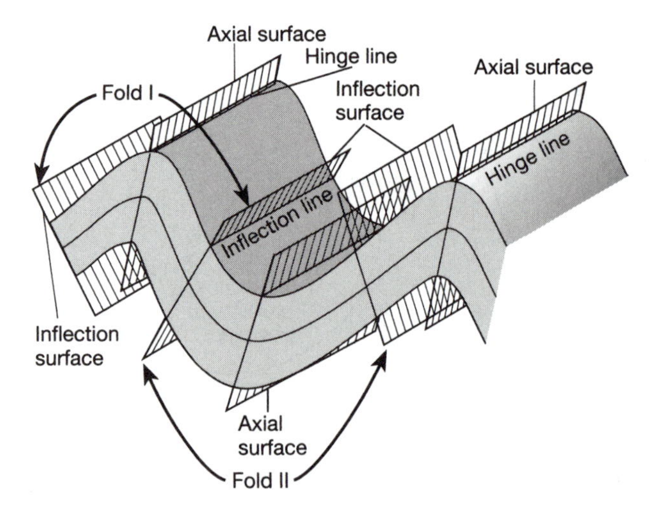

FIGURE 10.8 Folds in multilayers. A train of folds, showing the inflection surfaces, each of which contains the inflection lines of all the folded surfaces on one limb of a nested set of folds, and the hinge or axial surfaces, each of which contains all the hinge lines in a single nested set of folds.

stereographic projection from the measurement of two axial surface traces on two differently oriented exposure surfaces, since two intersecting lines always define a plane. The axial surface trace corresponds to the hinge line only if the hinge line is parallel to the exposure surface. For cylindrical folds, the intersection of any plane parallel to the axial surface with any part of the folded surface is parallel to the fold axis.

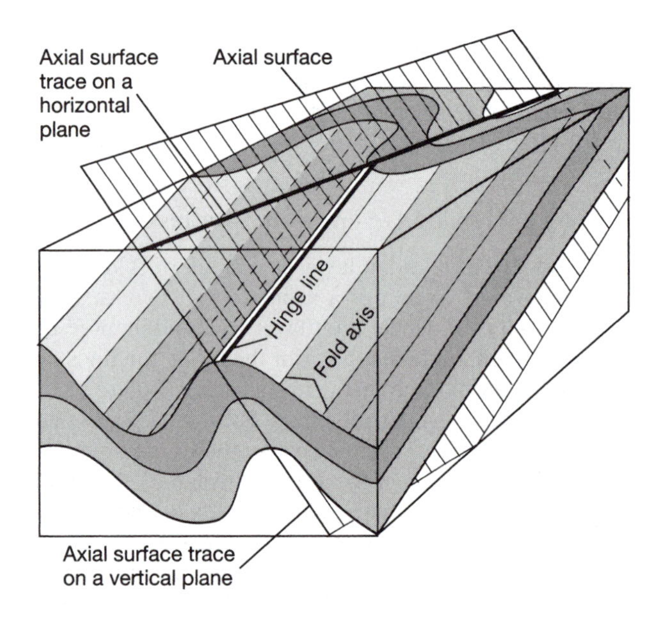

FIGURE 10.9 Block diagram of folds showing the distinction between the axial surface trace on a vertical and a horizontal surface, the hinge line, and the fold axis.

If the folded layers are sedimentary beds and if we can determine their relative ages, then we distinguish anticlines from synclines. **Anticlines**[1] are folds in which the older layers are on the concave side of a bedding surface, and the younger layers are on the convex side. **Synclines**[2] are folds in which the younger layers are on the concave side of a bedding surface, and the older layers are on the convex side (Figure 10.10*A*). Thus these terms are only used if the relative ages of the folded layers are known. If they are not known, then we describe a convex-up fold as an antiform and a concave-up fold as a synform (see Section 10.1-i).

Most anticlines are convex up (antiformal), and most synclines are concave up (synformal) (Figure 10.10*A*), although this geometry is not universal. In areas of complex deformation where the entire stratigraphy has been overturned, anticlines actually may be synformal and synclines may be antiformal (Figure 10.10*B*).

10.2 FOLD SCALE AND ATTITUDE

i. The Scale of Folds

Scale is a measure of the size of a fold in a layer or stack of layers. There are two components of the scale: the amplitude A and the wavelength λ (Figure 10.11). We define them with reference to the enveloping surfaces and the median surface (Figure 10.11). The **enveloping surfaces** are the two surfaces that bound the fold train developed in a single folded surface. The **median surface** includes all the inflection lines of a fold train in a single surface. The **amplitude** of any fold is the distance from the median surface to either of the enveloping surfaces measured parallel to the axial surface. The **wavelength** is the distance measured parallel to the median surface, between one point on a fold and the geometrically similar point on a neighboring fold, for example, from one antiformal hinge to the next or from one synformal hinge to the next.

ii. The Attitude of Folds

The orientation in three dimensions of a fold or train of folds is an important factor in any geologic study of folded rocks. Accordingly, an extensive nomenclature has been based on the attitude of folds. We express the attitude of a fold by the trend and plunge (see Appendix A1.1) of the hinge line or fold axis and the strike and dip of the axial surface. A fold is **upright** if the dip of the axial surface is close to vertical; it is **steeply, moderately**, or **gently** inclined as the dip angle progressively de-

[1]From the Greek words *anti* meaning "against," and *klinein*, meaning "to slope."

[2]The Greek word *syn* means "with, to."

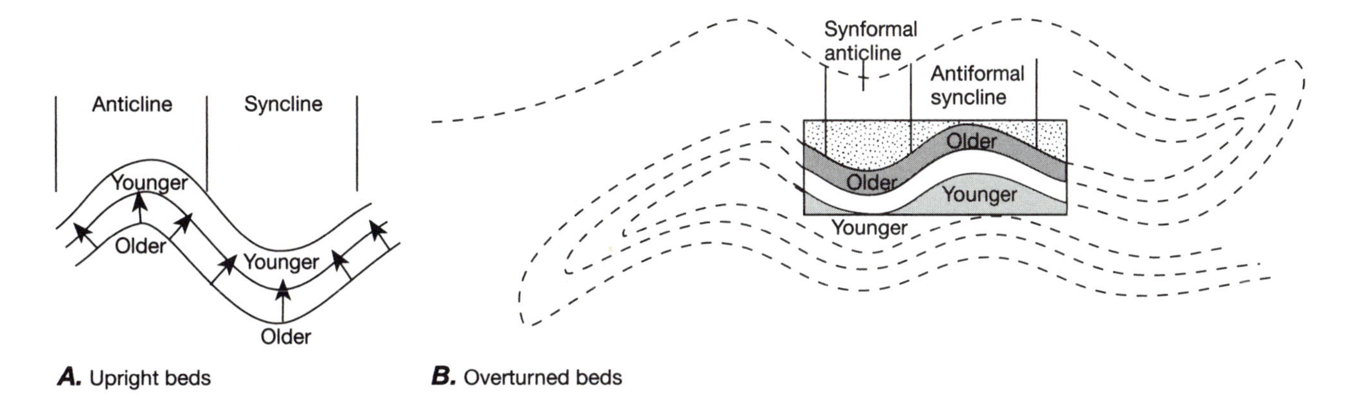

FIGURE 10.10 Distinction between an anticline and a syncline in a fold cross section. Arrows point from oldest to youngest beds, that is, in the stratigraphic-up direction. Dashed structure in *B* suggests one way in which large sections of strata could be overturned.

creases; and it is **recumbent** if the axial surface is close to horizontal. Depending on the plunge of the hinge, a fold is **horizontal, subhorizontal; gently, moderately,** or **steeply plunging**; and **subvertical** or **vertical**. A **reclined** fold is one whose hinge plunges down the dip of the axial surface. Thus a fold could be upright horizontal; upright moderately plunging; recumbent; and so forth. In Figure 10.12, the central triangular diagram displays graphically the conventional definitions of these terms, and the surrounding diagrams give examples of the various categories.

No folds are indefinite in length; all eventually die out along the hinge by decreasing in amplitude or terminating against a fault. Where upright or inclined horizontal folds die out, the hinge line must plunge. If a fold hinge plunges at both ends and the hinge line is at least a few times as long as the half-wavelength, it is a **doubly plunging** fold (see folds in the northwest part of Figure 10.2*A*). As the length of the hinge becomes comparable to the half-wavelength of the fold, the fold is called a **dome** or a **basin**, depending on whether it is antiformal or synformal, respectively.

Several other common terms specify relative orientations of the limbs of folds. A **homocline**[3] comprises a surface, such as bedding, that has a uniform nonhorizontal attitude over a regional scale with no major fold hinges (Figure 10.13*A*). A **monocline**[4] is a fold pair that has two long horizontal limbs connected by a relatively short inclined limb (Figure 10.13*B*). A **structural terrace** is a fold pair with two long planar inclined limbs connected by a relatively short horizontal limb (Figure 10.13*C*). An inclined or recumbent fold in which one limb is overturned—that is, rotated more than 90° from its original horizontal position (Figure 10.13*D*)—is sometimes called

[3]From the Greek *homo* meaning "same" and *klinein*, meaning "to slope."

[4]From the Greek words *mono* meaning "single, only" and *klinein*, which means "to slope."

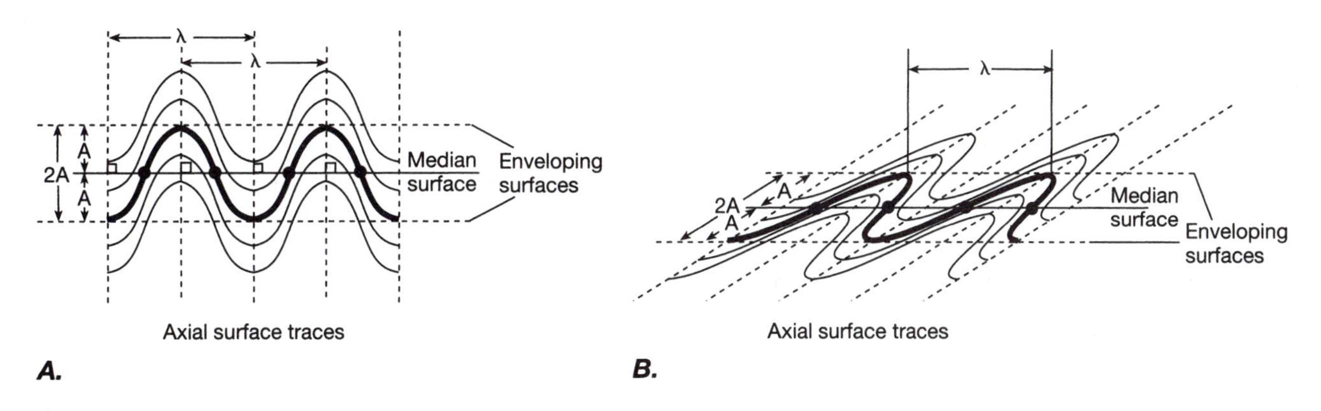

FIGURE 10.11 The scale of folding is defined by the wavelength (λ) and the amplitude (*A*) for *A*. Symmetric folds and *B*. Asymmetric folds.

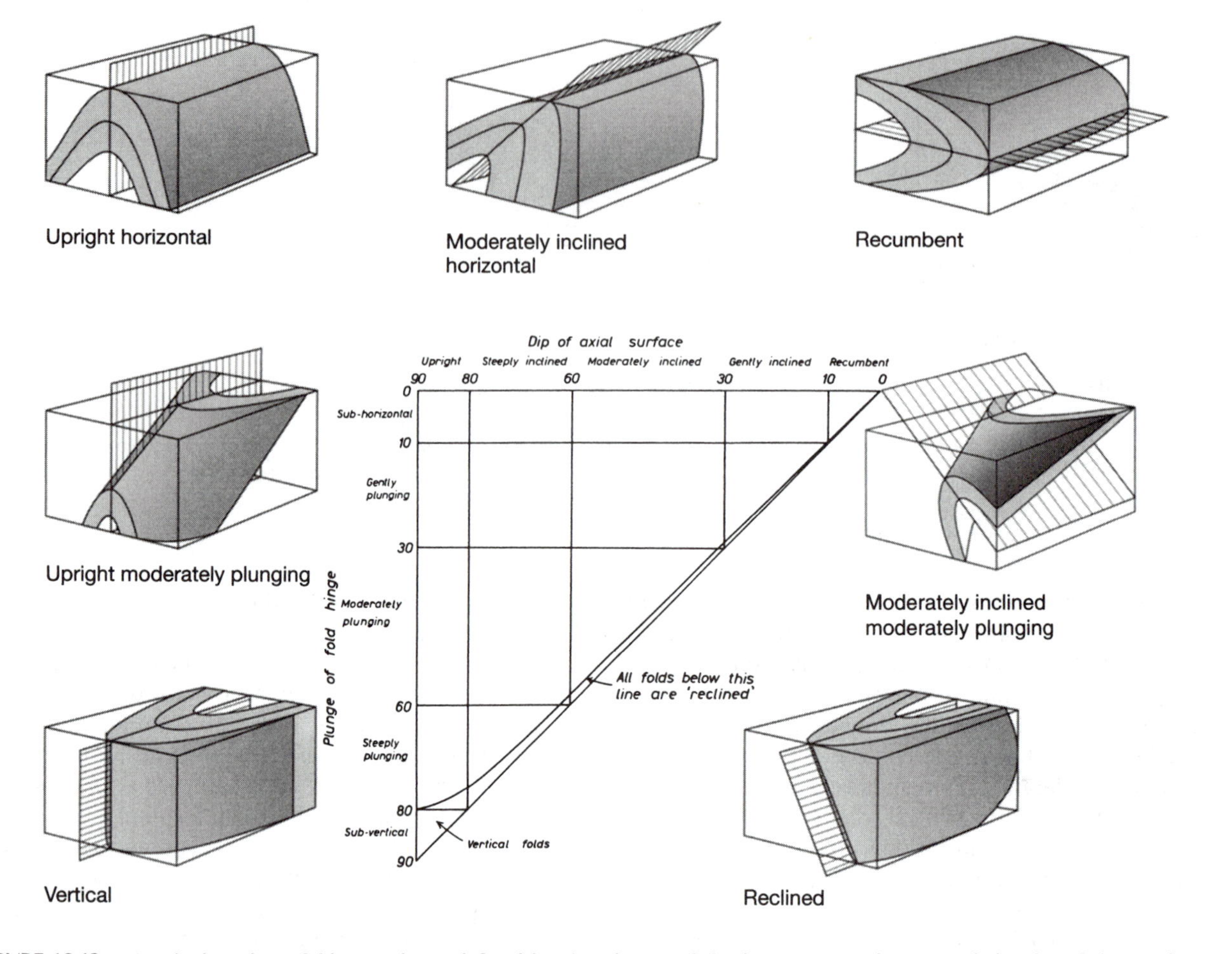

FIGURE 10.12 Graph describing fold attitude as defined by the plunge of the hinge (vertical axis) and the dip of the axial surface (horizontal axis). The graph shows the ranges of angles associated with each term. The surrounding diagrams show folds in varying attitudes, corresponding to the categories in the graph. (After Fleuty 1964)

an **overturned fold**. Note that the term *overturned* refers only to one limb of the fold, not to the whole fold. Thus an overturned anticline (Figure 10.13*D*) is not the same as an upside down, or synformal, anticline (Figure 10.10*C*).

10.3 THE ELEMENTS OF FOLD STYLE

The **style** of a fold is the set of features that describe its form. It is analogous, for example, to the architectural style of a building. Over years of working with folds, certain features have emerged as being particularly useful in describing folds and understanding how they develop. We refer to these features as the "elements of fold style," which we summarize in Table 10.1. In this section, we briefly define and discuss these elements. In Section 10.5, we apply these definitions to describe the most common fold styles that appear in deformed rocks. These are the main characteristics that we return to in Chapters 13 and 18 in our attempt to understand the significance of differences in fold geometry for the conditions of deformation within the Earth.

We must first define two angles that describe the amount by which a surface has been folded (Figure 10.14). The **folding angle** ϕ is the angle between the normals to the folded surface constructed at the two inflection points of a fold. It is the angle through which one limb has rotated, relative to the other. The more commonly used **interlimb angle** $\imath$ is the angle between the tangents to the two fold limbs constructed at the inflection points. It measures the dihedral angle between the two limbs and is the supplement of the folding angle (that is, $\imath = 180° - \phi$).

i. Style of a Folded Surface

1. Cylindricity The degree to which a fold approximates the geometry of a cylindrical fold (Section 10.1) is a feature that characterizes different styles of folding. The cylindricity is represented qualitatively on a stereo-

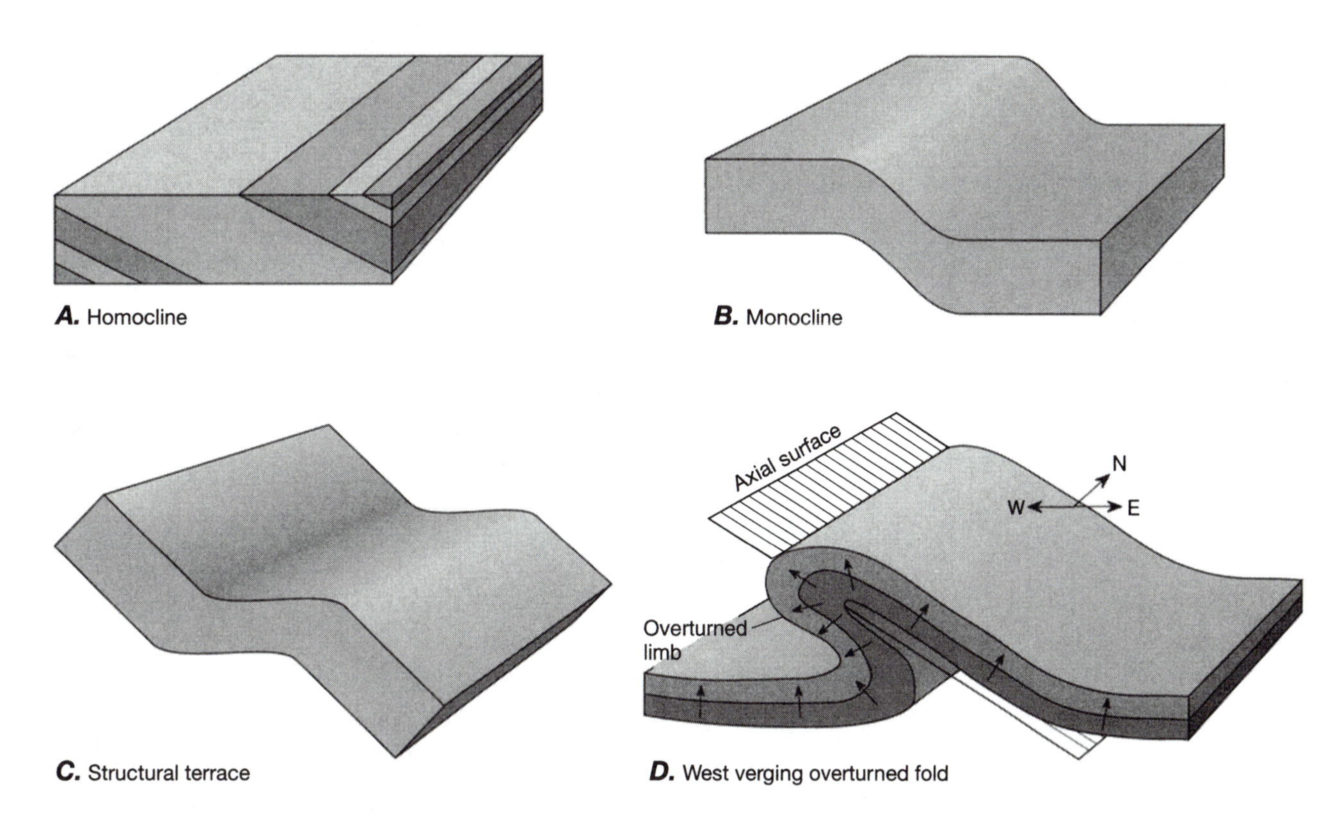

FIGURE 10.13 Structural terms describing the orientation of fold limbs. In *D*, the arrows show the stratigraphic-up direction on the sedimentary beds.

graphic projection by how closely the poles to planes around a fold fit a great circle distribution (Figure 10.6*C*). The distance along the hinge for which the cylindrical geometry is maintained, measured as a proportion of the half-wavelength, is also a significant characteristic of the fold style. A multilayer fold can be described as cylindrical if the attitudes from all surfaces in the multilayer conform to the geometry of a cylindrical fold (Figure 10.6*C*). The term **cylindroidal** is sometimes used to describe a fold that closely approximates an ideal cylindrical geometry.

TABLE 10.1 Elements of Fold Style

Style of a Folded Surface
1. Cylindricity 2. Symmetry 3. Aspect ratio 4. Tightness 5. Bluntness
Style of a Folded Layer (Ramsay's Classification)
1. Relative curvature: dip isogon pattern 2. Orthogonal thickness 3. Axial trace thickness
Style of a Folded Multilayer
1. Harmony 2. Axial surface geometry

2. Symmetry A folded surface forms a **symmetric fold** if, in profile, the shape on one side of the hinge is a mirror image of the shape on the other side and if adjacent limbs are identical in length (Figures 10.14*A* and 10.11*A*). For symmetric folds in single layers and multilayers, the mirror plane of symmetry is the axial plane. It is also the perpendicular bisector of the median surface between the inflection points, and it bisects the folding angle ϕ and the interlimb angle $\imath$.

Asymmetric folds in profile have no mirror plane of symmetry, and the limbs are of unequal length (Figures10.11*B* and 10.14*B*, *C*). The sense of asymmetry of a fold changes depending on whether we view the fold from one direction along the hinge or from the other. By convention, we specify the sense of asymmetry on a plunging fold when looking *down the plunge* of the hinge line. An asymmetric fold is a clockwise fold or a **z-fold** if the short limb has rotated clockwise with respect to the two long limbs, and the short limb with its two adjacent long limbs defines a Z shape (Figure 10.14*B*). An asymmetric fold is a **counterclockwise fold** or an **s-fold** if the short limb has rotated counterclockwise with

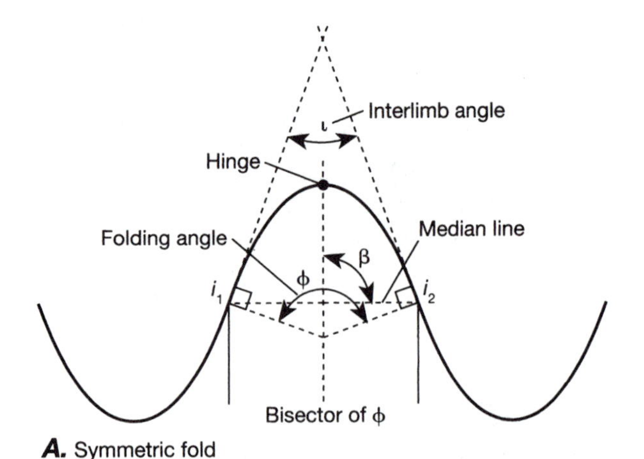

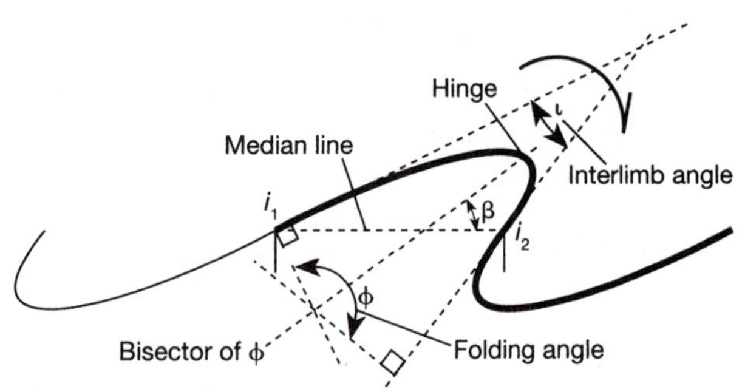

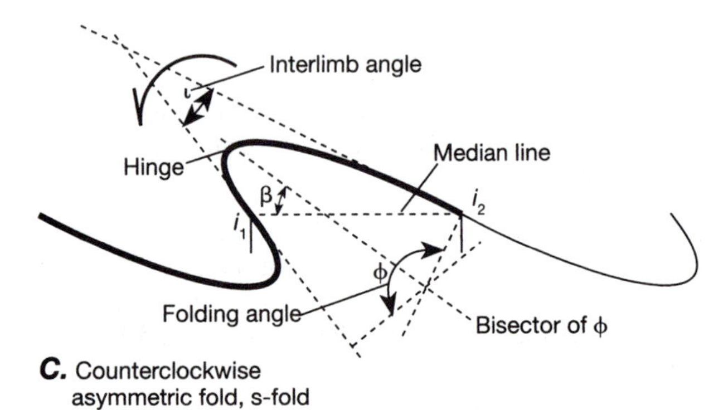

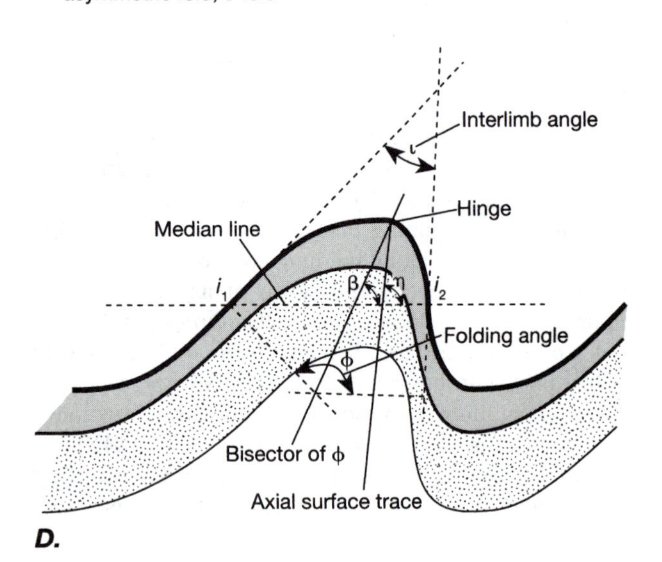

FIGURE 10.14 The folding angle, the interlimb angle, and the symmetry of folds. The folding angle ϕ is the angle between the normals to the folded surface constructed at the inflection points. It is the angle through which one limb has been rotated relative to the other by the folding. The interlimb angle is the angle between the tangents to the folded surface constructed at the inflection points. It is the supplement to the folding angle. *A*. For a symmetric fold, the fold profile from the hinge to i_l is the mirror image of the profile from the hinge to i_2. The mirror plane is the bisector of both the folding angle and the interlimb angle, and it is the perpendicular bisector of the median line $i_l i_2$. β equals 90°. *B* and *C*. Asymmetric folds for which there is no mirror plane of symmetry, the bisector of the interlimb and the folding angles does not bisect the median line, and $\beta \neq 90°$. *D*. An asymmetric multilayer fold is characterized by the inclination β of the folding angle bisector as well as by the inclination η of the axial surface with respect to the median surface. These two angles in general are not the same.

respect to the two long limbs and the short limb with its two adjacent long limbs defines an S shape (Figure 10.14*C*).

If the fold hinge is horizontal, the geographic direction of viewing must be part of the description of the asymmetry; for example, the fold is counterclockwise (or an s-fold) looking north. For an inclined fold with a horizontal to gently plunging hinge, however, the sense of asymmetry is often given by the **vergence**.[5] The vergence of an asymmetrical fold is the direction of "leaning" either of the bisector of the folding angle for a single folded surface or of the axial surface for a fold comprising one or more layers, that is, a nested stack of folded surfaces. The direction of vergence is the up-dip direction on either of these surfaces. For a multilayer fold, the axial surface is not in general the same as the folding angle bisector (Figure 10.14*D*). Thus in Figure 10.13*D*, for example, the vergence of the fold is to the west, or the fold verges to the west or is west-vergent.

Small symmetric folds, especially if they are within the core of a larger fold, are sometimes called **m-folds**.

3. Aspect ratio The **aspect ratio** *P* is the ratio of the amplitude *A* of a fold measured along the axial surface, to the distance *M* measured between the adjacent inflection points that bound the fold (Figure 10.15). For a periodic fold train, in which successive folds have the same wavelength λ (Figure 10.11), *M* is the half-wavelength ($\lambda/2$). Folds of increasing aspect ratio have a wide, broad, equant, tall, or elongate aspect, as defined in Table 10.2.

4. Tightness The **tightness** of folding is defined by the folding angle ϕ or the interlimb angle $\imath$ (Figure 10.16). As the degree of folding increases, the folding angle in-

[5]From the German word *vergenz*, meaning "overturn."

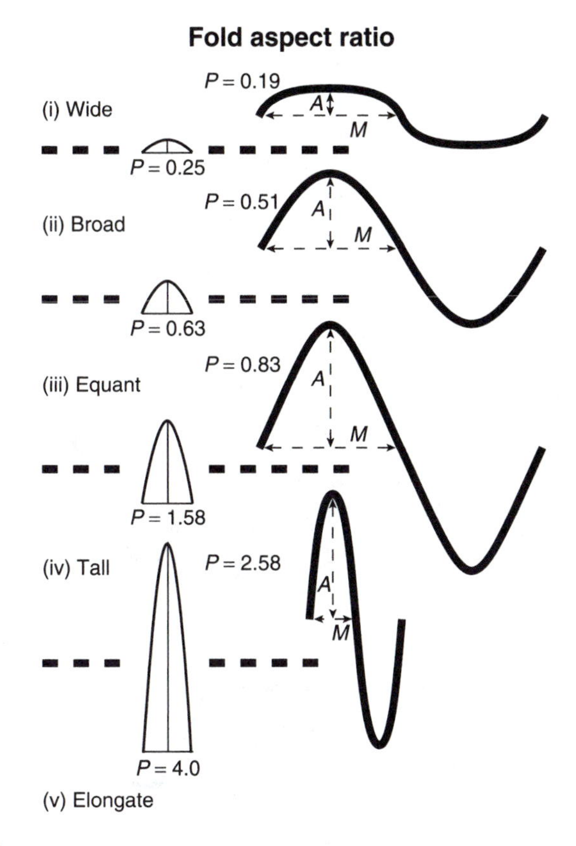

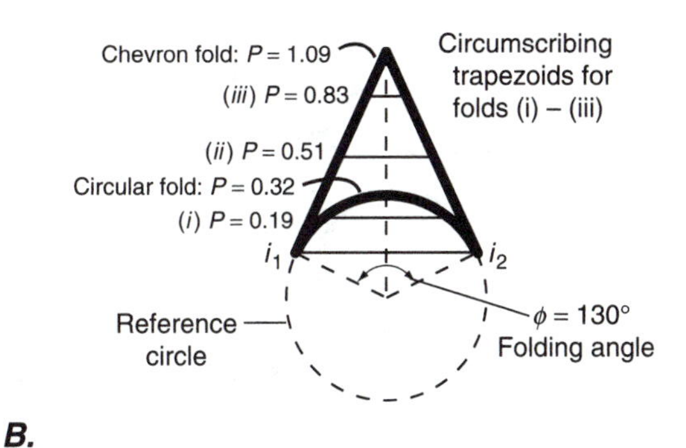

FIGURE 10.15 *A*. Examples of symmetric folds with various aspect ratios $P = A/M$ (see Table 10.2). Folds i–iii have the same folding angle $\phi = 130°$ (see *B* and Table 10.3) and the same half-wavelength *M*. For fold iv, the half-wavelength is one third of that for the other folds, and the folding angle is $\phi = 170°$. Fold forms along the heavy dashed lines show the aspect ratios at the boundaries between the named fields. An "elongate" fold form is not illustrated. *B*. The trapezoids that circumscribe folds i–iii in *A* and the associated reference circle tangent to the limbs at the inflection points i_1 and i_2. Single-hinged folds can only have aspect ratios between chevron folds (with planar limbs, with $P = 1.09$ for this folding angle) and circular folds (circular limbs and hinge, with $P = 0.32$ for this folding angle). Folds with aspect ratios less than the circular fold have a larger radius of curvature at the closure than the reference circle, and thus are double-hinged (see fold i in *A*).

TABLE 10.2 Aspect Ratio

Descriptive Term	Aspect Ratio P: $P = A / M$	Aspect Ratio P: Log P
Wide	$0 \le P < 0.25$	$-\infty \le \log P < -0.6$
Broad	$0.25 \le P < 0.63$	$-0.6 \le \log P < -0.2$
Equant	$0.63 \le P < 1.58$	$-0.2 \le \log P < 0.2$
Tall	$1.58 \le P < 4$	$0.2 \le \log P < 0.6$
Elongate	$4 \le P < \infty$	$0.6 \le \log P < \infty$

Source: Modified after Twiss 1988.

creases and the interlimb angle decreases. Folds are described as *gentle*, *open*, *close*, *tight*, *isoclinal*, *fan*, or *involute*, as defined in Table 10.3. Isoclinal folds fall on the boundary between acute folds ($\phi/2 < 90°$) and obtuse folds ($\phi/2 > 90°$).

5. Bluntness The **bluntness** b measures the radius of curvature[6] of the fold at its closure relative to that of a reference circle tangent to the limbs of the fold at the inflection points (Figure 10.17; cf. Figure 10.15*B*). It is defined by:

$$b = \begin{Bmatrix} \dfrac{r_c}{r_0} & \mathit{for}\ r_c \le r_0 \\ 2 - \left(\dfrac{r_0}{r_c}\right) & \mathit{for}\ r_c \ge r_0 \end{Bmatrix} \tag{10.1}$$

where r_c is the radius of curvature at the fold closure, represented by the radius of the circle that just fits the shape of the closure, and r_0 is the radius of the reference circle. Folds are described as *sharp*, *angular*, *subangular*, *subrounded*, *rounded*, or *blunt* (Table 10.4; Figure 10.17). A bluntness of $b = 0$ describes folds that have perfectly sharp hinges ($r_c = 0$); $b = 1$ describes perfectly circular folds, which, for both acute and obtuse folds, consist of a single circular arc; $b = 2$ describes a double-hinged fold with a flat closure ($r_c = \infty$). Thus all folds must have a bluntness between 0 and 2. For double-hinged folds, a complete description requires the bluntness of the hinges to be described in addition to the bluntness of the closure.

[6]For a curve defined as $y = f(x)$, the radius of curvature R is:

$$R = \frac{\left[1 + \left(\dfrac{dy}{dx}\right)^2\right]^{3/2}}{\left|\dfrac{d^2y}{dx^2}\right|}$$

The curvature is defined as $\kappa = 1/R$. For a symmetric fold, with the y axis parallel to the axial surface and perpendicular to the hinge, and with the x axis perpendicular to the axial surface and parallel to the median line, the slope at the hinge is $(dy/dx) = 0$, so the radius of curvature and the curvature at the hinge are, respectively,

$$R = 1/|d^2y/dx^2| \quad \kappa = |d^2y/dx^2|$$

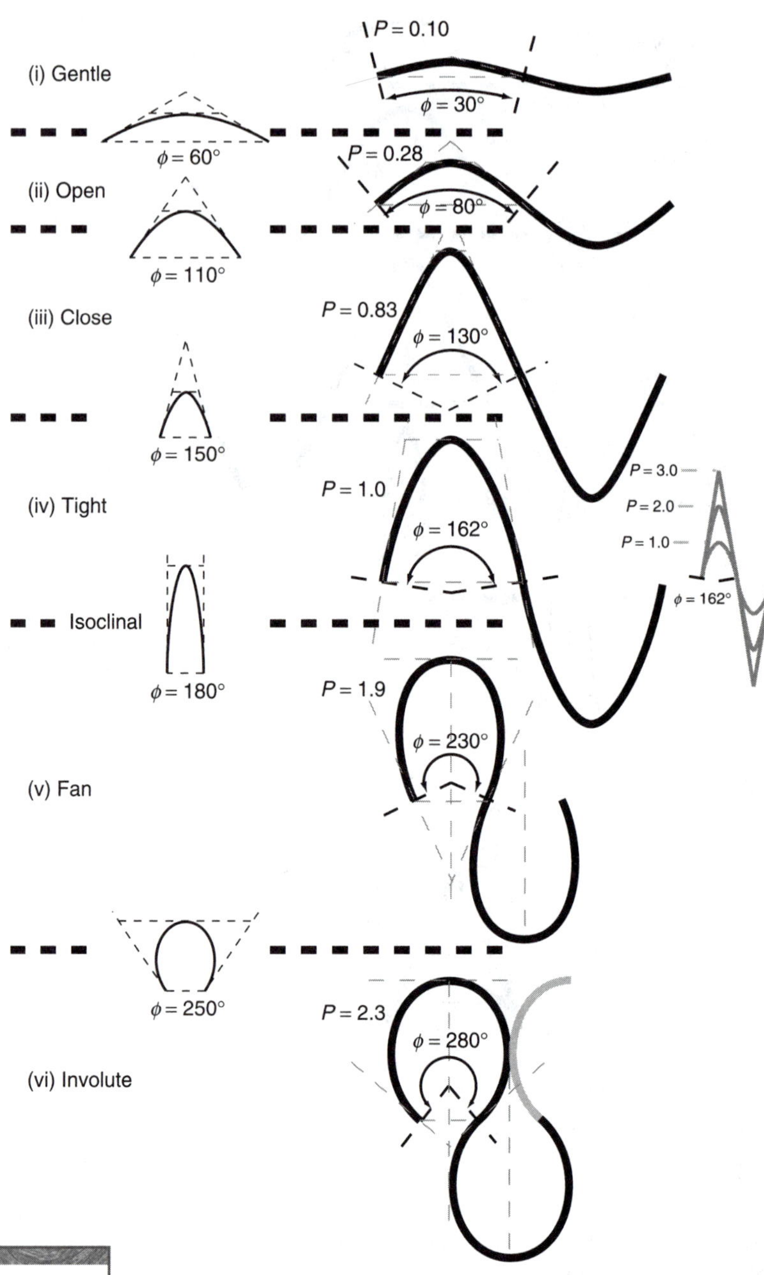

FIGURE 10.16 Examples of symmetric folds with various degrees of tightness, as defined by the folding angle ϕ in Table 10.3. Folds i–iv have the same half-wavelength M; for folds v and vi, the half-wavelength is half and a third, respectively, of that length. Fold forms along the heavy dashed lines show the folding angles at the boundaries between the named fields. The aspect ratio P is constrained to some extent by the folding angle, but it is not unique, as illustrated by the small gray fold shapes on the right side of iv, which all have the same folding angle. The gray fold with the smallest P is identical to the main fold in iv except for scale. In a fold train, the limbs of an obtuse fold cannot extend beyond the plane of symmetry of the adjacent fold, as illustrated by fold vi.

TABLE 10.3 Tightness of Folding

Descriptive Term	Folding Angle ϕ°	Interlimb Angle $\imath^\circ$
Acute		
Gentle	$0 < \phi < 60$	$180 > \imath > 120$
Open	$60 \leq \phi < 110$	$120 \geq \imath > 70$
Close	$110 \leq \phi < 150$	$70 \geq \imath > 30$
Tight	$150 \leq \phi < 180$	$30 \geq \imath > 0$
Isoclinal	$\phi = 180$	$\imath = 0$
Obtuse		
Fan	$180 < \phi \leq 250$	$0 > \imath \geq -70$
Involute	$250 < \phi \leq 360$	$-70 > \imath \geq -180$

Source: Modified after Fleuty 1964.

ii. The Style of a Folded Layer: Ramsay's Classification[7]

The style of a folded layer is determined by comparing the fold styles of the two surfaces of the layer. The comparison is conveniently made using three geometric parameters, which are defined relative to a given pair of parallel lines that are tangent, respectively, to the inner (concave) and outer (convex) surfaces of the layer on

[7]After J. G. Ramsay (1967), who first outlined this classification.

TABLE 10.4 Bluntness of Folds

Descriptive Term	Bluntness
Sharp	$0.0 \leq b < 0.1$
Angular	$0.1 \leq b < 0.2$
Subangular	$0.2 \leq b < 0.4$
Subrounded	$0.4 \leq b < 0.8$
Rounded	$0.8 \leq b \leq 1$
Blunt	$1 < b \leq 2$

Source: After Twiss 1988.

the fold profile (Figure 10.18). The inclination of the fold surface at the point of tangency is given by α, the angle between the tangent line and the line normal to the axial surface trace. The three geometric parameters are as follows: (1) The **dip isogon**, which is the line across the layer connecting two points of equal dip on opposite surfaces of the layer; (2) the orthogonal thickness t_α, which is the perpendicular distance between the two parallel tangents; and (3) the axial trace thickness T_α, which is the distance between the two tangents measured parallel to the axial surface trace. The two measures of layer thickness t_α and T_α are related by $t_\alpha = T_\alpha \cos \alpha$.

The elements of style for a folded layer are defined according to how these geometric parameters vary

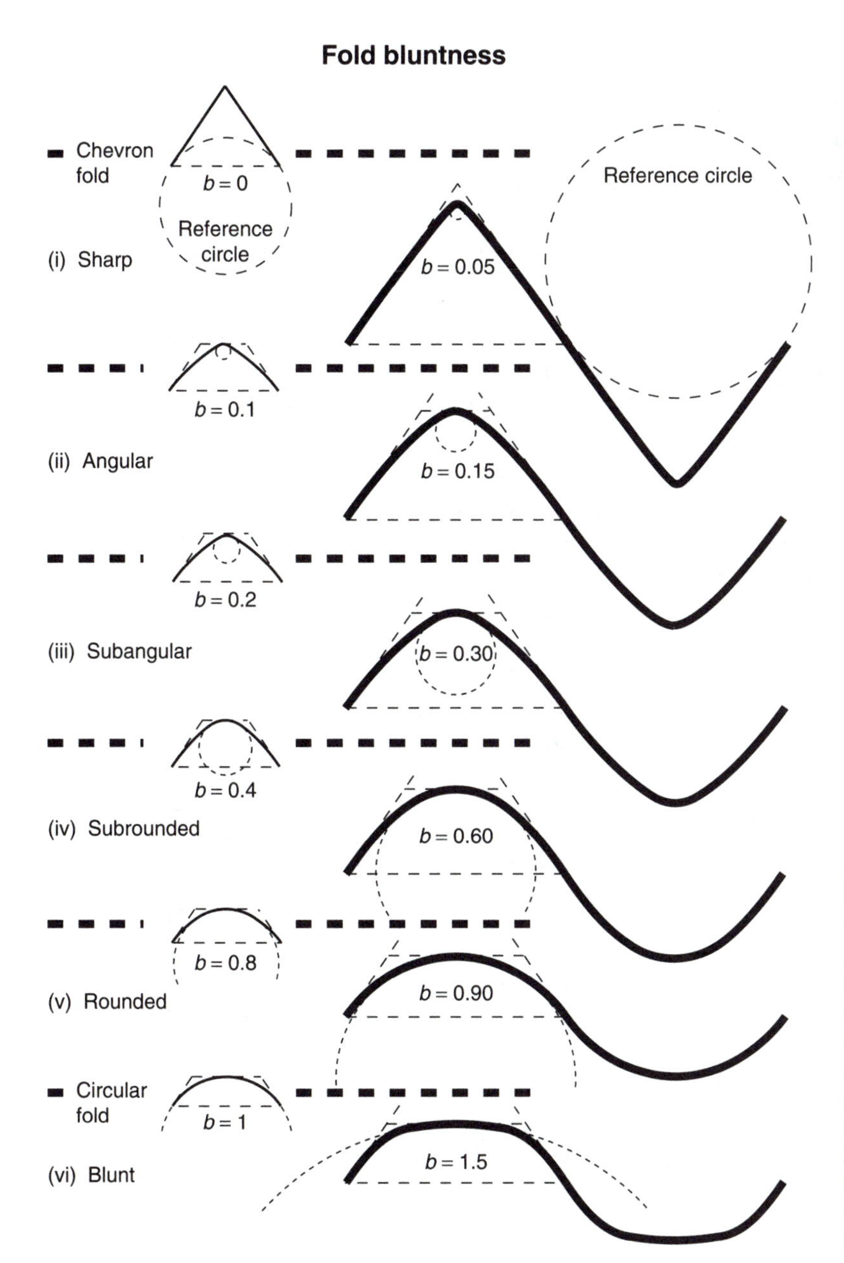

FIGURE 10.17 Examples of symmetric folds with various values of bluntness b as defined in Table 10.4 and by Equation (10.1). Fold forms along the heavy dashed lines show the bluntness at the boundaries between the named fields. Folds have a constant tightness (folding angle $\phi = 110°$), and folds in each column have a constant half-wavelength M. Aspect ratio P and folding angle ϕ are not unique, but P is constrained by that for chevron folds with planar limbs and by circular folds ($b = 1$) with circular limbs and hinge. All folds with $b \geq 1$ are necessarily double-hinged folds, as shown for fold vi.

across the fold from the hinge to the limbs, or with increasing values of the surface inclination α.

1. The relative curvature or dip isogon variation The relative curvature of the convex and concave surfaces is revealed by constructing a set of dip isogons at regular intervals from the hinge to the limb, each of which connects points of identical inclination on the inner (concave) and outer (convex) surfaces (Figure 10.19). If the dip isogons converge toward the inner side of the fold, the curvature of the inner surface is greater than that of the outer surface; if they diverge toward the inner surface, the opposite is true.

For most folded layers, the relative curvature is consistent and defines three styles of folds. Dip isogons that converge toward the inner side of the fold characterize **class 1 folds** (Figure 10.19), which are subdivided into subclasses 1A, 1B, and 1C, as described in item 2, which follows. Parallel dip isogons, which also are parallel to the axial surface, characterize **class 2 folds** (Figure 10.19). Folds of this style are referred to as **similar folds** because adjacent fold surfaces are ideally identical (similar) in form. Dip isogons that diverge toward the concave side of the fold characterize **class 3 folds** (Figure 10.19). The relative curvature is most obvious in the hinge zone, which generally allows folds to be classified by visual inspection.

2. Variation in the orthogonal thickness The variation of the orthogonal thickness from hinge to limb is characteristic of different styles of folds and is the basis for the subdivision of class 1 folds (Figures 10.19 and 10.20*A*).

For **class 1A folds**, the orthogonal thickness increases from hinge to limb (Figures 10.19 and 10.20). For **class 1B folds**, the orthogonal thickness is constant from hinge to limb (Figures 10.19 and 10.20). These folds also

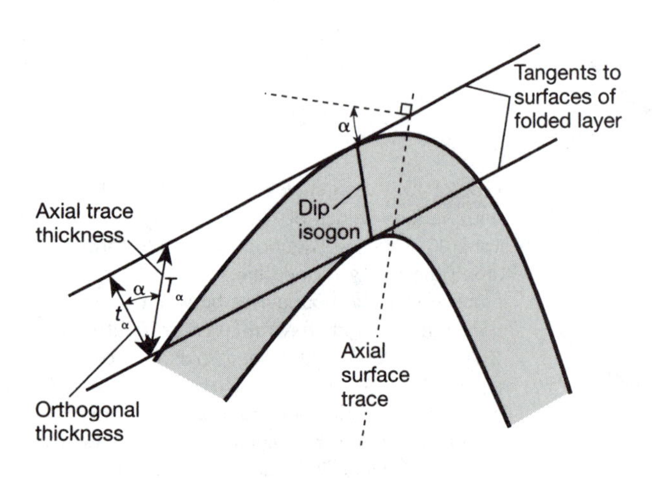

FIGURE 10.18 Definition of the layer inclination α, the dip isogon, the orthogonal thickness t_α, and the axial trace thickness T_α used to define the style of folded layers.

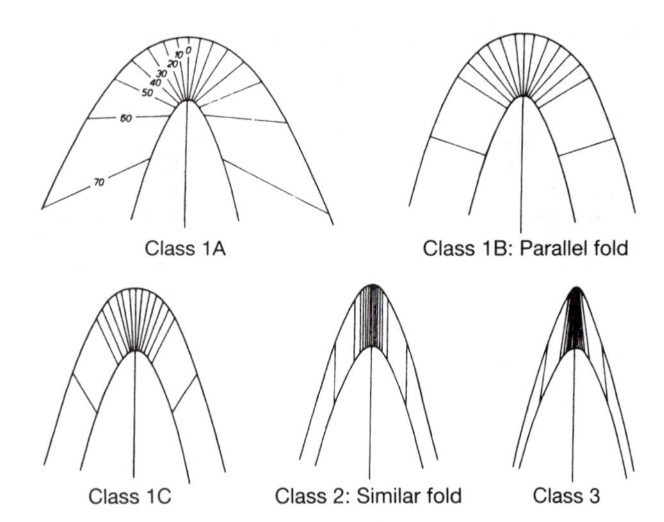

FIGURE 10.19 Ramsay's classification of folded layers (see Table 10.5). (After Ramsay 1967)

are commonly referred to as parallel folds because t_α is a constant all around the fold. **Concentric folds** are parallel folds whose inner and outer surfaces both have a bluntness of $b = 1$. Thus they are folds defined by two circular arcs having a common center. For **class 1C folds**, the orthogonal thickness decreases from hinge to limb (Figures 10.19 and 10.20*A*). The orthogonal thickness also decreases from hinge to limb for class 2 and class 3 folds (Figure 10.20*A*).

3. Variation in axial trace thickness The axial trace thickness also distinguishes the three classes of fold style. From hinge to limb, that is with increasing α, the axial trace thickness increases in class 1 folds, is constant in class 2 folds, and decreases in class 3 folds (Figure 10.20*B*).

Table 10.5 summarizes the characteristics of the different fold classes in terms of dip isogon geometry and the variations in orthogonal thickness and axial trace thickness. The characteristics of thickness shown in Figure 10.20 demonstrate that fold classes 1B and 2 are idealized geometries that form the boundaries between the other classes of folds. Thus some combinations of geometries in classes 1A and 1C closely approach class 1B style. Similarly, some combinations of geometries of classes 1C and 3 folds closely approach class 2 style. Not all possible folds are included in this classification, but most of the commonly observed geometries are included, and the characteristics of other styles can be presented on graphs such as in Figure 10.20.

iii. The Style of a Folded Multilayer

A multilayer fold comprises a stack of layers folded together. Its fold style can be defined in terms of the harmony of the folding and the axial surface geometry.

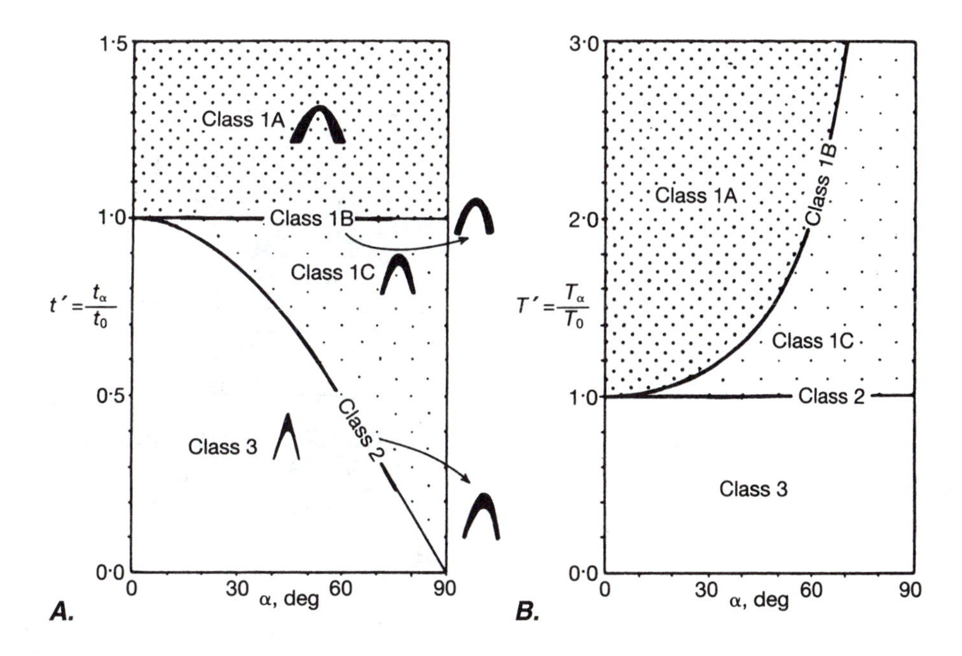

FIGURE 10.20 The classification of folded layers according to the thickness distribution in the layer with increasing α (that is, from hinge to limb). *A.* Fold classes distinguished by the normalized orthogonal thickness $t' = t_\alpha/t_o$, where t_o is the orthogonal thickness at the hinge where $\alpha = 0$. *B.* Fold classes distinguished by the normalized axial trace thickness $T' = T_\alpha/T_O$, where T_O is the axial trace thickness at the hinge where $\alpha = 0$.

1. Harmony of folding In profile, all multilayer folds must die out at both ends along the axial surface trace (Figure 10.21), unless the fold ends at a surface of discontinuity such as the Earth's surface. The depth of folding (D) is the distance along the axial surface trace over which the folding persists. The harmony H is a scale-independent measure of the rate at which the fold dies out along the axial surface trace and is equal to the ratio of the depth of folding (D) to the half-wavelength ($\lambda/2$):

$$H = 2D/\lambda \tag{10.2}$$

A **harmonic** fold is continuous along its axial trace for many multiples of the half-wavelength (Figure 10.21*A*). A **disharmonic** fold dies out within a couple of half-wavelengths or less (Figure 10.21*B*).

In general, because multilayer folds die out along the axial surface trace, dip isogons must form closed contours between two adjacent hinges (Figure 10.22*A*). As the fold amplitude increases, reaches a maximum, and then decreases along the axial surface trace, the dip isogons converge, are roughly parallel, and then diverge. Thus, in principle, all three of Ramsay's fold styles must occur in every multilayer fold. Although the strict definitions of the fold class nomenclature are more difficult to apply to multilayer fold classification, the dip isogon pattern still reveals important characteristics of the folds.

TABLE 10.5 Style of a Folded Layer

Class	Dip Isogon Geometry (progressing from convex to concave surface of layer)	Orthogonal Thickness (progressing from hinge to limb)	Axial Trace Thickness (progressing from hinge to limb)
1	Convergent		Increases
1A	Convergent	Increases	Increases
1B	Convergent	Constant	Increases
1C	Convergent	Decreases	Increases
2	Parallel	Decreases	Constant
3	Divergent	Decreases	Decreases

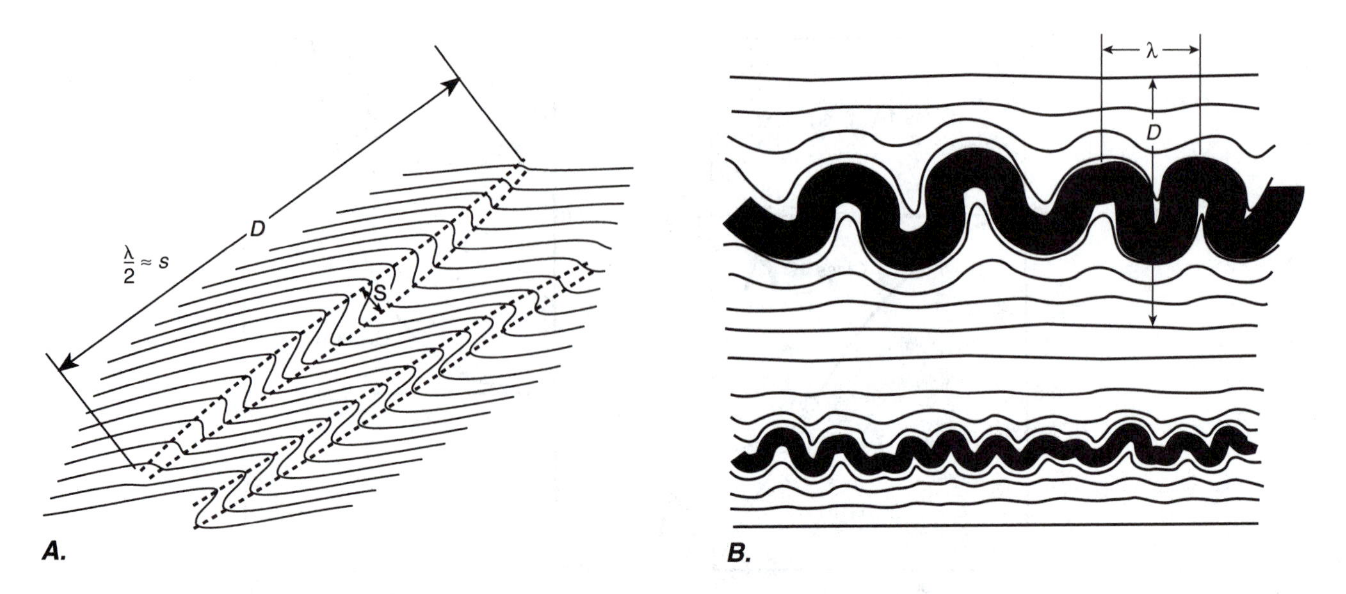

FIGURE 10.21 Harmony of folded multilayers. *A*. Harmonic folds affect layers for many times the half-wavelength along the axial surface trace in the profile plane. They have a large ratio H of the depth of folding D to half the wavelength λ ($H = 2D/\lambda$). The half-wavelength may conveniently be approximated by the spacing of adjacent axial surfaces S. This fold style is approximately a multilayer class 2 fold. *B*. Disharmonic folds die out within a very few layers along the axial surface trace in the profile plane and thus have a small depth to half-wavelength ratio. Nearby layers fold independently of one another.

For harmonic folds, the average convergence or divergence of the dip isogons is very small, so the folds approximate a class 2 (similar) geometry. Dip isogons constructed for each layer, however, may vary smoothly from layer to layer (Figure 10.22*A*) or change radically, in some cases alternating from convergent to divergent in succeeding layers (Figure 10.22*B*, *C*). In the latter cases, the harmony is determined by the trend of the dip isogons averaged over several adjacent layers.

For disharmonic folds (Figure 10.21*B*), the dip isogons converge or diverge very strongly along the axial surface trace. For multilayer folds that approximate class 1B (parallel) folds, for example, the radius of curvature decreases toward the concave side of the fold, and the dip isogons converge strongly (Figure 10.23*A*). Where the radius of curvature approaches zero, the fold must die out rapidly along the axial surface at a **decollement**[8] or sole fault (Figure 10.23). In fold and thrust belts this decollement commonly corresponds to the basal thrust fault into which thrusts converge (Figure 10.2*B*; see also Chapter 5).

2. Axial surface geometry Throughout our discussion so far, we have assumed that the axial surface is planar. Many folds, in fact, display parallel or subparallel axial surfaces that are planar or only slightly curved. It is not unusual, however, for folds to have a nonplanar axial surface; such folds are commonly called **convolute folds.** In some cases the axial surface itself describes a cylindrical fold (see Figures 13.36, 13.37, and 13.38), whereas in others it is more irregular. The convolution generally is the result of deformation of earlier folds by one or more subsequent generations of folding. We discuss the geometry of such superposed folding in Section 13.9).

Some folds of a single generation develop with axial surfaces that have widely disparate orientations or that split into two or more surfaces. Such folds are usually called **polyclinal**[9] **folds.**

10.4 THE ORDER OF FOLDS

Folds characteristically develop simultaneously at different scales, so that large folds include smaller-scale folds in their limbs and hinge zones. We generally distinguish these different scales of related folds by their **order**, with the largest-scale folds being first-order folds, and successively smaller-scale folds being higher-order numbers (Figure 10.24). First-order folds are generally regional-scale features. Folds observed on the outcrop scale are commonly second- or higher-order folds. Higher-order folds are sometimes called **parasitic folds**. The median surface of a set of high-order folds defines the folds of the next lower order. Thus the median surface of a train of third-order folds defines the second-order fold train, and the median surface of second-order

[8]After the French word *décollement* meaning "unsticking" or "detachment."

[9]After the Greek words *poly*, which means "many," and *klinein* meaning "to slope."

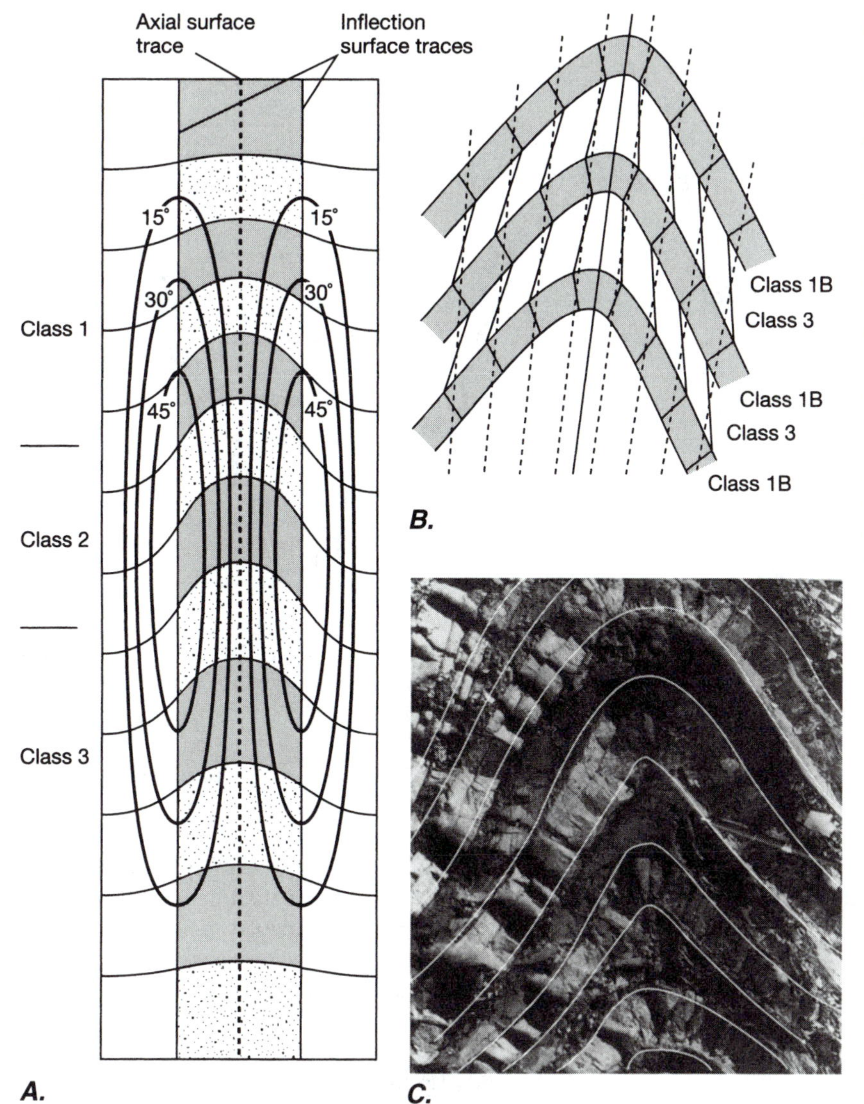

FIGURE 10.22 Profiles of multilayer folds showing dip isogon patterns in successive layers. *A.* The dip isogon pattern in a multilayer folds that dies out in both directions along the axial surface trace. In the shaded fold, isogons show regions of convergence (class 1), parallelism (class 2), and divergence (class 3). *B.* Diagram of a fold in which the dip isogons are alternately converging (class 1B folds) and diverging (class 3 folds) in successive layers. The average isogon pattern is approximately parallel to the axial surface, giving an approximately class 2 geometry. *C.* Folds in an interlayered chert-shale sequence that approximates the geometry shown in *B.* Lines drafted on the photo emphasize bed contacts (Franciscan complex, Marin Headlands, northern California).

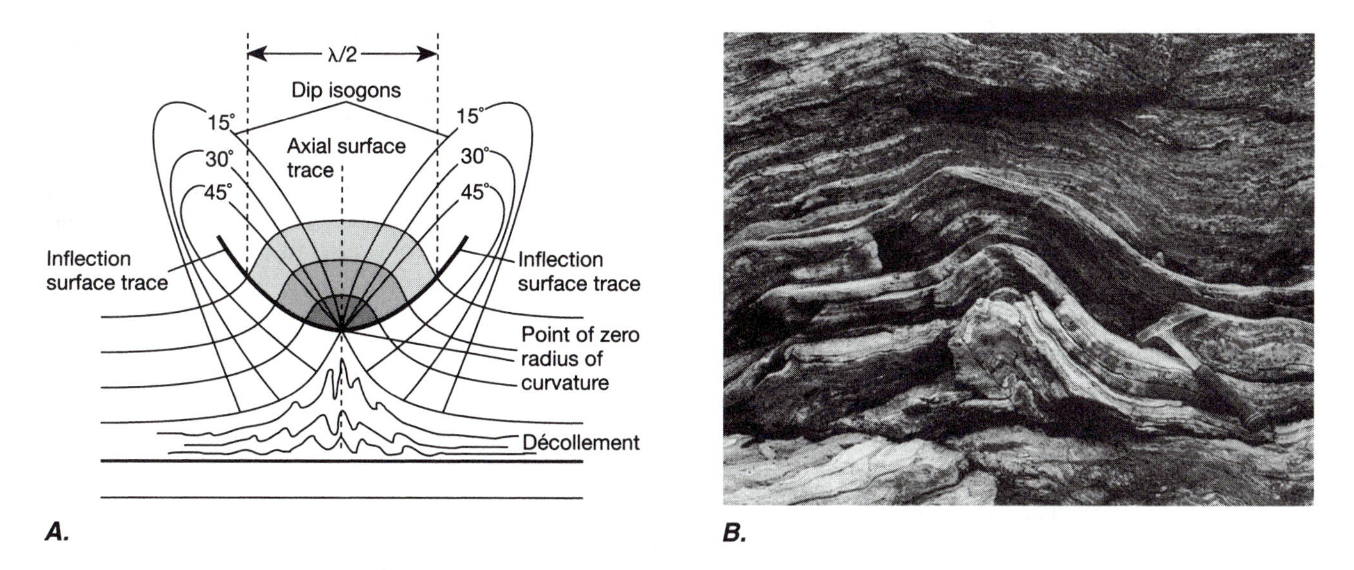

FIGURE 10.23 Disharmonic nature of class 1B folding resulting in a surface of disharmony or decollement. *A.* Diagram of a class 1B fold showing the surface of disharmony, the decollement. The half-wavelength is measured on the surface having the maximum amplitude. B. A class 1B fold formed during late-stage deformation in a banded gneiss.

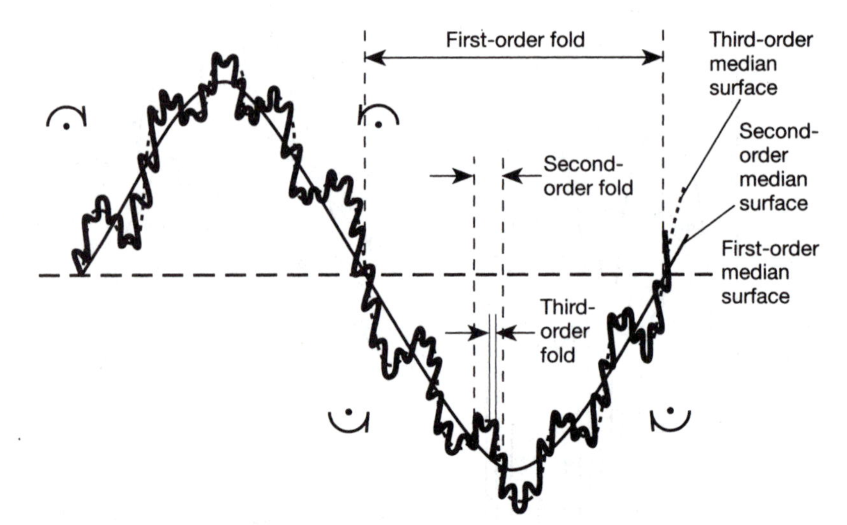

FIGURE 10.24 Illustration of different orders of folding. A fold train showing three orders of folds. The median surface of third-order folds defines the second-order folds, and the median surface surface of the second order folds defines the first order folds. The asymmetry of the second-order folds changes across the axial surface of the first-order folds, and the asymmetry of the third-order folds changes across the axial surfaces of the second-order folds. Because of the different limb lengths of the second-order folds, the predominant asymmetry of the third-order folds is different on opposite sides of a first-order axial surface.

folds describes the first-order fold shape (Figure 10.24; see Figure 13.21).

The asymmetry of higher-order folds changes across the axial surface of the next lower-order fold, as seen in Figure 10.24, and this feature is a very convenient field mapping tool for identifying the presence and location of low-order folds. The style and attitude of higher-order folds generally are very close to that of lower-order folds. This correspondence, known as **Pumpelly's rule,**[10] is also a valuable aid in deducing the geometry of large structures.

10.5 COMMON STYLES AND STRUCTURAL ASSOCIATIONS OF FOLDING

Some combinations of style elements occur so commonly together in deformed rocks that these fold styles have been given names. Moreover, certain common styles of folds are characteristic of particular tectonic settings. In this section we describe some of the more common of these associations.

i. Parallel Folds

This style of fold is strictly defined as class 1B for either single or multilayer folds. In standard usage, however, the term applies to class 1A and class 1C folds whose geometry is very close to the class 1B (Figure 10.20). Folds of this style characterize the geometry of fold and thrust belts, which lie on the margins of orogenic belts (Figure 10.2).

Rocks of these deformed belts are mostly unmetamorphosed to lightly metamorphosed layered sediments. Generally, the folds are approximately cylindrical over distances along the hinge that are large compared with the wavelength. Hinge lines are horizontal to gently plunging, and in the outer regions near the foreland they tend to have upright axial surfaces (Figure 10.12), wide aspect ratio (Figure 10.15), and gentle to open limbs (Figure 10.16). In the inner part of the belt closer to the hinterland, the aspect ratio tends to increase, limbs are tight or isoclinal, and the axial surfaces become inclined or recumbent, with vergence toward the foreland. Hinges are rounded in some belts but angular in others (Figure 10.17).

At a depth comparable to their dominant wavelength, the parallel folds of these belts die out at a sole fault or decollement, as required for the geometry of class 1B multilayer folds (Figure 10.2*B* and 10.23). This decollement tends to rise to progressively higher stratigraphic levels toward the foreland in a series of steps or ramps that alternately parallel and cross-cut the bedding. Some of these folds develop as the thrust sheet slides up these ramps and are called fault-ramp folds (see Figures 5.5, 5.6, 5.8, 5.14, 5.15, and 5.18).

The structure of fold and thrust belts in map view is exemplified by that of the Appalachian Valley and Ridge province (Figure 10.2*B*). The folds are continuous for up to tens or hundreds of kilometers. They typically die out as plunging structures or against tear faults (Figure 5.13), and the shortening accommodated by a fold that dies out is taken up by either adjacent folds or thrust faults. The higher aspect ratio folds are toward the interior of the range, with the aspect ratio and tightness of folds decreasing toward the foreland. In the Appalachian plateau, for example, the folding angle of the dominant folds is typically only a few degrees.

ii. Similar Folds

As strictly defined, similar folds have the geometry of class 2 single and multilayer folds. In common usage, however, the term is applied to a range of fold styles that

[10]The rule is named for Raphael Pumpelly, the geologist for the U.S. Geological Survey who first proposed this relationship, which he recognized from mapping the metamorphic rocks of the Green Mountains in western Massachusetts in 1894.

are very close to the class 2 style but that range from class 1C to class 3 (Figure 10.20). These folds are typical of the regionally metamorphosed central core zones of orogenic belts (Figure 10.1*B-D*). They vary in attitude, and many are recumbent, although upright and reclined folds are not usual. The folds are approximately cylindrical, although the distance along the hinge for which the cylindrical geometry is consistent is highly variable. Asymmetric folds are typical. The folds tend to have large aspect ratios (Figure 10.15) with close to isoclinal limbs (Figure 10.16) and angular to subangular hinges (Figure 10.17). Fold axial surfaces often are convolute and themselves describe fold systems. A planar alignment of platy minerals (a foliation; see Chapter 11) commonly is parallel to the axial surface of the folds.

Folds of this style that are large scale, recumbent, and isoclinal are called **fold nappes**[11] (Figure 10.1*B*, *C*). In some cases, the overturned limbs of these folds become sheared out so that the fold is further displaced by faulting (an example is the Morcles nappe, shown in Figure 10.1*B*), thus becoming a **thrust nappe**.

Folds in salt domes and glaciers tend also to be similar folds. In both settings, the folds are generally harmonic, tight to isoclinal, with subangular to angular hinges. Folds in salt domes are steeply reclined with their axes parallel to the margins of the structure, whereas in glaciers the folds tend to be gently plunging recumbent features (see Section 13.10).

iii. Other Styles of Folds

Chevron and **kink folds** (Figure 10.25) are cylindrical, harmonic, multilayer class 2 folds having angular to sharp hinges, equant aspect ratio, and gentle to close limbs. Chevron folds are symmetrical (Figure 10.25*A*, *C*), and kink folds are asymmetrical (Figure 10.25*B*). Both fold styles commonly develop in schists and other rocks having a strong planar preferred orientation of abundant platy minerals. They also develop in finely laminated rocks such as those composed of interbedded layers of sandstone or chert with shale. In such cases, the multilayer class 2

[11]The term *nappe* derives from the French word *nappe*, which means "cover sheet" or "tablecloth," and refers to any allochthonous sheet-like body of rock that has moved on a shallowly dipping surface. A nappe may originate as a recumbent isoclinal fold or as a thrust fault.

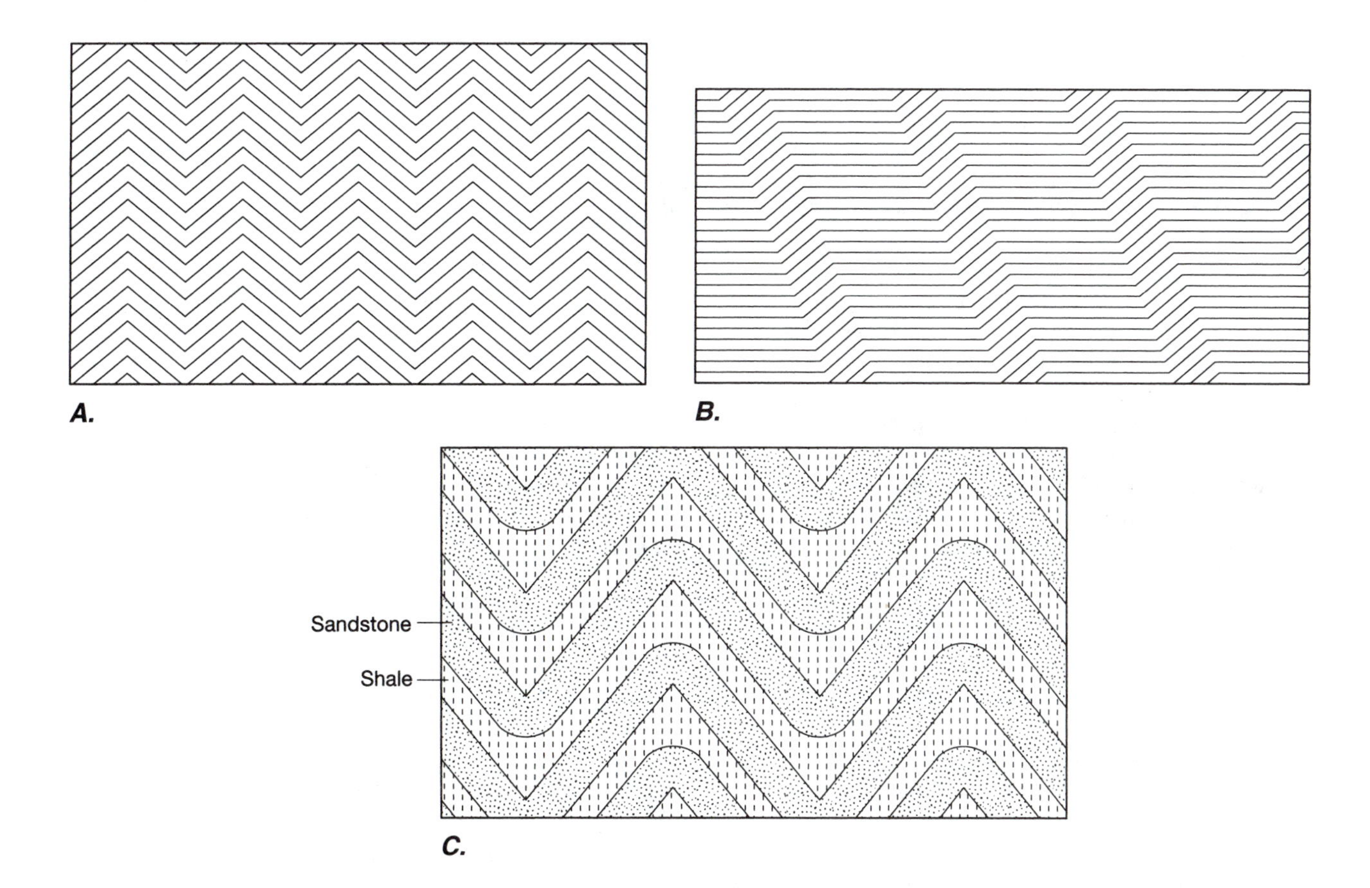

FIGURE 10.25 *A*. Chevron folds, *B*. Kink folds, *C*. Chevron folds in a sequence of alternating layers such as sandstone and shale.

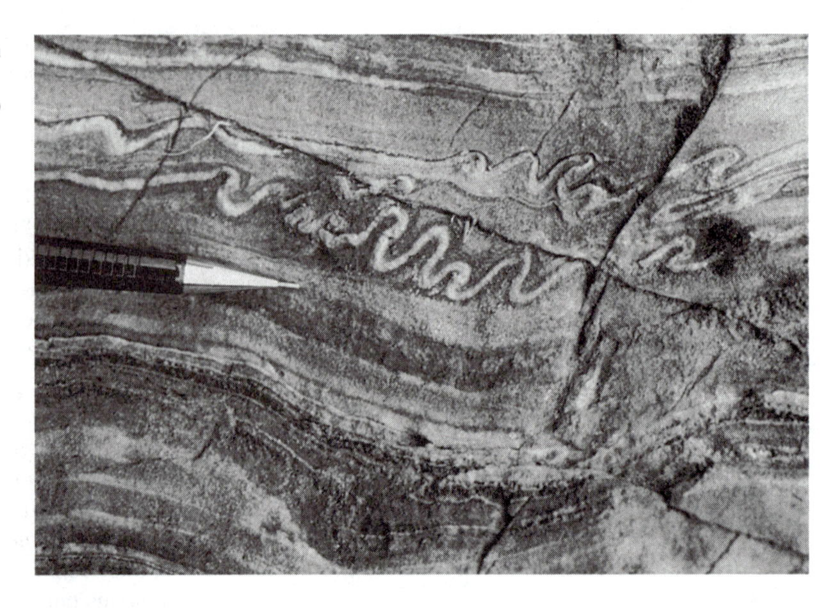

FIGURE 10.26 Ptygmatically folded layers in a banded marble, Bishop Creek roof pendant, Sierra Nevada, California. (Photo by Robert Varga)

geometry is provided by alternations between class 1 and class 3 folds in the sandstone and shale layers, respectively (Figures 10.25*C* and 10.22*C*).

Ptygmatic[12] folds (Figure 10.26) are disharmonic folds that develop in individual layers. The folds tend to be equant in aspect with close to fanning limbs, rounded to subrounded hinges, and class 1B or 1C layer geometry. They typically develop in layers, dikes, or veins in metamorphic rocks (Figure 10.26) and in sandstone layers or dikes in some sedimentary sequences.

[12]After the Greek word *ptygma*, meaning "fold."

REFERENCES AND ADDITIONAL READINGS

Allmendinger, R. W., R. Smalley Jr., M. Bevis, H. Caprio, and B. Brooks. 2005. Bending the Bolivian orocline in real time. *Geology* 33(11): 905–908.

Balk, R. 1949. Structure of the Grand Saline Salt Dome, Van Zandt County, Texas. *Bull. Amer. Assoc. Petrol. Geol.* 33: 1791–1829.

Bertelsen, A. 1960. Structural studies in the Precambrian of western Greenland. *Meddel. om Grœnl.* 123: 1–222.

Buxtorf, A. 1916. Prognosen und Befund beim Hauensteinbasis-nd Grenchenbergtunnel und die Bedeutung der letzteren für die Geologie des Juragebirges. *Verh. Naturforsch. Ges. Basel* 27: 185–254.

Collet, L. W. 1927. *The Structure of the Alps*. London, Edward Arnold & Co., 289 pp.

Debelmas, J., A. Escher, and R. Trumpy. 1983. Profiles through the western Alps. In N. Rast and F. M. Delany, eds., *Profiles of Orogenic Belts*, AGU-GSA Geodynamics Series volume 10: 83–97.

Eskola, P. 1949. The problem of mantled gneiss domes *Quart. J. Geol. Soc. Lond.* 104, Part 4: 461-476.

Fleuty, M. J. 1964. The description of folds. *Proc. Geol. Assoc.* 75: 461–492.

Hall, L. M., and P. Robinson. 1982. Stratigraphic-tectonic subdivisions of Southern New England. In P. St.-Julien and J. Béland, eds., *Major Structural Zones and Faults of the Northern Appalachians*, Geol. Assoc. of Canada Special Paper 24: 15–42.

Haller, J. 1956. Die Strukturelemente Ost Grœnlands zwischen 74° und 78° N., *Meddelelser om Grœnlands*, 154: 1–27.

Haller, J. 1957. Gekreutzte Faltensysteme in Orogenzonen. *Schweiz. Min. Petr. Mitt*, 37: 11–30.

Harris, H. D., A. G. Harris, W. DeWitt Jr., and K. C. Bayer. 1981. Evaluation of southern eastern overthrusts beneath Blue Ridge–Piedmont thrust. *Bull. Amer. Assoc. Petrol. Geol.* 65: 2497–2505.

Hudleston, P. J. 1973. Fold morphology and some geometrical implications of theories of fold development. *Tectonophysics* 16: 1–46.

Jackson, M. P. A., and C. J. Talbot. 1989. Anatomy of mushroom-shaped diapirs. *J. Struct. Geol.* 11: 211–230.

Milnes, A. G. 1974. Post-nappe folding in the western Lepontine Alps. *Eclogae Geologicae Helvetiae* 67: 333–348.

Muehlberger, W. R., P. S. Clabaugh, and M. L. Hightower. 1962. Palestine and Grand Saline salt domes, eastern Texas. In E. H. Rainwater and R. P. Zingula, eds., *Geology of the Gulf Coast and Central Texas and Guidebook of Excursions*, 1962 Annual Meeting of the Geol. Soc. Am.; Houston Geological Society, 266–277.

Oliver, J. E. 1980. Seismic exploration of the continental basement: Trends for the 1980's. In *Continental Tectonics*, Washington, DC, National Academy of Sciences, 117–126.

Pierce, G. G. 1947. *Cymric Oil field, Calif. Oil Fields, Div. Oil. and Gas*, Volume 33, No. 2, 7–15. Reported in *AAPG/SEPM Guidebook to California Oil Field Geology*, 223–231 (1951).

Plafker, G., and H. C. Berg. 1994. Overview of the geology and tectonic evolution of Alaska. In G. Plafker and H. C. Berg, eds., *The Geology of Alaska. Boulder, Colorado.* Geol. Soc. of Amer., The Geology of North America, volume G-1, 989–1021.

Ramberg, H. 1981. *Gravity, Deformation and the Earth's Crust.* Academic Press, 452 pp.

Ramsay, J. G. 1967. *Folding and Fracturing of Rocks.* New York, McGraw Hill.

Rodgers, J. 1970. *The Tectonics of the Appalachians.* New York, Wiley, 271 pp.

Rutten, M. G. 1969. *The Geology of Western Europe.* New York, Elsevier Publishing Co., 520 pp.

Shelton, J. S. 1966. *Geology Illustrated.* San Francisco and New York, W. H. Freeman and Co., 434 pp.

Suppe, J., and J. Namson. 1979. Fault-bend origin of frontal folds of the western Taiwan fold and thrust belt. *Petroleum Geology of Taiwan*, No. 16. 1–18.

Sussman, A. J., and A. B. Weil, eds. 2004. *Orogenic Curvature: Integrating Paleomagnetic and Structural Analyses*, Boulder, CO, Geol. Soc. Amer. Special Paper 383, 272 pp.

Thompson, J. B. Jr., P. Robinson, T. N. Clifford, and N. J. Trask Jr. 1968. Nappes and gneiss domes in west-central New England. In E.-A. Zen, W. S. White, J. B. Hadley, and J. B. Thompson Jr., eds., *Studies of Appalachian Geology: Northern and Maritime.* New York and London, Interscience Publishers, 203–218.

Trusheim, F. 1960. Mechanism of salt migration in northern Germany. *Amer. Assoc. Petroleum Geologists Bull.* 44: 1519–1540.

Twiss, R. J. 1988. Description and classification of folds in single surfaces. *J. Struct. Geol.* 10(6): 607–623.

Van der Voo, R. 2004. Paleomagnetism, oroclines, and growth of the continental crust. *GSA Today* 14(12): 4–9.

Vendeville, B. C. 2002. A new interpretation of Trusheim's classic model of salt-diapir growth. *Gulf Coast Assoc. Geological Soc. Trans.* 52: 943–952.

Weiss, L. E. 1972. *The Minor Structures of Deformed Rocks, a Photographic Atlas.* New York, Springer-Verlag, 431 pp.

Whitten, E. H. T. 1966. *Structural Geology of Folded Rocks.* Chicago, Rand McNally, 678 pp.

Chapter 11

FOLIATIONS AND LINEATIONS IN DEFORMED ROCKS

A **foliation**[1] is a planar structure in a rock that is **homogeneously**[2] distributed throughout the volume. Examples of foliations include sedimentary bedding; the planar alignment of sedimentary clasts; the planar structure defined by the parallel alignment of platy minerals in rocks such as schist (Figure 11.12*A*), slate (Figures 11.8 and 11.11), shale, or volcanic rock; the parallel alignment of flattened mineral grains (Figure 11.12*B*) and conglomerate pebbles (Figure 11.19*A*); and compositional banding defined by the concentration of particular minerals into layers, common in ultramafic rocks (Figure 11.2*A*), gneisses (Figure 11.2*B*), and some volcanic rocks.

A **lineation** is a homogeneously distributed linear structure. Lineations are **surficial** if they are present along discrete surfaces, and they are **penetrative** if they occur throughout the volume of a rock. Examples of surficial lineations include sedimentary groove casts in a bedding surface and the parallel alignment of mineral fibers that develop along some fault surfaces. Examples of penetrative lineations include the hinges of pervasive small crenulations in a foliation, the preferred alignment of elongate mineral grains such as amphiboles or quartz, or the linear alignment of elongate clusters of grains of a particular mineral such as quartz or mica.

Foliations and lineations are primary if they originate by primary sedimentary or igneous processes. Primary sedimentary processes such as sediment transport and deposition produce, for example, linear tool marks, a preferred orientation of sedimentary clasts, and bedding. Primary igneous processes, such as flow and crystallization; result in the preferred orientation of bubbles and pumice fragments or compositional streaks and bands. Foliations and lineations are secondary if they originate by secondary processes such as tectonic deformation or metamorphism. We sometimes use terms such as *sedimentary foliation*, *igneous lineation*, or *tectonite foliation* to specify the inferred origin of a foliation. Because the origin of a structure is an interpretation, however, it is an inappropriate basis for classification, and we therefore define foliation and lineation in strictly descriptive terms.

11.1 TECTONITES

In this chapter we discuss foliations and lineations that are characteristic of **tectonites**, which are rocks whose structure is a product of deformation and which are commonly, but not necessarily, metamorphosed. Tectonites are characteristic of orogenic belts and are common along fault zones, including major plate boundaries. Most tectonite foliations and lineations are secondary in origin and develop as a result of the deformation, although some may be inherited primary features. Although most tectonites have both foliations and lineations, if the tectonite structure is dominated by a foliation, we call it an **S-tectonite**,[3] and if is dominated by a lineation, we call it an **L-tectonite.**

[1]Derived from the Latin *folium*, which means "leaf." There is no universal agreement on the definition of the term *foliation*. Some authors use it in a more specific sense than we have adopted.

[2]A feature that is homogeneously distributed in a body has the same characteristics in any arbitrary volume of the body.

[3]The *S* comes from the German word *schiefer*, meaning "schist."

Generally we consider a structure to be homogeneously distributed, or **penetrative**, if the spacing or the scale of the structure in a rock is very small compared to the size of the rock volume under consideration. The terms *foliation* and *lineation* can be applied only to structures that are penetrative within a volume having a dimension on the order of tens of centimeters. Planar features that have an average spacing on the order of meters, for example, are not foliations but may be fractures, faults, shear zones, etc. The spacing of many foliations, such as the planar structure of slates, is so small as to be resolvable only under a microscope. If a lineation occurs on a penetrative planar structure such as a foliation, then the lineation also is penetrative.

Several other terms are commonly used to describe penetrative planar features in rocks. The term **S-surface** is generally synonymous with foliation. It refers to any penetrative planar feature of a rock and thus includes sedimentary bedding, schistosity, as well as axial surfaces of folds, which may be simply geometrical constructs rather than actual physical features in the rock. Bedding is commonly designated S_0, and other penetrative planar features such as foliations and axial surfaces are labeled S_1, S_2, etc., where the subscripts generally indicate the sequence in which the different features developed.

Rock cleavage, or simply **cleavage,** is the tendency of a rock to break or cleave along surfaces of a specific orientation. All cleavages are foliations, and the two terms often are used to describe the same structure. *Foliation* is a more general term than *cleavage,* however, because it includes planar geometric features that do not necessarily result in a cleavage. The planar alignment of slightly flattened grains, for example of quartz in a quartzite or olivine in a peridotite, or the compositional banding in a gneiss, would define a foliation but could provide so small a mechanical anisotropy that it would not result in a cleavage. The terms **layer** and **banding** describe planar tabular features in rocks that are distinguished by differences in composition or possibly texture from adjacent rock. The terms are commonly used in descriptions of plutonic igneous rocks and high-grade metamorphic gneisses. We outline a morphological classification for foliations in tectonites in Figure 11.1 and Sections 11.2 to 11.5.[4] The classification is based on either or both of the shape and the arrangement of components of the rock. This approach is preferable to the use of numerous older terms that are poorly defined, are imprecise, or have a genetic connotation (see Section 11.7).

Many foliations have a structure characterized by alternating **domains** that show marked differences in preferred orientation of mineral grains, in structure, or in composition. The domains may be laminar to lenticular in shape, and the spacing of the domains is sufficiently small that the structure is still considered penetrative. Foliations defined by domains having a spacing of 10 μm or more are **spaced foliations** (Figure 11.1; see also. Figures 11.2 to 11.10). Foliations having a finer domainal structure or no domainal structure at all are **continuous foliations** (Figure 11.1; see also Figures 11.11 and 11.12).

Foliation and cleavage			
	Spaced	Compositional	Diffuse
			Banded
		Disjunctive	Stylolitic
			Anastomosing
			Rough
			Smooth
		Crenulation	Zonal
			Discrete
	Continuous	Fine	Microcrenulation
			Microdisjunctive
			Microcontinuous
		Coarse	Mineral grain
			Discrete

FIGURE 11.1 Morphological classification scheme for foliations. (Modified from Gray 1977 and Powell 1979)

Spaced foliations are categorized on the basis of four features: (1) domain shape, (2) domain spacing, (3) distinguishing characteristics of individual domains such as mineral composition or the preferred orientation of mineral grains, and (4) the proportion of the rock occupied by the different types of domains. We recognize three categories of spaced foliation: **compositional, disjunctive,** and **crenulation** foliations (Figure 11.1).

Figure 11.18 and Sections 11.8 and 11.9 present a morphological classification of lineations in tectonites. We divide tectonite lineations into two major categories, **structural lineations** and **mineral lineations**, according to whether they are defined primarily by geometrical structures or by mineral grains or aggregates (Figure 11.18). Both types of lineation may be either surficial or penetrative, as defined in the chapter introduction.

11.2 COMPOSITIONAL FOLIATIONS

Compositional foliations (Figure 11.2) are marked by layers or laminae of different mineralogical composition. A planar alignment of platy or needle-shaped crystals may also be present, but the rock has at most a weak tendency to cleave parallel to the foliation. We subdivide these structures based on the mineralogical variation and the spacing and relative thicknesses of the compositional layers. **Diffuse foliations** are defined by widely spaced weak concentrations of a mineral in a rock of predominantly one lithology. They are common in ultramafic

[4] The classification is modified from a morphological classification of cleavage presented by Gray (1977) and Powell (1979).

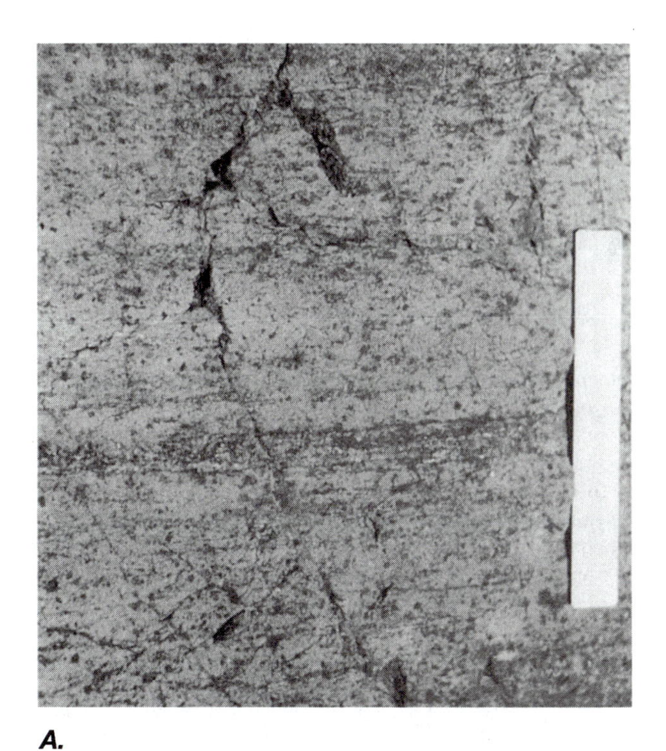

A.

B.

FIGURE 11.2 Compositional foliations. *A.* Diffuse compositional foliation in a dunite, China Mountain, eastern Klamath Mountains, northern California. The rock is composed mainly of olivine. Concentrations of pyroxene crystals in sparse layers defines the foliation. Scale is 6 inches (≈15.25 cm) long. (Courtesy N. Lindsley-Griffin) *B.* Banded compositional foliation in a high-grade metamorphic gneiss, Wopmay orogen, northwest Canada. (Courtesy M. St. Onge)

rocks such as dunites in which sparse layer concentrations of pyroxene crystals define a weak compositional layering (Figure 11.2*A*). Diffuse foliations are also common in deformed granites in which concentrations of mafic minerals define the foliation. **Banded foliations** are defined by relatively closely spaced compositional layers that are mineralogically distinct and are of comparable abundance. They are common in high-grade metamorphic gneisses (Figure 11.2*B*).

11.3 DISJUNCTIVE FOLIATIONS

Disjunctive[5] foliations (Figures 11.3 through 11.8) contain thin domains, called **cleavage domains** or **seams**, marked by concentrations of oxides and strongly aligned platy minerals. The cleavage domains are separated by

[5]From the Latin word *disjunctus*, meaning "disjoined or detached."

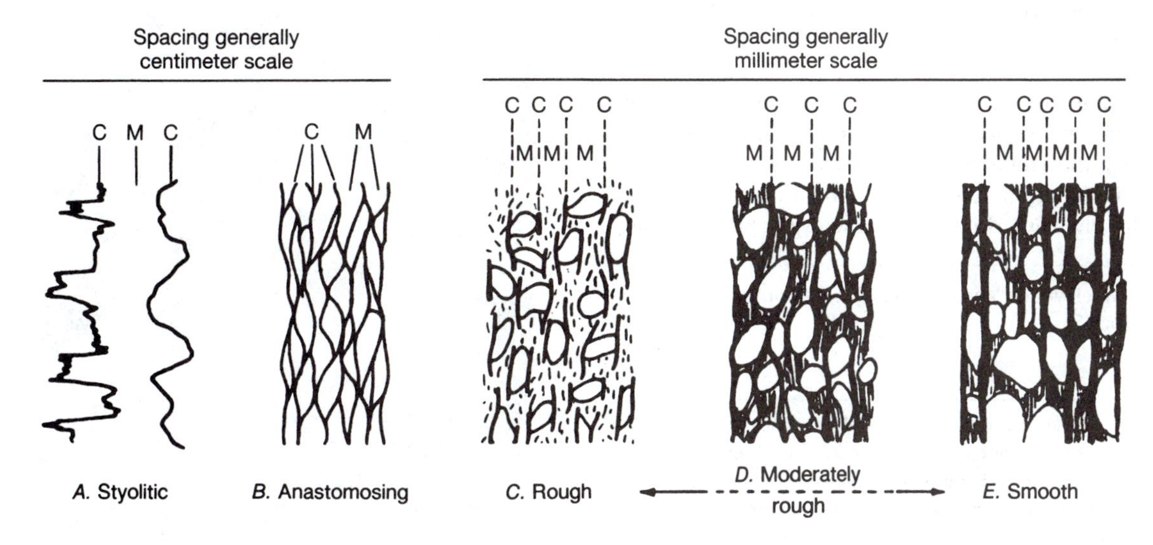

FIGURE 11.3 Sketches showing characteristics of the various types of disjunctive foliation. C marks cleavage domains; M marks microlithons. Note the change in the scale of the spacing from centimeters in *A* and *B* to millimeters in *C* through *E*. (From Powell 1979)

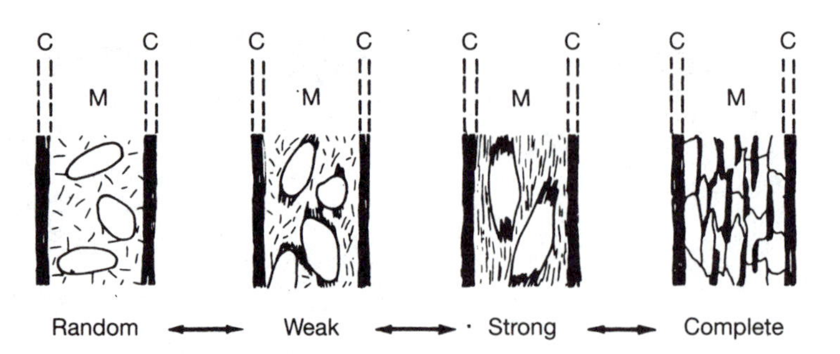

FIGURE 11.4 Preferred orientation within a microlithon bounded on either side by cleavage domains. C marks cleavage domains; M marks microlithons. *Random fabric*: the large grains have no preferred orientation, and the fine platy minerals in the matrix also are not oriented. *Weak fabric*: coarse mineral grains show a slight elongation and a weak preferred orientation of their long axes; mica "beards" weakly developed at the ends of the coarse mineral grains; platy minerals in the matrix show a weak preferred orientation. *Strong fabric*: the coarse mineral grains show distinct elongation and a strong alignment of their long axes; mica "beards" are well developed and oriented; and platy minerals in the matrix are strongly aligned. *Completely oriented fabric*: detrital grain shapes are not preserved; mineral grains are elongated and show a strong preferred orientation; the fabric is transitional to a continuous foliation. (From Powell 1979)

tabular to lenticular domains called **microlithons** in which platy minerals may be less abundant or more randomly oriented (Figures 11.3 and 11.4). Disjunctive foliations commonly form in previously unfoliated rocks such as limestones or mudstones, although they may also develop in some foliated rocks cross-cutting an earlier foliation.

We subdivide disjunctive foliations into four groups—**stylolitic**, **anastomosing**, **rough**, or **smooth**—based on the smoothness or regularity of the cleavage domains (Figure 11.3). This order corresponds to a general increase in smoothness of cleavage domains and a decrease in spacing, as well as a tendency toward stronger preferred mineral orientations within the microlithons. The fabric in the microlithons, particularly of the platy minerals, may range from randomly oriented to completely aligned, and it provides the basis for further subdivision (Figure 11.4).

Stylolitic[6] foliation consists of long continuous, but very irregular, cleavage domains commonly with a distinct tooth-like geometry in cross section (Figures 11.3*A* and 11.5*A*). Individually, such domains are called **stylolites**. This foliation type is common in limestones in which the cleavage domains characteristically are thin,

[6]From the Greek words "stylos" meaning "stalk" or "pen" and *lithos* meaning "rock."

A.

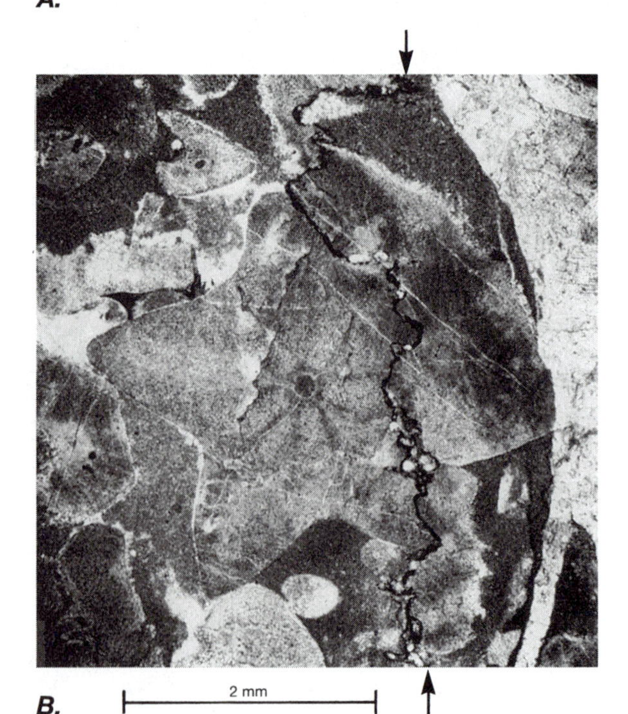

B.

FIGURE 11.5 Stylolitic foliation. A. Stylolitic foliation in limestone layers. Stylolites are the irregular dark seams approximately parallel to the pen. *B*. Stylolite truncating a pentacrinoid fossil in a limestone. The stylolite is the roughly vertical irregular black seam (see arrows). The fossil is shaped like a five-pointed star, but two of the points on the right side are truncated by the stylolite and mostly have been removed by solution. (Thin section courtesy of G. Protzman Nakamura; photo by E. Rodman)

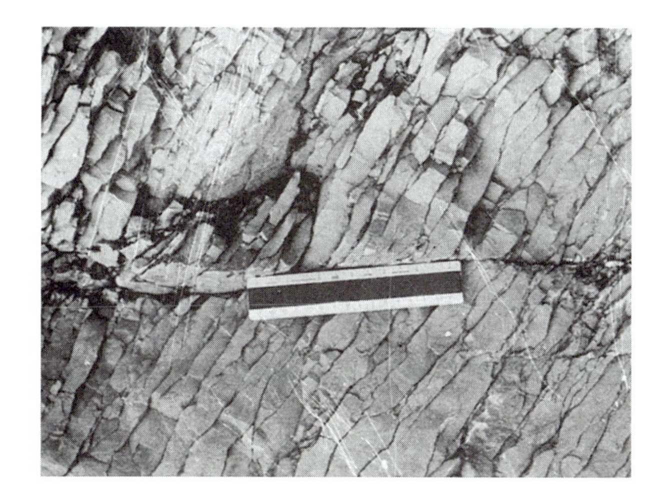

FIGURE 11.6 Anastomosing foliation in a limestone. Bedding is parallel to the ruler. (From Alvarez et al. 1978)

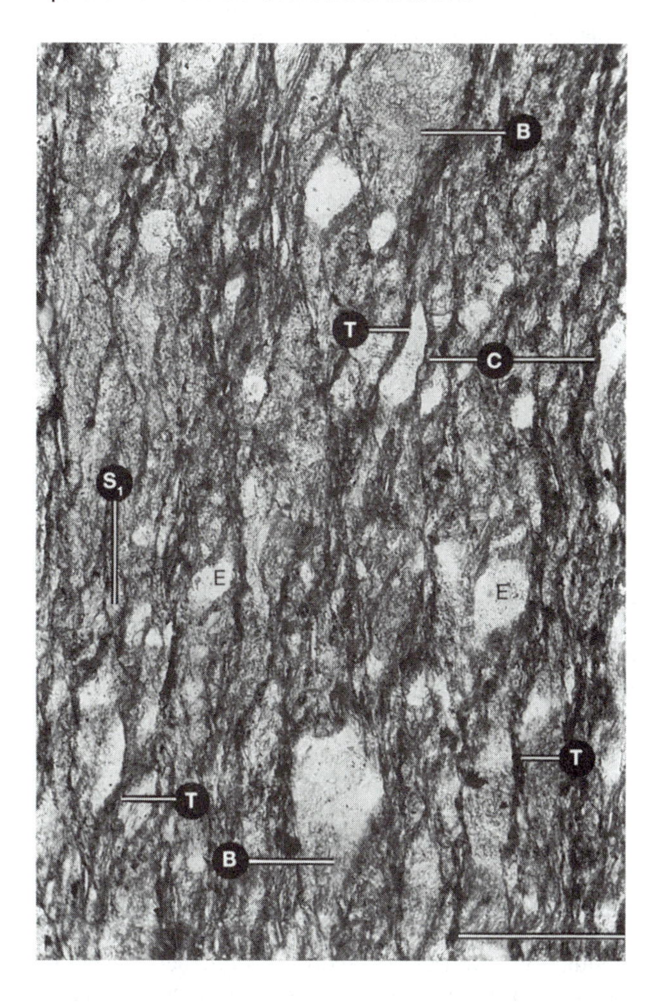

FIGURE 11.7 Rough foliation (S_1) in a deformed wacke. Dark seams are the cleavage domains composed of insoluble residues. At C, remnant detrital sand grains are truncated against cleavage domains. At T, thin plate-like quartz grains result from solution of the grains along cleavage domains (see the discussion in Sections 14.2(iii) and 14.3(ii)). B marks mica beard overgrowths on detrital grains. The scale bar is 1 mm. (From Gray 1978)

dark, clay-rich seams, and the microlithons generally show no preferred orientation. The spacing of the cleavage domains is commonly from 1 to 5 centimeters or more. In some limestones, stylolites may be seen to truncate fossil fragments (Figure 11.5*B*).

Anastomosing foliation is distinguished by long, continuous, wavy cleavage domains that form an irregular network outlining lenticular microlithons (Figures 11.3*B* and 11.6). Such foliations are common in limestones and in phyllites and schists. The spacing of the cleavage domains tends to be smaller than for stylolitic foliation, averaging perhaps 0.5 to 1 centimeter. The cleavage domains contain concentrations of platy minerals with a strong preferred orientation parallel to domain boundaries. The fabric within the microlithons is commonly random to weak.

Rough foliation typically develops in rocks containing abundant sand-size material. The cleavage domains are short discontinuous concentrations of highly oriented platy minerals that bound or envelop the coarse grains (Figures 11.3*C* and 11.7). The spacing of the cleavage domains generally is less than a millimeter. Within microlithons, the preferred orientations of mineral grains may vary widely from random to strongly oriented (Figure 11.4).

Smooth foliation represents the end-member of the spectrum from irregular to planar cleavage domains (Figure 11.3*D*, *E*) and is characteristic of some slates. Cleavage domains are long, continuous, and smooth with concentrations of highly oriented platy minerals (Figure 11.8). The cleavage domain spacing is generally less than a millimeter. Fabric development within the microlithons commonly ranges from random to completely oriented (Figure 11.4). With decreasing domain spacing, this type of foliation is transitional with microdomainal, fine, continuous foliations that are characteristic of some slates (Figure 11.8), as described in Section 11.4.

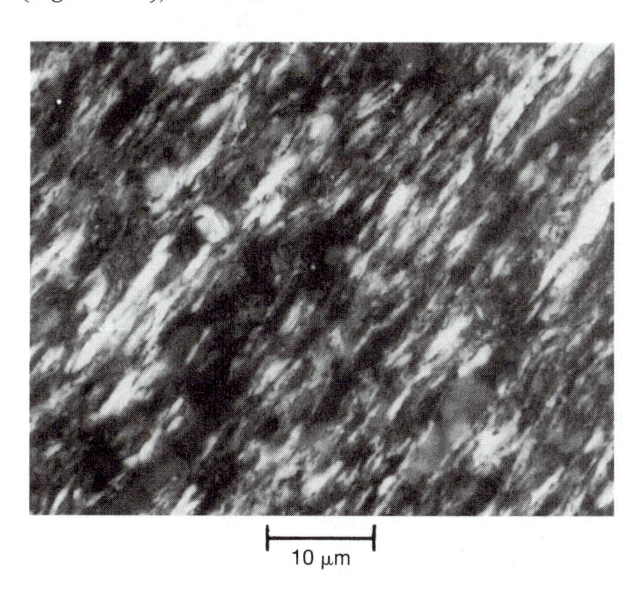

10 μm

FIGURE 11.8 Smooth foliation in a slate.

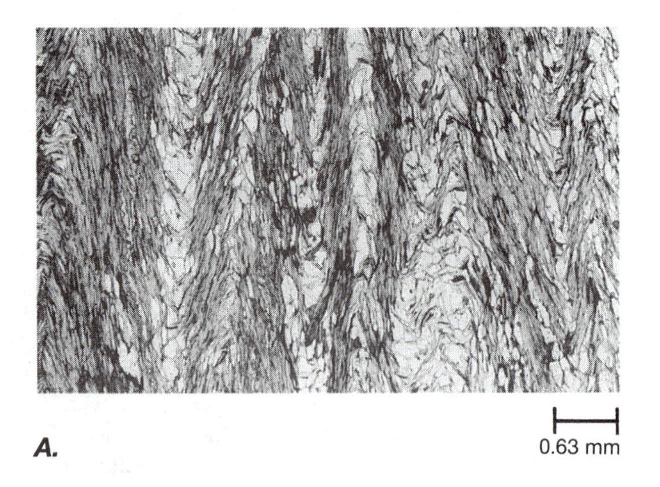

A.

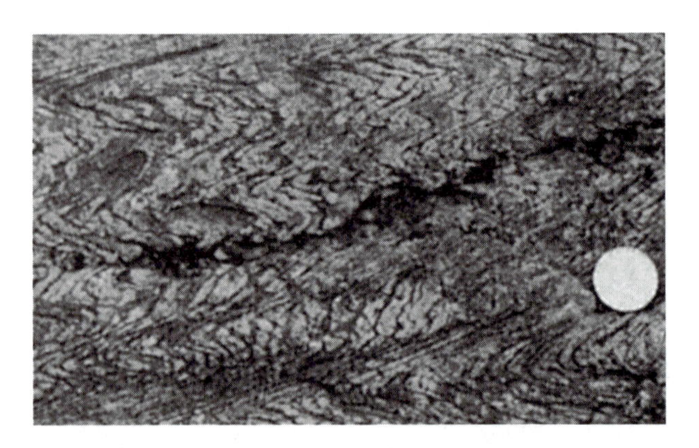

B.

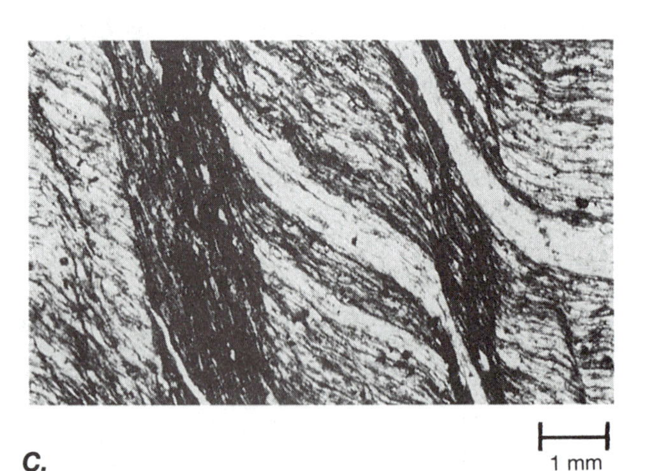

C.

FIGURE 11.9 Zonal crenulation foliations. Note that the laminations and the preferred orientation of the platy minerals vary continuously from microlithon to cleavage domain and that within the cleavage domain, the laminations and platy minerals are not strictly parallel to the new cleavage domain. *A.* If the crenulations are symmetric, both limbs define crenulation cleavage domains, and the hinge zone is preserved in the microlithons. Compositional differentiation leaves limbs of crenulations (dark bands) rich in mica and poor in quartz, and hinge zones rich in quartz and poor in mica. (Glen 1982) *B.* Asymmetrical crenulation foliation in schistose metagreywacke. Coin diameter is about 2.5 centimeters. Rotmell, Grampian Highlands, Scotland. (Courtesy of J. Treagus) *C.* Asymmetric crenulations in a quartz-rich schist. A loss of quartz from the cleavage domain results in a compositional differentiation of the domains. Cleavage domain on the right is definitely zonal; cleavage domain on the left borders on being discrete. (Gray 1977)

11.4 CRENULATION FOLIATIONS

Crenulation foliations are formed by harmonic wrinkles or chevron folds that develop in a pre-existing foliation. The new foliation cuts across the old foliation and is defined by either both limbs of symmetric crenulations (Figure 11.9*A*) or the long limbs of asymmetric crenulations (Figure 11.9*B*, *C*). The old foliation is preserved in the microlithons, either as the hinges of symmetric crenulations (Figure 11.9*A*) or as the short limbs of asymmetric crenulations (Figure 11.9*B*, *C*). The microlithon width is comparable to the half-wavelength (Figure 11.9*A*) or the wavelength (Figure 11.9*B*) of the crenulations.

The orientation pattern of platy minerals in the cleavage domain provides a further subdivision of crenulations. In a **zonal crenulation foliation**, the platy minerals in the new cleavage domain are oriented at a small angle to the domain and form a continuous variation of orientations from the platy minerals in the microlithons (Figure 11.9). The microlithon boundaries are gradational. There is commonly a compositional difference between cleavage domains and microlithons, with the proportion of platy minerals being relatively high in the cleavage domains and low in the microlithons.

In a **discrete crenulation foliation**, the orientation of platy minerals in the new cleavage domains is parallel to the domains and sharply discordant with the orientations of platy minerals in the microlithons (Figure 11.10). The crenulations are preserved in the microlithons (Figure 11.10). The cleavage domains are generally narrow and may, but do not necessarily, correspond to limbs of crenulations in the microlithons. Differences in mineralogy between the two domains are similar to those of zonal crenulation foliations.

All variations between these two end-members of crenulation foliation can be observed. In fact, it is not uncommon to find both morphologies in the same sample.

11.5 CONTINUOUS FOLIATIONS

Continuous foliations are defined either by domains with spacing less than 10 μm (Figure 11.11*A*, *B*) or by a nondomainal structure (Figure 11.12*A*, *B*). They are di-

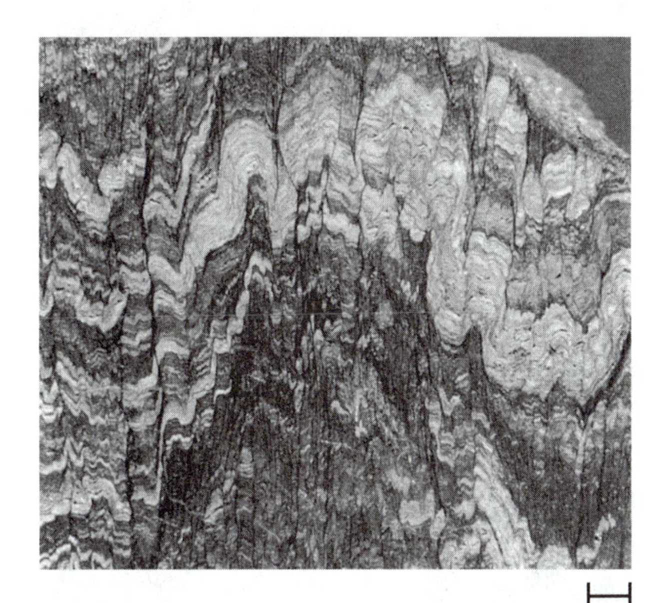

FIGURE 11.10 Discrete crenulation foliation in a calcareous slate. Note the sharp discontinuity in orientation marking the boundary between cleavage domain and microlithon. The orientation of the platy minerals in the cleavage domain is parallel to the domain boundary. Calaveras formation, Sierra Nevada foothills.

visible by grain size into **fine** and **coarse continuous foliations** (Figure 11.1), as exemplified, respectively, by slates and schists. Fine continuous foliations may be either microdomainal or microcontinuous. The **microdomainal** fine foliations may be microcrenulation (Figure 11.11*A*) or microdisjunctive (Figure 11.11*B*), and they have the same characteristics as their macroscopic counterparts, except that the microdomain spacing is less than 10 μm. A **microcontinuous** fine foliation is characterized by the parallel alignment of all platy or inequant grains in a rock, and it lacks any domainal structure. The terms *microdomainal* and *microcontinuous* are impractical to use as field classification terms because in fine-grained rocks only an electron microscope can reveal the distinction between the structures.

Coarse continuous mineral foliations are characterized by the complete orientation of homogeneously distributed platy minerals (Figure 11.12*A*) or by the alignment of flattened mineral grains (Figures 11.12*B* and 3.7). They have no domainal structure, which would be easily revealed by the coarse grain size. Coarse continuous discrete foliations are defined by the preferred orientation of deformed objects distributed within the rock. Objects such as pebbles in a conglomerate (Figure 11.19*A*), alteration spots in a slate (Figure 11.19*B*), or ooids in a limestone (see Figure 12.6*B*) can be flattened to define a discrete foliation parallel to the plane of flattening, as well as lengthened to define a discrete lineation parallel to the direction of maximum lengthening (see Figure 11.19*C*; Section 11.8).

11.6 THE RELATIONSHIP OF FOLIATIONS TO OTHER STRUCTURES

i. Relationship with Folds

Secondary foliations occur so commonly parallel or subparallel to the axial surfaces of folds that the association is almost axiomatic. Such foliations are called **axial surface foliations** or **axial plane cleavages**. The orientation of such foliations characteristically changes progressively from one side of the fold to the other, or **fans** across the fold, and is actually parallel to the axial surface only at the hinge surface. Foliation fans are **convergent** or **divergent**, depending on whether, in passing from the convex to the concave side of a fold, the foliation orientations converge toward one another (Figure 11.13*A*, layers I and III) or diverge from one another (Figure 11.13*A*, layer II).

It is important to distinguish foliation fans from fans of dip isogons on folds. The terminology and the geometry are the same for both, and diagrams of the two features look similar. The foliation fan, however, is an actual physical structure that can be observed in the rock (Figures 11.13*B*), whereas the dip isogon fan is a geometric construction (Figures 10.18 and 10.19). In general, the two features are not parallel.

The extent of fanning of an axial surface foliation is commonly dependent on the composition of the rock in which the foliation is developed. Foliations tend to be most strongly convergent across folds in rocks containing only small proportions of platy minerals such as sandstones, and they are least convergent or divergent in rocks rich in platy minerals such as schists and slates. The orientation of the foliation commonly changes significantly at a lithologic contact (Figure 11.13*A*, *B*). We call this feature a **refracted foliation** or **refracted cleavage** by analogy with the bending, or refraction, of a light ray as it passes obliquely across an interface between two media. The analogy has no significance, however, beyond the similarities of geometry.

The relationship of foliations subparallel to the axial plane of folds is so consistent that it can be used in field mapping to help determine the geometry of the folding. In an area that has been subjected to only one generation of folding, it also can be a valuable indicator of whether sedimentary beds are overturned or right side up, because a given surface of axial foliation may cut a particular folded surface only once (Figure 11.14*A*). Thus the relative orientations of bedding and foliation (Figure 11.14*B*) permit us to determine the general location and direction of the fold closures, as shown in Figure 11.14*A* (cf. Figure 11.13*A*, *B*). An interpretation of Figure 11.14*B* that has the fold closing to the left at the top of the photo and to the right at the bottom would be incorrect, because it would require the foliation plane to cut a single bedding surface more than once.

A useful rule of thumb by which to remember this relationship is to imagine an arrow drawn along the foliation

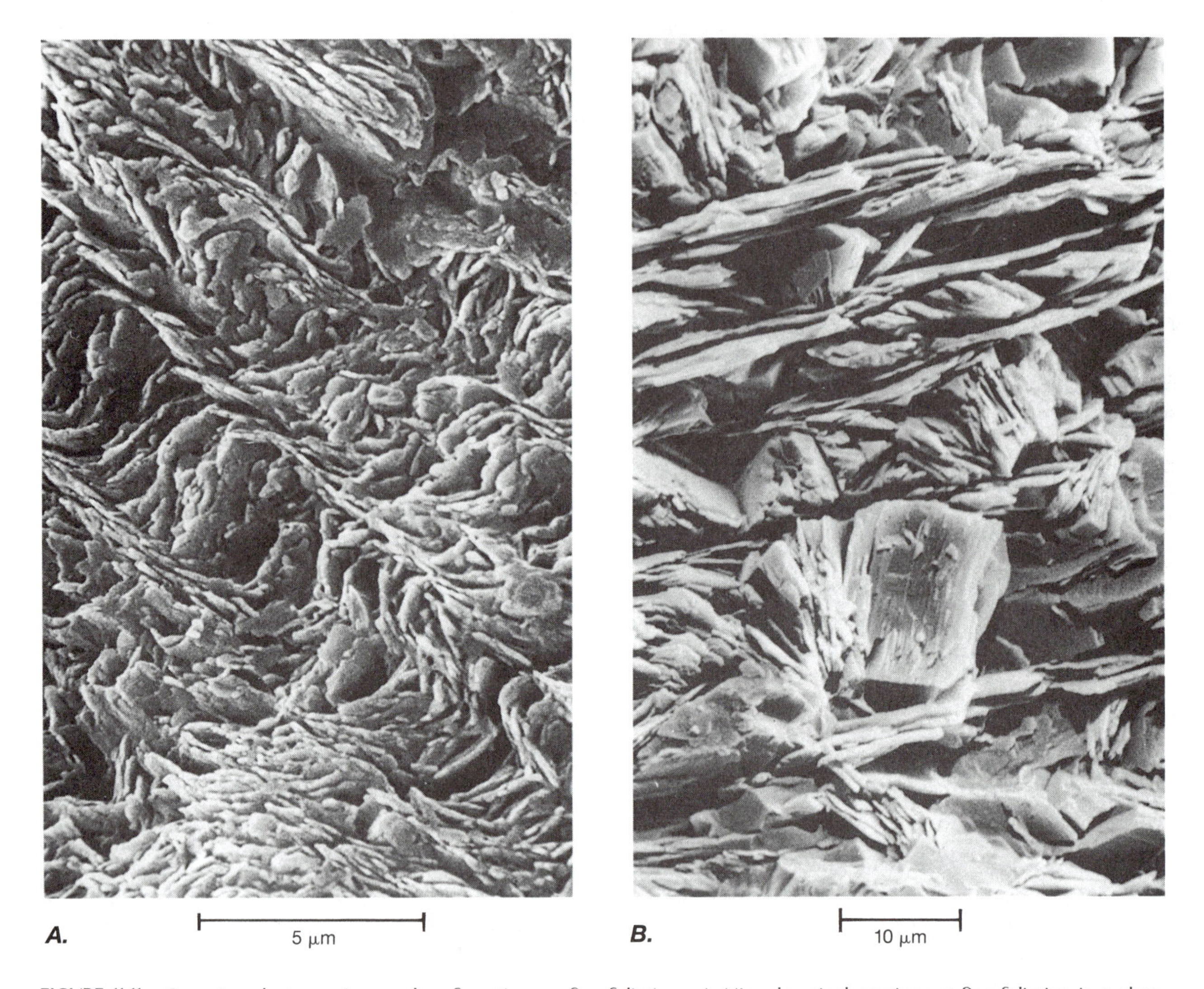

FIGURE 11.11 Scanning electron micrographs of continuous fine foliations. *A*. Microdomainal continuous fine foliation in a slate with an asymmetric microcrenulation structure. (From Weber 1981) *B*. A microdisjunctive continuous fine foliation in a slate. (From Weber 1981)

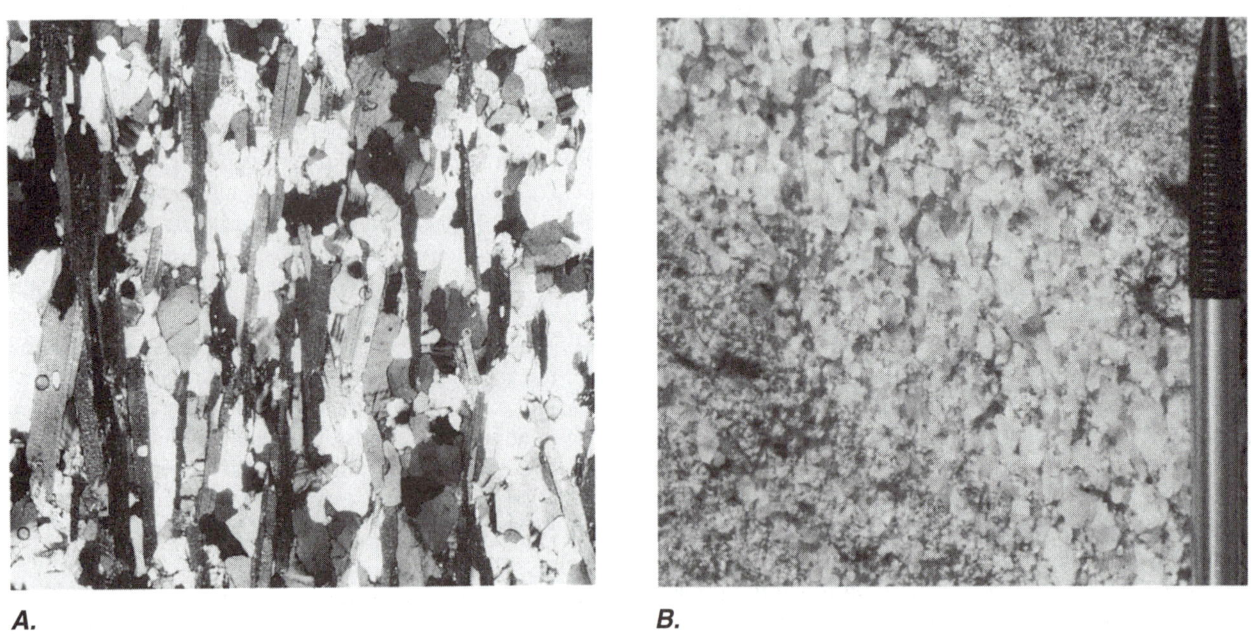

FIGURE 11.12 Coarse continuous foliation showing a strictly continuous structure. *A*. Photomicrograph of a schist with the foliation defined by mica. (Photo by E. Rodman) *B*. A grain-shape foliation parallel to the pencil in a very coarse grained marble layer.

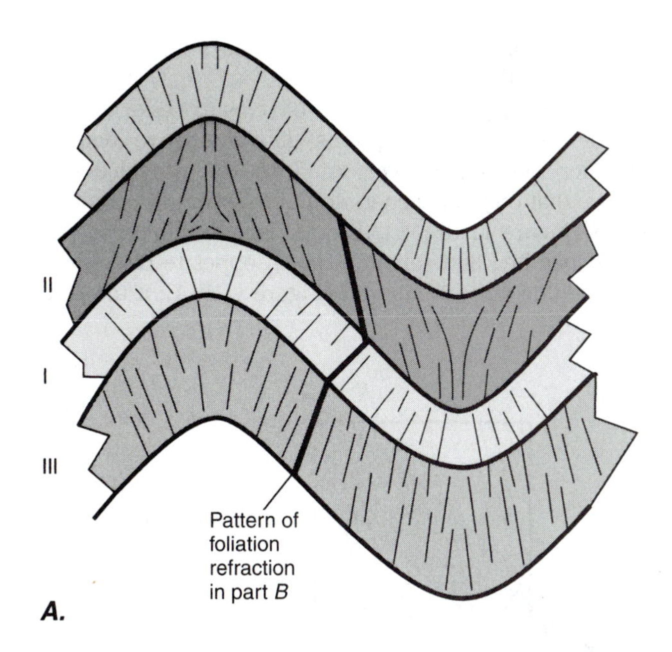

FIGURE 11.13 Convergent and divergent foliation fans on folds and the refraction of foliation across lithologic contacts. *A.* The typical pattern of foliations in a folded sequence of sandstone (I), shale or slate (II), and siltstone (III). The foliation pattern is convergent in the sandy (I) and silty (III) layers and divergent in the slate (II). Foliation orientation is "refracted" at the contacts between the layers. The heavy line across the middle limb emphasizes comparable foliation orientations shown in *B*. *B.* A sandstone (I), shale (II), siltstone (III) sequence showing the refraction of the foliation at the contacts. The photo illustrates the orientations of the beds shown in the middle limb of the diagram in *A*. (Photo courtesy of P. Stringer)

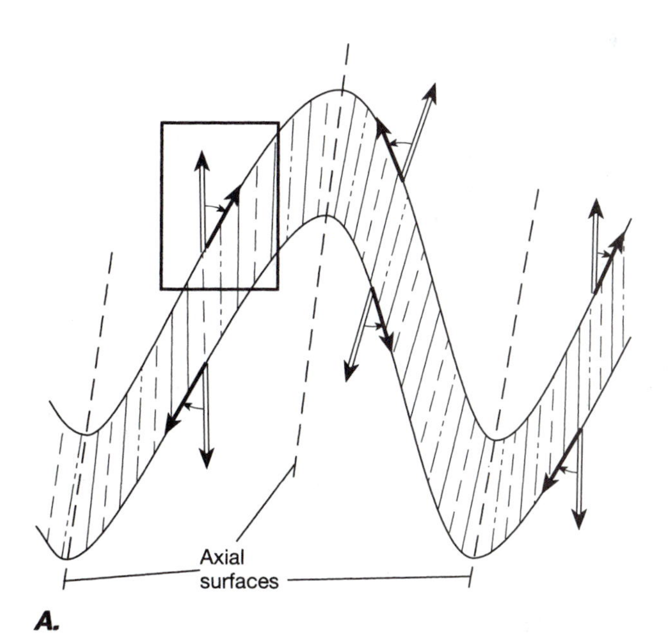

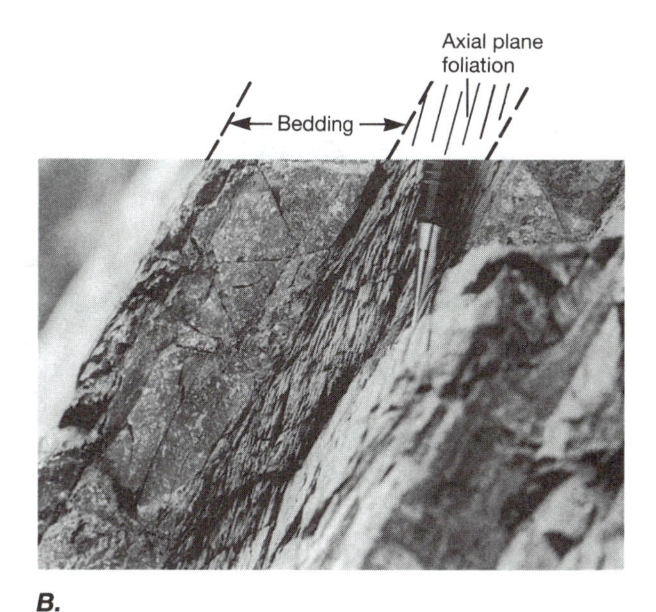

FIGURE 11.14 The use of bedding-foliation relationships to deduce the location of fold closures and axial surfaces. After one generation of folding, a foliation plane cannot cut a given bedding surface more than once. This is not true if there have been two or more generations of folding (see Figure 11.15*D*). *A.* Folded bedding with an axial foliation. Hollow arrows point along the foliation. Solid arrows point along bedding planes. The sense of rotation of the hollow arrow through the acute angle from the foliation toward the bedding (solid arrow) changes across an axial surface. The box outlines the foliation-bedding relationship shown in *B*, and the fold indicates the correct inference for the direction of fold closure. *B.* Bedding-foliation relationship in interbedded sandstone and shale. The foliation is obvious in the shale layer. The bedding-foliation relationship in the photograph indicates the fold closes upward to the right and downward to the left as shown in *A*. Marathon thrust belt, Texas. (Photo courtesy of R. J. Varga)

surface (Figure 11.14*A*) and then rotated through the acute angle between the foliation and the bedding. The sense of rotation is the same as the sense of asymmetry that a higher-order fold would have at the same location (cf. Figure 10.24*B*): clockwise (z) on the left side of an antiformal fold and counterclockwise (s) on the right side (Figure 11.14*A*). The sense of rotation changes across an axial surface. Thus it can be used to map the locations of axial surfaces and to infer the direction of closure of the larger fold, even if the exposure does not permit direct observation of these features.

Inferring the location of the fold closures as described in the preceding paragraphs allows one to deduce whether the bed is overturned or not, provided one knows that only one generation of folds has affected the rocks (Figure 11.15*A*–*C*). Another rule of thumb is helpful: If bedding and foliation dip in opposite directions, the bedding must be upright (Figure 11.15*A*); if the bed-

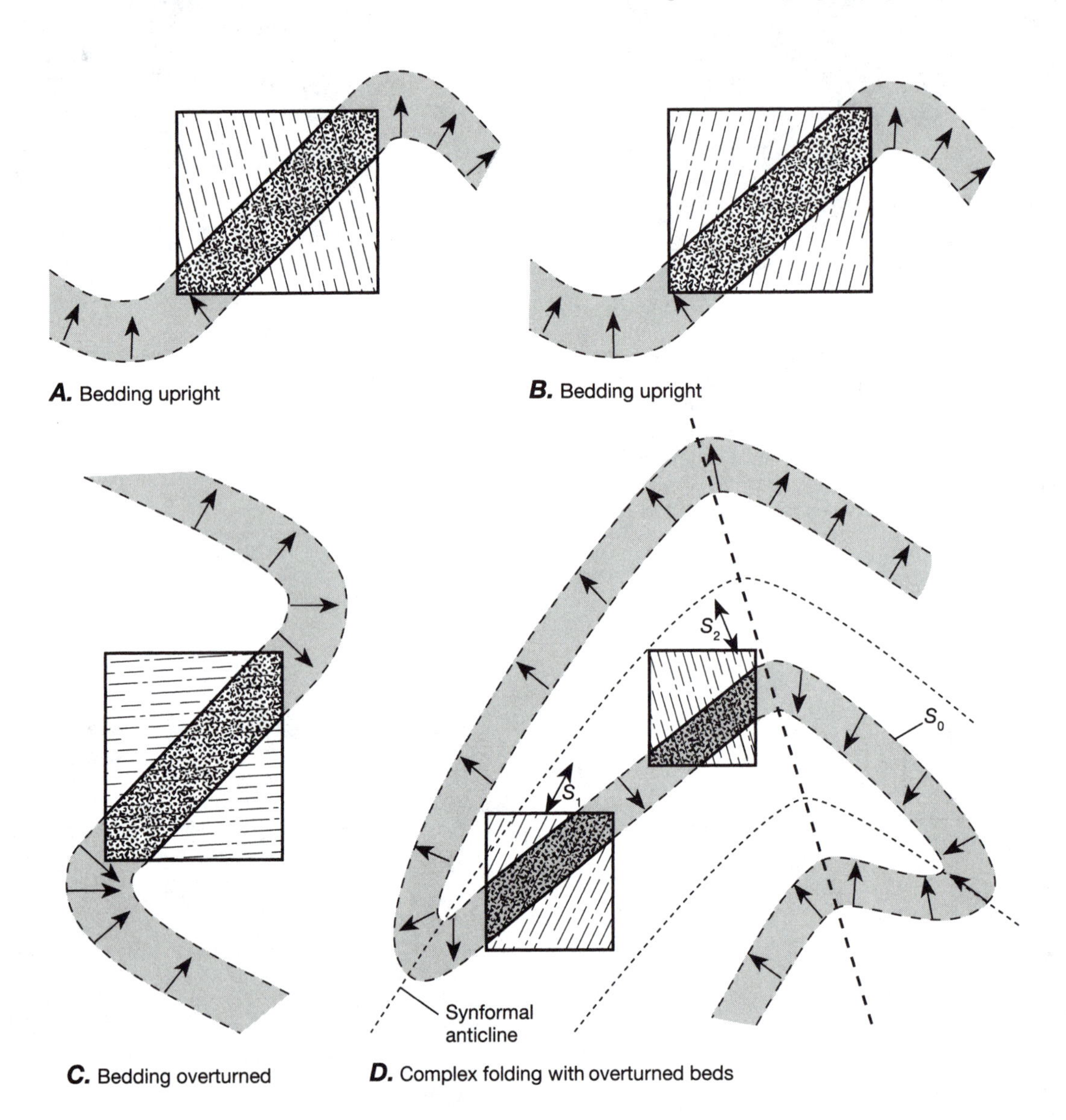

FIGURE 11.15 Use of bedding-foliation relationships to deduce the stratigraphic-up direction in simply folded layers. The boxes outline the part of the structure that is assumed to be observable in the field. The dashed and lightly shaded continuation of the folded layer shows the unobservable portion. The arrows in the folded layer indicate the stratigraphic-up direction. *A*. Bedding is upright if bedding and foliation dip in opposite directions. *B*. Bedding is upright if bedding and foliation dip in the same direction, and bedding has the shallower dip (cf. Figure 11.14*B*). *C*. Bedding is overturned if bedding and foliation dip in the same direction and bedding has the steeper dip. *D*. For two or more generations of folding, the bedding-foliation relationships do not give reliable results for the stratigraphic-up direction. In this example, note that the foliation could be S_1, parallel to the first-generation axial surface (lower box), or S_2, parallel to the second-generation axial surface (upper box), and in both cases the bedding is overturned. Comparison of the upper box with *A* and the lower box with *B* shows that the technique does not work in this case.

ding and the foliation dip in the same direction, the bedding is upright if it has a shallower dip than the foliation (Figure 11.15*B*), and it is overturned if it has a steeper dip than the foliation (Figure 11.15*C*).

This method for determining the stratigraphic-up direction does not work if multiple generations of folding have affected the rocks. In this case, the folding is complex, and different foliations may develop in association with the different fold generations. In places on a synformal anticline, for example, the boxes in Figure 11.15*D*, neither the S_1 nor the S_2 foliation provides the correct indication that the bedding is overturned (cf. Figure 11.15*A*, *B*). Thus the method must be applied with caution. When deformation is complex, we must rely on "way-up indicators" or geopetal structures, such as graded bedding and a variety of sedimentary structures preserved in the rocks themselves.

Rocks that display multiple generations of deformation commonly have two foliations, which may be the same or different types and may or may not be equally well developed. The earlier foliation commonly becomes folded with a second foliation developed subparallel to the axial surface of the second-generation folds. This relationship is most obvious in the hinge zone of the second-generation folds. On the limbs, the two foliations may be parallel and completely indistinguishable so that the rock appears to contain only one foliation. If multiple foliations are present, it is important to recognize and distinguish them while mapping, because they provide information about different parts of the deformation history and must be separated in the analysis of the structure of the area.

ii. Relationship with Ductile Shear Zones

In ductile shear zones, the rocks may contain two foliations, labeled S and C, both of which have developed during a single deformation (Figure 11.16). Such rocks are called **S-C tectonites**. The **S-foliation**[7] is a continuous coarse foliation defined by the preferred orientation of mica grains and commonly by elongate quartz grains; its predominant orientation is oblique to the ductile shear

[7] Again, the *S* derives from the German word *schiefer*. See footnote 3.

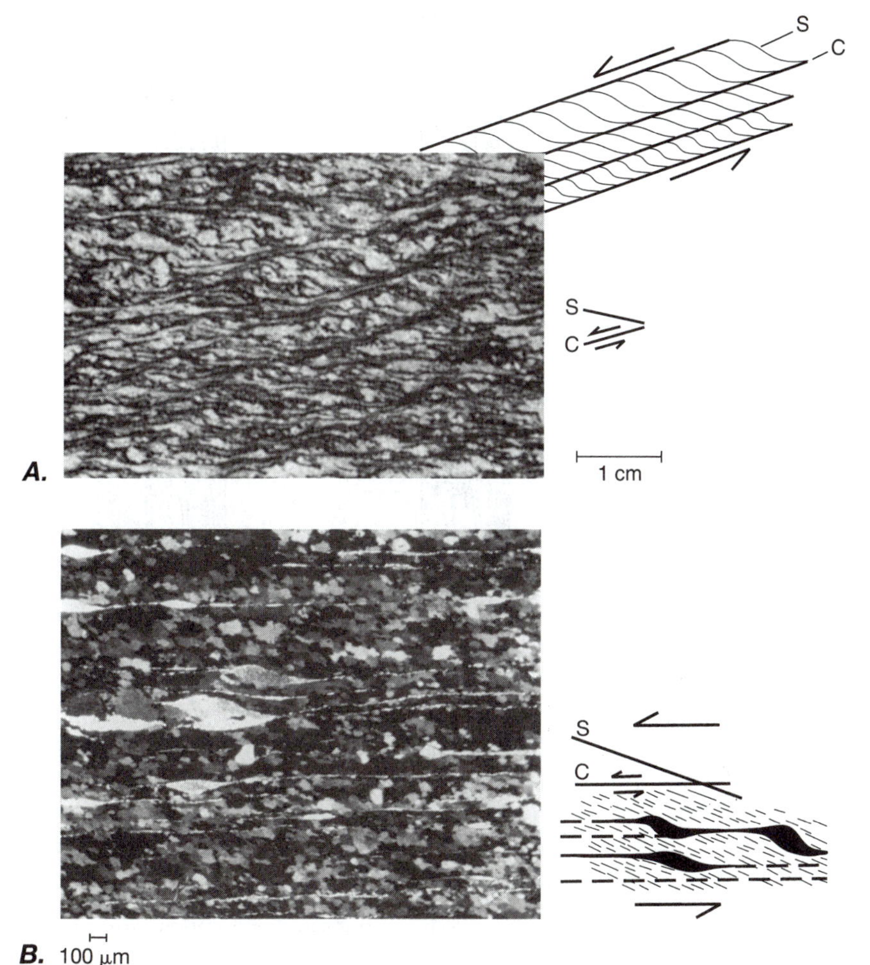

FIGURE 11.16 Foliations in an S-C tectonite. *A*. Type I S-C tectonite in a ductile shear zone in a granodiorite. The S-foliation is a continuous coarse mica foliation that curves toward an orientation parallel to the spaced C-foliation, which in turn is parallel to the shear zone boundaries. This sense of curvature is the same as the general shear sense and indicates a sinistral shear. The idealized geometry is shown in the adjacent diagram (Simpson and Schmid 1983). *B*. Type II S-C tectonite in a quartz-rich mylonite. The S-foliation is defined by the grain-shape foliation of the quartz and the preferred orientation of large mica porphyroclasts ("mica fish"). The C-foliation is the shear plane defined by the trails of fine micas commonly connected to the tips of the porphyroclasts. The sense of curvature from the mica porphyroclast tips to the mica trails is the same as the sense of shear on C and indicates sinistral shear. The idealized relationships are shown in the diagram, in which the micas are shown by heavy black lines and the quartz grain shape foliation by thin lines.

zone. The **C-foliation**[8] is a set of shear bands in the rock that develop subparallel to the boundaries of the shear zone and may have fibrous crystals (slickenfibers; see Section 11.9) lying on and subparallel to the foliation surfaces, indicating it was a shear surface during the deformation. If platy minerals are relatively abundant, a type I S-C tectonite develops in which the C-foliation crosscuts an S-foliation that has a sigmoidal shape between adjacent C surfaces (Figure 11.16*A*). If platy minerals are relatively sparse, as in some micaceous quartzites, a type II S-C tectonite develops. The S-foliation is defined by the preferred orientation of the large mica grains, called mica porphyroclasts or "fish," and by a grain shape foliation in the quartz. The C-foliation is defined by thin seams of very fine-grained mica connected to the ends of the mica fish. In both cases, micas in the S-foliation curve toward parallelism with C, and the sense of curvature defines the shear sense on the shear zone: Counterclockwise indicates sinistral shear, clockwise indicates dextral shear. With large amounts of shearing, the S- and C-foliations may become essentially parallel and indistinguishable, and a new foliation, labeled C′, may develop with characteristics similar to the C-foliation, but oriented at a low angle to the shear zone boundaries.

Although the S-C morphology is similar to some examples of crenulation foliation (e.g., Figure 11.9*C*), it is important to recognize that the S- and C-surfaces form during the same deformation, whereas crenulation foliations result from the superposition of two separate deformations.

A **transposition foliation** results from a superposition of a tectonite foliation on an earlier compositional layering, for example bedding or a compositional foliation. With progressive deformation, the compositional layering becomes isoclinally folded and dismembered (Figure 11.17), so that the folds are no longer recognizable except for scattered **rootless folds,** which are isolated isoclinal fold hinges that have axial surfaces parallel to the foliation and are not connected to any other hinges. The earlier layering is transformed into a discontinuous banding parallel to the new foliation.

Recent advances in radiometric dating, principally of the U-Pb, Rb-Sr, and K-Ar decay systems, have made it possible to date very small quantities of material. It has become possible to determine quantitatively the ages of minerals in separate parts of complex shear zones. This work promises to revolutionize our ideas about the relative ages of mineral-forming events in these structures and thus contribute to a significant refinement of understanding of the orogenic history in the regions where they are found.[9]

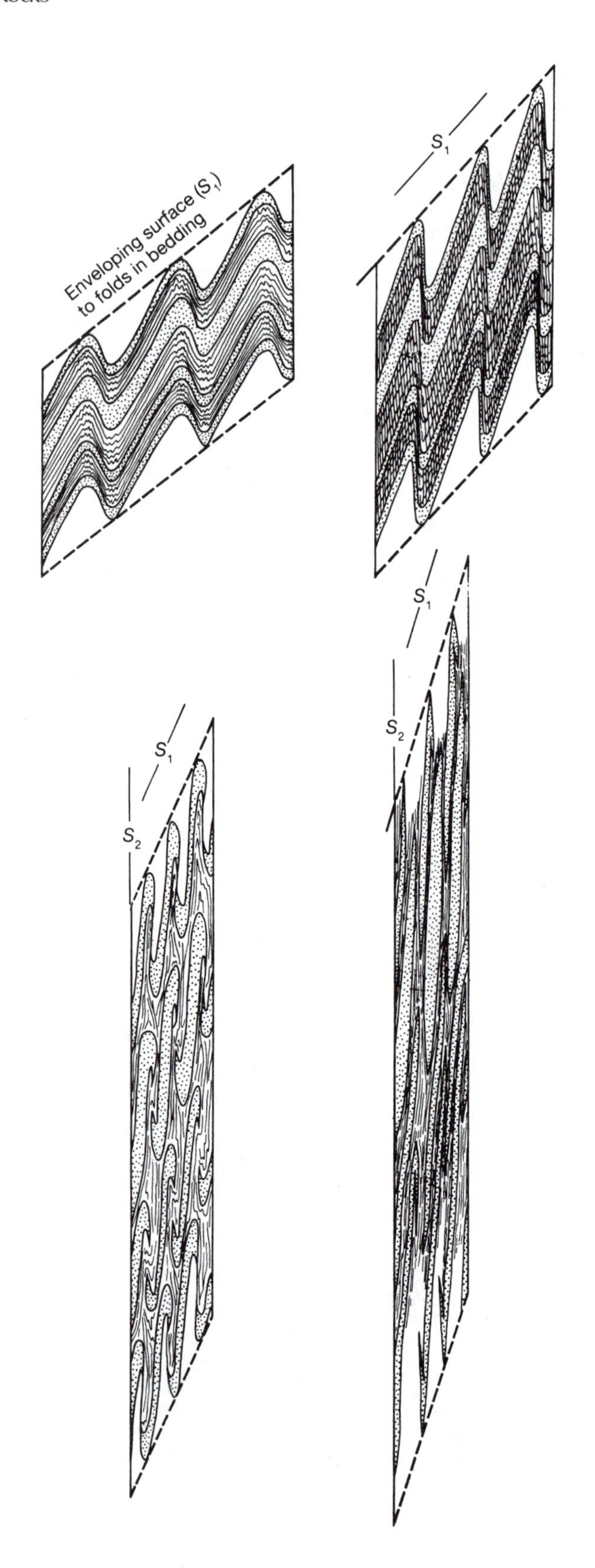

FIGURE 11.17 A possible sequence in the development of a transposition foliation. S_1 is the enveloping surface to the folds in bedding. S_2 is the transposition foliation. (From Turner and Weiss 1963)

[8]The *C* derives from the French word *cisaillement*, which means "shear."

[9]See references at the end of this chapter.

11.7 SPECIAL TYPES OF FOLIATION AND NOMENCLATURE

Many terms for various types of foliations exist in the geologic literature. Some terms are strictly descriptive and are therefore useful in referring to specific morphologic features. Others have genetic connotations that may not be correct. We strongly recommend abandoning the use of genetic terms for descriptive purposes because many of these terms are not well defined and their use leads to potentially misleading assumptions as to the origin of the structures. Interpretation, of course, has a valid place in any scientific investigation, but the use of interpretive terms for descriptive purposes inevitably leads to confusion. We review some common terms here and indicate what we believe to be the equivalents in the morphological classification presented in Sections 11.2 through 11.6.

Four terms—*slaty cleavage, phyllitic cleavage, schistosity,* and *gneissic foliation*—are not as specific as categories in the morphological classification given earlier, but they remain useful terms for general and field description. The first three of these essentially describe a continuum in grain size for foliations in rocks containing abundant platy minerals. The last pertains to rocks in which platy minerals are not abundant.

Slaty cleavage refers to fine continuous foliations characteristic of slates. Slates are very fine-grained, low-grade metamorphic rocks that contain abundant sheet silicates (generally clays, chlorites, and micas). They may also contain subordinate amounts of silty and carbonaceous material. The foliation may be either continuous or microspaced, but in the latter case, the microdomain spacing certainly cannot be recognized in the field. The foliation provides a very strong cleavage to the rock, along which the rock breaks easily and tends to weather preferentially. Rocks with slaty cleavage traditionally have been a valuable source of materials for such uses as roofing slates and blackboards.

Phyllitic cleavage resembles slaty cleavage except that the grain size of the rock is slightly coarser. It characterizes phyllites, which are low-grade (greenschist), fine-grained, metamorphic rocks containing abundant micas, chlorite, or both. In hand samples, the surface of the foliation has a sheen to it, and individual sheet silicate flakes may just be resolvable with a good hand lens. The foliation is generally intermediate between fine and coarse continuous foliation, although some phyllitic cleavages may be smooth disjunctive foliations. The foliation strongly affects the rock's weathering pattern.

Schistosity refers to the foliation found in coarse-grained, mica-rich, medium- to high-grade metamorphic rocks. Chlorite, biotite or muscovite defines the foliation, and the mineral grains are coarse enough to be visible with the unaided eye. It may appear as an anastomosing to smooth disjunctive foliation or as a coarse continuous foliation. It provides a strong cleavage to the rock.

Gneissic foliation refers to foliations that develop in gneisses, which are coarse-grained, high-grade metamorphic rocks in which platy minerals are sparse or absent. The term includes compositional foliations, as well as coarse continuous foliations defined by the alignment of sparse platy minerals, flattened mineral grains, or needle-shaped mineral grains. The foliation generally provides at best a weak cleavage.

The terms discussed in the next three paragraphs all have a genetic connotation and will not be used in this book. We include them for the sake of completeness and, we hope, for strictly historical interest and use in understanding the older literature.

Flow cleavage is a loosely defined term that seems to have been applied to continuous axial surface foliations interpreted to have been the result of a large amount of ductile deformation in the rocks. It was commonly, and erroneously, interpreted to represent the orientation of flow (shear) planes in the rock during ductile deformation.

Fracture cleavage refers to a variety of disjunctive foliations or discrete crenulation foliations. The term has most often been applied to disjunctive foliations in which the microlithon has little or no fabric and the cleavage domains are thin and can have the superficial aspect of a penetrative set of fractures, especially on weathered surfaces. The term is misleading because the fractures that are observed are in general secondary structures that form along previously developed foliation planes.

Shear cleavage, solution cleavage, and **strain-slip cleavage** are terms that have been used to describe a variety of spaced foliations. Solution cleavage refers to disjunctive foliations, especially at the more irregular end of the scale. Shear cleavage and strain-slip cleavage both refer to crenulation foliations. None of these terms is well defined and none should be used descriptively. If solution or shearing has been independently demonstrated, the term *solution cleavage* or *shear cleavage* is acceptable as an interpretative term.

11.8 STRUCTURAL LINEATIONS

Figure 11.18 shows a morphological classification of lineations in deformed rocks. We discuss the two main subdivisions, structural lineations and mineral lineations, in this and the next section.

Structural lineations are defined by the preferred orientation of a linear structure contained within a rock. They include **discrete lineations,** which are formed by the deformation of discrete objects such as ooids, pebbles, fossils, and alteration spots; and **constructed lineations,** which are formed from planar features constructed or deformed during the deformation and include the intersection of two foliations, crenulation hinge lines, boudin lines, structural slickenlines, and mullions (Figure 11.18).

Lineations in tectonites (surficial or penetrative)	Structural	Discrete	Pebbles Ooids Fossils Alteration spots
		Constructed	Intersections Hinge lines Boudin lines Mullions Structural slickenlines
	Mineral	Polycrystalline	Rods Mineral clusters Mineral slickenlines Nonfibrous overgrowths
		Mineral grain	Acicular habit grains Elongated grains Mineral fibers Fibrous vein filling Slickenfibers Fibrous overgrowths

FIGURE 11.18 Morphological classification scheme for lineations.

i. Discrete Lineations

Ductile deformation of the rock may distort discrete objects in the rock into well-aligned elongate shapes. Discrete lineations of this nature include stretched pebble conglomerates (Figure 11.19*A*), deformed oolitic limestone (see Figure 12.6*B*), and slates with alteration spots (Figure 11.19*B*). In these cases, objects that were roughly spherical before deformation are deformed into ellipsoidal shapes whose long and intermediate axes (***a*** and ***b***, respectively, in Figure 11.19*C*) may define a foliation, and whose long axes define a lineation. The true orientation of the lineation is apparent only on planes that contain the ***a*** axis of the ellipsoid (Figure 11.19*C*). Although other sections through an ellipsoid are generally elliptical in shape, the long axis of such an ellipse is not the true lineation.

Alteration spots, also called reduction spots, are volumes in rock, most commonly slates, that are distinguished mainly by color differences caused by chemical alteration of some of the rock's components (Figure 11.19*B*). They may be initially spherical features that develop in the sediment shortly after deposition and become deformed into an ellipsoidal shape that defines a lineation.

ii. Constructed Lineations

A variety of lineations fall into the category of constructed lineations, and they have in common the characteristic that the structures originated during deformation of the rock.

The intersection of two planar elements such as two foliations, one of which may be bedding, defines an **in-**

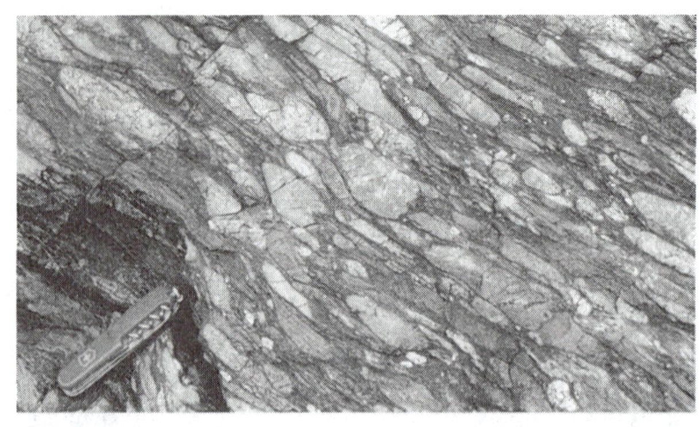

A.

B.

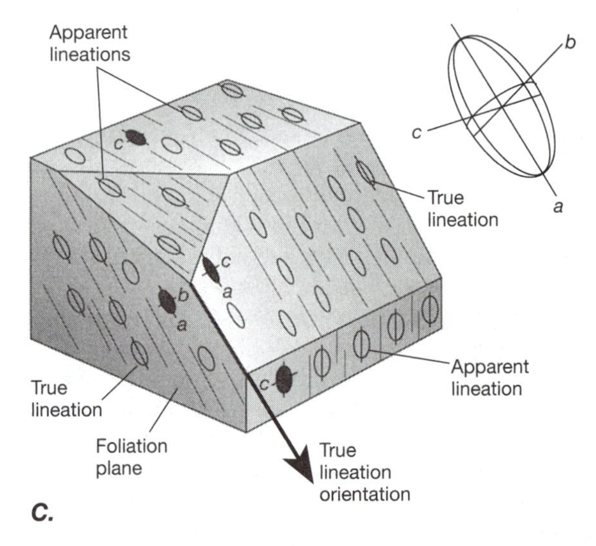

FIGURE 11.19 Discrete lineations. *A*. A stretched pebble conglomerate showing quartzite pebbles flattened parallel to the foliation and elongated to define a lineation. (Photo courtesy of M. G. Miller) *B*. Alteration spots in a slate. The view is parallel to the foliation, showing the maximum and minimum axes of the ellipsoids (*a* and *c*; see also *C*). (Photo courtesy of C. Simpson) *C*. True and apparent lineations associated with ellipsoidal structures. The true lineation orientation is shown on any plane containing the *a* axis (longest axis) of the ellipsoid. Planes of other orientations show elliptical sections through the ellipsoidal structures that do not define the true orientation of the lineation.

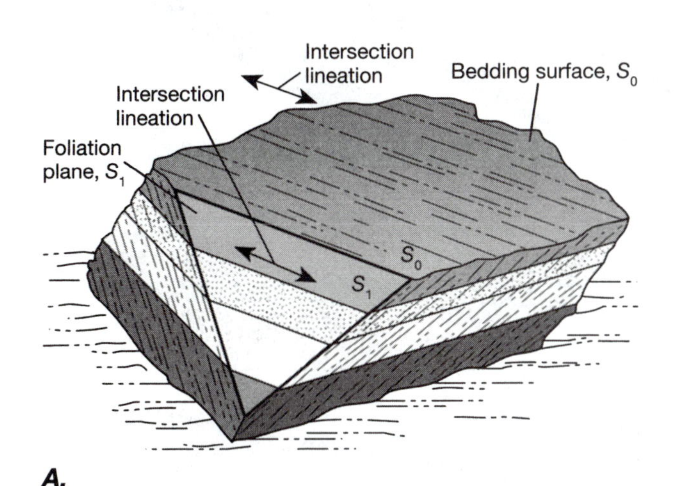

FIGURE 11.20 Intersection lineations. *A*. The intersection of a foliation and a bedding surface. The trace of the secondary foliation S_1 on the bedding S_0 as well as the trace of S_0 on S_1 are essentially the same lineation. *B*. Pencil cleavage in argillite, an intersection lineation defined by the intersection of two foliations, one of which may be bedding. Cleavage of the rock along both foliations produces elongate prisms, or pencils, of rock. (Photo courtesy of R. J. Varga)

tersection lineation. If one foliation is defined by platy minerals or flattened mineral grains, the lineation appears on the intersecting foliation as the parallel alignment of the edges of platy mineral grains or of the long axes of the flattened mineral grains (Figure 11.20). If one foliation is bedding or a spaced foliation in which the cleavage domains and the microlithons differ in mineralogy, the intersection lineation may appear on the intersecting foliation as streaks of different composition. Two intersecting foliations can produce a lineation called a **pencil cleavage** if the rock tends to cleave along both foliations, producing elongate rhombic prisms or "pencils" (for example, the intersection of S_1 with the S_0 surface in Figure 11.20*B*).

Fold hinge lineations are defined by the preferred orientation of microfold or crenulation hinges developed in foliations. The crenulations may, but need not, be associated with a crenulation foliation. On a regional scale, the orientation of hinges of outcrop-sized folds may be treated as a regionally penetrative lineation.

Boudins[10] are linear segments of a layer formed when the layer has been pulled apart along periodically spaced lines of separation called **boudin lines** (Figure 11.21*A*). Boudins are most easy to recognize in an exposure that is at a high angle to the boudin line (Figure 11.21). The process of forming boudins is called **boudinage**.

Boudins display a wide variety of shapes, many of which are summarized schematically in Figure 11.21. Shapes transitional between the idealized shapes shown are common. **Pinch-and-swell** structures (Figure 11.21*B*) are periodic oscillations in the thickness of a bed, with the "pinches" becoming thinner as the amount of lengthening of the bed increases (left to right in Figure 11.21*B*). **Tapered boudin blocks** may be connected by a neck (Figure 11.21*C*) or by a thin selvage of the original layer (Figure 11.21*D*), or they may actually be separated form one another (Figure 11.21*E*); these three diagrams represent progressively larger amounts of lengthening parallel to the layer. If the boudins are connected by necks, the boudin line is also referred to as a **neck line** (Figure 11.21*A*). Boudins also may be shaped like rectangular **blocks**, for which the separations in the layer can appear like extension fractures (Figure 11.21*F*). Boudins may also form by shearing along a surface oblique to the layering, in which case the cross section is shaped like parallelograms (Figure 11.21*G–I*). Sheared boudins may be in contact with one another (Figure 11.21*G*), they may be connected to one another by only a thin selvage of the original layer (Figure 11.21*H*), or they may even be totally disconnected (Figure 11.21*I*). These three diagrams again represent increasing lengthening parallel to the layer. Shearing is commonly accompanied by local rotation of the boudin blocks.

For tapered boudins, any fine laminations in the boudinaged layer itself also bend and thin into the area where the boudin block is necked or pinched off. The material

[10]From the French word *boudin* meaning "blood sausage." The term refers to the similarity between a set of boudins and a row of sausages laid out side by side (see Figure 11.21*A*). In a plane at a high angle to the boudin lines, a set of boudins can also resemble a chain of link sausages (Figure 11.21*C, D*), but, as pointed out by Pollard and Fletcher (2005), this analogy ignores their shape in the third dimension and has caused confusion as to the origin and significance of the term.

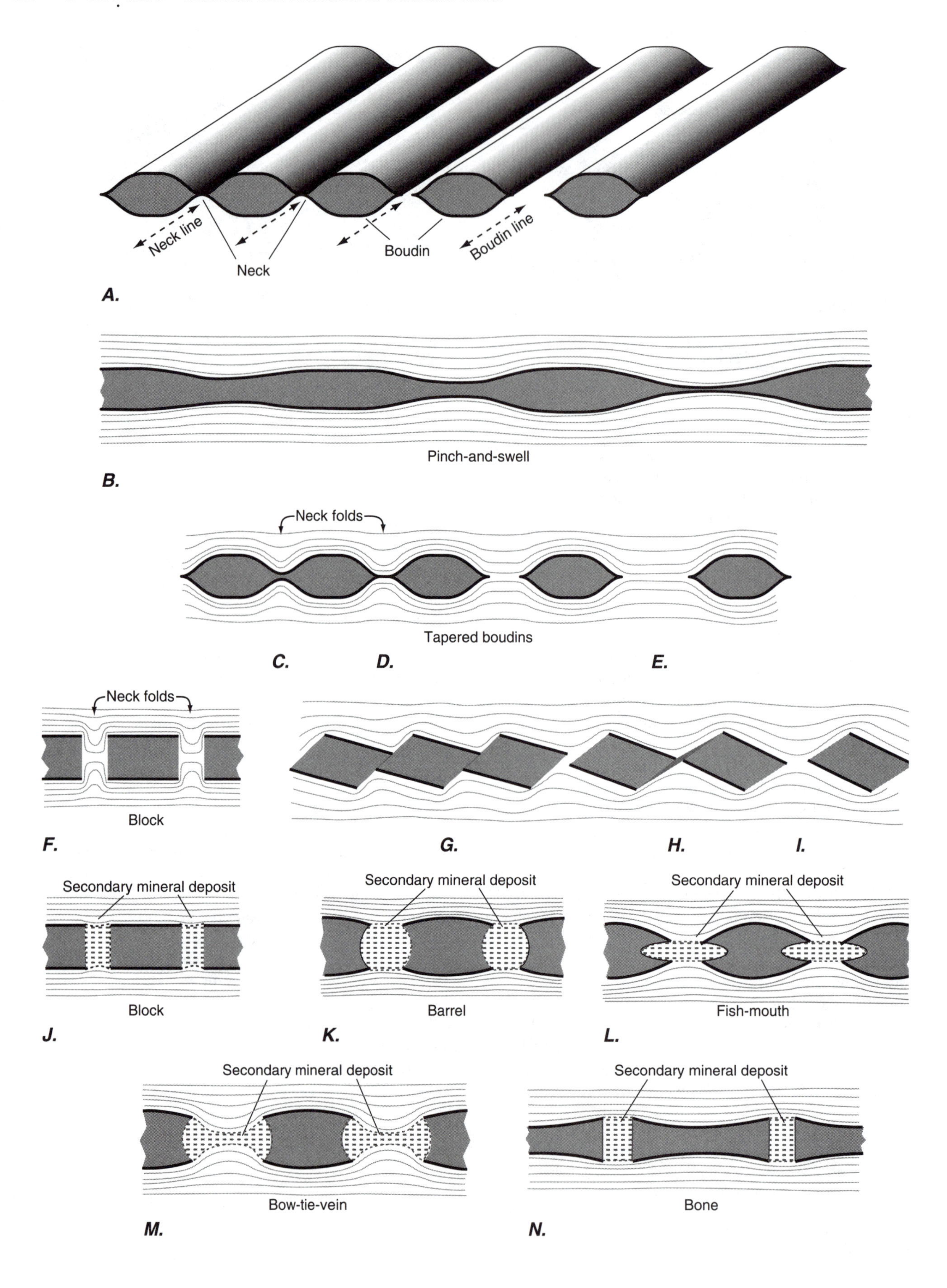
Neck line
Neck
Boudin
Boudin line
A.
Pinch-and-swell
B.
Neck folds
Tapered boudins
C.
D.
E.
Neck folds
Block
F.
G.
H.
I.
Secondary mineral deposit
Block
J.
Secondary mineral deposit
Barrel
K.
Secondary mineral deposit
Fish-mouth
L.
Secondary mineral deposit
Bow-tie-vein
M.
Secondary mineral deposit
Bone
N.

FIGURE 11.21 Boudins. (In part after Goscombe et al. 2004) *A*. Diagram of a boudinaged layer showing the relation between the boudins, boudin lines, necks, and neck lines. Dashed arrows are parallel to the neck lines or boudin lines. *B*. Pinch-and-swell structure. From left to right, the "pinch" becomes thinner as the amount of lengthening of the layer increases. *C–E*. Boudins connected by (*C*) a neck or (*D*) a thin selvage of the original layer or (*E*) completely disconnected from one another; these three diagrams reflect increasing amounts of lengthening of the layer. *F*. Rectangular block-shaped boudins in which the extensional spaces between the boudin faces are filled with neck folds of the surrounding rock. *G–I*. Boudins shaped like parallelograms with shear along the boudin faces. Boudin blocks (*G*) remain in contact; (*H*) are connected by a thin selvage of the original layer; or (*I*) are completely separated; these three diagrams again reflect increasing lengthening of the layer. *J–N*. Boudins for which all (*J*) or part (*K–N*) of the extension is accommodated by precipitation of a secondary mineral along the extensional face of the boudin. (*J*) Block-shaped boudins; (*K*) barrel-shaped boudins; (*L*) fish-mouth boudins; (*M*) bow-tie-vein boudins; and (*N*) bone-shaped boudins.

that fills the spaces between separated boudins is commonly just the surrounding rock within which the original layer was embedded. In cases where the rock on either side of the boudinaged layer is finely laminated, the laminations near the necks commonly form disharmonic folds, called **neck folds**. These folds conform to the shape of the boudin blocks at the interface and die out a short distance away (e.g., see neck folds labeled in Figure 11.21*C*, *D*, *F*, and other diagrams in Figure 11.21). The space between the boudin faces may also be filled with a secondary mineral, commonly quartz or calcite (Figure 11.21*J–N*). A variety of boudin shapes can develop, including block-shaped boudins (Figure 11.21*J*), **barrel boudins** (Figure 11.21*K*), **fish-mouthed boudins** (Figure 11.21*L*), **bow-tie-vein boudins** (Figure 11.21*M*), and **bone-shaped boudins** (Figure 11.21*N*).

Most boudin lines define a pronounced lineation. Some, however, display much scatter in orientation, and others may occur in two intersecting sets, with tablet-shaped boudins called **tablet boudinage** or **chocolate-tablet structure**, because of its resemblance to the tablets of a chocolate bar (cf. Figure 12.7*F*).

Mullions are linear fluted structures developed within a rock or at lithologic interfaces. The name derives from the resemblance of the geologic structure to the vertical fluted architectural structures, called mullions, that separate windows in Gothic cathedrals. They are characterized in cross section by convex surfaces with intervening cusps (Figure 11.22*A*) or by alternating convex and concave surfaces (Figure 11.22*B*). Characteristically they have a cross-sectional dimension of a few to several tens of centimeters and an indefinite linear dimension.

The surfaces of **fold mullions** are defined by parting along the cylindrically folded surfaces of layers or foliations. At a boundary between two thick layers of very different competence, such as a sandstone and a shale, the mullions appear in the more competent member as cylindrical surfaces convex toward the incompetent rock, joined by cusps that point into the competent rock (Figure 11.22*A*). This type of mullion is restricted to the bedding surface.

Irregular mullions are long fluted structures showing an irregular cross section that conforms in general with neither bedding nor foliation (Figure 11.22*B*). The surfaces of the mullions may be covered with a thin film of mica, and the surface of one mullion fits exactly against that of its neighbors. Some structures of this nature are fault mullions and are the result of irregularities in the fracture surface (Figure 3.8*B*). Other irregular mullions, such as in Figure 11.22*B*, are not well understood.

Structural slickenlines (Figure 3.8*A*, *B*) are grooves and ridges that appear on slickensides, the fine-grained polished surfaces that commonly develop along faults.[11] Larger-scale undulations in the fault surface (Figure 3.8*B*) are referred to as **fault mullions**.

11.9 MINERAL LINEATIONS

Mineral lineations consist of a preferred orientation of either single elongate mineral grains or elongate polycrystalline aggregates (Figure 11.18). **Mineral grain lineations** are formed by the parallel alignment of individual acicular[12] mineral grains such as amphibole, by grains of minerals that have been stretched into an elongate shape, or by mineral fibers that have grown in a preferred orientation. **Polycrystalline mineral lineations** are formed by the preferred orientation of elongate clusters of grains of a particular mineral measuring at least a few grains in diameter. A preferred orientation of a crystallographic axis of the mineral is commonly associated with both types of lineation, although it need not be parallel to the lineation orientation. Mineral lineations may occur as surficial lineations on lithologic contacts, foliation

[11]Slickensides are surfaces, although a number of authors use the term to refer to the lineations in the surface. Three types of lineations develop on slickensides: structural slickenlines, mineral slickenlines, and slickenfibers. The first is described here; the other two are described in the next section. We discuss the mechanisms by which these lineations form in Sections 14.2 and 14.6.

[12]From the Latin word *aciculus*, meaning "needle-like."

surfaces, or fault surfaces and as penetrative lineations in the rock.

i. Polycrystalline Mineral Lineations

A variety of structures fall into the general category of polycrystalline mineral lineations.

Rods are polycrystalline mineral lineations formed by rod-shaped concentrations of a particular mineral, commonly quartz (Figure 11.23). The rods may appear in cross section to be isolated cylindrical masses, rootless fold hinges, or layers or fold limbs that have been boudinaged. Thus they could in some cases be classified also as constructed structural lineations. They vary from approximately one to several tens of centimeters in diameter and typically are present parallel to a foliation plane and to the local orientation of fold hinges.

In many metamorphic rocks, **mineral cluster lineations** form small elongate concentrations or clusters of individual minerals on the scale of a millimeter to several centimeters (Figure 11.24). The texture of the minerals in the clusters is no different from that in any other part of the rock. The lineations may be quite subtle, such as the small polycrystalline trains of muscovite or quartz in a quartz-feldspar-muscovite schist, or they may be strikingly obvious as the lineations defined by elongate clusters of quartz, feldspar, and biotite in a gneiss (Figure 11.24). The lineations generally lie in a foliation plane, but in the case of a so-called **pencil lineation**, the rock is an L-tectonite, with a fabric dominated by a strong mineral cluster lineation and no evident foliation.

Mineral slickenlines appear as streaks developed on slickensides in fault zones (Figure 11.25). The streaks are probably the remnants of mineral grains or aggregates sheared out in the slickenside material, but the grain size is so small that individual mineral grains generally cannot be identified even with a hand lens. These lineations may not always be distinguishable from structural slickenlines such as those shown in Figure 3.8*A*, *B*, and the two types of lineation commonly are present together.

Nonfibrous overgrowths are concentrations of one mineral, commonly quartz, around inclusions or grains of another mineral such as garnet or pyrite. Both nonfibrous and fibrous overgrowths are commonly referred to col-

A.

B.

FIGURE 11.22 Mullions. *A*. Fold mullions in a sandstone at the contact with a shale (now eroded away). The mullion surface is restricted to the bedding surface; it is not a closed structure in cross section. (From Hobbs et al. 1976) *B*. Irregular mullions showing the irregular cross section and the strongly cylindrical structure of the lineation. The mullion surfaces may be coated with a thin film of mica. (Weiss 1972, figure 65A)

FIGURE 11.23 Quartz rod lineations. Rods are generally parallel to local fold hinges, and they may be isolated fold hinges or boudinaged fold limbs. In some cases, therefore, they could be classified as structural lineations as well. Northern Snake Range, Nevada.

lectively by the genetic term **pressure shadows**. Such overgrowths may define a polycrystalline mineral lineation if the overgrowths are elongate and have a preferred orientation. Mineral grains in the overgrowth do not necessarily have a dimensional or crystallographic preferred orientation, so the lineation is defined strictly by the dimensional orientation of the overgrowth (see Figure 11.26*B*).

ii. Mineral Grain Lineations

Three types of mineral grain lineations commonly exist in rocks and are formed respectively by acicular (needle-shaped) minerals, by elongate mineral grains, and by mineral fibers. The grain shape may, but need not, be simply related to the crystallography.

Some mineral grains, such as amphiboles or sillimanite, naturally grow with a prismatic or acicular habit. If their long axes have a preferred orientation, such minerals define an **acicular habit lineation**. If one crystallographic axis (e.g., the c axis in amphiboles and sillimanite) parallels the long axis of each mineral grain, the lineation is parallel to a crystallographic preferred orientation.

FIGURE 11.24 Mineral cluster lineation in a quartz-feldspar-biotite schist defined by elongate concentrations of quartz and feldspar and of biotite. Northern Snake Range, Nevada.

FIGURE 11.25 Mineral slickenlines on the slickenside of a fault surface. Lake Meade shear zone, southern Nevada.

Under some conditions, **elongate grain lineations** may form in a rock by deformation of pre-existing equant mineral grains into aligned elongate forms. Such mineral grains approximate triaxial ellipsoids in shape, and the lineation is parallel to the longest axis of the ellipsoids (cf. Figure 11.19*C*). These lineations are similar to the discrete lineations described in Section 11.8. Crystallographic axes commonly are aligned as well, but that alignment need not be parallel to the morphologic alignment of the mineral grains.

Mineral fiber lineations are formed by very elongate crystal grains of a particular mineral, commonly quartz, calcite, chlorite, or serpentine. The structure and composition of the mineral fibers are so distinct from the rock in which they occur that it is clear the fibers have grown in the rock during deformation. They occur packed densely together in fibrous sheets or bunches in which all the fibers are strongly aligned in either a linear or curvilinear arrangement (Figure 11.26*A*).

Mineral fiber lineations are commonly found as surficial lineations as both **fibrous vein fillings** in veins and slickenfiber lineations along fault planes (Figures 3.8*C* and 11.26*A*). In both cases, the mineral fibers have a very strong preferred orientation, which is commonly at a high angle to the vein wall and at a low angle to the fault surface. Mineral fiber lineations may also occur as a penetrative lineation where they occur in strongly oriented **fibrous overgrowths** on crystals or particles throughout the rock (Figure 11.26*B*). These lineations are common in low-grade metamorphic rocks. The growth of the fibers may result from several different mechanisms, and the correct interpretation of the fibers depends on understanding the mechanism by which they formed. We discuss the mechanisms in Chapter 14.

If the fibers in any of these mineral fiber lineations are strongly curved (Figure 11.26*A*), it may be difficult to define a unique lineation for the rock. Nevertheless, the study of any of these mineral fibers can yield significant information about the history of the deformation during which they grew.

The very strong linearity of the fibers need not reflect a comparable preferred orientation of their crystallographic axes. Many mineral fiber lineations display nearly random distribution of crystallographic axes, although most quartz and calcite fiber lineations have a strong crystallographic preferred orientation.

A.

B.

FIGURE 11.26 Mineral fiber lineations. *A*. Curvilinear serpentine slickenfibers on a fault surface. Feather River peridotite, northern California. *B*. Quartz fiber overgrowths on a pyrite grain in phyllite. French Pyrenees. (From Etchecopar and Malavielle 1987)

11.10 ASSOCIATIONS OF LINEATIONS WITH OTHER STRUCTURES

Lineations rarely are the only structure in an area, and the way they relate to other structures is important in understanding the structural history. The fabric of some rocks is completely dominated, at least locally, by a lineation. Pencil gneisses, for example, have a strong pencil lineation. Many lineations, however, are parallel to and lie within foliations or other planar features, and many are geometrically related to fold axes. A given area commonly contains different types of lineations, which may all have the same orientation or which may have different orientations.

i. Lineations and Foliations

Some lineations are defined at least in part by foliations, and of course these types must be parallel to that foliation. Intersection lineations (Figure 11.20*A*), including pencil cleavage (Figure 11.20*B*), must be parallel to the surfaces that define them.

Other lineations are defined by features that characteristically lie in a foliation. Acicular mineral grains may be distributed within a plane, defining a foliation, and they also may have a preferred orientation in the plane, defining an acicular habit lineation. Fold hinge lineations, fold mullions (Figure 11.22*A*), and in many cases rods (Figure 11.23) depend on folding for their linear character. If the folds are associated with an axial surface foliation, then these lineations also must be parallel to that foliation. Discrete lineations and mineral cluster, acicular habit, and elongated grain lineations also commonly lie in a foliation defined by platy minerals.

Some lineations, such as boudin lines, mineral fiber lineations, and structural and mineral slickenlines, are not defined by a foliation and do not contribute to the definition of one. Whether such lineations parallel a foliation or not depends on the geometry of the deformation.

Lineations, of course, may develop on surfaces other than foliations. Slickenlines and slickenfibers are commonly found on fault surfaces, and slickenfibers may be found on bedding surfaces in some circumstances, especially associated with folds in which slip on the bedding planes has occurred. Fold hinge lineations, boudin lines, and fold mullions must be parallel to the lithologic layers in which they develop. Intersection lineations involving lithologic layering, of course, must lie in the plane of the layering.

ii. Lineations and Folds

The relationship between folds and lineations can be of major importance in deciphering the structural geometry of an area and in interpreting the conditions under which the structures formed. Some lineations, such as fold hinge lineations, fold mullions (Figure 11.22*A*), and commonly rods (Figures 11.23), are generally parallel to the regional distribution of fold hinges. An intersection lineation defined by a folded surface and the axial foliation to the folds also parallels the fold axis if the folding is close to cylindrical (see Section 10.3(i)). Mineral lineations also are commonly parallel to fold hinges.

Because lineations are generally smaller-scale structures than folds, and because small-scale structures commonly reflect the geometry of large-scale structures, it may be easier to map the geometry of fold hinges by mapping the orientation of the appropriate lineations. The parallelism of a particular lineation with the hinges of a particular generation of folds, however, must be established independently.

Lineations such as boudin lines, acicular mineral grains, elongate mineral grains, and mineral cluster lineations, as well as discrete lineations and overgrowth lineations, may be found either parallel or perpendicular to fold axes. Some lineations, such as acicular habit lineations, have also been observed to be parallel to fold axes in hinge zones but perpendicular to them on the limbs. Slickenfiber lineations on folded bedding surfaces are usually perpendicular to the associated fold hinge and are most strongly developed on the limbs and faint to nonexistent in the hinge zone.

Less often, lineations are found at arbitrary angles to fold axes. Such a geometry is usually the result of the deformation of earlier lineations, as discussed in more detail in Chapter 13. Other possibilities cannot be dismissed, however, and each situation requires individual investigation.

Lineations can commonly be used to infer the distribution and geometric characteristics of the deformation in an area. We postpone discussion of these topics until after Chapter 12, where we introduce the concept of strain as a measure of deformation (see Sections 14.6 and 14.7).

REFERENCES AND ADDITIONAL READINGS

Alsop, G. I., R. E. Holdsworth, K. J. W. McCaffrey, and M. Hand (eds.). 2003. *Flow Processes in Faults and Shear Zones*. London, Geological Society, Special Publications, 224.

Alvarez, W., T. Engelder, and P. A. Geiser. 1978. Classification of solution cleavage in pelagic limestones. *Geology* 6: 263–266.

Christensen, J. N., J. L. Rosenfeld, and D. J. DePaolo. (1989). Rates of tectonometamorphic processes from rubidium and strontium isotopes in garnet. *Science* 244 (June 23): 1465–1469.

Christensen, J. N., J. Selverstone, J. L. Rosenfeld, and D. J. Depaolo. (1994). Correlation by Rb-Sr geochronology of garnet growth histories from different structural levels within the Tauern Window, eastern Alps. *Contributions to Mineralogy and Petrology* 118: 1–12.

Cloos, E. 1957. Lineation: A critical review and annotated bibliography. *Geol. Soc. Amer. Mem.* 18.

Engelder, T., and S. Marshak. 1985. Disjunctive cleavage formation at shallow depths in sedimentary rocks. *J. Struct. Geol.* 7(3/4): 327–343.

Etchecopar, A., and J. Malavielle. 1987. Computer models of pressure shadows: A method for strain measurement and shear sense determination. *J. Struct. Geol.* 9(5/6): 667–677.

Fleuty, M. J. 1975. Slickensides and slickenlines. *Geol. Mag.* 112: 319–322.

Glen, R. A. 1982. Component migration patterns during the formation of a metamorphic layering, Mount Franks area, Willyama Complex, N.S.W., Australia. *J. Struct. Geol.* 4: 457–468.

Goscombe, B. D., C. W. Passchier, and M. Hand. 2004. Boudinage classification: End-member boudin types and modified boudin structures. *J. Struct. Geol.* 26(4): 739–763.

Gray, D. R. 1977. Morphologic classification of crenulation cleavage. *J. Geol.* 85: 229–235.

Gray, D. R. 1978. Cleavages in deformed psammitic rocks from southeastern Australia: Their nature and origin. *Geol. Soc. Amer. Bull.* 89: 577–590.

Lister, G. S., and A. W. Snoke. 1984. S-C mylonites. *J. Struct. Geol.* 6: 617–638.

Müller, W. 2003. Strengthening the link between geochronology, textures and petrology. *Earth and Planet. Sci. Lett.* 206: 237–251.

Pollard, D. D., and R. C. Fletcher. 2005. *Fundamentals of Structural Geology*. Cambridge, UK, Cambridge University Press.

Powell, C. McC. 1979. A morphological classification of rock cleavage. *Tectonophysics* 58: 21–34.

Simpson, C., and S. Schmid. 1983. An evaluation of criteria to deduce the sense of movement in sheared rocks. *Geol. Soc. Amer. Bull.* 94: 1281–1288.

Turner, F., and L. Weiss. 1963. *Structural Analysis of Metamorphic Tectonites*. New York, McGraw-Hill.

Vance, D., W. Müller, and I. M. Villa (eds.). 2003. Geochronology: Linking the isotope record with petrology and textures. *Geol. Soc. Lond. Spec. Pub.* 220, 272 pp.

Weber, K. 1981. Kinematic and metamorphic aspects of cleavage formation in very low grade metamorphic slates. *Tectonophysics* 78: 291–306.

Chapter 12

GEOMETRY OF HOMOGENEOUS STRAIN

We describe folds, foliations, and lineations, which are structures that form through ductile deformation, in Chapters 10 and 11. To further our understanding of the origin and significance of these structures, we need to develop a means by which we can precisely describe the deformation that they record in the rocks. To this end, we must develop more carefully the concept of strain. We introduced some aspects of strain in Chapter 9, but we need a more thorough and systematic understanding. This will enable us to evaluate theoretically the models proposed for the formation of ductile structures, and to test these models against observations of natural deformation. We concentrate most of our analysis on two-dimensional strain or **plane strain**, because it is easier to diagram and is initially easier to grasp. For plane strain, all changes in size and shape of a body occur on a principal plane through the body, with no deformation occurring normal to that plane. Although two-dimensional strain is often used in analysis of geologic deformation, in many cases ignoring the third dimension can lead to misinterpretation and misunderstanding, so it is actually important to take account of the third dimension, and we discuss the generalizations to three dimensions where appropriate.

The **strain** of a body is simply the change in size and shape that the body has experienced during deformation. A change in size for two-dimensional strain is understood to mean simply a change in area; for three-dimensional strain, it is a change in volume. Figures 12.1*A* and *B* show how an original square of material changes size but maintains the same shape, and Figures 12.1*A* and *C*, and 12.1*A* and *D* show how the square can change shape but remain the same size. The strain is **homogeneous** if the changes in size and shape for each small part of the body are geometrically similar to those for the body as a whole, as illustrated by the grid cells in Figure 12.1*B–D*. Thus for any homogeneous strain, straight lines and flat planes remain straight and flat, respectively, and parallel lines and planes remain parallel (see the shaded bands in Figure 12.1*B–D*). The strain is inhomogeneous (Figure 12.1*E*, *F*) if the changes in size and shape of small parts of the body are not geometrically similar to that of the body as a whole and if those changes vary from place to place within the body, as illustrated by the grid cells in Figure 12.1*E*, *F*. Lines that are originally straight become curved, planes become curved surfaces, and parallel lines and planes generally do not remain parallel after deformation (see the shaded bands in Figure 12.1*E*, *F*).

Folding is an example of a strain that is necessarily inhomogeneous, because planes and lines, such as the surfaces of rock layers, in general do not remain planar, straight, or parallel. Within very small volume elements, however, such as the grid cells in Figure 12.1*E*, *F*, the local strain is approximately homogeneous. Thus we can describe an inhomogeneous strain in a structure as a variation from place to place of locally homogeneous strains. We discuss how big such a "small" volume element must be in Section 12.6.

In discussing the geometry of strain, we often refer to geometric objects such as lines, planes, circles, and ellipses. These features are so-called **material objects** if they are always defined by the same set of material particles, such as the shaded bands, the circles and ellipses, and the dashed lines in the diagrams of Figure 12.1. A bedding plane, for example, is a material plane because no matter how it moves and deforms, it is always defined by the same set of material particles. A coordinate plane defined by two reference axes, on the other hand, is a not a material plane because as a rock body deforms, its material particles can move through the coordinate

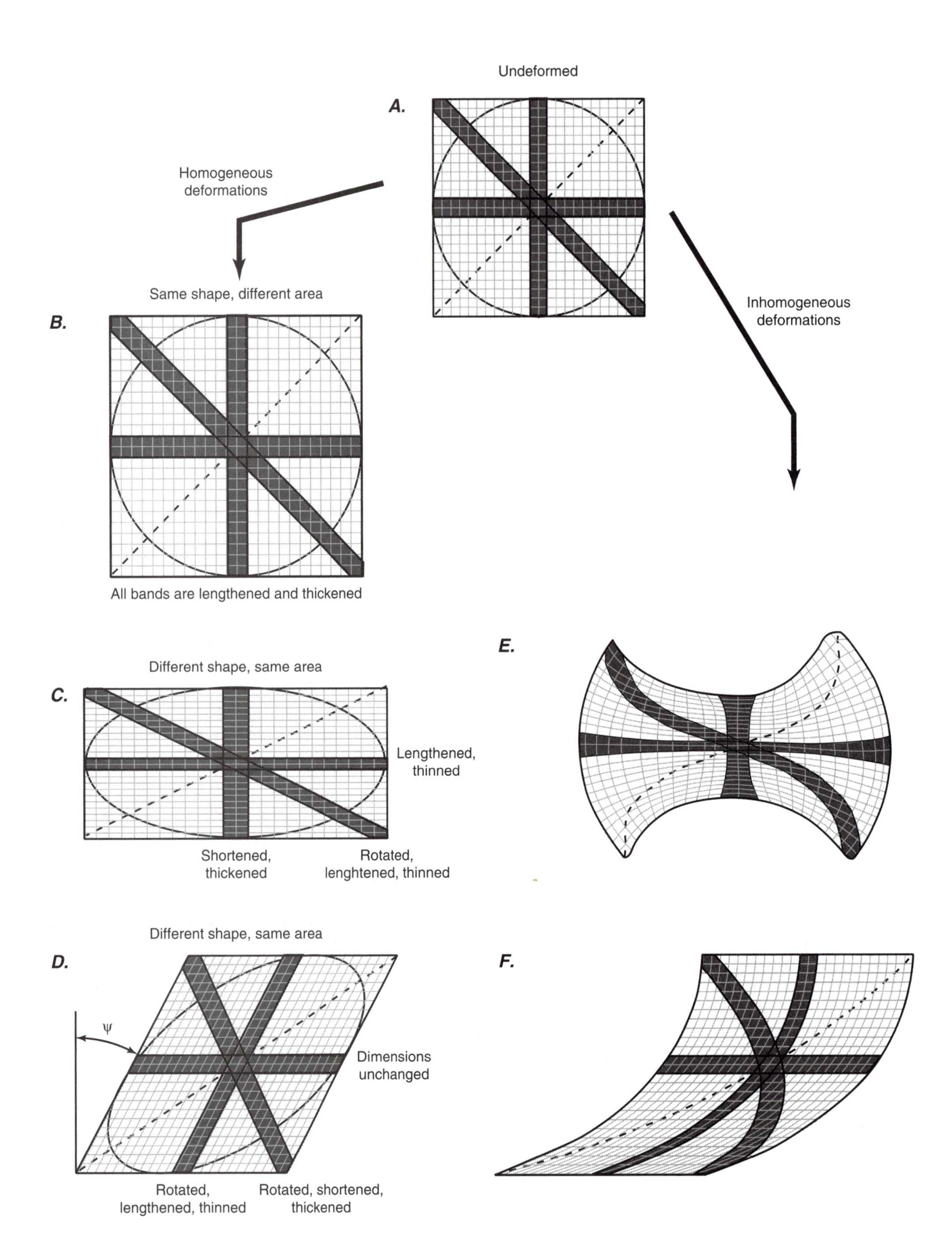
Undeformed
A.
Homogeneous deformations
Inhomogeneous deformations
Same shape, different area
B.
All bands are lengthened and thickened
Different shape, same area
C.
Lengthened, thinned
Shortened, thickened
Rotated, lenghtened, thinned
E.
Different shape, same area
D.
ψ
Dimensions unchanged
Rotated, lengthened, thinned
Rotated, shortened, thickened
F.

FIGURE 12.1 The components of homogeneous strain of a square (*A*), including change of area (*B*) and change of shape (*C, D*), compared with inhomogeneous strain (*E, F*). In homogeneous strain (*B, C, D*), straight lines remain straight; parallel lines remain parallel, and every small square cell deforms in the same way as the large square. In inhomogeneous strain (*E, F*), straight lines become curved, parallel lines do not remain parallel, and the deformation of the small square cells is heterogeneous and is not the same as the deformation of the larger square. *A*. An undeformed square with an inscribed material circle, three orientations of shaded bands representing material layers, and a material line (dashed) perpendicular to one of the bands. *B*. Homogeneous area increase of the square in *A* preserves the initial square shape. The inscribed figures change size but not shape. *C*. Homogeneous pure shear of the square in *A* to form a rectangle having the same area as the original square. The material circle changes into an ellipse; bands are either shortened and thickened or lengthened and thinned depending on their orientation; the diagonal band and the dashed line initially normal to it are each rotated so they are no longer perpendicular. *D*. Homogeneous simple shear of the square in *A* through the shear angle ψ. The resulting parallelogram has the same height and base length and thus the same area as the initial square. The circle is deformed into an ellipse; all bands, except those parallel to the shear plane (horizontal), are rotated and either shorten and thicken or lengthen and thin; the amount of rotation depends on orientation, and the dashed line is no longer perpendicular to the diagonal band. *E*. Inhomogeneous deformation of the square in *A* producing curved and non-parallel lines from initially straight and parallel lines. Deformation of the cells is highly heterogeneous. Area need not be conserved. *F*. Inhomogeneous simple shear of the square in *A*, resulting in a figure having the same height and base length and thus the same area as the original square. The shapes of the small cells change from one row to the next and are not the same as that of the deformed figure.

plane. Thus different sets of material particles occupy the coordinate plane at different times. This distinction is important in the following discussion.

12.1 MEASURES OF STRAIN

Strain in a body can be described as a change in the size of a body plus a change in its shape. A change in size is just a change in area for a two-dimensional body or a change in volume for a three-dimensional body, which can be described as changes in length measured parallel to the reference coordinates (two coordinates in two dimensions or three coordinates in three dimensions). A change in length parallel to a given coordinate is defined by a **linear strain**. Thus describing a change in size ultimately comes down to describing linear strains. A change in shape can be described by changes in angles, and a change in angle is defined by a **shear strain**. We discuss linear and shear strains in turn.

i. Linear Strain

The area of any given body is proportional to the product of two characteristic lengths of the body. For example, the area of a rectangle with edges parallel to the x and y coordinate axes that have lengths l_x and l_y, respectively, is $A = l_x\, l_y$. Similarly, the area of an ellipse with major and minor axes parallel to the x and y coordinate axes is given in terms of the radii parallel to those axes that have lengths r_x and r_y by $A = \pi r_x\, r_y$. The argument is easily extended to the volume V of a rectangular block that has edges of lengths l_x, l_y, and l_z parallel to the x, y, and z coordinate axes: $V = l_x\, l_y\, l_z$. Similarly, the volume of an ellipsoid that has semiaxes of lengths r_x, r_y, and r_z parallel to the x, y, and z coordinate axes is $V = [\frac{4}{3}\, \pi r_x r_y r_z]$. Thus the description of any change in size, that is, a change in area or in volume, for these two- or three-dimensional objects requires specification of the changes in the lengths of line segments in the x, y, (and z) coordinate directions. Comparable results apply to any two- or three-dimensional object.

We look, therefore, at how we describe the change in length of a line segment. Table 12.1 line A shows a line segment that has an undeformed length L. In line B, that line segment has been shortened, and in line C it has been lengthened. The lengths of the line segments are listed in the Length column, and the change in length from the undeformed state is listed in the Length Change column. The change in length, however, is an inadequate measure of the deformational state of a line segment, because for a given change in length, the intensity of the change is much greater for a short line segment than for a long one. For example, we intuitively understand that to stretch a 12-centimeter rubber band by 4 centimeters imposes a much more intense deformation on the rubber band than to stretch a 12-meter rubber band by 4 centimeters. To normalize for the original line length, therefore, the lengthening is expressed as a proportion of the original line length, and such normalized measures of the change in length describe the linear strain. Several measures of linear strain are used, but two of the most common, which we focus on in this book, are the stretch s_n and the extension e_n, where the subscript n indicates that the stretch or extension is measured in a direction parallel to a specified unit vector **n**, which in this case is parallel to the line segment.

The **stretch** s_n is the ratio of the deformed length l of a material line segment to its undeformed length L:[1]

$$s_n \equiv \frac{l}{L} \tag{12.1}$$

[1]We often use uppercase letters when referring to the undeformed state and lowercase letters when referring to the deformed state.

TABLE 12.1 Extensional Strain of a Material Line

	Length	Length Change ΔL	Stretch $s_n = l/L$	Extension $e_n = (l - L)/L$
A. Undeformed	$L = 12$	$\Delta L = 0$	$s_n = 1$	$e_n = 0$
B. Shortened	$l = 8$	$\Delta L = l - L < 0$ $\Delta L = 8 - 12$ $\Delta L = -4$	$0 < s_n < 1$ $s_n = 8/12$ $s_n = 0.67$	$e_n < 0$ $e_n = (8 - 12)/12$ $e_n = -0.33$
C. Lengthened	$l = 16$	$\Delta L = l - L > 0$ $\Delta L = 16 - 12$ $\Delta L = 4$	$s_n > 1$ $s_e = 16/12$ $s_e = 1.33$	$e_n > 0$ $e_n = (16 - 12)/12$ $e_n = 0.33$

In Table 12.1, the column labeled Stretch shows the characteristics of the stretch for an undeformed line segment (line A), a line segment that is shortened (line B), and a line segment that is lengthened (line C). Numerical values of the stretches are calculated for each of the line segments pictured at the left of the table. The stretch is always a positive number, and it is less than one ($0 < s_n < 1$) if the line segment has been shortened and greater than 1 ($s_n > 1$) if it has lengthened.

The **extension** e_n of a material line segment, which we introduce at the beginning of Chapter 9, is the ratio of its change in length, ΔL, to its initial length L, where the change in length is the final length minus the initial length:[2]

$$e_n \equiv \frac{l - L}{L} \equiv \frac{\Delta L}{L} \qquad (12.2)$$

In Table 12.1, the column labeled Extension shows the possible values for the extension for an undeformed line segment (line A), a shortened line segment (line B), and a lengthened line segment (line C). Numerical values of the extension are calculated for each line segment pictured at the left of the table. The extension is always negative ($e_n < 0$) if the line segment has been shortened, and it is always positive ($e_n > 0$) if it has been lengthened.

Comparing Equations (12.1) and (12.2) shows that these two measures of linear strain are related by:

$$e_n = \frac{l}{L} - \frac{L}{L} = s_n - 1 \qquad s_n = e_n + 1 \qquad (12.3)$$

Two other measures of linear strain—the quadratic elongation and the natural, or logarithmic, strain—are commonly used (see Box 12-1). We will find these mea-

[2]Note that with the sign conventions we have adopted, a positive value for extension measures a lengthening, whereas a positive value for stress measures a compression. We thus end up with a positive stress causing a negative extension. This incompatibility does not arise with the engineering sign convention for stress (see Box 7-4), which is one reason it is generally used in analytic applications of continuum mechanics.

BOX 12-1 Other Measures of Linear Strain

Two other measures used for linear strain are the **quadratic elongation** and the **natural strain**. The quadratic elongation is simply the square of the stretch, and it is often given the symbol[1] λ. It is a common measure in the analysis of finite strain, and it could have been used in Section 12.9, for example, instead of the explicit squares of the stretches that we used.

The **natural strain** $\bar{\varepsilon}_n$, also called the **logarithmic strain**, is the integral of all the infinitesimal increments of extension required to make up the deformation, where for each increment, the infinitesimal change in length is $d\ell$ and the reference length is taken to be the instantaneous deformed length ℓ:

$$\bar{\varepsilon}_n \equiv \int_L^{\ell_f} \frac{d\ell}{\ell} = \ln\left(\frac{\ell_f}{L}\right) = \ln s_n \qquad (12\text{-}1.1)$$

where L is the initial length, ℓ_f is the final length, and ln indicates the natural logarithm (the logarithm to base e).

Notice that the natural strain is the natural logarithm of the stretch. The natural strain is sometimes convenient for discussion of strain history (see Figures 12.23 and 12.25). It also provides a symmetric measure of shortening and lengthening. For example, for a line segment stretched to twice its initial length and one shortened to half its initial length, $s_n = 2$ and 0.5, and $e_n = 1$ and 0.5, but $\bar{\varepsilon}_n = 0.693$ and -0.693, respectively. The time derivative of the natural strain is also used as a measure of the strain rate (see Box 16-1, Equation (16-1.4)).

[1]Some authors, however, use this symbol to designate the stretch.

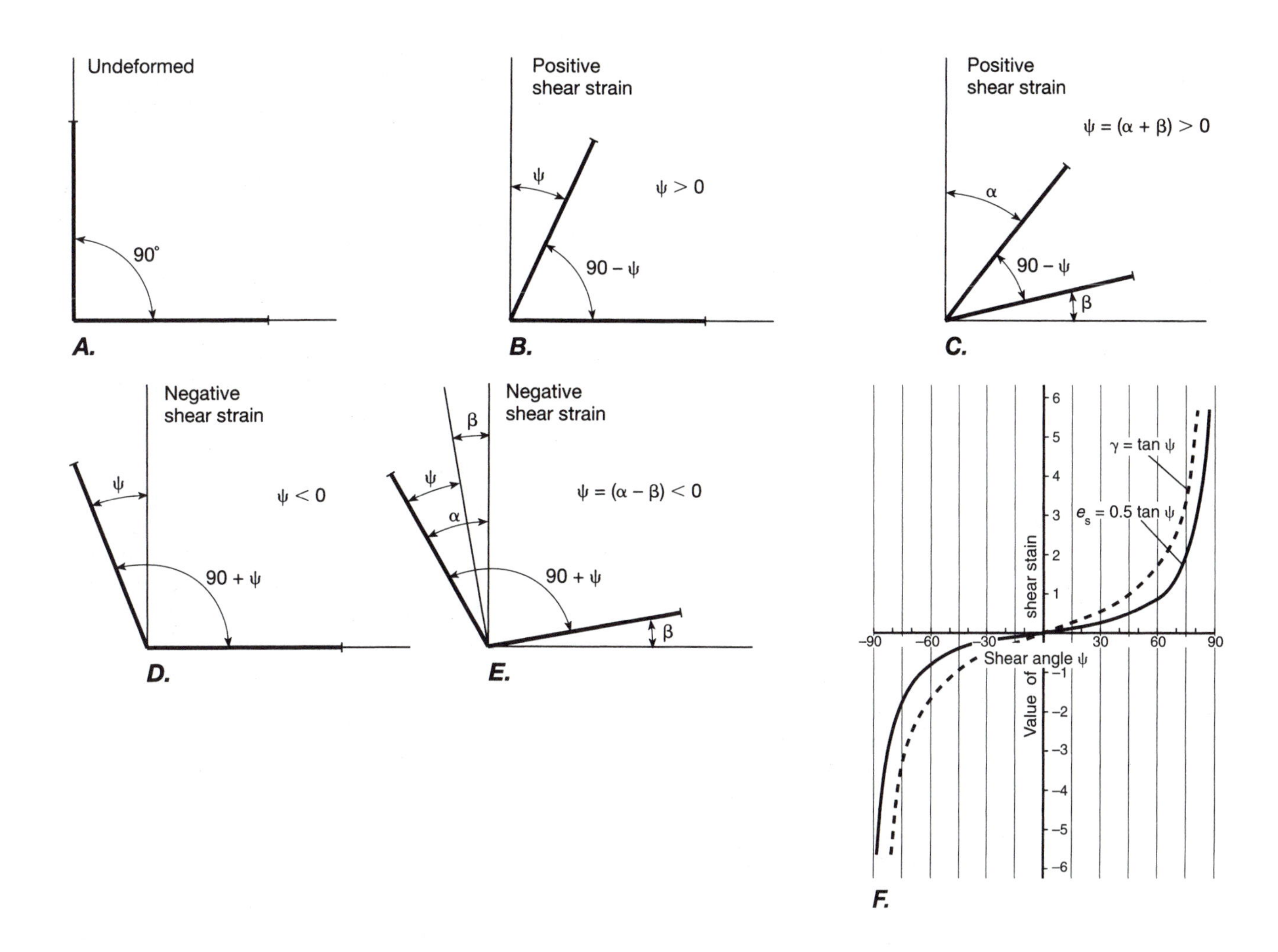

FIGURE 12.2 The tensor shear strain $e_s = 0.5 \tan \psi$ and the engineering shear strain $\gamma = \tan \psi$ of a material line, where ψ is the shear angle. *A*. The undeformed state: Shear of a material line is defined with reference to another material line initially normal to the first. *B* and *C*. Definition of a positive shear strain: $(90^\circ - \psi) < 90^\circ$. *D* and *E*. Definition of a negative shear strain: $(90 - \psi) > 90^\circ$. *F*. Tensor and engineering shear strains as a function of shear angle ψ. *Note:* The sign of the shear strain is determined by the sum of the displacement gradients $(\partial U_1/\partial X_3 + \partial U_3/\partial X_1)$, where X_1 and X_3 are the horizontal and vertical coordinate axes, and the deformation is defined by the displacement components U_1 and U_3 of material points parallel to those axes, respectively (see Box 12-3, Figure 12-3.2). The first and second of these displacement gradients define the shear of the two lines initially parallel to X_3, and X_1, respectively (see Figure 12-3.2). Thus $\partial U_1/\partial X_3$ is positive in *B* and *C*, because U_1 increases with increasing X_3; it is negative in Figure 12.2*D* and *E*. $\partial U_3/\partial X_1$ is zero in *B* and *D* and positive in *C* and *E*. Thus the sum of the displacement gradients is positive in *B* and *C*, and negative in *D*. In *E*, $\partial U_1/\partial X_3$ is more negative than $\partial U_3/\partial X_1$ is positive, so the sum is negative.

sures useful in some instances in later chapters, but in this chapter and for most of our discussions, we use the stretch or the extension.

We consider how the definition of linear strain leads to the definition of volumetric strain in Section 12.2(iii).

ii. Shear Strain

A body can change shape without changing size. For example, a square can deform into a parallelogram of the same area, or a circle can deform into an ellipse of the same area (Figure 12.1, the figures in *A* deform to those in *C* or *D*). In three dimensions, a cube can deform into a parallelepiped of the same volume, or a sphere can deform into an ellipsoid of the same volume. Changes in shape are described by the changes in the angle between pairs of lines that are initially perpendicular. The change in angle is called the **shear angle** ψ (Figure 12.1*D*), and the **shear strain** is defined by:

$$e_s \equiv 0.5 \tan \psi \quad \gamma \equiv \tan \psi \tag{12.4}$$

Here, e_s is the **tensor shear strain**, and γ is the **engineering shear strain**. They differ from each other by a factor of 2 ($\gamma = 2e_s$; $e_s = 0.5\gamma$). The factor of one half in the tensor shear strain is an unavoidable characteristic of the tensor components, but it is sometimes convenient to use the engineering definition to avoid having to carry that factor along in calculations.

For two material line segments originally oriented along the positive coordinate directions (Figure 12.2*A*), a decrease in angle between the two lines is considered a positive shear strain (Figure 12.2*B*, *C*), and an increase in angle is a negative shear strain (Figure 12.2*D*, *E*). Both γ, and e_s increase from 0 in the unstrained state to ∞ where $\psi = 90°$ (Figure 12.2*F*).

12.2 THE STATE OF STRAIN

We know the **state of strain at a point** if, for any material line of any orientation, we can determine its extension, and we can determine its shear strain with respect to any other line initially perpendicular to it. Any two-dimensional homogeneous strain always deforms a material circle into an ellipse called the **strain ellipse**, which completely defines the state of strain. (Figures 12.1 and 12.3*A*); in three-dimensional strain, a material sphere deforms into an ellipsoid called the **strain ellipsoid**.

i. Strain in Two Dimensions: The Strain Ellipse and Principal Axes of Strain

The stretch, extension, and shear strain all have a simple geometric interpretation related to the strain ellipse. Assume that a material circle in the undeformed state has a radius $R = 1$ (Figure 12.3*A*). After the deformation, any given radius of the circle R is transformed into a specific radius r of the strain ellipse whose length varies with orientation. Although R and r are lines made up of the same material points, they differ in length and orientation because of the deformation. If we superimpose the original unit circle on the strain ellipse (Figure 12.3*A*), we can see how much any radius of the strain ellipse has been shortened or lengthened from its original length $R = 1$. Using the definitions of the stretch (Equation (12.1)) and the extension (Equation (12.2)) and the fact that $R = 1$, we find that

$$s_n = \frac{r}{R} = r \qquad e_n = \frac{r-R}{R} = \frac{\Delta R}{R} = \Delta R \quad (12.5)$$

Thus for the deformation of the unit circle, the radius of the strain ellipse is the stretch, and the difference between the radius of the ellipse and that of the unit circle is the extension.

The shear strain of a line is determined with reference to another line initially normal to it. On a circle, the line T drawn perpendicular to any given radius R at its end point is tangent to the circle (Figure 12.3*B*). After deformation, the lines T and R are transformed into the lines t and r, respectively. Although t and r are no longer perpendicular in the deformed state, t is still tangent to the ellipse at the end point of the radius r. Accordingly, any radius and the associated tangent to the strain ellipse define the angle between two material lines that were perpendicular in the undeformed state. The change in angle ψ is thus easily constructed (Figure 12.3*B*), and it is a measure of the shear strain for that pair of lines. The shear strain of R relative to T is exactly the same as the shear strain of T relative to R, and the shear strain is therefore said to be *symmetric*.

We can define the geometry of any ellipse by the orientations and lengths of the two principal axes (Figure 12.4). For the strain ellipse, these axes are called the **principal axes of strain**. In two-dimensional strain analysis, the plane of the strain contains the maximum and minimum principal axes of the three-dimensional strain ellipsoid. By convention, we label the two **principal extensions**[3] $\hat{e}_1$ and $\hat{e}_3$ (cf. Equation (12.2)), and the two associated **principal stretches** $\hat{s}_1$ and $\hat{s}_3$ (cf. Equation (12.1)), such that $\hat{e}_1$ and $\hat{s}_1$ are measures of the maximum lengthening, whereas $\hat{e}_3$ and $\hat{s}_3$ are measures of the maximum shortening:

$$\hat{e}_1 \geq \hat{e}_3 \qquad \hat{s}_1 \geq \hat{s}_3 \quad (12.6)$$

In addition to the lengths of the principal axes, one angle is necessary to specify the orientation of the two orthogonal principal axes relative to a known reference coordinate system in two dimensions, such as the geographic coordinates East and North for a horizontal plane of strain. Thus, we need three numbers to describe the state of strain in two dimensions.

Tangents to the ellipse at the ends of the principal radii are perpendicular to the radii (Figure 12.4), and these are the only points on the ellipse where this is true. Because any tangent to a circle is necessarily perpendicular to the radius at the point of tangency, the principal radii and the tangents at their end points must have been perpendicular before deformation. Thus the shear strains for material lines parallel to the principal axes of the strain ellipse must all be zero. (Note the similarity to the stress ellipse and the fact that shear stresses on the principal planes of stress are zero).

ii. Strain in Three-Dimensions: The Strain Ellipsoid and Principal Axes of Strain

These concepts of two-dimensional strain can easily be extended to three dimensions. A sphere is deformed by

[3]As in our notation for stress, we again use the hats, or circumflexes, on the symbols to indicate principal values.

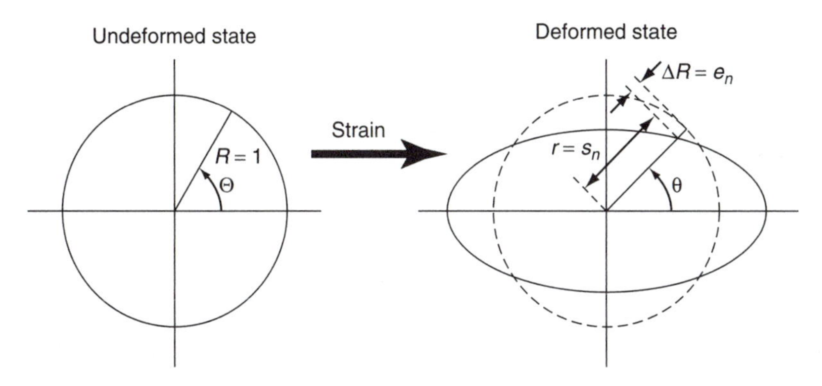

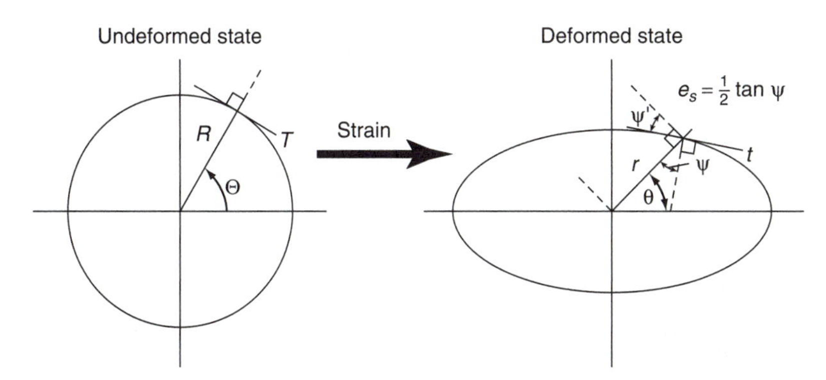

FIGURE 12.3 The relationship of the stretch, extension, and shear strain to the geometry of the strain ellipse. *A*. A homogeneous strain transforms the unit circle into an ellipse. An undeformed radius $R = 1$ is transformed into a deformed radius r, which has a different length and orientation. The stretch is the length of the radius of the ellipse, and the extension is the difference in radius between the initial unit circle and the ellipse. *B*. The shear strain is determined from the change in angle between a radius and a tangent at the end of the radius. The two lines R and T are perpendicular to each other on the circle but r and t are not, in general, perpendicular on the ellipse. The change in angle between the two lines is the shear angle that defines the shear strain for that pair of lines. The change in angle ψ of r relative to t is the angle between r and the perpendicular to t. The change in angle ψ' of t relative to r is the angle between t and the perpendicular to r. The two shear angles ψ and ψ' are equal.

a homogeneous deformation into an ellipsoid, the **strain ellipsoid**, whose geometry is defined by the orientation and lengths of the three mutually perpendicular **principal axes of strain** (see Figure 12.11). Parallel to the principal axes of the strain ellipsoid, the extensions (Equation 12.2) and stretches (Equation (12.1)) are a maximum, minimax,[4] and minimum, which by convention we designate, respectively:

$$\hat{e}_1 \geq \hat{e}_2 \geq \hat{e}_3 \qquad \hat{s}_1 \geq \hat{s}_2 \geq \hat{s}_3 \tag{12.7}$$

Because three numbers are needed to specify the orientations of these three principal axes relative to a known reference coordinate frame, such as geographic East, North, and Up, we need a total of six numbers to specify the state of strain in three dimensions.

Tangents to the ellipsoid at the ends of the three principal radii are perpendicular to the radii, and these are the only points on the ellipsoid where this is true. Because these radii and tangents must have been perpendicular before deformation, the shear strains for those radii and tangents all must be zero.

[4] $\hat{e}_2$ and $\hat{s}_2$ are each a minimax because each is a minimum in the ($\hat{e}_1$, $\hat{e}_2$)- or ($\hat{s}_1$, $\hat{s}_2$)-plane, respectively, and a maximum in the ($\hat{e}_2$, $\hat{e}_3$)- or ($\hat{s}_2$, $\hat{s}_3$)-plane, respectively, which is perpendicular to the first plane in each case.

iii. The Area and Volume Strains

Having defined measures for the linear strain, it is straightforward to define measures for the change in area and in volume. For a two-dimensional description of a deformation, if A is the undeformed area and a is the deformed area, then the area stretch s_A and the area extension e_A are defined by

$$s_A \equiv \frac{a}{A} \qquad e_A \equiv \frac{a - A}{A} = s_A - 1 \tag{12.8}$$

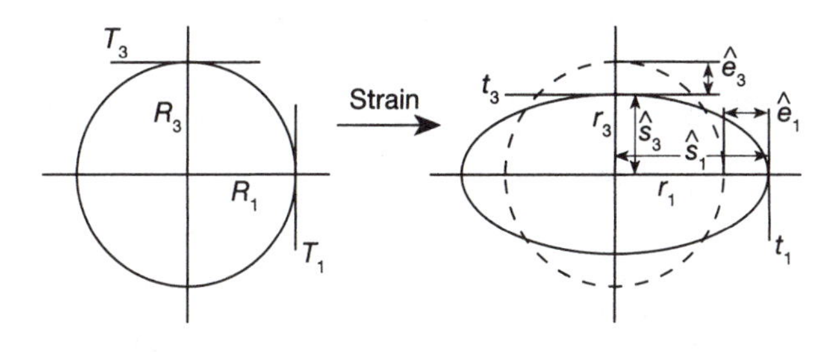

FIGURE 12.4 Representation of the principal stretches and the principal extensions on the strain ellipse formed from the unit circle. The shear strains are zero for the material lines parallel to the principal axes of strain because the tangents at the ends of the principal radii are perpendicular to those radii both before and after deformation.

For example, consider a rectangle whose edges are initially of lengths L_1 and L_3. If the rectangle undergoes a deformation for which its edges are parallel to the principal strain axes, the rectangle will remain a rectangle (because there is no shear strain for material lines parallel to the principal axes); and the deformed lengths of its edges are ℓ_1 and ℓ_3. The area stretch from the first Equation (12.8), Equation (12.1), and the second Equation (12.3), and the area extension from the second Equation (12.8) and the second Equation (12.3) are

$$s_A = \frac{\ell_1 \ell_3}{L_1 L_3} = \hat{s}_1 \hat{s}_3 = (\hat{e}_1 + 1)(\hat{e}_3 + 1) \qquad (12.9)$$
$$e_A = s_A - 1 = (\hat{e}_1 + 1)(\hat{e}_3 + 1) - 1 = \hat{e}_1 + \hat{e}_3 + \hat{e}_1 \hat{e}_3$$

If the strain under consideration is very small, as is generally the case for elastic deformation, then products of two extensions are extremely small and can be ignored, to a first approximation.[5] Under these circumstances, we find a particularly simple expression for the area extension

$$e_A \cong \hat{e}_1 + \hat{e}_3 \quad \text{if } \hat{e}_k << 1 \ (k = 1,3) \qquad (12.10)$$

It is straightforward to generalize these results to the volumetric strain. We define measures that we refer to as the volumetric stretch (s_V) and the volumetric extension[6] (e_V). If the undeformed volume is V and the deformed volume is v, then

$$s_V \equiv \frac{v}{V} \qquad e_V \equiv \frac{v - V}{V} = s_V - 1 \qquad (12.11)$$

We consider a rectangular block whose edges are initially of lengths L_1, L_2, and L_3. If we deform the block in such a way that the principal strain axes are parallel to the edges of the block, the block will remain rectangular. The deformed lengths of its edges are ℓ_1, ℓ_2, and ℓ_3. The volumetric stretch from the first Equation (12.11) is

$$s_V = \frac{\ell_1 \ell_2 \ell_3}{L_1 L_2 L_3} = \hat{s}_1 \hat{s}_2 \hat{s}_3 \qquad (12.12)$$

where the numerator is the deformed volume of the block and the denominator is the undeformed volume, and where we used Equation (12.1) to obtain the second relation. Using Equations (12.11) and (12.12) we can express the volumetric extension in terms of the linear extensions as follows:

$$s_V = \hat{s}_1 \hat{s}_2 \hat{s}_3 = (\hat{e}_1 + 1)(\hat{e}_2 + 1)(\hat{e}_3 + 1) \qquad (12.13)$$
$$e_V = s_V - 1 = (\hat{e}_1 + 1)(\hat{e}_2 + 1)(\hat{e}_3 + 1) - 1$$

$$e_V = \hat{e}_1 + \hat{e}_2 + \hat{e}_3 + \hat{e}_1 \hat{e}_2 + \hat{e}_2 \hat{e}_3 + \hat{e}_1 \hat{e}_3 + \hat{e}_1 \hat{e}_2 \hat{e}_3 \qquad (12.14)$$

For very small strains, we can ignore second- and third-order terms in the volumetric extension, so Equation (12.14) reduces to

$$e_V \cong \hat{e}_1 + \hat{e}_2 + \hat{e}_3 \quad \text{if} \quad |\hat{e}_k| << 1 \ (k = 1{:}3) \qquad (12.15)$$

The first Equation (12.13) expressing the volumetric stretch in terms of the three principal stretches is a relation that is not restricted to small strains. For that reason, Equation (12.13) is more general than Equation (12.15).

For a deformation that is two-dimensional ($\hat{s}_2 = 1$) and constant volume ($s_V = 1$), we must have from the first Equation (12.13)

$$\hat{s}_1 \hat{s}_3 = 1 \quad \hat{s}_1 = \frac{1}{\hat{s}_3} \quad \hat{s}_3 = \frac{1}{\hat{s}_1} \quad \text{if} \quad \hat{s}_2 = s_V = 1$$

iv. The Inverse Strain Ellipse

In the foregoing discussion, we measured stretches and shear strains with reference to line lengths and right angles in the undeformed state (Table 12.1; Figure 12.2). In analyzing large strains such as are common in ductilely deformed rocks, however, it often is more convenient to measure the stretches and shear strains with reference to line lengths and right angles in the strained state. This analysis requires a different strain ellipse called the **inverse strain ellipse**, which is the ellipse in the undeformed state that is transformed into a circle in the deformed state (Figure 12.5); stated another way, it is the ellipse in the undeformed state that is formed by the inverse deformation from a circle in the deformed state. The lengths of its principal axes are the inverse of the principal axes of the regular strain ellipse, and the material lines A and B that are parallel to the principal axes of inverse strain in the undeformed state become the lines a and b that are parallel to the principal axes of regular strain in the deformed state. For the purposes of our descriptive discussion, however, we deal mostly with the strain ellipse. In Section 12.9, however, we show how the inverse strain ellipse arises naturally from a consideration of the deformed state.

v. Why Study Strain?

All this discussion of circles and ellipses may seem academic and far removed from the study of real rocks. It is not, however, because structures that are initially ap-

[5]Rocks generally fracture under room pressure at extensions on the order of 0.01 (or 1%). Products of two extensions (second-order products) of this magnitude are thus on the order of 0.0001. Thus by ignoring second-order products, we limit the precision of the volumetric extension to no better than 0.0001, or one part in ten thousand.

[6]The volumetric extension in many works is given the symbol Δ. In this book, we reserve the symbol Δ to indicate the change in a variable.

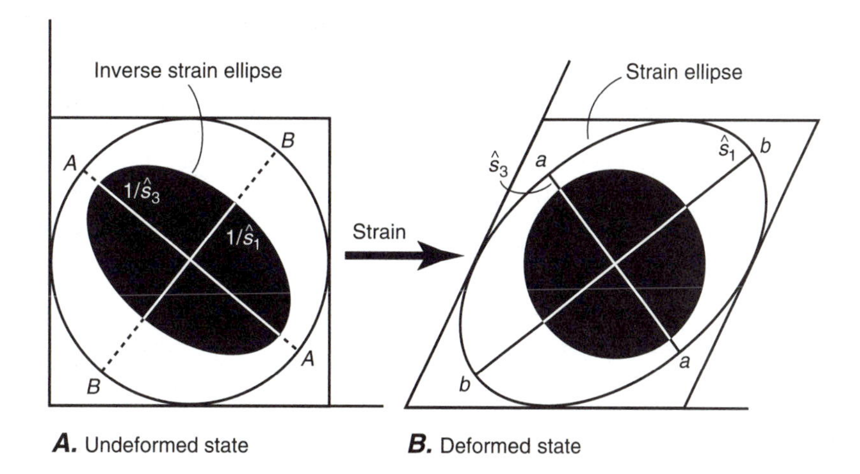

FIGURE 12.5 Definition of the inverse strain ellipse and its relationship to the strain ellipse. From the undeformed state (*A*), the inverse strain ellipse (solid black) is transformed into a circle (solid black) in the deformed state (*B*). The circle in the undeformed state (*A*) is transformed into the strain ellipse in the deformed state (*B*). Material lines A and B, which are parallel to the principal axes of the inverse strain ellipse in *A*, are rotated and changed in length by the deformation to material lines a and b, respectively, which are parallel to the principal axes of the strain ellipse in *B*. In general, A and B are not parallel to a and b, respectively. Note that none of the principal axes of strain or inverse strain are parallel to the diagonals of the initial square or its deformed equivalent parallelogram.

proximately circular or spherical are relatively common in some rock types. Where these rocks have been deformed, those structures provide a fascinating record of the distribution of strain throughout the rock. Ooids, for example, are small, almost spherical, pellet-like bodies common in limestones (Figure 12.6*A*), and they deform with the rock to record the shape and orientation of the strain ellipsoid (Figure 12.6*B*). Radiolaria and foraminifera, which are tiny spherical or disk-shaped fossils found in cherts or limestones, and alteration spots in slates (Figure 11.19*B*) also may serve as strain indicators. Other fossils, such as cephalopods and brachiopods, as well as pebbles and cobbles in conglomerates (Figure 11.19*A*) can provide information about the strain, even though they are not originally spherical and may have an original preferred dimensional orientation in the undeformed rock (see Figures 14.1*C*; 15-*A*.1). We discuss the significance of strain for interpreting the origin of structures in

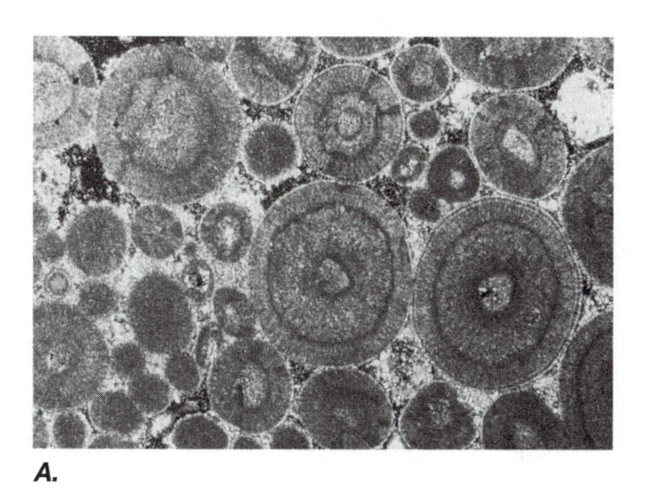

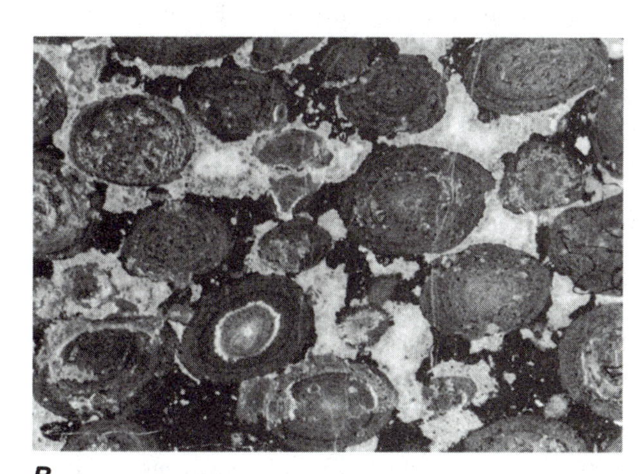

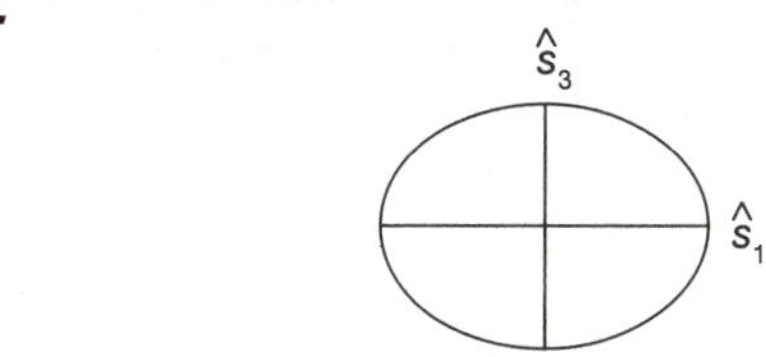

FIGURE 12.6 Ooids serve as strain markers in deformed limestone. *A*. An undeformed oolitic limestone. (Sample courtesy of G. Protzman-Nakamura; photo by E. Rodman) *B*. A deformed oolitic limestone. The ratio of the principal stretches is $\hat{s}_1/\hat{s}_3 \approx 1.4$. The larger ooids are approximately 4 millimeters in diameter. (Sample courtesy of John Christie; photo by Mary Graziose)

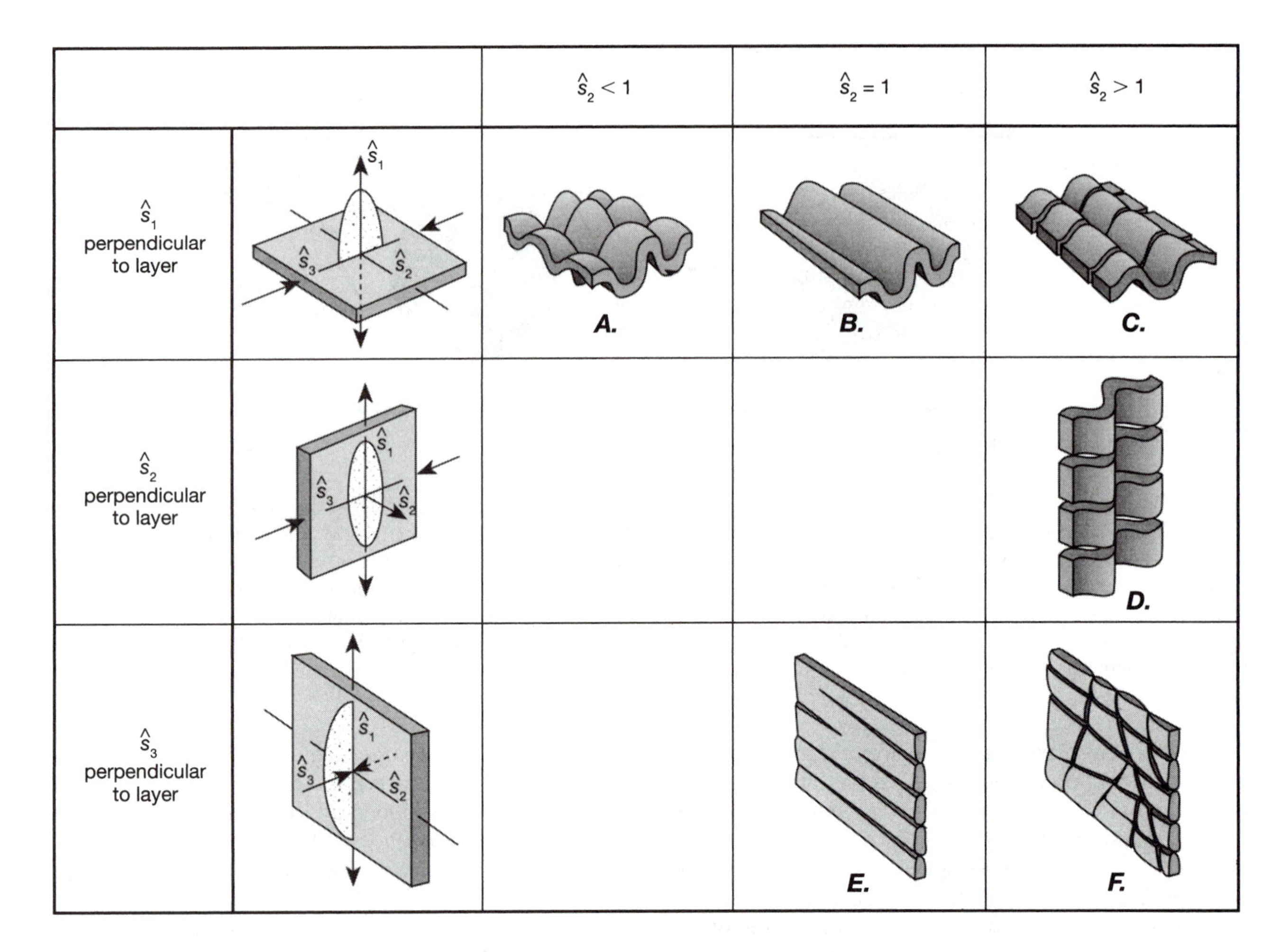

FIGURE 12.7 Structures that could develop in a stiff (competent) layer embedded in a soft (incompetent) matrix depend on the orientation of the layer relative to the principal stretches, and on the value of $\hat{s}_2$. In this diagram, we assume that lengthening the layer causes boudinage, shortening causes folding, and deformation is at constant volume, so that $\hat{s}_1 > 1$, $\hat{s}_3 < 1$, and $\hat{s}_2$ can be greater than, equal to, or less than 1, depending on the other two principal values and the constant volume condition $\hat{s}_1\hat{s}_2\hat{s}_3 = 1$. (After Ramsay 1967)

Chapters 13 and 14 and the measurement and observation of strain in deformed rocks in Chapter 15.

Some structures, such as folds and boudins, also record components of the strain. Consider, for example, a stiff layer embedded in an easily deformed matrix, such as a layer of limestone in a shale matrix. A variety of structures can develop, depending on the orientation of the strain ellipsoid relative to the stiff layer (Figure 12.7). A set of folds develops if the layer is parallel to a principal axis of shortening and normal to an axis of lengthening (Figure 12.7*A–D*). Boudins develop if the layer is parallel to a principal axis of lengthening (Figure 12.7*C–F*). Two interfering sets of folds form if the layer is parallel to two principal directions of shortening and normal to an axis of lengthening (Figure 12.7*A*). Folds develop with a boudinaged hinge line if the layer is perpendicular to a principal axis of lengthening, and the two principal axes parallel to the layer are axes of lengthening and shortening, respectively (Figure 12.7*C*, *D*). Finally, tablet boudinage develops if the layer is parallel to two principal axes of lengthening and perpendicular to one of shortening (Figure 12.7*F*). Thus the structures in such a layer can record components of the strain to which it has been subjected. The orientation of the layer relative to the principal stretches, however, is a major factor in determining what components of the strain can be recorded and thus what structures can develop, and the principal stretches do not have to be parallel or perpendicular to the layer.

12.3 SPECIAL STATES OF STRAIN

Strains with special characteristics are referred to by particular names, which we review in this section.

i. General Strains

Triaxial strain or **triaxial deformation** is a three-dimensional strain for which there is deformation parallel to all three of the principal strain axes. Thus none of the principal stretches equals 1 and none of the principal extensions equals 0:

$$\hat{s}_1 \neq 1, \hat{s}_2 \neq 1, \hat{s}_3 \neq 1 \quad \hat{e}_1 \neq 0, \hat{e}_2 \neq 0, \hat{e}_3 \neq 0 \quad (12.16)$$

This means that none of the principal axes of the strain ellipsoid is the same length as it was in the undeformed state (see Figure 12.11*A*, *B*).

Plane strain or **plane deformation** is a strain of a three-dimensional body for which there is no deformation parallel to the intermediate principal axis (Equation (12.9)):

$$\ell_2 = L_2 \quad \hat{s}_2 = 1 \quad \hat{e}_2 = 0 \quad (12.17)$$

The second two equations follow from the first and from Equations (12.1) and (12.2), respectively. Because of these conditions, a plane strain can always be analyzed as a two-dimensional deformation (see Figures 12.9*A*, *B*; 12.10*A*, *B*; and 12.11*C*).

Constant-volume strain or **constant-volume deformation** is a deformation of a three-dimensional body for which the volumetric stretch s_V is 1 and the volumetric extension is 0:

$$\begin{gathered} s_V = \hat{s}_1\hat{s}_2\hat{s}_3 = 1 \\ e_V = s_V - 1 = (\hat{e}_1 + 1)(\hat{e}_2 + 1)(\hat{e}_3 + 1) - 1 = 0 \end{gathered} \quad (12.18)$$

If the constant-volume strain is also a plane strain (see Figure 12.9*B*), then using the second Equation (12.17) in the first Equation (12.18) shows that

$$s_A = \hat{s}_1\hat{s}_3 = 1 \quad \hat{s}_1 = \frac{1}{\hat{s}_3} \quad \textit{if } s_V = 1 \text{ and } \hat{s}_2 = 1 \quad (12.19)$$

These equations show that the lengthening parallel to $\hat{s}_1$ exactly compensates for the shortening parallel to $\hat{s}_3$ so that the area remains constant. For very small extensions, the plane strain constant-volume condition results in a simple relation between the maximum and minimum principal extensions. From Equation (12.15) and the third Equations (12.17) and (12.18),

$$\begin{gathered} e_A = \hat{e}_1 + \hat{e}_3 = 0 \qquad \hat{e}_1 = -\hat{e}_3 \\ \text{if } e_V = 0, \hat{e}_2 = 0, \text{ and } |\hat{e}_k| << 1 \ (k = 1,3) \end{gathered} \quad (12.20)$$

ii. Homogeneous Strains

Various simple geometries of homogeneous strain are given specific names, which are useful to know. In the following discussion, we review some of the most common of these strain geometries, describing successively one-, two-, and three-dimensional deformations.

Pure strain is any strain for which the principal axes of strain are constant in orientation relative to the reference coordinate system (see, for example, Figures 12.8, 12.9, and 12.11*A*, *B*; Figures 12.10 and 12.11*C*, however, do not show a pure strain). Thus the principal axes of strain and the principal axes of inverse strain are parallel. All the strain geometries described in this subsection, except for simple shear (Figures 12.10 and 12.11*C*), belong to this general class of deformations. These strains are considered pure because they involve no rotational component (see Box 12-3 and Section 12.4).

Uniaxial strain (Figure 12.8) is characterized by having two of the principal stretches equal to 1. The third principal stretch may be either greater than 1 (uniaxial extension, or uniaxial lengthening, Figure 12.8*A*)

$$\hat{s}_1 \geq 1 \quad \hat{s}_2 = \hat{s}_3 = 1 \quad (12.21)$$

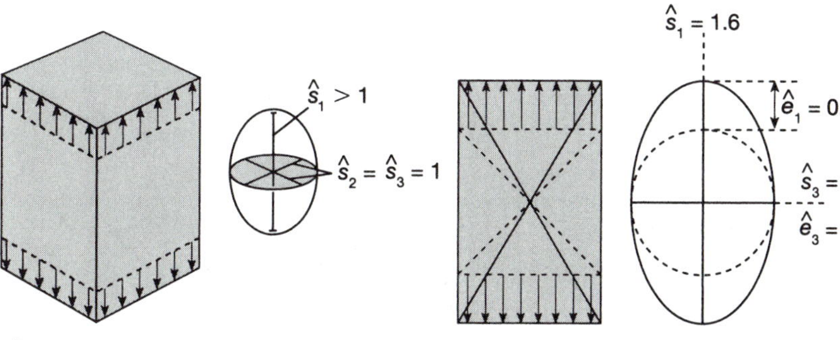

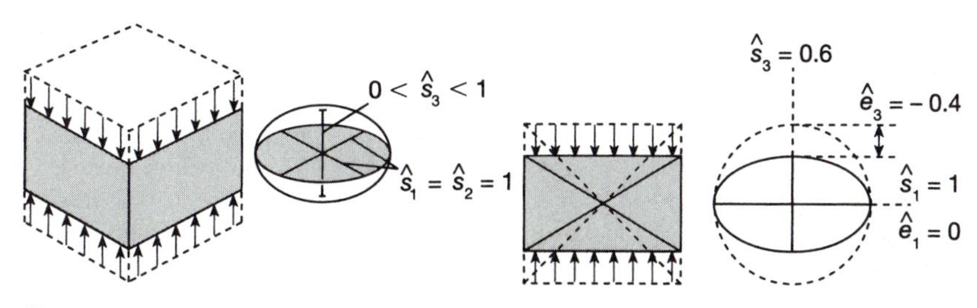

FIGURE 12.8 Uniaxial strain: Two principal stretches are both equal to 1. Dashed lines indicate the undeformed state, solid lines the deformed state.

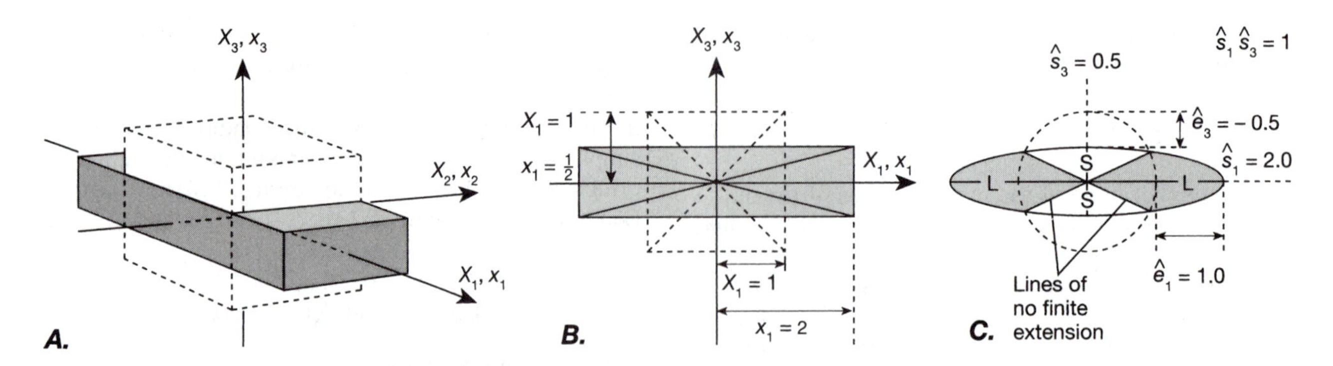

FIGURE 12.9 Pure shear: a constant-volume plane strain in which the principal axes of strain are not rotated by the deformation. *A*. Pure shear of a cube into a rectangular prism (shaded). *B*. Pure shear of a two-dimensional square to form a rectangle (shaded). The diagonals of the square are material lines that are rotated and stretched to become the diagonals of the rectangle. *C*. Pure shear of a unit circle to form an ellipse. The lines of no finite extension divide the strain ellipse into sectors (labeled *S*) in which all radii are shortened, and sectors (labeled *L*) in which all radii are lengthened. (These sector boundaries are not the same as the diagonal lines of the rectangle in *B*).

or less than 1 (uniaxial shortening, Figure 12.8*B*)

$$\hat{s}_1 = \hat{s}_2 = 1 \qquad 0 < \hat{s}_3 < 1 \tag{12.22}$$

Volume is not conserved. Lines perpendicular to the unique axis of stretch are unchanged in length. Lines in all other orientations are lengthened in uniaxial extension and shortened in uniaxial shortening.

Pure shear (Figure 12.9) is a constant-volume plane strain (Figure 12.9*A*):

$$\begin{matrix} \hat{s}_1 > 1 & \hat{s}_2 = 1 & \hat{s}_3 < 1 \\ s_V = 1 & \hat{s}_1 = 1/\hat{s}_3 & \end{matrix} \tag{12.23}$$

Material lines parallel to the principal axes of strain do not rotate, and they experience no shear strain. Material lines of all other orientations in the plane of strain, the $(\hat{s}_1, \hat{s}_3)$-plane, are rotated toward $\hat{s}_1$ (Figure 12.9*B*). The principal axes of strain have the same orientation for all magnitudes of strain and are parallel to the principal axes of inverse strain. Two orientations of line in the plane of strain have the same length as their initial length; these are the lines of no finite extension. They divide the ellipse into sectors within which all radial lines are either shortened (sector S in Figure 12.9*C*) or lengthened (sector L), depending on their orientation.

Simple shear (Figures 12.10 and 12.11*C*) is a constant-volume plane strain whose characteristics re-

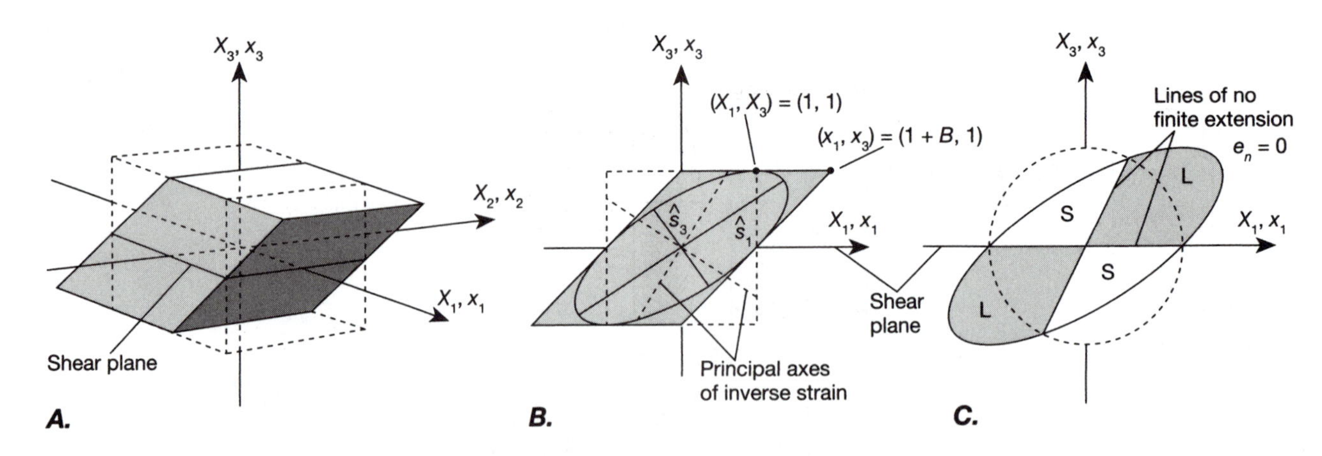

FIGURE 12.10 Simple shear: a constant-volume plane strain in which the displacement of all material particles is strictly parallel to the shear plane. Dashed lines indicate the undeformed state, solid lines the deformed state. *A*. Simple shear of a cube. *B*. Simple shear in two dimensions of a square. The principal axes of inverse strain in the undeformed state are dashed; the principal axes of strain in the deformed state are solid. Material lines parallel to the axes of inverse strain are rotated by the deformation into parallelism with the principal axes of strain. *C*. Lines of no finite extension in the strain ellipse divide the ellipse into sectors of shortened (*S*) and lengthened (*L*) radii of the ellipse.

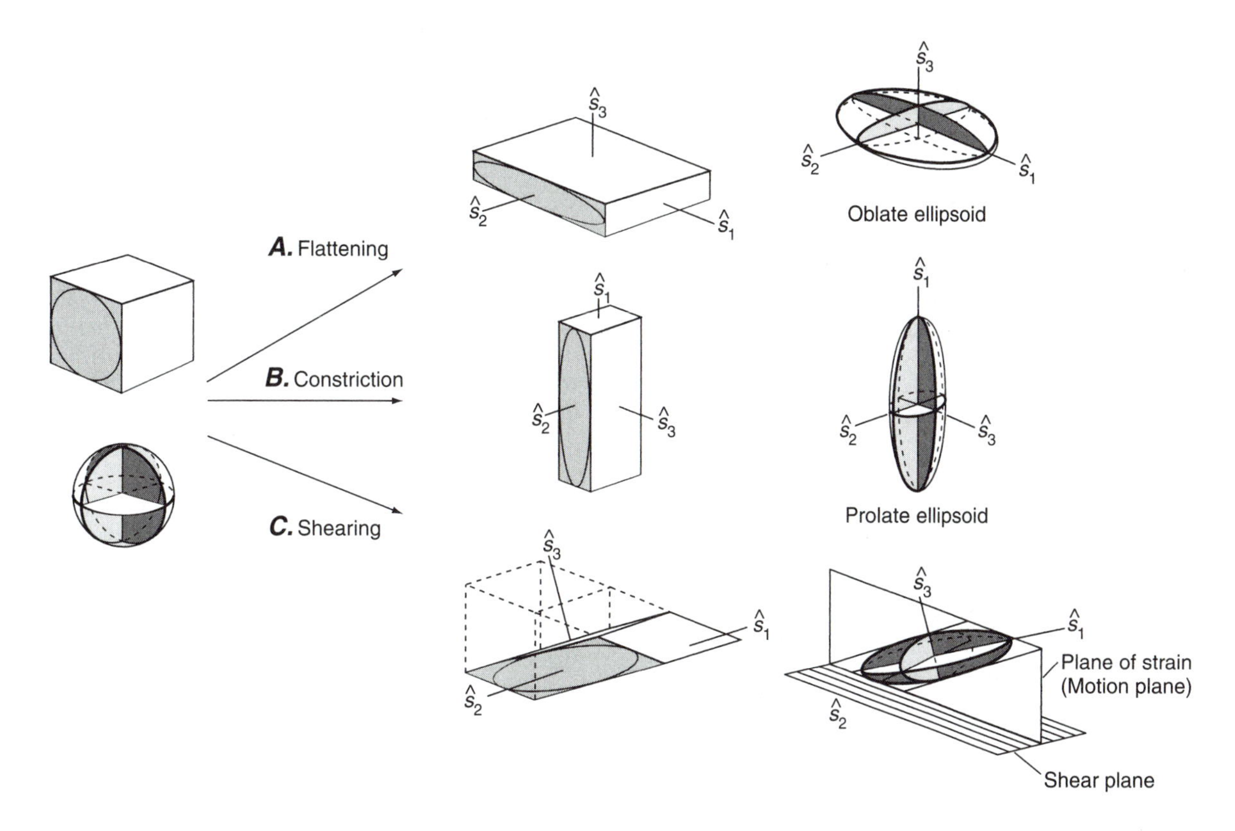

FIGURE 12.11 Types of homogeneous deformation imposed on a cube, a circle, and a sphere; $\hat{s}_1$, $\hat{s}_2$, and $\hat{s}_3$ are the directions of maximum, intermediate, and minimum stretch. The shaded faces of the blocks show the deformation in the $(\hat{s}_1, \hat{s}_3)$-plane. A circle in that plane is deformed into the strain ellipse whose maximum and minimum axes are parallel to $\hat{s}_1$ and $\hat{s}_3$. A sphere is deformed into strain ellipsoids whose principal axes are the principal directions of stretch $\hat{s}_1 \geq \hat{s}_2 \geq \hat{s}_3$. The strain ellipses on the shaded sides of the deformed blocks are the darkly shaded sections through the strain ellipsoids. *A*. Flattening. Two orthogonal dimensions are lengthened and one is shortened. A sphere is deformed into a pancake-shaped (oblate) ellipsoid. *B*. Constriction. Two orthogonal dimensions are shortened and one is lengthened. A sphere is deformed into a prolate (cigar-shaped) ellipsoid. *C*. Shearing. Particle motion is parallel to the shear plane, which contains $\hat{s}_2$, and to the motion plane, which contains $\hat{s}_1$ and $\hat{s}_3$. The displacement gradient is a maximum normal to the shear plane and zero normal to the motion plane. No deformation occurs parallel to the $\hat{s}_2$ direction. The cube is deformed into a rhomboid. A sphere becomes an ellipsoid with axes inclined relative to the shear plane.

semble the shearing of a deck of cards (compare the deformation in Figure 12.1 that carries the square in *A* into the parallelogram in *D*):

$$\begin{gathered} \hat{s}_1 > 1 \qquad \hat{s}_2 = 1 \qquad \hat{s}_3 < 1 \\ s_V = 1 \qquad \hat{s}_1 = 1/\hat{s}_3 \end{gathered} \tag{12.24}$$

Thus, for a homogeneous deformation, the side of the card deck changes from a rectangle to a parallelogram (Figure 12.10*A*). Displacement of all material particles is parallel to the shear plane, the (x_1, x_2)-plane in Figure 12.10*A*, as well as the plane of strain, the (x_1, x_3)- and $(\hat{s}_1, \hat{s}_3)$-planes, also referred to as the motion plane. In the plane of strain, all material lines are rotated except those parallel to the shear plane. There are two orientations of no finite extension in the plane of strain, one of which is always parallel to the shear plane. These lines divide the strain ellipse into sectors of shortened radii (S in Figure 12.10*C*) and lengthened radii (L).

Note that Equations (12.24) are identical to Equations (12.23), although the deformations are not the same because for simple shear, the orientations of principal strain axes change with the magnitude of shear, and the principal axes of strain and of inverse strain are not parallel (Figure 12.5). We discuss the differences between these two geometries of deformation in more detail in Section 12.4.

Uniform dilation is a pure volumetric strain with no change in shape of the deforming body:

$$\hat{s}_1 = \hat{s}_2 = \hat{s}_3 \neq 1 \tag{12.25}$$

A square or a cube is transformed into a body that is the same shape but is either larger (uniform expansion; $s_V > 1$) or smaller (uniform contraction; $0 < s_V < 1$). The same statement, of course, applies to both a circle and a sphere (Figure 12.1*A*, *B*). The stretch has the same value in all directions, as does the extension, and the shear strains are zero in all directions; that is, $\psi = 0$ for all orientations of line. All material lines change length, but none changes orientation.

A **flattening** (Figure 12.11*A*) is a pure triaxial strain that deforms a cube of material into a plate-like rectangular prism. One dimension of the original cube is shortened and the other two are both lengthened, although not necessarily by the same amount. Volume is not necessarily conserved, so the volumetric stretch s_V is not necessarily 1. A sphere is deformed into an oblate (pancake-shaped) ellipsoid:

$$\hat{s}_1 \geq \hat{s}_2 > 1 \qquad \hat{s}_3 < 1 \tag{12.26}$$

Simple flattening is a flattening for which the stretches in the two principal lengthening directions are equal. Thus the oblate strain ellipsoid is axially symmetric about its short axis and has a circular pancake shape. The orientations of the two principal lengthening directions are not unique because all directions normal to the short axis have the same stretch and extension:

$$\hat{s}_1 = \hat{s}_2 > 1 \qquad \hat{s}_3 < 1 \tag{12.27}$$

A **constriction** (Figure 12.11*B*) is a pure triaxial strain that deforms a cube of material into an elongate rectangular prism for which one dimension of the cube is lengthened and the other two dimensions are shortened, although not necessarily by the same amount. Volume is not necessarily conserved, so the volumetric stretch s_V is not necessarily 1. A sphere is deformed into a prolate (cigar-shaped) ellipsoid:

$$\hat{s}_1 \geq 1 \qquad 1 > \hat{s}_2 \geq \hat{s}_3 \tag{12.28}$$

Notice that in two dimensions, or in plane strain, the distinction between flattening and constriction cannot be made (see the shaded faces of the blocks in Figure 12.11*A*, *B*) because the intermediate stretch $\hat{s}_2$, which distinguishes these two geometries, either is not represented or, for plane strain, equals 1 in both cases.

Simple constriction or **simple extension** is a constriction for which the two principal axes of shortening are equal. Thus the strain ellipsoid is axially symmetric about its long axis and has the shape of a cigar with a circular cross section. The two principal shortening directions do not have a unique orientation because all directions normal to the long axis have the same stretch and extension:

$$s_1 \geq 1 \qquad 1 > \hat{s}_2 = \hat{s}_3 \tag{12.29}$$

These states of strain are all special cases of the infinite variety of possible states. They have no special qualities that make them uniquely applicable to the interpretation of rock deformation, but they are singled out for special names because the geometry of each is simple and well defined.

The geometry and orientation of the strain ellipse at the end of an arbitrary deformation can always be expressed as the sum of three components: (1) a pure strain that has stretches parallel to the axes of inverse strain (Figure 12.12*A*, *B*; compare Figure 12.5); (2) a rigid rotation of the body that brings the principal axes of strain into the proper orientation (Figure 12.12*C*); and (3) a rigid translation of the body that brings it into the proper location (Figure 12.12*D*). These components of the deformation, in principle, can be applied in any order. The strain ellipse resulting from a simple shear strain (Figure 12.10), for example, can be reproduced by the sum of a pure shear (Figure 12.9) parallel to the axes of inverse strain, a rotation of the principal axes, and a translation (Figure 12.12). The strain ellipses for other geometrically more complex deformations can be similarly reproduced.

Such descriptions of the final state of strain, however, only relate the final deformed state to the initial undeformed state. They do not imply anything about the intermediate states of strain through which the material passes in reaching the final state, and they do not define the different strain histories experienced by different orientations of material lines and planes. Defining this re-

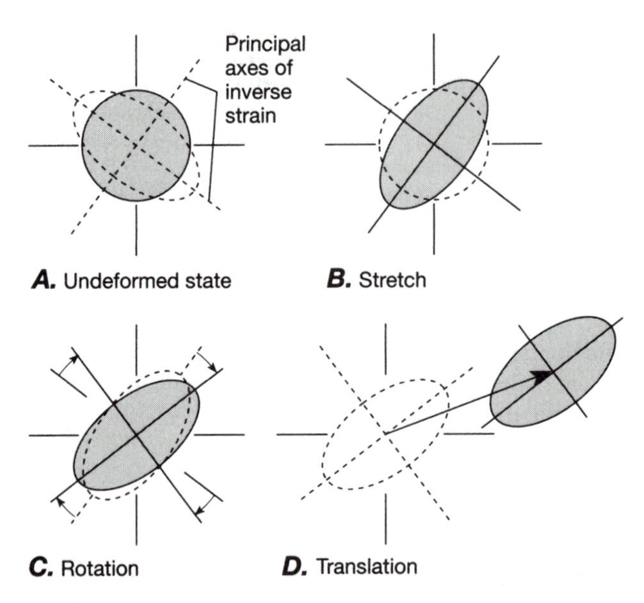

FIGURE 12.12 Decomposition of an arbitrary homogeneous strain into a pure strain, a rigid rotation, and a rigid translation. These components may be applied in any sequence. *A*. The undeformed state, showing the unit circle, the inverse strain ellipse (dashed), and the principal axes of inverse strain. *B*. Stretches are imposed parallel to the principal axes of inverse strain to reproduce the final shape of the strain ellipse. The inverse strain ellipse becomes a circle. *C*. Rigid-body rotation brings the principal axes into the correct final orientation. *D*. Rigid-body translation brings the body into the correct final location.

lationship requires a consideration of the motion by which the strain accumulates, which illuminates the distinction among different strain geometries. We consider this in the following section.

12.4 PROGRESSIVE DEFORMATION

We refer to the nonrigid motion of a body that carries the body from its initial undeformed state to its final deformed state as a **progressive strain** or **progressive deformation**. We can describe the motions of all material particles in the body by describing the deformed position of the particles as a function of their original position and of time (see Box 12-3). The strain states through which the body passes during a progressive deformation define the **strain path**. The state of strain of a body at any given time is the end point of the strain path at that time and thus the net result of all the previous deformations the body has undergone.

In rocks, we generally can observe only the final strained state and must infer the initial undeformed state. Thus in our discussion so far, we have simply related the deformed state to the undeformed state, without implying anything about the intermediate strain states that develop during the deformation. All states of strain, however, are the result of a progressive deformation, and the history of that deformation is also of great geological interest. Although the final geometry and orientation of the strain ellipse provide no information about the particular strain path that the body experienced, in some cases clues to the history are recorded by features in deformed rocks. Understanding the consequences of different strain paths can provide insight that is useful in interpreting strain and the history of deformation in rocks.

Structures such as folds, boudins, foliations, and lineations develop in rock in response to progressive deformations. Folds and boudins develop in material layers in the rock, such as sedimentary layers, cross-cutting veins, or dikes. Most spaced foliations are also defined by material surfaces. Therefore, in order to understand the relationship between such structures and the principal axes of strain, we investigate what happens to material lines of various orientations during different progressive plane deformations.

We can conceptualize the geometry of a progressive deformation by stopping it, marking a material circle on the body, and allowing the deformation to continue for a unit increment of time. The ellipse formed from that circle represents the increment of strain for that increment of time and is therefore called the **incremental strain ellipse**, or, preferably, the **instantaneous strain ellipse** if the time increments are very small. Thus we define the instantaneous extension ε_n, the instantaneous shear strain ε_s, and the instantaneous stretch ζ_n in terms of the instantaneous length of a material line ℓ, its infinitesimal change in length $d\ell$, and the infinitesimal shear angle $d\psi$ of two instantaneously perpendicular lines:

$$\varepsilon_n \equiv \frac{d\ell}{\ell} \qquad \varepsilon_s \equiv 0.5 \tan d\psi \qquad \zeta_n \equiv \frac{\ell + d\ell}{\ell} \tag{12.30}$$

(Note that the integral of the instantaneous extension is the natural, or logarithmic, strain; Box 12-1, Equation (12-1.1)). The lengths of the principal radii of the instantaneous strain ellipse, which are the radii parallel to the principal axes, are proportional to the principal instantaneous stretches $\hat{\zeta}_1 \geq \hat{\zeta}_3$. If the instantaneous strain ellipse has the same shape and orientation for every infinitesimal increment in time, the motion of the material particles is called a **steady motion**.

To illustrate the effects of different motions on material lines, consider two special steady motions: progressive pure shear and progressive simple shear. The paths traced out by individual particles during these progressive deformations are shown in Figure 12.13*A*, *B*, respectively. For these examples, the instantaneous strain ellipse has the geometric properties of either pure shear (Figure 12.9) or simple shear (Figure 12.10) for each increment of strain through time. Although a simple shear

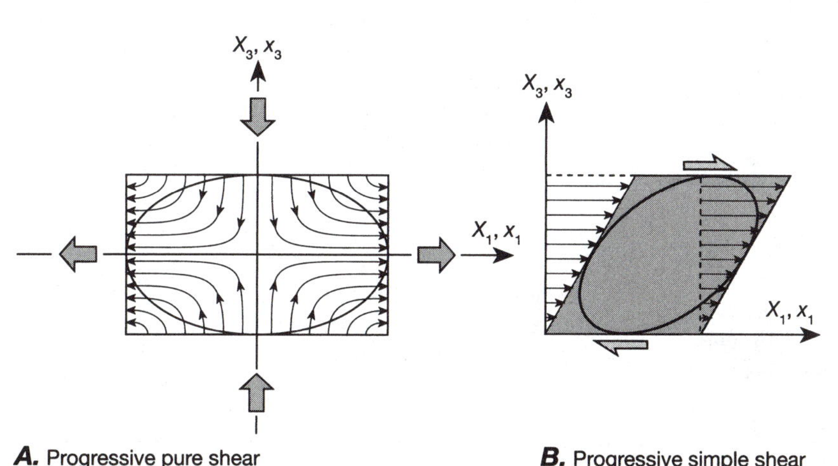

FIGURE 12.13 Particle motions during two progressive deformations. *A*. Particle motions during progressive pure shear. The lines with the arrowheads are parallel to the velocity vectors of the particles in the body. *B*. Particle motions during progressive simple shear are all strictly parallel to the shear direction in the shear plane (X_l direction). The velocity varies linearly with distance normal to the shear plane (X_3 direction).

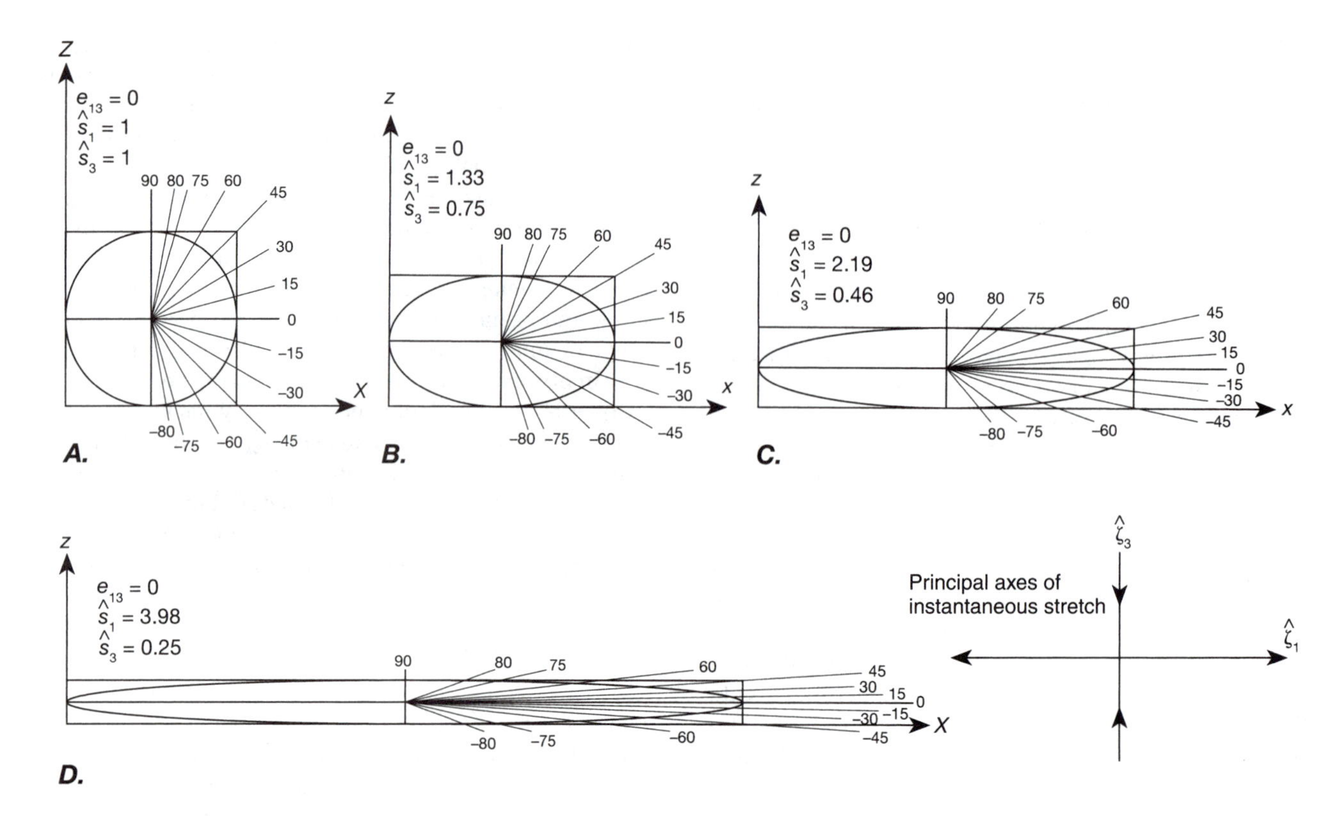

FIGURE 12.14 States of strain during a steady progressive pure shear. The axes at the bottom right of the figure indicate the constant orientation of the principal axes of instantaneous stretch. Material lines are labeled by the angle they make with X_l in the undeformed state. The lines 0 and 90 are the only ones that do not rotate during the deformation, and they are always parallel to the principal axes of strain. The magnitudes of the principal stretches in each diagram are the same as for the corresponding diagram in Figure 12.15.

strain ellipse may be equivalent to a pure shear strain ellipse plus a rigid rotation (see Figure 12.12), it is evident from these particle paths that a progressive simple shear is not equivalent to a progressive pure shear plus a rigid rotation.

Figures 12.14 and 12.15 illustrate the consequences of progressive pure shear and progressive simple shear, respectively. In each figure, *A* is the undeformed state, showing a sheaf of material lines. In Figure 12.14*A*, the material lines are oriented at regular angular intervals, and each line is labeled with the angle it originally makes with the *X* axis. In Figure 12.15*A*, the material lines are parallel to the axes of inverse strain for the state of strain in the diagram labeled with the corresponding letter. For example, the material lines C and C′ in *A*, are parallel to the principal axes of inverse strain for the strain state shown in *C*. These lines are rotated by the deformation into the orientations shown by *c* and *c*′, which become parallel to the principal axes of strain in *C*. In both figures, *B*–*D* shows the evolution of both the strain ellipse and the orientations of the same material lines as appear in *A*. The corresponding diagrams in Figures 12.14 and 12.15 show the same states of strain, although the orientations of the principal axes are different (see Figure 12.12).

A comparison of Figures 12.14 and 12.15 shows the following significant differences in behavior:

1. With respect to the coordinate axes (x, z), the principal axes of strain do not rotate in progressive pure shear, but in progressive simple shear they do. Thus the former is an **irrotational progressive deformation**, and the latter is a **rotational progressive deformation**. The difference in behavior of the principal strain axes is reflected by the **vorticity** of the deformation, which is a measure of the average rate of rotation of material lines of all orientations about each coordinate axis.[7] In the case of the plane strain in Figures 12.14 and 12.15, vorticity can occur only about the y axis.

   In Figure 12.14, for example, material lines oriented symmetrically relative to the x axis have exactly opposite rates of rotation. For example, the material lines in the upper-right quadrant rotate in exactly the opposite sense and rate to symmetrically oriented lines in the lower-right

[7]Technically, the vorticity is a vector $\boldsymbol{\omega}$ related to the antisymmetric part $\mathbf{w}$ of the velocity gradient tensor. See footnote 6, Section 17.7 for details.

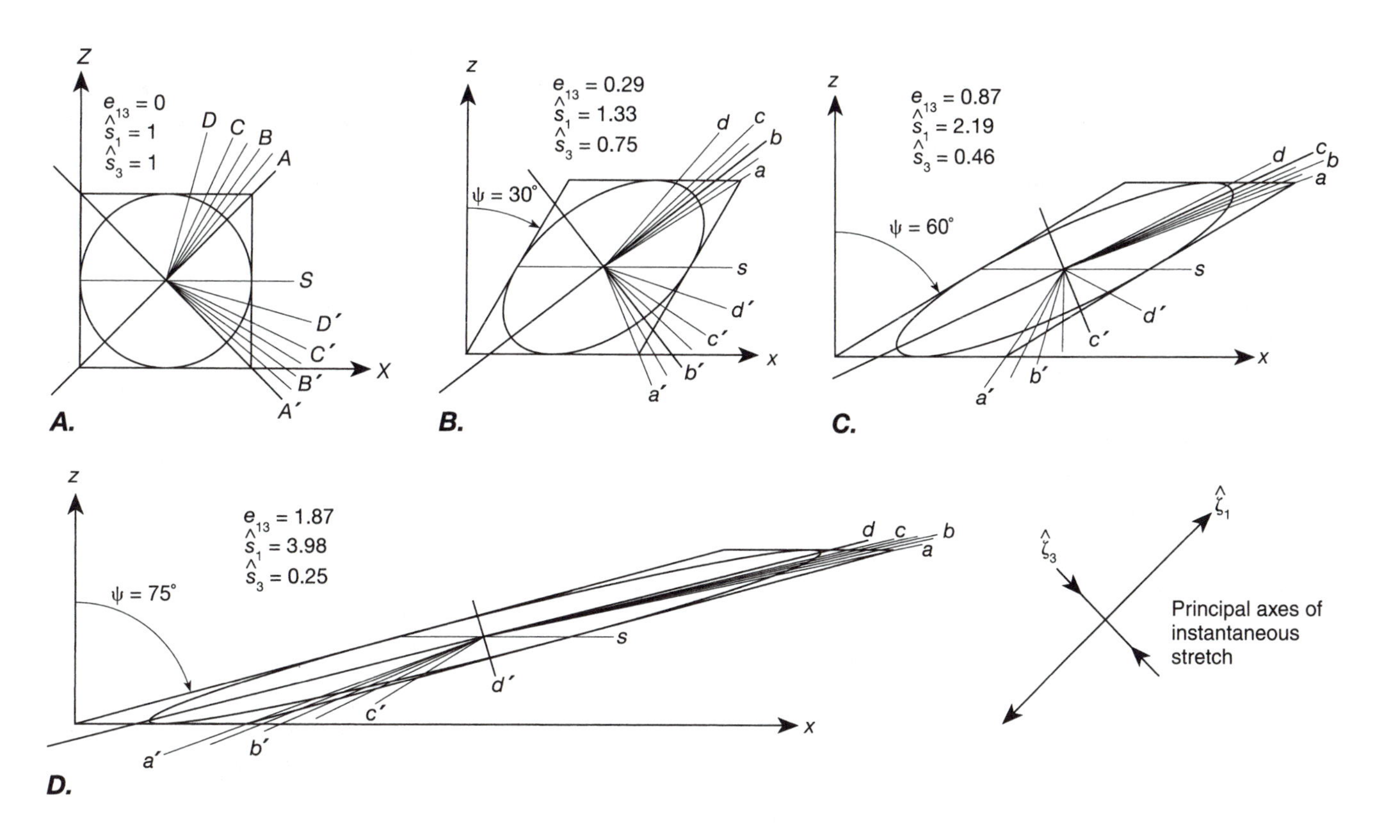

FIGURE 12.15 States of strain during steady progressive simple shear. The axes at the bottom right of the figure indicate the constant orientation of the principal axes of instantaneous stretch. The pairs of material lines in the undeformed state labeled (B and B'), (C and C') and (D and D') are parallel to the principal axes of inverse strain for the strain states shown in B, C, and D, respectively. These pairs of material lines take on the orientations in the deformed states indicated by the lines labeled in the equivalent lowercase letters, and each pair becomes parallel to the principal axes of strain in the diagram labeled with the same letter as the line pair. Thus the material lines rotate past the principal axes of strain, which themselves are not material lines. S and s indicate a material line parallel to the shear plane. This is the only orientation of line for which the orientation and length are constant throughout the deformation.

quadrant. The average rotation rate taken over all orientations of material line is therefore zero, and thus the vorticity for an irrotational deformation is zero. In contrast, all the material lines in Figure 12.15 (except lines parallel to *s*; see item 3 in this list) rotate in the same sense, so the average rate of rotation is nonzero, and thus the vorticity for a rotational deformation is nonzero.

2. In progressive pure shear, the principal axes of finite strain are always parallel to, or coaxial with, the principal axes of instantaneous strain. The deformation is therefore a **coaxial** progressive deformation. In progressive simple shear, the principal axes of finite strain rotate with respect to those of instantaneous strain, and this characteristic defines a **noncoaxial** progressive deformation. Note that for progressive simple shear, the principal axes of instantaneous strain always are oriented at a 45° angle to the shear plane.

   The terms *irrotational* and *coaxial* are not synonymous, nor are *rotational* and *noncoaxial*. The difference is in the reference frame from which the rotation is determined. A deformation is rotational or irrotational depending on how the principal axes of finite strain behave with respect to the coordinate system, which is always somewhat arbitrarily defined by the observer. A deformation is coaxial or noncoaxial depending on how the principal axes of finite strain behave with respect to the principal axes of instantaneous strain. This reference frame is intrinsic to the geometry of the deformation itself and is therefore not arbitrary. Thus the description of a progressive deformation as coaxial or noncoaxial is somewhat more fundamental than the description as rotational or irrotational, especially in geologic situations in which the best choice of an external reference coordinate system is not obvious.

3. In progressive pure shear, all material lines rotate during the deformation except those parallel to the principal axes of strain. Lines in adjacent quadrants of the principal coordinates have opposite senses of

rotation such that the lines all rotate toward parallelism with the $\hat{s}_1$ (maximum principal stretch) direction. Note that the term *irrotational* refers only to the behavior of the principal axes of strain and to the average rotation of all material lines, not to the rotation of any specific material line. In progressive simple shear, all lines rotate during the deformation, except those parallel to the shear plane (labeled *s* in Figure 12.15), and all have the same sense of rotation. The rotation rate of any line decreases with decreasing angle between the line and the shear plane.

4. In progressive pure shear, the lines that rotate most rapidly are those at an angle of 45° to the principal axes of the instantaneous strain ellipse. In progressive simple shear, the lines that rotate most rapidly are normal to the shear plane, and these lines also are oriented at a 45° angle from the principal axes of instantaneous strain. Lines parallel to the shear plane, however, do not rotate at all, and they too are 45° from the principal axes of instantaneous strain.

5. In progressive pure shear, the same pair of material lines remains parallel to the principal axes of strain throughout the deformation. In progressive simple shear, material lines rotate through the principal axes of strain. This characteristic shows that the principal axes of strain are not in general material lines. During progressive simple shear, material lines that are parallel to the principal axes at any time were originally orthogonal in the undeformed state. During the deformation, however, any such pair of lines is sheared out of orthogonality, then back into orthogonality when they become parallel to the principal axes, and finally out of orthogonality again (e.g., see the deformation of the lines C and C′ in Figure 12.15).

6. In both progressive pure shear and progressive simple shear, the stretch of material lines depends on their orientation. Some lines experience a history only of shortening, others experience only lengthening, and still others experience initial shortening followed by lengthening and can end up being either shorter or longer than they were originally. The pattern of variation determines what types of structures can develop. We discuss this further in the next section.

If the deformation stops at any time, the final state of strain can always be related to the initial state in Figure 12.14 by a pure shear strain or in Figure 12.15 by a simple shear strain. The converse of this statement, however, is not true: If a final state of strain can be related to the initial state by either a pure shear strain or a simple shear strain, it does not follow that the final state of strain was the result of a progressive pure shear or a progressive simple shear, respectively. An infinite number of strain paths lead from an undeformed state to a deformed state, and the final state of strain does not, by itself, provide sufficient information for any of the paths to be distinguished. It is very important to remember this when interpreting the strain in rocks.

From the foregoing discussion, it is evident that if a progressive deformation is noncoaxial, the principal axes of finite strain rotate relative to those of instantaneous strain, and that the principal axes of instantaneous strain are constant in orientation only if the deformation is steady. It should not be surprising, therefore, that the principal axes of finite strain are not in general parallel to the principal axes of stress. In fact, we see in Chapter 16, where we discuss the relationships between stress and strain, that for steady motions of homogeneous isotropic materials, the principal stress axes are parallel to the principal axes of instantaneous strain or of strain rate. Because most natural deformations are probably not steady, even this relationship may not be accurate for interpreting deformation that we observe in rocks. Thus as a general rule, *structures should always be interpreted in terms of the principal axes of strain. Only under very special circumstances can reliable inferences be made about the orientations of the principal stress axes.*

12.5 PROGRESSIVE STRETCH OF MATERIAL LINES

If the unit circle is superimposed on the associated finite strain ellipse, the radii to the intersection points define lines of no finite extension ($e_n = 0$), which are lines that are the same length as they were in the undeformed state ($s_n = 1$). These lines divide the ellipse (Figure 12.16*A*) into sectors labeled L in which radii are longer than they were originally ($s_n > 1$) and sectors labeled S in which the radii are shorter ($0 < s_n < 1$).

We can also examine a similar superposition of the unit circle on the instantaneous strain ellipse. For generality, we show in Figure 12.16*A*, *B* the finite and instantaneous principal strains in a relative orientation that can occur in nature only if the instantaneous principal axes have changed orientation during the deformation. The intersection of the circle with the instantaneous ellipse defines a pair of lines that instantaneously are not changing length ($\zeta_n = 1$). These lines divide the instantaneous strain ellipse (Figure 12.16*B*) into sectors labeled $\dot{\text{L}}$ and $\dot{\text{S}}$. In the $\dot{\text{L}}$ sectors, lines are becoming longer, and thus the instantaneous stretch is larger than one ($\zeta_n > 1$) and the finite stretch is increasing ($(ds_n/dt) > 0$). In the $\dot{\text{S}}$ sectors, the lines are becoming shorter, and thus the instantaneous stretch is between zero and one ($0 < \zeta_n < 1$) and the finite stretch is decreasing ($(ds_n/dt) < 0$).

The sector boundaries on the instantaneous strain ellipse (Figure 12.16*B*) are not in the same orientation as those on the finite strain ellipse (Figure 12.16*A*), and be-

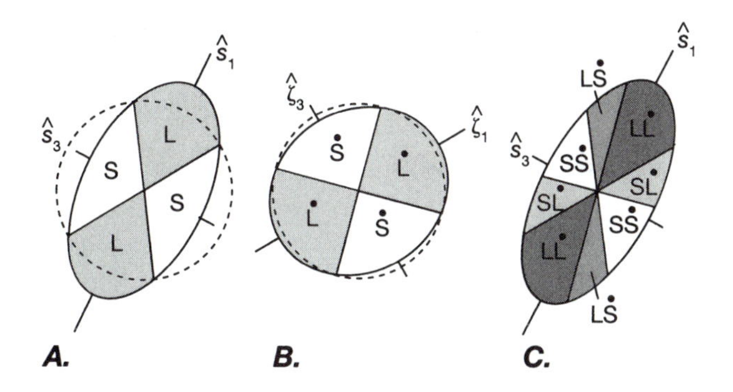

FIGURE 12.16 Geometry of finite and instantaneous strain ellipses for a deformation in which the instantaneous strain is superposed on a pre-existing homogeneous strain. For generality, we have chosen an orientation of the finite strain ellipse that can be formed only from an unsteady deformation, characterized by an instantaneous strain ellipse whose principal axes change orientation during the deformation. *A*. The strain ellipse, showing lines of no finite extension that define sectors in which radial material lines have been lengthened (*L*) or shortened (*S*) by the deformation. The unit circle is shown dashed. *B*. The instantaneous strain ellipse, showing lines of no rate of extension that divide the ellipse into sectors in which radial material lines are being lengthened (L̇) (positive rate of change of stretch) and sectors in which radial material lines are being shortened (Ṡ) (negative rate of change of stretch). *C*. The combination of the two sets of sectors from *A* and *B* on the strain ellipse defines sectors in which radial material lines have different combinations of stretch and rate of stretch.

cause material lines in general rotate during a deformation they can pass from one sector into another. Thus the finite strain ellipse can be divided into sectors in each of which the material lines have a different history of stretching (Figure 12.16*C*). The different possible histories are illustrated in Figure 12.17, where shortening of material lines is represented as folding or imbrication, and lengthening of material lines is represented as boudinage. In sectors labeled SṠ, lines are shorter than the original length and are now shortening; they have a continuous history of shortening (Figure 12.17*A*). In sectors labeled LṠ, lines are longer than the original length, indicating an initial history of lengthening, but they are now shortening (Figure 12.17*B*); with continued deformation, they may end up shorter than their initial length and therefore evolve into the SṠ sector (see Figure 12.17*E*). In sectors LL̇, lines have lengthened and are lengthening; they have a history of continuous lengthening (Figure 12.17*C*); and in sectors SL̇, lines are shorter than their initial length, indicating an initial history of shortening, although they are presently lengthening (Figure 12.17*D*). With continued deformation they may end up longer than their initial length and therefore evolving into the LL̇ sector. Thus SṠ sectors may be subdivided according to whether or not the lines had an initial history of lengthening that preceded the shortening, which has resulted in the lines presently being shorter than the initial length (compare Figure 12.17*A*, *B*). Similarly, LL̇ sectors may be subdivided according to whether or not the lines had an initial history of shortening that preceded the lengthening, which has resulted in the lines presently being longer than the initial length (compare Figures 12.17*C*, *D*: see lighter gray portions of LL̇ sectors in Figure 12.18*A*, *B*).

Thus, depending on the orientation of the material line with respect to the instantaneous and finite strain axes, the same deformation can produce folds, boudinage, boudinaged folds, or folded and imbricated boudins. The distributions of such sectors for progressive pure shear and for progressive simple shear are shown in Figure 12.18*A*, *B*, respectively. The main difference in these distributions of sectors about the principal axes of strain is the absence of an SL̇ sector for progressive simple shear subparallel to the shear plane. Thus the sectors of the strain ellipse for progressive pure shear have an overall orthorhombic symmetry, whereas the sectors for progressive simple shear have an overall monoclinic symmetry.[8] When these aspects of the deformation are taken into account, an arbitrary deformation cannot be reproduced by the sequence of operations indicated in Figure 12.12. It is evident, for example, that the rotation of the sectors with the strain ellipse produced by progressive pure shear (Figure 12.18*A*) does not reproduce the sectors in the strain ellipse formed by progressive simple shear (Figure 12.18*B*), even though the strain ellipses themselves are the same shape and differ only by a rigid rotation.

In principle, then, it should be possible to distinguish some features of the strain history, such as coaxial and noncoaxial progressive deformations, by examining the relationship between the deformational structures in the rock and their orientations. For example, if veins are intruded into a rock in a variety of orientations, subsequent deformation could cause veins to form folds and/or boudins depending on their orientation relative to the principal stretches. The observed distribution of these structures defines the sectors of the finite strain ellipse (Figure 12.18*C*). In practice, however, the sector patterns are difficult to establish. The distribution of orientations of deformed layers is usually not ideal (Figure 12.18*D*), and layers can shorten and thicken without folding or can

[8]Remember that in three dimensions, orthorhombic symmetry is characterized by three mutually orthogonal two-fold axes of rotational symmetry and three mirror planes of symmetry, each perpendicular to one of the two-fold axes (see Section 17.8(i); Figure 17.33*C*). Monoclinic symmetry is characterized by one two-fold axis of symmetry with one mirror plane of symmetry perpendicular to it (see Section 17.8(i); Figure 17.33*D*). In Figure 12.18*A*, the axes of orthorhombic symmetry are the two principal axes in the plane of the page and the third axis, the intermediate principal axis, normal to the page; in Figure 12.18*B* the unique axis of monoclinic symmetry is the intermediate principal axis of strain normal to the page.

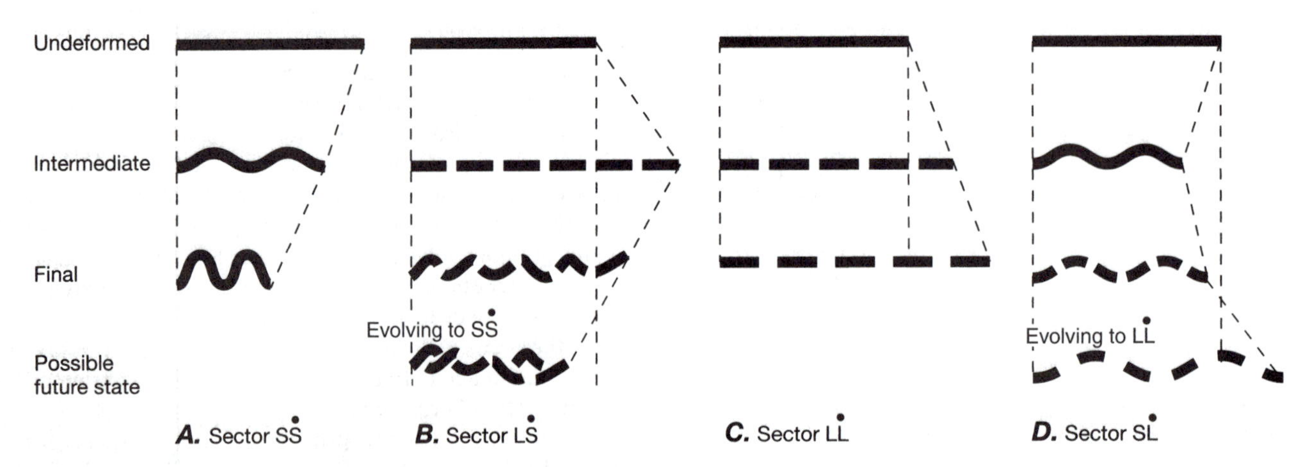

FIGURE 12.17 The histories of progressive deformation for competent layers oriented within the different sectors shown in Figure 12.16*C*. The undeformed, intermediate, and final states are points along the deformation path. *A*. Sectors $S\dot{S}$: shorter and being shortened. The layer is continuously folded. *B*. Sectors $L\dot{S}$: longer and being shortened. The layer was initially boudined and subsequently shortened, which caused folding and imbrication of the boudins. The "final" overall length is greater than the initial length but continued shortening could make it less, thereby transferring the line into the $S\dot{S}$ sector. *C*. Sectors $L\dot{L}$: longer and being lengthened. The layer is continuously boudinaged. *D*. Sectors $S\dot{L}$: shorter and being lengthened. The layer is initially folded and subsequently boudinaged. The "final" overall length is smaller than the original length, but continued lengthening could make it longer, thereby transferring the line into the $L\dot{L}$ sector. *E*. Boudins that have been shortened after formation, illustrating the deformational history in *B*.

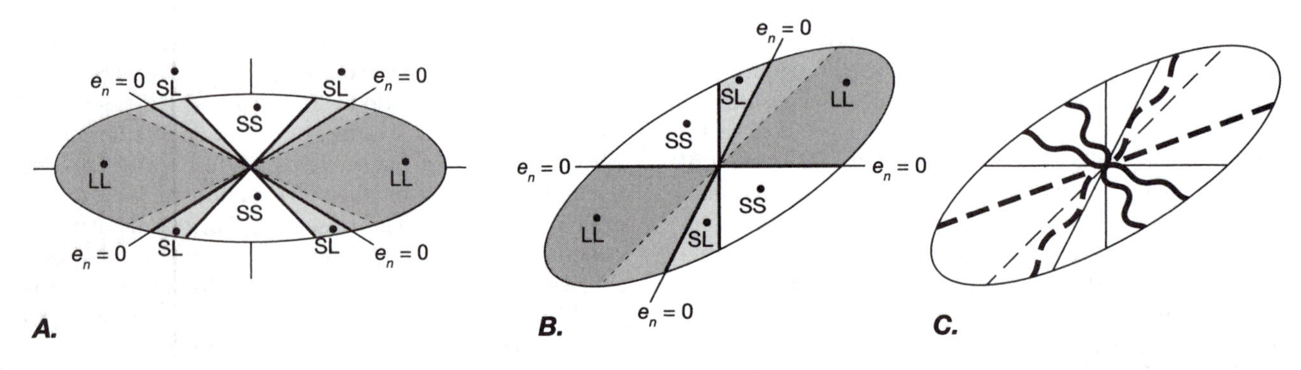

FIGURE 12.18 Distribution of sectors of stretch and stretching. Material lines in the lighter gray parts of the $L\dot{L}$ sectors (separated by the dashed line) have an initial history of shortening followed by lengthening and thus have rotated into the $L\dot{L}$ sectors from the adjacent $S\dot{L}$ sectors (see bottom of Figure 12.17*D*). *A*. Progressive pure shear. *B*. Progressive simple shear. This case differs from progressive pure shear mainly in the lack of symmetry of the ($S\dot{L}$) sectors about the principal axes of strain. *C*. Structures developed in stiff (competent) layers in a soft (incompetent) matrix consistent with the sectors for progressive simple shear. *D*. Folding of a layer (left) and simultaneous boudinage of a perpendicular layer (horizontal above pencil) during deformation of a marble.

lengthen and thin without boudinage. Despite its limited practical application, this analysis demonstrates the important fact that no single type of structure, such as folding or boudinage, is uniquely indicative of a particular geometry of deformation. Both may form in a single rock during a single deformation if different planar features have the appropriate orientations.

12.6 HOMOGENEOUS AND INHOMOGENEOUS DEFORMATION

So far in this chapter, we have restricted most of our discussion to homogeneous strains. As we noted at the beginning of the chapter, if we are interested in the inhomogeneous distribution of strain, such as in the formation of a fold, we assume the deformed body can be divided into volumes that are sufficiently small for the deformation to be described as locally homogeneous. The variation of these local strains across the body describes the inhomogeneous strain distribution. For any real material, we must realize that the description of a deformation as homogeneous at any particular scale is the result of averaging the deformation over volumes that are large compared with the scale of inhomogeneities that are of no immediate interest but small compared with the scale at which the inhomogeneity of the strain is of interest.

Figure 12.19, for example, shows the so-called deck-of-cards model for forming a passive shear fold (see also Figures 13.8 and 13.10). The deformation is accomplished by a discontinuity in the shear displacement at the card surfaces, with no deformation at all of the individual cards. On the scale of a fold limb, however, the deformation in this example can be regarded as homogeneous simple shear, and it produces the average strain ellipse shown on each fold limb in Figure 12.19. Thus the description of the strain as homogeneous results from averaging the strain over a region that is large compared with the thickness of the cards but small compared with the wavelength of the fold. In other words, the homogeneity depends on scale.

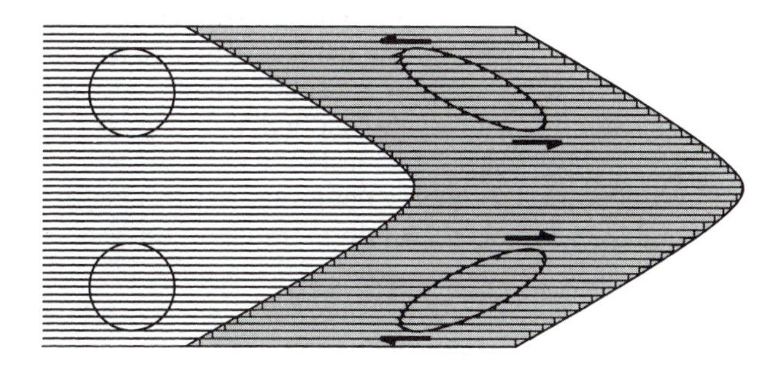

FIGURE 12.19 Deck-of-cards model of passive shear folding. On each card, the arcs of the undeformed circle are displaced so that they approximate the shape of the strain ellipse.

The variety of scales at which we could consider a deformation to be homogeneous is illustrated schematically in Figure 12.20. In Figure 12.20*A*, the block of folded rock measures about 1 km in length. The scale of the whole block is small compared with the dimension of a mountain belt but large compared with the wavelength of the folds. Thus at the scale of the mountain belt, the strain can be considered homogeneous within the block, and it is represented by the strain ellipse shown beside the block.

When we look at a scale comparable to the 100-meter fold wavelength, however, the strain is no longer homogeneous (Figure 12.20*B*). We then describe the inhomogeneous distribution of strain in a folded layer in terms of the variation in local strain, which is considered homogeneous on a scale, for example, of about a meter. That scale is small compared with the wavelength of the fold but large compared with the inhomogeneities in strain that might be present, for example, if the layer were sandstone containing a spaced foliation.

When we shift scales again, down to the level of the 10-millimeter spacing of the foliation domains (Figure 12.20*C*), we again find that the local strain distribution associated with the formation of the foliation is inhomogeneous from microlithons to cleavage domains. In this case, the local strain is determined by averaging over a volume that is small relative to the spacing of the foliation domains but large relative to the grain size.

Another shift in scale brings us down to the 100-μm scale (Figure 12.20*D*), where the local strain is inhomogeneous from grain to grain but can be considered homogeneous within individual grains. Here, the homogeneity is the result of averaging the deformation over a volume that is small relative to the spacing of the cleavage domains but large compared with the scale of crystal lattice imperfections.

Thus we can consider the strain to be "homogeneous" on a scale that is small compared with the particular structure within which we want to determine the strain distribution but large compared with the scale of inhomogeneities in which we are not interested and over which we want to average the deformation.

12.7 THE REPRESENTATION OF THREE-DIMENSIONAL STRAIN STATES AND PROGRESSIVE STRAINS

i. Finite Strain and the Flinn Diagram

We have concentrated largely on plane strains in the foregoing discussion because of the simplicity of diagramming and understanding the geometry. The deformation that we can expect in real geological situations, however, is unlikely to be that simple, and we cannot ignore three-dimensional deformation if we want to understand the orientation of structures in deformed rocks.

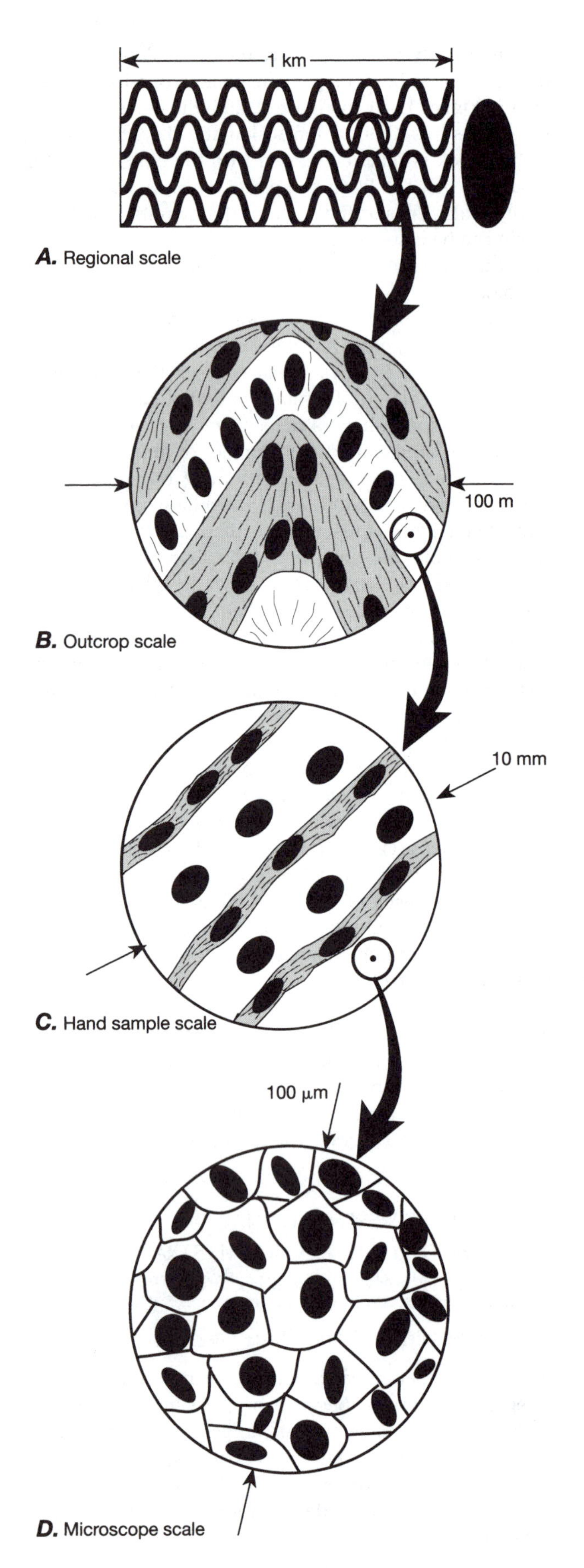

FIGURE 12.20 Scales of homogeneous and inhomogeneous strain. In each diagram, the volume over which the strain is averaged to form a locally homogeneous strain ellipse can be viewed at a smaller scale at which the strain distribution is inhomogeneous.

It is often useful to compare various states of triaxial strain in order to show, for example, how they are related to one another in heterogeneously deformed rocks or to illustrate the sequence of strain states that represents a particular progressive deformation. We can easily make such a comparison by plotting the information on a **Flinn diagram**[9] (Figure 12.21) on which the ordinate and abscissa are the ratios a and b of the principal stretches defined by:

$$a \equiv \frac{\hat{s}_1}{\hat{s}_2} \qquad b \equiv \frac{\hat{s}_2}{\hat{s}_3} \tag{12.31}$$

The origin of the coordinate axes for the Flinn diagram is generally taken to be (1, 1) because the definition of the relative values of the principal stretches ($\hat{s}_1 \geq \hat{s}_2 \geq \hat{s}_3$; second Equation (12.7)) means that both a and b cannot be less than 1. All three principal stretches are included in the plot, so it provides a representation of triaxial deformations.

Because the Flinn diagram is a plot of the ratios of the principal stretches, any component of isotropic volumetric deformation is not accounted for on the plot (an isotropic volumetric deformation preserves the shape of any deforming object). An isotropic volumetric deformation would change all the stretches by the same factor, and it therefore would be eliminated by taking the ratios of those stretches. Thus the Flinn diagram plots the shape of a strain ellipsoid but not the size. This is generally not a problem, because it is rare to be able to determine the volumetric component of the deformation anyway. One often knows the original shape of a strained object such as a fossil, but one rarely knows the original size, and without knowing the original size, we cannot determine the volumetric component of the deformation. For this reason, the study of geologic strains rarely includes the volumetric strain, and it is common to assume the deformation occurred at constant volume, although that assumption is not really justified and may be very wrong, as we discuss in Section 15.2(ii).

An unknown volumetric deformation can have important implications for the interpretation of the geometry of a strain ellipsoid, so first we discuss the Flinn diagram for constant-volume strain, and then we look at how a volumetric strain would affect the interpretation.

Any strain ellipsoid plots at a particular point (a, b)

[9]Introduced by the British geologist Derek Flinn (1962).

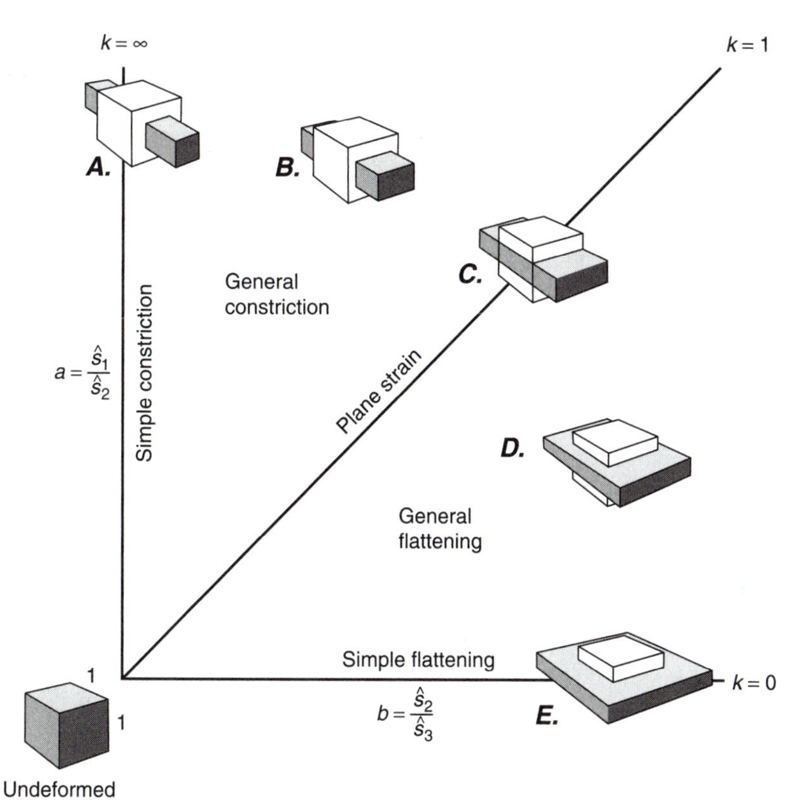

FIGURE 12.21 Flinn diagram for constant volume deformation showing the three lines ($k = 0$, $k = 1$, and $k = \infty$) and the two fields ($0 < k < 1$ and $1 < k < \infty$) of finite strain ellipsoids. (After Hobbs, Means, and Williams 1976)

on the Flinn diagram, and the slope k of the line from the origin (1, 1) to that point is the (rise / run) on the graph

$$k \equiv \frac{a-1}{b-1} = \frac{\hat{s}_1\hat{s}_3 - \hat{s}_2\hat{s}_3}{(\hat{s}_2)^2 - \hat{s}_2\hat{s}_3} \tag{12.32}$$

where we used Equation (12.31) to find the second Equation (12.32). For a constant volume deformation (first Equation (12.18) and Equation (12.19)), the value of k provides a useful way of classifying the types of ellipsoids (Figure 12.21). The three values for $\hat{s}_2 = (\hat{s}_1, 1, \hat{s}_3)$ substituted in turn into the right side of Equation (12.32) define three lines, with slopes $k = (0, 1, \infty)$, respectively, where we also use Equation (12.19) for the plane strain case ($\hat{s}_2 = 1$). These lines divide the graph into two fields, general flattening and general constriction, separated by the line $k = 1$ for plane strain. Ellipsoids of different characteristics plot along each line and within each field, as follows:

$k = 0$: simple flattening ($\hat{s}_1 = \hat{s}_2 > 1 > \hat{s}_3$) with oblate uniaxial ellipsoids (pancake shaped);

$0 < k < 1$: general flattening ($\hat{s}_1 > \hat{s}_2 > 1 > \hat{s}_3$) with oblate triaxial ellipsoids;

$k = 1$: plane strain ($\hat{s}_1 > \hat{s}_2 = 1 > \hat{s}_3$), which define two-dimensional deformations;

$1 < k < \infty$: general constriction ($\hat{s}_1 > 1 > \hat{s}_2 > \hat{s}_3$) with prolate triaxial ellipsoids; and

$k = \infty$: simple extension ($\hat{s}_1 > 1 > \hat{s}_2 = \hat{s}_3$) with prolate uniaxial ellipsoids (cigar-shaped).

Note that simple flattening and simple extension are both three-dimensional generalizations of pure shear (Figure 12.9*A*) for which the intermediate principal stretch $\hat{s}_2$ is equal, respectively, to the maximum principal stretch $\hat{s}_1$ or the minimum principal stretch $\hat{s}_3$.

ii. Nonzero Volumetric Strain and the Flinn Diagram

Although volumetric strain itself cannot be represented on the Flinn diagram, its value does affect the location of the line for plane strain. Because plane strain geometry ($\hat{s}_2 = 1$) must always separate the field of constriction ($\hat{s}_2 < 1$) from the field of flattening ($\hat{s}_2 > 1$), the location of the plane strain boundary determines the areas of the diagram in which a given plotted point represents a flattening or a constrictional strain ellipsoid. For example, if isotropic volume loss is a component of the deformation, then the value $\hat{s}_2 = 1$ must include both a component of shortening from the volumetric deformation, which does not show up on the Flinn diagram, and an exactly offsetting lengthening from the nonvolumetric part of the strain, which does show up on the Flinn diagram. Thus a plane strain ellipsoid is shifted away from the line $k = 1$, and that line therefore does not separate constriction from flattening if the deformation is not at constant volume.

In order to determine the line for plane strain when volumetric strain is not zero, we substitute the plane strain condition $\hat{s}_2 = 1$ into Equation (12.31), which gives

$$a = \hat{s}_1 \qquad b = \frac{1}{\hat{s}_3} \qquad \text{if } \hat{s}_2 = 1 \tag{12.33}$$

Solving the second equation for $\hat{s}_3$ and introducing the values for the three principal stretches from Equations (12.33) into Equation (12.12) gives the relation between a and b for plane strain in terms of the volumetric stretch

$$s_V = \frac{a}{b} \quad \text{or} \quad a = s_V b \quad \text{if } \hat{s}_2 = 1 \tag{12.34}$$

The second Equation (12.34) shows that the plane strain plots as a line on the Flinn diagram that goes through the point $(a, b) = (0, 0)$ with a slope of s_V. Figure 12.22 shows a set of plane strain lines for different values of the volumetric stretch. If there is no change in volume ($s_V = 1$), the line goes through both $(a, b) = (0, 0)$ and $(a, b) = (1, 1)$, and thus it is just the constant-volume result we obtained in the preceding subsection. The point $(a, b) = (0, 0)$ is a point that in fact has no physical significance, because the origin of the diagram is actually $(a, b) = (1, 1)$, but it does define the mathematical common intercept point of the plane strain lines for different volumetric stretches (Figure 12.22). The intercepts of these lines with either the a or the b axis can be found from Equation (12.34) by finding the value for a when $s_V > 1$ and $b = 1$, and the value for b when $s_V < 1$ and $a = 1$. Thus plane strain lines intersect the b axis, for which $a = 1$, at values of $1/s_V$, and they intersect the a axis, for which $b = 1$, at values of s_V (Figure 12.22).

It is sometimes more convenient to plot the Flinn diagram on a logarithmic scale. To show the effects of volumetric strain on the plane strain line, we take the natural logarithm of the second Equation (12.34) (the base-10 logarithm could also be used):

$$\ln a = \ln s_V + \ln b \quad \text{if } \hat{s}_2 = 1 \tag{12.35}$$

The result is a logarithmic Flinn diagram, on which the axes are $\ln a$ and $\ln b$, and the origin is at $(\ln a, \ln b) = (0, 0)$, since $\ln(1) = 0$ (Figure 12.23). Equation (12.35) shows that on this form of the Flinn diagram, the plane strain line maintains a constant slope of 1, and the volumetric stretch determines the intercept. Each line in Figures 12.22 and 12.23 represents the plane strain line for a different volumetric stretch, as labeled, and each line thus separates the field of constrictional strain above from the field of flattening strain below.

The danger of interpreting strain measurements without knowing the volumetric stretch is evident from Figure 12.23. A strain ellipsoid that plots at point A, for example, would be in the flattening field for $s_v \geq 1$ but in the constrictional field for $s_v \leq 0.8$. Similarly, a strain ellipsoid that plots at point B would be in the flattening

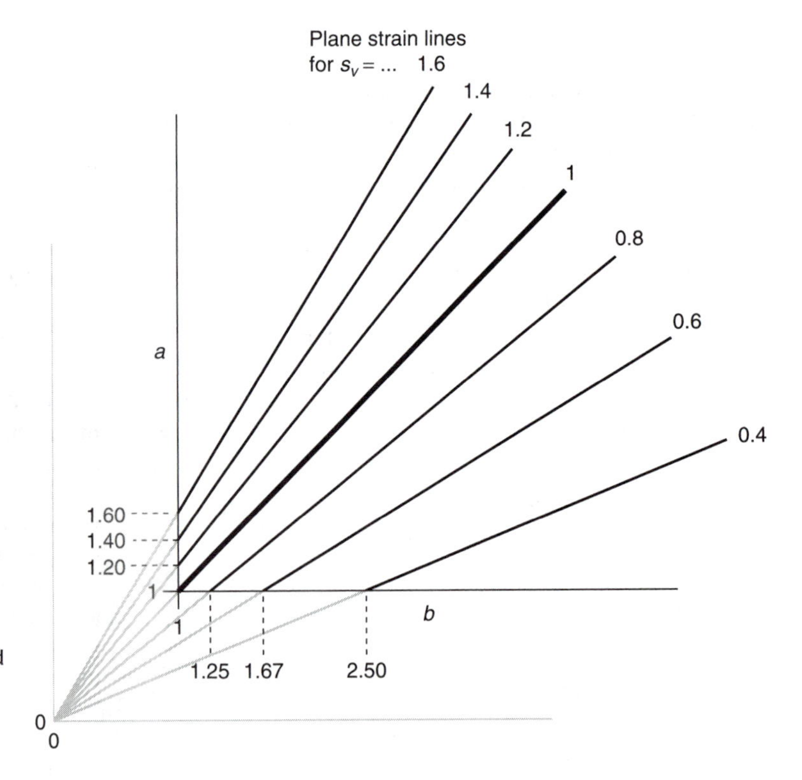

FIGURE 12.22 Flinn diagram plots of the line for plane strain at various values of the volumetric stretch, as labeled at the end of each line. Black axes are the standard axes for the Flinn diagram with the origin at $(a, b) = (1, 1)$. The axes and extensions of the plane strain lines are gray where they plot at physically meaningless values for a and b. The gray extensions of the plane strain lines show their common origin at the mathematical point $(a, b) = (0, 0)$.

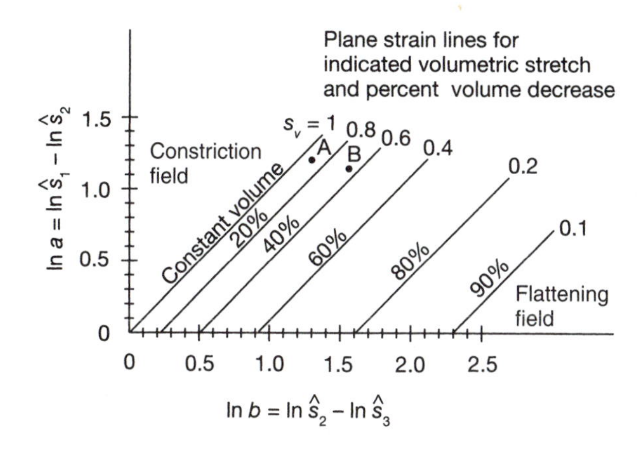

FIGURE 12.23 Logarithmic Flinn diagram for various amounts of volumetric stretch s_V (Equation (12.12)) showing the plane strain boundary lines (Equation (12.34)). Each line divides the field of flattening from the field of constrictional strain for the particular volumetric stretch.

field for $s_v \geq 0.8$ but in the constrictional field for $s_v \leq 0.6$. Thus plotting strain ellipses on the Flinn diagram without knowing the volumetric stretch can be misleading, and the common assumption of constant-volume deformation for rocks can lead to incorrect interpretations.

iii. Three-Dimensional Progressive Strains and Strain Facies

The Flinn diagram lends itself well to the representation of triaxial strain paths, which define the sequence of strain states through which a body passes in a progressive deformation, although it makes no distinction between strains produced by coaxial and noncoaxial progressive deformations. For example, progressive pure shear and progressive simple shear both plot along the line $k = 1$ because they are both progressive, constant-volume, plane deformations. We confine our attention to strain paths produced by steady deformations, for which the instantaneous strain ellipsoid is constant throughout the deformation, that is, its ellipticity and orientation do not change. Steady coaxial deformations and all progressive plane strains produce strain paths that plot on the Flinn diagram as straight lines of constant slope k. Steady triaxial deformations that are noncoaxial, however, may produce curved paths under certain conditions.

We can imagine a variety of steady progressive deformations by making combinations of coaxial and noncoaxial deformations. For example, imagine that any of the strain states diagrammed in Figure 12.21 was produced by a coaxial deformation with principal instantaneous stretches perpendicular to the faces of the initially undeformed cube. We can generalize the progressive deformation by adding an instantaneous simple shear (Figure 12.10*A*) such that the shear plane—(x_1, x_2) in Figure 12.10*A*—is parallel to one of the faces of the cube, and the shear direction (x_1 in Figure 12.10*A*) is parallel to one of the edges on that face. Since there are three orientations of cube faces and each face has two orientations of edges, there are six ways we could add a simple shear to each coaxial deformation. The multiplicity of possible deformations becomes too large to consider for the scope of this book.

As an example of deformations generalized in this way, however, we consider just the deformations that could occur in a vertical strike-slip shear zone in which a plane strain component of lengthening or shortening can occur perpendicular to the shear zone boundary, and volume is conserved by a compensating shortening or lengthening in a vertical direction. In Figure 12.24 the same plane strain is shown in two different orientations to represent constant-volume coaxial lengthening (divergence) normal to the shear zone boundary with compensating vertical shortening (Figure 12.24*A*,*B*) or a coaxial shortening normal to the shear zone boundary (convergence) with compensating vertical lengthening (Figure 12.24*A*,*C*). The addition of the horizontal simple shear component parallel to the front face of the cube then results in divergent strike-slip (Figure 12.24*D*) or convergent strike-slip (Figure 12.24*E*) deformation, commonly referred to as "transtension" and "transpression," respectively (see Box 12-2 for a discussion of this unfortunate terminology). Although each component of this deformation is meant to represent an infinitesimal deformation, they are diagrammed as large finite deformations for the purposes of clarity of illustration. Different strain paths result from varying the relative amounts of coaxial and noncoaxial components of the deformation, which is reflected by the relative velocity vector between opposite sides of the shear zone. The angle α between the shear zone boundary and that relative velocity vector defines the geometry of the deformation, with $\alpha = 0$ indicating pure strike-slip, and $\alpha = +90°$ or $-90°$ indicating, respectively, pure convergent or divergent deformation (Figures 12.24 and 12.26).

Strain paths for these generalizations of deformation are shown on a logarithmic Flinn diagram in Figure 12.25. Depending on whether the $\hat{s}_1$, $\hat{s}_2$, or $\hat{s}_3$ principal axis is vertical, the paths are drawn as solid, long-dashed, or short-dashed, lines, respectively. Thus it is apparent that there are three distinct deformation geometries distinguished by the orientations of the principal stretches, and that for a set of geometries with $|\alpha| < \sim 20°$, the deformation geometry changes dramatically part way along the progressive strain path. This change occurs where the strain path on the Flinn diagram touches one of the axes, that is, either a or b becomes equal to 1, and it occurs because two of the principal axes evolve to become equal, and then continue to evolve so that they exchange orientations. For convergent strike-slip, the $\hat{s}_2$ axis becomes the $\hat{s}_3$ axis and vice versa; and for divergent strike-slip, the $\hat{s}_2$ axis becomes the $\hat{s}_1$ axis, and vice versa. The

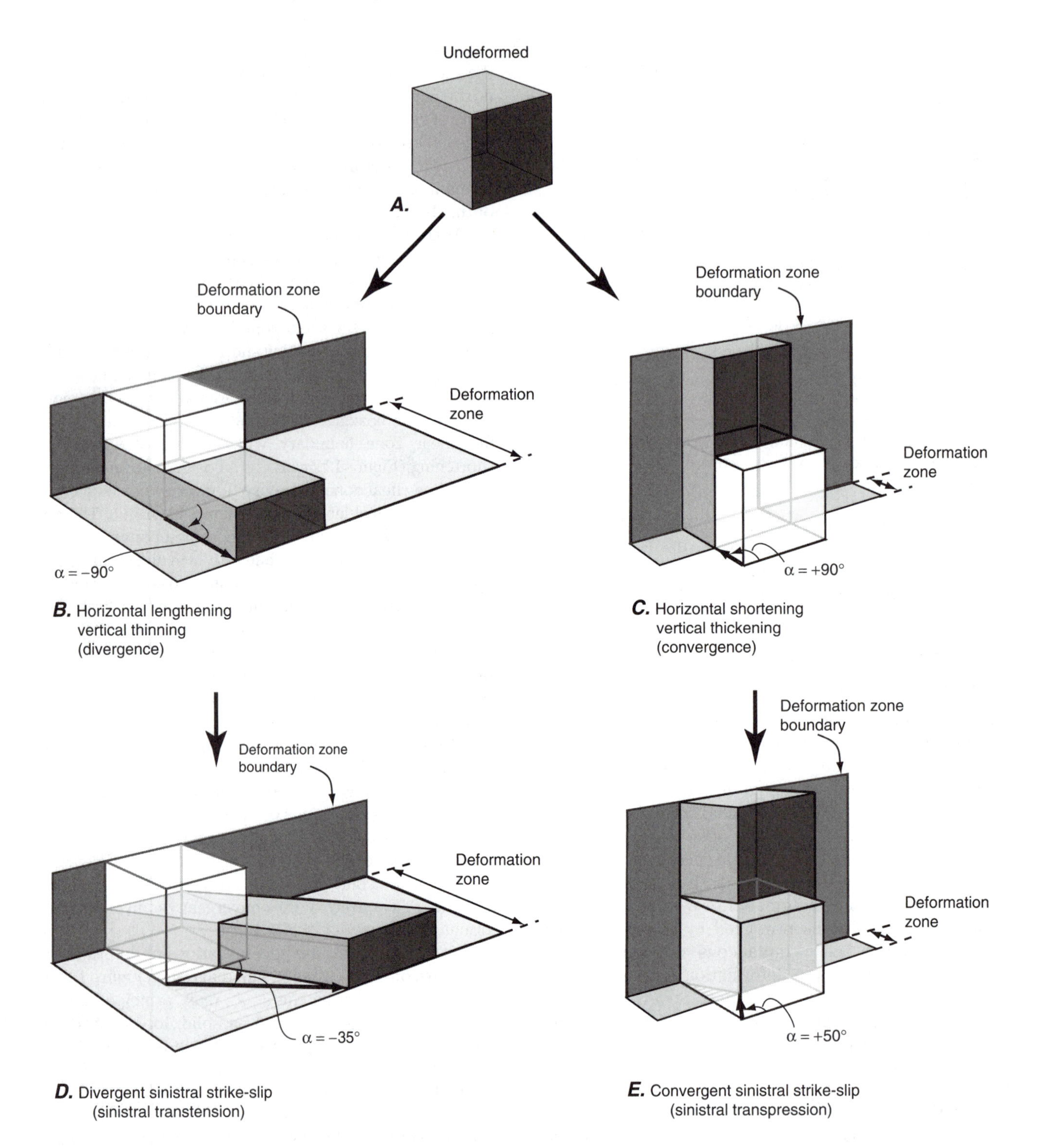

FIGURE 12.24 Divergent and convergent strike-slip deformation (transtension and transpression, respectively). Strain in shear zones can be a combination of simple shear parallel to the shear zone boundary plus lengthening or shortening normal to the boundary. The angle α defines the orientation of the relative velocity of the opposite boundaries of the shear zone. The two components of the deformation that are added together are both infinitesimal but are shown, for purposes of illustration, as large finite increments added together. *A*. Undeformed cube. *B*. Original cube is subjected to a horizontal lengthening normal to the shear zone and a vertical shortening (divergence). *C*. Original cube is subjected to a horizontal shortening normal to the shear zone and a vertical lengthening (convergence). *D*. Sinistral strike-slip shearing is added to the divergence (*B*) to give a divergent sinistral strike-slip (transtension). *E*. Sinistral strike-slip shearing is added to the convegence (*C*) to give a convergent sinistral strike-slip (transpression).

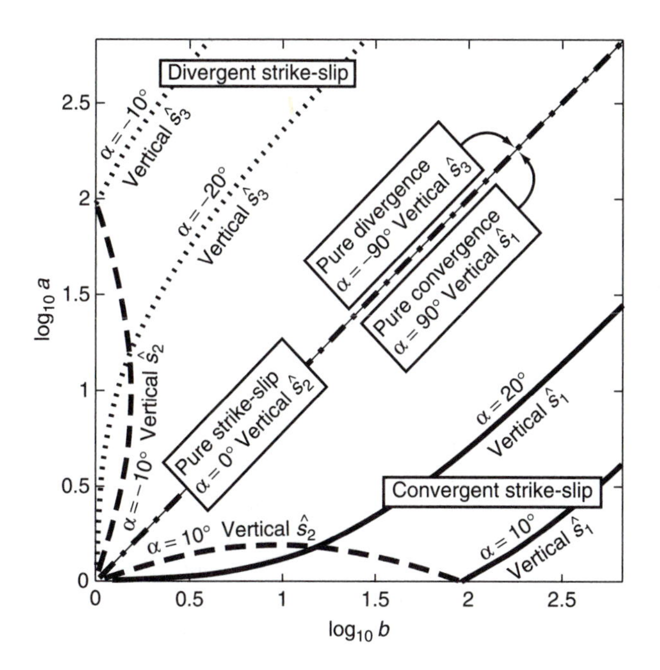

FIGURE 12.25 Logarithmic Flinn diagram of strain paths in shear zones. Convergent and divergent strike-slip strain paths are distinguished by the angle α between the shear zone boundary and the vector for the relative velocity of the opposite sides of the shear zone. Solid lines indicate that $\hat{s}_1$ is vertical; long dashed lines indicate that $\hat{s}_2$ is vertical; short dashed lines indicate that $\hat{s}_3$ is vertical. Note for values of $|\alpha| < \sim 20°$ the evolving strain path hits either the ordinate or the abcissa, at which point the geometry of the deformation changes from $\hat{s}_2$ vertical to either $\hat{s}_1$ vertical (for convergent strike-slip) or $\hat{s}_3$ vertical (for divergent strike-slip). The solid lines and the short dashed lines plot on the nonlogarithmic Flinn diagram as straight lines.

point at which the strain path hits the axis decreases toward zero as $|\alpha|$ approaches approximately 20°. The diagonal line represents the strain path for any type of constant-volume progressive plane strain, as represented by the three labels along the path indicating that any of the three principal axes can be vertical. Note that on a nonlogarithmic Flinn diagram, the solid and the short-dashed strain paths would all plot as straight lines.

A **strain facies** is characterized by three main factors: the geometry of the finite strain, as plotted on a Flinn diagram; the orientations of the principal strain axes in space; and the characteristics of structures that develop and their relationship to those principal strains. For this example of divergent or convergent strike-slip zones, we can identify three strain facies, defined according to which of the three principal stretches is vertical, or equivalently by the orientations of the foliation—the ($\hat{s}_1$, $\hat{s}_2$)-plane—and the stretching lineation—the $\hat{s}_1$ axis (Figure 12.26). The gray arrows in Figure 12.26 indicate paths of constant α, such as the strain paths in Figure 12.25, and the heavy curved lines in Figure 12.26 define the same transitions in strain facies as defined by the points where the strain paths in Figure 12.25 touch the axes of the graph. We discuss strain facies and their relation to characteristic structures in Chapter 15 after we describe how structures such as folds, foliations, and lineations are related to strain (Chapters 13 and 14).

In realistic conditions in the Earth, steady deformations over long periods of time are probably the exception, and strain paths on the Flinn diagram in general are curved and may even cross from the constrictional field into the flattening field, or vice versa.

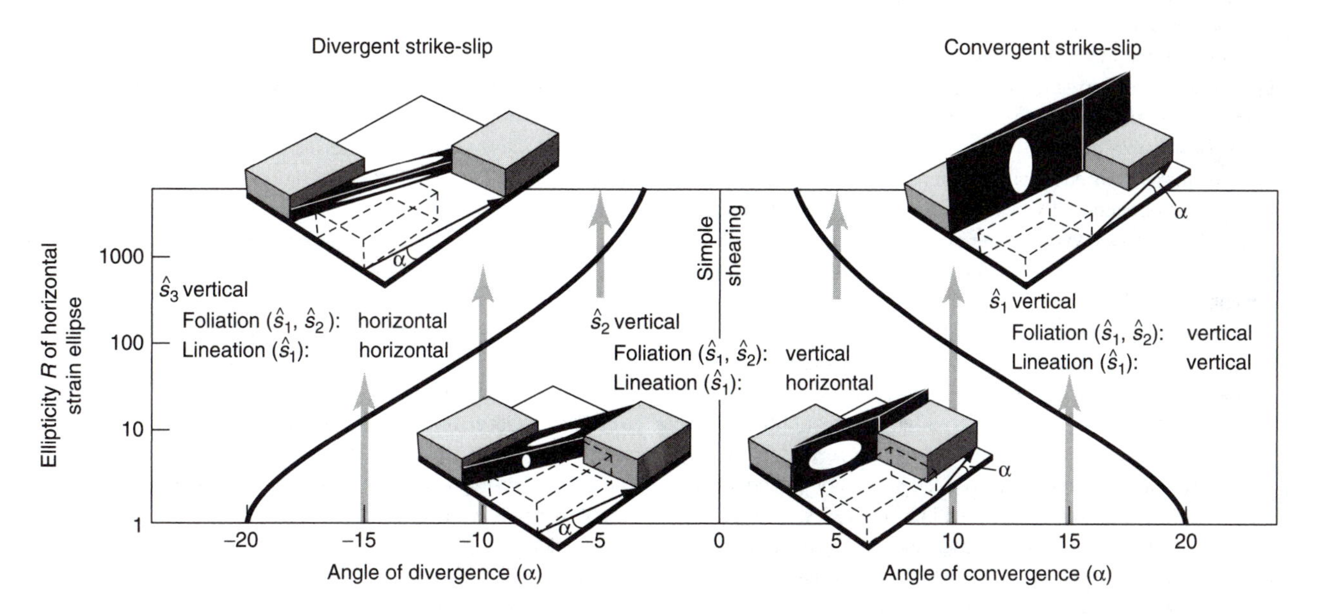

FIGURE 12.26 Fields of different strain facies (block diagrams) for convergent and divergent strike-slip deformation. Strain facies are defined by the characteristic orientations of the plane of flattening ($\hat{s}_1$, $\hat{s}_2$) and the principal stretching direction ($\hat{s}_1$). (After Teyssier and Tikoff 1999)

BOX 12-2 Terminology of Strain Compared with Stress: Beware!

It is common for geologists and geophysicists to confuse the terminology for stress and strain. This confusion occurs not only among students just learning the field, but also among professionals publishing research results in the field. Thus, even the formal literature is not always a reliable source from which to learn the terminology correctly. In some cases, some of the confusion may arise because the terminology itself is ambiguous. In other cases, however, the confusion reflects sloppiness in usage, which then promotes misunderstanding and further confusion. "Careless terminology is . . . dangerously close to confused thought."[2] We include this box, therefore, in the hope of dispelling some of these problems. As a summary, Table 12-2.1 lists and defines in separate columns words that are appropriately applied to strain and words that are appropriately applied to stress. Terms that could be confused are arranged in the same row. Table 12-2.2 lists antonym pairs, again in different columns for strain terms and stress terms, and again with terms that could be confused between strain and stress arranged in the same row.

Stress is a quantity that can never be observed directly. The measurement of stress always involves the direct measurement of some intermediate quantity and the implicit or explicit application of some equation, such as a constitutive relation between stress and strain that relates the intermediate quantity to the stress. For example, the determination of stress commonly involves the measurement of a displacement in an elastic material. The displacement is converted to a measure of strain, and the stress is then determined from the equations of elasticity that relate stress to strain for that material, either by direct calculation or by calibration. Examples of this type of relation include the elastic displacement of a spring scale like your bathroom scale, the elastic strain of an electrical resistance load cell, or the elastic distortion of the sensing element in an oil pressure gauge. Thus stress can never be a primary observation but must always be the result of some assumption of a relationship between stress and some other observable quantity.

We make a categorical statement, therefore, that should always be remembered when trying to interpret deformation structures in rocks: **We can never measure the orientations or magnitudes of the principal stresses *directly* from observation of deformed rocks.** Our observations are limited to the consequences of deformation, and they allow us to draw direct conclusions only about the orientations and magnitudes of the principal axes of finite or instantaneous strain. Any attempt to go beyond that to infer stress orientations or magnitudes always implies that we have some understanding of how the stress is related to the kinematics or to the strain. Such relations are well established in a few cases, such as the correlation of certain steady-state fabric or microfabric characteristics with differential stress, which we discuss in Box 17-2; in other cases, inferences of stress are based on implicit assumptions that can lead one into misinterpretation.

[2]Marrett and Peacock (1999); this paper, and Twiss and Unruh (1998) have informed much of the discussion in this box.

Stress terminology, when used in a descriptive sense, is fundamentally a genetic terminology, which means that we are imposing an interpretation on observations that, in the first instance, should be merely descriptive. Thus any descriptive discussion of structures observed in rocks should always be couched in kinematic or strain terminology, not in stress terminology. Evidence for strain can be observed directly; inferences of stress are always interpretive.

An example of terms that are commonly used as antonyms, at worst inappropriately and at best ambiguously, are the terms *extension* and *compression* (cf. Table 12-2.2). Unfortunately, both words are used in more than one sense. In common usage, *extension* means a lengthening, although its technical definition is the change in length divided by the initial length, which is positive for lengthening and negative for shortening. In either case, however, *extension* is unambiguously a strain word. Confusion is introduced, however, through use of the term *extensional stress*, which defines a specific relationship among the principal stresses, $\hat{\sigma}_1 = \hat{\sigma}_2 > \hat{\sigma}_3 > 0$ (see Section 7.3, Figure 7.12*E*). The meaning of *compression* is also ambiguous. In common usage, it is defined as "a reduction in volume that results from applying pressure," which conflates strain (volume decrease) with stress (pressure). In technical parlance, however, it is usually a stress word—we speak of a compressive stress but never of a compressive strain. In seismology, however, it is usual to use the terms *compression wave* and *rarefaction wave* as antonyms. Because *rarefaction* is unambiguously a strain word, the usage gives a connotation of strain to compression; an unambiguous alternative would be contraction wave. Thus using *extension* and *compression* as opposites is at best ambiguous and at worst inappropriate. In either case, such usage introduces confusion, and these terms should not be used as if they refer to opposite aspects of the same physical concept (see Table 12-2.2 for appropriate antonyms). It is preferable, whenever possible, to use the verbs *lengthen* and *shorten*, or the gerunds *lengthening* and *shortening*. These terms are general in nature and have no specific quantitative definition, but their meaning is completely unambiguous.

This same confusing ambiguity resides in the pair of terms *transtension* and *transpression*, which are kinematic terms describing strike-slip deformation with added components, respectively, of lengthening or shortening normal to the strike-slip fault (see Figure 12.24). *Transtension* is an ambiguous contraction (no pun intended!) of *transcurrent*, a kinematic word, with either *tension* or *extension*, the first of which is a stress word and the second a strain word. Similarly *transpression* is a contraction of *transcurrent* and *compression*. The latter word, as we discussed in the preceding paragraph, is itself ambiguous, being a stress word in most technical language but also used in some cases with at least implications of strain. Thus these terms have ambiguous connotations that introduce the potential for considerable confusion. Unfortunately, they are entrenched in the literature, but it should be clearly understood that they refer to the fault kinematics, not to the conditions of stress. Alternative terms that have been sug-

TABLE 12-2.1 Strain versus Stress Terminology

Strain Terms:	Definition / Comments	Stress Terms:	Definition / Comments
Strain	General term referring to a variety of different measures of change of length and change of shape; also the tensor defined at a point from which one can calculate the linear and angular deformation for a line of any orientation through that point	Stress	General term referring to the force per unit area; also a tensor defined at a point from which one can calculate the normal and shear components of force per unit area on a plane of any orientation through the point
Extension, normal strain	A measurement of linear strain; change of length divided by the initial length; can have positive or negative values; extension is also used as a synonym for lengthening	Normal stress	The component of a surface stress on a plane that acts perpendicular to the plane
Stretch	A measurement of linear strain; final length divided by the initial length; values are always positive		
Elongation	A measurement of linear strain not used in this book; can have positive or negative values		
Quadratic elongation	A measurement of linear strain equal to the square of the stretch; has only positive values		
Shear strain	Change in angle (ψ) between two initially perpendicular material line segments	Shear stress	The component of a surface stress on a plane that acts parallel to the plane
Tensor shear strain	0.5 tan ψ		
Engineering shear strain	$\gamma = \tan\psi$; twice the tensor shear strain		
Pure shear	A constant volume, plane, irrotational strain with $\hat{s}_1 = 1/\hat{s}_3$, and $\hat{s}_2 = 1$; progressive pure shear is a coaxial deformation for which the principal instantaneous strain axes are parallel to the finite strain axes	Pure shear stress	$\hat{\sigma}_1 = -\hat{\sigma}_3$; $\hat{\sigma}_2 = 0$; the term *pure shear* should never be used by itself to describe a state of stress
Simple shear	A constant volume, plane, rotational strain with $\hat{s}_1 = 1/\hat{s}_3$, and $\hat{s}_2 = 1$; extension normal to the shear plane is zero (no lengthening or shortening); progressive simple shear is a noncoaxial deformation for which the principal instantaneous strain axes are 45° to the shear plane; $\hat{s}_2$ is parallel to the shear plane	Deviatoric stress	A stress state for which the normal stress on the plane of maximum shear stress is zero; deviatoric normal stress components are the applied normal stress reduced by the mean normal stress; in two dimensions, the deviatoric stress is like a pure shear stress: the center of the Mohr circle is at the origin, $\hat{\sigma}_1^{(Dev)} = -\hat{\sigma}_3^{(Dev)}$
		Effective stress	A stress state for which the effective normal stresses are the applied normal stress reduced by the pore fluid pressure; normal effective stresses can be positive or negative
		Differential stress	The difference between the maximum and minimum principal stresses; a scalar invariant of stress that can have only positive values

gested are listed in Table 12-2.2, although most of them are either more cumbersome or less mellifluous. Notwithstanding the remote probability of replacing these entrenched terms, we choose in this book to use the unambiguous kinematic terms *divergent strike-slip* and *convergent strike-slip*.

An example of an inappropriate genetic stress term commonly used in a descriptive sense is *tension fracture*. This term is applied to a fracture that has opened with displacement

(continued)

BOX 12-2 Terminology of Strain Compared with Stress: Beware! *(continued)*

TABLE 12-2.2 Antonym Pairs: Strain versus Stress Terminology

Strain Antonyms		Comment	Stress Antonyms		Comment
Lengthening	Shortening	Very useful unambiguous but general terms not associated with a specifically defined quantitative measure			
Extension	Contraction	Extension in this sense is synonymous with lengthening, although it also has a more technical definition (see Table 12-2.1); refers to lengthening or shortening in a specific direction, e.g., vertical extension, horizontal contraction Some object to this use of contraction , claiming the word implies an isotropic volumetric deformation (an antonym of expansion), as opposed to a linear deformation as is intended here	Tension, Tensile stress	Compression, Compressive stress	A normal surface stress comprising a pair of equal and opposite tractions, that act,respectively, away from, or toward, each other across the surface In common language, compression is defined as a decrease in volume resulting from the application of pressure, and thus is more ambiguous than its technical usage
Extensional deformation	Contractional deformation	Synonymous with extension and contraction	Extensional stress	Axial compression, Confined compression	Used for a stress state with the unique axial stress, respectively, less compressive or more compressive than the equal radial stresses
Constriction	Flattening	Triaxial strains characterized by the unique axis being, respectively, a lengthening—forming an oblate (cigar-shaped) strain ellipsoid—or a shortening—forming a prolate (pancake-shaped) strain ellipsoid			The words *extension* and *extensional* by themselves should never be used to describe a state of stress.
			Deviatoric tension	Deviatoric compression	The two-dimensional deviatoric stress by definition plots as a circle centered on the origin of the Mohr diagram, so half of the normal stresses are always deviatoric tension, and half are always deviatoric compression.

perpendicular to the plane of the fracture, even though, for the most part, these fractures do not form under an applied tensile stress. A better genetic term would be *effective tension fractures* because we infer that they open under the influence of high pore fluid pressure, which creates an effective tensile stress, but such a term is rather cumbersome and still involves interpretation. Better would be to simply use the descriptive strain term *extension fracture*.

Perhaps one of the sources of the confusion is that we often tend to think of stress as being the causative factor in a mechanical process and strain as being the result. The manner in which the concept of stress is commonly introduced tends to make it seem the more appropriate quantity to use in interpreting structures. We have used this emphasis ourselves, for example, in Chapters 8 and 9 in discussing brittle fracture: The implication of this presentation, Section 8.7 notwithstanding, is that stress causes fracture, and different states of stress cause different types of fracture. However, it is just as correct to think of a displacement imposed on the boundaries of a block as causing a stress within the block as it is to think of a stress imposed on the boundaries of a block causing a strain within the

TABLE 12-2.2 *(continued)*

Strain Antonyms		Comment	Stress Antonyms		Comment
Divergent	Convergent	When used as kinematic terms with reference to a specific structure, these terms imply a component of horizontal lengthening or shortening normal to the structure	Tensile	Compressive	
Releasing, Extensional, Divergent	Restraining, Contractional, Convergent	Applied to fault bends, especially in strike-slip faults, across which there is a component, respectively, of lengthening or shortening normal to the bent segment	Tensile	Compressive	
Dilation (dilatation), Expansion, Expanding	Contraction, Shrinkage Shrinking.	Terms describing isotropic volumetric deformation *Contraction* is also used to describe linear strain as a synonym for *shortening*	(Isotropic Tension)	Pressure	The state of isotropic tension is so rare that a specific term has not been coined to identify it
Transtension	Transpression	Terms referring to a geometry of strain on a vertical fault consisting of a component of horizontal shear with a component of lengthening or shortening, respectively, normal to the fault. The terms have an ambiguous connotation, however, in that they combine kinematic terminology (transcurrent fault) with stress terminology (tension or compression), and thus are best avoided			
Divergent strike-slip Extensional strike-slip Transextension, Oblique extension, Oblique divergence	Convergent strike-slip Contractional strike-slip Transcontraction, Oblique contraction, Oblique convergence	Possible alternatives to transtension and transpression. The implicit presumption is that the component of lengthening or shortening is oriented perpendicular to the strike-slip fault in question			

block. These two possibilities are different mechanical conditions representing two different, but equally viable, boundary conditions on the block. For example, if a triaxial strain is imposed on a body of rock, the orientations of fractures that develop to accommodate the deformation (Figure 8.17*B*) are different from those that develop if a triaxial stress is imposed on the body (Figure 8.17*A*). In the former case, the deformation, not the stress, must be considered the cause of the process, and the Coulomb fracture criterion, which was developed for stress boundary conditions, does not work (see Section 8.7).

Stress, of course, has a valid place in the discussion of rock deformation. Its use, however, should be confined to situations in which the stress is known from external constraints on the system or is explicitly assumed. Stress terminology is appropriate for the genetic interpretation of structures, including theoretical mechanical models of formation, which require explicit and well-defined relations between stress and strain rate, and for the discussion of laboratory experiments for which the applied forces are well known. Field descriptions of structures or tectonic environments, however, should always employ kinematic or strain terminology.

12.8 TENSOR REPRESENTATIONS OF STRAIN

Our discussion of strain so far has emphasized the geometric aspects of its representation, using the strain ellipsoid and its principal axes, which represent the principal stretches. Because of the similarity of this representation with the stress ellipsoid and the principal stresses, it should come as no surprise that the strain, like the stress, is actually a second-rank tensor. In fact, the analogy with the stress extends even further: In order to know the state of strain at a point, we must be able to calculate the extension (or the stretch) and the shear strain for any orientation of line through that point (cf. the opening statement of Section 7.1(vii)). To make that calculation, it is sufficient to know the extensions for three initially, mutually perpendicular material lines through the point, and the shear strains of each of those lines relative to the other two. This implies that nine components are needed to define the three-dimensional state of strain at a point, and in this section, we introduce these quantities and the tensor notation by which they are generally represented.

i. The Strain Tensor

Figure 12.27 shows the two main components of strain, the volumetric component (Figure 12.27*A*) and the shear component (Figure 12.27*B*). These diagrams look rather complex at first glance, but each one is just a straightforward application of measures of strain we have already discussed, and it is mostly the repetition of those measures for each of the three coordinate planes that makes for the apparent complexity. Let us look at each diagram, one measure at a time.

We consider the increase in volume of an undeformed cube (Figure 12.27, dark shading) with edges parallel to an orthogonal coordinate system (X_1, X_2, X_3). We consider first the edge parallel to X_1, which has an undeformed length L_1. After the increase in volume, the length of this edge increases by ΔL_1 so the extension is (Figure 12.27*A*):

$$e_{11} = \frac{\Delta L_1}{L_1} \tag{12.36}$$

The first subscript on e_{11} indicates that the line is initially parallel to X_1, and the second subscript indicates that the change in length is also parallel to X_1. Similar relationships define the extensions e_{22} and e_{33} for material lines initially parallel to X_2 and X_3 respectively (Figure 12.27*A*).

If the large cube in Figure 12.27*A* is then subjected to components of shear, material lines initially parallel to the mutually orthogonal axes X_1, X_2, and X_3 are, after de-

A. Volumetric component of strain

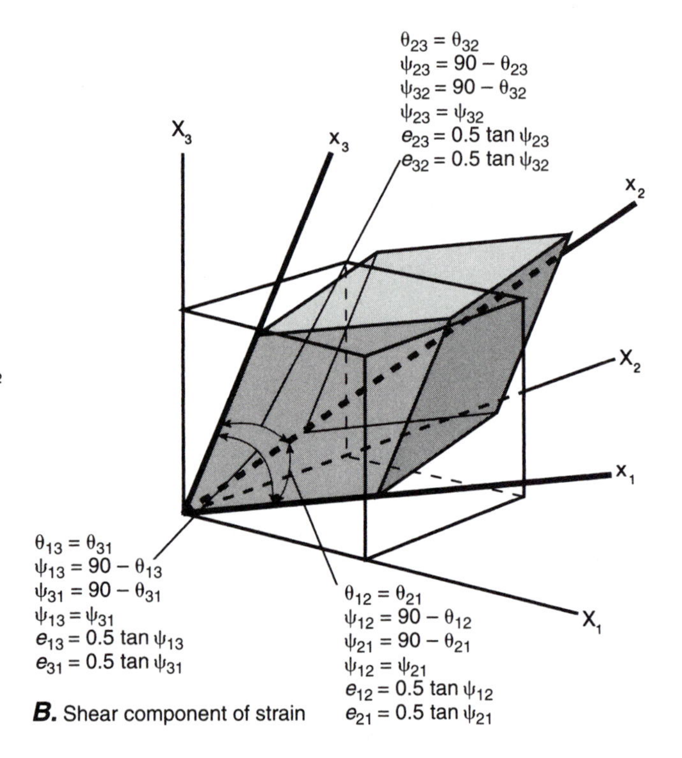

B. Shear component of strain

FIGURE 12.27 Geometric significance of the strain tensor components in three dimensions. *A*. Volumetric part of the strain. The small cube increases in volume to the larger cube by the equal lengthening of all sides of the cube. *B*. Shear part of the strain. The shear strain describes the change in shape from a cube into a parallelepiped (shaded). x_1, x_2, and x_3 are parallel to the deformed edges of the parallelepiped. All the tensor shear strain components are defined by three independent angles $\psi_{12} = \psi_{21}$, $\psi_{13} = \psi_{31}$, and $\psi_{23} = \psi_{32}$.

formation, parallel to the nonorthogonal axes x_1, x_2, and x_3, respectively (Figure 12.27*B*). Let us consider first the shear of a material line that is initially parallel to the X_1 axis. We must describe the shear of this line relative to material lines parallel to both X_2 and X_3. Consider first the shear of X_1 relative to X_2. After deformation, these material lines are parallel to the nonorthogonal axes x_1 and x_2. The initial 90° angle between these lines becomes sheared to the angle θ_{12} and thus the shear angle ψ_{12} and the associated shear strain e_{12} are:

$$\psi_{12} = 90° - \theta_{12} \qquad e_{12} = 0.5 \tan \psi_{12} \qquad (12.37)$$

A similar analysis shows that the material lines parallel to X_1 and X_3 are sheared so that in the deformed configuration, they are parallel to x_1 and x_3, respectively. The angle between x_1 and x_3 is θ_{13}, and the shear angle and associated shear strain are:

$$\psi_{13} = 90° - \theta_{13} \qquad e_{13} = 0.5 \tan \psi_{13} \qquad (12.38)$$

In each case the first subscript indicates that the shear strain is for the line initially parallel to X_1, and the second subscript indicates that the shear strain is determined relative to a line initially parallel to X_2 and to X_3, respectively (Figure 12.27*B*). The comparable strain components for the material line segment initially parallel to X_2 are e_{21} and e_{23}; for the material line segment initially parallel to X_3, they are e_{31} and e_{32} (Figure 12.27*B*).

Thus there are a total of nine strain components.[10] The extension and the two shear strain components for each of the three material lines are written in separate rows, forming an ordered array.

$$\begin{bmatrix} e_{11} & e_{12} & e_{13} \\ e_{21} & e_{22} & e_{23} \\ e_{31} & e_{32} & e_{33} \end{bmatrix} \qquad (12.39)$$

PRINCIPAL DIAGONAL

The components on the principal diagonal of the array, which have both subscripts the same, are the extensions (the volumetric extension is shown in Figure 12.27*A*). The off-diagonal components, which have two different subscripts, are the shear strains (Figure 12.27*B*). This array of strain components represents the **strain tensor**, which provides enough information for us to calculate the extension and shear strain for a line segment of any specified orientation (see Box 12-3 and Section 12.9).

The strain tensor is symmetric about the principal diagonal, because for a given pair of material lines initially parallel to X_1 and X_2, for example, the shear angle (ψ_{12}) of X_1 with respect to X_2 is the same as the shear angle (ψ_{21}) of X_2 with respect to X_1 (Figure 12.27*B*). Thus

$$e_{12} = e_{21} \qquad e_{23} = e_{32} \qquad e_{31} = e_{13} \qquad (12.40)$$

and therefore there are only six independent strain components in three-dimensional strain. Thus the strain, like the stress, is a symmetric second-rank tensor.

For plane strain, we have

$$e_{22} = 0 \qquad e_{21} = e_{23} = 0 \qquad (12.41)$$

and by Equations (12.40) and the second set of Equations (12.41), we must have

$$e_{12} = e_{32} = 0 \qquad (12.42)$$

Thus if we drop from Equation (12.39) all terms that necessarily become zero for plane strain, the plane strain tensor is represented in two dimensions by only four strain components, three of which are independent:

$$e_{kl} = \begin{bmatrix} e_{11} & e_{13} \\ e_{31} & e_{33} \end{bmatrix} \qquad e_{13} = e_{31} \qquad (12.43)$$

Thus in order to describe the state of plane strain, we need only the extension and one shear strain for each of the two material lines that originally are parallel to X_1 and X_3, respectively.

If we describe the strain relative to the principal coordinates, which are parallel to the principal axes of the strain ellipsoid, the representation of the strain tensor reduces to a particularly simple form in which the extensions are the principal values and the shear strains are zero. For three- and two-dimensional strains, respectively

$$\hat{e}_{kl} = \begin{bmatrix} \hat{e}_1 & 0 & 0 \\ 0 & \hat{e}_2 & 0 \\ 0 & 0 & \hat{e}_3 \end{bmatrix} \qquad \hat{e}_{kl} = \begin{bmatrix} \hat{e}_1 & 0 \\ 0 & \hat{e}_3 \end{bmatrix} \qquad (12.44)$$

It is very important to remember that in general the principal axes of finite strain are not parallel to the principal axes of stress. This becomes evident, for example, in the comparison of the orientations of the principal stresses and principal finite strains in folds, which we discuss in Section 18.7 (for example, compare Figures 18.17*B* and 18.18).

ii. Invariants of the Strain Tensor

Like the stress tensor, the strain tensor in two dimensions has two scalar invariants that define two quantities that

[10]The strain components described here are a first approximation that applies if the strains are very small—i.e., they are components of the so-called infinitesimal strain tensor. Finite strains involve additional nonlinear terms that make the analysis more complex (see Section 12.9). The volumetric strain, however, is correctly defined (see Equation (12.45)), and the discussion about the strain tensor is correct.

measure different aspects of the magnitude of the strain; in three dimensions there are three invariants. The most important, for our purposes here, is the first invariant, which is the sum of the diagonal components of the tensor matrix. We have shown in Section 12.2 that for the infinitesimal strain, this sum is just the area strain in two dimensions (Equation (12.10)), and the volume strain in three dimensions (Equation (12.15)), and it is the same for the components of infinitesimal strain in any given coordinate system x_k ($k = 1, 3$ in two dimensions; $k = 1{:}3$ in three dimensions):

$$e_A = \hat{e}_1 + \hat{e}_3 = e_{11} + e_{33}$$
$$e_V = \hat{e}_1 + \hat{e}_2 + \hat{e}_3 = e_{11} + e_{22} + e_{33} \qquad (12.45)$$
$$\text{if } \hat{e}_k \text{ and } e_{kl} << 1$$

(Compare the definition of the scalar invariants of the stress tensor in the first Equation (7.33) and Equation (7-2.4)).

iii. Other Tensor Measures of Strain

Because the stretch and the extension are different measures of the same deformation and are simply related to each other (Equation (12.3)), we can also define a tensor in which the components describing the linear deformation are the stretches rather than the extensions. This is sometimes referred to as the **deformation tensor** s_{kl}, and in three dimensions its components in a general coordinate system and the principal coordinate system are, respectively:

$$s_{kl} \equiv \begin{bmatrix} s_{11} & e_{12} & e_{13} \\ e_{21} & s_{22} & e_{23} \\ e_{31} & e_{32} & s_{33} \end{bmatrix} \qquad \hat{s}_{kl} = \begin{bmatrix} \hat{s}_1 & 0 & 0 \\ 0 & \hat{s}_2 & 0 \\ 0 & 0 & \hat{s}_3 \end{bmatrix} \qquad (12.46)$$

Notice that the components of the deformation tensor are quantities we have already defined: The diagonal components are the stretches, and the off-diagonal components are the same shear strains as appear in the strain tensor. Eliminating all components with a 2 in the subscripts gives the two-dimensional matrix for the deformation tensor.

The instantaneous extension ε_n and the instantaneous stretch ζ_n, which we introduced in Section 12.4, are also tensors whose representations ε_{kl} and ζ_{kl} are essentially similar to the matrices for the strain and deformation tensors, respectively, that we have discussed here. The instantaneous strain ellipse is always a measure of a small increment of strain that accumulates during a progressive deformation in any given small increment of time. Because the time increment is essentially infinitesimal, the instantaneous strain is effectively the same as the strain rate, which we discuss in more detail in Box 16-1.

BOX 12-3 A More Quantitative View of Strain

A homogeneous strain is described mathematically by a homogeneous transformation of a material body from the undeformed state to the deformed state. Such a transformation applied to any material point is represented mathematically by a linear relationship between the coordinates of the point in the undeformed state (X_1, X_3) and its coordinates in the deformed state (x_1, x_3). By convention, we use uppercase letters to describe the undeformed state and lowercase letters to describe the deformed state. If we restrict our analysis to plane deformation, the general form of such a transformation is

$$x_1 = AX_1 + BX_3 + C \qquad x_3 = DX_1 + EX_3 + F \qquad (12\text{-}3.1)$$

where A, B, C, D, E, and F are constants. The parts of the transformation defined by C and F are the same for all particles, and thus these constants describe a rigid-body translation. If any or all of these constants vary with time, then these equations describe the motion of the material particles.

The equations say that given the original location of any material particle in the undeformed state (X_1, X_3), we can calculate its final location in the deformed state (x_1, x_3). The equations may be solved for X_1 and X_3, so that given the deformed location of a material particle (x_1, x_3), we can also calculate its original location (X_1, X_3). The resultant equations define the **inverse transformation**:

$$X_1 = ax_1 + bx_2 + c \qquad X_3 = dx_1 + ex_3 + f \qquad (12\text{-}3.2)$$

where

$$a \equiv \frac{E}{AE - BD} \qquad b \equiv \frac{-B}{AE - BD} \qquad c \equiv \frac{BF - CE}{AE - BD}$$
$$d \equiv \frac{-D}{AE - BD} \qquad e \equiv \frac{A}{AE - BD} \qquad f \equiv \frac{DC - AF}{AE - BD} \qquad (12\text{-}3.3)$$

and where, again, c and f describe a rigid-body translation.

As examples of such a transformation and its inverse, the following equations describe a pure shear, which transforms a square with sides parallel to the principal coordinates into a rectangle (Figure 12.9B):

$$x_1 = AX_1 \qquad x_3 = \frac{1}{A} X_3$$
$$X_1 = \frac{1}{A} x_1 \qquad X_3 = Ax_3 \qquad (12\text{-}3.4)$$

A simple shear, which transforms a square into a parallelogram (Figure 12.10B), and its inverse are described by

$$x_1 = X_1 + BX_3 \qquad x_3 = \frac{1}{A} X_3$$
$$X_1 = \frac{1}{A} x_1 - Bx_3 \qquad X_3 = x_3 \qquad (12\text{-}3.5)$$

When the constants A in Equation (12-3.4) and B in Equation (12-3.5) are linear functions of time, the motions are steady, and these equations describe progressive pure shear and progressive simple shear, respectively (see Section 12.4).

With Equations (12-3.2), it is easy to show that a homogeneous deformation transforms a circle into an ellipse. A circle of unit radius in the undeformed state is represented by the equation

$$(X_1)^2 + (X_3)^2 = 1 \qquad (12\text{-}3.6)$$

If we substitute for X_1 and X_3 from Equations (12-3.2), we find the locus in the deformed state of all material particles that lie on the circle in the undeformed state. Because a rigid-body translation does not contribute to the strain, we assume $c = f = 0$. Then, making the substitution, we find

$$(a^2 + d^2)(x_1)^2 + 2(ab + de)x_1x_3 + (b^2 + e^2)(x_3)^2 = 1 \qquad (12\text{-}3.7)$$

Equation (12-3.7) is the equation of an ellipse with its principal axes tilted with respect to the coordinate axes, and it is, in fact, the strain ellipse.

The components of the strain tensor are related to the displacement vectors for the material particles. A displacement vector connects the position of a particle in the undeformed state to its position in the deformed state. The vector and its components (U_1, U_3) parallel to the X_1 and X_3 coordinate axes are (Figure 12-3.1A)

$$\mathbf{U} \equiv \mathbf{x} - \mathbf{X} \qquad (12\text{-}3.8)$$

$$U_1 = x_1 - X_1 \qquad U_3 = x_3 - X_3 \qquad (12\text{-}3.9)$$

When a material deforms, the displacement vectors for two neighboring material points are different. If they were the same, the "deformation" would be a rigid-body translation. The difference in these displacement vectors therefore describes the deformation. Thus we consider two neighboring points *A* and *B* that are displaced by the deformation to *a* and *b*, respectively. The displacement vectors for the two points are $\mathbf{U}^{(A)}$ and $\mathbf{U}^{(B)}$, and the difference between them is $d\mathbf{U}$ (Figure 12-3.1B). The material line segment $d\mathbf{X}$ connecting *A* to *B* is deformed into $d\mathbf{x}$ connecting *a* to *b*. The change in that line segment due to the deformation $\Delta d\mathbf{X}$ (where Δ is the symbol implying the change in) is also described by the vector $d\mathbf{U}$ (Figure 12-3.1B). Thus,

$$d\mathbf{U} \equiv \mathbf{U}^{(B)} - \mathbf{U}^{(A)} = \Delta d\mathbf{X} \equiv d\mathbf{x} - d\mathbf{X} \qquad (12\text{-}3.10)$$

The relationship between the first and last terms in this equation is just the differential of Equation (12-3.8).

We can consider the components dX_1 and dX_3 of the line segment $d\mathbf{X}$ to be two material line segments that are initially perpendicular to each other and parallel to the coordinate axes X_1 and X_3, respectively (Figure 12-3.2). If we restrict our analysis to infinitesimal strain, characterized by the conditions

$$dU_1 << 1 \text{ and } dU_3 << 1$$

then the displacement associated with each of these line segments due to the deformation can be expressed using Equation (12-3.10) and the chain rule of differentiation for $d\mathbf{U}$

$$\Delta d\mathbf{X} = \Delta dX_1 + \Delta dX_3 = d\mathbf{U} = \frac{\partial \mathbf{U}}{\partial X_1} dX_1 + \frac{\partial \mathbf{U}}{\partial X_3} dX_3 \qquad (12\text{-}3.11)$$

We can write the changes ΔdX_1 and ΔdX_3 in each of the line segments in terms of the components of the displacement vector $\mathbf{U}$ by (Figure 12-3.2):

$$\begin{aligned} \Delta dX_1 &= \frac{\partial \mathbf{U}}{\partial X_1} dX_1 = \frac{\partial U_1}{\partial X_1} dX_1 + \frac{\partial U_3}{\partial X_1} dX_1 \\ \Delta dX_3 &= \frac{\partial \mathbf{U}}{\partial X_3} dX_3 = \frac{\partial U_1}{\partial X_3} dX_3 + \frac{\partial U_3}{\partial X_3} dX_3 \end{aligned} \qquad (12\text{-}3.12)$$

For the material line segments dX_1 and dX_3, the extensional strains are labeled e_{11} and e_{33}, respectively. Each strain component is the change in length divided by the initial length, as defined in Equation (12.2). For dX_I, for example, the change in length is $(\partial U_1/\partial X_1)dX_1$ and the initial length is dX_I (Figure 12-3.2). Similar relations hold for dX_3. Thus

$$\begin{aligned} e_{11} &\equiv \frac{\Delta dX_1}{dX_1} = \frac{1}{dX_1}\left[\frac{\partial U_1}{\partial X_1} dX_1\right] = \frac{\partial U_1}{\partial X_1} \\ e_{33} &\equiv \frac{\Delta dX_3}{dX_3} = \frac{1}{dX_3}\left[\frac{\partial U_1}{dX_1} dX_3\right] = \frac{\partial U_3}{\partial X_3} \end{aligned} \qquad (12\text{-}3.13)$$

The shear strain of dX_1 relative to dX_3 and vice versa are labeled e_{13} and e_{31}, respectively, and are defined in Equation (12.4) to be half the tangent of the shear angle $\psi_{13} = \psi_{31} = \psi = \alpha + \beta$ (Figure 12-3.2). For very small strains, the angles α and β must also be very small, that is, $\alpha << 1$ and $\beta << 1$.

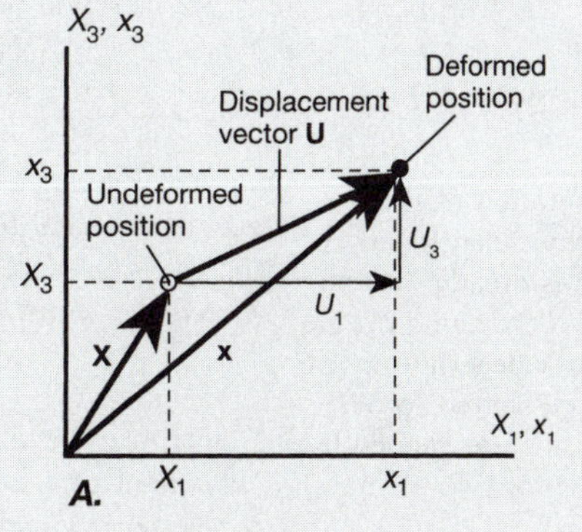

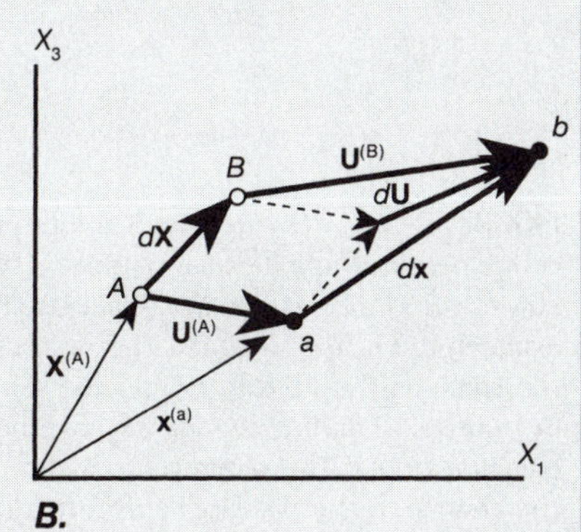

FIGURE 12-3.1 The displacement vector. *A.* The displacement vector U connects the position X of a material particle in the undeformed state to its position x in the deformed state. *B.* If the material is deformed, the displacement vectors for two neighboring points are different. Point *A* is deformed to the position *a*; *B* is deformed to the position *b*. The difference in the displacement vectors $d\mathbf{U} = \mathbf{U}^{(B)} - \mathbf{U}^{(A)} = d\mathbf{x} - d\mathbf{X}$ describes the deformation of the material.

(continued)

BOX 12-3 A More Quantitative View of Strain *(continued)*

Using this fact and the standard trigonometric identity for the tangent of the sum of two angles we get

$$\tan\psi \equiv \tan(\alpha + \beta) = \frac{\tan\alpha + \tan\beta}{1 - \tan\alpha\tan\beta} \approx \tan\alpha + \tan\beta \quad (12\text{-}3.14)$$

where we can use the approximation on the right because the product ($\tan\alpha \tan\beta$) is negligibly small. Recall that the tangent of an angle ϕ is defined on a right triangle as the length of the side opposite the angle ϕ divided by the length of the adjacent side. For infinitesimal strains, the side opposite the angle α is approximately $(\partial U_1/\partial X_3)dX_3$, and the adjacent side is dX_3 (Figure 12-3.2). Similar relationships hold for the angle β. Thus we have

$$\tan\alpha \approx \frac{1}{dX_3}\left[\frac{\partial U_1}{\partial X_3}dX_3\right] = \frac{\partial U_1}{\partial X_3}$$
$$\tan\beta \approx \frac{1}{dX_1}\left[\frac{\partial U_3}{\partial X_1}dX_1\right] = \frac{\partial U_3}{\partial X_1} \quad (12\text{-}3.15)$$

Then using the definition of the shear strain (first Equation (12.4)) with Equations (12-3.14) and (12-3.15) gives

$$e_{13} = e_{31} = 0.5\tan\psi \approx 0.5\left(\frac{\partial U_1}{dX_3} + \frac{\partial U_3}{\partial X_1}\right) \quad (12\text{-}3.16)$$

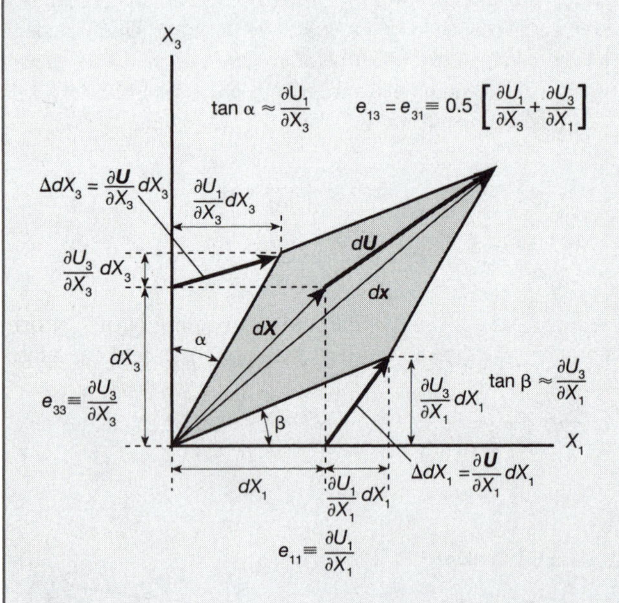

FIGURE 12-3.2 The geometrical interpretation of the components of infinitesimal strain for two-dimensional strain. For clarity, the infinitesimal strain is greatly exaggerated in the diagram. The vectors *dX*, *dx*, and *dU* are the same as the vectors having the same labels that appear in Figure 12-3.1*B*, but *dX* and *dx* have been shifted to a common origin. The strain components thus are defined by the change in the displacement vector *dU* for two neighboring points.

These relationships for the extensions and shear strains associated with the material line segments dX_1 and dX_3 are the components of the infinitesimal strain tensor. In shorthand component notation, we summarize Equations (12-3.13) and (12-3.16) by

$$e_{k\ell} \approx 0.5\left(\frac{\partial U_k}{\partial X_\ell} + \frac{\partial U_\ell}{\partial X_k}\right) \quad (k, \ell = 1,2,3) \quad (12\text{-}3.17)$$

This expression for $e_{k\ell}$ remains exactly the same if k and ℓ are interchanged, which shows that

$$e_{k\ell} = e_{\ell k} \quad (12\text{-}3.18)$$

Thus the strain tensor is a symmetric tensor (compare Equations (12.40) and (12.41), and the discussion of shear strain in Section 12.2 and Figure 12.3*B*) and is the symmetric part of the displacement gradient tensor $\partial U_k/\partial X_\ell$.

The antisymmetric part of the displacement gradient tensor can be shown to be the infinitesimal rotation tensor defined by

$$r_{k\ell} \equiv 0.5\left(\frac{\partial U_k}{\partial X_\ell} - \frac{\partial U_\ell}{\partial X_k}\right) \quad (k,\ell = 1,2,3) \quad (12\text{-}3.19)$$

The antisymmetric character of $r_{k\ell}$ is evident from this equation, because interchanging the subscripts k and ℓ gives the relationship:

$$r_{k\ell} = -r_{\ell k} \quad (12\text{-}3.20)$$

Adding Equations (12-3.17) and (12-3.19) shows that the sum of the infinitesimal strain and the rotation tensors is just the displacement gradient tensor, so the symmetric and antisymmetric parts of the displacement gradient tensor are complementary parts of that tensor. Components on the principal diagonal of the matrix $r_{k\ell}$, for which $k = \ell$, must be zero, as shown by setting $k = \ell$ in Equation (12-3.19). Thus, Equation (12-3.20) is true even if both subscripts are the same, because, for example, the equation $r_{11} = -r_{11}$ can be true only if $r_{11} = 0$. In two-dimensional strain,

$$r_{k\ell} = \begin{bmatrix} 0 & r_{13} \\ r_{31} & 0 \end{bmatrix}$$

and because of Equation (12-3.20), there is only one independent off-diagonal component $r_{13} = -r_{31}$. Thus from Equations (12-3.15) and (12-1.19) we can see that

$$r_{13} = 0.5\left(\frac{\partial U_1}{\partial X_3} - \frac{\partial U_3}{\partial X_1}\right) = 0.5(\tan\alpha - \tan\beta) \quad (12\text{-}3.21)$$

For very small angles, the tangent of the angle is approximately equal to the angle measured in radians. Thus for the very small angles that characterize shear in infinitesimal strain, we can write

$$r_{13} = 0.5(\alpha - \beta) \quad (12\text{-}3.22)$$

Because the deformation causes dX to have a clockwise component of rotation of α (Figure 12-3.2) and a counterclockwise component of rotation of β, the difference between these angles is the net amount by which $d\mathbf{X}$ rotates. Thus r_{13} is half the net rotation of the material line segment $d\mathbf{X}$.

The displacement components (U_1, U_3) can be expressed solely in terms of the coordinates of the material point in the undeformed state by substituting Equations (12-3.1) into (12-3.9), assuming the rigid translations are zero ($C = F = 0$):

$$U_1 = (A - 1)X_1 + BX_3 \qquad U_3 = DX_1 + (E - 1)X_3 \quad (12\text{-}3.23)$$

Using Equations (12-3.23) in (12-3.17), we find the values of the strain components in terms of the constants that define the motion of the material particles:

$$\begin{bmatrix} e_{11} & e_{13} \\ e_{31} & e_{33} \end{bmatrix} = \begin{bmatrix} (A - 1) & 0.5(B + D) \\ 0.5(D + B) & (E - 1) \end{bmatrix} \quad (12\text{-}3.24)$$

As indicated in the paragraph preceding Equation (12-3.11), the relationships given here are correct only for very small strains. The analysis of large strains is considerably more complex, although this geometric interpretation of the strain components remains intuitively useful.

For a line segment of arbitrary orientation in the undeformed state, given by the angle θ with respect to the $\hat{X}_1$ principal coordinate axis, it can be shown that the extension and the shear strain for infinitesimal plane strain are given in terms of the principal extensions by

$$e_n = \hat{e}_1 \cos^2\theta + \hat{e}_3 \sin^2\theta \quad (12\text{-}3.25)$$

$$e_s = (\hat{e}_1 - \hat{e}_3)\sin\theta\cos\theta$$

The first of these equations is identical in form to the first Equation (7.44) that we found for the normal stress. The second equation is identical in form to the second Equation (7.44) for the shear stress component. This similarity derives from the fact that the infinitesimal strain and the stress tensors are symmetric second-rank tensors, so their mathematical characteristics are identical. Thus these equations can be transformed easily into equations for the Mohr circle of the form of Equation (7.46) or (7.49) (see derivations in Section 7.4), so a Mohr circle can be defined for the infinitesimal strain just as for the stress.

Strains that precede brittle fracture are generally infinitesimal, as are tectonic strains that can be recorded by geodetic surveys, including those using GPS (global positioning satellites). The record of permanent deformation that we can observe in rocks, however, is a record of finite strain, for which the theoretical development is considerably more complex. Nevertheless, in Section 12.9, we derive the comparable relations for the finite stretch (Equations (12.53) and (12.54)) and finite shear strain (Equations (12.67) and (12.69)) of a line segment, and we show that these also lead to a representation of finite strain as a Mohr circle.

12.9 FINITE STRAIN OF AN ARBITRARY LINE SEGMENT AND THE MOHR CIRCLE

The state of strain at a point is determined completely by the principal stretches (or principal extensions) and their orientations. By definition, then, we must be able to use those quantities to calculate the stretch and the shear strain for any other orientation of line through the point. We show in this section how that can be done for finite strain, and in the process we will find that we can define two Mohr circles for the state of finite strain. It should come as no surprise that a Mohr circle can be plotted for strain, because, like the stress, the strain is a symmetric second-rank tensor, and the mathematics describing these two quantities is therefore very similar, although they describe very different physical quantities.

We consider a square that is deformed into a rectangle, with the square oriented so that its sides are parallel to the principal stretches (Figure 12.28*A*). We use uppercase letters to refer to the coordinate axes and lengths in the undeformed state and lowercase letters to refer to coordinate axes and lengths in the deformed state. The two sets of coordinate axes, however, are exactly the same in our example. The initial lengths of the sides of the square parallel to the $\hat{X}_1$ and $\hat{X}_3$ coordinate axes are L_1 and L_3, respectively. The square is deformed into a rectangle whose sides have lengths l_1 and l_3 so that the principal stretches that characterize the deformation (Figure 12.28*B*) are:

$$\hat{s}_1 = \frac{l_1}{L_1} \qquad \hat{s}_3 = \frac{l_3}{L_3} \quad (12.47)$$

Within that square, we can consider a line N of any orientation Θ relative to the $\hat{X}_1$ principal axis and calculate the stretch s_n (Figure 12.28), as well as the shear strain $\gamma = \tan\psi$ for that line relative to a second line P perpendicular to N (Figure 12.29*A*, *B*).

i. Stretch of an Arbitrary Line Segment

First we look at the stretch. The line of length N is the diagonal of a rectangular box shaded gray in Figure 12.28*A* whose sides have lengths N_1, and N_3 parallel to $\hat{X}_1$ and $\hat{X}_3$ respectively. As a result of the deformation, that gray box becomes deformed into the gray box shown in Figure 12.28*B*; the side N_1 is lengthened to n_1, whereas the side N_3 is shortened to n_3. Thus the lengths of the sides of the gray box in the undeformed and the deformed state are related by the principal stretches:

$$\hat{s}_1 = \frac{n_1}{N_1} \qquad \hat{s}_3 = \frac{n_3}{N_3} \quad (12.48)$$

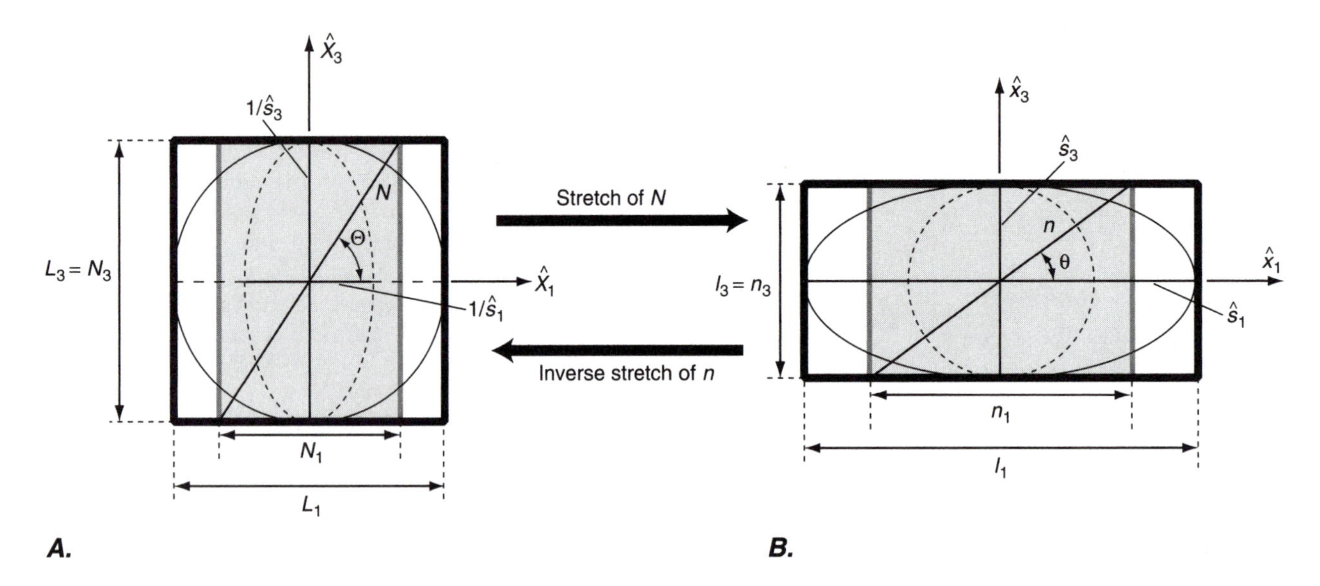

FIGURE 12.28 Stretch of an arbitrary line segment. *A*. Before deformation, the line segment *N* within the square makes an angle of Θ with the $\hat{X}_1$ axis. *N* is the diagonal of the shaded rectangle. *B*. After deformation, the original square becomes a rectangle, and the line segment *N* becomes the line segment *n*, which makes an angle θ with the $\hat{x}_1$ axis. The original shaded rectangle becomes the shaded rectangle with the diagonal *n*.

The lengths of the undeformed and deformed lines N and n are related to the lengths of the sides of the undeformed and deformed gray boxes by the Pythagorean theorem:

$$N^2 = N_1^2 + N_3^2 \qquad n^2 = n_1^2 + n_3^2 \tag{12.49}$$

The square of the stretch of the line N, using Equations (12.48), is then

$$s_n^2 = \frac{n^2}{N^2} = \frac{n_1^2 + n_3^2}{N_1^2 + N_3^2} \tag{12.50}$$

We use the square of the stretch just for the convenience of avoiding square roots of sums of quantities, which would make the mathematics very messy. This is one way in which the analysis of finite strain differs from infinitesimal strain. In the limiting case of infinitesimal strain, the second-order quantities can be adequately approximated by first-order terms.

The angle Θ between $\hat{X}_1$ and the undeformed line N and the angle θ between $\hat{x}_1$ and the deformed line n are given by

$$\sin\Theta = \frac{N_3}{N} \qquad \cos\Theta = \frac{N_1}{N}$$
$$\sin\theta = \frac{n_3}{n} \qquad \cos\theta = \frac{n_1}{n} \tag{12.51}$$

We want to calculate the stretch of the line s_n from the principal stretches and the orientation of the line. We could do this by solving Equations (12.48) for n_1 and n_3 and substituting the results into Equation (12.50), or by solving Equations (12.48) for N_1 and N_3 and substituting those results into Equation (12.50). Which we choose to do depends on which variables we want to keep in the final equations. As our observations are always made on some strained state of the rock, it is convenient to have the final equations expressed in terms of quantities we can observe in the strained state, and to that end, we use Equations (12.48) to eliminate N_1 and N_3 from Equations (12.50):

$$s_n^2 = \frac{n_1^2 + n_3^2}{\left(n_1/\hat{s}_1\right)^2 + \left(n_3/\hat{s}_3\right)^2} = \frac{n^2}{\left(n_1/\hat{s}_1\right)^2 + \left(n_3/\hat{s}_3\right)^2}$$

It simplifies the mathematics to take the reciprocal of this equation, which gives:

$$\frac{1}{s_n^2} = \frac{1}{\hat{s}_1^2}\frac{n_1^2}{n^2} + \frac{1}{\hat{s}_3^2}\frac{n_3^2}{n^2} \tag{12.52}$$

We substitute the appropriate relations from Equations (12.51) into Equation (12.52) to find

$$\frac{1}{s_n^2} = \frac{1}{\hat{s}_1^2}\cos^2\theta + \frac{1}{\hat{s}_3^2}\sin^2\theta \tag{12.53}$$

Here we see that a particularly simple relation arises that gives the inverse square of the line stretch in terms of the principal inverse squared stretches and functions of the angle θ between n and $\hat{x}_1$, which is measured in the de-

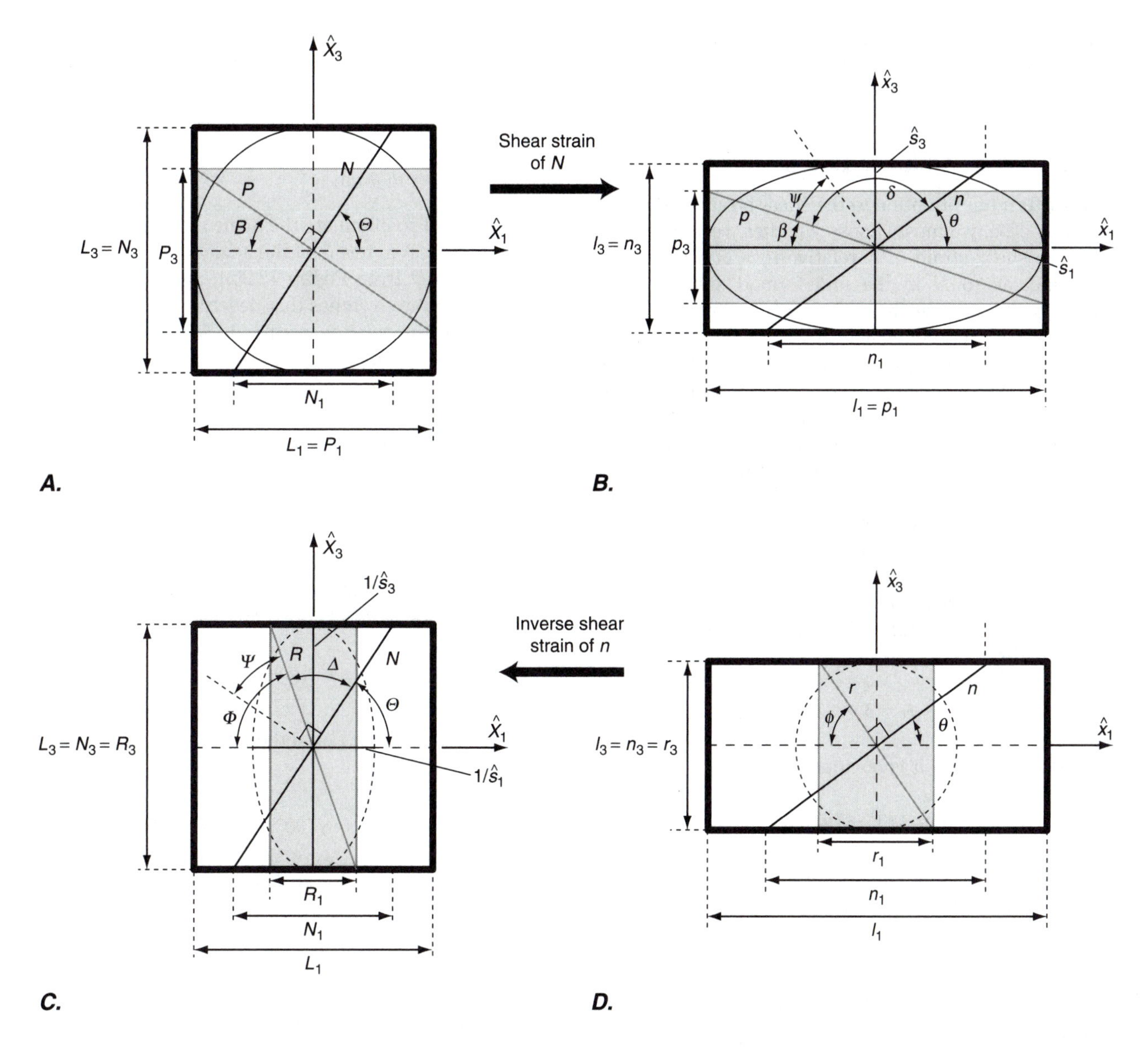

FIGURE 12.29 Shear of an arbitrary line segment. *A*. The undeformed state with line segment N and a line P perpendicular to it. P is the diagonal of the shaded rectangle. *B*. The deformed state. N and P are transformed to n and p, respectively. ψ is the shear angle of the line segment n relative to p. *C*. The undeformed state showing the inverse strain ellipse (dashed) and the inverse shear angle Ψ of the line segment N relative to R. These lines are perpendicular in the deformed state. *D*. The deformed state showing the line segment n and a line r perpendicular to it.

formed state. As we noted in Section 12.2, the inverses of the principal stretches are just the principal axes of the inverse strain ellipse (Figures 12.5 and 12.28*A*), and this equation shows why this strain ellipse turns out to be useful. The simplicity of using stretches raised to the second power accounts for the use of these second-order quantities in the normal continuum theory of finite strain. We also see that the inverse squares of the stretches are related by an equation that is similar in form to the first Equation (7.44), which gives the normal stress on a plane in terms of the principal stresses and the orientation of the plane.

Had we chosen to use Equation (12.48) to eliminate n_1 and n_3 instead of N_1 and N_3 from Equation (12.50), we would have found the relation

$$s_n^2 = \hat{s}_1^2 \cos^2 \Theta + \hat{s}_3^2 \sin^2 \Theta \qquad (12.54)$$

The relation still involves the squares of the stretches but not their inverses, and it still has the same general form as Equation (12.53) and the first Equation (7.44). The angle involved, however, is the angle between the line N and the $\hat{X}_1$ axis in the *undeformed* state, which generally is not what we have to work with when interpreting deformed rocks. This equation, however, is useful for forward modeling when we know the geometry of the undeformed state and want to calculate the effect of a known deformation.

ii. Shear of an Arbitrary Line Segment

We now turn our attention to the shear strain of the line N. In this case, we must choose whether we want to express the shear strain of N relative to a line P initially perpendicular to N in the undeformed state (Figure 12.29*A*) or the inverse shear strain of the line n relative to a line r that is perpendicular to n in the deformed state (Figure 12.29*D*). Again, it is a question of what variables we want in the final equation. We want to find the equation for the shear strain of a line in terms of the principal stretches, $\hat{s}_1$ and $\hat{s}_3$, and the angle θ that defines the orientation of the line in the deformed state. This gives us results consistent with Equation (12.53). If we consider the shear strain of N relative to P (Figure 12.29*A*), we will find the equations we want.

The initial right angle between N and P marked in Figure 12.29*A* ends up, after the deformation, being the angle δ in the deformed state (Figure 12.29*B*), and the shear angle ψ is then just

$$\psi = \delta - 90° \quad (12.55)$$

From Figure 12.29*B*, it is obvious that

$$\delta = 180° - (\theta + \beta) \quad (12.56)$$

where θ defines the orientation of n relative to $\hat{x}_1$ in the deformed state. Combining Equations (12.55) and (12.56) gives

$$\psi = 90° - (\theta + \beta) \quad (12.57)$$

Now we have to use a little algebra and trigonometry to manipulate the equations into a form that will be recognizable and useful. But hang in there; it's not too bad! We first set down a few standard trigonometric identities, in terms of arbitrary angles α and μ, that we will find useful:

$$\tan(90° - \alpha) = \cot \alpha \qquad \cot \alpha = \frac{1}{\tan \alpha}$$
$$\tan \alpha = \frac{\sin \alpha}{\cos \alpha} \qquad \cot \alpha = \frac{\cos \alpha}{\sin \alpha} \quad (12.58)$$
$$\tan(\alpha + \mu) = \frac{\tan \alpha + \tan \mu}{1 - \tan \alpha \tan \mu}$$

The engineering shear strain (second Equation (12.4)) is determined, using Equation (12.57):

$$\gamma \equiv \tan \psi = \tan(90° - (\theta + \beta))$$
$$= \cot(\theta + \beta) = \frac{1}{\tan(\theta + \beta)} \quad (12.59)$$

where we used the first two Equations (12.58). Applying the last Equation (12.58) to the last Equation (12.59) gives

$$\tan \psi = \frac{1}{\left[\dfrac{\tan \theta + \tan \beta}{1 - \tan \theta \tan \beta}\right]} = \frac{1 - \tan \theta \tan \beta}{\tan \theta + \tan \beta} \quad (12.60)$$

Now we want to eliminate the angle β from the equation by expressing it in terms of the angle θ. We know how B is related to Θ from Figure 12.29*A*, and from that relationship and the stretches that describe the deformation, we can find the relation we need. The stretches define how the sides of the gray box in Figure 12.29*A* deform into the gray box in Figure 12.29*B*, in the same way that they describe the deformation of the gray box in Figure 12.28*A–B* (Equation (12.48)):

$$\hat{s}_1 = \frac{p_1}{P_1} \qquad \hat{s}_3 = \frac{p_3}{P_3} \quad (12.61)$$

From the angular relations shown in Figure 12.29*A*, we see that

$$B + \Theta + 90° = 180° \qquad B = 90° - \Theta \quad (12.62)$$

Taking the tangent of B and then using the first Equation (12.58) we find

$$\tan B = \tan(90° - \Theta) = \cot \Theta \quad (12.63)$$

For the left side of Equation (12.63), we have from Figure 12.29*A*, B and Equations (12.61):

$$\tan B = \frac{P_3}{P_1} = \frac{\left(p_3/\hat{s}_3\right)}{\left(p_1/\hat{s}_1\right)} = \frac{\hat{s}_1}{\hat{s}_3}\frac{p_3}{p_1} = \frac{\hat{s}_1}{\hat{s}_3} \tan \beta \quad (12.64)$$

And similarly, for the right side of Equation (12.63), using Equation (12.48), we have:

$$\cot \Theta = \frac{N_1}{N_3} = \frac{\left(n_1/\hat{s}_1\right)}{\left(n_3/\hat{s}_3\right)} = \frac{\hat{s}_3}{\hat{s}_1}\frac{n_1}{n_3} = \frac{\hat{s}_3}{\hat{s}_1} \cot \theta \quad (12.65)$$

Substituting the results from Equations (12.64) and (12.65) into Equation (12.63) then gives

$$\frac{\hat{s}_1}{\hat{s}_3} \tan \beta = \frac{\hat{s}_3}{\hat{s}_1} \cot \theta \qquad \tan \beta = \frac{\hat{s}_3^2}{\hat{s}_1^2} \cot \theta \quad (12.66)$$

The second Equation (12.66) is the relationship we were looking for that expresses β in terms of θ and the principal stretches, and we can then substitute it for $\tan \beta$ in

Equation (12.60), use the second trigonometric identity in (12.58), and divide the numerator and denominator by $\hat{s}_3^2$ to get

$$\tan\psi = \frac{1 - \tan\theta\left(\hat{s}_3^2/\hat{s}_1^2\right)\cot\theta}{\tan\theta + \left(\hat{s}_3^2/\hat{s}_1^2\right)\cot\theta}$$

$$= \frac{\left(1/\hat{s}_3^2\right) - \left(1/\hat{s}_1^2\right)}{\left(1/\hat{s}_3^2\right)\tan\theta + \left(1/\hat{s}_1^2\right)\cot\theta}$$

Now using Equations (12.58) to express tangent and cotangent functions in terms of sines and cosines and simplifying the fractions gives

$$\tan\psi = \frac{\left(\frac{1}{\hat{s}_3^2} - \frac{1}{\hat{s}_1^2}\right)\sin\theta\cos\theta}{\frac{1}{\hat{s}_3^2}\sin^2\theta + \frac{1}{\hat{s}_1^2}\cos^2\theta}$$

We recognize that the denominator in this equation is just the expression we derived for the inverse stretch in Equation (12.53), so we can write the final result as

$$\frac{\gamma}{s_n^2} \equiv \frac{\tan\psi}{s_n^2} = \left(\frac{1}{\hat{s}_3^2} - \frac{1}{\hat{s}_1^2}\right)\sin\theta\cos\theta$$
$$\frac{\gamma}{s_n^2} = \frac{\hat{s}_1^2 - \hat{s}_3^2}{\hat{s}_1^2\hat{s}_3^2}\sin\theta\cos\theta \qquad (12.67)$$

The second expression for the shear strain in Equation (12.67) is written in terms of the difference in the inverse squared stretches and has a form exactly like the second Equation (7.44), which gives the shear stress on a plane in terms of the difference in principal stresses and the orientation of the plane. The left side of the equation is the engineering shear strain modified by the inverse square of the stretch of the line. This mixture of shear strain with stretch is inconvenient, in that the equation does not give the straight shear strain. However, the right side of the equation is a form that we have encountered before and that can be used to define a Mohr circle. It is therefore a convenient form to deal with.

We could choose to find the inverse shear strain $\tan\Psi$ (Figure 12.29*C*) of the line n relative to a perpendicular line r in the deformed state (Figure 12.29*D*). We would do an exactly analogous derivation using the angles in Figure 12.29*C*, *D* rather than those in Figure 12.29*A*, *B*. Expressing the result in terms of the original angle Θ between the line N and the $\hat{X}_1$ axis, and defining the inverse shear strain by

$$\Gamma \equiv \tan\Psi \qquad (12.68)$$

we would find:

$$\Gamma s_n^2 \equiv \tan\Psi\, s_n^2 = (\hat{s}_1^2 - \hat{s}_3^2)\sin\Theta\cos\Theta \qquad (12.69)$$

We find the relationship between the shear strain γ and the inverse shear strain Γ by dividing the two sides of the last Equation (12.67) by the corresponding sides of Equation (12.69), using Equations (12.51), and then using the definitions in the first Equation (12.50) and Equations (12.48) to find

$$\gamma = \frac{s_n{}^2}{\hat{s}_1\hat{s}_3}\Gamma \qquad \Gamma = \frac{\hat{s}_1\hat{s}_3}{s_n{}^2}\gamma \qquad (12.70)$$

iii. Mohr Circle for Finite Strain

The two Equations (12.53) and (12.67) allow us to calculate the inverse stretch and shear strain of a line that, in the deformed state, makes an angle θ with the $\hat{x}_1$ (and $\hat{s}_1$) axis. The two Equations (12.54) and (12.69) allow us to calculate the stretch and inverse shear strain of a line that, in the undeformed state, makes an angle Θ with the $\hat{X}_1$ (and $1/\hat{s}_1$) axis. Both pairs of equations have the same forms as Equations (7.44) for normal and shear stress. The same derivation for the Mohr circle leading from Equations (7.44) to (7.46) and to (7.49) can be done for the stretch and shear strains. Thus a Mohr circle, which allows a graphical solution of finite strain problems, can be defined for the finite strain with reference to either the deformed state or the undeformed state. The Mohr diagram of quantities referred to the deformed state requires that we plot the inverse squared stretch on the abscissa and the ratio of shear strain to squared stretch on the ordinate (Figure 12.30*A*). The Mohr diagram of quantities referred to the undeformed state requires that we plot the squared stretch on the abscissa and the product of inverse shear strain and squared stretch on the ordinate (Figure 12.30B).

We do not reproduce the derivations but simply record the final equations. For the finite strain Mohr circle referred to quantities measured in the deformed state, the parametric equations for the circle in terms of the angle θ are:

$$\left(\frac{1}{s_n^2}\right) = 0.5\left(\frac{1}{\hat{s}_3^2} + \frac{1}{\hat{s}_1^2}\right) - 0.5\left(\frac{1}{\hat{s}_3^2} - \frac{1}{\hat{s}_1^2}\right)\cos 2\theta \qquad (12.71)$$

$$\frac{\gamma}{s_n^2} \equiv \frac{\tan\psi}{s_n^2} = 0.5\left(\frac{1}{\hat{s}_3^2} - \frac{1}{\hat{s}_1^2}\right)\sin 2\theta \qquad (12.72)$$

These equations are comparable to Equations (7.46), except that Equation (12.71) has a minus sign instead of a plus sign in front of the second term on the right. This difference results in the requirement that an angle θ measured counterclockwise in physical space from the $\hat{x}_1$ (or $\hat{s}_1$) axis to the line segment, as in Figure 12.29*B*, *D*, is

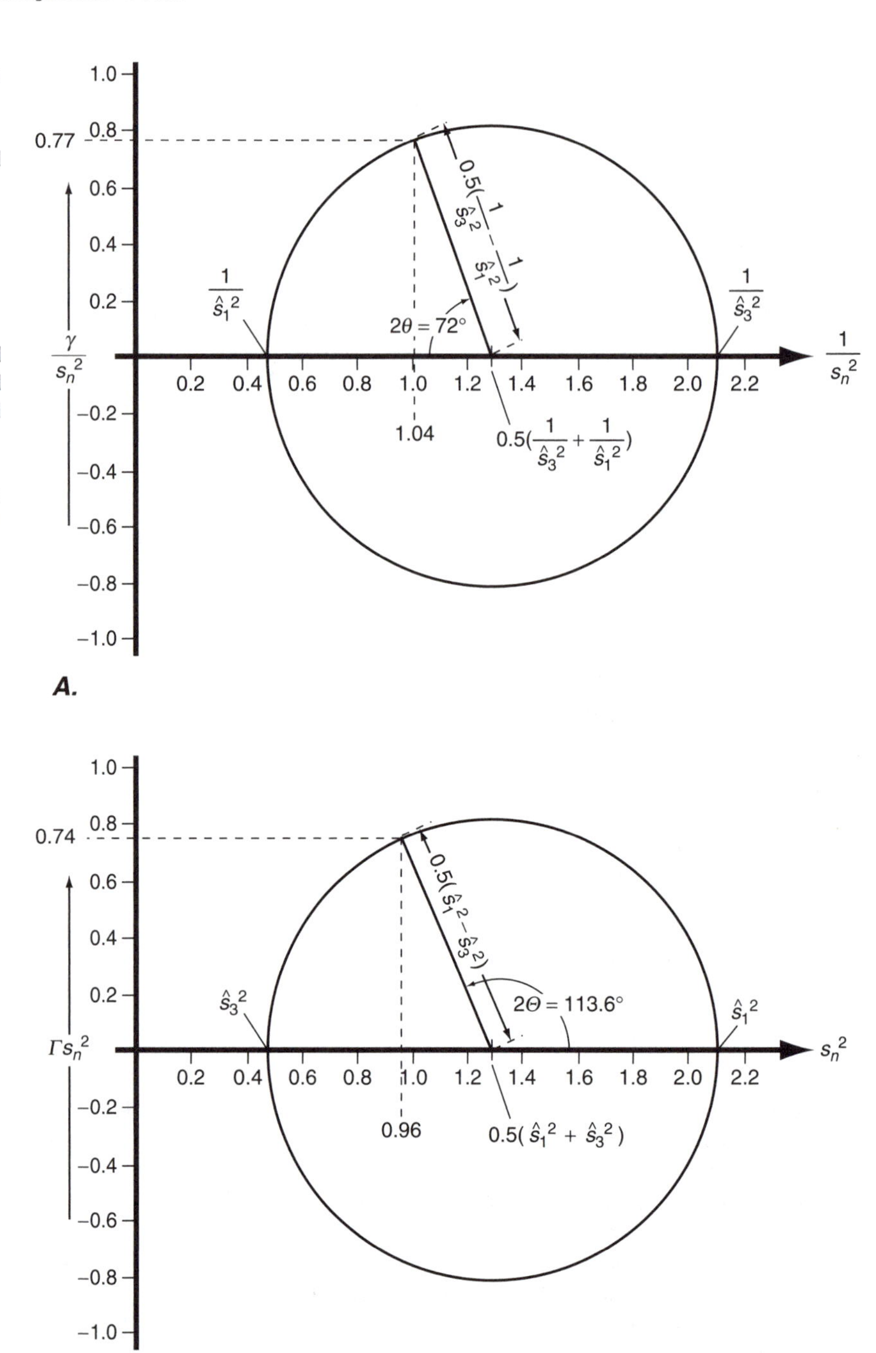

FIGURE 12.30 Mohr circles for the finite strain illustrated in Figure 12.29. This deformation is a constant-volume two-dimensional deformation, so $\hat{s}_1 = 1/\hat{s}_3$ and $\hat{s}_3 = 1/\hat{s}_1$. *A.* Finite strain referred to quantities measured in the deformed state. Note that the angle θ measured in physical space counterclockwise from $\hat{x}_1$ to the line segment n is plotted on the Mohr diagram as 2θ measured clockwise from the minimum on the inverse squared stretch axis $(1/\hat{s}_1^2)$. *B.* Finite strain referred to quantities measured in the undeformed state. The Mohr circle has the same size and position on the abscissa because the stretch quantities plotted on this abscissa are the inverse of those plotted in *A* and because $\hat{s}_1 = 1/\hat{s}_3$ and $\hat{s}_3 = 1/\hat{s}_1$.

plotted on the Mohr circle as an angle 2θ measured clockwise from the radius at the point $(1/\hat{s}_1^2)$ on the abscissa (Figure 12.30*A*), which is the smallest principal value on this Mohr circle. This differs from the construction for the stress Mohr circle. Recall that for the stress Mohr circle, a counterclockwise angle θ in physical space between the maximum compressive stress $\hat{\sigma}_1$ and the normal to a plane is plotted as a counterclockwise angle 2θ on the Mohr diagram from the largest principal value on the circle. Thus care must be taken to plot the angles in the appropriate sense and to interpret the orientations in physical space correctly. This difference arises simply because in the current case we are plotting the inverse of the squared stretch on the abscissa.

The direct equation for the Mohr circle is found by combining Equations (12.71) and (12.72) to eliminate the parameter θ. The derivation proceeds in the same manner as that for deriving Equation (7.49) from Equation (7.46), and the result is an equation of the same form as Equation (7.49):

$$\left[\frac{1}{s_n^2} - \frac{1}{2}\left(\frac{1}{\hat{s}_3^2} + \frac{1}{\hat{s}_1^2}\right)\right]^2 + \left[\frac{\gamma}{s_n^2}\right]^2 = \frac{1}{2}\left[\frac{1}{\hat{s}_3^2} - \frac{1}{\hat{s}_1^2}\right]^2 \tag{12.73}$$

Comparable equations to Equations (12.71) through (12.73) for the Mohr circle referred to the undeformed state are

$$s_n^2 = \left(\frac{\hat{s}_1^2 + \hat{s}_3^2}{2}\right) + \left(\frac{\hat{s}_1^2 - \hat{s}_3^2}{2}\right) \cos 2\,\Theta \quad (12.74)$$

$$\Gamma\, s_n^2 = \left(\frac{\hat{s}_1^2 - \hat{s}_3^2}{2}\right) \sin 2\,\Theta \quad (12.75)$$

$$\left[s_n^2 - \frac{\hat{s}_1^2 + \hat{s}_3^2}{2}\right]^2 + [\Gamma\, s_n^2]^2 = \left[\frac{\hat{s}_1^2 - \hat{s}_3^2}{2}\right]^2 \quad (12.76)$$

In this case, Equations (12.74) and (12.75) are identical in form to Equations (7.46), so the Mohr circle is plotted using the same conventions as for the stress Mohr circle.

Equations (12.72) and (12.75) give positive shear strains for any line segment whose angle θ from $\hat{x}_1$, or Θ from $\hat{X}_1$, respectively, is $0° \leq (\theta \text{ or } \Theta) = 90°$; these equations give a negative shear strain if $90° \leq (\theta \text{ or } \Theta) \leq 180°$. This sign convention is not consistent with the definition for the sign of the shear strain we gave in Section 12.1(ii) (Figure 12.2), where the reference axes are not the principal axes. This inconsistency is similar to the difference between the Mohr circle and the tensor sign conventions for the shear stress that we encountered in Chapter 7 (Sections 7.1(iv) and 7.5(ii); see Box 7-4, Table 7-4.1). In using the Mohr circles for finite strain, therefore, care must be taken to interpret the sign of the shear strain correctly.

For the deformation in Figure 12.29, the Mohr circle for finite strain referred to the deformed state (Equations (12.71) to (12.73)) is plotted in Figure 12.30*A*, and the Mohr circle for finite strain referred to the undeformed state (Equations (12.74) to (12.76)) is plotted in Figure 12.30*B*.

12.10 APPLICATIONS OF STRAIN ANALYSIS

Understanding deformation in rocks unequivocally requires an understanding of strain, as will become obvious from the discussions in subsequent chapters. The distribution of strain in folded layers helps distinguish the mechanisms by which folds form, as we show in Chapter 13. Understanding the relation between strain and foliations and lineations is also critical for interpreting those structures, as discussed in Chapter 14. In Chapter 15 we present a number of examples of the determination of strain in deformed rocks and discuss how this information is essential to understanding the deformation processes. In Chapter 16 we discuss the relationships between stress, strain, and strain rate. These relationships enable us to model rock deformation numerically with a computer and thereby to understand behavior of ductile deformation in different tectonic environments. In Chapter 17 we discuss the microscopic and submicroscopic mechanisms by which rocks flow, and we review the relationships that result from ductile deformation between strain and the crystallographic preferred orientations of mineral grains. Understanding these relationships provides another important tool for understanding the deformation in ductilely deformed rocks. Chapter 18 presents a discussion of scale models and numerically calculated models of rock deformation and thus is an application of the relationships between stress and strain rate that we discuss in Chapter 16. With these models, we can determine the distribution of strain in naturally evolving structures. For example, we compare the strain in the numerically calculated models of folding to the results from the kinematic models that we present in Chapter 13. This allows us to evaluate the kinematic models and to gain a more sophisticated understanding of how folds form. Finally, in Chapter 20, we discuss the application of strain measurements to investigating the processes of orogeny, or mountain building, including the use of stress–strain rate equations to calculate a model of orogeny.

The material in this chapter therefore provides a foundation for much of the rest of this book and therefore is essential to understand. The chapter contains a lot of information to digest at once, so we encourage readers to return to the relevant sections of this chapter to review the various concepts as they are encountered in the discussions in subsequent chapters and are applied to understanding deformation in the Earth.

REFERENCES AND ADDITIONAL READINGS

Eringen, A. C. 1967. *Mechanics of Continua*. New York, Wiley.

Flinn, D. 1962. On folding during three dimensional progressive deformation. *Quart. J. Geol. Soc. London* 118: 385–428.

Hobbs, B. E., W. D. Means, and P. F. Williams. 1976. *An Outline of Structural Geology*. New York, Wiley.

Marrett, R., and D. C. P. Peacock. 1999. Strain and stress. *J. Struct. Geol.* 21: 1057–1063.

Means, W. D. 1976. *Stress and Strain: Basic Concepts of Continuum Mechanics for Geologists*. New York, Springer-Verlag.

Passchier, C. W. 1990. Reconstruction of deformation and flow parameters from deformed vein sets. *Tectonophysics* 180(2–4): 185–199.

Ramsay, J. G. 1967. *Folding and Fracturing of Rocks*. New York, McGraw-Hill.

Ramsay, J. G. 1976. Displacement and strain. *Phil. Trans. Roy. Soc. London* A283: 3–25.

Ramsay, J. G., and M. I. Huber. 1983. *The Techniques of Modern Structural Geology. Vol. 1: Strain Analysis.* New York, Academic Press.

Teyssier, C., and B. Tikoff. 1999. Fabric stability in oblique convergence and divergence. *J. Struct. Geol.* 21(8/9): 969–974.

Tikoff, B., and H. Fossen. 1999. Three-dimensional reference deformations and strain facies. *J. Struct. Geol.* 21(11): 1497–1512.

Tickoff, B., and S. F. Wojtal. 1999. Displacement control of geologic structures. *J. Struct. Geol.* 21(8/9): 959–967.

Twiss, R .J., and J. R. Unruh. 1998. Analysis of fault slip inversions: Do they constrain stress or strain rate? *J. Geophys. Res.* 103(B6): 12,205–12,222.

Wood, D. S., and P. E. Holm. 1980. Quantitative analysis of strain heterogeneity as a function of temperature and strain rate. *Tectonophysics* 66: 1–12.

Chapter 13

KINEMATIC ANALYSIS OF FOLDS

The geometries of folds that are observed in naturally deformed rocks are described in Chapter 10. We would like to understand the significance of these various geometric forms in terms of the mechanism of folding. To this end, we discuss in this chapter **kinematic models** of folding and the implications of these models for the distribution of strain in the folded layer. Kinematic models specify the motion of the deforming body but in general do not relate that motion to the mechanical properties of the folded layer or to the stress. We first discuss various two-dimensional models for folding single layers (Sections 13.1 through 13.4) by which we can account for much of the geometric variation included in Ramsay's classification of folded layers (Figure 10.19). We then discuss two-dimensional models of multilayer folding, including flexural and passive shear folding of multilayers, which incorporate qualitatively the effects of the mechanical properties of the layers (Section 13.5). Section 13.6 presents models for the formation of kink and chevron folds. Folding often is a consequence of the propagation of faults through a layered sequence of rocks, and in Section 13.7 we describe kinematic models for fault-bend and fault-propagation folding.

These kinematic models for folding allow us to account for many of the fold geometries that we observe in nature. The models, however, are not unique, as evidenced, for example, by the various models of folding leading to class 1B folds. Thus a particular fold geometry cannot be used to infer uniquely the kinematics of fold formation. To try to distinguish among the various possible kinematic models, therefore, we look at the distribution of strain in the folded layers and, for some cases, at the pattern into which linear features on the bedding planes are folded, for each of the basic models of folding.

In Sections 13.8 through 13.10, we discuss models for some three-dimensional aspects of folding, including the relationship between drag folds and the slip direction on associated faults (Section 13.8), the geometry of superposed folds (Section 13.9), and models of folding during diapiric flow (Section 13.10). Our discussion of folding continues in Chapter 15, where we compare the characteristics of these kinematic models of folding with observations of natural folds.

The mechanical properties of the rocks during folding have a profound effect on the style of fold and the distribution of strain that develop in the folded layer. In this chapter, we account only qualitatively for differences in mechanical properties in terms of the competence of different layers of rock. A rock is said to be **competent** if, relative to other types of rock, it flows very slowly under a given stress or if it requires a very high stress to force it to flow at a given rate. A rock is said to be **incompetent** if, relative to other types of rock, it flows very rapidly at a given stress or if it requires a very small stress to force it to flow at a given rate. To include these factors more quantitatively in the models of folding, we must be able to define the relation between stress and the rate at which a ductile material is able to flow, which is the subject of Chapter 16.

We can then conclude our discussion of folding in Chapter 18 (Sections 18.3, 18.6, and 18.7), where we consider **scale models** and **mechanical models** of folding. These models incorporate the laws of mechanics; the particular relationship between stress, strain rate, and strain for the material in the folding layers; and the specific

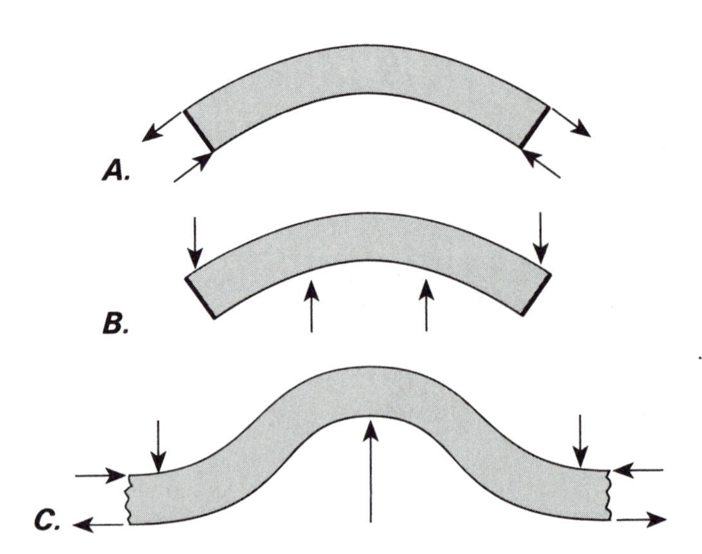

FIGURE 13.1 Flexure of a plate or a layer by bending. Bending is caused by application of pairs of torques. A torque is a force that is applied normal to a lever arm and tends to make a body rotate. *A*. Bending moments created by pairs of equal and opposite forces applied parallel to the plate. *B*. Four-point loading: torques are created by forces applied perpendicular to the plate. *C*. One possible distribution of forces required to bend an infinite layer into a single localized fold.

mechanical properties of the folding layers. The geometries of the folds that develop are therefore self-consistent consequences of the deformation and the material properties. We can then evaluate the applicability of the kinematic models introduced in this chapter for interpreting the formation of folds.

13.1 FLEXURAL FOLDING OF A LAYER

Class 1B folds are a common feature of many fold belts (Section 10.5(i)). The geometry of this class of folds may be explained by three different kinematic models, which we refer to as **orthogonal flexure**, **flexural shear**, and **volume-loss flexure**. Collectively these models are called **flexural folding**. In all three models, the orthogonal thickness of the layer remains constant during folding, thereby producing class 1B folds. Thus the class of the fold alone cannot be used to distinguish the exact kinematic model by which the fold formed. These folding models differ according to whether the convex side of a fold is lengthened or remains constant and whether its concave side is shortened or remains constant. As a result, the distribution of strain in the folded layer characterizes the different models of orthogonal flexure. We discuss orthogonal flexure and flexural shear in this section, but because the volume-loss mechanism can produce several geometries of fold, we consider it in a separate section (Section 13.3).

Flexural folding of layers of rock can result from either bending or buckling, which are two different ways of applying forces to the layers. **Bending** of a layer results from the application of pairs of forces that produce equal and opposite torques that bend the layer into a fold. In pure bending, there is no net tension or compression averaged over the layer, either parallel or perpendicular to it. Three possible systems of applied forces that can provide such torques are shown in Figure 13.1. Of these, only Figure 13.1*C* is geologically reasonable because only this example results in a localized fold in a horizontal layer of indefinite extent. In this diagram, the vertical force acting upward in the middle of the layer could represent the uniform fluid pressure of an intrusive magma body applied over a segment of strata. The resulting uplift of the strata into a localized fold is called a **laccolith**. This vertical force could also represent the effect of a normal fault in basement rocks that bends the overlying strata into a monoclinal fold (Figure 13.2).

Buckling results from the application of compressive stresses parallel to a competent layer (Figure 13.3*A*, *B*).

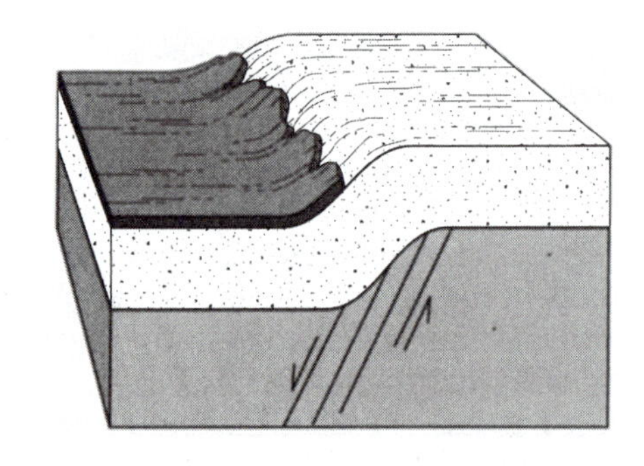

FIGURE 13.2 Monoclinal fold developed by bending. *A*. Diagram showing the formation of a monocline in sediments overlying a normal fault in basement rocks. *B*. Photograph of the Rattlesnake Mountain monocline, Wyoming. (Photo courtesy of Robert J. Varga)

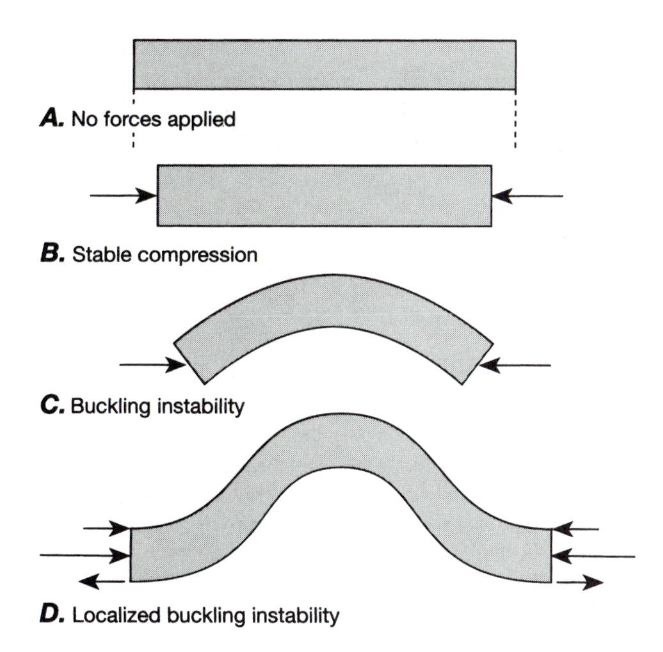

FIGURE 13.3 Flexure of a plate or layer by buckling. *A*. An unloaded plate. *B*. The plate is shortened by equal and opposite compressive forces applied to opposite ends of the plate. *C*. Forces increase until an instability develops, causing buckling. *D*. Combination of compression and torques that produce a localized fold.

If the compressive stress is sufficiently large, the layer becomes unstable and buckles into a fold, either under compressive stresses alone (Figure 13.3*C*) or in association with additional torques (Figure 13.3*D*). Of the two sets of forces shown in Figure 13.3, only those in Figure 13.3*D* result in a localized fold in a horizontal layer of indefinite extent. This type of buckling may be important in fold and thrust belts in which the compressive stress that drives the thrusting causes the layers to buckle, thereby shortening and thickening the thrust sheet (cf. Figure 10.2*B*). Folds in such belts, however, can also result from the sliding of thrust sheets up thrust ramps (Sections 5.2–5.3) and thus may form by a combination of bending and buckling (see Section 13.7). Buckling also is of prime importance in the formation of ptygmatic folds.

The diagrams showing the distribution of strain and rotated lineations in flexural folds are collected in Figures 13.6 (for bending) and 13.7 (for buckling) to allow easy comparison of the different models. The initial segment of the layer that becomes deformed into a half-wavelength fold, extending from an anticlinal hinge (line *A*) through an inflection line (line *I*) to a synclinal hinge (line *S*), is shown as an undeformed layer in Figure 13.6*A* and as a layer that is homogeneously shortened by the initial layer–parallel compression in Figure 13.7*A*. We assume plane strain during folding with the intermediate principal stretch ($\hat{s}_2 = 1$) everywhere parallel to the hinge. Thus all the strain appears in the profile plane of the fold, which contains the two principal stretches $\hat{s}_1$ and $\hat{s}_3$.

i. Orthogonal Flexure

Kinematic Model. A layer may respond to either bending or buckling loads by **orthogonal flexure** (Figure 13.4), a kinematic model for which lines perpendicular to the layer before folding (Figure 13.4*A*) remain so after folding (Figure 13.4*B*). In a profile plane perpendicular to the fold axis, the surface of the layer on the convex side of a fold is lengthened, and the surface on the concave side is shortened. Thus the layer-parallel stretch must vary across the thickness of the layer from the convex to concave side of the fold from greater than 1 to less than 1. There must therefore be a surface within the layer along which the stretch is 1, implying that there is no layer-parallel length change during the folding; this surface is called the **neutral surface**. The orthogonal thickness of the layer remains constant around the fold.

Orthogonal flexure should be characteristic of folds with low curvature in competent layers that are resistant to ductile deformation. As the curvature increases to high values, however, the orthogonality condition cannot be maintained.

Strain. During bending by orthogonal flexure, the $\hat{s}_1$ axes ($\hat{s}_1 > 1$) along the convex side of the fold profile are parallel to the surface and normal to the hinge line (Figure 13.6*B*(i)). On the concave side of the profile, the $\hat{s}_1$ axes form a convergent fan normal to the folded surface. Thus the profile plane and the top and bottom surfaces of the layer are all principal planes of strain.

The final strain in a buckle fold formed by orthogonal flexure is the sum of the initial shortening (Figure 13.7*A*(i)) and the strain distributions shown in Figure 13.6*B*, and it depends on the relative amounts of strain contributed by each.

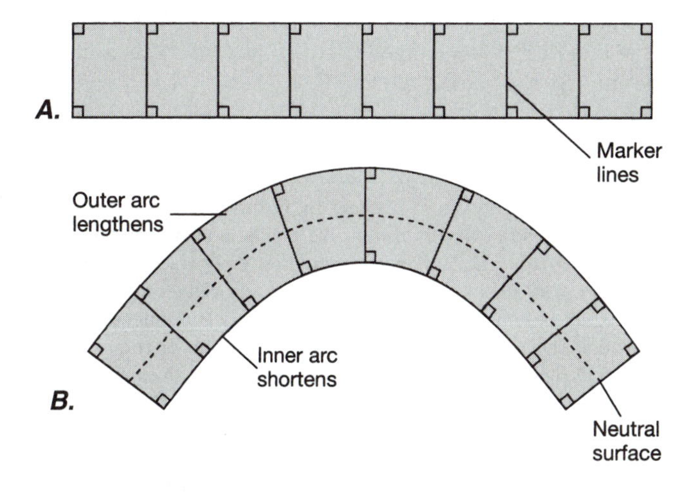

FIGURE 13.4 Geometry of orthogonal flexure. *A*. An undeformed layer with fiducial material lines normal to the layer. *B*. After folding, the fiducial lines remain normal to the layer.

For buckling by orthogonal flexure, the initial homogeneous layer–parallel shortening is counteracted on the convex side of the fold by the lengthening that develops during folding; on the concave side, the initial layer–parallel shortening is enhanced by the shortening that accumulates during folding (Figure 13.7*B*(i)). If the initial shortening is greater than the subsequent lengthening, the $\hat{s}_1$ axis is normal to the layer surface around the entire fold profile forming a convergent fan, and no neutral surface exists within the layer at all. If the lengthening due to the folding exceeds the initial shortening (Figure 13.7*B*(i)), then a neutral surface develops through part of the layer (edge of the shaded zone in Figure 13.7*B*(i); cf. Figure 9.17*C*).

Lineation Rotation. The geometry of a lineation on a folded surface depends on the strain in the surface. On a neutral surface, the folding rotates the lineation into different orientations, but the angle between the lineation and the fold axis *f* is everywhere constant and unchanged from the original angle (Figure 13.6*B*(ii)), because there is no strain in the plane of the neutral surface. Thus on a stereonet, the lineation plots along the arc of a small circle of angle β about the fold axis (Figure 13.6*B*(iii)).

On the top surface of the layer, however, the stretch normal to the fold axis involves a lengthening where it is on the convex side of a fold and a shortening where it is on the concave side (Figure 13.6*B*(i), (iv)). Thus the strain in the surface is inhomogeneous. Lineations lying in the surface rotate toward the axis of maximum stretch in the surface and away from the axis of minimum stretch. The top surface of the layer is lengthened most at the hinge on the convex side of a fold (line *A* in Figure 13.6*B*(i), (iv)) and is shortened most at the hinge on the concave side of a fold (line *S* in Figure 13.6*B*(i), (iv)). It is not stretched at all at the inflection line ('*I*' in Figure 13.6*B*(i), (iv)). Thus the angle between the fold axis *f* and the lineation (Figure 13.6*B*(iv)) is a maximum ($\beta'_A > \beta$) along the line *A*; it is unchanged ($\beta'_I = \beta$) at *I*, and it is a minimum ($\beta'_S < \beta$) along the line *S*.

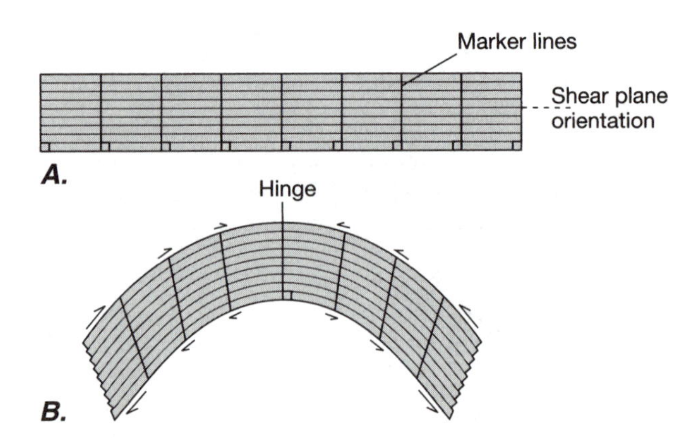

FIGURE 13.5 Geometry of flexural shear folding. *A.* Undeformed layer with fiducial material lines normal to the layer. Lines parallel to the layer indicate the orientation of potential shear planes. *B.* Fiducial lines are sheared out of perpendicular on shear planes. The shear magnitude decreases from limb to hinge and changes sense across the hinge surface.

On the stereonet, the lineations therefore fall between the small circles for the maximum and minimum angles (Figure 13.6*B*(v)). In general, then, we expect measurements of folded lineations from folds formed during bending by orthogonal flexure to plot within a small circle band concentric about the fold axis (Figure 13.6*B*(v)).

The pattern of lineations folded by buckling show little difference from those folded by bending. Before the initiation of buckling, the layer-parallel shortening homogenously rotates a lineation on the layer surface to a smaller angle with the incipient fold axis ($\beta_A = \beta_I = \beta_S = \beta' < \beta$; Figure 13.7*A*(i), (ii)) (see the difference between the small circles at angles β (dashed small circle) and β'_I (the middle solid small circle) in Figure 13.7*B*(iii). Thus the initial angle between lineation and hinge line is smaller for buckling than it would be for bending, but otherwise the patterns resulting from the different models of flexure remain the same (Figure 13.7*B*(iii)).

FIGURE 13.6 Models of bending by flexural folding showing the strain distribution in the profile planes and on the top surface of the fold. Each part of the figure shows the original layer in *A* folded into a half-wavelength anticline-syncline pair that extends from the hinge of the anticline (line *A*) through the inflection line (line *I*) to the trough of the syncline (line *S*). The mirror image of this half-wavelength is added to show a full-wavelength fold. The top surface of the layer in *A* includes a lineation that makes a constant angle $\beta_A = \beta_I = \beta_S = \beta$ with the incipient fold axis *f*, where the subscripts on the angle β identify the line parallel to the fold axis from which the angle is measured. In each of the fold models, the first column shows the total strain in the profile plane that has accumulated from the initial state shown in *A*. The second column shows the top surface of the fold, from the anticlinal hinge to the synclinal hinge, flattened out so that the strain in the surface and lineation orientations relative to the fold axis are preserved. The bar beneath each diagram shows the original undeformed length of the surface. The third column shows the geometry of the lineation rotation on a lower-hemisphere equal-area net. The orientation of the lineation is marked by *A* at the antiformal hinge, *I* at the inflection line, and *S* at the synformal hinge. *A.* The undeformed layer that becomes folded into the half-wavelength of the fold extending from the anticlinal hinge to the synclinal hinge. *B–D.* The results of different modes of bending, as labeled on the figure.

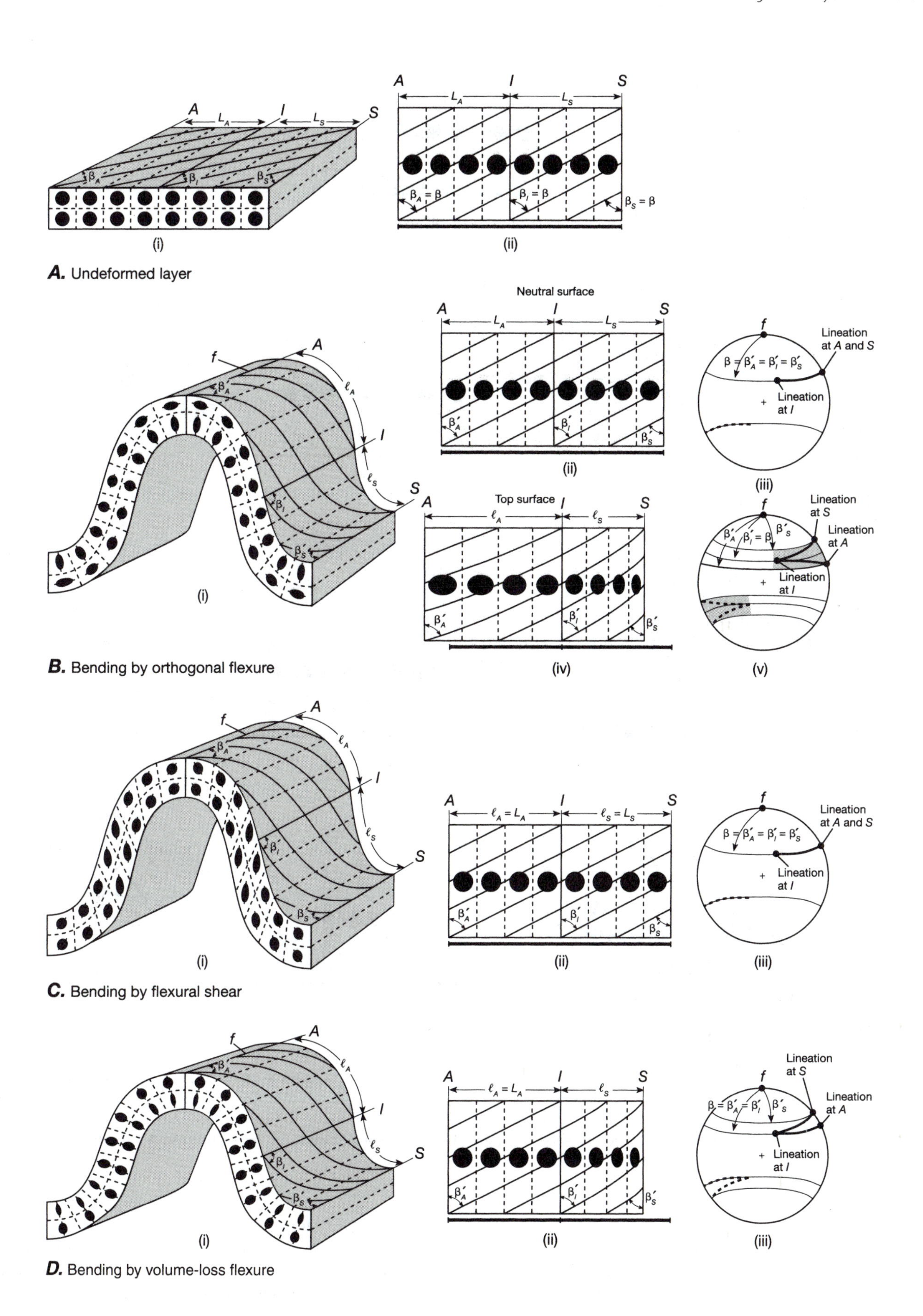
A. Undeformed layer
B. Bending by orthogonal flexure
C. Bending by flexural shear
D. Bending by volume-loss flexure
Neutral surface
Top surface
Lineation at A and S
Lineation at I
Lineation at S
Lineation at A

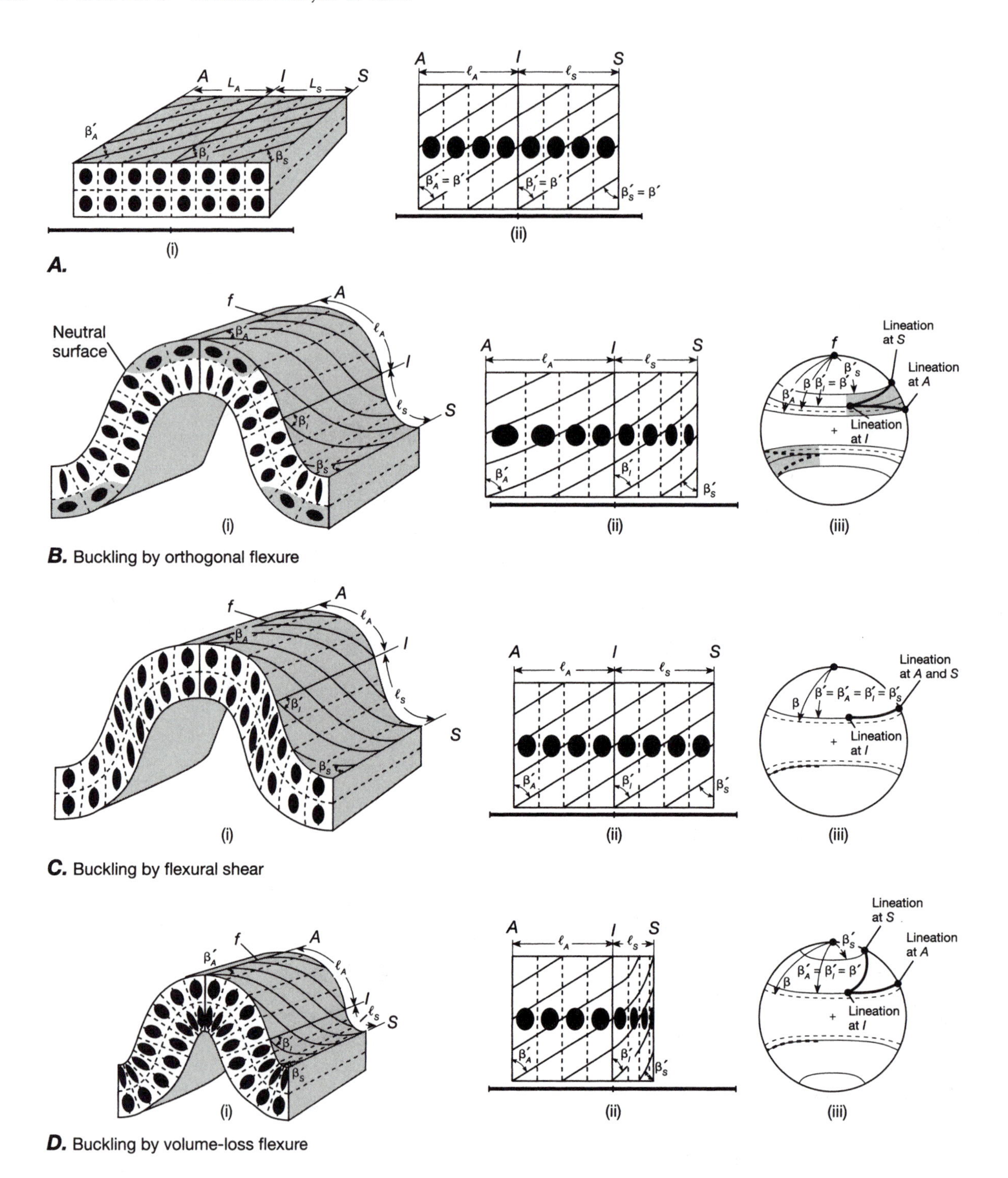

FIGURE 13.7 Models of buckling by flexural folding and the strain distribution in the profile planes and on the folded surfaces. See Figure 13.6 caption for details. *A*. The portion of the layer to be folded, shown after homogeneous shortening but just before the onset of buckling. The portion of the layer becomes the half-wavelength of the antiform-synform pair in the following diagrams. The initial dimensions before shortening are the same as in Figure 13.6*A*. *B–D*. The results of different modes of buckling.

ii. Flexural Shear

Kinematic Model. A layer can also respond to bending or buckling by **flexural shear**, also called **flexural flow** (Figure 13.5). Folding is accommodated by simple shear parallel to the layer, and there is no lengthening and shortening, respectively, of the convex and concave sides of the fold as there is in orthogonal flexure.

Flexural shear folding is analogous to the bending of a deck of cards (Figure 13.5*A*) in that the card surfaces represent the shear planes parallel to the layer, and all the deformation occurs by sliding parallel to those shear planes. The material on the convex side of a shear plane shears toward the fold hinge relative to that on the concave side (Figure 13.5*B*). The sense of shear on the limbs of such a fold changes across the fold axial surface, and the magnitude of the shear decreases toward the hinge. The thickness of the body, measured perpendicular to the shear planes, is constant. Lines originally perpendicular to the surface of the layer before folding, however, do not remain perpendicular except exactly at the hinge. During folding of a card deck, the length of individual cards is constant. Similarly, in flexural shear folding, any length measured in the profile plane parallel to the shear planes is constant, so neither the convex nor the concave surface of the layer changes length.

Flexural shear folding might occur instead of orthogonal flexure if the layer is less competent and undergoes ductile deformation more readily or if the layer has a strong planar mechanical anisotropy,[1] such as fine interbedding of chert and shale or a strongly developed schistosity parallel to the layer.

Strain. Bending by flexural shear folding involves internal shearing of the layer on shear planes that are parallel to the layer surface (Figure 13.5). The distribution of strain in the fold profile is illustrated in Figure 13.6*C*(i). The profile plane is a principal plane of strain, but the maximum and minimum principal stretches, $\hat{s}_1$ and $\hat{s}_3$, are neither parallel nor perpendicular to the layer, so the surfaces of the layer are not principal planes of strain (Figure 13.6*C*(i)). The $\hat{s}_1$ axis makes an angle of 45° or less with the layer, with the maximum angle near the hinge surface at *A* and *S*, where the shear strain across the entire layer decreases to zero, and the sense of shear reverses from one side of the hinge surface to the other. The amount of shear increases along a limb with increasing folding angle to a maximum at the inflection surface. With increasing shear, the $\hat{s}_1$ axis rotates to lower angles with the shear planes, which are parallel to the layer surface (Figure 13.6*C*(i)). As a result, $\hat{s}_1$ axes tend to form a curved pattern across the fold profile that can vary from a divergent fan in the hinge zone to a convergent fan along the limbs (Figure 13.6*C*(i)). There is no unique neutral surface, because all surfaces parallel to the surfaces of the layer are in effect neutral surfaces, as there is no layer-parallel stretching in any of those surfaces.

If buckling occurs by flexural shear, the resulting strains added to the layer-parallel shortening result in a more distinct convergent fan of $\hat{s}_1$ axes around the fold (Figure 13.7*C*(i)). Compared with pure flexural shear (Figure 13.6*C*), the axes of maximum stretch are oriented at higher angles to the layer.

Lineation Rotation. Surfaces parallel to the layer are planes of simple shear. They are therefore surfaces of no strain that are parallel to a circular section through the strain ellipsoid (Figure 13.6*C*(ii)) and not parallel to a principal plane. Thus on any of these surfaces, the original angle β between a lineation and the fold axis f remains unchanged by folding (Figure 13.6*C*(ii)). On a stereonet the orientations of folded lineations plot along a single small circle centered about the fold axis (Figure 13.6*C*(iii)), just as for the neutral surface in the orthogonal flexure fold (Figure 13.6*B*(iii)).

For buckling by flexural shear, the behavior of lineations is the same as for bending, with the only difference being that the angle β between the lineation and the fold axis f is reduced to β' by the initial layer–parallel shortening (Figure 13.7*C*(ii), (iii)).

13.2 PASSIVE SHEAR FOLDING OF A LAYER

i. Kinematic Model

Passive shear folding, also called **passive flow folding** or simply **flow folding**, occurs in an incompetent layer, which acts simply as a marker that records the deformation and thus exerts no influence on the process of folding. Deformation takes place by inhomogeneous simple shear on shear planes that cross-cut the layer. The amount and sense of shear vary systematically across the shear planes to produce the folded geometry. This process results in class 2, or similar, folds.

The behavior of a deck of cards is useful to illustrate this kinematic model of folding. In this case, the shear planes represented by the cards cut across the layer (Figure 13.8), as opposed to the flexural shear model (Figure 13.5) for which they are parallel to the layer. Along a given axial surface, the shape of the folded surface on the convex side of the layer is identical to the shape on the concave side, and thus the curvatures of the two sides of the layer also are identical. Consequently, the hinge lines that define the axial surface, or hinge surface, of a particular multilayer fold must lie on the same shear plane, and the shear planes are parallel to the axial surface (Figure 13.8). Because there is no deformation within any given shear plane, that is, none of the cards in the card deck changes size or shape, any fold axis that is parallel to the intersection of a shear plane with the folding layer is straight, and the associated fold is cylindrical. The axial trace thickness of the layer, which is measured parallel to the shear planes, is constant around the fold. These geometric characteristics are exactly those of class 2, or similar, folds.

[1]The mechanical properties of a mechanically anisotropic material are different in different directions in the material. For example, with a planar anisotropy, the properties parallel to the plane are different from those perpendicular to the plane.

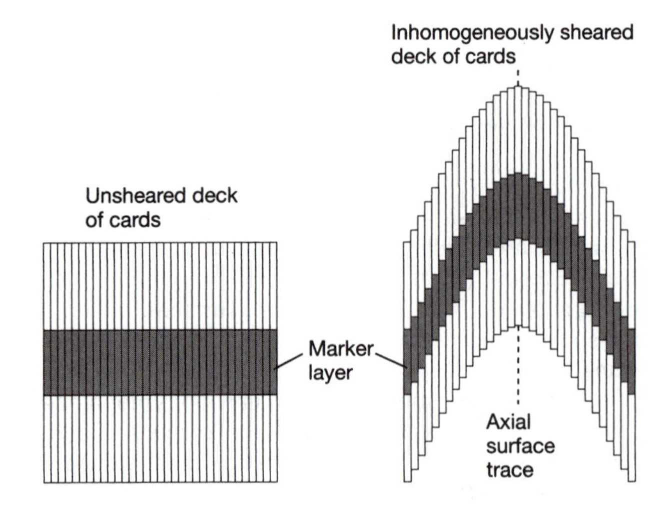

FIGURE 13.8 Passive shear folding. *A*. A block of material, including a marker layer (shaded black), cut by a set of shear planes perpendicular to the layer. *B*. Folding by inhomogeneous simple shear is approximated by "deck-of-cards" shear. The axial surface of the fold is parallel to the shear planes. The thickness of the layer parallel to the shear planes is constant. The shape of the fold is exactly the same on the convex and the concave sides of the folded layer.

In passive shear folds, the fold hinge and the fold axis must parallel the intersection of the shear planes with the original layer orientation (Figure 13.9). The shear planes can be oriented at any nonzero angle to the layer, and the shear direction within the shear planes can be in any orientation except parallel to the layer being folded. As long as there is a component of shear across the layer, a fold can form. Thus the orientation of the fold axis or hinge is not related to the direction of shear.

Natural fold geometries that come close to class 2 folds are characteristic of deformation in high-grade metamorphic rocks (Figure 10.1*B*, *C*, *D*), in salt domes (Section 13.10), and in glaciers, consistent with the inference that this fold class characterizes the deformation of incompetent materials. The model of passive shear folding certainly requires incompetent behavior, but as we show in the next two sections, it is not the only mechanism that produces folds having a geometry very close to that of class 2.

ii. Strain

The distribution of strain across the profile of a passive shear fold reflects the inhomogeneous simple shear that produces the fold (Figure 13.10*A*, *B*; cf. Figure 12.19). The shear planes are parallel to the axial surface, and the shear sense changes across that surface. Near the axial surface, where shear is a minimum, the angle between $\hat{s}_1$ axes and the surface is 45°. The acute angle between $\hat{s}_1$ and the shear plane decreases with increasing shear strain (Figure 13.10*B*), so that the $\hat{s}_1$ axes rotate toward parallelism with the shear plane and hence with the axial surface of the fold, especially in the limbs where the shear is the highest. In principle, however, the $\hat{s}_1$ axes can never actually become parallel with the shear planes, and the long axes of the strain ellipses diverge across the profile of a passive shear fold (Figure 13.10*B*).

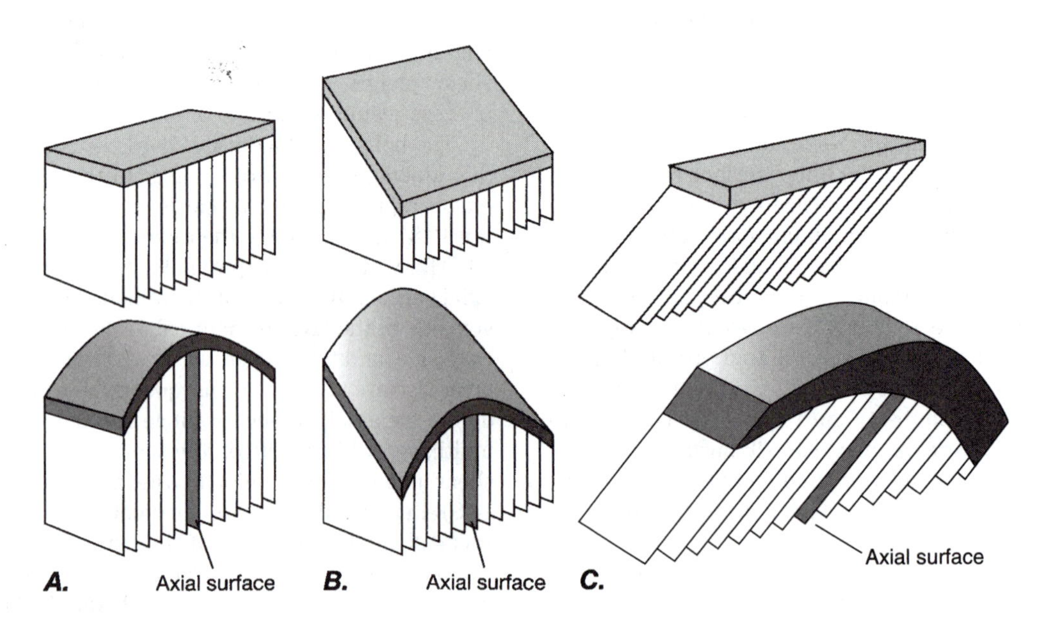

FIGURE 13.9 The orientation of the fold hinge for a passive shear fold is determined by the intersection of the shear planes with the original orientation of the layer to be folded. In *A–C*, the top diagram shows the relationship between the shear planes and the original orientation of the layer; the bottom diagram shows the layer after folding. The shear directions could have any orientation in the shear plane except parallel to the surface being folded. The orientation of the fold hinge does not indicate the direction of shear.

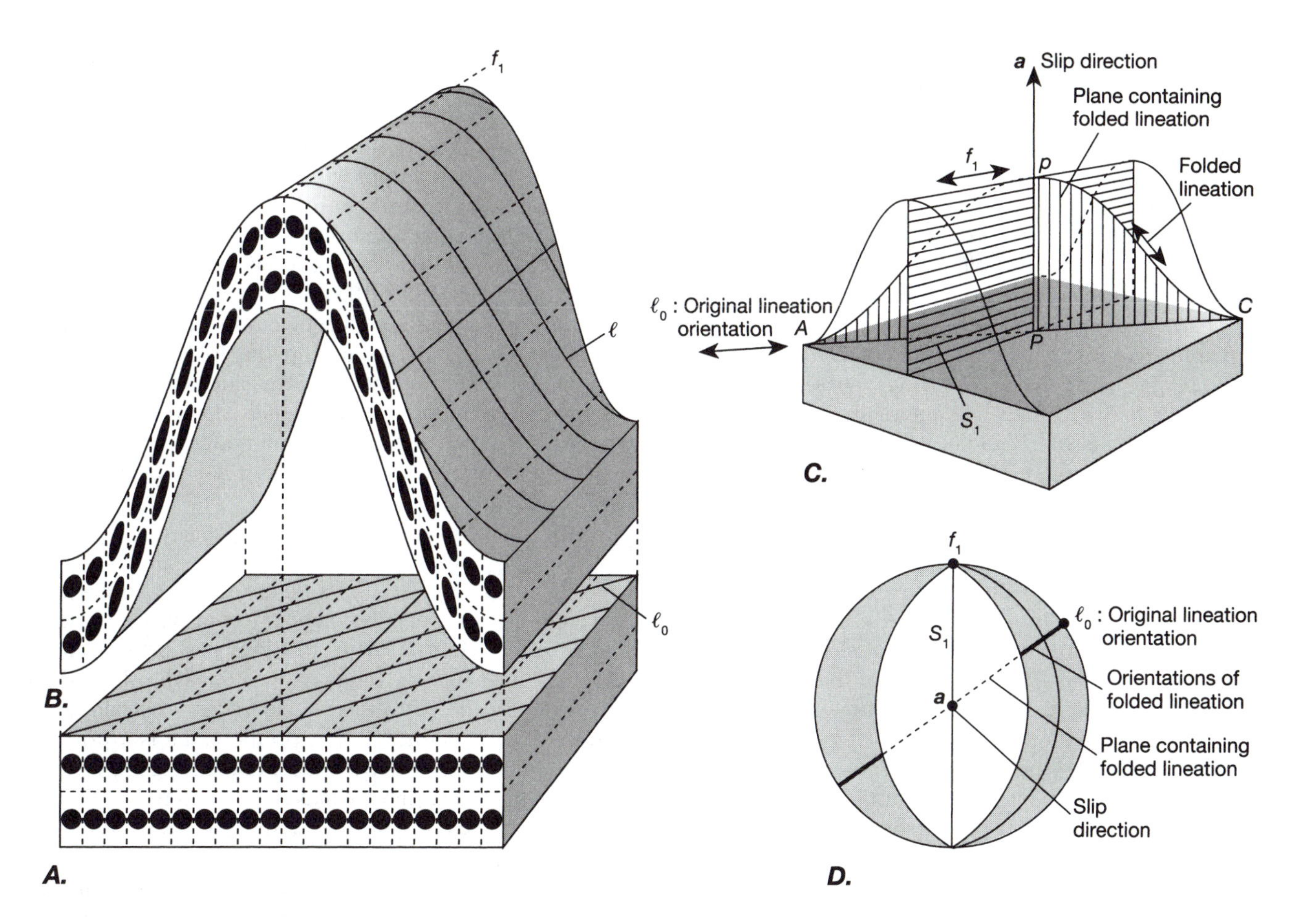

FIGURE 13.10 Passive shear folding. *A.* Initial layer to be folded into the full-wavelength antiform-synform pair. *B.* Inhomogeneous simple shear on a set of parallel shear planes intersecting a layer produces a class 2 fold. An original lineation ℓ_0 is folded into the form shown by the lines ℓ. The $\hat{s}_1$ directions of the strain ellipses on the profile plane of the fold diverge across the fold. Sections of the strain ellipsoids on the shear planes are circular sections. *C.* The original lineation AC is folded by passive shear on planes parallel to S_1 in the direction parallel to *Pp*. The folded lineation lies on the folded surface and in the plane *ApC*, which contains the original orientation of the lineation *AC* and the shear direction *Pp* = a. (After Ramsay 1967) *D.* A stereonet diagram showing the orientation of the fold axis f_1, the slip direction a, the shear plane and axial surface of the fold S_1, the orientation of the undeformed lineation ℓ_0, and the range of orientations of the folded lineation, which lie along a great circle representing the plane that contains ℓ_0, and a. Gray shading shows the location of orientations of great circles passing through the fold axis f_1 representing the folded orientations of the layer on the stereonet.

In simple shear, the $\hat{s}_2$ direction lies in the shear plane perpendicular to the shear direction, but $\hat{s}_2$ is not in general parallel to the fold axis unless it is initially parallel to the layer being folded (Figure 13.9). Thus the $\hat{s}_1$ and $\hat{s}_3$ axes do not in general lie in the profile plane of the fold. The shear planes always contain a circular section of the strain ellipsoid.

iii. Lineation Rotation

Passive shear deforms a linear feature into a range of orientations, all of which lie in a plane (Figure 13.10*B*, *C*). The orientations of lineations in a folded surface all plot on a stereonet along the great circle that contains the original lineation orientation ℓ_0 and the shear direction **a** (Figure 13.10*C*, *D*). The lineation is folded from its original orientation in both directions toward the slip direction. The slip direction also must lie in the shear plane, which is the axial surface. Thus on a stereonet, the intersection of the plane containing the lineation orientations with the plane of the axial surface defines the slip direction (Figure 13.10*C*, *D*) if the fold formed by passive shear. Both of these planes can be determined by field measurements.

13.3 VOLUME-LOSS FOLDING OF A LAYER

i. Kinematic Model

Volume-loss folding is a mechanism by which folds can form or be amplified by the gradual removal of material from particular zones in a folded layer. The loss generally results from solution, a process also called **solution folding**. The volume-loss mechanism, however, does not result in a unique class of folds, because the fold geometry depends on the orientation of the zones of volume loss relative to the layer. Folds may form with class 1B, 1C, or 2 geometry. Volume loss from discrete zones may result in the offset of beds, which gives an erroneous appearance of shearing along the zones; in fact, however, no shearing is required.

Three ideal fold geometries can result from volume-loss folding (Figure 13.11). Removing wedges of material that are symmetrical about a line normal to the layer surface results in a class 1B fold in which both the concave and convex surfaces are continuous (Figure 13.11*A*). Removing wedges of material that are symmetrical about a line oblique to the layer produces a fold with the approximate geometry of a class 1C fold, but the concave surface of the fold shows discontinuous offsets at the surfaces where volume loss has been concentrated, which could be misinterpreted as evidence of shearing (Figure 13.11*B*). For both models, the length of the convex side of the fold is unchanged by the loss of material, but the concave side of the fold is shortened.

Volume loss from parallel zones of constant thickness oriented oblique to an initial irregularity or gentle fold in the bedding can amplify a pre-existing fold or irregularity, although it cannot produce a fold from a flat layer (Figure 13.11*C*). To this extent, it is a discontinuous approximation to the geometry of homogeneous flattening, a process that we discuss further in the next section. The result of this geometry of volume loss is a fold that approaches a class 2 style. The discontinuous offsets in both the convex and concave surfaces suggest a fold formed by shearing on discrete shear surfaces, although no shearing is required.

Figure 13.12 shows an example of a fold that has been amplified by solution of material with a geometry comparable to that shown in Figure 13.11*B*. In the two photographs that make up Figure 13.12*A*, the bedding (left photo) and the solution surfaces at a high angle to the bedding (right photo) are visible especially near the hinge zone of the fold. Figure 13.12*B* shows the fold restored to a more open configuration by cutting along the main solution planes and opening up the fold to restore a smooth boundary on the concave side, effectively reversing this part of the volume-loss deformation. The empty wedge-shaped gaps in the photo illustrate the volume of material removed along major solution surfaces.

ii. Strain

Layer bending by volume-loss flexure results from solution along seams concentrated on the concave side of the fold (Figure 13.12*A*, *B*). Thus the layer is shortened in a direction normal to the hinge on the concave side of the fold, whereas the convex side undergoes no strain and forms the neutral surface. The profile plane is a principal plane of strain, with the maximum shortening $\hat{s}_3 < 1$ normal to the seams and the maximum stretch $\hat{s}_1 = 1$ parallel to the seams (Figure 13.6*D*(i)). If the seams are initially at a high angle to the layer surface, then $\hat{s}_1$ axes in the fold form a convergent fan around the fold. Only if the seams are normal to the layer, as in Figure 13.6*D*(i), is the layer surface a principal plane.

For buckling by volume-loss flexure, the initial strains associated with layer-parallel shortening normal to the hinge are preserved on the convex side of the layer

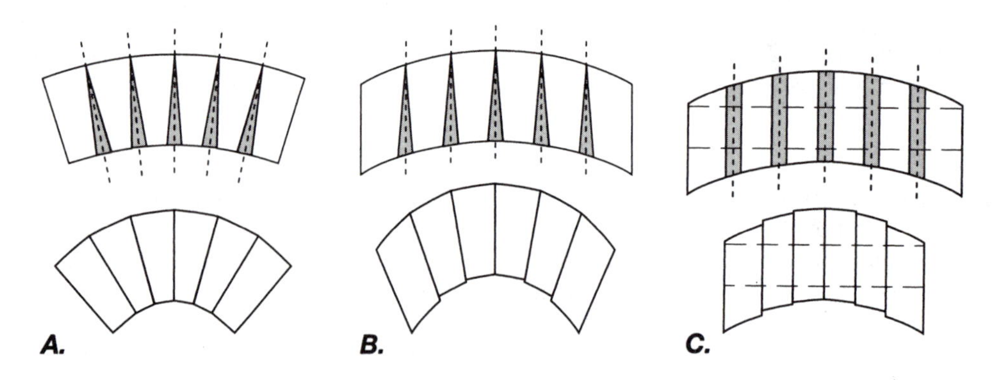

FIGURE 13.11 Volume-loss folding. The upper diagrams show the initial folds. The shaded areas are removed during intensification of the folds. (After Groshong 1975) *A*. Wedge-shaped areas of volume loss are symmetrical about lines (dashed) that are normal to the surfaces of the layer. Both the convex and concave surfaces of the resulting fold are continuous. *B*. Wedge-shaped areas of volume loss are symmetrical about lines (dashed) that are not normal to the surfaces of the layer (in this case they are parallel to one another). The convex surface of the resulting fold is continuous, but the concave surface has offsets that suggest shearing of the layer along the surfaces of volume loss. *C*. Lath-shaped areas of volume loss are parallel to one another and in general oblique to the layer. Both convex and concave surfaces of the fold show offsets that suggest shearing comparable to passive shear folding, even though displacement is strictly normal to the lath boundaries.

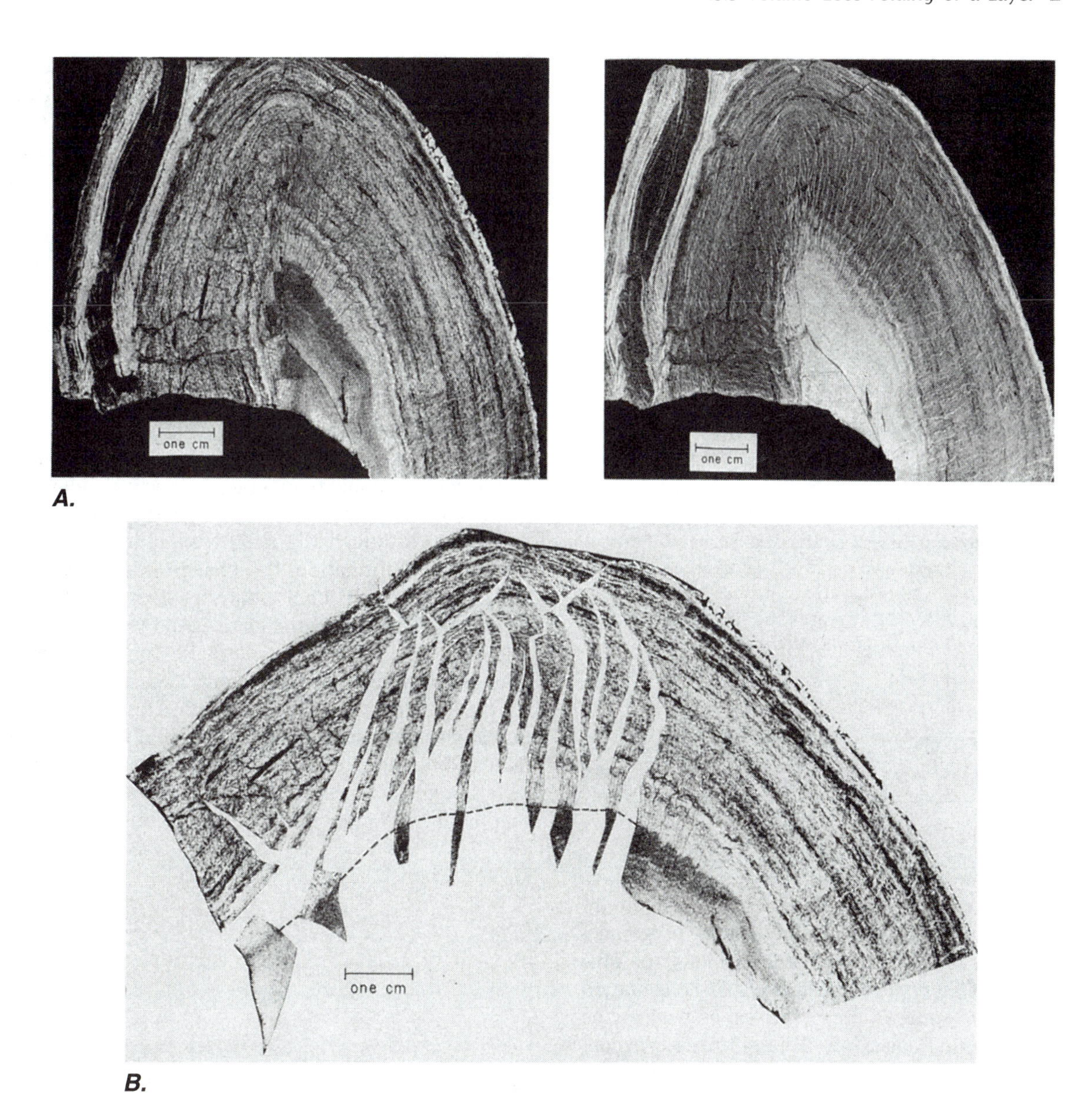

FIGURE 13.12 Partial unfolding of a fold tightened by volume-loss (solution) folding. (After Groshong 1975) *A*. Negative photos of acetate peels emphasizing the layering (left) and the foliation (right). *B*. Fold in *A* restored to the condition of having a smooth surface on the concave side (dashed line) by opening the fold along major solution seams. The volume of material lost is indicated by the blank areas in the photo along which the fold has been opened.

where the volume loss is zero or near-zero, with the result that $\hat{s}_1$ is normal to the layer (Figure 13.7*D*(i)). On the concave side of the fold, the initial shortening strains are increased by the folding, and $\hat{s}_1$ forms a convergent fan across the fold (Figure 13.7*D*(i)).

iii. Lineation Rotation

For a fold in which bending is accommodated by volume loss, the original angle on the convex surface between lineation and fold axis *f* is preserved, because the surface is not strained ($\beta'_A = \beta'_I = \beta$; Figure 13.6*D*(ii)). On the concave surface, the shortening normal to the fold axis rotates the lineation to smaller angles with the fold axis *f*, reaching a minimum angle ($\beta'_s < \beta$) at the hinge where the shortening is a maximum (Figure 13.6*D*(ii)). On a stereonet, lineations from the convex side of the fold plot along a small circle of angle β about the fold axis *f* and those from the concave side plot at smaller angles (Figure 13.6*D*(iii)). Thus, in general, the lineations plot within a small circle band between the angles β and β'_s from the fold axis.

For buckling, the geometry of the folded lineation is the same as for bending, except that the angle between the lineation and the fold axis *f* is decreased from β to β' by the initial layer-parallel shortening (Figure 13.7*D*(ii), (iii)).

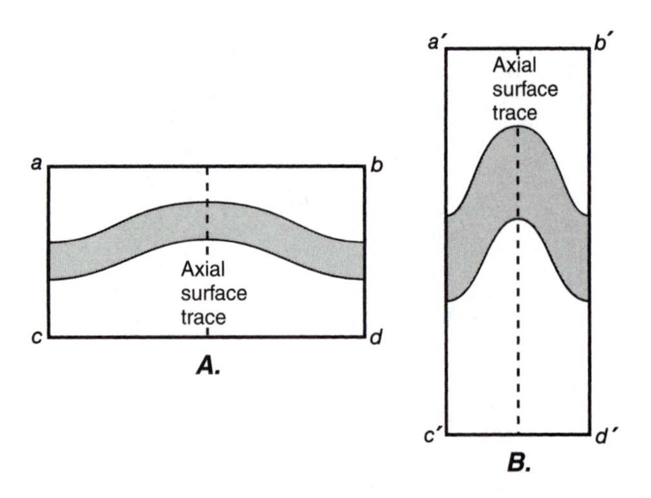

FIGURE 13.13 Intensification of a class 1B fold by homogeneous flattening normal to the axial surface. *A*. The initial fold within the rectangle *abcd* is a gentle class 1B fold. *B*. Progressive homogeneous flattening changes rectangle *abcd* into rectangle *a′b′c′d′*, increases the fold amplitude, and changes the fold into a class 1C fold.

13.4 HOMOGENEOUS FLATTENING OF FOLDS IN A LAYER

i. Kinematic Models

With the models of folding considered so far, we have succeeded in producing folds with a class 1B geometry (flexural folding, volume-loss folding), class 1C geometry (volume-loss folding), and class 2 geometry (passive shear folding). Other kinematic models of deformation also can produce some of these classes, as well as the other fold classes in Ramsay's fold classification. We consider here the effects of two-dimensional homogeneous flattening (Figures 12.1*A*, *C* and 12.11*A*) superimposed on folds formed by the mechanisms discussed in Sections 13.1–13.3. This geometry of deformation could be important in the formation of structures, because, for example, flexural folding can only accommodate a limited amount of shortening before the folds are so tight that they cannot take up any further shortening. The model for passive shear folding, moreover, does not permit any shortening whatsoever normal to the shear planes.

It is impossible to create a fold in a perfectly flat layer by a homogeneous flattening, because in any homogeneous deformation, such a layer remains planar with parallel surfaces (Figure 12.1*A*, *C*). Homogeneous flattening can, however, amplify an initial irregularity and change the geometry of a fold. An initial class 1B fold in a layer is contained within the rectangle *abcd* (Figure 13.13*A*). During homogeneous flattening with the maximum shortening direction normal to the axial surface, any part of the layer not exactly parallel to the principal direction of shortening is rotated away from that direction. In the fold hinges, where the layer is parallel to the principal shortening direction, the layer thickens in proportion to the change in length of the vertical sides *ac* and *bd* (Figure 13.13*B*). In the limbs of the fold, when the layer rotates to a high angle from the principal shortening direction, its thickness decreases in a manner analogous to the decrease in length of the horizontal sides *ab* and *cd* (Figure 13.13). Thus, after deformation, the layer is no longer of constant orthogonal thickness but is thicker in the hinge zone than in the limbs. The dip isogons still converge, so the curvature on the concave side of the fold is still greater than on the convex side. The resultant fold is a class 1C fold (Figure 13.13*B*).

Consider now an initial fold of class 2 style contained within a rectangle *abcd* (Figure 13.14*A*). If the axial surface trace is initially parallel to the vertical sides *ac* and *bd*, it remains parallel to these sides throughout the deformation. Although the axial trace thickness changes during the deformation, the change is everywhere the same and is proportional to the change in length of sides *ac* and *bd*. Thus the initial class 2 fold remains a class 2 fold during a homogeneous flattening for which the maximum shortening direction is normal to the axial surface.

If a fold undergoes a homogeneous flattening with the maximum shortening direction parallel to its axial surface and perpendicular to its hinge (Figure 13.15*A*), the layer thickness decreases in the hinge area and increases on the limbs (Figure 13.15*B*). Dip isogons still converge,

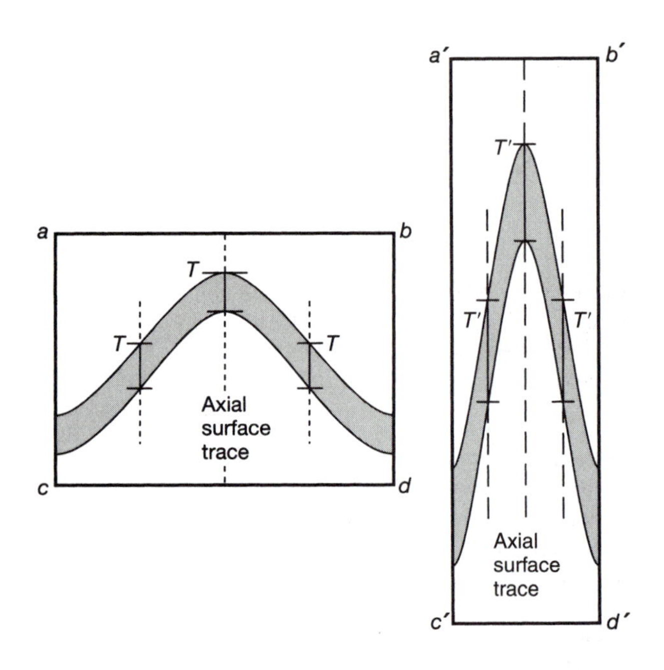

FIGURE 13.14 Intensification of a class 2 fold by homogeneous flattening normal to the axial surface. *A*. The initial rectangle *abcd* containing a class 2 fold. *B*. The deformed rectangle *a′b′c′d′*. The axial trace thickness *T* is changed by the deformation to *T′* but remains equal all around the fold. The fold remains class 2.

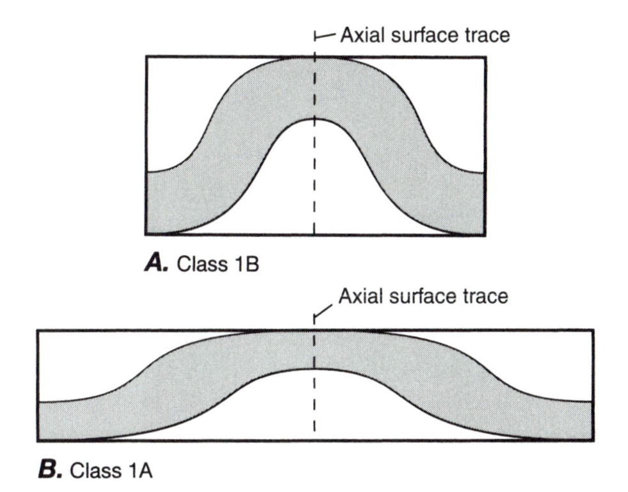

FIGURE 13.15 Homogeneous flattening of a class 1B fold parallel to the axial surface and perpendicular to the fold axis. *A*. A box containing a class 1B fold. *B*. The homogeneously deformed box showing that the fold is transformed into a class 1A fold.

but the orthogonal thickness increases from hinge to limb. The resulting geometry is a class 1A fold.

ii. Strain

If homogeneous flattening is superimposed on a fold with the maximum shortening direction perpendicular to its axial surface, $\hat{s}_1$ axes rotate toward the plane of homogeneous flattening, which, by assumption in this case, is the axial surface of the fold. We consider the effects on the strain for two kinematic models of folding: bending by orthogonal flexure and passive shear.

Figure 13.16 illustrates the effect on the strain distribution of superposed flattening normal to the axial surface of a simplified orthogonal flexure fold of constant curvature formed by bending. Note that a fold of constant curvature is not possible in nature because in a fold train, a discontinuity in strain would occur at the inflection surface where the strain in the profile plane must be zero (compare Figure 13.16*A* with 13.6*B*(i)). Flattening of the initial fold (Figure 13.16*A*) by 20% (Figure 13.16*B*) produces orientations of $\hat{s}_1$ axes somewhat similar to those in a flexural shear fold (compare Figure 13.16*B* with Figures 13.6*C*(i) and 13.7*C*(i)). The flattened fold is class 1C, however, and the flexural shear fold is class 1B. Flattening by 50% (Figure 13.16*C*) rotates the $\hat{s}_1$ axes subparallel to the axial surface and the fold approaches class 2 geometry (see Figure 13.17). Note that, in general, the dip isogons do not parallel the $\hat{s}_1$ axes of the strain ellipse (Figure 13.16*B*), although the difference in orientation may not be obvious if the amount of flattening is large (Figure 13.16*C*). Other types of flexural folding show similar effects.

Figure 13.17 shows the effects of superposed flattening normal to the axial surface of a class 2 fold formed by passive shear. Although the fold remains class 2, the $\hat{s}_1$ axes rotate toward the axial surface. The divergent fanning of $\hat{s}_1$ across the fold becomes less pronounced so that at 50% shortening it is difficult in practice to detect any fanning at all.

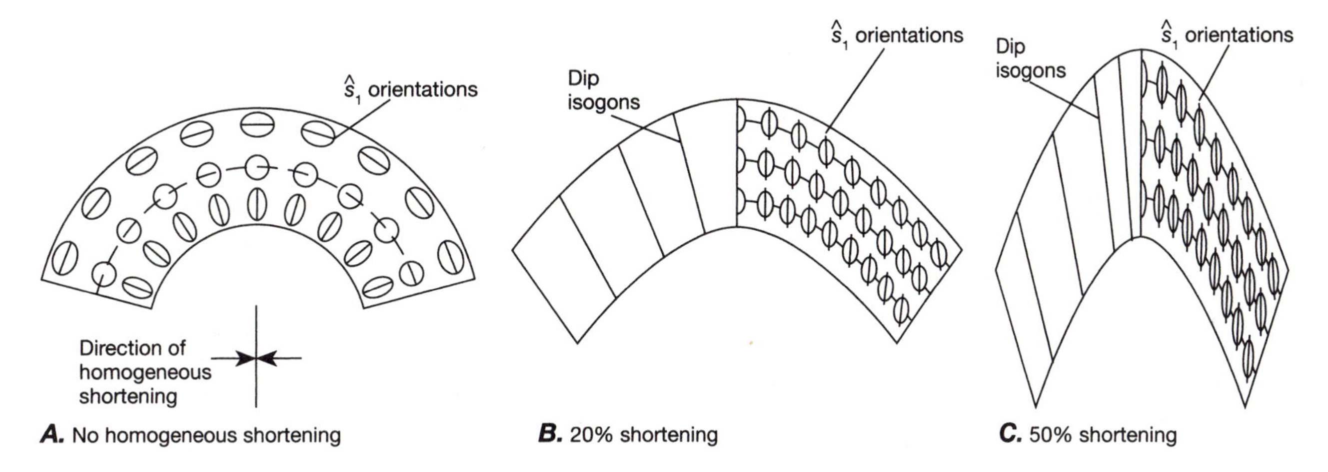

FIGURE 13.16 Homogeneous flattening normal to the axial surface of an orthogonal flexure fold formed in a layer by bending. In *B* and *C*, dip isogons are plotted on the left limbs and strain distributions on the right, with the line through the ellipses parallel to $\hat{s}_1$. (From Hobbs 1971) *A*. Bending of a plate by orthogonal flexure. The fold is class 1B. This simple fold form characterized by constant curvature would give rise to strain discontinuities at the inflection surface if it were part of a fold train, and thus it is an oversimplified model. A more realistic model, in which the strain diminishes to zero at the inflection surface, is shown in Figure 13.6*B*(i). *B*. 20% homogeneous shortening normal to the axial surface of the fold in *A*. The fold is class 1C. *C*. 50% homogeneous shortening normal to the axial surface of the fold in *A*. The fold is class 1C.

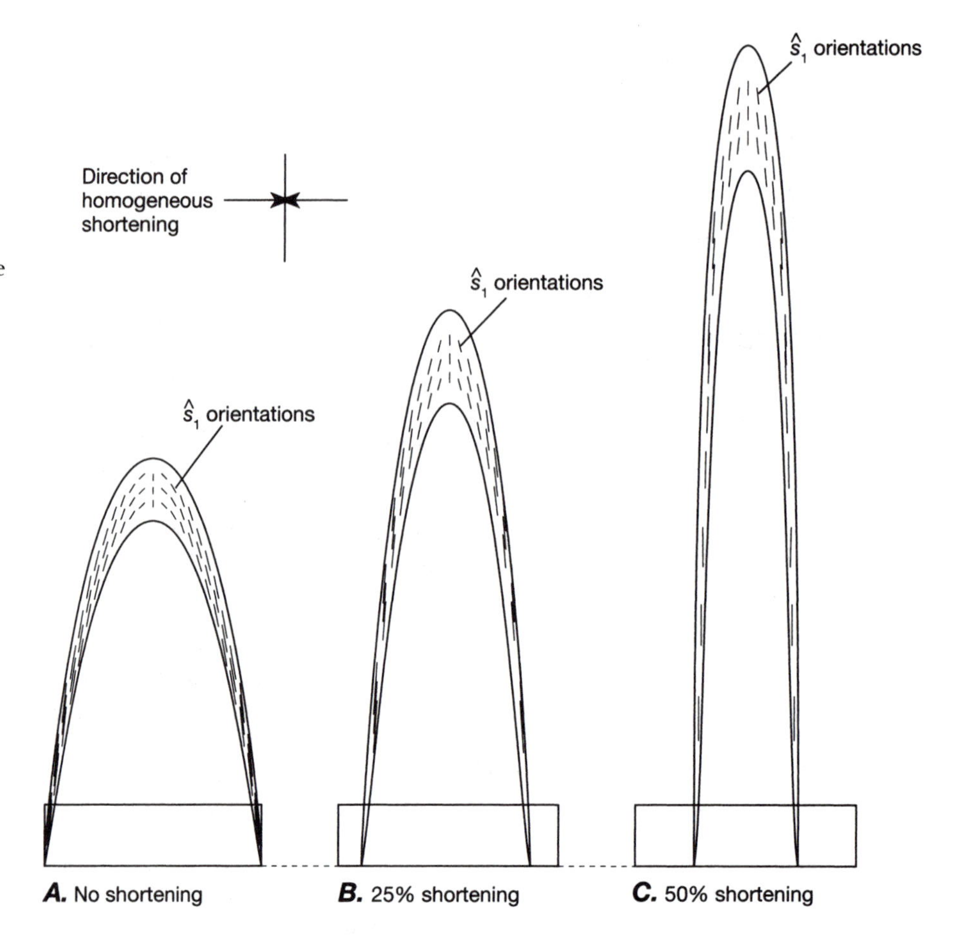

FIGURE 13.17 Homogeneous flattening normal to the axial surface of a class 2 passive shear fold. The lines in the folded layer show the orientations and relative magnitudes of $\hat{s}_1$; the undeformed length of the line is shown at the hinge of the fold in *A*. The initial dimensions of the undeformed plate are shown as the blank rectangle. (After Hobbs 1971) *A*. A class 2 fold formed by passive shear of the undeformed plate. *B*. A class 2 fold formed by 25% shortening of the fold in *A*. *C*. A class 2 fold formed by 50% shortening of the fold in *A*.

iii. Lineation Rotation

Superposed homogeneous flattening destroys the simple relationship between a flexure-folded lineation and a fold axis. With pronounced flattening, lineations that initially lie along a small circle about the fold axis, in general, are reoriented so that their distribution approaches a great circle, which represents a planar distribution.

Superposed homogeneous flattening of a lineation rotated by passive shear folding preserves its distribution in a plane, although the plane is rotated toward the superposed plane of flattening. The construction for determining the orientation of the slip direction of the passive shear folding (Figure 13.10) is still valid. The slip direction, however, is also rotated by the superposed flattening, unless it is parallel to one of the principal axes of the flattening strain.

Thus the geometry, strain distribution, and lineation orientations of a strongly flattened class 1B fold approach those of a class 2 fold. With incomplete or inexact data, the lineation distribution on a flattened class 1B fold could be mistaken for a class 2 distribution and erroneous conclusions may be drawn regarding the mechanism of folding.

13.5 FOLDING OF MULTILAYERS

i. Flexural and Passive Shear Folding of Multilayers: The Effects of Layer Competence

Most natural folding involves multilayered sequences of rocks that have a more complex folding geometry than single layers. An important cause of this complexity is the difference in mechanical properties that can exist between adjacent rock layers. We take qualitative account of this factor in our kinematic models of fold formation by considering the mean competence for the whole multilayer and the contrast in competence among individual layers (Figure 13.18). First we consider a simple fold model involving many layers of essentially the same high competence (high mean competence, low competence contrast). Then we consider the effect of alternating thin incompetent and thick competent layers (high mean competence, high competence contrast). Finally, we consider the effect of increasing the relative proportions of incompetent to competent material in the multilayer (decreasing mean competence, high competence contrast).

If the competence contrast is zero, then the multilayer behaves as a single layer according to the models dis-

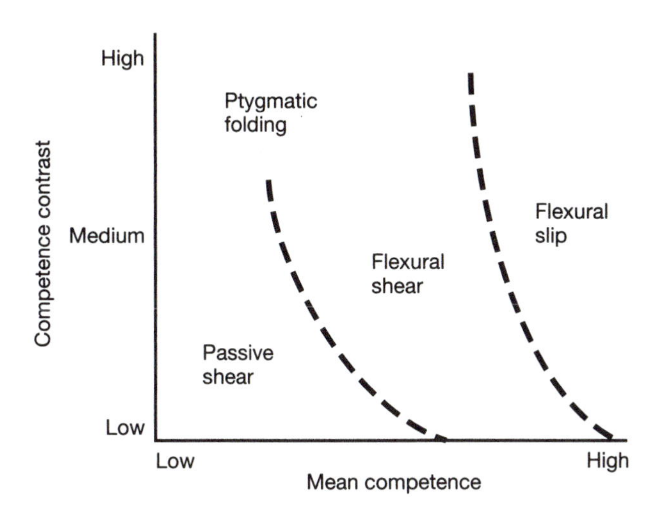

FIGURE 13.18 The dependence of the kinematic model of multilayer folding on the mean competence of the multilayer and the contrast in competence between adjacent layers. (Modified after Donath and Parker 1964)

cussed in Sections 13.1 through 13.4. Even for sequences of layers of different competence, however, the package of layers may behave like a single unit with an effective thickness greater than any one of the individual layers. In that case, if a neutral surface develops within the package, layers on the convex side of the neutral surface may be stretched and thinned at the hinges, giving them a class 1A geometry, another mechanism by which this fold geometry may be produced.

A stack of layers of can respond to either bending or buckling by **flexural-slip folding** if the layers have essentially the same high competence (high mean competence) and if the friction between the layers is relatively low, allowing them to slide freely (effectively a high competence contrast between the layer interfaces and the layer interiors) (Figures 13.18 and 13.19). If each layer folds by orthogonal flexure, the concave side of each layer is shortened, and the convex side is lengthened. Thus across a bedding surface, the layer on the convex side must slip toward the fold hinge relative to the layer on the concave side (Figure 13.19*A*). This relative slip between layers is greatest on the limbs and decreases to zero at the hinge line, where it changes shear sense. The geometry of deformation is similar to flexural shear folding (Figure 13.5) except that in flexural shear folding, the shear is distributed uniformly across the folding layers, whereas in flexure-slip folding it is concentrated along the interfaces between layers. This type of folding produces a class 1B fold.

Sliding of the layers past one another commonly results in the development of linear striations or mineral fibers (slickenside lineations or slickenlines) on the bedding surfaces perpendicular to the fold axis. The lineations are best developed on the limbs where the slip is a maximum, and they do not develop at all at the hinge (Figure 13.19*A*). The lineations such as ℓ', ℓ'', and ℓ''' labeled on the fold in Figure 13.19*A* plot on a stereonet along a great circle perpendicular to the fold axis f (Figure 13.19*B*).

If some degree of flexural shear occurs during folding (moderate mean competence, moderate competence contrast), then some of the potential slip between layers can be taken up by shear within the layer, and the amount of interlayer slip decreases. If all the slip is distributed

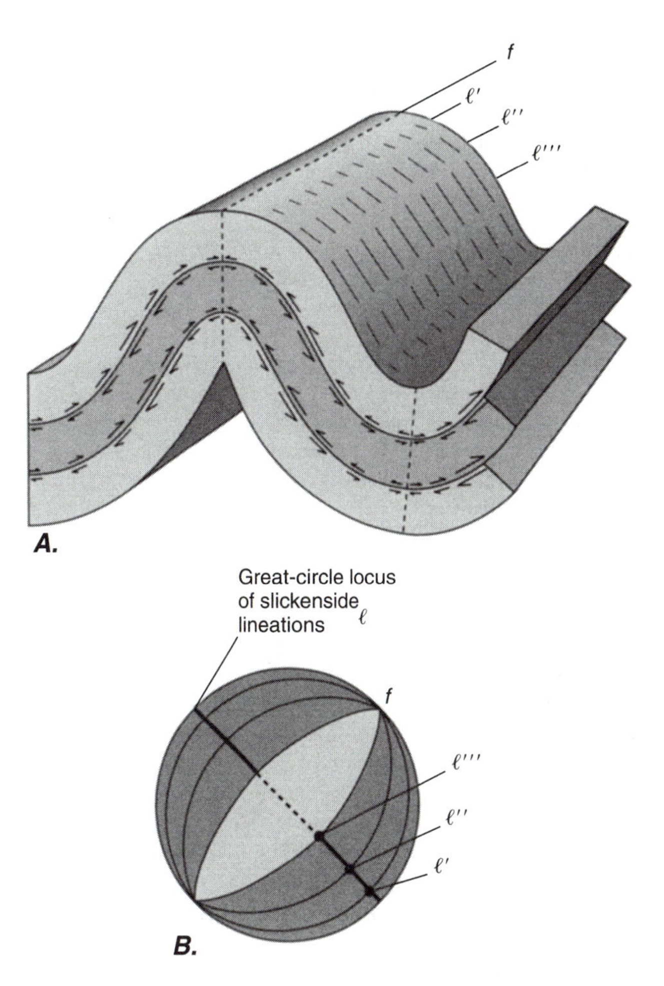

FIGURE 13.19 Flexure slip folding in a multilayer. *A*. Fold formed from an originally planar multilayer, showing relative displacement on layer surfaces. Layers on the convex side of a surface slip toward the hinge line relative to those on the concave side. The shear magnitude decreases to zero at the hinge surface and is of opposite sense on opposite sides of the surface. The lines on the surface of the layer indicate the orientation of slickenside lineations, and their lengths indicate relative amounts of slip. *B*. A stereonet diagram showing the range of orientations of the folded surfaces (great circles in the darkly shaded region) and the orientations of the lineations in those surfaces. The lineations lie on the solid black portion of the great circle normal to the fold hinge *f*. The labeled lineations correspond to those shown in *A*.

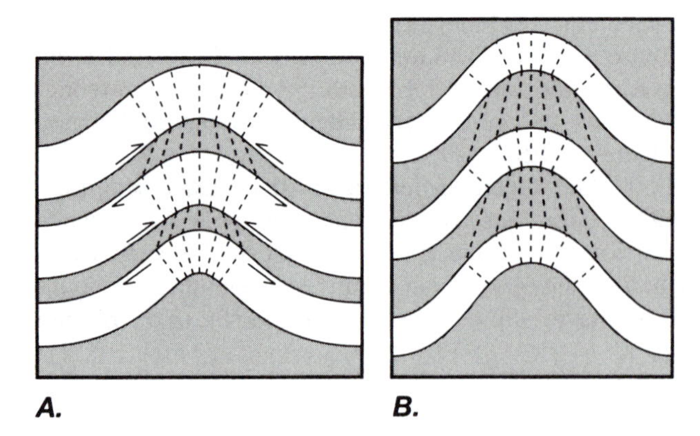

FIGURE 13.20 Flexural folding of interbedded competent and incompetent layers. *A.* A multilayer comprising three competent layers (unshaded) separated by thin incompetent layers (shaded). The dashed lines are dip isogons. Flexural folding of the competent layers is accommodated in the incompetent layer by shearing on the limbs, indicated by the arrow pairs, and flattening roughly normal to the axial surface in the hinge of the fold. *B.* Flexural folding of a multilayer in which the incompetent layers are comparable in thickness to the competent layers. The dip isogons (dashed lines), averaged across the multilayer, are parallel to the axial surface, indicating the fold is a multilayer class 2 fold. In detail, the multilayer fold comprises class 1B folds in the competent layers alternating with class 3 folds in the incompetent layers.

within the layers (moderate mean competence, zero competence contrast), the result is simply a multilayer flexural shear fold with class 1B geometry.

Many folds, however, consist of interlayered competent and incompetent lithologies of comparable thicknesses (moderate mean competence, high competence contrast). The competent layers as a group deform by flexural slip folding, and the interlayer slip is taken up by deformation in the incompetent layers (Figure 13.20*A*). A multilayer class 1C fold develops, as indicated by the fact that the dip isogons, averaged across the layers in the multilayer fold, converge toward the concave side of the fold, but not as strongly as for a class 1B fold. On the limbs of the fold, the incompetent layers are strongly sheared, whereas in the hinge zone they are simply flattened.

As the thickness of the incompetent layers increases relative to the competent layers (decreasing mean competence), the requirement that the adjacent competent layers nest tightly against one another becomes less stringent. Competent layers still fold by the flexural folding mechanisms discussed for single layers, and they still dominate the development of the fold. The curvatures of the adjacent surfaces of two competent layers, however, need not be the same because the incompetent layer between the two competent layers flows in whatever manner is required to accommodate the difference in geometry. This flow generally involves layer-parallel shearing on the limbs and flattening in the hinge zone (Figure 13.20*B*). Thus the dip isogons, averaged across the multilayer, are not strongly convergent and could be parallel or even divergent. In an incompetent layer, the fold has a smaller radius of curvature on its convex side than on its concave side, the dip isogons diverge, and the axial trace thickness T_{α} decreases from hinge to limb (Figure 13.20*B*). These features characterize single-layer class 3 folds, for which other models so far have not accounted.

Thus in an interlayered sequence of competent and incompetent layers, the class of fold that develops in the individual layers generally alternates between class 1B folds in the competent layers and class 3 folds in the incompetent layers. The pattern of the dip isogons averaged over a number of layers therefore may be convergent, parallel, divergent, or irregular, and this pattern determines the actual class of the multilayer fold. For the first three possibilities, the fold forms a multilayer class 1C, 2, or 3 fold, respectively.

If the incompetent layers are much thicker than the competent layers (low mean competence, high competence contrast), they dominate the large-scale deformation. The spacing between the competent layers is so large that flexural folding of one competent layer does not affect the adjacent one, and disharmonic ptygmatic folds develop (Figures 13.18 and 13.21). Although high-order folds in the individual competent layers are class 1B or 1C, the geometry of the lower-order multilayer folds are dominated by ductile flow of the incompetent layers and are class 2.

If the entire multilayer is made up of incompetent material with negligible difference in competence from layer to layer (low mean competence, low competence contrast), and if the layers do not slip past one another on their interfaces, then the multilayer is mechanically homogeneous and the layers simply act as passive markers of the deformation. Under these circumstances the material should deform by passive shear (Figures 13.8 and 13.10) with homogeneous flattening (Figure 13.14), rather than by bending or buckling, thereby forming class 2 folds.

Thus flexural folding in multilayers requires competent layers and a planar mechanical anisotropy that is provided, for example, by low friction interfaces or thin incompetent interlayers (high mean competence, high competence contrast). Passive shear folding in multilayers requires that an incompetent material dominates the mechanical behavior and that the effect of any competent layers is negligible (low mean competence, high to low competence contrast).

ii. Strain in Multilayer Folds

The pattern of strain in folded multilayers depends on the mechanisms by which the individual layers accom-

A.

B.

FIGURE 13.21 Folding of a multilayer in which the incompetent layers are much thicker than the competent layers. Fold geometry is dominated by flow in the incompetent layers. *A*. A diagram of ptygmatic folds in thin competent layers in a fold whose geometry is dominated by flow of the incompetent material. *B*. Photograph of black amphibolitic layers ptygmatically folded in a metasedimentary rock from the Matterhorn Peak roof pendant, Sierra Nevada, California. The geometry is similar to that shown in *A*.

modate the folding. If the competence contrast in the multilayer is negligible, then the strain distribution in the multilayer will conform to one of the patterns we have already discussed for flexural folding (Figures 13.6 and 13.7), for passive folding (Figure 13.10), or for those folding models with a component of homogeneous flattening added (Figures 13.16 and 13.17).

For the intermediate case, however, of an interlayered sequence of competent and incompetent rocks with layers of comparable thicknesses of each, the competence contrast is high and the mean competence is moderate. The competent layers fold by flexural folding, and these folds control the geometry of the folds in the incompetent layers, which develop into class 3 folds (Figures 13.20*A*, *B* and 10.22*B*, *C*). This requires strong shearing in the limbs of the incompetent layer and dominant flattening normal to the axial surface in the hinge zone. The distribution of $\hat{s}_1$ axes that results from buckle folding of such a multilayer is illustrated in Figure 13.22. A general pattern of convergent $\hat{s}_1$ axes develops across the competent layer folds (Figure 13.22). The fold limbs in the incompetent layer undergo strong shearing parallel to the layer as well as flattening approximately normal to it. In the hinge area, the incompetent layer is strongly flattened where it is compressed between the closing limbs of the adjacent competent layer, and there is some shear along the contact on the concave side of the fold. The resulting distribution of strain shows a divergent fan geometry of $\hat{s}_1$ axes in the incompetent layer (Figure 13.22).

Competent
Incompetent
Competent
$\hat{s}_1$ axis orientations

FIGURE 13.22 Buckling of interlayered competent and incompetent layers of comparable thickness. The short lines indicate the orientation of the $\hat{s}_1$ axes in the multilayer fold.

13.6 FORMATION OF KINK AND CHEVRON FOLDS

Folds with straight limbs and sharp hinges are **kink folds** if they are asymmetric and **chevron folds** if they are symmetric (see Section 10.5(iii); Figure 10.25). They develop in strongly layered or laminated sequences that have a dominant planar mechanical anisotropy, and they accommodate a component of shortening parallel to the layering or laminations.

i. Kink Folds

Kink folds occur in pairs with one short limb connecting two longer limbs (Figure 13.23). A **kink band** is the short

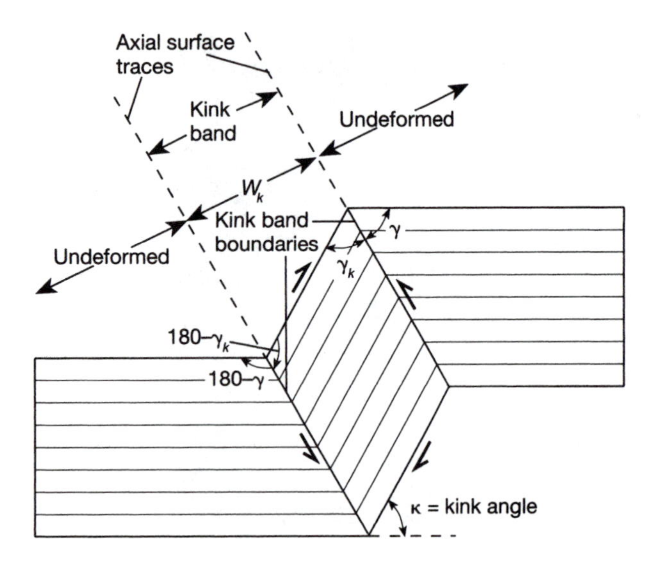

FIGURE 13.23 Geometry of a kink band, illustrating terminology. κ is the kink angle, w_k is the width of the kink band measured normal to the boundary, γ and γ_k are the angles between the kink band boundary (the axial surface) and the undeformed and deformed layers, respectively.

limb between the two axial surfaces, which are the **kink band boundaries**. In the kink band, laminations are deformed and are rotated with respect to the undeformed material by an angle κ called the kink angle. We describe four different kinematic models of kink band formation, each of which involves a component of shearing parallel to the laminations as well as preservation of continuity of the laminations across the kink band boundaries. The models differ from one another in the way in which the kink grows and in the geometry of the resulting deformation.

In two models (Figure 13.24), the kink develops by migration of the kink band boundary into the undeformed material. Folding by the migration of axial surfaces is different from any of the kinematic models of folding we have considered so far, although passage of the kink band boundary through undeformed material is accompanied by shearing of the material parallel to the laminations. Laminations in the undeformed and the kinked parts of the material maintain equal angles with the kink band boundary ($\gamma = \gamma_k$; Figures 13.23 and 13.24), and the line lengths parallel to the laminations, as well as the cross-sectional area, remain constant during kinking. In Figure 13.24*A*, the kink nucleates along a line (*AB*) normal to the laminations and grows by the rotation of the right

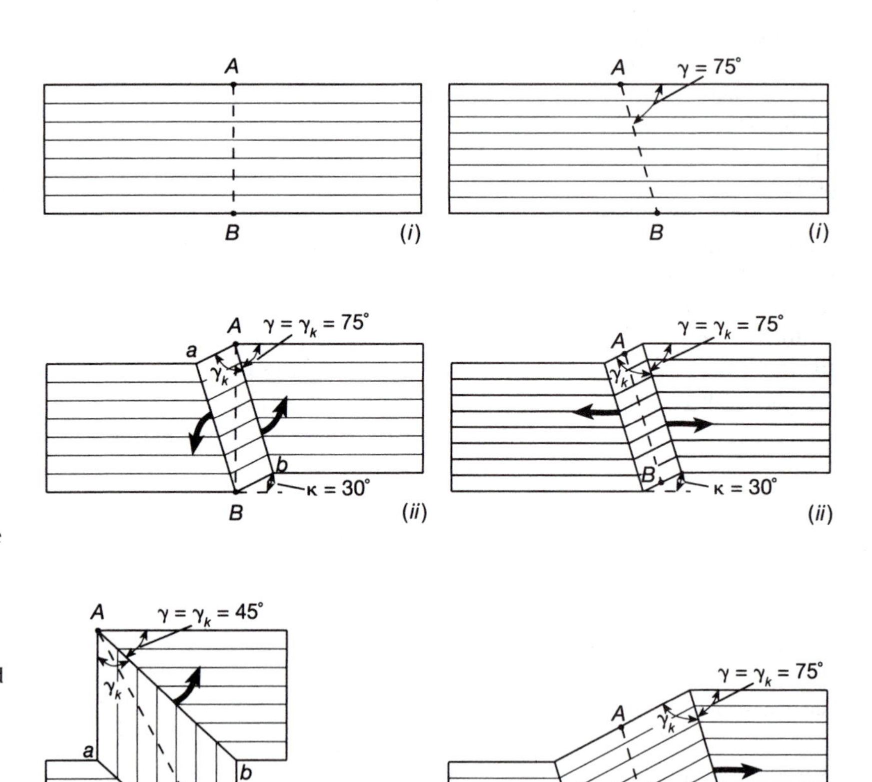

FIGURE 13.24 Kinematic models for the growth of a kink band in which the kink band grows by migration of the kink band boundary through the material. (After Weiss 1980) *A*. The kink band nucleates along the dashed line *AB* (*i*). It grows by rotation of the kink band boundary *Ab* counterclockwise about the fixed material point *A*, and by rotation of the opposite boundary *Ba* counterclockwise about the fixed material point *B* (*ii*) to (*iii*). As the kink band grows, the kink angle κ increases. The angles γ and γ_k both decrease during kink band growth, but they remain equal. *B*. The kink band nucleates along the dashed line *AB* (*i*). The kink band grows by the migration of the kink band boundaries in opposite directions into the undeformed material (*ii*) to (*iii*). As the kink band grows, the kink angle κ remains constant. The angles γ and γ_k remain equal and constant during kink band growth.

boundary (Ab) counterclockwise about A and of the left boundary (aB) counterclockwise around B, while A and B remain fixed points. The kink angle κ increases continuously with kink growth. In Figure 13.24B, the kink nucleates along a line (AB) oblique to the laminations, and the two margins migrate in opposite directions while maintaining the same orientation. The kink angle κ is fixed by the angle γ between the laminations and AB, because in order to preserve lamination line length during kinking, the angle between the laminations and the kink band boundary must be the same in the unkinked and the kinked material ($\gamma = \gamma_k$), and it does not change with growth of the kink band. In both cases, the deformation proceeds at constant volume.

In the two other models (Figure 13.25), the kink band boundaries do not migrate but rather mark the fixed boundaries of a shear zone. As the kink develops, the kink angle κ increases, but the angles γ and γ_k are not equal: γ remains constant whereas γ_k decreases. In Figure 13.25A, kinking produces a deformation equivalent to homogeneous simple shear parallel to the kink band boundaries, and therefore it is essentially like the passive shear model. The width w_k of the kink band measured normal to the kink band boundary is constant, and the laminations are deformable. The laminations first rotate toward an orientation perpendicular to the kink band boundary (Figure 13.25A(ii)), becoming shorter and thicker. With further rotation, the laminations lengthen and thin. The cross-sectional area of the kink band remains constant throughout.

In Figure 13.25B, folding involves shearing parallel to the laminations, and the laminations do not deform,

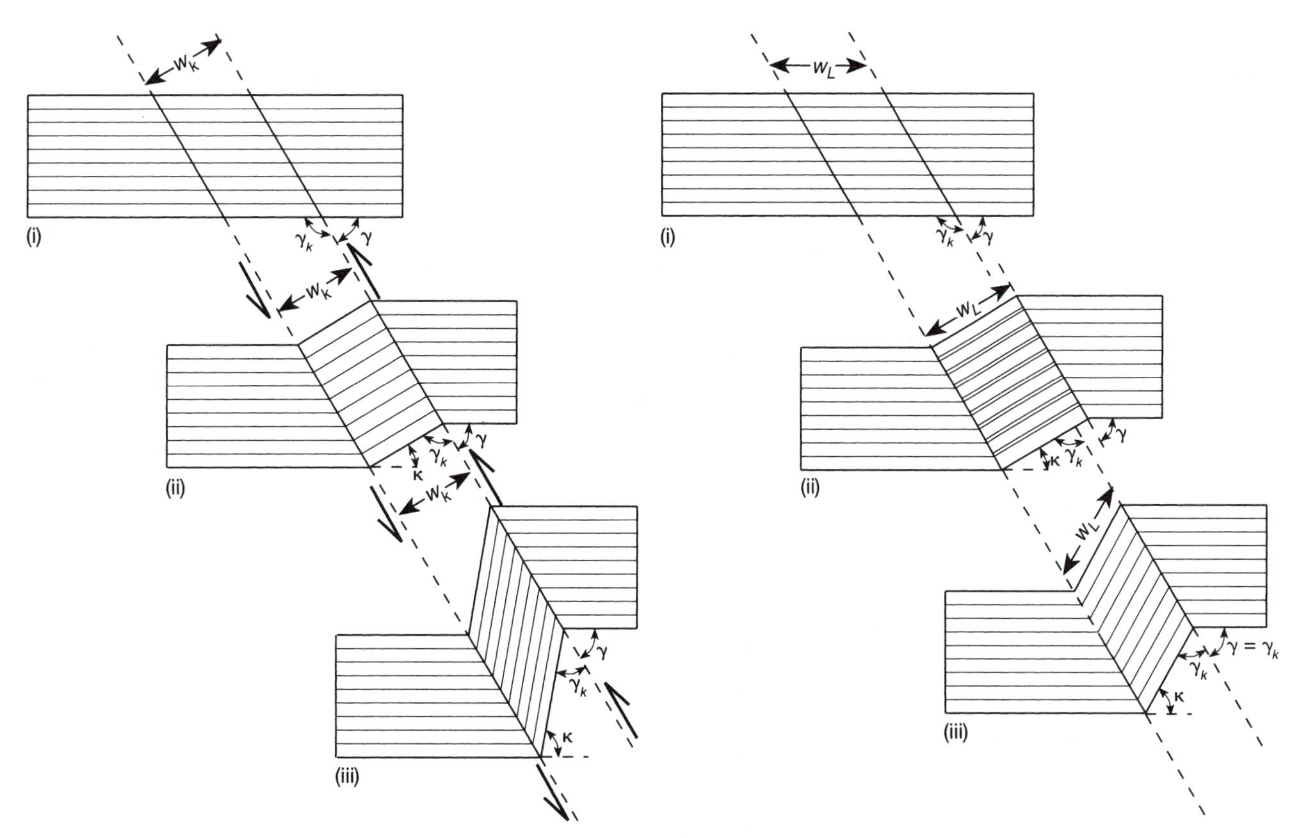

FIGURE 13.25 Models for formation of a kink band in which the kink band boundary remains fixed in the material. (After Ramsay 1967) *A*. Kink band growth by simple shear parallel to the kink band boundaries. The length of the laminations first decreases (*i*) to (*ii*) and their thickness increases until the laminations are perpendicular to the kink band boundary. With further rotation, their length increases and their thickness decreases. The width w_k of the kink band remains constant. As the kink band develops, the kink angle κ increases, γ remains constant, and γ_k decreases. Thus γ and γ_k are not in general equal. *B*. Formation of a kink band by rigid rotation of the laminations. The length of the laminations w_L in the kink band remains constant during kink band formation. Thus from (*i*) to (*ii*), where the laminations become perpendicular to the kink band boundary, the width of the kink band increases and the volume increases by the opening of spaces between the laminations. From (*ii*) to (*iii*) the width decreases and the spaces between laminations close. Thus the deformation is not at constant volume. As the kink angle κ increases, γ remains constant, and γ_k decreases until $\gamma_k = \gamma$ at which point the spaces between laminations are completely closed and further kinking is impossible.

maintaining constant length w_L and thickness. As the kink develops, the laminations rotate toward an orientation perpendicular to the kink band boundary (Figure 13.25*B*(ii)). The width of the kink band, measured normal to the kink band boundary, increases, and gaps open up between the laminations. With further rotation, the width of the kink band decreases again, and the gaps between laminations close. γ_k decreases progressively during the kinking, and when it reaches the value of γ, no further kinking is possible (Figure 13.25*B*(iii)). Because of the opening and closing of the gaps between the laminations, the cross-sectional area is not constant during the kinking.

Experiments on kink band formation indicate that kink bands do not develop along planes of high shear stress. Because the third and fourth models assume the kink band to be a zone of shear, the experiments suggest these models may not be appropriate for describing natural deformation. Evidence points most strongly to the operation of the first and second models, either singly or together, in kink band formation. Some natural kink bands, however, show evidence of an increase in volume during deformation, such as growth of later minerals between separated layers. This indicates that in some cases, at least, the fourth model represents a component of the kinking mechanism. Model *A* in Figure 13.25 may account for some kink formation in high-grade metamorphic rocks.

ii. Chevron Folds

Two kinematic models exist to account for the formation of chevron folds. In the first model, chevron folds develop where kink bands of conjugate orientation intersect (Figure 13.26). Transformation of the entire undeformed body into one completely filled with chevron folds requires a shortening of 50% (Figure 13.26*C*). Although this mechanism has been observed to operate during the experimental deformation of phyllites, in naturally deformed rocks the observed shortening resulting from kink folding rarely exceeds 25%, which is insufficient to form chevron folds by this mechanism.

Chevron folds can also develop by a process that is similar to flexural shear folding (Section 13.1(ii)). In this case, however, because the idealized laminations of the model are infinitesimally thin, the radius of curvature of the hinge does not have to change along the axial surface as it does for flexural shear folding of beds of finite thickness (cf. Figure 13.27). The result is a class 2 chevron fold formed by a flexural shear mechanism.

iii. Kink or Chevron Folding of Layered Sequences

Our idealized models have assumed the kinked material to be made of infinitesimally thin laminations and that shearing on the laminations results in a homogeneously distributed deformation. Such a condition is most closely approached in nature by foliated rocks such as slates, phyllites, or schists. Kink and chevron folds, however, also occur in thinly bedded rocks such as interbedded chert and slate. Whether a symmetrical chevron fold or an asymmetrical kink fold forms depends on whether the direction of maximum shortening is, respectively, parallel or oblique to the layering. The geometry of a chevron fold formed in a layered sequence having low friction bedding planes is illustrated in Figure 13.27. Formation of a class 2 fold by class 1B folding of the individual layers requires the opening of voids between the layers at the hinge. When such voids are filled with secondary mineral deposits, they are called **saddle reefs**. If the com-

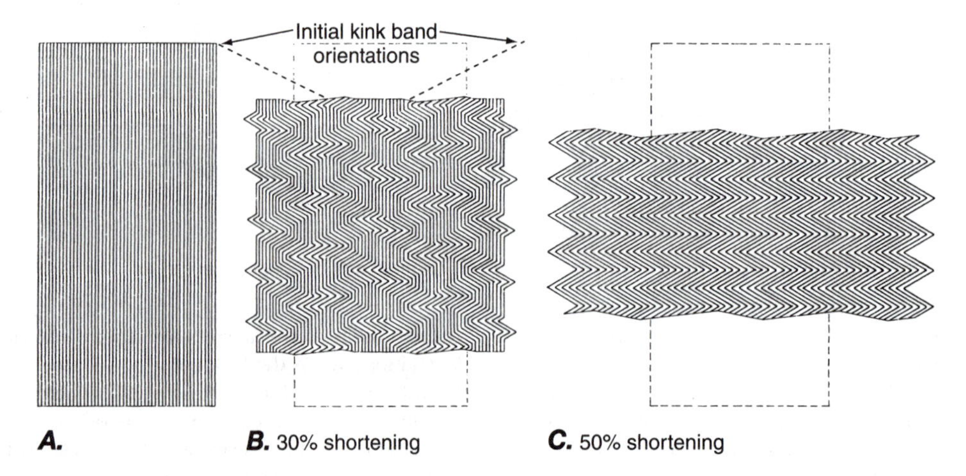

FIGURE 13.26 Development of chevron folds by kinking. (From Patterson and Weiss 1966) *A*. An undeformed block with a strong planar mechanical anisotropy. *B*. Shortening of a block parallel to the plane of weakness results in the formation of two sets of kink bands that have conjugate orientations. Chevron folds develop at the zones of interference between conjugate kink bands. *C*. As the widths of the conjugate kink bands increase, the area of interference, where the chevron folds develop, also increases until the entire block is filled with chevron folds.

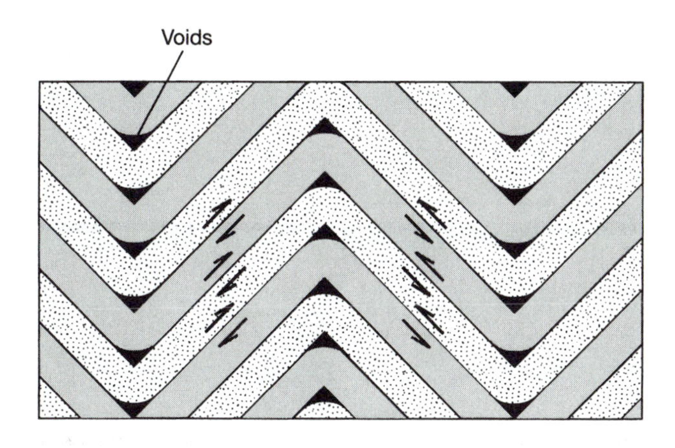

FIGURE 13.27 Chevron folding of layers of finite thickness by the flexural slip mechanism introduces voids in the hinge zone (black areas).

petent layers are separated by incompetent material instead of low friction interfaces, the incompetent material may flow from the limbs to the hinge zone to accommodate the mismatch in the fold form of the competent layers. This process once again produces multilayer class 2 fold geometry by alternate class 1B and class 3 folds in the competent and incompetent layers, respectively, as described in Section 13.5.

Our kinematic models do not explain why and under what conditions different folding mechanisms should operate. For example, they cannot resolve the question of why in some cases rounded folds form whereas in other cases chevron folds develop. In fact, we have not even explained why folds form at all instead of the layers simply becoming shorter parallel to the layer and thicker perpendicular to it. To approach these questions, we must consider the mechanics of fold formation, which involves the mechanical properties of the material and the relations between the stress and the deformation. We discuss the mechanics in Chapter 16 and Section 18.1 and the application to folding in Sections 18.6 and 18.7.

iv. Strain in Kink and Chevron Folds

For kink bands formed by any of the mechanisms of Figures 13.24 and 13.25, the acute angle between the maximum finite stretch axis $\hat{s}_1$ and the kink band boundary is less than 45° (Figure 13.28; cf. Figure 12.15). The kink angle κ of the layers, however, is not the angle of shear ψ of those layers in the kink band. κ measures the change in angle between two material lines, the laminations and the kink band boundary, but those two material lines are not perpendicular in the undeformed state. In Figure 13.28*A*, the heavy solid and dashed lines through the circle are perpendicular, so it is the change in angle between these two lines that defines the shear angle ψ (Figure 13.28*B*).

If kink bands form by either of the mechanisms shown in Figure 13.24, the kink is symmetric about the kink band boundary ($\gamma = \gamma_k$; Figure 13.28), and material lines parallel to the laminations, as well as those parallel to the kink band boundary, all are the same length after kinking as they were before. Thus these two directions must be lines of no finite extension, and the principal axes of the strain ellipse must bisect the angle between these lines. $\hat{s}_1$ therefore bisects the angle between the laminations in the kink band and the boundary of the kink band.

For the model of kinking in Figure 13.25 on the other hand, shearing accumulates progressively in the kink band, so the orientations of the principal axes are determined by the geometry of simple shear and are not uniquely related to the laminations or kink band boundary (cf. Figure 12.15). The mechanism illustrated in Figure 13.25*B* has the added complication of including a widening, and thus a volume increase, of the kink band until the laminations are perpendicular to the boundary, followed by a narrowing and volume decrease of the kink band until the kink band returns to its original width, at which point the kink is symmetric ($\gamma = \gamma_k$). This deviation from simple shear affects the orientations of the principal strain axes (cf. Figure 12.24; see also Figure 15.17).

13.7 FAULT-BEND AND FAULT-PROPAGATION FOLDING OF A MULTILAYER

Folds associated with thrust faults can be distinguished according to whether they result from folding above a fault ramp, from folding associated with the propagation of a thrust fault, or from buckling above a decollement. Fault-propagation folds develop ahead of a fault tip line to accommodate the deformation associated with the slip on the fault. Much research has been devoted to understanding the geometries of the folds formed in these various situations, and in this section, we review briefly two of the most successful kinematic models that have been used to account for such structures. Both models are two-dimensional and thus do not account for the flow of material into or out of the plane of the model. Such material motion in the third dimension does, in fact, occur, but it is generally a second-order phenomenon, so the two-dimensional models are useful.

i. The Kink Fold Model

In Chapters 4 and 5 concerning normal and thrust faults, we describe how a change in dip of the fault, for example, at a fault ramp or a bend in the fault surface, results in folding of the hanging wall block where it rides over the bend. Such folds associated with normal faults include the rollover anticlines (Figure 4.3) and fault-bend folds (Figure 4.5). Folds associated with thrust faults include fault-bend folds (Figures 5.5 and 5.6) and the complex anticlinal stacks above thrust duplexes (Figures 5.14

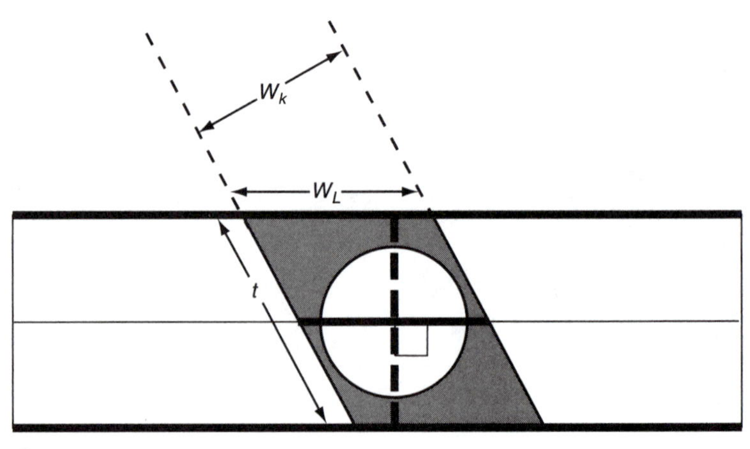

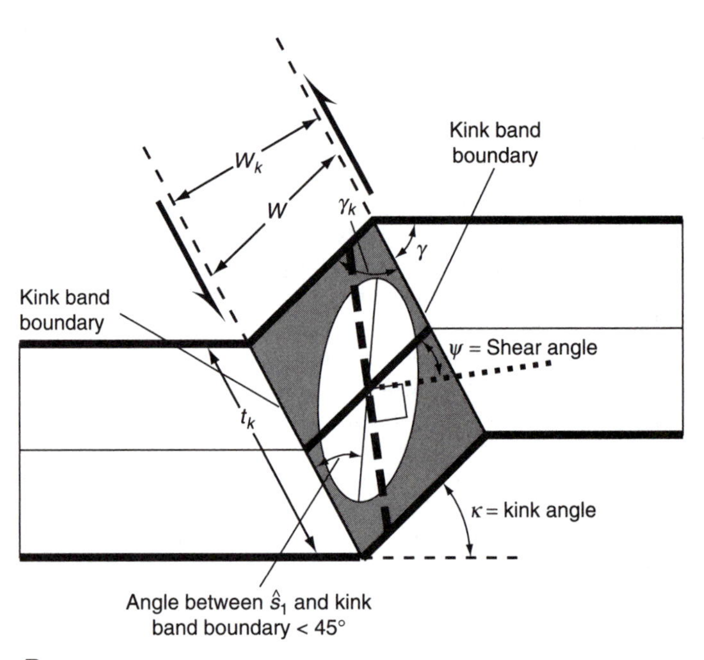

FIGURE 13.28 Strain orientation in a kink fold. *A.* The undeformed layer with a material circle (white) in the shaded area to be kinked. *B.* The kinked layer with the strain ellipse. The shear strain angle ψ of the kinked layer surface is not the same as the kink angle κ. The axis of maximum stretch $\hat{s}_1$ makes an angle of $<45^\circ$ with the kink band boundary. If the kink is symmetric ($\gamma = \gamma_k$), the thickness of the layer measured parallel to the kink band boundary does not change ($t = t_k$), and the width of the kink measured parallel to the layer in the kink band is the same before and after the kinking ($w = w_L$). Thus these two directions are directions of no finite stretch, and the $\hat{s}_1$ axis must bisect the angle γ_k between them.

and 5.15). We can explain the fold geometry in many natural examples of fault-bend folding by using some fairly simple assumptions concerning the kinematics of layers being displaced over fault bends. Such constraints have proved extremely useful in developing models to interpret some of the structures in fold and thrust belts and normal faulted terranes, as well as to create models for the development of fault-propagation folds.

The kinematic analysis requires the following assumptions:

1. No gaps are introduced as a result of slip along the fault plane;
2. Fault bends are sharp;
3. The orthogonal thicknesses of layers in the deformed block are preserved;
4. The lengths of layers in the deformed block are preserved; and
5. Layers that have not been transported across a fault bend are undeformed.

These assumptions imply that the resulting folds are kink folds (cf. Figure 13.24*B*) characterized by straight limbs, sharp hinges with a bluntness of zero, axial surfaces that bisect the angle between the kinked and unkinked layers, layer-parallel shear in the kink band, and a multilayer class 2 (similar) fold geometry. Note that because of its sharp hinge, a kink fold is the one geometry for which flexural folding can produce a class 2 fold. In nature, the model flexural shear that is distributed uniformly through a layer in reality may be approximated by localized flexural slip along layer boundaries. Nevertheless, because

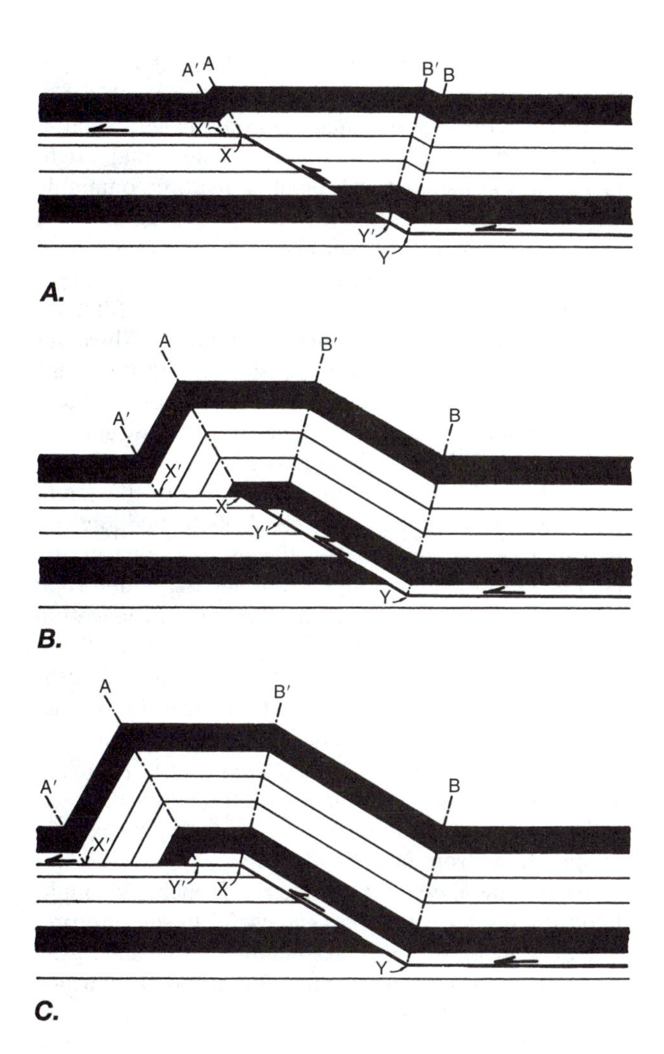

FIGURE 13.29 Kink model for the development of a fault-bend fold at a simple fault ramp. (From Suppe 1983) *A.* Initiation of the fault-bend fold by symmetrical kinking ($\gamma = \gamma_k$; cf. Figure 13.24*B*). *B.* Hinge surfaces A and B are fixed to the footwall at the top and bottom of the ramp, respectively. Hinge surfaces A′ and B′ are fixed to the hanging wall block and migrate with it. *C.* Hinge surface B′ reaches the top of the ramp and becomes fixed to that point in the footwall block, at which point hinge surface A becomes fixed in the hanging wall block. Hinge surface B remains fixed to the bottom of the ramp in the footwall block. Hinge surfaces A and A′ migrate with the hanging wall block.

of these restrictive assumptions, the relation between the fold geometry and the geometry of the fault and fault ramp can be quantitatively derived, and thus the evolution of these fault-bend fold models can be strictly defined. As a result, precise models can be constructed to interpret field data.

Fault-Bend Folding. As an example, Figure 13.29 illustrates the two phases in the development of a fault-bend fold at a simple ramp connecting horizontal decollements. At the initial increment of displacement on the fault, two kink bands form, with kink band boundaries A and A' for one and B and B' for the other. In the first phase of development, axial planes A' and B' are fixed at X' and Y', respectively, in the hanging wall block, and they migrate with the block as displacement accumulates on the fault. Axial planes A and B are fixed at X and Y, respectively, in the footwall block. Thus as displacement continues, material in the hanging wall block migrates through the axial surfaces A and B, and the kink folds grow (cf. Figure 13.24*B*). This phase of development continues until the point Y' in the hanging wall block, to which axial plane B' is attached, reaches the point X at the top of the ramp. At this instant, the fold reaches its maximum amplitude, the axial surface B' becomes fixed at the point X in the footwall block, and the axial surface A becomes fixed to the point Y' in the hanging wall block. With further displacement on the fault, the second phase of development begins. Axial planes A and A' are fixed in the hanging wall block and migrate with it. Axial planes B and B' are fixed with respect to the footwall block at the bottom and top of the ramp, respectively. Material in the hanging wall block migrates through these fixed axial planes, becoming sheared as it passes through B, and unsheared as it passes through B'. The assumed geometry of kink folding leads to a unique equation between the cutoff angle of the footwall bedding against the fault ramp and the folding angle of the front limb of the fold.

More complex models can also be treated, such as the development of fault-bend folds above imbricated thrust faults and duplexes (Figure 13.30). Note that the dips of the layers change abruptly at the axial planes, and that, in this case, the number of stepwise increases in dip at the front and back of the fold indicates the number of fault imbrications at depth. Thus under favorable circumstances the analysis of dip domains on fault-bend folds at the surface can help constrain the geometry of complex fault structures at depth. Details of these techniques are beyond the scope of this discussion.

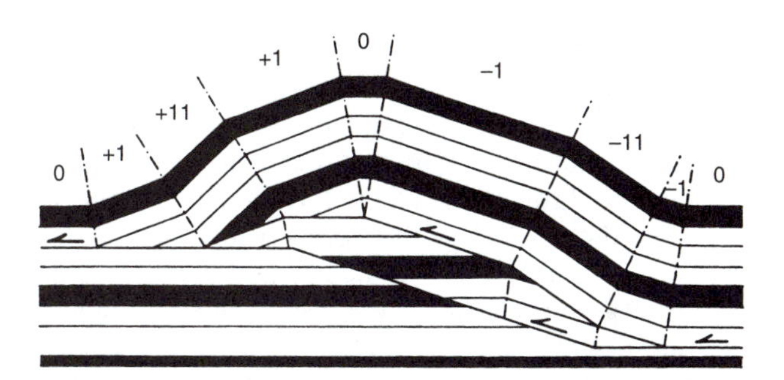

FIGURE 13.30 Kink model for a fault-bend fold resulting from imbrication of a thrust fault at a simple ramp and initiation of a duplex structure at depth. In this case, the number of increments of dip at the front and back of the fold indicate the number of imbrications on the fault. (From Suppe 1983)

This same model applied to normal faults (Figure 4.5) predicts the existence of fault-bend anticlines and synclines that reflect the geometry of the normal fault surface at depth. If we examine the left-hand part of Figure 4.5*B* that includes only the main fault that cuts the surface and the connecting flat, we see a simple model of a listric normal fault with a roll-over anticline. We might expect from the geometry of the kink fold that the deformation associated with folding would be accommodated in brittle rocks by shearing on a set of synthetic reverse faults parallel to both the main fault and the axial surfaces of the kink fold above the flat. Nature, however, does not necessarily pay attention to our neat kinematic models, as indicated in Figure 4.3. Here, the deformation in the hanging wall block required by the change in orientation of the listric normal fault is accommodated by a set of synthetic normal faults, which suggest a counterclockwise rotation of domino-like blocks (cf. Figures 4.20 and 4.23*B*). So models are just hypotheses and are only of use if they can be shown to reflect what actually happens in nature.

Fault-Propagation Folding. When a fault tip line propagates through a body of rock, the slip that occurs on the fault must be accommodated ahead of the tip line by some form of deformation. The kink model can be used to model this deformation as a fault-propagation fold (Figure 13.31). The model requires the fault tip to propagate at a velocity p that is twice the velocity of slip on the fault s ($p/s = 2$). Where the fault tip turns upward from a horizontal decollement to propagate across the layering, a pair of kink folds form with kink band boundaries A and A' for one kink and B and B' for the other (Figure 13.31*A*). The axial plane A' terminates at the tip line of the fault but is not parallel to the fault ramp. Thus it migrates through the material as the fault tip propagates. The kink band between A' and A accommodates the slip ahead of the fault. Axial plane B is fixed relative to the footwall block at the bend in the fault, and displacement on the fault causes material in the hanging wall block to migrate through B. Axial plane B' joins axial plane A at the same stratigraphic level where the fault tip is located. Below this stratigraphic level, folding is complete because further displacement is taken up by slip on the fault, not by folding. Axial planes A and B' also migrate through the material as the fault tip propagates, but the axial plane that forms below the junction of A and B' remains fixed in the hanging wall block and is displaced with it. This model also results in a unique equation between the cutoff angle of the footwall bedding at the fault ramp and the interlimb angle of the front fold limb that is different from the equation for a fault-ramp fold. The kinematics assumed for these kinks requires that the $\hat{s}_1$ axis bisect the angle between the layering and the kink band boundary, as discussed in Section 13.6(iv).

If folding becomes impossible at some point in this process (because, for example, of the resistance of a particular layer to tightening of the fold), the fault may break out and propagate rapidly through the stratigraphic section. If it cuts through between the axial planes A' and A, it leaves a tight syncline in the footwall block and a tight anticline in the hanging wall block, a feature commonly observed in nature. Such folds are often termed **drag folds** and are commonly ascribed to drag of one side of the fault against the other, rather than to fault-propagation folding that precedes the passage of the fault tip (compare Figure 5.10). The sense of curvature of the fault-propagation folds, however, is consistent with the sense of drag that would be expected from the slip on the fault.

Comparison of Figures 13.29 and 13.31 show similarities in the folds formed by fault-bend folding and fault-propagation folding. For fault-bend folding, the maximum possible cutoff angle between footwall bedding and the fault that is permitted by the assumptions of the model is 30°. For higher angles, the necessary deformation violates the assumptions. For fault-propagation folds, the model permits cutoff angles as high as 60°. In general, for a given cutoff angle, the interlimb angle for fault-propagation folds is smaller (and thus the folding angle is larger) than is possible for common fault-bend folding, so the origin of a fault-related fold in principle can be determined. For most cutoff angles that are smaller than 30°, tight folds result from fault propagation and open folds from fault-bend folding.

These kinematic and geometric models for fault-related folding have proved very useful in the interpretation of the deep structure in a number of fold and thrust belts. Such interpretations must be based on surface mapping, well data, seismic data, and regional stratigraphic data. The interpretations, however, are not unique. If any of the assumptions that are required for the model, which are very restrictive, are violated in natural deformation, the model does not provide reliable constraints on the reconstruction, and the distinction between the fault-bend and fault-propagation folds may become blurred.

These kink models assume the fault tip cuts up through the bedding to form a ramp from a horizontal decollement. But fault ramps may nucleate within a bed and then propagate both up and down to join existing decollements, or they may propagate down from an upper decollement to join a lower decollement. Moreover, the p/s ratio does not have to be 2, and it does not even have to be constant along the fault, as is assumed for the model in Figure 13.31. These possibilities lead to other predictions of the relationship between folding and fault ramps.

Beyond these sources of nonuniqueness, there is the observation that beds do not necessarily form kink folds. They may deform by shear that is not parallel to the layers, they may thicken or thin by homogeneous or even inhomogeneous deformation, or the volume of part of the section may be changed by solution of material. Some aspects of nonlayer-parallel shear can be included in the model, but most other types of deformation do not yield unique geometric constraints. In those cases, inconsis-

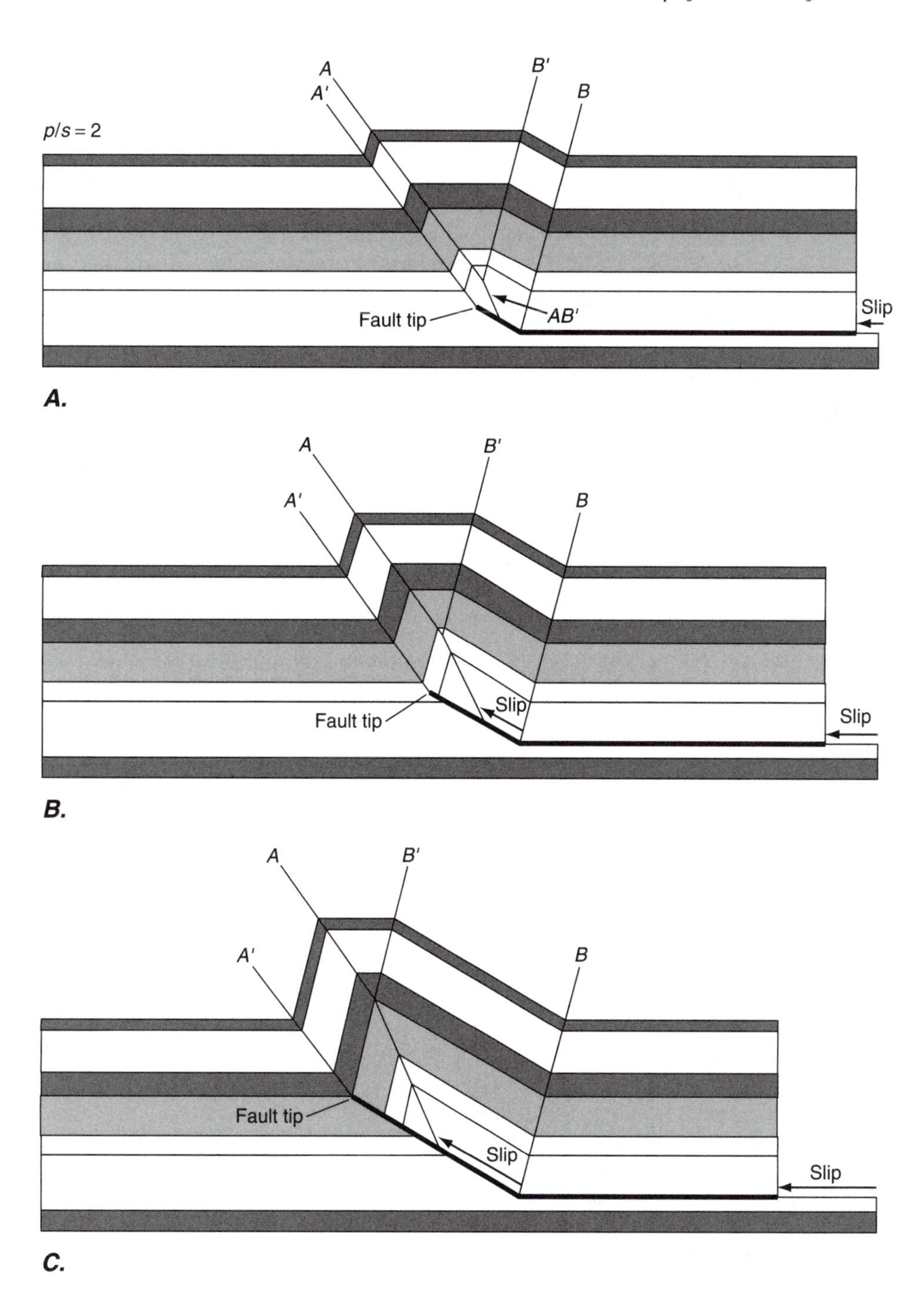

FIGURE 13.31 Kink model for the development of a fault-propagation fold above the tip of a propagating thrust fault. In any given increment, the fault tip propagates twice as far (p) as the fault slips (s) ($p/s = 2$). Hinge surfaces A and B′ intersect at the same stratigraphic level occupied by the fault tip. Hinge surface B is fixed to the bottom of the ramp in the footwall block. (From Suppe and Medwedeff 1990, with permission of Birkhäusser Verlag AG)

tencies in the reconstructions can point to cases in which the assumptions of the model do not apply.

ii. The Trishear Model of Fault-Propagation Folding

The **trishear model** of fault-propagation folding accounts for a variety of structures that cannot be produced by the kink model, but the success is at the expense of the precise and unique geometric reconstructions that the kink model permits. In the trishear model, the slip discontinuity on a fault becomes distributed into a triangular zone of shear in front of the fault having an apex at the propagating fault tip (Figure 13.32). The trishear model derives its name from the shape in two dimensions of this triangular shear zone that develops ahead of the

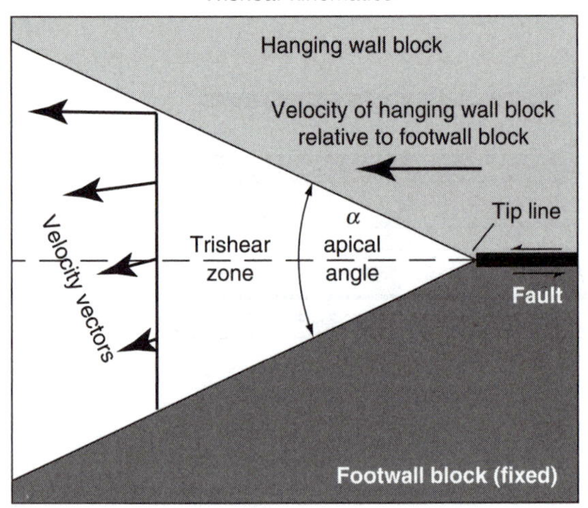

FIGURE 13.32 Characteristics of the trishear model for fault-propagation folding. Discontinuous slip on the fault is distributed from the tip line throughout the triangular trishear zone (unshaded). Velocities in the hanging wall block and along a line in the trishear zone normal to the fault plane are shown relative to a fixed footwall block. The symmetry of the trishear zone about the fault plane orientation and the velocity distribution shown result in a constant volume deformation. (Modified from Allmendinger 1998, figure 1b, p. 641, by permission of American Geophysical Union)

propagating fault tip line. In ductile rocks, deformation in this trishear zone develops by ductile flow. In brittle rocks, a **damage zone** forms in this trishear zone within which the rocks are damaged by distributed shear fracturing. Within this trishear zone, the forelimb of the asymmetric fault-propagation fold develops. This model is especially useful in modeling competent massive lithologies that are not strongly layered and thus do not form kink folds. It can also model asymmetric fault-propagation folds having rounded hinges and forelimbs in which the layers vary in dip and change in thickness. All of these characteristics are commonly present in fault-related folds but are not well represented by the kink model.

The basic kinematics of the model are illustrated in Figure 13.32. The figure shows a fault on the right side of the diagram separating the two fault blocks, and the triangular trishear zone spreading out to the left from the fault tip. The triangular zone must be symmetrical about the fault if a constant-volume deformation is to be maintained. In the trishear zone, along a line normal to the fault plane, the velocity relative to the footwall block varies in magnitude and orientation. Material adjacent to the hanging wall block must have the same velocity as the material in that block outside the shear zone. Toward the bottom boundary, the relative velocity in the shear zone must rotate toward parallelism with the bottom boundary and must approach zero at the boundary. These conditions are required to maintain continuity and constant volume in the trishear zone.

Shearing in the trishear zone is heterogeneous, because near the apex, the deformation is concentrated in a narrower width of the zone and thus is more intense than farther away (see the deformed material circles that approximate strain ellipses in Figure 13.34*A*, *B*). Thus, depending on the distance from the fault tip line, layers caught up in a trishear zone vary in the amount of rotation they experience and in whether they are shortened and thickened or lengthened and thinned. Initially, beds shorten and thicken as they rotate toward perpendicular to the fault orientation, and subsequently they lengthen and thin as they rotate further and become overturned. As a result, the layers in the forelimb of the fault-propagation fold have variable dips and variable thicknesses, which is not true for the kink model. If the fault tip propagates, it propagates into the trishear zone, causing the zone boundaries to migrate through the hanging wall and/or the footwall blocks. This migration adds to the heterogeneity of the deformation, because at any given time, material that has just crossed a boundary is less sheared than material that has spent more time within the trishear zone (see Figure 13.34*B*, *C*, and the associated discussion in the following paragraphs).

Folds formed in the trishear zone and subsequently cut by the propagating fault tip appear in the hanging wall and/or footwall blocks as folds truncated against the fault (Figures 13.33 and 13.34; compare Figure 5.10). These are often termed **drag folds**, implying the folding has resulted from the drag of one block against the other during fault slip. In the trishear model, however, the drag is

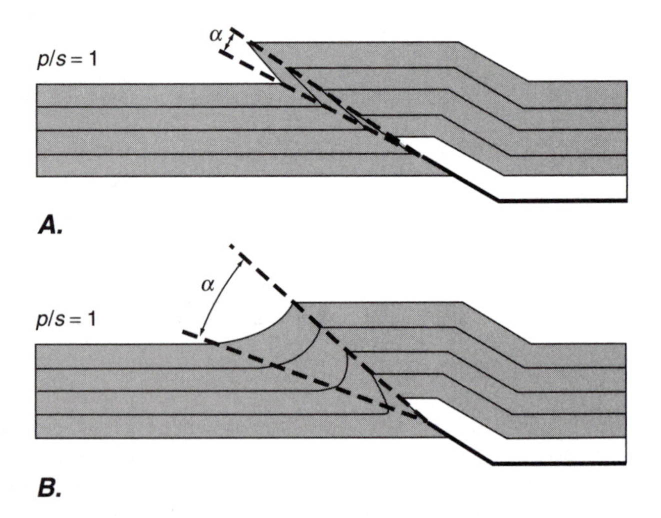

FIGURE 13.33 Trishear model for fault-propagation folding showing the effect of different apical angles α on folding within the trishear zone. Smaller angles concentrate the deformation in a smaller area making the folding more extreme. The fault propagates up from a horizontal decollement to form a ramp above which is a fault-ramp anticline. The back limb of the anticline forms a kink fold; the forelimb develops in the trishear zone, $p/s = 1$ in these models. (Modified after Hardy and Ford 1997, figures 8a, d; p. 847, by permission of American Geophysical Union) *A*. $\alpha = 5°$. *B*. $\alpha = 20°$.

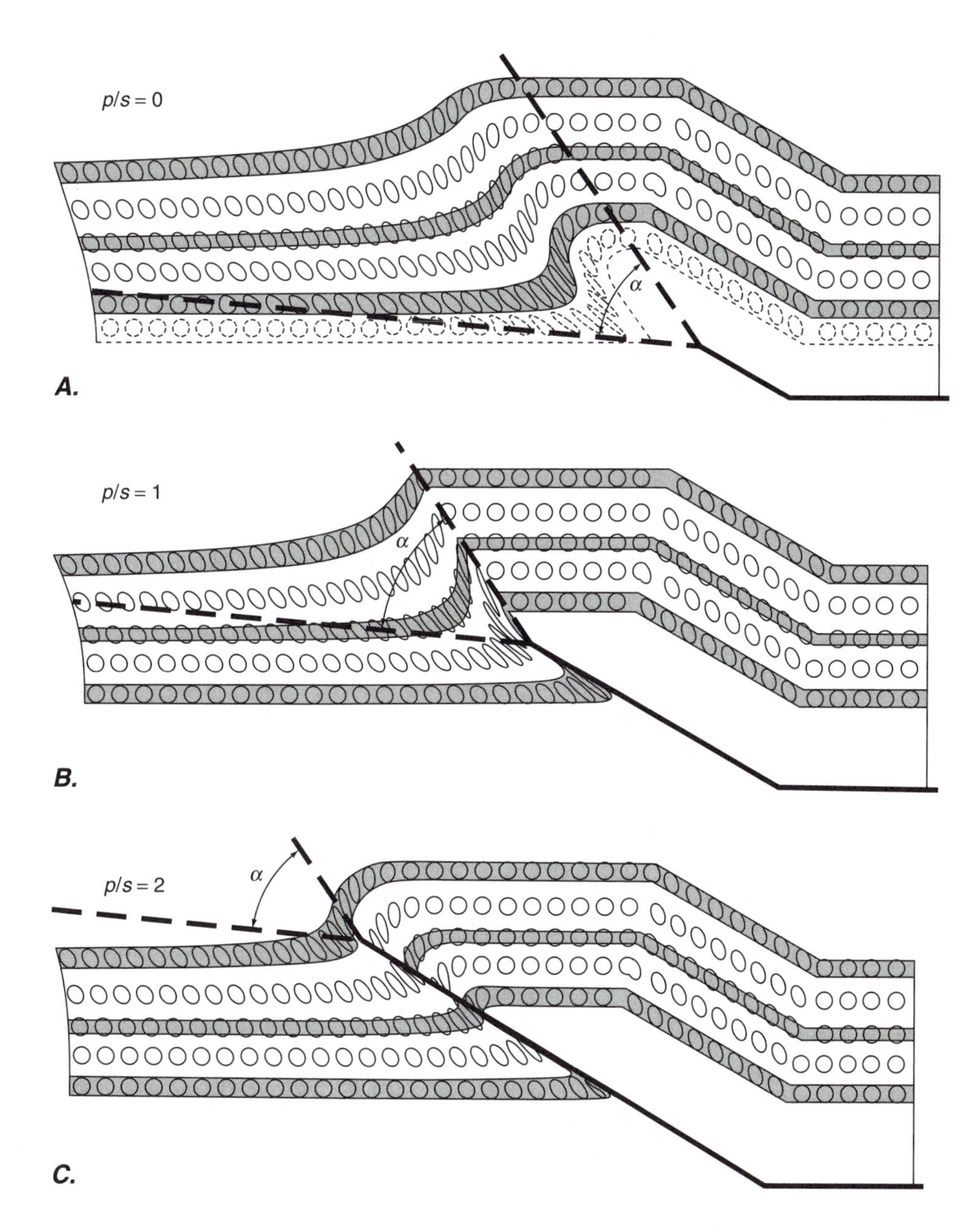

FIGURE 13.34 Trishear model for fault propagation folding showing the effect of different ratios (p/s) of fault tip propagation to fault slip. The fault tip propagates up from a horizontal decollelment to form a ramp. The bottom layer in *A* and the dashed strain ellipses have been sketched in to indicate how the fold sharpens to a kink at the bottom boundary of the trishear zone. Material circles are deformed into ellipses except where the strain is inhomogeneous on the scale of the diameter of the circle. (Modified after Allmendinger 1998, figures 2a, b, d; and Hardy and Ford 1997, figures 12a, b; by permission of American Geophysical Union) *A*. $p/s = 0$. The fault tip is stationary relative to the footwall block. *B*. $p/s = 1$. The fault tip is stationary relative to the hanging wall block. *C*. $p/s = 2$. The fault tip propagates upward relative to both footwall and hanging wall blocks. Compare with Figure 13.31, for which the p/s ratio is the same.

really a distributed shear in the trishear zone that precedes the faulting. Nevertheless, the sense of shear suggested by these cutoff trishear folds gives a correct indication of the sense of motion on the fault.

Within the confines of this model, two variables allow a wide range of fold geometries to develop: α, the apical angle of the trishear zone, and p/s, the ratio of the distance the fault tip propagates, p, to the associated amount of slip on the fault s (see Figures 13.32 and 13.34).

Variations in the apical angle determine the area over which the fault slip is distributed and thus the magnitude of the shear within the trishear zone (Figure 13.33). For smaller apical angles (Figure 13.33*A*), the shear is more highly concentrated, and the amount of rotation of the short limb of the fault-propagation fold is higher compared with larger apical angles (Figure 13.33*B*).

Material in the hanging wall and in the footwall can move into and out of the trishear zone during the fault slip as the zone boundaries migrate, in a manner that depends on the p/s ratio. The smaller the p/s ratio, the longer any given material point spends within the trishear zone, and thus the greater the deformation and the tighter the folding. If the ratio $p/s = 0$ (Figure 13.34*A*), the fault tip does not propagate relative to the footwall block. The lower boundary of the trishear zone is fixed in the footwall block, and material in the hanging wall block progressively moves across the trishear zone boundary into the shearing zone. The result is a fault-propagation fold with a blunt antiformal hinge and a sharper synformal hinge, which is actually kinked where the layers cross the lower trishear zone boundary.

If $p/s = 1$ (Figure 13.34*B*), the fault tip and the upper trishear zone boundary are fixed in, and propagate with, the hanging wall block. In this case, the fault-propagation fold has a kink along the upper trishear zone boundary, and the lower boundary migrates through the footwall block creating hinges that vary in bluntness along the axial surface (cf. Figure 13.33). The propagating fault leaves footwall synclines beneath the fault that are commonly referred to as drag folds but which, according to this model, form before the fault cuts through the forelimb of the fold.

The fault tip may also propagate at a rate greater than the slip rate on the fault, resulting in $p/s > 1$ (Figure 13.34*C*). In these cases, the upper boundary of the trishear zone migrates through the hanging wall block, and the lower boundary migrates through the footwall block, leaving behind fault-propagation folds with blunt hinges and cutoff hanging wall anticlines and footwall synclines. A sudden and large increase in the p/s ratio results in a breakout, which forms when a fault suddenly propagates across the stratigraphic section, cutting through any previously developing fault-propagation fold.

Many more-complicated folds can be produced by assuming a sudden change or a variation with time of the apical angle and/or the p/s ratio, or by incorporating fault ramps into the structure. Moreover, the model can be applied equally well to both thrust and normal faults. The main strength of the trishear model is its ability to reproduce a wide variety of the structures actually observed in the field. The main difficulty with its use is that the fold forms must be calculated on a computer by summing a series of small deformation increments; they cannot conveniently be determined analytically or geometrically, as can be done for the kink model.

13.8 DRAG FOLDS AND HANSEN'S METHOD FOR SLIP-LINE DETERMINATION

When subjected to ductile shear, rock layers commonly form asymmetric folds whose sense of asymmetry reflects the sense of shear of the deformation. Such folds are commonly called **drag folds**, the implication being that the velocity gradient in the shear zone has dragged the layer into a fold. Characteristically they are noncylindrical, asymmetric, and disharmonic. Because hinge orientations depend on the original orientation of the layer relative to the shear plane and on local inhomogeneities in the flow, they can vary widely, and the hinges need not be linear (see, for example, the folds in the salt bed in Figure 13.41*A*). Thus the hinge orientations do not indicate the slip direction. The hinges may form parallel to the shear plane; otherwise, subsequent deformation tends to rotate them toward parallelism with both the shear plane and the slip direction. Geometries of flow more complex than simple shearing, however, can rotate fold hinges toward being either parallel or perpendicular to the direction of flow (see Figures 14.20 and 20.31 and associated discussions).

The sense of asymmetry of any drag fold, whether its hinge is curved or straight, must be consistent with the sense of shear in the zone. This relationship of fold asymmetry to shear sense and the variable orientation of the fold axes are the basis of the Hansen method of determining the slip direction.[2] All hinge orientations are plotted on a stereonet, each labeled with its appropriate shear sense, which by convention is determined looking down the plunge of the hinge line. The hinge lines should lie approximately along a great circle defining the orientation of the shear plane (Figure 13.35*A*, *B*). Fold hinges that are subparallel to the slip line but are oriented on opposite sides of the line must have opposite shear senses. Thus the two hinges of opposite shear sense closest to being parallel to the slip direction (hinges 1 and 3 in Figure 13.35) define the separation angle, which constrains the possible slip-line orientation. The shear sense of the folds defines the sense of shear in the shear zone (Figure 13.35*B*).

13.9 SUPERPOSED FOLDING

In complexly deformed areas such as the central core regions of orogenic belts, folded layers of rock commonly display a geometry that indicates that earlier folds have been folded by one or more sets of later folds. Such multiple foldings are referred to as **superposed folding**, and the different sets of folds are called **generations of folds**. A first generation of folds is refolded by a second generation and by all subsequent generations.

In discussing superposed folds, we need a convenient notation to describe the successive surfaces and hinges formed. The terminology is illustrated in Figure 13.36. Surfaces are labeled S. Bedding is S_0, and the axial surfaces of first-, second-, and higher-generations of fold, which are assumed to form as planar surfaces, are designated S_1, S_2, and so on. Fold hinges are labeled f. The fold hinges of successive generations are f_1, f_2, and so on. It is also useful to include with the fold hinge symbol a designation of the surface being folded. Thus the fold

[2] A complete discussion of the application and pitfalls of this method can be found in Hansen (1971).

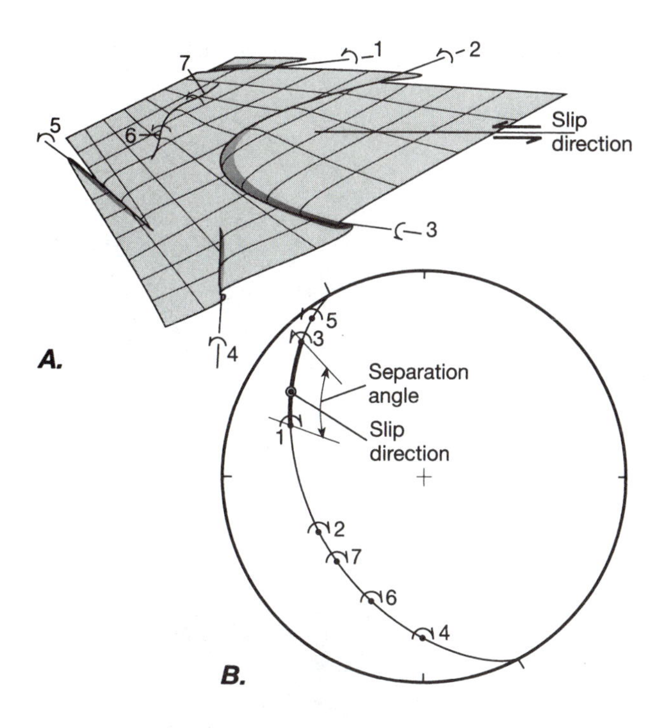

FIGURE 13.35 Hansen's method for slip line determination in folded layers subparallel to the shear plane. *A*. Fold hinges numbered 1 through 7 have a variety of orientations. The sense of asymmetry (clockwise or counterclockwise) is determined looking down the plunge of the fold hinge. All parts of all the fold hinges, however, are consistent with the same sense of shear in the shear zone. Because fold hinges 2 and 3 are oriented on opposite sides of the slip direction, they have opposite senses of asymmetry. *B*. On a stereonet, the hinge orientations plot parallel or subparallel to the shear plane (the great circle), and the asymmetry of the folds changes sense across the separation angle, which must contain the slip direction. The sense of fold asymmetry defines the shear sense of the deformation.

hinge $f_1^{S_0}$ means a first-generation fold hinge in the S_0 (bedding) surface. Second-generation folds develop in the already folded bedding S_0 and in the first-generation axial surfaces S_1. Thus second-generation fold hinges in these surfaces are labeled $f_2^{S_0}$ and $f_2^{S_1}$, respectively (Figure 13.36).

Understanding of the geometry of idealized superposed folds has allowed complex sequences of superposed structures to be unraveled. The details of the procedures are beyond the scope of this book, but we briefly review several types of superposed fold patterns that are characteristic of rocks affected by multiple generations of folding. The type of pattern depends in part on the mechanism of folding and in part on the relative orientations of the deformation for the first and second generations.

i. The Geometry of Superposed Flow Folding

If two generations of folding have affected a body of rock, where at least the second generation is a flow (or passive) folding, the basic pattern of orientations of fold hinges and axial surfaces can be analyzed according to fairly simple geometric rules. In general, the youngest generation of folding has planar axial surfaces. The axial surfaces of older generations are folded by all younger generations. After two generations of folding, for example, second-generation folds are developed in both the bedding S_0 and the earlier-generation axial surface S_1, and both have the same planar second-generation axial surfaces S_2 (Figure 13.36). Earlier-generation fold axes commonly behave like passive linear features and rotate during later generations of folding. The rotated axes develop predictable patterns that depend on the initial fold axis orientation and the geometry of the later deformation. We discussed some of these lineation patterns in Sections 13.1–13.4 (Figures 13.6, 13.7, and 13.10). Although the youngest-generation axial surfaces are commonly planar, the associated fold axes develop in a range of orientations depending on the initial orientation of the surface being folded (compare fold axes in Figure 13.9*A*, *B*). They are related to one another only in that all the youngest-generation fold axes must lie within the youngest-generation axial surface.

We examine the result of superposition of a second generation of folding that occurs by passive shear and that is similar in scale to the first generation. The resulting outcrop patterns of the superposed folds, called **interference patterns**, have characteristic styles that depend on two angles: the angle between the first-generation axial surface S_1 and the second-generation slip direction $\mathbf{a}_2$, and the angle between the first-generation fold axis f_1, and the second-generation axial surface S_2. These angles may vary from near 0° (small) to 90° (large).

The interference patterns are shown in the right-hand diagrams in Figure 13.37*A*, *B*, *C*. The first diagram in each part shows the geometry of first-generation (f_1) folding.

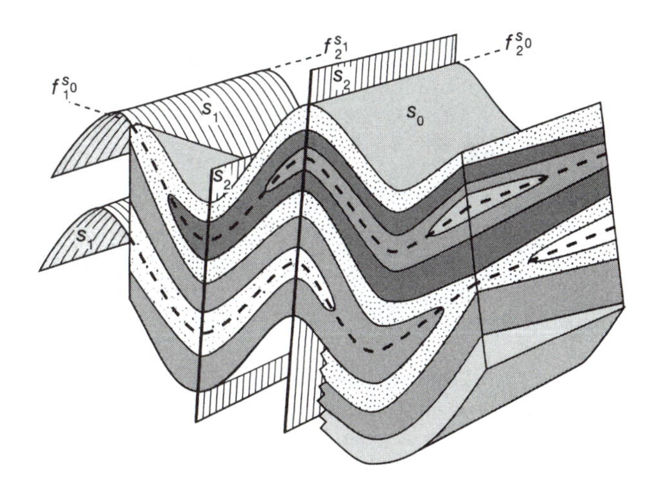

FIGURE 13.36 Geometric elements of a refolded fold. The first-generation folds are folds in S_0 with fold axis $f_1^{S_0}$ and axial surface S_1. The first-generation folds are refolded by second-generation folds having an axial surface S_2. Second-generation fold axes develop in S_0 ($f_2^{S_0}$) and in S_1 ($f_2^{S_1}$).

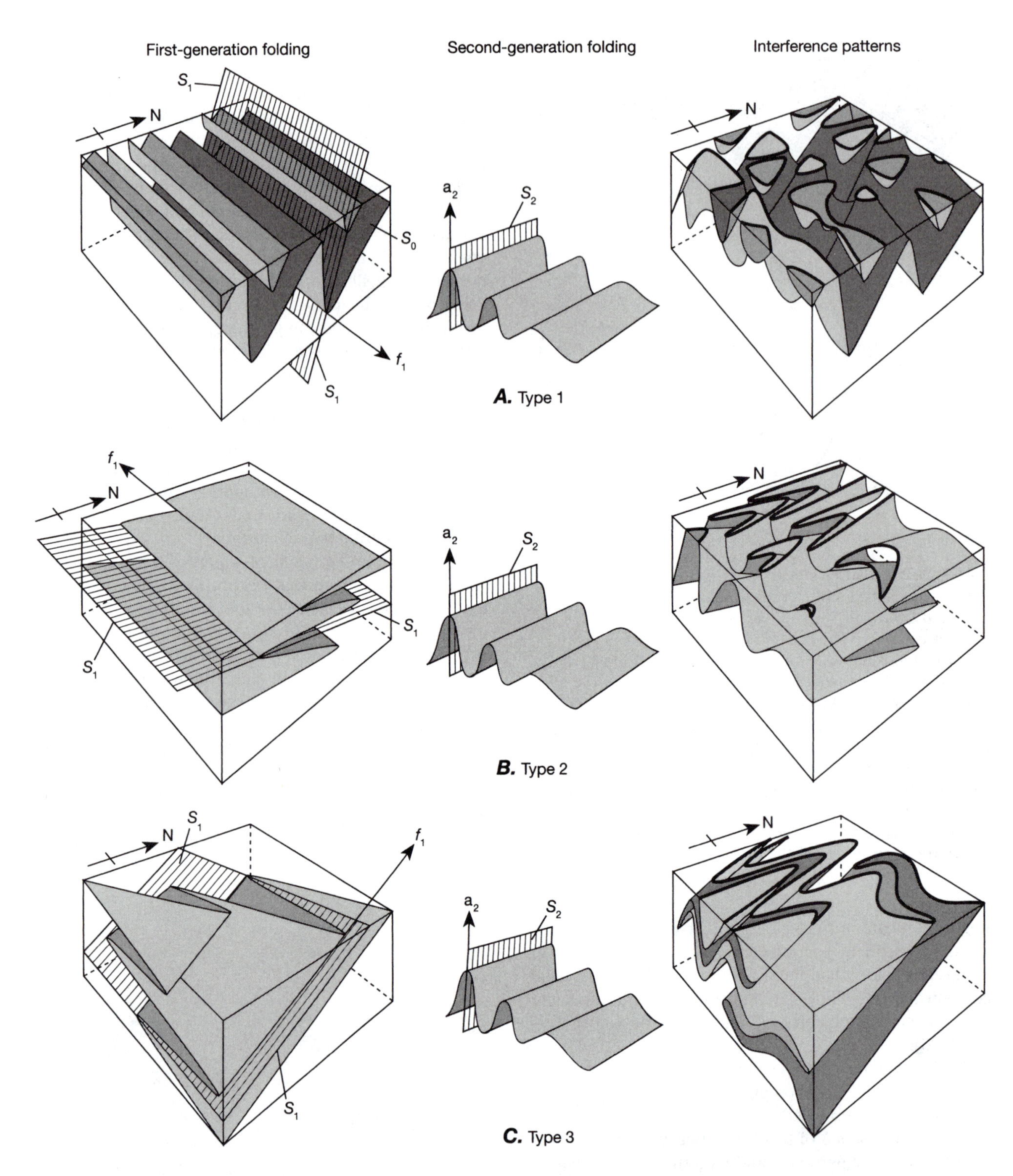

FIGURE 13.37 Fold interference patterns. In each case, the left diagram shows first-generation folds in S_0 and the axial surface S_1. The middle diagram shows the geometry of second-generation folding f_2 as it would appear in an initially horizontal surface. The right-hand diagram shows the superposition of the second-generation folding on the folds in the left-hand diagram, with the surface eroded down to a flat plane to reveal the characteristic outcrop patterns, which are shown in heavy lines. (Modified from Ramsay 1967) *A*. Type 1 interference folds, showing superposition of folds and dome-and-basin interference patterns. a_2 is at a small angle to S_1, and f_1 is at a high angle to S_2. *B*. Type 2 interference pattern, showing arrow-head or mushroom-shaped patterns. a_2 is at a high angle to S_1, and f_1 is at a high angle to S_2. *C*. Type 3 interference pattern, showing the wavy outcrop pattern of the S_1 axial surface. a_2 is at a high angle to S_1, and f_1 is at a small angle to S_2.

The second diagram shows the geometry of the second-generation folding as it would appear in an initially horizontal surface. It is the same cylindrical fold train for *A*, *B*, and *C*. When this folding geometry is superposed on the first-generation folds, the result is the refolded folds that appear on the right side of the figure. The intersection of these superposed fold styles with a horizontal surface of erosion produces characteristic interference patterns, which are emphasized with heavy lines in the diagrams on the right side of the figure. Note that the interference patterns depicted are simply the end members of a continuous gradation of patterns.

Type 1 interference folds (Figure 13.37*A*) are characterized by complete closures of the outcrop pattern of individual S_0 layers (Figure 13.38*A*). This pattern reflects the presence of domes, basins, and intervening saddles in the folded surface. It develops when the $\mathbf{a_2}$ direction is contained between the limbs of the first-generation folds and thus is generally at a small angle to S_1, and when the angle between f_1 and S_2 typically is large.

Type 2 interference folds (Figure 13.37*B*) are characterized by arrow head-, crescent-, and mushroom-shaped outcrop patterns of the folded surfaces (Figure 13.38*B*). These patterns develop when the $\mathbf{a_2}$ direction is not contained between the limbs of the first-generation folds and is therefore generally at a high angle to S_1 and when the angle between f_1 and S_2 typically is large, as for the type 1 pattern.

Type 3 interference folds (Figure 13.37*C*) are characterized by an undulating axial surface trace of first-generation folds (Figure 13.38*C*). This pattern develops if the slip direction $\mathbf{a_2}$ is not contained between the limbs of the first-generation folds and is therefore typically at a high angle to S_1, as for type 2 interference patterns, and when the angle between f_1 and S_2 is small.

ii. Geometry of Superposed Buckle Folding in a Single Layer

Flexural refolding by buckling behaves differently from refolding by passive shear because the mechanical strength of the layer being refolded has a significant effect on the manner in which the second-generation folds develop. We consider the simplest case of folding a single competent layer in an incompetent matrix in which the competent layer governs the form of the fold. We designate the first- and second-generation foldings by the symbols f_1 and f_2, respectively. We assume that the foldings f_1 and f_2 have approximately the same scale and that the f_2 maximum shortening direction is at a high angle to that for the f_1 folding.

Under these conditions, the superposed folding can occur as one of four distinct modes (Figure 13.39), depending on the initial tightness of the f_1 folding (refer to Table 10.3). If the f_1 folding angle is gentle, less than 45°, the interference is characterized by a mode 1 pattern of

A.

B.

C.

FIGURE 13.38 Natural examples of interference fold patterns. *A*. Type 1 style showing domes and basins in a gneiss. (From Skjernaa 1989) *B*. Type 2 style developed in banded marble, Northeastern Vermont, USA. *C*. Type 3 style developed in interlayered silicates and marble, Ruby Mountains, Nevada, USA.

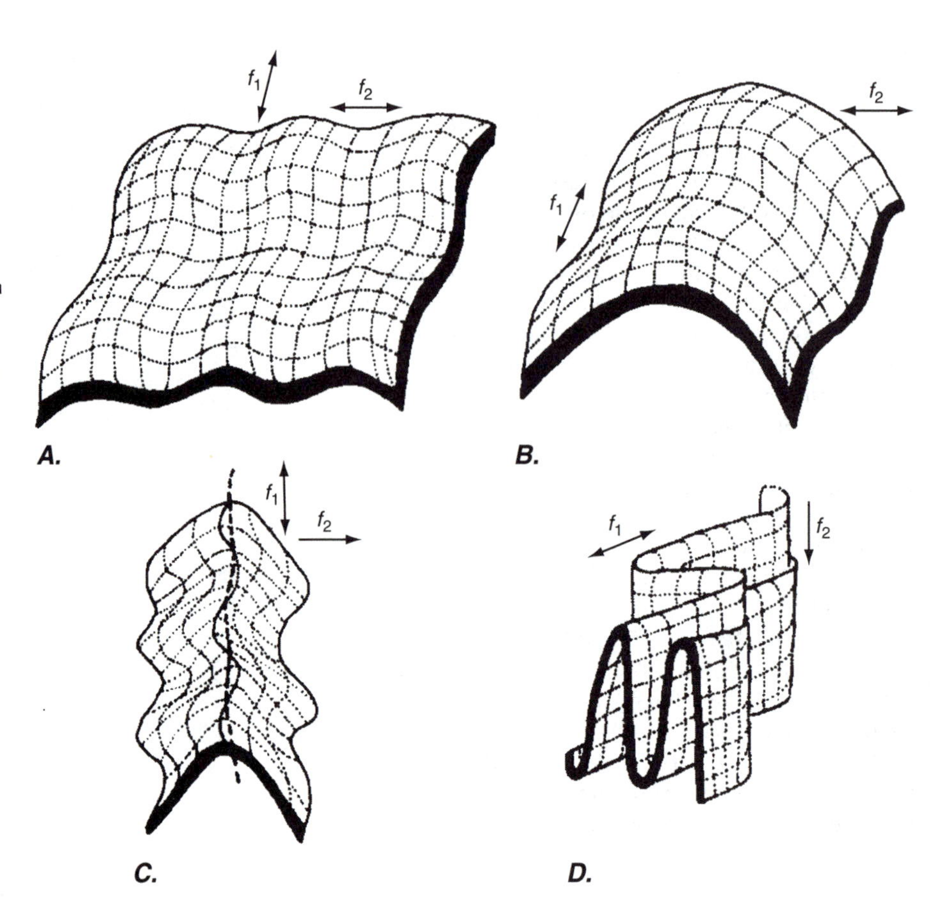

FIGURE 13.39 Modes of buckle fold interference for a single competent layer in an incompetent matrix. The shortening direction for the second folding is at a high angle to that for the first. (After Ghosh et al. 1993, Fig. 1, with permission from Elsevier) *A.* Mode 1 refolding: First generation is a gentle fold with a folding angle <45°. *B.* Mode 2 refolding: First-generation folding is a gentle to open fold with a folding angle between 45° and 90°. *C.* Mode 3 refolding: First-generation folding is mostly a close fold with folding angle between 90° and 150°. The dashed line shows the material particles originally defining the f_1 hinge line. *D.* Mode 4 refolding: First-generation folding is tight to isoclinal with folding angle between 150° and 180°.

domes and basins (Figure 13.39*A*). If the f_1 folding angle is gentle to open, between 45° and 90°, f_2 folds develop with a smaller wavelength than f_1 and form continuous corrugations across the earlier fold (Figure 13.39*B*). If the f_1 folds are mostly close folds, with folding angle between 90° and 150°, mode 3 interference develops with f_2 folds along the same axial surface changing from antiformal to synformal across the f_1 hinge and axial surface (Figure 13.39*C*). The original f_1 hinge is replaced by a sinuous hinge that occupies a different set of material points from the original f_1 hinge. Finally if the f_1 folds are tight to isoclinal, with folding angle between 150° and 180°, mode 4 interference develops a sinuously folded f_1 axis with f_2 hinges on opposite limbs of the f_1 folds being nested together with either constant antiformal or synformal geometry along any given second-generation axial surface (Figure 13.39*D*).

The geometry of these interference structures is affected by the curvature of the f_1 fold hinge and by the folding of multilayers rather than just the simple single layer, so the f_1 folding angles do not constitute rigidly defined boundaries for the different modes. The interference pattern on a planar outcrop depends in part on the relative orientation of the outcrop plane and the folded layer. If the outcrop plane is generally subparallel to the enveloping surface of the folded layer, however, modes 1 and 2 tend to form type 1 interference patterns (Figure 13.37), and modes 3 and 4 tend to produce type 3 interference patterns.

iii. Interpretation of Fold Generations

It is important to remember that to describe a second generation of folds as being superposed on a first generation implies only a sequence of deformational events. It says nothing about the interval of time between those events, and it does not necessarily imply that all folds of a particular generation developed at the same time everywhere. Moreover, the same number of fold generations does not necessarily appear everywhere in a deformed area, so any possible correlation of deformational events from place to place is probably more reliable if it is based on the particular style of folding rather than on the local generation number. These are very important restrictions on the interpretation of superposed folding. The deformation associated with two generations of folding could be associated with two distinct orogenic events separated by tens or hundreds of millions of years, or they could be the result of separate phases of a single orogenic event. In the latter case, different generations could represent separate chronological phases of a single orogeny, such as collision followed by strike-slip faulting. Alternatively, the different generations could represent spatial changes in the geometry of deformation that occur in a single event along the flow line for the rocks, such as the change in salt domes from convergent flow at the base, to upward shear flow in a pipe, to divergent flow in the head, as discussed in the following section. Only by very careful age dating of mineral grains that can be shown to have crystallized during a particular deformational event can such ambiguities be resolved.

13.10 DIAPIRIC FLOW

Diapirs[3] are rounded to elongate dome-shaped structures in which the rocks in the core have risen ductilely and pierced through the overlying rock. They are generally driven by buoyant forces caused by a density inversion, with the deeper lower-density rock rising up through overlying higher-density rock. The net effect is a lowering of the potential energy of the system, which makes it more stable. On horizontal sections through the structures, they have a circular to elliptical outline. The process of forming diapirs, called **diapirism,** is an extremely important one in geology. In continental crust, it is associated with the formation of salt domes, metamorphic gneiss domes, and igneous plutons. In the mantle, large-scale solid-state convective flow can occur by this process, as is evidenced by the overlying hot spots on the ocean floor, such as Hawaii or Iceland, or even in the continental crust, such as the Yellowstone hot spot in the northwestern United States.

[3]From the Greek word *diapiro*, meaning "I pierce, I penetrate," to be distinguished from "diaper," a baby's underclothing, from the Greek *diaspros*, meaning "pure white," which they generally are, at least at the start.

Salt diapirs were the first such structures to be recognized and are the best understood, in part because their economic importance in forming oil traps and in providing sources for salt and sulfur has led to much research to understand their origin. They are widespread in areas such as the north German Plain, western Iran, the Gulf Coast region of the United States and Mexico (Figure 4.16), the countries around the Caspian Sea—including Azerbaijan, Russia, Turkmenistan, and Kazakhstan—in the Mediterranean and Red Seas, as well as west central Africa, and the Canadian Arctic. Salt is deposited in oceanic basins in which circulation is very restricted so that evaporation can concentrate salts in solution until they precipitate. In a rifted margin tectonic setting, such as the Red Sea, salt deposits accumulate after ocean water first enters the rift but before the rift widens into an open ocean. Thus salt commonly lies at the base of a section of denser marine sediments. Salt can also be deposited in a restricted closing ocean basin, such as the Permian salt of the North German Plain and the Miocene salt of the Mediterranean Sea.

Salt forms diapirs because it is significantly lower in density (2.2×10^3 kg/m^3) than the sedimentary rocks that overlay the salt layers (generally $2.3–2.8 \times 10^3$ kg/m^3), creating a density inversion that drives the rise of the salt, and because salt flows very easily in the solid state, even at normal shallow crustal and surface temperatures. Diapirs begin as anticlinal or domal uplifts and evolve into walls, columns, bulbs, or mushroom shapes (Figure 13.40). The diapirs may even become detached from the original source layer of salt. As the salt moves upward, it pierces through the overlying sediments, which become bent upward along the margins of the diapir. These upturned sediments, truncated against the impermeable

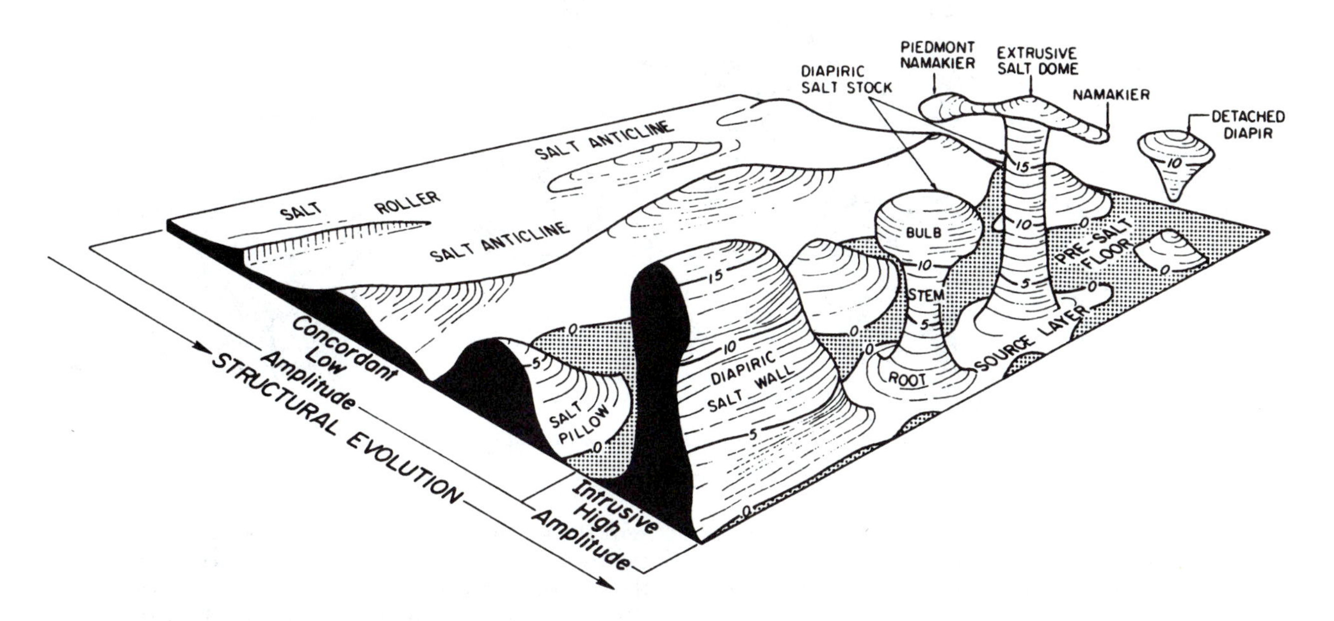

FIGURE 13.40 Common forms of salt intrusions. Diapirs originate from a layer of salt and then rise as salt pillows, salt stocks, or a salt wall, depending on the extent of intrusion. (From Jackson and Talbot 1989)

salt, provide the excellent traps for hydrocarbons, which makes salt domes very important economically. The tops of salt domes usually have been dissolved away by groundwater (see the top of Figure 13.41*C*) and are characterized by **caprock,** consisting of broad, subhorizontal, insoluble residues from the salt and by brecciated fragments of the overlying rock. A basin commonly forms above the diapir because of the solution. The sinking of the surrounding sediments to compensate for the rise of the salt commonly produces a rim syncline surrounding the diapir.

When we study the structures in rocks, our only clue to the geometry of the original deformation is usually the geometry of features such as folds. In salt diapirs, however, we have an unusual opportunity to examine the folding that results from a fairly well understood pattern of flow. Such examples provide us with models that we can use to understand other deformational environments in which a comparable style of folding is produced.

Models (see Section 18.4; Figures 18.5–18.7) suggest that horizontal radial flow converging toward the rising salt column initially forms a set of circumferential folds

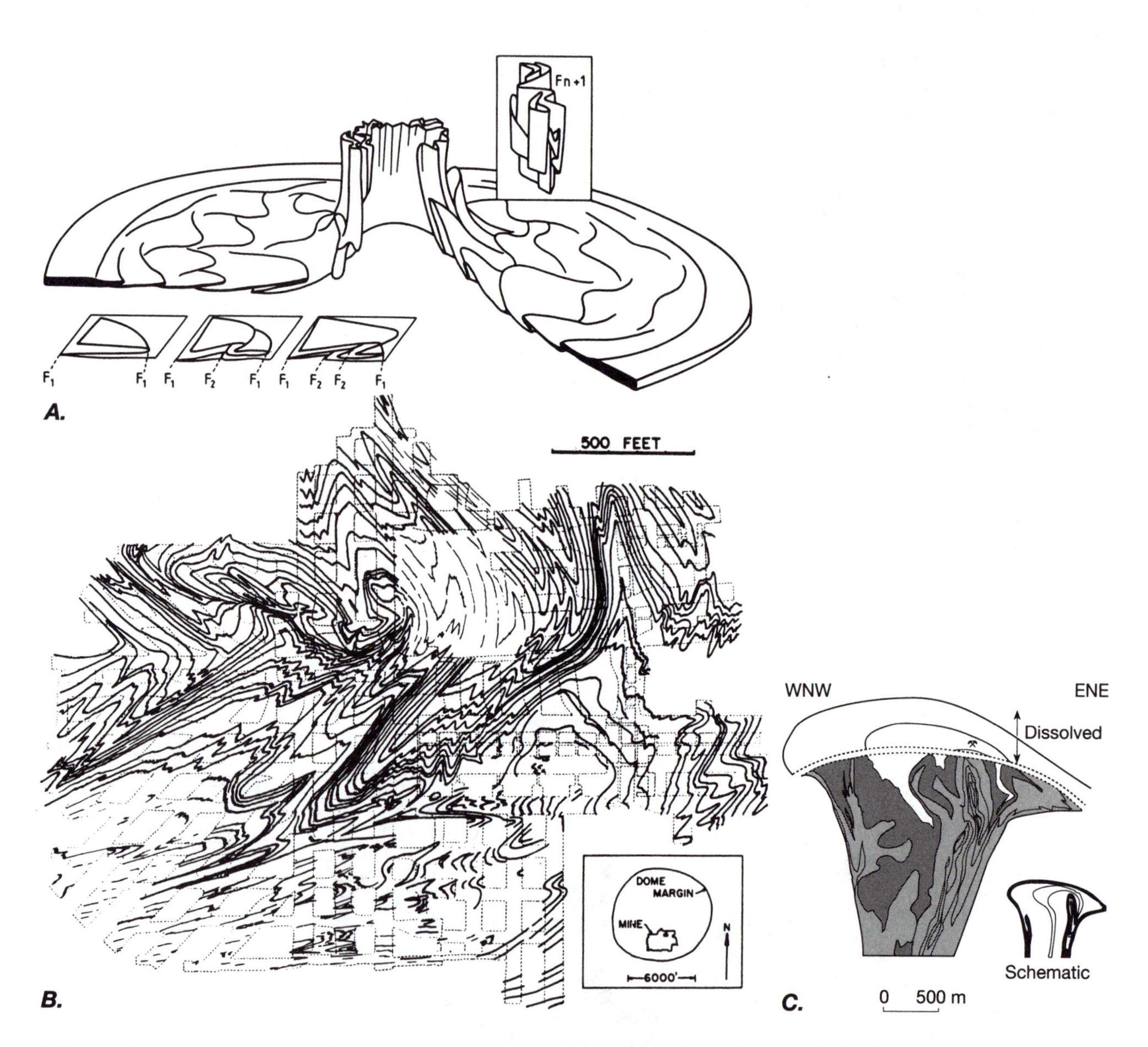

FIGURE 13.41 Internal structure of salt domes. *A.* Diagram of the base of a salt dome showing the evolution of folds and refolded folds that result from constrictional flow of salt from the layer into the stock of the salt dome. (From Talbot and Jackson 1987, figure 13) *B.* Generalized map of part of the Grand Saline salt dome in Texas, USA, showing the characteristic vertical folds and sheath folds. (From Muehlberger et al. 1962 and Muehlberger 1968) *C.* Cross sections of the internal structure of the Hänigsen salt dome, northeast of Hanover, West Germany, showing complex folding that results from flow within the rising diapir and from lateral spreading of the bulb. (From Jackson and Talbot 1989)

whose hinges become progressively rotated into a radial orientation (Figure 13.41*A*). Minor shifts in the flow geometry produce refolding of earlier generations of folds (lower inset, Figure 13.41*A*). As the salt moves up into the stem, the folds rotate into a vertical plunge, parallel to the main axis of the salt dome. A map of layers in the Grand Saline salt dome in Texas shows a complex geometry of class 2 refolded folds and sheath folds that have subvertical to vertical hinges (Figure 13.41*B*; compare upper inset in Figure 13.41*A*). In the bulbs and mushroom caps of the domes, lateral spreading of the salt and drag along the margins causes complex refolding of the salt layers and the possible entrainment of adjacent sediments into the folds (Figure 13.41*C*).

This example shows how different generations of folds can form from a change in the flow regime along the flow path. It also illustrates that different generations of folds can form in various places at the same time. By analogy, therefore, one must be cautious about interpreting the tectonic significance of different generations of folds that are characteristic of the metamorphic zones in orogenic belts.

Shale diapirs are present in some areas where folding of unconsolidated sediments has taken place or where rapid sedimentation and compaction have generated high fluid pressures in unlithified shales and caused them to move upward through the overlying rocks. The general form of these structures resembles that of salt domes. In some cases, the shale diapirs even reach the surface, where they form "mud volcanoes."

Serpentinite diapirs have been described from areas such as the California coast ranges and the forearc regions of active subduction zones, particularly the Marianas forearc. The serpentines are highly fragmented and contain some fragments of high-pressure low-temperature metamorphosed rocks. These rocks are believed to have formed in the subduction zone itself.

Mantled gneiss domes are domical bodies of gneissic rock found in the highly metamorphosed core zones of orogenic belts (Figure 13.42; compare Figure 18.7). They commonly display foliation parallel to the walls of the body, and they are surrounded, or "mantled" by a sheath of metamorphosed sedimentary rocks. Thus the folds associated with these bodies may be the result of diapiric flow of gneiss intruding into overlying rocks during intense regional metamorphism. In other cases, however, the structure may be the result of highly complex refolding of folds of crystalline basement and overlying metasediments or possibly the result of both processes acting in combination.

Rocks from the Earth's mantle are exposed on the surface either in peridotite massifs or ophiolites, both of which are composed predominantly of olivine with subordinate ortho- and clino-pyroxene and spinel organized into layers of different mineral abundances (see Figure 11.2*A*). These layers display complex fold structures reminiscent of those in the Grand Saline salt dome (Figure 13.41), suggesting an origin as a **mantle diapir**. Diapirism is probably common in the mantle, particularly beneath midocean ridges and under hotspot volcanoes such as Hawaii and Iceland. In these cases, the density inversion is caused by the higher temperature of the deeper material, which may rise either from the upper mantle or even from the deep mantle near the core-mantle boundary. One can imagine, therefore, that similar complex fold structures characterize much of the Earth's mantle.

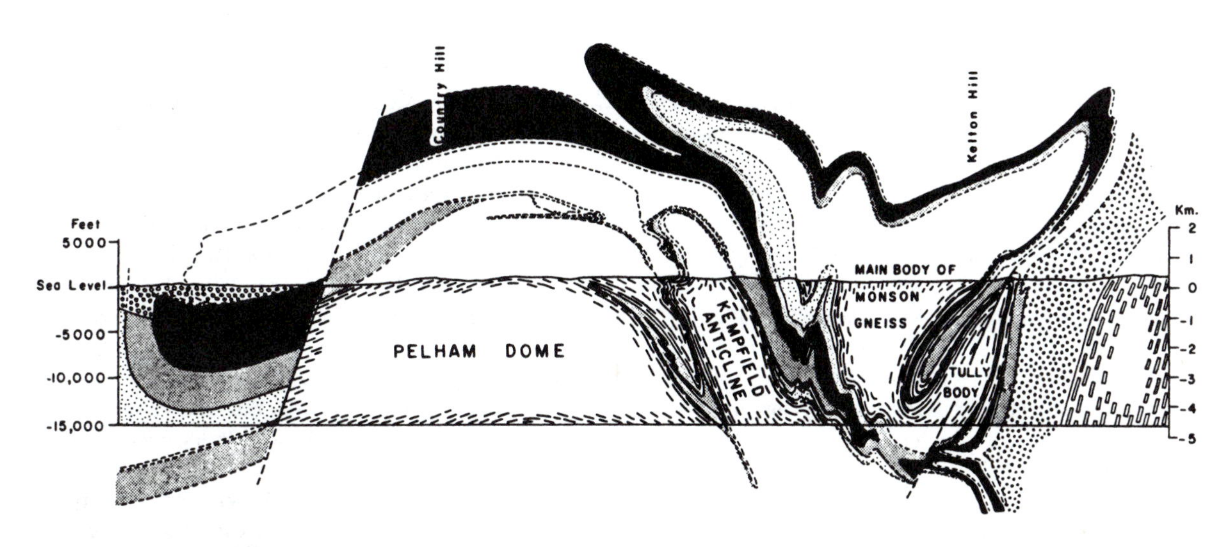

FIGURE 13.42 Cross section of a gneiss dome with the mantle of deformed metasediments from the Bronson Hill Anticlinorium of west-central New England, USA. The metasedimentary rocks were strongly deformed into recumbent nappes before deformation associated with the emplacement of the gneiss domes. (After Thompson et al. 1968)

REFERENCES AND ADDITIONAL READINGS

Allmendinger, R. W. 1998. Inverse and forward numerical modeling of trishear fault-propagation folds. *Tectonics* 17(4): 640–656.

Anastasio, D. J, E. A. Erslev, D. M. Fisher, J. P. Evans. 1997. Special issue; Fault-related folding. *J. Struct. Geol.* 19(3–4): 243–602.

Chester, J. S., J. M. Logan, and J. H. Spang. 1991. Influence of layering and boundary conditions on fault-bend and fault-propagation folding. *Geol. Soc. Amer. Bull.* 103(8): 1059–1072.

Donath, F. A., and R. B. Parker. 1964. Folds and folding. *Geol. Soc. Amer. Bull.* 75: 45–62.

Erslev, E. A. 1991. Trishear fault-propagation folding. *Geology* 19(6): 617–620.

Erslev, E. A., and K. R. Mayborn. 1997. Multiple geometries and modes of fault-propagation folding in the Canadian thrust belt. *J. Struct. Geol.* 19(3–4): 321–335.

Ghosh, S. K., N. Mandal, D. Khan, and S. K. Deb. 1992. Modes of superposed buckling in single layers controlled by initial tightness of early folds. *J. Struct. Geol.* 14: 381–394.

Ghosh, S. K., N. Mandal, S. Sengupta, S. K. Deb, and D. Kahn. 1993. Superposed buckling in multilayers. *J. Struct. Geol.* 15(1): 95–111.

Groshong, R. H. 1975. Strain, fractures and pressure solution in natural single-layer folds. *Geol. Soc. Amer. Bull.* 86: 1363–1376.

Gurnis, M. 2001. Sculpting the Earth from inside out. *Scientific American*. March 2001: 42–46.

Hansen, E. 1971. *Strain Facies*. New York, Springer-Verlag, 207 pp.

Hardy, S., and M. Ford. 1997. Numerical modeling of trishear fault propagation folding. *Tectonics* 16(5): 841–854.

Hobbs, B. E. 1971. The analysis of strain in folded layers. *Tectonophysics* 11: 329–375.

Hoeppener, R., M. Brix, A. Volbrecht. 1983. Some aspects on the origin of fold-type fabrics—theory, experiments and field applications. *Geologische Rundschau* 72: 421–450.

Jackson, M. P. A., and C. J. Talbot. 1989. Anatomy of mushroom-shaped diapirs. *J. Struct. Geol.* 11: 211–230.

Jin, G., and R. H. Groshong, Jr. 2006. Trishear kinematic modeling of extensional fault-propagation folding. *J. Struct. Geol.* 28(1): 170–183.

Mitra, S., and V. S. Mount. 1998. Foreland basement-involved structures. *Amer. Assoc. of Petrol. Geologists* 82(1): 70–109.

Muehlberger, W. R. 1968. Internal structures and mode of uplift of Texas and Louisiana salt domes. *Special Paper 88*, Geological Society of America, 359–364.

Muehlberger, W. R., P. S. Clabaugh, and M. L. Hightower. 1962. Palestine and Grand Saline salt domes, eastern Texas. In E. H. Rainwater and R. P. Zingula, eds., *Geology of the Gulf Coast and Central Texas: Guidebook of Excursions*, Houston Geological Society.

Narr, W., and J. Suppe. 1994. Kinematics of basement-involved compressive structures. *Amer. J. Sci.* 294: 802–860.

Patterson, M. S., and L. Weiss. 1966. Experimental deformation and folding in phyllite. *Geol. Soc. Amer. Bull.* 77: 343–374.

Ramsay, J. G. 1967. *Folding and Fracturing of Rocks*. New York, McGraw-Hill, 568 pp.

Skjernaa, L. 1975. Experiments on superimposed buckle folding. *Tectonophysics* 27: 235–270.

Skjernaa, L. 1989. Tubular folds and sheath folds: definitions and conceptual models of their development, with examples from the Grapesvare area, northern Sweden. *J. Struct. Geol.* 11(6): 689–703.

Suppe J., and D. Medwedeff. 1990. Geometry and kinematics of fault-propagation folding. *Eclogae Geologicae Helvetiae* 83(3): 409–454.

Suppe, J. 1983. Geometry and kinematics of fault bend folding. *Amer. J. Sci.* 283(7): 684–721.

Suppe, J. 1985. *Principles of Structural Geology*. Prentice-Hall, Inc., Englewood Cliffs, NJ, 537 pp.

Talbot, C. J., and M.P.A. Jackson. 1987. Internal kinematics of salt diapirs. *Amer. Assoc. Petrol. Geol. Bull.* 71(9): 1068–1093.

Tavani, S., F. Storti, and F. Salvini. 2006. Double-edge fault-propagation folding: Geometry and kinematics. *J. Struct. Geol.* 28(1): 19–35.

Thiessen, R. L., and W. D. Means. 1980. Classification of fold interference patterns: A reexamination. *J. Struct. Geol.* 2: 311–316.

Thompson, J. B, Jr., P. Robinson, T. N. Clifford, N. J. Trask, Jr. 1968. Nappes and gneiss domes in west-central New England. In E-A. Zen et al., eds., *Studies of Appalachian Geology: Northern and Maritime*. New York, Interscience Publishers: 203–218.

Turcotte, D. L., and G. Schubert. 1982. *Geodynamics: Application of Continuum Physics to Geological Problems*. New York, John Wiley & Sons, 450 pp.

Weiss, L. E. 1980. Nucleation and growth of kink bands. *Tectonophysics* 65: 1–38.

Whitney, D. L., C. Teyssier, and C. S. Siddoway. 2004. Gneiss domes in orogeny. *Geol. Soc. Amer. Special Paper 380*. Boulder, CO, Geological Society of America, 378 pp.

Zehnder, A. T., and R. W. Allmendinger. 2000. Velocity field for the trishear model. *J. Struct. Geol.* 22: 1009–1014.

Chapter 14

ANALYSIS OF FOLIATIONS AND LINEATIONS

The diverse foliations and lineations described in Chapter 11 form in many different ways, although the precise mechanism by which they form is not always clear. In this chapter, we discuss the principal mechanisms by which foliations and lineations can develop, which include strain of material objects in the rock, mechanical rotation, solution and precipitation, recrystallization and chemical reaction, and fracturing and mechanical wear. We then apply these ideas to the interpretation of the various morphological foliations and lineations defined in Chapter 11 (Figures 11.1 and 11.18).

A common field assumption is that foliations are parallel to the plane of flattening of the finite strain ellipsoid ($\hat{s}_1$, $\hat{s}_2$) and that lineations are parallel to the axis of maximum principal stretch ($\hat{s}_1$). Although in principle the parallelism for many cases cannot be exact, in practice it is often close enough to provide a useful indicator of the orientation of the principal strain axes. It is important, however, to understand the conditions under which this relationship does or does not hold. Part of our purpose in this chapter is to discuss these mechanisms of formation and how they relate to strain.

14.1 MATERIAL AND NONMATERIAL FOLIATIONS AND LINEATIONS

At the outset of our discussion, we must make a distinction between **material** and **nonmaterial** foliations and lineations, because the behavior of these two features relative to the strain ellipsoid is fundamentally different.

i. Material Foliations and Lineations

The essential characteristics of material foliations and lineations are that they are defined by a group of subparallel planes or lines each composed of a specific set of material points, and that their orientations are determined by the passive motion of these sets of material points as part of the deforming continuum (see the beginning of Chapter 12).

Structures interpreted as material foliations include features defined by compositional layers such as compositional foliations (Figure 11.2), mica films, compositionally differentiated cleavages (Figures 11.7; 11.9A, C; and 11.10), and possibly pervasive sets of microfractures. Spaced foliations and micro-spaced continuous foliations generally belong in this category. Material lineations include most constructed lineations (Figure 11.18), such as hinge lines, lines of intersection of material foliations (Figure 11.20*B*), boudin lines (Figure 11.21), and mullions (Figure 11.22).

ii. Nonmaterial Foliations

The essential characteristics of nonmaterial foliations and lineations are that they are defined by a dimensional preferred orientation of tabular or elongate objects within a material, and that the preferred orientation is not tied to the passive motion of material planes and lines during the deformation. Thus, although the objects that define the nonmaterial foliations and lineations are themselves material features of the rock, their shape and alignment are not, and the alignment behaves differently from specific material planes and lines.

Foliations and lineations in this category include those defined by the dimensional preferred orientation of passively strained objects such as deformed ooids (Figure 12.6*B*), pebbles (Figure 11.19*A*), spherical fossils such as radiolaria or foraminifera, alteration spots (Figure 11.19*B*), mineral grains (Figure 11.12*B*), and mineral grain clusters (Figure 11.24). Deformation of such features defines a foliation or lineation that is parallel, respectively, to the $(\hat{s}_1, \hat{s}_2)$-plane or the $\hat{s}_1$ axis of the finite strain ellipsoid. The principal planes and axes of the strain ellipsoid are not material planes and lines, as illustrated, for example, by a progressive noncoaxial deformation like simple shear. During simple shear, material lines rotate through the principal strain axes (see Section 12.4 and Figure 12.15). Thus the foliations or lineations defined by the principal axes of strain cannot be material planes or lines. The rigid rotation of planar or linear particles embedded in a deforming continuum also can result in preferred orientations that are not tied to material planes or lines.

The orientations of material and nonmaterial foliations and lineations coincide, however, after a coaxial deformation such as progressive pure shear (Figure 12.14), and in these circumstances, there is no distinction in the strain interpretation of the two.

iii. Rotation of Material Foliations and Lineations During Deformation

The relationship between strain and material foliations and lineations depends on the geometry of the deformation. Three conceivable associations of material foliations and lineations with strain are possible, depending on the relative orientation of the foliation and lineation and the nature of the strain (see Chapter 12, Figures 12.14 and 12.15):

1. If a material foliation forms parallel to the plane of flattening—the $(\hat{s}_1, \hat{s}_2)$-plane—or if a material lineation forms parallel to the maximum principal stretch $(\hat{s}_1)$, it can maintain that orientation throughout the progressive deformation only if the deformation is coaxial.
2. If a material foliation or lineation forms in the same orientation as described in 1, a progressive simple shear (a noncoaxial deformation) would rotate these structures out of their initial orientation and toward the shear plane or the shear direction, which are the so-called **fabric attractors** of the flow field. At the same time, however, the plane of flattening $(\hat{s}_1, \hat{s}_2)$ also rotates toward the shear plane, and the $\hat{s}_1$ axis rotates toward the shear direction. The principal planes and axes of strain, however, rotate at a different rate from the material foliation or lineation (see Figure 12.15). After a very large deformation, the angle between the $(\hat{s}_1, \hat{s}_2)$-plane and a material foliation, or between the $\hat{s}_1$ axis and a material lineation, decreases to the point that it may not be resolvable from field data.
3. If a material foliation or lineation forms in an orientation not parallel to the principal axes of the finite strain ellipsoid, in general it is rotated during progressive deformation. After a very large deformation, either coaxial or noncoaxial, the foliation or lineation may end up subparallel to the $(\hat{s}_1, \hat{s}_2)$-plane or the $\hat{s}_1$ axis, respectively.

Coaxial deformation is a very special geometry of deformation. Accordingly, the conditions under which it can occur are much more rare than those for noncoaxial deformation. Noncoaxial deformation is the most common condition in nature. Because material foliations and lineations can only be strictly parallel to a principal plane or axis of strain for coaxial deformation (see 1), we expect from arguments 2 and 3 that, in general, material foliations and lineations will only approximate such an orientation. For large strains, however, the deviation from parallelism becomes very small, generally below the precision of field measurement.

14.2 MECHANISMS OF FORMATION OF FOLIATIONS AND LINEATIONS AND THEIR RELATIONSHIPS TO STRAIN

i. Strain of Embedded Material Objects

A rock is made up of mineral grains,[1] and it commonly contains fossils or other deformable objects. When a rock deforms, embedded objects in the rock that initially are nearly spherical—such as ooids, some radiolarian tests, and alteration spots—are also deformed, becoming flattened and elongated into an ellipsoidal shape (Figure 14.1*A*). Their shapes provide one of the clearest and most abundant indicators of the nature of finite strain in deformed rocks. If the deformation is locally homogeneous, the parallel alignment of the planes of flattening $(\hat{s}_1, \hat{s}_2)$ then defines a nonmaterial foliation, and the parallel alignment of the maximum stretch $(\hat{s}_1)$ defines a nonmaterial lineation (Figure 14.1*A*), commonly referred to as a **stretching lineation**.

Other features, such as equant mineral grains or clusters of mineral grains and the clasts in a conglomerate, may have neither an initially spherical shape nor a random distribution of orientations. Nevertheless, a locally homogeneous strain deforms these features to produce foliations and lineations (Figure 14.1*B*, *C*).

Three limitations prevent such foliations and lineations from being ideal indicators of finite strain. First, in the undeformed state, most such features in rocks are spheroidal, that is, not exactly spherical, so that the deformed shapes are not true strain ellipsoids. Second, many features have an initial preferred orientation, the effect of which is never completely eliminated by defor-

[1] Excepting, of course, coal, opal, and volcanic glass.

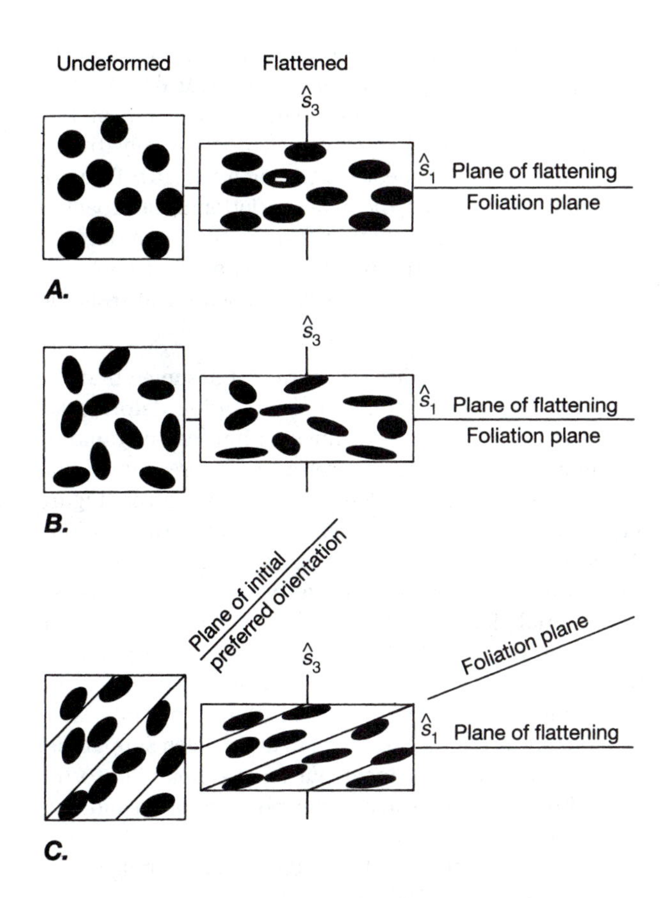

FIGURE 14.1 Foliations formed by deformation of discrete objects in a rock. (After Ramsay 1976) *A*. Flattening of initially spherical objects produces a foliation parallel to the plane of flattening—the $(\hat{s}_1, \hat{s}_2)$-plane. *B*. Flattening of randomly oriented elliptical objects produces a foliation statistically parallel to the plane of flattening. *C*. Flattening of initially elliptical objects with an initial preferred orientation produces a foliation in an orientation different from the plane of flattening.

mation; thus the principal axes of the deformed features are not the true principal axes of strain (Figure 14.1*C*). Third, because many "passive" markers are of different composition, and thus different competence, from the surrounding matrix, they might not deform by exactly the same amount as the matrix. Thus even if they were initially perfectly spherical objects, the strain they would record might not be the same as the net strain in the rock, and the foliation or lineation defined by their preferred orientation might not be parallel to the net strain axes. Moreover, if the markers are more competent than the surrounding matrix, their preferred orientation could be affected by a component of rigid rotation, which we discuss later.

Despite these problems, the inaccuracies become minor or even undetectable at sufficiently large deformations. Thus, *the nonmaterial foliations and lineations defined by deformed spheroidal features commonly approximate the orientation of the total finite strain ellipsoid.*

ii. Grain or Particle Rotation

Several lines of evidence indicate that rotation of mineral grains is a common phenomenon during deformation. Many rocks contain "snowball" garnets (Figure 3.19*E*) and curved fibrous overgrowths on undeformed mineral grains (Figure 11.26*B*). Detrital mica grains that are parallel to bedding in undeformed sediments are oriented parallel to a tectonite foliation in the equivalent deformed rocks. Mica grains in some crenulation foliations are rotated or bent into parallelism with the new foliation. Experimental deformation of rocks containing randomly oriented micas has produced a preferred orientation under conditions for which rotation is the only possible mechanism of reorientation. We infer, therefore, that systematic rotation of mineral grains into a preferred orientation is an important mechanism of formation of foliations and lineations. Because such a preferred orientation does not necessarily deform and rotate like a material plane or line in the rock, it is a nonmaterial foliation or lineation.

We can imagine three simple models to account for such rotations: (1) a mineral grain can rotate as a rigid particle embedded in a ductile matrix—the **Jeffrey model**[2] (Figure 14.2*A*); (2) a mineral grain can act as a strictly passive material marker that deforms with the surrounding rock—the **March model**[3] (Figure 14.2*B*); or (3) a mineral grain can deform internally by ductile shearing restricted to crystallographic slip planes, with an accompanying rotation of the grain that is required to keep its deformation compatible with the surrounding matrix—the **Taylor-Bishop-Hill**[4] model (Figure 14.2*C*). We discuss the effects of each model in turn on the formation of a foliation or lineation during simple shear, an example of a noncoaxial deformation. We discuss the effects of a coaxial deformation at the end of the section.

The Jeffrey Model. The consequences of the Jeffrey model for the formation of foliations and lineations are apparent when a rigid particle is suspended in a fluid undergoing progressive simple shear. If it is a plate-like particle, it rotates continually about an axis parallel to the shear plane and perpendicular to the shear direction

[2]After the British physicist G. B. Jeffrey, who in 1923 investigated theoretically the motion of rigid grains suspended in a deforming viscous fluid.

[3]After the German physicist A. March, who in 1932 analyzed theoretically the development of preferred orientation in deformable rods and plates.

[4]Named for the metallurgists G. I. Taylor, J. F. W. Bishop, and R. Hill, who developed the theory to explain the formation of crystallographic preferred orientations in ductilely deformed metals.

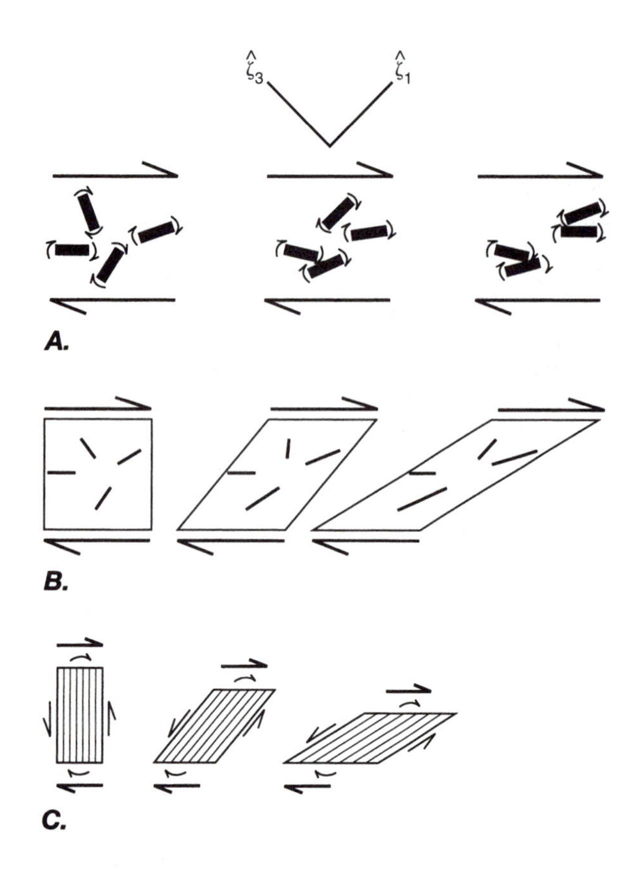

FIGURE 14.2 Mechanisms of rotation of crystal grains in a rock undergoing a noncoaxial deformation. *A*. The Jeffrey model: rotation of rigid elongate mineral grains in a ductile matrix. *B*. The March model: rotation of mineral grains that act as passive markers during the deformation. *C*. The Taylor-Bishop-Hill model: rotation of a mineral grain required to maintain compatibility between the imposed deformation and the ductile shear on a single set of slip planes within the crystal.

(Figure 14.2*A*). Although it never comes to rest, the plate rotates most slowly when it is parallel to the shear plane, because in this orientation, the torque applied to the plate by the shearing fluid is a minimum. The plate rotates most rapidly when it is perpendicular to the shear plane, because in this orientation, the applied torque is a maximum.

For a suspension of initially randomly oriented particles that rotate independently of one another in a shearing matrix, particles tend to pile up in orientations of low rotation rate (subparallel to the shear plane) and tend to be cleared out of orientations of high rotation rate (at a high angle to the shear plane). The resulting preferred orientation is not stable, however, because the particles never stop rotating, and it continually oscillates both in orientation and intensity. Overall, however, the result is a preferred orientation defining a nonmaterial foliation subparallel to the shear plane (Figure 14.2*A*). For elongate rigid particles, the Jeffrey model predicts that during progressive simple shear, the particles should develop a weak preferred orientation close to parallel with the shear plane and perpendicular to the shear direction.

During progressive simple shear, the plane of flattening of finite strain ($\hat{s}_1$, $\hat{s}_2$) rotates progressively toward parallelism with the shear plane (Figure 12.15). Thus for very high shear strains, the plane of flattening approaches parallelism with the foliation formed by this model, although in principle, the parallelism can never be exact. This model also predicts that a weak lineation also should develop subparallel to $\hat{s}_2$.

The March Model. The March model assumes that the particles defining the foliation or lineation deform passively as material planes or lines within the surrounding continuum. The rotation therefore is different from that of rigid particles embedded in a viscous matrix (Figure 14.2*B*). For example, during simple shear of the medium, the passive particles do not rotate continually, but the rotation rate decreases toward zero as the orientation of the material planes and lines approaches parallelism with the shear plane and the shear direction, respectively (Figure 14.2*B*; cf. Figure 12.15). Thus material planes and lines cannot rotate past the shear plane, and the resulting concentration of passive planar particles subparallel to the shear plane or linear particles subparallel to the shear direction results in a preferred orientation and thus a foliation or a lineation.

As the magnitude of the shear strain increases, the plane of flattening of the finite strain ellipsoid ($\hat{s}_1$, $\hat{s}_2$) rotates progressively toward the shear plane, and the maximum stretch axis ($\hat{s}_1$) rotates toward the shear direction. For any given orientation, however, the ($\hat{s}_1$, $\hat{s}_2$)-plane rotates more slowly than the parallel material planes, so material planes and lines continually rotate through the ($\hat{s}_1$, $\hat{s}_2$)-plane (cf. Figure 12.15). At very high shear strains, however, the angles between the foliation and the ($\hat{s}_1$, $\hat{s}_2$)-plane and between a lineation and the $\hat{s}_1$ direction become small, so that foliations become subparallel to the plane of flattening and lineations become subparallel to $\hat{s}_1$. At low shear strains, however, the angles are not negligible.

In some cases, particles that are rigid may nevertheless be constrained to lie along material planes in the continuum, such as stable boundaries of grains (i.e., grains that are not recrystallizing). These particles could rotate with the material planes and thus conform to the March model of rotation rather than the Jeffrey model. If, for example, mica flakes or amphibole needles behave rigidly but rotate with the stable grain boundaries of adjacent ductile mineral grains such as quartz, then their preferred orientation would be predicted by the March model. The preferred orientations increase in strength with increasing strain.

The Taylor-Bishop-Hill Model. Under some conditions, ductile shear on internal crystallographic slip planes of a mineral grain can result in rotation of the grain. For example, Figure 14.2*C* shows a hypothetical crystal grain that can deform by shear parallel to only one slip plane in one slip direction. Slip on this slip system, however, does not conform to the externally imposed geometry of shear, so a combination of shear on the slip system with a rotation of the grain must occur to make the internal and external deformations compatible. When the internal slip system coincides in orientation with the externally imposed shear, the crystal does not rotate further. This mechanism tends to rotate crystals into a preferred orientation (see Section 17.7(i), Figures 17.25 and 17.32), and if the crystallographic axes are associated with a platy or elongate crystal habit, a nonmaterial foliation or lineation is defined.

For example, muscovite generally forms thin platy crystal grains in which the dominant crystallographic slip plane is in the plane of the plates. According to the Taylor-Bishop-Hill model, if such crystal grains deform in a noncoaxial simple shear, they rotate toward parallelism with the plane of shear, and the resulting foliation would tend to be parallel to the shear plane. Because the ($\hat{s}_1$, $\hat{s}_2$)-plane of the finite strain ellipsoid approaches the orientation of the shear plane at very high shear strains, an approximate parallelism of the foliation to the plane of flattening could develop at high shear strains, although in principle, the parallelism would never be exact.

Unfortunately, it is difficult if not impossible to distinguish whether the rotation that produced a particular preferred orientation observed in rocks resulted from the March mechanism, the Jeffrey mechanism, the Taylor-Bishop-Hill mechanism, or some combination of these. The only factor that simplifies the interpretation of such foliations and lineations is that if the strains are very large, all these mechanisms produce preferred orientations that approach parallelism with the plane of flattening of the finite strain ellipsoid.

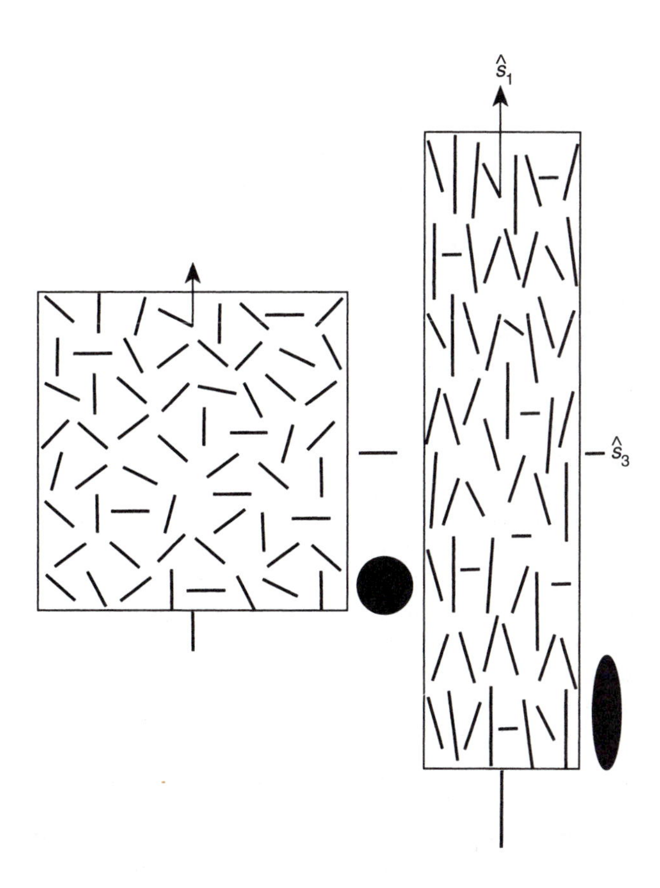

FIGURE 14.3 Rotation of passive particles in a ductile matrix undergoing coaxial deformation (the March model). If the deformation is homogeneous flattening, the particles rotate toward the plane of flattening and statistically define a foliation parallel to that plane. (After Ramsay 1976, figure 7, p. 12)

Effects of Particle Rotation During Coaxial Deformation. During a coaxial deformation such as homogeneous flattening (Figures 12.11*A* and 12.1*A*, *C*), any of these three rotation mechanisms result in the rotation of platy or elongate particles toward parallelism, respectively, with the plane of flattening of the finite strain ellipsoid—the ($\hat{s}_1$, $\hat{s}_2$)-plane—or the axis of maximum stretch ($\hat{s}_1$). In these circumstances, the resulting foliation or lineation is actually parallel, not just subparallel, to the appropriate principal stretch axes. Figure 14.3, for example, shows this effect for grains that rotate passively during the deformation according to the March model. During homogeneous constriction (cf. Figure 12.11*B*), elongate grains rotate toward parallelism with the maximum stretch direction ($\hat{s}_1$) to produce a lineation.

iii. Solution and Precipitation

The production of some foliations and lineations during rock deformation depends in part on the mobility of mineral constituents through the rock. The mechanisms involved include the breakdown of minerals by solution and chemical reaction, the migration of the chemical components through the rock, and the formation of new mineral grains by precipitation and recrystallization. Many spaced disjunctive and crenulation foliations result at least in part from such processes. The details of the process define how the resulting foliation or lineation is oriented relative to the principal axes of finite strain.

The dissolved material migrates through the rock, probably by grain boundary diffusion over short distances, or by transport in a fluid flowing through pores or fractures over larger distances. The dissolved material commonly reprecipitates locally, such as in microlithons (Figure 11.9*A*), where it accommodates a local dilation, or increase in volume; as overgrowths on pre-existing minerals or particles in the rock (Figures 11.7 (at "B") and 11.26*B*); as slickenfibers on shear surfaces (Figures 3.8*C* and 11.26*A*); and as fibrous or massive deposits in veins. In some cases, however, the bulk composition of the rock may be permanently changed by removal or introduction of one or more chemical components.

Two major factors affect the dissolution of minerals. First, any deformed mineral grain is more soluble than an undeformed one because of its higher internal strain energy (see Chapter 17). Second, a crystal subjected to a differential stress tends to dissolve more readily at surfaces on which the normal stress component is a maximum. This phenomenon is commonly called **Riecke's principle,**[5] and the process is called **pressure solution** or **solution transfer**. A corollary of Riecke's principle is that minerals tend to precipitate at surfaces on which the normal stress component is a minimum.

According to Riecke's principle, surfaces along which pressure solution occurs should be oriented perpendicular to the maximum compressive stress $\hat{\sigma}_1$, and those at which precipitation of minerals occurs should be perpendicular to the minimum compressive stress $\hat{\sigma}_3$. In a noncoaxial deformation, these orientations are in general not parallel to the principal axes of finite strain. Once such features have formed, however, they probably behave as material planes in the deforming rock and may rotate into a different orientation. The interplay between the orientation in which material foliations and lineations form, and the orientations toward which they may subsequently rotate makes interpretation of the mechanism and evolution of such structures particularly difficult.

iv. Recrystallization and Chemical Reaction

Recrystallization is the creation of new crystal grains out of old ones. Two kinds of recrystallization are important in structural geology. With **coherent recrystallization,** either old deformed grains are progressively transformed into new undeformed grains as a grain boundary migrates through the old crystal lattice, or old grains are subdivided into many new grains by the rotation of small internal domains called **subgrains**. The crystal structure and the composition of the old and new grains are the same, although new grains have different lattice orientations than the old. With **reconstructive recrystallization**, the old crystal structure breaks down—for example, during a chemical reaction—and a new structure forms, generally with a different composition. The distinction between the solution/precipitation process and reconstructive recrystallization is not always clear.

Both types of recrystallization can change the shape, arrangement, or preferred orientation of grains. A foliation or lineation can develop by solution or chemical reaction, for example, either by selective destruction of old grains, leaving only grains with a particular orientation, or by the production of new grains that grow in a preferred orientation. Phyllites have been found that contain both relict sedimentary mica grains and newly crystallized micas. The relict grains are characterized by large irregular grain shapes; the new grains are smaller with very regular grain boundaries and a strong preferred orientation that defines the new foliation. If recrystallization is extensive, it can obliterate all clues as to the nature and preferred orientation of the original grains.

Mimetic growth[6] is the growth of new crystals that nucleate on older crystals of similar structure in an orientation governed by the orientation of the older crystal. In this way, the growth of new crystals can enhance any pre-existing preferred orientation of the old crystals.

An existing foliation in a rock can also control the orientation of new mineral grains because growth parallel to the foliation may occur more easily than across it. Micas, for example, grow most rapidly parallel to their cleavage plane, and new mica grains that nucleate with cleavage planes parallel to a foliation grow more rapidly than micas in other orientations, thereby enhancing the pre-existing foliation.

Without some external controlling factor, reconstructive recrystallization generally does not produce a preferred orientation and can even destroy one. Recrystallization in association with deformation, however, can produce a preferred orientation or enhance an existing one. Because these processes are very commonly associated, the fabrics that result are strongly influenced by the interaction between these processes.

v. Brittle Mechanisms

During brittle faulting, fault zones commonly become filled with fault gouge, also called **cataclasite** (Figure 3.5*B*), which results from the mechanical comminution of the rock in a zone along the shear fracture. Surfaces of the shear fracture develop a fine-grained polished rind, referred to as a **slickenside**, which generally has a strong lineation called **slickenlines**. Some slickensides show evidence of recrystallization that evidently can occur during local frictional heating during faulting. Deformation in the gouge occurs by cataclastic flow, which involves frictional sliding and further fracturing of the rock particles (cf. Section 8.4(ii), Figure 8.10).

Slickenlines result from a number of processes, including the gouging of the surface by **asperities**[7] (Figure 3.8*A*), the mechanical abrasion of larger grains into streaks of pulverized rock (Figure 11.25), or the precipitation of mineral fibers (slickenfibers) from local solutions (Figures 3.8*C* and 11.26*A*).

Foliations can develop in cataclasite. Although the mechanisms are not well understood, these cataclastic foliations may form initially as *R* Riedel shears (Figure 8.7) and may involve mechanisms similar to those that form slickensides. If this interpretation were correct, the foliations would have an initial orientation at a low angle to the shear plane and thus would not directly reflect the orientation of the finite strain ellipsoid. Slickenlines

[5]After the nineteenth-century German physicist E. Riecke.

[6]After the Greek word *mimetikos*, meaning "imitation."

[7]The Latin word *asper* means "rough."

are undoubtedly parallel to the shear direction on the shear plane and therefore also are not directly related to the orientation of the finite strain ellipsoid.

14.3 INTERPRETATION OF THE MORPHOLOGICAL TYPES OF FOLIATION

In this section we review briefly some possible interpretations for the origin of the common morphological types of foliation that we introduced in Chapter 11. The type of foliation that develops is often dependent on the composition of the rock. For example, stylolitic foliations are mostly restricted to limestones and marbles, although they also form in other rocks, principally calcareous or argillaceous sandstones; a quartz-rich sandstone characteristically develops a rough to smooth disjunctive foliation, but never a fine continuous foliation; similarly, a crenulation foliation generally forms in rocks containing a high proportion of platy minerals, although it may also develop in finely laminated rocks.

i. Compositional Foliations

Compositional foliations are material foliations that result from original compositional banding in the rock or in some cases from metamorphic differentiation such as can occur with the formation of some crenulation foliations. In the first case, their initial orientation would have nothing to do with the subsequent finite strain ellipsoid, but after very high strains, the compositional foliations would be rotated to low angles with the finite strain plane of flattening. Compositional foliations formed by metamorphic differentiation could have an initial orientation related to the finite strain ellipsoid, which we discuss further in subsections (ii) and (iii) on disjunctive and crenulation foliations.

ii. Disjunctive Foliations

There is abundant evidence that the formation of disjunctive foliations involves the operation of solution processes. Stylolite cleavage domains may truncate fossils, which shows that part of the fossil has been dissolved (Figure 11.5*B*). The material filling these stylolites, largely clay minerals, iron oxides, and carbonaceous matter, is partly the insoluble residue from limestone solution but also is partly the result of local growth of new minerals. In some deformed sandy argillites that have a rough disjunctive foliation, truncation of detrital sand grains against cleavage domains ("C" in Figure 11.7) resulted from solution, not shear displacement along the cleavage. Detrital grains, such as quartz, that were originally equant ultimately may be nearly completely dissolved away into thin plate-like grains parallel to, and partly defining, the foliation ("T" in Figure 11.7). An insoluble residue of platy minerals, insoluble oxides, and carbonaceous material remains in the cleavage domains after the solution and removal of components such as quartz or calcite. In principle, the amount of shortening could be determined by measuring the amount of insoluble material in the cleavage domain and comparing that to its concentration in the original rock. Mineralogical and chemical alteration, however, can make such estimates highly uncertain.

We discuss possible models for the formation of disjunctive foliations in limestones and in argillites and the relationship these models imply between the orientations of the foliations and the finite strain ellipsoid.

For stylolites with a distinct tooth-like morphology (see the left diagram in Figure 11.3*A*), the teeth of the stylolite are generally at a high angle to the average stylolite surface (on "slickolites," however, the spikes are at a low angle to the solution surface; see Figure 14.16*C*). The morphology of this type of stylolite suggests that solution occurs at the ends of the teeth and that the sides must be parallel to the direction of relative displacement across the stylolite. Thus we infer that the sides of the teeth should be parallel to the direction of maximum shortening (minimum stretch $\hat{s}_3$) of the finite strain ellipsoid, and that shortening has occurred roughly perpendicular to the stylolite. To maintain this geometry, the principal axes of strain could not have rotated significantly, relative to the stylolite, during the shortening. An alternative interpretation, based on Riecke's principle, is that if the stylolite formed by pressure solution, we would expect it to be oriented normal to the maximum compressive stress $\hat{\sigma}_1$.

Strictly speaking, the two interpretations could be consistent only if the principal axes of stress and of instantaneous strain were parallel, a rheologically simple but reasonable assumption (see Section 16.6(i)), and if the deformation were coaxial so that the principal axes of instantaneous and finite strain would be parallel. In that case, the principal axes of stress and of finite strain also would be parallel. The two interpretations could also be approximately consistent if the total strain were small, so that even if the deformation were noncoaxial, the principal axes of instantaneous and finite strain would still be nearly parallel.

Other morphologies of disjunctive foliation in limestone (see Figures 11.3*A*, *B*; 11.5; and 14.4) show a progression from stylolitic foliation, through anastomosing and rough foliations, to smooth foliation, with a corresponding decrease in the spacing and an increase in the smoothness of the cleavage domains. Figure 14.4*A*, *B* shows that this progression correlates with a progressive increase in the amount of shortening at a high angle to the foliation. In this figure, foliations are at a high angle to the bedding. Shortening of the limestone bedding by solution is recorded by layers of chert in the limestone, which are much less soluble than the limestone and thus must accommodate the shortening by imbrication (Figure 14.4*A*). The amount of imbrication of the chert beds is a minimum measure of the shortening in the

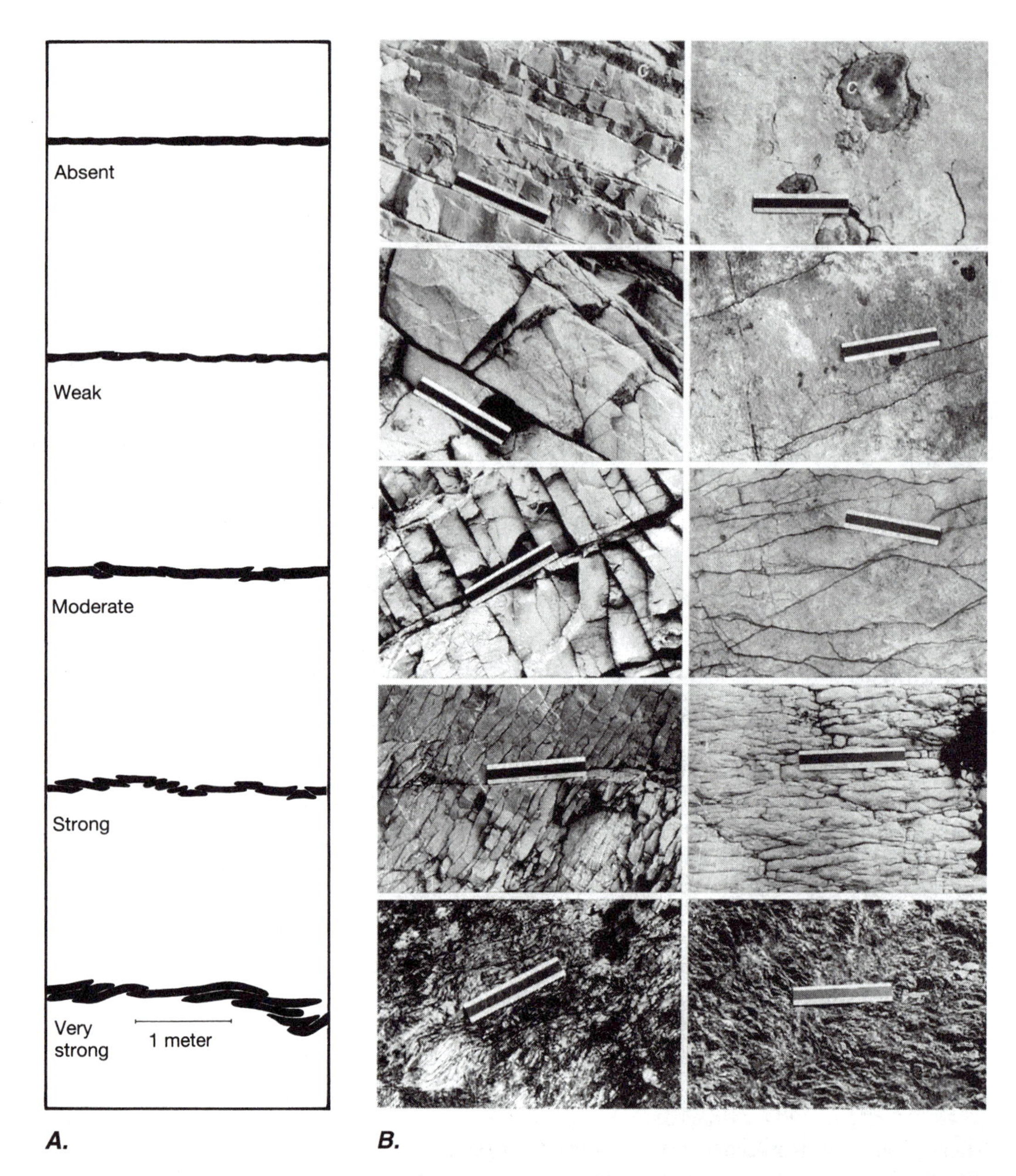

FIGURE 14.4 Development of disjunctive foliation in limestone correlated with the amount of shortening strain accommodated by solution. *A.* Observed progressive deformation of a bed of insoluble chert that becomes imbricated as the limestone shortens by solution along disjunctive foliation surfaces. The amount of imbrication is a minimum measure of the amount of shortening associated with solution. *B.* Foliation morphologies on surfaces perpendicular to bedding (left column; rulers parallel to bedding trace) and parallel to bedding (right column; rulers parallel to foliation trace) in a sequence (top to bottom) of increasingly deformed limestones. The shortening indicated by the diagrams of chert beds in *A* is associated with the adjacent foliation morphology shown in *B*. Scale in each photo is 6 in. (≈15.7 cm) long. *C.* Spaced disjunctive foliation in a limestone. The foliation diverges around insoluble chert nodules and converges toward regions of maximum imbrication of the nodules. (*A* and *B* from Alvarez et al. 1978; *C* from Alvarez and Engelder 1976, figure 3)

surrounding limestone parallel to bedding and at a high angle to the foliation. In weakly deformed rocks (top of Figure 14.4*A*, *B*), the foliation comprises a widely spaced irregular stylolitic foliation. Moderately deformed rocks (middle of Figure 14.4*A*, *B*) show a more closely spaced anastomosing foliation. The strongly deformed rocks (bottom of Figure 14.4*A*, *B*) show a closely spaced subplanar foliation. These observations are consistent with the interpretation that the foliation occurs subparallel to the plane of flattening of the finite strain ellipsoid, although an orientation normal to the maximum compressive stress, as predicted by Riecke's principle, is also possible.

A similar conclusion can be drawn from the example shown in Figure 14.4*C*. In the undeformed limestone, chert nodules are present in distinct layers, and nowhere are the nodules imbricated. The chert nodules are much less soluble than the limestone, so shortening in the limestone is accommodated by the imbrication of the chert nodules within the individual layers. Thus measurement of the amount of overlap of the imbricated nodules gives a minimum amount of shortening parallel to the bedding, which neglects any original spacing between nodules. The limestone is not strained or dissolved immediately adjacent to individual nodules, because the nodule has not strained. Shortening of the limestone by solution thus becomes locally concentrated in zones of imbrication of the nodules, and as a result, the distribution of strain near the nodule layer is highly heterogeneous. The foliation diverges around unshortened chert nodules and converges toward the highly imbricated zones (Figure 14.4*C*). This pattern is consistent with the distribution of orientations of the plane of flattening ($\hat{s}_1$, $\hat{s}_2$) of the finite strain ellipse that would develop in such an inhomogeneous deformation.

A model for the formation of disjunctive foliations in argillite is summarized in Figure 14.5. This model assumes plane coaxial deformation, with the stretch parallel to the coordinate axis x_2 always equal to 1. An undeformed argillite may have a foliation parallel to bedding, called a **bedding plane fissility**, which is formed by platy minerals being deposited with the mineral plane parallel to the depositional (bedding) surface. Subsequent compaction parallel to x_1 (Figure 14.5*A*) results in a volume decrease due to the loss of pore space, and thus the initial strain is an oblate ellipsoid (pancake-shaped; $\hat{s}_1 = \hat{s}_2 = 1$) with the ($\hat{s}_1$, $\hat{s}_2$)-plane parallel to the bedding (x_2, x_3).

Initial tectonic shortening parallel to x_3 is accommodated by both further compaction and solution, both volume-loss processes, thereby transforming the oblate total strain ellipsoid into a prolate one (cigar-shaped, $\hat{s}_1 = 1 > \hat{s}_2 > \hat{s}_3$) (Figure 14.5*B*). This deformation gradually develops a new tectonic disjunctive foliation that appears first as an anastomosing foliation and then a pencil cleavage defined by the intersection of the new disjunctive foliation with the primary foliation and bedding (cf. Figure 11.20*B*). Although the lineation is technically a stretching lineation parallel to the $\hat{s}_1$ axis (which is parallel to the x_2 coordinate axis in the figure), the deformation takes place largely by compaction and solution, and little if any real lengthening takes place (i.e., $\hat{s}_1 \approx 1$).

Further tectonic shortening parallel to x_3 occurs by constant-volume deformation, which involves lengthening normal to bedding (parallel to x_1) (Figure 14.5*C*). As this process continues, the primary foliation is gradually destroyed, and the tectonic foliation becomes increasingly planar and regular. The lengthening normal to bedding eventually makes this direction the maximum principal stretch $\hat{s}_1 > 1$, and a new stretching lineation develops parallel to the new orientation of $\hat{s}_1$. The earlier pencil lineation orientation is preserved as an intersection lineation between the tectonic foliation and the bedding, and it is parallel to what has become the intermediate principal stretch $\hat{s}_2 = 1$ (parallel to x_2; Figure 14.5*D*), which is commonly parallel to a set of fold axes. The progression of the strain states is indicated on a logarithmic Flinn diagram in Figure 14.5*E*.

The onset of noncoaxial deformation, such as is required by fold formation, would rotate material foliation surfaces, which would alter, if only slightly, the relationship between the foliation and the finite strain ellipsoid.

iii. Crenulation Foliations

Crenulation foliations are common features that may form by several kinematic mechanisms. Different mechanisms lead to different relationships between the orientation of the foliation and that of the principal strain axes. Thus the interpretation of strain orientations from crenulation foliations is tricky.

We distinguish zonal from discrete crenulation foliations (Figures 11.1, 11.9, 11.10) and symmetric from asymmetric crenulations (Figure 11.9*A*, *C*). Symmetric zonal crenulations probably form in response to shortening parallel to a pre-existing foliation S_1 (Figure 14.6*A*, *B*). The old foliation is rotated toward low angles to the axial surfaces of the new crenulations, which form the new foliation S_2. Solution of material from the limbs may strengthen the S_2 foliation. Such symmetrical crenulations, which may have rounded or sharp hinges (Figure 14.6*B*, *C*), commonly occur in the core of a lower-order fold, where the crenulation foliation is subparallel to the axial surface of the larger fold. In this model, the direction of minimum principal stretch $\hat{s}_3$ (maximum shortening) is normal to the crenulation foliation.

Models for the formation of asymmetric crenulations and the resulting foliations include (1) shortening oblique to a pre-existing foliation, (2) shear across an earlier foliation, and (3) rotation and asymmetric deformation of initially symmetric crenulations. Solution of material from cleavage domains may accompany the deformation. We discuss these models in turn.

Asymmetric crenulations can form by shortening at a low angle to the initial foliation (Figure 14.7). The axial

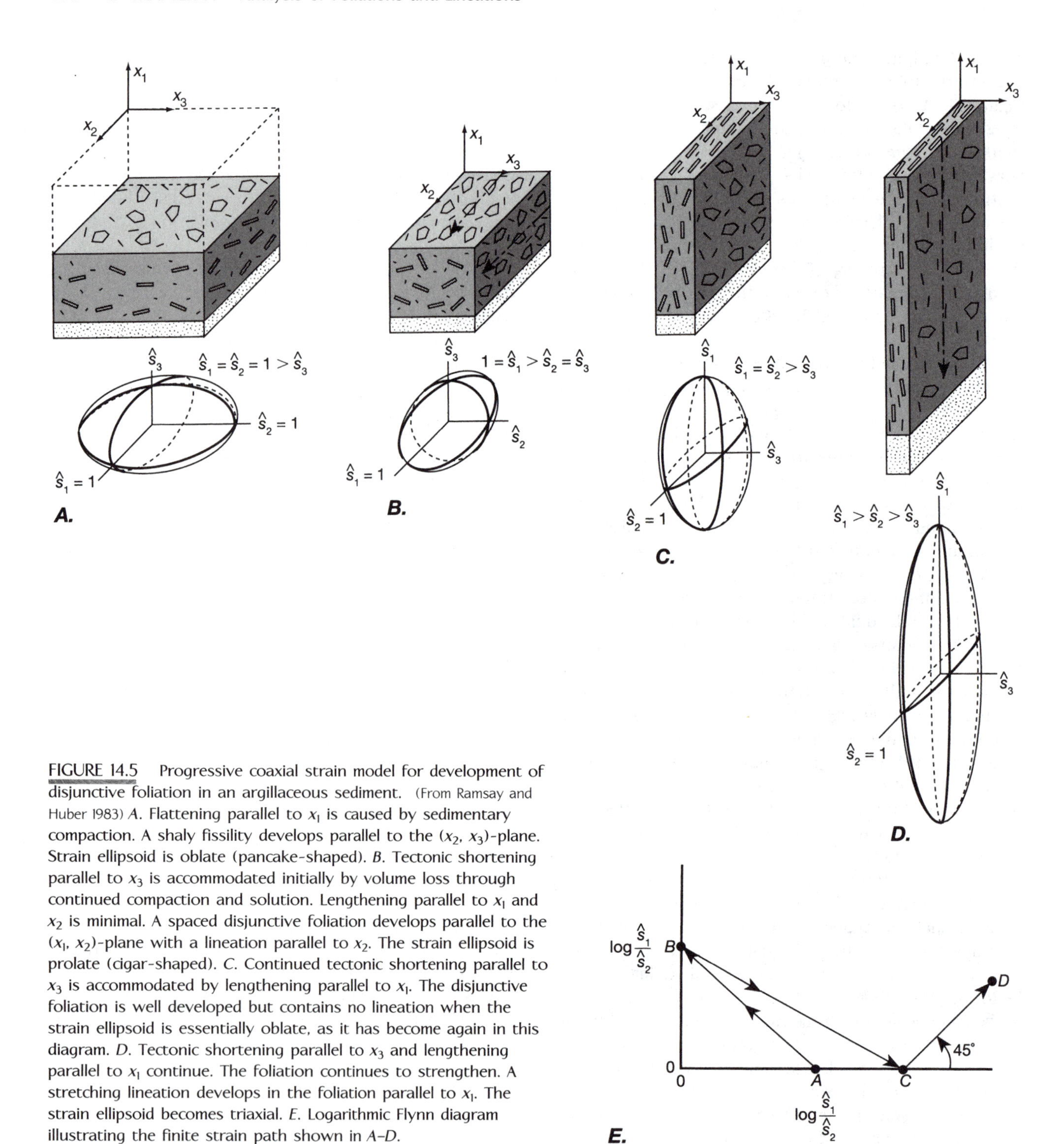

FIGURE 14.5 Progressive coaxial strain model for development of disjunctive foliation in an argillaceous sediment. (From Ramsay and Huber 1983) *A*. Flattening parallel to x_1 is caused by sedimentary compaction. A shaly fissility develops parallel to the (x_2, x_3)-plane. Strain ellipsoid is oblate (pancake-shaped). *B*. Tectonic shortening parallel to x_3 is accommodated initially by volume loss through continued compaction and solution. Lengthening parallel to x_1 and x_2 is minimal. A spaced disjunctive foliation develops parallel to the (x_1, x_2)-plane with a lineation parallel to x_2. The strain ellipsoid is prolate (cigar-shaped). *C*. Continued tectonic shortening parallel to x_3 is accommodated by lengthening parallel to x_1. The disjunctive foliation is well developed but contains no lineation when the strain ellipsoid is essentially oblate, as it has become again in this diagram. *D*. Tectonic shortening parallel to x_3 and lengthening parallel to x_1 continue. The foliation continues to strengthen. A stretching lineation develops in the foliation parallel to x_1. The strain ellipsoid becomes triaxial. *E*. Logarithmic Flynn diagram illustrating the finite strain path shown in *A–D*.

surfaces of the kinks form at a high angle to the original foliation and are parallel to a new crenulation foliation S_2. S_2 is defined by the short limbs of the crenulations in which the old foliation S_1 rotates into subparallelism with the new foliation. There is no net shearing of the body parallel to the new foliation, and the axis of maximum shortening (minimum principal stretch $\hat{s}_3$) is normal to the new foliation.

Asymmetric crenulation foliations can also form by the development of **shear bands**, which are bands of concentrated shear that cut across a pre-existing foliation and define new cleavage domains. A crenulation foliation inferred to have formed by this mechanism is sometimes called a **shear band foliation**, a genetic term. If the maximum instantaneous shortening direction $\hat{\zeta}_3$ is initially at a low angle to a pre-existing foliation S_1, shear

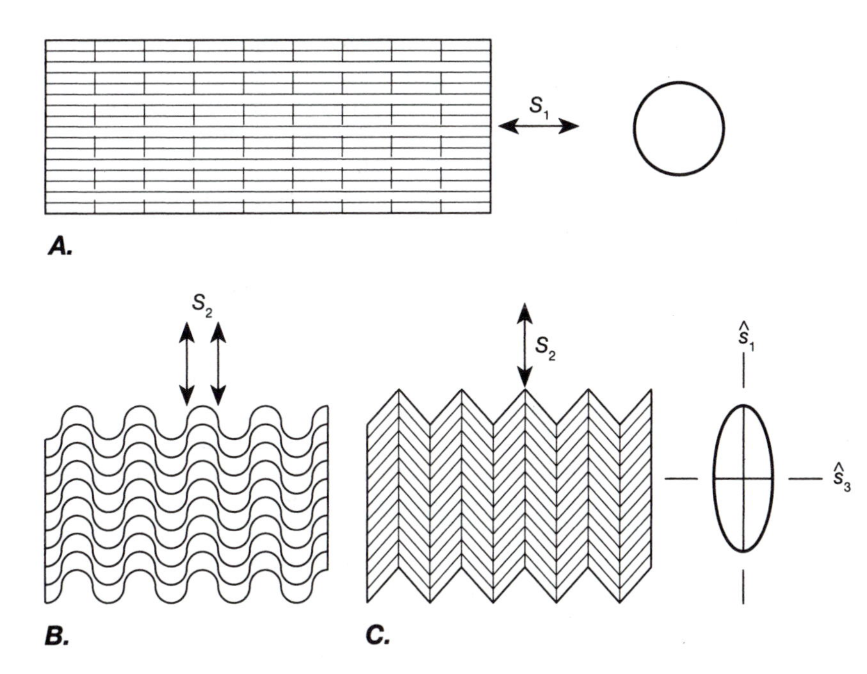

FIGURE 14.6 Production of a symmetrical crenulation foliation S_2 by buckling of S_1. Strain ellipses reflect the macroscopic strain. *A*. Initial foliation S_1 is parallel to the direction of maximum shortening. *B*. Progressive shortening produces symmetrical crenulations by buckling. The limbs become the cleavage domains for the new crenulation foliation S_2. *C*. Chevron style of crenulation. The axial surfaces define the new foliation S_2.

bands can form in which the old foliation undergoes a large rotation to form close to tight crenulations (Figure 14.8*A*, *B*). If the maximum instantaneous stretching direction $\hat{\zeta}_1$ is at a low angle to S_1, shear bands form in which S_1 only rotates a small amount, forming open to gentle crenulations (Figure14.9*A*, *B*). This geometry of shear band foliation is sometimes called an extensional crenulation foliation, implying that the maximum extension is subparallel to the original foliation. In both cases, the pre-existing foliation S_1 rotates toward parallelism with the shear band, thereby defining the new crenulation foliation S_2 (Figures 14.8*A*, *B* and 14.9*A*, *B*). The rotation may be enhanced by volume loss in the shear band, which accommodates additional shortening parallel to $\hat{s}_3$ and thins the shear band (Figures 14.8*C* and 14.9*C*).

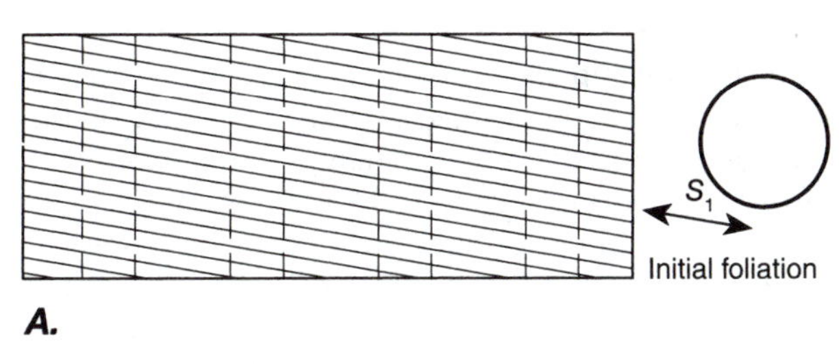

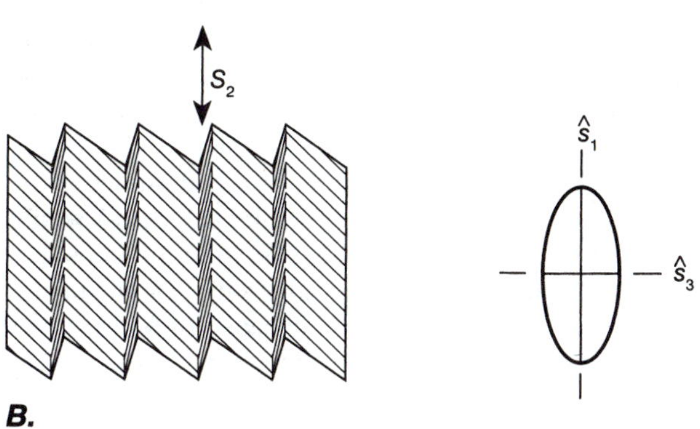

FIGURE 14.7 Formation of an asymmetric crenulation foliation S_2 by buckling of S_1. Strain ellipses reflect the macroscopic strain. *A*. The initial foliation S_1 is at a low angle to the direction of maximum shortening. *B*. Asymmetric crenulation forms by buckling of S_1. The short limbs of the crenulations, in which the original foliation has rotated to a low angle with the crenulation axial surface, define the new crenulation foliation S_2. Hinges may be sharp or rounded.

The shear bands form in orientations of high shear strain relative to the principal axes of instantaneous strain, but there is no general and unique relationship between the principal axes of finite strain and the foliation. With progressive deformation, the shear bands behave as a material foliation, so at very high strains, they may rotate into low angles with the finite strain plane of flattening (see the total strain ellipses in Figures 14.8 and 14.9). If this occurs, a second generation of shear bands could develop in an orientation of higher instantaneous shear strain.

An extensional crenulation foliation resulting from sequential superposed deformations forming first S_1 and then S_2 should not be confused with an S-C fabric in which both S and C develop in a single deformational event (see Figure 11.16 and Section 11.6(ii)). In some cases, however, the distinction is difficult to make. For example, during the formation of an S-C fabric in a shear zone, with increasing strain S rotates progressively toward parallelism with C. When the two become essentially parallel, shear bands, labeled C′, may develop at a low angle to the main shear zone. The morphology of the shear band foliation may be difficult to distinguish from the original S-C morphology. Fortunately, the sense of

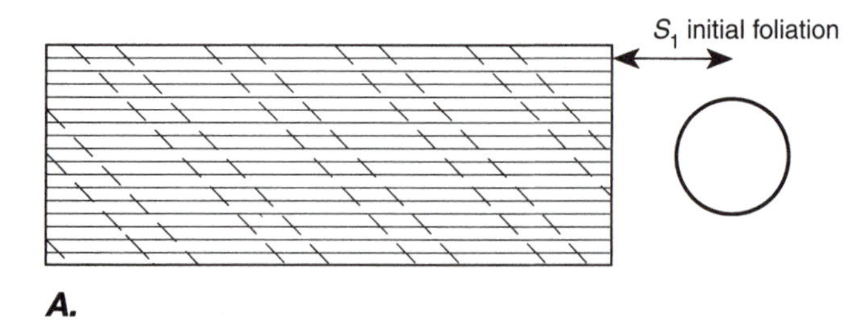

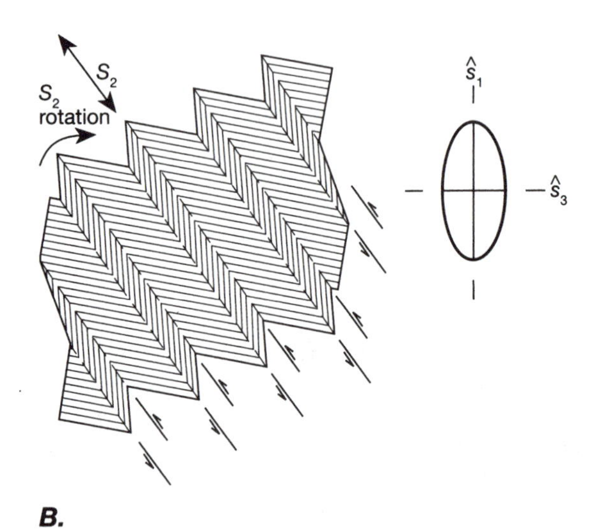

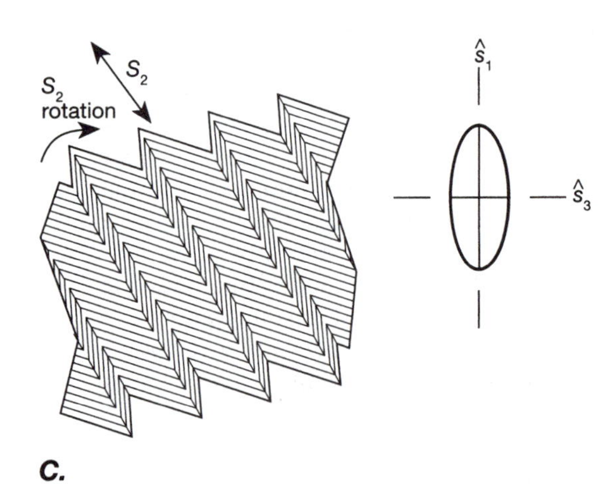

FIGURE 14.8 Production of an asymmetric crenulation foliation S_2 by the formation of shear bands crossing a pre-existing foliation. Strain ellipses reflect the average macroscopic strain. *A.* Undeformed state with foliation initially parallel to the direction of maximum instantaneous shortening $\hat{\zeta}_3$. *B.* Simple shear within shear bands produces asymmetric close to tight crenulations. Shear bands rotate toward higher angles with the axis of maximum finite shortening $\hat{s}_3$. The shear bands become the S_2 cleavage domains. Rounded hinges are also common. *C.* Volume loss in the cleavage domains strengthens the crenulation foliation S_2.

shear on the shear zone may be inferred from the shear band foliation using the same criterion as is applied to the original S-C relationship (see Figure 11.16), because even though the S-C and C′ foliations form sequentially, they are related to the same deformational event.

An asymmetric crenulation also could develop from a symmetric one by preferential solution of components from one set of limbs. This mechanism might occur, for example, if a layer containing an initial symmetrical crenulation (Figure 14.10*A*) is buckled into a lower-order fold by the flexural shear mechanism. Subsequent rotation of the crenulations in the limbs of a lower-order fold orients one set of crenulation limbs at a higher angle to the overall maximum shortening direction than the other set, and solution from those limbs and deposition in the adjacent limbs could accommodate both shortening of the rock normal to the axial plane of the low-order fold and shear parallel to its limbs (Figure 14.10*B*, *C*). The result could be interpreted as a consequence of apparent shear along the new crenulation foliation S_2, which would not be an accurate interpretation of the actual deformation. No unique relationship between foliation and strain arises from this mechanism, although the foliation probably would end up at a low angle to the plane of flattening.

Crenulation foliations commonly exhibit a compositional banding emphasizing the distinction between the cleavage domains and microlithons. Cleavage domains tend to be enriched in platy minerals and depleted in quartz compared to both the microlithons and the uncrenulated rock. In some cases, the microlithons are enriched in quartz, suggesting solution of quartz in the cleavage domains and precipitation in the microlithons. The banding shown in Figure 11.9*A*, *C*, for example, could result from preferential solution of more highly deformed quartz in the cleavage domains. It has also been suggested that quartz might dissolve more readily at the interface with a mica cleavage plane, and this effect might be enhanced, according to Riecke's principle, by a high compressive stress normal to the interface. If that were true, then in the limbs of the crenulations the maximum compressive stress $\hat{\sigma}_1$ would be at a high angle to the quartz-mica interfaces, and quartz would readily dissolve. In the microlithons, however, the quartz-mica interfaces would be subparallel to $\hat{\sigma}_1$ and at a high angle to the minimum compressive stress $\hat{\sigma}_3$, and on those surfaces, the quartz would tend to precipitate. This mechanism could account for the migration of quartz from cleavage domain to microlithon.

Discrete crenulation foliations develop from zonal crenulations by recrystallization of the platy minerals in the cleavage domains. New platy mineral grains replace the old grains and grow parallel to the cleavage domain, thereby producing a discontinuity between it and the microlithon.

In general, then, asymmetric crenulation foliations are difficult to interpret in terms of the orientation of the principal axes of strain, although if the strains are suffi-

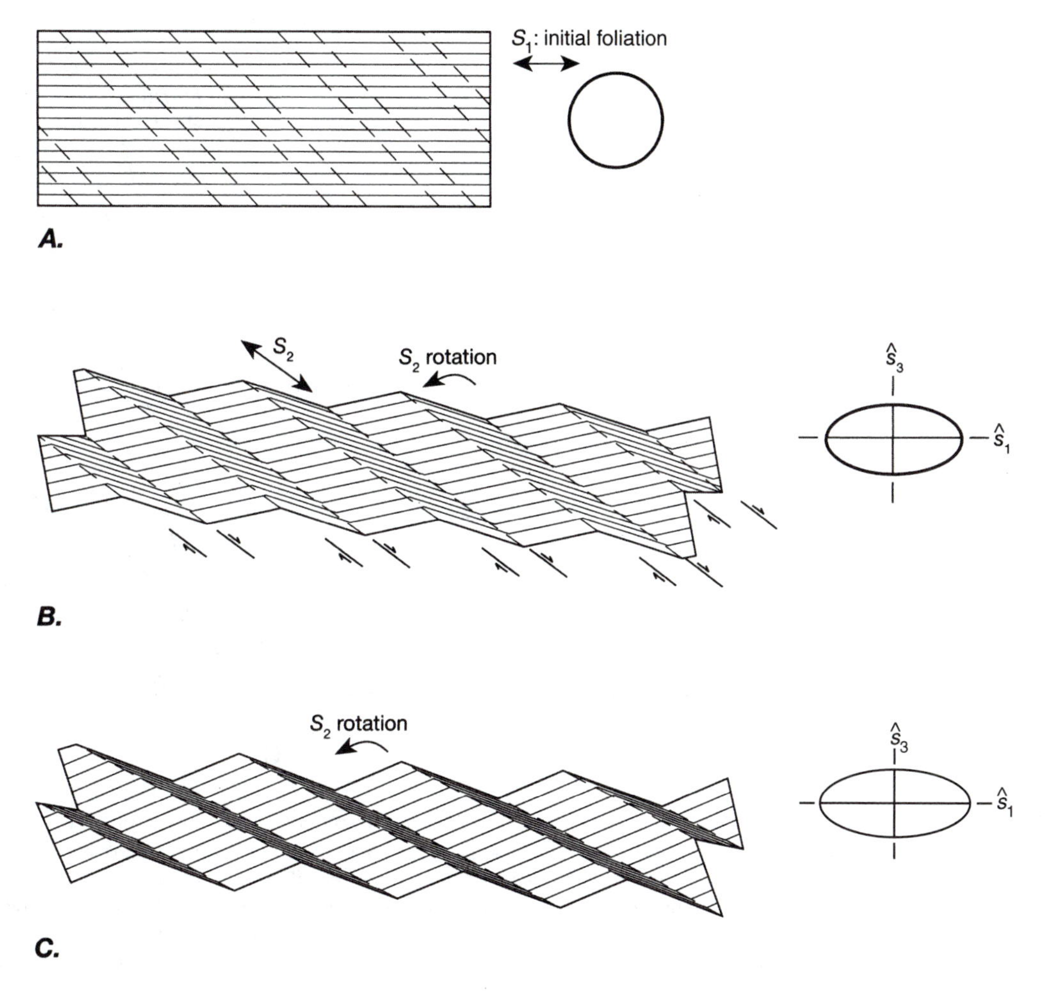

FIGURE 14.9 Production of an asymmetric extensional crenulation foliation S_2 by formation of shear bands that cross a pre-existing foliation. Strain ellipses indicate the average macroscopic strain. *A.* Undeformed state with S_1 foliation initially parallel to the direction of maximum instantaneous lengthening $\hat{\zeta}_1$. *B.* Extension is accommodated by the formation of shear bands producing open to gentle asymmetric crenulation foliation S_2. Rounded hinges are also common. *C.* The crenulation foliation is enhanced by volume loss from the S_2 cleavage domains.

ciently large, the foliations tend to rotate toward parallelism with the plane of flattening ($\hat{s}_1$, $\hat{s}_2$).

iv. Continuous Foliations

Although the continuous fine foliation known as "slaty cleavage" gives the impression of continuity down to the finest scales visible with the naked eye or a hand lens, microscopic and scanning electron microscopic examination of most slates reveals a microdomainal structure having the characteristics of crenulation or disjunctive foliations (see Figure 11.11). The evolution of slaty cleavage shown in Figure 11.11 is an example of the effect of progressive reconstructive recrystallization of clay minerals. The initial foliation is a microzonal crenulation foliation (Figure 11.11*A*). With increasing amounts of recrystallization of the platy clay minerals, the structure of the foliation changes to a microdiscrete crenulation foliation, and gradually to a microdisjunctive foliation with a very strong fabric in the microlithons but no remnant of the initial crenulations (Figure 11.11*B*). These slates are still low-grade metamorphic rocks.

Such studies suggest that microcontinuous slates, as well as coarse continuous foliations (see Figure 11.12*A*), develop as the products of recrystallization of earlier crenulation or disjunctive foliations. The processes of deformation, rotation, solution, and recrystallization can ultimately wipe out evidence of the initial structure (see Section 14.4). In rare circumstances, the origin of a coarse continuous foliation as an earlier crenulation foliation is preserved within porphyroblastic minerals such as garnets that overgrew and included the earlier foliation before it was transformed by recrystallization.

Although the complicated nature of the process does not lead to a simple prediction of the relationship between strain and foliation, crenulation and disjunctive

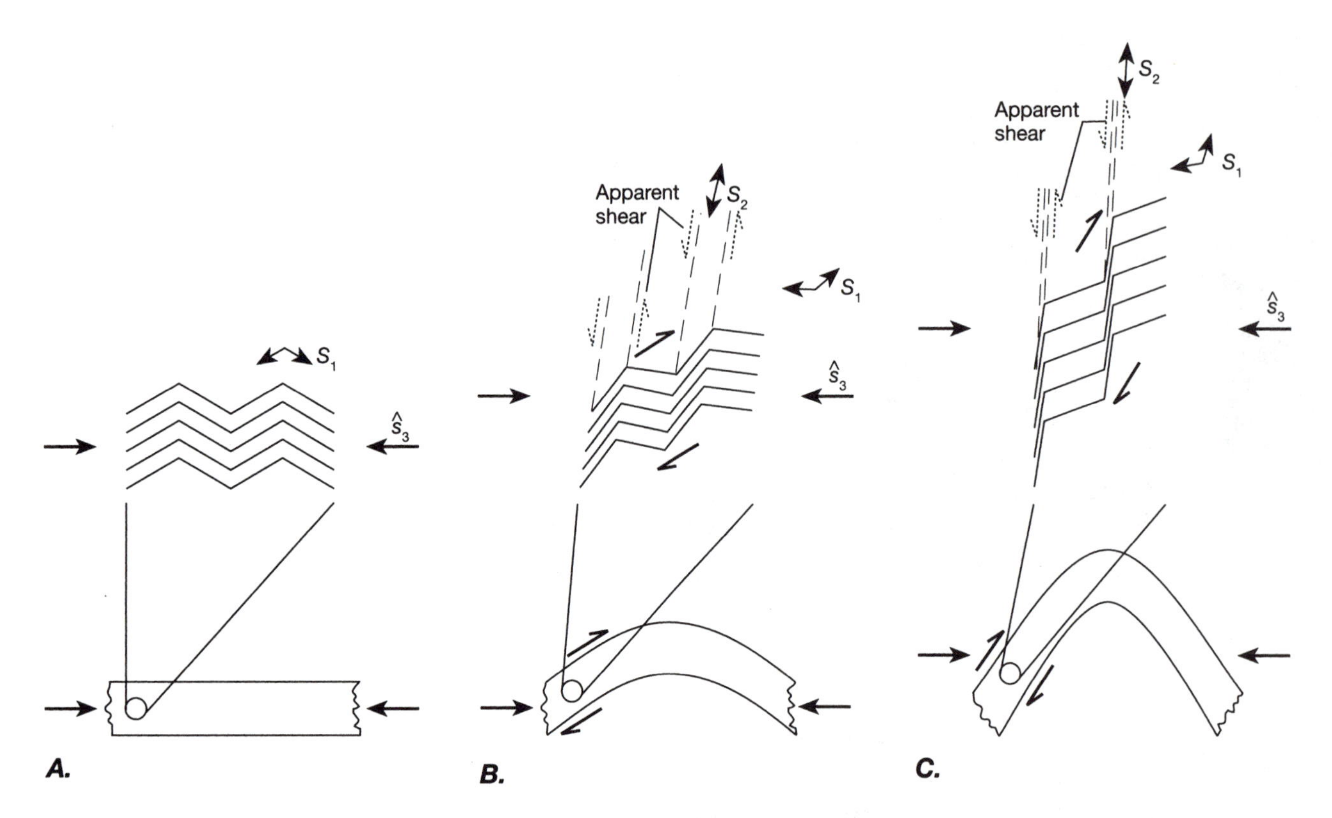

FIGURE 14.10 Asymmetric crenulations produced from symmetric crenulations by rotation during lower-order folding accompanied by solution in the crenulation limbs that lie at a high angle to the shortening direction and precipitation in the adjacent limbs. (Modified after Williams and Schoneveld 1981) *A*. Initial symmetric crenulations. *B* and *C*. Solution of material in the crenulation limbs that are at a high angle to $\hat{s}_3$ and precipitation in the adjacent limbs accommodates overall shortening and relative shear of the crenulated S_1 laminations. The set of crenulation limbs at a high angle to $\hat{s}_3$ become the cleavage domains of the S_2 foliation, and the other set of limbs becomes the microlithons. The apparent shear on S_2, indicated by dashed arrows, is an artifact.

foliations probably behave as material foliations that tend to evolve toward parallelism with the finite strain plane of flattening. Field studies in regions where the strain can be determined independently indicate consistently that, within the resolution of field measurements, which is probably on the order of about 5°, the foliation is parallel to the plane of flattening of the strain ellipsoid (see Sections 15.2 and 15.3).

Flattened grains of minerals such as calcite (Figure 11.12*B*) or quartz (Figure 3.7), define a continuous nonmaterial foliation, either fine or coarse. Since the grain shape reflects the strain in the rock, the foliation probably tracks closely the plane of flattening of the finite strain ellipsoid. An initial grain-shape preferred orientation is unlikely to be significant for such minerals, but recrystallization may reset the record of strain recorded by such grains.

Continuous coarse discrete foliations defined by deformed discrete objects such as pebbles (Figure 11.19*A*), alteration spots (Figure 11.19*B*), or ooids (Figure 12.6*B*), are nonmaterial foliations that tend to be parallel to the plane of flattening of finite strain, because the strain of the objects reflects the strain in the rock. Deviations from this orientation are caused by an initial shape preferred orientation (Figure 14.1), but the deviation decreases with increasing finite strain.

14.4 STEADY-STATE FOLIATIONS

Different types of foliation are not independent of one another and in fact may develop from one another, as is evident from Figures 11.9, 11.11, and 11.17. Thus crenulation foliations develop from the deformation of an earlier foliation, and with progressive deformation and recrystallization a crenulation foliation can evolve into a continuous foliation. Continuous foliations in turn can become crenulated (Figure 11.9) and evolve into a crenulation foliation.

Such sequences of development raise the question of whether a foliation ever reaches a final stage of evolution and, if so, under what circumstances that could occur. In some regions, two or more cycles of foliation evolution have been deciphered, each of which includes the crenulation of an initial foliation, whether of sedimentary or tectonic origin, followed by the formation of a crenulation foliation, commonly with compositional differentiation accentuating the difference between cleavage domains and microlithons, followed in turn by increasing recrystallization and the development of a new continuous foliation. Such cycles may in fact result from changing tectonic conditions such as the temperature of deformation or the geometry of the imposed deformation, or both. Thus the cycles do not necessarily characterize a steady flow of the rock under stable conditions, for which a stable morphology of foliation might in principle develop.

The spatial and temporal variation of conditions of deformation are probably common, such that the realistic identification of different generations of foliation and the correlation of particular generations from one area to another are highly suspect. The designation of different generations of foliation by numerical subscripts such as S_1 and S_2 therefore may only have very local significance. If applied over a large area, such a system of notation could indicate an erroneous conclusion concerning the correlation of different foliations and the simplicity of the deformational history.

14.5 FOLIATIONS AND SHEAR PLANES

The relationship between foliations and shear planes was for many years a source of confusion and misinterpretation in structural geology. In part this situation stemmed from the classical so-called passive shear model for producing similar folds, which requires inhomogeneous simple shear on planes parallel to the axial surface (see Section 13.2, Figure 13.8). Because a foliation is commonly subparallel to the axial surface of similar style folds, it was interpreted as the plane of simple shear. There is considerable evidence, however, that foliations are usually parallel or subparallel to the plane of flattening of the strain ellipsoid (see Section 14.3), an interpretation generally incompatible with foliations being parallel to shear planes. Under what circumstances, then, can shear strain accumulate across foliation planes? Under what circumstances can foliations be planes of simple shearing? And how reliable is the interpretation that shear has occurred on a foliation plane?

Any material foliation is a plane across which shear strain accumulates during a noncoaxial deformation. This fact is indicated by the initially perpendicular pairs of material lines such as a and a' through d and d' in Figure 12.15. In general, any two such lines are sheared into a nonperpendicular orientation, and they return to being perpendicular only for the instant that they are parallel to the principal axes of strain. At that instant, the total or finite shear strain for that pair of lines is zero. The instantaneous shear strain for those lines, however, is not zero because the principal axes of the instantaneous strain are not parallel to those of the total or finite strain. This difference in orientation means that at a given instant, two material lines can be parallel to the principal axes of finite strain and thus show no net shear strain and yet be shearing relative to one another because they are not parallel to the principal axes of instantaneous strain. Thus shear strain can accumulate across any material foliation plane throughout the deformation. This result is simply a consequence of the geometry of deformation of material lines and planes, but it does not imply that such foliations are planes of simple shearing.

Many foliations, both material and nonmaterial, are defined by strong planar mechanical anisotropies in the rock. Such anisotropies can have a significant effect on the geometry of ductile deformation, just as they do in the case of brittle fracture (see Figures 8.13–8.16). As an example, let us consider a state of pure shear stress (Figure 7.12*G*) or a deviatoric stress (Figure 7.12*H*) that is associated with a progressive simple shear in which the shear plane is the plane of maximum shear stress. The principal axes of instantaneous strain, $\hat{\zeta}_1$ and $\hat{\zeta}_3$, are parallel to the principal axes of stress, $\hat{\sigma}_1^{(Dev)}$ and $\hat{\sigma}_3^{(Dev)}$, respectively (Figure 14.11*A*). A foliation that is either a material plane or parallel to the plane of flattening of the finite strain ellipse rotates toward the shear plane as strain increases. If the yield stress for shear on the foliation is lower than the yield stress for shear across the foliation on the plane of maximum shear stress, then the foliation could become the shear plane when it rotates sufficiently close to the maximum shear stress.

Figure 14.11 illustrates this situation. On the Mohr diagrams of deviatoric stress, the higher yield criterion (solid lines) is for ductile shear across the foliation plane and the smaller yield criterion (dashed lines) is for ductile shear along the foliation plane. The surface stress on any plane cannot exceed the yield stress for shear along that plane, thus limiting the possible diameter of the Mohr circle. With the foliation at a high angle δ to the simple shear plane (δ is measured as the angle between the normals to the planes; Figure 14.11*A*), stress on the foliation is subcritical, and ductile shear occurs along the plane of maximum shear stress at a yield stress given by the solid line in Figure 14.11*B*. With increasing shear strain, the angle of the foliation decreases to δ' (Figure 14.11*C*). At this point the stress on the foliation and the stress on the plane of maximum shear stress are both at the critical value. With further decrease in the foliation angle to δ'' (Figure 14.11*D*), the differential stress must decrease so that the surface stress on the foliation does not exceed the yield stress. Thus the stress must relax to a value below the yield point for ductile flow on the original shear plane. In this situation, the foliation becomes the active shear plane even though it did not form as one.

If the foliation were initially parallel to the plane of flattening (Figure 14.12*A*), as soon as shearing on the foliation begins, the strain ellipse is rotated to a different orientation that is no longer parallel to the foliation

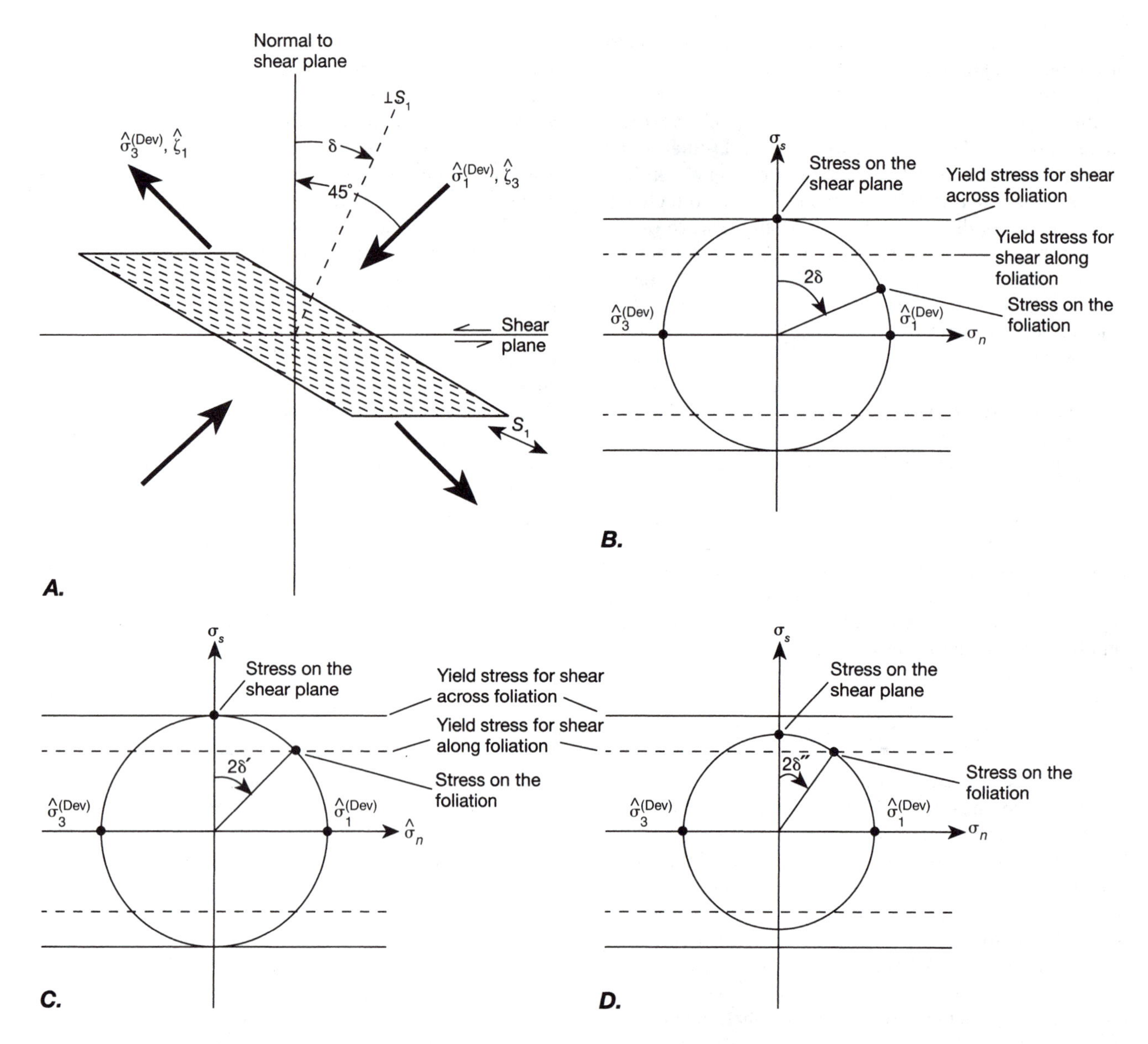

FIGURE 14.11 The transformation of a foliation plane from a passive plane to an active shear plane. In the Mohr diagrams, the solid lines are the von Mises criterion for ductile flow across the foliation; the dashed lines are the von Mises criterion for ductile flow on the foliation. *A.* During progressive simple shear, a material foliation S_1 rotates toward the shear plane: δ decreases progressively. *B.* The Mohr circle for deviatoric stress at large foliation angles. Shearing does not occur on the foliation because the stress on the foliation plane is below the yield stress for flow parallel to those planes. *C.* The foliation angle has decreased to the value δ', at which stress on the foliation plane is at the yield stress, resulting in shear parallel to the foliation plane. *D.* Further decrease in the foliation angle to δ'' requires the differential stress to decrease so that the stress on the foliation does not exceed the yield stress. Stress on the original shear plane drops below the yield stress, and shearing on that plane ceases; shearing on the foliation continues.

(Figure 14.12*B*). Note that the sense of rotation is initially in an opposite sense to that normally expected for the finite strain ellipse in simple shear.

The inference that shear has occurred on foliation surfaces commonly relies on the observation of displaced lithologic boundaries, which raises the problem of distinguishing between separation and displacement, as in the case of interpreting displacement on faults (cf. Figures 3.23 and 3.24). Beyond this geometric effect, however, several mechanisms exist that result in apparent shear of lithologic boundaries where in fact none has occurred, and we discuss these next.

First, if a lithologic layer is cut obliquely by a disjunctive foliation along which there has been significant amounts of solution, the effect can be an apparent displacement or shearing of the layer on the

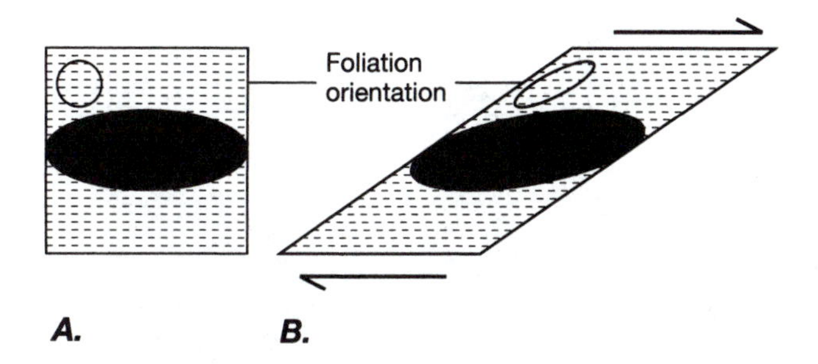

FIGURE 14.12 Shear parallel to the plane of flattening of the finite strain ellipse causes the ellipse to change shape so that the shear plane is no longer parallel to the plane of flattening. *A*. Finite strain ellipse (solid black) parallel to a foliation. The circle in the upper left records the subsequent deformation. *B*. Shear parallel to the plane of flattening in *A* causes the principal axes of finite strain initially to counterrotate away from the shear plane in a sense opposite to the sense of shear. The small ellipse in the upper left records the amount of shear.

foliation, whereas in fact the displacement may have been strictly normal to the foliation (Figure 14.13, layers 1, 2, and 5). Second, if a thin layer is initially ptygmatically folded in association with deformation of the surrounding rock, subsequent preferential solution of one set of limbs of these folds along a disjunctive foliation can produce an apparent shear of the layer along the foliation (Figure 14.13, layers 3 and 4, and Figure 14.14). Again, however, the displacement may have been strictly normal to the foliation. Note that in Figure 14.13*C*, one might infer opposite senses of shear on the foliation to explain the apparent displacement, even for layers such as 2, 3, and 4 that had the same initial orientation. Thus both senses of apparent shear could result from solution of layers with the same initial orientation. Figure 14.14 shows an example in which the selective solution of ptygmatic fold limbs is present in varying stages of advancement.

Although some foliation planes are demonstrable shear planes, such as C planes in shear zones (Figure 11.16) on which slickenfibers may develop, the association of a foliation plane with the dominant shear plane for the deformation cannot be assumed, and even evidence of apparent shear displacement on the foliation planes must be examined critically.

14.6 INTERPRETATION OF MORPHOLOGICAL TYPES OF LINEATION

Relatively few studies have been published specifically regarding the relationship between strain and lineations, although evidence for that relationship often results from studies of strain in deformed rocks. We review here what we might expect from our understanding of the mechanism by which various lineations form.

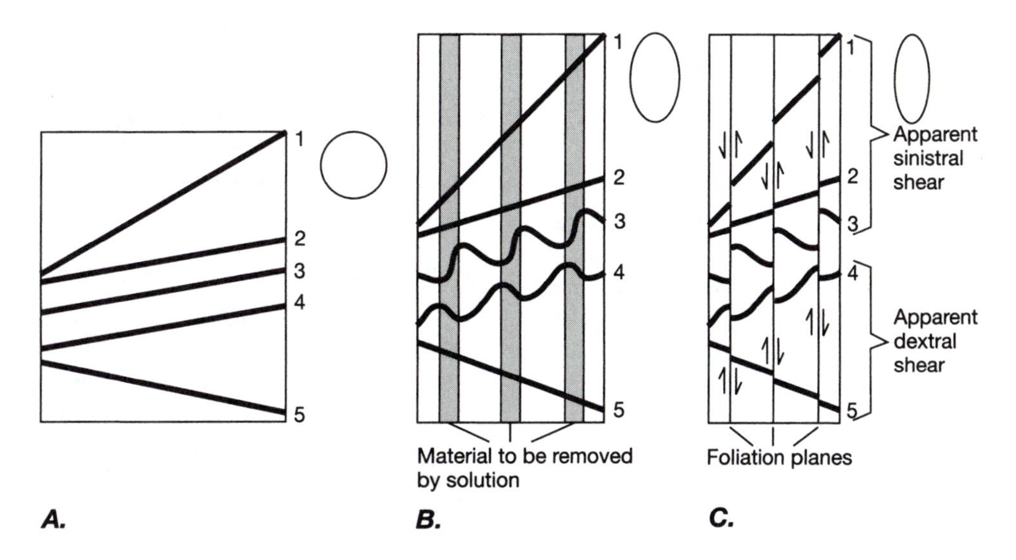

FIGURE 14.13 Apparent shear along a disjunctive foliation resulting from solution along the surfaces of the foliation. Displacement is strictly normal to the foliation. The strain ellipses show the bulk strain. *A*. The undeformed state. Layers 1, 2, and 5 are incompetent and behave passively during the deformation. Layers 3 and 4 are competent. *B*. After homogeneous constant volume flattening (pure shear), the competent layers 3 and 4 have buckled. The shaded areas indicate where solution will remove material, producing a disjunctive foliation. *C*. After deformation by solution along the foliation planes, layers 1, 2, and 3 suggest sinistral shear on the foliation, and layers 4 and 5 suggest dextral shear. The apparent shear direction is determined by the orientation of the layer with respect to the foliation (compare layers 1 and 2 with 5) or by which limb of the fold train is systematically removed by solution (compare layer 3 with layer 4).

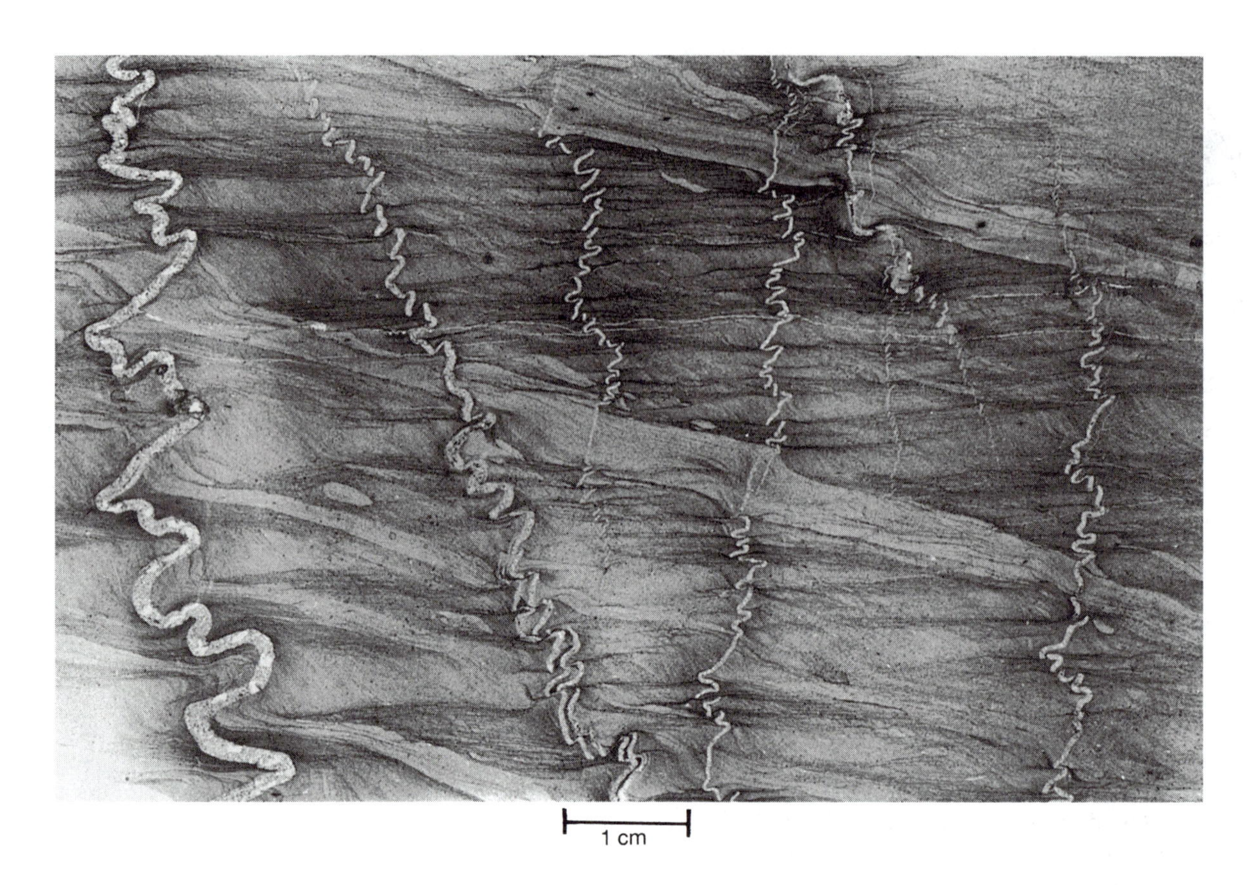

FIGURE 14.14 Photograph showing various stages in the progressive solution of limbs of ptygmatic folds. Compare with layer 3 in Figure 14.13*C* (cf. Figure 11.10). (From Gray 1979)

i. Discrete Structural Lineations

Initially spherical discrete objects that are distributed throughout the rock and that have deformed with it generally reflect the strain in the rock (Figures 11.19 and 12.6). Thus the long axes of these objects define a nonmaterial discrete **stretching lineation** that is closely parallel to the $\hat{s}_1$ axis of the finite strain ellipsoid. Any deviation of the discrete objects from an initially spherical shape and any associated initial preferred orientation can alter this relationship (cf. Figure 14.1), but as strain increases, the difference becomes smaller.

ii. Constructed Structural Lineations

Constructed lineations such as intersections of foliation and bedding, fold hinge lines, and boudin lines are material lineations that must lie within the plane of the bedding in which they form. It is always tempting to infer that boudin lines form normal to the direction of maximum principal stretch $\hat{s}_1$, and fold hinges form normal to the direction of minimum principal stretch $\hat{s}_3$. These relationships only hold, however, if the layers in which these structures form are parallel to the appropriate strain axis, as when tectonic shortening is parallel to bedding (cf. Figure 12.7). But the principal stretch axes associated with shearing, for example, need have no particular relationship to layers in the rocks, so in general, these structures simply record the component of stretch parallel to the layer in which they form (cf. Figure 12.18*C*, *D*). Thus boudin lines and fold hinges need not be parallel to any particular axis of strain. If these structural lineations behave as material lines in a deforming continuum, however, they would tend to rotate toward the direction of maximum stretch $\hat{s}_1$ as strains become large.

The Formation of Boudins. Boudins form during deformation if there is a component of lengthening parallel to a competent layer in an incompetent matrix (Figure 11.21). Ductile extension of the competent layer cannot keep pace with that of the incompetent matrix, and the layer tends to pull apart into boudins. As the difference in competence increases from small to large, the form of the boudins changes from a pinch and swell structure (Figure 11.21*B*), to separated boudins with pronounced necks (Figure 11.21*C*, *D*, *E*), to boudins with sharp ends that may actually be fractures (Figure 11.21*F*, *G*, *H*, *I*). Between the boudins is a region of locally low stress, which therefore becomes a favorable site for precipitation of minerals such as quartz or calcite (Figure 11.21*J*–*N*).

In simplified terms (Figure 14.15), the deforming incompetent matrix exerts a shear stress σ_s on both sur-

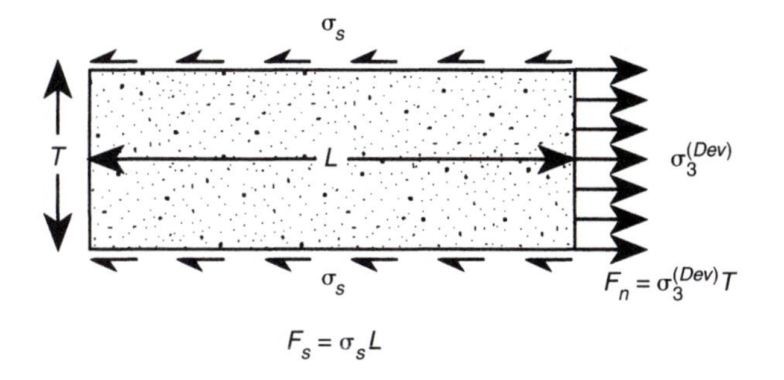

FIGURE 14.15 Formation of boudins. The length of each boudin is determined by the strength of the layer and the balance of forces created by the shear stresses σ_s on the surfaces of the layer and the deviatoric tensile stress $\sigma_3^{(Dev)}$ within and parallel to the layer (cf. Figure 11.21).

faces of the embedded competent layer, and the magnitude of the shear stress depends on the flow properties of the matrix and the rate of deformation. This shear stress provides a total shear force $2F_s$ that is parallel to the layer and increases with the length L of the layer on which the shear stress acts. The minimum principal deviatoric stress $\sigma_3^{(Dev)}$ within the layer is a tensile stress (cf. Figure 7.12*H*) that acts across the thickness T of the competent layer and provides a force F_n that balances the shear force.

$$F_s = \sigma_s L \qquad F_n = \sigma_3^{(Dev)}\, T \tag{14.1}$$

$$2F_s = F_n \tag{14.2}$$

The critical length L^* at which $\sigma_3^{(Dev)}$ reaches the strength $\sigma_3^{*(Dev)}$ of the competent layer (which is a measure of its competence) is the characteristic length of the boudin in cross section. This length thus defines the periodicity of failure sites in the layer. We find the length by substituting Equations (14.1) into Equation (14.2) and using the strength of the layer for the critical normal stress:

$$L^* = \frac{\sigma_3^{*(Dev)}}{2\sigma_s}\, T \tag{14.3}$$

Equation (14.3) indicates that boudin length should increase with increasing layer thickness and with increasing strength (competence) $\sigma_3^{*(Dev)}$ of the layer, and boudin length should decrease with increasing shear stress σ_s on the layer boundaries. Although the value of the shear stress depends in part on the competence of the matrix, its value can only be determined by solving the complete problem of flow (see Sections 16.1 and 18.1, Box 18-1); this is beyond the level of the current discussion.

The progression from pinch-and-swell structure (Figure 11.21*B*) to tapered boudins (Figure 11.21*C, D, E*) to block-shaped boudins (Figure 11.21*F, G, H, I, J*) represents a progressive increase in the competence contrast between the layer and the matrix. The evolution of the block boudin structure into barrel, fish-mouth, bow-tie-vein, or bone shapes (Figure 11.21*J, K, L, M, N*) is probably determined by the relative competences of the matrix, the competent layer, and the secondary mineral, as well as the rate at which the secondary mineral can accumulate to accommodate a portion of the layer lengthening. If precipitation of the secondary mineral can keep pace with the lengthening and the competence contrast between the layer and the matrix is high, the block structure in Figure 11.21*J* forms. If precipitation cannot keep up with lengthening or if the competence contrasts between the matrix and both the layer and the secondary mineral are modest, then the boudinage results in barrel (Figure 11.21*K*) and fish-mouth (Figure 11.21*L*) boudins, the latter reflecting either a larger amount of lengthening or a lower competence contrast between the matrix and both the layer and the secondary mineral filling. If the secondary mineral filling is less competent than the layer, lengthening becomes concentrated in the secondary mineral material, leading to the bow-tie-vein boudins (Figure 11.21*M*). If the layer is less competent than the secondary mineral filling, then lengthening is concentrated in the layer, forming the bone-shaped boudins (Figure 11.21*N*).

Structural Slickenlines. The structural slickenlines that form on fault surfaces (Figure 14.16*A, B*) are parallel to the shear plane and to the slip direction in that plane; therefore they cannot be strictly parallel to a principal direction of strain. If a shear zone is of finite width and deforms by either cataclastic or ductile flow, however, the direction of maximum stretch $\hat{s}_1$ within the deforming shear zone rotates toward progressively smaller angles with the shear direction as the shear strain increases.

Fault surfaces are never perfectly flat, but they contain minor irregularities or protrusions called **asperities** (see Figure 17.2). If asperities are sufficiently strong and resistant to abrasion and fracture, they can scratch and gouge the opposite surface of the fault, thereby giving rise to one type of structural slickenline (Figure 14.16*A*; cf. Figure 3.8*A*). Scratch and gouge marks end where an asperity breaks off the opposite surface of the fault. The length of these lineations is a lower-bound measure for the displacement on the fault surface.

Small ridges can develop where the fracture plane is deflected behind a hard asperity. A corresponding groove, of course, must form on the opposite surface. Similarly, ridge-in-groove lineations, or fault mullions, form if the fault surface is an irregular, rather than planar, surface, and if the irregularities are linear and parallel to the slip direction (Figure 14.16*B*; cf. Figure 3.8*B*). In both cases, the length of the lineation is not necessarily related to the amount of displacement on the fault because the ridge and matching groove form as part of the fracture propagation process, not as a result of the displacement on the fault. Thus they cannot be used to constrain the displacement magnitude.

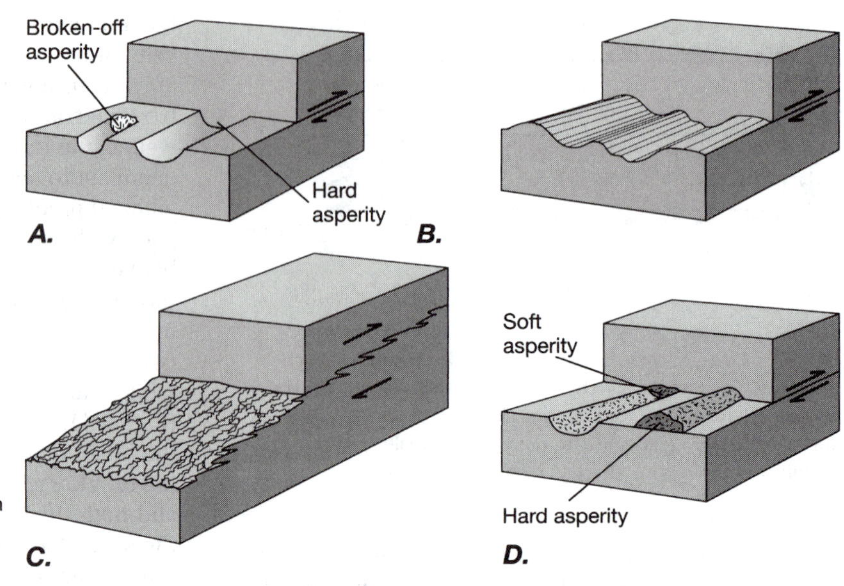

FIGURE 14.16 Structural and mineral slickenlines. (After Means 1987) *A*. Structural slickenlines formed by scratching and gouging of one side of the fault by hard asperities in the other side. *B*. Ridge-in-grove structural slickenlines (fault mullions) formed by linear irregularities in the fault surface that parallel the slip direction. *C*. Spikes on a slickolite. The slickolite is a solution surface subparallel to the direction of displacement across which a component of convergence is accommodated. The spikes are irregularities in the solution surface that parallel the slip direction and are comparable in origin to the teeth on stylolites. *D*. Mineral streak lineations form from the wearing down and smearing out of mineral grains and soft asperities or from the collection of gouge behind a hard asperity.

On fault surfaces across which there is a component of shortening, that component may be accommodated by solution along the fault, especially in limestone or marble. The mechanism is similar to that resulting in stylolites, except that the solution surface, called a **slickolite surface**, makes a small angle with the displacement direction rather than being approximately normal to it as for stylolites. The counterpart of the tooth structure on stylolites (Figure 11.3*A*, left diagram, and Figure 11.5) is a **spike** on a slickolite surface (Figure 14.16*C*). Spikes are oriented at a low angle to the surface, and they define a lineation that indicates the direction of relative shear. There is, however, no general strain interpretation for the orientation of slickolite planes or spikes.

iii. Polycrystalline Lineations

Rods are commonly parallel to local fold axes, which must lie in the plane of the folded layer and thus in principle need have no specific relationship to the orientation of the principal axes of finite strain (cf. Figure 12.7*B*, *D*). In practice, mineral rods usually occur in strongly sheared rocks, and the lineation is subparallel to the shear direction and the axis of maximum stretch $\hat{s}_1$.

Mineral cluster lineations generally are parallel to local fold axes in deformed metamorphic rocks, but their origin may be closer to that of discrete structural lineations in that the mineral clusters may reflect the orientation of the $\hat{s}_1$ axis of the finite strain ellipsoid and thus be a stretching lineation. This association should be demonstrated in the field, however, rather than assumed a priori.

Mineral slickenlines defined by streaks on a slickenside are the result of the smearing out of mineral grains and soft asperities (Figure 14.16*D*). They may also accumulate behind hard asperities, and they commonly form in combination with scratch and gouge lineations.

iv. Mineral Grain Lineations

Acicular and elongate mineral grain lineations may form parallel to one of the principal axes, and they are commonly parallel to the $\hat{s}_1$ direction. This relationship is not universal, however, and the parallelism of such lineations with the strain axes should be demonstrated rather than assumed.

v. Mineral Fibers

Mineral fiber lineations occur as slickenfibers on fault surfaces (Figures 3.8*C*, 11.26*A*, and 14.17*A*, *B*), fibrous vein fillings (Figure 14.17*C*, *D*), or fibrous overgrowths on grains or particles ("beards" at B in Figure 11.7; see also Figures 11.26*B* and 14.18). In almost all these cases, the fibers grow in an orientation parallel to the direction of displacement at the time of fiber growth. In some cases, however, the fibers grow normal to the crystal faces of a mineral grain such as pyrite (Figure 14.18*C*). The mechanism for this case is not well understood, but the different fiber orientations at different crystal faces make it easy to identify, and the suture line between the differently oriented fibers records the displacement history of the corner of the grain.

The continuity and morphology of the mineral fibers across the vein or shear surface imply that fiber growth accompanied and kept pace with gradual displacement and crack opening. Transport of the mineral constituents to the ends of the fibers occurs either by diffusion along grain boundaries or, more probably, through a fluid phase in which the components are dissolved. Slickenfibers on

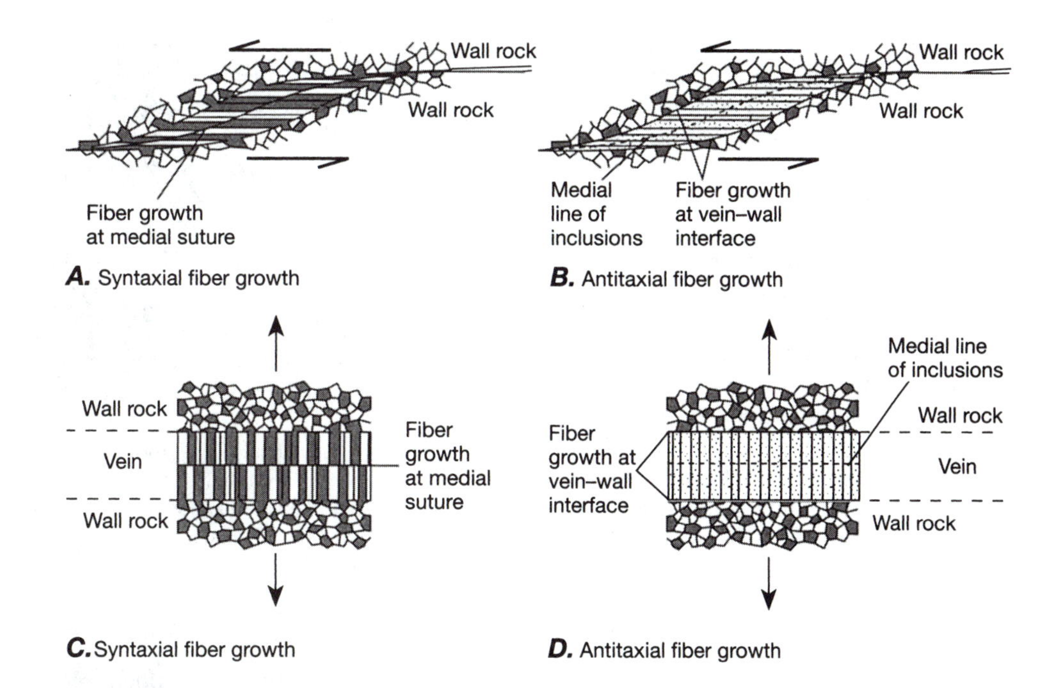

FIGURE 14.17 Comparison of syntaxial and antitaxial mineral fiber lineations in faults and veins. Syntaxial growth occurs if the mineral making up the fibers is also a common mineral in the host rock. If the mineral fibers are different from minerals in the host rock, antitaxial growth occurs. The arrows indicate the direction of displacement. (After Durney and Ramsay 1973; Ramsay and Huber 1983) *A*. Syntaxial growth of slickenfiber lineations on a fault surface. Growth occurs along the medial suture. *B*. Antitaxial growth of slickenfiber lineations on a fault surface. Growth occurs along the interface between the fibers and the wall of the fault. *C*. Syntaxial fiber growth in a vein occurs at the medial suture in the vein. *D*. Antitaxial fiber growth occurs at the interface between fibers and wall rock.

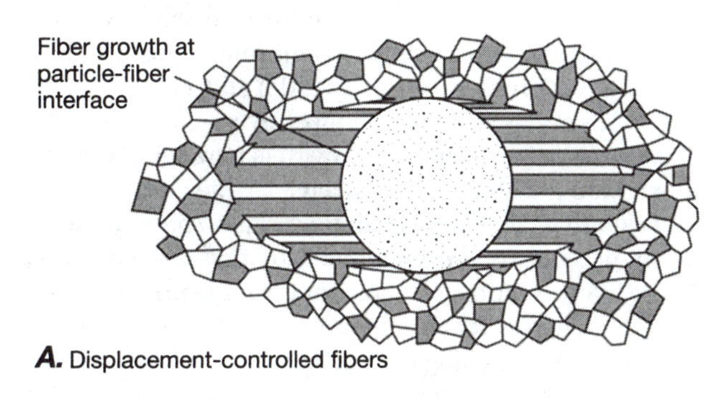

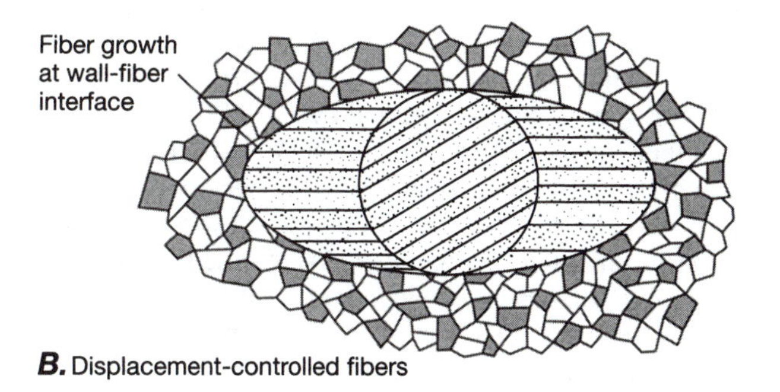

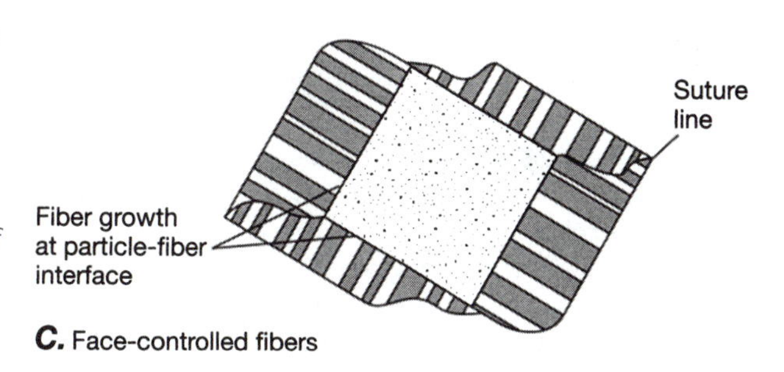

FIGURE 14.18 Fibrous overgrowths or pressure shadows. *A*. If the fiber mineral is similar to host rock minerals, and different from the particle mineral, fibers grow in optical continuity with similar mineral grains in the wall rock. Growth occurs at the fiber-particle interface. *B*. If the fiber mineral is the same as the particle but different from the minerals in the host rock, fibers grow in optical continuity with the particle, here illustrated by a twinned calcite grain. Growth occurs at the interface between the fiber and the wall rock. *C*. Face-controlled orientation in a fibrous overgrowth on a pyrite cube. The mineral fibers grow perpendicular to the crystallographic faces of the pyrite. Growth occurs at the fiber-pyrite interface, and the suture line between differently oriented groups of fibers indicates the displacement of the corner of the grain.

fault surfaces thus imply slow aseismic creep, rather than large rapid displacements that would produce earthquakes.

Four types of displacement-controlled fiber growth structure are recognized that reflect the process by which growth has taken place.

In **syntaxial growth**, fibers tend to grow in optical continuity with mineral grains of the same composition. Thus for this structure to develop, the fiber mineral must be a mineral that is present in the host rock. For slickenfibers (Figure 14.17*A*) and vein fillings (Figure 14.17*C*), the fibers extending from mineral grains in opposite walls meet at a medial suture at which there is both a structural and an optical discontinuity. The suture is the site of the latest growth of the fibers.

In fibrous overgrowths around a particle, the fibers can grow syntaxially on mineral grains in the host rock, with fiber growth occurring at the fiber-particle interface (Figure 14.18*A*), or they can grow syntaxially on the particle itself, as illustrated for a particle of twinned calcite in Figure 14.18*B*. In this case fiber growth occurs at the interface between fiber and wall rock.

Antitaxial growth occurs when the fiber mineral is absent or uncommon in the host rock. In slickenfibers (Figure 14.17*B*) and vein fillings (Figure 14.17*D*), a medial suture may contain inclusions of host rock, but the fibers are optically and structurally continuous across the suture. Fiber growth occurs along the margins of the vein or fault where there is a discontinuity in mineral composition.

Composite growth occurs if fibers of two different minerals grow, one of which is common and one rare or absent in the host rock. The fiber structure in veins may then show a central antitaxial band of the mineral that is rare in the host with bands of syntaxial fibers on either side. The fibers in both types of band grow at the interface between the antitaxial and syntaxial bands.

Fibers may also grow by the **crack-seal mechanism**, which involves repeated microfracturing across the fiber at random locations along its length, followed by deposition of optically continuous overgrowths that heal the fracture. Evidence for the mechanism includes abundant subplanar arrays of microscopic fluid inclusions crossing the fibers at the healed cracks (Figure 14.19). The resulting fibers are sometimes called **stretched crystals**. They can occur if the fiber mineral is the same as the dominant mineral in the host rock. Crystal fibers are structurally and optically continuous across the whole vein, and they often connect and are optically continuous with two fragments of crystal grain on opposite sides of the vein that apparently were originally a single grain. Because an increment of growth may occur at any place along the fiber, the orientation and shape of the fiber do not necessarily record the history of the displacement—only the net result.

Although we have discussed only straight fibers here, curved fibers on fault planes (Figure 11.26*A*) as well as in veins and overgrowths (Figure 11.26*B*) are relatively common. The curvature of the fibers records a component of rotation to the deformation during fiber growth.

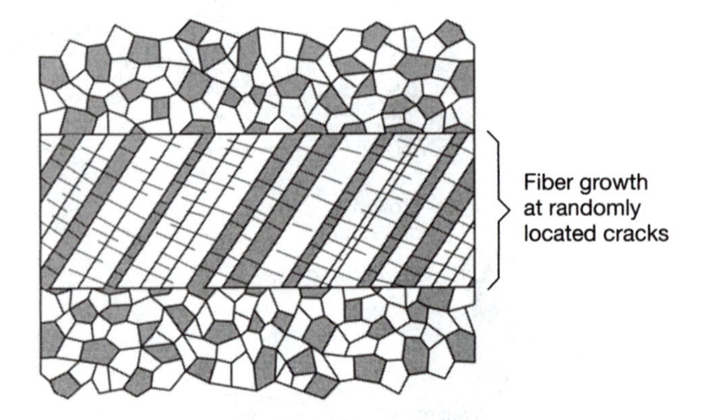

FIGURE 14.19 Crack-seal growth forming a mineral fiber lineation of stretched crystals in a vein. Growth does not occur at any particular surface but by repeated cracking followed by deposition of fiber mineral to close the crack. (After Ramsay and Huber 1983)

vi. Fibrous and Nonfibrous Overgrowths

Oriented overgrowths, or "pressure shadows," commonly form on mineral grains and particles if, during deformation, these particles behaved rigidly relative to the surrounding ductile matrix. The particle boundary and the matrix tend to separate along a surface at a high angle to the maximum instantaneous stretch $\hat{\zeta}_1$, where a local zone of abnormally low minimum compressive stress $\hat{\sigma}_3$ develops. Minerals in solution diffuse to the low stress area and precipitate as either nonfibrous or fibrous overgrowths (e.g. in Figures 11.7 (at "B") and 11.26*B*). In the case of fibers in veins and in most overgrowths, the displacement at the surface of growth is parallel to the axis of maximum instantaneous stretch $\hat{\zeta}_1$. Riecke's principle suggests that fiber growth in a pressure shadow should occur parallel to the minimum compressive stress $\hat{\sigma}_3$. These interpretations are compatible if the principal axes of instantaneous strain and stress are parallel, which is the case for a rheologically linear material (cf. Section 16.1(ii)). If the deformation is noncoaxial, however, the crystal and the existing overgrowths may rotate so that the $\hat{\zeta}_1$ axis is not parallel to the long axis of the overgrowth (Figure 11.26*B*).

In the case of slickenfibers on fault surfaces, the fibers grow in minor dilational irregularities in the fault surface (Figure 14.17*A*, *B*) and are parallel to the direction of relative displacement across the fault. In this case the irregularities in the fault probably cause a local perturbation in the large-scale stress, so the minimum principal stress $\hat{\sigma}_3$ is locally subparallel to the displacement direction across the dilational zone and to the mineral fibers.

Because they grow during a progressive deformation, mineral fibers preserve information about the deformation history, and in some cases, they can be used to distinguish coaxial from noncoaxial deformations. In veins, straight fibers generally indicate a coaxial deformation because they imply that the same material lines have remained parallel to the principal axes of instantaneous stretch $\hat{\zeta}_1$ throughout the deformation (Figures 14.17*C*, *D*). Conversely, curved mineral fibers indicate noncoaxial deformation, because they imply that the earliest fibers rotated progressively away from the direction of growth, which remained parallel to $\hat{\zeta}_1$. Detailed interpretation, however, requires a knowledge of whether the fibers form by syntaxial or antitaxial growth (Figure 14.17*C*, *D*) or by crack-seal growth (Figure 14.19), because the growth mechanism affects the fiber pattern that develops.

14.7 LINEATIONS ON FOLDS

Lineations are commonly associated with folds and in particular are likely to be oriented parallel to the fold hinges, but the significance of such lineations is not always clear. Some may be stretching lineations that are subparallel to the $\hat{s}_1$ direction. This could occur if the fold hinges formed or were rotated toward $\hat{s}_1$, as could occur, for example, as a result of large shear strain or if the rock is shortened normal to the flow direction (Figure 14.20; cf. Figure 12.7*D*).

A mineral lineation parallel to a fold axis and to $\hat{s}_1$ may be the result of dissolution of mineral grains. In this case it would be possible to have $\hat{s}_1 = 1 \geq \hat{s}_2 > \hat{s}_3$, for which no lengthening would have occurred parallel to the $\hat{s}_1$ direction (Figure 14.5*A*, *B*).

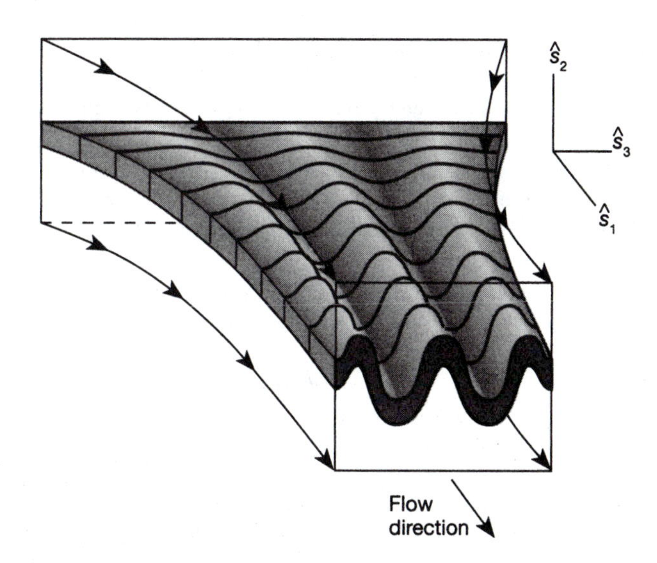

FIGURE 14.20 Convergent flow can result in the axis of maximum principal stretch being parallel to the flow direction and in the formation of fold hinges parallel to the maximum stretching direction and stretching lineations.

In other cases, the lineation may be constrained to develop parallel to a bedding plane. If so, it may form parallel to the maximum stretch in the bedding plane. It need not have any direct relationship with the principal axes of strain and may even change its geometric relation with the fold axis from hinge to limb.

REFERENCES AND ADDITIONAL READINGS

Alvarez, W., T. Engelder, and P. A. Geiser. 1978. Classification of solution cleavage in pelagic limestones. *Geology* 6: 263–266.

Alvarez, W., T. Engelder, and W. Lowrie. 1976. Formation of spaced cleavage and folds in brittle limestone by dissolution. *Geology* 4: 698–701.

Chapple, W. M., and J. H. Spang. 1974. Significance of layer parallel slip during folding of layered sedimentary rocks. *Geol. Soc. Amer. Bull.* 85: 1523–1534.

Durney, D. W. 1972. Solution transfer, an important geological deformation mechanism. *Nature* 235: 315.

Durney, D. W., and J. G. Ramsay. 1973. Incremental strains measured by syntectonic crystal growth. In K. A. DeJong and R. Scholten, eds., *Gravity and Tectonics*. John Wiley and Sons, New York, 67–96.

Gratier, J. P. 1983. Estimation of volume changes by comparative chemical analyses in heterogeneously deformed rocks (folds with mass transfer). *J. Struct. Geol.* 5(3/4): 329–339.

Gray, D. R. 1978. Cleavages in deformed psammitic rocks. *Geol. Soc. Amer. Bull.* 89A: 5677–5690.

Gray, D. R. 1979. Geometry of crenulation folds and their relationship to crenulation cleavage. *J. Struct. Geol.* 1(3): 187–205.

Gray, D. R., and D. W. Durney. 1979. Investigations on the mechanical significance of crenulation cleavage. *Tectonophysics* 58(1–2): 35–80.

Hobbs, B. E. 1971. The analysis of strain in folded layers. *Tectonophysics* 11: 329–375.

Jeffrey, G. B. 1923. The motion of ellipsoidal particles immersed in a viscous fluid. *Roy. Soc. Lond. Proc.*, Ser. A, 102: 161–177.

Mancktelow, N. S. 1979. The development of slaty cleavage, Fleurieu peninsula, South Australia. *Tectonophysics* 58: 1–20.

Marlow, P. C., and M. A. Etheridge. 1977. Development of a layered crenulation cleavage in mica schists of the Kanmantoo Group near Macclesfield,

South Australia. *Geol. Soc. Amer. Bull.* 88: 873–882.

Means, W. D. 1987. A newly recognized type of slickenside striation. *J. Struct. Geol.* 9: 585–590.

Passchier, C. W., and R. A. J. Trouw. 1996. *Microtectonics.* Springer-Verlag, New York, 289 pp.

Platt, J. P. 1979. Extensional crenulation cleavage. *J. Struct. Geol.* 1: 95–96.

Platt, J. P., and R. L. M. Vissers. 1980. Extensional structures in anisotropic rocks. *J. Struct. Geol.* 2(4): 397–410.

Ramsay, J. G. 1967. *Folding and Fracturing of Rocks.* McGraw-Hill, New York, 568 pp.

Ramsay, J.G. 1976. Displacement and strain. *Phil. Trans. Roy. Soc. Lond.* A 283: 3–25.

Ramsay, J. G., and M. I. Huber. 1983. *The Techniques of Modern Structural Geology*, vol. 1: *Strain Analysis*, Academic Press Ltd., London, 307 pp.

Ramsay, J. G., and M. I. Huber. 1987. *The Techniques of Modern Structural Geology*, vol. 2, *Folds and Fractures*, Academic Press, New York, 695 pp.

Reed, L. J., and E. Tryggvason. 1974. Preferred orientations of rigid particles in a viscous matrix deformed by pure shear and simple shear. *Tectonophysics* 24(1/2): 85–98.

Rees, A. I. 1979. The orientation of grains in a sheared dispersion. *Tectonophysics* 55(3/4): 275–288.

Weber, K. 1981. Kinematic and metamorphic aspects of cleavage formation in very low grade metamorphic slates. *Tectonophysics* 78: 291–306.

Williams, P. F. 1976. Relationships between axial plane foliations and strain. *Tectonophysics* 39: 305–328.

Williams, P. F. and C. Schoneveld. 1981. Garnet rotation and the development of axial plane crenulation cleavage. *Tectonophysics* 78: 307–334.

Willis, D. G., 1977. Kinematic model of preferred orientation. *Geol. Soc. Amer. Bull.* 88: 883–894.

Chapter 15

OBSERVATIONS OF STRAIN IN DEFORMED ROCKS

15.1 MEASURING STRAIN IN ROCKS

Our description of the geometry of strain during progressive deformation would be of limited application if it were not possible to measure in deformed rocks at least some of the parameters that characterize the strain. An extensive literature exists on various techniques for measuring strain in rocks and for determining the progressive deformation that the rocks have experienced.[1] We outline a few of the more common and straightforward methods here, both to illustrate the types of data that can be used and to impart some intuitive understanding of the concepts involved. More details on these methods are presented in the appendix to this chapter.

We limit our discussion of techniques for determining strain to two dimensions, because the geometry and the principles are most easily understood in two dimensions. Extension to three dimensions adds little to conceptual understanding but increases the practical difficulty considerably.

Techniques for measuring two-dimensional strain must be applied with caution. The intersection of a plane of any orientation with an ellipsoid is always an ellipse. Thus there is no way to determine, from a single plane, whether the observed maximum and minimum axes of that ellipse are parallel to the maximum and minimum principal stretches, or even whether they lie within one of the principal planes. Careful measurements on more than one arbitrary orientation of plane can be used to determine the ratios of the three principal stretches and their orientations. The measurements and the required calculations, however, are difficult and time-consuming. In some cases, however, the existence of particular structures enables us to infer the orientation of one or more of the principal axes or the principal planes. In such cases a two-dimensional analysis can be used with more confidence, and the problem of determining the ratios of the principal stretches in three dimensions also is simplified.

One of the frustrating problems we encounter in trying to determine the strain in rocks is that for many strain markers, the original shape of the marker is known but the original dimensions are unknown. In particular, we can often determine the deformed lengths of material lines parallel to the principal axes of strain, $\hat{\ell}_1$ and $\hat{\ell}_3$, and know that before the deformation those lines were both the same length L (Figure 15.1). We do not know, however, what that initial length L was, so we cannot determine the absolute value of the principal stretches $\hat{s}_k = \hat{\ell}_k/L$. Thus the volumetric strain is a quantity that can be determined only rarely and under very special circumstances. We can describe the shape of the strain ellipse, however, by determining the ratio of the principal axes, which is the ellipticity R:

$$R \equiv \frac{\hat{s}_1}{\hat{s}_3} = \frac{\hat{e}_1 + 1}{\hat{e}_3 + 1} = \frac{\hat{\ell}_1/L}{\hat{\ell}_3/L} = \frac{\hat{\ell}_1}{\hat{\ell}_3} \tag{15.1}$$

Equation (15.1) shows that R does not involve the original lengths L.

The two-dimensional state of strain is characterized by the components

$$e_{11},\ e_{13}\ (= e_{31}),\ e_{33} \tag{15.2}$$

[1]Ramsay and Huber (1983) provide an excellent summary, as well as an extensive bibliography, up to 1983.

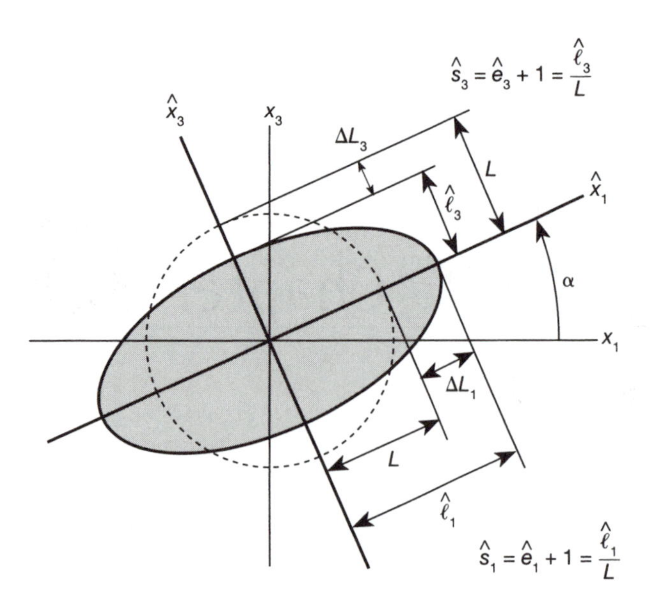

FIGURE 15.1 Three independent quantities, $\hat{s}_1$, $\hat{s}_3$, and α, with which we can define the shape and orientation of the strain ellipse. If L is unknown, we cannot determine the principal stretches independently but only their ratio $R = \hat{s}_1/\hat{s}_3 = \hat{\ell}_1/\hat{\ell}_3$.

The equality of the two shear strain components reflects the fact that the strain tensor is symmetric,[2] as discussed in Section 12.2(i) and 12.8(i) (Equation (12.40)). Thus only three independent measurements in the $(\hat{s}_1, \hat{s}_3)$ principal plane—which is also the $(\hat{e}_1, \hat{e}_3)$-plane—are required to define the shape, size, and orientation of the strain ellipse.

The three simplest independent measurements to deal with are the lengths $\hat{\ell}_1$ and $\hat{\ell}_3$ parallel to the principal coordinates $\hat{x}_1$ and $\hat{x}_3$, respectively, and the orientation given by the angle α between the maximum principal stretch and a reference axis x_1 in the plane (Figure 15.1). If we assume that the two-dimensional deformation was at constant area, the radius L of the undeformed circle that has the same area as the deformed elliptical object is:

$$L = \sqrt{\hat{\ell}_1 \hat{\ell}_3} \quad \text{for constant area} \tag{15.3}$$

The ratios of the deformed to the undeformed lengths define the two principal stretches $\hat{s}_1$ and $\hat{s}_3$, which define $\hat{e}_1$ and $\hat{e}_3$ (Figure 15.1). With these values, we can calculate the three independent components of strain e_{kl} (Equation (15.2)) in any coordinate system of known orientation relative to the principal coordinates. We cannot be sure, in general, that the deformation was two-dimensional and at constant area, however, so the original radius L of the feature is unknown. Thus we can determine only the *ratio* of the maximum and minimum stretches (Equation (15.1)), along with the orientation α, which means we can determine only two of the three quantities necessary to define the strain. Because the volumetric (or area) strain is usually indeterminate, our description is incomplete.

i. Deformed Objects of Initially Circular Cross Section

The simplest way to determine the strain in a plane is to observe directly the deformed shape of an object that is known to have been circular in that plane before deformation. Common examples of initially spherical objects include ooids and spherulites (Figure 12.6), alteration spots (Figure 11.19*B*), and some radiolaria and foraminifera shells. Some structures are only circular in one particular cross-sectional plane, so the strain in that plane can be determined easily, although the plane does not necessarily coincide with a principal plane of strain. Such features include circular disc-shaped segments of cylindrical crinoid stems and scolithus tubes, which are sediment-filled cylindrical worm holes. The crinoid stem discs are commonly deposited flat on a bedding plane, and scolithus tubes are initially oriented normal to bedding with a circular cross section in the bedding plane. Thus the bedding plane is the plane within which the strain can be determined from these features.

The initial shape of natural objects is never perfectly round, deformation is not always perfectly homogeneous, and our measurements are not as precise as we would like. Thus it is always necessary to measure a large number of deformed objects and to take some average ellipticity R and some average orientation of s_{max} as the best description of the state of two-dimensional strain in the plane. In Appendix 15-A.1, we give details of the R_f–ϕ method for determining strain from deformed objects that initially were approximately elliptical in shape.

ii. Deformed Linear Objects

In some cases, the stretch parallel to a deformed linear object, such as a needle-shaped crystal or a linear fossil, may be evident either as folding ($s_n < 1$) or as boudinage ($s_n > 1$) of the object. In these cases, measurement of the stretch in different directions enables us to determine the strain. This method has been applied, for example, to deformed acicular crystals such as tourmaline, amphibole, and rutile (Figure 15.2*A*, *B*; cf. Figure 12.18*C*) and to deformed cigar-shaped fossils such as belemnites. Ptygmatic folding and boudinage of layers, such as quartz–feldspar veins in a schist, can also be used to estimate two-dimensional strain if the deformation is measured in a principal plane of strain.

[2]In tensor form, the strain is characterized by a two-by-two symmetric matrix. In general coordinates (from Equation (12.43)):

$$e_{kl} = \begin{bmatrix} e_{11} & e_{13} \\ e_{31} & e_{33} \end{bmatrix} \quad \text{where} \quad e_{13} = e_{31}$$

The second equation expresses the symmetry of the matrix, and because of it, there are only three independent components in the matrix representing this strain tensor.

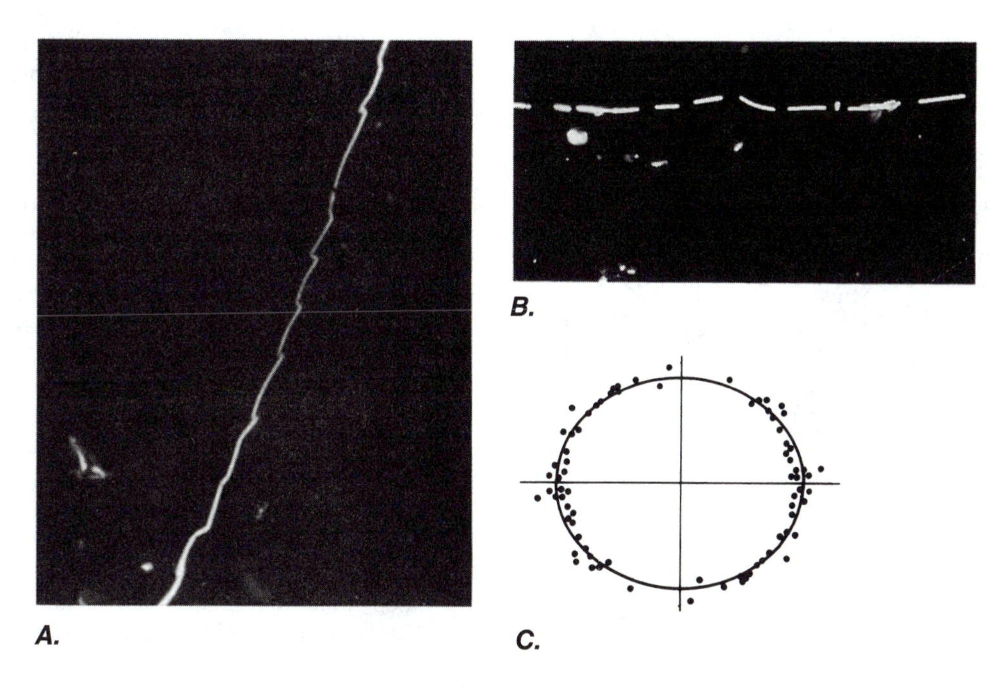

FIGURE 15.2 Examples of stretched linear objects. *A.* Needles of rutile included within a grain of quartz and folded by shortening during ductile deformation of the quartz. *B.* Needle of rutile stretched within a ductilely deformed grain of quartz. *C.* Strain ellipse determined from the stretching of rutile needles embedded in quartz crystals. The magnitude of the stretch of each needle is plotted along a radius parallel to the needle orientation to produce each data point in the plot. (From S. Mitra 1976)

To determine the stretch of the object, we take the original length to be the arc length of folds, or the sum of the segment-lengths of boudins. We take the final length to be the length of the fold wave train measured along the median surface or, for boudins, the total of the segment lengths plus the spacings between them. If the objects have themselves undergone no ductile deformation but only changed length by folding or boudinage, then the actual original length is measurable and can be used to determine the actual area (or volumetric) strain. Such measurements probably provide a minimum estimate of the stretch, however, because homogeneous ductile deformation of the linear object may have lengthened and thinned it or shortened and thickened it so that the folding or boudinage may not record all the stretching that occurred in the rock.

If the linear objects have a wide variety of orientations, then determination of the stretches in the different orientations can define the strain ellipse by plotting the values of the stretches as radii in the directions that correspond to the orientations of the linear objects measured. In principle, as noted earlier, we need to measure the stretches in only three independent directions to calculate a unique strain ellipse. In practice, however, random errors and uncertainties require that a much larger number of measurements be made. Figure 15.2*C*, for example, shows the strain in a quartz grain, determined from the deformation of embedded rutile needles (Figure 15.2*A*, *B*). Each point is the stretch of a particular needle, plotted in the orientation of that needle, relative to a prescribed reference axis. A more general application of these principles is the nearest neighbor center-to-center technique, which we discuss in Appendix 15-A.2.

Some fossils have a well-defined characteristic dimension, such as the spacing between segments of certain species of graptolite (see Figure 15.4). Because an original dimension is well known, these fossils provide one of the few structures from which absolute magnitudes of the strain can be determined (see Section 15.2).

iii. Sheared, Initially Orthogonal Pairs of Lines

Many fossils have features that are perpendicular to one another. Examples include the hinge line and symmetry plane of brachiopod shells (Figure 15.3) and the body segments and symmetry plane of trilobites. When such fossils are deformed, the angle between the initially orthogonal lines changes, making it possible to determine the shear strain. If only shear strains can be measured, however, it is impossible, even in principle, to determine the actual magnitudes of the principal stretches, because the shear strain depends only on the difference between the principal stretches, not on their actual magnitudes.[3]

[3]In the Mohr circle for finite strain, the shear strain depends on the diameter of the circle, and thus on the difference between the inverse squares of the principal stretches (cf. Equations (12.67) and (12.68)), but it does not depend on their magnitudes, which define where the center of the circle lies along the inverse-stretch axis.

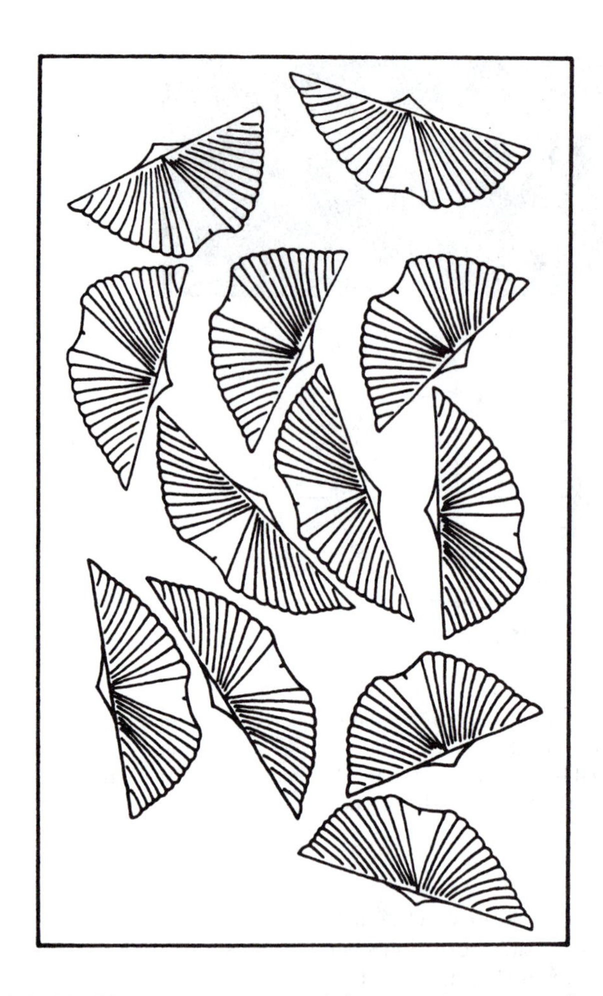

FIGURE 15.3 The hinge line and the original symmetry plane of brachiopod shells are initially perpendicular, and they therefore permit determination of the shear strain in the deformed state. (After Ramsay and Huber 1983)

Thus this method of determining strain can never provide a measure of the volumetric strain.

Theoretically, measurement of two shear strains in different directions is enough to determine the orientations and the ratio R of the principal stretches. In practice, of course, we rely on multiple determinations to estimate the average strain. We discuss a method for determining the strain from shear strain measurements in Appendix 15-A.3.

15.2 RELATIONSHIP OF STRAIN TO FOLIATIONS AND LINEATIONS

Determination of the relationship of strain to foliations or lineations requires that there be a good strain marker in the same rock that contains the foliation or lineation. This restriction makes it possible to study the relationship only in special circumstances. Understanding the relationship is important, however, because it helps determine how a foliation forms during deformation and because, if the relationship is simple, foliations and lineations may be useful as strain indicators where other more obvious methods are not available.

i. Strain in Slates

The relationship between strain and both foliation and lineation, has been most extensively studied in slates. These rocks have a very well-developed foliation, but they generally are not so highly metamorphosed that fossils and other strain markers are obliterated. If the slate is microdomainal in structure, the spacing of the foliation domains is very small compared with the size of most strain markers. Thus the strain markers effectively average the strain over many domains. Studies of alteration spots (see Figure 11.19*B*) show that the plane of flattening of the strain ellipsoid is parallel to the slaty foliation within a few degrees. Similar results have been reported from essentially every study of slaty foliation in which the independent determination of strain has been possible. The result is an empirically established parallelism between slaty foliation and the plane of flattening of the finite strain ellipsoid.

Determination of the state of strain associated with disjunctive foliations is more difficult than with slates. For stylolitic foliations, the spacing between domains is usually larger than most common strain markers, so the averaging inherent in strain determinations with slate is rarely possible. We discuss the association between the amount of bedding-parallel shortening and the morphology of a disjunctive foliation in a limestone in Section 14.3(ii) (Figure 14.4). Such data, while suggestive, are insufficient for a complete determination of the orientation and ellipticity of the strain ellipsoid.

The similarity between the microstructure of slaty foliations and that of other foliations (especially crenulation foliations), and the evidence that these foliations commonly transform into continuous foliations through recrystallization, suggest that the relationship between slaty foliations and strain could be applied to other crenulation foliations and continuous foliations. Our understanding of the mechanisms of foliation formation, however, reminds us to be cautious in making this interpretation. Some mechanisms do not lead necessarily to parallelism between foliation and the plane of flattening, and foliations that have acted as a plane of simple shearing cannot have exactly this orientation (see Section 14.5, Figure 14.12).

ii. Volumetric Strain in the Martinsburg Slate

Figure 15.4 presents results from a unique study of the formation of foliation in the Martinsburg slate[4] in the central Appalachian Valley and Ridge province. The strain indicators used were fossil graptolites, which were originally composed of chitinous material and are preserved

[4]The Martinsburg slate is Ordovician in age.

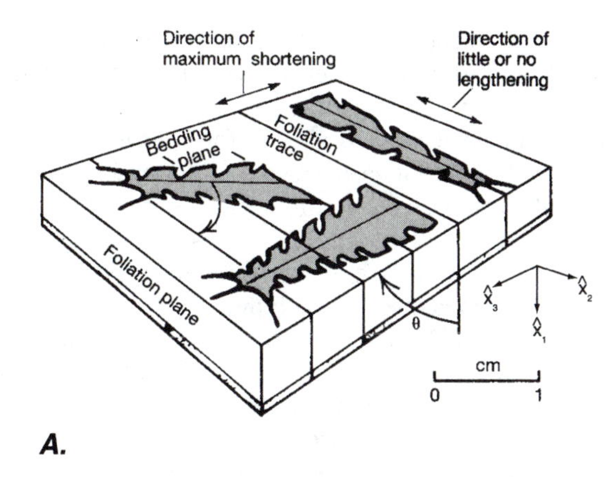

A.

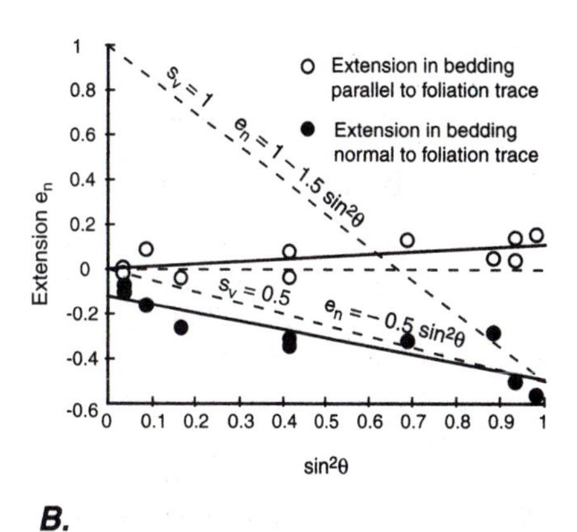

B.

FIGURE 15.4 The use of graptolites as an absolute measure of strain. The undeformed spacing of thecae, or segments, on the graptolites is known and relatively constant for a given species. *A.* Graptolites lying in the bedding plane are shortest and thickest where the stipe (the long dimension of the graptolite) is normal to the trace of the foliation, and they are longest and thinnest where the stipe is parallel to the trace of the foliation. (From Wright and Platt 1982) *B.* Extension in the bedding plane measured parallel to the foliation trace (open circles) and perpendicular to the foliation trace (solid circles) plotted against the square of the sine of the angle θ between bedding and foliation. The solid lines are the least-squares fits to the data. The theoretical relationships for plane strain ($\hat{s}_2 = 1$; $\hat{e}_2 = 0$) are shown by the dashed lines; the horizontal dashed line represents zero extension in the bedding plane parallel to the foliation trace; the two inclined dashed lines represent extension in the bedding plane normal to the foliation trace, respectively, for 50% volume loss ($s_v = \hat{s}_1 \hat{s}_2 \hat{s}_3 = 0.5$) with $(\hat{s}_1, \hat{s}_2, \hat{s}_3) = (1, 1, 0.5)$ and for constant volume ($\hat{s}_v = 1$) with $(\hat{s}_1, \hat{s}_2, \hat{s}_3) = (2, 1, 0.5)$. The latter model is definitively rejected by the data. The extension e_n parallel to a unit vector n is given by $e_n = \sum_{k=1}^{3} \hat{e}_k n_k^2$, where $\hat{e}_k = \hat{s}_k - 1$. For n in the bedding plane normal or parallel to the foliation trace, $n_k = [\cos\theta, 0, \sin\theta]$ or $[0, 1, 0]$, respectively. (Data replotted from Wright and Platt 1982).

as carbon films on bedding surfaces. It is unlikely that this material ever had much strength, so the graptolites probably record accurately the total strain in the rock. For certain graptolite species, the original spacing between segments (thecae) is constant and well known, so measurement of this spacing on deformed fossils can be used to determine absolute values for the stretches. In most studies of strain in rocks, only the ratio of the stretches can be determined. The graptolites lie parallel to the bedding plane (Figure 15.4*A*) and therefore record only the part of the strain that is represented by a bedding parallel section through the strain ellipsoid. However, because of folding, the bedding planes have a wide range of orientations with respect to the foliation, and they thereby provide information about the three-dimensional geometry of the strain relative to the foliation.

The trace of the foliation on the bedding plane is an intersection lineation, generally parallel to the fold axis, which provides a convenient reference orientation (Figure 15.4*A*). The open circles in Figure 15.4*B* show that the extension parallel to the foliation trace is always very small regardless of the angle θ between bedding and foliation. The solid circles show that the extension normal to the foliation trace approaches zero in a direction parallel to the foliation plane ($\theta = 0°$) and decreases to a minimum normal to the foliation plane ($\theta = 90°$). These data therefore demonstrate that there is almost no extension in any direction parallel to the foliation and that, normal to the foliation, the shortening is a maximum (stretch is a minimum). This result can be achieved only by a loss of volume. The trend defined by the solid circles is approximated by a theoretical curve indicating a volume loss for the rock of approximately 50 percent, and the volume loss was accomplished almost entirely by shortening perpendicular to the foliation.

This astounding result means that half of the original volume of the rock has been lost during deformation and the development of foliation! Other evidence, such as the partial solution of shells, indicates that most of this volume loss occurred by solution of material, although other processes, such as a decrease in porosity, loss of surface water on clay particles, and dehydration reactions of clay to micas, may have contributed.

This study proves conclusively that solution is a mechanism of major importance for low-temperature deformation and for the associated formation of foliations. It also indicates that solution can produce foliations oriented parallel to the plane of flattening of the finite strain ellipsoid. Moreover, this study demonstrates that if we assume a deformation is constant-volume simply because we cannot determine the volumetric strain, we may be making a serious error.

15.3 MEASUREMENT OF STRAIN IN FOLDS

Studies of strain distribution throughout folds indicate that folding is a more complex process than is assumed in simple kinematic models. Although the strain distributions do not have a unique interpretation, they constrain the possible folding mechanisms.

i. A Study of Orthogonal Flexure

Figure 15.5 shows an example of strain analysis in a close fold that approximates class 1B in geometry and that developed in a layer of limestone pebble conglomerate. Evidence of deformation includes deformed pebbles, pressure solution seams, twinned calcite in veins, extensional veins filled with calcite, stylolites parallel to bedding associated with compaction, and tectonic stylolites at high angles to bedding. Extensional veins are present on the limbs of the fold, and pressure solution films and twinned calcite crystals are most prominent along the concave side of the fold. Other extensional veins of calcite are found normal to the fold axis, indicating extension parallel to the fold axis.

The strain distribution around the fold is shown in Figure 15.5*A*, where measurements of deformed pebbles in each of the sectors are combined to define the average strain ellipse in that sector. In each sector of the hinge zone, between 10 and 50 clasts were measured; a few hundred were measured in each limb.

Figure 15.5*B–E* shows the strain distribution around the hinge zone predicted by four different kinematic models of folding. The limb strains are not shown. The model for orthogonal flexure by bending (Figure 15.5*B*; see also Figures 13.4 and 13.6*B*) shows a distribution of $\hat{s}_1$ orientations similar to that observed in the natural fold,

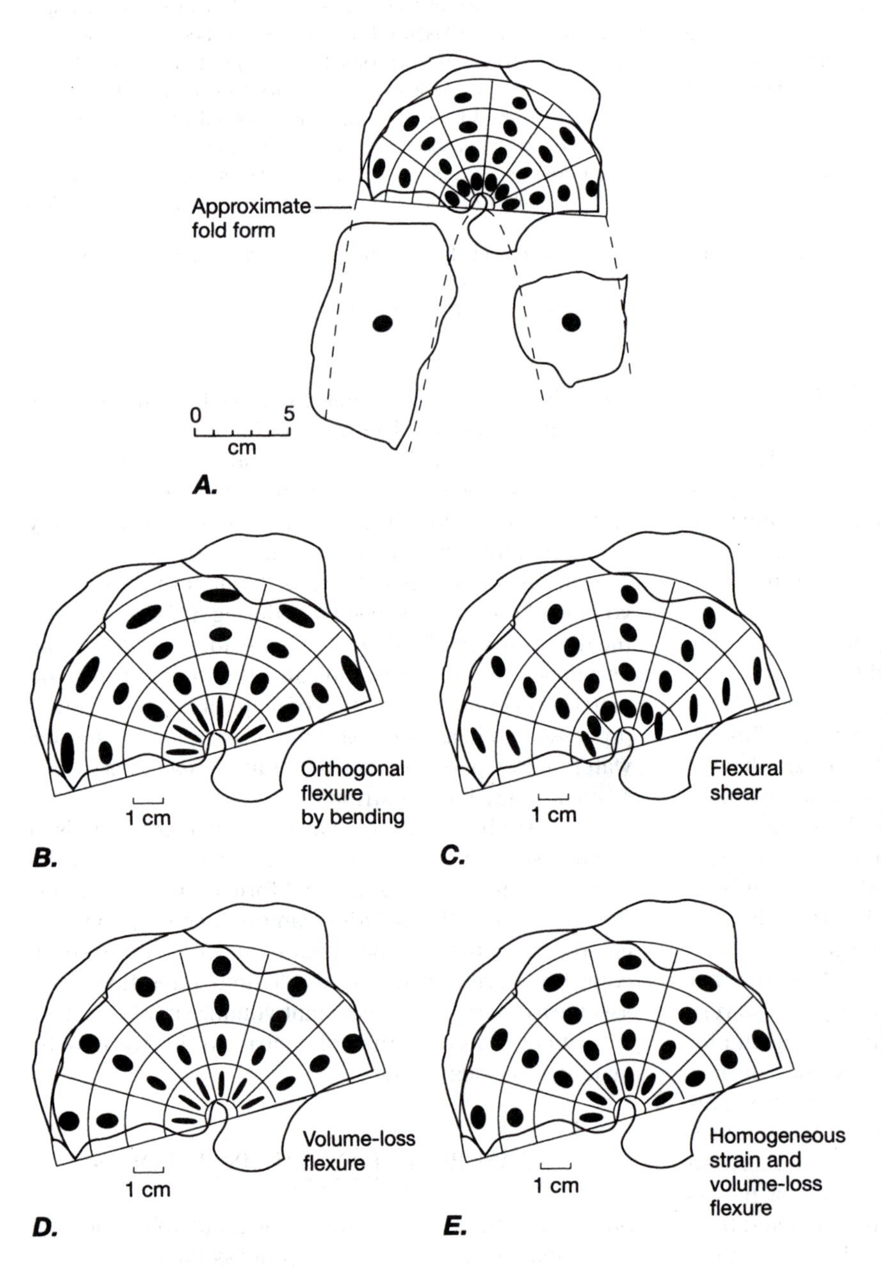

FIGURE 15.5 The strain distribution determined on a natural fold in a limestone pebble conglomerate, compared with the theoretical distribution of strain for different idealized models of folding. (From Hudleston and Holst 1984) *A.* Distribution of the strain ellipses in the profile plane of the natural fold. A large number of measurements of deformed clasts within each sector of the fold were combined for each strain determination. *B.* Strain distribution for orthogonal flexure by bending. *C.* Strain distribution for flexural-shear model of folding. *D.* Strain distribution for volume-loss flexural folding. The outer arc of the fold is assumed to remain constant in length. *E.* Strain distribution resulting from a homogeneous flattening strain normal to the bedding, as would result from compaction, followed by volume-loss flexural folding.

but the magnitudes of the principal stretches are markedly different. Models for flexural shear (Figures 15.5*C*; see also Figures 13.5 and 13.6*C*) and for volume-loss flexure from pressure solution (Figure 15.5*D*; see also Figures 13.6*D* and 13.11*A*, *B*) do not at all reproduce the essential features of stain in the natural fold and therefore are inappropriate models. Figure 15.5*E* shows a more complex model in which a homogeneous flattening normal to the bedding is followed by folding through volume loss. This model most nearly reproduces the orientation and magnitude of the strains in the natural fold.

In the limbs of the natural fold, the strain is small and homogeneous over a relatively large area. The models for orthogonal flexure by bending (see Figure 13.6*B*) or by volume loss (see Figure 13.6*D*) are consistent with very small limb strains, but the models for flexural-shear folding (see Figure 13.6*C*) or for an initial component of flattening normal to bedding are inconsistent.

Thus the best model for the hinge area is homogeneous flattening normal to bedding followed by volume-loss flexure, but the best models for the limbs are either orthogonal or volume-loss flexure. The discrepancy suggests either that the mechanism of folding was more complex than assumed by the models or that the strained clasts do not record the total strain. If measured values for the strain are too low, the model for orthogonal flexure by bending is probably the most appropriate, although the distribution of solution seams suggests some folding by volume loss. Such ambiguities are common in the interpretation of rocks, and the uncertainty inherent in studying the uncontrolled experiments performed by nature is a problem with which the geologist must always contend.

ii. The South Mountain Fold

A classic study of the relationship of strain to folds and foliation is the investigation of deformed oolitic limestones (see Figure 12.6) in the South Mountain fold at the western edge of the Blue Ridge Mountains in Maryland, eastern USA. Figure 15.6*A* is a map and cross section showing the regional setting of the fold, and Figure 15.6*B* is a generalized profile of the fold showing the relationships among the fold geometry, the strain distribution, and the foliation geometry. The $(\hat{s}_1, \hat{s}_2)$-plane of the strain ellipsoid inferred from the ooids is parallel to the foliation, and the orientations form a convergent fan around the fold. A stretching lineation is parallel to the $\hat{s}_1$ axes.

The convergent fan of $\hat{s}_1$ axes is consistent with a flexural-folding mechanism by volume loss (Figure 13.6*D*) and/or by buckling following layer-parallel shortening (Figure 13.7*A*, *B*). It is not consistent with passive-shear folding (Figure 13.10*A*). The maximum extensions indicated on the cross section by the shapes of the strain ellipses vary irregularly around the fold. The deformation of the ooids may not reflect the total deformation in the rock, and additional deformation might have been contributed, for example, by solution of the matrix. Thus, studying the ooids alone might result in underestimation of the total strain associated with folding.

iii. Foliation Patterns in Folds

The patterns of foliation orientation in folded layers generally show considerable diversity, much of which reflects the strain distribution. In the absence of independent strain measurements, the comparison of theoretically predicted strain distributions with such foliation patterns suggests a parallelism between foliation and the flattening plane of finite strain. In the remainder of this subsection, we describe several examples of such inconclusive but suggestive studies of folding.

In folded, interlayered, competent and incompetent beds, the foliation generally is refracted from one orientation to another across the bedding planes such that the angle between bedding and foliation at any given point on the fold is higher in competent beds than in incompetent beds (Section 11.6(i) and Figure 11.13). In places where the composition and, presumably, the competence vary gradually, the orientation of the foliation also changes gradually.

Figure 13.22 shows a model strain distribution having exactly the same pattern as is commonly observed in nature. In this case, the competent layers buckle predominantly by flexural shear, and the relative slip between the competent layers is accommodated by layer-parallel shear within the incompetent layers. The differences in the geometry and the amount of strain between competent and incompetent layers lead to the strong refraction of the $\hat{s}_1$ orientations across the bedding, especially in the limbs of the folds. In the incompetent layers, the strong divergent foliation fan near the hinges and the low intensity of foliation development on the concave side of the hinge zone (Figure 11.13) are also mirrored by the orientation of $\hat{s}_1$ axes and by the fact that the ellipticity of the strain ellipse is very small in this region.

The orientation of mica flakes in two folds of substantially different geometry is shown in Figures 15.7 and 15.8. The first fold is an asymmetric fold of class 1C geometry in a relatively competent layer embedded in a less competent matrix (Figure 15.7*A*). The orientations of mica flakes (Figure 15.7*B*) are unusual in the lower limb, where the dots represent areas of low preferred orientation, and in a zone through the upper limb, where some micas seem discordant with the predominant preferred orientation. Figure 15.7*C* shows three fold models consisting of initial flexural-shear folding followed by varying amounts of homogeneous flattening perpendicular to the enveloping surface. The orientations of the $\hat{s}_1$ axes are plotted in each model. The distribution of mica flake orientations is most closely mimicked by the orientations of $\hat{s}_1$ in model 2 (Figure 15.7*C*). The model is not unique, however; for example, it does not consider the possibility of any homogeneous shortening parallel to the layer. Thus the similarity of patterns does not prove that the micas are oriented

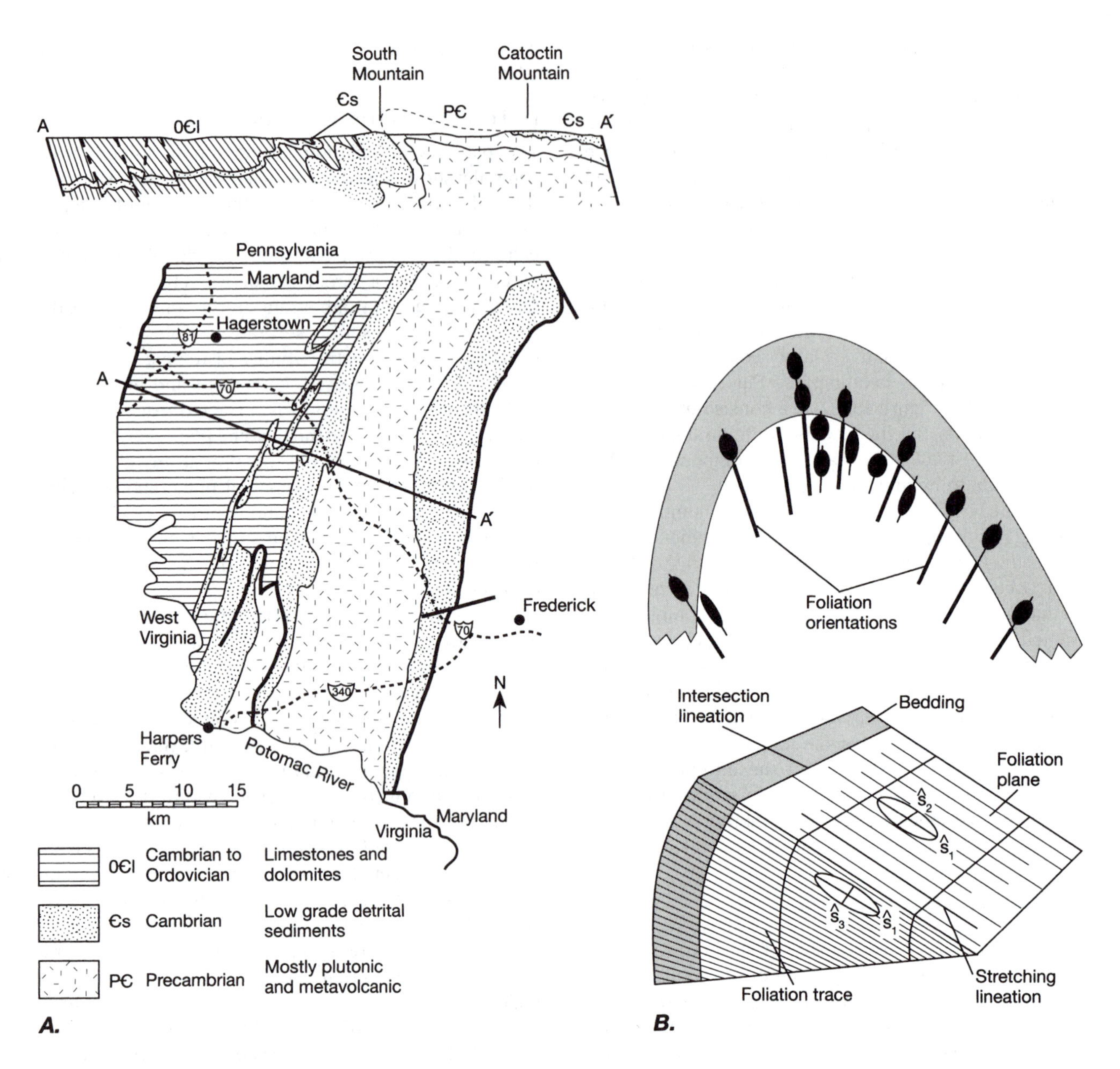

FIGURE 15.6 Strain across the South Mountain fold, Appalachian Mountains, western Maryland, USA. *A*. Cross section and generalized geologic map of the region of the South Mountain fold. (After Cleaves et al. 1968) *B*. Variation of strain across the South Mountain fold and its relationship to the foliation orientation. Fold profile is a composite section produced by down-structure projection. (After Cloos 1947, 1971)

parallel to the plane of flattening of the strain ellipsoid, although it is consistent with such an interpretation.

The second fold is one with class 2 geometry formed in a shear zone in a layered granulite (Figure 15.8*A*). A stretching lineation is oriented oblique to the hinge line. On a section through the fold normal to the lineation, the micas in the hinge zone form a divergent fan. In the limbs, they approach parallelism with the folded layer (Figure 15.8*B*). Figure 15.8*C* shows orientations of $\hat{s}_1$ for several fold models consisting of inhomogeneous simple shear (passive shear folding; cf. Figure 13.10*A*) followed by varying amounts of homogeneous flattening normal to the axial surface. The pattern of $\hat{s}_1$ axes that most closely reproduces the pattern of the mica orientations is shown as model 2 (Figure 15.8*C*). Although the similarity in patterns is striking, the model is again not unique. Thus the correspondence is suggestive—but not conclusive—evidence that the mica foliation is parallel to the plane of flattening of the finite strain ellipsoid.

Although the models presented are consistent with the data, they are based on untested assumptions, and other models might work equally well. Thus, in trying to inter-

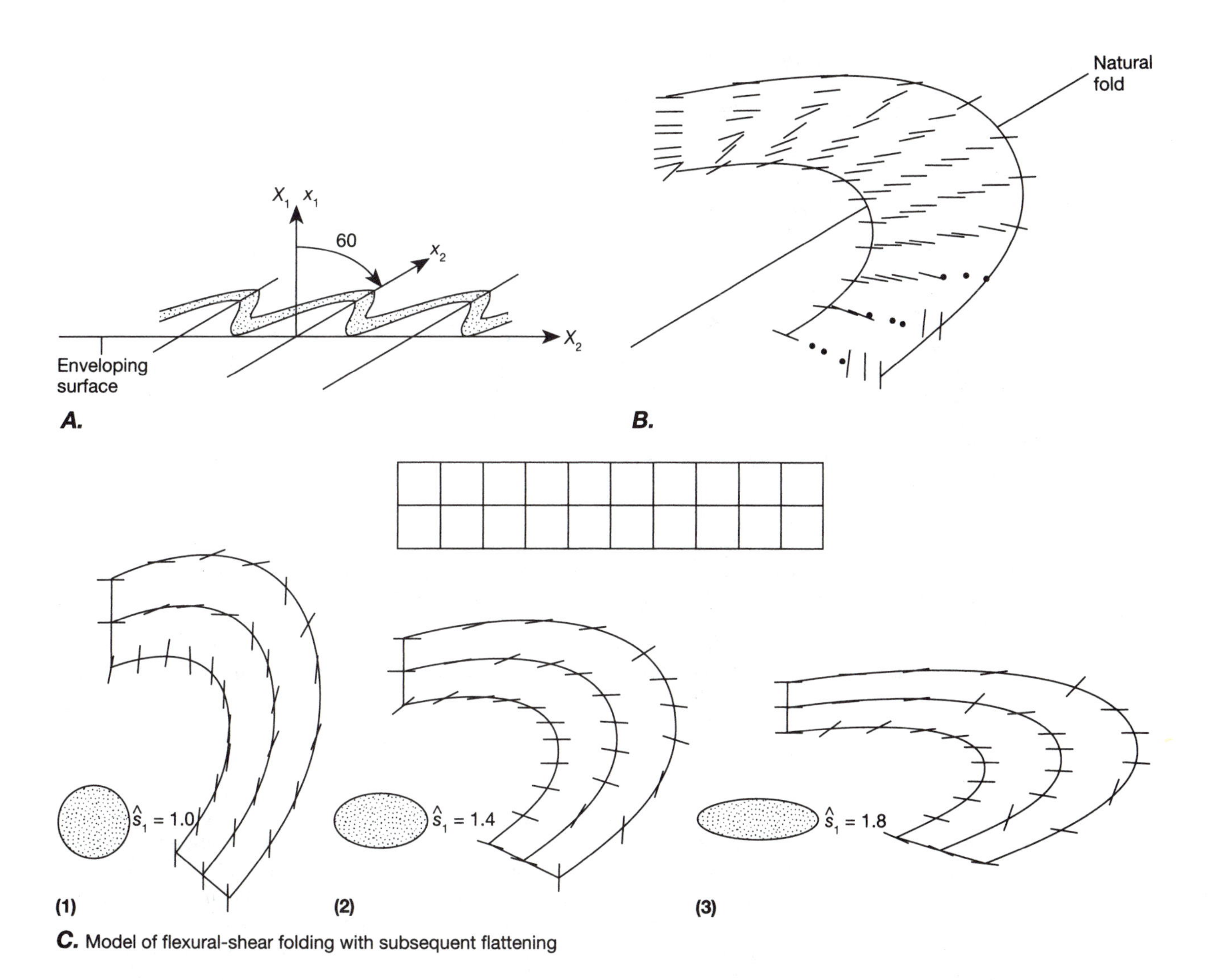

FIGURE 15.7 Correspondence between mica orientations in a class 1C fold and orientations of $\hat{s}_1$ in geometrically comparable theoretical folds. (After Hobbs 1971) *A*. The fold that is analyzed is one of the antiformal folds in an asymmetric fold train developed in a layer of quartz schist. *B*. Observed orientations of mica in the fold. *C*. Orientations of $\hat{s}_1$ for a fold model combining flexural shear folding with different amounts of subsequent homogeneous flattening normal to the enveloping surface. The rectangular bar shows the undeformed layer. Strain determinations for the different models are made at the corners of the squares. The homogeneous flattening component is indicated by the ellipses and the values of the maximum stretch. The orientations of $\hat{s}_1$ axes in model 2, for which the homogeneous flattening has a maximum stretch of $\hat{s}_1 = 1.4$, most successfully reproduce the distinctive orientations of the micas observed in the natural fold, although the model is not unique.

pret these structures, ideally we must try to test the assumptions, not simply accept them as being appropriate.

iv. Folds in the Cambrian Slate Belt of Wales

We often tend to think that folds reflect an overall shortening of the crust, and it is easy to assume that the fold axes form parallel to the $\hat{s}_2$ direction, with the maximum shortening direction $\hat{s}_3$ normal to the axial plane and $\hat{s}_1$ in the axial plane normal to the fold axis. This interpretation may be reasonable for symmetric parallel folds in foreland fold and thrust belts, for example, but in other situations the validity of such an interpretation cannot necessarily be presumed.

Strain studies of alteration spots in the Cambrian slate belt in Wales (Figure 15.9*A*) illustrate the point. Figure 15.9*B* shows a longitudinal section along the hinge surface of one of the folds. The upper heavy line shows the orientation of the hinge line in the axial surface, and the distance between the two lines represents the layer thickness in the fold hinge. Note that the diagram covers a section more than 20 kilometers in length. The hinge goes through several culminations and depressions of

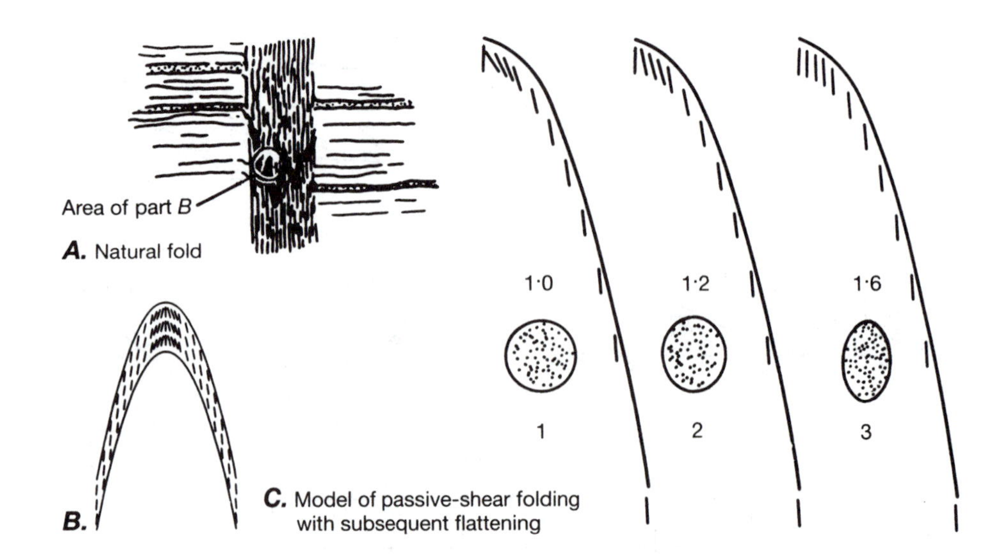

FIGURE 15.8 Correspondence between the mica orientations in a class 2 fold and the $\hat{s}_1$ orientations in theoretical folds of comparable geometry. (After Hobbs 1971) *A.* The similar fold is developed in layers in a quartzo-feldspathhic schist, within a shear zone that cuts through the schist. *B.* The distribution of mica flakes around the fold profile. *C.* Orientations of $\hat{s}_1$ for three fold models combining passive shear (inhomogeneous simple shear) with various superposed components of homogeneous flattening normal to the axial plane. The homogeneous flattening component is indicated by the ellipses and the values of the maximum stretch. Short lines are parallel to the orientations of $\hat{s}_1$ around the fold. Model 2, for which the homogeneous flattening has a maximum stretch of $\hat{s}_1 = 1.2$, most nearly reproduces the pattern defined by the mica orientations in *B*.

varying amplitudes along its length. The strain ellipses illustrated were measured in a vertical plane normal to the axial surface. Wherever there is a culmination in the hinge line, layer thickness increases and the strain ellipse shows an increase in ellipticity; maximum ellipticities correspond to maximum culminations. Conversely, hinge depressions correspond to the minimum layer thicknesses and the least elliptical strain ellipses. The $\hat{s}_1$ axes of the strain ellipsoids are vertical, and the orientation of the fold hinge is only locally parallel to the $\hat{s}_2$ direction, deviating in some places as much as 25°. In higher-grade metamorphic terranes, such deviations may become even more extreme.

Figure 15.9*C* shows a kinematic model that accounts for the associated variations in strain, layer thickness, and hinge line undulations. In the model, folding reflects an overall shortening of the body of rock perpendicular to the axial surface. The shortening is accommodated by vertical extension. The strain is inhomogeneous across the axial planes, leading to formation of the folds, and inhomogeneous along the axial surface, leading to the formation of culminations and depressions in the fold hinge. Where the shortening normal to the axial surface is greatest, the vertical extension is also the greatest, resulting in a thickening of the layers and a culmination in the fold hinge. Where shortening is least, the vertical extension is also small, resulting in a minimum in the layer thickness and a depression in the fold hinge.

These observations reinforce the conclusion that the orientations of many foliations and lineations are related to the orientation of the strain ellipsoid in the rocks. Our theoretical understanding of the mechanisms involved in the formation of foliations and lineations can help us make sense of the observed relationships, but in many cases it is not adequate to enable us to predict them. We need more studies that define the relationships among strain, folds, and the various types of foliations and lineations in rocks, and we need a more detailed understanding of the mechanisms by which these structures form (see Chapter 18).

15.4 STRAIN IN SHEAR ZONES

An increment of slip on a single shear zone accommodates a local strain in the shear zone for which the maximum and minimum instantaneous stretches ($\hat{\zeta}_1$, $\hat{\zeta}_3$) are at 45° to the shear zone. These principal axes lie in a plane called the motion plane, which is normal to the shear zone, and which contains the slip direction (Figure 15.10*A*). The bulk strain for a large body of rock containing an isolated shear zone is much smaller than the finite strain within a shear zone. Figure 15.10*B* shows how the local shear angle $\psi_{(sz)}$, when averaged over a larger volume of rock, results in a much smaller effective shear angle $\psi_{(bulk)}$. The principal axes of the bulk instantaneous strain for the volume of rock containing the

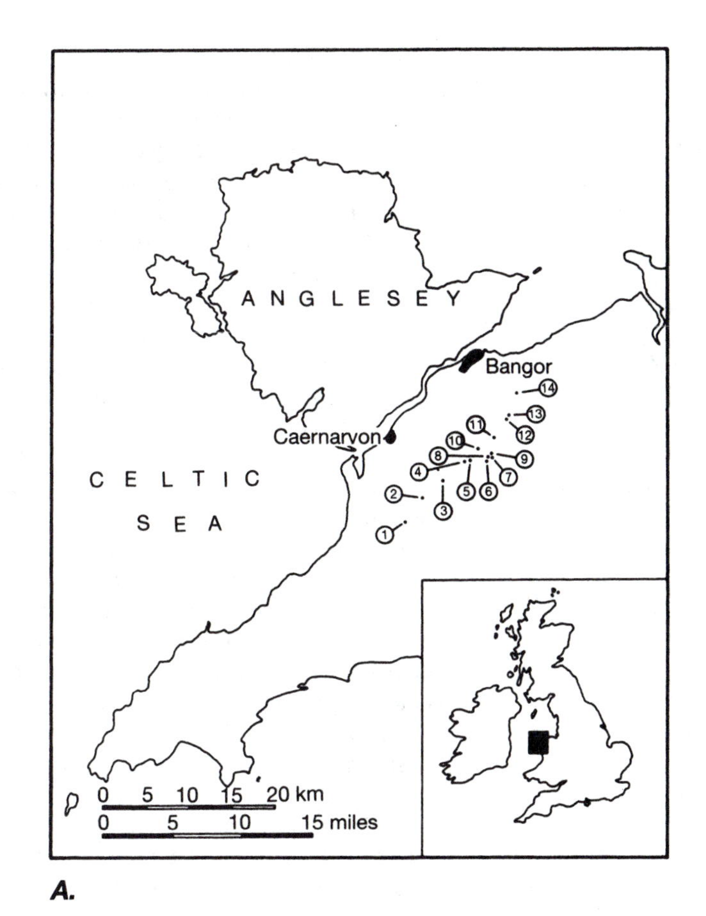

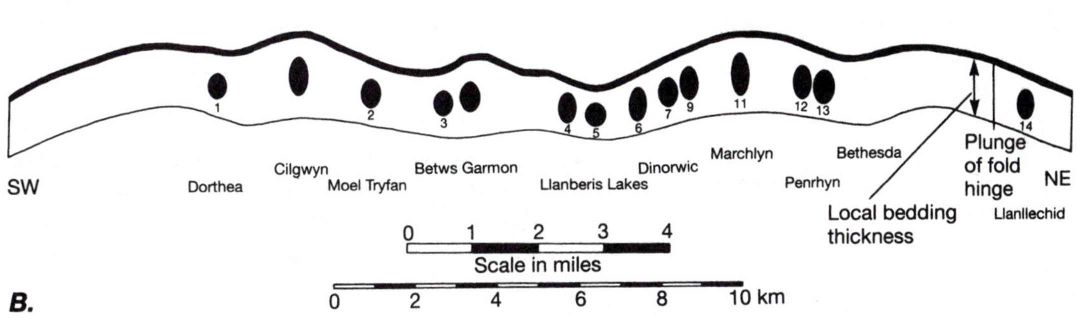

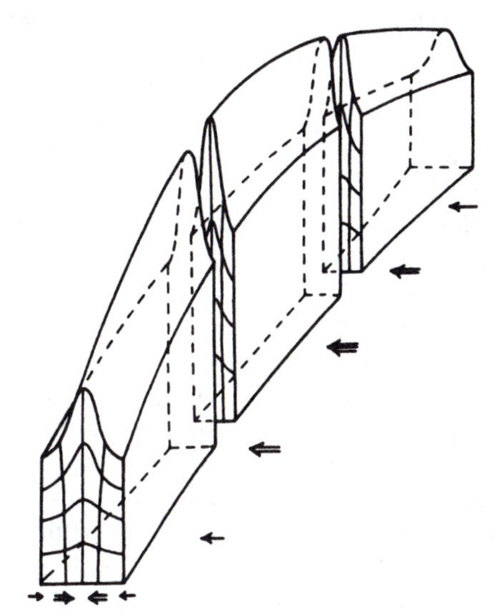

FIGURE 15.9 Deformation in the Cambrian slate belt in Wales. *A.* Location map for the slate belt in Wales. Dots labeled by circled numbers show where strain measurements were made. (From Wood 1973) *B.* Longitudinal profile of the hinge surface of a fold, showing the form of the hinge line (top line) and the variation in layer thickness (distance between top and bottom lines) along the surface. The strain ellipses show the principal axes of strain in a plane perpendicular to the hinge surface. Where the ellipticity is a maximum, the fold hinge goes through a culmination; where the ellipticity is a minimum, the fold hinge goes through a depression. (From Wood 1973) *C.* Model to account for the variation in strain and the associated changes in the fold hinge orientation. For a constant-volume plane strain deformation, the most extreme horizontal shortening must be associated with a maximum in vertical extension. (From Ramsay 1967)

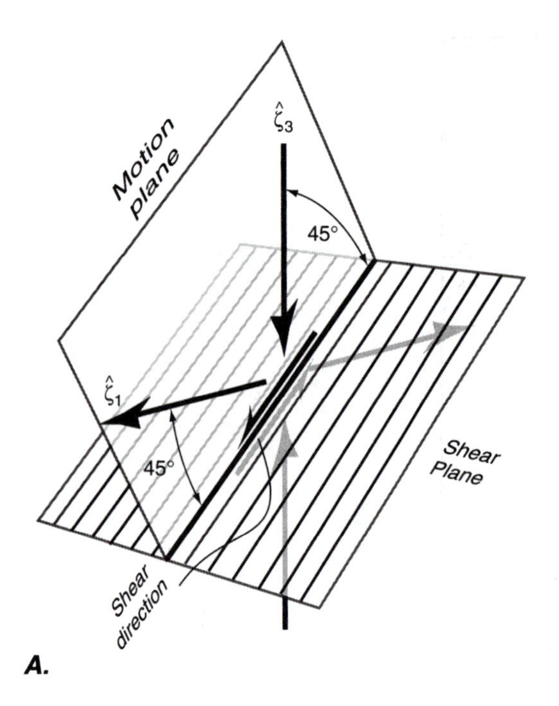

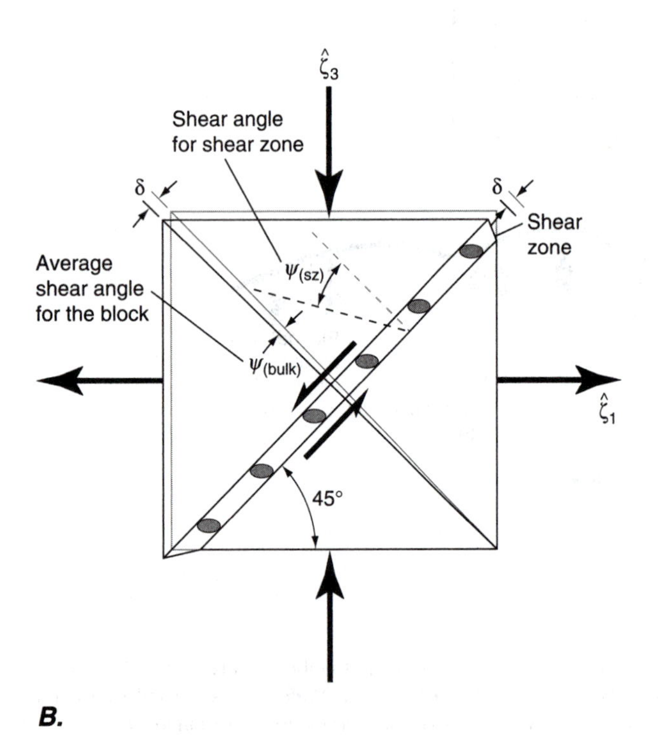

FIGURE 15.10 Orientation of the maximum and minimum axes of instantaneous stretch, $\hat{\zeta}_1$ and $\hat{\zeta}_3$, relative to shear zones. *A.* The maximum and minimum principal stretches lie in the motion plane at an angle of 45° to both the shear plane and the slip direction. The motion plane is normal to the shear plane and contains the slip direction. *B.* A finite increment of shear in a shear zone $\psi_{(sz)}$ contributes a very small average shear strain $\psi_{(bulk)}$ to the bulk strain of the block of rock containing the shear zone. The principal axes of bulk instantaneous stretch are oriented at 45° to both the shear plane and the slip direction.

shear zone are also oriented at 45° to the shear zone (Figure 15.10*B*). These are geometrically necessary relations that entail no assumptions or interpretation.

If, for example, a body of rock is deformed by a conjugate pair of shear zones that make an angle of 60° with each other, the principal axes of bulk instantaneous stretch for slip on each individual shear zone are at 45° to the shear zone and thus have different orientations, as shown in Figure 15.11*A*, *B*. The bulk principal instantaneous stretch axes are defined by the sum of the instantaneous strains that accumulate on each of the shear zones individually. If slip on each fault of the conjugate set is the same, then the sum of those shear strains averaged over the volume of rock containing the shear zones gives principal axes of bulk instantaneous stretch that bisect the angle between the conjugate shear zones, regardless of what that angle is (Figure 15.11*C*). If the slip is not the same on each fault, then the orientation of the principal instantaneous stretch axes depends on the relative amounts of shear that accumulate on the individual fractures. Thus the orientation of the bulk instantaneous stretches can vary between that shown in Figure 15.11*A* and that in Figure 15.11*B* as the relative amounts of slip vary from zero on one fracture to zero on the other.

For a body of rock cut by a multitude of shear zones, the bulk finite strain is just the sum of the averaged strains for each shear zone within the volume of rock. This type of averaging and summing of the strain associated with the discrete fault displacements allows us to treat the bulk deformation in such a situation as if it were a continuum deformation.

i. Strain from Brittle Deformation

The inference of strain from areas of brittle deformation applies the idea of a continuum measure of deformation (strain) to an area in which the deformation accumulates by discontinuous slip on a discrete set of faults. Implicitly, therefore, we must average the sum of the deformations contributed by the individual faults, over a volume that is large relative to the spacing of the faults. This is simply an application of the idea that a deformation can be considered homogeneous if the volume over which the strain is averaged is large relative to the discontinuities on which the deformation accumulates (see Figures 12.19 and 12.20).

Strain can be inferred from the systematics of fault length and displacement that we discussed in Sections 3.3(vi) and 3.4(i). We discuss this approach further in Box 15-1. The orientations of the principal strains and their relative magnitudes can also be inferred from the pattern of slip orientations on a set of diversely oriented faults.

Fractures associated with a major fault commonly have a wide variety of orientations on which shear occurs and slickenside lineations, or slickenlines, are developed. Such lineations are parallel to the direction of slip on the different shear planes. A bulk continuum deformation associated with the main fault is accommo-

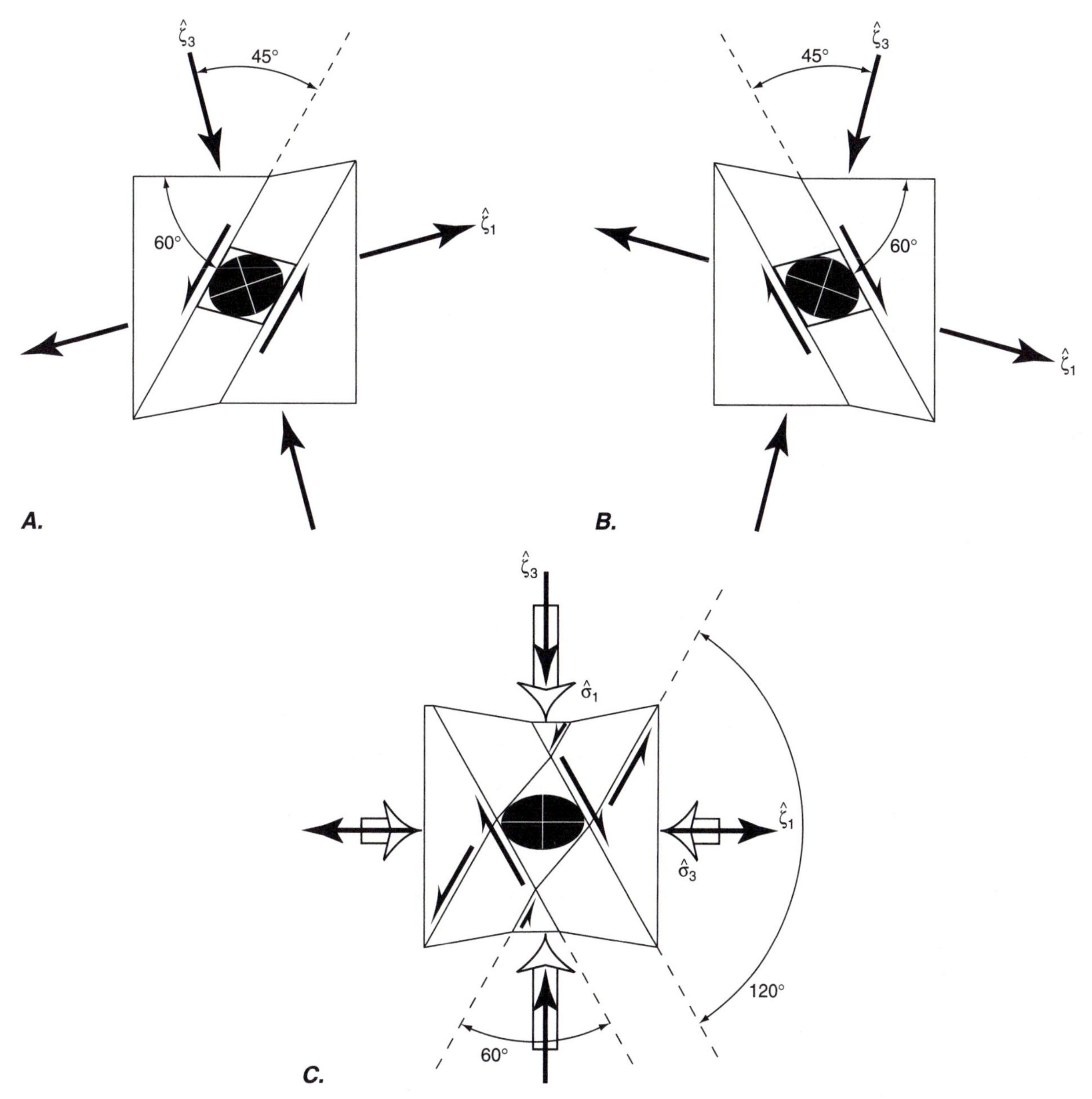

FIGURE 15.11 Strain associated with shear on conjugate shear zones. *A.* Counterclockwise shear on a shear zone dipping 60° to the left. Maximum and minimum principal axes of bulk instantaneous stretch are oriented at 45° to both the shear plane and the slip direction. *B.* Clockwise shear on a shear zone dipping 60° to the right. Maximum and minimum principal axes of bulk instantaneous stretch are oriented at 45° to both the shear plane and the slip direction. *C.* Simultaneous shear on the conjugate shear zones is approximated by superposing the shear in *B* on that in *A*. The diagram uses finite shear strain to represent instantaneous shear strain for the sake of diagrammatic clarity. If the amount of bulk shear contributed by each shear zone is the same, the principal axes of bulk instantaneous stretch (black arrows) bisect the angles between the conjugate shear zones. Different amounts of shear on the two shear zones, however, would result in principal axes that are not parallel to the bisectors. A common interpretation of conjugate shear zones is that the maximum compressive stress $\hat{\sigma}_1$ bisects the acute angle between the zones, and the minimum compressive stress $\hat{\sigma}_3$ bisects the obtuse angle (wide arrows). This assumes that the Coulomb fracture criterion is applicable.

dated by discrete shear on each of the brittle fractures and associated slip directions in this collection, which are distributed throughout the volume of deforming rock. Because this increment in strain is in general very small, we can equate it with the instantaneous strain. Thus we can use this collection of shear-plane/slip-direction data to infer the orientation and relative magnitude of the bulk instantaneous strain within the volume. In technical parlance, we can "invert" the shear-plane/slip-direction data to find the bulk instantaneous strain ellipsoid.[5] The inversion method requires two assumptions: (1) We assume the bulk instantaneous strain is homogeneous within the volume from which we collect the shear-plane/slip-direction data, and (2) we assume the slip direction on each plane is parallel to the direction of maximum resolved shear strain of the bulk instantaneous strain. Under these conditions, the slip directions vary systematically with both the orientation of the shear plane relative to the principal axes of bulk instantaneous strain and the relative magnitudes of the principal bulk instantaneous strains.

Patterns of shear-plane/slip-direction data for three different instantaneous strain ellipsoids are shown in the **tangent lineation diagrams** in Figure 15.12. To understand the meaning of these diagrams, imagine that a shear plane is plotted as a great circle on a lower-hemisphere spherical projection (Figure 15.12*A*, *B*). The pole to this plane is the plot of the orientation of the line perpendicular to the plane that passes through the center of the plotting hemisphere, and it marks the point on the hemisphere at which the plane would be tangent if it were moved to the outside of the hemisphere (Figure 15.12*B*). The slip direction in that plane is then plotted as an arrow parallel to the slickenline in the tangent plane, where the arrow gives the direction of motion of the material outside the hemisphere relative to the material inside the hemisphere. The arrow thus records the direction of motion of the footwall block for inclined faults and of the far side of the fault, relative to the center of the plotting hemisphere, for vertical faults. The resulting pattern of arrows (Figure 15.12*C–E*) thus defines how the material outside the hemisphere would move past the hemisphere on planes tangent to the surface of the hemisphere.

Two features of these patterns emerge immediately: (1) The arrows converge toward, or diverge from, the principal instantaneous strain axes, allowing the orientation of these axes to be identified from the arrow pattern; and (2) the particular pattern of arrow orientations varies with the relative magnitudes of the principal instantaneous extension axes, which can be defined by the ratio

$$D \equiv \frac{(\hat{\varepsilon}_2 - \hat{\varepsilon}_3)}{(\hat{\varepsilon}_1 - \hat{\varepsilon}_3)} \qquad (15.4)$$

The theory requires that this ratio be defined in terms of the principal instantaneous extensions $\hat{\varepsilon}_k$, but these are parallel to, and simply calculated from, the principal instantaneous stretches $\hat{\varepsilon}_k = \hat{\zeta}_k - 1$ (see Equations (12.30)).

We can calculate what the tangent lineation pattern ought to be for a given orientation of the instantaneous strain ellipsoid and the value of D, so we can write a computer program to do the inverse calculation that finds the ellipsoid that best accounts for a particular set of data. By inverting data collected within shear zones using such a program, we can map the distribution of instantaneous strain within those shear zones. This procedure can also be adapted for use with seismic focal mechanism solutions, for which we know the slip directions in each of two perpendicular nodal planes, one of which must be the actual fault plane (see Appendix Box A2-4). This allows us to use earthquakes to map the instantaneous strain ellipsoids within seismically active shear zones.

Figure 15.13 shows an example from an alignment of small earthquakes (magnitude < 3.0) from a right stepover in a major right lateral fault through the Coso Range in the Eastern California Shear Zone, east of the southern Sierra Nevada (Figure 15.13*A*; note that "up" on the map is South, to correspond with the block diagram in Figure 15.13*D* in which the view is toward the South; compare the image of the same area in Figure 19-1.2*A* in which the view direction is also toward the South). The fault is part of the Eastern California Shear Zone (see Figure 19.30). Different solutions are defined by the focal mechanisms in two depth ranges, one between 0 and 5 kilometers deep, and the other between 5 and 8 kilometers deep. The shallower set of focal mechanisms gives a solution (Figure 15.13[6] *B*, *D*) with $\hat{\varepsilon}_3$ subvertical $\hat{\varepsilon}_1$, subhorizontal nearly east–west, and $\hat{\varepsilon}_2$ subhorizontal (compare the theoretical tangent lineation pattern in Figure 15.12*D*), which indicates crustal thinning with nearly east–west lengthening. D is indistinguishable from 0.5, indicating plane strain.[7] Below 5-kilometer depth, the focal mechanisms give a solution (Figure 15.13[6] *C*, *D*) with $\hat{\varepsilon}_2$ subvertical and with $\hat{\varepsilon}_1$ subhorizontal having almost the same orientation as the shallower solution, and $\hat{\varepsilon}_3$ subhorizontal, which indicates horizontal shearing. D again is essentially 0.5, indicating plane strain. These results are consistent with a blind, dextral, strike-slip fault below 5-kilometer depth, with normal faulting accommodating the edge effects above the tip line of the blind fault (Figure 15.13*D*). Indeed, the surface at Wild Horse Mesa shows abundant normal faulting.

Brittle faults are often used to infer the orientations of the principal stresses to which the rock was subjected dur-

---

[5]In forward modeling, one would assume the orientation of the principal bulk instantaneous stretches and their relative magnitudes and calculate the orientation of the slip direction on planes of any desired orientation. The "inverse" of forward modeling is the process of using data on the orientation of the shear planes and the slip directions contained in them, to infer the orientation of the principal bulk instantaneous stretches that best account for the data.

[6]Note in Figure 15.13, that the primitive circle on the stereonet diagrams is horizontal with South at the *top* of the circle, and that the block diagram is viewed looking South.

[7]For constant volume, $\hat{\varepsilon}_1 + \hat{\varepsilon}_2 + \hat{\varepsilon}_3 = 0$, and for plane strain, $\hat{\varepsilon}_2 = 0$. Combining these relationships with Equation (15.4) shows that $D = 0.5$ for constant-volume plane strain.

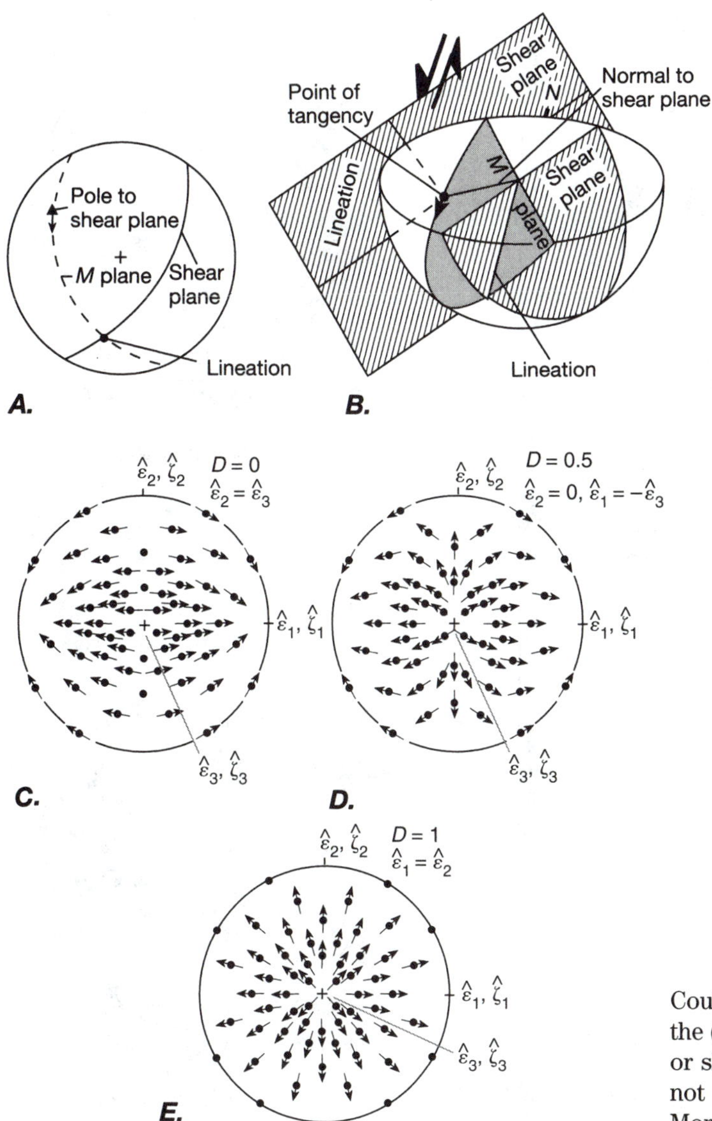

FIGURE 15.12 Construction and patterns of tangent lineation diagrams for shear-plane/slip-direction data. A tangent lineation diagram is a means of plotting on a lower hemisphere spherical projection both a shear plane and the slip direction in that shear plane, using only one symbol. *A* and *B*. To plot shear-plane/slip-direction data on a lower hemisphere spherical projection, we construct the great circle for the *M*-plane (the motion plane), which contains both the pole (the normal) to the shear plane and the orientation of the lineation in the shear plane. The *M*-plane is thus the plane normal to the shear plane that contains the lineation. The pole to the shear plane is the point on the plotting hemisphere where the shear plane would be tangent to the outside of the hemisphere. At the pole to the shear plane, we draw an arrow tangent to the *M*-plane great circle. The arrow is therefore parallel to the lineation in the tangent shear plane, and we choose to have it point in the direction of relative shear of the footwall block. *C*, *D*, and *E*. Patterns of tangent lineations for constant orientation of the principal instantaneous strain axes, but three different shapes of instantaneous strain ellipsoid. The shape is defined by the shape parameter *D*, which is defined in terms of the principal instantaneous extensions (Equation (15.4)): in C, $D = 0$, $\hat{\varepsilon}_2 = \hat{\varepsilon}_3$ (instantaneous simple constriction); in D, $D = 0.5$, $\hat{\varepsilon}_2 = 0$; $\hat{\varepsilon}_1 = -\hat{\varepsilon}_3$ (instantaneous plane strain); and in E, $D = 1$, $\hat{\varepsilon}_1 = \hat{\varepsilon}_2$ (instantaneous simple flattening).

ing brittle deformation. This interpretation relies on the Coulomb fracture criterion (Figures 8.3*B*, *D*, *E* and 8.4*A*), which suggests that for conjugate faults the maximum compressive stress $\hat{\sigma}_1$ should bisect the acute angle between the conjugate faults, the intermediate compressive stress $\hat{\sigma}_2$ should be parallel to the intersection line of the faults, and the minimum compressive stress should bisect the obtuse angle (Figure 15.11*C*). For two sets of faults to be firmly identified as conjugate faults, the angle between them should be between about 40° and 90°, they must have opposite senses of shear, and there must be good evidence—such as mutual cross-cutting relations—that the two orientations of fault were active at the same time.

The Coulomb fracture criterion, however, is not applicable to all shear fractures (see the discussion in Section 8.7). For example, the shear fractures predicted by Coulomb theory can accommodate deformation only in the $(\hat{\sigma}_1, \hat{\sigma}_3)$-plane. If there is a component of lengthening or shortening parallel to $\hat{\sigma}_2$, the Coulomb criterion cannot predict the orientation of the fractures that form. Moreover, shearing can occur on pre-existing fractures, faults, bedding planes, or other planes of weakness that have no relationship to the orientations predicted by Coulomb theory (cf. Figure 8.13).

Inversions of shear-plane/slickenline data are often interpreted as defining the orientation and relative values of the principal stresses, the assumption being that the slip directions on the faults are parallel to the direction of maximum resolved shear stress on those planes. The data, however, are fundamentally a reflection of the displacement on the faults, not of the stress, and thus they are better understood as inversions for the instantaneous extension axes (see Appendix Box A2-4). Equating the stress axis orientations with those of the instantaneous extension involves the implicit assumption that the rocks are mechanically isotropic. Assuming that *D* (Equation (15.4)) also defines the relative values of the principal stresses further requires the assumption that the rocks deform on a large scale with a linear deformation law (see Chapter 16). Neither assumption is necessarily valid.

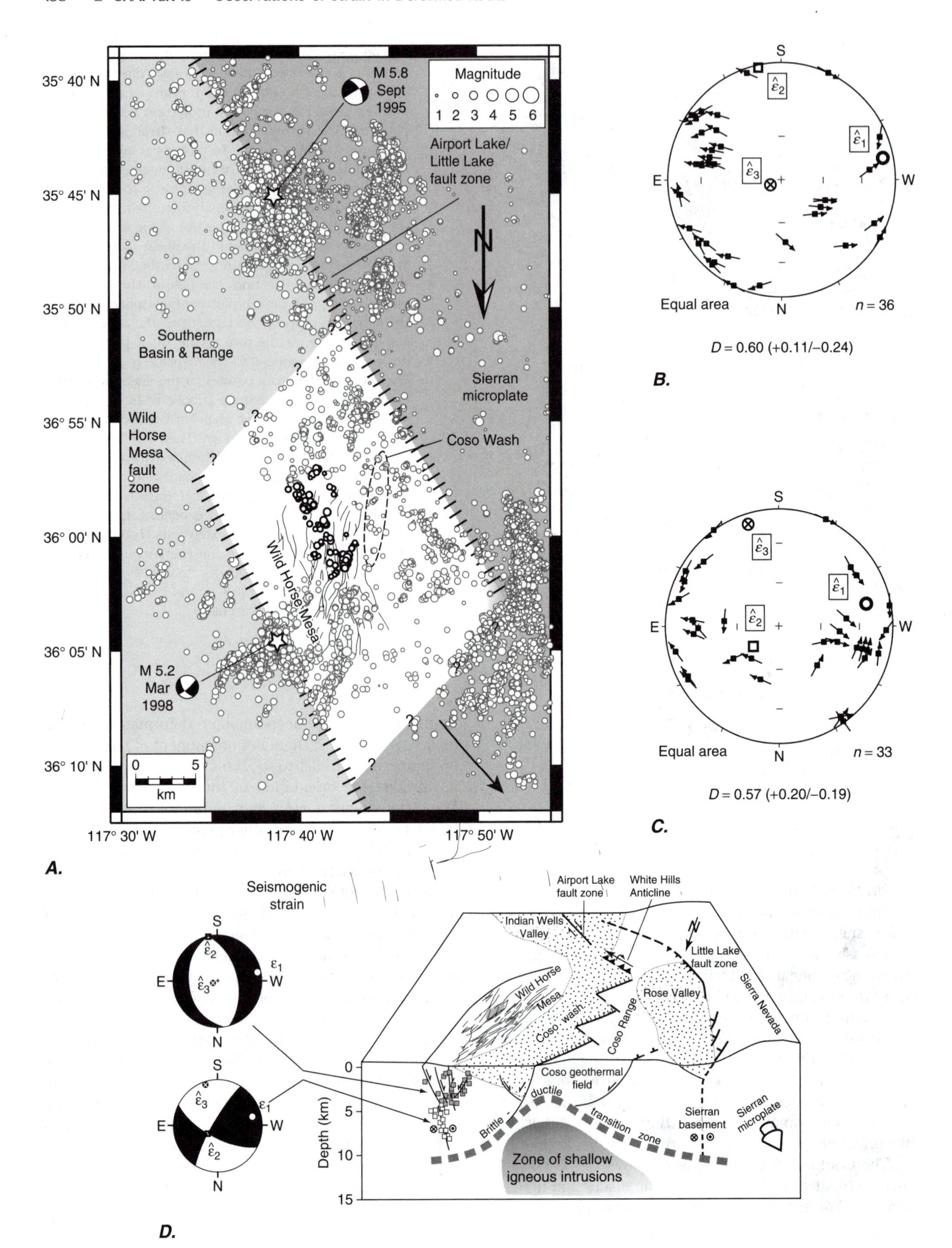
35° 40' N
35° 45' N
35° 50' N
36° 55' N
36° 00' N
36° 05' N
36° 10' N
117° 30' W
117° 40' W
117° 50' W
M 5.8
Sept
1995
Magnitude
1 2 3 4 5 6
Airport Lake/
Little Lake
fault zone
N
Southern
Basin & Range
Sierran
microplate
Wild
Horse
Mesa
fault
zone
Coso Wash
Wild Horse Mesa
M 5.2
Mar
1998
0 5
km
A.
S
E
W
N
Equal area
n = 36
D = 0.60 (+0.11/–0.24)
B.
S
E
W
N
Equal area
n = 33
D = 0.57 (+0.20/–0.19)
C.
Seismogenic
strain
ε1
Airport Lake
fault zone
White Hills
Anticline
Indian Wells
Valley
Little Lake
fault zone
Wild Horse
Mesa
Coso wash
Coso Range
Rose Valley
Sierra Nevada
Coso geothermal
field
Brittle
ductile
transition
zone
Sierran
basement
Sierran
microplate
Zone of shallow
igneous intrusions
Depth (km)
0
5
10
15
D.

FIGURE 15.13 Inversion solutions for the principal instantaneous extensions from seismic focal mechanisms under Wild Horse Mesa in the Coso Range, Eastern California Shear Zone, east of the southern Sierra Nevada. *A*. Map of the right stepover in the right lateral fault through the Coso Range. The bold circles are the epicenters for the seismic events under Wild Horse Mesa used for the analysis, which are among the thousands of events distributed throughout the region. The faults plotted in Wild Horse Mesa are mostly normal faults. "Beachball" diagrams (see Figure A2.3) show the predominantly strike-slip nature of large recent earthquakes on the faults bounding the stepover. Note that north is *down* on this map, to correspond with the block diagram in *D*. (Modified from Lewis et al. 2007, figure 3) *B*. Solution for the principal instantaneous extension axes from focal mechanisms between 0- and 5-kilometer depth, plotted on the tangent lineation diagram that shows the preferred shear-plane/slip-direction set for each focal mechanism. (Data from Lewis et al. 2007) *C*. Solution for the principal instantaneous extension axes from focal mechanisms between 5- and 8-kilometer depth plotted on the tangent lineation diagram that shows the preferred shear-plane/slip-direction set for each focal mechanism. (Data from Lewis et al. 2007) *D*. Interpretive block diagram of the local structure. Earthquake hypocenters are projected onto the vertical plane and shaded differently for the two different depth ranges. "Beachball" diagrams show the seismogenic instantaneous strain, which is essentially a vertical thinning for the shallow events and a horizontal shearing for the deeper events. (Modified from Lewis et al. 2007, figure 7)

BOX 15-1 Brittle Strain Inferred from Fault Systematics#

In order to describe the deformation associated with faulting, it is useful to be able to calculate the average strain of a volume of rock due to the slip on all the faults contained within the volume. As a simple example, we consider in Figure 15.10*B* the average shear strain in a block of rock that results from a finite displacement δ across a narrow shear zone through the rock. In this case, the length of the diagonal of the block normal to the shear zone is ℓ. Thus the shear strain averaged over the whole block is

$$e_s = \frac{1}{2} \tan \psi_{\text{bulk}} = \frac{1}{2} \frac{\delta}{\ell} \tag{15-1.1}$$

If we multiply the numerator and denominator of this result by the area of the fault over which the displacement δ occurs, we find how the shear strain is calculated from a volume average:

$$e_s = \frac{1}{2} \frac{\delta\, \ell h}{\ell\, \ell h} = \frac{1}{2} \frac{A\, \delta}{V} \tag{15-1.2}$$

where h is the thickness of the block normal to the diagram, ℓ is the length of the fault, so $A = \ell h$ is the area of the fault over which the displacement occurs, and $V = \ell^2 h$ is the volume of rock containing the fault. This equation tells us that for this particular example, the shear strain is half of the fault area per unit volume times the displacement. The quantity $A\,\delta$ is called the **geometric moment** M_0 of the fault†

$$M_0 \equiv A\, \delta \tag{15-1.3}$$

Thus, in terms of the geometric moment, the average shear strain in the volume due to slip on the fault is given for this example by

$$e_s = \frac{1}{2} \frac{M_0}{V} \tag{15-1.4}$$

#We are indebted to Randall Marrett for his contributions to this box.

†The product of the shear modulus times the geometric moment μM_0 is called the **seismic moment**, which is a common measure of the magnitude of an earthquake.

In this discussion, we are interested in determining the extension that we would measure along a traverse in a block of crust that has been deformed by slip on a large number of faults. We can show that the extension due to the *i*th fault along a traverse parallel to the unit vector **t** is

$$e_{(t)}^{(i)} = \frac{A_{(i)} \cos \theta_{(i)}\, \delta_{(i)} \cos \phi_{(i)}}{V} \tag{15-1.5}$$

which can be rewritten as

$$e_{(t)}^{(i)} = \frac{A_{(i)} \cos \theta_{(i)}}{(V/T)} \frac{\delta_{(i)} \cos \phi_{(i)}}{T} \tag{15-1.6}$$

where $\theta_{(i)}$ is the angle between the traverse direction and the normal to the fault plane, $\phi_{(i)}$ is the angle between the traverse direction and the slip direction on the fault, $A_{(i)}$ is the area of the fault over which the slip $\delta_{(i)}$ occurs, V is the volume of the rock over which we average the strain, and T is the length of the traverse. The second term in Equation (15-1.6) ($\delta_{(i)} \cos \phi_{(i)}/T$) is the component of the slip in the direction of the traverse divided by the length of the traverse. For small strains for which the undeformed traverse length T_O is approximately equal to the final traverse length T, this term is approximately equal to the standard definition of extension (see Equation 12.2). The first term in Equation (15-1.6) is the ratio of the fault area projected onto a plane normal to the traverse direction ($A_{(i)} \cos \theta_{(i)}$) to the cross sectional area of the averaging volume (V/T), where the cross section is also normal to the traverse direction. This term, therefore, is just the probability that a fault in the volume will be intersected by a linear (one-dimensional) sampling traverse. Thus the contribution of each fault to the extension is weighted by the probability of the fault being included in the sampling, which is related to the size of the fault. Equation (15-1.5) for the extension is comparable to Equation (15-1.2) for the shear strain, except that the extension does not include the factor of $^1/_2$ that is characteristic of the shear strain, and A and δ are each modified by a cosine function that takes account of the orientation of the traverse direction relative to the orientations of the fault and the slip direction.

(continued)

BOX 15-1 Brittle Strain Inferred from Fault Systematics *(continued)*

From Equation (15-1.5), the total extension is just the weighted sum of the extensions contributed by each fault cut by the traverse line.

$$e_{(t)}^{(tot)} = \sum_{i=1}^{N} e_{(t)}^{(i)} = \sum_{i=1}^{N} \frac{A_{(i)} \cos\theta_{(i)}\, \delta_{(i)} \cos\phi_{(i)}}{V} \quad (15\text{-}1.7)$$

We pursue this analysis for a simple case, for which we assume all the faults and all the displacements have a similar orientation, so the angles θ and ϕ are the same for each of the faults. We also approximate $A_{(i)}$ in Equations (15-1.5) through (15-1.7) in terms of the linear dimension of the fault $L_f^{(i)}$:

$$A_{(i)} \approx (L_f^{(i)})^2 \quad (15\text{-}1.8)$$

With these simplifications, the contribution of the ith fault to the total extension along the traverse line can be written from Equation (15-1.5)

$$e_{(t)}^{(i)} = \frac{(L_f^{(i)})^2}{V} \delta_{(i)} \cos\theta \cos\phi \quad (15\text{-}1.9)$$

The dimension of the volume V over which we average must be large relative to the dimension of the largest fault and should be on the order of T^3, where T is the length of the traverse, although the actual depth to which the volume extends must be smaller than the thickness of the brittle crust.

Measuring all the displacements on all the faults appears to be a formidable task, because the number of faults increases rapidly for shorter and shorter faults according to a power law (see Equation (3.2)). Although smaller faults have smaller displacements (Equation (3.1)), there are many more of the smaller faults than the large ones (Equation (3.2)), and it is not obvious whether the contribution of the many small faults to the total extension is significant compared to the contribution of the few large faults. Moreover, given the usual incomplete exposure, the measurement of displacements on all the faults along a traverse would in general be impossible. Fortunately, we are not completely stuck.

We can use the systematics of fault displacement (Section 3.3(vi)) and of fault length distributions (Section 3.4(i)) to estimate the total displacement without having to undertake the daunting task of measuring it all. The displacement on a fault is related to the linear dimension of the fault by a power law (Equation (3.1); Figure 3.26):

$$L_f^p = B\delta \quad \text{or} \quad L_f = (B\delta)^{1/p} \quad (15\text{-}1.10)$$

We can substitute the second Equation (15-1.10) into Equation (15-1.9) to find,

$$e_{(t)}^{(i)} = \frac{B^{2/p}}{V} \cos\theta \cos\phi\, \delta_{(i)}^{(p+2)/p} \quad (15\text{-}1.11)$$

Fault lengths have a power-law distribution given by (Equation (3.2); Figure 3.28*B*)

$$N(L_f) = K\, L_f^{-m} \quad (15\text{-}1.12)$$

By substituting L_f from Equation (15-1.10) into Equation (15-1.12), we find that the distribution of displacements must also be described by a power law given by

$$N(\delta) = K(B\delta)^{-m/p} \qquad \log(N(\delta)) = \log(KB^{-m/p}) - \frac{m}{p}\log\delta \quad (15\text{-}1.13)$$

$N(\delta)$ is the cumulative number of faults having a displacement greater than or equal to δ. Thus the fault with the largest δ (designated by $\delta_{(max)}$) is fault for which $N = 1$. Substituting this constraint into the first Equation (15-1.13) allows us to evaluate the constants in the equation.

$$1 = K(B\delta_{(max)})^{-m/p} \qquad KB^{-m/p} = \delta_{(max)}^{m/p} \quad (15\text{-}1.14)$$

Thus, we can rewrite the first Equation (15-1.13) as

$$N(\delta) = \delta_{(max)}^{m/p}\delta^{-m/p} \quad (15\text{-}1.15)$$

The number of faults having a component of displacement between δ and $\delta + d\delta$ is just the increment dN in the cumulative distribution, which we obtain from Equation (15-1.15)

$$dN = -\frac{m}{p}\delta_{(max)}^{m/p}\,\delta^{(-m-p)/p}\,d\delta \quad (15\text{-}1.16)$$

We can now calculate the increment in the extension due to slip on faults in the volume V if we treat the finite increments of extension associated with the ith fault (Equation (15-1.11)) as infinitesimal increments of a continuous function.

$$de_{(t)} = \frac{B^{2/p}}{V}\cos\theta\cos\phi\,\delta^{(p+2)/p}\,dN \quad (15\text{-}1.17)$$

For convenience, let

$$D \equiv \frac{B^{2/p}}{V}\cos\theta\cos\phi \quad (15\text{-}1.18)$$

Using Equations (15-1.16) and ((15-1.18) in Equation ((15-1.17) gives

$$de_{(t)} = -D\frac{m}{p}\delta_{(max)}^{m/p}\,\delta^{(2-m)/p}\,d\delta \quad (15\text{-}1.19)$$

We now take the indefinite integral to find the extension as a function of δ

$$e_{(t)} = -D\frac{m}{p}\delta_{(max)}^{m/p}\int \delta^{(2-m)/p}\,d\delta$$

$$e_{(t)} = -D\frac{m}{p}\delta_{(max)}^{m/p}\left[\frac{p}{p+2-m}\delta^{(p+2-m)/p} + C\right] \quad (15\text{-}1.20)$$

where C is the constant of integration.

We evaluate C by requiring, from Equation (15-1.11) that

$$\text{when} \quad \delta = \delta_{(max)} \quad \text{then} \quad e_{(t)} = e_{(t)}^{(1)} = D\,\delta_{(max)}^{(p+2)/p} \quad (15\text{-}1.21)$$

$$e_{(t)}^{(1)} = D\,\delta_{(max)}^{(p+2)/p} = -D\frac{m}{p}\delta_{(max)}^{m/p}\left[\frac{p}{p+2-m}\delta_{(max)}^{(p+2-m)/p} + C\right]$$

$$C = -\delta_{(max)}^{(p+2-m)/p}\frac{p}{m}\left[\frac{p+2}{p+2-m}\right] \quad (15\text{-}1.22)$$

Then from Equations (15-1.20) and (15-1.22) we have the extension contributed by all faults having a displacement greater than or equal to δ, that is, between δ and $\delta_{(max)}$.

$$e_{(t)}(\delta) = D\left[\frac{p+2}{p+2-m}\right]\delta_{(max)}^{(p+2)/p} - D\frac{m}{p+2-m}\delta_{(max)}^{m/p}\,\delta^{(p+2-m)/p} \quad (15\text{-}1.23)$$

The total extension is obtained by setting $\delta = 0$ in Equation (15-1.23), giving

$$e_{(t)}^{(tot)} = D\left[\frac{p+2}{p+2-m}\right]\delta_{(max)}^{(p+2)/p} = e_{(t)}^{(l)}\left[\frac{p+2}{p+2-m}\right] \quad (15\text{-}1.24)$$

This gives us the remarkable result that we can estimate the total extension contributed by all the faults along a traverse if we know only the extension contributed by the largest fault! We express the extension contributed by all faults having a displacement between δ and $\delta_{(max)}$ as a fraction of the total extension by dividing Equation (15-1.23) by the first Equation (15-1.24), which gives

$$\frac{e_{(t)}(\delta)}{e_{(t)}^{(tot)}} = 1 - \frac{m}{p+2}\left(\frac{\delta}{\delta_{(max)}}\right)^{(p+2-m)/p} \quad (15\text{-}1.25)$$

Before comparing these results with measurements made on actual fault systems, we must clarify the relation between the parameter m in these equations and that inferred from measurements on actual fault systems. The derivation above, through its use of the three-dimensional strain tensor and its relation to the geometric moment, implicitly assumes that Equation (15-1.12) reflects a three-dimensional sampling of fault lengths. Thus the parameter m in these equations should more appropriately be written m_3 to distinguish it as the parameter determined from three-dimensional sampling. For two-dimensional sampling of fault lengths on a map, for example, the slope determined from the plot of the measured data is m_2, and for one-dimensional sampling along a linear traverse, the slope would be m_1. These three parameters are related by

$$m_3 = m_2 + 1 = m_1 + 2 \quad (15\text{-}1.26)$$

(see the discussion of sampling dimension on the exponent in the power law given in Box 2-1(ii)).

Figure 3.26 shows that for a variety of different data sets, the value of p is approximately

$$p \approx 1.0 \quad (15\text{-}1.27)$$

Figure 3.28*B* shows data on the distribution of fault lengths determined from a two-dimensional sampling, for which the best-fit slope of the linear part of the curve gives

$$m_2 = 1.52 \quad \text{or} \quad m_3 = 2.52 \quad (15\text{-}1.28)$$

where we used the first Equation (15-1.26).

The parameter p is determined from multiple data sets that, as a whole, span about 6 orders of magnitude (Figure 3.26). The value of m_2 from Figure 3.28*B*, however, is determined from data that are log-linear over less than 1.5 orders of magnitude. Such a small range of data on a log–log plot limits the accuracy with which the parameters can be determined.

Figure 15-1.1 shows a plot of the displacement distribution for a population for faults in the Basin and Range province of

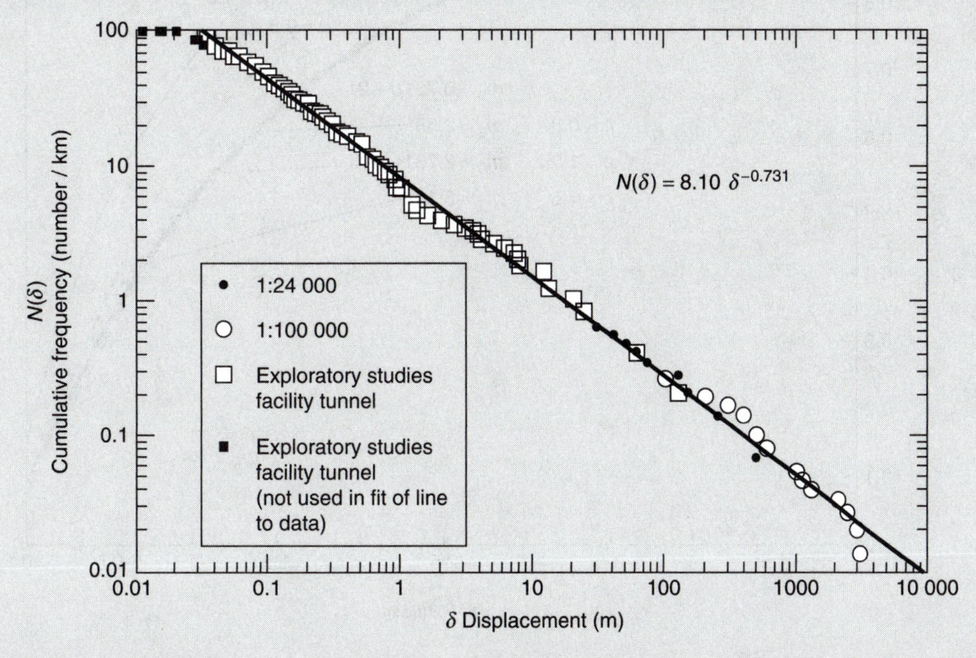

FIGURE 15-1.1 Plot of fault displacement vs cumulative frequency determined at different scales, as indicated in the key, along linear transects through the Paintbrush Group at Yucca Mt. in the Basin and Range province of southern Nevada, USA. Transects were oriented approximately perpendicular to the dominant strike of the faults. The line is fit only to the data from the Exploratory Studies Facility tunnel, and the fit of the rest of the data to this line demonstrates the consistency of the data over nearly 5 orders of magnitude. (Modified after Marrett et al. 1999, with permission of the Geological Society of America)

(continued)

BOX 15-1 Brittle Strain Inferred from Fault Systematics *(continued)*

the western United States in the vicinity of Yucca Mountain in southern Nevada, which were determined from one-dimensional sampling along a transect approximately normal to the dominant fault strike through the area. The data show a strikingly linear log-log relationship over approximately five orders of magnitude, implying that the assumption of a power-law distribution for the displacement is very robust, and that the parameters in Equation (15-1.13) are unusually well determined. This relation therefore provides an excellent constraint on the equation parameters. The slope of the plot is s_1, and from Figure 15-1.1

$$\log(N(\delta)) = \log(KB^{-s_1}) - s_1 \log \delta \quad \text{where} \quad s_1 = 0.731 \qquad (15\text{-}1.29)$$

By comparing the slopes in the second Equation (15-1.13) and Equation (15-1.29), we see that the parameter s_1 determined from a one-dimensional sampling of the fault displacements is related to the parameters determined from a three-dimensional sampling by

$$s_1 = \frac{m_1}{p} = \frac{m_3 - 2}{p} = 0.731 \qquad (15\text{-}1.30)$$

which gives

$$m_3 = 0.731\,p + 2 \qquad (15\text{-}1.31)$$

Equations (15-1.27) and (15-1.31) define well-constrained estimates for the values of p and m_3. Using Equation (15-1.31), we see the exponent on the variable δ in Equations (15-1.23) and (15-1.25) is constant:

$$\frac{p + 2 - m}{p} = \frac{p + 2 - 0.731p - 2}{p} = 0.269 \qquad (15\text{-}1.32)$$

In Figure 15-1.2, we plot Equation (15-1.25) for three values of p. For each value of p, we plot the curves for m_3 defined by Equation (15-1.31) and for $m_3 = 2.5$. If Equation (15-1.31) correctly defines the value of m_3, then Equation (15-1.32) shows that, regardless of the value of p, the plot of the extension will be almost identical (thick-line curves in Figure 15-1.2), differing only by the values of the constants in Equation (15-1.25). The result for $(p, m_3) = (1.0, 2.731)$ is the most reliable, however, because these constants are consistent with the data in Figures 3.26 and 15-1.1, which determine the values

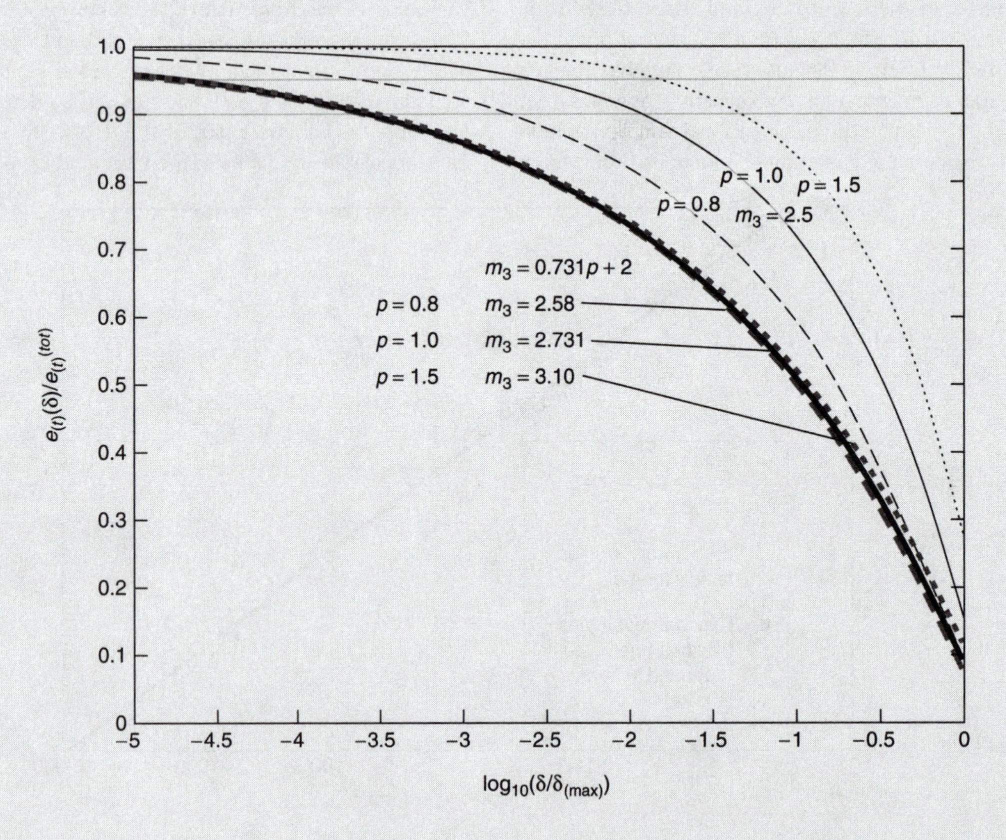

FIGURE 15-1.2 Cumulative extension as a function of displacement from Equation (15-1.23), plotted as fractions, respectively, of total extension and maximum displacement. Curves for $p = 0.8$, 1, and 1.5 are plotted for values of m taken from the second Equation (15-1.28) and Equation (15-1.31). The bold solid curve is for the preferred values of $(p, m) = (1.0, 2.731)$, and it indicates that to account for 95% of the extension, all faults having displacements from the maximum down to approximately 4.7 orders of magnitude smaller than the maximum displacement must be included.

of the parameters over the widest range. The curves for $m_3 = 2.5$ (thin-line curves, Figure 15-1.2) illustrate the consequences of different choices for the parameter values.

Figure 15-1.2 shows that, for the preferred parameter values, in order to account for 95% (or 90%), of the extension in the volume, we must account for all the faults having displacements from the maximum down to approximately 4.7 (or 3.6) orders of magnitude smaller than the maximum displacement. All the curves in Figure 15-1.2, however, show that small faults must contribute a significant fraction of the deformation in faulted crust.

The actual value of the parameters m, p, and B clearly has a very large effect on the conclusions we draw from Equations (15-1.24) and (15-1.25), and the accurate estimate of extension and the extension ratio from these equations requires that we use accurate numbers for these parameters. Thus the values of these parameters are the subject of much research and discussion.

Equation (15-1.24) is not in fact the best way to estimate the total extension in a region, because it ignores the fall-off from a power law of the distributions at the upper and lower limits of the data (Figures 3.26, 3.28*B*, 15-1.1). The faults at the lower limit of displacements contribute an insignificant amount to the total extension, as Figure 15-1.2 illustrates. Thus the deviation from the correct extension resulting from using Equation (15-1.24) for the lower end of the displacement distribution is insignificant. But the deviation from the power law distribution at the upper end of the displacement distribution would result in Equation (15-1.24) overestimating the extension in an area. A better estimate can be found by numerically summing the displacements for the largest faults, and then estimating the contribution from all the smaller faults by integrating the continuous function for the remainder of the power-law distribution of displacements.

The approach we use here to estimate the extension also is applicable to approximating other important aggregate characteristics of faulted rock. The fracture surface area is a significant aggregate characteristic for understanding chemical processes in pore fluids in the rock if the processes are sensitive to the area of the fluid-rock interface. Fracture porosity and permeability are important aggregate characteristics for the modeling of the flow of fluids such as water and oil through fractured rocks. Seismic shear wave anisotropy is an aggregate property that can depend on the fracture characteristics of the rock. Because this anisotropy can be measured remotely by seismology, understanding how the aggregate property derives from the fracture characteristics provides the possibility for making a remote evaluation of the fractured state of the rock.

Understanding these aggregate properties of rocks is important in a number of fields. An accurate estimation of these properties from the scale invariance of faults and fractures and their characteristics, however, requires a knowledge of the distribution of these characteristics and how to use that distribution to calculate the aggregate properties. Thus this topic is a subject of active research.

ii. Strain in Ductile Shear Zones

In massive crystalline rocks that have been deformed at low to intermediate metamorphic temperatures, the deformation commonly accumulates by concentrated shear along well-defined ductile shear zones. A schistosity develops in association with these shear zones, and with increasing distance from the boundary into the center of the shear zone, the massive rock characteristically becomes increasingly foliated. Figure 15.14*A* shows an example of such a shear zone developed in a massive gneiss. At the boundaries, the schistosity is very weakly developed and is oriented at an angle of approximately 45° to the boundary. Progressing in toward the central part of the shear zone, the schistosity becomes increasingly strongly developed, and the angle it makes with the shear zone boundary decreases.

Figure 15.14*B* shows another example of such a shear zone, but in this case the rock is a granite that contains an aplite dike and xenoliths, which record the magnitude of the shear strain. In the undeformed part of the granite, the thickness of the dike is 32 centimeters, there is no schistosity, and the xenoliths are roughly equidimensional. Near the boundary of the shear zone, the dike is only slightly deflected, a weak schistosity is present, oriented at an angle of approximately 45° to the boundary, and the xenoliths are slightly elongate. Within the central part of the shear zone, the dike is very strongly rotated and its thickness is reduced to 2 centimeters, the schistosity is strongly developed and oriented at a low angle to the shear zone boundary, and the xenoliths are stretched out into long, thin shapes. The dike and the xenoliths show that the strain increases from zero at the boundary of the shear zone to very high values at the center and that the angle between the foliation and the shear zone boundary decreases toward zero as the magnitude of the shear strain increases.

Simple Shear in Ductile Shear Zones. The simplest model to explain these features is an inhomogeneous progressive simple shear within the shear zone, with the magnitude of the shear strain varying from zero at both shear zone boundaries, to a maximum in the central part of the zone (Figure 15.15). If we assume that the schistosity is oriented parallel to the $(\hat{s}_1, \hat{s}_2)$-plane (the plane of flattening) of the finite strain ellipsoid, and thus is perpendicular to $\hat{s}_3$, then the schistosity should reflect the characteristic sigmoidal trajectories of the $\hat{s}_1$ axes of the strain ellipses that result from such a deformation (Figure 15.15). At the boundaries of the shear zone where the

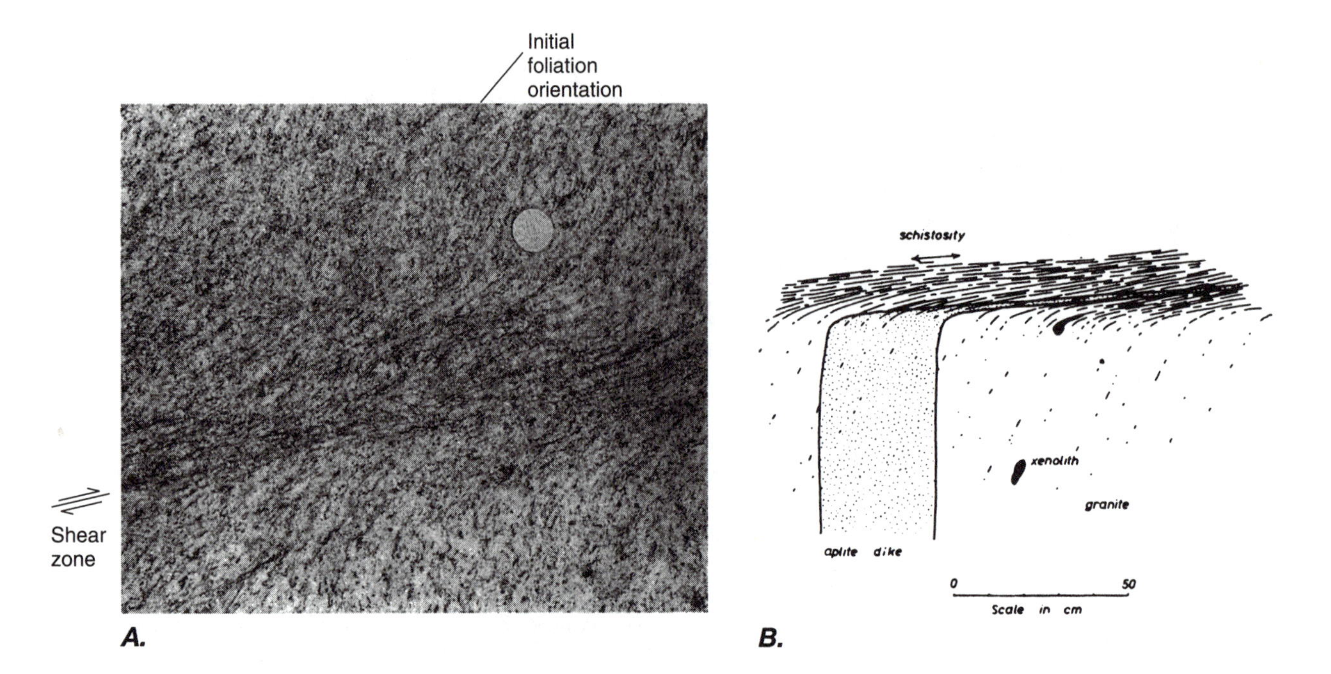

FIGURE 15.14 Deformation features of ductile shear zones. *A*. Development of schistosity in a small ductile shear zone in a metagabbro. (From Mukhopadhyay and Haimanot 1989) *B*. Deformation of an aplite dike and mafic xenoliths in a granite cut by a ductile shear zone normal to the dike. (From Ramsay and Graham 1970)

strain approaches zero, the orientation of the $\hat{s}_1$ axes is at an angle of 45° to the boundary (compare the maximum principal axis line *A* in Figure 12.15*A*). As the central part of the zone is approached, the magnitude of the shear strain increases and the angle between the shear zone boundary and the $\hat{s}_1$ axes decreases in a predictable way. At very high shear strains, the $(\hat{s}_1, \hat{s}_2)$-plane approaches parallelism with the shear zone boundary, and the maximum stretching direction approaches parallelism with the shear direction (compare the maximum principal axes in Figure 12.15*B–D*).

The shear zone in Figure 15.14*A* shows the characteristic sigmoidal pattern of the schistosity, which corresponds with the trajectories of the $\hat{s}_1$ axes of finite strain in the model (Figure 15.15). The angle between the foliation and the shear zone boundary in the gneiss, however, is considerably smaller than in the model shown in Figure 15.15, implying a significantly larger shear strain. The shearing of the xenoliths and the tabular dike in Figure 15.14*B* are also consistent with the model, except that the amount of shearing must be even greater than in Figure 15.14*A* to account for the near-parallelism of the foliation with the shear zone boundary and the extreme thinning of the dike. This implies that the shear strain in the shear zone must be extremely high. In both cases (Figure 15.14), the degree of development of schistosity also correlates with the inferred magnitude of the shear strain.

Volume Change in Ductile Shear Zones. Although progressive simple shear is a useful model for ductile shear

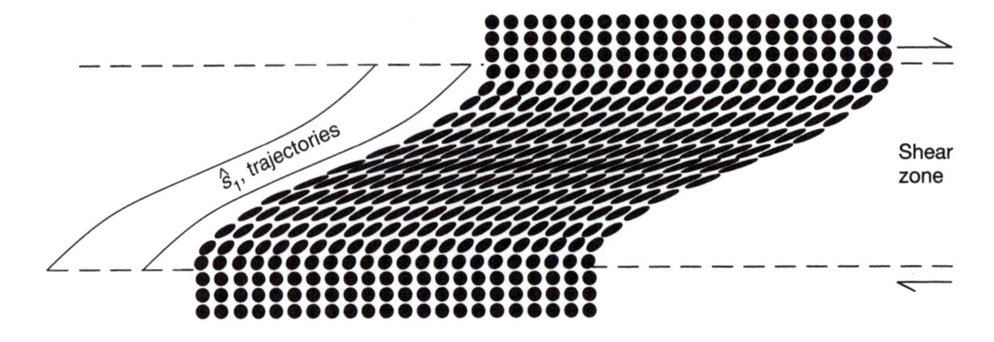

FIGURE 15.15 Orientation and magnitude of finite strain ellipses and trajectories of $\hat{s}_1$, across a ductile shear zone resulting from progressive simple shear. (From Ramsay and Huber 1983)

zones, it is by no means unique, and it may actually be wrong. Other models having different geometries of deformation must be considered as well. If we assume foliations are approximately parallel to the plane of flattening ($\hat{s}$, $\hat{s}_2$) and lineations are stretching lineations approximately parallel to $\hat{s}_1$, these other models suggest different interpretations of the observed geometry of natural shear zones such as are shown in Figure 15.14.

A more complex deformation model of a shear zone includes a component of shortening normal to the shear zone boundary, which could occur, for example, by heterogeneous volume loss. Figure 14.13 illustrates how a volume loss concentrated along individual planar features can produce a geometry similar to that of a brittle shear zone. Similarly, shortening normal to a shear zone boundary, accommodated by distributed heterogeneous volume loss, can produce a geometry similar to that of a ductile shear zone. Compare, for example, the undeformed line in Figure 15.16*A* after it has been deformed by simple shear (Figure 15.16*B*) with the same line after a heterogeneous volume loss (Figure 15.16*C*). The combination of a heterogeneous volume loss with a heterogeneous simple shear would provide a still more complex model of a shear zone, for which the resulting deformation is illustrated in Figure 15.16*D*. Faced with just the evidence in Figure 15.14, it is difficult to argue that the models shown in Figure 15.16*C*, *D* are not just as viable to explain the observations as the models in Figures 15.15 and 15.16*B*, but the inference of the strain involved would be very different for the different models.

The possibility that deformation in a shear zone may include not only simple shear but also a component of shortening or lengthening normal to a shear zone suggests a model to explain why, in some cases, gash fracture planes commonly do not bisect the angle between conjugate shear zones, as required by the stress interpretation (Figure 9.16). If we assume that gash fractures develop normal to the direction of maximum instantaneous stretch $\hat{\zeta}_1$, rather than normal to the minimum compressive stress $\hat{\sigma}_3$ as we assumed for Figure 9.16, then the observed variety of gash fracture orientations has a simple explanation. The addition of a lengthening normal to the shear zone causes $\hat{\zeta}_1$ to rotate toward a higher angle to the shear zone boundary and therefore results in gash fractures oriented at a smaller angle to the boundary (Figure 15.17*A*). Conversely, adding a shortening normal to the shear zone boundary rotates $\hat{\zeta}_1$ into a smaller angle with the boundary and thus results in gash fractures that have a higher angle to that boundary (Figure 15.17*B*).

Constant-Volume Deformation in Ductile Shear Zones. We can also consider a constant-volume deformation for which a component of shortening or lengthening perpendicular to the shear zone is compensated by lengthening or shortening, respectively, parallel to the shear zone boundary and normal to the shear direction, as shown in Figures 12.24 and 12.26. (In these figures, a convergence angle of $\alpha = 0°$ corresponds to the model of heterogeneous simple shear discussed for Figures 15.15 and 15.16*B*). We assume that foliations are parallel to the ($\hat{s}_1$, $\hat{s}_2$)-plane and lineations are parallel to $\hat{s}_1$, and we discuss the implications of convergent or divergent shear in terms of the orientations of the foliations and lineations. If the deformation is a convergent shear, the foliation orientation is stable and approaches parallelism with the shear zone boundary at high strains. The orientation of a stretching lineation, however, is unstable if the angle of convergence is $0 < \alpha < 20°$, because after sufficient strain, the $\hat{s}_1$ axis switches from subparallel to the shear direction to perpendicular to it (Figure 12.26). For larger convergent angles, $\alpha > 20°$, the stretching lineation is always perpendicular to the shear direction and parallel to the shear zone boundary. If the deformation is a divergent shear, the orientation of the stretching lineation is stable and approaches parallelism with the shear

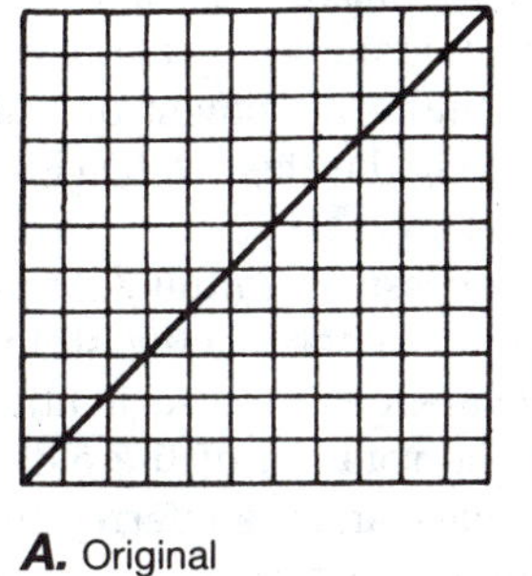

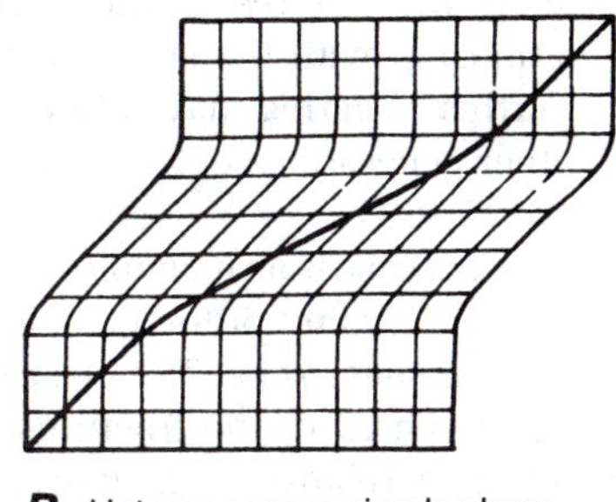

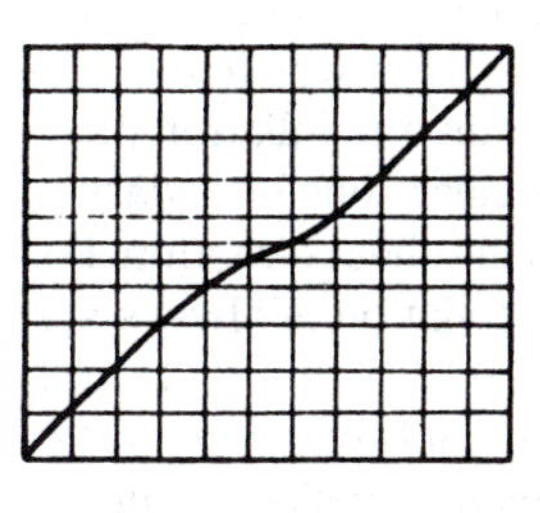

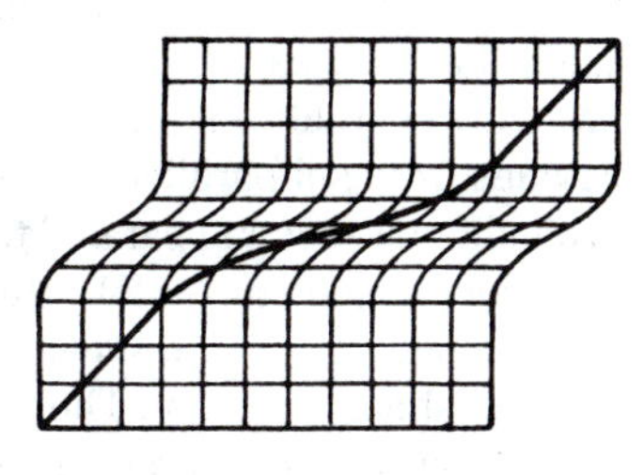

FIGURE 15.16 Displacement fields suggesting ductile shear zones. (From Ramsay 1980, figure 3) *A.* Undeformed state. *B.* Shear zone formed by a progressive inhomogeneous simple shear. *C.* Apparent ductile shear zone formed by a distributed inhomogeneous volume loss. *D.* Shear zone formed by the superposition of the displacement fields in *B* and *C*.

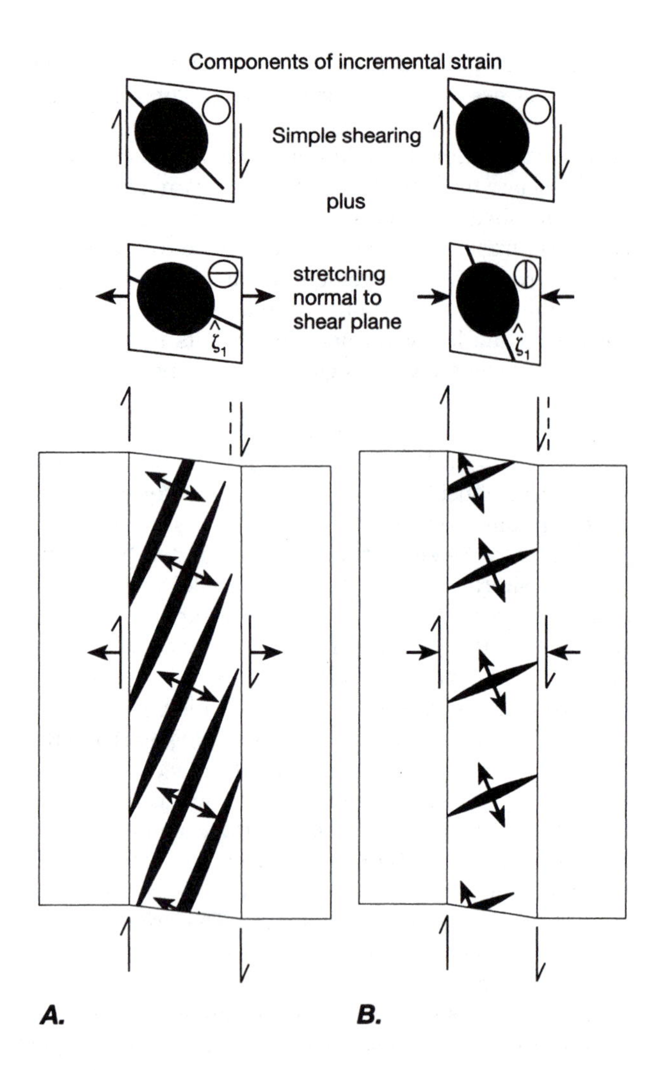

FIGURE 15.17 Assuming that gash fractures form normal to the orientation of the maximum instantaneous stretch $\hat{\zeta}_1$, the addition to the instantaneous shear of an instantaneous component of inhomogeneous volume increase *A* or decrease *B* normal to the shear zone can account for the various observed orientations of gash fracture relative to the shear zone. (Modified from Ramsay and Huber 1983)

direction, regardless of the amount of shear (Figure 12.26). For angles of convergence for which $0 > \alpha > -20°$, however, after sufficient strain, the foliation plane switches from subparallel to the shear zone boundary to perpendicular to it and parallel to the shear direction. For divergence angles $\alpha < -20°$, the foliation is always stable perpendicular to the shear zone boundary and parallel to the shear direction. These models could be particularly important in interpreting exhumed shear zones in high-grade metamorphic rocks.

15.5 DEFORMATION HISTORY

In order for us to infer the geometry of progressive deformation (that is, the deformation path), we require information from the rocks about the deformation history (see Sections 12.4 and 12.5). Determination of the finite strain ellipsoid is insufficient, because the finite strain is the net result of the whole preceding history of the deformation, and it does not define the deformation path.

Distinguishing between coaxial and noncoaxial deformation also requires some knowledge of the deformation history. To infer a coaxial geometry for the deformation, for example, we must be able to show that the principal axes of the finite strain ellipsoid have remained parallel to the same material lines throughout the deformation or, equivalently, that the principal axes of the finite strain ellipsoid and the instantaneous strain ellipsoid are parallel (see Sections 12.4 and 12.5).

Features that record different portions of the strain history can help us distinguish between coaxial and noncoaxial deformation. For example, features that are introduced into the rock at different times during its deformation (such as dikes, veins, and metamorphic segregations) record only the strain that accumulates after their introduction. An example of such a record can be found in brittle-ductile shear zones in which *en echelon* gash fractures are developed. We assume that the fractures open initially in an orientation either perpendicular to $\hat{\sigma}_3$, the axis of minimum compressive stress, or perpendicular to $\hat{\zeta}_1$, the axis of maximum instantaneous stretch (Figure 15.18*A*). The two directions may be the same, but that is not necessary, especially if the material is not mechanically isotropic. As shear strain accumulates by a combination of ductile flow and fracturing, the older parts of the gash fractures rotate, while the fractures continue to grow at the tips in a direction perpendicular to either $\hat{\sigma}_3$ or $\hat{\zeta}_1$ (Figure 15.18*B*). The result is a sigmoidal-shaped gash fracture. With sufficient rotation, extension fracturing on the original set of fractures ceases, and a second generation of fractures develops in the same original orientation as the first generation with respect to the stress or instantaneous strain axes. Cross-cutting relationships between the material filling the first- and second-generation fractures can clearly indicate the sequence of fracture development. After growth ceases at the tips of the old fractures, the tips may be rotated, along with the rest of the fracture, out of the unique relationship with the stress or instantaneous strain axes. Further shearing may result in the development of still other generations of fractures along the shear zone (Figure 15.18*C*).

This case is an example of successively forming structures recording different amounts of the ductile strain along a shear zone. The clear evidence of the sequential development of the fractures, the rotation of the older fractures relative to the newer ones, and the inferred relationship between the direction of fracture growth and the local stress or instantaneous strain axes indicate that material lines are rotated past the orientations of the principal axes of instantaneous strain. Thus the geometry of strain accumulation must have been noncoaxial, at least within the shear zone. The curvature and the accumulated widths of the gash fractures reflect a minimum

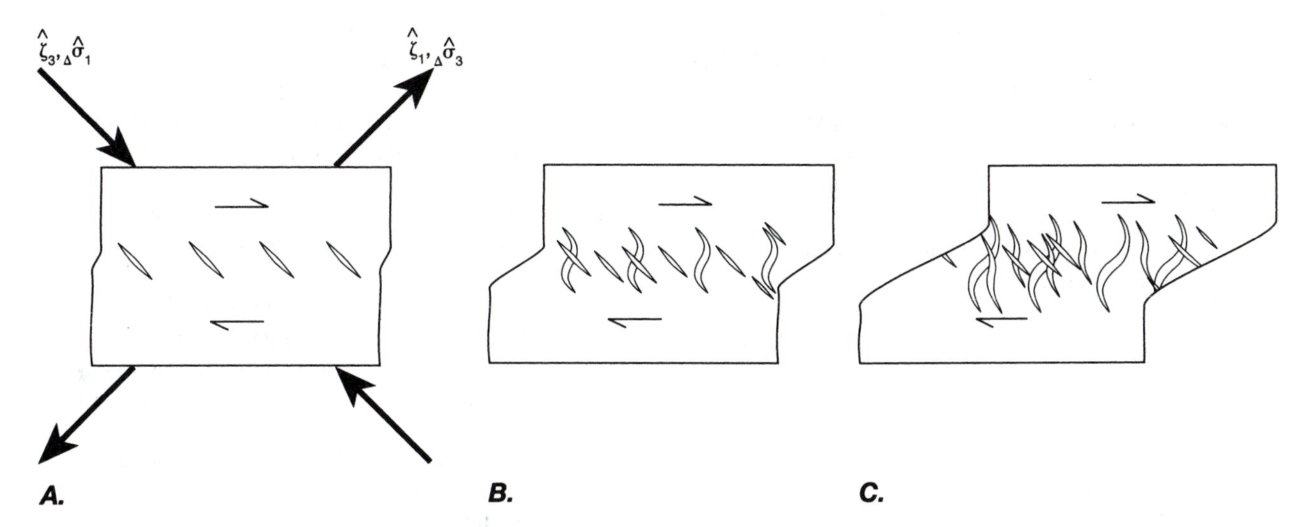

FIGURE 15.18 The pattern of growth of *en echelon* gash fractures in a shear zone undergoing both fracturing and ductile shearing. (From Durney and Ramsay 1973, figure 15) *A*. Initial development of gash fractures normal to $\hat{\zeta}_1$ and to $\hat{\sigma}_3^{(Dev)}$. The heavy arrows indicate the direction of principal instantaneous stretches and deviatoric stress. *B*. Growth of rotated gash fractures ceases, and a second generation of fractures develops in the original orientation, in places cross-cutting the older fractures. *C*. Growth of the rotated second generation of fractures ceases, and a third set begins to develop.

amount of shear strain that has occurred on the shear zone. The averaging of local noncoaxial deformation over the volume of rock containing the shear zone to give a bulk deformation and the possibility of the addition of stretches from multiple different shear zones do not preclude the possibility that the bulk strain could be coaxial (see, for example, Figure 15.11*C*).

A different technique for constraining the deformation history depends on the observation of mineral fibers that grow in the rock during the deformation (Sections 14.6(v) and 14.6(vi), Figures 14.17–14.19). Where fibers form the filling of extension fractures or veins or the overgrowths on particles in the rock and where fiber growth is displacement-controlled, the orientations of the fibers record the orientation history of the maximum instantaneous stretch axis $\hat{\zeta}_1$. Consider again a shear zone in which gash fractures form and are filled by mineral fibers. Because the deformation in this case is noncoaxial, the material planes defined by the fracture wall rotate with respect to the shear zone boundaries and the principal axes of instantaneous strain (see Figure 12.15). The fibers consistently grow parallel to the maximum instantaneous stretch axis $\hat{\zeta}_1$, however, which is constant in orientation with respect to the shear zone boundaries. Thus the fibers progressively change orientation with respect to the fracture wall and end up being curved. The pattern of curvature for a given noncoaxial deformation depends on whether fiber growth is syntaxial or antitaxial (Figure 14.17). The first orientation of fibers in later-generation gash fractures is perpendicular to the fracture wall and parallel to the last orientation of fibers in the earlier-generation fractures. This relationship confirms the interpretation that the fibers continually grow parallel to the maximum instantaneous stretch axis $\hat{\zeta}_1$.

Thus whether in extension fractures, shear fractures, or oriented overgrowths, fibers preserve a record of the history of displacement, and the length of the fibers records the total displacement across the fracture or overgrowth during the time of fiber growth. Curved fibers indicate the existence of a component of rotation during deformation relative to the instantaneous strain axes and thereby imply a noncoaxial deformation.

REFERENCES AND ADDITIONAL READINGS

Alvarez, W., T. Engelder, and P. A. Geiser. 1978. Classification of solution cleavage in pelagic limestones. *Geology* 6: 263–266.

Alvarez, W., T. Engelder, and W. Lowrie. 1976. Formation of spaced cleavage and folds in brittle limestone by dissolution. *Geology* 4: 698–701.

Cleaves, E. T., J. Edwards, Jr., and J. D. Glaser. 1968. *Geologic Map of Maryland*. Maryland Geological Survey.

Cloos, E. 1947. Oolite deformation in the South Mountain fold, Maryland. *Geol. Soc. Amer. Bull.* 58: 843–918.

Cloos, E. 1971. *Microtectonics Along the Western Edge of the Blue Ridge. Maryland and Virginia*. Johns Hopkins University Press, Baltimore.

Durney, D. W., and J. G. Ramsay. 1973. Incremental strain measured by syntectonic crystal growths. In K. A. DeJong and R. Scholten, eds., *Gravity and Tectonics*. Wiley, New York: 67–96.

Fry, N. 1979. Random point distributions and strain measurement in rocks. *Tectonophysics* 60: 89–105.

Gray, D. R., and D. W. Durney. 1979. Investigations on the mechanical significance of crenulation cleavage. *Tectonophysics* 58: 35–79.

Hobbs, B. E. 1971. The analysis of strain in folded layers. *Tectonophysics* 11: 329–375.

Hudleston, P. J., and T. B. Holst. 1984. Strain analysis and fold shape in a limestone layer and implications for layer rheology. *Tectonophysics* 106(3/4): 321–347.

Kostrov, V.V. 1974. Seismic moment and energy of earthquakes, and seismic flow of rock. *Izv. Acad. Sci. USSR Phys. Solid Earth, Engl. Transl.*, no. 1, p. 23-44.

Lewis, J. C., R. J. Twiss, C. J. Pluhar, and F. C. Monastero. 2007. Multiple constraints on divergent strike-slip deformation along the eastern margin of the Sierran microplate, SE California. In A. Till, et al., eds., *Exhumation Along Major Continental Strike-Slip Fault Systems*, Geol. Soc. Amer. Special Paper, in press.

Lisle, R. J. 1985. *Geological Strain Analysis: A Manual for the R_f/ϕ Technique*. Pergamon, Oxford.

Marrett, R. 1996. Aggregate properties of fracture populations. *J. Struct. Geol.* 18(2/3): 169–178.

Marrett, R., and R. W. Allmendinger. 1991. Estimates of strain due to brittle faulting: Sampling of fault populations. *J. Struct. Geol.* 13: 735–738.

Marrett, R., O. J. Ortega, and C. M. Kelsey. 1999. Extent of power-law scaling for natural fractures in rock. *Geology* 27(9): 799–802.

Mitra, S. 1976. A quantitative study of deformation mechanisms and finite strain in quartzites. *Contrib. Mineral. Petrol.* 59: 203–226.

Mitra, G. 1978. Microscopic deformation and flow laws in quartz within the South Mountain anticline. *J. Geol.* 86: 129–152.

Molnar, P. 1983. Average regional strain due to slip on numerous faults of different orientations. *J. Geophys. Res.* 88: 6430–6432.

Mukhopadhyay, D. K., and B. W. Haimanot. 1989. Geometric analysis and significance of mesoscopic shear zones in the Precambrian gneisses around the Kolar schist belt, south India. *J. Struct. Geol.* 11(5): 569–581.

Passchier, C. W., and R. A. J. Trouw. 1996. *Microtectonics*. Springer-Verlag, New York, 289 pp.

Ramsay, J. G. 1967. *Fracturing and Folding of Rocks*. McGraw-Hill, New York.

Ramsay, J. G. 1980. Shear zone geometry, a review. *J. Struct. Geol.* 2: 83–89.

Ramsay, J. G., and R. H. Graham. 1970. Strain variation in shear belts. *Can. J. Earth Sci.* 7: 786–813.

Ramsay, J. G., and M. I. Huber. 1983. *The Techniques of Modern Structural Geology*. Vol. 2: *Folds and Fractures*. Academic Press, London.

Teyssier, C., and B. Tikoff. 1999. Fabric stability in oblique convergence and divergence. *J. Struct. Geol.* 21(8/9): 969–974.

Turcotte, D. L. 1997. *Fractals and Chaos in Geology and Geophysics*, second edition. Cambridge University Press, New York, 398 pp.

Wellman, H. W. 1962. A graphical method for analyzing fossil distortion caused by tectonic deformation. *Geol. Mag.* 99: 348–352.

Wood, D. S. 1973. Patterns and magnitudes of natural strain in rocks. *Phil. Trans. Roy. Soc. Lond.* A274: 373–382.

Wright, T. O., and L. B. Platt. 1982. Pressure dissolution and cleavage in the Martinsburg shale. *Amer. J. Sci.* 282: 122–135.

Appendix 15-A

Common Techniques for Measuring Strain

In this appendix, we present some details of three different methods of measuring two-dimensional strain in deformed rocks. None of these methods provides a complete definition of the strain, because in all cases the volumetric strain cannot be defined. To obtain a measure of the volumetric strain, the initial dimension of the deformed objects must be known, but this requirement can be met only rarely. In general, therefore, we must be satisfied with determining only the ellipticity R and the orientation of the strain ellipse, but not the actual size.

15-A.1 STRAIN FROM DEFORMED INITIALLY ELLIPTICAL OBJECTS: THE R_f–ϕ METHOD

Some rocks, such as deformed conglomerates, contain deformed objects that were not initially spherical. If the undeformed objects were approximately elliptical in the plane on which deformation is to be measured, it is possible to separate the initial ellipticity from that imposed by the strain and thereby to estimate the strain. The technique is based on calculating the theoretical distribution of final ellipticities and orientations that result from imposing different strains on objects that have a known initial ellipticity and orientation (Figures 15-A.1, 15-A.2; see also Figure 14.1).

The final ellipticity R_f and the final orientation ϕ of a deformed object depend on the initial ellipticity R_i, on the initial orientation θ of the undeformed object, and on the ellipticity R_s of the imposed strain ellipse (see, for example, ellipses, *a*, *b*, and *c* in Figure 15-A.1). We choose the direction of the maximum principal axis of strain as the reference direction for our examples. Generally, we do not know this direction a priori, but any other direction can be used as a reference, and the analysis determines the orientation of $\hat{s}_1$.

As a simple example, we first consider the case in which the undeformed objects all have the same ellipticity $R_i = 2$ and are randomly oriented (Figure 15-A.1*A*), and we investigate the effect of imposing strains for which the ellipticity of the strain ellipse is $R_s = 1.5$ (Figure 15-A.1*B*) and $R_s = 3$ (Figure 15-A.1*C*). If the long axis of an object is parallel to the direction of maximum lengthening $\hat{s}_1$ of the imposed strain (ellipse *b* Figure 15-A.1*A*), then the final ellipticity R_f of the deformed object is a maximum (ellipse *b* Figure 15-A.1*B*, *C*), given by[1]

$$R_{f(\max)} = R_i R_s \qquad (15\text{-A.1})$$

If, on the other hand, the long axis of an object is parallel to the maximum shortening direction $\hat{s}_3$ of the imposed strain (ellipse *c* in Figure 15-A.1*A*), the final ellipticity R_f of the deformed object is a minimum (ellipse *c* in Figure 15-A.2*B*, *C*) given by

$$R_{f(\min)} = \begin{Bmatrix} R_i/R_s & \textit{if } R_s < R_i \\ R_s/R_i & \textit{if } R_s > R_i \end{Bmatrix} \qquad (15\text{-A.2})$$

For any other initial orientation of the object, the final deformed ellipticity is intermediate between these two values.

The graphs on the right in Figure 15-A.1 are the R_f–ϕ graphs that show how the final ellipticity R_f varies with the final orientation ϕ of the deformed ellipses for the different ellipticities of strain R_s. The heavy line in each graph is the curve for the initial ellipticity of $R_i = 2.0$, which characterizes the ellipses in the diagrams on the left. It shows the variation of the final ellipticity R_f as a function of the final orientation ϕ of the deformed objects. The R_i–curve varies with increasing strain from a straight line for the constant initial ellipticity R_i at zero strain (Figure 15-A.1*A*), to an open curve (Figure 15-A.1*B*), and to a closed curve (Figure 15-A.1*C*). The transition from an open to a closed curve occurs when the ellipticity of the strain equals the initial ellipticity of the deforming objects, $R_s = R_i$. The open curve means deformed ellipses of any orientation can occur (Figure 15-A.1*B*), whereas the closed curve indicates that orientations of the deformed ellipses are restricted to orientations of $|\phi| < 45°$ (Figure 15-A.1*C*). The other curves in the R_i family are for other initial ellipticities. The entire family of R_i curves takes on shapes that vary with the ellipticity of the different values of imposed strain.

In nature it is unlikely, of course, that any set of objects would have identical initial ellipticities. Thus we

[1]These equations were first obtained by Ramsay (1967), sections 5.5–5.6. See Lisle (1985) for a detailed description of the whole method.

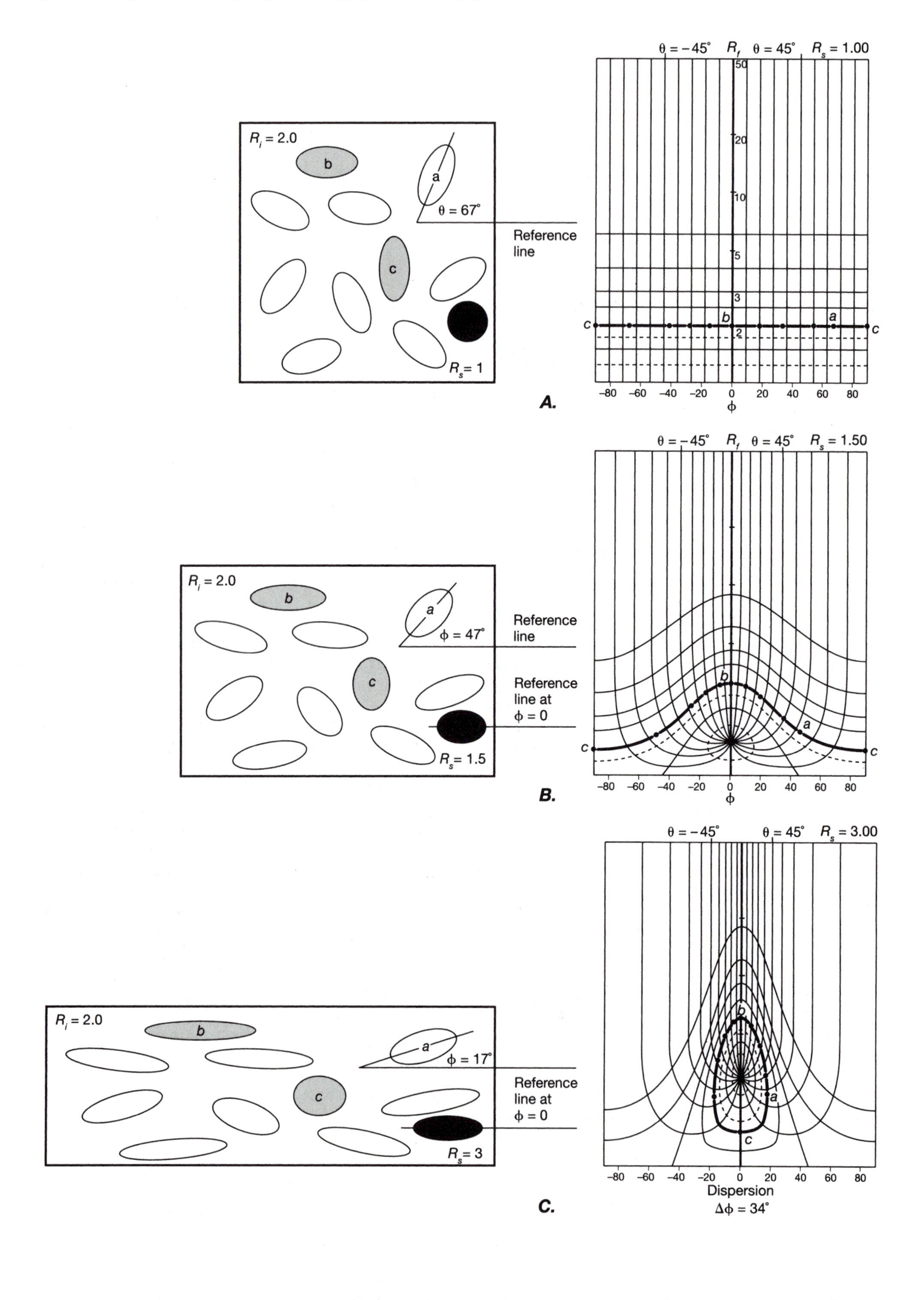

R_i = 2.0
b
a
θ = 67°
c
R_s = 1
Reference line
θ = −45° R_f θ = 45° R_s = 1.00
50
20
10
5
3
2
c
b
a
c
−80 −60 −40 −20 0 20 40 60 80
ϕ
A.
R_i = 2.0
b
a
ϕ = 47°
c
R_s = 1.5
Reference line
Reference line at ϕ = 0
θ = −45° R_f θ = 45° R_s = 1.50
b
a
c
c
−80 −60 −40 −20 0 20 40 60 80
ϕ
B.
R_i = 2.0
b
a
ϕ = 17°
c
R_s = 3
Reference line at ϕ = 0
θ = −45° θ = 45° R_s = 3.00
b
a
c
−80 −60 −40 −20 0 20 40 60 80
Dispersion
Δϕ = 34°
C.

consider as a second simple example the deformation of a group of objects having a constant initial orientation θ but variable initial ellipticities R_i (Figure 15-A.2). The heavy line in each of these R_f–ϕ graphs is one of a family of θ–curves. It shows for the objects in the diagrams on the left how the final ellipticities vary as a function of their final orientations for the particular initial orientation of $\theta = 63°$. In the undeformed state (Figure 15-A.2*A*), the line is a straight line at constant ϕ (63° in this example) indicating that all ellipses have the same initial orientation. The shape of the θ–curve changes with increasing strain. For any given strain ellipticity R_s, there is a family of θ–curves that characterize different initial orientations. The pattern of the family of θ–curves changes with increasing strain (compare Figure 15-A.2*A*, *B*, *C*). For an initially random distribution of orientations, the θ–curves for $\theta = +45°$ and $\theta = -45°$ must enclose half the data between them.

Because natural objects do not have the simple distribution of shapes or orientations considered in our two examples, we can expect natural data to be scattered on an R_f–ϕ plot (as shown, for example, in Figure 15-A.3*A*). For a random initial distribution of ellipticities and orientations, the final data should be symmetrically distributed about some line of constant ϕ that divides the data into two equal groups. That value of ϕ defines the orientation of the maximum principal axis of strain. Moreover, because all the orientations initially are randomly distributed within a 180° angle, 50 percent of the data should lie within the 90° angle that is bisected by the maximum principal strain axis. On the R_f–ϕ diagrams, therefore, 50 percent of the data should lie between the lines for $\theta = 45°$ and $\theta = -45°$ (Figure 15-A.1).

R_f–ϕ data measured from a large number of deformed ooids in a limestone are plotted in Figure 15-A.3*A*, where ϕ is measured relative to a convenient reference orientation. The data cluster is roughly symmetric about the line $\phi \approx 30°$, implying an initial random orientation of the markers and an orientation of the maximum principal stretch axis $\hat{s}_1$ at 30° from the reference line. The value of the strain ellipticity R_s is defined by finding the value of R_s that defines the family of θ–curves for which the $\theta = \pm 45°$ lines divide the data into two equal groups. The R_i curve that forms a tight envelope about most of the data defines the maximum initial ellipticity that is common among the deformed ooids. Because of the scatter in the data, the fitting of the curves provides only an approximate solution. A curve that shows a reasonable fit to the shape of the data cluster is for $R_s \approx 1.7$, $R_i \approx 1.6$ (Figure 15-A.3*B*). These values therefore provide an estimate of the strain and the maximum initial ellipticity of the ooids.

We can also determine the ellipticity of the strain ellipse R_s and the maximum initial ellipticity R_i analytically by combining Equations (15-A.1) and (15-A.2) to find

$$R_s^2 = \begin{Bmatrix} R_{f(\max)}\, R_{f(\min)} & \textit{if } R_s > R_i \\ R_{f(\max)} \,/\, R_{f(\min)} & \textit{if } R_s < R_i \end{Bmatrix} \qquad (15\text{-A.3})$$

$$R_i^2 = \begin{Bmatrix} R_{f(\max)} \,/\, R_{f(\min)} & \textit{if } R_s > R_i \\ R_{f(\max)}\, R_{f(\min)} & \textit{if } R_s < R_i \end{Bmatrix} \qquad (15\text{-A.4})$$

We take $R_{f(\max)}$ and $R_{f(\min)}$ to be the maximum and minimum values of R_f, measured on a set of deformed objects, ignoring isolated and widely scattered data points such as are found in Figure 15-A.3*A*.

If there was an initial preferred orientation to the objects being measured, the analysis is more difficult. The R_f–ϕ data then are not symmetric about any line of constant ϕ. Nevertheless, if the initial orientations are normally distributed about a preferred orientation, then we can find a θ–curve for some value of strain ellipticity R_s that still splits the data symmetrically into two equal groups. In principle, then, we can find a value of R_s, a value of ϕ, and an associated curve for constant θ that best fits a given data set, as well as an R_i–curve that most closely bounds the data. These values define the ellipticity R_s of the imposed strain ellipse, the orientation ϕ, of the maximum imposed stretch axis $\hat{s}_1$, the initial preferred orientation θ relative to $\hat{s}_1$, and the maximum initial ellipticity R_i of the deformed objects.

FIGURE 15-A.1 The effects of initial orientation θ of elliptical objects in the undeformed state on the final ellipticity R_f and final orientation ϕ of the objects in the deformed state. *A*. The undeformed state in which elliptical objects that have $R_i = 2$ are randomly oriented. The orientation θ is shown for ellipse *a* relative to a reference line, which for this example we take to be parallel to the maximum principal stretch in the deformed state. Ellipses *b* and *c* are in special orientations; their major and minor axes are parallel to the principal stretches in the deformed state. The R_f–ϕ plot shows that all these ellipses plot along one line, which is the $R_i = 2$ curve from the family of R_i curves for the undeformed state ($R_s = 1$). *B*. A deformed state for which the strain ellipticity $R_s = 1.5 < R_i$. The final ellipticity R_f is a maximum for ellipse *b* and a minimum for ellipse *c*. For $R_s < R_i$, the major axes of the deformed objects are dispersed within $\phi = \pm 90°$ of the maximum principal stretch. The graph shows the R_f–ϕ plot for $R_s = 1.5$. All the deformed ellipses plot along a single curve for which $R_i = 2$. This curve extends from $\phi = -90°$ to $\phi = +90°$, as do all R_i curves for $R_i > R_s$, reflecting the dispersion of the major axes about the direction of the maximum principal stretch. *C*. A deformed state for which $R_s = 3 > R_i$. The final ellipticity R_f is a maximum for ellipse *b* and a minimum for ellipse *c*. For $R_s > R_i$, the major axes of the deformed objects are dispersed within $\phi = \pm 45°$ of the maximum principal stretch, and in this case the dispersion is $\pm 17°$. The graph shows the R_f–ϕ plot for $R_s = 3.0$. All the deformed ellipses plot along a single curve for which $R_i = 2$. This curve is a closed curve (as are all R_i curves for $R_i < R_s$,) that includes a range of ϕ values from $-17°$ to $+17°$, reflecting the dispersion of the major axes about the direction of the maximum principal stretch. (R_f–ϕ curves from Lisle 1985)

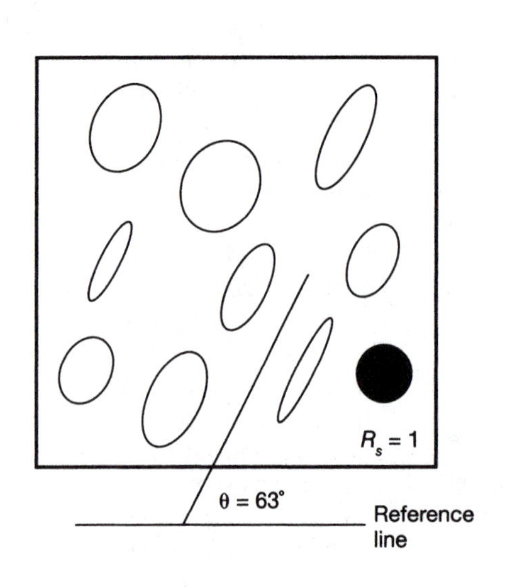

$R_s = 1$
$\theta = 63°$
Reference line

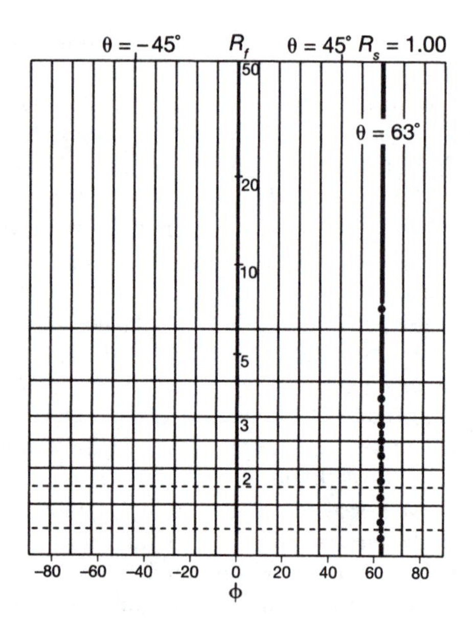

$\theta = -45°$
R_f
$\theta = 45°$
$R_s = 1.00$
$\theta = 63°$
50
20
10
5
3
2
-80 -60 -40 -20 0 20 40 60 80
ϕ

A.

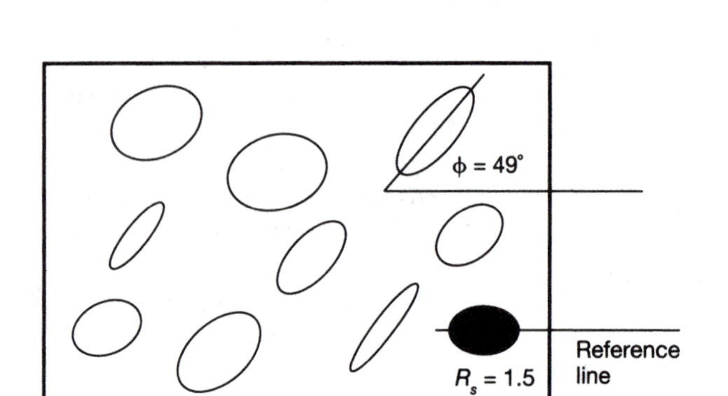

$\phi = 49°$
Reference line
$R_s = 1.5$

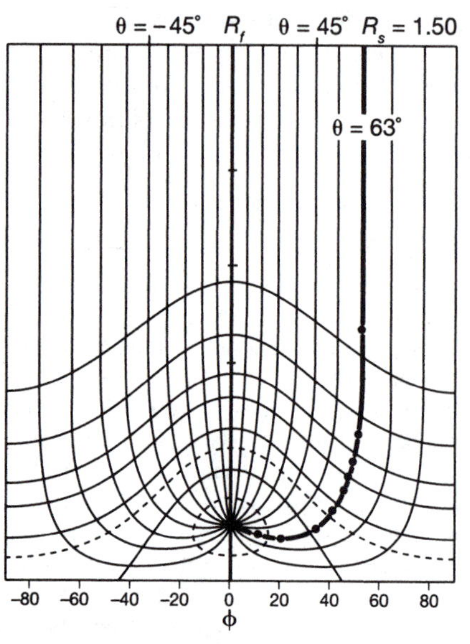

$\theta = -45°$
R_f
$\theta = 45°$
$R_s = 1.50$
$\theta = 63°$
-80 -60 -40 -20 0 20 40 60 80
ϕ

B.

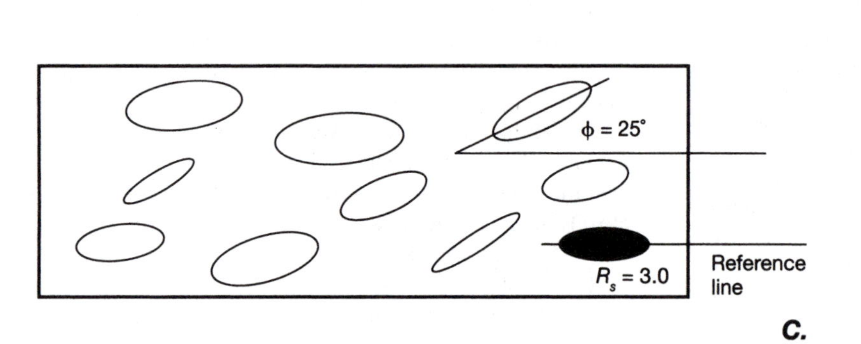

$\phi = 25°$
Reference line
$R_s = 3.0$

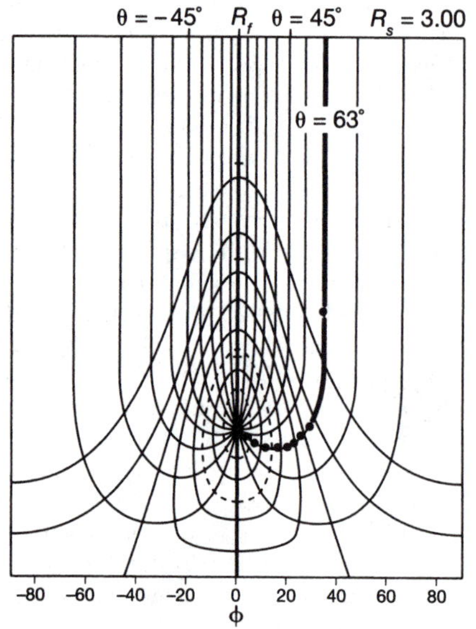

$\theta = -45°$
R_f
$\theta = 45°$
$R_s = 3.00$
$\theta = 63°$
-80 -60 -40 -20 0 20 40 60 80
ϕ

C.

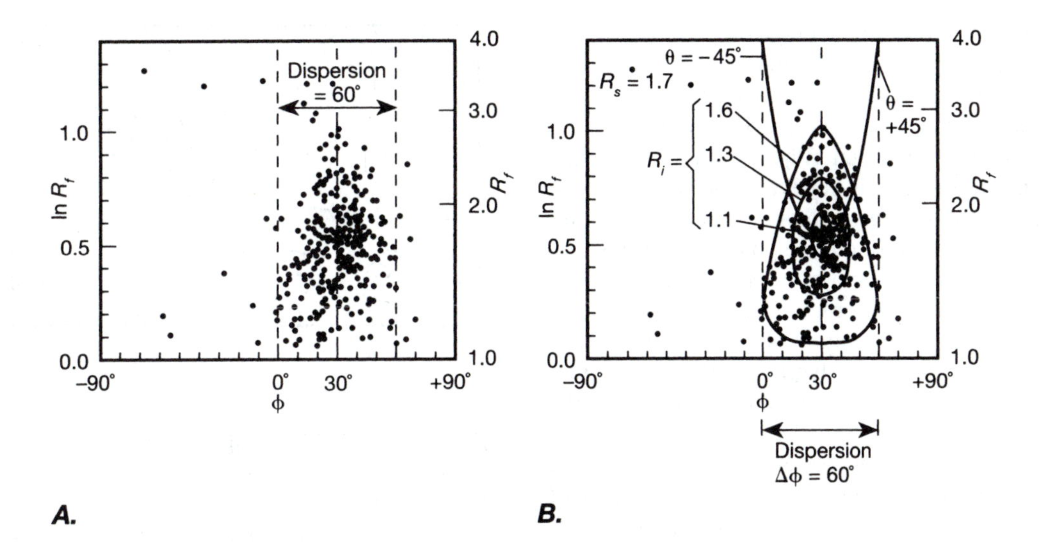

FIGURE 15-A.3 Interpretation of an R_f–ϕ plot for ooids in a deformed oolite. (Modified from Ramsay and Huber 1983, figure 5.11) *A*. The approximate symmetry of points about the value of ϕ at which R_f is a maximum means that initially the orientations of the elliptical ooids were nearly randomly distributed. Thus the value of ϕ at the symmetry line defines the orientation of $\hat{s}_1$ relative to the chosen reference line and divides data for an originally random distribution into two equal halves. *B*. Inference of the orientation of $\hat{s}_1$ ($\phi = 30°$), the strain ellipticity ($R_s = \sim 1.7$), and the initial ellipticities ($R_i < 1.6$) of the deformed objects for the data in *A*. For the best value of imposed strain R_s, the lines for $\theta = +45°$ and $\theta = -45°$ should separate data from an initially random distribution into two equal halves, and an R_i curve should form a tight envelope around the data. The dispersion of ϕ for the data is an indication of the largest initial ellipticities R_i of the undeformed objects. For the solution shown, the $\theta = +45°$ line actually lies too high to divide the data into two equal halves. This indicates that the estimate for R_s is slightly high.

The R_f–ϕ technique has been applied to deformed objects such as ooids, conglomerate pebbles, breccia fragments, quartz grains, porphyroblasts, mineral aggregates, amygdules, pumice fragments in tuff, the cross sections of worm burrows, and alteration spots in slates. Because of the abundance of rocks containing such objects, it is one of the most useful and reliable methods for determining tectonic strain.

FIGURE 15-A.2 The effect of varying initial ellipticity R_i of undeformed objects that all have the same initial orientation θ. *A*. The undeformed state ($R_s = 1$) with $\theta = 63°$ and R_i varying between 1.1 and 7.3. The R_f–ϕ plot shows the initial distribution of the data. *B* and *C*. Two deformed states with $R_s = 1.5$ and $R_s = 3.0$, respectively. The principal axes of the deformed objects do not remain parallel, and those with the smallest initial ellipticity change orientation the most. The R_f–ϕ plots show how the data plot asymmetrically relative to the direction of maximum principal stretch ($\phi = 0°$) along a curve of constant θ, and they indicate that the θ curves are different for different values of strain ellipticity. The $\theta = \pm 45°$ curves divide data from a random initial distribution of orientations, into two equal portions (compare Figure 15-A.1). (R_f–ϕ curves from Lisle 1985)

15-A.2 STRAIN FROM NEAREST-NEIGHBOR CENTER-TO-CENTER ANALYSIS: THE FRY METHOD

If the particles in an undeformed rock were initially distributed such that the distances between the centers of nearest neighbors were statistically constant and isotropic, then in a deformed rock, the distances between these centers provides a measure of the accumulated strain. Analysis of these distances constitutes the nearest-neighbor center-to-center technique. It can be applied, for example, to sand grains in a sandstone, to conglomerate pebbles in a conglomerate, and to ooids in an oolitic limestone. It can also be used to compare the bulk strain, with the strain recorded by the shapes of individual sand grains, conglomerate pebbles, and ooids.

Two particles are nearest neighbors if the center-to-center line does not cross any other particles. During deformation, the center-to-center lines behave as material lines and are rotated and stretched according to the shape and orientation of the strain ellipse. Thus this technique is in effect an application of the measurement of stretches of deformed linear objects having a wide variety of initial orientations. Measurements of the orientation and length of each nearest-neighbor center-to-center line about a given particle make it possible to determine the strain ellipse. In order to average out irregularities in the initial fabric and in the deformation, we must measure such sets of lines about many particles.

A graphical technique called the Fry method is useful for determining the strain ellipse from a large number of points. In essence, it involves plotting the length and orientation of a large number of center-to-center lines relative to a single reference point. The surface on which the strain is determined is commonly a photographic enlargement of an oriented thin section. Figure 15-A.4*A*, *B* shows the deformation of a polycrystalline aggregate with the centers of each grain marked. The strain is indicated by the adjacent strain ellipse. Analysis of the strain in Figure 15-A.4*B* by the Fry method proceeds as follows: On a sheet of tracing paper, make a base pattern of points by marking the centers for a large number of particles (Figure 15-A.4*C*). On a second transparent overlay, copy the base pattern and mark a convenient central reference point a'. Place the reference point a' of the overlay over each of the points on the base pattern in succession, maintaining a constant relative orientation of the two sheets of paper. In Figure 15-A.4*D*, the reference point a' on the overlay has been placed over the point b on the base pattern. For each such placement, trace onto the overlay the locations of the points on the base pattern. The final plot (Figure 15-A.4*E*) will show an elliptical central area, empty or nearly empty of points, or an elliptical concentration of points that defines the shape and orientation of the strain ellipse. The technique works because there is a minimum possible distance in any direction between two nearest-neighbor particles, and the variation with azimuth of this minimum distance shows up as the empty elliptical area in the plot. This procedure can be greatly simplified by scanning the base pattern into a computer and using a drafting program, in effect to shift the "overlay" and transfer the points from the base pattern onto the "overlay" multiple times.

Any initial ellipticity and preferred orientation of the particles, however, will be included in the final ellipse and will give an erroneous impression of the actual strain.

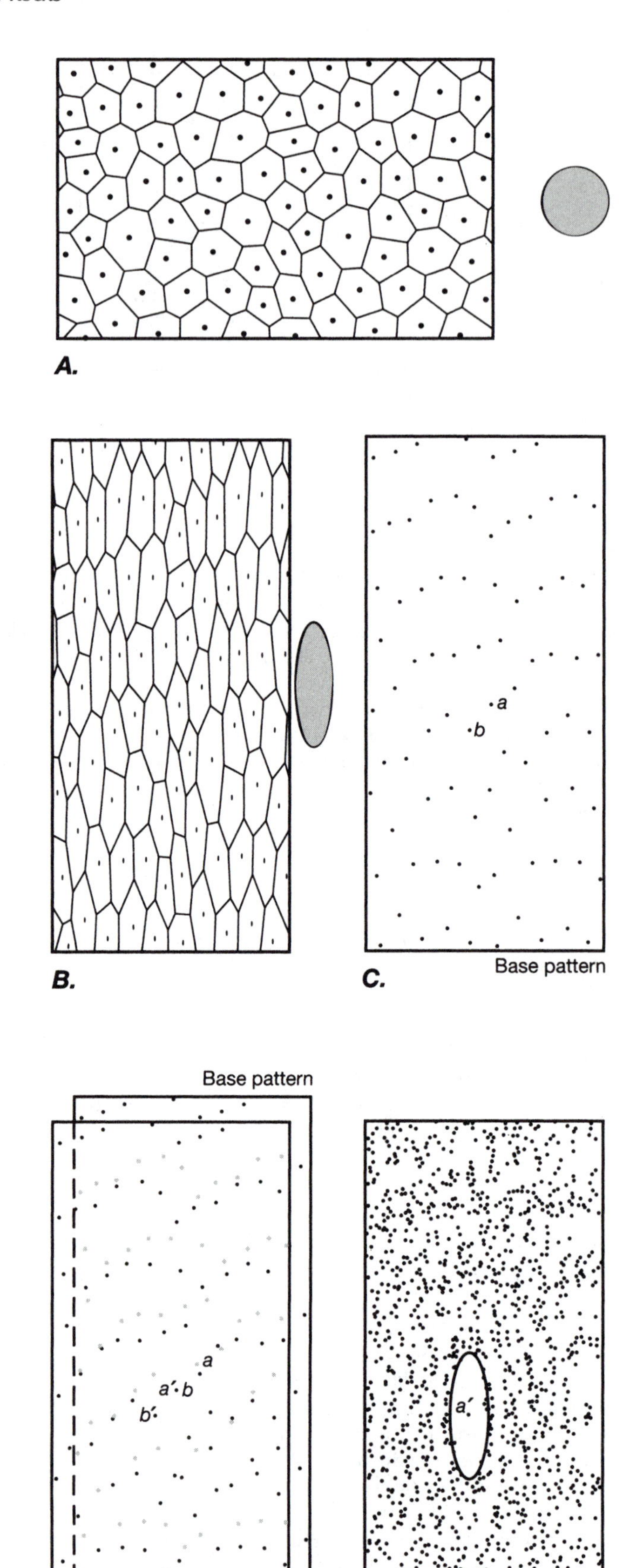

FIGURE 15-A.4 Strain determination using the nearest-neighbor center-to-center technique of Fry (1979). *A.* Undeformed polycrystalline aggregate. The dots are the centers of the grains. *B.* Deformed polycrystalline aggregate with strain ellipse. *C.* Base pattern of grain centers. Point *a* is chosen as the reference point. *D.* Transparent overlay, with the base pattern copied on it, is placed over the base pattern such that the reference point *a'* on the overlay is superimposed on point *b* in the base pattern. The base pattern is then copied onto the overlay. *E.* Placing the reference point *a'* on the overlay over a large number of points in the base pattern and copying the base pattern onto the overlay each time defines the strain ellipse about the reference point.

15-A.3 STRAIN FROM SHEARED ORTHOGONAL LINES: THE WELLMAN METHOD

For cases in which we can measure the angle between two material lines that we know to have been initially perpendicular (see Figure 15.3), we can determine the shape and orientation of the strain ellipse, but not the size. The volumetric strain is fundamentally indeterminate. The clearest method for determining the strain is the graphical construction technique known as the Wellman method. It is based on the geometric fact that any angle inscribed in a semicircle is necessarily a right angle (Figure 15-A.5*A*). Thus two perpendicular lines (*L* and *M*) of any given orientation, constructed from the opposite ends of any diameter of a circle, intersect on the circumference of the circle. After a homogeneous strain, any semicircle based on a diameter becomes a semi-ellipse, but the deformed lines (ℓ and m) must still intersect on the circumference of the ellipse, *regardless of the diameter on which they are constructed* (Figure 15-A.5*B*) because any such pair of lines *L* and *M* were initially perpendicular and initially intersected on the circle. All the perpendicular pairs of chords in the initial circle are sheared except one, and the lines of this pair are parallel to the principal stretches. Thus any possible shear strain is represented by some pair of chords inscribed within a semi-ellipse.

Reversing the argument, we see that if two deformed lines (ℓ and m, Figure 15-A.5*C*) were initially perpendicular, and if we construct lines parallel to ℓ and m (Figure 15-A.1*D*) from both end points of an arbitrary line AB representing a diameter of the strain ellipse, then these lines will intersect at p and p' on the circumference of the ellipse.

This geometry enables us to construct the shape of the strain ellipse if we are given a sufficiently large number of deformed objects. For each object (Figure 15-A.5*C*), the orientation of the initially orthogonal pair of lines is determined relative to a convenient reference line. To represent a diameter of the ellipse, choose a line AB (Figure 15-A.5*D*) of convenient length parallel to the reference line. Because shear strain data constrain only the ratio of the principal strains and not their magnitudes, the length of the line, and thus the size of the ellipse to be constructed, is arbitrary. From both end points of this diameter, construct the orientations of the pair of lines to determine their intersections (Figure 15-A.5*D*). Each pair of orientations generates two intersections that must lie on the strain ellipse. When points are constructed for a large number of strained objects of different orientations, the shape and orientation of the ellipse become apparent. The ellipticity, given by the ratio of maximum to minimum principal axes, is then the ratio of the maximum to minimum stretch, and the principal axes of the constructed ellipse define the orientation of the principal stretches.

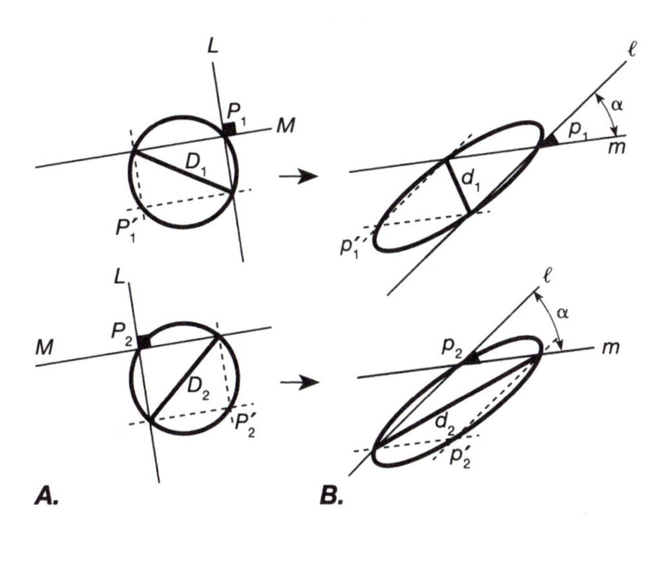

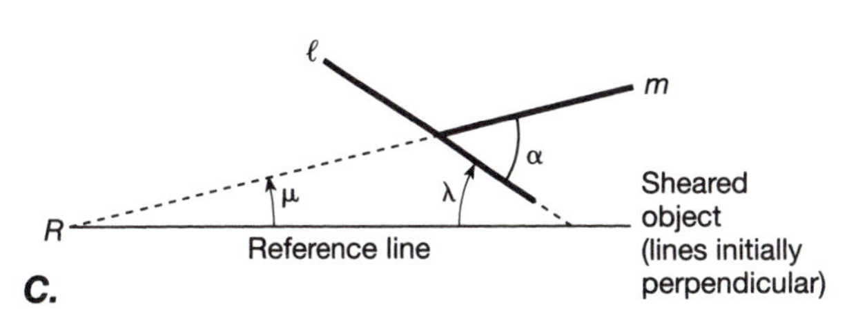

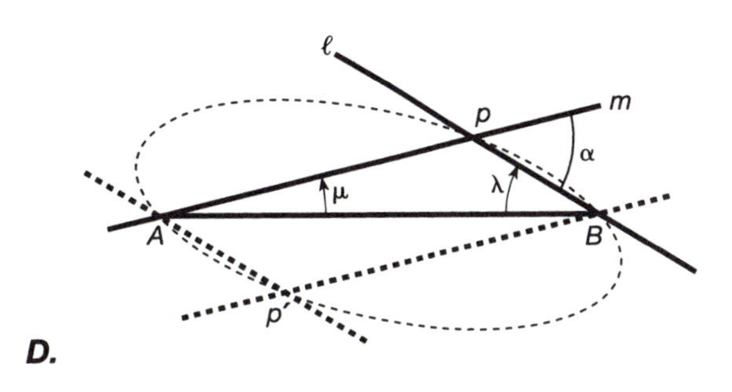

FIGURE 15-A.5 The use of measurements of shear strain to determine the ellipticity and orientation of the strain ellipse. *A* If the orientations of two perpendicular lines *L* and *M* are constructed from opposite end points of any diameter of a circle, such as D_1 or D_2, their points of intersection P_1, P_1', P_2, and P_2' must lie on the circle. *B*. Deformation changes the circle into an ellipse and shears the perpendicular lines. However, if the deformed orientations of two lines ℓ and m that were initially perpendicular are constructed from opposite end points of a line representing an arbitrary diameter of the strain ellipse, the points of their intersections p_1, p_1', p_2, and p_2' must lie on the ellipse. *C*. Sheared material lines ℓ and m were initially perpendicular. Angles λ and μ define the orientations of these lines relative to an arbitrarily defined reference line *R*. *D*. The line orientations from *C* are constructed at opposite ends of a reference line *AB* parallel to *R*, which represents a diameter of the strain ellipse. Two intersections (p and p') are defined, both of which must lie on the strain ellipse. The size of the ellipse is not constrained by data of this type.

PART III

RHEOLOGY

In our investigation of deformation in the Earth so far, we have discussed the relationship between stress and the fracture of rocks, and we have discussed the application of strain to the understanding of ductile deformation in rocks. Except for a brief introduction for the discussion of brittle fracturing in Chapter 9, however, we have not discussed the relationship between stress and strain. Broadly speaking, this relationship is the subject of Part III, which comprises the next three chapters.

Ultimately, if we are to have a deeper understanding of the origin of deformational structures in rocks, we must be able to use the relationships between stress and strain to make mathematical models of the deformation. With such models, we can calculate how structures develop. Our models, of course, must be based on the real behavior of rocks, and rock deformation experiments are essential in obtaining this information. Thus in Chapter 16, we discuss some of these models and describe the experimental evidence for how ductile rocks behave.

In Chapter 17 we take a closer look at the microscopic and submicroscopic mechanisms and structures that give rise to the ductile deformation of rocks. By this means, we gain an understanding of the underlying physical principles of ductile deformation. This knowledge enables us to evaluate extrapolations from laboratory conditions to conditions in the Earth, as well as to understand and interpret the microscopic structures associated with ductile flow.

Models of deformation in the Earth may be both physical and mathematical. Both types of models, however, must obey the fundamental conservation laws of physics. We introduce these laws at the beginning of Chapter 18. We then discuss scale models and the theory behind them, because they can provide physical analogues to deformation in the Earth. We illustrate the application of such models to investigations of the behavior of geometrically complex structures. We then consider mathematical models of deformation in the Earth. The equations describing the behavior of different types of material can be combined with the fundamental conservation, or balance, laws of physics to enable us to calculate mathematically the motion and deformation of a body of rock given the conditions to which it is subjected. In this way, we can determine how material properties affect the formation of structures. As examples of the application of this approach to modeling, we discuss some mathematical models of folding.

Models, of course, must always be simplifications and idealizations of reality. The accepted philosophy (known as Occam's razor[1]) is to employ models of minimum complexity, only incorporating those features necessary to account for the observations of interest. To the extent that the characteristics of various models correspond with our observations of the Earth, we accept the model as an acceptable representation of the conditions and processes that occur in the Earth. One must always keep in mind, however, that such models can never be more than simplifications and approximations to the real world.

[1]After William of Occam (≈1285–1349), one of the most influential philosophers of the fourteenth century, who prominently argued the principle.

Chapter 16

MACROSCOPIC ASPECTS OF ROCK DEFORMATION: RHEOLOGY AND EXPERIMENT

In previous chapters we explored the many different ways that rocks deform and the wide variety of structures that result—brittle fracturing resulting in extension fractures; brittle and localized ductile deformation resulting in faults and ductile shear zones; and distributed ductile deformation resulting in folds, foliations, and lineations. As a means of trying to understand how these structures formed, and therefore how we can use them to interpret the history of deformation in the Earth, we have discussed a variety of models for their formation. Most of these are kinematic models for which we prescribe a priori the geometry of the deformation, like we did for the different models of fold formation (Chapter 13). The models therefore are limited by our preconceived notions of how the material might behave. It would be preferable if we could start simply from a knowledge of the material properties and the conditions to which the body of material is subjected and then calculate how the material actually would behave. This would provide us with a much deeper understanding of the origin of the structures we observe in nature. This subject is referred to as the **dynamics** or the **mechanics** of rock deformation, and it requires that we know the mathematical equations that describe the relationships among factors such as stress, strain, strain rate, temperature, and pressure for the rocks. Such equations are known as **constitutive equations**, because their nature depends on the constitution of the material. They allow us to develop numerical models so that we can use computers to calculate how rocks deform under a given set of conditions. We can then determine how structures such as folds develop, and we can compare our model results with observations of natural structures to see how well our calculations reproduce nature. By this means, we try to understand the various factors that determine the final characteristics of the structures that we observe.

We begin our investigation of the dynamics of rock deformation by examining how rocks actually behave when they deform under a variety of conditions. This information comes primarily from laboratory experiments. Pieces of rock are placed into an apparatus in which the confining pressure, temperature, differential stress, and strain rate can be controlled. Other factors also are controlled in the experimental setup, such as grain size, composition, and chemical environment of the sample. The dependence of the mechanical behavior of rock on the different factors under experimental control can then be determined. From this information, the constitutive equations can be determined, and these equations allow the calculation of the rock behavior in many different circumstances.

The constitutive equations describe the bulk, or macroscopic, behavior of the rock. They ignore the local inhomogeneities and anisotropies associated, for example, with the fact that a rock is a polycrystalline material made up of a multitude of different crystal grains and usually of a mixture of a number of different minerals. In effect, the equations average the behavior of a body of rock over a volume that is large compared to the scale of these inhomogeneities but small compared to the size of the body that we want to investigate. Thus we assume, for these purposes, that rocks can be treated as a continuum (see Section 1.1).

Different materials behave differently under the same state of stress. For example, under a uniaxial compression, one type of material might shorten slightly and then stabilize, whereas another might flow continuously like putty. We must express such differences mathematically if we are to be able to calculate the mechanical behavior of a material. We first discuss, therefore, a few simple mathematical idealizations of material behavior that we can use to describe the relationship between stress and either strain or strain rate for different types of materials. Primarily, we emphasize one-dimensional constitutive relationships, because these contain the essence of the mechanical behavior without the complexity required by the more complete three-dimensional equations. (Three-dimensional relations for some types of behavior are discussed briefly in Box 16-4.)

16.1 CONTINUUM MODELS OF MATERIAL BEHAVIOR

i. Elastic Materials

One of the simplest and perhaps most familiar stress-strain equations is for a linear elastic solid. Such a material deforms by an amount proportional to the applied stress, but when the stress is released, the material returns to its original undeformed state. The deformation therefore is said to be **recoverable**. The relationship is a linear equation, which states either that a normal stress σ_n is proportional to the amount of extension e_n or that a shear stress σ_s is proportional to the amount of shear strain e_s (see Equation (9.2))

$$\sigma_n = Ee_n \qquad \sigma_s = 2\mu e_s \tag{16.1}$$

The constant E is Young's modulus; μ is the shear modulus or the modulus of rigidity. The graphical form of the equations is shown in Figure 16.1*A*.

These equations are identical in form to Hooke's law, which describes the behavior of a spring (Figure 16.1*B*): The applied force F is equal to the product of a spring constant k_1 and the displacement of the spring Δx. For this reason, a linear elastic solid is sometimes called a Hookean solid.

A variation of stress with time (Figure 16.1*C*) consisting of the application of stress at t_1, a linear increase to t_2, a constant value to t_3, and a linear decrease to zero at t_4 results in a similarly shaped strain-time history (Figure 16.1*D*).

Equations (16.1) may be rewritten to express the strain in terms of the stress (second Equation (9.2); compare Equation (9.5) for the three-dimensional form, which involves another material parameter, the Poisson ratio ν). Thus we may consider that the stress causes the strain or that the strain causes the stress (see Section 9.12). We discuss in Chapters 8 and 9 the importance of stress and elastic deformation in the formation of fractures and faults in the Earth's crust.

ii. Viscous Materials

At room temperature and pressure, a rock clearly reacts to stress very differently from a material such as water. We cannot expect, therefore, that an equation for elastic behavior, which applies very well to cold rocks, would be very successful in accounting for the behavior of water. Rocks also can flow under appropriate conditions, even in the solid state, although not nearly as easily as water. Nevertheless, to describe this type of behavior, we clearly need a different type of constitutive equation.

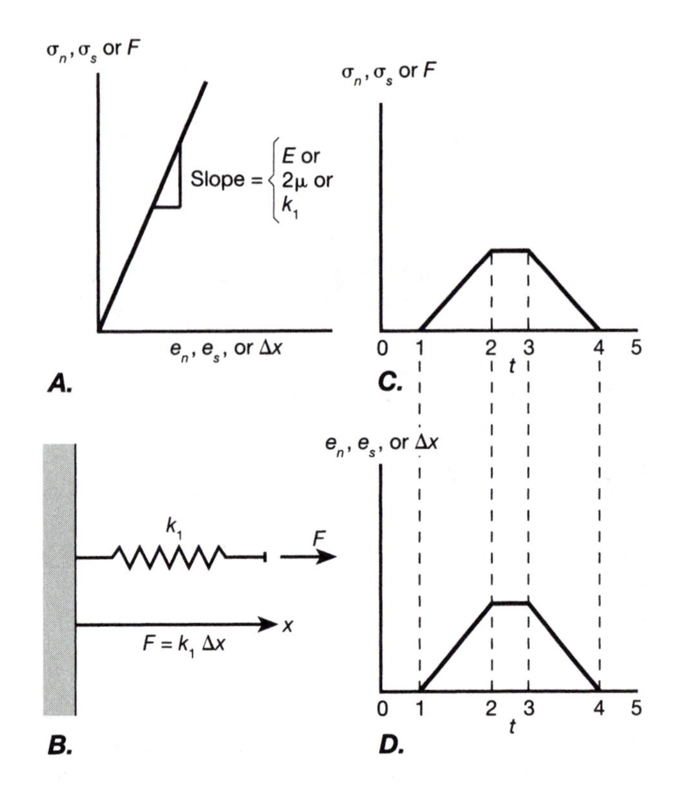

FIGURE 16.1 Characteristics of a linear elastic (Hookean) solid. *A*. A plot of normal stress, shear stress, (or force) versus extension, shear strain, (or displacement) for an elastic solid (or a spring) where the slope is E (Young's modulus), 2μ (twice the shear modulus), or k_l (the spring constant). *B*. A mechanical analogue to elastic behavior consists of a spring with spring-constant k_l attached to a rigid wall, subjected to a force F, and displaced a distance Δx beyond its unloaded length. *C*. A stress-versus-time history imposed on the elastic material. *D*. The extension-time behavior resulting from the imposed stress history.

If a deviatoric stress is applied to a fluid, it begins to flow. When the stress is removed, the flow stops, but the fluid does not return to its undeformed configuration, and the deformation therefore is said to be **nonrecoverable**. The larger the applied stress, the faster the fluid flows, suggesting a relationship between stress and strain rate.

This type of behavior is most simply idealized as a linearly viscous, or Newtonian, constitutive equation. The appropriate one-dimensional equation for constant-volume deformation relates the normal deviatoric stress $\sigma_n^{(Dev)}$ to the instantaneous extension rate $\dot{\varepsilon}_n$, or the shear stress σ_s to the instantaneous shear strain rate $\dot{\varepsilon}_s$

$$\sigma_n^{(Dev)} = 2\eta\dot{\varepsilon}_n \qquad \sigma_s = 2\eta\dot{\varepsilon}_s \tag{16.2}$$

The instantaneous extension rate is the rate of change

in length divided by the instantaneous deformed length (see Equation (16-1.3)). The instantaneous shear strain rate is half the rate at which the tangent of the instantaneous shear angle changes, where the instantaneous shear angle is the infinitesimal change in angle between two material lines that are instantaneously normal to one another (Equation (16-1.5)). These strain rates are nothing more than the strains defined by the instantaneous strain ellipse (Equation (12.30), Section 12.4) divided by the infinitesimal increment of time over which they accumulate. (For a detailed discussion of strain rates, see Box 16-1.) The proportionality constant η is called the **coefficient of viscosity** of the fluid.

The linear equation for viscous behavior is illustrated in Figure 16.2*A*, which is similar to the graph for elastic behavior except that the abscissa is the *strain rate* rather than the strain. When the stress goes to zero, the strain rate goes to zero, but the strain does not. Thus the deformation is permanent, and the amount of strain that can accumulate in such a material has no relationship to the magnitude of the stress. Any stress can produce an arbitrarily large strain given sufficient time.

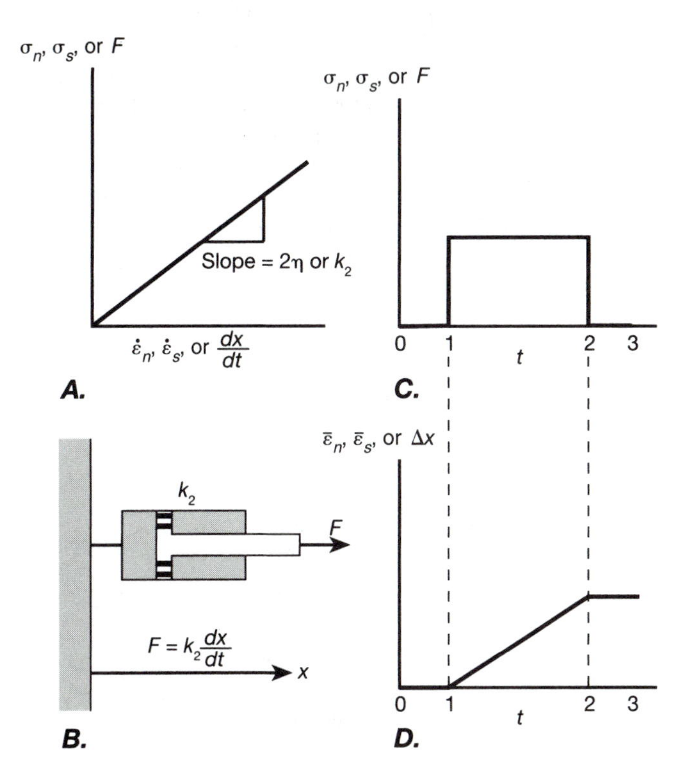

FIGURE 16.2 Characteristics of a linear viscous (Newtonian) fluid. *A*. A plot of stress (or force) versus instantaneous strain rate (or displacement rate) for a viscous material (or dash pot) has a slope 2η (or k_2), the coefficient of viscosity (dash pot constant). *B*. A mechanical analogue for viscous behavior is a dash pot consisting of a porous piston in a cylinder containing a viscous fluid, attached to a rigid support, and subjected to a force *F*. *C*. A stress-versus-time history imposed on the fluid. *D*. The natural-strain-versus-time history resulting from the imposed stress history.

A mechanical analogue for viscous behavior is a device called a dash pot, which consists of a fluid-filled cylinder containing a porous piston (Figure 16.2*B*). When a force is applied across the system the motion of the piston is governed by the rate at which fluid can flow through the pores in the piston. The greater the force applied, the faster the piston moves. This device is familiar in its use to prevent screen doors from slamming and as shock absorbers on motor vehicles.

A force suddenly applied to the dash pot at time t_1 and suddenly removed at time t_2 (Figure 16.2*C*) produces a linear increase of displacement between those times and leaves a permanent displacement when the force is removed (Figure 16.2*D*). Under some conditions of high temperature and pressure, rocks may deform according to a viscous constitutive equation (see Box 17-1).

iii. Plastic Materials

Under many conditions, including common experimental conditions in the laboratory, the model of the linearly viscous fluid does not describe well the observed behavior of polycrystalline solids, such as rocks and metals. Such materials commonly undergo no permanent deformation if the applied stress is smaller than a characteristic **yield stress**, but they flow readily at or slightly above the yield stress. Materials exhibiting this type of behavior are **plastic materials**.

We idealize plastic behavior mathematically by assuming that there is no deformation at all (the material is rigid) below the yield stress and that during the deformation the stress cannot rise above the yield stress except during acceleration of the deformation. This model describes a **perfectly plastic material** or a **rigid-plastic** material. The stress for ductile flow is a constant, and the constitutive equation is the **von Mises yield criterion** (see Figure 8.8):

$$|\sigma_s| \leq K \tag{16.3}$$

This equation requires that the magnitude of the shear stress σ_s cannot be greater than the yield stress *K*, which is a characteristic of the material. Thus for this ideal model, the stress does not determine the strain rate, and the constitutive equation is merely a limit on the possible values of the stress (Figure 16.3*A*). In a Mohr diagram, *K* is the radius of the Mohr circle at the yield point of the material.

A mechanical analogue for this type of behavior is the idealized frictional resistance to the sliding of a block on a surface (Figure 16.3*B*). Consider that the load *F* increases linearly from t_1 to the yield stress at t_2, remains constant to t_3, and then decreases linearly to t_4. The velocity of the block is zero until the applied force exceeds the force of friction, equivalent to the yield stress. At that point, the block starts to move, but the velocity of the block cannot be determined from the magnitude of the applied force, because that force never exceeds the frictional resistance except during acceleration. Thus the force history (Figure

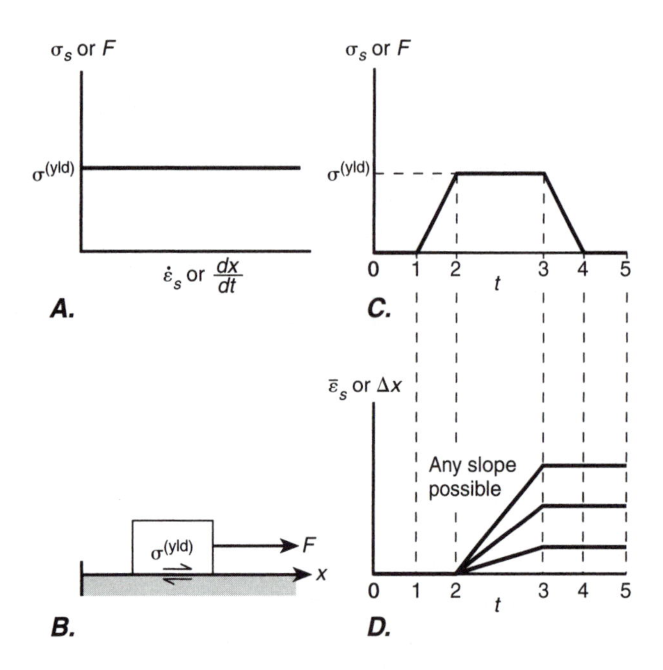

FIGURE 16.3 Characteristics of a perfectly plastic (Saint Venant) material. *A*. Characteristics of a plastic material and its mechanical analogue on a plot of stress (or force) versus instantaneous strain rate (or velocity). *B*. A mechanical analogue to plastic flow is the frictional resistance to sliding of a block on a plane, assuming static and dynamic coefficients of friction are equal. No displacement occurs until the applied force exceeds the frictional resistance. Once sliding has started, the applied force cannot rise above the frictional resistance regardless of the velocity of sliding, except during acceleration. *C*. A stress-versus-time history imposed on a plastic material. The maximum possible stress is the yield stress. *D*. The natural-strain-versus-time history resulting from the imposed stress. Any slope on the graph is possible because the rate of change of strain is not a function of the stress.

16.3*C*) produces an undefined rate of displacement (Figure 16.3*D*). The stress and strain rate behave in an analogous manner in a perfectly plastic material.

iv. Power-Law Materials

The rigid-plastic behavior is an idealization of material behavior that is convenient for some calculations but is not strictly observed in real materials. Many materials, including polycrystalline solids such as rocks, glaciers, metals, or ceramics, generally show that the strain rate is proportional to the differential stress raised to some power n, which is usually between 3 and 5

$$\dot{\varepsilon}_n = A(\sigma^{(Dif)})^n \tag{16.4}$$

where A may depend on temperature and various other factors (see Section 16.4). This type of behavior is called a **power-law rheology.** As the stress exponent n increases, the strain rate becomes more sensitive to changes in stress. This behavior is illustrated in Figure 16.4, in which the strain rate is plotted against stress for power-law flow models with the stress exponent $n = 1$, 3, 5, and 7. As the exponent increases, the power-law behavior progressively approaches the idealized model of perfect plasticity, indicated by the thin dashed line in Figure 16.4 (compare Figure 16.3*A*). We discuss the power-law behavior of rocks in greater detail in Section 16.4.

v. Other Continuum Models for Material Behavior

The simple elastic, viscous, plastic, and power-law models considered in the preceding subsections (i)-(iv) are useful, but they do not describe adequately all the important material behaviors. We will not delve deeply into most of these more complex models, but it is easy to gain an intuitive appreciation for some of them.

Figures 16.5 through 16.8 show four different rheologic models that result from combining the simple models in series (Figures 16.5 and 16.7) or in parallel (Figures 16.6 and 16.8). In each figure, *A* shows the relationships between stress and strain or strain rate; *B* shows the mechanical analogue for the behavior; *C* shows a stress-time history imposed on the material that is useful for demonstrating the major characteristics of the material behavior; and *D* shows the strain-time response of the material to the imposed stress. We use the natural strain because it relates directly to the strain rate that appears in the constitutive relation (see Box 16-1).

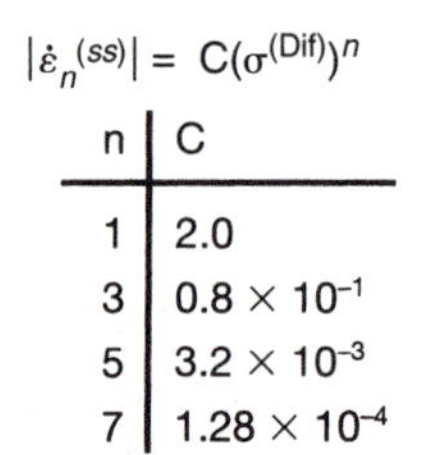

$|\dot{\varepsilon}_n^{(ss)}| = C(\sigma^{(Dif)})^n$

n	C
1	2.0
3	0.8×10^{-1}
5	3.2×10^{-3}
7	1.28×10^{-4}

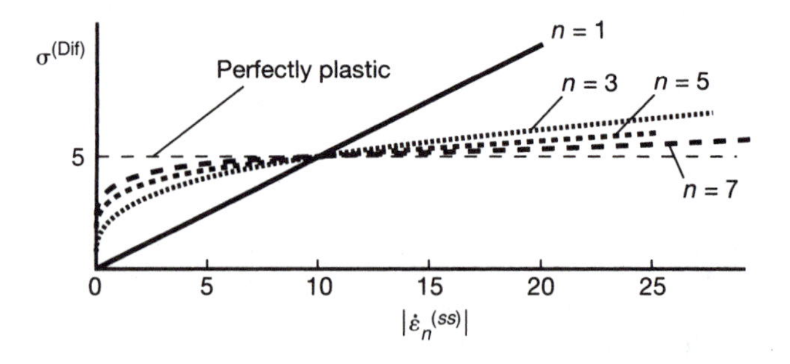

FIGURE 16.4 Relationship between a general power-law rheology and a perfectly plastic rheology. Curves for power-law creep are plotted for stress exponents of 1, 3, 5, and 7. The constant in the equation is adjusted to give the same strain rate for each curve at a stress of 5.

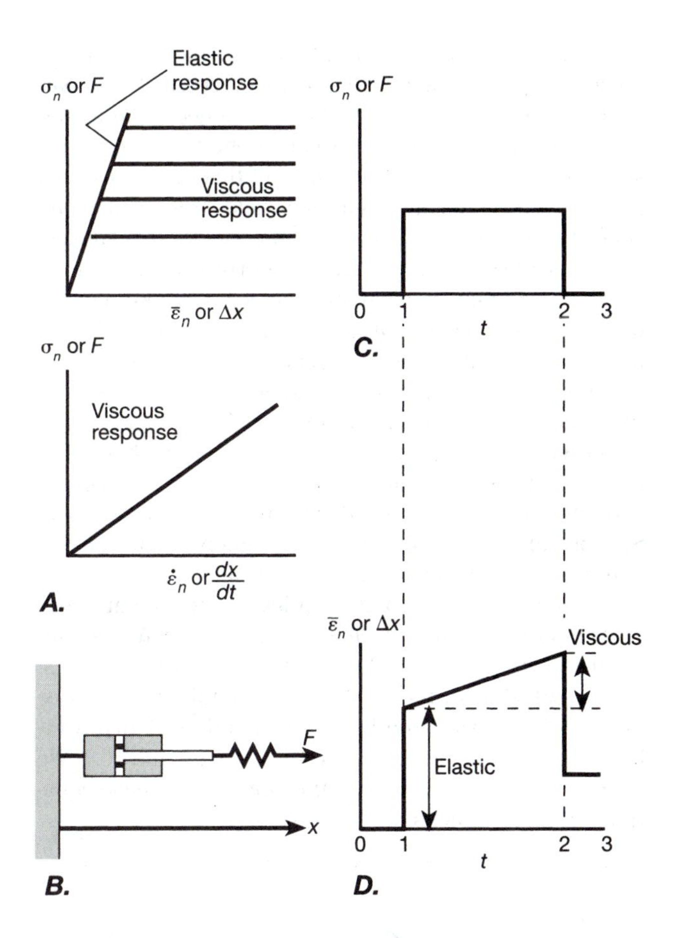

FIGURE 16.5 The characteristics of a viscoelastic (Maxwell) material. *A.* The behavior of a viscoelastic material shown on graphs of stress versus both finite natural strain (extension) (upper diagram) and instantaneous strain rate (lower diagram). The graph of stress versus finite natural strain shows that the magnitude of the elastic response depends on the magnitude of the stress. The graph of stress versus instantaneous strain rate shows the dependence of the viscous strain rate on the stress. *B.* The mechanical analogue comprising a dash pot in series with a spring. *C.* A stress-versus-time history imposed on the material. *D.* The natural-strain-versus-time response to the imposed stress includes an instantaneous recoverable elastic deformation and a nonrecoverable viscous deformation.

Visco-elastic materials are characterized by displaying both viscous and elastic properties. A **Maxwell material** (Figure 16.5) (also called an **elasto-viscous** or **elastico-viscous** material) behaves like an elastic spring and a viscous dash pot connected in series. The permanent strain accumulates viscously, and it begins as soon as a stress is applied. For high viscosities this material behaves like an elastic material for loads of short duration, but it behaves like a viscous material for loads of long duration (see discussion of Maxwell visco-elastic deformation in Section 18.7(v)). For an instantaneously imposed constant strain, the initial elastic response is gradually converted into permanent viscous deformation, and the associated stress decays with time. The Maxwell material is useful in modeling the response of the Earth's crust. The crust is observed to react elastically when it is subjected to a rapid, short-term loading and unloading, such as during the propagation of seismic waves or during the loading and unloading associated with the stick-slip of the earthquake cycle. The crust gradually flows, however, if the load is maintained for long periods, such as during the establishment of isostatic equilibrium of a mountain belt.

Another visco-elastic material, referred to as a **firmo-viscous**, **Kelvin**, or **Voigt** material (Figure 16.6) is elastic to the extent that the equilibrium strain is a linear function of the applied stress and is recoverable, but the strain rate is governed by a viscous response. An everyday application of the mechanical analogue is an auto-

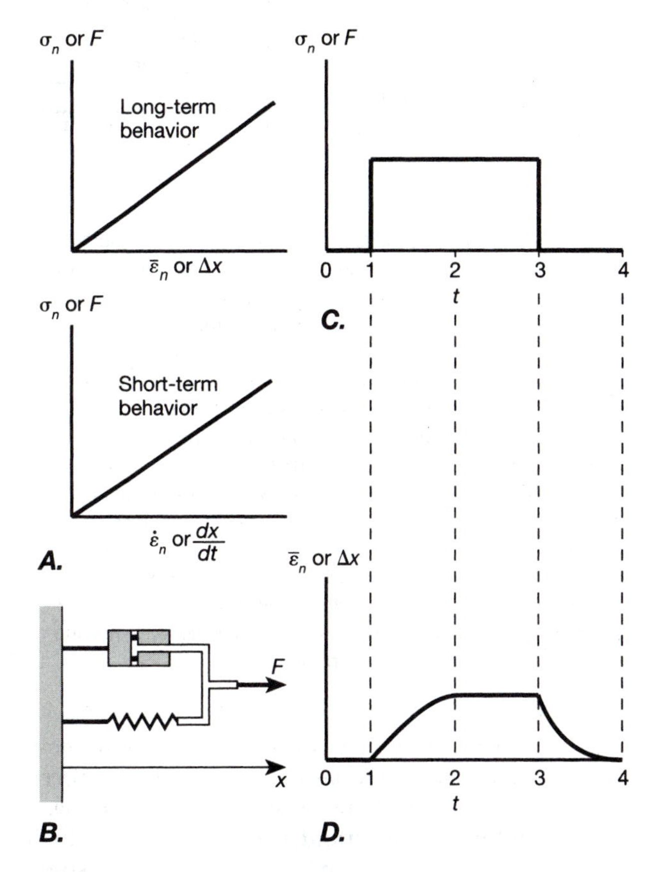

FIGURE 16.6 The characteristics of a firmo-viscous (Kelvin or Voight) material. *A.* The long- and short-term behaviors of a firmo-viscous material are revealed on plots of stress versus finite natural strain (long-term elastic dependence) (upper diagram) and stress versus instantaneous strain rate (short-term viscous dependence) (lower diagram). *B.* The mechanical analogue consists of a dash pot and a spring connected in parallel. *C.* The stress-versus-time history imposed on the material. *D.* The natural-strain-versus-time response to the imposed stress shows that the instantaneous strain rate depends on both the initial stress and the strain.

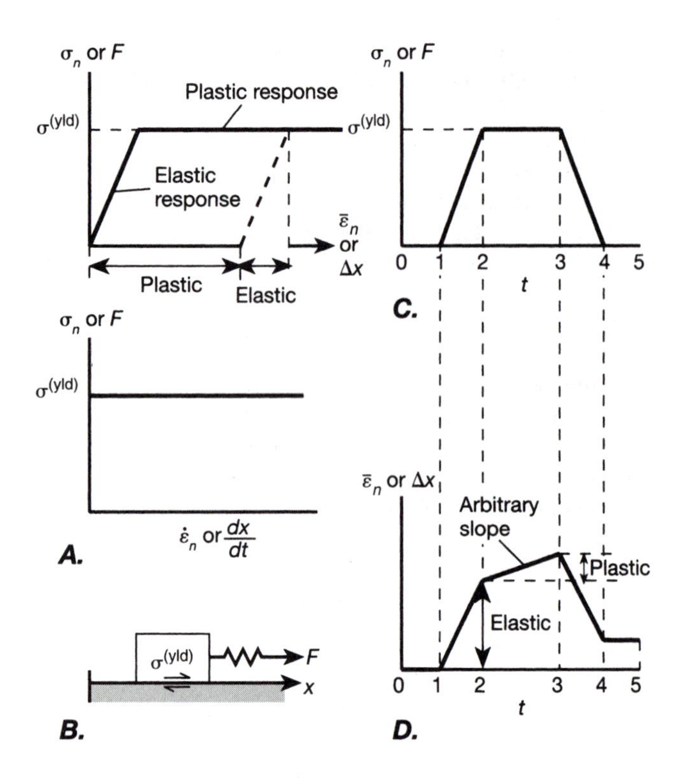

FIGURE 16.7 Characteristics of an elastic-plastic (Prandtl) material. *A.* The behavior of an elastic-plastic material showing the stress as a function of both the finite natural strain (extension) (upper diagram) and the instantaneous strain rate (lower diagram). The stress versus the finite natural strain shows the elastic response below the yield stress. The graph of stress versus instantaneous strain rate shows the independence of the plastic strain rate from the stress. *B.* A mechanical analogue comprising a friction block connected in series with a spring. *C.* A stress-versus-time history imposed on the material. *D.* The natural-strain-versus-time history resulting from the imposed stress. Below the yield stress, the material deforms elastically. At the yield stress, plastic deformation occurs at an undetermined rate. When the stress is removed, the elastic portion of the strain recovers, leaving a permanent strain equal to the plastic portion.

mobile suspension system consisting of a spring connected in parallel with a dash pot (a shock absorber), which damps out the elastic oscillations of the spring.

An **elastic-plastic (Prandtl)** material (Figure 16.7) also shows a combination of recoverable elastic strain and permanent deformation, analogous to the motion governed by a friction block and a spring connected in series. The permanent deformation is plastic, however, and it does not begin until the yield stress is reached. Release of the stress results in the disappearance (recovery) of the elastic deformation, but the plastic part of the deformation is permanent (nonrecoverable). Elastic-plastic behavior approximates the behavior of some crystals at high temperatures, as discussed in Section 16.4.

A **visco-plastic (Bingham)** material (Figure 16.8) displays linear viscous behavior only above a yield stress such as characterizes plastic materials, and its mechanical analogue is a dash pot and a friction block connected in parallel (in this case, they can also be connected in series; some models also add an elastic property by placing a spring in front). A natural example of this behavior is wet paint, which is fluid but has a small yield stress, which prevents it from running off a wall after a thin layer is applied.

Other possible models can be derived from combining the simple mechanical analogues in different configurations. The potential usefulness of any such model depends on whether its response corresponds to that observed for some particular material. All models, with the exceptions of plasticity (Figures 16.3 and 16.7) and the power-law rheology (Figure 16.4), have the limitation that they are superpositions of the linearly proportional responses of a spring or a dash pot to an imposed load. Such linear models have proved extremely useful in many applications in geology and engineering, but materials do not have to behave in mathematically convenient ways. Thus nonlinear elasticity theory is necessary to account for the elastic properties of rocks under very high pressures, and some of the processes of ductile flow of crystalline solids are inherently nonlinear and result in power-law rheology, as we see, for example, in Section 16.4. We discuss the generalization of the simplest of these equations to three dimensions in Box 16-4.

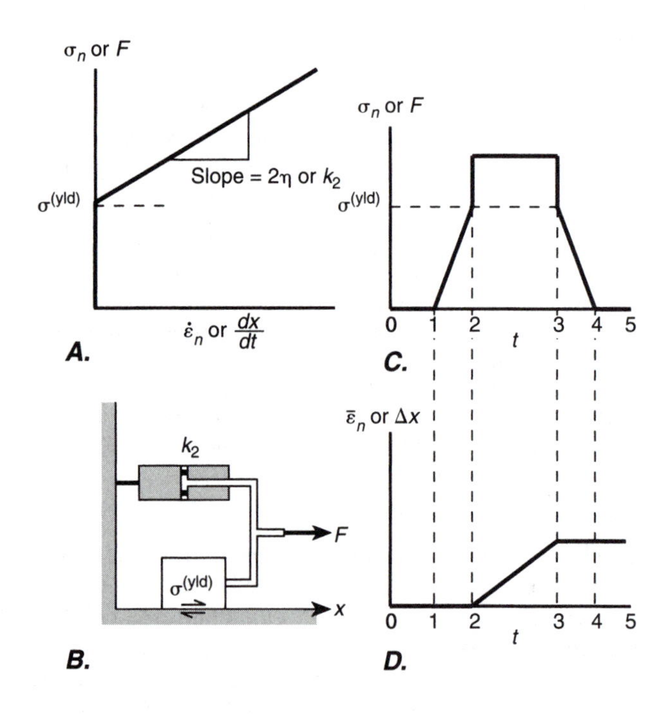

FIGURE 16.8 Characteristics of a visco-plastic (Bingham) material. *A.* The graph of stress versus instantaneous strain rate behavior of a visco-plastic material. Above the yield stress, the material behaves like a viscous fluid. *B.* The mechanical analogue consists of a dash pot and a friction block connected in parallel. *C.* A stress-versus-time history imposed on the visco-plastic material. *D.* The natural-strain-versus-time history resulting from the imposed stress.

BOX 16-1 Measures of Strain Rate

There are two strain rates that are used in the literature: one based on the rate of change of the finite strain, the other based on the rate of change of the instantaneous strain. Unfortunately, the symbols representing the different rates are not standardized. In fact, the difference between the two measures of strain rate is sometimes ignored in the literature, and it is not always clear which of the two rates is intended. The difference is significant, however, because using the wrong strain rate in a constitutive equation can give inaccurate results. Thus the reader must be careful about assuming which strain rate is meant. We demonstrate the difference between the two rates here.

Using the index (i) to indicate the initial deformed state at some time t, and the index (f) to indicate the final deformed state at a small increment of time Δt later, and using L for the undeformed length and ℓ for the deformed length, the rate of change of the finite extension is given by

$$\dot{e}_n = \frac{de_n}{dt} = \lim_{\Delta t \to 0} \frac{e_n^{(f)} - e_n^{(i)}}{\Delta t}$$

$$\dot{e}_n = \lim_{\Delta t \to 0} \frac{1}{\Delta t}\left[\left(\frac{\ell_f - L}{L}\right) - \left(\frac{\ell_i - L}{L}\right)\right]$$

$$\dot{e}_n = \lim_{\Delta t \to 0} \frac{1}{\Delta t}\left[\frac{\ell_f - \ell_i}{L}\right]$$

$$\dot{e}_n = \frac{d\ell/dt}{L} = \frac{d}{dt}\left[\frac{\ell}{L}\right] = \dot{s}_n \qquad (16\text{-}1.1)$$

We drop the subscript i or f on ℓ when it is clear that ℓ refers to the instantaneous deformed length. Notice that the rate of change of the finite extension is also the rate of change of the finite stretch s_n.

The finite shear strain rate is

$$\dot{e}_s = 0.5 \lim_{\Delta t \to 0} \frac{\tan \psi^{(f)} - \tan \psi^{(i)}}{\Delta t}$$

$$= 0.5 \lim_{\Delta t \to 0} \frac{[u_f/L] - [u_i/L]}{\Delta t} = 0.5 \lim_{\Delta t \to 0} \frac{\Delta u}{L\Delta t}$$

$$\dot{e}_s = \frac{0.5(du/dt)}{L} \qquad (16\text{-}1.2)$$

where u is the displacement normal to L associated with the shear of the line L through the angle ψ (Figure 16-1.1).

The instantaneous extension rate is just the instantaneous extension (first Equation (12.30)) that occurs in the time Δt:

$$\dot{\varepsilon}_n = \lim_{\Delta t \to 0} \frac{\varepsilon_n}{\Delta t} = \lim_{\Delta t \to 0} \frac{1}{\Delta t}\left[\frac{\ell_f - \ell_i}{\ell_i}\right] = \frac{d\ell/dt}{\ell} \qquad (16\text{-}1.3)$$

The instantaneous extension rate $\dot{\varepsilon}_n$ is also the rate of change of the natural strain $\bar{\varepsilon}_n$ defined in Equation (12-1.1):

$$\frac{d\bar{\varepsilon}_n}{dt} = \frac{d}{dt}\left(\ln\left(\frac{\ell_f}{L}\right)\right) = \frac{L}{\ell_f}\frac{d\ell_f/dt}{L} = \frac{d\ell/dt}{\ell} = \dot{\varepsilon}_n \qquad (16\text{-}1.4)$$

as well as the rate of instantaneous stretch, defined in the third Equation (12.30).

The rate of instantaneous shear strain is

$$\dot{\varepsilon}_s = 0.5 \lim_{\Delta t \to 0} \frac{\tan \Delta\psi}{\Delta t} = 0.5 \lim_{\Delta t \to 0} \frac{\Delta u}{\ell\,\Delta t} = \frac{0.5(du/dt)}{\ell} \qquad (16\text{-}1.5)$$

where $\Delta\psi$ is the very small shear angle that measures the instantaneous shear of a pair of instantaneously perpendicular lines, and Δu is a very small instantaneous displacement for a line of length ℓ, and where $\Delta\psi$, ℓ and Δu are comparable in geometry to $\psi^{(i)}$ and $u^{(i)}$ in Figure 16-1.1.

Thus the two strain rates differ in that for the rates of finite extension and shear strain, $\dot{e}_n$ and $\dot{e}_s$, the reference length is the original length L, whereas for the rates of instantaneous extension and shear strain, $\dot{\varepsilon}_n$ and $\dot{\varepsilon}_s$, the reference length is the instantaneous deformed length ℓ. The different rates are related by the stretch s_n:

$$\dot{\varepsilon}_n = \frac{d\ell/dt}{\ell} = \frac{d\ell dt}{L}\frac{L}{\ell} = \frac{\dot{e}_n}{s_n} \qquad (16\text{-}1.6)$$

$$\dot{\varepsilon}_s = \frac{du/dt}{\ell} = \frac{du/dt}{L}\frac{L}{\ell} = \frac{\dot{e}_s}{s_n} \qquad (16\text{-}1.7)$$

Thus at small strains, the difference between the rates is negligible because s_n is close to 1. At large strains, however, the difference becomes significant.

The strain rate tensor $\dot{e}_{k\ell}$ is a generalization of the instantaneous strain rates $\dot{e}_n$ and $\dot{e}_s$. The tensor components of this strain rate are the components of the instantaneous strain tensor $\varepsilon_{k\ell}$ for a unit increment of time. These components

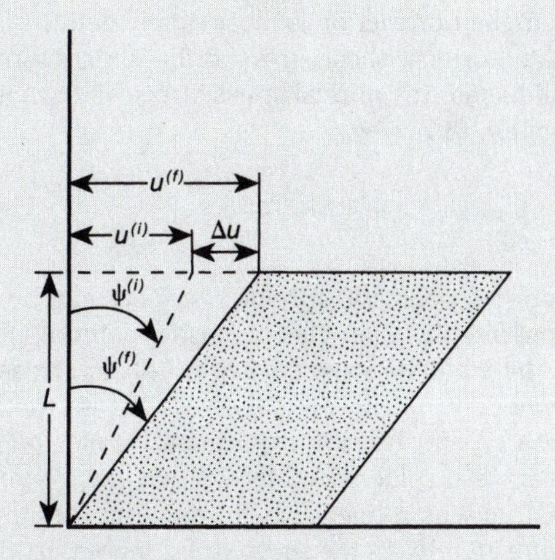

FIGURE 16-1.1 The finite shear strain in progressive simple shear is defined in terms of the increment of displacement $\Delta u = u^{(f)} - u^{(i)}$ per unit length L, per unit time, where the superscripts f and i indicate final and initial displacements.

BOX 16-1 Measures of Strain Rate *(continued)*

can be shown to be the symmetric part of the velocity gradient tensor, which is defined by:

$$\dot{\varepsilon}_{k\ell} = \dot{\varepsilon}_{\ell k} = 0.5\left[\frac{\partial v_k}{\partial x_\ell} + \frac{\partial v_\ell}{\partial x_k}\right] \tag{16-1.8}$$

where k and ℓ independently can take on any of the values 1, 2, or 3, and where v_k (or v_ℓ) are the components of the velocity of material points in the deforming continuum. Because the velocity is the displacement rate, these equations are comparable to Equations (12-3.17). The tensor components for which k and ℓ are equal, for example, $k = \ell = 1$, are rates of instantaneous extension. Those for which k and ℓ are unequal, for example, $k = 1$ and $\ell = 3$, are rates of instantaneous shear.

Use of the finite extension rate $\dot{e}_n$ or the finite shear strain rate $\dot{e}_s$ or, in general, $\dot{e}_{k\ell}$ can be justified when the mechanical behavior of the material depends on the total amount of strain accumulated in the body from the undeformed state, for in this case the undeformed state has a physical significance as a reference state for the behavior of the material.

If the behavior of the material is independent of the total strain, then the undeformed state is indistinguishable from the deformed state, as, for example, in the deformation of a viscous fluid such as water. In these circumstances, there can be little physical meaning in a strain rate that uses some arbitrarily defined state of the material as a reference state, and it is physically more relevant to refer the strain rate at any given time to the state of the material at that time. Thus the instantaneous extension rate $\dot{\varepsilon}_n$ or the instantaneous shear strain rate $\dot{\varepsilon}_s$ or, in general, $\dot{\varepsilon}_{k\ell}$ is the most appropriate measure to use.

For many materials, the current behavior is affected by the strain that has accumulated within a constant span of time before the present. Such materials are said to have a fading memory. The primary creep of many polycrystalline materials is an example of this behavior. Once steady-state creep is established, however, any state of strain is indistinguishable from any other state of strain, and the instantaneous strain rate, such as $\dot{\varepsilon}_n$, is most appropriate for describing the mechanical behavior.

Although experiments on ductile flow of rocks have concentrated mostly on the high-temperature steady-state deformation, the instantaneous strain rate is not universally employed, and in many cases, published reports of experimental deformation are not even explicit about which definition of strain rate is used.

16.2 EXPERIMENTS ON FRICTION AND CATACLASTIC FLOW: IMPLICATIONS FOR FAULTING

i. Static Rock Friction

The coefficient of friction is most simply defined by the ratio of the critical shear stress at the initiation of frictional sliding, to the normal stress on the sliding surface (cf. Equation (8.7)):

$$\bar{\mu} \equiv \frac{|\sigma_s^*|}{\sigma_n} \tag{16.5}$$

The frictional behavior of rock interfaces can be modeled most simply as an elastic-plastic material (Figure 16.7*B*); the yield stress is the critical shear stress, and it is dependent on the normal stress across the interface. In a typical friction experiment, a plot of shear stress versus displacement shows that the system deforms initially as a linear elastic material (points a to b in Figure 16.9*A*). As the shear stress increases, the response departs from the linear elastic curve (points b to c in Figure 16.9*A*) and defines a maximum at point c, which is characterized by the coefficient of static friction. With further displacement, the shear stress drops slightly and comes to a stable value at a shear stress characterized by the coefficient of sliding, or dynamic, friction (d in Figure 16.9*A*).

Experiments on the coefficient of friction on rock interfaces show that at low values of normal stress, roughly below 5 MPa, such as would apply for most civil engineering situations, the value of the maximum, or static, friction is highly variable and depends strongly on the initial roughness of the surfaces in contact. Figure 16.9*B* shows a compilation of low-pressure static friction data as a plot of critical shear stress versus the normal stress across the sliding surface.

At higher normal stresses, at least up to about 1700 MPa, which would be appropriate for depths characteristic of faults in the brittle crust, the static friction is essentially independent of surface roughness, and with few exceptions, it is also independent of the rock type (Figure 16.9*C*). Interfaces containing clay minerals such as montmorillonite or vermiculite, in particular, give anomalously low values of friction. Nevertheless, the friction for most rocks at these higher normal stresses can be summarized by two equations, which have come to be known as **Byerlee's law** after the researcher who proposed them (Figure 16.9*C*):

$$\begin{aligned} \sigma_s &= 0.85\,\sigma_n \text{ [MPa]} \quad \text{for} \quad 5 \text{ MPa} < \sigma_n \leq 200 \text{ MPa} \\ \sigma_s &= 50 + 0.6\,\sigma_n \text{ [MPa]} \quad \text{for} \quad \sigma_n \geq 200 \text{ MPa} \end{aligned} \tag{16.6}$$

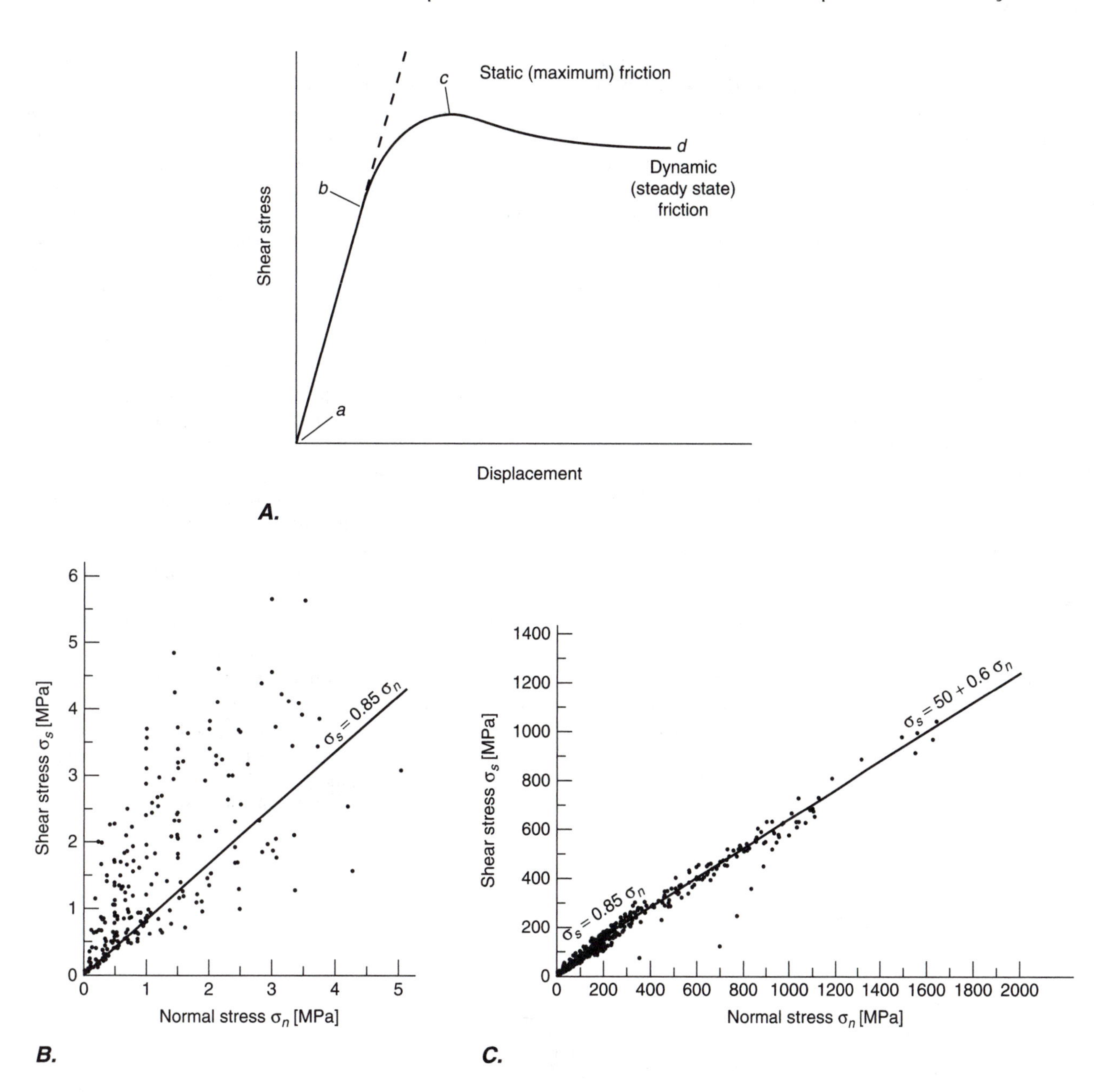

FIGURE 16.9 Experimental results for static rock friction. *A*. Typical shear stress versus displacement curve for a friction experiment. The line between points *a* and *b* is the linear elastic response. Point *c*: Shear stress reaches a maximum defining the static (or maximum) friction. Point *d*: Shear stress falls off to a steady-state value characteristic of dynamic friction. *B*. Experimental data for normal stress below 5 MPa for a variety of rock types, including igneous, metamorphic, and sedimentary rocks. The plotted line is for reference and comparison with *C*, not a fit to the plotted data. *C*. Experimental data for normal stresses above 5 MPa showing the two lines that constitute Byerlee's law of rock friction. Data are from a large variety of rock types, including granite (fractured and ground surfaces), granodiorite, quartz monzonite, gabbro, dunite, mylonite, gneiss, limestone, sandstone (fractured and saw-cut surfaces), and gouges composed of plaster, chlorite, serpentinite, illite, kaolinite, halloysite, monmorillonite, and vermiculite. (After Byerlee 1978)

ii. Dynamic Rock Friction and Cataclastic Flow

Displacement along a fault generally involves the frictional behavior of a layer of crushed rock, called gouge or cataclasite, that accumulates between the solid sliding surfaces, and so the sliding behavior of such a system is dominated by the cataclastic flow of this material (see Sections 3.2(i) and 8.4(ii), Figures 3.5*B* and 8.10, and Table 3.1). Thus an understanding of this process contributes to the continuing effort to understand the mechanics of earthquakes and the associated deformation. This understanding is particularly important in earthquake-prone regions, such as along plate boundaries like the subduction zones around the Pacific rim and the Alpine-Himalayan zone, and transform faults such as the San Andreas fault in California, the Dead Sea fault system in the Middle East, the North Anatolian fault in Turkey, and the Altyn Tagh fault system in China.

To investigate the rheology of cataclastic flow, experiments are performed on sliding interfaces that have been produced either by fracture or by cutting and various degrees of polishing. Finely crushed rock may be introduced between the rock interfaces to simulate fault gouge. Sliding at constant displacement rate, or less commonly, at constant applied stress, determines the effects on mechanical behavior for different values of the differential stress, displacement rate, normal stress, interface surface roughness, grain size of the gouge, and thickness of the gouge layer.

A diagnostic feature of cataclastic deformation is that the differential stress required for deformation increases markedly with increasing confining pressure. This behavior indicates a frictional mechanism, which is characterized by an increase in the force of friction with increasing normal stress across the plane of sliding (Equations (16.6); see Equation (8.7), Section 8.4(ii)).

Figure 16.10 shows some characteristic results of experiments at room temperature on a 1-millimeter-thick layer of simulated granite gouge sheared between two granite blocks. Figure 16.10*A* shows displacement-versus-time curves for creep experiments carried out at different constant applied shear stresses (given at the end of each curve in MPa). The curves exhibit two patterns of velocity with time, shown by the slope of the curves: Either there is a continual decrease of velocity with time, or there is a continual increase leading to catastrophic unstable slip. Note that only relatively small changes in stress separate stable sliding from unstable sliding. The instability leading to a catastrophic slip event is comparable to an earthquake, which is why understanding this behavior is of such great interest.

Figure 16.10*B* shows the result of a constant-displacement-rate test for the same experimental configuration. After almost 3 millimeters of displacement, the displacement was halted and then restarted. The behavior is very similar to the elastic-plastic (Prandtl) material (top graph, Figure 16.7*A*) except the stress peaks at the yield point and then decays with further displacement to a constant steady-state value (compare Figure 16.9*A*). After stopping and then restarting the slip, the initial peak in the frictional resistance is reproduced.

The transition from static to dynamic friction occurs as the system slips over a characteristic slip distance L_c. The final value of dynamic friction depends on the velocity of sliding; it is lower for higher sliding velocities. During periods of no slip, if an applied normal stress is maintained across the shear surface, the coefficient of static friction in the fault gouge increases with time, leading to a strengthening of the shear zone.

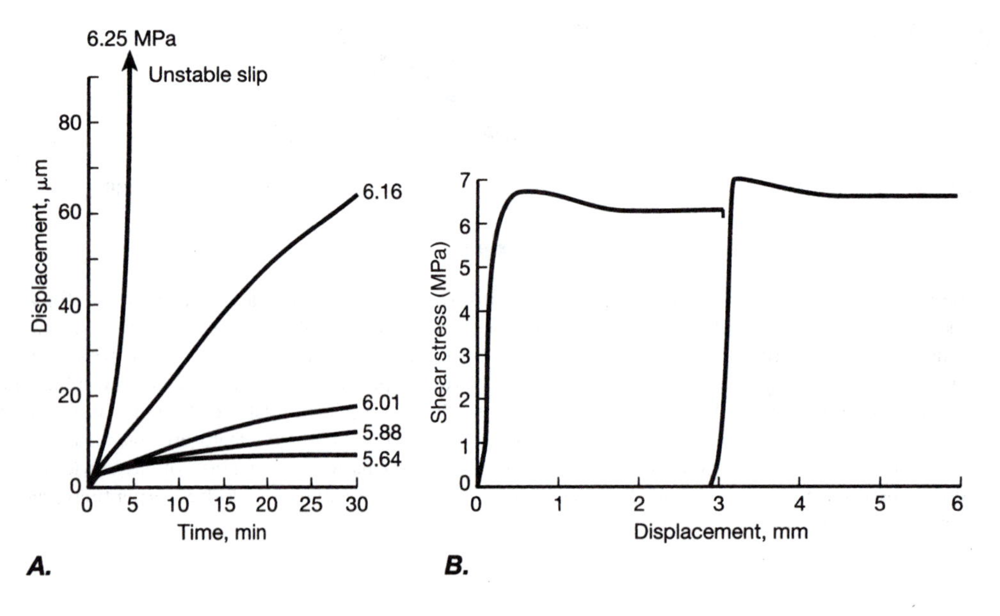

FIGURE 16.10 Experimental characteristics of the cataclastic flow of fault gouge. (From Dieterich 1981) *A*. Displacement versus time record for a 1-millimeter-thick layer of simulated gouge of 250-μm grain size, sheared between two granite blocks whose surfaces were finished with #60 abrasive. The various curves are for applied shear stresses ranging from 5.64 to 6.25 MPa. The slope at any point on any of the curves defines the displacement rate. *B*. Shear stress versus displacement record for a constant strain-rate experiment using the same configuration as in *A*. At 3-millimeter displacement, the displacement was stopped and then restarted.

The boundary between the stable sliding and the unstable, or stick-slip, sliding is a function of both temperature and pressure, as shown by the experimental mapping of this boundary for granite in Figure 16.11. Although the boundary is not sharply defined, there is clearly a high-temperature limit to the stick-slip field at any given pressure, which most likely marks the onset of ductile behavior in the rock. If the temperature is sufficiently low, at a constant temperature and increasing pressure there is a transition from stable sliding to stick-slip behavior. The variation of pressure with temperature in the Earth's crust is plotted as a line in Figure 16.11, assuming normal gradients of 30 MPa/km and 20°C/km. A depth scale is marked along this line. The high-temperature side of the transition zone crosses this line twice, first at about a crustal depth of 2.5 km and again at about 17 km. Using these two curves, we would predict that stick-slip behavior should occur in the crust between about 2.5 and 17 km, and this should be the seismogenic zone within which most earthquakes occur. In fact, along strike-slip faults, most earthquakes are observed to occur between depths of a few km and 15 to 20 km. Thus this boundary between stable sliding and stick-slip is consistent with observations of the earth.

The details of frictional behavior are more complicated than suggested here. Specifically, the friction depends not only on the rate of sliding, but also on the state of the frictional surface, the so-called **rate- and state-dependent friction**. These details can account for a number of critical features of fault behavior, and we give a brief introduction in Box 16-2.

iii. Relationship to Fault Movement

These experimental results resemble the behavior of natural faults. Stress buildup on a fault can lead to the sudden rapid displacement associated with an earthquake, which is comparable to the slip events in stick-slip behavior. This process implies a dynamic instability caused

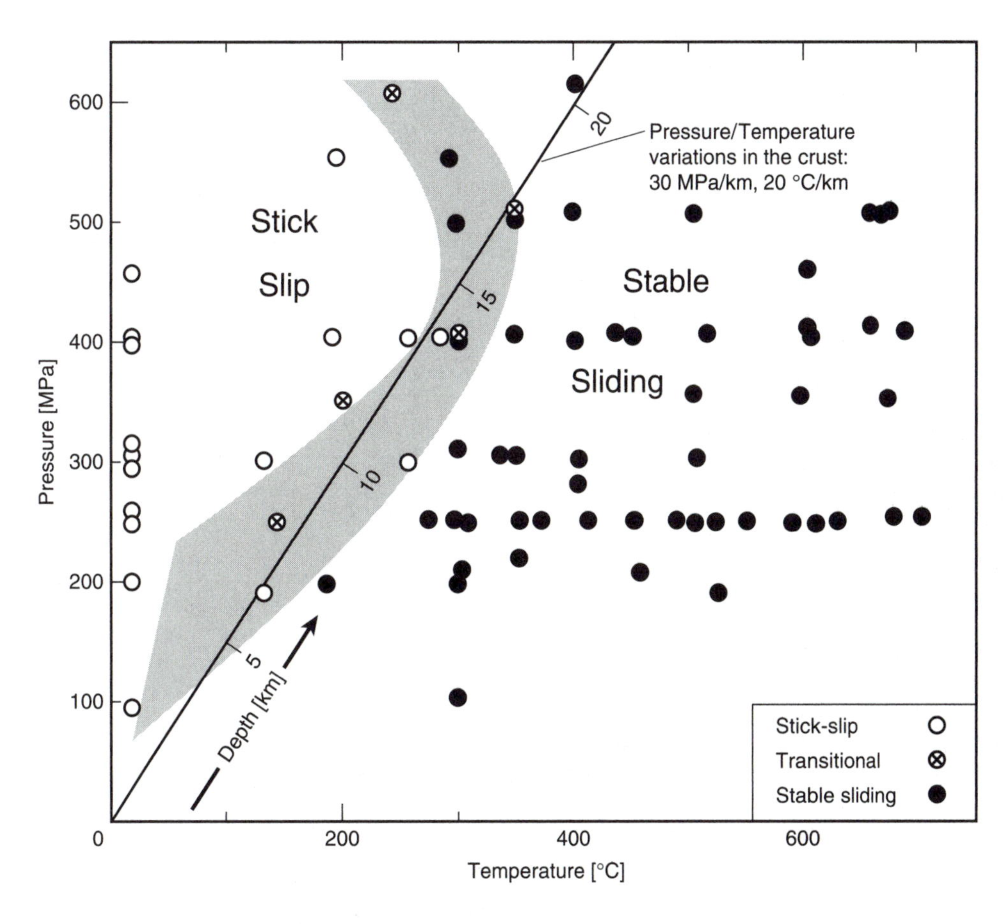

FIGURE 16.11 A mapping of the stick-slip/stable sliding boundary for frictional behavior in granite as a function of temperature and pressure. The straight line from the origin shows the variation of pressure with temperature through normal crust, assuming a pressure gradient of 30 MPa/km and a temperature gradient of 20°C/km. The depth scale is marked along this line. The high-temperature side of the stable sliding/stick-slip transition crosses the crustal pressure-temperature curve at about 2.5 km and again at about 17 km, suggesting that stick-slip behavior, and thus earthquakes, should occur between these depths. (Data after Scholz 1990, figure 2.21)

by decreasing strength with increasing displacement. The fact that faults with a substantial movement history still have a significant shear strength, however, requires a static strengthening mechanism so that resistance to slip can increase during periods of no displacement.

Figure 16.12*A* shows a simple mechanical analogue that reproduces the expected behavior of a fault system. It consists of a spring having a spring-constant (stiffness) k, and a slider block for which the coefficient of static friction is higher than that of sliding friction. The sliding interface of the block corresponds to the fault, and the spring corresponds to the behavior of rock surrounding the fault that transmits the stresses to the fault surface. We assume the area of the base is one square unit, a convenience that allows us to consider the applied force to be equal to the shear stress (force per unit area) across the base.

Assume the free end of the spring is displaced to the right at some slow rate v. As the spring stretches, the force exerted on the block by the spring increases, and the frictional shear stress on the base of the block builds up until it reaches the level of static friction. At that point, the block begins to slide, and the coefficient of friction starts decreasing, reaching the stable steady-state value for the coefficient of sliding friction, after a characteristic displacement L_C. As soon as the block slides faster than the imposed displacement rate v, the spring starts to relax, and according to Hooke's law (Figure 16.1*A*, *B*; see also Equation (16.1)), the force exerted by the spring on the slider must decrease. The behavior of the system then depends on the relative rates at which the spring force and the frictional resistance decrease with increasing displacement of the block (Figure 16.12*B*, *C*).

If the force exerted by the spring (dashed line, Figure 16.12*B*) decreases less rapidly with displacement than the frictional resistance (the solid line in Figure 16.12*B*), then the vertical space between the two lines represents the excess force exerted by the spring beyond what is required to counteract the frictional resistance. This excess force accelerates the block. The farther the block moves, the greater the excess force becomes, and the greater is the acceleration. Thus the sliding becomes unstable and a catastrophic slip event occurs. After the block has moved the characteristic distance L_C, the frictional resistance reaches the value of sliding friction and becomes constant, whereas the spring continues to relax. Acceleration of the block continues until the force exerted by the spring is less than that necessary to maintain the sliding. At that point, the excess of the frictional resistance over the spring force decelerates the block until it stops, and the frictional resistance returns to the value of static friction. Sliding cannot resume until the imposed displacement rate v stretches the spring sufficiently to once again exceed the force of static friction. The resulting periodic slip events are reminiscent of the stick-slip motion that characterizes the earthquake cycle on active faults, with the intermittent slip events corresponding to earthquakes (see Figure 8.11).

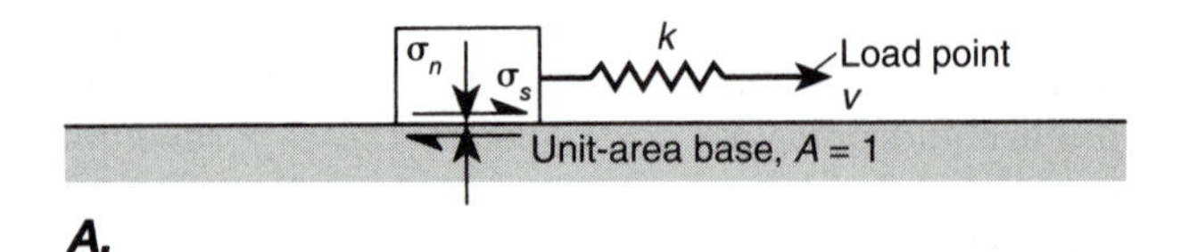

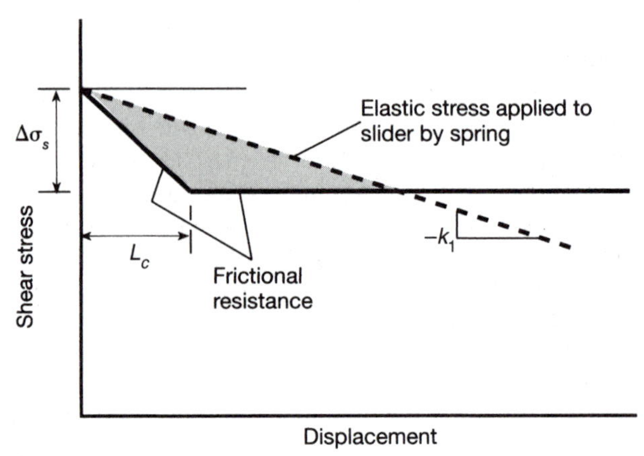

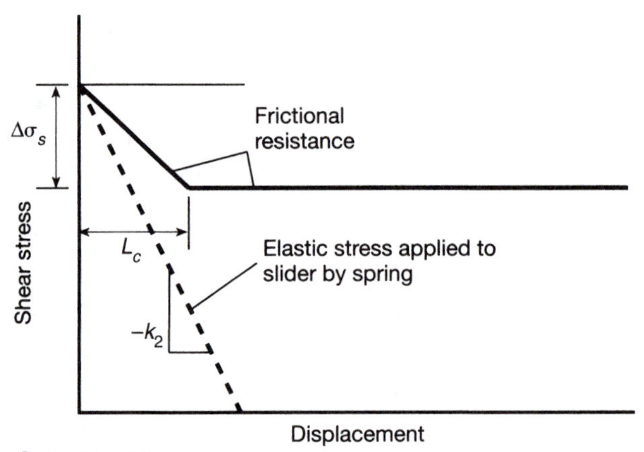

FIGURE 16.12 An analogue for the behavior of a fault system. (After Dieterich 1979) *A.* A mechanical analogue consisting of a spring, with spring constant k, connected in series to a frictional slider, for which the coefficient of friction decreases from the static value to the sliding value over a finite amount of displacement L_c, the characteristic length. If a constant velocity is imposed at the load point for the system, the slider block can respond either by stick-slip or by stable sliding. *B.* The instability leading to stick-slip is shown on a stress versus displacement record. During the slip event, the frictional resistance decreases more for each small increment of displacement than does the elastic stress from the spring, leaving the excess elastic stress available to accelerate the system, and thereby causing the slip instability. *C.* If slip is by stable sliding, the frictional stress tends to decrease less for each small increment of displacement than the elastic stress, which means that the elastic stress on the block cannot exceed the frictional resistance, and it is maintained by the constant displacement rate imposed at the load point for the system.

If, on the other hand, the force applied by the spring (dashed line, Figure 16.12*C*) decreases more rapidly with increasing displacement than the frictional resistance (solid line, Figure 16.12*C*), then the sliding of the block is stable because the spring can never exert an excess force on the block to accelerate it. With any infinitesimal increment of sliding, the spring force decreases below that necessary to maintain continued sliding, and only the continually imposed displacement rate v keeps the spring stretched sufficiently to just counteract the frictional resistance.

The potential for stick-slip behavior, and presumably for earthquakes, thus depends not only on the characteristics of the fault surface, represented in the model by the frictional properties of the bottom surface of the slider block, but also on the stiffness of the rock bodies on either side of the fault, represented by the spring with its stiffness k. A detailed understanding of the mechanics of cataclastic flow and friction may hold the key to understanding the mechanical behavior of faults and the earthquakes that occur along them. We discuss more details of the mechanics of this system in Box 16-2.

BOX 16-2 The Rate- and State-Dependent Friction Law

The mechanics of earthquakes is predominantly a function of the frictional behavior of pre-existing faults and fractures in a body of rock. Although brittle fractures do form and propagate, that behavior is of secondary importance in determining the behavior of the system. The behavior of a frictional system is governed by second-order frictional effects that are superimposed on the dominant frictional strength of the system. These effects are related to the fact that any surface has a degree of roughness to it, so that two surfaces in contact are in fact supported by contacts at the high points along both surfaces, the so-called asperities (see Figure 17.2). The actual area of contact that supports the surfaces is thus considerably less than the nominal area of each surface.

During experiments that investigate the sliding between two surfaces in frictional contact, when the sliding velocity undergoes a sudden increase, two effects are observed. First, there is an instantaneous increase in the friction proportional to the logarithm of the velocity; and second there is a gradual logarithmic decay in the friction to a new steady-state value. The steady-state value is reached as the surfaces slip over a characteristic distance L_c. This relation is expressed by the **rate- and state-dependent friction law**:

$$\sigma_s^* = \mu\sigma_n^{(Eff)} = \left[\mu_0 + a\ln\left(\frac{V}{V_0}\right) + b\ln\left(\frac{V_0\theta}{L_c}\right)\right]\sigma_n^{(Eff)} \quad (16\text{-}2.1)$$

where σ_s^* is the shear stress required to maintain sliding, and the coefficient of friction μ is:

$$\mu = \mu_0 + a\ln\left(\frac{V}{V_0}\right) + b\ln\left(\frac{V_0\theta}{L_c}\right) \quad (16\text{-}2.2)$$

The terms involving a and b describe the second-order effects mentioned at the beginning of this box. V is the slip velocity, which provides the rate-dependence part of the law; V_0 is an arbitrary reference velocity (for example, if there is a sudden jump in velocity, V_0 can be taken to be the initial velocity and V the final velocity); μ_0 is the steady-state coefficient of friction at $V = V_0$ (see Equations (16-2.4) through (16-2.6)); and L_c is the characteristic slip distance. θ is a "state variable" that provides the state-dependence part of the law, which we discuss further in the following paragraphs (see Equation (16-2.3)). It has units of time, and it defines a characteristic of the state of the sliding surface on which the magnitude of the frictional resistance depends.

This is an empirical equation that describes the experimental observations, and to that extent, it is like any of the other constitutive equations we have discussed. In particular, it is not based on a specific model of the physical mechanisms that give rise to the behavior. These mechanisms, of course, are always a matter of interest in interpreting an empirical constitutive equation. For example, we discuss mechanistic models for some of the constitutive equations in Box 17-1. We examine more closely the characteristics of the state variable θ and then describe the characteristics of the frictional behavior that are embodied in this equation.

The variation of θ with time is given by the empirical relation

$$\frac{d\theta}{dt} = 1 - \frac{\theta V}{L_c} \quad (16\text{-}2.3)$$

Equation (16-2.3) shows that when $\theta = L_c/V$, then $d\theta/dt$ is zero, and therefore the state variable is constant—the condition of steady state. When $\theta > L_c/V$, then $d\theta/dt$ is negative, so the value of θ decreases with time; conversely, when $\theta < L_c/V$, then $d\theta/dt$ is positive, so the value of θ increases with time. The closer θ is to L_c/V, the smaller $d\theta/dt$ is, and thus the more slowly θ changes. Equation (16-2.3) therefore says that θ always asymptotically approaches the value L_c/V and therefore always evolves toward a constant value as long as the velocity V remains constant. The state variable θ therefore defines the way the friction evolves as the surfaces slip over the characteristic distance L_c. When there is a sudden change in velocity, θ defines how long the system takes to re-establish a new steady-state value of friction.

Steady-state, indicated by the superscript (ss), is defined by the condition that θ does not change with time. From Equation (16-2.3):

$$\text{At steady state: } \frac{d\theta}{dt} = 0 \quad \text{implying: } \theta^{(ss)} = \frac{L_c}{V^{(ss)}} \quad (16\text{-}2.4)$$

$$\text{Thus, when } V^{(ss)} = V_0 \quad (16\text{-}2.5)$$

$$\text{then from Equation (16-2.2) } \mu^{(ss)} = \mu_0 \quad (16\text{-}2.6)$$

(continued)

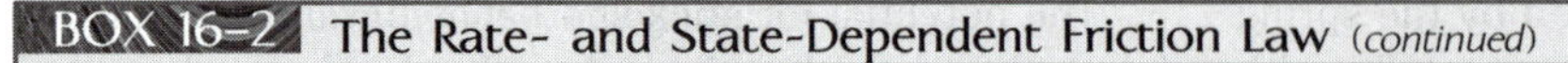

BOX 16-2 The Rate- and State-Dependent Friction Law *(continued)*

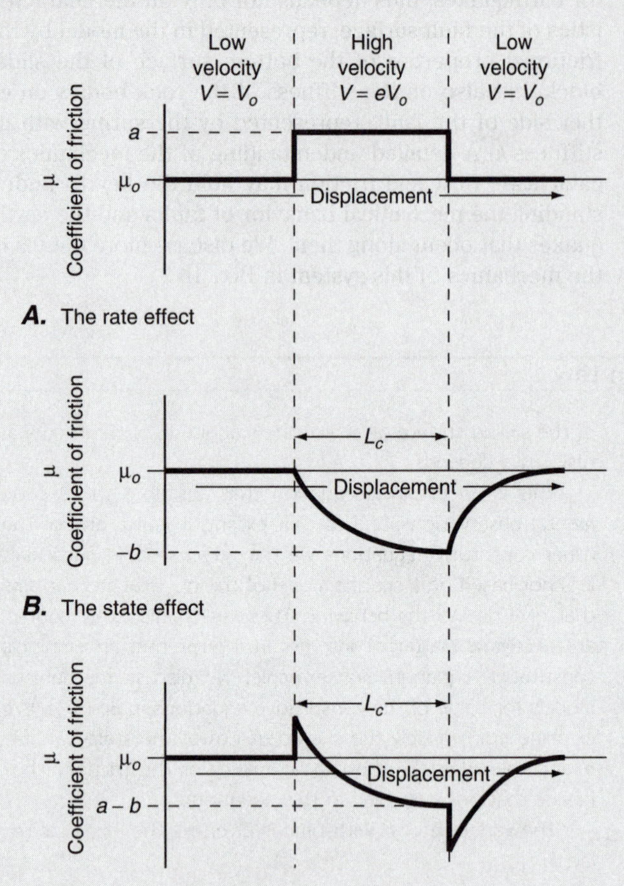

FIGURE 16-2.1 Variation of the coefficient of friction μ with slip as described by the rate- and state-dependent friction law (Equations (16-2.2) and (16-2.3)) for a velocity jump from an initial steady-state velocity V_0 to a higher steady-state velocity $V = eV_0$ (e is the base of the natural logarithm 2.718). After steady state is established at the higher velocity, the velocity jumps back to V_0, which results in an evolution of the coefficient of friction that mirrors the initial jump up. (After Scholz 1990, figure 2.20) *A.* The direct rate effect. When velocity jumps up by a factor of e, the coefficient of friction shows an instantaneous jump of magnitude a. *B.* The evolving state effect. After the instantaneous jump in velocity, the coefficient of friction declines gradually by an amount b, asymptotically approaching its new steady-state value. L_c is the characteristic distance for the system. *C.* The combined effect. The net change for the e-fold instantaneous increase in velocity is a decline in the coefficient of friction by an amount $(a–b)$.

Equation (16-2.4) shows that at a steady-state slip velocity $V^{(ss)}$, the state variable $\theta^{(ss)}$ is the time it takes for the system to slip the characteristic distance L_c. That distance is interpreted to be the slip distance over which the population of asperity contacts on the surface completely changes, so $\theta^{(ss)}$ is just an average contact lifetime for asperities at steady-state slip. In laboratory experiments, L_c is very small, of the order of 10 μm. When applied to large faults, however, it may be considerably larger, but less than 10 meters, which is the size of the smallest observed earthquakes.

Figure 16-2.1 illustrates the behavior of friction, as described by the rate- and state-dependent friction law (Equations (16-2.2) and (16-2.3)), when the sliding velocity suddenly jumps from $V = V_0$ to $V = eV_0$ ($e = 2.718$, the base of the natural logarithm), and then after a steady state is reached, when the velocity suddenly jumps back to $V = V_0$. With the instantaneous increase in velocity, there is an instantaneous increase in the friction of magnitude a (Figure 16-2.1*A*), as defined by the second term on the right in Equation (16-2.1) (the rate effect). There follows a time-dependent decrease in the friction (the state effect), associated with the asymptotic approach of θ toward its new steady-state value of L_c/V (Equation (16-2.4). At this point, the magnitude of the decrease in friction is $-b$, as defined by the last term in Equation (16-2.2) (Figure 16-2.1*B*). The net effect on the friction is shown by (Figure 16-2.1*C*), and the difference in steady-state friction at the two velocities is determined by the combined parameter $(a–b)$:

$$\Delta\mu^{(ss)} = (a-b)\ln\left(\frac{V^{(ss)}}{V_0}\right) \quad or \quad \Delta\mu^{(ss)} = (a-b) \qquad for\ V^{(ss)} = eV_0 \qquad (16\text{-}2.7)$$

where $\Delta\mu^{(ss)} = [\mu^{(ss)}]_{V=V^{(ss)}} - [\mu^{(ss)}]_{V=V^{(ss)}=V_0}$, and these two terms on the right are evaluated from Equation (16-2.2) using, respectively, $V = V^{(ss)}$ and $V = V^{(ss)} = V_0$, and using Equation (16-2.4) to evaluate $\theta = \theta^{(ss)}$.

Whether the sliding becomes unstable or not depends on the sign of $(a–b)$, the magnitude of the velocity jump, the temperature, and the effective normal stress. For the spring-and-slider-block model illustrated in Figure 16.12*A*, the difference in shear stress required to maintain the jump in sliding velocity is determined by the two levels of steady-state friction, and from Equation (16-2.1), it is given by

$$\Delta\sigma_s = \Delta\mu^{(ss)}\,\sigma_n^{(Eff)} = (a-b)\ln\left(\frac{V^{(ss)}}{V_0}\right)\sigma_n^{(Eff)} \qquad (16\text{-}2.8)$$

If $(a–b) > 0$, then the interface becomes stronger with increasing velocity. This behavior is inherently stable because it never results in an excess force that can accelerate the system. Additional force constantly must be applied through the spring to compensate for the relaxation of the spring during slip of the block and to keep the system sliding.

If, on the other hand, $(a–b) < 0$, the interface becomes weaker as the velocity increases. In this case, the size of the velocity jump and the magnitude of the effective normal stress determine whether the response will be stable or unstable (Figure 16-2.2). The boundary between stability and instability in the spring-and-slider-block system (Figure 16.12*A*) occurs when the slope of the friction evolution line equals the slope defining the response of the spring as determined by the stiffness (the spring constant k). This equality occurs when the jump in

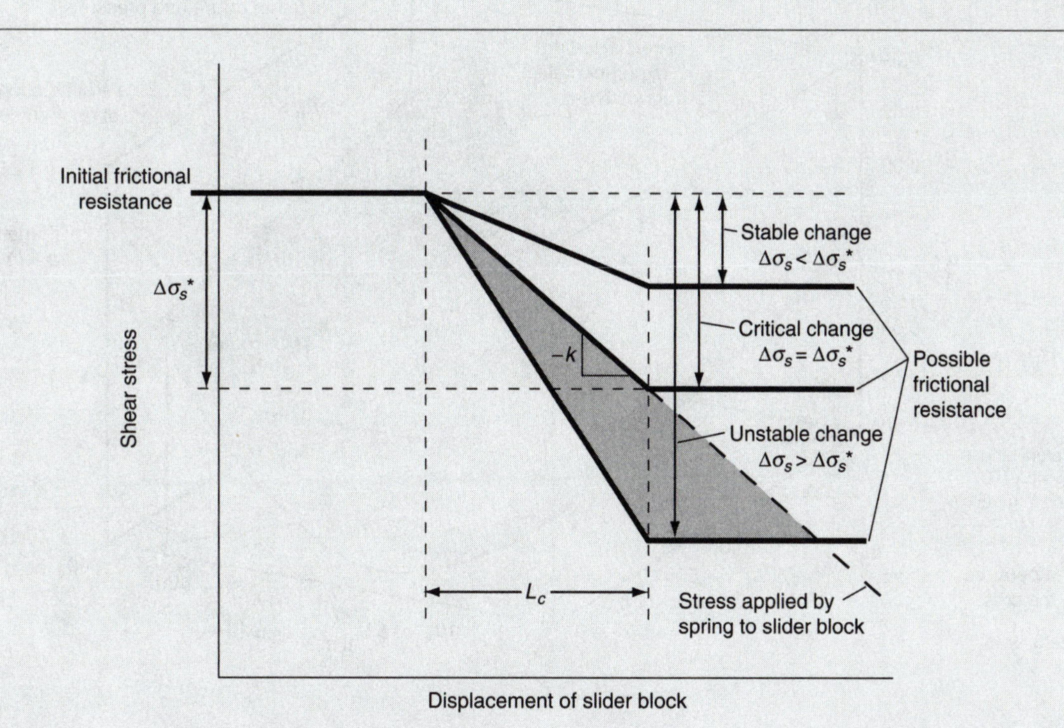

FIGURE 16-2.2 For a given spring stiffness k and critical slip distance L_c, a jump in velocity can be stable, critical, or unstable depending on the magnitude of the velocity increase and on the value of the effective normal stress, where the critical condition is defined by Equation (16-2.10).

shear stress between the two steady-state frictions is a critical magnitude, which, as illustrated in (Figure 16-2.2), is defined by the slope $-k$, to be:

$$\Delta\sigma_s^* = \Delta\mu^{(ss)}\sigma_n^{(Eff)} = -kL_c \qquad (16\text{-}2.9)$$

where the superscript asterisk indicates the critical value, and k is the stiffness defined by the slope of the stress versus displacement curve for the spring. Using the second Equation (16-2.8) in Equation (16-2.9), we find

$$\ln\left(\frac{V^{(ss)}}{V_0}\right)\sigma_n^{(Eff)} = \frac{-kL_c}{(a-b)} \qquad (16\text{-}2.10)$$

From a state of stable sliding at velocity V_0, the critical condition given in Equation (16-2.10) can be reached by a sudden increase in the velocity to a critical velocity $V^{(ss)} = V^*$ or by a sudden increase in the effective normal stress acting on the slider block, to a critical value $\sigma_n^{(Eff)} = \sigma_n^{(Eff)*}$. These critical values for V^* and for $\sigma_n^{(Eff)*}$ are given, respectively, by

$$\ln\left(\frac{V^*}{V_0}\right) = \frac{-kL_c}{(a-b)\sigma_n^{(Eff)}} \quad \text{for } \sigma_n^{(Eff)} = \text{constant} \qquad (16\text{-}2.11)$$

$$\sigma_n^{(Eff)*} = \frac{-kL_c}{(a-b)\ln(V^{(ss)}/V_0)} \quad \text{for } V^{(ss)}/V_0 = \text{constant} \qquad (16\text{-}2.12)$$

If, for a given effective normal stress on the slider block, the jump in velocity is smaller than the critical value given by Equation (16-2.11), then over the characteristic sliding distance L_c, decrease in stress from frictional resistance is smaller than the decrease from the spring, so the frictional resistance remains higher than the stress applied by the spring (*'Stable change'*, Figure 16-2.2). To maintain the sliding velocity, the stress applied through the spring must be constantly increased to compensate for the decrease in the spring stress associated with the relaxation of the spring as the block slides, and the sliding is stable. If the velocity jump were larger than the critical value, however, then over the characteristic sliding distance L_c, the decrease in stress from the frictional resistance of the system is larger than the decrease from the spring, so the frictional resistance remains lower than the stress applied by the spring (*'Unstable change'*, Figure 16-2.2). The spring would therefore exert an excess force on the system (shaded area in Figure 16-2.2), which would accelerate the sliding, and the system would be unstable.

This behavior can be seen in the experimental data plotted in Figure 16.10A, where the slope of the experimental displacement versus time curves show the velocities of slip. From the lowest curve on the graph, a small increase in shear stress leaves the system in the stable sliding mode, whereas a sufficiently large increase in shear stress results in a catastrophic unstable slip event. Thus, for a given velocity jump, the change in shear stress would be subcritical and the system stable if the normal stress were below the critical value given Equation (16-2.12). The same velocity jump, however, would be unstable if the effective normal stress exceeded that critical value. This is consistent with the plot in Figure 16.11, which shows a

(continued)

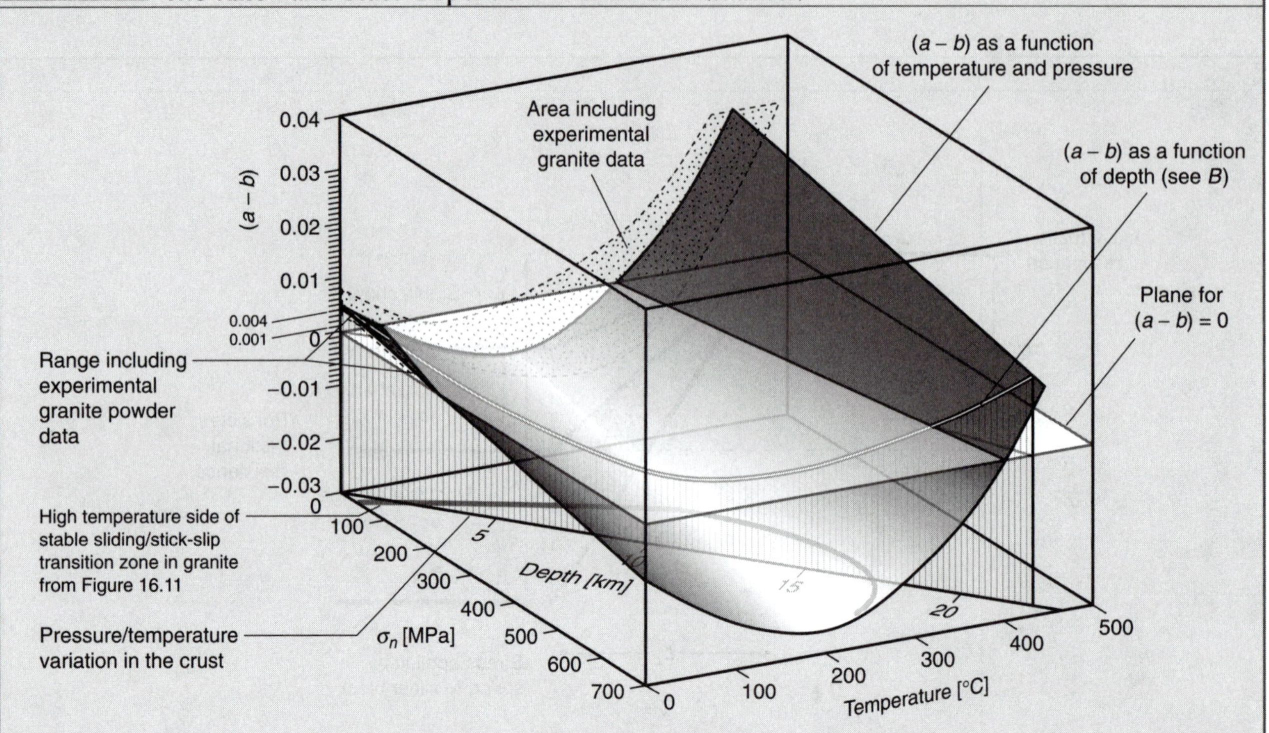

A.

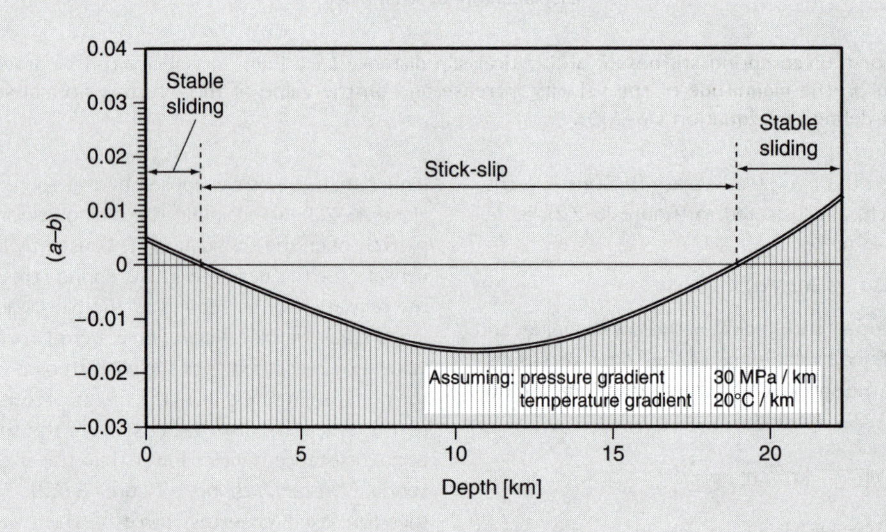

B.

FIGURE 16-2.3 Dependence of the combined parameter (*a–b*) on temperature, pressure, and depth in the crust. *A.* The back plane of the block diagram shows the dependence of (*a–b*) on temperature for granite. The outlined area shaded with the dot pattern shows the approximate variation of the experimental measurements, and the solid line through the middle is the best fit to the data. The variation of (*a–b*) with pressure is shown on the left side of the block diagram. Experimental measurements outlined with the dotted line were made over a range of normal stress from about 50 to 175 MPa, giving a range of (*a–b*) of roughly 0.004 to 0.001, and they tightly constrain the best-fit line. The pressure curve is used to project the temperature dependence along the normal stress axis to generate the concave upward curved surface. The line on the bottom plane of the block diagram shows the variation of pressure and temperature with depth in the Earth's crust, assuming a pressure increase of 30 MPa/km and a temperature increase of 20° C/km. The intersection of a vertical plane based on this line (shaded with vertical lines) with the curved surface showing the projected variation of (*a–b*) defines a curve that shows the variation of (*a–b*) with depth in the crust. The high-temperature side of the stable sliding/stick-slip transition in granite from Figure 16.11 is plotted on the bottom plane of the block. The prediction of the extent of the seismogenic zone from Figure 16.11 is consistent with that from this figure. (Experimental data from Scholz 1998) *B.* Dependence of (*a–b*) on depth in the crust showing the plane shaded with vertical lines from *A.* The value of (*a–b*) is positive in the very shallow crust above about 2 kilometers and in the deep crust below about 19 kilometers, which therefore should be regions of stable sliding. The value of (*a–b*) is negative in the depth range between about 2 and 19 kilometers, which therefore should be the region of stick-slip behavior in which earthquakes occur, the so-called seismogenic zone.

BOX 16-2 *(continued)*

transition at low temperatures from stable sliding to stick-slip with increasing pressure.

An analysis of the dynamics of the system, which goes beyond what we can present here, shows that at the critical state, the system can undergo oscillations in the sliding velocity. On the unstable side of the boundary, the amplitude of the oscillations progressively increases to an unstable slip, and on the stable side, the amplitude progressively decreases to zero.

In the static case, there is no distinction between V and V_0, so $V/V_0 = 1$, and the second term in Equation (16-2.2) is zero. In the second term of Equation (16-2.3), V/L_c in the dynamic system is equivalent to $1/t$ in the static system (the units of the two quantities are identical), and as t becomes very large, $d\theta/dt$ approaches 1. Thus θ becomes equivalent to the time t, which means that in the static case, the average contact time of asperities is just time itself. The argument of the third term in Equation (16-2.2) then approaches t/t_0, where V_0/L_c for the dynamic case is equivalent to an inverse reference time $1/t_0$ (the units are identical) in the static case. Thus for the static case, Equation (16-2.2) becomes

$$\mu = \mu_0 + b \ln\left(\frac{t}{t_0}\right) \qquad (16\text{-}2.13)$$

which shows that the coefficient of friction increases with the logarithm of the contact time, consistent with the observed healing behavior of a static fault.

The test for the rate- and state-dependent friction model is whether it correctly describes the behavior of brittle faults in the Earth. Figure 16-2.3 presents plots of the variation of the combined parameter $(a–b)$ for granite as a function of temperature (Figure 16-2.3*A*, back plane of the block) and for granite powder as a function of normal stress (Figure 16-2.3*A*, left plane of the block). We can use the normal stress data to suggest how the temperature-dependent curve might project along the normal stress axis (Figure 16-2.3*A*, curved surface). The pressure increases with depth in the crust generally at about 30 MPa/km, and the steady-state temperature increases at about 20° C/km. If we plot this pressure/temperature gradient for the crust on the bottom plane of the block in Figure 16-2.3*A*, then a vertical plane based on this line (Figure 16-2.3*A*, vertically lined plane) intersects the curved surface of $(a–b)$ values along a curve that represents the variation of $(a–b)$ with depth. This plane and the $(a–b)$ curve within it are shown in Figure 16-2.3*B*. With increasing depth $(a–b)$ changes from positive above about 2 kilometers to negative between about 2 and 19 kilometers, and back to positive below 19 kilometers. This variation implies a shift with increasing depth from stable sliding at shallowest depths to unstable sliding (stick-slip) in the middle depth range and back to stable sliding below about 19 kilometers. The realm of unstable sliding thus should define the depth range of the seismogenic zone within which earthquakes can occur. Earthquakes on strike-slip faults, for example, generally occur between a few kilometers and 15 to 20 kilometers deep, so the range of instability projected from the experimental data for granite is generally consistent with the observed distribution. These results are also consistent with the experimentally-determined transition from stable sliding to stick-slip as a function of pressure and temperature as shown in Figure 16.11. In particular, the high-temperature side of the transition zone, which is shown on the bottom plane of the block in Figure 16-2.3*A*, predicts a very similar range of stick-slip behavior in the crust to the curve in Figure 16-2.3*B*.

The rate- and state-dependent friction model embodies a range of complex behaviors and thus is a rich source for understanding a variety of characteristics of earthquakes. It is an active area of research, but further details are beyond the scope of this presentation.

16.3 EXPERIMENTAL INVESTIGATION OF DUCTILE FLOW

Experimental investigations of ductile behavior of rocks are generally conducted using an apparatus that deforms a sample of rock under high pressures and temperatures. The sample is held inside an electrical-resistance furnace within a pressure vessel. Pressure is applied to the sample by a gas, liquid, or ductile solid. The temperature is maintained by controlling the power to the furnace. The sample is deformed by squeezing it between opposing pistons. The samples are generally small cylinders of rock ranging from several millimeters to a few centimeters long, with a diameter between one-third and one-fourth of the length. They are subjected to confined compression (Figure 7.12*D*) at confining pressures ranging from a few hundred to a few thousand megapascals (1 MPa = 10 bars; 100 MPa = 1 kilobar) and at temperatures ranging between room temperature and the melting temperature of the material.

The experiments generally involve the slow continuous deformation, or **creep**, of the specimen under varied conditions. For technical and practical reasons, the lower bound for experimental strain rates is roughly 10^{-7} s^{-1}, although some techniques can push that strain rate down another order of magnitude or so. At this rate a specimen 1 centimeter long shortens by 10^{-7} centimeter each second, or less than 1 percent per day. Shortening the sample by only 10 percent from 10 millimeters to 9 millimeters, for example, takes approximately 11.6 days. Many experiments are performed at strain rates that are two or three orders of magnitude higher than that, at 10^{-5} to 10^{-4} s^{-1}.

Most estimates of geologic strain rates are on the order of 10^{-12} to 10^{-16} s^{-1}, which are many orders of magnitude lower than strain rates commonly used in the laboratory. The contrast between the strain rates achievable in the laboratory and natural strain rates raises concerns about the applicability of the experimentally determined constitutive equations to the much lower natural strain rates, because the range of applicability of a constitutive equation determined at laboratory strain rates may not extend down over many orders of magnitude to geologic strain rates. Different mechanisms may become predominant at the very low strain rates (see Figure 17.1).

Experiments are performed at either constant stress (creep experiments) or constant strain rate. The so-called constant-stress experiments usually are done at constant load, and the axial stress actually decreases slightly during the experiment because the cross-sectional area of the sample increases as the sample shortens. Similarly, the constant-strain-rate experiments usually are done at the constant displacement rate of the deforming piston, which results in an increase of the instantaneous strain rate during the experiment because the reference length of the sample is the instantaneous length, which decreases with time. More sophisticated experiments use computer control to continually adjust the load or the displacement rate to maintain the intended stress or strain rate at a constant value.

For investigating the ductile flow of rocks, experiments at constant stress are preferable to those at constant strain rate because the flow mechanism depends more directly on stress than on strain rate, and the material constants are more directly determinable from the experimental data (see Box 16-3). Constant-strain-rate experiments are generally simpler to perform, however, and for steady-state deformation, they give similar information (Box 16-3).

The ductile behavior of different materials is best compared at an equal **homologous temperature**, defined by the ratio T/T_m, where T is the temperature of the material, and T_m is its melting temperature, both expressed as an absolute temperature.[1] The melting temperature is a rough measure of the strength of the bonds binding a crystalline material together. Thus the behavior of many different materials tends to be the same at the same homologous temperatures, even if their absolute melting temperatures are very different. For example, at the high homologous temperature of 0.95, the mechanical behaviors of ice and olivine have similar characteristics, even though that temperature for ice is 259 K (−14° C), whereas for olivine it is approximately 2017 K (1744° C). The mechanical behavior of a material at different pressures is also the same at the same homologous temperature if the pressure effect on the melting temperature is included in the calculation of the homologous temperature. Thus the use of the homologous temperature has proven to be very useful in comparing the behavior of various rocks and minerals with one another.

The results of a constant-stress experiment are often plotted on a graph of strain versus time (Figure 16.13*A*,*B*). Deformation at homologous temperatures below about 0.5 is called **cold working**. Creep curves generally show a continuously decreasing creep rate (Figure 16.13*A*) and the creep is called **logarithmic creep** because the total strain increases with the logarithm of time.

High-temperature creep at homologous temperatures above about 0.5 is called **hot working**. A typical creep curve for a constant-stress experiment has several characteristic parts to it (Figure 16.13*B*). The first part is an essentially instantaneous recoverable elastic strain that accompanies the initial, ideally instantaneous, loading of the material. If the applied stress is above the yield stress, then the material begins creeping immediately upon application of the load. The creep rate initially is relatively high, but it steadily declines as the experiment proceeds. This phase of the curve is called **primary creep**, and the phenomenon of decreasing creep rate at constant stress is called **work hardening** or **strain hardening**. In this phase of the deformation, the material becomes less ductile with increasing strain.

Eventually the creep rate stabilizes at a constant value, called **steady state** or **secondary creep**. Steady-state creep is the part of the experiment that most interests geologists, because at constant stress it could continue indefinitely. Thus it presumably represents the long-term deformation processes that occur within the Earth.

In many experiments, the steady-state regime gives way to **tertiary creep**, during which the strain rate accelerates and ultimately the sample fractures. Tertiary creep is most common during high-stress, low-temperature experiments, and/or extensional stress experiments. In high-temperature experiments, it most likely results from the change in stress or strain rate associated with the changing geometry of the sample during creep. Thus it probably is an artifact of the method of experimentation rather than a fundamental characteristic of the deformation process.

For constant-strain-rate experiments the curve showing the differential stress versus time can be divided into comparable portions (Figure 16.13*C*, *D*). In this case, the buildup of elastic strain is not instantaneous because of the constant strain rate. Thus the stress builds up in proportion to the accumulating elastic strain. When the stress reaches the yield point, ductile deformation begins.

In experiments at low homologous temperatures (Figure 16.13*C*), the stress continues to increase indefinitely, at least to the strength limits of the sample (or the

[1]Absolute temperature is measured in units of kelvins and indicated by the symbol K; an increment of 1K is the same as 1°C (Celsius or Centigrade), but the freezing point of water is 0°C = 273 K.

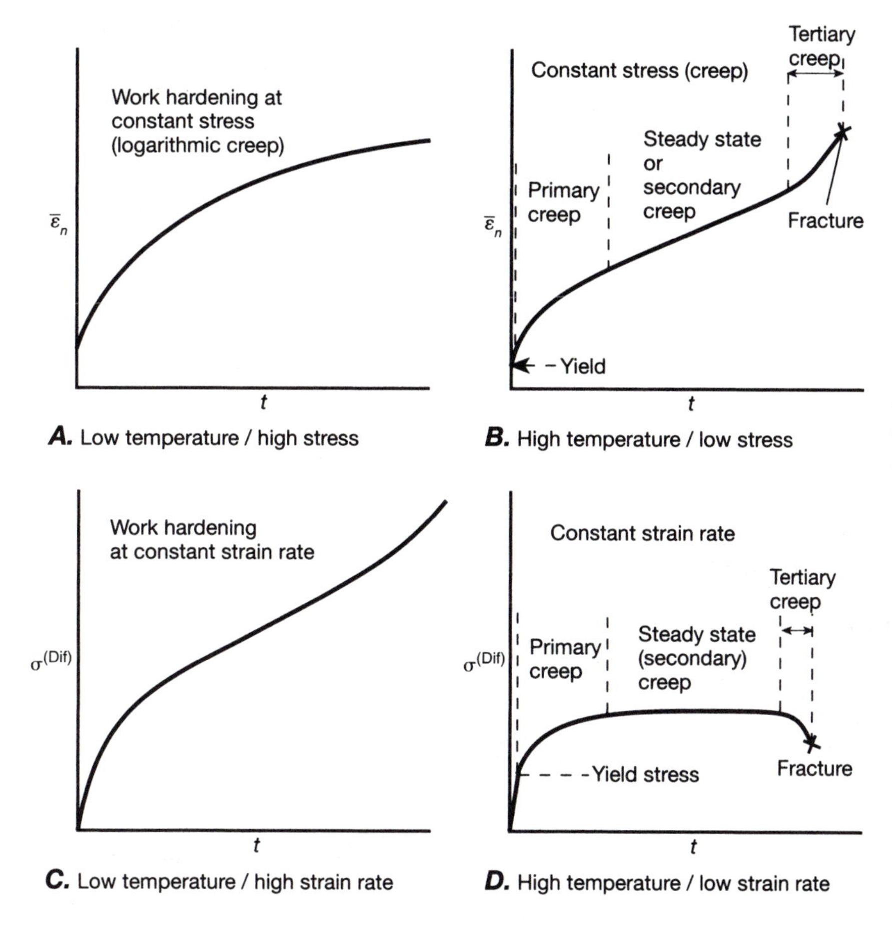

FIGURE 16.13 Characteristic creep curves for ductile deformation of a polycrystalline material. *A.* A curve showing logarithmic creep at constant stress, which is characteristic of work hardening during a low-temperature experiment. *B.* A creep curve for a constant-stress, high-temperature experiment showing the three different stages of creep. *C.* A creep curve for a constant-strain-rate, low-temperature experiment showing work-hardening behavior. *D.* A creep curve for a constant-strain-rate, high-temperature experiment showing the three different stages of creep.

apparatus!). In these cases, work hardening is the predominant mode of deformation.

In experiments at high homologous temperatures (Figure 16.13*D*), primary creep commences at the yield point, and the rate of increase of the stress gradually declines to zero. Where the stress becomes constant, the creep rate has reached the steady-state value. Steady-state creep continues for an indefinite period until the onset of tertiary creep, accompanied by a decline in the stress.

In practice, the yield point may be difficult to identify exactly because there is a gradational transition from purely elastic to predominantly ductile strain, and because in creep experiments, the apparatus cannot increase the stress rapidly enough to approximate an instantaneous loading. The shape of the curves may also be affected by the practice of maintaining constant load or constant displacement rate rather than constant stress or strain rate.

16.4 STEADY-STATE CREEP

Below depths of roughly 15 to 20 kilometers in the Earth, the homologous temperature generally exceeds 0.5. Deformation of rocks under these conditions is characterized by creep that can reach a steady state, and therefore large amounts of deformation can accumulate in the rocks, producing the complex folds, lineations, and foliations that we observe in nature. Thus we assume that the steady-state constitutive equations, or flow laws, that we determine in the laboratory can be used to model the high-temperature ductile deformation in the Earth and the large strains that result.

Many experiments have been performed to investigate the dependence of steady-state creep rate on differential stress ($\sigma^{(Dif)} \equiv \hat{\sigma}_1 - \hat{\sigma}_3$; Equation (7.38)) and temperature, as well as other parameters such as pressure, grain size, and chemical environment (see Section 16.5). For the purposes of discussion, it is convenient to divide the results into low-, moderate-, and high-stress deformation regimes. The boundaries vary for different rocks, but for olivine, for example, which is the most abundant mineral in the upper mantle, the moderate regime for the differential stress is roughly between 20 and 200 MPa, and it is of the same order of magnitude for most other minerals and rocks.

We discuss the moderate-stress regime first, because this regime is the most intensively investigated and is commonly presumed to be the most applicable to deformation in the Earth. The high-stress regime is probably not widely applicable to deformation in the Earth, although it may occur in some areas in association with faulting. The low-stress regime also may be very important for understanding deformation in the Earth. Because at these low stresses deformation proceeds very slowly, however, it is difficult to investigate experimentally and therefore has been ignored by many researchers.

i. The Moderate-Stress Regime: Power-Law Creep

The constitutive equation that accounts for most moderate-stress steady-state deformation observed in the laboratory is the power-law equation, so-called because the absolute value of the steady-state instantaneous strain rate[2] $|\dot{\varepsilon}_n^{(ss)}|$ (see Equations (16-1.3) to (16-1.5)) is related to the differential stress[3] $\sigma^{(Dif)}$ (Equation (7.38)) raised to a power n. The equation is written to give either the strain rate as a function of the stress, for constant-stress experiments, or the stress as a function of the strain rate, for constant-strain-rate experiments:[4]

$$|\dot{\varepsilon}_n^{(ss)}| = A_1(\sigma^{(Dif)})^n \exp\left[\frac{-E^*}{RT}\right]$$
$$\sigma^{(Dif)} = K_1|\dot{\varepsilon}_n^{(ss)}|^{1/n} \exp\left[\frac{E^*}{nRT}\right] \tag{16.7}$$

where A_1 and K_1 are constants, n is the stress exponent, E^* is the activation energy per mole for the creep process, and R is the Boltzman constant per mole (also called the "gas constant"). The constants A_1 and K_1 are related by

$$K_1 = \left[\frac{1}{A_1}\right]^{1/n} \tag{16.8}$$

The constants A_1, E^*, and n (alternatively, K_1, E^*, and n) have characteristic values for any particular material. The experimental techniques that allow these different constants to be determined are discussed in Box 16-3. It is clear from the first Equation (16.7) that both the temperature and the stress have a large effect on the strain rate.

Creep is a thermally activated process, which means there is an energy barrier that inhibits the creep mechanism. The dependence of the rate of a thermally activated process on an exponential term such as in Equations (16.7) is a characteristic of such processes, and the activation energy E^* is a measure of the energy barrier. At low temperatures, a high stress is required to overcome the energy barrier to produce a particular strain rate. At sufficiently high temperatures, however, random thermal fluctuations can provide the energy needed to surmount the barrier, so the strain rate can be maintained by a lower stress. Thus, in general, an increase in temperature increases the strain rate for a constant stress (the first Equation (16.7)) or lowers the stress required to produce a given strain rate (the second Equation (16.7)). This effect is accounted for by the rapid increase, with increasing temperature, of the exponential term in the first Equation (16.7), and by the corresponding rapid decrease of the exponential term in the second Equation (16.7).

Figure 16.14 shows the results of a typical series of constant-strain-rate experiments, in this case on dunite (polycrystalline olivine), at three different temperatures. The stress reaches a nearly constant magnitude with increasing strain, indicating steady-state conditions, which are the only conditions to which Equations (16.7) apply. The steady-state stress decreases markedly with increasing temperature. These experiments may be applicable to the behavior of the upper mantle.

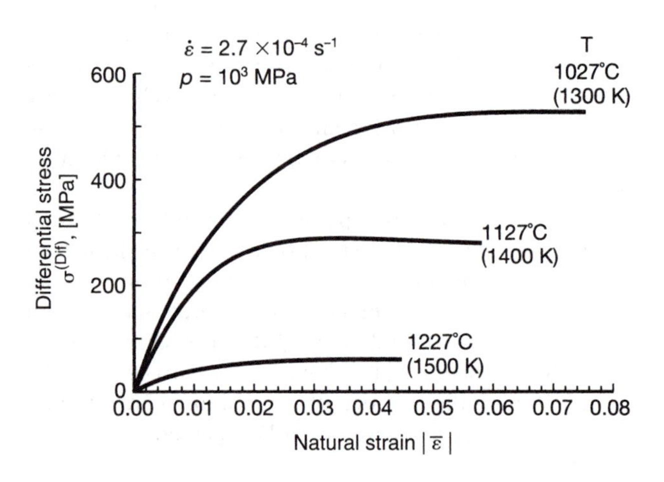

FIGURE 16.14 Effect of temperature on steady-state creep stress for coarse-grained dunite (average grain size $d = 1$ mm). Experiments were performed at constant strain rate, pressure, and temperature for three different temperatures, as labeled. (From Borch and Green 1989)

[2]We use the right superscript (ss) to indicate steady state.

[3]Remember that from its definition in Equation (7.38), the differential stress is the maximum minus the minimum principal stress, so it is always positive regardless of whether stresses are tensile or compressive. Thus we must use the absolute value for the rate of instantaneous strain.

[4]We use the standard notation $\exp[-x] \equiv e^{-x} = 1/e^x$ and $\exp[x] \equiv e^x$, where e is the base of the natural logarithm: $e = 2.718\ldots$. The exponentials in Equations (16.7) are the **Arrhenius factors,** which are characteristic of thermally activated processes. They are related to the theoretical probability that random thermal fluctuations can provide enough energy to surmount the energy barrier for a process.

The yield stress is very difficult to identify. In fact, the initial slopes of the stress-strain curves are all different, indicating that they result from a mixture of elastic strain, primary-creep strain, and the nonrigid behavior of the experimental apparatus. The yield stress of the material, therefore, is probably below the stress at which the curve turns substantially toward horizontal, which ideally would be identified as the yield stress.

Figure 16.15 shows the effect of different strain rates at two different constant temperatures. Decreasing the strain rate by an order of magnitude results in a drop in the stress under these conditions by a factor of slightly more than 0.5. This behavior implies stress exponents of $n = 3$ for the 1300°K experiments (Figure 16.15*A*) and $n = 2.35$ for the 1400°K experiments (see Box 16-3 and Figure 16-3.1). The best fit for a much more extensive set of experiments on olivine is actually about n = 3.5 (compare Figure 16.16). For many polycrystalline materials, including silicates, metals, and oxides, the stress exponents have values roughly between 3 and 5, and some can be as high as 7. Even higher values of the stress exponent are occasionally observed, but these are usually in the high-stress regime, where a different form of stress dependence is more appropriate (see Section 16.4(ii)).

Comparing Figures 16.14 and 16.15 with the stress-strain curve in Figure 16.7*A* indicates that the behavior of a power-law material at a given strain rate, temperature, and pressure approaches that of the elastic-plastic model for a continuum.

A compilation of steady-state creep data for olivine from several different investigators is plotted in Figure 16.16. The strain rate is plotted as a normalized strain rate to eliminate the effect of the different temperatures at which experiments were performed. In the range of moderate stresses, the data are fit reasonably well by a straight line with a slope of $1/n = 0.33$ (see the second Equation (16.7)), which gives $n = 3$.

ii. The High-Stress Regime: The Exponential Creep Law

At high stresses, roughly $\sigma^{(Dif)} > 200$ MPa, the strain rate becomes increasingly sensitive to differential stress as the stress increases. If we attempt to explain the data with the power-law creep model (Equation (16.7)), we find the stress exponent n increases with increasing stress. The data are better fit by a different creep model, called the exponential creep law. It has the form

$$|\dot{\varepsilon}_n^{(ss)}| = A_2 \exp[\beta\sigma^{(Dif)}] \exp\left[\frac{-E^*}{RT}\right] \qquad (16.9)$$

where A_2, β, and E^* are constants whose values are characteristic of particular materials. In this case, the steady-state strain rate depends exponentially on the differential stress, which means that it depends on e raised to the power $(\beta\sigma^{(Dif)})$. This type of creep behavior is evident from the steady-state data in the high-stress regime of Figure 16.16, where the data define a gradual curve, indicating the exponential-law behavior.

Data from the moderate- and high-stress regimes are sometimes jointly fit by a hyperbolic sine creep law

$$|\dot{\varepsilon}_n^{(ss)}| = A_3\,[\sinh(\alpha\sigma^{(Dif)})]^n \exp\left[\frac{-E^*}{RT}\right] \qquad (16.10)$$

which behaves like power-law creep in the moderate-stress regime ($\sigma^{(Dif)} << 1/\alpha$) and like exponential-law creep in the high-stress regime ($\sigma^{(Dif)} >> 1/\alpha$).[5] The solid line fit to the data in Figure 16.16 is an example of the hyperbolic sine law.

iii. The Low-Stress Regime: Power-Law Creep with Low *n*

At differential stresses roughly below 20 MPa, the constitutive equation is similar to Equation (16.7), but in this case the stress exponent n commonly has a value between 1 and 2, which is distinctly lower than the values observed for moderate stresses.

Figure 16.17 shows the steady-state creep behavior of the Solenhofen limestone over a range of more than two orders of magnitude in differential stress and at a variety of temperatures. The low-stress regime occupies most of the graph and shows power-law creep with a stress exponent of about $n = 1.7$. In the moderate-stress regime, the data are consistent with power-law creep with a stress exponent of about $n = 5$. One set of data extends into the high-stress regime and shows exponential-law creep. A similar transition to a low stress exponent in olivine has been observed at very fine grain sizes (see the discussion of grain size dependence in Section 16.5(ii)). The data on the deformation of single crystals of olivine at the lowest stresses in Figure 16.16 suggest that even in single crystals, at the lowest stresses there is a transition to a lower stress exponent.

5

$$\sinh(z) = 0.5(e^z - e^{-z})$$

$$\sinh(z) \approx 0.5e^z \quad \text{for } z >> 1$$

$$\sinh(z) \approx -0.5e^{-z} \approx -0.5(1-z) \quad \text{for } z << 1$$

where for $z << 1$, we obtained the second approximation by using the first two terms of the Maclaurin series expansion for the exponential term. Thus if $z = \alpha\sigma^{(Dif)}$, then raising the $\sinh(\alpha\sigma^{(Dif)})$ term to a power n gives an exponential law for $\alpha\sigma^{(Dif)} >> 1$ and a power law for $\alpha\sigma^{(Dif)} << 1$.

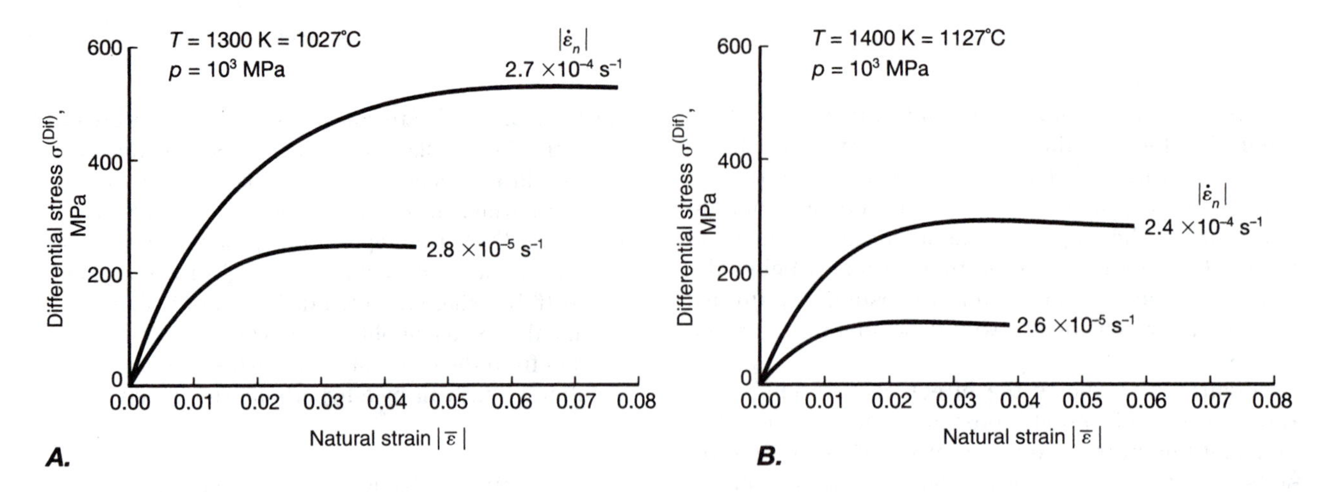

FIGURE 16.15 The effect of different constant strain rates on the steady-state stress for dunite at constant temperature and pressure. (Data from Borch and Green 1989) *A*. Two experiments at $T = 1300$ K, $p = 10^3$ MPa for strain rates of roughly 10^{-4} s^{-1} and 10^{-5} s^{-1}. *B*. Two experiments at $T = 1400$ K, $p = 10^3$ MPa for strain rates of roughly 10^{-4} s^{-1} and 10^{-5} s^{-1}.

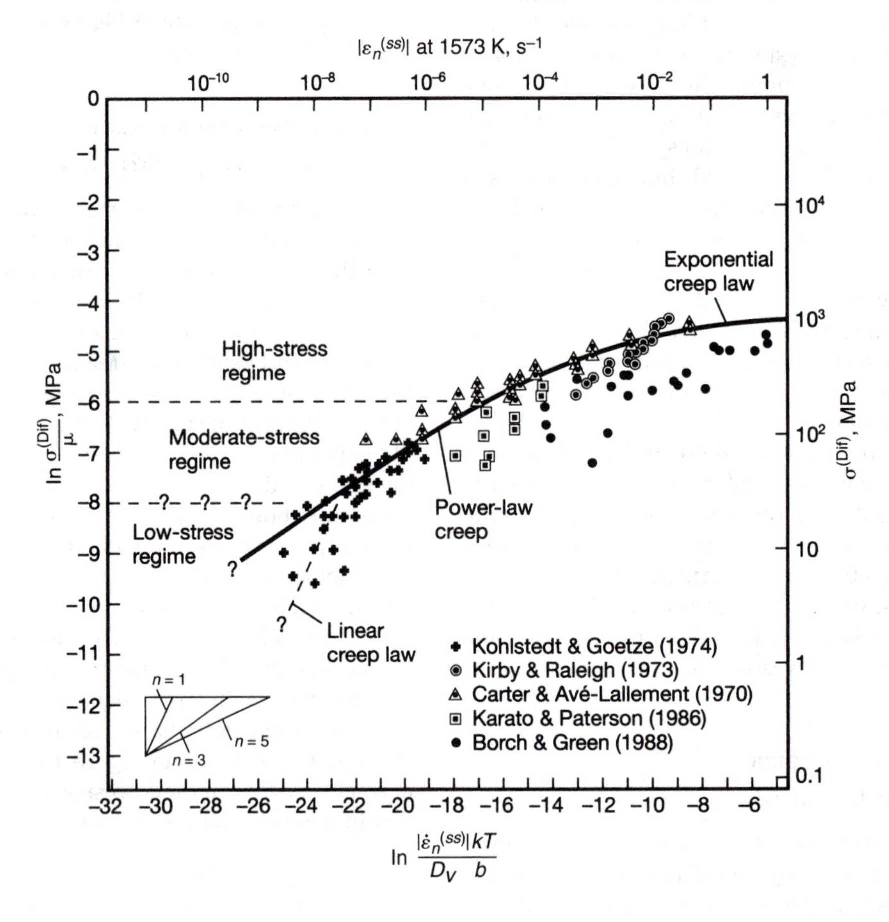

FIGURE 16.16 Normalized creep data for olivine showing regimes of high, moderate, and low stress. The more recent data sets of Karato and Paterson and of Borch and Green indicate slightly lower flow stresses than the older sets, probably reflecting improvements in the experimental technique. The data of Kohlstedt and Goetze are for single crystals. The line that is fit to the older data is a hyperbolic sine creep law. Plotting the nondimensional normalized strain rate $|\dot{\varepsilon}_n^{(ss)}|\ kT/D_v\mu b$, where $D_v = D_0 \exp[-E^*/RT]$, eliminates the effect of the different temperatures of the experiments in the diagram. In this expression, k is the Boltzman constant, D_v is the coefficient of volume self-diffusion, μ is the shear modulus, and b is the magnitude of the Burgers vector, a lattice constant (see Section 17.4(i), Figure 17.10). For olivine, we used $\mu = 7.91 \times 10^4$ MPa; $k = 1.38 \times 10^{-29}$ MPa m^3 K^{-1}; $D_v = (10^{-1} \mathrm{m}^2\ \mathrm{s}^{-1}) \exp[(-5.44 \times 10^5\ \mathrm{J\ mole}^{-1}) / (8.31\ \mathrm{J\ mole}^{-1}\ \mathrm{K}^{-1})\ (T\ \mathrm{K})]$; $b = 6.98 \times 10^{-10}$ m. The creep rates on the top scale are for 1573 K (1300°C), which is not necessarily the temperature at which the data were collected. (Modified after Twiss 1977)

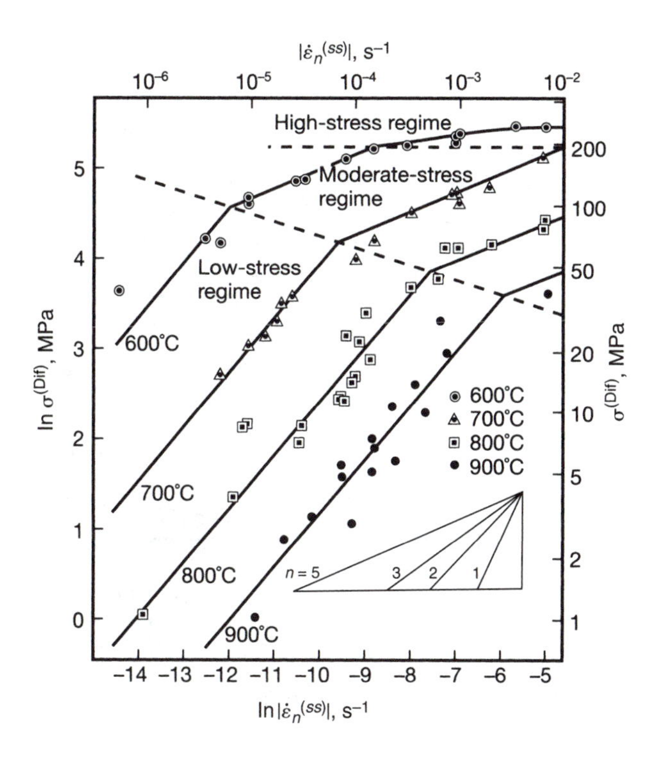

FIGURE 16.17 Creep data for Solenhofen limestone showing three different creep laws in the high-, moderate-, and low-stress regimes. Data are from constant-strain-rate tests, and points represent the strain rate and the stress at 10 percent strain, which is at or close to steady state. Different lines are plotted for each temperature because the strain rate is not normalized for temperature differences as in Figure 16.16. (From Schmid et al. 1977)

These different creep laws reflect the different processes that dominate at the different levels of stress, and they suggest that we should use caution in applying constitutive equations determined at laboratory creep rates to the rates characteristic of geologic processes. It is important, therefore, to understand the mechanisms that give rise to these different creep laws so that we may better evaluate the extrapolation to geologic conditions. We discuss the creep mechanisms in more detail in Chapter 17.

16.5 THE EFFECTS OF PRESSURE, GRAIN SIZE, CHEMICAL ENVIRONMENT, AND PARTIAL MELT ON STEADY-STATE CREEP

i. The Effect of Pressure

The effect of pressure on steady-state creep is relatively small at crustal depths within the Earth, so it is commonly ignored. Within the mantle, however, where pressures are much higher, the effect becomes important. An increase in pressure decreases the steady-state strain rate at constant stress or increases the steady-state stress at constant strain rate.

Figure 16.18 illustrates the pressure effect with two curves for steady-state creep of olivine at constant strain rate and temperature and at two different pressures. The pressure effect is commonly accounted for by modifying the first Equation (16.7) to read

$$|\dot{\varepsilon}_n^{(ss)}| = A_4\,(\sigma^{(Dif)})^n \exp\left[\frac{-(E^* + pV^*)}{RT}\right] \qquad (16.11)$$

where p is the pressure, V^* is the activation volume per mole, and $E^* + pV^* = H^*$ is the activation enthalpy per mole. V^* is generally interpreted as the volume of crystal affected by the activation process. As the pressure increases, the value of the exponential term decreases, and accordingly so does the strain rate at constant stress. For olivine, for example, the activation volume is approximately 2.7×10^{-5} m^3/mole (see Box 16-3, Figure 16-3.3). At a depth of 30 km, the pressure is roughly 10^9 N m^{-2} ($= 10^3$ MPa $= 1$ GPa),[6] so the pV^* term (2.7×10^4 N m mole^{-1} $= 2.7 \times 10^4$ J mole^{-1}) is less than 5% of the activation energy $E^* = 5.4 \times 10^5$ J mole^{-1}. For a temperature of 1000 K, a constant strain rate, and a stress exponent of $n = 3.5$, the differential stress at a depth of 30 kilometers would be only about 2.5 times that at the surface. For the same temperature and depth, for a constant differential stress, the strain rate would be about 0.04 times that at the surface. Such differences are within

[6] 1 megapascal (1 MPa) = 10^6 Newtons per square meter (N m^{-2}); 1 gigapascal (1 GPa) = 10^9 N m^{-2}; 1 N m = 1 Joule (1 J).

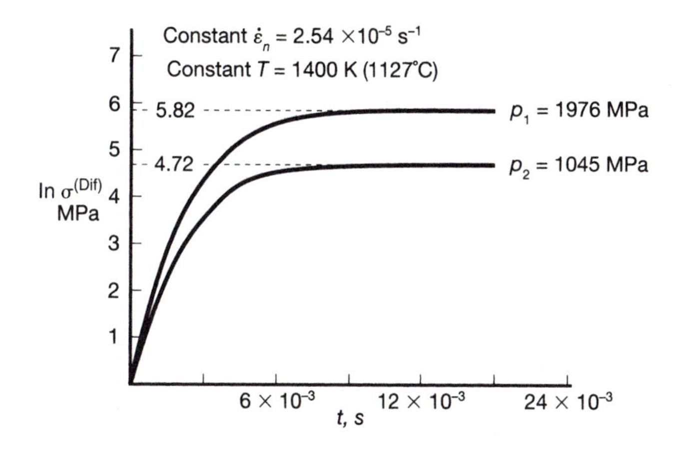

FIGURE 16.18 Effect of pressure on the steady-state creep stress for dunite. Experiments were run at constant strain rate, temperature, and pressure for two different pressures. (From Borch and Green 1987)

the variability of experimental data (see Figure 16.16, for example).

Experiments have shown that the dependence of the creep rate of polycrystalline materials on temperature and pressure is predictable if the temperature is normalized by the melting temperature for the mineral in the rock that controls the deformation rate:

$$|\dot{\varepsilon}_n^{(ss)}| = A_5\,(\sigma^{(Dif)})^n \exp\left[\frac{-gT_m}{T}\right] \tag{16.12}$$

where g is a dimensionless constant characteristic of a particular material. The pressure dependence is implicit in the equation because the melting temperature T_m depends on the pressure (it generally increases with increasing pressure). This form of the constitutive equation, in fact, takes account of complexities in the pressure dependence of the material behavior that are not adequately modeled by Equation (16.11).

ii. The Effect of Grain Size

When coarse-grained polycrystalline solids, such as rocks, are deformed at low homologous temperature in the moderate- to high-stress regime, the yield stress tends to decrease slightly with increasing grain size. This effect is illustrated in Figure 16.19*A* for limestones and marbles, in which the room-temperature yield stress varies inversely as the square root of the grain diameter. A similar effect is observed at higher temperatures in the moderate-stress regime, where during power-law creep, the steady-state flow stress for fine-grained Solenhofen limestone is considerably higher than it is for coarser-grained marbles.

The opposite effect, however, is observed for fine-grained materials in the low-stress regime, where the stress exponent is less than 2. Here the steady-state stress at a constant strain rate decreases rapidly with decreasing grain size. This effect has been observed experimentally both in fine-grained Solenhofen limestone below a mean grain diameter of 10.5μm (Figure 16.19*B*) and in a synthetic dunite below a mean grain diameter of about 27μm (Figure 16.19*C*). For constant temperature and strain rate, a log-linear fit to the data below these grain sizes gives

$$\log \sigma^{(Dif)} = B + C \log d \qquad or \qquad \log[\sigma^{(Dif)}\, d^{-C}] = B \tag{16.13}$$

where d is the mean grain diameter and (B, C) are constants. From (Figure 16.19*B, C*),

$$\begin{aligned} &\text{for limestone} \quad [B, C] = [0.04, 1.23] \\ &\text{for dry dunite} \quad [B, C] = [0.85, 1.21] \end{aligned} \tag{16.14}$$

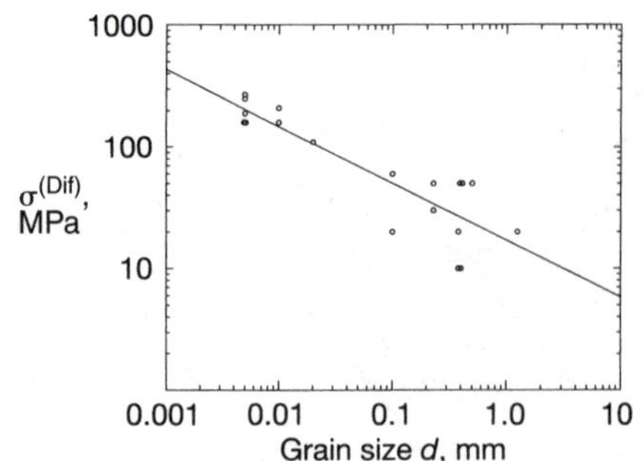

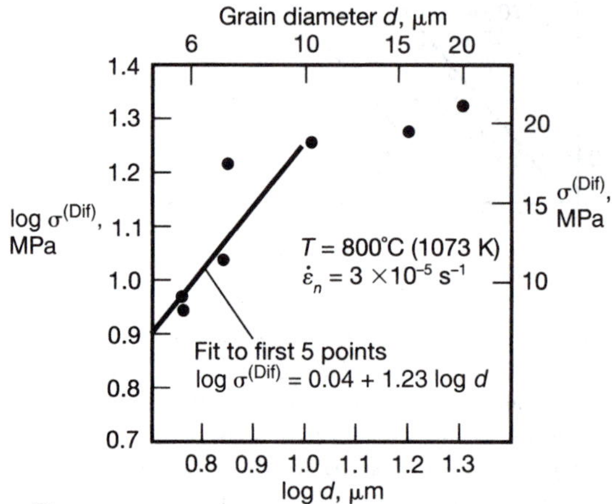

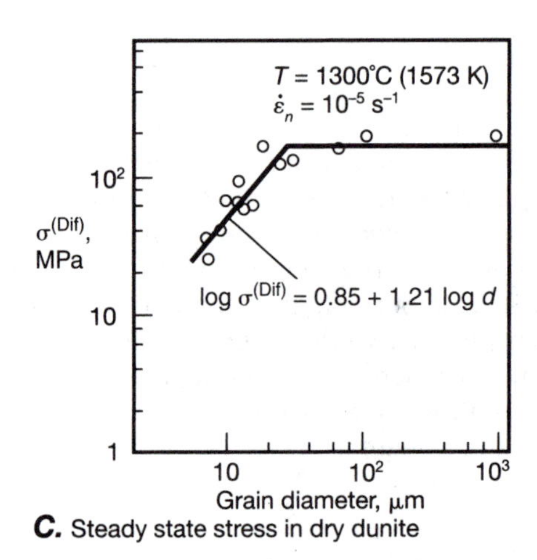

FIGURE 16.19 Grain-size dependence of creep properties. *A.* Grain-size dependence of the yield stress for a variety of calcite rocks of different grain size at room temperature and 100 MPa confining pressure. (From Olsson 1974) *B.* Grain-size dependence of the steady-state creep stress during constant-strain-rate experiments for fine-grained Solenhofen limestone. (From Schmid et al. 1977) *C.* Grain-size dependence of the steady-state creep stress during constant-strain-rate experiments for synthetic dunite. (From Karato and Paterson 1986)

In order to account explicitly for the grain-size dependence in the creep equation, we must modify Equation (16.11) by adding a grain-size-dependent term:

$$|\dot{\varepsilon}_n^{(ss)}| = A_6 \, d^{-b} \, (\sigma^{(Dif)})^n \exp\left[\frac{-(E^* + pV^*)}{RT}\right] \quad (16.15)$$

The second Equation (16.13) implies that the product $(\sigma^{(Dif)} \, d^{-C})$ is a constant for constant strain rate and temperature. In Equation (16.15), if we keep the temperature and strain rate constant, the product of the stress and the grain-size terms $(d^{-b} \, (\sigma^{(Dif)})^n)$ must also be constant. By raising the former product to the power n and equating the result to the latter product, we find

$$(\sigma^{(Dif)} \, d^{-C})^n = (\sigma^{(Dif)})^n \, d^{-b}$$

which implies

$$b = Cn \quad (16.16)$$

Taking $n = 1.7$ and 1.4 appropriate for the low-stress regimes of Figures 16.17 and 16.16, respectively, and using values for the constant C in Equations (16.14), Equation (16.16) gives values of about $b = 2.1$ and 1.7 for the grain-diameter exponent in Equation (16.15) for limestone and dry dunite. On theoretical grounds, we expect values of b between 2, for creep dominated by volume diffusion, and 3, for creep dominated by grain boundary diffusion (see Box 17-1). The data in Figure 16.19*B*, *C* are consistent with the lower end of this range. Other experiments have found a grain-size exponent of $b \approx 3$ and a stress exponent close to $n \approx 1$ for conditions applicable to the upper mantle. Similar values are commonly observed for many polycrystalline materials in comparable conditions.

Comparing the forms of Equations (16.11) and (16.15) shows that Equation (16.15) includes an expression of the functional dependence of the factor A_4 in Equation (16.11) on grain size, for the equations are equivalent if we write:

$$A_4(d) = A_6 d^{-b} \quad (16.17)$$

iii. The Effects of Chemical Environment

The chemical environment of deformation has a profound effect on the rheology of rocks; in particular, the partial pressures (or fugacities[7]) of water and oxygen have major effects on the rheology. These effects can be expressed as a functional dependence of the constant A_4 in Equation (16.11) on the various environmental factors in the same way that the grain-size dependence was expressed in Equation (16.17).

At elevated temperatures and pressures, water dissolves in very small amounts (parts per billion) in the lattices of silicates. In at least some minerals, such as quartz, feldspar, and olivine, it reduces the activation energy for creep, thereby significantly reducing the yield stress for creep. This so-called **water weakening** or **hydrolytic weakening** is notable in quartz, which is extremely strong when it is dry but becomes relatively weak when water is dissolved in the crystal lattice. The effect is pressure-dependent because the solubility of water in the silicate lattices increases with increasing pressure. To account for the concentration of water in the silicate lattice, Equation (16.15) is modified to read

$$|\dot{\varepsilon}_n^{(ss)}| = A_7 \, d^{-b} f_{H_2O}^{r} (\sigma^{(Dif)})^n \exp\left[\frac{-(E^* + pV^*)}{RT}\right] \quad (16.18)$$

where f_{H_2O} is the fugacity of water, which can be related to the concentration of water molecules, commonly expressed as the number of water molecules per 10^6 Si atoms, and r is a constant exponent, which for olivine has a value of $r \approx 1.2$. The presence of water also seems to lower the activation energy for creep, although in some experiments this effect is within the margins of error. Because the presence of water tends to decrease the melting temperature of a silicate, the change in creep behavior may be accounted for through Equation (16.12).

Because of the common occurrence of water in crustal rocks and some mantle rocks, it seems likely that it has a strong influence on deformation in fault zones, in crustal metamorphic terranes, and in some mantle environments. For example, under oceanic ridges, the amount of water in mantle rocks is limited, and when partial melting occurs to produce the mid-oceanic ridge basalts, that water will be concentrated in the melt phase, leaving relatively dry, and therefore relatively strong, olivine behind. In the mantle wedge above a subduction zone, however, where water released from the down-going slab is expected to be relatively abundant, the mantle can be expected to be considerably weaker.

Silicates, of course, all are compounds of oxygen. Experiments have demonstrated that the partial pressure, or fugacity, of oxygen also strongly affects the creep properties. Figure 16.20 shows, for example, that as the partial pressure of oxygen increases in a series of constant-strain-rate experiments on quartz, the creep strength decreases dramatically. A similar dependence

[7]The fugacity of a chemical component may be thought of in simple terms as an effective partial pressure of the component.

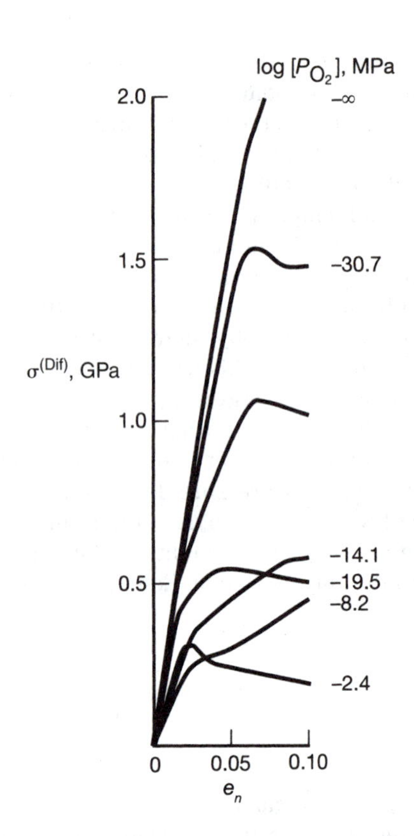

FIGURE 16.20 Constant-strain-rate experiments for quartz, showing the effect of the partial pressure of oxygen on the creep strength. Experiments were done at 10^{-5} s^{-1}, 800°C, and a confining pressure of 1.64 GPa (note: 1 GPa = 10^3 MPa). Curves are labeled with the value of log [P_{O_2}], the logarithm to base 10 of the oxygen partial pressure. (From Ord and Hobbs 1986)

has been demonstrated for olivine. To express this dependence, Equation (16.18) can be modified to

$$|\dot{\varepsilon}_n^{(ss)}| = A_8\, d^{-b} f_{H_2O}^r f_{O_2}^q (\sigma^{(Dif)})^n \exp\left[\frac{-(E^* + pV^*)}{RT}\right] \tag{16.19}$$

where f_{O_2} is the fugacity of oxygen and the constant exponent $q \leq 0.17$. For the mantle, the range of oxygen fugacities is expected to be small, so the oxygen fugacity term is commonly combined with the constant A_8.

The findings that the chemical environment can have important effects on the deformation properties of a mineral severely complicate the application of laboratory experiments to the Earth. Not only must we know the temperature, pressure, stress, and grain size, but we must also know the chemical environment of the mineral before a creep law can be applied with confidence. In many of the older published creep experiments, the chemical environment has not been carefully controlled, although more attention has been paid to this problem in recent experiments. The different environments associated with different experimental setups may explain part of the variation in results reported by different laboratories for comparable materials (see Figure 16.16). We can expect that ultimately, experimental determination of the constitutive equations that fully account for the chemical environment will enable us to determine even more about the conditions within the Earth and to explain with greater precision the processes that occur and the structures that we observe in deformed rocks.

iv. The Effect of Partial Melt

It seems intuitive that if a body of rock undergoes partial melting, then the presence of the liquid phase mixed among the solid crystals would affect the deformational properties of the body, because in general a liquid phase is much weaker than a solid crystalline phase of the same material. The effect depends on the fraction of the grain boundary area of the crystal aggregate that is occupied by the melt, for that determines the fraction of the shear stress in the material that is supported by the weaker melt. Thus the effect is most pronounced if the melt wets the boundaries of the mineral grains in the rock and least pronounced if the melt collects in pores at grain corners. This is an evolving area of research, but for melt fractions less than about 12 percent, the effect can be incorporated into the general power-law creep Equation (16.11) as an approximate empirical functional dependence of the factor A_4 on melt fraction ϕ. Including this modification into Equation (16.19) gives

$$|\dot{\varepsilon}_n^{(ss)}| = A_9\, d^{-b} f_{H_2O}^r f_{O_2}^q \exp[\alpha\phi](\sigma^{(Dif)})^n \exp\left[\frac{-(E^* + pV^*)}{RT}\right] \tag{16.20}$$

where α is a constant and the melt fraction is $\phi \leq 0.12$. For basaltic melt in olivine aggregates, the melt effectively wets the grain boundaries, and α has a value between 25 and 30 in the low-stress regime and between 30 and 45 in the moderate-stress regime. Partial melts are important in high-grade metamorphic terranes and in the mantle beneath midocean ridges and beneath the volcanic arcs associated with subduction zones. Thus understanding the details of this effect is an important part of understanding the deformation of the Earth's crust and upper mantle.

In summary, then, by comparing the basic power-law creep Equation (16.11) with Equation (16.20), we see that the factor A_4 in Equation (16.11) depends on grain size, chemical environment, and melt fraction, in a manner that can be written

$$A_4(d, f_{H_2O}, f_{O_2}, \phi) = A_9\, d^{-b} f_{H_2O}^r f_{O_2}^q \exp[\alpha\phi] \tag{16.21}$$

BOX 16-3 Experimental Determination of the Material Constants in the High-Temperature Creep Equation

In order to define the rheological equations for a given material, we first choose an appropriate form of constitutive equation whose general properties fit those observed for the material. For polycrystalline solids during high-temperature steady-state creep, one of the Equations (16.7), (16.11), (16.12), or (16.15) generally accounts well for the observations. We then use experimental data to determine the values of the constants in the constitutive equation that characterizes the particular material.

Our approach depends on whether the experiments are to be constant-stress experiments or constant-strain-rate experiments. Because the latter are more common among rock deformation experiments, we describe the procedure for obtaining the material constants n, E^*, V^*, b, and A_6 in Equation (16.15) with constant-strain-rate experiments.

We rearrange Equation (16.15) to write the differential stress as the dependent variable

$$\sigma^{(Dif)} = K_6\, d^{b/n} |\dot{\varepsilon}_n^{(ss)}|^{1/n} \exp\left[\frac{H^*}{nRT}\right]$$
$$H^* \equiv E^* + pV^* \quad K_6 \equiv \left[\frac{1}{A_6}\right]^{1/n} \qquad (16\text{-}3.1)$$

where H^* is the activation enthalpy. If we take the natural logarithm of both sides and rearrange, we can isolate different variables as the independent variable on the right side of the equation. Each of the following equations is arranged so that the terms in braces are either intrinsically constant or held constant during a particular set of experiments, and the independent variable is in the last term on the right:

$$\ln \sigma^{(Dif)} = \frac{1}{n}\left\{-\ln K_6 + b \ln d + \left[\frac{H^*}{RT}\right]\right\} + \left[\frac{1}{n}\right] \ln|\dot{\varepsilon}_n^{(ss)}| \qquad (16\text{-}3.2)$$

$$\ln \sigma^{(Dif)} = \frac{1}{n}\left\{-\ln K_6 + b \ln d + \ln|\dot{\varepsilon}_n^{(ss)}|\right\} + \left[\frac{H^*}{nR}\right]\frac{1}{T} \qquad (16\text{-}3.3)$$

Variables Held Constant	Graph: Ordinate	Graph: Abcissa	Slope	Equation	Figure
T, p, d	$\ln \sigma^{(Dif)}$	$\ln\|\dot{\varepsilon}_n^{(ss)}\|$	$1/n$	(16-3.2)	16-3.1
$\dot{\varepsilon}_n^{(ss)}, p, d$	$\ln \sigma^{(Dif)}$	$1/T$	H^*/nR	(16-3.3)	16-3.2
$\dot{\varepsilon}_n^{(ss)}, T, d$	$\ln \sigma^{(Dif)}$	p	V^*/nRT	(16-3.4)	16-3.3
$\dot{\varepsilon}_n^{(ss)}, T, p$	$\ln \sigma^{(Dif)}$	$\ln d$	b/n	(16-3.5)	16.18B
T, p, d	$\ln\|\dot{\varepsilon}_n^{(ss)}\|$	$\ln \sigma^{(Dif)}$	n		
$\sigma^{(Dif)}, p, d$	$\ln\|\dot{\varepsilon}_n^{(ss)}\|$	$1/T$	$-H^*/R$	(16-3.6)	
$\sigma^{(Dif)}, T, d$	$\ln\|\dot{\varepsilon}_n^{(ss)}\|$	p	V^*/RT		
$\sigma^{(Dif)}, T, p$	$\ln\|\dot{\varepsilon}_n^{(ss)}\|$	$\ln d$	b		

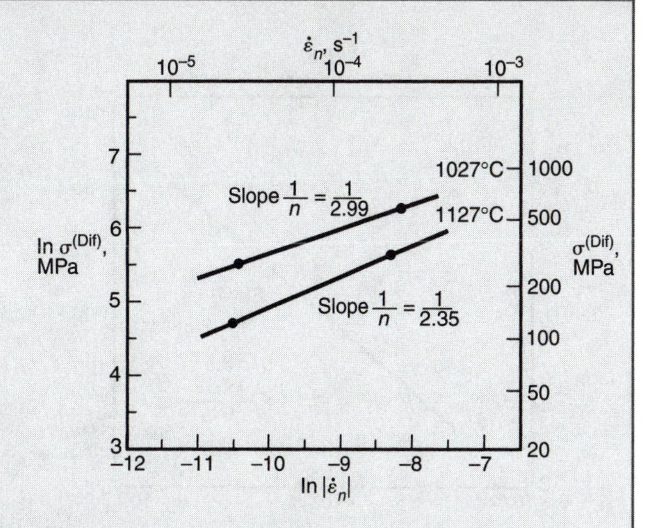

FIGURE 16-3.1 Steady-state differential stress plotted against instantaneous strain rate on natural logarithmic scales for experiments on dunite from Figure 16-15. The slope provides values for $1/n$ and thus determine the inverse stress-exponent. For an accurate determination, many more experiments than are shown here are required (compare Figure 16-17). (After Borch and Green 1987)

$$\ln \sigma^{(Dif)} = \frac{1}{n}\left\{-\ln K_6 + b \ln d + \ln|\dot{\varepsilon}_n^{(ss)}| + \left[\frac{E^*}{RT}\right]\right\} + \left[\frac{V^*}{nRT}\right]p \qquad (16\text{-}3.4)$$

$$\ln \sigma^{(Dif)} = \frac{1}{n}\left\{-\ln K_6 + \left[\frac{H^*}{RT}\right] + \ln|\dot{\varepsilon}_n^{(ss)}|\right\} + \frac{b}{n}\ln d \qquad (16\text{-}3.5)$$

The experimental technique involves holding constant all variables written between the braces so that in effect there is a linear relationship between $\ln \sigma^{(Dif)}$ and the one independent variable on the far right of each equation. For each type of experiment, plotting the dependent variable (left side of each equation) versus the independent variable (far right of each equation) on a graph allows us to measure the slope of the data distribution, which determines the value of the coefficient of the independent variable. The procedures for Equations (16-3.2) through (16-3.5) are summarized in the first four rows of the table at left.

The stress exponent n is determined directly as the inverse of the slope of Equation (16-3.2). Data for dunite from Figure 16.15 are plotted in Figure 16-3.1. The slopes for these data give $n = 2.35$ and 2.99, values that are slightly low compared to the preferred value of 3 to 3.5. In practice, however, many more experiments are required to determine the constant accurately (see, for example, the moderate-stress regime in Figure 16.16). The activation enthalpy H^* can be determined from the slope H^*/nR of Equation (16-3.3) because n is determined independently and R is a constant. Data for dunite from Figure 16.14 are plotted in Figure 16-3.2, which, for $n = 3.0$ to

(continued)

BOX 16-3 Experimental Determination of the Material Constants in the High-Temperature Creep Equation *(continued)*

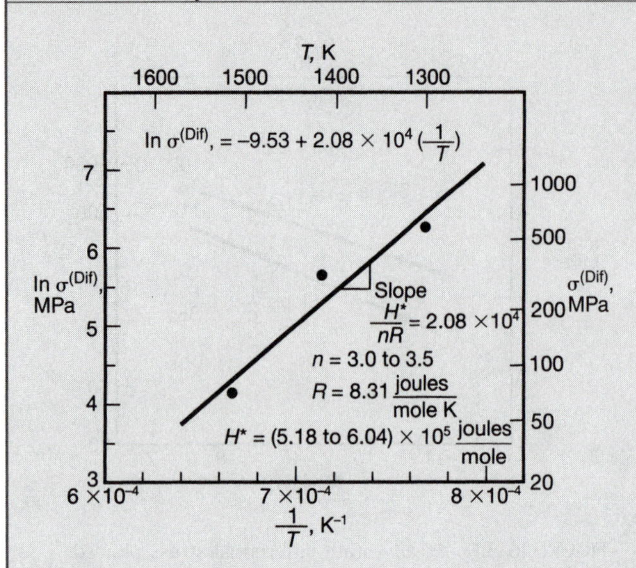

FIGURE 16-3.2 The natural logarithm of the steady-state differential stress plotted against inverse temperature for experiments on dunite from Figure 16-14. The slope provides a value for H^*/nR and thus determines the activation enthalpy, if n is known (as from Figure 16-3.1). For an accurate determination, many more experiments than are shown here are required. (After Borch and Green 1987)

3.5, gives an activation enthalpy between 518 and 604 kJ/mole. The result depends strongly on the value of n but is consistent with the preferred value of about 540 kJ/mole. Similarly, the activation volume V^* can be determined from the slope of Equation (16-3.4) because n is determined independently, R is a known constant, and T is held at a known constant value for the experiment. The data for dunite from Figure 16.18 are plotted in Figure 16-3.3. They indicate that the activation volume is approximately $V^* = (4.1 \text{ to } 4.8) \times 10^{-5}$ m^3/mole for n between 3 and 3.5. Again, this value is inferred from only two pressures, which is far too few for reliable results; the accepted activation volume is smaller than this value by a factor of between 0.5 and 0.1. Finally, the grain size exponent b is determined from the slope of Equation (16-3.5) because n is already known. For dunite, Figure 16.19C shows the slope to be $b/n = 1.21$ (the constant C in Equation (16.13) and the second Equation (16.14); see Equation (16.16)), and because in this set of experiments, n was determined to be about 1.4, we find $b = 1.7$. Knowing all the material parameters in the creep equation except K_6 allows the value of that constant to be determined from the value of the intercept on any of the graphs. Using the third Equation (16-3.1), we can then also determine the value of A_6 in Equation (16.15).

Note, however, that the determination of all of these material constants is dependent on the value of n. Thus any error in determining n is propagated into the error in the determination of H^*, V^*, and b. The same material constants can be determined independent of one another from experiments performed at constant stress, which makes these experiments much preferable to the constant-strain-rate experiments. For constant-stress experiments, the constitutive equation is rearranged so that $\ln|\dot{\varepsilon}_n^{(ss)}|$ appears as the dependent variable on the left of the equation. For example, taking the logarithm of both sides of Equation (16.15) and using the second Equation (16-3.1) gives

$$\ln\left|\dot{\varepsilon}_n^{(ss)}\right| = \left\{\ln A_6 - b \ln d + n \ln \sigma^{(Dif)}\right\} - \left[\frac{H^*}{R}\right]\frac{1}{T} \qquad (16\text{-}3.6)$$

The sixth row in the preceding table shows how a plot of $\ln|\dot{\varepsilon}_n^{(ss)}|$ versus $1/T$ provides the value of $-H^*/R$ directly, independent of n and of the error involved in determining that parameter (compare with Equation (16-3.3)). Equation (16-3.6) can be rearranged to place the variables $\ln \sigma^{(Dif)}$, p, and $\ln d$ in the position of the independent variable so that the slope of the resulting graph determines the values of n, V^*/RT, and b, respectively (see the fifth, seventh, and eighth rows of the table). In each of these cases, the slope of the graph is independent of n, making the constant-stress experiments a more direct method of determining the material parameters than the constant-strain-rate experiments.

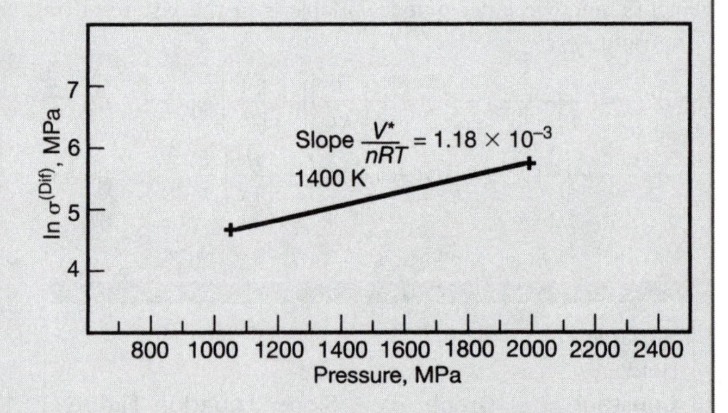

FIGURE 16-3.3 The natural logarithm of the steady-state differential stress plotted against pressure for dunite (see Figure 16-18). The slope determines the value of V^*/nRT and thus provides a value for the activation volume if n is known (as from Figure 16-3.1). (From Borch and Green 1987)

16.6 APPLICATION OF EXPERIMENTAL RHEOLOGY TO NATURAL DEFORMATION

i. Interpretation of Structures of Ductile Deformation with Relation to Finite Strain, Instantaneous Strain Rate, and Stress

It is common to find discussions of the origin of deformational structures couched in terms of the orientations of the principal stresses. We conclude, however, that in general it is preferable to discuss these origins in terms of the orientations of the principal finite strains. The logic for this conclusion is the following: Deformational structures provide information about the finite strain, which is the sum of all the deformations that a given body of rock has experienced, and in general, they do not provide information about the strain rate or the instantaneous strain. The constitutive equations that are relevant to describe the flow of rocks, however, such as Equations (16.2), (16.4) (16.7), (16.9), (16.11), (16.12), (16.15), (16.18), (16.19), and (16.20), all show that the stress is related to the strain rate, which is the instantaneous strain per unit time. Even in this case, the principal axes of stress are strictly parallel to the corresponding principal axes of the strain rate only for mechanically isotropic materials. The principal axes of strain rate, or instantaneous strain, and those of the finite strain, however, are not in general parallel. Thus we cannot directly infer the orientation of the principal stresses from an observation of the finite strain recorded by deformational structures. For example, one cannot assume that the maximum compressive stress is parallel to the maximum shortening direction of the finite strain. We conclude, therefore, that structures should generally be interpreted in terms of their relation to the orientations of the principal axes of finite strain, and it is incorrect to presume that they can be directly interpreted in terms of the orientations of the principal stresses.

Only if the entire deformation history of the rock were strictly coaxial, and only if the rock is and always was mechanically isotropic, could the principal axes of finite strain be strictly parallel to those of stress. It is unlikely that a deformation would remain coaxial throughout its history, however, and most rocks are not strictly isotropic. It is safest, therefore, to restrict the interpretation of structures to the observable characteristics, which are the finite strains, and not to make simplified assumptions regarding the deformation path or the mechanical properties of the material at the time of deformation in order to discuss the orientations of the principal stresses.

This is not merely an academic issue. For example, partly on the basis of an implicit assumption that the principal axes of stress parallel those of finite strain defined from fold structures along the San Andreas fault, it has been argued that the maximum compressive stress is nearly perpendicular to the fault plane. This would imply that the shear stress on the fault is very small and that therefore the fault must be very weak, and much effort has been expended in trying to understand how the mechanics of such a fault would work. It turns out, however, that the folds have been significantly rotated during a noncoaxial, convergent strike-slip deformation, and that once initiated, the folds themselves created a significant anisotropy in the deforming rock body. As a result, the principal axes of finite strain represented by the fold structures do not reflect the orientation of the principal axes of stress associated with the faulting. Thus this evidence does not indicate an abnormally weak fault, so the mechanics of its sliding are probably not the major problem that this misinterpretation had suggested. Correctly understanding the mechanics of the San Andreas fault is important in understanding the earthquake hazard for places like the San Francisco Bay area and the Los Angeles area, so this issue has important societal ramifications.

In comparing the results of experimental deformation of rocks with the continuum models we discussed in Section 16.1, we see that when the stress exponent n is close to 1, the material behaves at steady state like a linearly viscous (Newtonian) material, showing a linear relationship between the stress and the strain rate (Figure 16.2). When n has a value of 3 or more, on the other hand, the steady-state behavior shows a stress–strain-rate curve that resembles plastic behavior (compare Figure 16.4 with Figures 16.3 and 16.7), with the yield stress in effect being exponentially dependent on the inverse temperature (e.g., the second Equation (16.7)).

The rheological equations discussed in the preceding sections of this chapter are strictly one-dimensional. A more general description of material behavior requires that the constitutive equations be formulated for three-dimensional stress and strain. The resulting equations are considerably more complex than the one-dimensional equations, and we discuss some of these generalizations in Box 16-4.

Most experiments investigating rock rheology have used either monomineralic rocks such as quartzite, dunite, or marble, or single crystals of particular minerals, such as quartz, olivine, calcite, or feldspar. A few experiments, however, have used polymineralic rocks such as granite (see Table 16.1). Unfortunately, it has proved very difficult to predict accurately either the behavior of a polymineralic rock from data for its constituent minerals or the rheologic effect of a change in mineralogy. If a rock consists predominantly of one mineral, then we may assume that the rheology of the whole rock is governed by that mineral. A small percentage of a significantly weaker mineral, however, may strongly affect the ductile behavior of a rock. The difficulty in predicting this phenomenon fundamentally restricts our attempts to model accurately the behavior of Earth materials.

TABLE 16.1 Examples of Material Constants for Steady-State Power-Law Flow of Selected Rocks in the Moderate-Stress Regime

	Constants in Equation (16.7)		
Material	Stress exponent n	E^* [kJ mole^{-1}]	K_1 [MPa^{-n} s^{-1}]
Rock salt	5.3	102	6.29
Marble[a] (20–100 MPa)	7.6	418	5.07×10^{-28}
Marble[a] (<20 MPa)	4.2	427	1.98×10^{-9}
Quartzite	2.4	156	6.31×10^{-6}
Quartzite (wet)	2.3	154	2.52×10^{-4}
Granite	3.2	123	1.26×10^{-9}
Granite (wet)	1.9	137	2.0×10^{-4}
Quartz diorite	2.4	219	1.26×10^{-3}
Diabase	3.4	260	2.02×10^{-4}
Albite rock	3.9	234	2.59×10^{-6}
Anorthosite	3.2	238	3.27×10^{-4}
Dunite (dry)	3.0	540	4.0×10^{6}
Dunite (wet)[b]	3.0	420	1.9×10^{3}

Source: Modified after Kirby (1983) and Ranalli and Murphy (1987).
[a] Marble data from Schmid et al. (1980); the applicable range of differential stress $\sigma^{(Dif)}$ is listed.
[b] From Chopra and Paterson (1981); Karato and Paterson (1986).

ii. Experimental versus Geologic Strain Rates

It is important to investigate how the common experimental strain rates of 10^{-7} s^{-1} or higher compare with natural geologic strain rates and how we can evaluate the applicability of the experimentally derived constitutive equations to geologic conditions.

As a benchmark for geologic strain rates, we can consider the strain rates associated with the motions of tectonic plates. The spreading rates of modern oceanic ridges average about 5 cm/yr. If we assume that this displacement is accommodated by shear distributed linearly across the asthenosphere, and that the asthenosphere is approximately 200 km thick, then it is simple to calculate the shear strain rate:

$$\frac{5 \text{ cm/y}}{200 \text{ km}} = \frac{1.6 \times 10^{-7} \text{ cm/s}}{2 \times 10^{7} \text{ cm}} = 0.8 \times 10^{-14} \text{ s}^{-1} \tag{16.22}$$

It is difficult to change this number by a factor of 10 without making unreasonable assumptions. Thus a strain rate of 10^{-14} s^{-1} seems a reasonable order-of-magnitude approximation for mantle deformation associated with convection and plate tectonics. Comparison of the total strains measured in deformed rocks with the radiometrically or stratigraphically constrained times available for the deformation gives comparable orders of magnitude, although higher rates occur in faults and ductile shear zones where deformation is much more highly concentrated. In general, we expect geological deformation to proceed at strain rates between 10^{-11} s^{-1} and 10^{-16} s^{-1}. (The fastest of these rates has been documented for deformation of a 1- to 2-m-thick calcite mylonite within a thrust zone in the Helvetic Alps.) Thus geologic strain rates are generally 4 to 10 orders of magnitude lower than the rates practical for laboratory experiments, and differential stresses are also much lower by amounts that depend on the stress exponent n in the relevant constitutive equation.

How, then, can we tell whether the constitutive equations we determine in the laboratory are pertinent to geologic deformation, when they must be extrapolated over many orders of magnitude of strain rate? How can we be sure there are no other constitutive equations that dominate the material behavior at very low stresses and strain rates typical of geologic conditions? There are two methods for attacking these problems.

First, one can try to understand the mechanisms by which ductile deformation occurs and to identify microscopic and submicroscopic structures in the rock or in the crystal lattices that characterize the operation of these mechanisms. If similar microstructures are found in naturally deformed rocks, then we may conclude that the same deformation mechanisms operated under the two different sets of conditions and that therefore the laboratory-derived constitutive equations are applicable to geologic conditions. We discuss this method in more detail in Chapter 17 (see particularly Sections 17.6 and 17.7).

Second, one can compare observed natural deformation rates with the predictions based on laboratory-determined constitutive equations. For example, when the continental ice sheets of the last Pleistocene glacial advance melted about 10,000 years ago, a huge load was rapidly removed from the glaciated continents. The isostatic depression that had resulted from the ice load started to rebound by flow of mantle material toward the depression, causing uplift of the Earth's surface. This uplift raised and preserved a series of beaches along the shores of the rising continents. We can date these beaches using, for example, radiocarbon dating methods, and measure their current elevations relative to one another and to sea level. Thus we can determine the history of uplift and how it varied at different distances from the original ice sheet. We can estimate the stress that drives the flow in the mantle from the knowledge of the original area and thickness of the ice. The pattern of flow in the mantle in response to those stresses is governed by the rheology of the mantle, so we can compare the observed uplift histories with calculated histories based on different constitutive models for mantle flow. The constitutive model for mantle flow that most closely accounts for the observed uplift data provides an indication of the most likely flow law to apply to the mantle. This comparison thus provides a test for the relevance of laboratory-derived constitutive equations.

One such analysis of the data[8] indicates that the mantle behaves as a linearly viscous fluid, which implies the stress exponent in the constitutive equation is indistinguishable from $n = 1$. The inferred viscosity of the mantle is about $\eta \approx 10^{20}$ Pa s.[9] This viscosity is high compared to the viscosities of more familiar geologic fluids such as basaltic magma ($\eta \approx 3800$ Pa s to 7 Pa s, for temperatures roughly between 1100°C and 1400°C), and water ($\eta = 10^{-3}$ Pa s). Given the size of the Earth, however, and the great lengths of time available, even this high a viscosity is sufficiently low to allow flow in the solid mantle. The inferred stress exponent of $n = 1$, however, is controversial in part because it implies that the constitutive equation derived from most laboratory experiments on olivine, which give a stress exponent of about $n = 3$, cannot be used to model flow in the mantle. A transition from power-law creep to Newtonian viscosity may occur at a depth of 200 to 250 kilometers in the mantle, and the asthenosphere above about 200 kilometers may have a significantly lower viscosity than the rest of the mantle of about 10^{19} Pa s. Thus, although experimental data provide some of the best information available on the rheology of rocks, we must be cautious in applying those results to the Earth and must seek out other independent sources of constraint.

Experimentally determined values of n, E^*, and K_1 (the first Equation (16.7)) for a variety of rock types are summarized in Table 16.1. These data are only approximate, because there is much variability among experiments on the same material, but they give an order-of-magnitude estimate of how rocks might behave in the Earth.

The variation of mechanical properties with depth is of major interest in understanding the tectonic behavior of the Earth's lithosphere. To determine this variation, we must account for the increase of both temperature and pressure and for the change in mineral composition of the rocks. The pressure effect for lithospheric depths (100 km or less) is relatively small, and for a first approximation it can be ignored. The temperature increase, however, has a major effect, and the inferred geothermal gradient can vary significantly from one tectonic setting to another. Figure 16.21*A* shows simplified geotherms for regions below midoceanic ridges, hot continental areas such as the Basin-Range province of the western United States, oceanic areas underlain by mature lithosphere (more than 50 million years old), and cold continental areas (such as continental shield areas). The stress required to maintain a strain rate of 10^{-14} s^{-1} during power-law creep for a variety of rock types (see Table 16.1) is plotted in Figure 16.21*B* for the hot continental geotherm and in Figure 16.21*C* for the cold continental geotherm. Notice that strength drops dramatically with depth, and that rocks rich in quartz (quartzite, granite) generally are weaker than those rich in plagioclase (diabase, quartz diorite).

At shallow depths, the low temperatures would require a high stress to drive ductile creep, but creep is unlikely, because fracture and cataclastic flow can occur at much lower stress. However, the differential stress required for these brittle processes increases with increasing pressure (see, for example, Figures 16.9 and 8.10), and the differential stress required for ductile creep decreases with increasing temperature (see Figure 16.14 and the second Equation (16.7)). Because both pressure and temperature increase with increasing depth, there must be a transition region where brittle deformation gives way to ductile deformation. This is the so-called **brittle-ductile transition** (cf. Section 8.4, Figure 8.8). The fact that earthquakes occur no deeper than 15 to 20 kilometers along the strike-slip San Andreas fault of California, for example (see Chapter 6), is probably because the brittle-ductile transition occurs at that depth.

The actual variation of strength with depth thus depends on the geothermal gradient, the deformation mechanism, and the rock type. Figure 16.22*A*, *B* shows two possible distributions of crustal strength based on the assumption of Coulomb-type brittle behavior at low temperature and pressure, and power-law creep at high

[8] Cathles 1975.

[9] The SI unit of viscosity is the Pascal-second (Pa s), or (N s)/m^2; the cgs unit is the poise, or (dyne s)/cm^2. 1 Pa s = 10 poise.

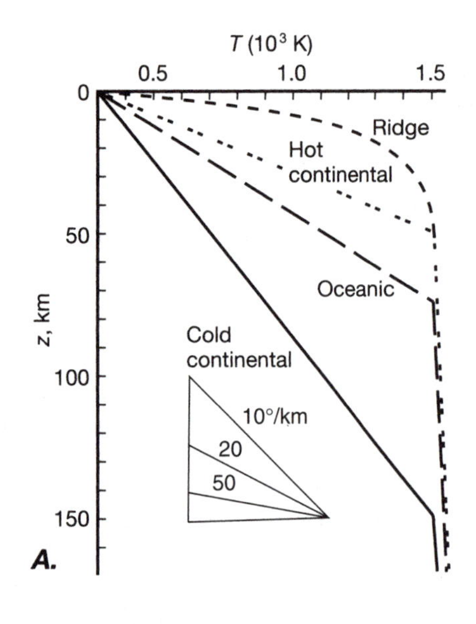

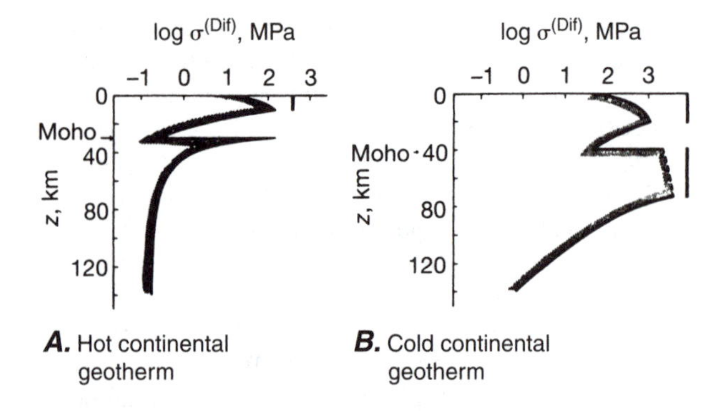

FIGURE 16.22 Distributions of strength with depth for two different models of the lithosphere. (From Ranalli and Murphy 1987) *A.* Model of a quartz/granite crust in a hot continental geotherm with the Moho at a depth of 30 kilometers. *B.* Model of a quartz/granite crust in a cold continental geotherm with the Moho at 40 kilometers.

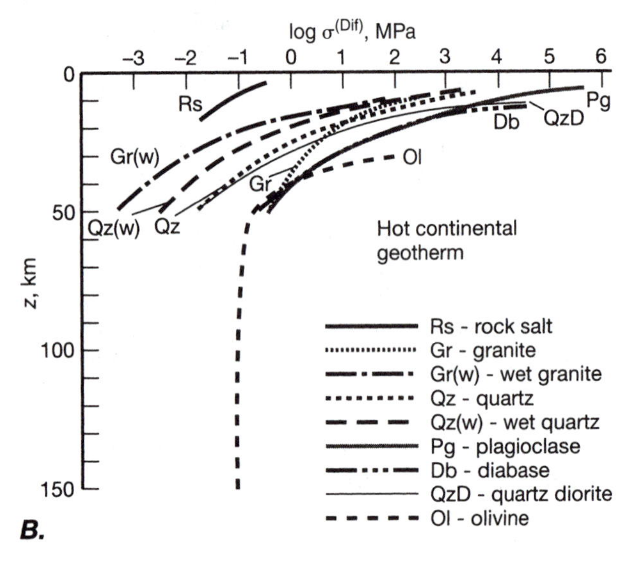

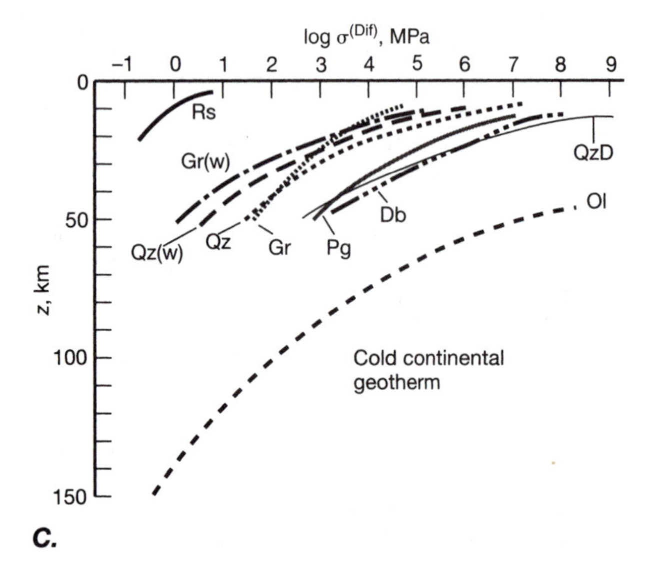

temperature and pressure (Table 16.1). Figure 16.22*A* reflects a hot continental geotherm (Figure 16.21*A*) with a quartz/granitic crust and a crust-mantle boundary (Moho) at 30 km. Figure 16.22*B* is for a cold continental geotherm (Figure 16.21*A*) with a 40-kilometer-thick quartz/granitic crust. Models assuming other distributions of rock type with depth would result in different distributions of crustal strength with depth, which could include distributions with two strength maxima and an intervening weak zone within the crust. The most significant factors, however, are the maximum in strength at the brittle-ductile transition in the crust, and a second maximum at the Moho where the rock composition changes from quartzo-feldspathic rocks to peridotite dominated by olivine.

These models of crustal strength differ considerably from the model of a homogeneous elastic crust assumed by Hafner (Figures 9.19 through 9.21), and to that extent, Hafner's models must be regarded as oversimplifications. On the other hand, as noted in the preceding discussion, models such as those in Figure 16.22 are also simplifications, as they do not account for the possibility that models based on laboratory data that are extrapolated over many orders of magnitude in strain rate

FIGURE 16.21 Variation of power-law creep rheology with depth for various rock types listed in Table 16.1. *A.* Simplified linearized geotherms for a midocean ridge, hot continental lithosphere, oceanic lithosphere, and cold continental lithosphere. The inset shows the slopes for different geothermal gradients as labeled. *B* and *C.* Steady-state creep stress required to maintain a creep rate of 10^{-14} s^{-1} for *B* the hot continental geotherm in *A* and for *C* the cold continental geotherm in *A.* (After Ranalli and Murphy 1987)

may well be unreliable. In addition, these models do not take into account possible chemical stratification of the crust, changes in the mineralogy caused by metamorphic reactions, or the potential for the effect on rheology of metamorphic reactions occurring during deformation. We try to formulate progressively more refined models as the detail of the phenomena we want to understand increases, as knowledge of conditions in the Earth improves, and as experimental results become more refined.

BOX 16-4 Constitutive Equations in Three Dimensions

Although one-dimensional constitutive equations are useful for explaining the different models of material behavior, the complete constitutive relations must take into account all three dimensions, and the equations accordingly become more complex. We summarize here the three-dimensional constitutive relations for isotropic elastic, viscous, plastic, and power-law continua. Isotropic materials are the simplest because they have the same mechanical properties in all directions. In anisotropic materials, the response of the material is different in different directions, and the constitutive equations become even more complex.

i. Elastic Behavior

For three-dimensional elasticity, we need six equations, one relating each of the independent stress components to the strain components, or (conversely) one relating each of the independent strain components to the stress components. For such materials, two elastic moduli, called the **Lamé parameters** λ and μ, are required to describe the mechanical behavior (anisotropic materials require more than two constants):

Normal Stress Components	Shear Stress Components	
$\sigma_{11} = \lambda e_V + 2\mu e_{11}$	$\sigma_{12} = 2\mu e_{12}$	
$\sigma_{22} = \lambda e_V + 2\mu e_{22}$	$\sigma_{13} = 2\mu e_{13}$	(16-4.1)
$\sigma_{33} = \lambda e_V + 2\mu e_{33}$	$\sigma_{23} = 2\mu e_{23}$	

The equations for normal stress components are on the left, and those for shear stress components are on the right. There are two terms in each equation for the normal stresses because a normal stress generally produces a change in both volume and shape. The first term accounts for the volume change in which e_V is the volumetric extension ($e_V = e_{11} + e_{22} + e_{33}$; see Equations (9.6), (12.15), (12.45)). The second term accounts for the change in shape. Shear stresses, on the other hand, produce only changes in shape and, therefore, are related to the strain by only one elastic constant. If the geologic tensor sign convention is used, the elastic moduli all have negative values because a positive (compressive) stress produces a negative (shortening) strain.

The six equations can be summarized in compact notation by the single equation

$$\sigma_{k\ell} = \lambda e_V \delta_{k\ell} + 2\mu e_{k\ell} \qquad (16\text{-}4.2)$$

where Equations (16-4.1) are obtained from Equations (16-4.2) by using different combinations of the subscript values chosen from (k = 1, 2, or 3) and (ℓ = 1, 2, or 3) and where $\delta_{k\ell}$ is a tensor, called the Kronecker delta, whose components equal 1 for $k = \ell$ and equal 0 for $k \neq \ell$.

$$\delta_{kl} \equiv \begin{bmatrix} 1 & 0 & 0 \\ 0 & 1 & 0 \\ 0 & 0 & 1 \end{bmatrix} \qquad (16\text{-}4.3)$$

The Lamé parameter μ in Equations (16-4.1) and (16-4.2) is the same as the shear modulus in the second Equation (16-1). Young's modulus E and Poisson's ratio ν, both of which appear in Equation (9.5) and the first of which appears in the first Equation (16.1), are related to the Lamé parameters by

$$E = \frac{\mu(3\lambda + 2\mu)}{\lambda + \mu} \qquad \nu = \frac{\lambda}{2(\lambda + \mu)} \qquad (16\text{-}4.4)$$

Thus one could rewrite Equations (16-4.1) to express λ and μ in terms of E and ν. It is strictly a matter of convenience which pair of elastic moduli are used to describe the behavior.

Although most minerals in rocks are anisotropic, the average elastic constants for the rock are isotropic if the mineral grains are randomly oriented. A rock with a preferred orientation of mineral grains such as a schist, however, would be mechanically anisotropic. The preferred orientation of olivine crystals in the upper mantle results in an anisotropy of elastic properties that shows up as a different velocity of seismic wave propagation for different orientations of ray path through the rock. This anisotropy can be detected in seismic refraction studies (see Appendix 2, especially Box A2-1) and can be used to constrain the geometry of flow in the upper mantle (see Section 17.7(i)). For the sake of simplicity in modeling deformation in the Earth, however, such anisotropies are commonly ignored.

ii. Viscous Behavior

To describe constant-volume viscous behavior in three dimensions, we must specify how each of the six independent components of the stress and of the instantaneous strain rate are related. We need, therefore, six independent equations. These six can be summarized as one equation:

$$\sigma_{k\ell}^{(Dev)} = 2\eta\dot{\varepsilon}_{k\ell} \qquad (16\text{-}4.5)$$

where $\dot{\varepsilon}_{k\ell}$ is the rate of instantaneous-strain tensor (see Box 16.1) and $\sigma_{k\ell}^{(Dev)}$ is the deviatoric stress tensor defined by relations comparable to Equation (7.37)

$$\sigma_{k\ell}^{(Dev)} \equiv \sigma_{k\ell} - \bar{\sigma}_n\delta_{k\ell} \qquad (16\text{-}4.6)$$

(continued)

BOX 16-4 Constitutive Equations in Three Dimensions *(continued)*

where $\bar{\sigma}_n$ is the mean normal stress defined in three dimensions by

$$\bar{\sigma}_n \equiv \frac{\sigma_{11} + \sigma_{22} + \sigma_{33}}{3} \quad (16\text{-}4.7)$$

(Compare the definition for mean normal stress in two dimensions given by the first Equations (7.32) and (7.33).) The subscripts k and ℓ in Equations (16-4.5) and (16-4.6) each take on values 1, 2, or 3, and all possible combinations of these subscript values produce nine equations. In Equation (16-4.6) the Kronecker delta $\delta_{k\ell}$ (Equation (16-4.3)) ensures that the mean normal stress $\bar{\sigma}_n$ is subtracted only from the stress component for which $k = \ell$, that is, only from the normal stress components. Three of the six equations for the shear stresses are redundant, however, because of the symmetry of the stress and strain-rate tensors (see Equations (7.56) and (12.40)).

The volumetric strain rate associated with the flow of a fluid is commonly very small and, accordingly, is generally ignored. By using the deviatoric stress in the constitutive equation, we ignore any effect of the pressure, which is appropriate if the volumetric deformation is negligible. Equations (16-4.5) are similar in form to the equations for elasticity (Equation 16-4.2) except that the strain rate appears in the equations instead of the strain, and the use of the deviatoric stress means that the mean normal stress is not explicitly constrained and that the deformation must be at constant volume (see Box 18-1(ii)).

iii. Plastic Behavior

In three dimensions, the constitutive equation for perfectly plastic behavior is written in terms of the deviatoric stress $\sigma_{k\ell}^{(Dev)}$ because, again, the mean normal stress does not affect the behavior of the material. Thus the yield criterion is that the second invariant of the deviatoric stress tensor I_2 is equal to a constant K^2:

$$I_2 \equiv \sum_{k=1}^{3} \sum_{\ell=1}^{3} (\sigma_{k\ell}^{(Dev)})(\sigma_{k\ell}^{(Dev)}) = K^2 \quad (16\text{-}4.8)$$

The double summation implies summation of all terms created by all possible combinations of values of k and ℓ. In principal coordinates, this relation, expressed in terms of the components of the full stress tensor, can be written

$$I_2 = \frac{1}{3}\left[(\hat{\sigma}_1 - \hat{\sigma}_2)^2 + (\hat{\sigma}_2 - \hat{\sigma}_3)^2 + (\hat{\sigma}_3 - \hat{\sigma}_1)^2\right] = \tau_{(oc)}^2 \quad (16\text{-}4.9)$$

$\tau_{(oc)}$ is the shear stress on the octahedral planes, which are the planes that are equally inclined to all three principal stress axes (and that therefore are the orientations of the surfaces of an octahedron based on the principal axes). The normal stress on those planes is $\bar{\sigma}_n$, the mean normal stress. In two dimensions I_2 is directly related to the square of the radius of the Mohr circle (see the second Equation (7.32)), so this yield criterion is simply a generalization of the von Mises yield criterion (Figure 8.8) that the maximum shear stress, $(\hat{\sigma}_1 - \hat{\sigma}_3)/2$, be equal to a constant.

The strain rate is constrained by the requirement

$$\dot{\varepsilon}_{k\ell} = \frac{\sqrt{E_2}}{K}\sigma_{k\ell}^{(Dev)} \quad (16\text{-}4.10)$$

where E_2 is the second invariant of the strain rate tensor $\dot{\varepsilon}_{k\ell}$ defined in a similar manner to I_2 in Equation (16-4.8). Equation (16-4.10) requires the principal axes of the strain rate to be parallel to those of the stress, but the complete strain rate is indeterminate because E_2 is not constrained by the relationship. In fact, by forming the second invariant of both sides of the equation, we simply recover Equation (16-4.8), which is therefore implicit in Equation (16-4.10).

iv. Power-Law Behavior

A generalized tensor form of the power-law equation can be written in terms of the second invariant of the deviatoric stress I_2

$$|\dot{\varepsilon}_{k\ell}^{(ss)}| = A_1 I_2^{(n-1)/2} \sigma_{k\ell}^{(Dev)} \exp\left[\frac{-E^*}{RT}\right] \quad (16\text{-}4.11)$$

$$I_2 = \sum_{i=1}^{3} \sum_{j=1}^{3} (\sigma_{ij}^{(Dev)})(\sigma_{ij}^{(Dev)}) \quad (16\text{-}4.12)$$

In the case of confined compression, for which $\hat{\sigma}_2 = \hat{\sigma}_3$, $\sqrt{I_2}$ reduces to a constant times the differential stress $\sigma^{(Dif)}$ and in principal coordinates for $(k, \ell) = (1,1)$, Equations (16-4.11 and 16-4.12) reduce to the first Equation (16.7).

REFERENCES AND ADDITIONAL READINGS

Amin, K. E., A. K. Mukherjee, and J. E. Dorn. 1970. A universal law for high-temperature, diffusion-controlled transient creep. *J. Mech. Phys. Solids* 18: 413–426.

Borch, R. S., and H. W. Green II. 1989. Deformation of peridotite at high pressure in a new molten salt cell: Comparison of traditional and homologous temperature treatments. *Phys. of the Earth and Planet. Int.* 55: 269–276.

Byerlee, J. 1978. Friction of rocks. *Pure and Appl. Geophys.* 116(4–5): 615–626.

Carter, N. 1976. Steady state flow of rocks. *Rev. of Geophys. and Space Phys.* 14: 301–360.

Carter, N. L., and H. G. Avé-Lallement. 1970. High-temperature flow of dunite and peridotite. *Geol. Soc. Amer. Bull.* 81: 2181-2202.

Carter, N. L., and S. H. Kirby. 1978. Transient creep and semibrittle behavior of crystalline rocks. *Pure Appl. Geophys.* 116: 807–839.

Cathles, L. M. III. 1975. *The Viscosity of the Earth's Mantle.* Princeton University Press, 386 pp.

Chopra, P. N., and M. S. Paterson. 1981. The experimental deformation of dunite. *Tectonophysics* 78: 453–473.

Dieterich, J. H. 1979. Modeling of rock friction 1. Experimental results and constitutive equations. *J. Geophys. Res.* 84(B5): 2161–2168.

Dieterich, J. 1981. Constitutive properties of faults with simulated gouge. In N. L. Carter, M. Friedman, J. M. Logan, and D. W. Stearns, eds., *Mechanical Behavior of Crystal Rocks*. Amer. Geophys. Union Geophysical Monograph 24: pp. 103–120.

Dieterich, J. 1994. A constitutive law for rate of earthquake production and its application to earthquake clustering. *J. Geophys. Res.* 99(B2): 2601–2618.

Garofalo, F. 1965. *Fundamentals of Creep and Creep Rupture of Metals*. MacMillan, New York: 258 pp.

Green, H. W. II, and R. S. Borch. 1987. The pressure dependence of creep. *Acta Metall.* 35(6): 1301–1305.

Griggs, D. T. 1967. Hydrolytic weakening of quartz and other silicates. *J. Roy. Astron. Soc.* 14: 19–31.

Gu, J.-C., J. R. Rice, A. L. Ruina, and S. T. Tse. 1984. Slip motion and stability of a single degree of freedom elastic system with rate and state dependent friction. *J. Mech. Phys. Solids* 32(3): 167–196.

Hager, B. H. 1991. Mantle viscosity: A comparison of models from post-glacial rebound and from the geoid, plate driving forces, and advected heat flux. In R. Sabadini, K. Lambeck, E. Boschi, eds., *Glacial Isostasy, Sea Level and Mantle Rheology*, D. Reidel Publishing Company, Dordrecht-Boston: 493–513.

Herwegh, M., J. H. P. de Bresser, and J. H. ter Heege. 2005. Combining natural microstructures with composite flow laws: An improved approach for the extrapolation of lab data to nature. *J. Struct. Geol.* 27(3): 503–521.

Hirth, G., and D. L. Kohlstedt. 1996. Water in the oceanic upper mantle: Implications for rheology, melt extraction and the evolution of the lithosphere. *Earth Planet. Sci. Lett.* 144: 93–108.

Hirth, G., and D. L. Kohlstedt. 2003. Rheology of the upper mantle and mantle wedge: A view from the experimentalists. In J. Eiler, ed., *Inside the Subduction Factory*, American Geophysical Union.

Ismat, Z., and G. Mitra. 2001. Folding by cataclastic flow at shallow crustal levels in the Canyon Range, Sevier orogenic belt, west-central Utah. *J. Struct. Geol.* 23: 355–378.

Johnson, A. M. 1970. *Physical Processes in Geology*. Freeman, Cooper & Co., San Francisco, 577 pp.

Karato, S.-I., and H. Jung. 2003. Effects of pressure on high-temperature dislocation creep in olivine. *Phil. Mag.* 83: 401–414.

Karato, S.-I., and M. S. Paterson. 1986. Rheology of synthetic olivine aggregates: Influence of grain size and water. *J. Geophys. Res.* 91(B8): 8151–8176.

Karato, S.-I., and P. Wu. 1993. Rheology of the upper mantle: A synthesis. *Science* 260: 771–778.

Kirby, S. H. 1983. Rheology of the lithosphere. *Rev. of Geophys. and Space Phys.* 21: 1458–1487.

Kirby, S. H., and C. B. Raleigh. 1973. Mechanisms of high temperature, solid-state flow in minerals and ceramics and their bearing on the creep behavior of the mantle. *Tectonophysics* 19: 165.

Kohlstedt, D. L., Q. Bai, Z.-C. Wang, and S. Mei. 2000. Rheology of partially molten rocks. In N. Bagdassarov, D. Laporte, and A. B. Thompson, eds., *Physics and Chemistry of Partially Molten Rocks*, Kluwer Academic Publishers: 3–28.

Kohlstedt, D. L., and C. Goetze. 1974. Low-stress high-temperature creep in olivine single crystals. *J. Geophys. Res.* 79: 2054.

Mei, S., and D. L. Kohlstedt. 2000. Influence of water on plastic deformation of olivine aggregates 1. Diffusion creep regime. *J. Geophys. Res.* 105(B9): 21457–21469.

Mei, S., and D. L. Kohlstedt. 2000. Influence of water on plastic deformation of olivine aggregates 2. Dislocation creep regime. *J. Geophys. Res.* 105(B9): 21471–21481.

Miller, D. D. 1998. Distributed shear, rotation, and partitioned strain along the San Andreas fault, central California. *Geology* 26: 867–870.

Mount, V. S., and J. Suppe. 1987. State of stress near the San Andreas fault; implications for wrench tectonics. *Geology* 15(12): 1143–1146.

Mukherjee, A. K., J. E. Bird, and J. E. Dorn. 1969. Experimental correlations for high temperature creep. *Trans. ASM* 62: 155–179.

Olsson, W.A.. 1974. Grain size dependence of yield stress in marble. *J. Geophys. Res.* 79: 4859–4862.

Ord, A., and B.E. Hobbs. 1986. Experimental control of the water-weakening effect in quartz. In H. C. Heard and B. E. Hobbs, eds., *Mineral and Rock Deformation: Laboratory Studies—The Paterson Volume*, Geophysical Monograph 36, Amer. Geophys. Union, Washington, DC: 51–72.

Ord A., B. E. Hobbs, and K. Regenauer-Lieb. 2004. A smeared seismicity constitutive model. *Earth Planets Space* 56(12): 1121–1133.

Paterson, M. S. 2001. Relating experimental and geological rheology. *Int. J. Earth Sci.* (*Geologische Rundschau*) 90: 157–167.

Peltier, W. R.. 1998. Global glacial isostasy and relative sea level: Implications for solid Earth geophysics and climate system dynamics. In P. Wu, ed., *Dynamics of the Ice Age Earth*: 17–54.

Pfiffner, O. A., and J. G. Ramsay. 1982. Constraints on geological strain rates: Arguments from finite strain states of naturally deformed rocks. *J. Geophys. Res.* 87: 311–321.

Ranalli, G., and D. C. Murphy. 1987. Rheological stratification of the lithosphere. *Tectonophysics* 132: 281–296.

Rice, J. R., and A. L. Ruina. 1983. Stability of steady frictional slipping. *Trans. ASME, J. Appl. Mech.* 50: 343–349.

Ross, J. V., H. G. Avé Lallemant, N. L. Carter. 1979. Activation volume for creep in the upper mantle. *Science* 203: 261–263.

Schmid, S. M., J. N. Boland, and M. S. Paterson. 1977. Superplastic flow in fine-grained limestone. *Tectonophysics* 43: 257–291.

Schmid, S. M., M. S. Paterson, and J. N. Boland. 1980. High temperature flow and dynamic recrystallization in Carrara marble. *Tectonophysics* 65: 245–280.

Scholz, C. H. 1990. *The Mechanics of Earthquakes and Faulting*. Cambridge University Press, New York. 439 pp.

Scholz, C. H. 1998. Earthquakes and friction laws. *Nature* 391: 37–42.

Scholz, C. H. 2000. Evidence for a strong San Andreas fault. *Geology* 28(2): 163–166.

Takeuchi, S., and A. S. Argon. 1976. Review: Steady state creep of single phase crystalline matter at high temperatures. *J. Mat. Sci.* 11: 1542–1566.

Tullis, J. 1979. High temperature deformation of rocks. *Rev. Geophys. & Space Phys.* 17: 1137–1154.

Twiss, R. J. 1976. Structural superplastic creep and linear viscosity in the Earth's mantle. *Earth and Planet. Sci. Letts.* 33: 86–100.

Xu, Y., M. E. Zimmerman, and D. L. Kohlstedt. 2002. Deformation behavior of partially molten mantle rocks. In G. D. Karner, N. W. Driscoll, B. Taylor, and D. L. Kohlstedt, eds., *MARGINS Theoretical and Experimental Earth Science Series. Volume 1: Rheology and Deformation of the Lithosphere at Continental Margins*, Columbia University Press.

Chapter 17

MICROSCOPIC ASPECTS OF DUCTILE DEFORMATION: MECHANISMS AND FABRICS

In preceding chapters, we considered rock deformation from the continuum point of view, which assumes that rock is homogeneous and has no discontinuities of structure or mechanical properties. In fact, rocks are made of crystal grains, and the grain boundaries are discontinuities in the crystallographic orientation, and often in the structure, composition, and mechanical properties of the grains. Thus, if we assume the material is homogeneous, in effect we can only describe the behavior averaged over a volume that is large compared to the grain size. The description of the flow of rocks at the continuum level requires a phenomenological approach, for which we must determine the average mechanical properties of the material from experiments. In this chapter, however, we explore the deformation processes at the microscopic and submicroscopic scales. Processes at these scales ultimately result in the apparent continuum deformation at the macroscopic level. By studying these processes at the microscopic scale, however, we gain a deeper understanding of how the average mechanical properties arise. Thus we explore just how a solid crystalline material can flow to produce the large-scale structures, such as folds, that we observe in the field and to permit the convection in the mantle that ultimately drives the relative motion of lithospheric plates. If we can understand these mechanisms, we will be better able to predict the behavior of rocks under a variety of conditions, especially those that are unavailable to experimental observation, and better able to interpret the structures in deformed rocks.

We therefore wish to answer several questions: What mechanisms permit solid rocks to flow? Under what conditions do these mechanisms operate? What rheology (that is, what relationship between strain rate and stress) is associated with each of these mechanisms? What microscopic and submicroscopic structures can we identify in the rock that reflect the deformation mechanisms that produced them? What can we infer from these structures about the conditions of deformation?

Macroscopic flow of rocks can result from a number of different mechanisms, most of which involve either the motion of linear crystal defects called **dislocations** or the motion of **point defects** that result in so-called **diffusive mass transport**. Exponential-law creep and power-law creep result from dislocation motion in the high- and moderate-stress regimes, respectively (see Figures 16.16 and 16.17). At lower stresses, deformation is accommodated by diffusion, either along grain boundaries (so-called **solution creep** and **Coble creep**) or through the volume of crystals (**Nabarro-Herring creep**). At high stresses, especially at upper crustal depths where temperature and pressure are low, cataclastic flow dominates the rheology. We discuss these mechanisms in greater detail in the subsequent sections of this chapter.

Each mechanism may dominate the rheology of a rock at a particular range of physical conditions such as temperature, confining pressure, and differential stress. The conditions of stress and temperature at which some of the flow mechanisms dominate can be plotted on a so-called **deformation mechanism map**, or more simply a **deformation map**. Figure 17-1 shows typical deformation maps of dislocation and diffusion mechanisms for quartz, calcite, and olivine. Other deformation mechanisms, including cataclastic flow (see Table 3-1 and Sections 8.4(ii), 16.2(ii), and 17.1(iv), are not plotted on these deformation maps. Stress is normalized to a dimensionless number by dividing the differential stress $\sigma^{(Dif)}$ by

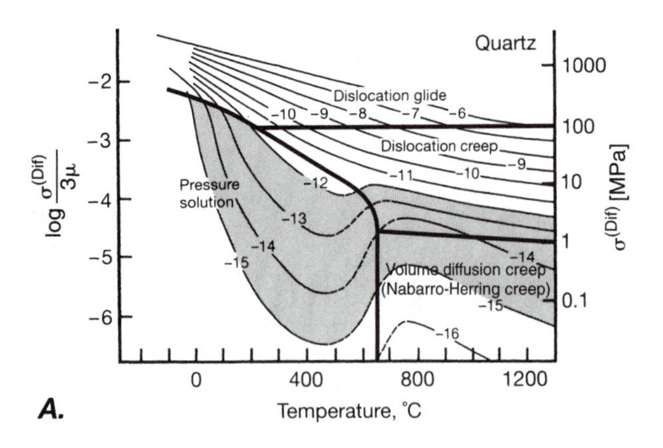

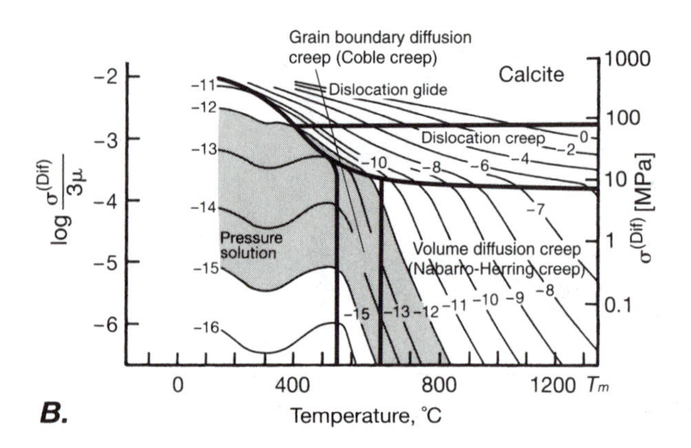

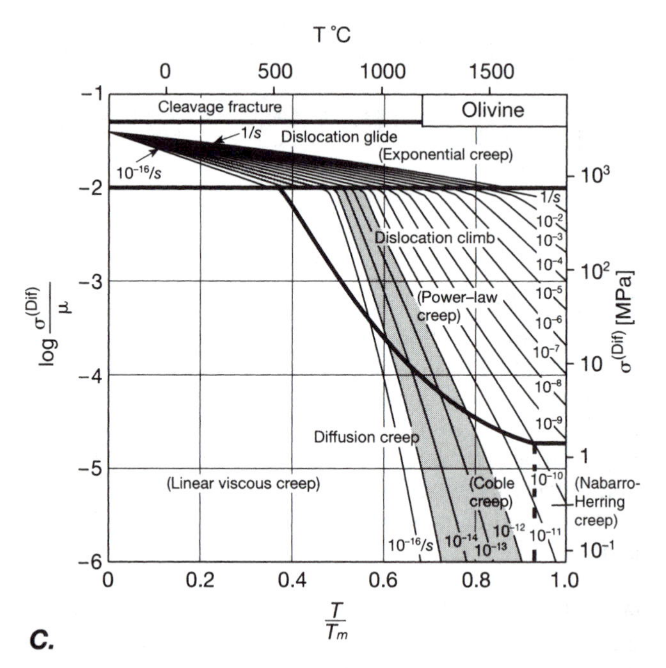

FIGURE 17.1 Deformation mechanism maps. Representative plots of normalized stress (differential stress divided by shear modulus) versus temperature, showing the fields in which different mechanisms dominate the strain rate, and the boundaries between the fields at which the strain rates from the two mechanisms are equal. Fine lines are contours of strain rate. Expected geological strain rates (10^{-12} s^{-1} to 10^{-15} s^{-1}) are shaded. *A.* Deformation map for wet quartzite with 100 μm grain size. Stress scale at right is evaluated for the shear modulus μ at 900° C. (From Rutter 1976) *B.* Deformation map for marble with 100 μm grain size. Stress scale at right is evaluated for the shear modulus μ at 500° C. (From Rutter 1976) *C.* Deformation map for olivine with 100 μm grain size. Temperature scale is normalized by the melting temperture T_m to give homologous temperature. (From Ashby and Verrall 1977)

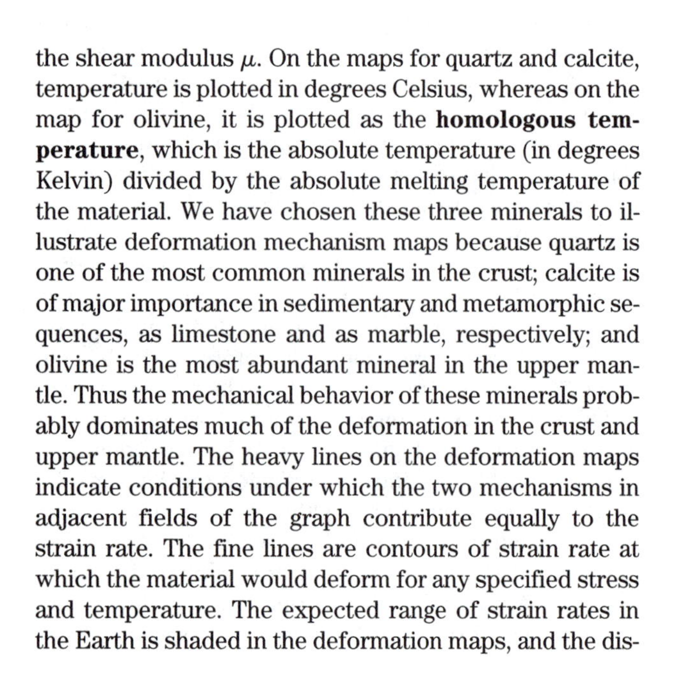
the shear modulus μ. On the maps for quartz and calcite, temperature is plotted in degrees Celsius, whereas on the map for olivine, it is plotted as the **homologous temperature**, which is the absolute temperature (in degrees Kelvin) divided by the absolute melting temperature of the material. We have chosen these three minerals to illustrate deformation mechanism maps because quartz is one of the most common minerals in the crust; calcite is of major importance in sedimentary and metamorphic sequences, as limestone and as marble, respectively; and olivine is the most abundant mineral in the upper mantle. Thus the mechanical behavior of these minerals probably dominates much of the deformation in the crust and upper mantle. The heavy lines on the deformation maps indicate conditions under which the two mechanisms in adjacent fields of the graph contribute equally to the strain rate. The fine lines are contours of strain rate at which the material would deform for any specified stress and temperature. The expected range of strain rates in the Earth is shaded in the deformation maps, and the distribution of the shaded band across the fields dominated by each of these mechanisms indicates that they can all be important at various conditions in the Earth.

These maps give some idea of the relationships among differential stress, temperature, and the dominant mechanisms of flow, although they are approximate and are undoubtedly incorrect in detail. Evidence from natural rocks suggests, for example, that for quartz and calcite, dislocation creep (see Section 17.5) may dominate at lower stresses and temperatures than suggested by these maps. Moreover, the maps do not include all known deformation mechanisms, nor do they account for all conditions that rocks encounter in the natural environment. For example, hydrolytic weakening (see Section 16.5(iii), which is not accounted for in these maps, reduces the stress and temperature at which dislocation creep can dominate the deformation process in quartz and olivine. Moreover, the maps are constructed for a particular grain size and therefore do not indicate the effect of changing grain size (see Section 16.5(ii). Thus the maps should be considered at best only a qualitative illustration of the relative importance of the different deformation mechanisms.

Most mechanisms of deformation produce characteristic microscopic and submicroscopic structures in the rock as a whole, as well as within the individual mineral grains. These structures impart a **fabric** to the rock, which is defined by the geometrical organization of the structures in the rock.[1] We distinguish between **macrofabric** and **microfabric**. The **macrofabric** comprises the preferred orientations of fold axes and axial surfaces, foliations, lineations, joints, veins, and similar macroscopic structures. The **microfabric** includes the preferred orientations of crystallographic axes of mineral grains, the preferred orientations of the long, short, and intermediate dimensions of elongate or tabular grains, and the **texture**, which is defined by the characteristics of the shape and arrangement of crystal grains in the rock. On the submicroscopic scale, the microfabric also includes the crystal substructure, including the arrangement of linear and planar imperfections in the crystal lattices. To be considered part of a fabric, a structure must be a penetrative feature of a volume of rock that is large compared to the scale of the structure. Such a structure is thereby termed a **fabric element** of the rock. From the association between these fabrics and particular deformation mechanisms, we can infer from observation of naturally deformed rocks the mechanisms that operated during the deformation.

In the remainder of this chapter, we discuss these mechanisms of deformation and the structural effects that these mechanisms leave in the rocks. Section 17.1 introduces the basic concepts of elastic deformation and then deals with low-temperature deformation mechanisms that share the characteristic of being highly sensitive to the magnitude of the confining pressure. Such pressure-sensitive deformation is characteristic of mechanisms that involve friction. Section 17.2 describes another low-temperature deformation mechanism, twin gliding, that is particularly common in rocks with a mineral composition dominated by calcite ($CaCO_3$). The remainder of the chapter deals with mechanisms of high-temperature deformation and their effects on the structure of rocks. The fundamental mechanisms include the migration of point defects (diffusion, Section 17.3) and the motion of linear defects (dislocation creep, Section 17.4) through the crystal lattices. By themselves, these mechanisms generally cannot sustain an indefinite flow of the rock, and other processes must be involved to allow this to occur. We review these other processes and the results of their interaction in Section 17.5. Most of these deformation mechanisms produce some sort of rock microfabric, and we pay particular attention to the microfabrics associated with high-temperature deformation processes. Section 17.6 discusses microstructural features that result from these mechanisms of deformation, and Section 17.7 discusses the preferred crystallographic orientation fabrics that result from dislocation mechanisms. Finally, in Section 17.8, we introduce the very general concepts of symmetry that govern the relationships among externally applied constraints, the associated deformation, and the internal structure of deformed rocks.

[1]Different authors use the term *fabric* in slightly different ways. We adopt a broad definition of the term.

17.1 MECHANISMS OF LOW-TEMPERATURE DEFORMATION

i. Elastic Behavior

Elastic behavior reflects the interatomic or interionic forces and the potential energy of the atoms or ions in a crystal lattice. The position of any ion in a lattice is one of minimum potential energy that results from the balance of attractive force from the net ionic charges and repulsive force from the nuclei of the atoms. For a simple crystal such as salt, NaCl, the oppositely charged ions Na^+ and Cl^- are attracted to each other by long-range electrostatic forces associated with the net charge on the whole ion. The positively charged nuclei of both ions, however, repel each other, but that repulsive force is not effective at large distances because of the shielding of the electron cloud. The equilibrium spacing of the ions in the crystal lattice[2] occurs at the potential energy minimum, where these two forces balance each other.

During elastic deformation the ions are forced out of their positions of minimum potential energy. Young's modulus of elasticity (see Section 9.1) is a measure of the increase in potential energy for a given applied stress. Removal of the stress causes the lattice to relax to its original position of minimum potential energy. Thus elastic deformation disappears when the stress is removed.

[2]It is worth remembering that a mineral is simply a three-dimensional network of ions arranged such that ionic charges balance, and the ions are at an equilibrium spacing, which is the so-called "ionic radius" of the various ions. Silicate minerals fundamentally consist of SiO_4 tetrahedra that are linked in various ways by different amounts of sharing of the oxygens of neighboring tetrahedra. For example, olivine consists of stacked free SiO_4 tetrahedra. SiO_4 tetrahedra link to form chains in pyroxene and amphibole, sheets in micas and serpentine, and three-dimensional networks in feldspar and quartz. The linked silicate structures are in turn bound together by other ions necessary to preserve charge balance; these ions must be the proper size to fit into spaces in the three-dimensional crystal structure. The nature and strength of the various bonds determine the resistance of the mineral to deformation.

ii. Friction

The physical mechanism associated with friction depends on the nature of the contact between the two surfaces. No surface is perfectly flat, and the contact between two surfaces, therefore, is actually supported by the high spots on the surface, called **asperities** (Figure 17.2), which characteristically have a fractal size distribution for natural surfaces (see the discussion of fractals in Box 2-1(i). The actual contact area is much smaller than the nominal area of the interface. Thus the stress is concentrated in the asperities and is much higher than the nominal applied stress. The behavior of the asperities therefore determines the characteristics of friction.

The asperity model of a frictional interface can account for the observed behavior of friction in fault gouge. The normal stress, concentrated on the supporting asperities, causes them to creep, presumably by some ductile mechanism, thereby progressively increasing the area of contact between the surfaces. The more interlocked the surfaces become as a result of the creep of the asperities, the higher the frictional resistance to sliding. Thus, increasing the sliding velocity decreases the frictional resistance because a given set of asperities are in contact for a shorter time, which decreases the degree to which they can become interlocked. Shearing off of asperities and formation of a thin layer of gouge also decreases the interlocking of the two surfaces and causes a decrease in friction (see Section 16.2(ii) and Box 16-2).

iii. Granular Flow

Granular flow involves the rolling and sliding of rigid particles past one another. It characterizes the deformation of either unlithified sedimentary layers or slurries, and it occurs only at low effective confining pressures.

When particles move past one another, they cannot remain in a close packed arrangement (see, for example, a diagram of close packing in two dimensions in Figure 17.9*A*). As such particles are displaced, the space between them increases, so the volume of the body must increase as well. High effective confining pressure suppresses granular flow because an increase in volume requires work to be done against the effective confining pressure. Moreover, sliding of particles past one another involves friction, and frictional resistance to sliding increases with increasing normal stress across the sliding surface. Thus the higher the effective confining pressure, the more work is expended in dilation and frictional sliding, and the less efficient the granular flow mechanism becomes.

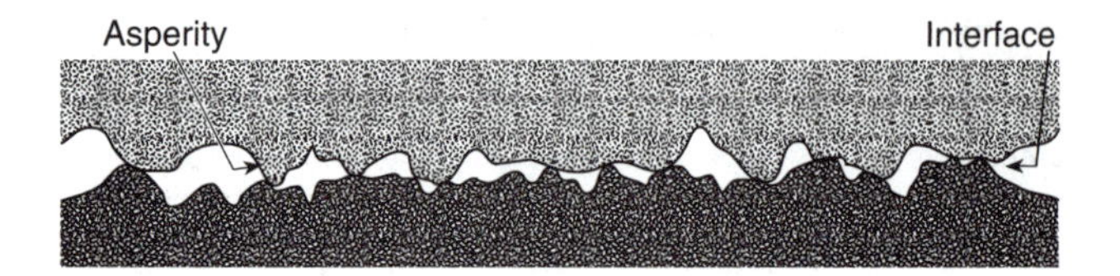

FIGURE 17.2 Asperities support the contact between two surfaces and control the characteristics of friction.

High pore fluid pressure lowers the effective confining pressure by reducing the normal stress supported by grain-to-grain contacts (see Section 8.5). Thus even at considerable depths, a pore fluid pressure approaching the minimum compressive stress would permit granular flow.

Granular flow is important in soft-sediment deformation, such as the slumping of sediment layers on the ocean floor, and it is probably the mechanism of deformation in at least some parts of the accretionary prisms that accumulate above subduction zones. The effects of granular flow are difficult to detect in rock microfabrics. If the rocks are deformed (for example, folded) but the mineral grains themselves show no evidence of deformation, granular flow may have occurred. Subsequent crystalline flow and recrystallization, however, would make it impossible to recognize the earlier operation of granular flow from the microfabric.

iv. Cataclastic Flow

Cataclastic flow is a process of deformation that involves continuous brittle fracturing of grains in a rock, with attendant frictional sliding and possibly some rolling of the fractured particles past one another (see Sections 3.2(i), 8.4(ii), and 16.2(ii). Although it results in a macroscopically coherent and permanent deformation, it differs from granular flow in that small-scale fracturing is an integral part of the process. It occurs instead of granular flow when the work required to fracture grains is less than that required to slide or roll them past one another. Thus cataclastic flow occurs where effective confining pressures are too high to permit pure granular flow. A continuous gradation exists between the two processes, determined by the different relative amounts of fracturing, sliding, and rolling. Cataclastic flow differs rheologically from ductile flow (discussed in Sections 17.3 through 17.5) in that the stress required to maintain a given strain rate increases strongly with increasing effective confining pressure.

Cataclastic fabrics are characterized by the sharp angular shapes of the clasts and grains, by pervasive cracks, by a broad spectrum of grain and clast sizes that characteristically has a fractal distribution (see the discussion of fractals in Box 2-1(i), and commonly by an absence of any foliation (Figure 3.5*B*; Table 3.1). Cataclastic flow produces a progressive decrease in the grain size and a progressive increase in the volume of the cataclastic material through the continual comminution of larger grains into smaller (see Box 2-1(iii)) and the fracturing and abrasion of the sides of the fault zone. For individual rock types, a correlation exists between displacement on a fault and the thickness of gouge in the fault, and in principle, a correlation also exists between grain size and total strain.

Pseudotachylites (Figure 3.6; Table 3.1) are unfoliated rocks of glass or cryptocrystalline material that are commonly associated with cataclasites and are present as veins along some fault zones. We interpret the glass to be a quenched melt of the rock that forms locally by frictional heating when the rock is dry and slip rates are sufficiently high. These conditions are most likely during fault-slip events associated with earthquakes. The resulting melt intrudes through small fractures in the adjacent rock before being quenched.

Although the process by which cataclasites form seems clear, our understanding of the rheology of the process is based largely on experimental work. Theoretical models for the rheology associated with cataclastic flow are in only the early stages of development (see Sections 16.2(ii) and (iii), and Box 16-2).

17.2 TWIN GLIDING

Many minerals have a crystal structure capable of forming a twin across particular crystallographic planes. The twinned structure may be related to the original structure by a mirror reflection across the twin plane, such as in calcite, or by rotation of the original structure about an axis normal to the twin plane, such as the albite twin in plagioclase. Twins may form during either crystallization or deformation.

Twin gliding is a process whereby twins are produced from an original structure by a simple shearing parallel to the twin plane. Calcite twins commonly form in this manner, as shown in Figure 17.3*A–C*. Figure 17.3*A* shows the untwinned calcite structure with the dots representing

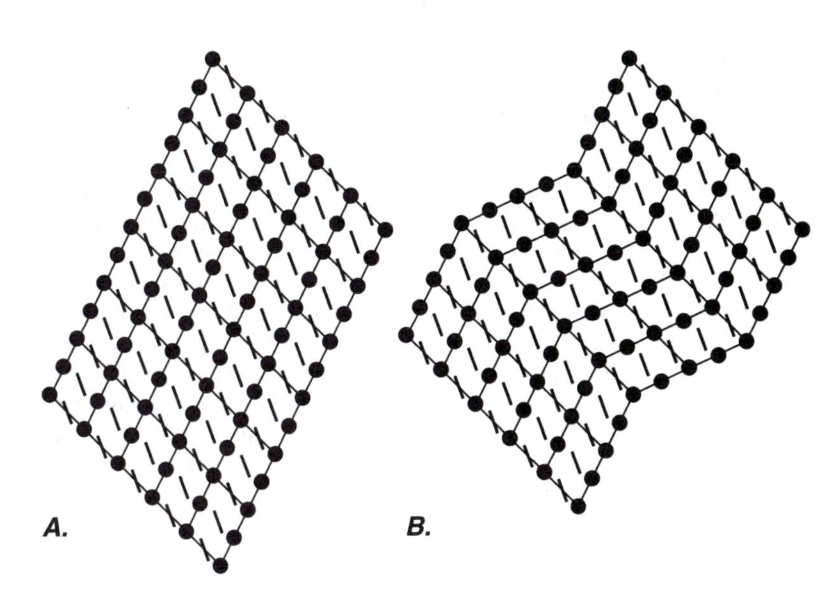

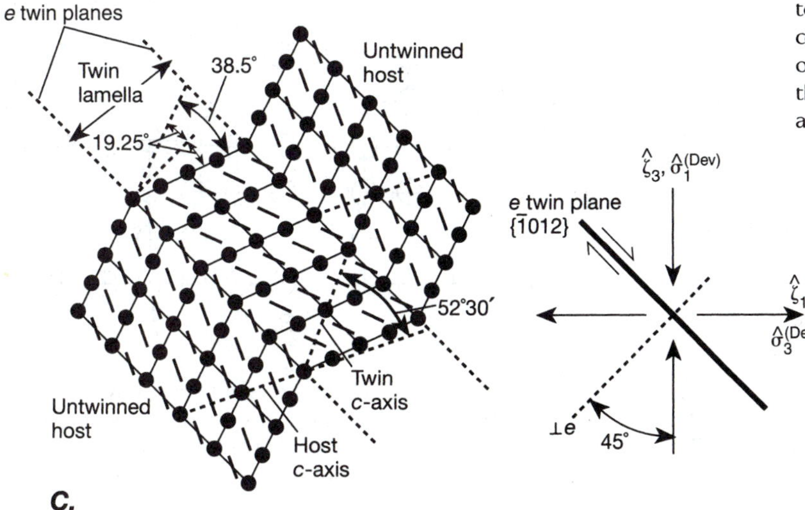

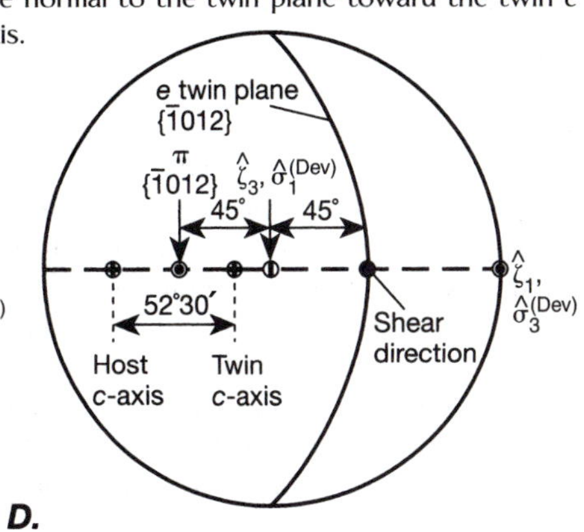

FIGURE 17.3 The geometry of twinning in calcite. *A*. Structure of an untwinned calcite lattice. The short lines are parallel to the CO_3^{-2} complexes. *B*. Calcite lattice after the middle band has been subjected to a simple shear parallel to the *e* twin plane. The orientation of the CO_3^{-2} complexes shows that this structure is not an exact reflection twin of the original structure. *C*. A minor rotation of the CO_3^{-2} complexes has produced a perfect reflection twin across the shear plane. The maximum shear strain may be calculated from the maximum displacement permitted by the twin geometry:

$$e_s = \frac{1}{2} \tan \psi = \frac{1}{2}[2 \tan 19.25] = 0.349$$

D. Geometry of calcite twinning shown in *C* plotted in the plane of diagrams *A–C* (left) and on a stereonet (lower hemisphere equal-angle projection). A great circle joins the *c* axis of host, the *c* axis of the twin, and the pole (π) to the twin plane. The axis of maximum compressive deviatoric stress is assumed to lie on this great circle at a 45° angle measured from the normal to the twin plane toward the twin *c* axis.

the Ca^{+2} ions and the short lines representing the edges of the CO_3^{2-} planes. Figure 17.3*B* shows the calcite structure after a simple shear deformation of the Ca^{+2} lattice parallel to the *e* twin plane $\{\bar{1}012\}$ (see footnote to Table 17.1). The resulting structure is not quite a twin because of the orientation of the CO_3^{2-} radicals. A minor rotation of the CO_3^{2-} planes, however (Figure 17.3*C*), brings the structure into a perfect reflection twin.

In calcite, this twin gliding occurs at a relatively low shear stress (10 MPa) on the twin plane. Twinning an entire calcite crystal grain allows the accumulation of a maximum shear strain of $e_s = 0.35$. Greater strains must be accommodated by another mechanism. Thus the mechanism does not give rise to steady-state creep and is not depicted on the deformation map for calcite (Figure 17.1*B*).

The ease of twinning in calcite, however, makes twin gliding a significant deformation mechanism at low temperatures in limestones and marbles. Rocks thus deformed have abundant twins, most of which are in an orientation of high instantaneous shear strain and presumably high shear stress. Because total strains associated with this mechanism typically are not large, the twin planes do not rotate far from the orientation in which they formed. Accordingly, the twin planes and the slip directions of twinning can be used to infer the orientations of the principal instantaneous strains and, by further assumption, of the principal stresses associated with deformation of the rocks (Figure 17.3*D*). The dispersion of twin plane orientations can also be used to constrain the magnitude of the differential stress (see Box 17-2).

17.3 DIFFUSION AND SOLUTION CREEP

Diffusion is the slow migration of one material through another that results from the random thermal motions of atoms and that is driven by a gradient in chemical activity, such as is provided by a concentration gradient or by a nonzero differential stress. In **diffusion creep**, rock deformation takes place by the migration of atoms of the material through the solid material itself from areas of high compressive stress to areas of low compressive stress. Diffusion creep may result from the diffusion of point defects through a crystal lattice, the diffusion of atoms or ions along grain boundaries, or the diffusion of dissolved components in a fluid along the grain boundaries. In the process of **self-diffusion**, atoms may be transferred through a crystal lattice from a crystal face oriented so it is under high compressive stress to another face oriented so it is under a lower compressive stress. Diffusion may also greatly enhance the rate of strain by aiding the motion of linear crystal defects (dislocations; see Section 17.4(ii)) and by accommodating the shape changes of mineral grains required for grain boundary sliding.

Point defects in crystal lattices are formed by both interstitials and vacancies. **Interstitials** are extra atoms that are stuffed between the normal lattice sites of a crystal (Figure 17.4*A*). The lattice is deformed around an interstitial because it must expand to accommodate the defect. A **vacancy** is a lattice site that is unoccupied by any atom (Figure 17.4*B*). The lattice around a vacancy is also deformed because it tends to collapse into the void. In general, an interstitial is a higher energy defect than a vacancy, and therefore it has a lower probability of forming. The concentration of stable vacancies in a lattice increases with increasing temperature. As a result, vacancies are more common and more important in high-temperature ductile flow.

If an atom jumps from its lattice site into an adjacent vacant site, it fills that site but leaves a vacant site behind, in effect causing the vacancy to jump in the opposite direction from the atom (Figure 17.5). Thus the movement of vacancies in one direction is equivalent to the movement of atoms in the opposite direction. The same applies, of course, to ions, the only difference being the nature of the bonding in the crystal.

i. Nabarro-Herring Creep (Volume Diffusion)

The motion of vacancies is a thermally activated process, which means that vacancies make random jumps from site to site through the lattice with a frequency that depends on the temperature (see footnote 4, Section 16.4). If differential stress is applied to a crystal, vacancies are created at the surface of the crystal where the compressive stress is a minimum (Figure 17.6*A*), and they are destroyed at the surface where the compressive stress is a maximum (Figure 17.6*B*). The resulting concentration gradient of the vacancies causes a diffusive flux (Figure 17.6*C*): The vacancies move from the surface of low com-

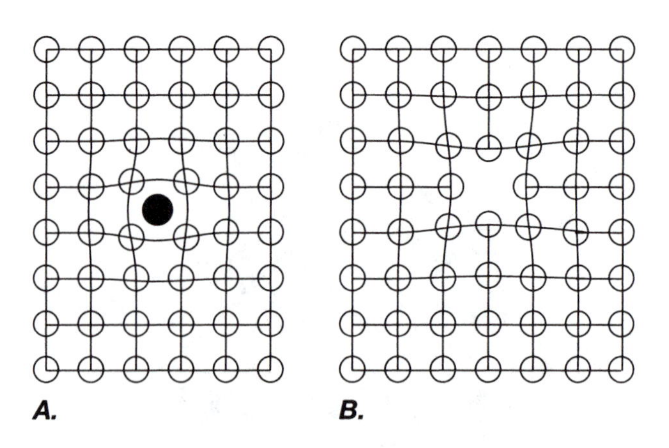

FIGURE 17.4 Point defects in a crystal lattice. *A*. An interstitial. The black circle represents an extra atom in the crystal, causing distortion of the lattice. *B*. A vacancy. The lattice tends to collapse in around the vacancy, thereby distorting the structure.

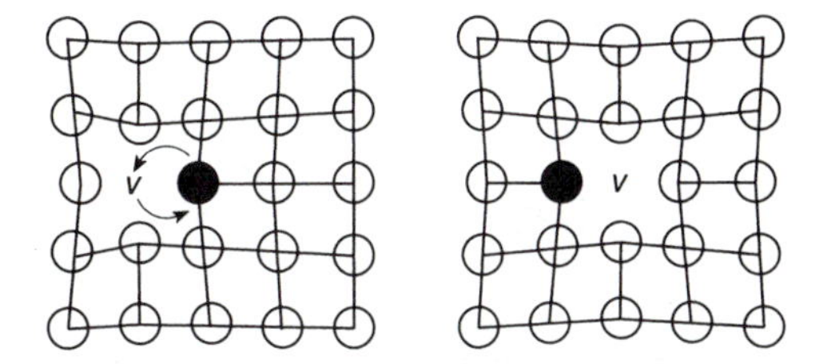

FIGURE 17.5 The motion of a vacancy (indicated by "v") from one lattice site to an adjacent site is opposite to the motion of the atom from the adjacent site into the vacant site. Thus material diffuses in a direction opposite to that of the vacancies.

pressive stress toward the surface of high compressive stress. The flux of atoms is opposite to that of the vacancies, and a crystal grain changes shape by atoms moving away the high-stressed face and accumulating on the low-stressed face, thereby shortening the crystal parallel to the maximum compressive stress and lengthening it parallel to the minimum compressive stress (Figure 17.7).

The deformation that results is called **Nabarro-Herring** creep, and it produces a linear relationship between the strain rate and the differential stress (Box 17-1), comparable to a Newtonian, or linearly viscous, fluid (see Section 16.1(ii)). Figure 17.1 indicates that it is an effective deformation mechanism only at high temperatures and low stresses, and in olivine it is effective only near the melting temperature (homologous temperature near 1, Figure 17.1*C*).

ii. Coble Creep (Grain Boundary Diffusion)

Diffusion of atoms occurs not only through the volume of crystals, but also along grain boundaries. Diffusion along grain boundaries is more rapid than through the volume, largely because the activation energy for grain boundary diffusion is roughly two thirds that for volume diffusion. Thus, although the diffusion path for an atom or ion around the sides of a grain may be longer than the path directly through the volume, the higher rate of grain boundary diffusion can make it the more efficient mechanism of the two.

The process is the same as for volume diffusion: Atoms diffuse away from surfaces subjected to high compressive stress and accumulate on surfaces of low compressive stress. The resulting **Coble creep** also exhibits a linear stress–strain rate relationship. Coble creep is effective at lower temperatures than Nabarro-Herring creep because of its lower activation energy. The two fields are separated by a line of constant temperature (Figure 17.1).

iii. Solution Creep

During **solution creep**, mineral grains dissolve more readily at faces under high compressive stress. The dissolved components then diffuse through the fluid phase on the grain boundaries and precipitate on surfaces of low compressive stress. Evidence for solution creep and for the mobility of components of the rock comes from the study of folds (Section 13.3), foliations (Sections 14.2(iii), 14.3(ii) and (iii), and 15.2), and lineations (Sections 14.2(iii), and 14.6(v) and (vi)).

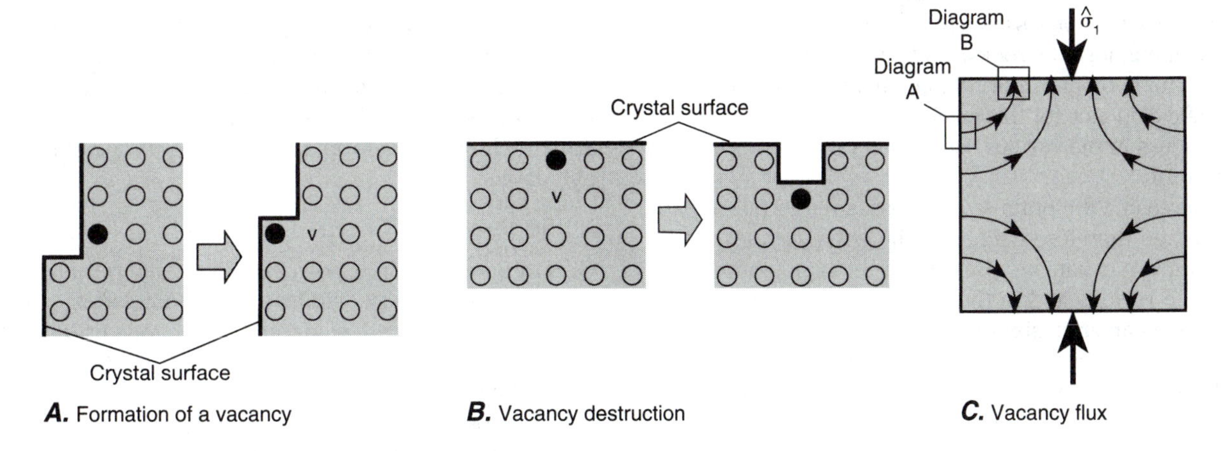

FIGURE 17.6 Nabarro-Herring creep caused by volume diffusion of vacancies in response to a uniaxial compressive stress. *A*. The process of creation of a vacancy at a crystal surface of minimum compressive stress. The solid lines mark the surface of the crystal, which is never perfectly planar at the atomic scale. The solid circle marks the ion whose position changes during the process to create the vacancy (marked "v"). The surface thus gets gradually built out, lengthening the crystal normal to the compressive stress. *B*. The process of destruction of a vacancy at a crystal surface of maximum compressive stress. Symbols have the same meaning as in *A*. The accumulation of vacancies on the crystal surface gradually results in the displacement of the surface, and thus shortening the crystal parallel to the maximum compressive stress. *C*. Under a uniaxial compressive stress, vacancies diffuse toward the surface of highest normal stress along the indicated paths through the crystal, with a resulting flux of atoms in the opposite direction.

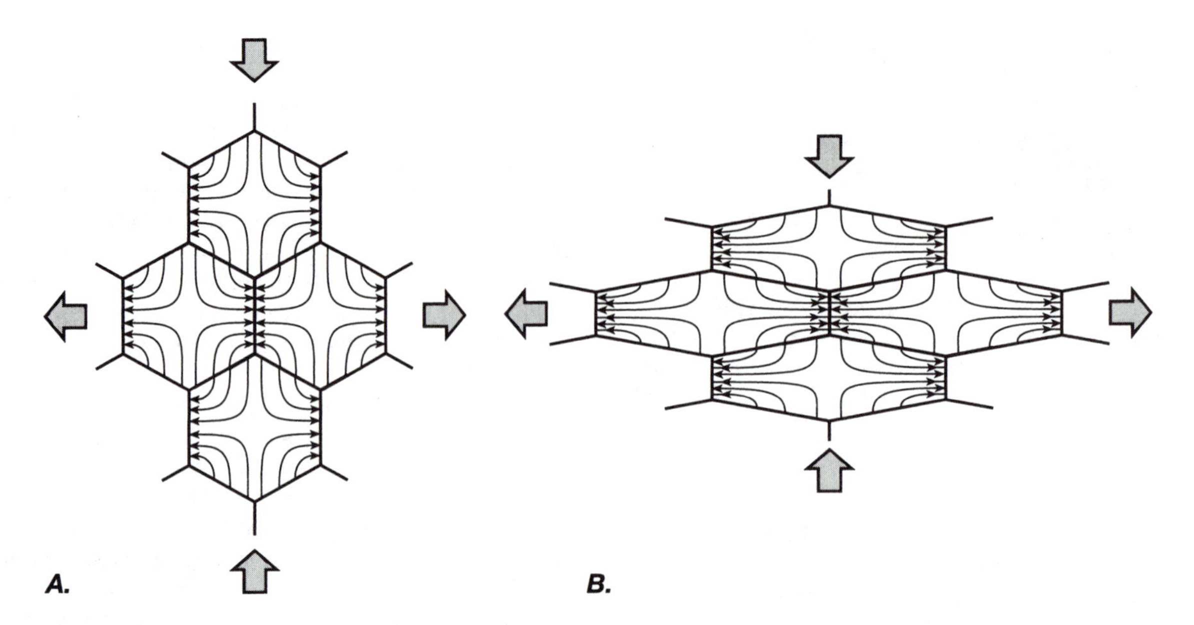

FIGURE 17.7 The change in grain dimension accomplished by Nabarro-Herring creep increases the diffusive path length and leads to a slowing down of the diffusion creep. A similar geometry results from grain boundary diffusion.

Solution creep probably is very similar to Coble creep, the principal difference being that the diffusivity is higher along fluid-filled grain boundaries than along dry grain boundaries. Like the other diffusion creep mechanisms, solution creep probably has a Newtonian viscous rheology. On the deformation mechanism maps for calcite and quartz, the rheology estimated for solution creep restricts (Figure 17.1*B*) or eliminates (Figure 17.1*A*) the field of Coble creep. It is the dominant deformation mechanism at temperatures of several hundred degrees Celsius or less, such as would be found in low-grade metamorphic rocks, and at the strain rates expected for geologic deformation. This result accords with the field evidence for this mechanism of deformation that we discuss in the various sections cited in the preceding paragraph.

Increasing the grain size increases the diffusive path length and therefore decreases the efficiency of this deformation mechanism (Box 17-1). In high-grade metamorphic regions and in the mantle, it seems very likely that the mean grain size of deforming rocks is larger than the 100 μm (0.1 mm) assumed for the maps in Figure 17.1. Thus the actual range of conditions at which diffusion mechanisms operate in the Earth may be considerably smaller than shown. For a tenfold increase in grain size, strain rates decrease by a factor of 100 for Nabarro-Herring creep and by a factor of 1000 for Coble and solution creep. (This result derives from the dependence of the creep rate on the grain size d; see Section 16.5(ii) and Equations (16.15), (17-1.1), and (17-1.2)). In such a case, the dislocation creep field would expand at the expense of the diffusion creep fields, and the Nabarro-Herring field would expand into the Coble and solution creep fields.

During any diffusive mass transfer process, grains tend to shorten parallel to the maximum compressive stress and to lengthen parallel to the minimum compressive stress (Figure 17.7). As the shape of the grains changes, the lengths of the diffusive paths increase, and consequently the strain rate decreases. At high temperatures, however, grain boundary migration becomes important (see Section 17.5), and this process could maintain a roughly equant grain shape, thereby allowing diffusion creep to continue as a steady-state process.

Solution creep, of course, requires the presence of an aqueous fluid in the rocks. This is common in metamorphic rocks, but if deformation occurs under dry conditions, then solution creep is impossible, and the diffusion creep field in the deformation maps would be considerably larger.

iv. Superplastic Creep

Grain boundaries in a polycrystalline material also can be important because sliding on the boundaries can accommodate strain. If voids are not to open along the boundaries, however, the shapes of the grains must change slightly as they slide past one another, and this deformation can be accommodated by diffusion (Figure 17.8). **Superplastic creep** is a phenomenon observed in some metals and inferred for some rocks. It results from coherent grain boundary sliding in which deformation occurs without the opening of gaps or pores between adjacent crystal grains. It is characterized by rapid strain rates at low stresses, compared to other mechanisms of ductile deformation, and by a power-law rheology with a stress exponent between 1 and 2.

Figure 17.8 shows one possible mechanism of superplastic creep for four isolated grains in a two-dimensional

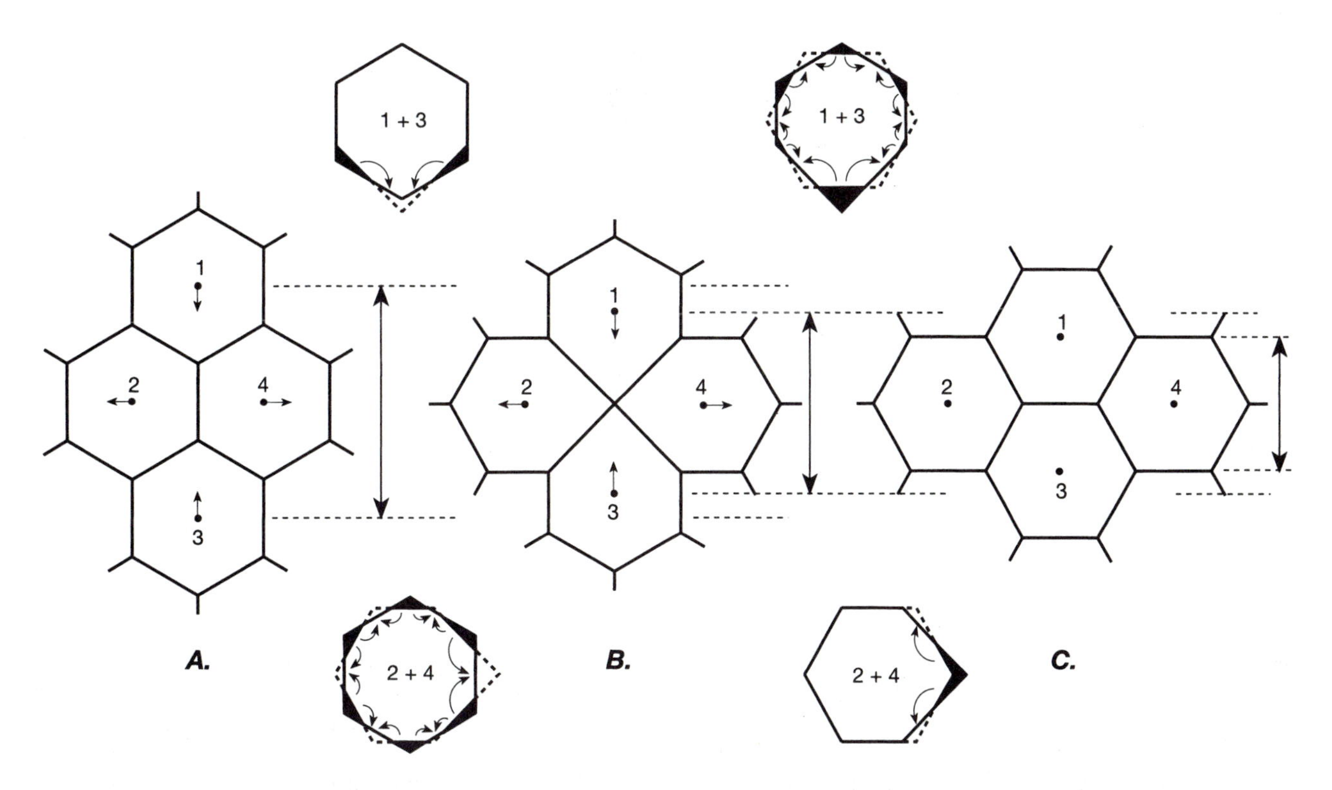

FIGURE 17.8 Model of superplastic creep by diffusion-accommodated coherent grain boundary sliding. The curved arrows in the isolated grains indicate the diffusive paths by which material is transferred in order to change the shape of the grains from *A* to *B* and *C*. Solid lines show the initial grain shape; dashed lines the final shape. Black shading indicates the part of the old crystal that must diffuse to fill in the area outlined by the dashed lines. This change of shape permits coherent grain boundary sliding by which neighboring grains become separated and non-neighboring grains become neighbors, a "neighbor switching" event. (After Ashby and Verrall 1973)

polycrystal. The changes in grain shape required to allow the grain boundary sliding are indicated by the isolated grains in the diagram, where the material shaded in black must diffuse along paths indicated by the curved arrows, to fill in the areas outlined by the dashed lines. The straight arrows at the grain centroids indicate the direction of motion of the grains as a result of grain boundary sliding. In the diagram, grains 2 and 4 start out as neighbors, and after the grain boundary sliding, grains 1 and 3 have become neighbors. Like other diffusion creep mechanisms, this mechanism results in a linear viscous rheology. However, because the average diffusive path lengths are shorter than for Coble or Nabarro-Herring creep, and because the grains slide along their boundaries, the strain rate is theoretically about five times higher than for Nabarro-Herring or Coble creep. In this mechanism, the grains remain approximately equant so that large strains can be accommodated (Figure 17.8; compare Figure 17.7). The effect of this process on the deformation maps of Figure 17.1, would be to increase the strain rate at any given temperature within the diffusion creep field and to increase the area of this field at the expense of dislocation creep field.

Other mechanisms may operate during superplastic creep to accommodate grain boundary sliding. If the grains change shape by the motion of dislocations, which we discuss in the following sections, a power-law rheology is predicted with a stress exponent of 2. It is possible for both the diffusion and the dislocation mechanisms to be active at the same time in accommodating grain boundary sliding.

17.4 LINEAR CRYSTAL DEFECTS: THE GEOMETRY AND MOTION OF DISLOCATIONS

The ductile deformation of a crystal causes a change in its shape but not in its crystalline structure. We can visualize how this process might occur in a perfect crystal by considering the model illustrated in Figure 17.9. Deformation could proceed by shift of lattice points along a crystallographic glide plane, as illustrated in the progression from Figure 17.9*A* to *C*. Repeating such shifts on parallel planes in the crystal would allow accumulation of large homogeneous strains (Figure 17.9*D*). Each step, however, would require a very large stress, because all atoms above the glide plane must be lifted at the same time out of their potential energy wells and moved over the adjacent atoms lying below the glide plane (Figure 17.9*B*) before they can settle back into an equivalent low-

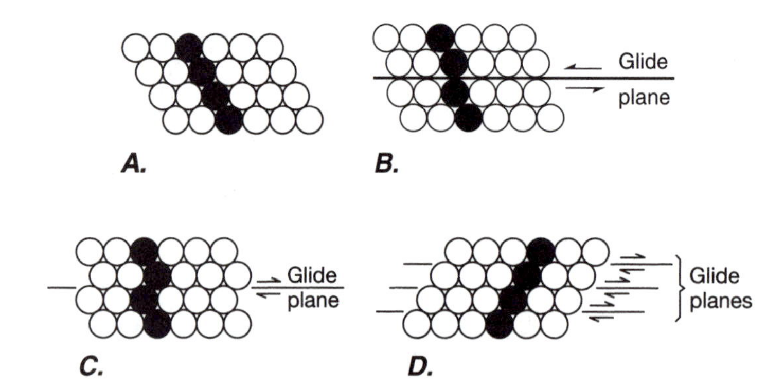

FIGURE 17.9 The theoretical process of slip in a perfect crystal involves the sliding of an entire lattice plane one lattice distance over the adjacent lattice plane. Black circles identify the same set of atoms in each diagram. *A.* Low-energy close-packed array of atoms. *B.* Transitional structure showing the dilation of the lattice as atoms slide out of low-energy well. *C.* Low-energy close-packed array after slip of one lattice distance along the glide plane. *D.* Homogeneous deformation due to sliding on all adjacent glide planes.

energy lattice site (Figure 17.9*C*). Because of this large stress, this model requires the theoretical yield strength of a perfect crystal to be very large.

The observed yield strengths for real crystals, however, are roughly one order of magnitude smaller than theoretical strengths for perfect crystals, which suggests the presence of crystal defects. For the two-dimensional crystal in Figure 17.9, shearing only one lattice point at a time on the glide plane, instead of an entire row, requires a much smaller stress. This process, however, introduces a defect into the lattice that separates sheared from unsheared parts of the lattice. In three dimensions, such defects are linear features called **dislocations**. The propagation of a dislocation across an entire glide plane has the net effect of shearing the entire glide plane one lattice spacing as in Figure 17.9*C*. The motion of dislocations through a crystal lattice is probably the most important mechanism for producing ductile deformation in crystalline materials.

i. Dislocation Geometry

The two principal types of linear defects that occur in crystal lattices are **edge dislocations** and **screw dislocations**. Both types of dislocation mark the boundary between sheared and unsheared crystal. They are distinguished by whether the sheared part of the crystal has moved perpendicular or parallel to the dislocation line, respectively.

A perfect crystal is represented in detail as a set of atoms in a regular three-dimensional array (Figure 17.10*A*). We produce dislocations by shearing part of the lattice (the shaded volumes at the corners of the block in Figure 17.10*A*) along the glide plane relative to the rest of the crystal. A dislocation develops within the crystal lattice along the interior edge of each sheared volume, so it lies on the glide plane at the boundary between the sheared and the unsheared areas of the glide plane. To produce an edge dislocation, we shear the shaded volume by one lattice spacing perpendicular to its interior edge (Figure 17.10*B*). The shearing leaves an extra half-plane of atoms above or below the glide plane at the boundary between the sheared and unsheared lattice, and the edge of that half-plane is the edge dislocation line. To produce a screw dislocation, we shear the shaded volume by one lattice spacing parallel to its interior edge (Figure 17.10*C*). This shearing leaves the initially planar arrays of lattice points shifted into a continuous spiral whose axis is the screw dislocation line. In both cases, the shearing requires breaking the bonds in lattice planes that cross the glide plane, and reconnecting the half-planes above the glide plane with the half-planes below the glide plane that are one lattice spacing away from the original connection.

The slip vector that is characteristic of a dislocation is called the **Burgers vector**.[3] It is parallel to the glide plane, has a length of one lattice spacing, and is perpendicular to an edge dislocation (Figure 17.10*B*) and parallel to a screw dislocation (Figure 17.10*C*). Any counterclockwise circuit of lattice steps around a dislocation—for example, starting from below and left of the dislocation and consisting of an equal number of lattice steps to the right, up, left, and down—will not end up at the starting point. The vector from the end of a counterclockwise circuit to the starting point is the characteristic Burgers vector for the dislocation, and its orientation defines the sign of the dislocation. Both edge and screw dislocations can develop on a given glide plane with two different signs; for screw dislocations, these are right-handed or left-handed screws, depending on the sense of the helix of lattice planes around the dislocation line. Dislocations of opposite sign on the same glide plane are shown for

[3]After the Dutch-American materials scientist J. M. Burgers.

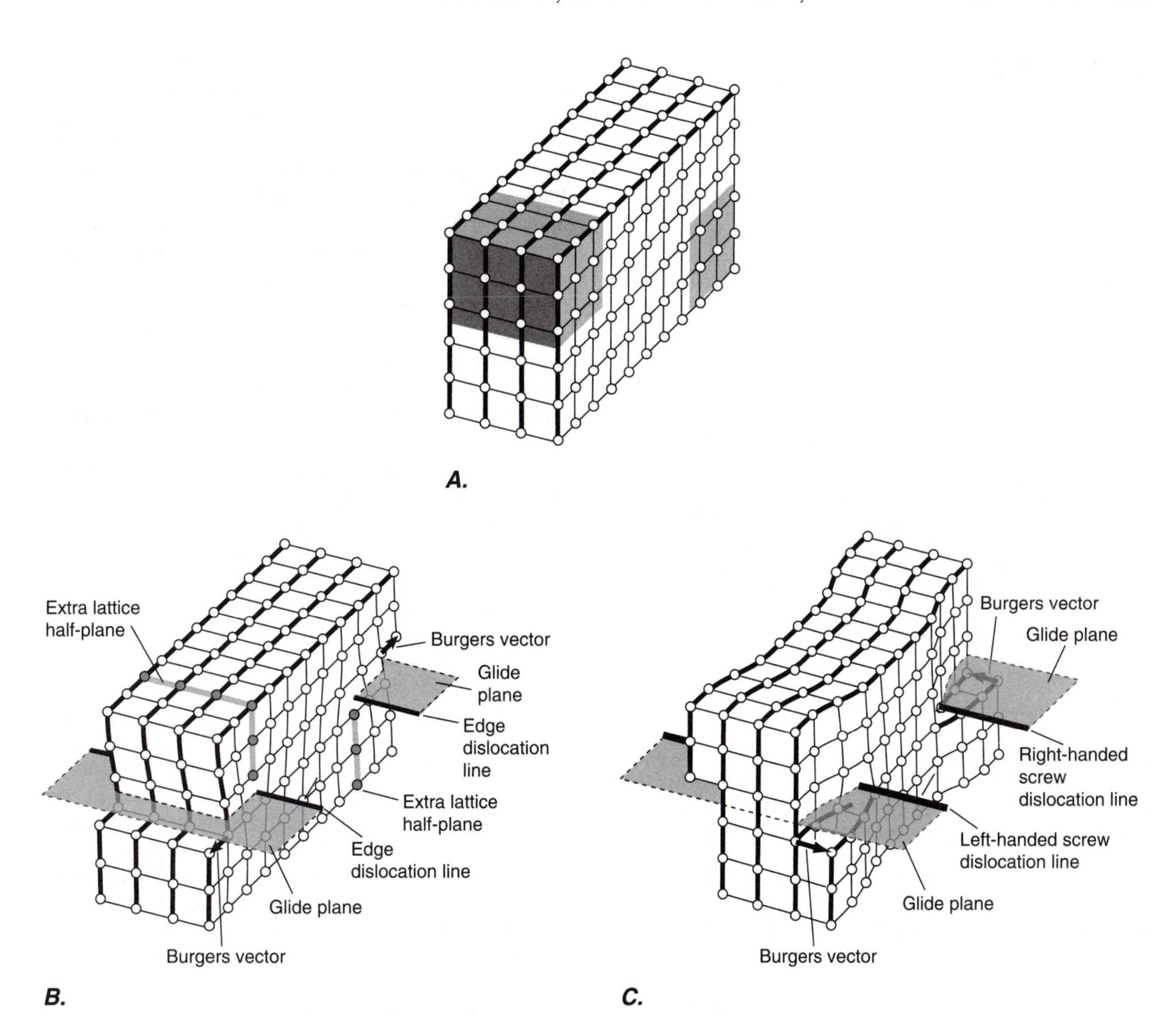

FIGURE 17.10 The geometry of edge and screw dislocations. The two examples of each type of dislocation in *B* and *C* are dislocations of opposite sign. The Burgers vector is shown for each dislocation. *A.* A perfect crystal lattice with balls representing atoms and bars the bonds between them. Shaded corners at opposite ends of the block indicate volumes to be sheared to produce either edge (*B*) or screw (*C*) dislocations. *B.* Two edge dislocations of opposite sign at opposite ends of the block produced by shearing the shaded part of the lattice in *A* one lattice spacing perpendicular to the boundary between sheared and unsheared areas of the glide plane (shaded). The edge dislocation is at the interior edge of the extra half-plane of atoms that is created by the shearing and separates sheared from unsheared areas of the glide plane. The atoms and lattice plane defining the extra half-plane are shaded gray. The Burgers vector is perpendicular to each edge dislocation line, and the vectors are of opposite sign for the two edge dislocations shown. *C.* Screw dislocations of opposite sign produced by shearing the shaded part of the crystal lattice in *A* one lattice dimension in a direction parallel to the boundary between sheared and unsheared areas of the glide plane. The crystal lattice planes form a continuous helical surface around the dislocation line with either a right-handed or a left-handed screw. The Burgers vector is parallel to each screw dislocation line, and the vectors are of opposite sign for the two dislocations shown.

edge dislocations in Figure 17.10*B* and for screw dislocations in Figure 17.10*C*. Note that in each case, the Burgers vectors **b** have opposite directions.

Dislocations need not be straight lines through the lattice. Curvature of a dislocation results from offsets in the dislocation line. The offsets can be short segments of edge dislocation type, called **jogs**, in a screw dislocation, or short segments of screw dislocation type, called **kinks**, in an edge dislocation. For example, in Figure 17.11*A*, the slipped region of the glide plane (shaded) is bounded by a curved dislocation composed of segments of edge dislocations offset by kinks and segments of

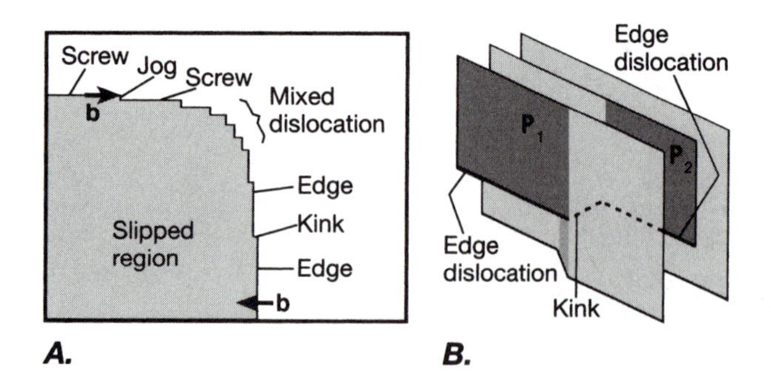

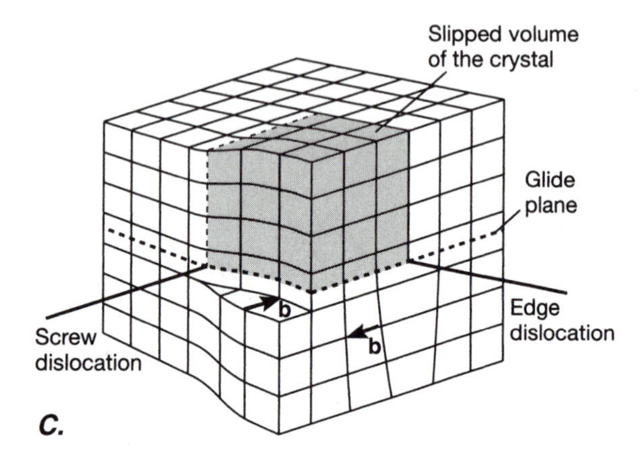

FIGURE 17.11 The nature of a curved dislocation. *A.* Curvature of an edge dislocation within the slip plane can be ascribed to an increasing density of kinks along the edge. At moderate angles to the Burgers vector, the dislocation line takes on mixed character, and at low angles it becomes a screw dislocation with edge-like jogs along it. *B.* Geometry of a kink in an edge dislocation. The portions of the planes that are shaded and labeled P_1 and P_2 are the portions that are extra half-planes in the crystal lattice. The edge dislocation is offset by the kink. The kink has the character of a short segment of screw dislocation, as indicated by tracing a clockwise circuit around the kink segment from plane P_1 to P_2. *C.* The shaded volume of the crystal has slipped relative to the unshaded portion. The boundaries on the glide plane are an edge and a screw dislocation, as shown in *A.*

screw dislocation offset by jogs. Where neither type of dislocation is dominant, the dislocation is of mixed character. Figure 17.11*B* shows the detailed geometry of a kink in an edge dislocation. Jogs also permit both types of dislocations to curve out of their glide planes (see Figure 17.15*C*). The lattice geometry for a curved dislocation that changes character from screw to edge dislocation as it changes orientation is illustrated at a corner of a glide plane in Figure 17.11*C*. Notice that for this particular geometry of curved dislocation, the Burgers vectors are of opposite sign for the screw and the edge dislocations (compare Figure 17.10*B*, *C*; see Figure 17.12*B*).

Because a dislocation marks the boundary on a glide plane between slipped and unslipped portions of the crystal lattice, it cannot simply stop inside the crystal. It must either continue to the edge of the crystal where it ends at a step in the lattice, or it must form a closed loop within the crystal. Figure 17.12 shows schematically the structure of a rectangular dislocation loop, with the lattice planes below the glide plane shown in thick gray lines and those above the glide plane shown in thin black lines. Within the closed dislocation loop, the crystal lattice above the glide plane has all slipped over the glide plane in the same direction relative to the lattice below the glide plane. Thus the closed dislocation loop must have portions that are edge dislocation (perpendicular to the Burgers vector) and portions that are screw dislocation (parallel to the Burgers vector) (Figure 17.12). A curved dislocation would also have portions that are mixed in character where the dislocation line is oblique to the Burgers vector (Figure 17.11*A*).

The dislocation loop has opposite Burgers vectors on opposite sides of the loop (Figure 17.12*B*). Where the slip vector is perpendicular to the boundary of the slipped region, the dislocation is an edge, with a positive Burgers vector where the extra half-plane lies below the glide plane and a negative Burgers vector where it lies above the glide plane (Figure 17.12*B*; compare Figure 17.10*B*). Where the slip vector is parallel to the boundary, the dislocation is a screw, with a positive Burgers vector for a right-handed screw and a negative Burgers vector for a left-handed screw (Figure 17.12*B*; compare Figure 17.10*C*). Notice that if two dislocations of the same type but opposite sign on the same glide plane were to come together, the dislocations would annihilate each other. For example, if two edge dislocations of opposite sign come together, the extra half-plane below the glide plane for the positive dislocation would join the extra half-plane above the glide plane for the negative dislocation to form a complete lattice plane with no dislocation (Figures 17.10*B* and 17.12*B*). Similarly, if two screw dislocations of opposite sign on the same glide plane come together, they annihilate each other, leaving perfect crystal (Figures 17.10*C* and 17.12*B*). Thus any dislocation loop on a single glide plane (Figure 17.12*B*) could shrink down to nothing and disappear, leaving only perfect crystal.

In the diagrams illustrating dislocation geometry, we have used a cubic lattice. In many materials (including silicates), however, the chemical compositions and the crystal structures are much more complex. For such structures, the principles of dislocation geometry remain the same except that instead of individual atoms and planes of atoms, we must consider unit cells and planes of unit cells of the crystal structure. In complex crystals the Burgers vector is the spacing of the unit cell rather than the interatomic spacings implied by the diagrams of simple structures.

Although dislocations are too small to be directly observable with an optical microscope, several techniques provide direct evidence for them. In some materials they can be made observable by "decorating" them with another material. For example, when natural olivine is held at elevated temperature in an oxidizing atmosphere for a

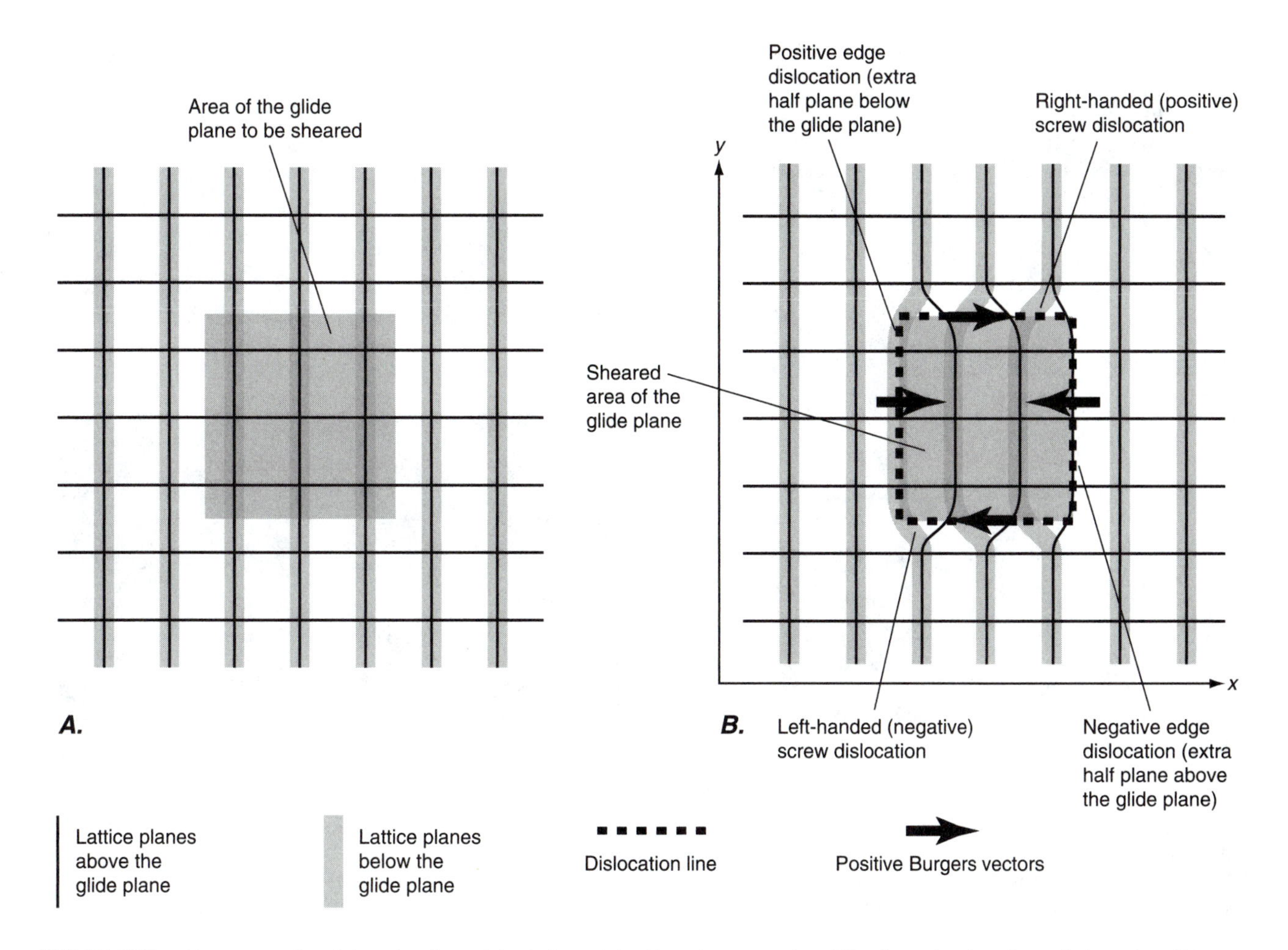

FIGURE 17.12 Geometry of a dislocation loop (shaded area) looking down on the glide plane. Lattice planes below the glide plane are drawn in heavy gray lines; lattice planes above the glide plane are drawn in thin black lines. *A*. Perfect crystal lattice showing the area of the glide plane (shaded) to be sheared to create the dislocation loop. *B*. Geometry of the dislocation loop, with the dislocation line shown as short dashed lines. Notice that the lattice planes above the glide plane are sheared to the right, and those below are sheared to the left. The Burgers vectors for dislocations on opposite sides of the loop have opposite signs.

short period of time, oxygen diffuses rapidly along the dislocations and oxidizes some of the iron in the olivine. The small lines of oxides are then visible under an ordinary petrographic microscope, revealing the dislocation structure (Figure 17.13*A*).

When a polished surface of a crystal is chemically etched, the strained material near an emerging dislocation reacts more vigorously with the etchant than the unstrained material, thereby producing a pit in the surface. Such etch pits are readily observable with a reflecting microscope (Figure 17.13*B*).

Dislocations in very thin foils of material can be imaged directly in a transmission electron microscope. The image results from the interaction of the electron beam with the strain field around a dislocation. Under ideal conditons, the characteristics of this interaction can even be used to determine the orientation of the Burgers vector of a dislocation and thus identify its nature, but in normal electron micrographs the two dislocation types cannot be distinguished. Figure 17.13*C* shows an electron beam image of dislocations in quartz.

ii. The Motion of Dislocations

Ductile deformation can occur in crystal lattices by the propagation of enormous numbers of dislocations through the lattice. We now look at the ways in which dislocations produce ductile flow and at some of the simple interactions that can occur between and among dislocations as they move through the crystal.

Figure 17.14 illustrates how **glide** of an edge dislocation through a crystal produces ductile deformation. Figure 17.14*A* shows a perfect lattice with one of the glide planes marked. Columns of atoms are numbered 1 to 8, and the letters *A* and *B* indicate the parts of the column above (*A*) and below (*B*) the glide plane. A shear stress

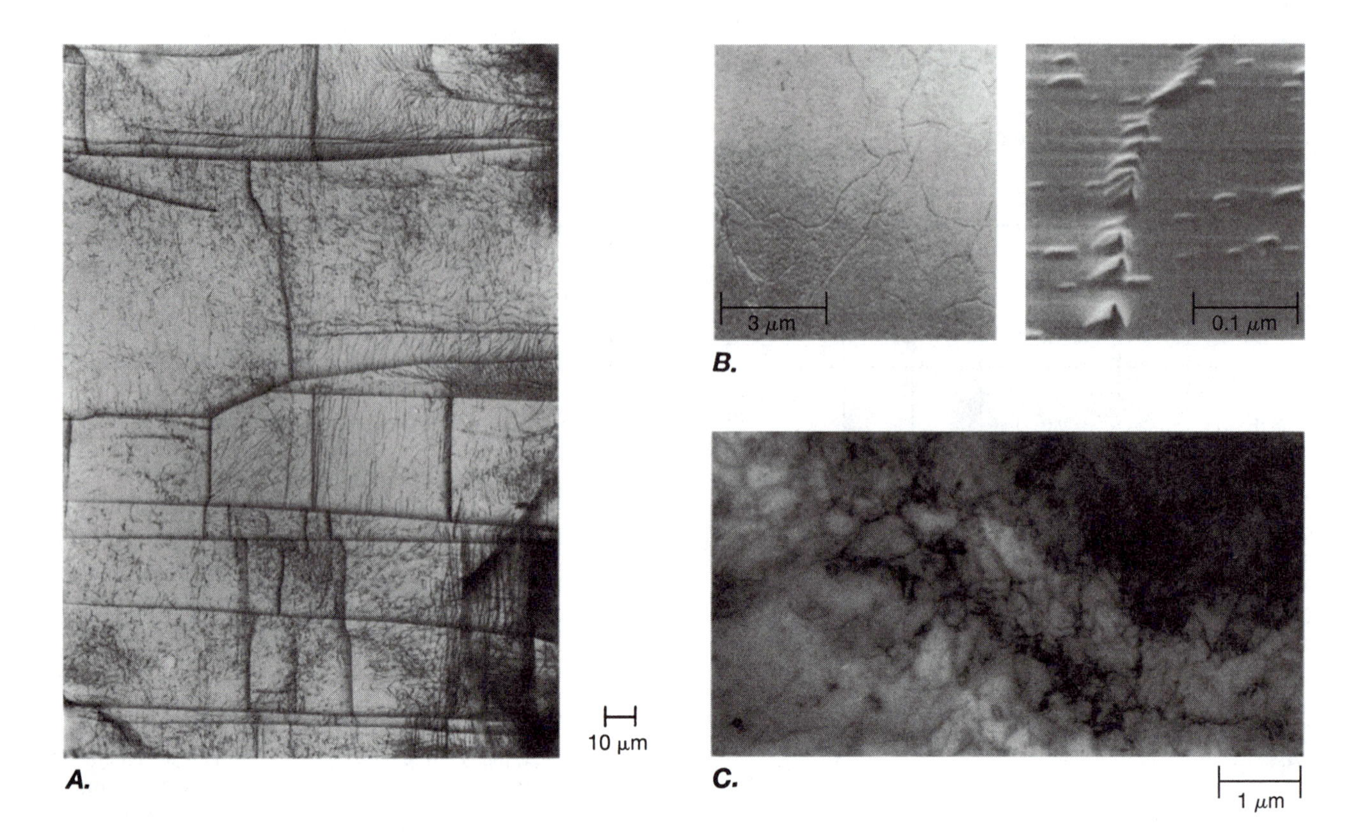

FIGURE 17.13 Images of dislocations in silicates. *A.* Dislocations in olivine (fine black lines) decorated by oxidation and imaged under the petrographic microscope. Rectangular cells outlined by dark boundaries are subgrains. (Courtesy Jin Zhen-Ming) *B.* Etch pits in quartz showing locations on the crystal surface where dislocations intersect the surface. (From Ball and White 1977) *C.* Electron microscope image of dislocations in quartz (fine black lines). (From Twiss 1976)

applied to the crystal (Figure 17.14*B*) breaks the bonds crossing the glide plane in columns 1 and 2 (Figure 17.14*C*); column 1A then forms a new bond with column 2B. Column 2A becomes an extra half-plane forming an edge dislocation in the glide plane. Continued shear stress (Figure 17.14*D*) causes column 3B to break from 3A and then to join with 2A, leaving 3A as the extra half-plane and moving the dislocation one lattice space to the right. Several more such bond-switching events (Figure 17.14*E*) result in the dislocation gliding completely out of the crystal leaving a step of one Burgers vector width on the surface of the crystal (Figure 17.14*F*). Propagation of many dislocations through the crystal from one side to the other on the set of parallel glide planes results in solid-state flow, that is, a permanent change in the shape of the crystal without the destruction of the solid crystalline structure.

The model in Figure 17.14 is only two-dimensional, so in fact a whole row of bond-switching events must accompany each step made by the dislocation line. Because only one row of bonds is broken and re-formed for each Burgers vector length that a dislocation moves, the motion of a dislocation requires much less stress than shearing a whole sheet of atoms all at once, as in Figure 17.9*A–C*. In fact, the stress required for advancing a whole dislocation line is minimized by propagating kinks or jogs down the dislocation line, thereby requiring the breaking of only one bond at a time.

A screw dislocation and an edge dislocation having the same Burgers vector are perpendicular to each other, and they propagate in perpendicular directions under the same shear stress. For example, in Figure 17.11*A*, the edge dislocation propagates to the right, whereas the screw propagates up, leaving behind a region of crystal above the glide plane that has slipped one lattice spacing to the right. Under a given shear stress, parallel dislocations of opposite sign propagate in opposite directions. Thus, depending on the orientation of the shear stress and the signs of the dislocations, a dislocation loop tends either to expand out to the edges of the crystal or to shrink down until it disappears, with dislocation segments of opposite signs on opposite sides of the loop in effect annihilating each other.

A set of parallel crystallographic glide planes with an associated slip vector constitutes a **slip system** in a crystal. In general, the most closely packed planes and the directions with the shortest lattice spacings tend to be the most easily activated slip systems. Table 17.1 lists the

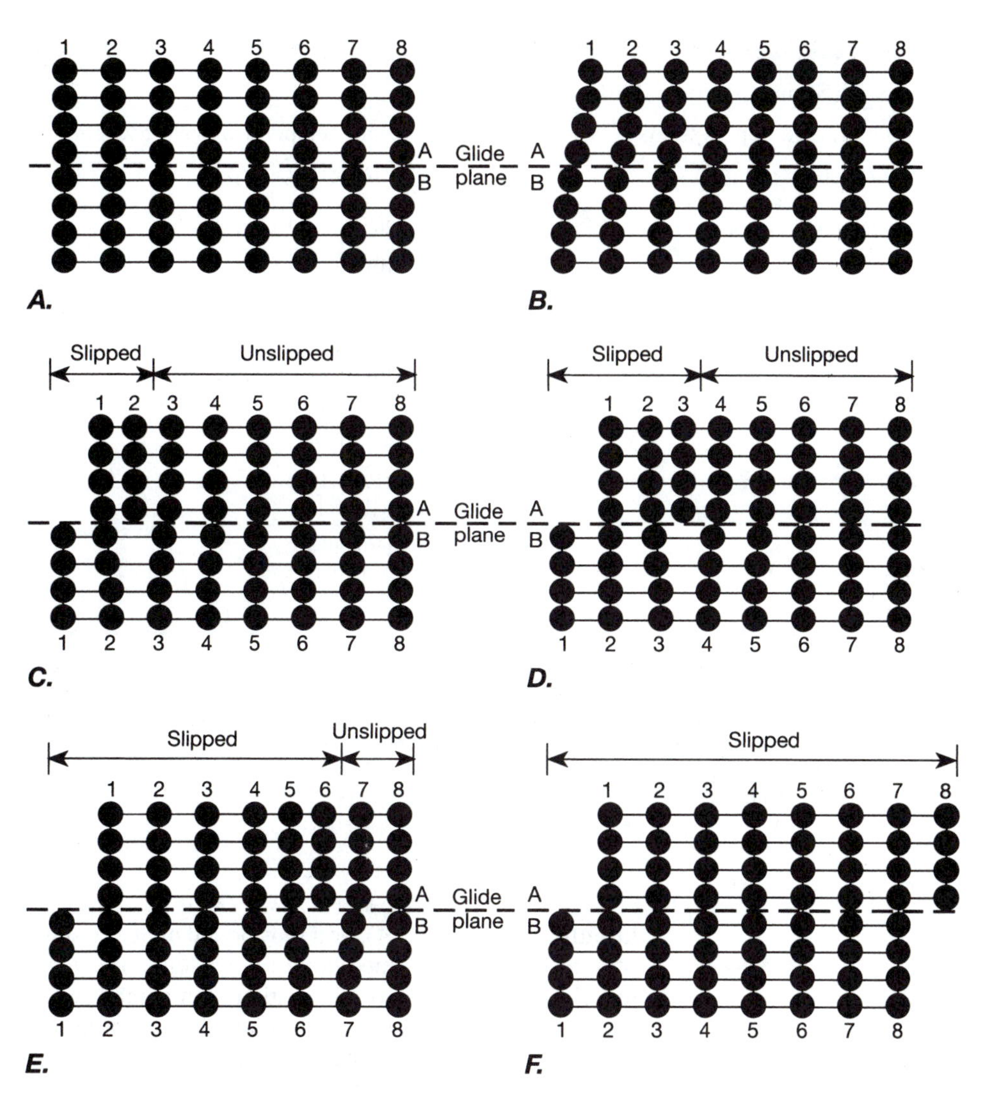

FIGURE 17.14 The glide of a dislocation through the crystal lattice is accomplished by switching the bonds of neighboring atoms across the glide plane. *A* through *F* illustrate the process of forming an edge dislocation and propagating it through the crystal to produce one Burgers vector of offset in the crystal. Crystallographic planes in the undeformed crystal are numbered 1 through 8 and the letters A and B refer to the parts of the planes above and below the slip plane, respectively. (From Spry 1969)

principal slip systems of some common minerals. Whether a slip system is activated depends on the magnitude of the critical shear stress necessary to move a dislocation and on the magnitude of the shear component of the applied stress on the glide plane in the direction of slip.

Edge dislocations glide only on planes that contain both the dislocation line and the Burgers vector. They can leave their original glide plane by a process called **climb**. In order for an edge dislocation to climb, atoms must either be added to, or taken away from, the edge of the extra half-plane. The edge dislocation in Figure 17.15*A*, for example, climbs down one lattice spacing (the half-plane lengthens) if an atom from a neighboring site jumps onto the bottom of the half-plane, leaving a vacancy behind (Figure 17.15*A* to *B*), which can then diffuse away. The edge dislocation climbs up if a vacancy diffuses to the bottom of the half-plane, and the plane shortens (Figure 17.15*B* to *A*). By the process of climb, edge dislocations act as sources or sinks for vacancies in the lattice. Jogs are created in the dislocation line if different segments of the dislocation climb by different amounts, as shown by the irregular edge of the isolated extra half-plane in Figure 17.15*C*.

Screw dislocations need not glide on a single plane because the Burgers vector and the dislocation line are parallel and thus do not define a unique plane. The glide planes for a screw dislocation belong to a family of planes, each of which must be parallel to the Burgers vector (Figure 17.16). Thus it is possible for a segment of a screw to change glide planes, which is referred to as **cross-slip**. Figure 17.16 shows that when a segment of a screw dislocation undergoes cross slip, it leaves behind a segment of edge dislocation that connects it to the rest of the screw.

We describe in Section 17.4(i) how dislocations of opposite sign, either edge or screw, on the same glide plane can annihilate each other by gliding toward each other. Two edge dislocations of opposite sign can also annihilate each other by climbing toward each other. For example, if two edge dislocations of opposite sign bound opposite sides of an extra lattice plane within the crys-

TABLE 17.1 Dominant Slip Systems of Some Common Minerals

Mineral	Low-Temperature Slip Systems[a]	High-Temperature Slip Systems[a]
Calcite	$\{10\bar{1}1\}\langle\bar{1}012\rangle$: $\{r_1\}\langle r_1 \cap f_2\rangle$ **Twinning:** $\{\bar{1}012\}\langle 10\bar{1}1\rangle$: $\{e_1\}\langle e_2 \cap a_2\rangle$	$\{\bar{2}021\}\langle 1\bar{1}02\rangle$: $\{f_1\}\langle r_2 \cap f_3\rangle$ $\{\bar{2}021\}\langle\bar{1}2\bar{1}0\rangle$: $\{f_1\}\langle r_3 \cap f_1\rangle$
Quartz	$(0001)\langle 11\bar{2}0\rangle$:$(base)\langle a\rangle$	$\{10\bar{1}0\}[0001]$: $\{m_1\}[c]$ $\{10\bar{1}0\}\langle 1\bar{2}10\rangle$: $\{m_1\}\langle a\rangle$ $\{10\bar{1}0\}\langle 1\bar{2}13\rangle$: $\{m_1\}\langle c + a\rangle$ $\{10\bar{1}1\}\langle 1\bar{2}10\rangle$: $\{r\}\langle a\rangle$ $\{10\bar{1}1\}\langle 2\bar{1}\,\bar{1}3\rangle$: $\{r\}\langle c \pm a\rangle$
Micas		$(001)\langle 110\rangle$ $(001)[100]$
Halite	$\{110\}\langle 1\bar{1}0\rangle$	$\{110\}\langle 1\bar{1}0\rangle$ $(001)\langle 1\bar{1}0\rangle$
Olivine	$(100)[001]$ $\{100\}[001]$	$\{110\}[001]$ $(010)[100]$ $\{0k\ell\}[100]$

[a] In labeling slip systems, we give the Miller indices of the slip plane first, followed by the components of the slip direction vector. If a specific plane and direction are indicated, they are written in parentheses and in square brackets, respectively: (plane) [direction]. If a set of symmetrically equivalent slip systems is indicated, one plane of the set is written in braces and the corresponding direction is written in angle brackets: {plane} <direction>. The letter symbols following the Miller indices for the slip systems are standard abbreviations for the different planes and directions. Some directions are indicated by the intersection of two planes $[p_1 \cap p_2]$.
Source: Data after Nicolas (1988) and Nicolas and Poirier (1976).

tal lattice (Figure 17.17*A*), or if they bound opposite sides of a missing lattice plane within the crystal lattice (Figure 17.17*B*), then the two dislocations can climb toward each other and annihilate each other. The extra lattice plane is either eliminated by diffusion (Figure 17.17*A*) or the missing lattice plane is filled in by diffusion (Figure 17.17*B*) to leave perfect crystal, and no dislocations remain. Because the elimination of dislocations from the lattice lowers the internal strain energy of the crystal, dislocations of opposite sign tend to attract one another unless driven apart by a shear stress on the glide plane or a tensile or compressive stress across the extra lattice plane.

Dislocations of the same sign on the same glide plane, however, tend to repel one another. The strain about a single edge dislocation is contractional on one side and extensional on the other, leading to compressional and tensile stresses on opposite sides of the glide plane

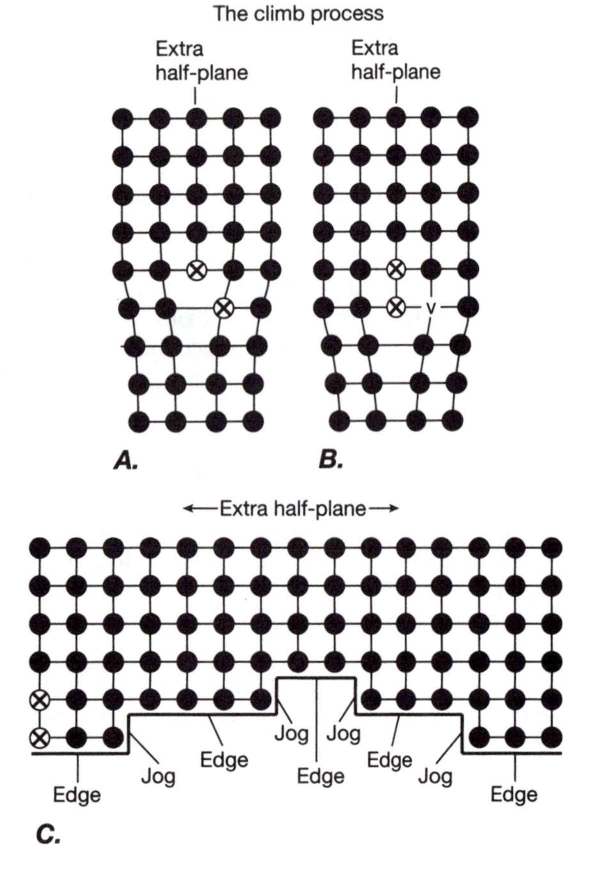

FIGURE 17.15 Climb of an edge dislocation. The dislocation climbs downward if an atom from a neighboring site jumps onto the half-plane (A to *B*) leaving a vacancy behind that can then move away by diffusion. The dislocation climbs upward if a vacancy from a neighboring site jumps onto the half-plane (*B* to *A*). *C*. The extra half-plane associated with an edge dislocation along which there have been different amounts of climb, leaving jogs in the dislocation line. The view shows the third dimension of the extra half-plane labeled in *B*.

around the dislocation (Figure 17.17*C*). As two edge dislocations approach each other, their strain fields overlap and add together (Figure 17.17*D*). The tendency of the lattice to adopt a low-energy configuration provides an effective force that drives the two dislocations apart.

Edge dislocations of the same sign on different parallel-slip planes are attracted into **dislocation walls** (Figure 17.17*E*) in which the dislocations are aligned above one another. This is a low-energy configuration because the contractional strain associated with one side of one dislocation is in part canceled out by the extensional strain of the adjacent dislocation. A dislocation wall is also called a **low-angle boundary** or a **tilt boundary**, because it is a boundary between two parts of the crystal lattice that are tilted at a low angle with respect to each other (Figure 17.17*E*). They commonly form the boundaries of **subgrains** (Figure 17.13*A*), which are small volumes within individual crystal grains that commonly differ in orientation by 1° to 2° from neighboring subgrains. The greater the density of dislocations in a tilt boundary, the greater the angle of tilt.

Dislocations are easily lost from a lattice by gliding or climbing out of the lattice or by annihilation within the lattice, as described three paragraphs above and in Sec-

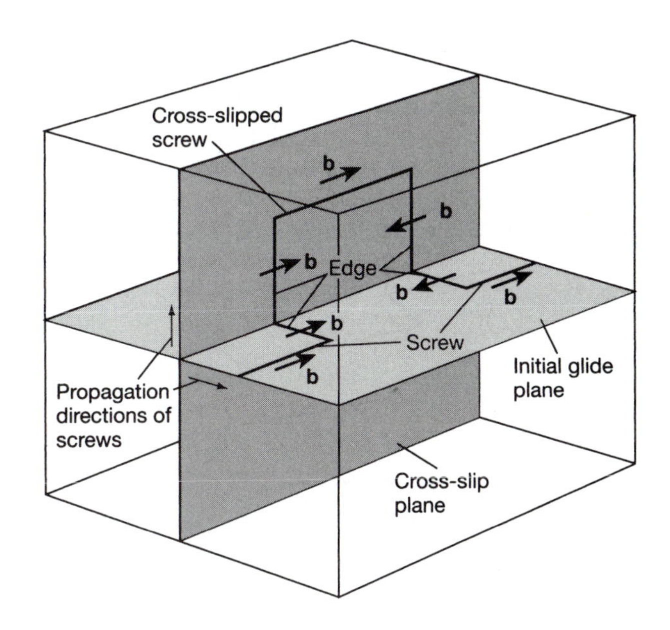

FIGURE 17.16 Cross slip occurs when a screw dislocation changes slip planes, and the process leaves segments of edge dislocation connecting the cross-slipped segment with the rest of the screw dislocation. In this diagram, the left and right portions of the screw have continued gliding on the original slip plane, whereas the central portion has cross slipped.

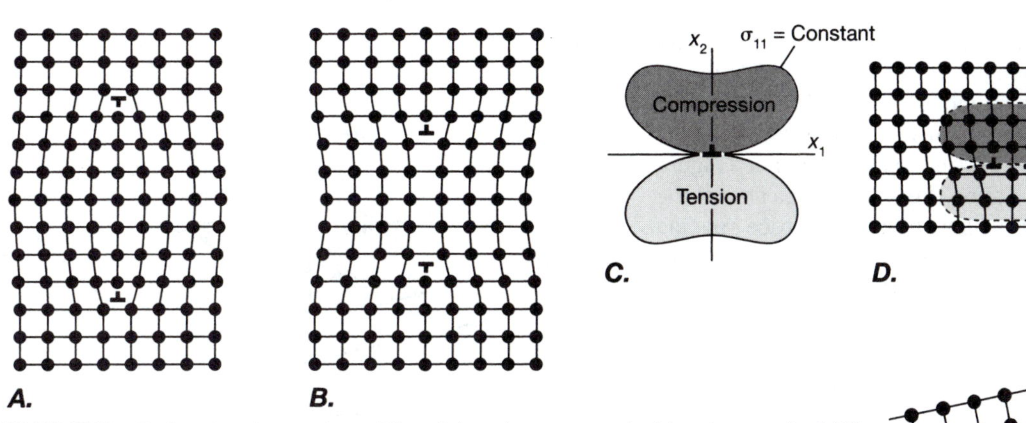

Low-angle boundary

E.

FIGURE 17.17 Dislocation interactions. Edge dislocations are marked by the standard "T" symbol, where the stem of the "T" extends toward the extra half-plane. *A* and *B*. Dislocations of opposite sign tend to attract each other and annihilate, thereby eliminating imperfections in the crystal and lowering the strain energy. In *A*, if dislocations climb toward each other, they eliminate the extra lattice plane between them. In *B* if they climb toward each other, they complete the missing lattice plane between them. *C*. The form of a contour of constant stress σ_{11} about an edge dislocation. Note that one side of the dislocation is in compression and the other side is in tension. (From Hirth and Lothe 1968) *D*. The local distortion of the crystal lattice, and thus the local strain energy per unit volume, increases as two dislocations of the same sign on the same glide plane get closer together. The equilibrium configuration corresponds to the lowest possible energy state. Thus these dislocations effectively repel each other because the local strain energy per unit volume is lowered as the dislocations move apart. *E*. Edge dislocations tend to accumulate in a wall because it is a low-energy configuration. If two parallel dislocations of the same sign occur one above the other, the strain energy per unit volume for the lattice between the dislocations is reduced because the contractional strain associated with one dislocation partly cancels the extensional strain associated with the other. The accumulation of a large number of dislocations in this relationship forms a dislocation wall—also called a low-angle or a tilt boundary—which separates two parts of a crystal lattice with slightly different orientations.

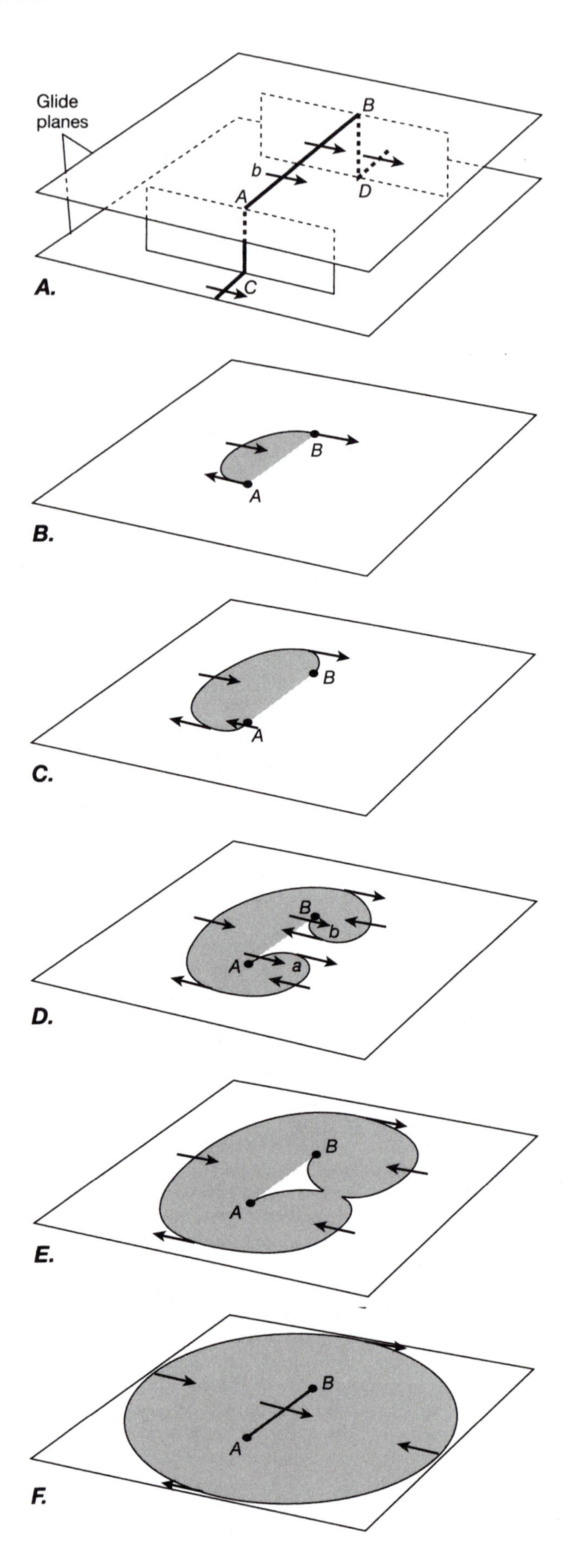

FIGURE 17.18 The operation of a Frank-Read source for dislocation generation. Burgers vectors for the edge and screw segments of the curved dislocation are shown in each diagram. *A*. A segment AB of an edge dislocation has climbed from the lower glide plane to the upper one, generating two dislocations AC and BD, which are not on the glide plane and therefore cannot move easily. They pin the segment AB at its ends. *B*. The dislocation segment AB bows out in response to an applied shear stress, generating screw dislocation segments of opposite sign extending from A and from B. *C*. The edge part of the loop propagates parallel to its Burgers vector, while the screw parts of the loop propagate perpendicular to their Burgers vector, causing the dislocation to pivot about the pinning points A and B. *D*. The parts of the dislocation in lobes a and b, which are screw segments of opposite sign, approach each other. *E*. The lobes a and *b* meet and annihilate. *F*. The dislocation snaps back to the position AB, leaving a dislocation loop expanding in the glide plane.

tion 17.4(i). Thus, in order for them to be able to produce an arbitrary amount of ductile deformation, there must be some mechanism for generating dislocations. A common mechanism for dislocation generation is called a **Frank-Read** source. Assume that the segment *AB* of the dislocation in Figure 17.18*A* is pinned at both ends *A* and *B*. This might happen, for example, if the segment had climbed or cross slipped out of its original slip plane and the jogs were unable to move easily. With increasing shear stress on the material, the dislocation tends to bow out more and more from the pinning points (Figure 17.18*B*, *C*, *D*). In each figure, the shaded area shows the region of the crystal in which slip of one lattice spacing has occurred starting from the configuration in Figure 17.18*A*. The Burgers vectors indicate the signs of the pure edge and pure screw portions of the curved dislocation (compare the dislocation loop in Figure 17.12*B*). As the shear stress increases further, the dislocation pivots around each pinning point (Figure 17.18*C*) because dislocations of opposite sign are driven to glide in opposite directions, and edge and screw dislocations are driven to glide in directions perpendicular to each other. The two lobes of the dislocation at *a* and *b* (Figure 17.18*D*) are of opposite sign, and when they come together they annihilate (Figure 17.18*E*). By this process all the material within the dislocation loop has slipped one lattice spacing, and a dislocation segment is left between the original pinning points *A* and *B*. The dislocation loop that has been created continues to expand in response to the applied stress (Figure 17.18*F*), and the segment *AB* can bow out again an unlimited number of times to create unlimited numbers of dislocations. The same mechanism works for both edge and screw dislocations. With many sources such as this operating throughout a crystal, unlimited amounts of ductile deformation can be produced.

17.5 MECHANISMS OF DISLOCATION CREEP

Propagation of dislocations through crystal lattices is one of the most common mechanisms of ductile deformation of crystalline materials. Different mechanisms that can limit the rate of dislocation motion give rise to different rates of deformation in response to the same applied stress, thereby accounting for some of the observed differences in rheology.

For a crystal to undergo an arbitrary constant-volume deformation, a condition called the **von Mises condition** must be satisfied. This condition should not be confused with the Von Mises criterion for plastic failure named for the same person (see Section 8.4(i), Equation (8.6)). The von Mises condition for a crystal to undergo any arbitrary constant-volume deformation is that it have five independent slip systems. This condition arises because in three dimensions, we need six independent numbers to describe the state of strain, either the three principal strains (Equation (12.7); Figure 12.11) and their orientations, or the six independent components of the strain tensor (Equations (12.39) and (12.40)). If one imposes the condition of constant-volume deformation ($s_V = 1$ or $e_V = 0$), then we have one relationship among the three stretches or the three extensions (Equations (12.18)), so then we need only five independent numbers to define a constant-volume strain. Five independent slip systems are required to produce these five independent components of strain, whence the von Mises condition. In a polycrystal, if the strain is allowed to be inhomogeneous from one crystal grain to another but is still required to be coherent (that is, the grain boundaries do not separate), then four, or even three, independent slip systems may be enough to accommodate an arbitrary strain of the material, especially if grain boundary sliding can occur.

i. Dislocation Glide

At low temperatures and/or high stresses, ductile deformation of crystalline solids occurs predominantly by glide of dislocations through the lattice, a process called **cold working**. Resistance to the motion of the dislocations comes in part from the lattice itself, because of the necessity of breaking bonds for the dislocations to move. In silicates, the bonds are largely covalent and therefore strong, and the stress necessary to overcome the bond strength is high. Resistance also comes from obstacles that occur in the path of a gliding dislocation, such as other dislocations or impurities.

When several slip systems operate simultaneously, dislocations gliding on different slip systems inevitably intersect. This process introduces kinks and jogs into the dislocation lines that impede their motion. As strain increases, the number of dislocations increases, and the complexity of interactions produces tangles of dislocations that impede dislocation motion even more. The higher the density of dislocations, the more difficult it becomes for any dislocation to move. Thus higher stress is required to force the dislocations to glide and to produce more dislocations. This process is the origin of the **work-hardening** behavior described in Section 16.3. Figure 17.13*C* shows an example of dislocation tangles that resulted from deformation of quartz.

If the stress is held constant during cold working, the creep rate gradually declines and may approach zero because of the work hardening. If the strain rate is kept constant, the stress increases with increasing strain until either the material fractures or a steady-state saturation stress is reached that is sufficiently high to force the dislocations to glide out of the crystal. The steady-state condition is only approached at high strains, for example, $e_n > 1$. Thus the transient creep stage is very long, and in many low-temperature experiments a steady state is not observed. Dislocations may become so closely packed together that they act like Griffith cracks (see Section 8.8(i), or grain boundaries may open to form Griffith cracks, leading to the brittle failure of the sample before steady state can be established.

A theoretical model based on kinetic theory enables us to calculate the rate of glide of dislocations through a crystal lattice when the glide rate is controlled by obstructions. The theory predicts a strain rate dependent on an exponential function of the differential stress (Box 17-1). This rheology coincides with the experimentally observed behavior of crystalline solids in the high-stress regime.

The glide mechanism dominates the creep rate of minerals at very high stresses across a wide range of homologous temperature (Figure 17.1). Because the temperature dependence of creep is the same for the exponential law and the power law, the boundary between the two mechanisms is dependent only on the normalized differential stress (Figure 17.1).

ii. Recovery by Dislocation Climb

At homologous temperatures above approximately 0.5, work hardening is counteracted by recovery processes that allow the rearrangement of dislocations. One of the most important recovery mechanisms is climb of edge dislocations. Edge dislocations can climb toward others of opposite sign and annihilate them, they can climb into low-energy walls, tangles can become untangled through climb, and the climb of jogs on dislocations can allow the dislocations to glide more freely. Glide of the dislocations is still an important factor in ductile deformation, but climb is the rate-limiting process, because it is the slowest mechanism in the process.

Thus the steady-state creep observed at homologous temperatures above about 0.5 is a balance between the processes of work hardening, which tend to decrease dislocation mobility, and those of recovery, which tend to enhance their mobility. The mutual repelling force between dislocations, called the **backstress**, counteracts the effect of the applied stress on dislocations and thus inhibits dislocation glide. At steady state, the backstress is essentially equal to the applied stress, and new dislocations cannot be produced and cannot move unless others are eliminated, either by annihilation, or by climb or cross-slip into walls or out of the crystal.

A number of different models have been proposed to explain steady-state creep based on assumptions as to what the rate-limiting step is in the creep or recovery process. All of these models lead to equations of the power-law type (Box 17-1 and Equation (16.4)), which is consistent with the inference that the ductile deformation observed in experiments is controlled by these dislocation processes.

iii. Dynamic Recrystallization

Dynamic recrystallization, which in geology is sometimes referred to as **syntectonic recrystallization**, is a process by which new crystal grains form from old grains during the deformation process.[4] Two mechanisms are recognized by which dynamic recrystallization occurs: boundary migration recrystallization and subgrain rotation recrystallization.

Boundary migration recrystallization is recrystallization by means of the migration of a grain boundary separating highly strained from unstrained crystals of the same mineral. The material in the highly strained crystal crosses the grain boundary and is added to the unstrained crystal. Thus the grain boundary migrates into the more highly strained region and leaves unstrained crystal behind it. This process is comparable to a recovery process because it lowers the dislocation density, and therefore the strain energy, of deformed grains. In this case, however, deformed grains are completely replaced by new undeformed grains. The new grains in turn become deformed and are subsequently replaced. This process can be an important means of maintaining steady-state deformation at stresses much lower than normal recovery processes would require. Figure 17.19*A*, *B* shows the effects of boundary migration recrystallization in quartz. The highly serrated grain boundaries indicate high mobility, and such boundaries can bulge rapidly out into a volume of strongly strained crystal to nucleate a new recrystallized grain.

During **subgrain rotation recrystallization**, subgrain boundaries in highly strained regions of a crystal accumulate an increasing number of dislocations, which causes a progressive rotation of the subgrain lattice with respect to the surrounding crystal. Eventually, at roughly 10 degrees of relative rotation, the boundary becomes saturated with dislocations. Any further rotation changes the boundary into the high-angle grain boundary of a newly recrystallized grain, which can no longer be represented by a dislocation wall. Thus subgrain rotation recrystallization is not distinct from recovery, but rather results from the operation of recovery mechanisms such as dislocation climb. Figure 17.19*C* shows an old large grain of quartz that is filled with subgrains. The misorientation of the subgrains increases progressively from the center of the grain (region A) toward the boundary (region C) where the subgrains become indistinguishable from the new recrystallized grains that surround the old quartz grain.

Grain boundaries are high-angle boundaries characterized by large misorientations or structural discontinuity between adjacent grains. Such boundaries consist of a patchwork of areas of partial fit and areas of no fit between the lattices of adjacent grains. High-angle boundaries are very mobile because it is easy for atoms or ions to transfer one at a time across them. Subgrain boundaries, however, are relatively immobile because they can move only by the coordinated glide of all the dislocations in the subboundary wall.

No models of dislocation creep rigorously account for the effects of recrystallization by boundary migration or subgrain rotation. Empirically, steady-state creep during boundary migration recrystallization displays a higher creep rate than steady-state creep during recovery. In both cases, however, the material displays a power-law rheology. The increased creep rate presumably results from the higher rate of primary creep relative to steady-state creep (compare Figure 16.13*B*), which must occur in each new recrystallized grain immediately after its formation. The continual resupply of such new grains by dynamic recrystallization maintains a significant portion of the crystal grains in the rock in the primary creep regime.

At the strain rates characteristic of geologic deformation, dislocation creep is probably important in the deformation of quartz at high grades of metamorphism (Figure 17.1*A*), especially if the dislocation creep field is enlarged by higher creep rates associated with recrystallization; by the presence of water, which causes hydrolytic weakening; and by larger grain sizes, which suppress diffusion creep. This is consistent with the development of crystallographic preferred orientations, which are commonly observed in quartz tectonites (see Section 17.7). The deformation map for olivine (Figure 17.1*C*) indicates that power-law creep probably is also a major creep mechanism in upper mantle deformation. This conclusion is again consistent with the evidence for crystallographic preferred orientation of olivine in peri-

[4]As used here the term *recrystallization* refers to processes that leave the chemical composition of the affected mineral grains essentially unchanged. Such recrystallization is "dynamic" if it occurs during deformation and "static" if deformation is not occurring. Metamorphic petrologists commonly use the same word to describe the result of chemical reactions during which the reactant minerals break down and new minerals form. The term **neomineralization** can also be used for the latter process. The meaning is clear from the context, although failure to recognize the difference in usage can cause confusion.

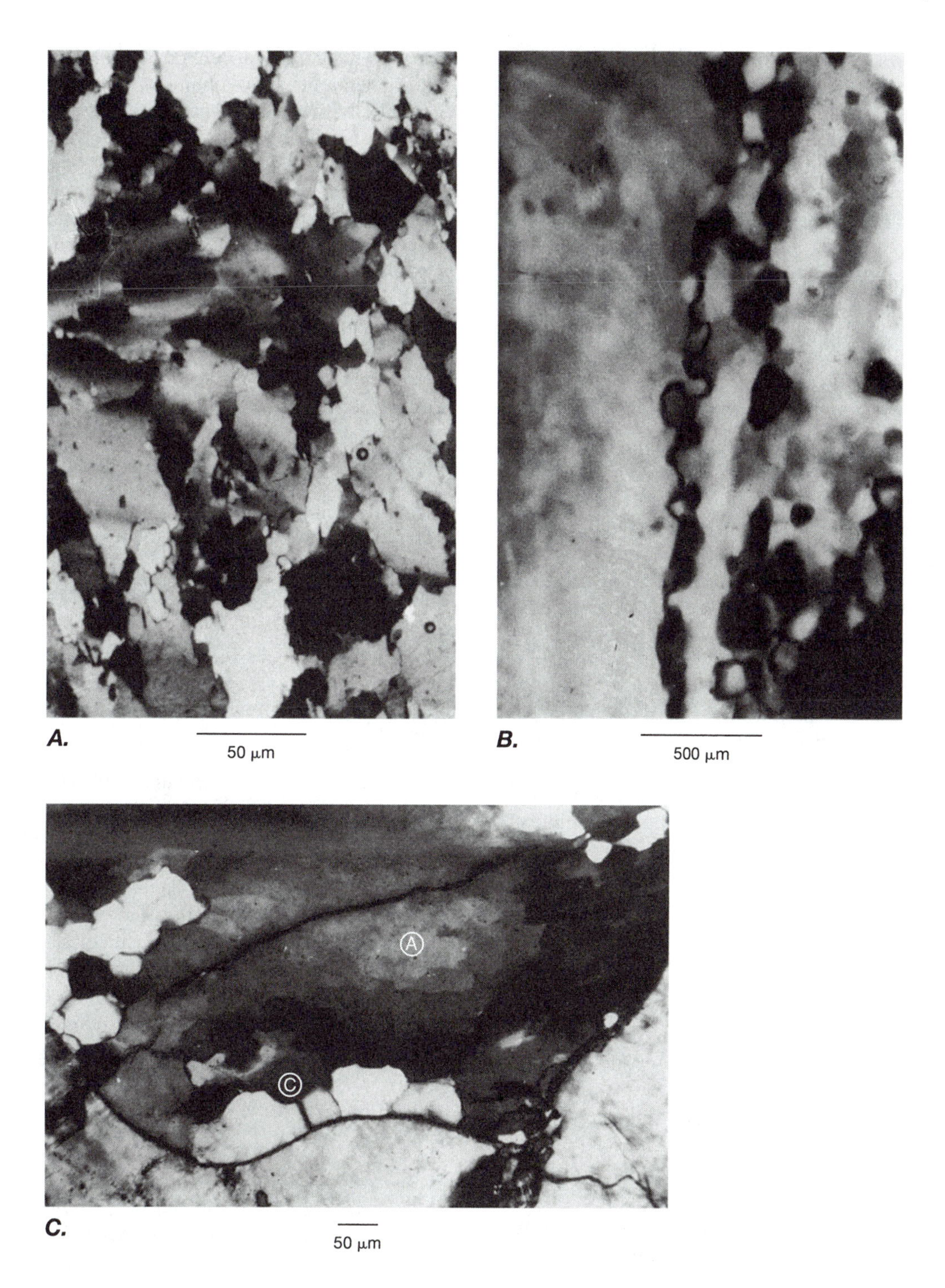

FIGURE 17.19 Mechanisms of dynamic recrystallization in quartz. Photomicrographs under crossed nicols. *A*. The highly serrated crystal boundaries in this sample of polycrystalline quartz are characteristic of high mobility and migration of the grain boundaries. (From Kohlstedt and Weathers 1980) *B*. Boundary migration recrystallization in polycrystalline quartz begins with the migration of a high-angle boundary to form a lobe into the adjacent crystal. (From Bell and Etheridge 1976) *C*. Subgrain rotation recrystallization. The misorientation of the subgrains, indicated by the shade of gray, increases progressively from the core of the grain at A toward the rim at C, where there is little or no distinction between recrystallized grains and highly misoriented subgrains. (From Bell and Etheridge 1976)

dotite tectonites as well as with the interpretation of the observed anisotropy of seismic velocities in the upper mantle as resulting from crystallographic preferred orientation produced by dislocation creep mechanisms (see Section 17.7). Calcite apparently should not deform much by dislocation creep (Figure 17.1*B*), although the large grain size and strong crystallographic preferred orientation of many marbles suggests the boundaries of the dislocation creep field in this map may be at too high a stress.

iv. Harper-Dorn Creep

The creep rates controlled by recovery and recrystallization (described earlier) all have power-law rheologies. For olivine (Figure 17.1*C*), the field of the deformation map occupied by these mechanisms covers most of the expected range of stresses for mantle deformation at high temperatures, especially considering that this field would expand at the expense of the diffusion creep field for the larger grain sizes characteristic of the mantle. Thus power-law creep may dominate in much of the upper mantle, at least. This conclusion is not consistent with geophysically derived inferences of a linear rheology for the entire mantle (see Section 16.6(ii)).

The resolution of this problem may lie in the existence of yet another dislocation deformation mechanism called **Harper-Dorn** creep. The exact mechanism is not well understood, but it is a dislocation mechanism for which the dislocation density is constant and independent of the differential stress, which leads to a linear relationship between strain rate and stress (Box 17-1). Like diffusion creep, it is effective at low stresses, and it produces a linear viscous rheology. In contrast to diffusion creep, however, it can provide higher strain rates than other mechanisms in coarse-grained materials, roughly greater than 0.5 mm. Because of these features, it is tempting to speculate that Harper-Dorn creep may be an important creep process in the mantle. Although it has so far only been definitely identified in some metals, the similarity between other deformation mechanisms in metals and silicates suggests that it also might be applicable to the Earth. Thus the dislocation microstructures associated with dislocation creep (see Sections 17.6 and 17.7), are not necessarily incompatible with a linear rheology, but the suggestion is currently very speculative and in need of more research. If Harper-Dorn creep does occur in rocks, the deformation maps in Figure 17.1 would have to be significantly revised, especially in the low-stress areas.

v. Annealing

The microstructure of a polycrystalline material can continue to evolve at high temperature even after the differential stress is removed, because recovery and recrystallization processes continue to operate. In the absence of a differential stress, **static recovery** and **static recrystallization**, referred to collectively as **annealing**, are driven by the associated decrease in the internal strain energy per unit volume of the material.

Static recovery involves the rearrangement and annihilation of dislocations by climb and cross slip. Static recrystallization may be subdivided into **primary** and **secondary recrystallization. Primary recrystallization**, like dynamic recrystallization, reduces the internal strain energy of the material by replacing deformed grains having a high dislocation density with new grains that are essentially strain-free. Grain boundaries become straight, and for materials such as quartz and olivine that have nearly isotropic grain boundary energies, the grain boundaries at triple grain junctions tend to meet at angles of 120° (Figure 15-A.4*A*). During **secondary recrystallization**, small crystal grains are preferentially eliminated by the exaggerated growth of a few larger grains. Such grains tend to have highly irregular boundaries, and grains that on a planar section are apparently isolated, in fact, can be part of a single grain that is highly convoluted in three dimensions. The growth of very large grains decreases the internal energy of a material because the grain boundary energy per unit volume is reduced.

vi. Evaluation of Relative Creep Rates

Different deformation mechanisms give rise to different rheological equations (Box 17-1), and accordingly, the strain rate for each mechanism is different. This effect is illustrated in Figure 17.20, where the logarithm of the normalized creep rate is plotted against the logarithm of the normalized stress for olivine. The rheologies for Nabarro-Herring diffusion creep (Equation (17-1.1)) and for power-law creep by dislocation glide and recovery (Equation (17-1.6)) are plotted using material constants appropriate for olivine and four different grain sizes for the diffusion creep mechanism. Because the graph is a log-log plot, the slope of each line is equal to the exponent on the stress in the corresponding rheological equation (Equations (17-1.1) and (17-1.6)).

The stress exponent is 3.5 for the power-law creep and 1 for Nabarro-Herring creep. Because of the different slopes, the lines cross, producing a so-called **cross-over stress**. At stresses above the cross-over stress, the highest strain rate is provided by the power-law creep, but below that stress, the linear Nabarro-Herring creep provides a higher strain rate. The variation of this cross-over stress with homologous temperature (the heavy dots in Figure 17.20) defines the boundary on the deformation maps between the different deformation mechanisms (Figure 17.1*C*). In principle, both mechanisms can operate on either side of the cross-over stress, and the total strain rate is the sum of strain rates contributed by each mechanism. Except for a small stress range immediately around the cross-over stress, however, the strain rate contribution of one mechanism or the other dominates the behavior of the material.

It is apparent in Figure 17.20 that the cross-over stresses for realistic mantle grain sizes of roughly 1 mm

BOX 17-1 Rheologies Inferred from Mechanisms of Ductile Deformation

One method for trying to understand the mechanisms of ductile deformation is to make models of how the process might function and to use those models to derive an equation describing the theoretical rheology. The model, of course, is based on observations of the physical phenomenon. One test of the applicability of the model is whether the resulting theoretical rheology compares with the observed rheology. In this box, we give descriptive sketches of various models of ductile deformation and compare the rheological equations derived from them with observation.

The model for Nabarro-Herring creep is that deformation occurs by a flux of vacancies through the crystal lattice. Thus the strain rate depends on the vacancy density and the rate of vacancy migration. The vacancy density in turn depends on temperature, and it varies at different surfaces of the crystal grain depending on the magnitude of the normal component of stress at the surface. Under nonhydrostatic stress, therefore, the vacancy concentration at differently stressed surfaces is different, setting up a concentration gradient down which the vacancies diffuse. As mentioned in Sections 17.3 and 17.5, the strain rate is related to the vacancy flux, giving for the theoretical rheologic equation:

$$|\dot{\varepsilon}_n^{(ss)}| = \frac{6V_v^* D_0^{(v)}}{RT} d^{-2} \sigma^{(Dif)} \exp\left[\frac{-H_v^*}{RT}\right] \quad H_v^* \equiv E_v^* + pV_v^* \qquad (17\text{-}1.1)$$

where $\dot{\varepsilon}_n^{(ss)}$ is the steady-state instantaneous strain rate, $\sigma^{(Dif)}$ is the differential stress, d is the grain diameter, T is the absolute temperature, $D_0^{(v)}$ is the diffusion constant for diffusion of vacancies through the volume of the crystal, and R is the Boltzmann constant for a mole (gram molecular weight) of material (the so-called "gas constant"). H_v^* is called the activation enthalpy for diffusion of vacancies through the volume. It is defined as the activation energy for vacancy diffusion E_v^*, which is the energy needed to make a vacancy jump from one lattice site to another, plus the pressure p times the activation volume V_v^*, which is the local volume increase in the crystal lattice required to allow a vacancy to jump from one lattice site to another. Thus the pV_v^* term expresses the work against the external pressure needed for a vacancy to jump from one lattice site to another.

Equation (17-1.1) has several interesting features. First, it has a form very close to Equation (16.15) with $n = 1$, except that in Equation (16.15) the constant A_6 does not include an inverse temperature dependence. Second, with the stress exponent $n = 1$, this creep mechanism produces the rheology of a Newtonian viscous fluid. Third, the strain rate depends inversely on the square of the grain diameter, which means that as the grain size increases, this mechanism rapidly becomes less efficient.

A similar approach for Coble creep and for solution creep differs mainly in that the diffusion path must be along the grain boundaries rather than through the volume. The theoretical model for Coble creep leads to the rheological equation:

$$|\dot{\varepsilon}_n^{(ss)}| = \frac{6V_b^* D_0^{(b)} \delta}{RT} d^{-3} \sigma^{(Dif)} \exp\left[\frac{-H_b^*}{RT}\right] \qquad (17\text{-}1.2)$$

with a similar form resulting for solution creep. This equation is very similar to that for Nabarro-Herring creep with three exceptions. First, $D_0^{(b)}$, H_b^*, and V_b^* are the diffusion constant, the activation enthalpy, and the activation volume, respectively, for grain boundary diffusion rather than volume diffusion. Second, the parameter δ, which is the thickness of the grain boundary, appears in the coefficient. Third, the strain rate depends on the inverse third power of the grain diameter, as compared with the inverse second power for Nabarro-Herring creep. This difference means that as grain size increases, the efficiency of Coble creep decreases more rapidly than for Nabarro-Herring creep. Nevertheless, this diffusive mechanism also provides a Newtonian viscous rheology because the stress exponent is 1.

One model for the mechansim of dislocation creep, known as Weertman creep, assumes the rate of dislocation glide is limited by the rate at which attached dislocation segments not in the glide plane can climb to keep up with the gliding dislocations. The strain rate then depends on the density of dislocations ρ, the magnitude of the Burgers vector b, which defines the amount of slip associated with each dislocation, and the climb velocity of the dislocations v_c:

$$|\dot{\varepsilon}_n^{(ss)}| = \beta b \rho v_c \qquad (17\text{-}1.3)$$

where β is a geometrical constant. Theoretically the dislocation density should vary as the square of the differential stress. That is,

$$\rho = \alpha(\sigma^{(Dif)})^2/(\mu b)^2 \qquad (17\text{-}1.4)$$

where α is a constant, μ the shear modulus, and b is the magnitude of the Burgers vector (cf. Equation (17-2.1)). The climb velocity depends on the differential stress and the coefficient of vacancy diffusion through the volume, which in turn is a thermally activated quantity:

$$v_c = \frac{C}{RT} \sigma^{(Dif)} D_0^{(v)} \exp\left[\frac{-H^*}{RT}\right] \qquad (17\text{-}1.5)$$

where C is a constant of proportionality. Substituting Equations (17-1.4) and (17-1.5) into Equation (17-1.3) gives a power-law relation that has the form

$$|\dot{\varepsilon}_n^{(ss)}| = \frac{\beta_0}{RT} (\sigma^{(Dif)})^3 \exp\left[\frac{-H^*}{RT}\right] \qquad \beta_0 \equiv \alpha\beta C D_0^{(v)}/\mu^2 b \qquad (17\text{-}1.6)$$

This equation is identical in form to Equation (16.11) with $n = 3$, with the exception of the inverse temperature dependence of the first term. Still other assumptions can be made about the rate-limiting mechanism that gives rise to a stress exponent of $n = 4.5$, and empirical values are commonly between 3 and 5.

For Harper-Dorn creep, Equation (17-1.4) does not apply, because the dislocation density is apparently independent of stress. The resulting rheological equation gives a linear relationship between strain rate and stress.

A theoretical model based on kinetic theory has been used to calculate the rate of glide of dislocations through a crystal lattice when the glide rate is controlled by obstructions. The equation has the form

$$|\dot{\varepsilon}_n^{(ss)}| = A \exp\left[1 - \frac{\sigma^{(dif)}}{\sigma_0}\right] \exp\left[\frac{-E^*}{RT}\right] \qquad (17\text{-}1.7)$$

where A is a constant and σ_0 is the differential flow stress at 0 K. The dependence of the creep rate on an exponential function of the stress is comparable to the experimentally derived Equation (16.9), which suggests the model of dislocation glide may be appropriate.

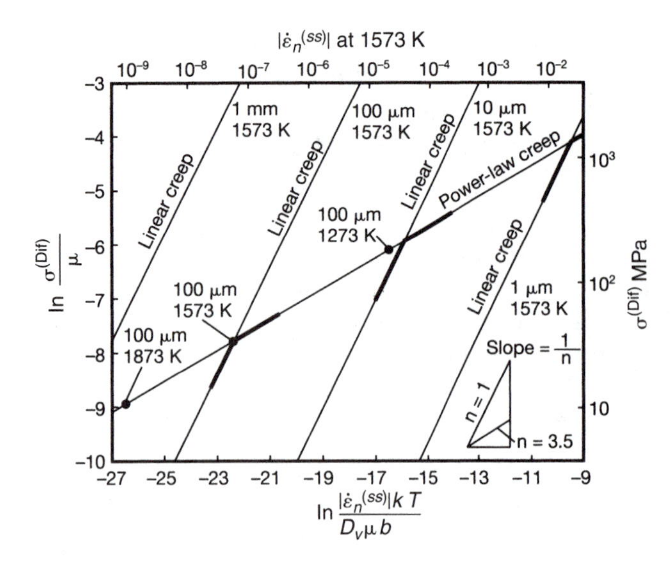

FIGURE 17.20 Plot of normalized stress versus normalized strain rate for olivine, showing how the strain rates for power-law creep and for diffusion creep are related. The heavy obtuse angles mark the stresses where the dominant deformation mechanism changes for grain sizes of 1 μm, 10μm, and 100μm at 1573 K. The heavy line segments indicate the line that provides the highest strain rate at the given conditions, and the apex of the angle is the cross-over point where the dominant deformation mechanism changes. The three solid dots indicate how the cross-over stress changes from the middle point with a change in temperature of $\pm$300 K for a 100-μm grain size. The vertical scale on the right gives the differential stress, and the horizontal scale at the top gives the absolute strain rate for 1573 K (1300° C). The resultant equations for the linear and power-law creep curves are, respectively:

$$\ln \frac{|\dot{\varepsilon}_n^{(ss)}|kT}{D_V\mu b} = \left[-9.92 + \ln \frac{D_b}{D_V}\right] - 2\ln\frac{d}{b} + \ln\frac{\sigma^{(Dif)}}{\mu}$$

$$\ln \frac{|\dot{\varepsilon}_n^{(ss)}|kT}{D_V\mu b} = 4.87 + 3.5\ln\frac{\sigma^{(Dif)}}{\mu}$$

The ratio D_b/D_V appears in the linear creep law because the plotted strain rate is normalized by the coefficient of volume self-diffusion D_V, whereas the appropriate quantity for normalizing the linear diffusion creep should be the coefficient of grain boundary self-diffusion D_b. This term introduces the temperature dependence of the plot for linear creep. $D_V = 0.1 \exp[-Q_V/RT]$; $D_b = 0.1 \exp[-Q_b/RT]$; $Q_V = 5.4 \times 10^5$ joule; $Q_b = 2.9 \times 10^5$ joule; other constants are listed in the caption to Figure 16.16. The plot is based on the results of Karato et al. (1986).

or larger are at such low strain rates that they could never be examined directly in an experiment. Only by increasing the temperature several hundred degrees or decreasing the grain sizes to about 1 to 10 μm are the cross-over stresses accessible experimentally, and these conditions are unlikely to be representative of the mantle. Our inference as to which mechanism actually should dominate under mantle conditions, therefore, depends strongly on our ability to identify the appropriate mechanisms and to determine accurately the rheological equations. Only then can we extrapolate reliably to mantle conditions. The uncertainty of the appropriate values for the material constants in the equations—to say nothing of the possibility that there are other deformation mechanisms that we do not know or understand sufficiently—makes the interpretation of the appropriate rheology for the mantle still a debated issue.

17.6 MICROSTRUCTURAL FABRICS ASSOCIATED WITH DISLOCATION CREEP

Creep and recrystallization result in the development of characteristic fabrics in mineral grains and in rocks, including dislocation microstructures, grain textures, and crystallographic preferred orientations. These fabrics provide important clues to the nature of the deformation.

Crystals deformed by cold working, when examined in thin section between crossed polarizers, commonly exhibit a characteristic undulatory extinction that reflects deformation-imposed curvature in the lattice (Figure 17.21*A*). This curvature represents the excess of dislocations of one sign over those of the other. If the temperature is high enough for recovery to occur, either during or after the deformation, these dislocations may climb and be annihilated in pairs or form low-energy walls, which may be evident as subgrain boundaries or a series of tilt boundaries of the same sign (Figure 17.21*B*).

Deformation lamellae are features that develop in quartz (see Box 17-2(ii)) and olivine; they indicate deformation has occurred at relatively high strain rate or high differential stress and at temperatures in the low range for dislocation creep (Figure 17.22). They appear in thin section as thin discontinuous planar features in a mineral grain and have a slightly different index of refraction from the surrounding crystal. They are commonly formed by pile-ups and tangles of dislocations on

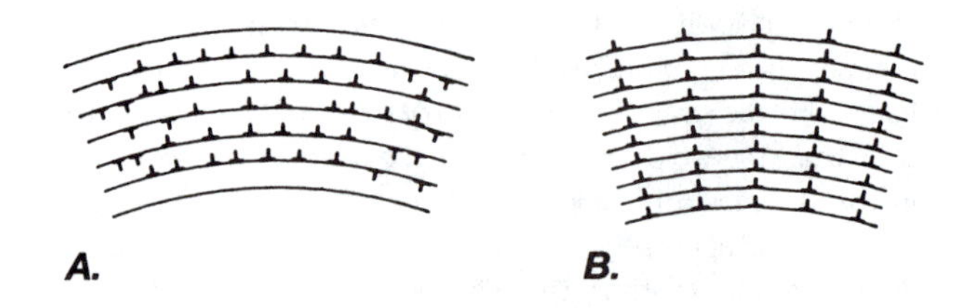

FIGURE 17.21 A dislocation model for undulatory extinction and polygonization. (From Spry 1969) *A*. Glide planes in the crystal contain dislocations of both signs, but more of one sign than the other, giving rise to the curvature of the lattice planes. *B*. Recovery during annealing allows dislocations to annihilate and to array themselves into low-energy walls, thereby segmenting the crystal into subgrains separated by low-angle tilt walls, so that each subgrain has a small misorientation with respect to its neighbors.

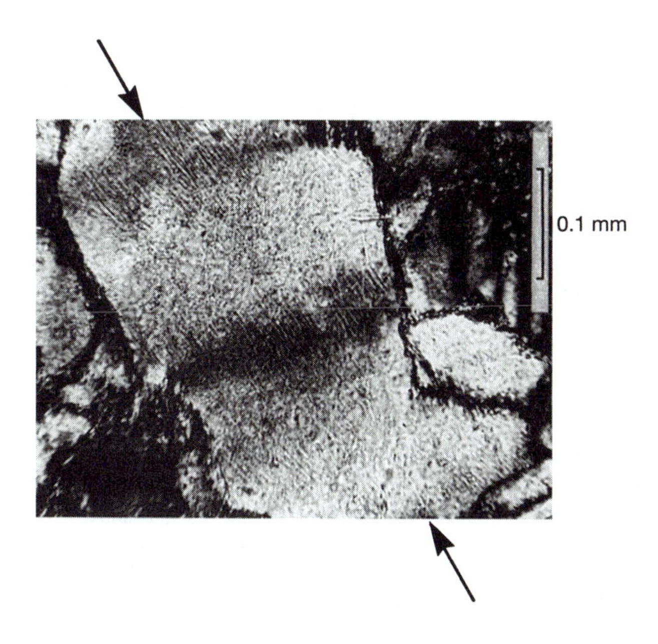

FIGURE 17.22 Deformation lamellae in quartz seen under the optical microscope. (From McLaren and Hobbs 1972)

the active glide planes in the crystal and are evidence of a low rate of recovery.

Ribbon-shaped grains, especially of quartz, that have been severely flattened and stretched (Figure 7.23), indicate that the rocks were deformed at moderate temperatures below that required for recrystallization.

Subgrains in crystals are another indication that dislocation creep mechanisms have operated. Commonly these subgrains are visible in the petrographic microscope as a slight difference in extinction position of areas within a single crystal grain (Figure 17.19*C*). The misorientation of many subgrains, however, is too small to be optically visible, and techniques such as decorating the dislocations (Figure 17.13*A*), etching polished crystal surfaces (Figure 17.13*B*), or imaging the crystal in the transmission electron microscope (Figure 17.13*C*) must be used.

Highly serrated grain boundaries (Figure 17.19*A*, *B*) are evidence that the grain boundaries have been mobile and therefore that the homologous temperature of deformation was relatively high. Mobility of the grain boundaries promotes dynamic recrystallization and grain growth and allows the grains to maintain a more equant shape during deformation. If deformation ceases while temperatures are still high, annealing occurs and mobile grain boundaries tend to become straight, which is a minimum energy configuration (Figure 15-A.4). Straight grain boundaries, however, are not a unique characteristic of postdeformational annealing. They also develop during dynamic subgrain rotation recrystallization and as a result of diffusional mechanisms of deformation.

The effects of dynamic recrystallization are easiest to recognize where there has been a dramatic reduction in grain size but at least some original grains are still preserved (Figure 3.7). Boundary migration recrystallization may produce highly irregular and sutured grain boundaries (Figure 17.19*B*). Subgrain rotation recrystallization results in increasing misorientation of subgrains from the core to the boundary of large relict grains (Figure 17.19*C*).

Piezometers, or stress measures (Box 17-2), are steady-state fabric characteristics that change with the magnitude of the applied differential stress. If they are preserved in the mineral grains, we can use them to infer the paleostresses that caused the ductile deformation. Because dislocations multiply until the backstress equals the applied stress, the dislocation density[5] at steady-state creep is a function of the applied differential stress (Figure 17.24*A* and Box 17-2). The size of subgrains and of dynamically recrystallized grains formed during steady-state dislocation creep is also dependent on the magnitude of the applied stress, and the size decreases with increasing stress (Figure 17.24*B*, *C* and Box 17-2). Piezometrically determined stresses as high as 100 to 200 MPa have been found in crustal rocks associated with mylonites in ductile shear zones. At temperatures that put the deformation into the dislocation creep regime, such stresses must be associated with geologically rapid strain rates. For quartz, at temperatures as low as 400° C, the strain rate at 100 MPa differential stress is roughly 10^{-10} s^{-1} (Figure 17.1*A*), which is a very high strain rate for ductile flow in a geologic process.

[5]Dislocation density is measured in units of total dislocation length per unit volume, which reduces to units of inverse length squared. Common units are cm^{-2}. Undeformed crystals may have free dislocation densities of 10^3 to 10^5 cm^{-2}; deformed crystals may have densities up to 10^{10} cm^{-2} (Figure 17.24*A*). For the larger figure, the total length of dislocations in a cubic centimeter is comparable to the total length of blood vessels in the human body. If they were strung end to end, they would reach two and a half times around the world!

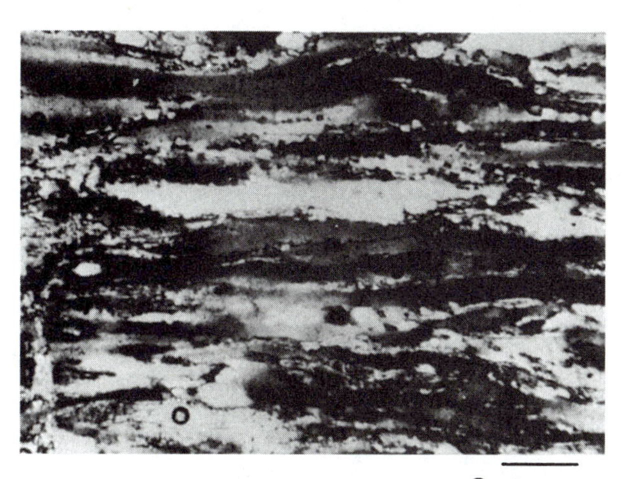

FIGURE 17.23 Ribbon quartz in a strongly deformed quartzite. (From Kohlstedt and Weathers 1980)

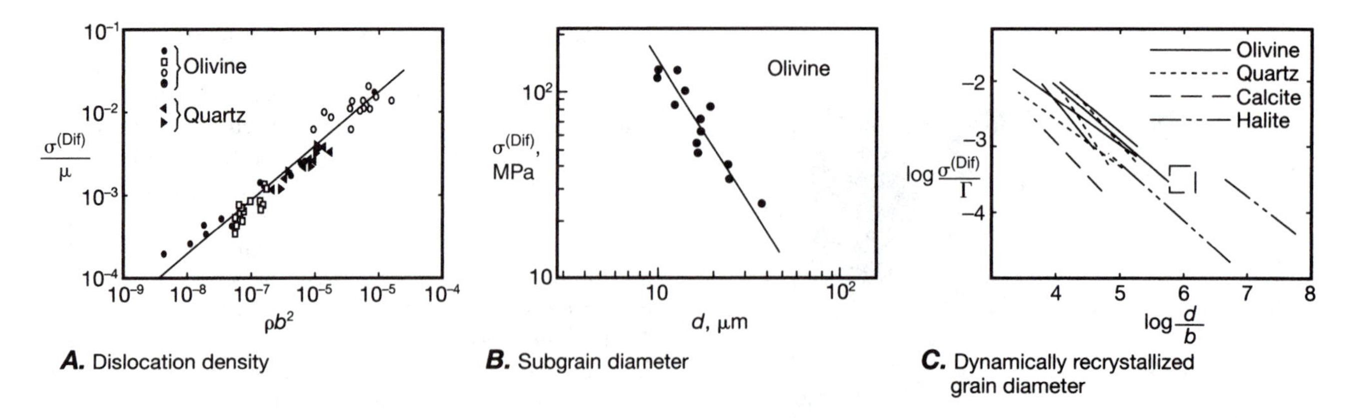

FIGURE 17.24 Dependence of stress-sensitive fabric elements on differential stress. A. Variation of free dislocation density with differential stress for olivine and quartz showing the relationship given in Equation (17-2.1). (From Kohlstedt and Weathers 1980) *B.* Variation of the mean subgrain diameter with the differential stress at steady state illustrating the relationship given in Equation (17-2.2). (From Karato et al. 1980) *C.* Variation of mean dynamically recrystallized grain diameter with differential stress illustrating the relationship given in Equation (17-2.3). Data include relations for wet and dry olivine and quartz, and for calcite (box and dashed line) and halite recrystallizing by migration and rotation recrystallization. The normalization modulus is $\Gamma = \mu/(1 - \nu)$, where μ is the shear modulus and ν the Poisson ratio. (From Schmid 1982)

BOX 17-2 Inferring the Orientation and Magnitude of Paleostresses in Deformed Rocks

In applying our understanding of the rheology of rocks to the interpretation of structures in the Earth, we face a dilemma: The orientations and magnitudes of the stresses as well as the strain rates associated with the observable ductile deformation are usually impossible to determine. Without one of these pieces of information, it is impossible to use the rheological equations, and inferences of rheology are speculative or, at best, very approximate. Under some circumstances, however, it is possible to use the structures in deformed mineral grains or the characteristics of the grain fabrics to infer the orientations and/or magnitudes of the stresses that were associated with the deformation. Such measurements are extremely important in allowing us to check laboratory and theoretical models against actual conditions of deformation in the Earth.

The orientations of calcite twin lamellae and of quartz deformation lamellae can be used to determine the orientations of the principal instantaneous strain axes. In Section 15.4(i), we also described the use of fault-slip data, such as shear-plane/slickenline data, to determine the orientation and relative magnitudes of the principal instantaneous strains. In the literature, such determinations are often referred to as "stress determinations," although identifying the instantaneous strains with stresses requires assumptions that may not be accurate.

The dispersion of calcite twin lamellae can be related to the magnitude of the differential stress during deformation and the microstructural characteristics of dislocation density, subgrain diameter, and dynamically recrystallized grain size, all are observed empirically to vary monotonically with the magnitude of the differential stress at steady state. These relations can be justified on theoretical grounds, and with adequate calibration, all can be used to infer the magnitude of the paleostresses.

In applying these techniques, therefore, it is important to be clear about exactly what we are measuring and how we infer stress from these measurements, because commonly measurements that are referred to as "stress measurements" in the literature are in fact measurements of strain, and the inference of stress from those measurements relies on specific, but often only implicit, assumptions.

We discuss the methods using calcite, quartz, and microstructural fabrics in turn.

i. Calcite

Calcite *e* twin lamellae (Figure 17.3) are planar microstructures that accommodate a shear of the crystal lattice parallel to the twin plane (Section 17.2). The total deformation of the crystal then equals the net effect of all the shear strains that accumulate on all the twin lamellae in the crystal. The twin lamellae tend to form in orientations of high shear strain. Such planes have a restricted range of orientations relative to the principal instantaneous strain axes $\hat{\zeta}_k$. The only crystals that can develop twins, therefore, are those that are oriented such that twin planes and twin shear directions are in optimal orientations relative to the principal axes of instantaneous strain. Thus the preferred orientations of the planar twin lamellae and of the crystallographic axes of the crystal in which they occur can be used to infer the orientation of the principal axes of instantaneous strain (Figure 17.3*D*).

During the twinning process, the *c* axis of the twin is flipped into an orientation that is closer to the axis of minimum stretch

$\hat{\zeta}_3$ (maximum shortening) than the c axis of the host grain (Figure 17.3*A*, *C*, *D*). By measuring the orientations of the twin plane as well as the c axes in both the twin and the host, we can use the geometry of Figure 17.3*D* to infer the orientation of the principal instantaneous strains associated with that twinning deformation. The preferred orientation of these principal strain axes, determined for a large number of calcite grains in a sample, defines the best estimate of the principal strain axes for the deformation of the sample. It is generally assumed that the strains are small enough that the twin planes have not been rotated significantly by a finite deformation, and that the principal axes of finite and instantaneous strain are not much different. Because observed shear strains are generally much less than the theoretical maximum of about $e_s = 0.35$ for a crystal that is 100 percent twinned (Figure 17.3), that assumption is probably reasonable.

It is commonly assumed that the principal axes of stress are parallel to the inferred orientations of the principal instantaneous strain axes, for both individual crystals and a rock sample as a whole. The rationale for this assumption is that deformation twins in calcite can develop only if the crystal is oriented relative to the principal stresses such that the resolved shear stress on a twin plane in the direction of shear required for twinning exceeds the yield shear stress for twinning, which is about 10 MPa. The direction of twin shearing in the twin plane is therefore assumed to be an orientation of high resolved shear stress, and therefore the construction for the orientation of the principal instantaneous strain axes also provides a solution for the principal stress axes.

If a differential stress of 20 MPa is applied to a volume of limestone or marble, the maximum possible shear stress is 10 MPa, which is the yield stress for calcite twinning. Thus twinning can occur only in those calcite grains oriented so that the twin plane is exactly parallel to the plane of maximum resolved shear stress and the twinning direction is exactly parallel to the direction of maximum resolved shear stress. For such a situation, the inferred $\hat{\sigma}_1$ axes for twinned grains should form a tight maximum. For higher differential stresses, the maximum possible shear stress is also higher, and grains that are not in the ideal orientation would also be able to twin. Thus the inferred directions for $\hat{\sigma}_1$ would spread out into a larger maximum. The size of the maximum therefore can be used to infer the approximate magnitude of the differential stress.

ii. Quartz

Quartz deformation lamellae (Figure 17.22) are planar microstructures that are oriented subparallel to dominant crystallographic slip planes in the quartz lattice. They are characterized by a high density of tangled dislocations that accumulate either because the strain rate is too high or the temperature is too low, for dislocation climb to efficiently disperse the tangles. The orientations of the lamellae define planes on which shearing of the crystal lattice has been accommodated. To that extent, quartz deformation lamellae are similar in significance to the calcite twin lamellae discussed in the preceding subsection. They differ, however, in that their geometry is not as rigorously defined by the crystal lattice: In particular, the lamella plane is not necessarily an exact crystallographic plane, the amount of possible shear that can be accommodated on a lamella is not limited by the crystal structure, the slip direction is not clearly defined, and a separate c axis orientation for the lamella cannot be determined. Thus the method of analysis is necessarily different.

Figure 17-2.1 illustrates the use of quartz deformation lamellae in the study of a small fold in a quartzite layer shown in Figure 17-2.1*D*. The orientations of the lamella planes and the host-grain c axes were determined for a large number of grains in samples taken from the three different limbs of the fold, which are plotted respectively in Figure 17-2.1*A*, *B*, and *C*.

Poles to lamellae are plotted in the top set of stereograms, and they tend to form two maxima that may spread out into a partial small-circle girdle. The principal axes of instantaneous stretch are taken to be parallel to the symmetry axes for these distributions; two of these axes, therefore, bisect the acute and obtuse angles between the lamellae pole maxima (see Section 17.8).

The identity of these principal axes is defined by constructing arrow diagrams (bottom stereograms in Figure 17-2.1*A*, *B*, *C*). For each measured lamella, an arrow is plotted along the great circle spanning the acute angle from the host-grain c axis to the lamella pole. These arrows tend to point toward the axis of maximum instantaneous stretch $\hat{\zeta}_1$ and away from the axis of minimum instantaneous stretch $\hat{\zeta}_3$. The arrows shown in Figure 17-2.1*A*–*C* are only those for which the lamella pole is within 10° to 25° of the c axis, because they show this relationship most clearly. These lamellae are subparallel to the basal plane (0001)—that is, they lie at a high angle to the c axis. This is an empirical method whose effectiveness has been confirmed by comparing results from quartz deformation lamellae and calcite twin lamellae from the same rock.

The relationship between arrows and principal instantaneous stretch axes must reflect a general consistency in the orientation of slip directions in the lamellae relative to the c axis. A possible explanation is as follows: A lamella forms only if a crystal were oriented, relative to the principal instantaneous stretch axes ($\hat{\zeta}_1$, $\hat{\zeta}_3$), such that the component of shear strain would be high on two crystallographic slip systems each of which had a slip direction close to the ($\hat{\zeta}_1$, $\hat{\zeta}_3$) plane. Figure 17-2.2 shows one such orientation in which slip could occur on the basal slip system (0001)$[a_1]$ or $[-a_3]$, and the rhombohedral slip system $(r_1)[c + a_1]$ or $[c - a_3]$. Slip on these two slip systems would produce a sub-basal lamella that lies between the two slip planes in an orientation of high resolved shear strain (near 45° from $\hat{\zeta}_1$ and $\hat{\zeta}_3$), and that has a net slip direction subparallel to the ($\hat{\zeta}_1$, $\hat{\zeta}_3$) plane. The ($\hat{\zeta}_1$, $\hat{\zeta}_3$) plane (labeled (X—X') in Figure 17-2.2) also contains the c axis and

(*continued*)

BOX 17-2 Inferring the Orientation and Magnitude of Paleostresses in Deformed Rocks *(continued)*

the lamella pole. Thus an arrow constructed from the *c* axis to the lamella pole would tend to point toward $\hat{\zeta}_1$ and away from $\hat{\zeta}_3$ (Figure 17-2.2). Given the multiplicity of possible slip systems in quartz (Table 17.1) and their angular relations (Figure 17.29*B*), a wide diversity in arrow orientations could arise, but a subset of the lamellae that are sub-basal in orientation would give the most consistent results (Figure 17-2.2).

Generally, it is implicitly assumed that the principal axes of stress ($\hat{\sigma}_3$, $\hat{\sigma}_1$) are parallel, respectively, to those of instantaneous stretch ($\hat{\zeta}_1$, $\hat{\zeta}_3$). Although this assumption may be a reasonable approximation, it is not, strictly speaking, a necessary relationship, especially if the material is anisotropic, which rocks and minerals generally are. Accurate information on stress orientation is only possible from deformation lamellae if the initial distribution of *c* axes for grains in the rock is isotropic, and if the total strain is limited to relatively low values—perhaps 20 percent or less. Nevertheless, applying this assumption to the fold in Figure 17-2.1*D* leads to a reasonable interpretation.

Figure 17-2.1*D* shows the folded layer with the arrows indicating the directions of the principal stresses inferred from the fabric diagrams in Figure 17-2.1*A*, *B*, *C*, and the dashed lines representing the average orientation of the lamellae. Figure 17-2.1*E* show the same layer unfolded to the point that the maximum compressive stresses are all parallel. Apparently the lamellae were formed early in the deformation (compare Figure 18.16*A*) and subsequently were rotated by the folding. The fact that the lamellae do not represent the stresses expected later in the folding process suggests that the magnitude of the stress decreased as the folding progressed. These conclusions are consistent with the mechanical models discussed in Section 18.7(i), and they indicate a complication of interpreting strain or stress orientations inferred from fabric data: Later deformation may rotate the inferred principal axes out of their original orientations.

iii. Microstructural Paleopiezometers

Paleopiezometers are structural characteristics of deformed rocks that vary with the magnitude of the applied differential stress under which they formed and that therefore provide a means of determining the magnitude of the paleostress. For rocks that have deformed at steady state, three different elements of the microstructural fabric can be used as paleopiezometers. With increasing differential stress $\sigma^{(Dif)}$, the free dislocation density ρ increases, and the mean diameter of subgrains d_s and of dynamically recrystallized grains d_r decreases according to the following relationships:

Fabric Element	Piezometric Equation	
Dislocation density	$\sigma^{(Dif)} = \alpha \mu b \rho^{1/2}$	(17-2.1)
Subgrain diameter	$\sigma^{(Dif)} = K_1 \mu b d_s^{-r}$	(17-2.2)
Dynamically recrystallized grain diameter	$\sigma^{(Dif)} = K_2 \mu b d_r^{-p}$	(17-2.3)

In these equations, μ is the elastic shear modulus; b is the magnitude of the Burgers vector; and α, K_1, K_2, r, and p are all constants that must be determined experimentally. Experimental calibrations of these relationships for various minerals are shown, respectively, in Figure 17.24 *A*, *B*, *C*. α is approximately equal to 1, and r and p generally have values of $r \approx 1$, $p \approx 0.7$, although values of p as high as 1.4 have been reported.

Although dislocation densities decrease rapidly during annealing at high temperature after the stress is removed, the dislocation substructure can be preserved if deformation occurs at low temperature, if cooling is rapid, or if cooling occurs under constant differential stress.

The subgrain diameter responds more slowly to changes in stress than the dislocation density because it requires larger strains to form subgrains and because boundaries are less mobile than individual dislocations during annealing. Subgrains therefore provide a more stable piezometer.

The constants in Equation (17-2.3) for the two mechanisms of dynamic recrystallization are different. Subgrain rotation recrystallization produces smaller grains than boundary migration recrystallization, at least in calcite and halite (see Figure 17.24*C*). Thus care must be taken to ensure that the piezometric equation applied to observations of naturally deformed mineral grains is appropriate for the mechanism of recrystallization that occurred.

The highest differential paleostresses inferred for tectonic processes in the Earth, using the different piezometers discussed in this subsection, have come from mylonites in ductile shear zones. Magnitudes on the order of 100 to 200 MPa are indicated from crustal shear zones such as the Moine thrust in northern Scotland, the Arltunga complex in central Australia, metamorphic core complexes in the western Cordillera of the United States, and ultramafic bodies of mantle origin that have been thrust up over crustal rocks in orogenic belts. The application of piezometry to olivine in xenoliths brought up from depths of 100 to 200 kilometers suggests the existence of a wide range of stresses in the mantle, ranging from the high stresses mentioned above, down to stresses on the order of a few megapascals. The higher stresses are probably associated with the local process—such as diapirism—that resulted in the eruption of the lavas containing the xenoliths, and the lower stresses may represent the regional flow stresses in the upper mantle. The difficulty of distinguishing steady-state microstructures from those resulting from annealing makes the determinations of the lower stress values somewhat suspect, although determining the complete distribution of grain sizes rather than just the mean grain size has shown some promise in distinguishing the history and mechanism of the grain size change.

The application of these piezometers to rocks is still not very precise. Experimental calibration of the constants in the equations is not adequate for most minerals, and experimental evidence shows that the presence of water dissolved

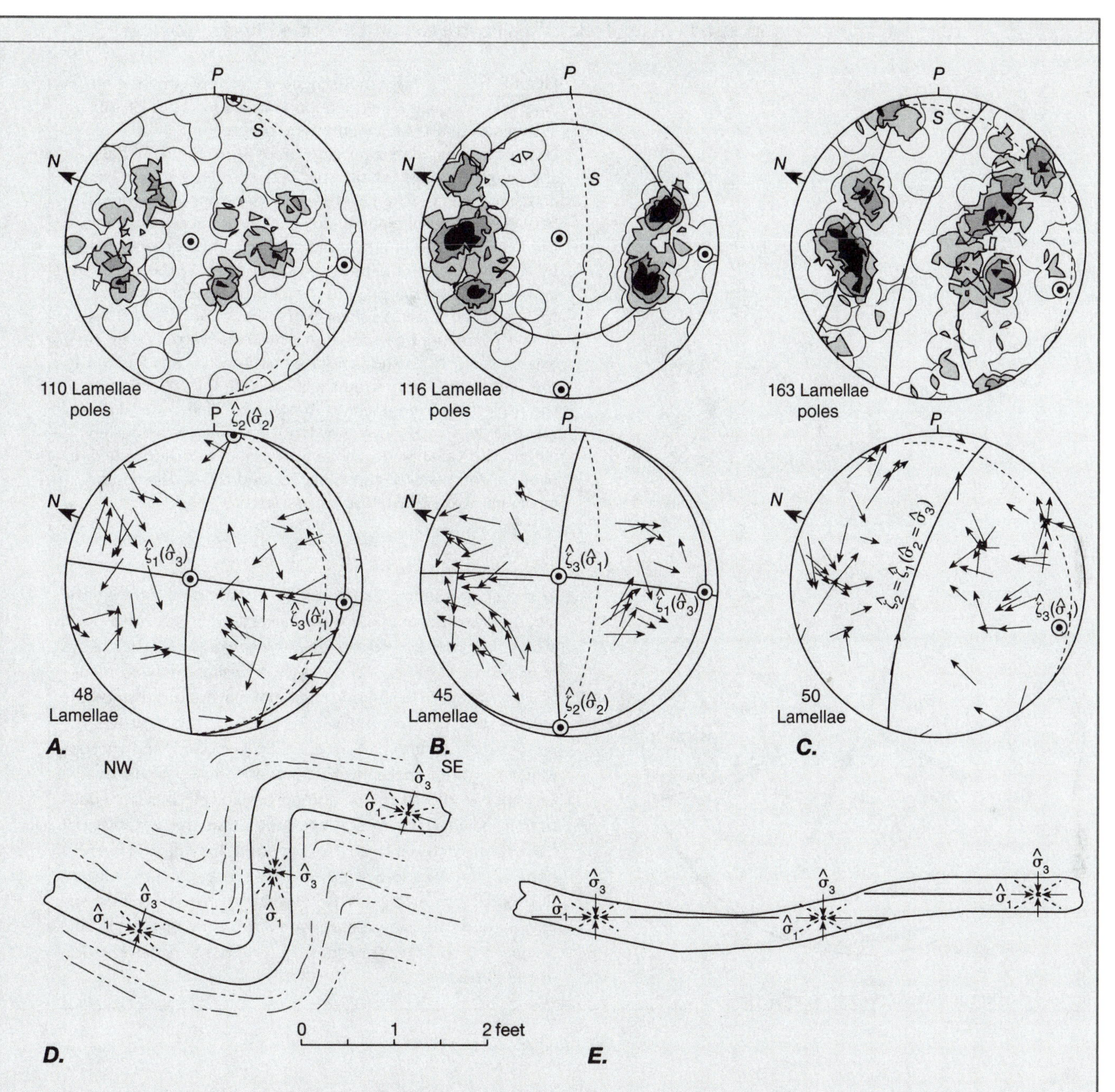

FIGURE 17-2.1 Principal instantaneous strain axes deduced from quartz deformation lamellae in a minor fold, with interpretation of principal stress axes. (After Scott et al. 1965) *A*, *B*, and *C*. *Top*: poles to deformation lamellae with bedding orientation (dashed great circle; N is north direction), small circle distribution of lamella poles, and symmetry axes of the fabric (circled dots); *bottom*: arrow diagrams with arrows drawn along great circles connecting the optic *c* axis (tail) with the lamella pole (head). Only arrows subtending angles between 10 and 25° are shown. Solid great circles are symmetry planes of the lamella fabric. The dashed great circle is the bedding orientation. The principal axes of instantaneous strain are inferred to be parallel to the symmetry axes of the lamella fabric, and the arrows tend to point away from the maximum shortening axis $\hat{\zeta}_3$ and toward the maximum lengthening axis $\hat{\zeta}_1$. The maximum and minimum compressive stresses $\hat{\sigma}_1$ and $\hat{\sigma}_3$ are assumed to be parallel to $\hat{\zeta}_3$ and $\hat{\zeta}_1$, respectively. The point *P* on each stereonet is the orientation of the line normal to the cross sections in *D* and *E*. Lower-hemisphere equal-area projections onto the horizontal plane. *A*. Contours 0.9, 2.7, 4.5, and 6.4 percent per 1 percent area. *B*. Contours 0.9, 2.6, 4.4, and 6.1 percent per 1 percent area. *C*. Contours 0.6, 1.8, 3.1, and 4.3 percent per 1 percent area. *D*. Profile of the fold showing the interpreted stress axes in the fold limbs. *E*. The quartzite layer unfolded to the point that the interpreted axes of maximum compressive stress are all parallel.

(*continued*)

BOX 17-2 Inferring the Orientation and Magnitude of Paleostresses in Deformed Rocks *(continued)*

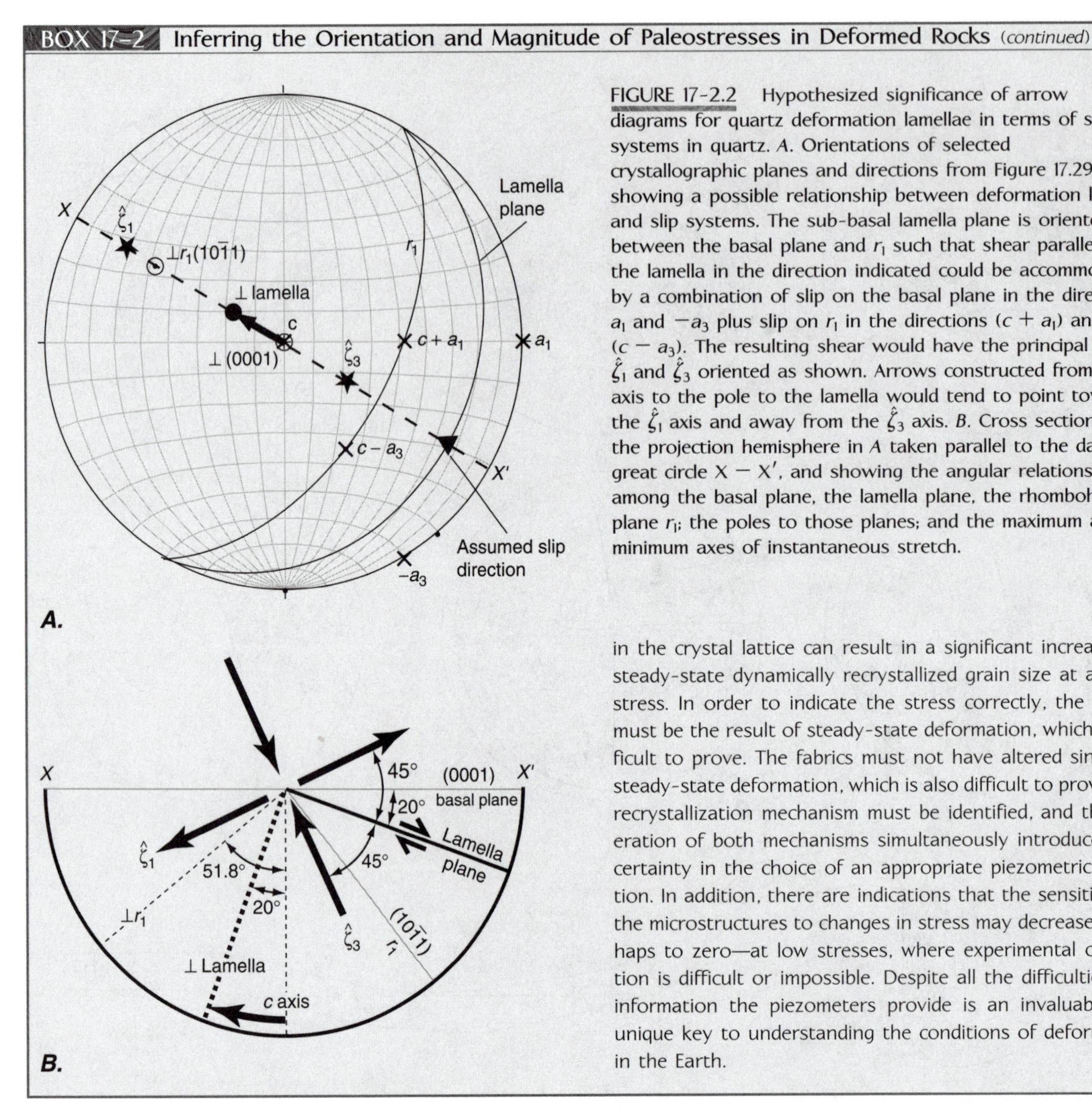

FIGURE 17-2.2 Hypothesized significance of arrow diagrams for quartz deformation lamellae in terms of slip systems in quartz. *A.* Orientations of selected crystallographic planes and directions from Figure 17.29*B*, showing a possible relationship between deformation lamella and slip systems. The sub-basal lamella plane is oriented between the basal plane and r_1 such that shear parallel to the lamella in the direction indicated could be accommodated by a combination of slip on the basal plane in the directions a_1 and $-a_3$ plus slip on r_1 in the directions $(c + a_1)$ and $(c - a_3)$. The resulting shear would have the principal axes $\hat{\zeta}_1$ and $\hat{\zeta}_3$ oriented as shown. Arrows constructed from the *c* axis to the pole to the lamella would tend to point toward the $\hat{\zeta}_1$ axis and away from the $\hat{\zeta}_3$ axis. *B.* Cross section of the projection hemisphere in *A* taken parallel to the dashed great circle X − X′, and showing the angular relationships among the basal plane, the lamella plane, the rhombohedral plane r_1; the poles to those planes; and the maximum and minimum axes of instantaneous stretch.

in the crystal lattice can result in a significant increase the steady-state dynamically recrystallized grain size at a given stress. In order to indicate the stress correctly, the fabrics must be the result of steady-state deformation, which is difficult to prove. The fabrics must not have altered since the steady-state deformation, which is also difficult to prove. The recrystallization mechanism must be identified, and the operation of both mechanisms simultaneously introduces uncertainty in the choice of an appropriate piezometric equation. In addition, there are indications that the sensitivity of the microstructures to changes in stress may decrease—perhaps to zero—at low stresses, where experimental calibration is difficult or impossible. Despite all the difficulties, the information the piezometers provide is an invaluable and unique key to understanding the conditions of deformation in the Earth.

17.7 PREFERRED ORIENTATION FABRICS OF DISLOCATION CREEP

Ductile deformation by dislocation creep produces characteristic preferred orientations of mineral crystallographic axes. The pattern of preferred orientation that develops for a particular mineral in an initially randomly oriented polycrystalline aggregate depends on the slip systems that are active in that mineral and on the geometry and magnitude of the externally applied deformation. The number and orientation of the slip systems that operate in the particular mineral may change with different conditions of temperature and/or stress, thereby producing different fabrics under the different conditions. Diffusion creep does not lead to preferred orientations because it is not associated with any preferred plane and orientation in the crystal.

Coaxial deformations produce fabrics that are symmetric with respect to the principal axes of finite strain, whereas noncoaxial deformations tend to produce asymmetric fabrics (see the discussion of symmetry principles in Section 17.8). The strength of the preferred orientations increases with increasing strain, in some cases approaching a limiting value. Thus we should be able to deduce constraints on the geometry, magnitude, and conditions of deformation from observations of natural preferred orientation fabrics and a knowledge of the ductile behavior of particular minerals.

i. Formation of a Crystallographic Preferred Orientation

The formation of a crystallographic preferred orientation during ductile deformation of a rock requires some mechanism by which individual crystals rotate toward a preferred orientation as a result of slip on their particular slip systems. Two models have been proposed for this process. One, called the **Taylor-Bishop-Hill theory**, requires that a measure of the rotation rate of crystals, called the **spin** (defined in the following paragraph), be the same as that of the imposed large-scale homogeneous deformation. The model assumes that at least five independent slip systems are active in the crystals, satisfying the von Mises condition. The other model, which we will refer to as the **misfit minimization model**, is based on determining the rotation of individual crystals that is required to minimize the geometric misfit in an aggregate of crystals that can deform only on a limited number of slip systems. Both models predict aspects of observed crystallographic preferred orientations, and the actual process of crystallographic rotation probably includes both mechanisms. We discuss the basic aspects of the two mechanisms of rotation and then describe characteristics of natural crystallographic fabrics found in olivine and in quartz.

In a deforming continuum, the **spin** at a point is a measure of the average rate of rotation, or angular velocity, of all material lines through that point.[6] The angular velocities of the material lines in a continuum undergoing either progressive pure shear or progressive simple shear are shown for a two-dimensional deformation in Figure 17.25*A*, *B*, respectively. The length of the arrow at the tip of each material line indicates the magnitude of the angular velocity of that line. The symmetry of these arrows in Figure 17.25*A* shows that for progressive pure shear the average of these angular velocities, and therefore the spin, is zero. For progressive simple shear, however (Figure 17.25*B*), the spin is clearly not zero, that is, the material is rotating about the point.

To describe the essential idea of the Taylor-Bishop-Hill theory, we consider the simplified model of a crystal grain of random orientation embedded in a continuum undergoing a macroscopic progressive pure shear. We assume the crystal has only one slip direction on one slip plane oriented normal to the crystallographic *c* axis and that the deformation in the crystal grain is homogeneous (Figure 17.25*C*, *D*). In the theory, the spin of the crystal is identical to that of the surrounding continuum, that is, it must be zero in this case (Figure 17.25*A*). Slip on the one slip system in the crystal produces a progressive simple shear such that the strain ellipse in the crystal is identical in shape to that of the macroscopic pure shear.

Simple shearing, however, has a nonzero spin (Figure 17.25*B*). The average spin in the crystal can be made zero, if a compensating rigid-body angular velocity is added to the progressive simple shear (Figure 17.25*E*, *F*). This angular velocity maintains the principal axes of instantaneous strain in the crystal parallel to the principal axes of macroscopic strain, and in so doing, it rotates the slip plane toward a limiting orientation normal to the shortening direction, and rotates the *c* axis toward $\hat{\zeta}_3$, the axis of maximum instantaneous shortening. Note that the crystallographic axes, such as the *c* axis, are not material lines and are not rotated by shear on the slip planes. The rate of rotation decreases toward zero as the limiting orientation is approached.

If we apply this model to all grains of a polycrystalline aggregate and ignore the constraints of the von Mises condition (see Section 17.5), we predict a symmetrically distributed maximum of *c* axes around the direction of maximum instantaneous shortening $\hat{\zeta}_3$ and, because this example is a coaxial deformation, around the maximum finite shortening axis $\hat{s}_3$ as well. The concentration of the maximum increases with increasing strain.

Under similar conditions, a progressive simple shear of the continuum should cause the *c* axis of the crystal

---

[6]Like the infinitesimal displacement gradient tensor (Box 12-3), the velocity gradient tensor can be divided into a symmetric and an antisymmetric part. The symmetric part is the instantaneous strain rate $\dot{\varepsilon}_{k\ell}$, and the antisymmetric part is the spin $w_{k\ell}$:

$$\frac{\partial v_k}{\partial x_\ell} = \dot{\varepsilon}_{k\ell} + w_{k\ell} \quad \dot{\varepsilon}_{k\ell} \equiv 0.5\left[\frac{\partial v_k}{\partial x_\ell} + \frac{\partial v_\ell}{\partial x_k}\right]$$

$$w_{k\ell} \equiv 0.5\left[\frac{\partial v_k}{\partial x_\ell} - \frac{\partial v_\ell}{\partial x_k}\right] \qquad \text{(i)}$$

$$\dot{\varepsilon}_{k\ell} = \dot{\varepsilon}_{\ell k} \quad w_{k\ell} = -w_{\ell k} \qquad \text{(ii)}$$

Equations (ii) define the symmetric character of $\dot{\varepsilon}_{k\ell}$ and the antisymmetric character of $w_{k\ell}$. The second Equation (ii) shows that there are only three independent components to the spin tensor in three dimensions, w_{12} $(= -w_{21})$, w_{13} $(= -w_{31})$, and w_{23} $(= -w_{32})$, because for $k = \ell$, Equation (ii) requires $w_{k\ell} = 0$ (in two dimensions, there is only one independent component). These three independent components define a vector called the **vorticity** $\boldsymbol{\omega}$ whose components are:

$$[\omega_1, \omega_2, \omega_3] \equiv [(w_{32} - w_{23}), (w_{13} - w_{31}), (w_{21} - w_{12})] \qquad \text{(iii)}$$

$$= \left[\left(\frac{\partial v_3}{\partial x_2} - \frac{\partial v_2}{\partial x_3}\right), \left(\frac{\partial v_1}{\partial x_3} - \frac{\partial v_3}{\partial x_1}\right), \left(\frac{\partial v_2}{\partial x_1} - \frac{\partial v_1}{\partial x_2}\right)\right]$$

Thus the vorticity is the curl of the velocity field at a point, $\boldsymbol{\omega} = \nabla \times \mathbf{v}$. It can be shown that the three components of the vorticity vector at a point are the average of the rates of rotation, or angular velocities, of all material lines of all orientations through that point, about each of the three reference axes, respectively (compare the infinitesimal rotation tensor in Box 12-3). It also can be shown that the vorticity is the angular velocity of the principal axes of the instantaneous strain rate tensor $\dot{\varepsilon}_{k\ell}$. Some authors refer to the spin tensor as the vorticity tensor and give the word *spin* a specialized definition.

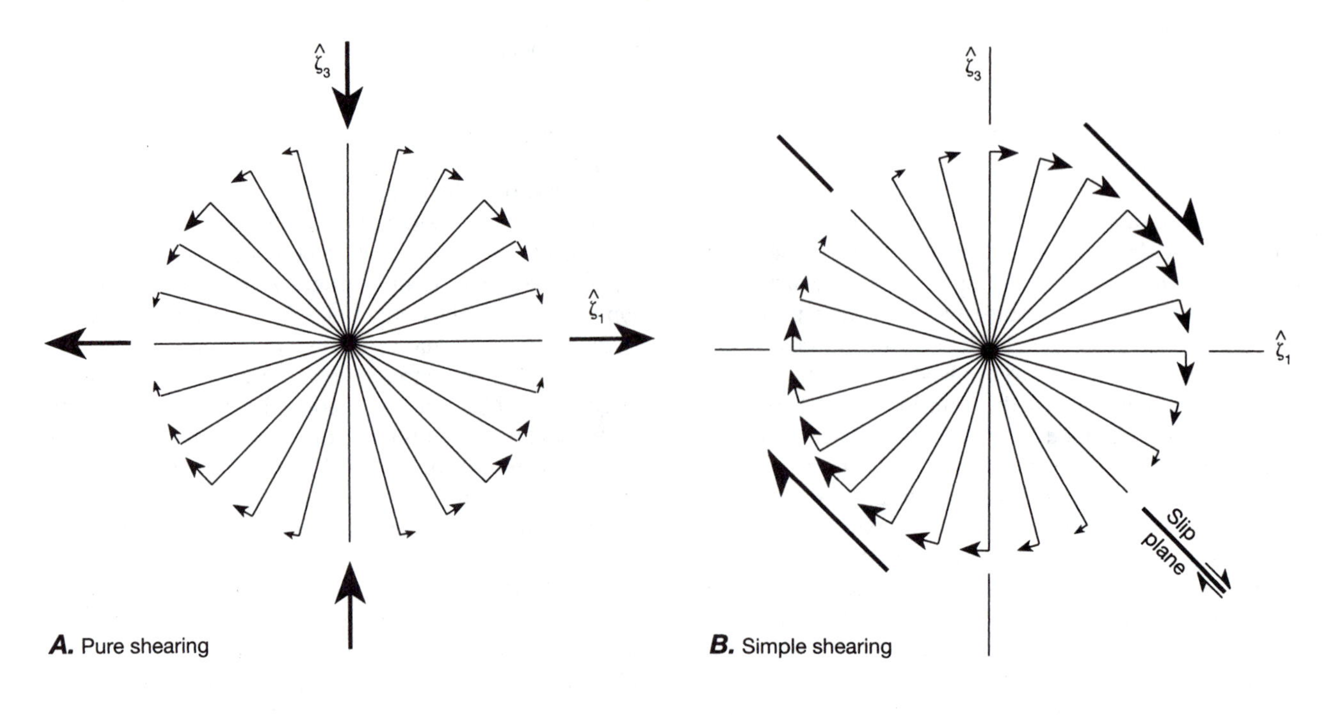

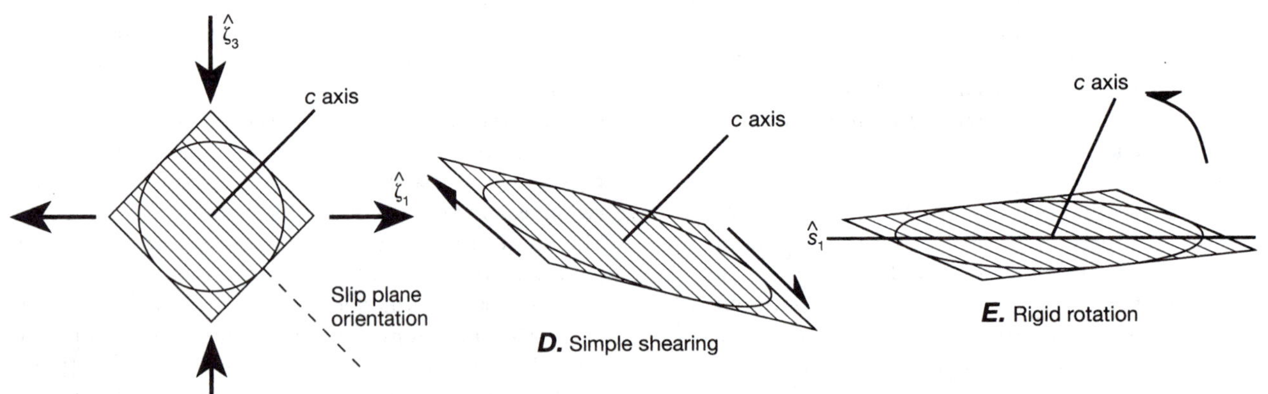

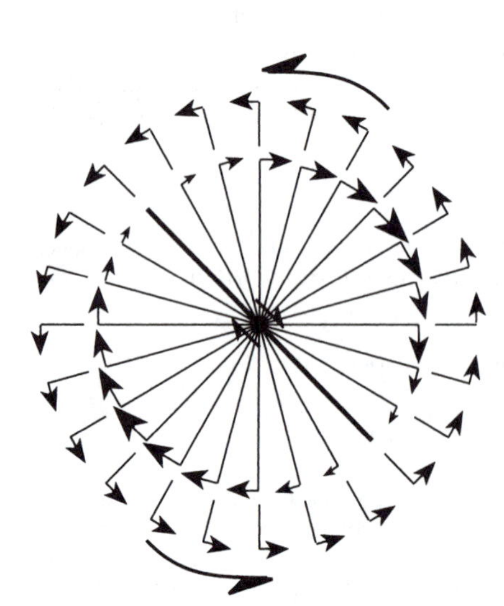

F. Simple shearing and rigid rotation rate to give zero spin

FIGURE 17.25 The model for producing a preferred crystallographic orientation by dislocation glide in crystals with a single crystallographic slip system, assuming the net spin of individual crystal grains must equal the spin of the macroscopic deformation. *A.* During progressive pure shear, the angular velocities of material lines, indicated by the curved arrows at the ends of the lines, are distributed symmetrically about the principal axes of instantaneous strain, and the spin is therefore zero. *B.* For progressive simple shear, the angular velocities of the material lines are not symmetric about the instantaneous strain axes, and the average of these rotational velocities, the spin, is not zero. *C.* The initial orientation of the crystal. The set of thin parallel solid lines indicates the orientation of the crystallographic slip planes; the slip direction is in the plane of the diagram; the heavy line is the orientation of the optic *c* axis. *D.* Macroscopic pure shearing as in *A* can only be accommodated in the crystal by simple shearing on the single slip system. The crystal deforms by simple shear to a strain equal to the imposed macroscopic pure shear strain. This shear does not rotate either the non-material *c* axis, which remains normal to the crystallographic shear planes, or the material lines parallel to the slip planes, but it does rotate all other orientations of material lines in the crystal, such as the crystal boundaries. The spin associated with the simple shear is not zero, as shown in *B*. *E.* The rigid-body angular velocity of the crystal reduces the spin to zero and keeps the principal axes of finite strain in the crystal parallel to the macroscopic instantaneous strain axes as in *A*. The angular velocity rotates the *c* axis toward the direction of maximum instantaneous shortening. *F.* The rigid-body angular velocity of the crystal grain (indicated on the outer circle) is sufficient so that when it is added to the angular velocities from progressive simple shear (inner circle; compare *B*), the resulting angular velocities of the material lines are the same as for pure shearing (*A*)—that is, the spin is zero. The rigid-body angular velocity reorients the *c* axis, causing it to rotate toward the principal shortening direction.

to rotate toward the perpendicular to the macroscopic shear plane. In this limiting orientation, the crystallographic shear plane is parallel to the macroscopic shear plane, and the instantaneous strain rate and spin in the crystal are the same as in the surrounding continuum. The rate of rotation of the crystal decreases asymptotically toward zero as it approaches the limiting orientation. Thus in simple shear, a crystallographic preferred orientation should develop that places the crystallographic slip plane parallel to the macroscopic shear plane and the crystallographic slip direction parallel to the direction of macroscopic shear.

For the coherent deformation of a polycrystalline aggregate, however, crystals must satisfy the von Mises condition of having five independent slip systems. The analysis and the preferred orientation patterns consequently become more complex. The crystallographic fabrics that develop are affected by different independent slip systems that can operate, by their relative ease of operation, by the deformation path, and by any initial crystallographic fabric in the material. The basic principles described in this subsection, however, still apply for understanding the production of crystallographic preferred orientation fabrics.

Most minerals do not possess five independent slip systems that actually operate during ductile deformation. Having to choose five to permit an arbitrary deformation of the crystals may not be realistic and may lead to preferred orientations that do not develop in nature.

Crystallographic rotations can also result from the requirement that the geometric misfit of a heterogeneously deforming polycrystal be a minimum. This means that there should be no gaps between the crystals and that the gaps that do appear in the model deformation therefore should be minimized. The essence of the misfit minimization theory is illustrated in Figure 17.26. Figure 17.26*A* shows an idealized undeformed polycrystal. The lines within each grain show the orientation of the single permitted slip direction. The von Mises condition is ignored. Figure 17.26*B* shows the polycrystal as it would look if the deformation were homogeneous. Figure 17.26*C* shows the closest approximation to the homogeneous deformation that can be attained, given that each crystal has only one slip system. Rotations and translations of the crystals are introduced to minimize the gaps and overlaps resulting from the heterogeneity of the actual crystal slip. These rotations result in the reorientation of the crystal lattices and the development of a preferred orientation. The remaining gaps and overlaps are ignored, the implicit assumption being that other deformation mechanisms (including solution and diffusion) can operate in nature to keep the deformation coherent. This model also predicts a concentration of crystallographic slip planes subparallel to the plane of flattening in coaxial deformation and subparallel to the shear plane in simple shear.

No simple model has emerged from experimental results to predict the effects of either dynamic or static recrystallization on preferred orientation fabrics. In general, the orientation of new grains produced during static recrystallization seems controlled by the orientation of the original grain, although the crystallographic relationship is not a simple one. Dynamic recrystallization, however, does not seem to alter preferred orientation fabrics significantly.

Whatever the detailed mechanisms for rotating crystals into a crystallographic preferred orientation during deformation of polycrystalline material such as rock, however, empirical observations of natural fabrics seem to support the following general rules:

For both coaxial and noncoaxial progressive deformations, there is a progressive increase toward 90° in the angle between the principal

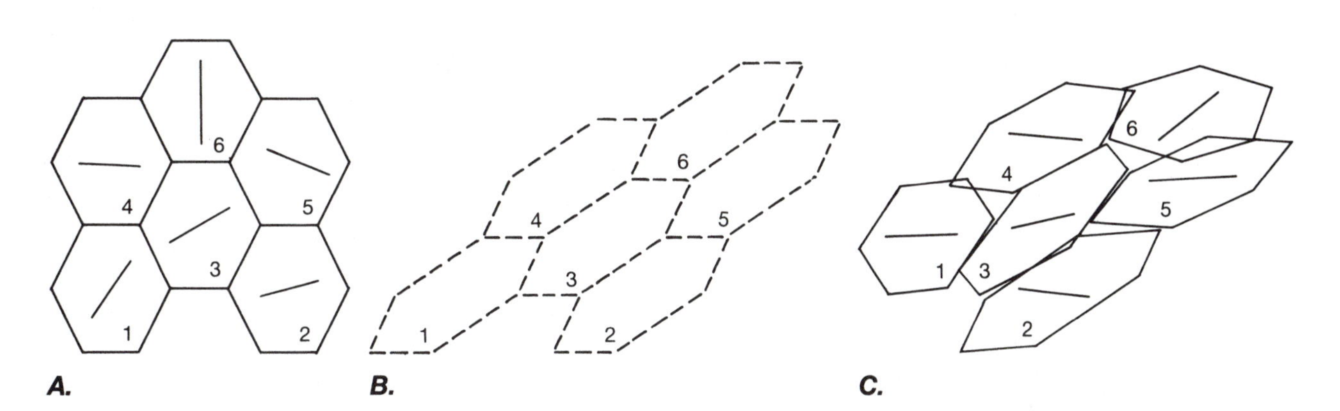

FIGURE 17.26 The model for producing a preferred crystallographic orientation by dislocation glide in crystals with a single crystallographic slip system, assuming the geometric misfit of crystal grains in a polycrystalline aggregate is minimized. (From Etchecopar and Vasseur 1987) *A*. The undeformed polycrystal. Lines in each crystal show the orientation of the slip plane. The slip direction is in the plane of the diagram. *B*. The form the crystals would take if the deformation were homogeneous. *C*. The best approximation to the homogeneous form of the deformation given that crystals can slip on only one slip system and that the area of misfit and overlap must be a minimum.

shortening axis of finite strain $\hat{s}_3$ and the active crystallographic slip plane.

During coaxial macroscopic progressive deformation, active crystallographic slip planes tend to rotate toward being perpendicular to the $\hat{s}_3$ axis of finite strain. The $\hat{s}_3$ axis does not rotate with respect to the axes of instantaneous strain. Rock fabrics tend to have orthorhombic or axial symmetry (see Section 17.8(i), Figure 17.33*B*, *C*).

During a noncoaxial macroscopic progressive deformation, active crystallographic slip planes tend to rotate toward parallelism with the macroscopic shear plane, and active crystallographic slip directions tend to rotate toward parallelism with the macroscopic shear direction. The $\hat{s}_3$ axis of finite strain tends to rotate with respect to the principal axes of instantaneous strain toward 90° to the macroscopic shear plane (see Figure 12.15). Thus the angle between $\hat{s}_3$ and the crystallographic slip plane tends to increase toward 90°. Rock fabrics in general have monoclinic symmetry (see Section 17.8(i), Figure 17.33*D*).

If more than one slip system is active in a given mineral, crystallographic preferred orientations will develop associated with the slip on each slip system, according to the preceding principles.

We illustrate these general principles by examining the slip systems and crystallographic preferred orientation fabrics for olivine and quartz in the following two subsections.

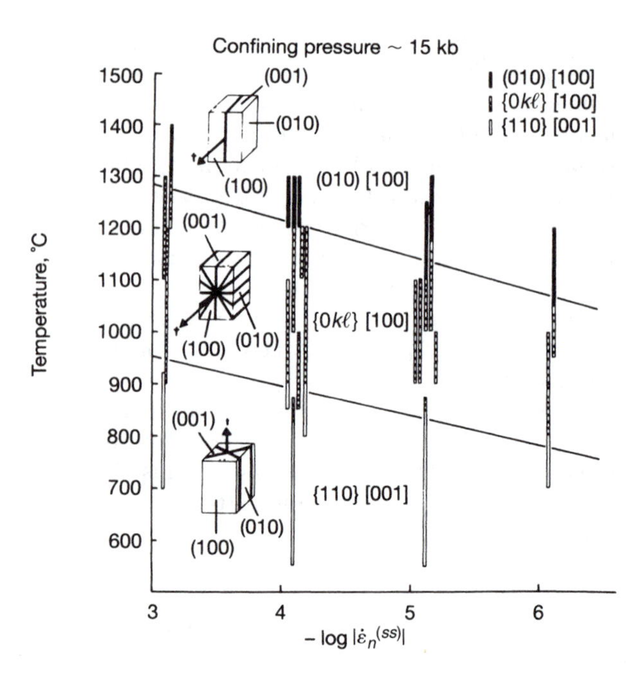

FIGURE 17.27 Experimentally determined slip systems that operate in olivine as a function of temperature and strain rate. The different patterns on the vertical bars indicate the conditions under which the different slip systems have been observed. Parentheses ( ) indicate the Miller indices of particular crystallographic slip planes; braces { } indicate a complete set of symmetrically related crystallographic slip planes; and brackets [ ] indicate the coordinates of particular crystallographic slip directions. The blocks represent an olivine crystal on which are indicated the active slip planes and slip directions (arrows). (From Carter and Avé Lallement 1970)

ii. Olivine Fabrics

The dominant slip systems in dry olivine change with temperature and strain rate, as shown by the crystal diagrams and the experimental data plotted in Figure 17.27 (see also Table 17.1).

At low temperatures, the dominant slip system is {110}[001], which means the slip planes are the {110} planes, which intersect both the *a* and *b* crystallographic axes and are parallel to the *c* axis of the olivine crystal, and the slip direction is [001], which is parallel to the crystallographic *c* axis (see lower diagram in Figure 17.27; the caption to this figure gives an explanation of the conventions for designating the shear plane and slip direction of a slip system). At intermediate temperatures, the dominant slip system is {0*kℓ*}[100], where *k* and *ℓ* can have any integer value. For this slip system, called **pencil glide**, any plane parallel to [100] (the crystallographic *a* axis) is a slip plane with the slip direction parallel to [100] (middle diagram in Figure 17.27). At the highest temperatures, the dominant slip system is (010)[100], which means slip occurs on planes parallel to the *a* and *c* crystallographic axes in the [100] (or *a*) direction (upper diagram in Figure 17.27).

The dominant high-temperature slip system may be significantly modified by the presence of water in solution in the crystal lattice. As the water content increases from zero to high (on the order of 1200 H/10^6 Si), the exact value depending on the applied stress, the slip system changes from (010)[100] to (001)[100] and eventually to (100)[001]. At high stresses for a wide range of water contents, the dominant slip system becomes (010)[001]. Thus, depending on the conditions of deformation, a variety of different crystallographic preferred orientations should develop.

Examples of preferred orientation fabrics for the *b* axes, that is, the [010] fabrics, in naturally deformed olivine are shown in Figure 17.28. In Figure 17.28*A*, the *b* crystallographic axes [010] form a maximum normal to the foliation plane but are somewhat spread out in a girdle also normal to the foliation. If the foliation is interpreted to be the plane of flattening of the finite strain ellipsoid, the ($\hat{s}_1$, $\hat{s}_2$)-plane, with the lineation parallel to $\hat{s}_1$, the symmetry of the [010] fabric with respect to the foliation indicates a coaxial deformation (see Section 17.8). Under these conditions, the observed fabric is what we would expect if the high-temperature slip system

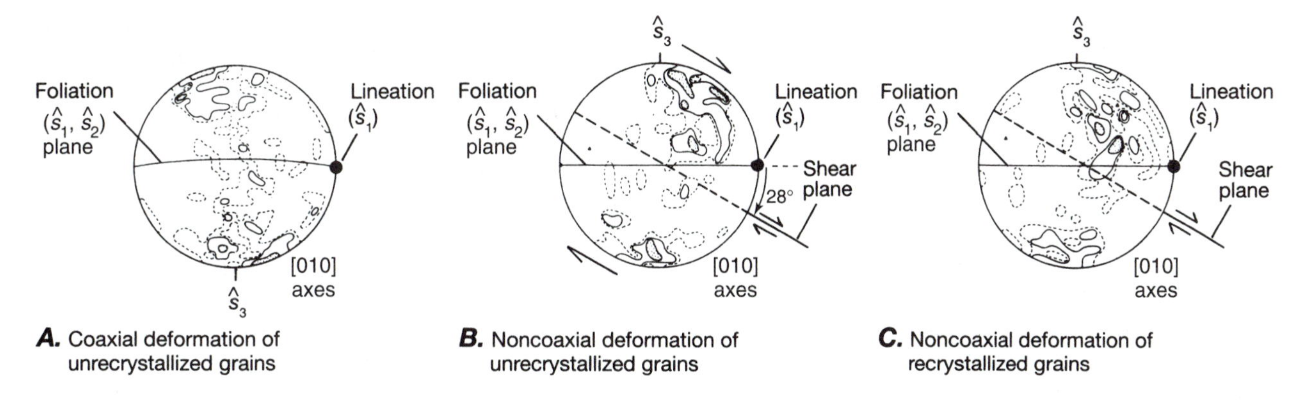

FIGURE 17.28 [010] olivine fabrics from naturally deformed dunites. Lower hemisphere, equal area projections; contours at 1, 2, 4, and 8 percent per 0.45 percent area. (From Nicolas and Poirier 1976) *A.* [010] olivine fabric from 100 unrecrystallized grains from the Baldissero ultramafic body (Ivrea zone, western Alps). A girdle normal to the foliation plane contains a maximum normal to the foliation. The symmetry indicates coaxial deformation. If the foliation is taken to be the ($\hat{s}_1$, $\hat{s}_2$)-plane of the finite strain ellipsoid (the plane of flattening) and the lineation is the $\hat{s}_1$ direction, then this pattern is consistent with (010)[001] slip producing the maximum, and {0*kℓ*}[100] slip producing the girdle. The inference is that deformation was at moderate temperatures for ductile deformation in olivine. *B* and *C.* [010] olivine fabrics from the Lanzo ultramafic body (Ivrea zone, western Alps). *B.* [010] fabric in 100 unrecrystallized olivine grains. *C.* [010] fabric in 100 dynamically recrystallized olivine grains. Both show a diffuse maximum inclined to the foliation, indicating noncoaxial deformation, consistent with dominant glide on (010)[100] during deformation having a strong component of simple shear. (From Boudier 1976)

(010)[100] were active (Figure 17.27), because then the crystals would rotate so that the slip plane (010) would tend to become perpendicular to the shortening direction $\hat{s}_3$ and the *b* crystallographic axis [010] would become concentrated parallel to the maximum shortening direction (compare Figure 17.25*C*, *D*, *E*). The fact that there is also an indication of a distribution of *b* axes in a girdle normal to the foliation is consistent with a contribution from slip on the pencil glide slip systems {0*kℓ*}[100], because an orientation of the crystal so that any one of these possible slip planes is normal to the direction of maximum shortening $\hat{s}_3$ would result in the *b* axes lying along a great-circle girdle that contained the $\hat{s}_3$ direction. This distribution indicates that deformation occurred under the moderate temperature conditions that lead to the activation of the pencil glide slip systems (Figure 17.28). Evidence that both these slip systems were active indicates that deformation occurred under conditions transitional between the fields in which each of these systems dominates.

Olivine fabrics produced experimentally at high temperatures during coaxial deformation are consistent with the natural fabric in Figure 17.28*A*. In these experiments, the unrecrystallized and recrystallized grains showed the same fabric.

The [010] fabrics in Figure 17.28*B*, *C* are for unrecrystallized and recrystallized grains, respectively, in a naturally deformed peridotite. They indicate deformation under noncoaxial conditions because the [010] (*b* axis) maximum is inclined to the foliation, not normal to it. For unrecrystallized grains (Figure 17.28*B*) the [010] maximum is diffuse but at a high angle to the inferred macroscopic shear plane, which is indicated by the dashed great circle. Thus the fabric is consistent with the slip plane in the crystals (010) rotating toward being parallel to the macroscopic plane of shear. The foliation plane indicates the orientation of the plane of flattening, the ($\hat{s}_1$, $\hat{s}_2$)-plane, and the lineation indicates the orientation of the $\hat{s}_1$ direction in that plane. Thus the principal finite strain axes are inclined to the macroscopic shear plane, and the orientation of $\hat{s}_1$ relative to the macroscopic shear plane defines the sense of shear that produced the fabric (compare with Figure 12.15).

The angle between the $\hat{s}_1$ direction, indicated by the stretching lineation, and the shear plane, indicated by the normal to the *b* axis maximum, should decrease with increasing strain (Figure 12.15), as long as the fabric is not affected by dynamic recrystallization. This angle therefore provides a minimum measure of the shear strain magnitude. In Figure 17.28*B*, this angle is about 28°, which falls roughly between the strains in Figure 12.15*B* (37°) and *C* (24°). In the same rock, the newly recrystallized grains show a pattern (Figure 17.28*C*) similar to, but even more diffuse than, the unrecrystallized grains (Figure 17.28*B*). The interpretation of fabrics in this manner can provide essential information concerning the coaxial or noncoaxial nature of the deformation and a rough estimate of the magnitude of the shear in naturally deformed rocks.

Olivine fabrics of this nature have been invoked to explain seismic anisotropy in fossil subduction zones and in the mantle. An anisotropy of seismic velocity is commonly observed in mantle rocks above depths of about 200 to 250 kilometers. Olivine crystals are seismically anisotropic, with the highest velocity parallel to the crystallographic a direction [100] and the slowest parallel to the c direction [001]. If we assume the high-temperature creep mechanism (010)[100] has operated in olivine in the mantle (Figure 17.27), the a slip direction [100] in the olivine crystals should align approximately with the shear direction of the deformation. Thus the fast seismic direction can be associated with a preferred orientation of olivine a axes, and the seismic anisotropy can be used to infer the direction of shear in the mantle. This interpretation is complicated by the experimental evidence that water dissolved in the olivine can change the dominant slip system. For mantle environments in which significant amounts of water were present, this could invalidate the assumption that the dry high-temperature slip system was dominant. With further work, however, the seismic anisotropy of the mantle may become a means of mapping the mantle flow.

The inference that the seismic velocity anisotropy indicates a crystallographic preferred orientation of olivine in the mantle also suggests that above a depth of 200 to 250 km creep in the mantle is dominated by dislocation mechanisms. These mechanisms give rise to crystallographic preferred orientations and result in a power-law rheology. Below that depth, deformation is dominated by diffusion mechanisms that do not produce crystallographic preferred orientations and give a linear Newtonian rheology (see Section 16.6(ii).

iii. Quartz Fabrics

Quartz is a more difficult mineral to understand than olivine because of its higher symmetry and its greater number of possible slip systems, and possibly because the conditions in the crust under which quartz is deformed span the range of conditions over which the preferred slip systems in the mineral change. The major slip planes and directions that have been identified in quartz on the basis of experimental deformation of single crystals are shown in Figure 17.29 (see also Table 17.1). At lower temperatures and higher strain rates, slip most commonly occurs on the basal plane, which is normal to the c axis, parallel to one of the a axis directions $(0001)\langle 11\bar{2}0\rangle$. At higher temperatures and lower strain rates, slip occurs on one of the prism planes m parallel to an a axis or the c axis $\{10\bar{1}0\}\langle 1\bar{2}10\rangle$, [0001] and on the rhomb planes r parallel to the a and the $c + a$ directions $\{1\bar{1}01\}\langle 11\bar{2}0\rangle$, $\langle 11\bar{2}3\rangle$. Numerous other slip systems also have been observed, including slip on the prism planes m parallel to $c + a$, and on second-order pyramidal planes $\{2\bar{1}\bar{1}1\}$ parallel to $c + a$.

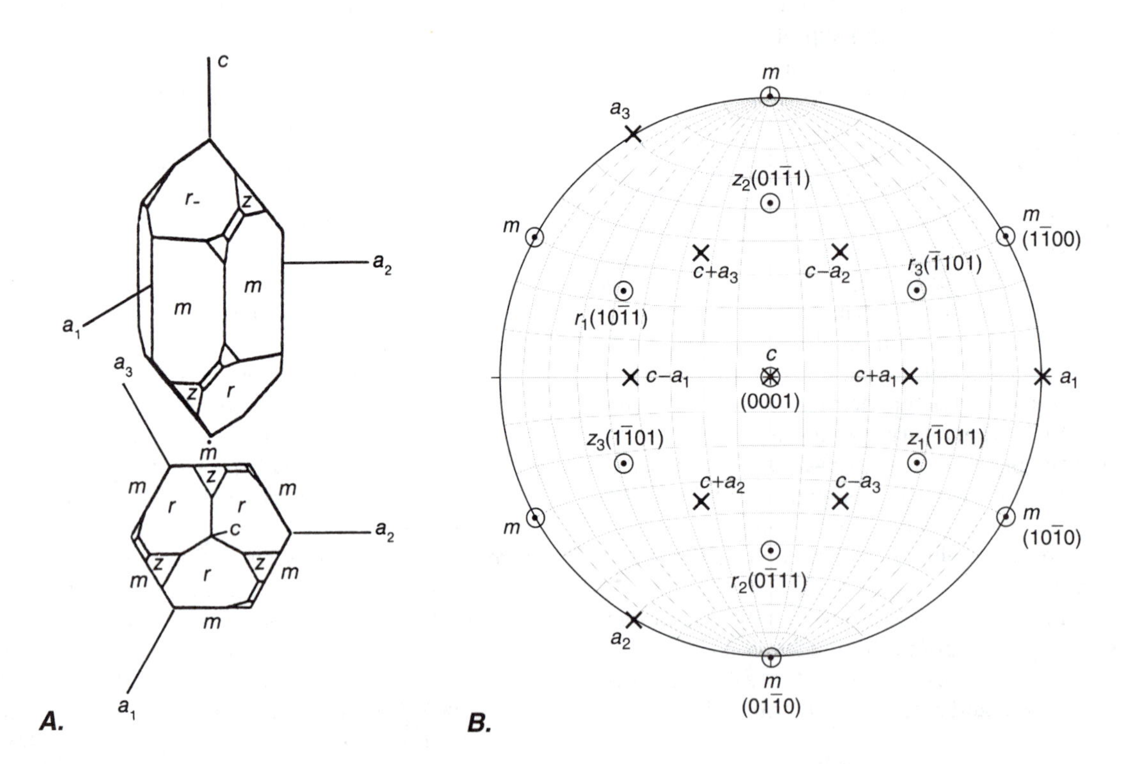

FIGURE 17.29 Crystal faces and directions in quartz showing the most important slip planes and slip directions. (Nicolas and Poirier 1976) *A.* Crystal morphology with crystallographic axes and faces labeled. Here m indicates the prism planes; r and z indicate, respectively, the positive and negative rhombohedral planes; the plane normal to c is the basal plane. *B.* Upper hemisphere equal-angle projection of the poles to crystallographic planes (dots) and the crystallographic directions (x) in quartz, oriented as shown in the lower of the two diagrams in *A.* (Note: the upper-hemisphere projection is common in mineralogy, whereas the lower-hemisphere projection is the standard in structural geology).

In natural rocks, however, fabric data that include both the c and a axes show that there is a dominant tendency for an a axis to be oriented parallel to the direction of slip, indicating that it is the dominant slip direction. This direction is common to several slip planes, such as the basal plane, the prism planes m, and the positive and negative rhombohedral planes r and z. Thus the preferred orientation of c axes depends on the geometry of the deformation and the particular slip planes that are active.

Figure 17.30 shows idealizations of the common types of quartz c axis and a axis fabrics showing orthorhombic symmetry, along with examples of natural c axis fabric patterns. Diagrams *A* through *E* and *G* through *J* are superimposed on a Flinn diagram illustrating the states of strain associated with the various patterns (compare Figure 12.21). The ratios of the principal strains for the natural fabrics (the large black dots in Figure 17.30) were determined largely from the shapes of deformed grains in the rocks. The idealized c axis diagrams are so-called **fabric skeletons**, which are constructed by drawing lines on contoured fabric data through the maxima and along the lines of maximum concentration. The idealized a axis fabrics are shown as schematic contour diagrams. Although a axis patterns provide additional information useful for interpreting the kinematic significance of the fabric patterns, their determination requires X-ray techniques and sophisticated computer analysis. Determining these complete fabrics is difficult and time-consuming and has been done in only a few studies.

The symmetry of the fabrics in Figure 17.30 with respect to the principal stretches indicates the deformation was coaxial (see Section 17.8). For simple flattening ($k = 0$), c axes tend to lie on a small-circle girdle about the maximum shortening direction $\hat{s}_3$, and a axes also define a small-circle girdle about $\hat{s}_3$ having a larger opening angle (Figure 17.30*A*). In the general flattening field ($0 < k < 1$), patterns are transitional between simple flattening and plane strain (Figure 17.30*B*). For plane strain ($k = 1$), the c axes define a crossed girdle pattern (Figure 17.30*C*).

There are two types of crossed girdle patterns. The **type I crossed girdle pattern** appears like a distorted small-circle girdle about the maximum shortening direction $\hat{s}_3$, connected by a partial great-circle girdle in the ($\hat{s}_2$, $\hat{s}_3$)-plane. The a axes fall predominantly into two maxima symmetric about the direction of maximum extension $\hat{s}_1$ (Figure 17.30*C*). Less common is the **type II crossed girdle pattern**, which approximates a pair of crossing great-circle distributions. This pattern has been associated with both plane strain (Figure 17.30*C*) and a general extension (Figure 17.30*D*) for which the pattern represents a transition to the pattern for simple extension.

In simple extension, the c axes tend to define a **cleft girdle pattern**, which consists of a small circle with large opening angle centered on the direction of maximum extension $\hat{s}_1$ (Figure 17.30*E*). The associated a axis pattern evolves toward a small-circle distribution with a small opening angle also centered about the $\hat{s}_1$ direction (Figure 17.30*E*).

Another c axis pattern common in high-grade metamorphic rocks is a single maximum parallel to the foliation and normal to the lineation (Figure 17.30*F*). The a axis pattern mimics the distribution of the a axes in a quartz single crystal.

These fabrics consistently show that the a directions of the crystals have a strong preferred orientation symmetrical relative to the principal stretches. The associated c axis orientations then consistently place one of the principal slip planes containing the a axis in an orientation of high shear strain. This geometry implies that a is the dominant direction of slip and that the slip systems tend to become oriented so as to accommodate the imposed deformation. Most of the orthorhombic c axis fabric diagrams are more easily understood after we discuss the geometrically simpler case of simple shear. For Figure 17.30*F*, however, the preferred crystallographic orientation accommodates coaxial plane strain by slip in the a direction on two prism planes symmetrically oriented with respect to $\hat{s}_1$ and $\hat{s}_3$.

In noncoaxial deformation, the c axis pattern becomes asymmetric with respect to the foliation in the rock, and therefore presumably with respect to the plane of flattening ($\hat{s}_1$, $\hat{s}_2$). Figure 17.31 shows idealized c axis fabric skeletons and a axis contour diagrams, along with some examples of c axis fabrics from naturally sheared rocks. The diagrams are plotted so that the foliation plane is the vertical great circle from left to right across each stereogram, and the shear sense on the shear plane is dextral. The progression from *A* to *E* represents an increase in the angle Ω between the maximum principal axes of the instantaneous stretch $\hat{\zeta}_1$ and the finite stretch $\hat{s}_1$. As the amount of shear increases from zero to infinity, the angle Ω increases from 0° to 45°. This increase reflects the fact that during simple shear, the maximum instantaneous stretching direction $\hat{\zeta}_1$ remains at a constant 45° to the shear plane, whereas with increasing shear strain, $\hat{s}_1$ starts at 45° to the shear plane and progressively rotates toward parallelism with the shear plane.

For simple shear, the initial c axis fabric skeleton is a type I crossed girdle pattern characteristic of coaxial plane strain (compare Figures 17.31*A* with 17.30*C*). Ω is essentially zero, and the shear plane is at 45° to the center of the distorted small-circle girdle. With increasing strain, c axes tend to concentrate toward a great-circle girdle that is roughly normal to the shear plane. The inclination of the c axis great-circle girdle relative to the foliation, which marks the ($\hat{s}_1$, $\hat{s}_2$)-plane, indicates the direction of relative shear on the shear plane. Thus, well-defined quartz fabrics, asymmetric with respect to the foliation, can be used to infer the sense of shearing in the rocks (see Section 17.8). This type of information, if gathered across an entire region, is useful in inferring the tectonic history.

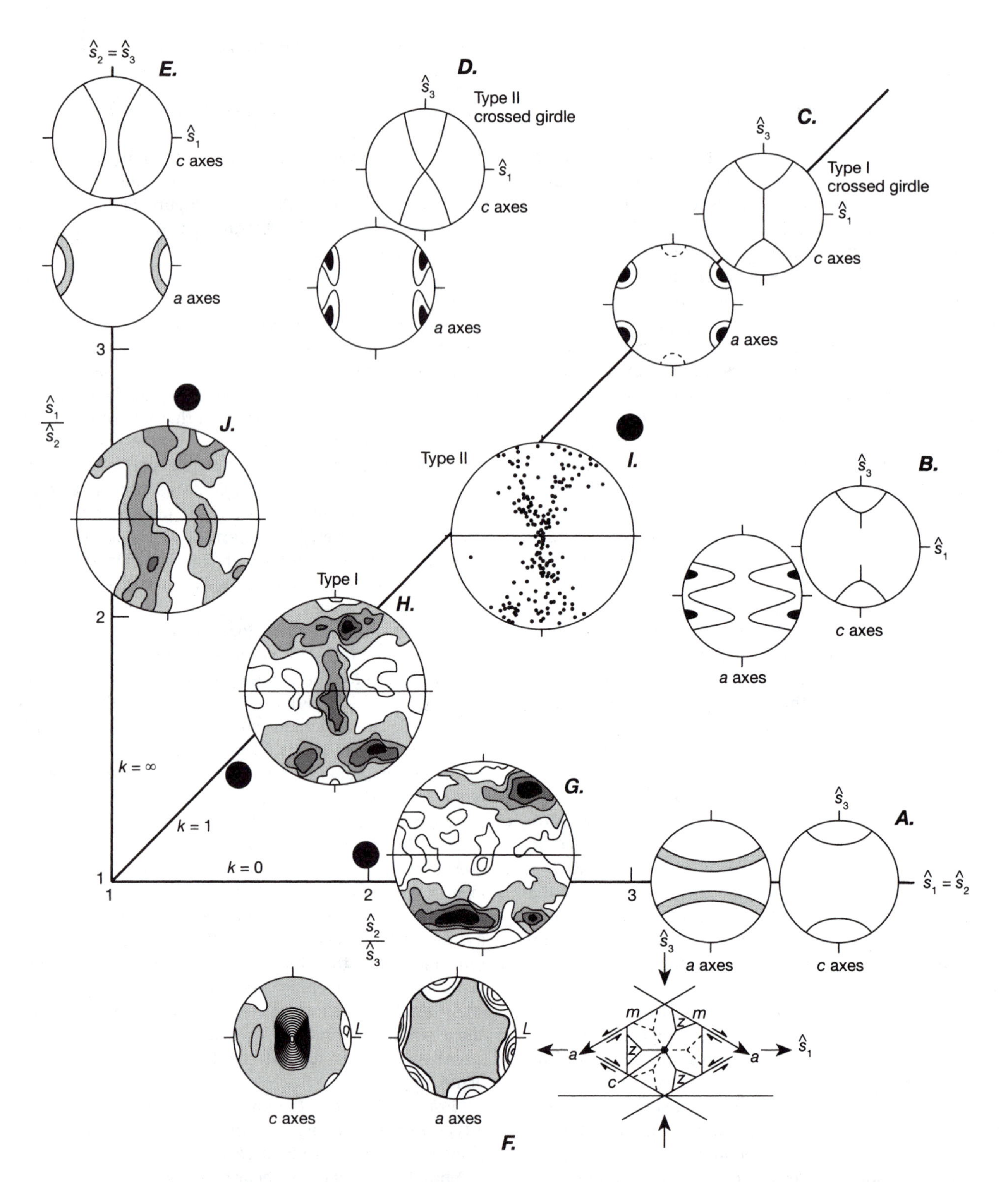

FIGURE 17.30 Common quartz fabrics for coaxially deformed rocks: natural *c* axis fabrics, idealized *c* axis fabric skeletons, and schematic contours of *a* axis fabrics as a function of the value of *k* on a Flinn diagram (see Figure 12.21). The large black dots plotted on the Flinn diagram indicate the principal stretch ratios for the natural fabrics, which have been independently determined from deformed grain shapes. In each diagram, the foliation plane strikes east-west and is vertical. *A.* Simple flattening ($k = \infty$). B. General flattening ($0 < k < 1$). C. Plane strain ($k = 1$). The type I cross girdle fabric consists of a small-circle girdle with a connecting partial great-circle girdle; the type II cross girdle fabric consists of two crossing great-circle girdles. *D.* General extension ($1 < k < \infty$) showing a type II crossed girdle pattern transitional from plane strain to simple extension. *E.* Simple extension ($k = \infty$). *F.* Quartz *c* axis fabric consisting of a single maximum parallel to the foliation and perpendicular to the lineation, the associated *a* axis fabric, and the interpretation of the active slip systems. Such fabrics are common in high-grade metamorphic rocks. G. Natural quartz *c* axis fabric comparable to *A*. *H.* Natural quartz *c* axis fabric comparable to *C*. *I.* Natural quartz *c* axis fabric comparable to *D*, although the inferred strain ratios imply the deformation is in the field of general flattening, whereas the fabric in *D* plots in the field of general constriction, suggesting the inferred strain ratios may be in error. *J.* Natural quartz *c* axis fabric comparable to *E*. (*A* through *E*: after Schmid and Casey 1986; *F*: from Schmid and Casey 1986, sample "Gran 133"; *G* through *J*: from Price 1985, samples 9, 46, 62, and 72, respectively).

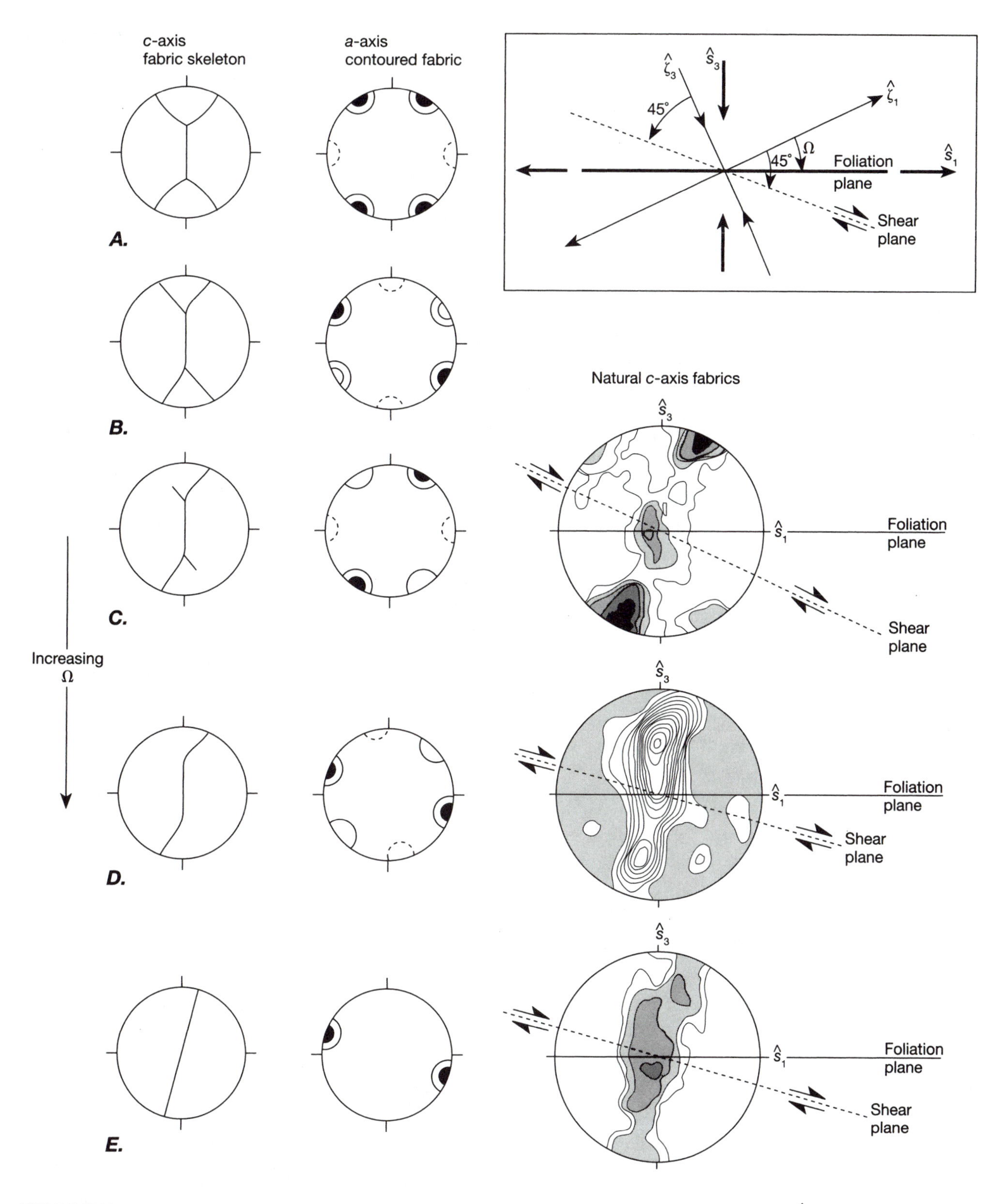

FIGURE 17.31 Common quartz fabrics associated with noncoaxial deformation: natural *c* axis fabrics, idealized *c* axis fabric skeletons, and schematic contours of *a* axis fabrics as a function of increasing amounts of noncoaxial deformation (*A* through *E*) as measured by increasing values of the angle Ω between the maximum principal axes $\hat{\zeta}_1$ and $\hat{s}_1$ of the instantaneous and finite strain ellipses. (Fabric skeletons, schematic *a* axis fabrics, and natural fabric in *D* from Schmid and Casey 1986; natural fabrics in *C* and *E* from Price 1985)

The origin of these c axis preferred orientation patterns can be understood in terms of the orienting mechanism shown in Figure 17.25*C–E*. In simple shear, the crystallographic slip direction tends to become aligned with the direction of macroscopic shear, and the crystallographic slip planes tend to become oriented parallel to the plane of simple shear. Because slip in the a crystallographic direction appears to be dominant in naturally deformed quartz, different c axis maxima can be associated with slip on different specific slip systems in the a direction. Thus in Figure 17.32*A*, maximum I is associated with slip on the prism planes $\{m\}<a>$ because with c in this orientation, one m-plane is parallel to the shear plane, and the a axis in that m-plane is parallel to the shear direction (Figure 17.32*B*). Similarly, maxima II (Figure 17.32*A*) are associated with slip on the rhombohedral planes $\{r\}<a>$ and $\{z\}<a>$ (Figure 17.32*C*); and maximum III (Figure 17.32*A*) is associated with slip on the basal planes, which are normal to the c axis, in one of the a directions—that is, $(0001)<a>$ (Figure 17.32*D*).

From this association of c axis maxima with different slip systems, we can see that the c axis patterns having orthorhombic symmetry are consistent with a coaxial deformation accommodated on slip systems symmetrically oriented relative to the principal axes of strain. Thus, for example, from Figure 17.32*A* we interpret the crossed girdle patterns (Figures 17.32*E*, 17.30*D*, *I*) to indicate that coaxial deformation was accommodated by slip on all three slip systems (Figure 17.32*B–D*), with shear in two directions symmetrically oriented with respect to both $\hat{s}_1$ and $\hat{s}_3$.

Deformation experiments have successfully reproduced some aspects of these quartz fabrics. Single c axis maxima and small-circle girdle distributions have been produced experimentally under conditions of simple flattening (Figure 17.30*B*). This is consistent with the active slip system being on the basal plane in the a direction. Instead of the single maximum III in Figure 17.32*A* associated with simple shear, however, the axial symmetry of the experimental deformation results in the slip planes aligning tangent to an open cone about the $\hat{s}_3$ axis. Thus the c axes lie on a small-angle cone about that axis, in effect spinning the maximum III about the $\hat{s}_3$ axis, which is normal to the foliation plane. The transition from a

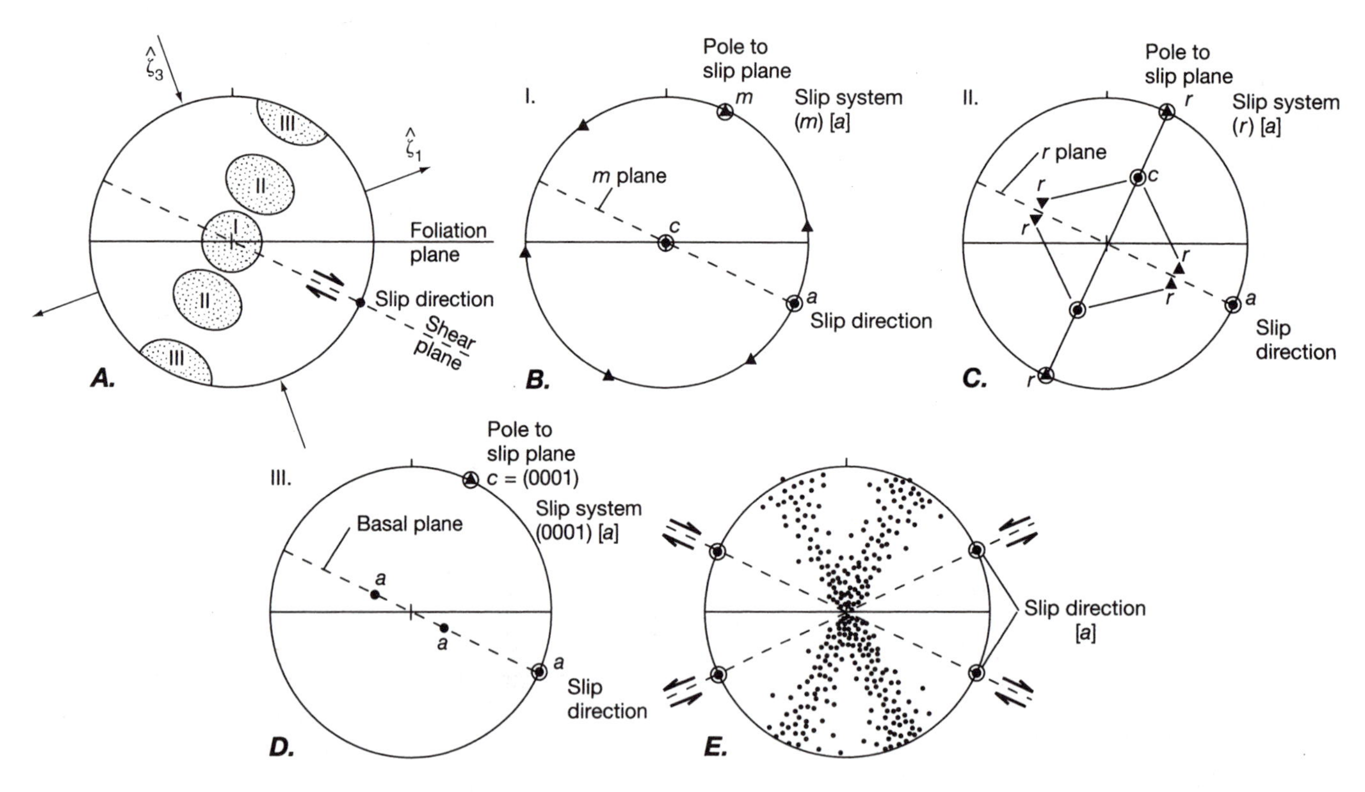

FIGURE 17.32 Slip systems associated with particular c axis maxima assuming that the active slip plane must be parallel to the shear plane and that a is the crystallographic slip direction. Crystallographic axes are shown as dots, poles to crystallographic planes as triangles. Circled points indicate directions and poles relevant to the active slip system. *A*. Locations of three common c axis maxima, I, II, and III, in noncoaxial deformations, relative to the foliation and shear planes. *B*. Crystallographic orientation for maximum I associated with prismatic slip $(m)[a]$ (compare Figure 17.30*F*). *C*. Crystallographic orientation for maximum II associated with rhombohedral slip $(r)[a]$ and $(z)[a]$. *D*. Crystallographic orientation for maximum III associated with basal slip $(0001)[a]$. *E*. Hypothetical type II crossed girdle c axis pattern resulting from coaxial deformation with active crystallographic slip systems illustrated in *A* through *D*. The plotted c axis pattern results from the a axes tending to align parallel to the circled dots. (After Schmid and Casey 1986)

maximum to a small-circle girdle with progressively increasing angle occurs as temperature increases and strain rate decreases. This transition is consistent with a shift from predominantly basal plane slip parallel to a (maximum III in Figure 17.32*A*, *D*) to mixed basal plane and either rhomb (r, z) or prism slip, both parallel to a, with an increasing dominance of the rhomb slip system (maximum II in Figure 17.32*A*, *C*) and the prism slip system (maximum I in Figure 17.32*A*, *B*) over the basal slip system. Experiments in simple shear with a component of axial shortening have also succeeded in producing the asymmetric fabrics characteristic of natural noncoaxial deformation.

The fabrics that develop depend in large part on which slip systems require the least critical resolved shear stress to be activated. This critical shear stress for the different slip systems varies with temperature and the dissolved water content of the crystal, but the specific shear stress values and how they depend on temperature and water content is not well known. Nevertheless, by using reasonable values for the different slip systems, investigators making theoretical calculations using the Taylor-Bishop-Hill model have produced many of the characteristics of quartz c-axis fabrics shown in Figures 17.30 and 17.31.

For simple shear, the Taylor-Bishop-Hill model and the misfit optimization model yield markedly different results. The Taylor-Bishop-Hill model predicts a c axis fabric skeleton comparable in many respects to that in Figure 17.31*B*, *C*, but the fabric does not evolve to the single girdle of Figure 17.31*E*. The misfit minimization model, however, predicts the formation of a single girdle inclined to the ($\hat{s}_1$, $\hat{s}_2$)-plane, similar to the fabric skeleton of Figure 17.31*E*.

Thus although the theoretical models successfully reproduce various aspects of naturally observed quartz c axis fabrics, they are not completely successful. We need better knowledge of the relative ease of slip on the different slip systems and of the effect of conditions such as temperature and dissolved water content, as well as more sophisticated models that incorporate the effects of both spin and geometric fit. We can anticipate that such developments will make it possible to interpret quartz fabric data more accurately and will allow us greater insight into the deformation conditions under which the fabrics were produced.

17.8 SYMMETRY PRINCIPLES IN THE INTERPRETATION OF DEFORMED ROCKS

One of the main goals of structural geologists, particularly in studying deformed metamorphic rocks, is to try to infer the geometry of the flow field that has produced the structures and fabrics that we observe. Among the most general concepts that can help interpret the kinematic significance of structures in rocks are symmetry principles, which can be applied to any type of process. They permit us to deduce constraints on the symmetry of unknown factors, such as the causes of a deformation process, from the symmetry of related quantities, such as the observable effects of the deformation, without knowing anything about the mechanisms that are involved in the process. Among effects of a deformation is the fabric of the rock, which we can observe and measure. Thus the question we address in this section is: What can the symmetry of fabrics in deformed rocks tell us about the deformation that produced the fabric?

Application of symmetry principles to the interpretation of rock fabrics has a long and controversial history in structural geology, starting in the 1920s and 1930s with the pioneering work of the German structural geologists Bruno Sander and W. Schmid. Structural geologists are still trying to understand the relationship between both structures and fabrics in rocks and the flow field that produced them (at one time this was a major focus of interest, particularly before the mechanisms of fabric development began to be understood). The generality of the symmetry principles still provide a useful constraint on our interpretation of rock fabrics.

We cannot hope to provide a detailed review of the subject here, but we aim in this section at least to introduce the concepts and illustrate at a basic level the use of symmetry arguments in the interpretation of fabrics in rocks. We rely substantially on the definitive paper by Paterson and Weiss (1961) that reviewed the literature and cleared up the misconceptions that existed.

We begin by reviewing the symmetries that can characterize rock fabrics.

i. Symmetries Common in Deformation and Fabrics

A symmetry operation, when applied to a physical system such as a deformed rock, is an operation such as a rotation or a reflection that leaves the system looking exactly the same as it looked before the operation was applied. Such an operation is called a **symmetry element** of the system. For our purposes, then, the term "symmetry elements" refers to axes of rotational symmetry and to mirror planes of symmetry. Two symmetry elements are considered common if they describe the same symmetry and have the same orientation in space.

There are five symmetry classes that can characterize the fabric symmetry of a deformed rock, and we summarize their characteristics, progressing in order from the highest to the lowest symmetries.

1. **Isotropic**, or **spherical**, **symmetry** (Figure 17.33*A*) is the symmetry of a sphere and thus is characterized by an infinite number of mirror planes and an infinite number of infinite-fold axes of rotational symmetry. A hydrostatic pressure and a volumetric strain that involves no shear strain are examples of fields having isotropic symmetry. The stress or strain ellipsoids in these cases reduce to a

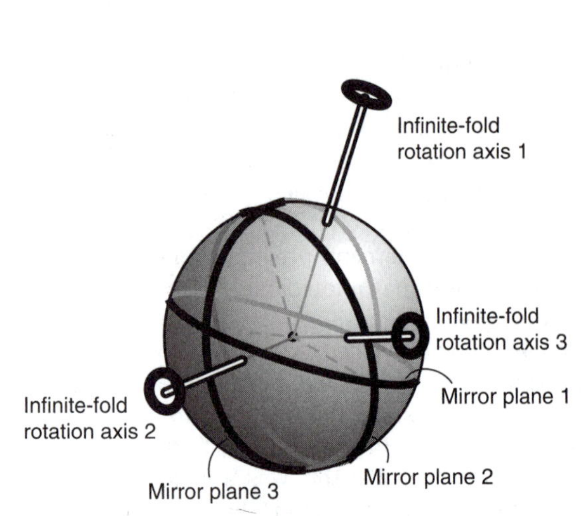

A. Isotropic symmetry

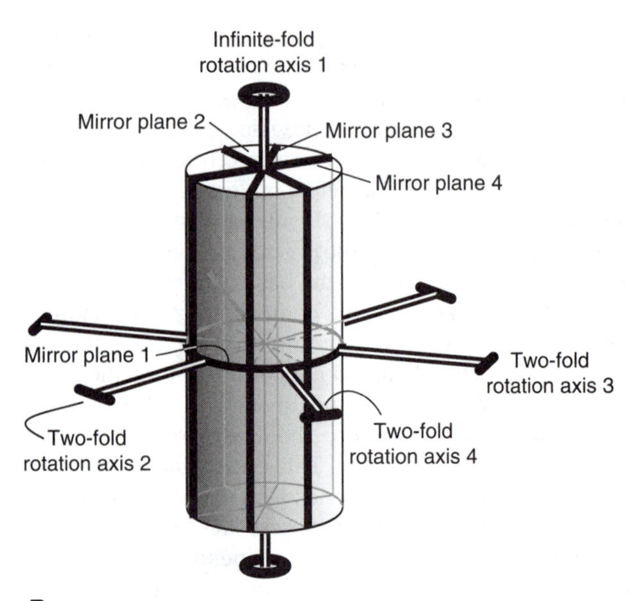

B. Axial symmetry

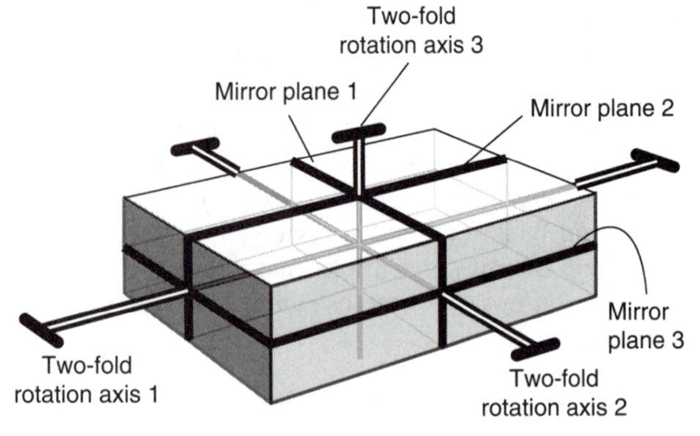

C. Orthorhombic symmetry

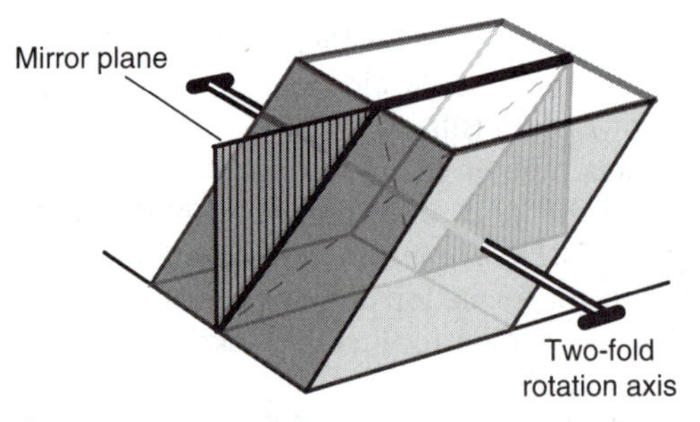

D. Monoclinic symmetry

FIGURE 17.33 Classes of symmetry relevant to the fabrics of deformed rocks. Bold black lines mark the intersection of mirror planes with the geometric figures. Double bold lines are axes of rotational symmetry. Each rotation axis is perpendicular to a mirror plane. The ends of the infinite-fold rotation axes are marked by a doughnut; the ends of the two-fold rotation axes are marked by "T-handles." *A.* Isotropic symmetry: the symmetry of a sphere. Only three mirror planes and the perpendicular infinite-fold rotation axes are shown, selected arbitrarily from the infinite number of possible mirror planes and perpendicular infinite-fold rotation axes. *B.* Axial symmetry: the symmetry of a cylinder. Of the infinite number of mirror planes that parallel the infinite-fold rotation axis, only three with their perpendicular two-fold rotation axes are shown. *C.* Orthorhombic symmetry: the symmetry of a rectangular block, characterized by three mutually orthogonal mirror planes of symmetry, with three two-fold axes of rotation, each perpendicular to a mirror plane. *D.* Monoclinic symmetry: the symmetry of a rhombic parallelepiped characterized by a single mirror plane with a single two-fold axis of rotation perpendicular to it.

perfect sphere. A rock with an equant grain shape, no crystallographic preferred orientation, and no lineation or foliation has isotropic fabric symmetry.

2. **Axial symmetry** (Figure 17.33*B*) is the symmetry of a cylinder. It is characterized by one infinite-fold axis of rotational symmetry (the cylinder axis) that is perpendicular to one mirror plane of symmetry, and by an infinite number of two-fold axes of rotational symmetry (parallel to the radii of the cylinder), each of which is perpendicular to a mirror plane of symmetry (planes parallel to the cylinder axis that bisect the cylinder) that contains the infinite-fold rotation axis. A rock characterized by a lineation and no foliation has axial fabric symmetry. The infinite-fold axis of rotational symmetry is parallel to the lineation.

3. **Orthorhombic symmetry** (Figure 17.33*C*) is the symmetry of a rectangular block having three

orthogonal edges of unequal length. There must be three orthogonal mirror planes of symmetry (each bisecting the block and parallel to one of the faces), and three two-fold axes of symmetry, one normal to each of the mirror planes (and parallel to the lines of intersection of the three mirror planes). The symmetry of the stress, the strain, and the strain rate can be no lower than orthorhombic, which is the symmetry of a triaxial ellipsoid. The mirror planes and the two-fold rotation axes of symmetry in these cases are identical in orientation to the principal planes and principal axes of the ellipsoid. A rock characterized by a lineation that lies in a foliation has orthorhombic fabric symmetry, for which the three mirror planes of symmetry are (1) perpendicular to the lineation (and thus the foliation), (2) parallel to the lineation and perpendicular to the foliation, and (3) parallel to both the lineation and the foliation.

4. **Monoclinic symmetry** (Figure 17.33*D*) is the symmetry of a rhombic parallelepiped such as is formed by the simple shear of a cube parallel to one of its faces and edges. It is characterized by a single mirror plane of symmetry with a single two-fold axis of symmetry normal to it. In the case of the sheared cube, the mirror plane is normal to the simple shear plane and parallel to the shear direction. A rock with an *S-C* fabric has monoclinic symmetry, with the mirror plane perpendicular to the line of intersection of the *S* and *C* fabric planes.
5. **Triclinic symmetry** has no mirror planes of symmetry and no axes of rotational symmetry. This symmetry commonly results from the superposition of unrelated deformations on a rock for which fabric elements from both deformations are preserved. Because the deformations are not related, the orientations of the strain axes associated with the two individual deformations have no simple relationship. Thus the fabrics that result have no symmetry elements in common, in which case the overall fabric symmetry is triclinic.

ii. The Symmetry Principle

The main principle of use to us here is attributed to the nineteenth-century physicist Pierre Curie[7] and can be stated as follows:

The symmetry elements that characterize the effects of a physical process must include the symmetry elements common among all the independent causes that govern the process and may include additional symmetry elements. However, any symmetry element not included in the effect symmetry must be absent from the symmetry of at least one of the independent causes.

Put another way, this principle states:

The symmetry of the effects can be the same or higher than the combined symmetry of the causes, but it cannot be lower.

Conversely, the combined symmetry of the causes can be the same or lower than the symmetry of the effects, but it cannot be higher.

The converse statement provides the constraints that we can apply to understanding the causes of a deformation. An illustration of an effect having a higher symmetry than the combined causes would be a uniform temperature increase (isotropic effect symmetry) that could result from a simple shearing deformation of a mechanically isotropic solid (monoclinic cause symmetry).

The symmetry principle raises the issue of how one objectively identifies cause and effect, and this is not always obvious or even possible. For our current purposes, however, we can consider that the independent causes are the *independent* constitutive variables and the constitutive material parameters. These material parameters are just the mechanical properties of the material that appear as constants in the constitutive equations. The effects are the *dependent* constitutive variables, as well as the structural and fabric elements in the rock that are associated with a deformation.

The constitutive variables could include such quantities as the temperature, the stress, the strain, the strain rate, or the velocity gradient. Determining which of the constitutive variables are dependent and which are independent depends on the externally applied constraints on the deforming system known as the boundary conditions (see Section 18.1(iii)). The material parameters could be constants such as the elastic moduli, the viscosity, or the yield strength. The fabric elements include such features as the crystallographic preferred orientation of different minerals, the foliations, and the lineations of a deformed rock. The symmetry elements that are common to all such fabric elements define the symmetry of the total rock fabric.

Thus an application of the symmetry principle to our specific area of interest could be stated:

The symmetry of the fabric of a deformed rock can be no lower, but can be higher, than the symmetry of all the combined independent *factors that govern the deformation process.*

Conversely, the symmetry of the combined independent factors that govern a deformation process can be the same or lower than the fabric symmetry, but it cannot be higher.

[7]Paterson and Weiss (1961).

We examine how this symmetry principle can aid in inferring constraints on the kinematics of the deformation from the rock fabrics that resulted from the deformation. We show in the next subsection that application of this principle allows us to eliminate certain kinematic interpretations from consideration but does not permit a unique interpretation without additional assumptions.

iii. Symmetry in the Interpretation of Deformation Fabrics

In discussing some of the c axis fabrics for olivine and quartz in Section 17.7, we made reference to the symmetry of the fabrics. We examine these fabrics again here in the light of the symmetry principle.

The symmetry of the fabric of a deformed rock generally is defined by just a few of the fabric elements, such as the foliation, the lineation, and the crystallographic preferred orientation of one of the major minerals. Determination of additional fabric elements, such as the crystallographic preferred orientations of other minerals, generally does not change the fabric symmetry. The symmetry of the fabric can be deduced from the distribution of fabric data plotted on a stereogram.

The olivine fabric in Figure 17.28*A* consists of a foliation plane, a diffusely defined girdle of [010] crystallographic axes oriented approximately normal to the foliation plane, an [010] maximum in that girdle normal to the foliation, and a stretching lineation in the foliation plane normal to the [010] girdle. From the plot in Figure 17.28*A*, we can conclude that the fabric symmetry is essentially orthorhombic. Of the three mirror planes of symmetry, one is parallel to the foliation, one is parallel to the [010] girdle, and one is nearly parallel to the primitive circle of the stereogram, and these planes are mutually orthogonal. The three two-fold axes of symmetry are respectively perpendicular to each of the mirror planes; one of these axes is parallel to the lineation, and one is parallel to the [010] maximum.

What does the symmetry principle allow us to deduce about the kinematics of deformation from the orthorhombic fabric symmetry? The principle states that this fabric symmetry must include all the symmetry elements common to the combined causes, that is, to the material properties and the independent physical fields that govern the process, but that it may contain other symmetry elements in addition. Thus we must conclude that the symmetry of the combined causes, which could include, for example, the material properties and either the stress or the velocity gradients, can be no higher than the orthorhombic symmetry of the fabric but that it could have a lower symmetry.

Further interpretation must rest on some assumptions. Let us assume that the material properties are isotropic. Let us further assume that the deformation, defined by the velocity gradients, is a cause of the process. By the symmetry principle, the combined causes can have at most the orthorhombic symmetry of the fabric. Because the material properties are already assumed to have isotropic symmetry, the most restrictive conclusion we can then draw is that the deformation must have no higher symmetry than the same orthorhombic symmetry as the fabric, but it could be lower.

This conclusion only eliminates axial and isotropic symmetry for the deformation and permits either orthorhombic or monoclinic symmetry. Thus the symmetry principle would be consistent with the fabric being formed by either a coaxial (orthorhombic) or a noncoaxial (monoclinic) deformation. A similar analysis applies to the quartz fabrics in Figure 17.30*F–J*.

We can eliminate some of the nonuniqueness in the interpretation of the orthorhombic olivine [010] axis and quartz c axis fabrics (Figures 17.28*A* and 17.30*F–J*) if we apply independent information about the significance of the fabric elements. We can assume that the foliation plane is the plane of flattening of the finite strain ($\hat{s}_1$, $\hat{s}_2$) and that the lineation is parallel to the maximum principal finite stretch $\hat{s}_1$. Furthermore, because we understand the mechanism by which crystallographic preferred orientations form (Figures 17.25*C–E* and 17.32), and we know which slip system(s) are active, we can infer that the principal axes of instantaneous strain $\hat{\zeta}_k$ are parallel to the axes of orthorhombic symmetry for the observed crystallographic preferred orientation. The fabrics in Figures 17.28*A* and 17.30*F–J* therefore show that the principal axes of finite and instantaneous strain are coaxial, so we can conclude that the fabrics result from a coaxial deformation.

In the case of the olivine [010] axis fabrics in Figure 17.28*B*, *C* and the quartz c axis fabrics in Figure 17.31*C*, *D*, *E*, the fabric symmetry is monoclinic, because the maxima and the girdle defined by the crystallographic axes are inclined (not perpendicular) to the foliation and lineation. Because all the symmetry elements common to the independent causes of the process must also be found among the symmetry elements of the effect (the fabric), the symmetry of the causes can be only the same or lower, that is, monoclinic or triclinic. Given the assumptions we have made, we can conclude that the deformation must have been noncoaxial. Thus a monoclinic fabric symmetry provides a more restrictive constraint on the symmetry of the independent governing factors than an orthorhombic fabric symmetry.

Our assumptions about the isotropic symmetry of the material properties still leave a potential nonuniqueness in the interpretation of the deformation. The lower the symmetry of one of the causes, the less restriction exists on the symmetry of the other causes, because the combined causes can have no higher symmetry than the lowest symmetry of any one cause. Thus, for example, in the case of the monoclinic fabric, if the symmetry of the material properties is less than isotropic, the symmetry of the deformation could be higher than monoclinic, as long as the symmetry of the combined causes is no higher than the monoclinic symmetry of the fabric. Such possibilities therefore require caution in using symmetry principles to interpret rock fabrics.

iv. Symmetry in the Interpretation of Deformation Geometry

An example of the usefulness of symmetry arguments in understanding physical processes is the simple shear of a mechanically isotropic material. An isotropic material has the same mechanical properties in all directions. Assume a homogeneous stress is imposed on the material, so the stress is the cause of the process, and the velocity gradient is an effect.

What are the symmetry elements of the combined causes? They are the symmetry elements common to the isotropic mechanical properties and the orthorhombic stress. Thus the symmetry of the combined causes is the orthorhombic symmetry of the stress.

What can we infer about the velocity gradient from the cause symmetry? Because the symmetry of the velocity gradient (the effect), can be no lower than the symmetry of the combined causes, we conclude that the velocity gradient must have at least orthorhombic symmetry. Thus it is impossible for such a physical situation to result in a simple shear, because the symmetry of the velocity gradient in simple shear is monoclinic. This symmetry is lower than the orthorhombic symmetry of the combined causes, and the symmetry principle states that is not possible. Thus a homogeneous stress applied to a mechanically isotropic material can never cause a simple shearing deformation.

The only simple way to obtain a simple shearing in an isotropic material is to impose the simple shearing a priori (see Figure 12.13*B*), or in other words, to make a monoclinic velocity gradient the cause of the process and the stress an effect.

What would be the symmetry of the combined causes in this case? The monoclinic velocity gradient and the isotropic mechanical properties only have the monoclinic symmetry elements of the velocity gradient in common. This monoclinic symmetry must then appear in the symmetry of the effects.

What is the symmetry of the effects? It is just the orthorhombic symmetry of the stress, which therefore must include the monoclinic symmetry elements of the causes. Thus, the two-fold axis of rotation and the mirror plane that characterize the monoclinic symmetry of the velocity gradient, must be among those symmetry elements that characterize the symmetry of the stress. This, then, is an example of a process in which the symmetry of the effect (orthorhombic stress) is higher than that of the combined causes (monoclinic velocity gradient).

Of course, if the material properties were not isotropic, then their symmetry would be lower than in the cases we consider above, and this could lower the symmetry of the combined causes. In that case, the symmetry of the effect could also be lower, and a situation could be envisioned in which a stress imposed on an anisotropic material could cause a simple shear.

We further discuss cause and effect and the significance of boundary conditions in Section 18.1(iv).

REFERENCES AND ADDITIONAL READINGS

Ashby, M. F., and R. A. Verrall. 1973. Diffusion-accommodated flow and superplasticity. *Acta Metall.* 21: 149–163.

Ashby, M. F., and R. A. Verrall. 1977. Micromechanisms of flow and fracture, and their relevance to the rheology of the upper mantle. *Phil. Trans. Roy. Soc. Lond.* 288A: 59–95.

Ball, A., and S. White. 1977. An etching technique for revealing dislocation structure in deformed quartz grains. *Tectonophysics* 37(4): T9–T14.

Bell, T. H., and M. A. Etheridge. 1976. The deformation and recrystallization of quartz in a mylonite zone, central Australia. *Tectonphysics* 32: 235–269.

Boudier, F. 1976. Le Massife Lherzolitique de Lanzo, Étude Structurale et Pétrologique, Thèse Doctorat d'Etat Nantes, 167 pp.

Carter, N. L. 1976. Steady state flow of rocks. *Rev. Geophys. and Sp. Phys.* 14: 301–360.

Carter, N. L., and H. Avé Lallement. 1970. High temperature flow of dunite and peridotite. *Geol. Soc. Amer. Bull.* 81: 2181–2202.

Carter, N. L., and M. Friedman. 1965. Dynamic analysis of deformed quartz and calcite from the dry creek ridge anticline, Montana. *Amer. J. Sci.* 263: 747–785.

Carter, N. L., and S. H. Kirby. 1978. Transient creep and semibrittle behavior of crystalline rocks. *Pure and Appl. Geophys.* 116: 807–839.

Dell'Angelo, L. N., and J. Tullis. 1989. Fabric development in experimentally sheared quartzites. *Tectonophysics* 169: 1–21.

Etchecopar, A. 1977. A plane kinematic model of progressive deformation in a polycrystalline aggregate. *Tectonophysics* 39: 121–139.

Etchecopar, A., and G. Vasseur. 1987. A 3-D kinematic model of fabric development in a polycrystalline aggregate: Comparisons with experimental and natural examples. *J. Struct. Geol.* 9: 705–717.

Etheridge, M. A., and J. C. Wilkie. 1981. An assessment of dynamically recrystallized grain size as a paleopiezometer in quartz-bearing mylonite zones. *Tectonophysics* 78: 475–508.

Friedman, M. 1964. Petrofabric techniques for the determination of principal stress directions in rocks. In W. R. Judd, ed., *State of Stress in the Earth's Crust*. American Elsevier, New York: 450–552.

Green, H. W., II, and S. V. Radcliffe. 1972. Deformation processes in the upper mantle. In H. C. Heard, I. Y. Borg, N. L. Carter, and C. B. Raleigh, eds., *Flow and Fracture of Rocks*, Geophys. Monog. Series 16, Amer. Geophysi. Union, Washington, DC: 139–156.

Herwegh, M., J. H. P. de Bresser, and J. H. ter Heege. 2005. Combining natural microstructures with composite flow laws: An improved approach for the extrapolation of lab data to nature. *J. Struct. Geol.* 27(3): 503–521.

Hirth, J. P., and J. Lothe. 1968. *Theory of Dislocations*, McGraw-Hill, New York, 780 pp.

Hobbs, B. E., W. D. Means, and P. F. Williams. 1976. *An Outline of Structural Geology*, John Wiley & Sons, New York, 571 pp.

Jung, H., and S. Karato. 2001. Effects of water on dynamically recrystallized grain-size of olivine: *J. Struct. Geol.* 23: 1337–1344.

Karato, S. F. 1984. Grain size distribution and rheology of the upper mantle. *Tectonophysics* 104: 155–176.

Karato, S.-I., M. S. Paterson, and J. D. FitzGerald. 1986. Rheology of synthetic olivine aggregates: Influences of grain size and water. *J. Geophys. Res.* 91: 8151–8176.

Karato, S.-I., M. Toriumi, and T. Fujii. 1980. Dynamic recrystallization of olivine single crystals during high-temperature creep. *Geophys. Res. Letts.* 7(9): 649–652.

Katayama, I., H. Jung, and S. Karato. 2004. New type of olivine fabric from deformation experiments at modest water content and low stress. *Geol.* 32(12): 1045–1048,

Kirby, S. H. 1983. Rheology of the lithosphere. *Rev. of Geophys. and Sp. Phys.* 21: 1458–1487.

Kirby, S. H., and A. K. Kronenberg. 1987. Rheology of the lithosphere: Selected topics. *Rev. of Geophys. and Sp. Phys.* 25: 1217–1244.

Knipe, R. J. 1989. Deformation mechanisms—Recognition from natural tectonites. *J. Struct. Geol.* 11(1/2): 127–146.

Kohlstedt, D. L., and M. S. Weathers. 1980. Deformation-induced microstructures, paleopiezometers, and differential stresses in deeply eroded fault zones. *J. Geophys. Res.* 85: 6269–6285.

Lister, G. S., and B. E. Hobbs. 1980. The simulation of fabric development during plastic deformation and its application to quartzite: The influence of deformation history. *J. Struct. Geol.* 2: 355–370.

Lister, G. S., and M. S. Paterson. 1979. The simulation of fabric development during plastic deformation and its application to quartzite: fabric transitions. *J. Struct. Geol.* 1: 283–297.

Lister, G. S., M. S. Paterson, and B. E. Hobbs. 1978. The simulation of fabric development in plastic deformation and its aplication to quartzite: The model. *Tectonophysics* 45: 107–158.

Lister, G. S., and P. F. Williams. 1983. The partitioning of deformation in flowing rock masses. *Tectonophysics* 92: 1–33.

McLaren, A. C., and B. E. Hobbs. 1972. Transmission electron microscope investigation of some naturally deformed quartzites. In *Flow and Fracture of Rocks*, the Griggs volume; Amer. Geophys. Union Monograph 16, Amer. Geophys. Union, Washington DC: 55–66.

Mercier, J.-C. C. 1985. Olivine and pyroxenes. In H.-R. Wenk, ed., *Preferred Orientation in Deformed Metals and Rocks: An Introduction to Modern Texture Analysis*, Academic Press, New York: 407–430.

Nicolas, A., F. Boudier, and A. M. Boullier. 1973. Mechanisms of flow in naturally and experimentally deformed peridotites. *Amer. J. Sci.* 273: 853–876.

Nicolas, A., and J. P. Poirier. 1976. *Crystalline Plasticity and Solid State Flow in Metamorphic Rocks*, John Wiley and Sons, New York, 444 pp.

Paterson, M. S., and L. E. Weiss. 1961. Symmetry concepts in the structural analysis of deformed rocks. *Geol. Soc. Amer. Bull.* 72: 843–882.

Price, G. P. 1985. Preferred orientations in quartzites. In H.-R. Wenk, ed., *Preferred Orientation in Deformed Metals and Rocks: An Introduction to Modern Texture Analysis*, Academic Press, New York: 385–406.

Rutter, E. H. 1976. The kinetics of rock deformation by pressure solution. *Phil. Trans. Roy. Soc. Lond.* A283: 203–217.

Rybacki, E. and G. Dresen. 2004. Deformation mechanism maps for feldspar rocks. *Tectonophysics* 382: 173–187.

Schmid, S. M. 1982. Microfabric studies as indicators of deformation mechanisms and flow laws operative in mountain building. In K .J. Hsü, ed., *Mountain Building Processes*, Academic Press, New York: 95–110.

Schmid, S. M., and M. Casey. 1986. Complete fabric analysis of some commonly observed quartz c-axis patterns. In B. E. Hobbs and H. C. Heard, eds., *Mineral and Rock Deformation: Laboratory Studies*, The Paterson Volume. Geophysical Monograph 36, Amer. Geophys. Union, Washington DC: 263–286.

Scott, W. H., E. C. Hansen, and R. J. Twiss. 1965. Stress analysis of quartz deformation lamellae in a minor fold. *Amer. J. Sci.* 263: 729–746.

Spry, A. 1969. *Metamorphic Textures*. Pergamon, New York, 350 pp.

Tullis, J. A. 1979. High temperature deformation of rocks and minerals. *Rev. Geophys. and Space. Phys.* 17(6): 1137–1154.

Tullis, J., J. M. Christie, and D. T. Griggs. 1973. Microstructures and preferred orientations of experimentally deformed quartzites. *Geol. Soc. Amer. Bull.* 84: 297–314.

Twiss, R. J. 1976. Some planar deformation features, slip systems and submicroscopic features in synthetic quartz. *J. Geol.* 84: 701–724.

Twiss, R. J. 1986. Variable sensitivity piezometric equations for dislocation density and subgrain diameter and their relevance to olivine and quartz. In H. C. Heard and B. E. Hobbs, eds., *Mineral and Rock Deformation: Laboratory Studies—The Paterson Volume*, Amer. Geophys. Union Monograph 36, Amer. Geophys. Union, Washington, DC: 247–261.

Weertman, J., and J. R. Weertman. 1964. *Elementary Dislocation Theory*. MacMillan, New York, 213 pp.

Wenk, H. R., ed. 1985.; *Preferred Orientation in Deformed Metals and Rocks: An Introduction to Modern Texture Analysis*, Academic Press, New York, 610 pp.

Wenk, H. R. 1985. Carbonates. In H. R. Wenk, ed., *Preferred Orientation in Deformed Metals and Rocks: An Introduction to Modern Texture Analysis*, Academic Press, New York: 361–384.

Wilson, C. J. L. 1973. The prograde microfabric in a deformed quartzite sequence, Mount Isa, Australia. *Tectonophysics* 19: 39–81.

Chapter 18

SCALE MODELS AND QUANTITATIVE MODELS OF ROCK DEFORMATION

A principal reason for studying the rheology of rocks under various conditions is to understand how the flow of rocks contributes to the tectonic evolution of the Earth. In order to gain insight into these processes, we make mechanical models of deformation that we can compare with actual observations. If the model reproduces the observed characteristics, we infer that the model can provide insight into the natural conditions of deformation.

In Chapters 13, 14, and 15, we consider predominantly kinematic models of ductile deformation for which we prescribe the motion. Such models do not attempt to account for the motions of the material as being a consequence of their rheology, but rather just prescribe the motions a priori. Models that do incorporate the rheology, however, can help us evaluate whether the kinematic models are mechanically possible and reasonable and give us a deeper level of understanding of the origin of the structures.

In this chapter, we discuss the use of both scale models and mathematical models to study the mechanics of rock deformation, and as examples of these methods, we use investigations of fold-and-thrust belts, normal faulting in the brittle crust, intrusion of diapirs, plastic slip lines and faulting, and the relationships among rheology, fold geometry, and foliation orientations.

Scale models are actual physical models of parts of the Earth. The models are constructed with materials whose properties scale in such a way that their behavior over short times and small distances reproduces the behavior of rocks over long periods of time and large distances. The correct choice of the model materials requires a knowledge of the mechanical behavior of the rocks, expressed in terms of constitutive equations, and a knowledge of **scaling theory** to guide us in the selection of materials that could be used to model the Earth. It is difficult if not impossible to make the model correctly scaled in every respect, however, so usually the choice of an appropriate material and model geometry requires compromises that limit to some degree the correspondence of the model to the Earth.

Mathematical models describe behavior in terms of sets of equations that define the mechanical behavior of particular materials (constitutive equations), the appropriate physical conservation laws that all systems must obey (in particular, the conservation of mass, momentum, angular momentum, and energy), and the specific physical constraints to which the body of material is subjected (the boundary and initial conditions). We solve these sets of equations to find the distribution of unknown quantities such as the velocity, stress, strain rate, displacement, and strain throughout a deforming body. In principle, we can solve the equations either analytically or numerically. **Analytic solutions** are general solutions to the equations that express the values of the unknown quantities as a mathematical function of the specific material properties and of the position within the body. Thus the solution can be evaluated for a wide variety of specific conditions, and we can determine from the general solution how changes in conditions, such as the viscosity, affect the solution. Analytic solutions, however, can be found only if the geometry of the body and the geometry of the boundary conditions are both relatively simple, and therefore they generally apply only for small strains. **Numerical solutions** are found with a computer and provide the approximate values of the unknown quantities at a specific set of points throughout the body, for a specific set of material properties and boundary conditions. The solutions, therefore, are not general; we get one solution for one specific set of conditions. To investigate a different set of conditions, for example, a different material viscosity, we have to solve the equations all over again to get another specific solution. The advantage of numerical solutions for investigating geological deformations, however, is that we can analyze geometrically complex bodies and large and complex deformations. The formulation of a mathematical model for a geological system becomes more complex and difficult as the system becomes more realistic.

The literature on mechanical models in geology is very large. In this chapter, therefore, we can only introduce some fundamental ideas behind scale and mathematical models of rock deformation and give a few examples of the results from each type of model. Our intent is to provide some insight into how the models are constructed and to demonstrate the value of mechanical models in understanding geologic structures.

18.1 CONSTRAINTS ON PHYSICAL MODELS

Regardless of whether we decide to model the deformation of the Earth using scale models or mathematical models, the model behavior is governed by three sets of equations: Conservation laws are the fundamental laws of physics that all systems must obey; constitutive equations define the mechanical properties of the particular material to be deformed; and boundary conditions define what we want to happen along the boundaries of the model. We discuss each of these sets of equations briefly.

i. Conservation or Balance Laws

In any body undergoing deformation, every point of the body has to obey the same fundamental conservation laws of physics. For the strictly mechanical problems that we are concerned with here, these conservation laws include the conservation of mass, the conservation of momentum, and the conservation of angular momentum. For systems involving temperature gradients and heat flux, such as thermal convection, we also need to consider the conservation of energy, but we will not deal with that complexity.

For any scale model, of course, the material automatically obeys these laws, and we do not need to consider them explicitly. For a mathematical model, however, the expression of these laws as mathematical equations is essential. The equations expressing the balance laws, together with the constitutive equations, constitute a set of equations in which there are as many equations as there are unknowns, which is an essential condition for solving the equations (see Box 18-1).

The **conservation of mass** requires that mass be neither created nor destroyed, and this condition is automatically satisfied if we assume constant-volume deformation and a homogeneous mass density.

The **conservation of momentum** is Newton's second law. It states that the net force on a body is equal to the mass m multiplied by the acceleration a:

$$\sum_i F_{(i)} = ma \tag{18.1}$$

where $F_{(i)}$ includes all the possible forces acting on a body.

Geological deformations are generally so slow that the acceleration is negligible.[1] Under these circumstances, the right side of Equation (18.1) becomes zero, and Newton's second law reduces to the **equilibrium equation**, which requires all the forces on a body be balanced, that is, they must sum to zero:

$$\sum_i F_i = 0 \tag{18.2}$$

These equations can be put in a form that expresses the conservation of momentum at a point in a continuum, and this form involves first-order partial derivatives of the stress with respect to the spatial coordinates (see Box 18-1).

The **conservation of angular momentum** requires that the net torque on the body be equal to its moment of inertia multiplied by its angular acceleration:

$$\sum_i M_i = \sum_i F_i(d_i) = I\omega \tag{18.3}$$

where M_i are the moments about a point, F_i are the forces that act on the moment arms of length d_i about a point, I is the moment of inertia, and ω is the angular acceleration. Again, applying this to geological situations, we can assume the angular acceleration is zero, giving

$$\sum_i M_i = \sum_i F_i(d_i) = 0 \tag{18.4}$$

We show in Section 7.1(vii) that applying the balance of moment of momentum at a point in a continuum (Equation (7.27)) leads to the requirement for symmetry of the shear stress components (Equations (7.29), (7.31), (7.56)).

ii. Constitutive Equations

In order to model the deformation of the Earth or any part of it, we must have some idea of how the strain rate and the stress in the rock are related. We describe what we know about this relationship with a **constitutive equation**. In Section 16.1 we introduce the one-dimensional constitutive equations that describe the behavior of elastic materials (Equation (16.1); Figure 16.1), viscous ma-

[1]The obvious exceptions are earthquakes and rapid landslides.

terials (Equation (16.2); Figure 16.2), plastic materials (Equation (16.3); Figure 16.3), and power-law materials (Equation (16.4); Figure 16.4), and we give graphical illustrations for a number of more complex behaviors (Figures 16.5–16.8). We discuss the three-dimensional form of the first four types of behavior in Box 16-4. We also discuss the experimentally determined power-law creep equations for rocks (e.g., Equations (16.7), (16.11), (16.12), (16.15), (16.18), (16.19), (16.20)). These are all examples of constitutive equations.

Whether we want to model deformation in the Earth with scale models or with mathematical models, we need to know what type of behavior to model. For scale models, for example, it might not be reliable to model the power-law creep of rocks with a material that behaved as a linearly viscous material, and even if we had a putty that follows a power-law flow equation, we would need to know the form of the power-law equation for the rock so that we could figure out the appropriate scaling constants. We discuss scaling theory in detail in the next section.

For mathematical models, the constitutive equations are an essential part of the system of equations that we must solve to model a particular system in the Earth, for they are the means of incorporating the material behavior into the mathematical model.

iii. Boundary and Initial Conditions

No physical system, whether in the Earth, a scale model in the laboratory, or a mathematical model we solve with a computer, can be of infinite extent. All systems must have boundaries, and the conditions that exist at those boundaries affect how the material within those boundaries behaves. If we build a scale model of a deforming orogen in a box, for example, we must consider what should happen along the boundaries of the box. Should we grease the sides of the box to try to make the shear stress as close to zero as possible? Or should we line the sides with sandpaper to try to make the material in the box stick to the sides? Whatever we do defines the boundary conditions for the model and affects how the model behaves.

In mathematical modeling, we face the same problem. The model must have boundaries beyond which we do not want to solve for the deformation; but what should we specify to happen along those boundaries? It turns out that specifying how the material behaves along the boundaries, that is, the boundary conditions, is essential to finding solutions to the mathematical equations (see Box 18-1).

It may not always be clear what boundary conditions are appropriate for a model, because in many cases, the boundaries of a model do not correspond to a boundary in the Earth on which either the stress or the velocity is necessarily constant or even easily specified a priori. In the real Earth, conditions on an arbitrarily selected boundary most likely evolve with time and with the deformation of the material. Thus it is not always clear what boundary conditions we should specify for a model. Under those conditions, any boundary conditions we may choose are a distortion of the real situation in the Earth, and we must be concerned with what the effects of simple boundary conditions are on the behavior of the model and thus on how well the model represents the Earth. It may be possible to make the model sufficiently large that the effects of the boundary are minimal in the parts of the model far from the boundary that are of interest, but that is not always possible.

If we are modeling the time-dependent evolution of a part of the Earth, we need to start the deformation in the model at some initial time. Because we do not have a solution for the problem at the beginning, however, we must impose a condition that defines the behavior of the model at the start of the process from which its subsequent behavior evolves; this is called an **initial condition**. It is essentially like a boundary condition, except that it is a boundary in the time coordinate rather than in the spatial coordinates.

iv. Cause and Effect in Physical Processes

There seems to be a common prejudice toward assuming that stress is always a principal cause of the process of flow in rock and that the deformation rate is an effect. In mathematical terms, the cause(s) should be the independent variable(s), which can be varied by external factors, such as the choice of the researcher doing the modeling, and the effect is the dependent variable, whose value is determined from the equations by the value of the independent variable(s). Beyond that, however, the balance and constitutive equations are completely neutral as to which variables are independent (the causes) and which are dependent (the effects). In the flow of a continuum, for example, the stress can be considered a cause, and the deformation rate is then an effect, that is, we calculate the deformation rate from a knowledge of the stress. Conversely, the deformation rate just as easily can be considered a cause, and the stress is then an effect, in which case we calculate the stress from a knowledge of the deformation rate. In all cases, the material properties must be considered one of the causes of the flow of a continuum, because the physical characteristics of the material that is deforming necessarily affect the characteristics of the flow. The force of gravity must also be considered a cause where it is significant, because everything on Earth is subjected to this force.

We must look to the boundary conditions to distinguish between cause and effect. The boundary conditions, after all, define what is imposed on the body. We can require the stresses on the surface of a body to be constant (stress boundary conditions), in which case the stress is the cause of the strain rate within the body. On the other hand, we can specify the velocity on the boundaries of the body, in which case the strain rate (velocity gradients) is the cause of the stress field within the body. Of course, we can also specify stress on some parts of the boundary and velocity on the other parts (mixed boundary conditions), in which case the cause is neither all one nor all the other.

In some cases, the assumption of velocity or displacement boundary conditions seems to provide a simpler explanation of observed structures than the assumption of stress boundary conditions, implying that deformation rate is the cause of the process. An example is the origin of gash fracture orientations, which we explained in terms of the stress in Section 9.8 (Figure 9.16) and in terms of the displacements in Section 15.4(ii) (Figure 15.17). The latter provides a simpler model for the observed variety of gash fracture orientations and argues that they might better be interpreted as indicators of the strain in the rocks rather than the stress.

In interpreting structures, therefore, it is important not to be prejudiced toward one mode of thinking, even if intuitively it seems somehow more "natural." Intuition is often very useful, but it also can lead one astray. As a rule of thumb, we recommend that unless there is a definite reason for assuming otherwise, it is safest to assume that velocity boundary conditions are more appropriate and that an imposed deformation is the cause of the stress. Thus the interpretation of deformational structures in rocks is usually best done in terms of the deformation field in which they developed, rather than in terms of the stress field.

18.2 THE THEORY OF SCALE MODELS

For many geologic situations, it is useful to make a scale model whose behavior on a scale of meters and hours is equivalent to what occurs in the Earth, the prototype system, on a scale of kilometers and millions of years. The main problem is to know how to construct a model so that its behavior reliably represents that of the prototype.

In order to describe completely the mechanical relationship between a scale model and the prototype we need the three independent scale factors for length (λ), time (τ), and mass (μ). We define them by:

$$\lambda = \frac{L_m}{L_p}, \quad \tau = \frac{t_m}{t_p}, \quad \mu = \frac{m_m}{m_p} \tag{18.5}$$

where subscripts m and p refer to the model and prototype, respectively, and L is length, t is time, and m is the mass of the material in question. Models are **geometrically similar** if all linear dimensions of the model are λ times the equivalent dimension in the prototype. They are **kinematically similar** if the time required for the model to undergo a change in size, shape, or position is τ times the time for the prototype to undergo a geometrically similar change.

The scaling of all the mechanical quantities of interest can be derived from these basic three scale factors, and thus the properties of the scale model are tightly constrained. Table 18.1 lists some of the different mechanical quantities, their units, and the scale factors derived for them.

Scaling laws require that the geometry and time span of a deformation must scale appropriately from the prototype to the model. Thus the response to each type of scaled force in the model must be geometrically and kinematically similar to the prototype's response to natural forces. The scale factor that must apply to each type of force is defined in terms of the three independent scale factors as $(\mu\lambda/\tau^2)$ (Table 18.1). If each force in the model is related to the corresponding one in the prototype by this same scale factor, the model and the prototype are **dynamically similar**.

In Chapter 7, we distinguish two kinds of forces: body forces and surface forces. Of the body forces on mechanical systems, we need consider only the inertial force (F_i), which is the force associated with acceleration, and the gravitational force (F_g). We ignore others, such as electrostatic and magnetic forces. Surface forces arise from the resistance of the material to deformation, and the relationship between force and deformation is expressed by a constitutive equation such as those we consider in Chapter 16 (e.g., Equations (16.1), (16.2), (16.4)). Thus to achieve dynamic similarity, the behavior of the model and of the prototype must be characterized by the same constitutive equation, and the material constants in those equations must be appropriately scaled.

The scale factors for the relevant material constants in terms of the three independent scale factors in Equation (18.5) can be derived from the appropriate constitutive equation. For linearly viscous materials, for example, the shear stress is directly proportional to the shear strain rate (second Equation (16.2); compare Equations (18-1.14) and (18-1.18)):

$$\sigma_s = 2\eta\dot{\varepsilon}_s \tag{18.6}$$

where η is the viscosity coefficient. Solving for η, we find

$$\eta = \frac{\sigma_s}{2\dot{\varepsilon}_s} \tag{18.7}$$

As strain is a dimensionless quantity, the unit of strain rate is inverse time. Thus the units of viscosity are

$$[\text{stress/strain rate}] = [(F/A) \,/\, (1/t)] = [Ft/L^2]$$

The ratio of the viscosity in the model to that in the prototype defines the scaling factor for viscosity, which we

TABLE 18.1. Scale Factors for Selected Variables in Mechanics

Quantity	Symbol	Units	Scaling Ratios	Scale Factor
1. Length	L	L	$\frac{L_m}{L_p}$	λ
2. Time	t	t	$\frac{t_m}{t_p}$	τ
3. Mass	m	m	$\frac{m_m}{m_p}$	μ
4. Area	A	L^2	$\frac{A_m}{A_p} = \frac{(L_m)^2}{(L_p)^2}$	λ^2
5. Volume	V	L^3	$\frac{V_m}{V_p} = \frac{(L_m)^3}{(L_p)^3}$	λ^3
6. Density	ρ	$\frac{m}{V}$	$P \equiv \frac{\rho_m}{\rho_p} = \frac{m_m(L_p)^3}{m_p(L_m)^3}$	$P = \frac{\mu}{\lambda^3}$
7. Velocity	v	$\frac{L}{t}$	$\frac{v_m}{v_p} = \frac{L_m/t_m}{L_p/t_p} = \frac{L_m t_p}{L_p t_m}$	$\frac{\lambda}{\tau}$
8. Acceleration	a	$\frac{L}{t^2}$	$\frac{a_m}{a_p} = \frac{L_m/t_m^2}{L_p/t_p^2} = \frac{L_m t_p^2}{L_p t_m^2}$	$\frac{\lambda}{\tau^2}$
9. Force	F	$\frac{mL}{t^2}$	$\frac{F_m}{F_p} = \frac{m_m(L_m/t_m^2)}{m_p(L_p/t_p^2)} = \frac{m_m L_m t_p^2}{m_p L_p t_m^2}$	$\frac{\mu\lambda}{\tau^2}$
10. Stress	σ	$\frac{F}{L^2}$	$\Sigma \equiv \frac{\sigma_m}{\sigma_p} = \frac{F_m/A_m}{F_p/A_p} = \frac{F_m L_p^2}{F_p L_m^2}$	$\Sigma = \frac{\mu}{\lambda\tau^2}$
11. Viscosity	η	$\frac{Ft}{L^2}$	$H \equiv \frac{\eta_m}{\eta_p} = \frac{\sigma_s^{(m)}/\dot{\varepsilon}_s^{(m)}}{\sigma_s^{(p)}/\dot{\varepsilon}_s^{(p)}} = \frac{\sigma_s^{(m)}\dot{\varepsilon}_s^{(p)}}{\sigma_s^{(p)}\dot{\varepsilon}_s^{(m)}}$	$H = \Sigma\tau = \frac{\mu}{\lambda\tau}$

find in terms of the three independent scale factors in Equation (18.5) (see Table 18.1):

$$\frac{\eta_m}{\eta_p} = \frac{\sigma_s^{(m)}/\dot{\varepsilon}_s^{(m)}}{\sigma_s^{(p)}/\dot{\varepsilon}_s^{(p)}} = \frac{F_m t_m/L_m^2}{F_p t_p/L_p^2} = \frac{F_m t_m L_p^2}{F_p t_p L_m^2}$$

$$= \frac{(m_m L_m/t_m^2) t_m L_p^2}{(m_p L_p/t_p^2)\ t_p L_m^2} = \frac{m_m L_p t_p}{m_p L_m t_m}$$

Thus

$$\frac{\eta_m}{\eta_p} = \frac{\mu}{\lambda t} \tag{18.8}$$

Ratios for other material properties can be defined in a similar manner. Thus in principle, once we have chosen aspects of the model that define the three independent scale factors μ, λ, and τ, the viscosity of the model material is defined, and if we want to make a dynamically similar model, we must use a material with that viscosity; we have no options.

In some circumstances, it may be convenient to replace one of the fundamental scale factors λ, τ, and μ with a different one defined in terms of the one it replaces. For example, we will see in the next section that it is convenient to consider the scale factor for stress Σ as one of the independent scale factors in place of the scale factor for mass μ, which is permissible because the scale factor for stress ($\Sigma = \mu/\lambda\tau^2$, Table 18.1) includes μ. We can only ever specify a priori the values of three of the scale factors, and those three must be independent. The remaining scale factors are then determined, thus constraining the characteristics of the scale model.

Although construction of scale models may seem straightforward in principle, numerous problems complicate the process. It is impossible to devise a model in which all factors can be scaled correctly, and compromises and approximations are necessary. The greater the number of physical phenomena to be modeled, the more difficult it is to scale all relevant factors correctly.

To have an accurate mechanical scale model, it is necessary to know the rheology of the prototype rocks. All too often, we can only guess this fundamental characteristic. Even if we can devise an appropriate scale model, we still must find a usable material that satisfies the scaling requirements.

Any model is of finite size, and therefore it must include boundary conditions that do not necessarily reflect the prototype geologic environment. Boundary conditions for any restricted part of the Earth are often difficult if not impossible to define precisely in any case. Thus although the technique of scale modeling can provide some fascinating insights into the behavior of geologic systems, it is not without drawbacks, and its limitations must be kept in mind in interpreting model behavior.

In the following three sections, we analyze the results of scale model experiments of different geologic situations. Examples of other models illustrated previously include Figures 8.6, 8.7, and 9.17.

18.3 SCALE MODELS OF FOLDING

Many investigators have made scale models of folding using plasticine, silicone putty, and similar materials. As with any model, the important question is whether we are justified in interpreting the model structures to be equivalent to those observed in rocks. We can use the theory of scale models to answer this question, and we illustrate the analysis by considering a model of fold development in the Jura Mountains.[2]

We discuss the Jura in the chapters on strike-slip faults and folds (see the maps in Figures 5.22 and 6.10 and the detailed cross section in Figure 10.2*B*). The structure of the mountains consists of a series of large folds developed in sedimentary rocks lying north of the main Alpine orogenic zone (Figures 5.22 and 18.1). The folds are in general of class I geometry, and they are underlain by a decollement.

The width of the Jura is about 30 km, or 3×10^6 cm. A reasonable model would have a corresponding cross-sectional length of 30 cm. Thus the geometric scale factor is

$$\lambda = \frac{L_m}{L_p} = \frac{30 \text{ cm}}{3 \times 10^6 \text{ cm}} = 10^{-5} \qquad (18.9)$$

[2]In this discussion we follow the works of Bucher (1956) and Dennis and Häll (1978).

The time required for the Jura folds to form was approximately one million years. Suppose we want to be able to run a model experiment within 9 hours, or approximately 10^{-3} year (9 hrs ÷ 24 hrs/day ÷ 365 days/yr = 1.03×10^{-3} yr). The scale factor for time would be

$$\tau = \frac{t_m}{t_p} = \frac{10^{-3} \text{ yr}}{10^6 \text{ yr}} = 10^{-9} \qquad (18.10)$$

Finally, the model materials that we are likely to be able to use do not have a great range in densities compared to rocks. As a first approximation, we can assume that the ratio of model to prototype densities is approximately 1, whereby

$$P = \frac{\rho_m}{\rho_p} \cong 1 = \frac{\mu}{\lambda^3} \qquad \mu \cong \lambda^3 \qquad (18.11)$$

We thus have values for the three independent scale factors, and we can derive values of all the other mechanical scale factors. We first inquire what the scale factor for forces should be. Using Equations (18.9) through (18.11) in the scale factor for force (Table 18.1) we find:

$$\frac{F_m}{F_p} = \frac{\mu\lambda}{\tau^2} = \frac{\lambda^4}{\tau^2} = \frac{10^{-20}}{10^{-18}} = 10^{-2} \qquad (18.12)$$

Thus each type of force on the model must be 0.01 times the corresponding force on the prototype.

This requirement immediately presents a problem because both the prototype and the model are subjected to the same gravitational force, which implies the scale factor for force should be 1. There are two possible ways to deal with this contradiction. First, if the gravitational force is not a significant factor in fold development compared to the surface forces, then we can ignore the gravitational forces and the constraint they impose on the force scale factor. Second, the force of gravity must be

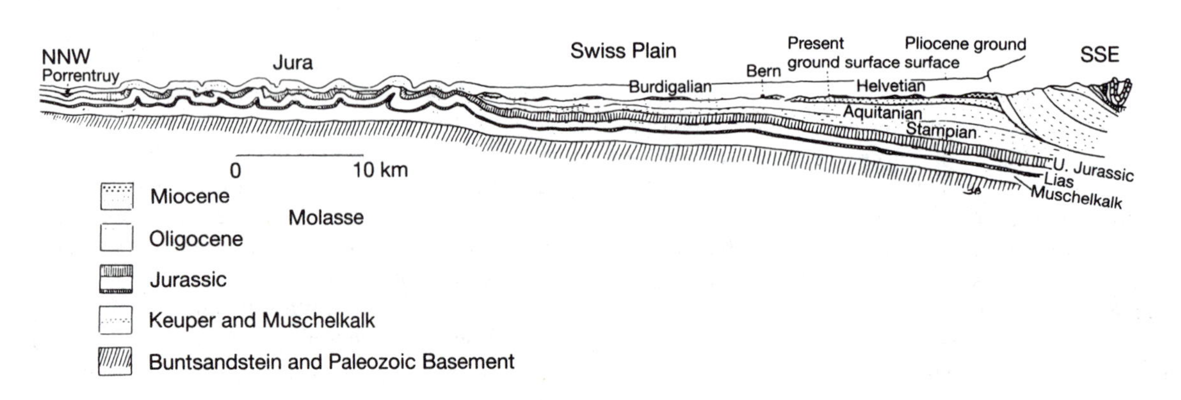

FIGURE 18.1 Cross section through some of the Jura folds showing the fold style, the underlying decollement, and the relationship to the front of the Alps. (After Dennis and Häll 1978)

of major significance in gravitational collapse of a topographic high, in any convective process such as the rise of diapiric structures and gneiss domes, or in isostatic adjustments. If gravitational force is the dominant factor compared to surface forces, then it is possible to scale the inertial force up in the model by using a centrifuge so that by comparison, the gravitational force becomes negligible. The controllable and variable inertial force then can be substituted for the uncontrollable gravitational force. We discuss this possibility further in the next section.

For the Jura model we assume that gravity plays a negligible role in the development of the folds. Furthermore, because the deformation proceeds very slowly, accelerations are very small, and therefore inertial forces also are negligible. Assuming constant-volume deformation, the density is not a factor in the constitutive relations either. For this reason, the scale factors for mass or density are not directly pertinent to our problem, and we may choose a model material of any convenient density without affecting the correlation between model and prototype.

Thus we may choose the stress ratio Σ as a third independent scale factor. To model the natural behavior of rocks, however, we must know what that behavior is. Experiments on ductile deformation of limestone show that it behaves as a power-law material at high stress and that at low stress and fine grain size its rheology approaches that of a viscous fluid (Figure 16.17). None of the experiments, however, evaluated the rheology of solution creep, which the structures in the rock suggest was significant in the Jura folding, and which in principle could increase the range of stresses and grain sizes for which limestone would deform viscously (see Figure 17.1*B*).

Let us assume, therefore, that the rocks deformed viscously and that the viscosity was between 10^{12} and 10^{16} Pa s.[3] Using the scale factor for viscosity (Table 18.1), the scale factor for time (Equation (18.10)), and this range of values for the prototype viscosity, we find that

$$\frac{\eta_m}{\eta_p} = \Sigma \mathcal{T} \tag{18.13}$$

$$\eta_m = 10^{-9}\,\eta_p\,\Sigma = 10^3\,\Sigma \text{ to } 10^7\,\Sigma \tag{18.14}$$

Equation (18.14) shows that it is possible to maintain dynamic similarity for the model over a large range of model viscosities simply because it is easy to vary the stresses applied to the model, and thus the stress scale factor, over a wide range. This result indicates that for this model in particular, and for fold models in general for which gravitational forces are insignificant, the viscosity of the particular material used to model the rock deformation is not important. As long as sufficient stress can be applied to deform the material at the appropriate rate, dynamic similarity can be maintained.

It is convenient to choose a model material that flows on time scales of hours but not on time scales of minutes. Such a property permits relatively large deformations within reasonable lengths of time at easily attained stresses. At the same time, such materials are easy to handle while constructing the model, and they permit the model to be cut apart after deformation for examination without destroying the geometry of the experiment. An appropriate viscosity range is between 10^4 and 10^7 Pa s; materials that are approximately visco-elastic and have viscosities within this range include silicone putty and stitching wax.[4]

The rocks in the Jura include shale interbeds between the dominant limestone layers. The rheological properties of shale have not been well studied, but field evidence indicates they are considerably less competent than the limestones. The models shown in Figures 18.2 and 18.3 were constructed from layers of stitching wax separated by layers of grease to approximate the properties of the interlayered limestone and shale.

Figure 18.2 shows the results of one model experiment on the Jura Mountains. The details of the experimental setup are shown in Figure 18.2*A*. The interlayered sequence of stitching wax and grease was compressed by the gravitationally induced collapse and spreading of a large block of stitching wax. The resulting deformation (Figure 18.2*B*) shows a striking resemblance to the deformation depicted in the cross section of the Jura shown in Figure 18.1.

The style of folds generally observed at the elevated temperatures characteristic of high grades of metamorphism, indicates that large contrasts in the competence of different rock types are at a minimum or disappear altogether. The model in Figure 18.3 was designed to reproduce this style of deformation. It consists of a rigid block adjacent to a multilayer of warm and softened stitching wax. The block and multilayer are overlain by a thin covering of interlayered stitching wax and grease. Deformation of the warm wax was driven by the advancing rigid block, which caused the wax to shorten and thicken and then to spread laterally as a result of gravitational collapse. The resulting structure is a recumbent anticline that has spread out underneath the covering layers. The lower limb of the fold is highly attenuated, and the style of folding has

[3] A Pa s (Pascal second = (N s)/m^2 or Newton second per square meter) is the unit of viscosity in the SI (Systeme Internationale), which is the mks (meter-kilogram-second)) system of measurement.

[4] A pure form of wax, usually beeswax, used as a dry thread lubricant in complex sewing projects, such as quilting or leatherwork.

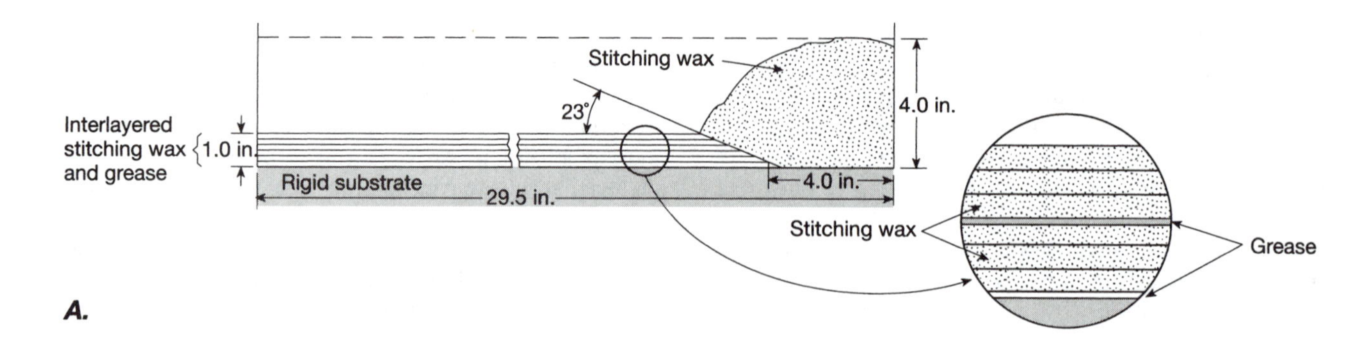

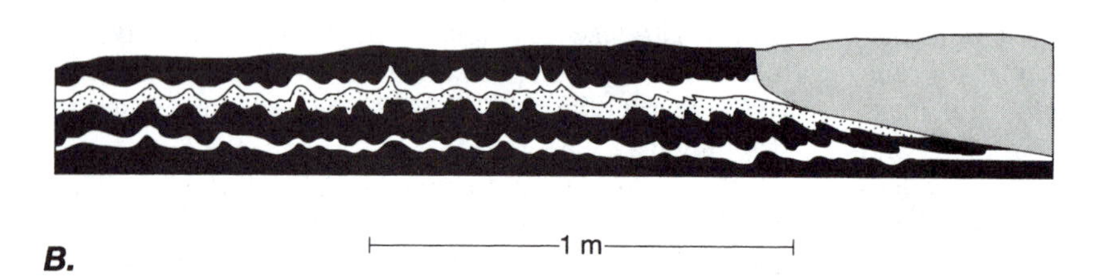

FIGURE 18.2 Scale model of the formation of the Jura Mountains using interlayered stitching wax and grease. (From Bucher 1956) *A*. Details of the experimental setup showing the sequence of layers of stitching wax and grease and the block of stitching wax that compressed the layered sequence as it flowed out and flattened during gravitational collapse. *B*. The result of one of the deformation experiments showing folding comparable to that observed in the cross section of the Jura (see Figure 18.1)

much in common with major recumbent folds that are observed in the metamorphic cores of many orogenic belts (see, for example, Figure 10.1*B*, *C*, *D*). The advancing fold has also shortened and folded the layers that lie ahead of it.

The experiment does not properly model orogenic deformation, however. The model does not fulfill all the requirements for dynamic similarity with a model orogenic zone because the gravitational forces, on which the deformation depends, are not correctly scaled. Moreover, the boundary conditions represented by the motion of the rigid block are unlikely to duplicate the conditions actually associated with orogeny (compare, however, the numerical model in Figure 20.34). The results of the experiment are intriguing because of their similarity to real geologic structures, but they do not provide a reliable model of deformation in the Earth, because the model is not correctly scaled, particularly because of the lack of accommodation of gravity, which is an important force in orogenic development. Thus we need to turn to the question of scaling of gravity in models.

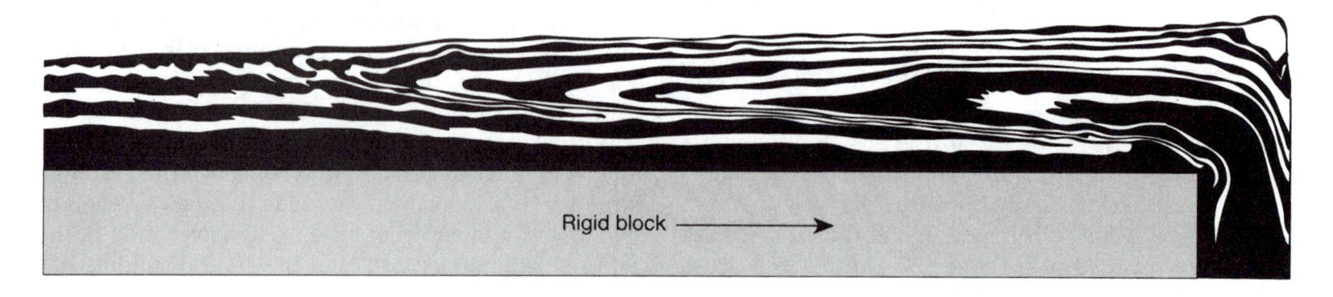

FIGURE 18.3 Stitching-wax model of the formation of a recumbent anticline. A short thick multilayer of warmed stitching wax with no interlayers of grease was compressed by a rigid block representing colder more competent material. Overlying the warm wax and rigid block are several thin layers of wax with grease interbeds. Deformation is driven by motion of the block into the warmed wax and by gravitational collapse of the resulting uplift. (After Bucher 1956)

18.4 SCALE MODELS OF GRAVITY-DRIVEN DEFORMATION

In many model experiments, both the model and the prototype are deformed under the same gravitational acceleration. Thus using the scale factors for acceleration from Table 18.1, we must have:

$$\frac{g_m}{g_p} = \frac{\lambda}{\tau^2} = 1 \qquad \tau = \lambda^{1/2} \tag{18.15}$$

where g_m and g_p are the gravitational accelerations for model and prototype, respectively.

We consider first the requirements for modeling gravity-driven brittle deformation in the Earth's crust, which is governed by the Coulomb fracture criterion (Equation (8.2)). Because time is not a variable in the Coulomb fracture criterion, the scale factor τ is not important for the scale model. Of the two material constants in the fracture criterion, the coefficient of internal friction μ_c is dimensionless and thus should be the same in both prototype and model (we introduce a subscript c for the coefficient of internal friction to distinguish it from the scale factor for mass). The cohesion c_0 has dimensions of stress and so must be scaled accordingly. For a deformation driven by gravitational forces, such as extensional normal faulting, we use the stress scale factor, from Table 18.1 with the second Equation (18.15) to find

$$\Sigma \equiv \frac{\sigma_m}{\sigma_p} = \frac{\mu}{\lambda\tau^2} = \frac{\mu}{\lambda^2} = \frac{\mu}{\lambda^3}\,\lambda = P\lambda \tag{18.16}$$

where we used the scale factor for density P from Table 18.1. Thus if the density scale factor is roughly 1, the strength of the materials must scale with the length. Cohesions for rocks are generally less than 50 MPa, and many are more than an order of magnitude less. Thus taking $\lambda = 10^{-5}$ from Equation (18.9), we see that even to model the strongest rocks requires a model material with negligible cohesion. Consequently, we model the brittle deformation of the Earth's crust using dry sand!

Figure 18.4 shows one experiment in which layers of dry sand were deformed to 50% extension by the uniform stretching of a rubber substrate. The results show horst-and-graben structure bounded by conjugate normal faults, listric normal faults bounding tilted fault blocks, and domains extending across several fault blocks in which bedding is tilted uniformly in one direction—all features found in the Basin and Range province (compare Figures 4.3, 4.6, 4.12, and 4.14).

If both time and the force of gravity are significant factors in the formation of a structure, as is the case for gravity-driven viscous deformation, special problems arise in the development of a scale model. If we choose a reasonable length scale factor for modeling geologic systems—say $\lambda = 10^{-5}$—Equation (18.15) requires that the time scale factor be 3.16×10^{-3}. Thus a geologic event that would occur over a period of 10^6 years would require 3160 years to occur in a dynamically similar scale model. Clearly it is not practical to do such experiments.

The use of a centrifuge to alter the scaling factors for inertial forces and for gravitational forces can provide a means around this problem. Deformation in the prototype is so slow that inertial body forces are very much smaller than the gravitational body forces and therefore can be ignored. If we place the model in a centrifuge where accelerations equivalent to several thousand times the acceleration of gravity can be attained, then for the model, the gravitational body forces are very much smaller than

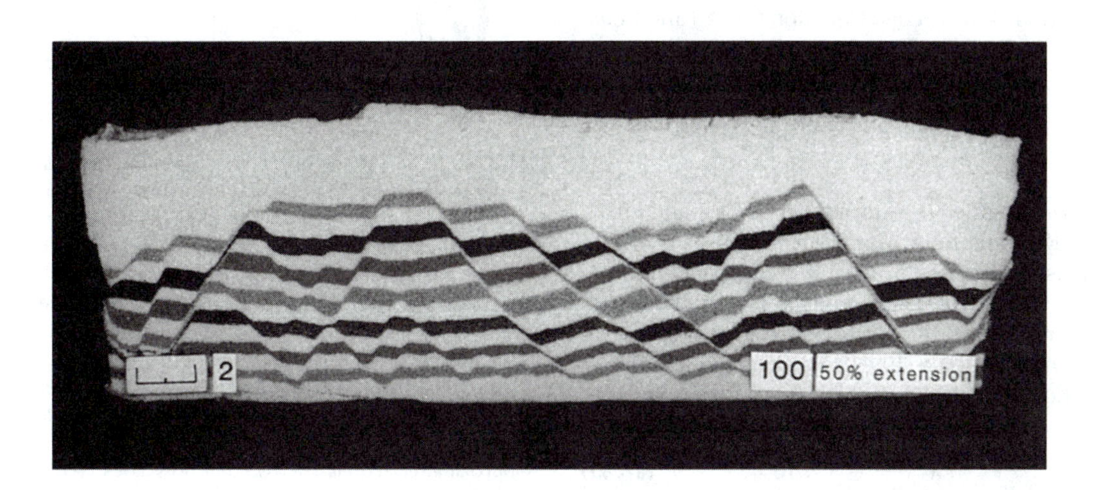

FIGURE 18.4 Model of deformation in the brittle crust made by using layered dry sand deformed to 50% extension by uniform stretching of a rubber substrate. Note horst-and-graben structure, listric normal faults, and tilted fault blocks. (From McClay and Ellis 1987)

the inertial body forces and therefore can be ignored. If one or both quantities in the ratio defining a scale factor are unimportant for the behavior of the system, then scaling that quantity from prototype to model is immaterial and the constraints imposed by the scale factor for that particular type of force can be ignored. Applying this line of argument to centrifuged models, then, the gravitational forces are negligible in the model, and the inertial forces are negligible in the prototype. We can therefore ignore the constraint implied by Equation (18.15), and we may treat λ and τ as independent scale factors.

Assuming the rocks behave viscously, the significant constraint is that the body forces and the viscous forces must be scaled by the same scale factor. The body force per unit mass is provided by the mechanical acceleration a_m in the model and by the gravitational acceleration g_p in the prototype. Thus, using the superscripts (v), (i), and (g) to identify viscous, inertial, and gravitational forces, respectively, we must have

$$\frac{F_m^{(v)}}{F_p^{(v)}} = \frac{F_m^{(i)}}{F_p^{(g)}} \tag{18.17}$$

Notice that the term on the right is the ratio of inertial forces in the model to gravitational forces in the prototype. To get an expression for the units of the viscous force in terms of the coefficient of viscosity, we look at Equation (18.6), from which we write

$$\frac{F^{(v)}}{L^2} = 2\eta\frac{1}{t} \qquad F^{(v)} = \frac{2\eta L^2}{t} \tag{18.18}$$

and for the units of inertial and gravitational forces in terms of the accelerations, we write

$$F^{(i)} = ma = (\rho L^3)a \qquad F^{(g)} = mg = (\rho L^3)g \tag{18.19}$$

Using the forms of the second Equation (18.18) and Equations (18.19) in Equation (18.17), we find

$$\frac{\eta_m(L_m)^2/t_m}{\eta_p(L_p)^2/t_p} = \frac{[\rho_m(L_m)^3]a_m}{[\rho_p(L_p)^3]g_p}$$

We rearrange terms and use the definitions for λ and τ from Equations (18.5) and for P from Equation (18.11) to write this relationship in terms of the three independent scale factors λ, τ, and P. We then introduce values of $P = 0.5$, which is approximately the correct ratio of densities for silicone putty and rock, $\lambda = 10^{-5}$ from Equation (18.9), $\tau = 10^{-9}$ from Equation (18.10), and $(a_m/g_p) = 10^3$ to find

$$\frac{\eta_m}{\eta_p} = P\lambda\tau\frac{a_m}{g_p} = 5 \times 10^{-12} \tag{18.20}$$

The ratio of mechanical to gravitational accelerations is not a formal scale factor because it is a ratio of two different quantities. Its value is determined by the capabilities of the centrifuge. If we assume the prototype viscosities in an orogenic zone to be in the range 10^{14} to 10^{18} Pa s, then Equation (18.20) gives approximately

$$10^3 \leq \eta_m \leq 10^7 \text{ Pa s} \tag{18.21}$$

which is in the range of viscosities available from materials like silicone putty and stitching wax.

Diapiric structures are common in a number of geologic environments, including sedimentary basins under-

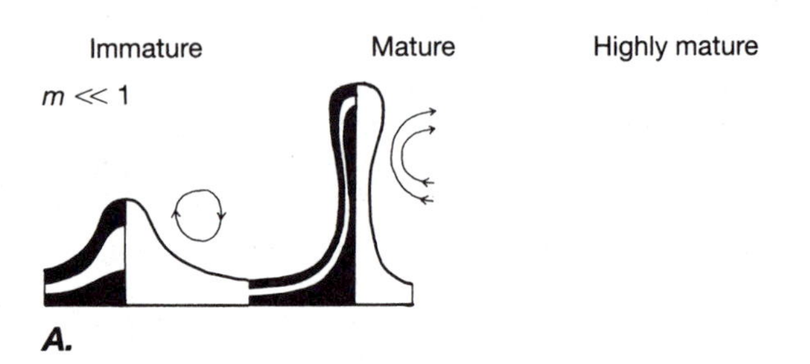

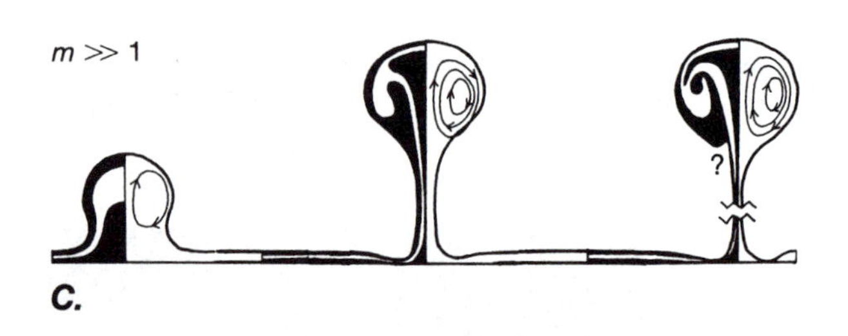

FIGURE 18.5 Styles of diapiric structures revealed by model experiments. Diapir shape and structure are a function of the ratio m of the viscosity of the high-density material to the viscosity of the low-density material and of the maturity of diapiric development and are determined largely by the toroidal flow that develops around the diapir due to drag on the diapir as it intrudes the overburden. (From Jackson and Talbot 1989) *A.* If the viscosity of the diapir is much greater than the surroundings, the toroidal flow occurs mainly in the high-density material, and columnar diapiric stocks develop. *B.* If the viscosities of diapir and surroundings are approximately equal, the toroidal flow lines cross the diapir boundary, and mushroom-shaped diapirs develop. *C.* If the viscosity of the diapir is much less than the surroundings, the toroidal flow is restricted to the interior of the diapir, and bulb-shaped diapirs develop.

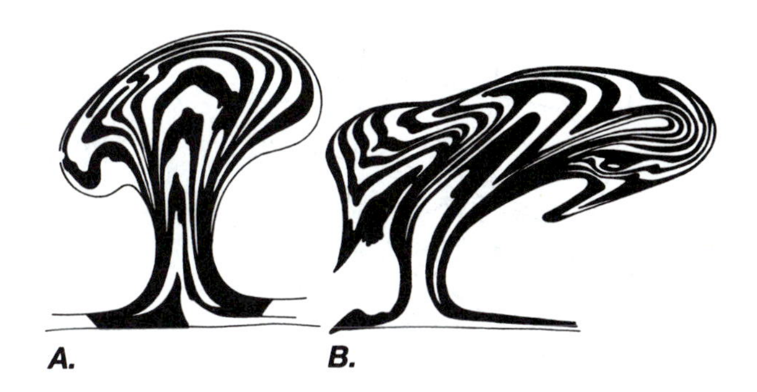

FIGURE 18.6 Internal structure of model diapirs deformed in a centrifuge, showing the folding of initially planar layers in the low-density material. (From Jackson and Talbot 1989)
A. Asymmetric diapir with a bulb shape on the right and a mushroom shape on the left. *B*. Tilted highly mature mushroom-shaped diapir.

lain by thick deposits of lower-density salt, which rises through the sediments to form salt diapirs, metamorphic environments where granitic gneisses rise diapirically through overlying denser metamorphic rocks, and the mantle where thermal or compositional density inversions lead to the diapiric rise of plumes. Salt diapirs are of particular interest because they form economically important traps for oil.

Salt commonly is present in thick layers formed as evaporites during the incipient opening (Gulf Coast, Red Sea) or advanced closing (Mediterranean Sea) of an ocean basin. Salt is a low-density substance (2.2×10^3 kg/m^3) with a very low strength at room temperature (see Table 16.1; Figure 16.21*B*, *C*). The sediments that commonly overlie the salt, such as sandstone, shale, and limestone, have densities significantly greater than the salt (2.3 to 2.8×10^3 kg/m^3). Thus when thick salt layers lie beneath a sufficiently thick overburden, the salt tends to flow slowly up, forming diapirs (see Section 13.10). Figure 13.40 illustrates a variety of shapes that are observed in salt diapirs (see also Figure 13.41).

Many workers have modeled diapiric intrusions in a centrifuge using silicon putties, modeling clay, and similar materials, and they have successfully reproduced many features found in nature. These models show that the shear between the rising diapir and the surrounding material induces a toroidal flow along the margins of the diapir (Figure 18.5). The effective viscosity ratio $m = \eta_{(hi\ \rho)}/\eta_{(lo\ \rho)}$ between the high- and low-density materials has a strong effect on the shape of the diapir, because it governs the location of the toroidal flow (Figure 18.5). If the diapiric material is much more viscous than its higher-density surroundings, then $m << 1$, and the toroidal flow is concentrated in the high-density material, resulting in the diapir adopting a fairly columnar shape (Figure 18.5*A*). If the viscosities are approximately equal, then $m \approx 1$, and the toroidal flow lines cross the margin between the diapir and the surrounding material, deforming the margin to form a mushroom-shaped diapir (Figure 18.5*B*). If the diapir has a much lower viscosity than the surrounding material, then $m >> 1$ and the toroidal flow is concentrated within the diapir, which therefore adopts a bulb shape (Figure 18.5*C*). The mushroom shape can also develop as a result of lateral spreading of the diapir if it reaches a level above which it cannot rise (see Figure 18.7). Figure 18.6 shows some of the complexity of internal structure that develops from initially horizontal layering in centrifuged silicone putty diapirs. These models can be compared with the diagrams in Figures 13.40 and 13.41, showing the structure of naturally occurring salt diapirs.

Gneiss domes are a common feature of many high-grade metamorphic belts. They have been interpreted as the result of the diapiric rise of lower-density gneisses into a higher-density cover (Figure 13.42). Centrifuged models designed to simulate orogenic zones (Figure 18.7) show a striking resemblance to the structures associated with mantled gneiss domes. These studies support the belief that gravity-driven deformation is a significant factor in the development of major structures in orogenic zones.

18.5 PLASTIC SLIP-LINE FIELD THEORY AND FAULTING

i. Plastic Slip-Line Field Theory

Faults in the Earth are characterized by the facts that the velocity of the material is parallel to the fault, and that a discontinuous jump in the velocity occurs across the

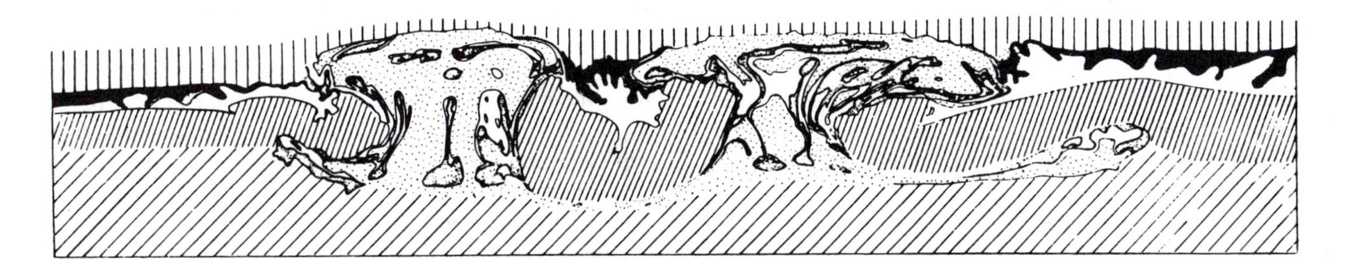

FIGURE 18.7 Centrifuged scale model of an orogenic zone showing the consequences of gravity-driven deformation. (Ramberg 1981)

fault. We can model this behavior with a rigid-plastic, isotropic, homogeneous material undergoing plane strain. For such a material under these conditions, it can be shown that the constitutive equations for plasticity (see Box 16-4(iii), Equations (16-4.8) and (16-4.9)) and the conditions of equilibrium (Equation (18.2); compare Equations (18-1.11) and (18-1.12)), provide an analytic solution that defines two families of lines called **slip lines**, which by convention are designated α and β for the lines with right- and left-lateral senses of slip, respectively (Figures 18.8*A* and 18.9). The mathematical development of the theory is beyond the scope of this book but can be found in a variety of references.[5] The results, however, are not difficult to understand. Each line in one family is orthogonal to all the lines in the other family, and each line is everywhere tangent to the maximum shear stress and the maximum shear strain rate in the material. Thus the maximum and minimum principal stresses ($\hat{\sigma}_1$, $\hat{\sigma}_3$) and the corresponding maximum shortening and lengthening principal axes of instantaneous stretch ($\hat{\zeta}_3$, $\hat{\zeta}_1$) bisect the right angles between any two intersecting slip lines (Figure 18.8*A*).

The components of displacement and velocity parallel and normal to the slip lines must obey a few simple rules. Extensional strain parallel to any of the slip lines must be zero, and this requires the component of displacement or velocity parallel to a slip line to be constant along that slip line (Figure 18.8*B*). It also means that the component of displacement or velocity normal to any slip line must be constant across the slip line, because the slip lines are mutually orthogonal (Figure 18.8*C*). The component of displacement and velocity parallel to a slip line, however, may be different on opposite sides of that slip line without violating these restrictions (Figure 18.8*D*), and any such discontinuity must be the same along the entire length of that particular slip line. Thus a velocity discontinuity that is imposed by the boundary conditions propagates through the material along the slip lines that emanate from the point of discontinuity. It is this last property that makes the slip lines relevant to the interpretation of faults.

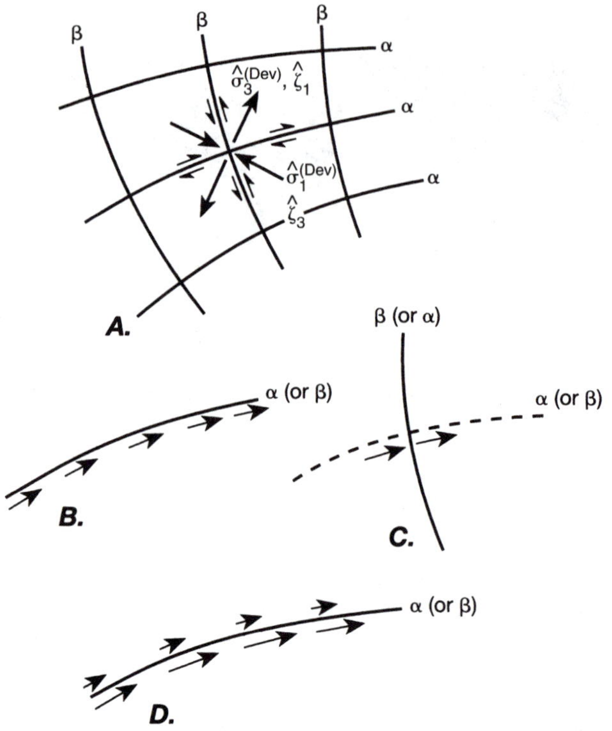

FIGURE 18.8 Characteristics of plastic slip-line fields. (After Backofen 1972) *A*. Slip lines form two sets of curves, α (dextral shear) and β (sinistral shear), respectively, that are everywhere mutually orthogonal. Each line is tangent to the orientation of maximum resolved shear stress and maximum resolved instantaneous shear strain, so the principal axes of both quantities bisect the angles between the two sets of slip lines. *B*. Along any one slip line, the tangential component of the velocity and displacement must be constant. *C*. The component of velocity and displacement normal to a slip line must be the same on both sides of the slip line. *D*. The component of velocity and displacement parallel to a slip line can be discontinuous across the slip line, so the slip line can behave like a fault.

The geometry of the slip lines depends on the geometry of the deforming system. As an example, we consider the slip lines resulting from a rigid indenter impinging on the edge of a plastic layer (Figure 18.9). In Figure 18.9*A*, the plastic material is constrained between two rigid boundaries that support no shear stress and that are symmetrically placed on either side of the indenter. In Figure 18.9*B*, the indenter impinges on the straight edge of a plastic layer that in effect fills half the two-dimensional plane. Figure 18.9*C* shows the indenter impinging near the corner of a plastic plate that in effect fills a quarter of the two-dimensional plane. The heavily stippled area in each diagram is a region of "dead" material that in Figure 18.9*A*, *B* is fixed to the indenter and does not deform. In Figure 18.9*C*, however, the "dead zone" is not a steady-state feature, but rather it must slip to the right across the top of the indenter and become deformed as the indenter advances into the plate. The slip lines are the lines along which the material shears during the deformation.

In the case of Figure 18.9*C*, the presence of a corner in the plate near the indenter dominates the geometry of the slip lines, causing them to develop asymmetrically. The β slip lines that extend toward the free surface that is perpendicular to the indented surface are the predominantly active slip lines, and the lower-right corner of the plate is extruded down and to the right. To the left of the α slip line that makes an angle of 77.8° with the contact face of the indenter, the plastic plate is under a compressive stress.

[5]See, for example, Johnson et al. (1970) and Odé (1960).

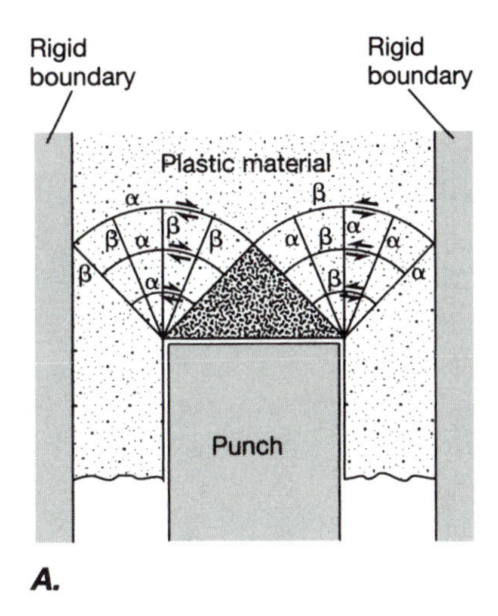

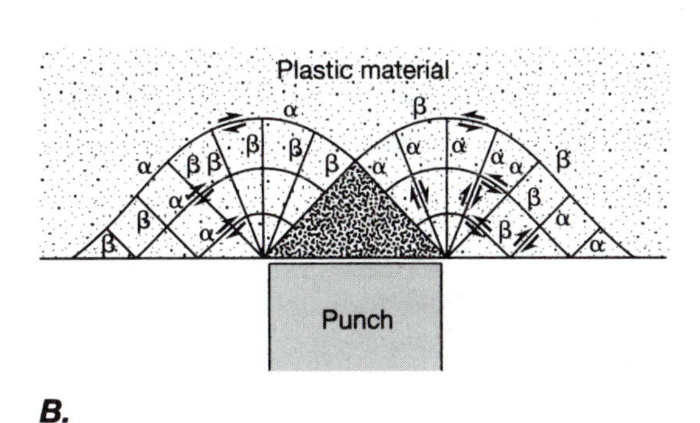

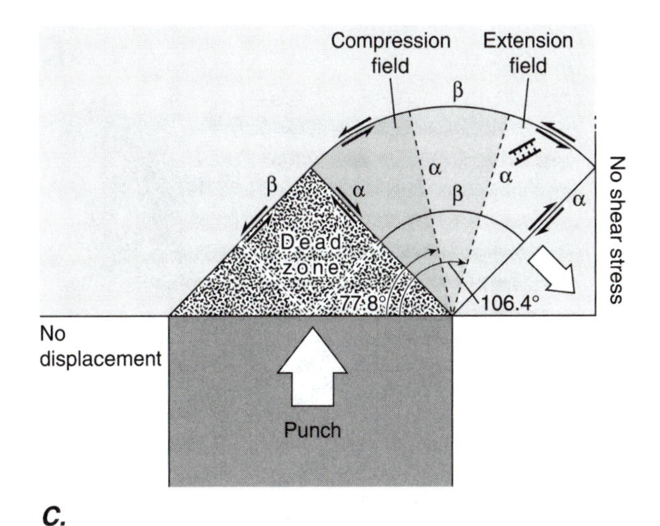

FIGURE 18.9 Geometry of plastic slip-line fields in an isotropic, homogeneous, rigid-plastic material deformed by a rigid punch. The stippled area represents the plastic material. The heavily stippled triangular area is the "dead zone" in which no deformation occurs. *A*. The plastic material is constrained by rigid boundaries that support no shear stress symmetrically placed on either side of the punch. *B*. The punch impinges on the straight edge of the plastic material, which is unconstrained on either side of the punch. *C*. The punch impinges near the corner of a plastic plate, where the side adjacent to the surface of contact with the indenter is a free surface that supports no shear stress. The presence of this free boundary leads to the asymmetric development of the plastic slip lines. The lower-right corner of the indented plate shifts down and to the right. (*A*, *B* after Backofen 1972; *C* after Molnar and Tapponnier 1977)

Between this slip line and the α slip line at 106.4° from the contact face of the indenter, the minimum principal stress is tensile but the mean stress is compressive; and to the right of this line, even the mean stress is tensile. In principle, the indenter can be composed either of the more rigid or the more ductile material. In the latter case, the pattern of faulting is different from that shown in Figure 18.9.

Because discontinuities in the tangential component of velocity are permitted across the slip lines, they have been used to interpret patterns of strike-slip faulting in the Earth's crust that result from tectonic collisions between a less-ductile continental block, the indenter, and a more-ductile one, the plastic material.

ii. Scale Model for Faulting in Southeast Asia: Analysis of Model Scaling

As an example of the interpretation of major transcurrent faults in terms of a plastic slip-line field, we discuss the formation of the system of major strike-slip faults in central Asia including the Altyn Tagh, Kansu, Kunlun, Kang Ting, and Red River faults (see Figure 6.2*B*). This system of faults has been compared to plastic slip lines such as those in Figure 18.9*C*, with India acting as the indenter during its collision with Asia. A scale model has been constructed to model the tectonic evolution of southeast Asia (Figures 18.10 and 18.11). India is represented by the rigid indenter 5 centimeters in width, and Asia by a layer of plasticine 11 centimeters thick and 30 centimeters square. One side of the plasticine layer perpendicular to the indented surface is left unconstrained to represent the eastern edge of the Asian plate along the line of subduction zones in the western Pacific. The deformation is constrained to be a plane strain by a pair of stiff plates above and below the plasticine layer that are lubricated to eliminate shear stresses. We first evaluate the geometric, kinematic, and dynamic similarity between this model and the Earth, and then we briefly discuss the implications of the model for the tectonic evolution of southeast Asia.

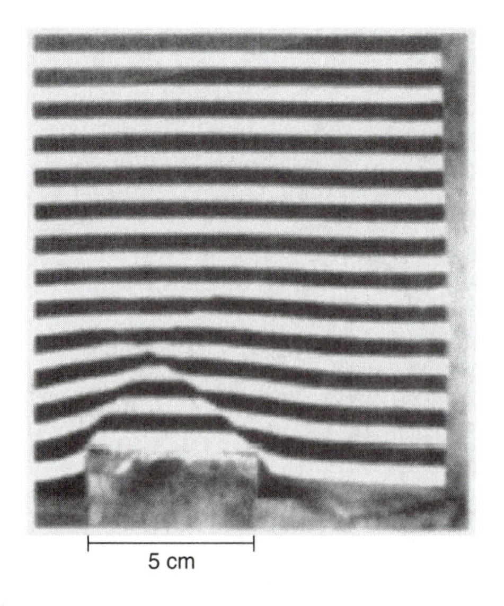
5 cm

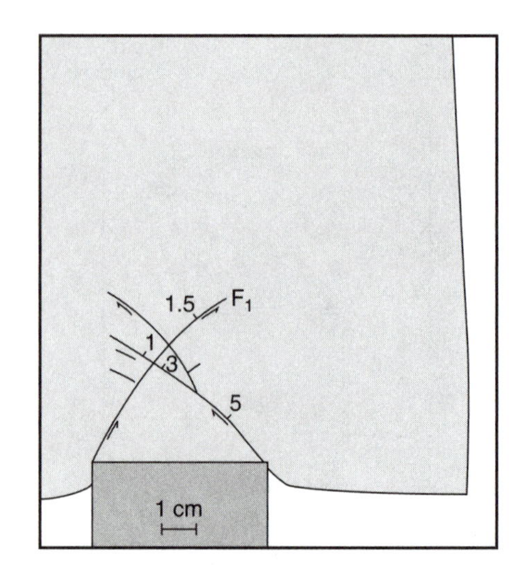
1.5
F1
1
3
5
1 cm

A.

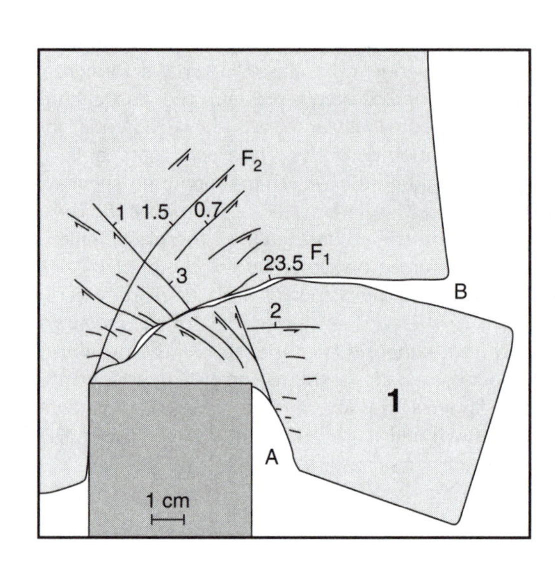
F2
1
1.5
0.7
23.5
F1
3
2
B
1
A
1 cm

B.

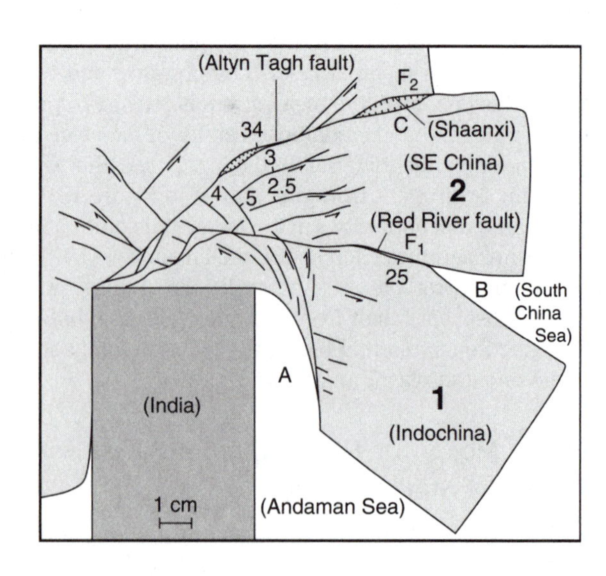
(Altyn Tagh fault)
F2
34
C
(Shaanxi)
3
(SE China)
4
5
2.5
2
(Red River fault)
F1
25
B
(South China Sea)
A
1
(India)
(Indochina)
1 cm
(Andaman Sea)

C.

FIGURE 18.10 Photos (left) and labeled line drawings (right) of a plasticene scale model of the collision of India with the southern margin of Eurasia. The rigid indenter, representing India, impinges on a plasticine plate representing Eurasia. The contact zone is near a corner of the plate, with the adjacent side a free boundary (compare Figure 18.9C). Small numbers indicate cumulative displacement (in millimeters) on the major faults. Deformation is constrained to plane strain by essentially rigid lubricated plates above and below the layer of plasticine. (After Tapponnier et al. 1982) *A*. Slip lines initially develop symmetrically from the corners of the indenter, defining a triangular zone at the head of the indenter. *B*. The free boundary to the right leads to the preferential development of the β (sinistral) slip lines, which propagate out to intersect the free boundary. Block 1 is extruded toward the free boundary, and its rotation opens gaps labeled "A" and "B". *C*. The triangular zone advances with the indenter, becomes deformed, and slides to the right across its top. Slip line F_1 is deformed with the material in the triangular zone and is rotated into an orientation unfavorable for further slip. A new β slip line F_2 forms and propagates out to the free boundary, cutting out block 2, which extrudes toward the free boundary from in front of the indenter, carrying block 1 along with it. Extension occurs along F_2 in the areas surrounded by hatchured lines, including the area labeled "C" near the end of F_2, and the gap between block 1 and the indenter, labeled "A" continues to grow. Features labeled in parentheses indicate the geographic names from east Asia to which they bear an approximate resemblance (see Figure 18.11).

The scaling parameters of the model are determined by the 5 cm width of the indenter representing the roughly 2000 km width of the Indian block, giving a length scale of

$$\lambda \equiv \frac{L_m}{L_p} = \frac{5\text{ cm}}{2000\text{ km}}$$

$$= \frac{5\text{ cm}}{2 \times 10^3\,[\text{km}] \times 10^3\,[\text{m/km}] \times 10^2\,[\text{cm/m}]}$$

$$\lambda = 2.5 \times 10^{-8} \qquad (18.22)$$

The thickness of the plasticine plate does not scale at all with the crust or lithosphere, but the constraint of plane strain deformation means the third dimension of the model is inconsequential. The time scale is determined by the roughly 2 h duration of the experiment representing the 40 My since India began colliding with Asia. Thus

$$\tau \equiv \frac{t_m}{t_p} = \frac{2[\text{h}]}{40[\text{My}]} = \frac{2[\text{h}]}{4 \times 10^7\,[\text{y}] \times 365\,[\text{d/y}] \times 24\,[\text{h/d}]}$$

$$\tau = 5.7 \times 10^{-12} \qquad (18.23)$$

Thus the velocity should scale as (Table 18.1, item 7):

$$v \equiv \frac{v_m}{v_p} = \frac{\lambda}{\tau} = \frac{2.5 \times 10^{-8}}{5.7 \times 10^{-12}} = 4.4 \times 10^3 \qquad (18.24)$$

Reconstructions using seafloor magnetic anomalies have indicated that there has been 2500 to 3500 km of convergence of northeast India with Asia since 40 to 50 Ma, giving an average convergence rate of between 5 and 8.75 cm/y, or between 5.7×10^{-4} and 1×10^{-3} cm/h. Equation (18.24) thus requires a model velocity of between 2.5 and 4.4 cm/h. The model used the slower rate of 2.5 cm/h. Thus the requirements for geometric and kinematic similarity are satisfied by the model.

We have only defined two independent scale factors for the model (λ and τ), leaving us free to choose one other scale factor. The constraint of plane strain means that we ignore any thickening of the plate, although we know that this is not strictly applicable to southeast Asia, as the Tibetan plateau is one of the largest pieces of high-elevation real estate in the world. Accepting this as a compromise for a reasonable first approximation, however, we can assume that gravitation and its interaction with topography are negligible factors. The extremely low velocities also mean we can ignore acceleration. Ignoring these two factors means that mass and density do not constrain the similarity between model and prototype. We can then define the third independent scale factor based on the material properties of the Earth and the model. This scaling, however, is difficult because of our limited knowledge of the material properties for the Earth and the magnitude of the stresses that exist during a collisional event. Given the uncertainties, we make independent estimates for the scale factors for both stress and viscosity and then see to what extent they are consistent.

The plasticine used for the model has a power-law rheology with a stress exponent of $n = 7.5$ (see Equation (16.4)) and a yield stress of about 10^5 Pa at a strain rate of $10^{-7}\,\text{s}^{-1}$. The strain rate for the experiment must have significant spatial variation, but if we average the convergence rate of 2.5 cm/h across a distance that is roughly twice as big as the width of the indenter (compare Figure 18.9*C*), or 10 cm in the model, we find an average strain rate of

$$\dot{\varepsilon}_n = \frac{2.5\,[\text{cm/h}] \times (1[\text{h}]/60[\text{min}]) \times (1[\text{min}]/60[\text{s}])}{10[\text{cm}]}$$

$$= 6.9 \times 10^{-5}\,\text{s}^{-1} \qquad (18.25)$$

Local strain rates are likely to be higher. This strain rate is 690 times higher than that at which the yield condition is assessed, but because of the power-law rheology with

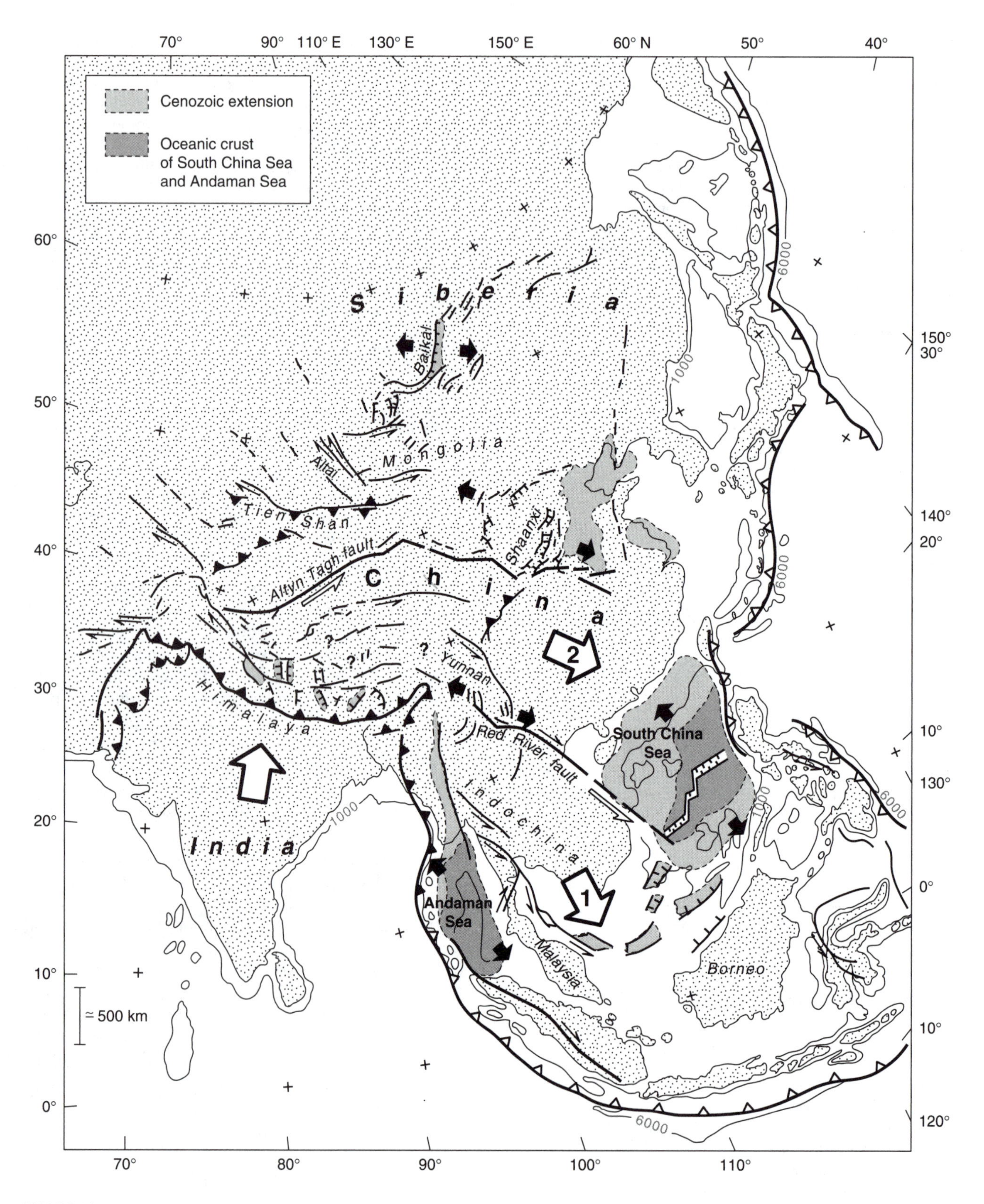

FIGURE 18.11 Map showing major strike-slip faults and areas of extension of southeast Asia, interpreted in terms of the model in Figure 18.10. (From Tapponnier et al. 1982)

a stress exponent of 7.5, an increase in strain rate by this factor would represent an increase in stress by a factor only of about 2.4 ($2.4^{7.5} \approx 2.39^{7.5} \approx 690$, using Equation (16.4)). Thus for a large range of strain rates, the yield stress provides a reasonable order-of-magnitude estimate for the stress in the model (compare Figure 16.4 for $n = 7$).

Stress in the Earth, as inferred from deformation textures in rocks (see Box 17-2(iii)), may be expected to be as little as ~1 MPa for steady-state flow in the mantle to as much as ~100 MPa, for shearing on major crustal faults. We could also infer from Figure 16.21*B* that a rough range of stresses in the crust for a variety of rocks at midcrustal depths of about 25 km and a strain rate of $10^{-14}\ \mathrm{s}^{-1}$ would be about 0.1 to 100 MPa. Thus, converting the largest range of these stresses to Pascals and using the yield stress for plasticine, a reasonable range for the stress scale factor Σ can be estimated to be (Table 18.1, item 10):

$$\Sigma \equiv \frac{\sigma_m^{(Dif)}}{\sigma_p^{(Dif)}} = \left\{ \frac{10^5\ \mathrm{Pa}}{10^5\ \mathrm{Pa}} \rightarrow \frac{10^5\ \mathrm{Pa}}{10^8\ \mathrm{Pa}} \right\} = \{1 \rightarrow 10^{-3}\} \tag{18.26}$$

Turning to the viscosities, we realize that a power-law material does not have a single viscosity as defined by Equation (18.7) because the dependence of strain rate on stress is not linear. The viscosity depends on the stress or the strain rate at which it is measured (see the discussion in Section 18.7, Equation (18.37)). Nevertheless, we can estimate an effective viscosity of the crust for a strain rate of $10^{-14}\ \mathrm{s}^{-1}$ by using a stress range of 0.1 to 100 MPa in Equation (18.7):

$$\eta_p = \frac{\sigma_s}{2\dot{\varepsilon}_s} \cong \frac{0.5\ \sigma^{(Dif)}}{2\dot{\varepsilon}_n} \tag{18.27}$$

$$\eta_p = \left\{ \frac{10^5\ \mathrm{Pa}}{4 \times 10^{-14}\ \mathrm{s}^{-1}} \rightarrow \frac{10^8\ \mathrm{Pa}}{4 \times 10^{-14}\ \mathrm{s}^{-1}} \right\}$$

$$\eta_p = \{2.5 \times 10^{18} \rightarrow 2.5 \times 10^{21}\}\mathrm{Pa\ s} \tag{18.28}$$

In Equation (18.27), we used $\sigma_s^{(\max)} = 0.5(\hat{\sigma}_1 - \hat{\sigma}_3) = 0.5\ \sigma^{(Dif)}$ and $\dot{\varepsilon}_s^{(\max)} = 0.5(\dot{\hat{\varepsilon}}_1 - \dot{\hat{\varepsilon}}_3)$. We also used Equation (12.15), which applies to the instantaneous strain rates as well (cf. Equation (18-1.5)), and with the restriction $\dot{\varepsilon}_V = \dot{\hat{\varepsilon}}_2 = 0$, this gives $\dot{\varepsilon}_s^{(\max)} = \dot{\hat{\varepsilon}}_1 = \dot{\varepsilon}_n^{(\max)}$. (Had we used the more precisely defined Equation (18.37) for the relation between the coefficient of viscosity and the differential stress, the numerical factor would have been 3.3 instead of 2.5—hardly a significant difference, given the range of estimated values). For plasticine, we estimate a range for the effective viscosity for two strain rates, first by using the yield conditions for plasticine noted in the text above Equation (18.25), and second by using the strain rate from Equation (18.25) and the associated stress determined from the power-law relation for the plasticine with a stress exponent of $n = 7.5$ (Equation (16.4)):

$$\eta_m = \frac{\sigma_s}{2\dot{\varepsilon}_s} \cong \frac{0.5\ \sigma^{(Dif)}}{2\dot{\varepsilon}_n}$$

$$= \left\{ \frac{10^5\ \mathrm{Pa}}{4 \times 10^{-7}\ \mathrm{s}^{-1}} \rightarrow \frac{2.4 \times 10^5\ \mathrm{Pa}}{4 \times (6.9 \times 10^{-5})\mathrm{s}^{-1}} \right\}$$

$$\eta_m = \{2.5 \times 10^{11} \rightarrow 8.7 \times 10^8\}\mathrm{Pa\ s} \tag{18.29}$$

Thus the possible extreme values for the viscosity scale factor (Table 18.1, item 11) are

$$H \equiv \frac{\eta_m}{\eta_p} = \left\{ \frac{2.5 \times 10^{11}\ \mathrm{Pa\ s}}{2.5 \times 10^{18}\ \mathrm{Pa\ s}} \rightarrow \frac{8.7 \times 10^8\ \mathrm{Pa\ s}}{2.5 \times 10^{21}\ \mathrm{Pa\ s}} \right\}$$

$$H = \{1.0 \times 10^{-7} \rightarrow 3.5 \times 10^{-13}\} \tag{18.30}$$

We can determine the consistency of the scale factors we have determined for stress and for viscosity by using the relationship between them (Table 18.1, item 11). This relationship allows us to determine an independent value for the viscosity scale factor H from the scale factors for stress Equation (18.26) and for time Equations (18.23):

$$H = \Sigma\tau = \{1 \rightarrow 10^{-3}\} \times 5.7 \times 10^{-12}$$

$$H = \{5.7 \times 10^{-12} \rightarrow 5.7 \times 10^{-15}\} \tag{18.31}$$

We can compare this result with the result in Equation (18.30). Taken together, our estimates of the viscosity scale factor from Equations (18.30) and (18.31) vary over more than seven orders of magnitude, and they overlap only between about 10^{-12} and 10^{-13}. The range estimated from the rheology (Equation (18.30)) is generally several orders of magnitude larger than the range estimated indirectly from the stresses and the time scale factor (Equation (18.31)). Thus if the model and the Earth could be considered dynamically similar, it would only be at the extreme high end for the range of stress we have chosen and the extreme low end for the effective viscosity. This difference in the viscosity scale factors H and the fact that the power-law stress exponent for the plasticine is 7.5, whereas for the Earth it is most likely to be 3 or less, should make us cautious about accepting the dynamic similarity between this model and the Earth. Nevertheless, the geometric and kinematic similarities between the model and the geology are striking, as we discuss next, so it is probable that the model provides some useful insight into the large-scale tectonics of the region.

iii. Plastic Slip-Line Model for Faulting: Results of the Model

Figure 18.10 shows the progressive development of slip lines in the plasticine model. As the indenter first pushes into the plasticine layer, faults initiate at the corners of the indenter and propagate toward and beyond each other to create a triangular zone at the head of the indenter (Figure 18.10*A*). With continued penetration of the indenter (Figure 18.10*B*), the faults then develop asymmetrically, with the β plastic slip line, the sinistral fault F_1, becoming dominant and propagating from the tip of the triangular zone out to the free boundary (compare Figure 18.9*C*). Displacement on this fault ultimately results in block 1 extruding 25 to 35 mm away from the advancing indenter and rotating clockwise by nearly 25°. Although subsidiary faults form in block 1, particularly near the indenter, the motion of block 1 is mostly a rigid-body motion. The size of the block depends on the width of the indenter and the distance from the free lateral boundary to the edge of the indenter. The triangular zone shears across the top of the indenter toward the free boundary, and the fault F_1 is caught up in the deformation of this material and rotated into an orientation unfavorable for further slip. Finally, F_1 is abandoned and another major β slip line, fault F_2, initiates from the far corner of the indenter (Figure 18.10*C*). The pattern of deformation is then substantially repeated, with a large block, block 2, being extruded and rotated clockwise as slip accumulates on F_2. Subsidiary faults also develop in block 2 near the indenter, although most of the block remains undeformed. Block 1 continues to extrude along with block 2 and is further rotated to about 40°. Displacement on the α slip lines (dextral faults) remains small relative to the massive displacements on the (sinsitral) β slip lines F_1 and F_2.

Gaps open up along the major faults like pull-apart basins on large strike-slip faults. While F_1 is active, a large wedge-shaped gap, labeled "B" (Figure 18.10*B*, *C*), opens along the free lateral margin as block 1 rotates away from the adjacent undeformed material, and a large gap, labeled "A" also opens between block 1 and the advancing indenter (Figure 18.10*C*) that widens continuously during the slip on both F_1 and F_2. These gaps indicate areas of extensional deformation.

Figure 18.11 shows some of the major tectonic structures of southeast Asia interpreted in the light of this scale model. The large open arrows on the Indian continent and on the Indochina and south China blocks indicate the directions of motion of these blocks, corresponding to the kinematics of the scale model (compare Figure 18.10*C*). The major faults in Asia that seem comparable in kinematic importance to the faults F_1 and F_2 in the model are the Red River fault between Indochina and south China and the Altyn Tagh fault system between south China and Mongolia. The latter fault system, which includes the Kansu fault (Figure 6.2) and other contiguous faults to the east, is 2500 km long and dominates recent deformation in the area. The Indochina block is bounded to the north by the Red River fault and therefore corresponds to block 1 in the model (Figure 18.10*C*). The south China block is bounded on the south and north, respectively, by the Red River and Altyn Tagh faults, and therefore corresponds to block 2 in the model. Subsidiary faults such as the Kang Ting and Kunlun faults in south China (Figure 6.2*B*) correspond to the subsidiary left-lateral faults that develop in block 2 just ahead of the indenter (Figure 18.10*C*).

The Andaman Sea west of Malaysia and Indochina (Figure 18.11) is a region of dextral strike-slip faulting and extensional seafloor spreading. It corresponds with the gap that opens between the indenter and the adjacent plasticine, labeled "A" in Figure 18.10*C*, and that accommodates the extrusion of block 1 and the movement of the indenter. The South China Sea, located northeast of Indochina (Figure 18.11), also involves extension of the crust and is comparable to the formation of the wedge-shaped gap labeled "B" in the model at the end of F_1 (Figure 18.10*C*). A comparable area of extension exists in the Shaanxi Graben system near the end of the Altyn Tagh fault system (Figure 18.11), and this is comparable to the extensional deformation in the model labeled "C" near the end of the fault F_2 (Figure 18.10*C*).

Current shear on the Red River fault is dextral, although this is apparently a reversal in shear sense from the initial sinistral displacement. This reversal of shear does not show up in the model but can be understood if Indochina did not continue to extrude with the south China block as the Altyn Tagh fault became dominant. In that case the south China block would extrude past the Indochina block, resulting in a reversal of shear sense on the Red River fault from sinistral shear when Indochina was being extruded, to dextral shear when extrusion of south China became dominant. This suggests the actual boundary conditions for the Indochina block are not as free as they are in the model.

The asymmetry of the fault systems and of the displacement magnitude apparent in the model also occurs in the geology. Dextral offsets on the northwest-trending faults in the Tien Shan and Altai regions are no greater than tens of kilometers, whereas Tertiary offsets on the major sinistral Red River and Altyn Tagh faults are on the order of hundreds of kilometers. In the model, 10 mm of displacement corresponds to 400 km of displacement in the prototype, so the model is generally consistent with the observed geology.

Thus the model shows many similarities with the existing geology, and it provides a testable scenario for the evolution of some of the main Tertiary structures in Asia. It also suggests that continued advance of the indenter beyond that in Figure 18.10*C* would repeat the cycle in which fault F_2 would be abandoned and a new sinistral fault would propagate from the left corner of the indenter to the free boundary, cutting out another continental block from the undeformed plate adjacent to block 2. The future trace of such a fault extending from the northwest corner of India through Tien Shan, Baikal to the Sea of Okhotsk may already be evident in the active strike-slip and extensional structures (Figure 18.11).

The model should not be taken to provide a full picture of the tectonics of Asia, however. It imposed plane strain on the material, thereby preventing any vertical deformation, eliminating the possibility of crustal thickening by thrust faulting, and emphasizing the strike-slip faulting. Yet we know the Tibetan plateau is an area of very high elevation. The model also does not account for variations of rheology with depth and the effects that has on the geology. Nevertheless, it provides important insight into the mechanics of large-scale tectonic deformation and the formation of large continental transcurrent faults (see Figure 20.35 for a different interpretation of Tibetan tectonics).

18.6 ANALYTIC SOLUTION FOR THE VISCOUS BUCKLING OF A COMPETENT LAYER IN AN INCOMPETENT MATRIX

As an example of an analytic mechanical model for the formation of a geologic structure, we consider a model for the formation of folds by viscous buckling of a layer having a thickness h and a high viscosity η_L embedded in an infinitely thick viscous matrix having a lower viscosity η_M. The measure of viscosity is one example of a precise method of defining the competence; a relatively high-viscosity material is competent, and a relatively low-viscosity material is incompetent (see Section 13.5(i); Figure 13.18).

The geometry of the problem is set out in Figure 18.12. The layer is given an initial deflection, exaggerated in the figure for the sake of clarity, which is necessary for the folding to start and which approximates normal irregularities found in nature. We wish to calculate how the layer will behave if a known force **P** is applied parallel to the layer and if the resistance **K** of the matrix is proportional to the rate of deflection of the layer. For the boundary conditions, we specify that the deflection caused by deformation is zero along the entire length of the layer before the deformation begins, and it must always be zero at $x = 0$ and $x = L$, where L is the unknown wavelength we want to determine. Thus the layer is pinned at 0 and L, and these points are nodes of the folding. In addition, we ignore any possible shortening and thickening of the layer in response to the force **P**.

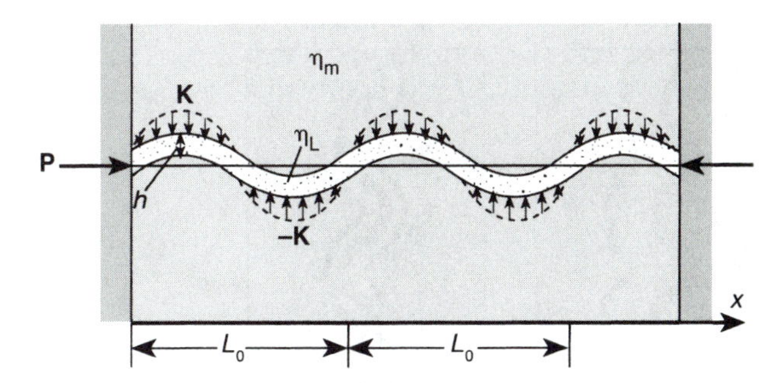

FIGURE 18.12 Model for the theoretical analysis of the folding of viscous materials. A competent viscous layer of viscosity η_L and thickness h is compressed parallel to the layer by a force P. It is embedded in an incompetent viscous matrix of viscosity η_M that resists the folding of the layer with a force K that is proportional to the vertical rate of shear induced by the folding layer. L_O is the wavelength of the initial deflection in the layer.

The formulation of a problem of viscous deformation is reviewed briefly in Box 18-1. We focus here on the solution to this model, which gives the magnitude of the displacement u normal to the layer for an initial wavelength L_O having an initial displacement perturbation δ_0. The solution is a function of the time t, of the growth coefficient a, and of the initial wavelength L_0, and it varies with the distance x along the layer:

$$u = \delta_0 \exp[t/a]\sin(2\pi x/L_0) \qquad (18.32)$$

The fold growth coefficient a is a function of the stress in the layer, or equivalently of the imposed strain rate, as well as of the viscosity of the layer (its competence) and the relative viscosities of the layer and the matrix (the competence contrast). The time constant, or characteristic time, for a fold to reach an amplitude of e[6] times the initial perturbation is $t_c = a$, although the time dependence applies only for the small increment of deformation before the initial geometry of the layer is appreciably changed. The predominant wavelength L is the one that grows the fastest, and it is therefore the one for which the function a is a minimum. It can be shown that this condition leads to Biot's relation for L:

$$L = 2\pi h\left[\frac{\eta_L}{6\eta_M}\right]^{1/3} \qquad (18.33)$$

where η_L and η_M are the viscosities of the layer and the matrix, respectively, and h is the layer thickness. Thus this equation states that if layer-parallel shortening is imposed on a single competent layer that has a variety of scales of initial perturbations from a planar geometry, and that is embedded in a less-competent matrix, the layer will develop a fold train with a dominant wavelength L that depends strongly on the layer thickness h and less strongly on the competence contrast between layer and matrix (the cube root dependence on η_L/η_M).

This solution is applicable to small amplitude folds only, for which the folding angle is less than about 20° to 30° and for which the true wavelength of the fold λ and the length along the layer L for the same wavelength are approximately the same. With large amounts of shortening, the wavelength decreases as the folding angle in-

[6] e is the base of natural logarithms, $e = \exp(1) = 2.71828\ldots$

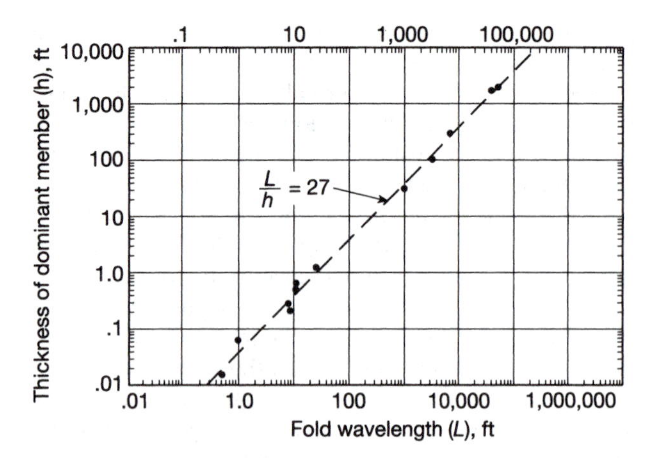

FIGURE 18.13 Dependence of wavelength of folding on the thickness of the folded layer determined from natural folds. The line is for a wavelength to thickness ratio $L/h = 27$. (From Currie et al. 1962)

creases, the difference between λ and L increases, and this solution no longer applies.

Despite this limitation, the relationship defined by Equation (18.33) seems to hold in nature. Figure 18.13 shows the relationship between layer thickness and fold wavelength measured from natural folds. It shows that the ratio of wavelength to layer thickness L/h is essentially a constant, which is the relationship predicted by Equation (18.33) if the viscosity ratio (i.e., the competence contrast) is relatively constant.

In principle it should be possible, therefore, to measure the wavelength of a train of folds and the thickness of the folded layer and determine the ratio of the viscosities of the layer and the medium. For all the folds plotted in Figure 18.13, for example, the wavelength-to-thickness ratio L/h is approximately 27, implying that the corresponding ratio of the layer viscosity to the matrix viscosity η_L/η_M is approximately 476. The wavelength, however, is not a very sensitive function of the viscosity ratio. Thus if η_L/η_M increases by a factor of 10, then L/h increases only by a factor of a little less than 2.2. The result is also not reliable because the folds plotted in Figure 18.13 are multilayer folds, whereas the analysis was done for single layers. Ptygmatic folds would be preferable for this type of analysis, because they develop in isolated competent layers in an incompetent matrix, thus closely resembling the geometry of the theoretical model.

Another major limitation of this model of folding is the implicit assumption that the layers of rock actually deform as a viscous material. Our discussion of rock rheology in Chapter 16, however, shows that this assumption may not be valid. Furthermore, the predicted form of the folds would be no different if we had assumed the rocks were elastic materials, because the constitutive equations for viscous fluids and elastic solids are very similar in form. For the buckling instability of a relatively stiff elastic layer in a less stiff elastic matrix, the wavelength L is related to the layer thickness h, and the ratio of the elastic moduli of layer and the matrix B_L/B_M by[7]

$$L = 2\pi h\left[\frac{B_L}{6B_M}\right]^{1/3} \qquad B \equiv \frac{E}{1-\nu^2} \tag{18.34}$$

where B_L and B_M are the elastic moduli of the layer and the matrix, respectively, E is Young's modulus, and ν is the Poisson ratio.

Figure 18.14A illustrates this relationship for elastic materials with three rubber strips, one twice the thickness of the other two, embedded in a less stiff gelatin.

[7]After Biot in Johnson (1970).

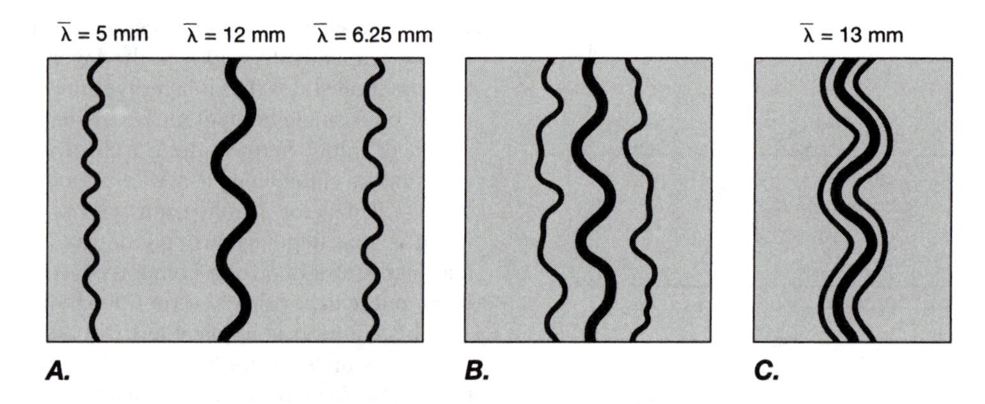

FIGURE 18.14 Folding of elastic rubber strips embedded in a softer elastic gelatin matrix. (After Currie et al. 1962) A. Widely separated strips 0.4 millimeter thick (two strips) and 0.8 millimeter thick (center strip). The average wavelength in the thicker strip ($\lambda \approx 12$ mm) is approximately twice the average wavelength in the strips that are half as thick ($\lambda \approx 5.6$ mm). B. With less separation, the folding of adjacent strips begins to interfere. C. With still less separation, the multilayer behaves as a single layer forming folds with a wavelength larger than any of the individual strips but smaller than a single strip with a thickness equal to the combined thickness of all the strips.

The wavelength for the thicker strip is very close to twice the wavelength for the strips that are half as thick, as predicted by Equation (18.34).

Figure 18.14*B*, *C* show the effects of folding of a multilayer in which the competent layers are close enough together that the folding patterns interfere and the wavelength of the folding is modified. The folds in Figure 18.14*C* have a wavelength characteristic of a single layer with a thickness greater than the thickest layer but less than the combined thickness of all three layers. Experiments such as these serve to test theoretical predictions and provide an intuitive grasp of the factors that affect the form of folds.

Because Equation (18.34) is identical in form to Equation (18.33), it is impossible to distinguish elastic from viscous effects on the basis of the form of folds alone. Despite this ambiguity, the theory provides insight into the process of fold formation and the probable influence of initial geometry and material properties on the final fold geometry.

18.7 NUMERICAL MODELS OF BUCKLING AND THE EFFECTS OF DIFFERENT RHEOLOGIES

In this section, we consider numerical models—principally of the buckling of a single competent layer in an incompetent matrix—using both viscous and power-law rheology (see also Section 13.1, Figure 13.3). Because we want to calculate the development of the model through time into the realm of large-scale deflections of the layer, we cannot use the general analytic solution (Equation (18.33)) for the viscous rheology because it only applies to small deflections. Moreover, we cannot even obtain an analytic solution for the power-law rheology. Thus we must use numerical solutions, and therefore we must specify the viscosities for each solution.

i. Linear Viscous Rheology and Folding

We examine first a numerical model for the folding of a linearly viscous layer in a lower viscosity matrix. Figure 18.15 shows the geometry of the model. An initial irregularity of wavelength L and amplitude A_0, chosen to give initial dips between 1° and 2°, is built into the layer of thickness h. Such geometric irregularities are included in the model layer to approximate natural irregularities at a wavelength near the theoretically dominant one (Equation (18.33)). Irregularities are inherent in natural systems, and unless they are introduced into models, the perfection of the model geometry would cause the system to be artificially stable and folds would never form. Because of the symmetry of the geometry, calculations need only be performed for points within the area *ABCD* if L is close to the dominant wavelength. The boundary conditions require that the lines *AB* and *CD* both approach the stationary axis $x = 0$ at a constant velocity and that there be no shear stresses on *BC*. The time in-

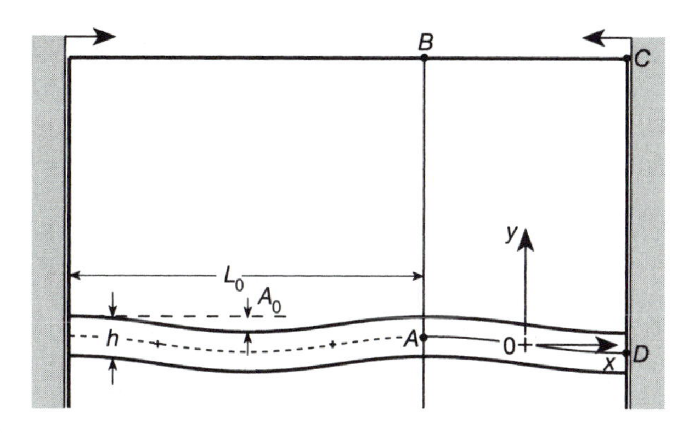

FIGURE 18.15 Model for studying numerically the folding of a viscous layer in a less viscous medium. A layer of thickness h has an initial irregularity of wavelength L_0 and amplitude A_0 built into it. L_0 is close to the predominant wavelength for the system (Equation (18.33)). Calculations need only be done for the section of the model outlined by points *ABCD* because of the geometrical symmetry of the model. The boundary conditions require the planes *AB* and *DC* to approach the origin $x = 0$ at the same constant rate. (From Dieterich and Carter 1969)

crements for each iteration are chosen so that the average extensional strain for each increment parallel to the initial layer is $e_n = -0.02$.

Figure 18.16 shows one model for the development of a fold for which $L/h = 12$ and the viscosity ratio $\eta_L/\eta_M = 42.1$. The model produces a class 1B fold, and the figure shows how the orientations of the axes of maximum compressive stress $\hat{\sigma}_1$ change across the structure and through time. In the hinge zones, the maximum compressive stress is parallel to the layer on the concave sides of folds where layer-parallel shortening occurs, and it is roughly perpendicular to the layer on the convex sides where layer-parallel lengthening occurs. In the limbs, the maximum compressive stress tends to rotate with the limbs as the folding angle increases until limb dips become steep, at which point it returns toward its original orientation and tends to be at high angles to the bedding. The orientations of $\hat{\sigma}_1$ within the layer can be compared with those of Figure 9.17*C*, which were observed experimentally in an elastic bar of gelatin with no surrounding medium.

The magnitudes of the stresses also vary across the fold and throughout the course of the deformation. On the convex side of the hinge zone, $\hat{\sigma}_1$ drops to very small values early in the deformation and remains small throughout. On the concave side of the hinge zone, the compressive stresses are high throughout the folding process. Within the limbs, $\hat{\sigma}_1$ starts at high values but decreases by a factor of 10 by the time the folding angle approaches 180°. This change reflects the fact that the competent layer bears a large proportion of the force applied to the system when the layer is parallel to the shorten-

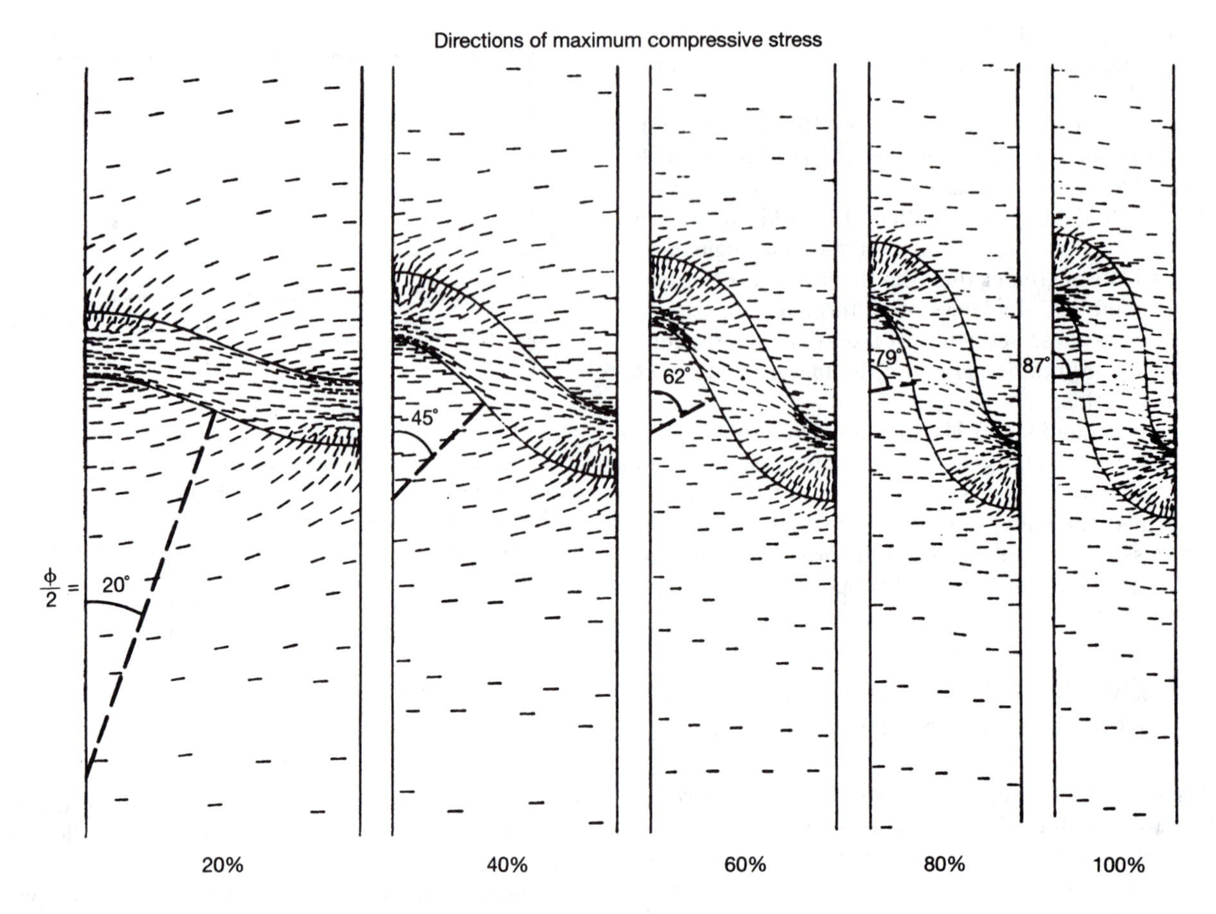

FIGURE 18.16 Numerical modeling of progressive folding of a layer with a high competence compared to the matrix, where the ratio of the layer viscosity to the matrix viscosity is $\eta_L/\eta_M = 42.1$, a high competence contrast. The wavelength to thickness ratio is $L/h = 12$. The short lines indicate the orientation of $\hat{\sigma}_1$, the maximum compressive stress. For each frame, the half folding angle $\phi/2$ is indicated, and the overall shortening is shown as a natural strain at the bottom of each frame. (From Dieterich and Carter 1969)

ing direction, but its strengthening effect decreases as the limbs rotate to higher angles. (We reached this same conclusion in Box 17-2(ii) from studying the quartz deformation lamellae in a folded layer; compare Figure 18.16*A* with Figure 17-2.1*E*).

The orientations of the principal stretches can also be calculated from the numerical models of folding. Figure 18.17 shows the distribution of $\hat{s}_1$ axes throughout the system for the same fold model as shown in Figure 18.16. The maximum principal stretch axes $\hat{s}_1$ form a convergent fan in the competent layer. Within the incompetent matrix, the $\hat{s}_1$ axes form a divergent fan on the convex side of the hinge in the layer (Figure 18.17, areas labeled "O") and are parallel to the axial surface or slightly convergent on the concave side of the folded layer (areas labeled "I").

Figure 18.18 shows the distribution of minimum compressive stress axes, $\hat{\sigma}_3$, for the fold in the last frame of Figure 18.16. It is instructive to compare the orientations of $\hat{\sigma}_3$ axes (Figure 18.18) and of $\hat{s}_1$ axes (Figure 18.17*B*) with each other and with the orientations of foliations from natural folds of a competent material such as sandstone in an incompetent matrix such as shale (Figure 11.13). Comparing Figure 18.17*B* with Figure 18.18 shows that there are significant differences in the orientations of the maximum finite stretch $\hat{s}_1$ and the minimum compressive stress $\hat{\sigma}_3$. In the hinge zone, the pattern is significantly different, and in the limbs, the amount of refraction of $\hat{s}_1$ is much higher than that of $\hat{\sigma}_3$. Overall, the pattern shown by the maximum principal stretch axes $\hat{s}_1$ shows a closer resemblance to the natural foliation pattern than does the pattern of minimum compressive stress axes $\hat{\sigma}_3$. This similarity lends credence to the hypothesis that foliations that are subparallel to the axial surface mark the orientation of the plane of flattening ($\hat{s}_1$, $\hat{s}_2$) of the finite strain ellipsoid. It also suggests that the refraction of foliations across lithologic boundaries reflects differences in the mechanical properties of the adjacent layers, which give rise to the different amounts and orientations of strain across the lithologic contact.

A change in the viscosity contrast has a profound effect on the geometry of folds and the orientation of $\hat{s}_1$. Figure 18.19 shows the fold produced by a model for which $L/h = 9$ and the viscosity ratio $\eta_L/\eta_M = 17.5$. In this model, homogeneous shortening and thickening of

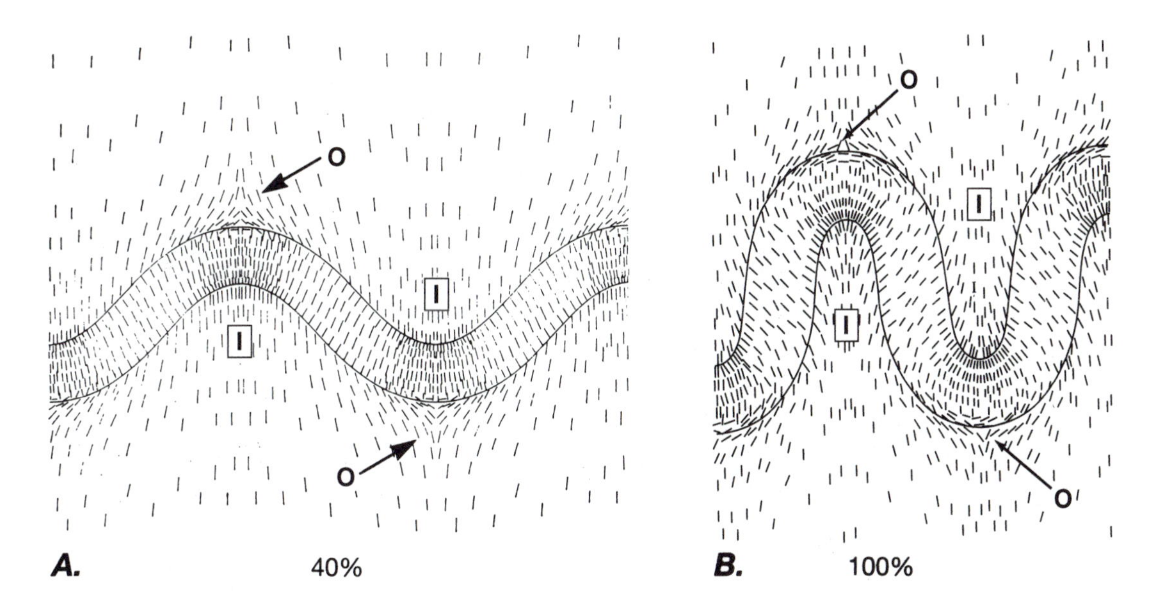

FIGURE 18.17 Orientations of $\hat{s}_1$, the maximum stretch, for the folds in the same model shown in Figure 18.16 for (*A*) 40% overall shortening and (*B*) 100% overall shortening. $\hat{s}_1$ directions are divergent on the convex sides of the folded layer (regions marked "O") and parallel to slightly convergent on the concave sides of the folded layer (regions marked "I"). Within the layer, the $\hat{s}_1$ directions are convergent in the limbs, and in the hinge zone $\hat{s}_1$ may be convergent (*A*) or vary from divergent to convergent across the layer (*B*). (From Dieterich 1969)

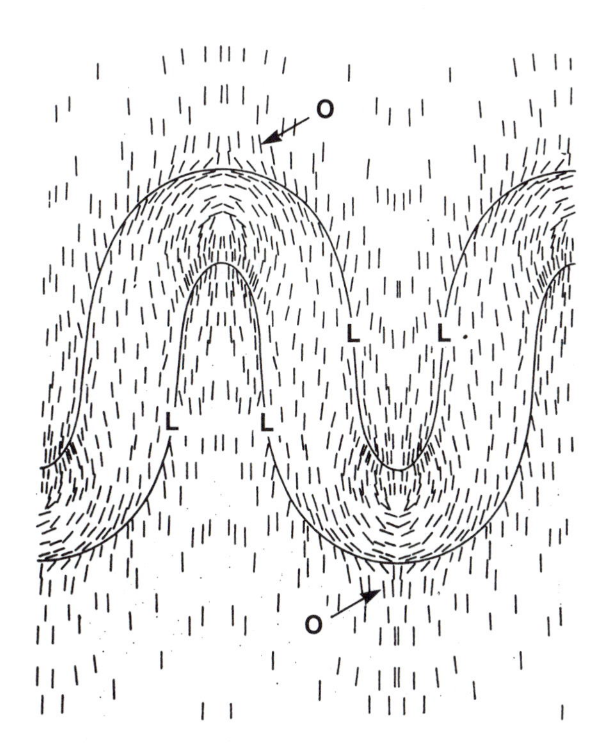

FIGURE 18.18 Orientation of the directions of minimum compressive stress $\hat{\sigma}_3$ in the same fold shown in the last frame of Figure 18.16. The pattern of orientations is similar to that of $\hat{s}_1$ in Figure 18.17*B* except that the divergent pattern is not so pronounced in the matrix on the convex side of the folded layer (regions labeled "O"), and there is little refraction of orientations across the layer boundaries in the limbs (regions marked "L"). (From Dieterich 1969)

the layer occurred along with the folding, especially during the early stages of folding. The result is a fold of class 1C geometry (see Section 10.3(ii) and Figures 10.19, 10.20, and 13.13). The refraction of the orientations of $\hat{s}_1$ at the layer boundary is considerably less than in the first model for folds having similar limb dips (compare Figure 18.17 with Figure 18.19). This model also demonstrates that buckle folding need not produce layer-parallel extensional strains on the convex side of the hinge zone in the layer. Note that in Figure 18.19*B*, the concave sides of the fold develop into a cusp, and the convex side forms a lobe comparable to a fold mullion (Figure 11.22*A*). The cusps point into the material that was more competent at the time of deformation. Because the relative competence for the same two rock types can actually change with changing conditions of deformation (see Figure 18.20 and discussion that follows), the cusps can be a useful criterion for determining relative competence in naturally deformed rocks.

Although the assumption of linear viscosity for the model materials might not correspond well to the rheology of real rocks, the correspondence between the calculated structures and observations in naturally deformed rocks suggests that the assumption may not be bad as a first approximation.

ii. Power-Law Rheology and Folding

The assumption of a power-law rheology for models that have the geometry shown in Figure 18.15 provides an alternative approximation to the rheology of real rocks, at

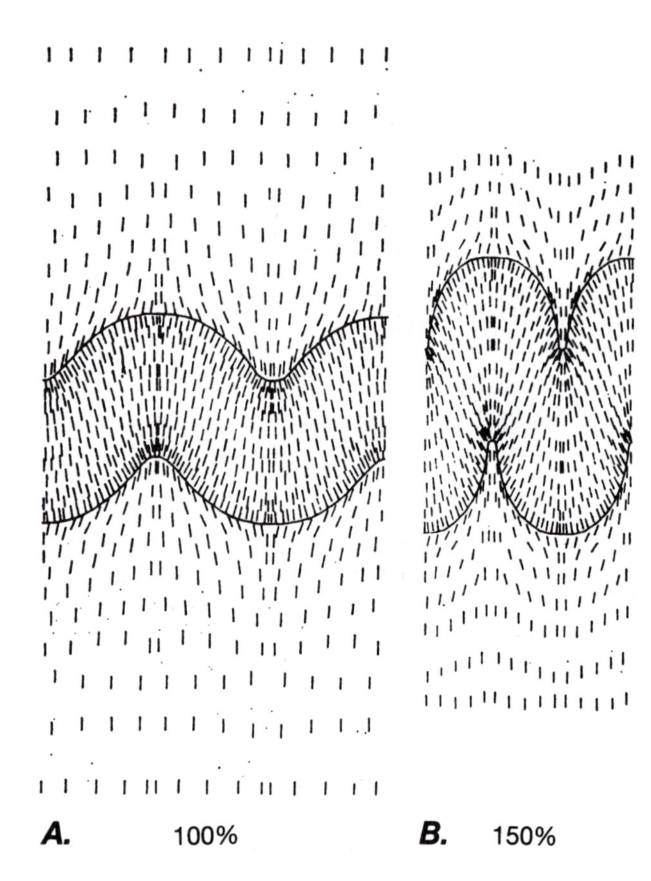

FIGURE 18.19 Numerical modeling of progressive folding of a layer in a matrix, where the ratio of the layer viscosity to the matrix viscosity is $\eta_L/\eta_M = 17.5$, a low competence contrast. The wavelength to thickness ratio is $L/h = 9$. Short lines show the orientations of $\hat{s}_1$. The shortening measured as natural strain is 100% in *A* and 150% in *B*. (From Dieterich 1969)

least for some conditions of ductile deformation. This model assumes experimentally determined power-law rheologies of wet quartzite for the layer and of marble for the matrix, as defined by:

wet quartzite:

$$|\dot{\varepsilon}_n^{(ss)}| = 1.77 \times 10^{-2}\,(\sigma^{(Dif)})^{2.8} \exp\left[\frac{-2.30 \times 10^5}{RT}\right] \tag{18.35}$$

marble:

$$|\dot{\varepsilon}^{(ss)}| = 1.20 \times 10^{-4}\,(\sigma^{(Dif)})^{8.3} \exp\left[\frac{-2.59 \times 10^5}{RT}\right] \tag{18.36}$$

where the stresses are in megapascals, the activation energies are in joules/mole, and the strain rates are in s^{-1}. (Note that the constants used here are different from those listed in Table 18.1. This discrepancy reflects the differences that occur in various experimental evaluations of these constants.)

The materials that obey a power-law rheology are not characterized by a single viscosity, because the ratio of stress to strain rate (Equation (18.7)) varies with the magnitude of the stress. Nevertheless it is convenient to express the strain rate under any particular set of conditions in terms of an **effective viscosity**, which is defined as the ratio of the stress to the strain rate. In terms of the differential stress and the axial strain rate, the effective viscosity $\eta^{(Eff)}$ is

$$\eta^{(Eff)} = \frac{\sigma^{(Dif)}}{3|\dot{\varepsilon}_n|} \tag{18.37}$$

This is a more precise definition than the approximation used in Equation (18.27).[8] The effective viscosity is de-

[8]This relationship is derived from the constitutive equation for linearly viscous materials, Equations (18-1.14), by assuming an axial symmetry to the deformation. For the convenience of having the same subscripts for the components of the principal stresses and the parallel components of the instantaneous strain rates, in this footnote we refer both quantities to the same principal coordinate system $(\hat{x}_1, \hat{x}_2, \hat{x}_3)$. The maximum compressive stress and the maximum instantaneous shortening rate are taken to be parallel to the $\hat{x}_3$ axis and are designated $\hat{\sigma}_{33}$ and $\hat{\dot{\varepsilon}}_{33}$, respectively. The axial symmetry of the deformation then requires $\hat{\sigma}_{11} = \hat{\sigma}_{22}$ and $\hat{\dot{\varepsilon}}_{11} = \hat{\dot{\varepsilon}}_{22}$. We also assume a constant-volume deformation as defined by the first Equation (18-1.7) and the first Equation (18-1.8), which, with the axial shortening geometry, gives

$$\hat{\dot{\varepsilon}}_{11} = -\frac{1}{2}\,\hat{\dot{\varepsilon}}_{33} \tag{i}$$

We form the second scalar invariant for the deviatoric stress I_2 (Equations (16-4.9)) because this quantity can be directly related to the differential stress, and we use the constitutive Equation (18-1.14), to find

$$I_2 \equiv \sum_{k=1}^{3}\sum_{\ell=1}^{3} (\sigma_{k\ell}^{(Dev)})(\sigma_{k\ell}^{(Dev)}) = \sum_{k=1}^{3}\sum_{\ell=1}^{3} (2\eta\dot{\varepsilon}_{k\ell})(2\eta\dot{\varepsilon}_{k\ell}) \tag{ii}$$

We use the constant-volume deformation condition (18-1.7) and (18-1.8) to subtract $\dot{\varepsilon}_V/3 = 0$ from both $\dot{\varepsilon}_{kl}$ in Equation (ii). We can, for convenience, express this relation in principal coordinates so all off-diagonal terms are zero. We then use the definition of the deviatoric stress (Equations (16-4.6) and (16-4.7); see also Equation (7.37)) and Equation (18-1.5) and expand the squared terms. After some algebraic manipulation the result can be written as

$$[(\hat{\sigma}_{11} - \hat{\sigma}_{22})^2 + (\hat{\sigma}_{22} - \hat{\sigma}_{33})^2 + (\hat{\sigma}_{33} - \hat{\sigma}_{11})^2] = 4\eta^2[(\hat{\dot{\varepsilon}}_{11} - \hat{\dot{\varepsilon}}_{22})^2 + (\hat{\dot{\varepsilon}}_{22} - \hat{\dot{\varepsilon}}_{33})^2 + (\hat{\dot{\varepsilon}}_{33} - \hat{\dot{\varepsilon}}_{11})^2] \tag{iii}$$

Applying the constraints for an axial constant-volume deformation as defined in Equation (i), this equation reduces to

$$2(\hat{\sigma}_{11} - \hat{\sigma}_{33})^2 = 4\eta^2\left[2\left(\frac{3}{2}\,\hat{\dot{\varepsilon}}_{33}\right)^2\right] \tag{iv}$$

$$\sigma^{(Dif)} = \eta 3\dot{\varepsilon}_n \tag{v}$$

To get the last equation from the previous one, we simplified the numerical factors, took the square root of both sides of the equation, used the definition of the differential stress Equation (7.38), and set $\hat{\dot{\varepsilon}}_{33} = \dot{\varepsilon}_n$ for the axial deformation geometry. Equation (18.37) follows from this last equation.

pendent on the stress and the temperature, as can be seen by substituting for the strain rate in Equation (18.37) from either Equation (18.35) or Equation (18.36). For a power-law material at a constant stress and temperature the effective viscosity equals that of a linearly viscous material flowing at the same rate under the same stress. It is a mechanically exact measure of the competence of the different rocks under consideration, but in this case it is important to realize that the competence changes with stress as well as with temperature.

In Figure 18.20 we plot the effective viscosity ratio $\eta_{qtz}^{(Eff)}/\eta_{mbl}^{(Eff)}$ for wet quartzite and marble, as a function of temperature, for a strain rate of 10^{-14} s^{-1}. At temperatures below about 550°C, $(\eta_{qtz}^{(Eff)}/\eta_{mbl}^{(Eff)}) \geq 1$, and the quartzite has the higher effective viscosity (greater competence). Because marble has a higher activation energy, however, the marble becomes the more competent material above that temperature. This behavior has interesting implications for the possible development of structures in metamorphic rocks, as we discuss below in association with Figure 18.22.

Figure 18.21 shows a plot of the $\hat{\sigma}_3$ axes, the $\hat{s}_1$ axes, and the equivalent viscosities for one numerical experiment at a temperature of 375° C and an imposed strain rate of 10^{-14} s^{-1}, which gives an apparent viscosity ratio from Figure 18.20 of $(\eta_{qtz}^{(Eff)}/\eta_{mbl}^{(Eff)}) = 10$. The initial dip of the limbs is 10°, and the time increments were chosen to give increments of natural strain (Equation (12-1.1)) of $\bar{\varepsilon}_n = 0.05$.

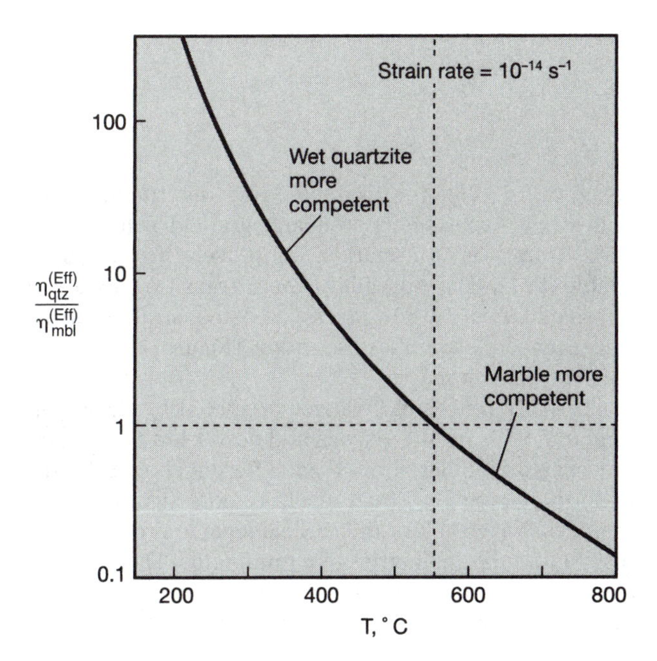

FIGURE 18.20 A plot of the ratio of effective viscosity of wet quartzite to marble $(\eta_{qtz}^{(Eff)}/\eta_{mbl}^{(Eff)})$ as a function of temperature at a strain rate of 10^{-14} s^{-1}. The relative competence reverses at a temperature of about 550°C (dashed lines). (From Parrish et al. 1976)

The geometry of the fold formed is a class 1B fold. The $\hat{s}_1$ axes form a divergent fan in the marble on the convex side of the hinge zone and a convergent fan within the quartzite layer. The difference between the $\hat{\sigma}_3$ and the $\hat{s}_1$ orientations is similar to the case of viscous folding (compare Figure 18.21*B* with Figures 18.17*B* and 18.18). The effective viscosity increases significantly as the stress decreases, so it does not remain constant around the structure or through time (Figure 18.21). The strain rate is highest in the hinge zone, and because of the power-law rheology, the effective viscosity there is the lowest. The lower effective viscosity in turn tends to concentrate the strain rate even more into the hinge zone. The strain rate in the limbs is relatively low and the effective viscosity correspondingly high. Thus the power-law rheology tends to produce folds with sharper hinges and more planar limbs than the linearly viscous rheology. Compare the folds in Figure 18.21 with those of Figures 18.16 through 18.18, also of class 1B geometry, which formed in linearly viscous materials. For the power-law rheologies, the angular variation in the fans of $\hat{s}_1$ is not as large as for the linear rheology—a difference particularly apparent at high strains. The difference in the geometry is subtle, however, even though the effective viscosity ratio for the power-law materials is one-fourth the viscosity ratio for the viscous materials. Given the uncertainties in the rheologies of natural rocks, it would be difficult to infer a reliable rheology for a natural fold from the geometry of the fold and the orientation of $\hat{s}_1$ axes.

Because of the temperature dependence of the rheologies and of the effective viscosity ratio, an increase in the temperature of metamorphism could change the mechanical behavior of the system, thereby affecting the geometry of the folds that develop. As an example, Figure 18.22 shows the model of the quartz layer in the marble matrix subjected to a 20% shortening at 375°C (top diagram), followed by continued shortening up to 200% at 550°C. At the higher temperature, the effective viscosity ratio of quartzite to marble is 1, indicating that the mechanical difference between these two materials is very small. Thus this example is a well-defined mechanical model equivalent to the kinematic model discussed in Section 13.4(i) (Figure 13.13), for which initial gentle folding was followed by homogeneous flattening normal to the axial surface. The initial class 1B fold (top diagram, Figure 18.22) is flattened into class 1C geometry which, beyond about 120% shortening, closely resembles a class 2 geometry.

Figure 18.23 shows the effect of different temperatures on the same model. In this case all folds initially formed by shortening to 40% at 375°C$(\eta_{qtz}^{(Eff)}/\eta_{mbl}^{(Eff)} = 10)$ and then were shortened to a total of 100% at different temperatures and thus different effective viscosity ratios. With increasing temperature from 450° C to 550° C, the geometry of the folds is class 1C, with characteristics ranging from nearly class 1B to nearly class 2. In this temperature range the fold geometry reflects the behavior of the more competent quartzite layer. Above 550° C, how-

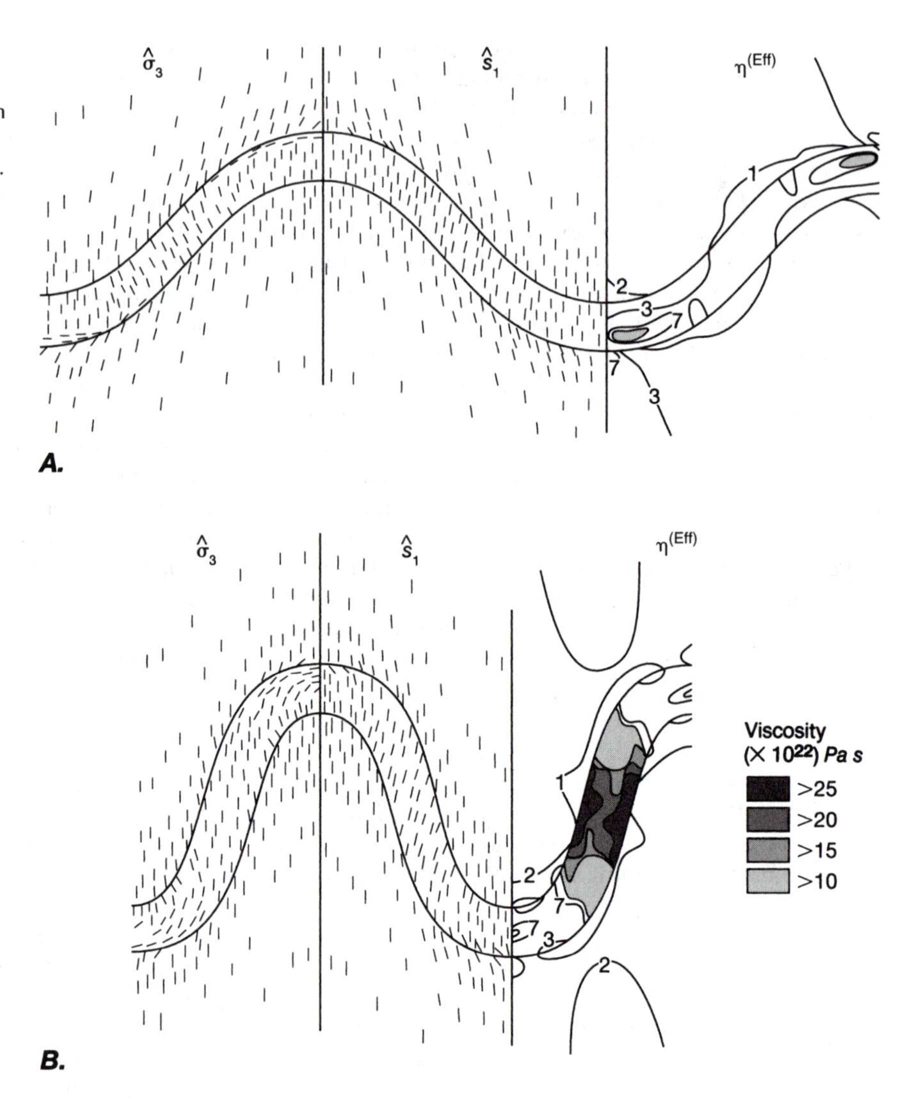

FIGURE 18.21 Numerically calculated folds formed under the power-law rheologies of a layer of wet quartzite in a matrix of marble at 375° C and shortened at an initial rate of 10^{-14} s^{-1}. In the left panel, short lines are parallel to $\hat{\sigma}_3$; in the middle panel, they are parallel to $\hat{s}_1$. The right panel is a contour plot of the magnitude of the effective viscosity $\eta_{qtz}^{(Eff)}$. Shortening strains are 40% in *A* and 80% in *B*. (From Parrish et al. 1976)

ever, the quartzite layer becomes less competent than the marble and ceases to control the geometry of the folding. In these cases, we see the quartzite layer develop class 3 folds, with the curvature at the hinges dominated by the behavior of the marble. This fold geometry resembles that of the incompetent layer in Figures 10.22*B*, *C*; 13.20; and 13.22.

These numerical fold models give a sound mechanical basis for the relationship between rheology and fold geometry (Section 13.5). The degree to which they represent natural conditions, however, is uncertain. Linear viscosity may be an unrealistic rheology for real rocks, although at low stresses it might be reasonable for solution folding. This mechanism, however, is not being modeled in these examples, because these models assume no volume loss. On the other hand, power-law rheologies derived experimentally at high strain rates may not be appropriate for modeling of real rocks at geologically realistic strain rates, which are many orders of magnitude smaller. The rheology adopted for marble in Equation (18.21), for example, was determined from high-stress experiments on coarse-grained Yule marble. Data from Carrara marble, moreover, are different (Table 18.1). The possibility for a transition to lower stress-exponents at low stress, as demonstrated for the fine-grained Solenhofen limestone (Figure 16.17), also was not considered.

The differences in behavior predicted by these particular models for the viscous and power-law rheologies, however, are not large, although different choices for the material parameters could result in more significant differences. Nevertheless, the models seem to provide some insight into the geometries of natural folds. These results also indicate that assuming a linear viscous rheology is probably a reasonable first-order approximation.

iii. Anisotropy

The models described so far generally produce subrounded to rounded folds (Figure 10.17), but they do not

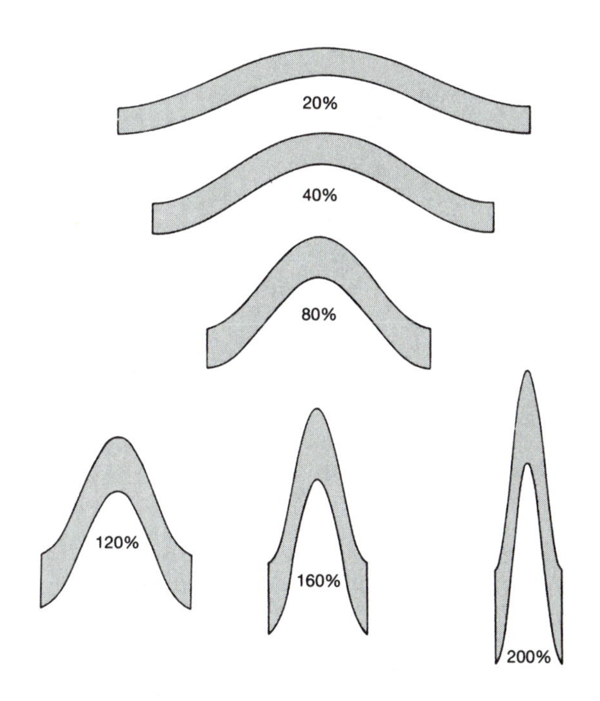

FIGURE 18.22 Fold geometries that result from power-law rheologies for a quartzite layer in a marble matrix shortened to 20% natural strain at 375° C ($\eta_{qtz}^{(Eff)}/\eta_{mbl}^{(Eff)} = 10$), and then shortened up to 200% at 550° C ($\eta_{qtz}^{(Eff)}/\eta_{mbl}^{(Eff)} = 1$). Numbers indicate the percentage of shortening.
(From Parrish et al. 1976)

seem to account for chevron folds with sharp hinges and planar limbs. Such folds are characteristic of rock sequences from a wide variety of conditions, ranging from unconsolidated sediments to high-grade metamorphic rocks. The rocks in which chevron folds develop, however, all have in common a strong planar anisotropy, which may control the geometry of developing folds. Certainly the geometry of chevron folds is inconsistent with a linearly viscous rheology, so their existence argues strongly that another rheology must be responsible.

Figure 18.24 shows three models comparing the effects of power-law rheology with those of anisotropy. The unfolded layer in each case started out with a rectangular grid of material lines, with the three columns of rectangles in each of the hinge zones being half the width of the rectangles in the limb to give higher resolution in areas of higher strain and strain rate. The subsequent orientation and spacing of the lines is therefore an index of the internal deformation of the layer. In all these models, the viscosity or the effective viscosity of the layer is 100 times larger than that of the matrix.

Figure 18.24*A* shows a layer with a power-law rheology having a stress exponent of 10. Such a rheology provides a **strain-rate softening** mechanism, for which the effective viscosity decreases with increasing strain rate (compare Figure 18.21). Deformation is concentrated in the high-strain-rate, low-effective-viscosity hinge zones

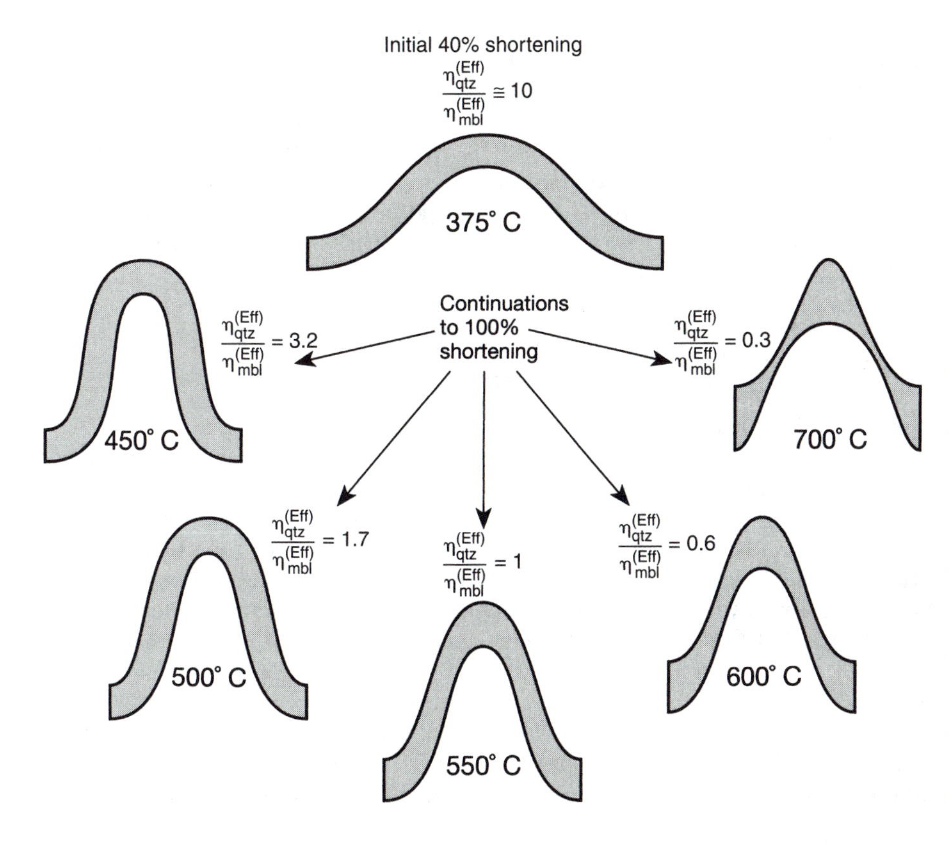

FIGURE 18.23 Power-law rheology folding for a quartzite layer in a marble matrix. Folds were deformed to 40% overall shortening at 375° C ($\eta_{qtz}^{(Eff)}/\eta_{mbl}^{(Eff)} = 10$) and shortened further to 100% average shortening (measured as natural strain) at the temperature and effective viscosity ratio indicated for each fold. Note the change in fold style from class 1 in the folds for which $\eta_{qtz}^{(Eff)}/\eta_{mbl}^{(Eff)} \gtrsim 1$, to class 3 for which the quartzite layer becomes less competent than the marble matrix ($\eta_{qtz}^{(Eff)}/\eta_{mbl}^{(Eff)} < 1$). (From Parrish et al. 1976)

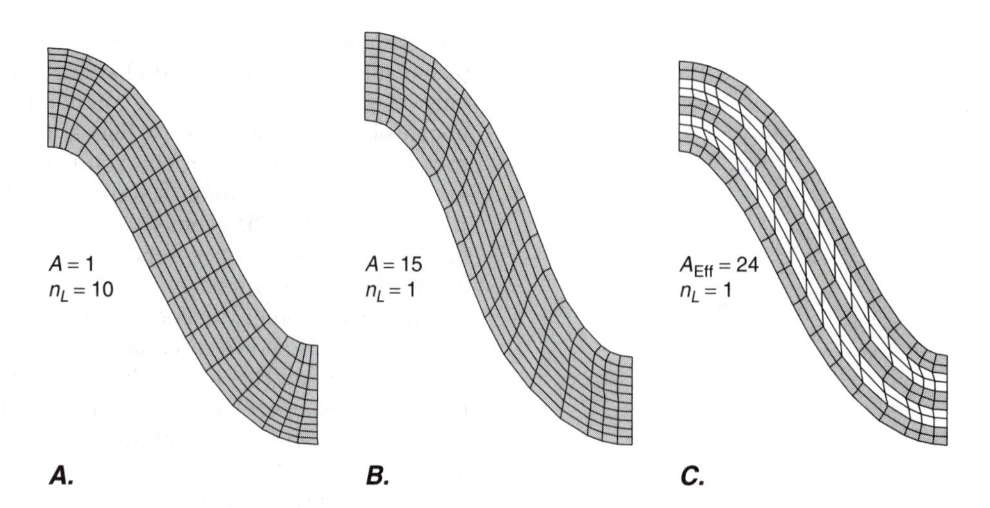

FIGURE 18.24 Buckle folds characterized by straight limbs and hinge zones that vary approximately from subrounded to subangular from the convex to the concave sides of the layer. In all cases, the initially rectangular cells in the three columns in the hinge zone were initially half the width of the cells in the limbs; $L_O/h = 12$, $A_O/h = 0.1$, $\eta_L/\eta_M = 100$, and shortening 40% (variables are defined in Figures 18.12 and 18.15). The matrix is not shown. (After Lan and Hudleston 1996) *A*. Power-law exponent $n_L = 10$; isotropic medium. *B*. Power-law exponent $n_L = 1$; anisotropic medium with the ratio of viscosity for normal stress to the viscosity for the shear stress $A = 15$. *C*. Multilayer composed of three competent layers (gray) separated by two incompetent layers (white). $\eta_L/\eta_M = 100$ for the competent layers compared with both the matrix and the incompetent layers; rheology is linear viscous ($n_L = 1$) and isotropic for the individual layers and matrix. The effective anisotropy for the multilayer viscosities is approximately $A = 24$.

and is essentially zero in the limbs. The convex side of the hinge zones undergoes layer-parallel lengthening, and the concave side undergoes layer-parallel shortening. Overall, there is a modest thickening of the layer in the hinge zone relative to the limb.

Figure 18.24*B* shows a fold with anisotropic material properties, with the viscosity for the normal stress 15 times the viscosity for the shear stress, as defined by the ratio $A = 15$. The model in Figure 18.24*C* imparts a mechanical anisotropy by using alternately high- and low-viscosity layers (gray and white, respectively) that individually are linearly viscous and isotropic but that differ in viscosity from each other by a factor of 100, with the low-viscosity layers being the same viscosity as the matrix (compare the elastic model in Figure 18.14*C*). The overall effective anisotropy is characterized by the ratio $A = 24$, that is, the effective viscosity for the normal stress is 24 times the effective viscosity for the shear stress. In both these cases of anisotropic viscosity, there is little deformation in the hinge areas and significant layer-parallel shearing in the limbs that reaches a maximum at the inflection point. For the interlayered medium (Figure 18.24*C*), the shear is concentrated in the lower-viscosity layers, but the average across the entire layer is similar to the anisotropic continuum (Figure 18.24*B*). Thickening of the hinge zone is small.

Even these models, however, do not produce the true chevron fold style, suggesting that other mechanisms may be involved. Mechanisms that could lead to even more concentration of deformation in the hinge zone include **strain softening**, in which the effective viscosity of the material decreases with increasing strain. Both dynamic recrystallization and the formation of a crystallographic preferred orientation could cause the stress required for steady-state creep to decrease with increasing strain. In both cases, because the strain and the strain rate are largest in the hinge area, the material there would tend to become progressively less resistant to deformation than elsewhere, and this in turn would cause an even greater concentration of strain and an increase in strain rate. The result would be a fold in which almost all the bending is concentrated in a small sharp hinge area, leaving the limbs almost planar.

iv. Comparison of Kinematic Models with Numerical Models of Folding

In the numerical models we have discussed so far, illustrated in Figures 18.17, 18.19, 18.21, and 18.24, the kinematic behavior of the layer during folding is determined by the constitutive equations, the balance laws, the boundary conditions, the material properties, and the temperature, which strongly affects the material properties. The conditions assumed for some of these models make them comparable to some of the kinematic models of folds illustrated in Figures 13.6 and 13.7, for which we imposed a preconceived kinematic behavior. We want now to evaluate how accurate and useful the kinematic models in fact are. We first determine which of the two types of models could be compared and then evaluate the limitations of the kinematic models as indicated by the mechanically more complete numerical models.

Volume is conserved in the numerical models, so the kinematic models for volume-loss folding (Figures 13.6*D* and 13.7*D*) are not pertinent to the comparison. The boundary conditions imposed on the numerical models lead to the buckling mode of folding, and thus the folds in Figure 13.6 are not comparable, as they are assumed to be formed in the bending mode. Thus a comparison of the numerical fold models is relevant only with the kinematic fold models in Figure 13.7*B*, *C*. These kinematic fold models in turn compare most closely in folding angle ϕ with the numerical models in Figures 18.17*B*, 18.19*B*, and 18.21*B* and with all three models in Figure 18.24. Thus we can make at least a qualitative comparison of the fold form and the distribution of $\hat{s}_1$ orientations among these sets of models. In making the comparison we must take account of the higher resolution for the strain distribution provided by the numerical models than is shown for the kinematic models.

The kinematic models (Figure 13.7*B*, *C*) are both class 1B folds. Of the models assuming a linear viscous rheology, the folds in Figure 18.17 are class 1B, but those in Figure 18.19 are class 1C. The class 1C geometry is attributable to the low viscosity ratio between the layer and matrix of 17.5 (low competence contrast), which leads to a large amount of layer shortening and thickening in the early stages of folding and to nearly homogeneous flattening during later stages of folding. For the linear anisotropic rheologies (Figure 18.24*B*, *C*), very little deformation occurs in the hinge zone, and although there is some slight hinge-zone thickening, the folds are very close to class 1B. For the power-law rheologies, the fold form is very close to class 1B, although some thickening does occur in the hinge zone, especially for the model in Figure 18.24*A*, where the strain rate is highest and the effective viscosity is the lowest and where deformation therefore tends to be concentrated (compare Figure 18.21*B*, right panel).

The strain distribution in the numerical model based on linear viscous rheology (Figure 18.17*B*) has aspects similar to buckling by both orthogonal flexure (Figure 13.7*B*) and flexural shear (Figure 13.7*C*). In particular the layer-parallel orientation of $\hat{s}_1$ on the convex side of the hinge zone (Figure 18.17*B*) is similar to that resulting from orthogonal flexure (Figure 13.7*B*) but quite distinct from the orientations for flexural shear (Figure 13.7*C*). In the limbs of the numerical model (Figure 18.17*B*), however, the angle between $\hat{s}_1$ and the bedding is not as high as in orthogonal flexure (Figure 13.7*B*), but instead is more akin to the angle in the limbs of the flexural shear folds (Figure 13.7*C*). Thus the kinematic models do not represent accurately the behavior of these linear viscous models.

The strain distribution in the numerical models based on the power-law rheologies depends on the particular rheology and the effective viscosity ratio between the layer and the matrix. For the model in Figure 18.21*B*, the stress exponent is 2.8 (Equation (18.35)), and the effective viscosity of the layer is only 10 times that of the matrix (Figure 18.20 for 375° C). The orientations of the $\hat{s}_1$ principal axis are qualitatively similar to the kinematic model for buckling by flexural shear (Figure 13.7*C*), especially in the limbs. On the convex side of the hinge, the numerical model shows a small effect of layer-parallel lengthening, which is consistent with the orientation of the minimum principal stress $\hat{\sigma}_3$ in that part of the fold. Nevertheless, the $\hat{s}_1$ axes are nearly parallel to the axial surface throughout most of the hinge zone, and in the limbs the inclination of $\hat{s}_1$ to the layer boundaries suggests a component of layer-parallel shearing.

In contrast to the model in Figure 18.21*B*, the model in Figure 18.24*A* is characterized by a stress exponent of 10 and an effective viscosity of the layer equal to 100 times that of the matrix. These differences result in a fold that much more nearly reflects the characteristics of buckling by orthogonal flexure (Figure 13.7*B*). The hinge zone is characterized by layer-parallel shortening on the concave side and layer-parallel lengthening on the convex side. The limbs have experienced very little deformation, so that the lines initially orthogonal to the layer surface remain in that orientation after folding, comparable to the fold in Figure 13.7*B*. In the kinematic model, however, we assumed a certain amount of initial layer-parallel shortening throughout the fold, which is not apparent in the power-law rheology model of Figure 18.24*A*. Thus we find very different kinematic behaviors for power-law rheology depending on the stress exponent and the effective viscosity ratio between the layer and the surrounding medium.

The two folding models based on an anisotropic rheology (Figure 18.24*B*, *C*) are generally consistent with buckling by flexural shear (Figure 13.7*C*). Very little deformation occurs in the hinge zone, and in particular there is no layer-parallel lengthening on the convex side of the zone; the limbs are dominated by layer-parallel shear. For the multilayered structure, the shearing is concentrated in the less competent layers, but the average across the multilayer is similar to, although somewhat larger than, the anisotropic continuum (Figure 18.24*B*).

Based on these results of numerical modeling, we can conclude that folding is probably not well represented by either pure orthogonal flexure or pure flexural shear. This result should come as no great surprise, because our kinematic models are simple postulated end-members of possible kinematic behaviors, and natural processes generally do not conform to our ideals of such simple behavior.

The models we have examined are only a small sampling of the rheologies, effective viscosities, and competence contrasts that are possible, but even these have produced a variety of fold geometries and distributions of strain in the folded layer. A few general conclusions seem warranted: Folds that have a class 1B geometry are indicative of a high competence contrast; folds with a lobe-and-cusp class 1C geometry such as in Figure 18.19*B* are indicative of a small but significant competence contrast between layer and matrix; folds with sharp hinges and planar limbs are incompatible with linear viscosity and indicate a mechanism involving strain softening or strain-rate softening (like a power-law rheology) or both. At a more detailed level of analysis, however, subtle differences in the fold geometry can reflect major differences in the rheology, and fold geometry and strain distribution may not provide unique constraints on the rheological properties. Taking into account the irregularities inherent in natural systems, it is not clear that fold geometry and strain distribution can provide more than a general constraint on the rheology, although additional independent information such as the variation of fabric or texture around a fold may help constrain the folding mechanism further.

v. Other Constitutive Models for Folding

In the models we investigate in the foregoing subsections of this section (18.7), the assumption is implicit that a quarter wavelength is sufficient to model the folding, where that wavelength is chosen to be the one that is amplified most rapidly during buckling, according to Equation (18.33). This assumption implies that the fold train is an exactly periodic series of identical folds and that the fold hinges do not migrate along the layer. Neither of these assumptions is necessarily correct. Moreover, these models are based on the assumption of either a linear viscous or a power-law rheology that is constant through time, with the exception of the stepwise change in rheology illustrated in Figures 18.22 and 18.23. But fold trains observed in nature commonly are not exactly regular and periodic, and that is the motivation for looking at more complex models to see whether that characteristic simply reflects initial irregularities in the natural system or if it could be an important indication of the material properties at the time of folding. It is not difficult to devise rheological models that are more complex (see Figures 16.5 through 16.8), and some of these may provide a better model for rock rheology. It is possible that the rheology also could evolve through time, as in strain-softening behavior.

Natural material behavior might be more realistically represented, for example, by a rheology that couples elastic behavior either with viscous behavior, as exemplified by a Maxwell visco-elastic material (Figure 16.5), or with power-law behavior (similar to the elastic-plastic model in Figure 16.7). We look particularly at the Maxwell visco-elastic rheology. For a Maxwell material, the total strain rate $\dot{\varepsilon}_T$ equals the elastic strain rate $\dot{\varepsilon}_\mu$ plus the viscous strain rate $\dot{\varepsilon}_\eta$:

$$\dot{\varepsilon}_T = \dot{\varepsilon}_\mu + \dot{\varepsilon}_\eta = \frac{1}{2\mu}\dot{\sigma}_s + \frac{1}{2\eta}\sigma_s \qquad (18.38)$$

The second Equation (18.38) comes from the time derivative of the second Equation (16.1), assuming small strains, and from the second Equation (16.2).

The extent to which a material appears to behave as either a solid or a fluid depends in part on the time scale of the observation. For example, rocks that appear solid may actually be observed to flow if the time scale of observation is sufficiently long, like millions to tens of millions of years. Whether a material appears to behave as a fluid or a solid is reflected by the value of a nondimensional ratio called the **Deborah number**, *De*:[9]

$$De \equiv \frac{\textit{intrinsic relaxation time of the material}}{\textit{time scale of observation}}$$

$$\equiv \frac{t_r}{t_p} \qquad (18.39)$$

If the relaxation time for a material t_r is short compared to the time t_p over which the deformation is observed, then *De* is small, and the material appears to behave like a fluid. On the other hand, if the relaxation time for a material is very long compared to the time scale of observation, then *De* is large, and the material appears to behave like a solid.

For a Maxwell visco-elastic material, the intrinsic relaxation time for the material is the time it takes for the stress that results from a constant, instantaneously applied total strain, to relax to $1/e$ of its initial value, where e is the base of natural logarithms ($e = 2.71828$). That time is given by the ratio $t_r = \eta/\mu$, where η is the coefficient of viscosity of the viscous element (Figure 16.5*A* (bottom panel), 16.5*B*) and μ is the elastic shear modulus for the elastic element (Figure 16.5*A* (top panel), 16.5*B*).[10] For the time scale of observation, we choose

[9]Named after the prophetess Deborah of the Torah and the Old Testament who, after the Israelites defeated the Canaanites in battle, sang that "The mountains flowed before the Lord" (Judges 5:5). Her words have been interpreted to imply that the mountains flowed on the time scale of the Lord's observation (essentially infinite), but not on the scale of mortal observation (see Reiner, 1964). This inferred reference to the relation between flow and the time scale of observation is apparently the origin for the name of this ratio. Other translations of the biblical text use verbs such as "melted" (the King James version) and "quaked" rather than "flowed," which we simply mention without further comment, the correct translation and interpretation of biblical text being far beyond our area of expertise.

[10]If a total strain is imposed instantaneously and then held constant, $\dot{\varepsilon}_T = 0$. To find the stress as a function of time, use this condition with the second Equation (18.38), rearrange, and integrate to find

$$\int_{\sigma_0}^{\sigma_t} \frac{d\sigma_s}{\sigma_s} = \int_0^t -\frac{\mu}{\eta}\,dt$$

where σ_0 is the stress associated with the instantaneously imposed strain. Integrating and then expressing the resulting natural logarithm in terms of an exponential we find

$$\frac{\sigma_t}{\sigma_0} = \exp\left[-\left(\frac{\mu t}{\eta}\right)\right]$$

Thus the time it takes for σ_t to decay to $(1/e)$ times the initial stress σ_0 is defined if the right side of the equation equals $1/e$, for which

$$\frac{\mu t}{\eta} = 1 \quad \text{or} \quad t = \frac{\eta}{\mu}$$

the characteristic time for the fold amplitude to grow to e times the initial deflection during the process of buckling. From Equation (18.32), that time for viscous buckling is equal to the fold growth coefficient $t = t_p = a$, where a is a function of strain rate, layer competence, and the competence contrast between layer and matrix. Thus for visco-elastic folding, the appropriate definition of the Deborah number is

$$De = \frac{\eta}{a\mu} \tag{18.40}$$

Numerical experiments have shown that if the Deborah number is less than one, that is, the time scale for viscous buckling a is large compared to the viscous relaxation time of the material, then the material is dominated by homogeneous shortening. No buckling instabilities develop, and the geometry of the fold train that develops is determined by the passive amplification of the initial perturbations and imperfections in the layer. On the other hand, if the Deborah number is greater than one, then the time scale for the viscous relaxation of stress is large compared to the time required to amplify the folds formed by viscous buckling. The fold train that results will be dominated by buckling instabilities and by the selective growth of a preferred fold wavelength as defined by Equation (18.33). Large Deborah numbers are the result of lower values of the elastic modulus μ, higher values of the viscosity coefficient η, higher strain rates, and higher competence contrast.

The effect of increasing competence contrast $R = \eta_L/\eta_M$ resulting in higher Deborah numbers is illustrated in Figure 18.25. The lowest competence contrast $R = 20$ shows the passive amplification of initial perturbations with a superimposed weak selection of a dominant wavelength, giving a rather irregular fold train. The highest competence contrast of $R = 200$ shows the strong predominance of a selected wavelength with evidence of the initial perturbation essentially eliminated.

More sophisticated analyses than we can present here have shown that for different conditions, buckling instability in such a system can lead to the formation of a single-fold wavelength, as given by Equation (18.33), or to two dominant wavelengths that are amplified independently and are the result of the coupling between the elastic and viscous characteristics of the material. These result in a fold train that appears like a first-order set of folds with a second-order set superimposed on the first (compare Figure 10.24).

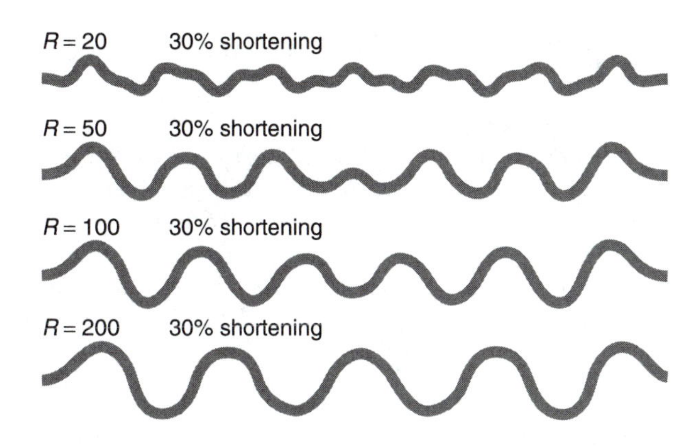

FIGURE 18.25 The effect of competence contrast R on the selection of the predominant wavelength in a visco-elastic material at a strain rate of $\dot{\varepsilon}_n = 10^{-14}$ s^{-1} at a bulk shortening of 30%. R is assumed to have the same value for the ratios of viscosity, Young's modulus, and shear modulus (η_L/η_M, E_L/E_M, and μ_L/μ_M) of the layer relative to the matrix. For the model with $R = 20$, the material constants for the layer are $\eta_L = 2 \times 10^{23}$ Pa s, $E_L = 3.5 \times 10^{10}$ Pa, $\mu_L = 1.4 \times 10^{10}$ Pa; for the matrix, these constants are $\eta_M = 1 \times 10^{22}$ Pa s, $E_M = 1.75 \times 10^9$ Pa, and $\mu_M = 0.7 \times 10^9$ Pa. For the other models, the matrix constants are decreased to give the appropriate value of R. (From Hobbs et al. 2000)

The models we review in this section do not include the possibility of the evolution of the rheology with time, which could result from the changes in the texture and the fabric of the rock with progressive deformation. Such nonlinear effects can lead to strain-weakening behavior and the development of localized packets of folds rather than periodic fold trains of indefinite length.

BOX 18-1 Formulation of a Mathematical Model with Application to the Problem of Viscous Deformation

In making a mathematical model, we wish to calculate the distribution of stress, strain, strain rate, velocity, and displacement throughout a body by specifying only the mechanical properties of the body and the stresses or the displacements (or velocities) that are imposed on its boundaries. To solve any set of equations, we need as many equations as there are unknowns. In Section 18.1, we indicated that we need three different sets of equations to solve a problem in mechanics: conservation laws (or balance equations), constitutive equations, and boundary conditions. We use the example of a linearly viscous material to show here that these equations are required to give a complete set of equations that have as many equations as there are unknowns. We need to use the generality of tensor notation in this discussion, which we present in Section 7.5(i).

(continued)

BOX 18-1 Formulation of a Mathematical Model with Application to the Problem of Viscous Deformation *(cont*

i. Conservation Laws

The motions and the deformation of a body are subject to the fundamental conservation laws of physics, which must apply to all materials in all situations, no matter what the problem. As discussed in Section 18.1, these conservation laws include the conservation of mass, the conservation of momentum, and the conservation of moment of momentum (angular momentum).

Conservation of Mass. The conservation of mass requires that the rate of change of mass of a body be zero. Thus the only way the mass density (mass per unit volume) can change with time is if the volume containing the mass changes size. This is expressed at a point in a continuum by

$$\frac{\partial \rho}{\partial t} + \rho \sum_{k=1}^{3} \frac{\partial v_k}{\partial x_k} = 0 \qquad (18\text{-}1.1)$$

or in expanded form,

$$\frac{\partial \rho}{\partial t} + \rho\left(\frac{\partial v_1}{\partial x_1} + \frac{\partial v_2}{\partial x_2} + \frac{\partial v_3}{\partial x_3}\right) = 0 \qquad (18\text{-}1.2)$$

where the first term is the rate of change of density. The summation in the second term is called the divergence of the velocity, and it equals the volumetric strain rate. Thus the second term expresses the rate at which the density can change due to volumetric deformation.

To show that the summation term in Equation (18-1.1) is the volumetric strain rate, we note that the volumetric strain is defined from Equation (12.15) by

$$e_V \cong \hat{e}_1 + \hat{e}_2 + \hat{e}_3 \qquad \text{if } |\hat{e}_k| << 1\ (k = 1{:}3) \qquad (18\text{-}1.3)$$

The instantaneous extensions ε_n, defined in the first Equation (12.30), are just the very small extensions that accumulate in an infinitesimal instant of time. Thus the instantaneous extensions satisfy the condition expressed in the second relationship of Equation (18-1.3), and we can write Equation (18-1.3) in terms of the instantaneous strain, as

$$\varepsilon_V = \hat{\varepsilon}_1 + \hat{\varepsilon}_2 + \hat{\varepsilon}_3 = \varepsilon_{11} + \varepsilon_{22} + \varepsilon_{33} \qquad (18\text{-}1.4)$$

where the second relationship follows because the sum of the normal components of symmetric tensor in any coordinate system is a scalar invariant (second Equation (12.45); compare Equations (7.32), (7.33), and (7-2.4)). Taking the time derivative of the second Equation (18-1.4) gives

$$\dot{\varepsilon}_V \equiv \dot{\varepsilon}_{11} + \dot{\varepsilon}_{22} + \dot{\varepsilon}_{33} \qquad (18\text{-}1.5)$$

Equation (16-1.8) defines the instantaneous strain rate in terms of the velocity gradients:

$$\dot{\varepsilon}_{k\ell} = \dot{\varepsilon}_{\ell k} = 0.5\left[\frac{\partial v_k}{\partial x_\ell} + \frac{\partial v_\ell}{\partial x_k}\right] \qquad (18\text{-}1.6)$$

Using Equation (18-1.6) with $k = \ell$ in Equation (18-1.5) gives the volumetric deformation rate in terms of the velocity gradient components:

$$\dot{\varepsilon}_V = \frac{\partial v_1}{\partial x_1} + \frac{\partial v_2}{\partial x_2} + \frac{\partial v_3}{\partial x_3} = \sum_{k=1}^{3} \frac{\partial v_k}{\partial x_k} \qquad (18\text{-}1.7)$$

Thus if a deformation is a constant-volume deformation, the summation term in Equation (18-1.7) is zero, and the conservation of mass Equation (18-1.1) becomes

$$\sum_{k=1}^{3} \frac{\partial v_k}{\partial x_k} = 0 \Rightarrow \frac{\partial \rho}{\partial t} = 0 \qquad (18\text{-}1.8)$$

Thus the conservation of mass is satisfied if density at a point is constant, that is, its rate of change is zero.

Conservation of Momentum. The conservation of momentum, Newton's second law of motion, applied at each point in a deformable body, is expressed by the equation

$$\sum_{k=1}^{3} \frac{\partial \sigma_{k\ell}}{\partial x_k} + \rho g_\ell = \rho a_\ell \qquad (18\text{-}1.9)$$

which is called the **equation of motion**. Because ℓ can independently take on any of the values 1, 2, or 3, Equation (18-1.9) defines a set of three equations, which, if we expand the subscript notation, take the form:

$$\begin{aligned}
&\frac{\partial \sigma_{11}}{\partial x_1} + \frac{\partial \sigma_{21}}{\partial x_2} + \frac{\partial \sigma_{31}}{\partial x_3} + \rho g_1 = \rho a_1 \\
&\frac{\partial \sigma_{12}}{\partial x_1} + \frac{\partial \sigma_{22}}{\partial x_2} + \frac{\partial \sigma_{32}}{\partial x_3} + \rho g_2 = \rho a_2 \\
&\frac{\partial \sigma_{13}}{\partial x_1} + \frac{\partial \sigma_{23}}{\partial x_2} + \frac{\partial \sigma_{33}}{\partial x_3} + \rho g_3 = \rho a_3
\end{aligned} \qquad (18\text{-}1.10)$$

Each term in each equation has units of force per unit volume. The right side of the equation is the inertial force term—that is, the mass per unit volume ρ multiplied by the acceleration a_ℓ. The first summation term on the left side of the Equation (18-1.9) is called the divergence of the stress and is the sum of three partial derivatives of the stress tensor. Each partial derivative has units of [stress/length] = [force/(area × length)] =[force/volume]. These terms are the forces per unit volume arising from the resistance of the material to deformation. The sum is the net force per unit volume resulting from stress components acting parallel to the coordinate axis x_ℓ, giving us three equations, for $\ell = 1, 2$, and 3, respectively. The second term on the left side of Equation (18-1.9) is the force of gravity per unit volume (mass per unit volume times acceleration), where g_ℓ is the acceleration due to gravity. Thus each equation expresses the balance of momentum parallel to one of the three coordinate directions x_ℓ, as indicated by the value of the subscript ℓ.

We can interpret the partial derivatives of stress for the simple example of fluid flow through a pipe parallel to x_l. We take $\ell = 1$ and thus look at the first Equation (18-1.10). The first partial derivative in the sum is the pressure gradient that drives the flow; the second two partial derivatives are the forces per unit volume from the shear stresses along the sides of the pipe that resist the flow.

Geological deformations are generally so slow that the acceleration is negligible and can be set equal to zero (with the obvious exceptions of earthquakes and rapid landslides). Under these circumstances, the right sides of Equations (18-1.9) and (18-1.10) become zero, and Newton's second law reduces to the **equilibrium equation**, which requires all the forces on a body be balanced. If we ignore the force of gravity, which generally we only can do for deformation on a relatively small scale, the only forces we must consider are those provided by the resistance of the material to deformation, and these are expressed by the stresses in the body. At each point in the body, the equilibrium equation then takes the form

$$\sum_{k=1}^{3} \frac{\partial \sigma_{k\ell}}{\partial x_k} = \frac{\partial \sigma_{1\ell}}{\partial x_k} + \frac{\partial \sigma_{2\ell}}{\partial x_2} + \frac{\partial \sigma_{3\ell}}{\partial x_3} = 0 \qquad (18\text{-}1.11)$$

where, again, ℓ takes on the values 1, 2, and 3 to give three equations:

$$\frac{\partial \sigma_{11}}{\partial x_1} + \frac{\partial \sigma_{21}}{\partial x_2} + \frac{\partial \sigma_{31}}{\partial x_3} = 0$$
$$\frac{\partial \sigma_{12}}{\partial x_1} + \frac{\partial \sigma_{22}}{\partial x_2} + \frac{\partial \sigma_{32}}{\partial x_3} = 0$$
$$\frac{\partial \sigma_{13}}{\partial x_1} + \frac{\partial \sigma_{23}}{\partial x_2} + \frac{\partial \sigma_{33}}{\partial x_3} = 0 \qquad (18\text{-}1.12)$$

For many geological problems, these are acceptable approximations. For large-scale deformations, however, such as the gravitational collapse of overthickened crust, the diapiric rise of salt domes or gneiss domes, and mantle convection, the gravitational force cannot be ignored. We use the simplest equations (Equation (18-1.11) or (18-1.12)), however, to illustrate the formulation of a mechanical model.

Conservation of Moment of Momentum (Angular Momentum). We showed in Section 7.1 that applying the balance of moment of momentum at a point in a continuum (Equation (7.27)) leads to the requirement for symmetry of the shear stress components (Equations (7.29) and (7.56)):

$$\sigma_{12} = \sigma_{21} \qquad \sigma_{13} = \sigma_{31} \qquad \sigma_{23} = \sigma_{32} \qquad (18\text{-}1.13)$$

We now take stock of the number of variables we must solve for and the number of equations we have to solve for them:

Variable	Number of Variables	Equation	Number of Equations
ρ	1	(18-1.1)	1
v_k	3	(18-1.12)	3
σ_{kl}	9	(18-1.13)	3
Total	**13**		**7**

Thus we have thirteen unknowns and only seven equations to determine them. Obviously the problem is underdetermined, and we need six more equations if we are to be able to find a solution. This should come as no surprise, because intuitively we understand that the deformational behavior of a body must depend on the mechanical properties of the material of which the body is composed, and so far our equations have specified nothing about the particular material properties. This is the function of the constitutive equations.

ii. Constitutive Equations

In Box 16-4 we discuss the three-dimensional constitutive equations that describe the behavior of elastic, viscous, plastic, and power-law materials. These equations express the stress as some function of the strain or strain rate, or vice versa. We develop here the equations from which we calculate the mechanical behavior of a body undergoing viscous, constant-volume deformation.

For a viscous material, the deviatoric stress is directly proportional to the strain rate:

$$\sigma_{k\ell}^{(Dev)} \equiv \sigma_{k\ell} - \overline{\sigma}_n \delta_{k\ell} = 2\eta \dot{\varepsilon}_{k\ell} \qquad (18\text{-}1.14)$$

or

$$\sigma_{k\ell} = \overline{\sigma}_n \delta_{k\ell} + 2\eta \dot{\varepsilon}_{k\ell} \qquad (18\text{-}1.15)$$

where the first Equation (18-1.14) is the general definition of the deviatoric stress components and δ_{kl} is called the Kronecker delta, whose components equal 1 for $k = \ell$ and 0 for $k \neq \ell$:

$$\delta_{k\ell} \equiv \begin{bmatrix} 1 & 0 & 0 \\ 0 & 1 & 0 \\ 0 & 0 & 1 \end{bmatrix} \qquad (18\text{-}1.16)$$

and where

$$\overline{\sigma}_n \equiv \frac{\sigma_{11} + \sigma_{22} + \sigma_{33}}{3} \qquad (18\text{-}1.17)$$

When written out in full, with the subscripts k and ℓ each taking on the values 1, 2, and 3, Equation (18-1.15) gives six independent equations that relate the six independent components of the stress to the six independent components of the instantaneous strain rate:

Normal Stress Components	Shear Stress Components
$\sigma_{11} = \overline{\sigma}_n + 2\eta\dot{\varepsilon}_{11}$	$\sigma_{12} = 2\eta\dot{\varepsilon}_{12}$
$\sigma_{22} = \overline{\sigma}_n + 2\eta\dot{\varepsilon}_{22}$	$\sigma_{13} = 2\eta\dot{\varepsilon}_{13}$
$\sigma_{33} = \overline{\sigma}_n + 2\eta\dot{\varepsilon}_{33}$	$\sigma_{23} = 2\eta\dot{\varepsilon}_{23}$ (18-1.18)

where Equation (18-1.13) and the first Equation (18-1.6) show that the nine components of the stress and strain-rate tensors in fact reduce to only six independent components each. These equations provide the six additional equations we need to solve the problem except that they introduce six new variables, the instantaneous strain rate $\dot{\varepsilon}_{k\ell}$. The second Equation (18-1.6), however, gives us six equations that define the instantaneous strain rate in terms of the velocity, so using those relationships, we can write the constitutive equations in terms of the velocity

$$\sigma_{k\ell} = \overline{\sigma}_n \delta_{k\ell} + \eta \left[\frac{\partial v_k}{\partial x_\ell} + \frac{\partial v_\ell}{\partial x_k} \right] \qquad (18\text{-}1.19)$$

(continued)

BOX 18-1 Formulation of a Mathematical Model with Application to the Problem of Viscous Deformation *(continued)*

which in expanded form reads:

Normal Stress Components / Shear Stress Components

$$\sigma_{11} = \bar{\sigma}_n + 2\eta \frac{\partial v_1}{\partial x_1} \qquad \sigma_{12} = \eta\left[\frac{\partial v_1}{\partial x_2} + \frac{\partial v_2}{\partial x_1}\right]$$

$$\sigma_{22} = \bar{\sigma}_n + 2\eta \frac{\partial v_2}{\partial x_2} \qquad \sigma_{13} = \eta\left[\frac{\partial v_1}{\partial x_3} + \frac{\partial v_3}{\partial x_1}\right]$$

$$\sigma_{33} = \bar{\sigma}_n + 2\eta \frac{\partial v_3}{\partial x_3} \qquad \sigma_{23} = \eta\left[\frac{\partial v_2}{\partial x_3} + \frac{\partial v_3}{\partial x_2}\right] \qquad (18\text{-}1.20)$$

These six constitutive equations therefore complete our system of equations, so that with them, we have as many equations (13) as we have unknowns (see the table near the end of Subsection *i*).

These constitutive equations actually define the components of the deviatoric stress (see Equation (18-1.14)), which means that the mean normal stress is not determined by the constitutive equations. This is evident if we try to calculate the mean normal stress using the definition Equation (18-1.17) and the constitutive equations for the normal stress from Equations (18-1.20):

$$\bar{\sigma}_n \equiv \frac{\sigma_{11} + \sigma_{22} + \sigma_{33}}{3} = \frac{1}{3}(\bar{\sigma}_n + \bar{\sigma}_n + \bar{\sigma}_n) + \frac{2}{3}\eta\left(\frac{\partial v_1}{\partial x_1} + \frac{\partial v_2}{\partial x_2} + \frac{\partial v_3}{\partial x_3}\right)$$

$$\bar{\sigma}_n = \bar{\sigma}_n + \frac{2}{3}\eta\left(\frac{\partial v_1}{\partial x_1} + \frac{\partial v_2}{\partial x_2} + \frac{\partial v_3}{\partial x_3}\right)$$

$$\frac{\partial v_1}{\partial x_1} + \frac{\partial v_2}{\partial x_2} + \frac{\partial v_3}{\partial x_3} = \sum_{k=1}^{3} \frac{\partial v_k}{\partial x_k} = 0 \qquad (18\text{-}1.21)$$

This is just the condition for a constant-volume deformation (Equation (18-1.7)). Thus the constitutive equations we have adopted do not constrain the mean normal stress, and they require the volumetric strain rate to be zero.

iii. The Governing Equations

We can reduce the number of equations and unknowns that we have to deal with in the solution by combining the equations of equilibrium (Equation (18-1.11)) with the constitutive Equations (18-1.19). Taking the partial derivative of Equation (18-1.19) with respect to x_k, summing over the index k, and setting the result equal to zero as required by Equation (18-1.11), we obtain

$$\sum_{k=1}^{3} \frac{\partial \sigma_{k\ell}}{\partial x_k} = 0 = \frac{\partial \bar{\sigma}_n}{\partial x_l} + \eta \sum_{k=1}^{3}\left[\frac{\partial^2 v_k}{\partial x_k \partial x_\ell} + \frac{\partial^2 v_\ell}{\partial x_k^2}\right] \qquad (18\text{-}1.22)$$

The first term under the summation on the right is zero because from Equation (18-1.21):

$$\sum_{k=1}^{3} \frac{\partial^2 v_k}{\partial x_k \partial x_\ell} = \frac{\partial}{\partial x_\ell} \sum_{k=1}^{3} \frac{\partial v_k}{\partial x_k} = 0$$

Thus the final differential equations that we have to solve are

$$\frac{\partial \bar{\sigma}_n}{\partial x_\ell} = -\eta \sum_{k=1}^{3} \frac{\partial^2 v_\ell}{\partial x_k^2} \qquad \sum_{k=1}^{3} \frac{\partial v_k}{\partial x_k} = 0 \qquad (18\text{-}1.23)$$

where the second equation is Equation (18-1.21). The first equation represents three equations for $\ell = 1, 2,$ and 3, respectively. Note that although we eliminated the stress components from the equations by combining the constitutive equation with the equation of equilibrium, we still have the mean normal stress as an unknown, along with the three velocity components v_k. Equations (18-1.23) then provide a set of four differential equations to solve for the four unknowns $\bar{\sigma}_n$ and v_k.

iv. Boundary Conditions

Solving the differential Equations (18-1.23) requires integration. The process of integration introduces arbitrary constants that permit a whole set of acceptable solutions to the problem. We can obtain a specific solution if we can evaluate the arbitrary constants, and we can do so only if we specify additional constraints for the problem called **boundary conditions**. For example, we may want to determine the stresses, velocities, and strain rates throughout a body when the boundary is subjected to specified stresses. Alternatively, we may want to calculate the stresses, velocities, and strain rates throughout a body when the boundary is subjected to specified velocities. The stresses or velocities imposed on the boundary constitute the boundary conditions for the problem. The solution for the unknowns within the body must be consistent with the conditions prescribed on the boundary. Thus the constitutive equations, the balance laws, and the boundary conditions are the three essential parts of finding a numerical solution to a problem of deformation.

v. Solution

If an analytic solution can be obtained for the equations, we can calculate the velocity and mean normal stress as a function of position in the body (Equation (18-1.23)), and we can then determine the stress from Equations (18-1.20) and the strain rate from Equation (18-1.6). From this solution, we obtain the displacement by multiplying the velocity by a small increment of time. That displacement, however, changes the geometry of the body (for example, a flat layer may become folded), so that even if an analytic solution can be obtained for the initial step, the new more complicated geometry that results after the first increment of the deformation makes an analytic solution difficult if not impossible to obtain for the remainder of the deformation. Thus we generally cannot follow the development of a geometrically complex structure through time with analytic solutions.

With numerical techniques, however, we can obtain approximate solutions for the displacement field and the stress distribution at the end of each successive increment of time for a specified network of points across the body. We use iterative techniques by which we make an initial guess at a solu-

BOX 18-1 Formulation of a Mathematical Model with Application to the Problem of Viscous Deformation *(continued)*

tion for the stress and the velocity at the points on the network and then repeatedly adjust that guess to get progressively closer to the values that satisfy the equations. When the adjustments required at each iteration are small enough to be ignored, we multiply the velocity of those points by a small time increment to find the displacement and thus determine the new shape of the body. This new shape then becomes the starting point from which we solve for the velocity and stress at the network points, and this solution then defines the increment of displacement that occurs during the next increment of time. We thus model the actual smooth evolution of the system through time as a series of stepwise changes. If the time increments are not too large, the complex time-dependent behavior of a system can be followed with reasonable accuracy.

In order to study such problems as the deformation in a collisional orogen, the driving forces of plate tectonics, or convective motions in the mantle, we should include the effects of temperature and heat transfer. The additional variables that are introduced require additional constitutive equations governing the flow of heat, as well as the balance law for the conservation of energy. The solutions quickly become very complex but, with the increasing speed and memory of electronic computers, they are not intractable.

The detailed development of the mathematics and the solutions, both analytic and numerical, are beyond the scope of this book. We give a general idea of the value of the method, however, by discussing briefly several models for the formation of folds in Sections 18.6 and 18.7.

REFERENCES AND ADDITIONAL READINGS

Backofen, W. A. 1972. *Deformation Processing*, Addison-Wesley, Reading, Mass., 326 pp.

Balk, R. 1949. Structure of Grand Saline salt dome, Van Zandt County, Texas. *Amer. Assoc. Petro. Geol. Bull.* 33: 1791–1829.

Biot, M. A. 1957. Folding instability of a layered viscoelastic medium under compression. *Proc. Roy. Soc. Lond.* A242: 111–454.

Biot, M. A. 1961. Theory of folding of stratified viscoelastic media and its implications in tectonics and orogenesis. *Geol. Soc. Amer. Bull.* 72: 1595–1618.

Biot, M. A. 1965. *Mechanics of Incremental Deformations.* John Wiley, New York, 504 pp.

Brun, J.-P. 2002. Deformation of the continental lithosphere: Insights from brittle-ductile models. In S. DeMeer et al., eds., *Deformation Mechanisms, Rheology and Tectonics: Current Status and Future Perspectives*, Geol. Soc. Lond., Sp. Pub. 200: 355–370.

Bucher, W. H. 1956. Role of gravity in orogenesis. *Geol. Soc. Amer. Bull.* 67: 1295–1318.

Currie, J. B., H. W. Patnode, and R. P. Trump. 1962. Development of folds in sedimentary strata. *Geol. Soc. Amer. Bull.* 73: 655–673.

Dennis, J. G., and R. Häll. 1978. Jura-type platform folds: A centrifuge experiment. *Tectonophysics* 45: T15–T25.

Dieterich, J. H. 1969. Origin of cleavage in folded rocks. *Amer. J. Sci.* 267: 155–165.

Dieterich, J. H., and Carter, N. L. 1969. Stress-history of folding. *Amer. J. Sci.* 267: 129–154.

Hobbs, B. E., H.-B. Mühlhaus, A. Ord, Y. Zhang, and L. Moresi. 2000. Fold geometry and constitutive Behaviour. In M. W. Jessell and J. L. Urai, eds., *Stress, Strain and Structure: A Volume in Honour of W. D. Means. Journal of the Virtual Explorer*, 2. http://virtualexplorer.com.au/2000/Volume2/www/contribs/hobbs/index.html

Hubbert, M. K. 1937. Theory of scale models as applied to the study of geologic structures. *Geol. Soc. Amer. Bull.* 48: 1459.

Hudleston, P. J. and L. Lan. 1994. Rheological controls on the shapes of single-layer folds. *J. Struct. Geol.* 16: 1007–1021.

Jackson, M. P. A., and C. J. Talbot. 1989. Anatomy of mushroom-shaped diapirs. *J. Struct. Geol.* 11(1/2): 211–230.

Johnson, A. M. 1970. *Physical Processes in Geology.* Freeman Cooper & Co., San Francisco, 576 pp.

Johnson, A. M., and R. C. Fletcher. 1994. *Folding of Viscous Layers.* Columbia University Press, New York, 461 pp.

Johnson, K. M., and A. M. Johnson. 2002. Mechanical analysis of the geometry of forced-folds. *J. Struct. Geol.* 24(3): 401–410.

Keller, J. V. A., S. H. Hall, and K. R. McClay. 1997. Shear fracture pattern and microstructural evolution in transpression fault zones from field and laboratory studies. *J. Struct. Geol.* 19(9): 1173–1187.

Lan, L., and P. J. Hudleston. 1995. The effects of rheology on the strain distribution in single layer buckle folds. *J. Struct. Geol.* 17(5): 727–738.

Lan, L., and P. J. Hudleston. 1996. Rock rheology and sharpness of folds in single layers *J. Struct. Geol.* 18(7): 925–931.

McClay, K. R., T. Dooley, R. Gloaguen, P. Whitehouse, and S. Khalil. 2001. Analogue modelling of extensional fault architectures: Comparisons with natural rift fault systems. In K. C. Hill and T. Bernecker, eds., *Eastern Australasian Basins Symposium 2001: A Refocused Energy Perspective for the Future, Petroleum Exploration Society of Australia Special Publication*, v.1: 573–584.

McClay, K. R., T. Dooley, P. Whitehouse, L. Fullarton, and S. Chantraprasert. 2004. 3D analogue models of rift systems: Templates for 3D seismic interpretation. In R. J. Davies, J. A. Cartwright, S. A. Stewart, M. Lappin, and J. R. Underhill, eds., *3D Seismic Technology; Application to the Exploration of Sedimentary Basins: Memoirs of the Geological Society of London*, v. 29: 101–115.

McClay, K. R., and P. G. Ellis. 1987. Analogue models of extensional fault geometries. In M. P. Coward, J. F. Dewey, and P. L. Hancock, eds., *Continental Extensional Tectonics*, Geol. Soc. Lond. Sp. Pub. 28, Blackwell Scientific Publications, Oxford: 109–125.

Molnar, P., and P. Tapponnier. 1977. Relation of the tectonics of eastern China to the India-Eurasia collision: Application of slip line field theory to large-scale continental tectonics. *Geology* 5(4): 212–216.

Odé, H. 1960. Faulting as a velocity discontinuity. In D. T. Griggs and J. Handin, eds., *Rock Deformation*, Geol. Soc. Amer. Memoir 79.

Parrish, D. K., A. L. Krivz, and N. L. Carter. 1976. Finite-element folds of similar geometry. *Tectonophysics* 32: 183–207.

Ramberg, H. 1963. Fluid dynamics of viscous folding. *Amer. Assoc. Petrol. Geol. Bull.* 47: 484–515.

Ramberg, H. 1981. *Gravity, Deformation and the Earth's Crust.* Academic Press, New York, 452 pp.

Reiner, M. 1964. The Deborah number. *Physics Today*: 62.

Tapponnier, P., G. Peltzer, A. Y. Le Dain, R. Armijo, and P. Cobbold. 1982. Propagating extrusion tectonics in Asia: New insights from simple experiments with plasticine. *Geology* 10(12): 611–616.

Trusheim, F. 1960. Mechanism of salt migration in northern Germany. *Bull. Amer. Assoc. Petrol. Geol.* 44: 1519–1540.

Zhang, Y., N. S. Mancktelow, B. E. Hobbs, A. Ord, and H. B. Mühlhaus. 2000. Numerical modelling of single-layer folding: Clarifcation of an issue regarding the possible effect of computer codes and the influence of initial irregularities. *J. Struct. Geol.* 22: 1511–1522.

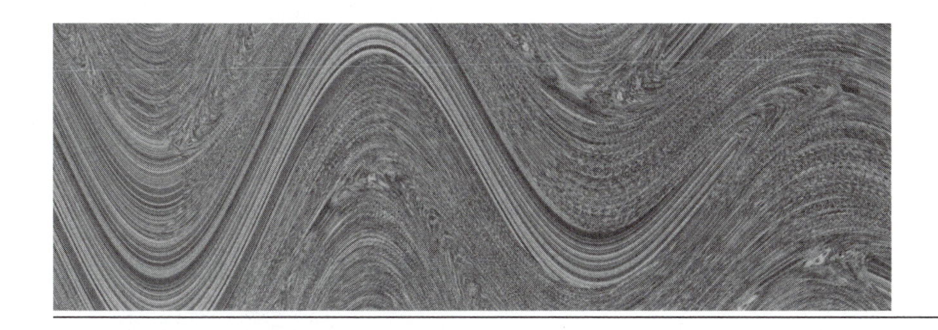

PART IV

REGIONAL ASSOCIATIONS OF STRUCTURES

At this point we have described the basic types of structures observed in the Earth, and in trying to understand their origin and significance, we have considered kinematic and mechanical models for their formation and discussed mechanisms by which rocks deform and the structures develop. In Chapter 1, we describe briefly the major tectonic features of the Earth, and we give a few examples of regional associations of structures in Chapters 4–6 and 9–11. In Part IV, the final chapters of the book, we turn our attention to a more systematic attempt to integrate our understanding of small-scale structures with large-scale tectonic processes. In particular, we discuss the regional associations of structures, how they characterize various tectonic environments, and how they can be used to constrain and understand the larger-scale tectonic processes. The literature on these subjects is vast, and within the scope of this book, we cannot cover these topics in detail. We hope, however, that the brief outline will at least stimulate curiosity; the reader can find a path to more information through the selected bibliographies at the end of each chapter.

Plate margins that are active today provide natural examples of the associations of structures that characterize the different principal tectonic environments. From studying these, we can gain an understanding of the characteristics we should look for to identify these different environments in the geologic record. To this end, in Chapter 19 we present a brief discussion of the associations of structures that form at active plate margins.

Orogenic belts have traditionally been the subject of much attention in structural geology and tectonics, because it is predominantly in these belts that the structures discussed in the foregoing chapters are exposed, and it is here that the record of the structural and tectonic evolution of the Earth is preserved, especially for times earlier than about 200 Ma. In Chapter 20, therefore, we review the structural characteristics of orogenic belts, discuss the application of small-scale structures to the understanding of the process of orogeny, and present a first-order model of the process of collision at a subduction zone and its relation to the structural evolution of orogens.

Chapter 19

DEVELOPMENT OF STRUCTURES AT ACTIVE PLATE MARGINS

The margins of the tectonic plates are the locations where two plates move relative to each other. They can move away from each other, creating new plates at the boundary, they can slide horizontally past each other, or one plate can slide underneath the other. Because of this relative motion, these are the sites at which most of the deformation of crustal rocks currently accumulates (Figure 19.1). It therefore seems logical to assume that significant deformation observed in ancient crustal rocks probably originated in similar tectonic settings. By studying the deformation currently accumulating at plate margins, therefore, we can begin to understand the conditions under which a wide variety of structures formed. This study not only helps us to interpret the current tectonic activity, but also provides a guide to the interpretation of ancient structural associations. In this chapter we look at how the different plate margins give rise to different associations of structures and how those structures are related to the tectonic and deformational processes that we can observe.

We begin with divergent margins on continents (Section 19.1). Our discussion focuses on the formation of structures as these margins evolve from continental rifts to rifted passive continental margins on the shores of ocean basins. Rifting of continents leads eventually to the formation of new ocean basins, with active rifting occurring at the midocean ridges. In Section 19.2, we discuss the deformation processes and the resulting structures in these oceanic environments, as revealed by geophysical and in situ observation of active oceanic rifts, as well as observations of oceanic crust, inferred to have formed at extinct rifts, that has subsequently been uplifted and exposed on land. Major strike-slip faults occur in both oceanic and continental crust, and we discuss the associated structures in Section 19.3. Many of these faults are transform faults that form important parts of plate margins; other large strike-slip faults are particularly important in continental crust at collisional margins. The structural development along both types of strike-slip fault is similar, and they are distinguishable only from their tectonic setting. In Section 19.4, we discuss the development of structures at active convergent margins, including the structurally complex accretionary prisms. Finally, in Section 19.5 we briefly review the types of collision zones to which convergent margins give rise, although a more detailed discussion of the complex structure of orogenic belts that develop from such collision zones is left to Chapter 20.

19.1 DIVERGENT MARGINS ON THE CONTINENTS: CONTINENTAL RIFTING

Divergent margins are locations where two plates move apart from each other. The structures are dominated by normal faulting, which accommodates horizontal lengthening and vertical thinning of the Earth's crust. Other structures also are present, however. We examine first the structures occurring along active rifts in continents, structures along new continental margins bounding ocean basins, and then the structures characteristic of continental margins that once were active rifts but have become passive rifted margins as the continents drifted away from the site of active rifting.

i. New Rifts in Continental Crust

We briefly discuss continental regions of the world dominated by extensional normal faulting in Section

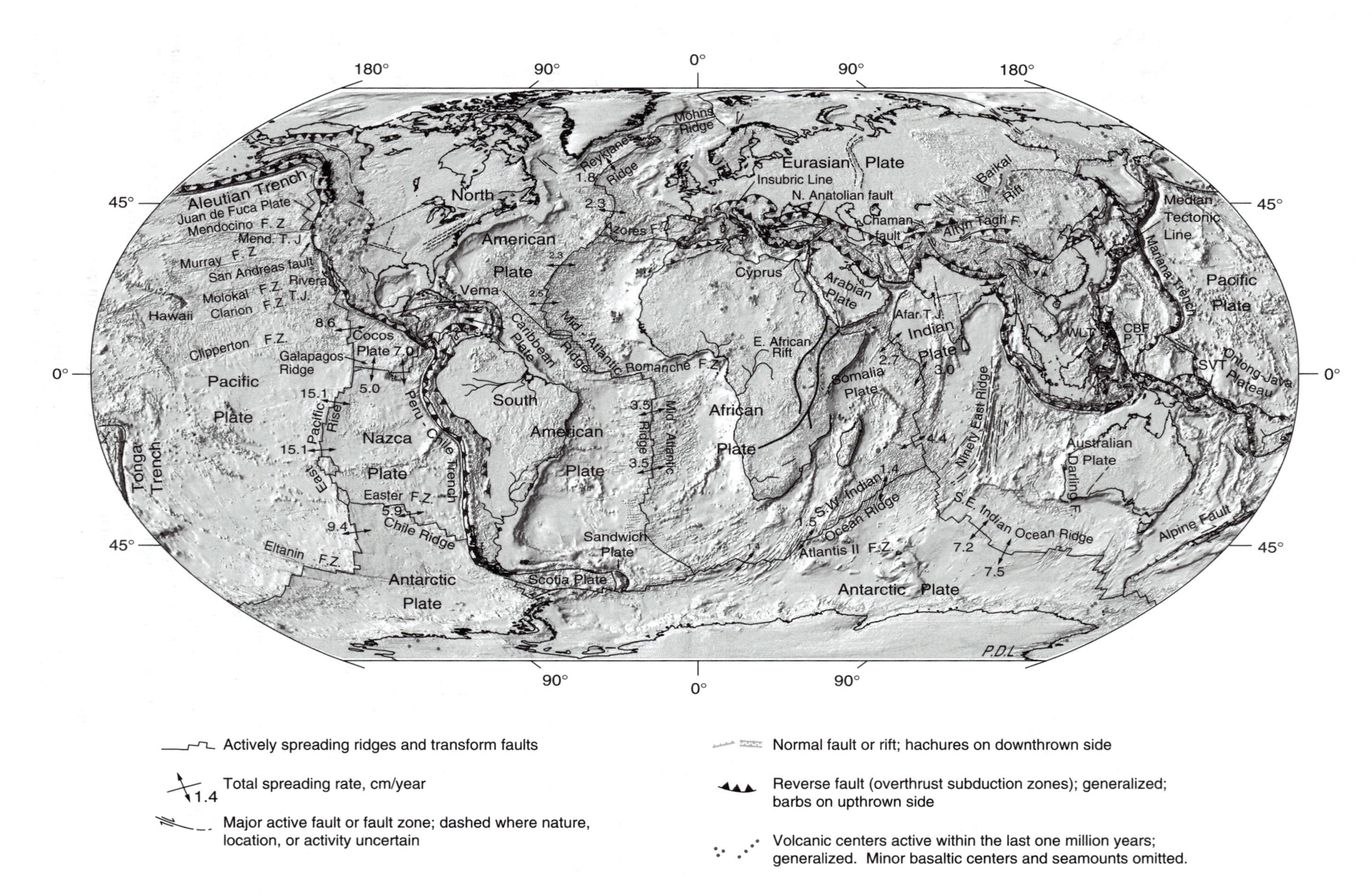

FIGURE 19.1 Map of the world showing major oceanic fracture zones, active transform faults, and active and inactive continental strike-slip faults. On Philippine Plate, CBF is the Central Basin fault, P.T. is the Philippine trench; south of the Ontong-Java plateau, SVT is the Solomon-Vanuatu trench. (After Menard and Chase 1971, NASA, 2002 (see Chapter 1 for reference))

4.3(ii) (Figure 4.9), but not all of these regions are necessarily areas of rifting marking the incipient development of oceanic basins. Evidence for active continental rifting is clearly displayed in the African continent along its eastern margin at the Red Sea and along the intracontinental East African rift system. The East African rift apparently formed by rifting of a series of pre-existing domes that are tens to hundreds of kilometers in diameter (Figures 19.2 and 19.3*A*), although many domes and basins not associated with rifts also occur scattered throughout the continent (Figure 19.2). The entire rift system is comparable in size to the Basin and Range province in the western United States (cf. Figure 4.9), although it is debatable whether the Basin and Range is of similar origin.

The structures developed during the earliest phases of continental rifting are illustrated in Figure 19.3*B*. On both the western and eastern rifts (Figure 19.3*A*; see Figure 19.2), structures predominantly include normal faults that consist of a series of half-grabens formed by faults that are arcuate in plan view and listric in the third dimension. Half-grabens overlap each other and are separated by complex transfer structures (Figure 19.3*B*, *C*).

ii. Young Rifted Continental Margins and New Ocean Basins

The Red Sea represents a more advanced stage in the process of continental rifting and is in fact connected to the East African rift at the Afar ridge-ridge-ridge triple junction near the southwest corner of the Arabian Peninsula (Figures 19.1 and 19.2). Parts of the Red Sea are still underlain by thinned continental crust, and other parts, particularly in the south, actually have oceanic crust forming along the central ridge axis. The continental margins of the Red Sea have been interpreted as both symmetric and asymmetric structures, a difference possibly reflecting evolution in the ideas of rifted margin formation. Figure 19.4 shows two interpretations of approximately the same cross section. An earlier cross section (Figure 19.4*B*) assumes symmetrical spreading. A more recent interpretation (Figure 19.4*C*) invokes an asymmetry in the structure of the rift that is characterized by an east-dipping low-angle detachment fault that cuts into the upper mantle (Figure 19.4*C*) with an upper plate on the east side. The rift margins of the Red Sea near the Afar triangle are volcanic-rich. The trend of active volcanism, shown by the concentrations of Tertiary volcanics in the Arabian Peninsula east of the Red Sea (Figure 19.4*A*), is more northerly than that of the rift itself, however, which follows an old northwest-trending late-Precambrian zone of weakness (see schematic section in Figure 19.7*A*).

Within the Red Sea itself, thick deposits of salt have accumulated, and are currently accumulating, on the seafloor. Dense brines are present in the Red Sea, thought to be the product of repeated incursion and evaporation of sea water by intermittent opening and closing of the southern end of the Red Sea at the Afar triple junction (triangle in Figure 19.4). Salt crystallizing from these brines has resulted in the accumulation of salt deposits approximately one kilometer thick, which then have been covered by sediments originating from the erosion of the margins of the adjacent continents and from the settling of marine biogenic debris.

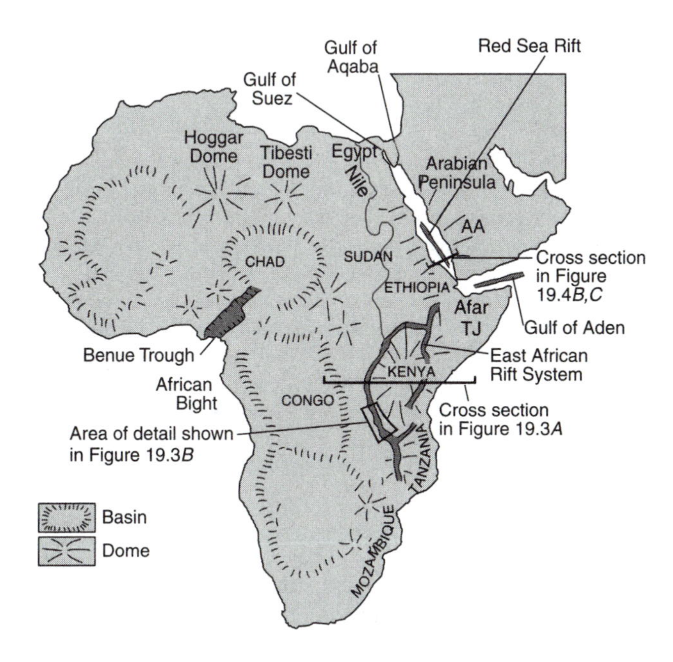

FIGURE 19.2 Map of Africa, showing domes and basins, the East African rift, the Benue trough, and other adjacent tectonic features. Location of cross sections are shown. AA = African-Arabian dome. Afar TJ = Afar triple junction. (After Clifford and Gass 1970; Burke and Whiteman 1972, 1973)

iii. Mature Rifted Continental Margins

The rifting of a continent to form an intervening ocean basin leaves passive rifted continental margins along the rim of the ocean basin (Figure 19.5). The structure of these margins reflects the original rifting process and the subsequent accumulation of sediments after the continental margin has drifted away from the site of active rifting. Various margins of the world's current ocean basins preserve a record of the different rifting processes, characterized by a highly variable abundance of volcanic rocks (Figure 19.5).

Volcanic-rich margins display thick sequences of volcanic flows, thought to be mostly basaltic, that dip gently down toward the ocean basin and form "seaward-dipping reflector sequences" recognized in many seismic reflection profiles (Figure 19.6). Seaward of volcanic-rich margins, huge anomalously thick oceanic volcanic plateaus are common, which may represent the voluminous outpouring of magma above a rising mantle diapir. For example, Iceland ('I' on Figure 19.5) is a 100,000-km^2 volcanic plateau astride the Mid-Atlantic Ridge. Volcanic

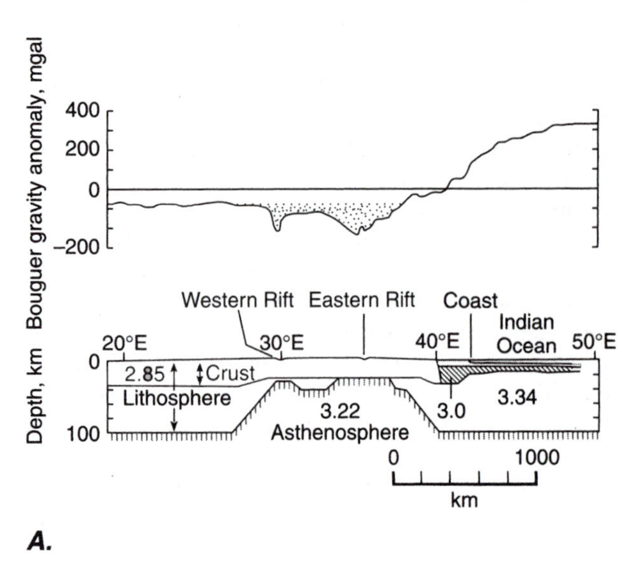

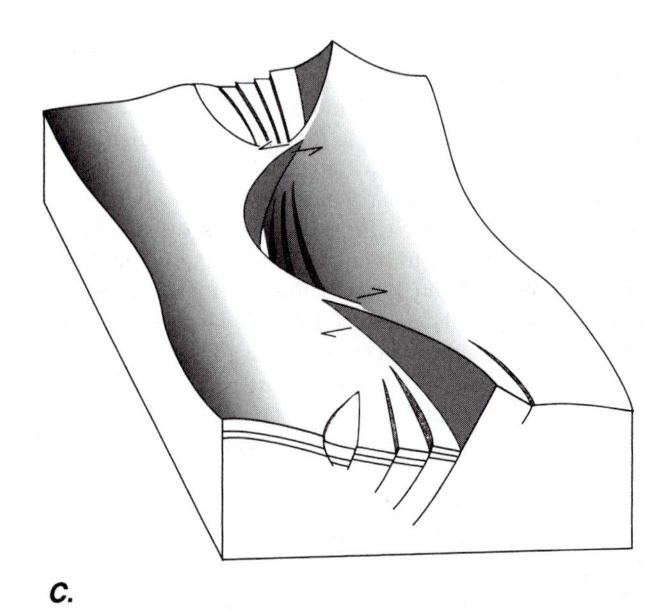

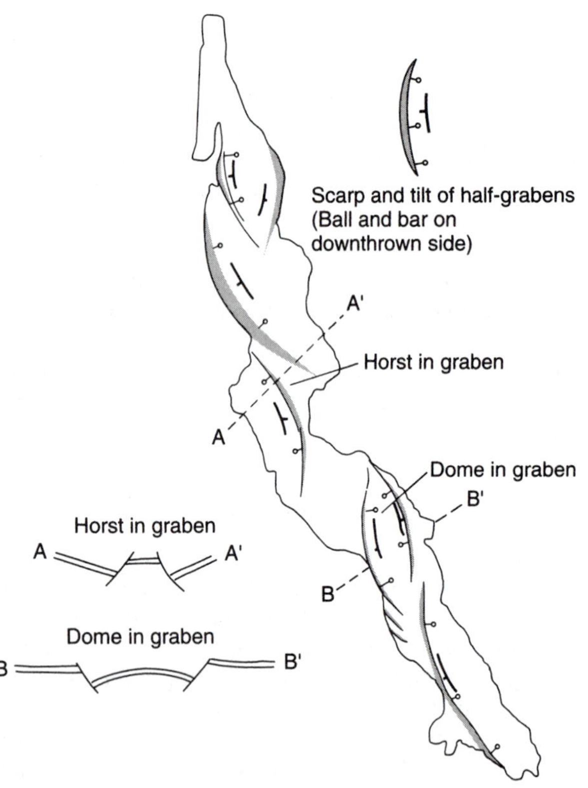

FIGURE 19.3 The East African rift. *A*. Generalized cross section of East Africa showing the Bouguer gravity anomaly (top) and a density model (bottom; gm/cm^3) based on gravity and seismic data. Note the similarity in width with the Basin and Range province. (After Fairhead 1989) *B*. Map and selected cross sections of Lake Tanganyika (outlined), showing alternating and overlapping half-grabens. Cross section A-A′ shows a central horst bounded by two normal faults. Cross section B-B′ shows a dome developed in a graben. Not to scale. (After Rosendahl et al. 1986) *C*. Three-dimensional sketch of the geometry of faulting shown in *B*. (After Rosendahl et al. 1986)

ridges extend from it to the volcanic-rich margins of Greenland and northwest Europe.

Volcanic-poor margins, on the other hand, display abundant normal faulting, similar to that of the southern Red Sea (see schematic section in Figure 19.7*A*). The Spanish Galicia margin is an example of a nonvolcanic sediment-starved margin characterized by a low-angle detachment that dips beneath the Iberian continent (Figure 19.7*B*). Its structure has been documented by studies including dredging, geophysical surveys, and drilling by the Ocean Drilling Project. These studies have revealed the presence of the listric faults and a diapir of serpentinized peridotite.

The margins of eastern North America and western Africa are good examples of mature rifted continental margins. Figure 19.8 shows schematic cross sections of these margins. The Central Scotian (Figure 19.8*A*) and the Moroccan margins (Figure 19.8*B*) were essentially op-

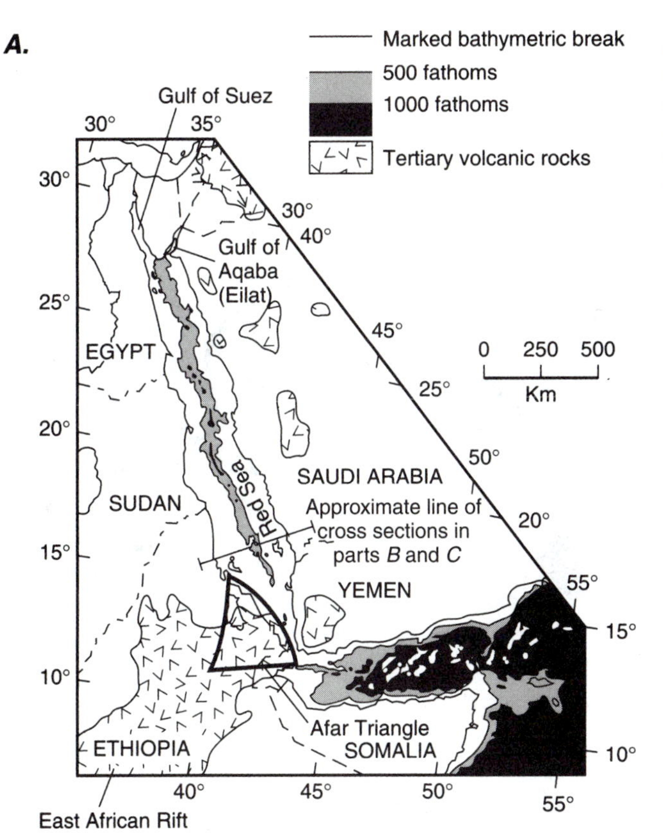

FIGURE 19.4 Structure of the southern Red Sea. *A*. Map of the Red Sea and neighboring region. (After Lowell and Genik 1974; Dixon et al. 1989) *B*. Cross section of the southern Red Sea, assuming symmetrical rifting. (After Lowell and Genik 1974) C. Cross section of the Red Sea, assuming asymmetrical rifting. (After Voggenreiter et al. 1988)

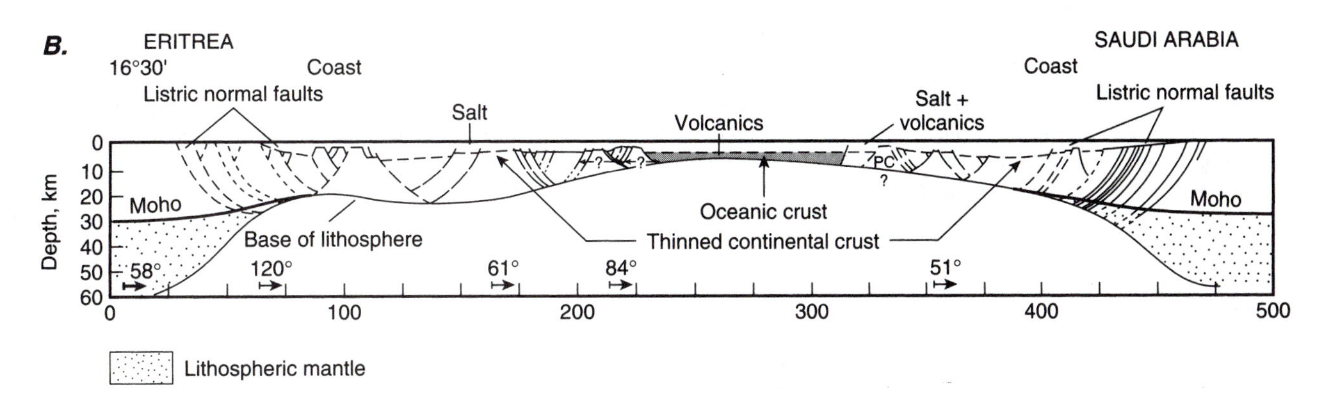

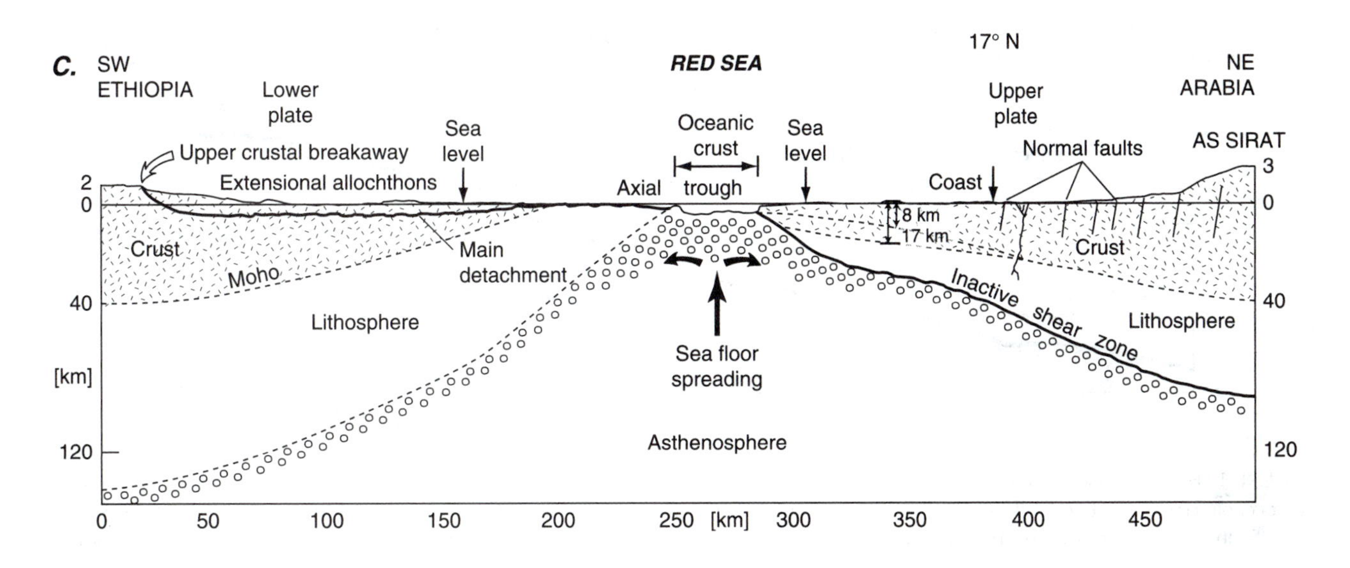

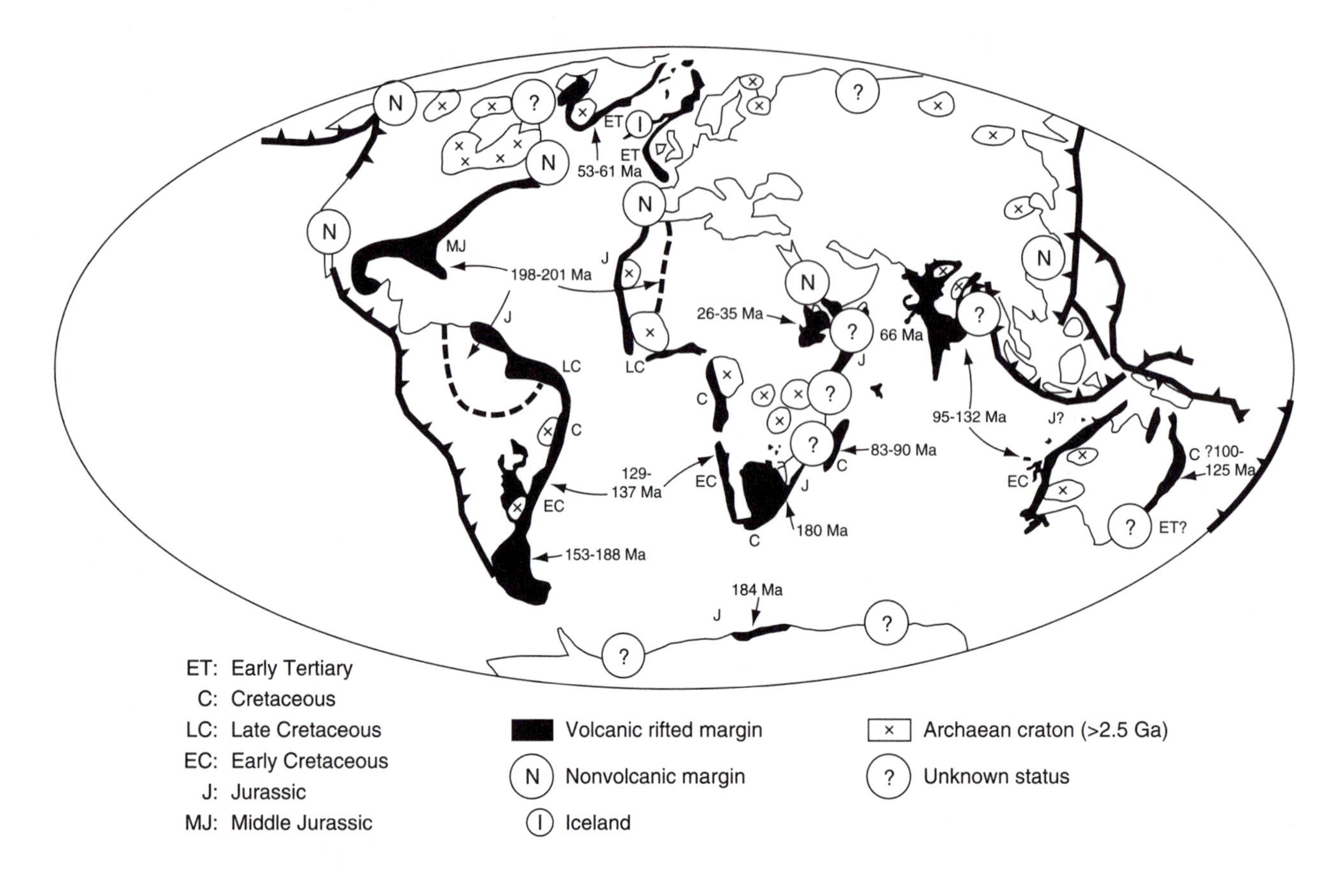

FIGURE 19.5 Generalized world map showing areas of volcanic-rich and volcanic-poor rifted margins, as well as regions where the type of margin is uncertain. Numeric ages are from dated basalts along the margins. Heavy dashed lines in northern South America and northwest Africa indicate the limits of mantle-plume-related volcanics associated with the break-up of the continents. (Modified after Menzies et al. 2002)

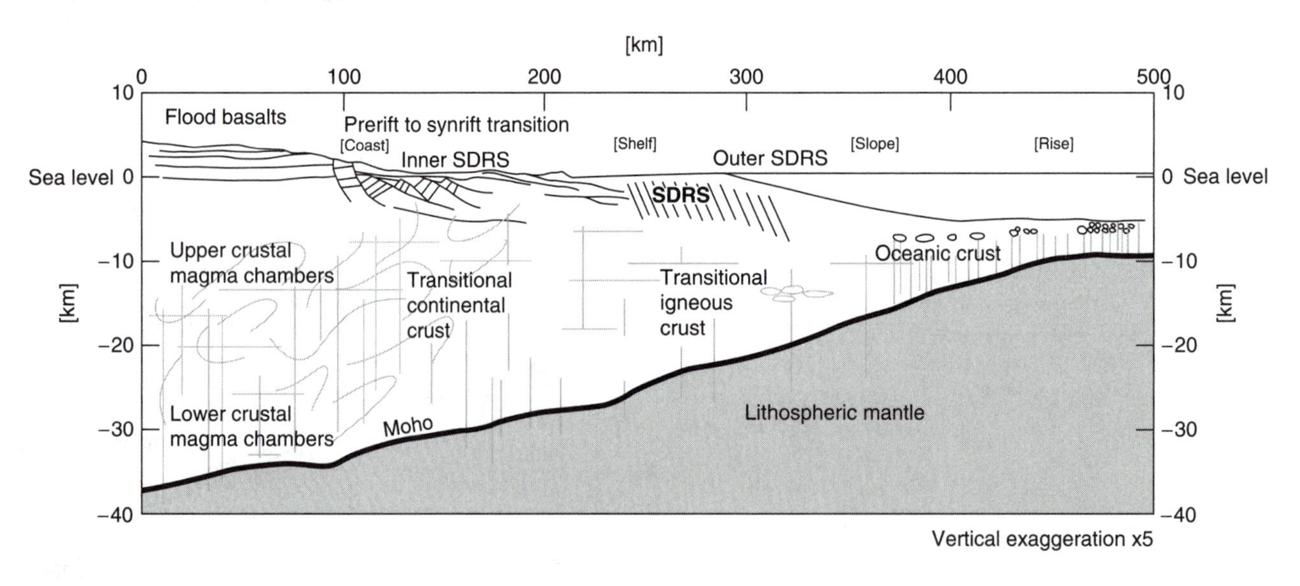

FIGURE 19.6 Schematic cross section of a volcanic-rich rifted margin, showing characteristic features from continental interior to oceanic crust: a subaerial flood basalt province, early listric normal faults, an extensive region of seaward-dipping reflectors (SDRS) in seismic reflection profiles, and a magma-rich oceanic crust. Vertical exaggeration approximately 4:1. (After Menzies et al. 2002)

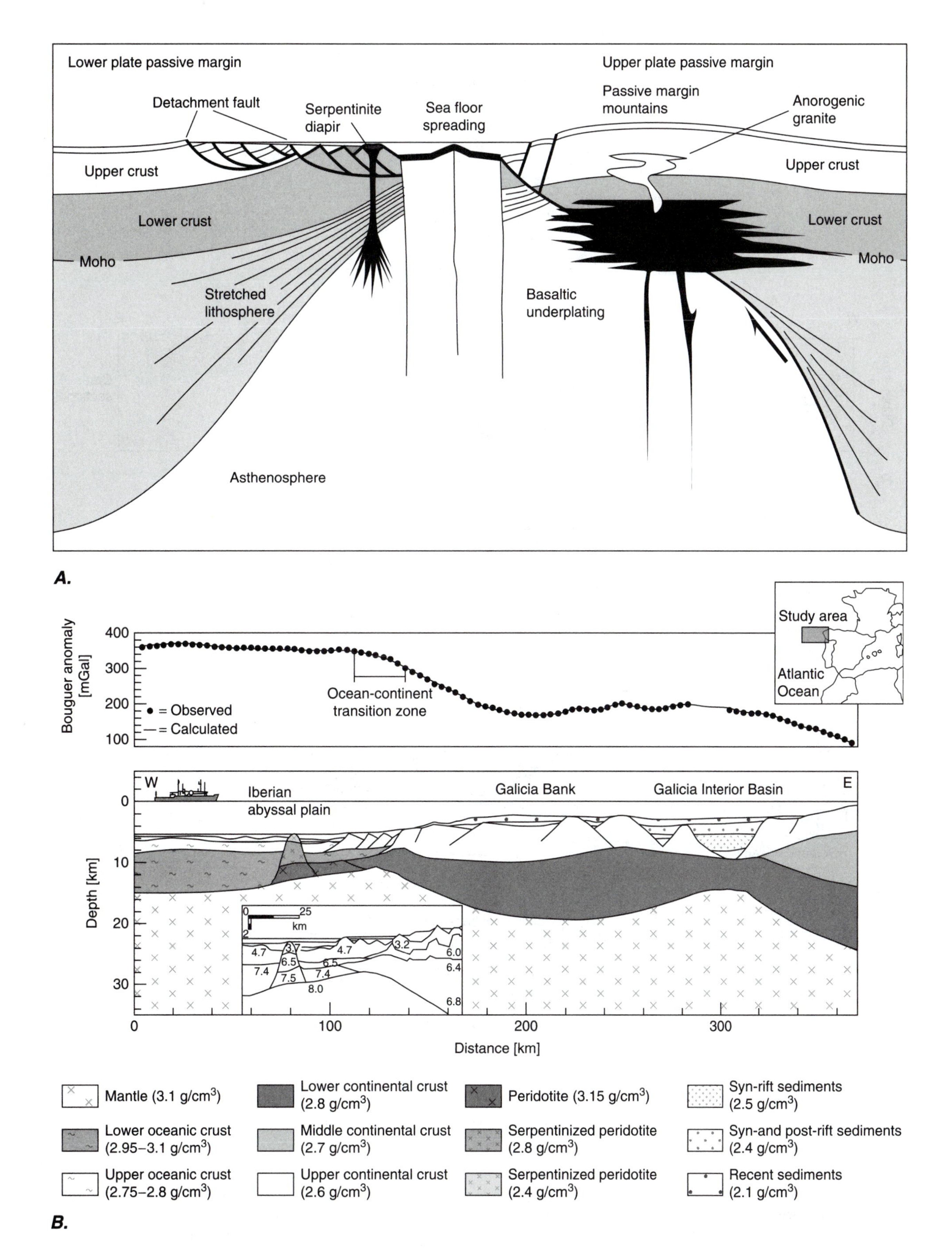

FIGURE 19.7 *A.* Schematic cross section of a volcanic-poor rifted margin, showing upper and lower plates, listric detachment faults, stretched lithosphere, possible basaltic magmas near the Moho in the upper plate, giving rise to anorogenic granites, and serpentinite diapirs in the lower plate. (After Lister et al. 1991; Reston et al. 1996) *B.* Structure of the non-volcanic Galicia Atlantic margin off northwest Spain. Vertical exaggeration approximately 1.5:1. Note the presence of listric normal faults and serpentinite diapir. Lower inset shows seismic p-wave velocity structure. (From Carbó et al. 2004)

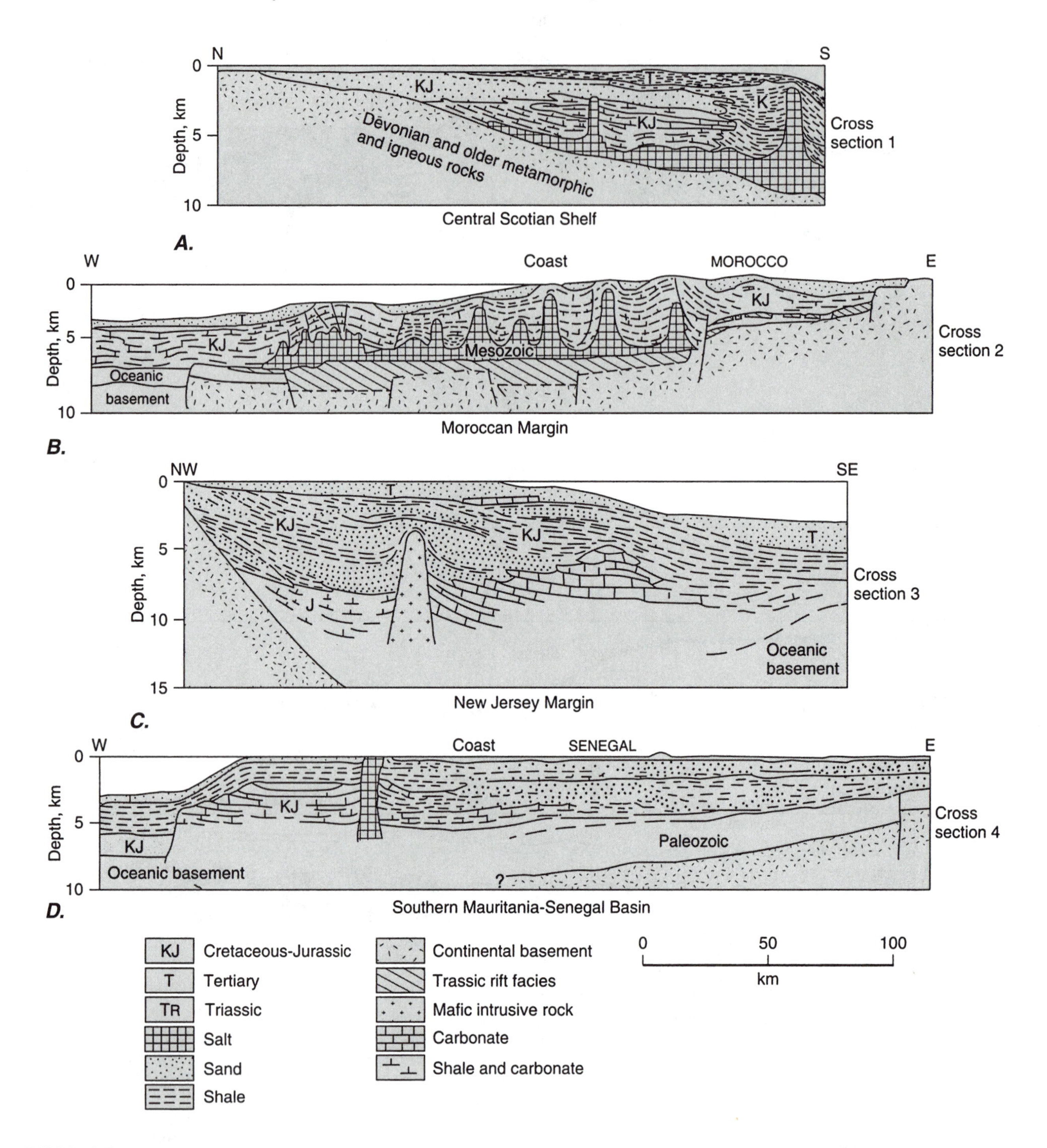

FIGURE 19.8 Structure of the conjugate North American and West African margins. *A*. The Central Scotian shelf and *B*. the Moroccan margin were opposite each other. *C*. The New Jersey margin and *D*. the Southern Mauritania-Senegal Basin were also opposite each other. Vertical exaggeration 5:1. (After Schlee 1980)

posite each other as were the New Jersey margin (Figure 19.8*C*) and Southern Mauritania-Senegal basins (Figure 19.8*D*). Thus the Scotian-Moroccan margin pair and the New Jersey–Mauretania/Senegal margin pair constitute so-called conjugate margins, which are the originally adjacent sides of the same rifted continent. Characteristics in common include the old basement, continental rift facies rocks of Triassic age, a layer of salt in the Scotia-Morocco conjugate, but not in the New Jersey–Mauritania-Senegal regions, and marine deposits. These cross sections do not address the issues of whether the rifting was symmetric or asymmetric or whether listric detachment faults are present. They do demonstrate, however, the overall similarity of the sequence of rocks deposited along two conjugate margins that are now widely separated from each other. A similarity in stratigraphic section can just imply a similarity in the rifting process and need not imply the initial proximity of similar sections.

iv. Models of the Rifting Process

Based on the evidence from current rifting processes that we review in the preceding Subsections 19.1(i) through (iii), most workers believe that continental rifting can begin by either active rifting or passive rifting. Active rifting of the continent begins with uplift during the formation of a structural dome in a continental platform (Figure 19.9*A*), accompanied by a significant volume of alkalic igneous eruptions. The dome formation presumably reflects the rise of a mantle diapir at depth, and the alkalic volcanics are derived from deep (100 km or more) partial melting of the rising diapir. The radial extension caused by the doming is accommodated by three rifts forming a "Y" intersection at the top of the dome (compare Figure 4.7 for comparable faulting above a small-scale dome). Ultimately, these three rifts either develop into a ridge-ridge-ridge (RRR) triple junction in oceanic crust, which separates three different continental masses (Figure 19.9*B*), or they develop into a single rift that leaves a sharp angle in the separated continental margins, with a failed rift that intersects the apex of the angle where it forms a re-entrant in the coast of one of the continents (Figure 19.9*C*). The failed rift opens sufficiently to develop a deep sediment-filled trough trending from the triple junction into the continental interior, but it never develops fully into an ocean basin (Figure 19.9*C*).

Evidence for this process is preserved in modern rifted continental margins, which characteristically show sharp changes in trend reflecting the geometry of an original RRR triple junction. The sharp corner on the southwest of the Arabian Peninsula, for example, preserves the geometry of intersection of the Red Sea rift with the Gulf of Aden rift at the Afar triple junction (Figures 19.1 and 19.2). The Benue Trough, is an example of a failed rift of Cretaceous age that intersects the sharp change in trend of African margin defining the African Bight (Figure 19.2). It was active at approximately the same time as the separation of South America from Africa. The two intersecting margins of the bight mark the original location of the other two arms of the RRR triple junction, which remained active to produce the adjacent part of the south Atlantic Ocean. Similar sharp bends in the margins around the Atlantic Ocean probably formed in a comparable manner. The Benue Trough may be a modern counterpart to Proterozoic aulacogens that are present in many continental interiors (see Section 1.7(ii), Figure 1.19).

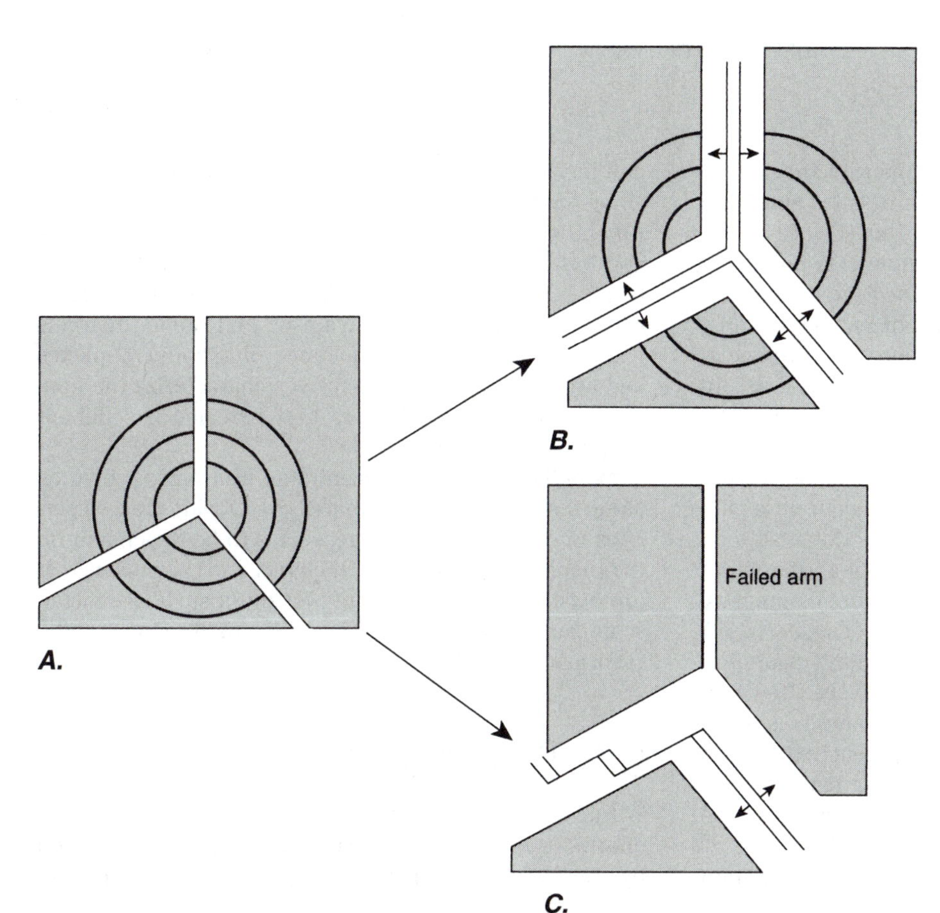

FIGURE 19.9 Evolution of a rift starting as a three-armed graben on a domal uplift. *A* to *B*. Breakup of a continent into three masses with the development of a ridge-ridge-ridge triple junction. *A* to *C*. Breakup of a continent into two masses with a predominantly ridge boundary, a predominantly transform boundary, and a failed arm of a ridge forming a trough that extends into one continent at the concave angle (often called a "bight") in the continental margin.

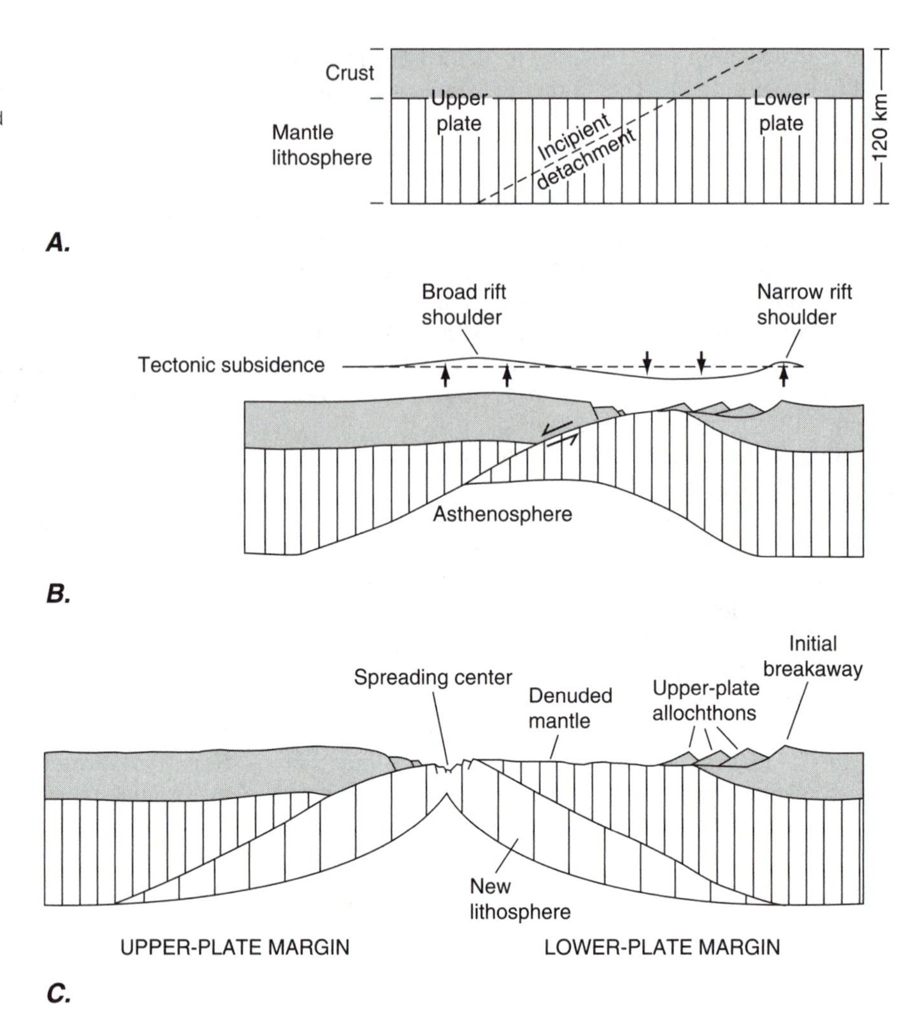

FIGURE 19.10 Detachment model for the evolution of an asymmetic passive continental margin. (After Froitzheim and Manatschek 1996)

Passive rifting of a continent begins with subsidence along the incipient rift as the plates are pulled apart and the continental crust and lithosphere thin (Figure 19.10). This stretching and thinning of the continent is commonly accommodated by an asymmetric listric detachment fault complex. Subsidence and deposition of sediments in the resulting depression are followed by mantle upwelling to redress the isostatic imbalance (Figure 19.10*A*, *B*). In most cases, this upwelling mantle partially melts, and the magma thus produced intrudes the crust or extrudes on the surface, eventually to form oceanic crust (Figure 19.10*C*). Where mantle melting does not occur or is minimal, the structure of the rifted margin is dominated by the listric detachment fault system and the formation of structures comparable to metamorphic core complexes. Passive rifting is more common along segments of rift that lie between the domes, and active rifting characterizes the areas over the domes. As the passively rifted margin develops, subsidence within the rift area is isostatically compensated by a broad uplift apparent in the upper plate of the listric detachment and a more narrow uplift of the lower plate in the area of the breakaway of the listric detachment fault system (Figure 19.10*B*).

These models of rifting can account for the formation of new ocean basin separating the rifted margins of the continents, but the structure and the rocks deposited along the rifted margins reveal the differences in history. Alkalic extrusives followed by ocean margin sediments and basaltic extrusives characterize active rifting; sediments covered by later basaltic lavas characterize passive rifting with subsequent volcanism; and faulted margins with structures similar to metamorphic core complexes and little or no basaltic extrusives characterize the nonvolcanic margins, which are best revealed if sediment supply was also minimal.

Whether rifting is concentrated in a narrow belt, as along the Red Sea, or distributed over a wide area, as perhaps in the Basin and Range, seems to be dependent on the rheological structure of the crust and the presence of pre-existing crustal zones of weakness such as crustal-scale faults or ancient sutures, which tend to localize the deformation.

19.2 DIVERGENT MARGINS IN OCEAN BASINS

Within the ocean basins, midoceanic ridges constitute an enormous system of active structures that are dominated by simultaneous normal faulting and magmatic activity (Figure 19.1; see Figures 1.6 and 1.8). We focus our dis-

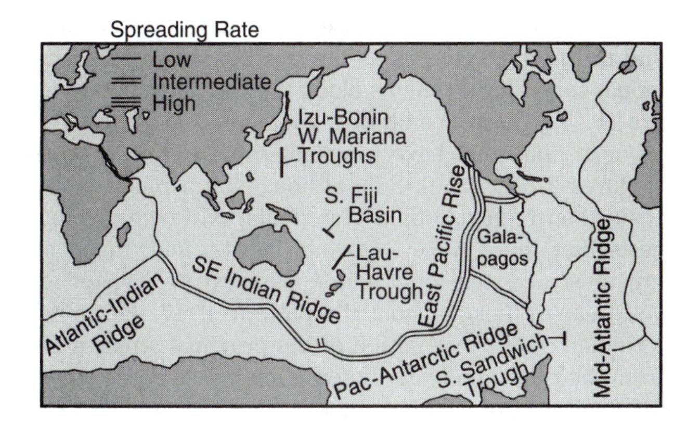

FIGURE 19.11 Map of world's ridges and "back-arc" spreading centers—active spreading centers in back-arc basins, classified according to spreading rate. Back-arc spreading centers—Izu-Bonin zone, Mariana trough, S. Fiji Basin, Lau Havre trough west of the Tonga-Kermadec island arc, South Sandwich trough—are all slow-spreading. (Modified after Macdonald 1982; Moores and Twiss 1995, figure 5.18; Lagabrielle et al. 1998; Perfit et al. 1994, p. 375)

cussion on those features that directly reflect the structural processes occurring at the spreading center.

i. Large-Scale Topography of Oceanic Ridges

Beyond the continental margins, oceanic crust is dominated by rock of basaltic composition, which has been produced at the spreading centers along the oceanic ridge system. Minor amounts of spreading can also occur in back-arc basins, but we focus on the predominant oceanic ridge system. Figure 19.11 shows a map of the active oceanic ridges, together with back-arc spreading centers, with different symbols employed for the three different categories of spreading rate. Most ridges stand about 2.5 kilometers above the flanking abyssal plane (Figure 19.12). As oceanic lithosphere moves away from the spreading center, it cools and undergoes an associated thermal contraction. The vertical component of that contraction results in subsidence of the ocean floor with time, which eventually, after 60 –80 My, approaches a steady-state depth of approximately 5–6 km (Figure

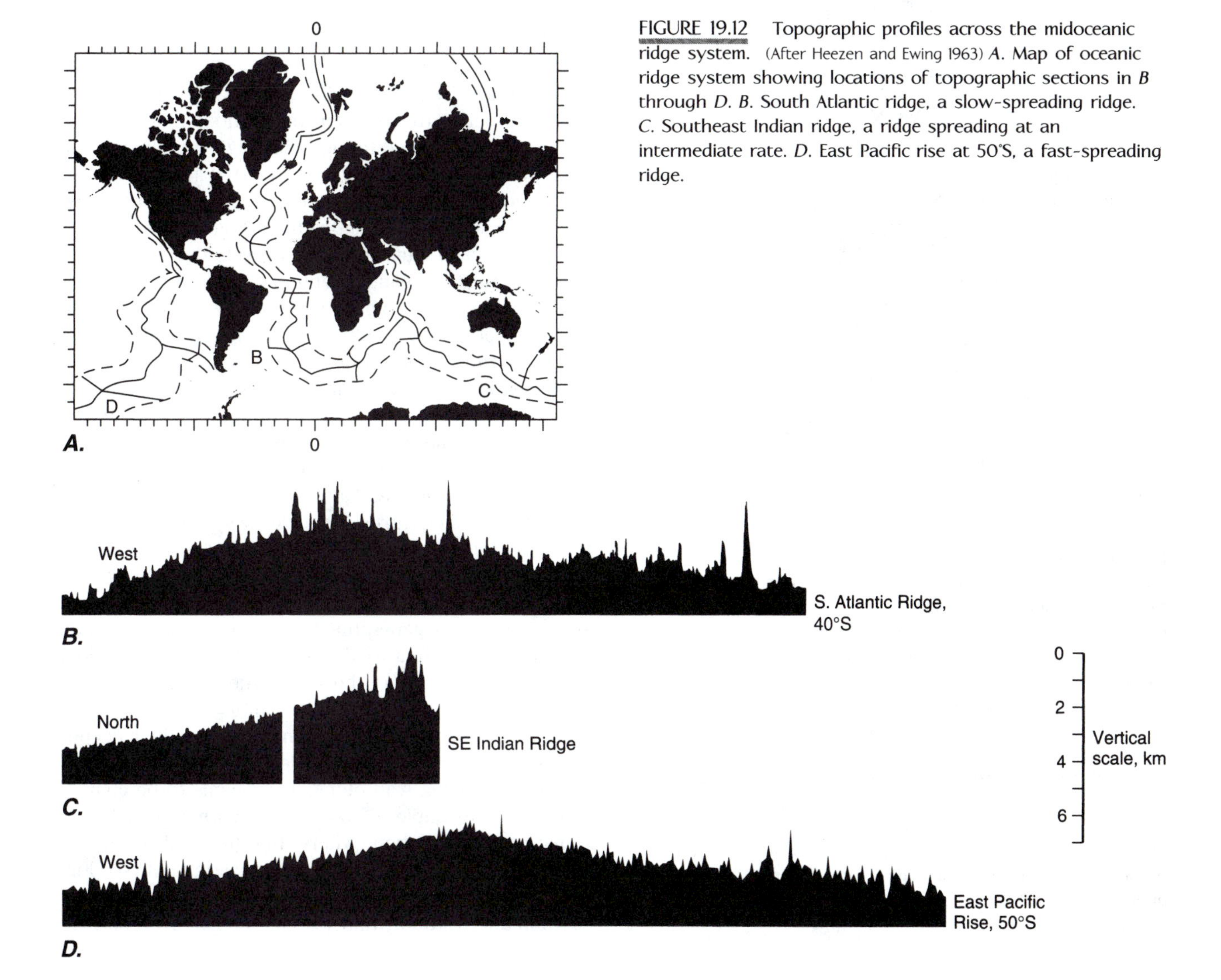

FIGURE 19.12 Topographic profiles across the midoceanic ridge system. (After Heezen and Ewing 1963) *A*. Map of oceanic ridge system showing locations of topographic sections in *B* through *D*. *B*. South Atlantic ridge, a slow-spreading ridge. *C*. Southeast Indian ridge, a ridge spreading at an intermediate rate. *D*. East Pacific rise at 50°S, a fast-spreading ridge.

19.13*A*). The cooling also results in the isotherms in the lithosphere moving downward. Because the base of the lithosphere is defined by a particular isotherm, the lithosphere becomes thicker, and after 60–80 My it approaches a steady-state thickness of about 100 km (Figure 19.13*B*).

ii. Ridges Interpreted as Crustal-Scale Extension Fractures

All ridges exhibit discontinuities at a variety of length scales. The most obvious first-order discontinuities at scales of tens to hundreds of kilometers are transform faults at which ridge segments are offset. We discuss these in more detail in Section 19.3. Other smaller discontinuities are present as well, including such second- and third-order features as the overlapping tips of individual spreading segments along the spreading axis (Figure 19.14). These tips show a geometry similar to overlapping joints that have propagated toward each other (Figures 2.10*F* and 9.12*C*), although on a much larger scale than regular joints. The similarity, however, suggests that these ridge segments can be interpreted as crustal-scale extension fractures that have propagated toward each other and that the pattern of overlap is determined by the orientation of the principal stresses far from the ridge and by the interaction between the stress field and the propagating cracks, as we discussed in Section 9.5(viii).

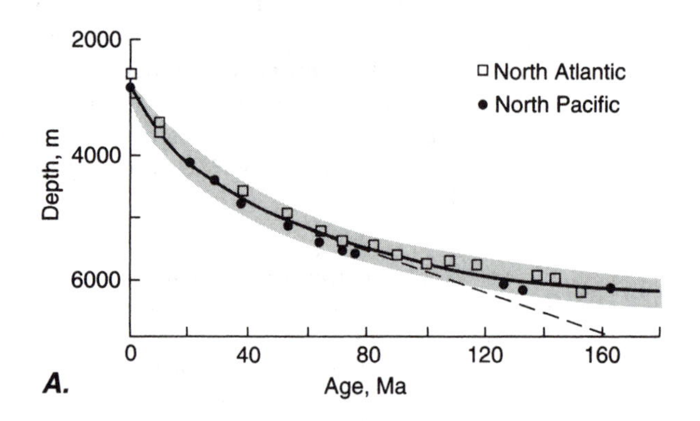

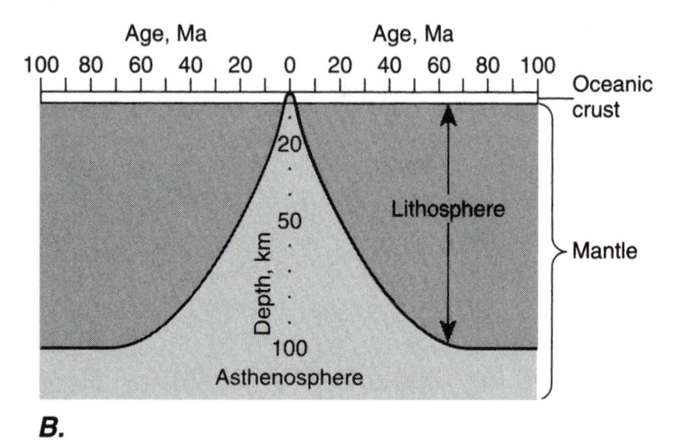

FIGURE 19.13 Topographic effects of cooling oceanic lithosphere. *A*. Plot of mean ocean depth versus age for the North Pacific and North Atlantic. The shaded area shows an estimate of scatter in the original points used to produce the curve. The solid line shows the theoretical topography, assuming that the lithosphere reaches a limiting thickness at about 80 Ma. The dashed line shows the theoretical topography for a lithosphere whose thickness increases indefinitely with the square root of time ($t^{1/2}$). (After Sclater et al. 1980) *B*. Model for thickening of the oceanic lithosphere with age. The lithosphere reaches a maximum thickness of approximately 100 kilometers after about 80 My, corresponding to the depth/age relationship of *A*. The vertical exaggeration varies with spreading rate, but it is approximately 10X for a spreading rate of 5 cm/y. (After Sclater et al. 1981)

iii. Structures at Oceanic Spreading Centers

For descriptive purposes, oceanic ridges are commonly separated into three categories according to whether the spreading rates are low (1–5 cm/yr), intermediate (5–9 cm/yr), or high (9–20 cm/yr) (Figure 19.11). The classification is justified by differences in topography, distribution of rock types, and structure, as well as other physical characteristics.

Ridges with low spreading rates (Figure 19.15*A*) are characterized by a deep axial graben with depths up to 1.5–3 km below the adjacent ocean floor that is superimposed on the broad topographic high that defines the crest of the ridge itself (Figure 19.12*B*). Within the graben itself, elongate shield volcanoes are intermittently present as shown diagrammatically in Figure 19.15*A*; they tend to occur in clusters. This suggests that the magma supply at these spreading centers is not continuous.

Typically, the graben is characterized by rough faulted topography that is approximately symmetrical about the ridge axis and that consists of an imbricate series of listric normal faults dipping toward the axial trough. Extrusive layers of lava dip away from the spreading center, and dikes dip toward it. These orientations of lavas and dikes are consistent with the rotation of normal fault blocks about a horizontal axis due to sliding on the listric faults as spreading carries the rocks away from the spreading axis (see Figure 19.19*B*). Spreading is also accommodated in part by magmatic addition to the crust as magma chambers at depth, dikes above that, and extrusive lavas on top, generating the characteristic ophiolitic crustal sequence that we describe in the next subsection.

In some situations, however, the extension is accommodated almost entirely by faulting with the formation of a listric low-angle detachment along which deeper levels of the crust or even mantle are pulled out from beneath the hanging wall block. This seems to be characteristic, for example, of the "inside corner" of a ridge-transform intersection where the material from the spreading ridge abuts the active transform fault (see Figure 19.27). Here, the topography is high relative to the surrounding seafloor, and lower crust or even serpen-

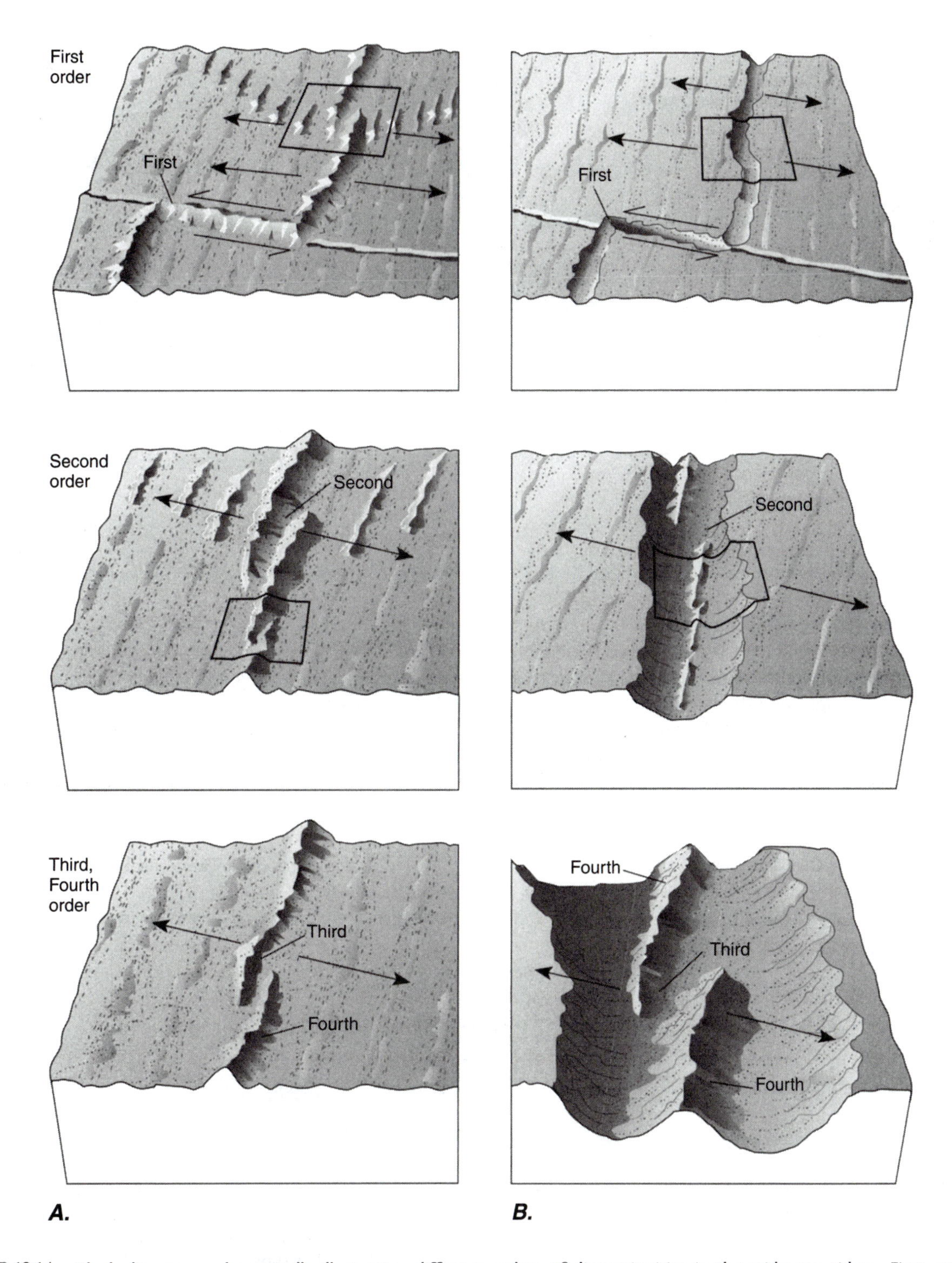

FIGURE 19.14 Block diagrams, schematically illustrating different orders of discontinuities in the midocean ridges. First-order discontinuities are transform faults that continue beyond the ridge intersections as fracture zones. Second-order discontinuities on fast-spreading centers are overlapping rifts formed by propagating ridges that are preserved as V-shaped trains of ridge tips. Third- and fourth-order discontinuities are lesser overlaps, offsets in volcanoes, or slight deviations in strike. (After Macdonald and Fox 1990, p. 74) *A*. Fast-spreading centers. *B*. Slow-spreading centers.

tinized peridotite from the mantle lithosphere is exposed at the seafloor (Figure 19.16). The "outside corner" of the ridge-transform intersection, where spreading material from the ridge abuts the inactive segment of the transform, shows a listrically faulted upper oceanic crustal sequence. In all cases, the flanks, or footwall blocks, of the graben are progressively uplifted along the faults relative to the graben bottom.

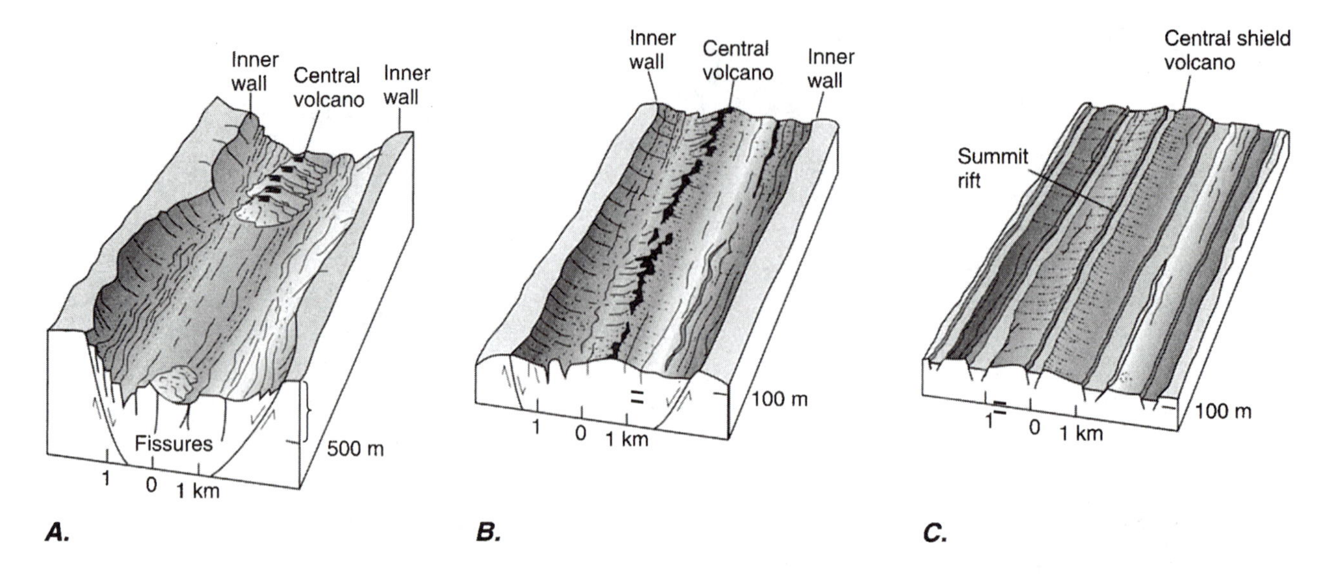

FIGURE 19.15 Diagrams showing the topography and structure of the axial zone on ridges of varying spreading rate. Note that the block diagrams have a vertical exaggeration of 2:1. (After Macdonald 1982) *A*. Slow-spreading ridge (1–5 cm/yr). *B*. Intermediate-spreading ridge (5–9 cm/yr). *C*. Fast-spreading ridge (9–20 cm/yr).

Ridges having intermediate spreading rates display topography intermediate between the steep topography of slow-spreading ridges and the subdued topography of fast-spreading ridges (Figures 19.15*B* and 19.12*C*). A central graben is present, but it is generally not more than 100 to 200 m deeper than the adjacent ocean floor. Central volcanoes are nearly continuous, commonly with extension fractures arranged *en echelon* along their crests (Figure 19.15*B*).

Ridges with high spreading rates typically display no axial valley (Figure 19.15*C*; compare Figure 19.12*D*). The volcanic edifice at the center of the rift is more continuous

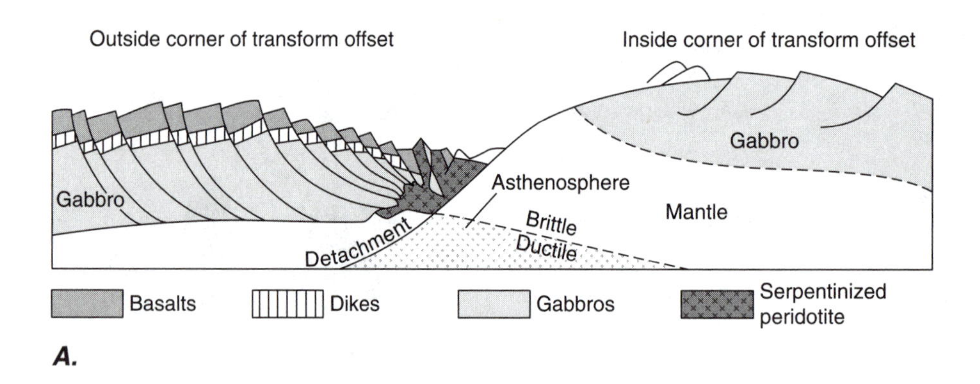

FIGURE 19.16 Schematic cross sections showing structural models of slow (*A–C*) and fast (*D*) spreading ridges. *A*. Model of faulted ridge with complete crustal sequence on the left flank and eroded mantle-crustal sequence on the right. With respect to the ridge segments offset along a transform fault, the "inside corner" of the ridge abuts the active part of the transform fault between the two ridge segments, and the "outside corner" abuts the inactive part of this transform. Mantle is exposed by faulting on the detachment dipping steeply to the left during periods of minimal to zero magma supply to the spreading ridge. *B*. Model of asymmetrically faulted ridge where oceanic crustal sequence is incomplete and tectonized serpentinized mantle rocks form the oceanic crust. *C*. Model of symmetrically faulted ridge with incomplete oceanic crustal sequence and tectonized serpentinized mantle forming the oceanic crust. *D*. Schematic block diagram of fast-spreading ridge showing volcanic ridges, an axial summit caldera, a thin axial magma chamber, and a wider zone of very hot or partially molten rock. Enriched mid-ocean ridge basalts (E MORB) have higher abundances, relative to normal mid-ocean ridge basalts (N MORB), of the highly incompatible elements such as Rb, Ba, Th, U, Nb, the light rare-earth elements, and the radiogenic isotopes of Sr and Pb. (*A–C* after Lagabrielle et al. 1998; *D* after Perfit et al. 1994)

(continued)

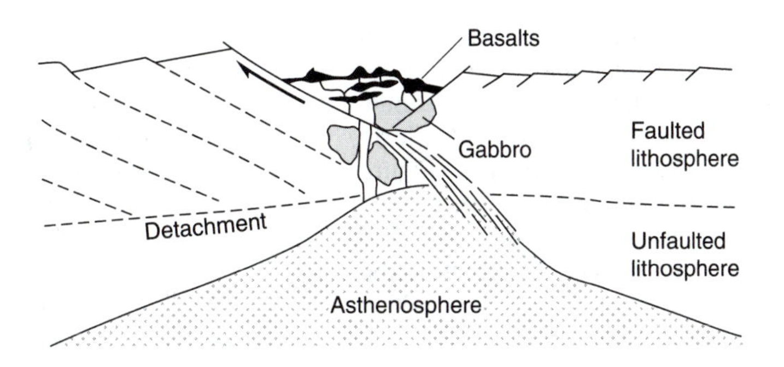

B.

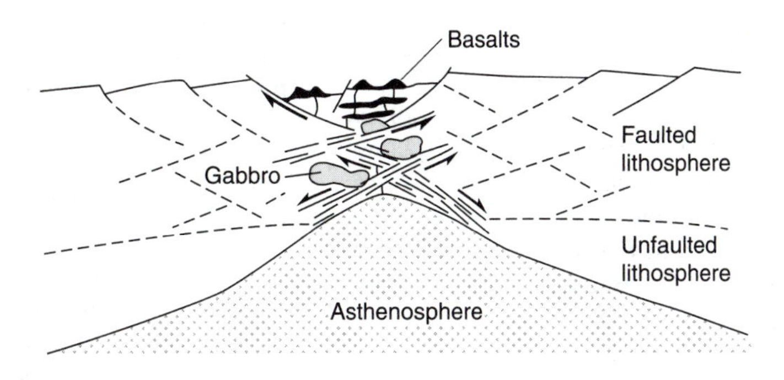

C.

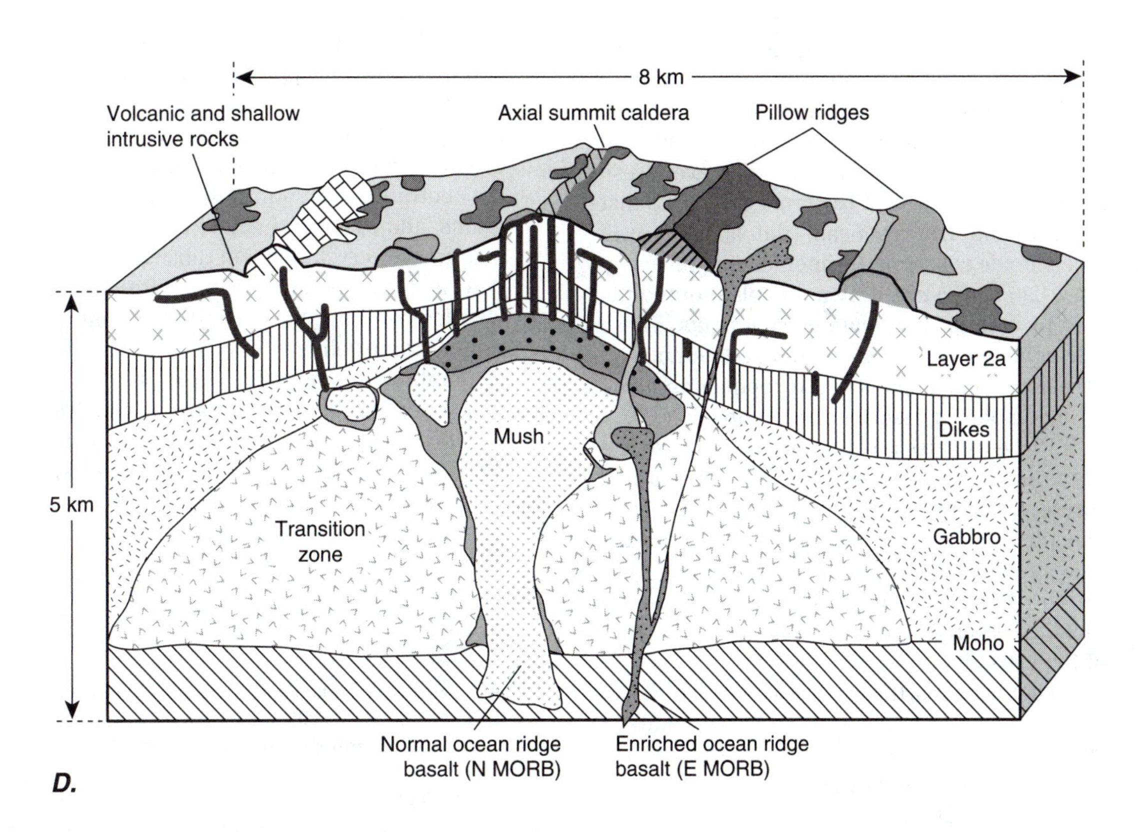

D.

FIGURE 19.16 *(continued)*

and reminiscent of Hawaiian volcanoes, with a small summit ridge or graben. The topography is smooth, in contrast with that of slow-spreading rifts. Fast-spreading ridges display little or no faulting. Thus lavas tend to form horizontal layers, and dikes are nearly vertical (Figure 19.16*D*), although in a few cases, the lavas tilt toward the spreading center and dikes tilt away. This structure presumably reflects a greater buildup of lava adjacent to the spreading center, with an associated isostatic sinking where the buildup is greatest, resulting in a tilt of the lava contacts toward the spreading center. These characteristics are consistent with a more or less continuous supply of magma as the ridge spreads, with individual successive lava flows thinning away from the central feeder zone.

The differences in structural characteristics of spreading ridges are intimately connected to the supply of magma to the crust. On slow-spreading ridges, magma supply is continual but intermittent, although on some very slow-spreading ridges, no magmatic rocks at all are present. Magma chambers are apparently exceedingly rare, although clearly they must exist episodically, and mantle rocks become exposed on the seafloor by normal faulting on low-angle detachments during the amagmatic periods.

Figure 19.16 shows three models of lithosphere evolution at the axis of a slow-spreading ridge. Figure 19.16*A* represents a possible cross section near a ridge-transform intersection. Here, the spreading center tends to be starved of magma, probably because the proximity to the cooler lithosphere that abuts the ridge across the transform fault suppresses the generation of magma (Figure 19.16 *A–C*). Extension is accommodated almost entirely by the formation of a listric low-angle detachment, and the footwall block consists of serpentinized or even fresh peridotite from deeper levels of the crust or mantle that have been extracted from beneath the hanging wall block. On the outside corner of the intersection, the hanging wall block is characterized by a complete ophiolitic sequence cut by a set of imbricate normal faults. These low-angle detachment faults that accommodate magma-starved extension are analogous to the continental metamorphic core complexes described in Section 4.3(iii).

Figure 19.16*B* illustrates a region of low to moderate rate of magma production. Some volcanic and plutonic rocks are present, but much of the crust is formed of tectonized lithosphere. The model suggests an asymmetric normal fault situation. In Figure 19.16*C*, a more symmetric distribution of listric normal faults is envisaged that deforms the serpentinized lithosphere, which makes up most of the crust.

At fast-spreading ridges, the geophysical data indicate that the layers of volcanic extrusive and dike complex are underlain at the ridge by a lens-like axial magma chamber sitting on top of a region of very hot or partially molten rock (Figure 19.16*D*). A prominent model for evolution of ridges invokes the existence of episodic magma chambers along the axis of a ridge. A pulse of magma in one location will cause the development of a swell in the topography and propagation of a spreading center along the axis until the magma is exhausted. Two adjacent pulses of magma along a ridge axis will result in overlapping centers. In this way, spreading proceeds by injection of dikes presumably in a plane perpendicular to $\hat{\sigma}_3$, similar to hydrofracturing (see Sections 8.5, 9.2(ii)).

iv. Ophiolites

The investigation of active ridges is hampered by the depth of the ocean. Thus much of our information comes from geophysical and remote-sensing surveys. As a result of this work, however, we now recognize that there are places where the oceanic crust has been thrust up and exposed on land, and investigation of these exposures, called **ophiolite sequences**, contributes significant detail to our understanding of the structure and processes at oceanic ridges.

In basic outline, an ideal ophiolite sequence includes the following sequence of rocks, from bottom to top (Figure 19.17*A*):

A. At the base of the sequence is tectonized peridotite, which in composition is commonly harzburgite (olivine + orthopyroxene) but also lherzolite (olivine + orthopyroxene + clinopyroxene),[1] with subordinate dunite (which consists of >95% olivine) and chromitite (which contains concentrations of chrome spinel). Compositionally these rocks are relatively uniform, and the Mg/Fe ratio in olivines and pyroxenes is approximately 9:1. Structurally they are tectonites that are characterized by a pronounced foliation defined by preferred orientation of elongate mineral grains, by laminar concentrations of subordinate pyroxene and spinel grains, and by fabrics and textures that reveal a history of recrystallization during ductile deformation. The microscopic fabric is characterized by strong crystallographic preferred orientation of olivine. Layers of pyroxene, chromite, or dunite are commonly folded into isoclinal (class 2 or class 3) folds, and in places such folds are refolded. These rocks are sometimes called "Alpine peridotites" because of their characteristic occurrences in the Alps. We infer that they represent the upper-mantle lithosphere.

[1]Olivine is made of stacked individual silicate tetrahedra and compositionally is an Mg-Fe silicate. Pyroxenes are made of chains of silicate tetrahedra: Orthopyroxenes have orthorhombic crystal symmetry and consist essentially of Mg-Fe silicates; clinopyroxenes have monoclinic crystal symmetry and can have a wide range of compositions. Many are Ca-Mg-Fe^{2+} silicates, but many other cations can substitute for these, including Na, Mn, Fe^{3+}, and Al, among the more important ones.

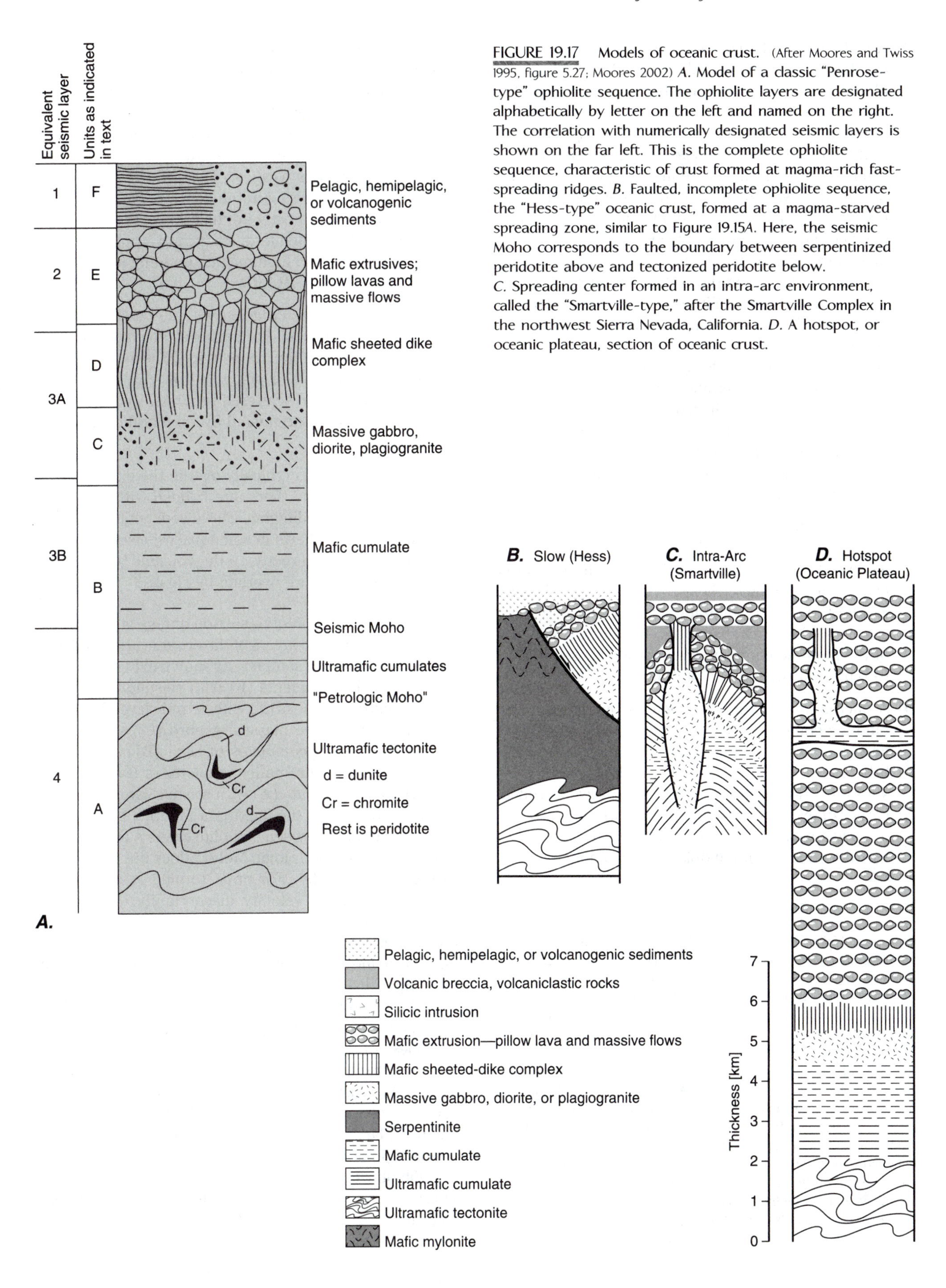

FIGURE 19.17 Models of oceanic crust. (After Moores and Twiss 1995, figure 5.27; Moores 2002) *A*. Model of a classic "Penrose-type" ophiolite sequence. The ophiolite layers are designated alphabetically by letter on the left and named on the right. The correlation with numerically designated seismic layers is shown on the far left. This is the complete ophiolite sequence, characteristic of crust formed at magma-rich fast-spreading ridges. *B*. Faulted, incomplete ophiolite sequence, the "Hess-type" oceanic crust, formed at a magma-starved spreading zone, similar to Figure 19.15*A*. Here, the seismic Moho corresponds to the boundary between serpentinized peridotite above and tectonized peridotite below. *C*. Spreading center formed in an intra-arc environment, called the "Smartville-type," after the Smartville Complex in the northwest Sierra Nevada, California. *D*. A hotspot, or oceanic plateau, section of oceanic crust.

B. A mafic-ultramafic stratiform plutonic complex overlies the tectonite or Alpine peridotite. This plutonic complex commonly is rich in olivine and pyroxene at the base, and grades upward into units rich in plagioclase and, near the top, even quartz or hornblende. These rocks also are compositionally layered, but their layers are more regular than those of the underlying tectonite. The mineral grain textures are characteristic of crystallization from a magma, rather than of recrystallization during metamorphism or ductile flow. The peridotite at the base of this layer and the constituent mineral compositions are more iron-rich than the underlying Alpine peridotites. Layering and compositions suggest that many of these rocks formed by accumulation of crystals at the bottom of a magma chamber by gravity settling. Thus the lower contact of these rocks marks the true crust-mantle boundary (the "petrologic Moho"[2]) because it is the contact between the solid ductilely deformed mantle and the oceanic crust formed from magmatic rocks.

C. The top of the plutonic complex generally is composed of coarse- to fine-grained vari-textured gabbro, diorite, and leucocratic quartz diorite (often called plagiogranite). These rocks are probably the result of in situ crystallization, rather than the gravity-settling that characterizes the stratiform part of the plutonic complex. We interpret these rocks to represent the top of a plutonic body.

D. A finer-grained mafic dike complex overlies the plutonic complex in many instances. This complex consists of dikes intruded into plutonic rocks at the base, areas of 100% dikes in the middle, called a **sheeted dike complex**, where dikes have intruded earlier dikes, and at the top, dikes intruded into extrusive rocks, commonly pillow basalts (Figure 19.18*A*). In areas of multiple dike injection, dikes intrude either along the margin or into the center of pre-existing dikes (Figure 19.18*B*). Where a later dike splits an earlier dike in half, presumably it is because the older dike is still hot and therefore weak in the central region but cooler and stronger along its margins, and a new fracture develops where the rock is weakest. This process is repeated many times, resulting in the development of dike-like bodies that mostly are half-dikes, bounded on each side by chilled margins having a polarity from margin into the chilled magma pointing in the same direction (Figure 19.18*C*).

E. Overlying the dike complex with a gradational contact is a sequence of extrusive volcanic rocks, generally basaltic in composition. These rocks consist of massive or pillowed flows, with a few sills or dikes, and scattered breccias. We infer that these sequences originated by submarine extrusion of lavas.

F. Pelagic sediments typically overlie the extrusive sections, here and there interbedded with metal-rich chemical sediments. These sediments include thinly bedded cherts in some places; elsewhere limestones rest directly on the volcanic basement and themselves are overlain by cherts.

This idealized ophiolite section agrees well with the seismically identified layers of the oceanic crust and with evidence from direct samples. As shown in Figure 19.17*A*, we identify the sediments (F) with seismic layer 1, the volcanic extrusive rocks (E) with seismic layer 2, the dike complex (D) and the massive gabbro (C) with seismic layer 3A, the mafic cumulates of layer (B) with seismic layer 3B, and the ultramafic cumulates of layer (B) plus the ultramafic tectonite (A) with seismic layer 4. Implicit in this correlation is the important fact that the igneous-tectonite contact between ophiolite layers A and B, sometimes called the "petrologic Moho," does not correspond to the seismic Moho. For crust formed at magma-rich ridges, the seismic velocity discontinuity defining the seismic Moho lies between seismic layers 3 and 4 within the plutonic complex (ophiolite layer B), between the mafic (olivine-poor) cumulate rocks above and the ultramafic (olivine-rich) cumulate rocks below (Figure 19.17*A*). The sharpness of this petrologic transition from poor to rich olivine compositions is a measure of the sharpness of the seismic Moho. In many ophiolites, this transition ranges in thickness from 50 to a few hundred meters. In crust formed at magma-poor, faulted ridges, such as portrayed on Figure 19.17*B*, however, the seismic boundary is between fresh peridotite below and serpentinized peridotite above.

Ophiolites also display variations reminiscent of those observed in the oceanic crust. Some contain complete untilted sequences such as in Figure 19.17*A* (e.g.,

[2]The term *Moho* is a contraction of "Mohorovičic′ discontinuity," which is also sometimes referred to as the "M-discontinuity." The discontinuity is a worldwide sharp increase in seismic velocity with increasing depth that occurs at a depth of about 5–8 km under the oceanic crust and 35–50 km under continental crust. It was first identified in 1909 by the Croatian seismologist Andrija Mohorovičic′ and was subsequently named after him. Since then various contractions of the name have come into use, presumably because "Mohorovičic′" is such a mouthful. Although the term fundamentally refers to the seismic velocity boundary between the crust and mantle, it has come to be used for other definitions of the crust/mantle boundary. Thus, if it is defined by the boundary between magmatic rocks and tectonites—the "petrologic Moho,"—it may have a slightly different location than if it is defined by the jump in seismic velocity, which corresponds to the change from mafic to ultramafic rocks—the "seismic Moho" (see Figure 19.17*A*).

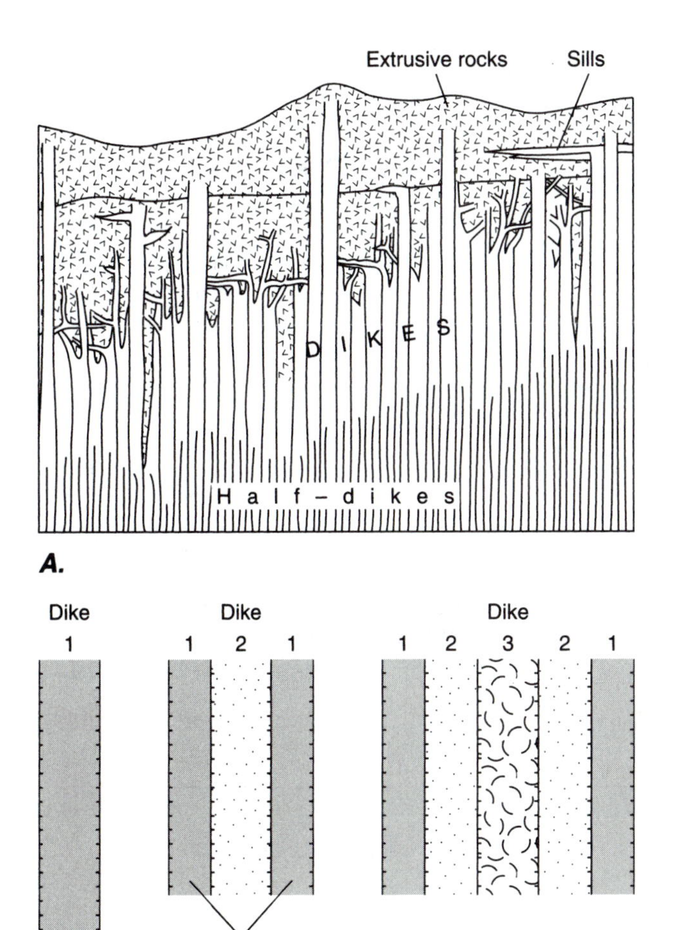

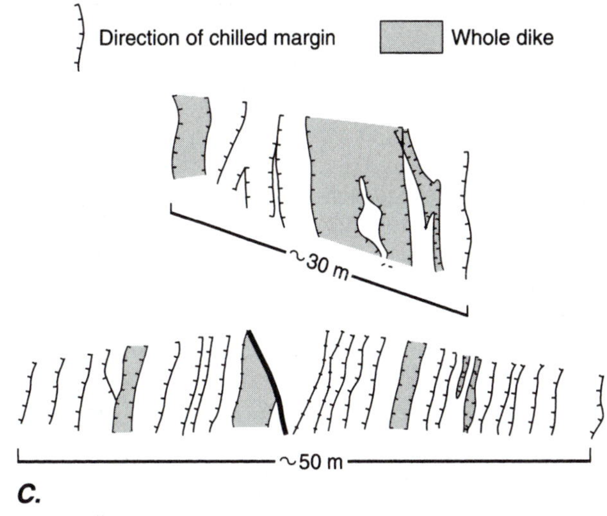

FIGURE 19.18 Sheeted dike complexes. *A*. Troodos sheeted dike complex and its relationship to the overlying extrusive rocks. *B*. Three steps showing the development of dike-within-dike relationships by successive intrusion of new dikes up the center of an earlier dike. The final result is areas of dike intrusions characterized by chilled margins all of the same polarity. *C*. Representative cross sections through the Troodos sheeted dike complex showing the polarity of the chilled margins. (*A* after Wilson 1959; *B* and *C* after Moores and Vine 1971)

Semail complex, Oman) that suggest formation at magma-rich spreading centers. Others display incomplete ophiolite sequences, suggesting formation at slow-spreading magma-starved ridges (Figure 19.17*B*) (e.g., several ophiolites in the Alps, Italy, the Balkan peninsula, and Europe). Others display graben structures that may be either fossil axial graben or faulted regions of fast-spreading oceanic crust (Troodos complex, Cyprus, which is an example of a complete ophiolite as shown in Figure 19.17*A*; see Figure 19.19). Others display incomplete sequences, reminiscent of those from slow-spreading magma-starved ridges (Figure 19.17*B*) (e.g., several ophiolites in the Alps, Italy, the Balkan Peninsula, and Europe). The "Smartville-type" ophiolite illustrated in Figure 19.17*C* represents a sequence of extrusive and intrusive mafic rocks, probably formed in an island arc setting, subsequently intruded by a sheeted dike complex. Such a complex may have formed in a rifted arc setting. The "Oceanic plateau" type of oceanic crust represents crust formed by extrusion of thick lavas over a preexisting oceanic crust or development of thick oceanic crust at a magma-rich spreading center.

v. Subaerial Exposures of Oceanic Spreading Centers

It is difficult to map midocean ridges in detail because the observations must be made by either geophysical remote sensing or direct observation from a deep-ocean submersible. Thus the places in the world where oceanic crust has been thrust up to be subaerially exposed provide a wealth of detailed data on the structure and the process of extension that complement and constrain the models of oceanic spreading centers. These regions are the ophiolite complexes mentioned in Subsection 19.2(iv). One good example is the Troodos complex, Cyprus (see Figure 19.19). The complex includes pillow lavas, sheeted dikes, and plutonic rocks. The Solea graben is a prominent feature that has an axis trending north-northeast. In the north, the graben is covered by the uppermost lavas; south from that are exposed progressively-deeper levels of the crust, reaching into the plutonic rocks of Mount Olympus in the south. Relationships in the Solea graben indicate the presence of a complete ophiolite sequence that has been modified by listric normal faults that dip

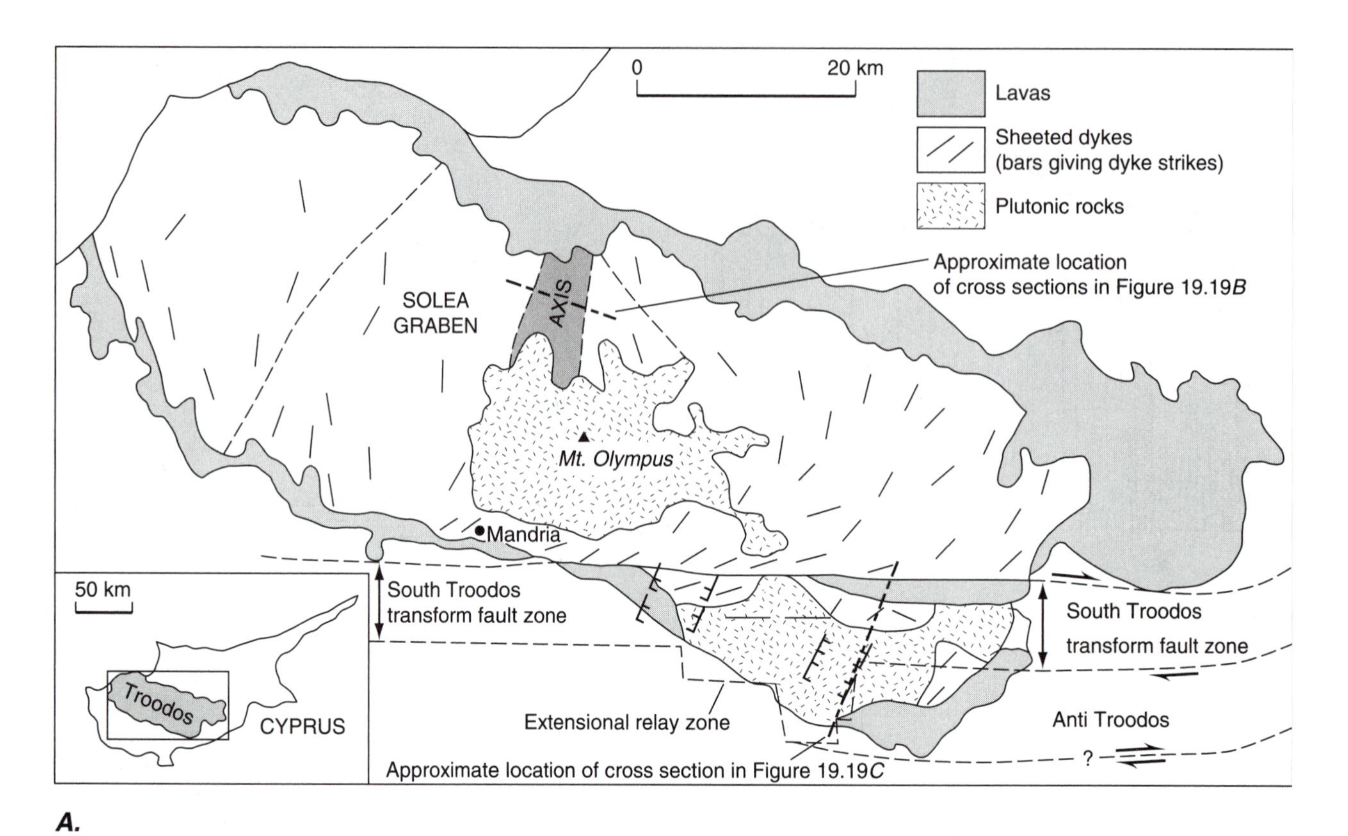

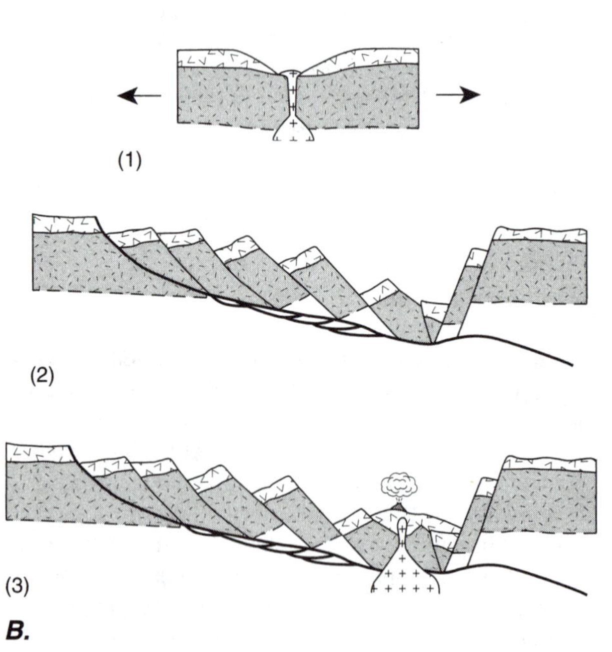

FIGURE 19.19 Troodos Complex, Cyprus. *A*. Simplified geologic map of the Troodos complex, showing the principal features of the spreading geometry. The Solea graben (the area within the dashed lines) is a possible extinct ridge axis. It extends through the plutonic rocks and connects with the South Troodos Transform fault zone. The anti-Troodos area is the plate south of the transform fault. (After Gass et al. 1994) *B*. Schematic cross sections showing structural/ igneous relations along the Solea graben axis. (1) Production of a complete ophiolite section preceded (2) listric normal faulting, which (3) was then intruded and overlain by later magmas. The complex may represent a slow-spreading ridge or a fast-spreading ridge subsequently ruptured by faulting and subsequent vulcanism. See text for discussion. (After Moores and Twiss 1995, figure 5.29) *C*. Schematic cross section through the Arakapas transform fault, Troodos complex, Cyprus. (After Gass et al. 1994)

(continued)

toward the central graben axis. Rotation of hanging wall blocks on these faults leaves lavas tending to dip away from the graben axis and dikes tending to dip toward it. These rocks have subsequently been intruded and overlain by younger igneous rocks (Figure 19.19*B*). These characteristics suggest that the Solea graben is the axial graben, either of a relatively magma-rich slow-spreading center or of a fast-spreading center, which terminates southward at a transform fault. The Troodos complex provides an opportunity to map and study in detail the relations between igneous and tectonic activity that occur at a spreading center, where they are clearly exposed and easily accessible. Such detail cannot be obtained from deeply submerged oceanic rifts, and thus they provide unequaled insight into the mid-ocean ridge processes.

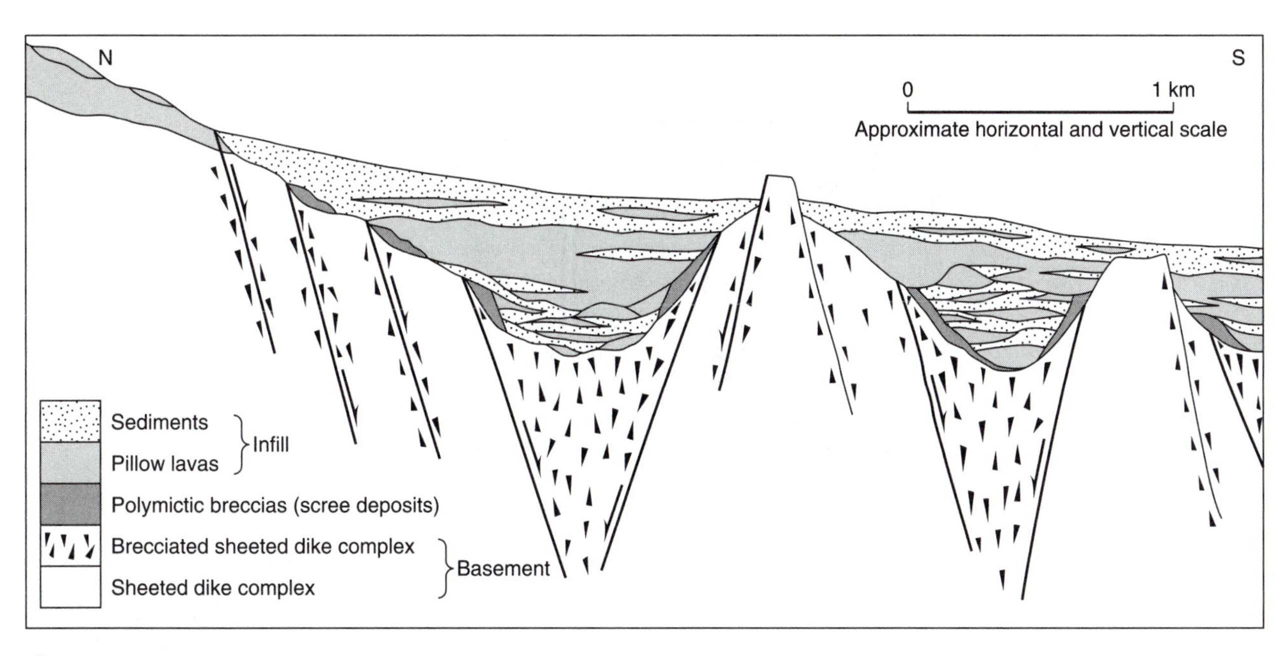

FIGURE 19.19 *(continued)*

19.3 MAJOR STRIKE-SLIP FAULTS: TRANSFORM FAULTS AND MEGASHEARS

Strike-slip faults are important structures in the Earth's crust. Many of them are transform faults that form parts of the plate boundaries, but others are faults of major scale that develop at convergent margins, commonly during collisions. These latter are sometimes referred to as continental megashears. We discuss them all in this section, however, because the structures associated with major strike-slip faults, whether transform faults or megashears of collisional origin, are basically indistinguishable, and the distinction is only apparent from their tectonic environment.

i. Oceanic Transform Faults

Transform faults are plate margins characterized by strike-slip faulting where adjacent plates move horizontally past each other and, ideally, where lithosphere is neither created nor destroyed. **Fracture zones** are prominent breaks in the oceanic crust that appear as conspicuous linear topographic features on the ocean floor and that mark discontinuities in the linear magnetic anomaly stripes. They include both active transform faults that are located between two spreading ridge segments (Figure 19.20*A*) or between a ridge segment and a subduction zone (Figure 19.20*B*, *C*) and the inactive extensions of these faults into the interior of a single plate. Across the inactive parts of a fracture zone, the older side of the crust was initially part of the active transform fault before it passed the adjacent ridge segment and became juxtaposed against newly created oceanic crust. At that point, both sides of the fracture zone belong to the same plate and move with the same velocity. Thus strike-slip activity does not occur on the inactive parts of a fracture zone, which therefore preserve the structures created in the older crust when it was part of the active transform fault. Only relatively minor vertical adjustments occur across the inactive parts of a fracture zone.

Figure 19.1 shows the location of major transform faults and associated fracture zones in the world. Three types of transform fault are possible, depending on the nature of the plate boundaries that are connected by the fault (Figures 19.20 and 19.21).

A **ridge-ridge transform fault** connects two segments of a ridge or divergent plate margin (Figures 19.1 and 19.20*A*), and they constitute the overwhelming majority of transform faults (Figure 19.1). As a result, they have been the subject of most of the studies of transform faults. Our understanding of transform fault structure and processes, which we discuss in greater detail in Subsection 19.3(ii) (see Figures 19.22–19.26), reflects this bias. Indeed such faults are present on the average approximately every 100 km along the strike of a ridge axis. In most cases, each transform fault maintains a constant length between ridge segments, which is the part of the fault that is actively shearing; beyond this part, the faults are inactive fracture zones (Figure 19.20*A*). Examples of such transform faults are especially prominent in the equatorial Atlantic and Indian Oceans, but they are present on other ridges as well (Figure 19.1).

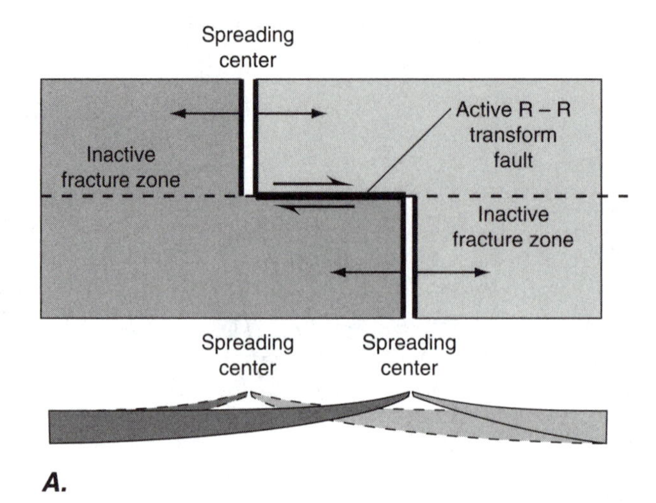

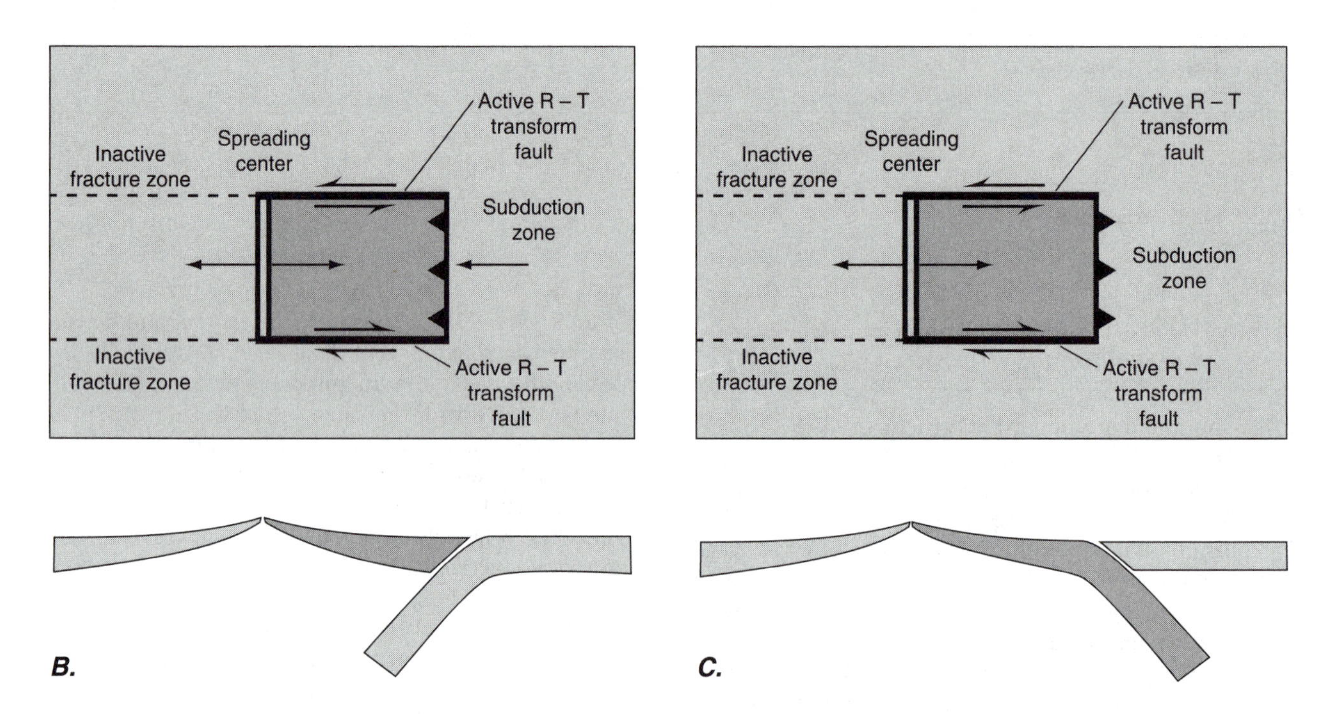

FIGURE 19.20 Idealized oceanic transform faults. *A.* Ridge-ridge transform fault showing map view and cross section in which the structure on the near side of the transform is in solid lines and that on the far side is in dashed lines. *B.* Ridge-trench transform fault in the over-riding plate. *C.* Ridge-trench transform fault in the down-going plate.

Ridge-trench transform faults connect a ridge to a trench; there are two types, depending on the polarity of the trench relative to the transform fault (Figures 19.1 and 19.20*B*, *C*). One connects a ridge on the over-riding plate with the convergent boundary (Figure 19.20*B*); the other connects a ridge on the down-going plate with the convergent boundary (Figure 19.20*C*).

Three types of **trench-trench transform faults** occur, distinguished by the polarities of the two convergent margins connected by the transform. The subduction zones of the connected trenches can dip toward each other (Figure 19.21*A*), they can dip away from each other (Figure 19.21*B*), or they can dip in the same direction (Figure 19.21*C*). Other types of transform fault end at triple junctions, such as the San Andreas transform fault in California, which terminates at the Mendocino triple junction in the north and at the Rivera triple junction in the south (Figure 19.1).

ii. Structures of Oceanic Ridge-Ridge Transform Faults

Most of our knowledge about oceanic transform faults comes from ridge-ridge transform faults. Little information is available on the other types of faults outlined in the preceding subsection. In all known cases, the faults

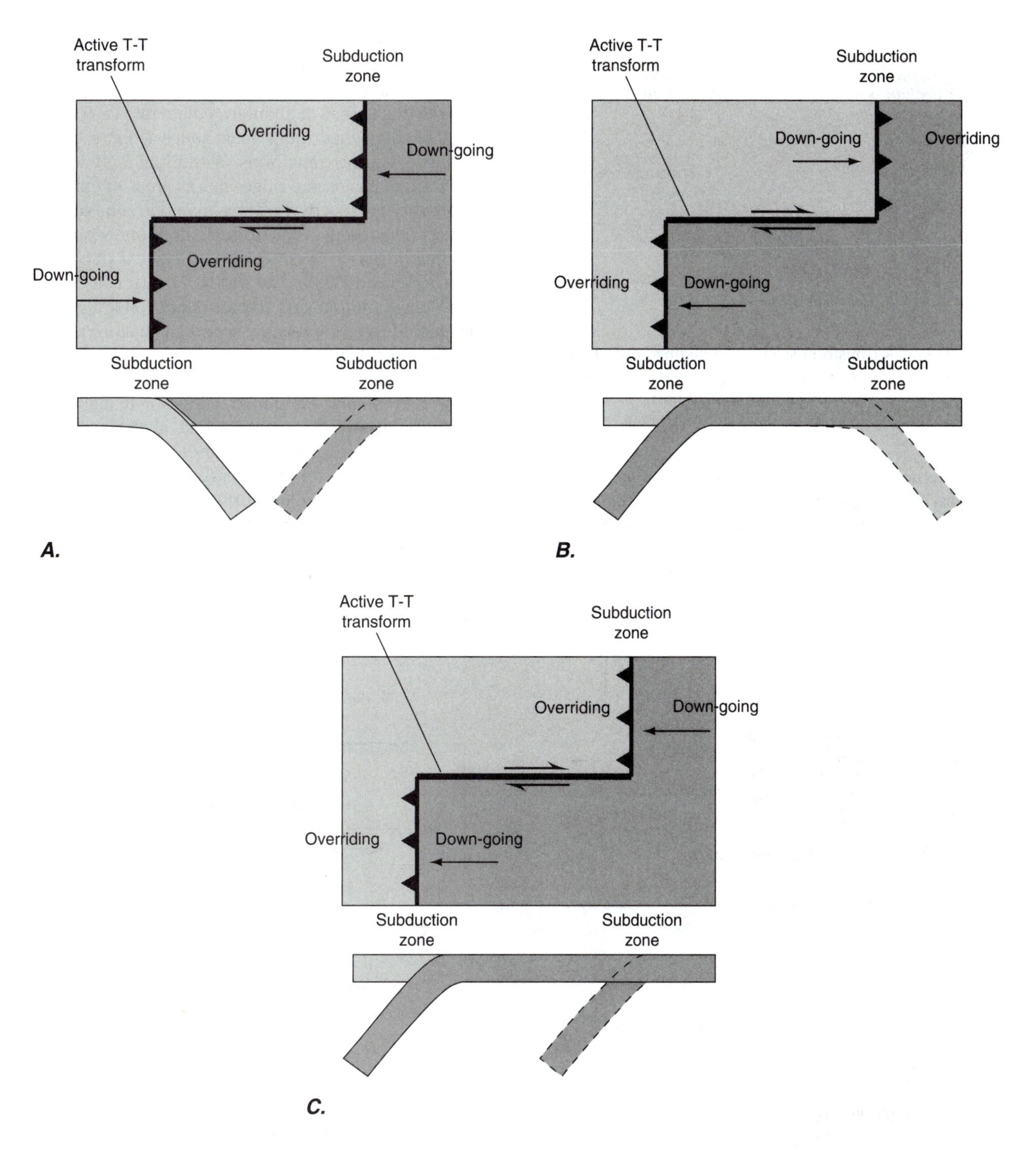

FIGURE 19.21 Trench-trench transform faults. *A*. Subduction zones dip toward each other. *B*. Subduction zones dip away from each other. *C*. Subduction zones are parallel.

are characterized by sharp topographic discontinuities, where steep ridges or scarps form the sides of narrow, deep basins. Fault scarps are especially apparent on ridge-ridge transform fracture zones (Figure 19.22; compare Figure 19.19*C*) and are most pronounced along long transform fracture zones where there is a large age difference across the fault. The younger, hotter lithosphere has an equilibrium elevation that is higher than the older, colder lithosphere (Figure 19.22), and the higher-temperature lithosphere occurs on the high side of a topographic scarp. Because of this relationship, the high side of the fault changes sides midway between the ridge segments.

Like any other strike-slip fault, a transform fracture is not a single plane through the Earth's crust, but rather

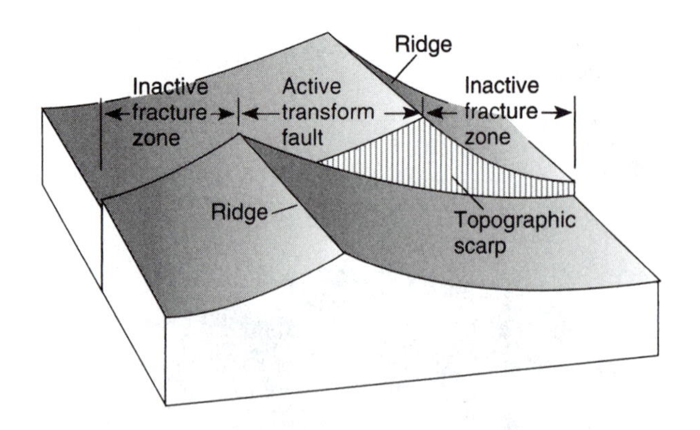

FIGURE 19.22 A transform scarp is formed because of the contrast in age and temperature of the lithosphere across the fracture zone. Young hot lithosphere stands topographically higher than older cooler lithosphere.

it is a fracture zone that can be many kilometers wide. The width and structural complexity of ridge-ridge transform faults increases as the spreading rate increases. Transform faults connecting slow-spreading ridges (< 5 cm/yr; Figure 19.23*A*), have a pronounced valley, within which is a single rather narrow fault zone, generally less than one kilometer in width, along which most of the displacement takes place. Transform faults connecting ridge segments that spread at an intermediate rate (5–9 cm/yr), display a somewhat wider fault zones (Figure 19.23*B*) than slow-shearing transforms. Within the fault zone, an alternation of basins and ridges marks areas of extension and intrusion, respectively. More than one zone of active slip may be present. Transform faults connecting fast-spreading ridges (9–18 cm/yr) are wide zones of complex faulting (Figure 19.23*C*). The width of the transform zone may be tens of kilometers to more than 100 km, and the zones include short spreading segments that are not necessarily perpendicular to the spreading direction and that may propagate across the zone.

Careful dredging, combined with seismic data, direct observation from submersibles, and analysis of fossil transform zones in subaerially exposed ophiolites gives a complex picture of a transform fracture zone. A schematic model for some of the main features of such zones is shown in Figure 19.24 as vertical sections parallel (Figure 19.24*B*) and perpendicular (Figure 19.24*C*) to the fault. The normal oceanic crustal sequence, from

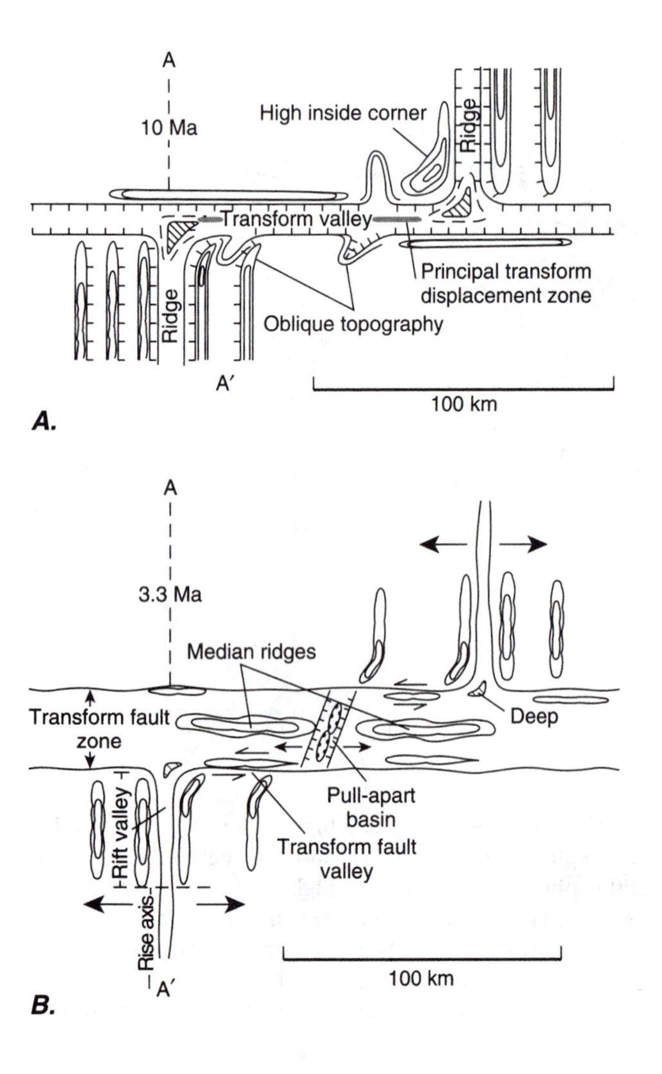

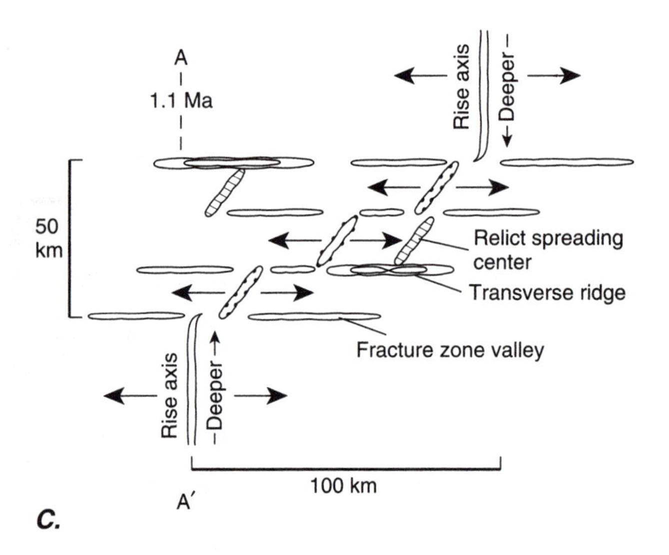

FIGURE 19.23 Schematic maps of transform faults connecting spreading centers that spread at different rates. (After Fox and Gallo 1984) *A*. Low spreading rate (< 5 cm/yr): the presence of a transform valley and of a principal transform displacement zone along which most of the current displacement is concentrated. *B*. Intermediate spreading rate (5–9 cm/yr): a wider transform zone that includes parallel ridges and valleys and one or more pull-apart basins within the transform zone. *C*. High spreading rate (>9 cm/yr): a complex zone, tens of kilometers wide, that includes transverse ridges and valleys with numerous active or inactive pull-apart basins or spreading centers and distributed shearing.

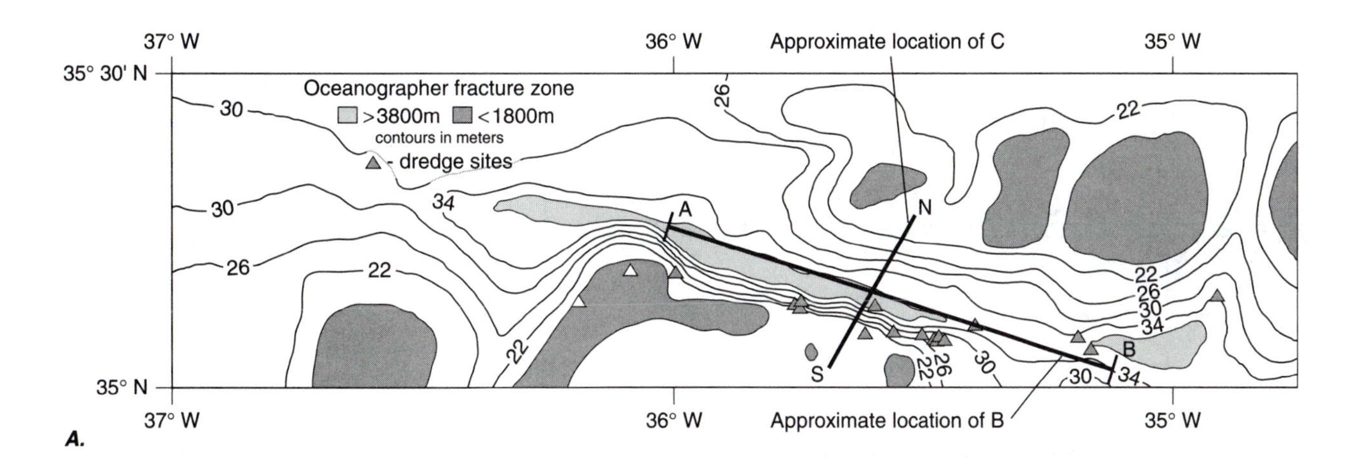

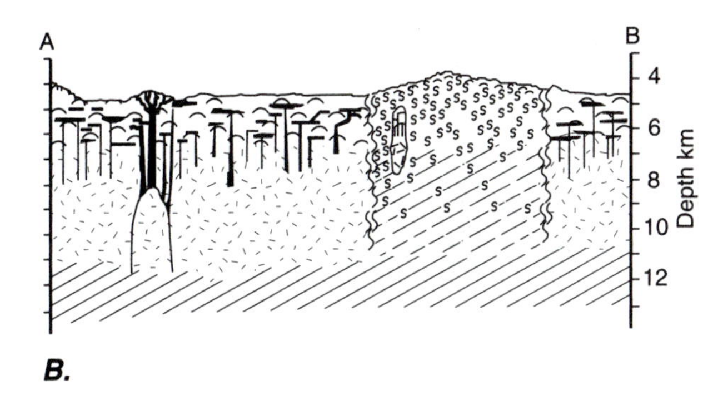

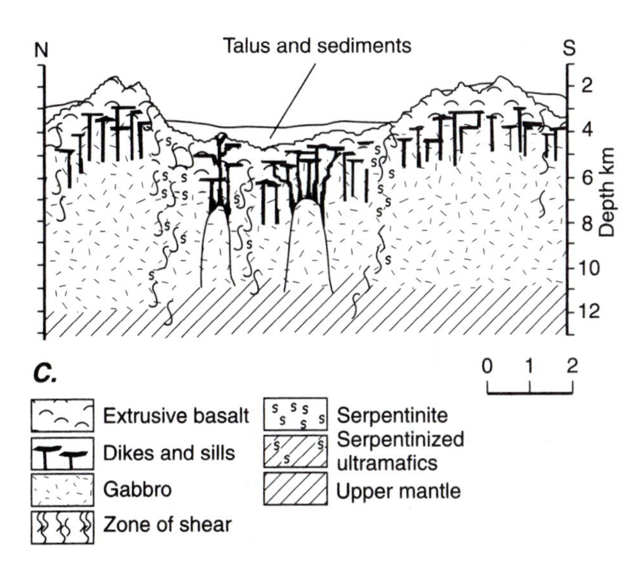

FIGURE 19.24 The possible structure of a transform fault zone illustrated by the slow-spreading Oceanographer transform fault, showing possible geology based on dredge samples. (After Fox et al. 1976) *A.* The location of the Oceanographer Transform fault showing the location of dredge samples. Labeled lines show the approximate locations of the generalized longitudinal (A–B) and transverse (N–S) cross sections, respectively, along and across the length of the studied fault shown in *B* and *C*. *B.* Longitudinal cross section. *C.* Transverse cross section.

top to bottom, of pillow basalts and extrusives, dikes and sills, gabbro, and peridotite is partially metamorphosed and cut by zones of deformation, which are commonly serpentinized. Serpentinite diapirs intrude the crust in places, and fault slices and horses of these different rock types occur in the fault zone, out of place with respect to the normal sequence. Surficial talus deposits are common along the scarps, and young extrusives may cover the deformed rocks and even the talus deposits. A large area within the fracture zone may be underlain by serpentinite.

The geologic and topographic complexity along transform faults is interpreted to result from the geometry of plates moving on the surface of a sphere. A detailed analysis, which is beyond the current discussion (see Moores and Twiss, *Tectonics*, Chapter 4), shows that to accommodate the relative motion of adjacent plates, components of shortening or lengthening perpendicular to the transform fault plane must develop. This kinematic fact requires transform faults to continually change their configuration and kinematic character and leads to the complexity of structure observed.

Another reason to expect complexity along transform fracture zones arises from the nature and timing of the processes that occur along a transform fault, shown schematically in Figure 19.25. Normal oceanic crust is formed by magmatic intrusion and extrusion at each ridge, and it is carried away from the ridge by the spreading. Rocks erupted at a ridge segment adjacent to a transform fault, however, become deformed within the transform fault zone if they are on the inside corner of the ridge-transform intersection, adjacent to the active transform. Talus breccias accumulate on the low side of the transform fault scarp and cover the older deformed extrusives. The talus deposits may subsequently become involved in the deformations themselves. As these rocks move past the opposite ridge segment, the deformed crust and talus deposits are intruded and/or covered by new magma. Beyond this ridge segment, however, shearing on the transform fracture zone ceases, so that the newest lavas remain undeformed. Thus transform fracture zones

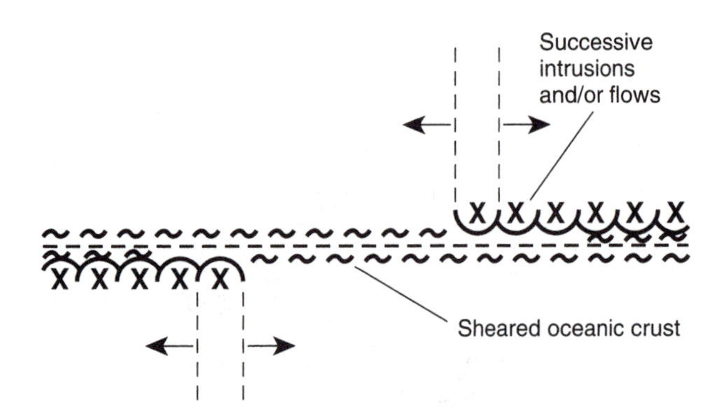

FIGURE 19.25 Diagram illustrating the complex relation between intrusion and deformation along a transform fault. The active transform fault between ridge segments shows deformed oceanic crust. The portion of the fracture zone beyond the ridges shows complex intrusion into and extrusion over previously deformed crust.

can be expected to display complex intrusive and extrusive contacts between deformed and undeformed igneous and sedimentary rocks.

Two offset ridge segments connected by a transform fault may change length in a coordinated fashion, such that as one of the segments lengthens, the adjacent segment shortens. Such ridge segments are referred to as **propagating ridges** (see Section 19.1, Figure 19.14). As a result, the connecting transform fault must migrate, transferring crustal material from one plate to another and creating a wide and structurally complex transform fault zone. In some cases, such propagating ridge segments advance and retreat in an oscillating manner (Figure 19.26), such that as one segment advances, the other retreats, and vice versa. Such ridge segments are said to be **dueling ridges**. Crust that is transferred from one plate to another when a transform fault migrates past, changes its direction of motion relative to the adjacent ridge segments (see Figure 19.26*B*, *C*, block *a*). If a piece of crust is transferred back and forth from one plate to another as a transform fault oscillates first one way and then back, it could become an anomalously old fragment

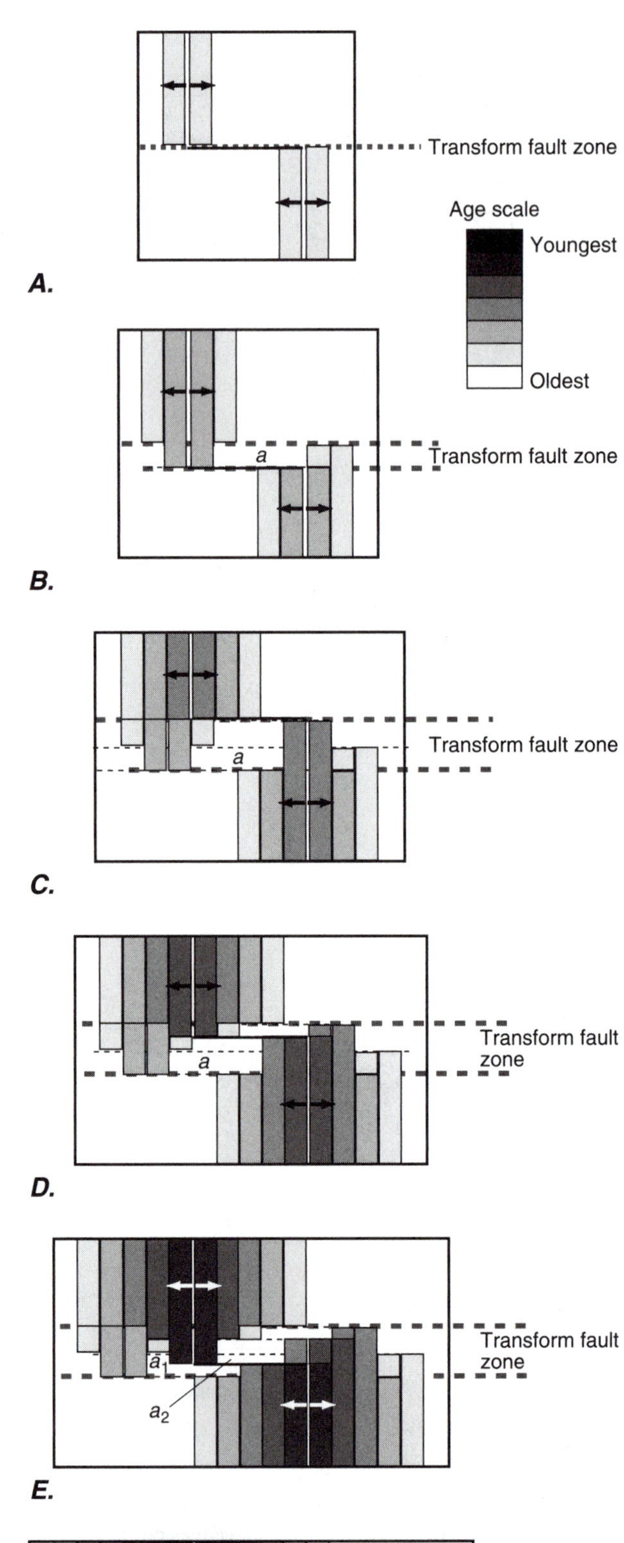

FIGURE 19.26 Effects on a transform fault zone of "duelling ridge segments." The tips of the adjacent ridge segments advance and retreat, sweeping the transform fault back and forth across the ocean floor to form a wide fault zone. For the different steps, starting from *A*, ridge tips and the active transform have propagated in the following directions: *B*: south; *C*: north; *D*: south; *E*: south; *F*: north. White are oldest rocks and black are youngest; active transform faults connect the tips of the active spreading centers are solid lines; thin dashed lines indicate previously active transform faults; thick gray dashed lines indicate the boundaries of the transform fault zone. Ages of blocks of rock within the transform fault zone are very jumbled, with the oldest rocks (white shading) juxtaposed in places against the youngest (darkest shading) and different ages mixed together.

of crust very close to a ridge segment. For example, in Figure 19.26*B*, block a is transferred to the opposite plate (Figure 19.26*C*) and then is split into blocks a_1 and a_2 on opposite plates (Figure 19.26*E*). Subsequently, these blocks become part of the same plate again (Figure 19.26*F*). All the white blocks near the active ridge segments (Figure 19.26*F*) are anomalously old crust. The result of such a process could explain the large width and complexity of many transform fault zones and could account for complex age relationships within some transforms, including the fact that in places, such as in the equatorial Atlantic Ocean, old rocks have been found very near the midoceanic ridge crest along transform fault zones.

Near transform faults, where the crust is thinner than farther from the fault, even fast-spreading ridges tend to develop an axial valley. A possible explanation for this observation is that near such faults, the hot asthenosphere that rises under the spreading segments of the plate boundary encounters anomalously cool lithosphere along the side bounded by the transform fault. The enhanced cooling of the rising hot asthenosphere that results would depress the amount of partial melting and impede the separation of melt from solid mantle. Less magma would be available to form the magmatic crust, thereby resulting in a thinner crust. For a similar reason, the lower lithospheric temperatures adjacent to transform faults could promote the formation of axial valleys by making the thermal structure of the adjacent fast-spreading ridge resemble that of a slow-spreading ridge (Figure 19.27).

Near a transform fault, topographic lineaments, faults, and dikes characteristically change orientation from being almost perpendicular to the transform to being oriented at roughly 45° to the fault zone (Figure 19.27). We can account for these changes in terms of the bulk stress field associated with the ridge system if we assume an Anderson model for the relationship between the principal stresses and the faults (see Figure 9.18). The stress field far from a transform fault then should have the orientation characteristic for the normal faults that parallel the ridge axis, with the maximum compressive stress $\hat{\sigma}_1$ vertical, the intermediate compressive stress $\hat{\sigma}_2$ horizontal and parallel to the ridge, and the minimum compressive stress $\hat{\sigma}_3$ horizontal and perpendicular to the ridge (Figure 19.27; compare Figure 9.18*A*). At the transform fault, however, the orientation of the principal stresses should be characteristic of strike-slip faults, with $\hat{\sigma}_2$ vertical, $\hat{\sigma}_1$ horizontal at an angle of 30° to 45° to the fault, and $\hat{\sigma}_3$ horizontal at an angle of 45° to 60° to the fault. The horizontal stresses also must be oriented to give the appropriate shear stress on the fault (Figure 19.27; cf. Figure 9.18*C*). If dikes are injected perpendicular to the minimum compressive stress $\hat{\sigma}_3$ and parallel to the ($\hat{\sigma}_1$,$\hat{\sigma}_2$)-plane, the inferred stress fields would result in vertical dikes parallel to the ridge axis away from the transform fault and vertical dikes oriented at 30° to 45° to the transform fault adjacent to the fault, which is the pattern illustrated in Figure 19.27.

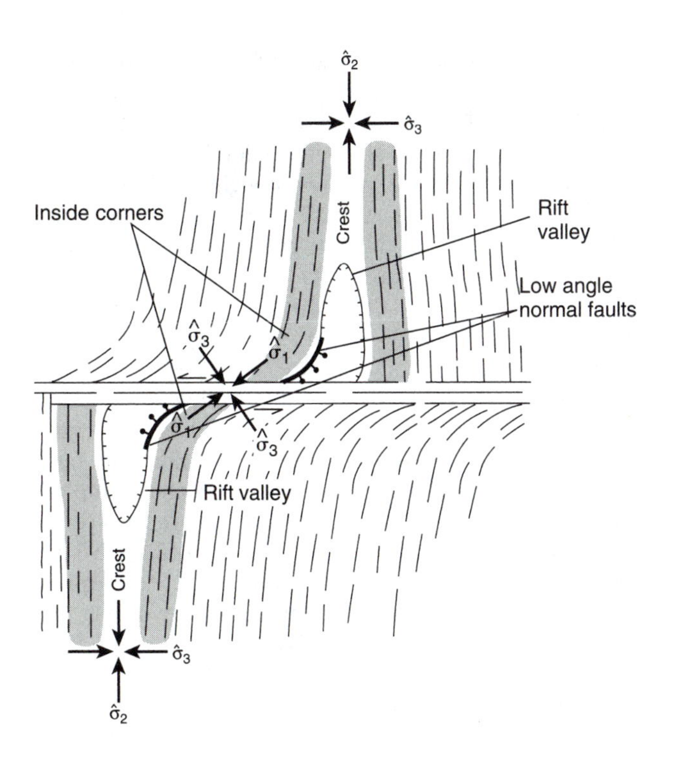

FIGURE 19.27 Diagrammatic plan view of a fast-spreading ridge showing development of an axial rift valley near a transform fault. Also shown is the change in the orientation of the structural grain, defined by topography, faults, and dikes; and the orientations of the principal stress directions far from, and near to, the transform fault.

On ridge-ridge transform faults, the "inside corner" of the intersection of the ridge and the transform is the corner between the two ridge segments. Many inside corners, particularly on slow-spreading ridges, are topographically high relative to the surrounding seafloor and have mantle rocks exposed at the surface. These mantle exposures are the footwall blocks to a low-angle normal fault that strikes parallel to the spreading center, as illustrated schematically in Figure 19.27. The removal of the crust to expose the mantle indicates the presence of **oceanic metamorphic core complexes** in these regions (see Figure 4.15).

iii. Subaerial Exposures of Oceanic Transform Faults

Fossil transform faults have been recognized in several ophiolites. In particular, the Troodos complex in Cyprus, and possibly the Bay of Islands complex in Newfoundland. The ophiolites of Italy, Iran, and Turkey, may also preserve structures formed in oceanic ridge-ridge transform faults. An oceanic trench-trench transform fault may be preserved in the complex deformed ophiolitic rocks of the northwest Sierra Nevada–Klamath Mountains of northern California. Study of such exposed features can lend insight into the nature of transform faults covered by deep oceans.

The South Troodos Transform fault provides a good example of an on-land exposure of a ridge-ridge transform fault. Figure 19.19*A*, *C* shows the region of this fault. It comprises a zone of sheared ophiolitic rocks, including a few graben-like structures thought to represent an extensional relay zone. A small piece of the plate south of the transform fault, the Anti Troodos plate, is present in the southern exposure of the complex. Relationships indicate that a swing in dike direction toward parallelism with the transform fault east of the Solea graben axis resulted from tectonic rotation after formation of the dikes with a northerly strike. (This postspreading deformation has resulted in a different orientation of dikes than would have been predicted from Figure 19.27.) Along the fault zone, steep faults cut brecciated sheeted dike complex, and form troughs filled with "polymictic" breccias (ones with clasts derived from many different rock types), interlayered with pillow lavas (Figure 19.19*C*). The presence of the lavas indicates magmatic activity during formation of the fault. Relationships indicate that the fault was a left-step fault, with displacement mainly dextral along the fault zone between the Troodos and the Anti Troodos plate. The South Troodos Transform Fault zone is probably the best-known on-land exposure so far of an oceanic transform fault.

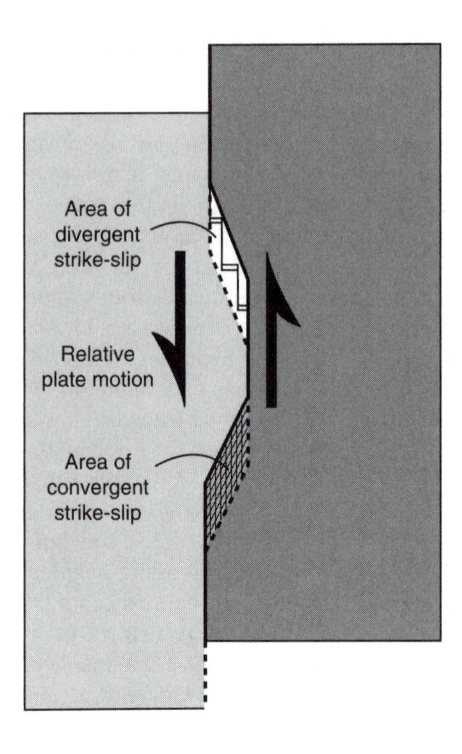

FIGURE 19.28 Diagram showing development of convergent and divergent strike-slip along irregular trace of continental transform fault zone. See text for discussion.

iv. Continental Transform Faults

Transform faults that cut continental crust provide another opportunity to study on land the processes associated with transform faulting. These faults are similar to oceanic transform faults in that they form plate boundaries that connect two spreading centers, two subduction zones, or a spreading center and a subduction zone; their motion is essentially strike-slip; and seismicity on the faults is limited to shallow depths of generally less than about 20 km. The involvement of continental crust, however, affects the character of the faults and adds to their complexity, making it difficult to draw a strong parallel to the oceanic situation. Nevertheless, the study of continental transform faults is important in its own right because of our interest in being able to interpret ancient structures in continental rocks, not to mention our need to understand the seismic hazard they present to populated areas.

Transform faults in continents can be expected to follow pre-existing zones of weakness or discontinuity, such as have been created by older faults, which themselves need not necessarily have formed as strike-slip faults. Such fault zones need not be parallel to the current relative plate motion. As a result, extensional or contractional strike-slip duplexes could form, depending on the relative orientations of the fault and plate motion. Figure 19.28 shows schematically a situation in which a fault zone is not everywhere parallel to the relative velocity vector for the two plates. The relative plate motion in principle should cause the plates to converge within the cross-hatched area, thus producing contractional structures such as folds and thrust faults, whereas the motion should cause the plates to diverge within the white area, thus producing extensional structures, such as normal faults, pull-apart basins, and short ridge segments separated by transform faults.

Prominent active continental transform systems include the San Andreas fault system in California-Mexico, the Alpine fault of New Zealand, and the Dead Sea fault zone in the Middle East. These faults are several hundred kilometers in length and are characterized by multiple strands of faults that in places distribute the faulting over a width of as much as 100 km. Individual faults may develop mylonite or gouge zones that are 1 km or more thick. We look at some of the complex structures that can develop along such faults by discussing briefly the San Andreas system, one of the best-studied continental transform faults in the world.

v. The San Andreas–Gulf of California Transform System

We discuss aspects of the southern California section of the San Andreas–Gulf of California transform fault system in Sections 6.3(ii) and 6.4; Figures 6.12 and 6.17). The entire system is a zone approximately 3000 km long, reaching from the Mendocino trench-transform-transform triple junction off northwestern California to the Rivera ridge-trench-transform triple junction near the mouth of the Gulf of California (Sea of Cortez; Figure 19.29). The transform system consists of two main parts. The San Andreas strike-slip fault zone and associated faults extend from Cape Mendocino in the northwest to the Imperial Valley in the southeast (Figures 19.29 and 19.30). A second part, southeast of the San Andreas fault zone, extends from the head to the mouth of the Gulf of California and consists of a series of long transform faults and very short ridge segments that accommodate the divergent strike-slip, or oblique rifting, in the Gulf of California (Figure 19.29).

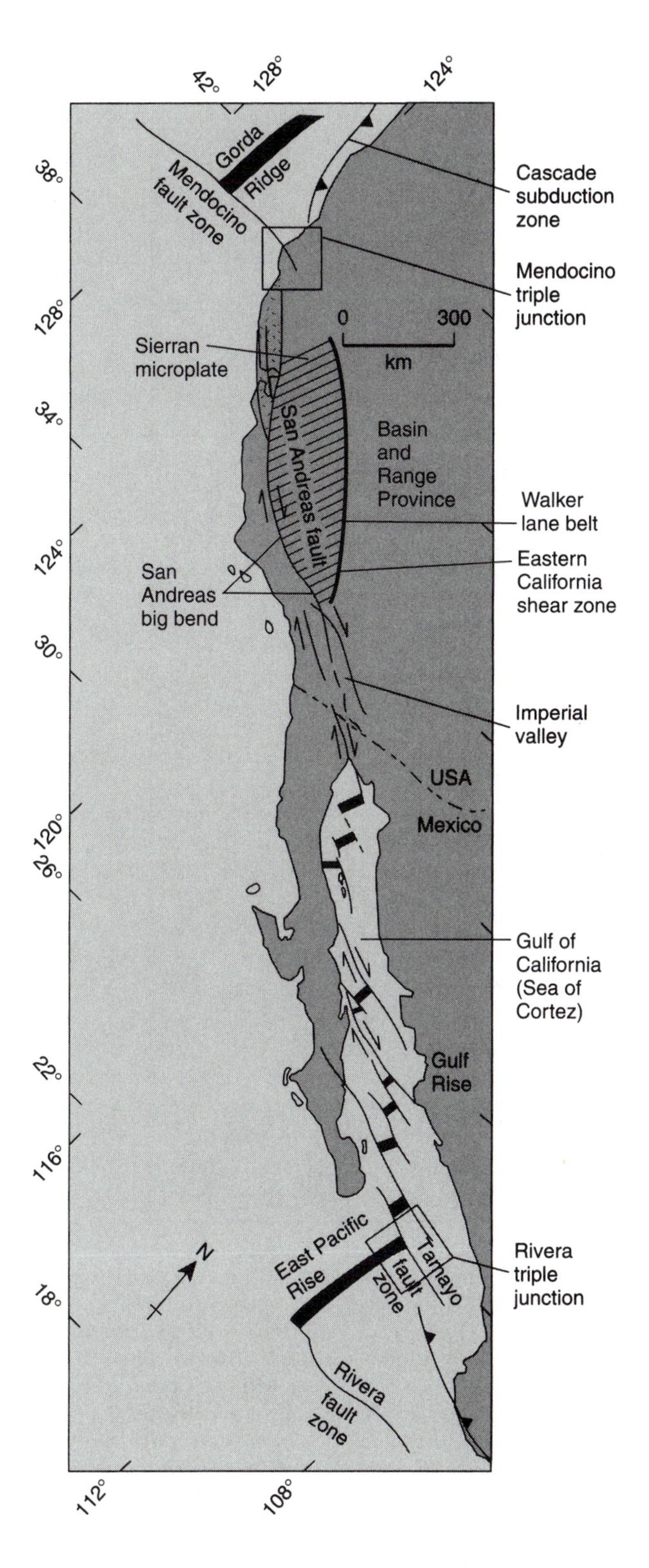

A splay off the San Andreas fault system called the Eastern California shear zone is connected with the main fault system south of the big bend in the San Andreas fault and trends north along the eastern front of the Sierra Nevada, where it is known as the Walker Lane Belt (Figures 19.29 and 19.30). The crust between the main San Andreas fault and the Eastern California shear zone is a subsidiary microplate, the "Sierran microplate" (Figures 19.29 and 19.30), whose northern boundary in northern California and southern Oregon is diffuse and poorly defined. The microplate is moving at about 11–14 mm/y relative to the North American plate, and thus the Eastern California shear zone/Walker Lane Belt accommodates at least 22% of the Pacific–North American plate motion.

The total displacement on this entire system is estimated to be approximately 1000 km, based on calculations from global plate motions, and as much as 600 km, based on displacement of geologic features along the part of the fault system that is on land. Some strands of the fault, however, also are present off the coast of California, where assessing the displacement from the geology is difficult or impossible.

The dominant style of faulting is certainly strike-slip faulting. In such an extensive fault system, however, it is not surprising that a wide range of structures develops in association with the strike-slip faulting. We highlight just a few examples of these complexities.

One of the most prominent departures from a strict strike-slip fault occurs in the southern part of the San Andreas system where the left-lateral Garlock and Big Pine faults intersect the San Andreas system (Figure 19.30). Here the San Andreas system makes a prominent left bend, commonly referred to as the "big bend." This bend is in a contractional orientation for the right-lateral slip on the San Andreas system. The convergence across this segment of the fault zone is taken up by thrust faults in the prominent Transverse Range of mountains, represented in Figure 19.30 by the nearly east-west thrust faults to the west of the bend.

The motion of the Sierran microplate relative to North America can be described as a counterclockwise rotation about a pole located west of the southern California coast. The eastern margin of the Sierran microplate is marked by the Walker Lane Belt, which is connected to the San Andreas fault system through the Mojave Desert by the Eastern California shear zone (Figure 19.30). This boundary lies approximately along a small circle about

FIGURE 19.29 San Andreas–Gulf of California transform system, showing Sierran microplate. (After Macdonald et al. 1979; Unruh et al. 2003)

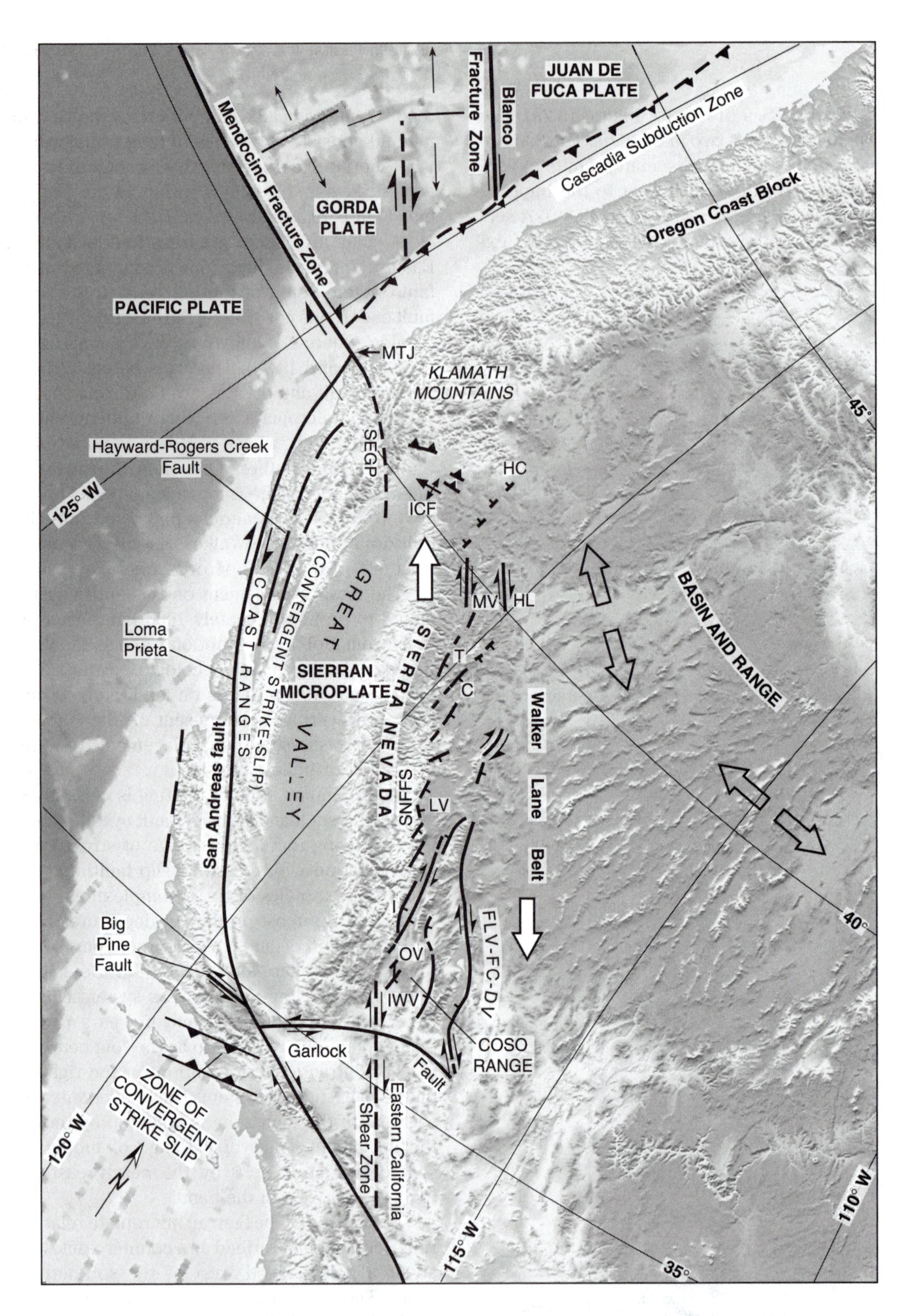

FIGURE 19.30 Generalized digital relief map of the southwest United States, showing the San Andreas fault zone, the Sierran microplate, a zone of diffuse extension in the western Basin and Range province, the central Nevada block, and central Utah, zones of transpression along the western edge of the Sierran microplate, and in the Transverse Ranges near Los Angeles. Rates of motion are based on Global Positioning System measurements and are shown relative to the Colorado Plateau. Map projection is chosen so that small circles about the Sierran–North American pole of rotation plot as straight lines from top to bottom on the map. C: Carson Valley; FLV-FC-DV: Fish Lake Valley–Furnace Creek–Death Valley fault zone; HC: Hat Creek graben; HL: Honey Lake fault; ICF: Inks Creek fold belt; IWV: Indian Wells Valley; LV: Long Valley; MTJ: Mendocino triple junction; MV: Mohawk Valley; OV: Owens Valley; SEGP: subducted southern edge of the Gorda plate; SNFFS: Sierra Nevada frontal fault system; T: Lake Tahoe basin. (After Unruh et al. 2003; Bennett et al. 2003; Unruh et al. 1992, 2002)

the Sierran–North American relative pole of rotation. Figure 19.30 is a map projection chosen such that these small circles plot as straight lines from top to bottom on the map. The frontal fault system along the east side of the Sierra Nevada consists of a series of *en echelon* normal faults, arranged along the small-circle boundary and striking approximately 30° to 45° clockwise from the boundary. Although these faults are predominantly normal faults, their *en echelon* arrangement along the transcurrent Sierran–North American plate boundary shows that they form a closely spaced series of releasing faults that predominantly accommodate the dextral strike-slip motion of the plate boundary. In that sense, they serve a similar kinematic function, at a very large scale, to gash fractures (cf. Figure 9.16*A*), except that they are normal faults instead of extension fractures.

Thus a single nominally strike-slip fault system can in detail involve a wide variety of structures, including both thrust and normal faults, depending on the local geometry of the faults relative to the plate motion vector. Any accurate understanding of these faults can be achieved only by interpreting them within the larger tectonic setting. In Box 19-1, we provide a closer look at the structures in a contractional bend in the San Andreas fault, and a releasing stepover in the southern Walker Lane belt.

vi. Major Continental Strike-Slip Faults

Transform faults account for only some of the major continental strike-slip faults that are known. Active strike-slip faults with large displacements are present in a number of regions, chiefly associated either with active subduction systems or regions of continental collision. Except for the tectonic setting, there seems to be no way to distinguish continental transform faults from these other major strike-slip faults.

Examples of active major strike-slip faults associated with subduction zones are shown in Figure 19.1 and include the Median Tectonic Line in Japan, the Denali fault system of Alaska, and the Atacama fault in Chile. In each of these cases, the strike-slip fault occurs at a subduction zone where convergence is oblique, and it lies parallel to the subduction zone in the over-riding plate. One hypothesis to account for this phenomenon suggests that where subduction is oblique ($> \sim 30°$), the work expended to accommodate the slip is minimized if the slip becomes partitioned into a pure thrust component at the subduction zone and a strike-slip component on a vertical fault behind a small strip at the edge of the overriding plate.

Examples of major strike-slip faults associated with continental collision are also shown on Figure 19.1 and include the Insubric Line in the Alps, the north and east Anatolian faults in Turkey, the Chaman fault in Afghanistan and Pakistan, and the Altyn-Tagh and Red River faults in eastern Asia. Each of these faults, most of them currently active, characteristically shows hundreds of kilometers of strike-slip displacement. Most workers now agree that they represent tectonic features linked with the complex interactions of continental blocks in the continent-continent collision of Africa, the Arabian Peninsula, and India with Eurasia. As such, they are considered in more detail in the context of these collisions in Section 19.5.

BOX 19-1 Structures of Convergent and Divergent Strike-Slip along the Boundaries of the Sierran Microplate

i. The Loma Prieta Fault

The San Andreas fault goes through a subtle contractional bend just south of San Francisco along a segment of the fault system called the Loma Prieta fault. This strand of the fault diverges from the local plate motion by as much as 18°, and the convergence across this bend has resulted in the uplift of the Santa Cruz Mountains in this vicinity. After a large earthquake on this fault in 1989 (the magnitude 7.1 Loma Prieta earthquake), the well-located aftershocks showed that the fault in this region is predominantly a blind reverse-oblique slip fault that dips steeply southwest at about 70° (Figure 19-1.1). Three main segments of the blind fault (Southern, Central, and Northern) define a restraining bend relative to the plate motion (Figure 19-1.1*B*). The Southern Segment is almost parallel to the plate motion; the Central Segment and, to a lesser extent, the Northern Segment are oriented counterclockwise from the plate motion and thus must accommodate a component of shortening across the faults. The upper tip line of the blind fault is at a depth of about 4–6 km (Figure 19-1.1*A*,*C*). Above the tip line is a complex of faults that are predominantly thrust to oblique dextral thrust faults arranged *en echelon* above the main fault and striking counterclockwise relative to it (Figure 19-1.1.*C*). These shallow thrust faults can be explained as edge effects characterized by the rotation of the principal stresses in the region above the blind fault tip.

ii. The Coso Range in the Southern Walker Lane Belt

Near the southern end of the Walker Lane Belt, just north of the Garlock fault between Owens Lake and the north end of Indian Wells Valley ("IWV" in Figure 19.30), the Coso Range (Figure 19-1.2*A*) lies in a releasing step-over of the Eastern California Shear Zone (Figure 19.30). It is an area of active normal faulting and abundant recent bimodal volcanic activity, including basaltic flows and rhyolitic domes. There, the slip at the northern end of the Eastern California shear zone on the Airport Lake fault is transferred across a releasing step-over via the Coso Wash fault onto the Owens Valley fault to the

(continued)

BOX 19-1 Structures of Convergent and Divergent Strike-Slip along the Boundaries of the Sierran Microplate *(continued)*

north of the range (Figure 19-1.2*B*). The step-over is underlain by a granitic intrusive body at a depth of about 8 kilometers that provides the heat source to the Coso geothermal field (see Figure 15.13*D*).

Surface mapping and seismic reflection studies show that the Coso Wash fault zone is a set of *en echelon* listric normal faults that accommodate extensional strike-slip motion and that sole at a depth of about 5 kilometers into the brittle-ductile transition overlying the granitic intrusives (see Figure 15.13*D*). Underlying Wildhorse Mesa to the east of the Coso Wash fault is an imbricate set of normal faults that terminate at a depth of about 5 km near the tip line of a blind strike-slip fault. The strike-slip fault extends from a depth of about 5 km down to nearly 10 km, where it probably terminates against the continuation of the Coso Wash fault along the brittle-ductile transition (Figure 15.13*D*). The complex has been interpreted to be a modern metamorphic core complex in the process of formation (compare Figure 15.13*D* with Figures 4.13 and 4.15*B*).

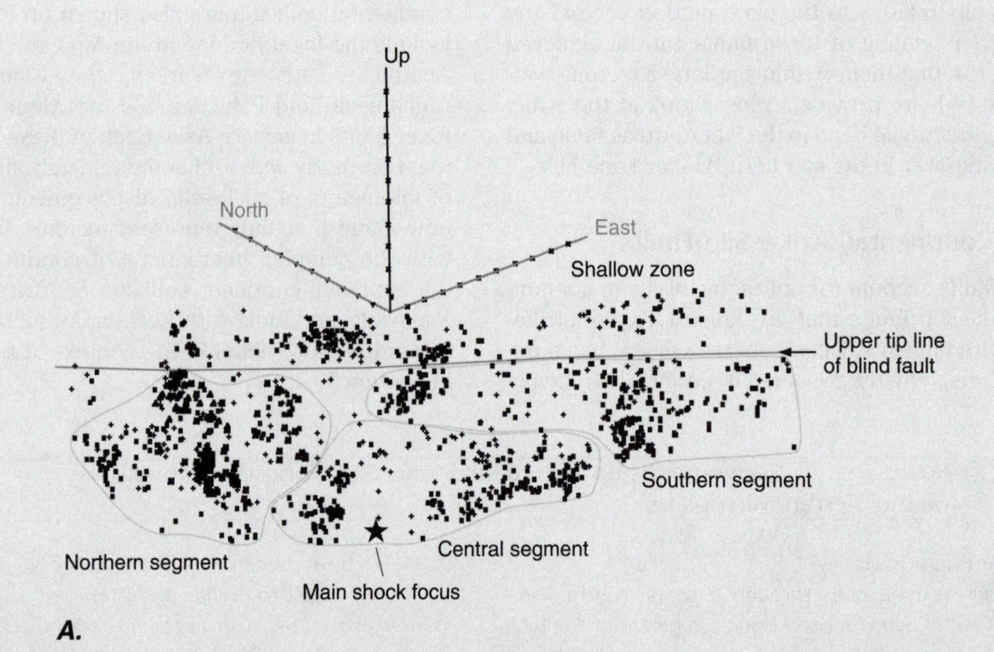

FIGURE 19-1.1 Contractional bend in the Loma Prieta fault, part of the San Andreas fault south of San Francisco (see Figure 19.30), as illuminated by the aftershocks of the 1989 Loma Prieta earthquake. (After Twiss and Unruh 2007) *A*. View looking in the direction [041, 25] normal to the mean fault plane. The Southern, Central, and Northern segments of the main fault are outlined and the upper tip line of the main fault is shown. Faults in the shallow zone above the upper tip line are mostly oblique reverse faults. *B*. View of main fault zone aftershocks (not including the Shallow Zone) looking [azimuth, plunge] = [170, 51], approximately down the dip of the main fault segments. The contractional bend defined by the Central Segment is clearly shown and is particularly obvious if the figure is viewed at a low angle to the page in a direction parallel to one of the alignments. The Southern Segment consists of two parallel faults. Wide arrows show the approximate relative plate motion projected onto the plane of the diagram. *C*. View toward [070, 25] of the idealized geometry of the Loma Prieta fault zone, as approximated by planar fault segments. The three segments of the main fault are shown as three planes of irregular outline, which approximates the distribution of aftershocks (cf. *A*). The faults in the shallow zone are shown in more detail. The three main fault segments are projected onto the three coordinate planes bounding the figure; shallow faults are projected only onto the horizontal coordinate plane. The *en echelon* arrangement of the shallow faults is clearly shown by the projection onto the horizontal plane. The relative plate motion is shown by the large gray arrows on the bottom horizontal plane.

(continued)

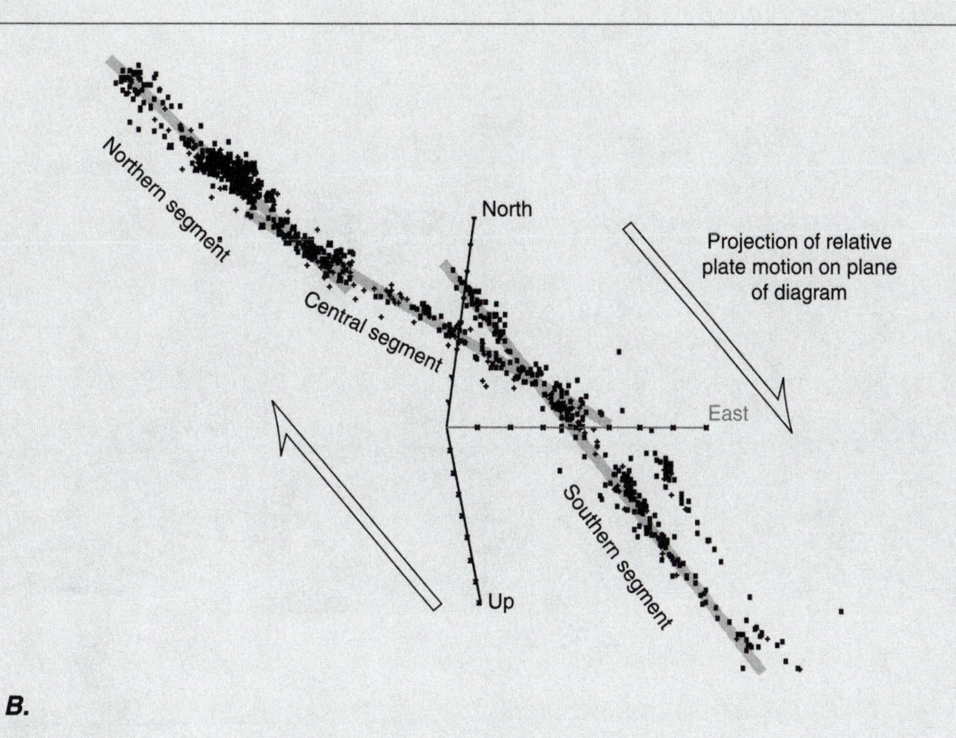

B.

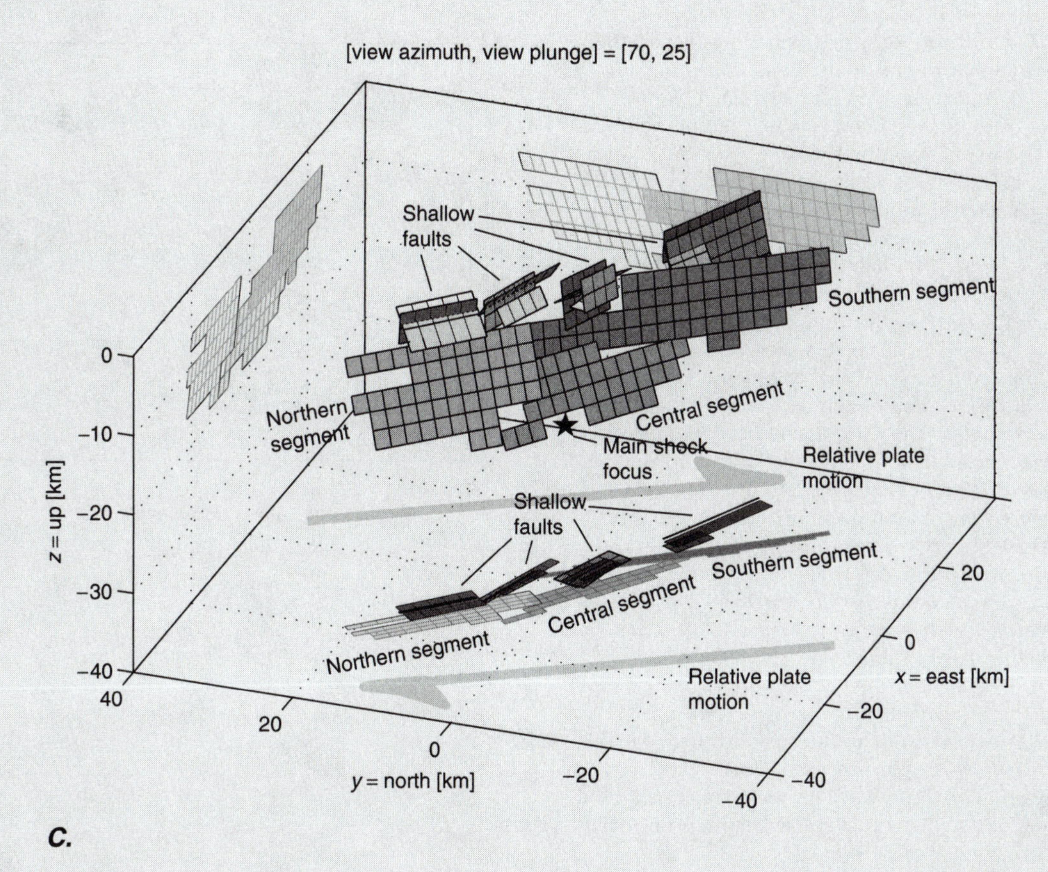

C.

FIGURE 19-1.1 *(continued)*

BOX 19-1 Structures of Convergent and Divergent Strike-Slip along the Boundaries of the Sierran Microplate *(continued)*

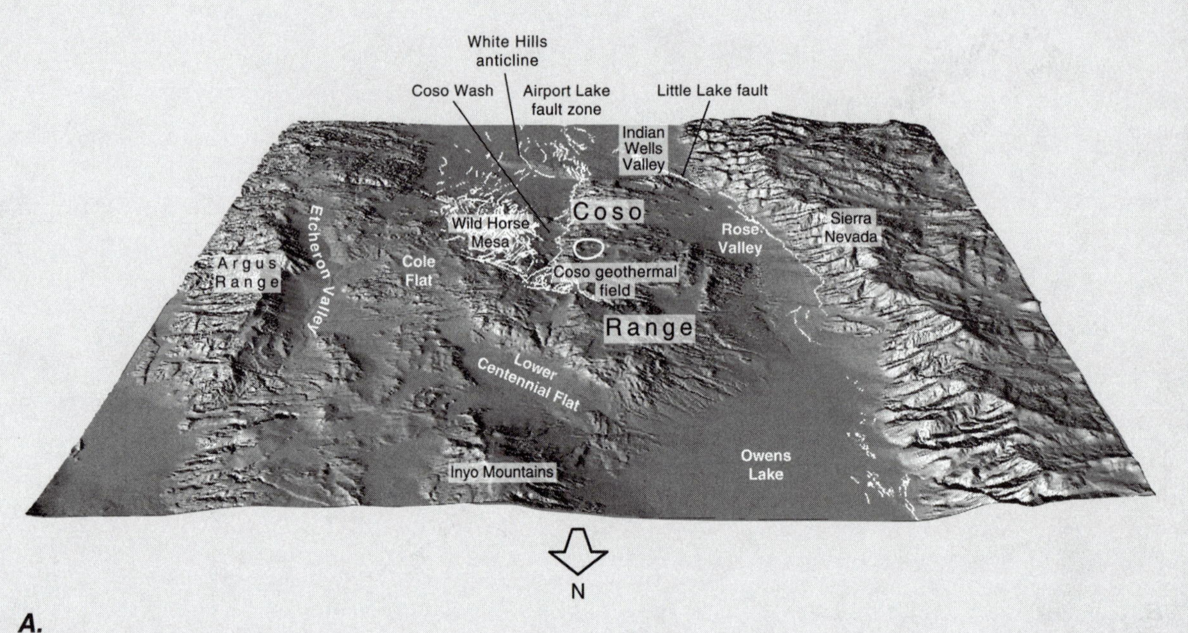

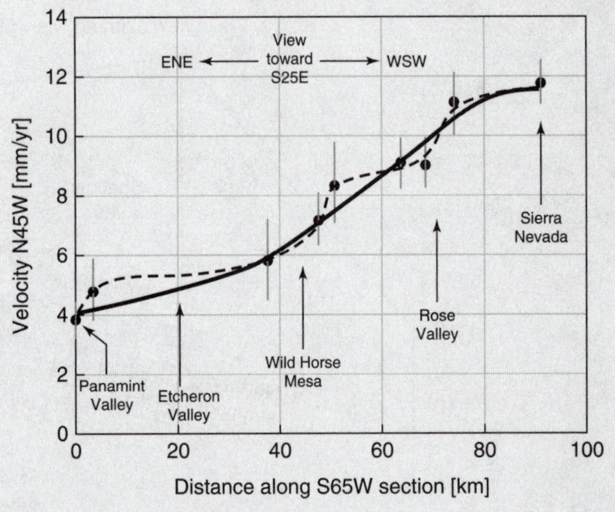

FIGURE 19-1.2 Structure of a releasing step-over in the Eastern California shear zone at the Coso Range (compare Figure 15.13*A*, *D*). *A*. Oblique view of hill-shaded digital elevation map of the Coso Range (center) looking south, bounded on the west (right) by the Sierra Nevada and on the east (left) by the Argus Range. White lines are faults active in the Quaternary in the Coso Range, Indian Wells Valley, and Rose Valley. Illumination is from a source 45° above horizontal with an azimuth of 045°; no vertical exaggeration. (Image courtesy of E. Cowgill; DEM data from the National Elevation Dataset (http://ned.usgs.gov/); fault trace data courtesy of J. Unruh, compiled by Unruh from work by Unruh, Duffield and Bacon (1981), Jennings (1994), Whitmarsh (1997), and Benner (2004)) *B*. Distribution of slip determined by GPS along a S65W traverse across the Coso Range near the middle of the topographic image in *A*. The view is looking S25E, so west is to the right and east to the left. The arrows indicate the approximate location of geographic features labeled in *A*, with the exception of the Panamint Valley, which is to the east of the Argus Range. Error bars on the data points are 1 standard deviation. The solid line is a smooth fit that lies within the error bars of all the data. The dashed line is a better, but more speculative, fit that suggests stepwise jumps in the velocity, presumably across active faults. The velocity jump to the west (right) of Wild Horse Mesa corresponds with the Coso Wash fault on the west side of Coso Wash. The step near Rose Valley, however, does not correlate with any major faults exposed on the surface, although it could result from slip on a blind fault. (After Monastero et al. 2005)

19.4 CONVERGENT MARGINS

At convergent or consuming plate margins, the adjacent plates move toward each other and the motion is accommodated by one plate over-riding the other. These margins, also called **subduction zones**, characterize some 45,000 kilometers of plate boundary on the Earth. They are either **oceanic** or **subcontinental** depending on whether the crust of the over-riding plate immediately above the convergent margin is oceanic or continental crust (Figure 19.31). The crust on the over-ridden, or subducted, plate is usually oceanic crust, because where it is continental crust, a **collision zone** develops, forming an orogenic mountain belt that impedes or terminates the normal subduction process. These zones are discussed briefly in Section 19.5, and in more detail in Chapter 20.

i. Physiographic Features

Subduction zones are characterized by an association of physiographic features, diagrammed in Figure 19.32*A*, *B*. The ocean floor generally shows a linear depression, the **trench**, where the subducted plate bends over an **outer swell** and starts its descent into the mantle. The trench may be partially or even completely filled with sediment, depending on the supply of sediment available from the over-riding plate. On the over-riding plate, an arcuate chain of volcanoes develops, generally called a **volcanic arc**. If it is built on oceanic crust, it is referred to as an **oceanic island arc**, or just an **island arc**, as the volcanoes generally rise above sea level to create a chain of islands (Figure 19.32*A*, *C*; cf. Section 1.5(i); if it is built on continental crust, it is called a **continental arc** (Figure 19.32*B*, *D*). **Continental arcs** are also called **Andean-style continental margins** (Figure 19.32*D*; cf. Section 1.8(iv), reflecting the modern tectonic situation of the Andes along western South America. Often, we use the terms *trench*, *island arc*, and *continental arc* to refer to these features of convergent plate margins.

Many areas of active subduction have discontinuous volcanic chains (Figure 19.32*C*, *D*). This is true of both continental and oceanic arcs such as the Andes, the Philippines, and the Izu-Bonin arcs. In the Andes, the nonvolcanic areas are characterized by shallow dips of the down-going lithospheric slab, but such low-angle slabs do not necessarily characterize other volcanic-free regions.

An **arc-trench gap** lies between the trench and the active arc, and it is generally a basin that collects sediment from the volcanic arc, underlain toward the trench, by an **accretionary prism**, which comprises deformed sediments deposited from the arc and scraped off the down-going plate (see Subsection 19.4(iv).

Behind an oceanic arc is a **back-arc basin**, in which active spreading may take place. Back-arc basins may contain **remnant arcs**, or inactive volcanic ridges, that may have been split off the main arc by the back-arc spreading process. Back-arc spreading also takes place in some continental arcs, including the Aegean Sea and early spreading of the Basin and Range province. In addition, the Sea of Japan represents a back-arc basin that separates a former continental arc, the Japan arc, from the main area of the Asian continent. Clearly, as with continental rifting, pronounced back-arc spreading in a continental arc ultimately will result in oceanic crust and result in an island arc essentially indistinguishable from an oceanic arc. Japan is a good case of a continental arc that has become an oceanic arc through formation of a back-arc basin.

In the remainder of this section, we review some of the structural features of oceanic and continental convergent margins, many of which are broadly similar for both types. There are many complexities, however, and it is difficult to make sweeping statements that apply to all margins.

ii. Structure of Island Arcs

Island arcs consist of an arcuate chain of active volcanoes developed on a basement that varies in composition from place to place. The structure of these volcanic chains is diverse, and generalizations are difficult. Many western Pacific arcs overlie a basement of oceanic rocks not older than Eocene, whereas others, such as Japan, are built on continental rocks as old as early Paleozoic, now separated from the main continent by the Sea of Japan. Many arcs display extensional structures in the region of active volcanism, although a few display contractional structures. Most arcs consist of a single chain of volcanoes, but some, such as southwest Japan, have two volcanic chains, each of which erupts a distinctive composition of lava. The spacing of volcanoes along the arc tends to be fairly regular, averaging approximately 70 kilometers (Figure 19.32).

The back-arc region of oceanic arcs is characterized by extensional tectonics. Several back-arc regions have zones of active seafloor spreading. In some cases, the spreading is focused at a well-defined spreading center, similar to midoceanic ridges; in other cases, it is more diffuse.

iii. Structure of Continental Arcs

Continental arcs develop at the edges of continents where subduction zones dip beneath the continent. The archetype of continental volcanic arcs is the Andes Mountains, which include a chain of active volcanoes that extends some 5000 km from north to south. They sit on a high plateau as much as 6000 m in elevation, underlain by a complex basement of late Precambrian–Mesozoic rocks. Individual peaks are as high as 6900 m. Along a considerable length of the Andes, there is little or no accretionary prism (see Figure 19.31). Other active continental arcs include the North American Cordillera in southern Alaska, and from Vancouver Island, Canada, to Cape Mendocino, California, at the northern end of the

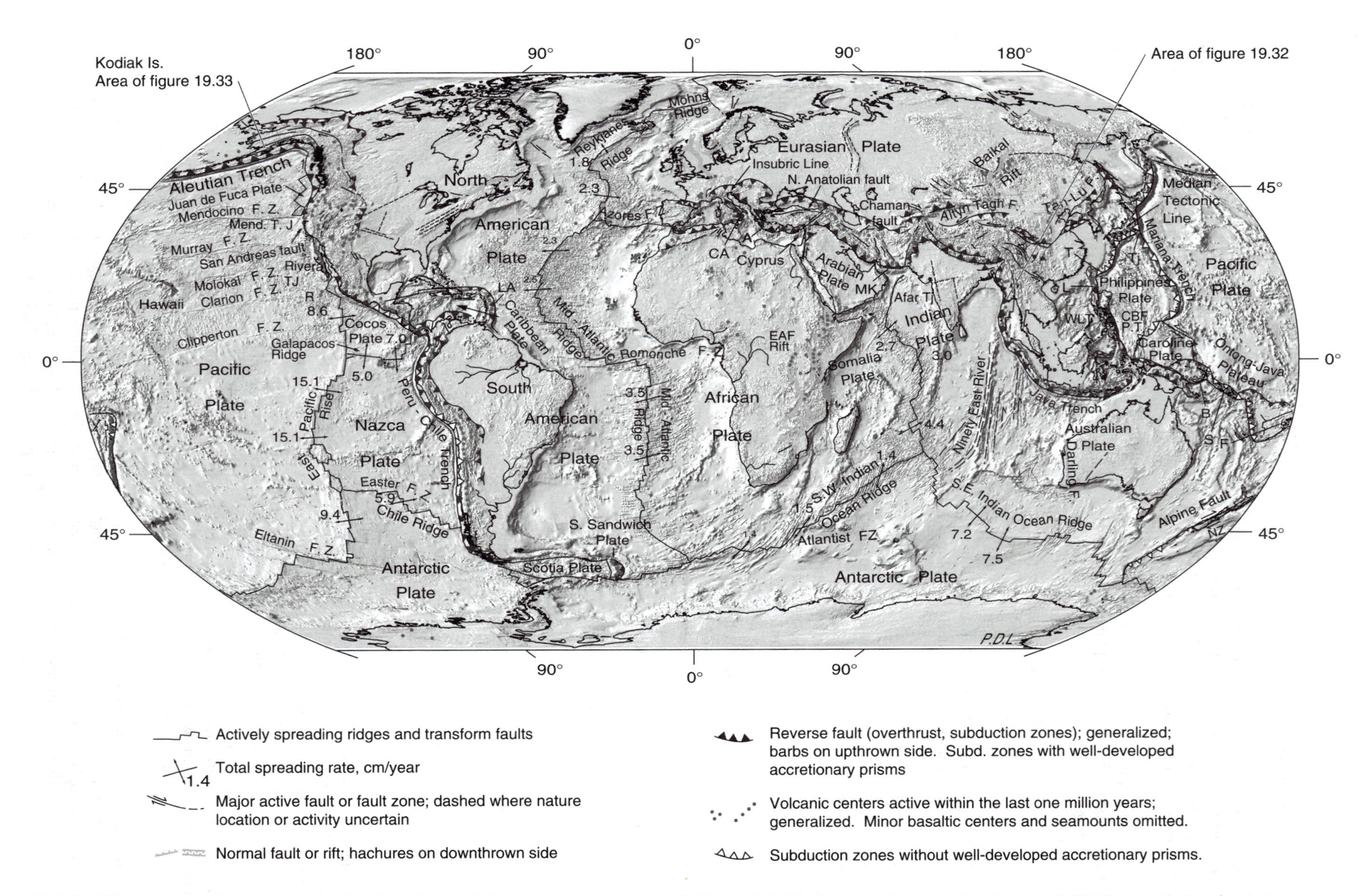

FIGURE 19.31 Generalized map showing plate boundaries of the world. A: Aegean arc; B: Bismarck Archipelago; C: Cascades: CA; Calabria; F: Fiji Plateau; H: Honshu; L: Luzon; LA: Lesser Antilles; MK: Makran; NZ: New Zealand; R: Rivera triple junction; S: Solomon Islands; T: Taiwan; Tj: Boso triple junctions. (After Moores and Twiss 1995, figure 7.1; distinction of subduction zones with and without well-developed accretionary prisms from von Huene and Scholl 1993)

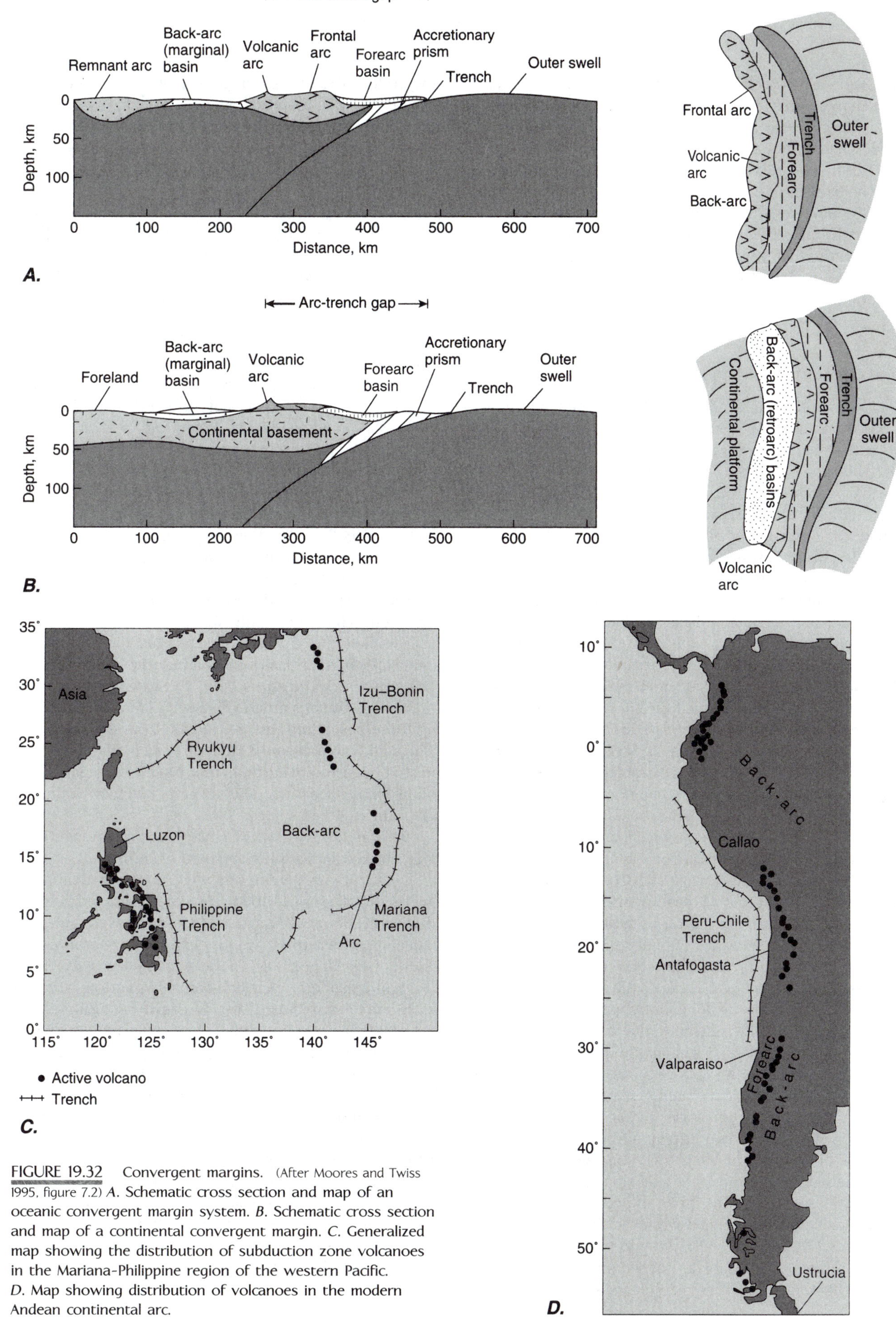

FIGURE 19.32 Convergent margins. (After Moores and Twiss 1995, figure 7.2) *A*. Schematic cross section and map of an oceanic convergent margin system. *B*. Schematic cross section and map of a continental convergent margin. *C*. Generalized map showing the distribution of subduction zone volcanoes in the Mariana-Philippine region of the western Pacific. *D*. Map showing distribution of volcanoes in the modern Andean continental arc.

San Andreas, as well as in Central America. The volcanoes in these arcs also sit on a complex basement ranging in age from early Paleozoic to mid-Cenozoic. Areas of eroded and inactive volcanic chains are marked by large batholithic complexes from Alaska to Central America.

The back-arc regions of continental arcs characteristically contain a sedimentary basin. A few of these basins, principally along the Andes, have been deformed by thrust faults that verge toward the continental interior. This is not a ubiquitous feature, however, as other continental arcs, such as those mentioned in the preceding paragraph, do not display such a feature.

iv. Accretionary Prisms

For both oceanic and continental convergent margins, the inner wall of the trench is the topographic break in slope that marks the boundary between the down-going plate and the over-riding plate, and thus the location of the main subduction thrust fault. Behind (toward the volcanic arc from) the trench inner wall, and thus over-riding the down-going plate, the **forearc region** may be underlain by a broad and thick wedge of mostly deformed sedimentary rocks known as the **accretionary prism** (Figure 19.33), although many forearcs lack well-developed accretionary prisms, as indicated in Figure 19.31 and discussed near the end of this subsection. Where these wedges occur, the trench inner wall marks the deformation front, behind which the sediments on the over-riding plate are deformed, and in front of which the sediments on the down-going plate are undeformed.

Seismic reflection studies of active accretionary prisms show that at least to the resolution of the imaging method, the prisms seem to be dominated by imbricate thrust faulting (Figure 19.33; see also Figures 19.34 and 19.36). These faults commonly all have the same vergence (Figure 19.33*A*), but in some places, an antithetic set of faults is present (Figure 19.33*B*). The differences between the structures of these prisms may depend in part on the shape of the buttress ("basement" in Figure 19.33) against which the accretionary prism is built. If the accretionary prism abuts against basement rocks at an interface that forms a high angle to the subduction zone, the thrust faults may be dominantly vergent toward the trench (Figure 19.33*A*). If the contact between the accretionary prism and the basement dips shallowly toward the subduction zone, an antithetic set of thrusts faults may develop above the basement contact (Figure 19.33*B*; see Figure 9.24*A*). In this case, the accretionary prism may actually develop two critical surface slopes on either side of a topographic high, with one side sloping toward the trench and the other side sloping away.

For the most part, the rocks in accretionary prisms are either sediments eroded from the over-riding plate and carried by turbidity currents into the bottom of the trench, or deep sea sediments carried into the trench on the down-going plate. These sediments are carried back toward the arc as the down-going plate is subducted. The basal thrust fault beneath the accretionary prism propagates out into the undeformed sediments, thereby adding the sediments above the fault to the accretionary prism, a process referred to as **offscraping**. In some regions, seamounts on the down-going plate may even be incorporated into the accretionary prism by this mechanism.

In some cases, rocks and sediments that have been subducted beneath the accretionary prism are subsequently attached to the over-riding plate at depth by the progressive propagation of thrust fault ramps into the rocks of the down-going plate, forming a thrust duplex structure. This process, shown in Figure 19.33*C*, is called **underplating**. If the supply of sediment to the trench is very small, or if most of the sediment in the trench is subducted, little or no accretionary prism can develop (Figure 19.33*D*), and the trench inner wall is characterized by normal faults that cut down into basement rocks.

Sediments may also be ponded in trench-slope basins formed by active imbricate thrust faults that deform the interior of the accretionary prism and sole into the subduction decollement (Figures 19.33 and 19.34). Sediments in these basins may then become deformed as the imbricate thrusts propagate through them (Figure 19.34).

The upper part of the arc-trench gap is a gently sloping region that commonly forms a wide sedimentary basin called the **forearc basin**. This basin develops above an irregular basement that consists of the crustal rocks on which the arc is built, and it may include part of the accretionary prism (Figure 19.33). Sediments filling these basins are derived mostly from the active arc or from the arc basement rocks. They commonly are deposited by turbidity currents that travel either parallel to the arc along the basin axis or perpendicular to the arc down the regional slope.

Deformational structures in accretionary prisms have been studied in samples obtained from drill cores from active accretionary prisms as well as by examination of emergent portions of subduction complexes and inferred fossil accretionary prisms. Such studies reveal an abundance of structures of more than one generation, including pervasive shear fractures or faults, slaty cleavage, and irregularly cleaved sediments, variously called **scaly clay, scaly argillite,** or **argille scagliose** (the Italian term), as well as small-scale folds and boudins in the sedimentary layering. Folding is evident on both large and small scales, as illustrated by data from the Aleutian Trench near Kodiak Island, Alaska (Figure 19.35*A*). The map in Figure 19.35*B* shows folding with a wavelength on the order of 1 kilometer that has been mapped in both the bedding and the spaced cleavage. The cross section in that figure shows the form of the fold as revealed by the orientations of the spaced cleavage. Figure 19.35*C* shows much smaller-scale folding observed in outcrop. Some folding is evident on many seismic profiles, chiefly folds with subhorizontal axes and axial surfaces that trend parallel to the plate boundary and dip moderately

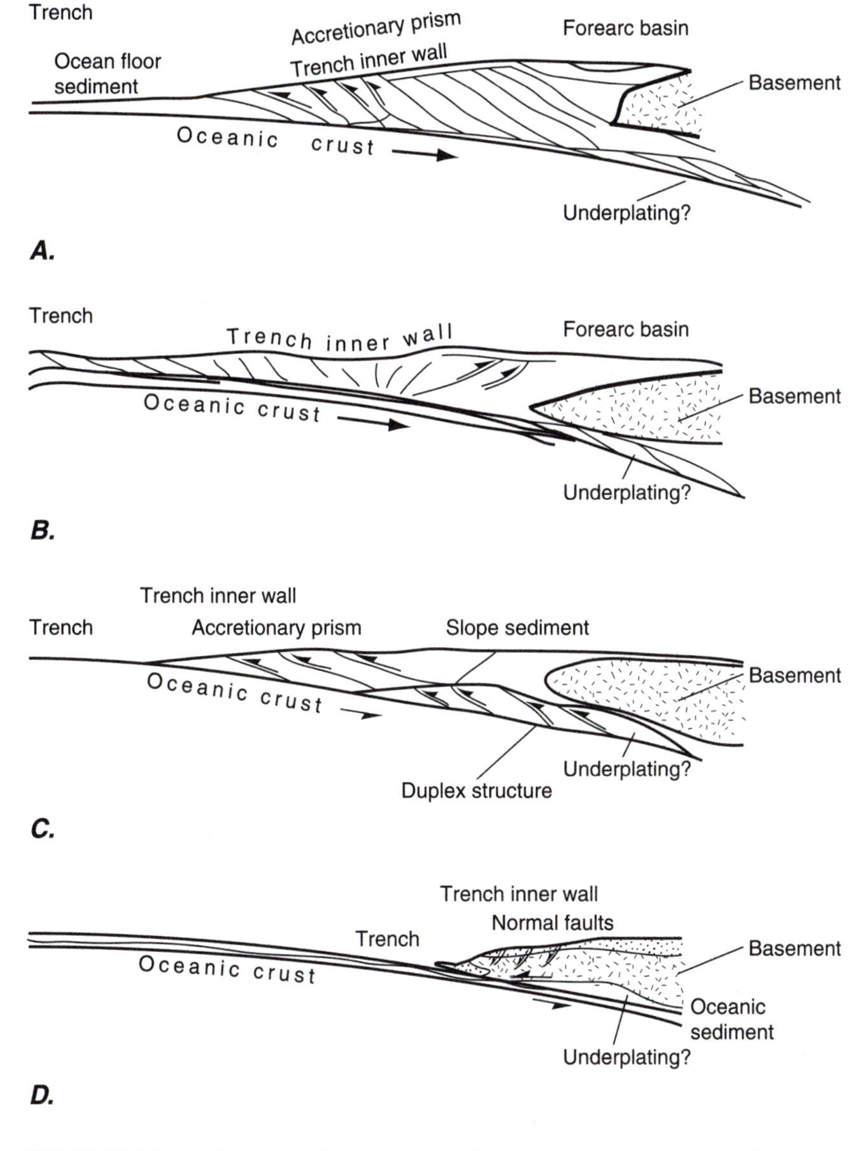

FIGURE 19.33 Variations in the structure of convergent margins depending on shape and size of accretionary prisms. *A.* Well-developed accretionary prism, with thrust verging toward down-going plate. *B.* Well-developed accretionary prism with thrust verging in two directions. *C.* Small accretionary prism, with large zone of underplating. *D.* Nonaccreting margin, with sediment veneer on upper plate. Essentially all sediment is subducted either to underplating region or deeper into the mantle. (*A, C, D* after von Huene and Scholl 1993; *B* after Moores and Twiss 1995, figure 7.10)

to steeply in the direction of subduction, although in places there seems to be little order. Evidence for soft-sediment deformation is abundant. In some cases, the deformation in the accretionary prism is so intense that any pre-existing stratigraphic continuity is destroyed. Such chaotic deposits are called **melanges** and are discussed in more detail in Subsection 19.4(vi).

Many trenches show relatively undeformed sediments on the down-going plate that extend for several kilometers beneath deformed rocks of the trench inner wall. In some cases, the sediments on the down-going plate are underplated to the bottom of the accretionary prism (Figures 19.33 and 19.36). In other cases, sediments in the accretionary prism may be added to the down-going plate in a process called **subduction erosion**, if the main subduction thrust propagates up into the over-riding accretionary prism, or even into the arc basement, effectively transferring the cut-off material onto the down-going plate. This process may be marked by the presence of normal faults in the inner trench walls (Figures 19.33*D* and 19.37*D*) and faults in accretionary prisms where sediments above the decollement are truncated against the decollement. Sediments that adhere to the down-going plate may descend into the mantle to become involved in the generation of arc magmas or may even pass deeper to be recycled into the mantle.

These observations lead to a model of the accretionary prism as a dynamic volume of lithified and unlithified sediment that acts like a critical Coulomb wedge (see Section 9.11(v) and Box 9-2; Figures 9.24*B* and 9.26*B*) and that continually deforms in response to the relative motion at the plate boundary and to the addition and subtraction of material. All of these processes can lead to changes in the shape of the accretionary wedge and thus to internal deformation of the wedge as it reacts to maintain the critical Coulomb taper. The accretionary prism is thus characterized by a complex association of deformational structures. If the accretionary wedge lengthens by the addition of off-scraped material to the front of the wedge or thins because of subduction erosion, the surface slope will decrease, and to maintain the slope the wedge must shorten and thicken internally by thrust faulting, folding, and underplating (Figure 19.37*A*). Addition of sediment to the top surface of the wedge increases the surface slope. Addition of material to the bottom of the wedge by underplating uplifts the wedge and therefore also increases the surface slope. In this case, the critical Coulomb taper is maintained by thinning the wedge through internal normal faulting and by lengthening the wedge through propagation of the decollement out in front of the existing wedge (Figure 19.37*B*).

Continual underplating by duplex formation at the base of the wedge and thinning and lengthening of the wedge by normal faulting at shallower levels could result

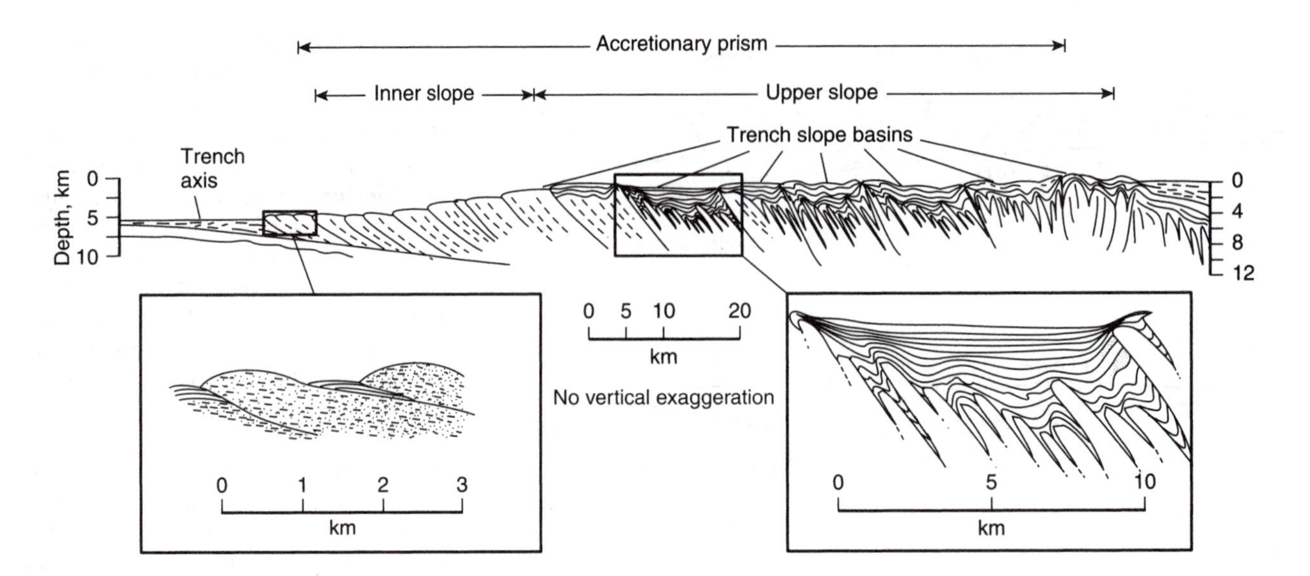

FIGURE 19.34 Structure of the Sunda arc, a typical well-developed accretionary prism. The inner trench slope and forearc basin region are underlain by an accretionary prism of imbricate thrusts and trench-slope basins. See Figure 19.31 for location. Accreted sediments are thrust over the slope sediments (insets). Compared with the inner trench slope area, the thrust faults are steeper under the forearc basin and involve the older sediments in the faulting. (After Moores and Twiss 1995, figure 7.7; Moore et al. 1980, pp. 18–20)

in the eventual uplift of rocks from deep in the wedge to be exposed at the surface (Figure 19.37*C, D*). Indeed, in ancient accretionary prisms, such as the Franciscan complex of northern California and the accretionary terranes of Japan, this relationship is just what we observe; the grade of metamorphism of the rocks in the accretionary complex increases with increasing distance from the front of the wedge. Note that this process involves both thrusting and normal faulting in different parts of the same accretionary wedge at the same time during plate convergence.

These scenarios have assumed the slope of the decollement remains constant, but the dip angle of the subduction fault can also change, complicating the predictions for the deformation required to maintain a critical Coulomb wedge (see Equation (9.19)).

Not all subduction zones have well-developed accretionary prisms, and in fact, nearly half the total length of trenches (21,000 km) may be nonaccreting, in which case old rocks of the island arc or continental basement can be traced out to the lower trench slope. Only thin layers of sediment accumulate in basins on the basement rocks, and very little if any sediment occurs in the trench. Although a lack of sediment supply may be part of the reason for the absence of an accretionary prism, subduction of all available sediment is apparently also a significant factor (Figure 19.33*D*). The arc basement is commonly cut by normal faults downthrown on the trench side (Figure 19.33*D*) and although thrust faulting may be present at depth, there is little evidence for it at the surface.

These processes of sediment supply, sediment subduction, underplating, subduction erosion, uplift, and subsidence in accretionary prisms, not to mention metamorphism and the evolution of pore fluid pressure, make it extremely difficult to predict the evolution of an accretionary wedge under the various possible sets of con-

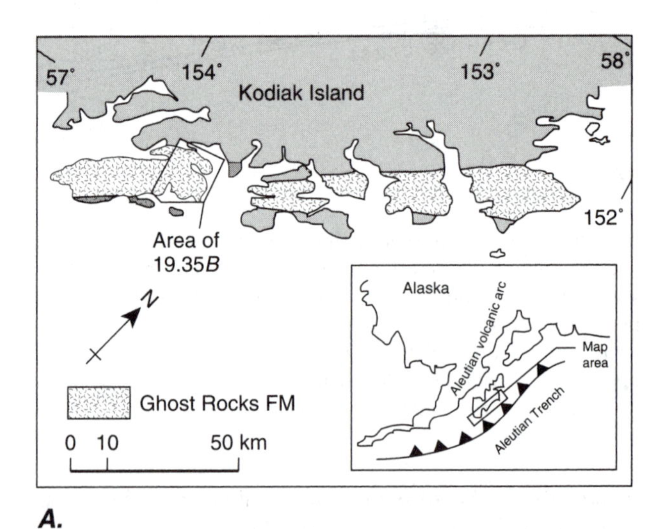

FIGURE 19.35 Fold styles in accreted forearc sediments of the Aleutian arc, Kodiak Island, Alaska. See Figure 19.31 for location. (After Moores and Twiss 1995, figure 7.11; Byrne 1982, p. 235) *A*. Location map of Kodiak Island, Alaska. *B*. Map and cross section of a portion of the accretionary prism showing a large-scale fold. *C*. Small-scale fold styles that formed in partially lithified sediments. Diagram 4 shows spaced cleavage cutting across the axial surface of earlier folds.

(continued)

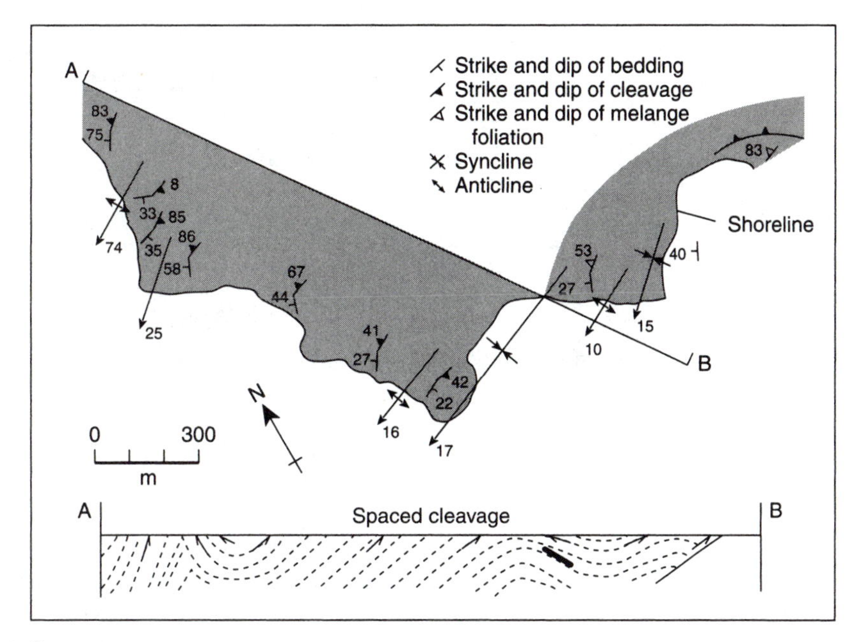

B.

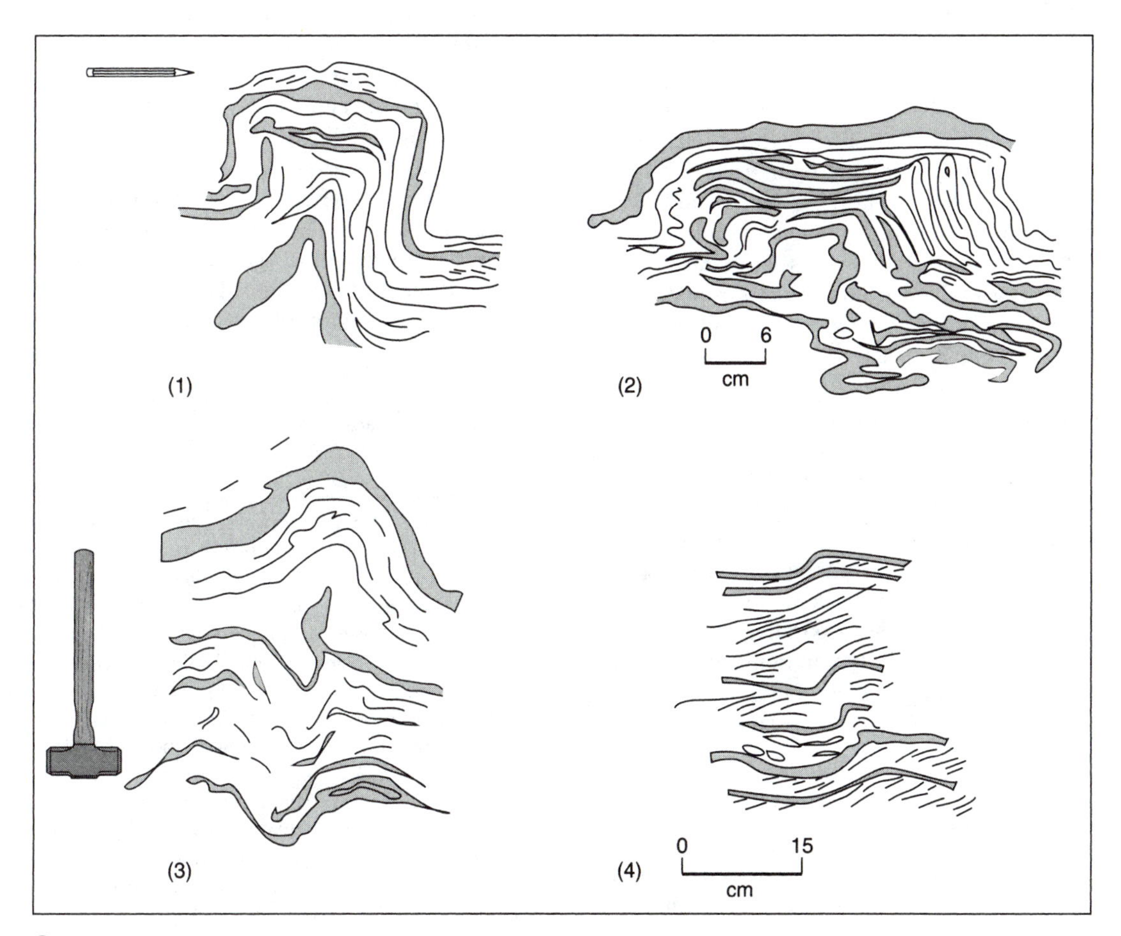

C.

FIGURE 19.35 *(continued)*

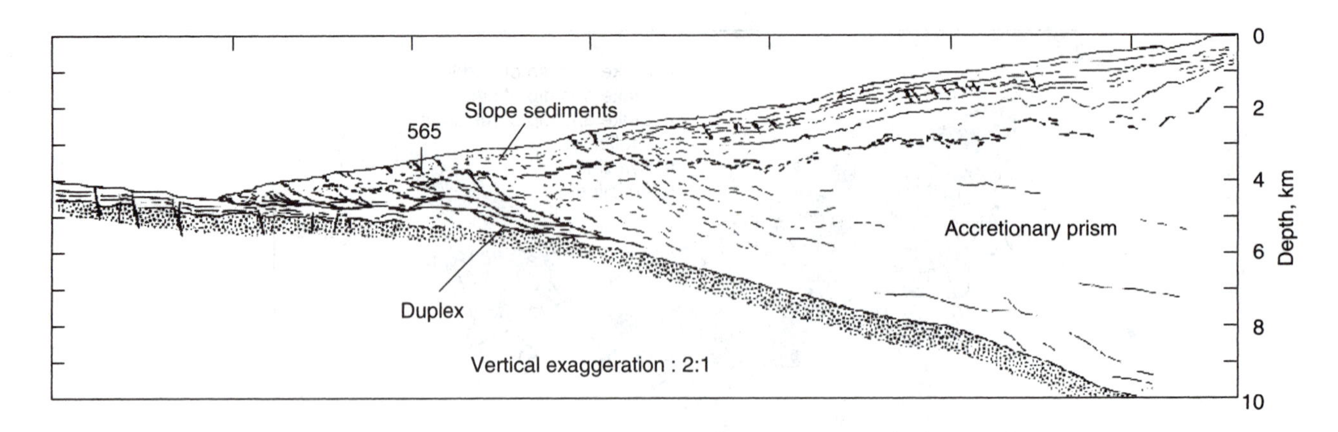

FIGURE 19.36 Underplating of the accretionary prism off Costa Rica by duplex formation, based on a seismic reflection profile. (After Shipley et al. 1992)

ditions. As yet, we do not have a sufficiently complete understanding.

v. Partitioned Strike-Slip Faulting at Convergent Margins

At spreading centers, the plate boundary is organized into ridge segments and transform faults, and these structures are oriented so the relative plate motion is perpendicular to the ridge segments and parallel to the transform faults. We do not observe oblique opening of a simple plate boundary. This phenomenon can be accounted for in terms of the minimum work principle, with less work being required to separate plates whose boundary is partitioned into orthogonal ridge and transform segments, than to open and shear an oblique segment.

A similar principle appears to apply at subduction zones, although the results are different. If the angle between the relative convergent plate velocity and the subduction zone is highly oblique, roughly $< 60°$, the slip becomes partitioned at the subduction zone into a near dip-slip component on the subduction fault and a strike-slip component on an adjacent fault that cuts the over-riding plate between the trench and the volcanic arc (Figure 19.38). This partitioning creates a prism of the forearc on the over-riding plate that slips parallel to the trench at a velocity such that the subduction beneath the prism is approximately dip-slip. The relative motion across the subduction zone at depth behind the forearc prism, however, must be oblique.

vi. Chaotic Deposits and Melanges

Many orogenic belts contain so-called **chaotic deposits** or **melanges**,[3] which are rocks composed predominantly of sediments in which the normal stratigraphic relations, such as regular bedding and consistent stratigraphic sequences, have been disrupted or destroyed. Their presence is associated with modern and inferred ancient forearc regions. A chaotic deposit can be the product of either a submarine landslide, in which case it is called a **sedimentary melange** or an **olistostrome**,[4] or very large pervasive strains, in which case it is called a **tectonic melange**. Exposures may be such that the origin of the deposit as either sedimentary or tectonic is difficult to decipher from field evidence.

Olistostromes are found within an otherwise normal stratigraphic sequence generally consisting of shale, sandstone, or both. They characteristically contain a disordered arrangement of blocks of one lithology embedded in a matrix of another. For example, Figure 19.39 shows a cross section of a region in the northern Apen-

[3]After the French word *mélange* meaning "mixture."

[4]After the two Greek words, *olistos*, which means "sliding," and *stroma*, which means "bed."

FIGURE 19.37 Dynamic processes in the formation of an accretionary prism. *A*. At the toe of the accretionary wedge, if the slope is too low, shortening and thickening occur by formation of imbricate thrusts synthetic to the subduction zone, as well as some antithetic thrusts. Underplating occurs by duplex formation and folding at deeper levels. *B*. Underplating continues uplifting older rocks and steepening the slope at the rear of the wedge. The oversteepened slope drives gravitational collapse by listric normal faulting. Stippled pattern shows rocks that have been above 1000 MPa pressure. *C*. Growth of the accretionary prism through continued uplift by underplating from below, compensated for by gravitational collapse on listric normal faults above. Deeper rocks are uplifted, and the accretionary prism grows out over the subduction zone. *D*. Continuation of these processes forms complex normal fault geometry near the surface, nappes of high-pressure rocks on listric normal faults at depth and extension of the toe of the accretionary prism on late thrust faults. (After Platt 1986)

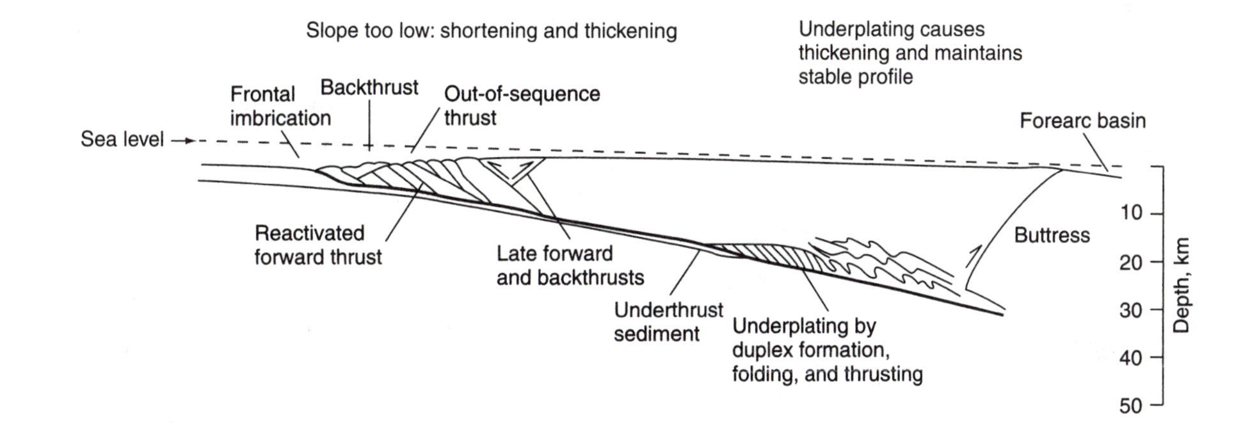
Slope too low: shortening and thickening
Underplating causes thickening and maintains stable profile
Frontal imbrication
Backthrust
Out-of-sequence thrust
Sea level
Forearc basin
Reactivated forward thrust
Late forward and backthrusts
Buttress
Underthrust sediment
Underplating by duplex formation, folding, and thrusting
10
20
30
40
50
Depth, km

A.

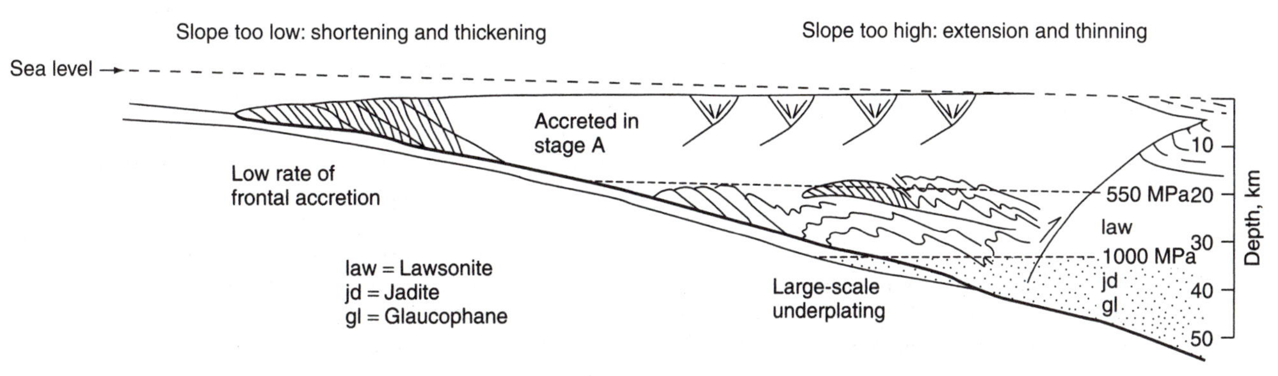
Slope too low: shortening and thickening
Slope too high: extension and thinning
Sea level
Accreted in stage A
Low rate of frontal accretion
10
550 MPa
20
law
30
1000 MPa
law = Lawsonite
jd = Jadite
gl = Glaucophane
Large-scale underplating
jd
gl
40
50
Depth, km

B.

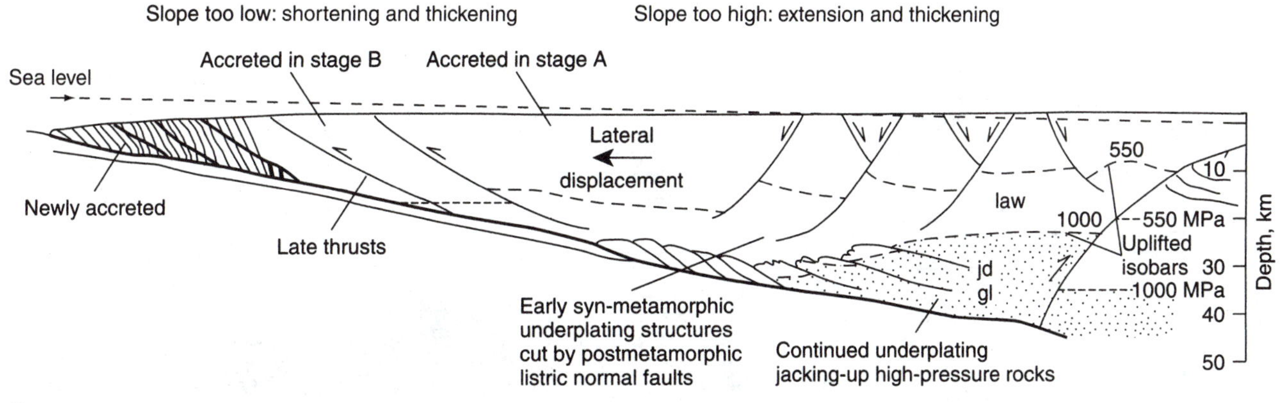
Slope too low: shortening and thickening
Slope too high: extension and thickening
Sea level
Accreted in stage B
Accreted in stage A
Lateral displacement
550
10
Newly accreted
law
Late thrusts
1000
550 MPa
Uplifted isobars
jd
gl
30
1000 MPa
40
Early syn-metamorphic underplating structures cut by postmetamorphic listric normal faults
Continued underplating jacking-up high-pressure rocks
50
Depth, km

C.

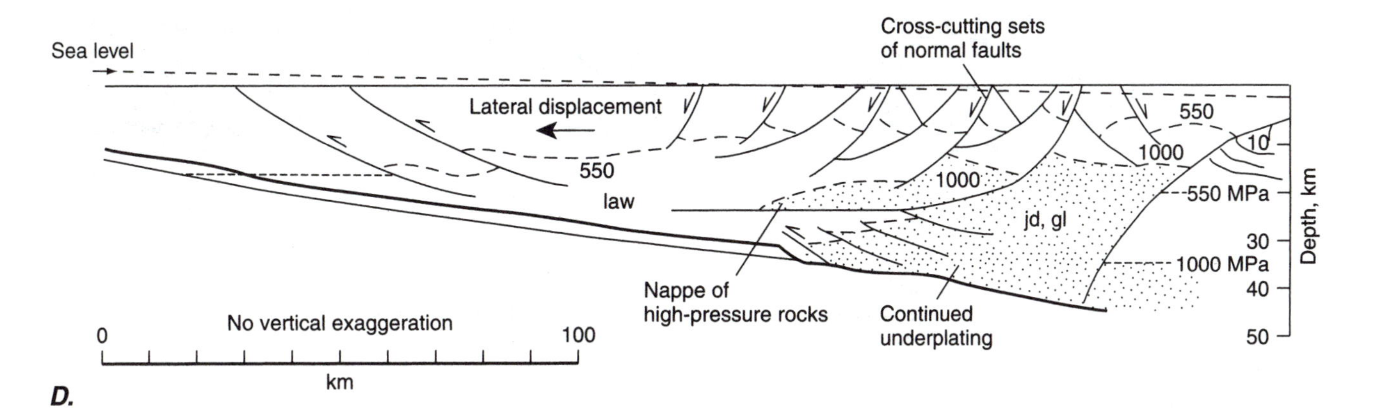
Sea level
Cross-cutting sets of normal faults
Lateral displacement
550
10
1000
550
1000
law
550 MPa
jd, gl
30
1000 MPa
40
Nappe of high-pressure rocks
Continued underplating
No vertical exaggeration
0
100
km
50
Depth, km

D.

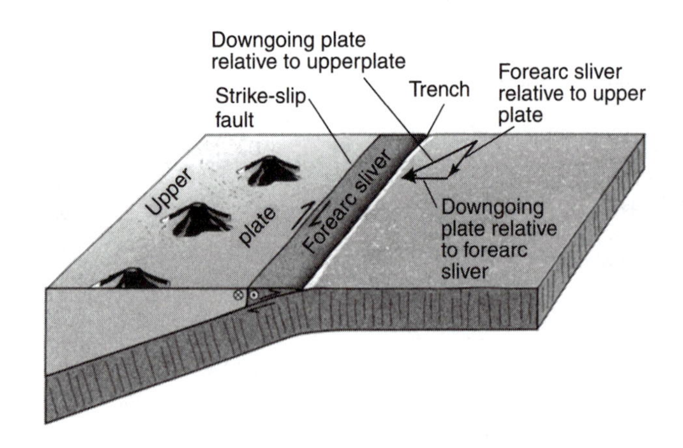

FIGURE 19.38 Block diagram showing a model of processes in obliquely subducting zones. Partitioning of oblique displacement into strike-slip and dip-slip components leads to the development of a forearc sliver bounded on opposite sides by a strike-slip fault and the trench. (From Moores and Twiss 1995, figure 7.28)

nines of Italy where olistostromes of ophiolitic lithologies are present in a sequence of shale and calcareous turbidites. Many olistostromes contain mappable blocks of an individual lithology that occur on a scale of meters to kilometers, called **olistoliths** (Figure 19.39). The olistostromes probably formed by landsliding from a submarine topographic high of oceanic crust, possibly a ridge along a fracture zone. Other topographic highs that might serve as the source of such slides are topographic scarps along passive continental margins, over-steepened trench walls, and the flanks of midplate volcanic islands such as the Hawaiian Islands. Very large landslides from midplate volcanic islands are widespread; some researchers have estimated that as much as 6% of the ocean floor area is covered with such deposits.

Tectonic melanges, on the other hand, clearly are the result of a large and pervasive deformation of tectonic origin. Most tectonic melanges are chaotic mixtures of diverse rock types in an irregularly foliated matrix (Figure 19.40) and are characteristically found in parts of the accretionary complexes of subduction zones. Blocks of ophiolitic lithologies, such as peridotite, volcanic rock, gabbro, and pelagic chert or limestone, are found in many melanges, and blocks of metamorphic rocks and shallow- or deep-water terrigenous sediments also are common. These blocks can range in size from a few millimeters to as much as several kilometers. The matrix is generally a scaly argillite or a serpentinite, both of which have an anastomosing disjunctive foliation along which the rock breaks to form small scale-like fragments.

Melanges exist in mountain belts as old as late Precambrian (e.g., Anglesey, North Wales; Damaran belt, Namibia). In some regions, such as the Coast Ranges of California, Turkey, and Iran, they form vast terranes thou-

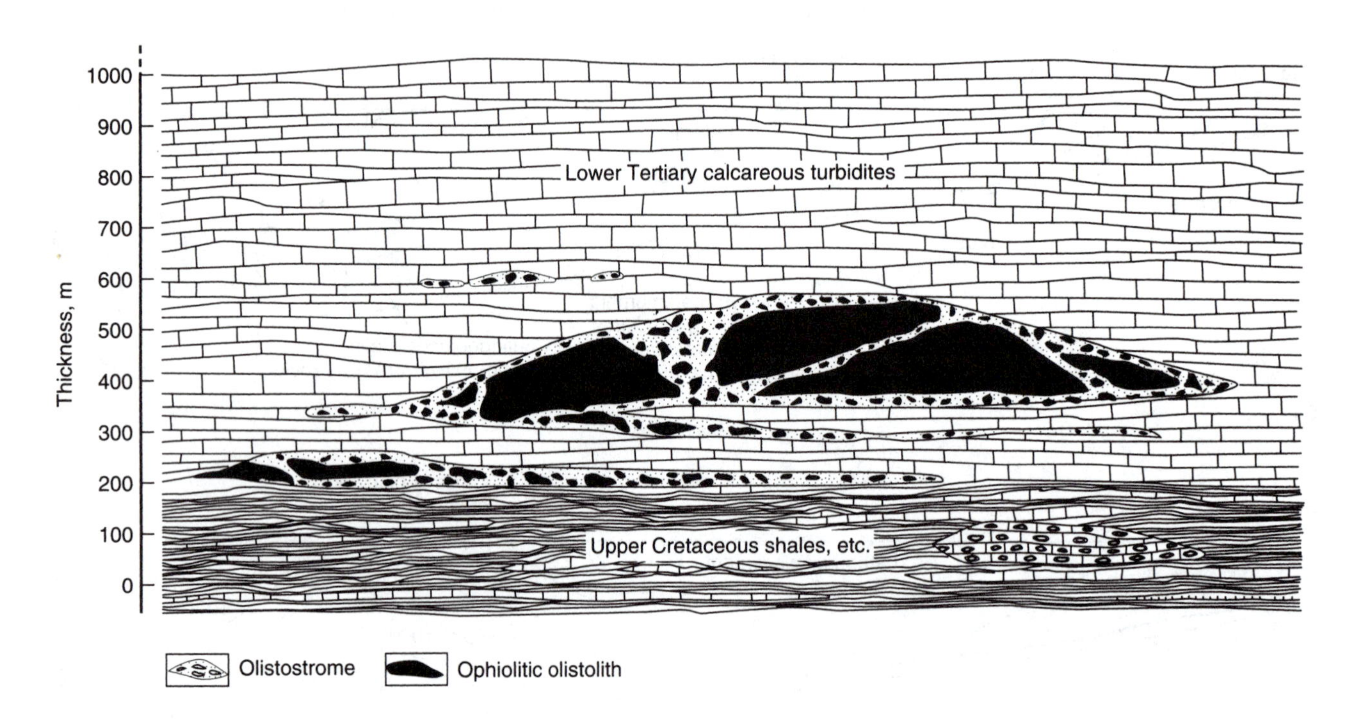

FIGURE 19.39 Schematic cross section, illustrating olistostrome occurrences in the northern Apennines of Italy. Chaotic masses of ophiolitic debris in a pelitic matrix are interbedded with upper Cretaceous–lower Tertiary shales and calcareous turbidites. The larger olistostromes contain several olistoliths of ophiolitic lithologies. Estimated vertical exaggeration about 2×. (After Moores and Twiss 1995, figure 7.19; Abbate et al. 1970)

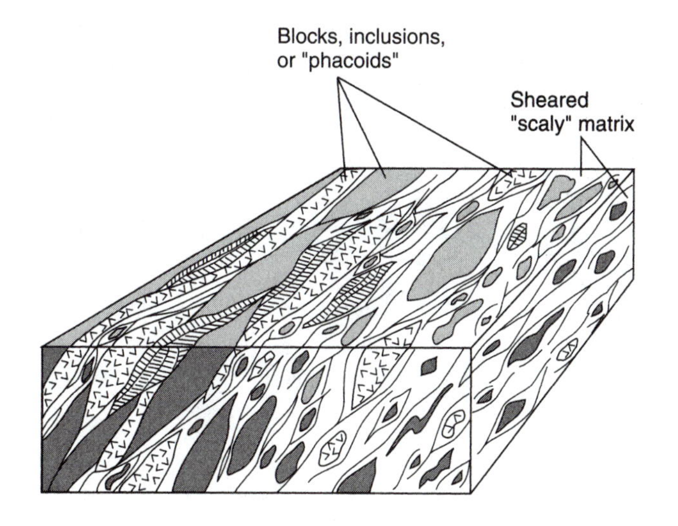

FIGURE 19.40 Block diagram of typical melange showing diverse elongate blocks and irregularly foliated (scaly) matrix. Scale could be anything from centimeters to kilometers. Matrix could be sheared clay, sand clay mixture, or serpentinite. Blocks could be sedimentary, metamorphic, or ophiolitic lithologies. (After Moores and Twiss 1995, figure 7.20; Cowan 1985)

sands of square kilometers in area. Elsewhere they are more limited in extent. Melanges also have been identified in a number of modern accretionary prisms. Cores obtained from ocean drilling in areas such as the Aleutians and the Middle America accretionary prisms have retrieved scaly clay, suggesting at least one origin for the material.

The chaotic nature of the rocks in melanges and their general lack of stratigraphic continuity make them difficult to map and interpret (see Figure 19.40). Fossil or radiometric age information obtained from blocks in the melange at best provides a maximum possible age for the incorporation of the block into the melange, for the block must have been formed before it became part of the melange. Melange formation, however, could have started before the rock was formed, or it could have started long after the rock was formed.

Figure 19.41 shows a number of sites in a subduction zone setting where melanges could form. Sedimentary melanges could form along the margins of forearc or slope basins or along the inner wall of the trench as a result of gravity sliding and disruption. Tectonic melanges could form by disruption associated with dewatering. Dewatering of the down-going rocks could also result in

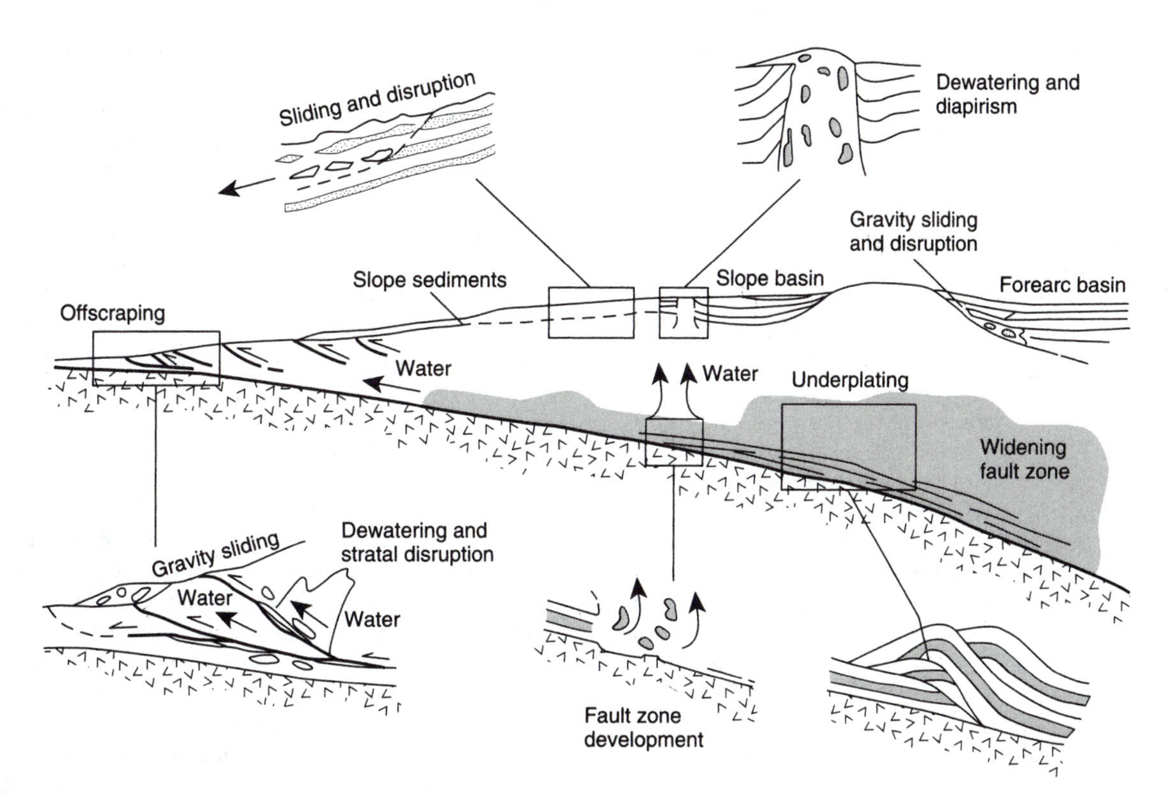

FIGURE 19.41 Generalized cross section showing possible origins of chaotic deposits. Gravity sliding could be along trench slope or in margins of basins. Dewatering of sediments in offscraping or underplating regions could release water and disrupt rock units, which then could be transported upwater with escaping water. Melanges could also form by progressive widening of fault zone. (After Moores and Twiss 1995, figure 7.21; Cowan 1985)

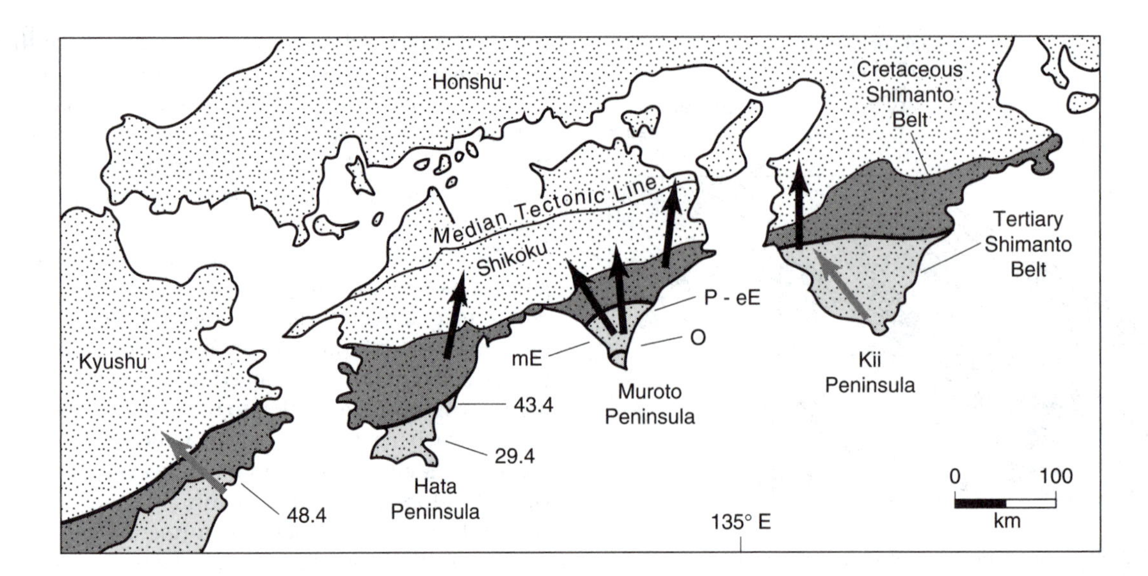

FIGURE 19.42 Directions of underthrusting from paleokinematic data compiled from the accretionary complex along the southwest coast of Japan. Data from the Muroto Peninsula are from faults with slickenlines, where cross-cutting relations show that the north-directed underthrusting is the oldest, and the northwest directed underthrusting reflects a shift in the plate motions at the margin at about early Eocene ($< \sim$55Ma) to Oligocene ($> \sim$25Ma) time. Numbers along the coast are K-Ar ages for very fine-grained cleavage-forming phyllosilicates. Letters refer to paleontologically determined ages. mE: middle Eocene; O: Oligocene; P-eE: Paleocene to early Eocene. (From Lewis et al. 2002, figure 2D)

diapiric activity, which could bring to the surface chaotically disrupted rocks. Widening fault zones in a zone of underplating could also result in melange formation. Once formed, a melange can redeform easily because of the mechanical weakness of such an incoherent deposit. Thus areas of extensive melange exposure, such as the Coast Ranges of California or the Apennines of Italy, also display abundant landsliding. The relative role of original or subsequent disruption on an individual exposure often is difficult, if not impossible, to decipher.

Because the deformation in a tectonic melange is generally driven by the relative motion across a subduction zone, the structures in the melange should preserve a record of that motion. Analysis of these structures can then provide crucial constraints on the kinematics of the plate that has been consumed at the subduction zone. For example, a kinematic analysis comparable to that described for brittle shear zones in Section 15.4(i) has been applied to sets of small faults in the melange along the south coast of Honshu, Japan. The faults were divided into two distinct groups based on the character of the faulting, one set being active before lithification of the sediment, and the other active after lithification. Each group gives a distinct orientation to the direction of subduction at the subduction zone (Figure 19.42). The ages of the faults are constrained by fossils in the rocks they cut, and thus the change in subduction direction can be correlated with changes in the orientation of seafloor magnetic anomalies. The different subduction slip directions were used to constrain the location of the poles of relative rotation that define the plate motions at the subduction zones and thus provide a test and a predictive model for the reconstruction of past plate motions in the western Pacific. This approach to interpreting the deformation only works if the direction of subduction is not partitioned into two components, one on the subduction zone and the other on a strike-slip fault parallel to the subduction zone. In this example, the obliquity of both inferred subduction directions (Figure 19.42) was sufficiently small to allow the analysis.

19.5 ACTIVE COLLISIONS

A consequence of the evolution of consuming margins is that, eventually, a down-going plate will carry some continental or island arc crust into a subduction zone, unless it consists completely of oceanic crust. At that point, a collision zone replaces the consuming margin. A collision is a plate interaction for which thick low-density crust on the down-going plate is subducted beneath thick low-density crust on the over-riding plate. The thick crust may be continental crust, or it may be the crust beneath an island arc or oceanic plateau. The buoyancy of the low-density crust being dragged into a higher-density mantle counteracts the driving forces of subduction. Thus subduction ceases and a rearrangement of plate motions occurs, which could include the generation of new plate margins.

Several possible types of collision involving continental crust or island arc crust exist, and they can be grouped according to whether one or two subduction zones are involved (Figure 19.43). If only one subduction zone is present, a passive continental margin on the down-going plate can collide with either an active continental margin (Figure 19.43*A*) or the forearc of an active island arc (Figure 19.43*B*) on the over-riding plate. If two subduction zones dip the same way, the back-arc region of an island arc on the down-going plate can collide with

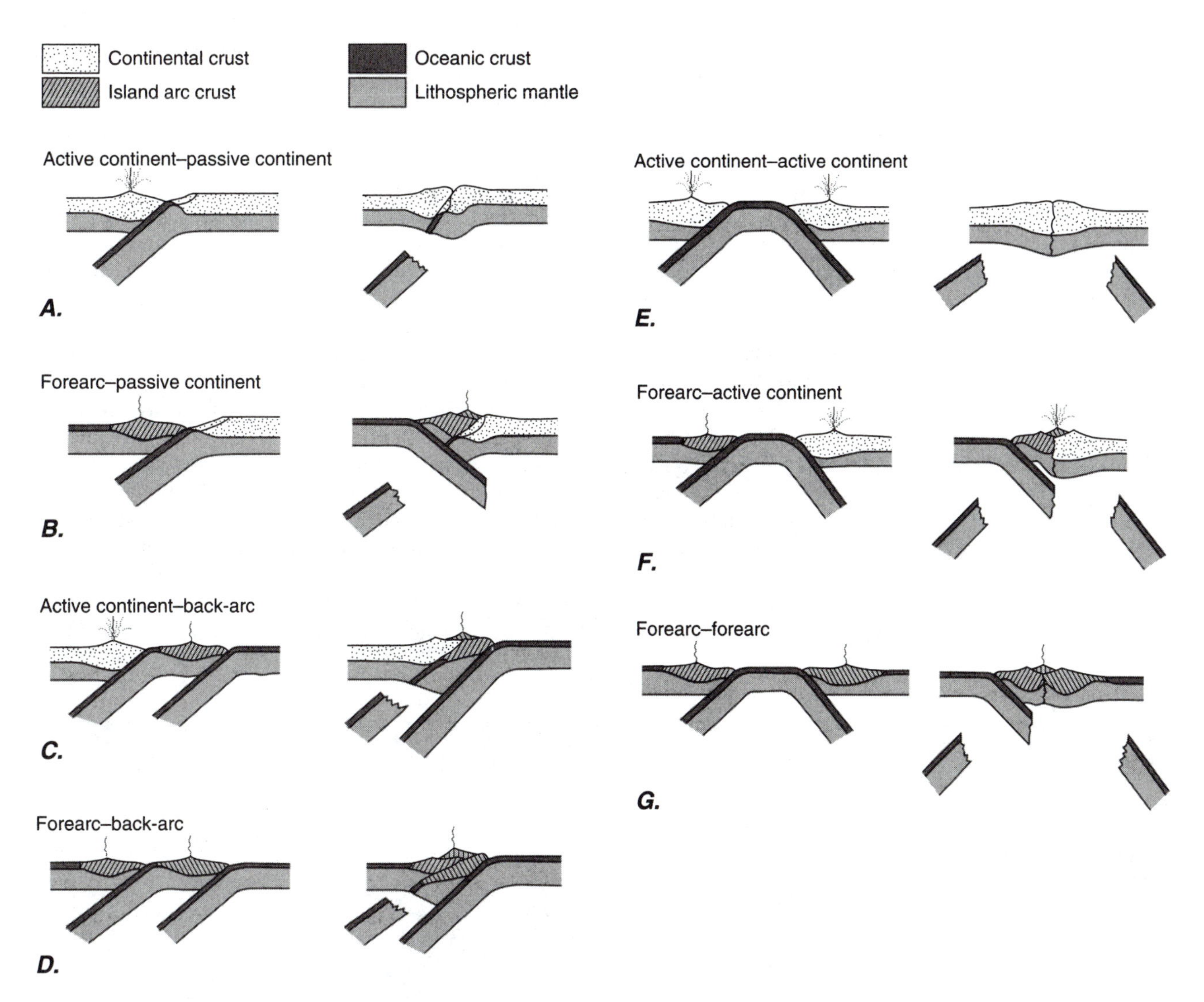

FIGURE 19.43 Schematic diagrams showing the theoretically possible types of collision involving one and two subduction zones. The left column shows the precollision geometry and the right column shows the postcollision geometry of the crustal blocks and the down-going plate(s). Following collision in all cases, the original down-going lithospheric plate is detached from the surface plate and the subduction geometry is reorganized. (After Moores and Twiss 1995, p. 213) *A.* Collision of an active continental margin with a passive continental margin. Following the collision, subduction ceases and must be accommodated elsewhere in the plate tectonic system. *B.* Collision of a forearc margin with a passive continental margin. After collision, the subduction polarity flips and the down-going slab dips under the collision zone. *C.* Collision of a back-arc margin with an active continental margin. After collision, the remaining active subduction zone dipping under the island arc can accommodate the required increase in subduction rate. *D.* Collision of a forearc margin with a back-arc margin. After collision, the required increase in subduction rate can be taken up by the remaining active subduction zone. *E.* Collision of two active continental margins as the intervening plate is subducted at opposite dipping subduction zones. After collision, subduction ceases and must be accommodated elsewhere in the plate tectonic system. *F.* Collision of a forearc margin with an active continental margin with the subduction of the intervening plate under both margins. After collision, a new subduction zone could form dipping under the collision zone. *G.* Collision of a forearc margin with a forearc margin with the subduction of the intervening plate under both margins. After collision, a new subduction zone could form dipping in either direction underneath the collision zone.

either an active continental margin (Figure 19.43*C*) or the forearc of another island arc (Figure 19.43*D*) on the over-riding plate. Finally, if the two subduction zones dip in opposite directions, two over-riding plates collide in three possible combinations of forearc and active continental margin (Figure 19.43*E*, *F*, *G*). Thus just from these simple models, we can envision two types of continent-continent collision (Figure 19.43*A*, *E*), three types of arc-continent collision (Figure 19.43*B*, *C*, *F*), and two types of arc-arc collision (Figure 19.43*D*, *G*).

A termination of subduction following collision must lead to a shift in either the location or the polarity of local subduction or to an alteration in the plate tectonic geometry and kinematics, as illustrated by the right-hand diagrams in Figure 19.43. For a continent-continent collision (Figure 19.43*A*, *E*), the convergence must be accommodated elsewhere in the plate tectonic system, and this in turn could lead to a worldwide reorganization of plate kinematics. For the forearc-passive continental margin collision (Figure 19.43*B*), the oceanic crust can begin subducting under the collision zone, giving a flip in the polarity of subduction. For the parallel double-subduction zone configuration, collision terminates subduction at one of the subduction zones, and the convergence across the system could be taken up by an increase in the subduction rate at the surviving subduction zone (Figure 19.43*C*, *D*). For the opposing double-subduction zone configuration, continued convergence could be accommodated locally by the initiation of a new subduction zone just outboard of the collision zone that carries oceanic crust under the collision zone (Figure 19.43*F*, *G*).

The distinction between the different collision types is subtle. A key clue, however, is the structural sequence exhibited in the collision zone. Deciphering the stratigraphy and tectonic sequence in a deformed belt will give some indication as to the types of margins involved, and the identity of the overriding plate.

Collisions typically result in horizontal shortening and vertical thickening of the crust by underthrusting at the subduction zone, with the consequent formation of ductile fold nappes (see Figures 10.1*B*, *C*), major thrust nappes (see Figure 10.2), and imbricate thrust zones (see Figures 5.11, 5.12, and 5.20*B*, *C*). Ophiolite belts are commonly found within continents (Figure 19.44), particularly within orogenic belts. The models of collisions involving continents suggest that such ophiolites represent remnants of the ocean basins that once separated two colliding crustal masses and that have become trapped within the collision zone. Most ophiolites have been emplaced by collision of a passive continental margin with an intra-oceanic subduction zone (Figure 19.45*A*, *B*). Typically, these ophiolite remnants are from the over-riding plate and are thrust over the down-going shelf sediments. The lower contact of the ophiolite is thus a tectonized zone characterized by subhorizontal thrust faults that have transported the ophiolites and underlying tectonic slices or melange large distances toward the interior of

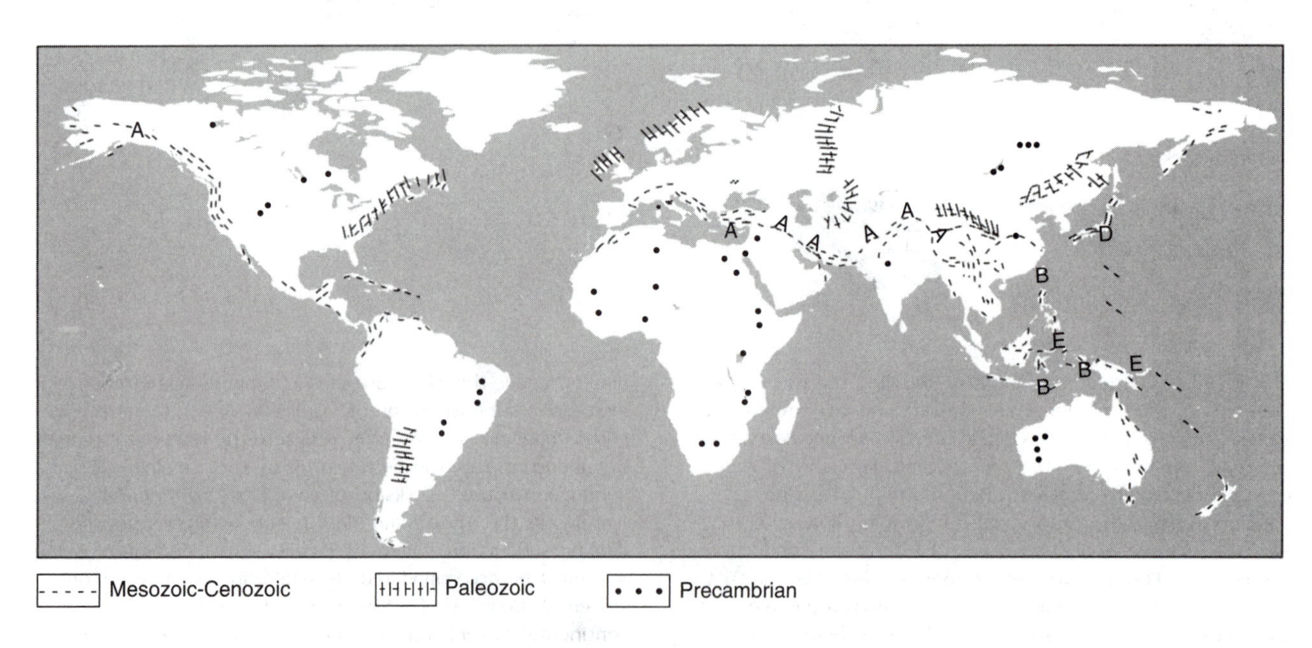

FIGURE 19.44 Map of world showing ophiolite belts of various ages, and zones of active collision indicated by various letters. The letters in active collision zones correspond to the type of collision characterized by the respective diagrams in Figure 19.43. These ongoing collisions include the Alpine-Himalayan belt (A), the West Luzon arc–East China collision in Taiwan (B), the Sangihe-Halmahera collision south of the Philippines (E), the New Britain–Trobriand trough collision in Papua New Guinea (E), the Indonesian-Australian collision (B), the Alaska active continental margin–Yakutat block collision in Alaska-Canada (A), and the Izu-Bonin-Japan collision (D). (After Moores and Twiss 1995; Gass 1982; Moores 2002; see Moores and Twiss 1995, chapter 9)

the continent. A few ophiolites are emplaced by incorporation of part of a down-going oceanic plate into the over-riding plate, which can happen if an antithetic thrust fault develops in the down-going plate (Figure 19.45*C*, *D*). Thus a belt of ophiolites marks the location of a former ocean basin and is one indication of an important tectonic **suture.** The great variety and number of ophiolite belts in the world indicates that many collisions have taken place and that collision is a tectonic process of primary importance. With so many possibilities for collision geometry, we can expect the interpretation of the geologic record to be quite difficult.

In addition to the two-dimensional consequences of collisions outlined in Figure 19.43, there are interactions in the third dimension resulting in the transport of major crustal blocks subparallel to the collisional orogen and thus perpendicular to the cross sections by which these areas are usually depicted. Many collisions involve not only thrust faults and fold nappes, but also major strike-slip faults. For example, Figure 19.46 shows a map of the Alpine-Himalayan belt, which involves an active ongoing collision between the Eurasian continent and the continents of Africa, including the Arabian Peninsula, and India. Prominent on the map are strike-slip fault zones and small plates that result from adjustments to ongoing collision of continental crust on the over-riding plate. Typically, these faults accommodate the lateral movement of crustal blocks on the over-riding plate away from an impinging promontory of continental crust on the down-going plate. Thus, for example, the motion of the small Turkish and Aegean plates (Figure 19.46) are a result of the movement of material on the over-riding Eurasian plate out of the way of the colliding Arabian promontory on the down-going plate where it is forming the Caucasus Mountains. Similarly, several strike-slip faults in Asia around the margin of India are structures that accommodate the impinging of India, which is on the down-going plate, into the over-riding Southeast Asian continental mass (see Section 18.5(iii)).

During many collisions, thickened crust results from thrusting of one crustal block over another or from some more complex process, as suggested in Figure 19.43. Eventually such crustal thickening carries continental crustal rocks to sufficient depth that partial melting of the lower crust occurs, producing granitic magma. The Himalaya-Tibet region is an active zone of continent-continent collision that exemplifies this process. Figure 19.47 shows a generalized cross section of this region in

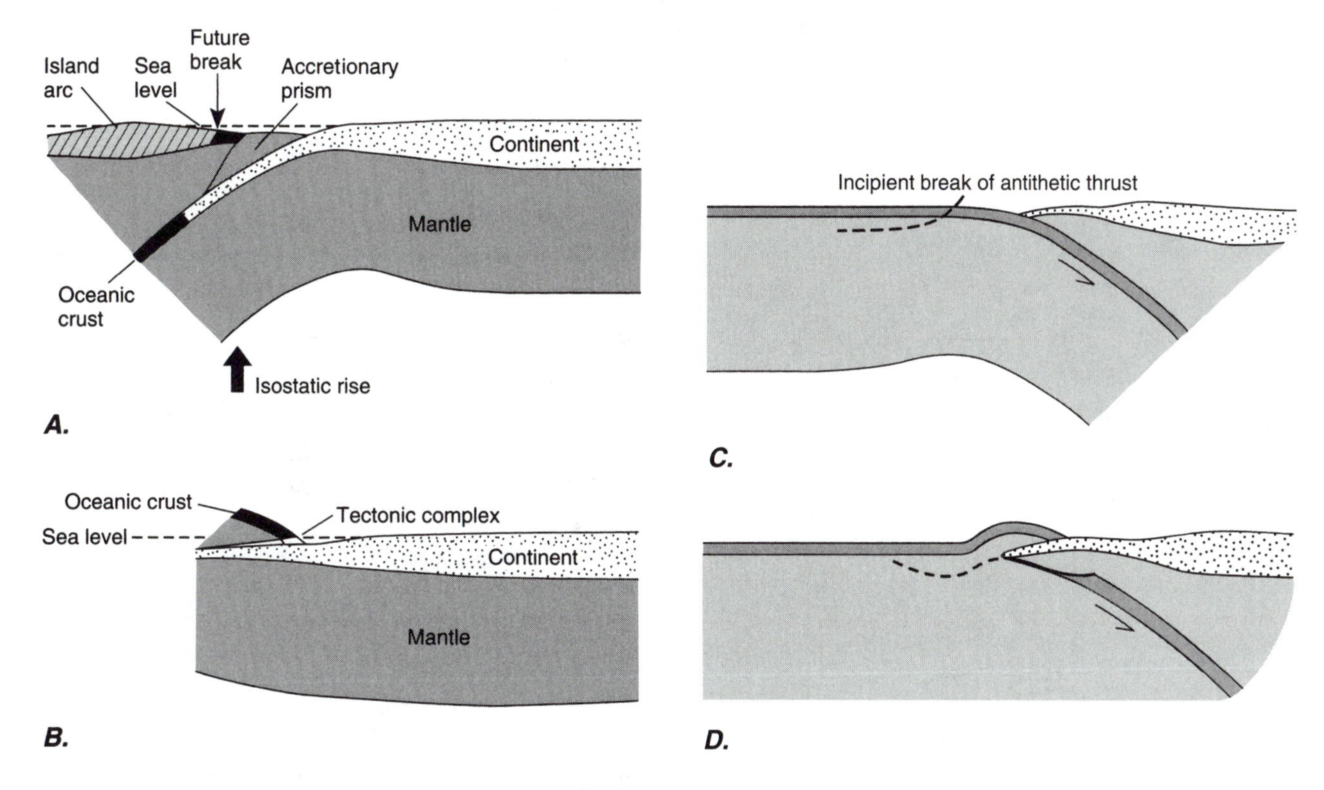

FIGURE 19.45 Emplacement of ophiolite complexes. (After Moores and Twiss 1995, pp. 239–240) *A*. Collision of continent with a subduction zone dipping away from the continental margin. Shoving the buoyant crust into the mantle eventually chokes off the subduction. *B*. Subsequent isostatic rise leaves a remnant of the former over-riding plate on the continental margin. *C*. Incorporation of ophiolite complex onto a continental margin by development of an antithetic thrust in the down-going plate. *D*. Continental margin wedges under a portion of the oceanic plate, emplacing the ophiolite over the continental margin.

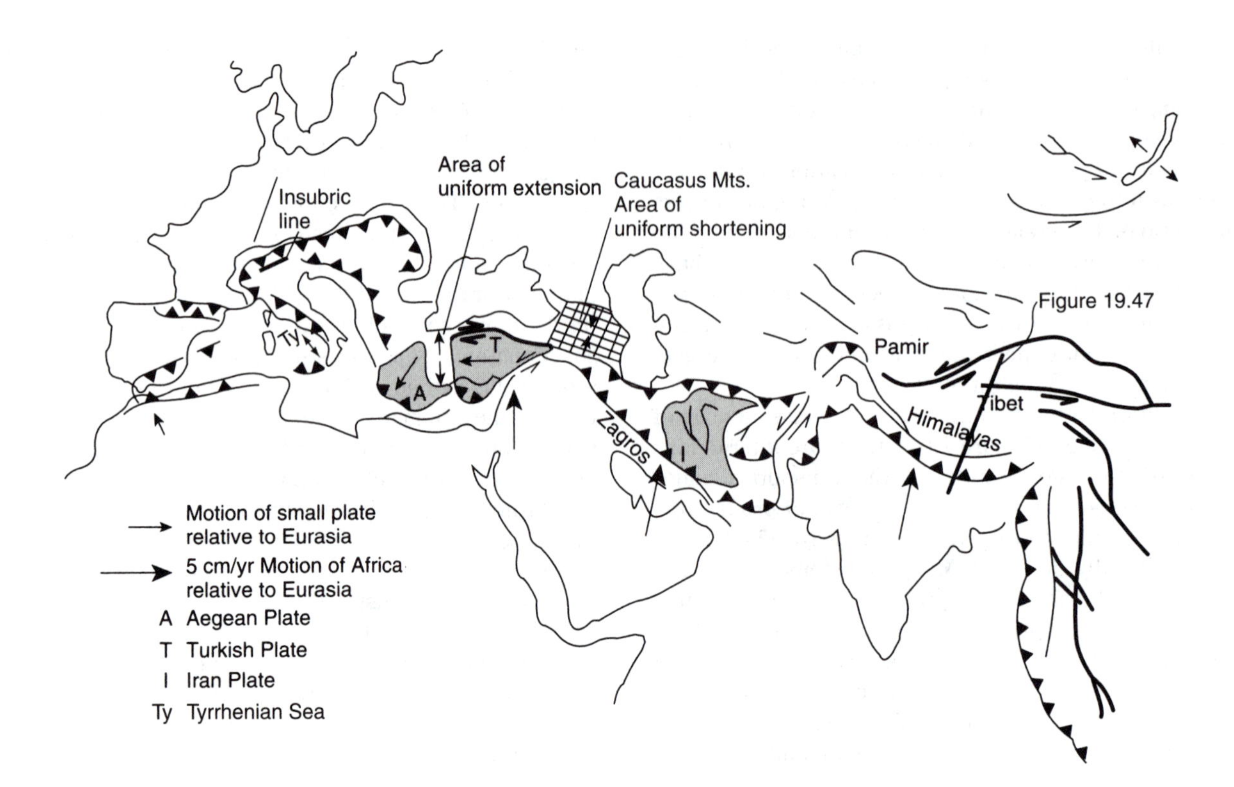

FIGURE 19.46 Generalized map of Alpine-Himalayan system, showing principal thrust regions and three mini-plates, the Aegean, Iranian, and Turkish plates. Small arrows indicate relative motion along faults or motion of small plates. Large arrows indicate overall convergence rates of Eurasia with respect to the African, Arabian, and Indian plates. (After Moores and Twiss 1995; Dewey 1977; McKenzie 1970)

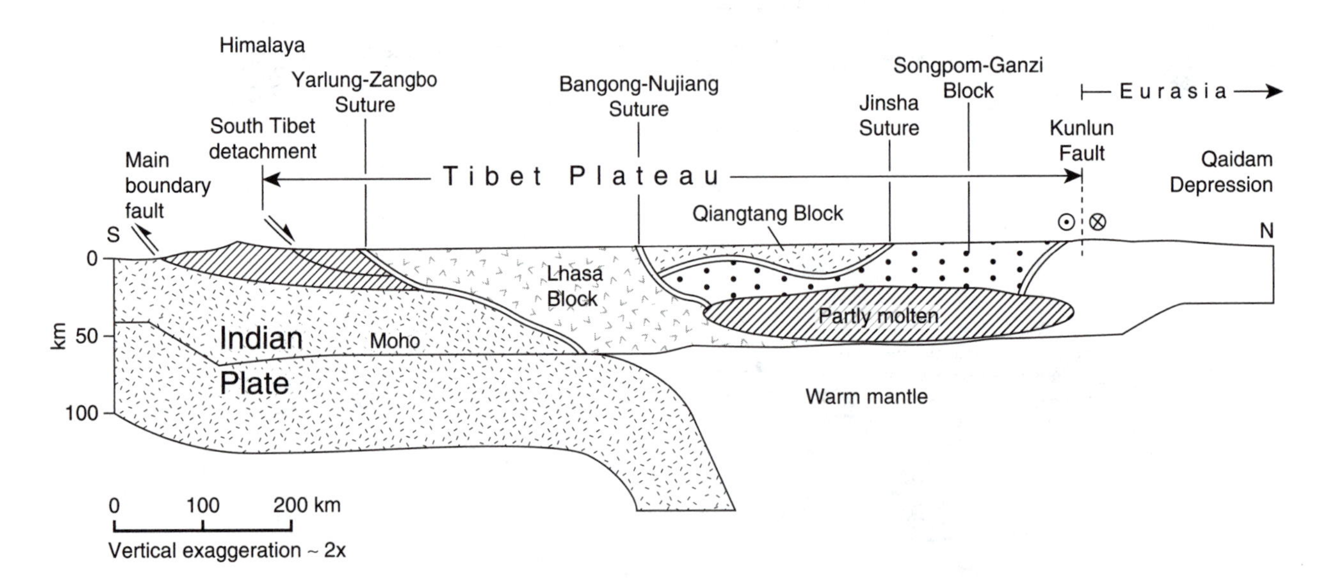

FIGURE 19.47 Generalized cross section of the Himalaya-Tibet region (see Figure 19.46 for location). The Indian Plate dips northward beneath Tibet; the Himalaya are composed of rocks from the now-subducted passive margin formerly along the northern margin of the Indian continent. The Lhasa, Qiangtang, and Songpan-Ganzi blocks are microcontinents or intraoceanic island arcs caught up in the Indian-Eurasian collision. The Bangong-Nujiang suture is of mid-Jurassic age; the Main Boundary fault is the main lower active thrust fault of the Himalaya; the South Tibet detachment zone is a north-dipping detachment (normal) fault active during southward-thrusting and uplift of the Himalaya All the sutures contain ophiolites. (Modified after Haines et al. 2003)

which the several sutures in the Tibetan crust testify to the complexity of the collisional history, as opposed to the simple models shown in Figure 19.43. Figure 19.47 also shows the presence of Indian crust and mantle lithosphere beneath Tibet, as well as zones of fluids, presumably partially molten rock, in the thickened crust beneath Tibet, which have a profound effect on the evolution of the Tibetan Plateau.

REFERENCES AND ADDITIONAL READINGS

i. Rifted Margins

Allmendinger, R. W., T. A. Hauge, E. C. Hauser, C. J. Potter, S. L. Klemperer, K. D. Nelson, P. Knuepfer, and J. Oliver. 1987. Overview of the COCORP 40° N Transect, western United States: The fabric of an orogenic belt. *Geol. Soc. Amer. Bull.* 98: 364–372.

Bertotti, G., Y. Y. Podlachikov, and A. Daehler. 2000. Dynamic link between the level of ductile crustal flow and style of normal faulting of brittle crust. *Tectonophysics* 320: 195–218 (doi: 10.1016/S0040-1951(00)00045-7).

Block, L., and L. H. Royden. 1990. Core complex geometries and regional scale flow in the lower crust. *Tectonics* 9: 557–567.

Brun, J.-P. 2002. Deformation of the continental lithosphere: Insights from brittle-ductile models. In S. DeMeer et al., eds., *Deformation Mechanisms, Rheology and Tectonics: Current Status and Future Perspectives*, Geol. Soc. Lond., Sp. Pub. 200: 355–370.

Buck, W. R. 1991. Modes of continental lithospheric extension. *J .Geophys. Res.* 96: 20161–20178.

Burk, C. A. 1968. Buried ridges within continental margins, N.Y. *Acad. Sci. Trans.* 30: 135–160.

Burke, K. 1980. Intracontinental rifts and aulacogens, in NAS-NRC. *Continental Tectonics* 42–49.

Burke, K., and A. J. Whiteman. 1973. Uplift, rifting, and the breakup of Africa. In D. H. Tarling and S. K. Runcorn, eds., *Implications of Continental Drift to the Earth Sciences*, Academic Press, vol. 2: 735–755.

Burke, K., and J. T. Wilson. 1976. Hot spots on the Earth's surface, *Scientific American*, 235(2): 46-57; also in J.T. Wilson (ed.), *Continents Adrift and Continents Aground*. Readings from *Scientific American*. W. H. Freeman Co., San Francisco, Calif., United States (USA), 230 pp.

Carbó, A., M. Muñoz, M. Druet, P. Llanes, and J. Álvarez. 2004. New gravity map of the western Galicia margin: The Spanish exclusive economic zone project. *EOS, Trans. Amer. Geophys. Union* 85(52): 565–568.

Christensen, N. I., and Salisbury, M. H. 1975. Structure and constitution of the lower oceanic crust. *Rev. Geophys. Space Phys.* 13: 57–86.

Clifford, T. N., and I. G. Gass. 1970. *African Magmatism and Tectonics*. Oliver & Boyd, Edinburgh, 461 pp.

Coffin, M. F., and Eldholm, O. 1994. Large igneous provinces: Crustal structure, dimensions, and external consequences. *Rev. Geophys.* 32: 1–36.

Coleman, R. G. 1974. Geologic background of the Red Sea. In C. A. Burk and C. L. Drake, eds., *Continental Margins*, Springer, New York: 743–753.

Coward, M. P, J. F. Dewey, and P. L. Hancock. 1987. *Continental Extensional Tectonics*, Geol. Soc. Lond. Spec. Paper 28.

Deffeyes, K. S. 1970. The axial valley, a steady state feature of the terrain. In H. Johnson and B. L. Smith, eds, *The Megatectonics of Continents and Oceans*, Rutgers University Press, New Brunswick, NJ: 194–222.

Detrick, R. S., P. Buhl, E. Vera, J. Mutter, J. Orcutt, J. Madsen, and T. Brocher. 1987. Multi-channel seismic imaging of a crustal magma chamber along the East Pacific Rise. *Nature* 326: 35–41.

Dixon, T. H., E. R. Ivins, and B. J. Franklin. 1989. Topographic and volcanic asymmetry around the Red Sea: Constraints on rift models. *Tectonics* 8: 1193–1216.

Eaton, G. P. 1980. Geophysical and geological characteristics of the crust of the Basin and Range province. In Assembly of Mathematical and Physical Sciences (U.S.). Geophysics Study Committee, *Continental Tectonics*, Washington, D.C., National Academy of Sciences: 96–113.

Fairhead, J. D. 1986. Geophysical controls on sedimentation within the African rift systems. In Frostick, L. E., R. W. Renaut, I. Reid, and J. J. Tiercelin (eds.), *Sedimentation in the African Rifts*, Geol. Soc. Lond., Spec. Publ. 25: 19–27.

Froitzheim, N., and G. Manatschal. 1996. Kinematics of Jurassic rifting, mantle exhumation, and passive margin formation in the Austroalpine and Penninnic nappes, eastern Switzerland. *Geol. Soc. Amer. Bull.* 108: 1120–1133.

Gass, I. G., C. J. MacLeod, B. J. Murton, A. Panayiotou, K. O. Simonian, and C. Xenophontos. 1994. The geology of the southern Troodos transform fault zone. *Geol. Surv. Dept. Mem.* 9. Ministry of Agriculture and Natural Resources, Geological Survey Department, Nicosia, Cyprus (CYP). 218 pp.

Hamilton, W. 1978. Mesozoic tectonics of the western United States. In Howell, D. G. and K. McDougall, (eds.) *Pacific Coast Paleogeography Symposium, no.2,*

Pacific Section, Society of Economic Paleontologists and Mineralogists, Los Angeles, CA, United States (USA): 33–70.

Heezen, B., and M. Ewing. 1963. The mid-oceanic ridge, in *The Sea* vol. 3, M. N. Hill, ed., Interscience, New York: 388–410.

Hey, R. N. 1977. A new class of pseudofaults and their bearing on plate tectonics: A propagating rift model. *Earth Planet. Sci. Lett.* 37: 321–325.

JOIDES. 1991. *North Atlantic Rifted Margins Detailed Planning Group Report*, 100 pp.

King, P. B. 1977. *Evolution of North America*, second edition, Princeton University Press, Princeton, NJ.

Lachenbruch, A. 1976. Dynamics of a passive spreading center. *J. Geophys. Res.* 81: 1883–1902.

Lagabrielle, Y., D. Bideau, M. Cannat, J. A. Karson, and C. Mével. 1998. Ultramafic-mafic plutonic rock suites exposed along the mid-Atlantic ridge (10°N–30°N): Symmetrical-asymmetrical distribution and implications for seafloor spreading processes. In W. R. Buck, P. T. Delaney, J. A. Karson, and Y. Lagabrielle, eds., *Faulting and Magmatism at Mid-Ocean Ridges*, Amer. Geophys. Union Monograph 106: 153–176.

Lister, G., M. A. Etheridge, and P. A. Symonds. 1991. Detachment models for the formation of passive continental margins. *Tectonics* 10: 1038–1064.

Lister, G., M. A. Etheridge, and P. A. Symonds. 1986. Detachment faulting and the evolution of passive continental margins. *Geology* 14: 246–250.

Lowell, J. D., G. J. Genik, T. H. Nelson, P. M. Tucker. 1975. Petroleum and plate tectonics of the southern Red Sea. In A. G. Fischer and S. Judson, eds., *Petroleum and Plate Tectonics*, Princeton University Press, Princeton, NJ: 129–153.

Macdonald, K. C. 1982 Mid-ocean ridges: Fine scale tectonic, volcanic and hydrothermal processes within the plate boundary zone. *Ann. Rev. Earth Planet. Sci.* 10: 155–190.

Macdonald, K. C. 1989a. Tectonic and magmatic processes on the East Pacific Rise. In E. L. Winterer, D. M. Hussong, and R. W. Decker, eds., *Geology of North America*, vol. N, East Pacific Ocean and Hawaii, Geol. Soc. Amer., Boulder, CO: 93–110.

Macdonald, K. C. 1989b. Anatomy of the magma reservoir. *Nature* 339: 178–179.

Macdonald, K. C., and P. J. Fox. 1990. The mid-ocean ridge. *Scientific American*, June: 72–79.

Macdonald, K. C., J.-C. Sempere, P. J. Fox, and R. Tyce. 1987. Tectonic evolution of ridge axis discontinuities by the meeting, linking, or self-decapitation of neighboring ridge segments. *Geology* 15(11): 993–997.

May, P. R. 1971. Pattern of Triassic-Jurassic diabase dikes around the North Atlantic in the context of predrift position of the continents. *Geol. Soc. Amer. Bull.* 82: 1285–1292.

Mayer, L. 1986. Topographic constraints on models of lithospheric stretching of the Basin and Range province, western United States. *Geol. Soc. Amer. Spec. Pap.* 208: 1–14.

McClain, J. S. 1981. On the long-term thickening of the oceanic crust. *Geophys. Res. Lett.* 8: 1191–1194.

Menard, H. W., and T. E. Chase. 1970. Fracture zones. In A. E. Maxwell, ed., *The Sea*, Vol. 4, part 1: New Concepts of Sea Floor Evolution: 421–455.

Menzies, M. A., S. L. Klemperer, C. J. Ebinger, and J. Baker. 2002. Characteristics of volcanic rifted margins. In M. A. Menzies, S. L. Klemperer, C. J. Ebinger, and J. Baker, eds., *Volcanic Rifted Margins*, Geol. Soc. Amer. Spec. Pap. 362: 1–14.

Montadert, L., D. G. Roberts, O. de Charpal, and P. Guennoc. 1979. Rifting and subsidence of the northern continental margin of the Bay of Biscay. *Initial Reports Deep-Sea Drilling Project* 48: 1205–1260.

Moores, E. M. 1982. Origin and emplacement of ophiolites. *Rev. Geophys. and Space Phys.* 20: 735–760.

Moores, E. M. 2002. Pre-1 Ga (pre-Rodinian) ophiolites: Their tectonic and environmental implications. *Geol. Soc. Amer. Bull.* 114: 80–95.

Moores, E. M., and R. J. Twiss. 1995. *Tectonics*. Freeman, New York, 415 pp.

Moores, E. M., and F. J. Vine. 1971. The Troodos complex, Cyprus, and other ophiolites as oceanic crust: Evaluation and implications. *Trans. Roy. Soc.* 286A: 144–166.

Nur, A., and A. Ben-Avrahem. 1982. Oceanic plateaus, the fragmentation of continents, and mountain building. *J. Geophys. Res.* 87: 3644–3662.

Perfit, M. R., D. J. Fornari, M. C. Smith, J. F. Bender, C. H. Langmuir, and R. Haymon. 1994. Small scale spatial and temporal variations in midoceanic ridge crest magmatic processes. *Geology* 22: 375–379.

Rabinowitz, P. D. 1974. The boundary between oceanic and continental crust in the western North Atlantic. In C. A. Burk and C. L. Drake, eds., *Continental Margins*, Springer, New York: 67–84.

Reston, T. J., C. M. Krawczyk, and D. Klaeschen. 1996. The S reflector west of Galicia (Spain): Evidence from prestack depth migration for detachment faulting during continental breakup. *J. Geophys. Res.* 101: 8075–8091.

Reston, T. J., J. Pennell, A. Stubenrauch, I. Walker, and M. Perez-Gussinye. 2001. Detachment faulting, mantle serpentinization, and serpentinite mud volcanism beneath the Porcupine Basin, southwest of Ireland. *Geology* 29: 587–590.

Rosenbaum, G., K. Regenauer-Lieb, and R. Weinberg. 2005. Continental extension: From core complexes to rigid block faulting. *Geology* 33(7): 609–612.

Rosendahl, B., D. J. Reynolds, P. M. Lorber, C. F. Burgess, J. McGill, D. Scott, J. J. Lambiase, and S. J. Derksen. 1986. Structural expressions of rifting: Lessons from

Lake Tanganyika, Africa. In Frostick, L. E., R. W. Renaut, I. Reid, and J. J. Tiercelin (eds.), *Sedimentation in the African Rifts*, Geol. Soc. Lond., Spec. Publ. 25: 29–43.

Schlee, J. 1980. A comparison of two Atlantic-type continental margins. U.S. Geol. Surv. Prof. Paper 1187, 74 pp.

Sclater, J. G., and J. Francheteau. 1970. The implications of terrestrial heat flow observations on current tectonic and geochemical models of the crust and upper mantle of the Earth. *Geophys. J.*, Royal Astronomical Society, 20: 509–542.

Sclater, J. G., C. Jaupart, and D. Galson. 1980. The heat flow through oceanic and continental crust and the heat loss of the Earth. Rev. Geophys. Space Phys. 18: 269–311.

Sheridan, R. E., J. A. Grow, and K. D. Klitgord. 1988. Geophysical data. In R. E. Sheridan and J. A. Grow, eds., *Geology of North America*, v. I-2, The Atlantic Continental Margin: U.S.: 177–196.

Smith, R. B. 1978. Seismicity, crustal structure, and intraplate tectonics of the interior of the western Cordillera. *Geol. Soc. Amer. Mem.* 152: 111–144.

Stewart, J. H. 1972. Initial deposits in the Cordilleran geosyncline: Evidence of a late Precambrian (<850 m.y.) continental separation. *Geol. Soc. Amer. Bull.* 83: 1345–1360.

Stewart, J. H. 1978. Basin-range structure in western North America: A review. *Geol. Soc. Mem.* 152: 1–31.

Stewart, J. H. 1980. *Geology of Nevada*, Nevada Bureau of Mines & Geology, Spec. Publ. 4, 136 pp.

Talwani, M. 1965. Crustal structure of the mid-ocean ridges, 2: Computed model from gravity and seismic refraction data. *J. Geophys. Res.* 70: 341–452.

Tapponnier, P., and J. Francheteau. 1978. Necking of the lithosphere and the mechanics of slowly accreting plate boundaries. *J. Geophys. Res.* 83: 3955–3970.

Varga, R .J., and E. M. Moores. 1985. Spreading structure of the Troodos ophiolite, Cyprus. *Geology* 13: 846–850.

Varga, R. J., and E. M. Moores. 1990. Intermittent magmatic spreading and tectonic extension in the Troodos ophiolite: Implications for exploration for black smoker-type ore deposits. In J. Malpa, E. M. Moores, A. Panayiotou, and C. Xenophontos, eds., *Ophiolites, Ocean Crust Analogues: Proceedings of the Symposium Troodos* 87: 53–64.

Vine, F. J. 1968. Magnetic anomalies associated with mid-ocean ridges. In R. A. Phinney, ed., *The History of the Earth's Crust*, Princeton University Press, Princeton, NJ: 73–90.

Voggenreiter, W., H. Hötzl, and J. Mechie.1988. Low-angle detachment origin for the Red Sea Rift System? *Tectonophysics* 140: 51–75.

Williams, H., and R. K. Stevens. 1974. The ancient continental margin of eastern North America. In C. A. Burk and C. L. Drake, eds., *The Geology of Continental Margins*, Springer, New York: 781–796.

Wilson, R. A. M. 1959. *The Geology and Mineral Deposits of the Xeros-Troodos Area, Cyprus*, Geological Survey Memoir 1, 179 pp.

ii. Transform Margins

Abbate, E., V. Bortolotti, and G. Principi. 1980. Appenine ophiolites: A peculiar oceanic crust. *Ofioliti*, Special Issue, Tethyan Ophiolites, v. I: 59–97.

Anderson, D. L. 1971. The San Andreas fault. *Scientific American*, 225(5): 52–68.

Argus, D. F., and R. G. Gordon. 1991. Current Sierra Nevada–North America motion from very long baseline interferometry: Implications for the kinematics of the western United States, *Geology* 19: 1085–1088.

Atwater, T., and P. Molnar. 1973. Relative motion of the Pacific and North American plates deduced from seafloor spreading in the Atlantic, Indian, and south Pacific Oceans. In R. L. Kovach and A. Nur, eds., *Proceedings of the Conference on Tectonic Problems of the San Andreas Fault System*, Stanford University Publications 13, Stanford, CA: 136–148.

Atwater, T., and J. Stock. 1998. Pacific–North America plate tectonics of the Neogene southwestern United States: An update. In W. G. Ernst and C. A. Nelson, eds., *Integrated Earth and Environmental Evolution of the Southwestern United States*, Bellweather, Columbia, MD: 393–419.

Benner, D. C. 2004. A neo-tectonic study of seismogenic deformation and surface faulting in Rose Valley and the western Coso Range, Inyo County, California. M.S. thesis, 2005, Geology Department, University of California at Davis, 250 pp.

Bennett, R. A., B. P. Wernicke, N. A. Niemi, and A. M. Friedrich. 2003. Contemporary strain rates in the northern Basin and Range province from GPS data. *Tectonics* 22(2): 1008 (doi:10.1029/2001TC001355).

Bowman, D., G. King, and P. Tapponnier. 2003. Slip partitioning by elastoplastic propagation of oblique slip at depth. *Science* 300: 1121–1123.

Carter, R. M., and R. H. Norris. 1976. Cainozoic history of southern New Zealand: An accord between geological observations and plate-tectonic predictions. *Earth and Planet. Sci. Lett.* 31: 85–94.

DeMets, C., R. G. Gordon, D. F. Argus, and S. Stein. 1990. Current plate motions. *Geophys. J. Int.* 101: 425–478.

Duffield, W. A., and Bacon, C. R. 1981. Geologic map of the Coso Volcanic field and adjacent areas, Inyo County, California: U.S. Geological Survey Miscellaneous Investigations Series Map I-1200, scale 1:50,000.

Ernst, W. G., Y. Seki, H. Onuki, and M. C. Gilbert. 1970. Comparative study of low-grade metamorphism in the California Coast Ranges and the Outer

Metamorphic Belt of Japan. *Geol. Soc. Amer. Mem.* 124, 276 pp.

Fox, P. J., E. Schreiber, H. Rowlett, and K. McCamy. 1976. The geology of the oceanographer fracture zone: A model for fracture zones. *J. Geophys. Res.* 81: 4117–4128.

Fox, P. J., R. S. Detrick, and G. M. Purdy. 1980. Evidence for crustal thinning near fracture zones: Implications for ophiolites. In A. Panayiotou, ed., *Proceedings of the International Ophiolite Symposium, Cyprus.* Geol. Surv. Dept.: 161–168.

Fox, P. J., and D. G. Gallo. 1984. A tectonic model for ridge-transform-ridge plate boundaries: Implications for the structure of oceanic lithosphere. *Tectonophysics* 104: 205–242.

Grindley, G. W. 1974. New Zealand. In A. M. Spencer, ed., *Mesozoic-Cenozoic Orogenic Belts*, Geological Society of London Special Publication No. 4: 387–417.

Jennings, C. W. 1994. Fault activity map of California and adjacent areas: California Department of Conservation, Division of Mines and Geology, Geologic Data Map No. 6, scale 1:750,000.

Karson, J. 1986a. Lithosphere age, depth and structural complications from migrating transform faults. *J. Geol Soc. London* 153: 785–788.

Karson, J. 1986b. Variations in structure and petrology in the Coastal complex, Newfoundland: Anatomy of an oceanic fracture zone. *Geol. Soc. London Spec. Pub.* 13: 131–144.

Macdonald, K. D., K. Kastens, S. Miller, and F. N. Spiess. 1979. Deep-tow studies of the Tamayo transform fault. *Mar. Geophys. Res.* 4: 37–70.

Menard H. W., and T. E. Chase. 1970. Fracture zones. In A. E. Maxwell, ed., *The Sea*, Wiley, New York, 4: 321–443.

Monastero, F. C., A. M. Katzenstein, J. S. Miller, J. R. Unruh, M. C. Adams, and K. Richards-Dinger. 2005. The Coso geothermal field: A nascent metamorphic core complex. *Geol. Soc. Am. Bull.* 117: 1534–1553.

Moores, E. M. 1982. Origin and emplacement of ophiolites. *Rev. Geophys. Space Phys.* 20: 735–760.

Moores, E. M., and F. J. Vine. 1971. The Troodos Massif, Cyprus, and other ophiolites as oceanic crust: Evaluation and implications. *Phil. Trans. Roy. Soc. London* 278A: 443–466.

Powell, R. E., R. J. Weldon II, and J. C. Matti, eds. 1993. *The San Andreas Fault System: Displacement, Palinspastic Reconstruction, and Geologic Evolution.* Geol. Soc. of Amer. Memoir 178.

Robinson, P. T., B. T. R. Lewis, M. F. Flower, M. H. Salisbury, and H.-U. Schmincke. 1983. Crustal accretion in the Gulf of California, an intermediate-rate spreading axis. In *Initial Reports of the Deep Sea Drilling Project* 65: 739–752.

Simonian, K., and I. G. Gass. 1978. Arakapas fault belt, Cyprus, a fossil transform fault. *Geol. Soc. Amer. Bull.* 89: 1220–1230.

Twiss, R. J., and J. R. Unruh. 2007. Structure, deformation, and strength of the Loma Prieta fault as inferred from the 1989–1990 Loma Prieta aftershock sequence. *Geol. Soc. Amer. Bull.* (in press).

Unruh, J. R., A. Streig, and S. Sundermann. 2003a. Mapping and characterization of neotectonic structures in a releasing stepover, northern Coso Range, eastern California. CD of Presentations, Eighth Annual U.S. Navy Geothermal Program Office Technical Symposium, University of California, Davis, 29–30 May.

Unruh, J. R., and E. M. Moores. 1992. Quaternary blind thrusting in the southwestern Sacramento Valley, California. *Tectonics* 11: 192–203.

Unruh, J. R., E. Haukkson, F. C. Monastero, R. J. Twiss, and J. C. Lewis. 2002. Seismotectonics of the Coso Range—Indian Wells Valley region, California; Transtensional deformation along the southeastern margin of the Sierran microplate. In Glazner, A. F., J. D. Walker, J. M. Bartley, eds., *Geologic Evolution of the Mojave Desert and Southwestern Basin and Range*, Geol. Soc. Amer. Mem. 195: 277–294.

Unruh, J. R., J. Humphrey, and A. Barron. 2003b. Transtensional model for the Sierra Nevada frontal fault system, eastern California. *Geology* 31(4): 327–330.

Whitmarsh, R. W. 1998. Geologic map of the Coso Range. DOI: 10.1130/1998-whitmarsh-coso. http://gsamaps.gsajournals.org/gsamaps/map/10.1130/1998-whitmarsh-coso

iii. Convergent Margins

Abbate, E., V. Bortolotti, and P. Passerini. 1970. Olistostromes and olistoliths. *Sedimentary Geol.* 4: 521–557.

Abbate, E., V. Bortolotti, and G. Principi 1980. Appennine ophiolites: A peculiar oceanic crust. *Ofioliti Special Issue, Tethyan Ophiolites*, v. 1, Western Area, G. Rocci, ed., 59–96.

Bard, J. P. 1983. Metamorphism of an obducted island arc: Example of the Kohistan sequence (Pakistan) in the Himalayan collided range. *Earth. Planet. Sci. Lett.* 65: 133–144.

Byrne, T. 1982. Structural evolution of coherent terranes in the Ghost Rocks formation, Kodiak Island, Alaska. *Geol. Soc. Lond. Spec. Pap.* 10: 229–242.

Cowan, D. G. 1985. Structural styles in Mesozoic and Cenozoic melanges in the western Cordillera of North America. *Geol. Soc. Amer. Bull.* 96: 451–462.

Coward, M. P., B. F. Windley, R. D. Broughton, I. W. Luff, M. G. Petterson, C. J. Pudsey, D. C. Rex, and M. Asif Khan. 1986. Collision tectonics in the NW Himalayas. In M. P. Coward and A. C. Ries, eds., *Collision Tectonics*, Geol. Soc. Lond. Spec. Publ. 19: 203–219.

Curray, J., and D. G. Moore. 1974. Sedimentary and tectonic processes in the Bengal Deep-Sea fan and geosyncline. In C. A. Burk and C. L. Drake, eds., *The*

Geology of Continental Margins, Springer, New York: 617–628.

Davis, D., J. Suppe, and F. A. Dahlen. 1983. Mechanics of fold-and-thrust belts and accretionary wedges. *J. Geophys. Res.* 88(B2): 1153–1172.

Dewey, J. F. 1976. Plate tectonics. In J. T. Wilson, ed., *Continents Adrift and Continents Aground*, Scientific American, New York: 34–45.

Eiler, J., ed., 2003. *Inside the Subduction Factory*, Amer. Geophys. Union Monograph 138, 311 pp.

Dickinson, W. R. 1970. Relations of andesites, granites, and derivative sandstones to arc-trench tectonics. *Rev. Geophys. Space Phys.* 8: 813–860.

Ernst, W. G. 1974. Arcs and subduction zones. In W. R. Dickinson, ed., *Geological Interpretations from Global Tectonics, with Applications for California Geology and Petroleum Exploration*, San Joaquin Geological Society, Bakersfield, CA: 5-1–5-7.

Ernst, W. G., Y. Seki, H. Onuki, and M. C. Gilbert. 1970. Comparative study of low-grade metamorphism in the California Coast Ranges and outer metamorphic belt of Japan. *Geol. Soc. Amer. Mem.* 124, 276 pp.

Hayes, D. E. 1974. Continental margin of western South America. In C. A. Burk and C. L. Drake, eds, *Geology of Continental Margins*, Springer, New York: 581–590.

Isacks, B., and P. Molnar. 1971. Distribution of stresses in the descending lithosphere from a global survey of focal-mechanism solutions of mantle earthquakes. *Rev. Geophys. Space Phys.* 9: 103–174.

Karig, D. E. 1974. Evolution of arc systems in the western Pacific. *Ann. Rev. Earth. Planet. Sci.* 2: 51–76.

Karig, D. E. 1972. Remnant arcs. *Geol. Soc. Amer. Bull.* 83: 1057–1068.

Kennett, J. P., A. R. McBirney, and R. C. Thunell. 1977. Episodes of Cenozoic volcanism in the circum-Pacific region. *J. Volcanol. and Geothermal Res.* 2: 145–163.

Lewis, J. C., and T. B. Byrne. 2001. Fault kinematics and past plate motions at a convergent plate boundary; Tertiary Shimanto Belt, Southwest Japan. *Tectonics* 20(4): 548–565.

Lewis, J. C., T. B. Byrne, and X. Tang. 2002. A geologic test of the Kula-Pacific ridge capture mechanism for the formation of the West Philippine Basin. *Geol. Soc. Amer. Bull.* 114(6): 656–664.

Lawver L., and J. W. Hawkins. 1978. Diffuse magnetic anomalies in marginal basins: Their possible tectonic and petrologic significance. *Tectonophysics* 45: 323–339.

Lehner, F., H. Douslt, G. Bakker, P. Allenbach, and T. Gueneau. 1983. Active margins, part 3. In A. W. Bally, ed., *Seismic Expressions of Structural Styles* 3, Amer. Assoc. Petrol. Geol., Tulsa, 3.4.2: 45–81.

Maxwell, J. C. 1974. Anatomy of an orogen. *Bull. Geol. Soc. Amer.* 85: 1195–1204.

Marsh, B. D. 1979. Island-arc volcanism. *Amer. Sci.* 67: 161–172.

Marsh, B. D., and I. S. E. Carmichael. 1974. Benioff zone magmatism. *J. Geophys. Res.* 79: 1196–1206.

McCaffrey, R. 1992. Oblique plate convergence, slip vectors, and forearc deformation. *J. Geophysical Res.* 97: 8905–8915.

McCaffrey, R. 1994. Global variability in subduction thrust-forearc systems. *Pure Appl. Geophys.* 142: 173–224.

Meissner, R. O., E. R. Flueh, F. Stibane, and E. Berg. 1976. Dynamics of the active plate boundary in southwest Colombia according to recent geophysical measurements. *Tectonophysics* 35: 115–137.

Michael, A .J. 1990. Energy constraints on kinematic models of oblique faulting; Loma Prieta versus Parkfield-Coalinga. *Geophys. Res. Lett.* 17(9): 1453–1456.

Moore, G. F., J. R. Curray, D. G. Moore, and D. E. Karig. 1980 Variations in geologic structure along the Sunda fore arc, NE Indian Ocean. In D. E. Hayes, ed., *The Tectonic and Geologic Evolution of Southeast Asian Seas and Islands*, Amer. Geophys. Union Monograph 23: 145–160.

Moore, J. C., ed. 1986. *Structural Fabrics in Deep Sea Drilling Project Cores from Forearcs*. Geol. Soc. Amer. Mem. 166, 160 pp.

Moores, E. M., and Twiss, R. J. 1995. *Tectonics*, W. H. Freeman & Co., New York, 415 pp.

Mwrozoski, C. L., and D. E. Hayes. 1980. A seismic reflection study of faulting in the Mariana fore arc. In D. E. Hayes, ed., *The Tectonic and Geologic Evolution of Southeast Asian Seas and Islands*, Amer. Geophys. Union Monograph 23: 223–234.

Oxburgh, E. R., and D. L. Turcotte. 1970. Thermal structure of island arcs. *Geol. Soc. Amer. Bull.* 82: 1665–1688.

Petterson, M. G., and B. F. Windley. 1985. Rb-Sr dating of the Kohistan arc-batholith in the Trans-Himalaya of north Pakistan, and tectonic implications. *Earth Planet. Sci. Lett.* 74: 45–57.

Platt, J. P. 1986 Dynamics of orogenic wedges and the uplift of high-pressure metamorphic rocks. *Geol. Soc. Amer. Bull.* 97: 1037–1053.

Raymond, L. A., ed. 1984. Melanges, their nature, origin and significance. *Geol. Soc. Amer. Spec. Pap.* 198, 170 pp.

Scott, R., and L. Kroenke. 1980. Evolution of back arc spreading and arc volcanism in the Philippine sea: Intrepretation of Leg 59 DSDP results. In D. E. Hayes, ed., *The Tectonic and Geologic Evolution of Southeast Asian Seas and Islands*, Amer. Geophys. Union Monograph 23: 283–292.

Shreve, R. L., and M. Cloos. 1986. Dynamics of sediment subduction, melange formation and prism accretion. *J. Geophys. Res.* 91: 10,229–10,245.

Silver, E. A., M. J. Ellis, N. A. Breen, and T. H. Shipley. 1985. Comments on the growth of accretionary wedges. *Geology* 13: 6–9.

Tatsumi, Y., and S. Eggins. 1995. *Subduction Zone Magmatism*. Blackwell, Oxford, 211 pp.

Toksoz, M. N. 1976. Subduction of the lithosphere. In J. T. Wilson, ed., *Continents Adrift and Continents Aground*, Scientific American, New York: 112–123.

Turcotte, D. L., and G. Shubert. 2002. *Geodynamics*, second ed. Cambridge University Press, New York, 456 pp.

Twiss, R. J., and E. M. Moores. 1992. *Structural Geology*, W. H. Freeman and Co., New York, 532 pp.

Uyeda, S. 1976. Some basic problems in the trench-arc-back arc system. In M. Talwani and W. Pitman, eds., *Island Arcs, Deep Sea Trenches, and Back-Arc Basins*, Maurice Ewing Series 1, American Geophysical Union, Washington, DC, USA: 1–14.

Uyeda, S. 1978. *The New View of the Earth*, W. H. Freeman and Co., San Francisco, 217 pp.

Von Huene, R., and D. W. Scholl. 1991. Observations at convergent margins concerning sediment subduction, subduction erosion, and the growth of continental crust. *Rev. Geophys.* 29: 279–316.

Von Huene, R., and D. W. Scholl. 1993. The return of sialic material to the mantle indicated by terrigeneous material subducted at convergent margins. *Tectonophysics* 219: 163–175.

Weissel, J. K. 1981. Magnetic lineations in marginal basins of the western Pacific. In F. J. Vine et al., conveners, *Phil. Trans. Roy Soc. Lond.* 300A: 223–247.

Zhao, W.-L., D. M. Davis, F. A. Dahlen, and J. Suppe. 1986. Origin of convex accretionary wedges: Evidence from Barbados. *J. Geophys. Res.* 91: 10,246–10,258.

iv. Collisions

Addicott, W., and P. W. Richards, coordinators, 1982. Plate tectonic map of the Circum-Pacific region, Pacific basin sheet. Amer. Assoc. Petrol. Geol. Tulsa, OK.

Bain, J. H. C. 1973. A summary of the main structural elements of Papua New Guinea. In P .J. Coleman, ed., *Western Pacific Island Arcs*, University of Western Australia Press, Nedlands, WA: 147–161.

Baranawski, J., J. Armbruster, L. Seeber, and P. Molnar. 1984. Focal depths and fault plane solutions of earthquakes and active tectonics in the Himalaya. *J. Geophys. Res.* 89: 6918–6928.

Bird, P. 1978. Finite element modelling of lithosphere deformation: The Zagros collision orogeny. *Tectonophysics* 50: 307–336.

Byrne, T. B., and C.-S. Liu, eds. 2002. *Geology and Geophysics of an Arc-Continent Collision, Taiwan*. Geol. Soc. Amer. Spec. Pap. 358, 215 pp.

Cardwell, R. K., B. L. Isacks, and D. E. Karig. 1980. The spatial distribution of earthquakes, focal mechanism solutions, and subducted lithosphere in the Philippine and northeastern Indonesian islands. Amer. Geophys.Union Monograph 23: 1–36.

Carney, J. N., and A. MacFarlane. 1982. Geological evidence bearing on the Miocene to recent structural evolution of the New Hebrides arc. *Tectonophysics* 87: 147–175.

Casey, J. F., and J. F. Dewey. 1984. Initiation of subduction zones along transform and accreting plate boundaries, triple-junction evolution, and forearc spreading centres—Implications for ophiolitic geology and obduction. In I. G. Gass, S. Lippard, and A. Shelton, eds., *Ophiolites and Oceanic Lithosphere*, Geol. Soc. Lond., Spec. Publ. 13, Blackwell: 269–287.

Cloos, M. 1993. Lithospheric buoyancy and collisional orogenesis: Subduction of oceanic plateaus, continental margins, island arcs, spreading ridges, and seamounts. *Geol. Soc. Amer. Bull.* 105: 715–737.

Couch, R., R. Whitsett, B. Huehn, and L. Briceno-Guarupe. 1981. Structures of the continental margin of Peru and Chile. *Geol. Soc. Amer. Mem.* 154, 703–728.

Coward, M. P., and A. C. Ries, eds. 1986. *Collision Tectonics*, Geol. Soc. Spec. Publ. No. 19, Palo Alto, 415 pp.

Dewey, J. F. 1977. Suture zone complexities, a review. *Tectonophysics* 40: 53–67.

Dewey, J. F. 1980. Episodicity, sequence, and style at convergent plate boundaries. In D. W. Strangway, ed., *The Continental Crust and Its Mineral Deposits*, Spec. Pap. Geol. Assoc. Can. 20: 533–576.

Dewey, J. F., and J. M. Bird. 1970. Mountain belts and the new global tectonics. *J. Geophys. Res.* 75: 2625–2647.

Eaton, J. P. 1966. Crustal structure in northern and central California from seismic evidence. *Calif. Div. Mines Geol. Bull.* 190: 419–426.

England, P. 1982. Some numerical investigations of large-scale continental deformation. In K. J. Hsü, ed, *Mountain Building Processes*, Academic Press, New York: 129–141.

Gansser, A. 1980. The significance of the Himalayan suture zone. *Tectonophysics* 62: 37–53.

Gass, I. G. 1982. Ophiolites. *Scientific American* Aug: 122–131.

Haines, S. S., S. L. Klemperer, L. Brown, J. Guo, J. Mechie, R. Meissner, R. Ross, and W. Zhao. 2003. INDEPTH III seismic data: From surface observations to deep crustal processes in Tibet. *Tectonics.* 22(1)(doi:10.1029/2001TC001305. P. 1-1-1-18).

Helwig, J. 1976. Shortening of continental crust in orogenic belts and plate tectonics. *Nature* 260: 768–780.

Hilde, T. W. C., S. Uyeda, and L. Kroenke. 1977. Evolution of the western Pacific and its margin. *Tectonophysics* 38: 145–166.

Krogstad, E. J., S. Balakrishnan, D. K. Mukhopadhyay, V. Rajamani, and G. N. Hanson. 1989. Plate tectonics 2.5 billion years ago: Evidence at Kolar, south India. *Science* 243: 1337–1340.

Liou, J. G., C.-Y. Lan, J. Suppe, and W. G. Ernst. 1977. *The East Taiwan Ophiolite, Its Occurrence, Petrology, Metamorphism, and Tectonic Setting.* Mining Research and Service Organization, Taipei, Taiwan, 208 pp.

McCaffrey, R., and G. A. Abers. 1991. Orogeny in arc-continent collision: The Banda arc and western New Guinea. *Geology* 19: 563–566.

McCaffrey, R., E. A. Silver, and R. W. Raitt. 1980. Crustal structure of the Molucca Sea collision zone, Indonesia. Amer. Geophys. Union Monograph 23: 161–179.

McKenzie, D. P. 1972. Active tectonics in the Mediterranean region. *Geophys. J. Roy. Astron. Soc.* 30: 109–185.

Monger, J. W. H., and E. Irving. 1978. Northward displacement of north-central British Columbia. *Nature* 285: 289–294.

Moores, E. M. 1982. Origin and emplacement of ophiolites. *Rev. Geophys. and Space Phys.* 20: 735–760.

Moores, E. M. 2002. Pre-Ga (pre-Rodinian) ophiolites: Their tectonic and environmental implications. *Geol. Soc. Amer. Bull.* 114: 80–95.

Ni, J., and M. Barazangi. 1984. Seismotectonics of the Himalayan collision zone: Geometry of the underthrusting Indian plate beneath the Himalaya. *J. Geophys. Res.* 89: 1145–1163.

Nur, A., and Z. Ben-Avrahem. 1982. Oceanic plateaus, the fragmentation of continents and mountain building. *J. Geophys. Res.* 87: 3644–3661.

Pigram, C. J., and H. L. Davies. 1987. Terranes and the accretion history of the New Guinea orogen. *BMR J. Australian Geol. and Geophys.* 10: 193–211.

Silver, E. A., L. D. Abbott, K. S. Kirchoff-Stein, D. L. Reed, B. Bernstein-Taylor, and D. Hilyard. 1991. Collision propagation in Papua New Guinea and the Solomon Sea. *Tectonics* 10: 863–874.

Stöcklin, J. 1974. Possible ancient continental margins of Iran. In C. Burk and C. L. Drake, eds, *Geology of Continental Margins*, Springer, New York: 873–888.

Tapponnier, P. 1977. Evolution tectonique du systeme alpine en Mediterranee: poinçoinnement et ecrasement rigid-plastique. *Bull. Soc. Geol. France* 7(XIX): 437–460.

Tapponnier, P., and P. Molnar. 1976. Slip-line field theory and large-scale continental tectonics. *Nature* 264: 319.

Vink, G. E., W. J. Morgan, and W.-L. Zhao. 1984. Preferential rifting of continents: A source of displaced terranes. *J. Geophys. Res.* 89: 10,072–10,076.

Whittington, H. B. 1973. Ordovician trilobites. In A. Hallam, ed., *Atlas of Paleobiogeography*, Elsevier, New York: 13–18.

Chapter 20

ANATOMY AND TECTONICS OF OROGENIC BELTS

20.1 INTRODUCTION

Orogenic[1] belts coincide, not accidentally, with some of the Earth's great mountain chains. Since time immemorial, mountains have held the attention of humans. In times past, they were features of terror, homes of gods or demons, to be traveled through as quickly as possible. More recently, however, their image has become more benign; they have attracted the curiosity of geologists, provided challenge to adventurers, and inspired poets and philosophers. Since the late 1960s, orogenic belts have taken their place among the many features of Earth and its history that we have recognized to be inevitable consequences of plate tectonic activity. This activity includes subduction of one plate beneath another, during which "collisions" between crustal masses, such as two continents, a continent and an island arc, or a continent and an oceanic plateau, can occur. Because all oceanic crust older than about 200 Ma (early Jurassic) has been subducted, orogenic belts are the prime repositories of information regarding plate tectonic interactions for the first 95% of Earth history. Studying these features gives us an opportunity to decipher part of the tectonic history of the Earth that is preserved nowhere else.

In Sections 20.2 through 20.8 we discuss the structural characteristics of the major parts of orogenic belts, illustrating them with examples from some of the major mountain belts of the world. Our examples come chiefly from the North American Cordillera, the Appalachian-Caledonian, and the Alpine-Himalayan systems, but also from the North American Precambrian regions and the Urals and Altaids of Central Asia (see Figure 20.1). In Sections 20.9 through 20.13, we focus on the tectonic aspects of orogeny and on models for the orogenic process, and we discuss the structural evidence that relates the two.

Figure 20.2 shows generalized maps of the North American Cordillera (Figure 20.2*A*), the Alpine-Iranian portion of the Alpine-Himalayan orogen (Figure 20.2*B*), the Appalachian–Caledonian orogen on a predrift reconstruction (Figure 20.2*C*), the Ural-Altaids (Figure 20.2*D*), and the Canadian Shield (Figure 20.2*E*). The Cordillera and the Alpine-Himalayan orogens include plate margins that are still active. The major plates involved in the North American Cordilleran orogen are the North American, Pacific, Rivera, Cocos, and Juan de Fuca plates; those involved in the Alpine-Himalayan orogens include the Eurasian, African, Arabian, and Indo-Australian plates. Other smaller plates are involved in each orogen (compare Figures 19.29, 19.30, and 19.47). The Appalachian-Caledonian orogen was formed by the collision of Africa and Europe (or Africa and Baltica) with North America (or Laurentia[2]) in late Paleozoic time, and it was fragmented into its separate parts by subsequent opening of the Atlantic Ocean in Mesozoic-Cenozoic time. The Ural-Altaid orogen formed by progressive accretion throughout the Paleozoic of a series of oceanic island arcs, culminating with the sandwiching of the entire orogenic complex between the Angara and Russian cratons. The North American Precambrian regions (including the Canadian Shield) show a complex agglomeration of Archean and Proterozoic accreted and collisional terranes.

[1]From the Greek words *oros*, which means "mountain," and *genesis*, which means "origin, birth," see footnote 13, Section 1.7(ii).

[2]Baltica and Laurentia are the names given, respectively, to the Neoproterozoic-Paleozoic European and North American continents.

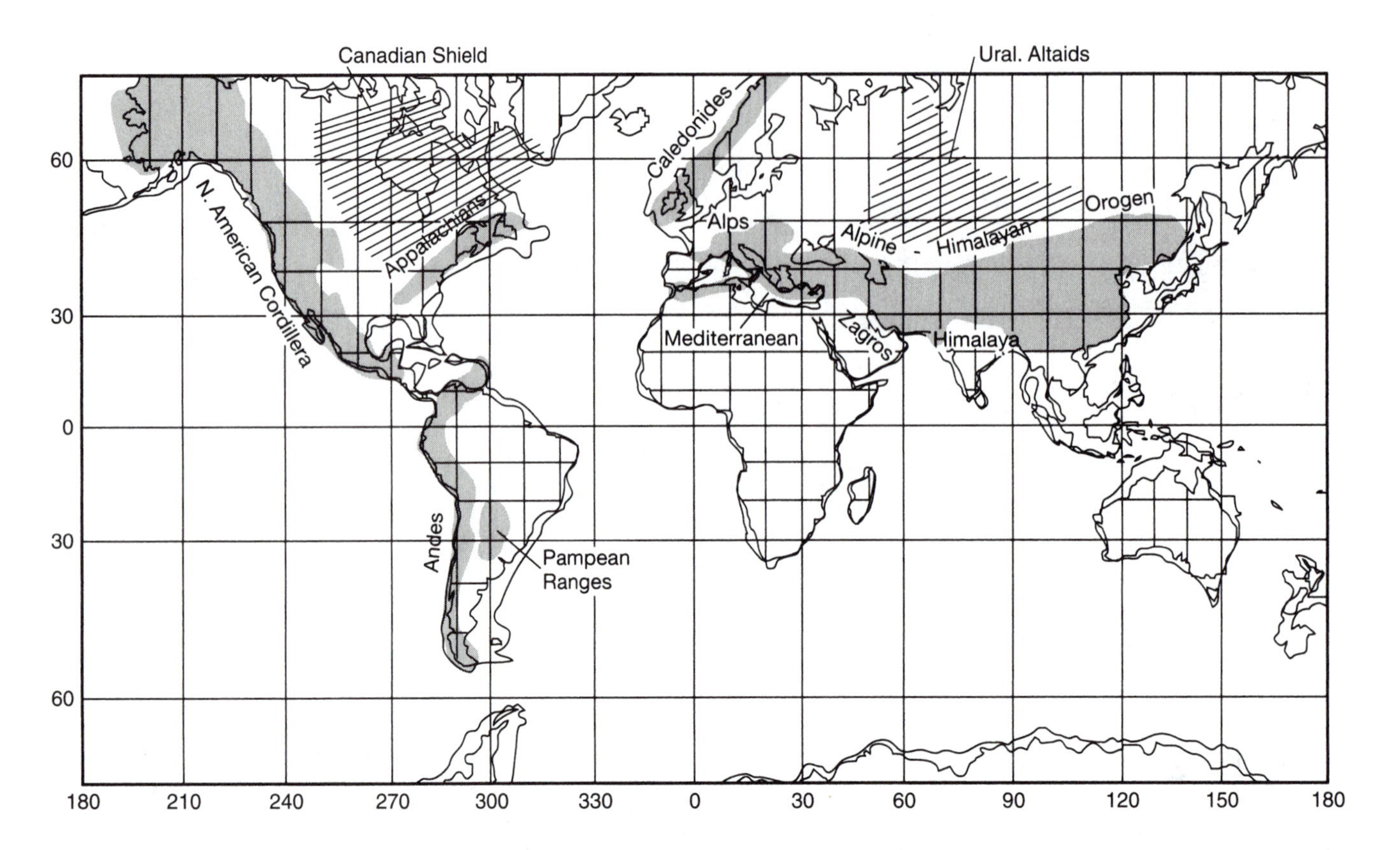

FIGURE 20.1 World map, showing the location of major orogenic belts referred to in this chapter: North American Cordillera, Appalachians, Caledonides, Alpine-Himalayan belt, Andes, Canadian shield, and Ural-Altaids.

From the maps of Figure 20.2 and the discussion of the types of collision in Section 19.5, it should be clear that no single map or cross section could provide a universal model of an orogenic belt. Nevertheless, these and other orogens have a number of features in common. Orogens generally display a rough bilateral structural symmetry, which in some orogens includes the presence of a foreland or undeformed plate on either side, as in the case of the Urals and the Appalachian-Caledonian system before opening of the Atlantic Ocean basin. Many, but not all, orogens are flanked by sedimentary basins called foredeeps or foreland basins, and by fold and thrust belts, which lie adjacent to an internal crystalline core zone of metamorphosed and deformed sedimentary and volcanic rocks, mafic-ultramafic complexes, and granitic plutons. The core zones include sutures marked by ophiolitic rocks, a slate belt, or other regions of diverse and separate stratigraphic and structural history. These common features of orogens permit us to construct a schematic composite cross section (Figure 20.3) that provides the basis for our discussion of these important features in this chapter. Not all orogens have all of these features, of course, and the cross section of any portion of one of these orogens will differ from that of another part of the same or another orogen.

The deep structure of orogens is best revealed by seismic reflection profiles, which show that the Moho is flat beneath some orogenic belts and is offset beneath others. Offsets in the Moho are associated with thrust faults that cut through the entire crust, and they may be associated with the remnants of subducted lithosphere.

20.2 THE FOREDEEP OR FORELAND BASIN

Along most orogenic belts, the boundary between the deformed rocks and the undeformed continental platform contains a thick series of syn-orogenic clastic sediments derived from a rising thrust–faulted source area in the adjacent mountains. These sediments are deposited in a **foredeep** or **foreland basin** (Figures 20.3 and 20.4) (a **molasse basin** in the terminology of Alpine geology), and they reach thicknesses of as much as 8 to 10 kilometers near the mountain front. Generally the coarseness of the basin fill decreases away from the mountain front: Conglomerates pass into sandstones and then into shales, which in turn may pass into carbonate marine shelf sediments (Figure 20.4).

Environmental indicators from different basins suggest two possible modes of basin development. Along the eastern side of the Cordilleran belt in the western United

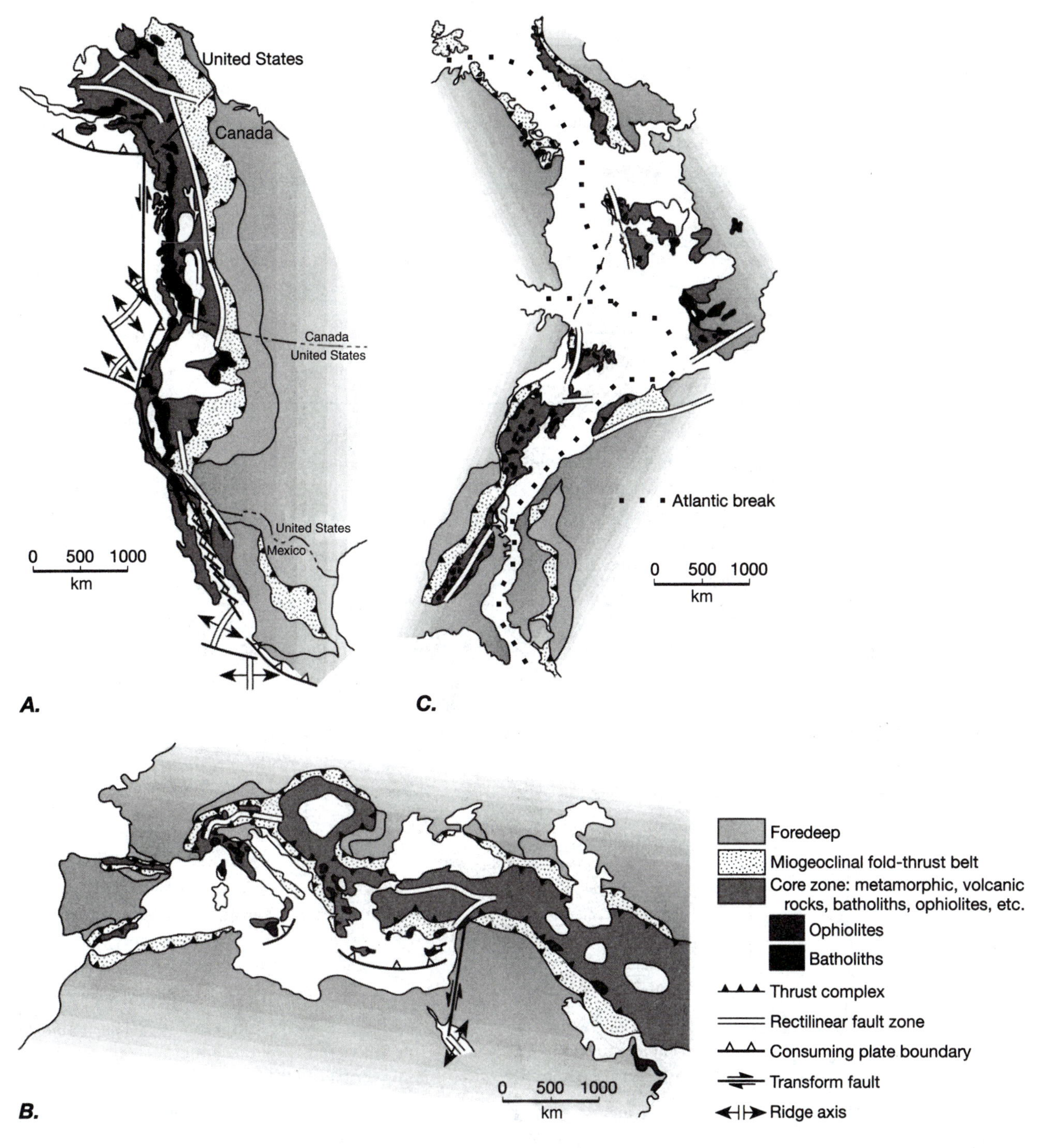

FIGURE 20.2 Generalized maps of five orogenic belts, at approximately the same scale, showing major tectonic features to be compared with model cross section (Figure 20.3). *A.* Generalized map of North American Cordillera. (Modified after King 1977) *B.* Generalized map of the Alpine-Iranian, or western, segment of the Alpine-Himalayan orogen. (Modified after Burke et al. 1977) *C.* Generalized map of the Appalachian-Caledonian orogenic belt (including the West African orogen), on a predrift reconstruction of the continents around the Atlantic Ocean. (Modified after Williams 1984) *D.* Generalized map of the Ural-Altaids. (Modified after Sengör and Natal'in 1996; Savilieva et al. 2002) *E.* Map of Precambrian regions of North America, including the Canadian shield, showing the main orogenic belts and their dates. Baltica (Europe) shown to the right in pre-Atlantic opening position. CB-Cape Smith Fold Belt; MRV-Minnesota River Valley gneiss. (Modified after Hoffman 1988; Percival et al. 2004)

(continued)

FIGURE 20.2 *(continued)*

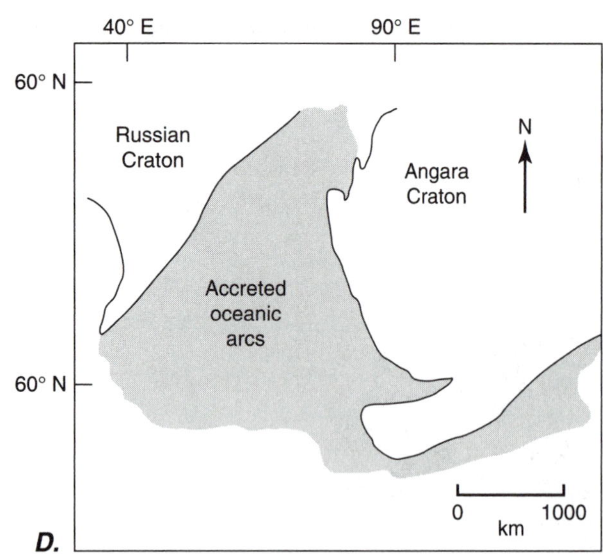

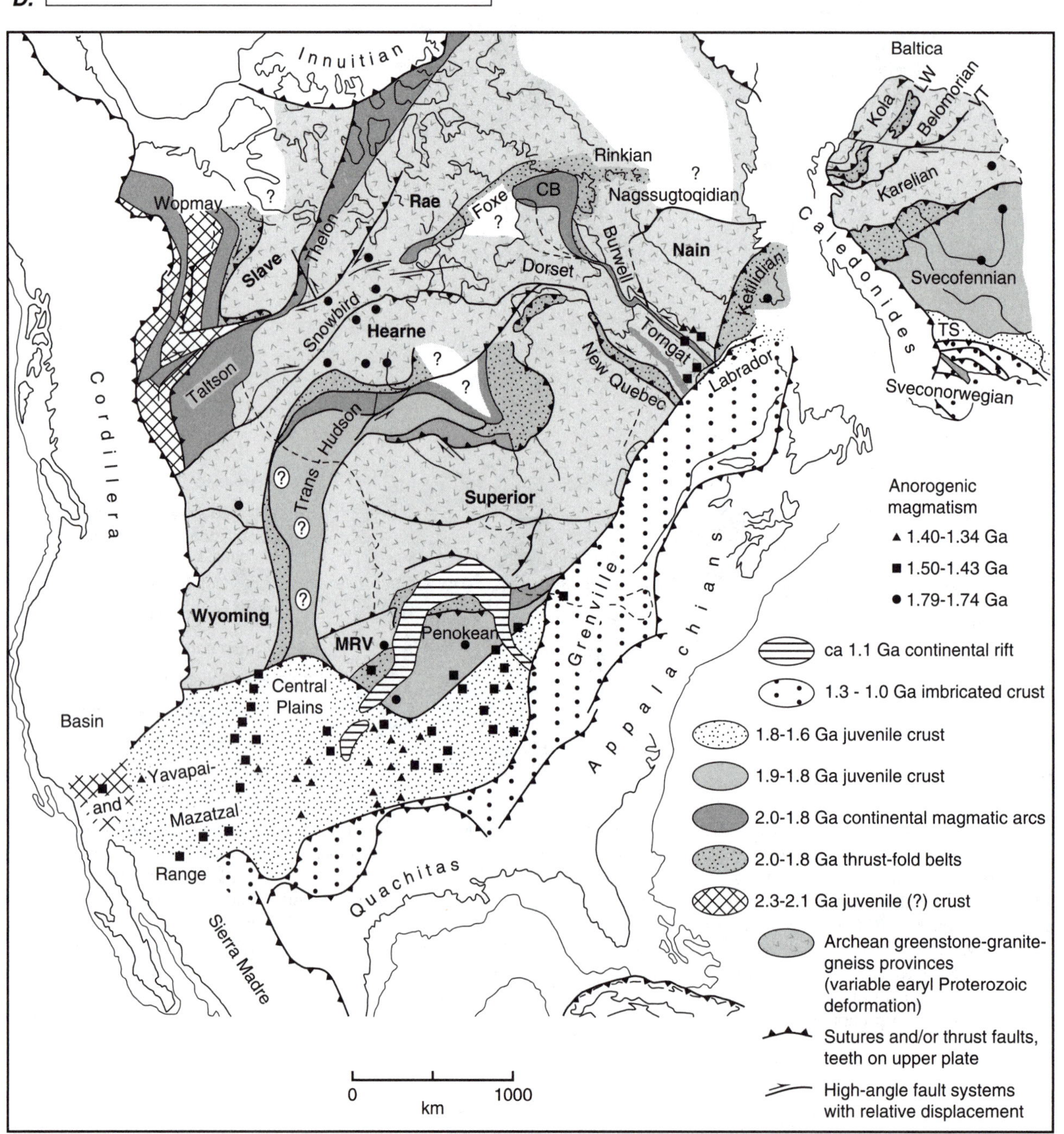

E.

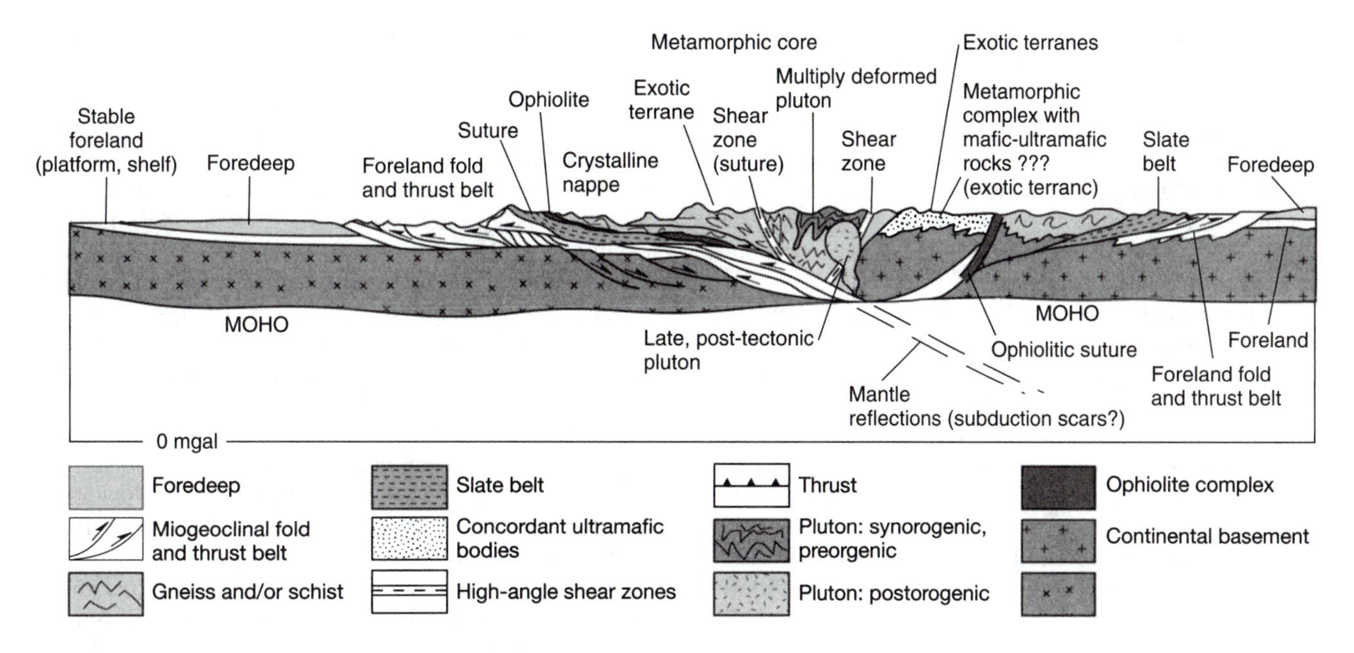

FIGURE 20.3 Cross section across a model composite orogenic belt. (Generalized after Hatcher and Williams 1986; Moores and Twiss 1992)

States, for example, the Cretaceous stratigraphy begins with deep-water sandstones and shales that pass upward into shallow-water deposits, suggesting that the basin formed relatively abruptly and was gradually filled. Shallow-water sandstones and continental deposits expand through time from west to east across the basin (Figure 20.4), indicating a gradual shallowing of the basin. By contrast, though the Ordovician foredeep deposits of the central and southern Appalachians contain deep-water deposits, the Devonian foredeep trough of the northern Appalachians, exposed in New York State, formed slowly enough that sedimentation generally kept pace with subsidence, and deep-water sediments are rare. In the northern part of the Appalachians in eastern Canada, however, there is no foredeep developed at all. The differences in development of foredeep deposits may be related to the strength of the continental lithosphere relative to the thickness and weight of the overthrusting mass exposed toward the center of the orogenic belt.

In some sequences, the clasts in the conglomerates and sandstones reveal an **unroofing sequence**, in which stratigraphically younger deposits contain debris from successively deeper levels in the mountains. Such a sequence reflects the progressive uplift and erosion of the adjacent mountains, referred to as **exhumation** or **unroofing**.

The amount of deformation in most of the foredeep rocks is generally slight, indicating either that they were

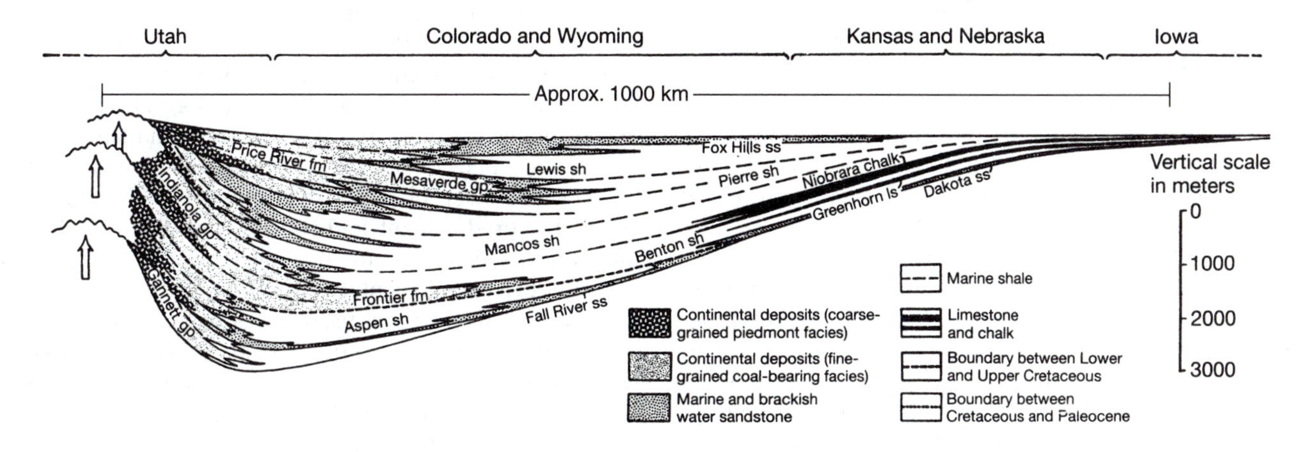

FIGURE 20.4 Diagrammatic cross section of a Cretaceous foredeep on the east side of the Cordillera from Utah to Iowa. Such basins are also called molasse basins or foreland basins. Note thickening and coarsening of sediments toward the western source area. (After King 1959)

mostly deposited after the main phase of deformation in the interior of the orogenic belt or that they were far enough away from the main orogenic deformation to be unaffected by it. There are exceptions to this generalization, however. For example, north of the Alps, the Jura folds (see Figures 20.26, 6.10, and 10.2*B*) deform strata of the foredeep deposits as well as the underlying platform, or foreland, rocks, and thus this deformation must postdate the foredeep deposition.

Closer to the core of the mountain belt, however, most foredeep deposits are caught up in the folding and thrusting that characterize the adjacent foreland fold and thrust belt (see Section 20.3). The folds are generally open, multilayer, class 1B folds having folding angles of 90° or less. The wavelengths, which often exceed 1 km, are controlled by the thick competent layers in the stratigraphy. Folding increases in intensity toward the orogenic core, as evidenced by higher folding angles and larger aspect ratios. The folds change toward multilayer class 1C geometry, and in some instances they are inclined or even overturned, with a vergence toward the stable foreland. Allochthonous masses of the sediments are found in a few places in the foredeep basins, emplaced presumably by gravity sliding of surficial sheets of sediment.

In some regions, such as the central Rocky Mountains of the United States (Figure 20.2*A*) or the Pampean ranges of the southern Andes (Figure 20.1), exceptions to this general pattern are present. In these regions, low to moderately dipping thrust faults emplace basement rocks, on which the foredeep sediments were deposited, on top of the outer foredeep deposits or on the shelf deposits of the foreland (Figure 20.2*A*, *B*). Seismic reflection profiles over several uplifts indicate that the thrust penetrates to the base of the continental crust (e.g., see the Wind River thrust in Figure 5.3).

20.3 THE EXTERNAL THRUST COMPLEX: FORELAND FOLD AND THRUST BELT, SLATE BELT, OPHIOLITES, AND SUTURES

Behind the foredeep basins, closer to the core of the orogenic belt, lies a **foreland fold and thrust belt**. This belt consists predominantly of folded and thrust faulted sedimentary rocks of the **passive margin sequence** (sometimes called the **miogeocline**), which are pre-orogenic deposits of sandstones, shales, and carbonates. These sediments accumulated on the passive continental margin before it was involved in a collision and have been pushed away from the orogenic core and out over the foreland basin or the stable foreland (also called the *shelf* or *platform*; Figures 20.2 and 20.3). The foredeep deposits are overthrust by the miogeoclinal rocks at the front of the main fold and thrust belt, and some foredeep deposits may be incorporated into the hanging wall blocks of some the thrusts. The miogeoclinal sediments generally thicken toward the core of the orogenic belt, and the folds tend to become tighter. Fold style changes from multilayer class 1B to multilayer class 1C and from upright to overturned with a vergence toward the foreland, as with folds in foredeep deposits (see, for example, the South Mountain fold on the east side of the Appalachian Valley and Ridge province, Figure 15.6). Typical examples of these structural regions occur in the Appalachian Valley and Ridge province (Figures 5.11*A* and 5.12*A*); the Cordilleran overthrust belt, especially north of the Basin and Range province (Figures 5.11*B* and 5.12*B*); the Caledonian fold–thrust belt of Scandinavia; the southern Himalaya (Figure 5.11*C*); and the Zagros mountains of Iran (Figure 5.21*B*, *C*). The Jura Mountains north of the Alps also comprise a fold and thrust belt, although they are anomalous in that they consist essentially of platform and foredeep deposits (Figures 6.10, 10.2*B* and 18.1) rather than the more usual miogeoclinal rocks.

In Section 5.3(iii), we described the general structural features of foreland fold and thrust belts. The fundamental characteristic is the presence of a **sole fault**, or **basal decollement**, that separates the folded and faulted rocks of the thrust sheet from the underlying undeformed basement and that rises through the stratigraphic section toward the foreland to give a wedge-shaped geometry to the thrust sheet. Above the sole fault, there may be several decollements, all of which ultimately are branches off the main sole fault and, like the sole fault, tend to rise through the stratigraphic section toward the foreland (Figures 5.12, 5.15, and 5.20*B*, *C*). Each fault characteristically adopts a ramp-flat geometry that cuts steeply up through competent layers such as sandstone or limestone and forms bedding-parallel faults in incompetent layers such as shale, gypsum, or salt. Duplex structures are common (Figures 5.14 and 5.15).

Movement on such faults accommodates horizontal shortening and vertical thickening of the thrust wedge and creates fault-ramp folds. Once created, these folds may accommodate further deformation by tightening beyond the folding angles imposed by the ramp–flat geometry of the fault. Other folds may form above a flat decollement to accommodate shortening and thickening of the thrust wedge. Some thrust faults may develop when folding becomes too tight to accommodate more shortening, and faults cut up from the decollement through the steep or overturned limb of a fold (Figure 5.10). The predominant fold style is class IB to IC (Figure 10.19), which of geometrical necessity must be associated with a decollement (Section 10.5(i)). Many of these folds are asymmetric, in most cases having a vergence away from the orogenic core (Figure 10.13*D*). The fold wavelength is related to the thickness of the folded competent layers (Figures 18.13 and 18.14).

In many fold and thrust belts, age relationships generally show that thrusts and folds near the orogenic core and those shallower in the thrust stack are older than those near the foreland and deeper in the thrust stack.

This decrease in age of deformation toward the foreland is sometimes called a **prograding deformation** (see Section 5.4). **Out-of-sequence thrusts**, which are new thrust faults that form behind the frontal thrust, are common, however. The simple model of thrust wedges (Section 9.11(v); Box 9-2) requires that a thrust wedge maintain a critical taper, which implies there must be a prograding deformation as internal shortening and thickening of the thrust wedge occurs. This shortening and thickening takes place by folding or by out-of-sequence thrusting.

Faults that form early in the deformation are commonly folded when a new fault propagates out under the older fault forming a new decollement, and folding occurs above this younger fault (Figure 5.15). This folding of the older faults makes continued slip on them increasingly difficult, and they eventually become inactive. Such deformed faults may subsequently be cut by later out-of-sequence faults.

In some cases, imbricate thrust faults dominate the deformation of the thrust wedge, as in the southern Appalachian Valley and Ridge province (Figures 5.11*A* and 5.12*A*). In other cases, the formation of folds characterizes the deformation, as in the northern Valley and Ridge (Figures 5.11*A* and 10.2*A*).

Most major sole faults remain above the strong crystalline basement and within the miogeoclinal sedimentary sequence, which characteristically contains abundant layers of weak rocks such as shales, gypsiferous layers, or salt. This style of deformation, where the basement remains undeformed by the thrusting, is known as **thin-skinned tectonics**. This style is not universal, however, and in many orogenic belts, such as the northern Appalachians, the Grenville orogen, the Alps, and the Himalaya, some of the thrust faults penetrate into the lower crust and, based on seismic reflection profiles, even into the mantle. In addition, in the Alps, the thrust faults were subsequently deformed into major nappe structures.

In the inner parts of foreland fold and thrust belts closer to the orogenic core, fault slices of crystalline rocks become incorporated into the thrust sheets (Figure 20.3). In some cases, such as the western Alps, these rocks formed the basement on which the miogeoclinal rocks were deposited; they are called **external massifs** (Figure 20.5). In other cases the crystalline rocks bear no obvious relationship to the sediments of the thin-skinned belt. Regardless of their origin, these basement crystalline rocks are variably deformed and in some cases contain both mylonite zones and folds that reflect faults and folds in the associated cover. Such blocks could be easily incorporated into the thrust sheets if earlier normal faults associated with rifting were reactivated as thrust faults.

Observations of major thrust sheets indicate that they have a wedge shape (Section 9.11(v)). If major thrust systems are effectively branches off a subduction zone that cut through the sediments of a down-going passive continental margin, the polarity of the system is determined by the subduction zone, and the thrust fault must rise through the stratigraphy in the direction of the foreland (as shown in Figures 5.12 and 5.15, for example). Moreover, the sedimentary sections involved in foreland fold and thrust belts are miogeoclinal sections that become thinner with increasing distance from the orogenic core. Thus there is a gentle but significant upward slope to the basement of the sedimentary section from the hinterland, or orogenic core, toward the foreland, or the stable continental interior. This slope is accentuated by, and in part attributable to, the rifted and thinned character of the continental basement at a passive continental margin (Figures 5.12 and 1.23); thus the sedimentary section is thickest at the outer edge of the continental margin where the continental basement is thinnest. Where these sections are deformed, they ubiquitously exhibit transport from the margins toward the foreland, and the basement slope helps direct the thrust faults upward in this direction.

Most rocks in foreland fold-thrust belts are unmetamorphosed, although in some regions, low-grade metamorphism is present. Clay minerals, for example, are recrystallized to chlorites and micas, coal is anthracite-grade, and the magnetic vectors of the rocks are reset. Slaty cleavage is characteristic of argillaceous sediments and may be cut by a second generation of spaced foliation formed by solution during deformation. In spite of these metamorphic effects, fossils are fairly abundant in many areas, the stratigraphy is relatively easy to work out, and the rocks are amenable to correlation from one thrust block to another.

Moving from the front of the fold and thrust belt further toward the interior of the orogenic belt several changes take place. The character of the sediments involved in the deformation changes from shallow-water sediments to deeper-water fine-grained clastic sediments. In some areas, called **slate belts**, the rocks are characterized by a monotonous predominance of relatively unfossiliferous shales and slates. The monotony of the stratigraphy, the lack of fossils, and the poor exposure of the easily eroded shales and slates make stratigraphic analysis and correlations difficult and imprecise. The rocks in these regions apparently were laid down off the edges of continental margins as continental rise or abyssal deposits or in an offshore volcanic environment. It is possible to distinguish between these different provenances by stratigraphic and petrologic analysis of the sediments.

Because slate belts are closer to the orogenic core, the grade of metamorphism tends to increase relative to outer zones, reaching high zeolite or low greenschist facies. Ductile deformation becomes increasingly prominent, and the style of folding becomes dominated by class 1C to class 2 folds, reflecting a more ductile sedimentary pile and a decrease in the predominant influence of competent layers on the folding of the sedimentary section

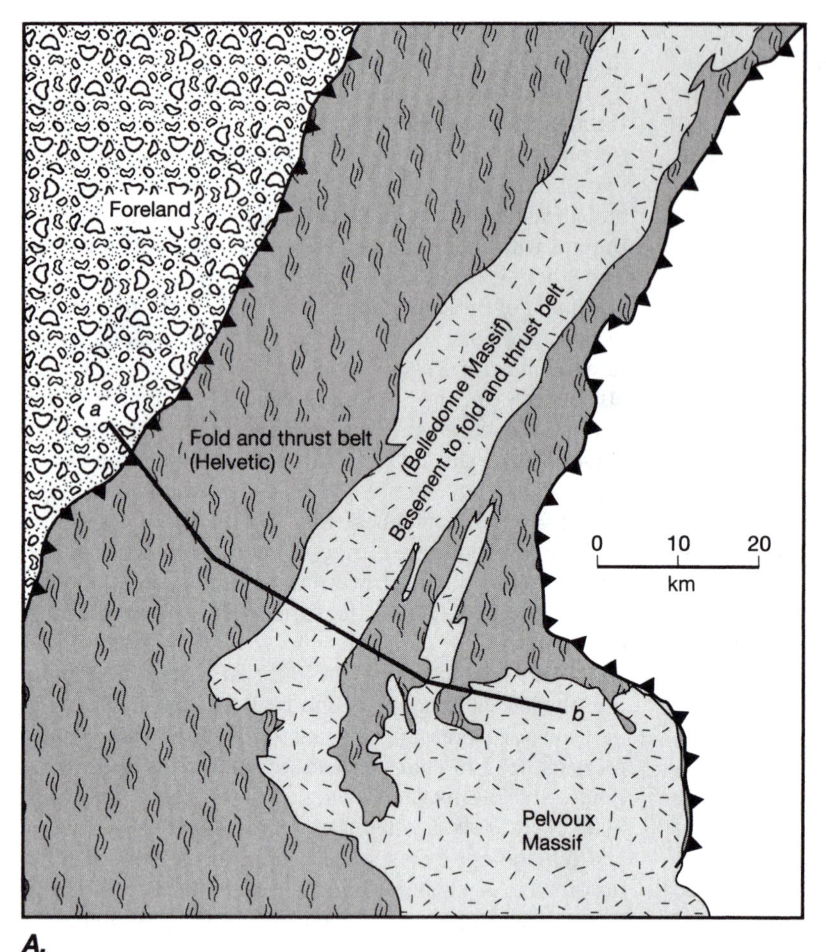

FIGURE 20.5 External massifs of the western Alps, resulting from the involvement of basement rock in the fold and thrust belt, which itself has been involved in major nappe formation. (Modified after Ramsay 1963) *A*. Generalized map of the Belledone and Pelvoux massifs, France. *B*. Cross section along line a-b. *C*. Index map, reduced version of Figure 20.2*B* showing the location of *A* (see map of the Alps in Figure 20.26).

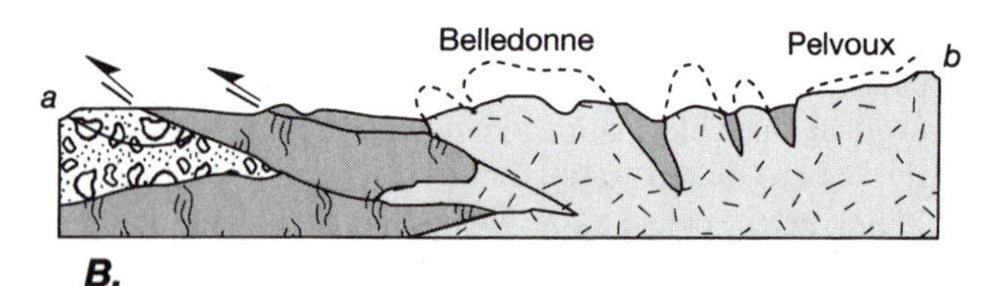

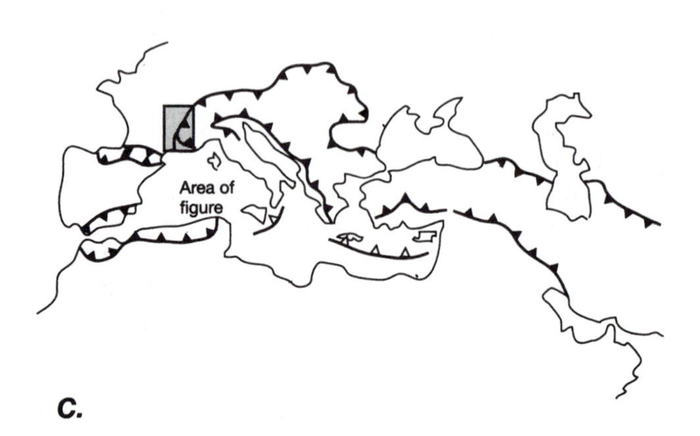

(Section 13.5). In places, multiple generations of folding are present. Folds become more inclined, even recumbent, and they can form huge fold nappes. This change in fold style is attributable in part to the higher temperature indicated by the increase in metamorphic grade and in part to a change in lithology from predominantly sandstone and limestone to the more ductile shale and its metamorphic equivalent slate.

As the rocks are more highly deformed and more recrystallized, they tend to exhibit pervasive continuous foliation such as slaty cleavage and phyllitic foliation, which in places may be overprinted by later spaced foliations. Faults are present, but in many cases the lack of distinctive markers or piercing points makes them difficult to interpret.

The slate belt of Wales in the United Kingdom is one of the best-studied examples (Figure 20.6). There, a thick sequence of late Precambrian and lower Paleozoic deep-sea sediments and associated volcanic rocks and melange display a pervasive sequence of upright folds and a penetrative cleavage, with the maximum extension axis oriented down-dip in the cleavage. Thrust faults are recog-

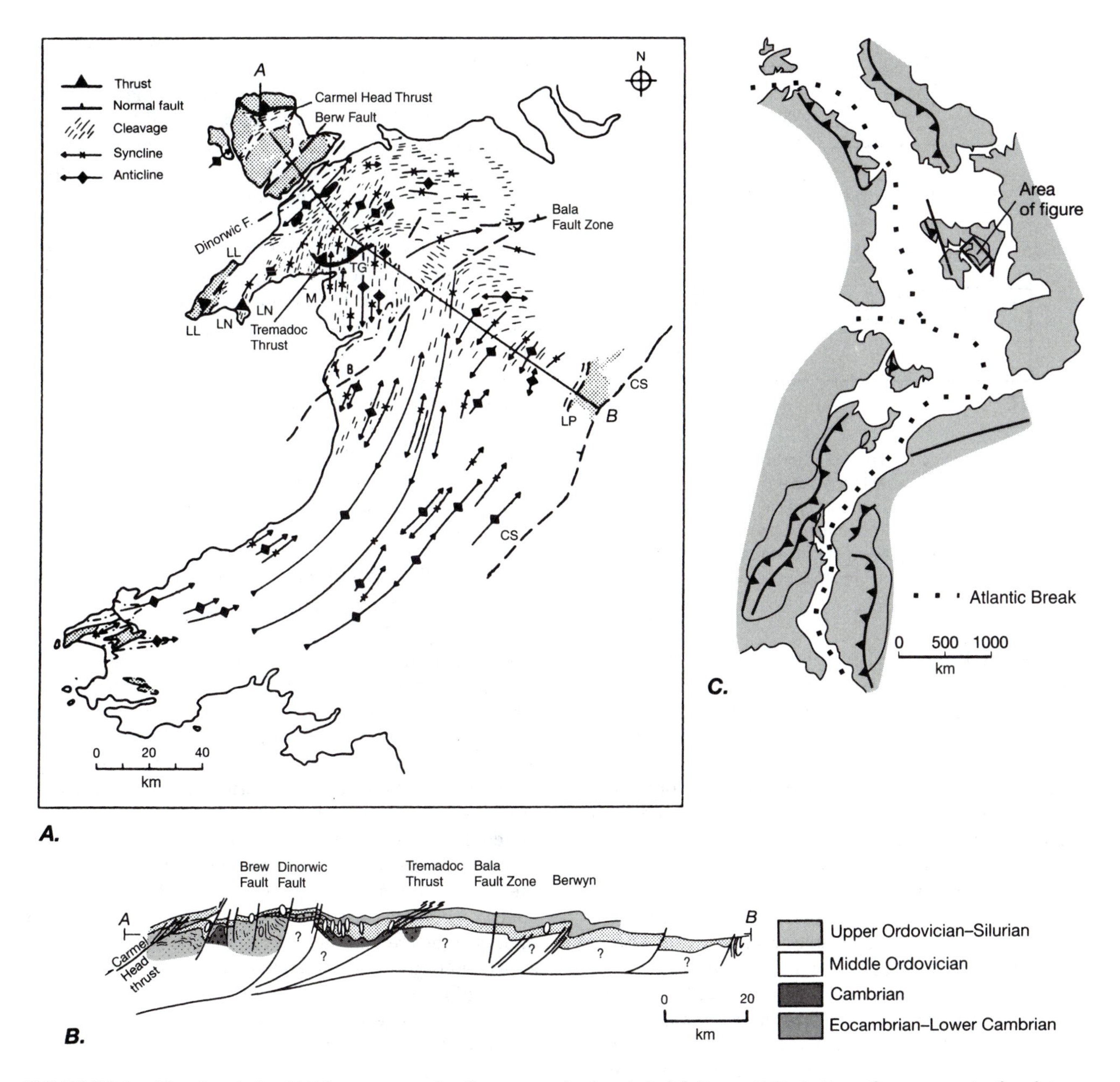

FIGURE 20.6 The slate belt of Wales, an example of an orogenic slate belt (cf. Figure 15.9). *A*. Map of structures in the slate belt showing the cross section line. *B*. Cross section. *C*. Index map, a reduced version of Figure 20.2C showing the location of *A*. (*A* and *B* modified after Coward and Siddans 1971)

nized in a few areas, but rarely can they be traced for long distances. As shown in the cross section (Figure 20.6*B*), the structures are thought to be part of a series of imbricate thrusts above a basal decollement, but the question marks on the section emphasize the uncertainty in this interpretation.

Most orogenic belts contain ophiolites. Where best exposed, the thicknesses of well-preserved complexes resemble that of the oceanic crust (see Figure 19.17). Ophiolites occur predominantly in two ways: Small tectonic slices or blocks of incomplete sequences are present in many accretionary prisms formed during the subduction of an oceanic plate; and complete ophiolitic sequences commonly occur as large subhorizontal thrust sheets hundreds of kilometers in dimension. Belts of many ophiolites of similar age such as in the Alpine-Iranian or the Ural-Altaid orogenic belts (Figure 20.2) may extend for thousands of kilometers. Individual ophiolite thrust sheets commonly overlie a tectonic complex of thrust slices of platform, slope rise, and abyssal sediments, or a melange complex (Figure 20.7*A*). Palinspastic restoration of these tectonic slices (Figure 20.7*B*) suggests that

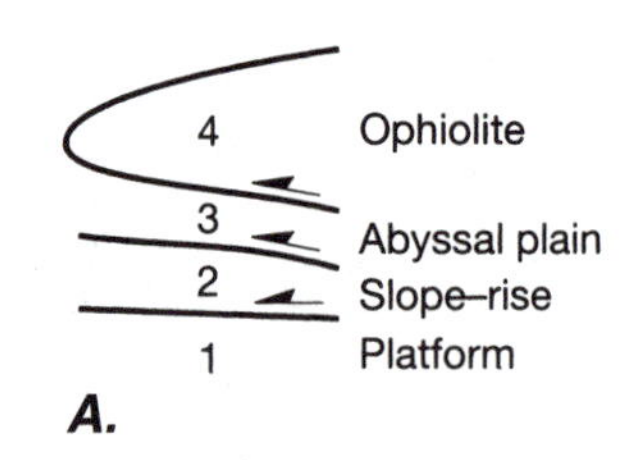

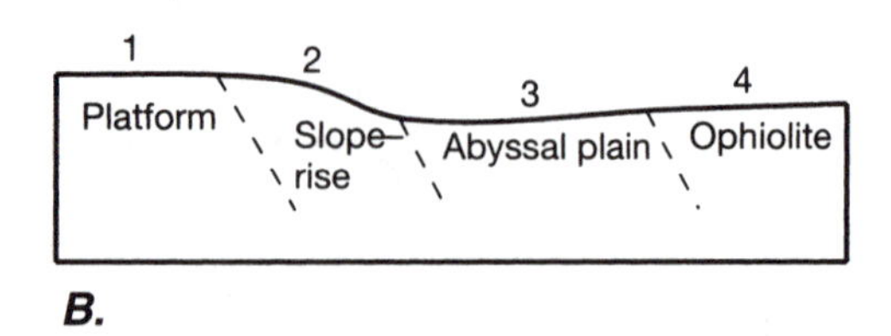

FIGURE 20.7 Reconstruction of the orogenic emplacement of ophiolites (see Figure 19.17 for a diagram of the ophiolite sequence). A. Diagram illustrating typical tectonic stacking of thrust sheets of different provenance beneath an ophiolite. *B*. Diagram illustrating palinspastic restoration of the thrust sheets in *A* to their original relative positions.

successively higher thrust sheets originated from positions progressively further from the foreland, with the ophiolite complex representing the highest and most oceanic thrust sheet, another example of a prograding thrust fault system.

The structural relationships of these ophiolite complexes indicate that they represent thrust sheets of oceanic crust and mantle, originally rooted in the oceanic mantle. Many geologists think that these well-preserved ophiolitic thrust complexes are the result of a continental margin on a down-going plate colliding with a subduction zone. The ophiolite may represent the leading edge of the over-riding plate, which is oceanic crust immediately adjacent to the subduction zone. Alternatively, it may represent a slice of oceanic crust from the down-going plate that has been underplated onto the bottom of the over-riding plate and has been followed by underplated thrust sheets sliced progressively off the down-going continental margin as it enters the subduction complex (Figure 20.7). The ophiolite could also have been emplaced initially on the over-riding plate, either a continental margin or an island arc complex, by a thrust fault antithetic to subduction before its collision with the down-going continental margin. In these cases, the presence of such ophiolites indicates not only thrusting and shortening of the down-going continental margin, but also an unknown amount of intra-oceanic thrusting that has been accommodated by the subduction zone. Thus ophiolites mark a zone of unknown amounts of thrust displacement where possibly vast amounts of oceanic lithosphere and overlying sediment have disappeared down the subduction zone. Because of this uncertainty, it is intrinsically impossible to balance cross sections across such zones, unlike structures in foreland fold and thrust belts (Section 3.5).

A **suture** is a boundary between two pieces of continental crust that initially were separated by an ocean basin and that have collided as a result of the subduction of that intervening ocean basin. Ophiolites, therefore, mark sutures because they are remnants of the old oceanic crust that have been caught up and preserved in a collision zone. Other features that can mark the location of sutures include melanges (Section 19.4(vi), Figures 19.40 and 19.41), wide ductile shear zones (Figures 3.4 and 3.19), markedly different geology on opposite sides of a shear zone, paired metamorphic belts (discussed in Section 20.4), and island arc sequences present within the core of a mountain belt. Although ophiolitic thrust sheets have received less attention than the more familiar and easily analyzed fold and thrust complexes, they may be more important in interpretation of the history of an orogenic belt.

20.4 THE CRYSTALLINE CORE ZONE: METAMORPHISM

The rocks in the central portions of most orogenic belts are metamorphic rocks. Although a comprehensive discussion of metamorphism is beyond the scope of this book, a brief review is pertinent to our discussion of the structure and tectonics of orogens. The temperature and pressure conditions at which various metamorphic mineral assemblages are in equilibrium have been determined by extensive laboratory experiments and field studies (Figure 20.8). These mineral assemblages define the various metamorphic grades and different mineral zones within those grades, and these different grades and mineral zones reflect the different temperature and pressure conditions under which the rocks recrystallized. The distribution of metamorphic zones in most orogenic belts is roughly symmetrical, so that the highest-grade rocks underlie the central portions of the belt, and unmetamorphosed rocks are found on the flanks.

Based on the different mineral assemblages, we can distinguish metamorphism that has occurred under conditions of high temperature and low pressure (Buchan type), normal pressure and temperature (Barrovian type), and high pressure and low temperature (blueschist type) (Figure 20.8). Barrovian metamorphism reflects a normal continental geothermal gradient and is widespread in all mountain belts. For this reason it is also known as classic regional metamorphism. The temperature-depth profiles implied by these types of metamorphism are the result of different tectonic environments.

High-temperature–low-pressure (Buchan) metamorphism implies that temperatures are elevated above a normal geothermal gradient. This condition develops most commonly in situations in which excess heat has been advected into the shallow continental crust by mag-

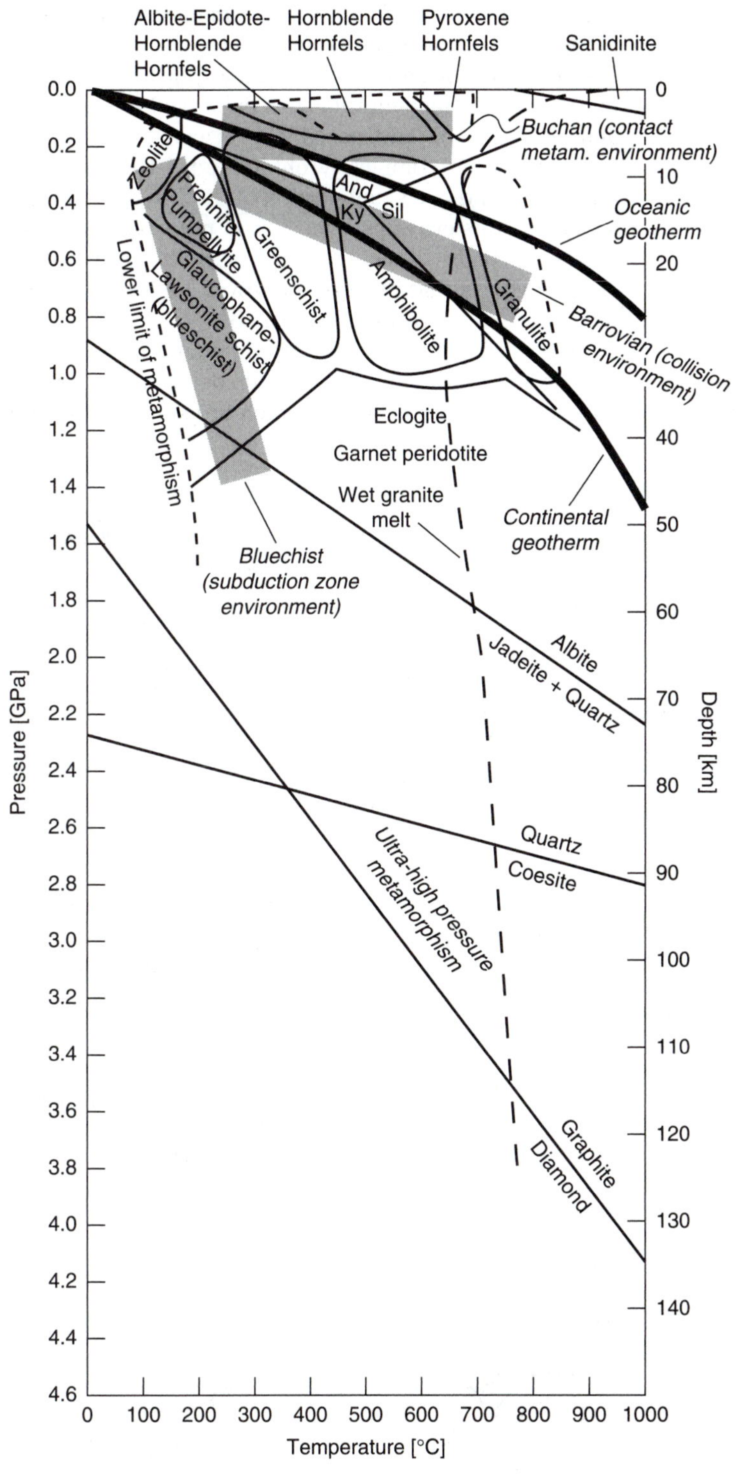

FIGURE 20.8 Petrogenetic grid showing the pressure-temperature domains of major metamorphic facies and critical metamorphic reactions. The gray bands show the approximate locations of metamorphic facies series for conditions of high-pressure–low-temperature (blueschist) metamorphism characteristic of a subduction zone environment; intermediate pressure and temperature (Barrovian) metamorphism characteristic of normal temperature gradients in a collisional environment; and low-pressure–high-temperature (Buchan) metamorphism in a shallow contact aureole environment. Also shown is the generalized region of ultra-high-pressure metamorphism below the quartz-coesite transition. The depth scale is approximate and assumes a density of 3100 kg/m^3, which is too high a density for the crust, and too low a density for the mantle. Heavy black lines show approximate average steady-state geotherms for continental and oceanic crust. (Modified after Turner 1968; Liou 2000; Carswell and Zhang 2000; Krabbendam and Dewey 1998)

matic intrusions. Thus we find Buchan metamorphism developed in a volcanic arc environment as contact metamorphic aureoles around shallow-level igneous intrusions. Less commonly, the crust may be abnormally heated by magmas from partial melting of the mantle, rising and ponding against the bottom of the crust. The Buchan metamorphism may be pre-orogenic, in which case it may be overprinted by subsequent syn-orogenic Barrovian metamorphism; alternatively, it may be post-orogenic, in which case it overprints all earlier metamorphic phases.

Blueschist metamorphism implies that temperatures in the Earth are depressed significantly below their normal values. This situation occurs in subduction zones where cold shallow rocks are carried to large depths more rapidly than thermal diffusion can reestablish normal temperatures. Blueschist metamorphism tends to occur relatively early in the orogeny and is rarely preserved in most mountain belts. Where present, it commonly is overprinted by younger Barrovian metamorphism. Blueschist and Buchan metamorphism, where present together as adjacent parallel belts, may constitute a **paired metamorphic belt** indicative of the former existence of a subduction zone (a high-pressure–low-temperature environment) adjacent to a volcanic arc (a high-temperature–low-pressure environment).

Figure 20.8 also shows the region of so-called **ultra-high-pressure (UHP) metamorphism,** defined as metamorphism that occurs in the stability field of coesite, a high-pressure polymorph of quartz. Some UHP metamorphic rocks even contain diamond, indicating that continental rocks have been carried down to mantle depths of more than 100 km, presumably by subduction, and have subsequently been uplifted to become exposed back at the surface. The coesite- or diamond-bearing assemblages characteristically are tectonic blocks included in quartzo-feldspathic metamorphic rocks, eclogites, or garnet-bearing peridotite assemblages. UHP metamorphism was discovered in the late 1990s in several mountain belts, particularly in the western Alps and the Dabie Shan mountains in northeast China. It has also been identified in such diverse localities as the Betic Cordillera and Riff orogens around Gibraltar in the western Mediterranean, the Hercynian belt of central Europe, the Scandinavian Caledonides, Indonesia, ophiolites of the Himalaya, the late Precambrian orogen of Mali, and the Paleozoic orogens of central Asia. Thus UHP metamorphism, remarkable as it may seem, is by no means an anomalous feature of orogenic core zones.

The Alps provide a good example of the regional relationships among the various metamorphic types (Figure 20.9). The regions of "burial metamorphism" are characterized by zeolite facies metamorphism, or even lower grades, that result simply from the burial of rocks by subsequent deposition. The blueschist metamorphism is 70 Ma to 85 Ma in age and is overprinted in places by the 15–25 Ma Barrovian regional metamorphism. Elsewhere only Barrovian metamorphism is preserved in the rocks, and evidence of any possible previous metamorphic event has been completely obliterated. UHP metamorphism is found in a small portion of the Dora-Maira massif of the western Alps, where it is preserved in small lenses or pods of eclogite and garnet schist within granitic gneiss that preserves a lower-pressure regional metamorphic overprint (Figure 20.9).

The distribution of temperature and pressure domains in metamorphic rocks can be determined by mapping metamorphic isograds, which are surfaces defined

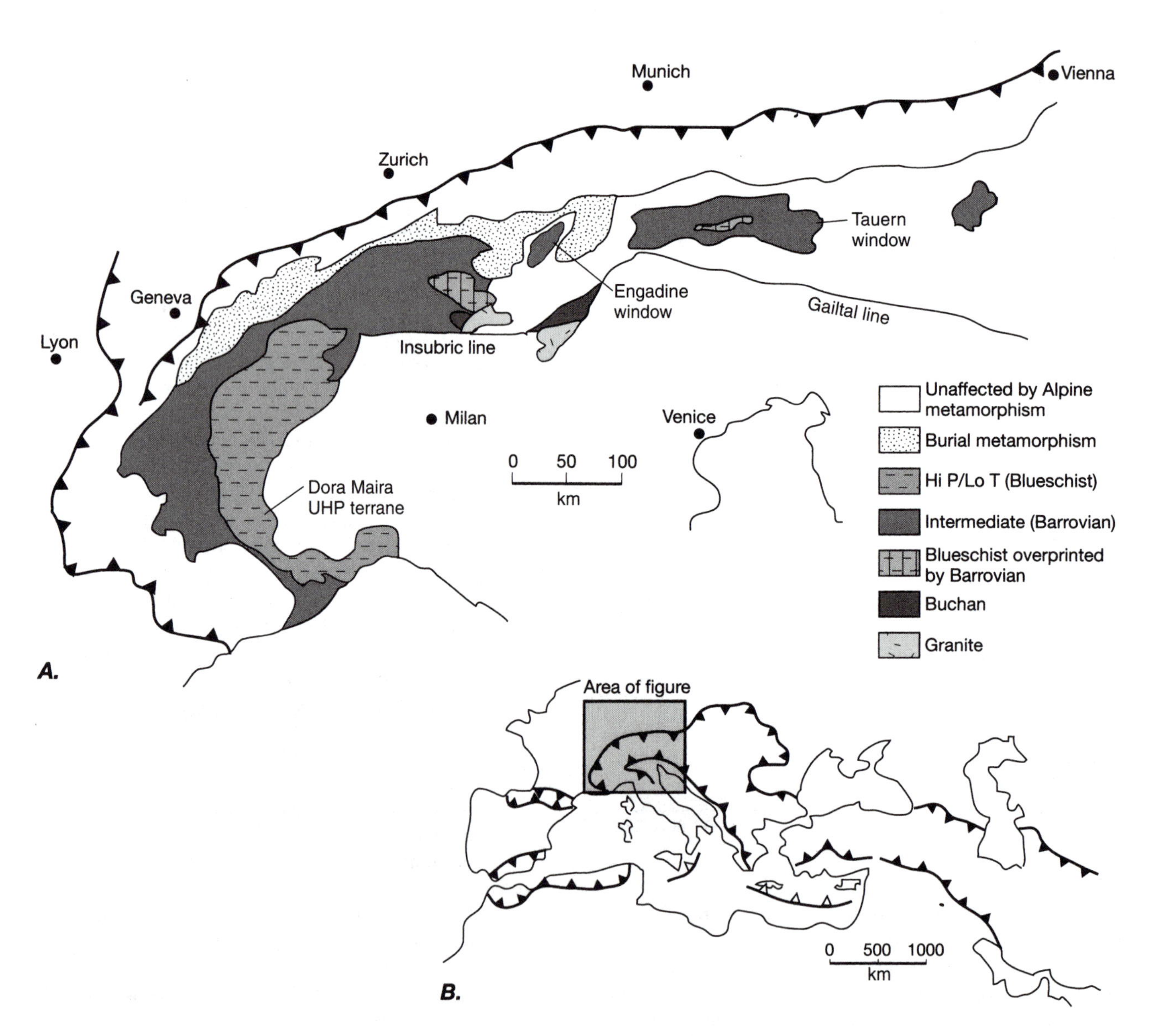

FIGURE 20.9 Distribution of metamorphic facies in the Alpine orogenic belt. A. Generalized metamorphic map of the Alps (compare Figures 20.26 and 5.22). Note the scarcity of granitic rocks. (Redrawn after Frey et al. 1974) *B.* Index map, showing the location of *A*.

by the appearance or disappearance of diagnostic metamorphic minerals in the rock. These boundaries mark the location of critical metamorphic reactions that occurred at specific temperature and pressure conditions.

In areas of good exposure, careful mapping can reveal the local three-dimensional shape of these isograd surfaces. Where they are undeformed or only mildly deformed, they appear to be gently curved surfaces that intersect the Earth's surface at small angles. For example, Figure 20.10 shows a map of the structure of part of the northern Appalachians (Figure 20.10*A*) and of the Barrovian metamorphic zones in the same region (Figure 20.10*B*). In Figure 20.10*A* the upper, central, and lower nappes form a subhorizontal stack along which there is a north-trending antiform-synform pair and a concentration of gneiss domes. These folds in turn are warped into a series of culminations and depressions about west-northwest–trending axes, giving rise to type 1 and type 2 interference patterns (cf. Figures 13.37*A*, *B* and 13.38*A*, *B*) that are particularly evident in the boundary between the upper and central nappes (Figure 20.10*A*). The grade of metamorphism increases continuously from the upper

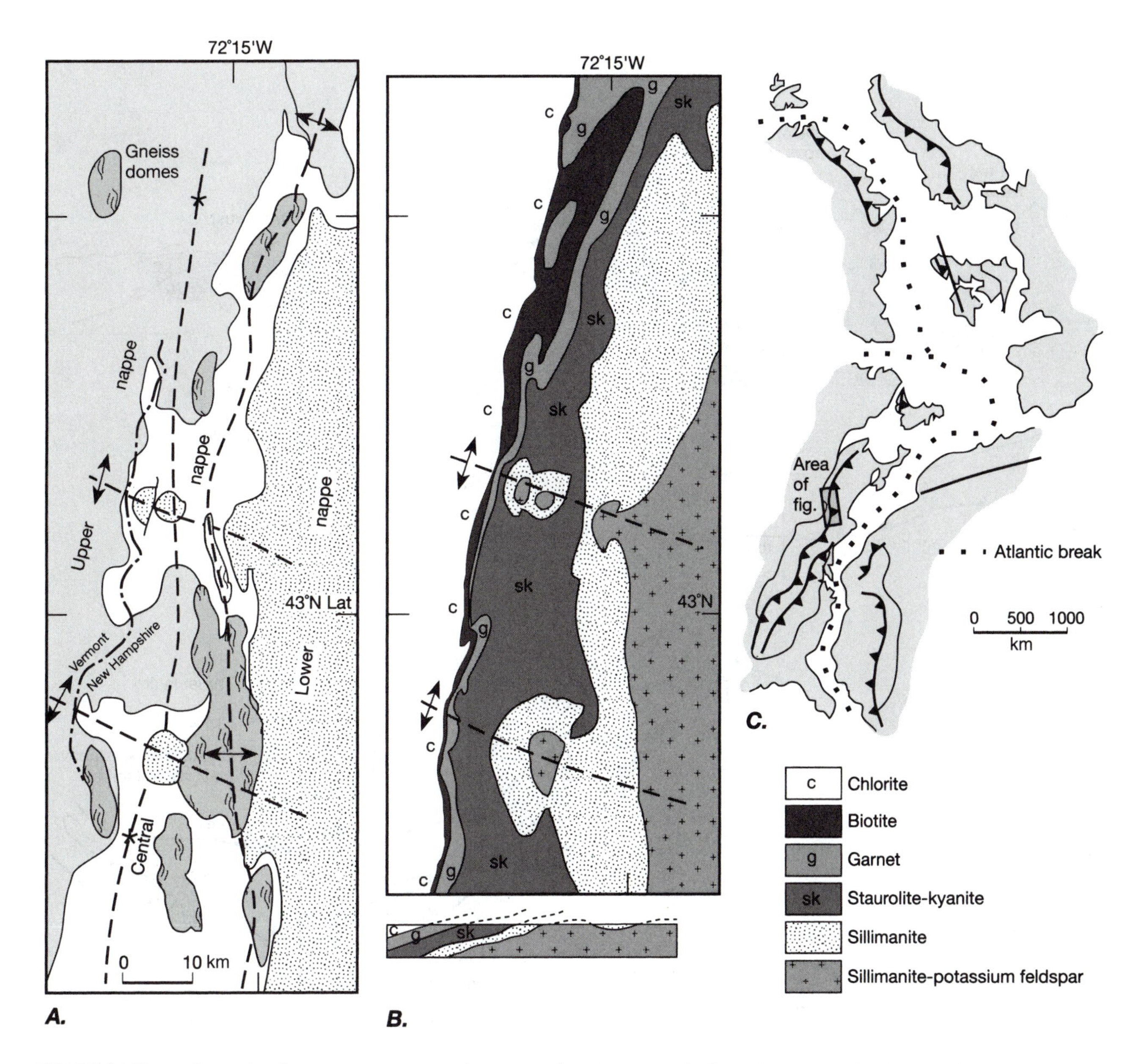

FIGURE 20.10 Relationship between structure and metamorphism, an example from the New England Appalachians. *A*. Tectonic map showing the distribution of major nappes and gneiss domes. *B*. Map and schematic cross section of mineral isograds of Barrovian metamorphism. Note that in general the metamorphic grade increases with progressively deeper tectonic levels. (After Thompson et al. 1968: 212–213) *C*. Index map, a reduced version of Figure 20.2*C*, showing location of *A* and *B*.

to the lower nappe (Figure 20.10*B*) through the Barrovian mineral zones of the Greenschist and Amphibolite metamorphic facies (chlorite, biotite, garnet, kyanite, sillimanite; cf. Figure 20.8). These metamorphic isograds are gently folded about the same two directions as the nappes, giving the interference patterns on the isograd map (Figure 20.10*B*). Thus, although the isograds do cut across the nappe boundaries, they share in the two gentle foldings of the tectonic units, suggesting that emplacement of the nappes ceased before or during peak metamorphism and that subsequently the nappes and isograds together were gently folded.

In other regions, metamorphic zones clearly have been displaced or even inverted. In the central Himalaya, for example (Figure 20.11), the metamorphic grade increases continuously through Barrovian mineral zones from the bottom to the top of the structure, revealing an inverted metamorphic gradient. Such an inversion could develop above a subduction zone where a cold downgoing slab cools off the lower portions of the over-riding plate (Figure 20.12*A*). It also could indicate a tectonic inversion of the isotherms following metamorphism, as would occur on the inverted limb of a recumbent fold (Figure 20.12*B*).

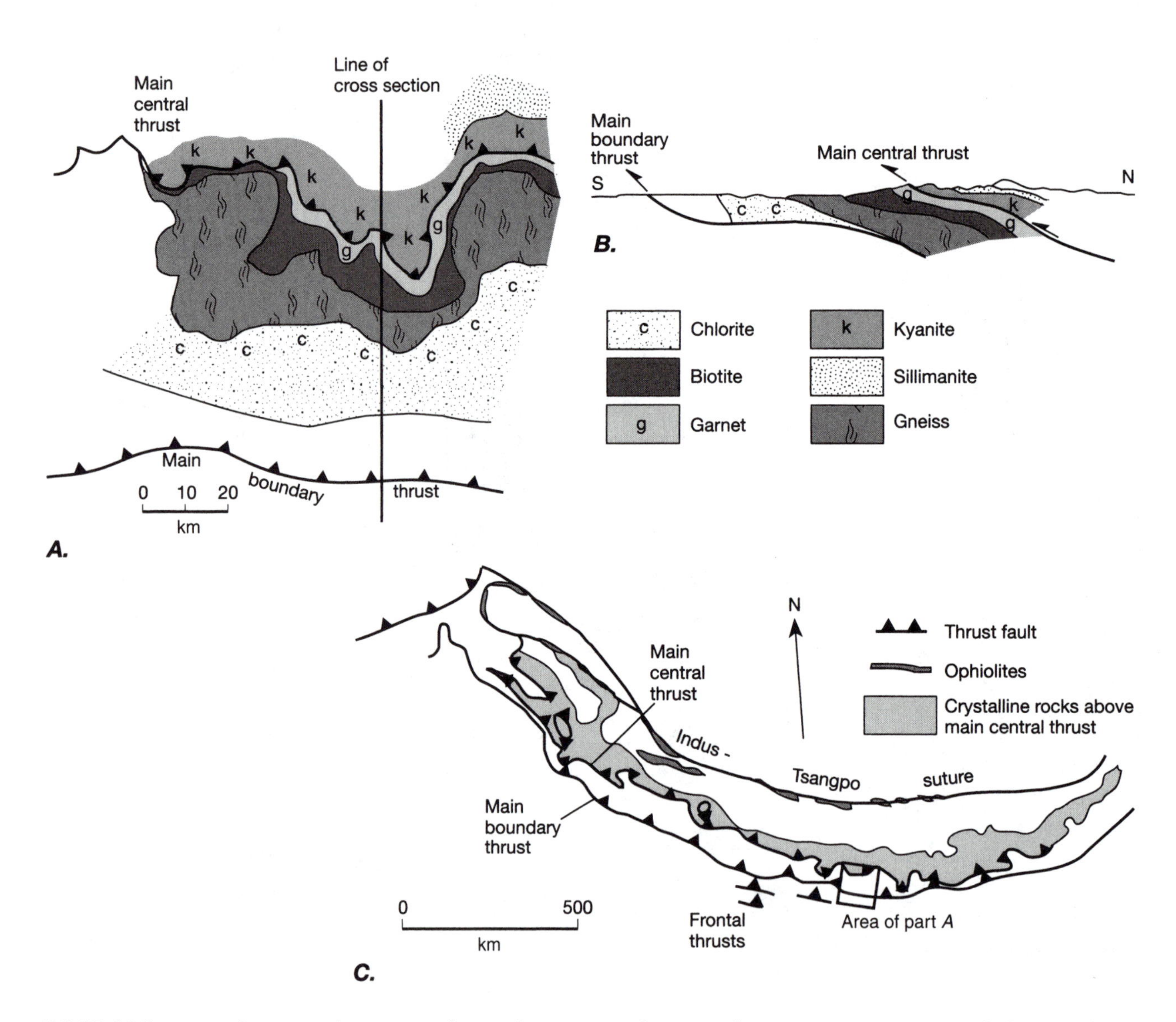

FIGURE 20.11 Inverted metamorphic zones in the Himalaya. *A*. Map of metamorphic zones. (After LeForte 1975) *B*. Cross section showing tectonic units and metamorphic isograd inversion across the same region. (After Gansser 1974) *C*. Location map, simplified from Figure 5.11C, showing the location of *A*.

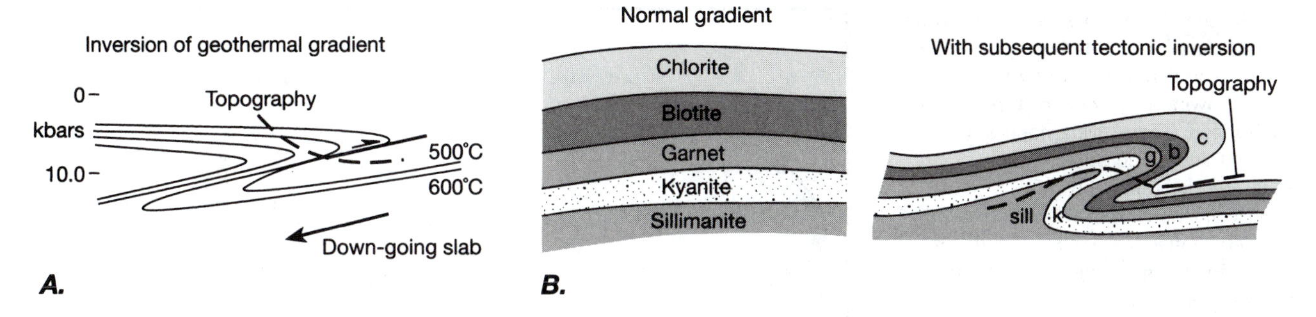

FIGURE 20.12 Possible origin of inverted metamorphic zones. *A*. Inversion of isotherms during thrusting, showing the possible temperature distribution during metamorphism. (After LeForte 1975) *B*. Development of a recumbent nappe of metamorphic rocks subsequent to metamorphism.

20.5 THE CRYSTALLINE CORE ZONE: STRUCTURE AND LITHOLOGY

The center or "core" of an orogenic belt is made up of high-grade metamorphic and plutonic "crystalline" rocks that have deformed extensively by ductile flow. The resulting structures include complex multiple-deformation structures and thrust or fold nappes, which can be very large, covering as much as 100,000 to 250,000 km^2. The core is invariably thrust out over the rocks of the foreland fold and thrust belt (Figure 20.3).

Multiple generations of folding in core zones produce a variety of fold interference structures (Figures 13.37 and 13.38). A structural sequence that frequently emerges from the geometrical analysis of such areas includes one or more generations of recumbent isoclinal folds, refolded by a generation of upright more open folds, and finally deformed by a generation of either smaller-scale kink and chevron folds or of ductile shear zones (Figure 20.13).

In rocks that preserve both fine-scale layering and large-scale stratigraphy, small-scale (high-order) folds mimic the orientations and styles of the larger (lower-order) folds. Detailed studies of high-order folds in critical outcrops, therefore, can be used to infer the basic geometry of regional deformation. In plutonic igneous rocks where no original layering is present, the deformed rocks become foliated, and mylonite zones of high ductile shear strain define boundaries of areas where deformation has been less intense.

The rocks in core zones of mountain belts are of diverse origins. In this section, we describe the principal lithologic components of most orogenic core zones.

i. Metamorphosed Sedimentary Rocks and their Basement

In some cases, such as in some Alpine crystalline nappes (Figure 20.14), the crystalline rocks represent metamorphosed deep-water sedimentary rocks and their thinned continental crystalline basement. In such regions, former crystalline basement rocks represented by coarsely crystalline gneisses commonly form the cores of nappe structures, and these are surrounded by an envelope or sheath of younger metasedimentary rocks. Figure 20.14 shows an example of such structures from the Penninic zone of the western Alps, where the Adula, Tambo, and Suretta nappes form a stack of thrust sheets, each consisting of basement rocks with their metasedimentary cover (compare Figure 20.20).

In the deeper structural levels of an orogenic belt, metamorphic temperatures can approach or even exceed the granite solidus (see Figure 20.8). The resulting highly ductile rocks (Figure 20.15*A*) are gravitationally unstable, and they rise diapirically (Figure 20.15*B*), forming

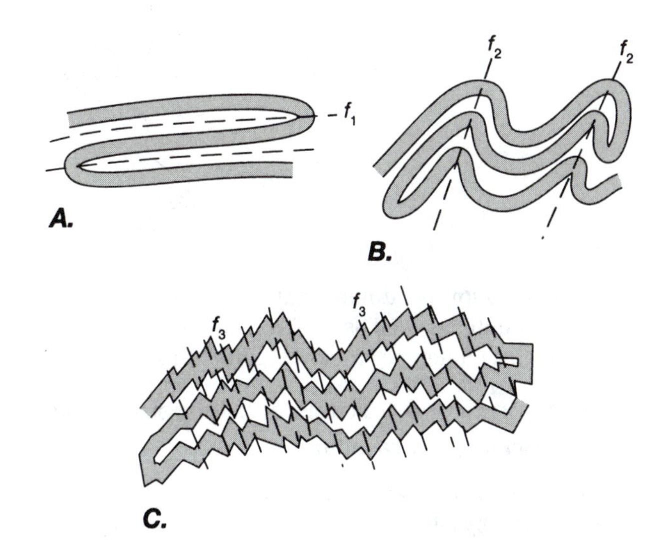

FIGURE 20.13 Diagrammatic cross sections illustrating the progressive sequence of folding proceeding from: *A*. first-generation isoclinal folds, to *B*. more upright second-generation folds superposed on the earlier deformation, to *C*. a third-generation kinking superposed on all earlier foldings.

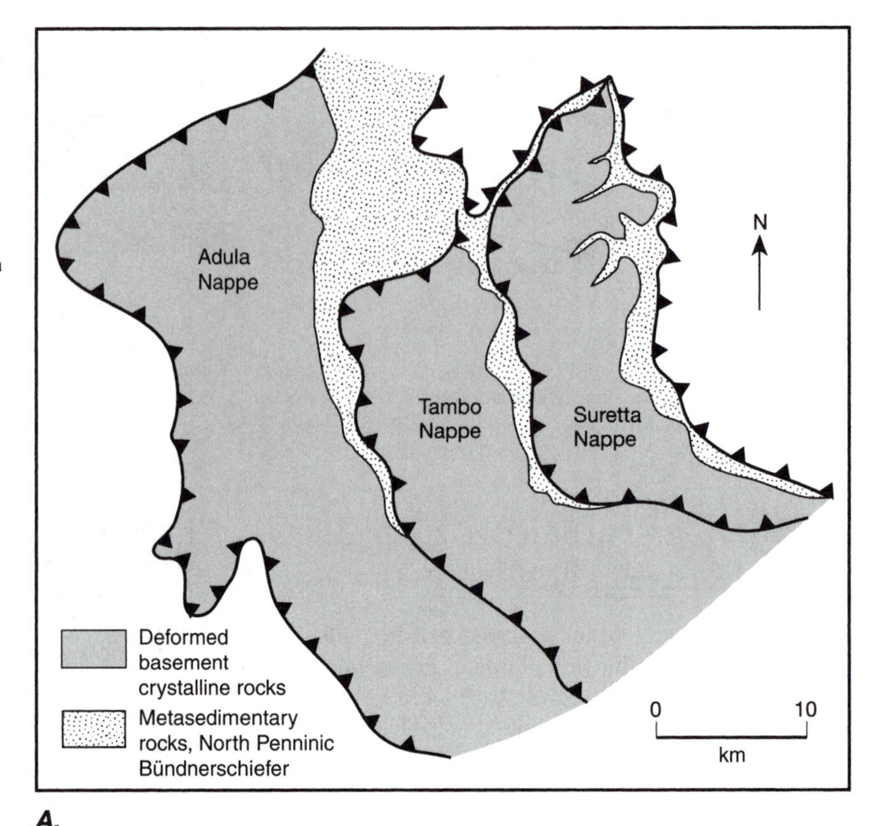

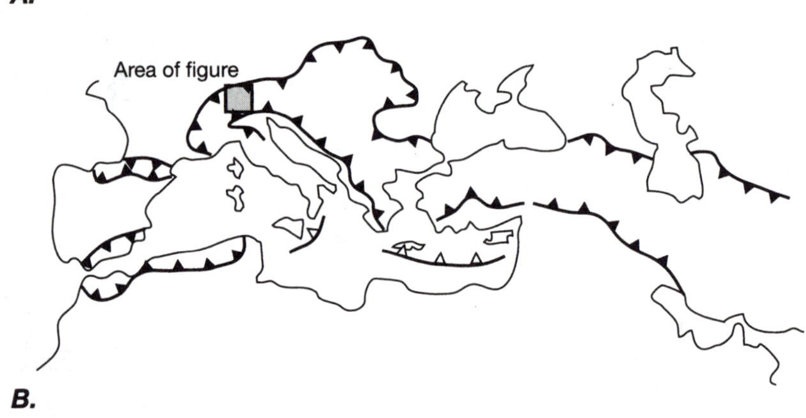

FIGURE 20.14 A stack of nappes in an orogenic core zone. *A*. Generalized map of part of the Penninic (core) zone of the Alps, showing map view of three major subhorizontal crystalline nappes, the Adula, Tambo, and Suretta nappes, separated by thin septa of metasedimentary rocks (compare Figure 20.20*B*). (After Spicher 1980) *B*. Index map, a reduced version of Figure 20.2*B* showing the location of *A*.

huge **mantled gneiss domes** that may contain a core of intrusive granite, as well as gneiss, mantled by an envelope of metasedimentary rocks (Figure 20.15*C*).[3]

ii. Metamorphosed Volcanic and Igneous Rocks and Associated Metasediments

Orogenic core zones generally contain large areas of rock characterized by a lack of pronounced or continuous layered stratigraphy and by an abundance of intrusive rocks. Metamorphism is commonly intense in these rocks, reaching amphibolite or even granulite conditions (see Figure 20.8). The vast areas of these rocks therefore tend to be massive or banded amphibolite or granulite in which the original stratigraphic and intrusive relationships become next to impossible to determine.

The Appalachians in New England and southern Canada provide some especially well-documented examples of these rocks. As shown in Figure 20.16, the rocks there include metasedimentary rocks, metavolcanic rocks, and a number of gneiss domes. The rocks display complex, multiply folded structures, generally with patterns and numbers of deformational phases similar to

[3]Gneiss domes appear to be gradational into domes formed by multiple folding. Indeed, it is possible in some cases that little or no piercing, i.e., diapiric activity, of the gneiss through the overlying rock has taken place. The contact between the gneiss and metasedimentary rocks generally is a ductile shear zone, however.

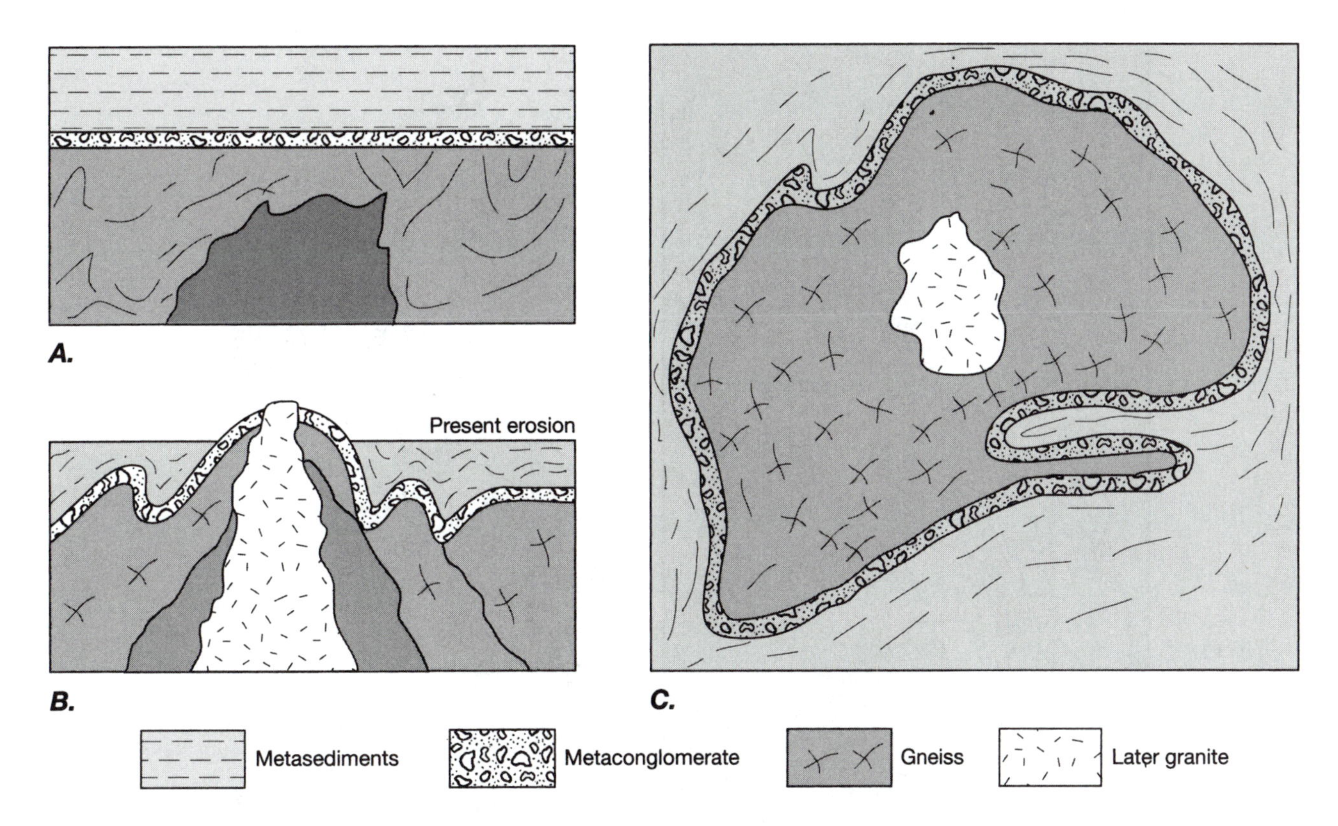

FIGURE 20.15 Development of mantled gneiss domes. (Redrawn after Eskola 1949) *A*. Deposition of a sedimentary sequence unconformably on metamorphosed sediments and intrusive rocks. *B*. Deformation of entire sequence in *A* during a new deformation, followed by intrusion of a new granitic body. *C*. Generalized map of the resulting gneiss dome after exposure by erosion.

those outlined in the introduction to this section (Figure 20.13).

Such metavolcanic rocks are understood to be a juxtaposition of continental rise-slope deposits, with volcanic rocks and volcanogenic sediments that formed in continental volcanic arcs, oceanic arcs, or midoceanic volcanic complexes. The juxtaposition is a consequence of collisions at one or more subduction zones, as illustrated in Figure 19.43.

iii. Metamorphosed Ophiolitic Sequences

Ophiolite belts are a feature of many orogenic core zones, and in many cases they have participated in the regional deformation and metamorphism. When this has happened, the pseudostratigraphy of the ophiolite complexes (see Figure 19.17) develops complex structures including large-scale recumbent or multiply refolded folds. At high grades of metamorphism and large amounts of deformation, pillow lavas become massive greenschists or amphibolites; mafic dikes and plutonic complexes become massive or banded amphibolites; mantle peridotites become serpentinized at lower grades of metamorphism and are dehydrated back into peridotite at higher grades, with the original mantle fabric overprinted or even obliterated by the later deformation.

iv. Lower Continental Crust and Mantle

In some regions of collisional orogenic belts, where overthrusting and subsequent uplift has been extreme, highly metamorphosed quartzo-feldspathic gneisses are found overlying peridotite. The Ivrea zone of the Alps is a good example of this situation (Figure 20.17; compare Figures 20.20*A* and 5.22). These regions probably represent the contact between the lower continental crust and the underlying mantle, that is, the original Moho. Like ophiolitic peridotites, both the peridotites and the gneisses of continental assemblages typically display an old metamorphic assemblage and deformation fabric overprinted by a younger one. Some subcontinental peridotites contain irregular layers and veinlets of gabbroic rocks that pass into bodies or dikes, suggesting that partial melting has occurred. The pressure-temperature conditions inferred from the gabbro and peridotite mineralogy suggest that this melting took place during emplacement of the complex beneath the continental crust during continental stretching or rifting.

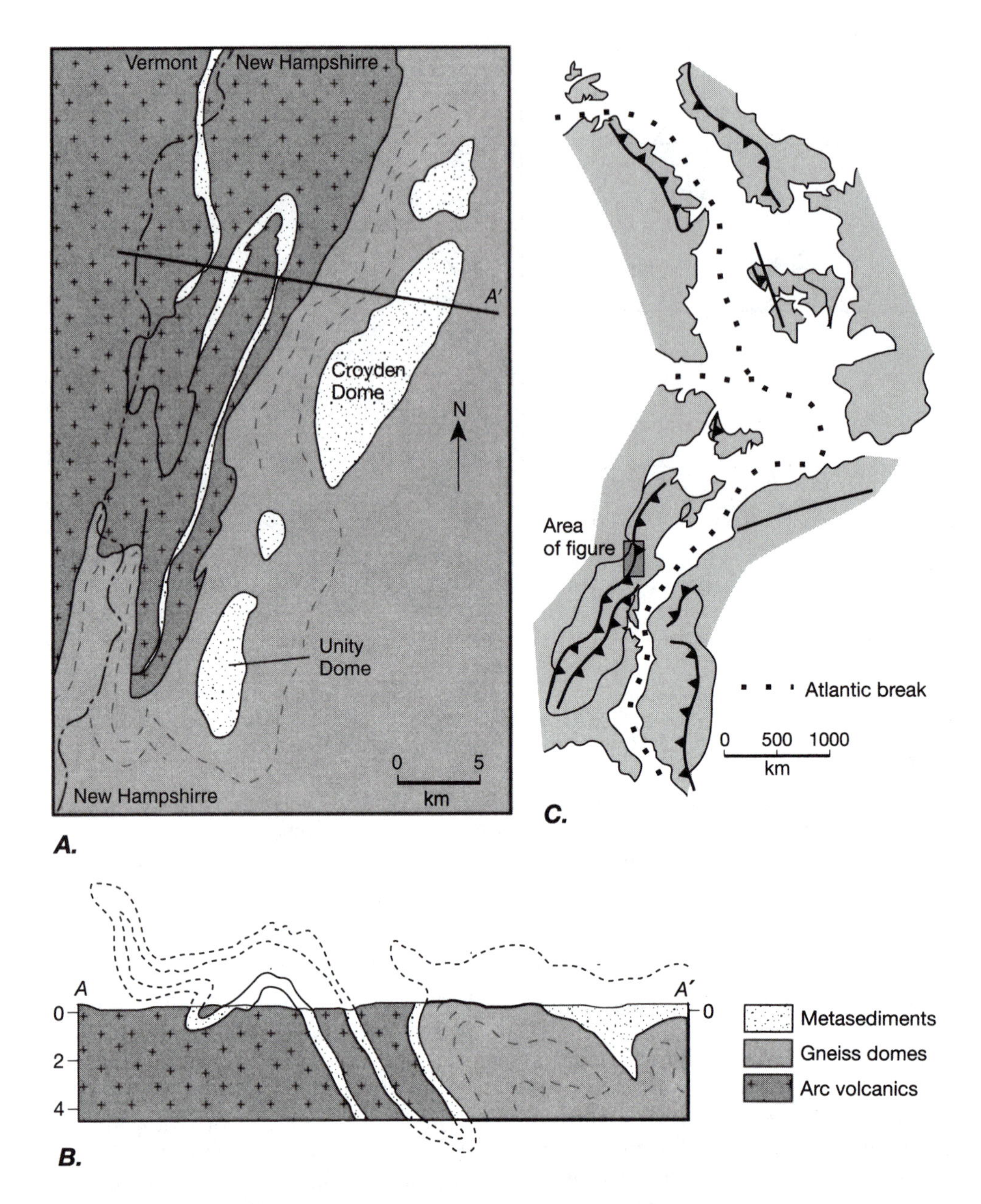

FIGURE 20.16 A typical multiply folded "eugeoclinal" core zone, the northern Appalachians in Vermont. Of particular interest is the metavolcanic sequence, possibly an early Paleozoic island arc complex that collided with North America in Ordovician time. (*A* and *B* modified after Thompson 1968) A. Map. B. Cross section. *C.* Index map, a reduced version of Figure 20.2C showing location of *A.*

v. Gneissic Terranes with Abundant Ultramafic Bodies

Some orogenic core zones include terranes characterized by amphibolite or granulite-facies gneisses and schists that contain numerous small, discontinuous ultramafic bodies, most less than 1 kilometer long. These bodies typically consist of fresh peridotite, pyroxenite, or dunite or their serpentinized equivalents. The southern Appalachians provides an especially good example of such bodies, as shown in Figure 20.18, but similar terranes are present in the Alps and the Caledonides, as well. The bodies characteristically are elongate parallel to the regional structural grain and display an internal fabric concordant with the regional fabric.

The interpretation of these terranes is a difficult tectonic problem. In many cases, we understand neither the mechanism by which the ultramafic rocks are incorporated into the metamorphic terranes, nor the protolith[4]

[4]After the Greek words *protos*, which means "first," and *lithos*, which means "rock." In other words, the first, or original, rock.

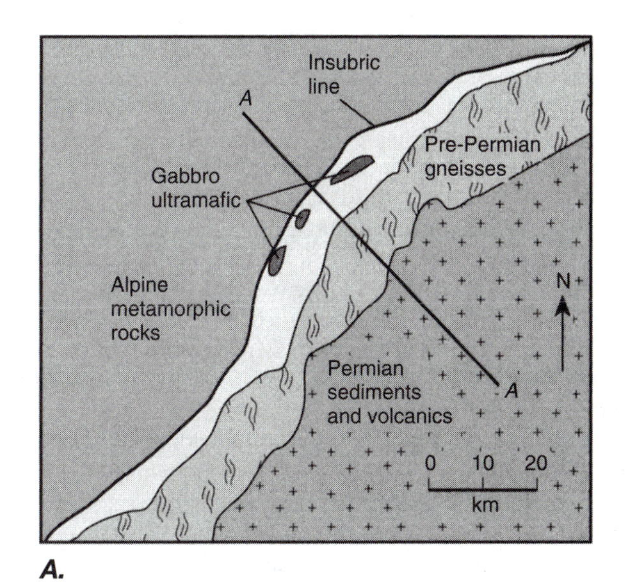

A.

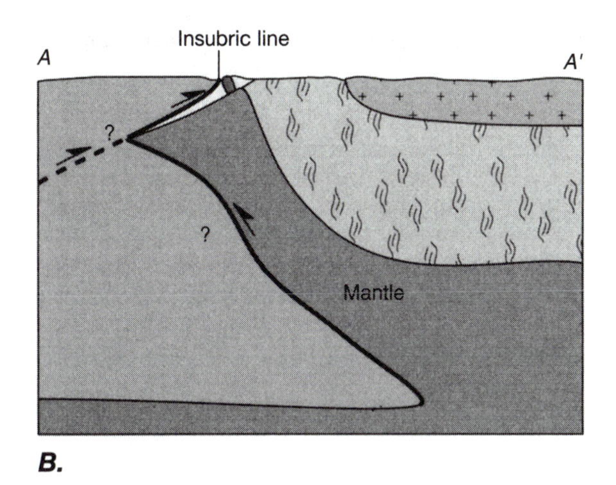

B.

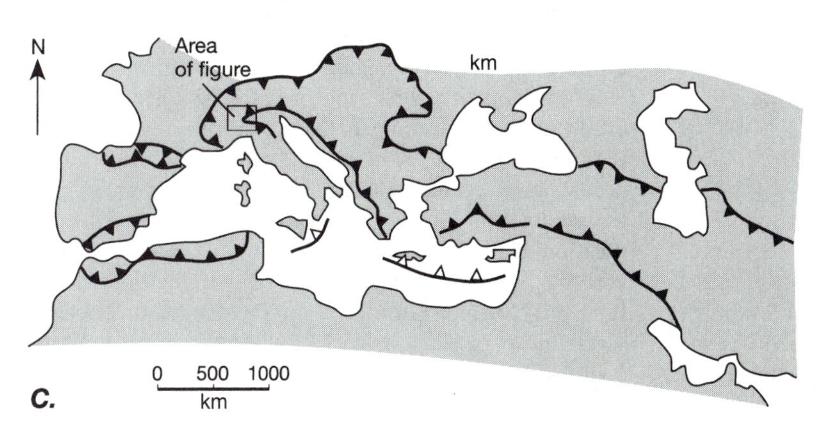

C.

FIGURE 20.17 Exposure of lower continental crust and mantle, Ivrea zone, southern Alps (see map of the Alps, Figures 5.22 and 20.26). *A.* Generalized map. *B.* Cross section based on geology and geophysics showing mantle thrust over the northern continental edge, as well as back thrust in the opposite direction. See the alternative interpretation of the cross section in Figure 20.20*A*. (After Zingg and Schmid 1979) *C.* Index map, a reduced version of Figure 20.2*B*, showing the location of *A*.

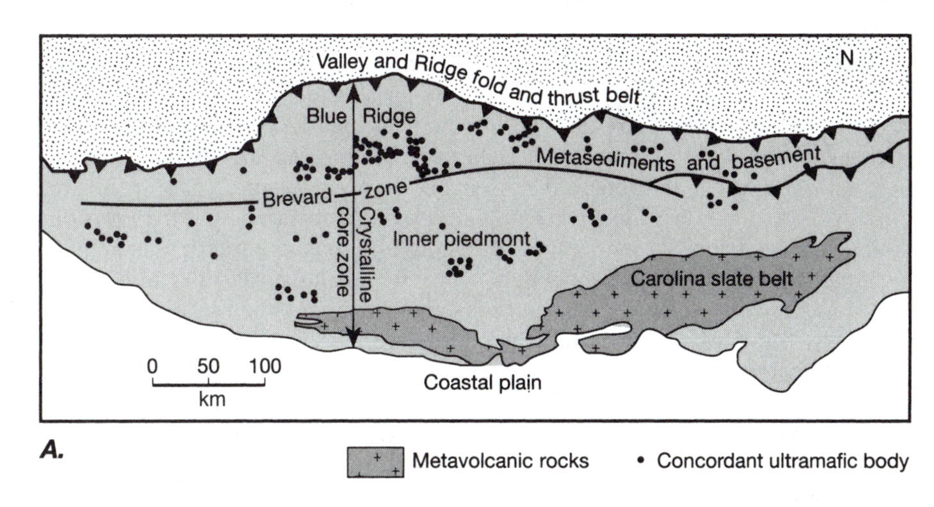

A.

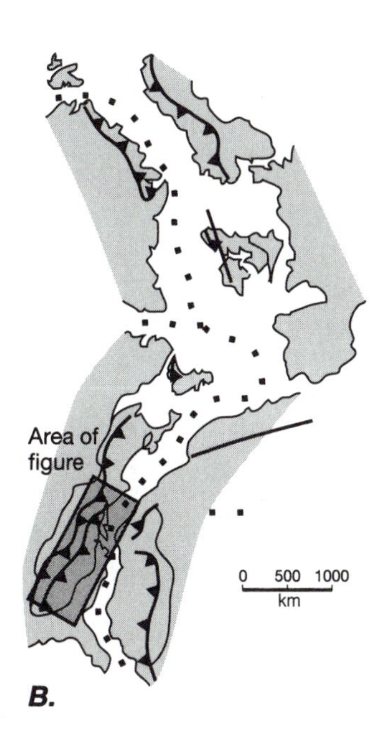

B.

FIGURE 20.18 *A.* Concordant ultramafic bodies in the crystalline core zone of the southern Appalachians. (After Misra and Keller 1978) *B.* Index map, a reduced version of Figure 20.2*C*, showing the location of *A*.

of the metamorphic rocks. We can suggest four possible origins for these enigmatic regions:

1. They may be exposures of deep crustal levels where fragments of the subcontinental or subisland arc mantle were somehow incorporated into the crust during deformation.
2. They may be remnants of ophiolitic or mantle slabs that have been completely disrupted after emplacement, by extreme deformation and metamorphism that produced the gneissic terranes.
3. They may be zones of melange formed at shallow levels and carried to deeper levels during orogenic development.
4. They may be mafic and ultramafic igneous rocks intruded into continental margin sediments during rifting. Mafic and ultramafic rocks tend to be less ductile than gneisses or schists, so during ductile deformation, the more ductile rocks flow relatively easily, and the less ductile rocks tend to become boudinaged. Thus any original continuity of layers can become completely lost.

vi. Granitic Batholiths

Granitic batholiths are large areas of plutonic igneous rocks, generally of dioritic to granitic composition, that occur in many mountain belts, in some cases dominating the area of the orogenic core. The North American Cordillera provides a good example, where vast areas of batholithic rocks extend discontinuously from northern Alaska to northern Mexico (Figures 20.2*A*). The Alps, on the other hand, have only minor amounts of granitic intrusives. This difference must reflect a fundamental difference in tectonic environment, which is not well understood. It may be related to the amount of subduction that took place prior to continent-continent collision. The Alps are the product of closure of a narrow ocean basin, where not enough subduction occurred to develop abundant magmatic activity such as is typical of circum-Pacific convergent margins. Most batholiths are not a single body, but comprise tens to hundreds of individual plutons, each a few tens to hundreds of square kilometers in area. They exhibit wide differences in rock type, degree of deformation, and apparent depth of emplacement.

Granitic rocks are either I-type or S-type granites, where the designations "I" and "S" imply derivation by partial melting, respectively, of an originally igneous or sedimentary source. I-type granites are characteristically hornblende-biotite quartz diorites in composition. They are thought to have formed by partial melting of a hydrous mantle or a previous crystallized igneous rock. S-type granites are richer in K and characteristically contain both biotite and muscovite and, more rarely, garnet. They may be derived from the partial melting of sedimentary rocks.

The timing of batholithic activity relative to deformation in the orogen also is variable. In a number of mountain belts, the age of granitic rocks overlaps with periods of deformation. Older granitic plutons commonly exhibit evidence of this deformation, such as the development of foliation and/or folds parallel to regional trends, whereas younger intrusive rocks do not. Thus they can be considered to be pre-orogenic if they intruded prior to the main deformation, syn-orogenic if they intruded during deformation, or post-orogenic if they intruded after deformation.

The timing of emplacement of individual plutons is also a subject of some debate. Some plutons may intrude as a single unit. Others consist of an amalgamation of multiple small intrusions into the same volume of crust, and the intrusions apparently can span a time of several million years. Because of their similarity, the rocks are often mapped as a single pluton, but careful examination of the intrusive history clearly shows they can be more complex.

The dominant rock type in batholiths is different in different mountain ranges. Quartz diorite or granodiorite is dominant in the western U.S. Cordillera and the Andes, whereas granite or quartz monzonite is more characteristic of the Appalachians and Caledonides. The composition of the granitic rock displays a crude correlation with the type of country rock and the timing of intrusion. Both pre-orogenic batholiths and those intruding oceanic sedimentary or volcanic rocks tend to be poorer in K than post-orogenic batholiths and those invading continental rocks.

The batholiths of the Sierra Nevada of California and Nevada (Figure 20.19*A*) provide a good example of the variability of batholithic history. They have an intrusive history ranging in age from approximately 270 Ma to 70 Ma (Figure 20.19*B*, *C*). Some plutons in the Sierra Nevada are deformed, whereas others are not. Granitic rocks in the Sierra Nevada cluster into three age groups: early-mid Jurassic, late Jurassic, and mid-late Cretaceous (Figure 20.19*C*). As the principal deformation in this region was mid-late Jurassic (175 Ma to 140 Ma) these groups generally correspond to pre-orogenic, syn-orogenic, and post-orogenic intrusives. As mentioned in the preceding paragraphs, individual plutons apparently accumulated incrementally over several million years (Figure 20.19*D*).

We can gain some insight into the variations one sees in granitic batholithic terranes by considering a plate-tectonic model. Pre-orogenic I-type granitic bodies may be the product of normal igneous activity at a consuming margin that occurs before the collision in which the granitic bodies are deformed. The granites are Na-rich and are associated with volcanic rocks of similar composition, which are also deformed in the later collision. The syn-orogenic and post-orogenic S-type granites may result from partial melting of the lower part of the con-

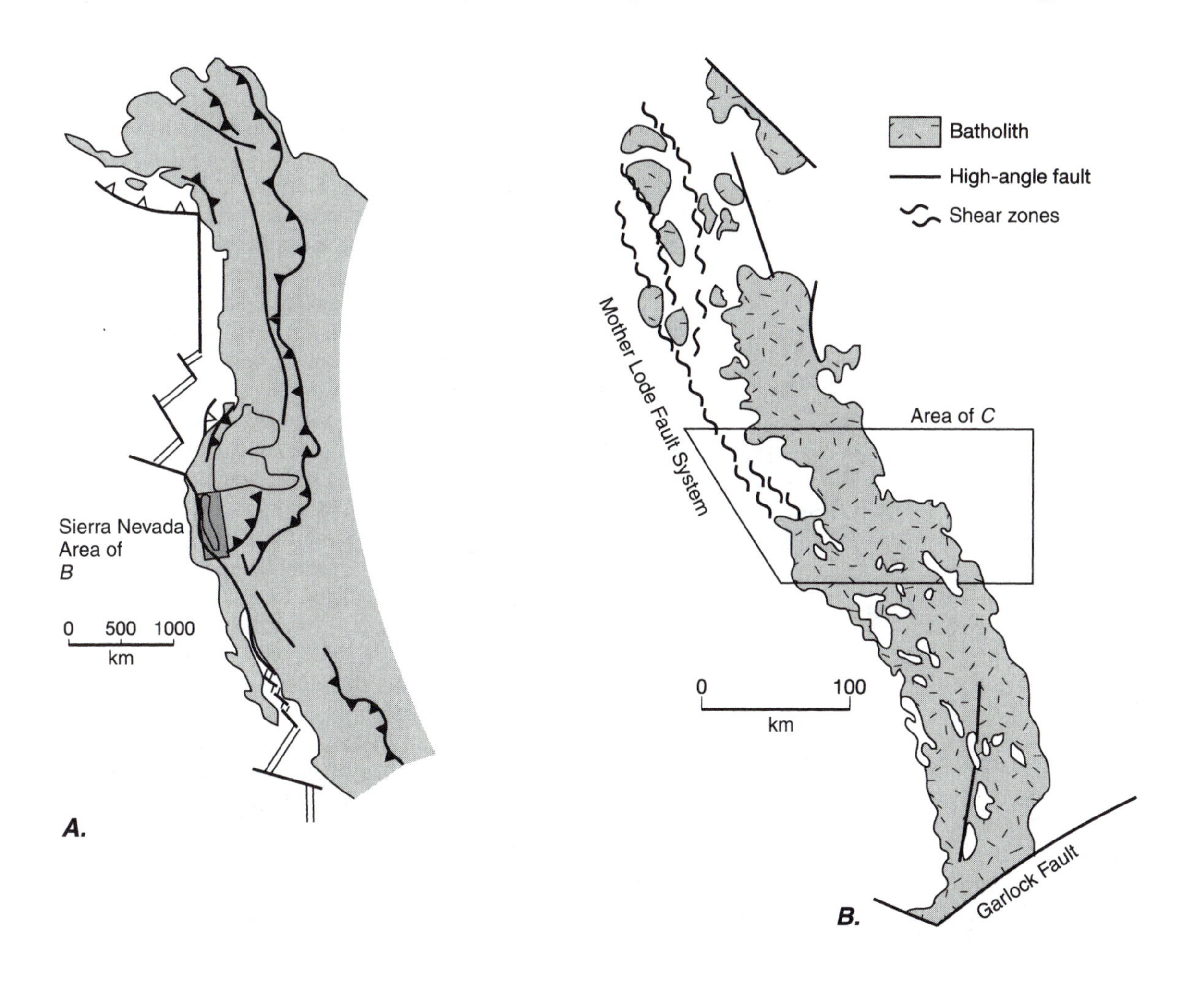

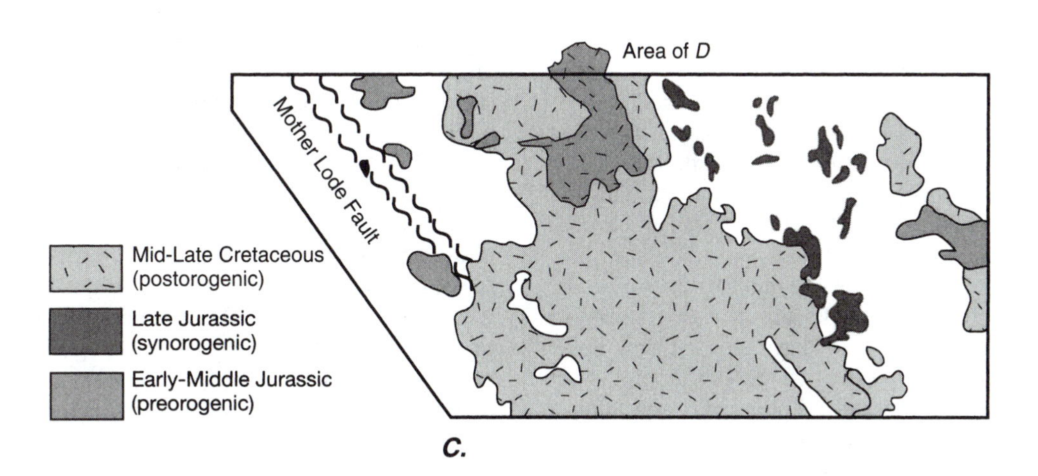

FIGURE 20.19 Batholiths in the North American Cordillera. *A*. Index map, a reduced version of Figure 20.2*A*, showing the location of the Sierra Nevada shown in *B*. *B*. Generalized map of batholithic rocks in the Sierra Nevada, California (see *A* for location). (After Bateman 1981) *C*. More detailed map of part of central Sierra Nevada, showing radiometric age relations of granitic plutonic rock (see *B* for location). *D*. Map of the Tuolumne zoned pluton, with age range of various units indicated (see *C* for location). (Redrawn after Glazner et al. 2004)

(continued)

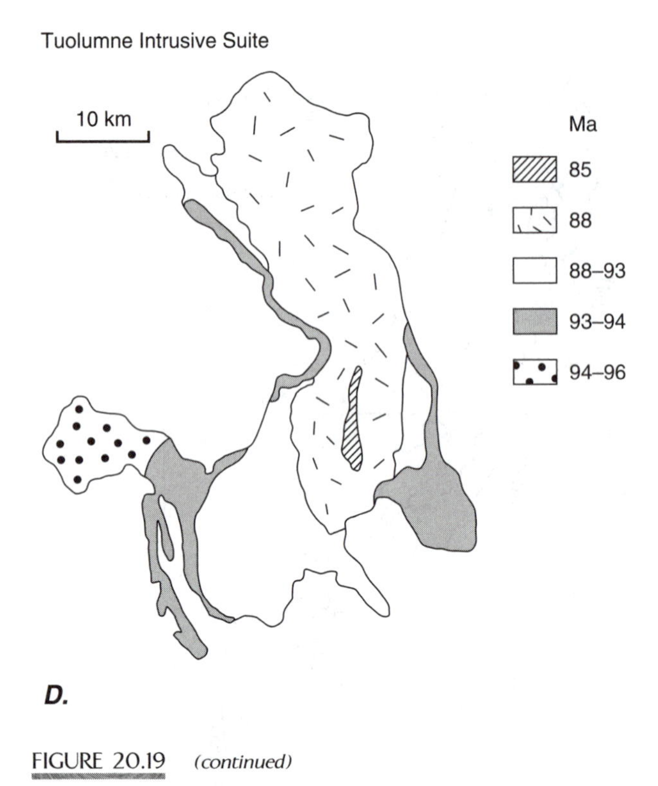

FIGURE 20.19 *(continued)*

tinental crust that has been thickened by formation of a mountain root during collision. They are K-rich, and possibly even Al-rich, granitic rocks. Post-orogenic alkalic granites form after all orogenic phases have ceased and may reflect early stages of an episode of subsequent continental rifting.

vii. The Deep Structure of Core Zones

In many instances, the crust thickens markedly under orogenic belts, from a normal continental thickness of about 35–40 km to a thickness of about 60 km. The area where the Moho is depressed by the crustal thickening is referred to as the **root** of the mountain belt, and it is generally in isostatic equilibrium with the elevated topography of the mountains.

In cross sections of many orogenic belts, the major folds are recumbent, and faults dip at a low angle. In places, however, these features have steeper dips. Here, fold axial surfaces are vertical to steeply dipping with horizontal to steeply plunging fold axes (see Figure 10.12), thrust faults are nearly vertical, pronounced down-dip lineations occur, and, in places, ductile strains become very large. Some of these zones are also the sites of major shear zones as discussed in Section 20.7. Generally, structures and tectonic units are not continuous across the region. Understanding the deep structure and its origin in orogenic core zones and their roots has been a major challenge for structural geology that even yet is not entirely resolved (for more discussion, see Section 20.12).

In the Alps it has been shown that the steep-dip region, which lies just north of the Insubric Line in Figure 20.20*A*, *B*, is in fact a limb of a huge second-generation fold that deforms originally subhorizontal nappes and thrust sheets (see Figure 10.1*B* for an older interpretation of a western cross section of the Alps that does not extend to as great a depth). Because these folds have a vergence that is opposite to the vergence of the nappe structures, they are referred to as **back folds**. In the Alps, this region has been referred to as the "root zone." This rather outmoded term is a different usage from the "root" of a mountain belt defined at the beginning of this subsection. It derives from the time when these areas were viewed as the source area from which large-scale nappes and folds emerged as a result of extreme compression and to which they were connected, or "rooted" (see right-hand end of the model in Figure 18.3).

We now understand these back-fold structures to be the result of collision and the consequent development of a crustal-scale zone of thrusting conjugate to the subduction zone, which may be enhanced by underplating of material onto the over-riding plate and by continued shortening during isostatic uplift (Figure 20.20*A*, *B*; see Section 20.12(i)). An alternative model for the origin of these structures is that they may develop from a change in direction of dip of a subduction zone, as illustrated schematically in Figure 20.21*A*.

The depth to which surface structures descend seems to vary from one mountain belt to another. In the Alps, seismic reflection and refraction evidence suggest that the structures involved in the back fold apparently extend to deep levels and that the Moho is offset across the orogenic belt (Figures 20.20 and 20.21*A*). In other regions, the Moho is smooth, and seismic reflection data suggest that the crystalline core zone is allochthonous. For example, in the southern Appalachians, the crystalline core overlies a series of flat-lying reflectors that may be the little-deformed equivalents to the sediments of the Valley and Ridge province (Figure 20.22; cf. Figure 20.2*C*). This result implies that the entire deformed belt may be allochthonous and may have been displaced hundreds of kilometers over an autochthonous continental basement. Similar seismic results from elsewhere along this orogenic belt suggest that along much of its length, the entire core zone is allochthonous. Such a detachment involving an entire orogenic belt, sometimes called an **orogenic float,** has profound implications for the nature of the movements and forces that caused it.

20.6 EXTENSIONAL DEFORMATION AND LOW-ANGLE DETACHMENTS

In many mountain belts, major areas of extensional deformation have been recognized, characterized by low-angle detachment faults that mark a large discontinuous jump in metamorphic grade. Across these faults,

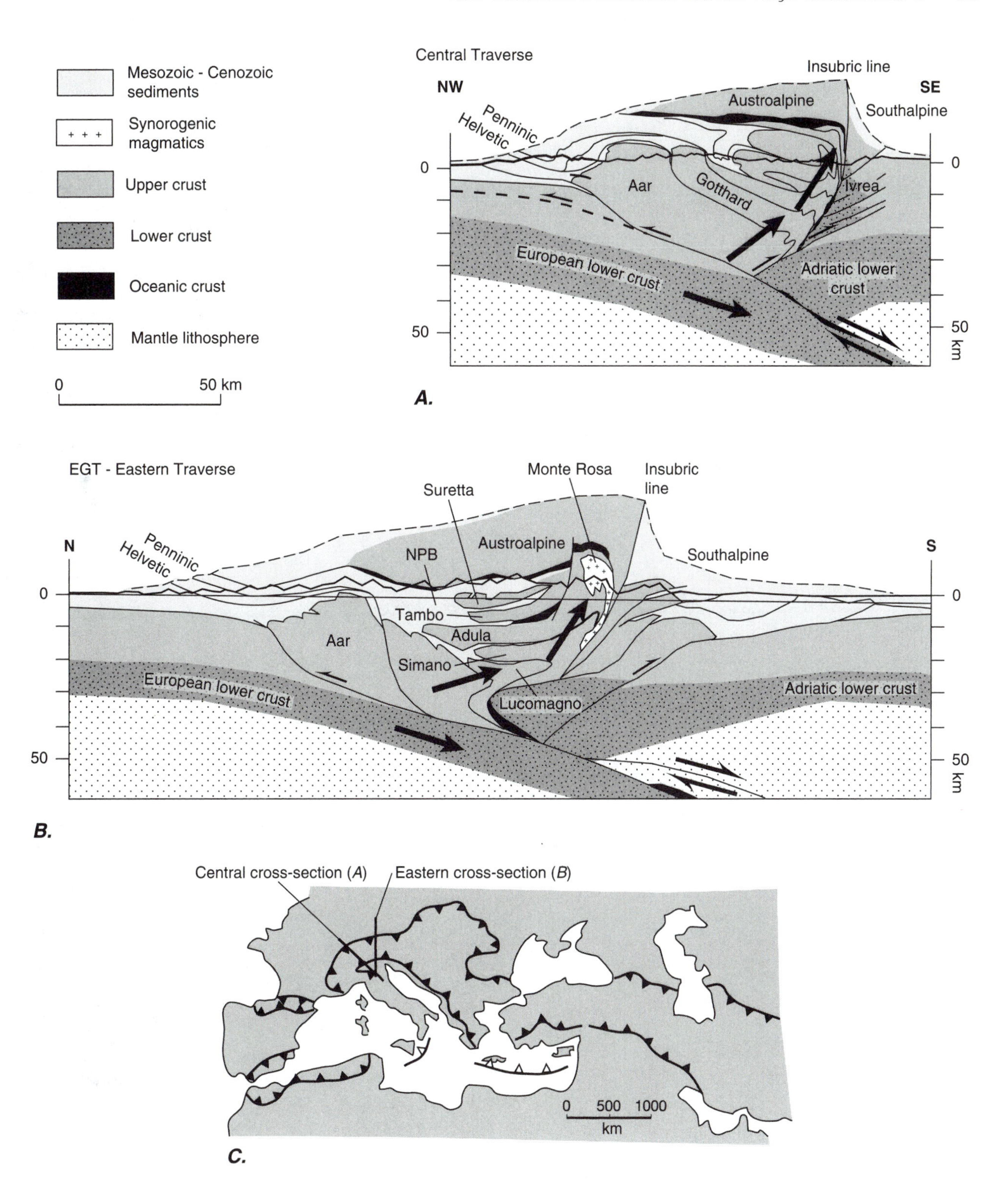

FIGURE 20.20 Crustal-scale cross sections of the Swiss Alps based on seismic and structural data. Bold jagged line indicates current topography; the cross section projected above topography is based on estimates of eroded material. The European lower crust and lithospheric mantle dip south beneath the Adriatic lower crust and lithospheric mantle. *A*. Central cross section: The Ivrea zone is interpreted to be the Adriatic lower crustal material entrained in the back-folding (cf. Figure 20.34). *B*. The eastern cross section shows the bivergent nature of the Alpine orogen. NPB: North-Penninic Bündnerschiefer, a metamorphosed clastic sequence that probably originated as late Mesozoic–early Tertiary turbidites. *C*. Index map, a reduced version of Figure 20.2*B*, showing location of sections in *A* and *B*. (*A* and *B* from Pfiffner et al. 2000, Figure 2*A*, *B*, with permission from the American Geophysical Union)

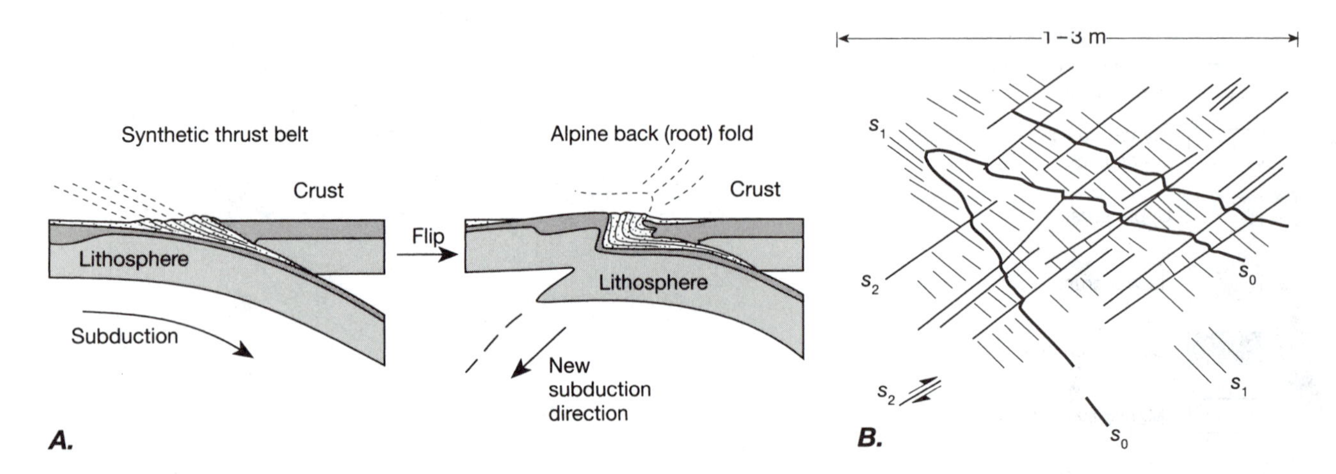

FIGURE 20.21 Hypothesis for the relationship among thrust belts, back folds, and subduction (see also Figure 20.34). (After Roeder 1972) *A*. Model for formation of multiply deformed structures such as the Alpine back fold by formation of synthetic thrust faults during collision of continents and subsequent back-folding after a "flip" of subduction polarity. *B*. Outcrop-scale structure from within the "root zone" showing two episodes of deformation, marked by cleavage s_1 and kink-bands s_2, and interpreted as reflecting deformation during synthetic thrusting and back-folding, respectively.

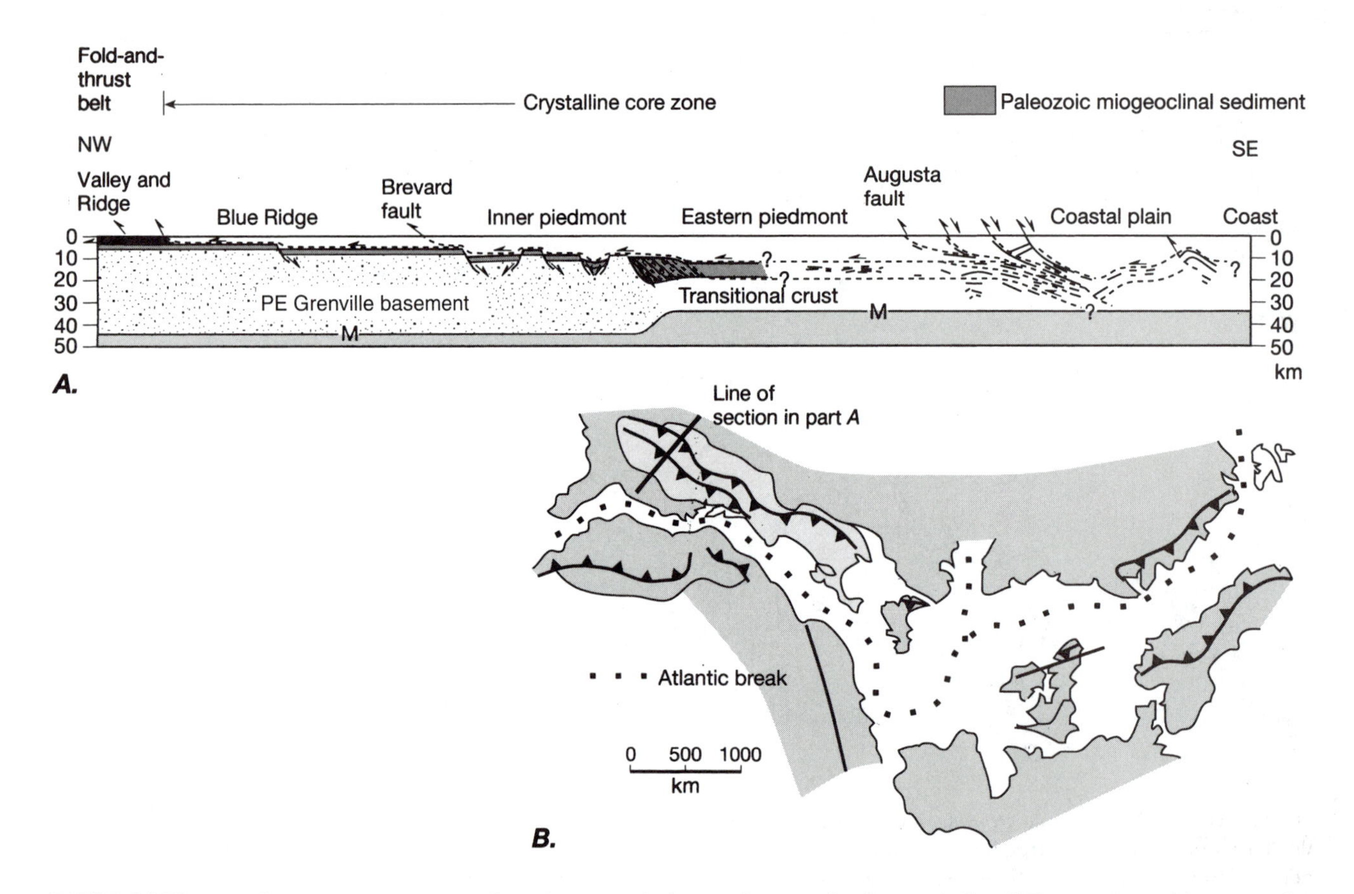

FIGURE 20.22 A. Schematic cross section of southern Appalachians, showing the "orogenic float." The continental basement is believed to extend beneath the crystalline core of mountain belt, including the Blue Ridge and Piedmont provinces. The basal decollement may either extend beneath the Coastal Plain or "root" beneath the eastern Piedmont. (After Cook et al. 1981) *B*. Index map, a reduced version of Figure 20.2C, showing the location of *A*.

low-grade metamorphic rocks are emplaced over high-grade metamorphic rocks, the latter of which may reach grades as high as ultra-high-pressure metamorphism. Such rocks have been exhumed from depths as far down as 120 km (see Figure 20.8). In some belts, much of the extension occurs parallel to the trend of the belt; in others the extension is at a high angle to the orogen.

In the eastern Alps, for example, the over-riding Austroalpine nappes have been tectonically stripped off the underlying Penninic core zone rocks, with the extension carrying the overlying rocks eastward, subparallel to the orogen toward the Pannonian basin, along a complex of roughly north-south normal faults and east-west conjugate strike-slip faults (Figures 20.23 and 20.26). The underlying metamorphosed Penninic nappes are now exposed in the Tauern window.

In the Norwegian Caledonides, on the other hand, the extension on mostly top-to-the-west, low-angle detachment faults (Nordfjord-Sogn detachment, Jotun detachment, Laerdal-Gjende fault zone) has been at a high angle to the orogen, exposing the underlying Western Gneiss region along the coast of Norway north of Bergen (Figure 20.24). The hanging wall blocks to these detachment faults include greenschist- to amphibolite-grade gneisses, Caledonian nappes, and younger, overlying, weakly metamorphosed rocks deposited in Devonian sedimentary basins (Hornelen Basin). Footwall rocks of the Western Gneiss region are predominantly Proterozoic granodioritic and granitic gneisses that have been metamorphosed to the eclogite and ultra-high-pressure facies and subsequently have been largely retrograded to amphibolite grade. Rocks preserving the highest-pressure metamorphic grades occur in the western part of the region and commonly contain coesite, a high-pressure form of quartz (see Figure 20.8). Some occurrences of diamond have also been found.

A similar structure is present in the Himalaya. The South Tibetan detachment fault is a north-dipping detachment structure that has been active at the same time as the main Himalaya south-vergent thrusting (Figure 20.25).

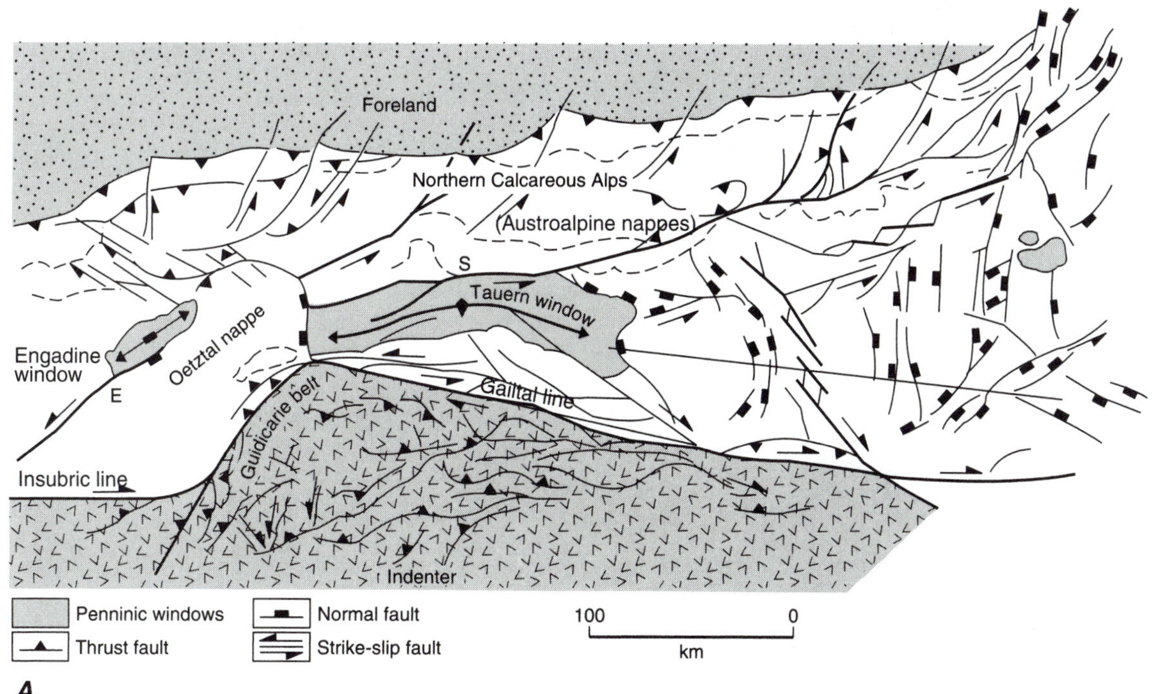

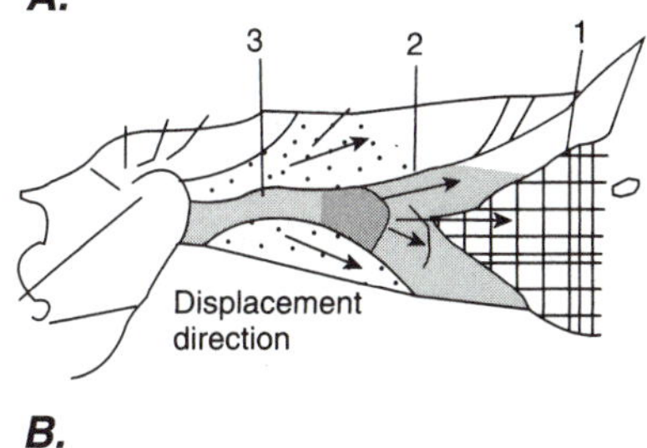

FIGURE 20.23 Lateral extension in the eastern Alps. *A*. Details of the faults and windows in the orogenic core zone of the eastern Alps. Penninic rocks in the Tauern window have been exhumed by a combination of uplift, erosion, and the tectonic removal of the overlying rocks on a complex of roughly north-south normal faults and east-west conjugate strike-slip faults (see the map of the Alps in Figure 20.26). *B*. Schematic diagram of the eastern Alps based on the data in *A* showing three main tectonic wedges (numbered 1–3) bounded by conjugate strike-slip faults. Arrows indicate the direction of displacement of the rocks relative to the orogen. (From Ratschbacher et al. 1991)

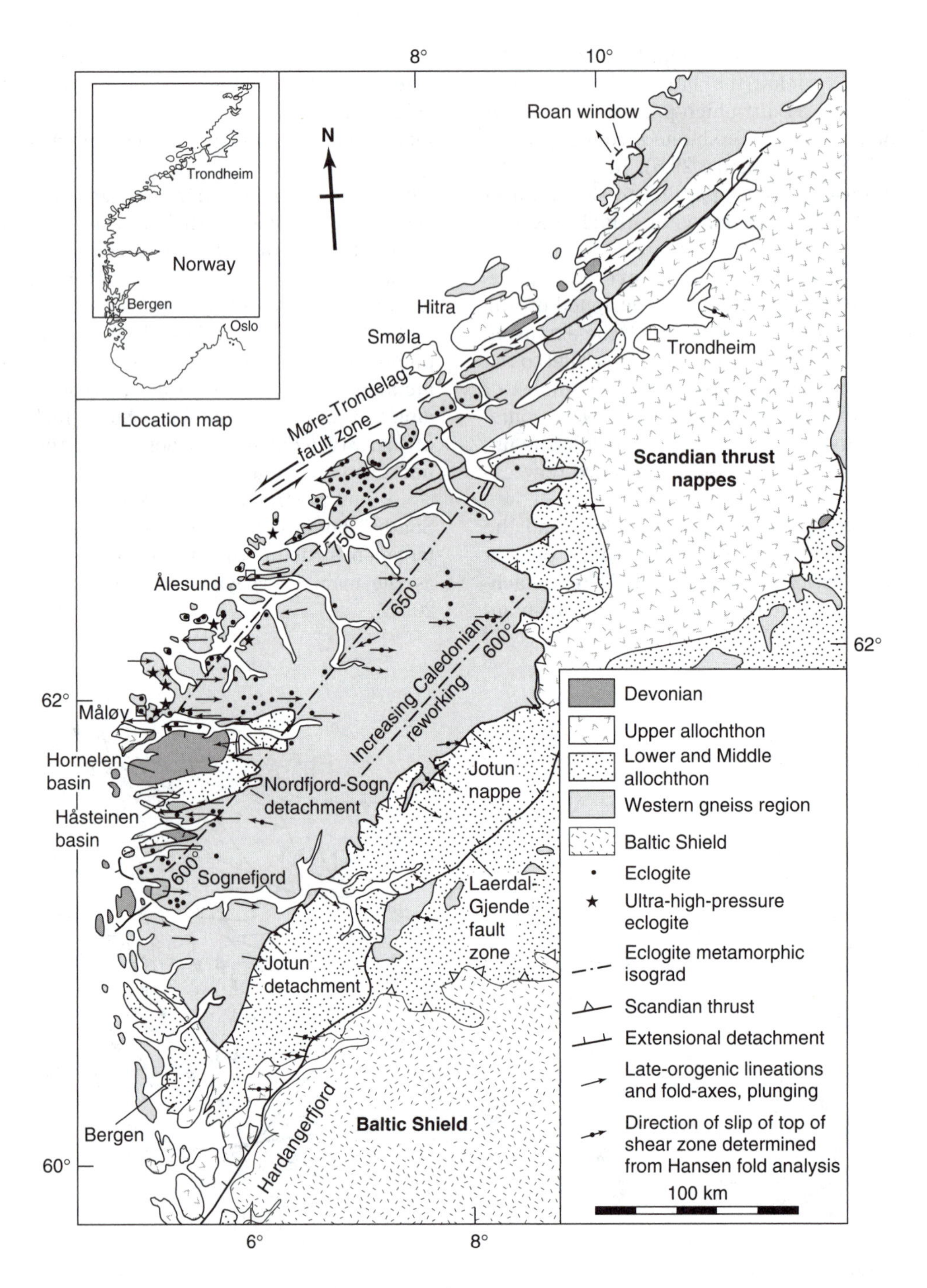

FIGURE 20.24 Low-angle detachment faulting exposing the Western Gneiss region of the Norwegian Caledonides showing exposures of eclogite and ultra-high-pressure eclogite metamorphic rocks and isotherms of eclogite metamorphism. Arrows show orientations of lineations and the fold axes used to determine the shear direction of the tops of shear zones by Hansen's method (see Section 13.8, Figure 13.35). The Nordfjord-Sogn detachment fault in the southwest separates the eclogite-bearing gneisses from the overlying allochthonous units that are metamorphosed mostly to the greenschist facies. The Jotun detachment was initially an east-directed thrust fault but was reactivated as a west-directed low-angle normal fault. (After Krabbendam and Dewey 1998; Hansen et al. 1967)

These observations have raised important questions about the process of orogenesis: How can large amounts of extensional deformation occur in what seemingly should be an environment of shortening deformation associated with continental collision? How do continental rocks become buried to depths of more than 100 km and then return to be exposed at the surface? We discuss some possible models for these processes in Section 20.12.

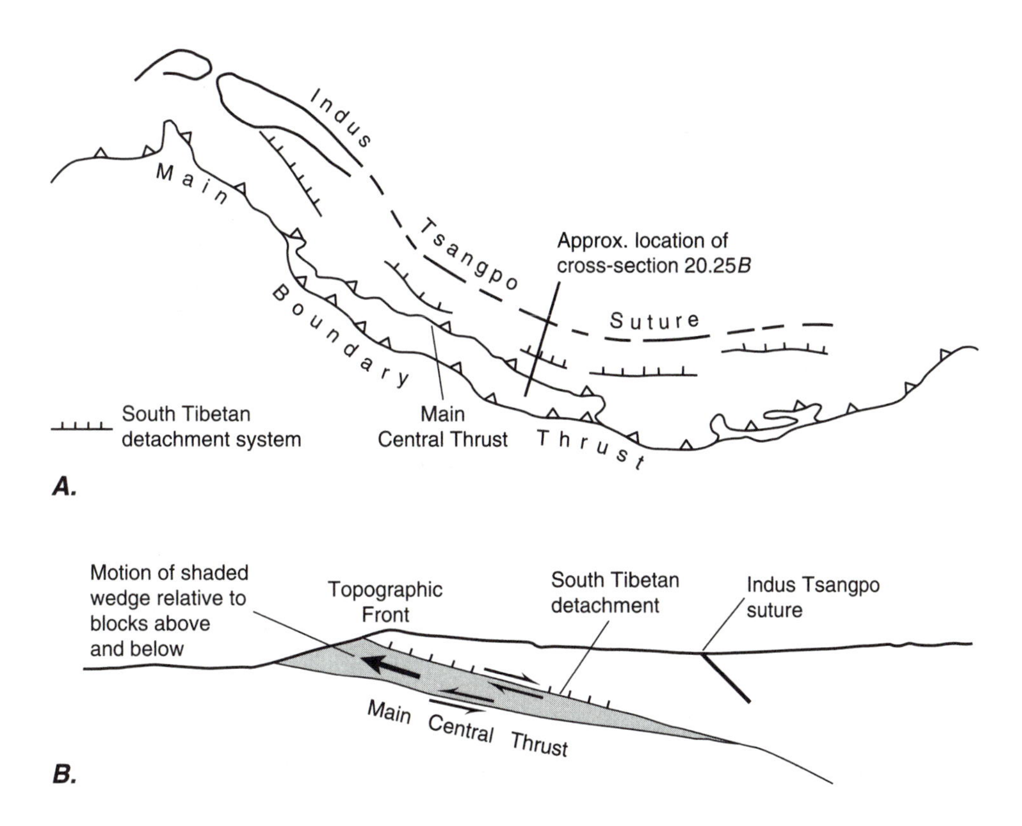

FIGURE 20.25 The South Tibetan detachment system. (After Burchfiel et al. 1992) *A*. Schematic map of Himalaya, showing the Main Boundary fault, Main Central thrust, South Tibetan detachment system, and Indus-Tsangpo suture. *B*. Schematic cross section showing South Tibetan detachment system active at the same time as the Main Central thrust. The shaded block moves up and to the left relative to rocks above and below.

20.7 HIGH-ANGLE FAULT ZONES

High-angle fault zones, which in general are moderately to steeply dipping and transect all the other features, are present in nearly every mountain belt (e.g., Figure 20.2). They vary from a few hundred meters to 10 kilometers in width, although most are relatively narrow and are marked by a band of well-developed mylonite. They extend typically for tens or hundreds of kilometers roughly parallel to the axis of the orogen, forming a major structural boundary along it. In Sections 18.5, 19.3(vi), and 19.5, we briefly discuss examples of such high-angle continental faults, including the Altyn Tagh fault in northern Tibet, the North Anatolian fault in northern Turkey, and the Insubric Line in the southern Alps (see Figure 19.1). Here, however, we review some of the characteristics of such fault zones.

Some shear zones offset geologic features that can be identified on both sides of the fault, so that estimates of the displacement are possible. In such regions, both dip-slip and strike-slip components of displacement have been identified. The lineations observed in such zones tend to be approximately parallel to the direction of demonstrable displacement.

Other fault zones separate regions for which the predominant metamorphic ages on either side of the fault may be radically different, or the geologic or paleogeographic history may be distinct. Such structures are illustrated by the Indus-Tsangpo fault zone north of the Himalaya (Figure 20.25*A*) and the Insubric-Guidicarie-Gailtal Line in the Alps (Figure 20.26; compare Figures 20.9 and 20.23). In such cases, no correlation can be made across the fault, and such zones probably represent **sutures**, where former oceanic crust has been subducted, leaving only remnants preserved along the sutures as ophiolites (see Figure 20.3).

Rocks within such shear zones are highly recrystallized and even mylonitic and often possess a different metamorphic grade, generally lower, but in places higher, than those in the surrounding region. In most places, diverse lithologies are present as discontinuous lenses—horses (see Figure 3.27)—within the fault zone. Mineral lineations are common, as are minor folds with the fold axes parallel to the mineral lineations. These lineations can be horizontal or steep, depending on the geometry of the deformation within the fault zone, which could be strike-slip or dip-slip, with or without a component of convergence or divergence normal to the fault (see Figure 12.26).

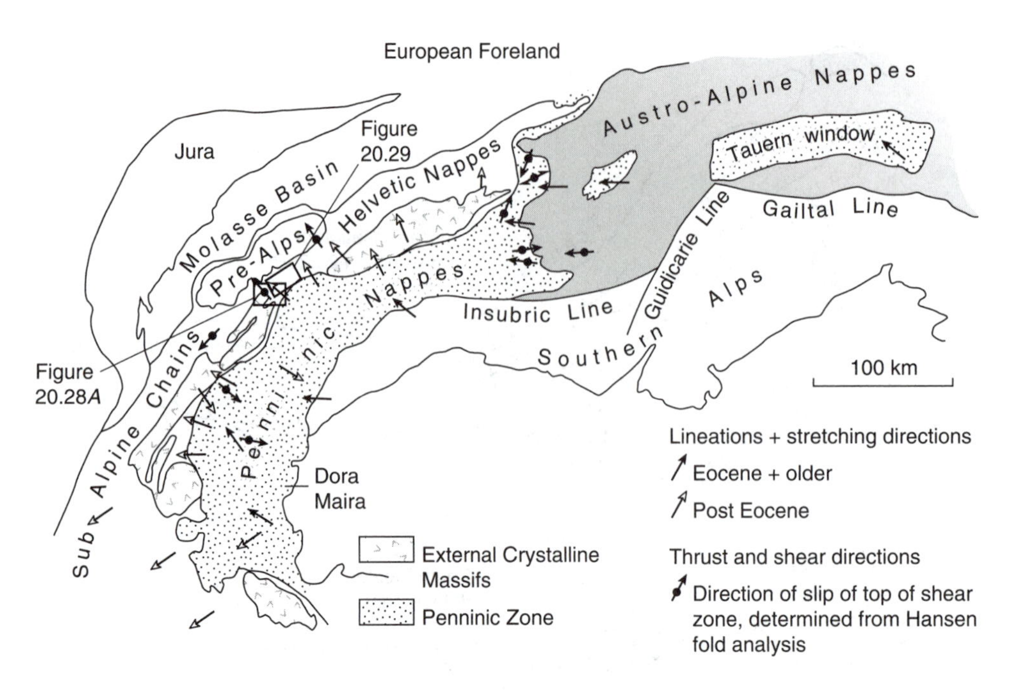

FIGURE 20.26 Generalized tectonic map of the Alps showing stretching lineations and the shear directions in shear zones of the upper layers relative to lower layers, as deduced from Hansen's method (see Section 13.8, Figure 13.35). Arrows pointing from the foreland in toward the core may reflect back-thrusting. (After Platt 1986; Hansen et al. 1967)

The timing of the deformation in these faults is variable. Some zones are late faults that sharply transect the pre-existing structures. Others, however, show deformational structures that are gradational into the structure of the surrounding rocks. Fold axial surfaces, for example, may be progressively deflected from the regional attitude into attitudes parallel to the zones. Such relationships make the timing of the deformation difficult to interpret.

Most of the steep shear zones that are well documented display a complex history of repeated movement with several different senses. Three examples serve to illustrate this characteristic. The Brevard zone of the southern Appalachians (Figures 20.18 and 20.22) extends for more than 500 km along strike. It appears to have an early history of thrust motion, followed by dextral strike-slip motion. The Insubric Line of the Alps (Figures 5.22, 20.9; 20.20*A*, *B*; and 20.26) also provides a good example of one of these zones. It is a steep fault zone that separates the Ivrea zone to the south from the Penninic zone of the Alps to the north. It also appears to have an early thrust history with a subsequent dextral strike-slip and normal faulting, so the displacement history is complex and the cross section varies along its length. The Mother Lode fault system of the Sierra Nevada, California (Figure 20.19*B*, *C*), also represents an example of such faults. This steeply east-dipping system of faults in part represents the traces of highly deformed thrust faults displaying dominantly west-over-east dip-slip motion that subsequently has been folded and reactivated during an episode of east-over-west back-folding, followed by recent dextral strike-slip and normal faulting. Both the Mother Lode fault system and the Insubric Line contain the remnants of a suture in their respective orogens.

Recent developments in radiometric dating have made it possible to separate various phases of deformation along fault zones by dating minerals that crystallized during those separate phases. This analysis uses such techniques as the K-Ar decay scheme (generally the $^{40}Ar/^{39}Ar$ method) for dating of clay minerals or micas, and various thermoluminescent methods[5] for dating of fault gouge itself.

20.8 MINOR STRUCTURES AND STRAIN IN THE INTERPRETATION OF OROGENIC ZONES

Our efforts to understand the origin of orogenic cores and their relationship to plate tectonic events prompts such questions as: In what directions were the rocks transported during the deformation? What is the distri-

[5]Thermoluminescence refers to the light emitted by a material when it is heated. If a material is exposed to radiation from, for example, the decay of radioactive elements, the radiation creates atoms that have excited metastable electron states. Heating allows the electrons to return to their normal states, and they emit visible light in the process. The longer the exposure to the radiation, the more atoms with excited electron states are created, and the more intense is the emitted light when the material is heated. The sensitivity of the material is calibrated by exposure to a known radiation source, and the natural radiation levels can be determined from the measurement of the concentration of radioactive elements in the material.

bution of strain through the rocks? What large-scale pattern of flow gave rise to the observed structures and strain distributions, and how can these features be explained in terms of tectonic processes? These questions bear on the way that orogenic cores evolve, and answering them ultimately may aid in correlating core zone evolution with plate kinematics.

In this section, we present examples of how the analysis of minor structures helps understand the history of deformation in orogenic core zones. Our focus here is on kinematic analysis of folds (discussed in Chapter 13), foliations (discussed in Chapter 14), and mineral fibers (discussed in Section 14.6(v) and 15.4(i); and strain analysis (Chapter 15) and kinematic analysis of crystallographic preferred orientations (discussed in Section 17.7). We cannot aim here to give a comprehensive account of the application of structural analysis to the study of the kinematics of orogenic zones. Rather we try to illustrate useful research approaches through brief accounts of some examples in which such analysis has proved fruitful.

i. Kinematic Analysis of Folds

Analysis of folding, in some cases, can provide good constraints on the kinematics of regional deformation. For example, the application of the Hansen method of determining the slip direction (Section 13.8) to deformed rocks in the Alps shows that in the outer areas of the thrust nappes, the shear direction is complex. In some places, it is transverse to the orogenic belt, as one might generally expect. In the central region of the orogenic core, however, many shear directions tend to be parallel to the axis of the orogenic belt (Figure 20.26). These results are similar to thrust and shear directions inferred from stretching lineations throughout the Alps (Figure 20.26). Similar results have been found in the Norwegian Caledonides (see Figure 20.24) and the northern Appalachians. These longitudinal shear directions could be accounted for by flow models in which localized collision of irregularities in continental margins results in lateral flow away from the collision zone. Such models also have been proposed for the Himalayan-Tibet region, as discussed in Subsection 20.12(x) (see Figure 20.35).

ii. Kinematic Interpretation of Foliations

Because in general, foliations are approximately parallel to the plane of flattening of the finite strain ellipsoid (Sections 14.2, 15.2, 15.3, and 15.4(ii)), they can be used to infer a shear sense in a fault zone if the orientation of the shear zone is also known. The intersection of the foliation plane and the shear plane is a line approximately perpendicular to the direction of shear, and the sense of rotation (clockwise or counterclockwise) through the acute angle from the foliation plane to the shear plane defines the sense of shear on the shear plane (Figure 20.27*A*).

This relationship is confirmed, for example, by studies of the foliation developed in sediments in southern Alaska that have been deformed in the accretionary prism above the subduction zone (Figure 20.27*B*). Earthquake focal mechanisms and Pacific plate reconstructions determine the relative plate motion independently, and it indeed lies perpendicular to the intersection of the foliation with the thrust plane. Such a relationship between deformation structures and plate kinematics at a subduction zone may not hold, however, if the subduction direction is very oblique to the subduction zone. In those circumstances, the shear associated with subduction may be partitioned into an orthogonal slip component at the subduction zone and a parallel slip component on a strike-slip fault within the arc complex (see Figure 19.38).

iii. Finite Strain Analysis

The Morcles nappe is one of the Helvetic nappes along the northwestern edge of the Swiss Alps, southeast of the Lake of Geneva (Figures 20.9 and 20.26). It is one example in which the regional distribution of strain has been determined (Figure 20.28). The map (Figure 20.28*A*) shows the distribution of the ($\hat{s}_1$, $\hat{s}_2$) axes with the horizontal projection of $\hat{s}_1$ plotted with the correct trend. Note that $\hat{s}_1$ is commonly perpendicular to the fold axes (indicated by the arrows) and to the general trend of the orogen and thus is parallel to the expected direction of displacement of the nappe relative to the European foreland, the so-called **direction of tectonic transport**. In places, however, $\hat{s}_1$ is parallel to the fold axes and perpendicular to the direction of tectonic transport. The exposure of the Morcle nappe (Figures 20.28*A*) is in effect an inclined section through the nappe with the deepest parts of the nappe exposed in the west and the shallowest parts exposed in the east. Thus projection of the exposures onto a vertical plane gives a cross section of the nappe (Figure 20.28*B*). This cross section shows that the strains are highest near the base of the nappe where the $\hat{s}_1$ axis is oriented at small angles to the subhorizontal thrust. Higher in the nappe, the strain magnitude decreases progressively, and the plunge of the $\hat{s}_1$ axes becomes steeper.

The total strain accumulated throughout the deformation history is recorded in the final finite strain ellipsoid. This history includes a component of initial flattening associated with sedimentary compaction, followed by multilayer buckle folding during which the limestones behaved as the competent members, followed by inhomogeneous simple shearing associated with the emplacement of the nappe. Thus the finite strain ellipsoids do not record just the process of nappe emplacement, and the discrimination of the different phases of deformation requires additional independent information that cannot be inferred from just the finite strains.

iv. Instantaneous Strain Analysis: Kinematic Interpretation of Mineral Fibers

Mineral fibers in oriented overgrowths on pyrite concretions have been used to infer the extension history over a region of the Swiss Alps (Figure 20.29) that overlaps

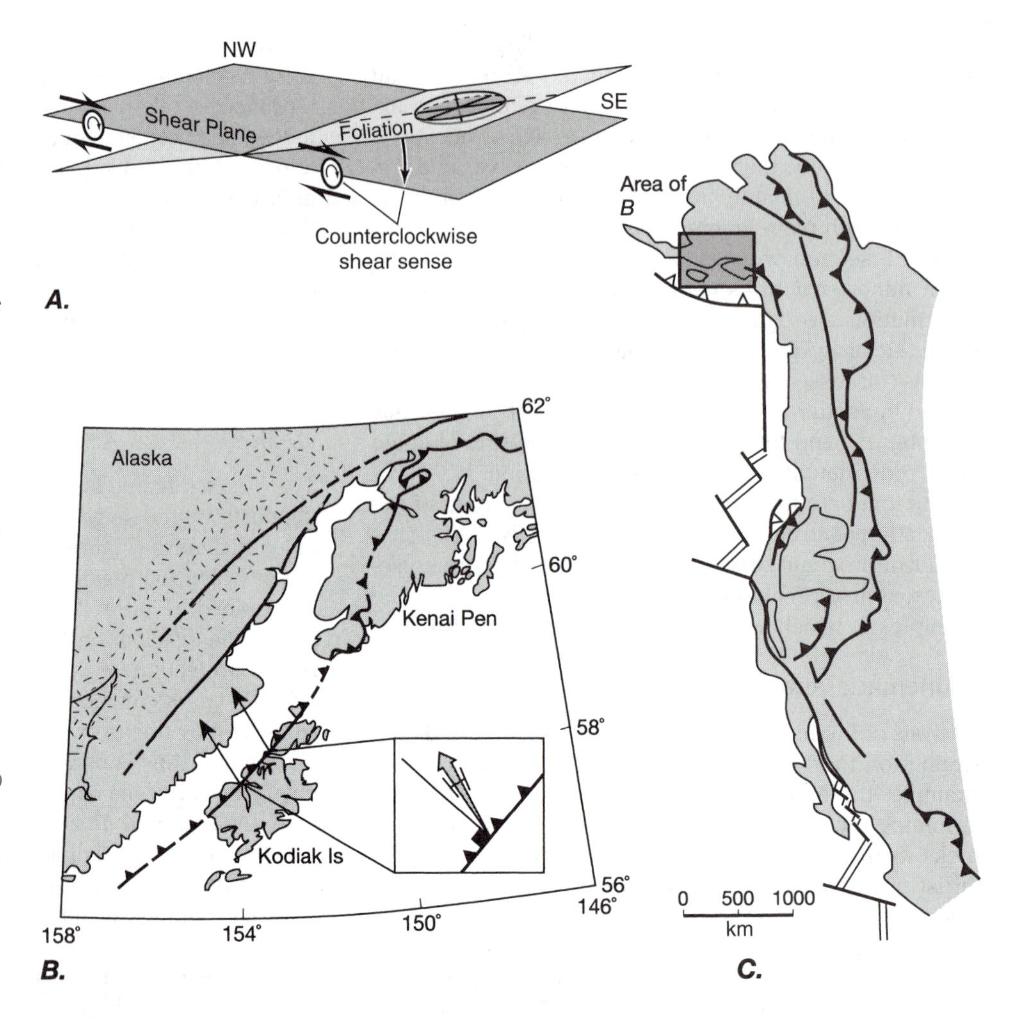

FIGURE 20.27 Relationship between foliation and the shear plane. *A*. The foliation and the shear plane intersect in a line that is perpendicular to the direction of shearing. The sense of rotation from the foliation plane through the acute angle to the shear plane defines the shear sense on the shear plane, which in this example is clockwise as viewed in the diagram. *B*. An example application of this relationship in the field, Kodiak Island, Alaska. The generalized map shows arrows indicating the relative plate motion deduced from analysis of Pacific plate motions. The inset shows that relative motion (thick arrow) and the deduced shear direction from foliation-shear plane relationship (thin line). See *C* for location. (After Moore 1978) *C*. Index map, a reduced version of Figure 20.2*A*, showing the location of *B*.

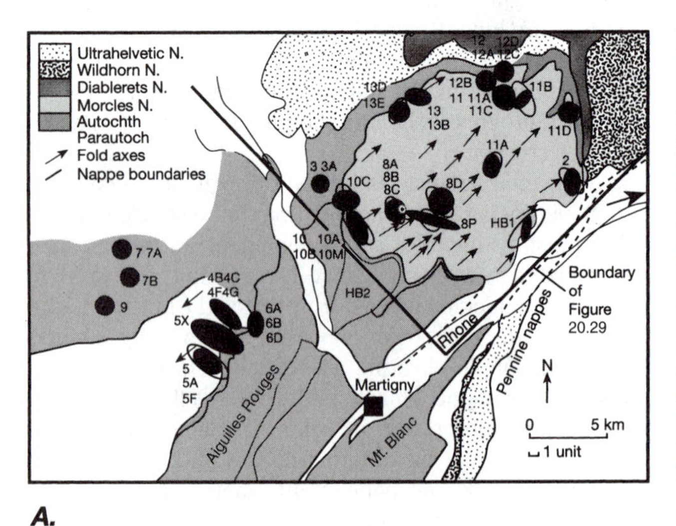

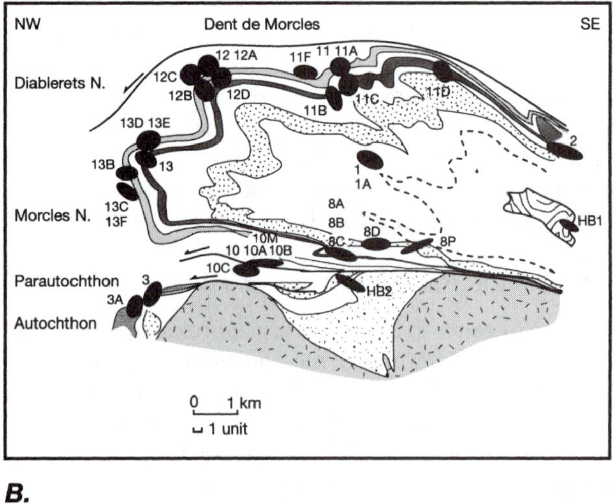

FIGURE 20.28 Distribution of strain in the Morcle Nappe in the Alps of western Switzerland. See Figure 20.26 for location. (From Siddans 1983, figures 2 and 3) *A*. Geologic map of the Morcles Nappe and its surroundings showing horizontal sections through the strain ellipsoid (solid black ellipses) and the $(\hat{s}_1, \hat{s}_2)$ section of the finite strain ellipsoids projected onto the horizontal with $\hat{s}_1$ oriented parallel to its correct bearing (open ellipses). *B*. Composite down-plunge projection through the Morcles Nappe showing the distribution of the $(\hat{s}_1, \hat{s}_3)$ section through the finite strain ellipsoids.

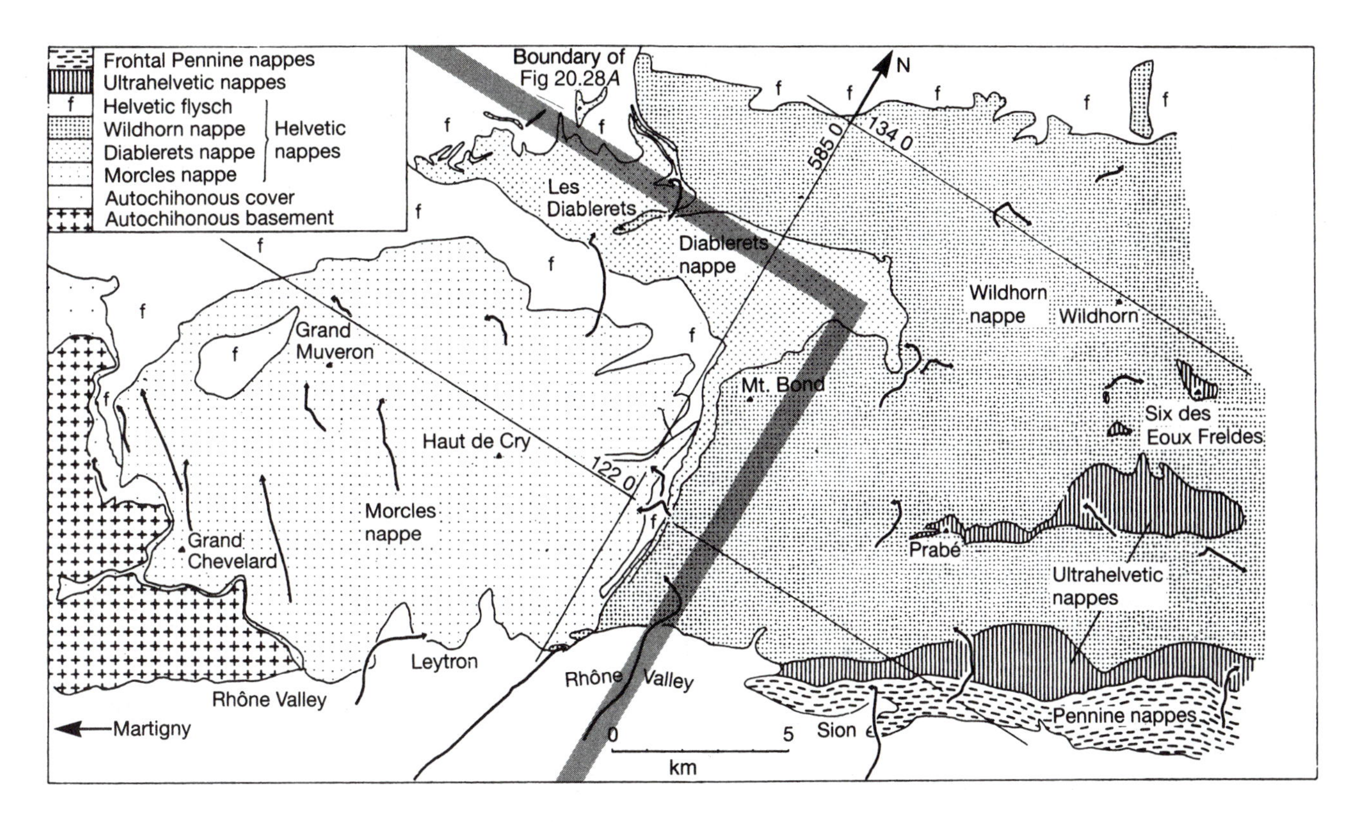

FIGURE 20.29 History of the maximum instantaneous extension in the western Helvetic nappes of the Swiss Alps as deduced from fibrous overgrowths on pyrite crystals. Lines are parallel to the directions of maximum incremental extension $(1 + \zeta_1)$ and line lengths in any given direction indicate magnitudes of the extension in that direction. The lines do not indicate the displacement of material points in the rock. Triangular dots are at the youngest end of the line and are plotted at the location where the data were measured. See Figure 20.26 for location. (Durney and Ramsay 1973, figure 22)

with Figure 20.28. The lines plotted at the triangular dots show the history of the orientation of the maximum instantaneous extension $\hat{\zeta}_1$ at that point, with the youngest increment of the extension history plotted at the dot. The length of any segment of the line is proportional to the magnitude of the extension in that direction. Note that the lines do not indicate the direction and amount of displacement of material points in the rocks.

Lower in the nappe (in the west), the extension history recorded by the fibers is simple, and the instantaneous extension directions $\hat{\zeta}_1$ indicated by the fiber directions (Figure 20.29) are roughly parallel to the maximum extension direction $\hat{s}_1$ of the finite strain ellipsoids (Figure 20.28*A*). Structurally higher in the Morcles nappe (toward the east), and still farther east in much of the overlying Wildhorn nappe, the fibers indicate a change in the orientation of the instantaneous extension $\hat{\zeta}_1$ during the deformation. Early in the deformation, $\hat{\zeta}_1$ is oriented roughly northwest-southeast in the Morcle nappe and approximately north-south in the Wildhorn nappe, directions that are at a high angle to the orogenic core zone (cf. Figure 20.26); late in the deformation, however, especially in the Wildhorn nappe, the fibers record a shift to an orientation that shows significant components of east-west extension, roughly parallel to the orogenic core. This consistent pattern over a large area suggests a fundamental change in the geometry of the deformation, which we discuss further in Section 20.11(i). Such details of deformation history have not yet been incorporated, however, into a unified model of the emplacement of these nappes during the orogeny.

Thus the information about the history of the instantaneous strain obtained from the fibers helps to separate the different phases of deformation, for which only the net result is recorded by the finite strain data, and thus it provides more stringent constraints on any model that attempts to account for the emplacement of these masses of rock.

v. Kinematic Analysis of Crystallographic Preferred Orientations

Along the basal thrusts of ophiolites, the ultramafic rocks or underlying metamorphic rocks exhibit a foliation oriented approximately parallel to the thrust surface. Fabrics of olivine and orthopyroxene in the ultramafic rocks and of quartz in the underlying metamorphic rocks in some cases develop an orientation that can be related to sense of shear (see Sections 17.7 and 17.8). If these basal thrusts represent fossil plate boundaries, as suggested in

Sections 5.5 and 19.5 (see Figures 5.20*C* and 19.43), these fabrics could be used to constrain the relative plate motions along the boundaries.

Similar fabrics are developed in quartz-rich or calcite-rich mylonitic rocks along many major thrust faults, and these have been used to constrain the models of emplacement of the overlying thrust sheets.

20.9 TECTONICS, TOPOGRAPHY, AND EROSION

The relationship between tectonics and topography has long been a subject of debate. Since the development of plate tectonic theory, however, it has become clear that tectonic forces associated with collisions are of major importance in the vertical uplift associated with the formation of the topographic highs we know as mountain belts. Before discussing the evolution of topography, however, it is important to define exactly what we mean by "uplift," because there are three distinct but related concepts. **Surface uplift** is the vertical displacement of the Earth's topographic surface with respect to the geoid. (We could also use mean sea level as a reference, if account were taken for eustatic sea level fluctuations due, for example, to the accumulation or melting of continental ice caps and glaciers.) **Rock uplift** is the vertical displacement of a given volume of rock with respect to the geoid. **Exhumation** is the vertical displacement of a given volume of rock with respect to the topographic surface. Thus exhumation is just the difference between rock uplift and surface uplift, that is,

$$\text{exhumation} = \text{rock uplift} - \text{surface uplift}$$

A similar relationship holds for the rates of these different variables. If the surface uplift equals the rock uplift for a given volume of rock, then the rock volume does not approach the surface and the exhumation is zero.

Rock and surface uplift can occur together as a result of isostatic uplift or tectonic crustal thickening. Processes that produce such thickening include underplating at a subduction zone or horizontal shortening and vertical thickening of a portion of the crust through thrust faulting or folding. In such cases, exhumation occurs only if material is removed from the surface by erosion or by tectonic denudation on low-angle normal faults that in effect lengthen and thin the crust. Thus, when

$$\text{exhumation} = \text{erosion} + \text{tectonic denudation}$$

then

$$\text{rock uplift} - \text{surface uplift} = \text{erosion} + \text{tectonic denudation}$$

and the rock uplift rate must exceed the surface uplift rate.

Rock uplift without appreciable surface uplift can occur by convective rise of rock in a diapiric structure such as a gneiss dome, or during a return flow of material up a subduction zone. Such a flow can occur if material that is dragged down the subduction zone is blocked by a constriction and forced to flow back up. In these cases, the exhumation equals the rock uplift.

As a rule, high topography is supported isostatically by a thickened root of low-density crustal material that extends down into the underlying higher-density mantle. In effect, the crust floats in the mantle much as an iceberg floats in water. The height of the iceberg above the water surface is supported below the surface by the displacement of higher-density water with the lower-density ice. In orogenic belts, isostatic surface uplift can be caused in two ways:

1. Thrust faulting and folding during orogeny thicken the crust and cause a surface uplift that is supported isostatically by a deepening crustal root.
2. A decrease in density of the mantle immediately below the crust causes isostatic uplift of the surface.

Such a density decrease can occur by **delamination** of the lower crust or the mantle lithosphere, a process whereby dense lower crust or mantle lithosphere separates from the shallower crust and sinks into the Earth's interior, to be replaced by hotter, less-dense asthenospheric mantle. A comparable decrease in mantle density can also occur when rifting thins the lithosphere and hot, less-dense mantle (asthenosphere) is emplaced immediately beneath the continental crust. It is this process that results in surface uplift to form mountains along rifted margins such as those in East Africa. Topography can also be supported to some extent by the bending strength of the crust or by dynamic stresses that arise from flow in the mantle. The scale of orogenic zones, however, is generally large enough that these processes are secondary to the isostatic support of topography.

A major factor in the surface uplift that results in the topography of mountainous areas is the intimate interplay between erosion and tectonically driven rock uplift. Tectonic denudation also can be very significant; we discuss this process further in Section 20.12(ii). Erosion rates in mountainous areas are on the order of tenths of a millimeter per year to several millimeters per year, and they tend to increase with increasing topographic relief. Rock uplift rates for orogenic zones are of the same order. The higher the topography, the faster the erosion, and the faster the erosion, the more rock uplift and exhumation can occur.

High-grade Barrovian metamorphic rocks that have come from depths of 20 to 40 kilometers (see Figure 20.8) are now exposed at the surface in most mountain ranges. In some places, ultra-high-pressure metamorphic minerals are found in rocks now at the surface (e.g., Figures 20.9 and 20.24), showing that the rocks have been carried down to depths exceeding 100 kilometers and subsequently have been returned to the surface. Thus vast thicknesses of rock must have been removed to expose these deep metamorphic rocks. In order to form the familiar mountains of collisional orogenic belts such as the Alpine-Himalayan belt, it is apparent that during an orogenic cycle, the rock uplift rate must initially exceed the rate of erosion, so the surface elevation increases. As the surface uplift increases, however, the erosion rate increases. Thus in principle a steady state eventually can develop for which the surface uplift rate is zero and the erosion rate, and thus the exhumation rate, approximately equals the rock uplift rate. Eventually, the tectonic forces driving orogeny decrease, and the rock uplift rate decreases below the erosion rate, resulting in a negative surface uplift rate and thus a gradual decline in the topographic elevation. This state is exemplified by much of the core zone in the Appalachian Mountains, which are little more than gently undulating hills in those places unaffected by recent glaciation.

In some collisional belts, however, high-grade metamorphic rocks are not exposed, implying that the exhumation rate must have been relatively low. Examples include the Canadian Appalachians, the Altaids of western Asia, and the Urals. Because the topography of these areas is not exceptionally high, the implication is that not only was there a smaller difference between the rock uplift and surface uplift rates, but that both rates must have been considerably lower than in other orogenic belts where high-grade metamorphic rocks are exposed. The reasons for this contrast are unclear, but they may have to do with the extent to which the down-going continent underthrusts the over-riding continent in a continent-continent collision and thereby drives the uplift.

Our understanding of the relationship of tectonics and topography is making rapid progress. New means of dating erosional surfaces provide us the ability to measure long-term erosion and uplift rates, and new means of measuring ongoing surface motions, such as using the Global Positioning System (GPS) and Interferometric Synthetic Aperture Radar (InSAR), provide data on current motions. These techniques are bringing new insight into the relationship between present plate motions and their interactions with the continental crust.

20.10 TECTONICS AND METAMORPHISM

The sequence of metamorphic assemblages found in metamorphosed terranes is a reflection of the temperature-depth profile at the time of metamorphism. Radiometric dating techniques can provide a date for the time at which individual minerals in a metamorphic rock passed through the **closure temperature** for a specific radioactive decay scheme. Below the closure temperature, the elements and isotopes involved in the decay are effectively frozen into the mineral grain, thereby starting the radiogenic clock. The closure temperature is different for different minerals and for different decay schemes, so as a rock cools, the time since it passed through the different closure temperatures can be preserved if the appropriate minerals are present. The combination of such mineral assemblage studies and radiometric age determinations has provided constraints on the evolution of pressure and temperature with time experienced by a specific body of rock, the so-called **PTt paths**. These paths for a given volume of rock in turn provide constraints on the tectonic history.

Figure 20.30 shows a set of pressure-temperature paths (black lines labeled A through F) plotted on a background

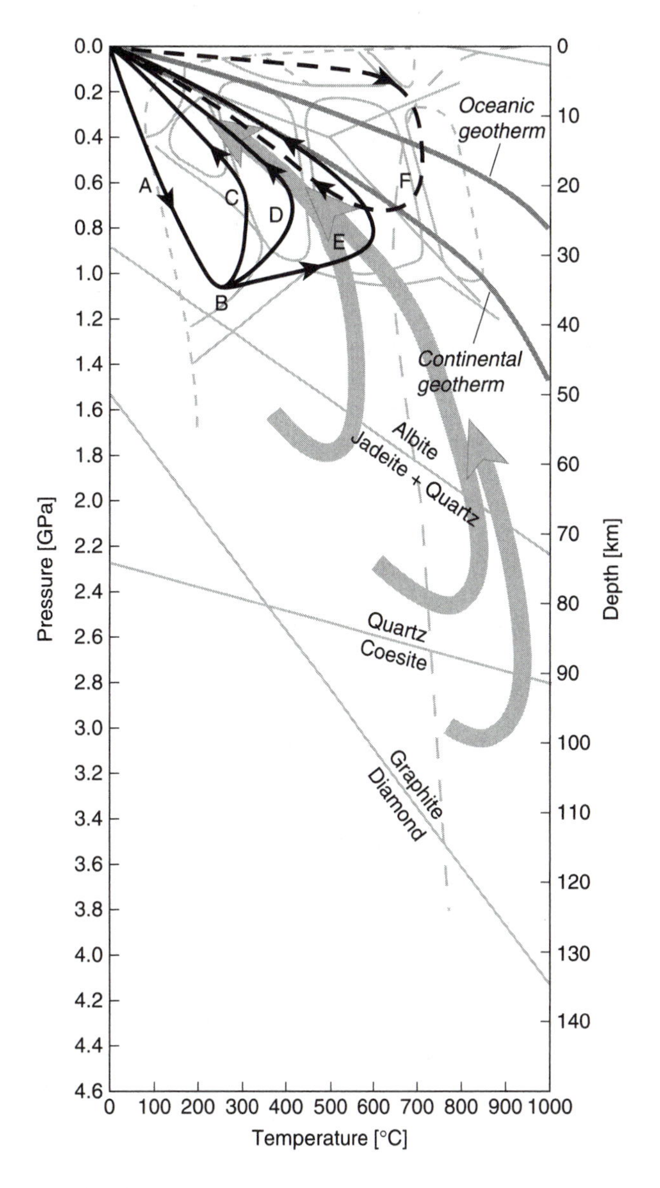

FIGURE 20.30 Diagram showing several schematic pressure-temperature-time (PTt) paths (black solid and dashed lines labeled A through F) for rocks undergoing metamorphism; time increases along the path in the direction of the arrows, but no time scale is implied. Thick gray arrows indicate PTt paths for very high- and ultra-high-pressure metamorphic rocks from the Western Gneiss region of the Norwegian Caledonides. See the text for discussion. Petrogenetic grid and pressure-depth scale from Figure 20.8. (After England and Thompson 1984; Krabbendam and Dewey 1998; Wakabayashi 2004)

of the petrogenetic grid (compare Figure 20.8). These paths, which are idealizations of paths that have been determined from metamorphic rocks, essentially trace the burial and exhumation cycles for individual volumes of rock. Time increases nonlinearly along each path in the direction of the arrows, but is not specifically plotted here. Paths A–E show cycles characterized by an initial rapid pressure increase along path A from the upper-left corner of the graph to point B, which is associated with rapid burial at temperatures that are markedly lower than the steady-state continental geotherm. The rapid burial is followed for paths C, D, and E by pressure decrease, indicating exhumation, and varying degrees of gradual heating. Peak temperatures for each path are reached after the peak in pressure and during the initial phase of pressure decrease. Subsequent cooling with a further pressure decrease completes the cycle.

Such paths are typical of subduction zone and continent-continent collision environments. For example, path A, which implies rapid burial to the maximum depth at B, could occur in a subduction zone where cooler rocks are subducted to great depths more rapidly than they can heat up by thermal conduction. The rock might follow path C if exhumation begins very soon after the rock reaches the maximum depth, in which case there would not be time for temperatures to equilibrate to the values for the steady-state geothermal gradient at that depth. This could occur, for example, if the rock were part of a return flow up the subduction zone, almost retracing its downward path. Alternatively, the rock might follow path D or E if it is exhumed more slowly, as might happen if it was underplated onto the over-riding plate during a continent-continent collision and was only gradually exhumed. In case E, the rock undergoes minimal exhumation for a lengthy period while the temperature in the rock increases and approaches the steady-state geotherm. Calculations for the time it would take rocks to heat up to the steady-state geotherm from point B along a path such as E during only small amounts of pressure decrease suggest that the peak in temperature may come 10 million years or more after the peak in pressure.

Other PTt paths can show a cycle with the opposite sense of evolution (curve F dashed in Figure 20.30). In this case, a rapid heating at a low, relatively constant pressure, that is, in an environment of abnormally high geothermal gradient, is followed by a pressure increase during burial and then a cooling during exhumation. Such rapid heating events are generally the result of advection of heat by magmatic intrusion into the shallow crust, although the crust can also be abnormally heated by magmatic underplating of the crust or delamination of the mantle lithosphere from the crust, resulting in a replacement of lithospheric mantle by upwelling hot asthenospheric mantle. The increase in pressure subsequent to the heating is generally of tectonic origin when, for example, the heated shallow crust is over-ridden and rapidly buried by thrust sheets during a collisional event. Subsequently, the rocks cool toward a steady-state continental geotherm, and then pressure and temperature decline along the steady-state geotherm during exhumation. This type of history has been inferred for back-arc areas, which are initially heated by magmatic intrusions from the arc or underplated by magmas during back-arc extension. A subsequent collision at the arc shortens and thickens the orogen, burying the high-temperature rocks, which are then gradually exhumed as the orogen matures. A wide variety of PTt paths have been found, and we can imagine many different tectonic scenarios that might account for these paths. The interpretations are not unique but must be constrained by other available geologic observations.

The assemblage of minerals that make up a metamorphic rock can vary, depending on the pressure and temperature conditions through which the rock passes, the time it spends at the different conditions (the PTt path), and the relative rates of the metamorphic reactions. The reaction rates determine how quickly the different minerals grow and, once grown, whether they survive conditions outside their stability fields. Thus path A-B-C might give rise to surface exposure of blueschist metamorphism, path A-B-D to blueschist overprinted by Barrovian metamorphism, and path A-B-E to Barrovian metamorphism. Path F would give Buchan metamorphism overprinted by regional Barrovian metamorphism.

Only parts of such PTt paths may be preserved in the mineral assemblage due to the replacement of earlier-formed minerals by later ones, a process referred to as **overprinting**. In general, retrograde metamorphism during decreasing temperature-pressure conditions is less efficient at overprinting earlier metamorphic minerals than prograde metamorphism during increasing temperature-pressure conditions. The reason for this difference is in part because prograde metamorphic reactions tend to be dehydration reactions. Because the water released as a product of these reactions tends to migrate out of the rock, the retrograde hydration reactions must occur under relatively dry conditions, and without sufficient water, the reactions cannot run to completion. Moreover, the reaction rates are the highest at the highest temperatures, so the reactions that create the high-grade mineral assemblages run to completion most rapidly. As the rocks cool down, the retrograde reactions occur at lower temperatures and thus tend to progress more slowly, so the overprinting by lower-grade assemblages may not be complete.

The wide gray arrows in Figure 20.30 show inferred PTt exhumation paths for high- to ultra-high-pressure metamorphic rocks (e.g., from the Norwegian Caledonides, the Alps, and other places where such rocks are exposed). These rocks have been carried down during a continent-continent collision to exceptional depths for crustal rocks and have been exhumed so rapidly that the mineralogic evidence for the high pressure has survived. For the deepest rocks in the Norwegian Caledonides, radiometrically determined ages have determined an exhumation rate from pressures of approximately 3 GPa to 1 GPa to be about 6 to 9 mm/y. Exhumation rates for ultra-high-pressure metamorphic rocks in the Alps have been determined to be in excess of 20 mm/y. An exhumation rate of 10 mm/y can lead to exhumation of rocks from depths of 100 km up to the surface in only 10 million years. The processes that provide such astounding rates of exhumation are not well understood (see the discussion in Section 20.12(ii)).

These considerations of metamorphic mineral assemblages and PTt paths are of major importance in constructing and testing models of orogeny, which is the focus of our discussion in Section 20.12.

20.11 SIMPLE MODELS OF OROGENIC DEFORMATION

One of our objectives in conducting a structural study of an area is to understand the relationship between local small-scale structures and large-scale tectonic processes, up to and including plate tectonics. What is the structure of the orogenic core zones at depth? How far down do the structures extend that we observe at the surface? What are the tectonic environments in which these structures form, and how can we infer these environments from the characteristics of the structures? The significance of the local structures for interpreting the larger scale can emerge only from the integration of detailed studies over large areas, which requires much time-consuming fieldwork and analysis. After decades of study by hundreds if not thousands of geologists in many mountain ranges all over the world, and with the use of numerical modeling on high-speed computers, models for the processes of orogeny are starting to emerge. The models, however, are still oversimplified, and much work remains to be done. The difficulty of relating details of structural geology to large-scale tectonic motions remains a major fascinating problem of orogenic belts.

It is easiest to interpret field observations from areas of good exposure because the record is more complete. The models of regional tectonics derived from such areas can then provide the framework for interpreting the structure of less-well-exposed regions. Because the Alps, the Caledonides, and the Himalaya have been scraped clean by continental glaciation in the recent geologic past, exposure is exceptionally good, and because these orogens are relatively accessible, much of the work on these problems has come from studies in these areas.

i. Simple Patterns of Ductile Flow and the Direction of Tectonic Transport

The pattern of flow of rock during ductile deformation determines the orientation and distribution of strain in the rocks and the types and orientations of the structures that form. These flow patterns relate the strain and the associated structures to the tectonic motions that generate the deformation. The flow pattern is best described by the **streamlines**, which are everywhere tangent to the velocity vectors of material points. In studying the tectonics of orogenic zones, the velocities are generally specified relative to the adjacent stable continental foreland, which is considered fixed. In this case, the orientation of the streamlines is commonly referred to as the **direction of tectonic transport** (compare Figure 20.26). We consider three idealizations of the patterns of flow that could occur: convergent flow, divergent flow, and shear flow. The formation of structures such as folds in the rocks is the result of inhomogeneities in the flow field, which are not considered explicitly in these models.

In convergent flow (Figure 20.31*A*), all streamlines converge in the downstream direction, and for the deformation to proceed at a constant volume, the velocity must increase in the downstream direction. Any volume of rock is subjected to a coaxial deformation, and the strain path along a given streamline plots in the constrictional strain field of the Flinn diagram (Figure 12.21). Thus the $\hat{s}_1$ (maximum lengthening) direction of the finite strain ellipsoid is oriented essentially parallel to the streamlines (Figure 20.31*A*). We therefore expect stretching lineations and fold hinges (cf. Figure 14.20) formed during the deformation to be parallel to the streamlines. Because in general material lines rotate toward the $\hat{s}_1$ direction during a coaxial deformation (compare Figure 12.14), we also expect older lineations defined by material lines to rotate toward parallelism with the streamlines.

In divergent flow (Figure 20.31*B*), all streamlines diverge from one another in the downstream direction, and the material velocity must decrease in that direction if material volume is to be conserved. The deformation is again coaxial, but the strain path along a given streamline in this case lies in the flattening strain field of the Flinn diagram (Figure 12.21), and the $\hat{s}_3$ (maximum shortening) direction of the finite strain ellipsoid is essentially parallel to the streamlines. Stretching lineations and fold axes formed during the deformation are perpendicular to the streamlines, and lineations defined by material lines are rotated toward the $\hat{s}_1$ direction or toward the plane of flattening, the ($\hat{s}_1$, $\hat{s}_2$)-plane, and therefore toward being perpendicular to the streamlines.

In shear flow (Figure 20.31*C*), the streamlines are parallel, and the velocity of the material does not change in the downstream direction for a constant-volume deformation. The velocity does change, however, with distance perpendicular to the streamlines, and that velocity gradient is a maximum parallel to the movement plane and is zero parallel to the shear plane. The deformation is noncoaxial. Progressive simple shear is one example of such a shear flow (Figures 12.13*B* and 12.15), and it illustrates the fact that both the $\hat{s}_1$ direction and material lines rotate progressively toward the streamlines. In principle, however, neither $\hat{s}_1$ nor rotated material lines ever

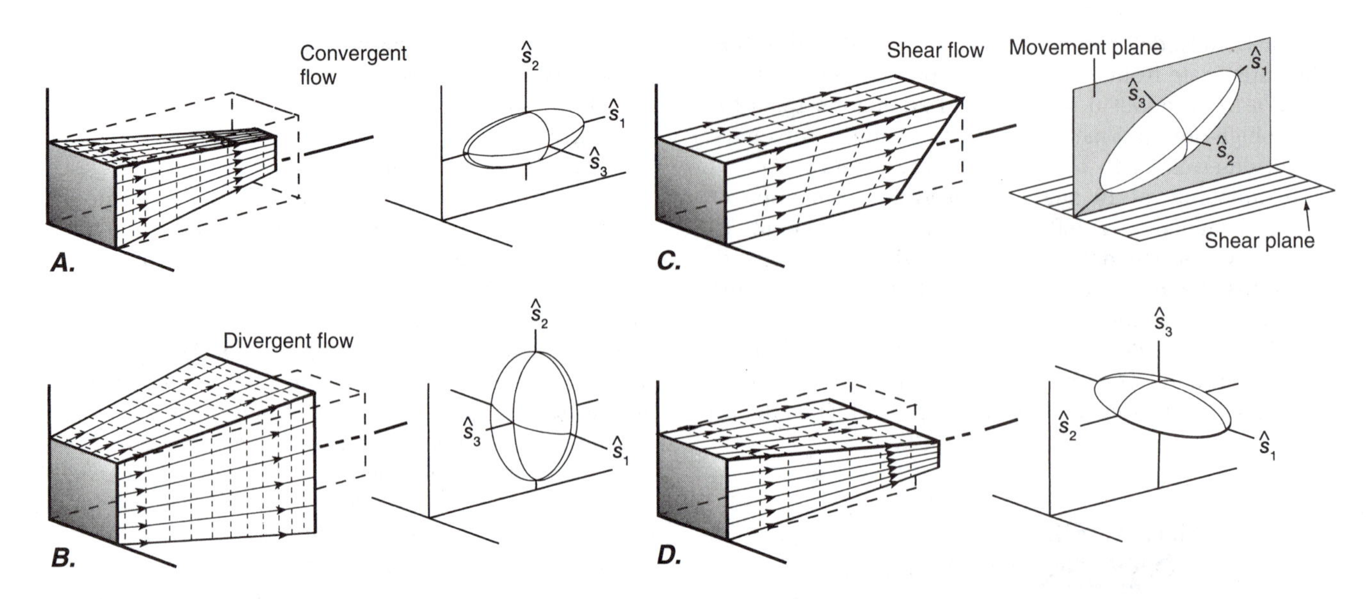

FIGURE 20.31 Streamlines for different flow geometries showing the finite strain ellipsoids associated with each flow field. Note that the orientations of the principal axes of finite strain are not uniquely related to the orientation of the streamlines. (After Hansen 1971) *A.* Convergent flow: The streamlines all converge in the downstream direction, and velocity increases downstream. *B.* Divergent flow: The streamlines all diverge in the downstream direction, and velocity decreases downstream. *C.* Shear flow: The streamlines are all parallel, and velocity does not change in the downstream direction. The velocity increases linearly from the bottom to the top of the block. *D.* Combined convergent and divergent flow.

become exactly parallel with the streamlines, although for extreme values of shear, the difference may become negligible.

These simple types of flow also can be combined to give more complex flows. For example, the streamlines may converge in one plane but diverge in a plane normal to the first (Figure 20.31*D*). Moreover, a shear flow may be combined with any of the flows involving convergent or divergent flow, with major consequences for the orientation of the strain ellipsoid (see for example Figures 12.24–12.26) .

The preceding discussion shows that there is no simple relationship between the streamlines (the direction of tectonic transport) and the orientations of the principal axes of finite strain, fold axes, or other material lineations. This is because streamlines reflect two components of the motion, a rigid translation and a deformation. Structures, on the other hand, can only record deformation. Rigid translation is characterized by a complete absence of gradients in the velocity field, that is, the velocity does not change from one place to another. Moreover, the orientation of the velocity vectors depends in part on what is chosen as a fixed reference coordinate system. For example, the velocity in a reference frame fixed to the platform of the down-going plate would be different from the velocity in a reference frame fixed to the over-riding plate. Deformation, however, is defined only by the velocity gradients, that is, how the velocity changes from one point to the next along the different coordinate directions. Because the change in velocity is referenced to the deforming material itself, the choice of external coordinate frame is immaterial.

As these simple examples in Figure 20.31 show, velocity gradients parallel to the streamlines can have a positive sign (as in convergent flow; velocity increases in the downstream direction) or a negative sign (as in divergent flow; velocity decreases in the downstream direction), and velocity gradients also can be perpendicular to the streamlines (as in shear flow). More complex combinations of these simple patterns are also possible (see Figure 12.26). Each gives rise to very different geometries of deformation and different orientations of finite strain ellipsoids. Thus the orientation and shape of the finite strain ellipsoid do not uniquely define the orientation of the streamlines or the direction of tectonic transport.

The slip line of a shear flow is generally expressed in terms of the direction of motion of the block on one side of a shear zone relative to the block on the other side. For example, we can always specify the direction of shear in a ductile thrust fault zone by giving the direction in which the hanging wall block moves relative to the footwall block. This direction is only the direction of tectonic transport, however, if the footwall block is fixed relative to the stable continental platform that serves as the reference for the streamlines. Commonly the equivalence of the reference frames for shearing and for the streamlines is implicitly assumed. It may be a reasonable assumption, but it is not a necessary one, and the validity of the assumption must be evaluated in terms of the regional geologic context. A shear flow, moreover, can be modified by the addition of one of the other flow types (see Figure 12.26), in which case the orientation of the principal finite strain axes may not be diagnostic of the direction of tectonic transport, even if the assumption is correct.

ii. Simple Models of Ductile Nappe Emplacement

In Section 9.11 we discussed the problem of the emplacement of thrust sheets, with particular attention to foreland fold and thrust belts in which the thrust wedge behaves like a Coulomb-type of material (Section 9.11(v)), and high pore fluid pressure plays an important role in reducing frictional resistance at the base of the thrust wedge. We also mentioned the possibility that Coulomb fracture and friction may be irrelevant mechanisms to invoke if the thrust sheet deforms in a ductile manner (Section 9.11(vi)). For example, the emplacement of the Helvetic nappes of the Alps (Figures 20.26, 20.28, and 20.29) may have been dominantly by ductile flow. It is to this possibility that we now turn our attention.

We consider three simple models that have been proposed to explain the emplacement of ductile fold or thrust nappes: gravity glide, horizontal compression, and gravitational collapse. Each of these models predicts a different distribution of the principal finite strain axes, so in principle we should be able to distinguish these different mechanisms of emplacement. Although these models are overly simplistic and thus may not provide a realistic representation of nappe formation, they illustrate how different mechanisms lead to different patterns of strain that can provide useful guides to interpretation.

The critical differences among these models show up on cross sections through the thrust sheet constructed parallel to the direction of maximum shearing, which is the slip direction of the hanging wall block relative to the footwall block. We represent a portion of the thrust sheet in two dimensions by a rectangular block resting on a thrust fault at the base of the block (Figure 20.32*A*).

The gravity-glide model of nappe motion assumes that a nappe may be emplaced by gravitational forces that cause the nappe to flow down a gently inclined base by ductile shearing within the nappe (Figure 20.32*B*). In the simplest case, the nappe neither shortens nor lengthens parallel to the slip direction. The resulting flow is an example of inhomogeneous progressive simple shear (compare the homogeneous case in Figure 20.31*C*). The base of the nappe is a zone of intense shear strain. With progressively shallower depths in the nappe, the shear strain decreases, reaching zero near the surface (Figure 20.32*B*). Thus at the top of the nappe, the $\hat{s}_1$ axis of the finite strain ellipse is oriented at 45° to the shear plane, but the ellipticity of the strain ellipse is vanishingly small (Figure 20.32*B*; compare Figure 12.15*A*). With increasing depth in the nappe, the orientation of the $\hat{s}_1$ axis rotates increasingly toward lower angles with the shear plane, reaching the minimum angle at the base of the nappe, where the ellipticity of the strain ellipse is a maximum (compare Figure 12.15*D*).

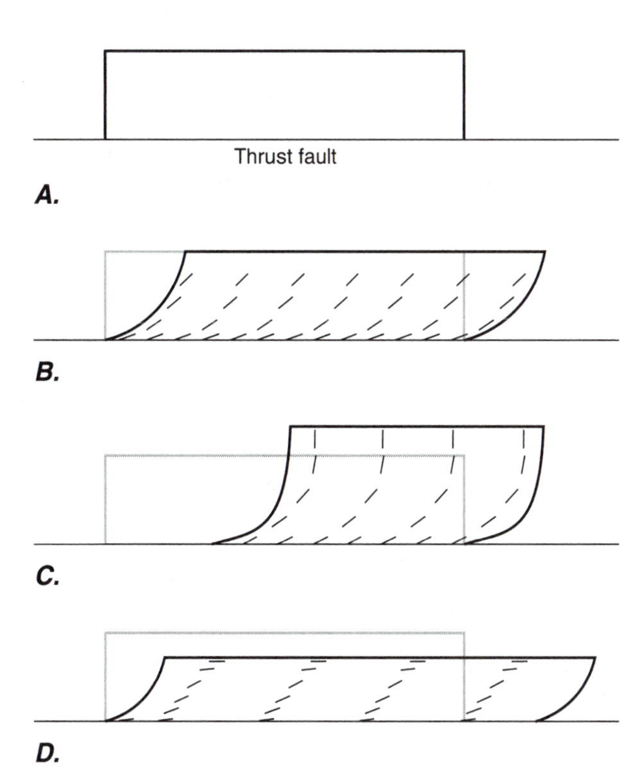

FIGURE 20.32 Idealized distributions of maximum finite stretch $\hat{s}_1$ in thrust nappes deformed according to different possible mechanisms of emplacement. (After Sanderson 1982) *A*. A portion of a nappe before deformation. *B*. Gravitational glide: The nappe deforms by simple shear parallel to the basal fault. *C*. Horizontal shortening, or a "push from behind": The nappe deforms by a combination of simple shear with horizontal shortening parallel to the basal fault and vertical thickening. *D*. Gravitational collapse: The nappe deforms by a combination of simple shear with horizontal lengthening parallel to the basal fault and vertical shortening.

The second model assumes that the nappe is emplaced by a horizontal compression applied to the rear of the nappe, the so-called "push from behind" (Figure 20.32*C*). The result is a ductile shearing over the base, similar to that described for the first model, on which is superimposed a shortening of the nappe parallel to the slip line. For simplicity, we assume the shortening is homogeneous throughout the nappe. The flow is then a combination of inhomogeneous shear flow and divergent flow in the plane of the diagram (compare Figures 20.31*B*, *C*). Thus at the top of the nappe where the shear flow is zero, the horizontal shortening and vertical thickening cause the $\hat{s}_1$ axes to be vertical. With increasing depth approaching the basal shear zone, the shortening component of the strain added to the simple shear causes the $\hat{s}_1$ axes to be oriented at a higher angle to the shear plane than for the case of simple shear alone.

The third model assumes that the nappe is emplaced by a process of gravitational collapse (Figure 20.32*D*), which involves the ductile horizontal spreading and vertical thinning of the nappe in a manner analogous to the flow of a continental ice sheet. In this model, for both the nappe and the ice sheet, the flow is driven by the tendency for the surface slope of the body to flatten out to horizontal. We assume for simplicity that the flattening of the nappe is homogeneous within the illustrated portion of the cross section. The resulting deformation is a combination of inhomogeneous shear flow (compare Figure 20.31*C*) and flow that converges downstream toward the thrust plane (comparable to a vertical cross section parallel to the streamlines through Figure 20.31*A*, with no convergence normal to the plane of the section). At the top of the thrust sheet where the shear flow is zero, the vertical flattening of the nappe causes the $\hat{s}_1$ axes to be horizontal and parallel to the streamlines (Figure 20.32*D*; cf. Figure 12.13*A*). Where the shear flow is nonzero, the vertical flattening tends to rotate the $\hat{s}_1$ axes toward lower angles with the shear plane than would occur for simple shear alone (Figure 20.32*D*; cf. Figure 14.12). The net result is a sigmoidal pattern of $\hat{s}_1$ orientations with the smallest angles between $\hat{s}_1$ and the shear plane near the top and bottom of the nappe and the highest angles near the center (Figure 20.32*D*).

Figure 20.33 shows a plasticine model of gravitational collapse characterized by vertical thinning, horizontal lengthening, and shear along the base, with a complex rolling under of the top of the nappe at the front. Such a process could explain the major fold noses and inverted limbs of fold nappes and the superposed crenulation cleavage such as is found in the inverted limb of the Morcles nappe. Note that in this physical model, the shearing decreases toward the surface but does not go to zero except perhaps within a very small distance from the surface, which is contrary to the assumptions on which the model in Figure 20.32*D* was constructed.

FIGURE 20.33 Strain distribution in a plasticine model of a nappe undergoing gravitational collapse with shear along the base. Note the top of the nappe at the front gets rolled under the advancing front of the nappe, a process that would explain the development of recumbent folds at the fronts of nappes. (After Merle 1986)

These models, of course, are highly simplified, particularly in that they ignore end effects, they assume a highly idealized geometry for the nappes, they ignore the effects of surrounding rock, and they assume the nappes undergo an idealized two-dimensional ductile deformation. Nevertheless, they provide a basis on which to build an interpretation of field data. The strain in the Morcles nappe, for example (Figures 20.28 and 20.29), is reasonably consistent with the models for gravity glide or possibly gravitational collapse, but not with the push-from-behind model (Figure 20.32). The history of the maximum instantaneous strain orientation $\hat{\zeta}_1$ for the Morcles nappe shown in Figure 20.29 is consistent with a model of nappe emplacement in which top-to-the-north shear flow with slip lines oriented away from the orogenic core (Figure 20.31C) dominates in the lower part of the Morcles nappe throughout the emplacement history. Higher in the Morcles nappe and in the Wildhorn nappe (or farther east in the orogen on the map), however, the flow geometry changed late in the emplacement history. The initial shear flow was followed by a flow consistent with horizontal divergent flow (e.g., Figure 20.31*B*) with streamlines at a high angle to the orogenic core and $\hat{\zeta}_1$ oriented roughly horizontal and subparallel to the orogenic core. It may be a component of shearing parallel to the orogenic core during this latter phase of the deformation that is reflected by the longitudinal slip lines indicated by the Hansen fold analysis (see Figure 20.26; Section 20.8(i)). The inhomogeneity of the strain and the complicated history of the deformation (Section 20.8), however, make comparison with such simple models very imprecise.

Other models of nappe formation that are more complex have been investigated using scale-model experiments. One hypothesis proposes that the emplacement of some nappes is driven by convective overturn of crustal rocks in an orogen. Models of the process produced by centrifuging layered blocks of different putty-like materials to induce density-driven flow (Section 18.4) show striking similarities to cross sections of some deformed orogens (e.g., Figure 18.7). The similarity does not prove the hypothesis, however, because the hypothesis is not necessarily unique. Some nappes are composed of high-pressure metamorphic rocks for which the density is higher than the surrounding rocks; they are unlikely, therefore, to have been emplaced by a density-driven convective overturn. Comparable deformation, moreover, might occur through the dynamics of a continent-continent collision, the gravitational collapse of an orogenic root, or a combination of the two, which we discuss in the following subsection. Such models, however, can provide possible ways of interpreting the field data and can point to specific field observations by which the different models might be distinguished.

20.12 A TWO-DIMENSIONAL PLATE TECTONIC MODEL OF OROGENY

The following summary presents a simplified model for the complex process of orogeny. We use the diagrams in Figure 20.34, showing a numerical model of a continent-continent collision, as just one example of a class of models that illustrates some of the main features of orogeny as we currently understand it. The deformation in these diagrams is depicted by a "lagrangian grid," that is, a grid of material lines that deforms with the material. We fully recognize, however, that the assumptions on which this particular numerical model is based (see the figure caption) are highly oversimplified and not necessarily representative of the conditions we would expect for the behavior of rock under these circumstances. Nevertheless, this model displays a number of characteristics that help us understand field and laboratory results from the actual rocks. This concept of orogeny is highly influenced by the geology of the Alps in particular, because extensive fieldwork in this mountain range, as well as in the Caledonides, has provided us with the most detailed information available about how orogens have evolved. It may therefore help us understand other orogens in which less work has been done or in which poor exposure precludes even the possibility of accumulating such detailed data.

The model of orogeny that we consider is only for a continent-continent collision. If it applies at all to the Alps or other complex orogens, it is only to the continent-continent collision, which is likely to be the last phase of the orogeny, as in the case of the Alps. Every orogen, of course, has its own unique history that will be reflected in the details of its structure. The Alps, for example, have an extensive history of orogeny that begins in the mid-Cretaceous (at approximately 95 Ma) and involves the incorporation into the orogen of a number of continental slivers, microcontinents, and intervening sedimentary and oceanic basins. Finally in the Oligocene (at approximately 38 Ma), the southern margin of Europe entered the subduction zone, starting a continent-continent collision that continued into the Pliocene (approximately 3 Ma). Thus the whole orogen evolved over a span of roughly 90 million years, of which only the last 35 million years or so was a continent-continent collision.

We discuss this model, however, in the hope that the first-order processes are common to many orogens and that models such as this eventually will help us unravel and understand the structural characteristics of orogeny.

i. Collision, Underplating, Uplift, Nappe Formation, Back-Folding/Thrusting, and Rotation

The transition from subduction to continent-continent collision occurs when a block of continental crust enters a subduction zone (Figure 20.34*A*), marking the onset of a collisional orogeny. The evolution of the collisional orogenic belt is dominated by underplating, which occurs as the subduction zone fault steps down into the upper layers of the down-going continental crust, slicing off a piece and effectively adding it to the bottom of the orogenic pile in the over-riding plate. Although the numerical model (Figure 20.34) shows the underplating as a continuous deformation, field evidence shows it must in fact occur as a series of discrete steps (compare Figure 20.20), which this numerical model cannot reproduce. The numerical model also includes erosion of material from topographic highs, with the rate of erosion increasing with increasing topographic elevation.

Continued subduction brings new parts of the down-going continental crust to the bottom of the orogenic pile, and nappe emplacement consists of transferring the rocks in the nappe from the down-going to the over-riding plate and then dragging the remaining down-going plate underneath the nappe. With each successive step of the main subduction thrust into the down-going plate, a new slice of the down-going continent is added to the bottom of the orogenic pile on the over-riding plate. Each slice forms a new thrust nappe that is emplaced over the down-going continent as the continent is dragged further down the subduction zone. Continued underthrusting and underplating uplifts the material in the over-riding plate. During the early phase of the collision, subduction drags original suture material down underneath each newly separated nappe. The suture material includes subducted oceanic sediments and possibly imbricate slices of the down-going oceanic crust that preceded the continental margin down the subduction zone (Figure 20.34*A*). It is represented in successive frames of Figure 20.34 by the light gray material. The result is a stack of thrust nappes that are surrounded by sheaths of the suture material, with the oldest nappes on top (Figure 20.20). The top nappe would probably be the oceanic crust that preceded the continental crust down the subduction zone. It would be underplated onto the bottom of the overriding plate and would then be preserved as an ophiolite that would mark the original suture between the overriding and downgoing continental crust.

As a continental margin follows oceanic crust into the subduction zone, rocks deposited in offshore abyssal environments are subducted first, followed by rocks from the continental rise and slope, and finally the continental shelf. Thus the rocks in the stack of nappes should progress from top to bottom from oceanic crust (ophiolite) to abyssal and to shelf environments, and the lowest nappes ultimately could include continental basement (compare Figure 20.7). In the numerical model, the interleaving of thrust slices and suture material cannot be represented, but the uplift, deformation, erosion, and rotation in the core zone can be seen by following the white suture material in the sequential frames in Figure 20.34. Ultimately, the suture material, which was initially dragged

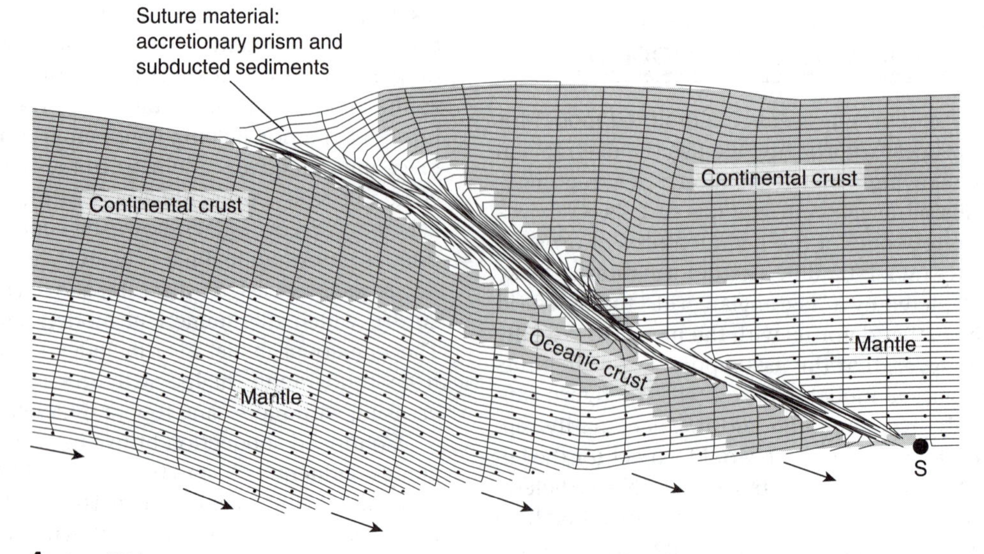

A. $\Delta x = 50$ km

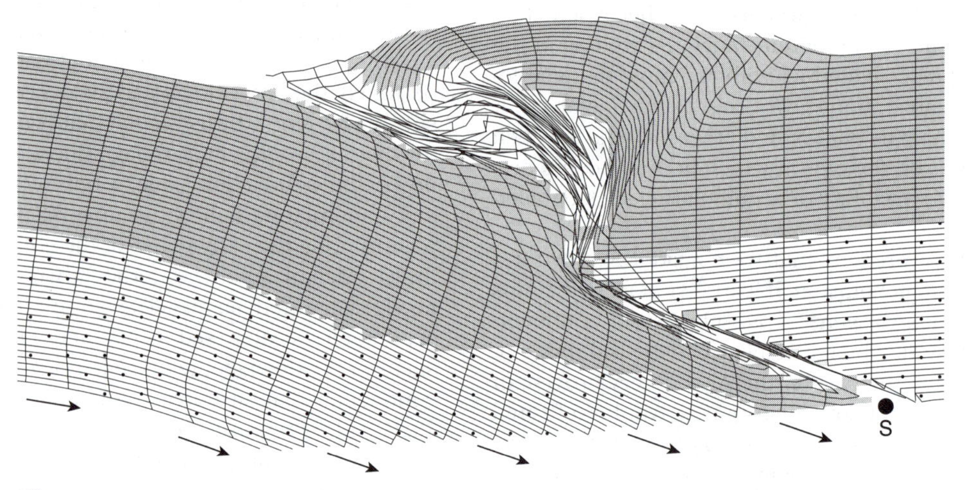

B. $\Delta x = 100$ km

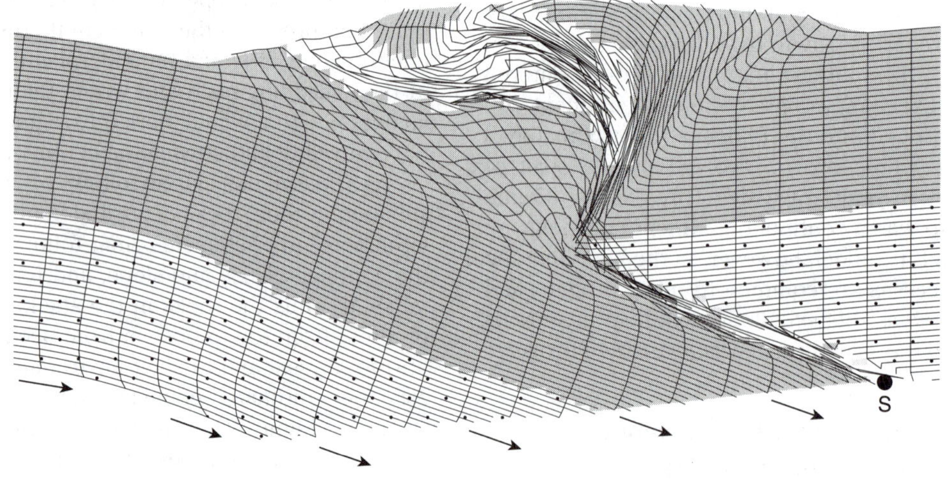

C. $\Delta x = 150$ km

FIGURE 20.34

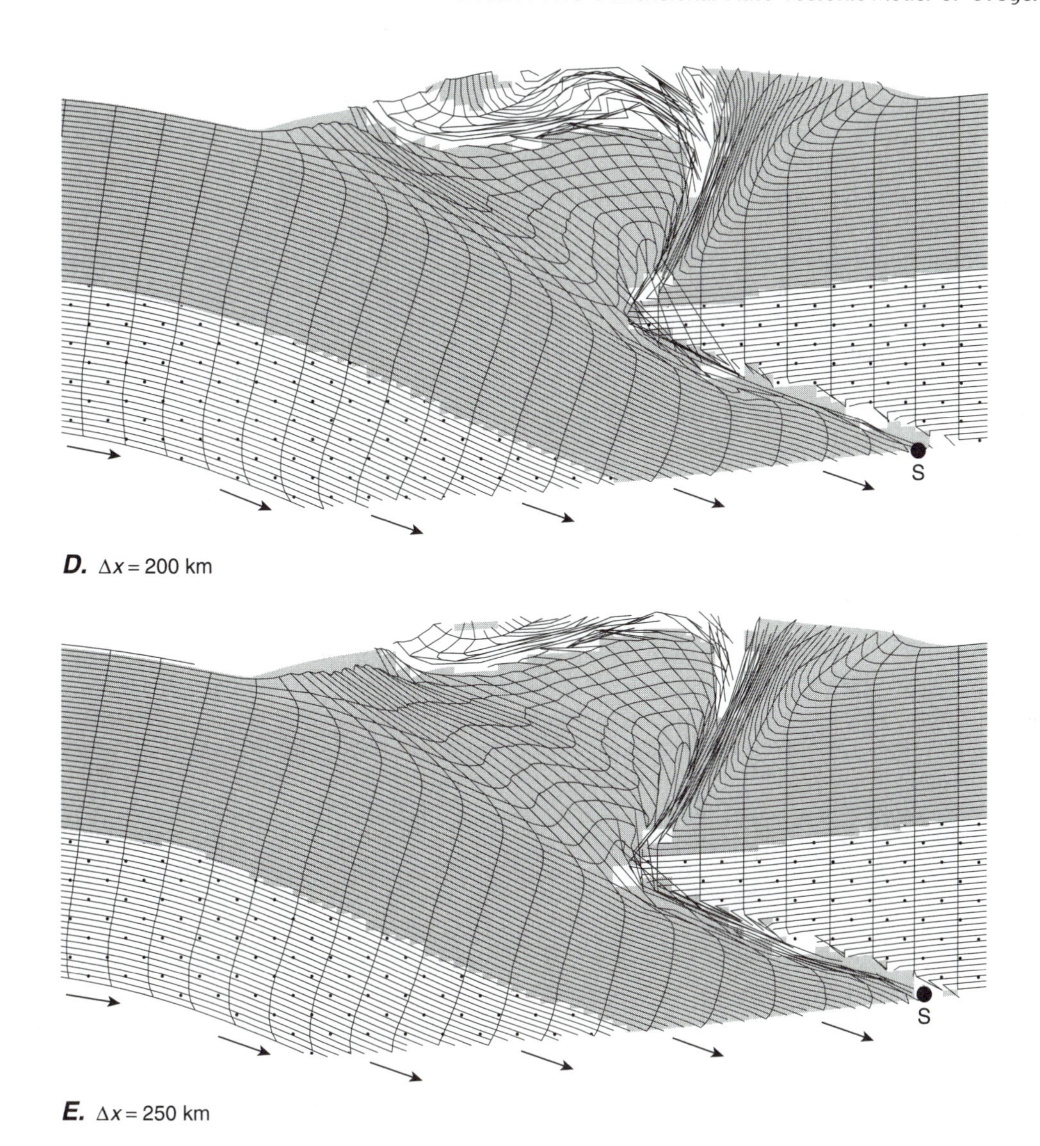

FIGURE 20.34 *(continued)* Numerical model of a continent-continent collision using Coulomb-type materials. Dark gray is strong continental crust; white is weak sediments of the accretionary prism and subduction zone ("suture material"). The grid is a lagrangian system of lines that deforms with the material. Arrows represent velocity boundary condition of material leaving the model. S indicates the point of velocity discontinuity. The model extends to an initial depth of 60 km and assumes a constant convergence rate, a moderate erosion rate that increases with the topographic elevation, and a rigid mantle in the over-riding plate that is only allowed to bend. The model assumes only one collision, an oversimplification of most collisional orogens that include one or more island arcs sandwiched between the two continents. In addition, the model does not include thermal equilibration or sedimentation and cannot model discontinuous deformation such as the formation of faults.
(From Pfiffner et al. 2000, Figure 5a-e, with permission from the American Geophysical Union)

down into the subduction zone and metamorphosed at depth (Figure 20.34*A*), ends up lying subhorizontally near the top of the orogen (Figure 20.34*E*).

In the Alps, the stack of nappes that result from this process is evident in Figure 20.20*B* as the ophiolite sheet (oceanic crust), and the Suretta, Tambo, Adula, Simano, and Lucomagno nappes. Each nappe beneath the ophiolite is surrounded by an envelope of the North Penninic Bundnerschiefer (NPB in Figure 20.20*B*), also referred to as the Schistes Lustrés or early Alpine flysch (compare Figure 20.14), which are the metamorphosed oceanic sediments of the suture zone. After all the suture zone material has been uplifted above the main subduction zone thrust, the thrust continues to cut down into the downgoing continental crust, but the material dragged under each of these younger nappes is continental crust. This process creates the nappes of the basement massifs such as the Belledonne, Pelvoux, Gotthard and Aar massifs of the Alps (Figures 20.5 and 20.20; these are among the external crystalline massifs of Figure 20.26). In this model, the highest nappes are the oldest, although in fact out-of-sequence thrusting is possible.

The underplating also results in the formation of a zone of back-folding and back-thrusting, which is essentially a steep thrust zone antithetic to the subduction zone. Development of this back-fold–thrust zone leads to the bivergent symmetry of the orogen (compare Figures 20.3 and 20.34*E*), although the asymmetry of the subduction zone is still preserved in the steep dip of the back-fold–thrust zone relative to the shallower dip of the subduction zone. The deformation of the orogenic pile that results from the continuing underplating is characterized by ductile nappe formation at depth, as suggested by the fold-like structures that develop in the lagrangian grid at depth in the orogen (Figure 20.34*C–E*). Folds such as we now see exposed, for example, in the wall of Grandjeans Fjord on the east Greenland side of the Caledonian mountains (Figure 10.1*C*), reflect this type of deformation. Ductile fold nappes may also develop adjacent to the back-thrust–fold zone and may involve parts of the stack of nappes depicted in the cross sections of the Alps (Figures 20.20 and 10.1*B*). The Monte Rosa nappe in the Alps (Figures 10.1*B* and 20.20) may represent this phase of deformation. Rock uplift along the back-fold–thrust zone entrains the leading edge of the over-riding continental plate, and with exhumation it can bring rocks from the base of the over-riding crust up to the surface (Figure 20.34*D–E*). We find such exposures now in the Ivrea zone in the southern Alps (Figures 5.22, 20.20*A* and 20.26).

ii. Exhumation

The exhumation processes of erosion and tectonic denudation remove rock from the shallower levels of the orogen to expose rock uplifted from greater depths. In the numerical model (Figure 20.34), erosion plays the dominant role in the removal of material from the top of the orogenic pile and the consequent lowering of the topographic slope. The rate of model erosion is assumed to increase as the topographic elevation of the surface increases. In reality, however, the topographic high created by the underplating can also result in gravitational collapse of the orogenic pile, which also lowers the topographic slope. This collapse occurs by the formation of low-angle normal faults that tectonically strip material off the topographic high and place low-grade to unmetamorphosed rocks on top of high- to very high-grade rocks. This process is similar to the development of metamorphic core complexes (compare Section 4.3(iii)). Such normal faults, which in effect lengthen and thin the top of the orogen, therefore can develop at the same time as the folding and thrust faulting that shorten and thicken the orogen at depth. Faulting cannot be modeled by the numerical techniques used to produce Figure 20.34, but subhorizontal lengthening and vertical thinning at the topographic high of the orogen is evident by comparing the spacings of the lagrangian grid in Figure 20.34*A*, *B* (in subsequent panels of the figure, the deformation has become too complex to see this effect clearly, although it is important).

As mentioned in Sections 20.6 and 20.9, tectonic denudation must be a major contributor to exhumation in real mountain belts. In the Alps, for example, a reconstruction of the now-eroded thickness of the Austroalpine nappes that were part of the over-riding plate (Figure 20.20*A*, *B*) indicates that they were insufficiently thick to account for the high-pressure metamorphism of the underlying and now-exposed rocks. Thus other material must have contributed to the thickness of the orogenic pile, and erosion alone cannot account for the exhumation of rocks from depths as large as are required by the metamorphic mineral assemblages. Material must have been removed from within the orogenic pile, which can only be done on low-angle normal faults.

Underplating, rock uplift, and erosion and tectonic denudation in effect create a circulating flow in the orogen against the backstop of the over-riding continental mass. This flow allows the exhumation of metamorphic rocks and nappe structures from deep within the orogen where they initially formed (Figure 20.20). The rock uplift and exhumation is greatest near the back-fold–thrust zone (Figure 20.34), which is therefore where the highest-grade metamorphic rocks are exposed to form the core zone of the orogen.

In principle, a steady state can be reached in which the flux of material into the orogenic pile by underplating of material from the down-going continental crust is balanced by the erosion and tectonic denudation of material off the top of the orogen. If the flux of underplated material exceeds the rate of exhumation, then surface uplift will occur. If the rate of exhumation exceeds the flux due to underplating, the surface topography will decline. As long as underplating continues, however, regardless of the motion of the surface, rock uplift within the orogen will occur. Rock uplift may also occur during gravitational collapse of the orogen, as described in Subsection 20.12(vii). This process is similar to that described in Chapter 19 for the evolution of accretionary prisms (Figure 19.37).

The original suture zone material can thus end up at the surface in nappe structures whose basal thrusts have been rotated from the original dip parallel to the subduction zone (Figure 20.34*A*) into a dip toward the foreland of the down-going plate (Figure 20.34*E*). With continued underplating and exhumation (only slightly beyond the state in Figure 20.34*E*) these nappes may become completely detached from the associated material in the back-fold–thrust zone. The slices of oceanic crust in the original suture material eventually are exposed in the orogen as ophiolites, which provide a distinctive marker for the suture between the material from the over-riding plate and the material from the down-going plate.

iii. Root Formation

Only a portion of the down-going crust is underplated; the rest is subducted, and it either produces the root to the orogenic zone, which may be sufficiently deep to result in ultra-high-pressure metamorphism (Figure 20.8), or it is entrained into the deep mantle. The model in Figure 20.34 does not describe the deformation of rocks initially below 60 km depth, so it does not provide a reliable model of either the formation of the orogenic root or the process of carrying continental crust into the realms of ultra-high-pressure metamorphism and back. It is possible that before convergence completely shuts down, the buoyancy of the down-going continental crust could cause it to deform and pile up into a sub-orogenic root while the mantle lithosphere continues subducting. This could produce a substantially larger and deeper root than is depicted in Figure 20.34.

iv. Buoyancy and the Cessation of Collision

Ultimately, the buoyant force arising from the increasing volume of low-density continental material carried down into the high-density mantle counteracts the driving forces of subduction, and the plate convergence ceases. Presumably, the accommodation of the plate motions may be taken up by a global reorganization of the directions and magnitudes of the relative plate velocities and even in a reorganization of some of the plate boundaries. The oceanic part of the down-going slab may detach from the continental part and continue sinking by itself into the mantle, to be replaced by hotter upwelling asthenosphere. With the cessation of subduction, the orogenic root, which in part has been held down dynamically by the forces driving subduction, is out of isostatic equilibrium, and it begins to rise.

v. Orogenic Heating

Slow heating of the crustal rock in the orogenic zone, through radioactive decay and conduction, gradually tends to change the low geothermal gradient associated with subduction of a cold slab, toward a more normal steady-state continental geotherm. The increasing temperature makes the rock increasingly ductile, and the deepest rocks in the orogen may even reach temperatures at which partial melting can form post-orogenic granites (see the wet granite melting curve in Figure 20.8).

If the deep crust has a basaltic composition, it may undergo metamorphism to eclogite (clinopyroxene plus garnet), which is sufficiently dense that it may delaminate from the upper crust and sink into the mantle, to be replaced by hotter upwelling asthenosphere. This process may further increase temperatures in the crust and lead to increased isostatic uplift by replacing dense eclogitic lower crust with less-dense and hotter asthenosphere. This process has been suggested to explain the uplift during the past 5 million years of the Sierra Nevada in the Cordilleran chain in California.

vi. Metamorphism and the Preservation of Ultra-High-Pressure Rocks

The plate tectonic interpretation of the metamorphic evolution of rocks in orogens is complex, because the rates at which rock is carried down a subduction zone and uplifted again are comparable to the rates at which they can heat up or cool off. In the initial stages of collision, the rate of exhumation must be considerably less than the rate of tectonic thickening to account for the thickened crust in orogenic belts such as the Zagros and the Himalaya-Tibet areas. Thus we might expect a given volume of rock to be buried quickly. The rate of exhumation would then determine the extent and character of the subsequent metamorphic overprint. If exhumation is rapid, the temperature increase would be minimized and the mineral assemblage in equilibrium with deep, cool conditions could be preserved (curve ABC, Figure 20.30). If uplift is initially very slow, the temperature would have time to reequilibrate to the steady-state continental geotherm, and an overprint of Barrovian metamorphism could dominate the final mineral assemblage in the rocks (curve ABE, Figure 20.30). Although it is not difficult to invent plate tectonic scenarios that would account for such differences, such hypotheses are not unique, and it is difficult to prove that any given scenario is what actually happened.

We know that some terranes that were initially metamorphosed at high-pressure–low-temperature blueschist conditions have been extensively overprinted by later higher-temperature Barrovian metamorphism, because the rocks make it back to the surface as Barrovian metamorphic mineral assemblages, some of which still preserve remnant minerals from the colder conditions. Other terranes, in some cases covering many tens of square kilometers such as the Franciscan terrane in northern California, preserve high-pressure–low-temperature assemblages that are little altered from that peak metamorphic condition.

Two mechanisms have been proposed to account for the preservation of high-pressure–low-temperature minerals. The rocks may return to the surface from blueschist metamorphic conditions (see Figure 20.8) by some sort of return flow within a subduction zone. Such a mechanism, however, does not seem able to account for large terranes of high-pressure–low-temperature mineral assemblages such as the Franciscan terrane. Alternatively, the rocks might return to the surface by rapid underplating in combination with exhumation by erosion and active tectonic denudation on low-angle normal faults. This process has been proposed for active accretionary prisms (Section 19.4(iv)). Major low-angle normal faults also are characteristic of metamorphic core complexes (see Figure 4.13) and have been recognized in many orogenic mountain belts (see Section 20.6), so this mechanism undoubtedly contributes to the exhumation of high-pressure–low-temperature metamorphic rocks in orogens as well.

The ultra-high-pressure metamorphic rocks that have been found exposed at the surface in orogens present a more difficult problem. The process by which these continental crustal rocks are carried down to depths of as much as 120 to 140 km and then returned to the surface is still poorly understood, especially considering that normal orogenic roots only extend to depths of 60 to 70 km. Dragging the crustal rocks down to the extreme depths undoubtedly occurs as part of the subduction process during a collision. The mechanism by which they rise back to the surface from such crustal depths, however, is a problem. Exhumation must have been rapid enough to preserve the ultra-high-pressure metamorphic assemblages as the rock traversed regions in which lower-grade metamorphic conditions prevailed.

Field relations in some areas suggest the rocks have moved upward as a slice a few kilometers in thickness, bounded above and below by ductile shear zones or faults, but the details of the mechanism driving the uplift are not clear. This process has been attributed to expulsion or extrusion (see the discussion in Section 20.9). It may be that exhumation of extremely deep crustal rocks requires a two-stage process. A deeply subducted continental fragment might be held at depth by the dynamic forces driving subduction. When these forces are relaxed, initial uplift to more normal depths for orogenic roots would occur by a density-driven upflow, essentially collapsing and spreading the ultra-deep root. Further uplift might then occur through underplating during a subsequent collision of a large continent. Tectonic denudation and erosion would then exhume the rocks, eventually exposing them at the surface. Again, this is a possible scenario, but proving that any such model represents the actual history is much more difficult than dreaming up the hypothesis. Much work remains to be done in solving these problems.

vii. Gravitational Collapse of the Orogenic Root and the Topographic High

Although the topographic high of the orogen and its root in the mantle may be in isostatic equilibrium as a whole, within the orogenic pile the stress is nonhydrostatic. At shallow levels, the stress results predominantly from the surface slope of the orogenic pile, and at deeper levels it is predominantly from the slope of the interface between the root and the surrounding mantle. If the rocks become sufficiently hot and weak, these stresses tend to produce a gravitational collapse and horizontal spreading of the orogen. The topographic high collapses down to lower elevations, and the root flows upward to decreasing depths. The averaged streamlines for the whole gravitationally collapsing orogenic pile would resemble those depicted in Figure 12.13*A*.

Collapse of the topographic high occurs through the development of low-angle normal faults that strip material off the top of the orogenic core zone and spread it laterally toward the foreland. Such faults are shown in Figure 20.23, exposing the Penninic nappes in the Tauern window in the Alps, and in Figure 20.24, exposing the Western Gneiss region in the Norwegian Caledonides, as well as in Figure 20.25, showing the South Tibetan detachment fault; the early Tertiary denudation faults in parts of the North American Cordillera are probably another example. These faults help to expose deeper orogenic rocks with their high grades of metamorphism and ductile flow structures.

Collapse of the root occurs by a vertical divergent flow (Figure 20.31*B*) that turns horizontal toward the edges of the orogen, resulting in large recumbent ductile fold nappes, an approximately horizontal plane of flattening of the finite strain ellipsoid, and a subhorizontal foliation. Horizontal seismic reflectors are a common feature of the deep parts of orogenic belts and may be a reflection of subhorizontal foliations associated with such a flow field. This process of orogenic gravitational collapse also could contribute in the Alps to the emplacement of the Helvetic nappes on previously formed thrusts, as well as to the emplacement of the basement massifs, such as the Aar and Gotthard massifs (Figures 20.5 and 20.20*A*, *B*). Gravitational collapse could also be a driving mechanism for the deformation in the foreland fold and thrust belts on either side of the orogenic welt (compare the scale model in Figure 18.2 with the cross section of the Jura Mountains in Figure 18.1).

viii. Relation to Observed Generations of Deformation

Numerous proposals have been made to account for the different generations of deformation observed in orogenic core zones, although the data are not adequate to support any of them clearly. As a volume of rock proceeds through the different deformational environments associated with underplating and nappe emplacement, uplift, and gravitational collapse, it is not difficult to imagine that the structures would preserve a record of superposed generations of folding. Different generations of deformation also could relate to multiple phases of collision in a complex collision zone.

In principle, the details of these superposed generations of deformation should provide clues to the sequence of deformational environments experienced by the rock and thereby a means to refine the details of the model and to understand some of the complexities that develop in real orogens. At our current state of understanding, however, it has been difficult to make a unique association between small-scale structures observed in outcrop and the large-scale tectonic events. This remains one of the biggest challenges in understanding the structural geology of orogenic belts.

ix. Foreland Thin-Skinned Fold and Thrust Belts

Structural evidence from active decollement-style fold and thrust belts in the southwest Pacific and from the Zagros Mountains in Iran indicate that these belts develop during the subduction of a passive continental margin and are synthetic to the direction of subduction (Figure

5.20*C*). These observations imply that the formation of fold and thrust belts should be a process that occurs only when continental crust arrives at a subduction zone on a down-going plate. The deformation results from the propagation of a set of imbricate thrust faults up-section and toward the foreland of the down-going plate from a decollement that ultimately is rooted in the subduction-zone thrust (Figures 5.12, 5.20*C*, and 5.21). Crustal thickening, with subsequent gravitational collapse of the orogenic topographic high also may help drive the formation of the fold and thrust belt, as may the gravitational collapse of the root of the orogen (Figure 5.20*B*). Splays off the back-fold–thrust zone propagate out into the foreland of the over-riding crust, creating the fold and thrust belt on the "backside" of the orogenic welt and thus the bilateral symmetry of many orogens.

This model is by no means universally accepted as a general explanation of foreland fold and thrust belts. A major difficulty is the presence in the eastern Andes of a foreland fold and thrust belt antithetic to the present subduction of the Nasca plate. The Andes are generally thought by many to have formed on an over-riding plate above a subduction zone that, since the Permian, has consistently dipped to the east under the continental margin. Thus the Andean orogeny is not ascribed to a collision of the South American continent with a west-dipping subduction zone, as would be implied by the model of orogeny presented in the preceding subsections of Section 20.12. In this view, the observed fold and thrust belts did not form during a collision and did not form synthetic to a subduction zone. Instead, they are generally assumed to have formed antithetic to the subduction zone as a result of a compression along the leading edge of the over-riding plate derived from tractions associated with subduction of the down-going plate.

It is probable, however, that the tectonic history at the Andean convergent margin has been considerably more complex. Geologically distinct oceanic terranes have been described from the northern (Colombia to Ecuador) and southern (Patagonian) Andes, and oceanic volcanic arc rocks characterize the Coastal Ranges of Chile. Possibly, the South American continent has been involved in collisions with oceanic island arcs of the type depicted in Figure 19.43*B*, *C*, or *F*, one or more of which may have involved a west-dipping subduction zone (Figure 19.43*B*). If this were the case, then the Andean fold and thrust belts referred to in the preceding paragraph could have formed on the foreland of subducted continental crust, consistent with the model we described in this subsection.

Indeed, exactly this scenario has been invoked to explain some features of the western North American Cordillera. Thus we suspect that even the Andes will turn out to be consistent with the model presented in this subsection. In this view, the fold and thrust belts are the result of the South American continent colliding with a west-dipping subduction zone and with the subduction zone thrust cutting down in a series of imbricate thrusts into the down-going continental sediments and splaying eastward toward the foreland. This shortening in the foreland may also in part result from the partial gravitational collapse of the orogenic pile that accumulated above the collision zone (see Figures 18.2 and 20.34*E*) and that is responsible in part for the high elevation of the Andes.

x. Details and Complexities

The details of the deformation and its distribution in orogens ultimately must depend on factors such as the relative mechanical strengths of the crust and mantle in the two plates, the heterogeneities in structure and in mechanical properties in both the over-riding and the down-going plates, the temperature profiles in the two colliding crustal blocks, the rate of subduction, and the details of the subduction geometry, including irregularities and misfits in the margins of the colliding continental blocks. At the current state of research, modeling can only hint at the possibilities, but it is unlikely to reproduce all the details of any given orogenic belt. Continuum models such as that shown in Figure 20.34 are not capable of modeling discontinuous deformation such as occurs on faults. Collision often involves a series of colliding continental fragments or island arcs that precede the main continent-continent collision and result in more than one suture zone in the orogen. The resulting structure would be considerably more complex than can be accounted for by the model in Figure 20.34.

We have emphasized the two-dimensional cross-sectional view of orogeny, but we know that orogen-parallel movements are common. The convergence vector at most subduction zones is not perpendicular to the plate boundary, leading to a component of strike-slip motion parallel to the subduction zone. The outlines of colliding continental masses are generally irregular, and convergence of these irregularities may be accommodated by orogen-parallel extrusion of rock in the colliding zone. These complexities are possible sources of the orogen-parallel shearing and extension found both within orogens (Figures 20.24, 20.26, and 20.29) and along the steep strike-slip zones that commonly flank the orogenic cores, such as the Insubric Line in the Alps (Figure 20.26).

Despite decades of assiduous study of orogenic core zones, it has been extraordinarily difficult to make any precise associations between particular structures and specific tectonic events. We infer that peak temperature of metamorphism was more or less synchronous with at least a major part of the deformation and that peak temperature conditions probably postdated any single collision by a few tens of millions of years (Section 20.10), because generally there is no evidence that minerals defining foliations have recrystallized after deformation.

Although the stacking up of nappes presumably occurred in the active subduction zone, it remains unclear whether or how the other generations of structures observed in the core zones can be associated with underplating, isostatic uplift, or gravitational collapse of the orogenically thickened crust or even other collisional events. Perhaps as absolute dating techniques improve, it will become possible to relate more precisely the age of formation of an individual structure or fabric-forming event to the inferred relative plate motion for the same time, but currently the problem remains unresolved.

Tibet may provide one example where the probable formation of large ductile fold nappes in the mid to lower crust can be related to tectonic events that we can see currently happening. In Section 19.5 (Figure 19.47), we discussed briefly the inference that the lower crust beneath Tibet may be partially molten. The lower and midcrust would therefore be expected to be very hot and weak. Along the east side of the Tibetan plateau immediately south, and to a lesser extent immediately north, of the Sichuan Basin (Figure 20.35), the normally steep plateau boundary is absent, and instead there is a rather gradual large-scale topographic ramp. One hypothesis to account for this feature is that it results from the gravitational collapse of the high topography of the plateau by flow of midcrustal material eastward out from under the plateau, as indicated by the arrows in Figure 20.35. Such a flow would undoubtedly produce large crystalline nappes such as are observed in crystalline cores of orogenic zones (e.g., see Figure 10.1*C*).

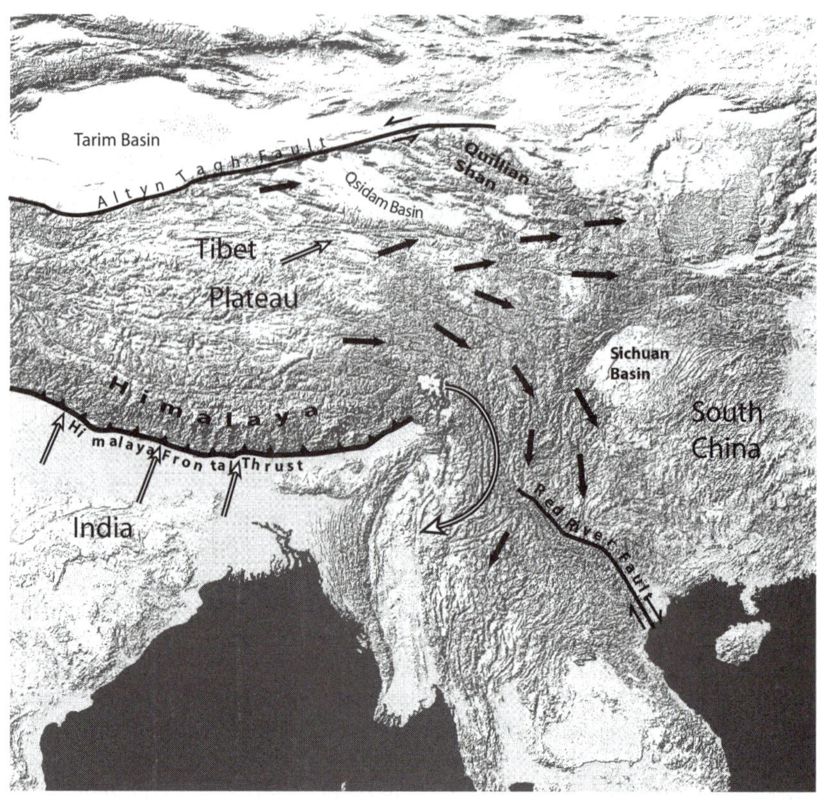

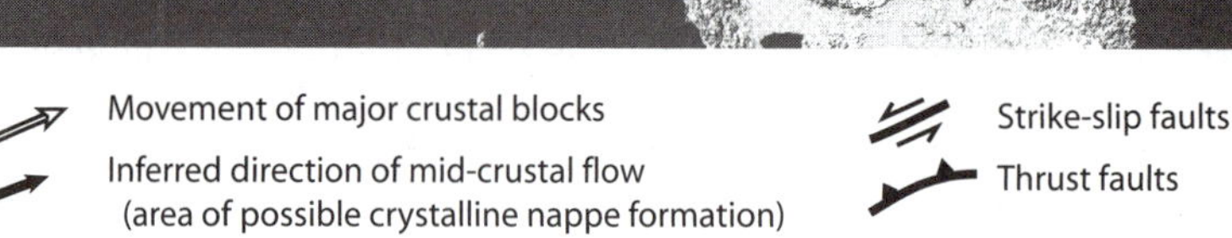

FIGURE 20.35 Oblique-view hill-shaded digital elevation map of the Tibetan plateau and the surrounding region showing topographic ramps at the eastern edge of the plateau south and north of the Sichuan basin. The ramps have a comparatively gentle topographic gradient, which is interpreted to result from the flow of rock out from under the plateau at mid- to lower-crustal levels, as indicated by the solid arrows. Such a flow is probably forming large fold nappes in the rock (cf. Figures 6.23, 18.10, and 18.11). (After Clark and Royden 2000; Burchfiel 2004; hill-shaded DEM image from http://seamless.usgs.gov/)

The ongoing collision of India with Asia is a major focus of study, as is the deep structure of the Himalayas and the Tibetan plateau. India is on a down-going plate and is currently colliding with Asia, moving relatively in a north-northeast direction. Evidently this collision has uplifted the Himalayan range and the Tibetan plateau to the north. Moreover, the rocks in the Tibetan plateau contain several sutures marking the locations of earlier collisional events. Thus Tibet may provide a location where we can study the tectonic environment associated with a large-scale mid- to deep-crustal flow subparallel to a major orogen and at a high angle to the direction of current tectonic convergence. Apparently, the plate motion that results in a major orogenic mountain range need not be obvious from a knowledge of the flow that is forming deep ductile nappes at the same time.

Such a geometry of flow is suggested for part of the Alps, both by slip lines determined using Hansen's method (Figure 20.26), the orientations of stretching lineations, and the maximum instantaneous stretching direction of the last phase of deformation inferred from the analysis of mineral fiber overgrowths (Figure 20.29). The Tibetan model suggests another possible explanation for this geometry of deformation. Although it is not clear this model would apply directly to the Alps, where a large plateau behind the main orogenic zone has not developed, the lateral escape of material into the Pannonian basin and the tectonic denudation of the Tauern window (Figure 20.23) may reflect this type of process. Thus even in a case for which we can infer the relationship between nappes and tectonics, many problems in understanding these structures remain unresolved.

20.13 THE "WILSON CYCLE" AND PLATE TECTONICS

Observations concerning the development of orogenic belts that have accumulated over the past century or so of geologic research suggested a pattern that was termed the "orogenic cycle."[6] Although the advent of the plate tectonic theory has swept away many of the old concepts, the same observations must nevertheless be accounted for by any new model. The characteristics of an "orogenic cycle" include:

1. Accumulation in separate areas, of thick deposits of both shallow-water (miogeoclinal) and deep-water (eugeoclinal) marine sediments, the latter in association with intrusions or extrusions of mafic or intermediate magmatic rocks.
2. Commencement of deformation in the foreland fold and thrust belt together with the emplacement of ophiolitic rocks and subsequent uplift of the ophiolite and the deformed sediments beneath it.
3. Continued deformation in the fold and thrust belt and in the core zone, metamorphism, deformation, intrusion of granitic batholiths, uplift, and tectonic denudation of the overthickened mountain core. In marginal foredeep basins, the deposition of syn-orogenic sediments (the "molasse" of Alpine geology).
4. Partial deformation of post-orogenic continental sediments in the outer foredeep during continuing deposition on the flanks of the orogen, continuing uplift of the orogenic region, and tectonic denudation of shallow levels of the orogenic core.
5. Block faulting, development of fault-bounded basins, and intrusion of scattered alkalic dikes or intrusive bodies.

In 1966, just as the idea of seafloor spreading was gaining widespread acceptance, the Canadian tectonicist J. T. Wilson proposed that the Atlantic Ocean had closed and reopened. This early idea spawned the concept of the **Wilson cycle,** which embeds the process of orogeny that we discussed in Section 20.12 in a grand tectonic "cycle" consisting of continental rifting and the formation of an ocean basin, followed by subduction and the closing of the ocean basin, and ending in collision and orogeny followed by reopening, and so on.

The idea of the Wilson cycle can accommodate the observations of events leading to orogeny that were outlined in Section 20.12. One possible, but by no means unique, scenario for the Wilson cycle is shown schematically in Figure 20.36:

[6]The term *cycle* is used rather loosely, because rarely, if ever, is there evidence of exactly the same sequence of events repeating in the same orogen.

1. The rifting of a continent and opening of a new ocean basin produces gradually subsiding passive continental margins. These margins accumulate thick deposits of shallow-water (miogeoclinal) sediments characterized by sandstones and limestones with subordinate shales, similar to the continental margins along the present Atlantic Ocean. Farther offshore, the eugeoclinal oceanic rocks accumulate in the ocean basin, comprising turbidite deposits of interlayered sandstones, siltstones, and shales along the continental rises, deep-water shales in the abyssal plains, and basaltic volcanics erupted at midocean spreading centers (Figure 20.36*A*, *B*). Eventually, when the pattern of plate motions shifts and the spreading pattern changes, a subduction zone develops in the ocean basin, probably along pre-existing fractures such as transform faults, oceanic fracture zones, or ridge-parallel faults (Figure 20.36*C*). The resulting island arcs produce deposits of volcanic rocks and volcanogenic sedimentary rocks that complete the oceanic, or eugeoclinal, suite.
2. A passive continental margin on the down-going plate collides with the subduction zone (Figure 20.36*D*), emplacing a piece of oceanic crust and island arc crust (an ophiolite) onto the subducted continental margin, juxtaposing the eugeoclinal and miogeoclinal suites, and initiating the formation of a foreland fold and thrust belt in the sediments of the passive margin. This collision probably is not synchronous along the entire margin but migrates along strike with time, depending on the relative geometries of the subduction zone and the continental margin.
3. Following the first collision, the polarity of the subduction zone flips (cf. Figure 19.43*B*), producing a continental arc (like the current Andean-style continental margin) along the now-deformed former passive continental margin (Figure 20.36*E*) and initiating subduction of the remainder of the ocean basin. The consuming continental margin depicted in Figure 20.36*E* need not be a part of the originally rifted continent, because the convergent relative velocities need not be the exact opposite of the original divergent relative velocities. Pre-orogenic plutons intrude the deformed continental margin. Deep basins in the trench, along the continental margin, or deep-sea fans derived from the continental margin in the narrowing ocean constitute the deep-water "orogenic sediments." Eventually, the second passive continental margin arrives at the subduction zone and begins a continent-continent collision (Figure 20.36*F*, *G*), which sutures the formerly separate pieces of continent. All the rocks in the suture region are deformed; the thrusting of the continental crust on

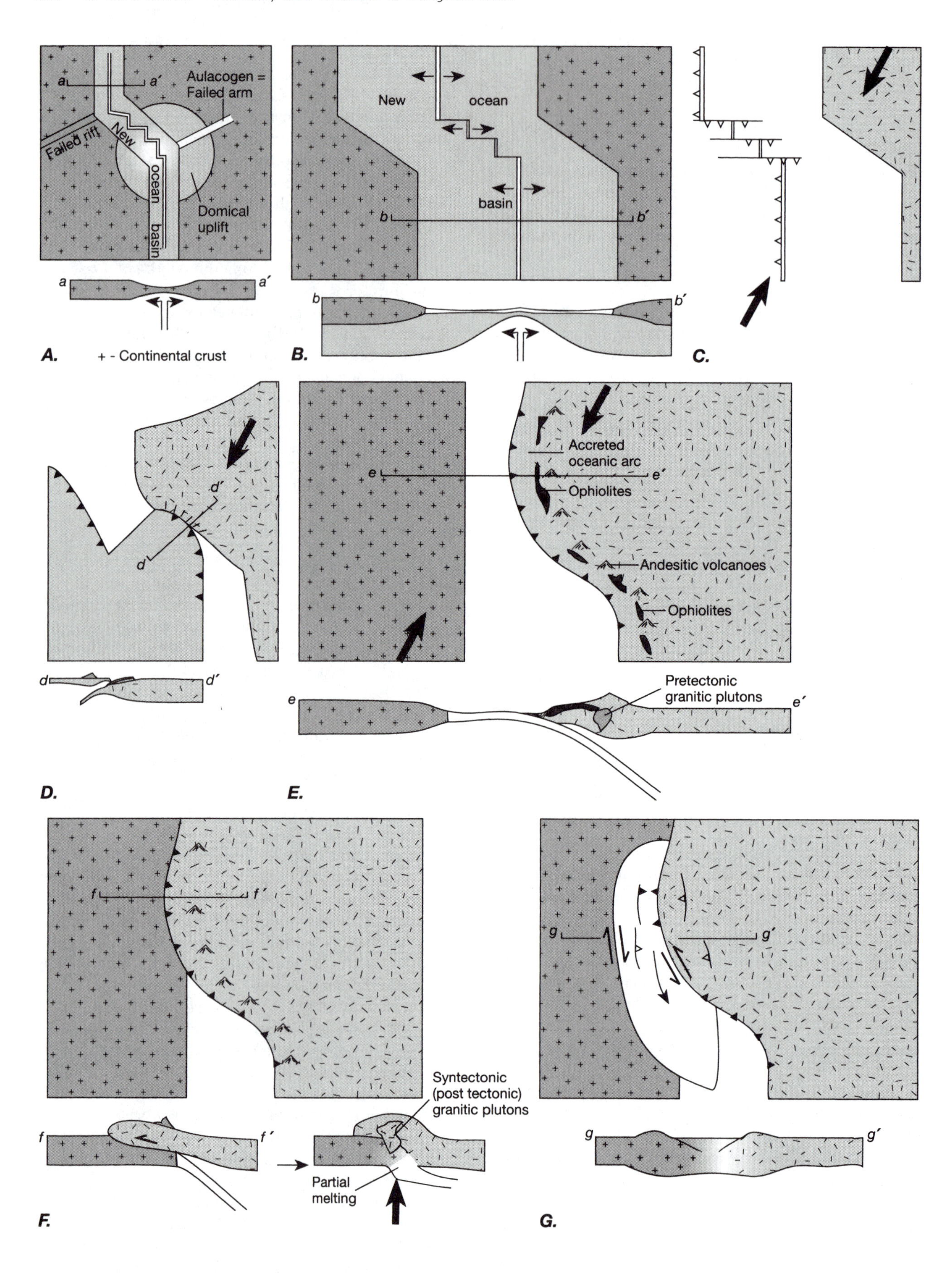
a
a′
Aulacogen =
Failed arm
Failed rift
New
ocean
basin
Domical
uplift
a
a′
A.
+ - Continental crust
New
ocean
basin
b
b′
b
b′
B.
C.
d′
d
d
d′
D.
e
e′
Accreted
oceanic arc
Ophiolites
Andesitic volcanoes
Ophiolites
e
Pretectonic
granitic plutons
e′
E.
f
f′
f
f′
Syntectonic
(post tectonic)
granitic plutons
Partial
melting
F.
g
g′
g
g′
G.

FIGURE 20.36 Diagrammatic sketch maps and cross sections illustrating possible development of a plate-tectonic Wilson cycle to account for traditional observations of the history of formation of orogenic belts. *A*. Rifting of the continental margin and formation of domical uplifts and aulacogens along failed arms. *B*. Development of mature Atlantic-style ocean with passive margins. *C*. Change of relative plate motion, commencement of closure of ocean, development of subduction zones along pre-existing fracture zones, ridge-transform fault intersections, or ridge parallel faults in the oceanic crust. *D*. Collision of a hypothetical intra-oceanic arc system with margin of the continent on the right, with emplacement of ophiolites. *E*. "Flip" of subduction direction, development of Andean-style margin on continent to the right, convergence of continent on the left with the subduction zone. *F*. Collision of two continents with subduction of Atlantic-style continental margins under the Andean-style margin, formation of mountain root. *G*. Adjustment of continental margin collision zone by strike-slip movement or renewed rifting.

the down-going plate beneath the continental crust on the over-riding plate causes uplift and creates a deep root. Partial melting near the base of the root produces late-stage intrusive or extrusive magmatic rocks. Strike-slip faulting takes place as the irregular edges of the continents adjust to each other.

4. Uplift of the collision zone by underplating and isostatic rise creates a source for post-orogenic sediments and deforms earlier sediments. Strike-slip faults continue as the zone accommodates further post-collisional convergence (Figure 20.35*G*).
5. Finally, the suture zone may be torn apart by another rifting event, which causes normal faulting and the intrusion and extrusion of alkalic to basaltic magmas.

The Wilson cycle as outlined here is based largely on Appalachian-Caledonian tectonic history, although recent investigations have shown that history to be much more complex. It is highly unlikely that two rifted continental margins will come back together in exactly the same place from which they rifted, so the idea of a "cycle" cannot be understood too strictly. Rather, given the changes in plate motion observed as a normal consequence of multi-plate tectonic evolution, it is much more probable that different parts of the margins on the two rifted continents, or even other margins of the same or other continents, will approach each other during the closing of the ocean basins and that the margins will not fit together well. Indeed, many continental collisions probably have occurred between two continental margins that were never previously close to one another. Subduction directions are generally oblique, at least to some extent, and misfits along colliding continental margins will adjust by lateral motion of crustal blocks. Thus we can expect to find that strike-slip motion in orogenic regionships is as important as the more obvious contractional motion. Finally, the Wilson cycle does not account for the agglomeration of exotic terranes, which is a common feature of orogenies.

The Wilson cycle may capture, imperfectly, an important aspect of tectonic cycles that operate on a global scale, specifically the assembly and breakup of supercontinents. The late Paleozoic assembly of Pangaea and the Mesozoic dispersal of its fragments are well known. Preceding this latest supercontinent cycle, however, are the assembly of the supercontinent of Rodinia about 1000 Ma, its breakup about 800 Ma to 600 Ma, and the assembly of Gondwanaland approximately 700 Ma to 500 Ma as an early phase of the final assembly, around 250 Ma, of Pangaea. To account for this cyclicity, several geologists have speculated on the existence of **superplumes**, which are inferred from very large seismic velocity anomalies in the mantle. These anomalies are postulated to represent convective upwellings (anomalously low seismic velocity) or downwellings (anomalously high seismic velocity) that involve the entire mantle and that are more than 1000 km in diameter. The model for assembly of supercontinents holds that relatively cold subducted plates tend to accumulate at the 600-km-deep phase transition in the mantle. Over a long time period, a sufficiently large volume of plate material accumulates so that it eventually triggers a catastrophic "avalanche" of cold slab material into the lower mantle. The resulting converging flow in the upper mantle pulls together a number of formerly separate continents into a supercontinent. The upwelling of a subsequent hot superplume, possibly from interaction between the core and the lower mantle, leads to the subsequent breakup of the supercontinent. Although little is known of the geometry of convection cells in the mantle, the breakup of the lithosphere tends to follow old faults, sutures, or other lines of weakness that we observe in the crust, even though it may not be located directly over an upwelling limb of a mantle convection cell.

Based on seismic velocity anomalies, two major upwellings are thought to exist currently, one beneath Africa and the other beneath the southwest Pacific. A counterpart downwelling, or "cold superplume," is postulated to exist beneath Asia, thus perhaps giving rise to the continuing convergence of the Indian-Australian plate with the Eurasian plate, despite the Alpine-Himalayan continent-continent collision that extends from western Europe to southeast Asia.

According to another hypothesis, the Pacific Ocean between Asia, Australia, Antarctica, and North and South America, is now contracting as a result of the opening of the Atlantic and Indian Oceans. These continents resulted from the breakup of the supercontinent Rodinia about 800 Ma to 600 Ma, and they may in the future reassemble, with parts of their boundaries that were external to the Rodinian supercontinent becoming the sites of major sutures within a new supercontinent, in effect turning the former supercontinent inside-out. How this hypothesis fits with the inference of the hot superplume below the southwest Pacific and the cold superplume below Asia is not clear and remains a question for future research.

Thus the expanded concept for the Wilson cycle may involve a grand scheme of supercontinent formation and breakup over a cycle time of hundreds of millions of years that involves many island arcs and midplate volcanics, in addition to the main enveloping continents. This postulated complexity is supported by the abundance of "suspect" or "exotic" terranes present in most major orogenic belts. We discuss the analysis of these terranes briefly in Section 20.14, which follows.

20.14 TERRANE ANALYSIS

Many, if not most, of the world's orogenic belts include a composite of distinct terranes that originate not only from the continent(s) or arc(s) involved in the final collision but also from areas that are clearly "exotic" to the main crustal blocks or whose relationship to those blocks is "suspect." We call such terranes **exotic** or **suspect terranes.**

An exotic terrane is thus an area surrounded by sutures[7] and characterized by rocks having a stratigraphy, petrology, or paleolatitude that is distinctly different from that of neighboring terranes or continents. In a collisional orogen, it is a remnant of crust that has had a different history from the subducted oceanic crust or from the main crustal blocks that have collided. Some mountain belts, such as the North American Cordillera, are characterized by collisions between a single continent and numerous exotic terranes rather than between two continents. Some of these terranes apparently are similar to the areas of anomalous oceanic crust discussed in Section 1.5(ii).

A unique method of analysis, referred to as **terrane analysis**, is required to understand these exotic or suspect terranes and their role in the formation of an orogen. Application of that analysis has begun to shed new light on the history of these complex regions. The object of terrane analysis is to work out when adjacent terranes were apart and when they came together, as revealed by a detailed comparison of the geologic histories. Figure 20.37 illustrates schematically the method of analysis with a hypothetical map (Figure 20.37*A*) and a strati-

[7]A **suture** is defined as a zone that marks the boundary between two adjacent continental terranes that were initially separated by an ocean basin. Thus it is a site at which that ocean basin has been subducted (see Sections 20.12 and 20.13).

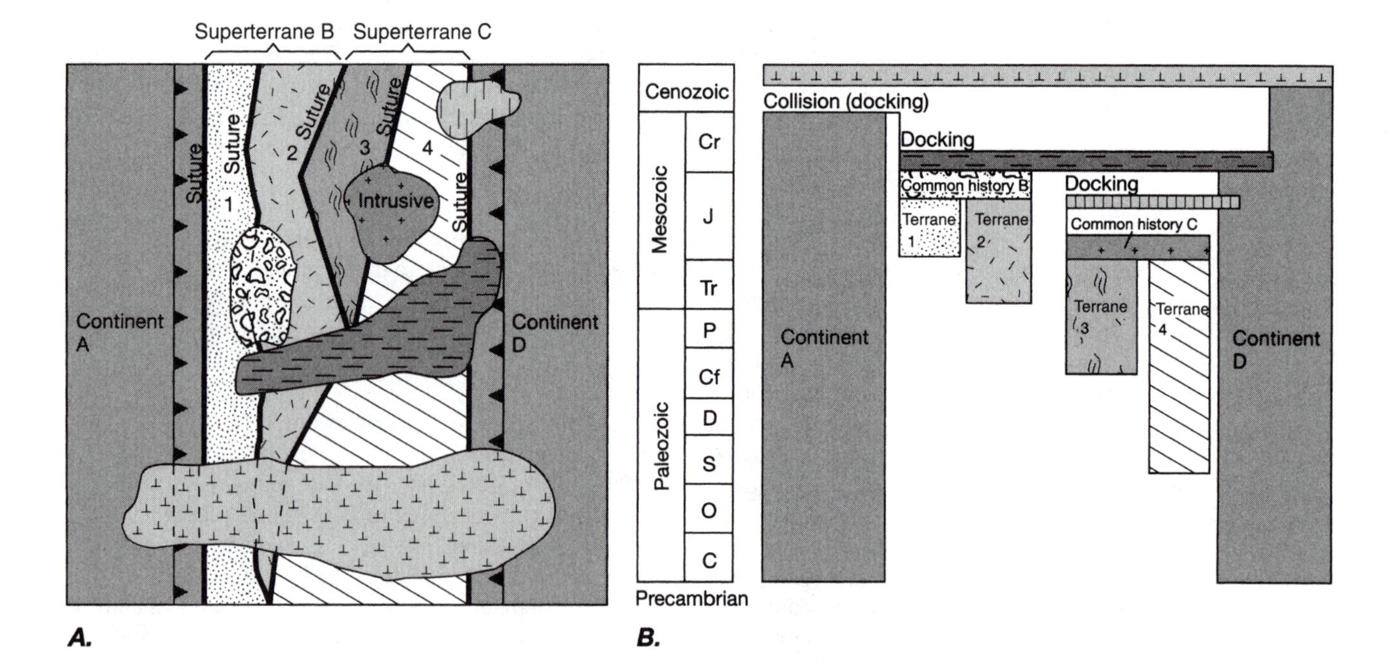

FIGURE 20.37 Analysis of exotic terranes. *A*. Schematic map of two continents separated by several exotic terranes. *B*. Stratigraphic-tectonic diagram illustrating ages of rocks in individual terrains and ages of "docking" and of common histories. See the text for discussion.

graphic-tectonic diagram (Figure 20.37*B*). Four terranes, numbered 1 through 4, are separated from one another and from continents A and D by sutures. By plotting the stratigraphy from each terrane in a column, with adjacent columns for adjacent terranes (Figure 20.37*B*), it is possible to determine when the terranes began a common history and thereby to constrain when they collided, or "docked." The time of collision or docking of two terranes or a terrane and a continent is given by the maximum age of either cross-cutting intrusives or sediments unconformably overlying both terranes. Apparent polar wander (APW) paths (see Appendix 2, Section A2.4) of two separate terranes should be different until the docking, after which they should exhibit a single APW path.

Terranes 3 and 4 contain rocks of Carboniferous to Cenozoic and Siberian to Cenozoic age, respectively. They share a common history "C" from Jurassic time onward, as indicated by a date on the pluton ("+" pattern) that intrudes the suture. In mid-Jurassic time, they collided with continent D, as indicated by the sedimentary unit (vertical-ruled) that overlies that suture. Rocks in terrane 1 are of Jurassic through Cenozoic age, and in terrane 2, they are of Triassic through Cenozoic age. The diagram shows that they share a common history "B" beginning in late Jurassic, when a rock unit (dot-and-blob pattern) was deposited across the suture between them. Terranes 1 and 2 docked or collided with the already amalgamated terranes 3 and 4 and continent D in early Cretaceous time, as indicated by the sedimentary unit overlying all the sutures between terranes 2-3, 2-4, 3-4, and 4-D (dark gray with horizontally ruled pattern). Continent A collided with the composite of the terranes and continent D in Cenozoic time, when the sedimentary deposits covering all the previously accreted terranes (inverted T pattern) began also to cover continent A.

The recognition of exotic terranes in orogens around the world requires a major modification of the Wilson cycle and of our ideas about how mountain belts develop. Given that areas of anomalously thick oceanic crust can develop in the ocean basins (Section 1.5(ii)), the process of subduction must sweep them into the subduction zone ahead of any continent that rides on the same plate. Before collision, these terranes have their own individual geologic history, but after collision, they become part of the over-riding crustal block. Thus the argument about whether orogeny is an ongoing or highly episodic event may be settled in favor of both sides, in that the docking of exotic terranes or island arc collisions may be a recurring process during subduction, but the collision of major continents is only a single episode. Thus the motions of the major crustal blocks in a collision may only represent a part of the activity that constructs an orogenic belt, and many other relative plate motions may be represented by the numerous sutures among different exotic terranes.

The thickness of exotic terranes is also quite variable. Seismic evidence indicates that although some are of full crustal thickness, most are thin, not more than a few kilometers thick, and thus qualify as complex multi-lithologic thrust sheets, rather than as true crustal blocks. Thus the amount of material added by terranes to individual continents remains uncertain.

REFERENCES AND ADDITIONAL READINGS

Anhausser, K., C. R. Anhausser. 1984. C. R.: Structural elements of Archaean granite-greenstone terranes as exemplified by the Barberton Mountain Land, southern Africa. In A. Kröner and R. Greiling, eds, *Precambrian Tectonics Illustrated*, Schweizerbartsche, Stuttgart: 57–78.

Anonymous. 1950. *Der Bau der Erde, 1:140,000,000*. Justus Perthes, Gotha, Thuringia, Germany.

Bally, A. W. 1980. *Basins and Subsidence—A Summary*. Amer. Geophys. Union, GSA Geodynamic Series 1: 5–20.

Bateman, P. C. 1981. Geologic and geophysical constraints on models for the origin of the Sierra Nevada Batholith, California. In W. G. Ernst, ed., *The Geotectonic Development of California; Rubey Volume I*, Prentice-Hall, Englewood Cliffs, NJ: 71–86.

Burbank, D., and Anderson, R. S. 2001. *Tectonic Geomorphology*. Blackwell, Malden, MA, 274 pp.

Burchfiel, B. C. 1983. The continental crust. In R. Siever, ed., *The Dynamic Earth*, Scientific American.

Burchfiel, B. C. 2004. 2003 Presidential address: New technology, new geological challenges. *GSA Today* 14(2): 4–9.

Burchfiel, B. C., C. Zhiliang, K. V. Hodges, L. Yuping, L. H. Royden, D. Changrong, and X. Jiene. 1992. The South Tibetan detachment system, Himalayan Orogen: extension contemporaneous with and parallel to shortening in a collisional mountain belt., Geol. Soc. Amer. Spec. Paper 269, 41 pp.

Burke, K. 1980. Intracontinental rifts and aulacogens In *Geophysics Study Committee, Continental Tectonics*, National Academy of Sciences: 42–50.

Carswell, D. A., and R. Y. Zhang. 2000. Petrographic characteristics and metamorphic evolution of ultrahigh-pressure eclogites in plate-collision belts. In W. G. Ernst and J. G. Liou, eds., *Ultrahigh-Pressure Metamorphism and Geodynamics in Collision-Type Orogenic Belts: Final Report of the Task Group III-6 of the International Lithosphere Project*, International Book Series, v.4: 39–56;

International Lithosphere Program, Contribution No. 344, Geol. Soc. Amer., Boulder, CO.

Chopin, C., and H.-P. Schertl. 2000. The UHP unit in the Dora-Maira massif, western Alps. In W. G. Ernst and J. G. Liou, eds., *Ultrahigh-Pressure Metamorphism and Geodynamics in Collision-Type Orogenic Belts: Final Report of the Task Group III-6 of the International Lithosphere Project*, International Book Series, v.4: 133–148; International Lithosphere Program, Contribution No. 344. Geol. Soc. Amer., Boulder, CO.

Clark, M. K., and L. H. Royden. 2000. Topographic ooze: Building the eastern margin of Tibet by lower crustal flow. *Geology* 28: 703–706.

Clark, S. P. Jr., B. C. Burchfiel, and J. Suppe, eds. 1988. Processes in continental lithospheric deformation. Geol. Soc. America Spec. Paper 218, 212 pp.

Continental tectonics / Geophysics Study Committee, Geophysics Research Board, Assembly of Mathematical and Physical Sciences, National Research Council. 1980. *Continental Tectonics*: National Academy of Sciences, Washington, D.C., 197 pp.

Cook, F. A., L, D. Brown, S. Kaufman, J. E. Oliver, and T. A. Petersen. 1981. COCORP seismic profiling of the Appalachian orogen beneath the coastal plain of Georgia. *Geol. Soc. Amer. Bull.* 92(10): I 738–I 748.

Coward, M. P., and A. W. B. Siddans. 1979. The tectonic evolution of the Welsh Caledonides. In A. L. Harris, C. H. Holland, and G. E. Leake, eds., *The Caledonides of the British Isles Reviewed*, Geol. Soc. Lond. Sp. Pub. 8: 187–199.

Dalziel, I. W. D., and R. D. Forsythe. 1985. *Andean Evolution and the Terrane Concept: Circum-Pacific Council for Energy and Mineral Resources.* Earth Science Series, No. 1: 565–581.

Davis, D., J. Suppe, and F. A. Dahlen. 1983. Mechanics of fold-and-thrust belts and accretionary wedges. *J. Geophys. Res.* 88: 1153–1172.

Dewey, J. F. 1977. Suture zone complexities: A review. *Tectonophysics* 40: 53–67.

Durney, D. W., and J. G. Ramsay. 1973. Incremental strain measured by syntectonic crystal growths. In K. A. De Jong and R. Scholten, eds. *Gravity and Tectonics*, Wiley, New York: 67–96.

England, P. C., and A. B. Thompson. 1984. Pressure-temperature-time paths of regional metamorphism. I. Heat transfer during the evolution of regions of thickened continental crust. *J. Petrol.* 25: 894–928.

Ernst, W. G., and J. G. Liou, eds. 2000. *Ultrahigh-Pressure Metamorphism and Geodynamics in Collision-Type Orogenic Belts: Final Report of the Task Group III-6 of the International Lithosphere Project*, International Book Series, v.4, 293 pp.; International Lithosphere Program, Contribution No. 344. Geol. Soc. Amer., Boulder, CO.

Ernst, W. G., and J. G. Liou. 2000. Overview of UHP metamorphism and tectonics in well-studied collisional orogens. In W. G. Ernst and J. G. Liou, eds., *Ultrahigh-Pressure Metamorphism and Geodynamics in Collision-Type Orogenic Belts. Final Report of the Task Group III-6 of the International Lithosphere Project*, International Book Series, v.4: 3–19. Geol. Soc. Amer., Boulder, CO.

Eskola, P. 1949. The problem of mantled gneiss domes. *Quart. J. Geol. Soc. London* 104: 461–475.

Francheteau, J. 1983. The oceanic crust. In *The Dynamic Earth*, Scientific American.

Frey, M., J. C. Hunziker, W. Frank, J. Bocquet, G. V. Dal Piaz, E. Jager, and E. Niggli. 1974. Alpine metamorphism of the Alps, a review. *Schweizerische Mineralogische und Petrografische Mitteilung = Bulletin Suisse de Mineralogie et Petrographie*, 54(2–3): 247–290.

Galson, D. A., and S. Mueller, eds. 1984. Proceedings of the First Workshop on the European Geotraverse (EGT), the Northern Segment, Munsch. European Science Foundation, Strasbourg, France.

Galson, D. A., and S. E. Mueller. 1986. The European geotraverse, part 2. *Tectonophysics* 128(3/4): 163–396.

Gansser, A. 1964. *Geology of the Himalayas.* Wiley-Interscience, New York.

Glazner, A. F., J. M. Bartley, D. S. Coleman, W. Gray, and R. Z. Taylor. 2004. Are plutons assembled over millions of years by amalgamation from small magma chambers? *GSA Today* 14(4/5): 4–11 (Doi:10.1130/1052-5173(2004)014<0004:APAOMO> 2.0.CO;2).

Goodwin, A. M. 1981. Archean plates and greenstone belts. In A. Kroner, ed., *Precambrian Plate Tectonics*, Elsevier.

Hansen, E. C. 1971. *Strain Facies.* Springer Verlag, New York, 207 pp.

Hansen, E. C., W. H. Scott, and R. S. Stanley. 1967. Reconnaissance of slip line orientations in parts of three mountain chains. *Carnegie Inst. Washington Year Book* 65: 406–410.

Harris, L. D., and K. C. Bayer. 1979. Sequential development of the Appalachian orogen above a master decollement—A hypothesis. *Geology* 7: 568–572.

Hatcher, R., and R. T. Williams. 1986. Mechanical model for single thrust sheets. *Geol. Soc. Amer. Bull.* 97: 975–985.

Hoffman, P. F. 1988. United plates of America, the birth of a craton: Early Proterozoic assembly and growth of Laurentia. *Ann. Rev. Earth Planet. Sci.* 16: 543–603.

Hoffman, P. F., J. F. Dewey, and K. C. Burke. 1974. Aulacogens and their genetic relation to geosynclines, with a Proterozoic example from Great Slave Lake, Canada. In R. H. Dott, Jr., and R. H. Shaver, eds., *Modern and Ancient*

Geosynclinal Sedimentation, Soc. Econ. Paleont. Mineral. Spec. Publ. 19: 38–55.

Jackson, M. P. A. 1984. Archean structural styles in ancient gneiss complex of Swaziland, southern Africa. In A. Kröner and R. Greiling, eds., *Precambrian Tectonics Illustrated*. Stuttgart, Schweizerbartsche: 1–18.

Jordan, T. E., B. L. Isacks, R. W. Allmendinger, J. A. Brewer, V. A. Ramos, and C. J. Ando. 1983. Andean tectonics related to geometry of subducted Nazca plate. *Geol. Soc. Amer. Bull.* 94: 341–361.

Kay, M. 1951. *North American Geosynclines*, Geol. Soc. Amer. Mem. 48.

King, P. B. 1977. *Evolution of North America*, second edition. Princeton University Press, Princeton, 197 pp.

Krabbendam, M., and J. F. Dewey. 1998. Exhumation of UHP rocks by transtension in the Western Gneiss region, Scandanavian Caledonides. In R. E. Holdsworth, R. A. Strachan, and J. F. Dewey, eds., *Continental Transpressional and Transtensional Tectonics*, Geol. Soc. Sp. Pub. No. 135. Geol. Soc. Lond.: 159–181.

Kröner, A., and R. Greiling, eds. 1984. *Precambrian Tectonics Illustrated*, Schweizerbartsche, Stuttgart, 419 pp.

Laubscher, H. P. 1982. Detachment, shear, and compression in the central Alps. *Geol. Soc. Amer. Mem.* 158: 191–211.

LeForte, P. 1975. Himalayas: The collided range. Present knowledge of the continental arc. *Amer. J. Sci.* 275A: 1–44.

Liou, J. G. 2000. Petrotectonic summary of less intensively studied UHP regions. In W. G. Ernst and J. G. Liou, eds., *Ultra-High Pressure Metamorphism and Geodynamics in Collision-type Orogenic Belts*. Bellwether Geol. Soc. Amer. Boulder, CO: 20–38.

Merle, O. 1986. Patterns of stretch trajectories and strain rates in spreading-gliding nappes. *Tectonophysics* 124: 211.

Miller, H., S. Mueller, and G. Perrier. 1983. Structure and dynamics of the Alps: A geophysical inventory. In H. Berckhemer and K. Hsu, eds., *Alpine Mediterranean Geodynamics*, Geol. Soc. Amer., Amer. Geophys. Union, Geodynamics Series 7: 75–204.

Misra, K. C., and F. B. Keller. 1978. Ultramafic bodies in the southern Appalachians: A review. *Amer. J. Sci.* 278: 389–418.

Moore, J. C. 1978. Orientation of underthrusting during latest Cretaceous and earliest Tertiary time, Kodiak Islands, Alaska. *Geology* 6: 209–213.

Moores, E. M., and R. J. Twiss. 1995. *Tectonics*. Freeman, New York, 415 pp.

National Research Council (U.S.A.). Ad Hoc Panel to Investigate the Geological and Geophysical Research Needs and Problems of Continental Margins. 1979. *Continental Margins: Geological and Geophysical Research Needs and Problems*. National Academy of Sciences, no. 2793, 302 pp.

Nisbet, E. G. 1987. *The Young Earth: An Introduction to Archaean Geology*. Boston: Allen & Unwin, 402 pp.

Oldow, J. S., A. W. Bally, and H. G. Avé Lallemant. 1990. Transpression, orogenic float, and lithospheric balance. *Geology* 18: 991–994.

Percival, J. A., W. Bleeker, F. A. Cook, T. Rivers, G. Ross, and C. vanStaal. 2004. PanLITHOPROBE Workshop IV: Intraorogen correlations and comparative orogenic anatomy. *Geoscience Canada* 31: 23–39.

Pfiffner, O. A., S. Ellis, and C. Beaumont. 2000. Collision tectonics in the Swiss Alps: Insight from geodynamic modeling. *Tectonics* 19(6): 1065–1094.

Platt, J. P. 1986. Dynamics of orogenic wedges and the uplift of high-pressure metamorphic rocks. *Geol. Soc. Amer. Bull.* 97: 1037–1053.

Platt, J. P. 1993. Exhumation of high-pressure rocks: A review of concepts and processes. *Terra Nova* 5: 119–133.

Press, F., and R. Siever. 1986. *Earth*, fourth edition, Freeman, New York.

Price, R. A., and R. D. Hatcher, Jr. 1983. Tectonic significance of similarities in the evolution of the Alabama-Pennsylvania Appalachians and the Alberta–British Columbia Canadian Cordillera. *Geol. Soc. Amer. Mem.* 158: 149–160.

Ramsay, J. G. 1963. Stratigraphy, structure and metamorphism in the western Alps. *Geol. Assoc. Proc. London* 74: 357–392.

Ratschbacher, L. O., W. Frisch, H.-G. Linzer, and O. Merle. 1991. Lateral extrusion in the Eastern Alps, Part II: Structural analysis. *Tectonics* 10: 256–272.

Roeder, D. H. 1973. Subduction and orogeny. *J. Geophys. Res.* 78: 5005–5024.

Roeder, D. H., O. E. Gilbert, Jr., and W. D. Witherspoon. 1978. Evolution and macroscopic structure of Valley and Ridge thrust belt, Tennessee and Virginia. *Studies in Geology* 2, 25 pp.

Sanderson, D. J. 1982. Models of strain variation in nappes and thrust sheets: A review. *Tectonophysics* 88: 201–233.

Savilieva, G. N., A. Y. Sharaskin, A. A. Saviliev, P. Spadea, A. N. Pertsev, and I. I. Babarina. 2002. *Ophiolites and Zoned Mafic-Ultramafic Massifs of the Urals: A Comparative Analysis and Some Tectonic Implications*. Amer. Geophys. Union Geophys. Monogr. 132: 135–153.

Schaer, J.-P., and J. Rodgers, eds. 1987. *The Anatomy of Mountain Ranges*. Princeton University Press, Princeton, 298 pp.

Seeber L., J. G. Armbruster, and R. C. Quittmeyer. 1981. Seismicity and continental subduction in the Himalayan arc. In H. K. Gupta and F. M. Delany, eds., *Zagros-Hindu Kush-Himalaya Geodynamic Evolution*, Washington D.C. Amer. Geophys. Union, Geophys. Soc. Amer. Geodynamics Series 3: 215–242.

Sengör, A. M. C., and B. A. Natal'in. 1996. Turkic-type orogeny and its role in the making of the continental crust. *Ann. Rev. Earth and Planet. Sci.* 24: 263–337, Annual Reviews, Palo Alto, CA, USA.

Siddans, A. W. B. 1983. Finite strain patterns in some Alpine nappes. *J. Struct. Geol.* 5(3/4): 441.

Smithson, S. B., P. N. Shive, and S. K. Brown. 1977. Seismic velocity, reflection, and structure of the crystalline crust. In J. G. Heacock, ed., *The Earth's Crust, Its Nature and Physical Properties*, Amer. Geophys. Union Monograph 20: 254–270.

Smithson, S. B., J. Brewer, S. Kaufman, J. Oliver, and C. Hurich. 1978. Nature of the Wind River thrust, Wyoming, from COCORP deep-reflection data and from gravity data. *Geology* 6: 648–652.

Spicher, A. 1980. Tektonische Karte der Schweiz, Schweizer Geologisches Kommission.

Stanley, S. 1993. *Earth and Life through Time.* W. H. Freeman and Co., New York.

Stanton, R. L. 1972. *Ore Petrology.* McGraw-Hill, New York.

Thompson, J.B., Jr., P. Robinson, T. N. Clifford, and N. J. Trask, Jr. 1968. Nappes and gneiss domes in west-central New England. In Zen, E-A, W. S. White, J. B. Hadley, and J. B. Thompson, Jr. 1968. *Studies of Appalachian Geology: Northern and Maritime*, New York, Interscience Publishers: 203–219.

Trendall, A. P. 1968. Three great basins of Precambrian banded iron formation deposition: A systematic comparison, *Geol. Soc. Amer. Bull.* 79: 1527–1544.

Turner, F. J. 1968. *Metamorphic Petrology, Mineralogical and Field Aspects.* McGraw Hill, New York, 403 pp.

Uyeda, S. 1978. *The New View of the Earth*, W. H. Freeman and Co., New York, 217 pp.

Vail, P. R., R. M. Michum, Jr., and S. Thompson. *AAPG Mem.* 19: 84.

Wakabayashi, J. 2004. Tectonic mechanisms associated with *P-T* paths of regional metamorphism: alternatives to single-cycle thrusting and heating. *Tectonophysics* 392(1–4): 193–218.

Williams, H. 1984. Miogeoclines and suspect terranes of the Caledonian-Appalachian Orogen; tectonic patterns in the North Atlantic region. *Can. J. Earth Sci.* 21: 887–901.

Windley, B. F. 1993. *The Evolving Continents*, third edition, Wiley, New York.

Windley, B. F., F. C. Bishop, and J. V. Smith. 1981. Metamorphosed layered igneous complexes in Archean granulite-gneiss belts. *Ann. Rev. Earth and Planet. Sci* 9: 175–198, Annual Reviews, Palo Alto, CA, USA.

Yin, A. 2006, Cenozoic tectonic evolution of the Himalayan orogen as constrained by along-strike variation of structural geometry, exhumation history, and foreland sedimentation. *Earth Science Reviews.* 76: 1–131.

Yin, A., and M. Harrison, eds. 1996. *The Tectonic Evolution of Asia. Rubey Colloquium.* Cambridge University Press, Cambridge, UK.

Zingg, A., and R. Schmid. 1979. Multidisciplinary research on the Ivrea zone. *Schweische Mineralogische und Petrografische Mitteilung* 59: 189–197.

Appendix 1

THE ORIENTATION AND REPRESENTATION OF STRUCTURES

Investigations in structural geology require familiarity with basic techniques of observation and of reporting and displaying three-dimensional information such as the orientations of bedding planes or fold axes. Geologists therefore have developed standardized methods for measuring the orientations of planes and lines in space and for displaying the information in graphical form. Geologic maps and cross sections display data in a geographic framework; histograms and spherical projections portray only orientation data without regard to geographic location. Most discussions of structural geology and tectonics assume the reader is familiar with all these techniques, and laboratories in structural geology traditionally cover these topics in detail. In this appendix, we review only briefly the essentials necessary for understanding the discussion in the main text.

A1.1 THE ATTITUDE OF PLANES AND LINES

Many of the structures observed in rock outcrops are approximately planar or linear features. Planar features include bedding, fractures, fault planes, dikes, unconformities, and planar preferred orientations of micas. Linear features include grooves and streaks on a surface, intersections of two planar features, fold hinge lines, and linear preferred orientations of mineral grains. We can describe features that are not planar, such as folded surfaces and folded linear features, respectively, by specifying the range of orientations of tangent planes or tangent lines around the structure.

Describing the **attitude** of a plane or a line—that is, defining its orientation in space—is fundamental to the description of structures. We specify the attitudes of both planes and lines by specifying two angles, which we define in the following two paragraphs. For the attitude of a plane, these angles are called, respectively, the strike and the dip; for the attitude of a line, they are the trend and plunge.

The **strike line** for any orientation of plane is the line in the plane that is horizontal. The **strike** of a plane, then, is the angle measured from geographic north to the strike line (Figure A1.1*A*). It has a unique orientation for any given orientation of the plane except for a horizontal plane, because all lines lying in a horizontal plane are of course horizontal. We specify the strike as either a bearing or an azimuth. The bearing is an angle between 0° and 90° measured from north in the quadrant east or west of north (in some cases it is measured east or west of south). Thus N25E and N56W are bearings. The azimuth is the angle between 000° and 360° increasing clockwise from north. Azimuths are commonly written as three-digit numbers to distinguish them from the dip or plunge angle, defined in the next paragraph, which are never more than 90° and thus never need more than two digits. Thus 025 and 304 are the azimuth readings for the two bearings listed above. Strikes that differ by 180° represent the same orientation, for example, N50E and S50W or 050 and 230.

The **dip line** is the line that lies in the planar structure perpendicular to the strike line, and it is the line of steepest descent on the plane. The **dip angle**, or simply the **dip**, is the angle between 0° and 90°, measured in a vertical plane from the horizontal down to the dip line (Figure A1.1*A*). It is the largest possible acute angle between the horizontal plane and the inclined plane. For a given strike, a particular value of the dip angle identifies two planes that slope in opposite directions. To

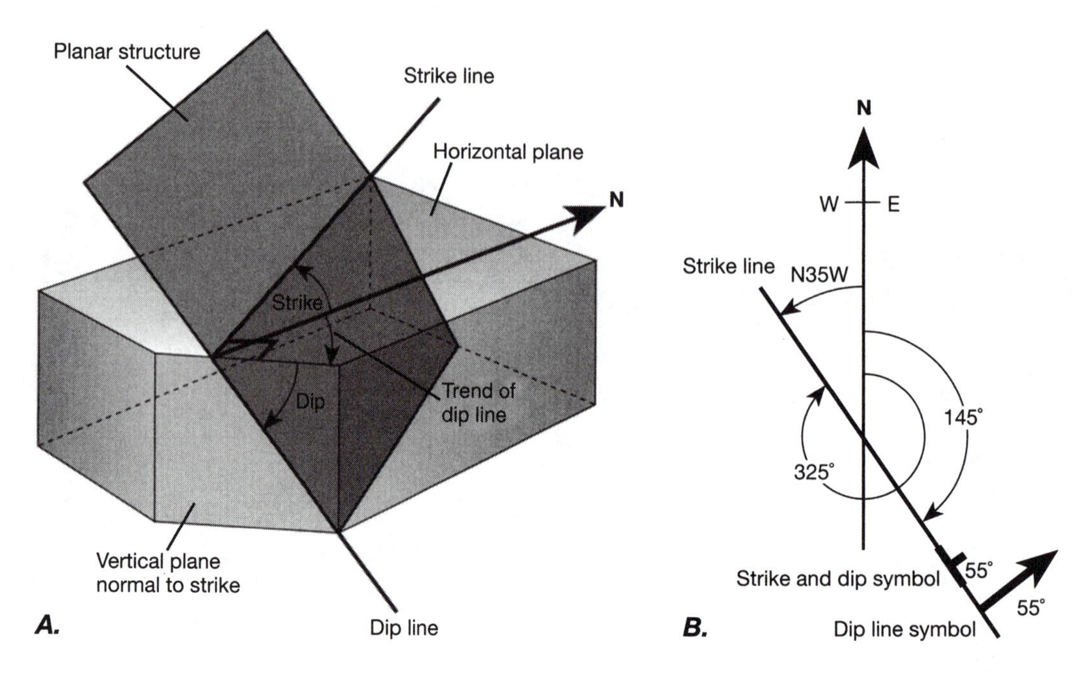

FIGURE A1.1 The strike and dip of a planar structure. *A*. The strike line of a planar structure is a line defined by the intersection of a horizontal plane with the planar structure. The strike is the angle between north and that horizontal line in the planar structure. The dip angle is measured between horizontal and the planar structure in a vertical plane normal to the strike line. The dip direction in this diagram is to the northeast. *B*. The strike shown here is defined by a bearing of N35W or an azimuth of either 325° or 145°. The attitude is indicated on a map by a T-shaped symbol with the cross-bar parallel to the strike line and the short stem parallel to the dip direction. The dip angle is written beside the stem.

distinguish between them, we specify the quadrant (NE, SE, SW, NW) or the approximate compass direction (N, E, S, W) of the down-dip direction. For example, for the same strike, dips of 36NE and 36SW identify differently oriented planes. Dips of 36NE and 36N are two ways of writing the same measurement.

We generally specify the orientation of a plane by writing the strike, dip, and dip direction in that order. Thus (N35W; 55NE), (325; 55NE), and (145; 55NE) all specify the same attitude of plane (Figure A1.1*B*). In some cases, the attitude of a plane is specified by the strike and dip angles alone, where according to the convention of the **right-hand rule**, if one faces in the direction of the azimuth specifying the strike, the plane must dip down to the right. For the plane specified above, this convention would define the attitude as (325; 55). Other possible conventions specify the orientation of a plane by the trend and plunge (defined after Equation (A1.2)) of the dip line alone, which for the aforementioned plane would be (035; 55). Because for a nonzero dip angle there is only one possible horizontal line normal to the dip line, the strike line is uniquely determined by the orientation of the dip line.

For computational purposes, it is commonly most convenient to specify the orientation of a plane by the Cartesian components of a unit vector normal to the plane (see Box 7-1). With the positive directions of the Cartesian coordinates (x, y, z) oriented (East, North, Up), respectively, and the strike and dip designated, respectively, by (α, δ) using the right-hand rule, a plane has a downward-plunging unit normal whose components are

$$[n_x, n_y, n_z] = [-\sin\delta\,\cos\alpha,\ \sin\delta\,\sin\alpha,\ -\cos\delta] \quad (A1.1)$$

For the plane oriented (035, 55), for example, the downward-pointing unit normal has components

$$[n_x, n_y, n_z] = [-0.671,\ 0.470,\ -0.574] \quad (A1.2)$$

These components define the unit normal that intersects the lower hemisphere of the plotting sphere that we describe in Section A1.2.

The **trend** of a linear structure is the angle between geographic north and the **trend line**, which is the orthogonal projection of the linear structure onto the horizontal plane and thus the horizontal line that lies in the vertical plane containing the linear structure. It is unique except for an exactly vertical linear structure, which has no trend (Figure A1.2*A*). As with the strike, the trend can be specified by quadrant or azimuth, N56E and 056 be-

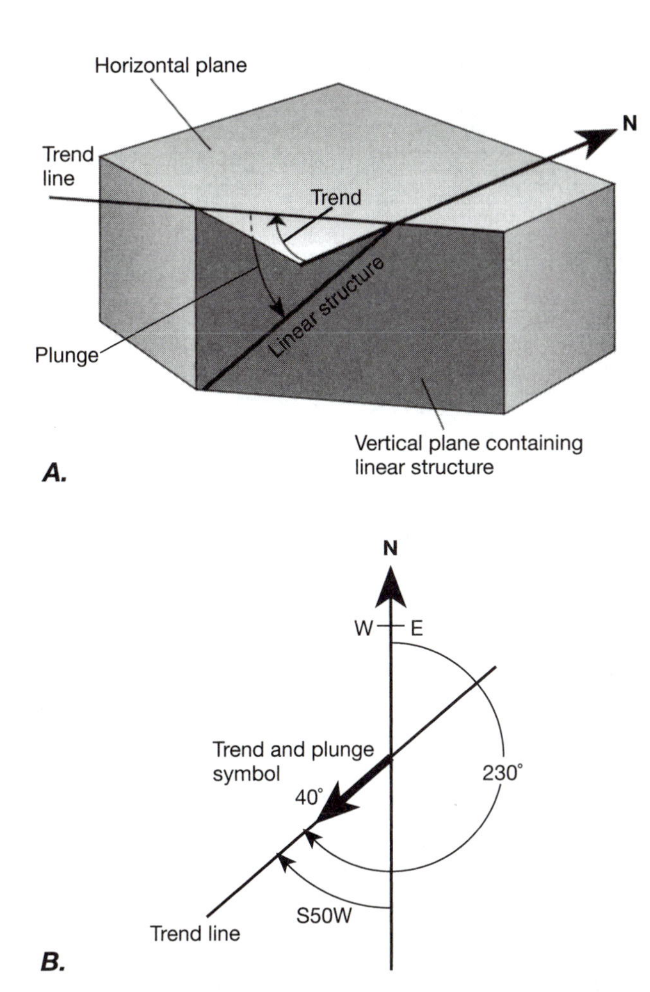

FIGURE A1.2 The trend and plunge of a linear structure. *A*. The trend is the angle between north and the horizontal line that is the orthogonal projection of the linear structure onto the horizontal plane. The plunge is the angle between horizontal and the linear structure, measured in the vertical plane that contains the structure. *B*. The trend shown here is given in the down-plunge direction as a bearing of S50W or as an azimuth of 230°. On a map, linear features are plotted as an arrow parallel to the trend in the down-plunge direction. The plunge angle is written beside this arrow.

ing equivalent trends. The **plunge** is the angle between the horizontal and the linear structure itself, measured in a vertical plane containing the linear structure (Figure A1.2*A*). By common convention, the trend specifies the down-plunge direction of the line, in which case trends that differ by 180° define a different orientation of line. Some lines, however, have directionality to them that can be represented by an arrow, such as slickenlines on a fault that specify the direction of slip of the hanging wall block. In this case, the trend specifies the direction of the arrow, and then the plunge can have either a positive or a negative value. If the linear structure points up, then the plunge is negative and is measured up from the horizontal in a vertical plane containing the linear structure.

According to one convention, we write the plunge first and then the trend, for example, [40; 230] or [40; S50W], which helps to distinguish attitudes of lines from those of planes. The opposite convention of writing [trend, plunge] is also common, however; [230; 40] and [S50W; 40] are equivalent measures of the previous attitude. If the trend is not taken to define the down-plunge direction of the line, then the plunge must be modified with the compass direction or quadrant toward which the line plunges down, for example, [40SW; N50E] or equivalently [N50E; 40SW]. All these measurements define the same orientation of line (Figure A1.2*B*).

For computational purposes, it is usually most convenient to define the orientation of a line by the Cartesian components of a unit vector parallel to the line. If the trend and plunge are, respectively [τ, ρ], and if the Cartesian coordinates (x, y, z) are defined as before, then a line has Cartesian components:

$$[l_x, l_y, l_z] = [\cos\rho \sin\tau, \cos\rho \cos\tau, -\sin\rho] \quad \text{(A1.3)}$$

For the line oriented [230, 40], for example, the Cartesian components are

$$[l_x, l_y, l_z] = [-0.587, -0.492, -0.643] \quad \text{(A1.4)}$$

If we specify the strike and dip of a plane (α, δ) using the right-hand convention, then the trend and plunge [τ, ρ] of the downward-pointing unit normal to the plane are given by

$$\tau = \alpha - 90° \quad \text{(A1.5)}$$

$$\rho = 90° - \delta \quad \text{(A1.6)}$$

Substituting Equations (A1.5) and (A1.6) into Equation (A1.3) and using standard trigonometric equivalences gives the right side of Equation (A1.1).

A1.2 GRAPHICAL PRESENTATION OF ORIENTATION DATA

Often it is desirable to present orientation data in such a way that the distribution of orientations is emphasized independently of the geographic location of the data. For example, it may be useful to know whether there is a pattern of preferred orientation of beds, joints, or linear features in an area, regardless of how the data are distributed across a map. The types of diagrams most frequently used to present such information are histograms, rose diagrams, and spherical projections.

Orientation **histograms**[1] are plots of one part of the orientation data, such as strike azimuth, against the

[1]In Greek, *histos* means "mast" or "web," and *grammē* means "line."

frequency of orientations that are found within particular azimuth intervals. The frequency may be plotted as a percentage of all observations or as the number of observations within each interval. The plots characteristically consist of a series of rectangles, where the width of the rectangle represents the orientation interval and its height represents the frequency.

Rose diagrams are essentially histograms for which the orientation axis is transformed into a circle to give a true angular plot. The intervals of azimuth are plotted as pie-shaped segments of a circle in their true orientation, and the length of the radius is proportional to the frequency of measurements having that orientation. The use of the true angle conveys an intuitive sense of the orientation distribution. Rose diagrams are used for displaying such features as the direction of sediment transport and the strike of vertical joints (see, for example, Figure 2.9*A*).

Both histograms and rose diagrams can present only one aspect of the attitude of planar or linear features, such as the strike or trend, respectively. In cases where the dip of the planar features is always essentially vertical or the plunge of linear features is always essentially horizontal or some other constant angle, this limitation is of little consequence. If the dip or plunge of the feature is an important variable in defining its attitude, however, the best method of plotting orientation data is the spherical projection.

When orientation data are plotted on a **spherical projection**, all planes and lines are considered to pass through the center of the plotting sphere, and their attitudes are then defined by their intersections with the surface of the sphere. The intersection of a plane with the plotting sphere is a great circle; the intersection of a line with the plotting sphere is a point (Figure A1.3). Using the full sphere, however, is actually redundant; a hemisphere is all that is needed. Applications for structural geology and tectonics generally use the hemisphere below the horizontal plane (the lower hemisphere), whereas applications in mineralogy usually employ the upper hemisphere. In both cases, the hemisphere is projected onto a flat plane to permit the convenient two-dimensional graphical presentation of data.

There are a variety of methods of projecting a sphere onto a plane, although two projections are most often encountered in structural geology and tectonics. The first may be referred to as a **stereographic projection** or as an **equal-angle projection**; the other is a **Lambert projection** or an **equal-area projection**. They differ only in the way the hemisphere is projected onto a plane called the image plane.

For both types of projections, the image plane for the projection is tangent to the hemisphere at point T (Figure A1.4) and parallel to the plane containing the edge of the hemisphere. For the equal-angle projection (Figure A1.4*A*), the highest point on the sphere opposite the image plane, the zenith point Z, is the projection point. A line, whose orientation is represented by a point on the lower hemisphere, is projected to the image plane by constructing a projection line from the zenith point Z through the point on the hemisphere (P or Q) to the image plane (P' or Q'). Thus any orientation of line has a unique projection on the image plane. A plane whose orientation is represented by a great circle on the lower hemisphere is projected to the image plane by constructing projection lines that extend from the zenith point through all points along that great circle to the image plane. The locus of those points on the image plane defines the projection of the great circle, which again is unique for any given orientation of plane.

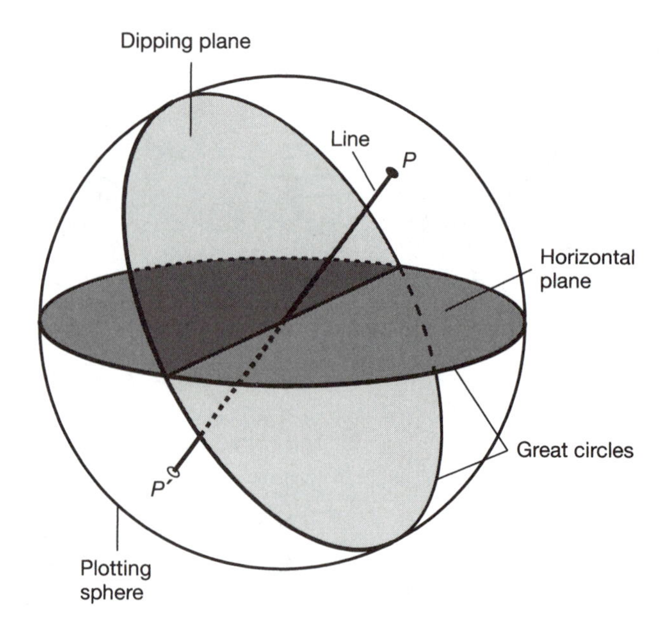

FIGURE A1.3 Plotting orientation data on a plotting sphere. Planar and linear features are considered to pass through the center of the sphere and to intersect its surface. Thus the attitude of a plane is defined by the great circle that is the intersection of the plane with the plotting sphere. A horizontal plane and a dipping plane are shown. The attitude of a line is defined by two points (P and P') that are the intersections of the line with the plotting sphere.

The advantage of this type of projection is that angles between lines on the hemisphere are not distorted by the projection. Moreover, circles on the hemisphere remain circles on the projection, although the center of the circle on the hemisphere does not project to the center of the circle on the image plane. All great circles are also arcs of circles on the image plane. Areas that are equal on the hemisphere, however, do not in general project as equal areas on the image plane.

An equal-area projection is constructed by using the point of tangency T between the hemisphere and the image plane as a center of rotation (Figure A1.4*B*). The pro-

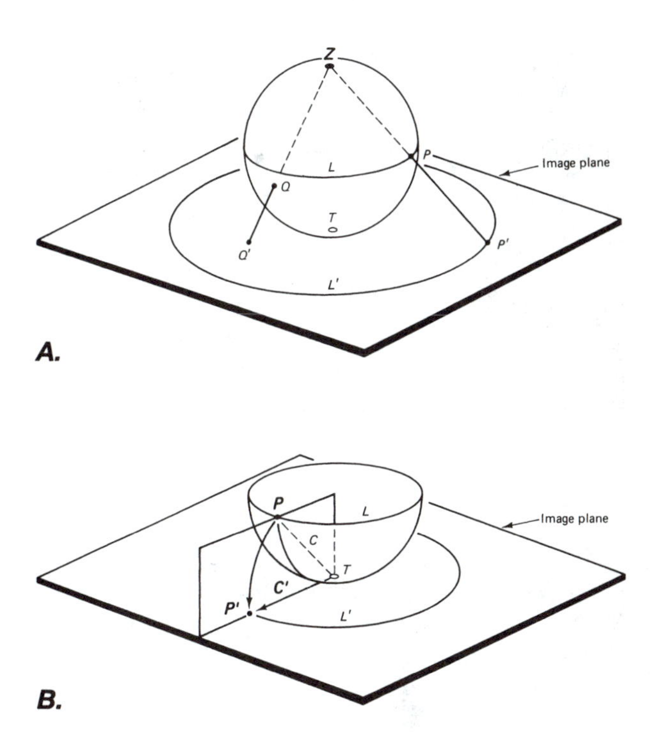

FIGURE A1.4 Projection of the lower half of the plotting sphere onto the image plane. (From Suppe, J. 1985. *Principles of Structural Geology*. Prentice-Hall, Englewood Cliffs, NJ, figure 2.9) *A*. The principle of stereographic, or equal-angle, projection. The zenith point Z is the projection point. Points P and Q on the plotting hemisphere are projected to points P' and Q' on the image plane by a line passing from Z through P and Q, respectively, to the image plane. A great circle is described on the image plane by projecting each point of the great circle on the plotting hemisphere to the image plane. *B*. The principle of Lambert, or equal-area, projection. The point P is projected to the image plane by constructing a cord C of the sphere from T (the tangent point of the image plane) to P and rotating that cord in a vertical plane about T down to C' in the image plane. The end of the rotated cord P' is the projected point. The great circle is projected by rotating each point of the great circle on the plotting hemisphere down to the image plane in a similar manner.

jection of a point on the hemisphere is determined by constructing a chord C from T to the point P on the hemisphere and rotating this chord in a vertical plane about T down to C' in the image plane. The end of the chord P' in the image plane defines the projection of the point. Projections of great circles are constructed, in principle, by rotating all points on the great circle from the hemisphere down to the image plane in a similar manner. This projection has the advantage that any two equal areas on the hemisphere project as equal areas on the image plane. This type of projection, therefore, is used to present data when the statistical concentration of points is important to the interpretation, because those concentrations are not distorted by this projection, as they are for the equal-angle projection. The shapes of areas on the hemisphere are not preserved by the projection, however, so angular relationships are distorted, although they can still be determined if the angles are measured along a great circle.

A1.3 GEOLOGIC MAPS

Geologic maps are the basis of all studies of structure and tectonics. They are two-dimensional representations of an area of the Earth's surface on which are plotted a variety of data of geologic interest. A geologic map emphasizes the distribution of lithologies, their ages, and the structures in the rocks, whereas a tectonic map may plot the distribution of units of similar tectonic significance. The data plotted may include the distribution of the different rock types, soil types and surficial deposits, the location and nature of the contacts between the rock types, and the location and attitude of structural features such as bedding, faults, or folds. These data are based on observations from many outcrops and on the judicious inference of relationships that are not directly observable.

All geologic maps are smaller than the area they represent. Exactly how much smaller is represented by the **scale** of the map, which is the ratio of the distance on the map to the equivalent distance on the ground. A scale of 1:25,000 (one to twenty-five thousand), for example, indicates that 1 unit of distance on the map (such as a centimeter or an inch) represents a horizontal distance of 25,000 of the same unit on the ground. Because the scale is a ratio, it applies to any desired unit of measurement. The scales of most maps used in structural and tectonic work range between 1:1,000 and 1:100,000, though other scales are also used. In particular, maps of large regions such as states, provinces, countries, and continents are published at scales between 1:500,000 and 1:20,000,000.

As the scale of a map changes, the size of the features and the amount of detail that can be represented on the map also change. If a map is of a very small region (an area a few meters in dimension, for example), then correspondingly small features and great detail can be portrayed. If the map represents a large region (an area hundreds of kilometers in dimension, for example), then only large features and little detail can be shown.

Unfortunately, the word *scale* is used in two different and opposite ways, which can lead to confusion. With regard to geologic features, the term refers to the dimensions of the feature. Thus small-scale features have a characteristic dimension roughly in the range of centimeters to perhaps hundreds of meters. Large-scale features have a characteristic dimension of roughly hundreds of meters to thousands of kilometers. With regard to maps, however, scale refers to the distance on a map divided by the equivalent distance on the ground. Thus small-scale maps (such as 1:100,000) cover larger areas than large-scale maps (such as 1:1,000). Confusion arises because small-scale features are portrayed on large-scale maps, and vice versa.

Because the Earth's surface is very nearly spherical, any planar map of the surface is a distortion of true shape. For small areas, even up to standard 1:24,000 or 1:25,000 quadrangles (approximately 17 km in a north-south direction and, in midlatitudes, 11 km to 15 km east to west), this distortion is minor and usually is ignored. For larger regions, however, distortions become significant. For example, Greenland is considerably smaller in area than South America, notwithstanding its appearance on most Mercator map projections. Many different types of projections of the spherical surface onto a plane are used. Each represents a compromise between minimizing the distortion of the shape of the region and minimizing the distortion of its area.

A1.4 CROSS SECTIONS: PORTRAYAL OF STRUCTURES IN THREE DIMENSIONS

A geologic map provides the basis for detailed understanding of the structural geometry of an area. The map, however, is only a two-dimensional representation of three-dimensional structures. Cross sections complement the information on maps by showing the variation of structure with depth, usually as it would appear on a vertical plane that cuts across the area of a geologic map. Fundamentally, cross sections are extrapolations to depth of data available at the surface, but they may also be constrained by data on the regional stratigraphy or lithology, by direct observation from drill holes or mines, and by geophysical data (see Appendix 2). Without such independent constraints, cross sections are highly interpretive, because they are based on the assumptions that the structure at depth is a simple projection of the structure observed at the surface and that the attitudes and geometry do not change along the line of projection. The validity of that assumption varies a great deal, depending on the characteristics of the local geology.

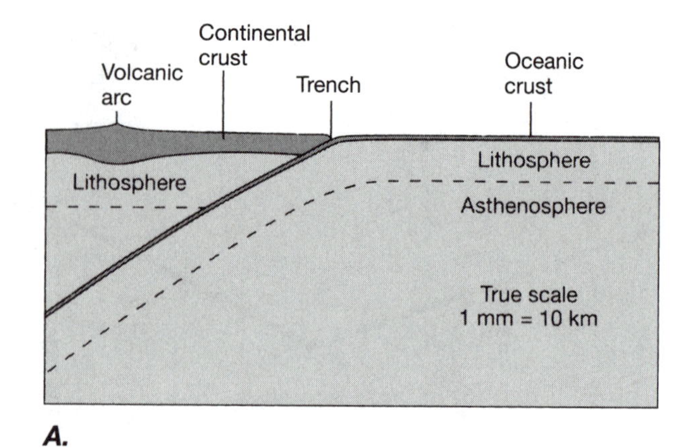

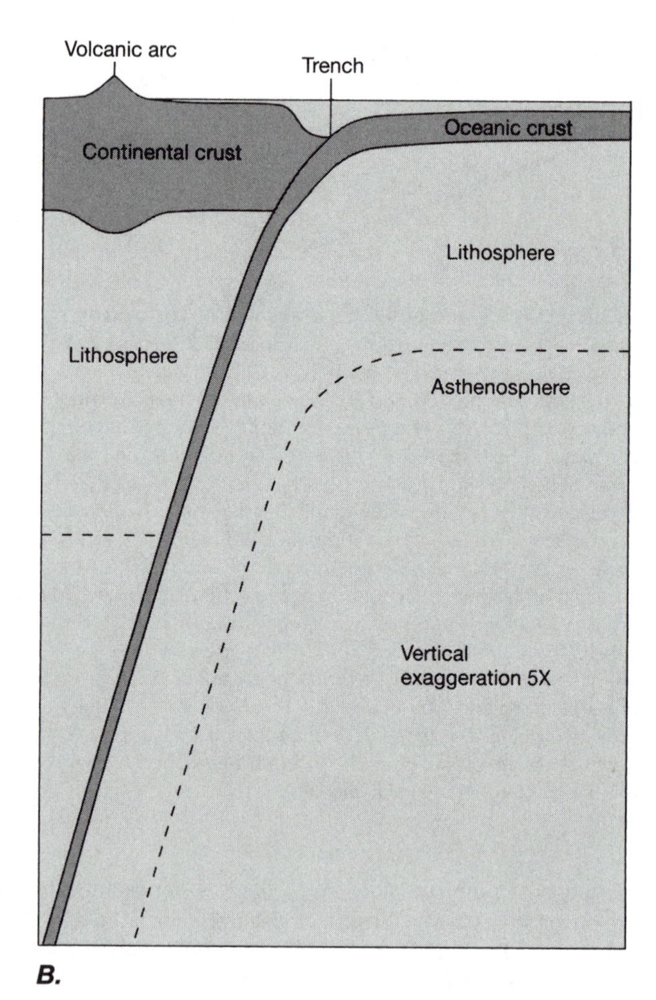

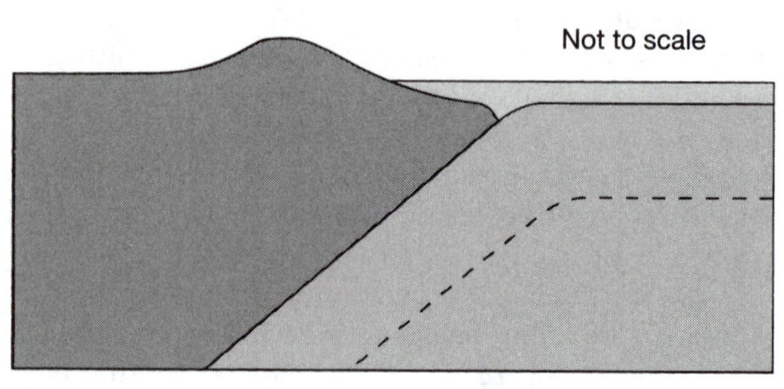

FIGURE A1.5 The effect of vertical exaggeration on cross sections. *A*. True-scale cross section of a subduction zone in which the horizontal and vertical scales are equal (1 mm = 10 km). Note that topographic variations are almost imperceptible. *B*. A cross section of the same subduction zone as in *A*, here shown at a vertical exaggeration of 5×, that is, the vertical scale is 5 times the horizontal scale. Note the exaggeration of the dip of the subducted slab and the distortion of the thickness of layers of crust and lithosphere depending on the magnitude of the dip. *C*. Schematic cross section that combines vertical exaggeration for the topography with no vertical exaggeration for structures below the surface. This mixing of vertical scales precludes the construction of an accurate cross section.

Cross sections ideally are oriented normal to the dominant strike of planar structures in an area. In this orientation, the dip of those structures is most accurately represented on the section. In some cases, however, if the structure is complicated with a variety of attitudes, no one orientation of cross section adequately represents all the structures, and the apparent dip of any plane whose strike is not perpendicular to the cross-section line is less than the true dip of that structure.

In order to show an undistorted view of structures at depth, we must take the vertical scale ratio equal to the horizontal scale ratio. In some cases, however, the features of interest are best shown by using **vertical exaggeration**, for which the vertical scale ratio is larger than the horizontal scale ratio. For example, the relatively small changes in topography and in stratigraphic thickness that commonly occur over large distances can be shown clearly only with vertical exaggeration. As a result, vertically exaggerated cross sections are standard in marine geology and stratigraphy.

The habitual use of vertical exaggeration to portray certain features, however, gives a false impression of the nature of those features. Figure A1.5, for example, shows cross sections of a subduction zone and an adjacent volcanic island arc. The true scale section is shown in Figure A1.5*A*. Note that the topographic variation is almost impossible to see! A properly vertically exaggerated section is shown in Figure A1.5*B*. The vertical exaggeration necessary to emphasize the topography also exaggerates the surface slopes so that their appearance is not at all representative of actual slopes. The effect of vertical exaggeration is just as dramatic on the dip of planar features (compare the angle of the subducting lithospheric slab in Figure A1.5*A* with that in Figure A1.5*B*). Published informal cross sections often give an even more confusing view by combining vertically exaggerated topography with no vertical exaggeration below the surface, as shown in Figure A1.5*C*.

Vertical exaggeration makes dipping structures appear much steeper than they are. The effect is much stronger on shallowly dipping planes than on steeply dipping ones, and the effect is that at high values of the vertical exaggeration, the distinction between shallow and steep true dips effectively disappears.

Another effect of vertical exaggeration is to cause beds of the same thickness but different dip to appear to differ in thickness. In Figure A1.5*B* the subducted lithospheric slab and oceanic crust appear to be thinner where they are dipping than where they are horizontal, an effect caused simply by exaggeration of the vertical dimension.

Thus vertical exaggeration causes distortions of the geometry of features that seriously alter the way they look. Because much of structural geology involves visualizing the true shape of features in three dimensions, the use of vertical exaggeration with structural cross sections should be avoided whenever possible.

REFERENCES AND ADDITIONAL READINGS

Compton, R. R. 1985. *Geology in the Field.* Wiley, New York, 391 pp.

Marshak, S., and G. Mitra. 1988. *Basic Methods of Structural Geology.* Prentice-Hall, Englewood Cliffs, NJ, 446 pp.

Appendix 2

GEOPHYSICAL TECHNIQUES

Geophysical data such as seismic reflection profiles and gravity measurements are essential to the interpretation of many large-scale structures, and a structural geologist must understand how these data are obtained and interpreted in order to understand their uses and limitations.

Although mapping rocks that are exposed at the surface provides good information about the three-dimensional structure near the surface, it cannot reveal the structure of areas covered by alluvium, deep soils, vegetation, or water such as lakes, seas, and oceans. Nor can surface mapping provide information about the structure deep within the crust and in the mantle. Information about the shapes of major faults at depth, the presence of magma chambers at depth, the location of the crust-mantle boundary, the thickness and nature of the lithosphere, and the structure of the deeper mantle can come only from the interpretation of geophysical measurements, especially from seismic, gravity, and magnetic measurements. We review briefly the application of these aspects of geophysics to large-scale structure and tectonics because they have become essential and because a structural geologist must at least be aware of the techniques and their limitations. Adequate coverage of these topics, however, would require at least a separate book, and we encourage students to take appropriate courses in geophysics.

A2.1 SEISMIC STUDIES

Seismic waves are oscillations of elastic deformation that propagate away from a source. Waves from large sources such as major earthquakes and nuclear explosions can be detected all around the world. Small explosions are often used as sources to investigate structure at a more local scale. Body waves, which can travel anywhere through a solid body, are of two kinds: compressional (contractional) (P) waves, for which the particle motion is parallel to the direction of propagation, and shear (S) waves, for which the particle motion is normal to the direction of propagation. The designations "P" and "S" refer to primary and secondary waves, so named because of the normal sequence of arrival of the waves as revealed on a seismogram. P waves travel faster than S waves and so arrive at a detector first. They are therefore easier to recognize and to measure accurately. For this reason, and because explosions generate mostly P waves, they are the waves predominantly used to investigate the structure of the Earth's crust. The propagation of seismic waves may be described by seismic rays, which are lines everywhere perpendicular to the seismic wave fronts. Three types of seismic studies are particularly important in structure and tectonics: seismic refraction, seismic reflection, and first-motion studies.

Seismic refraction can be used to infer the structure of the Earth by studying the arrival times of those seismic rays that propagate through boundaries where the seismic velocity changes. As a seismic ray travels across a boundary where the seismic velocity increases, the ray is refracted, or bent, away from the normal to the boundary (see Box A2-1, Figure A2-1.1). Thus, in the Earth, where seismic velocity generally increases with depth, the ray paths tend to be concave upward. The travel time of rays from the source to different receivers is plotted against the distances of the receivers from the source. Such plots make it possible to determine the ray velocity in the deepest layer through which the ray travels. By

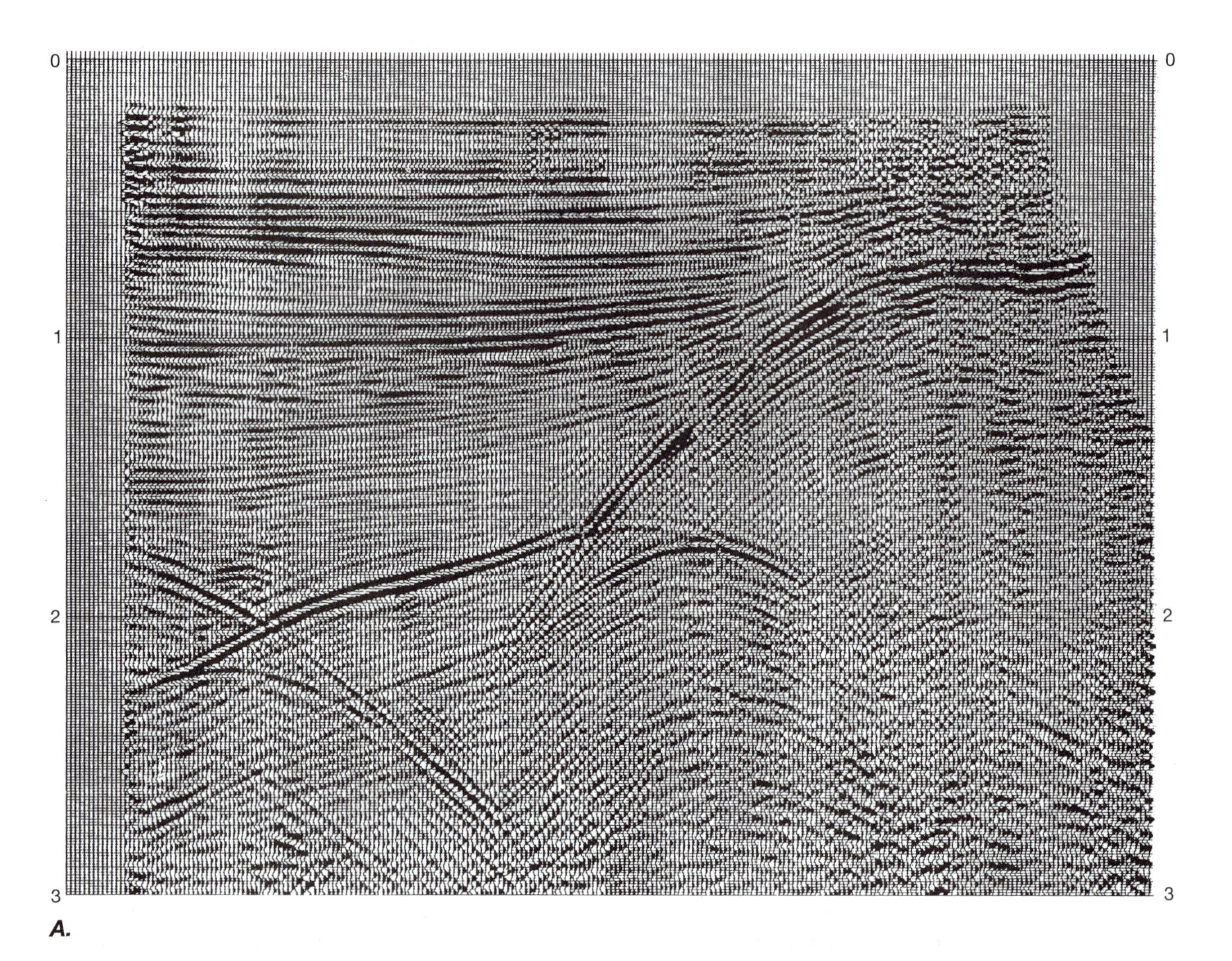

FIGURE A2.1 Seismic reflection profiles. Individual seismic records are the wavy vertical lines plotted side by side along the distance axis. The vertical axis is the two-way travel time, which increases downward. Peaks in each record are shaded black to show up reflectors that can he traced from one record to the next. Good horizontal reflectors are particularly evident at two-way travel times of less than about 1.7 seconds in the left half of the profile. (From Lindseth 1982) *A*. Unmigrated seismic profile. *B*. Migrated seismic profile.

measuring travel times for rays that penetrate to greater and greater depths, the investigator can determine the velocity structure of the Earth.

The velocities of P and S waves depend on the density and the elastic constants of the rock. Thus knowing how seismic P and S velocities vary with depth provides information about the distribution of the density and elastic properties in the Earth, and locating where changes in seismic velocity occur reveals where the rock type changes.

Although seismic refraction studies give a good "reconnaissance" view of the structure of a large area, the technique has several disadvantages. The presence of low-velocity layers cannot be detected (Box A2-1). Deep structures ordinarily can be detected only at distances from the source that are greater than the depth. The properties of the Earth are averaged over large distances, so details of structure are lost. And nonhorizontal or discontinuous layers and complex structure are difficult or impossible to resolve.

In **seismic reflection** studies, the structure of the Earth is revealed by reflections of P waves off internal boundaries. Seismic signals are recorded by as many as several hundred to several thousand geophones at a time, often set out in a linear array, and the resulting data are analyzed by computer (Boxes A2-2 and A2-3). Each seismogram shows the temporal sequence of arrivals of seismic waves at a particular geophone. Each seismogram is plotted along a vertical line so that time increases downward, and the seismograms from all the geophones are arranged side by side along the horizontal distance axis, according to the distance of each geophone from a common origin. The distance from zero-time to the individual peaks, or events, that make up a seismogram, records the two-way travel time for each reflected wave, which

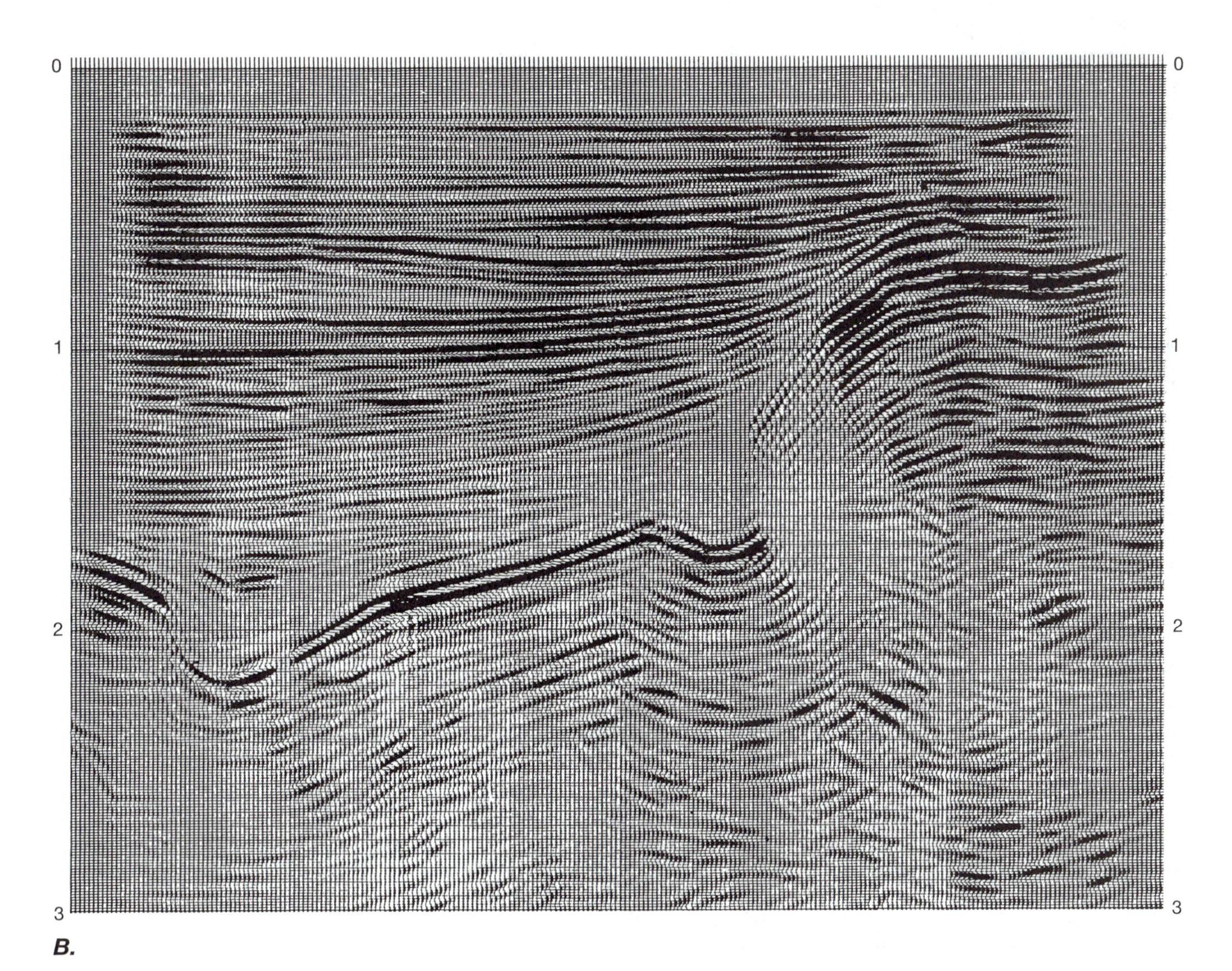

FIGURE A2.1 *(continued)*

is the time required for a wave to travel from a shot point on the surface to a reflector and back up to a geophone on the surface. The resulting plot is called a **record section** (Figure A2.1*A*). Peaks in each seismogram are shaded black so that a high-amplitude signal shows up as a black linear feature across the record section, tracing the arrival of individual seismic waves as they travel across the geophone array. A continuous line across the record section indicates a continuous reflector at depth.

Sophisticated computerized digital processing of the seismic records, including the very important processes of stacking and migration, allow complex structures to be resolved (Figure A2.1*B*) and therefore yield an incomparable image of the subsurface structure. The **stacking** of seismic records is a method of enhancing the signal-to-noise ratio by adding together reflections that occur at different angles from the same subsurface point (Box A2-2, Figure A2-2.1). **Migration** is a technique that allows the true locations of reflectors to be determined. All reflection signals on a seismogram plot along the time axis at the specific receiver distance as though they were vertical reflections (Figure A2.1*A*). Any given reflection, however, could have come from any point on an arc of constant travel time around the receiver, and the process of migration corrects the seismograms to determine the actual location of the reflector (Figure A2.1*B*; Box A2-3, Figure A2-3.1).

We can illustrate the benefits of migration by comparing an unmigrated seismic record (Figure A2.2*A*), a migrated profile (Figure A2.2*B*), and the true geologic cross section (Figure A2.2*C*). Migration eliminates the artifacts and errors in the unmigrated record. The actual geologic section, however, can be determined only if the two-way travel time can be converted into depth, which requires knowledge of the way velocity varies with depth. Note that if the seismic velocity changes with depth, the record section distorts the vertical scale because it is a time axis, not a true distance axis.

Despite their obvious value, the utility of reflection seismic studies is limited by their expense. Producing one

BOX A2-1 Seismic Refraction

The time required for seismic rays to travel directly from a source to different detection stations distributed around the source is affected by the particular paths the seismic rays take, and these in turn are determined by the structure and thus the seismic velocity of the material along each path. If a seismic ray travels obliquely across a boundary from a low- to high-seismic-velocity material, it is refracted away from the normal to the boundary (Figure A2-1.1). If the ray travels from high- to low-velocity material, it is refracted toward the normal to the boundary.

Travel-time measurements can be interpreted to reveal the variation of seismic wave velocity with depth. The principle is illustrated in Figure A2-1.2, which shows the location of a seismic source and an array of detectors. Some of the ray paths shown stay within the crust; others travel in part through the higher-velocity upper mantle. A time-distance plot indicates the arrival times of those different rays at the detectors. Because the seismic velocity in the mantle is higher than that in the crust, mantle rays reach distant detectors before crustal rays. For the layered structure shown, the difference in arrival times at the different detectors reflects the speed of the rays through the deepest layer along the ray path. Thus the slopes of the two lines on the time-distance plot are the inverse of the velocities in the crust and mantle, respectively.

If a layer that has a lower seismic velocity occurs at depth between rocks that have higher seismic velocities, rays are bent toward the normal to the boundary on entering the layer and away from the normal on leaving (Figure A2-1.1). Seismic rays, therefore, can never reach their maximum depth in a low-velocity layer, and the seismic velocity of that layer—and therefore its very existence—cannot be detected on a time-distance plot.

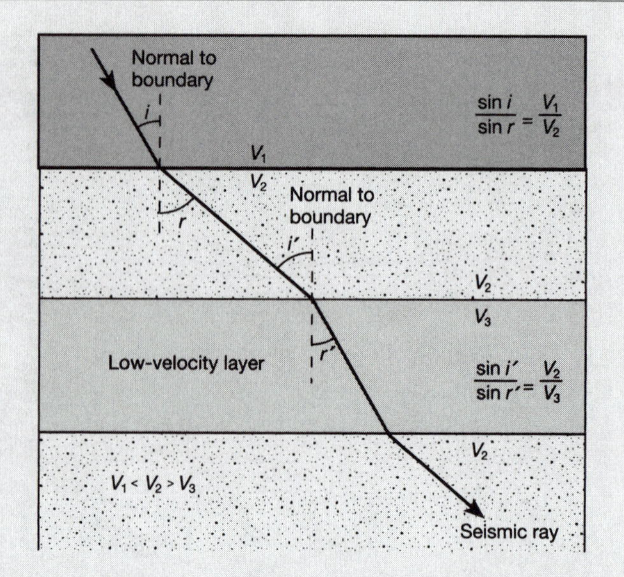

FIGURE A2-1.1 Refraction of seismic rays. Refraction is away from the normal to the boundary if the ray travels from a low- to a high-seismic-velocity material (here V_1 to V_2 or V_3 to V_2). Refraction is toward the normal to the boundary if the ray travels from a high- to a low-seismic-velocity material (V_2 to V_3).

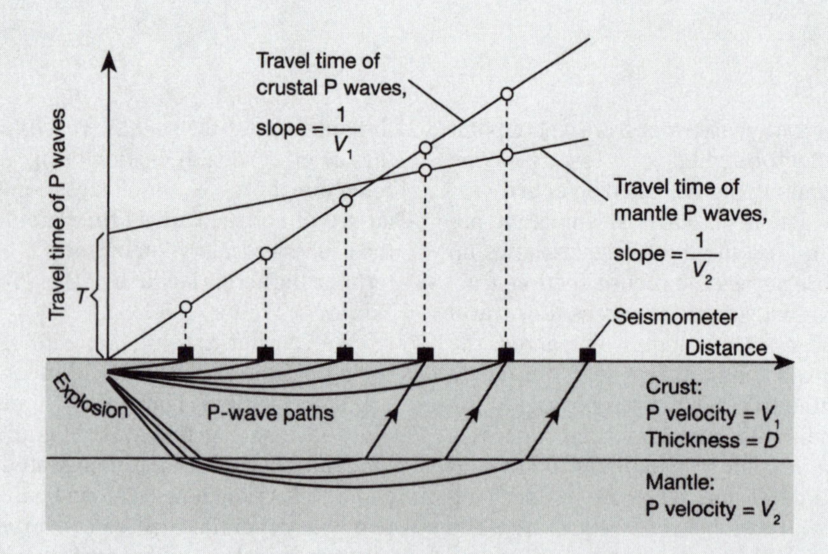

FIGURE A2-1.2 Illustration of the principle of seismic refraction in a two-layer structure in which velocity increases with depth. The diagram shows ray paths for P waves through the structure. The travel-time plot indicates the arrival times of the rays at the different detectors. (After Press and Siever 1986, p. 414)

BOX A2-2 Stacking of Seismic Records

Figure A2-2.1 illustrates the principle involved in the common depth-point stacking of seismic records. If explosions are detonated at shot points S_1, S_2, S_3, and S_4, reflections from the same point P on a horizontal subsurface boundary will be received at geophones G_1, G_2, G_3, and G_4, respectively. The same is true for all other horizontal reflectors below the surface point p. The corresponding shot points and geophones (S_i and G_i) are equidistant from the point p above P. If the travel times are corrected for the difference in length of the ray paths, the records can be added together, or stacked. The time-corrected signals from the reflections at P reinforce one another, and the signals from random noise tend to cancel out, thereby increasing the signal-to-noise ratio. The result is an enhanced seismogram showing the reflections as they would appear if the shot point and receiver were both at p.

In practice, data are gathered from a large linear array of shot points and geophones. Each geophone records many reflections from different depths, and the stacking is done by computer to produce enhanced seismograms at each point in the profile. Figure A2.1A is an example of a stacked seismic profile, which shows abundant horizontal reflectors at shallow depths.

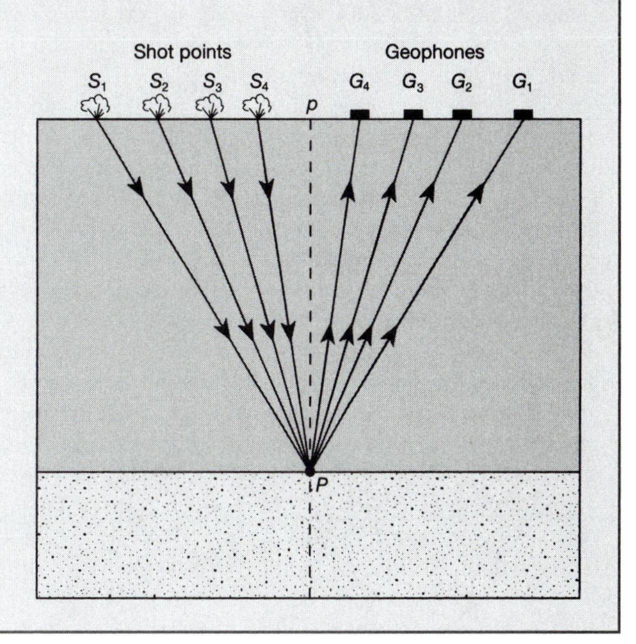

FIGURE A2-2.1 The principle of stacking of seismic records. Explosions are set off at shot points S_1, S_2, S_3, and S_4. The rays reflected from the same point P on a reflector at depth are received, for each of the explosions, at geophones G_1, G_2, G_3, and G_4, respectively. Adjusting the arrival times of these reflections for the different lengths of ray path allows the four records to be added together, which enhances the signal from the reflection at P and cancels out random noise. The effect is an increase in the signal-to-noise ratio.

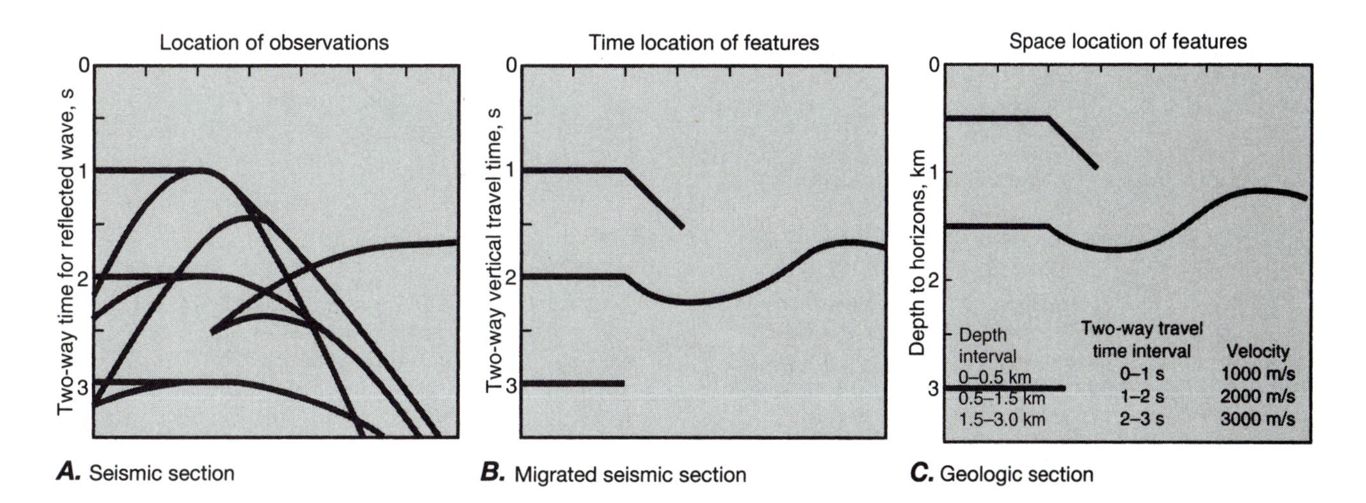

FIGURE A2.2 Diagrams illustrating the effects of migration on seismic reflection profiles. (After Sheriff 1978, p. 210) *A*. An unmigrated seismic section with multiple intersecting curved reflections. *B*. The same section as in *A* after migration. The ambiguities and artifacts of the unmigrated section are all removed. *C*. The corresponding geologic section. The depth scale is different from the two-way travel-time scale because seismic velocity varies with depth.

BOX A2-3 Migration of Seismic Records

Although horizontal or very shallowly dipping reflectors are common in undeformed sedimentary basins (shallow parts of Figure A2.1*A*, *B*), much of the structure of interest in structural and tectonic investigations is a great deal more complex (deeper parts of Figure A2.1*A*, *B*). For example, beds with significant and variable dips and discontinuous beds (possibly truncated by a fault) are common. Such structures give rise to distortions and artifacts in seismic profiles (Figure A2.1*A*), which must be corrected by **migration**.

We describe the principle of migration by using an example for which the seismic velocity of the material is constant, and the source and detector are at the same surface point p (Figure A2-3.1*A*). A particular reflection that apparently plots at P below the detector could come from a boundary that is tangent to any point on a circular arc of constant two-way travel time having radius pP around p—for example, from P'. On two adjacent reflection seismograms (Figure A2-3.1*B*), a reflection apparently plots at P_1 beneath p_1 and at P_2 beneath p_2. It would therefore appear that the reflector had the dip of the line P_1P_2. In fact, however, the reflector must be the common tangent to the two constant-travel-time arcs of radius p_1P_1 about p_1 and p_2P_2 about p_2. The reflections must therefore come from points P_1' and P_2'. Thus an uncorrected profile shows erroneous locations and dips for dipping reflectors, and the reflection points P_1 and P_2 must be migrated along their respective constant-travel-time arcs to the correct locations at P_1' and P_2'. The higher the true dip, the greater the distortion. Vertical reflectors plot on unmigrated seismic profiles as an alignment of reflections having a 45° dip.

If the seismic source and the detector are not at the same point, the arc of constant two-way travel time becomes an ellipse, and it is further distorted if the velocity is not constant. These are complications that must be accounted for in any analysis of a real seismic record, although the principle remains the same.

Another problem occurs if a reflector is discontinuous. The end of the reflector acts as a diffraction point, which takes energy from any angle of incidence and radiates it in all directions as though the point were a new source (point D in Figure A2-3.2). The signals recorded by the nearby detectors—for example, at p_1, p_2, and p_3—then plot on an uncorrected seismic profile along a parabolic arc (the dotted line in Figure A2-3.2) at P_1, P_2, and P_3, respectively (see also Figures A2.1*A*, A2.2*A*). On this plot, the travel time between p_i and P_i is the same as that between p_i and the diffraction point D, and that time increases as the distance of the detector p_i from the end of the reflector increases. Thus the possible locations of the diffraction point that could generate the signal P_i recorded at a given receiver p_i must lie along an arc of constant two-way travel time about the receiver p_i, and these arcs are shown as

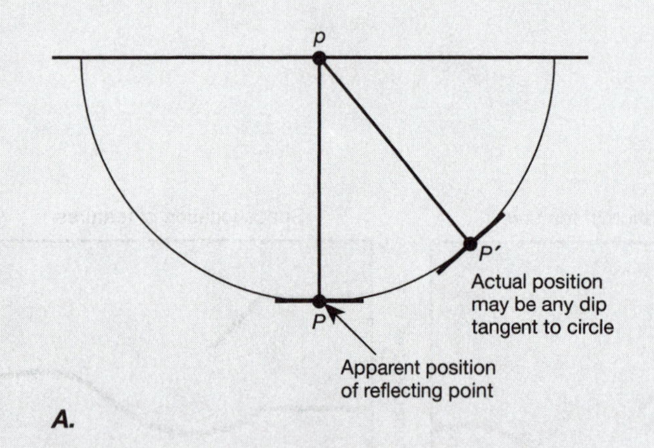

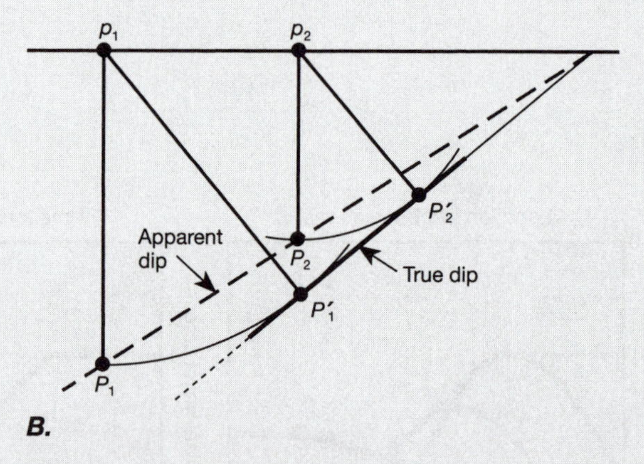

FIGURE A2-3.1 The migration of seismic signals corrects the seismic records to give the true location and dip of reflectors. In this example, seismic velocity is considered constant, and the shot point and receiver are both located at the same point. (After Lindseth 1982) *A*. A reflection received at p appears on the seismic record at a two-way travel time that plots at P. In fact, the signal could come from any reflector, such as P', that is tangent to the semicircular arc of radius pP around p. That arc is the locus of constant-two-way travel time. *B*. Reflections detected at p_1 and p_2 plot vertically below each point at P_1 and P_2, respectively, giving the reflector the apparent dip and location of the line P_1P_2. The true location and dip of the reflector, however, must be given by the line $P_1'P_2'$, which is the common tangent to the constant two-way travel-time arcs about P_1 and P_2, respectively. Note that P_1' is the actual location of the reflector below p_2.

the dashed arcs in Figure A2-3.2. The true location of the diffraction point is at D, the common intersection of the arcs constructed for several detectors. Thus migrating each signal along its arc to the common point identifies the true location of the diffraction point D.

In practice, the process of migration consists of taking each individual event on a given reflection seismogram and, along its constant-travel-time arc, adding that event to any other seismogram intersected by the arc. Thus in Figure A2-3.2, the event at P_3 is added to the p_2 seismogram at P_3', to the p_1 seismogram at P_3'', and to the p_0 seismogram at D. Similarly, the event at P_2 is added to the p_3 seismogram at P_2'', to the p_1 seismogram at P_2', and to the p_0 seismogram at D; and the event at P_1 is added to the p_2 seismogram at P_1' and to the p_0 seismogram at D. All these additions result only in random noise unless they coincide with one another; thus if migrated signals are added to the same point on a record, they tend to cancel one another out. The only exception is for the additions at D, and there, all the events reinforce one another to become a large amplitude signal. The resulting seismic record section is then a series of seismograms, each of which consists of the original record altered by the addition of all the events that migrate to that record. With this procedure, reflecting boundaries appear as coherent traces of events across the section in their correct location, and diffracted signals sum together at the location of the diffraction point. The other additions to the different seismograms tend to cancel each other out and do not produce coherent patterns on the seismic profile.

Determining the constant two-way travel-time arc, of course, requires determining the velocity structure. The amount of computation required to migrate every event in a profile to every seismogram intersected by its constant-travel-time arc is prodigious; in practice, it can be handled only by a computer.

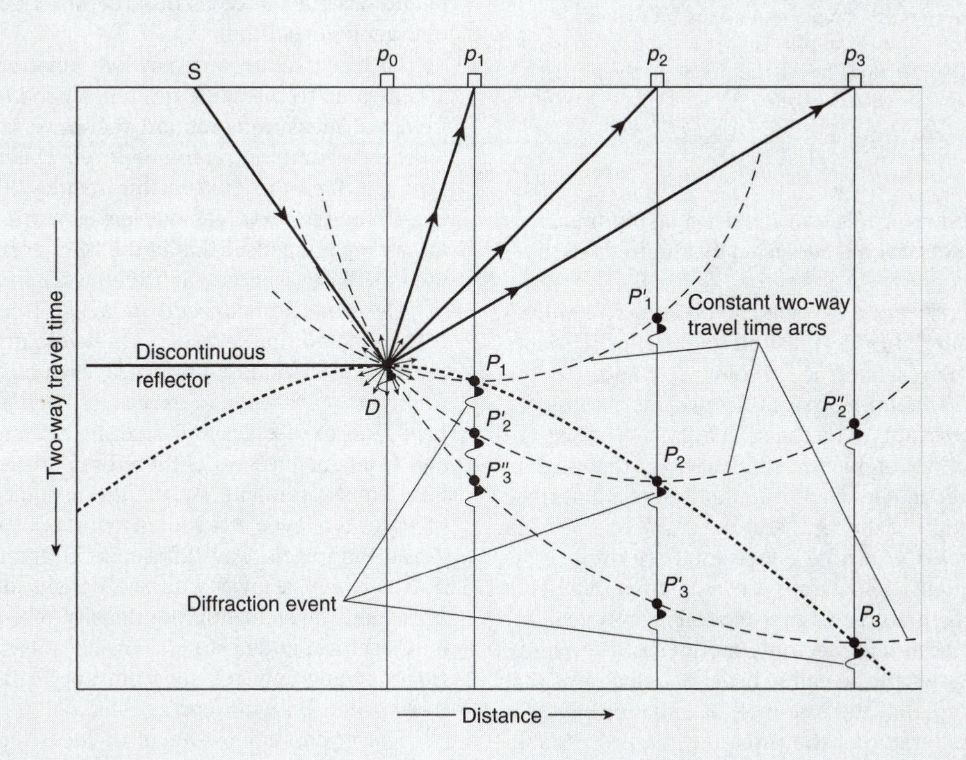

FIGURE A2-3.2 The end of a discontinuous reflector (D) acts as a diffraction point and radiates seismic energy in all directions for any angle of incidence. The farther the receiver is from D, the later the diffracted ray arrives. Thus the diffracted energy arrives at receivers p_1, p_2, and p_3, for example, at times that fall along a parabolic arc at P_1, P_2, and P_3, respectively. The three constant-travel-time arcs constructed about the three receivers with radii p_1P_1, p_2P_2, and p_3P_3, respectively, must intersect at the location of the diffraction point. Thus we migrate each event along its constant-travel-time arc, adding the event to any intersected seismogram (e.g., points P_1', P_2', P_2'', P_3', P_3''). At the common point of intersection of all the arcs D, the addition of events creates a large-amplitude event that indicates the true location of the diffraction point.

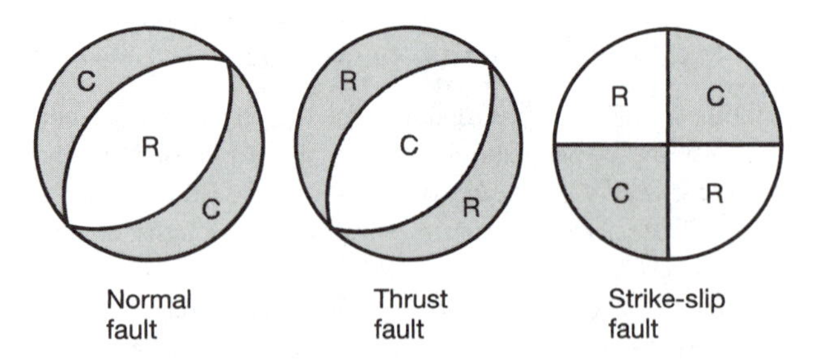

FIGURE A2.3 Equal-area projections showing the radiation pattern of contractional first motions (C) and rarefaction first motions (R) for the three main types of faults, informally referred to as "beachball diagrams." Planes separating the sectors are nodal planes, one of which must be the fault plane. The planes whose orientations are plotted on such diagrams are all considered to pass through the center of the plotting sphere (see Appendix 1, Section A1.2). All orientations of lines that plot within a given sector are the orientations of rays that leave the source with the first-motion characteristic of that sector. Material on each side of each nodal plane moves toward the contraction sectors, defining the sense of shear on those planes. T axes (maximum lengthening) bisect the nodal planes in the shaded sectors; P axes (maximum shortening) bisect the nodal planes in the unshaded sectors (see Box A2-4).

seismic reflection profile can involve making hundreds of shots, each of which is recorded by hundreds to thousands of geophones.

The first arrival of a P wave can be either a contraction of the material (a decrease in volume) or its opposite, a rarefaction (an increase in volume). Whether a contraction or a rarefaction arrives first is revealed by the direction of first motion on the seismogram. The pattern of contraction or rarefaction first motions that radiate out from a sudden slip event on a fault is characteristic of the orientation of the fault and the sense of slip (Box A2-4). **First-motion studies**, therefore, are used to determine the orientation of, and sense of slip on, faults at depth. Regional patterns of first motions can reveal, for example, the large-scale tectonic motions of the plates. Figure A2.3 shows the so-called beachball diagrams that plot on a stereogram the first-motion radiation patterns that are characteristic of the three basic types of faulting: normal, thrust, and strike-slip.

A2.2 ANALYSIS OF GRAVITY ANOMALIES

Gravity measurements are perhaps the second most important source of geophysical data (after seismic data) used in structural geology and tectonics. A **gravity anomaly**[1] is the difference between a measured value of the acceleration of gravity, to which certain corrections are applied, and the reference value for the particular location. The reference value is determined from an internationally accepted formula that gives the gravitational field for an elliptically symmetric Earth. Because gravity anomalies arise from differences in the density of rocks, the goal in structural geology is to relate these differences in density to structural features. If no density contrasts exist, then the structure can have no effect on the gravitational field, and gravity anomalies cannot aid in the interpretation of that structure.

The structure at depth is interpreted by matching the gravity anomaly profile observed, generally along a linear traverse, with the anomaly profile calculated from an assumed model of the structure. The model is adjusted until the model anomaly profile shows a satisfactory fit to the observed anomaly profile. Although the model can never be unique, it is usually constrained by surface mapping, possibly by seismic data, and by requirements that the model be as simple as possible, consistent with expectations for the types of structure likely in the particular geologic setting.

To calculate an anomaly, we must correct the measured value to the same reference used for the standard field. All measurements are therefore corrected to sea level as a common reference level. This altitude correction, the **free-air correction**, results in an increase in most land-based values but leaves surface observations at sea unchanged. If this is the only correction applied, the calculated anomaly is called a **free-air anomaly**.

The **Bouguer correction** is also frequently applied. It is assumed that between sea level and the altitude of the measurement is a uniform layer of continental crustal rock that represents an excess of mass piled on the surface. The excess gravitational attraction that would result from such a layer is therefore subtracted from land-based measurements. At sea, it is assumed that the layer of water is a layer deficient in mass, because water is less dense than rock. The difference in gravitational attraction between a layer with the density of water and an equivalent layer having the density of continental crust is therefore added to sea-based measurements. The Bouguer anomaly results from application of both the free-air and Bouguer corrections.

The gravitational effects of local topography differ measurably from those of the uniform layer assumed for the Bouguer correction, however. Thus a refinement of the simple Bouguer anomaly, called the complete Bouguer anomaly, requires a terrain correction to account for these local effects.

Thus, the Bouguer anomaly compares the mass of existing rocks at depth below sea level to the mass of standard continental crust whose elevation is at sea level. Bouguer anomalies are generally strongly negative over areas of high topography, indicating that there is a deficiency of mass below sea level compared to standard continental crust. They are strongly positive over ocean

[1]From the Greek word *anomalia*, which means "irregularity" or "unevenness."

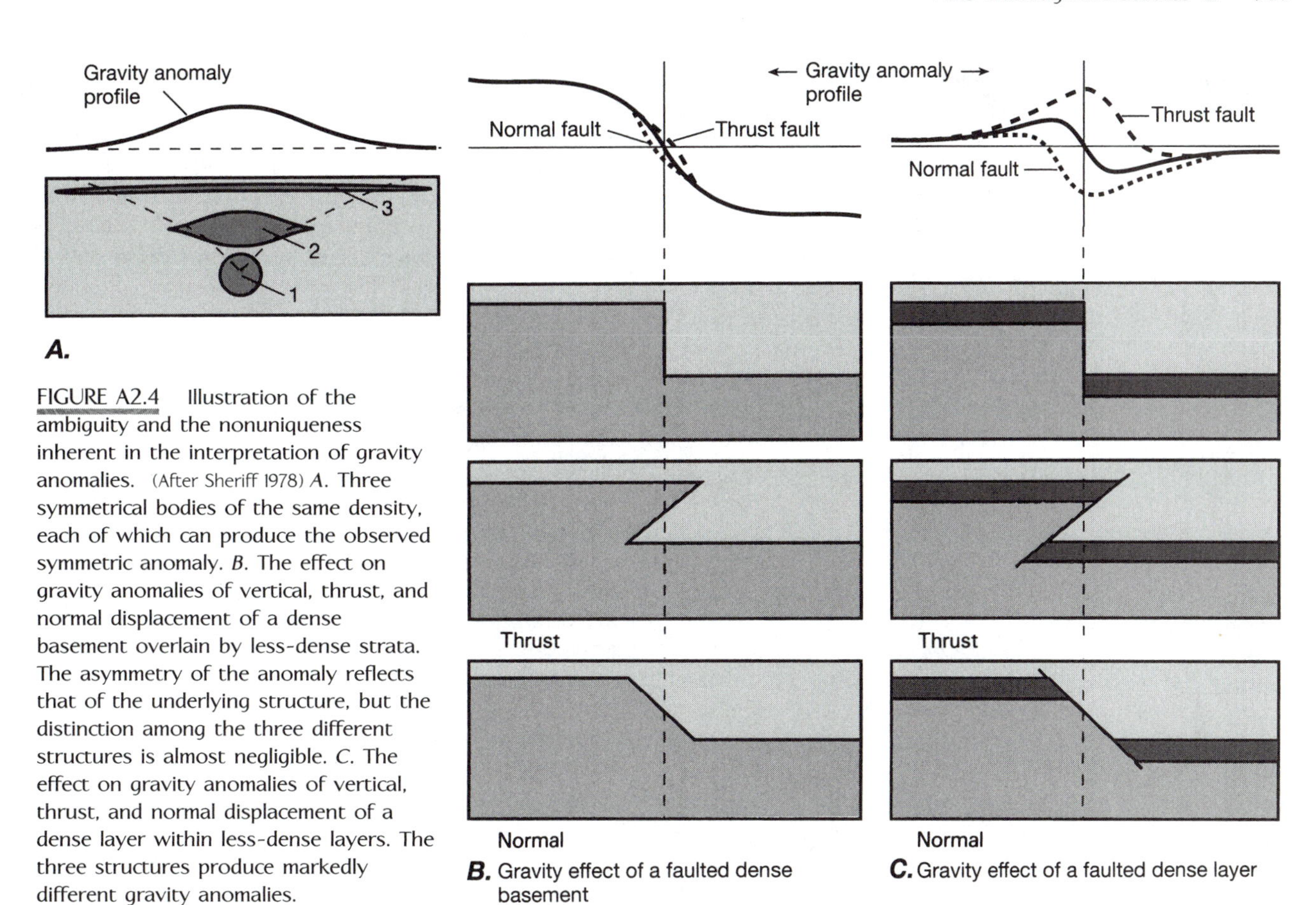

FIGURE A2.4 Illustration of the ambiguity and the nonuniqueness inherent in the interpretation of gravity anomalies. (After Sheriff 1978) *A*. Three symmetrical bodies of the same density, each of which can produce the observed symmetric anomaly. *B*. The effect on gravity anomalies of vertical, thrust, and normal displacement of a dense basement overlain by less-dense strata. The asymmetry of the anomaly reflects that of the underlying structure, but the distinction among the three different structures is almost negligible. *C*. The effect on gravity anomalies of vertical, thrust, and normal displacement of a dense layer within less-dense layers. The three structures produce markedly different gravity anomalies.

basins, indicating that there is an excess of mass below the ocean bottom compared to standard continental crust.

The area under a gravity anomaly profile provides a unique measure of the total excess or deficiency of mass at depth, and the shape of the profile constrains the possible distribution of the anomalous mass. The interpretation of mass distribution is not unique, however, because a given anomaly profile can be produced by a wide range of density differences and distributions. Figure A2.4*A*, for example, shows three symmetric bodies of the same density, each of which produces the same symmetric gravity anomaly. Figure A2.4*B*, *C* illustrates how the faulting of different density distributions affects the gravity anomaly profile. If a low-density layer overlies a thick higher-density layer and the structure is faulted (Figure A2.4*B*), the gravity anomaly profile is asymmetric, but the different geometries of faulting have only a minor effect on the anomaly shape. If the denser material is in a relatively thin layer (Figure A2.4*C*), the gravity anomaly profile is again asymmetric, though the shape is different from that in Figure A2.4*B*, and the effect of different fault geometry is significant. Thus although the anomaly shape imposes constraints on the possible structure, to be reliable, gravity models should be based on additional structural and geophysical information.

A2.3 GEOMAGNETIC STUDIES

A magnetic field is a vector quantity that has both magnitude and direction. For the Earth's magnetic field, the magnitude can be specified by the magnitudes of the horizontal and vertical components of the field. The orientation is specified by the declination and inclination, which are essentially the trend and plunge of the field line (see Appendix 1, Section A1.1), though the inclination also includes the polarity, which defines whether the magnetic vector points up or down. Studies of the Earth's magnetic field include the study of magnetic anomalies and of paleomagnetism.

Magnetic anomalies are measurements of the variation of the Earth's magnetic field relative to some locally defined reference. There is no international standard reference field from which anomalies are measured, because the Earth's magnetic field is not constant and changes significantly even on a human time scale.

Regional maps of magnetic anomalies are made by using both aerial and surface measurements. The principal use of continental magnetic anomaly maps is to infer the presence of rock types and structures that are covered by other rocks, sediments, or water. In some cases, the presence of particular rock types at depth can be inferred on the basis of characteristic patterns on a mag-

BOX A2-4 First-Motion Radiation Pattern from a Faulting Event

The first-motion pattern of seismic radiation from a fault-slip event can be understood by referring to the two-dimensional model shown in Figure A2-4.1. The undeformed state is shown in Figure A2-4.1A; it is represented by two squares drawn on opposite sides of an east-west line that represents the future location of a fault. Gradual prefaulting deformation of the rock (Figure A2-4.1B) deforms the squares into parallelograms, shortening the northwest-southeast–oriented dimensions of the squares (such as *AD* and *CF*) and lengthening the northeast-southwest–oriented dimensions (such as *BC* and *DE*). North-south and east-west dimensions remain unchanged.

An earthquake occurs when cohesion on the fault plane is lost, and sudden slip returns each parallelogram separately to its undeformed condition (Figure A2-4.1C). During faulting, the outer points *A*, *B*, *E*, and *F* remain stationary, while the points *C* and *D* on the fault separate into the respective pairs C_N and D_N, and C_S and D_S.

In this process, the northwest-oriented dimensions suddenly become longer (for example, D_N moves away from *A*, and C_S moves away from *F*), creating a rarefaction for the first motion. The northeast-oriented dimensions, however, suddenly become shorter (for example, C_N moves closer to *B* and D_S moves closer to *E*), creating a contraction for the first motion. Again, the north-south and east-west dimensions remain unchanged. Thus contractional first motions radiate outward in the northeast and southwest quadrants, and rarefaction first motions radiate outward in the northwest and southeast quadrants. The quadrants are separated by nodal planes, which are the fault plane and the plane normal to it, for dimensions do not change in these directions during faulting, and the amplitude of the seismic wave that propagates out in these directions is therefore zero.

The first-motion radiation pattern of contractions and rarefactions thus enables us to identify two nodal planes, one of which must be the fault plane. It also indicates the sense of slip on either plane that would generate that pattern: Relative to the line of intersection of the two nodal planes, slip is toward the quadrant of contractional first motions (C) and away from the quadrant of rarefactional first motions (R) on either side of either nodal plane. The actual fault plane often can be identified by taking into account geologic considerations or by studying the location of aftershocks that occur along the fault plane.

The same principle works in three dimensions, and the nodal planes can be identified from first motions by using a worldwide array of seismometers that in effect form a three-dimensional array surrounding the earthquake. The axis of maximum shortening, the P axis, bisects the nodal planes in the quadrants of rarefactional first motions, and the axis of maximum lengthening, the T axis, bisects the nodal planes in the quadrants of contractional first motions (Figure A2-4.1C). It is a potential source of confusion that the axis of maximum shortening (P) is located in the quadrant of rarefactional first arrivals (R), and the axis of maximum lengthening (T) is located in the quadrant of contractional first arrivals (C).

In Chapter 15 we show that the minimum instantaneous extension $\hat{\varepsilon}_3$ (maximum instantaneous shortening) parallels the P axis in the quadrant of rarefactional first motions, and the maximum instantaneous extension $\hat{\varepsilon}_1$ (maximum instantaneous lengthening) parallels the T axis in the quadrant of contractional first motions. The maximum and minimum compressive stresses, $\hat{\sigma}_1$ and $\hat{\sigma}_3$, are commonly equated with the P and T axes, respectively, although, strictly speaking, this is not correct, because, for example, the angle between $\hat{\sigma}_1$ and the plane of fracture ideally is about 30° rather than the 45° between the P axis and the fracture plane (see Chapter 15).

netic anomaly map. For example, the extension of rocks of the Canadian shield beneath thrust faults of the Canadian Rocky Mountains can be inferred from the extension of the shield magnetic pattern beneath the thrust front.

Marine magnetic surveys have resulted in the well known maps of the patterns of magnetic anomaly stripes symmetric about oceanic spreading axes, which have been so fundamental to the development of plate tectonic theory. When correlated with the magnetic reversal time scale, these maps can be interpreted to give a map of the age of ocean basins and the rate of seafloor spreading.

Magnetic anomalies also can be used in a manner similar to gravity anomalies to infer structure at depth, except that the magnetic anomalies are due to differences in magnetic properties of the rocks rather than to differences in their densities. Modeling of magnetic information is more complex, because a given anomaly in total field intensity can result from either differences in intensity of magnetization of the rocks or different orientations of the magnetic vector. Models of structure based on magnetic anomalies suffer from the same lack of uniqueness as models based on gravity anomalies and for similar reasons.

By a variety of processes that include crystallization, cooling, sedimentation, and chemical reaction in the Earth's magnetic field, rocks can become magnetized in a direction parallel to the ambient field and can preserve that magnetism even if the rocks are rotated to new orientations. Studies of paleomagnetism involve measuring the orientation of the magnetic field preserved in rocks

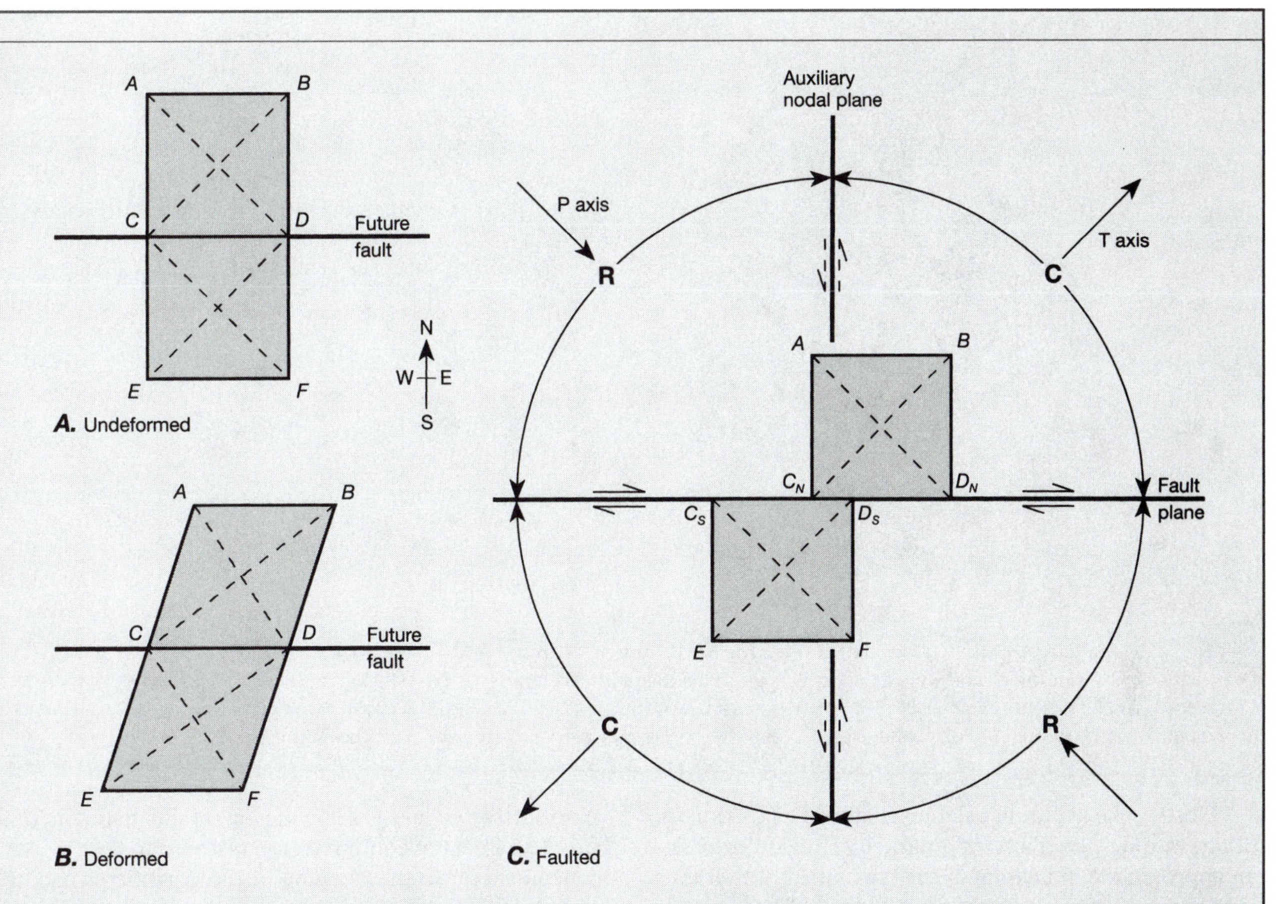

FIGURE A2-4.1 A two-dimensional model for the mechanism of first-motion radiation patterns. *A*. Undeformed state represented by squares on either side of a future fault. *B*. Deformed state before faulting. North-south and east-west dimensions of the squares are unchanged, but northeast-southwest dimensions (such as *BC* and *DE*) are lengthened and northwest-southeast dimensions (such as *AD* and *CF*) are shortened. *C*. Faulted state: Sudden slip on the fault generates an earthquake. North-south and east-west dimensions of the squares are still unchanged, but northeast-southwest dimensions (such as BC_N and D_SE) are suddenly shortened, and northwest-southeast dimensions (such as AD_N and C_SF) are suddenly lengthened. Thus first motions are contractions for rays leaving the source in the quadrants marked *C* and are rarefactions for rays leaving the source in the quadrants marked *R*. The fault plane and the plane normal to it are nodal planes along which no change in dimension occurs, and the amplitude of the first motion is therefore zero. P is the axis of maximum shortening, and T is the axis of maximum lengthening.

and comparing it to the orientation of the present field. If the original horizontal plane in the sample is known, these measurements can be interpreted to indicate the declination and inclination of the Earth's field at the time of magnetization. Horizontal or gently dipping unaltered sediments and volcanic rocks provide the most reliable paleomagnetic measurements, but more deformed or metamorphosed rocks and plutonic rocks are sometimes useful. Rocks that have been tilted since magnetization are generally assumed to have tilted about a horizontal axis, so they are restored to the original horizontal by rotation about an axis parallel to the strike of the bedding.

The Earth's magnetic field, when averaged over a sufficiently long time, like ten thousand years, is approximately symmetric about the axis of rotation, and the inclination of the field lines varies systematically with latitude from vertically down at the north pole through horizontal at the equator to vertically up at the south pole. Because this relationship is assumed to have been constant throughout geologic time, the paleo-declination determined for a sample in its original horizontal attitude indicates the amount of rotation a rock has undergone about a vertical axis, and the paleo-inclination relative to the original horizontal indicates the latitude at which the sample was magnetized. Such measurements can therefore define the changes in latitude and the rotations about a vertical axis resulting from the large-scale tectonic motions that rocks have experienced since magnetization. They can provide no information, however, on the changes in longitude associated with these motions.

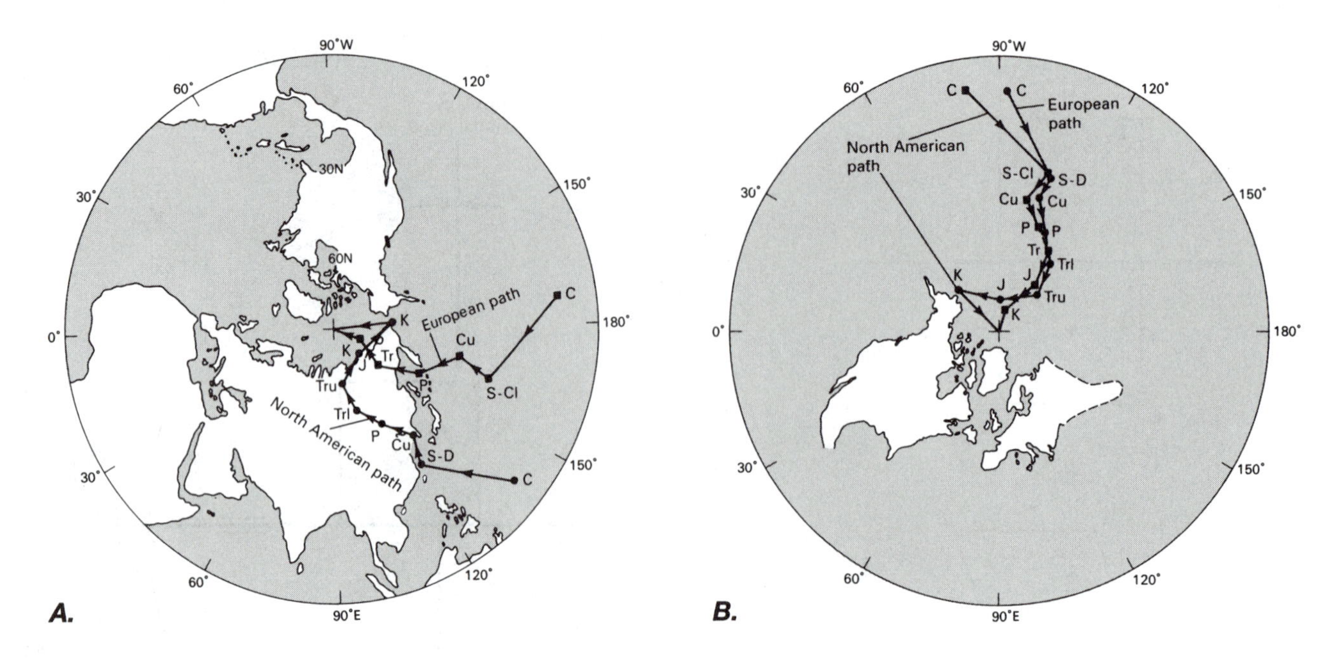

FIGURE A2.5 Apparent polar wander path (APW) for North America (circles) and Europe (squares). C: Cambrian; S: Silurian; D: Devonian; Cl and Cu: lower and upper Carboniferous; P: Permian; Tr, Trl, and Tru: Triassic, lower Triassic, and upper Triassic; K: Cretaceous. (After Press and Siever. 1986. *Earth*. W. H. Freeman, New York, p.425; McElhinny 1973) *A*. Polar wander paths for the continents in their present positions. *B*. Polar wander paths for the continents before the opening of the Atlantic Ocean.

Plotting the apparent paleomagnetic pole position for different time periods from a particular region provides an approximate indication of the movement of that area with respect to the Earth's geographic pole. These results are usually presented in the form of apparent polar wander maps, such as the map of paleopole positions for North America and Europe during the Phanerozoic (Figure A2.5*A*). If the continents are restored to their relative positions before the opening of the Atlantic Ocean, the apparent polar wander paths coincide approximately from Silurian through Triassic, indicating the period of time the continents were joined (Figure A2.5*B*).

REFERENCES AND ADDITIONAL READINGS

Fowler, C. R. M. 1990. *The Solid Earth: An Introduction to Global Geophysics*. Cambridge University Press, New York, 472 pp.

Lindseth, R. O. 1982. Digital processing of geophysical data: A review. Continuing Education Program, Society of Exploration Geophysicists, Teknica Resource Development Ltd., Calgary, Alberta, Canada.

Sheriff, R. E. 1978. A first course in geophysical exploration and interpretation. Boston: International Human Resources Development Corporation.

Appendix 3

UNITS AND CONSTANTS

Basic SI (Système Internationale; mks) Units			
Symbol	Name	Description	Equivalents
1 m	meter	distance	1000 mm = 100 cm = 39.37 in.
1 kg	kilogram	mass	1000 g = 2.205 lbm
1 s	second	time	31,536,000s = 1 y

Table of SI Multiples*

Multiple	Name (prefix)	Symbol	Multiple	Name (prefix)	Symbol
10^{1}	deca	da	10^{-1}	deci	d
10^{2}	hecto	h	10^{-2}	centi	c
10^{3}	kilo	k	10^{-3}	milli	m
10^{6}	mega	M	10^{-6}	micro	μ
10^{9}	giga	G	10^{-9}	nano	n
10^{12}	tera	T	10^{-12}	pico	p
10^{15}	peta	P	10^{-15}	femto	f
10^{18}	exa	E	10^{-18}	atto	a
10^{21}	zetta	Z	10^{-21}	zepto	z
10^{24}	yotta	Y	10^{-24}	yocto	y

*The prefix is placed before any basic unit of the SI system to indicate the multiple of that unit.

Other Systems of Units			
Symbol	Name	Description	Equivalents
CGS UNITS			
1 cm	centimeter	distance	10^{-2} m = 10 mm
1 g	gram	mass	10^{-3} kg
1 s	second	time	
ENGLISH UNITS			
1 in.	inch	distance	2.540 cm
1 ft	foot	distance	12 in. = 0.3048 m
1 lbm	pound mass	mass	0.4535 kg
1 s	second	time	

Table of Units and Constants Used in the Book and Some Common Equivalents*

Symbol	Name	Comment	Units/dimensions	Equivalents
FORCE				
1 N	Newton	SI (mks) unit of force	kg m s^{-2}	0.225 lbf
1 MN	Meganewton		10^6 kg m s^{-2}	10^6 N
1 dyne	Dyne	cgs unit of force	gm cm s^{-2}	10^{-5} N
1 lbf	Pound force	English gravitational unit of force = weight of 1 pound mass in standard Earth gravitational field	32.2 lbm ft s^{-2}	
PRESSURE or STRESS				
1 Pa	Pascal	SI (mks) unit of pressure or stress	N m^{-2}	
1 MPa	Megapascal		MN m^{-2} = N mm^{-2}	10^6 Pa = 10 b
1 GPa	Gigapascal		GN m^{-2}	10^9 Pa = 10^3 MPa = 10 kb
1 b	Bar	cgs unit of pressure or stress	10^6 dynes cm^{-2}	10^{-1} MPa
1 kb	Kilobar		10^9 dynes cm^{-2}	10^3 b = 10^2 MPa
1 atm	Atmosphere	pressure exerted by a column of mercury (Hg) 760 mm (29.92126 in) high	14.696 lbf/in^2	101.325 kPa $\sim 10^5$ Pa = 10^{-1} MPa = 1 b
ENERGY				
1 J	Joule	SI (mks) unit of energy	N m	
1 cal	Calorie (15 °C)	cgs unit of energy; definitions vary (by <0.1%), depending on the temperature of the water used as a standard. (In nutrition, '1 calorie' is generally 1 kcal (10^3 cal) as defined here)	Heat needed to raise 1 g of water at 15°C by 1°C	4.1855 J
1 Btu (BTU)	British thermal unit (59 °F)	English unit of energy; definitions vary (by <0.5%) depending on the temperature of the water used as a standard	Heat needed to raise 1 lb avoirdupois of water at 59°F by 1°F	1054.804 J
POWER				
1 W	Watt	SI (mks) energy per unit of time	J s^{-1}	N m s^{-1} = kg · m^2 · s^{-3}
TEMPERATURE				
1 °C	Degree Celcius	10^{-2} of the temperature range between the freezing and boiling points of water at 1 atm pressure, taken to be 0 °C and 100 °C respectively absolute zero = −273.15 °C	°C	°C = K − 273.15 °C = (5/9)(°F − 32)
1 K	Degree Kelvin	1 K is the same interval of temperature as 1 °C. absolute zero = 0 K	K	K = °C + 273.15
1 °F	Degree Farenheit	(1/180) of the temperature range between the freezing and boiling points of water, taken to be 32 °F and 212 °F, respectively. 1 °F = (5/9) of 1 K and of 1 °C; absolute zero = –459.67°F	°F	°F = (9/5) °C + 32

GRAVITY		Mean acceleration of gravity at Earth's surface = $9.81\ m\ s^{-2}$		
1 gu	Gravity unit	SI (mks) unit of gravitational acceleration	$10^{-6}\ m\ s^{-2}$	10^{-1} mgal
1 gal	Gal	1 galileo; cgs unit	$1\ cm\ s^{-2}$	10^{4} gu
1 mgal	Milligal	cgs unit commonly used to measure gravity anomalies	$10^{-3}\ cm\ s^{-2}$	$10^{-5}\ m\ s^{-2} = 10^{-3}\ gal = 10\ gu$

FUNDAMENTAL CONSTANTS				
1 mol	Mole (gram molecular weight)	The number (Avogadro's number) of particles (atoms, ions, molecules, or formula units) in a given mass of a substance, defined such that the mass in grams is numerically equal to the number defining the molecular weight of the substance.	$6.022\ 10^{23}$ particles	
k	Boltzman constant	Constant relating temperature to energy	$1.3806\ 10^{-23}\ J\ K^{-1}$	$1.3806\ 10^{-23}\ Pa\ m^3\ K^{-1} = 1.3806\ 10^{-29}\ MPa\ m^3\ K^{-1}$
R	Gas constant	Boltzman constant per mole	$8.3145\ J\ K^{-1}\ mole^{-1}$	$k \times 6.022\ 10^{23}\ J\ K^{-1}\ mole^{-1}$
G	Gravitational constant	Defines the gravitational attraction F between two masses m and M with a distance R between their centers of mass, such that $F = GmM/R^2$	$6.673\ 10^{-11}\ m^3\ kg^{-1}\ s^{-2}$	
e	e	Base of natural logarithms	2.7182818284 . . .	
π	pi	Dimensionless ratio of the circumference c to the diameter d of a circle: c/d	3.1415926535 . . .	

MATERIAL CONSTANTS				
1 Pa s	Pascal second	SI (mks) unit of viscosity; ratio of an applied shear stress to the resulting engineering shear strain rate	$1\ N\ s\ m^{-2}$	$1\ kg\ m^{-1}\ s^{-1} = 10$ poise
1 poise	Poise	cgs unit of viscosity; ratio of an applied shear stress to the resulting engineering shear strain rate	$1\ dyne\ s\ cm^{-2}$	$1\ gm\ cm^{-1}\ s^{-1} = 10^{-1}\ Pa\ s$
E	Young's modulus	Elastic modulus for isotropic materials; ratio of an applied normal stress to the extension parallel to the applied stress; same units as stress	MPa	$= \frac{\mu(3\lambda + 2\mu)}{\lambda + \mu} = 2\mu(\nu - 1)$
μ	Shear modulus; modulus of rigidity; one of the two Lamé parameters (see λ)	Elastic modulus for isotropic materials; ratio of an applied shear stress to the resulting engineering shear strain; same units as stress	MPa	$= \frac{E}{2(1 + \nu)}$
ν	Poisson's ratio	Elastic modulus for isotropic materials; absolute value of the ratio of extension normal to an applied compressive stress to the extension parallel to the applied stress	Dimensionless	$= \frac{\lambda}{2(\lambda + \mu)} = \frac{E - 2\mu}{2\mu}$
λ	One of the two Lamé parameters (see μ)	Elastic modulus for isotropic materials for volumetric deformation; ratio of the mean normal stress to the associated volumetric extension; same units as stress	MPa	$= \frac{-\mu(E - 2\mu)}{E - 3\mu} = \frac{E\nu}{(1 + \nu)(1 - 2\nu)}$

*The SI ('Systeme Internationale') or mks (meter-kilogram-second) units are now considered the worldwide standard. We give equivalences in the cgs (centimeter-gram-second) system and the English system mostly for reference to the older literature and to the U.S. engineering literature. (Source for some of the definitions: *Wikipedia*, http://www.wikipedia.org/).

Index

A
Abyssal plains, 17
Accretionary prisms, 615, 618–622, 619*f*
deformation structures, 618–619
by duplex formation, 619*f*, 622*f*
dynamic formation of, 622
fold styles, 620*f*, 621*f*
Sunda arc, 620*f*
underplating, 618, 622*f*, 623*f*
ac fractures, 58, 58*f*
Acicular habit lineation, 315, 316
ac joints, 59
Aegean arc, 616*f*
Afar triangle (Somalia), Red Sea rift margins, 583–584, 585*f*
Africa
East African rift, 583, 584*f*
tectonic features (map), 583*f*
African Bight, 589
Alaska
earthquake of 1964, 70
oroclines, 273, 274*f*
Yakutat block collision, 628
Allochthon, 115, 118, 118*f*
Allochthonous rocks, 115, 117
Alpine fault (New Zealand)
termination, 146, 148*f*
transform fault, 608
Alpine-Himalayan belt
collisions, 628*f*, 629
orogenic belts, 640*f*, 641*f*
system, structures of, 630*f*
Alpine peridotites, 596, 598
Alps
Austro-Alpine nappes, 131–132, 132*f*
crustal-scale cross section, 661*f*
crystalline nappes, 654*f*
exhumation rates, 673
external massifs, 646*f*
folds, 275*f*
Helvetic nappes, 667
Ivrea zone, 655
lateral extension, 663*f*
lineations and shear directions, 666*f*
maximum instantaneous extension, 669*f*
metamorphic facies, 650, 650*f*
Morcle Nappe, 667, 668*f*, 676*f*
Penninic nappes, 682
root zone, 660, 662*f*
suture zone material, 679
Alteration spots, 310, 310*f*
as strain markers, 424
Altyn Tagh fault (China)
strike-slip faults, 558–560, 558*f*, 611
transcurrent faults, 136*f*
Amontons' second law, 212, 219
Amplitude of folds, 280
Analytic solutions
functions of, 543
viscous buckling of competent layer in incompetent matrix, 561–563, 561*f*, 562*f*, 563*f*
Anastomosing foliations, 299*f*, 301, 301*f*
Anatolian faults (Turkey), 611
Andaman Sea, seafloor spreading, 560
Andean-style margins, 27, 28*f*, 29, 29*f*, 615
Anderson's theory, faulting, 254–255, 255*f*
Andes
oroclines, 273
orogenic belts, 640*f*
Andesitic flow, columnar jointing, 41*f*
Angara graben (Siberia), pull-apart basin, 143*f*
Angle
interlimb, 282, 284, 286
of internal friction, 212
of repose, 260
Angular momentum, conservation of, 544, 574
Anisotropy
effect on fracture, 221–224, 222*f*, 223*f*, 224*f*
numerical model of folding, 568–570, 570*f*
Annealing, 516
Anorogenic magmatic suites, 22
Anorthosite, Proterozoic terrane, 22
Anticlines
defined, 280
fault-ramp, 95*f*
fold train, 276*f*
compared to synclines, 281*f*
rollover, 66
thrust faults, 119, 121*f*
Antiformal stack, 125*f*
Antiforms, 276
Antitaxial growth, mineral fibers, 420
Antithetic faults, 70, 87, 87*f*, 94*f*, 95, 108, 108*f*, 126*f*
Apennines (Italy), olistostromes, 624, 624*f*, 626
Appalachian Mountains
eugeoclinal core zone, 656*f*
fold and thrust belts, 644
metamorphism, 651*f*
orogenic belts, 640*f*, 641*f*, 642*f*
orogenic float, 660, 662*f*
thrust fault displacement, 131
thrust faults, 116*f*, 122*f*
ultramafic bodies, 657*f*
Appalachian plateau (New York), joints, map of, 46*f*

Note: *f* following page number indicates figure; *t* following page number indicates table.

Appalachian Valley and Ridge
folds, 276*f*, 292
Martinsburg slate fracture experiments, 221–222, 222*f*
Martinsburg slate volumetric strain, 426–427, 427*f*
oroclines, 273
Apparent lineations, 310, 310*f*
Apparent polar wander (APW) paths, 689
Applied stress, versus local stress, 226–227
Aquathermal pressuring, 238
Archean terranes, 19–22
age of rocks, 19
rock types, 19, 20*f*
structural features, 19, 20*f*
sutures, 21, 21*f*
tectonic conditions, 20–22
Arches, Phanerozoic regions, 25, 26*f*
Arc-trench gap, 615
Ardmore Basin (Oklahoma)
flower structure, 142*f*
positive flower structure, 142*f*
Area strain, 325–326
Argille scagliose, 618
Argillite, disjunctive foliations, 407, 408*f*
Arltunga complex (Australia), paleostresses, 522
Arrhenius factors, 478
Aseismic ridges, 14*f*, 17
Asia, active tectonics in, 137*f*
Aspect ratio, fold style, 284, 285*t*
Asperities, 66, 404, 417
and friction, 498, 498*f*
Asthenosphere, 8, 8*f*, 11, 29*f*
and midoceanic ridges, 594*f*, 595*f*
Asymmetric crenulation foliations, 407–411, 409*f*, 410*f*, 411*f*
Asymmetric folds, 283–284, 293
Atacama fault (Chile), 611
Atlantic-style margins, 27, 28*f*
Attitude
of folds, 280–282, 282*f*
of lines, 693
of planes, 693
Aulacogens, Proterozoic terranes, 22, 24*f*, 589
Austro-Alpine nappes (Alps), 131–132, 132*f*
Autochthon, 115, 118, 118*f*
Autochthonous rocks, 115, 117
Axial compression
Mohr diagram, 173, 174*f*
stress, 173, 174*f*
Axial extension, Mohr diagram, 173, 174*f*
Axial plane
cleavages, 303
folds, 279, 280*f*
Axial surface
folds, 279–280, 280*f*
foliations, 303, 305*f*
trace, 279, 280*f*, 288*f*
Axial symmetry, 536, 536*f*
Azimuth, 693

B
Back-arc basins, 591, 591*f*, 615, 617*f*
Back-arc margins, 30, 30*f*
Back folds, 660, 680
Backstress, 514
Back thrusts, 126, 126*f*, 128
Bacon, Sir Francis, inductive (Baconian) method, 5–6
Balanced cross sections, 88
Banding
banded foliations, 299, 299*f*
defined, 298
Bangong-Nujiang suture (Tibet), 630*f*
Barberton Mountain Land (Africa), Archean terrane, 19, 20*f*
Barrel boudins, 312*f*, 313
Barrovian metamorphism, 648, 649, 649*f*, 650*f*, 681
Basal decollement, 644
Basal ductile flow, thrust sheet emplacement, 259
Basal friction
thrust sheet emplacement, 259
Basalts
oceanic spreading centers, 594*f*, 595*f*
oceanic transform faults, 605*f*
in ophiolite sequences, 597*f*, 598
Basin
cratonic, 25
foreland, 640
molasse, 640
pull-apart, 140
Basin and Range province (North America)
and Anderson's theory, 255
continental crust thickness, 17
continental rifts, 27
Great Basin, structure of, 101*f*
location/relief map, 610*f*
metamorphic core complexes, 99*f*, 100–102
normal faults, 92, 98, 99*f*, 100
Basins
Africa (map), 583*f*
fold as, 281
oceanic. *See* Ocean basins
Phanerozoic regions, 25, 26*f*
Batholiths, orogenic belts, 658–660, 659*f*–660*f*
Bay of Islands complex (Newfoundland), transform faults, 607–608
bc fractures, 58, 58*f*
bc joints, 59
Beachball diagrams, seismogenic strain, 438*f*, 439
Bearing, 693
Bedding-foliations, and folds, 305*f*, 306–307, 306*f*
Bedding joints, 40
Bedding plane fissility, 407, 408*f*
Belemnites, 424
Belledon massifs (France), 646
Bending, 363
flexural shear, 368
orthogonal flexure, 365
volume-loss flexure, 372
Bends
contractional/restraining, 139*f*, 140
extensional/releasing/dilatant, 139*f*, 140
strike-slip faults, 85, 86*f*, 139–140, 139*f*
Benue Trough (Nigeria), failed rift, 589
Big Bend (San Andreas fault), 144–145, 145*f*, 609, 610*f*
Bight, 589*f*
Bingham materials, behavior of, 464, 464*f*
Bird River complex (Canada), Proterozoic dike system, 22, 23*f*
Bishop Creek (Nevada), ptygmatic folds, 294
Bismarck Archipelago, 616*f*
Blind fault, 83
Blocks, boudins, 311, 312*f*
Blue Ridge Mountains (Maryland), South Mountain fold, 429, 430*f*
Blueschist metamorphism, 648, 649, 649*f*, 650*f*
Bluntness, fold style, 285, 287*f*
Body forces, 153
Bone-shaped boudins, 312*f*, 313
Boso triple junction, 616*f*
Boudins
boudinage, 311
defined, 311
formation of, 311, 416–417, 417*f*
shapes of, 311–313, 312*f*
as strain markers, 328, 328*f*, 424–425
Bouger gravity anomaly, 584*f*
Boundary conditions
physical models, 545–546
scale models, 547–548
viscous deformation model, 575
Boundary migration recrystallization, 514, 515*f*
Bow-tie-vein boudins, 312*f*, 313
Box method, fractal dimension measure, 49, 51*f*
Branch lines, 83–84, 85*f*
Breccia series, faults, 63, 64*t*, 92–93
Brittle deformation
defined, 11, 35
faults, 84–85
inferred from fault systematics, 439–443
strain measures from, 434–443, 437*f*, 438*f*
See also Fracture(s); Joint(s)
Brittle-ductile transition, 110*f*, 489–490, 490*f*
Brittle fracture
experiments on brittle rock, 210
failure of, 210–211
Griffith theory. *See* Griffith theory of fracture
Bronson Hill Anticlinorium (New England), gneiss dome, 397*f*
Buchan metamorphism, 648–649, 649*f*, 650*f*
Buckling
analytic solution, viscous buckling, 561

flexural folding, 364–365, 365*f*, 368*f*
interference for single layer, 393–394, 394*f*
kinematic models, 365–366, 369, 372–373*f*
kinematic and numerical models compared, 570
multilayer folds, 379, 379*f*
numerical models of, 563
orthogonal flexure, 366, 368*f*
volume-loss folds, 368*f*
Burgers vector, 504–506, 505*f*, 506*f*, 512
Burial
joint formation, 241, 243–245
metamorphism, 650, 650*f*
sediment, mechanical properties, 244*t*
Bushveld Complex (South Africa), Proterozoic dike system, 22, 23*f*
Byerlee's law, 466

C
Calabria, 616*f*
Calcite
deformation mechanism map, 496*f*
paleostress orientation/magnitude, inferring, 520–521
twin gliding, 499–500, 499*f*, 520–521
Caldera, normal faults, 98, 98*f*
Caledonides
fold and thrust belts, 644
low-angle detachment fault, 663, 664*f*
orogenic belts, 640*f*, 641*f*, 642*f*
California-style margins, 29, 29*f*
Cambrian slate belt (Wales), strain measurement, 431–432, 433*f*
Cambrian units, 119, 119*f*
Canadian Rockies
oroclines, 273
tectonic wedging, 126*f*
thrust fault displacement, 131
Canadian Shield
Archean sutures, 21, 21*f*
orogenic belts, 640*f*, 641*f*, 642*f*
Caprock, 396
Cartesian coordinate system, 154
Cascades, 616*f*
Cataclasites
cataclasis process, 219, 404
and dynamic rock friction, 467–469, 468*f*
and fault depth, 63
and faults, 63–65, 63*t*, 64*t*, 65*f*, 92, 404
foliations in, 404–405
pseudotachylites, 63, 64*t*, 65, 65*f*, 499
as self-similar, 63
types of, 63, 64*t*
Cataclastic flow, 498
defined, 498
and deformation, 498–499
and dynamic rock friction, 467–469, 468*f*
experiments on, 468
Cate Creek window (Tennessee), thrust fault, 120*f*, 131
Causation, versus correlation, 7
Central Scotian Shelf, 584, 588, 588*f*
C-foliations, 307*f*, 308, 308*f*
Chaman fault (Afghanistan/Pakistan), 611
Chaotic deposits, 622–626
formation sites, 625–626
olistostromes, 622, 624, 624*f*
origin of, 625*f*
tectonic melanges, 622, 624–625, 625*f*
Chevron folds, 293, 293*f*, 382–383
kinematic models, 382–383, 382*f*
layered sequences, 382–383
strain in, 383
Chief Mountain klippe (Tennessee), 120*f*
China Mountain (California), diffuse foliations, 299*f*
Chocolate-tablet structure, 313
Chromite, Proterozoic terrane, 22, 23*f*
Class 1, 1A, 1B, 1C folds, 288, 289*f*, 289*t*, 291*f*
Class 2 folds, 288, 288*f*, 289*f*, 289*t*
Class 3 folds, 288, 288*f*, 289*f*, 289*t*
Clastic dikes
defined, 55
joint formation evidence, 243
Cleavage
defined, 298
flow, 309
fracture, 309
and lineations, 311, 311*f*
phyllitic, 309
shear/solution/strain-slip, 309
slaty, 309
Cleavage domains, and foliations, 299–300, 299*f*, 300*f*, 301, 302*f*, 303, 309
Cleavage plane, stress on, 222–224, 223*f*
Cleft girdle pattern, 531, 532*f*
Climb, edge dislocations, 509–510, 510*f*, 513–514
Closure, folds, 277, 277*f*
Closure temperature, 671
Coarse continuous foliations, 303, 304*f*
Coast Range Ophiolite (CRO) (California), antithetic thrusting, 126*f*
Coaxial progressive deformation, 335, 403
Coble creep, 495, 501
Coefficient of internal friction, 212
Coefficient of viscosity, 461
Coherent recrystallization, 404
Cohesion, Couloumb fracture criterion, 212
Cold working, 476, 513
Collisions, 626–631
active, world maps, 628*f*
cessation of, 681
collision margins versus convergent margins, 27
collision zone, 615
crustal effects, 628–629
India with Asia, 684
ophiolite complexes, 628–629, 629*f*
orogeny, 677–684
types of, 627*f*, 628
Wilson cycle, 685–687, 686*f*
Columnar joints
features of, 40, 41*f*
formation of, 246
hexagonal shape, 246
Comminution, fractal model, 51–52, 51*f*
Competent rock, defined, 363
Completely oriented fabric, 300*f*
Composite growth, mineral fibers, 420
Compositional foliations, 298–299, 299*f*
development of, 405
Compression, defined, 346
Compressional (P) wave, 701
reflection of, 701
refraction of, 705
Compressive stress, 158, 158*f*
defined, 348
Griffith fracture criteria, 229
Concentric folds, 288
Confined compression, 173
Confining pressure
Coulomb fracture criterion, 212–216, 213*f*
and fracturing, 216–220
and frictional sliding, 219
in Griffith theory, 229
and mode of deformation, 218–219, 218*f*
and shear failure, 216–219, 218*f*
Congruous steps, fault displacement, 72, 73*f*, 74
Conical folds, 278, 279*f*
Conjugate fans, 109*f*
Conjugate normal faults, 226*f*
Conjugate shear fractures, 57, 57*f*, 210*f*, 216, 218
Conjugate shear planes, 212, 213*f*
Connecting fault, 87*f*
Conservation laws
conservation of angular momentum, 544, 574
conservation of mass, 544, 573
conservation of momentum, 573–574
viscous deformation model, 573–574
Conservative boundaries, plate movement, 11, 12
Constant volume deformation, Flinn diagram, 341, 341*f*
Constant-volume strain, 328*f*, 329
Constitutive equations
functions of, 3, 459, 545
in three-dimensions, 491–492
viscous deformation model, 574–575
See also Material behavior
Constriction
defined, 348
homogeneous deformation, 331*f*, 332
Constructed lineations, 309, 310–313, 311*f*, 312*f*
boudins, 416–417
development of, 416–418
structural slickenlines, 417–418, 418*f*
Continental arcs, structure of, 615, 618

Continental crust, 17–19
Moho discontinuity, 18*f*, 19
Phanerozoic regions, 25–30
Precambrian shields, 19–25
seismic model, 17–18, 18*f*
structure of, 17–19
thickness variation, 17–18, 17*f*, 18*f*
Continental ice sheets, melting, and plate bending, 237
Continental margins, 27–30
bight, 589*f*
convergent, 615–626
faulting along, 102–106, 105*f*, 106*f*
modern study of, 27
rifted. *See* Rifted continental margins
types of, 27–29
Continental rifts, features of, 27
Continental shelves, world map, 14*f*
Continent-continent collision, orogeny model, 677–684
Continua, defined, 2
Continuous foliations
coarse, 303, 304*f*, 412
development of, 411–412
features of, 298, 298*f*, 299*f*, 302–303
fine, 303, 304*f*
Continuum mechanics, study of, 2
Contraction, defined, 348, 349*t*
Contractional, defined, 349*t*
Contractional bends, 139*f*, 140
Contractional duplexes, 140
Contractional (P) wave. *See* Compressional wave
Convergent, defined, 349*t*
Convergent boundaries, plate movement, 11
Convergent foliation fans, 303, 305*f*
Convergent margins, 615–626
accretionary prisms, 618–622
chaotic deposits, 622–626
versus collisional margins, 27
continental arcs, 615, 618
cross-section, 617*f*
formation of, 16, 27, 29
island arcs, 615
partitioned strike-slip faulting at, 622
physiographic features, 615
Convergent strike-slip deformation, 343, 344*f*, 345*f*
Convolute folds, 290
Cordillera (North America)
batholitic rocks, 658, 659*f*
fold and thrust belts, 644
normal faults, 99*f*
orogenic belts, 640*f*, 641*f*, 643*f*
paleostresses, 522
thrust faults, 122*f*, 126–127, 126*f*
Core complex, metamorphic, 100, 589, 592, 602, 680
Core of Earth
composition of, 8
layers of, 8, 8*f*
Core zone, crystalline, 648, 653
Correlation, versus causation, 7
Coso Range (California)
seismogenic strain, 438*f*, 439
strike-slip structures, 610*f*, 611–614, 614*f*
Couloumb fracture criterion
Anderson's theory of faulting, 254–255, 255*f*
Berea sandstone, 213
for confined compression, 212–216, 213*f*
defined, 212
equation, 213–214
and frictional sliding, 219, 219*f*
limitations of, 225–226
principal stresses, 214
shear fracture formation, 251
See also Normal faults; Strike-slip faults; Thrust faults
Counterclockwise folds, 283–284
Crack-seal growth, mineral fibers, 420, 420*f*
Cratonic platforms, 25, 26*f*
Cratons, 22
Creep
Coble, 495, 501
defined, 475
diffusion, 500
dislocation, 513–535
exponential-law, 479
Harper-Dorn, 516
logarithmic, 476
Nabarro-Herring, 495, 500–501, 501*f*, 516–517
power-law, 478, 479
primary/secondary/tertiary, 476–477, 477*f*
solution, 495, 501–502, 502*f*
steady-state. *See* Steady-state creep
superplastic, 502–503, 503*f*
Creep rate, relative evaluation of, 516
Crenulation foliations
asymmetric, 407–411, 409*f*, 410*f*, 411*f*
development, theories of, 407–411
discrete, 302, 302*f*, 303*f*, 407, 410
extensional, 409–410
features of, 302, 302*f*, 303*f*
symmetric, 407, 409*f*
zonal, 302, 302*f*, 407
Crest, folds, 277, 278*f*
Crinoid stems, as strain markers, 424
Critical Coulomb wedge model
compared to natural thrust wedges, 262
tapered thrust sheets, 260–263, 261*f*
Crossed girdle patterns, 531, 532*f*
Cross joints, 40–41, 59
Cross-over stress, 516, 518*f*
Cross sections
balanced, 88
displacement, determination of, 132–133
geometric constraints, 107, 107*f*
normal faulting, 110–111, 110*f*
thrust faults, 118*f*, 123*f*, 125*f*, 132
Cross-slip, dislocations, 509, 511*f*
Crust of Earth, 9–19
composition of, 8
hypsometric diagrams, 9, 11*f*
stress in. *See* Stress in Earth
structural distributions, 9, 10*f*
See also Continental crust; Oceanic crust
Crystallographic preferred orientation
formation of, 525–528, 526*f*, 527*f*
kinematic analysis, 669–670
misfit minimization model, 525, 527
Taylor-Bishop-Hill theory, 525, 526*f*
Crystals
dislocations, 504–512
ductile deformation of, 503–504, 504*f*
interstitials, 500, 500*f*
point defects, 495, 497, 500, 500*f*
recrystallization, 514–516, 515*f*
in slip system, 508–509, 510*f*
stretched, 420
subgrains, 404, 511, 519
See also Mineral(s)
Culminations, 123
folds, 277, 278*f*
Curvature, of folds, 276–277, 277*f*
Cutoff lines, 68*f*, 69, 79, 79*f*
Cylindrical folds
features of, 278, 278*f*, 282–283
stereographic projection, use of, 279
Cylindroidal, use of term, 283

D
Damage zones
faults, 86–88, 87*f*
trishear zones, 388
Dante's Domes (Mid-Atlantic ridge), metamorphic core complex, 104*f*
Darwin, Charles, deductive (Darwinian) method, 5–6
Dash pot, 461
Dasht-E Bayaz fault (Iran), extensional duplex, 144, 144*f*
Dead Sea, transform fault, 608
Death Valley National Monument (California, Nevada), Titus Canyon megabreccia, 65*f*
Deborah number, 572, 576–577
Deck-of-cards-model, passive shear fold, 339, 339*f*, 369–370, 370*f*
Decollements
basal, 644
features of, 121*f*, 123
and multilayer folds, 276*f*, 290, 291*f*
Deductive (Darwinian) method, 5–6
Deformation
brittle deformation, 12
brittle strain inferred from, 439–443
confining pressure, 212–220
defined, 1
deformation maps, 495–496, 496*f*
ductile. *See* Ductile deformation
dynamics/mechanics of, 459
elastic, 231–232, 497
frictional behavior, 466–475

gravity-driven, scale model, 551–553, 551*f*, 552*f*, 553*f*
history, information from, 446–447
laboratory study, 214, 216, 216*f*
material behavior, models of, 460–464
microscopic structures related to, 497
orogenic belts, 673–676
progressive. *See* Strain
rheology, applications, 487–492
steady-state creep, 476–486
and study of geology, 2, 6–7
symmetry elements. *See* Deformation symmetries
viscous, mathematical model, 573–576
vorticity of, 334–335
Deformation lamellae, 518–519, 519*f*
quartz, 521–522, 523*f*, 524*f*
Deformation mechanism map, 496*f*
deformation map, 495
Deformation symmetries, 535–539
axial, 536, 536*f*
of deformation fabrics, 538–539
of deformation geometry, 539
isotropic/spherical, 535–536, 536*f*
monoclinic, 536*f*, 537
orthorhombic, 536–537, 536*f*
symmetry principal, 537–538
triclinic, 537
Deformation tensors, 352
Delamination, 670
Denali fault (Alaska), 611
Depressions, 123
folds, 277, 278*f*
Detachment faults
of metamorphic core complex, 101–102, 103*f*, 104*f*
normal faults, 93, 94*f*, 108*f*, 109*f*
Detachments, 123, 123*f*
Deviatoric compression, defined, 348
Deviatoric stress
defined, 347*t*, 348
Mohr diagram, 174*f*, 175
Devil's Postpile National Monument (California), columnar joints, 41*f*
Dextral faults. *See* Right-lateral (dextral) faults
Diagonal joints, 41
Diapirism, 395
Diapirs, 119, 121*f*, 395–399
formation of, 395–396
forms of, 395*f*
internal structure of, 396*f*
mantle, 397
model experiments, 552–553, 552*f*, 553*f*
normal faults, 97*f*
serpentine, 397
shale, 397
Diatremes, 98
Differential stress
defined, 347*t*
Mohr diagram, 174*f*, 175
Diffuse foliations, 298–299, 299*f*
Diffusive mass transport, 495
Diffusion
and deformation, 500
grain boundary, 501, 502*f*
self-diffusion, 500
volume diffusion, 500–501, 501*f*
Diffusion creep, 500
relation to power-law creep, 518*f*
Dikes
clastic, 55, 243
at oceanic spreading centers, 592, 594*f*, 595*f*, 596
oceanic transform faults, 605*f*, 607
in ophiolite sequences, 597*f*, 598
Proterozoic, 22, 23*f*
sheeted dike complex, 598, 599*f*, 601*f*
Dilatant bends, 139*f*, 140
Dilatant faults, 66
Dilation, defined, 228, 349*t*
Dip, dip angle, 693
Dip isogons
folds, 287–288, 288*f*, 290*f*, 291*f*
homogeneous flattening, 375, 375*f*
Dip line, 693
Dip joints, 40
Dip separation, 79, 79*f*
Dip-slip faults, 61–62, 62*f*, 63*f*, 86*f*
inclined. *See* Normal faults
Direction of tectonic transport, 667, 673
Discrete crenulation foliations, 302, 302*f*, 303*f*, 407, 410
Discrete lineations
development of, 416
features of, 309, 310, 310*f*
Disharmonic folds, 289–290, 290*f*, 294, 294*f*
Disjunctive foliations
development, theories of, 405–407, 406
features of, 299–302, 299*f*, 300*f*, 301*f*
Dislocation(s), 504–512
Burgers vector, 504–506, 505*f*, 506*f*, 512
climb, 509–510, 510*f*, 513–514
cross-slip, 509, 511*f*
curved, 505–506, 506*f*
defined, 495, 504
dislocation walls, 511, 511*f*
edge, 504, 505*f*, 508–512
Frank-Read source, 512, 512*f*
geometry of, 504–507, 505*f*, 506*f*, 507*f*
glide of, 507–510, 509*f*, 512*f*, 513
interactions, 511*f*
motion of, 507–512
and recrystallization, 514–516, 515*f*
screw, 504–506, 505*f*, 508–509
viewing of, 506–507, 508*f*
Dislocation creep, 513–535
creep rates, 516, 518
crystallographic preferred orientation, formation of, 525–528, 526*f*, 527*f*
Harper-Dorn, 516–518
microstructural fabrics related to, 518–520
olivine fabrics, 528–530, 528*f*, 529*f*
quartz fabric, 530–535, 530*f*, 532*f*, 533*f*
Dislocation density, 519–520, 520*f*
measurement of, 519
Dislocation walls, 511
Displacement, faults. *See* Fault displacement
Displacement vector, 353, 354*f*
Dissolution, minerals, 403–404
Divergent, defined, 349*t*
Divergent boundaries, plate movement, 11, 12
Divergent flow, 673, 675, 682
Divergent foliation fans, 303, 305*f*
Divergent margins
ocean basins, 590–601
rifted continental margins, 581–590
Divergent strike-slip deformation, 343, 344*f*, 345*f*
Domes
Africa (map), 583*f*
dome-in-graben, 584*f*
fold as, 281
mantled gneiss, 654
normal faults, 96, 97*f*, 98, 102
Phanerozoic regions, 25
salt. *See* Diapirs
turtle-back, 102
Dora Maira, 650*f*, 666*f*
Doubly plunging folds, 276*f*, 281
Down-scale, 2
Drag folds, 388
faults, 68*f*, 69, 92
Hansen's method, slip-line determination, 390, 391
Dry Creek fault (Nevada), 101*f*
Ductile deformation, 12, 271–272
cataclastic flow, 498–499
creep, 500–503
criteria for recognizing, 271–272
defined, 271
diffusion, 500
dislocation creep, 513–518
dislocations, 504–512
experimental investigations, 475–477
friction, 498
granular flow, 498
mechanisms of, 495–518
rheologies inferred from, 517
steady-state creep, 476–486
twin gliding, 499–500, 499*f*
Ductile failure, 210
Ductile nappe emplacement, 675–676, 675*f*
Ductile shear fault zones, 62*f*
and fault displacement, 75
and foliations, 307–308, 307*f*, 308*f*
simple shear, 443–445, 444*f*
strain measures from, 443–446, 444, 445*f*, 446*f*
volume change, 444–446, 445*f*, 446*f*
Dueling ridge segments, 606–607, 606*f*
Duplexes
accretionary prisms, 619*f*, 622*f*
extensional, 109*f*, 110, 144*f*
faults, 85, 86*f*
strike-slip, 140–141, 140*f*, 141*f*
thrust, 124, 125*f*, 127–128, 128*f*
Dynamic recrystallization, 514, 515*f*, 519

Dynamic rock friction
and cataclastic flow, 467–469, 468*f*
static to dynamic transition, 468–469, 469*f*

E
Earth
crust of. *See* Crust of earth
as dynamic planet, 1
eons of, 12*f*
geology as study of, 1–7
heat/temperature of, 8–9
interior layers of, 8–9, 8*f*
stress in. *See* Stress in Earth
topographic elevations, 11, 11*f*
Earthquakes
and stick-slip sliding, 219
stress determination method, 234, 236
Eclogite, 19
Edge dislocations, 504, 505*f*, 508–512
climb, 509–510, 510*f*
interactions, 511*f*
Effective stress
defined, 220, 347*t*
Mohr diagram, 174*f*, 175
Effective viscosity, 566
Elastic crust, 237
Elastic deformation, 231–232
definition, 231
at low-temperature, 497
Elastic material behavior, 460, 460*f*
elastic-plastic (Prandtl) material, 464, 464*f*
elasto-viscous (Maxwell) material, 463–464, 464*f*
firmo-viscous (Kelvin or Voight) material, 463–464, 463*f*
three-dimensional elasticity, 491
Elongate grain lineations, 315
Elongation
defined, 347*t*
quadratic, 322, 347*t*
Ely-Black Rock fault (Nevada), 101*f*
Emperor Seamounts (Hawaii), aseismic ridges, 14*f*, 17
En echelon arrays
fringe fractures, 247, 247*f*
gash fractures, 252, 252*f*, 447*f*
pinnate fractures, 252
San Andreas fault, 611, 612*f*
strike-slip faults, 135, 138, 139, 140
transfer zones, 124, 126*f*
Engadine window, 131–132, 132*f*
Engineering Mohr circle sign convention, 187
Engineering shear strain, 323*f*, 324, 347*t*
Engineering tensor sign convention, 187
Envelope, shear fracture, 212. *See* Coulomb fracture criterion
Enveloping surfaces, folds, 280, 281*f*
Epeiric seas, 25
Epistemology, defined, 6
Equal angle projection, 696–697, 697*f*
Equal area projection, 696–697, 697*f*
Equilibrium equation, conservation of momentum, 544

Etch pits, 506
Evaporite, thrust sheet mechanics, 259
Exfoliation joints. *See* Sheet joints
Exhumation
defined, 670
and orogeny, 643, 680
tectonics of, 680
Exotic terranes
analysis of, 688–689, 688*f*
features of, 688
Expansion, defined, 349*t*
Exponential-law creep, 479
Extension
defined, 346, 347*t*, 348
normal faults, estimation of, 111–112, 111*f*
simple, 332
uniaxial, 329
volumetric 326, 329
Extensional bends, 139*f*, 140, 140*f*
Extensional, defined, 349*t*
Extensional crenulation foliations, 409–410
Extensional deformation
defined, 348
orogenic belts, 660, 663–664, 663*f*
Extensional duplex, 109*f*, 110, 144*f*
Extensional strain, of material line, 322*f*
Extensional stress, defined, 346, 348
Extension fractures, 37–42
defined, 38, 210
experiments on brittle rock, 210, 210*f*
and fault displacement, 73–74, 74*f*, 87, 87*f*
joints, 38–41
oblique extension, 37
pinnate (feather), 41, 42*f*
plumose structure on, 251
propagation of, 37, 38*f*
compared to shear fractures, 251–252
spacing of, 250–251, 251*f*
External forces, types of, 153
External massifs, 645, 646*f*

F
Fabric element, 497
Fabrics
cataclastic, 498
coaxial, 525–535, 538
crossed girdle patterns, 531, 532*f*
of dislocation creep, 518–520
fabric skeletons, 531
microfabric/macrofabric, 497
noncoaxial deformation, 531, 533*f*
symmetry principles. *See* Deformation symmetries
Faceted spurs, 69, 69*f*, 92*f*
Facies, metamorphic, 649*f*
Failure
brittle fracture, 209–210
ductile, 209, 216
Fans
foliation, 303, 305*f*
imbricate. *See* Imbricate fans
listric, 109*f*

Fault(s), 62–63
categories of, 61–62, 62*f*, 63*f*, 68*f*
defined, 61
displacement. *See* Fault displacement
Earth's crust, view of, 64*f*
fault blocks, 61–63
fault zone, 62*f*, 64*f*, 65
and fractures, 42*f*, 57, 57*f*, 252–253
graphic depiction of. *See* Fault geometry
high and low-angle, 61
impact on geologic units, 66–69
indications of, 66–69, 68*f*
normal, 91–112
oceanic transform, 601–608
origin of term, 61
physiographic criteria for, 69–70
reverse. *See* thrust
ramps/jogs/duplexes, 85, 86*f*
rocks of, 64*t*, 63–65
as shear fractures, 214–216
shear sense criteria, 71
slip vectors/components, 61–62, 62*f*
strike-slip, 135–149
termination lines, 82–85, 84*f*
terminology related to, 63*t*
textures/structures of, 63–66
three-dimensional view, 81, 82*f*, 85*f*
thrust, 115–133
transcurrent, 135–136
transform, 135–136*f*
See also specific types of faults
Fault benches, 69
Fault-bend folds, 118*f*, 119, 385–386, 385*f*
Fault blocks, 61–63
footwall block, 61, 62*f*, 63*f*
hanging wall block, 61, 62*f*, 63*f*
Fault displacement, 70–80
asymmetric step formation, 71–72, 73*f*, 74*f*
complete determination of, 71
cross section estimation, 132–133
damage zones, 86–88, 87*f*
frictional sliding, 219–220, 469–471, 470*f*
geological maps, 77, 78*f*, 79*f*
kinematics of, 70–77
length systematics, 80, 81*f*, 83*f*
mechanical analogue, 470*f*
motion detection methods, 70–71
nonunique constraints on, 79–80, 79*f*
normal faults, 95, 95*f*
partial determination, large-scale structures, 77–79
partial determination, small-scale structures, 71–77
and piercing points, 71, 72*f*
relative and absolute, 70
secondary fracturing, 73–74, 74*f*, 86–88
separations, 79–80, 79*f*, 80*f*
and slickenfibers, 72, 73*f*, 420
strike-slip faults, 78*f*, 138*f*, 140, 143*f*, 148–149
structure contour maps, 77, 78*f*
and synthetic and antithetic faults, 70
thrust faults, 119, 131–133

Fault geometry, 81–88
duplexes. *See* Duplexes
extension, estimation of, 111–112, 111*f*
fault-length scaling, 81–82
normal faults, 107–110, 107*f*, 108*f*, 109*f*, 110*f*
strike-slip faults, 139
thrust faults, 121*f*, 125*f*, 128–129
transfer faults, 100, 100*f*
transfer zones, 86, 86*f*
Faulting
Anderson's theory, 254–255, 255*f*
along continental margins, 102–106, 105*f*, 106*f*
cross-sections, 110–111, 110*f*
stress with depth distribution, 255–258, 256*f*, 257*f*
Fault mullions, 66, 67*f*, 313
Fault-propagation folds, 121, 121*f*, 383, 386–390
kink folds, 386–387, 387*f*
trishear model, 387–390, 388*f*, 389*f*
Fault-ramp fold, 118*f*, 119
Fault scarps, 69, 69*f*, 93*f*
fault line scarps, 69, 69*f*
trench-trench transform faults, 603, 604*f*
Fault trace, 82, 84*f*, 85*f*, 92*f*
thrust faults, 118*f*, 131
Feather fractures, 41, 42*f*
Feather River (California), mineral fiber lineations, 316
Feldspar, 77
Fensters, thrust faults, 118, 118*f*, 131
Fiber growth
antitaxial, 420
composite, 420
crack-seal, 420
syntaxial, 420
Fibrous overgrowths, 316, 316*f*, 420–421
types of, 419*f*, 420
Fibrous vein fillings, 315–316, 316*f*
Fiji Plateau, 616*f*
Fine continuous foliations, 303, 304*f*
Finite shear strain rate measure, 465, 465*f*
Finite strain
analysis, 667
ellipse, 336–337, 337*f*
Flinn diagram, 340–345
Mohr diagrams, 359–361, 360*f*
orogenic belt deformations, 667
Firmo-viscous (Kelvin or Voight) material, behavior of, 463–464, 463*f*
First motion studies, 710
First-rank tensors, 184
Fish-mouthed boudins, 312*f*, 313
Fiskenaesset complex (Greenland), folding pattern, 19, 21*f*
Flanks, 277, 277*f*
Flat jack method, stress determination, 234, 235*f*
Flattening, homogeneous deformation, 331*f*, 332
Flexural folding, 364–369, 366*f*
bending, 364, 364*f*, 367
buckling, 364–365, 365*f*, 368*f*
flexural shear, 366*t*, 367*f*, 368–369, 368*f*
orthogonal flexure, 365–366, 365*f*, 367*f*, 368*f*
Flexural shear folds, 366*t*, 367*f*, 368–369, 368*f*
kinematic model, 368–369
and strain, 369
strain measurement study, 431*f*
Flexural slip folds, in multilayer, 377–378, 377*f*
Flinn diagram, 340–345
components of, 340
for constant volume deformation, 341
nonzero volumetric strain, 341–343, 342*f*, 343*f*
strain facies, 345, 345*f*
triaxial strain paths, 343–345
Floor fault, 109*f*, 110*f*, 125*f*
Flow cleavage, 309
Flower structures
negative, 140*f*, 141, 142*f*
seismic profiles, 142*f*
strike-slip faults, 140–141, 140*f*, 141*f*
Flow folding. *See* Passive shear fold
Fold(s)
attitude of, 280–282, 282*f*
block diagram, 280*f*
chevron, 293, 293*f*, 382–383
class, 286
and diapiric flow, 395–399
drag, 388, 390
fault-bend folds, 118*f*, 119, 385–386
and fault displacement, 79, 80*f*
fault-propagation, 121, 121*f*, 383, 386–390
flexural folds, 364–369
folded layers, parts of, 279–280
fold nappes, 293
fold systems, 276
fold train, 276, 276*f*
and foliations, 303–307, 305*f*, 306*f*
and fractures, 57–58, 58*f*, 253–254, 253*f*, 254*t*
geometry, 274
homogeneous flattening, 374–376
as inhomogeneous strain, 319
kinematic analysis of, 667
kink, 293, 379–385
and lineations, 316–317, 420, 421*f*
multilayer, 279, 280*f*, 288–291, 376–379
normal faults, 92, 108*f*
order of development, 290–292, 292*f*
and orogenic belts, 273, 274*f*, 292
parallel, 292
parasitic, 290
passive shear folds, 369–371
ptygmatic, 294, 294*f*, 416*f*
Pumpelly's rule, 292
scale models, folding, 548–550, 548*f*, 550*f*
scale of, 274*f*, 280
similar, 292–293
single, parts of, 274, 276–279, 277*f*
as strain markers, 328, 328*f*
strain measures from, 427–432, 428*f*, 430*f*, 431*f*, 432*f*
style of. *See* Fold style
superposed, 390–394
thrust faults, 121, 121*f*, 123, 127, 131
volume-loss folds, 371–374
Fold and thrust belts
Appalachian, 120*f*, 122*f*, 123*f*, 276*f*, 644
Cordilleran, 122*f*, 123*f*, 126–127*f*, 644
foreland, 123, 123*f*, 643*f*, 644–645, 646*f*
Idaho-Wyoming, 128
Jura, 276*f*, 644
models, 128*f*, 129*f*, 643*f*
of orogenic belts, 644–645, 646*f*
root zone model, 129*f*
subduction model, 129, 129*f*, 130*f*
tectonic wedging, 126–127, 126*f*
thin-skinned, 645
Fold axis, 278, 278*f*, 279*f*
Fold hinge lineations, 311, 316
Folding angle, 282, 284*f*
Fold mullions, 313, 314*f*
Fold style, 282–290
and accretionary prisms, 620*f*, 621*f*
angles of folded surface, 282, 284*f*
aspect ratio, 284, 285*t*
asymmetric folds, 283–284
bluntness, 285, 287*f*
chevron, 293, 382–383*f*
class 1, 1A, 1B, 1C folds, 288, 289*f*, 289*t*, 291*f*
class 2 folds, 288, 288*f*, 289*f*, 289*t*
class 3 folds, 288, 288*f*, 289*f*, 289*t*
clockwise, 283
concentric folds, 288
conical, 278, 279*f*
counterclockwise folds, 283–284
cylindrical, 278, 278*f*, 282–283
elements of, 283*t*
harmonic and disharmonic, 289–290, 290*f*
noncylindrical, 278–279
ptygmatic, 294
Ramsay's classification, 286–288, 288*f*, 289*t*
similar, 288, 288*f*, 369, 370*f*, 371*f*, 374*f*, 376*f*, 378*f*. *See* class 2 folds
symmetric, 283, 287*f*
tightness, 284–285, 286*f*
Foliations
brittle mechanisms, 404
classification scheme, 298*f*
and cleavage, 309
compositional, 298–299, 299*f*, 405
continuous, 298, 298*f*, 299*f*, 302–303, 304*f*, 411–412
crenulation, 302, 302*f*, 303*f*, 407–411
defined, 297, 298
diffuse, 298–299, 299*f*
disjunctive, 299–302, 299*f*, 300*f*, 301*f*, 405–407

Foliations (*continued*)
and ductile shear zones, 307–308, 307*f*, 308*f*
examples of, 297
and folds, 303–307, 305*f*, 306*f*
formation of, 58, 333
Jeffrey model, 401–402, 402*f*
kinematic interpretation of, 667
and lineations, 316
March model, 401, 402, 402*f*
material, 399–400, 405
mineral dissolution, 403–404
mylonitic rocks, 65
nonmaterial, 399–400
recrystallization, 404
rotation of grains/particles, 401–402, 403*f*
shear band, 408–409
and shear planes, 413–415, 414*f*, 415*f*, 668*f*
shear zone, 75–77, 76*f*
spaced, 298, 298*f*, 299*f*
S-surface, 298
steady-state, 412–413
and strain, 400–401, 401*f*
strain measures from, 429–432, 430*f*, 431*f*, 432*f*
Taylor-Bishop-Hill model, 401, 402–403, 402*f*
and tectonic stresses, 245
of tectonites, 297–298
Footwall, 61
Footwall block, 61, 62*f*, 63*f*, 79*f*
Foraminifera, as strain markers, 327, 424
Force
components of, 152*t*
defined, 153
intensity of, 151, 152, 152*t*, 153*f*
opposing, 156, 157*f*, 158
resultant force R, 154
surface, 153
and traction, 153, 156, 157*f*
types of, 153
Forearc basin, 618*f*
Forearc region, 618
Foredeep/foreland basins, 122*f*, 123, 125*f*
orogenic belts, 640, 643–644, 643*f*
Foreland fold and thrust belts, 644–645, 646*f*
Form lines, joints, 45
Forward modeling, 436
Fossils, as strain markers, 327
Fractal geometry, 46–52
box method, 49, 51, 51*f*
cataclastic rocks, 65
defined, 43
fractal limits, 49
fractile dimensions, 48–49, 50*f*, 51*f*
fragmentation (comminution), model for, 51–52, 51*f*
joint pattern description, 49–51, 50*f*
von Koch curve, 46–47, 47*f*
Fracture(s)
defined, 37
extension, 37–42
and faults, 42*f*, 57, 57*f*, 252–253
and folds, 57–58, 58*f*, 253–254, 253*f*, 254*t*
formation, timing of, 54–57, 55*f*
gash, 41–42, 42*f*
and igneous intrusions, 58–59
intersections, 55, 56*f*
and joint formation, 243–250
mechanics of. *See* Fracturing
microfractures, 45
modes I/II/III propagation, 37, 38*f*
orientation of, 43–44, 43*f*
pattern and distribution, study of, 45, 50*f*
scale and shape, 44–45, 44*f*
self-similar geometry, 37, 46
shear, 37, 38*f*
spacing of, 45, 45*f*
surface features, 52–54, 53*f*
veins, 42
See also specific types of fractures
Fracture angle, 211, 215*f*
Fracture cleavage, 309
Fracture criteria. *See* Griffith fracture criteria
functions of, 209
shear fracture. *See* Coulomb fracture criteria
for tension fractures, 210–212, 211*f*
Fracture plane angle, 211, 215*f*, 222*f*
Fracture zones
oceanic crust, 15–16, 15*f*, 601, 602*f*
oceanic transform faults, 604–606, 605*f*, 606*f*
Fracturing, 209–230
anisotropy, effects on, 221–224, 222*f*, 223*f*, 224*f*
confining pressure, effects of, 216–220
Couloumb fracture criterion, 212–216, 213*f*, 215*f*, 216*f*, 225–226
experiments on Berea sandstone, 214–216, 215*f*, 216*f*
experiments on brittle rocks, 209–210
Griffith theory, 226–230
intermediate principal stress, 224–225
Mohr diagrams, 211–213, 211*f*, 213*f*, 216–225, 221, 223*f*, 224*f*
and pore fluid pressure, 220–221, 238, 243–245, 243*f*, 245
scale effects, 225
temperature effects, 225
tension fracture criterion, 210–212, 211*f*
Fragmentation (comminution), fractal model, 51–52, 51*f*
Frank-Read source, dislocation generation, 512, 512*f*
Free air anomaly, ocean basins, 14
Friction, 466–475
Amontons' second law, 212, 219
asperity model, 498, 498*f*
dynamic rock, 467–469, 468*f*, 469*f*
and fault movement, 469–471, 470*f*
internal, coefficient of, 212
rate-and state-dependent law, 469, 471–475, 472*f*, 473, 474*f*
static rock, 466–467, 467*f*
static to dynamic transition, 468–469, 469*f*
thrust sheet emplacement, 259, 262
Frictional sliding
Coulomb criterion, 219, 219*f*
and fault movement, 219–220, 469–471, 470*f*
and pore fluid pressure, 220, 221*f*
stable sliding, 219, 220, 220*f*
stick-slip sliding, 219, 220, 220*f*
Fringe faces, 52, 53*f*
Fringe fractures, and joint formation, 247*f*
Frontal ramps, 93, 119
Fry method, strain measure, 453–455, 454*f*

G
Gabbros
at oceanic spreading centers, 594*f*, 595*f*, 598
oceanic transform faults, 605*f*
Galicia margin (Spain), volcanic-poor margins, 584, 587*f*
Garlock fault system (California), fault system, 144–145, 145*f*, 147, 609, 610*f*
Garnets, 76, 401
Gash fractures, 41–42, 42*f*
and faults, 252, 252*f*
shear zone deformation, 445–446, 446*f*, 447*f*
strain interpretation of, 446*f*
stress interpretation of, 252, 252*f*
Generations of folds
interpretation of, 394–395
superposed folds, 390–394
Geologic sign convention
Mohr diagram, 158–159, 158*f*, 187
tensor sign convention, 183*f*, 184–186, 187
Geology, 1–7
and scientific method, 4–7
structural, 1–7
tectonics, 1–2
Geomagnetic studies, 709–712
Geometric models, functions of, 3
Geometry
fault, 81–88
fractal, 46–52
Glide, of dislocations, 507–510, 509*f*, 512*f*, 513
Glide horizons, 125*f*
Gneisses
foliations, 309
high-grade, Archean terranes, 19, 20*f*
mantled gneiss domes, 397, 397*f*, 553, 654, 655*f*
mylonitic, 65
orogenic belts, 656, 658
terranes with ultramafic bodies, 656–658, 657*f*
Gondwanaland
cratonic platforms, 25
cyclicity, 687

Gouge
fault displacement, 74, 75*f*
fault surface, 67*f*
See also Cataclasites
Grabens
cross-section, 107, 107*f*
dome-in-graben structure, 584*f*
horst-and-graben structure, 94*f*, 95–96, 584*f*
normal faults, 94*f*, 95–96, 96*f*, 97*f*
at oceanic spreading centers, 592, 594*f*
Grain boundaries
diffusion, 501, 502*f*
postdeformational shapes, 519
Grain size, and steady-state creep, 482–483, 482*f*, 502, 502*f*
Grampian Highlands (Scotland), zonal crenulation foliations, 302*f*
Grandjeans-Fjord (Greenland), folds, 275*f*
Grand Saline salt dome (Texas), 396*f*, 397
Granitic batholiths, orogenic belts, 658–660, 659*f*–660*f*
Granular flow, and deformation, 498
Granulites, structure of, 17–18
Graptolites, as strain measure, 426–427, 427*f*
Gravitational sliding, thrust sheet emplacement, 259–260, 260*f*
Gravity-driven deformation, scale model, 551–553, 551*f*, 552*f*, 553*f*
Gravity profile, thrust faults, 117*f*
Great Valley Sequence (California), antithetic thrusting, 126*f*
Greenstone belts, Archean terranes, 19, 20*f*
Griffith fracture criteria, 229
Griffith theory of fracture, 226–230
for brittle facture, 209
confining pressure in, 229
Griffith cracks, 226–227, 227*f*, 228*f*, 229–230
intermediate principal stress in, 229–230
longitudinal splitting, 227–228
pore fluid pressure in, 229
shear fractures, 228–229, 228*f*, 229*f*
tensile and compressive stress, 229
tension fractures, formation of, 227, 227*f*
Growth faults, 102, 104, 106*f*
Gulf Coast (United States), growth faults, 102, 104, 105*f*, 106*f*

H
Hackle, 52
Hackle fringe, 52, 53*f*
Hackle plume
features of, 52–54, 53*f*
and joint formation, 246–247, 247*f*
Haig Brook window (Tennessee), 120*f*
Half-graben, normal faults, 94*f*, 95, 110*f*
Hanging wall, 61
Hanging wall anticline, 121*f*
Hanging wall block
features of, 61, 62*f*, 63*f*, 69, 79*f*
normal faults, 93*f*, 95, 108*f*
thrust faults, 118–119
Hänigsen salt dome (Germany), 396*f*
Hansen's method, slip-line determination, 390, 391*f*, 667
Harmonic folds, 289–290, 290*f*
Harper-Dorn creep, 516–518
Head-to-tail rule, vector addition, 154–155, 154*f*
Heat of Earth, production of, 8–9
Heave, and faults, 62
Helvetic nappes, 666*f*, 667, 668*f*, 669*f*
Hess-type oceanic crust, 597*f*
High-angle faults, 61
fault zones, 665–666
High-grade gneissic regions, Archean terranes, 19, 20*f*
High-stress regime for ductile flow, 479
Himalaya Mountains
and Anderson's theory, 255
earthquakes, 122*f*
faults, 122*f*
Himalaya-Tibet region, 630*f*
and India-Asia collision, 684
inverted metamorphic zones, 652*f*
orogenic belts, 681
South Tibetan detachment system, 665*f*
thrust wedges, 260, 261*f*
Hinge line, 71, 72*f*
folds, 277, 277*f*, 279, 280*f*, 292
Hinge surface, 279, 280*f*
Hinge zone, 277, 277*f*
Hinterland, thrust faults, 122*f*, 123, 125*f*, 129
Homoclines, 281, 283*f*
Homogeneous flattening, 331*f*, 332, 374–376
kinematic model, 374–376, 375*f*
and strain, 375
Homogeneous strain, 319–321
defined, 319
geometries of, 329–332, 331*f*
scales of, 339, 340*f*
of square, 320*f*, 321
Homologous temperature, 476, 496
Honshu (Japan), 616*f*, 626
Hookean solid, behavior of, 460, 460*f*
Hope fault (New Zealand), horsetail splay, 146, 148*f*
Horizontal folds, 281, 282*f*
Horizontal separation, 79
Horizontal stress in Earth, 239–241
and joint formation, 243–244
nontectonic, 239, 239*f*
tectonic, 239–241, 239*f*
Horses (fault slices), 67, 82*f*
thrust faults, 124, 125*f*, 127
Horsetail splay
features, 146
secondary faults, 87, 87*f*
strike-slip faults, 146, 146*f*, 148*f*
Horst, normal faults, 94*f*, 95
Horst-and-graben structures, 94*f*, 95–96, 584*f*
Hotspot, oceanic plateau, 597*f*
Hot working, 476
Hydraulic fracturing (hydrofrac) method, 234, 236*f*
Hydrofracture
joint formation, 243–245, 243*f*
See also Pore fluid pressure
Hydrolic weakening, 483
Hydrolytic weakening, 483
Hydrostatic stress, Mohr diagram, 173, 174*f*
Hypotheses, in scientific method, 4–5
Hypsometric diagrams, 9, 11*f*

I
Igneous intrusions, and fractures, 58–59
Igneous rocks, orogenic belts, 654–655
Imbricate fans
normal faults, 109*f*
splay faults, 84, 85*f*
strike-slip faults, 146*f*
thrust faults, 123*f*, 124, 128*f*, 131
Imbricate faults, 93, 94*f*, 109*f*
Incompetent rock, defined, 363
Incongruous steps, fault displacement, 72, 73*f*, 74*f*
Incremental strain, ellipse, 333. *See* instantaneous strain
ellipse, 333
India, collision with Asia, 684
Indonesian-Australian collision, 628*f*
Inductive (Baconian) method, 5–6
Infinitesimal strain, 351, 354–355, 354*f*
Inflection line, folds, 276, 277*f*, 278*f*
Inflection surface, multilayer folds, 279, 280*f*
Inhomogeneous strain, 319–321
defined, 319
folding as, 319
scales of, 339, 340*f*
of square, 320*f*, 321
Initial condition, 545
Instantaneous extension strain rate measure, 465
Instantaneous strain analysis, 667, 669, 669*f*
Instantaneous strain ellipse, 333, 337
Insubric Line (Alps), 611
Interference patterns
superposed folds, 391–393, 392*f*
Types 1 through 3, 393, 393*f*
Interlimb angle, 282, 284*f*
Intermediate principal stress
and fracturing, 224–225
in Griffith theory, 229–230
Internal forces, examples of, 153
Internal friction
angle of, 212. *See* Angle of internal friction
coefficient of, 212. *See* coefficient of internal friction
Griffith theory and, 229
International Ocean Drilling Project (IODP), 27, 234
Intersection lineations, 311, 311*f*

Intersections, joints, origin of, 248–250, 249*f*, 250*f*
Interstitials, 500, 500*f*
Intrusive contacts, features of, 66, 67*f*
Invariants, scalar
strain, 351
stress, 172, 180
Inverse strain ellipse, 326, 327*f*, 332*f*
Inversion tectonics, 126
Inverted metamorphic zones, 652, 652*f*, 653*f*
Irregular mullions, 313, 314*f*
Irrotational progressive deformation, 334
Island arc-deep sea trench, 14*f*, 16, 16*f*
Island arcs, structure of, 615
Isolated lens, fault displacement, 87
Isopach maps, fault displacement, 77, 78*f*
Isotherms, inversion of, 653*f*
Isotropic symmetry, 535–536, 536*f*
Isotropic tension, 349
Ivrea zone, 655, 657*f*, 661*f*, 666
Izu-Bonin-Japan collision, 628
Izu-Bonin Trench, 617*f*
Izu-Bonin zone, back-arc basins, 591*f*

J
Japan Sea-style margins, 30, 30*f*
Jeffrey model, foliations and lineations, 401–402, 402*f*
Jogs, 505, 506*f*
See also Bends
Joint(s), 38–41
bedding, 40
classification, 38
columnar, 40, 41*f*
cross, 40–41
defined, 38
fractal geometry, pattern descriptions, 49–51
joint sets, 38, 39*f*
joint system, 38–39, 39*f*
joint zone, 38, 39*f*
master, 45
oblique and diagonal, 41
orientation of, 40–41
origin of. *See* Joint formation
rose diagrams, 43*f*, 44
sheet (exfoliation), 39, 41
strike and dip, 40
systematic and nonsystematic, 38, 39*f*
terminations of, 44–45, 44*f*
See also specific types of joints
Joint formation, 243–250
during burial, 241, 243–244
columnar joints, 246
extension fractures, spacing of, 250–251, 251*f*
hydrofracture and depth of burial, 243–245, 243*f*
joint curvature, 247–248, 247*f*
joint intersections, origin of geometries, 248–250, 249*f*, 250*f*
and plumose structure, 246–247
and pore fluid pressure, 244–245, 248
sheet joints, 246
stress paths, 241
tectonic stresses, 245–246
uplift and erosion, 243–245
during uplift/erosion, 241, 243–245
Jura Mountains (Switzerland)
fold and thrust belts, 644
folding, scale model, 548–550, 548*f*, 550*f*
folds, 276*f*, 548*f*
tear faults, 141, 143*f*

K
Kalahari craton (Africa), Archean terranes, 19, 20*f*
Kang Ting faults (China), 560
Kansu fault (China), 560
Kelvin material, behavior of, 463–464, 463*f*
Keystone thrust fault (Nevada), 115, 116*f*
Kinematic models
chevron folds, 382–383, 382*f*
flexural shear folds, 368–369
folding, compared to numerical model, 570–572
functions of, 3, 107, 363
homogeneous flattening, 374–376, 375*f*
kink folds, 380*f*–381*f*, 384–385
normal faults, 107–111, 107*f*, 108*f*, 109*f*, 110*f*
orthogonal flexure, 365–366
passive shear folds, 369–370, 369–371
strike-slip faults, 146–148, 149*f*
thrust faults, 127–128, 128*f*
volume-loss folds, 371–374, 372*f*
Kinematics
crystallographic preferred orientations, 669–670
defined, 3
fault displacement indicators, 70–77
of folds, 667
of foliations, 667
instantaneous strain analysis, 667, 669
Kink(s), 505, 506*f*
Kink folds, 293, 293*f*, 379–385
fault-bend folding, 385–386
fault-propagation folding, 386–387
kinematic model, 380*f*–381*f*, 384–385
kink bands, 379–380, 381*f*, 383
layered sequences, 382–383
strain in, 383, 384*f*
Kink fractures, and joint formation, 247, 247*f*
Klamath Falls (Oregon), normal faults, 92*f*
Klamath Mountains (California)
diffuse foliations, 299*f*
transform faults, 607–608
Klippe, thrust faults, 118, 118*f*, 120*f*, 131
Kodiac Island (Alaska), accreted forearc fold styles, 618, 620*f*
Kunlun faults (China), 560

L
Lake Meade (Nevada), shear zone, 315
Lakes, caldera structure, 98*f*
Lambert projection. *See* Equal angle projection
Lamé parameters, 491
Laminations, zonal crenulation foliations, 302*f*
Lateral ramps, 85, 86*f*, 93, 119, 121*f*
Lau Havre trough, back-arc basins, 591*f*
Laws, in scientific method, 5
Layer, defined, 298
Left-lateral (sinistral) faults, 62, 63*f*, 138*f*
Lengthening, 348*t*
Length systematics, fault displacement, 80, 81*f*
Lesser Antilles, 616*f*
Lewis thrust fault (Canada-United States)
duplex structure, 124, 125*f*
fault displacement, 131
lateral ramps, 119, 120*f*
Limbs, folds, 277, 277*f*
Line
attitude of, 693
boudin, 311
branch, 82
cut-off, 66
hinge, 277
neck, 311
tip, 82
Lineaments, joint systems, 39, 40*f*
Linear crystal defects. *See* Dislocations
Linear strain, 321–323
defined, 321
measures of, 321–322
Lineations
and boudins, 311–313, 312*f*
classification scheme, 310*f*
and cleavage, 311, 311*f*
constructed, 310–313, 311*f*, 312*f*, 416–418
defined, 297
discrete, 310, 310*f*, 416
examples of, 297
faults, 67*f*, 72, 138
and folds, 316–317, 421, 421*f*
and foliations, 316
formation of, 58
Jeffrey model, 401–402, 402*f*
March model, 402, 402*f*
mineral, 298, 313–316, 315*f*, 316*f*, 418–420
mineral dissolution, 403–404
and mullions, 313, 314*f*
mylonitic rocks, 65
polycrystalline, 313–315, 315*f*, 418
recrystallization, 404
rotation of grains/particles, 401–402, 403*f*
slickensides, 72, 92, 314, 315*f*, 316, 404
and strain, 400–401
strain measures from, 426–427
stretching, 400, 401*f*, 416
structural, 298, 309–313
Taylor-Bishop-Hill model, 402–403
of tectonites, 297–298
Listric fans, 109*f*
Listric normal faults, 92, 94*f*, 104, 106*f*
displacement, geometric model, 107–108, 108*f*

with ramp and flat, 109*f*
stress distribution effects, 257
Lithoprobe project, 20–21
Lithospheric mantle, 8, 8*f*, 9–11, 12, 29*f*
elastic part, 237
plate bending, causes of. *See* Stress in Earth
Little Shuteye Pass (California), sheet joints, 41*f*
Local stress
versus applied stress, 226–227
and joint intersections, 248, 250*f*
Logarithmic creep, 476
Logarithmic strain, 322
Loma Prieta fault (San Andreas system), 610*f*, 611, 612*f*
Longitudinal splitting
experiments on brittle rock, 210, 210*f*
Griffith theory, 227–228
Low-angle boundary, 511, 511*f*
Low-angle faults, 61
detachment faulting, 663, 664*f*
thrust faults, 119, 120*f*
Low-stress regime, 49
L-tectonites, 297
Luzon, 616*f*, 617*f*
Lyons sandstone, Paintbrush tuff, 50*f*

M
Macrofabric, 497
Macroscopic behavior of rock, constitutive equations, 459
Mafic rocks
Archean terranes, 19, 20*f*
distribution of Proterozoic complexes, world map, 23*f*
minerals of, 15
Magnetic anomalies, 709
Makran, 616*f*
Mantled gneiss domes, 397, 397*f*, 553, 654, 655*f*
Mantle diapir, 397
Mantle of Earth
composition of, 8
layers of, 8, 8*f*
Maps, geologic, 697
Marathon thrust belt (Texas), bedding foliation, 305*f*
March model, foliations, and lineations, 401, 402, 402*f*
Mariana Trench, 617*f*
Mariana trough, back-arc basins, 591*f*
Marias Pass (Montana), ramps, 120*f*
Martinsburg slate (Pennsylvania)
fracture experiments, 221–222, 222*f*
volume strain, 426–427, 427*f*
Mass, conservation of, 544, 573
Massif, external, 645
Master joints, 45
Material behavior, 460–464
Deborah number, 572, 576–577
elastic material, 460, 460*f*
elastic-plastic (Prandtl) material, 464, 464*f*
plastic material, 461–462, 462*f*
power-law materials, 462, 462*f*
recoverable/nonrecoverable deformation, 460
visco-elastic (Maxwell) materials, 463–464, 463*f*, 572, 576, 576*f*
visco-plastic (Bingham) material, 464, 464*f*
viscous material, 460–461, 461*f*
Material foliations, 399–400, 405
Material lineations, 399
Material lines. *See* Material objects
Material objects, defined, 319
Material planes. *See* Material projects
Material points, 156
Material properties, 153
Material surface, 156
Mathematical models
boundary conditions, 545
functions of, 543–544
viscous deformation problem, 573–576
Matrix
stress components, 181–182
symmetric, 186
Matterhorn Peak (California), multilayer folding, 379*f*
Mauritania-Senegal Basin, 588, 588*f*
Maximum shear stress, 172
Maxwell materials, behavior of, 463–464, 463*f*, 572, 576, 576*f*
McConnell thrust fault (Alberta), ramp, 119*f*
Mean normal stress, 172
Mechanically anisotropic rocks, 221
Mechanically isotropic rocks, 221
Mechanical models
functions of, 3, 363–364
viscous buckling of competent layer in incompetent matrix, 561–563, 561*f*, 562*f*, 563*f*
Mechanics, defined, 3
Median surface of folds, 280
Median Tectonic Line (Japan), 611
Megabreccia, faults, 63, 64*t*, 65, 92
Mega-mullion structure, 102, 104*f*
Melanges
defined, 619
tectonic, 622, 624–625, 625*f*
Mesosphere, 8, 8*f*
Metamorphic core complexes, 99*f*, 100–102, 102*f*
components of, 102
detachment faults, 101–102, 104*f*
origin of, 102*f*
Metamorphic facies, 648–652, 649*f*
defined, 19
Metamorphic rocks, 117, 648
Metamorphism, 648–653
Barrovian (normal pressure/temperature), 648, 649, 649*f*, 650*f*, 681
Blueschist (high pressure/low temperature), 648, 649, 649*f*, 650*f*
Buchan (high-temperature-low-pressure), 648–649, 649*f*, 650*f*
burial, 650, 650*f*
inverted metamorphic zones, 652, 652*f*, 653*f*
pressure/temperature domains, 649*f*
relationship to structure, 651–652, 651*f*
and tectonics, 671–673
ultra-high-pressure (UHP), 649–650
Metavolcanic rocks, orogenic belts, 654–655
M-folds, 284
Mica, 77
Microbreccia, 63, 64*t*
Microcontinuous fine foliations, 303, 304*f*
Microdomainal fine foliations, 303, 304*f*
Microfabric, 497
Microfaults, 61
Microfractures, 45
Microlithons, 299*f*, 300, 300*f*
Midoceanic ridges. *See* Oceanic ridges
Migmatite, 18
Mimetic growth, crystals, 404
Mineral(s)
dating of, 308
dissolution of, 403–404
in fault zones, 66, 71
in joints/fractures, 53, 54, 57
mineral streaks, 66, 71
of porphyroblasts, 76–77
of porphyroclasts, 77
recrystallization of, 404
rotation during deformation, 401–402, 402*f*, 403*f*
in shear zone, 75, 76*f*
slip systems of, 510*t*
in veins, 42
See also Crystals
Mineral fibers, slickenfibers, 66, 72, 73*f*
Mineral lineations, 298, 313–316, 315*f*, 316*f*, 418–420
fibrous overgrowths, 316, 316*f*, 419*f*, 420–421
mineral cluster, 314, 315*f*, 418
mineral fiber, 315–316, 315*f*, 418–420
mineral grain, 315–316, 316*f*, 418
mineral slickenlines, 418, 418*f*
nonfibrous overgrowths, 314–315, 420–421
polycrystalline, 313–315
Miogeoclines, 644
Misfit minimization model, 525, 527
Missing sections, faults, 68, 68*f*
Models
analytical, 543
geometric, 3
kinematic, 3
mathematical, 543–544
mechanical, 3
numerical, 543–544
physical, 544–546
scale, 543, 546–561
in scientific method, 4–7
See also specific types of models
Moderate stress regime, 478
Moderately inclined folds, 280, 282*f*
Moderately plunging folds, 281, 282*f*

Modes I/II/III propagation, fractures, 37, 38*f*
Modulus, of rigidity, 95–96. *See* Shear modulus
  shear, 136
  Young's, 136
Moho, use of term, 598
Moho discontinuity
  brittle-ductile transition, 490, 490*f*
  continental, 18–19, 18*f*
  defined, 598
  in ophiolite sequences, 597*f*, 598
  and orogenic belts, 640
Mohr diagrams
  applications related to stress, 169–175
  circle sign convention for stress, 165*t*, 168*f*, 187
  diagram features, 169, 169*f*
  finite strain, 359–361, 360*f*
  fracturing, 211–213, 211*f*, 213*f*, 216–225, 223*f*, 224*f*
  geologic sign convention for stress, 158–159, 158*f*, 187
  Mohr circle, features of, 168–169*f*, 178, 178*f*
  stress tensor, 188–189
  three-dimensional stress, 168, 178–180
  two-dimensional stress, 168–173, 175–177
Moine thrust (Scotland), paleostresses, 522
Molasse basins, 640, 643*f*
Moment of momentum. *See* Angular momentum
Momentum, conservation of, 544, 573–574
Monoclines, 281, 283*f*
Monoclinic symmetry, 536*f*, 537
  strain ellipse, 337
Montague Island (Alaska), fault, 70, 71*f*
Morcle nappe, 667, 668*f*, 676
Moroccan Margin, 584, 588, 588*f*
Mount Crandell thrust fault (Canada)
  fault displacement, 131
  roof thrust, 124, 125*f*
Mullions
  fault, 66, 67*f*, 313
  fold, 313, 314*f*
  and lineations, 313, 314*f*
Multilayer folds, 288–291, 376–379
  buckling, 379, 379*f*
  and decollements, 290, 291*f*
  flexural slip folding, 377–378, 377*f*
  layer competence, 376–378, 377*f*, 378*f*, 379*f*
  parts of, 279, 280*f*
  profiles of, 290*f*
  strain in, 378–379
  style of, 288–290, 290*f*
Muscovite, crystal formation, 403
Muskox complex (Canada), Proterozoic dike system, 22, 23*f*
Mylonitic rocks
  and faults, 64*f*, 64*t*, 65, 66*f*, 93, 100
  transform faults, 608
  types of, 65

N
Nabarro-Herring creep, 495, 500–501, 501*f*, 516–517
Nappe
  defined, 293
  folds, 293
  gravitational collapse, 676*f*
  salt, 104, 105*f*, 110
  thrust, 293
  thrust faults. *See* Thrust sheets
Natural strain, 322
Nearest-neighbor center-to center (Fry) method, strain measure, 453–455
Neck folds, 312*f*, 313
Neck line, 311, 312*f*
Negative flower structure, 140*f*, 141, 142*f*
Neomineralization, defined, 514
Neutral surface, 365, 365*f*, 368*f*
New Britain-Trobriand collision (Papua New Guinea), 628*f*
New Jersey Margin, 588, 588*f*
Niagaran dolomite, 50*f*
Noncoaxial deformation, 531, 533*f*
Noncoaxial progressive deformation, 335
Noncollisional orogenic belts, 27
Noncylindrical folds, 278–279
Nonfibrous overgrowths, mineral lineations, 314–315, 420–421
Nonmaterial foliations, 399–400
Nonmaterial lineations, 399
Nonrecoverable deformation, 460
Nonsystematic joints, 38, 39*f*
Nordfjord-Sogn detachment fault, 664*f*
Normal faults, 62–63*f*, 68*f*, 85*f*, 91–112
  along continental margins, 102–106, 105*f*, 106*f*
  displacement, 95, 95*f*
  extension, determination of, 111–112, 111*f*
  folds, 92, 108*f*
  kinematic models, 107–111, 107*f*, 108*f*, 109*f*, 110*f*
  metamorphic core complexes, 99*f*, 100–102, 102*f*
  regional systems of, 98–100, 99*f*, 101*f*
  separations of stratigraphy, 91–92, 93*f*
  shape of, 93
  structural associations of, 96–98, 97*f*
  surface features, 92–93
  systems of, 94*f*
  world map of, 98*f*
Normal stress, defined, 347*t*
Normal traction component, 156
North America, crustal thickness, 17*f*
North Kootenay Pass, ramps, 120*f*
Nose of fold, 277, 277*f*
Numerical models
  anisotropic material and folding, 568–570, 570*f*
  folding, compared to kinematic model, 570–572
  functions of, 543–544
  linear viscous rheology and folding, 563–565, 564*f*, 565*f*, 566*f*
  power-law rheology and folding, 565–568, 567*f*, 568*f*, 569*f*
  visco-elastic (Maxwell) rheology and folding, 572

O
Oblique fractures
  features of, 58, 58*f*
  oblique extension fractures, 37
Oblique joints, 41
Oblique ramps, 85, 85*f*, 119, 120*f*
Oblique-slip faults, 61–62, 62*f*, 63*f*, 85*f*
Occam's Razor, 4, 458
Ocean basins, 14–17
  age of, 13*f*
  features, world maps, 14*f*, 582*f*
  free air anomaly, 14
  layered model, 14–15, 14*f*
  plate interiors, 17
  plate margins, 15–16
  transform faults. *See* Oceanic transform faults
Ocean Drilling Project (ODP), 27, 584
Oceanic crust
  age of, 13, 13*f*
  continental margins, 27–30
  fracture zones, 15–16, 15*f*, 601, 602*f*
  Hess-type, 597*f*
  lithosphere, cooling of, 591–592*f*
  Penrose-type, 597*f*
  ridges. *See* Oceanic ridges
  seafloor spreading, 11–12
  Smartville-type, 597, 597*f*, 599
Oceanic metamorphic core complexes, 607
Oceanic plateaus, 14*f*, 17
Oceanic ridges, 590–600
  back-arc basins, 591, 591*f*
  as crustal-scale extension fractures, 592, 593*f*
  midoceanic, features of, 15, 15*f*, 591–593, 591*f*
  midoceanic discontinuities, 593*f*
  ophiolite sequences, 596–599, 597*f*
  spreading center, Troodos Complex example, 599–600, 600*f*
  spreading center structures, 592–596, 594*f*, 595*f*
  spreading center subaerial exposures, 599–600, 600*f*
  spreading rates, 592–594, 593*f*, 594*f*, 596
  topographic profile, 591–592, 592*f*
  world map, 591*f*
Oceanic transform faults, 601–608
  boundaries, oceanic, 15–16, 15*f*
  fracture zone complexity, 604–606, 605*f*, 606*f*
  ridge-ridge, 601–607, 602*f*
  ridge-trench, 602, 602*f*
  spreading rates, 604*f*
  subaerial exposures, 607–608
  trench-trench, 602, 603*f*
Oceanographer Transform fault, 604–605, 605*f*

Offscraping, 618
Offsets. *See* Step-overs
Offset streams, 69, 70*f*
Olistoliths, 624
Olistostromes, 622, 624, 624*f*
Olivine
 deformation map, 496*f*
 dislocations in, 506–507, 508*f*
 fabrics, formation of, 528–530, 528*f*, 529*f*
 in ophiolite sequences, 596
Ooids, as strain markers, 327, 327*f*, 424
Ophiolites, 14, 397, 598–599
 collisions and complexes, 628–629, 629*f*
 development of, 647–648, 648*f*
 orogenic belts, 647–648, 648*f*
Ophiolite sequences, 596–599, 597*f*
 models of, 597*f*
 of orogenic belts, 655
 rocks of, 596–598, 597*f*
Order of folds, 290–292, 292*f*
Orientation data, graphical representation of, 695
Orientation of planes
 Mohr circle, 169–171, 170*f*, 179
 stress across, 159–161, 160*f*
Ornach-Nal faults, north-south trends, 147*f*
Oroclines, 273, 274*f*
Orogenic belt(s), 26–27
 core zones, deep structure, 660, 661*f*
 cross-sections of, 643*f*
 crystallographic preferred orientations, kinematic analysis, 669–670
 deformations. *See* Orogenic belt deformation
 and exotic terranes, 688–689, 688*f*
 and folds, 273, 274*f*, 292
 foredeep/foreland basins, 640, 643–644, 643*f*
 foreland fold and thrust belt, 644–645, 646*f*
 gneissic terranes, 656, 658
 granitic batholiths of, 658–660, 659*f*–660*f*
 igneous rocks of, 654–655
 metamorphism, 648–653
 metavolcanic rocks of, 654–655
 mineral fibers, kinematic interpretation, 667, 669
 ophiolites, 647–648, 648*f*
 ophiolitic sequences of, 655
 orogenic cycle, features of, 685
 peridotites of, 655
 rock types of, 26
 sedimentary rocks of, 653–654
 slate belt, 645–647, 647*f*
 sutures, 648
 tectonic conditions, 27
 thrust faults, 123, 129
 types of, 27
 world map of, 640*f*
Orogenic belt deformations, 273
 direction of tectonic transport model, 673–675
 ductile nappe emplacement model, 675–676
 extensional deformation, 660, 663–664, 663*f*
 finite strain analysis, 667
 folds, kinematic analysis, 667
 foliations, kinematic analysis, 667
 high-angle fault zones, 665–666
 low-angle detachment faulting, 663, 664*f*
Orogenic float, 660, 662*f*
Orogeny, 677–684
 buoyant force, 681
 collisional, types of, 627*f*
 continent-continent collision, numerical model of, 677
 exhumation, 680
 fold-and-thrust belts, 682–683
 and generations of deformation, 682
 limitations in study of, 683–684
 orogenic heating, 681
 root collapse, 682
 root formation, 680–681
 subduction to collision process, 677–680, 678*f*, 679*f*
 ultra-high-pressure metamorphic rocks, preservation of, 681–682
 and Wilson cycle, 685–688
Orthogonal flexure, 365*f*, 367*f*, 368*f*
 bending, 366, 367*f*
 buckling, 366, 368*f*
 kinematic model, 365–366
 strain measure, 428–429, 428*f*
Orthogonal thickness, folds, 288, 288*f*, 289*f*
Orthorhombic symmetry, 536–537, 536*f*
 strain ellipse, 337
 in three dimensions, 337
Outcrop scale, meaning of, 2
Outer swells, 615
Out-of-sequence thrusts, 128, 645
Outward normal units, 183
Overburden
 and horizontal tectonic stress, 240–241, 240*f*
 standard state stress, 256
 stress determination method, 237
Overcoring, 233–234, 235*f*
Overgrowths
 fibrous, 315
 nonfibrous, 314
Overprinting, 672
Overthrusts. *See* Thrust faults; Thrust sheets
Overturned folds, 282, 283*f*
Oxygen, and steady-state creep, 483–484, 484*f*

P
Paired metamorphic belt, 649
Paleomagnetism, fault displacement detection, 77, 79
Paleopiezometers, paleostress orientation/magnitude, inferring, 522–524
Palinspastic restoration
 extension, estimation of, 112
 thrust faults, 125*f*, 132
Palm tree structure, 141, 142*f*
Paradigms, shifts in, 5
Parallel folds, 292
Parallelogram rule, vector addition, 154–155, 154*f*
Parasitic folds, 290
Passive margins, 27, 28*f*, 644
Passive rifting, 590, 590*f*
Passive shear fold, 369–371
 deck-of-cards-model, 339, 339*f*, 369–370, 370*f*
 fold hinge orientation, 370*f*
 kinematic model, 369–370
 and strain, 370–371, 371*f*
Patton Bay Fault (Alaska), uplift contours, 70, 71*f*
Pelagic sediments
 formation of, 14
 in ophiolite sequences, 597*f*, 598
Pelvoux massifs (France), 646
Pencil cleavage, 311, 311*f*
Pencil glide, 528
Pencil lineations, 314
Penetrative lineations, 297, 298
Penetrative (structure), 298
Penninic nappes (Alps), 131–132, 132*f*
Penrose-type oceanic crust, 597*f*
Perdido fold belt (Gulf Coast), 105*f*, 110
Perfectly plastic material, 461, 462*f*
Peridotites
 in ophiolite sequences, 596, 598
 of orogenic belts, 655
Peru-Chile Trench, 617*f*
Petrologic Moho, in ophiolite sequences, 597*f*, 598
Phanerozoic regions, 25–30
 continental margins, 27–30
 continental rifts, 27
 cratonic platforms, 25, 26*f*
 interior lowlands, 25
 orogenic belts, 26–27
Philippine Trench, 617*f*
Phyllitic cleavage, 309
Physical models, 544–546
 and boundary conditions, 545–546
 cause and effect in, 545–546
 conservation laws, 544
 constitutive equations, 459, 544–545
 and initial conditions, 545
Piercing points, fault displacement, 71, 72*f*
Piezometers
 defined, 519
 paleostress orientation/magnitude, inferring, 522–524
Pinch-and-swell structures, 311, 312*f*, 417
Pine Mountain thrust fault (Tennessee), tear faults, 119, 120*f*
Pinnate (feather) fractures, 41, 42*f*
 and shear fractures, 251
Pinning points, 88
Plagioclase, 22

Plagiogranite, in ophiolite sequences, 597*f*, 598
Plane, attitude of, 693
Plane of flattening, 399
Plane strain, defined, 319, 329
Plastic, use of term, 272, 461
Plastic material behavior, 461–462, 462*f*
  elastic-plastic (Prandtl) material, 464, 464*f*
  three-dimensional, 492
  visco-plastic (Bingham) material, 464, 464*f*
Plastic slip-line field theory, 553–561
Plate boundaries
  conservative, 9. *See* transform
  convergent, 9, 615, 626
  divergent, 9, 581, 590
  transform, 9, 601
Plate tectonics
  development of study, 2
  stress and plate motion, 237
Platform
  cratonic, 25
Pluck holes, 74, 75*f*
Plumose structure, 52–54, 53*f*
  on extension fractures, 251
  and joint formation, 246–247
Plunge, 695
Plutonic complex, in ophiolite sequences, 597*f*, 598
Plutonic rocks, 117
Point defects, 495, 497, 500, 500*f*
Poisson effect
  joint formation, 243–245
  Poisson expansion, 232–233, 239
Poisson ratio
  defined, 232
  elasticity equation, 239
Polar wander maps, 712
Polyclinal folds, 290
Polycrystalline mineral lineations, 313–315, 315*f*, 418
Pore fluid pressure
  aquathermal pressuring, 238
  defined, 210
  effective stress and, 173
  extension fractures, spacing of, 250–251
  and fracturing, 220–221, 238, 243–245, 243*f*, 245
  and frictional sliding, 220, 221*f*
  geological importance of, 220–221
  in Griffith theory, 229
  hydraulic fracturing and, 234
  and joint formation, 244–245, 248
  pore fluid, origins of, 220–221
  thrust sheet emplacement, 259, 262–263
Porphyroblasts
  and fault displacement, 75–77
  minerals of, 76–77
Porphyroclasts, 65, 66*f*
  and fault displacement, 75–77
  minerals of, 77
  tail types, 75, 76*f*
Positive flower structure, 141, 142*f*
Postgrading deformation, 645
Power-law creep
  and depth of rock, 490*f*
  relation to diffusion creep, 518*f*
  with low n, 479, 480*f*
  steady-state creep, 478–479, 480*f*, 490
Power-law materials behavior
  power-law rheology, 462, 462*f*, 565–568, 567*f*, 568*f*, 569*f*
  three-dimensional, 492
Prandtl material, behavior of, 464, 464*f*
Precambrian shields, 19–24
  Archean terranes, 19–21
  formation of, 19
  Proterozoic terranes, 22–25
Preferred orientation fabrics of dislocation creep, 524
Pressure
  changes and stress, 238
  confining. *See* Confining pressure
  and steady-state creep, 481–482, 481*f*
Pressure shadows, 76, 315, 419*f*, 420
Pressure solution, 404
Pressure-temperature-time (PTt) paths, 671, 671*f*
  exhumation paths, 671*f*, 672
Primary creep, 476–477, 477*f*
Primary recrystallization, 516
Principal axes, 163, 164*f*
  of instantaneous strain, 333
  of strain, 324–325, 325*f*, 334*f*, 335*f*
  of stress, 163
Principal coordinates, 163, 164*f*, 185
Principal diagonal, 182, 182*f*
Principal planes, 163, 164*f*
Principal stress
  Couloumb fracture criterion, 214
  intermediate, effect of, 224–225
  Mohr circle, 169, 169*f*
  three-dimensions, 166, 178–179
  two-dimensions, 163–166, 164*f*, 189–190, 190*f*
Principal stretches, 324
Principal values, 185
Profile, folds, 277, 278*f*
Prograding deformation, 645
Progressive deformation, 333, 336
  coaxial, 335
  irrotational, 334
  noncoaxial, 335
  rotational, 334
Progressive pure shear, 333–336, 333*f*, 334*f*, 338*f*
Progressive simple shear, 333–334, 333*f*, 335*f*, 336, 338*f*
Progressive strain, 333–339
  defined, 333
  geometry of, 333*f*
Propagating ridges, 606
Proterozoic terranes, 22–25
  age of rocks, 19
  aulacogens, 22, 24*f*
  deformed belts, 22
  rock types, 22, 22–23, 23*f*
  tectonic conditions, 24–25
Protomylonite, 65
Pseudotachylites, 63, 64*t*, 65, 65*f*, 499
P shears, strike-slip faults, 138–139
Ptygmatic folds, 294, 294*f*, 416*f*
Pull-apart, fault displacement, 87
Pull-apart basins, 141, 143*f*
Pumpelly's rule, 292
Pure shear
  crystallographic preferred orientation, 525, 526*f*
  defined, 330, 330*f*, 347*t*
  progressive, 333–336, 333*f*, 334*f*
Pure shear stress
  defined, 347*t*
  Mohr diagram, 173, 174*f*
Pure strain, 329, 331–332, 332*f*
P waves, 495. *See* Compressional waves
Pyrenees (France), mineral lineations, 316*f*

Q
Quadratic elongation, defined, 347*f*
Quartz
  deformation lamellae, 521–522, 523*f*, 524*f*
  deformation map, 495, 496*f*
  fabric, formation of, 530–535, 530*f*, 532*f*, 533*f*
  paleostress orientation/magnitude, inferring, 521–522
Quartzites
  mylonitic, 66*f*
  Proterozoic terrane, 22
Quartzose sandstones, 22
Quetta-Chaman fault (Pakistan), north-south trends, 145, 147*f*

R
R_f–ϕ method, for strain measurement, 449
Radar interferometry, fault movement survey, 70
Radiation pattern, first motion from faulting event, 710
Radiolaria, as strain markers, 327, 424
Radiometric dating, age of minerals, 308
Railroad Valley (Nevada)
  drag folds, 92
  graben, 96, 96*f*
Ramp(s)
  fault geometry, 85, 86*f*, 118
  normal faults, 93, 95*f*
  ramp anticlines, 118*f*, 119
  thrust faults, 118–119, 118*f*
Ramp-flat, fault surface, 119
Ramsay's classification, fold style, 286–288, 288*f*, 289*t*
Random fabric, 300*f*
Rapakivi granite, Proterozoic terrane, 22
Rarefaction waves, 346
Rate-and state-dependent friction law, 469, 471–475, 472*f*, 473*f*, 474*f*
Rattlesnake Mountain (Wyoming), monoclinal folds, 364*f*
Reclined folds, 281, 282*f*

Reconstructive recrystallization, 404
Recoverable deformation, 460
Recoverable strain, 231
Recrystallization, 514–516, 515*f*
  annealing, 516
  boundary migration, 514, 515
  coherent, 404
  dynamic, 514, 515*f*, 519
  foliations/lineations formation, 404
  primary/secondary, 516
  process of, 404
  reconstructive, 404
  static, 516
  subgrain rotation, 514, 515*f*
  syntectonic, 514
Recumbent folds, 281, 282*f*
Red River fault (China)
  dextral shear, 560
  strike-slip faults, 558–560, 558*f*, 611
Red Sea
  continental rifting, 583–584, 584*f*, 585*f*, 589, 590
  seafloor spreading, 585*f*
  southern, structure of, 585*f*
Reduction spots, 310, 310*f*
Reelfoot Rift (United States), aulacogens, 22, 24*f*
Re-entrant (syntaxis), 122*f*, 123
Reflected seismic waves, 705, 706
Refracted cleavage, 303
Refracted foliations, 303, 305*f*
Refracted seismic waves, 704
Relative displacement, and faults, 61
Relay faults, 86
Relay ramps, 86
Relay zones, 86
Releasing, defined, 349*t*
Releasing bends, 139*f*, 140
Remnant arcs, 615
Remote stress
  defined, 248
  and joint intersections, 248, 250*f*
Repeated sections, faults, 67–68, 68*f*
Restraining, defined, 349*t*
Restraining bends, 139*f*, 140
Resultant force R, 154
Reverse faults
  62–63*f*
  See also Thrust faults
Rheology, 487–492
  deformation, experimental applications, 487–492
  ductile deformation mechanisms, 517
  linear viscous rheology and folding, 563–565, 564*f*, 565*f*, 566*f*
  material behavior, 460–464
  power-law, 462, 462*f*, 565–568, 567*f*, 568*f*, 569*f*
  power-law rheology and folding, 565
  rock types used, 487, 488*f*
  steady-state creep, 476–486
  visco-elastic (Maxwell) rheology and folding, 572
Rhine Graben (Europe), normal faults, 99
Ribbon quartz, 519, 519*f*
Rib marks
  features of, 52–53, 53*f*
  and joint formation, 247, 247*f*
Ridge-in-groove lineation
  features of, 66, 67*f*, 71
  shear fractures, 251
Ridge-ridge-ridge (RRR) triple junction, 589, 589*f*
Ridge-ridge transform faults, 601–607, 602*f*
  dueling ridge segments, 606–607, 606*f*
  fast-spreading ridges, 607, 607*f*
  propagating ridges, 606
  structures of, 602–607, 604*f*, 605
Ridge-trench transform faults, 602, 602*f*
Riecke's principle, 404, 405
Riedel (R) shears
  conjugate (R') type, 139, 217*f*
  and Coulomb fracture criterion, 225–226
  defined, 216
  fault displacement, 74, 74*f*
  strike-slip faults, 138–139
Rifted continental margins, 27, 28*f*, 581–590
  evolution of, 589*f*
  mature, 583, 586*f*, 587*f*, 588, 588*f*
  passive rifting, 590, 590*f*
  ridge-ridge-ridge (RRR) triple junction, 589, 589*f*
  rifting process, models, 589–590, 589*f*, 590*f*
  volcanic-poor, 584, 586*f*, 587*f*
  volcanic-rich, 583, 586*f*
  young and ocean basins, 583, 585*f*
Rift Valley (Africa), continental rifts, 27
Right-lateral (dextral) faults, 62, 63*f*
Rigid-body motion, defined, 1
Rigidity. *See* Modulus of rigidity
Rigid-plastic material, 461
Right-hand rule for vectors, 154, 154*f*
Ring faults, 98, 98*f*
Ripple marks, 52–53, 53*f*
Rivera triple junction, 616*f*
Rock cleavage. *See* Cleavage
Rock uplift, 670, 680
Rodinia, 687, 688
Rods lineations, 314, 315*f*
Rollover anticlines, 66, 67*f*, 107, 109*f*
Roof fault, 109*f*, 124, 125*f*
Rootless folds, 308
Root zone, 660, 662*f*
Rose diagrams, 696
  fractures, 57, 57*f*
  joints, 43*f*, 44
Rotational faults, 63*f*, 95
  domino block model, 108, 109*f*
Rotational progressive deformation, 334
Rotations, minerals during deformation, 401–402, 402*f*, 403*f*
Rough foliations, 299*f*, 301, 301*f*
R shears, R′ shears. *See* Riedel shears
Ruby Mountains (Nevada), interference folds, 393*f*
Rutile, as strain marker, 424, 425*f*
Ryukyu Trench, 617*f*

S
Saddle reefs, 382–383
Sag ponds, 70
Salient, 122*f*, 123
Salt domes. *See* Diapirs
Salt nappe, 104, 105*f*, 110
San Andreas fault (California), 608–614
  and Anderson's theory, 255
  Big Bend, 144–145, 145*f*, 609, 610*f*
  brittle-ductile transition, 489
  fault slip history, 219, 220*f*
  kinematic model, 146–147, 149*f*
  Loma Prieta fault, 610*f*, 611, 612*f*
  offset stream, 70*f*
  shoreline, 77*f*
  Sierran microplate, 609–614, 610*f*
  size/location of, 608–609*f*
  strike-slip fault, 136*f*, 137*f*, 146–147, 148, 148*f*, 609–611
  as transform fault, 148, 609
  Transverse Ranges, 609, 610*f*
Sangihe-Halmahera collision, 628*f*
Satellites
  fault movement survey, 70
  as geological tool, 2
Scalar
  defined, 154, 156
  scalar components, 155, 155*f*
  as zero rank tensor, 184
Scalar invariants of stress, Mohr diagram, 172–173, 172*f*, 180
Scale
  of folds, 274*f*, 280
  and fracturing, 225
  of maps, 697
Scale-invariant geometry, 46
Scale models, 543, 546–561
  boundary conditions, 547–548
  diapiric structures, 552–553, 552*f*, 553*f*
  dynamically similar to prototype, 546
  faulting, Southeast Asia, 555–561, 556*f*, 558*f*
  of folding, 548–550, 548*f*, 550*f*
  functions of, 363–364, 543
  geometrically similar to prototype, 546
  gravity-driven deformation, 551–553, 551*f*, 552*f*, 553*f*
  kinematically similar to prototype, 546
  modeling material, 549, 550*f*, 551, 551*f*, 553
  plastic slip-line field theory, 553–561
  scale factors, 547*t*
  similarities in, 546
Scaling theory, 543
Scaly argillite, 618
Scaly clay, 618
Scarps. *See* Fault scarps
Schist, with foliations, 304*f*, 309
Schuppen zones, thrust faults, 123*f*, 124
Scientific method, 4–7
  correlation versus causation, 7
  deductive (Darwinian) method, 5–6
  hypotheses to laws process, 4–5
  inductive (Baconian) method, 5–6
Scissor faults, 62, 95, 141

Scolithus tubes, as strain markers, 424
Screw dislocations, 504–506, 505*f*, 508–509
S-C tectonites, foliations in, 307–308, 307*f*
Seafloor spreading
  evidence of, 11–12
  Red Sea, 585*f*
  See also Oceanic ridges
Sea level
  fault motion detection, 70
  first and second-order cycles, 25*f*
Sea of Japan, back-arc basin, 615
Seaward-dipping reflectors (SDRS), 583, 586*f*
Secondary creep. *See* Steady-state creep
Secondary recrystallization, 516
Secondary shears, 216
  See also Riedel shears
Second-rank tensors, 181, 184–185
Sedimentary melange, 622
Sedimentary rocks, orogenic belts, 653–654
Seismic focal mechanism. *See* P-wave first motion studies
Seismic rays, 701
  diffraction of, 706
  P-wave first motion studies, inference of strain (or stress) from, 710. *See* Seismic focal mechanism
  reflection of, 704, 704*f*
  refraction of, 705, 706
  migration of, 706
  stacking of, 705
Seismic reflection profile, listric normal fault, 94*f*
Seismic studies, 701–708
Self-diffusion, 500
Self-similar geometry
  elements of, 46
  fractures, 37
Separations
  fault displacement, 79–80, 79*f*, 80*f*
  normal faults, 91–92, 93*f*
Serpentine diapirs, 397
Serpentinites, 102
  oceanic crust, 597*f*
  oceanic transform faults, 605*f*
S-folds, 283–284
S-foliations, 307–308, 308*f*
Shaanxi Graben (China), 560
Shale diapirs, 397
Shansi Graben (China), normal faults, 98
Shear
  pure, 330, 330*f*
  simple, 330–331, 330*f*
  strike-slip faults, 146, 148*f*
Shear angle, 320*f*, 321, 323
Shear bands
  defined, 408
  shear band foliations, 408–409
Shear cleavage, 309
Shear couple, 157*f*, 158
Shear displacement, fractures. *See* Extension fractures
Shear failure
  and confining pressure, 216–219, 218*f*
  failure envelopes, 218, 218*f*
  von Mises criterion, 217–218, 218*f*
Shear fracture envelope, 212
Shear fractures
  in compression, 228–229, 229*f*
  conjugate shear, 57, 57*f*, 210*f*, 216, 218
  Coulomb fracture criterion, 212–216, 213*f*, 215*f*, 216
  defined, 61
  experiments on brittle rock, 210, 210*f*
  compared to extension fractures, 251–252
  Griffith theory, 228–229, 228*f*, 229*f*
  pinnate fractures of, 251
  propagation of, 37, 38*f*
  Riedel. *See* Riedel (R) shears
  sinistral simple, 217*f*
  stresses, on fracture plane/cleavage plane, 224*f*
  See also Faults
Shear modulus. *See* Modulus, shear
Shear planes, 331*f*
  and foliations, 413–415, 414*f*, 415*f*, 668*f*
Shear sense indicators, 71
Shear strain, 323–325
  arbitrary line segment, 357*f*, 358–359
  defined, 321, 323, 347*f*
  geometric shape change, 350*f*
  positive and negative, 323*f*
  tensor and engineering, 323*f*, 324
Shear strength, and fracture, 222–224, 222*f*
Shear stress
  defined, 347*t*
  planes of maximum, 171–172, 172*f*, 179, 181*f*
Shear (S) wave, 701
Shear traction, 153
Shear zones
  fault displacement, 75–77, 76*f*
  strain measures from, 432–446, 435*f*
  strain paths in, 343, 345*f*
Sheath folds, 75, 76*f*
Sheeted dike complex, 598, 599*f*, 601*f*
Sheet joints, 39, 41
  formation of, 246
Shortcut fault, 109*f*
Shutter ridges, strike-slip faults, 138
Sidewall ramps, 85, 86*f*, 119, 120*f*, 121*f*
Sierra Nevada (California)
  batholiths, 658, 659*f*
  location/digital relief map, 610*f*
Sierran microplate (San Andreas fault), 609–614, 610*f*
Sign conventions, stress, 187, 187*t*
  geological, Mohr circle, 156, 158*f*, 187, 187*t*
  geologic tensor, 182, 187, 187*t*
Sign convention, traction, 156
Sigsbee escarpment (Gulf Coast)
  growth fault, 105*f*, 106*f*
  salt nappe, 104, 105*f*, 110
Sills, oceanic transform faults, 605*f*
Similar folds, 292–293
Simple constriction, homogeneous deformation, 331*f*, 332
Simple flattening, homogeneous deformation, 331*f*, 332
Simple shear, 330–331, 330*f*
  crystallographic preferred orientation, 525, 526*f*
  defined, 330, 347*t*
  progressive, 333–334, 333*f*, 335*f*, 336
Sinai Peninsula (Red Sea), fractural topograph, 40*f*
Sinistral faults. *See* Left-lateral (sinistral) faults
Slates
  fracture experiments, 221–222, 222*f*
  slate belts, 645–647, 647*f*
  strain measures from, 426–427
Slaty cleavage, 309
Slickenfibers
  composition of, 66
  and fault displacement, 72, 73*f*, 420
  mineral fiber lineations, 418–420, 419*f*
Slickenlines
  formation of, 404, 417–418, 418*f*
  mineral, 418*f*
  structural, 417–418, 418*f*
Slickensides
  fault planes, 66, 67*f*, 71
  fractures, 54, 66, 67*f*
  lineations, 72, 92, 314, 315*f*, 316, 404
  mineral, 314, 315*f*
  strike-slip faults, 138
  structural, 313
Slickolites, 73, 73*f*, 405, 418, 418*f*
Sliding. *See* Frictional sliding
Slip
  and faults, 61–63*f*, 108
  See also Strike-slip faults
Slip-lines
  determination, Hansen's method, 390, 391*f*, 667
  geometry of, 554–555, 555*f*
  plastic slip-line field theory, 553–561
Slip system
  in crystal, 508–509, 510*f*
  crystallographic orientation, 534, 534*f*
  pencil glide, 528
Smartville Complex (California), 597*f*
  Smartville-type oceanic crust, 597*f*, 599
Smooth foliations, 299*f*, 301, 301*f*
Snake Range (Nevada)
  folds, 275*f*
  mineral lineations, 315*f*
Snowball garnets, 401
Solea graben (Troodos complex), 599, 600*f*, 608
Sole faults (detachments), 123, 123*f*
  foreland fold and thrust belts, 644–645
  See also Decollements
Solomon Islands, 616*f*
Solution cleavage, 309
Solution creep, 495, 501–502, 502*f*
Solution folding. *See* Volume-loss folding
Solution transfer, 404

South China Sea, seafloor spreading, 560
Southeast Asia
  faulting, scale model, 555–561, 556*f*, 558*f*
  strike-slip faults (map), 558
South Fiji Basin, back-arc basins, 591*f*
South Mountain (Maryland), strain and foliation study, 429, 430*f*
South Sandwich trough, back-arc basins, 591*f*
South Yorkshire coal fields (United Kingdom), fault-length systematics, 83*f*
Spaced foliations, 298, 298–302, 298*f*, 299*f*
  compositional, 298–299, 299*f*
  crenulation, 302, 302*f*, 303*f*
  disjunctive, 299–302, 299*f*, 300*f*, 301*f*
Spherical projections, 696
Spherulites, as strain markers, 424
Spikes, on slickolites, 418, 418*f*
Spin, 525
Splay faults, 82*f*
  imbricate fans, 84, 85*f*
Spring-and-slider block system, 472–473, 473*f*
S-surface, foliations, 298
Stable sliding, 219, 220, 220*f*
Stain path, 333
Standard state stress, 256
State-dependent friction law, 469, 471–475, 472*f*, 473*f*, 474*f*
State of strain, 324
Static recovery, 516
Static recrystallization, 516
Static rock friction, 466–467, 467*f*
  static to dynamic transition, 468–469, 469*f*
Staurolite, 76
Steady motion, defined, 333
Steady-state creep, 476–486, 477*f*
  chemical environment effects, 483–484, 484*f*
  exponential creep law, 479
  grain size effects, 482–483, 482*f*
  at high stresses, 479
  high-temperature creep equation, material constants in, 485–486
  at low stresses, 479–481
  at moderate stresses, 478–479
  partial melt effects, 484
  power-law creep, 478–479, 480*f*, 490
  power-law creep with low *n*, 479, 480*f*
  pressure effects, 481–482, 481*f*
  temperature effects, 478–481, 478*f*
Steady-state foliations, development of, 412–413
S-tectonites, 297
Step-overs
  contractional/restraining, 139*f*, 140
  extensional/releasing/dilatant, 139*f*, 140
  strike-slip faults, 85, 139–140, 139*f*
Stick-slip sliding, 219, 220, 220*f*
Stillwater complex (United States), Proterozoic dike system, 22, 23*f*
Stillwater Range (Nevada), normal fault, 92*f*
Stitching wax, modeling material, 549, 550*f*
Strain
  area, 325–326
  constant-volume, 328*f*, 329
  constriction, 331*f*, 332, 339
  defined, 319, 347*t*
  ellipse, 324, 325*f*, 327*f*, 332, 332*f*, 333, 337*f*
  ellipsoid, 231, 324–325, 325*f*, 331*f*
  extensional, 322, 322*t*
  finite, Mohr circle, 359–361
  flattening, 331*f*, 332, 339
  Flinn diagram, 340–345
  and foliations, 400–401, 401*f*
  homogeneous, 319–321, 320*f*, 329–333, 331*f*, 339, 340*f*
  inhomogeneous, 319–321, 320*f*, 339
  inverse strain ellipse, 326, 327*f*, 332*f*
  linear, 321–323
  and lineations, 400–401
  measurement of. *See* Strain measures
  natural, 321, 322
  plane, 319, 329
  principal axes, 324–325, 325*f*, 334*f*, 335*f*
  progressive, 333–339
  pure, 329, 331–332, 332*f*
  quantitative view of, 352–355
  shear, 323–325, 350*f*
  shear of arbitrary line segment, 358–359
  sign convention, linear, 321, 322*t*
  sign convention, shear, 323, 323*f*
  state of, 324
  strain facies, 345, 345*f*
  versus stress terminology, 346–349
  and stretch, 325*f*, 355–358
  study, rationale for, 326–328
  tensor representations. *See* Strain tensors
  in three-dimensions, 324–325, 328–329, 343–345
  triaxial, 328–329, 332
  in two-dimensions, 324, 324*f*, 423
  uniaxial, 329–330, 329*f*
  uniform dilation, 331–332
  volume, 325–326, 350*f*
Strain facies, Flinn diagram, 345, 345*f*
Strain hardening, 476
Strain logarithmic. *See* Strain, natural
Strain measures
  from brittle deformation, 434–443, 437*f*, 438*f*
  from deformed initially circular cross-section, 424
  from deformed linear objects, 424–425
  from ductile shear zones, 443–446, 444, 445*f*, 446*f*
  finite shear strain rate, 465, 465*f*
  finite strain analysis, 667
  in folds, 427–432, 428*f*, 430*f*, 431*f*, 432*f*
  from foliation patterns, 429–432, 430*f*, 431*f*, 432*f*
  graptolites as absolute measure, 426–427, 427*f*
  from initially elliptical objects, 424, 449–453, 450*f*, 452*f*, 453*f*
  from initially orthogonal pairs of lines, 425–426, 426*f*
  instantaneous extension rate, 465
  instantaneous strain analysis, 667, 669, 669*f*
  limitations of, 423–424, 424*f*
  and lineations, 426–427
  nearest-neighbor center-to center (Fry) method, 453–455, 454*f*
  orthogonal flexure study, 428–429, 428*f*
  from sheared initially orthogonal pairs of lines, 425–426, 426*f*
  from sheared orthogonal lines (Wellman method), 455, 455*f*
  in shear zones, 432–446, 435*f*
  in slates, 426–427
  volumetric, 426–427, 444–446, 445*f*, 446*f*
Strain rate
  experimental versus geologic, 488–491
  measures of, 465–466
  strain-rate softening, 569–570
Strain shadows. *See* Pressure shadows
Strain-slip cleavage, 309
Strain softening, 570
Strain tensors, 350–355
  defined, 351
  deformation tensor, 352
  infinitesimal, 351, 354–355
  instantaneous extension/stretch, 352
  in three dimensions, 350–351, 350*f*
  in two dimensions, 351–352
Stratigraphic separation, 79
Streamlines, 673–674, 674*f*
Strength, defined, 209
  shear, 221
  tensile, 210. *See* Tensile strength
Stress, 151–208
  applied, 226–227
  concept/components of, 152*t*, 159–161, 160*f*
  critical, 223, 223*f*
  defined, 151, 347*t*
  distribution and faulting, 255–258
  effective, 220
  fracture formation. *See* Fracturing
  local, 226–227, 248
  measurement limitations, 346
  Mohr diagram, 158–159, 168–173, 175–177, 188–189
  notation, 191*t*
  numerical example of, 161–163
  paleostresses in deformed rocks, 520–524
  across planes of orientation, 159–161, 160*f*, 169–171, 170*f*, 179
  remote, 248
  scalar invariants, 172–173, 180
  stable, 223, 223*f*
  standard state, 256
  versus strain terminology, 346–349

Stress (*continued*)
stress components sign conventions, 187
surface, 151, 152*t*, 156–159, 157*f*
three-dimensional state. *See* Three-dimensional stress
two-dimensional state. *See* Two-dimensional stress
types of, 158, 173, 175
yield stress, 218, 218*f*
Stress components sign convention, 187*t*
Stress determination methods, 233–236
drill holes method, 234
earthquake first-motion studies, 234, 236
flat jack method, 234, 235*f*
hydraulic fracturing (hydrofrac) method, 234, 236*f*
overcoring, 233–234, 235*f*
Stress in Earth, 233–243
aquathermal pressuring, 238
determining. *See* Stress determination methods
horizontal/vertical motions, 237
normal vertical/horizontal stress, 238–239, 238*f*, 239*f*
and origin of joints. *See* Joint formation
overburden, 237
and plate motion, 237
pressure effects, 238
stress maps, 241, 242*f*
tectonic horizontal stress, 239–241
thermal effects, 237–238
Stress ellipse
features of, 163, 164*f*
Stress maps, 241, 242*f*
Stress tensor, 180–191
Mohr diagrams, 188–189
stress components matrix, 181–182
symmetric tensor, 186
tensor sign convention, 182–186, 183*f*, 187
two-dimensions, 186, 188–189
Stretched crystals, 420
Stretches
arbitrary line segment, 355–358, 356*f*
defined, 321–322, 347*t*
of object, determining, 425, 425*f*
principal, 324
progressive, of material lines, 336–339, 338*f*
Stretching lineations, 400, 401*f*, 416
Strike, 693
Strike joints, 40
Strike line, 693
Strike separation, 79, 79*f*, 80*f*
Strike-slip deformation, divergent and convergent, 343, 344*f*, 345*f*
Strike-slip faults, 135–149
balancing of, 149
bends, 85, 86*f*, 139–140, 139*f*
displacement, 78*f*, 138*f*, 140, 146*f*, 148–149
duplexes, 86, 140–141, 140*f*, 141*f*
en echelon arrays, 135, 138, 139, 140
fault systems, 136, 136*f*, 137*f*, 145*f*
features, 61–63, 62*f*, 136–139
isopach map, 78*f*
kinematic models of, 146–148, 148*f*, 149*f*
left-lateral (sinistral), 62–63*f*, 138*f*
major faults, examples of, 611
partitioned, at convergent margins, 622
right-lateral (dextral), 62–63*f*
San Andreas fault, 136*f*, 137*f*, 146–147, 148, 148*f*, 609–611
shape of, 139, 139*f*
single, 139–140
Southeast Asia (map), 558
step-overs, 85, 139–140, 139*f*
structures related to, 85–86, 86*f*, 138*f*, 139, 146
tear, 119, 135, 141, 141*f*, 143*f*
terminations, 144–146, 147*f*, 148*f*
transcurrent, 135–136, 141
transfer, 86, 119, 135
transform, 135–136, 141
Strong fabric, 300*f*
Structural geology, 1–7
development of, 2
study of, 2
Structural lineations, 298, 309–313
constructed, 310–313
defined, 309
discrete, 310
Structural slickenlines, 313
Structural terraces, defined, 281, 283*f*
Structure contour maps, fault displacement, 77, 78*f*
Style of folds. *See* Fold style
Stylolites, 72–73
Stylolitic foliations, 299*f*, 300–301, 300*f*, 405
Subduction erosion, 619
Subduction zones
continent-continent collision, 677–680, 678*f*, 679*f*
plate movement, 11, 12, 237
trench-trench transform faults, 602, 603*f*
See also Convergent margins
Subgrain(s), 404, 511, 519
Subgrain rotation recrystallization, 514, 515*f*
Subhorizontal folds, 281
Subsurface stress, components of, 152*t*
Subvertical folds, 281
Sunda arc, accretionary prisms, 620*f*
Supercontinents
Gondwanaland, 25, 687
Pangaea, 687
Rodinia, 687, 688
and Wilson cycle, 687–688
Superplastic creep, 502–503, 503*f*
Superplumes, 687
Superposed folds, 390–394
buckle folding, single layer, 393–394, 394*f*
fold generations, interpretation of, 394–395
geometry of, 391, 391*f*
interference patterns, 391–393, 392*f*
Surface forces, 153
Surface loads, and plate bending, 237
Surface stress, 151, 152*t*, 156–159, 157*f*
Mohr circle, 169–171, 169*f*, 179
Surface uplift, 670
Surficial lineations, 297
Suspect terranes, 688
Sutures
Archean, 20–21*f*
defined, 648, 688
Himalaya-Tibet region, 630*f*
orogenic belts, 648
structures marking location of, 648
Symmetric crenulation foliations, 407, 409*f*
Symmetric folds, 283, 286*f*, 287*f*, 293
Symmetric matrix, 186
Symmetric tensor, 185
stress, 182
strain, 324, 350
Symmetry and deformation. *See* Deformation symmetries
Symmetry element, 535
Symmetry principle, 537–538
Synclines
compared to anticlines, 281*f*
defined, 280
fault-ramp, 95*f*, 109*f*, 121*f*
fold train, 276*f*
Synforms, 276
Syntaxial growth, mineral fibers, 420
Syntaxis, 122*f*, 123
Syntectonic recrystallization, 514
Synthetic faults, 70, 87, 87*f*, 94*f*, 95, 109*f*
Systematic joints, 38, 39*f*

T
Tablet boudinage, 313
Taiwan fold and belt thrust, 129, 130*f*
thrust wedges, 260, 261*f*
Talus deposits, oceanic transform faults, 605, 605*f*
Tangent lineation diagrams, 436, 437*f*, 438*f*
Tapered boudin blocks, 311, 312*f*, 313
Tapered thrust sheets, 260–263, 261*f*
Tauern window (Alps), Austro-Alpine nappes, 131–132, 132*f*
Taylor-Bishop-Hill model
crystallographic preferred orientation, 525, 526*f*
foliations and lineations, 401, 402–403
Tear faults
features, 119, 120*f*
strike-slip faults, 135, 141, 141*f*, 143*f*
thrust sheets, 123–124, 124*f*
Tectonic melanges, 622, 624–625, 625*f*
Tectonic plate boundaries
movements at, 11, 12
world map of, 616*f*
Tectonic plates, transform versus transcurrent faults, 136

Tectonics
and metamorphism, 671–673
study of, 9–11
relationship to topography/erosion, 670–671
See also Plate tectonics
Tectonic stress, and joint formation, 245–246
Tectonic transport, direction of, 667
Tectonic wedging, 126–127, 126*f*
Tectonites
foliations of, 297–298
L, 297
lineations of, 297–298
S, 297
S-C tectonites, 307–308, 307*f*
Temperature
and creep, 478–481, 478*f*, 485–486
and fracturing, 225
homologous, 476–477, 496
and joint formation, 243–244
orogenic heating, 681
and pore fluid pressure, 238
thermal expansion as stress, 237–238
Tensile strength, 211
theoretical, 226
Tensile stress, 158, 158*f*
defined, 348
depth factor, 241
Griffith fracture criteria, 229
and joint formation, 244–245
Tension fractures
experiments on brittle rock, 210, 210*f*
fracture criterion, 210–212, 211*f*
Griffith theory, 227, 227*f*
tension fracture envelope, 211, 211*f*
Tensors
first-rank, 184
quantities, types of, 184
second-rank, 181, 184–185
See also Strain tensors; Stress tensors
Tensor shear strain, 323*f*, 324, 347*f*
Termination(s)
joints, 44–45, 44*f*
strike-slip faults, 144–146, 147*f*, 148*f*
Termination lines, faults, 82–85, 84*f*
Terrace, structural, 281
Terrain correction, 708
Terrane analysis, 688–689
of exotic terranes, 688–689, 688*f*
Terrestrial bodies, 9
Tertiary creep, 476–477, 477*f*
Theory, in scientific method, 5
Thermoluminescence, defined, 666
Thin-skinned tectonics, 645
Three-dimensional strain, 324–325, 328–329, 343–345
Three-dimensional stress, 166–168
geometry of, 188*f*
Mohr diagram, 168, 178–180
stress components, 167*f*, 168
stress ellipsoid, 166, 167*f*, 168
Three-dimensions constitutive relations, constitutive equations, 491–492
Throw, and faults, 62
Thrust faults, 62–63*f*, 68*f*, 86*f*, 115–133
displacement, 119, 131–133
fault geometry, 121*f*, 125*f*, 128–129
with folds, 121, 121*f*, 123, 127, 131
kinematic models, 127–128, 128*f*
local, 119
rocks of, 115, 117
shape of, 116*f*, 117–119, 118*f*, 123*f*
stratigraphic features, 115–117, 116*f*
thrust systems, 122*f*, 123–127
Thrust nappes, 293
maximum finite stretch, 675*f*
Thrust sheets, 115, 118*f*, 119, 123–124, 124*f*, 131–132, 132*f*
ductile nappe emplacement, 675–676, 675*f*
emplacement, simplified model, 258–267
plastic behavior, 263, 266
slip domains, 266–267
tapered, critical Coulomb wedge model, 260–263, 261*f*
thrust wedge measured geometry, 262*f*
thrust wedges cross sections, 261*f*
Thrust duplex, 124, 125*f*
Tibet
digital elevation map, 684*f*
Himalaya-Tibet region, 630*f*
and India-Asia collision, 684
South Tibetan detachment system, 665*f*
Tibetian plateau, continental crust thickness, 17
Tightness, fold style, 284–285, 286*f*
Tilt boundary, 511, 511*f*
Time, deep time, 1
Tip lines, 83–84, 84*f*
Titus Canyon (Death Valley), megabreccia, 65*f*
Tonga-Kermadec island arc, back-arc basins, 591*f*
Tool marks, fault displacement, 74, 75*f*
Traction
components of, 152*t*
defined, 151, 156
magnitude across surface, 156, 156*f*
numerical example of, 161–163
opposing, 157*f*, 158
Transcontinental Arch (United States), cratonic basins, 25, 26*f*
Transcurrent faults, of strike-slip faults, 135–136*f*
Transfer faults
fault geometry model, 100, 100*f*
features, 135
of strike-slip faults, 86, 119, 135
Transfer zones, faults, 86, 86*f*, 100, 124, 126, 126*f*
Transformation equations
stress components, 161
tensors, 185
vector components, 155, 161
Transform faults
continental, 608
features, 135–136*f*, 601
oceanic. *See* Oceanic transform faults
plate movement at boundary, 11, 12
San Andreas, 148, 608–614
world map, 14*f*
Transform margins, 29, 29*f*
Transposition foliation, 308, 308*f*
Transpression, defined, 343, 344*f*, 346, 349*t*
Transtension, defined, 343, 344*f*, 346, 349*t*
Transverse Ranges (San Andreas fault), 609, 610*f*
Trenches, accretionary prisms, 615, 619, 619*f*
Trench-trench transform faults, 602, 603*f*
Trend, 694
Trend line, 694
Triaxial strain, 328–329, 332
Triaxial stress
and Coulomb fracture, 214
Mohr diagram, 173, 174*f*
Triclinic symmetry, 537
Trishear model, fault-propagation folds, 387–390, 388*f*, 389*f*
Troodos Complex (Cyprus)
oceanic spreading, 599–600, 600–601*f*
sheeted dike complex, 599, 599*f*, 601*f*
transform faults, 607–608
Trough lines, folds, 277, 278*f*
True lineations, 310, 310*f*
Tulip structure, 140*f*, 141, 142*f*
Turtle-back domes, 102
Twin gliding, 499–500, 499*f*
calcite, 499–500, 499*f*, 520–521
and deformation, 500
Two-dimensional strain, 324, 324*f*, 423
Two-dimensional stress, 163–166
Mohr circle stress problems, 193–208
Mohr diagram, 168–173, 175–177
Mohr equations, derivation of, 175–177
principal stresses, 163–166, 164*f*, 189–190, 190*f*
stress components, 152*f*
stress ellipse, 163, 164*f*
stress tensor components, 186, 188–189
Types I and II crossed girdle patterns, 531, 532*f*

U

Ultra-high-pressure (UHP) metamorphism, 649–650
Ultramafic rocks
Archean terranes, 19, 20*f*
distribution, world map, 23*f*
of gneissic terranes, 656–658, 657*f*
minerals of, 15
Ultramylonite, 65
Underplating
accretionary prisms, 618, 622*f*, 623*f*
and back-folding, 680
and exhumation, 680
Underthrusting
and collisions, 628
Japan, coast of, 626*f*
Uniaxial strain, 329–330, 329*f*

Uniaxial stress
elastic deformation, 231–232, 232*f*
Mohr diagram, 173, 174*f*
Uniaxial tension, Mohr diagram, 173, 174*f*
Uniform dilation, 331–332
Unroofing sequence, 96, 643
Uplift, joint formation, 241, 243–245
Upright folds, 280, 282*f*
Ural-Altaids, orogenic belts, 640*f*, 641*f*

V
Vacancy, 500, 501*f*
Vectors, 154–155
addition, 154
defined, 154
scalar components, 155, 155*f*
vector quantity, 153
Veins, composition of, 42
Velocity inversions, 17
Vergence
asymmetric folds, 284
back thrusts, 126, 126*f*, 128
thrust faults, 119, 124, 126, 131
Vertical folds, 281, 282*f*
Vertical separation, 79
Vertical stresses in Earth, 238–239, 238*f*
and joint formation, 244–245
Virgation, 122*f*, 123
Viscosity
coefficient of, 461
effective, 566
Viscous material behavior, 460–461, 461*f*
firmo-viscous (Kelvin or Voight) material, 463–464, 463*f*
three-dimensional, 491–492
visco-elastic (Maxwell) materials, 463–464, 463*f*, 572, 576, 576*f*
visco-plastic (Bingham) material, 464, 464*f*
viscous deformation, mathematical model, 573–576
Voight material, behavior of, 463–464, 463*f*
Volcanic arcs, 615
Volcanic structures
caldera, 98, 98*f*
midoceanic ridges, 594, 595*f*, 596
Volume diffusion, 500–501, 501*f*
Volume-loss folds, 371–374, 372*f*, 373*f*
buckling, 368*f*
kinematic model, 371–374, 372*f*
Volume strain, 325–326, 350*f*
strain measures from, 426–427, 427*f*, 444–446, 445*f*, 446*f*
Von Koch curve, 46–47, 47*f*
Von Mises condition, 513, 527
Von Mises criterion, 217–218, 218*f*, 461
Vorticity of deformation, 334–335

W
Wales, slate belt, 646–647, 647*f*
Walner lines, 54
Water
hydraulic weakening, 483
pore fluid. *See* Pore fluid pressure
and steady-state creep, 483–484
Water, bodies of
lakes, caldera structure, 98*f*
offset streams, 69, 70*f*
pull-apart basins, 141, 143*f*
sag ponds, 70
See also entries under Ocean and Oceanic
Waterton field thrust (Canada), duplex structure, 124, 125*f*
Wavelength of folds, 280
Waves, seismic, 701
Weak fabric, 300*f*
Weddell Sea-Ross Sea (Antarctica), normal faults, 99
Weertman creep, 517
Wellman method, strain measure, 455, 455*f*
Wells fault (Nevada), 101
Whipple Mountain (California)
cataclasite, 65*f*
metamorphic core complex, 101–102, 103*f*
Wild Horse Mesa (California), seismogenic strain, 438*f*, 439
Wilson cycle, 685–688, 686*f*
Windows (fensters), thrust faults, 118, 118*f*, 131
Wind River thrust fault (Wyoming), gravity profile, 117*f*
Wing cracks, 87, 87*f*
Wopmay orogen (Canada), banded foliations, 299
Work hardening, 476, 477*f*, 513
Wrench fault, 135

Y
Yakutat block collision (Alaska), 628
Yield stress, 218, 218*f*, 461
Young's modulus
defined, 232
geologic sign convention, 232, 232*f*
Y shears, strike-slip faults, 139
Yucca Mountain (Nevada), Paintbrush tuff, 50*f*

Z
Zagros Mountains (Iran)
crush zone, 130*f*
fold and thrust belts, 644
orogenic belts, 681
Z-folds, 283
Zonal crenulation foliations, 302, 302*f*, 407